Handbook of
Industrial and Hazardous Wastes Treatment

9/2007

Handbook of Industrial and Hazardous Wastes Treatment

Second Edition, Revised and Expanded

edited by

Lawrence K. Wang

Zorex Corporation, Newtonville, New York, New York, and Lenox Institute of Technology, Lenox, Massachusetts, U.S.A.

Yung-Tse Hung

Cleveland State University, Cleveland, Ohio, U.S.A.

Howard H. Lo

Cleveland State University, Cleveland, Ohio, U.S.A.

Constantine Yapijakis

The Cooper Union, New York, New York, U.S.A.

Consulting Editor

Kathleen Hung Li

NEC Business Network Solutions, Irving, Texas, U.S.A.

MARCEL DEKKER

MARCEL DEKKER, INC. NEW YORK • BASEL

The first edition of this book was published as *Handbook of Industrial Waste Treatment, Volume 1*, edited by Lawrence K. Wang and Mu Hao Sung Wang (Marcel Dekker, Inc., 1992).

Library of Congress Cataloging-in-Publication Data
A catalog record for this book is available from the Library of Congress.

ISBN: 0-8247-4114-5

This book is printed on acid-free paper.

Headquarters
Marcel Dekker, Inc., 270 Madison Avenue, New York, NY 10016, U.S.A.
tel: 212-696-9000; fax: 212-685-4540

Distribution and Customer Service
Marcel Dekker, Inc., Cimarron Road, Monticello, New York 12701, U.S.A.
tel: 800-228-1160; fax: 845-796-1772

Eastern Hemisphere Distribution
Marcel Dekker AG, Hutgasse 4, Postfach 812, CH-4001 Basel, Switzerland
tel: 41-61-260-6300; fax: 41-61-260-6333

World Wide Web
http://www.dekker.com

The publisher offers discounts on this book when ordered in bulk quantities. For more information, write to Special Sales/Professional Marketing at the headquarters address above.

Current printing (last digit):

10 9 8 7 6 5 4 3 2 1

PRINTED IN THE UNITED STATES OF AMERICA

Preface

Environmental managers, engineers, and scientists who have had experience with industrial and hazardous waste management problems have noted the need for a handbook that is comprehensive in its scope, directly applicable to daily waste management problems of specific industries, and widely acceptable by practicing environmental professionals and educators.

Many standard industrial waste treatment and hazardous waste management texts adequately cover a few major industries, for conventional in-plant pollution control strategies, but no one book, or series of books, focuses on new developments in innovative and alternative environmental technology, design criteria, effluent standards, managerial decision methodology, and regional and global environmental conservation.

This handbook emphasizes in-depth presentation of environmental pollution sources, waste characteristics, control technologies, management strategies, facility innovations, process alternatives, costs, case histories, effluent standards, and future trends for each industrial or commercial operation, such as the metal plating and finishing industry or the photographic processing industry, and in-depth presentation of methodologies, technologies, alternatives, regional effects, and global effects of each important industrial pollution control practice that may be applied to all industries, such as industrial ecology, pollution prevention, in-plant hazardous waste management, site remediation, groundwater decontamination, and stormwater management.

In a deliberate effort to complement other industrial waste treatment and hazardous waste management texts, this handbook covers new subjects as much as possible.

Many topics, such as industrial ecology, pollution prevention, in-plant hazardous waste management, stormwater management, photographic processing wastes, soap and detergent wastes, livestock wastes, rubber processing wastes, timber industry wastes, power plant wastes, and metal finishing wastes, are presented in detail for the first time in any industrial waste treatment book. Special efforts were made to invite experts to contribute chapters in their own areas of expertise. Since the field of industrial hazardous waste treatment is very broad, no one can claim to be an expert in all industries; collective contributions are better than a single author's presentation for a handbook of this nature.

This handbook is to be used as a college textbook as well as a reference book for the environmental professional. It features the major industries and hazardous pollutants that have significant effects on the environment. Professors, students, and researchers in environmental, civil, chemical, sanitary, mechanical, and public health engineering and science will find valuable educational materials here. The extensive bibliographies for each industrial waste treatment or practice should be invaluable to environmental managers or researchers who need to trace, follow, duplicate, or improve on a specific industrial hazardous waste treatment practice.

A successful modern industrial hazardous waste treatment program for a particular industry will include not only traditional water pollution control but also air pollution control, noise control, soil conservation, site remediation, radiation protection, groundwater protection, hazardous waste management, solid waste disposal, and combined industrial-municipal waste treatment and management. In fact, it should be a total environmental control program. Another intention of this handbook is to provide technical and economical information on the development of the most feasible total environmental control program that can benefit both industry and local municipalities. Frequently, the most economically feasible methodology is combined industrial-municipal waste treatment.

We are indebted to Dr. Mu Hao Sung Wang at the New York State Department of Environmental Conservation, Albany, New York, who coedited the first edition, and to Ms. Kathleen Hung Li at NEC Business Network Solutions, Irving, Texas, Consulting Editor for this new edition.

Lawrence K. Wang
Yung-Tse Hung
Howard H. Lo
Constantine Yapijakis

Contents

Contributors

Donald B. Aulenbach Rensselaer Polytechnic Institute, Troy, New York, U.S.A.

Adel Awad Tishreen University, Lattakia, Syria

Renbi Bai National University of Singapore, Singapore

Charles J. Banks University of Southampton, Southampton, England

Thomas Bechtold Leopold-Franzens-University, Innsbruck, Austria

Thomas W. Bober* Eastman Kodak Company, Rochester, New York, U.S.A.

Michal Bodzek Silesian University of Technology, Gliwice, Poland

Trevor J. Britz University of Stellenbosch, Matieland, South Africa

Eduard Burtscher Leopold-Franzens-University, Innsbruck, Austria

Chein-Chi Chang District of Columbia Water and Sewer Authority, Washington, D.C., U.S.A.

J. Paul Chen National University of Singapore, Singapore

Thomas J. Dagon* Eastman Kodak Company, Rochester, New York, U.S.A.

Harvey E. Fowler Eastman Kodak Company, Rochester, New York, U.S.A.

Sudhir Kumar Gupta Indian Institute of Technology, Bombay, India

Sunil Kumar Gupta Indian Institute of Technology, Bombay, India

Mohd Ali Hassan University Putra Malaysia, Serdang, Malaysia

Yung-Tse Hung Cleveland State University, Cleveland, Ohio, U.S.A.

Tsuyoshi Imai Yamaguchi University, Yamaguchi, Japan

Volodymyr Ivanov Nanyang Technological University, Singapore

Işik Kabdaşh Istanbul Technical University, Istanbul, Turkey

Venera Z. Latypova Kazan State University, Kazan, Russia

Kathleen Hung Li NEC Business Network Solutions, Irving, Texas, U.S.A.

Howard H. Lo Cleveland State University, Cleveland, Ohio, U.S.A.

Simo Olavi Pehkonen National University of Singapore, Singapore

*Retired.

Hana Salman Tishreen University, Lattakia, Syria

Svetlana Yu. Selivanovskaya Kazan State University, Kazan, Russia

Swee-Song Seng National University of Singapore, Singapore

Thomas T. Shen Independent Environmental Advisor, Delmar, New York, U.S.A.

Yoshihito Shirai Kyushu Institute of Technology, Kitakyushu, Japan

Kuan-Yeow Show Nanyang Technological University, Singapore

Nadezda Yu. Stepanova Kazan Technical University, Kazan, Russia

Suresh Sumathi Indian Institute of Technology, Bombay, India

Joanna Surmacz-Gorska Silesian University of Technology, Gliwice, Poland

Jerry R. Taricska Hole Montes, Inc., Naples, Florida, U.S.A.

Joo-Hwa Tay Nanyang Technological University, Singapore

Stephen Tiong-Lee Tay Nanyang Technological University, Singapore

Robert Leo Trotta Sullivan County Division of Public Works, Monticello, New York, U.S.A.

Olcay Tünay Istanbul Technical University, Istanbul, Turkey

Masao Ukita Yamaguchi University, Yamaguchi, Japan

Dominick Vacco Eastman Kodak Company, Rochester, New York, U.S.A.

Corné van Schalkwyk University of Stellenbosch, Matieland, South Africa

Lawrence K. Wang Zorex Corporation, Newtonville, New York, and Lenox Institute of Water Technology, Lenox, Massachusetts, U.S.A.

Zhengjian Wang University of Southampton, Southampton, England

Joseph M. Wong Black & Veatch, Concord, California, U.S.A.

Shahrakbah Yacob University Putra Malaysia, Serdang, Malaysia

Lei Yang National University of Singapore, Singapore

Constantine Yapijakis The Cooper Union, New York, New York, U.S.A.

Shuaiwen Zou National University of Singapore, Singapore

1

Implementation of Industrial Ecology for Industrial Hazardous Waste Management

Lawrence K. Wang
Lenox Institute of Water Technology, Lenox, Massachusetts, U.S.A., and Zorex Corporation, Newtonville, New York, U.S.A.

Donald B. Aulenbach
Rensselaer Polytechnic Institute, Troy, New York, U.S.A.

1.1 INTRODUCTION

Industrial ecology (IE) is critically reviewed, discussed, analyzed, and summarized in this chapter. Topics covered include: IE definitions, goals, roles, objectives, approach, applications, implementation framework, implementation levels, industrial ecologists' qualifications, and ways and means for analysis and design. The benefits of IE are shown as they relate to sustainable agriculture, industry, and environment, zero emission and zero discharge, hazardous wastes, cleaner production, waste minimization, pollution prevention, design for environment, material substitution, dematerialization, decarbonation, greenhouse gas, process substitution, environmental restoration, and site remediation [1–46]. Case histories using the IE concept have been gathered by the United Nations Industrial Development Organization (UNIDO), Vienna, Austria [39–41]. This chapter presents these case histories to illustrate cleaner production, zero discharge, waste minimization, material substitution, process substitution, and decarbonization.

1.2 DEFINITIONS OF INDUSTRIAL ECOLOGY

Industry, according to the Oxford English Dictionary, is "intelligent or clever working" as well as the particular branches of productive labor. Ecology is the branch of biology that deals with the mutual relations between organisms and their environment. Ecology implies more the webs of natural forces and organisms, their competition and cooperation, and how they live off one another [2–4].

The recent introduction of the term "industrial ecology" stems from its use by Frosch and Gallopoulos [10] in a paper on environmentally favorable strategies for manufacturing. Industrial ecology (IE) is now a branch of systems science for sustainability, or a framework for designing and operating industrial systems as sustainable and interdependent with natural systems. It seeks to balance industrial production and economic performance with an emerging understanding of local and global ecological constraints [10,13,20].

A system is a set of elements inter-relating in a structured way. The elements are perceived as a whole with a common purpose. A system's behavior cannot be predicted simply by analysis of its individual elements. The properties of a system emerge from the interaction of its elements and are distinct from their properties as separate pieces. The behavior of the system results from the interaction of the elements and between the system and its environment (system + environment = a larger system). The definition of the elements and the setting of the system boundaries are "subjective" actions.

In this context, industrial systems apply not only to private sector manufacturing and service, but also to government operations, including provision of infrastructure. A full definition of industrial systems will include service, agricultural, manufacturing, military and civil operations, as well as infrastructure such as landfills, recycling facilities, energy utility plants, water transmission facilities, water treatment plants, sewer systems, wastewater treatment facilities, incinerators, nuclear waste storage facilities, and transportation systems.

An industrial ecologist is an expert who takes a systems view, seeking to integrate and balance the environmental, business, and economic development interests of the industrial systems, and who will treat "sustainability" as a complex, whole systems challenge. The industrial ecologist will work to create comprehensive solutions, often simply integrating separate proven components into holistic design concepts for possible implementation by the clients.

A typical industrial ecology team includes IE partners, associates, and strategic allies qualified in the areas of industrial ecology, eco-industrial parks, economic development, real estate development, finance, urban planning, architecture, engineering, ecology, sustainable agriculture, sustainable industry systems, organizational design, and so on. The core capability of the IE team is the ability to integrate the contributions of these diverse fields into whole systems solutions for business, government agencies, communities, and nations.

1.3 GOAL, ROLE, AND OBJECTIVES

An industrial ecologist's tasks are to interpret and adapt an understanding of the natural system and apply it to the design of man-made systems, in order to achieve a pattern of industrialization that is not only more efficient, but also intrinsically adjusted to the tolerances and characteristics of the natural system. In this way, it will have a built-in insurance against further environmental surprises, because their essential causes will have been designed out [29].

A practical goal of industrial ecology is to lighten the environmental impact per person and per dollar of economic activity, and the role of the industrial ecologist is to find leverage, or opportunities for considerable improvement using practical effort. Industrial ecology can search for leverage wherever it may lie in the chain, from extraction and primary production through final consumption, that is, from cradle to rebirth. In this regard, a performing industrial ecologist may become a preserver when achieving endless reincarnations of materials [3].

An overarching goal of IE is the establishment of an industrial system that recycles virtually all of the materials. It uses and releases a minimal amount of waste to the environment. The industrial systems' developmental path follows an orderly progression from Type I, to Type II, and finally to Type III industrial systems, as follows:

1. Type I industrial systems represent an initial stage requiring a high throughput of energy and materials to function, and exhibit little or no resource recovery. It is a once flow-through system with rudimentary end-of-pipe pollution controls.

2. Type II industrial systems represent a transitional stage where resource recovery becomes more integral to the workings of the industrial systems, but does not satisfy its requirements for resources. Manufacturing processes and environmental processes are integrated at least partially. Whole facility planning is at least partially implemented.
3. Type III industrial systems represent the final ideal stage in which the industrial systems recycle all of the material outputs of production, although still relying on external energy inputs.

A Type III industrial ecosystem can become almost self-sustaining, requiring little input to maintain basic functions and to provide a habitat for thousands of different species. Therefore, reaching Type III as a final stage is the goal of IE [11]. Eventually communities, cities, regions, and nations will become sustainable in terms of natural resources and the environment.

According to Frosch [9]:

> "The idea of industrial ecology is that former waste materials, rather than being automatically sent for disposal, should be regarded as raw materials – useful sources of materials and energy for other processes and products. The overall idea is to consider how the industrial system might evolve in the direction of an interconnected food web, analogous to the natural system, so that waste minimization becomes a property of the industrial system even when it is not completely a property of a individual process, plant, or industry."

IE provides a foundation for sustainable industrialization, not just incremental improvement in environmental management. The objectives of IE suggest a potential for reindustrialization in economies that have lost major components of their industrial base. Specifically, the objective of industrial ecology is not merely to reduce pollution and waste as traditionally conceived, it is to reduce throughput of all kinds of materials and fuels, whether they leave a site as products, emissions, or waste.

The above objectives of IE have shown a new path for both industrial and developing countries. Central objectives of an industrial-ecology-based development strategy are making economies profoundly more efficient in resource use, less dependent upon nonrenewable resources, and less polluting. A corollary objective is repair of past environmental damage and restoration of ecosystems. Developing countries that recognize the enormous opportunity opened by this transformation can leapfrog over the errors of past industrialization. They will have more competitive and less polluting businesses [21].

1.4 APPROACH AND APPLICATIONS

The IE approach involves (a) application of systems science to industrial systems, (b) defining the system boundary to incorporate the natural world, and (c) seeking to optimize that system.

Industrial ecology is applied to the management of human activity on a sustainable basis by: (a) minimizing energy and materials usage; (b) ensuring acceptable quality of life for people; (c) minimizing the ecological impact of human activity to levels natural systems can sustain; (d) conserving and restoring ecosystem health and maintaining biodiversity; (e) maintaining the economic viability of systems for industry, trade, and commerce; (f) coordinating design over the life cycle of products and processes; and (g) enabling creation of short-term innovations with awareness of their long-term impacts.

Application of IE will improve the planning and performance of industrial systems of all sizes, and will help design local and community solutions that contribute to national and global solutions. For small industrial systems applications, IE helps companies become more competitive by improving their environmental performance and strategic planning. For medium-sized industrial systems, IE helps communities develop and maintain a sound industrial base and infrastructure, without sacrificing the quality of their environments. For large industrial systems, IE helps government agencies design policies and regulations that improve environmental protection while building business competitiveness.

Several scenarios [20] offer visions of full-blown application of IE at company, city, and developing country levels. Lists of organizations, on-line information sources, and bibliographies in the book provide access to sources of IE information.

1.5 TASKS, STEPS, AND FRAMEWORK FOR IMPLEMENTATION

Pratt and Shireman [25] propose three simple but extraordinarily powerful tasks, over and over again, for practicing industrial ecological management:

1. *Task 1, Eco-management*: Brainstorm, test, and implement ways to reduce or eliminate pollution;
2. *Task 2, Eco-auditing*: Identify specific examples of materials use, energy use, and pollution and waste reduction (any form of throughput);
3. *Task 3, Eco-accounting*: Count the money. Count how much was saved, then count how much is still being spent creating waste and pollution, and start the cycle over.

The above three tasks are essentially eco-management, eco-auditing, and activity-based eco-accounting, which are part of an inter-related ecological management framework. Pratt and Shireman [25] further suggest a way to implement the three tasks by going through a series of perhaps 14 specific steps, spiraling outward from the initial Step 1, "provide overall corporate commitment," to the final Step 14, "continue the process," which flows back into the cycle of continuous improvement:

Step 1: Provide overall corporate commitment.
Step 2: Organize the management efforts.
Step 3: Organize the audit.
Step 4: Gather background information.
Step 5: Conduct detailed assessment.
Step 6: Review and organize data.
Step 7: Identify improvement options.
Step 8: Prioritize options.
Step 9: Implement fast-track options.
Step 10: Analyze options.
Step 11: Implement best options.
Step 12: Measure results.
Step 13: Standardize improvement.
Step 14: Continue the process.

Each of the components within the "three tasks" does not necessarily fall into discrete categories. For clarity of presentation, each of the tasks is divided into steps. Table 1 shows that these steps overlap and are repeated within this systematic approach. The names of tasks and steps have been slightly modified by the current author for ease of presentation and explanation.

Table 1 Implementation Process for Applying Industrial Ecology at Corporate Level

Task 1: Eco-management		Task 2: Eco-auditing		Task 3: Eco-accounting	
Step 1	Overall corporate commitment	*Step 3*	Organize the audit	*Step 5*	Conduct detailed assessment
Step 2	Organize management efforts	*Step 4*	Gather background information	*Step 12*	Measure results
Step 7	Identify improvement options	*Step 5*	Conduct detailed assessment		
Step 8	Prioritize options	*Step 6*	Review and organize data		
Step 9	Implement fast-track options	*Step 7*	Identify improvement options		
Step 10	Analyze options	*Step 12*	Measure results		
Step 11	Implement best options				
Step 13	Standardize improvements				
Step 14	Continue the process				

As shown in Table 1, the company must initially provide the overall corporate commitment (Step 1) and organize the management efforts (Step 2) in Task 1 that will drive this implementation process forward (and around). Once the industrial ecological implementation process is initiated by the eco-management team in Task 1 (Steps 1 and 2), the eco-auditing team begins its Task 2 (Steps 3–7) with background and theory that support an industrial ecology approach, and the eco-accounting team begins its Task 3 (Step 5) to conduct detailed assessment. The eco-management team must then provide step-by-step guidance and directions in Task 1 (Steps 7–11) to identify, prioritize, implement, analyze, and again implement the best options. Subsequently, both the eco-auditing team (Task 2, Step 12) and the eco-accounting team (Task 3, Step 12) should measure the results of the implemented best options (Task 1, Step 11). The overall responsibility finally to standardize the improvements, and to continue the process until optimum results are achieved (Task 1, Steps 13, 14) will still be carried out by the eco-management team.

1.6 QUALIFICATIONS OF INDUSTRIAL ECOLOGISTS

The implementation process for applying industrial ecology at the corporate level (as shown in Table 1) may sound modest in its concept. In reality, each step in each task will face technical, economical, social, legal, and ecological complexity, and can be accomplished only by qualified industrial ecologists.

Accordingly, the most important element for industrial ecology implementation will be drawing on in-company expertise and enthusiasm as well as outside professional assistance. The qualified industrial ecologists retained for their service must have their respective knowledge in understanding the rules and regulations, assessing manufacturing processes and wastes, identifying various options, and measuring results. Because it is difficult to find a single industrial ecologist who has all the required knowledge, several experts in different areas are usually assembled together to accomplish the required IE tasks.

The team of qualified industrial ecologists assembled should have a clear sense of the possibilities and methodologies in the following professional areas specifically related to the problem:

1. Industrial or manufacturing engineering of the target industrial system;
2. Energy consumption and material balances for environmental auditing;
3. Cleaner production, materials substitution, and dematerialization;
4. Zero emission, decarbonization, waste minimization, and pollution prevention;
5. Sustainable agriculture and sustainable industry;
6. Industrial metabolism and life-cycle analyses of products;
7. Site remediation and environmental restoration;
8. Ecological and global environmental analyses;
9. Accounting and economical analyses;
10. Legal, political affairs, and IE leverage analyses.

An IE team may not be required to have all of the above expertise. For example, the expertise of site remediation may not be required if the industrial system in question is not contaminated by hazardous substances. The expertise of global environmental analyses may not be needed if the IE level is at the company level, instead of at the regional or national level.

1.7 WAYS AND MEANS FOR ANALYSIS AND DESIGN

Each task and each step outlined in Table 1 for implementation of an industrial ecology project can not be accomplished without understanding the ways and means for IE analysis and design. Indigo Development, a Center in the Sustainable Development Division of RPP International [13] has identified seven IE methods and tools for analysis and design: (a) industrial metabolism; (b) urban footprint; (c) input–output models; (d) life-cycle assessment; (e) design for environment; (f) pollution prevention; and (g) product life extension. Ausubel [2] and Wernick et al. [45] suggest that searching for leverage will be an important tool for IE implementation.

The United Nations Industrial Development Organization [39–41] and Ausubel and Sladovich [4] emphasize the importance of cleaner production, pollution prevention, waste minimization, sustainable development, zero emission, materials substitution, dematerialization, decarbonization, functional economic analysis, and IE indicators. These ways and means for analysis and design of industrial ecology are described separately herein.

1.8 SUSTAINABLE AGRICULTURE, INDUSTRY, AND ENVIRONMENT

Because IE is a branch of systems science of sustainability or a framework for designing and operating industrial systems as sustainable living systems interdependent with natural systems, understanding and achieving sustainable agriculture and industry will be the most important key to the success of sustainable environment.

An industrial ecologist may perceive the whole system required to feed planet Earth, preserve and restore its farmlands, preserve ecosystems and biodiversity, and still provide water, land, energy, and other resources for a growing population. The following is only one of many possibilities for achieving sustainable agriculture and industry: utilization of large volumes of carbon dioxide gases discharged from industrial and commercial stacks as a resource for decarbonation, pollution control, resource development, and cost saving [22,24,39–42].

Meeting the challenges involved in sustainable systems development, which can be either technical or managerial, will require interdisciplinary coordination among many technical, economic, social, political, and ecological research disciplines.

1.9 ZERO EMISSION, ZERO DISCHARGE, CLEANER PRODUCTION, WASTE MINIMIZATION, POLLUTION PREVENTION, DESIGN FOR ENVIRONMENT, MATERIAL SUBSTITUTION, DEMATERIALIZATION, AND PROCESS SUBSTITUTION

1.9.1 Terminologies and Policy Promotion

The terms of zero emission, zero discharge, cleaner production, waste minimization, pollution prevention, design for environment, material substitution, and dematerialization are all closely related, and each is self-explanatory. The US Environmental Protection Agency (USEPA), the United Nations Industrial Development Organization (UNIDO) and other national and international organizations at different periods of time have promoted each [8,19,23,30–34,39–46].

Design for environment (DFE) is a systematic approach to decision support for industrial ecologists, developed within the industrial ecology framework. Design for environment teams apply this systematic approach to all potential environmental implications of a product or process being designed: energy and materials used; manufacture and packaging; transportation; consumer use, reuse or recycling; and disposal. Design for environment tools enable consideration of these implications at every step of the production process from chemical design, process engineering, procurement practises, and end-product specification to postuse recycling or disposal. It also enables designers to consider traditional design issues of cost, quality, manufacturing process, and efficiency as part of the same decision system.

1.9.2 Zero Emission

Zero emission has been promoted by governments and the automobile industry in the context of energy systems, particularly in relation to the use of hydrogen as an energy source. Recent attention has focused on electric cars as zero-emission vehicles and the larger question of the energy and material system in which the vehicles are embedded. Classic studies about hydrogen energy may be found in a technical article by Hafele et al. [12]. The term "zero emission" is mainly used in the field of air emission control.

1.9.3 Zero Discharge

Zero discharge is aimed at total recycling of water and wastewater within an industrial system, and elimination of any discharge of toxic substances. Therefore, the term "zero discharge" is mainly used in water and wastewater treatment plants, meaning total water recycle. In rare cases, total recycling of air effluent within a plant is also called "zero discharge." Wastewater recycling is important, not only for environmental protection, but also for water conservation in water shortage areas, such as California, United States. Several successful IE case histories are presented to show the advantages of zero discharge:

Total Wastewater Recycle in Potable Water Treatment Plants

The volume of wastewater produced from a potable water treatment plant (either a conventional sedimentation filtration plant or an innovative flotation filtration plant) amounts to about 15% of a plant's total flow. Total wastewater recycle for production of potable water may save water and cost, and solve wastewater discharge problems [15,35–38].

Total Water and Fiber Recycle in Paper Mills

The use of flotation clarifiers and fiber recovery facilities in paper mills may achieve near total water and fiber recycle and, in turn, accomplish the task of zero discharge [16].

Total Water and Protein Recycle in Starch Manufacturing Plants

The use of membrane filtration and protein recovery facilities in starch manufacturing plants may achieve near total water and protein recycle and, in turn, accomplish the task of zero discharge [39–41].

Cleaner production, waste minimization, pollution prevention, designs for benign environmental impacts, material substitution, and dematerialization are all inter-related terms. Cleaner production is formally used and promoted by UNIDO (Vienna, Austria) [39–40], while waste minimization and pollution prevention are formally used and promoted by USEPA and U.S. state government agencies. Design for minimal environmental impact is very similar to cleaner production, and is mainly used in the academic field by researchers. Cleaner production emphasizes the integration of manufacturing processes and pollution control processes for the purposes of cost saving, waste minimization, pollution prevention, sustainable agriculture, sustainable industry, and sustainable environment, using the methodologies of material substitution, dematerialization, and sometimes even process substitution. Accordingly, cleaner production is a much broader term than waste minimization, pollution prevention, sustainability, material substitution, process substitution, and so on, and is similar to design for benign environmental impact. Furthermore, cleaner production implementation in an industrial system always saves money for the plant in the long run. Considering that wastes are resources to be recovered is the key for the success of an IE project using a cleaner production technology.

1.10 CASE HISTORIES OF SUCCESSFUL HAZARDOUS WASTE MANAGEMENT THROUGH INDUSTRIAL ECOLOGY IMPLEMENTATION

Several successful IE case histories are presented here to demonstrate the advantages of cleaner production for hazardous wastes management [40].

1.10.1 New Galvanizing Steel Technology Used at Delot Process SA Steel Factory, Paris, France

Galvanizing is an antirust treatment for steel. The traditional technique consisted of chemically pretreating the steel surface, then immersing it in long baths of molten zinc at 450°C. The old process involved large quantities of expensive materials, and highly polluting hazardous wastes. The cleaner production technologies include: (a) induction heating to melt the zinc, (b) electromagnetic field to control the molten zinc distribution, and (c) modern computer control of the process. The advantages include total suppression of conventional plating waste, smaller inventory of zinc, better process control of the quality and thickness of the zinc coating, reduced labor requirements, reduced maintenance, and safer working conditions. With the cleaner production technologies in place, capital cost is reduced by two-thirds compared to the traditional dip-coating process. The payback period was three years when replacing existing plant facilities.

1.10.2 Reduction of Hazardous Sulfide in Effluent from Sulfur Black Dyeing at Century Textiles, Bombay, India

Sulfur dyes are important dyes yielding a range of deep colors, but they cause a serious pollution problem due to the traditional reducing agent used with them. The old dyeing process involved four steps: (a) a water soluble dye was dissolved in an alkaline solution of caustic soda or sodium carbonate; (b) the dye was then reduced to the affinity form; (c) the fabric was dyed; and (d) the dye was converted back into the insoluble form by an oxidation process, thus preventing washing out of the dye from the fabric. The cleaner production technology involves the use of 65 parts of starch chemical Hydrol™ plus 25 parts of caustic soda to replace 100 parts of original sodium sulfide. The advantages include: reduction of sulfide in the effluent, improved settling characteristics in the secondary settling tank of the activated sludge plant, less corrosion in the treatment plant, and elimination of the foul smell of sulfide in the work place. The substitute chemical used was essentially a waste stream from the maize starch industry, which saved them an estimated US$12,000 in capital expenses with running costs at about US$1800 per year (1995 costs).

1.10.3 Replacing Toxic Solvent-Based Adhesives With Nontoxic Water-Based Adhesives at Blueminster Packaging Plant, Kent, UK

When solvent-based adhesives were used at Blueminster, UK, the components of the adhesive, normally a polymer and a resin (capable of becoming tacky) were dissolved in a suitable organic solvent. The adhesive film was obtained by laying down the solution and then removing the solvent by evaporation. In many adhesives, the solvent was a volatile organic compound (VOC) that evaporated to the atmosphere, thus contributing to atmospheric pollution. The cleaner production process here involves the use of water-based adhesives to replace the solvent-based adhesives. In comparison with the solvent-based adhesives, the water-based adhesives are nontoxic, nonpolluting, nonexplosive, nonhazardous, require only 20–33% of the drying energy, require no special solvent recovery systems nor explosion-proof process equipment, and are particularly suitable for food packaging. The economic benefits are derived mainly from the lack of use of solvents and can amount to significant cost savings on equipment, raw materials, safety precautions, and overheads.

1.10.4 Recovery and Recycling of Toxic Chrome at Germanakos SA Tannery Near Athens, Greece

Tanning is a chemical process that converts hides and skins into a stable material. Tanning agents are used to produce leather of different qualities and properties. Trivalent chromium is the major tanning agent, because it produces modern, thin, light leather suitable for shoe uppers, clothing, and upholstery. However, the residual chromium in the plant effluent is extremely toxic, and its effluent concentration is limited to 2 mg/L. A cleaner production technology has been developed to recover and reuse the trivalent chromium from the spent tannery liquors for both cost saving and pollution control. Tanning of hides is carried out with chromium sulfate at pH 3.5–4.0. After tanning, the solution is discharged by gravity to a collection pit. In the recovery process, the liquor is sieved during this transfer to remove particles and fibers originating from the hides. The liquor is then pumped to a treatment tank where magnesium oxide is added, with stirring, until the pH reaches at least 8. The stirrer is switched off and the chromium precipitates as a compact sludge of chromium hydroxide. After settling, the clear liquid is decanted off. The remaining sludge is dissolved by adding concentrated sulfuric acid until a pH of 2.5 is reached. The liquor now contains chromium sulfate and is pumped back to a storage tank for reuse. In the conventional chrome tanning processes, 20–40% of the chrome used was discharged into wastewaters as hazardous substances. In the new cleaner production process, 95–98% of the spent trivalent chromium can be recycled for reuse. The required capital investment for the Germanakos SA plant was US$40,000. Annual saving in tanning agents and pollution control was $73,750. The annual operating cost of the cleaner production process was $30,200. The total net annual savings is $43,550. The payback period for the capital investment ($40,000) was only 11 months.

1.10.5 Recovery of Toxic Copper from Printed Circuit Board Etchant for Reuse at Praegitzer Industries, Inc., Dallas, Oregon, United States

In the manufacture of printed circuit boards, the unwanted copper is etched away by acid solutions as cupric chloride. As the copper dissolves, the effectiveness of the solution falls and it must be regenerated, otherwise it becomes a hazardous waste. The traditional way of doing this was to oxidize the copper ion produced with acidified hydrogen peroxide. During the process the volume of solution increased steadily and the copper in the surplus liquor was precipitated as copper oxide and usually landfilled. The cleaner production process technology uses an electrolytic divided cell, simultaneously regenerating the etching solution and recovering the unwanted copper. A special membrane allows hydrogen and chloride ions through, but not the copper. The copper is transferred via a bleed valve and recovered at the cathode as pure flakes of copper. The advantages of this cleaner production process are: improvement of the quality of the circuit boards, elimination of the disposal costs for the hazardous copper effluent, maintenance of the etching solution at optimum composition, recovery of pure copper for reuse, and zero discharge of hazardous effluent. The annual cost saving in materials and disposal was US$155,000. The capital investment cost was $220,000. So the payback period for installation of this cleaner production technology was only 18 months.

1.10.6 Recycling of Hazardous Wastes as Waste-Derived Fuels at Southdown, Inc., Houston, Texas, United States

Southdown, Inc., engages in the cement, ready-mixed concrete, concrete products, construction aggregates, and hazardous waste management industries throughout the United States. According to Southdown, they are making a significant contribution to both the environment and energy conservation through the utilization of waste-derived fuels as a supplemental fuel source. Cement kiln energy recovery is an ideal process for managing certain organic hazardous wastes. The burning of organic hazardous wastes as supplemental fuel in the cement and other industries is their engineering approach. By substituting only 15% of its fossil fuel needs with solid hazardous waste fuel, a modern dry-process cement plant with an annual production capacity of 650,000 tons of clinker can save the energy equivalent of 50,000 barrels of oil (or 12,500 tons of coal) a year. Southdown typically replaces 10–20% of the fossil fuels it needs to make cement with hazardous waste fuels.

Of course, by using hazardous waste fuels, the nation's hazardous waste (including infections waste) problem is at least partially solved with an economic advantage.

1.10.7 Utilization and Reduction of Carbon Dioxide Emissions at Industrial Plants

Decarbonization has been extensively studied by Dr L. K. Wang and his associates at the Lenox Institute of Water Technology, MA, United States, and has been concluded to be technically and economically feasible, in particular when the carbon dioxide gases from industrial stacks are collected for in-plant reuse as chemicals for tanneries, dairies, water treatment plants, and municipal wastewater treatment plants [22,23,42]. Greenhouse gases, such as carbon dioxide, methane, and so on, have caused global warming over the last 50 years. Average temperatures across the world could climb between 1.4 and 5.8°C over the coming century. Carbon dioxide emissions from industry and automobiles are the major causes of global warming. According to the UN Environment Program Report released in February 2001, the long-term effects may cost the world about 304 billion US dollars a year in the future. This is due to the following projected losses: (a) human life loss and property damages as a result of more frequent tropical cyclones; (b) land loss as a result of rising sea levels; (c) damages to fishing stocks, agriculture, and water supplies; and (d) disappearance of many endangered species. Technologically, carbon dioxide is a gas that can easily be removed from industrial stacks by a scrubbing process using any alkaline substances. However, the technology for carbon dioxide removal is not considered to be cost-effective. Only reuse is the solution. About 20% of organic pollutants in a tannery wastewater are dissolved proteins that can be recovered using the tannery's own stack gas (containing mainly carbon dioxide). Similarly, 78% of dissolved proteins in a dairy factory can be recovered by bubbling its stack gas (containing mainly carbon dioxide) through its waste stream. The recovered proteins from both tanneries and dairies can be reused as animal feeds. In water softening plants using chemical precipitation processes, the stack gas can be reused as precipitation agent for hardness removal. In municipal wastewater treatment plants, the stack gas containing carbon dioxide can be reused as neutralization and warming agents. Because a large volume of carbon dioxide gases can be immediately reused as chemicals in various in-plant applications, the plants producing carbon dioxide gas actually may save chemical costs, produce valuable byproducts, conserve heat energy, and reduce global warming problem [47].

By reviewing these case histories, one will realize that materials substitution is an important tool for cleaner production and, in turn, for industrial ecology. Furthermore, materials substitution is considered a principal factor in the theory of dematerialization. The theory asserts that as a nation becomes more affluent, the mass of materials required to satisfy new or growing economic functions diminishes over time. The complementary concept of decarbonization, or the diminishing mass of carbon released per unit of energy production over time, is both more readily examined and has been amply studied by many scientists. Dematerialization is advantageous only if using fewer resources accompanies, or at least leaves unchanged, lifetime waste in processing, and wastes in production [43].

It is hoped that through industrial ecology investigations, strategies may be developed to facilitate more efficient use of material and energy resources and to reduce the release of hazardous as well as nonhazardous wastes to our precious environment. Hopefully, we will be able to balance industrial systems and the ecosystem, so our agriculture and industry can be sustained for very long periods of time, even indefinitely, without significant depletion or environmental harm. Integrating industrial ecology within our economy will bring significant benefits to everyone.

REFERENCES

1. Allen, D.T.; Butner, R.S. Industrial ecology: a chemical engineering challenge. Chem. Engng. Prog. **2002**, *98* (11), 40–45.
2. Ausubel, J.H. The virtual ecology of industry. J. Ind. Ecol. **1997**, *1* (1), 10–11.
3. Ausubel, J.H. Industrial ecology: a coming of age story. Resources **1998**, *130* (14) 28–31.
4. Ausubel, J.H.; Sladovich, H.E. *Technology and Environment*; National Academy of Science: Washington, DC, 1989.
5. Ayres, R.U.; Ayres, L.W. *Industrial Ecology: Towards Closing the Materials Cycle*; Edward Elgar Publishing: Cheltenham, UK, 1996.
6. Cox, B. High-mileage precept still just a high-priced concept. Times Union, Automotive Weekly, February 22, 2001; 16 pp.
7. AIChe. Society merges technology and ecology. Chem. Engng. Prog. **2001**, *97* (4), 13–14.
8. Evers, D.P. Facility pollution prevention. In *Industrial Pollution Prevention Handbook*; Freeman, H.M., Eds.; McGraw-Hill: New York, 1995; 155–179.
9. Frosch, R.A. Toward the end of waste: reflections on a new ecology for industry. Daedalus **1996**, *125* (3), 199–212.
10. Frosch, R.A.; Gallopoulos, N.E. Strategies for manufacturing. Scientific American **1989**, 144–152.
11. Graedel, T.E.; Allenby, B.R.; Comrie, P.R. Matrix approaches to abridged life cycle assessment. Environ. Sci. Technol. **1995**, *29*, 134A–139A.
12. Hefele, W.; Barner, H.; Messner, S.; Strubegger, M.; Anderer, J. Novel integrated energy systems: The case of zero emissions. In *Sustainable Development of the Biosphere*; Clark, W.C., Munns, R.E., Eds.; Cambridge University Press: Cambridge, UK, 171–193.
13. Indigo Development. *Creating Systems Solution for Sustainable Development Through Industrial Ecology*; RPP International: Oakland, California, elowe@indigodev.com, June 5, 2000.
14. Klimisch, R.L. *Designing the Modern Automobile for Recycling. Greening Industrial Ecosystems*; Allenby, B.R., Richards, D., Eds.; National Academy Press: Washington, DC.
15. Krofta, M.; Wang, L.K. *Development of Innovative Floatation Processes for Water Treatment and Wastewater Reclamation*, National Water Supply Improvement Association Conference, San Diego, August 1988, 42 pp.
16. Krofta, M.; Wang, L.K. Total closing of paper mills with reclamation and deinking installations. Proceedings of the 43rd Industrial Waste Conference, Purdue University: W. Lafayette, IN, 1989; 673 pp.
17. Lovins, A.B.; Lovins, L.H. *Supercars: The Coming Light-Vehicle Revolution*, Technical report, Rocky Mountain Institute: Snowmass, CO, 1993.
18. Lovins, A.B.; Lovins, L.H. Reinventing the wheels. *Atlantic Monthly* **1995**, *January*.
19. Lowe, E.; Evans, L. Industrial ecology and industrial ecosystems. J. Cleaner Prod. **1995**, *3*, 1–2.
20. Lowe, E.A.; Warren, J.L.; Moran, S.R. *Discovering Industrial Ecology: An Executive Briefing and Sourcebook*; Battelle Press: Columbus, OH, 1997. ISBN 1-57477-034-9.
21. Lowe, E.A. *Creating Systems Solutions for Sustainable Development Through Industrial Ecology: Thoughts on an Industrial Ecology-Based Industrialization Strategy*, Indigo Development Technical Report, RPP International: 26 Blachford Court, Oakland, California, USA, 2001.
22. Nagghappan, L. *Leather Tanning Effluent Treatment*; Lenox Institute of Water Technology: Lenox, MA. Master Thesis (Wang, L.K., Krofta, M., advisors), 2000; 167 pp.
23. NYSDEC. *New York State Waste Reduction Guidance Manual*; NYS Department of Environmental Conservation: Albany, NY, 1989.
24. Ohrt, J.A. *Physicochemical Pretreatment of a Synthetic Industrial Dairy Waste*. Lenox Institute of Water Technology: Lenox, MA. Master Thesis (Wang, L.K.; Aulenbach, D.B., advisors), 2001; 62 pp.
25. Pratt, W.B.; Shireman, W.K. *Industrial Ecology: A How-to-Manual: The Only 3 Things Business Needs to Do to Save the Earth*, Technical Manual. Global Futures Foundation: Sacramento, CA, 1996, www.globalff.org.
26. Renner, M. *Rethinking the Role of the Automobile*. Worldwatch Institute: Worldwatch Paper 84: Washington, DC, 1988.

27. Rittenhouse, D.G. Piecing together a sustainable development strategy. Chem. Engng. Prog. **2003**, *99* (3), 32–38.
28. Swan, C. *Suntrain Inc. Business Plan*. Suntrain Inc.: San Francisco, CA, 1998.
29. Tibbs, H. Industrial ecology: an environmental agenda for industry. Whole Earth Rev. **1992**, *Winter*, 4–19.
30. US Congress. *From Pollution to Prevention: A Progress Report on Waste Reduction*. US Congress, Office of Technology Assessment, US Government Printing Office: Washington, DC, 1992; OTA-ITE-347.
31. USEPA. *Waste Minimization Issues and Options*. US Environmental Protection Agency: Washington, DC, 1986; 530-SW-86-04.
32. USEPA. *Waste Minimization Benefits Manual, Phase I*. US Environmental Protection Agency: Washington, DC, 1988.
33. USEPA. *Pollution Prevention Benefits Manual, Phase II*. US Environmental Protection Agency: Washington, DC, 1989.
34. USEPA. *Facility Pollution Prevention Guide*. US Environmental Protection Agency, Office of Solid Waste: Washington, DC, 1992; EPA/600/R-92/083.
35. Wang, L.K. Recycling and reuse of filter backwash water containing alum sludge. Water Sewage Works **1972**, *119* (5), 123–125.
36. Wang, L.K. Continuous pilot plant study of direct recycling of filter backwash water. J. Am. Water Works Assoc. **1973**, *65* (5), 355–358.
37. Wang, L.K. Design and specifications of Pittsfield water treatment system consisting of air flotation and sand filtration. Water Treatment **1991**, *6*, 127–146.
38. Wang, L.K.; Wang, M.H.S.; Kolodzig, P. Innovative and cost-effective Lenox water treatment plant. Water Treatment **1992**, *7*, 387–406.
39. Wang, L.K.; Cheryan, M. *Application of Membrane Technology in Food Industry for Cleaner Production*. The Second International Conference on Waste Minimization and Cleaner Production. United Nations Industrial Development Organization: Vienna, Austria, 1995; Technical Report No. DTT-8-6-95, 42 pp.
40. Wang, L.K.; Krouzek, J.V.; Kounitson, U. *Case Studies of Cleaner Production and Site Remediation*. United Nations Industrial Development Organization: Vienna, Austria, 1995; Training Manual No. DTT-5-4-95, 136 pp.
41. Wang, L.K.; Wang, M.H.S.; Wang, P. *Management of Hazardous Substances at Industrial Sites*. United Nations Industrial Development Organization: Vienna, Austria, 1995; Technical Report No. DTT-4-4-95, 105 pp.
42. Wang, L.K.; Lee, S.L. *Utilization and Reduction of Carbon Dioxide Emissions: An Industrial Ecology Approach*. The 2001 Annual Conference of Chinese American Academic and Professional Society (CAAPS), St. Johns University, New York, NY, USA, April 25, 2001.
43. Wernick, I.K.; Herman, R.; Govind, S.; Ausubel, J.H. Materialization and dematerialization measures and trends. Daedalus **1993**, *125* (3), 171–198.
44. Wernick, I.K.; Ausubel, J.H. *Industrial Ecology: Some Directions for Research*; The Rockefeller University: New York, 1997. ISBN 0-9646419-0-7.
45. Wernick, I.K.; Waggoner, P.E.; Ausubel, J.H. Searching for leverage to conserve forests: the industrial ecology of wood products in the US. Journal of Industrial Ecology **1997**, *1* (3), 125–145.
46. Wernick, I.K.; Ausubel, J.H. National Material Metrics for Industrial Ecology. In *Measures of Environmental Performance and Ecosystem Condition*; Schuize, P., Ed.; National Academy Press: Washington, DC, 1999; 157–174.
47. Wang, L.K.; Pereira, N.C.; Hung, Y. *Air Polution Control Engineering*; Human Press, Totowa, NJ, 2004.

2
Bioassay of Industrial Waste Pollutants

Svetlana Yu. Selivanovskaya and Venera Z. Latypova
Kazan State University, Kazan, Russia

Nadezda Yu. Stepanova
Kazan Technical University, Kazan, Russia

Yung-Tse Hung
Cleveland State University, Cleveland, Ohio, U.S.A.

2.1 INTRODUCTION

Persistent contaminants in the environment affect human health and ecosystems. It is important to assess the risks of these pollutants for environmental policy. Ecological risk assessment (ERA) is a tool to estimate adverse effects on the environment from chemical or physical stressors. It is anticipated that ERA will be the main tool used by the U.S. Department of Energy (US DOE) to accomplish waste management [1]. Toxicity bioassays are the important line of evidence in an ERA. Recent environmental legislation and increased awareness of the risk of soil and water pollution have stimulated a demand for sensitive and rapid bioassays that use indigenous and ecologically relevant organisms to detect the early stages of pollution and monitor subsequent ecosystem change.

Aquatic ecotoxicology has rapidly matured into a practical discipline since its official beginnings in the 1970s [2–4]. Integrated biological/chemical ecotoxicological strategies and assessment schemes have been generally favored since the 1980s to better comprehend the acute and chronic insults that chemical agents can have on biological integrity [5–8]. However, the experience gained with the bioassay of solid or slimelike wastes is as yet inadequate.

At present the risk assessment of contaminated objects is mainly based on the chemical analyses of a priority list of toxic substances. This analytical approach does not allow for mixture toxicity, nor does it take into account the bioavailability of the pollutants present. In this respect, bioassays provide an alternative because they constitute a measure for environmentally relevant toxicity, that is, the effects of bioavailable fraction of an interacting set of pollutants in a complex environmental matrix [9–12].

The use of bioasssay in the control strategies for chemical pollution has several advantages over chemical monitoring. First, these methods measure effects in which the bioavailability of the compounds of interest is integrated with the concentration of the compounds and their intrinsic toxicity. Secondly, most biological measurements form the only way of integrating the effects on a large number of individual and interactive processes. Biomonitoring methods are often cheaper, more precise, and more sensitive than chemical analysis in detecting adverse conditions in the environment. This is due to the fact that the biological response is very integrative and accumulative in nature, especially at the higher levels of biological organization. This may lead to a reduction in the number of measurements both in space and time [12].

A disadvantage of biological effect measurements is that sometimes it is very difficult to relate the observed effects to specific aspects of pollution. In view of the present chemical-oriented pollution abatement policies and to reveal chemical specific problems, it is clear that biological effect analysis will never totally replace chemical analysis. However, in some situations the number of standard chemical analyses can be reduced, by allowing bioeffects to trigger chemical analysis (integrated monitoring), thus buying time for more elaborate analytical procedures [12].

2.2 GENERAL CONSIDERATIONS

According to USEPA, the key aspect of the ERA is the problem formulation phase. This phase is characterized by USEPA as the identification of ecosystem components at risk and specification of the endpoints used to assess and measure that risk [13]. Assessment endpoints are an expression of the valued resources to be considered in an ERA, whereas measurement endpoints are the actual measures of data used to evaluate the assessment endpoint.

Toxicity tests can be divided according to their exposure time (acute or chronic), mode of effect (death, growth, reproduction), or the effective response (lethal or sublethal) (Fig. 1) [11]. Other approaches to the classifications of toxicity tests can include acute toxicity, chronic toxicity, and specific toxicity (carcinogenecity, genotoxicity, reproduction, immunotoxicity, neurotoxicity, specific exposure to skin and other organs). For instance, genotoxicity reveals the risks for interference with the ecological gene pool leading to increased mutagenecity and/or carcinogenecity in biota and man. Unlike normal toxicity, the incidence of genotoxic effect is thought to be only partially related to concentration (one-hit model).

A toxicity test may measure either acute or chronic toxicity. Acute toxicity is indicative for acute effects possibly occurring in the immediate vicinity of the discharge. An acute toxicity test

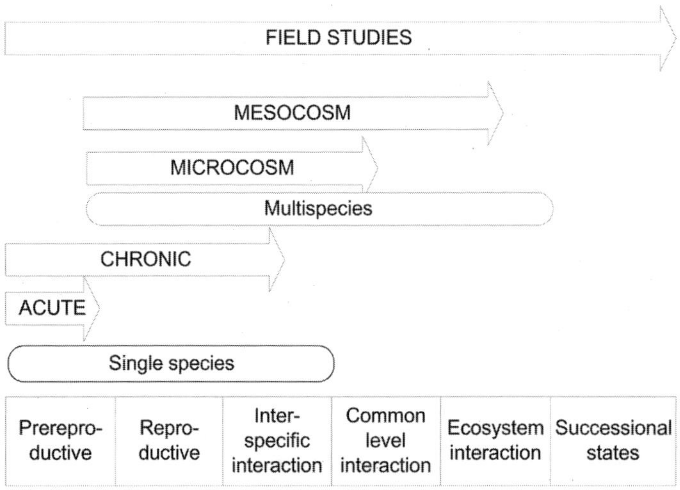

Figure 1 Classification of toxicity tests in environmental toxicology.

is defined as a test of 96-hours or less in duration, in which lethality is the measured endpoint. Acute responses are expressed as LC_{50} (lethal concentration) or EC_{50} (effective concentration) values, which means that half of the organisms die or a specific change occurs in their normal behavior. Sometimes in toxicity bioassays the NOEC (no observed effect concentration) can be used as the highest toxicant concentration that does not show a statistically significant difference with controls. The EC_{10} can replace the NOEC. This is a commonly used effect parameter in microbial tests [14–17]. At the EC_{10} concentration there is a 10% inhibition, which might not be very different from the NOEC concentration, but the EC_{10} does not depend on the accuracy of the test.

Acute toxicity covers only a relatively short period of the life-cycle of the test organisms. Chronic toxicity tests are used to assess long-lasting effects that do not result in death. Chronic toxicity reflects the extent of possible sublethal ecological effects. The chronic test is defined as a long-term test in which sublethal effects, such as fertilization, growth, and reproduction are usually measured in addition to lethality. Traditionally, chronic tests are full life-cycle tests or a shortened test of about 30 days known as an "early-stage test." However, the duration of most EPA tests have been shortened to 7 days by focusing on the most sensitive early life-cycle stages. The chronic tests produce the highest concentration percentage tested that caused no significant adverse impact on the most sensitive of the criteria for that test (NOEC) as the result. Alternative results are the lowest concentration tested that causes a significant effect (lowest observed effect concentration; LOEC), or the effluent concentration that would produce an observed effect in a certain percentage of test organisms (e.g., EC_{10} or EC_{50}). The advantage of using the LC or EC over the NOEC and LOEC values, is that the coefficient of variation (CV) can be calculated. In some case, since toxicity involves a relationship with the effect concentration (test result; the lower the EC, the higher the toxicity), all test results are converted into toxic units (TU). The number of toxic units in an effluent is defined as 100 divided by the EC measured (expressed as a dilution percentage). Two distinct types of TUs are recognized by the EPA, depending on the types of tests involved (acute: $TU_a = 100/LC_{50}$; chronic $TU_c = 100/NOEC$). Acute and chronic TUs make it easy to quantify the toxicity of an effluent, and to specify toxicity-based effluent quality criteria.

However, the effect of a harmful compound should be studied with respect to the community level, not only for the organism tested. Tests with several species are realized in microcosm and mesocosm studies. Mesocosms are larger with respect to both the species number and the species diversity and are often performed outdoors and under natural conditions.

Choice of method is the most important phase if reliable data are to be obtained successfully. A good toxicity test should measure the right parameters and respond to the environmental requirements. When selecting from among available test organisms, the investigator should choose species that are relevant to the overall assessment endpoints, representative of functional roles played by resident organisms, and sensitive to contaminants. In addition, the test should be fast, simple, and repetitive [1,11,18]. The selection of ecotoxicological test methods also depends on the intended use of the waste and the entities to be protected. Usually a single test cannot be used to detect all biological effects, and several biotests should therefore be used to reveal different responses. The ecological relevance of the single species tests has been criticized, and the limits associated with these tests representing only one trophic level have to be acknowledged.

Biological toxicity tests are widely used for evaluating the toxicants contained in the waste. Most toxicity bioassays have been developed for liquid waste. Applications of bioassays in wastewater treatment plants fall into four categories [19]. The first category involves the use of bioassays to monitor the toxicity of wastewaters at various points in the collection system, the major goal being the protection of biological treatment processes from toxicant action.

These screening tests should be useful for pinpointing the source of toxicants entering the wastewater treatment plant. The second category involves the use of these toxicity assays in process control to evaluate pretreatment options for detoxifying incoming industrial wastes. The third category concerns the application of short-term microbial and enzymatic assays to detect inhibition of biological processes used in the treatment of wastewaters and sludges. The last category deals with the use of these rapid assays in toxicity reduction evaluation (TRE) to characterize the problem toxic chemicals. In addition to the abovementioned categories, we could point out another one: whole effluent testing (WET) in accordance with International (National) Environmental Policy.

Ecotoxicological testing of the pollutants in solid wastes should be considered in the following cases: supplementary risk assessment of contaminated waste; assessment of the extractability of contaminants with biological effects in cases where the waste can affect the groundwater; ecotoxicological assessment of the waste intended for future utilization as soil fertilizer, conditioner, amendment (for example, compost from organic fraction of municipal solid waste, sewage sludge, etc.); control of the progress in biological waste treatment.

All the tests used for estimation of solid waste toxicity can be divided into two groups: tests with water extracts (elutriate toxicity tests) and "contact" toxicity tests. The majority of the assays (e.g., with bacteria, algae, Daphnia) for testing toxicity have been performed on water extract. The water path plays a dominant role in risk assessment. Water may mobilize contaminants, and water-soluble components of waste contaminants have a potentially severe effect on microorganisms and plants, as well as fauna. Owing to their low bioavailability, adsorbed or bound species of residual contaminants in waste represent only a low risk potential. However, mobilized substances may be modified and diluted along the water path. Therefore investigations of water extracts may serve as early indicators [9]. Meanwhile, owing to the different solubility of each contaminant in the water, water extracts represent only a part of contamination. Water elutriation could underestimate the types and concentrations of bioavailable organic contaminants present [20,21]. Evaluation of results requiring sample extraction appears extremely difficult. The evaluation of toxicity with extracts sometimes ignores the interactions that may occur in contacts with substances in a solid phase. Therefore "contact" tests involve the use of organisms in contact with the contaminated solids. Such tests have been standardized and used for soils, for example, using higher plants [9,22,23]. During the past few years some applications of bacterial contact assays have been suggested [17,21,24–27]. We also present the bioassays that have been used for estimation of toxicity of liquid and solid wastes.

2.3 MICROBIAL TESTS

Microbial toxicity tests are known to be fast, simple, and inexpensive. These properties of the tests have resulted in their ever-increasing use in environmental control, assessment of pollutants in waste, and so on. Toxicity test methods based on the reaction of microbes are useful in toxicity. In particular they can be a very valuable tool for the toxicity classification of samples from the same origin. Microbial tests can be performed using a pure culture of well-defined single species or a mixture of microbes. The variables measured in toxicity tests may be lethality, growth rate, change in species diversity, decrease in degradation activity, and energy metabolism or activity of specific enzymes. The results are generally expressed as the dose–response concentration and the EC_{50} or EC_{10} value [11,15,17,28,29].

2.3.1 Tests Based on Bioluminescence

One of the commonly used tests is the bioluminescence-measuring test. It is based on the change of light emission by *Vibrio fischeri* (*Photobacterium phosphoreum*) when exposed to toxic chemicals. The bioluminescence is directly linked to the vitality and metabolic state of the cells, therefore a toxic substance causing changes in the cellular state can lead to a rapid reduction of bioluminescence. Thus a decrease in the light emission is the response to serious damage to metabolism in the bacterial cells. This test is a fast and reliable preliminary toxicity test and is comparable with other toxicity tests [11,29–31]. The procedure has been developed for the investigation of water, for example, wastewater, but can be applied without problems to the investigations of soil and waste extracts. Toxicity extracts can be determined using standard test methods such as the BioTox or Microtox methods [32]. The test criterion is the inhibition of light emission. The result is expressed as the G_L value (or lowest inhibitory dilution LID value). This is the lowest value for dilution factor of the extract which exhibits less than 20% inhibition of light emission under test conditions. In the case of individual toxicants the result is presented as EC_{50} or EC_{20}. This test is probably the most popular commercial test for assessing toxicity in wastewater treatment plants [19,33] and whole effluence testing. However, an expensive luminometer is required for the scoring of results. One of the reasons for the widespread application of this assay is the (commercial) availability of the bacteria in freeze-dried form, which eliminates the need for culturing of the test organisms [34–37].

A "direct contact test" has been developed for solid samples. A solid-phase assay eliminates the need for soil extracts and utilizes whole sediments and soils. In the current procedure the solid sample is suspended in 2% NaCl. Dilutions of the stock suspension are measured to determine the EC_{50} and EC_{10} at 5 and 15 minute contact times. For this the homogenized sample and photobacterial suspension mixture are incubated. The suspended solid material is then centrifuged out and light emission of the supernatant determined [24–26,32].

The bioluminescent direct contact flash test has been proposed as a modification of the direct contact luminescent bacterial test [24,38]. This method was developed for measuring the toxicity of solid and color samples, and involves kinetic measurements of luminescence started at the same time that the *V. fischeri* suspension is added to the sample. The luminiscence signal is measured 20 times per second during the 30 second exposure period.

2.3.2 Tests Based on Enzyme Activity

Enzyme activity tests can be used to describe the functional effects of toxic compounds on microbial populations. Many enzymes are used for toxicity estimation. The enzymes used to assess the toxicity of solid-associated contaminants (soils, composts, wastes) are phosphatase, urease, oxidoreductase, dehydrogenase, peroxidase, cellulase, protease, amidase, etc. Determining dehydrogenase activity is the most common method used in enzyme toxicity tests [11,29]. The method measures a broad oxidizing spectrum and does not necessarily correlate with the number of microbes, production of carbon dioxide, or oxygen demand. In ecological studies, correlations have been determined between dehydrogenase activity and the concentration of harmful compounds. Substrates for dehydrogenase activity are triphenil tetrazoliumchloride (TTC), nitroblue tetrazolium (NBT), 2-(*p*-iodophenyl)-3-(*p*-nitrophenyl)-5-phenyl tetrazoliumchloride (INT), and resasurine [21,29].

Toxi-Chromotest^{TM} is a commercial toxicity assay that is based on the assessment of the inhibition of β-galactosidase activity, measured using a chromogenic substrate and a colorimeter. A mutant strain of *Escherichia coli* is revitalized from a lyophilized state prior to the test [39]. The principle of the MetSoil^{TM} test is similar to that of the Toxi-Chromotest^{TM}.

The bacterial mutant is mainly sensitive to metals and should therefore be used in conjunction with another bacterial test. This microbiotest is commercially available and is designed specifically for testing soils, sediments, and sludges. Semiquantitative results are obtained after three hours [40].

The MetPAD™ test kit (Group 206 Technologies, Gainesville, Florida) has been developed for the detection of heavy metal toxicity. It has been used to determine the toxicity of sewage water and sludge, sediments, and soil [41]. The test is based on the inhibition of β-galactosidase activity in an *Escherichia coli* mutant strain. Performance of the test does not require expensive equipment and it is therefore easily applied as a field test.

The MetPLATE™ test (Group 206 Technologies, Gainesville, Florida) is a fast β-galactosidase activity microtiter plate test [40]. The test is specific for heavy metal toxicity. MetPLATE is in a 96-well microtitration plate format and is suitable for determination of toxicity characteristics such as median inhibitory concentrations. MetPLATE is based on the activity of β-galactosidase from a mutant strain of *E. coli* and uses chlorphenol red galactopyranoside as enzyme substrate. The test is suitable for sewage water as well as for sewage sludge, sediments, and soil. The MetPLATE test is more sensitive to heavy metals than the Microtox™ test, which is based on bioluminescence inhibition. However, this test does not react-sensitively to organic pollutants. The MetPAD and the MetPLATE tests are available in kit form.

The ECHA (Cardiff, England) Biocide Monitor™ is a qualitative test developed for environmental samples and is based on measurement of dehydrogenase activity [41,42]. This test is performed with a small plastic strip carrying an absorbent pad impregnated with a sensitive microorganism, nutrients, and an indicator of metabolic activity and growth. Solid samples are tested directly without extraction. Semiquantitative results are evaluated after 5–24 hours with this assay, which is available as a commercial kit.

A toxicity testing procedure using the inhibition of dehydrogenase enzyme activity of *Bacillus cereus* as test parameter has been developed [21]. This microbial assay includes direct contact of bacteria with solids over 2 hours and the following measurement of dehydrogenase enzyme activity on the base of resazurine reduction. It is the authors' opinion that this method can integrate the real situation in a more complex system much better than extracts. There are numerous results from different solid phases assayed with *B. cereus*. Experiments were conducted with several contaminants, which show differences in environmental behavior: Tenside and heavy metals (high adsorption, good solubility in water), para-nitrophenol (low adsorption, good solubility in water), polycyclic aromatic hydrocarbons (high adsorption, low solubility in water). For most of the substances, the contact assay shows higher sensitivity than elutriate testing, that is, the EC_{50} is lower (Table 1). Studies with soil samples spiked with organic compounds and copper indicate the higher sensitivity of solid-phase bioassay compared to water extract testing [17]. A comparison of the sensitivity of the *B. cereus* contact test and the *Photobacterium phosphoreum* solid-phase test demonstrates that the *B. cereus* test is more sensitive for copper. The test is the scientific tool to elucidate the importance of exposure routes for compounds in soils and solid wastes. However, the authors note that the problems in predicting ecological effects of contaminants (e.g., soil contaminants) exist.

Toxi-ChromoPad™ (EBPI, Ontario, Canada) is a simple method for evaluation the toxicity of solid particles [25,26,32,39]. The test is based on the inhibition of the synthesis of ™β-galactosidase in *E. coli* after exposure to pollutants. The method has been used to measure acute toxicity of sediment and soil and other solid samples. The test bacterial suspension is mixed with homogenized samples and incubated for 2 hours. A drop of the test solution is pipetted onto a fiberglass filter containing an adsorbed substrate. A color reaction indicates the synthesis of enzyme, while a colorless reaction indicates toxicity. It has previously been shown

Table 1 Comparison of the Results of *Bacillus cereus* Contact Assay and Elutriate Toxicity for Some Spiked Soils

Substance	EC_{50} for contact assay	EC_{50} for elutriate assay
Benzalkonium-chloride	500 mg/kg	Up to 2000 mg/kg no effect
Alkylphenolpolyethylene-glycolether	3700 mg/kg (EC_{30})	Up to 4200 mg/kg no effect
Sodium alkylbenzenesulfonate	130 mg/kg	450 mg/kg
p-Nitrophenol	250/750/1000 mg/kg: 40.7/86.3/95.0% inhibition	250/750/1000 mg/kg: 13.2/65.0/82.3% inhibition
bis-tri-*n*-butyltinoxide	250/500 mg/kg: 94.2/95.1% inhibition	250/500 mg/kg: 34.0/80.5% inhibition
Naphthol	450 mg/kg	1000 mg/kg
Catechol	20 mg/kg	400 mg/kg
Lubricant oil	1.15 Gew%	3.40 Gew%
Copper	200 mg/kg	Up to 500 mg/kg no effect

that inducible enzyme metabolism can be considered a sensitive indicator for detecting the effects of harmful compounds [43]. Moreover Dutton *et al.* [44] found that β-galactosidase *de novo* biosynthesis in *E. coli* was a more sensitive reaction to harmful compounds than enzymatic activity.

2.3.3 Tests Based on Growth Inhibition

Growth inhibition tests are available for determination of the toxicity of harmful compounds. *Pseudomonas putida* is a common heterotrophic bacteria in soil and water and the test is therefore suited for evaluation of the toxicity of sewage sludge, soil extracts, and chemicals [45]. The test criterion is the reduction in cell multiplication determined as the reduction in growth of the culture. According to the standard test ISO 10712 [46] *P. putida* is grown in liquid culture to give a highly turbid culture, which is then diluted by mixing with the sample solution. After incubation of the culture for 16 hours, growth is measured as turbidity during this period. Inhibition of an increase in turbidity in the samples is compared with that of the control using the following equation:

$$I = \frac{B_c - B_n}{B_c - B_o} \times 100$$

where I is the cell multiplication inhibition, expressed as a percentage, B_n is the measured turbidity of biomass at the end of the test period, for the nth concentration of test sample, B_c is the measured turbidity of biomass at the end of the test period in the control, and B_o is the initial turbidity measurement of biomass at time t_0 in the control.

The inhibition values (I) for each dilution should then be plotted against the corresponding dilution factor. The desired values of EC_{50}, EC_{20}, and EC_{10} are located at the intersection of the straight lines with lines parallel to the abscissa at ordinate values of 10, 20, and 50%. The evaluation may also be performed using an appropriate regression model on a computer.

Another growth inhibition test of *B. cereus* is used to determine the toxicity of chemicals and sediments [41]. This test is based on the measurement of an inhibition zone.

An agar plate method is presented by Liu *et al.* [47]. On an agar plate covered by a bacterial suspension, an inhibition zone is formed and measured around the spot where the toxic sample has been placed. The duration of the test depends on the growth of the bacterial species (from 3 to 24 hours). This assay is not available in a commercial kit but it is simple to perform as part of routine testing. Any bacterial strain can be used, but solid samples can only be tested as extracts.

2.3.4 Test Based on the Inhibition of Motility

The test based on motility inhibition of the bacterium *Spirillum volutants* is a very simple and rapid test for the qualitative screening of wastewater samples or extracts [48]. The organisms are observed under the microscope immediately after the addition of the test solution. The maintenance of a bacterial culture is necessary as in the previous type of assay.

2.3.5 Tests Based on Respiration Measurements

The assay microorganisms in Polytox are a blend of bacterial strains originally isolated from wastewater [48]. The Polytox kit (Microbiotest Inc., Nazareth, Belgium), specifically designed to assess the effect of toxic chemicals on biological waste treatment, is based on the reduction of respiratory activity of rehydrated cultures in the presence of toxicants. The commercially available kit is specifically designed for testing wastewaters. Quantative results can be obtained in just 30 minutes.

Respiration inhibition kinetics analysis (RIKA) involves the measurement of the effect of toxicants on the kinetics of biogenic substrate (e.g., butyric acid) removal by activated sludge microorganisms. The kinetic parameters studied are q_{max}, the maximum specific substrate removal rate (determined indirectly by measuring V_{max}, the maximum respiration rate), and K_S, the half-saturation coefficient [19]. The procedure consists of measuring with a respirometer the Monod kinetic parameters, V_{max} and K_S, in the absence and in the presence of various concentrations of the inhibitory compound.

2.3.6 Genotoxicity

Genotoxicity is one of the most important characteristics of toxic compounds in waste. The Ames test with *Salmonella* is the most widely used test for studying genotoxicity [49]. The test has been applied in genotoxic studies on waste, contaminated soil, sewage sludge, and sediments [11,19,50–52]. Specific *Salmonella typhimurium* strains with obligatory requirements for histidine are used to test mutagenicity. On histidine-free medium, colonies are formed only by those bacteria that have reverted to the "wild" form and can produce histidine. Addition of a mutagenic agents increases the reversion rate.

The SOS Chromotest^{TM} (Labsystems, Helsinki, Finland) is a test based on *E. coli* with an additional *lacZ* gene with SOS gene promoter *sfiA*. Under the influence of mutagenic agents, the DNA of the bacterial cells is damaged and an enzymatic SOS-recovering program and *stifA* gene promoter induce *de novo* transcription and synthesis of β-galactosidase. Commercial SOS Chromotests^{TM} are used for estimation of soil and sediment contaminants [41,42,53].

Genotoxicity may also be tested with a Mutatox^{TM} test (Azur Environmental Ltd., Berkshire, England), using a dark mutant strain of bioluminescent bacterium *V. fischeri* [54]. DNA-damaging substances are recognized by measuring the ability of a test sample to restore the luminescent state in the bacterial cells. The authors pointed to the sensitivity of the test to chemicals that damage DNA, bind DNA, or inhibit DNA synthesis.

Muta-Chromoplate is a modified version of the classical Ames test for the evaluation of mutagenicity. The bioassay uses a mutant strain of *S. typhimurium*. The reverse mutation is recorded as absence of bacterial growth after 5 days incubation [55].

2.3.7 Tests Based on Nutrient Cycling

Sometimes the risk of waste is estimated on the basis of nutrient cycling tests. As a rule such investigation is carried out for surface waste disposal or its land application. The carbon cycle is very sensitive to harmful compounds. Soil respiration is considered a useful indicator of the contaminants' effects on soil microbial activity [56–59]. The production of carbon dioxide can be followed as short-term and long-term respiration tests.

Many organisms take part in processes that release inorganic nitrogen as a result of the mineralization of organic matter, leading initially to the formation of NH_4^+ ions. In contrast, relatively few genera of autothrophic bacteria, such as *Nitrosomonas* and *Nitrobacter* acting in sequence, take part in the transformation of ammonium to nitrite and nitrate. Toxicity assays based on the inhibition of both *Nitrosomonas* and *Nitrobacter* have been developed for determining the toxicity of wastewater samples [19]. However, *Nitrosomonas* appears to be much more sensitive to toxicants than *Nitrobacter*. A rapid method for testing potential nitrification on the basis of ammonium oxidation in soil is under development at ISO [11]. This method is used to estimate the effects of toxicants contained in soil or sewage sludge [60,61].

Soil microbial processes, like mineralization of organic matter or soil respiration, can be relatively little affected by moderate levels of heavy metals, while the processes carried out by a few specialized organisms, that is, nitrogen fixation, are more sensitive [56–60,62]. Toxicity tests exist for both symbiotic and free-living nitrogen-fixing microorganisms. It is generally agreed that N_2 fixation is more sensitive than soil respiration to toxicants such as metals.

One of the most commonly used parameters in soil biology is microbial biomass. The level of microbial biomass is used for assessment of the effects of contaminants in sewage sludge or compost of municipal solid waste in short-term or long-term experiments [56–59,63–69].

2.4 TESTS WITH FAUNA SPECIES

2.4.1 Tests with Crustaceans

Throughout the last three decades, only one taxon has emerged (for reasons of practicality as well as of sensitivity) as the key group for standard ecotoxicological tests with invertebrates, namely the cladoceran crustaceans, and more particularly the daphnids. *Daphnia* tests are currently the only type of freshwater invertebrate bioassay that are formally endorsed by international organizations such as the US EPA, the EEC, and the OECD, and that are required by virtually every country for regulatory testing [70]. The reasons for the selection of daphnids for routine use in toxicity testing are both scientific and practical. Daphnids are widely distributed in freshwater bodies and are present throughout a wide range of habitats. They are an important link in many aquatic food chains (they graze on primary producers and are food for many fish species). They have a relatively short life-cycle (important for reproduction tests) and are relatively easy to culture in the laboratory. They are sensitive to a broad range of aquatic contaminants. Their small size means that only small volumes of test water and little benchspace are required. *Daphnia magna* and *D. pulex* are the most frequently used invertebrates in standard acute and chronic bioassays. *Ceriodaphnia* species are used extensively in the United States, mainly in short-term chronic bioassays [71].

A large number of papers have been published on the use of acute *Daphnia* toxicity tests, on a whole range of fundamental and applied toxicological problems. Excellent reviews of ecotoxicological testing with *Daphnia* have been written by Buikema *et al.* [72] and Baudo [73]. Standard protocols are introduced in Refs. 74–83. Acute bioassays with *Daphnia* sp. are among the most frequently used toxicity tests because, once a good laboratory culture is established, the tests are relatively easy to perform on a routine basis and do not require highly skilled personnel. Moreover, compared to acute toxicity tests with fish, acute *Daphnia* tests are cost-effective because they are shorter (48 vs. 96 hours) and the culture and maintenance of the daphnids requires much less space, effort, and equipment.

The acute *Daphnia* bioassay is recognized to be one of the most "standardized" aquatic toxicity tests presently available and several intercalibration exercises report a reasonable degree of intra- and interlaboratory reproducibility [84–87].

In addition to acute toxicity tests, two standard chronic toxicity test methods are widely accepted by various regulatory agencies: the seven-day *Ceriodaphnia* survival and reproduction test and the 21-day *Daphnia* reproduction test.

Cereodaphnia dubia was first identified in toxicity testing as *Cereodaphnia reticulata* [88] and subsequently as *Cereodaphnia affinis* [89]. The *Ceriodaphnia* survival and reproduction test is a cost-effective chronic bioassay for on-site effluent testing and is now one of the most used invertebrate chronic freshwater toxicity tests in the United States. The major arguments for introducing this method are that it is a more ecologically relevant test species in the United States (than *D. magna*), is easier to culture, and has an exposure period that is only one-third of that of the *D. magna* chronic test [88]. Owing to its ease of culturing, short test duration, low technical requirements, and high sensitivity, the seven-day *Ceriodaphnia* chronic test is a very attractive and relatively cost-effective bioassay, which can be performed by moderately skilled personnel. Key documents and standard protocols may be found in Refs. 71, 88, and 90. Different standard bioassays (Toxkit tests) are now available. In Daphtoxkit F^{TM} magna (Microbiotest Inc., Nazareth, Belgium) and pulex inhibition of mobility of *D. magna* and *D. pulex* is recorded after 24 and 48 hours exposure [91]. The test organisms are incorporated into commercial kits Daphtoxkit F^{TM} magna and Daphtoxkit F^{TM} pulex as dormant eggs and can be hatched on demand from the dormant eggs 3 to 4 days before testing [92,93]. IQ^{TM} Fluotox-test is presented by Janssen and Persoone [94]. The damaged enzyme systems (β-galactosidase) of the crustacean *D. magna* after exposure to toxic substances can be detected by their inability to metabolize a fluorescently marked sugar. Healthy organisms with unimpaired enzyme systems will "glow" under long-wave ultraviolet light, while damaged organisms will not. This microbiotest is commercially available and only takes a one-hour exposure. CerioFast^{TM} is a rapid assay based on the suppression of the feeding activity of *C. dubia* in the presence of toxicants [93,95,96]. After a one-hour exposure to the toxicant, the *C. dubia* is fed on fluorescently marked yeast and the fluorescence is observed under an epifluorescent microscope or long-wave ultraviolet light. The presence or absence of fluorescence in the daphnid's gut is used as a measure of toxic stress. This microbiotest is commercially available and only takes a few hours to complete.

The test organisms are exposed for 24, 48, and 96 hours to different concentrations of testing water. After the exposure period the number of dead organisms is counted. Each test sample container is examined and the number of dead organisms counted (looking for the absence of swimming movements). A test is regarded as valid if the mortality in the control is $<10\%$. Toxicity is calculated as:

$$T = \frac{N_0 - N_t}{N_0} \times 100\%$$

where T is toxicity in %, N_0 is the average quantity of test organisms at time 0, and N_t is the average quantity of test organisms at time t.

There are many procedures for calculating LC_{50}s. LC_{50} or EC_{50} values are calculated using the probit-derived method. A very simple procedure consists of plotting the calculated percent mortalities on a log concentration/% mortality sheet. The procedure for estimation of the LC_{50} is as follows:

1. Indicate the concentrations or dilutions used in the dilution series on the Y-axis.
2. Plot the calculated percent mortality on the horizontal line at the height of each concentration or dilution.
3. Connect the plotted mortality points on the graph with a straight line.
4. Locate the two points on the graph that are separated by the vertical 50% mortality line and read the LC_{50} at the intersect of the two lines. Expression and interpretation of the toxicity data of wastewaters; all median toxicity values are converted into toxic units (TU), that is, the inverse of the LC/EC_{50} expressed in %, according to the formula $TU = [1/L(E)C_{50}] \times 100$.

This expression is the dilution factor, which must be applied to the effluent so as to obtain a 50% effect, and is directly proportional to toxicity. The result of several toxicity tests is applied on the base of the most sensitive test species.

2.4.2 Tests with Protozoa

Dive and Persoone [97] advanced a number of arguments in favor of tests with protozoa: unicellular organisms combine all biological mechanisms and functions in one single cell; the generation time of protozoa is very short in comparison to metazoa; large numbers of organisms can be produced in a small volume; and unicellular organisms play a significant role in aquatic ecosystems, especially in the transformation and degradation of organic matter.

The standard *Colpodium campylum* toxicity test developed by Dive and colleagues [98,99] measures the inhibition of growth of this ciliate, cultured monoxenically on *E. coli*. The reduction of the number of generations is measured in increasing concentrations of the toxicant, and the effects are expressed as 24 hour IC_{50} values. This bioassay is relatively easy to learn, to carry out, and to interpret.

The microbiotest with ciliate protozoan *Tetrahymena thermophila* (Protoxkit F^{TM}, which only became available commercially recently) evaluates the growth inhibition of the unicellulars submitted for 20 hours to a toxicant [100]. The decreased multiplication of the ciliates is determined indirectly via the reduction in their food uptake, by optical density measurement in 1 cm spectrophotometric cells.

A test with *Paramecium caudatum* was suggested for estimation of the toxicity of inflowing municipal wastewater entering the treatment plant as well as of local wastewater during the process of channeling [18,101,102]. Use of *P. caudatum*, a typical representative of the organisms of activated sludge, permits us to foresee the impact of toxicants on the processing of the wastewater treatment plant. The test reaction is the death of the test organism when exposed to tested wastewater or waste extract for 1 hour. The toxicity is calculated as:

$$T\ (\%) = (N_f : N_i) \times 100$$

where N_f is the number of dead *P. caudatum* (the average from five replications), and N_i is the initial number of *P. caudatum*.

Another test organism suggested for the estimation of wastewater entering the treatment plant is *Euplotes patella* [103].

2.4.3 Tests with Cnidaria

The freshwater cnidarian *Hydra attenuata* was only recently exploited to assess the acute lethal toxicity of wastewaters [37,104]. The advantages of using *Hydra* for bioassay include its wide distribution in freshwater environments, thereby making it a representative animal for conducting environmental hazard assessment, as well as its robustness, which makes it easily manipulable, and easily reared and maintained in the laboratory. Upon exposure to bioavailable toxicants, *Hydra* undergoes profound morphological changes, which are first manifested by sublethal and then lethal effects. From their normal appearance, the animals progressively exhibit bulbed (clubbed) tentacles as an initial sign of toxicity, followed by shortened tentacles and body. After these sublethal manifestations, and if toxicity continues to prevail, *Hydra* reaches the tulip phase, where death then becomes an irreversible event. The postmortem stage is finally indicated by disintegration of the organism. Noting *Hydra* morphology during exposure allows for simple recording of (sub)lethal toxicity effects. *Hydra* assay demonstrates good sensitivity in detecting effluent toxicity [105].

2.4.4 Tests with Fish

Toxic characteristics of industrial wastewater in many countries are still assessed using fish [106–108]. The standardized procedure describes testing with different species in different life stages. For ethical reasons, as well as those linked to cost- and time-effectiveness, labor-intensiveness, analytical output, and effluent sample volume requirements, there is unquestionable value in searching for alternative procedures that would eliminate the drawbacks associated with fish testing. Investigators therefore use an *in vitro* cell system, which can greatly decrease the need for the *in vivo* fish model [37].

2.4.5 Tests with Invertebrates

Soil invertebrates are also good subjects for evaluating the possible harmful effects of toxic substances. There is a wide range of methods that involve soil invertebrates in toxicity testing. There are standard methods for earthworms (*Eisenia fetida*), collembola (*Folsomia candida*), and enchytraeide (*Enchytraeidae* sp.) [11,19,110,111]. When considering the use of invertebrates for ecological testing, the species should be selected with respect to how well it represents the community of organisms in question and how feasible is the culture of the species in the laboratory throughout the year.

As protozoa and nematodes live in pore water in the soil, most of the methods are adapted from toxicity tests designed for aquatic samples. Among the protozoa the tests with ciliates *Tetrahymena pyriformis*, *Tetrahymena thermophiia*, *Colpoda cucullus*, *Colpoda inflata*, *Colpoda steinii*, *Paramecium caudatum* and *Paramecium aurelia* have been developed [102,112–117]. It is the opinion of some authors that the sensitivity of infusorians is higher than that of microorganisms [115,116].

Bacteriovorus nematodes offer possibilities for toxicity testing because a large number of different species can be extracted from the soil and reared in the laboratory. Among the nematodes used are *Caenorhabditis elegans*, *Panagrellus redivivus*, and *Plectus acuminatus* [118–120]. The endpoint most often used has been mortality of the test organisms, expressed as the LC_{50}. Furthermore, fecundity, development, morphology, growth, population growth rate, and behavior have been used to assess toxic effects. Recently, assays for *C. elegans* that measure the induction of stress reporter genes have been developed [119]. The major problem in tests with nematodes and protozoans is extrapolation of the results for environmental risk assessment of hazardous compounds. Usually the tests are performed with artificial media; the composition of the media thus has a bearing on the results [11]. The survival, growth, and maturation of the nematode *P. redivivus* is evaluated such that three endpoints can be measured from this toxicity test: acute, chronic, and genotoxic [121]. This microbiotest is not available in commercial form, but the maintenance of these organisms is rather simple. Extracts form solid samples are prepared by a simple procedure, directly in the test media. A disadvantage of this 96 hour test is that qualified staff is needed to evaluate the results under the microscope.

Earthworms are often used for the assessment of toxicant effects due to their sensitivity to most of the factors affecting soil ecosystems, especially those associated with the application of agriculture chemicals. Earthworms respond to chemicals in several ways, for example, increase in body burdens, increase in mortality, and overall decrease in activities normally associated with viable earthworm populations [122]. Species recommended by standards ASTM (American Society for Testing and Materials) and OECD (Organization for Economic Cooperation and Development) are *Eisenia fetida* and *Eisenia andrei*, which commonly occur in compost and dung heaps, and can be easily cultured in the laboratory [11,123]. Another recommended species is *Limbricus terrestris* [124,125]. The ASTM standard test for soil toxicity with *E. fetida* is designed to assess lethal or sublethal toxic effects on earthworms in short-term tests. The sublethal effects examined can be growth, behavior, reproduction, and physiological processes, as well as observations of external pathological changes, for example, segmental constrictions, lesions, or stiffness. Callahan [122] has presented three different earthworm bioassays: the 48 hour contact test, 14-day soil test, and a neurological assay. The contact test is effective in detecting toxicity when the toxicant is water-soluble, and the soil test is effective in indicating the toxicity of a range of toxicants, both water-soluble and water-insoluble. Nerve transmission rate measurements have been found to be very efficient in picking up toxicity at lower concentrations and shorter exposure times. The contact test and the soil test appear to be adequate for toxicity assessment of pollutants in hazardous wastes.

In the past few years the use of rotifers in ecotoxicological studies have substantially increased. The main endpoints used are mortality, reproduction, behavior, cellular biomarkers, mesocosms and species diversity in natural populations [126]. Several workers have used *Brachionus calyciflorus* for various types of toxicity assessments. Thus, comprehensive evaluation of approximately 400 environmental samples for the toxicity assessment of solid waste elutriates, monitoring wells, effluents, sediment pore water, and sewage sludge was carried out by Persoone and Janssen [127]. The mortality of rotifers hatched from cysts is evaluated after 24 hours exposure. This microbiotest has been commercialized in a Rotoxkit F^{TM} [128,129].

2.5 ALGAE TESTS

Algae may also serve as test organisms in toxicity testing. In standard algal toxicity test methods published by various organizations such as APHA, ASTM, ISO, and OECD [130–133], a rapidly growing algal population in a nutrient-enriched medium is exposed to the toxicant for 3 or 4 days. *Selenastrum capricornutum* (renamed *Raphidocelis subcapitata*) and *Scenedesmus subspicatus* are the most frequently used, although others have also been used or recommended.

Increasing the simplicity and cost efficiency of algal tests has been an important research activity in recent years. [134,135] New tests procedures involve the application of flow cytometry, microplate techniques, and immobilized algae [135–140].

A miniaturized version of the conventional flask method with *S. capricornutum* has been developed by Blaise *et al.* [136]. In this assay the algae are exposed to the toxicant in 96-well microplates for a period of 96 hours, after which the cell density is determined using a hemocytometer or electronic particle counter. ATP content measurements [136] or chlorophyll fluorescence [141,142] have also been proposed as test criteria. Compared to the flask method, the main advantages of the microplate assay are: (a) the small sample volumes and reduced bench space requirements, (b) the use of disposable materials, (c) the large number of replicates, and (d) the potential for automation of the test set-up and scoring [136,143,144].

Another alternative algal assay with *S. capricornutum* that has recently been developed is the Algaltoxkit F[TM] (Microbiotest Inc., Nazareth, Belgium) [135,140,145]. One of the main features of this kit test is that no pretest culturing of algae is required as the algae are supplied in the form of algal beads that can be stored for several months. The algae are de-immobilized from the beads in order to test for growth inhibition by optical density measurement in "long-cell" test cuvettes.

2.5.1 Calculation of Percent Growth Inhibiton in Algae Tests

A growth curve for the algae test is drawn up by assessing the cell concentration (number of cells/mL) or optical density for each concentration of the sample being investigated and plotting against time. In order to evaluate the relationship between growth and concentration the EC_{50} is calculated for every period of time at which the biomass was measured during the test (24, 48, and 72 hours) according to OECD [133]. The effect is estimated by using the area under the growth curves as a measure of the growth (EC_{50}: measure of the effect on biomass = concentration at which the area under the growth curve comes to half of the area under the growth curve of the control). The curves are constructed using the average values of the replicates.

The area under the growth curve is calculated for each of the points in time as:

$$A = \frac{N_1 - N_0}{2} t_1 + \frac{N_1 + N_2 - 2N_0}{2}(t_2 - t_1) + \frac{N_{n-1} + N_n - 2N_0}{2} \times (t_n - t_{n-1})$$

where A is area, N_0 is the nominal number of cells or absorption measured at time t_0, N_1 is the number of cells measured or absorption measured at time t_1, N_n is the number of cells measured or absorption measured at time t_n, t_1 is the point in time at which the first measurement was made after the start of the test, and t_n is the time of the nth measurement after the start of the test.

The percent growth inhibition for each test concentration is calculated by

$$I_a = \frac{(A_c - A_a) \times 100}{A_c}$$

where I_a is the percent inhibition of concentration a, A_c is the area of the control growth curve, and A_a is the area of the growth curve of concentration a. The concentration–effect curves and the EC_{50} are determined by means of linear regression analysis.

2.6 PLANT TESTS

Plants constitute the most important components of ecosystems because of their ability to capture solar energy and transform it into chemical energy. Oxygen and the sugars produced by plants from solar energy and carbon dioxide are essential to all living organisms. The sensitivity of plants to chemicals in the environment varies considerably. Plants sensitive to harmful substances can be used as bioindicators. The plant tests used in environmental analysis can be classified into five groups: (a) biotransformation (detecting changes in the amounts of chemicals caused by plants); (b) food chain uptake (determining the amounts and concentrations of toxic chemicals that enter the food chains via plant uptake); (c) phytotoxicity (determining the toxicity and hazard posed by pollutants to the growth and survival of plants); (d) sentinel (monitoring the pollutants by observing toxicity symptoms displayed by plants); (e) surrogate (instead of animal or human assay).

Most attention has been devoted to phytotoxicity tests. Many plant species and numerous phytotoxic assessment endpoints have been used to characterize toxicant impacts on vegetation. Phytotoxicity can be determined as seed germination, root elongation, and seedling growth [22,23,57,58,146–151]. The tests can be carried out in pots or in petri dishes. The majority of plants commonly used in phytotoxicity tests have been limited to species of agricultural importance. A recent update of ASTM methodology for terrestrial plant toxicity testing lists nearly 100 plant taxa [152]. OECD has developed a plant bioassay [11]. This test is a simple test and includes at least one monocotyledon and one dicotyledon plant. The plant species recommended for growth experiments by OECD are listed in Table 2. Some test species are also recommended in ISO documents (Table 3). The test examines the reaction of growth of a plant species in the early stages of development. The efficiency of plant growth within 14 days is determined by establishing the average fresh mass after cutting the shoots above the soil surface. The calculation of the reduction in growth as a percentage of the average mass of the plants from the test samples compared to that of controls is then carried out, such that

$$\text{Percentage growth reduction} = \frac{C - T}{C} \times 100$$

where C is the average fresh mass in the control, and T is the average fresh mass of the plants from the diluted test waste or soil. The level of significance of any growth inhibition observed is computed using Student's t-test or Dunnett's t-test.

The other parameter for phytotoxicity assessment is emergence, calculated as

$$\text{Percentage emergence} = \frac{C_{\mathrm{e}} - T_{\mathrm{e}}}{C_{\mathrm{e}}} \times 100$$

Table 2 Plant Species Recommended for Assessment of Toxicity by OECD

Common name	Latin name
Ryegrass	*Lolium perenne*
Rice	*Oryza sativa*
Oat	*Avena sativa*
Wheat	*Triticum aestivum*
Sorghum	*Sorghum bicolor* (L.) *Moench*
Mustard	*Sinapis alba*
Rape	*Brassica napus*
Radish	*Raphanus sativus*
Turnip	*Brassica rapa*
Chinese cabbage	*Brassica campestris*
Vetch	*Vicia sativa*
Mungbean	*Phaseoli aureus*
Red clover	*Trifolium opratense*
Lettuce	*Lactuca sativa*
Cress	*Lepidium sativum*

Table 3 Test Species of the Plants Recommended for Phytotoxicity Assessment by ISO

Monocotyledonous		Dicotyledonous	
Common name	Latin name	Common name	Latin name
Rye	*Secale cereale* L.	Mustard	*Sinapis alba*
Ryegrass	*Lolium perenne* L.	Rape	*Brassica napus* (L.)
Rice	*Oryza sativa* L.	Radish	*Raphanus sativus* L.
Wheat, soft	*Triticum aestivum* L.	Chinese cabbage	*Brassica campestris* L.
Oat	*Avena sativa* L.	Birdsfoot fenugreek	*Trifolium ornithopodioides* L.
Spring or winter barley	*Hordeum vulgare* L.	Lettuce	*Lactuca sativa* L.
Sorghum	*Sorghum bicolor* (L.) Moench	Cress	*Lepidium sativum* L.
Sweetcorn	*Zea mays* L.	Tomato	*Lycopersicon esculentum* Miller
		Bean	*Phaseolus aureus* Roxb.

where C_e is the average number of emerged seeds in the control, and T_e is the average number of emerged seeds in the diluted test waste or soil.

Plant growth and germination tests are the most common techniques used to determine compost maturity and toxicity. A large number of studies have been carried out with different plant species such as ryegrass [153], barley [57,58,149], barley and radish [154], poplar [150], red maple, white pine, pin oak [155], and lettuce [151]. Furthermore, phytotoxicity parameters are used to ascertain whether the different kinds of waste (sewage sludge and municipal solid waste) are suitable for agricultural use or soil rehabilitation [57,58]. However, whether sewage sludge or municipal solid waste is used, it is convenient to submit it first to a process of composting to avoid risks associated with the presence of the phytotoxic substances. On the other hand, the fresh products are suitable for addition to soils with a view to their rehabilitation.

2.7 COMMERCIAL STANDARD BIOASSAYS: TOXKIT TESTS

The impact of xenobiotics on aquatic environments, including wastewaters, is generally determined by acute and chronic toxicity tests. However, because of the large inventory of chemicals, short-term bioassays are now being considered for handling this task. The major attraction of the new bioassays is that they bypass one of the major handicaps of toxicological testing, namely the necessity of continuous recruitment and/or culturing of live stock of test species in good health and in sufficient numbers [156]. On the basis of the information supplied by 35 ecotoxicological laboratories in Europe, Persoone and Van de Vel [157] performed a cost analysis of the three acute aquatic toxicity tests recommended by the OECD, and came to the conclusion that maintenance and culturing of live stocks makes up at least half of the expense of any of those bioassays. Maintenance and culturing of live organisms furthermore requires highly skilled personnel and the availability of temperature-controlled rooms provided with specific equipment.

In a review on "Microbiotests in aquatic ecotoxicology," Blaise [28] comments on 25 different test procedures with bacteria, protozoa, microalgae, invertebrates, and fish cell lines, worked out and used to date by different research laboratories. When examining each of these tests from the point of view of practical features according to five criteria (availability in kit format, portability, maintenance-free bioindicator, performance in microplates, minimal training and equipment requirement), Blaise comes to the conclusion that the Toxkit tests are the only types of bioassays that abide by all five features.

The first steps in bypassing of the biological, technological, and financial burden of live stock culturing or maintenance were made more than 20 years ago through the development of a "bacterial luminescence inhibition test" [34,35]; this bioassay is presently known and used worldwide as the Microtox® test. The revolutionary principle of this test is that it uses a "lyophilized" strain of a (marine) bacterium (*Photobacterium phosphoreum*). This makes the bioassay applicable anytime, anywhere, without the need for continuous culturing of the test species.

The second breakthrough in cost-effective toxicity screening was made through the development of "cyst-based" toxicity tests [158,159]. The new approach is based on the use of cryptobiotic stages (generally called cysts) of selected aquatic invertebrate species; the cysts are used as the "dormant" biological material from which live test organisms can easily be hatched. Like seeds of plants, "resting eggs" can be stored for long periods of time without losing their viability, and can be hatched "on demand" within 24 hours. The continuous availability of live test organisms through hatching of cysts eliminates all the problems inherent or related to continuous recruitment or culturing of live stocks, and solves one of the major bottlenecks in routine ecotoxicological testing. Commercial products for toxicity measurement on liquid and solid samples are already available (Table 4).

2.8 APPLICATION OF THE BIOASSAYS FOR ASSESSMENT OF TOXICITY OF SOLID WASTE

The ecological risk assessment of toxicants in waste requires reproducible and relevant test systems using a wide range of species. It is generally acknowledged by ecotoxicologists and environmental legislators that single species toxicity tests provide an adequate first step toward the ecological risk assessment of toxicants in soil and water [116,161].

2.8.1 Application of Single Species Bioassays

Use of tests based on luminescence is proposed by Carlson-Ekvall and Morrison [162] for estimation of the copper in the presence of organic substances in sewage sludge. The authors applied the Microtox toxicity test and Microtox solid-phase method and revealed that copper toxicity in sewage sludge can increase dramatically in the presence of certain organic substances (linear alkylbenzene sulfonate, caffeine, myristic acid, palmitic acid, nonylphenol, ethyl xanthogenate, and oxine) in sewage sludge. They attributed this effect to synergism and potentially the formation of lipid-soluble complexes. Based on the results of the toxicity found in this study they concluded that all organic substances tested in some way affected copper toxicity and measurements of total metal concentration in sewage sludge is insufficient for decision making concerning the suitability of sludge for soil amendment.

The Microtox test has been used for determination of toxicity of wastewater effluents, complex industrial wastes (oil refineries, pulp and paper), fossil fuel process water, sediments extracts, sanitary landfill, and hazard waste leachates [19].

The contribution of polycyclic aromatic hydrocarbons present in sewage sludge to toxicity measured with the ToxAlert® bioassay has been investigated by a Spanish group [163]. A ToxAlert® bioassay based on the inhibition of *V. fischeri* and chemical analysis using gas chromatography–mass spectrometry was applied to sludge extracts after purification by column chromatography. The toxicity data can be explained by the levels and composition of different

Table 4 Commercially Available Toxicity Tests

Testkit	Test organism and test process	References
Bacteria		
BioTox Kit	*Vibrio fischeri*, luminiscence	[38]
Microtox	*Vibrio fischeri*, luminiscence	[19,32–37,39]
Microtox Solid-Phase Test		
ECHA Biocide Monitor	*Bacillus* sp., inhibition of dehydrogenase activity	[29,41,42]
MetPAD	*E. coli*, mutant strain, inhibition of β-galactosidase activity	[19,29,41]
MetPLATE Kit	*E. coli*, inhibition of β-galactosidase activity	[19,40]
Toxi-Chromotest Kit	*E. coli* mutant strain, inhibition of β-galactosidase activity	[19,25,39,43,160]
MetSoil	*E. coli*, mutant strain, inhibition of β-galactosidase activity	[40]
Toxi-ChromoPad Kit	*E. coli*, inhibition of the *de novo* synthesis of β-galactosidase	[25,26,32,39]
Polytox	Blend of bacterial strains originally isolated from wastewater, reduction of respiratory activity	[19]
Muta-Chromoplate Kit	Modified version of Ames test	[55]
Mutatox	Dark mutant strain of *Photobacterium phosphoreum* (*V. fisheri*), genotoxicity	[54]
SOS-Chromotest Kit E	Mutant strain of *E. coli*, genotoxicity	[41,42,53]
Invertebtates		
Daphnotoxlit F magna	Cladoceran crustacean, *Daphnia magna*	[91]
Daphnotoxkit F pulex	Cladoceran crustacean, *Daphnia pulex*	[91]
IQ Toxicity Test Kit	*Daphnia magna*	[94]
Artoxkit F	Anostracan crustacean, *Artemia franciscana* (formerly *A. salina*)	[91]
Thamnotoxkit F	Crustacean *Thamnocephalus platyurus*	[161]
Rotoxkit F	Rotifer *Branchionus calyciflorus*	[130]
Protozoa		
Protoxkit F	Ciliate, *Tetrachymena thermophila*	[91]
Algae		
Algaltoxkit F	Algal growth test, *Selenastrum capricornutum*	[136]

polycyclic aromatic hydrocarbons in sewage sludge samples. It is the authors' opinion that the present approach can contribute to evaluating the toxicity of sewage sludge. Furthermore, these bioassays may help researchers in developing processes that produce ecologically sustainable soils [164].

Genotoxicity is one of the most important characteristics of toxic compounds in waste. For studying genotoxicity of waste, contaminated soil, sewage sludge, and sediments the conventional Ames test with *Salmonella* is usually used together with SOS-Chromotest™ and Mutatox™ [11,19,50–52,165–167].

2.8.2 Application of the Battery of Toxicity Tests

In many studies on solid waste in which ecotoxicological tests have been used, little attention has been given to such aspects as the selection of test species, sensitivity of the tests, and the simplicity and cost of the assays. Very few serious endeavors have been made to determine the minimum battery of the test required [10,168]. The potential toxicity of the product of composting pulp and paper sewage sludge has been determined using a battery of toxicity tests [11]. The tests were the bioluminescent bacteria test, the flash method, Mutatox™, MetPLATE™, MetPAD™, ToxiChromotest™, the reverse electron transfer (RET) test, and seed germination with red clover. Differences in sensitivity were found between the tested parameters. The high concentration of organic matter masked the toxicity effect due to the activation of bacterial metabolism and enzymatic reaction. Another disturbing factor was color, especially for the bioluminescence test. The flash method was found to be more sensitive than the traditional luminescent bacteria test and, in addition, the most sensitive test for solid samples.

A Russian group have suggested using a battery of biotests for toxicity estimation of ash from a power plant [169]. The ash of six power plants was intended for use in organo-mineral fertilizers. However, the presence of metals (Mn, Cu, Str, Ni, Mg, Cr, Zn, Co, Cd, Pb, Fe) required the performance of an investigation into their biological effects and safety. The battery included tests with the protozoan *Tetrachymena piriformis*, the water flea *Daphnia magna*, the algae *Scenedesmus quadricauda*, and barley seeds. It was established that the sensitivity of the tests varies. Results of the bioassays are presented in Table 5. The algae test and the water flea test were found to be more sensitive. It is the authors' opinion that a bioassay using such a battery of tests utilizing different kinds of organisms is needed for the estimation of biological effects of the ash and its suitability for agriculture.

A battery of toxicity tests has been used to study decontamination in the composting process of heterogenous oily waste [10]. This particular waste from an old dumping site was composted in three windrows with different proportions of waste, sewage sludge, and bark. Samples from the windrow having intermediate oil concentrations were tested with toxicity tests based on microbes (*Pseudomonas putida* growth inhibition test, ToxiChromotest, MetPLATE, and three different modifications of luminescent bacterial tests: BioTox, the bioluminescent direct contact test, and the bioluminescent direct contact flash test), Mutatox genotoxicity assay, enzyme inhibition (reverse electron transport), plants (duckweed growth inhibition and red clover seed germination), and soil animals (*Folsomia candida*, *Enchytraeus albidus*, and *Enchytraeus* sp.). The luminescent bacterial tests were used as prescreening tests. The bioassays were accompanied by chemical analysis. As a consequence of the investigation the authors concluded that the most sensitive tests, which also correlated with the oil hydrocarbon reduction, were the RET assay, the BioTox test, the bioluminescent direct contact test, the bioluminescent

Table 5 Bioassay of Water Extracts of the Ash Produced in Power Plants

	Value for the dilution factor of water extract, which exhibits 50% inhibition of the estimating function			
Power plant	Barley seeds	*Scenedesmus quadricauda*	*Daphnia magna*	*Tetrachymena piriformis*
Shaturskaya	–[a]	1 : 4	1 : 4	–
Azeiskaya	–	1 : 4	1 : 4	–
Kuzneckaya	–	1 : 2	1 : 2	1 : 0
CZKK	–	1 : 5	1 : 0	1 : 0
Irsha-Borodinskaya	1 : 0	1 : 5	1 : 3	1 : 0
Stupinskaya		1 : 5	1 : 5	1 : 0

[a]Indicates absence of toxic effect.

flash test, the red clover seed germination test, the test with soil arthropod *F. candida*, and the test with *Enchytraeus* sp. These tests represent different trophic levels and also assess the effects of solid samples and extracts. It is the authors' opinion that one test of each category should be used to assess the environmental impact of the composted product. The Mutatox assay can also be included in the battery to assess the disappearance of genotoxicity. Note that one biotest is sufficient if only process monitoring is concerned. The most suitable test for screening and monitoring during composting was the luminescent bacterial test, in particular flash modification.

An integrative approach, using toxicological and chemical analyses to screen toxic substances that could be added to the septic sludge obtained at the wastewater treatment plant was proposed by Robidoux et al. [170,171] to assist in the management of septic sludge. The necessity of the development of this ecotoxicological procedure was provoked by the temptation for producers of toxic substances to mix their hazard waste with chemical-toilet sludge shipments. At the first stage, four toxicity tests (Microtox, bacterial respiration, root elongation, and seed germination tests) were used to estimate the toxicity range of a "normal" sludge and for determination of the threshold limits criteria. These detection criteria can be used with relative efficiency and confidence to determine whether a sludge sample is contaminated or not. Taken individually, the seed germination test was the least discriminating toxicological method (detecting only 10% of the spiked samples). The bacterial respiration test was relatively better (detecting 72% of the spiked samples). As a whole, the battery of toxicity tests detected at least 93% of the spiked samples. Using a limited battery of two toxicity tests (Microtox and respiration test), the identification of contaminated chemical-toilet sludge can be detected with good efficiency and possibly greater reliability (more than 80% of spiked samples). An integrated ecotoxicological approach to screen for illicit discharge of toxic substances in chemical-toilet sludge received at a wastewater treatment plant is proposed by the authors based on chemical and toxicological analyses (Fig. 2). After sampling the sludge received at the wastewater plant, a 1 L sample is sent to the laboratory for toxicological characterization and Microtox and bacterial respiration analyses performed. A result below one of the following criteria would indicate "abnormal" sludge. For the Microtox assay, the two lower criteria suggested by these authors are: an IC_{50}-5 minute value of 0.20% (w/w), and IC_{50}-30 minute value of 0.10% (w/w). Microtox IC_{50} values higher than 0.51% w/w (5 minute) or 0.22% w/w (30 minute) would indicate that the sludge could be considered normal. For the bacterial respiration test an oxygen consumption rate less than 14.4 mg/L hour would be considered "abnormal." The sludge would be considered normal if its respiration test rate is higher than 49.2 mg/L hour. Results lying between the two criteria for each test would be considered dubious. The sludge in this latter range is "probably abnormal" and would necessitate an investigation and closer monitoring by the manager to avoid subsequent illicit discharge of contaminants. In the absence of additional incriminating information, the suspicious sludge otherwise should be considered "normal."

In Russia the disposal cost of waste depends on the class of hazard. For sewage sludge the ecotoxicological procedure has been outlined for its attribution to different classes of hazard (nonhazard, low hazard, moderate hazard, and hazard) [172,173]. This approach combines chemical analysis with bioassay. The data of chemical analysis are used for the determination of the class of hazard by a method of calculation. However, all compounds could not be taken into account. Therefore the bioassay of sewage sludge was added. The battery of biotests employed the protozoan *Paramecium caudatum*, the bacterium *Pseudomonas putida*, the higher plant *Raphanus sativus*, and water flea *Daphnia magna*. These organisms are relevant to overall assessment endpoints, representative of functional roles played by resident organisms, and sensitive to the contaminants present. In addition, they are characterized by rapid life-cycles,

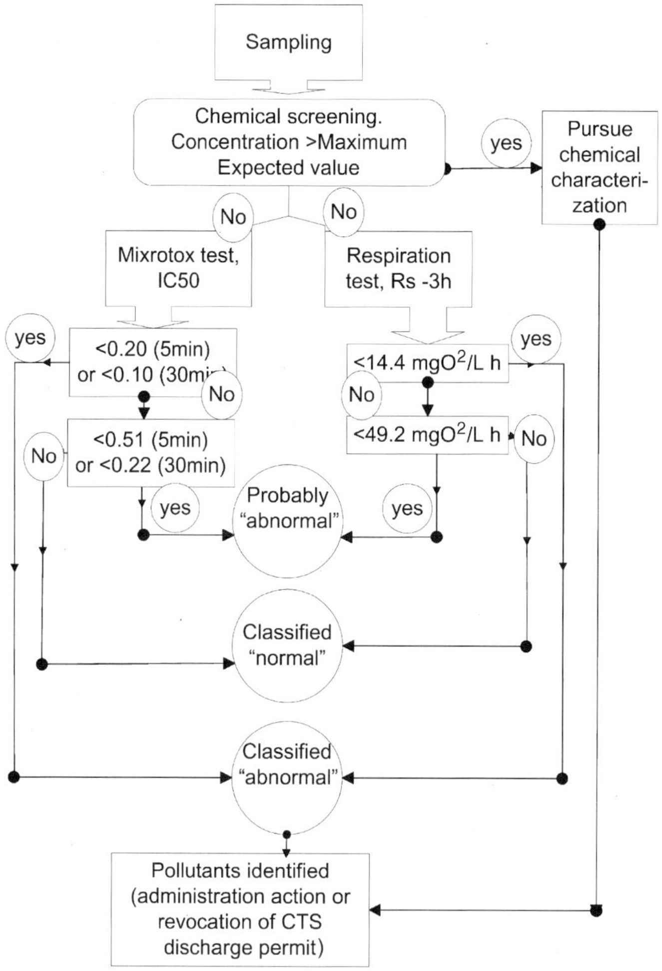

Figure 2 Proposed ecotoxicological procedure to screen for illicit discharge to toxic substances in chemical-toilet sludge.

uniform reproduction and growth, ease of culturing and maintenance in the laboratory, uniformity of population-wide phenotypic characteristics, and similar routes of exposure to those encountered in the field. The first stage consisted of the spiking of three samples of real sewage sludge with inorganic contaminants (metals) in such a manner as to create the samples on the bounds of the different classes of hazard. According to Russian legislation it is the metal content in the sewage sludge that defines the method of its disposal and its attribution to classes of hazards. Later the threshold limits criteria of these samples were established by determination of the lowest value for the dilution factor (LID_{10}) and the toxicity unit (TU) of the water extract, which exhibits less than 10% inhibition of the estimating function. Thus, the attribution of the sewage sludge samples to different classes of hazard includes the chemical analysis and the following calculation of the class of hazard and their simultaneous bioassay with the following attribution to the classes of hazard on the basis of the TU determined (Table 6). As a whole the sample of sewage sludge is attributed to the hazardous class by experimental and calculation methods (Fig. 3). In the following for the attribution of the real waste to the classes of hazard the

Table 6 Attribution of the Sewage Sludge to the Classes of Hazard in Relation to the Results of the Bioassays, Expressed as TU $[(LID_{10})^{-1} \times 100]$

	Class of hazard			
Indexes	Hazard	Moderate hazard	Low hazard	Nonhazard
Index of hazard (K) calculated on the base of the analytical data	$K > 1000$	$1000 \geq K > 100$	$100 \geq K > 10$	$10 \geq K$
Pseudomonas putida bioassay	$LID_{10} < 0.15$	$0.15 \leq LID_{10} < 4$	$LID_{10} \geq 4$	Nontoxic without dilution
Paramecium caudatum bioassay	$LID_{10} < 0.07$	$0.07 \leq LID_{10} < 0.6$	$ID_{10} \geq 0.6$	Nontoxic without dilution
Daphnia magna bioassay	$LID_{10} < 0.07$	$0.07 \leq LID_{10} < 0.43$	$ID_{10} \geq 0.43$	Nontoxic without dilution
Raphanus sativus bioassay	$LID_{10} < 1$	$1 \leq LID_{10} < 17$	$ID_{10} \geq 17$	Nontoxic without dilution

TU, toxic unit.

following procedure was carried out. After sampling of about 5 kg of the waste, the sample was divided into two parts. In one part, the pollutants were analyzed by chemical methods then the class of hazard calculated. The second part of the sample was analyzed by biological methods using bioassays with four test organisms. For this, the water extract (1 : 10) was produced, the series of dilutions obtained, and the toxicity measurement carried out.

The germination experiments (in quadruplicate) were carried out on filter paper in petri dishes. The corresponding water extracts (5 mL) (1/10) from the sewage sludge or soils were introduced into the dishes, with distilled water as the control in other dishes. Twenty-five radish seeds (*Raphanus sativus*) were then placed on the filter paper and the dishes placed in a germination chamber maintained at 20°C. The root lengths were measured after three days.

The tests with *Daphnia magna* were performed in 50 mL beakers. They were filled with 20 mL test solution and five animals (aged 6–24 hours) were added to each solution. For each dilution of the extract 2 × 5 daphnids were applied in parallel samples. The daphnids were incubated without feeding. After 96 hours the number of immobilized specimens was determined visually.

The toxicity tests with *Paramecium caudatum* were carried out in a special plate and examined under a Laboval microscope (Carl Zeiss, Jena). The test reaction was the death of the test organisms when exposed to 0.3 mL of test solution for 1 hour, using 10 individuals of *Paramecium*. Analysis was conducted five times simultaneously.

For toxicity testing with *Pseudomonas putida*, the inoculum, which has been adjusted to a specific turbidity, is added to the culture flask filled with the cultural medium and the test sample. Each dilution step should encompass three parallel batches. After an incubation period of 16 ± 1 hours at a constant temperature of 23°C in the dark, the measurement of turbidity, after homogenization by shaking, was carried out.

In all cases, the percent of inhibition ($I\%$) was determined by comparing the response given by a control solution to the sample solution. After that, an inhibition curve was fitted to

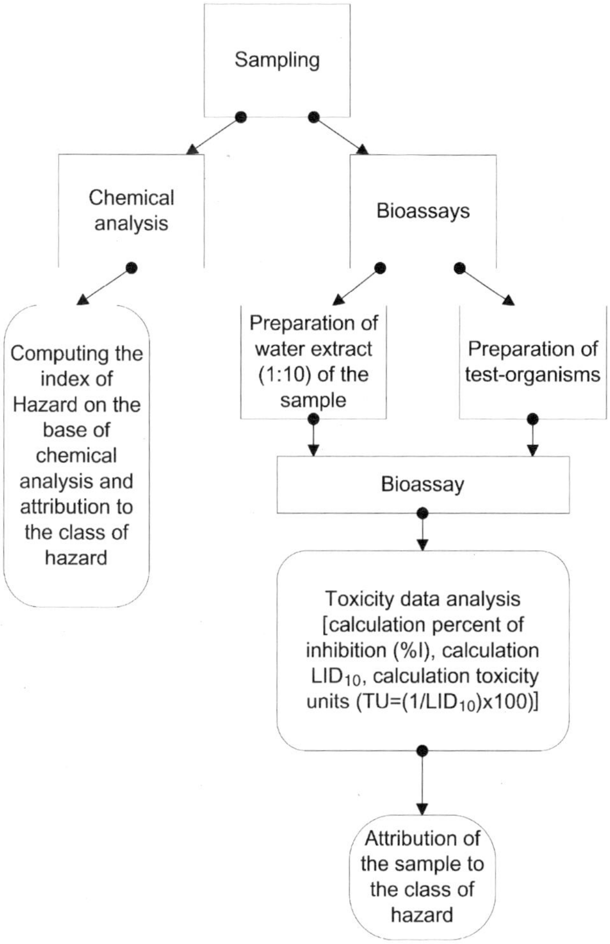

Figure 3 Proposed ecotoxicological procedure for assessment of solid waste toxicity and calculation of the classes of hazard.

calculate the 10% value for the dilution factor (LID_{10}) of the extract (Fig. 4). Acute toxicity was calculated in toxicity units (TU) according to the following formula:

$$TU = (\mathrm{LID}_{10})^{-1} \times 100$$

This toxicity unit reflects the total toxicity of all toxic substances in the sample.

The examples of the attribution of the real sewage sludge formed on different treatment plants to the classes of hazard are presented in Table 7. The same approach was adopted in Russia for the attribution of the waste as a whole to the classes of hazard. The only difference is the use of the test organisms representing water life (water flea, algae, protozoa) [174].

A similar regulation concerning solid waste is applied in Hungary. Evaluation of hazard of the waste and the establishment of fines are based on the results of ecotoxicological tests. Classification of wastes is based on the results of toxicological tests (algal test, *Selenustrum capricornutum*; seeding test, *Sinapis alba*; crustacean, *Daphnia magna*; fish, *Zebradanio rerio*;

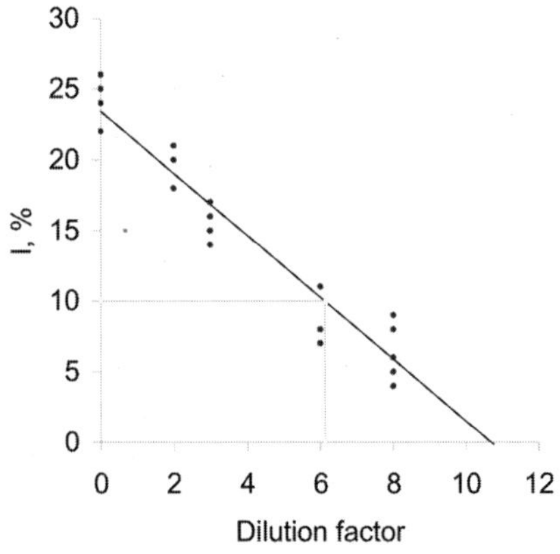

Figure 4 Example of the calculation of LID_{10}.

bacteria, *Azotobacter agile*, *Pseudomonas fluorescens*, Terravita mixed microflora) [175]. If at least one of abovementioned tests is positive in 10-fold dilution, the waste is valued for hazard.

The use of a battery of environmental bioassays for the management of hazardous wastes is applied in the Czech Republic [176]. This battery of environmental bioassays has included representatives of producers, consumers, and destructors: *D. magna* (possible substitution by *D. pulex*), acute, reproduction, chronic test; *Scenedesmus quadricauda* (*S. capricornutum*) as bottle test or in microwell plates; *Poecillia reticulate* (*Danio rerio*) acute, chronic, embryolarval tests; *S. alba* (*Lactuca sativa*) germination test, 72 hours.

2.9 APPLICATION OF THE BIOASSAYS FOR REGULATORY REQUIREMENTS OF EFFLUENTS

Chapman [177] described in his paper an historical aspect of biotesting application, and wrote that we could date toxicity tests back at least to Aristotle, who collected "bloodworms" (most probably chironomids) from freshwater muds downstream of where Athenians discharged their sewage and observed the responses of these animals when placed into salt water. Similar experimentation has occurred on an investigator-specific basis through to the present century [178]. Effluent toxicity testing in support of organized efforts to assess and control water pollution began in the 1940s; the first attempt at standardizing effluent toxicity tests occurred in the 1950s [179]. In 1985, whole effluent toxicity (WET) testing was formalized by the U.S. Environmental Protection Agency (U.S. EPA), with the intent: "To identify, characterize, and eliminate toxic effects of discharges on aquatic resources" [180]. Whole effluent toxicity testing is clearly a useful tool [181], but has a number of imperfections.

Among the objectives that strongly relate to the control function of wastewater biomonitoring are (1) the prevention/reduction of effects occurring in receiving water bodies; (2) to permit compliance testing as a part of the permit formulation; and (3) testing and steering the progress of technology based on improvement of effluent quality. Early warning of disasters and accident spills together with the prediction of effects occurring in receiving water bodies are mainly related to the alarm and the prediction function, respectively [177].

In order to use effluent toxicity data for pollution control purposes, it is necessary to test effluent samples that are representative of the characteristics of the effluent. Because an effluent

Table 7 Attribution of the Real Sewage Sludge Formed on Treatment Plants of Different Cities in Russia to the Classes of Hazard on the Basis of the Bioassay Data

	Toxicity expressed as TU of the water extract				
Type of sewage sludge	*Daphnia magna*	*Paramecium caudatum*	*Pseudomonas putida*	*Raphanus sativus*	Class of hazard
Raw mixture of primary and secondary sewage sludge (municipal treatment plant, Zelenodolsk)	0	0	0	0	Nonhazard
Anaerobically digested fresh sewage sludge (municipal treatment plant, Nabereznie Chelni)	0.7	0	4	33	Low hazard
Anaerobically digested sewage sludge, stored municipal treatment plant, Nabereznie Chelni)	0	0	10	0	Low hazard
Raw secondary sewage sludge (municipal treatment plant, Kogalim)	0.4	0	0	0	Moderate hazard
Raw primary sewage sludge (municipal treatment plant, Kogalim)	0.75	2.6	4.2	25	Low hazard
Mixture of primary and secondary sewage sludge, treated with filter press (municipal treatment plant, Kazan)	1	10	3.4	8.3	Moderate hazard
Mixture of primary and secondary sewage sludge, stored in landfill (municipal treatment plant, Kazan)	5	0	20	50	Moderate hazard
Mixture of primary and secondary sewage sludge (municipal treatment plant, Usadi)	0	0	0	0	Nonhazard

may vary significantly in quantity and toxicity either randomly or with regular cycles, the design of an appropriate sampling regime is difficult, as illsutrated in Figure 5. The variability of toxicity of samples from the Kazan municipal treatment plant is strongly dependent on the time and intervals of sampling.

Whole effluent toxicity test species are generally not the same as the resident species that the results of WET testing are aimed at protecting, particularly where nontemperate environments (e.g., tropical and Arctic environments) are concerned, or for estuaries [177]. Also, not all resident species have the same sensitivities to individual or combined contaminants in effluents. Further, differences exist between sensitivities and tolerances of WET species. Such differences are not unexpected; hence, it is desirable to use more than one toxicity test organism and endpoint to assess effluent toxicity.

Pontasch et al. [182] summarize the shortcomings of single species tests as follows:

1. They do not take into account interactions among species;
2. They utilize genetically homogeneous laboratory stock test populations;
3. They utilize species of unknown relative sensitivities;
4. They are mostly conducted under experimental conditions that lack similarity to natural habitats;
5. They utilize species that are not usually indigenous to the receiving ecosystem.

Indeed, toxicity assays are performed on a very limited set of species, and thus only represent a small fraction of the phylogenetic assemblages that characterize natural systems.

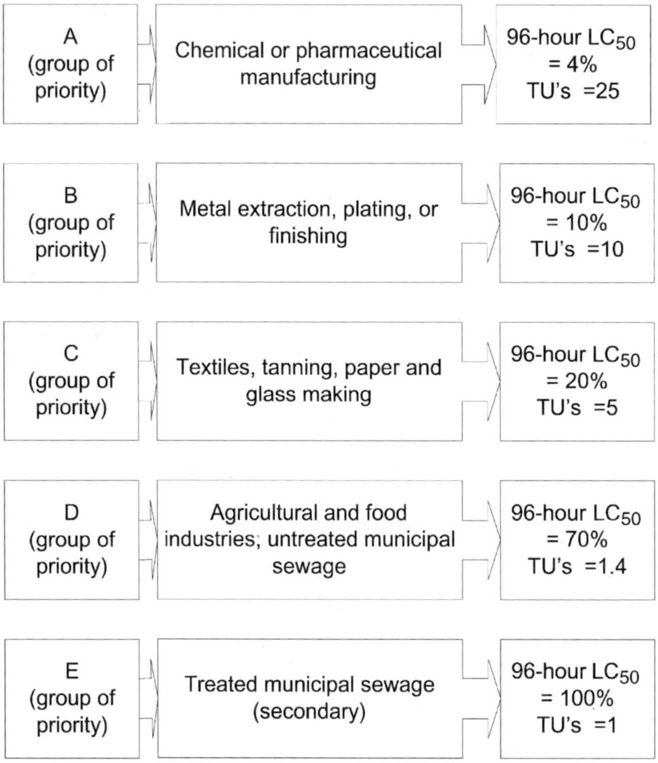

Figure 5 Irish industry specific criteria for whole effluent toxicity.

Test species currently used are those that are easily cultured and/or maintained in the laboratory. However, because of their much broader tolerances to natural environmental stressors, these biota may be poor predictors of the responses of organisms growing in a more delicately balanced and biologically inter-related environments. However, the search for the most sensitive taxon fortunately died a natural death when it was realized that different types of toxicants have different modes of action, and that no general toxicological relationship exists that is applicable to all categories of chemicals, and for all species. Yet, the question that has not yet been solved is how many species and what types of species need to be tested to adequately represent the whole range of indigenous biota of natural systems. For waste mixtures, the suite of biota must cover a much broader range of phylogenetic groups, unless it can be demonstrated that particular groups of biota are much less sensitive than others, and can be excluded from the battery [183]. In order to take the (often neglected) ecological realism in toxicity testing into consideration, the battery of bioassays are composed of test species belonging to the three trophic levels of aquatic food chains: producers, consumers, and decomposers. Bernard *et al.* [168] used *Scenedesmus subspicatus* (micro-algae) and *Lemna minor* (duckweed) for producers; *Brachionus calyciflorus* (rotifers) and *Daphnia magna* for consumers; and *Ceriodaphnia dubia* and *Thamnocephalus platyurus* (crustaceans), for the decomposers. *Vibrio fisheri* (bacteria) and *Spirostomum ambiguum* (ciliate protozoan) were used for testing such complex effluents as landfill leachates. Based on the results of their investigations the authors made recommendation for using a further such battery of tests: prokaryotes (*V. fisheri*), unicellular animal eukaryotes (*S. ambiguum*), unicellular plant eukaryotes (*S. subspicatus*) and one representative of either a multicellular plant, or various groups of animal eukaryotes.

In Russian legislation there is a requirement for using two test organisms from different trophic levels in a battery of recommended test species: decomposer, bacteria *V. fisheri* and *E. coli* (Toxi-Chromotest), unicellular animal eucaryotes *P. caudatum*; producers, unicellular plant eukaryotes *Chlorella vulgaris*, *Scenedesmus quadricauda*; and consumers, multicellular animal eukaryotes *D. magna*, *Ceriodaphnia dubia* (*affinis*). Sensitivity of test species depends on the wastewater composition, but sensitivity is often accorded in decreasing order as: *S. quadricauda* (*C. vulgaris*) → *P. caudatum* → *V. fisheri* → *D. magna* [*C. dubia* (*affinis*)]. In particular we would like to underline the ability of microalgae to increase biomass during wastewater testing. This effect is well known as stimulated, which necessitates eutrofication of the recipient water body. The use of ciliate *P. caudatum* has low sensitivity, but due to high expression (1 hour) is a very popular test, in particular for toxicity screening of wastewater in the sewage system before biological treatment. The test permits the most toxic wastewaters to be analyzed rapidly and cost-effectively.

2.9.1 Control of the Toxicity of Industrial Discharges

In most European countries, the control of toxicity of industrial discharges is carried out, to date, almost exclusively through quantitative chemical analysis of each compound for which a limit value has been set. Unfortunately, this practise is not very efficient from the point of view of protection of the aquatic ecosystem for the following two major reasons:

1. Chemical analyses are limited to a restricted number of compounds, which do not necessarily reflect the qualitative nor quantitative "overall" composition of the waste.
2. Wastes are very often complex mixtures of substances, each of which are present in a different concentration [156].

With regard to the first reason, it must be stressed that whereas each legislation prescribes explicitly that an industrial discharge should not affect the biota of the receiving waters, the practical implementation totally overlooks the (potential) toxic effects of compounds for which no limit values have been set, but which may make up a substantial part of the effluent. With regard to the second reason, it is virtually impossible to calculate the ultimate toxicity of a (complex) waste from the individual toxicities of each chemical present. A simple comparison to illustrate the latter statement is the impossibility "to predict" (at least with a certain degree of precision) the final color of a set of different dyes, to be mixed in different proportions. The only valid approach to determine the final color (i.e., in the case of hazard assessment: the ultimate toxicity) is the "experimental" way, namely by ecotoxicological testing [156].

Although ecotoxicological testing is the only valid approach to establish the real hazard of effluent discharges, it is seldom practiced in routine unless it is explicitly imposed by legislation, which is the case in only a few countries.

The data concerning the use of bioassays in the biomonitoring of liquid waste are presented in different reviews [12,19,184]. Hereafter we represent some information from these reviews.

2.9.2 Canada

Environment Canada recently developed an evaluation system based on effluent toxicity testing, capable of ranking the environmental hazards of industrial effluents [185]. This so-called Potential Ecotoxic Effects Probe (PEEP) incorporates the results of a variety of small-scale toxicity tests into one relative toxicity index to prioritize effluents for sanitation. In the index no allowance has been made for in-stream dilution, therefore the actual risk for environmental effects is not modeled. The tests performed on each effluent are the following: bacterial assay [*V. fisheri* (*P. phosphoreum*), *Microtox*], microalgal assay (*S. capricornutum*); crustacean assay (*C. dubia*); bacterial genotoxicity test (*E. coli*, SOS-test).

All test results are expressed as threshold values (LOECs), and subsequently transformed to toxic units (TUs). The entire scheme results in a total number of 10 TUs per effluent. The results are put through the following calculation to produce the PEEP index.

$$PEEP = \log_{10}\left[1 + n\left(\frac{\sum_{i=1}^{N} TU_i}{N}\right)Q\right]$$

where N is the total number of bioassays performed, n is the number of bioassays indicating toxicity, and Q is the flow rate of the effluent in m^3/hour.

Based on the correlation matrix of all bioassays data obtained with 37 effluents, it can be concluded that none of the bioassays produces data that are redundant. In other words, all bioassay procedures add to the information content of the PEEP index.

In the 37-effluent study, the effluents of the pulp and paper industry proved to be consistently far more toxic than those of other types of industries (PEEP > 5). The same study revealed that approximately 90% of the total toxic discharge is caused by the added toxicity of only three effluents of the 37. The effluent pipes for these are clearly considered the most rewarding for counteractive measures [12].

2.9.3 USA

In 1984, the U.S. Environmental Protection Agency (EPA) [186] recommended the use of "biological techniques as a complement to chemical-specific analysis to assess effluent discharges and express permit limitations." Already in 1985 [187] a guidance document had been produced on the use of effluent toxicity test results in the process of granting permits for discharge. The Organization for Economic Cooperation and Development (OECD) [188,189] in 1987 and 1991 fully adhered to the guidelines provided by the EPA. Discharging industries are required to provide quality-assured data on toxicity according to a tiered approach, where the in-stream dilution is the first screening level, and increasing toxicity requires more complicated and definitive testing with increasing numbers of species from different trophic levels, at increasing frequencies. The permit requirements are set to the level where there is a minimal risk for ecosystem damage outside the in-stream mixing zone. Inside the mixing zone some nonlethal effects are allowed to occur, depending on the types of organisms and their duration of residence in the dilution plume. The 1985 scheme was rather complicated with respect to determining the balance between the projected in-stream toxicity and uncertainty/reliability. Since new policies and regulations have been promulgated and a vast amount of knowledge and experience has been gained in controlling toxic pollutants, the testing and evaluation scheme was greatly simplified, while retaining its integrity, in 1991 [190]. Genotoxicity is addressed in a chemical-specific way with respect to human health only, based on the average daily intake (ADI) with drinking water and the ADI with fish consumption. The aspect of bioaccumulative capacity is also dealt with in a chemical-specific way.

The biological approach (whole effluent) to toxics control for the protection of aquatic life involves the use of acute and chronic toxicity tests to measure the toxicity of wastewaters. Whole effluent tests (WET) employ the use of standardized, surrogate freshwater or marine (depending on the mixture of effluent and receiving water) plants (algae), invertebrates, and vertebrates.

The evaluation strategy applied to the combined data on in-stream dilution and multiple data on effluent toxicity involves a comparison of the calculated concentration of the effluent in the receiving water under worst case conditions (RWS = receiving water concentration) with statistically derived "safe" concentrations of that specific effluent [the critical continuous concentration (CCC), based on chronic testing, and the critical maximum concentration (CMC), based on acute testing]. RWC, as well as CCC and CMC, are expressed as TUs. Action is taken when RWC > CCC or RWC > CMC. As a minimum input from toxicity testing it is required to perform acute toxicity tests on three different species quarterly for a period of at least one year. Additionally, some extrapolation to chronic toxicity has to be provided, or chronic toxicity has to be tested, depending on the rate of in-stream dilution. If the dilution is less than 1 : 100, chronic toxicity is required. If neither of the CCC or CMC are violated and the dilution is less than 1%, then it has to be demonstrated that combination effects will not occur in the receiving water (use up-stream dilution water in toxicity tests), and that the toxicity is nonpersistent (repeatedly test effluent/up-stream water samples after progressive storage under realistic conditions).

The EPA realized that setting water quality criteria with respect to toxic load, although playing an important role in assuring a healthy aquatic environment, has not been sufficient to ensure appropriate levels of environmental protection. The primary objective of the U.S. Clean Water Act (1987) is "... the restoration and maintenance of the chemical, physical, and biological integrity of the Nation's waters." To meet this objective, EPA rightfully states that water quality criteria should address biological integrity. Therefore, the Agency recommends that the water quality authorities begin to develop and implement biological criteria in their water quality standards. In order to verify the compliance of water bodies to their assigned standards, ecosystem monitoring is considered a necessity. In the guidance document on water quality based toxics control [190], it is explicitly stated that the chemical-specific and the whole effluent approaches for controlling water quality should eventually be integrated with ecological bioassessment approaches [12].

2.9.4 Argentina

As in many countries, the first attempts at understanding the effects of pollution on aquatic ecosystems in Argentina began within the academic and scientific community [191]. A systematic approach using toxicity tests with aquatic organisms is applied only in scientific laboratories.

2.9.5 Chile

The use of bioassays in environmental monitoring has not been developed in Chile [191]. In 1998 the Ministry of Agriculture started to set up a bioassay laboratory for evaluation of the presence of toxic substances in water for irrigation and animal consumption. This ministry is now in the process of implementation of EPA standardized crustacean and algal tests with *Daphnia* and *Selenastrum capricornutum*, respectively. There is no governmental wastewater bioassay monitoring.

In 1998, two bioassay methods were considered by the Chilean Regulation Institute (INN) as the first attempts for the introduction of microbioassays for routine testing in Chilean regulations: (1) the *Bacillus subtilis* growth inhibition test for toxicity evaluation of industrial effluents discharged into sewers, to detect interference with the BOD, is near endorsement; and (2) the assessment of acute toxicity in receiving waters using *D. pulex* is presently under discussion.

2.9.6 Columbia

The use of bioassays as an analytical tool for the assessment of environmental pollution is relatively new in Columbia. Even though the Ministry of Health established in Decree 1594 (1984) that environmental control agencies should propose acceptable LC_{50} values for 22 substances of ecotoxicological interest in order to protect fauna and flora, none of the entities has carried out this action up to mid-1998.

The control of toxic substances by means of bioassays at a governmental level has had little development. Even though there has been no great industrialization in this country, control of industrial contamination has centered on the implementation of treatment systems to remove organic material and bacteria. Consequently, although it is well-known that 85% of industrial effluents are discharged into continental waters and seas without any treatment, and that 74% of them are found around the Caribbean basin, currently proposed monitoring programs are centered on physico-chemical evaluation and the reduction of organic and bacteriological contamination [191].

2.9.7 Japan

In Japan, many chemicals are monitored at specific sites in rivers, lakes, and coastal areas, and data are published through the Japanese Environmental Agency. Environmental standards of water quality were revised in 1993 and over 50 chemicals were added to the list. Ecotoxicological monitoring is now considered to be very important for risk assessment of chemicals, and guidelines for ecotoxicological evaluation of chemicals are presently under examination at the level of the Japanese Government [192]. The methods that will be taken into consideration are in most cases in accordance with OECD Guidelines [79,133]. From the 10 toxicity tests described in the OECD Guidelines, the algal growth inhibition test, the *Daphnia* acute immobilization and reproduction test, and the fish toxicity test have been selected and the PNEC values from literature sources are compared with environmental concentrations. However, bioassays are not yet endorsed legally as a tool for environmental monitoring and hazard assessment in Japan. Toxic hazard is still only evaluated through chemical analysis.

2.9.8 France

In France, industrial effluents are regularly monitored for toxicity with daphnids. The toxicity data are used as a base for discharge taxation [193]. The Microtox test, chronic toxicity test, and a test on mutagenicity to the set of required bio-criteria are also used for wastewater monitoring [12,194].

2.9.9 Germany

German water authorities adopted a permit system for effluent emission where the requirements are based on fish toxicity [195]. Daphnia, algae, and luminescent bacteria are including for a screening additionally to the fish test. In this scheme the fish test (*Goldorfo*; *Leuciscus idus*) is still considered to be the only test producing definitive results.

The toxicity requirements are established per type of industry, in terms of the maximum number of times the effluents needs to be diluted to produce a no observed effect concentration (NOEC), defined as Gf for fish, Gd for daphnia, Ga for algae, and Gl for luminescent bacteria. Testing is limited to the exposure to only the appropriate Gx level, which should not produce any observed effect [the G-value corresponds with the dilution of the effluent, expressed as the lowest dilution factor $(1, 2, 4, \ldots)$ causing less than 10% mortality]. The level of maximum allowable toxicity per industrial branch is based on the level that is considered to be attainable with state-of-the-art process and/or treatment technology. Violating the toxicity requirements results in a levy, which makes state-of-the-art compliance a more economic option [12].

2.9.10 Ireland

In Ireland, compliance with toxicity limits for selected industries is ascertained by annual or biannual test on representative samples of effluent. The test species most commonly used is the rainbow trout (*Salmo gairdneri*). Control authorities normally require results from 96-hour tests. The toxicity values are expressed as the minimum acceptable proportion of effluent (as a percentage) in a test resulting in 50% fish mortality after 96 hours of exposure. The toxic units (TU) are defined as the maximum number of times an effluent may be diluted to produce the test criteria ($TU = 100/96$-hour LC_{50}, with LC_{50} expressed as the percentage of effluent in the test) (Fig. 5).

In order to encourage the optimum selection of sites for new industries, it is recommended that receiving waters at all times must provide a minimum of 20 dilutions in the immediate vicinity of the discharge for each toxic unit discharged. Flow measurements, mixing and dispersion studies are therefore a necessary addition to monitoring toxicity limits of effluents [12].

2.9.11 The Netherlands

For the control of water quality, the Netherlands government identified two pathways in a tiered procedure. The first path, the emission approach, requires dischargers to apply best available and/or best affordable technologies for the reduction of the environmental risk of their effluents with respect to good housekeeping, process control, choice of (raw) materials, and effluent pretreatment. Currently, this process is only iteratively guided by chemical-specific evaluation of effluent quality. In a combined effort, the Ministry of Housing, Spatial Planning and the Environment, together with the Ministry of Transport, Public Works and Water Management, are in the process of developing a whole effluent evaluation system that will complement the chemical-specific approach. The whole effluent evaluation method will only be applied to selected effluents (large quantities, high risk) to assist in formulating additional pollution reduction strategies. The method will be comprised of effluent tests on mutagenicity, persistence, chemical and biological oxygen demand (COD and BOD), acute and chronic toxicity, and bioaccumulation as intrinsic properties of the effluent (Fig. 6) [12,196,197].

Once effluent quality is considered to be acceptable, the water quality based approach will be followed, in which the remaining risks for effects in the receiving water are evaluated. In this framework, ambient water quality, inside and outside the mixing zone, will be verified against compound-specific water quality objectives, designated use requirements, the presence of actual toxicity (TRIAD) and biological integrity (biological water quality objectives). The results of the remaining risk evaluation may lead to the requirement of further risk reducing measures in the effluent. Additionally, the possibilities for setting permit limit requirements in the sense of whole effluent toxicity are also being evaluated.

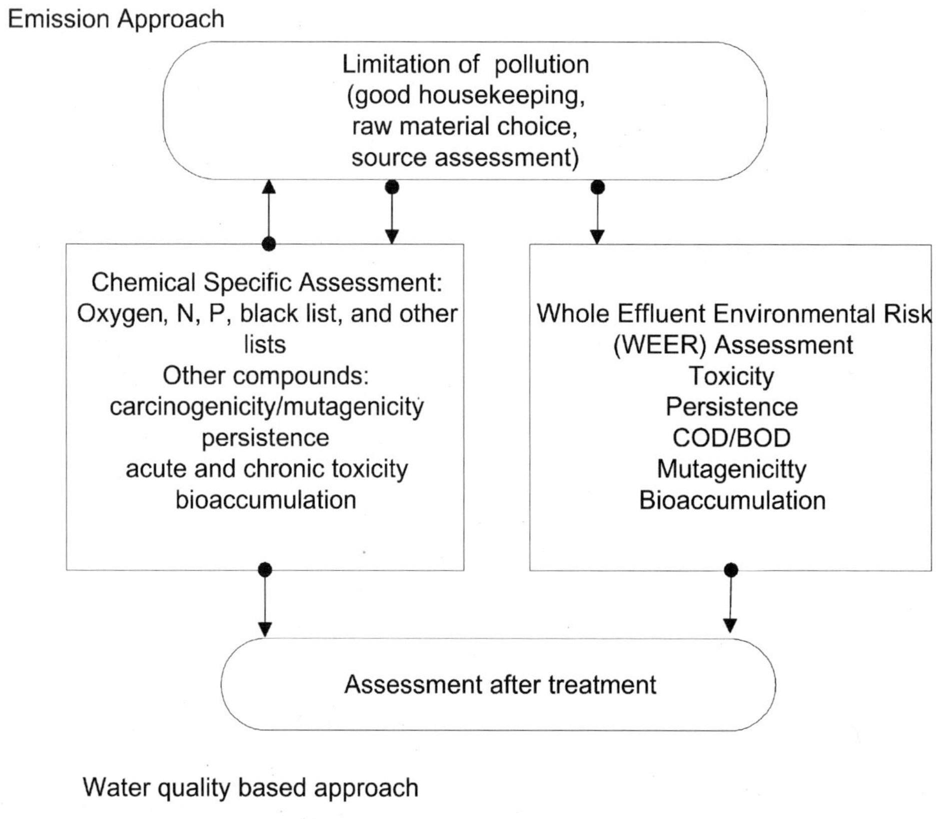

Figure 6 The Netherlands system for water quality control.

2.9.12 United Kingdom

The biological testing of waste water initially only consist of acute toxicity screening with luminescent bacteria (Microtox) and a 24 hour *Daphnia* lethality test for freshwater or a 24 hour *Oyster* larvae test for estuarine or marine waters to reveal the need for further testing [12,198,199]. The results of these tests classify the permit requirements for the effluent in four categories. The most stringent class requires the effluent to be monitored with three or four acute toxicity tests (Freshwater: 72/96 hour algal growth inhibition test with *Selenastrum*, 48 hour *Daphnia* lethality test, and a 96 hour fish lethality test with *Salmo trutta*, *Oncorhynchus mykiss*, or *Cyprinus carpio*. Marine/estuarine water: 72/96 hour algal growth inhibition test with *Pheodactylum* or *Skeletonema*, 48 hour *Oyster* embryo/larvae development test and a 96 hour fish lethality test with *Pleuronectes paltessa* or *Scopthalmus maximus*). The second stringent class prescribes effluent monitoring with one of the screening tests after verification with the abovementioned three or four acute tests. A third lower level of toxicity leaves the obligation for toxicity monitoring to one of the screening tests only, and at the fourth level no toxicity monitoring will be required. Measurements of chronic toxicity are not considered, neither are evaluations of accumulation, persistency, degradability, and genotoxicity.

2.9.13 Sweden

In Sweden, industrial effluents are to be characterized by chemical composition, toxicity, bioaccumulative capacity, and degradability [200]. The evaluation is performed according to the following tiered procedure:

Step 1

- Degradability is measured as BOD_7/COD;
- Acute toxicity is evaluated for fish, crustacea, algae, and higher plants (model organisms);
- Bioaccumulation capacity is estimated by extraction with an organic solvent, followed by the separation of the lipophilic compounds with thin layer chromatography. The migration distances give information on possible bioconcentration factors. The compounds of interest can be isolated from the TLC-plate and analyzed by GC/MS;
- Chemical analysis, including group variables like absorbable organic halogenids or total organic chlorine.

Step 2

- Degradability – added test with possibly a characterization (toxicity or bioaccumulation) or identification of the nondegradable fraction;
- Biological effects measurements – chronic toxicity and mutagenicity tests;
- Bioaccumulation and chemical evaluation – involve more and more elaborate analysis.

Step 3

- Step 3 is only prescribed in general terms, but should be tailored for the specific effluent on the basis of the results from tier 1 and 2.

2.9.14 Norway

Norway has a standardized test program for permit derivation, comprising the Ames mutagenicity test, acute and chronic toxicity tests and a biodegradation test. For monitoring purposes, it is advised to start screening the toxicity of an effluent with a comparatively large diversity of tests. The determination of precise concentration–effect relationships can then be restricted to the most sensitive types of organisms [12,201].

2.9.15 Poland

The first Polish toxicity standards were elaborated more than 25 years ago [202]:

- determination of acute toxicity to *Chlorella*;
- determination of acute toxicity to *Daphnia magna*;
- determination of acute toxicity to *Lebistes reticulatus.*

Those standards were modified and adapted to ISO standards 10 years ago. Standard bioassays are used mainly *D. magna*, luminescent bacteria (Microtox^{TM}, Lumistox) and Spirotox (protozoan *Spirostomum ambiguum*).

2.9.16 Estonia

In Estonia the monitoring of effluents is based on chemical analysis. The list of controlled water quality parameters depends on the type of industry. Bioassays are not used as a monitoring tool. However, according to HELCOM Recommendations No 16/5 "Requirement for discharging of waste water from the chemical industry" and No 16/10 "Reduction of discharges and emission from production of textiles," the toxicity effect of discharges into water bodies should be determined by (at least) two toxicity tests, which could be chosen out of the following four toxicity tests [203]:

- toxicity to fish;
- toxicity to algae;
- toxicity to invertebrates (*Daphnidae*);
- toxicity to bacteria.

2.9.17 Hungary

Chemical analyses are mainly used for detecting hazard of liquid and solid wastes [175]. Governmental orders and laws regulate the evaluation of hazard of effluent by toxicological tests. Waste control includes the determination of 30 chemical parameters, coliform count, and the result of ecotoxicological test (*D. magna* test). Category of toxicity:

- $>$100-fold dilution $\rightarrow$ strongly ecotoxic;
- 50 to 100-fold dilution $\rightarrow$ ecotoxic;
- 10 to 50-fold dilution $\rightarrow$ lightly ecotoxic;
- $<$10-fold dilution $\rightarrow$ non toxic.

2.9.18 Czech Republic

The official use of bioassays for the environmental management of hazardous wastes and chemicals are requested in Law No 157/98 (chemicals) and 132/97 (industrial and domestic wastes). As well as others law (protected areas, reservoirs conservation, etc.) there is also a system of regulations, directives of the Ministry of the Environment (mostly wastes, monitoring system, etc.), agriculture (drinking water, agricultural soil, and water for animals) and guidelines of local governments. The philosophy is led by the idea that every battery of environmental bioassays includes representatives of producers, consumers, and destructors [176].

Here is an overview of ecotoxicological bioassays cited in the above-mentioned legislation:

- *Daphnia magna* (*Thamnocephalus platyurus*);
- *Scenedesmus quadricauda (Selenastrum capricornutum)* as bottle test or in microwell plates;
- *Poecillia reticulate (Danio rerio)* acute, chronic, embryolarval tests;
- Microtox test (or equivalent);
- Activated sludge respiration test.

2.9.19 Slovenia

Routine ecotoxicological tests are regulated for some wastewaters by a law that controls emission and taxation. For wastewaters that flow into receptacles, the inhibition of the mobility of *Daphnia magna* Straus (Cladocera, crustacea) acute toxicity test is obligatory [79]. For some types of wastewaters flowing through biological purification plants the evaluation of aerobic biodegradability of organic compounds in an aqueous medium/static test (Zahn–Wellens method) is likewise obligatory [204].

2.9.20 Lithuania

According to wastewater requirements, the water quality of effluents should not be toxic on the basis of results of two acute toxicity tests. The following tests can be applied: toxicity to fish, toxicity to daphnia, toxicity to luminescent bacteria, toxicity to green algae [205].

2.9.21 Russia

The battery of tests used involves the conventional crustacean test (*D. magna*, *C. dubia*), conventional algal test (*S. quadricauda*, *C. vulgaris*), protozoan tests (*P. caudatum*, death and chemotaxis) and ToxiChromotest, which are applied for taxation of discharge water [206,207]. Researchers should choose no less than two of the bioassays. The level of the tax depends on the level of the toxicity of the discharged water and is calculated by multiplication of the basal tax by a special coefficient, which is determined on the basis of the dilution factor, expressed as toxic units (Table 8). Analysis of data from 13 regions of Russia [207] has shown that only 47% of all samples of wastewater were nontoxic and 5% of them were extremely toxic (Fig. 7). Analysis of toxicity of industrial discharges of the Tatarstan region has shown that toxicological load on the recipient water body is proportional of wastewater toxicity. Thus, about 0.03% of the water capacity of the River Kazanka is necessary for dilution of toxic wastewaters every day (Table 9).

Table 8 Classification of Effluents Based on Toxicity Assessment

Level of toxicity	Toxic unit	Multiplying coefficient
Low toxicity	1.1–16	1.3
Moderate toxicity	16–50	1.5
High toxicity	50–90	1.8
Extremely high toxicity	>99	2.0

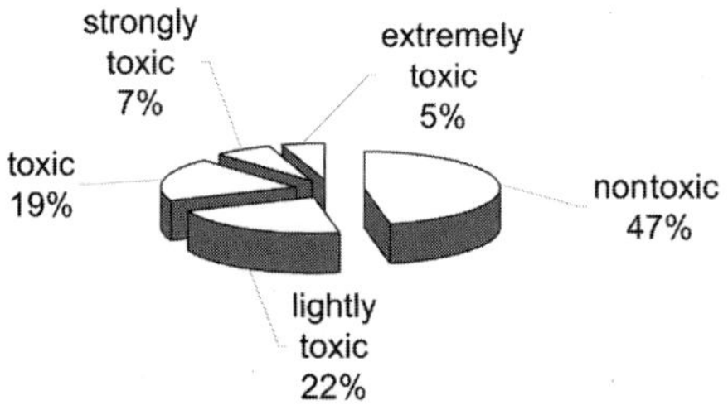

Figure 7 Wastewater classification on the basis of toxicological results (data analysis of 13 Russian regions).

Gelashvilly *et al.* [208] showed distribution of toxic input on the basis of types of industrial wastewaters for the Nizhni Novgorod region (Fig. 8). Toxicological contribution of different industrial wastewaters on recipient water bodies was calculated using the following equation:

$$T = TU \cdot Q \cdot t$$

Table 9 Toxicity Control of the Wastewaters Discharged into Recipient Water Bodies (Tatarstan Region, Russia)

Source of industrial discharge	TU_a	Capacity of discharge in m^3/day	Volume of natural water for decreasing acute toxicity of discharge (m^3/day)	Name of the recipient river
Kazan municipal treatment plant	1.3	572,000	743,600	Volga
Zainsk municipal treatment plant	3.75	8,000	30,000	Zai
Hennery "Yudinski"	12	240	2,880	Volga
Milk plant (Sabinsk)	7.2	280	2,016	Sabinka
Alcagol distilling plant (Usadi)	1	1,895	1,895	Kazanka
Optic-mechanical plant	6	253.44	1,774.1	Kazanka
Milk plant (Arsk)	6	200	1,200	Kazanka
Hennery (Laishevo)	1	800	800	Pond
Municipal treatment plant (Pestresi)	1	700	700	Miesha
Sanatorium "Krutushka"	4	150	600	Kazanka
Agricultural firm "Serp i Molot"	1	400	400	Kazanka
Municipal treatment plant (Laishevo)	1	300	300	Kama
Building plant (Kurkachi)	1	250	250	Kazanka
Kazan tuberculosis hospital	1.5	90	135	Kazanka
Sanatorium "Vasilievo"	1	90	90	Volga
Nutritive plant (Laishevo)	1	15	15	Kama

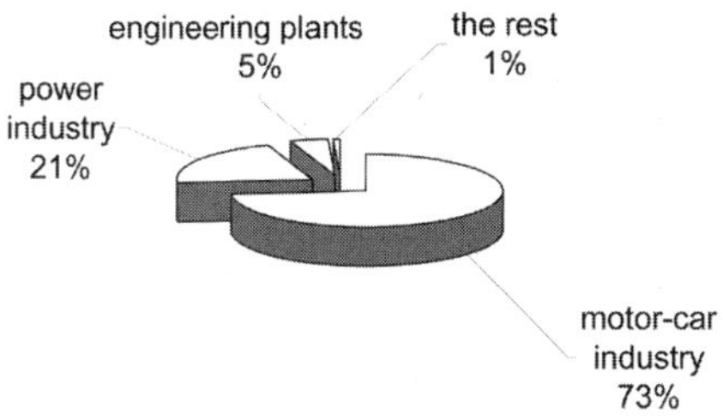

Figure 8 Toxic contribution of the different industrial wastewaters to the recipient water bodies of the Nizhni Novgorod region.

where T is toxicological load on the water body in m^3/year, TU is acute toxicity results of the most sensitive tests, Q is capacity of industrial discharge in m^3, and t is period of time (year).

The average toxicity of wastewaters decreased more than four-fold in the Nizhni Novgorod region as a result of the economic mechanism of tax collection.

2.10 CONCLUSION

Interest in bioassays as a tool in risk assessment of waste has definitely grown in recent years. There are many different toxicological bioassays, but most of them are developed for liquid waste or environments, with only a few standard tests being available for solid waste. What is more, very few have been widely adopted for routine toxicity evaluation. The main reason for this is difficulty in maintaining healthy laboratory cultures of the proposed organisms for long periods, resulting in a low degree of standardization of toxicity tests, which are thus not accepted by standardization and/or regulatory organizations. Another problem is the choice of suitable bioassays to produce a good toxicity test battery. The choice of toxicity method should be based on screening, regulatory requirements, or predictive hazard assessment. Each type of bioassay can have its own merits when properly used in the correct context.

Toxicity tests integrate interactions among complex mixtures of contaminants. They measure the total toxic effect, regardless of physical and chemical composition. As such, these tests are a useful tool. But they are not a perfect tool, particularly because they are commonly applied to conditions that do not reflect the test exposures. No single, perfect, universal tool exists; all tools have advantages, disadvantages, and assumptions [174]. Alone, bioassays cannot fulfill their stated purpose "to identify, characterize, and eliminate toxic effects of discharges on aquatic resources." However, together with other appropriate tools in a risk assessment framework (i.e., joint, not independent, applicability), toxicity testing is essential for ultimately achieving this purpose.

REFERENCES

1. Markwiese, J.T.; Ryti, R.T.; Hooten, M.M.; Michael, D.I.; Hlohowskyj, I. Toxicity bioassays for ecological risk assessment in arid and semiarid ecosystems. Rev. Environ. Contam. Toxicol. **2001**, *168*, 43–98.
2. Jouany, J.M. Ecologie et nuisances. Actual Pharmaceut. **1971**, *69*, 12–22.
3. Butler, G.C. *Principles of Ecotoxicology*; Wiley: Chichester, UK, 1978; SCOPE Vol. 12.
4. Ramade, F. *Ecotoxicologie*; Masson: Paris, France, 1979; Collection d'ecologie, No. 9.

5. Chapman, P.M. Sediment quality criteria from the sediment quality triad: an example. Environ. Toxicol. Chem. **1986**, *5*, 957–964.
6. Thomas, J.M.; Skalski, J.R.; Cline, J.F.; McShane, M.C.; Simpson, J.C.; Miller, W.E.; Peterson, S.A.; Callahan, C.A.; Green, J.C. Chemical characterization of chemical wastesite contamination and determination of its extent using bioassays. Environ. Toxicol. Chem. **1986**, *5*, 487–501.
7. Blaise, C.; Sergy, G.; Bermingham, N.; Van Coillie, R. Biological testing – development, application and trends in Canadian Environmental Protection Laboratories. Tox. Assess. Int. J. **1988**, *3*, 385–406.
8. Dutka, B. Priority setting of hazards in waters and sediments by proposed ranking scheme and battery of tests approach. German J. Appl. Zool. **1988**, *75*, 303–316.
9. Kreysa, G.; Wiesner, J., Eds. *Bioassays for Soils*; Schon & Wetzel GmbH: Frankfurt am M., Germany, 1995.
10. Juvonen, R.; Martikainen, E.; Schultz, E.; Joutti, A.; Ahtiainen, J.; Lehtokari, M. A battery of toxicity tests as indicators of decontamination in composting oily waste. Ecotox. Environ. Safe. **2000**, *47*, 156–166.
11. Kapanen, A.; Itavaara, M. Ecotoxicity tests for compost applications. Ecotox. Environ. Safe. **2001**, *49*, 1–16.
12. Zwart, D. *Monitoring Water Quality in the Future*; Ministry of Housing, Spatial Planning and the Environment, Department for International Relations: The Netherlands. Vol. 3: Biomonitoring, 1995; 83 pp.
13. USEPA. *Ecological Risk Assessment Guidance for Superfund: Process for Designing and Conducting Ecological Risk Assessments*; US Environmental Protection Agency: Washington, DC, 1997.
14. Haanstra, L.; Doelman, P. An ecological dose response model approach of short and long-term effects of heavy metals on arylsulphatase activity in soil. Biol. Fert. Soils **1991**, *11*, 18–23.
15. Van Beelen, P.; Doelmann, P. Significance and application of microbial toxicity tests in assessing ecotoxicological risks of contaminants in soil and sediment. Chemosphere **1997**, *34*, 455–499.
16. Tubbing, D.; Santhagens, L.R.; Admiraal, W.; Van Beelen, P. Biological and chemical aspects of differences in sensitivity of natural populations of aquatic bacterial communities exposed to copper. Environ. Toxic. Water Quality **1993**, *S8*, 191–205.
17. Ronnpagel, K.; Janssen, E.; Ahlf, W. Asking for the indicator function of bioassays evaluating soil contamination: Are bioassay results reasonable surrogates of effects on soil microflora? Chemosphere **1998**, *36*, 1291–1304.
18. Selivanovskaya, S.Y.; Petrov, A.M.; Egorova, K.V.; Naumova, R.P. Protozoa and metazoa communities treating a simulated petrochemical industry wastewater in rotating disc biological reactor. World J. Microb. Biot. **1997**, *18*, 197–204.
19. Bitton, G. Toxicity testing in wastewater treatment plants using microorganisms. In *Wastewater Microbiology*; Wiley-Liss: New York, 1994, 478 pp.
20. Ongley, E.D.; Birkholz, D.A.; Carey, J.H.; Samoiloff, M.R. Is water a relevant sampling medium for toxic chemicals? An alternative environmental sensing strategy. J. Environ. Qual. **1988**, *17*, 391–401.
21. Ronnpagel, K.; Liss, W.; Ahlf, W. Microbial bioassays to assess the toxicity of solid-associated contaminants. Ecotox. Environ. Saf. **1995**, *31*, 99–103.
22. ISO 11269-1. *Soil quality – Determination of the Effects of Pollutants on Soil Flora – Part 1: Method for Measurement of Inhibition of Root Growth*; International Organization for Standardization: Geneve, 1993; 9 pp.
23. ISO 11269-2. *Soil Quality – Determination of the Effects of Pollutants on Soil Flora – Part 2: Effects of Chemicals on the Emergence and Growth of Higher Plants*; International Organization for Standardization, Geneve, 1995, 7 pp.
24. Brouwer, H.; Murphy, T.; McArdle, L. A sediment-contact bioassay with *Photobacterium phosphoreum.* Environ. Toxicol. Chem. **1990**, *9*, 1353–1358.
25. Kwan, K.K.; Dutka, B.J. A novel bioassay approach: direct application of Toxi-Chromotest and SOS Chromotest to sediments. Environ. Toxic. Water Quality **1992**, *7*, 49–60.

26. Kwan, K.K. Direct assessment of solid phase samples using the Toxi-Chromotest kit. Environ. Toxic. Water Quality **1993**, *8*, 223–230.
27. Prokop, Z.; Holoubek, I. The use of a microbial contact toxicity test for evaluating cadmium bioavailability in soil. J. Soil Sediment **2001**, *1*, 21–24.
28. Blaise, C. Microbiotests in aquatic ecotoxicology: Characteristics, utility, and prospects. Environ. Toxic. Water Quality **1991**, *6*, 145–155.
29. Bitton, G.; Koopman, B. Bacterial and enzymatic bioassays for toxicity testing in the environment. Rev. Environ. Contam. Toxicol. **1992**, *125*, 1–22.
30. Dutka, B.J.; Nyholm, N.; Petersen, J. Comparison of several microbiological toxicity screening tests. Water. Res. **1983**, *17*, 1363–1368.
31. Ribo, J.M.; Kaiser, K.L.E. Photobacterium phosphoreum toxicity bioassay. I. Test procedures and application. Toxic. Assess. **1987**, *2*, 305–323.
32. Kwan, K.K.; Dutka, B.J. Evaluation of the Toxi-Chromotest direct sediment toxicity testing procedure and Microtox solid-phase testing procedure. B. Environ. Contam. Tox. **1992**, *49*, 656–662.
33. Day, C.; Dutka, B.J.; Kwan, K.K.; Batista, N.; Reynoldson, T.B.; Metcalfe-Smith, J.L. Correlations between solid-phase microbial screening assays, whole-sediment toxicity tests with macroinvertebrates and in situ benthic community structure. J. Great Lakes Res. **1995**, *21*, 192–206.
34. Bulich, A.A.; Green, M.M. *The use of luminescent bacteria for biological monitoring of water quality*. In Proceedings of the International Symposium on the Analysis and Application of Bioluminescence and Chemiluminescence; Schram, E., Philip, Eds.; Schram, State Printing and Publ. Inc., 1979; 193–211.
35. Bulich, A.A.; Isenberg, D.L. Use of the luminescent bacteria systems for rapid assessment in aquatic toxicology. Adv. Instrument **1980**, *35*, 35–40.
36. Bulich, A.A. A practical and reliable method for monitoring the toxicity of aquatic samples. Proc. Biochem. **1982**, *17*, 45–57.
37. Blaise, C. Canadian application of microbiotests to assess the toxic potential of complex liquid and solid media. In *New Microbiotests for Routine Toxicity Screening and Biomonitoring*; Persoone, G., Janssen, C., De Coen, W., Eds.; Kluwer Academic/Plenum Publishers, New York, 2000, 312.
38. Lappalainen, J.; Juovonen, R.; Vaajasaari, K.; Karp, M. A new flash method for measuring the toxicity of solid and colored samples. Chemosphere **1999**, *38*, 317–328.
39. Kwan, K.K. Direct sediment toxicity testing procedure using Sediment-Chromotest Kit. Environ. Toxil. Water Quality **1995**, *9*, 193–196.
40. Bitton, G.; Jung, K.; Koopman, B. Evaluation of a microplate assay specific for heavy metal toxicity. Arch. Environ. Con. Tox. **1994**, *27*, 25–28.
41. Ronco, A.E.; Sorbero, M.C.; Rossini, G.D.; Alzuet, P.R.; Dutka, B.J. Screening for sediment toxicity in the Rio Santiago basin: a baseline study. Environ. Toxic. Water Quality **1995**, *10*, 35–39.
42. Dutka, B.J.; McInnis, R.; Jurkovic, A.; Liu, D.; Castillo, G. Water and sediment ecotoxicity studies in Temulko and Rapel river basin, Chile. Environ. Toxic. Water Quality **1996**, *11*, 237–247.
43. Reinhartz, A.; Lampert, I.; Herzberg, M.; Fish F. A new, short-term sensitive, bacterial assay kit for the detection of toxicants. Toxic. Assess. **1987**, *2*, 193–206.
44. Dutton, R.J.; Bitton, G.; Koopman, B. Enzyme biosynthesis versus enzyme activity as a basis for microbial toxicity testing. Toxic. Assess. **1988**, *3*, 245–253.
45. Bringmann, G.; Kuhn, R. Limiting values for damaging action of water pollutions to bacteria *Pseudomonas putida* and green algae *Scenedesmus quadricauda* in cell multiplication inhibition test. Z. Wasser Abwasses-Forschung **1977**, *10*, 87–98.
46. ISO/DIS 10712. *Water Quality – Pseudomonas putida growth inhibition test (Pseudomonas Cell Multiplication Inhibition Test)*; International Organization for Standardization: Geneve, 1995; 14 pp.
47. Liu, D.; Chau, Z.K.; Dutka, B.J. Rapid toxicity assessment of water-soluble and water-insoluble chemicals using a modified agar plate method. Water Res. **1989**, *23*, 333–339.
48. Kilroy, A.; Gray, N.F. Treatability, toxicity and biodegradability test methods. Biol. Rev. **1995**, *70*, 243–275.

49. Maron, D.M.; Ames, B.N. Revised methods for the Salmonella mutagenicity test. Mutat. Res. **1983**, *113*, 174–210.
50. Donelly, K.C.; Brown, K.W.; Andersson, C.S.; Thomas, J.C.; Scott, B.R. Bacterial mutagenity and acute toxicity of solvent and aqueous extracts of soil samples from a chemical manufacturing site. Environ. Toxicol. Chem. **1991**, *10*, 1123–1131.
51. Ivanchenko, O.B.; Ilyinskaya, O.N.; Karamova, N.S.; Kostyukevich, I.I. Ecological and genetic description of the technogenic soils. Soil Sci. **1996**, *11*, 1394–1398.
52. Ivanchenko, O.B.; Karamova, N.S.; Schmidt, M.A.; Ilyinskaya, O.N. Toxicogenetic aspects in regulating the application of the sewage sludge. Toxicol. Rev. **2001**, *3*, 22–26 (in Russian).
53. Fish, F.; Lampert, I.; Halachmi, A.; Riesenfeld, G. The SOS Cromotest Kit: A rapid method for detection of genotoxicity. Toxic. Assess. **1987**, *2*, 135–147.
54. Kwan, K.K.; Dutka, B.J. Mutatox test: a new test for monitoring environmental mutagenic agents. Environ. Pollut. **1990**, *65*, 323–332.
55. Rao, S.S.; Lifshitz, R. The Muta-ChromoPlate Method for measuring mutagenicity of environmental samples and pure chemicals. Environ. Toxic. Water Quality **1995**, *10*, 307–313.
56. Garcia, C.; Hernandes, T. Effect of bromacil and sewage sludge addition on soil enzymatic activity. Soil Sci. Plant Nutr. **1996**, *42*, 191–195.
57. Pascual, J.A.; Auso, M.; Garcia, C.; Hernandez, T. Characterization of urban wastes according to fertility and phytotoxicity parameters. Waste Mgt. Res. **1997**, *15*, 103–112.
58. Pascual, J.A.; Garcia, C.; Hernandes, T.; Ayso, M. Changes in the microbial activity of the arid soil amended with urban organic wastes. Biol. Fert. Soil **1997**, *24*, 429–434.
59. Khan, M.; Scullion, J. Effect of soil on microbial responses to metal contamination. Environ. Pollut. **2000**, *110*, 115–125.
60. McGrath, S.P. Effects of heavy metals from sewage sludge on soil microbes in agricultural ecosystems. In *Toxic Metals in Soil-Plant Systems*; Ross, S.M., Ed.; John Wiley & Sons Ltd: New York, 1994; 247–274.
61. Johansson, M.; Stenberg, B.; Torstensson, L. Microbial and chemical changes in two arable soils after long-term sludge amendments. Biol. Fert. Soil **1999**, *30*, 160–167.
62. Wetzel, A.; Werner, D. Ecotoxicological evaluation of contaminated soil using the legume root nodule symbiosis as effect parameters. Environ. Toxic. Water Quality **1995**, *10*, 127–133.
63. Flieβbach, A.; Martens, R.; Reber, H.H. Soil microbial biomass and microbial activity in soils treated with heavy metal contaminated sewage sludge. Soil Biol. Biochem. **1994**, *26*, 1201–1205.
64. Hue, N.V. Sewage sludge. In *Soil Ammendments and Environmental Quality*; Rechcigl, J.E., Ed.; Lewis Publishers: Boca Raton, FL, 1995; 199–247.
65. Dar, G.H. Impact of lead and sewage sludge on soil microbial biomass and carbon and nitrogen mineralization. B. Environ. Contam. Tox. **1997**, *58*, 234–240.
66. Baath, E.; Diaz-Ravina, M.; Frostegard, A.; Campbell, C.D. Effect of metal-rich sludge amendments on the soil microbial communities. Appl. Environ. Microb. **1998**, *64*, 238–245.
67. Moreno, J.L.; Hernandez, T.; Garcia, C. Effects of cadmium-contaminated sewage sludge compost on dynamics of organic matter and microbial activity in an arid soil. Biol. Fert. Soil **1999**, *28*, 230–237.
68. Kelly, J.J.; Haggblom, M.; Robert, L.T. Effects of the land application of sewage sludge on soil heavy metal concentrations and soil microbial communities. Soil Biol. Biochem. **1999**, *31*, 1467–1470.
69. Chander, K.; Dyckmans, J.; Joergensen, R.; Meyer, B.; Raubuch, M. Different sources of heavy metals and their long-term effects on microbial properties. Biol. Fert. Soil. **2001**, *34*, 241–247.
70. Persoone, G.; Janssen, C.R. Freshwater invertebrate toxicity tests. In *Handbook of Ecotoxicology*; Calow, P., Ed.; Blackwell Scientific Publications, 1994; Vol. 1, Chap. 4, 51–65.
71. Horning, W.B.; Weber, C.I. *Short-term Methods for Estimating the Chronic Toxicity of Effluents and Receiving Waters to Freshwater Organisms*, US EPA, Report EPA/600/4-85/014, 1985; 161 pp.

72. Buikema, A.L.; Geiger, J.C.; Lee, D.R. *Daphnia* toxicity tests. In *Aquatic Invertebrate Bioassays*; Buibema, A.L., Cairns, J., Eds.; ASTM: Philadelphia, 1980; ASTM STP 715, 48–69.
73. Baudo, R. Ecotoxicological testing with *Daphnia*. In *Daphnia*; Peters, R.H., de Bernardi, R., Eds.; Mem. Ist. Ital. Idrobiol. **1987**, *45*, 461–482.
74. British Health and Safety Commission (BHSC). Testing for acute toxicity to *Daphnia*. In *Methods for the Determination of Ecotoxicity*; BHSC: London, 1982; 14–18.
75. ISO. *Water Quality – Determination of the Inhibition of Mobility of Daphnia magna Straus (Cladocera, Crustacea)*, 1st Ed.; Report 6341–1982; ISO: Geneva, 1982.
76. EPA, US Environmental Protection Agency. *Daphnid Acute Toxicity Test*; US EPA, Office of Toxic Substances, Document EG-1/ES-1: Washington, DC, 1982.
77. Association francaise de normalisation (AFNOR). Essais des eaux, determination de l'inhibition de la mobilite de *Daphnia magna*. Norme francaise homologuee, NF T90-301. Association francaise de normalisation: Paris, France, 1983.
78. American Society for Testing and Materials (ASTM). Standard practice for conducting static acute toxicity tests on waste water with *Daphnia*. In *Annual Book of ASTM Standards*; ASTM: Philadelphia, 1984; Vol. 11.01, D4229, 27–39.
79. Organization for Economic Cooperation and Development (OECD). *Daphnia spp., Acute Immobilization and Reproduction Test. OECD Guideline for Testing Chemicals*; Organization for Economic Cooperation and Development: Geneva, 1984; Vol. 202.
80. Peltier, W.H.; Weber, C.I. *Methods for Measuring the Acute Toxicity to Effluents of Freshwater and Marine Organisms*; US EPA/600/4-85/013, Environmental Protection Agency: Cincinatti, 1985; 216 pp.
81. Green, J.C.; Bartels, C.L.; Warren-Hicks; Parkhurst, B.P.; Linder, G.L.; Peterson, S.A.; Miller, W.E. *Protocol of Short-Term Toxicity Screening of Hazardous Waste Sites*; US EPA: Corvallis, OR, 1988.
82. Poirier, D.G.; Westlake, G.F.; Abernethy, S.G. *Daphnia Magna Acute Lethalithy Toxicity Test Protocol*; Ontario Ministry of Environment, Aquatic Toxicity Unit, Water Res. Br.: Rexbale, Ontario, Canada, 1988.
83. Environmental Protection Series (EPS). *Biological Test Method: Acute Lethality Test Using Daphnia spp.*; Report EPS 1/RM/11, Environment Canada, 1990; 57 pp.
84. Rue, W.J.; Flava, J.A.; Grothe, D.R. A review of inter- and intralaboratory effluent toxicity test method variability. In *Aquatic Toxicology and Hazard Assessment*; Adams, W.J., Chapman, G.A., Landis, W.G., Eds.; ASTM: Philadelphia, 1988; Vol. 10, ASTM STP 971, 190–203.
85. Grothe, D.R.; Kimerle, R.A. Inter- and intra-laboratory variability in *Daphnia magna* effluent toxicity test results. Environ. Toxicol. Chem. **1988**, *4*, 189–192.
86. Canton, J.H.; Adema, D.M.M. Reproducibility of short-term and reproduction toxicity experiments with *Daphnia magna* and comparison of the sensitivity of *Daphnia magna* with *Daphnia pulex* and *Daphnia cucullata* in short-term experiments. Hydrobiologia **1978**, *59*, 135–140.
87. Parker, W.R. *Results of an Interlaboratory Study on the Toxicity of Potassium Dichromate to Daphnia*; EPS Report; Environmental Protection Service, Environment Canada: Canada, 1983.
88. Mount, D.I.; Norberg, T.J. A seven-day life-cycle cladoceran toxicity test. Environ. Toxicol. Chem. **1984**, *3*, 425–434.
89. Cowgill, U.M.; Keating, K.I.; Takahashi, I.T. Fecundity and longevity of *Ceriodaphnia dubia/affinis* in relation to diet at two different temperatures. J. Crustacean Biol. **1985**, *5*, 420–429.
90. American Society for Testing and Materials (ASTM). Standard guide for conducting three-brood, renewal toxicity tests with *Ceriodaphnia dubia*. In *Annual Book of ASTM Standards*; ASTM: Philadelphia, 1989; Vol. 11.04, 360–379.
91. Persoone, G. Development and validation of Toxkit microbiotests with invertebrates, in particular crustaceans. In *Microscale Testing in Aquatic Toxicology: Advantages, Techniques and Practice*; Wells, P.G., Lee, K., Blaise, C., Eds.; CRC Press LLC: Boca Raton, FL, USA, 1998, 437–449.
92. Daphtoxkit F^{TM} magna. *Crustacean Toxicity Screening Test for Freshwater. Standard Operational Procedure*; Creasel: Deinze, Belgium, 1996; 16 pp.

93. Daphtoxkit F^{TM} pulex. *Crustacean Toxicity Screening Test for Freshwater. Standard Operational Procedure*; Creasel: Deinze, Belgium, 1996; 17 pp.
94. Janssen, C.R.; Persoone, G. Rapid toxicity screening tests for aquatic biota: I. Methodology and experiments with *Daphnia magna*. Environ. Toxicol. Chem. **1993**, *12*, 711–717.
95. Bitton, G.; Rhodes, K.; Koopman, B.; Cornejo, M. Short-term toxicity assay based on daphnid feeding behavior. Water Environ. Res. **1995**, *67*, 290–293.
96. Bitton, G.; Rhodes, K.; Koopman, B. Ceriofast^{TM}: an acute toxicity test based on *Ceriodaphnia dubia* feeding behavior. Environ. Toxicol. Chem. **1996**, *15*, 123–125.
97. Dive, D.; Persoone, G. Protozoa as test organisms in marine ecotoxicology: luxury or necessity? In *Ecotoxicological Testing for the Marine Environment*; Persoone, G., Jaspers, E., Claus, C., Eds.; State Univ. Ghent and Inst. Mar. Sci. Res.: Belgium, 1984; Vol. 1, 281–306.
98. Dive, D. Nutrition et croissance de Colpodium campylum. Contribution experimentale. Possibilites d'application en ecotoxicologie; Universite Sci. Techn.: Lille, France, 1981; 295 pp. These Doctorale.
99. Dive, D.; Blaise, C.; Le Due, A. Standard protocol proposal for undertaking the *Colpodium campylum* ciliate protozoan growth inhibition test. Z. Angew. Zool. **1991**, *78*, 1.
100. Protoxkit^{TM}. *Freshwater Toxicity Test With a Ciliate Protozoan. Standard Operational Procedure*; Creasel: Deinze, Belgium, 1998; 18 pp.
101. Selivanovskaya, S.Yu.; Petrov, A.M.; Egorova, K.V.; Naumova, R.P. Forming of immobilized communities treated the waste water from the organic compounds. Chem. Technol. Water **1995**, *17*, 618–622 (in Russian).
102. Petrov, A.; Stepanova, N.; Gabaydullin, A.; Shagidullin, R. Use of *Paramecium caudatum* for toxicity screening of a local industrial flow into the sewage system before biological treatment. In *International Symposium on New Microbiotests for Routine Toxicity Screening and Biomonitoring*, Brno, Czech Republic, 1–3 June, 1998; 83 pp.
103. Selivanovskaya, S.Yu.; Maslov, A.P.; Naumova, R.P. Toxicological testing of waste water subjected to biological treatment by means of ciliate infusoria. Chem. Technol. Water **1993**, *15*, 286–290 (in Russian).
104. Fu, L.J.; Staples, R.E.; Stahl, R.G., Jr. Assessing acute toxicities of pre- and post-treatment industrial wastewaters with *Hydra attenuata*: a comparative study of acute toxicity with the fathead minnow *Pimephales promelas*. Environ. Toxicol. Chem. **1994**, *13*, 563–569.
105. Kusui, T.; Blaise, C. Ecotoxicological assessment of Japanese industrial effluents using a battery of small-scale toxicity tests. In *Impact Assessment of Hazardous Aquatic Contaminants: Concept and Approaches*; Salem, R., Ed.; Ann Arbor Press: Michigan, USA, 1998; 161–181.
106. Organization for Economic Cooperation and Development (OECD). *Draft Update OECD Guideline or Testing of Chemicals – Fish Early-Life Stage Toxicity Test*; Organization for Economic Cooperation and Development: Paris, France, 1991; OECD guidelines, TGP/145, 22 pp.
107. Organization for Economic Cooperation and Development (OECD). *Draft Updated OECD Guideline for Testing of Chemicals 203 – Fish, Acute Toxicity Test*; OECD: Paris, France, 1991; TGP/148, 10 pp.
108. ASTM *Standard Guide for Conducting Early Life-Stage Toxicity Tests With Fish*; Annual Book of ASTM Standards; American Society for Testing of Materials: Philadelphia, USA, 1991; Vol. 11.04, 857–882.
109. ISO/FDIS 11267. *Soil Quality – Inhibition of Reproduction of Collembola (Folsomia candida) by Soil Pollutants*; 1998.
110. ISO WD 16387. *Soil Quality – Effects on Pollutants on Enchytraedae (Enchytraeus sp.): Determination of Effects on Reproduction*; 1999.
111. Juvonen, R.; Martikainen, E.; Schultz, E.; Joutti, A.; Ahtiainen, J.; Lehtokari, M. A battery of toxicity tests as indicators of decontamination in composting oily waste. Ecotox. Environ. Safe. **2000**, *47*, 156–166.
112. Alekperov, I.Ch.; Kasimov, R.U. Infusorian for biotesting of the boring sludge. Hydrobiological J. **1986**, *22*, 96–98 (in Russian).

113. Cronin, M.T.D.; Schultz, T.W. Structure–toxicity relationships for phenols to *Tetrahymena pyrifirmis*. Chemosphere **1999**, *32*, 1453–1468.
114. Sauvant, M.P.; Pepin, D.; Bohatier, J.; Groliere, C.A.; Guillot, J. Toxicity assessment of 16 inorganic environmental pollutants by six bioassays. Ecotox. Environ. Safe. **1997**, *37*, 131–140.
115. Campbell, C.; Warren, A.; Cameron, C.; Hope, S. Direct toxicity assessment of two soils amended with sewage sludge contaminated with heavy metals using protozoan (*Colpoda steinii*) bioassay. Chemosphere **1997**, *34*, 501–514.
116. Bogaerts, P.; Bohatier, J.; Bonnemoy, F. Use of the ciliated protozoan *Tetrahymena pyriformis* for the assessment of the toxicity and quantitative structure–activity relationships of xenobiotics: comparison with the microtox test. Ecotox. Environ. Safe. **2001**, *49*, 293–301.
117. Nicolau, A.; Dias, N.; Mota, M.; Lima, N. Trends in the protozoa in the assessment of wastewater treatment. Res. Microbiol. **2001**, *152*, 621–630.
118. Kammenga, J.E.; Koert, P.; Riksen, J.; Korthals, G.W.; Bakker, J. A toxicity test in artificial soil based on the life-history strategy of the nematode *Plectus acuminatys*. Environ. Toxicol. Chem. **1996**, *15*, 722–727.
119. Traunspurger, W.; Haitzer, M.; Hoss, S.; Beier, S.; Ahlf, W.; Steinberg, C. Ecotoxicological assessment of aquatic sediments with *Caenorhabditis elegans* (nematoda) – a method for testing liquid medium and a whole-sediment samples. Environ. Toxicol. Chem. **1997**, *16*, 245–250.
120. Gratzer, H.; Ahlf, W. Adjustment of a formulated sediment for sediment testing with *Caenorhabditis elegans* (nematoda). Acta Hydrochim. Hydrobiol. **2001**, *29*, 41–46.
121. Samoiloff, R.; Schulz, S.; Jordan, Y.; Arnott, E. A rapid simple long-term toxicity assay for aquatic contaminants using the nematode *Panagrellus redivivus*. Can. J. Fish. Aquat. Sci. **1980**, *37*, 1167–1174.
122. Callahan, C.A. Earthworms as ecotoxicological assessment tools. In *Earthworms in Waste and Environmental Management*; Edwards, C.A., Neuhauhauser, E.F., Eds.; SPB Academic: The Hague, 1998; 295–301.
123. Staint-Dernis, M.; Narbonne, J.F.; Arnaud, C.; Ribera, D. Biochemical responses of the earthworm *Eisenia fetida andrei* exposed to contaminated artificial soil: effects of lead acetate. Soil Biol. Biochem. **2001**, *33*, 395–404.
124. Pallant, E.; Hilster, L.M. Earthworm response to 10 weeks of incubation in a pot with acid mine spoil, sewage sludge and lime. Biol. Fert. Soil. **1996**, *22*, 355–358.
125. Berry, E.C.; Jordan, D. Temperature and soil moisture content effects on the growth of *Lumbricus terrestris (Oligochaeta: Lunbricidae)* under laboratory conditions. Soil Biol. Biochem. **2001**, *33*, 133–136.
126. Snell, T.W.; Janssen, C.R. Rotifers in ecotoxicology: a review. Hydrobiologia **1995**, 313–314, 231–247.
127. Persoone, G.C.; Janssen, C.R. Freshwater invertebrate toxicity tests. In *Handbook of Ecotoxicology*; Calow, P., Ed.; Blackwell Publishers: UK, 1993; 51–66.
128. Rotoxkit F^{TM} *Rotifer Toxicity Screening Test for Freshwater. Standard Operation Procedure*; Creasel: Deinze, Belgium, 1992; 22 pp.
129. Snell, T.W.; Janssen, C.R. Microscale toxicity testing with rotifers. In *Microscale Testing in Aquatic Toxicology: Advantages, Techniques and Practice*; Wells, P.G., Lee, K., Blaise, C., Eds.; CRC Press LLC: Boca Raton, FL, 1998; 409–422.
130. APHA. *Toxicity Testing with Phytoplancton. Standard Methods for the Examination of Water and Wastewater*, 17th Ed.; American Public Health Association: Washington, DC, 1989.
131. ASTM. *Standard Guide for Conducting Static 96h Toxicity Tests with Microalgae*, E1218-90; American Society for Testing and Materials: Philadelphia, PA, 1990.
132. ISO. *Algal Growth Inhibition Test*; Draft ISO Standard ISO/DIS 10253.2; International Organization for Standardization: Paris, France, 1987.
133. Organization for Economic Cooperation and Development (OECD). *Algal Growth Inhibition Test, OECD Guideline for Testing Chemicals*, No. 201; Organization for Economic Cooperation and Development: Geneva, Switzerland, 1984; Vol. 201.

134. Radetski, C.M.; Ferard, J.M.; Blaise, C. A semistatic microplate based phytotoxicity test. Environ. Toxicol. Chem. **1995**, *14*, 299–302.
135. Persoone, G. Development and first validation of a "Stock culture free" algal microbiotest: the Algaltoxkit. In *Microscale Testing in Aquatic Toxicology; Advantages, Techniques and Practice*; Wells, P.G., Lee, K., Blaise, C., Eds.; CRC Press Boca Raton, FL, 1998a; 311–320.
136. Blaise, C.; Legault, R.; Bermingham, N.; Van Coille, R.; Vasseur, P. A simple microplate algal assay for aquatic toxicity assessment. Tox. Assess. Int. J. **1986**, *1*, 261–281.
137. Bozeman, J.; Koopman, K.; Bitton, G. Toxicity testing using immobilized algae. Aquat. Toxicol. **1989**, *9*, 345–352.
138. Gala, W.R.; Giesy, J.P. Flow cytometric techniques to assess toxicity to algae. In *Aquatic Toxicology and Risk Assessment*; Landis, W., Vander Schalie, W.H., Eds.; ASTM: Philadelphia, PA, 1993; Vol. 13, 237–246.
139. Wren, M.J.; McCaroll, D. A simple and sensitive bioassay for the detection of toxic materials using a unicellular green alga. Environ. Pollut. **1990**, *64*, 87–91.
140. Amparado, R.F. *Development and Application of a Cost-Effective Algal Growth Inhibition Test with the Green Alga Selenastrum capricornutum (Printz)*; University of Gent: Belgium, 1995; 217 pp, Ph.D. Thesis.
141. Caux, P.Y.; Blaise, C.; Le Blanc, P.; Tache, M. A phytoassay procedure using fluorescence induction. Environ. Toxicol. Chem. **1992**, *11*, 549–557.
142. Willemsen, A.; Vaal, M.A.; de Zwart, D. *Microbiotests as Tools for Environmental Monitoring*; Report No 9, 607042005; National Institute of Public Health and Environmental Planning (RIVM): The Netherlands, 1995; 39 pp.
143. Blaise, C.; Sergy, G.; Bermingham, N.; Van Coillie, R. Biological testing-development, application and trends in Canadian Environmental Protection Laboratories. Tox. Assess. Int. J. **1988**, *3*, 385–406.
144. Blaise, C. Microbiotests in aquatic ecotoxicology: characteristics, utility and prospects. Environ. Toxic. Water Quality **1991**, *6*, 145–156.
145. Algaltoxkit F^{TM}. *Freshwater Test with Microalgae. Standard Operational Procedure*; Creasel: Deinze, Belgium, 1996; 28 pp.
146. Ostroumov, S.A. Some aspects of the estimation of biological activity of xenobiotics. Bulletin of Moscow University **1990**, *2*, 27–34.
147. Iannotti, D.A.; Grebus, M.E.; Toth, B.L.; Madden, L.V.; Hoitink, H.A.G. Oxygen respirometry to assess stability and maturity of composted municipal solid waste. J. Environ. Qual. **1994**, *23*, 1177–1183.
148. Kapustka, L.A.; Lipton, J.; Galbraith, A.; Cacela, O.; Leyeune, K. Metal and arsenic impacts to soils, vegetation communities and wildlife habitat in southwest Montana uplands contaminated by smelter emissions. 2. Laboratory phytotoxicity studies. Environ. Toxicol. Chem. **1995**, *14*, 1905–1912.
149. Boelens, J.; De Wilde, B.; De Baere, L. Comparative study on biowaste definition: Effects on biowaste collection, compost process and compost quality. Compost. Sci. Util. **1996**, *4*, 60–72.
150. Campbell, A.; Zhang, X.; Tripepi, R.R. Composting and evaluation a pulp and paper sludge for use as a soil amendment/mulch. Compost. Sci. Util. **1995**, *84*, 84–95.
151. Lau, S.; Fang, M.; Wong, J. Effects of composting process and flu ash amendment on phytotoxicity of sewage sludge. Arch. Environ. Con. Tox. **2001**, *40*, 184–191
152. Markwiese, J.T.; Ryti, R.T.; Hooten, M.M.; Michael, D.I.; Hlohowskyj, I. Toxicity bioassays for ecological risk assessment in arid and semiarid ecosystems. Rev. Environ. Contam. Tox. **2001**, *168*, 43–98.
153. Keeling, A.A.; Griffiths, B.S.; Ritz, K.; Myers, M. Effects of compost stability on plant growth, microbiological parameters and nitrogen availability in media containing mixed garden waste compost. Bioresource Technol. **1995**, *54*, 279–284.
154. Itavara, M.; Vilman, M. Venelampi, O. Windrow composting of biodegradable packaging materials. Compost. Sci. Util. **1997**, *5*, 84–92.

155. Maynard, A.A. Utilization of MSW compost in nursery stock production. Compost. Sci. Util. **1998**, *6*, 38–44.
156. Persoone, G. Ecotoxicology and water quality standards. In *River Water Quality – Ecological Assessment and Control*; Newman, P., Piavaux, A., Sweeting, R., Eds.; 1992; 751 pp.
157. Persoone, G.; Van de Vel, A. *Cost-Analysis of Five Current Aquatic Toxicity Tests*; Report EUR 11342 EN, Commission of the European Communities, 1988; 119 pp.
158. Persoone, G. Cyst-based toxicity tests. I. A promising new tool for rapid and cost-effective toxicity screening of chemicals and effluents. Zeitschr. Für Angewandte Zoologie **1991**, *78*, 235–241.
159. Janssen, C.R.; Vangheluwe, M.; Van Sprang, P. A brief review and critical evaluation of the status of microbiotests. In *New Microbiotests for Routine Screening and Biomonitoring*; Persoone, G., Janssen, C., Coen, W., Eds.; Kluwer Academic/Plenum Publishers: New York, 2000; 27–37.
160. Thamnotoxkit F^{TM} *Crustacean Toxicity Screening Test for Freshwater. Standard Operational Procedure*; Creasel: Deinze, Belgium, 1995; 23 pp.
161. Lokke, H. Ecotoxicological extrapolation: tool or toy? In *Ecotoxicology of Soil Organisms*; Donker, M.H., Eijsakers, H., Heimbach, F., Eds.; SETAC Special Publication. Lewis: Boca Raton, FL, 1994; 411–425.
162. Carlson-Ekvall, C.E.A.; Morisson, G.M. Toxicity of copper in the presence of organic substances in sewage sludge. Environ. Technol. **1995**, *16*, 243–251.
163. Perez, S.; Farre, M.; Garcia, M.J.; Barcelo, D. Occurrence of polycyclic aromatic hydrocarbons in sewage sludge and their contribution to its toxicity in the ToxAlert® 100 bioassay. Chemosphere **2001**, *45*, 705–712.
164. Sayles, D.; Achenson, C.M.; Kupferle, M.J.; Brenner, R.C. Land treatment of PAH contaminated soil: performance measured by chemical and toxicity assays. Environ. Sci. Technol. **1999**, *33*, 4310–4317.
165. Canna-Michaelidou, S.; Nicolaou, A.S.; Neopfytou, E.; Christodoulidou, M. The use of a battery of microbiotests as a tool for integrated pollution control: evaluation and perspectives in Cyprus. In *New Microbiotests for Routine Toxicity Screening and Biomonitoring*; Persoone, G., Janssen, C., De Coen, W., Eds.; Kluwer Academic/Plenum Publishers: New York, 2000; 39–48.
166. Ehrlichmann, H.; Manh, B.; Dott, W.; Eisentraeger, A. Development of a miniaturized *Salmonella typhimurium* reversion test with kinetic data acquisition. In *New Microbiotests for Routine Toxicity Screening and Biomonitoring*; Persoone, G., Janssen, C., De Coen, W., Eds.; Kluwer Academic/Plenum Publishers: New York, 2000; 503–510.
167. Ivanchenko, O.; Ilinskaya, O.; Kruglova, Z.; Petrov, A. Genotoxicity monitoring of environmental samples in Tatarstan, Russia. In *New Microbiotests for Routine Toxicity Screening and Biomonitoring*; Persoone, G., Janssen, C., De Coen, W., Eds.; Kluwer Academic/Plenum Publishers: New York, 2000; 511–516.
168. Bernard, C.; Persoone, G.; Colin, J.; Le Du-Delepierre, A. Estimation of the hazard of landfills through toxicity testing of leachates: determination of leachate toxicity with a battery of acute tests. Chemosphere **1996**, *33*, 2203–2230.
169. Pryadko, A.L.; Alekseeva, T.V. The using of biotesting for estimation of toxicity the ash of power plant. Hygiene and Sanitation **1992**, *3*, 69–71 (in Russian).
170. Robidoux, P.Y.; Lopes-Gastey, J.; Choucri, A.; Sunahara, G.I. Procedure to screen illicit discharge of toxic substances in septic sludge received at a wastewater treatment plant. Ecotox. Environ. Safe. **1998**, *39*, 31–40.
171. Robidoux, P.Y.; Lopes-Gastey, J.; Choucri, A.; Sunahara, G.I. Screening of illicit toxic substances discharged in chemical toilet sludge. Qual. Assur. **1999**, *6*, 23–44.
172. Selivanovskaya, S.Yu.; Latypova, V.Z. Substantiation of the system for experimental estimation of classes of hazard of sewage sludge and the selection of the way of its disposal. Ecol. Chem. **2001**, *10*, 124–134 (in Russian).

173. Semanov, D.A.; Ravzieva, G.M.; Chabibullin, D.I.; Latypova, V.Z.; Selivanovskaza, S.Yu. Comparative analysis of the approaches to the definition of toxicity classes of sewage sludge. Toxicol. Rev. **2001**, *3*, 2–6 (in Russian).
174. Ministry of Natural Resources in Russia. Criteria for the attribution of the hazardous solid waste to classes of hazard. Ministry of Natural Resources of Russia, 15.06.2001. Ecol. Consulting **2001**, *2*, 30–34 (in Russian).
175. Törökne, A. *State of Environmental Pollution and Toxicity Testing/Monitoring in Hungary*, International Workshop FITA 4 Programme Tallin, Estonia. September 10–11, 1999.
176. Marsalek, B. *Ecotoxicological Bioassays in the Czech Republic*, International Workshop FITA 4 Programme, Tallin, Estonia. September 10–11, 1999.
177. Chapman, P.M. Whole effluent toxicity testing – usefulness, level of protection, and risk assessment. Envir. Toxicol. Chem. **2000**, *19*, 3–13.
178. Anderson, B.G. Aquatic invertebrates in tolerance investigations from Aristotle to Naumann. In *Aquatic Invertebrate Bioassays*; Buikema, A.L. Jr., Cairns, J. Jr., Eds.; American Society for Testing and Materials: Philadelphia, PA, 1980; Vol. 3, 3–35.
179. American Public Health Association. *Standard Methods for the Examination of Water, Sewage and Industrial Wastes*; American Public Health Association, American Water Works Association, Water Pollution Control Federation: Washington, DC, 1955.
180. EPA (U.S. Environmental Protection Agency) *Regions 9 and 10 Guidance for Implementing Whole Effluent Toxicity Testing Programs*, Technical Report; Seattle, WA, 1996.
181. Grothe, D.R.; Johnson, D.E. Bacterial interferences in whole effluent toxicity tests. Environ. Toxicol. Chem. **1996**, *15*, 761–764.
182. Pontasch, K.W.; Niederlehner, B.R.; Cairns, J. Jr. Comparisons of single species, microcosms and field responses to a complex effluent. Environ. Toxicol. Chem. **1989**, *8*, 521–532.
183. Persoone, G.; Janssen, C.R. Field validation of predictions based on laboratory toxicity tests. In *Freshwater Field Tests for Hazard Assessment of Chemicals*; Hill, I.A., Helmbach, F., Leeuwangh, P., Matthiessen, P., Eds.; CRC Press, Inc.: Boca Raton, FL, 1994; 379–397.
184. Metcalf & Eddy, Inc. *Wastewater Treatment Engineering: Treatment, Disposal and Reuse*; Singapore, 1991; 102–108.
185. Costan, G.; Bermingham, N.; Blaise, C.; Ferard, J.F. Potential Ecotoxic Effects Probe (PEEP): a novel index to assess and compare the toxic potential of industrial effluents. Environ. Toxic. Water Quality **1993**, *8* (1).
186. EPA (U.S. Environmental Protection Agency). *Policy for the Development of Water Quality-Based Limitations for Toxic Pollutants*; US Environmental Protection Agency: Washington DC, 1984; EPA-49-FR-9016.
187. EPA (U.S. Environmental Protection Agency). *Technical Support Document for Water Quality-Based Toxics Control*; US Environmental Protection Agency, Office of Water: Washington, DC, 1985; EPA-440/4-85-032.
188. OECD. *The Use of Biological Tests for Water Pollution Assessment and Control*; Organisation for Economic Cooperation and Development: Paris, France, Environment Monographs, No. 11, 1987.
189. Hanmer, R.W. *Biological Testing of Complex Effluents in Wastewater Regulation: OECD Work and Implementation in the United States*, International Conference on River Water Quality – Ecological Assessment and Control, Palias des Congres: Brussels, December 16–18, 1991.
190. EPA (U.S. Environmental Protection Agency). *Technical Support Document for Water Quality-Based Toxics Control*; US Environmental Protection Agency, Office of Water: Washington, DC, USA, 1991; EPA/505/2-90-001.
191. Ronco, A.E.; Castillo, G.; Diaz-Baez, M.C. Development and application of microbioasays for routine testing and biomonitoring in Argentina, Chile and Colombia. In *New Microbiotests for Routine Screening and Biomonitoring*; Persoone, G., Janssen, C., Coen, W., Eds.; Kluwer Academic/Plenum Publishers: New York, 2000; 49–61.

192. Aoyama, I.; Okamura, H.; Rong, L. Toxicity testing in Japan and the use of Toxkit microbiotests. In *New Microbiotests for Routine Screening and Biomonitoring*; Persoone, G., Janssen, C., Coen, W., Ed.; Kluwer Academic/Plenum Publishers: New York, 2000; 123–133.
193. Garric, J.; Vindimian, E.; Ferard, J.F. *Ecotoxicology and Wastewater: Some Practical Applications*; Secotox: Amsterdam 1992. The Science of the Total Environment, Supplement, 1993; 1085–1103.
194. Vasseur, P.; Ferard, J.F.; Babut, M. The biological aspects of the regulatory control of industrial effluents in France. Chemosphere **1991**, *22* (5), 625–633.
195. Steinhäuser, K.G.; Hansen, P.D. *Biologische Testverfahren*; Gustav-Fisher Verlag: Stuttgart, 1992; 884 pp.
196. Tonkes, M.; Botterweg, J. *Totaal Effluent Milieubezwaarlijkheid.* RIZA-nota, AquaSense-rapport 93.0435, Rijksinstituut voor Integraal Zoetwaterbeheer en Afvalwaterbehandeling: Lelystad, The Netherlands, 1994; 157 pp.
197. Heinis, F.; Brils, J.M.; Klapwijl, S.P.; Poorter, L.R.M. In *New Microbiotests for Routine Screening and Biomonitoring*; Persoone, G., Janssen, C., Coen, W., Eds.; Kluwer Academic/Plenum Publishers: New York, 2000; 65–72.
198. Hunt, D.T.E.; Johnson, I.; Milne, R. *The Control and Monitoring of Discharges by Biological Techniques*; IWEM 91 Conf. paper. J. IWEM **1992**; *6*, 269–277.
199. Crawshaw, T. *Pre-Congress Workshop: SETAC Effluent Toxicity Program, Implementation, Compliance and Enforcement*; National Rivers Authority (NRA): Worthing, West Sussex, UK, March 28, 1993. SETAC, Lissabon 1993.
200. SNV. *Biological-chemical Characterization of Industrial Waste Water. Application When Granting Permits and Exercising Supervisory Authority for Activities Harmful to the Environment*; Swedish Environmental Protection Agency: Solna, Sweden, 1990.
201. Tapp, J.F.; Williams, B.R.H. *An Assessment of the Application of Acute Toxicity Testing for the Monitoring and Control of Oil Refinery Effluents*; Conservation Clean Air and Water Europe: Brussels, Belgium, CONCAWE Report No. BL/A/2894, 1986; 96 pp.
202. Nalecz-Jawecki, G. *Environmental (Water) Pollution in Poland. Ecotoxicological Bioassays in Poland*, International Workshop FITA 4 Programe, Tallin, Estonia, September 10–11, 1999.
203. Kahru, A.; Blinova, I. *Monitoring of Surface Water in Estonia*, International Workshop FITA 4 Programe, Tallin, Estonia. September 10–11, 1999.
204. Kolar, B. *The State of Art of Environmental Pollution, Toxicity Testing and Hazard Monitoring in Slovenia*, International Workshop FITA 4 Programe, Tallin, Estonia, September 10–11, 1999.
205. Manusadzianas, L. *General Requirements for Treated Wastewaters in Lithuania*, International Workshop FITA 4 Programe, Tallin, Estonia, September 10–11 1999.
206. Zmur, N.S. Monitoring problems of sources of natural water contaminations: conditions of decision making and some perspectives. Ecol. Chem. **1998**, *7*, 191–199 (in Russian).
207. Stepanova, N.; Latypova, V. Chemical structure and waste water toxicity: several results of economic experiment in Republic of Tatarstan. Ecol. Consulting **2001**, *3*, 17–20 (in Russian).
208. Gelashvilly, D.B.; Bezrukova, N.V.; Bezrukov, M.E. Ecotoxicological analysis of toxic load of industrial enterprises of Nizhni Novgorod to water bodies of river part of Cheboksarski reservoir. News of Samara Scientific Center of Russian Academy of Science **2000**, *2*, 244–251 (in Russian).

3
Treatment of Pharmaceutical Wastes

Sudhir Kumar Gupta and Sunil Kumar Gupta
Indian Institute of Technology, Bombay, India

Yung-Tse Hung
Cleveland State University, Cleveland, Ohio, U.S.A.

3.1 INTRODUCTION

The pharmaceutical industry manufactures biological products, medicinal chemicals, botanical products, and the pharmaceutical products covered by Standard Industrial Classification Code Numbers 2831, 2833, and 2834, as well as other commodities. The industry is characterized by a diversity of products, processes, plant sizes, as well as wastewater quantity and quality. In fact, the pharmaceutical industry represents a range of industries with operations and processes as diverse as its products. Hence, it is almost impossible to describe a "typical" pharmaceutical effluent because of such diversity. The growth of pharmaceutical plants was greatly accelerated during World War II by the enormous demands of the armed forces for life-saving products. Manufacture of the new products, particularly the antibiotics that were developed during World War II and later periods, exacerbated the wastewater treatment problems resulting from this industry. Industrialization in the last few decades has given rise to the discharge of liquid, solid, and gaseous emissions into natural systems and consequent degradation of the environment [1]. This in turn has led to an increase in various kinds of diseases, which has necessitated the production of a wide array of pharmaceuticals in many countries. Wastewater treatment and disposal problems have also increased as a result. From 1999 to 2000, the U.S. Geological Survey conducted the first nationwide reconnaissance of the occurrence of pharmaceuticals, hormones, and other organic wastewater contaminants (OWC) in a network of 139 streams across 30 states. The study concluded that OWC were present in 80% of the streams sampled. The most frequently detected compounds were basically of pharmaceutical origin, that is, coprostanol (fecal steroid), cholesterol (plant and animal steroids), *N,N*-diethyltoluamide (insect repellant), caffeine (stimulant), and triclosan (antimicrobial disinfectant), and so on [2].

3.2 CATEGORIZATION OF THE PHARMACEUTICAL INDUSTRY

Bulk pharmaceuticals are manufactured using a variety of processes including chemical synthesis, fermentation, extraction, and other complex methods. Moreover, the pharmaceutical industry produces many products using different kinds of raw material as well as processes;

Table 1 Classes of Pharmaceutical Products and Typical Examples [3]

Classes	Subclasses with typical examples
Medicinal	Antibiotics (e.g., penicillins, tetracyclines)
	Vitamins (e.g., B, E, C, A)
	Anti-infective agents (e.g., sulphonamides)
	Central depressants and stimulants (e.g., analgesics, antipyretics, barbiturates)
	Gastro-intestinal agents and therapeutic nutrients
	Hormones and substitutes
	Autonomic drugs
	Antihistamines
	Dermatological agents–local anesthetics (e.g., salicylic acid)
	Expectorants and mucolytic agents
	Renal acting and endema reducing agents
Biologicals	Serums/vaccines/toxoids/antigens
Botanicals	Morphine/reserpine/quinine/curare
	Various alkaloids, codeine, caffeine, etc.

hence it is difficult to generalize its classification. In spite of extreme varieties of processes, raw materials, final products, and uniqueness of plants, a first cut has been made to divide the industry into categories having roughly similar processes, waste disposal problems, and treatment methods. Based on the processes involved in manufacturing, pharmaceutical industries can be subdivided into the following five major subcategories:

1. Fermentation plants;
2. Synthesized organic chemicals plants;
3. Fermentation/synthesized organic chemicals plants (generally moderate to large plants);
4. Biological production plants (production of vaccines–antitoxins);
5. Drug mixing, formulation, and preparation plants (tablets, capsules, solutions, etc.).

Fermentation plants employ fermentation processes to produce medicinal chemicals (fine chemicals). In contrast, synthesized organic chemical plants produce medicinal chemicals by organic synthesis processes. Most plants are actually combinations of these two processes, yielding a third subcategory of fermentation/synthesized organic chemicals plants. Biological production plants produce vaccines and antitoxins. The fifth category comprises drug mixing, formulation, and preparation plants, which produce pharmaceutical preparations in a final form such as tablets, capsules, ointments, and so on.

Another attempt was made to classify the industry based on production of final product. The Kline Guide in 1974 defined the various classes of bulk pharmaceutical final products. Based on that, the NFIC–Denever (recently renamed NEIC, National Enforcement Investigation Center), Washington, D.C., classified the pharmaceutical industry into three major categories as depicted in Table 1 [3].

3.3 PROCESS DESCRIPTION AND WASTE CHARACTERISTICS

Pharmaceutical waste is one of the major complex and toxic industrial wastes [4]. As mentioned earlier, the pharmaceutical industry employs various processes and a wide variety of raw materials to produce an array of final products needed to fulfill national demands. As a result, a number of waste streams with different characteristics and volume are generated, which vary by plant, time, and even season, in order to fulfill the demands of some specific drugs. It has been reported that because of the seasonal use of many products, production within a given pharmaceutical plant often varies throughout the year, which changes the characteristics of wastewater by season [5]. Hence, it is difficult to generalize the characteristics of the effluent discharged from these industries.

Fermentation plants generally produce extremely strong and highly organic wastes, whereas synthetic organic chemical plants produce wastes that are strong, difficult to treat, and frequently inhibitory to biological systems. The production of antitoxins and vaccines by biological plants generates wastewater containing very high BOD (biochemical oxygen demand), COD (chemical oxygen demand), TS (total solids), colloidal solids, toxicity, and odor. The waste load from drug formulating processes is very low compared to the subcategory 1, 2, 3, bulk pharmaceutical manufacturing plants [3]. Characteristics of the waste produced and the process description of various types of pharmaceutical industries are described in the following sections.

3.3.1 Fermentation Plants

These plants use fermentation techniques to produce various pharmaceuticals. A detailed description of the fermentation process including formulation of typical broths, fermentation chemistry, and manufacturing steps of various medicines are given in the NEIC report [6]. Major unit operations involved in the fermentation process are generally comprised of seed production, fermentation (growth), and chemical adjustment of broths, evaporation, filtration, and drying. The waste generated in this process is called spent fermentation broth, which represents the leftover contents of the fermentation tank after the active pharmaceutical ingredients have been extracted. This broth may contain considerable levels of solvents and mycelium, which is the filamentous or vegetative mass of fungi or bacteria responsible for fermentation. One commercial ketone solvent has been reported as having a BOD of approximately 2 kg/L or some 9000 times stronger than untreated domestic sewage. One thousand gallons of this solvent were calculated as equivalent in BOD to the sewage coming from a city of 77,000 people. Similarly, amyl acetate, another common solvent, is reported as having a BOD of about 1 kg/L and acetone shows a BOD of about 400,000 mg/L [7–9]. The nature and composition of a typical spent fermentation broth is depicted in Table 2 [3].

3.3.2 Synthetic Organic Chemical Plants

These plants use the synthesis of various organic chemicals (raw materials) for the production of a wide array of pharmaceuticals. Major unit operations in synthesized organic chemical plants generally include chemical reactions in vessels, solvent extraction, crystallization, filtration, and drying. The waste streams generated from these plants typically consist of cooling waters, condensed steam still bottoms, mother liquors, crystal end product washes, and solvents resulting from the process [10]. The waste produced in this process is strong, difficult to treat, and frequently inhibitory to biological systems. They also contain a wide array of various chemical components prevailing at relatively high concentration produced from the production of chemical intermediates within the plant. Bioassay results on the composite waste from a plant in India approximated 0.3% when expressed as a 48 hour TLm. A typical example of untreated synthetic organic chemical waste for a pharmaceutical plant located in India is given in Table 3

Table 2 Characteristics of a Typical Spent Fermentation Broth [3]

Composition	
Total solids	1–5%
The total solids comprise	
Protein	15–40%
Fat	1–2%
Fibers	1–6%
Ash	5–35%
Carbohydrates	5–27%
Steroids, antibiotics	Present
Vitamin content of the solids	Thiamine, Riboflavin, Pyridoxin, HCl, Folic acid at 4–2,000 μg/g
Ammonia N	100–250 mg/L
BOD	5,000–20,000 mg/L
pH	3–7

BOD, biochemical oxygen demand.

[11]. Various types of waste streams were generated from this plant depending upon the manufacturing process. Waste was segregated into various waste streams such as strong process waste, dilute process waste, service water, and composite waste [12]. The strength and magnitude of various waste streams generated at the Squibb, Inc. synthetic penicillin and antifungal plant in Humaco, Puerto Rico, is given in Table 4.

Many other researchers have segregated the waste generated from a synthetic organic chemical pharmaceutical plant located in Hyderabad, India, into different wastewater streams such as floor washing, also known as condensate waste, acid waste, and alkaline waste [13–15]. This plant is one of the largest of its kind in Asia and is involved in the production of various drugs, such as antipyretics, antitubercular drugs (isonicotinic acid hydrazide), antihelminthic, sulfa drugs, vitamins, and so on. Tables 5 to 8 present the characteristics of each waste stream generated from a synthetic drug plant at Hyderabad, along with the characteristics of the combined waste streams. Wastewater from this plant exhibited considerable BOD variation among the various waste streams generated from the plant. The BOD of the condensate waste

Table 3 Characteristics of Untreated Synthetic Drug Waste [11]

Parameter	Concentration range (mg/L)
p-amino phenol, *p*-nitrophenolate, *p*-nitrochlorobenzene	150–200
Amino-nitrozo, amino-benzene, antipyrene sulfate	170–200
Chlorinated solvents	600–700
Various alcohols	2,500–3,000
Benzene, toluene	400–700
Sulfanilic acid	800–1,000
Sulfa drugs	400–700
Analogous substances	150–200
Calcium chloride	600–700
Sodium chloride	1,500–2,500
Ammonium sulfate	15,000–20,000
Calcium sulfate	800–21,000
Sodium sulfate	800–10,000

Table 4 Characteristics of Synthetic Organic Chemicals, Wastewater at Squibb, Inc., Humaco [12]

Waste	Flow, g/day		BOD (mg/L)	COD (mg/L)	BOD load (lb/day)		COD load (lb/day)	
	Avg.	Max.			Avg.	Max.	Avg.	Max.
Strong process	11,800	17,400	480,000	687,000	47,300	74,200	67,600	105,800
Dilute process	33,800	37,400	640	890	180	190	250	280
Service water	35,300	–	–	–	–	–	–	–
Composite	80,900	–	70,365	109,585	47,500	–	67,900	–

BOD, biochemical oxygen demand; COD, chemical oxygen demand.

was found to be very low compared to other wastes. Acidic waste contributed 50% of the total waste flow at 600 m^3/day and had a pH of 0.6. The combined waste had a pH of 0.8 (including acidic waste stream), whereas the pH of the waste without acidic waste stream was 9.3. The BOD to COD ratio of alkaline, condensate and combined wastewater was around 0.5–0.6, while for the acidic waste alone it was around 0.4, indicating that all these wastewaters are biologically treatable. The combined wastewater had average TOC, COD, and BOD values of 2109 mg/L, 4377 mg/L, and 2221 mg/L. Heavy metal concentration of the wastewater was found to be well below the limits according to IS-3306 (1974). Most of the solids present were in a dissolved form, with practically no suspended solids. The wastewater contained sufficient nitrogen, but was lacking in phosphorus, which is an essential nutrient for biological treatment. The 48-hour TL_m values for alkaline and condensate wastes showed 0.73–2.1% (v/v) and 0.9% (v/v),

Table 5 Characteristics of Alkaline Waste Stream of a Synthetic Drug Plant at Hyderabad [13,15]

Parameters	Ranges (max. to min.) From Ref. [15]	Ranges (max. to min.) From Ref. [13]
Flow (m^3/day)	1,400–1,920 (1,710)	1,710
pH	4.1–7.5	2.3–11.2
Total alkalinity as $CaCO_3$	1,279–2,140	624–5630
Total solids	1.29–2.55%	11825–23265 mg/L
Total volatile solids	13.1–32.6% of TS	1,457–2,389 mg/L
Total nitrogen (mg/L)	284–1,036 (TKN)	266–669
Total phosphorus (mg/L)	14–42	10–64.8
BOD_5 at 20°C (mg/L)	2,874–4,300	2,980–3,780
COD (mg/L)	5,426–7,848	5,480–7,465
BOD : COD	–	0.506–0.587
BOD : N : P	–	100 : (8.9–17.7) : (0.265–1.82)
Suspended solids (mg/L)	–	11–126
Chlorides as Cl^- (mg/L)	–	2,900–4,500

TS, total solids; TKN, total Kjeldhal nitrogen; BOD, biochemical oxygen demand; COD, chemical oxygen demand.

Table 6 Characteristics of Condensate Waste Stream of a Synthetic Drug Plant at Hyderabad [13,15]

Parameters	Ranges (max. to min.)	
	From Ref. [15]	From Ref. [13]
Flow (m^3/day)	1,570–2,225 (1,990)	1,570–2,225 (1,990)
pH	2.1–7.3	7–7.8
Total alkalinity as $CaCO_3$	498–603	424–520
Total solids	0.31–1.22%	2,742–4,150 mg/L
Total volatile solids	13.6–37.2% of TS	363–800 mg/L
Total nitrogen (mg/L)	120–240 (TKN)	120–131
Total phosphorus (mg/L)	2.8–5	3.1–28.8
BOD_5 at 20°C (mg/L)	1,275–1,600	754–1,385
COD (mg/L)	2,530–3,809	1,604–2,500
BOD : COD	–	0.4–0.688
BOD : N : P	–	100 : (10.9–16.71) : (0.28–3.82)
Suspended solids (mg/L)	–	39–200
Chlorides as Cl^- (mg/L)	–	700–790

TS, total solids; TKN, total Kjeldhal nitrogen; BOD, biochemical oxygen demand; COD, chemical oxygen demand.

respectively. Table 9 gives the characteristics of a typical pharmaceutical industry wastewater located at Bombay producing various types of allopathic medicines [16].

3.3.3 Fermentation/Synthetic Organic Chemical Plants

These plants employ fermentation techniques as well as synthesis of organic chemicals in the manufacturing of various pharmaceuticals. Typically, they are operated on a batch basis via fermentation and organic synthesis, depending upon specific requirements of

Table 7 Characteristics of an Acid Waste Stream of a Synthetic Drug Plant at Hyderabad [13]

Parameters	Ranges (max. to min.)
Flow (m^3/day)	435
pH	0.4–0.65
BOD_5 at 20°C (mg/L)	2,920–3,260
COD (mg/L)	7,190–9,674
BOD/COD ratio	0.34–0.41
Total solids (mg/L)	18,650–23,880
Total volatile solids (mg/L)	15,767–20,891
Suspended solids	Traces
Total nitrogen (mg/L)	352
Total phosphorus (mg/L)	9.4
Total acidity as $CaCO_3$	29,850–48,050
Chlorides as Cl^- (mg/L)	6,500
Sulfate as SO_4^{2-} (mg/L)	15,000

BOD, biochemical oxygen demand; COD, chemical oxygen demand.

Table 8 Characteristics of Combined Wastewater[a] of a Synthetic Drug Plant at Hyderabad [15]

Parameters	Range	Standard deviation
pH	2.9–7.6	–
BOD_5 at 20°C (mg/L)	1,840–2,835	2,221 ± 301
COD (mg/L)	4,000–5,194	4,377 ± 338
BOD/COD ratio	0.46–0.54	–
Total organic carbon (C) (mg/L)	1,965–2,190	2,109 ± 73
BOD exertion rate (k) constant[b]	0.24–0.36	0.28 ± 0.02

[a]Alkaline and condensate wastewater mixed in 1:1 ratio.
[b]BOD, biochemical oxygen demand; COD, chemical oxygen demand.

Table 9 Characteristics of Pharmaceutical Industry Wastewater Producing Allopathic Medicines [16]

Parameter	Range of concentration	Average concentration
pH	6.5–7.0	7
BOD (mg/L)	1,200–1,700	1,500
COD (mg/L)	2,000–3,000	2,700
BOD/COD ratio	0.57–0.6	0.55
Suspended solids (mg/L)	300–400	400
Volatile acids (mg/L)	50–80	60
Alkalinity as $CaCO_3$ (mg/L)	50–100	60
Phenols (mg/L)	65–72	65

various pharmaceuticals. Characteristics of the waste generated vary greatly depending upon the manufacturing process and raw materials used in the production of various medicines.

3.3.4 Biological Production Plants

These plants are mainly involved in the production of antitoxins, antisera, vaccines, serums, toxoids, and antigens. The production of antitoxins, antisera, and vaccines generates wastewaters containing animal manure, animal organs, baby fluid, blood, fats, egg fluid and egg shells, spent grains, biological culture, media, feathers, solvents, antiseptic agents, herbicidal components, sanitary loads, and equipment and floor washings. Overall, 1,80,000 G/day of waste is generated by biological production plants [17]. The various types of waste generated mainly include:

- waste from test animals;
- pathogenic-infectious waste from laboratory research on animal disease;
- toxic chemical wastes from laboratory research on bacteriological, botanical, and zoological problems;
- waste from antisera/antitoxins production;
- sanitary wastes.

Table 10 gives the characteristics of liquid waste arising in liver and beef extract production from a biological production pharmaceutical plant [18]. These wastes can be very high in BOD, COD, TS, colloidal solids, toxicity, color, and odor. The BOD/COD ratio of the

Table 10 Characteristics of Liquid Waste Arising in Liver and Beef Extract Production from a Biological Production Pharmaceutical Wastewater [18]

Constituents	Range	Mean
pH	5–6.3	5.8
Temperature (°C)	26.5–30	28
BOD_5 (mg/L)	11,400–16,100	14,200
COD (mg/L)	17,100–24,200	21,200
BOD/COD ratio	0.66–0.67	0.67
Total solids (TS) (mg/L)	16,500–21,600	20,000
Volatile solids (VS) (mg/L)	15,900–19,600	19,200
TKN (mg/L)	2,160–2,340	2,200
Crude fat (mg/L)	3,800–4,350	4,200
Volatile fatty acids (VFA) (mg/L)	1,060–1,680	1,460

BOD, biochemical oxygen demand; COD, chemical oxygen demand; TKN, total Kjeldhal nitrogen.

Table 11 Characteristics of Typical Spent Stream of Biologicals Production Plant at Greenfield, IN [20]

Parameter	Value
Flow (G/day)	15000
pH	7.3–7.6
BOD (mg/L)	1,000–1,700
Total solids (TS) (mg/L)	4,000–8,500
Suspended solids (mg/L)	200–800
Percentage suspended solids	5–10

BOD, biochemical oxygen demand.

waste is around 0.66. The waste contains volatile matter as 95% of TS present in the waste, containing easily degradable biopolymers such as fats and proteins. Table 11 presents the characteristics of spent streams generated from a typical biological production plant, Eli Lilly and Co., at Greenfield, IN [19,20].

3.3.5 Drug Mixing, Formulation, and Preparation Plants

Drug formulating processes consist of mixing (liquids or solids), palletizing, encapsulating, and packaging. Raw materials utilized by a drug formulator and packager may include ingredients such as sugar, corn syrup, cocoa, lactose, calcium, gelatin, talc, diatomaceous, earth, alcohol, wine, glycerin, aspirin, penicillin, and so on. These plants are mainly engaged in the production of pharmaceuticals primarily of a nonprescription type, including medications for arthritis, coughs, colds, hay fever, sinus and bacterial infections, sedatives, digestive aids, and skin sunscreens. Wastewater characteristics of such plants vary by season, depending upon the production of medicines to meet seasonal demands. However, the waste can be characterized as being slightly acidic, of high organic strength (BOD, 750–2000 mg/L), relatively low in suspended solids (200–400 mg/L), and exhibiting a degree of toxicity. During the period when cough and cold medications are prepared, the waste may contain high concentration of mono- and disaccharides and may be deficient in nitrogen [5]. A drug formulation plant usually operates a single shift, five days a week. Since drug formulating is labor-intensive, sanitary waste constitutes a larger part of total wastes generated, therefore waste loads generated from such plants are very low compared to other subcategories of bulk pharmaceutical manufacturing plants.

3.4 SIGNIFICANT PARAMETERS IN PHARMACEUTICAL WASTEWATER TREATMENT

Significant parameters to be considered in designing a treatment and disposal facility for pharmaceutical wastewater are given in Table 12. Biochemical oxygen demand measurements of the waste have been reported to increase greatly with dilution, indicating the presence of toxic or inhibitory substances in some pharmaceutical effluents. The toxicity impact upon various biological treatments by various antibiotics, bactericidal-type compounds, and other pharmaceuticals has been described in the literature [21–24].

Discharge permits for pharmaceutical manufacturing plants place greater attention on high concentrations of ammonia and organic nitrogen in the waste. Considerable amounts of TKN (total Kjeldhal nitrogen) have been found to still remain in the effluent even after undergoing a high level of conventional biological treatment. It has also been reported that the nitrogen load of treated effluent may sometimes exceed even the BOD load. This generates an oxygen demand, increased chlorine demand, and formation of chloramines during chlorination, which may be toxic to fish life and create other suspected health problems. The regulatory authorities have limited the concentration of unoxidized ammonia nitrogen to 0.02 mg/L in treated effluent.

Certain pharmaceutical waste may be quite resistant to biodegradation by conventional biological treatment. For example, various nitroanilines have been used in synthesized production of sulfanilamide and phenol mercury wastes and show resistance against biological attack. Both ortho and meta nitroaniline were not satisfactorily degraded even after a period of many months [25]. Other priority pollutants such as tri-chloro-methyl-proponal (TCMP) and toluene must be given attention in the treatment of pharmaceutical wastewater. With careful controls, *p*-nitroaniline can be biologically degraded, although the reaction requires many days for acclimatization [25,26].

Table 12 Parameters of Significance for the Pharmaceutical Industry Wastewater [3]

pH	Fecal coliform
Temperature	Manganese
BOD_5, BOD_{Ult}	Phenolics
COD	Chromium
Dissolved oxygen	Aluminum
TOC	Cyanides
Solids (suspended and dissolved)	Zinc
Oil and Grease	Lead
Nitrogen, (NH_4 and organic-N)	Copper
Sulfides	Mercury
Toxicity	Iron

BOD, biochemical oxygen demand; COD, chemical oxygen demand; TOC, total organic carbon.

3.5 WASTE RECOVERY AND CONTROL

Production processes used in the pharmaceutical/fine chemical, cosmetic, textile, rubber, and other industries result in wastewaters containing significant levels of aliphatic solvents. It has been reported that of the 1000 tons per year of EC-defined toxic wastes generated in Ireland, organic solvents contribute 66% of the waste [27]. A survey of the constituents of pharmaceutical wastewater in Ireland has reported that aliphatic solvents contribute a significant proportion of the BOD/COD content of pharmaceutical effluents. Organic solvents are flammable, malodorous, and potentially toxic to aquatic organisms and thus require complete elimination by wastewater treatment systems.

Pretreatment and recovery of various useful byproducts such as solvents, acids, sodium sulfate, fermentation solids, and fermentation beers comprise a very important waste control strategy for pharmaceutical plants. Such an approach not only makes expensive biological treatment unnecessary, but also gives economic returns in recovery of valuable byproducts [19,21,28–33].

In fermentation plants, the spent fermentation broth contains considerable levels of solvents and mycelium. As mentioned earlier, these solvents exhibit very high BOD strength and also some of the solvents are not biologically degradable; hence, if not removed/recovered, the latter places a burden on the biological treatment of the waste and destroys the performance efficiency of biological treatment. Intense recovery of these solvents in fermentation processes is thus recommended as a viable option to reduce flow into pharmaceutical effluents. The mycelium, which poses several operational problems during treatment, can be recovered for use as animal feed supplements. Separate filtration, drying, and recovery of mycelium has been recommended as the best method for its use as animal feed or supplements. Moreover, spent fermentation broth contains high levels of nutrients and protein, which attains a high value when incorporated into animal feeds. Large-scale fermentation solids recovery is practiced at Abbott Labs, North Chicago, IL; and has been conducted at Upjohn Co., Kalamazoo, Michigan, and at Abbott Labs, Barceloneta, Puerto Rico [3].

Spent beers contain a substance toxic to the biological system and exhibit considerable organic strength; hence, it needs to be removed/recovered to avoid the extra burden on the biological treatment. Large-scale recovery of antibiotic spent beers by triple-effect evaporators was carried out at Upjohn Co., Kalamazoo, Michigan, in the 1950s. Biochemical oxygen demand reduction with the triple-effect evaporation system was reported to be 96 to 98% for four different types of antibiotic spent beers. A similar practise had been adopted by pharmaceutical plants Pfizer (Terre Haute, IN) and Lederle Labs (Pearl River, NY) for the recovery of spent beers in the 1950s and 1960s, but these practises have been discontinued due to changing products or other conditions.

From 1972 to 1973, Abbott Labs in North Chicago, IL, recovered beers with a BOD_5 (five-day biological oxygen demand) load potential of 20,000 lb/day or greater. In the process, the spent beers were concentrated by multiple effect evaporators to 30% solids and the resulting syrup sold as a poultry feed additive. Any excess was incinerated in the main plant boilers. Abbott Labs reported that an average overall BOD reduction efficiency of the system up to 96% or more could be achieved.

Recovery of valuable products from penicillin, riboflavin, streptomycin, and vitamin B_{12} fermentation has been recommended as a viable waste control strategy when incorporated into animal feeds or supplements. Penicillin wastes, when recovered for animal feed, are reported to contain valuable growth factors, mycelium, and likewise evaporated spray-dried soluble matter [31,32,34].

Recovery of sodium sulfate from waste is an important waste control strategy within synthetic organic pharmaceutical plants. A sodium sulfate waste recovery system was employed in the Hoffmann–La Roche (Belvidere, NJ) plant, which manufactured synthetic organic pharmaceuticals. In 1972, the company reported 80 tons/day of sodium sulfate recovery [3]. The recovery and subsequent sale of sodium sulfate not only gave an economic return, but also reduced the influent sulfate concentration that may otherwise cause sulfide toxicity in anaerobic treatment of the pharmaceutical effluents.

To use water efficiently, the cooling and jacketing tower water must be segregated from the main waste streams and should be recycled and reused in cooling towers. Scavenging and recovery of high-level ammonia waste streams is recommended as a viable option of ammonia recovery for waste streams containing high concentrations of ammonia nitrogen.

The recovery of alcohol by distillation, concentration of organics, and use of waste activated sludge as a soil conditioner and fertilizer has also been reported [35].

Based on extensive experience in wastewater reduction and recovery experience at Bristol Labs (Syracuse, NY) and at the Upjohn Company (Kalamzoo, Michigan), the following practises have been recommended for waste control and recovery of byproducts in pharmaceutical industries [8,9,36,37]:

1. Install stripping towers for solvent removal (recover solvents wherever possible);
2. Conduct a program of sampling and testing solvents on wastewater flows;
3. Collect and incinerate nonreusable combustible solvents and residues;
4. Remove all mycelium;
5. Carefully program dumping of contaminated or spoiled fermentation batches;
6. Eliminate all possible leakage of process materials;
7. Separate clean waters from contaminated wastewaters;
8. Collect and haul selected high organic wastes to land disposal or equivalent;
9. Recycle seal waters on a vacuumed pump system;
10. Improve housekeeping procedures.

3.6 TREATMENT OF PHARMACEUTICAL WASTEWATER

The pharmaceutical industry employs a wide array of wastewater treatment and disposal methods [3]. Wastes generated from these industries vary not only in composition but also in magnitude (volume) by plant, season, and even time, depending on the raw materials and the processes used in manufacturing of various pharmaceuticals. Hence it is very difficult to specify a particular treatment system for such a diversified pharmaceutical industry. Many alternative treatment processes are available to deal with the wide array of waste produced from this industry, but they are specific to the type of industry and associated wastes. Available treatment processes include the activated sludge process, trickling filtration, the powdered activated carbon-fed activated sludge process, the anaerobic hybrid reactor. An incomplete listing of other treatments includes incineration, anaerobic filters, spray irrigation, oxidation ponds, sludge stabilization, and deep well injection. Based upon extensive experience with waste treatment across the industry, a listing of the available treatments and disposals is summarized as follows [3]:

- Separate filtration of mycelium, drying and recovery of fermentation broth and mycelium for use as animal feed supplements.
- Solvent recovery at centralized facilities or at individual sectors, reuse and/or incineration of collected solvents.
- Special recovery and subsequent sale of sodium sulfate.
- Cooling towers for reuse of cooling and jacketing waters.

- Scavenging and recovery of high-level ammonia waste streams.
- Elimination of barometric condensers.
- Extensive holding and equalization of wastewater prior to main treatment.
- Extensive neutralization and pH adjustment.
- The activated sludge process including multiple-stage, extended aeration, the Unox pure oxygen system, aerated ponds, and other variations.
- The trickling filter process, including conventional rate filters, multiple-stage, high-rate systems, and bio-oxidation roughing towers.
- Treatment of selected waste streams by activated carbon, ion exchange, electro-membranes, chemical coagulation, sand and dual and multimedia filtration.
- Spray irrigation of fermentation beers and other pharmaceutical wastes.
- Collection of biological, synthetic organic and pathogenic waste for incineration or disposal by separate means such as steam cooking and sterilization of pathogenic wastes.
- Multiple effects evaporation–steam and/or oil, multiple hearth and rotary kiln incineration, and other special thermal oxidation systems.
- Incineration of mycelium and excess biological sludge. Incineration system may also receive pathogenic wastes, unrecoverable solvents, fermentation broths or syrups, semi-solid and solid wastes, and so on. The system can be further integrated with the burning of odorous air streams.
- Acid cracking at low pH.
- Excess biological sludge can be handled by flotation, thickening, vacuum filtration, centrifugation, degasification, aerobic and/or anaerobic digestion, lagooning, drying, converting to useable product, incineration, land spreading, crop irrigation, composting, or land filling.
- Chlorination, pasteurization, and other equivalent means of disinfecting final effluents. Disinfection is generally utilized inside vaccine-antitoxins production facilities, and in some cases dechlorination may be required.
- Extensive air stream cleaning and treatment systems.
- Municipal waste treatment.

The treatment options cited above are very specific to the type of waste. To have a clear understanding of the various unit operations used in the treatment and disposal of various types of wastes produced in the pharmaceutical industry, the treatment processes can be divided into the following three categories and subcategories:

1. physicochemical treatment process;
2. biological treatment process,
   (i) aerobic treatment,
   (ii) anaerobic treatment,
   (iii) two-stage biological treatment,
   (iv) combined treatment with other waste;
3. integrated treatment and disposal facility for a particular plant wastewater.

3.6.1 Physicochemical Treatment

Physicochemical treatment of pharmaceutical wastewater includes screening, equalization, neutralization/pH adjustment, coagulation/flocculation, sedimentation, adsorption, ozone and hydrogen peroxide treatment. Detailed descriptions of the various physicochemical treatment processes are described in the following sections.

Extensive Holding and Equalization of Waste

As explained earlier, waste produced from the pharmaceutical industry varies in composition and magnitude depending upon various factors, that is, raw materials, manufacturing processes, process modifications, and specific demand of seasonal medicines, and so on. Such variation in the quality and quantity of the wastewater may cause shock as well as underloading to the various treatment systems, which leads to malfunctioning or even failure of treatment processes, particularly biological treatment. To avoid these operational problems, extensive holding and equalization of wastewater is extremely important. Use of an equalization basin has been reported effectively to control shock loading on further treatment units treating the pharmaceutical waste [5]. The retention time and capacity of the holding tank in such cases is designed based on the degree of variability in composition and magnitude of the wastewater.

Neutralization/pH Adjustment

Wastewater generated from the pharmaceutical industry varies greatly in pH, ranging from acidic to alkaline. For example, the pH of an alkaline waste stream from a synthetic organic pharmaceutical plant ranges from 9 to 10, whereas a pH of 0.8 has been reported for acidic waste streams [13,15]. Nevertheless, almost all types of waste streams produced from the pharmaceutical industry are either alkaline or acidic, and require neutralization before biological treatment. Thus, neutralization/pH adjustment of the waste prior to the biological system is a very important treatment unit for the biological treatment of pharmaceutical wastewater. The pH of the wastewater in this unit is adjusted by adding alkali or acid depending upon the requirement of the raw wastewater.

Coagulation/Flocculation

Coagulation and flocculation of the wastewater are carried out for the removal of suspended and colloidal impurities. The application of such treatment units greatly depends upon the suspended and colloidal impurities present in the raw wastewater. Coagulation and flocculation of pharmaceutical wastewater have been reported to be less effective at a pharmaceutical plant in Bombay that produces allopathic medicines [16]. The effects of various coagulants such as $FeSO_4$, $FeCl_3$, and alum on suspended solids and COD removal efficiency were evaluated. The wastewater used in the study contained an average BOD of 1500 mg/L; COD, 2700 mg/L; phenol, 65 mg/L and SS (suspended solids), 400 mg/L (Table 9). It was found that at the optimum doses of $FeSO_4$ (500 mg/L), $FeCl_3$, (500 mg/L), and alum (250 mg/L), the COD and SS removal efficiency was 24–28% and 70%, respectively. The study indicates that high doses of the coagulants were required, but the COD removal efficiency was marginal. Based on the above results, it was concluded that physicochemical treatment of effluent from this type of plant prior to biological treatment is neither effective nor economical [16]. A similar observation was made in a coagulation study of wastewater from the Alexandria Company for Pharmaceuticals and Chemical Industries (ACPCI) [38].

Air Stripping

Air stripping of pharmaceutical wastewater is a partial treatment used in particular for the removal of volatile organics from wastewater. M/S Hindustan Dorr Oliver, Bombay, in 1977 studied the effect of air stripping on the treatment of pharmaceutical wastewater and reported that a COD removal efficiency up to 30–45% can be achieved by air stripping. It was found that adding caustic soda did not appreciably increase the air stripping efficiency.

Ozone/Hydrogen Peroxide Treatment

Pharmaceutical wastewater contains various kinds of recalcitrant organics such as toluene, phenols, nitrophenols, nitroaniline, trichloromethyl propanol (TCMP), and other pollutants that exhibit resistance against biodegradation. Since these pollutants cannot be easily removed by biological treatment, biologically treated effluent exhibits a considerable oxygen demand, that is, BOD and COD, in the effluent. It has also been reported that activated carbon adsorption may not always be successful in removing such recalcitrant organics [39,40]. Economic constraints may also prohibit the treatment of pharmaceutical wastewater by activated carbon adsorption [41]. In such cases, ozone/hydrogen peroxide treatment may appear to be a proven technology for treating such pollutants from pharmaceutical wastewater.

The removal of organic 1,1,1-trichloro-2-methyl-2-propanol (TCMP), a common preservative found in pharmaceutical effluent, by ozone and hydrogen peroxide treatment has been studied [39]. Oxidation of TCMP was quite effective when it was contained in pure aqueous solutions, but almost nil when the same quantity of TCMP was present in pharmaceutical wastewater. Competitive ozonation of other organic solutes present inhibits the degradation of TCMP in pharmaceutical wastewater. Hence it has been concluded that for effective removal of TCMP by ozone/hydrogen peroxide, biological pretreatment of the wastewater for the removal of other biodegradable organics is crucial. It has been concluded that biological pretreatment of pharmaceutical wastewater before ozonation/hydrogen peroxide treatment should be utilized in order to increase the level of treatment.

3.6.2 Biological Treatment

The biological treatment of pharmaceutical wastewater includes both aerobic and anaerobic treatment systems. Aerobic treatment systems have traditionally been employed, including the activated sludge process, extended aeration activated sludge process, activated sludge process with granular activated carbon, or natural or genetically engineered microorganisms and aerobic fixed growth system, such as trickling filters and rotating biological contactors. Anaerobic treatment includes membrane reactors, continuously stirred tank reactors (anaerobic digestion), upflow filters (anaerobic filters), fluidized bed reactors, and upflow anaerobic sludge blanket reactors. Anaerobic hybrid reactors, which are a combination of suspended growth and attached growth systems, have recently become popular. Pharmaceutical/fine chemical wastewater presents difficult substrates for biological treatment due to their varying content of a wide range of organic chemicals, both natural and xenobiotic, which may not be readily metabolized by the microbial associations present in the bioreactors. Various processes dealing with the biological treatment of pharmaceutical wastewater are summarized in subsequent sections.

Activated Sludge Process

The activated sludge process has been found to be the most efficient treatment for various categories of pharmaceutical wastewater [14,15,19,42–46]. It has also been reported that this process can be successfully employed for the removal of tert-butanol, a common solvent in pharmaceutical wastewater that cannot be degraded by anaerobic treatment [44]. At a volumetric loading rate of 1.05 kg COD/m^3 day, HRT (hydraulic retention time) of 17 hours, and mixed liquor dissolved oxygen concentration of 1 mg/dm^3, the tert-butanol can be completely removed by the activated sludge process.

The activated sludge process has been successfully employed for the treatment of a wide variety of pharmaceutical wastewaters. The American Cynamid Company operated an activated sludge treatment plant to treat wastewater generated from the manufacture of a large variety of chemicals [19]. The activated sludge process has also been successfully employed for the treatment of wastewater in the chemical and pharmaceutical industry [42]. M/S Hindustan Dorr Oliver of Bombay studied the performance of the activated sludge process for the treatment of wastewater from its plant in 1977, and concluded that at an MLSS (mixed liquor suspended solids) concentration of 1800–2200 mg/L and aeration period of 24 hours, a COD removal efficiency of 50–83% can be achieved.

The performance of the activated sludge process for the treatment of wastewater from a synthetic drug factory, has been reported [14,15,45]. One of the biggest plants of its kind in Asia, M/S Indian Drugs and Pharmaceutical Ltd., Hyderabad, went into production in 1966 to make sulfa drugs such as sulfanilamides: antipyretics (phenacetin), B-group vitamins, antitubercular drugs (isonicotinic acid hydrazide) and antihelminthic, and so on.

When the performance of the activated sludge process was first studied for the treatment of simulated pharmaceutical wastewater, it was found that the wastewater was biologically treatable and that this process can be successfully employed for treating wastewater from pharmaceutical plants [45]. Based on Mohanrao's [14] recommendation, the performance of the activated sludge process for the treatment of actual waste streams generated from this plant, that is, alkaline waste, condensate waste, and a mixture of the two along with domestic sewage (1 : 2 : 1) as evaluated. Characteristics of various types of wastes used in the study are depicted in Table 13. The study demonstrated that condensate waste, as well as mixture, could be treated successfully, yielding an effluent BOD of less than 10 mg/L. However, the BOD removal efficiency of the system for the alkaline waste alone was found to be only 70%. The settleability of the activated sludge in all three units was found to be excellent, yielding a sludge volume index 23 and 45. The study indicated that biological treatability of the waste remained the same, although the actual waste was about 10 times diluted compared with the synthetic waste.

In 1984, the performance of a completely mixed activated sludge process for the treatment of combined wastewater was again evaluated. It was found that the activated sludge process was amenable for the treatment of combined wastewater from the plant, concluding that segregation and giving separate treatment for various waste streams of the plant would not be beneficial. The study was conducted at various sludge loading rates (0.14–0.16, 0.17–0.19, and 0.20–0.26 kg BOD/kg MLVSS (mixed liquor volatile suspended solids) per day and indicated that for the lower two loadings, effluent BOD was less than 50 mg/L, while for the other two higher loading

Table 13 Characteristics of Alkaline and Condensate Wastes Generated from a Synthetic Drug Plant at Hyderabad [14]

	Alkaline waste			Condensate waste		
Parameters	Min.	Max.	Avg.	Min.	Max.	Avg.
pH	8.6	9.4	–	7.0	7.6	–
BOD (mg/L)	1025	1345	1204	155	490	257
COD (mg/L)	2475	3420	2827	413	850	572
COD/BOD	2.41	2.54	2.3	2.66	1.73	2.2
Total solids (%)	0.53	0.66	0.63	0.12	0.14	0.13
Volatile solids (% of TS)	29.3	67.7	51.0	36.6	50.6	45.3
Total nitrogen (mg/L)	–	–	560	–	–	56
Total phosphorus (mg/L)	–	–	Nil	–	–	Nil

TS, total solids; BOD, biochemical oxygen demand; COD, chemical oxygen demand.

effluents BOD was less than 100 mg/L. The average TOC, COD, and BOD reductions were around 80, 80, and 99% respectively. The settleability of the activated sludge was found to be excellent with an SVI of 65–72 [15].

A similar study was conducted at Merck & Co. (Stonewall Plant, Elkton, Virginia) to assess the feasibility of the activated sludge process for treating wastewater generated from this plant. This plant is one of the six Merck Chemical Manufacturing Division facilities operated on a batch basis for fermentation and organic synthesis and has been in operation since 1941. A bench-scale study revealed that a food to microorganism (F/M) ratio from 0.15 to 0.25, MLVSS of 3500 mg/L, HRT 4 days, and minimum DO (dissolved oxygen) concentration of 3 mg/L was essential for meeting the proposed effluent limits and maintaining a viable and good settling sludge in the activated sludge process [46]. Based on these design criteria, a pilot plant and full-scale system were designed and studied. The old treatment plant consisted of an equalization basin, neutralization, primary sedimentation, roughing biofilter, activated sludge system, and rock trickling filter with final clarifiers. In the proposed study, the old activated sludge system, rock filter, and final clarifier were replaced with a new single-stage, nitrification-activated sludge system. A schematic diagram of the pilot plant is presented in Figure 1. The study demonstrated that BOD_5 removal efficiencies of the pilot and bench-scale plant were 94 and 98%, respectively. The TKN and NH_4-N removal were found to be 65 and 59%, respectively. It has also been observed that system operation was stable and efficient at F/M ratios ranging from 0.19 to 0.30, but prolonged operation at an F/M ratio less than 0.15 led to an episode of filamentous bulking.

The performance of the activated sludge process has been evaluated for the treatment of ACPCI (Alexandria Company for Pharmaceutical and Chemical Industry) effluent. These drug formulation and preparation-type plants are mainly involved in the production of a wide variety of pharmaceuticals, including analgesics, anthelmintics, antibiotics, cardiacs, chemotherapeutics, urologics, and vitamins. A study indicated that significant dispersed biosolids were found in the treated effluent when applying aeration for 6 hours. However, extending the aeration to 9–12 hours and maintaining the MLSS at levels higher than 2500 mg/L improved sludge

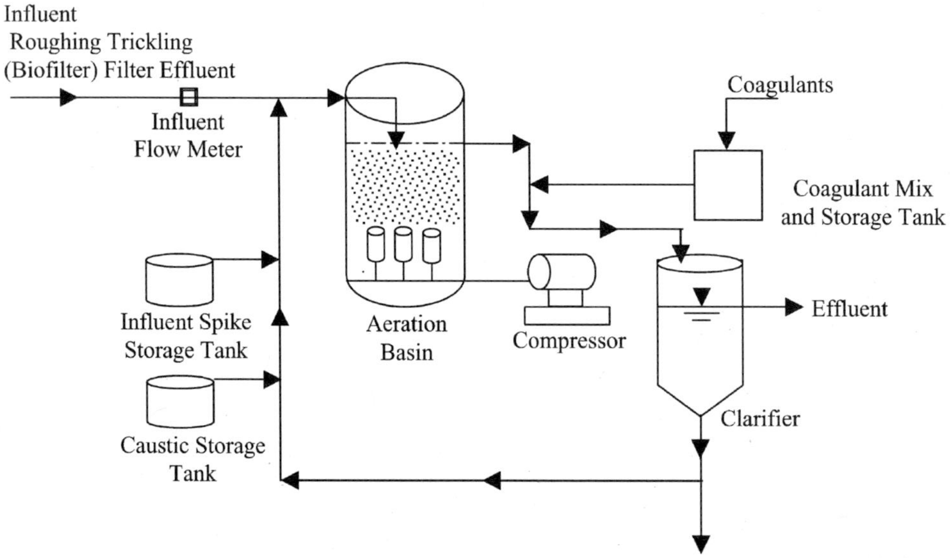

Figure 1 Schematic of the pilot plant at Merck and Co. Stonewall Plant in Elkton, VA.

settling and produced effluent with low SS. The study concluded that the activated sludge process is capable of producing effluent with BOD and SS values within the limits of the Egyptian standards. However, sand filtration was needed for polishing the treated effluent [38].

Powdered Activated Carbon Activated Sludge Process

Various researchers [47,48] have investigated the effect of powdered activated carbon (PAC) on the performance of the activated sludge process for the treatment of pharmaceutical wastewater. Various treatment units such as the activated sludge process (ASP), PAC-ASP, granular activated carbon (GAC), and a resin column were studied and compared in removing priority pollutants from a pharmaceutical plant's wastewater [47]. The wastewater generated from the plant contained 0-nitroaniline (0-NA), 2-nitrophenol (2-NP), 4-nitrophenol (4-NP), 1,1,2-trichloroethane (TCE), 1,1-dichloroethylene (DCE), phenol, various metals, and other organics. Characteristics of the wastewater collected from the holding pond are given in Table 14. The study concluded that there are treatment processes available that can successfully remove the priority pollutants from pharmaceutical wastewater. The treatment systems, ASP, PAC-ASP, and GAC, were all quite efficient in removing phenol, 2-NP and 4-NP, while the resin column was found unable to treat phenol. However, 2-NP and 4-NP can be treated to a certain extent (72 and 65%, respectively). The author further concluded that 1,1,2-dichloroethane and 1,1-dichloroethane can be treated successfully by all four treatment systems, but the efficiency of the resin column and GAC exceeded the other two system. In terms of TOC removal, ASP and PAC-ASP were found to be more efficient than either GAC or the resin column. However, the performance of the PAC-fed ASP was found to be most efficient. In terms of color removal, PAC, GAC, and the resin process were more efficient than ASP, whereas in terms of arsenic removal, GAC and resin column were found most efficient. The performance summary of various treatment systems is given in Table 15. In general, it may be concluded that the addition of PAC in the ASP produced a better effluent than the ASP.

Addition of PAC to the activated sludge process increases the soluble chemical oxygen demand (SCOD) removal from the pharmaceutical wastewater but no measurable effect in terms

Table 14 Characteristics of Wastewater from a Typical Pharmaceutical Industry [47]

Parameters	Average	Ranges (min.–max.)
Color	4,648	1,800–6,600
TSS (mg/L)	234	47–2,700
VSS (mg/L)	152	17–1,910
TOC (mg/L)	387	205–630
Arsenic (mg/L)	5.82	4–12
o-Nitraniline (ONA) (μg/L)	12,427	3,200–30,500
Phenol (μg/L)	1,034	<10 to 3,700
2-NP (μg/L)	1,271	<10 to 2,900
4-NP (μg/L)	635	<10 to 2,300
TCE (μg/L)	4,080	620–6,550
DCE (μg/L)	291	<10 to 1,060

TSS, total suspended solids; VSS, volatile suspended solids; 4-NP, 4-nitrophenol; 2-NP, 2-nitrophenol; TCE, 1,1,2-trichloroethane; DCE, 1,1-dichloroethylene; TOC, total organic carbon.

Table 15 Performance Efficiency of Various Systems for the Treatment of Pharmaceutical Wastewater [47]

	Removal efficiency (%)			
Parameter	ASP	PAC-ASP	GAC	Resin column
Color	46.3	94.9	96.9	92
TOC	72.4	89.7	43.9	15
Phenol	95.8	>99	95.4	Nil
2-Nitrophenol	93.8	>99.2	99.1	72.3
4-Nitrophenol	89.4	96.5	96.5	65.8
o-Nitraniline	58.6	94.1	99.9	96.7
Arsenic	20.6	42.8	73.9	62.5
1,1,2-trichloroethane	94.2	96.4	99.4	99.8
1,1-dichloroethylene	94.5	>96.6	95.5	96.6

ASP, activated sludge process; PAC-ASP, powdered activated carbon activated sludge process; GAC, granular activated carbon; TOC, total organic carbon.

of soluble-carbonaceous biochemical oxygen demand (S-CBOD) was observed [48]. Moreover, addition of PAC increased the sludge settleability, but the MLSS settling rate remained at a very low level (0.01 to 0.05 cm/min) and resulted in a viscous floating MLSS layer at the surface of the activated sludge unit and clarifier. This study concluded that a PAC-fed ASP cannot be recommended as a viable option for this plant wastewater until the cause of the viscous floating MLSS layer is identified and adequate safeguards against its occurrence are demonstrated. The relationship to estimate the dose of activated carbon required for producing a desired quality of the effluent is given in Eq. (1).

$$\frac{X}{M} = 3.7 \times 10^{-7} C_{\mathrm{e}}^{2.1} \tag{1}$$

where X is the amount of SCOD removal attributed to the PAC (mg/L), M is the PAC dose to the influent (mg/L), and C_e is the equilibrium effluent SCOD concentration (mg/L).

Extended Aeration

The performance of the ASP has been found to be more efficient when operating on an extended aeration basis. The design parameters of the process were evaluated for the treatment of combined wastewater from a pharmaceutical and chemical company in North Cairo that produced drugs, diuretics, laboratory chemicals, and so on [49]. The study revealed that at an extended aeration period of 20 hours, COD and BOD removal efficiency ranges of 89–95% and 88–98%, respectively, can be achieved. The COD and BOD values of the treated effluent were found to be 74 mg/L and 43 mg/L, respectively.

In contrast, the performance of an extended aeration system for the treatment of pharmaceutical wastewater at Lincoln, Nebraska, was poor. At an organic loading of 30 kg BOD/day and a detention period of 25 hours, the percentage BOD reduction ranged from 30 to 70%. The degree of treatment provided was quite variable and insufficient to produce a satisfactory effluent. The pilot plant study performed at various feeding rates of 1.5, 2.4, 3.0, 3.6 and 4.8 L/12 hours indicated that at feeding rate of 4.8 L/12 hours, the sludge volume index was 645 and suspended solids were being carried over in the effluent.

Oxidation Ditch

The performance of an oxidation ditch for treating pharmaceutical wastewater has been evaluated and described by many researchers [16,50]. Treatability of wastewater from a typical pharmaceutical industry at Bombay producing various types of allopathic medicines was studied in an oxidation ditch at HRTs ranging from 1 to 3 days, corresponding to an SRT (solid retention time) of 8–16 days. The average MLVSS concentration in the reactor varied from 3000 to 4800 mg/L during the investigation period. The study indicated that on average about 86–91% of influent COD and 50% of phenols could be removed by this process [16].

A pilot-scale oxidation ditch was evaluated for the treatment of pharmaceutical wastewater at a Baroda unit. The treatment system was comprised of neutralization followed by clarifier and oxidation ditch. Primary treatment of the wastewater using neutralization with lime followed by sedimentation in a clarifier demonstrated SS and BOD removal of 30–41% and 28–57%, respectively. The effluent from the clarifier was further treated in an oxidation ditch operating on an extended aeration basis. It was found that at loading of 0.1–0.5 lb BOD/lb MLSS/day, an MLSS concentration of 3000–4000 mg/L and aeration period of 22 hours, a BOD removal up to 70–80% could be achieved. The high COD of treated effluent indicated the presence of organic constituents resistant to biodegradation. Considering the high COD/BOD ratio of the wastewater, it has been suggested that the biological treatment should be supplemented with chemical treatment for this type of plant wastewater [50].

Aerated Lagoon

The performance studies of aerated lagoons carried out by many researchers [14,51] have demonstrated that lagoons are capable of successfully treating wastewater containing diversified fine chemicals and pharmaceutical intermediates.

A laboratory-scale study of alkaline and condensate waste streams from a synthetic drug factory at Hyderabad demonstrated that an aerated lagoon is capable of treating the wastewater from this industry [14]. The BOD removal rate K of the system was found to be 0.18/day and 0.155/day based on the soluble and total BOD respectively. Based on the laboratory studies, a flow sheet (Fig. 2) for the treatment of waste was developed and recommended to the factory.

Trickling Filter

The performance of a trickling filter has been studied by many researchers [14,38,49,51–53] and it was found that a high-rate trickling filter was capable of treating wastewater containing diversified fine chemicals and pharmaceutical intermediates to a level of effluent BOD less than 100 mg/L [51]. A similar conclusion was made in the performance study of a trickling filter for the treatment of wastewater from chemical and pharmaceutical units [53].

It has also been reported that wastewater from a pharmaceutical plant manufacturing antibiotics, vitamins, and sulfa drugs, can be treated by using a trickling filter [52]. One study evaluated the efficiency of a sand bed filter for the treatment of acidic waste streams from a synthetic organic pharmaceutical plant at Hyderabad. The acidic waste stream was neutralized to a pH of 7.0 and treated separately through a sand bed filter. The sand bed filter was efficient in treating the acidic waste stream to a level proposed for its discharge to municipal sewer [14].

The efficiency of the biological filter (trickling filter) for treatment of combined wastewater from a pharmaceutical and chemical company in North Cairo has been evaluated. The treatment system consisted of a biological filter followed by sedimentation. The degree of treatment was found quite variable. The COD and BOD removal efficiency of the trickling filter at an average OLR (organic loading rate) of 26.8 g BOD/m^2 day were found to be 43–88% and

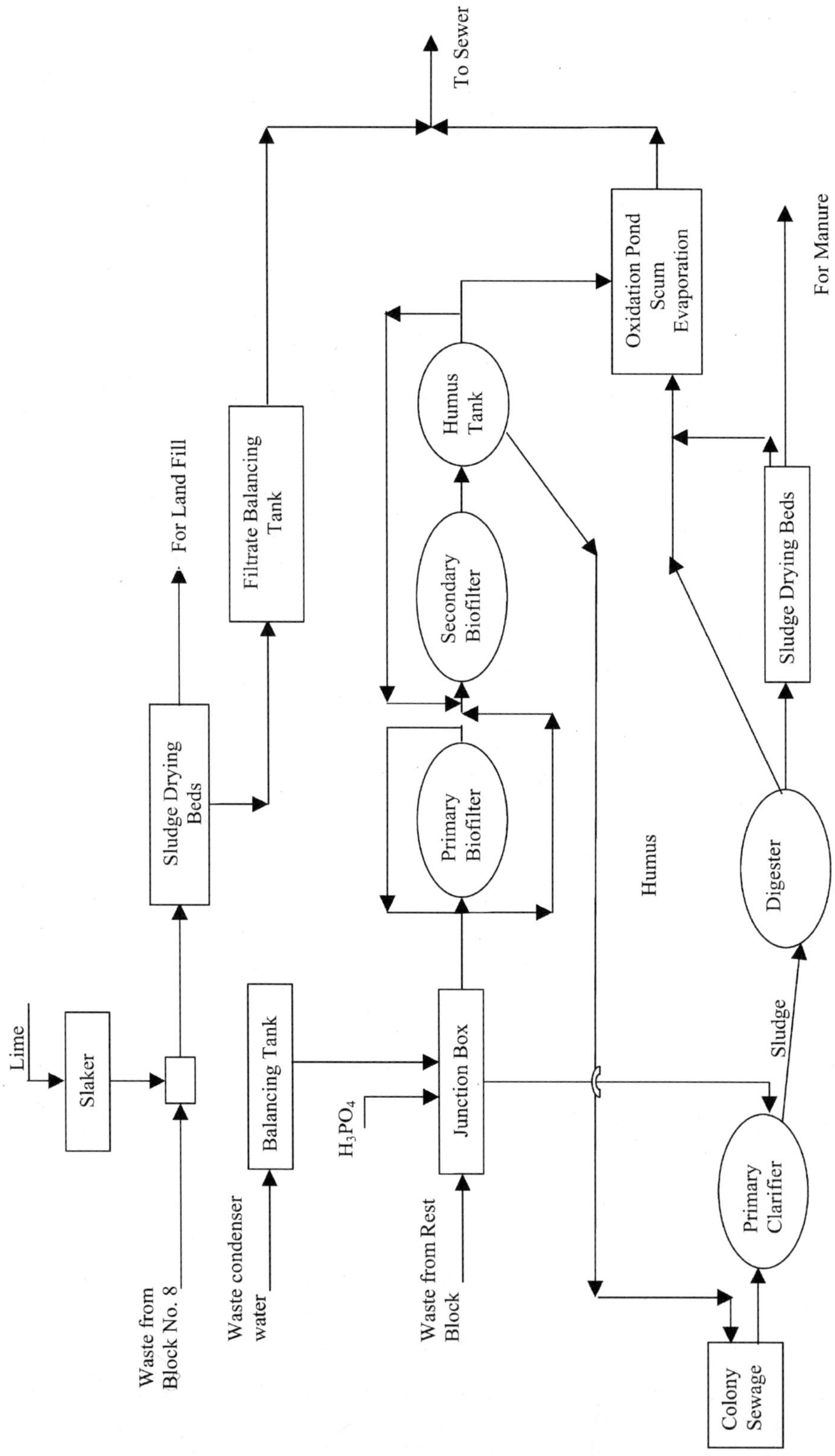

Figure 2 Flow sheet for treatment of synthetic drug waste.

58–87%, respectively. The study revealed that a biological filter alone was unable to produce effluents to a level complying with the national standards regulating wastewater disposal into the surface water [49].

Similar conclusions were made in the treatment of ACPCI effluent using a biofilter. The low performance efficiency and presence of dispersed biosolids in the effluent have made the trickling filter unsuitable for the treatment of this plant wastewater [38].

Anaerobic Filter

The anaerobic filter has been reported to be a promising technology for the treatment of wide varieties of pharmaceutical wastewater [4,10,54–59]. The performance of the anaerobic filter was first studied at a pharmaceutical plant in Springfield, Missouri [54]. The characteristics of the waste fed into the reactor are given in Table 16. The treatability study revealed that at an HRT of two days, an OLR ranging from 0.37 to 3.52 kg COD/m^3 day and influent COD concentration ranging from 1000 to 16,000 mg/L, COD removal efficiencies of 93.7 to 97.8% can be achieved. Moreover, the problem of sludge recycling and sludge disposal in the case of the anaerobic filter can be reduced to a great extent due to the much smaller biomass yield, that is, 0.027 g VSS (volatile suspended solids)/g COD removed. The shock loading study revealed that shock increase in organic loading did not result in a failure of the capability of the filter to treat the waste. This is a distinct feature of anaerobic filters, especially when dealing with pharmaceutical wastewater, which is supposed to cause shock loading due to frequent variation in composition as well as in magnitude of the waste load. In contrast, it has been reported that the

Table 16 Physical and Chemical Characteristics of Pharmaceutical Waste in Springfield, MO [54]

Parameters	Range
pH	7.5–10.1
COD (mg/L)	15,950–16,130
SS (mg/L)	28–32
TS (mg/L)	432–565
Alkalinity (mg/L as $CaCO_3$)	412–540
Nitrogen (mg/L)	
Ammonia	0–11.8
Organic	33.3–34.2
Phosphorus (mg/L)	
Ortho	0.4–0.5
Total	0.9–0.95
Heavy metals (mg/L)	
Lead	0.005–0.007
Copper	0.140
Zinc	0.018–0.11
Manganese	0.020–0.22
Iron	0.05–0.56
Cadmium	0.020–0.01
Calcium	9.7–58.7
Magnesium	7.5–14.7

COD, chemical oxygen demand; SS, suspended solids; TS, total solids.

anaerobic filter fed with pharmaceutical wastewater containing high ammonia nitrogen could not withstand a three-fold increase in OLR [55]. It has been further concluded that the amber color of the untreated waste can be removed through treatment, but due to poor degradability of the odor-producing toluene, the effluent maintained the tell-tale odor of toluene, indicating that it passed through the filter with little or no treatment.

The suitability of the anaerobic filter for treatment of wastewater from a chemically synthesizing pharmaceutical industry has been studied [10]. Characteristics of the strong waste stream used in the study are given in Table 17. The study revealed that at an HRT of 48 hours and COD concentration of 1000 mg/L, waste can be treated at least to a level of treatment generally occurring when employing aerobic treatment. Moreover, methane-rich biogas is generated in this treatment, which can be utilized later as an energy source. Thus the use of an anaerobic filter system would be a net energy producer rather than an energy consumer as in the case of current aerobic systems. In addition, the effluent from this system was found to contain far less color than the effluent from the existing system.

The performance of an anaerobic mesophilic fixed film reactor (AMFFR) and an anaerobic thermophilic fixed film reactor (ATFFR) for the treatment of pharmaceutical wastewater of a typical pharmaceutical plant at Mumbai were studied and compared [56]. The study revealed that at an OLR of 0.51 kg/m^3 day and HRT of 4.7 days, the COD removal efficiency of mesophilic was superior (97%) to the thermophilic reactor (89%). The effect of organic loading and reactor height on the performance of anaerobic mesophilic (30°C) and thermophilic (55°C) fixed film reactors have demonstrated that the AMFFR can take a load of several orders of magnitude higher, with higher removal efficiency compared to the ATFFR for pharmaceutical wastewater [56]. Wastewater used in the study was collected from an equalization tank of the pharmaceutical industry treatment plant at Bombay. The characteristics of the wastewater are given in Table 18. The start-up study has indicated that a starting-up period for the AMFFR (four months) was far less than the starting-up period for the ATFFR (six months). The gas production and methane percentage were also found to be higher in the AMFFR compared to the ATFFR. The effective height of the reactor was found to be in the range of 30–90 cm. Other researchers [10,54,55,58,59] have reported a similar effective height range of 15–90 cm. They have

Table 17 Characteristics of a Concentrated Waste Stream of Synthesized Organic Chemicals—Type Pharmaceutical Industry [10]

Parameters	Sample 1 (28-02-76)	Sample 2 (20-04-76)	Sample 3 (10-10-76)	Sample 4 (20-11-76)
pH	3.6	3.5	2.2	1.6
BOD_5 (mg/L)	Varies	–	–	–
COD (mg/L)	514,900	533,000	89,000	62,530
TS (mg/L)	37,740	38,520	13,090	5,190
TDS (mg/L)	37,650	38,420	13,030	5,180
TVSS (mg/L)	18,880	19,070	5,180	2,090
Dissolved volatile solids (mg/L)	18,800	18,980	5,120	2,080
TKN (mg/L)	19.3	25.8	23.0	33.6
NH_4^-N (mg/L)	BDL	BDL	BDL	BDL
SO_4^{2-} (mg/L)	–	–	75.0	183
Total phosphorous (mg/L)	BDL	BDL	BDL	BDL

BOD, biochemical oxygen demand; COD, chemical oxygen demand; TS, total solids; TDS, total dissolved solids; TVSS, total volatile suspended solids; TKN, total Kjeldhal nitrogen; BDL, below detectable limit.

Table 18 Characteristics of Wastewater from a Typical Pharmaceutical Industry at Bombay [56]

Parameters	Concentration range	Average
pH	5.5–9.2	7.2
COD (mg/L)	1,200–7,000	2,500
TSS (mg/L)	30–55	40
Total alkalinity as $CaCO_3$ (mg/L)	70–1,500	750
TVA (mg/L)	70–2,000	750
NH_4^+-N (mg/L)	80–500	200
PO_4^{3-}P (mg/L)	3.5–35	16
SO_4^{2-} (mg/L)	100–700	300
Chloride (mg/L)	500–1,200	900
Sulfide (mg/L)	2–8	5
Cobalt (mg/L)	0–0.6	0.2
Potassium (mg/L)	5–25	18
Lead (mg/L)	0.05–0.9	0.35
Iron (mg/L)	0.2–0.9	0.45
Zinc (mg/L)	0.05–0.15	0.09
Chromium (mg/L)	0.1–0.6	0.3
Mercury (mg/L)	0.15–0.50	0.25
Copper (mg/L)	0–0.10	0.1
Cadmium (mg/L)	0.07–0.25	0.10
Sodium (mg/L)	200–3,000	2,000
Manganese (mg/L)	0.1–0.4	0.2
Silicon (mg/L)	5–50	25
Magnesium (mg/L)	5–60	40
Tin (mg/L)	0.1–1.5	0.6
Aluminum (mg/L)	0.05–0.20	0.10
Barium (mg/L)	0.1–0.3	0.16
Arsenic (mg/L)	0.1–0.5	0.25
Bismuth (mg/L)	0.09–0.3	0.15
Antimony (mg/L)	0.50–3.0	1.4
Selenium (mg/L)	0.1–0.95	0.38

TVA, total volatile acid; COD, chemical oxygen demand; TSS, total suspended solids.

reported that rapid change in most of the characteristics occurs only in the lower portion of the reactor.

Two-Stage Biological System

The two-stage biological system generally provides a better quality of effluent than the single-stage biological system for the treatment of pharmaceutical wastewater. It has been reported that a single-stage biological system such as activated sludge process and trickling filter alone is not capable of treating the wastewater to the effluent limit proposed for its safe discharge to inland surface water [49]. However, the combined treatment using a two-stage aerobic treatment system is efficient in treating wastewater to a level complying with national regulatory standards. A performance study of a two-stage biological system for the treatment of pharmaceutical wastewater generated from Dorsey Laboratories Plant (drug mixing and formulation type plant) at Lincoln, Nebraska, was carried out and the following conclusions drawn:

- Shock organic and hydraulic loading created serious operational problems in the system. Bulking sludge and the inability to return solids from the clarifier to the aeration unit further complicated plant operation.
- Microscopic observations of the sludge flock showed the presence of filamentous organisms, *Sphaerotilus natans*, in high concentrations. The presence of these organisms was expected to be due to deficiency of the nitrogen in the wastewater.

To overcome the problem of sludge bulking, nitrogen was supplemented in the wastewater as ammonium sulfate, but operational problems continued even after nitrogen was added. Hence, to avoid shock loading on the treatment, the effluent treatment plant (ETP) was expanded. The expanded treatment system (Fig. 3) consists of a communicator, basket screen, equalization basin, biological tower, activated sludge process, disinfection, and filtration. The study indicated that the equalization basin and biological tower effectively controlled shock loading on the activated sludge process. Overall, BOD and COD removal of 96 and 88%, respectively, may be achieved by employing a two-stage biological system [5]. It has also been found that a two-stage biological system generally provides a high degree of treatment. However, bulking sludge causes severe operational problems in the extended aeration system and sand filter.

A two-stage biological treatment system consisting of anaerobic digestion followed by an activated sludge process was developed for the treatment of liquid waste arising from a liver and beef extract production plant. Being rich in proteins and fats, the waste had the following characteristics: pH, 5.8; COD, 21,200 mg/L; BOD, 14,200 mg/L; and TS, 20,000 mg/L. The treatability study of the waste in anaerobic digestion revealed that at an optimum organic loading rate of 0.7 kg COD/m^3 day and an HRT of 30 days, a COD and BOD removal efficiency of 89 and 91% can be achieved [18]. The effluent from anaerobic digestion still contains a COD of 2300 mg/L and BOD of 1200 mg/L. The effluent from anaerobic digestion was settled in a primary settling tank. At an optimum retention time of 60 minutes in the settling tank, the percentage COD and BOD removal increased to 94 and 95%, respectively. The effluent from the settling tank was then subjected to the activated sludge process. At an optimum HRT of 4 days, the COD and BOD removal increased to 96 and 97%, respectively. The effluent from the activated sludge process was settled for 1 hour in a secondary settling tank, which gave an increase in COD and BOD removal to 98 and 99%, respectively. The study therefore revealed that the combination of anaerobic–aerobic treatment resulted in an overall COD and BOD reduction of 98 and 99%, respectively. The final effluent had a COD of 290 mg/L and BOD of 50 mg/L, meeting the effluent standard for land irrigation.

The performance of two-stage biological systems was examined for the treatment of wastewater from a pharmaceutical and chemical company in North Cairo. A combined treatment using an extended aeration system (20 hour aeration) or a fixed film reactor (trickling filter) followed by an activated sludge process (11 hour detention time) was found efficient in treating the wastewater to a level complying with national regulatory standards. From a construction cost point of view, the extended aeration system followed by activated sludge process would be more economical than the fixed film reactor followed by activated sludge process. The flow diagrams of the two recommended alternative treatment processes for the treatment of this plant wastewater are depicted in Figure 4 and Figure 5, respectively [49].

Anaerobic treatment of high-strength wastewater containing high sulfate poses several unique problems. The conversion of sulfate to sulfide inhibits methanogenesis in anaerobic treatment processes and thus reduces the overall performance efficiency of the system. Treatment of high sulfate pharmaceutical wastewater via an anaerobic baffled reactor coupled

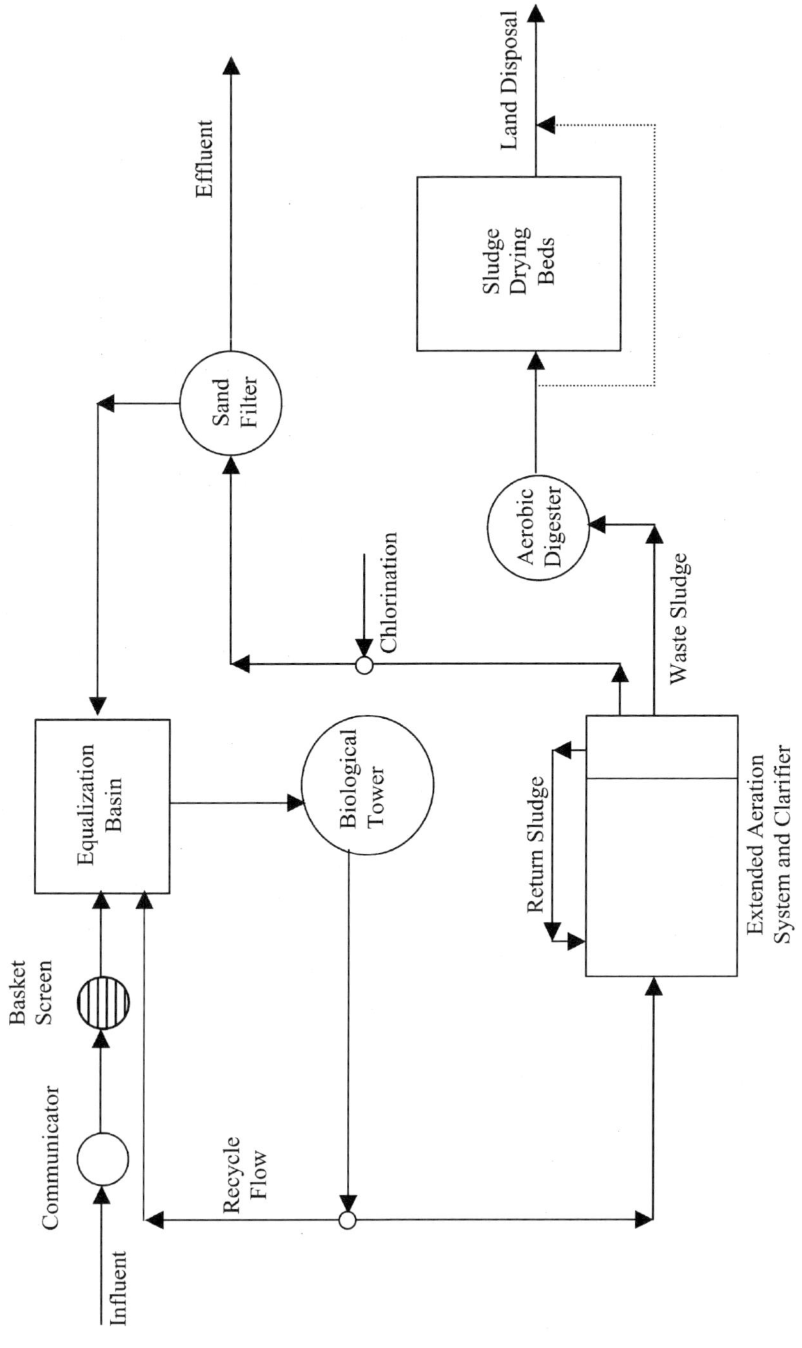

Figure 3 Flow diagram of wastewater treatment plant at Dorsey Laboratory.

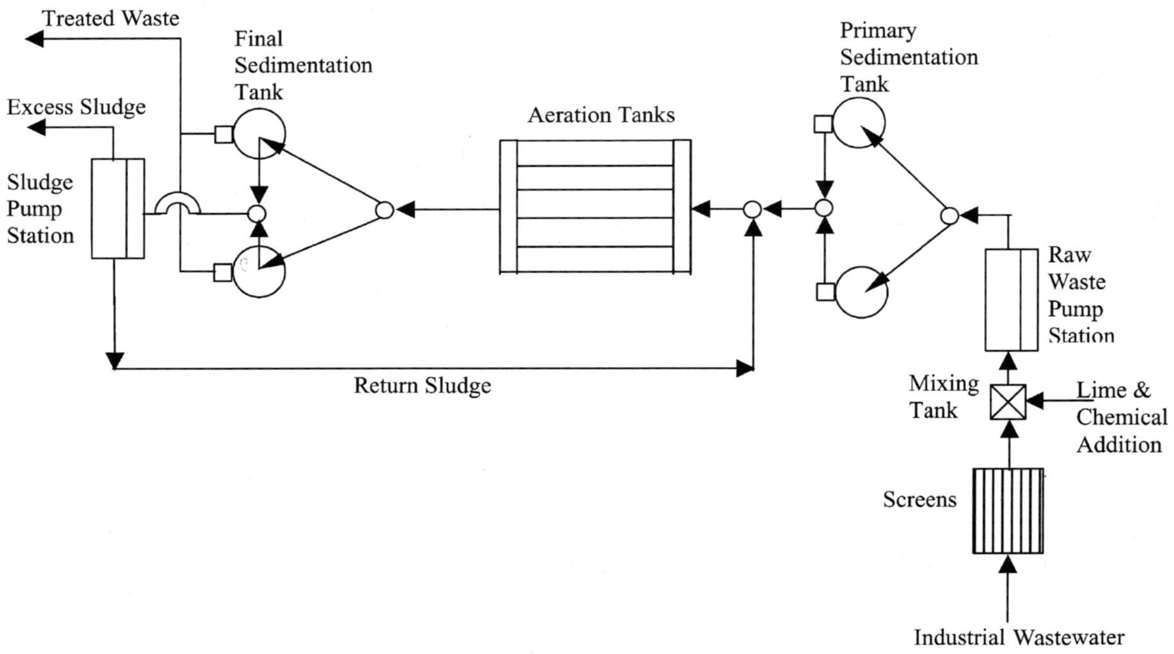

Figure 4 Flow diagram for treatment process using activated sludge, extended aeration.

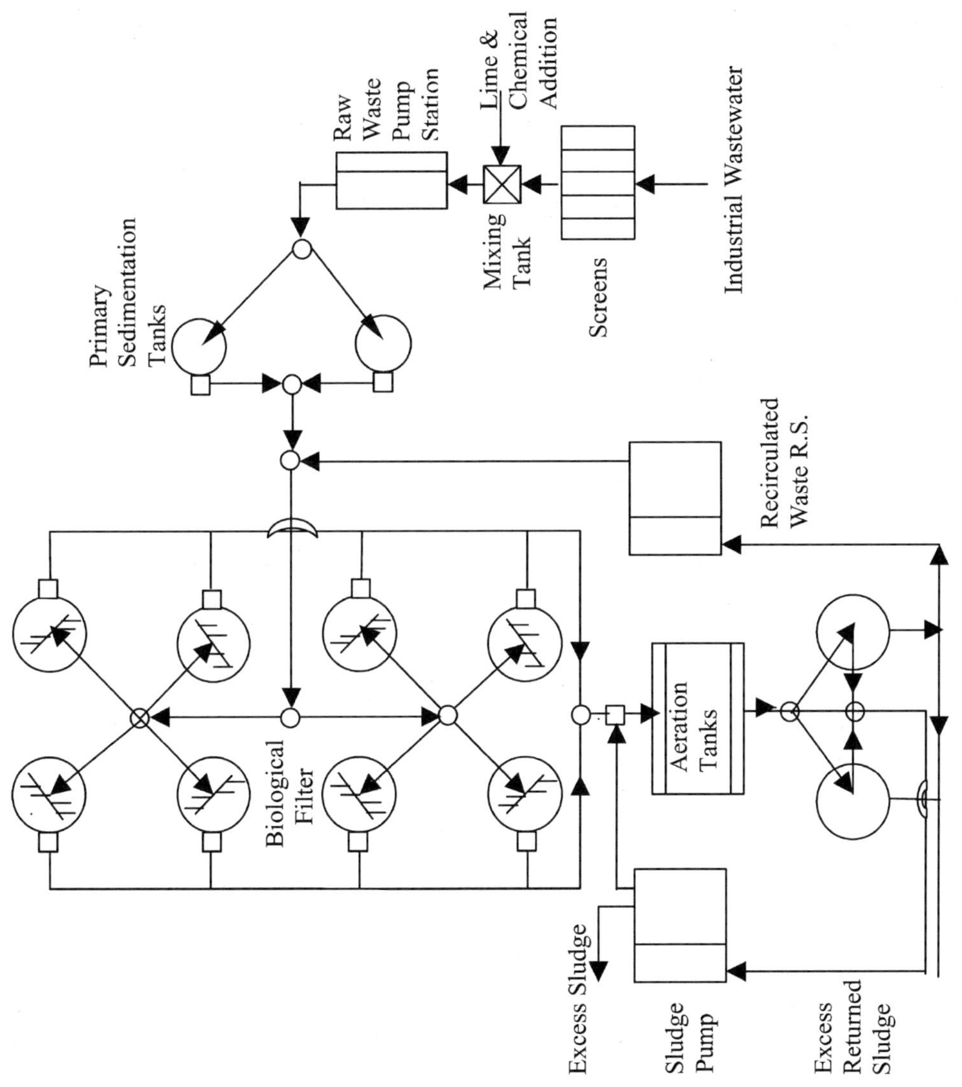

Figure 5 Flow diagram for treatment process using biological filters followed by activated sludge process.

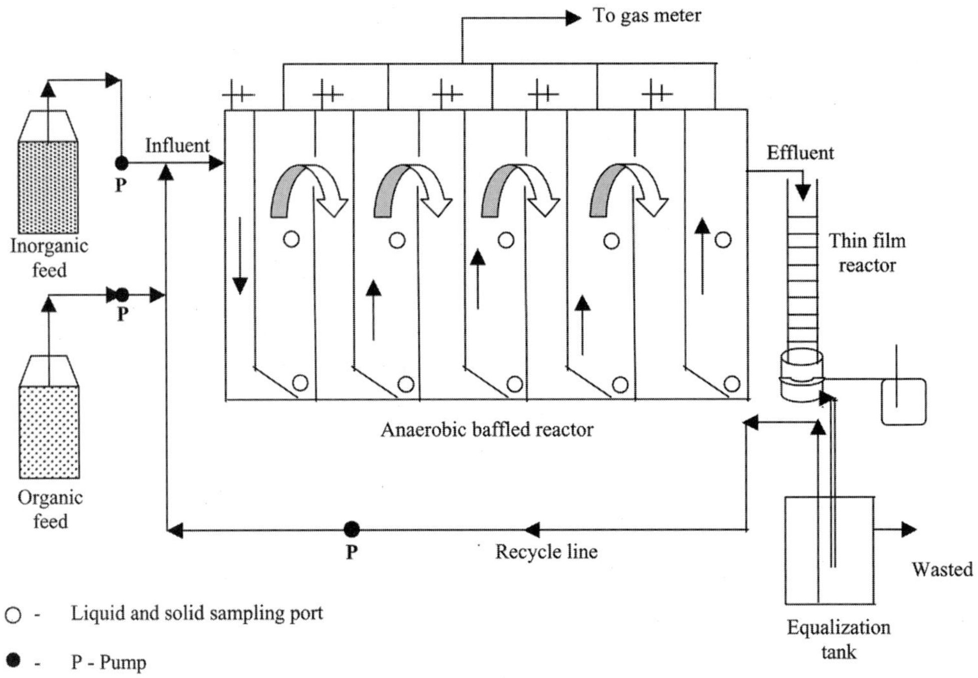

Figure 6 Schematic of anaerobic baffled reactor followed by thin film sulfide oxidizing reactor.

with biological sulfide oxidation was carried out and evaluated. The schematic view of the combined treatment system is given in Fig. 6. The wastewater used in the study contained isopropyl acetate, sulfate, and cellular product. The COD and sulfate concentration of the wastewater were 40,000 mg/L and 5000 mg/L, respectively. Treatment of the wastewater using an anaerobic baffled reactor alone was found effective at 10% dilution but at higher concentration, sulfide inhibition reduced the efficiency of both COD conversion and sulfate conversion. To reduce sulfide inhibition, the treated effluent was subjected to a thin film sulfide oxidizing reactor to facilitate biological oxidation of sulfide into elemental sulfur. The study indicated that at an influent concentration of 40% and HRT of 1 day, COD removal efficiencies greater than 50% can be achieved. The conversion of influent sulfate was greater than 95% with effluent sulfide concentration less than 20 mg/L [60]. Coupled anaerobic/aerobic treatment of high sulfate-containing wastewater effectively alleviated the sulfide inhibition of both methanogenesis and sulfate reduction. A thin film sulfide oxidizing reactor was also effective in converting the sulfide to elemental sulfur without adding excess oxygen, which made recycling of treated anaerobic effluent through the sulfide oxidizing reactor feasible. This indicates that biological sulfide oxidation could provide an alternative method to remove sulfide produced during anaerobic treatment, thereby alleviating sulfide inhibition by removing sulfur from the wastewater stream.

Anaerobic Hybrid Reactor

The anaerobic hybrid reactor is generally a combination of suspended growth and attached growth systems. Recently, this technology has become popular in the treatment of industrial wastewater, in particular in cases of high-strength wastewater. It has been reported that this reactor design presents a viable alternative to continuously stirred reactors, anaerobic filters, and anaerobic fluidized bed reactors for the high-rate treatment of pharmaceutical wastewater containing C_3 and C_4 aliphatic alcohol and other solvents [44]. The suitability of an anaerobic hybrid reactor for the treatment of synthetic pharmaceutical wastewater containing target solvents C_3 and C_4, tert-butanol, sec-butanol, and ethyl acetate was assessed at various organic loadings and varying influent concentrations. The study indicated that isopropanal, isobutanol, and sec-butanol can be almost fully degraded by using the anaerobic hybrid reactor. At OLR ranging from 3.5 to 4.5 kg COD/m^3 day and HRT of 2 days, the reactor achieved total and soluble COD removal efficiencies of 97 and 99%, respectively. However, the reactor was unable to degrade the tert-butanol, resulting in a decrease in soluble COD removal efficiency to 58%. A bacterial enrichment study with the tert-butanol as a sole substrate indicated that this is poorly degradable in anaerobic conditions. The observed recalcitrance of the tert-butanol in the present case contrasts with the findings of earlier researchers, who have listed these solvents as being amenable to anaerobic digestion [61,62]. Degradation of tert-butanol in the activated sludge process has been evaluated, and it was found that aerobic posttreatment/polishing of the anaerobically treated effluent of pharmaceutical wastewater is essential for removing the residual solvent [43]. The addition of a trace metals cocktail in the feed did not affect steady-state reactor performance, but was found beneficial in handling the influent compositional changes. Moreover, the methanogenic activity of the granular sludge fed with trace metals was found significantly higher than the granular sludge of the reference anaerobic hybrid reactor.

Combined Waste Treatment with Other Industrial Waste

The possibility of treatment of pharmaceutical wastewater combined with other industrial waste has been explored and evaluated [63]. One study carried out nitrification of high-strength nitrogenous wastewater (a concentrated stream from a urea plant) in a continuously stirred tank reactor. Pharmaceutical wastewater was used as an organic carbon source to maintain a COD/TKN ratio of 1. The reactor was operated at an HRT of 1.5–2.1 days and solid retention time (SRT) ranging from 10–62.5 days. Characteristics of the wastewater from the urea plant, pharmaceutical wastewater, and combined wastewater are depicted in Table 19. The study concluded that pharmaceutical wastewater may be used as a co-substrate to supply energy for nitrification of high-strength nitrogenous wastewater. Such treatment alternatives establish the advantages of a dual mechanism of treatment, that is, nitrification as well as oxidation of organic pollutants.

Table 19 Characteristics of Urea Plant, Pharmaceutical Plant, and Combined Wastewater [63]

Parameter	Urea plant	Pharmaceutical plant	Combined wastewater
pH	11.0–12.5	5.0–8.0	7.0–9.0
COD (mg/L)	3,520–4,850	1,100–5,500	1,010–1,290
Alkalinity as $CaCO_3$ (g/L)	1.005–1.010	0.30–2.0	4.4–5.45
PO_4^{3-}-P (mg P/L)	0.7–1.0	2.8–14.4	22.8
NH_4^-N (mg N/L)	38,000–45,000	30–50	500–550
Urea-N (mg N/L)	1,860–2,380	–	500

COD, chemical oxygen demand.

3.6.3 Integrated Treatment and Disposal Facilities for Specific Pharmaceutical Waste

The above-cited studies demonstrate the performance of a particular unit system for the treatment of specific type of waste stream. A particular unit system alone may not be able to treat the wastewater to a level of effluent standard prescribed for its safe disposal. Hence a number of pretreatments, such as screening, sedimentation, equalization, and neutralization, and post-treatment units such as secondary sedimentation, sludge thickening, digestion and disposal, disinfection, and so on, are extremely important for complete treatment. The effluent treatment and disposal facilities adopted by various types of pharmaceutical industries are described in the following sections.

Treatment of Synthetic Organic Bulk Pharmaceutical Waste

The Hoffman–La Roche plant in Belvedere, NJ, manufactures synthetic organic bulk pharmaceuticals, including dry vitamin powders, sulfa drugs, vitamin C, riboflavin, aromatics, and sodium sulfate salts. An integrated sodium sulfate recovery system was employed in this plant to recover sodium sulfate. The plant's waste control and treatment system includes screening, preclarifier, equalization with aeration (1 day detention time), pH adjustment/neutralization, flocculator-clarifier, activated sludge process, secondary settler, two oxidation ponds in series, sludge thickening, aerobic sludge digestion, sludge drying beds, and final chlorination. The treatment plant was initially designed for a design flow of 1 MGD (million gallons per day) with BOD_5 and TSS removal efficiency of the system at 97.4 and 98%, respectively. Effluent from this plant had a BOD_5 of 50 mg/L and TSS of 20 mg/L. In 1973, the raw waste load at the plant increased from 1 MGD to 1.6 MGD with BOD load of 30,000 lb · BOD/day or more, together with 8400 lb/day of TSS. By late 1973, the effluent load was about twice the design specification. Although data on the performance of the treatment plant for the current waste loads (1973, 1974) were lacking, the author has indicated a typical removal of BOD, COD, and TSS of 97.5, 90, and 90%, respectively.

Treatment of Fermentation/Synthetic Organic Bulk Pharmaceutical Waste

Pfizer, Inc. (Terre Haute, IN) is a fermentation/synthesized organic bulk pharmaceutical type plant mainly involved in the manufacture of streptomycin, terramycin, two undefined antibiotics, fumaric acid, benzoic acid, and so on. This plant employs a five-stage biological system with a retention time of process waste varying from 45 to 65 days. The treatment plant consists of a primary clarifier, two extended aeration (activated sludge) basins in series (12 days detention), secondary settling tank, two clari-digesters in parallel, two standard rate trickling filters in parallel, a high-rate bio-oxidation tower, final clarifier, two aerated stabilization ponds in series, stabilization pond, chlorination, aerobic sludge digester, sludge stabilization pond, land/crop application of stabilized sludges, and holding pond for spent cooling waters (1 day detention). The plant was designed for combined waste of 1.3 MGD of process waste and 5 MGD of spent cooling water flow. In 1972, Pfizer reported average BOD and TSS removal of 98 and 97.5%, respectively. From 1973 to 1974, the BOD and TSS removal were reported to be 99.1 and 97.8%, respectively. The treated effluent contained a BOD of 10–15 mg/L and TSS of 20–30 mg/L. The Pfizer system was capable of giving 50% phosphorous reduction. The TKN, NH_4-N and organic nitrogen removal were reported to be 75, 67, and 81%, respectively.

A similar plant, Clinton Laboratories (Clinton, IN) is mainly involved in producing a cephalosporin-type antibiotic. Major products include monensin sodium, keflex, and kefzol. The waste generated in this plant includes mycelia, general trash, concentrated chemical wastes, diluted chemical wastes, water process waste, sanitary sewage, and a clear water stream.

The control and treatment system in this plant mainly relies on the chemical destruction of waste rather than biological processes. The plant generates a raw waste load as high as 400,000 lb · BOD/day. From 1973 to 1974, the company reported a total waste flow of 3.5–4.3 MGD containing a BOD of 1710–1960 lb/day, COD of 3700–4000 lb/day, and TSS of 1040–1250 lb/day. The treatment system included the following units:

- concentration of waste streams to minimum volume;
- oversized strippers for solvent recovery;
- stripper system for waste preconditioning;
- Carver–Greenfield multistage, oil dehydration, steam evaporator system (fermentation waste);
- two John Zink thermal oxidation incineration systems (chemical wastes);
- Bartlett–Snow rotary kiln incinerator (plant trash and mycelium);
- small biological treatment plant (sanitary wastes);
- cooling water towers;
- scrubbing of air effluents from incinerators and waste heat boiler on Carever–Greenfield.

Both concentrated and dilute waste were sent to a pair of John Zink thermal oxidizers equipped with adjustable venturi scrubbers for removal of particulates prior to stack discharge. Water process waste originating primarily from fermentation sectors was sent to the Carver–Greenfield evaporation system. The evaporator utilized a multistep oil dehydration process and was equipped with a centrifuge, waste heat boiler, and a venturi scrubber. The Clinton Laboratory reported an overall BOD and COD reduction of 90 and 99%, respectively, depending upon the configuration used.

Treatment of Fermentation, Organic Synthesis Processing, and Chemical Finishing and Packaging Type Bulk Pharmaceutical Waste

Abbott Labs (Chicago, IL) has extensive fermentation, organic synthesis processing, and chemical finishing and packaging facilities and is engaged mainly in production of antibiotics, that is, erythromycin and penicillin, and hundreds of medicinal and fine chemicals. Characteristics of various types of wastes generated from this plant are depicted in Table 20. The typical units involved in the Abbott treatment works are as follows:

- waste screening and neutralization;
- two equalization basins (1.5 day detention);
- six activated sludge basins (100,000 gallon);
- degasification chambers for mixed liquors;

Table 20 Characteristics of the Abbott Laboratory Wastewater [3]

Parameters	Fermentation waste	Chemical waste	Combined waste
Flow (MGD)	0.312	0.262	0.575
pH	6.7	5.4	6.1
BOD (mg/L)	3620	2520	3120
TSS (mg/L)	1660	510	1140
TDS (mg/L)	3590	5690	4620

MGD, million gallons per day; BOD, biochemical oxygen demand; TSS, total suspended solids; TDS, total dissolved solids.

- two final settlers in parallel;
- pasteurization of final process effluent;
- chlorination of process plus cooling flows;
- evaporation/drying of spent fermentation broths;
- enclosure of treatment works;
- centrifuging of excess biological sludge;
- ducting of various odorous air streams to main plant boilers;
- incineration of sludge and odorous air streams in main boilers;
- recovery of select waste streams high in ammonia for bulk fertilizer sales;
- connection to municipal AWT plant.

Process waste averaging 0.6–0.7 MGD was sent to the activated sludge treatment system. Cooling water flows of 14–15 MGD were sent for chlorination before final discharge. This plant also employed a spent fermentation beer recovery system integrated with an expansive incinerator ducting system. Exhaust air from the drying of spent beer was collected into a specially designed duct system. This also collected the odorous stream from the fermentors, exhaust from degassing chambers, and exhaust from the enclosed activated sludge tank and sludge holding tanks. The combined air stream was then carried to the main plant boilers and incinerated therein. Treated effluent characteristics are given in Table 21. In 1972, overall BOD and TOC reductions were reported to be 94.6 and 86%, respectively. In 1973, the average BOD and TOC reductions were reported as 96.7 and 98%, respectively. The annual costs of the Abbott treatment works were U.S. $1.2 million, which was equivalent to U.S. $4.50–5.5 per 1000 gallons of process waste. In view of the state effluent limits of 4 mg/L BOD and 5 mg/L TSS for discharges into Lake Michigan by 1975, the treated effluent is scheduled for connection to the regional municipal AWT plant [29,30,33,64].

A treatment plant including the following units was recommended for handling the wastewater from drug formulation and packaging type bulk pharmaceutical waste [3]:

- possible separate handling of process and sanitary wastes;
- screening;
- equalization (2 days detention or more) with auxiliary aeration;
- activated sludge, multichamber, (approximately 24 hour detention);
- secondary settling;

Table 21 Characteristics of Treated Effluent from Abbott Laboratory Works and 1972 Effluent Standards [3]

Parameters	Treated effluent plus cooling water flow	1972 state standard
Flow (MGD)	15	–
pH	7.5	–
BOD (mg/L)	16	20
TSS (mg/L)	20	25
TDS (mg/L)	400	750
Phenolics (mg/L)	0.02	0.30
Mercury (mg/L)	0.0003	0.0005
Coliforms/100 mL	11	400

MGD, million gallons per day; BOD, biochemical oxygen demand; TSS, total suspended solids; TDS, total dissolved solids.

- sludge thickening;
- aerobic digestion of excess sludges with residues to landfill;
- chlorination of final effluent.

A similar system with minor modifications should be fairly adaptable to biological production type pharmaceutical plants.

3.7 OPERATIONAL PROBLEMS AND REMEDIAL MEASURES

Much research has focused on bulking of the sludge in the aerobic treatment of pharmaceutical wastewater [46,65–67]. The filamentous organism *Sphaerotilus natans* has been reported to be responsible for sludge bulking. The growth of these filamentous organisms was coupled with a deficiency of nitrogen in the wastewater and shock organic and hydraulic loading applied in the system. Another researcher identified the Type 021N microorganism as being responsible for sludge bulking [46]. Three microorganisms, Type 0092, *Microtrix parvicella*, and Type 0041, were also identified to be responsible for sludge bulking. It has been further noted that another factor responsible for the bulking of sludge is influent wastewater variability. Subsequently it has been concluded that all three organisms are correlated with filamentous bulking at low organic loading [66]. To deal with the problem of sludge bulking, the addition of nitrogen was recommended, but even after doing so, operational problems continued and the decision was made to expand the treatment facility to avoid shock organic and hydraulic loading in the reactor. It was further observed that the addition of PAC in the activated sludge process resulted in some improvement in sludge settleability; however, the MLSS settling rate remained at a very low level (0.01–0.05 cm/min). The study demonstrated that due to nitrification, the pH decreased, causing a viscous floating layer of MLSS formed on the surface of the aeration basin and clarifier that resulted in significant reductions in the MLSS and PAC concentration in the system.

Chlorination of mixed liquor has been recommended to address the problem of sludge bulking. It was expected that chlorination of the mixed liquor at dosages ranging from 3 to 7.5 lb Cl_2/1000 lb MLSS could control the problem of sludge bulking; however, chlorination had in fact severely affected the treatment process and stopped nitrification. To resolve this problem, it was suggested that the plant should always operate at an F/M ratio above 0.15 to avoid filamentous growth, and that any increase in filaments should be treated before intense chlorination [46]. Another study recommended that sludge bulking be controlled by operating the system at a dissolved oxygen (DO) concentration of MLSS greater than 3 mg/L. An optimal dissolved oxygen control strategy for an activated sludge system in treatment of pharmaceutical wastewater is described by Brandel [68].

Temperature has been shown to affect the performance of the activated sludge process [46]. Pilot plant results indicated that system efficiency was excellent as long as the aeration basin temperature was less than 38°C, whereas at temperatures exceeding 38°C, BOD_5 removal efficiency decreased considerably, accompanied with the cessation of nitrification. High temperatures resulted in killing of the nitrifiers and inhibited carbonaceous removal. Hence, a heat exchanger in the influent line has been suggested to bring down the wastewater temperature.

3.8 ENVIRONMENTAL PROTECTION AGENCY EFFLUENT LIMITATIONS FOR THE PHARMACEUTICAL INDUSTRY

The EPA has developed effluent limitations in terms of percentage reductions of raw waste loads or effluent concentration as shown in Table 22 [3]. Additional parameters that should receive

Table 22 EPA Effluent Limitations for Pharmaceutical Plants [3]

Parameter	Limit
Average daily based on max. monthly raw waste load	
BOD_5	92–95%
COD	80–82%
TSS	82.5%
Ammonia N	70–75%
pH	6–9
Fecal coliforms	Average, 200/100 mL Max. daily, 400/100 mL
For daily limitations = 2 to 3× average daily levels given above, suggested limits for metals, trace ions	
Iron, Zinc	1.0–1.5 mg/L
Mn, Cu	0.5–1 mg/L
Phenolics, total Cr	0.25–0.5 mg/L
Aluminum	1.0–2.0 mg/L
Sulfide (approx.)	0.5 mg/L
Lead	0.1–0.25 mg/L
Mercury (total plant)	45.36 g/day

BOD, biochemical oxygen demand; COD, chemical oxygen demand; TSS, total suspended solids.

attention at many bulk manufacturing plants include copper, cyanides, tin, cadmium, nickel, arsenic, chlorinated hydrocarbons, and pesticides.

In India, domestic and industrial wastewaters are required to meet the standards set out in the Environment (Protection) Third Amendment Rules (1993) and Water (Prevention and Control of Pollution) Act (1974). The tolerance limits for the disposal of industrial effluents into inland surface water are given in Table 23 [69].

3.9 SUMMARY AND CONCLUSIONS

The information included in this chapter on pharmaceutical wastewater encompasses only a fragment of the research in this area. Owing to extreme variability of pharmaceutical wastewater characteristics, treatability studies should be conducted on a case-by-case basis to identify and confirm the required design parameters. As discussed earlier, physico-chemical treatment such as air stripping and coagulation was not found effective and beneficial for this wastewater, but in many cases, sedimentation has been found effective. The treatability study of almost all kinds of waste streams has indicated that waste is biologically treatable. Hence, a combination of physical, chemical, and biological processes seem to be feasible for the treatment of pharmaceutical wastewater. A two-stage biological system or a combination of aerobic and anaerobic processes proved effective for some pharmaceutical wastewater. Keeping in mind the varying characteristics of pharmaceutical wastewater, the shock loading capacity of the treatment units must also be given much attention in identifying and evaluating the technical feasibility of the processes. After identifying the technical feasibility of the processes, the final selection should be made based on economic analysis.

Table 23 Schedule VI of Environment (Protection) Third Amendment Rules (1993) [69]

		Standards[a]			
Serial No.	Parameters	Inland surface water	Public sewers	Land for irrigation	Marine coastal areas
1	Color and odor	[b]	[b]	[b]	[b]
2	Suspended solids (mg/L), max	100	600	200	(a) For process waste water, 100 (b) For cooling water effluent 10% above total suspended matter of influent
3	Particle size of suspended solids	Shall pass 850 micron IS sieve	–	–	(a) Floatable solids, solids max. 3 mm (b) Settleable solids, max. 856 microns
4	pH value	5.5–9.0	5.5–9.0	5.5–9.0	5.5–9.0
5	Temperature	Should not exceed 5°C above the receiving water temperature	–	–	Should not exceed 5°C above the receiving water temperature
6	Oil and grease (mg/L), max	10	20	10	20
7	Total residual chlorine (mg/L), max	1	–	–	1
8	Ammonical nitrogen as (N) (mg/L), max	50	50	–	50
9	Total Kjeldahl nitrogen (as N) (mg/L), max	100	–	–	100
10	Free ammonia (as NH_3) (mg/L), max	5	–	–	5
11	Nitrate nitrogen	10	–	–	20
12	BOD_5 (mg/L), max	30	350	100	100
13	COD (mg/L), max	250	–	–	250
14	Arsenic (as As) (mg/L)	0.2	0.2	0.2	0.2
15	Mercury (as Hg) (mg/L), max	0.01	0.01	–	0.01
16	Lead (as Pb) (mg/L), max	0.1	0.1	–	2
17	Cadmium (as Cd) (mg/L), max	2	1	–	2
18	Hexavalent chromium (as Cr^{6+}) (mg/L), max	0.1	2	–	1
19	Total chromium (as Cr) (mg/L), max	2	2	–	2
20	Copper (as Cu) (mg/L), max	3	3	–	2

(continues)

Table 23 Continued

Serial No.	Parameters	Standards[a] Inland surface water	Public sewers	Land for irrigation	Marine coastal areas
21	Zinc (as Zn) (mg/L), max	5	15	–	15
22	Selenium (as Se) (mg/L), max	0.05	0.05	–	0.05
23	Nickel (as N_i) (mg/L), max	3	3	–	5
24	Cyanide (as CN) (mg/L), max	0.2	2	0.2	0.2
25	Fluoride (as F) (mg/L), max	2	15	–	15
26	Dissolved phosphates (as P) (mg/L), max	5	–	–	–
27	Sulfide (as S), (mg/L)	2	–	–	5
28	Phenolic compounds (as C_6H_5OH) (mg/L), max	1	5	–	5
29	Radioactive materials				
	(a) Alpha emitters (micro-Curie mg/L), max	10^{-7}	10^{-7}	10^{-8}	10^{-7}
	(b) Beta emitters (micro-Curie, mg/L), max	10^{-6}	10^{-6}	10^{-7}	10^{-6}
30	Bio-assay test after 96 hours in 100% effluent	90% survival of fish	90% survival of fish	90% survival of fish	90% survival of fish
31	Manganese (as Mn) (mg/L), max	2	2	2	2
32	Iron (as Fe) (mg/L), max	3	3	3	3
33	Vanadium (as V) (mg/L), max	0.2	0.2	–	0.2

[a]These standards shall be applicable for industries, operations, or processes other than those industries, operations, or process for which standards have been specified in Schedule I.
[b]All efforts should be made to remove color and unpleasant odor as for as practicable.

Based on extensive study and experience in treatment of pharmaceutical wastewater, the following specific conclusions may be drawn:

- Pretreatment of pharmaceutical industry wastewater such as air stripping and coagulation is not beneficial; however, sedimentation of treated effluent was found effective in further reduction of SS and COD of the effluent. Hence, the pretreatment of pharmaceutical wastewater is not advisable.
- In many cases, anaerobic filter treatment was found to successfully treat pharmaceutical industry wastewater. This can be an excellent alternative for conventional aerobic treatment, which is energy intensive and requires the disposal of sludge. The anaerobic filter, on the other hand, can produce energy in the form of biogas and does not require

sludge disposal. Moreover, the anaerobic filter is more resistant and capable of handling shock loading as compared to the aerobic system.

- All waste streams, with the exception of acid waste streams of a synthetic drug factory, must be treated collectively rather than treated separately, as the performance efficiency of combined waste has been proved to be better than that of waste treated separately. Moreover, the segregation of acid waste streams could result in the following benefits:
  - recovery of useful acids from the waste;
  - the volume of the waste needing neutralization has been reduced to 50% and has eliminated the necessity of adjusting the pH of the combined waste for biological treatment;
  - the burden on the biological treatment has been reduced.
- The problem of sludge bulking in the case of the activated sludge process can be controlled in the following ways:
  - chlorination of the mixed liquor;
  - operating the system at min^m DO concentration of 3 mg/L;
  - operating the system at higher organic loading.
- Treatment processes such as ASP, PAC-ASP, GAC, and resin columns can successfully remove priority pollutants from pharmaceutical wastewater.
- In general, the trickling filter and activated sludge were found to satisfactorily cope with the needs of wastewater treatment for the pharmaceutical industry.
- Addition of PAC in the activated sludge process was found beneficial in improving the effluent quality, but it cannot be recommended until the problem of viscous layer formation is solved.

3.10 DESIGN EXAMPLES

Example 1

A synthetic organic chemicals plant discharges mainly two types of waste streams, namely strong process waste and dilute process waste. The flow and BOD_5 of the waste streams are given in the following table.

Type of wastes	Flow (GPD)	BOD_5 (mg/L)
Strong process waste	11,800	480,000
Dilute process waste	33,800	640

GPD, gallons per day; BOD, biochemical oxygen demand.

In addition, the plant discharges 35,300 GPD service wastewater. If the total BOD load of the composite waste is 47,500 lb/day, estimate (i) the BOD_5 of the composite waste and domestic waste; and (ii) the BOD load of the each stream and their contribution to the total BOD load of the plant.

Solution

Determine the BOD_5 of the wastes. The first step is to find out the total flow of the composite waste by summing the flow of the various waste streams of the plant.

Total flow of the composite waste = 11,800 + 33,800 + 35,300 = 80,900 GPD

$$\text{BOD}_5 \text{ of the composite waste} = \frac{\text{Total BOD load (lb/day)} \times 453.6\ (\text{g/lb}) \times 1000\ (\text{mg/g})}{\text{Flow (GPD)} \times 3.785\ (\text{L/gal})}$$

$$= \frac{47{,}500 \times 453.6 \times 1000}{80{,}900 \times 3.785}$$

$$= 70{,}364.28\ \text{mg/L}$$

$$\text{BOD load of the strong process waste} = \frac{\text{Flow (GPD)} \times 3.785\ (\text{L/gal}) \times \text{BOD}_5\ (\text{mg/L})}{10^3\ (\text{mg/g}) \times 453.6\ (\text{g/lb})}$$

$$= \frac{11{,}800 \times 3.785 \times 480{,}000}{1000 \times 453.6}$$

$$= 47{,}262.43\ \text{lb/day}$$

$$\text{BOD load of the dilute process waste} = \frac{\text{Flow (GPD)} \times 3.785\ (\text{L/gal}) \times \text{BOD}_5\ (\text{mg/L})}{10^3\ (\text{mg/g}) \times 453.6\ (\text{g/lb})}$$

$$= \frac{33{,}800 \times 3.785 \times 640}{1000 \times 453.6}$$

$$= 180.50\ \text{lb/day}$$

$$\text{BOD load due to domestic waste} = 47{,}500 - (47{,}262.43 + 180.5)$$

$$= 57.05\ \text{lb/day}$$

$$\text{BOD}_5 \text{ of the domestic waste, mg/L} = \frac{\text{BOD load (lb/day)} \times 453.6\ (\text{g/lb}) \times 1000\ (\text{mg/g})}{\text{Flow (GPD)} \times 3.785\ (\text{L/gal})}$$

$$= \frac{57.07 \times 453.6 \times 1000}{35{,}300 \times 3.785}$$

$$= 193.75\ \text{mg/L}$$

Comment
The total BOD load of the plant is mainly due to strong process waste. Segregation of strong process waste can result in significant reduction in total BOD load of the plant.

Example 2

The five-days BOD at 20°C and flow of the various types of waste streams generated from a synthetic drug plant are given in the following table.

Type of wastes	Flow (m^3/day)	BOD_5 (mg/L)
Alkaline waste stream	1710	3500
Condensate waste stream	1990	1275
Acid waste stream	435	3090

BOD, biochemical oxygen demand.

Estimate the BOD_5 and subsequent BOD load of the composite waste. If the acid waste stream has to be segregated for the recovery of acids then find out (i) the BOD_5 of the combined waste excluding acid waste; and (ii) comment on the effect of segregation in BOD loading of the plant.

Solution

$$\text{BOD load of the alkaline waste} = \text{Flow } (m^3/\text{day}) \times BOD_5 \, (g/m^3) \times 10^{-3} \, (kg/g)$$
$$= 1710 \times 3500 \times 10^{-3} = 5985 \text{ kg BOD/day}$$

$$\text{Similarly, BOD load of the condensate waste} = 1990 \times 1275 \times 10^{-3}$$
$$= 3834.61 \text{ kg BOD/day}$$

$$\text{BOD load of the acid waste} = 435 \times 3090 \times 10^{-3} = 1344.15 \text{ kg BOD/day}$$

$$\text{Total BOD load of the composite waste} = 5985 + 3834.61 + 1344.15$$
$$= 11{,}163.76 \text{ kg BOD/day}$$

$$\text{Total flow of the composite waste} = 1710 + 1990 + 435 = 4135 \, m^3$$

$$BOD_5 \text{ of the composite waste} = \frac{\text{BOD load (kg/day)} \times 10^6 \text{ (mg/kg)}}{\text{Flow } (m^3/\text{day}) \times 10^3 \, (L/m^3)}$$
$$= \frac{11{,}163.76 \times 10^6}{4135 \times 10^3}$$
$$= 2699.82 \text{ mg/L}$$

$$\text{BOD load of alkaline and condensate waste} = 5985 + 3834.61$$
$$= 9819.61 \text{ kg BOD/day}$$

$$\text{Total flow of the alkaline and condensate waste} = 1710 + 1990$$
$$= 3700 \, m^3$$

$$BOD_5 \text{ of combined (alkaline and condensate)waste} = \frac{9819.61 \times 10^6}{3700 \times 10^3}$$
$$= 2653.95 \text{ mg/L}$$

Comment
Segregation of the acid waste stream has resulted in significant reduction in total BOD load of the plant, but the BOD_5 of the composite waste remains almost the same. Hence the acid waste stream can be segregated from the main stream without affecting the treatability of the waste.

Example 3

A primary sedimentation tank has been designed for the pretreatment of 0.312 MGD of fermentation waste generated from the pharmaceutical industry. The raw waste SS concentration is 1660 mg/L. At a detention time of 2 hours the effluent SS concentration is reduced to 260 mg/L. Determine (i) the SS removal efficiency of the sedimentation tank; and (ii) the quantity of sludge generated per day. Assume the specific gravity of sludge (S_{sl}) is 1.03, which contains 6% solids.

Solution

(A) SS removal efficiency of the tank can be obtained as follows

$$\text{SS removal efficiency} = \frac{(1660 - 260) \times 100}{1660}$$
$$= 84.34\%$$

(B) Determine the mass of dry solids removed per day

$$W_s = 0.312\ (\text{MGD}) \times 10^6\ (\text{gal/M}) \times 3.785\ (\text{L/gal}) \times (1660 - 260)\ (\text{mg/L}) \times 10^{-6}\ (\text{kg/mg}) = 1653.29\,\text{kg/day}$$

(C) Determine the volume of sludge produced per day

$$V_{sl} = \frac{W_s}{\rho_w \times S_{sl} \times P_s}$$

where W_s is the mass of dry solids removed per day, ρ_w is the density of the water, and P_s is the percentage of sludge solids.

$$V_{sl} = \frac{1653.29\ (\text{kg/day})}{1000 \times 1.03 \times 0.06}$$
$$= 26.752\,\text{m}^3\ (7067.96\,\text{GPD})$$

Example 4

Physicochemical treatment of a typical pharmaceutical plant generating 33,800 GPD wastewater has indicated that at optimum doses of $FeSO_4$ (500 mg/L), $FeCl_3$ (500 mg/L), and alum (250 mg/L), COD and SS removal of the effluent of 25 and 70%, respectively, can be achieved. Determine the quantities of various chemicals required per day. If 49% strength alum is to be used and 30 days supply is to be stored at the treatment facility, estimate the storage capacity required for the alum.

Solution

(A) The quantities of the various chemicals required per day can be obtained as follows:

$$\text{Quantity of } FeSO_4 \text{ required per day} = 500\ (\text{mg/L}) \times 10^{-6}\ (\text{mg/kg}) \times 33{,}800\ (\text{GPD}) \times 3.785\ (\text{L/gal})$$
$$= 63.97\,\text{kg/day}$$

$$\text{Quantity of } FeCl_3 \text{ required per day} = 500\ (\text{mg/L}) \times 10^{-6}\ (\text{mg/kg}) \times 33{,}800\ (\text{GPD}) \times 3.785\ (\text{L/gal})$$
$$= 63.97\,\text{kg/day}$$

$$\text{Quantity of alum required per day} = 250\ (\text{mg/L}) \times 10^{-6}\ (\text{mg/kg}) \times 33{,}800\ (\text{GPD}) \times 3.785\ (\text{L/gal})$$
$$= 31.98\,\text{kg/day}$$

(B) Determine the weight of alum per m^3 of 49% liquid alum.

$$\text{Weight per m}^3 = 0.49 \times 80\ (\text{lb/ft}^3) \times 16.0185\ (\text{kg/m}^3 \cdot \text{lb/ft}^3)$$
$$= 627.925\,\text{kg/m}^3$$

(C) Determine the storage capacity required for 30 days.

$$\begin{aligned}\text{Storage capacity} &= 31.98\ (\text{kg/day}) \times 30\ (\text{days})/627.925\ (\text{kg/m}^3)\\ &= 1.527\,\text{m}^3\ (1527\,\text{L})\end{aligned}$$

Example 5

Estimate the quantity of sludge produced in a chemical precipitation of 1710 m^3/day of pharmaceutical wastewater with SS concentration 560 mg/L. The addition of the $FeSO_4$ (500 mg/L), $FeCl_3$ (500 mg/L), and lime (600 mg/L) increases the SS removal efficiency of the primary sedimentation tank from 60 to 70%. Comment on the chemical precipitation process on the basis of sludge production. Assume $CaCO_3$ solubility = 15 mg/L, specific gravity of sludge = 1.03, and moisture content of the sludge = 95%.

Solution

(A) Determine the mass of SS removed per day without chemical addition

$$\begin{aligned}M_{ss1} &= 1710\ (\text{m}^3/\text{day}) \times 10^3\ (\text{L/m}^3) \times 0.6 \times 560\ (\text{mg/L}) \times 10^{-6}\ (\text{kg/mg})\\ &= 574.56\,\text{kg/day}\end{aligned}$$

(B) Determine the mass of SS removed per day after chemical addition

$$\begin{aligned}M_{ss1} &= 1710\ (\text{m}^3/\text{day}) \times 10^3\ (\text{L/m}^3) \times 0.7 \times 560\ (\text{mg/L}) \times 10^{-6}\ (\text{kg/mg})\\ &= 670.320\,\text{kg/day}\end{aligned}$$

(C) Volume of the sludge without chemical addition

$$\begin{aligned}V_{sl} &= \frac{W_s}{\rho_w \times S_{sl} \times P_s}\\ &= \frac{574.56\ (\text{kg/day})}{1000 \times 1.03 \times (1 - 0.95)}\\ &= 11.16\,\text{m}^3/\text{day}\ (2948.48\,\text{GPD})\end{aligned}$$

(D) Determine the quantity of sludge produced due to addition of chemicals. This can be calculated from the stochiometry of the chemical reactions taking place with the addition of these chemicals. The chemical reactions taking place are described below. When $FeSO_4$ and lime are added:

$$\underset{(278)}{FeSO_4 \cdot 7H_2O} + \underset{(100)}{Ca(HCO_3)_2} \Longleftrightarrow \underset{(178)}{Fe(HCO_3)_2} + \underset{(136)}{CaSO_4} + \underset{(7\times 18)}{7H_2O}$$

$$\underset{(178)}{Fe(HCO_3)_2} + \underset{(2\times 56)}{2Ca(OH)_2} \Longleftrightarrow \underset{(2\times 100)}{2CaCO_3} + \underset{(89.9)}{Fe(OH)_2} + \underset{(2\times 18)}{2H_2O}$$

$$\underset{(4\times 89.9)}{4Fe(OH)_2} + \underset{(32)}{O_2} + \underset{(2\times 18)}{2H_2O} \Longleftrightarrow \underset{(4\times 106.9)}{4Fe(OH)_3}$$

$$\underset{(56)}{Ca(OH)_2} + \underset{(44)}{H_2CO_3} \Longleftrightarrow \underset{(100)}{CaCO_3} + \underset{(2\times 18)}{2H_2O}$$

$$\underset{(56)}{Ca(OH)_2} + \underset{(100)}{Ca(HCO_3)_2} \Longleftrightarrow \underset{(2\times 100)}{2CaCO_3} + \underset{(2\times 18)}{2H_2O}$$

When $FeCl_3$ is also added:

$$\begin{array}{llll} 2FeCl_3 + 3Ca(OH)_2 & \Longleftrightarrow & 2Fe(OH)_3 + 3CaCl_2 \\ (2 \times 162)\ \ (3 \times 56) & & (2 \times 106.9)\ (3 \times 111) \end{array}$$

The addition of $FeSO_4$ mainly produces precipitable flocs of $CaCO_3$ and $Fe(OH)_3$. The quantity of $CaCO_3$ precipitated by addition of 500 mg/L of $FeSO_4$ can be estimated as:

$$\begin{aligned} \text{Quantity of } CaCO_3 &= 500\ (\text{mg/L}) \times (200/278) \\ &= 359.71\ \text{mg/L} \end{aligned}$$

Similarly, the quantity of $Fe(OH)_3$ precipitated by addition of 500 mg/L of $FeSO_4$

$$\begin{aligned} &= 500\ (\text{mg/L}) \times (106.9/278) \\ &= 192.27\ \text{mg/L} \end{aligned}$$

Amount of lime consumed in formation of $Fe(OH)_3$ flocs

$$\begin{aligned} &= 192.27\ (\text{mg/L}) \times (56/106.9) \\ &= 100.72\ \text{mg/L} \end{aligned}$$

Similarly, the quantity of $Fe(OH)_3$ precipitated by addition of 500 mg/L of $FeCl_3$

$$\begin{aligned} &= 500\ (\text{mg/L}) \times (106.9/162) \\ &= 329.94\ \text{mg/L} \end{aligned}$$

Amount of lime consumed in formation of $Fe(OH)_3$ flocs by addition of $FeCl_3$

$$\begin{aligned} &= 329.94\ (\text{mg/L}) \times (3 \times 56/2 \times 106.9) \\ &= 259.26\ \text{mg/L} \end{aligned}$$

Total amount of lime remaining

$$\begin{aligned} &= 600 - (100.72 + 259.26) \\ &= 240.02\ \text{mg/L} \end{aligned}$$

Amount of $CaCO_3$ precipitated by addition of lime

$$\begin{aligned} &= 240.02 \times (3 \times 100/2 \times 56) \\ &= 642.91\ \text{mg/L} \end{aligned}$$

Determine the total amount of $CaCO_3$ precipitated per day

$$\begin{aligned} &= 1710\ (\text{m}^3/\text{day}) \times 10^3\ (\text{L/day}) \\ &\qquad \times (359.71 + 642.91 - 15)\ (\text{mg/L}) \times 10^{-6}\ (\text{kg/mg}) \\ &= 1688.83\ \text{kg/day} \end{aligned}$$

Similarly, the total amount of $Fe(OH)_3$ precipitated per day

$$= 1710\ (\mathrm{m^3/day}) \times 10^3\ (\mathrm{L/day}) \times (192.27 + 329.94)\ (\mathrm{mg/L}) \times 10^{-6}\ (\mathrm{kg/mg})$$
$$= 892.98\ \mathrm{kg/day}$$

Total volume of sludge on dry basis per day

$$= 670.32 + 1688.83 + 892.98$$
$$= 3252.13\ \mathrm{mg/L}$$

Hence the volume of the sludge produced per day with chemical addition:

$$V_{\mathrm{sl}} = \frac{W_{\mathrm{s}}}{\rho_{\mathrm{w}} \times S_{\mathrm{sl}} \times P_{\mathrm{s}}}$$
$$= \frac{3252.13\ (\mathrm{kg/day})}{1000 \times 1.03 \times (1 - 0.95)}$$
$$= 63.15\ \mathrm{m^3/day}\ (16{,}684.28\ \mathrm{GPD})$$

Comment

The problem of sludge disposal increased to a greater extent in the case of chemical precipitation than in the sedimentation without the chemical.

Example 6

Estimate the food–microorganism ratio (F/M) and sludge age (solid retention time) of an activated sludge process designed to reduce the BOD_5 of the spent stream generated from a biological production plant from 1500 mg/L to 50 mg/L. The wastewater flow is $Q = 15{,}000$ GPD, aeration tank volume = 45 m^3, MLVSS = 3000 mg/L, and net biomass yield coefficient (Y_n) = 0.28 kg/kg. Also compute the performance efficiency of the plant.

Solution

$$\text{Total substrate removed (kg BOD/day)} = Q(\mathrm{GPD}) \times 3.785\ (\mathrm{L/gal}) \times (S_{\mathrm{i}} - S_{\mathrm{e}})\ (\mathrm{mg/L}) \times 10^{-6}\ (\mathrm{kg/mg})$$
$$= 15{,}000 \times 3.785 \times (1500 - 50) \times 10^{-6}$$
$$= 82.32\ \mathrm{kg\ BOD/day}$$

$$\text{Total MLVSS (kg MLVSS)} = \text{MLVSS (mg/L)} \times 10^{-6}\ (\mathrm{kg/mg}) \times V\ (\mathrm{m^3}) \times 10^3\ (\mathrm{L/m^3})$$

$$\text{Total MLVSS} = 3000 \times 10^{-6} \times 45 \times 10^3 = 135\ \mathrm{kg\ MLVSS}$$

$$\text{Total substrate applied per day} = 15{,}000 \times 3.785 \times 1500 \times 10^{-6}$$
$$= 85.16\ \mathrm{kg\ BOD/day}$$

$$\text{F/M ratio (day}^{-1}) = \frac{\text{Total substrate applied (kg BOD/day)}}{\text{Total MLVSS (kg MLVSS)}}$$
$$= (85.16/135) = 0.63\ \mathrm{day}^{-1}$$

$$\text{Net MLVSS produced (kg VSS/day)} = Y_{\mathrm{n}}\ (\mathrm{kg/kg}) \times \text{total substrate removed (kg/day)}$$
$$= 0.28 \times 82.32$$
$$= 23.05\ \mathrm{kg\ VSS/day}$$

$$\text{Sludge age (solid retention time)}\ (\theta_c) = \frac{\text{Total MLVSS}}{\text{Net VSS produced per day}} = \frac{135}{23.05} = 5.86\,\text{day}$$

$$\text{BOD removal efficiency} = \frac{(S_i - S_e) \times 100}{S_i} = \frac{(1500 - 50) \times 100}{1500} = 96.67\%$$

Example 7

Design a complete-mix activated sludge process for the treatment of 1710 m^3/day of settled condensate wastewater with BOD_5, 1500 mg/L generated from a synthetic organic chemical type of pharmaceutical industry. Assume the following conditions are applicable:

1. Effluent contains 25 mg/L biological solids, of which 65% is biodegradable;
2. MLSS concentration in the reactor = 5000 mg/L;
3. MLVSS (X) = 0.8 × MLSS;
4. Solid retention time, $\theta_c = 5$ days;
5. $BOD_5 = 0.68\ BOD_L$ (ultimate biological oxygen demand);
6. Return sludge concentration = 1%;
7. Effluent BOD_5 = 50 mg/L;
8. Maximum yield coefficient, $Y = 0.6$ mg/mg;
9. Decay constant, $K_d = 0.07\ \text{day}^{-1}$.

Solution

(A) Determine the influent soluble BOD_5 escaping the treatment:

(i) BOD_L of the biodegradable effluent solid

$$= 25\ (\text{mg/L}) \times 0.65 \times 1.42\ (\text{mg } O_2 \text{ consumed/mg cell oxidized}) = 23.075\,\text{mg/L}$$

(ii) BOD_5 of the effluent SS = 23.075 (mg/L) × 0.68 = 15.69 mg/L (say 15.7 mg/L)

(iii) Influent soluble BOD_5 escaping the treatment

$$= 50 - 15.7 = 34.3\,\text{mg/L}$$

(B) Efficiency of the process:

(i) Process efficiency based on soluble BOD_5

$$E_s = \frac{(1500 - 34.3) \times 100}{1500} = 97.71\%$$

(ii) Similarly, overall plant efficiency of the system

$$E_s = \frac{(1500 - 50) \times 100}{1500}$$
$$= 96.67\%$$

(C) Determine the capacity of the aeration basin

$$V \text{ (volume)} = \frac{Y\theta_c Q(S_i - S)}{X(1 + K_d\theta_c)}$$

where Y = maximum yield coefficient (mg/mg), θ_c = mean cell residence time (day), Q = flow (m^3/day), S_i = substrate concentration in the influent (mg/L), S = substrate concentration in effluent (mg/L), X = mass concentration of microorganism in reactor (mg/L), and K_d = endogenous decay coefficient (day^{-1}). On substituting the values, the above equation results in:

$$V = \frac{0.6 \text{ (mg/mg)} \times 5 \text{ (day)} \times 1710 \text{ (m}^3\text{/day)}(1500 - 50) \text{ (mg/L)}}{0.8 \times 5000 \text{ (mg/L)} \times [1 + 0.07 \text{ (day}^{-1}) \times 5 \text{ (day)}]}$$
$$= 1377.5 \text{ m}^3$$

Check for the F/M ratio and OLR

$$\text{HRT } (\theta) = V/Q$$
$$\theta = 1377.5 \text{ (m}^3)/1710 \text{ (m}^3\text{/day)}$$
$$= 0.805 \text{ day (19.33 hours)}$$
$$\text{F/M ratio} = (S_i/\theta X)$$
$$= \frac{1500 \text{ mg/L}}{0.805 \text{ (day)} \times 0.8 \times 5000 \text{ (mg/L)}}$$
$$= 0.466 \text{ day}^{-1}$$

$$\text{Amount of BOD}_5 \text{ consumed} = (1500 - 34.3) \text{ (mg/L)} \times 10^{-6} \text{ (kg/mg)} \times 1710 \text{ (m}^3\text{/day)} \times 10^3 \text{ (L/m}^3)$$
$$= 2506.35 \text{ kg BOD/day}$$

$$\text{OLR} = \frac{(1500-34.3) \text{ (mg/L)} \times 10^{-6} \text{ (kg/mg)} \times 1710 \text{ (m}^3\text{/day)} \times 10^3 \text{ (L/m}^3)}{1377.5 \text{ (m}^3)}$$
$$= (2506.35/1377.5)$$
$$= 1.82 \text{ kg BOD/m}^3\text{day}$$

(D) Sludge recycling

The recycling ratio (r) can be computed as follows:

$$r = \frac{X}{(X_r - X)}$$

where X_r = MLVSS in the recycled effluent

$$= \frac{0.8 \times 5000 \text{ (mg/L)}}{0.8 \times (10{,}000 - 5000) \text{ (mg/L)}}$$

$$= 0.5$$

Hence the recycling flow = 0.5 Q = 0.5 × 1710 (m^3/day) = 855 m^3/day.

(E) Sludge production

(i) Net VSS production = XV/θ_c

$$= \frac{0.8 \times 5000 \text{ (mg/L)} \times 10^{-6} \text{ (kg/mg)} \times 1377.5 \text{ (m}^3\text{)} \times 10^3 \text{ (L/m}^3\text{)}}{5 \text{ (days)}}$$

$$= 1102 \text{ kg/day}$$

(ii) Net SS production = 1102 (kg/day)/0.8

$$= 1377.5 \text{ kg/day}$$

(iii) Volume of the sludge produced

$$= \frac{1377.5 \text{ (kg/day)}}{1000 \text{ (kg/m}^3\text{)} \times 1.03 \times 0.01}$$

$$= 133.73 \text{ m}^3\text{/day}$$

(iv) VSS production per kg BOD_r (biological oxygen demand removed)

$$= \frac{1102 \text{ (kg/day)} \times 10^6 \text{ (mg/kg)}}{(1500 - 34.3) \text{ (mg/L)} \times 1710 \text{ (m}^3\text{/day)} \times 10^3 \text{ (L/m}^3\text{)}}$$

$$= 0.44 \text{ mg/mg}$$

(F) Oxygen requirement

(i) Theoretical O_2 required = (BOD_L removed) − (BOD_L of solids leaving)

$$= 1.47\ (1500 - 34.3) \text{ (mg/L)} \times 1710 \text{ (m}^3\text{)} \times 10^3 \text{ (L/m}^3\text{)}$$

$$\times 10^{-6} \text{ (kg/mg)} - 1.42 \times 1102 \text{ (kg/day)}$$

$$= 2119.49 \text{ kg/day}$$

(ii) Theoretical air requirement assuming that air contains 23.2% oxygen by weight and density of air = 1.201 kg/m^3

$$= \frac{2119.49 \text{ (kg/day)}}{0.232 \times 1.201 \text{ (kg/m}^3\text{)}}$$

$$= 7606.47 \text{ m}^3\text{/day}$$

(iii) Actual air requirement at an 8% transfer efficiency

$$= 7606.47 \text{ (m}^3\text{/day)}/0.08$$

$$= 95{,}084.65 \text{ m}^3\text{/day}$$

(iv) Check for the air requirement per unit volume

$$= 95{,}084.65\ (\text{m}^3/\text{day})/1710\ (\text{m}^3/\text{day})$$

$$= 55.60\,\text{m}^3/\text{m}^3$$

(v) Air requirement per kg of BOD_5 removed

$$= 95{,}084.65\ (\text{m}^3/\text{day})/2506.35\ (\text{kg/day})$$

$$= 37.94\,\text{m}^3/\text{kg BOD}_5 \text{ removed}$$

(G) Power requirement assuming the aerators are designed to give 2 kgO_2/kWh and the field efficiency is 70%.

$$\text{Power required} = \frac{2119.49\ (\text{kg/day})}{2\ (\text{kg/kWh}) \times 0.7 \times 24\ (\text{h/day})}$$

$$= 63.08\ (\text{kW}) \times 1.3410\ (\text{hp/kW})$$

$$= 84.59\,\text{hp (say 85 hp)}$$

Example 8

1710 m^3/day of alkaline waste stream with $BOD_5 = 3500$ mg/L is treated in an extended aeration system. The BOD removal efficiency of the system is 96%. If the volume of the aeration basin is 1780 m^3, estimate (i) detention time (hydraulic retention time, HRT) and organic loading rate (OLR). Also compute the BOD_5 of the treated effluent.

Solution

$$\text{HRT}(\theta)\ (\text{day}) = \frac{\text{Volume of the tank}, V(\text{m}^3)}{\text{Flow}, Q\ (\text{m}^3/\text{day})}$$

$$= 1780/1710 = 1.04\,\text{day} = 24.98\,\text{hours}$$

$$\text{OLR (kg BOD/m}^3\text{day)} = \frac{Q\ (\text{m}^3/\text{day}) \times 10^3\ (\text{L/m}^3) \times E \times S_i\ (\text{mg/L}) \times 10^{-6}\ (\text{kg/mg})}{V\ (\text{m}^3)}$$

$$= (1710 \times 10^3 \times 0.96 \times 3500 \times 10^{-6})/1780$$

$$= 3.23\,\text{kg BOD/m}^3\text{ day}$$

$$\%\ \text{BOD removal efficiency} = \frac{(S_i - S_e) \times 100}{S_i}$$

$$96 = \frac{(3500 - S_e) \times 100}{3500}$$

$$S_e = 140\,\text{mg/L}$$

Example 9

An extended aeration activated sludge process is designed to treat 2000 m^3/day of condensate waste generated from a synthetic organic chemical plant. The system is operating at an organic loading rate of 1.2 kg COD/m^3 day. If the BOD_5 of influent raw waste and treated effluent is 1275 mg/L and 76.5 mg/L, respectively, determine the HRT and performance efficiency of the system.

Solution

$$\text{OLR (kg BOD/m}^3\text{day)} = \frac{Q\ (\text{m}^3/\text{day}) \times 10^3\ (\text{L/m}^3) \times (S_i - S_e)\ (\text{mg/L}) \times 10^{-6}\ (\text{kg/mg})}{V\ (\text{m}^3)}$$

$$1.2 = [2000 \times 10^3 \times (1275 - 76.5) \times 10^{-6}]/V$$

$$V = 1997.5\ \text{m}^3$$

$$\text{HRT},\ \theta\ (\text{day}) = \frac{V\ (\text{m}^3)}{Q\ (\text{m}^3/\text{day})}$$

$$= 1997.5/2000 = 0.99\ \text{day, say 1 day (24 hours)}$$

$$\%\ \text{BOD removal efficiency} = \frac{(S_i - S_e) \times 100}{S_i}$$

$$= \frac{(1275 - 76.5) \times 100}{1275}$$

$$= 94\%$$

Example 10

Design an extended aeration process for the treatment of 1275 m^3/day of pharmaceutical wastewater with a BOD_5 of 3500 mg/L. Assume the following conditions are applicable:

- Effluent contains 20 mg/L biological solids of which 70% is biodegradable;
- MLSS concentration in the reactor = 6000 mg/L;
- MLVSS = 0.75 × MLVSS;
- Solid retention time, $\theta_c = 12$ days;
- $BOD_5 = 0.68\ BOD_L$;
- Return sludge concentration = 2%;
- Effluent BOD_5 = 30 mg/L;
- $Y = 0.65$ mg/mg;
- Decay constant, $K_d = 0.075\ \text{day}^{-1}$.

Solution

(A) Determine the influent soluble BOD_5 escaping the treatment:

(i) BOD_L of the biodegradable effluent solid

$$= 20\ (\text{mg/L}) \times 0.70 \times 1.42\ (\text{mg } O_2 \text{ consumed/mg cell oxidized})$$

$$= 19.88\ \text{mg/L}$$

(ii) BOD_5 of the effluent SS

$$= 19.88\ (mg/L) \times 0.68$$
$$= 13.52\ mg/L\ (say\ 13.5\ mg/L)$$

(iii) Influent soluble BOD_5 escaping the treatment

$$= 30 - 13.5$$
$$= 16.5\ mg/L$$

(B) Efficiency of the process:

(i) Process efficiency based on soluble BOD_5

$$E_s = \frac{(3500 - 16.5) \times 100}{3500}$$
$$= 99.5\%$$

(ii) Similarly, overall plant efficiency of the system

$$E_s = \frac{(3500 - 30) \times 100}{3500}$$
$$= 99.1\%$$

(C) Determine the capacity of the aeration basin

$$V = \frac{Y\theta_c Q(S_i - S)}{X(1 + K_d\theta_c)}$$
$$= \frac{0.65\ (mg/mg) \times 12\ (day) \times 1275\ (m^3/day) \times (3500 - 30)\ (mg/L)}{0.75 \times 6000\ (mg/L) \times [1 + 0.075\ (day^{-1}) \times 12\ (day)]}$$
$$= 4036.15\ m^3\ (say\ 4050\ m^3)$$

Check for the F/M ratio and OLR and HRT

$$HRT(\theta) = V/Q$$
$$\theta = 4050\ (m^3)/1275\ (m^3/day)$$
$$= 3.18\ day$$

$$F/M\ ratio = (S_i/\theta X)$$
$$= \frac{3500\ (mg/L)}{3.18\ (day) \times 0.75 \times 6000\ (mg/L)}$$
$$= 0.24\ day^{-1}$$

$$\text{Amount of } BOD_5 \text{ removed} = (3500 - 16.5)\ (mg/L) \times 10^{-6}\ (kg/mg) \times 1275\ (m^3/day) \times 10^3\ (L/m^3)$$
$$= 4441.46\ kg/day$$

$$OLR = \frac{(3500 - 16.5)\ (mg/L) \times 10^{-6}\ (kg/mg) \times 1275\ (m^3/day) \times 10^3\ (L/m^3)}{4050\ (m^3)}$$
$$= (4441.46/4050)$$
$$= 1.10\ kg\ BOD/m^3 day$$

(D) Sludge recycling

The recycling ratio (r) can be computed as follows:

$$r = X/(X_r - X)$$

$$= \frac{0.75 \times 6000 \text{ (mg/L)}}{0.75 \times (20{,}000 - 6000) \text{ (mg/L)}}$$

$$= 0.43$$

Hence the recycling flow $= 0.43\,Q = 0.43 \times 1275\ (\text{m}^3/\text{day}) = 548.25\ \text{m}^3/\text{day}$

(E) Sludge production

(i) Net VSS production $= XV/\theta_c$

$$= \frac{0.75 \times 6000 \text{ (mg/L)} \times 10^{-6} \text{ (kg/mg)} \times 4050 \text{ (m}^3) \times 10^3 \text{ (L/m}^3)}{12 \text{ (day)}}$$

$$= 1518.75\ \text{kg/day}$$

(ii) Net SS production $= 1518.75\ (\text{kg/day})/0.75$

$$= 2025\ \text{kg/day}$$

(iii) Volume of the sludge produced $= \dfrac{2025 \text{ (kg/day)}}{1000 \text{ (kg/m}^3) \times 1.03 \times 0.02}$

$$= 98.3\ \text{m}^3/\text{day}$$

(iv) VSS production per kg BOD_r

$$= \frac{1518.75 \text{ (kg/day)} \times 10^6 \text{ (mg/kg)}}{(3500 - 16.5) \text{ (mg/L)} \times 1275 \text{ (m}^3\text{/day)} \times 10^3 \text{ (L/m}^3)}$$

$$= 0.34\ \text{mg/mg}$$

(F) Oxygen requirement

(i) Theoretical O_2 required $= (\text{BOD}_L \text{ removed}) - (\text{BOD}_L \text{ of solids leaving})$

$$= 1.47\ (3500 - 16.5)\ (\text{mg/L}) \times 1275\ (\text{m}^3) \times 10^3\ (\text{L/m}^3) \times 10^{-6}\ (\text{kg/mg}) - 1.42 \times 1518.75\ (\text{kg/day})$$

$$= 4372.32\ \text{kg/day}$$

(ii) Theoretical air requirement assuming that air contains 23.2% oxygen by weight and density of air $= 1.201\ \text{kg/m}^3$

$$= \frac{4372.32 \text{ (kg/day)}}{0.232 \times 1.201 \text{ (kg/m}^3)}$$

$$= 15{,}692.11\ \text{m}^3/\text{day}$$

(iii) Actual air requirement at an 8% transfer efficiency

$$= 15{,}692.11\ (\text{m}^3/\text{day})/0.08$$
$$= 196{,}151.41\ \text{m}^3/\text{day}$$

(iv) Check for the air requirement per unit volume

$$= 196{,}151.41\ (\text{m}^3/\text{day})/1275\ (\text{m}^3/\text{day})$$
$$= 153.84\ \text{m}^3/\text{m}^3$$

(v) Air requirement per kg of BOD_5 removed

$$= 196{,}151.41\ (\text{m}^3/\text{day})/4441.46\ (\text{kg/day})$$
$$= 44.16\ \text{m}^3/\text{kg BOD}_5\ \text{removed}$$

(G) Power requirement assuming the aerators are designed to give 2 kgO_2/kWh and the field efficiency is 70%.

$$\text{Power required} = \frac{4372.32\ (\text{kg/day})}{2\ (\text{kg/kWh}) \times 0.7 \times 24\ (\text{h/day})}$$
$$= 130.13\ (\text{kW}) \times 1.3410\ (\text{hp/kW})$$
$$= 174.5\ \text{hp (say 175 hp)}$$

Example 11

A powdered activated carbon fed activated sludge process is designed to treat 15,000 GPD of pharmaceutical wastewater. The SCOD (soluble chemical oxygen demand) of the treated effluent is 590 mg/L. Determine the dose of PAC (powdered activated carbon) required for further reduction of effluent SCOD from 590 mg/L to 200 mg/L. Use the Freundlich equation [48]: $X/M = (3.7 \times 10^{-6})\ C_e^{2.1}$ to determine the dose of powdered activated carbon.

Solution

(A) SCOD concentration at equilibrium $C_e = 200$ mg/L
(B) Amount of SCOD removal attributed to the PAC, X (mg/L) $= 590 - 200 = 390$ mg/L
(C) The dose of activated carbon (M) can be determined by the Freundlich equation:

$$\frac{X}{M} = (3.7 \times 10^{-6})C_e^{2.1}$$

$$M = X/3.7 \times 10^{-6} \times C_e^{2.1}$$

$$M = 390/3.7 \times 10^{-6} \times 200^{2.1}$$
$$M = 1551\ \text{mg/L (1.55 g/L)}$$

(D) The dose of PAC per unit SCOD removed

$$\frac{X}{M} = \frac{1551\ (\text{mg/L})}{390\ (\text{mg/L})}$$

$$= 3.98\ \text{mg PAC/mg SCOD}_r$$

Example 12

The result of a pilot plant study of PAC-fed activated sludge process is given in the following table.

PAC dose (mg/L)	Effluent SCOD in control reactor (mg/L)	Effluent SCOD in PAC-fed reactor (mg/L)
208	825	459
827	825	265
496	670	314
1520	583	194

PAC, powdered activated carbon; SCOD, soluble chemical oxygen demand.

Using the Freundlich equation $(X/M) = kC_e^{1/n}$, find the values of constants k and n.

Solution

(A) The first step is to estimate the values of X/M against the equilibrium SCOD concentration. The SCOD removal attributed to PAC can be calculated by subtracting the effluent SCOD in the PAC-fed reactor from the effluent SCOD of the control reactor. The estimated values of X and X/M are given in the following table.

PAC dose (mg/L) (M)	Effluent SCOD (mg/L) C_e	SCOD removal by PAC (mg/L) X	Ratio (X/M)	log C_e	log(X/M)
208	459	366	1.76	2.66	0.25
827	265	560	0.68	2.42	−0.17
496	314	356	0.72	2.50	−0.14
1520	194	389	0.20	2.29	−0.70

(B) The second step is to plot the log(X/M) values against the various values of the log C_e as shown in Figure 7. By taking the log of both sides of the Freundlich equation we get a straight line whose intercept gives the value K and slope gives the value of $1/n$. The log of the Freundlich equation results in the following equation:

$$\log\left(\frac{X}{M}\right) = \log K + \left(\frac{1}{n}\right)\log C_e$$

The values of log(X/M) and log C_e have been calculated and given in the table above and plotted as shown in Figure 7. From the graph, the slope of the line gives a value of $1/n = 2.4218$, hence $n = 0.41$ and the intercept gives the value of $\log K = -6.1665$, hence $K = 6.81 \times 10^{-7}$.

Example 13

An aerated lagoon is to be designed to treat 15,000 GPD of spent stream generated from a biological production plant. The depth of the lagoon is restricted to 3.3 m and the HRT of the

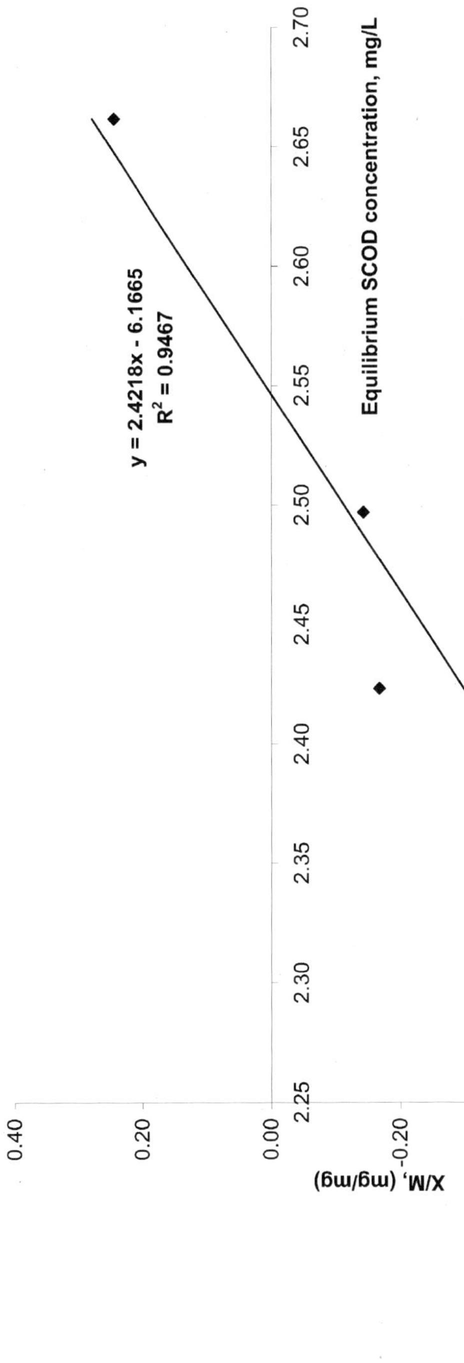

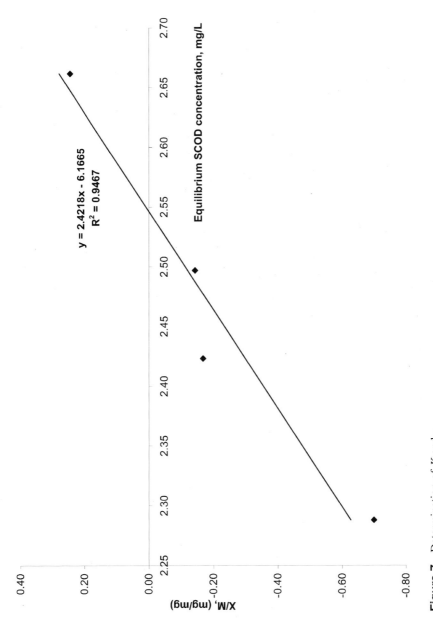

Figure 7 Determination of K and n.

lagoon is 4 days. Determine the surface area of the lagoon. If the wastewater enters the lagoon at a temperature of 65°C and the mean ambient temperature is 30°C, estimate the lagoon temperature assuming complete mixing condition and exchange coefficient $f = 0.5$ m/day. Also comment on the effect of temperature on process efficiency.

Solution

$$\text{Volume of the aerated lagoon (m}^3) = \text{flow}, V(\text{GPD}) \times 3.785\ (\text{L/gal}) \times 10^{-3}\ (\text{m}^3/\text{L}) \times \text{HRT}, \theta(\text{day})$$

$$= 15{,}000 \times 3.785 \times 10^{-3} \times 4$$

$$= 227.1\ \text{m}^3$$

$$\text{Surface area of the lagoon (m}^2) = \frac{\text{Volume, } V\ (\text{m}^3)}{\text{Depth, } D(\text{m})}$$

$$= (227.1/3.3) = 68.82\ \text{m}^2$$

The lagoon temperature can be obtained by the law of conservation of energy:

Total heat gain = Total heat loss

$$Q\ (\text{GPD}) \times 3.785\ (\text{L/gal}) \times 10^3\ (\text{m}^3/\text{L}) \times (T_i - T_e)\ (°\text{C}) = f\ (\text{m/day}) \times \text{A}\ (\text{m}^2) \times (T_w - T_a)\ (°\text{C})$$

For complete mixing condition

$$T_e = T_w$$

$$15{,}000 \times 3.785 \times 10^3 \times (65 - T_w) = 0.5 \times 68.82 \times (T_w - 30)$$

$$T_w = 65°\ \text{C}$$

Comment

The temperature of the aerated lagoon of more than 38°C has been reported to decrease the system efficiency. At this high temperature nitrifiers cannot survive. Hence, a heat exchanger on the influent line must be provided to reduce the high temperature of the raw wastewater.

Example 14

Design a flow-through aerated lagoon to treat 0.575 MGD of composite wastewater generated from a pharmaceutical plant. Assume that the following conditions and requirements apply.

1. Mean cell residence time, $\theta_c = 10$ days;
2. Depth of the lagoon = 3.3 m;
3. Kinetic coefficients: $Y = 0.6$ mg/mg, $K_s = 210$ mg/L, $k = 4.6$ day^{-1}, $K_d = 0.1$ day^{-1};
4. Influent BOD_5 after settling = 3100 mg/L;
5. Influent SS concentration = 1140 mg/L;
6. O_2 transfer capacity of the aerator in field = 1.22 kg O_2/kWh;
7. Power requirement for mixing = 0.6 hp/1000 m^3.

Solution

(A) Determine the size of the aerated lagoon based on θ_c (solid retention time)

(i) Volume of wastewater generated per day

$$= 0.575\ (\text{MGD}) \times (3.785 \times 10^3)\ (\text{m}^3/\text{Mgal})$$

$$= 2176.37\ \text{m}^3/\text{day}$$

(ii) Volume of the aerated lagoon required $= 2176.37\ (\text{m}^3/\text{day}) \times 10\ (\text{day})$

$$= 21{,}763.75\ \text{m}^3$$

(iii) Surface area of the lagoon $= 21{,}763.75\ (\text{m}^3)/3.3\ (\text{m})$

$$= 6595.07\ \text{m}^2$$

(B) Determine the soluble effluent BOD_5 using the kinetic data

$$S = \frac{K_s(1 + \theta K_d)}{\theta(Yk - K_d) - 1}$$

$$= \frac{210\ (\text{mg/L})[1 + 10\ (\text{day}) \times 0.1\ (\text{day}^{-1})]}{10\ (\text{day})[0.60\ (\text{mg/mg}) \times 4.6\ (\text{day}^{-1}) - 0.1\ (\text{day}^{-1})] - 1}$$

$$= 16.41\ \text{mg/L}$$

(C) Determine the O_2 requirement

(i) Estimate the concentration of biological solids produced

$$= \frac{Y(S_i - S)}{1 + K_d\theta_c}$$

$$= \frac{0.6\ (\text{mg/mg})\ (3100 - 16.41)\ (\text{mg/L})}{1 + 0.1\ (\text{day}^{-1}) \times 10\ (\text{day})}$$

$$= 925.08\ \text{mg/L}$$

(ii) Estimate the SS concentration in the lagoon before settling

$$= 925.08 + 1140$$

$$= 2065.08\ \text{mg/L}\ (2065.08\ \text{g/m}^3)$$

(iii) Estimate the amount of solids wasted per day

$$= 925.08\ (\text{g/m}^3) \times 10^{-3}(\text{kg/g}) \times 2176.37\ (\text{m}^3/\text{day})$$
$$= 2013.32\ \text{kg/day}$$

(iv) Estimate the amount of O_2 required $= 1.47\ Q\ (S_i - S) - 1.42\ P_x$

$$= 1.47 \times 2176.37\ (\text{m}^3/\text{day}) \times (3100 - 16.41)\ (\text{mg/L})$$
$$\times 10^{-3}\ (\text{kg/mg} \cdot \text{L/m}^3) - 1.42 \times 2013.32\ (\text{kg/day})$$
$$= 7006.33\ \text{kg/day}$$

(D) Determine the power required to meet the O_2 requirement:

$$\text{Power} = \frac{7006.33\ (\text{kg/day})}{(1.22\,\text{kgO}_2/\text{kWh}) \times 24\ (\text{h/day})}$$

$$= 239.29\,\text{kW}$$

$$\text{Power} = 239.29\ (\text{kW}) \times 1.3410\ (\text{hp/kW})$$
$$= 320.89\,\text{hp}$$

(E) Determine the power required for mixing:

(i) Lagoon volume $= 21{,}763.75\ (\text{m}^3) \times 35.3147\ (\text{ft}^3/\text{m}^3)$

$$= 768{,}580.3\,\text{ft}^3$$

(ii) Power required for mixing

$$= 768{,}580.3\,(\text{ft}^3) \times 0.6\left(\frac{\text{hp}}{1000\,\text{ft}^3}\right)$$

$$= 461.15\,\text{hp}$$

(F) Determine the horse-power rating of the aerator:

Horse power rating = 461.15 hp to fulfill the requirement of both mixing and O_2 supply

Example 15

Design a trickling filter to treat 33,800 GPD of pharmaceutical wastewater using the empirical method of Ten States for the data given below:

1. Influent BOD_5 of the raw wastewater, $S_i = 6000$ mg/L;
2. Efficiency of the filter, $E = 90\%$;
3. Depth of filter is restricted to 1.8 m.

Solution

(A) Determine the recirculation ratio required to give 90% efficiency:

$$E = \frac{(1 + R/Q)}{1.5 + (R/Q)}$$

$$0.90 = \frac{(1 + R/Q)}{1.5 + (R/Q)}$$

$$R/Q = 3.5$$

$$R = 3.5 \times 33{,}800\ (\text{GPD})$$

$$= 118{,}300\ \text{GPD}$$

(B) Determine the filter volume required by providing the maximum organic loading rate 1.2 kg/m^3 day:

$$V = [33{,}800 \text{ (GPD)} \times 3.785 \text{ (L/gal)} \times 0.9 \times 6000 \text{ (mg/L)}$$
$$\times 10^{-6} \text{ (kg/mg)}]/1.2 \text{ (kg/m}^3 \text{ day)}$$
$$= 575.70 \text{ m}^3$$

(C) Determine the size of filter required:

$$\text{Surface area required} = 575.70 \text{ (m}^3\text{)}/1.8 \text{ (m)}$$

$$(\pi/4)D^2 = 319.83 \text{ (m}^2\text{)}$$
$$D = 20.18 \text{ m (say 20.2 m)}$$

(D) Hydraulic loading including the recirculation

$$= \frac{(33{,}800 + 3.5 \times 33{,}800) \text{ (GPD)} \times (3.785 \times 10^{-3}) \text{ (m}^3\text{/gal)}}{(\pi/4) \times (20.2)^2 \text{ (m}^2\text{)}}$$

$$= 1.80 \text{ m}^3\text{/m}^2 \text{ day}$$

Example 16

Design a UASB (upflow anaerobic sludge blanket) reactor for treatment of 435 m^3/day of wastewater generated from a typical pharmaceutical plant. The COD removal efficiency of the reactor at HRT of 2 days and organic loading of 3.52 kg/m^3 is 94%. Assume the following design data are applicable:

1. Influent COD = 7000 mg/L;
2. Methane yield = 0.35 m^3/kg COD removed;
3. Solubility of methane = 0.028 m^3/m^3 effluent;
4. Biomass yield = 0.027 mg/mg;
5. MLVSS of the sludge bed = 70 kg/m^3;
6. MLVSS of the sludge blanket = 4 kg/m^3;
7. Depth of the sludge bed = 1.5 m;
8. Depth of the sludge blanket = 3.5 m.

Solution

(A) Determine the size of the reactor

(i) Volume of the reactor, $V = Q \times \theta$

$$V = 435 \text{ (m}^3\text{/day)} \times 2 \text{ (day)}$$
$$= 870 \text{ m}^3$$

(ii) Depth of the reactor, $H = 1.5 + 3.5$

$$H = 5 \text{ m}$$

(iii) Surface area required, $A = V/H$

$$A = 870\ (\text{m}^3)/5\ (\text{m})$$
$$= 174\,\text{m}^2$$

(iv) Diameter of reactor, $D = (4 \times A/\pi)^{1/2}$

$$D = (4 \times 174/\pi)^{1/2}$$
$$= 14.88\,\text{m (say 14.9 m)}$$

(B) Determine the organic loading rate

$$= (Q \times S_\text{i} \times E/V)$$
$$= 435\ (\text{m}^3/\text{day}) \times 7000\ (\text{mg/L}) \times 10^{-3}\ (\text{kg/mg} \cdot \text{L/m}^3) \times 0.94/870\ (\text{m}^3)$$
$$= 3.29\,\text{kg/m}^3\ \text{day} < 3.52\,\text{kg/m}^3\ \text{day, hence OK}$$

(C) Determine the upflow velocity, $v = H/\theta$

$$v = 5\ (\text{m})/[2\ (\text{day}) \times 24\ (\text{h/day})]$$
$$= 0.1\,\text{m/h}$$

(D) Determine the SRT

(i) Total COD removed per day $= 435\ (\text{m}^3/\text{day}) \times 7000\ (\text{mg/L}) \times 10^{-3}\ (\text{kg/mg} \cdot \text{L/m}^3) \times 0.94$

$$= 2862.3\,\text{kg/day}$$

(ii) Biomass produced per day $= 0.027\ (\text{mg/mg}) \times 2862.3\ (\text{kg/day})$

$$= 77.28\,\text{kg/day}$$

(iii) Total biomass in the reactor

$$= \text{Biomass in the sludge bed} + \text{biomass in the sludge blanket}$$
$$= 70\ (\text{kg/m}^3) \times \pi/4 \times (14.9)^2\ (\text{m}^2) \times 1.5\ (\text{m}) + 4\ (\text{kg/m}^3) \times \pi/4 \times (14.9)^2\ (\text{m}^2) \times 3.5\ (\text{m})$$
$$= 20{,}749.58\,\text{kg}$$

(iv) SRT = Total biomass in the reactor/biomass produced per day

$$= 20{,}749.58\ (\text{kg})/77.28\ (\text{kg/day})$$
$$= 268.5\,\text{day}$$

(E) $$\text{F/M ratio} = \frac{435\ (\text{m}^3/\text{day}) \times 7000\ (\text{mg/L}) \times 10^{-3}\ (\text{kg/mg} \cdot \text{L/m}^3)}{20{,}749.58\ (\text{kg})}$$

$$= 0.15\,\text{day}^{-1}$$

(F) Methane production

(i) Total quantity of methane generated $= 0.35\ (\text{m}^3/\text{kg COD}_\text{r}) \times 2862.3\ (\text{kg COD}_\text{r}/\text{day})$

$$= 1001.8\,\text{m}^3/\text{day}$$

(ii) Methane leaving as dissolved in the effluent $= 0.028\ (m^3/m^3) \times 435\ (m^3/day)$

$= 12.18\ m^3/day$

(iii) Usable methane $= 1001.8 - 12.18$

$= 989.62\ m^3/day$

(G) Specific gas production

(i) Specific gas production per m^3 per m^3 of reactor per day

$= 1001.8\ (m^3/day)/870\ (m^3)$
$= 1.15\ m^3/m^3/day$

(ii) Specific gas production per m^3 per m^3 of effluent

$= 1001.8\ (m^3/day)/435\ (m^3/day)$
$= 2.3\ m^3/m^3$

(H) Energy equivalent of biogas
$= 989.62\ m^3/day \times 10{,}000\ (kcal/m^3)$

$= 989.62 \times 10^4\ kcal/day$
$= 989.62 \times 10^4\ (kcal/day) \times 1.1633 \times 10^{-3}\ (kWh/kcal)$
$= 11{,}512.25\ kWh/day$

(I) Coal equivalent of biogas $= 989.62 \times 10^4\ (kcal/day)/4000\ (kcal/kg)$

$= 2474.05\ kg/day$
$= 2.47\ tonnes/day$

Example 17

Determine the size of the UASB reactor for the treatment of 1275 m^3/day of wastewater generated from a typical pharmaceutical plant with COD of 16,000 mg/L. The COD removal efficiency at an HRT of 4.7 days is 97%. If the following data and conditions are applicable, estimate (i) OLR and upflow velocity and (ii) methane yield and energy equivalent.

1. Overall depth = 4 m;
2. Percentage of methane in biogas = 65%;
3. Specific biogas production rate = 7.64 m^3/m^3 effluent.

Solution

(A) Determine the size of the reactor

(i) Volume of the reactor, $V = Q \times \theta$

$V = 1275\ (m^3/day) \times 4.7\ (day)$

$= 5992.5\ m^3$

(ii) Surface area required, $A = V/H$

$$A = 5992.5\ (\text{m}^3)/4\ (\text{m})$$
$$= 1498.125\ \text{m}^2$$

(iii) Diameter of reactor, $D = (4 \times A/\pi)^{1/2}$

$$D = (4 \times 1498.125/\pi)^{1/2}$$
$$= 43.67\ \text{m}$$

Because the diameter required is much more, therefore three parallel units must be designed with area of each unit as follows

$$\text{Area} = 1498.125\ (\text{m}^2)/3$$
$$= 499.375\ \text{m}^2$$

(iv) Diameter of each unit $= (4 \times 499.375/\pi)^{1/2}$

$$= 25.21\ \text{m}$$

(B) Determine the OLR

(i) Total COD removed/day $= 1275\ (\text{m}^3/\text{day}) \times 16{,}000\ (\text{mg/L}) \times 10^{-3}\ (\text{kg/mg} \cdot \text{L/m}^3) \times 0.97$

$$= 19{,}788\ \text{kg/day}$$

(ii) COD removed per unit $= 19{,}788\ (\text{kg/day})/3$

$$= 6596\ \text{kg/day}$$

(iii) $\text{OLR} = 6596\ (\text{kg/day})/499.375\ (\text{m}^2)$

$$= 13.21\ \text{kg COD/m}^3\ \text{day}$$

(C) Determine the upflow velocity $= H/\theta$

$$= 4\ (\text{m})/[4.7\ (\text{day}) \times 24\ (\text{h/day})]$$
$$= 0.035\ \text{m/h}$$

(D) Determine the methane yield

(i) Total quantity of methane generated

$$= 7.64\ (\text{m}^3/\text{m}^3\ \text{effluent}) \times 1275\ (\text{m}^3/\text{day}) \times 0.65$$
$$= 6331.65\ \text{m}^3/\text{day}$$

(ii) Methane yield = total methane generated/total COD removed

$$= 6331.65\ (\text{m}^3/\text{day})/19{,}788\ (\text{kg/day})$$
$$= 0.32\ \text{m}^3/\text{kg COD}_\text{r}$$

(E) Determine the energy equivalent

(i) Methane leaving as dissolved in the effluent

$$= 0.028\ (\text{m}^3/\text{m}^3) \times 1275\ (\text{m}^3/\text{day})$$
$$= 35.7\ \text{m}^3/\text{day}$$

(ii) Usable methane $= 6331.65 - 35.7$

$$= 6295.95\,\text{m}^3/\text{day}$$

(iii) Energy equivalent of biogas

$$= 6295.95\ (\text{m}^3/\text{day}) \times 10{,}000\ (\text{kcal}/\text{m}^3)$$
$$= 6295.95 \times 10^4\ \text{kcal/day}$$
$$= 6295.95 \times 10^4\ (\text{kcal/day}) \times 1.1633 \times 10^{-3}\ (\text{kWh/kcal})$$
$$= 73{,}240.79\ \text{kWh/day}$$

3.11 DISCUSSION TOPICS AND PROBLEMS

1. The BOD_5 and flow of various types of waste streams generated from Abbott Laboratory (a typical pharmaceutical plant) is given in the following table.

Type of waste stream	Flow (MGD)	BOD_5 (mg/L)
Chemical waste	0.262	2520
Fermentation waste	0.312	3620

In addition, the plant generates 470 m^3/day of domestic wastewater with $BOD_5 = 675$ mg/L. Calculate (a) BOD_5 of the composite waste and (b) total BOD load of the plant.
[*Answers*: (a) = 2684.09 mg/L; (b) 7091.18 kg BOD/day]

2. A synthetic organic chemical plant generates mainly two types of waste streams, i.e. strong process waste and dilute process waste. The BOD_5 of the 45,000 GPD of combined waste generated from the plant is 75,000 mg/L. If the BOD_5 and flow of the dilute process waste are 1200 mg/L and 33,800 GPD, respectively, estimate (a) BOD_5 of the strong process waste and (b) the BOD load of each waste stream and their contribution to the total BOD load of the plant.
[*Answer*: (a) = 297,717.73 mg/L; (b) BOD load of strong and dilute process waste = 12,620.85 kg/day (98.8%) and 153.52 kg/day (1.2%), respectively]

3. A primary settling tank is designed for the pretreatment of 0.575 MGD of wastewater with SS concentration of 1140 mg/L generated from a typical pharmaceutical plant. If the SS removal efficiency of the sedimentation tank is 60%, find (a) the effluent SS concentration and (b) the quantity of sludge generated. Assume the specific gravity of the sludge is 1.03 and that the sludge contains 5% solids.
[*Answer*: (a) = 456 mg/L; (b) 7636.89 GPD]

4. A typical pharmaceutical industry generates 15,000 GPD of wastewater with SS concentration 800 mg/L. The addition of alum (200 mg/L) and $FeCl_3$ (150 mg/L) reduces the SS concentration of effluent from 800 mg/L to 50 g/L. Determine the quantity of sludge generated per week. Assume the specific gravity of the sludge is 1.04 and that the sludge contains 3% solids.
[*Answer*: 3014.5 GPD]

5. The BOD removal efficiency of an activated sludge process treating 2000 m^3/day of condensate waste generated from a synthetic organic chemical plant is 94%. If the organic loading rate and BOD_5 of the raw waste are 3.171 kg/m^3 day

and 1275 mg/L, estimate (a) BOD_5 of the treated effluent and (b) hydraulic retention time.
[*Answer*: (a) 76.5 mg/L; (b) 9.12 h]

6. An activated sludge process is designed to treat 1950 m^3/day of pharmaceutical wastewater with BOD_5 concentration of 3250 mg/L. If the performance efficiency of the process based on BOD_5 removal is 85%, determine (a) the organic loading rate (OLR) and BOD_5 of the treated effluent; assuming the following data and conditions are applicable:
   (i) Aeration tank volume = 1500 m^3;
   (ii) Depth of the aeration tank = 2.5 m;
   (iii) MLVSS = 6000 mg/L.
   Also compute (b) the hydraulic loading rate and F/M ratio.
   [*Answer*: (a) OLR = 3.59 kg/m^3 day and BOD_5 = 487.5 mg/L; (b) HLR = 3.25 $m^3/m^2 \cdot$ day and F/M ratio = 0.7 day^{-1}]
7. Determine the F/M ratio and solid retention time of an extended aeration system designed for the treatment of 33,800 GPD of pharmaceutical wastewater. The BOD_5 of the raw wastewater and treated effluent are 5000 mg/L and 560 mg/L, respectively. Assume the following data and conditions are applicable:
   (i) HRT = 5 days;
   (ii) MLSS = 5600 mg/L;
   (iii) MLVSS/MLSS = 0.75;
   (iv) $Y_n = 0.45$ mg/mg;
   [*Answer*: F/M = 0.24 day^{-1} and SRT = 10.51 days]
8. The PAC-fed activated sludge process is designed to treat the alkaline waste stream generated from a synthetic organic chemical plant. The influent BOD_5 of the alkaline waste is 1275 mg/L, which can be treated to a BOD_5 of 275 mg/L by the activated sludge process. The addition of PAC at a dose of 500 mg/L gives a further reduction of effluent BOD_5 from 275 mg/L to 150 mg/L. Determine the constant K of the Freundlich equation given below $(X/M) = KC_e^{2.2}$. Also comment on the efficiency of the system before and after addition of PAC.
   [*Answer*: $K = 4.08 \times 10^{-6}$; performance efficiency of the system can be increased by approximately 10% by addition of PAC]
9. An aerated lagoon is designed to treat 435 m^3/day of acid waste stream with a BOD_5 of 3500 mg/L generated from a synthetic organic chemical plant. The depth of lagoon is restricted to 4 m and organic loading rate is 0.7 kg/m^3 day. Estimate (a) the surface area and hydraulic loading rate. If the performance efficiency of the lagoon is 97%, determine (b) the BOD_5 of the treated effluent.
   [*Answer*: (a) A = 527.44 m^2 and HLR = 0.82 m^3/m^2 day; (b) 105 mg/L]
10. An aerated lagoon is designed to treat 0.575 MGD of composite waste (including chemical and fermentation waste) with a BOD_5 of 3150 mg/L. The depth and HRT of the lagoon are restricted to 3.5 m and 5 days, respectively. Find (a) the surface area of the lagoon. If the temperature of composite waste entering into the lagoon is 60°C and mean ambient temperature is 15°C during winter, estimate (b) the lagoon temperature assuming complete mixing condition and exchange coefficient f = 0.54 m/day. Also comment (c) on the effect of wastewater temperature in the process efficiency of the lagoon.
    [*Answer*: (a) A = 3109.10 m^2; (b) T_w = 40.4°C; (c) the temperature of waste 60°C will result in the temperature in aerated lagoon being >38°C, which is found to reduce the process efficiency]

11. A trickling filter is designed to treat 435 m^3/day of acid waste stream generated from a synthetic organic chemical plant. The BOD_5 of the acid waste before and after the primary sedimentation is 3250 mg/L and 2850 mg/L, respectively. The efficiency of the filter at a recirculation ratio of 4.5 is 92%. If the depth of filter is restricted to 1.6 m and the value of the constant in Eckenfelder's equation is $n = 0.5$, determine the value of constant K_f assuming the hydraulic loading rate = 17.5 $m^3/m^2 \cdot day$.
    [*Answer*: $K_f = 3.12\ m^{1/2} \cdot day^{1/2}$]
12. A pharmaceutical wastewater with BOD_5 of 3000 mg/L is to be treated by a trickling filter. Design the filter for 15,000 GPD of wastewater to give the desired effluent BOD_5 of 50 mg/L. Use the NRC (US National Research Council) equation for the design of the filter. The following data and conditions are applicable:
    (i) Depth of filter = 1.7 m;
    (ii) Recirculation ratio = 2 : 1;
    (iii) Wastewater temperature = 20°C;
    (iv) Assume efficiencies of the two-stage filters are equal: $E_1 = E_2$.
    [*Answer*: (a) $E_1 = E_2 = 87\%$; (b) diameter of 1st stage filter $D_1 = 23$ m and 2nd stage filter $D_2 = 63.95$ m]
13. A UASB reactor is designed to treat 1275 m^3/day of wastewater with a BOD concentration of 2000 mg/L generated from a typical pharmaceutical industry. At an HRT of 1.5 days, the COD and BOD removal efficiencies of the reactor are 80 and 95%, respectively. Determine (a) the size of the reactor; (b) the total quantity of methane produced; and (c) the coal equivalent and energy equivalent. Assume that the following data and conditions are applicable:
    (i) Depth of the reactor is restricted to 4.5 m;
    (ii) Biogas yield = 0.6 $m^3/kg\ COD_r$;
    (iii) Methane content of biogas = 70%;
    (iv) Solubility of methane = 0.028 m^3/m^3 effluent;
    (v) Calorific value of methane = 10,000 $kcal/m^3$;
    (vi) Calorific value of coal = 4000 kcal/kg.
    [*Answer*: (a) Diameter = 23.26 m; (b) 1102.24 m^3/day; (c) 2.67 tons/day and 12,410 kWh/day]
14. The COD removal efficiency of a UASB reactor treating pharmaceutical wastewater is 96% at an organic loading rate of 0.5 kg $COD/m^3/day$. If the plant generates 33,800 GPD wastewater with a COD concentration of 1000 mg/L and the depth of reactor is restricted to 3 m, estimate (a) the size of the UASB reactor; (b) the HRT; and (c) the specific gas production rate assuming a methane yield of 0.3 $m^3/kg\ COD_r$.
    [*Answer*: (a) Diameter = 10.21 m; (b) 1.92 days; (c) 0.29 m^3/m^3 effluent and 0.15 $m^3/m^3/day$]

NOMENCLATURE

ACPCI	Alexandra Company for Pharmaceutical and Chemical Industry
AMFFR	anaerobic mesophilic fixed film reactor
ASP	activated sludge process
ATFFR	anaerobic thermophilic fixed film reactor

BOD	biochemical oxygen demand (mg/L)
BOD_5	5-day biochemical oxygen demand (mg/L)
BOD_r	biochemical oxygen demand removed (kg/day)
BOD_L	ultimate biochemical oxygen demand (mg/L)
COD	chemical oxygen demand (mg/L)
DCE	dichloroethylene
DO	dissolved oxygen (mg/L)
DOC	dissolved organic carbon (mg/L)
F/M	food to microorganism ratio (day^{-1})
gal	gallon
GAC	granular activated carbon
h	hour
HLR	hydraulic loading rate (m^3/m^3 day)
HRT	hydraulic retention time (day)
kg/day	kilogram per day
L/gal	liter/gallon
mg/L	milligram per liter
MGD	million gallons per day
MLSS	mixed liquor suspended solids (mg/L)
MLVSS	mixed liquor volatile suspended solids (mg/L)
NA	nitroaniline
NEIC	National Enforcement Investigation Center
NP	nitrophenol
OLR	organic loading rate (kg COD/m^3 day)
OWC	organic wastewater contaminants
PAC	powdered activated carbon
SCOD	soluble chemical oxygen demand (mg/L)
S-CBOD	soluble carbonaceous oxygen demand (mg/L)
SRT	solid retention time (day)
SS	suspended solids (mg/L)
SVI	sludge volume index
TCMP	tri-chloromethyl-propanol
TDS	total dissolved solids (mg/L)
TKN	total Kjeldahl nitrogen (mg/L)
TOC	total organic carbon (mg/L)
TS	total solids (mg/L)
UASB	upflow anaerobic sludge blanket
VSS	volatile suspended solids (mg/L)
V/V	volume/volume

REFERENCES

1. Mehta, G.; Prabhu, S.M.; Kantawala, D. Industrial wastewater treatment–The Indian experience. J. Indian Assoc. Environ. Management **1995**, *22*, 276–287.
2. Kolpin, D.W.; Furlong, E.T.; Meyer, M.T.; Thurman, E.M.; Zaugg, S.D.; Barber, L.B.; Buxton, H.T. Pharmaceuticals, hormones and other organic wastewater contaminants in U.S. streams, 1999–2000: A national reconnaissance. Environ. Sci. Technol. **2002**, *36*, 1202–1211.

3. Struzeski, E.J. Status of wastes handling and waste treatment across the pharmaceutical industry and 1977 effluent limitations. *Proceedings of the 35th Industrial Waste Conference*, Purdue University, West Lafayette, IN, 1980, 1095–1108.
4. Seif, H.A.A.; Joshi, S.G.; Gupta, S.K. Effect of organic load and reactor height on the performance of anaerobic mesophilic and thermophilic fixed film reactors in the treatment of pharmaceutical wastewater. Environ. Technol. **1992**, *13*, 1161–1168.
5. Andersen, D.R. Pharmaceutical wastewater treatment: A case study. *Proceedings of the 35th Industrial Waste Conference*, Purdue University, West Lafayette, IN, 1980, 456–462.
6. Struzeski, E.J. *Waste Treatment and Disposal Methods for the Pharmaceutical Industry*. NEIC-Report EPA 330/1-75-001; U.S. Environmental Protection Agency, Office of Enforcement, National Field Investigation Center: Denver, CO, 1975.
7. Molof, A.H.; Zaleiko, N. Parameter of disposal of waste from pharmaceutical industry. Ann. N. Y. Acad. Sci. **1965**, *130*, 851–857.
8. Syracuse, N.Y.; Mann, U.T. Effects of penicillin waste in Ley Creek Sewage Treatment Plant. Sewage Ind. Waste **1951**, *23*, 1457–1460.
9. Gallagher, A., *et al.* Pharmaceutical waste disposal. Sewage Ind. Waste **1954**, *26* (11), 1355–1362.
10. Seeler, T.A.; Jennet, J.C. Treatment of wastewater from a chemically synthesized pharmaceutical manufacturing process with the anaerobic filter. *Proceedings of the 33rd Industrial Waste Conference*, Purdue University, West Lafayette, IN, 1978, 687–695.
11. Patil, D.M.; Shrinivasen, T.K.; Seth, G.K.; Murthy, Y.S. Treatment and disposal of synthetic drug wastes. Environ. Health (India) **1962**, *4*, 96–105.
12. Lawson, J.R.; Woldman, M.L.; Eggerman, P.P. Squibb solves its pharmaceutical wastewater problems in Puerto Rico. Chem. Engng. Progress Symposium Series No. 107, 1971, "Water-1970", 1970, 401–404.
13. Murthy, Y.S.; Subbiah, V.; Rao, D.S.; Reddy, R.C.; Kumar, L.S.; Elyas, S.I.; Rama Rao, K.G.; Gadgill, J.S.; Deshmukh, S.B. Treatment and disposal of wastewater from synthetic drugs plant (I.D.P.L.), Hyderabad, Part I – Wastewater characteristics. Indian J. Environ. Health **1984**, *26* (1), 7–19.
14. Deshmukh, S.B.; Subrahmanyanm, P.V.R.; Mohanrao, G.J. Studies on the treatment of wastes from a synthetic drug plant. Indian. J. Environ. Health **1973**, *15*, 2.
15. Deshmukh, S.B.; Gadgil, J.S.; Subrahmanyanm, P.V.R. Treatment and disposal of wastewater from synthetic drugs plant (I.D.P.L.), Hyderabad, Part II – Biological treatability. Indian J. Environ. Health **1984**, *26*, 20–28.
16. Mayabhate, S.P.; Gupta, S.K.; Joshi, S.G. Biological treatment of pharmaceutical wastewater. Water Air Soil. Poll. **1988**, *38*, 189–197.
17. Howe, R.H.L.; Nicoles, R.A. Waste treatment for veterinary and plant science research and production at Eli Lilly Greenfield Laboratories. *Proceedings of the 14th Industrial Waste Conference*, Purdue University, West Lafayette, IN, 1959; 647–655.
18. Yeole, T.Y.; Gadre, R.V.; Ranade, D.R. Biological treatment of a pharmaceutical waste. Indian J. Environ. Health **1996**, *38* (2), 95–99.
19. Howe, R.H.L. Handling wastes from the billion dollar pharmaceuticals Industry. Waste Engng **1960**, *31*, 728–753.
20. Howe, R.W.L.; Coates, D.G. Antitoxin and vaccine wastes treated at Eli Lilly plant. Waste Engng **1955**, *26*, 235.
21. Lines, G. Liquid wastes from the fermentation industries. J. Water Pollut. Conf. **1968**, 655.
22. Genetelli, G.J.; Heukelekian, H.; Hunter, J.V. *Use of Research Techniques for Determination of Design Parameters*. Journal Series, New Jersey Agricultural Experiment Station, Rutgers – The State University of New Jersey, Department of Environmental Sciences: New Brunswick, NJ, 1967.
23. Nedved, T.K.; Bergmann, D.E.; Camens, A.A. Pharmaceutical laboratory BOD studies, a matter of philosophy. *2nd Symposium of Hazardous Chemicals and Disposal*, Indianapolis, IN, 1971.
24. Genetelli, E.J.; Heukelekian, H.; Hunter, J.V. A rational approach to design for a complex chemical waste. *Proceedings of the 5th Texas Water Pollution Association, Industrial Water and Waste Conference*, 1965; 372–396.

25. Young, J.C.; Affleck, S.B. Long-term biodegradability tests of organic industrial wastes. *Proceeding of the 29th Industrial Waste Conference*, Purdue University, West Lafayette, IN, 1974; 26.
26. Howe, R.H.L. Biological degradation of waste containing certain toxic chemical compounds. *Proceedings of the 16th Industrial Waste Conference*, Purdue University, West Lafayette, IN, 1961; 262–276.
27. O'Flaherty, T. The chemical industry. In *Environment and Development in Ireland*; Feehan, J., Ed.; Environmental Institute, University College: Dublin, 1991; 136–142.
28. Colovos, G.C.; Tinklenberg, N. Land disposal of pharmaceutical manufacturing wastes. Biotech. Bioeng. **1962**, *IV*, 153–160.
29. Quane, D.E.; Stumpf, M.R. Coal-fired boilers burn waste sludge and odors in an integrated pollution control system. *46th Annual Water Pollution Control Federation Conference*, Cleveland, OH, 1973.
30. Barker, W.G.; Schwarz, D. Engineering processes for waste control. Civil Engng Progress **1961**, *65*, 58–61.
31. Blaine, R.K.; Van Lanen, J.H. Application of waste-to-product ratios in fermentation industries. Biotech. Bioeng. **1962**, *IV*, 129–138.
32. Edmondson, K.H. Disposal of antibiotic spent beers by triple effect evaporation. *Proceedings of the 8th Industrial Waste Conference*, Purdue University, West Lafayette, IN, 1953; 46–58.
33. Barker, W.G.; Stumpf, H.R.; Schwarz, D. Unconventional high performance activated sludge treatment of pharmaceutical wastewater. *Proceedings of the 28th Industrial Waste Conference*, Purdue University, West Lafayette, IN, 1973.
34. Jackson, C.J. Fermentation wastes disposal in Great Britain. *Proceedings of the 21st Industrial Waste Conference*, Purdue University, West Lafayette, IN, 1966; 19–32.
35. McCallum, D., *et al.* Wastewater management in the pharmaceutical industry. *3rd International Conference on Effluent Treatment from Biochemical Industry*, Wheatland: Watford, England, 1980.
36. Paradiso, S.J. What to do with wastes when volumes overtake capacity. Industry and Power **1955**, *69*, 35–39.
37. Paradiso, S.J. Disposal of fine chemical wastes. *Proceeding of the 10th Purdue Industrial Waste Conference*, Purdue University, West Lafayette, IN, 1955; 49–60.
38. Hamza, A. Evaluation of treatability of the pharmaceutical wastewater by biological methods. In *Current Practices in Environmental Engineering*; Hamaza, A., Ed.; International Book Traders, Delhi, India. 1984; 37–44.
39. Gulyas, H.; Hemmerling, L.; Sekoulov, I. Moglichkeiten zur weitergehenden Entfernung organischer Inhaltsstoffe aus Abwassern der Altolaufbereitung. Z. Wasser Abwasser Forsch. **1991**, *24*, 253–257.
40. Delaine, J.; Gough, D. An evaluation of process for treatment of pharmaceutical effluents. *3rd International Conference on Effluent Treatment from Biochemical Industry*, Wheatland: Watford, England, 1980.
41. Gulyas, H.; von Bismarck, R.; Hemmerling, L. Treatment of industrial wastewaters with ozone/hydrogen peroxide. Water Sci. Technol. **1995**, *32* (7), 127–134.
42. Dryden, F.E.; Barrett, P.A.; Kissinger, J.C.; Eckenfelder, Jr., W.W. High rate activated sludge treatment of fine chemical wastes. Sewage Ind. Wastes **1956**, *29*, 193.
43. Henary, M.P. *Biological Treatment of Pharmaceutical Wastewater*. Ph.D. thesis, National University of Ireland, 1994.
44. Henary, M.P.; Donlon, B.P.; Lens, P.N.; Colleran, E.M. Use of anaerobic hybrid reactor for treatment of synthetic pharmaceutical wastewaters containing organic solvents. J. Chem. Technol. Biot. **1996**, *66*, 251–264.
45. Mohanrao, G.J.; Subramanyam, P.V.; Deshmukh, S.B.; Saroja, S. Water treatment at a synthetic drug factory in India. J. Water Pollut. Conf. **1970**, *42* (8), 1530–1543.
46. Donahue, R.T. Single stage nitrification activated sludge pilot plant study on a bulk pharmaceutical manufacturing wastewater. *Proceeding of the 38th Industrial Waste Conference*, Purdue University, West Lafayette, IN, 1984; 173–180.
47. Kincannon, D.F. Performance comparison of activated sludge, PAC activated sludge, granular activated carbon and a resin column for removing the priority pollutants from a pharmaceutical

wastewater. *Proceeding of the 35th Industrial Waste Conference*, Purdue University, West Lafayette, IN, 1980; 476–483.
48. Osantowski, R.A.; Dempsey, C.R.; Dostal, K.A. Enhanced COD removal from pharmaceutical wastewater using powdered activated carbon addition to an activated sludge system. *Proceeding of the 35th Industrial Waste Conference*, Purdue University, West Lafayette, IN, 1980; 719–727.
49. El-Gohary, F.A.; Abou-Elela, S.I.; Aly, H.I. Evaluation of biological technologies for the wastewater treatment in the pharmaceutical industry. Water Sci. Technol. **1995**, *32*, 13–20.
50. Gopalan, R., *et al.* Treatment and disposal of effluents from pharmaceutical and dyestuff industries in Baroda. *Proceedings of the Symposium on Environmental Pollution*, Central Public Health Engineering Research Institute, Zonal Laboratory, 1973; 88–94.
51. Vogler, J.F. Chemical and antibiotics waste treatment at William Island, West Virgina; Present chemical wastes treatment. Sewage Ind. Wastes **1952**, *24*, 485.
52. Reimers, A.E.; Rinace, U.S.; Poese, L.E. Trickling filter studies on fine chemicals plant waste. Sewage Ind. Wastes **1954**, *26*, 51.
53. Liontas, J.A. High rate filters treat mixed wastes at Sherp and Dohme. Sewage Ind. Wastes **1954**, *26*, 310.
54. Jennet, J.C.; Dennis, N.D. Anaerobic filter treatment of pharmaceutical waste. J. Water Pollut. Conf. **1975**, *47* (1), 104–121.
55. Elliot, S.F.; Jennet, J.C.; Rgand, M.C. Anaerobic treatment of synthesized organic chemical pharmaceutical wastes. *Proceedings of the 33rd Industrial Waste Conference*, Purdue University, West Lafayette, IN, 1978; 507–514.
56. Seif, H.A.A. *Comparative Study on Treatment of Pharmaceutical Wastewater by Anaerobic Mesophilic and Thermophilic Fixed Film Reactors*. Ph.D. thesis, CESE, IIT, Bombay, India, 1990.
57. Seif, H.A.A.; Joshi, S.G.; Gupta, S.K. Treatment of pharmaceutical wastewater by anaerobic mesophilic and thermophilic fixed film reactors. *First Symposium on Hazard Assessment and Control of Environmental Contaminants in Water*, Kyoto, Japan, 1991; 630–637.
58. Khan, A.N.; Siddiqui, R.H. Wastewater treatment by anaerobic contact filter. Indian J. Environ. Health **1976**, *18* (4), 282.
59. Young, J.C.; Dhab, M.F. The effect of media design on the performance of fixed bed anaerobic reactors. Water Sci. Technol. **1983**, *15*, 369.
60. Fox, P.; Venkatasubbiah, V. Coupled anaerobic/aerobic treatment of high-sulfate wastewater with sulfate reduction and biological sulfide oxidation. Water Sci. Technol. **1996**, *34* (5/6), 359–366.
61. Mormile, M.R.; Liu, S.; Sulfita, J.M. Anaerobic biodegradation of gasoline oxygenates: extrapolation of information to multiple sites and redox conditions. Environ. Sci. Technol. **1994**, *28*, 1727–1732.
62. Speece, R.E. Anaerobic biotechnology for industrial wastewater treatment. Environ. Sci. Technol. **1983**, *17*, 417–427.
63. Gupta, S.K.; Sharma, R. Biological oxidation of high strength nitrogenous wastewater. Water Res. **1996**, *30* (3), 593–600.
64. Otto, R.; Barker, W.G.; Schwarz, D.; Tjarksen, B. Laboratory testing of pharmaceutical wastes for biological control. Biotech. Bioeng. **1962**, *IV*, 139–145.
65. Storm, P.F. *Review of Bulking Episode at the Merck and Co., Inc., Elkton, Virginia Wastewater Treatment Pilot Plant*, Merck & Co., Inc.: Elkton, VA, 1981.
66. Storm, P.F.; Jenkins, D. Identification and significance of filamentous microorganism in activated sludge. *54th Annual Conference of the Water Pollution Control Federation*, Detroit, Michigan, 1981.
67. Jenkins, D. The control of activated sludge bulking. *52nd Annual Conference, California Water Pollution Control Association*, California, 1980.
68. Brandel, J.S. Pharmaceutical company's aeration system saves energy. Ind. Wastes **1980**, *26* (2), 16–19.
69. Arceivala, S.J., Ed. *Wastewater Treatment for Pollution Control*; Tata McGraw-Hill Publishing Company Limited: New Delhi, India, 1998.
70. Hindustan D.O. *Studies on Pharmaceutical Wastewater*; Park-Davies, Hindustan Dorr Oliver: Bombay, India.

4

Treatment of Oilfield and Refinery Wastes

Joseph M. Wong
Black & Veatch, Concord, California, U.S.A.

Yung-Tse Hung
Cleveland State University, Cleveland, Ohio, U.S.A.

The petroleum industry, one of the world's largest industries, has four major branches [1]. The production branch explores for oil and brings it to the surface in oilfields. The transportation branch sends crude oil to refineries and delivers the refined products to consumers. The refining branch processes crude oil into useful products. The marketing branch sells and distributes the petroleum products to consumers. The subject of this chapter is the treatment of liquid wastes from the production and refining branches.

4.1 OIL PRODUCTION

Each year more than 30 billion barrels of crude oil are produced in the world. The average worldwide and U.S. production rates are 83 million and 5.9 million barrels per day (bpd), respectively. Saudi Arabia produced the most crude in 1999, at more than 7.5 million bpd, followed by the former Soviet Union countries, at more than 7.3 million bpd (data taken from Oil & Gas J., December 18, 2000).

Oil production starts with petroleum exploration. Oil geologists study rock formations on and below the Earth's surface to determine where petroleum might be found. The next step is preparing and drilling an oil well. After completing the well, which means bringing the well into production, petroleum is recovered in much the same way as underground water is obtained.

4.1.1 Oil Drilling

There are three well-established methods of drilling [1]. The first oil crews used a technique called cable-tool drilling, which is still used for boring shallow holes in hard rock formation. Today, most U.S. crews use the faster and more accurate method of rotary drilling. On sites where the well must be drilled at an angle, crews use the directional drilling technique. Directional drilling is often used in offshore operations because many wells can be drilled directionally from one platform. Petroleum engineers are also testing a variety of drilling methods to increase the depth of oil wells and reduce the cost of drilling operations.

Cable-tool drilling works in much the same way as a chisel is used to cut wood or stone [1]. A steel cable repeatedly drops and raises a heavy cutting tool called a bit. Bits may be as long as 8 feet (2.4 m) with a diameter of 4 to 12.5 inches (10–31.8 cm). Each time the bit drops, it drives deeper and deeper into the earth. The sharp edges of the bit break up the soil and rock into small particles. From time to time, the workers pull out the cable and drill bit and pour water into the hole. They then scoop up the water and particles at the bottom of the hole with a long steel tool known as a bailer.

The rotary drilling method works like a carpenter's drill boring through wood [1]. The bit on a rotary drill is attached to the end of a series of connected pipes called the drill pipe. The drill pipe is rotated by a turntable on the floor of the derrick. The pipe is lowered into the ground. As the pipe turns, the bit bores through layers of soil and rock. The drilling crew attaches additional lengths of pipe as the hole becomes deeper.

The drill pipe is lowered and raised by a hoisting mechanism called the draw works, which operates somewhat like a fishing rod. Steel cable is unwound from the hoisting drum, then threaded through two sets of pulleys (blocks) – the crown block, at the top of the rig, and the traveling block, which hangs inside the derrick. The workers attach the upper end of the drill pipe to the traveling block with a giant hook. They can then lower the pipe into the hole or lift it out by turning the hoisting drum in one direction or the other.

During rotary drilling, a fluid called drilling mud is pumped down the drill pipe. It flows out of the openings in the bit and then back up between the pipe and the wall of the hole to just below the derrick. This constantly circulating fluid cools and cleans the bit and carries cuttings (pieces of soil and rock) to the surface. Thus, the crew can drill continuously without having to bail out the cuttings from the bottom of the well. The drilling mud also coats the sides of the hole, which helps prevent leaks and cave-ins. In addition, the pressure of the mud on the well reduces the risk of blowouts and gushers.

In cable-tool drilling and most rotary drilling, the well hole is drilled straight down from the derrick floor. In directional drilling, the hole is drilled at an angle using special devices called turbodrills and electrodrills. The motors that power these drills lie directly above the bit and rotate only the lowermost section of the drill pipe. Such drills enable drillers to guide the bit along a slanted path. Drillers may also use tools known as whipstocks to drill at an angle. A whipstock is a long steel wedge grooved like a shoehorn. The wedge is placed in the hole with pointed end upward. The drilling path is slanted as the bit travels along the groove of the whipstock.

4.1.2 Recovering Petroleum

Petroleum is recovered in two ways [1]. If natural energy provides most of the energy to bring the fluid to the surface, the recovery is called primary recovery. If artificial means are used, the process is called enhanced recovery.

In primary recovery the natural energy comes mainly from gas and water in reservoir rocks. The gas may be dissolved in the oil or separated at the top of it in the form of a gas cap. Water, which is heavier than oil, collects below the petroleum. Depending on the source, the energy in the reservoir is called solution-gas drive, gas-cap drive, or water drive. In solution-gas drive, the gas expands and moves toward the opening, carrying some of the liquid with it. In gas-cap drive, gas is trapped in a cap above the oil as well as dissolved in it. As oil is produced from the reservoir, the gas cap expands and drives the oil toward the well. In water drive, water in a reservoir is held in place mainly by underground pressure. If the volume of water is sufficiently large, the reduction of pressure that occurs during oil production causes the water to expand. The water then displaces the petroleum, forcing it to flow into the well.

Enhanced recovery can include a variety of methods designed to increase the amount of oil that flows into a producing well. Secondary recovery consists of replacing the natural energy in a reservoir. Water flooding is the most widely used method, which involves injecting water into the reservoir to cause the oil to flow into the well. Tertiary recovery includes a number of experimental methods of bringing more oil to the surface. These methods may include steam injection or burning some of the petroleum in the reservoir. The heat makes the oil thinner, enabling it to flow more freely into the well.

Oil leaving the producing well is a mixture of liquid petroleum, natural gas, and formation water. Some production may contain as much as 90% produced water [2]. This water must be separated from the oil, as pipeline specifications stipulate maximum water content from as low as 1% to 4%. The initial water–oil separation vessel in a modern treating plant is called a free-water-knockout [2]. Free water, defined as that which separates within five minutes, is drawn off to holding to be clarified prior to reinjection or discharge. Natural gas is also withdrawn from the free-water-knockout and piped to storage. The remaining oil usually contains emulsified water and must be further processed to break the emulsion, usually assisted by heat, electrical energy, or both. The demulsified crude oil flows to a stock tank for pipeline shipment to a refinery.

4.2 OIL REFINING

After crude oil is separated from natural gas, it is transported to refineries and processed into useful products. Refineries range in size from small plants that process about 150 barrels of crude oil per day to giant complexes with a capacity of more than 600,000 bpd [1]. As of January 1, 2002, there are 732 operating refineries in the world and 143 operating refineries in the United States. The worldwide and U.S. crude capacities are 81.2 and 16.6 million bpd, respectively [3]. Table 1 shows the distribution and crude capacities of operating refineries in the United States [3].

A petroleum refinery is a complex combination of interdependent operations engaged in separating crude molecular constituents, molecular cracking, molecular rebuilding, and solvent finishing to produce petroleum-derived products. Figure 1 shows an overall flow diagram for a generalized refinery production scheme [4].

In its 1977 survey, the U.S. Environmental Protection Agency (USEPA) identified over 150 separate processes being used in refineries [5]. A refinery may employ any number or a combination of these processes, depending upon the type of crude processed, the type of product being produced, and the characteristics of the particular refinery. The refining processes can generally be classified as separation, conversion, and chemical treatment processes [1].

Separation processes separate crude oil into some of its fractions. Fractional distillation, solvent extraction, and crystallization are some of the major separation processes.

Conversion processes convert less useful fractions into those that are in greater demand. Cracking and combining processes belong to the class of conversion processes. Cracking processes include thermal cracking and catalytic cracking, which convert heavy fractions into lighter ones. During cracking, hydrogenation may be used to further increase the yield of useful products. Combining processes do the reverse of cracking – they form more complex fractions from simple gaseous hydrocarbons. The major combining processes include polymerization, alkylation, and reforming.

Chemical treatment processes are used to remove impurities from the fractions. The method of treatment depends on the type of crude oil and on the intended use of the petroleum product. Treatment with hydrogen is a widely used method of removing sulfur compounds. Blending with other products or additives may be carried out to achieve certain special properties.

Table 1 Survey of Operating Refineries in the United States (State Capacities as of January 1, 2002)

State	No. of refineries	Crude capacity (b/cd)[a]
Alabama	3	148,225
Alaska	6	373,500
Arkansas	3	67,700
California	20	1,975,100
Colorado	2	88,000
Delaware	1	175,000
Georgia	1	6,000
Hawaii	2	149,000
Illinois	5	940,550
Indiana	2	433,500
Kansas	3	278,500
Kentucky	2	227,500
Louisiana	20	2,703,780
Michigan	1	74,000
Minnesota	2	360,000
Mississippi	2	318,000
Montana	4	175,100
New Jersey	3	557,000
New Mexico	3	97,600
North Dakota	1	58,000
Ohio	4	530,000
Oklahoma	5	438,858
Pennsylvania	5	761,700
Tennessee	1	175,000
Texas	25	4,440,500
Utah	5	160,500
Virginia	1	58,600
Washington	5	618,520
West Virginia	1	11,500
Wisconsin	1	33,250
Wyoming	4	130,000
Total	**143**	**16,564,483**

[a]b/cd = barrels per calendar day.
Source: *Oil & Gas J.*, Dec. 24, 2001.

In addition to these major processes, there are other auxiliary activities that are critical to the operation in a refinery. These auxiliary operations and the major refining processes are briefly described below, along with their wastewater sources [5].

4.2.1 Crude Oil and Product Storage

Crude oil, intermediate, and finished products are stored in tanks of varying size to provide adequate supplies of crude oils for primary fractionation runs of economical duration; to equalize process flows and provide feedstocks for intermediate processing units; and to store final products prior to shipment in adjustment to market demands. Generally, operating schedules permit sufficient detention time for settling of water and suspended materials.

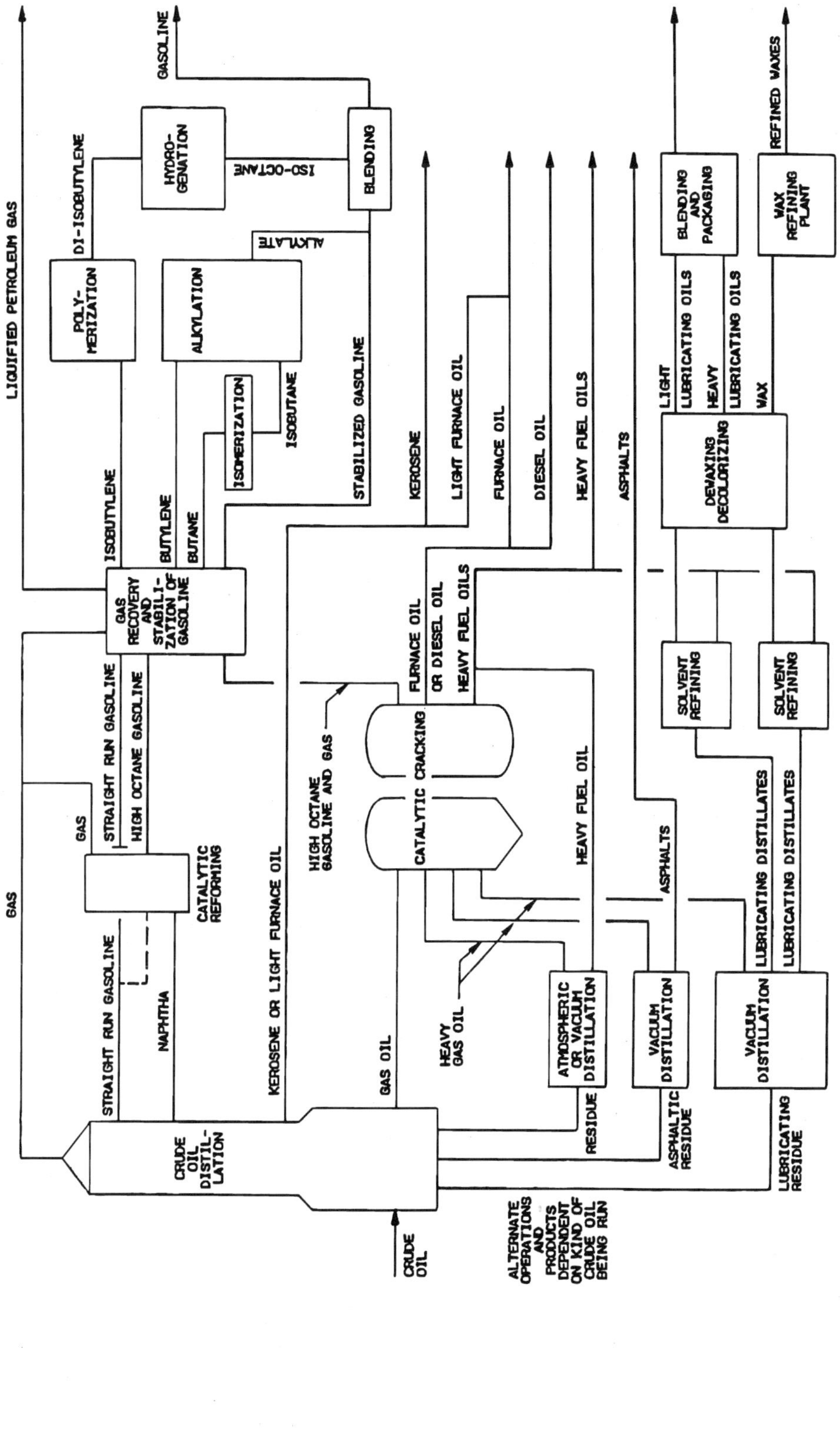

Figure 1 Generalized flowchart for petroleum refining. Crude oil is separated into different fractions and processed into many different products in a refinery. (From Ref. 4.)

Wastewater pollutants associated with storage of crude oil and products are mainly free oil, emulsified oil, and suspended solids. During storage, water and suspended solids in the crude oil separate. The water layer accumulates below the oil, forming a bottom sludge. When the water layer is drawn off, emulsified oil present at the oil–water interface is often lost to the sewers. This waste is high in chemical oxygen demand (COD) levels and, to a lesser extent, biochemical oxygen demand (BOD). Bottom sludge is removed at infrequent intervals. Waste also results from leaks, spills, salt filters (used for product drying), and tank cleaning.

Intermediate storage is frequently the source of polysulfide-bearing wastewaters and iron sulfide suspended solids. Finished product storage can produce high-BOD, alkaline wastewaters, as well as tetraethyl lead. Tank cleaning can contribute large amounts of oil, COD, and suspended solids and a minor amount of BOD. Leaks, spills, and open or poorly ventilated tanks can also be a source of air pollution through evaporation of hydrocarbons into the atmosphere.

4.2.2 Ballast Water Storage

Tankers that ship intermediate and final products discharge ballast water (approximately 30% of the cargo capacity is generally required to maintain vessel stability). Ballast waters have organic contaminants that range from water-soluble alcohol to residual fuels. Brackish water and sediments are also present, contributing high COD and dissolved solids loads to the refinery wastewater. These wastewaters are usually discharged to either a ballast water tank or holding ponds at the refinery. In some cases, the ballast water is discharged directly to the wastewater treatment system, and potentially constitutes a shock load to the treatment system.

4.2.3 Crude Desalting

Common to all types of desalting are an emulsifier and settling tank. Salts can be separated from oil by one of two methods. In the first method, water wash desalting in the presence of chemicals is followed by heating and gravity separation. In the second method, water wash desalting is followed by water–oil separation in a high-voltage electrostatic field acting to agglomerate dispersed droplets. A process flow schematic of electrostatic desalting is shown in Figure 2. Wastewater containing removed impurities is discharged to the wastewater system, and desalted crude oil flows from the upper part of the holding tank.

Much of the bottom sediment and water content in crude oil is a result of the "load-on-top" procedure used on many tankers. This procedure can result in one or more cargo tanks containing mixtures of seawater and crude oil, which cannot be separated by decantation while at sea, and are consequently retained in the crude oil storage at the refinery. Although much of the water and sediment are removed from the crude oil by settling during storage, a significant quantity remains to be removed by desalting before the crude is refined.

The continuous wastewater stream from a desalter contains emulsified oil (occasionally free oil), ammonia, phenol, sulfides, and suspended solids, all of which produce a relatively high BOD and COD concentration. It also contains enough chlorides and other dissolved materials to contribute to the dissolved solids problems in discharges to freshwater bodies. Finally, its temperature often exceeds 95°C (200°F), thus it is a potential thermal pollutant.

4.2.4 Crude Oil Fractionation

Fractionation is the basic refining process for separating crude petroleum into intermediate fractions of specified boiling point ranges. The various subprocesses include prefractionation and atmospheric fractionation, vacuum fractionation, and three-stage crude distillation.

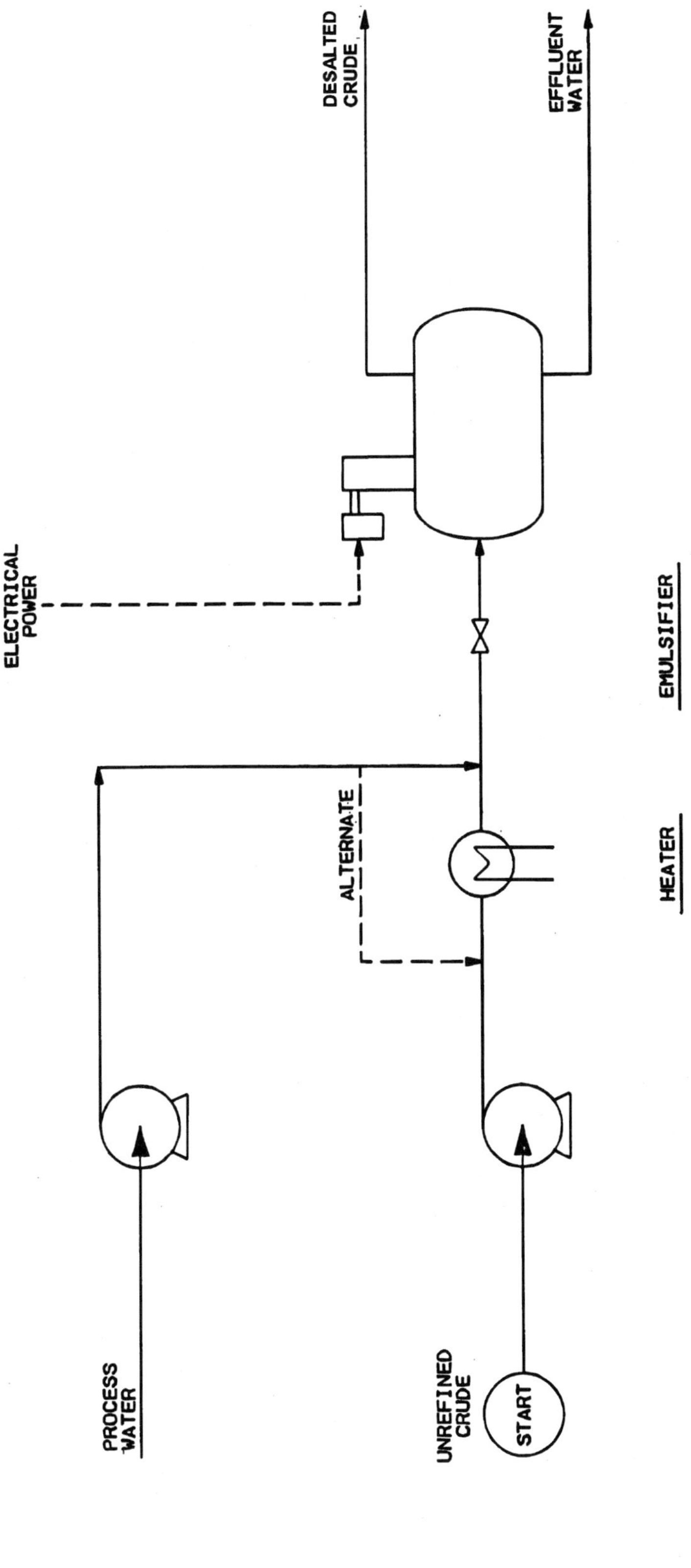

Figure 2 Crude desalting (electrostatic desalting). A high-voltage electrostatic field acts to agglomerate dispersed oil droplets for water–oil separation after water wash desalting. (From Ref. 5.)

Prefractionation and Atmospheric Distillation (Topping or Skimming)

Prefractionation is an optional distillation process to separate economic quantities of very light distillates from the crude oil. Lower temperatures and higher pressures are used than in atmospheric distillation. Some process water can be carried over to the prefractionation tower from the desalting process.

Atmospheric distillation breaks the heated crude oil as follows:

1. Light overhead (gaseous) products (C_5 and lighter) are separated, as in the case of prefractionation.
2. Sidestream distillate cuts of kerosene, heating oil, and gas oil can be separated in a single tower or in a series of topping towers, each tower yielding a successively heavier product stream.
3. Residual or reduced crude oil remains for further refining.

Vacuum Fractionation

The asphaltic residuum from atmospheric distillation amounts to roughly one-third (U.S. average) of the crude charged. This material is sent to vacuum stills, which recover additional heavy gas oil and deasphalting feedstock from the bottoms residue.

Three-Stage Crude Distillation

Three-stage crude distillation, representing only one of many possible combinations of equipment, is shown schematically in Fig. 3. The process consists of (1) an atmospheric fractionating stage, which produces lighter oils; (2) an initial vacuum stage, which produces well-fractionated, lube oil base stocks plus residue for subsequent propane deasphalting; and (3) a second vacuum stage, which fractionates surplus atmospheric bottoms not applicable for lube production, plus surplus initial vacuum stage residuum not required for deasphalting. This stage adds the capability of removing catalytic cracking stock from surplus bottoms to the distillation unit.

Crude oil is first heated in a simple heat exchanger, then in a direct-fired crude charge heater. Combined liquid and vapor effluent flow from the heater to the atmospheric fractionating tower, where the vaporized distillate is fractionated into gasoline overhead product and as many as four liquid sidestream products: naphtha, kerosene, and light and heavy diesel oil. Part of the reduced crude from the bottom of the atmospheric tower is pumped through a direct-fired heater to the vacuum lube fractionator. Bottoms are combined and charged to a third direct-fired heater. In the tower, the distillate is subsequently condensed and withdrawn as two sidestreams. The two sidestreams are combined to form catalytic cracking feedstocks, and an asphalt base stock is withdrawn from the tower bottom.

Wastewater from crude oil fractionation generally comes from three sources. The first source is the water drawn off from overhead accumulators prior to recirculation or transfer of hydrocarbons to other fractionators. This waste is a major source of sulfides and ammonia, especially when sour crudes are being processed. It also contains significant amounts of oil, chlorides, mercaptans, and phenols.

The second waste source is discharge from oil sampling lines. This should be separable, but it may form emulsions in the sewer.

A third waste source is very stable oil emulsions formed in the barometric condensers used to create the reduced pressures in the vacuum distillation units. However, when barometric condensers are replaced with surface condensers, oil vapors do not come into contact with water and consequently emulsions do not develop.

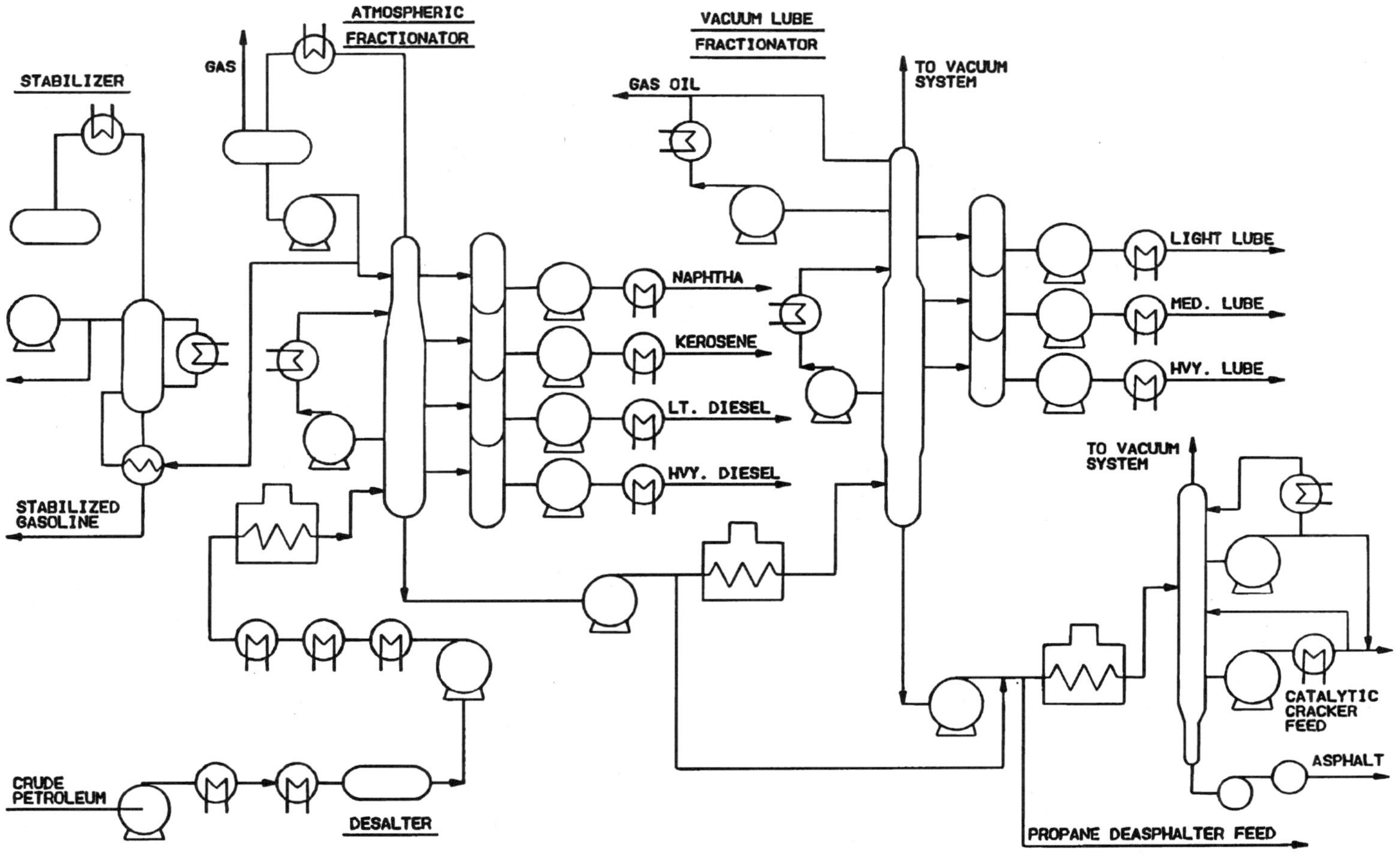

Figure 3 Crude fractionation (crude distillation, three stages). An atmospheric fractionating stage produces lighter oils. An initial vacuum stage produces lube oils. A second vacuum stage fractionates bottoms from the other stages to produce asphalt and catalytic cracker feed. (From Ref. 5.)

4.2.5 Thermal Cracking

Thermal cracking can include visbreaking and coking, in addition to regular thermal cracking. In each of these operations, heavy gas oil fractions (from vacuum stills) are broken down into lower molecular weight fractions such as domestic heating oils, catalytic cracking stock, and other fractions by heating, but without the use of catalyst. Typical thermal cracking conditions are 480–600°C (900–1100°F), and 41.6–69.1 atm (600–1000 psig). The high pressures result from the formation of light hydrocarbons in the cracking reaction (olefins, or unsaturated compounds, are always formed in this chemical conversion). There is also a certain amount of heavy fuel oil and coke formed by polymerization and condensation reactions.

The major source of wastewater in thermal cracking is the overhead accumulator on the fractionator, where water is separated from the hydrocarbon vapor and sent to the sewer system. This water usually contains various oils and fractions and may be high in BOD, COD, ammonia, phenol, sulfides, and alkalinity.

4.2.6 Catalytic Cracking

Catalytic cracking, like thermal cracking, breaks heavy fractions, principally gas oils, into lower molecular weight fractions. The use of catalyst permits operations at lower temperatures and pressures than with thermal cracking, and inhibits the formation of undesirable products. Catalytic cracking is probably the key process in the production of large volumes of high-octane gasoline stocks; furnace oils and other useful middle molecular weight distillates are also produced.

Fluidized catalytic processes, in which the finely powdered catalyst is handled as a fluid, have largely replaced the fixed-bed and moving-bed processes, which use a beaded or pelleted catalyst. A schematic flow diagram of fluid catalytic cracking (FCC) is shown in Fig. 4.

The FCC process involves at least four types of reactions: (1) thermal decomposition; (2) primary catalytic reactions at the catalyst surface; (3) secondary catalytic reactions between the primary products; and (4) removal of polymerization products from further reactions by adsorption onto the surface of the catalyst as coke. This last reaction is the key to catalytic cracking because it permits decomposition reactions to move closer to completion than is possible in simple thermal cracking.

Cracking catalysts include synthetic and natural silica-alumina, treated bentonite clay, fuller's earth, aluminum hydrosilicates, and bauxite. These catalysts are in the form of beads, pellets, and powder, and are used in a fixed, moving, or fluidized bed. The catalyst is usually heated and lifted into the reactor area by the incoming oil feed which, in turn, is immediately vaporized upon contact. Vapors from the reactors pass upward through a cyclone separator which removes most of the entrained catalyst. The vapors then enter the fractionator, where the desired products are removed and heavier fractions are recycled to the reactor.

Catalytic cracking units are one of the largest sources of sour and phenolic wastewaters in a refinery. Pollutants from catalytic cracking generally come from the steam strippers and overhead accumulators on fractionators, used to recover and separate the various hydrocarbon fractions produced in the catalytic reactors.

The major pollutants resulting from catalytic cracking operations are oil, sulfides, phenols, cyanides, and ammonia. These pollutants produce an alkaline wastewater with high BOD and COD concentrations. Sulfide and phenol concentrations in the wastewater vary with the type of crude oil being processed, but at times are significant.

Regeneration of spent catalyst in the steam stripper may produce enough carbon monoxide and fine catalyst particles to constitute an air pollution problem.

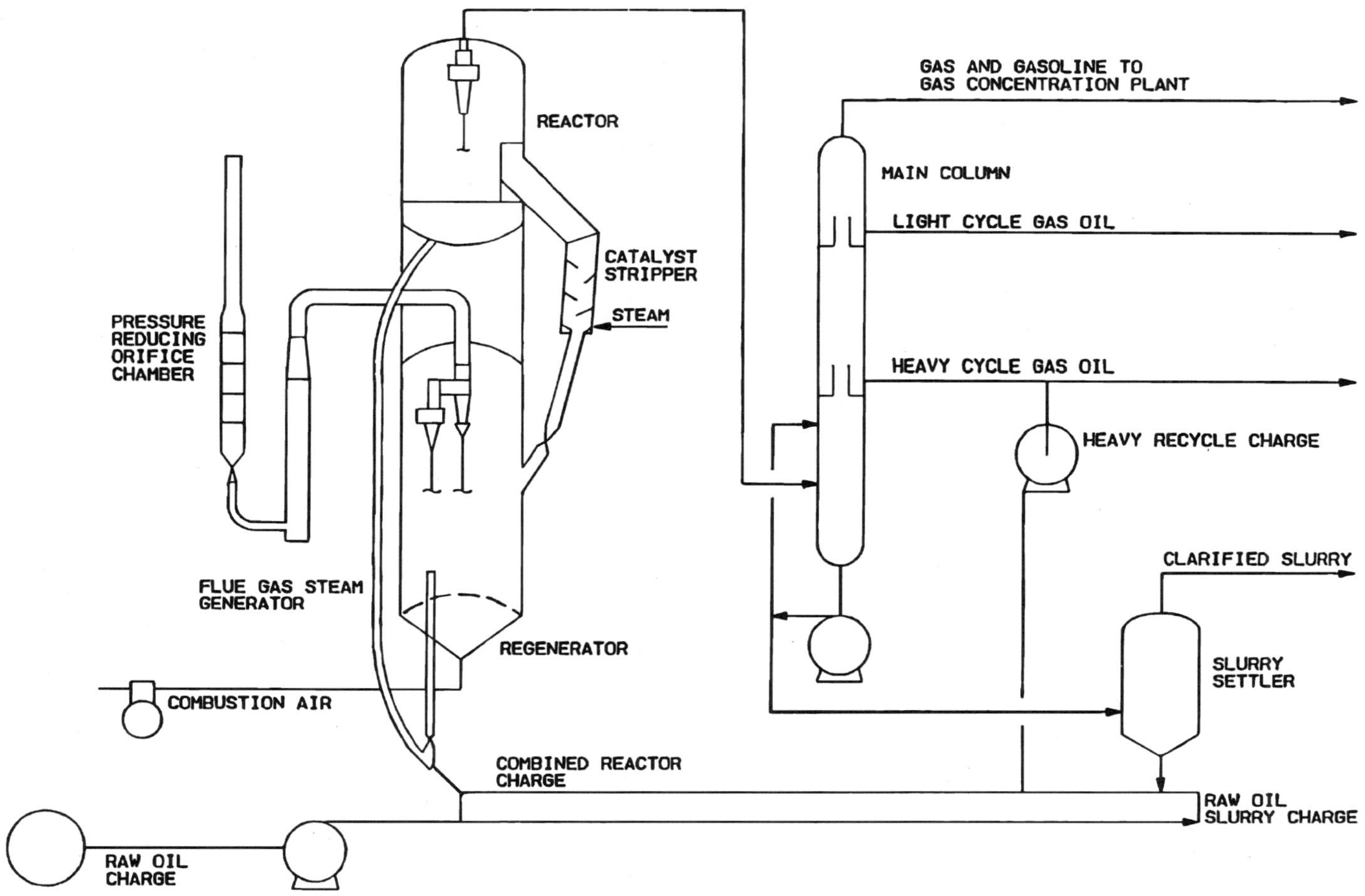

Figure 4 Catalytic cracking (fluid catalytic cracking). Heavy fraction gas oils are cracked (broken down) into lower molecular weight fractions in the presence of finely powdered catalyst, handled as a fluid. (From Ref. 5.)

4.2.7 Hydrocracking

This process is basically catalytic cracking in the presence of hydrogen, with lower temperatures and higher pressures than FCC. Hydrocracking temperatures range from 200 to 425°C (400–800°F) and pressures range from 7.8 to 137.0 atm (100–2000 psig). Actual conditions and hydrogen consumption depend upon the feedstock and the degree of hydrogenation required. The molecular weight distribution of the products is similar to catalyst cracking, but with reduced formation of olefins.

At least one wastewater stream from the process should be high in sulfides, as hydrocracking reduces the sulfur content of the material being cracked. Most of the sulfides are in the gas products that are sent to a treating unit for removal or recovery of sulfur and ammonia. However, some of the H_2S dissolves in the wastewater being collected from the separator and fractionator following the hydrocracking reactor. This water is probably high in sulfides and may contain significant quantities of phenols and ammonia.

4.2.8 Polymerization

Polymerization units convert olefin feedstocks (primarily propylene) into higher octane polymers. These units generally consist of a feed treatment unit (to remove H_2S, mercaptans, and nitrogen compounds), a catalytic reactor, an acid removal section, and a gas stabilizer. The catalyst is usually phosphoric acid, although sulfuric acid is used in some older methods. The catalytic reaction occurs at 150–224°C (300–435°F) and at a pressure of 11.2–137.0 atm (150–2000 psig). The temperature and pressure vary with the subprocess used.

Polymerization is a rather dirty process in terms of pounds of pollutants per barrel of charge, but because of the small polymerization capacity in most refineries, the total waste production from the process is small. Even though the process makes use of acid catalysts, the waste stream is alkaline because the acid catalyst in most subprocesses is recycled, and any remaining acid is removed by caustic washing. Most of the waste material comes from the pretreatment of feedstock, which removes sulfides, mercaptans, and ammonia from the feedstock in caustic and acid wastes.

4.2.9 Alkylation

Alkylation is the reaction of an isoparaffin (usually isobutane) and an olefin (propylene, butylene, amylenes) in the presence of a catalyst at carefully controlled temperatures and pressures to produce a high-octane alkylate for use as a gasoline blending component. Propane and butane are also produced. Sulfuric acid is the most widely used catalyst, although hydrofluoric acid is also used. Figure 5 shows a flow diagram of the alkylation process using sulfuric acid [6]. The reactor products are separated in a catalyst recovery unit, from which the catalyst is recycled. The hydrocarbon stream then passes through a caustic and water wash before going to the fractionation section.

The major discharges from sulfuric acid alkylation are the spent caustics from the neutralization of hydrocarbon streams leaving the alkylation reactor. These wastewaters contain dissolved and suspended solids, sulfides, oils, and other contaminants. Water drawn off from the overhead accumulators contains varying amounts of oil, sulfides, and other contaminants, but is not a major source of waste. Most refineries process the waste sulfuric acid stream from the reactor to recover clean acids, use it to neutralize other waste streams, or sell it.

Hydrofluoric acid (HF) alkylation units have small acid rerun units to purify the acid for reuse. HF units do not have a spent acid or spent caustic waste stream. Any leaks or spills that

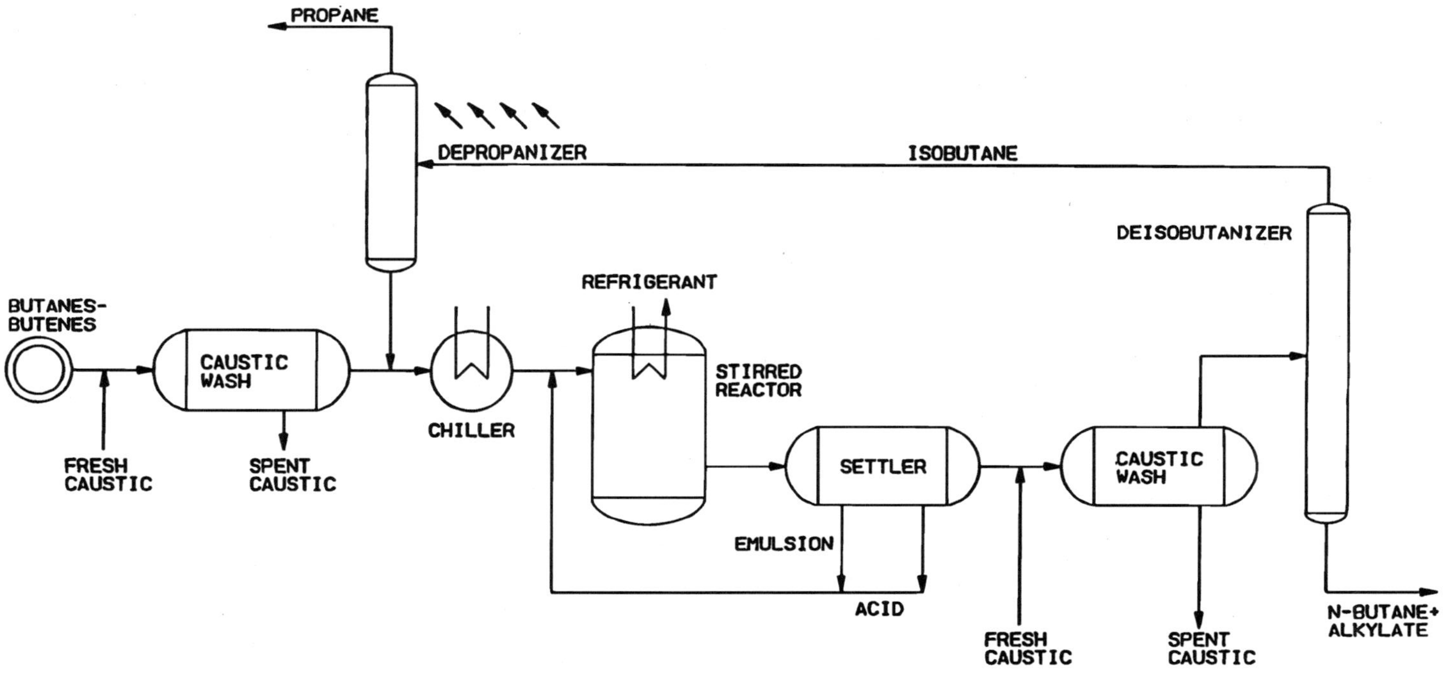

Figure 5 Alkylation process using sulfuric acid. Butanes and butenes react in the presence of a catalyst (sulfuric acid) to form an alkylate for use as a gasoline blending component. Propane and butane are also produced. (From Ref. 6.)

involve loss of fluorides constitute a serious and difficult pollution problem. Formation of fluorosilicates has caused line plugging and similar problems. The major sources of waste materials are the overhead accumulators on the fractionator.

4.2.10 Isomerization

Isomerization is a process technique for converting light gasoline stocks into their higher octane isomers. The greatest application has been, indirectly, in the conversion of isobutane from normal butane for use as feedstock for the alkylation process. In a typical subprocess, the desulfurized feedstock is first fractionated to separate isoparaffins from normal paraffins. The normal paraffins are then heated, compressed, and passed through the catalytic hydrogenation reactor, which isomerizes the *n*-paraffin to its respective high-octane isomer. After separation of hydrogen, the liquids are sent to a stabilizer, where motor fuel blending stock or synthetic isomers are removed as products.

Isomerization wastewaters present no major pollutant discharge problems. Sulfides and ammonia are not likely to be present in the effluent. Isomerization wastewaters should also be low in phenolics and oxygen demand.

4.2.11 Reforming

Reforming converts low-octane naphtha, heavy gasoline, and naphthene-rich stocks to high-octane gasoline blending stock, aromatics for petrochemical use, and isobutane. Hydrogen is a significant byproduct. Reforming is a mild decomposing process, as some reduction occurs in molecular size and boiling range of the feedstock. Feedstocks are usually hydrotreated to remove sulfur and nitrogen compounds prior to charging to the reformer, because the platinum catalysts widely used are readily poisoned.

The predominant reaction during reforming is dehydrogenation of naphthenes. Important secondary reactions are isomerization and dehydrocyclization of paraffins. All three reactions result in high-octane products.

One subprocess may be divided into three parts: the reactor heater section, in which the charge plus recycle gas is heated and passed over the catalyst in a series of reactions; the separator drum, in which the reactor effluent is separated into gas and liquid streams, the gas being compressed for recycle; and the stabilizer section, in which the separated liquid is stabilized to the desired vapor pressure. There are many variations in subprocesses, but the essential and frequently only difference is in catalyst involved.

Reforming is a relatively clean process. The volume of wastewater flow is small, and none of the wastewater streams has high concentrations of significant pollutants. The wastewater is alkaline, and the major pollutant is sulfide from the overhead accumulator on the stripping tower used to remove light hydrocarbon fractions from the reactor effluent. The overhead accumulator catches any water that may be contained in the hydrocarbon vapors. In addition to sulfides, the wastewater contains small amounts of ammonia, mercaptans, and oil.

4.2.12 Solvent Refining

Refineries employ a wide spectrum of contact solvent processes, which are dependent upon the differential solubilities of the desirable and undesirable feedstock components. The principal steps are countercurrent extraction, separation of solvent and product by heating and fractionation, and solvent recovery. Naphthenics, aromatics, unsaturated hydrocarbons, and sulfur and other inorganics are separated, with the solvent extract yielding high-purity products. Many of the solvent processes may produce process wastewaters that contain small amounts of the solvents employed. However, these are usually minimized because of the economic incentives for reuse of the solvents.

The major solvent refining processes include solvent deasphalting, solvent dewaxing, lube oil solvent refining, aromatic extraction, and butadiene extraction. These processes are briefly described below.

Solvent deasphalting is carried out primarily to recover lube or catalytic cracking feedstocks from asphaltic residuals, with asphalt as a byproduct. Propane deasphalting is the predominant technique. The vacuum fractionation residual is mixed in a fixed proportion with a solvent in which asphalt is not soluble. The solvent is recovered from the oil via steam stripping and fractionation, and is reused. The asphalt produced by this method is normally blended into fuel oil or other asphaltic residuals.

Solvent dewaxing removes wax from lubricating oil stocks, promoting crystallization of the wax. Solvents include furfural, phenol, cresylic acid-propane (DuoSol), liquid sulfur dioxide (Eleleanu process), B,B-dichloroethyl ether, methyl ethyl ketone, nitrobenzene, and sulfurbenzene. The process yields de-oiled waxes, wax-free lubricating oils, aromatics, and recovered solvents.

Lube oil solvent refining includes a collection of subprocesses improving the quality of lubricating oil stock. The raffinate or refined lube oils obtain improved viscosity, color, oxidation resistance, and temperature characteristics. A particular solvent is selected to obtain the desired quality raffinate. The solvents include furfural, phenol, sulfur dioxide, and propane.

Aromatic extraction removes benzene, toluene, and xylene (BTX) that are formed as byproducts in the reforming process. The reformed products are fractionated to give a BTX concentrate cut, which, in turn, is extracted from the napthalene and the paraffinics with a glycol base solvent.

Butadiene extraction accounts for some 15% of the U.S. supply of butadiene, which is extracted from the C4 cuts from the high-temperature petroleum cracking processes. Furfural or cuprous ammonia acetate is commonly used for the solvent extraction.

The major potential pollutants from the various solvent refining subprocesses are the solvents themselves. Many of the solvents, such as phenol, glycol, and amines, can produce a high BOD. Under ideal conditions the solvents are continually recirculated with no losses to the sewer. Unfortunately, some solvent is always lost through pump seals, flange leaks, and other sources. The main source of wastewater is from the bottom of fractionation towers. Oil and solvent are the major wastewater constituents.

4.2.13 Hydrotreating

Hydrotreating processes are used to saturate olefins, and to remove sulfur and nitrogen compounds, odor, color, gum-forming materials, and others by catalytic action in the presence of hydrogen, from either straight-run or cracked petroleum fractions. In most subprocesses, the feedstock is mixed with hydrogen, heated, and charged to the catalytic reactor. The reactor products are cooled, and the hydrogen, impurities, and high-grade product separated. The principal difference between the many subprocesses is the catalyst; the process flow is similar for essentially all subprocesses. Figure 6 shows a flow diagram of the hydrotreating process [2].

Hydrotreating reduces the sulfur content of product streams from sour crudes by 90% or more. Nitrogen removal requires more severe operating conditions, but generally 80% reductions or better are accomplished.

The primary variables influencing hydrotreating are hydrogen partial pressure, process temperature, and contact time. Higher hydrogen pressure gives a better removal of undesirable

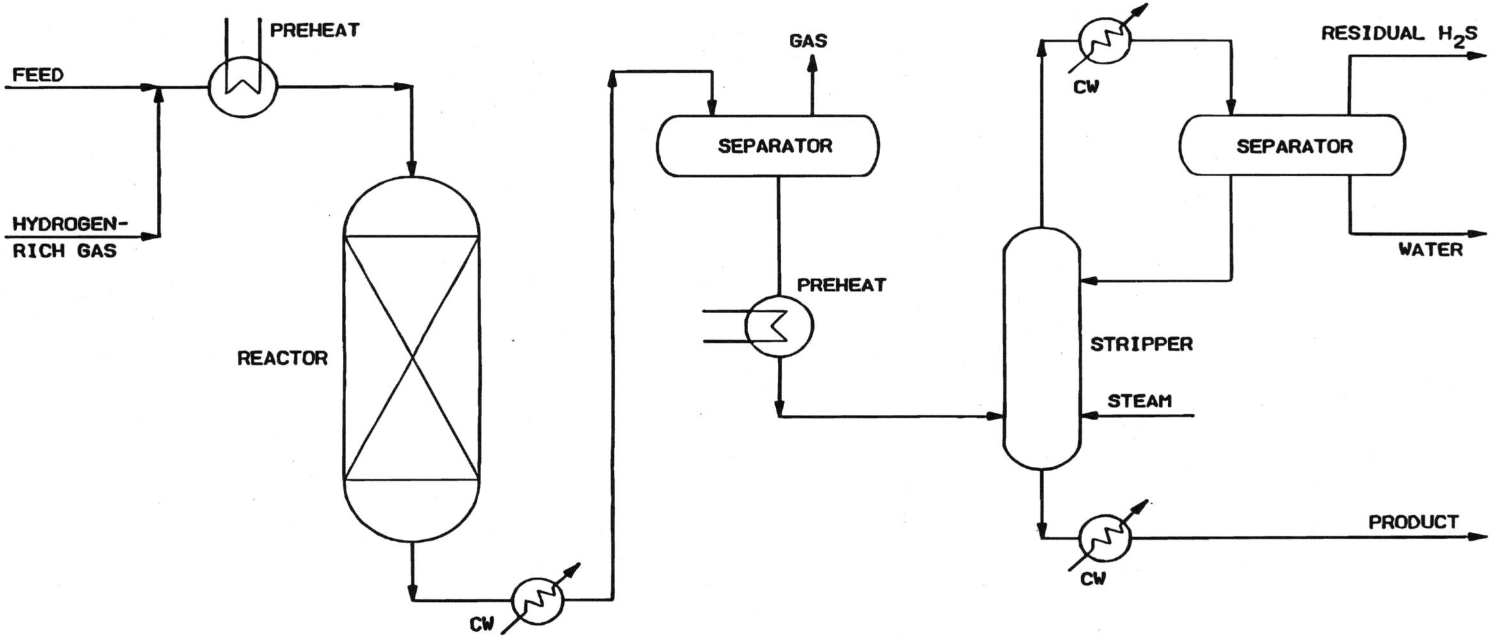

Figure 6 Hydrotreating process. Hydrogen reacts with hydrocarbon feed to remove sulfur from the stream. The formed hydrogen sulfide is steam-stripped from the product. (From Ref. 2.)

materials and a better rate of hydrogenation. Make-up hydrogen requirements are generally great enough to require a hydrogen production unit. Excessive temperatures increase the formation of coke, and the contact time is set to give adequate treatment without excessive hydrogen usage or undue coke formation. For the various hydrotreating processes, the pressures range from 7.8 to 205 atm (100 to 3000 psig). Temperatures range from less than 177°C (350°F) to as high as 450°C (850°F); most processing is carried out in the range 315–427°C (600–800°F). Hydrogen consumption is usually less than 5.67 cubic meters (200 scf) per barrel of charge.

The principal hydrotreating subprocesses used are as follows:

- pretreatment of catalytic reformer feedstock;
- naphtha desulfurization;
- lube oil polishing;
- pretreatment of catalytic cracking feedstock;
- heavy gas-oil and residual desulfurization;
- naphtha saturation.

The strength and quantity of wastewaters generated by hydrotreating depends upon the subprocess used and feedstock. Ammonia and sulfides are the primary contaminants, but phenols may also be present if the feedstock boiling range is sufficiently high.

4.2.14 Grease Manufacturing

Grease manufacturing processes require accurate measurements of feed components, intimate mixing, and rapid heating and cooling, together with milling, dehydration, and polishing in batch reactions. The feed components include soap and petroleum oils with inorganic clays and other additives.

Grease is primarily a soap and lube oil mixture. The properties of grease are determined in large part by the properties of the soap component. For example, sodium soap grease is water soluble and not suitable for water contact service. A calcium soap grease, on the other hand, can be used in water service. The soap may be purchased as a raw material or may be manufactured on site as an auxiliary process.

Only small volumes of wastewater are discharged from a grease manufacturing process. A small amount of oil is lost to the wastewater system through leaks in pumps. The largest waste loading occurs when the batch units are washed, resulting in soap and oil discharges to the sewer system.

4.2.15 Asphalt Production

Asphalt feedstock (flux) is contacted with hot air at 200–280°C (400–550°F) to obtain desirable asphalt product. Both batch and continuous processes are in operation at present, but the batch process is more prevalent because of its versatility. Nonrecoverable catalytic compounds include copper sulfate, zinc chloride, ferric chloride, aluminum chloride, phosphorus pentoxide, and others. The catalyst does not normally contaminate the process water effluent.

Wastewaters from asphalt blowing contain high concentrations of oil and have high oxygen demand. Small quantities of phenols may also be present.

4.2.16 Drying and Sweetening

Drying and sweetening is a broad class of processes used to remove sulfur compounds, water, and other impurities from gasoline, kerosene, jet fuels, domestic heating oils, and other middle distillate products. Sweetening is the removal of hydrogen sulfide, mercaptans, and thiophenes, which impart a foul odor and decrease the tetraethyl lead susceptibility of gasoline. The major sweetening operations are oxidation of mercaptans or disulfides, removal of mercaptans, and destruction and removal of all sulfur compounds. Drying is accomplished by salt filters or absorptive clay beds. Electric fields are sometimes used to facilitate separation of the product.

The most common waste stream from drying and sweetening operations is spent caustic. The spent caustic is characterized as phenolic or sulfidic, depending on which is present in the largest concentration; this in turn is mainly determined by the product stream being treated. Phenolic spent caustics contain phenol, cresols, xylenols, sulfur compounds, and neutral oils. Sulfidic spent caustics are rich in sulfides, but do not contain any phenols. These spent caustics have very high BOD and COD. The phenolic caustic streams are usually sold for the recovery of phenolic materials.

Other waste streams from the process result from water washing of the treated product and regeneration of the treating solution such as sodium plumbite (Na_2PbO_2) in doctor sweetening. These waste streams contain small amounts of oil and the treating material, such as sodium plumbite (or copper from copper chloride sweetening).

The treating of sour gases produces a purified gas stream and an acid gas stream rich in hydrogen sulfide. The H_2S-rich stream can be flared, burned as fuel, or processed for recovery of elemental sulfur.

4.2.17 Lube Oil Finishing

Solvent-refined and dewaxed lube oil stocks can be further refined by clay or acid treatment to remove color-forming and other undesirable materials. Continuous contact filtration, in which an oil–clay slurry is heated and the oil removed by vacuum filtration, is the most widely used subprocess.

Acid treatment of lubricating oils produces acid-bearing wastes occurring as rinse waters, sludges, and discharges from sampling, leaks, and shutdowns. The waste streams are also high in dissolved and suspended solids, sulfates, sulfonates, and stable oil emulsions.

Handling of acid sludge can create additional problems. Some refineries burn the acid sludge as fuel, which produces large volumes of sulfur dioxide that can cause air pollution problems. Other refineries neutralize the sludge with alkaline wastes and discharge it to the sewer, resulting in both organic and inorganic pollution. The best method of disposal is probably processing to recover the sulfuric acid, but this also produces a wastewater stream containing acid, sulfur compounds, and emulsified oil.

Clay treatment results in only small quantities of wastewater being discharged to the sewer. Clay, free oil, and emulsified oil are the major waste constituents. However, the operation of clay recovery kilns involves potential air pollution problems of hydrocarbon and particulate emissions. Spent clays are usually disposed of by landfill.

4.2.18 Blending and Packaging

Blending is the final step in producing finished petroleum products to meet quality specifications and market demands. The largest volume operation is the blending of various gasoline stocks (including alkylates and other high-octane components) and antiknock (tetraethyl lead), antirust, anti-icing, and other additives. Diesel fuels, lube oils, and waxes involve blending of various components and additives. Packaging at refineries is generally highly automated and restricted to high-volume consumer-oriented products such as motor oils.

These are relatively clean processes because care is taken to avoid loss of product through spillage. The primary source of waste material is from the washing of railroad tank cars or tankers prior to loading finished products. These wash waters are high in emulsified oil.

Tetraethyl lead was the major additive blended into gasolines in the past, and it must be carefully handled because of its high toxicity if it is still used. Sludges from finished gasoline storage tanks can contain large amounts of lead if tetraethyl lead is still used and should not be washed into the wastewater system.

4.2.19 Hydrogen Manufacture

The rapid growth of hydrotreating and hydrocracking has increased the demand for hydrogen beyond the level of byproduct hydrogen available from reforming and other refinery processes. The most widely used process for the manufacture of hydrogen in the refinery is steam reforming, which utilizes refinery gases as a charge stock. The charge is purified to remove sulfur compounds that would temporarily deactivate the catalysts.

The desulfurized feedstock is mixed with superheated steam and charged to the hydrogen furnace. On the catalyst, the hydrocarbons are converted to hydrogen, carbon monoxide, and carbon dioxide. The furnace supplies the heat needed to maintain the reaction temperature.

The gases from the furnace are cooled by the addition of condensate and steam, and then passed through a converter containing a high or low temperature shift catalyst, depending on the degree of carbon monoxide conversion desired. Carbon dioxide and hydrogen are produced by the reaction of the carbon monoxide with steam.

The gas mixture from the converter is cooled and passed to a hydrogen purifying system where carbon dioxide is absorbed into amine solutions and later driven off to the atmosphere by heating the rich amine solution in the reactivator.

Because some refining processes require a minimum of carbon dioxide in the product gas, the oxides are reacted with hydrogen in a methanation step. This reaction takes place in the methanator over a nickel catalyst at elevated temperatures.

Hydrocarbon impurities in the product hydrogen usually are not detrimental to the processes where this hydrogen will be used. Thus, a small amount of hydrocarbon is tolerable in the effluent gas.

The hydrogen manufacture process is relatively clean. In the steam reforming subprocess a potential waste source is the desulfurization unit, which is required for feedstock that has not already been desulfurized. This waste stream contains oil, sulfur compounds, and phenol. In the partial oxidation subprocess, free carbon is removed by a water wash. Carbon dioxide is discharged to the atmosphere at several points in the subprocess.

4.2.20 Utility Functions

Utility functions such as the supply of steam and cooling water generally are set up to service several processes. Boiler feed water is prepared and steam is generated in a single boiler house. Noncontact steam used for surface heating is circulated through a closed loop, whereby various quantities are made available for the specific requirements of the different processes. The condensate is nearly always recycled to the boiler house, where a certain portion is discharged as blowdown.

Steam is also used as a diluent, stripping medium, or source of vacuum through the use of steam jet ejectors. This steam actually contacts the hydrocarbons in the manufacturing processes and is a source of contact process wastewater when condensed.

Noncontact cooling water is normally supplied to several processes from the utilities area. The system is either a loop that utilizes one or more evaporative cooling towers, or a once-through system with direct discharge.

Cooling towers work by moving a predetermined flow of ambient air through the tower with large fans. A small amount of the water is evaporated by the air; thus, through latent heat transfer, the remainder of the circulated water is cooled.

Wastewater streams from the utility functions include boiler and cooling tower blowdowns and waste brine and sludge produced by demineralizing and other water treatment systems. The quantity and quality of the wastewater streams depend on the design of the systems and the water source. These streams usually contain high dissolved and suspended solids concentrations and treatment chemicals from the boiler and cooling tower. The blowdown streams also have elevated temperatures.

4.3 WASTE SOURCES AND CHARACTERISTICS

Wastes generated from oil fields include produced water, drilling muds and cuttings, and tank bottom sludges. These wastes are associated with the drilling, recovery, and storage of crude oil. Wastes from petroleum refineries generally include process wastewater, wastewater from utility operations, contaminated storm water, sanitary waste, and miscellaneous contaminated streams. These waste streams are usually discharged to a central wastewater treatment system; some of these streams, such as sour water, are pretreated first.

4.3.1 Oil Field Wastes

The most important wastes from oil fields are produced water and drilling muds. The characteristics of these waste streams are discussed below.

Produced water is the water brought to the surface with the oil from a production well. It is estimated that for every barrel of oil produced, on average 2–3 barrels of water are produced, ranging from a negligible amount up to values over 100 barrels of water per barrel of oil [7]. Once on the surface, the water and oil are separated. The oil is prepared for distribution, leaving the water to be disposed of by some means.

Produced water is typically saline. A great deal of data exist regarding the quality of the inorganic components of the produced water [8]. Table 2 is a summary of this information. To date very little information has been published regarding the concentration of the traditional pollutant parameters in the produced water. Table 3 presents the ranges of various water quality

Table 2 Inorganic Components in Oilfield Produced Water

Constituent	Concentration (mg/L)
Sodium	12,000–150,000
Potassium	30–4,000
Calcium	1,000–120,000
Magnesium	500–25,000
Chloride	20,000–250,000
Bromide	50–5,000
Iodide	1–300
Bicarbonate	0–3,600

Source: From Ref. 9.

Table 3 Produced Water Quality

Parameter	Range (mg/L)
Biochemical oxygen demand	50–1,400
Chemical oxygen demand	450–5,900
Phenols	0.7–7.6
Oil and grease	15–290
Ammonia nitrogen	4–206
Total suspended solids	35–300
Sulfides	0.2–800
pH	6.7–9.0

Source: From Ref. 14.

parameters measured in produced water from over 30 individual wells in several California oilfields [9]. Work done by Chevron showed that typical produced waters from the U.S. west coast and the Gulf of Mexico, after oil removal, had compositions ranging from 20,000 to 135,000 mg/L total dissolved solids, 45 to 130 mg/L ammonia (as N), and 0.1 to 3.0 mg/L phenols [10].

Drilling muds are fluids that are pumped into the bore holes to aid in the drilling process. Most are water based and contain barite, lignite, chrome lignosulfate, and sodium hydroxide [11], but oil-based drilling muds are still used for economic and safety reasons [12]. Used muds can be removed by vacuum trucks, pumped down the well annulus, or allowed to dewater in pits, which are then covered with soil or disposed of by land farming.

The main components of pollution concern in drilling muds include (1) oil itself, especially in oil fluids, (2) salts, and (3) soluble trace elements consisting of zinc, lead, copper, cadmium, nickel, mercury, arsenic, barium, and chromium associated with low grades of barite [13]. Owing to its variability, very little information has been published regarding the concentration of pollutants in spent drilling mud. Copa and Dietrich [14] conducted a wet air oxidation experiment on a sample of spent drilling mud taken from a storage lagoon. The material was a concentrated mud, having a suspended solids concentration of approximately 500 g/L. The original drilling mud contained emulsifying agents and oils, which inhibited dewaterability. The characteristics of the diluted (4 : 1) spent drilling mud are shown in Table 4.

Table 4 Characteristics of Spent Drilling Mud

Analyses	Concentration (diluted 4 : 1)
COD, soluble (mg/L)	5,720
BOD, soluble (mg/L)	2,625
TOC, soluble (mg/L)	2,010
Total solids (g/L)	113.4
Ash (g/L)	107.5
Suspended solids (g/L)	103.7
Suspended ash (g/L)	100.9
Specific filtration resistance ($cm^2/g \times 10^{-7}$)	155

Source: From Ref. 14.

4.3.2 Refinery Wastewater

The sources of wastewater generation in petroleum refineries have been discussed previously in this chapter. Table 5 presents a qualitative evaluation of wastewater flow and characteristics by fundamental refinery processes [5]. The trend of the industry has been to reduce wastewater production by improving the management of the wastewater systems. Table 6 shows wastewater loadings and volumes per unit fundamental process throughput in older, typical, and newer technologies [15]. Table 7 shows typical wastewater characteristics associated with several refinery processes [16].

In addition to those from the fundamental processes, wastewaters are also generated from other auxiliary operations in refineries. Figure 7 shows the various sources of wastewater and their primary pollutants in a refinery [17].

In the USEPA study to develop effluent limitation guidelines [7], refinery operations were grouped together to produce five subcategories based on raw waste load, product mix, refinery processes, and wastewater generation characteristics. These subcategories are described below.

1. *Topping* Includes topping, catalytic reforming, asphalt production, or lube oil manufacturing processes, but excludes any facility with cracking or thermal operations.
2. *Cracking* Includes topping and cracking.
3. *Petrochemical* Includes topping, cracking, and petrochemical operations.
4. *Lube* Includes topping, cracking, and lube oil manufacturing processes.
5. *Integrated* Includes topping, cracking, lube oil manufacturing processes, and petrochemical operations.

The term petrochemical operations means the production of second-generation petrochemicals (alcohols, ketones, cumene, styrene, and so on) or first-generation petrochemicals and isomerization products (BTX, olefins, cyclohexane, and so on) when 15% or more of refinery production is as first-generation petrochemicals and isomerization products.

All five subcategories of refineries generate wastewaters containing similar constituents. However, the concentrations and loading of the constituents (raw waste load) vary among the categories. The raw waste loads, and their variabilities, for the five petroleum refining subcategories are presented in Tables 8 to 12 [7].

In addition to the conventional pollutant constituents, USEPA made a survey of the presence of the 126 toxic pollutants listed as "priority pollutants" in refinery operations in 1977 [5]. The survey responses indicated that 71 toxic pollutants were purchased as raw or intermediate materials; 19 of these were purchased by single refineries. At least 10% of all refineries purchase the following toxic pollutants: benzene, carbon tetrachloride, 1,1,1-trichloroethane, phenol, toluene, zinc and its compounds, chromium and its compounds, copper and its compounds, and lead and its compounds. Zinc and chromium are purchased by 28% of all refineries, and lead is purchased by nearly 48% of all plants.

Forty-five priority pollutants are manufactured as final or intermediate materials; 15 of these are manufactured at single refineries. Benzene, ethylbenzene, phenol, and toluene are manufactured by at least 10% of all refineries. Of all refineries, 8% manufacture cyanides, while more than 20% manufacture benzene and toluene. Hence, priority pollutants are expected to be present in refinery wastewaters. The EPA's short-term and long-term sampling programs conducted later detected and quantified 22 to 28 priority pollutants in refinery effluent samples [5].

Table 5 Qualitative Evaluation of Wastewater Flow and Characteristics by Fundamental Refinery Processes

Production processes	Flow	BOD	COD	Phenol	Sulfide	Oil	Emulsified oil	pH	Temperature	Ammonia	Chloride	Acidity	Alkalinity	Suspended solids
Crude oil and product storage	XX	X	XXX	X		XXX	XX	O	O	O		O		XX
Crude desalting	XX	XX	XX	X	XXX	X	XXX	X	XXX	XX	XXX	O	X	XXX
Crude distillation	XXX	X	X	XX	XXX	XX	XXX	X	XX	XXX	X	O	X	X
Thermal cracking	X	X	X	X	X	X		XX	XX	X	X	O	XX	X
Catalytic cracking	XXX	XX	XX	XXX	XXX	X	X	XXX	XX	XXX	X	O	XXX	X
Hydrocracking	X			XX	XX				XX	XX				
Polymerization	X	X	X	O	X	X	O	X	X	X	X	X	O	X
Alkylation	XX	X	X	O	XX	X	O	XX	X	X	XX	XX	O	XX
Isomerization	X													
Reforming	X	O	O	X	X	X	O	O	X	X	O	O	O	O
Solvent refining	X		X	X	O		X	X	O			O	X	
Asphalt blowing	XXX	XXX	XXX	X		XXX								
Dewaxing	X	XXX	XXX	X	O	X	O							
Hydrotreating	X	X	X		XX	O	O	XX		XX	O	O	X	O
Drying and sweetening	XXX	XXX	X	XX	O	O	X	XX	O	X	O	X	X	XX

XXX = major contribution; XX = moderate contribution; X = minor contribution; O = insignificant; Blank = no data.
BOD, biochemical oxygen demand; COD, chemical oxygen demand.
Source: From Ref. 5.

Table 6 Waste Loadings and Volumes Per Unit of Fundamental Process Throughput in Older, Typical, and Newer Technologies

Fundamental process	Older technology				Typical technology				Newer technology			
	Flow (gal/bbl)	BOD (lb/bbl)	Phenol (lb/bbl)	Sulfides (lb/bbl)	Flow (gal/bbl)	BOD (lb/bbl)	Phenol (lb/bbl)	Sulfides (lb/bbl)	Flow (gal/bbl)	BOD (lb/bbl)	Phenol (lb/bbl)	Sulfides (lb/bbl)
Crude oil and product storage	4	0.001	—	—	4	0.001	—	—	4	0.001	—	—
Crude desalting	2	0.002	0.20	0.002	2	0.002	0.10	0.002	2	0.002	0.05	0.002
Crude fractionation	100	0.020	3.0	0.001	50	0.0002	1.0	0.001	10	0.0002	1.0	0.001
Thermal cracking	66	0.001	7.0	0.002	2	0.001	0.2	0.001	1.5	0.001	0.2	0.001
Catalytic cracking	85	0.062	50.0	0.03	30	0.010	20	0.003	5	0.010	5	0.003
Hydrocracking	Not in this technology				Not in this technology				5	—	—	—
Reforming	9	T	0.7	T	6	T	0.7	0.001	6	T	0.7	0.001
Polymerization	300	0.003	1.4	0.22	140	0.003	0.4	0.010	Not in this technology			
Alkylation	173	0.001	0.1	0.005	60	0.001	0.1	0.010	20	0.001	0.1	0.020
Isomerization	Not in this technology				Not in this technology				—	—	—	—

Solvent refining	8	—	3	T	8	—	3	T	8	—	3	T
Dewaxing	24	0.52	2	T	23	0.50	1.5	T	20	0.25	1.5	T
Hydrotreating	1	0.002	0.6	0.007	1	0.002	0.01	0.002	8	0.002	0.01	0.002
Deasphalting	—	—	—	—	—	—	—	—	—	—	—	—
Drying and sweetening	100	0.10	10	—	40	0.05	10	—	40	0.05	10	—
Wax finishing	—	—	—	—	—	—	—	—	—	—	—	—
Grease manufacturing	—	—	—	—	—	—	—	—	—	—	—	—
Lube oil finishing	—	—	—	—	—	—	—	—	—	—	—	—
Hydrogen manufacture	Not in this technology				Not in this technology				—	—	—	—
Blending and packaging	—	—	—	—	—	—	—	—	—	—	—	—

T = trace; — = data not available for reasonable estimate; BOD, biochemical oxygen demand.
gal/bbl = gallons of wastewater per barrel of oil processed.
lb/bbl = pounds of contaminant per barrel of oil processed.
Source: From Ref. 15.

Table 7 Typical Waste Characteristics

Characteristic	Spent caustic stream: Benzene sulfonation scrubbing	Orthophenylphenol washing	Alkylate washing	Polymerization
Alkalinity (mg/L)	33,800	18,400	46,250	209,330
BOD (mg/L)	53,600	18,400	256	8,440
COD (mg/L)	112,000	67,600	3,230	50,350
pH	13.2	9–12	12.8	12.7
Phenols (mg/L)	8.3	5,500	50	22.2
NaOH (wt %)	1	0.2–0.5		
Na_2SO_4 (wt %)	1.5–2.5			
Sulfates (mg/L)	3,760	2,440		
Sulfides (mg/L)			2	3,060
Sulfites (mg/L)	7,100	4,720		
Total solids (mg/L)	90,300	40,800		

Characteristic	Process waste: Crude Desalting	Catalytic cracking	Naphtha cracking	Sour condensates from distillation cracking, etc.
Ammonia (mg/L)	80			135–6,550
BOD (mg/L)	60–610	230–440		500–1,000
COD (mg/L)	124–470	500–2,800	53–180	500–2,000
Oil (mg/L)	20–516	200–2,600	160	100–1,000
pH (mg/L)	7.2–9.1			4.5–9.5
Phenols (mg/L)	10–25	20–26	6–10	100–1,000
Salt (as NaCl) (wt %)	0.4–25			
Sulfides (mg/L)	0–13			390–8,250 (H_2S)

Characteristic	Acid waste: Acid wash: alkylation	Acid wash: phenol still bottoms	Acid wash: orthophenylphenol	Sulfite wash: liquid OP-phenol distillation
Acidity (mg/L)	1,105–12,325		24,120	675
BOD (mg/L)	31	20,800	13,600	105,000
COD (mg/L)	1,251	248,000	23,400	689,000
Dissolved solids (mg/L)		340,500	81,300	176,800
Oil (mg/L)	131.5			
pH	0.6–1.9	1.0	1.1	3.8
Phenols (mg/L)		3,800	1,500	16,400
Sulfate (mg/L)			54,700	
Sulfite (mg/L)		34,800	2,920	74,000
Total solids (mg/L)		403,200	81,600	176,900

BOD, biochemical oxygen demand; COD, chemical oxygen demand.
Source: From Ref. 16,

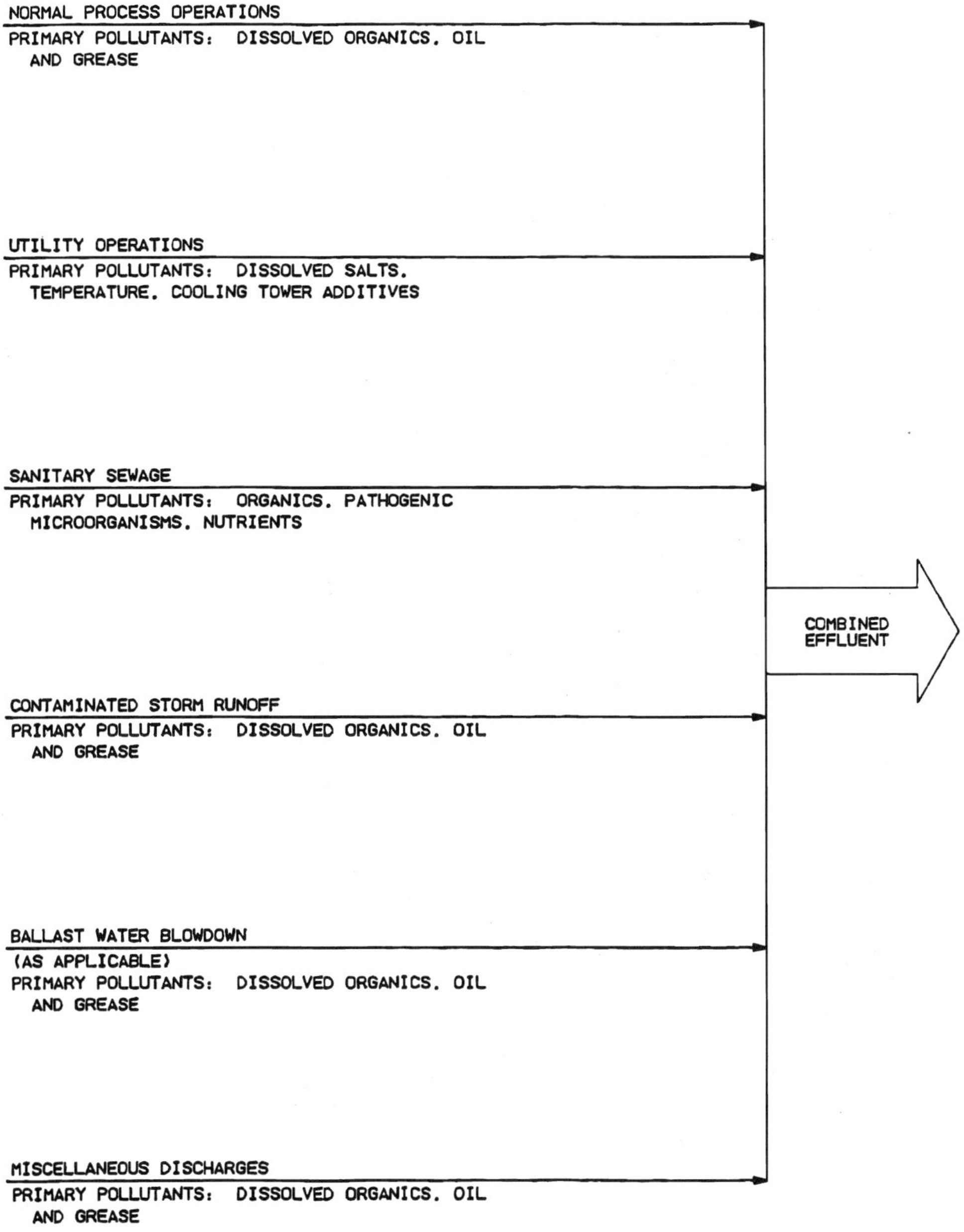

Figure 7 Components of pollutants by source. These principal pollutants are present in waste streams from each refinery operations/sources. (From Ref. 17.)

4.3.3 Refinery Solid and Hazardous Wastes

According to a USEPA survey, many of the more than 150 separate processes used in petroleum refineries generate large quantities of hazardous wastes. Typical wastes generated from refinery processes include bottom sediments and water from crude storage tanks, spent amines, spent acids and caustics, spent clays, spent glycol, catalyst fines, spent Streford solution and sulfur,

Table 8 Topping Subcategory Raw Waste Load Effluent from Refinery API Separator[a]

	Probability of occurrence, percent less than or equal to		
Parameters	10	50 (median)	90
BOD_5	1.29 (0.45)	3.43 (1.2)	217.36 (76)
COD	3.43 (1.2)	37.18 (13)	486.2 (170)
TOC	1.09 (0.38)	8.01 (2.8)	65.78 (23)
TSS	0.74 (0.26)	11.73 (4.1)	286 (100)
O&G	1.03 (0.36)	8.29 (2.9)	88.66 (31)
Phenols	0.001 (0.0004)	0.034 (0.012)	1.06 (0.37)
Ammonia	0.077 (0.027)	1.20 (0.42)	19.45 (6.8)
Sulfides	0.002 (0.00065)	0.054 (0.019)	1.52 (0.53)
Chromium	0.0002 (0.00007)	0.007 (0.0025)	0.29 (0.1)
Flow[b]	8.00 (2.8)	66.64 (23.3)	557.7 (195)

[a]Values represent kg/1000 m^3(lb/1000 bbl) of feedstock throughput.
[b]1000 m^3/1000^3 feedstock throughput (gallons/bbl).
BOD, biochemical oxygen demand; COD, chemical oxygen demand; TOC, total organic carbon; TSS, total suspended solids; O&G, oil and grease.
Source: From Ref. 7.

coking fines, slop oil, and storage tank bottoms. Most are hazardous wastes. Figure 8 shows a refinery schematic diagram indicating representative sources of solid wastes in refinery systems [18].

Also, the plant's utility systems often contribute to the volume of waste. Utility water systems generate raw water treatment sludge, lime softening sludge, demineralizer regenerants, and cooling tower sludge. These wastes may or may not be hazardous, depending on characteristics such as pH and metal concentrations. Figure 9 shows a refinery schematic

Table 9 Cracking Subcategory Raw Waste Load Effluent from Refinery API Separator[a]

	Probability of occurrence, percent less than or equal to		
Parameters	10	50 (median)	90
BOD_5	14.3 (5.0)	72.93 (25.5)	466.18 (163)
COD	27.74 (9.7)	217.36 (76.0)	2516.8 (880)
TOC	5.43 (1.9)	41.47 (14.5)	320.32 (112)
O&G	2.86 (1.0)	31.17 (10.9)	364.65 (127.5)
Phenols	0.19 (0.068)	4.00 (1.4)	80.08 (28.0)
TSS	0.94 (0.33)	18.16 (6.35)	360.36 (126.0)
Sulfur	0.01 (0.0035)	0.94 (0.33)	39.47 (13.8)
Chromium	0.0008 (0.00028)	0.25 (0.088)	4.15 (1.45)
Ammonia	2.35 (0.82)	28.31 (9.9)	174.46 (61.0)
Flow[b]	3.29 (1.15)	92.95 (32.5)	2745.6 (960.0)

[a]Values represent kg/1000 m^3 (lb/1000 bbl) of feedstock throughput.
[b]1000 m^3/1000 m^3 feedstock throughput (gallons/bbl).
BOD, biochemical oxygen demand; COD, chemical oxygen demand; TOC, total organic carbon; TSS, total suspended solids; O&G, oil and grease.
Source: From Ref. 7.

Table 10 Petrochemical Subcategory Raw Waste Load Effluent from Refinery API Separator[a]

	Probability of occurrence, percent less than or equal to		
Parameters	10	50 (median)	90
BOD_5	40.90 (14.3)	171.6 (60)	715 (250)
COD	200.2 (70)	463.32 (162)	1086.8 (380)
TOC	48.62 (17)	148.72 (52)	457.6 (160)
TSS	6.29 (2.2)	48.62 (17)	371.8 (130)
O&G	12.01 (4.2)	52.91 (18.5)	234.52 (82)
Phenols	2.55 (0.89)	7.72 (2.7)	23.74 (8.3)
Ammonia	5.43 (1.9)	34.32 (12)	205.92 (72)
Sulfides	0.009 (0.003)	0.86 (0.3)	91.52 (32)
Chromium	0.014 (0.005)	0.234 (0.085)	3.86 (1.35)
Flow[b]	26.60 (9.3)	108.68 (38)	443.3 (155)

[a]Values represent kg/1000 m^3 (lb/1000 bbl) of feedstock throughput.
[b]1000 m^3/1000 m^3 feedstock throughput (gallons/bbl).
BOD, biochemical oxygen demand; COD, chemical oxygen demand; TOC, total organic carbon; TSS, total suspended solids; O&G, oil and grease.
Source: From Ref. 7.

diagram indicating representative sources of solid waste in utility water systems [18]. Wastes generated from wastewater treatment systems include API/CPI separator sludge, dissolved-air flotation or induced-air flotation system floats, pond and tank sediments, and biosolids. Of these, only the biosolids from the biological wastewater treatment system may be nonhazardous. Figure 10 shows a refinery schematic diagram indicating representative sources of solids waste in wastewater treatment systems [18].

Table 11 Lube Subcategory Raw Waste Load Effluent from Refinery API Separator[a]

	Probability of occurrence, percent less than or equal to		
Parameters	10	50 (median)	90
BOD_5	62.92 (22)	217.36 (76)	757.9 (265)
COD	165.88 (58)	543.4 (190)	2288 (800)
TOC	31.46 (11)	108.68 (38)	386.1 (135)
TSS	17.16 (6)	71.5 (25)	311.74 (109)
O&G	23.74 (8.3)	120.12 (42)	600.6 (210)
Phenols	4.58 (1.6)	8.29 (2.9)	51.91 (18.5)
Ammonia	6.5 (2.3)	24.1 (8.5)	96.2 (34)
Sulfides	0.00001 (0.000005)	0.014 (0.005)	20.02 (7.0)
Chromium	0.002 (0.0006)	0.046 (0.016)	1.23 (0.43)
Flow[b]	68.64 (24)	117.26 (41)	772.2 (270)

[a]Values represent kg/1000 m^3 (lb/1000 bbl) of feedstock throughput.
[b]1000 m^3/1000 m^3 feedstock throughput (gallons/bbl).
BOD, biochemical oxygen demand; COD, chemical oxygen demand; TOC, total organic carbon; TSS, total suspended solids; O&G, oil and grease.
Source: From Ref. 7.

Table 12 Integrated Subcategory Raw Waste Load Effluent from Refinery API Separator[a]

	Probability of occurrence, percent less than or equal to		
Parameters	10	50 (median)	90
BOD_5	63.49 (22.2)	197.34 (69.0)	614.9 (215)
COD	72.93 (25.5)	328.9 (115)	1487.2 (520)
TOC	28.6 (10.0)	139.0 (48.6)	677.82 (237)
O&G	20.88 (7.3)	74.93 (26.2)	268.84 (94.0)
Phenol	0.61 (0.215)	3.78 (132)	22.60 (7.9)
TSS	15.16 (5.3)	58.06 (20.3)	225.94 (79.0)
Sulfur	0.52 (0.182)	2.00 (0.70)	7.87 (2.75)
Chromium	0.12 (0.043)	0.49 (0.17)	121.55 (42.5)
Ammonia	3.43 (1.20)	20.50 (7.15)	121.55 (42.5)
Flow[b]	40.04 (14.0)	234.52 (82.0)	1372.8 (480)

[a]Values given represent kg/1000 m^3 (lb/1000 bbl) of feedstock throughput.
[b]1000 m^3/1000 m^3 feedstock throughput (gallons/bbl).
BOD, biochemical oxygen demand; COD, chemical oxygen demand; TOC, total organic carbon; TSS, total suspended solids; O&G, oil and grease.
Source: From Ref. 7.

The amount and type of wastes generated in a refinery depend on a variety of factors such as crude capacity, number of refining processes, crude source, and operating procedures. A 130,000 bpd integrated refinery on the West Coast generates about 50,000 tons per year of hazardous waste (including recycled streams and unfiltered sludges). The major wastes are wastewater treatment plant sludge, spent caustics, Stretford solution and sulfur, and spent catalysts [19]. A much simpler 50,000 bpd refinery generates only 400 tons per year of hazardous waste. Major wastes in this refinery are wastewater treatment plant sludge (dewatered by pressure filtration), spent catalysts, and spent clay filter media [19].

4.4 ENVIRONMENTAL REGULATIONS

Three categories of regulatory limitations apply to wastewater discharge from industrial facilities such as oilfields and petroleum refineries [20]. The first category includes effluent limitations, which are designed to control those industry-specific wastewater constituents deemed significant from the standpoints of water quality impact and treatability in conventional treatment systems. In the United States, these limitations are the EPA Effluent Guidelines, issued under Public Law 92-500.

The second category includes pretreatment discharge requirements established both by the EPA and certain municipalities that treat combined industrial and domestic wastes in their publicly owned treatment works. These standards have not been updated by USEPA as of 2003.

The third category includes effluent limitations associated with maintaining or establishing desirable water uses in certain bodies of effluent-receiving waters, that is, water-quality-limiting segments as defined in Public Law 92-500. This last category became the overriding category in many locations in the United States when the EPA published its final surface water toxics control rule on June 2, 1989 [21]. These three categories of effluent limitations are discussed below.

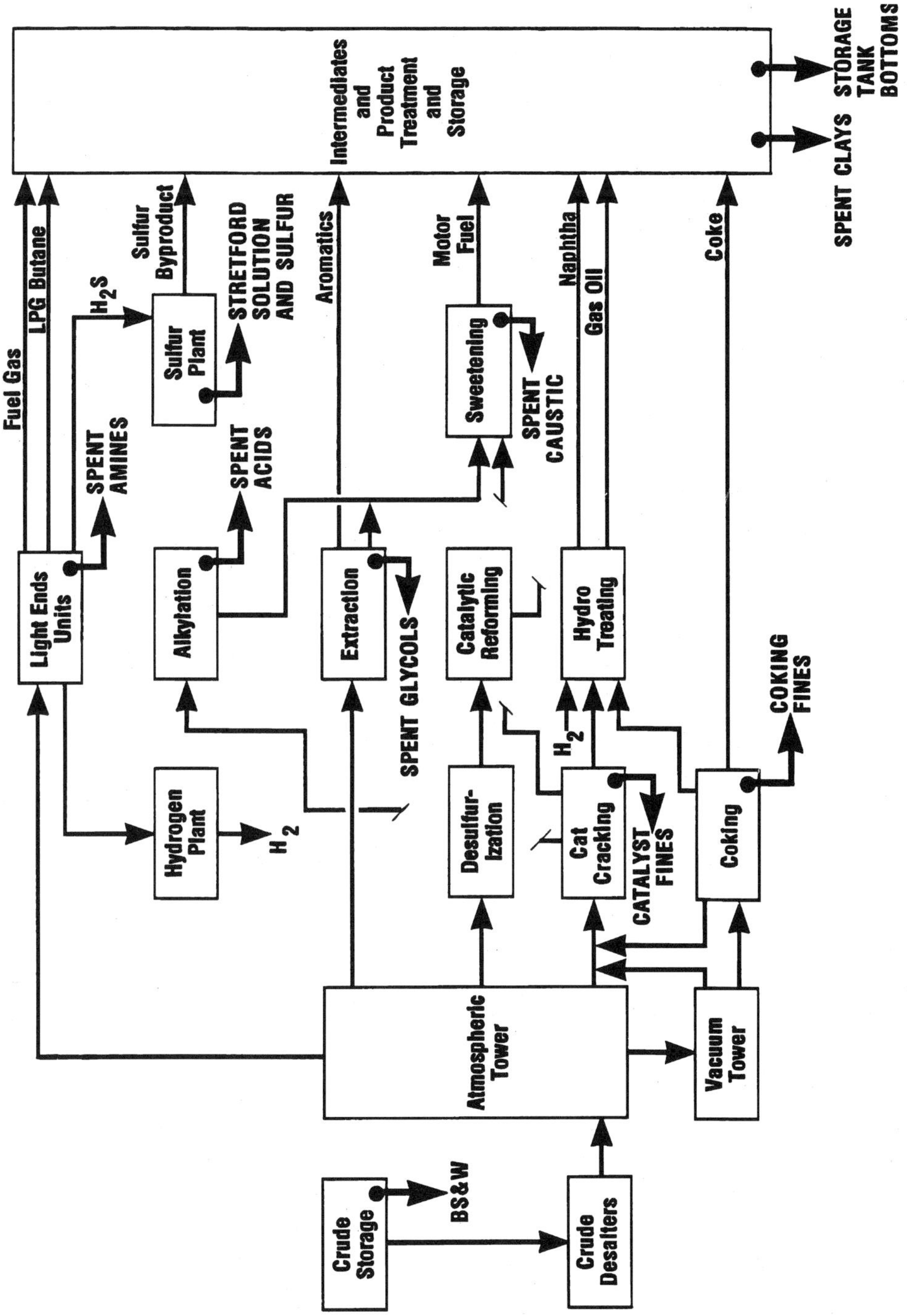

Figure 8 Refinery schematic diagram indicating representative sources of solid waste in refinery system. Most solid wastes from refineries are considered hazardous wastes in the United States. (From Ref. 18.)

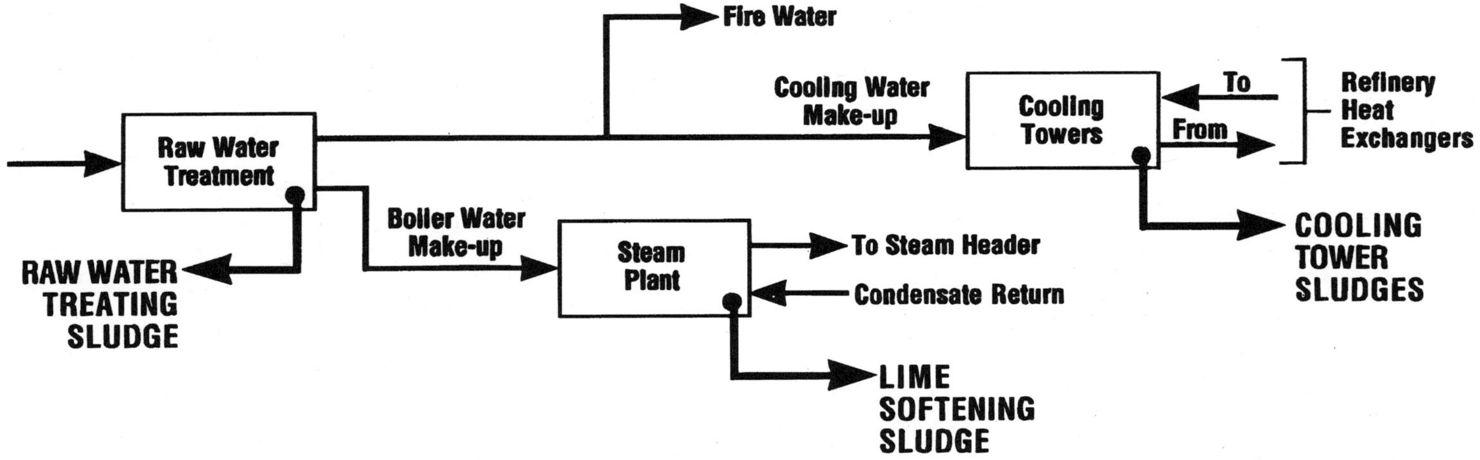

Figure 9 Refinery schematic diagram indicating representative sources of solid waste in utility water system. These wastes may not be classified as hazardous in the United States. (From Ref. 18.)

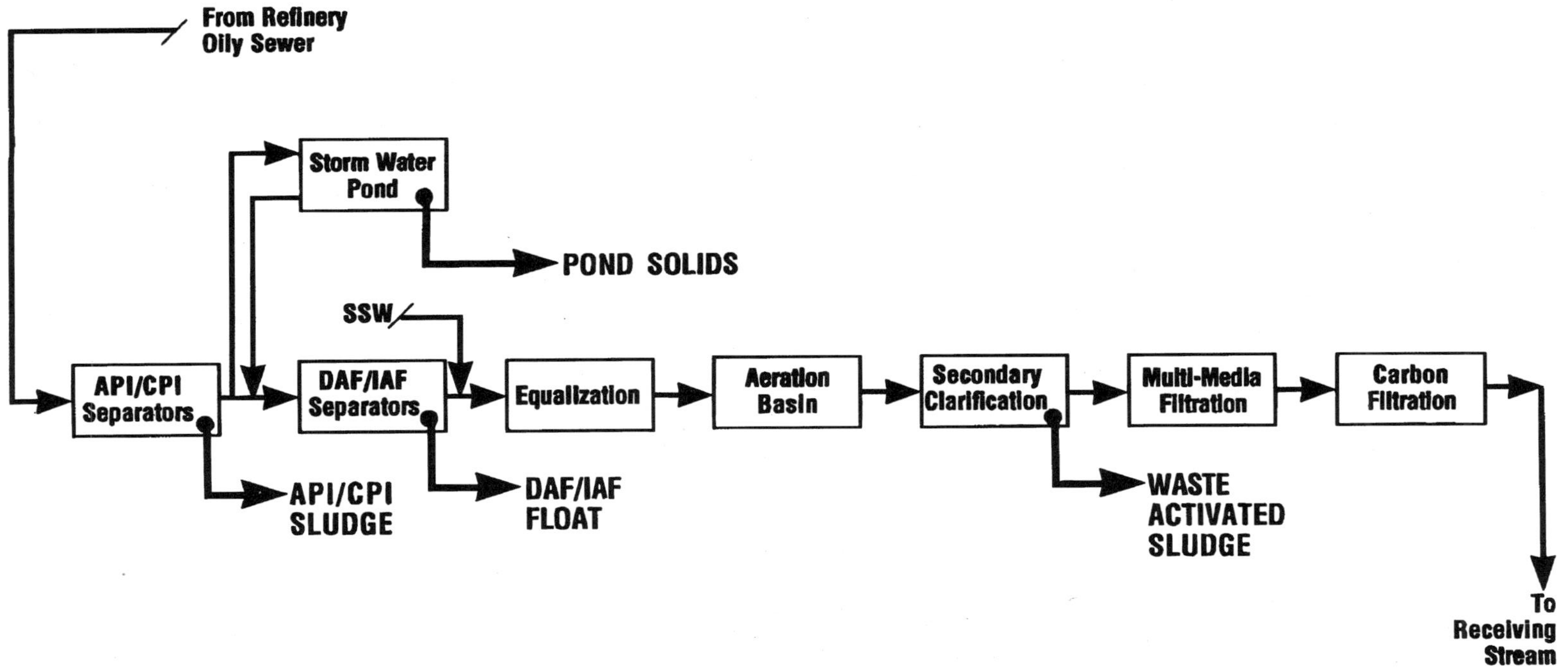

Figure 10 Refinery schematic diagram indicating representative sources of solid waste in wastewater treatment system. All wastes except waste activated sludge are classified as hazardous wastes because of their oil contents. (From Ref. 18.)

4.4.1 Effluent Guidelines for Industrial Point Source Categories

USEPA has established effluent limitations on wastewater constituents for various industrial categories. The EPA effluent limitations for Oil and Gas Extraction Point Source Category are under 40 CFR Part 435 (Code of Federal Register, 1988). The regulations differentiate between offshore, onshore, and coastal facilities. The limitation for onshore oil and gas facilities is no discharge of wastewater pollutants into navigable waters from any source associated with production, field exploration, drilling, well completion, or well treatment (produced water, drilling muds, drill cuttings, and produced sand). Owing to a challenge in court (API vs. EPA, 1981), the limitation was suspended for facilities located in the Santa Maria Basin of California.

For onshore facilities located in the continental United States and west of the 98th meridian for which the produced water has a use in agriculture or wildlife propagation when discharged into navigable waters, discharge of produced water is allowed if its oil and grease (O&G) concentration does not exceed 35 mg/L. Other wastes from these onshore facilities are not to be discharged to navigable waters.

The effluent limitations for offshore and coastal oil and gas facilities are identical. The main criteria for discharge are O&G concentrations. For produced water, the effluent limitations are 72 mg/L of O&G maximum for anyone day and 48 mg/L of O&G average for 30 consecutive days. For other industrial wastes from these facilities, the effluent limitations are no discharge of free oil.

The EPA promulgated Effluent Guidelines and Standards for the Petroleum Refining Industry under 40 CFR Part 419 on May 9, 1974, and published the most recent update to Part 419 on August 12, 1985 (Federal Register). Standards for direct dischargers are mass-limited, not concentration-limited, and are expressed in pounds per 1000 barrels of feedstock. The standards are further subdivided into five subcategories within the petroleum refining category, as described earlier in this chapter. The standards for each subcategory may in turn be modified by "size" and "process" factors. For example, in the topping subcategory, a plant of less than 24,000 bpsd of feedstock would have a size factor of 1.02 applied to the effluent limitations, and a plant of 150,000 bpsd or greater would have a size factor of 1.57 applied.

The EPA has established four different control technologies for the petroleum refining industry: best practicable control technology (BPT), best available technology economically achievable (BAT), best conventional pollutant control technology (BCT), and new source performance standards (NSPS). Table 13 shows the BPT and NSPS standards that must be met by the various subcategories (40 CFR Part 419). The limitations for BPT actually incorporate those of both BAT and BCT for this industry.

In addition to these effluent standards, the EPA has also established separate BPT, BAT, BCT, and NSPS standards for ballast water and BPT, BAT, and BCT standards for contaminated storm water (40 CFR Part 419). Once-through cooling water is allowed for direct discharge if the total organic carbon concentration does not exceed 5 mg/L.

4.4.2 Pretreatment Requirements

Presently there are no EPA pretreatment standards for the oil and gas extraction (oilfield) point source category. The EPA pretreatment standards for discharge from existing and new petroleum refining facilities to publicly owned treatment works include 100 mg/L each for oil and grease (O&G) and ammonia (as N). For new facilities a total chromium concentration of 1 mg/L for the cooling tower discharge part of the refinery effluent is also required (40 CPR Part 419).

In addition to meeting the EPA pretreatment standards, indirect dischargers are required to meet individual municipal pretreatment limits. Publicly owned treatment works establish limits

Table 13 Effluent Standards for Five Subcategories of the Petroleum Refining Point Source Category

	Effluent limitation (daily average for 30 consecutive days, in lbs/1000 bbl of feedstock)									
	Topping		Cracking		Petroleum		Lube		Integrated	
Parameters	BPT	NS	BPT	NS	BPT	NS	BPT	NS	BPT	NS
BOD_5	4.25	2.2	5.5	3.1	6.5	4.1	9.1	6.5	10.2	7.8
TSS	3.6	1.9	4.4	2.5	5.25	3.3	8.0	5.3	8.4	6.3
COD	31.3	11.2	38.4	21.0	38.4	24.0	66.0	45.0	70.0	54.0
O&G	1.3	0.70	1.6	0.3	2.1	1.3	3.0	2.0	3.2	2.4
Phenolic compounds	0.027	0.16	0.036	0.020	0.0425	0.027	0.065	0.043	0.068	0.051
Ammonia as N	0.45	0.45	3.0	3.0	3.8	3.8	3.8	3.8	3.8	3.8
Sulfide	0.024	0.012	0.029	0.017	0.035	0.022	0.053	0.035	0.056	0.042
Total chromium	0.071	0.037	0.088	0.049	0.107	0.068	0.160	0.105	0.17	0.13
Hexavalent chromium	0.0044	0.0025	0.0056	0.0032	0.0072	0.0044	0.160	0.0072	0.011	0.008

Note: pH (within the range of 6.0 to 9.0); BPT incorporates BAT and BCT; BPT, best practicable control technology; NS, new source performance standards; BOD, biochemical oxygen demand; TSS, total suspended solids; COD, chemical oxygen demand; O&G, oil and grease; BAT, best available technology economically achievable; BCT, best conventional pollutant control technology.

Source: From Ref. 40 CFR, Part 419, 1988.

to control pollutants that could be deleterious to conventional biological treatment systems or that could cause the municipality to violate receiving water standards. Table 14 shows the industrial effluent limits established by the City of San Jose, CA (San Jose Municipal Code, 1988). This city has also adopted an effluent toxicity requirement for industrial dischargers using the public sewer. Discharges are not to exceed a median threshold limit of 50%.

4.4.3 Water Quality Based Limitations

In the United States, as control of conventional pollutants has been significantly achieved, increased emphasis is being placed on reduction of toxic pollutants. The EPA has developed a water quality based approach to achieve desired water quality where treatment control based discharge limits have proved to be insufficient [22]. The procedure for establishing effluent limitations for point sources discharging to a water quality based segment generally involves the use of some type of mathematical model or allocation procedure to apportion the allowable

Table 14 Industrial Waste Pretreatment Limits for a Publicly Owned Treatment Works

Toxic substance	Max. allowable concentration (mg/L)
Aldehyde	5.0
Antimony	5.0
Arsenic	1.0
Barium	5.0
Beryllium	1.0
Boron	1.0
Cadmium	0.7
Chlorinated hydrocarbons, including but not limited to pesticides, herbicides, algaecides	Trace
Chromium, total	1.0
Copper	2.7
Cyanides	1.0
Fluorides	10.0
Formaldehydes	5.0
Lead	0.4
Manganese	0.5
Mercury	0.010
Methyl ethyl ketone and other water insoluble ketones	5.0
Nickel	2.6
Phenol and derivatives	30.0
Selenium	2.0
Silver	0.7
Sulfides	1.0
Toluene	5.0
Xylene	5.0
Zinc	2.6
pH, su	5.0 to 10.5

SU = standard unit
Source: From City of San Jose, CA, Municipal Code, 1988.

loading of a particular toxicant to each discharge in the segment. These allocations are generally made by the state regulatory agency [20].

State and regional regulatory agencies may also establish general effluent limitations for a particular water body to control the total discharge of toxic pollutants. Table 15 shows the discharge limits for toxic pollutants established by the San Francisco Bay Regional Water Quality Board (1986).

This agency has also adopted biomonitoring and toxicity requirements for municipal and industrial dischargers. Biomonitoring, or whole-effluent toxicity testing, has become a requirement for many discharges in the United States. As of 1988, more than 6000 discharge permits have toxicity limits to protect against chronic toxicity [22]. When a discharge exceeds the toxicity limits, the discharger must conduct a toxicity identification evaluation (TIE) and a toxicity reduction evaluation (TRE). A TRE is a site-specific investigation of the effluent to identify the causative toxicants that may be eliminated or reduced, or treatment methods that can reduce effluent toxicity.

4.5 CONTROL AND TREATMENT TECHNIQUES FOR OILFIELD WASTES

Major waste liquids arising from oil and gas production include produced water and drilling fluids and muds. These waste streams are handled and disposed of separately.

4.5.1 Produced Water Treatment and Disposal

Produced water (brine) disposal practices may be divided into the broad categories of surface discharge, subsurface discharge, evaporation, and reuse. Approximately 30 states produce some amount of oil or gas, and brine handling practises vary considerably because of variations in climate, geology, brine quantity and quality, and regulatory framework [23].

Table 15 Effluent Limitations for Selected Toxic Pollutants for Discharge to Surface Waters (All Values in $\mu g/L$)

	Daily average	
	Shallow water	Deep water
Arsenic	20	200
Cadmium	10	30
Chromium (VI)	11	110
Copper	20	200
Cyanide	25	25
Lead	5.6	56
Mercury	1	1
Nickel	7.1	71
Silver	2.3	23
Zinc	58	580
Phenols	500	500
PAHs	15	150

PAHs = Polynuclear aromatic hydrocarbons
Source: From Water Quality Control Plan, San Francisco Bay Basin, 1986.

Surface Discharge

Because onshore oil and gas facilities are not allowed to discharge wastes to navigable waters, surface discharge is only practiced at coastal facilities. In some states indirect surface discharge is practiced by simple dilution through an existing municipal or industrial wastewater treatment facility [23].

The main pollutant of concern for brine discharge is oil and grease (O&G). However, other pollutants may be important if they violate state-set water quality criteria for local water bodies. Michalczyk et al. [9] suggested a typical production water treatment system to meet the criteria of the California Ocean Plan. As shown in Fig. 11, treatment processes include equalization, oil removal by flotation, pH adjustment, and activated sludge. Experimental results obtained by Michalczyk *et al.* indicate that biological treatment effectively reduces BOD/COD and phenol in oilfield produced waters to acceptable levels, but nitrification can be inhibited by inorganic or biologically refractory organic compounds. Wang et al. [24] reported the use of hydrocarbon deterioration bacteria with gas lift processing to treat oily produced water. With oil content above 300 mg/L, COD of 250–480 mg/L, the treated water has 10 mg/L of oil and less than 120 mg/L of COD. A special group of bacteria named WS3 were selected for treatment testing after an elaborate screening process.

Palmer et al. [10] reported the results of two pilot field studies of treating oilfield produced water by biodisks in southern California. The TDS concentration of the produced water was 20,000 mg/L. The results indicate that dissolved organics and ammonium compounds can be

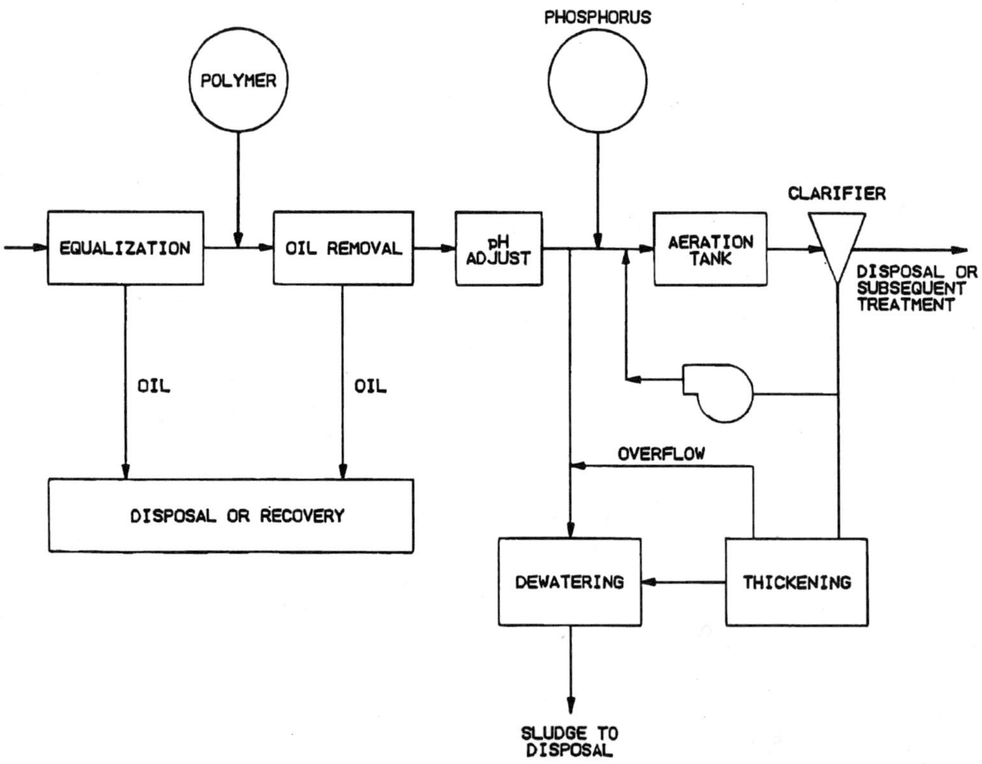

Figure 11 Produced water treatment system. Treatment is mainly for oil and organics removal. (From Ref. 9.)

removed by biological oxidation in a biodisk unit to meet California Ocean Plan criteria. Earlier, Beyer et al. [25] demonstrated the feasibility of biological oxidation by aerated lagoons to remove dissolved compounds such as ammonia and phenols from produced waters. Ali et al. [26] conducted laboratory and field tests to successfully demonstrate that a two-stage filtration process can effectively reduce oil and grease content in offshore discharged produced water. The first stage (Crudesorb) removes dispersed oil and grease droplets, and the second stage (polymeric resin) removes dissolved hydrocarbons, aliphatic carboxylic acids, cyclic carboxylic acids, aromatic carboxylic acids, and phenolic compounds.

Subsurface Discharge

Disposal of brine in subsurface wells is probably the most widely used control method, especially in the western and southern oil and gas producing states [23]. For this to be an effective disposal option, two conditions must be met: the natural aquifer must be naturally saline and must not leak to freshwater aquifers, and the reinjection pressure must not exceed the fracture pressure of the formation [9]. Produced water is usually pretreated to prevent equipment from being corroded and to prevent plugging of the sand at the base of the well. Pretreatment may include the removal of oils and floating material, suspended solids, biological growth, dissolved gases, precipitable ions, acidity, or alkalinity [27]. A typical system is shown in Fig. 12.

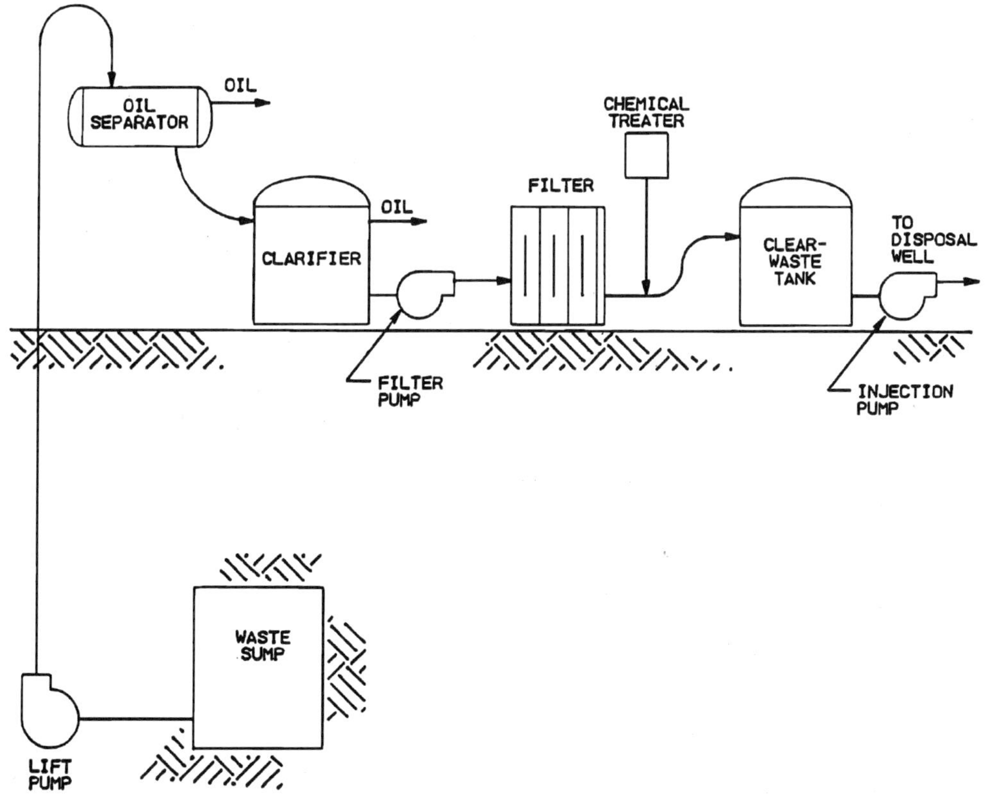

Figure 12 Typical subsurface waste disposal system. Waste is treated for oil removal, filtered, and chemically treated before subterranean injection. (From Ref. 27.)

In the United States, injection wells are classified into three categories: Class 1 wells are used to inject hazardous wastes; Class 2 wells are used to inject fluids brought to the surface in connection with the production of oil and gas or for disposal of salt water; Class 3 covers solution mining wells [28]. Class 1 wells are heavily regulated by the EPA. However, tougher rules for casing and cementing are being considered for Class 2 wells. After conducting a random sample of Class 2 wells in four states in 1987 and 1988, the General Accounting Office (GAO) claimed that federal and state regulations are not preventing brine injection wells from contaminating U.S. drinking water aquifers [29]. The GAO recommended that the EPA require all existing injection wells to be checked for leakage and require state agencies to examine permit applications for new injection wells more closely.

Evaporation

The use of open pits or ponds for evaporation of brine is widely practiced in southwestern states where evaporation exceeds precipitation [23]. For example, about 75% of all oil and gas waste fluids are disposed of by evaporation pits in New Mexico [30]. Evaporation ponds require large land areas, and they may contaminate groundwater. Today regulators view evaporation pits with disfavor because faulty pond design and operation have allowed salts to migrate into usable groundwater reservoirs [9].

Reuse

The most desirable disposal method is to reuse the produced water. Produced water can be treated and reinjected into the subsurface reservoir to cause the oil to flow into the well to increase yields (water flooding). The produced water is usually filtered to remove oil and suspended solids before injection. In a steam flooding project in the Far East, produced water was treated and used as feedwater to generate steam for enhanced oil recovery [31]. The treatment processes included induced air flotation, filtration, softening, and deaeration. Treatment technologies for reclaiming oilfield-produced water for beneficial reuse were evaluated by Doran et al. [32]. The investigators selected precipitating softening and high-pH reverse osmosis (RO) for pilot testing based on a literature review and benchscale softening tests that indicated hardness, boron, and silica removal could be simultaneously optimized. The results of a pilot study were used to perform a conceptual design and cost estimation for a 7000 m^3/day (44,000 bpd) treatment facility for converting produced water to drinking water or other reuse quality [32]. Depending on reuse quality requirements, the capital cost ranged from \$3.1 million to \$12.3 million, and the operating and maintenance (O&M) cost from \$0.28/m^3 to \$1.45/m^3 of water recovered.

Another potential use of the brine is for highway application. Sodium and calcium chlorides have been widely used in highway applications both for winter deicing and for road stabilization and dust control. Sack et al. [23] sampled and analyzed produced brines from 13 counties representing 8 different geological formations. A significant number of West Virginia brines were found to be of suitable quality for highway application.

4.5.2 Drilling Fluids Treatment and Disposal

Potential treatment and disposal methods for drilling fluids include (1) fluid ejection, (2) pit and solids encapsulation, (3) injection into safe formations, (4) removal to disposal sites off location, (5) incineration, (6) microorganism processing, and (7) distillation, liquid extraction, and chemical fixation [13].

Direct Ejection of Fluids

This disposal method only applies to water-based drilling fluids. The fluids may be spread directly over adjacent agricultural or forest land after adjustment of pH and ion content. Treatment may include coagulation, flocculation, filtration, and pH adjustment before spreading. A major consideration is chloride ion content. With higher chlorides, some transport of the fluid to a better disposal site may be necessary.

Pit and Solids Encapsulation

Pit encapsulation means constructing a reserve pit to contain the fluids and seal it at the end of drilling. Normal procedures involve slurry trenching for sidewalls, plowing in organic-treated bentonite for a bottom base, placing a synthetic liner in this excavation, and covering the liner with additional soil containing some bentonite for puncture protection. The pit is then filled with waste drilling fluid. At well completion, the fluid is allowed to evaporate. When the fluid is substantially dewatered, the pit is covered with a top layer of soil containing organic-treated bentonite. The location of the burial site is recorded.

Solids encapsulation is removing solids from a fluid by some form of polymer coating procedure. The coated solids are buried. One novel treatment method adds a microorganism "cocktail", along with nutrients, to a fluid containing suspended solids and high chloride ion content [13]. The microorganisms utilize chloride during growth and coagulate the solids. After sufficient aging, clean water can be pumped off, leaving the coagulated solids residue, which is buried. Chloride ion concentration is normally below 200 mg/L following aging.

Pumping into Safe Formation

Deep well injection of spent fluids is another possible alternative. The criteria for injection of drilling fluids are similar to those for injection of produced water discussed earlier in this chapter.

Removal to Designated Disposal Sites

The fluids can be hauled by vacuum trucks to an approved disposal site for such wastes. There are different classes of disposal sites. If regulatory agencies require that a fluid be disposed in a hazardous waste "secure" landfill, the cost would be very high.

Incineration

Incineration offers the complete destruction of oil and organic materials. However, it is very expensive and may cause air pollution. Incineration would be used when other less costly options are unavailable.

Microorganism Processing

Biological treatment may be used to degrade the oil and grease fractions in drilling fluids prior to solids separation. Marks et al. [33] conducted batch treatment tests for drilling fluids and production sludges and demonstrated that biological treatment is feasible. However, more biokinetic tests are required for further evaluation.

Distillation, Liquid Extraction, and Chemical Treatment

Several emerging processes may be applicable for treatment of oily drilling muds prior to disposal. One process being tested in Europe involves the use of an electric distillation kiln to break down solids-laden oil-based drilling muds [13]. Another process uses critical fluid to extract oil and organics from oily sludges so that they can be landfilled [34]. Copa and Dietrich [14] treated a sample of spent drilling mud with wet air oxidation. The COD content was reduced by 45 to 64% and the dewaterability of the mud was improved.

Chemical fixation is another possible process to handle drilling fluids. A typical process uses a mixture of potassium or sodium silicate with portland cement to turn a drilling fluid into a soil-like solid that may be left in place, used as a landfill, or even used as a construction material [13].

4.6 IN-PLANT CONTROL AND TREATMENT TECHNIQUES FOR REFINERY WASTES

The management of wastes from refineries includes in-plant source control, pretreatment, and end-of-pipe treatment. In-plant source control reduces the overall pollutant load that must be treated in an end-of-pipe treatment system. Pretreatment reduces or eliminates a particular pollutant before it is diluted in the main wastewater stream, and may provide an opportunity for material recovery. End-of-pipe treatment is the final stage for meeting regulatory discharge requirements and protection of stream water quality. These techniques are discussed in more detail in the following sections.

4.6.1 In-Plant Source Control

Source control means different things to different people. Here it means knowing the sources and the amounts of water and contaminants and continuously monitoring them, then reducing the amounts by in-plant operating and equipment changes.

There are many ways in a refinery to reduce the amount of wastewater flows and contaminants. These can include good housekeeping, process modifications, and recycle–reuse.

Good Housekeeping

Good housekeeping can play an important role in reducing unnecessary flows that must be treated downstream. Good housekeeping practises include minimizing waste when sampling product lines; shutting off washdown hoses when not in use; having a good maintenance program to keep the refinery as leakproof as possible; and individually treating waste streams with special characteristics, such as spent cleaning solutions [35].

Many more things can be done; here are just a few. The use of dry cleaning, without chemicals, aids in reducing water discharges to the sewer. Using vacuum trucks to clean up spills, then charging this recovered material to slop oil tanks, reduces the discharge of both oil and water to the wastewater system. Process units should be curbed to prevent the contamination of clean runoff with oily storm runoff and to prevent spills from spreading widely. Sewers should be flushed regularly to prevent the buildup of material, eliminating sudden surges of pollutants during heavy rains. Collection vessels should be provided whenever maintenance is performed on liquid processing units, to prevent accidental discharges to the sewers.

Housekeeping practises within a refinery can have substantial impact on the loads discharged to the wastewater facilities. Knowlton [36] reported how source control by good housekeeping helped a Chevron refinery meet new NPDES permit requirements. Good housekeeping practises to reduce wastewater loads require judicious planning, organization, and operational philosophy. They also require good communication and education for all personnel involved. A refinery newsletter is a good tool to communicate and educate refinery personnel on pollution control issues.

Process Modifications

Many new and modified refineries incorporate reduced water use and pollutant loading into their process and equipment design. Modifications include:

1. Substitution of improved catalysts, which require less regeneration and thus lower wastewater loads.
2. Replacement of barometric condensers with surface condensers or air fan coolers, reducing a major source of oil–water emulsion.
3. Substitution of air cooling devices for water cooling systems.
4. Use of hydrocracking and hydrotreating processes that produce lower wastewater loadings than existing processes.
5. Improved drying, sweetening, and finishing procedures to minimize spent caustics and acids, water washes, and filter solids requiring disposal.

Wastewater Recycle–Reuse

Wastewater reuse is a good way to reduce overall pollutant loadings. However, water quality is critical in water reuse. The contaminants present must be compatible with the reuse. For example, reuse waters with high solids content are not satisfactory for crude unit desalting. Stripped foul water containing low H_2S and ammonia and high concentrations of phenols has essentially no solids. It is suitable for crude unit desalter wash water if the phenols extracted by the crude are subsequently converted by hydroprocessing units into nonphenolic compounds [36]. Some other examples include:

1. Use of recycling cooling towers to replace a once-through cooling system.
2. Reuse of cooling tower blowdown as seal water on high-temperature pumps, where mechanical seals are not practicable.
3. Use of stripped sour water as low-pressure boiler makeup.
4. Reuse of wastewater treatment plant effluent as cooling water, as scrubber water, or as plant makeup water.
5. Putting high-pressure water in cokers through a gravity separator to remove floating oil and settleable coke fines.

4.6.2 Segregation and Pretreatment

The first step in good pretreatment practise is the segregation of major wastewater streams. This frequently simplifies waste treating problems as well as reducing treatment facility costs. Treatment at the source is also helpful in recovering byproducts that otherwise would not be economically recovered from combined wastes downstream [35]. Four major pretreatment processes that are applicable to individual process effluents or groups of effluents within a refinery are sour water stripping, spent caustics treatment, ballast water separation, and slop oil recovery. These are discussed below.

Sour Water Stripping

Many processes in a refinery use steam as a stripping medium in distillation and as a diluent to reduce the hydrocarbon partial pressure in catalytic or thermal cracking [37]. The steam is eventually condensed as a liquid effluent commonly referred to as sour or foul water. The two most prevalent pollutants found in sour water are H_2S and NH_3 resulting from the destruction of organic sulfur and nitrogen compounds during desulfurization, denitrification, and hydrotreating. Phenols and cyanides also may be present in sour water.

The purpose of sour water pretreatment is to remove sulfides (H_2S, ammonium sulfide, and polysulfides) before the waste enters the sewer. The sour water can be treated by stripping with steam or flue gas, air oxidation to convert sulfides to thiosulfates, or vaporization and incineration.

Sour water strippers are designed primarily for the removal of sulfides and can be expected to achieve 85–99% removal. If acid is not required to enhance sulfide stripping, ammonia will also be stripped, the percentage varying widely with stripping pH and temperature. Depending on pH, temperature, and contaminant partial pressure, phenols and cyanides can also be stripped with removal as high as 30%.

There are many different types of strippers, but most of them involve the downward flow of sour water through a trayed or packed tower while an ascending flow of stripping steam or gas removes the pollutants. The stripping medium can be steam, flue gas, fuel gas, or any inert gas. Owing to its higher efficiency, the majority of installed refinery sour water strippers employ steam as both a heating medium and a stripping gas [37]. Some of the steam strippers are provided with overhead condensers to remove the stripping steam from the overhead H_2S and NH_3. The condensed steam is recycled or refluxed back to the stripper. The results of a 1972 survey by the American Petroleum Institute suggested that, overall, refluxed strippers remove a greater percentage of H_2S and NH_3 than nonrefluxed strippers [5].

The operating conditions of sour water strippers vary from 0.1 to 3.5 atm (1–50 psig) and from 38 to 132°C (100–270°F). The sour water may or may not be acidified with mineral acid prior to stripping. H_2S is much easier to remove than NH_3. In pure water at 100°F, for example, the Henry's Law coefficient for NH_3 is 38,000 ppm/psia, whereas that for H_2S is 184 ppm/psia [37]. To remove 90% of the NH_3, a temperature of 110°C (230°F) or higher is usually employed, but 90% or more of the H_2S can be removed at 100°F.

Two-stage strippers are installed in some refineries to enhance the separate recovery of sulfide and ammonia. Acidification with a mineral acid is used to fix the NH_3 in the first stage and allow more efficient H_2S removal. In the second stage the pH is readjusted by adding caustics for efficient NH_3 removal. One example is the Chevron WWT process, which is essentially two-stage stripping with ammonia purification, so that the H_2S and NH_3 are separated. The H_2S goes to a conventional Claus sulfur plant and the NH_3 can be used as a fertilizer [20]. Figure 13 shows a schematic flow diagram of the Chevron WWT process.

Another way to treat sour water is air oxidation under elevated temperature and pressure. Compressed air is injected into the stream followed by sufficient steam to raise the reaction temperature to at least 88°C (190°F). Reaction pressure of 3.7 to 7 atm (50–100 psig) is required. Oxidation proceeds rapidly and converts practically all the sulfides to thiosulfates, and about 10% of the thiosulfates to sulfate [38]. Air oxidation, however, is much less effective than stripping in reducing the oxygen demand of sour waters, as the remaining thiosulfates can later be oxidized to sulfates by aquatic microorganisms. Air oxidation is sometimes carried out after sour water stripping as a sulfide polishing step.

Stripping of sour water is normally carried out to remove sulfides, hence the effluent may contain 50 to 100 ppm of NH_3, or even considerably more, depending on the influent ammonia

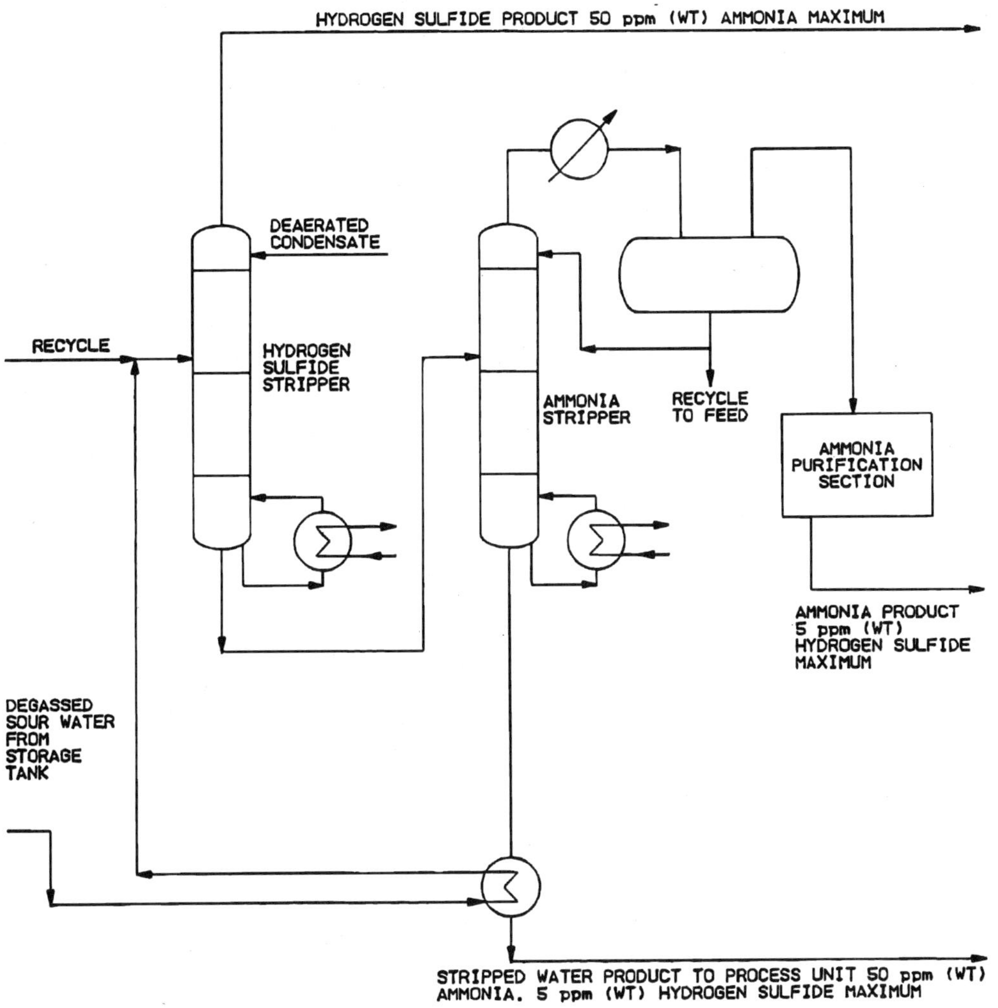

Figure 13 Chevron WWT process. Acid is used in first stage to enhance hydrogen sulfide removal. Caustic is used in second stage to enhance ammonia removal. (From Ref. 20.)

concentration. Values of NH_3 have been reported to be as low as 1 ppm, but generally the effluent NH_3 concentration is held to approximately 50 ppm to provide nutrient nitrogen for the refinery biological waste treatment system. Because of more stringent effluent requirements for NH_3, many refineries seek to improve the sour water stripping systems for NH_3 removal. This can be done by (1) increasing the number of trays, (2) increasing the steam rate, (3) increasing tower height, and (4) adding a second column in series. All these methods are now available to the refining industry [5].

Spent Caustics Treatment

Caustics are widely used in petroleum refineries. Typical uses are to neutralize and to extract acidic materials that may occur naturally in crude oil, acidic reaction products that may be produced by various chemical treating processes, and acidic materials formed during thermal and catalytic cracking such as H_2S, phenolics, and organic acids.

Spent caustics may therefore contain sulfides, mercaptides, sulfates, sulfonates, phenolates, naphthenates, and other similar organic and inorganic compounds [38]. Spent caustics can also be classified as phenolic and sulfidic [37]. Sulfidic spent caustics are rich in sulfides, contain no phenols, and can be oxidized with air. Phenolic spent caustics are rich in phenols and must be neutralized with acid to release and remove the phenols.

At least four companies process spent caustics to market the phenolics and the sodium hyposulfite. However, the market is limited and most of the spent caustics are very dilute, so the cost of shipping the water makes this operation uneconomic. Concentration can be increased by recycling spent caustics at the treater or recycling the spent caustics found in the water bottoms of intermediate product tanks [39].

Some refineries neutralize the caustic with spent sulfuric acid from other refining processes, and charge it to the sour water stripper. This removes the H_2S. The bottoms from the sour water stripper go to the desalter, where the phenolics can be extracted by the crude oil.

Spent caustics usually originate as batch dumps, and the batches may be combined and equalized before being treated and discharged to the refinery sewer. Spent caustics can also be neutralized with flue gas to form carbonates. Sulfides, mercaptides, phenolates, and other basic salts are converted by the flue gas (reaction time 16–24 hours) stripping. Phenols can be removed, then used as a fuel or sold. H_2S and mercaptans are usually stripped and burned in a heater. Some sulfur is recovered from stripper gases. The treated solution contains mixtures of carbonates, sulfates, sulfites, thiosulfates, and some phenolic compounds.

Ballast Water Separation

Ballast water normally is not discharged directly to the refinery sewer system because of the intermittent high-volume discharges [38]. The potentially high contents of salt, oil, and organics in ballast water would upset the treatment facilities if not controlled. Ballast water may also be treated separately by heating, settling, and at times filtration. The settling tank can also be provided with a steam coil for heating the tank contents to help break emulsions, and an air coil to provide agitation. The recovered oil, which may be considerable, is generally sent to the slop oil system.

Slop Oil Treatment

Separator skimmings, which are generally referred to as slop oil, require treatment before they can be reused because they contain an excess amount of solids and water. Solids and water contents of about 1% generally interfere with processing [38].

In most cases slop oils are easily treated by heating to 88°C (190°F) for 12 to 14 hours. At the end of settling, three definite layers exist: a top layer of clean oil; a middle layer of secondary emulsion; and a bottom layer of water containing soluble components, suspended solids, and oil. It may be advantageous or even necessary to use acid or specific chemical demulsifiers to break slop oil emulsions. The water layer has high BOD and COD contents, but also low pH (after acid treatment), and must be treated before it can be discharged. Slop oil can also be successfully treated by centrifugation or by precoat filtration using diatomaceous earth.

4.6.3 End-of-Pipe Treatment

Conventional refinery wastewater treatment technology is mainly concerned with removing oil, organics, and suspended solids before discharge. However, because of new stringent discharge requirements for specific toxic constituents as well as whole-effluent toxicity, specific advanced treatment processes are becoming a necessity for many refineries. This section describes the conventional treatment processes used in refineries. Specific advanced treatment processes are described in the next section.

Conventional refinery wastewater treatment processes can be categorized into primary, intermediate, secondary, and tertiary treatment processes [17]. Primary processes include API separators and parallel or corrugated plate interceptors (CPI) to remove free oil. Intermediate processes include dissolved air flotation (DAF) or induced air flotation (IAF) and equalization. Secondary processes include biological treatment processes in their different forms or combinations. These can include activated sludge, trickling filters, aerated lagoons, stabilization, and rotating biological contactors (RBC). Tertiary treatment processes include filtration and granular activated carbon (GAC) adsorption. Activated sludge enhanced with powdered activated carbon (PACT®), a combination of secondary and tertiary processes, is discussed in the next section.

API Separators

The API separator is a widely used gravity separator for removal of free oil from refinery wastewater. It can be installed either in the central wastewater treatment plant or as an upstream pretreatment process to remove gross quantities of free oil and solids.

The process involves removal of materials less dense than water (such as oil) and suspended materials that are more dense than water by settling. The API separator does not separate substances in solution, nor does it break emulsions. The effectiveness of a separator depends on the temperature of the water, the density and size of the oil globules, and the amounts and characteristics of the suspended materials. The susceptibility to separation (STS) test is normally used as a guide to determine what portion of the influent to a separator is amenable to gravity separation [38]. In terms of globule size, an API separator is effective down to globule diameters of 0.015 cm (15 microns).

The API has long been active in the study of oil–water separators. Its design recommendations are clearly and adequately set forth in the API manual [40]. The basic design of an API separator is a long rectangular basin, with enough detention time for most of the oil to float to the surface and be removed. Most API separators are divided into more than one bay to maintain laminar flow within the separator, making the separator more effective. They are usually equipped with scrapers to move the oil to the downstream end of the separator where it is collected in a slotted pipe or on a drum. On their return to the upstream end, the scrapers travel along the bottom moving the solids to a collection trough. Sludge can be dewatered and either incinerated or disposed of in hazardous waste landfills. To control volatile organic compound emissions to the atmosphere, U.S. refineries are required to install covers for oil–water separators (40 CFR Part 60).

Because of the limitations in gravity separator design, the lower limit of free oil in API separator effluent is usually around 50 mg/L. Removal of other contaminants in an API separator is highly variable. Table 16 shows typical removal efficiencies of oil separator units for several contaminants [17]. Chemical oxygen demand removal efficiencies range from 16 to 45%, and suspended solids removal ranges from 33 to 68%.

Parallel and Corrugated Plate Separators

Parallel and corrugated plate separators are improved types of oil–water separators with tilted plates installed at an angle of 45°. This increases the collection area many times while decreasing the overall size of the unit accordingly. As the water flows through the separator, the oil droplets coalesce on the underside of the plates and travel upward to where the oil is collected.

Table 16 Typical Efficiencies of Oil Separation Units

Oil content					
Influent (mg/L)	Effluent (mg/L)	Oil (percent removed)	Type of separator	COD (percent removed)	SS (percent removed)
300	40	87	Parallel plate	–	–
220	49	78	API	45	–
108	20	82	Circular	–	–
108	50	54	Circular	16	–
98	44	55	API	–	–
100	40	60	API	–	–
42	20	52	API	–	–
2,000	746	63	API	22	33
1,250	170	87	API	–	68
1,400	270	81	API	–	35

COD, chemical oxygen demand; SS, suspended solids.
Source: From Ref. 17.

Because of the coalescing action, these separators can separate oil droplets as small as 0.006 mm (6 microns) in diameter and produce effluent-free oil concentrations as low as 10 mg/L [27].

There is a broad range applications for tilted-plate separators. As little space is required, they can be installed to polish the effluent from existing API separators that are either overloaded or improperly designed, or they can be installed parallel with existing separators, reducing the hydraulic load and enhancing the oil removal capacity of the system.

Dissolved Air Flotation

Dissolved air flotation (DAF) is a process commonly used in refineries to enhance oil and suspended solids from gravity-separator effluent. In some refineries it is used as a secondary clarifier for activated sludge systems and as a sludge thickener. The process involves pressurizing the influent or recycled wastewater at 3–5 atm (40–70 psig) then releasing the pressure, which creates minute bubbles that float the suspended and oily particulates to the surface. The float solids are removed by a mechanical surface collector.

If a significant portion of the oil is emulsified, chemical addition with rapid-mix and flocculation chambers are a part of the flotation unit, breaking the emulsion and enhancing the separation. Chemicals normally used include salts of iron and aluminum and polyelectrolytes.

Dissolved air flotation in combination with flocculation can reduce oil content in refinery wastewater to levels approaching oil solubility [40]. According to Katz [41], DAF plus chemical aids for flocculation can be expected to reduce BOD and COD by 30–50% and to reduce total oil to the range 5–25 mg/L. Table 17 shows some data for oil removal from refinery wastewater [27]. Removal efficiencies range from 70 to 90%. The accepted design overflow rates for DAF units are between 60 and 120 L/min per square meter (1.5–3.0 gpm/sq ft) [17].

Dissolved air flotation equipment is available from a number of manufacturers. Packaged units of steel construction are available with capacities to 7.6 cu m/min (2000 gpm). The essential elements of the DAF system are the pressurizing pump, air injection facilities, pressurization tank or contact vessel, back-pressure regulating device, and the flotation chamber [40].

Three principal variations in the process design of DAF systems are full-flow, split-flow, and recycle operation (Fig. 14). Full-flow operation consists of pressurizing the entire waste

Table 17 Oil Removal by Dissolved Air Flotation in Refineries

Coagulant dosage (mg/L)	Oil concentration (mg/L)		Removal (percent)
	Influent	Effluent	
0	125	35	72
100 (alum)	100	10	90
130 (alum)	580	68	88
0	170	52	70

Source: From Ref. 27.

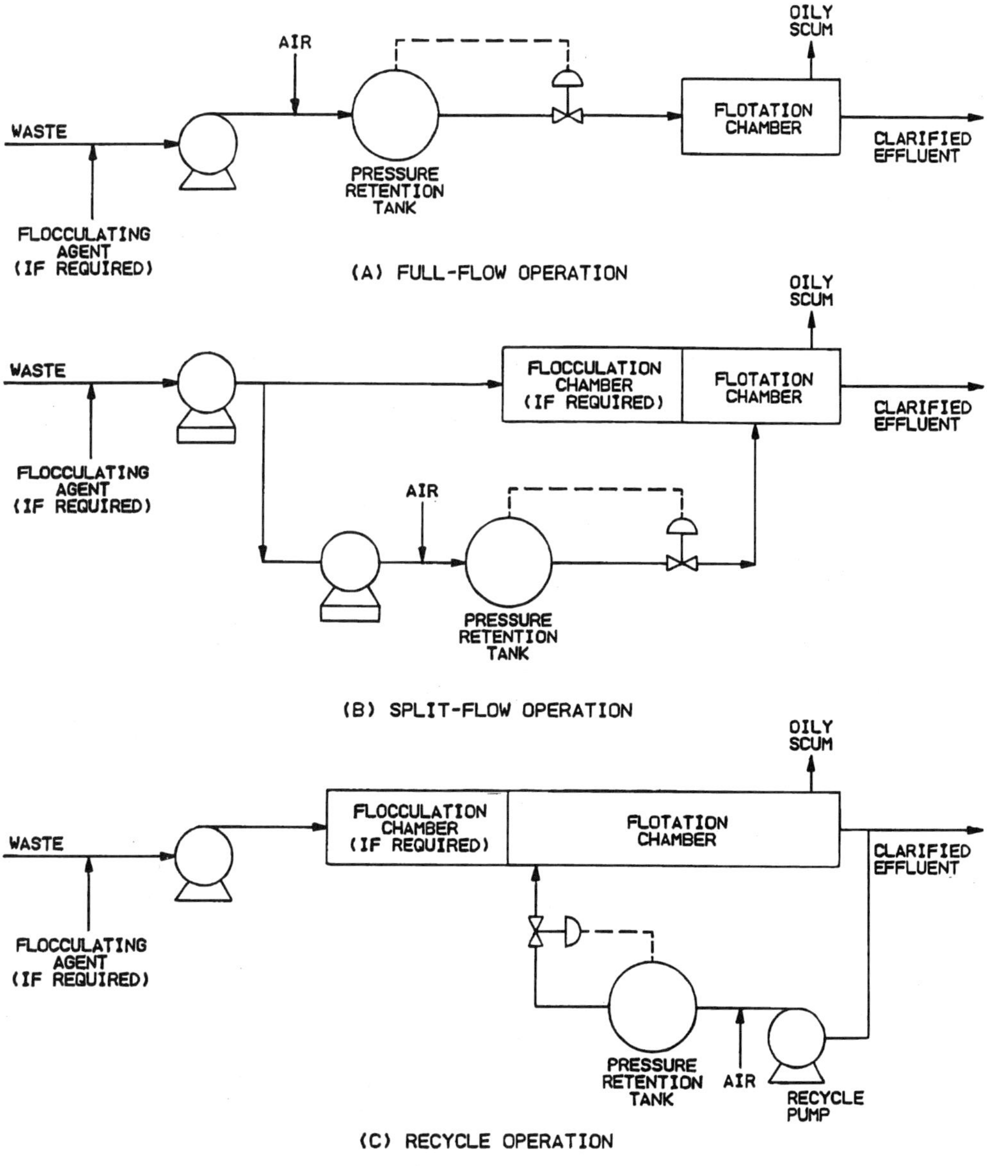

Figure 14 Variations in dissolved air flotation (DAF) design. (A) Full-flow operation; (B) split-flow operation; (C) recycle operation. (From Ref. 40.)

stream, followed by release of pressure and bubble formation at the inlet to the flotation chamber. Split operation consists of pressurizing only part of the waste flow and diverting the remainder directly into the flocculation or flotation chamber. Recycle operation consists of pressurizing a recycle stream of the clarified effluent. The recycled stream usually amounts to between 20 and 50% of the oily wastewater flow. The pressure is released and the bubble-containing recycle stream is mixed with the unit influent. Each of these variations has its advantages and disadvantages.

A relatively new design of a high-rate DAF unit uses a shallow bed system (Supracell) with only 3 minutes of retention time and operated at an overflow rate of 140 Lpm/sq m (3.5 gpm/sq ft) [42]. This unit has been used for industrial and municipal wastewater treatment and offers lower capital cost and headroom requirements. It was installed at a petrochemical complex in Texas as a secondary clarifier to improve the operation and the capacity of an existing activated sludge system [43]. In recent years, nitrogen has replaced air in covered DAF systems because of the potential for explosion. These systems are called dissolved nitrogen flotation (DNF) systems. The operations of DAF and DNF are similar.

Induced Air Flotation

The induced air flotation (IAF) system operates on the same principles as a pressurized DAF unit [27]. The air, however, is typically mixed into the effluent by a rotor-disperser mechanism. The rotor, which is submerged in the liquid, forces the liquid through the disperser openings, thereby creating a negative pressure. This pulls the air downward into the liquid, causing the desired gas–liquid contact. The liquid moves through the flotation cell(s), and the float skimmings pass over the overflow weirs on each side of the unit.

The advantages of an IAF unit include significantly lower capital cost and smaller space requirements than a DAF unit. On the other hand, it has a higher connected power requirement than a DAF unit. It also has a higher volume of float skimmings: the normal range is 3–7% of the incoming flow for IAF units and less than 1% for DAF units [27]. Induced air flotation units have been used in petroleum refineries and oilfields for removing free oil and suspended solids.

Equalization

The purpose of equalization is to dampen out surges in flows and loadings to maintain optimum conditions for subsequent treatment processes. This is especially necessary for a biological treatment plant, as high concentrations of certain materials will upset or completely kill the bacteria in the biotreater. Many wastewater discharges within refinery complexes are from washdowns, tank cleanings, batch operations, and inadvertent spills, necessitating a basin capable of receiving these waters and allowing their controlled release [17].

The equalization step in a refinery usually consists of one or more large ponds or tanks that may contain mixers to stir the wastes. Many refineries are planning to or have replaced ponds with steel tanks because of the requirements for groundwater protection. The use of covered and vented tanks also provides more positive control of odors from equalization systems when anaerobic conditions develop.

Equalization basins may be designed to equalize flow, concentrations, or both [27]. For flow equalization, the cumulative flow is plotted vs. time over the equalization period, which is usually 24 hours. The maximum volume above the constant-discharge line is the equalization volume required. The basin may also be sized to restrict the discharge to a maximum concentration of a critical pollutant. For example, if the maximum effluent from an activated sludge unit is 20 mg/L BOD_5, the maximum allowable effluent from the equalization basin may be computed and thereby provide a basis for sizing the unit. Novotny and England [44] suggested a formula for computing the required equalization time for the case of near-constant wastewater flow and a normal statistical distribution of wastewater composite analyses.

Activated Sludge

Activated sludge is the most common biological treatment process because of the high rate and degree of organic stabilization possible. It is widely used in treating refinery wastewater [5].

Activated sludge is an aerobic biological treatment process in which high concentrations (1500–3000 mg/L) of newly grown and recycled microorganisms are suspended uniformly throughout a holding tank to which raw wastewaters are added. Oxygen is introduced by mechanical aerators, diffused air systems, or other means. The organic materials in the waste are removed by the microbiological growth and stabilized by biochemical synthesis and oxidation reactions. The term activated stems from the fact that the microbial sludge is a floc that is highly active in adsorbing colloidal and suspended waste matter from the aqueous stream [25].

The basic activated sludge system consists of an aeration tank followed by a sedimentation tank. The microbial floc removed in the sedimentation tank is recycled to the aeration tank to maintain a high concentration of active microorganisms. Although the microorganisms remove almost all of the organic matter from the waste being treated, much of the converted organic matter remains in the system in the form of microbial cells. Because of their oxygen demand, these cells must be removed from the treated wastewater before discharge. Thus, final sedimentation and recirculation of biological solids are important elements in an activated sludge system.

Although refinery wastewaters are generally highly amenable to activated sludge treatments, the exact treatability of a refinery-petrochemical installation is a function of many factors such as the classification of the refinery, the type of crude charge, the age of the facility and nature of its collection system, the relative effluent volume attributed to utility blowdown, and the degree of in-plant control. For these reasons, activated sludge facilities vary from one installation to another [17]. Treatability studies using bench- and pilot-scale trials therefore are used to formulate the basic design criteria and predict treated effluent quality.

The three basic types of activated sludge systems are conventional, contact stabilization, and extended aeration systems [40]. Other types include high-purity oxygen systems and sequencing batch reactors, but these are not commonly used in refineries.

The conventional activated sludge system allows for absorption, flocculation, and synthesis in a single step. It usually employs long, rectangular aeration tanks that approximate plug flow conditions, or crossflow aeration tanks that approach complete mixing. The oxygen utilization rate is high in the beginning of the aeration tank, but decreases with aeration time or distance down the tank. Where complete treatment is required, the oxygen utilization rate approaches the endogenous value toward the end of the aeration tank. The conventional process can operate over a wide loading range, which is limited by flocculation, settling, and separation requirements of the microbial flocs.

The contact stabilization system provides for removal of the organics from the wastewater by contact with activated sludge (absorption) and transfer to a separate aeration tank for oxidation and synthesis. This process is applicable to wastes containing a large proportion of the BOD in suspended or colloidal form. The influent is first contacted with the activated sludge in an aeration basin of a relatively short retention period (15–30 minutes). This contact basin removes the suspended or colloidal content from the stream by absorption on the sludge floc. The mixed liquor flows to the settler-thickener where the clarified effluent overflows and the thickened sludge flows to a stabilization basin. A small part of the thickened sludge is discarded as waste. The recycled sludge is aerated in the stabilization basin for 1–5 hours. During this period, the adsorbed organics undergo synthesis and endogenous respiration and the sludge becomes stabilized. This process results in savings in total basin area as only the recycled sludge, not the whole waste stream, is subjected to long-time aeration. However, if the oxygen demand of the influent is due mostly to dissolved rather than suspended contaminants, the short retention period in the contact basin may not produce a satisfactory effluent [25].

The extended aeration system is one in which the synthesized cells undergo autooxidation, resulting in a minimum of solids disposal. Extended aeration is reaction-defined rather than a hydraulically-defined mode and can be designed as a plug flow or a complete mix system. Design parameters include a food/microorganism ratio (F/M) of 0.05–0.15, a sludge age of 15–35 days, and mixed-liquor suspended solids (MLSS) concentrations of 3000–5000 mg/L [27]. This process has low cell growth rates, low sludge yields, and high oxygen requirements compared with the conventional activated sludge process. The advantages are high-quality effluent and less sludge production.

The extended aeration process can be sensitive to sudden increases in flow due to resultant high-MLSS loadings on the final clarifier, but is relatively insensitive to shock loads in concentrations due to the buffering effect of the large biomass volume. Because of the long sludge age, nitrification can be incorporated into the design of the extended aeration process. Extended aeration in the form of loop-reactor or ditch systems has been used significantly in wastewater treatment during recent years.

Other variations of activated sludge such as deep shaft high-rate activated sludge and sequencing batch reactor (SBR) have been used for refinery wastewater treatment. A refined deep shaft process has been installed and in operation at the Chevron refinery in Burnaby, British Columbia, Canada, since 1996 [45]. In the course of a recent wastewater treatment upgrade, a BP refinery on the eastern Australian coast converted an existing lagoon to an SBR system [46].

The design organic load for most activated sludge systems ranges from 0.1 to 1.0 lb BOD_5/(day)(lb MLSS) [17]. Higher loadings can be imposed, but generally at the expense of poorer efficiency and higher organic levels in the treated effluent. Table 18 shows the performance of typical activated sludge systems in refineries based on loading and retention time [40].

One particularly important parameter for the influent to an activated sludge system in a refinery is oil and grease, which can lower floc density to a level where the sludge-settling properties are destroyed. A study conducted for USEPA [47] indicated that an activated sludge

Table 18 Performance of Typical Activated Sludge Systems

Type of waste	Sludge loading (lbs of BOD/day/lbs of sludge)	Detention time (hours)	Percentage of BOD reduction	Comment
Refinery	0.6	4–5	90–95	99% phenol removal
Refinery	0.3–0.4	3–4	90–95	98% phenol removal
Refinery	0.1–0.2	18–22	88–92	Minimal sludge production
Petrochemical	0.65–0.76	8–10	95–97	Sludge bulks for long periods

BOD, biochemical oxygen demand.
Source: From Ref. 40.

system will perform satisfactorily with continuous loading of hexane extractables of 0.1 lb/lb MLSS. It was recommended that the influent to the biological system should contain less than 75 mg/L hexane extractables and preferably less than 50 mg/L.

Aerated Lagoon

Aerated lagoons are low-rate biological systems in which a flow-through basin allows microorganisms in contact with the wastewater to reduce organic constituents biochemically. Unlike activated sludge there is no solids recycle. Retention times are usually between 3 and 10 days [38]. Oxygenation and mixing can be carried out with mechanical or diffused aeration units and through induced surface aeration. Depths of 3–4.3 m (10–14 ft) are used to accommodate the aeration equipment and minimize area requirements [40].

Aerated lagoons have been extensively used in refineries to treat wastewaters because of their ease of operation and maintenance, ability to equalize wastewater, and ability to dissipate heat when desirable. However, because of their inherent limitations, they are usually used upstream from waste stabilization ponds or as an interim treatment process that can be converted to an activated sludge system. BOD_5 reductions in completely mixed aerated lagoons may range from 40 to 60%, with little or no reduction in suspended solids [38]. Because of more stringent effluent discharge requirements, many lagoons have been converted to other more effective processes such as SBR, as discussed previously in the activated sludge section [46].

Waste Stabilization Ponds

A stabilization pond is a simple pond in which aeration is not mechanically enhanced. Its shallow depth allows the pond to function aerobically without mechanical aerators. Algae in the pond produce oxygen through photosynthesis, which is then used by the bacteria to oxidize the wastes. Because of the low loadings, little biological sludge is produced and the pond is fairly resistant to upsets due to shock loadings.

The stabilization pond is practical where land is plentiful and cheap. It has a large surface area and a shallow depth, usually not exceeding 2 m (6 ft). Stabilization ponds have a long retention, ranging from 11 to 110 days [38], depending on the land available as well as the design requirement.

Stabilization ponds have been successfully used in the treatment of refinery and petrochemical wastewaters. They are used either as the major treatment step or as a polishing process after other treatment processes. In the United States, because land is generally quite expensive, the use of waste stabilization ponds is limited [17].

Trickling Filter

The trickling filter is a packed bed of medium covered with biological slime growth through which the wastewater is passed. Wastewater is sprinkled onto the medium through a rotating distribution system above the bed. As the wastewater passes through the slime, organics and oxygen diffuse into the microbial mass where they are oxidized to carbon dioxide, water, and metabolic byproducts [17]. The trickling filter is followed by a clarifier to settle sloughed-off slimes. Recycle flow may be taken either before or after clarification.

Conventional trickling filters contain 6 to 10 cm (2.5 to 4 in) rocks and vary in depth from 1 to 2.5 m (3–8 ft). Hydraulic loadings are 20 Lpm/sq m (0.5 gpm/sq ft) or less. Plastic packings are employed in depths up to 12 m (40 ft), with hydraulic loadings as high as 240 Lpm/sq m (6.0 gpm/sq ft) [40]. Trickling filters are fixed reactors and are simple to operate. However, the reaction rate for treating soluble industrial wastewaters is relatively low, hence they are not economically attractive for high treatment efficiency (85% BOD reduction) of such wastewaters [27].

The petroleum industry uses them mostly as roughing devices to reduce the loading on activated sludge systems. In some cases, trickling filters are used to pretreat steam-stripped sour water before mixing it with other refinery wastewater streams for secondary treatment [48].

Rotating Biological Contactors

Rotating biological contactors (RBCs) have attracted widespread attention in the United States since 1969 [5]. RBCs generally consist of rows of plastic discs mounted on horizontal shafts that turn slowly keeping the disc about 40% immersed in a shallow tank containing wastewater as shown in Fig. 15. A 1 to 4 mm layer of slime biomass is developed on the media. This is equivalent to 2500–10,000 mg/L of MLSS in a mixed system [27]. Single RBC units are up to 3.7 m (12 ft) in diameter and 7.6 m (25 ft) long, containing up to 9300 square meters (100,000 sq ft) of surface in one section.

The RBC is a combination of fixed film reactor and mechanical aerator. The fixed film reactor is the disc upon which microorganisms attach themselves and grow. Aeration occurs while a section of disc is above water level. Microorganisms produce a film on the surface of the disc, remove organic matter from the wastewater, and accumulate on each disc. Excess biomass is stripped and returned to the wastewater stream by the shearing action of water against rotating discs. Waste biomass is held in suspension by the mixing action of the discs, and carried out of the reactor for removal in a clarifier. Treatment efficiency can be improved by increasing the number of RBCs in series, and by temperature control, sludge recycle, and chemical addition.

Advantages of RBCs include the ability to sustain shock loads because of high microorganism concentrations, ease of expansion because of modular design, and low power consumption, which may be particularly attractive for industrial application. Full-scale RBC installations in refineries have performances in removal of oxygen-demanding pollutants comparable to activated sludge systems [5].

Filtration

The use of filtration to polish biological treatment system effluent has become more popular in recent years because of more stringent discharge requirements. The 1977 EPA survey of petroleum refineries indicated that 27 of 259 plants used filtration as part of the existing treatment scheme and 16 others planned to install filtration systems in the near future [5]. Filtration can improve effluent quality by removing oil, suspended solids, and associated BOD and COD, and carryover metals that have already been precipitated and flocculated. Improved effluent filtration in one recent instance helped a Colorado refinery to meet the newly adopted discharge toxicity requirements [49].

Granular-medium filters are the predominant type of filtration systems used in refineries. The medium can be sand, dual medium of anthracite and sand, or multimedium of anthracite, sand, and garnet. As the water passes down through a filter, the suspended matter is caught in the pores. When the pressure drop through the filter becomes excessive, the flow is reversed to remove the collected solids. The backwash cycle occurs approximately once a day, depending on the loading, and usually lasts for 10 to 15 minutes. The normal surface loading rate is between 80 and 200 Lpm/sq m (2–5 gpm/sq ft). Coagulants such as iron and aluminum salts and polyelectrolytes can enhance suspended solids removal.

Several advanced filtration systems are finding applications in treating refinery wastewaters. Examples include the HydroClear filter (Zimpro, Rothschild, WI) and the Dynasand filter (Parkson Corporation, Fort Lauderdale, FL). The HydroClear filter employs a single sand medium (0.35–0.45 mm) with an air mix (pulsation) for solids suspension and regeneration of the filter surface. Filter operation enables periodic regeneration of the medium

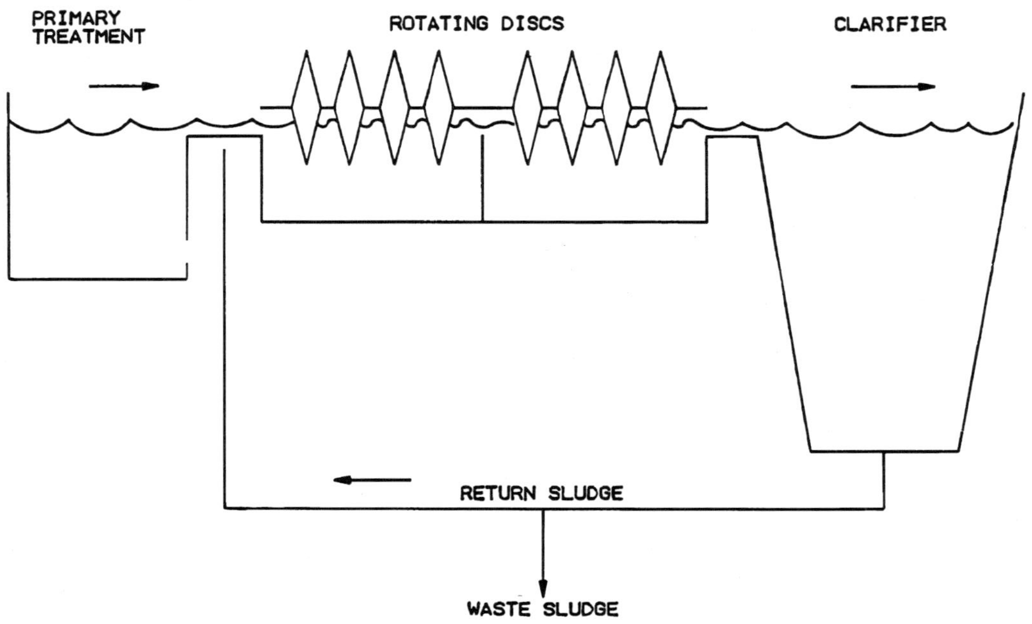

Figure 15 Rotating biological contactors. Plastic discs rotate slowly in a shallow tank. About 60% of each disk is above water surface for aeration. (From Ref. 5.)

surface without backwashing. The Dynasand filter is a continuous self-cleaning upflow deep-bed granular-medium filter [27]. The filter medium is cleaned continuously by recycling the sand internally through an airlift pipe and sand washer. The regenerated sand is redistributed on top of the bed, allowing for a continuous uninterrupted flow of filtered water and reject water.

Activated Carbon Adsorption

Activated carbon adsorption is most often employed for removal of organic constituents from water and wastewater. Granular activated carbon (GAC) or powdered activated carbon (PAC) may be used. Granular activated carbon columns can be used for secondary treatment of industrial wastewaters or for tertiary treatment to remove residual organics from biological treatment effluent. The primary use of PAC in wastewater treatment has been in the PACT® process (Zimpro), in which PAC is added to the activated sludge process for enhanced performance. This process is discussed in the next section of this chapter.

A GAC system is generally preceded by a filtration system to remove suspended solids to minimize plugging of the adsorption sites (pores). Filtered water flows to a bank of GAC columns arranged in series or parallel. As the water flows through the columns the pollutants are adsorbed onto the carbon, gradually filling the pores. The exhausted carbon is removed for regeneration in a furnace or disposed of in appropriate landfills. Figure 16 shows the process flowsheet for a GAC system with a regeneration system.

The adsorption of organics from the liquid to a solid phase is generally assumed to occur in three stages [50]. The first is the movement of the contaminant (adsorbate or solute) through a film surface surrounding the solid phase (adsorbant). The second is the diffusion of the adsorbate within the pores of the activated carbon. The final stage is the sorption of the material onto the surface of the sorbing medium. The overall rate of adsorption is controlled by the rate of diffusion of the solute molecules within the capillary pores of the carbon particles [27].

Adsorption can be divided into two types. Chemical adsorption results in the formation of a monomolecular layer of the adsorbate on the surface through forces of residual valence of the surface molecules. Physical adsorption results from molecular condensation in the capillaries of the solid. In general, substances of the highest molecular weight are most easily adsorbed [27].

Currently the use of full-scale GAC systems in the U.S. petroleum refining industry is very limited. Some refineries used GAC as the secondary treatment process but have discontinued the operations. Two examples are the Atlantic Richfield (Arco) system near Wilmington, CA, and the British Petroleum (BP) system in Marcus Hook, PA [17].

The Arco GAC system was designed to treat 50 MGD of combined storm runoff and process water during periods of rainfall when the treatment plant of Los Angeles County Sanitation District (LACSD) cannot accommodate the storm runoff from the refinery. The GAC system included 12 adsorber cells, a carbon handling system, and a multiple-hearth regeneration system. The design was based on COD removal of 85% at an average influent concentration of 250 mg/L. The operating results indicated that the effluent COD was in the range of the predicted level when the influent concentration did not exceed the design basis. However, the carbon consumption rate ranged from 0.30 to 0.35 kg COD removed per kg of carbon, rather than the 1.75 kg COD/kg carbon predicted. The system is no longer in operation primarily because the treatment requirements imposed by LACSD have been changed.

The BP refinery used a filtration/GAC system to treat API separator effluent before discharge. It consisted of three parallel adsorbers each containing 42,000 kg of carbon in beds 14 m (45 ft) deep. The design contact time was 40 minutes and theoretical carbon capacity was 0.3 kg TOC/kg carbon. The regeneration facility was a 1.5 m diameter, multiple-hearth furnace. After several years of operation, BP abandoned the GAC system and installed a biological

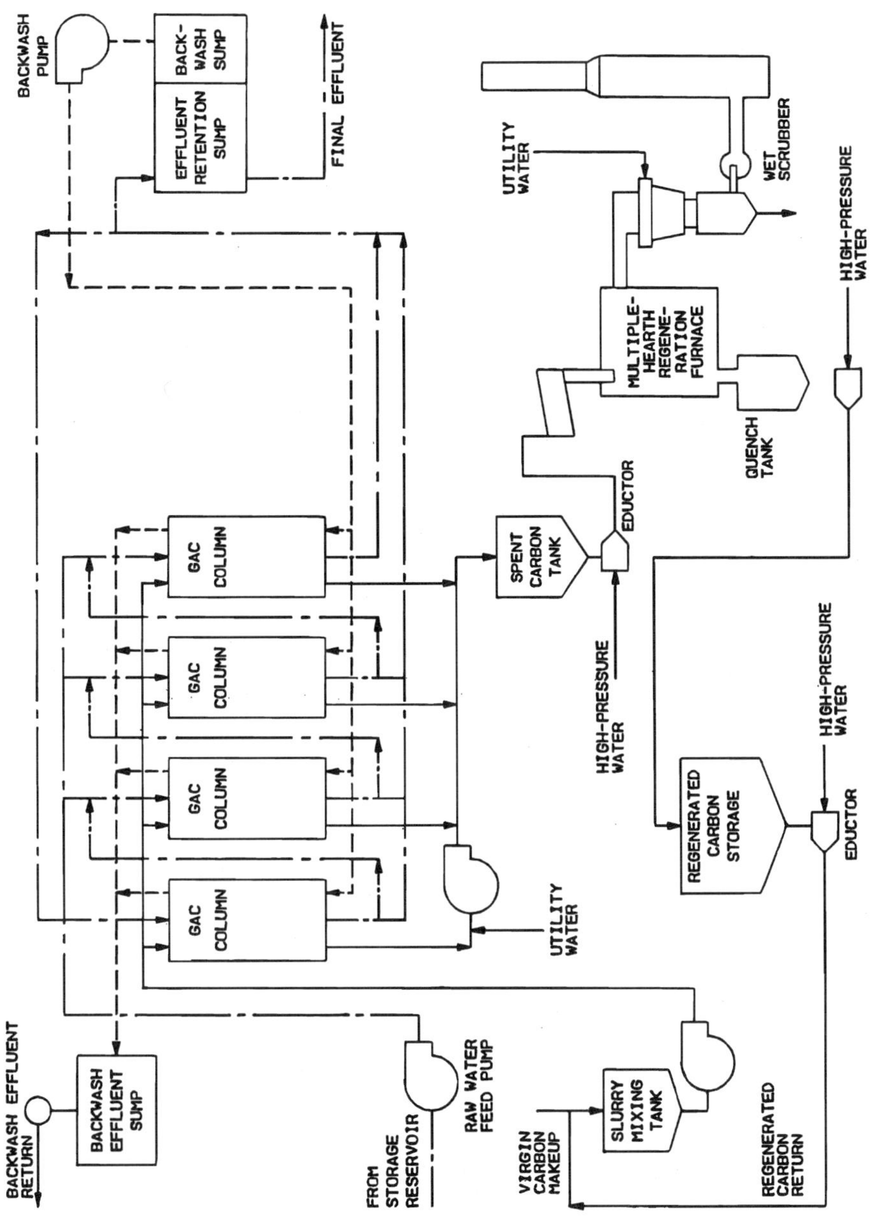

Figure 16 Process flowsheet of a GAC system with regeneration. In this complete GAC adsorption and regeneration system, four GAC columns can be operated in parallel or in series. Spent carbon is transferred to a multiple-hearth furnace for thermal regeneration. Regenerated carbon is mixed with virgin makeup and pumped back to the GAC columns. The GAC columns are backwashed periodically. (From Ref. 27.)

treatment system for secondary treatment because of operational problems including inadequate pretreatment of the API separator effluent in terms of O&G and soluble organics removal, buildup of anaerobic biological growths and oily materials in the carbon media, and a 40% decrease in adsorptive capacity of the regenerated carbon.

The use of GAC systems to follow biological treatment processes is a more promising application. Adding GAC as a polishing process may be necessary in the future in certain refineries to meet more stringent discharge requirements for toxic constituents. In pilot studies of GAC as a tertiary treatment process for refinery and petrochemical plants, carbon adsorption following biological treatment was particularly effective in reducing both BOD and COD to low levels; Table 19 shows the results for COD removal in some of these studies [51]. Activated carbon also removes a variety of toxic organic compounds from water and wastewater [52]. More discussions of GAC for control of whole effluent toxicity are presented in the next section.

4.6.4 Specific Advanced Treatment Processes

Many refineries in the United States are being required to control whole-effluent toxicity as well as specific toxic constituents to meet new wastewater discharge limits. There can be a variety of toxic constituents that may need to be controlled, depending on waste characteristics and local water quality objectives. The more common constituents in refinery wastewater include cyanide and heavy metals. The treatment processes for control of whole-effluent toxicity, cyanide, and heavy metals are discussed below.

Control of Whole Effluent Toxicity

Any treatment process that can remove the toxicity-causing constituents can reduce whole-effluent toxicity of a discharge. If the primary cause of effluent toxicity can be identified through the TIE or TRE procedures, specific treatment processes can be incorporated into the existing treatment system to control the toxicity. However, for a complex wastewater such as that from refinery and petrochemical facilities, the cause of toxicity may not be easily identified. The toxicity can be caused by a combination of constituents that exhibit synergistic or antagonistic effects.

The PACT® process has great potential for controlling whole-effluent toxicity in refinery wastewater. It involves the addition of powdered activated carbon (PAC) to the activated sludge process for enhanced performance [53]. Figure 17 shows the process flow diagram of the PACT® process. The addition of PAC has several process advantages: decreasing variability in effluent quality, removing nondegradable organics by adsorption, reducing inhibitions in industrial wastewater treatment (e.g., nitrification), and removing refractory priority pollutants [27].

Table 19 Carbon Pilot Plant Results for Petrochemical and Refining Wastewaters

Type of wastewater	Design Q (MGD)	Process application	Influent COD (mg/L)	Effluent COD (mg/L)	Percent removal
Petrochemical	3	Tertiary	150	49	67
Refinery	26	Tertiary	100	41	59
Refinery	28	Tertiary	300	50	83
Refinery	8	Tertiary	100	40	60
Petrochemical	29	Tertiary	150	48	68

Source: From Ref. 51.

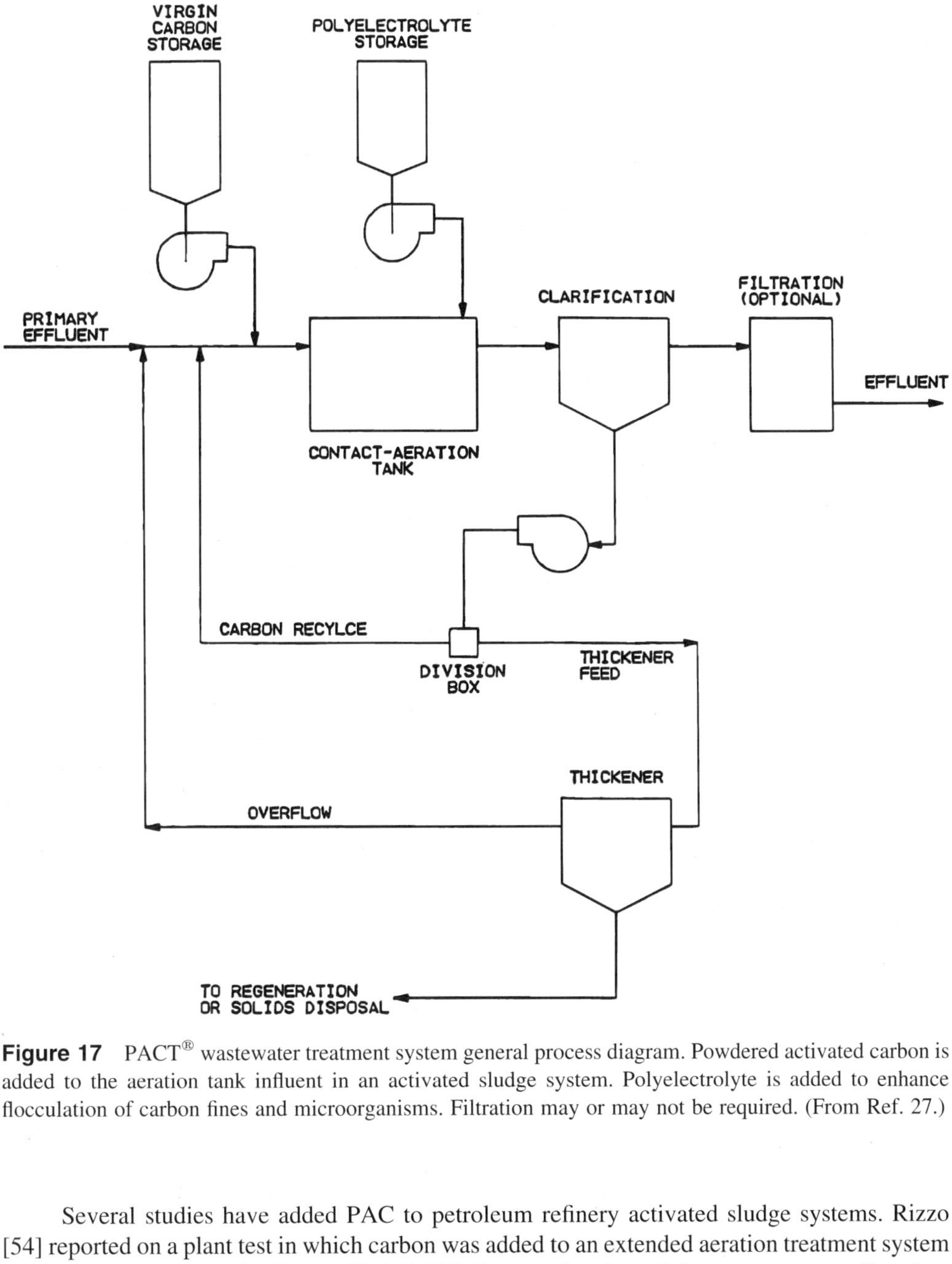

Figure 17 PACT® wastewater treatment system general process diagram. Powdered activated carbon is added to the aeration tank influent in an activated sludge system. Polyelectrolyte is added to enhance flocculation of carbon fines and microorganisms. Filtration may or may not be required. (From Ref. 27.)

Several studies have added PAC to petroleum refinery activated sludge systems. Rizzo [54] reported on a plant test in which carbon was added to an extended aeration treatment system at the Sun Oil Refinery in Corpus Christi, TX. Test results showed that even very small carbon dosages (9–24 mg/L) significantly improved removal of BOD, COD, and total suspended solids, as well as producing uniform effluent quality, a clearer effluent and eliminating foam. Grieves *et al.* [55] reported on a pilot plant study at the Amoco refinery in Texas City where PAC was added to the activated sludge process in 10 gal (37.9 L) pilot plant aerators. Significant amounts of soluble organic carbon (53%), soluble COD (60%), NH_3-N (98%), and phenolics were removed after 50 mg/L of PAC was added. The amounts removed increased with increasing carbon dosage.

Thibault et al. [56] reported on a field-scale test with aerator PAC levels of 1000 mg/L or more in an Exxon refinery. They found significantly improved effluent quality and noted improvement in shock loading resistance leading to process stability. An additional 10% of TOC and COD was removed.

Wong and Maroney [47] reported on a pilot-plant comparison of PACT® and extended aeration for toxicity reduction in wastewater from a West Coast refinery. The average PAC dosage used was approximately 70 mg/L in the influent. Flow-through bioassays were used to monitor the toxicity of the treated effluent. Although both PACT® and extended aeration performed similarly in COD removal, only the PACT® system yielded an effluent meeting the discharge requirements for whole-effluent toxicity. A full-scale PACT® system installed at this refinery has been operating satisfactorily. Similar results in toxicity reduction have been reported for wastewaters from other industries [57].

Butterworth [58] has presented case histories of how refineries have used GAC to achieve compliance with NPDES permit requirements for toxicity. There are five major refineries in the San Francisco Bay Area as of 2003. Because of the stringent toxicity requirements for direct discharge to the Bay, four of the major refineries have installed GAC systems to polish secondary treatment plant effluent prior to discharge (Chevron Texaco, Valero, Tesoro, and Shell Equilon). The one exception is the ConocoPhilips Refinery in Rodeo, which has a PACT® system for organics and toxicity removal. These GAC systems are designed mainly to reduce toxicity rather than COD. The toxicity of treated refinery effluent is believed to be caused mainly by naphthenic acid [59]. The spent GAC from the refineries is regenerated offsite by a contractor. The cost of GAC treatment in these refineries has been lower than anticipated because COD removal is not critical for meeting toxicity requirements, and thus the GAC beds can last much longer between regenerations.

Cyanide Control

Historically, refinery cyanide control was not a concern because cyanide levels in refinery effluent were usually much lower than those in wastewaters from metal finishing and plating industries. Regulatory agencies have now established new and more stringent cyanide effluent limits for most wastewater discharges. One example is the cyanide effluent limit of 0.025 mg/L (as total cyanide) in the San Francisco Bay imposed by the California Water Resources Control Board [60].

Fluid catalytic cracking (FCC) and coker units generate most of the cyanides in refineries [61]. Cracking organic nitrogen compounds liberates cyanide and other nitrogen compounds, such as ammonia and thiocyanates. Figure 18 shows a simplified FCC/coker gas fractionation system and the path the waste stream containing cyanide follows in a typical refinery [62]. The FCC/coker reactor gases, including cyanide and NH_3, go overhead on the fractionation column, where water is injected into the overhead line for corrosion control. This water is collected in an accumulator and pumped to a steam stripper along with other sour water to remove NH_3 and H_2S. Part of the cyanide is also removed. The remaining cyanide goes to the wastewater treatment system where simple cyanide is biodegraded and complexed cyanide may pass through the treatment plant and be discharged.

Because the complexed cyanide species that pass through biological treatment plants are usually very stable, common cyanide removal methods such as chlorination and precipitation do not reduce the effluent cyanide concentrations to below detection limits. Wong and Maroney [62] identified four potential end-of-pipe treatment processes to remove cyanide to very low levels: (1) a cyanide-selective ion exchange resin Amberlite IRA-958 developed by Rohm and

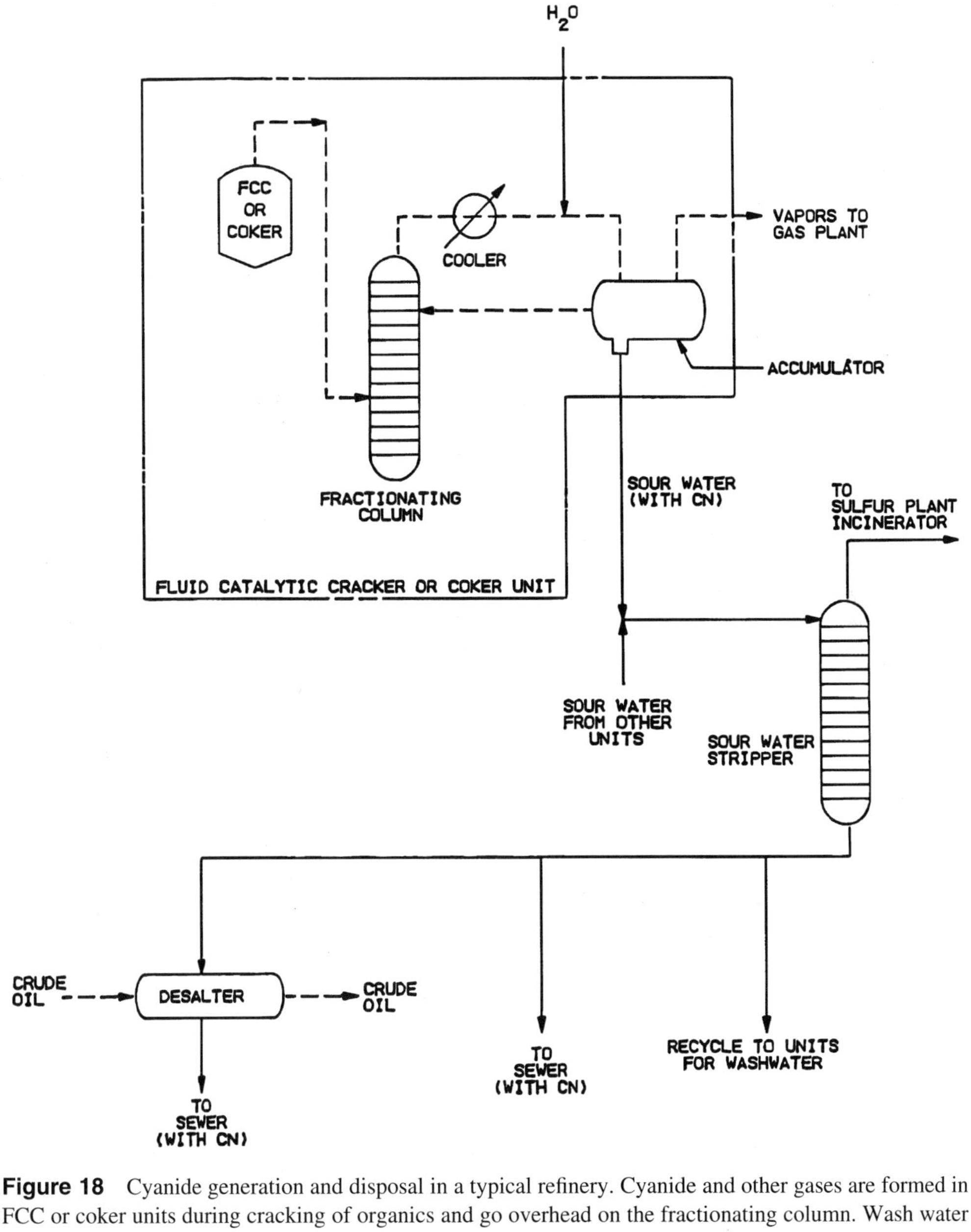

Figure 18 Cyanide generation and disposal in a typical refinery. Cyanide and other gases are formed in FCC or coker units during cracking of organics and go overhead on the fractionating column. Wash water dissolves these gases and becomes sour water. Part of the cyanide is removed by the sour water stripper and the rest goes to the sewer and eventually to the wastewater treatment system. (From Ref. 48.)

Haas Company (Philadelphia, PA), (2) reverse osmosis (RO), (3) adsorption/oxidation with PAC and copper, and (4) ultraviolet irradiation/ozonation (UV/O_3).

All these processes are very expensive for the purpose of removing a small amount of cyanide. The adsorption/oxidation process with PAC and copper could be easily incorporated into existing biological treatment systems; however, the concern of copper toxicity in the final effluent makes this process undesirable.

The most economical cyanide control method in a refinery appears to be upstream control using polysulfides. Sodium and ammonium polysulfide (APS) have been used to inhibit cyanide-induced corrosion in FCC and coker fractionation systems [63]. The polysulfide combines with cyanide, forming thiocyanate according to the reaction

$$CN^- + S_x^{-2} \rightarrow SCN^- + S_{x-1}$$

The thiocyanate is readily biodegradable and is innocuous in refinery effluent. Knowlton *et al.* [63] reported that one large refinery generated several hundred pounds per day of cyanide in its FCC and coker units. When APS solution was used to thoroughly scrub gases produced in the FCC and coker units, the cyanide content in the final effluent was consistently less than the detection limit. The polysulfide treatment method is effective at high temperatures and when the cyanide is still in the free form. However, careful design and operation control are critical to the success of implementing a polysulfide treatment system. Some refineries have reported severe fouling and plugging in the sour water strippers when APS was used [64].

Heavy Metals Removal

Heavy metals such as copper, zinc, lead, nickel, silver, arsenic, selenium, cadmium and chromium may originate from many sources within a refinery and may, in specific cases, require end-of-pipe treatment. Some agencies have set discharge limits that are beyond the capability of common metals removal processes such as lime precipitation and clarification to achieve. Other treatment processes such as iron coprecipitation and adsorption, ion exchange, and reverse osmosis may be required to achieve these low effluent concentrations [52].

The iron coprecipitation and adsorption process involves adding an iron salt such as ferric chloride to the wastewater. The iron hydrolyzes, forming an amorphous iron hydroxide floc. Metals adsorb onto the floc and are removed by clarification or filtration. Cationic metals (e.g., cadmium, zinc) are best adsorbed at high pH and anionic metals (e.g., arsenic, selenium) adsorb better at low pH. The process can remove metals to very low levels, in the ppb range [65]. It operates within the physiological pH range (6–9) and produces relatively few waste solids. This process can be incorporated at one or more points in an existing treatment plant if metals removal is required. The success of the process may depend on the forms of the metallic species and the extent of interference by organics in the wastewater. The iron coprecipitation process has been used successfully in several San Francisco Bay Area refineries to remove selenium to below 50 ppb in treated effluent. Based on bench- and pilot-scale tests in a refinery, a ferric chloride dosage of 50 mg/L as Fe was necessary to achieve the required 50 ppb selenium at all times [66]. The iron coprecipitation system in the Shell Equilon Refinery generates a large amount of iron sludge for disposal. An outside contractor uses an onsite belt filter press system to dewater the iron sludge before its offsite disposal as hazardous waste in California (Glaze, D.E., 2002, personal communication).

Ion exchange can be used to remove soluble heavy metals to very low levels [67]. Because it can remove all ionic species in water and thus chemical regeneration cost is high, its use has been more common for treatment of water or wastewater with low dissolved solids. Pretreatment is required to prevent excessive resin fouling. There are many ion-specific resins for removal of different metals [68]. However, several different resins are needed when different metals must be removed. One significant use of ion exchange wastewater treatment is for chromate-containing blowdown from recirculating cooling water systems [52]. With proper pH adjustment (to pH 4.0–5.0), the chromate is removed even in the presence of several hundred mg/L sulfate and chloride [69].

Reverse osmosis can remove dissolved metals to very low levels. It can also remove a variety of pollutants such as cyanide and residual organics from refinery wastewater. However, because it is an expensive process, it would be competitive only if removal of total dissolved solids is also required. It also requires extensive pretreatment to prevent membrane fouling and deterioration [52]. The pretreatment processes may include filtration to remove suspended solids, pH adjustment, softening, and activated carbon treatment to remove organics and chlorine. A major drawback of the RO process is the handling and disposal of the reject stream, which can amount to 20–30% of the influent flow.

4.6.5 Treatment Modifications Due to Newer Regulations

Since 1990, several new or revised U.S. environmental regulations, which significantly affect refinery wastewater treatment systems, have been promulgated. The most important ones include the revised Toxicity Characteristics (TC) rule, the Primary Sludge rule, and the Benzene NESHAP (National Emissions Standards for Hazardous Air Pollutants) rule. These regulations and their impacts on refinery wastewater facilities are briefly discussed below.

Revised TC Rule

The TC rule was revised to include 26 organic chemicals, including benzene and cresols. It broadened the definition of a characteristically toxic hazardous waste to include a large number of wastes that were previously not included. This rule came into effect on September 25, 1990. The presence of benzene, for example, renders a wastewater hazardous when benzene concentrations are greater than 0.5 mg/L [70]. Refinery waste streams typically contain benzene. The greatest impact of the revised TC rule is on ponds, lagoons, and impoundments that have been managing wastewater that was not previously considered to be hazardous [71]. These units become RCRA regulated surface impoundments if they receive TC hazardous wastewater, and would have had to be retrofitted with two liners and leachate collection by March 19, 1994. Because these units are usually very large, they are very costly to retrofit. Several alternatives are available to retrofitting these units. Some refineries replaced the ponds and lagoons with above-ground tanks. Other plants have installed air or steam stripping facilities to remove benzene before the wastewater enters these surface impounds. And yet others installed high-rate biological treatment systems to biodegrade benzene so that they can continue to use the ponds without retrofitting. This is economically feasible because benzene can be easily biodegraded.

One of the Bay Area refineries has installed a second above-ground biological treatment system to treat waste streams with higher benzene concentrations (Glaze, D.E., 2002, personal communication). The existing biotreater is pond-based with DAF clarifiers. Figure 19 shows a block flow diagram of the revised effluent treatment system [72]. The process train includes conventional refinery treatment processes, two different biological treatment systems, an iron coprecipitation system for selenium removal, and GAC for toxicity reduction.

Primary Sludge Listing

The Primary Sludge rule, effective May 2, 1991, lists primary petroleum refinery sludge, designated F037 and F038, as hazardous wastes [70]. It governs all sludges generated from the separation of oil/water/solids during the storage or primary treatment of process wastewaters and oily cooling waters. These include API separator sludge, DAF floats, and sludges from all surface impoundments prior to biological treatment. Surface impoundments that receive or generate these wastes must comply with minimum technology requirements (MTRs) within four years of the promulgation date. Examples of these MTRs are double liners, leachate collection,

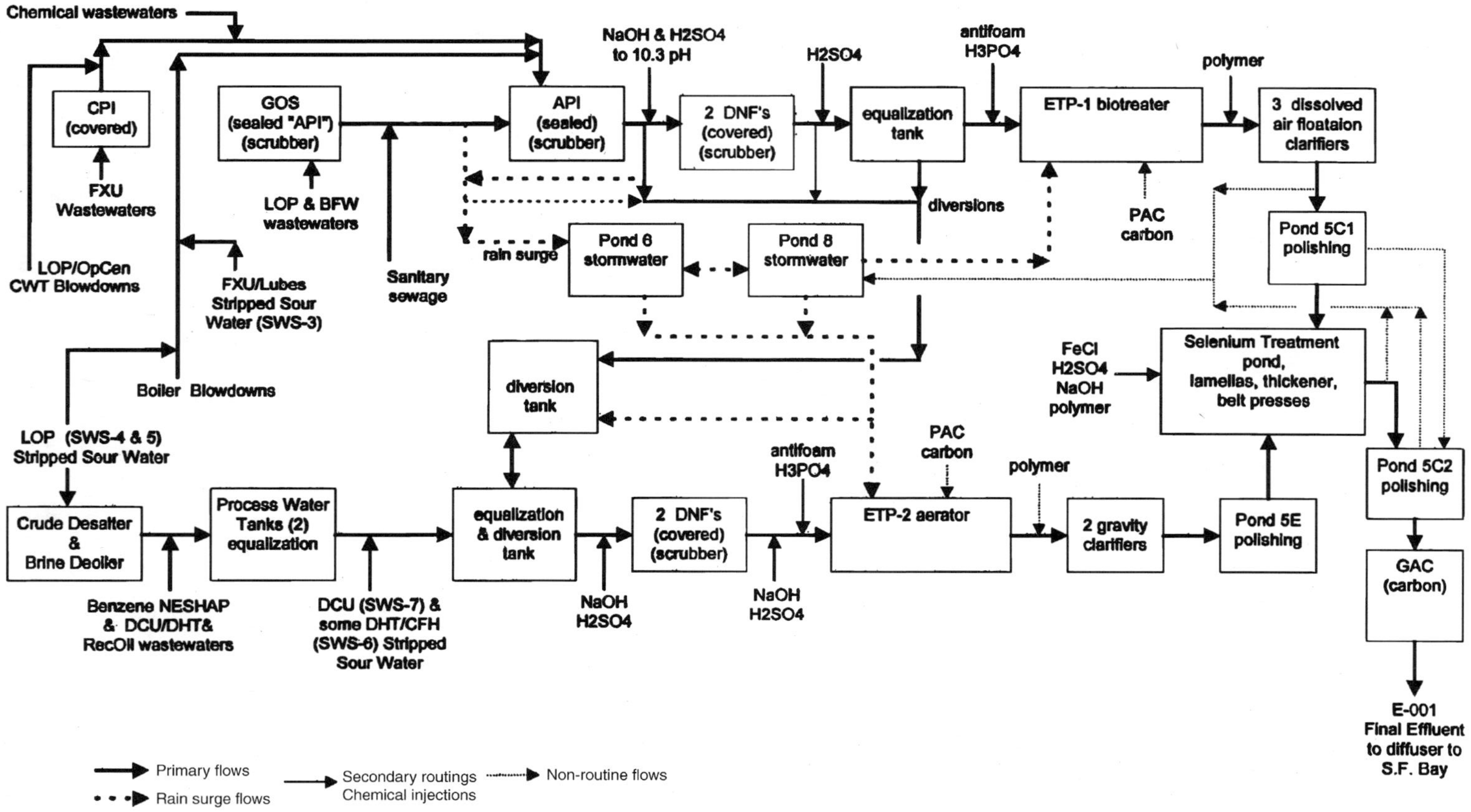

Figure 19 Bay area refinery effluent treating block flow diagram. This refinery has a complicated wastewater treatment scheme because of the toxicity characteristics rule to separate streams with higher benzene concentrations for treatment in aboveground biotreater. (From Ref. 72.)

and groundwater monitoring. Most refiners chose to reconfigure their wastewater collection and treatment systems by replacing impoundments with above-ground tanks, and by lining or enclosing process wastewater conveyance ditches. About 25 U.S. refineries practise sludge coking to dispose of oily, indigenous sludges [70]. In this process, the sludge is injected into coke drums during the quench cycle.

Benzene NESHAP

USEPA issued the NESHAP for benzene waste operations March 7, 1990, under the Clean Air Act. The compliance date was May 1992. It affects not only equipment leaks but also emissions of benzene in wastewater streams. Facilities with greater than 10 tonnes/year benzene in wastewater streams are affected. They must identify wastewater streams containing greater than 10 mg/L benzene and divert them to units that will reduce benzene to acceptable levels, that is, below 10 mg/L or by 98%. This rule affected most major refineries and olefins plants. Mobil Corp. spent $10 million on a benzene recovery project at its Chalmette, La., refinery. The refinery uses vacuum steam stripping to decrease benzene emissions by about 10 tonnes/year. One Gulf Coast petrochemical plant has also spent $10 million on a wastewater stripping facility, which reduced benzene levels from several thousand mg/L to less than 5 mg/L [70].

On March 5, 1992, USEPA delayed the effective date of the NESHAP until it clarified some confusing points raised by members of the petroleum industry. The final clarifying amendments to the benzene NESHAP were issued by USEPA on January 7, 1993 [72a].

4.6.6 Treatment for Recycle/Reuse and Zero Discharge

Petroleum refineries require a reliable supply of fresh water for steam generation, process cooling, product manufacturing, and other purposes. Because fresh water is becoming more valuable in many parts of the world, many locations have undertaken to reclaim and reuse waters for cooling, steam generation, and process use [73]. Bresnahan [74] presented two case studies that illustrate some of the technical challenges that were encountered when reusing water in refining and petrochemical complexes. One case was use of reclaimed municipal wastewater for most of the cooling towers at Mobil's Torrance Refinery in Los Angeles County, CA, which began in 1995. After working with chemical suppliers to formulate an appropriate treatment program together with optimization and continuous improvements, the reuse program has been operating successfully.

Another case involves the 300,000 bpd Chevron refinery in the San Francisco Bay Area. It is the largest user of potable water in the area [75]. Nearly half of the refinery's water demand (23,000 m^3/day) is used as makeup water in the cooling towers. The water utility identified the potential water reuse for this application in 1979. A pilot plant testing program was completed in 1987, which demonstrated that using lime/soda ash softening treatment on secondary effluent would produce a consistently high-quality reclaimed water for use as makeup water in the refinery's cooling towers. A full-scale plant (23,000 m^3/day) was completed in 1995. Figure 20 shows a process flow schematic of the reclamation plant. Secondary effluent from the WWTP is stored in a 6400 m^3 equalization tank. The influent is pumped to two 17 m diameter solids contact clarifiers after chemical treatment with lime/soda ash. The clarifier overflow is pH adjusted and filtered by four deep-bed, continuous-backwash sand filtration units. The filter effluent is disinfected by sodium hypochlorite for 90 minutes before being pumped to the refinery. The sludge from the clarifiers is thickened in two 10.7 m diameter thickeners and dewatered by a plate and frame filter press with 1.5 m plates.

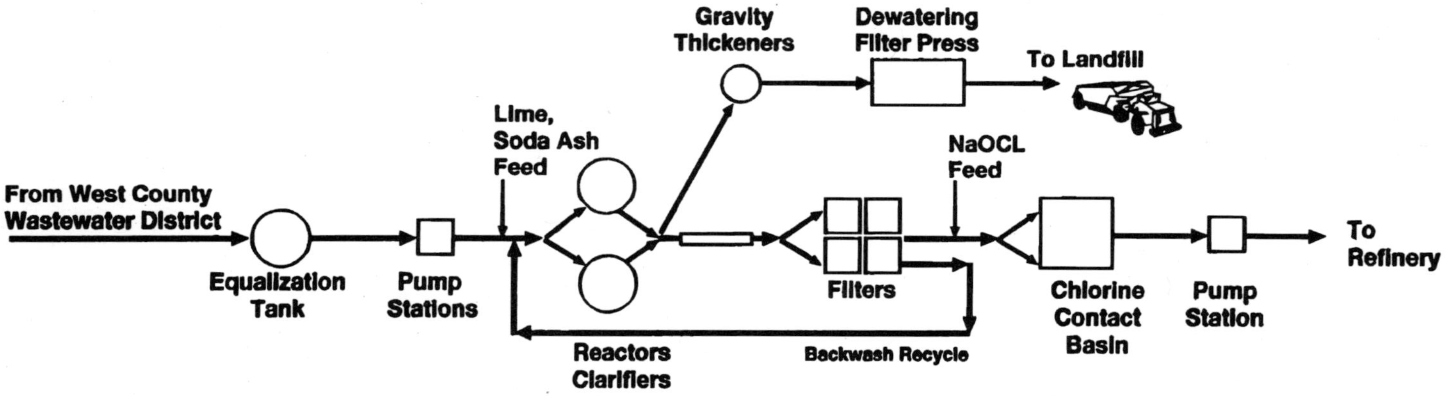

Figure 20 EBMUD North Richmond Reclamation Plant process flow schematic. The tertiary treated wastewater is reused in the Richmond Chevron refinery as cooling tower makeup. (From Ref. 75.)

The El Segundo, CA, Chevron refinery takes a further step in water reuse [76]. It receives 16,300 m^3/day of reclaimed water to feed its boilers. Microfiltration (MF) and RO are used to treat secondary effluent from the Hyperion Wastewater Treatment Plant to provide low-pressure boiler feedwater while a second pass RO is used to produce high-pressure boiler feedwater.

The concept of water reuse and zero liquid discharge in petroleum refineries has been proposed and debated for many years [77]. The principal drawback for zero liquid discharge is the generation of large amount of solid waste, mostly salt from the wastewater. It is this problem that caused USEPA to back off from zero liquid discharge in the 1970s, and it remains the primary deterrent today. However, there are two refineries in Mexico that have recently gone to zero discharge [78]. Wastewater from the refineries and nearby municipalities are treated with biological, physical/chemical processes, RO, brine concentrator evaporator and crystallizer to maximize water recycle to the refineries, minimize water makeup from the river and to attain zero liquid discharge. Figure 21 shows a process schematic diagram of the refinery wastewater recycle/zero liquid discharge system.

4.7 POLLUTION PREVENTION/HAZARDOUS WASTE MINIMIZATION

Refineries generate a large amount of hazardous wastes. As a result, they have been hit hard by environmental regulations and unfavorable public opinion, and Congress mandated in 1984 that refineries minimize waste [79]. In California, refiners turned to waste minimization, or pollution prevention, en masse in 1991 when the state's Source Reduction and Hazardous Waste

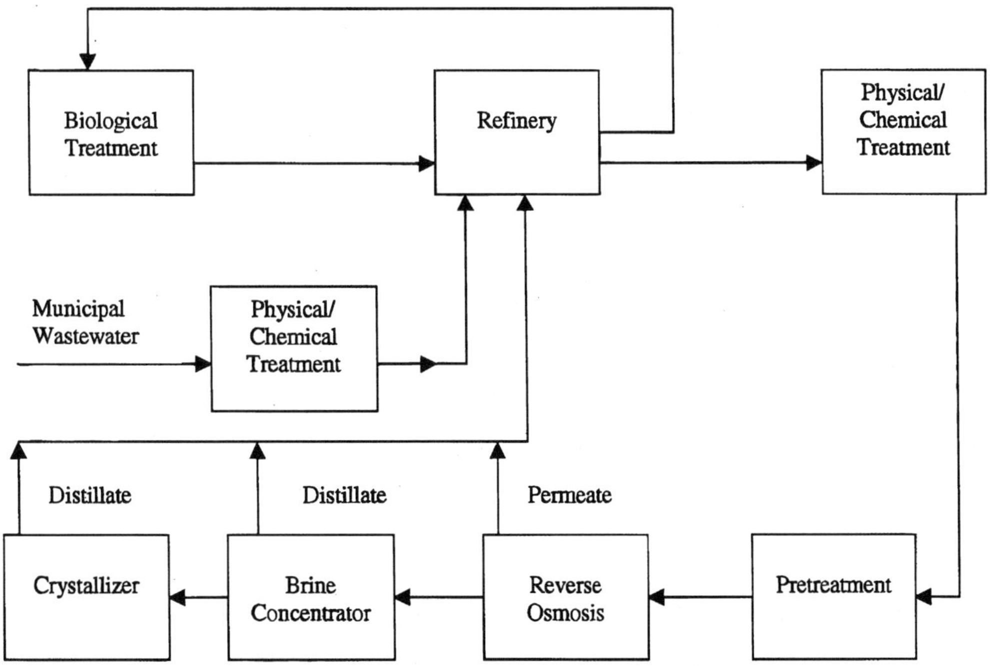

Figure 21 Refinery wastewater recycle/zero liquid discharge scheme. Pretreatment and reverse osmosis are used to recycle water, and brine concentrator and crystallizer are used to treat the rejects to achieve zero liquid discharge. (From Ref. 78.)

Management Review Act of 1989, commonly referred to as Senate Bill (SB) 14, went into effect. Inspired by USEPA and California regulations, other states have pursued similarly restrictive paths.

4.7.1 Pollution Prevention Program

A pollution prevention or waste minimization program usually consists of the following [79]:

- conducting a waste survey;
- screening waste streams for minimization opportunities;
- developing minimization options;
- screening minimization options;
- evaluating high-priority options;
- scheduling and implementing desirable options;
- evaluating and reviewing program performance periodically.

For a waste minimization program to succeed, refinery managers must provide the necessary staff and other resources to accomplish their goals. A team committed to the tasks is usually assembled. Because a refinery is a complex facility and there are numerous emission sources and waste streams to take into account, the team should consider and give the highest priority to:

- *Pollution prevention hierarchy* – The order of preference (highest to lowest) is source reduction, recycling, treatment, and secure land disposal;
- *Reduction of waste volume* – Volume reduction will usually reduce cost for handling, treatment, and disposal;
- Ease of implementation;
- Proven performance;
- Safety and health risks to employees and the public.

Some waste minimization approaches are proving to be more successful than others. Studying several refineries for waste minimization opportunities led to these eye-opening conclusions [79]:

- Housekeeping is the most cost-effective way to minimize waste;
- Solids that enter the refinery's wastewater treatment system automatically are classified as hazardous waste. Therefore, refiners can lower the volume of hazardous waste generated by keeping nonhazardous waste out of the treatment system;
- Raw materials (e.g., crude source) substitution is difficult because the choice of material is dictated by economics, availability, and the process units at the refinery;
- Process modifications can be implemented but may require considerable research and development;
- In-plant and offsite reuse of wastes plays a major role in waste minimization.

4.7.2 The 130,000 bpd Refinery Example

Take the example of a 130,000 bpd West Coast refinery that generates approximately 50,000 tons per year of hazardous waste [19]. Since 1984, this refinery has initiated waste management practices to handle:

- *Spent caustic*. At the end of 1990, 100% of the spent caustic was recycled onsite or offsite. The alkylation/dimersol and fluid catalytic cracking unit (FCCU) spent caustic is recycled to neutralize acidic wastewater. The virgin light-ends spent caustic is transported offsite for reuse at a paper manufacturing facility. The alkylation unit propane spent caustic is used in three other alkylation unit caustic washes. A Minalk Treating System was installed at the FCCU to replace an existing Merox Treating System to convert mercaptans to disulfide. This replacement reduces FCCU spent caustic generation by 1700 tons/year.

- *Stretford solution*. Since 1987, 100% of the waste Stretford solution has been shipped to a metals reclamation facility. Vanadium is reclaimed as vanadium pentoxide.
- *Wastewater treatment sludge*. In 1987, the refinery started a program to recycle this sludge to the coker within the refinery. At the end of 1990, 60% of the sludge was recycled to the coker. The remaining 40% was dewatered onsite at a belt filter press and then landfilled offsite or incinerated. Since 1986, the refinery has paved five plant areas to reduce the amount of dirt and debris washing into the sewer.
- *Catalysts, desiccants, and catalyst inerts*. In 1988, the refinery began to recycle nonhazardous catalysts, desiccants, and catalyst fines. It recycles electrostatic precipitator fines, Claus catalyst, and catalyst support inerts for use in cement manufacture. Two other catalysts, zinc oxide and iron chromate from the hydrogen plant, are reprocessed at smelters to recover the metals.

California SB 14 regulations required the refinery to further evaluate source reduction opportunities. The following are some of the measures identified by the refinery for further evaluation and implementation:

- Modification of the coker silo area to reduce dirt and debris to the wastewater treatment system;
- Reuse of the waste Stretford sulfur stream at a sulfuric acid manufacturer;
- Use of a transportable treatment unit to oxidize thiosulfate salts in the Stretford solution to allow them to be recycled in the Stretford process;
- Installation of sulfur de-entrainment devices in the Claus Plant sulfur condensers to allow them to be recycled in the Stretford process;
- Installation of asphalt lips around sewer to inhibit entry of dirt and debris;
- Evaluation of increasing the amount of wastewater treatment sludge recycled to the coker.

REFERENCES

1. Bass, D.M. Petroleum. In *The World Book Encyclopedia*; World Book, Inc.: Chicago, IL, 1989; 330–350.
2. Kemmer, F.N. *The Nalco Water Handbook*; McGraw-Hill: New York, 1979.
3. Stell, J. Survey of Operating Refineries in the U.S. Oil & Gas J. **2001**, December 24.
4. Shreve, R.N. *Chemical Process Industries*, 3rd Ed.; McGraw-Hill: New York, 1967.
5. U.S. Environmental Protection Agency. *Development Document for Effluent Limitations, Guidelines and Standards for the Petroleum Refining Point Source Category,* EPA-440/1-82-014, 1982.
6. Hengstebeck, R.J. *Petroleum Processing*; McGraw-Hill: New York, 1959.
7. U.S. Environmental Protection Agency. *Brine Disposal Treatment Practices Relating to the Oil Production Industry*, EPA-660/2-74-037, 1974.
8. U.S. Environmental Protection Agency. *Development Document for Interim Final Effluent Guidelines and Proposed New Source Performance Standards for the Oil and Gas Extraction Point Source Category*, EPA 440/1-76-005-a-Group II, 1976.

9. Michalczyk, B.L.; Pollock, T.E.; White, H.R. Treatment of oil field production waters. *Proceedings of the Industrial Wastes Symposium, 57th Annual WPCF Conference*, New Orleans, Louisiana, 1984, 457–471.
10. Palmer, L.L.; Beyer, A.H.; Stock, J. Biological oxidation of dissolved compounds in oilfield produced water by a field pilot biodisk. J. Petrol. Tech. **1981**, *June*, 1136.
11. Bates, M.H. Land farming of reserve pit fluids and sludges: Fates of selected contaminants. Wat. Res. **1988**, *22*, 793.
12. Vielvoye, R. Cleaning up cuttings. Oil & Gas J. **1989**, *Sept. 25*, 46.
13. Nesbitt, L.E.; Sanders, J.A. Drilling fluid disposal. J. Petrol. Tech. **1981**, *Dec.*, 2377–2381.
14. Copa, W.M.; Dietrich, M.J. Wet oxidation of oils, oil refinery sludges, and spent drilling muds. *Proceedings of Oil Waste Management Alternatives Symposium*, California Dept. of Health Services: Oakland, California, April, 1988.
15. U.S. Department of the Interior. *The Cost of Clean Water, Volume III, Industrial Waste Profile No. 5–Petroleum Refining*; Federal Water Pollution Control Administration, 1967.
16. Gloyna, E.F.; Ford, D.L. *The Characteristics and Pollutional Problems Associated With Petrochemical Wastes*; Federal Water Pollution Control Administration, Ada, OK, 1970.
17. Ford, D.L. Water pollution control in the petroleum industry. In *Industrial Wastewater Management Handbook*; Azad, H.S., Ed.; McGraw-Hill: New York, 1976.
18. Bryant, J.S.; Moores, C.W. Disposal of hazardous wastes from petroleum refineries. *Proceedings, 45th Purdue Industrial Waste Conference*, West Lafayette, IN, 1990: Lewis Publishers, Inc.
19. Wong, J.M. Hazardous waste minimization (SB 14) in California petroleum refineries. *Proceedings of 50th Purdue Industrial Waste Conference*, Lewis Publishers, Inc., Chelsea, MI, 1993.
20. Sittig, M. *Petroleum Refining Industry Energy Saving and Environmental Control*; Noyes Data Corp.: Park Ridge, NJ, 1978.
21. Truitt, R. POTWs feel the heat of toxics control. Water/Engng & Mgnt **1989**, *Sept.*, 14–15.
22. Thomas, N.A. Use of biomonitoring to control toxics in the United States. Wat. Sci. Technol. **1988**, *20*(10), 101–108.
23. Sack, W.A.; Eck, R.W.; Romano, C.R. Recovery of waste brines for highway applications. *Proceedings of 40th Purdue Industrial Waste Conference*, West Lafayette, IN, 1985.
24. Wang, W.D.; Li, X.M.; Chen, Y.; Zhang, S.T.; Jiang, Y. The technology of microbial treating drained water of oil field. *Proceedings, SPE Asia Pacific Improved Oil Recovery Conference*, Kuala Lumpur, Malaysia, 8–9 October, 2001.
25. Beyer, A.H.; Palmer, L.L.; Stock, J. Biological oxidation of dissolved compounds in oilfield-produced water by a pilot aerated lagoon. J. Petrol. Technol. **1979**, *Feb.*, 241–245.
26. Ali, S.A.; Henry, L.R.; Darlington, J.W.; Occapinti, J. New filtration process cuts contaminants from offshore produced water. Oil & Gas J. **1998**, *Nov.*, 73.
27. Eckenfelder, W.W., Jr. *Industrial Water Pollution Control*, 2nd Ed.; McGraw-Hill: New York, 1989.
28. McNally, R. Tougher rules challenge future for injection wells. Petrol. Eng. Int. **1987**, *July*, 28–30.
29. Anonymous. GAO finds brine still contaminates aquifers. Oil and Gas J. **1989**, *Oct. 16*, 38.
30. Waite, B.A.; Blauvelt, S.C.; Moody, J.L. *Oil and Gas Well Pollution Abatement Project*, ME No. 81495; Moody and Associates, Inc.: Meadville, PA, 1983.
31. Chen, J.C.T.; Stephenson, R.L. Cost effective treatment of oil field produced wastewater for 'Wet Stream' generation – a case history. *Proceedings, 39th Purdue Industrial Waste Conference*, West Lafayette, IN, 1984.
32. Doran, G.F.; Carini, F.H.; Fruth, D.A.; Drago, J.A.; Leong, L.Y.C. Evaluation of technologies to treat oil field produced water to drinking water or reuse quality. *Proceedings, 68th SPE Annual Western Regional Meeting*, Bakersfield, CA, 38830, 1998.
33. Marks, R.E.; Field, S.D.; Wojtanowicz, A. Biodegradation of oilfield production pit sludges. *Proceedings, 42nd Purdue Industrial Waste Conference*, West Lafayette, IN, 1987.
34. Moses, J.; Abrishamian, R. Case study: SITE program puts critical fluid solvent extraction to the test. Hazardous Waste Management Magazine **1988**, *Jan/Feb.*

35. U.S. Environmental Protection Agency. *Development Document for Proposed Effluent Limitations Guidelines and New Source Performance Standards for the Petroleum Refining Point Source Category*, Report EPA-440/1-73/014, 1973.
36. Knowlton, H.E. Source control in petroleum refineries. *Proceedings, National Petroleum Refiners Association Annual Meeting*, San Antonio, TX, March 19–21, 1978.
37. Beychok, M.R. *Aqueous Wastes from Petroleum and Petrochemical Plants*; John Wiley & Sons: London, 1967.
38. Hackman, E.E., III. *Toxic Organic Chemicals Destruction and Waste Treatment*; Noyes Data Corp.: Park Ridge, NJ, 1978.
39. Knowlton, H.E. Control refinery odors and effluent quality to meet environmental regulations. *Proceedings, National Petroleum Refiners Association Annual Meeting*, San Antonio, TX, March 24–26, 1985.
40. American Petroleum Institute. *Manual on Disposal of Refinery Wastes*; American Petroleum Institute: Washington, DC, 1969.
41. Katz, W.J. Adsorption – secret of success in separating solid by air flotation. *Wastes Eng.* **1959**, *July*.
42. Krofta, M.; Wang, L.K. Flotation technology and secondary clarification. *TAPPI J.* **1987**, *70*(4).
43. Krofta, M.; Guss, D.; Wang, L.K. Development of low-cost flotation technology and systems for wastewater treatment. *Proceedings, 42nd Purdue Industrial Waste Conference*, West Lafayette, IN, 1987.
44. Novotny, V.; England, A.J. Water Res. **1974**, *8*, 325.
45. Anonymous. A refined process. Water Qual. Int. **1998**, *1–2*, 37.
46. Hudson, N.; Doyle, J.; Lant, P.; Roach, N.; de Bruyn, B.; Staib, C. Sequencing batch reactor technology: The key to a BP refinery (Bulwer Island) upgraded environmental protection system: A low cost lagoon based retrofit. Water Sci. Technol. **2001**, *43*(3), 339.
47. U.S. Environmental Protection Agency. *The Impact of Oily Materials on Activated Sludge Systems*, Hydroscience, Inc., EPA Project No. 12050 DSH, March 1971.
48. Wong, J.M.; Maroney, P.M. Pilot plant comparison of extended aeration and PACT® for toxicity reduction in refinery wastewater. *Proceedings, 44th Purdue Industrial Waste Conference*, West Lafayette, IN, 1989.
49. Brown and Caldwell Consulting Engineers. Confidential report; Brown and Caldwell: Denver, CO, October 1989.
50. Adams, C.E., Jr.; Ford, D.L.; Eckenfelder, W.W., Jr. *Development of Design and Operational Criteria for Wastewater Treatment*; Enviro Press, Inc.: Nashville, Tennessee, 1981.
51. Ford, D.L.; Manning, F.S. Treatment of petroleum refinery wastewater. In *Carbon Adsorption Handbook*; Cheremisinoff, P.N., Ellerbusch, F., Ed.; Ann Arbor Science: Ann Arbor, MI, 1978.
52. Patterson, J.W. *Industrial Wastewater Treatment Technology*, 2nd Ed.; Butterworth: Boston, 1985.
53. Hutton, D.G.; Robertaccio, F.L. Waste water treatment process. U.S. Patent 3,904,518, September 9, 1975.
54. Rizzo, J.A. Case history: Use of powdered activated carbon in an activated sludge system. *First Open Forum on Petroleum Refinery Wastewaters*, Tulsa, OK, 1976.
55. Grieves, C.G.; Stenstrom, M.K.; Walk, J.D.; Grutsch, J.F. Effluent quality improvement by powdered activated carbon in refinery activated sludge processes. *API Refining Department, 42nd Midyear Meeting*, Chicago, IL, May 11, 1977.
56. Thibault, G.T.; Tracy, K.D.; Wilkinson, J.B. Evaluation of powdered activated carbon treatment for improving activated sludge performance. *API Refining Department, 42nd Midyear Meeting*, Chicago, IL, May 11, 1977.
57. Zimpro, Inc. CIBA-GEIGY meeting tough bioassay test. Reactor **1986**, *June*, 13–14.
58. Butterworth, S.L. Granular activated carbon as a toxicity reduction technology for wastewater treatment. Proc. Am. Chem. Soc. Spring Natl. Meet. Fuel Chem. Div. **1996**, *41*(1), 466.
59. Wong, D.C.L.; van Compernolle, R.; Nowlin, J.G.; O'Neal, D.L.; Johnson, G.M. Use of supercritical fluid extraction and fast ion bombardment mass spectrometry to identify toxic chemicals from a refinery effluent adsorbed onto granular activated carbon. Chemosphere **1996**, *32*, 621.

60. California Water Resources Control Board. *Water Quality Control Plan for the San Francisco Bay Basin*, Region 2, 1986.
61. Prather, B.; Berkemeyer, R. Cyanide sources in petroleum refineries. *Proceedings, 30th Purdue Industrial Waste Conference*, West Lafayette, IN, 1975.
62. Wong, J.M.; Maroney, P.M. Cyanide control in petroleum refineries. *Proceedings, 44th Purdue Industrial Waste Conference*, West Lafayette, IN, 1975.
63. Knowlton, H.E.; Coombs, J.; Allen, E. Chevron process reduces FCC/coker corrosion and saves energy. Oil Gas J. **1980**, *April 14.*
64. Kunz, R.; Casey, J.; Huff, J. Refinery cyanide: A regulatory dilemma. Hydrocarbon Proc. **1978**, *October.*
65. Merrill, D.T.; Manzione, M.A.; Peterson, J.J.; Parker, D.S.; Chow, W.; Hobbs, A.D. Field evaluation of arsenic and selenium removal by iron coprecipitation. J. WPCF **1986**, *58*, 18–26.
66. Nurdogan, Y.; Schroeder, R.P.; Meyer, C.L. Selenium removal from petroleum refinery wastewater. *Proceedings, 49th Purdue Industrial Waste Conference*, West Lafayette, IN, 1994.
67. U.S. Environmental Protection Agency (1981). Summary Report: Control and Treatment Technology for Metal Finishing Industry; Ion Exchange, EPA 625/8-81-007.
68. Peters, R.W.; Ku, Y.; Bhattacharyya, D. Evaluation of recent treatment techniques for removal of heavy metals from industrial wastewaters. AIChE Symposium Series **1985**, *81*(243).
69. Anderson, R.E. Some examples of the concentration of trace heavy metals with ion exchange resins. *Proceedings, Traces of Heavy Metals in Water-Removal Processes and Monitoring*, USEPA 902/9-74-001, 1974.
70. Rhodes, A.K. Recent and pending regulations push refiners to the limit. Oil Gas J. **1991**, *Dec. 16*, 39–46.
71. American Petroleum Institute. *Applying the Revised Toxicity Characteristic to the Petroleum Industry*; API: Washington, DC, 1991.
72a. Wong, J.M. Advanced Wastewater treatment for refineries. Proc. Petroleum Refinery/Petrochemical Wastewater Treatment and River Basin/Water Quality Management Workshop, Kaohsiung, Taiwan, ROC, December 4–7, 2001.
72. *Environmental Reporter*; The Bureau of National Affairs, Inc.: Washington, DC, January 15, 1993.
73. Wong, J.M. Petrochemicals. Water Environ. Res. **1998**, *71*(5), 828.
74. Bresnahan, W.T. Water reuse in oil refineries. *Proceedings, Natl. Assoc. Corrosion Eng. Corrosion '96 Conf.*, Houston, TX, 1996.
75. Wong, J.M. Water conservation and reuse in the petrochemical industries. *Proc. Industrial Water Conservation Workshop*, Taipei, Taiwan, ROC, April 11–12, 1996.
76. Anonymous. Effluent recycling plant expands to provide boiler feedwater to Chevron refinery. Civil Eng. **2000**, *70*(7), 14.
77. Diepolder, P. Is 'Zero Discharge' Realistic? Hydrocarbon Proc. **1992**, *October.*
78. Heimbigner, B. Water and wastewater treatment in petroleum refineries. Presented at the Plock Refinery, Poland, June, 1999.
79. Wong, J.M. Pollution prevention/waste minimization in california petroleum refineries. OCEESA J. **2002**, *19*, 1.

5
Treatment of Metal Finishing Wastes

Olcay Tünay and Işık Kabdaşlı
İstanbul Technical University, İstanbul, Turkey

Yung-Tse Hung
Cleveland State University, Cleveland, Ohio, U.S.A.

5.1 INDUSTRY OVERVIEW

In the metal finishing industry, metals or alloys are used as starting materials to manufacture a wide range of metal components. The metal finishing and metal fabricating industry is identified as SIC (Standard Industrial Classification) code 34. SIC code 34 is further divided into nine classes: SIC 341 to SCI 349. These subclasses indicate product groups [1].

Metal finishing involves the forming and shaping of metals and the altering of surface properties to enhance corrosion resistance, electrical conductivity or resistance, wear resistance, chemical and tarnish resistance, solderability, torque tolerance, ability to bond rubber and adhesives or organic coatings, hardness, reflectivity, and decorative appearance. A broad range of processes is employed in the metal finishing industry. Metal fabrication covers mostly mechanical operations such as cutting and forming. Surface treatment involves plating, conversion coating, anodizing, painting, heat treating, and many other operations. Degreasing, cleaning, pickling, and etching are supporting processes. The industry manufactures a wide range of metal components such as cans, hand tools, hardware, cutlery, and structural metal products. Many industries use metal finishing in their manufacturing processes. Metal finishing is an essential part of a number of industries including automotive, electronics, defense, aerospace, hardware, heavy equipment, appliances, telecommunication and jewelry. With this profile, the metal finishing industry is among the most common industrial activities in the United States and in many other countries as well. While production methods and applications are similar in all metal finishing plants, capacities vary widely. Metal finishing facilities are grouped into two major categories: captives and job shops. Captive facilities are part of a larger operation and perform metal finishing processes on in-house manufactured parts. The plants in this category tend to be larger in capacity than job shops. Job shops are independently owned small plants that rely on a variety of customers and work on the parts manufactured by others. Job shops may also be used as subcontractors by the captive facilities. This application tends to be more common [2]. Captive facilities are more specialized in their operations, while job shops are more flexible in operations to respond to the varying demands of customers.

Metal finishing industries use a variety of chemicals including solvents, acids, bases, surfactants, complex organic substances, and metal salts such as cadmium, nickel, and chromium. The industry is one of the largest users of toxic chemicals in the United States [3] and is a significant source of pollutants that are discharged to virtually all receiving media, including air, surface water, land and publicly owned treatment works (POTWs). A considerable part of the wastes is sent offsite for treatment or recycle. This profile makes the metal finishing industry the focus of environmental pollution control and prevention applications and subject to many limitations and standards. Thus, the metal finishing industry uses a greater variety of treatment technologies compared to other industries. Increased cost and sophistication of treatment methods have led to development of in-plant controls, as well as limitations as to chemicals for which control and treatment are considered to be inadequate and/or unreliable. Many of the chemicals, particularly solvents used in the metal finishing industry, have been phased out. A great variety of means and technologies has been employed for pollution prevention. Captive plants, being a part of greater enterprises, tend to be more proactive in their approach to environmental management. However, job shops having the ability to adapt themselves to varying conditions can be quite successful in implementing pollution prevention measures and may even develop original solutions for pollution control.

Pollution prevention measures and applications in the metal finishing industry are mostly associated with production methods and recovery options. Therefore, an evaluation of pollution prevention applications in the industry requires an acquaintance with the unit operations of production.

5.2 PROCESS AND OPERATIONS OF THE INDUSTRY

5.2.1 Overview of Processes

Metal finishing and other related manufacturing categories employ some 50 unit processes and operations. These operations are applied in many variations depending on production demands; however, the main characteristics of the operations or processes remain unchanged. Forty-five unit operations and processes of metal finishing and metal fabricating are listed in Table 1 [3]. Some processes or operations are important in terms of waste generation but are not included in the table because they either employ a group of unit processes to produce a specified product

Table 1 Unit Operations of Metal Finishing Industry [3]

• Electroplating	• Thermal cutting	• Hot dip coating
• Electroless plating	• Welding	• Sputtering
• Anodizing	• Brazing	• Vapor plating
• Conversion coating	• Soldering	• Thermal infusion
• Etching (chemical milling)	• Flame spraying	• Salt bath descaling
• Cleaning	• Sand blasting	• Solvent degreasing
• Machining	• Other abrasive jet machining	• Paint stripping
• Grinding	• Electrical discharge machining	• Painting
• Polishing	• Electrochemical machining	• Electrostatic painting
• Tumbling (barrel finishing)	• Electron beam machining	• Electropainting
• Burnishing	• Laser beam machining	• Vacuum metalizing
• Impact deformation	• Plasma arc machining	• Assembly
• Pressure deformation	• Ultrasonic machining	• Calibration
• Shearing	• Sintering	• Testing
• Heat treating	• Laminating	• Mechanical plating

such as printed circuit boards (PCB), or constitute integral parts of some processes such as rinsing. Therefore, rinsing, cooling and lubrication, and PCB manufacturing are also briefly explained. One additional process that is not a production process and therefore not shown in the table is fume and exhaust scrubbing. This process has been widely used in metal finishing plants and constitutes a significant source of pollution. Brief information about these processes and operations is given below [1–5]. After a short definition of the process, emphasis is placed on the chemicals used in the processes. A general scheme of the metal finishing processes indicating the sequence of application is given in Figure 1 [2].

5.2.2 Description of Processes and Operations

Electroplating

Electroplating is defined as coating a metal surface with another metal by applying electric current (electrolysis). Electroplating baths contain metal ions that are reduced to a metallic state on the metal being plated. The metal being plated serves as the cathode. The anode is mostly the metal providing plating by dissolution in the bath solution. If an inert metal serves as the anode, the bath solution is replenished by the addition of a salt of the metal as it is consumed by plating. Electroplating baths contain metal salts, complexing agents, pH buffers, as well as organic and organometallic additives. Table 2 presents the major constituents of commonly used electroplating baths [4]. A great variety of metals or alloys is used for plating; those that are commonly used include cadmium, copper, chromium, nickel, zinc, and precious metals.

Electroless Plating

Electroless plating, as the name implies, does not utilize an electric current for plating. Electroless plating can be carried out as autocatalytic and immersion plating. Autocatalytic plating is the process in which the metal ion in the solution is forced to convert into the metallic state and deposit onto the object to be plated by the use of reducing agents. The process is started by the catalytic action of the surface being plated. For this purpose the surface is pretreated, usually by the application of metal plating. Electroless plating can be applied to metal and nonmetal substrates. The process requires the use of specific chemicals in addition to reducing agents, such as complexing agents and stabilizers. Specific conditions such as pH and temperature also need to be satisfied. Hydrazine, dimethylamine borane (DMAB), hypophosphite, and formaldehyde are common reducing agents. Many types of complexing agents such as EDTA, Rochelle salt, tartrates, and citrates are used. Thiodiglycolic acid, mercaptobenzotiazone (MBT), thiourea, fluoride salts, heavy metals, thioorganic compounds, and cyanides are among the stabilizers used. HCl, H_2SO_4, NaOH, and ammonium hydroxide are used for pH adjustment. Copper and nickel electroless plating is commonly used for PCBs. Electroless plating of precious metals is also common.

Immersion plating is also carried out without the application of electric current. However, in this case the metal ion in solution is plated onto the base metal, not by forcing with reducing

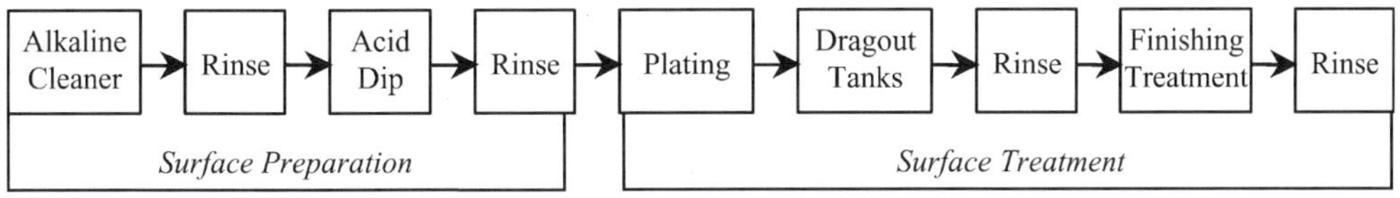

Figure 1 Sequence of the metal finishing operations. (From Ref. 2.)

Table 2 Common Electroplating Bath Compositions [4]

Electroplating baths	Composition
Brass and bronze	Copper cyanide Zinc cyanide Sodium cyanide Sodium carbonate Ammonia Rochelle salt
Cadmium cyanide	Cadmium cyanide Cadmium oxide Sodium cyanide Sodium hydroxide
Cadmium fluoroborate	Cadmium fluoroborate Fluoroboric acid Boric acid Ammonium fluoroborate Licorice
Copper cyanide	Copper cyanide Sodium cyanide Sodium carbonate Sodium hydroxide Rochelle salt
Copper fluoroborate	Copper fluoroborate Fluoroboric acid
Acid copper sulfate	Copper sulfate Sulfuric acid
Copper pyrophosphate	Copper pyrophosphate Potassium hydroxide Ammonia
Fluoride-modified copper cyanide	Copper cyanide Potassium cyanide Potassium fluoride
Chromium	Chromic acid Sulfuric acid
Chromium with fluoride catalyst	Chromic acid Sulfate Fluoride

agents, but spontaneously and as a thin film only, by the difference of electrode potential of the metal in solution and base metal. The metal in solution has a higher electrode potential, that is, it has a higher tendency to be reduced. Immersion plating baths contain alkalis and complexing agents. Sometimes nonalkaline heated baths are used, as in the case of copper plating on steel, or aluminum. Complex agents commonly used are lactic, glycolic, and malic acid salts, ammonia, and cyanide. Sulfuric and hydrofluoric acids are used for nonalkaline applications. Aluminum, copper alloys, zinc, and steel are plated by immersion plating. Commonly used metals for plating are cadmium, copper, nickel, tin, zinc, and precious metals.

Conversion Coatings

Conversion coating is one of the main categories of coating and it is widely applied in metal finishing processes. The purpose of conversion coating is to form a film of a substance that is bonded by the metal coating. This film converts the characteristics of the metal surface into a structure that is more resistant to external effects or amenable to further processing. Chromating, phosphating, passivation, metal coloring, and anodizing are among the common applications of the process.

Chromating or chromate conversion coating is realized through a chemical or electrochemical reaction of hexavalent chromium with the metal surface. Mineral acids and activators such as acetate, formate, phosphate, nitrate, and chloride are used in the process. Chromate coating provides corrosion resistance. It is widely applied to aluminum, copper, cadmium, and zinc. Among the precious metals, silver is often chromated.

Anodizing is a process in which the metal is covered by an insoluble metal oxide. Anodizing particularly refers to oxide coating to aluminum; however, the process is applied to other metals, particularly to zinc, magnesium, and titanium. It is an electrochemical process in which the metal being coated with an oxide layer constitutes the anode. A thin nonporous oxide layer forms on the metal. This layer provides corrosion and wear resistance as well as facilitating further coating operations. The electrolyte in the bath is an acid. Aluminum anodizing includes the use of chromic acid anodizing, sulfuric or boric-sulfuric acid anodizing. Magnesium anodizing solutions are mixtures of fluoride, phosphate, and chromic acids, or of potassium hydroxide, aluminum hydroxide, and potassium fluoride. Phosphoric and oxalic acids are also used for anodizing. Anodized metals are generally sealed. Nickel/cobalt acetate is widely used to seal anodic coatings. Dichromate seal is also very effective.

Passivation is a process by which protective films on the metal surface are formed by immersing the metal in an acid solution for oxidation. Strong oxidizing solutions like nitric acid and/or sodium dichromate are used. The oxide layer provides corrosion resistance. Passivation can be accomplished through anodizing or chromating.

Phosphating is a conversion coating by which nonmetallic, nonconductive surfaces composed of insoluble metal phosphate crystals are obtained. The main function of phosphating is to impart absorptivity to the surface and to provide a base for adhesion of paints, lacquers, and plastic coating. Phosphating is carried out by immersing the metal into a phosphoric acid solution. The phosphates of zinc, iron, and manganese are commonly used for phosphating. Strontium, cadmium, aluminum, chromium, as well as fluorine, boron, and silicon are common constituents. Nitrate, nitrite, chlorate, and peroxides are used as accelerators. Phosphating is performed on aluminum, cadmium, iron, magnesium, and zinc.

Coloring is a chemical conversion of a metal surface into an oxide or other metallic compound to produce a decorative finish. Coloring is commonly applied to copper, zinc, steel, and cadmium. A wide variety of solutions is used for coloring. Examples of coloring solutions are ammonium molybdate, ammonium polysulfide, copper carbonate, ferric chloride and potassium ferricyanide, potassium dichromate and nitric acid, potassium chlorate, and nickel sulfate.

Chemical Milling and Etching

Chemical milling is used for shaping or blanking the metal parts. Highly concentrated sodium hydroxide solutions are used in chemical milling. Chemical etching is used to remove relatively small amounts of metal from the surface of metals for surface conditioning or producing a pattern as in PCB manufacturing. Highly acidic solutions containing ferric chloride, ammonium persulfate, cupric chloride, sodium persulfate, and chromic acid can be used for the process.

Cleaning

Cleaning soil (oil and dirt) from metal surfaces is an essential step preceding many of the unit operations involved in metal finishing. Oil and dirt may be organic or inorganic in nature. Organic materials include saponifiable oils of animal and vegetable origin, mineral oils and waxes, and other organic contaminants such as inhibitors. Metal oxides, residues from the operations such as polishing, abrading, fluxing, and dust are the main inorganic materials. Cleaning operations consume large amounts of water and involve the use of a variety of chemicals. Some processes such as electroplating and electroless plating require a high degree of cleanliness, while others may not require the same degree of cleanliness. On the other hand, the composition and physical properties of the material being cleaned are important for the cleaning processes. As a result of these varying requirements, many types of cleaning processes have been developed. However, considering the basic character of the cleaning solution, four main cleaning groups can be defined: solvent cleaning, alkaline cleaning, electrocleaning, and acid cleaning. Solvent cleaning is described under solvent degreasing. Alkaline cleaning involves the use of builders such as sodium or potassium salts of phosphates, carbonates, silicates and hydroxides, surfactants, and, sometimes, antioxidants or inhibitors, complex formers, stabilizers, and small amount of solvents. Newly introduced nonemulsifying surfactants are very effective in separating the soil from surfaces. Strong alkaline cleaners may also contain cyanide. Alkaline cleaning is more effective for removing soil from surfaces.

Electrocleaning uses a strong alkaline with an electric current either reverse, periodic reverse, or direct, to remove soils and activate the surface. It is applied, generally, as a last step of cleaning. An electric current electrolyzes the water, evolving hydrogen and oxygen gases. Oxygen exerts a scrubbing effect on the surfaces. In reverse cleaning, the workpiece functions as the anode and the evolved oxygen assists in the removal of soil. In direct cleaning, the workpiece is the cathode and liberated hydrogen at the surface facilitates the scrubbing action. In the periodic reverse system, the workpiece is made alternately anodic and cathodic.

The three most common application modes of aqueous cleaning are immersion with mechanical agitation, immersion with ultrasonic agitation, and spray washing. Ultrasonic cleaning is a highly effective method. The method uses high-frequency sound waves, which locally exert high pressure and temperatures to loosen and remove the contaminants.

Sometimes semi-aqueous methods of cleaning can be used. Emulsion cleaning uses common organic solvents such as kerosene, mineral oil, and benzene, dispersed in an aqueous medium. Diphase cleaning is a two-layer system of water-soluble and water-insoluble organic solvents.

Machining and Grinding

Machining and grinding are the mechanical operations to shape and condition metals and their surfaces. These operations involve the use of natural and synthetic oils for cooling and lubrication.

Polishing, Tumbling, and Burnishing

Polishing is used to smooth out surface defects using polishing and buffing compounds. Metallic soaps, mineral oils, dispersing agents, and waxes are among the chemicals used.

Barrel finishing (tumbling) is used to remove burrs and scales. Several chemicals in addition to abrasives are used in the process. Oils, soaps, organic acids such as citric and maleic acids, sodium dichromate, as well as sodium cyanide are among the chemicals used.

Burnishing is a smooth finishing by displacement of small surface irregularities. Lubricants and soap solutions are used for cooling of burnishing tools. Light spindle oil, sodium cyanide (as a wetting agent), and rust inhibitors are used.

Impact Deformation, Pressure Deformation, and Shearing

Impact deformation, pressure deformation, and shearing are all mechanical operations. Oils, light greases, and pigmented lubricants are used for the deformation and shearing equipment.

Heat Treating

Heat treating aims to modify the physical properties of workpieces through the application of controlled heating and cooling cycles. Case-hardening produces a hard surface over a metal core. The surface is wear-resistant and durable. Quenching is realized using several types of solutions. Brine solutions contain sodium and calcium chloride and mineral acids. Water/oil emulsions contain soaps, alcohols, oils, emulsifiers, in addition to dissolved salts. Liquid carburizing and carbonitriding solutions contain sodium cyanide, detergents, and dissolved salts. High-temperature quenching baths contain sodium cyanide, boron oxide, sodium fluoride, dissolved salts, as well as manganese dioxide and silicon carbide. Molten lead is used for heat treatment of steel.

Thermal Cutting, Welding, Brazing, Soldering and Flame Spraying

Thermal cutting is accomplished using oxyacetylene oxygen, or electric arc tools. A rinsing step may follow the operations. Welding, brazing, and soldering operations are used to join metal parts, applying heat and pressure to melt the metal or filling material. The operations can be followed by quenching, cooling, or annealing in solutions or emulsified oils.

Flame spraying is the process of applying a metallic coating to a workpiece using powdered metal together with fluxes. This process is also followed by quenching, cooling, or annealing in a solution, or emulsified oils.

Sand Blasting

Sand blasting involves the use of abrasive grains pneumatically directed against workpieces to mechanically clean the surfaces. Rinsing may follow the operation.

Other Abrasive Jet Machining

Abrasive jet machining is a mechanical operation similar to sand blasting. Abrasive materials such as aluminum oxide, silicon carbide, and dolomite are used in alkaline or emulsified solutions.

Electrical Discharge Machining and Electrochemical Machining

Electrical discharge machining is applied to conductive materials using an electrode that creates an electrical spark. Dielectric fluids such as hydrocarbon-petroleum oils, kerosene, silicone oils, and ethylene glycol are used in the operation. Electrochemical machining is an electrolysis process in which the metal to be treated is the anode. Both aqueous and organic solvents are used as electrolytes.

Electron Beam Machining, Laser Beam Machining, Plasma Arc Machining, and Ultrasonic Machining

Electron beam machining and laser beam machining are thermoelectric processes. Plasma arc machining involves the use of high-temperature ionized gas at high velocity. Ultrasonic machining is the use of ultrasonic energy to machine hard and brittle material in a liquid.

Sintering and Laminating

Sintering is a process of forming a metal coating from powdered metal under pressure and heat. Laminating is bonding layers of metal, plastic, or wood by adhesives. A rinsing or cooling may follow the process.

Hot Dip Coating

Hot dip coating is achieved by coating a metal with another metal by immersion in a molten metal bath. The molten metal coats the part by forming an alloy at the interface of the two metals. Aluminum, lead, and tin can be used for hot dip coating; however, the most common application is zinc coating (galvanizing). Cleaning operations and fluxing precede the hot dip coating. In galvanizing, a zinc ammonium chloride flux is used. Sometimes coated pieces are quenched.

Sputtering, Vapor Plating, and Thermal Infusion

Sputtering is coating by bombarding metal ions in a gas discharging tube. Vapor plating is coating by decomposition of a metal compound on the heated surface of the base material. Thermal infusion is the application of a fused metal on heated ferrous material.

Salt Bath Descaling

There are several types of baths such as oxidizing, reducing, and electrolytic, for removing surface oxides and scale, in which workpieces are immersed. Salt baths are followed by quenching and acid dipping. The baths contain molten salts, sodium hydroxide, sodium hydride, and other chemical additives.

Solvent Degreasing

Solvent cleaning aims to remove oil and oily contaminants. Cold cleaning, diphase cleaning, and vapor phase cleaning are the three main methods applied. Cold cleaning is the application of unheated solvents of nonhalogenated type by wipe cleaning, soak cleaning, ultrasonic cleaning or steam gun stripping. Diphase cleaning systems use both water and solvent. After a water bath, solvent spray is applied to remove oil. Vapor phase cleaning is carried out in a tank where a halogenated solvent is heated to its boiling point. The parts to be cleaned are placed in the vapor zone. Solvent vapors condensed on the parts dissolve the oil and drip down the liquid phase. Air pollution devices such as coolers and condensators are placed above the vapor zone to minimize solvent emissions. Vapor degreasing involves the use of chlorinated solvents such as trichloroethylene, and perchloroethylene.

Paint Stripping

Paint stripping is the removal of an organic coating from a workpiece. Stripping solutions may contain caustic soda, wetting agents, detergents, emulsifiers, foam soaps, alcohol amines, ammonia, or solvents. Chlorinated solvents, polar solvents like acetone, methyl ethyl ketone, benzene, and toluene are commonly used. A rinsing step follows the process.

Painting, Electropainting, and Electrostatic Painting

Painting is the application of an organic coating such as paint, varnish, lacquer, shellac, and plastics. Spray painting is the most common method of painting; however, a variety of other methods such as dipping, brushing, and roll coating are used. Electrostatic painting is the application of electrostatically charged paint particles to an oppositely charged surface of the workpiece. It is followed by thermal fusing. Electropainting is carried out in a bath of aqueous emulsion of the paint, and the workpiece to be painted is made anodic or cathodic to collect the paint. Paint and other coating ingredients include a wide range of chemicals: pigments, resins, solvents, and other additives.

Vacuum Metalizing and Mechanical Plating

Vacuum metalizing is the process of coating a workpiece with a metal by flash heating the metal vapor in a high vacuum and condensing it on the workpiece surface. Mechanical plating is a process where cadmium, zinc, and tin powders are used for coating in barrels immersed in an acid solution by inert impact media. The plated parts are then rinsed.

Assembly, Calibration, and Testing

Assembly is the fitting together of previously manufactured parts or components. Calibration is the application of thermal, electrical, or mechanical energy to set or establish a reference point for a component or assembly. Testing is used to control the suitability or functionality of a finished or semi-finished product by the application of thermal, electrical, or mechanical means. Oils and fuels are used in nondestructive testing. Oily penetrants are used in dye-penetrant testing. Kerosene, ethylene glycol, and lubricating oils are among common penetrants.

Rinsing

Rinsing is the most common operation that follows many unit operations in metal finishing. Plating, cleaning, degreasing, and heat treating are the operations for which rinsing is an integral part of the operation. Rinsing may also be used after dry processes such as sand blasting. Rinsing is the major water-using process – a significant part of plant wastewater originates from rinsing. Many rinsing operations are continuous, which determines the bulk of continuous wastewater. The aim of rinsing is to remove contaminants of the preceding process. These contaminants may be oily, solid, but mostly aqueous. The drops or particles of contaminants that have the same or similar character of the solution or bath content on the workpiece are termed "dragout," which is mostly valuable. Therefore, rinsing is required to completely remove the contaminants, to achieve this with a minimum amount of water, and to enable recovery of dragout in a most efficient way. Of course, there is no ideal way to combine all these purposes, but what is applied is to select an optimum system depending on the process, that is, the value of the dragout, flow conditions, quality requirements of the subsequent processes, or of the finished product. Several rinsing systems may serve as an optimum in terms of general applications. Rinsing systems are based on several types of rinsing techniques: single running, countercurrent, in series, spray, dead, or economical. Single running rinse is the simplest and is an efficient method, but it consumes the largest amount of water. Countercurrent rinsing makes use of several tanks, but only the last tank receives fresh water, while preceding tanks are fed with the overflow of the following tank. Series or multistage static rinse is made up of several tanks in series, each having a separate feed. The conditions of each tank can be set up independently. Spray or fog rinse is the most efficient mode of continuous dilution rinsing. Dead or economy rinse follows the bath, receiving the most concentrated dragout, and is used to make up the preceding bath, or for recovery. Several methods of rinsing systems have been developed by combining these techniques.

Cooling and Lubrication

Cooling and lubrication is a common application employed in many operations such as machining operations, grinding, burnishing, and testing. Metalworking fluids are used for cooling and lubrication. Metalworking fluids are applied to the workpiece or cutting tool in order to cool the workpiece and/or tool, lubricate, wash away chips, to inhibit corrosion or surface oxidation, and to provide a good finish. Metalworking fluids can be air-blasted, sprayed, or applied by suction. Aqueous solutions contain an alkali such as borax, sodium carbonate, or trisodium phosphate. Emulsions are suspension of oil or paste in water. Oil–water or synthetic emulsions can be used. Some operations use a high oil to water ratio, for example, greater than 1/20, while others that require primarily lubrication rather than cooling, use oil. In addition to synthetic or petroleum-based oil content, metalworking fluids may contain chlorine, sulfur and phosphorus compounds, phenols, creosols, and alkalis.

Printed Circuit Board Manufacturing

Printed circuit board manufacturing is widely used in many fields, from electronics to transportation. The processes employed throughout the manufacturing are, to a great extent, metal finishing unit processes such as plating and etching. Wastewaters originating from PCB manufacturing also have a very similar character to those of metal finishing plants. Printed circuit board manufacturing is based on creating a circuit by sandwiching a conductive metal, usually copper, on or between layers of plastic or glass boards. There are three main production methods: subtractive process, additive process, and semi-additive process. Cleaning and surface preparation is the first step common to all processes. Cleaning and surface preparation includes the processes of scrubbing, alkaline cleaning, etching, and acid cleaning. Catalyst application involving the use of palladium and tin is essential for additive and semi-additive processes. Electroless copper plating in different patterns is common to all methods. Electroless copper plating uses copper salts, mostly copper sulfate, formaldehyde as the reducing agent, chelating compounds, mostly EDTA or tartrates, various polymers and amines, and sodium hydroxide. Copper electroplating used in subtractive and semi-additive processes can be carried out using a cyanide copper bath or other copper baths such as fluoroborate, pyrophosphate, and sulfate copper baths. Other processes employed in the manufacturing include solder plating, solder brightening, nickel and gold plating, and immersion plating.

Fume and Exhaust Scrubbing

Air pollution is a common problem in the metal finishing industry. Solvent wastes, metal ion bearing mists, metal fumes, acid mists and fumes are primary pollutant sources. Air pollution control is, in many cases, a mandatory application. Control of water-soluble contaminants and gases is usually carried out using wet collectors. Wet collectors can be simple spray chambers or packed-bed scrubbers. Scrubber blowdown contains the same pollutants as those of the sources and contributes to wastewater flow in varying proportions depending on plant characteristics.

5.3 ORIGIN AND SOURCES OF WASTEWATERS

Numerous wastewater sources exist in parallel with the high number of unit processes and other activities that result in wastewater generation in the metal finishing industry. However, these sources can be grouped to provide ease of evaluation. Each group involves sources of similar character in terms of wastewater generation pattern and relative amount of wastewater. The grouping also may help to evaluate the in-plant control. The main wastewater groups are:

- rinsing;
- bath dumps;
- used/spent process solutions;
- washing;
- spills, leaks;
- in-plant treatment and recovery.

The rinsing operation as defined in Section 5.2 produces large volumes of wastewater. Generally, pollutant concentrations in rinsing wastewaters are low, but due to high volumes, pollution loads originating from rinsing may become significant. High volumes of rinsing wastewaters have made these sources a focus of water reuse applications. In some cases, for example, plating bath rinses, well-defined and relatively uncontaminated pollutant content of rinse wastewater promotes material recovery usually in combination with water reuse applications. Rinsing is generally a continuous source discharging during working periods. Flow rate may be constant within a certain interval or follows a defined pattern for each source. Rinse wastewaters constitute the most significant group in determining the wastewater control system capacity.

The category of bath dumps covers plating baths, rinse baths, cleaning baths, and other baths. These baths are dumped periodically. While they represent only a minimal part of wastewater, they constitute the greatest part of pollution load. Therefore, they play a key role in determining wastewater control, recovery, and reuse applications as well as wastewater treatment sludge handling and disposal system characteristics. Used and spent solutions are of importance since they are the main wastewater sources of many processes and operations. They are similar to bath dumps in that they generate a small amount of wastewater, but with significant loads of pollution. Their pattern of dumping may not be well defined.

Washing is applied either for cleaning of equipment or floor washing. Washing is common to all operations even for zero discharge processes.

Spills, leaks, and drips of process solutions may be accidental or are sometimes sourced from process characteristics as in the case of testing. These, depending on the severity of the leak, contribute to wastewater and may pose serious problems for wastewater control. Many of the in-plant control and safety measures have been designed for the control of these sources.

In-plant treatment and recovery applications have become quite common and presently they are integral parts of many unit processes. However, some of these applications generate wastewater, sometimes contributing considerably to pollution loads. Filter backwash of painting systems, concentrates of membrane systems, and ultrafiltration applications are a few examples.

As explained in the previous section, air pollution control contributes to wastewaters and is mandatory for many operations. This source must be taken into account. It is rather difficult to relate this wastewater with operations. Although, the operations for which air pollution control are required are known, the application may cover a number of combinations depending on the finishing plant operations and their locations.

Water use characteristics, as well as main source groups of the unit operations of the metal finishing, are shown in Table 3 [3]. In this table, zero discharge operations have been determined by both operation characteristics and evaluations conducted onsite. Plants employing only these

Table 3 Water Usage by Unit Operations [3]

Unit operation	Major water usage	Minimal water usage	Zero discharge	Wastewater source groups: Rinsing	Washing	Area washing	Bath dumps	Spent solution dumps	Spills	Leaks	In-plant control
Electroplating	*			*	*		*				*
Electroless plating	*			*	*		*				*
Anodizing	*			*	*		*				*
Conversion coating	*			*	*		*				*
Etching (chemical milling)	*			*	*		*				*
Cleaning	*			*	*		*				*
Machining	*					*		*	*	*	*
Grinding	*					*		*	*	*	
Polishing		*				*					
Tumbling (barrel finishing)	*			*		*		*			
Burnishing		*		*		*		*			
Impact deformation		*				*					
Pressure deformation		*				*					
Shearing		*				*					
Heat treating	*			*	*		*		*	*	
Thermal cutting		*		*	*						
Welding[a]	*			*				*			
Brazing[a]		*		*				*			

(continues)

Table 3 Continued

				Wastewater source groups							
Unit operation	Major water usage	Minimal water usage	Zero discharge	Rinsing	Washing	Area washing	Bath dumps	Spent solution dumps	Spills	Leaks	In-plant control
Soldering[a]		*		*				*			
Flame spraying[a]		*		*				*			
Sand blasting	*			*							
Other abrasive jet machining		*			*			*	*	*	
Electrical discharge machining		*		*	*						
Electrochemical machining		*		*	*		*				
Electron beam machining			*								
Laser beam machining			*								
Plasma arc machining			*								
Ultrasonic machining			*		*						
Sintering				*		*					
Laminating		*									
Hot dip coating[a]		*						*			
Sputtering			*								
Vapor plating			*								
Thermal infusion			*	*				*	*		
Salt bath descaling	*					*			*	*	
Solvent degreasing		*		*				*	*	*	
Paint stripping	*					*					
Painting	*						*				
Electrostatic painting	*				*						
Electropainting	*					*					*
Vacuum metalizing			*		*						
Assembly		*				*					
Calibration			*		*				*	*	
Testing	*			*		*		*			
Mechanical plating	*										

operations may not have wastewater discharges unless they conduct wastewater-generating pre- or postoperations, that is, cleaning, testing. Printed circuit board manufacturing employs mainly plating, etching, and cleaning operations, and has the wastewater sources belonging to these operations.

The amount of wastewater generated from metal finishing shops and plants varies over a wide range. Establishment of production and wastewater generation ratio (production-based wastewater flow or unit flow) is quite difficult and cannot be defined on the basis of the whole industry. In some cases, however, unit flows can be determined on the basis of operations or groups of operations. This proves very useful in terms of in-plant control measures as well as legal enforcement. Unit wastewater flows generally apply to plating operations and are expressed as volume of wastewater per unit surface finished. Amount of daily wastewater generation rarely exceeds 2000 m^3 even for big and complex plants such as the automotive industry. Wastewater flows frequently remain in the range 0–400 m^3/day. However, daily flows are lower than 100 m^3 for workshop-sized plants [3].

5.4 CHARACTERISTICS OF WASTEWATERS

Metal finishing wastewaters originating from the sources depicted in previous sections contain the chemicals used and workpiece materials. Owing to the great variety of chemicals and materials used in the processing, wastewaters contain a long list of pollutants [3], which includes metals, organic wastes, suspended solids, cyanide, phosphate, fluoride, and ammonia. Ammonia, which generally does not reach high concentrations, is mostly mixed with metals and cyanide. If it exists in high concentrations, it needs to be controlled because of its interference with the removal of these parameters. Generally, ammonia has not been considered as a parameter for regulation. Cyanide, phosphate, and fluoride are well-known parameters having defined impacts on the environment and they are regulated parameters. Suspended solids is a traditional parameter controlling both the treatment performance and discharge of organic and inorganic toxic parameters. Metals constitute an important group, including over 30 metals. Some metals are considered nontoxic, such as iron, magnesium, manganese, and titanium. They are not regulated and are removed to a great extent through the metal removal processes. Toxic metals have been regulated by imposing stringent discharge limitations. Cadmium, chromium, copper, lead, nickel, zinc, and silver are found at significant concentrations in the metal finishing wastewaters, and have been commonly regulated.

Organic wastes vary in a wide spectrum as far as both their structure and their impacts on the environment are concerned. Two parameters have been widely used to represent and control organic wastes: oil and grease, and toxic organics. Oil and grease is a conventional parameter. Its source definition, treatment, and measurement can be readily carried out. Both conventional and advanced methods of oil and grease removal are available, providing very high efficiencies. Oil and grease removal, while providing the control of a significant part of organic matter content that is measurable by chemical oxygen demand (COD), contributes to some extent to the removal of toxic organics. Therefore, oil and grease is one of the most important standard and regulated parameters of metal finishing wastewaters.

The origins of toxic organics are quite complex. However, for ease of evaluation, three groups of material use can be considered. The first group is solvent use, which is the most important group in terms of toxicity. Solvent use is the primary source of priority pollutant content of wastewaters. Its control largely relies on in-plant control and separate treatment methods. Therefore, this source group is generally considered as a separate subcategory for water quality management. The second group of organic matter covers complex formers, mostly chelating agents, which are widely used in plating baths, being mixed with metals. They strongly interfere with metal removal, requiring specialized control and treatment methods. This group is also conceived as a separate subcategory. The third group involves all remaining organic matter sources. Metalworking fluids, emulsifying agents, soluble oils, antibacterial agents, and surfactants are among the important chemicals determining the toxic organic content of wastewaters. Control of this group is largely dependent on in-plant control, material recovery, and material changes rather than treatment. When treatment is inevitable, advanced methods of treatment rather than conventional methods need to be used either on the separated flows or on the total wastewaters. Regulation of organic pollutants is based on priority pollutant limitation, although they can be represented as a collective parameter of TTO (total toxic organics). A list of significant organic parameters is presented in Table 4 [6].

Table 4 Toxic Organics in Metal Finishing Wastewaters [6]

Acenaphthene	4-Chlorophenyl phenyl ether	Fluorene
Acrolein	4-Bromophenyl phenyl ether	Phenanthrene
Acrylonitrile	*bis* (2-Chloroisopropyl) ether	1,2,5,6-Dibenzanthracene
Benzene	*bis* (2-Chloroethoxy) methane	Indeno(1,2,3-cd) pyrene
Benzidine	Methylene chloride	2,3-Phylene pyrene
Carbon tetrachloride	Methyl chloride	Pyrene
Chlorobenzene	Methyl bromide	Tetrachloroethylene
1,2,4-Trichlorobenzene	Bromoform	Toluene
Hexachlorobenzene	Dichlorobromomethane	Trichloroethylene
1,2-Dichloroethane	Chlorodibromomethane	Vinyl chloride
1,1,1-Trichloroethane	Hexachlorobutadiene	Aldrin
Hexachloroethane	Hexachlorocyclopentadiene	Dieldrin
1,1-Dichloroethane	Isophorone	Chlordane
1,1,2-Trichloroethane	Naphthalene	4,4-DDT
1,1,2,2-Tetrachloroethane	Nitrobenzene	4,4-DDE (*p,p*-DDX)
Chloroethane	2-Nitrophenol	4,4-DDD (*p,p*-TDE)
bis (2-Chloroethyl) ether	4-Nitrophenol	Alpha-endosulfan
2-Chloroethyl vinyl ether	2,4-Dinitrophenol	Beta-endosulfan
2-Chloronaphthalene	4,6-Dinitro-o-cresol	Endosulfan sulfate
2,4,6-Trichlorophenol	*N*-nitrosodimethylamine	Endrin
Parachlorometa cresol	*N*-nitrosodiphenylamine	Endrin aldehyde
Chloroform	*N*-nitrosodi-*n*-propylamine	Heptachlor
2-Chlorophenol	Pentachlorophenol	Heptachlor epoxide
1,2-Dichlorobenzene	Phenol	(BHC-hexachloro-cyclohexane)
1,3-Dichlorobenzene	*bis* (2-Ethylhexyl) phthalate	Alpha-BHC
1,4-Dichlorobenzene	Butyl benzyl phthalate	Beta-BHC
3,3-Dichlorobenzidine	Di-*n*-butyl phthalate	Gamma-BHC
1,1-Dichloroethylene	Di-*n*-octyl phthalate	Delta-BHC
1,2-*trans*-Dichloroethylene	Diethyl phthalate	(PCB-polychlorinated biphenyls)
2,4-Dichlorophenol	Dimethyl phthalate	PCB-1242 (Arochlor 1242)
1,2-Dichloropropane	1,2-Benzanthracene	PCB-1254 (Arochlor 1254)
1,3-Dichloropropylene	Benzo(a)pyrene	PCB-1221 (Arochlor 1221)
2,4-Dimethylphenol	3,4-Benzofluoranthene	PCB-1232 (Arochlor 1232)
2,4-Dinitrotoluene	11,12-Benzofluoranthene	PCB-1248 (Arochlor 1248)
2,6-Dinitrotoluene	Chrysene	PCB-1260 (Arochlor 1260)
1,2-Diphenylhydrazine	Acenaphthylene	PCB-1016 (Arochlor 1016)
Ethylbenzene	Anthracene	Toxaphene
Fluoranthene	1,12-Benzoperylene	2,3,7,8-Tetrachlorodibenzo-*p*-dioxin

The above discussion indicates that segregation of wastewater is the basic approach for management, control, and treatment of the metal finishing wastewaters. Within this context, seven wastewater groups have been defined [3]. Of these, three groups have already been defined in relation to organic matter. These are oily wastewaters, complexed wastewaters, and solvents, all requiring in-plant control and pretreatment. Three additional groups requiring similar attention and pretreatment are cyanide wastewaters, hexavalent chromium wastewaters, and precious metal wastewaters. The remaining group represents the wastewaters containing mainly metals that can be handled by metal treatment techniques without a need for pretreatment, that is, the common metals wastewaters. Table 5 shows the sources of the main materials determining the wastewater groups on the basis of unit operations [5]. Table 5 also indicates the unit operations that are of importance as sources of toxic pollutants. Although every group of wastewater has its own characteristic parameters, all groups of wastewater contain virtually all pollutant parameters of metal finishing wastewaters. A brief evaluation of the general characteristics of wastewater groups is given below.

Common Metals

Basic pollutants contained in this wastewater group are the metals and the acids. Metal concentrations vary from several mg/L to several hundred mg/L. Concentrations of other pollutants are generally low, but their values are determined by flow segregation practice.

Precious Metals

Gold and silver are two important and characteristic parameters of the precious metals wastewater group, while other precious metals like palladium and rhodium are less often used. Owing to the high price of precious metals, this wastewater is wasted only after efficient recovery operations. Although silver concentration may reach up to one hundred mg/L, gold concentration never exceeds a few mg/L.

Complexed Metals

Electroless plating and immersion plating are two major sources of complexed metal wastewaters. Depending on other sources, many metals may occur in this wastewater, but the main metal content is determined by the electroless and immersion plating applications. In this context, copper, nickel, and tin are frequently encountered. Commonly used complex formers are given in Table 6 [3].

Cyanide

Cyanide wastewaters are subject to a more strict flow segregation. Although electroplating and heat treating are major sources, other sources such as electrical degreasing contribute to the cyanide wastewater flow. Cyanide concentrations may reach several thousand mg/L.

Hexavalent Chromium

Similar to cyanide wastewaters, hexavalent chromium wastewaters are carefully segregated. The number of sources may be higher than those of complexed metals and cyanide wastewaters. Chromium concentrations vary, but a very high concentration of hexavalent chromium can be encountered.

Table 5 Wastewater Group Determining Parameters by Unit Operations [5]

Unit operation	Metals	Hexavalent chromium	Cyanide	Oils	Toxic organics
Electroplating	*	*	*		*
Electroless plating	*	*			*
Anodizing	*	*			*
Conversion coating	*	*	*		*
Etching (chemical milling)	*	*	*		*
Printed circuit board	*				*
Manufacture	*	*	*	*	*
Cleaning	*			*	
Machining	*			*	
Grinding	*			*	
Polishing	*	*	*	*	
Tumbling (barrel finishing)	*		*	*	
Burnishing	*			*	
Impact deformation	*			*	
Pressure deformation	*			*	
Shearing	*		*	*	
Heat treating	*				
Thermal cutting	*				
Welding	*				
Brazing	*				
Soldering	*				
Flame spraying	*				
Sand blasting	*				
Other abrasive jet machining	*			*	
Electrical discharge machining	*		*		*
Electrochemical machining	*				
Laminating	*				
Hot dip coating	*			*	*
Salt bath descaling	*			*	*
Solvent degreasing	*			*	*
Paint stripping	*				*
Painting	*	*			*
Electrostatic painting	*				*
Electropainting	*			*	*
Assembly				*	
Testing	*	*			
Mechanical plating					

Oily Wastewaters

Treatment and control of oily wastewater depends on the concentration of oils. Table 7 shows the unit operations with oily waste characteristics [3]. In concentrated sources, oil concentrations may reach several hundred thousand mg/L, and even higher. Oily wastewater contains a significant part of toxic pollutants.

Table 6 Common Complexing Agents [3]

• Ammonia	• *O*-phenanthroline
• Ammonium chloride	• Oxine, 8-Hyroxyquinoline (Q)
• Ammonium hydroxide	• Oxinesulphonic acid
• Ammonium bifluoride	• Phthalocyanine
• Acetylacetone	• Potassium ethyl xanthate
• Citric acid	• Phosphoric acid
• Chromotropic acid	• Polyethyleneimine (PEI)
• Cyanide	• Polymethacryloylacetone
• DTPA	• Poly (*p*-vinylbenzyliminodiacetic acid)
• Dipyridyl	• Rochelle salts
• Disulfopyrocatechnol	• Sodium tripolyphosphate
• Dimethylglyoxime	• Succinic acid
• Disalicylaldehyde 1,2-propylenediimine	• Sulphosalicylic acid (SSA)
• Dimercaptopropanol (BAL)	• Salcylaldehyde
• Ethylenediaminetetraaceticacid (EDTA)	• Salicylaldoxime
• Ethylenebis (hydroxyphenylglycine) (EHPG)	• Sodium hydroxyacetate
• Ethylenediamine	• Sodium citrate
• Ethylenediaminetetra (methylenephosphoric acid) (EDTPO)	• Sodium fluoride
• Glyceric acid	• Sodium malate
• Glyolic acid	• Sodium amino acetate
• Hydroxyethylenediaminetriacetic acid (HEDTA)	• Tartaric acid
• Hydroxyethylidenediphosphonic acid (HEDP)	• Trisodium phosphate
• HEDDA	• Trifluoroacetylacetone
• Lactic acid	• Thenoyltrifluoroacetone (TTA)
• Malic acid	• Triaminotriethylamine
• Monosodium phosphate	• Tetraphenylporphin
• Nitrilotriacetic acid (NTA)	• Toluene dithiol
• *N*-Dihydroxyethylenephosphonic acid (NTPO, ATMP)	• Thioglycolic acid
• Nitrilotrimethylenediamine	• Thiourea

Solvents

The solvent group does not produce a wastewater flow. This waste type will be dealt with in Section 5.6.

5.5 WASTEWATER TREATMENT

5.5.1 Methodology

Treatment of metal finishing wastewaters is based on the wastewater groups as defined in Section 5.4. Segregating the wastewaters to form the wastewater groups is an essential part of treatment. Flow segregation facilitates some treatment applications, but the main reason for segregation is that mixing of wastewater groups causes inefficient treatment, even total failure of treatment applications. Moreover, mixture of wastewater groups results in health hazards and economic losses. In this context, wastewater treatment will be dealt with on a wastewater group basis. A general flow segregation and waste treatment scheme is illustrated in Figure 2 [3]. Precious metals wastewater treatment will not be addressed separately. Widely employed recovery technologies for this group are evaporation, ion exchange, and electrolytic recovery,

Table 7 Characterization of Oily Waste [3]

	Character of oily waste generated	
	Concentrated	Dilute
Cleaning		*
Machining	*	*
Grinding	*	*
Polishing		*
Tumbling (barrel finishing)		*
Burnishing	*	
Impact deformation	*	
Pressure deformation	*	
Shearing	*	
Heat treating	*	*
Welding	*	
Brazing	*	
Soldering	*	
Flame spraying	*	
Other abrasive jet machining	*	
Electrical discharge machining	*	
Salt bath descaling		*
Solvent degreasing	*	
Paint stripping		*
Assembly		*
Calibrating		*
Testing	*	

which are explained in the common metals treatment section. Recovery and disposal will be discussed in Section 5.6.

Some heavy metals may not be treated adequately by conventional schemes due to the stringent limitations imposed upon them. These metals may require special attention, flow segregation, and proper treatment for control. Cadmium and mercury are examples for these applications. The optimum pH for cadmium removal is too high and a standard application of hydroxide precipitation may not provide the required effluent concentrations. Therefore, cadmium-containing flows can be separated and treated by the methods mostly providing recovery, for example, evaporative recovery and ion exchange. Mercury frequently exists as the sole or main component of segregated flows and efficient treatment methods, for example, sulfide precipitation, may be applied for control.

5.5.2 Common and Complexed Metals Treatment

Hydroxide Precipitation

Hydroxide precipitation is the most common conventional treatment applied to metal finishing wastewaters to remove heavy metals, as well as many other particulate and soluble pollutants. The method is based on low solubility of metal hydroxides at alkaline pH values. As the metals are converted to the solid phase, they are separated from wastewater by physical means such as sedimentation, flotation, and filtration. Iron, copper, zinc, cadmium, beryllium, cobalt, mercury, manganese, and aluminum are among the metals that can be treated by hydroxide precipitation.

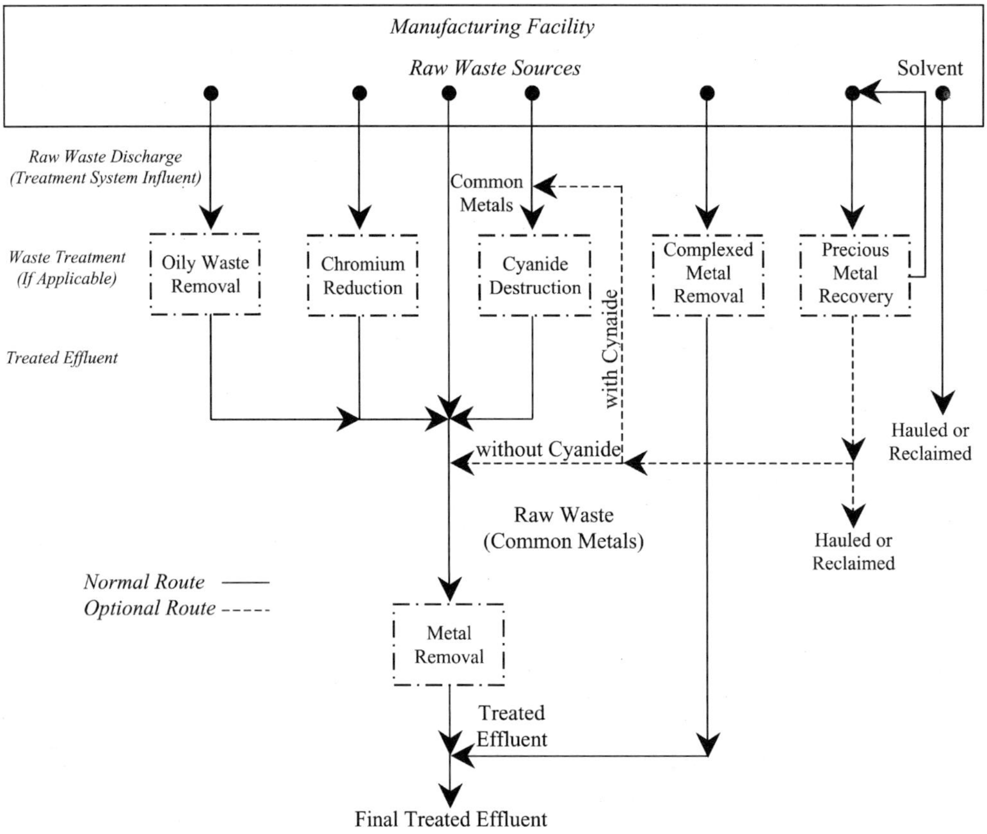

Figure 2 Wastewater groups and treatment schematic. (From Ref. 3.)

Chromium can be precipitated only in the trivalent state. The process is effective in removing other soluble and particulate pollutants, often by the aid of organic and inorganic coagulants.

Hydroxide precipitation has the advantage of removing many of the pollutant parameters existing in metal finishing wastewaters without pretreatment. The process operates at ambient conditions and its operation is easy and suited to automatic control. The most important advantage of the process is its low cost, particularly when lime is used for pH adjustment. Common use of the process has provided accumulation of data and experience on which precipitation strategies can be based and prediction of performance can be made within a certain range of precision. Existing information about the theoretical basis of the process as well as ongoing research in the field help modeling and planning of the process.

However, there are disadvantages to hydroxide precipitation. The first and most important one is the relatively high quantity of sludge produced. Secondly, the optimum pH values of the different metals, the pH at which a certain metal reaches minimum solubility, are different. Maximizing the removal of a certain metal, generally, results in significant reductions in the removal of other metals. Sometimes a two-stage treatment may be needed to provide high efficiencies for all the metals existing in the wastewater. On the other hand, solubilities of the metal hydroxides are not low enough to remove complexed metals. Furthermore, as with all precipitation–sedimentation processes, the performance of the process is determined by the solid separation step. Another drawback is the use of lime for pH adjustment. Lime is used as slurry, which is difficult to handle in terms of pumping, piping, and feeding. However, it is very cheap compared to other alkalis and, more importantly, it proves quite beneficial for further process steps.

Solubility of hydroxides is dependent on pH. This dependency is not only due to the hydroxide ion concentration that determines the solubility of metal hydroxides, but also to a larger extent due to the formation of metal hydroxo complexes. Concentration of hydroxo complexes is also determined by pH. Hydroxo complexes may be positively or negatively charged or neutral. Positively charged ones dominate at lower pH, namely the pH values lower than the minimum solubility of the metal. Negatively charged species prevail at higher pH values. Concentration of neutral species is, however, independent of pH. Polymer (polynuclear) type hydroxo complexes that have more than one metal atom also exist. Metal solubility is the sum of free metal ion and hydroxo complex concentrations. Figure 3 shows the solubility diagram of cadmium hydroxide [7]. As seen from the figure, solubility exhibits a shape similar to a parabola or an upside-down bell shape. This shape sometimes approximates to a triangle. In this shape, solubility decreases with increasing pH up to a minimum, then it increases with further increase in pH. The point of minimum solubility is termed "optimum pH." Some metals such as Fe^{3+} do not exhibit such a solubility diagram due to very low solubility product, and solubility monotonously decreases as the pH is increased (Fig. 4) [3]. For such metals, there is no optimum pH. In Figure 3, cadmium solubility is seen to be determined by Cd^{2+} and major

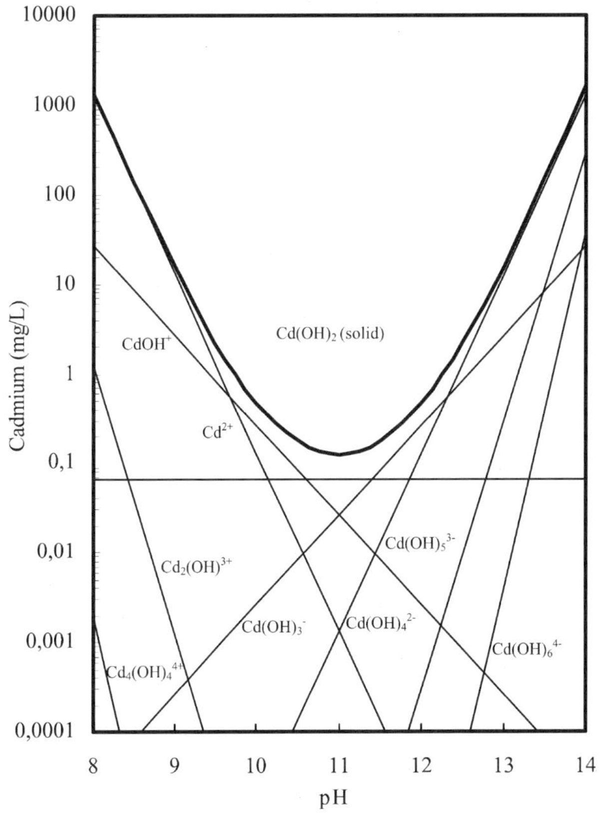

Figure 3 Cadmium solubility diagram. (From Ref. 7.)

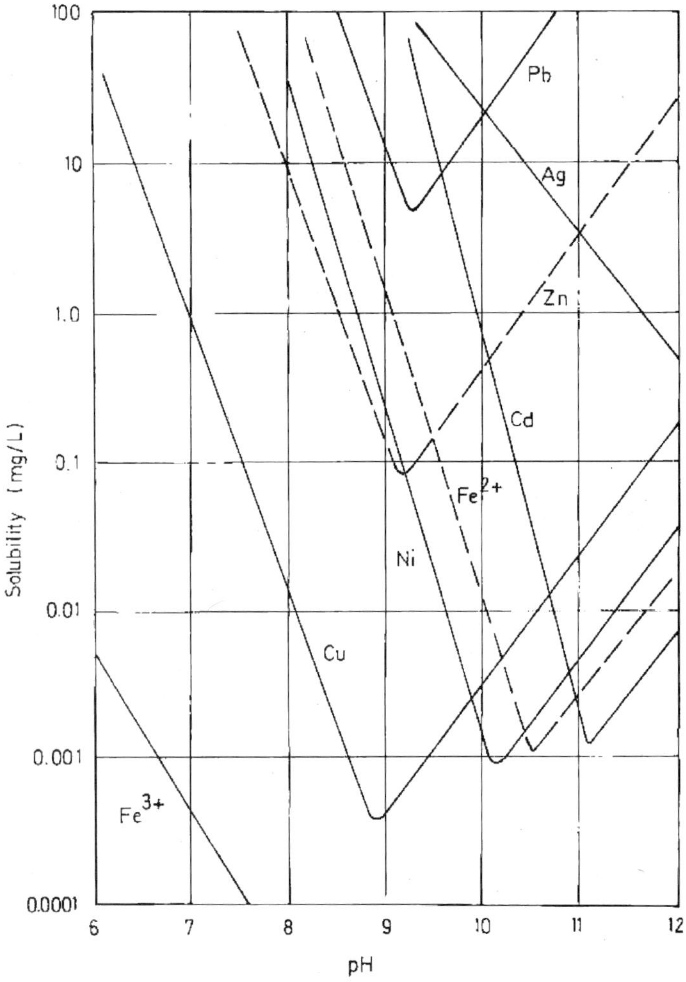

Figure 4 Metal hydroxide solubilities. (From Ref. 3.)

complexes $CdOH^+$ and $Cd(OH)_2^0$ as the pH approaches to optimum pH. $Cd(OH)_2^0$ is, as mentioned above, independent of pH and constitutes a limit to minimum solubility at all points near the optimum pH.

Beyond the optimum pH of 11.2, solubility increases due to increasing concentration of negatively charged complexes and is determined to a great extent by $Cd(OH)_3^-$ and $Cd(OH)_4^{2-}$. Optimum pH is quite important for many of the metals because their solubility changes abruptly around the optimum pH. In Figure 4, the optimum pH of zinc is read as 9.2 and the solubility at this pH is lower than 0.1 mg/L, which is a low enough value for direct discharge to many receiving bodies. However, one unit decrease in pH increases the solubility nearly a hundred-fold and zinc concentration reaches 9 mg/L at pH 8.2. However, some metals have a rather flat shape around optimum pH and solubility does not exhibit radical changes over pH ranges of one or even two units. Trivalent chromium is a good example for this case (Fig. 5) [8]. Figure 5 is obtained from the data given in the literature [9]. The shape of the solubility curve determines the precipitation strategy and minimum concentration obtainable by the hydroxide precipitation

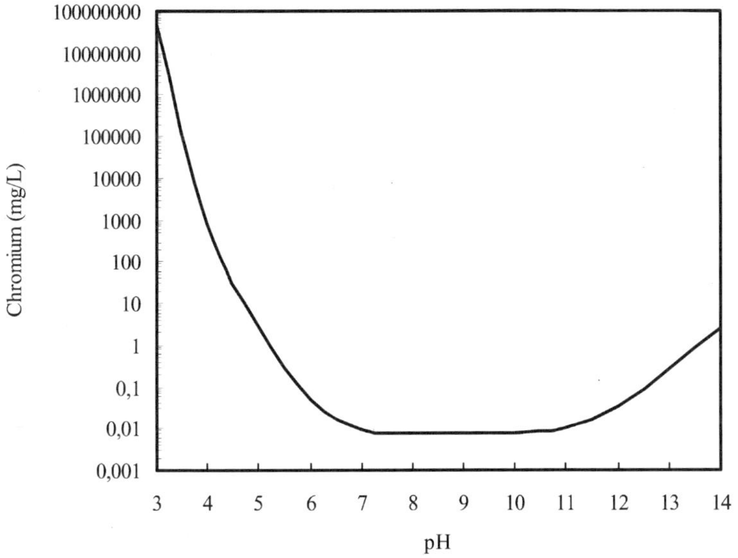

Figure 5 Chromium solubility diagram. (From Ref. 8.)

for the metal of concern. When more than one metal ion exist in solution, the situation becomes more complex and pH adjustment requires a detailed evaluation to reach minimum or acceptable concentrations for all metals. Several cases can be selected to illustrate on Figure 4 the selection of working pH for a mixture of two metals. In the first case, copper and nickel exist in solution and hydroxide precipitation is required to reduce the concentration of both metals below 1.0 mg/L. This is the easiest case to handle in that adjustment of pH to any value in the wide pH range between 7.0 and 12.0 meets the requirement.

In the second case, the metals existing in solution are copper and cadmium; final concentrations required for these metals are 0.1 mg/L for cadmium and 1.0 mg/L for copper. In this case, pH should be adjusted close to the optimum pH of cadmium because copper solubility is always below 1.0 mg/L between pH 7.0 and 12.0. If effluent concentrations of both metals are required to be below 0.05 mg/L for the above case in finding a proper pH, shifting the pH from the optimum pH of cadmium towards the optimum pH of copper is needed. A proper pH range satisfying the requirement is 10.6–11.1.

In another case, cadmium and zinc exist in solution and final concentrations of cadmium and zinc are required to be 0.1 and 1.0 mg/L, respectively. There is a very narrow pH interval in which this requirement can be met. A small shift of pH from 10.5 towards 11.0 will cause the zinc limit to be exceeded. In such a case, there is no solution other than a separate treatment. Then, zinc is first precipitated at its optimum pH, which has no effect on the cadmium concentration if the latter is not exceedingly high (over 100 mg/L). After separating the zinc hydroxide, increasing the solution pH to the optimum pH of cadmium provides the most effective cadmium removal, being well below the 0.1 mg/L limit. These examples indicate the importance of pH adjustment for hydroxide precipitation. pH adjustment depends on the case and degree of treatment requirement for the metals of concern. Determination of the operating pH requires a very careful examination of the theoretical basis of the process as the number of metals in the wastewater increases. The determination of the discharge standards, if the hydroxide precipitation is assumed to be the technological basis, in turn, takes account of these evaluations to set out achievable limits.

Many factors affect the performance of hydroxide precipitation [10]. A group of factors directly affects the solubility of metal hydroxides, among which, those of importance are:

- temperature;
- ionic strength;
- complex formation;
- formation of other solid phases;
- alkalinity; and
- buffers.

The effect of temperature on solubility can be calculated by the changes in equilibrium and stability constants of relevant species. The change in ambient conditions and wastewater temperatures do not significantly affect the process performance. Ionic strength is a measure of dissolved salt content of wastewater, and it increases as dissolved solids concentration increases. In general, solubility increases as the ionic strength increases. Solubility correction is carried out using activity coefficients, which can be calculated simply by Davies or Güntelberg approximation formulas up to ionic strengths of 0.5 M [11]. However, in these approaches, complex species cannot be accounted for in a precise manner. In one study, the effect of ionic strength on hydroxide precipitation performance was evaluated employing the Ion Pairing Model [12]. In this approach, complex species can be separately accounted for. Application of the model to nickel and copper indicated that at pH 10.0, differences in solubilities for zero and 0.9 M ionic strength remain within the range of 14–16%.

Complex formation is the most important factor causing radical increases in metal solubility. The effect of complex formers (ligands) on solubility depends on the hydroxide solubility of metal and on the stability constant of the metal–ligand complex. While some ligands such as short-chain fatty acids do not exert a significant effect on solubility, strong ligands may totally upset the process. The effect of complexing on hydroxide precipitation is dealt with in the subsequent subsection on complexed metal treatment. Formation of solid phases other than metal hydroxide is a common phenomenon and can be encountered in several forms. One of the most important occurrences is the formation of a solid phase of the metal being precipitated within the pH range of hydroxide precipitation. Carbonate may frequently precipitate instead of, or together with, hydroxide.

At high alkalinity values, zinc carbonate precipitates at pH values very close to the optimum pH of zinc hydroxide [10]. Hydroxy-carbonate species such as $NiCO_3 \cdot Ni(OH)_2 \cdot 4H_2O$ tend to precipitate within wide pH ranges depending on carbonate concentration [13]. Some anions commonly found in metal finishing wastewaters, such as chloride and sulfate, may precipitate metals, for example, silver and lead. Carbonate and sulfate and some organic anions precipitate calcium ions added as lime for pH adjustment. Alkalinity plays an important role in determining process performance and operation in more than one way. It may cause precipitation of metal carbonate or the lime added, and provides a good buffer for pH adjustment. Fine adjustment of pH, as noted above, is an important step in hydroxide precipitation. Fine pH adjustment, however, can only be realized if proper buffers exist. Metal finishing wastewaters often contain buffers, that is, organic acids, ammonia, phosphate, and carbonate, but difficulties arise if their amounts or their effective pH ranges are not adequate, unless a proper buffer is provided.

Another group of variables that affects process performance and/or operation is related to solids formation. Metal hydroxides initially may not be ideal crystals. Solubility decreases as the solid converts to ideal crystal by aging. The importance of crystallinity was emphasized and the conclusion drawn from experimental studies with copper (Cu^{2+}) that system performance was dependent on the least-soluble kinetically precipitated phase rather than thermodynamically favored [14]. The rate of formation of the solid phase depends on supersaturation, existence of particulate matter, and mixing. Nucleation generally does not pose a problem because wastewaters commonly contain particulate matter and/or are oversaturated in terms of metals. If very concentrated solutions such as immersion copper baths are precipitated, the solids initially formed may be very finely divided and hard to separate by gravity [15].

In cases of moderately oversaturated solutions with no particulate matter, sludge recycling may help formation of particles that can be readily coalesced and settled. The type of pH adjustment agent also affects the properties of the solids formed. Sodium hydroxide generally causes the formation of smaller particles, but upon coagulation they yield a very clear supernatant. The best solution to the solid formation and separation problem is the use of inorganic and organic coagulants. For this purpose, $FeCl_3$, alum, and various polyelectrolytes have been commonly used. While they provide an efficient solid separation, they enable or promote the removal of particulate and dissolved pollutants.

Suspended solids are common to many metal finishing wastewaters. Coarse particles may not always have high rates of sedimentation as in the case of oil or other organics. Coagulation helps to remove these particles effectively within reasonable settling durations. However, the main function of coagulation is to remove colloid material existing both in the wastewater and formed by hydroxide precipitation. Because hydroxide precipitation efficiency depends on physical separation of suspended solids, the degree of heavy metal and other pollutant removal is limited and determined by suspended solids removal. With a proper coagulation and settling of suspended solids, concentration in the effluent can be kept below 20 mg/L [3]. Organic and inorganic matter, soluble and emulsified oil, as well as micropollutants, can be removed to a significant extent by precipitation. Their removal efficiencies, and in turn, the basis of regulatory limitation by this process, rely almost totally on the capability of adsorption of the flocs that are formed by the addition of flocculants.

Experimental studies have shown that a variety of metal hydroxides may be used for preconcentration of polynuclear aromatic hydrocarbons and polychlorinated biphenyls, indicating the potential of coagulation to remove these pollutants [16]. Complex formers in wastewaters affect not only the solubility of metal ions, but also behave like stabilizers for inorganic coagulants, particularly for ferrous sulfate [17]. The presence of weak complexing agents such as tartrate, citrate, and ammonia was determined to have a slight effect on particle distribution tending to form smaller particles. Inorganic flocculants also affect the soluble metals that may be in equilibrium with the formed solid phase or that remain unaffected by hydroxide precipitation. Freshly precipitated alum and ferric chloride flocs are known to have a considerable capacity to adsorb/coprecipitate metal ions [18]. On the other hand, while lime also has the potential to remove some pollutants, it was claimed that its precipitates, $CaCO_3$ and $CaSO_4$, have a limited capacity to remove soluble nickel and zinc [13,19]. It has been reported that lead can be immobilized on calcium hydroxy-apatite [20].

All the abovementioned factors and variables involved in hydroxide precipitation give rise to difficulties in evaluating the process. Prediction of performance usually requires case-by-case theoretical and experimental evaluation. On the other hand, prediction of performance on a general basis is important for the design of projected plants, as well as for setting the technology-based discharge standards. The concept of solubility domain was proposed to provide a solution to this problem. This approach graphically represents the effect of all known or predictable variations and defines a domain for solubility rather than a single line drawn for a set of conditions [10]. Figure 6 illustrates the solubility domain presentation. This figure was obtained for zinc hydroxide and carbonate solubility using data in the literature [8]. The values of solubility product and complex stability constants vary over a wide range. The phase boundries shown in the figure represent the solubilities drawn for different set of equilibrium constants. The solubility domain boundary enveloping all the solubility diagrams represents minimum and

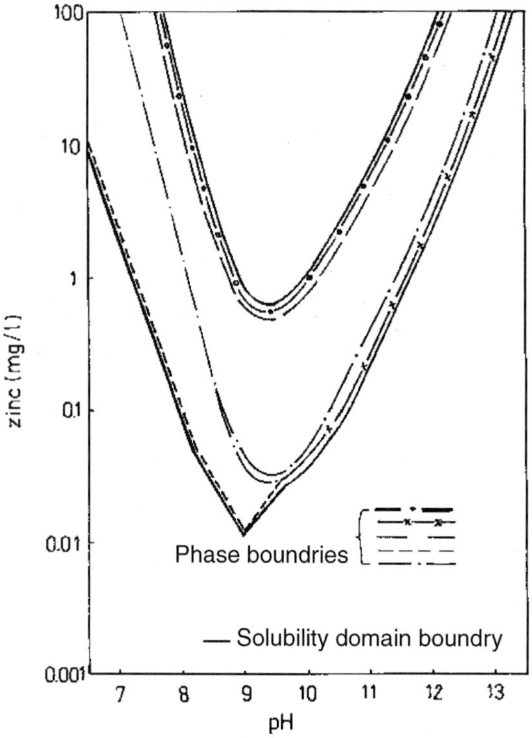

Figure 6 Illustration of solubility domain. (From Ref. 8.)

maximum solubilities for a given pH. In this approach, the domain can be confined in narrower bands as the theoretical and experimental studies provide better understanding and predictability to the process. This approach has been used by several researchers for elaborating the effects of solubility limiting phases, electrolyte composition, adsorption capacity of inorganic coagulants, and of other factors [13,14,19,21].

Hydroxide precipitation also removes some other parameters of importance. In addition to the parameters removed by the aid of coagulants, phosphate, fluoride, sulfate, and alkalinity are effectively removed by the process. Phosphate and fluoride can be precipitated by forming calcium salts. Therefore, their removal depends on the use of lime as well as on providing the conditions to maximize their removal. Calcium phosphate chemistry is quite complex [22]. Many calcium phosphate compounds have low solubility at high pH values. Although the thermodynamically favored species is hydroxy-apatite, other calcium phosphate salts, if they are kinetically dominating, may form. This complexity, together with the main target of maximizing metal removal, generally yields lower removal of phosphate than expected, or, than can be otherwise achieved. Effluent phosphate concentration was statistically determined to be about a mean value of 10 mg/L [3]. Fluoride removal is accomplished by the formation of sparingly soluble calcium fluoride. However, its removal is limited and effluent flouride concentrations of 10–20 mg/L are obtainable as a result of both the relatively high solubility of calcium flouride and the factors affecting its formation and solubility. Coprecipitation on precipitates, particularly on alum and magnesium hydroxide flocs, is effective in the removal [3,23].

Alkalinity and sulfate are precipitated by calcium. Alkalinity removal is quite high, particularly if pH is over 10. Formation of calcium carbonate may help the removal of some pollutants; however, it results in loss of buffer capacity and an increase in the amount of sludge. Solubility of calcium sulfate is higher and its removal depends on ionic strength, competing ions like carbonate and phosphate, as well as on complex formation with other metals. In concentrated wastewaters, sulfate concentrations achievable by calcium sulfate precipitation have been determined to be 1500–1700 mg/L [15].

Carbonate, Sulfate, and Chloride Precipitation

Carbonate precipitation can be applied within hydroxide precipitation, or separately. Advantages of carbonate precipitation over hydroxide precipitation include the potential of lower pH operation, easier settling, and improved removal efficiency. The main difference between carbonate precipitation and hydroxide precipitation is that the former's solubility is a function of carbonate concentration. Smaller metal solubilities can be provided by increasing the carbonate doses at lower pH values than those of the optimum pH of hydroxide precipitation. However, carbonate precipitation has its limitations. A major limitation is precipitation kinetics. Experimental studies conducted to determine carbonate precipitation performance with zinc, cadmium, and nickel have shown that the obtained residual metal concentrations were above the theoretical solubilities [24]. However, carbonate precipitation is advantageous and commonly applied at least for some metals such as lead, which have high hydroxide solubilities. Carbonate precipitation also provides recovery of such metals. In some cases, carbonate precipitation is inevitable. Experimental studies conducted with solutions containing 10 mg/L lead, 10^{-2} M sulfate, and 3×10^{-5} to 1.5×10^{-2} M carbonate between pH 3.9 and 11.3 have indicated that $PbSO_4(s)$, $PbCO_3(s)$, and $Pb_3(CO_3)_2(OH)_2(s)$ precipitates formed, while no $Pb(OH)_2(s)$ was detected [25]. Carbonate precipitation applied to battery manufacturing wastewaters yielded residual lead values of less than 0.2 mg/L [26]. Lead precipitation with carbonate in the presence of Fe^{3+} resulted in the same residual lead concentration [27]. The optimum pH for lead carbonate precipitation was found to be 8.0–9.0, yielding a residual lead concentration of 0.3 mg/L.

Sulfate precipitation of lead was determined to require at least twice the stoichiometric dose between pH values 3.5 and 5.5 to obtain lead concentrations as low as 2.0 mg/L [28].

Sulfate precipitation is commonly applied to barium removal. Barium concentrations around 1.0 mg/L are obtainable with stoichiometric sulfate doses [23,29]. Barium can also be precipitated as carbonate at pH values over 10 with higher efficiencies. Chloride precipitation, particularly, in combination with cyanide oxidation, is applied to the removal and recovery of silver [23].

Treatment of Complexed Metals

Complexed metals form a group of wastewater pollutants that contain complexing agents. Complexing agents prevent the metals from being precipitated. In fact, almost all groups of wastewaters from metal finishing operations contain inorganic and organic complex formers that may interact or interfere with many of the treatment methods. In some cases, metals can be effectively removed independently of how strong the metal binding is, as in the case of sulfide precipitation. Therefore, complexed metal definition is made relative to a reference. This reference is mainly hydroxide precipitation. If a wastewater containing complexing agents cannot be treated to remove metals by hydroxide precipitation within the limits of usual operation, it is considered to be complexed metal wastewater. While some weak complexing agents such as citrate and tartrate may not interfere with hydroxide precipitation [30], most of the ligands presented in Table 6 modify the precipitation performance to yield unacceptable effluent metal concentrations. Many of these complex formers are chelates, which bind the metals in more than one position to form stable structures. Because treatment of complexed metals tends to be expensive, they need to be segregated. The treatment methods applicable to complexed metals wastewaters can be classified as:

- pretreatment followed by hydroxide precipitation;
- modified hydroxide precipitation; and
- other methods of treatment.

Pretreatment of complexed metal wastewaters aims to retain hydroxide precipitation as the main or central treatment. Thus, the purpose of pretreatment is to destroy the complexing agents or to convert them into a form or into compounds such that they cannot interfere with the precipitation. One of the methods being used is the addition of complex breakers such as sodium dimethyldithiocarbamate (DTC). This method may be expensive and some complex breakers may have toxic effects on the environment. Another pretreatment method is chemical reduction, which is based on a complexing equilibrium. Many of the complex formers are weak acids, which convert to their acid forms at low pH, thus freeing the metal ion. If the metal is reduced at this pH, using chemical reduction agents, to an oxidation state in which it cannot re-combine with complexing agents, then it can be separated by conversion into a compound that is separable by precipitation or by using other means.

Oxidation is another method aimed at the destruction of complex formers, generally not to mineralization, that is, carbon dioxide and water, but to structures for which they have no capacity of complexing or at least to a much weaker extent than the initial organic matter. Strong oxidants such as chlorine or ozone are used for oxidation. Existing metals may serve as catalysts. Oxidation may be conducted at pH values, usually acid, beyond the range of hydroxide precipitation. The method may be useful in removing organic priority pollutants. Selective ion exchange, as in the case of ammonia removal by clinoptilolite, may also prove useful [31].

Modified hydroxide precipitation is based on the addition of chemicals to which complexing agents preferentially bind, allowing freed metal ions to be precipitated. The theory and application basis of the hydroxide precipitation of complexed metals is given in the literature [32]. The added chemicals are usually metals having affinity to the complexing agents at least to a comparable extent with those of the metals to be precipitated. The added metal generally does not have a greater capacity than the existing metals. Therefore, it cannot favorably compete with the existing metals at the usual pH values for their hydroxide precipitation, but increasing the pH shifts the equilibrium in favor of the added metal and causes the existing metals to be freed and precipitated as hydroxides. This application is therefore also known as high pH precipitation [3,32]. The mechanism is explained by the ligand-sharing effect of the added metal. Figures 7 and 8 illustrate the mechanism by which ligand sharing is carried out by calcium. These figures were drawn to present ligands of varying strength [33]. In Figure 7, succinic acid is seen to increase the cadmium concentration at pH 9.0, particularly, for high concentrations of succinic acid. The ligand-sharing effect of calcium at this pH does not seem to be effective. However, increasing the pH to 11.0 causes the succinic acid to be bound almost totally by calcium, and cadmium solubility returns to its noncomplexed state. Succinic acid is a relatively weak ligand and the pH elevation needed to completely overcome its effect is up to the usual optimum pH of cadmium. For this case, an extra pH increase is not needed, as long as calcium is used for ligand sharing. However, for NTA, a strong complex former, the pH increase needed for cadmium to turn back to normal solubility is seen to be one unit more than the optimum pH. The effect of calcium and increased pH on hydroxide precipitation of nickel is seen for the common strong ligands of NTA and EDTA in Figure 8 [33]. Calcium is one of the strongest ligand-sharing metals. The ability of other common metals for ligand sharing action has been investigated [34]. Calcium, ferrous and ferric ions, Mn^{2+} and Mg^{2+} were theoretically

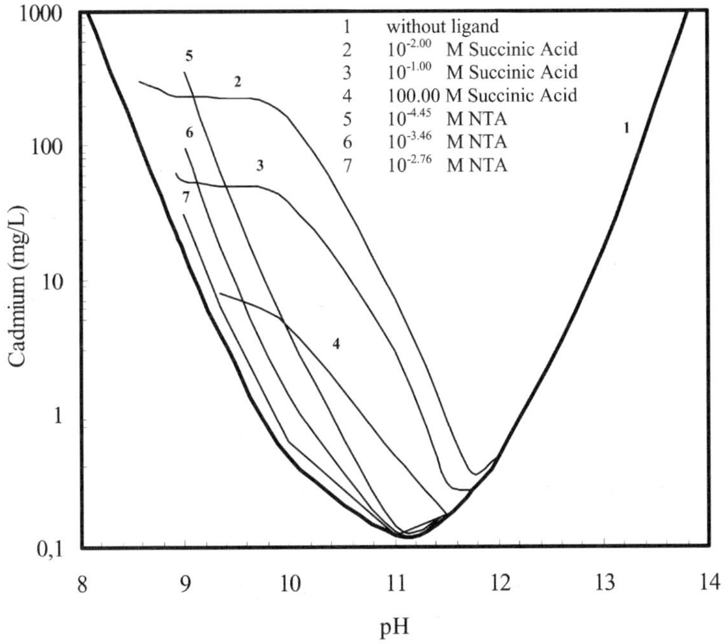

Figure 7 Cadmium–succinic acid and NTA. (From Ref. 33.)

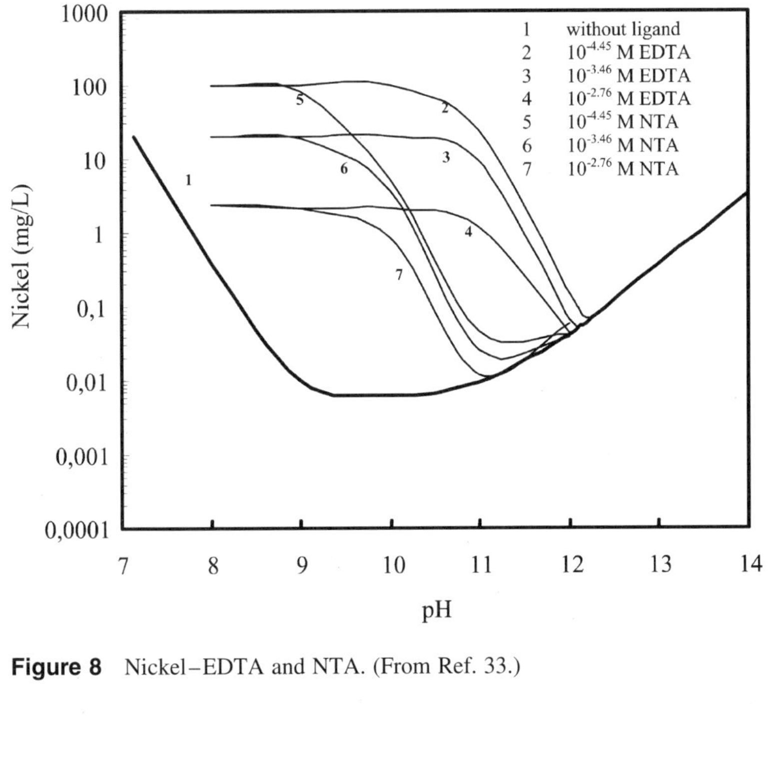

Figure 8 Nickel–EDTA and NTA. (From Ref. 33.)

and experimentally evaluated in terms of their effectiveness in precipitating complexed metals. Calcium is found to be the most effective, while Fe^{2+} and Mn^{2+} have limited capacity. Fe^{2+} was also reported to be useful in the destabilization of the Cu–EDTA complex [35]. However, as noted in the literature, Fe^{2+} and Mn^{2+} are readily oxidizable at alkali pH, thus losing their ability to bind complex formers [34].

The use of other methods that are less affected or not affected by the existence of complexed metals is the third alternative. Sulfide precipitation is the best precipitation method applicable to complexed metals [3]. Membrane processes may be used for the separation of complexed metals [3], and developing methods of adsorption and ion exchange may also prove useful for this purpose.

Sulfide Precipitation

Sulfide precipitation is based on low solubility of metal sulfides. Metal sulfide solubilities are much lower than those of hydroxides. Therefore, very efficient metal removal can be accomplished by the use of sulfide precipitation. Metal sulfide solubilities are also dependent on pH. Figure 9 presents the solubility of metal sulfides depending on pH [3]. The process is

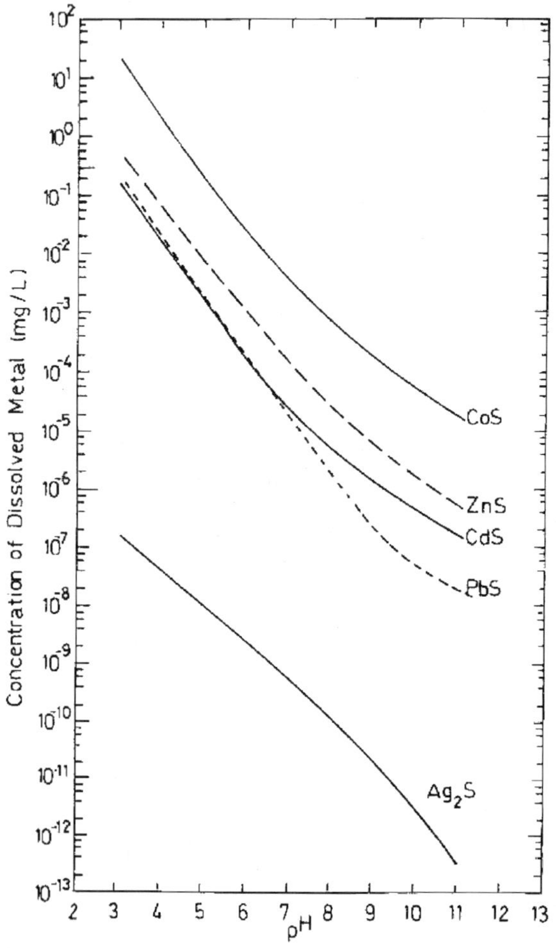

Figure 9 Metal sulfide solubilities. (From Ref. 3.)

conducted at alkaline pH values to obtain minimum solubilities, but more importantly to avoid toxic H_2S gas evolution. Sulfide can be added as soluble sulfide or insoluble sulfide. In the first method, soluble sulfides such as sodium sulfide, calcium polysulfide, or sodium hydrosulfide, are added while pH is maintained above 8.0. In the second method slightly soluble ferrous sulfide is added as a slurry. This slurry can be obtained by mixing $FeSO_4$ and sodium hydrosulfide. Most heavy metal sulfides have lower solubilities than FeS, therefore heavy metal sulfides precipitate as FeS dissolves. The reaction is performed at pH values above 8 and iron precipitates as hydroxide [3,36]. An important advantage of the process is the reduction of chromium by sulfide, which is oxidized to the elemental state. Chromium precipitates as hydroxide. The reaction for the case of FeS addition is:

$$H_2CrO_4 + FeS + 2H_2O \longrightarrow Cr(OH)_3 + Fe(OH)_3 + S^0 \qquad (1)$$

The process has other advantages. One of the most important advantages is the ability to precipitate the complexed metals without pretreatment and at normal operating conditions. The process does not require fine pH adjustment nor does it necessitate a two-stage treatment for different metals.

Sulfide precipitation has some disadvantages as well as strict operating rules. Sulfide precipitates have smaller particle sizes and poor settling characteristics. Thus, the use of coagulants is essential. The amount of sludge produced is greater than that produced by hydroxide precipitation. Handling and disposal of the sludge is difficult due to its hazardous nature. Excess sulfide is present in the effluent, which has H_2S evolution potential and must be removed by air oxidation in an additional tank.

The operation of sulfide precipitation begins with the adjustment of wastewater pH to alkaline values. Lime or sodium hydroxide can be used to increase the wastewater pH. After neutralization, hydroxide sludges form and can be separately removed. Sulfide is added in a controlled manner. Excess sulfide addition causes sulfide odors, and may also cause formation of colloidal precipitates, which are quite difficult to precipitate. Overdosing also increases the need for sulfide oxidation after precipitation. Sulfide level is controlled by oxidation reduction potential (ORP) measurement, or by using sulfide-specific electrodes.

Application of Precipitation and Separation

A summary of the technologies applied for precipitation is given below. Some of the technologies are also commonly used for the treatment of other wastewater groups.

Application of hydroxide precipitation is made using a conventional flash-mix, flocculation, and sedimentation scheme. Chemicals are then added to the flash-mixing basins. The detention time of the flash-mixing basins may vary from one minute to 15 minutes, depending on the solubility and reaction time of the chemicals, the time needed for pH adjustment, and the degree of mixing. For mixing, mechanical mixers are mainly used to provide mixing intensities over $350\ s^{-1}$ velocity gradients. The dosage of inorganic coagulants varies depending on the type and wastewater quality. Ferric chloride doses of 50 mg/L to several hundred mg/L are common. Polyelectrolyte feeding, generally a few mg/L depending on the type of polymer, is added to the flocculation basin. Detention times employed for flocculation vary between 10 and 45 minutes. Slow mixing, generally, is provided by turbine mixers for small units and batch operation, and axial paddle-type mixers for continuous flow. Sedimentation tanks can be rectangular or circular, with several hours of detention time, and are equipped with an effective sludge scraping and removal system. Hydraulic loading may begin from a low value of $20\ m^3/m^2 \cdot day$, particularly for oily wastewater. Sedimentation basins can be upgraded by installing inclined plates or similar equipment within the tank. Recently, the use of inclined plates or clarifiers, which have higher efficiencies and require much less area than conventional clarifiers, has been a common application. Centrifugal-type clarifiers have also been used.

Flotation can also be used for solid separation and is an effective process. It is based on reducing the average density of the flocs as rising air bubbles attach to them, and causing the flocs to float to the surface. Solids and oil have specific gravities near to one, so their settling, or rising velocities are low, requiring a long time for complete separation. Flotation removes these particles in a much smaller time. There are several types of flotation devices; in each type the method of generation of air bubbles differs. In foam flotation, air is blown through the wastewater in which flotation reagents have been added. Only the particles having water-repellant surface character adhere to the air bubbles. Other particles remain in solution. A froth layer forms on the surface, separating effectively and selectively the particles to be removed. Dispersed air flotation is based on producing bubbles of relatively large size by introducing the air at the bottom together with mechanical agitation. It is relatively less effective compared to dissolved air and vacuum flotation. Another type of flotation is nozzle air flotation (NAF), which uses a gas aspirator nozzle that draws air into the wastewater to form a mixture of air and water as it enters the flotation tank. Vacuum flotation is based on applying vacuum to a wastewater that has been saturated with air. Very finely dispersed air bubbles evolve under the vacuum. Saturation of the air is carried out as the wastewater enters the tank on the suction of a pump or in the tank. Dissolved air flotation (DAF) is the most commonly used method. It is a well-proven and very effective technology. The application is based on the saturation of wastewater with air at high pressure and discharging the flow to atmospheric pressure in the flotation tank. Saturation can be made on a recycle flow or by direct pressurization of the influent. Dissolved air floatation generates very fine air bubbles, which attach to solid or oil particles and rise to the surface. The floating material is continuously skimmed. The addition of coagulating chemicals is necessary to increase the separation of solids and oils. This is generally carried out in a coagulation–flocculation system preceding DAF, or chemicals can be fed into the entrance of the tank. Additional chemicals such as activators and collectors are also used to promote air adhesion and flotation. A detailed review of principles and applications of DAF is available in the literature [37]. The main design parameter of flotation is the air-to-solids (oil) (A/S) ratio. Optimum values of this parameter can be determined by experiments in the laboratory. A detailed review of the literature on flotation of metal hydroxide precipitation and results of experimental studies conducted with the use of collectors is given by Gopalratnam *et al.* [38]. A separate cadmium hydroxide sludge flotation by DAF and dispersed air flotation has been evaluated in another study [39].

Another efficient flotation method is electroflotation, which is based on the production of gas bubbles through electrolysis. Its application to plating industry wastewaters and oily wastewater as well as design and operating parameters have been discussed in the literature [40,41]. Adsorbing colloid flotation is a newly developed technology. Inorganic flocculants such as $FeCl_3$ and synthetic resins or other synthetic material, for example, potassium ethyl xanthogenate and zeolite fines, are used in flotation to further increase removal efficiencies [42–44].

Filtration is applied after settling to polish the effluent. Filtration removes suspended solids that do not settle in the clarifier and also provides a safety against any upset that may occur in the settling thank. Several types of granular bed filtration are commonly used. A filter may use a single medium such as sand, or dual or mixed media. Dual media consist of a fine bed of sand under a coarser bed of anthracite coal. A dual media filter provides higher efficiencies and permits higher flow rates. The coarse upper layer collects the greater part of the particles, and the sand layer performs a polishing function. The flow is usually top to bottom, but there are also upflow and horizontal filters. Filters are backwashed as the head increases. Sometimes surface wash is provided as auxiliary cleaning. Filters generally operate by gravity flow; however, pressure filters may be preferred when the flow is pumped for further treatment because they permit higher loadings. Addition of polyelectrolytes improves filter performance. Volumetric loads used for gravity sand and high-rate mixed media filters are 40–50 $L/min{\cdot}m^2$ and 80–120 $L/min{\cdot}m^2$, respectively. Buoyant media filtration has been considered as an alternative to conventional filtration [45]. Diatomaceous filters are sometimes used instead of granular bed filters. Diatomaceous earth filters provide a very high-quality effluent.

Membrane filtration is sometimes used for the removal of precipitated solids, with excellent removal efficiencies. Membrane filtration is preceded by pH adjustment. Proprietary chemicals are added before filtration to form a nongelatinous and stable precipitate. The filter module typically contains tubular membranes.

Evaporation

Evaporation is not a treatment process, but rather a concentration process. Concentration of wastewaters primarily serves in material recovery and also reduces the wastewater flow to be treated. Evaporated water can be reused after condensation. Evaporation is the process with the highest energy requirement. However, it is a proven and reliable technology and is widely used to avoid or reduce wastewaters with highly toxic character, as in the case of cadmium. Evaporation is generally feasible with multistage countercurrent rinse systems. It may also serve the dual purpose of exhaust fume scrubbing and evaporation. A common example for this application is chromium-containing wastewater. Evaporation results in a buildup of contaminants in the concentrate and also decomposition of some materials. If the degree of contamination prevents recovery, the concentrate needs to be treated by ion exchange or by other means.

Two basic modes of evaporation are atmospheric evaporation and vacuum evaporation. Atmospheric evaporation can be accomplished by heating the water until boiling, or spraying heated liquid onto a surface and blowing air over the surface. The most common type is a packed column on which heated water is sprayed. A fan draws the air through the packed bed. Air is humidified with vaporizing water as it is drawn upwards and exhausted to the atmosphere. Mechanical vapor recompression vaporization is similar to evaporation, but works with increased pressure on the vapor. Although there are some types with condensation, evaporated water is not reused.

Vacuum evaporation is applied to lower the boiling temperature. Low temperatures prevent the degradation of plating additives. The vapor is condensed and noncondensing gases are removed by a vacuum pump. Submerged tube type evaporators are commonly used. Another type is the climbing film evaporator. Vacuum evaporation can be double or multiple effect. In double-effect evaporators, the vapor from the first evaporator is fed to the second as the heat source. A second evaporator works at a lower pressure. Cold vaporization units use a similar principle, but with increased vacuum so that water evaporates at 10–20°C.

Ion Exchange

Ion exchange is essentially a sorption process, a surface phenomenon that involves the exchange of an ion held by electrostatic forces on the functional groups of the surface materials with another ion in the solution. The process takes places on the surface of ion exchange material, which is in contact with the solution. The ion exchange material, commonly called resin, may be synthetic or natural. The ions taken up from the solution have a greater preference than the ones existing on the resin and the existing ions are released to the solution as the preferred ones are captured by the resin. In the application, contaminants in the wastewater solution are exchanged with the harmless or acceptable ions of the resin. The ions released by the resin are generally hydrogen or sodium for cation exchange, and hydroxyl, sulfate, or chloride for anion exchange. The key to the process is selectivity, the preference of ions to be bound by the resins.

The exchange reaction is reversible in that it depends on the concentration of the ion in solution. In other words, a less preferred ion can be taken up by the resin if its concentration is higher than the preferred ones. This is used as the basis of regeneration of the ion exchangers. Two important characteristics of ion exchange are high removal efficiency and selectivity. High removal efficiency provides effluent qualities to meet discharge standards and/or reuse criteria of treated water, enabling water reuse. Selectivity is the basis of material recovery. Selectivity of an ion over other ions may reach a factor of several thousand. Selectivity is also important in determining removability, which is the minimum concentration of the preferred ion obtainable with ion exchange depending on the concentration of the less preferred ions in the solution.

Regeneration of the resins produces a concentrated solution of the ion or ions removed. The concentration may reach several thousand mg/L. This level may be adequate for material replacement in some baths or a further concentration or treatment may be needed for recovery. If the regenerant is not used for recovery, it is sent to the treatment system. The application of ion exchange is realized in columns containing beds of resins through which the solution to be treated passes. Pretreatment requirement for ion exchange involves the removal of suspended solids and substances that foul the resin. Fouling is the loss of exchange capacity of a resin, commonly by irreversible binding of some substance or by structural damage. Organic matter removal to prevent fouling may be carried out using activated carbon beds ahead of the ion exchange column. High temperatures and the presence of oxidizing agents including dissolved oxygen are limiting factors. A pH adjustment before feeding may be necessary to avoid precipitation of metals in the exchange bed. Some complexed metals may be fed after dilution. There are resins that operate on chelation. These have a very high selectivity to a specific cation, for example, copper, and they are not affected by the presence of other cations even if they exist at high concentrations. Strong base anion resins may absorb acid molecules. This property is used in the technology known as acid sorption. There are many types of resins for both anion and cation exchange that exhibit different characteristics and selectivities. A broad categorization is strong or weak acid and basic resins. While some metals are retained as cations, some are held in the anion exchangers in anion form such as hexavalent chromium. In the application, there are several systems. In general, more than one set of exchangers are used to provide high efficiency and reliability. Generally, standby units are used for cyclic regeneration, providing the regeneration of an exhausted bed while the other is in service. Acids and bases are also used for regeneration. The usable capacity of a resin for a particular metal depends on a number of factors. Determination of breakthrough characteristics is the basis of operation. Ion exchange is used as an end-of-pipe treatment, but its primary application is recovery of rinse waters and process chemicals. Among the many different uses of ion exchange in the metal finishing industry, recovery of precious metals, recovery of chromium, nickel, phosphate solutions, and sulfuric acid, and rinse water are quite common.

Extensive studies have been conducted on ion exchange, mostly as a part of water reuse and materials recovery [46–48]. Innovative processes have been developed by the use of specialty resins [49–51]. There is also a renewed interest in the use of natural ion exchange material, notably zeolites and sepiolite for metal removal [52–56].

Electrolytic Metal Recovery

Electrolytic metal recovery, or electrowinning, is the recovery of metals from solution using the electroplating process. It has been applied in many metal-plating processes such as PCB manufacturing, and rolling mills. Cadmium, tin, copper, solder alloy, silver, and gold are among the metals commonly recovered by electrowinning [4,57]. Electrowinning cannot be applied to chromium. It is applied to the static rinse following plating by circulating the bath solution through the electrowinning tank or to spent process baths prior to final treatment. Electrowinning involves an electrochemical cell where metal ions are reduced and deposited in metallic form onto the cathode, while oxygen is evolved at the anode. Two types of cathodes are used. In the first, parallel plate systems generate a solid metal that can be reclaimed by stripping off the metal from the cathode base or can be used as the anode in an electroplating bath. As the metal concentration in solution decreases, the efficiency of metal removal decreases drastically. Therefore, electrowinning solution requires further treatment before discharge. An ion exchange system may be used in combination with the electrowinning system for effluent treatment and recycle. However, the metal ions that are not removed by electrowinning accumulate in the ion exchanger. A second type of cathode used is the woven carbon or copper mesh cathode, providing much greater surface area than parallel plate cathodes. This system produces effluents with very low metal concentrations; however, the metal collected on the surface of woven material cannot be removed and recovered. Electrowinning can handle low cyanide concentrations, oxidizing the cyanide to carbon dioxide and nitrogen, but strong oxidizing agents such as nitric acid pose problems. Because electrowinning operation is different for each metal, electrowinning of a mixture of metals is difficult. Therefore, wastewater segregation, and sometimes concentration, are needed for effective operation.

Reverse Osmosis

Reverse osmosis (RO) is a pressure-driven membrane process. The RO process uses a semipermeable membrane that permits the passage of water but retains the dissolved salts. The advantage of the process is its ability to recover all additives in addition to metals [2]. Organic compounds may not be totally rejected by the membranes, necessitating the use of other methods such as activated carbon in combination with RO. The membranes are not resistant to solutions with high oxidation potential such as chromic acid and high pH cyanide solutions. The main use of RO in metal finishing is the rinse water treatment for material recovery and water reuse. A feasibility study has been conducted on the treatment and recycling of a wastewater originating from metal plating where different aspects of the application were discussed [58]. There are many applications of RO to metal finishing wastewaters such as watts nickel, bright nickel, and silver cyanide baths. Recent research indicates further development in the field [57–60].

Electrodialysis

Electrodialysis is also a membrane process using selective membranes and an electric potential to separate positive and negative ions. An electrical potential applied across the membrane causes the ions to migrate towards the electrodes, while the cation and anion permeable membranes separate and concentrate the ions. It is an effective method for material recovery and water recycle from the rinse water [57]. It requires suspended solids and pH control. It can work on dilute wastewaters, and is used to regenerate chrome-anodizing baths (through the oxidation of trivalent chromium at the surface of membrane), chromic acid and copper recovery. Wiśniewski and Wiśniewska have carried out a detailed review and discussed the application potential of the process [61,62].

Diffusion Dialysis

Diffusion dialysis (DD) uses ion exchange membranes placed to separate two flows that act countercurrently. Anion and cation exchanging membranes can be used [2]. In an anion exchange system, wastewater is fed into a compartment separated from another compartment by an anion exchange membrane. Pure water (strip stream) is fed into the second compartment countercurrently. Only anions and hydrogen ion pass through the membrane as a result of the concentration difference. The water becomes enriched with anions and hydrogen ions, creating an acid solution. In the cation mode, the system works similarly, but the membrane allows the passage of cations and hydroxyl ions, making a base solution. In Donnan dialysis, the strip stream is a mineral acid instead of water. Diffusion dialysis is used for alkali recovery from caustic cleaners, acid recovery from spent pickling liquors, and anodizing baths.

Crystallization

Crystallization is applied to concentrated solutions. The process makes use of lower solubility at reduced temperatures. Saturated or highly concentrated solutions are circulated through a refrigeration unit where crystallization takes place. It is used for onsite treatment of concentrated bath solutions [5].

Other Technologies

Many other technologies exist for metal removal. Some methods are newly developed or still being tested. A review of these technologies is now presented. Adsorption, in the form of activated carbon or using other synthetic or natural material, is used to remove metals. Activated carbon adsorption is used as a polishing step or treatment of diluted streams for metal removal. It has been applied to hydroxide and sulfide precipitation effluents. Excess sulfide removal has also been provided by the activated carbon [3,63]. Several types of activated carbon have been found to be successful for metal removal from complexed metal solutions [64]. Activated carbon prepared from coirpith and almond shells has been used for metal removal [65,66], and bone char has been used for cadmium adsorption [67]. Synthetic material such as starch graft copolymers has been found effective in removing metal ions from aqueous solution [68]. Among the other adsorbents used for metal removal are blast-furnace slag, sand, silica, coal, alumina, modified kaolinites, fly ash, tailings, sulfonated coal, and iron oxide coated sand [69–75]. Biosorbents prepared from microorganisms and algae have also been used for metal sorption [75–79]. In total, 67 sorbents have also been used for heavy metal removal [80].

Peat adsorption is a polishing process to remove diluted metals. Peat is a complex material with functional groups that provide metal bonding. Peat also adsorbs polar organic molecules [3]. Peat adsorption characteristics for copper and nickel have been presented in the literature [81].

Insoluble xanthates are used for the precipitation of metals. Xanthates are obtained using carbon disulfide and starch or alcohol in an alkali. Xanthates are essentially ion exchangers and remove dissolved metals from solution [3]. Starch xanthate has been used to remove uncomplexed metal, including hexavalent chromium, by reduction and increasing the pH, with high efficiencies. Cellulose and starch xanthate have been used in combination with chemical oxidation for the removal of silver, cadmium, and mercury [82], while potassium ethylxanthate has been used to remove copper from solution effectively [83]. Low-cost insoluble straw xanthate has been reported to remove various metal ions simultaneously [84], and agro-based starch xanthate has been used to remove copper and zinc [75].

Chemical reduction using sodium borohydride has been studied to recover nickel from solution [85]. Meanwhile, use of organic acids such as oxalic, malonic, fumaric, and maleic acids has been investigated as a means of metal precipitation [86].

The emulsion liquid membrane system is based on double emulsions such as water/oil/water (W/O/W) or oil/water/oil (O/W/O). In W/O/W, the aqueous phase is emulsified in an organic solvent. This emulsion is then dispersed in another aqueous phase to form a new emulsion. Separation occurs as the solute transfers from the outer aqueous phase to the inner aqueous phase within the organic phase [57]. Emulsion liquid membranes are a new means of metal removal. Lead and cadmium extraction from aqueous solution by emulsion liquid membrane systems employing lead and cadmium-di-2-ethylhoxyl phosphoric acid has been reported to be successful [87]. Modeling of an aqueous hybrid liquid membrane system for metal separation has also been made [88]. Meanwhile, electrochemical membrane systems have been used for the treatment of complexed copper [89]. Ion transfer technology is generally applied to chromic acid plating baths.

Cementation and metallurgical recovery are also employed for metal treatment and disposal. These methods will be briefly explained in Section 5.6.

Biological treatment can also be utilized for the removal of metals. Joint treatment of domestic and industrial wastewater indicated and proved the ability of biological treatment to absorb, concentrate, and remove metals [90]. Tannery wastewater treatment is a good example of chromium removal by the activated sludge process [91]. Special applications in this field, such as zinc removal in an anaerobic sulfate-reducing substrate process, have been made [92].

5.5.3 Cyanide Wastewater Treatment

Chlorine Oxidation

Oxidation of cyanide by chlorine is the most commonly used and effective method. The method can be operated in batch or continuous modes of operation. It is suited to automation and works at ambient conditions. It is a well-proven method generating a vast experience of operation [3].

Cyanide oxidation by chlorine is a two-step process. In the first step, cyanide is converted to cyanate; in the second step, cyanate is hydrolyzed to carbon dioxide and nitrogen gas [93].

In the first step, chlorine gas or hypochlorites reacts with cyanide to yield cyanogen chloride:

$$NaCN + Cl_2 \longrightarrow CNCl + NaCl \tag{2}$$

$$NaCN + NaOCl + H_2O \longrightarrow CNCl + 2NaOH \tag{3}$$

$$2NaCN + Ca(OCl)_2 + 2H_2O \longrightarrow 2CNCl + Ca(OH)_2 + 2NaOH \tag{4}$$

These reactions are not dependent on pH and are almost instantaneous. Cyanogen chloride is toxic and readily volatilizes. Therefore, it needs to be converted into a harmless compound immediately as it forms. Cyanogen chloride breaks down rapidly above pH 10.0 and 20°C. At a lower pH, conversion to cyanate is slow; therefore pH 8.0 is a minimum to cyanate cyanogen chloride formation. Conversion of cyanogen chloride to cyanate, which is a stable and much less toxic compound than cyanide, is carried out by alkaline hydrolysis as:

$$CNCl + 2NaOH \longrightarrow NaCNO + NaCl + H_2O \tag{5}$$

Completion of this reaction takes 10–30 minutes at pH 8.5–9.0. Increasing the pH to 10–11 provides the completion of the reaction in 5–7 minutes, while working on the safe side to avoid formation of cyanogen chloride. Therefore, chlorine oxidation of cyanide to cyanate is always carried out at pH higher than 9.5–10.0, and pH values above 10.5 are preferred.

In the second stage, cyanate is hydrolyzed to yield ammonia and carbon dioxide. The reaction proceeds at alkali pH, but is very slow unless free chlorine is present. The reaction takes several hours at pH 10.0. However, at pH 8.5–9.0, the reaction completes in reasonable durations.

$$3Cl_2 + 4H_2O + 2NaCNO \longrightarrow 3Cl_2 + (NH_4)_2CO_3 + Na_2CO_3 \tag{6}$$

Chlorine does not take part in the reaction, but accelerates the process. In spite of the presence of chlorine, the reaction still requires 1–1.5 hours to complete. In the presence of free chlorine, ammonia is rapidly converted to nitrogen gas.

$$3Cl_2 + 6NaOH + (NH_4)_2CO_3 + Na_2CO_3 \longrightarrow 2NaHCO_3 + N_2 \uparrow + 6NaCl + 6H_2O \quad (7)$$

As with all breakpoint chlorination applications, other products such as N_2O and NCl_3 are also formed. Another way of hydrolyzing the cyanate is to employ acid conditions ($pH < 2.5$).

$$2NaCNO + H_2SO_4 + 4H_2O \longrightarrow (NH_4)_2SO_4 + 2NaHCO_3 \quad (8)$$

The reaction is completed within 5 minutes. However, this method is generally not preferred due to the high cost and neutralization requirement of the effluent.

There are several limitations and interferences to chlorine oxidation. Wastewaters containing cyanide also contain heavy metals. Almost all metals are bonded by cyanide through complex formation. Some cyanide metal complexes are very strong and resist oxidation. While ferrocyanide can be oxidized easily, ferricyanide cannot be oxidized. Some of the other transition metals such as cobalt also resist chlorine oxidation. Nickel cyanide can be oxidized, but requires longer retention times; if complete oxidation of cyanide in two stages is realized, however, it does not pose a problem. Because nickel and cyanide rarely co-exist in the same bath, flow separation can also be practiced. If ammonia is present in the wastewater, chloramines form. Chlorine addition may be increased beyond the breakpoint to destroy chloramines. However, chloramines can also oxidize cyanide. Because the reaction rate of chloramines is slower, an additional 45–50 minutes should be allowed for chloramine oxidation of cyanide. At any rate, the oxidation rate of insoluble metal cyanides by chloramines is so slow that complete oxidation may not be achieved in reasonable durations. This may cause some cyanides to be collected in the sludge.

Application of the first stage of oxidation requires a minimum of 30 minutes of reaction time. Minimum pH should be kept between 9.5 and 10.0. An ORP level of 350–400 mV is maintained throughout the reaction. Monitoring of residual chlorine provides additional control of oxidation conditions and oxidant addition. A slow mixing in the reaction mixture should be provided. Theoretical chlorine requirement is 2.73 parts for each part of cyanide as CN. NaOH consumption is 1.125 parts per part of Cl_2. In practice, higher chlorine dosages may be required due to the existence of other chlorine-consuming substances, in particular organic matter and oxidizable metals like cuprous copper. In the second stage, pH is usually adjusted to 8.0, while ORP is maintained at 600 mV. Reaction time is a minimum of 1.5 hours. Slow mixing should also be provided. Theoretical chlorine requirements for this stage is 4.1 parts per part of cyanide. NaOH requirement is 1.125 parts per part of chlorine applied. If ammonia is initially present, 10 mg/L chlorine per mg/L of ammonia nitrogen must be additionally provided for breakpoint chlorination [3].

Ozone Oxidation

Ozone oxidation of cyanide has become a common application. Ozone oxidation is effective in destroying the strong cyanide complexes such as iron and nickel, but cobalt cyanide may resist ozone treatment. Ozonation is primarily used to oxidize cyanide to cyanate:

$$CN^- + O_3 \longrightarrow CNO^- + O_2 \quad (9)$$

Theoretically, 1.85 parts ozone is required to oxidize one part of cyanide. Oxidation of cyanide by ozone is a rapid process completing in about 15 minutes at pH 9.0–10.0. A small amount of copper highly accelerates the reaction. Conversion of cyanide to nitrogen and carbon dioxide by ozone is a slower process, but catalysts can be used to accelerate the process.

However, application of other processes such as bio-oxidation following the first-stage ozone oxidation may be preferred [5].

Oxidation by ozone with UV radiation is another application that is particularly useful for complex cyanides. The process is also effective in removing halogenated organics [3].

Permanganate Oxidation

Potassium permanganate can also be used to oxidize cyanide, although it is not widely practiced. The method is advantageous because hydroxide ion is a product of the reaction, keeping the medium alkaline throughout the reaction [94]. The permanganate oxidizes the cyanide only to cyanate.

$$2KMnO_4 + 3CN^- + H_2O \longrightarrow 3CNO^- + 2MnO_2 + 2KOH \tag{10}$$

Hydrogen Peroxide Oxidation

Use of hydrogen peroxide is also uncommon because it decomposes in dilute solutions and decomposition is accelerated by the presence of heavy metals. Copper-catalyzed hydrogen peroxide oxidation processes are, however, specially designed and patented processes. One process uses a stabilized 41% solution of hydrogen peroxide together with copper catalyst and formaldehyde. Hydrogen peroxide converts cyanide to cyanate as [95]:

$$CN^- + H_2O_2 \longrightarrow CNO^- + H_2O \tag{11}$$

Cyanate may only partly be converted to ammonia. Formaldehyde helps to reduce cyanide metal complexes. The process is applied to cyanide copper and cyanide cadmium bath and rinse waters and also provides metal precipitation [3,94]. Another process developed by the Degussa Corporation utilizes only hydrogen peroxide and copper sulfate [94]. In one study, the heterogeneous catalyst Ru/MgO was tested for the oxidation of cyanide using hydrogen peroxide and relevant parameters were defined for the process [96].

Precipitation of Cyanide

Free cyanide is strongly bound by ferrous ions as a stable complex. There is evidence that other metal cyanides may precipitate as their neutral complexes fall below pH 4, but the process is used to remove free cyanide. Ferrous sulfate or chloride is added at pH 8.5. Ferric ion reacts with ferrous ions to form an insoluble complex. The approximate composition of the two mixed iron cyanide precipitates is $KFe[Fe(CN)_6].H_2O$. The mixed iron complex precipitation completes in 15–30 minutes [95].

Sulfur Dioxide/Air Cyanide Destruction

Sulfur dioxide/air destruction of cyanide is known as the INCO process. The process is based on conversion of cyanide to cyanate with SO_2 and oxygen in air in the presence of elevated copper concentrations and at pH 8–10, according to the reaction:

$$CN^- + SO_2 + O_2 + Cu + H_2O \longrightarrow CNO^- + Cu + H_2SO_4 \tag{12}$$

where copper acts as a catalyst. The Noranda process uses only sulfur dioxide. These two processes were developed for the treatment of cyanide in gold mining wastewaters [95].

Electrolytic Decomposition

Wastes such as spent baths and alkaline descalers containing high concentrations of cyanide can be treated by electrolytic decomposition. Cyanide waste is subjected to anodic electrolysis at elevated temperatures and cyanide is broken down to carbon dioxide, nitrogen, and ammonia. The reaction may take a long time, particularly for the reduction of residual cyanide concentration to a few mg/L. The process can be supported by the addition of strong oxidizing agents or NaCl solution, which upon electrolysis forms chlorine or hypochlorite, providing additional oxidation. The process is less effective on wastes containing sulfate [3,23]. A detailed treatment of the process emphasizing the role of catalysts is available in the literature [97]. Processes employing Ti/Pt and stainless steel anodes and providing simultaneous cyanide removal and copper recovery are defined [98,99]. The use of Ti/Co_3O_4 electrodes for cyanide oxidation has also been studied [100].

Evaporation and Cyanide Recovery

Evaporation is typically applied to concentrated rinse waters and to cyanide plating baths. Evaporative recovery is applied as a closed loop for the cyanide plating process sequence [23]. However, energy price is a limiting factor for evaporation. Cyanide recovery can be carried out, mostly in the mining industry, by stripping off the cyanide as HCN at near-neutral or acid pH, depending on the form of cyanide, and absorbing the HCN in a basic solution, mainly in lime slurry [95].

Other Methods

Ion exchange and chlorine dioxide oxidation are the methods tested for cyanide removal [95,101]. Photo-oxidation of cyanide has also been studied as a promising process. Photosensitized cyanide oxidation over TiO_2 and TiO_2/ZnO and direct photo-oxidation using TiO_2 have been tested with high efficiencies [102–104].

Alkaline hydrolysis is another cyanide destruction method and it does not involve the use of chemicals other than caustic. pH is adjusted to 9.0–10.0 and temperature is increased to 165–185°C at a pressure of 6.3–7.0 kg/cm^2. Reaction time is about 1.5 hours [5].

5.5.4 Hexavalent Chromium Wastewater Treatment

Conventional treatment of hexavalent chromium wastewaters involves the reduction of hexavalent chromium to trivalent chromium and subsequent precipitation of trivalent chromium as hydroxide. Hexavalent chromium reduction is commonly accomplished using chemical reduction.

Hexavalent chromium is reduced to trivalent chromium at acid pH using chemical reduction agents. Sodium bisulfide, sodium metabisulfide, and ferrous sulfate are among the common reduction agents used for small plants. In larger plants, gaseous sulfur dioxide is more common and economical. Sulfur dioxide is hydrolyzed in water to yield sulfite.

$$3SO_2 + 3H_2O \longrightarrow 3H_2SO_3 \quad (13)$$

Sulfite ions are oxidized by the half reaction:

$$SO_3^{2-} + H_2O \longrightarrow 2e^- + 2H^+ + SO_4^{2-} \quad (14)$$

Ferrous ions are oxidized by the half reaction:

$$Fe^{2+} \longrightarrow Fe^{3+} + e^- \quad (15)$$

Hexavalent chromium reduction proceeds with the half reaction:

$$HCrO_4^- + 7H^+ + 3e^- \longrightarrow Cr^{3+} + 4H_2O \quad (16)$$

A combination of oxidation and reduction half reactions gives the reduction reaction for the chemical used. For the use of sulfite ions, the balanced reaction is:

$$3SO_3^{2-} + 2HCrO_4^- + 8H^+ \longrightarrow 2Cr^{3+} + 3SO_4^{2-} + 5H_2O \quad (17)$$

This indicates that 1.5 moles sulfite ion are used to reduce 1 mole of hexavalent chromium. The reducing agent requirement can be calculated from the relevant equation [3,105]. However, an excess of reducing agent is always needed. First, the reaction must go to completion, otherwise unreduced chromium passes through the hydroxide precipitation. To ensure that this occurs, at least 10 mg/L unreacted sulfite is required to remain after the reaction is completed. Secondly, there are reducible substances in the wastewater that consume the reducing agent. Dissolved oxygen and ferric ions are two such important substances consuming the reducing agent. Another occurrence highlighted in Reaction (17) is the consumption of hydrogen ions. This means that the reaction favors acid conditions. The reaction rate is also a function of pH. The rate of reduction reaction is so slow at $pH > 5.0$ that it is impractical to conduct the reduction. At pH 4, completion of the reaction takes about one hour, and is even more for SO_2 reduction, while at pH 2.0 the required reaction time is 5–10 minutes [3,105]. In addition to the reducing agents mentioned above, other reducers may be used. The use of scrap iron has been suggested in the literature [106].

Chromium reduction is a fully proven technology suited to automatic control, particularly when using SO_2 addition and operating at ambient conditions. Final hexavalent chromium concentrations of 0.05 mg/L are readily obtainable. Operation of the system for continuous flow conditions is commonly realized in two reactors connected in series and mixed slowly, each having 45 minutes retention time. pH is maintained between 2.0 and 3.0 in the first tank, while the second tank pH is kept around 2.0. SO_2 addition is made to maintain the ORP in the range 250–300 mV. Batch systems may also have two pH reduction stages [3].

Electrochemical chromium reduction is another technology used to treat hexavalent chromimum wastewaters. It uses an electric current applied to iron electrodes that dissolve to release ferrous ions to solution. Hexavalent chromium ions are reduced as ferrous ions are oxidized to ferric ions [3]. A neutral pH is sufficient. After the reaction is completed, the solution is passed to a clarifier. Ferric hydroxide and chromium hydroxide are settled generally without a need for alkali addition. The process also removes zinc, copper, and other heavy metals. The parameters affecting the process efficiency were experimentally investigated by Kongsricharoern and Polprasert [107].

Another process used for chromium reduction is high pH reduction. In this reaction, $FeSO_4$ is used for the reduction, and pH is maintained between 7.0 and 10.0. Addition of sodium sulfide was suggested to obtain additional chromium reduction and removal [108].

5.5.5 Oily Wastewater Treatment

Oily wastewater treatment may be applied to both combined and segregated wastewaters. However, effective removal of characteristic parameters is most efficiently carried out in segregated waste streams. This point is of importance for priority pollutant and total toxic organics removal. Two main parameters of the oily wastewaters are oil and grease, and toxic organics. The treatment scheme for oily waste is determined by the form of the oil and organic content. Oil can be found as free, emulsified, and soluble forms. The main treatment methods for these forms are summarized below.

Free oils separate easily by decantation. Gravity oil separators of various types can be used to separate and remove free oils. Detailed information about gravity separators can be found elsewhere [109].

Coalescing and centrifugation can be used for oily waste treatment. For diluted aqueous suspensions, use of aluminum silicate-based powders may also be considered for the adsorption of oil [3].

Emulsion breaking and skimming is a basic technology for the removal of emulsified oil. Chemical emulsion breaking aims to break the ion bond between the oil and water, and to coagulate the freed oils to form separable particles by gravity. A variety of chemicals can be used for this purpose. Acids, being effective emulsion breakers, are commonly used. Adjustment of pH to 3 or lower provides high efficiency without causing high sludge production. Cost of the process and need of pH correction after the process are limiting factors. Inorganic flocculants such as ferric chloride and alum function by both lowering the pH and providing effective coagulation. Although they are reasonably priced, they produce excessive sludge. Organic polymers do not cause sludge production, but when used alone their effectiveness needs to be evaluated in terms of type and dosage. Good mixing agitation is required for homogeneous dissipation of chemicals and for the process of emulsion breaking. Retention times of 15–20 minutes are commonly employed. At least two hours separation time should be allowed after emulsion breaking for effective oil removal. Skimming is the operation to skim off the floating oil layer, and can be carried out using several types of skimming devices, such as rotating drum and belt skimmers. Heat application reduces the separation time. Ultrafiltration may replace the emulsion breaking and skimming, particularly for the cases where recovery of oils is considered. Thermal emulsion breaking is similar to chemical emulsion breaking except with the use of heat instead of chemicals. This very effective but costly method is used for recovery rather than for treatment [3].

Flotation, particularly dissolved air flotation (DAF), is a very effective way of separating oil following emulsion breaking. Dissolved air flotation, as explained in the metal hydroxide separation applications section, is a very well established technology. Design and operating information about DAF application to oily wastewater can be found in the literature [110,111]. Electroflotation is particularly effective for oily waste treatment.

Electrocoagulation, an electrochemical method using iron or aluminum, has been suggested for oil–water emulsion treatment [112].

Ultrafiltration (UF) uses a semipermeable membrane that physically separates oil from oil–water emulsions. The separation is realized by pressurizing the liquid so that it permeates the membrane. The membrane works as a molecular screen. Pore sizes vary between 0.0025 and 0.01 microns. Water, solvent, and low molecular weight solutes pass through the membrane, while dispersed material and substances with high molecular weight (generally in the range of 1000–100,000 g) are retained. The fluid is pumped to the membrane unit under pressures of 0.7–3.5 kg/cm^2. Retained oil droplets are removed continuously. Because the retained particle sizes are much greater than the pore size of the membrane, clogging generally does not pose a problem. A wide range of membranes are available. Membranes may be tubular, hollow fiber, or spiral wound. Tubular membranes are preferred for small flows. A very broad range of capacities, beginning with a few $m^3/m^2 \cdot day$ to about 1000 $m^3/m^2 \cdot day$ are also available. The limitations of ultrafiltration are temperature and the substances that are hazardous to the membrane. Membranes can withstand up to 70°C. Oxidizing agents, some solvents, and organic compounds can cause dissolution of the membrane. Ultrafiltration is a proven technology with relatively lower capital and operating costs. It is highly efficient for the recovery of solids and liquids. In metal finishing it is used for the separation of oils, toxic organics, and solids. Oily wastewater of all kinds and electropainting wastes are major areas of use. It also has a very common application for the treatment and recovery of metalworking fluids [2,57].

Microfiltration employs larger pore sizes (0.1 micron or greater) and retains only fine solids and micro emulsions. It can be used in the recovery of aqueous cleaners and after hydroxide precipitation for solid separation. Cross-flow type units are most commonly used in metal finishing because of their self-cleaning ability [57].

Adsorption, chemical oxidation, and biological treatment are used as polishing treatments for oily waste, mostly aiming at soluble oil and toxic organics removal. Adsorption is a very effective technology with a wide area of application. Activated carbon obtained from resins of phenol formaldehydes or acrylic esters, and from a variety of other materials, can be used for adsorption. Activated carbon can be applied in powdered or granulated form. Powdered activated carbon (PAC) application is limited and it is generally used in combination with other methods such as biological treatment, due to operational difficulties. Granulated activated carbon is used in columns similar to filters. Suspended solids and emulsified oil must be treated efficiently before activated carbon applications to prevent frequent backwashing and poisoning of the carbon. Increased temperatures reduce the adsorption rate. Regeneration of activated carbon can be made thermally or by oxidation. If regeneration is not economically justifiable, the carbon is disposed of as solid or hazardous waste. Activated carbon efficiently removes organic matter except very low molecular weight and highly solute ones, and inorganic pollutants such as heavy metals, chlorine, and cyanide.

Air stripping is applied to aqueous waste to separate low concentration of volatile substances having Henry's law constants of 10 atm by passing air through the waste solution. volatile organic carbon (VOC) removal from wastewaters up to 10 mg/L concentration by air stripping is a commonly used technology. Steam stripping is similar to air stripping, but steam is used instead of air. It is a more versatile technology and used to separate and concentrate volatile substances existing in the solution in high concentrations. VOC concentration is a common application [57].

5.5.6 Metalworking Fluid Treatment and Recovery

Metalworking fluids are costly to replace and difficult to treat. They are considered as hazardous waste due to their content and concentrated nature. Therefore, they are intended to be recycled in-house, or offsite. In either case, their effective segregation is required. Spills and contamination to wastewater should be minimized. There are several technologies used for recycling and treatment of metalworking fluids. Gravity and vacuum filtration are used to separate solids. Generally, disposable filters are used [4].

Hydrocyclones are used to separate solids from the fluid. They work by the principle of cyclones. Fluid is directed to the top of the cyclone under pressure. Hydrocyclones cannot efficiently remove oil and particles below 5 μm. Centrifugation, employing forces of several thousand *g*, separates free oils as well as emulsified oil. Pasteurization is used, generally, in combination with centrifuging to clean up bacterial contamination. However, pasteurization is an energy-intensive process and may not be effective in total destruction of specific coolant bacteria [4].

Oil removal is an important step. Along with conventional means, use of ozone and salts has been reported to help demulsification of cutting oil [113]. Skimming is used to separate oil from the fluids. It is a gravity separation conducted in simple tanks with extended durations.

Skimming can be carried out manually or using automatic devices with oil-attracting belts, floating ropes, or mechanical separation devices [4].

Coalescing is the use of special surfaces having oleophilic properties. Oil in emulsion form clings to the surface and is skimmed off by special equipment into weirs as it coalesces. Coalescing uses mediums with high surface area such as tubes or corrugated plates [4]. One of the most effective means of oil separation is DAF. Ultrafiltration is commonly and effectively used for onsite recovery of metalworking fluids. The process is used after separation of free oil and suspended matters. Ultrafiltration is generally combined with ion exchange, microfiltration, or chemical treatment such as ozonation [114,115].

Ultrafiltration uses a minimum of chemicals. Equipment sizes are small. It is a reliable and easy-to-use technology.

The combinations of UF/RO and UF/membrane distillation have also been suggested for oily waste treatment [116,117]. High efficiencies have been obtained from UF treatment of drawing oils. Treatment of permeate with ion exchange has yielded high-quality effluent [115]. A review and evaluation of commercial ultrafiltration systems applied to automotive oily wastewaters can be found in the literature [118].

Biological treatment alternatives have also been tested for metalworking fluids. Pilot-plant studies were conducted on permeates of ultrafiltration of boron- and ethanolamine-containing coolants, for use as a carbon source for denitrification of electrochemical wastewater [119]. Anaerobic removal of COD in metal-cutting fluid wastewater was studied, and the biodegradable part of COD, which was determined to be 65% of the total, was removed efficiently [120].

5.6 HAZARDOUS WASTES, TREATMENT SLUDGE MANAGEMENT, AND MINIMIZATION

5.6.1 Hazardous Wastes

Hazardous Wastes Sources

The metal finishing industry is an important source of hazardous wastes. Hazardous wastes are defined in many ways. They are, in general, the materials or wastes that may pose a substantial hazard to human health and the environment and can be in any physical form. Sources of hazardous wastes include industrial processes, operations, and treatment works, as well as many other activities such as mining and material storage. The exact definition of hazardous wastes employs the use of criteria and lists as given in the regulations. Storage, collection, treatment, and disposal of hazardous wastes are subject to specific and stringent rules. Hazardous waste disposal requires special attention and is more expensive than that of other wastes. Any waste contaminated by hazardous wastes is also considered to be hazardous. Therefore, the identification, segregation, and proper management of hazardous waste is of primary importance. In this context, hazardous waste management in the metal finishing industry is an important task.

Many hazardous waste sources exist in the metal finishing industry. Source control and recovery applications can provide significant reductions in the amount of wastes, as in the case of metal plating bath recovery [4]. However, even some pollution prevention and recovery applications generate hazardous wastes. Important sources of hazardous wastes include:

- solvent wastes;
- spoiled or contaminated metalworking fluids;
- painting wastes;

- case-hardening spent baths and spent packing when pack carburizing is used;
- spent plating and processing baths and their wastes, such as anode sludges and hot dip dross;
- oils and suspended solids separated from cleaning baths;
- wastes originating from recovery and reuse applications: filter wastes or cartridges used for pretreatment of ion exchange or adsorption operations, evaporation residues, still bottoms, membrane filter concentrates, wasted treatment material such as resins, and adsorbants, etc.;
- scrap metals mixed with metalworking fluids or contaminated by other hazardous wastes;
- exhaust scrubber blow down;
- treatment sludges containing toxic metals.

A detailed list of hazardous wastes originating from metal finishing activities is outlined in Table 8 [121].

Solvent wastes, together with spent baths, are among the important hazardous waste sources. Pollution prevention measures for spent baths are described in Section 5.7. The other important hazardous waste sources are solvent wastes, metalworking fluids, paint wastes, case hardening, and operations involving the use of abrasives. Pollution prevention for these sources is summarized below.

Solvent Waste Management and Control

Solvents originate from several sources in metal finishing. Primary sources are vapor degreasing and other degreasing operations. In degreasing operations, as the solvent becomes too contaminated for further use, it is reclaimed or disposed of as hazardous waste.

Two important methods for reducing the amount of solvent waste at the source are maintaining solvent quality and minimizing evaporation losses [4]. To achieve these goals, the following measures can be taken: installation of tank lids, increase in freeboard space, and installation of freeboard chillers on vapor degreasers. Quality maintenance can be best carried out by preventing cross-contamination and water contamination to solvents and removing the sludge that collects in the bottom of the tank. Use of appropriate make-up solutions is important for keeping the ability of the solvent to neutralize acids. Instead of adding fresh solvent only, addition of specific substances and stabilizers, and providing balanced and working components is preferable. Other measures to be taken include solvent standardization in a plant and consolidating cold cleaning operations into a centralized degreasing operation. Solvent dragout reduction and good housekeeping are other applications to maximize solvent recovery.

Replacing solvents with less toxic ones is an effective way of reducing toxic and hazardous wastes. Replacement of aromatics with aliphatic solvents, use of terpenes, and soft surfactants are the alternatives. Aliphatic solvents, however, are not preferred, because they are flammable and have health hazards. Aqueous cleaners may replace solvents in some cases. Use of mechanical and thermal alternatives such as air blast and sand blasting may also help in reducing the use of solvent.

Replacement of chlorinated solvents with aqueous or semi-aqueous cleaners, petroleum hydrocarbons, hydrochlorofluorocarbons (HCFCs), and other organic solvents can be considered. Supercritical fluids or carbon dioxide are the new replacement methods. The use of HCFCs is restricted due to their ozone-depleting potential and health effects. Catalytic wet oxidation cleaning and absorbent media cleaning may be alternative methods to chlorinated solvents [122].

To facilitate solvent reuse and removal, solvents should be kept segregated and sorted. The main rules are to segregate chlorinated from nonchlorinated, aliphatic from aromatic,

Table 8 Hazardous Wastes of Metal Finishing Industry [119]

Hazardous waste no.	Hazardous waste	Hazard code
F001	Halogenated solvents used in degreasing: trichloroethylene, methylene chloride, tetrachloroethylene,1,1,1-trichloroethane, carbon tetrachloride, and chlorinated fluorocarbons; all spent solvent mixtures/blends used in degreasing containing, before use, a total of 10% or more (by volume) of one or more of the above halogenated solvents or those solvents listed in F002, F004, and F005; and still bottoms from the recovery of these spent solvents and spent solvent mixtures.	Toxic waste
F002	Halogenated solvents: tetrachloroethylene, methylene chloride, chlorobenzene, trichloroethylene, 1,1,1-trichloroethane, trichlorofluoromethane 1,1,2-trichloro-1,2,2-trifluoroethane, ortho-dichlorobenzene, and 1,1,2-trichloroethane; all spent solvent mixtures/blends containing, before use, a total of 10% or more (by volume) of one or more of the above halogenated solvents or those listed in F001, F004, or F005; and still bottoms from the recovery of these spent solvents and spent solvent mixtures.	Toxic waste
F003	The following spent nonhalogenated solvents: xylene, acetone, ethyl benzene, ethylacetate, ethyl ether, methyl isobutyl ketone, *n*-butyl alcohol, cyclohexanone, and methanol; all spent solvent mixtures/blends containing, before use, only the above spent nonhalogenated solvents; and all spent solvent mixtures/blends containing, before use, one or more of the above nonhalogenated solvents, and, a total of 10% or more (by volume) of one or more of those solvents listed in F001, F002, F004, and F005; and still bottoms from the recovery of these spent solvents and spent solvent mixtures.	Ignitable waste
F004	Nonhalogenated solvents: cresols and cresylic acid, and nitrobenzene; all spent solvent mixtures/blends containing, before use, a total of 10% or more (by volume) of one or more of the above nonhalogenated solvents or those solvents listed in F001, F002, and F005; and still bottoms from the recovery of these spent solvents and spent solvent mixtures.	Toxic waste
F005	Spent nonhalogenated solvents: toluene, methyl ethyl ketone, carbon disulfide, isobutanol, pyridine, benzene, 2-ethoxy-ethanol, and 2-nitropropane; all spent solvent mixtures/blends containing, before use, a total of 10% or more (by volume) of one or more of the above nonhalogenated solvents or those solvents listed in F001, F002, or F004; and still bottoms from the recovery of these spent solvents and spent solvent mixtures.	Toxic and ignitable waste
F006	Wastewater treatment: sludges from electroplating operations except from the following processes: (a) sulfuric acid anodizing of aluminum; (b) tin plating on carbon steel; (c) zinc plating (segregated basis) on carbon steel; (d) aluminum or zinc–aluminum plating on carbon steel; (e) cleaning/stripping associated with tin, zinc, and aluminum plating on carbon steel; and (f) chemical etching and milling of aluminum.	Toxic waste

Table 8 Continued

Hazardous waste no.	Hazardous waste	Hazard code
F007	Spent cyanide plating bath solutions from electroplating operations.	Reactive, toxic waste
F008	Plating bath residues from the bottom of plating baths from electroplating operations where cyanides are used in the process.	Reactive, toxic waste
F009	Spent stripping cleaning bath solutions from electroplating operations where cyanides are used in the process.	Reactive, toxic waste
F010	Quenching bath residues from oil baths from metal heat treating operations where cyanides are used in the process.	Reactive, toxic waste
F011	Spent cyanide solutions from salt bath pot cleaning from metal heat treating operations.	Reactive, toxic waste
F012	Quenching waste water treatment sludges from metal heat treating operations where cyanides are used in the process.	Toxic waste
F019	Wastewater treatment: sludges from the chemical conversion coating of aluminum except zirconium phosphating in aluminum can washing when such phosphating is an exclusive conversion coating process.	Toxic waste

chlorofluorocarbons from light solvents such as methylene chloride, and water from solvents. Solvent wastes should be labeled to indicate their identification and other properties, and should be preserved at a hazardous waste collection site.

Used solvents can be recycled onsite. If it proves feasible, onsite recycling reduces the amount of waste to be disposed of or hauled, lowers liability and possibly unit cost, enables closer control of reclaimed solvent purity, and reduces paperwork. However, disadvantages include a high capital cost, the requirement for trained personnel, the imposition of liabilities through occupational health, spills, and other risks. In some cases, the operation may result in low efficiencies and even loss of solvent. Gravity separation and filtration using reusable filters such as metal mesh or filter bags are used before distillation. The distillation or fractionation process is the main method for solvent recycling. Solids and high boiling point ($>200^\circ$C) liquids remain as residues. In same cases, distilled solvent may be further dried using a molecular sieve, ion exchange, or a desiccant. Solvent waste can also be burned as a fuel in boilers if its flash point is above 60°C. Solvents can be recycled offsite, can be recovered in central facilities, or can be used as waste exchange material and may be burned in appropriate media such as cement kilns [4]. The distillation process is explained below.

Distillation is the separation of two liquids having different vapor pressures. The process applies to situations in which the components to be separated exist in significant concentrations. The process is based on the separation of the more volatile component or components to the vapor phase by applying heat, then condensing the vapor phase to obtain the concentrate. When more than one component are volatilized, their separation may also be possible by fractionation. Distillation uses significant amounts of heat; therefore, it is an energy-intensive technology. In the metal finishing industry, the primary use for distillation is solvent recovery; however, it can be used in electroplating for the recovery of spent acids, etching solutions, and pickling solutions [57].

Metalworking Fluid Management and Control

Metalworking fluids become contaminated as they are used. A primary problem is contamination with tramp oil. The degree of contamination can be minimized through maintenance and periodic change of seal and wipers. Other maintenance measures include: use of stable cutting and grinding fluids, use of demineralized water for mixing, fluid concentration control, control of fluid chemistry, and regular cleaning of fluid through filtering or centrifugation. Recycling of deteriorated fluids reduces the amount to be disposed of or hauled, and minimizes replacement with new fluid material [4]. While small plants and shops prefer offsite recycling, larger plants can apply in-plant recycle using the methods described in the metalworking fluid treatment in the Section 5.5.

Paint Wastes

The main sources of paint wastes are empty paint containers, leftover paint, outdated paints, paint overspray, paint collected by air pollution equipment, spent stripping solutions, and equipment cleaning wastes.

Bulk purchasing and returning the bulk containers to paint manufactures where applicable reduces the amount of paint containers, leftover paint, and outdated paint. Measures should be taken for waste segregation. For small plants or applications, the purchase of paint in small amounts also minimizes the amount of residues. Segregation of paint wastes from solid wastes and trash reduces the amount of hazardous waste. Paint overspray forms the main loss of paint. Overspray is related to the efficiency of the spray method. As spraying efficiency increases, the amount of overspray reduces. While conventional air atomization provides 30–60% efficiency, pressure atomization and electrostatic atomization methods reach efficiencies of 70–95%. Operator training for best operation is also a very effective measure. Material substation to reduce the hazardous components is the most important step in the prevention of pollution. In this context water-based coatings, which use polymers such as alkyd, polyester, vinyl acetate, acrylic, and epoxies, are becoming more common. These eliminate the hazardous components of the paint. Water-based paints can be used with existing equipment and increase the efficiency of recovery by simply concentrating the water-captured overspray. Radiation-curable coating and powder coating are new methods of painting that reduce paint residues to very low levels. Preventive maintenance of the painting system (such as painting tools, and water curtains) plays an important role for almost every type of paint application in increasing the efficiency and thus reducing the amount of waste. Recycling of paint from cleaning wastes can be carried out using filtration, distillation, and reuse of the collected wastes in another batch of painting [4,57].

Other Wastes

Abrasives are used with an oil-based or water-based binder. Wastes from the use of abrasives are worn-out cloth wheels and similar abrasive application materials. These materials contain abrasives, binders, metal particles, and oxides. Use of water-based binders reduces or eliminates the need of additional cleaning such as alkaline soaking. Liquid spray compositions provide the optimum quantity to be used. Instead of natural material, which increases the amount of hazardous waste by grinding down together with metal fines, synthetic abrasives such as aluminum oxide can be used. Control of water in the mass finishing material decreases the attrition rate and provides longer use of abrasives [4].

Case hardening uses very concentrated salt baths as described in Section 5.2. Spent baths and solutions from this process are frequently classified as hazardous wastes, for example, nitriding and cyaniding process baths. Cleaner technologies are preferred for waste reduction. Energy hardening methods do not use chemicals and are becoming alternative solutions, in particular for ferrous metals [4].

5.6.2 Treatment of Sludge

Many residues and sludges originate from the treatment operation. Some residues are related to in-plant control and recovery applications. Among them are filter cartridges, cloth filters, spent adsorbates such as activated carbon or xanthates, membrane filter concentrates, regenerate solutions, unrecoverable electrolytic treatment residues, distillation and evaporation residues, filter sludges, oily wastes, and sludges separated from the baths [2]. These wastes and their collection, control, and disposal should be evaluated within the context of hazardous waste management. Among the residues of end-of-pipe treatment, two sources are of importance: oily waste from separators and chemical sludge produced by precipitation processes.

Oily waste originates from API separators, skimming tanks, and DAF units. These oily wastes need to be separately handled in consideration of their hazardous properties as well as their adverse effect on dewaterability as they are mixed with other sludges. Pollution prevention measures can significantly reduce the amount of oil to be handled in the treatment stage. Source-based reprocessing and reclamation is much easier and more economical. The recoverability potential of oily waste collected at the treatment plant is much weaker. However, separating and concentrating the oil fraction of these wastes are of importance in terms of their subsequent handling and onsite or offsite treatment and disposal. Separation and concentration can be carried out by the methods given for oily wastewaters and metalworking fluid concentration [3].

Chemical precipitation sludge is produced through either sulfide or hydroxide precipitation. Sulfide precipitation sludge contains wastewater suspended solids and metal sulfides. Sulfide sludge needs to be carefully handled and disposed of due to toxic H_2S evolution potential and other limitations addressed in the Section 5.5. Hydroxide precipitation sludge contains wastewater suspended solids, which are generally metal oxides and organics such as paint and additives, and precipitation sludge. Precipitation sludge is the primary source and contains metal hydroxides, oxides, metal carbonates, and – when lime is used as a precipitant – calcium carbonate, calcium sulfate, calcium phosphates, and calcium fluoride. Use of inorganic coagulants such as ferric chloride and alum contributes significantly to sludge production through formation of iron or aluminum hydroxides. Separation of the sludge is needed after withdrawal to avoid resolubilization and mixing with other residues. Separated sludge is sent to gravity thickeners where the solid content is increased 2–3 times. Thickener supernatant is returned to precipitation tanks. The thickened sludge is pumped to dewatering. Dewatering can be achieved using pressure filtration, vacuum filtration, centrifugation, and sludge drying beds. Centrifugation and pressure filtration are more common, particularly in small plants. Pressure filtration provides a higher degree of dryness. Use of sludge conditioners, mostly polymers, aids dewatering and achieving higher solids content. Sludge cakes are then disposed of or hauled [3].

Stabilization of hazardous sludges can be achieved by chemical fixation or physical fixation methods. Chemical fixation is altering the properties of hazardous waste by chemical reaction to render the waste less mobile or less toxic. An example is heating the chromate with carbon and oxygen. This yields Cr_2O_3, which is attached firmly to the carbon matrix. This product is stable and has a low solubility. Chemical and physical fixation often occur together. Polymeric inorganic silicates containing calcium and aluminum are widely used as a fixation matrix. Among the physical fixation methods, solidification with cement and silicate materials are more common. Portland cement (generally Type I, or Types II and IV for sulfate-containing wastes) binds water from sludge and reacts with some wastes and provides a solid matrix. Water-insoluble silicates (pozzolanic substances) and in some cases soluble silicates are used with a setting agent, which may be cement, gypsum, or lime, for solidification. Additives such as emulsifiers can be used to improve the product properties. Encapsulation, either coating of small particles (micro encapsulation), or imbedding larger aggregates (macro encapsulation) is an effective method. Thermoplastics, asphalt, and waxes can be used for encapsulation. Organic polymers produced in contact with solid wastes provide an inert matrix. Vitrification is the use of glass material to fix the hazardous wastes. Selection of the method depends on the type and effect of wastes, applicability of the method to the waste, complexity, and cost of the method. Thermal methods can also be employed to support the application of stabilization techniques [123,124]. A case study was conducted on metal hydroxide sediment sludge to investigate detoxification by microwave heating [125]. In addition to the above mentioned methods, new methods of sludge stabilization and recovery have been developed. Hydroxide precipitation sludge has been acid leached and metals were recovered by sulfide precipitation followed by solvent extraction [126]. Recovery of chromic acid from electroplating sludges by re-dissolving and electro-oxidation has also been realized [127]. Recovery of copper concentrates from galvanic copper sludges by leaching and precipitation–calcination has been experimentally studied [128], and a jigging process has been developed to recover ferrochromium metal from waste dumps [129].

Pollution prevention is quite effective in reducing the amount and toxicity of the sludges produced. Reduction of the metal content of the wastewater lowers the amount of sludge produced. Use of trivalent chromium instead of hexavalent chromium eliminates the chromium reduction step. Chromium reduction, conducted at very low pH, requires a significant amount of alkali for the adjustment of pH to alkaline levels of precipitation. When lime is used for precipitation, calcium sulfate precipitates with the metal hydroxides, increasing the sludge production. Segregation of complexed metals and cyanide wastewaters helps to avoid hazardous precipitates. Separate treatment of concentrated sources such as bath dumps provides optimum use of precipitation chemicals. However, there are alternatives to reduce sludge production. One of the most effective ways is to use caustic soda instead of lime. Switching to caustic soda, although it is more expensive, may provide significant reductions in sludge production depending on wastewater composition. Use of polyelectrolytes instead of, or in combination with, inorganic coagulants such as ferric chloride and alum helps to reduce the formation of additional metal hydroxide precipitates. Efficient dewatering, although not affecting the amount of dry matter, reduces the overall amount and volume of the sludge produced. For this purpose, air drying can be used in addition to efficient dewatering techniques [4].

5.6.3 Regulatory Framework

The Resource Conservation and Recovery Act (RCRA) of 1976, which amended the Solid Waste Disposal Act, addresses hazardous waste management activities. Regulations promulgated pursuant to subtitle C of RCRA (40 CFR Parts 260–299) establish the system of hazardous waste management and state the definitions, limitations, and rules concerning the applications. The hazardous wastes given in Table 8 are taken from the list of 40 CFR Chapter I, Part 261, as issued in Federal Register dated November 20, 2002 [121].

The Clean Air Act (CAA) lists hazardous air pollutants and controls the emissions of these substances by National Emission Standards (NESHAP). The United States Environmental Protection Agency (USEPA) has designed a NESHAP for the halogenated solvent degreasing/cleaning source category to reduce halogenated solvent emissions. The emission standards are based on maximum achievable control technology (MACT). Another standard related to the metal finishing category is the ozone standard.

5.7 POLLUTION PREVENTION

5.7.1 Pollution Prevention Approach

USEPA defines pollution prevention as any practice that reduces the amount of any hazardous substance, pollutant, or contaminant entering the waste stream or otherwise released to the environment prior to recycling, treatment, or disposal, and reduces the hazards to public health and the environment. The application of pollution prevention is based on source control. Source control generally is a process of minimizing the pollution and health hazards that originate from a certain production by taking appropriate measures, from selection of production method to good operating practices. Every effort for pollution prevention ultimately, at least legally, aims to avoid banned applications, to satisfy occupational health measures and, for water pollution control, to meet discharge standards within the economical constraints of the plant and of the market. An outline of discharge standard applications will be given later in this section.

Pollution prevention is a complex process involving several inter-related aspects of running a plant. They may be grouped as management-related issues, personnel practices and procedural measures, operational measures, and engineering measures. Management plays a key part in effective pollution prevention. Successful pollution prevention is initiated by the perspective and determination of plant owners or managers. One essential aspect of pollution prevention is the concept and application of total quality management. Pollution prevention measures, in fact, begin with the establishment of the plant: process selection, infrastructure, and personnel policy. Among the important tasks of managers are administration of pollution control and management, pursuing the pollution control issues, and even foreseeing developments in the area and making timely decisions. Personnel practices cover basic and periodical training, developing skills, and management initiatives. Procedural measures involve all kinds of supporting activities such as accounting, documentation, materials handling and storage, material tracking and inventory control, scheduling, as well as data and record keeping, handling, and processing. Operational measures are defined and implemented within the context of the engineering system and procedures [2,4]. These measures include:

- minimizing water use;
- monitoring operations;
- leak and spill prevention;
- preventive maintenance;
- emergency preparedness;
- hazardous waste containment;
- general housekeeping.

Minimization of water use in metal finishing is not limited to, but to a great extent is determined by the rinsing processes. Pollution prevention in rinsing operations will be dealt with separately. Leak and spill prevention is an important task in the metal finishing industry. This application is related mostly to wastes of a hazardous nature such as heat treatment solutions, solvents, and metalworking fluids. In this respect, it is also associated with hazardous waste containment.

In essence, engineering procedures constitute in-plant measures to reduce and control pollution. The applications vary over a wide range in terms of their targets and economic implications. The main applications are:

- waste stream segregation;
- process and system design;

- chemical substitution;
- equipment modification;
- alternative processes;
- onsite recycling and recovery;
- offsite recycling and recovery.

Waste stream segregation covers several different applications. The first is the segregation of domestic wastewaters and nonprocess (dilute waste streams) wastewaters originating from noncontact cooling, boilers, water conditioning, and so on, from process wastewaters. Segregation of regulated process waste streams can be fulfilled by making these separations; however, more than one regulated process wastewaters may exist in large and complex plants. The second application is the separation of flows that have undergone recycling and recovery operations. The third and the most important application, as emphasized in Section 5.5, is segregation of six wastewater groups and solvents in metal finishing. A further segregation of waste streams may be needed to facilitate treatment through applications such as flow proportioning, neutralization, and so on. Other options of in-plant control for the metal finishing industry are compiled and discussed in several USEPA reports [1,2]. Table 9 summarizes the application of these options for metal plating operations [2].

Much effort has been devoted to analyze and apply waste minimization approaches for the metal finishing industry. A review of waste prevention options and their applications with regard to zero discharge is given by Gallerani and Drake [130]. An evaluation of waste minimization techniques and economic analysis of waste minimization for electroplating plants has been presented by Lo and Tsao [131]. One study demonstrated an integrated waste minimization approach for the PCB manufacturing shop of Lawrence Berkeley Laboratory. In this study, source-reduction applications were assessed, and flow segregation for chelated effluents and their treatment with ferrous sulfate were realized. In another study, a closed-loop system employing a reverse osmosis–ion exchange combination together with source control applications was reported to provide 90% reduction in acid wastes and 99% metal removal [132]. A zero liquid discharge system was suggested for electroplating wastewaters and employing the boiler as a receiver of treated rinse waters as feed [45].

Three operation groups (aqueous cleaning, plating, and rinsing) are important to pollution prevention. Applications in these operation groups are summarized below.

5.7.2 Pollution Prevention in Aqueous Cleaning

Aqueous cleaning is accomplished with alkaline and acid solutions and with the use of the chemicals given in Section 5.2. Pollution prevention in cleaning solutions includes removal of sludge and use of dry methods. Frequent removal of the sludge from cleaning baths can reduce waste stream volume and increases the time interval between dumping episodes. Sludge removal can be carried out by pumping the solution to a hydrocyclone and/or a sludge retention tank where sludge is separated, and the solution is sent back to the cleaning tank. Alkaline cleaners can be regenerated with microfiltration. Dry methods of cleaning such as sand and bead blasting or pressurized gas can be used where applicable. In some cases, use of abrasives may replace the aqueous cleaning. Vacuum de-oiling, laser ablation, and supercritical fluid cleaning and plasma cleaning are new methods of cleaning that minimize or eliminate pollution [2,4].

Free oils in cleaning baths can be recycled using gravity oil separators. Acid pickling baths can be frequently and effectively recovered. As acid baths are used, they become diluted and ineffective. Sulfuric acid baths can be recycled by crystallizing and separating the ferrous sulfate. While the separated ferrous sulfate crystals can be sold after washing and drying, the

Table 9 Waste Minimization Options [2]

Waste minimization options	Examples	Applications	Limitations
General waste reduction practice	• Improved operating procedures • Dragout reduction • Rinse water use reduction	• Applicable to all conventional plating • Should be considered standard operating practice • Cost benefit typically outweighs any necessary expenditures	• Existing facilities might not be able to accommodate changes because of process configuration and/or space constraints
Alternative processes	• Thermal spray coatings • Combustion torch • Electric arc spraying • Plasma spraying • Physical vapor deposition • Ion plating/plasma-based • Ion implantation • Sputtering and sputter deposition • Laser surface cladding • Chemical vapor deposition	• Primarily repair operations, although they are now being incorporated into original manufacturing • Primarily high-technology applications that can bear extra costs • Expected to improved product quality and durability	• Technologies in varying states of development; commercial availability might be limited in certain cases • Expense often limits applications to expense parts (e.g., aerospace, military, and electronics) • Might require improved process controls, employee training, and automation
Process substitution	• Product changes • Input changes that eliminate toxic materials including – Cyanide – Chromium – Cadmium	• Applicable to most conventional plating operations • Captive shops/manufacturers might be able to explore product changes	• Job shops have little control over input decisions • Product changes might need to be evaluated in terms of customer preference • Product specifications might eliminate consideration of some process substitutes
Process solution maintenance	• Conventional maintenance methods • Advanced maintenance methods • Microfiltration • Ion exchange • Acid sorption	• Conventional methods applicable to all plating operations	• Advanced methods might require significant changes in process design, operation, and chemistry • Application limited for some plating process/technology combinations (e.g., microfiltration should not be used for copper or aluminum)

(continues)

Table 9 Continued

Waste minimization options	Examples	Applications	Limitations
	• Membrane electrolysis • Process monitoring and control		
Chemical recovery technologies	• Ion exchange • Evaporation • Electrowinning • Electrodialysis • Reverse osmosis	• Requires significant engineering, planning, and characterization of process chemistry	• Costs are highly variable for advanced methods
Offsite metals recovery	• Ion exchange • Filtration • Electrowinning • Electrodialysis • Reverse osmosis	• Metal-bearing wastewater treatment sludge • Spent solvents	• Waste materials must be acceptable to recyclers

separated solution can be used in the bath by adding concentrated acid. Hydrochloric acid can be completely recycled from baths by distilling off the acid. Distillation is also applicable to nitric acid [2,4].

5.7.3 Pollution Prevention in Plating Baths

Pollution Prevention Practice

As mentioned earlier, plating baths contain high concentrations of specific metals such as cadmium, chromium, copper, nickel, zinc, as well as cyanide and other toxic compounds. The baths are not dumped unless impurities build up or solution constituents are lost by dragout. An exhausted plating bath is highly concentrated and needs extensive treatment. The source reduction methods for spent plating baths include:

- monitoring bath composition/chemistry;
- chemical addition to reduce dragout;
- equipment modifications;
- removal of impurities from the baths;
- chemical substitution;
- replacement of process.

Proper control of bath operating parameters increases the lifespan of the solution. Increased bath temperatures reduce volume of dragout and allow the use of lower solution concentrations. When the process is carried out at higher temperatures, lower plating solution concentrations can be used to obtain results equivalent to those at higher concentrations at lower temperatures [2,4]. On the other hand, higher operating temperatures increase evaporation energy costs and worker exposure due to higher emission from the baths (e.g., cyanide baths). Reducing the concentration of the bath solution also decreases the tolerance of impurities.

Several chemical additives such as wetting agents and nonchelated process chemicals are used to assist in the plating process [2]. Wetting agents reduce dragout loss and can improve quality of finishing. On the other hand, some wetting agents can create foaming problems in process baths. Some baths' chemistries are also not compatible with wetting agents. Chelating compounds inhibit the precipitation of metals and can be replaced by nonchelating formulas in some baths. However, these applications require continuous monitoring and cleaning of the baths.

Equipment modifications provide decreasing contamination of plating baths, extending their lifespan and reducing waste generation. These applications include the use of the proper anode, use of purified water, and ventilation/exhausting systems [2]. Anodes with lower grades of purity can contaminate the plating baths. One modification is to place cloth bags around anodes to prevent impurities from entering the bath. De-ionized, distilled, or reverse osmosis water can be used for process bath make-up and rinsing operations instead of tap water. These applications can increase the potential for dragout recovery and decrease the frequency of bath dumping. Use of mist suppressants, foam blankets, and floating balls reduce evaporation and mist formation. Ventilation/exhaust systems, while removing hazardous mist and vapors, may help the recovery of mists.

The life of process solutions can be lengthened by removing impurities from the baths. Removal techniques include filtration, carbonate freezing, electrolysis, and carbon treatment [2]. Filtration is one of the most common applications for extending bath life. Cartridge or reusable filters, either in a tank or as external units, are used. Since carbonate build-up adversely affects cyanide bath operation, the build-up of salts can be prevented by applying carbonate freezing. Precipitation is used as an alternative to carbonate freezing for cyanide bath cleaning.

Precipitation using agents such as lime, sodium sulfide, hydrogen peroxide, and potassium permanganate can be employed to remove metal contaminants and phosphorus in cyanide and noncyanide baths. Low-current electrolysis (dummy plating) is another method for extending the bath life (especially for zinc and nickel baths). Organic impurities can be removed from the baths by carbon treatment, which operates on either a continuous or a batch basis.

Substituting toxic plating solutions with non- or less toxic constituents are important for pollution prevention in the plating. For instance, cyanide plating solutions can be replaced with less toxic cyanide-free solutions. Replacing cyanide solutions with noncyanide solutions can reduce hazardous wastes, safety risks to workers, and regulatory reporting requirements. They can also decrease environmental impact and corporate liability. Noncyanide-based alternatives are available for several plating processes. Table 10 presents an overview of the available alternatives [2]. On the other hand, noncyanide alternatives have limitations as also indicated in the table. Although use of noncyanide processes eliminates the cyanide oxidation step, many noncyanide processes pose difficulties in the treatment and often generate more sludge than that of the cyanide bath [2]. Another disadvantage of noncyanide substitutes is that some of the common recovery technologies are more difficult to use. Also, they require installation of more than one operation line to replace a single cyanide line. Noncyanide alternatives also require upgrading of cleaning or degreasing techniques to ensure high-quality plating, because they are often less tolerant of poor surfaces. These alternatives are also more expensive than cyanide-based conventional baths.

Cadmium exhibits superior corrosion resistance in land and marine environments, and has other specific engineering properties, but it is closely regulated because of its high toxicity. Common cadmium plating baths contain cyanide. Therefore, pollution prevention for cadmium plating baths is considered in two stages: the replacement of cyanide, and the elimination of both cyanide and cadmium. These alternatives are presented in Table 10. Similar replacement alternatives for brass, copper, copper strike, gold, silver, and zinc plating are also given in Table 10.

Hexavalent chromium is used in many plating operations. The most common applications of chromium are decorative and hard chromium coatings, aluminum conversion coatings, and copper bright dippings [2]. Chromium provides excellent coating, physical, and aesthetic properties. It can be operated in closed-loop systems. However, its high toxicity, high treatment cost, and mist evolution in the operation have led researchers to seek alternatives to the use of hexavalent chromium. Two alternative approaches are (1) the use of trivalent chromium instead of hexavalent chromium; and (2) use of other metals for coating [2]. These alternatives are shown in Table 11 [2]. The table also shows replacement processes for chromium.

The replacement alternatives to the nickel plating bath with nickel-free baths are also given in Table 11. Elimination of nickel is crucial for avoiding hazardous waste and serious health problems [2]. When nickel-containing wastewater is treated by hydroxide precipitation, the resulting sludge is classified as hazardous waste.

Electroless copper baths are commonly prepared using copper salts, EDTA as a chelating agent, and formaldehyde as the reducing agent. Formaldehyde, however, is a suspected carcinogenic material. Additionally, as mentioned in Section 5.5, complexed wastewaters do not respond to hydroxide precipitation. Therefore, a carbon technology called the Blackhole Process has been tested to avoid the use of these compounds, as shown in Table 11. The Blackhole Process is presented as an environmentally attractive technology using conventional plating equipment and aqueous black carbon dispersed in the solution [2]. It reduces water use, process steps, and waste treatment costs. It also decreases air pollution, and health and safety concerns.

Sulfuric acid anodizing and chromium plating alternatives have been developed to avoid the use of hexavalent chromium. These alternatives are outlined in Table 11 [2].

Table 10 Pollution Prevention in Cyanide-Based Plating [2]

	Alternative bath solution	Comments	Replacement process
Brass plating	• Brass pyrophosphate • Brass pyrophosphate-tartare • Zirconium nitride	✗ Insufficient color ✗ Poor appearance ✗ Narrow operating range ✗ Bath instability	• Electrocoating
Cadmium plating	Noncyanide alternatives • Cadmium neutral • Cadmium acid sulfate • Cadmium acid fluoroborate Noncyanide and noncadmium alternatives • Zinc–nickel acid or alkaline (A) • Zinc–cobalt acid or alkaline (B) • Zinc–iron acid or alkaline (C) • Tin–nickel acid or near neutral (D) • Tin–zinc acid or alkaline or neutral	✓ Good corrosion protection ✓ Satisfactory or good appearance Noncyanide alternatives ✗ Increased electrical contact resistance for zinc coating ✗ Reduced lubricity ✗ Decreased throwing power Noncyanide and noncadmium alternatives ✓ No silver required for black chromating (B, C) ✗ May contain chelators (A, B) ✓ Good hardness and wear resistance (D)	• 50/50 zinc–cadmium alloy using alternative deposition method • Ion vapor deposition of aluminum
Copper plating	• Copper alkaline (cupric copper) • Copper acid sulfate • Copper fluoroborate (F) • Copper pyrophosphate	✗ Higher surface cleaning requirement ✗ Higher operating costs of the bath ✓ Faster plating speed ✓ Lower sludge generation ✓ Cheaper treatment costs (except F)	• Ion plating • Plasma spraying • Sputter deposition
Copper strike	• Copper pyrophosphate (G) • High pH–nickel (H)	✗ Time consuming operation (G) ✗ Higher sludge generation (H)	
Gold plating	• Gold sulfide • Palladium • Cobalt-hardened	✗ New processes with more research requirement	• Ion plating • Sputter deposition
Silver plating	• Ammonium silver (J) • Silver methanesulfonate–potassium iodate (K)	✗ Not promising due to worker health and safety (J)	• Sputter deposition

(continues)

Table 10 Continued

	Alternative bath solution	Comments	Replacement process
	• Technic noncyanide silver (L) • RCA silver solution (M)	✗ Not tested on a commercial scale (K) ✗ More expensive and difficulty-to-polish surface (L) ✗ More expensive, toxic, sensitive to light (M)	
Zinc plating	• Zinc alkaline (N) • Zinc acid chloride (P)	✓ Cheapest of all the zinc platings (N) ✓ Lower sludge generation (N) ✓ Lower energy cost (P) ✓ Easy treatment of nontoxic electrode (P) ✗ Higher initial equipment investments (P)	• Autophoteric coatings

✓, advantage; ✗, disadvantage.

Table 11 Pollution Prevention in Noncyanide Platings [2]

	Alternative bath solution	Comments	Replacement process
Chromium plating	• Trivalent chromium (A) • Electroless nickel phosphorous (B) • Nickel–tungsten–boron alloy (C) • Nickel–tungsten–silicon carbide (D) • Tin–cobalt alloy (E) • Tin–nickel acid or near neutral (F)	✓ All alternatives eliminate mist problem ✗ Color of deposit is main concern (A) ✗ Not suitable for hard chromium (A) ✓ Good deposition on substrate without excessive buildup at the edges (B) ✗ Bath sensitivity to impurities and poorer physical properties (B) ✗ Short bath life, requirement of frequent disposal and replenishment (B) ✓ Reduced energy cost and more uniform deposition (C, D) ✗ Lack of maturity and increased chemical cost (C, D) ✓ High ductility and favorable resistance (C) ✓ Higher plating rate and cathode current efficiency (D) ✓ Better throwing power and wear resistance (D) ✗ Susceptibility to metallic and biological contamination (D) ✓ Adequate wear resistance and corrosion for indoor application (E) ✓ Excellent frictional resistance (F)	• Ion vapor deposition of aluminum • Physical vapor deposition • Metal spray coating
Nickel plating	• Yellow/white bronze (G) • Palladium (H)	✓ Better bactericide property (G) ✓ Better throwing power (G) ✗ More expensive (H)	
Electroless copper			• Carbon technology

(continues)

Table 11 Continued

	Alternative bath solution	Comments	Replacement process
Chromic acid anodizing	• Sulfuric acid anodizing	✓ Reduced energy cost ✓ Lower wastewater treatment cost	• Ion vapor deposition of aluminum
Blackening	• The Versy Black process	✓ No environmental and safety risks ✓ Easy to relieve and lacquer ✓ Excellent adhesion	
Chromating	• Cobalt/molybdenum (J) • Gardolene VP 4683 (K) • Oxide layer growth in high-temperature deionized water (L) • Nonchromate passivation zinc (M) • SANCHEM-CC (N) • Zirconium oxide (P)	✓ Much less hazardous (J) ✗ More expensive (J) ✗ Requirement of additional tank for the sealing process (J) ✓ Hexavalent or trivalent chromium free (K, M, N, P) ✓ Completely inorganic and nontoxic (L) ✓ Very diluted chemical solution (L) ✗ Requirement of many additional steps (L)	
Passivation	• Chelant-based solutions	✓ Biodegradable and nontoxic ✗ Wastewater treatment problems	

✓, advantage; ✗, disadvantage.

Stainless steel passivation is generally carried out using nitric acid and sodium dichromate. Owing to toxicity of nitric acid fumes, a chelant-based solution has been developed for use with passivation on steel. The chelants used are nontoxic and biodegradable (Table 11) [2].

The Versy Black process has been developed for blackening, which currently makes use of hazardous materials such as arsenic, lead, and antimony (Table 11) [2].

Recycling and Recovery in Plating Baths

There have been efforts to recycle and recover metals, as well as concentrated acid and alkaline solution, onsite [2]. These applications are quite beneficial, but have not yet been standardized. It is a very promising field that employs new processes. Some examples of these applications are outlined in Table 12 [2].

Alternative Methods of Metal Deposition

Methods for depositing metal coatings in traditional electroplating have inherent pollution problems. Several alternative technologies have been developed to eliminate the use of toxic components from plating processes and to coat a substrate with metal without using electrolytic solutions or plating baths. These alternative technologies can also reduce the amount of metal contained in the wastewater and sludge. Table 13 shows these alternatives that replace the conventional processes [2].

5.7.4 Pollution Prevention in Rinsing

Rinsing (see Section 5.2 for a description of purpose and requirements) is the most extensively studied system due to its importance in pollution prevention. There are many alternatives that have been applied for pollution prevention. These alternatives can be divided into two groups; alternative rinsing practices and rinse water recycling and recovery techniques.

Alternative rinsing practices are shown in Table 14 [4]. The applications given in the table are grouped into three categories [4]. However, almost all applications are in fact inter-related and one measure may help the other in one way or another. A brief description of the alternatives is presented below [2,4].

Dragout control is one of the most effective ways of reducing waste and minimizing water use in rinsing. It is also an inexpensive method. There are several alternatives to reduce the dragout. One is lowering the viscosity of plating solution. Viscosity reduction can be realized by reducing the chemical concentration of the process bath or increasing the temperature of the bath. These applications are related to bath composition and operation and have been dealt with in pollution prevention in plating baths. Speed of withdrawal of the workpieces from the bath can have a significant effect on dragout volume. As the speed is reduced, drain time increases and a thinner film of bath solution remains on the surface of the workpieces. However, a limitation of this application is that, in addition to loss of operating time, drying of the dragout causes spots on the surface of the workpieces, particularly in hot baths. Surfactants can be used to lower the surface tension of the plating solution. Nonionic surfactants can be used without being degraded by electrolysis. Lowering the surface tension of the liquid reduces the edge effect, the amount retained in crevices, holes, and at the edges as the workpiece is removed from the bath. Racking and rack design is an effective way to reduce the dragout. Workpieces should be arranged so that the solution drains freely and without dripping onto another piece. Although there are some rules for proper positioning, owing to the diversity of the shapes of workpieces, the best solution should be determined experimentally.

Table 12 Recovery and Reuse Applications in Plating [2]

	Applications	Methods
Copper plating	• Copper recovery from bath and rinse water	• Ion exchange
	• Copper recovery from rinse	• Electrodialysis
	• Copper recovery from rinse	• Electrolytic recovery
	• Copper recovery from acid copper or copper cyanide baths	• Reverse osmosis
Gold plating	• Gold recovery	• Ion exchange
	• Gold recovery	• Reverse osmosis
Silver plating	• Silver recovery from rinse	• Electrolytic recovery
	• Silver recovery from rinse	• Ion exchange
	• Silver recovery from rinse	• Polymer filtration
Zinc plating	• Plating solution recovery from rinse and bath	• Reverse osmosis
Chromium plating	• Chromium recovery from bath	• Porous pots
	• Chromium recovery from bath	• Membrane electrolysis
	• Chromium recovery from bath	• Ion exchange
Nickel plating	• Nickel recovery from rinse	• Electrodialysis reversal process
	• Nickel recovery from concentrated solutions, e.g., dragout tanks	• Electrolytic recovery • Ion exchange
	• Nickel recovery from rinse	
	• Nickel recovery from rinse	• Reverse osmosis
Electroless plating	• Metallic ions recovery	• Electrolytic recovery
	• The ionized chemicals recovery	• Electrodialysis
Chemical and electrical conversion	• Chromic acid regeneration	• Ion exchange
	• Sulfuric acid anodize generation	• Ion exchange
	• Sulfuric acid anodize generation	• Electrodialysis
	• Sulfuric acid anodize generation	• Acid retardation
Chromating	• Chromating acid generation	• Ion exchange
	• Chromating acid generation	• Electrochemical methods
Acid solutions	• Onsite	• Distillation • Acid sorption • Membrane electrolysis • Crystallization • Diffusion dialysis
Acid baths	• Onsite	• Reactive rinsing in alkaline and mild etching
Spent bath solutions	• Onsite	• Metal precipitation • Chromium treatment • Complexed metal treatment • Cyanide treatment

Table 13 Alternative Methods for Metal Deposition [2]

Thermal spray coatings
• Combustion torch/flame spraying
• Combustion torch/high velocity oxy fuel
• Combustion torch/detonation gun
• Electric arc spraying
• Plasma spraying
Vapor deposition technologies: physical vapor deposition
• Ion plating/plasma-based
• Ion implantation
• Sputtering and sputter deposition
• Laser surface alloying/laser cladding
Vapor deposition technologies: chemical vapor deposition

Racking must also take effective rinsing into consideration. Barrel plating creates more dragout. Use of correct barrels, proper hole size, and rotation of the barrel in the upright position are among the important measures reducing the amount of dragout. Cleaning and maintenance of the racks also helps to reduce the amount of dragout. Use of drainboards collecting the dripping dragout and delivering back to the plating tank is also an effective solution. Another alternative for reducing the dragout is the use of dragout tanks. Dragout tanks, also called dead or static rinse tanks, are placed between the plating bath and rinsing tank and operate without a continuous flow of water. Workpieces are immersed in the dragout tank before rinsing. As the rinse water becomes concentrated, it is used to make up the plating bath. If the plating bath evaporation rate is high, workpieces can be rinsed directly over the plating bath. Air knives are used to blow air to push the dragout into the process tank or rinse tank. This operation also helps drying.

Another approach for pollution prevention is the reduction of the amount of water through the application of alternative rinsing methods. Reduction of water can be provided by increasing

Table 14 Alternative Rinsing Practices [4]

Reduction in dragout of process chemicals:
• Speed of workpiece withdrawal
• Surface tension of the plating solution and physical shape and surface area of the workpiece
• Plating bath concentration
Source Reduction Options:
• Surfactant use
• Solution temperature
• Workpiece positioning
• Dragout recovery
System Design Considerations:
• Rinse tank design
• Multiple rinsing tanks
• Reactive rinsing
• Foy nozzles and sprays
• Automatic flow controls
• Rinse bath agitation

rinsing efficiency and controlling water flow. Improvement of rinsing efficiency can be achieved by modifying the contact time and/or agitation of the rinse tank with air or water. Water flow can be controlled using alternative methods of rinsing as described in Section 5.2. In addition, warm rinsing in the case of alkaline solutions and reactive rinsing can be applied. Reactive rinsing uses the liquors of acid pickling, which generally follows alkaline cleaners, in the alkaline rinse for neutralization. Use of de-ionized water can extend plating bath life by reducing impurity build-up. Rinse water flow controls can be carried out using well-calculated flow rates, flow restrictors, and level controls in the tanks. Conductivity and pH measurement provide additional water conservation. Use of multiple tanks, designed in several ways as described in Section 5.2, works to effectively reduce rinse water.

The most effective method of water conservation and of reducing pollution loads is the application of rinse water recycling and recovery techniques. These techniques have not only become common, but also a part of the technologies employed in metal finishing. Rinse water recycle can be made using closed- or open-loop systems. In a closed-loop system, rinse water effluent is treated and returned to the rinse. Only a small amount of waste is discharged from closed-loop systems. In the open-loop system, the application is similar to the closed-loop procedure except the final rinse, which is fed by fresh water to ensure high quality [2,4].

Chemical recovery from rinse water may also be feasible. The recovered material can be applied back into the process or, if it is unacceptable for its original use, recovered onsite or offsite for other applications. The main methods applicable to metal and chemical recovery are evaporation, electrolytic metal recovery, electrodialysis, and ion exchange. The methods employed for chemical solution maintenance are acid sorption, diffusion dialysis, ion exchange, ion transfer, microfiltration, and membrane electrolysis [2].

Recent research discusses the application of new and advanced processes for rinsewater reuse as well as material recovery. Reuse of spent rinse water from metal plating using reverse osmosis has been studied by laboratory experiments. High-quality effluent with 45 μS/cm conductivity was obtained. Ultrafiltration was suggested as a pretreatment for reverse osmosis [58]. Reverse osmosis was applied on a pilot scale in a metal finishing plant to global rinse wastewaters after pretreatments of cyanide oxidation, suspended solids removal, and pH adjustment. In total, 75–95% recovery of water with de-ionized water quality was obtained [59]. The application of electrochemical technology to closed-loop systems of the metal finishing industry has been assessed and discussed. A critical evaluation of the usefulness of different types of electrodes, electrochemical cell design, as well as an application for the removal of nickel from process liquids was provided by Janssen and Koene [133]. A complex system comprised of microfiltration, UV irradiation, carbon adsorption, nanofiltration, and ion exchange was applied for recycling of spent final rinse of an electroless plating operation. The system was used for reclaiming wastewater containing heavy metals but low in monovalent ions [134].

5.7.5 Overview of Regulations

The metal finishing industry is regulated by several federal, state, and local statutes for the control of its environmental impact. Three major federal laws regulating the releases of the industry are the Clean Air Act (CAA), the Clean Water Act (CWA), and the Resource Conservation and Recovery Act (RCRA). The limitations of RCRA and CAA in relation to the metal finishing industry were summarized in Section 5.6. The CWA regulates wastewater discharges, which are categorized as direct and indirect. Direct discharge is the discharge of wastewater into a body of water. Indirect discharge covers the discharges to POTWs. Indirect dischargers comply with local pretreatment standards, which may be more stringent than those in the federal effluent guidelines due to local quality requirements. Two CWA regulations affect

Table 15 New Source Performance Standards for 40 CFR 433 [6]

Pollutant or pollutant property	Maximum for any 1 day	Monthly average shall not exceed (mg/L)
Total cadmium	0.11 mg/L	0.07 mg/L
Total chromium	2.77 mg/L	1.71 mg/L
Total copper	3.38 mg/L	2.07 mg/L
Total lead	0.69 mg/L	0.43 mg/L
Total nickel	3.98 mg/L	2.38 mg/L
Total silver	0.43 mg/L	0.24 mg/L
Total zinc	2.61 mg/L	1.48 mg/L
Total cyanide	1.20 mg/L	0.65 mg/L
Total toxic organics (TTO)	2.13 mg/L	*
Oil and grease	52 mg/L	26 mg/L
Total suspended solids	60 mg/L	31 mg/L
pH	6–9	6–9

*No maximum monthly average TTO concentration regulated.

the direct discharges of metal finishing industry. These are the Effluent Guidelines and Standards for Metal Finishing (40 CFR Part 433) and the Effluent Guidelines and Standards for Electroplating [6]. The Effluent Guidelines and Standards for Metal Finishing are applicable to wastewaters generated by the processes of electroplating, electroless plating, anodizing, coating, chemical etching and milling, and PCB manufacturing. If any of those six operations are present, then the standards apply to discharges from all the unit processes of the metal finishing. Companies regulated by electroplating standards remain subject to electroplating standards as long as they do not make modifications in their operations. Otherwise, they become subject to more stringent metal finishing standards. All new companies are subject to the metal finishing standards. The metal finishing standards are given in Table 15 [6].

In some cases, effluent limitations and standards for the following industrial categories may be effective and applicable to wastewater discharges from metal finishing operations. In such cases, Part 433 limits shall not apply and the following regulations shall apply: Nonferrous Metal Smelting and Refining (40 CFR Part 421), Coil Coating (40 CFR Part 465), Porcelain Enameling (40 CFR Part 466), Battery Manufacturing (40 CFR Part 461), Iron and Steel (40 CFR Part 420), Metal Casting Foundries (40 CFR Part 464), Aluminum Forming (40 CFR Part 467), Copper Forming (40 CFR Part 468), Plastic Molding and Forming (40 CFR Part 463), Nonferrous Forming (40 CFR Part 471), and Electrical and Electronic Components (40 CFR Part 469).

The Part 433 standard does not apply to metallic platemaking and gravure cylinder preparation conducted within or for printing and publishing facilities and existing indirect discharging job shops and independent PCB manufacturers that are covered by 40 CFR Part 413.

REFERENCES

1. USEPA. *EPA Office of Compliance Sector Notebook Project. Profile of the Fabricated Metal Product Industry*, EPA/310-R-007, 1995.
2. USEPA. *Pollution Prevention for the Metal Finishing Industry, A Manual for Pollution Prevention Technical Assistance Providers*, EPA/742/B-97/005, 1997.
3. USEPA. *Development Document for Effluent Limitations Guidelines and Standards for the Metal Finishing, Point Source Category*, EPA/440/1-82/091-b, 1982.

4. USEPA. *Guides to Pollution Prevention, The Fabricated Metal Products Industry*, EPA/625/7-90/006, 1990.
5. Riikonen, N. *Industrial Wastewater Source Control. An Inspection Guide*; Technomic Publishing Co. Inc.: Lancaster, Basel, 1992.
6. USEPA. *Electronic Code of Federal Regulations*, e-CFR, 40 CFR – Chapter I – Part 433, December, 2002. http://www.access.gpo.gov/hara/cfr
7. Kabdaşlı, N.I. *Treatment of Complexed Heavy Metals by Hydroxide Precipitation Method*, M.Sc. Thesis, Istanbul Technical University (in Turkish), 1990.
8. Kabdaşlı, I.; Tünay, O. Phase diagrams for metal hydroxides and carbonates; Unpublished study, 2000.
9. Rai, D.; Sass, B.M.; Moore, D.A. Chromium (III) hydrolysis constants and solubility of chromium (III) hydroxide. Inorg. Chem. **1987**, *26*, 345–349.
10. Tünay, O.; Taşlı, R.; Orhon, D. Factors affecting the performance of hydroxide precipitation of metals. *Proceedings of 46th Industrial Waste Conference*, Purdue University, West Lafayette, IN, 1991, 467–478.
11. Stumm, W.; Morgan, J.J. *Aquatic Chemistry – Chemical Equilibria and Rates in Natural Waters*, 3rd Ed.; John Wiley and Sons, 1998.
12. Öztürk, Z. *The Effect of Ionic Strength on Hydroxide Precipitation of Heavy Metals*, M.Sc. Thesis, Istanbul Technical University (in Turkish) 2000.
13. Lin, X.; Burns, R.C.; Lawrance, G.A. Effect of electrolyte composition, and of added iron (III) in the presence of selected organic complexing agents, on nickel (II) precipitation by lime. Water Res. **1998**, *32*, 3637–3645.
14. Baltpurvins, K.A.; Burns, R.C.; Lawrance, G.A. Heavy metals in wastewater: modelling the hydroxide precipitation of copper (II) from wastewater using lime as the precipitant. Waste Mgmt **1996**, *16*, 717–725.
15. Tünay, O.; Kabdasli, I.; Orhon, D. Wastewater control in welding electrode manufacturing – a case study. Water Sci. Technol. **2000**, *42*, 309–313.
16. Hess, G.G.; McKenzie, D.E.; Hughes, B.M. Selective preconcentration of polynuclear aromatic hydrocarbons and polychlorinated biphenyls by in situ metal hydroxide precipitation. J. Chromatogr. **1986**, *366*, 197–203.
17. Crowe, C.W. Evaluation of agents for preventing precipitation of ferric hydroxide from spent treating acid. J. Petrol. Technol. **1985**, *37*, 4, 691–695.
18. Martinez, C.E.; McBride, M.B. Solubility of Cd^{2+}, Cu^{2+}, Pb^{2+}, and Zn^{2+} in aged coprecipitates with amorphous iron hydroxides. Environ. Sci. Technol. **1998**, *32*, 743–748.
19. Baltpurvins, K.A.; Burns, R.C.; Lawrance, G.A.; Stuart, A.D. Effect of electrolyte composition on zinc hydroxide precipitation by lime. Water Res. **1997**, *31*, 973–980.
20. Ma, O.Y.; Traina, S.J.; Logan, J.; Ryan, J.A. In situ lead immobilization by Apatite. Environ. Sci. Technol. **1993**, *27*, 1803–1810.
21. Baltpurvins, K.A.; Burns, R.C.; Lawrance G.A. Use of the solubility domain approach for the modeling of the hydroxide precipitation of heavy metals from wastewater. Environ. Sci. Technol. **1996**, *30*, 1493–1499.
22. Abbona, F.; Madsen, H.E.L.; Boistelle, R. The initial phases of calcium and magnesium phosphates precipitated from solutions of high to medium concentrations. J. Cryst. Growth **1986**, *74*, 581–590.
23. Patterson, J.W. *Industrial Wastewater Treatment Technology*, 2nd Ed.; Butterworth: London, 1985.
24. Petterson, J.W.; Scala, J.J.; Allen, H.E. Heavy metal treatment by carbonate precipitation. *Proceedings of 30th Industrial Waste Conference*, Purdue University, West Lafayette, IN, 1975, 132–150.
25. Marani, D.; Macchi, G.; Pagano, M. Lead precipitation in the presence of sulphate and carbonate: testing of thermodynamic predictions. Water Res. **1995**, *29*, 1085–1092.
26. Macchi, G.; Marani, D.; Pagano, M.; Bagnuolo, G. A bench study on lead removal from battery manufacturing wastewater by carbonate precipitation. Water Res. **1996**, *30*, 3032–3036.
27. Macchi, G.; Pagano, M.; Santori, M.; Tiravanti, G. Battery industry wastewater: Pb removal and produced sludge. Water Res. **1993**, *27*, 1511–1518.

28. Kabdaşlı, I.; Tünay, O.; Teymur, P. Lead removal by chemical precipitation. J. Turkish Water Poll. Cont. **1996**, *6*, 41–47 (in Turkish).
29. Kabdaşlı, I.; Tünay, O. Treatment of sulfate containing wastewaters using barium salts. Water Sci. Technol. **1993**, *28*, 257–265.
30. Ku, Y.; Peters, R.W. The effect of weak chelating agents on the removal of heavy metals by precipitation processes. Environ. Prog. **1986**, *5*, 147–153.
31. Jorgensen, S.E.; Libor, O.; Graber, K.L.; Barkacks, K. Ammonia removal by use of clinoptilolite. Water Res. **1976**, *10*, 213–224.
32. Tünay, O.; Kabdaşlı, N.I. Hydroxide precipitation of complexed metals. Water Res. **1994**, *28*, 2117–2124.
33. Kabdaşlı, I.; Tünay, O. Phase diagrams of heavy metal hydroxides in the presence of weak and strong ligands, Unpublished study, 2002.
34. Tünay, O.; Kabdaşlı, I.; Tasli, R. Pretreatment of complexed metal wastewaters. Water Sci. Technol. **1994**, *29*, 265–274.
35. Chang, L.Y. An industrial wastewater pollution prevention study: evaluation of precipitation and separation processes. Environ. Prog. **1996**, *15*, 28–37.
36. Bhattacharyya, D.; Jumawan, A.B., Jr.; Grieves, R.B. Separation of toxic heavy metals by sulfide precipitation. Separ. Sci. Technol. **1979**, *14*, 441–452.
37. Edzwald, J.K. Principles and applications of dissolved air flotation. Water Sci. Technol. **1995**, *31*, 1–23.
38. Gopalratnam, V.C.; Bennett, G.F.; Peters, R.W. Effect of collector dosage on metal removal by precipitation/flotation. J. Environ. Eng, ASCE **1992**, *118*, 923–948.
39. Zouboulis, A.I.; Matis, K.A. Removal of cadmium from dilute solutions by flotation. Water Sci. Technol. **1995**, *31*, 315–326.
40. Ibrahim, M.Y.; Mostafa, S.R.; Fahymy, M.F.M.; Hafez, A.I. Utilization of electroflotation in remediation of oily wastewater. Separ. Sci. Technol. **2001**, *36*, 3749–3762.
41. Oussedik, S.M.; Khelifa, A. Reduction of copper ions concentration in wastewaters of galvanoplastic industry by electroflotation. Desalination **2001**, *139*, 383.
42. Rubio, J.; Tessele, F. Removal of heavy metal ions by adsorptive particulate flotation. Miner. Eng. **1997**, *10*, 671–679.
43. Alexandrova, L.; Grigorov, L. Precipitate and adsorbing colloid flotation of dissolved copper, lead and zinc ions. Int. J. Miner. Process. **1996**, *48*, 111–125.
44. Lin, C.S.; Huang, S.D. Removal of Cu(II) from aqueous solution with high ionic strength by adsorbing colloid flotation. Environ. Sci. Technol. **1994**, *28*, 474–478.
45. Dietz, J.D. Evaluation of new and emerging waste-management technologies in the metal finishing industry. Plat. Surf. Finish. **1987**, *74*, 62–68.
46. Koene, L.; Janssen, L.J.J. Removal of nickel from industrial process liquids. Electrochim. Acta. **2001**, *47*, 695–703.
47. Tenório, J.A.S.; Espinosa, D.C.R. Treatment of chromium plating process effluents with ion exchange resins. Waste Mgmt **2001**, *21*, 637–642.
48. Yalcin, S.; Apak, R.; Hizal, J.; Afsar, H. Recovery of copper (II) and chromium (III, VI) from electroplating-industry wastewater by ion exchange. Separ. Sci. Technol. **2001**, *36*, 2181–2196.
49. Totura, G. Innovative uses of specialty ion exchange resins provide new cost-effective options for metals removal. Environ. Prog. **1996**, *15*, 208–212.
50. Sengupta, S. Electro-partitioning with composite ion-exchange material: an innovative in-situ heavy metal decontamination process. React. Funct. Polym. **1999**, *40*, 263–273.
51. Clearfield, A.; Bortun, A.I.; Khainakov, S.A.; Bortun, L.N.; Strelko, V.V.; Khryaschevskii, V.N. Spherically granulated titanium phosphate as exchanger for toxic heavy metals. Waste Mgmt **1998**, *18*, 203–210.
52. Rodríguez-Iznaga, I.; Gómez, A.; Rodríguez-Fuentes, G.; Benítez-Aguilar, A.; Serrano-Ballan, J. Natural clinoptilolite as an exchanger of Ni^{2+} and NH_4^+ ions under hydrothermal conditions and high ammonia concentration. Micropor. Mesopor. Mat. **2002**, *53*, 71–80.

53. Petruzzelli, D.; Pagano, M.; Tiravanti, G.; Passino, R. Lead removal and recovery from battery wastewaters by natural zeolite clinoptilolite. Solvent Extr. Ion Exc. **1999**, *17*, 677–694.
54. Ouki, S.K.; Kavannagh, M. Treatment of metals-contaminated wastewaters by use of natural zeolites. Water Sci. Technol. **1999**, *39*, 115–122.
55. Panayotava, M.; Velikov, B. Kinetics of heavy metal ions removal by use of natural zeolite. J. Environ. Sci. Heal. A **2002**, *37*, 139–147.
56. Brigatti, M.F.; Lugli, C.; Poppi, L. Kinetics of heavy-metal removal and recovery in sepiolite. Appl. Clay Sci. **2000**, *16*, 45–57.
57. Freeman, H.M. *Industrial Pollution Prevention Handbook*; McGraw-Hill Inc.: USA, 1995.
58. Qin, J.-J.; Wai, M.-N.; Oo, M.-H.; Wong, F.-S. A feasibility study on the treatment and recycling of a wastewater from metal plating. J. Membrane Sci. **2002**, *208*, 213–221.
59. Benito, Y.; Ruíz, M.L. Reverse osmosis applied to metal finishing wastewater. Desalination **2002**, *142*, 229–234.
60. Chai, X.; Chen, G.; Yue, P.-L.; Mi, Y. Pilot scale membrane separation of electroplating wastewater by reverse osmosis. J. Membrane Sci. **1997**, *123*, 235–242.
61. Wisniewski, J.; Wiśniewska, G. Water and acid recovery from the rinse after metal etching operations. Hydrometallurgy **1999**, *53*, 105–119.
62. Wiśniewski, J.; Wiśniewska, G. Application of electrodialysis and cation exchange technique to water and acid recovery. Environ. Protec. Eng. **1997**, *23*, 35–45.
63. Ku, Y.; Peters, R.W. Innovative uses for carbon adsorption of heavy metals from plating wastewaters: I. Activated carbon polishing treatment. Environ. Prog. **1987**, *6*, 119–124.
64. Leinonen, H.; Lehto, J. Purification of metal finishing waste waters with zeolites and activated carbons. Waste Mgmt Res. **2001**, *19*, 45–57.
65. Kadirvelu, K.; Thamaraiselvi, K.; Namasivayam, C. Adsorption of nickel (II) from aqueous solution onto activated carbon prepared from coirpith. Separ. Sci. Technol. **2001**, *24*, 497–505.
66. Toles, C.A.; Marshall, W.E. Copper ion removal by almond shell carbons and commercial carbons: batch and column studies. Separ. Sci. Technol. **2002**, *37*, 2369–2383.
67. Cheung, C.W.; Porter, J.F.; Mckay, G. Sorption kinetic analysis for the removal of cadmium ions from effluents using bone char. Water Res. **2001**, *35*, 605–612.
68. Zhang, L.-M.; Chen, D.Q. An investigation of adsorption of lead(II) and copper(II) ions by water-insoluble starch graft copolymers. Colloid Surface A **2002**, *205*, 231–236.
69. Dimitrova, S.V.; Mehandgiev, D.R. Lead removal from aqueous solutions by granulated blast-furnace slag. Water Res. **1998**, *32*, 3289–3292.
70. Yabe, M.J.S.; de Oliveira, E. Heavy metals removal in industrial effluents by sequential adsorbent treatment. Adv. Environ. Res. **2001**, *7*, 2, 263–272.
71. Héquet, V.; Ricou, P.; Lecuyer, I.; Le Cloirec, P. Removal of Cu^{2+} and Zn^{2+} in aqueous solutions by sorption onto mixed fly ash. Fuel **2001**, *80*, 851–856.
72. Gupta, G.; Torres, N. Use of fly ash in reducing toxicity of and heavy metals in wastewater effluent. J. Hazard. Mater. **1998**, *57*, 243–248.
73. Suraj, G.; Iyer, C.S.P.; Lalithambika, M. Adsorption of cadmium and copper by modified kaolinites. Appl. Clay Sci. **1998**, *13*, 293–306.
74. Kleiv, R.A.; Sandvik, K.L. Using tailings as heavy metal adsorbents – the effect of buffering capacity. Miner. Eng. **2000**, *13*, 719–728.
75. Bose, P.; Bose, M.A.; Kumar, S. Critical evaluation of treatment strategies involving adsorption and chelation for wastewater containing copper, zinc and cyanide. Adv. Environ. Res. **2002**, *7*, 179–195.
76. Xie, J.Z.; Chang, L.-H.; Kilbane II, J.J. Removal and recovery of metal ions from wastewater using biosorbents and chemically modified biosorbents. Bioresource Technol. **1996**, *57*, 127–136.
77. Schiewer, S.; Volesky, B. Ionic strength and electrostatic effects in biosorption of divalent metal ions and protons. Environ. Sci. Technol. **1997**, *31*, 2478–2485.
78. Zhao, M.; Duncan, J.R.; Van Hille, R.P. Removal and recovery of zinc from solution and electroplating effluent using azolla filiculoides. Water Res. **1999**, *33*, 1516–1522.

79. Butter, T.J.; Evison, L.M.; Hancock, I.C.; Holland, F.S.; Matis, K.A.; Philipson, A.; Sheikh, A.I.; Zouboulis, A.I. The removal and recovery of cadmium from dilute aqueous solutions by biosorption and electrolysis at laboratory scale. Water Res. **1998**, *32*, 400–406.
80. Mohammad, A.; Mohamed Najar, P.A. Physico-chemical adsorption treatments for minimization of heavy metal contents in water and wastewaters. J. Sci. Ind. Res. India **1997**, *56*, 523–539.
81. Ho, Y.S.; Wase, D.A.J.; Forster, C.F. Kinetic studies of competitive heavy metal adsorption by sphagnum moss peat. Environ. Technol. **1996**, *17*, 71–77.
82. Tiravanti, G.; Petruzzelli, D.; Passino, R. Low and non waste technologies for metals recovery by reactive polymers. Waste Mgmt **1996**, *16*, 597–605.
83. Chang, Y.-K.; Chang, J.-E.; Lin, T.-T.; Hsu, Y.-M. Integrated copper-containing wastewater treatment using xanthate process. J. Hazard. Mater. B **2002**, *94*, 89–99.
84. Kumar, A.; Rao, N.N.; Kaul, S.N. Alkali-treated straw and insoluble straw xanthate as low cost adsorbents for heavy metal removal – preparation, characterization and application. Bioresource Technol. **2000**, *71*, 133–142.
85. Tünay, O.; Alp, K.; Kabdasli, I.; Ceviker, S.; Eremektar, G. Heavy metal treatment by chemical reduction. Fresen. Environ. Bull. **1997**, *6*, 278–283.
86. Gyliené, O.; Šalkauskas, M. Precipitation of metal ions by organic acids as a mean for metal recovery and decontamination of wastewater. J. Radioanal. Nucl. Ch. **1998**, *229*, 123–127.
87. Raghuraman, B.J.; Tırmızı, N.P.; Kim, B.-S.; Wiencek, J.M. Emulsion liquid membranes for wastewater treatment: equilibrium models for lead- and cadmium-di-2-ethylhexyl phosphoric acid systems. Environ. Sci. Technol. **1995**, *29*, 979–984.
88. Kislik, V.; Eyal, A. Aqueous hybrid liquid membrane process for metal separation. Part I. A model for transport kinetics and its experimental verification. J. Membrane Sci. **2000**, *169*, 119–132.
89. Juang, R.-S.; Lin, L.-C. Treatment of complexed copper(II) solutions with electrochemical membrane processes. Water Res. **2000**, *34*, 43–50.
90. Huang, C.P.; Wang, J.M. Factors affecting the distribution of heavy metals in wastewater treatment processes: role of sludge particulate. Water Sci. Technol. **2001**, *44*, 47–52.
91. Kabdaslı, I.; Tunay, O.; Daymen, E.; Meric, S. The factors affecting chromium precipitation in leather tanning industry wastewater. Fresen. Environ. Bull. **1998**, *7*, 859–866.
92. Tuppurainen, K.O.; Väisänen, A.O.; Rintala, J.A. Zinc removal in anaerobic sulphate-reducing liquid substrate process. Miner. Eng. **2002**, *15*, 11, 847–852.
93. White, G. *Handbook of Chlorination*, 2nd Ed.; Van Nostrand: Princeton, NJ, 1986.
94. Berkowitz, J.B.; Funkhouser, J.T.; Stevens, J.I. *Unit Operations for Treatment of Hazardous Industrial Wastes*; Noyes Data Corporation: Park Ridge, NJ, 1978.
95. Smith, A.; Mudder, T.I. *The Chemistry and Treatment of Cyanidation Wastes* 1991, Mining Journal Books, Ltd., London, UK.
96. Pak, D.; Chang, W. Oxidation of aqueous cyanide solution using hydrogen peroxide in the presence of heterogeneous catalyst. Environ. Technol. **1997**, *18*, 557–561.
97. Cheng, S.C.; Gattrell, M.; Guena, T.; MacDougall, B. The electrochemical oxidation of alkaline copper cyanide solutions. Electrochim. Acta **2002**, *47*, 3245–3256.
98. Szpyrkowicz, L.; Kaul, S.N.; Molga, E.; DeFaveri, M. Comparison of the performance of a reactor equipped with a Ti/Pt and an SS anode for simultaneous cyanide removal and copper recovery. Electrochim. Acta **2000**, *46*, 381–387.
99. Szpyrkowicz, L.; Zilio-Grandi, F.; Kaul, S.N.; Rigoni-Stern, S. Electrochemical treatment of copper cyanide wastewaters using stainless steel electrodes. Water Sci. Technol. **1998**, *38*, 261–268.
100. Stavart, A.; Van Lierde, A. Electrooxidation of cyanide on cobalt oxide anodes. J. Appl. Electrochem. **2001**, *31*, 469–474.
101. Parga, J.R.; Cocke, D.L. Oxidation of cyanide in a hydrocyclone reactor by chlorine dioxide. Desalination **2001**, *140*, 289–296.
102. Peral, J.; Muñoz, J.; Domènech, X. Photosensitized CN^- oxidation over TiO_2. J. Photoch. Photobiol. A **1990**, *55*, 251–257.
103. Peral, J.; Domenech, X. Photocatalytic cyanide oxidation from aqueous copper cyanide solutions over TiO_2 and ZnO. J. Chem. Technol. Biot. **1992**, *53*, 93–96.

104. Dabrowski, B.; Zaleska, A.; Janczarek, M.; Hupka, J.; Miller, J.D. Photo-oxidation of dissolved cyanide using TiO_2 catalyst. J. Photoch. Photobiol. A **2002**, *151*, 201–205.
105. Taylor, C.R.; Qasim, S.R. More economical treatment of chromium-bearing wastes. *Proceedings of 37th Industrial Waste Conference*, Purdue University, West Lafayette, IN, 1982, 189–196.
106. Bowers, A.R.; Ortiz, C.A.; Cardozo, R.J. Iron process for treatment of Cr(VI) wastewaters. *Proceedings of 41st Industrial Waste Conference*, Purdue University, West Lafayette, IN, 1986, 465–473.
107. Kongsricharoern, N.; Polprasert, C. Electrochemical precipitation of chromium (Cr^{6+}) from an electroplating wastewater. Water Sci. Technol. **1995**, *31*, 9, 109–117.
108. Higgins, T.E.; Sater, V.E. Combined removal of Cr, Cd, and Ni from wastes. Environ. Prog. **1984**, *3*, 12–25.
109. USEPA. *Pretreatment of Industrial Wastes*, manual of practice No.FD-3 Facilities Development Water Pollution Control Federation 2626. Pennsylvania Avenue, NW, Washington DC, 20037, 1981.
110. Zouboulis, A.I.; Avranas, A. Treatment of oil-in-water emulsions by coagulation and dissolved-air flotation. Colloid Surf. A **2000**, *172*, 153–161.
111. Al-Shamrani, A.A.; James, A.; Xiao, H. Separation of oil from water by dissolved air flotation. Colloid Surf. A **2002**, *209*, 15–26.
112. Ögutveren, U.B.; Koparal, S. Electrocoagulation for oil–water emulsion treatment. J. Environ. Sci. Heal. A **1997**, *32*, 2507–2520.
113. Song, Y.-C.; Kim, I.-S.; Koh, S.-C. Demulsification of oily wastewater through a synergistic effect of ozone and salt. Water Sci. Technol. **1998**, *38*, 247–253.
114. Chang, I.-S.; Chung, C.-M.; Han, S.-H. Treatment of oily wastewater by ultrafiltration and ozone. Desalination **2001**, *133*, 225–232.
115. Lin, S.-H.; Lan, W.-J. Treatment of waste oil/water emulsion by ultrafiltration and ion exchange. Water Res. **1998**, *32*, 2680–2688.
116. Karakulski, K.; Kozlowski, A.; Morawski, A.W. Purification of oily wastewater by ultrafiltration. Separ. Technol. **1995**, *5*, 197–205.
117. Gryta, M.; Karakulski, K.; Morawski, A.W. Purification of oily wastewater by hybrid UF/MD. Water Res. **2001**, *35*, 3665–3669.
118. Kim, B.R.; Kalis, E.M.; Florkey, D.L.; Swatsenbarg, S.L.; Luciw, L.; Bailey, C.H.; Gaines, W.A.; Phillips, J.H.; Kosokowsky, G.B. Evaluation of commercial ultrafiltration systems for treating automotive oily wastewater. Water Environ. Res. **1998**, *70*, 1280–1289.
119. Schuch, R.; Gensicke, R.; Merkel, K.; Winter, J. Nitrogen and DOC removal from wastewater streams of the metal-working industry. Water Res. **2000**, *34*, 295–303.
120. Kim, B.R.; Zemla, J.F.; Anderson, S.G.; Stroup, D.P.; Rai, D.N. Anaerobic removal of COD in metal-cutting-fluid wastewater. Water Environ. Res. **1992**, *64*, 216–222.
121. USEPA. *Electronic Code of Federal Regulations*, e-CFR 40 CFR – Chapter I – Part 261, December 2002. http://www.access.gpo.gov/nara/cfr
122. USEPA. *Guide to Cleaner Technologies, Alternatives to Chlorinated Solvents for Cleaning and Degreasing*, EPA/625/R-93-016, 1994.
123. Manahan, S. *Hazardous Waste Chemistry, Toxicology and Treatment*; Lewis Publishers, 1990.
124. Cushnie, G.C., Jr. *Removal of Metals from Wastewater Neutralization and Precipitation*, Pollution Technology Review, No. 107; Noyes Publications, 1984.
125. Gan, Q. A case study of microwave processing of metal hydroxide sediment sludge from printed circuit board manufacturing wash water. Waste Mgmt. **2000**, *20*, 695–701.
126. Renard, D.E. Metal recovery from leached plating sludge. Plat. Surf. Finish. **1987**, *74*, 46–48.
127. Chmielewski, A.G.; Urbanski, T.S.; Migdal, W. Separation technologies for metals recovery from industrial wastes. Hydrometallurgy **1997**, *45*, 333–344.
128. Jandová, T.; Štefanová, T.; Niemczyková, R. Recovery of Cu-concentrates from waste galvanic copper sludges. Hydrometallurgy **2000**, *57*, 77–84.
129. Coetzer, G.; Giesekke, W.; Guest, R.N. Hexavalent chromium in the recovery of ferrochromium from slag. Can. Metall. Quart. **1997**, *36*, 261–268.

130. Gallerani, P.; Drake, D. Wastewater management for the metal finishing industry in the 21st century. Plat. Surf. Finish. **1993**, *80*, 10, *October*, 28–35.
131. Lo, S.-L.; Tsao, Y.-C. Economic analysis of waste minimization for electroplating plants. Water Sci. Technol. **1997**, *36*, 383–390.
132. Chang, L.-Y. A waste minimization study of a chelated copper complex in wastewater-treatability and process analysis. Waste Mgmt **1995**, *15*, 209–220.
133. Janssen, L.J.J.; Koene, L. The role of electrochemistry and electrochemical technology in environmental protection. Chem. Eng. J. **2002**, *85*, 137–146.
134. Wong, F.-S.; Qin, J.J.; Wai, M.N.; Lim, A.L.; Adiga, M. A pilot study on a membrane process for the treatment and recycling of spent final rinse water from electroless plating. Separ. Technol. **2002**, *29*, 41–51.

6
Treatment of Photographic Processing Wastes

Thomas W. Bober*, Dominick Vacco, Thomas J. Dagon*, and Harvey E. Fowler
Eastman Kodak Company, Rochester, New York, U.S.A.

6.1 THE PHOTOGRAPHIC PROCESS

6.1.1 Exposure

A photographic material consists of a base made of a sheet of plastic film, glass, or paper coated with a photographic emulsion (consisting of a polymeric material such as gelatin containing numerous fine crystals of light-sensitive silver halide). Silver halides used in photographic emulsions include silver bromide, silver chloride, silver iodide, or mixtures of these. Exposure of this photographic emulsion to light results in the formation of a "latent image," a four-to-ten atom speck of metallic silver on the silver halide crystal. In simplified terms, light striking molecules of silver halide (AgX) in the emulsion causes some of them to be reduced to metallic silver (Ag^0) atoms.

This can be represented by the following simplified equation:

$$\underset{\text{(silver halide)}}{AgX} + \text{light} \longrightarrow \underset{\text{(metallic silver latent image)}}{Ag^0} \qquad (1)$$

This latent image is invisible to the naked eye, because it consists of only a minute portion of the total amount of silver halide available in the emulsion. The number of atoms converted to silver depends on the *intensity* of the light and the *duration* (time) of exposure, as well as several other factors not discussed here.

The following discussion of development covers black and white images. Color development is examined in a later section. See additional references [1–4] for a more complete description of the photographic process.

6.1.2 Development

In order to be useful, this nearly invisible latent image must be enhanced or "amplified" by converting other surrounding silver halide molecules to silver metal, until the metallic image becomes fully visible. This is done chemically by a process step known as *development*. The latent image is *developed* by immersing the emulsion into a solution (the *developer*) containing a

*Retired.

mild chemical reducing agent (the *developing agent*) in water, usually at an alkaline pH. The developer may include other ingredients added to enhance the process. A chemical reaction takes place in which the developing agent furnishes an electron (e^-) to reduce the silver halide molecule (AgX) to a metallic silver atom (Ag^0), releasing halide ion (X^-) in the process. The reaction describing this step can be summarized as follows:

$$\underset{\text{(silver halide surrounding latent image)}}{AgX} + \underset{\text{(electron from developer)}}{e^-} \longrightarrow \underset{\text{(metallic silver image)}}{Ag^0} + \underset{\text{(free halide ion)}}{X^-} \tag{2}$$

This chemical reduction proceeds as an autocatalytic chain reaction, enhanced or catalyzed by the presence of already formed metallic silver atoms. Thus, the silver halide molecules in the immediate vicinity of the latent image react more quickly than those that are farther away.

Those sites in the latent image that were originally exposed to the most intense light will contain the largest concentration of metallic silver atoms. Therefore, development will proceed faster around these areas than around other locations where weaker light had exposed fewer silver atoms. The chain reaction thus proceeds at different rates within the silver halide crystal and among crystals, producing groups of dark metallic silver "grains," with the groupings sized roughly in proportion to the amount of light that struck the area. If the emulsion were allowed to remain in the developer for a long time, eventually the reaction would proceed to completion and all the silver halide crystals would be reduced to metallic silver, leaving a fully developed black surface. Thus, *time of development* is an important factor in this step, with time (among other factors) being carefully controlled to yield a photographic image of a desired "density," "contrast," "grain," "sharpness," "resolution" and other characteristics that are discussed further in most standard photographic texts. Another very important factor is *developer temperature.* As the temperature is increased, a silver image is produced in a shorter time, but may have less sharpness and contrast than if the reaction were conducted at a lower temperature. Other factors influencing the results include pH, the measure of acidity or alkalinity of a solution (developers are usually strongly alkaline, in a pH range of 9–12); *strength* of the developer solution, with more dilute solutions producing finer developed silver particles (i.e., less grain) but at a slower rate; the *type of emulsion* used (emulsions have varying thickness and are often combined in multiple layers, with each layer containing various additives to modify the process and achieve certain desired results); *agitation* or mixing (to remove unwanted development byproducts from the emulsion and allow fresh developer to access the silver halide crystals); and other factors. In the real world of the photoprocessing laboratory, *developer stability* is an important factor. Because the developing agent is a chemical reducing agent, it is easily oxidized by oxidizing agents, including air, and must be protected. Thus, certain advantages one could gain by running the process one way may be traded off for other advantages gained by operating it under slightly modified conditions.

A chemical known as a *preservative*, usually sodium sulfite, is added to protect the developer from unwanted aerial oxidation; it does this by sacrificing itself and being oxidized instead, generally to sulfate. This is summarized by the reaction equation

$$\underset{\text{(sulfite ion)}}{SO_3^{2-}} + \underset{\text{(oxygen from air)}}{\tfrac{1}{2}O_2} \longrightarrow \underset{\text{(sulfate ion)}}{SO_4^{2-}} \tag{3}$$

During development, unwanted byproducts form in the developer that gradually decrease the activity of the developing agent and begin to retard development. One of these is the free halide ion (chloride, bromide, or iodide) that is released to the solution when the developer converts the silver halide to metallic silver. Another is the oxidized (used) developing agent itself. When it reduces (i.e., gives up an electron to) the silver halide molecule, it in turn becomes oxidized (i.e., it has lost an electron). Thus, it can no longer enter into the development reaction and, in the case of black and white developer, becomes a useless, less-soluble molecule that impedes the action of its unused neighboring developer molecules. It can also form an objectionable brown scum or stain on the surface of the emulsion. Fortunately, a rapid chemical reaction occurs between the oxidized developing agent and the sulfite preservative to produce a sulfonated form of the spent developing agent that is less reactive and more soluble in water. This helps prevent stain or scum formation on the emulsion by keeping the spent developer in solution, and also helps reduce interference with the unused developing agent.

At some point in the development cycle, the amount of spent reaction products could build up to a point where they would begin to seriously interfere with developer activity if not controlled. This buildup of unwanted development products has historically been remedied by discarding a portion of the solution and reconstituting the remainder with fresh developer (known as "replenishment").

In order to produce the highest quality result with the desired characteristics in the final developed image, it is necessary to be able to stop the development reaction quickly. This can be done in one of several ways: quickly lowering the pH, rapidly reducing the concentration of developer, or abruptly decreasing the temperature to a low value (the latter way is usually not practical, too expensive, and therefore generally not used). pH and developer concentration can be lowered in one of two ways: with the use of an acid solution (in either an appropriately named *stop bath* or an acidic *fixer*), or by dilution with water (which is generally not as sudden or precise) using a water wash.

6.1.3 Stop Bath

The *stop bath* (or "stop") is generally an organic acid in water. Acetic acid is used most of the time because it is relatively inexpensive, nontoxic, commercially available, and has good buffering capability (i.e., the ability to remain at a relatively fixed acidic pH even though a certain amount of alkali may be added to it). As described above, it stops the action of the developing agent when the film is immersed in this bath. Today the stop bath is frequently combined with the next solution in the process, the fixer, to form a fixer in the pH range 4–6 (sometimes known as "acid fix") that combines the action of stopping development with that of fixing, described next. This economizes on the number of separate steps and, correspondingly, the number of processing tanks or trays needed to process the emulsion.

6.1.4 Fixing

After the latent silver halide image has been developed into a metallic silver image, the remainder of the emulsion still contains undeveloped silver halide. Because this is an opaque, pinkish-to-grayish material that will not pass light (and will eventually turn into metallic silver if exposed to light or heat for too long a time), it must be removed from the emulsion. The *fixer* is a solvent that will selectively dissolve silver halide molecules from the emulsion while leaving adjacent metallic silver atoms relatively untouched. It is generally composed of thiosulfate, $S_2O_3^{2-}$ (usually the sodium, potassium, or ammonium salt, depending on certain processing considerations). Fixer is also sometimes known in the photographic trade as "hypo," an abbreviated version of "hyposulfite," the now obsolete chemical term for thiosulfate. This is an inexpensive, nontoxic, commercially available, quick-acting solubilizing agent (also known as a "complexing agent" or "sequestrant," because it tends to form a stable chemical complex with the metal ion it dissolves). The thiosulfate fixer dissolves the unused silver halide from the emulsion, forming a tightly bound silver thiosulfate complex, $Ag(S_2O_3)_2^{3-}$. This is a stable, water-soluble chemical complex that has very low toxicity compared to free silver ion (Ag^+); thus, it is relatively safe for photographic personnel to handle and does not create toxicity problems if discharged to a sewer for biological treatment. (See Section 6.2.2, "Environmental Parameters," and Section 6.3.2, "Secondary (Biological) Treatment.")

6.1.5 Color Versus Black and White Processes

The above description adequately portrays the typical black and white negative process, in which the final desired image is a black metallic silver image on a transparent or white background. Color processes follow approximately the same steps, but with some important variations.

A black and white developing agent is a mild reducing agent, usually a rather simple organic molecule like hydroquinone, which has no further value to the photographic process once it has become spent (oxidized) in developing the silver atom. In contrast, the color developing agent is a more complicated molecule, usually a para-phenylenediamine-based compound, which comprises one-half of a dye-forming molecule.

The emulsion of a color film or paper is also more complex than its black and white counterpart. In addition to the profusion of fine silver halide crystals, the gelatin matrix also contains a dispersion of tiny globules of the other half of the dye-forming molecule, known as the "coupling agent" or *coupler.* Whereas the developer half of the final dye molecule is always the same regardless of color, the coupler half of the molecule is a different type of compound, depending on the particular color (cyan, magenta, or yellow) to be formed in that emulsion layer.

When the color developing agent reduces the silver halide molecule (AgX) to metallic silver (Ag^0) by supplying an electron (e^-), it in turn becomes oxidized by losing an electron, as described above. However, instead of becoming a waste product as in the case of the black and white developing agent, the oxidized color developing agent now seeks to join up (couple) with the nearest coupler molecule, to form a dye. This dye, being a large bulky molecule and relatively insoluble in water, has no tendency to migrate from its position in the matrix but remains in place. Thus, the image at this point consists of tiny globules of dye sharing space with clusters of developed silver grains, with both surrounded by a dispersion of transparent, unused coupler globules and undeveloped silver halide crystals in a gelatin matrix. The reactions can be described as follows:

$$\underset{\text{(silver halide surrounding latent image)}}{\text{AgX}} + \underset{\text{(electron from developer)}}{e^-} \longrightarrow \underset{\text{(metallic silver image)}}{\text{Ag}^0} + \underset{\text{(free halide ion)}}{X^-} \tag{4}$$

$$\text{Color developing agent} - \underset{\text{(electron lost to silver)}}{e^-} \longrightarrow \text{Oxidized color developing agent} \tag{5}$$

$$\text{Oxidized color developing agent} + \text{Coupler} \longrightarrow \text{Colored dye image} \tag{6}$$

The emulsion now contains both developed metallic silver and undeveloped silver halide, neither of which is wanted in the final product. Fixing at this point would remove only the silver halide but not the silver metal. Thus, it is necessary to convert the silver metal back to silver halide before both can be satisfactorily dissolved from the emulsion by the same fixing step.

6.1.6 Bleaching

The conversion of silver metal back to silver halide is accomplished by using a mild oxidizing agent known as a *bleaching agent* together with a water-soluble halide salt, such as potassium bromide, in a water solution. Together these are known as a photographic *bleach.* The bleaching agent is mild enough to not adversely affect the gelatin or dye in the emulsion, yet strong enough to take electrons from the silver metal in the presence of the halide, thus converting the silver back to silver halide. Typically, iron complexes such as iron EDTA (ethylenediamine tetraacetic acid), iron PDTA (propylenediamine tetraacetic acid) or less commonly ferricyanide [$Fe(CN)_6^{3-}$] are used since they can supply the proper bleaching activity without harming the emulsion. These are all relatively nontoxic salts that are cheap, commercially available, and safe to handle. Two are commonly used in food products: EDTA is found in bread, baked goods, and pharmaceuticals [5], whereas ferricyanide is used to prevent table salt and foods from caking and as a blue pigment in cosmetics, inks, and paints [5,6]. Ferricyanide was formerly the photographic bleaching agent of choice because it was relatively easy to recover and reuse, but was largely removed from the marketplace because of public concern over the word "cyanide." However, ferrocyanide, or hexacyanoferrate, is an extremely stable, complex iron salt with very low toxicity compared to simple cyanide.

The reaction between the bleach and the metallic silver halide can be described in simplified fashion as:

$$\underset{\text{(ferric iron salt)}}{Fe^{3+}} + \underset{\text{(metallic silver)}}{Ag^{0}} + \underset{\text{(halide ion)}}{X^{-}} \longrightarrow \underset{\text{(silver halide)}}{AgX} + \underset{\text{(ferrous iron salt)}}{Fe^{2+}} \quad (7)$$

After the metallic silver is converted back to silver halide, the entire emulsion can be fixed to remove all the silver, leaving only an image of finely divided colored dye globules in a transparent matrix.

6.1.7 Bleach-Fixes

In some processes, particularly paper processes, which are more easily bleached than film processes, the bleach and fixer can be combined into a single solution known as *bleach-fix* (also commonly known in trade jargon as a "blix" or a "bleach-fixer"). Some chemical synergy is achieved by mixing the two solutions; therefore, the concentration of each can be lowered slightly to achieve the same photographic effect. The gentle oxidizing action of the bleach component is insufficient to damage the fixer component; therefore, they are able to survive together for a reasonable period of time in a single solution. This single solution saves time in processing and simplifies the processing machine. It may also save money in shipping processing chemicals and may produce environmental benefits since it may result in less chemical usage. Bleach-fixes were first introduced commercially in the late 1960s for color paper, specifically for their environmental features as well as reduced processing steps. They remain as a predominant processing solution in color print processes today.

6.1.8 Washes

Before finally drying a photographic emulsion, residual chemicals from processing solutions and reaction byproducts must be eliminated to avoid future interactions that would limit the life of the product. In conventional processing, a final water wash is most often used for both color and black and white products. Also, at certain critical junctures in various processes, it may be desirable to introduce an intermediate wash to remove residual chemicals and/or alter the pH or chemical balance before entering the next solution. Water washes generally contain much lower concentrations of the same chemicals found in the preceding tank.

In the past, all water washes were usually discarded. In more recent years, because of environmental and energy concerns as well as economic considerations, many schemes for purifying and reusing wash waters were proposed and in some cases were successfully implemented to accomplish at least partial recycling and reuse. In some of these cases, when an additional chemical was needed to treat the water or rejuvenate a purification bed, disposal of the treatment chemical may have posed a separate problem. However, water conservation techniques such as countercurrent washes, and mechanical devices known as "water savers" to turn off the water when it is not needed, have generally been shown through practical experience to be much more economical and environmentally beneficial than more complicated recovery techniques.

6.1.9 Stabilizers

Historically, at the end of the process, the gelatin emulsion, having undergone a series of swelling and shrinking cycles as it passed from one processing solution to another, may have lost some of the hardness and physical strength it originally had, which could make it susceptible to scratching or damage. Also, some of the newly formed dyes in the emulsion needed to be further chemically protected against aging and light fading. Both of these tasks can be performed by treating the emulsion with a stabilizer, usually the last solution in a color process. However, through the use of modern forehardened emulsions and other chemical modifications, a number of newer processes have been reformulated to eliminate the need for an emulsion hardening agent in stabilizers.

Currently many processes are designed to save or eliminate water. "Washless minilab" processes are intended to provide processing for a customer in 1 hour or less, and may be located in department store, drugstore, or storefront locations not having sewers. Increasingly, such processes are also being found in professional and commercial photography houses. In these cases the stabilizer may also serve the function of a wash, by eliminating residual chemicals in the emulsion prior to the drying step, which is necessary for image stability upon long-term keeping.

Throughout much of the history of color films and papers, the most common and effective stabilizer was a water solution of formaldehyde, sometimes containing additional ingredients such as citric acid. However, in recent years, because of heightened medical concerns over the handling of formaldehyde plus its annoying lachrymatory odor, most processes today use alternative materials.

6.1.10 Solution Carryover and Replenishment

If each of the above steps could be carried out under ideal, pristine conditions, there would be few unwanted reactions or byproducts and therefore waste would be minimal. Unfortunately, in actual practice this is not the case.

Oxygen from the air is the primary cause of unwanted reactions. It slowly oxidizes components such as the developing agent and fixer upon long-term standing or solution agitation, both of which tend to promote dissolving of air, and during attempts to reuse solutions or recover silver. As previously mentioned, this oxidation necessitates adding preservatives such as sulfite and other ingredients needed to counter the effects of oxidation. These preservatives are also eventually consumed by oxidation, thereby forming byproducts of their own.

Solution carryover is the second major cause of chemical loss, since a solution is carried on the surface and within the saturated emulsion from one tank to the next, thereby losing the solution from the first tank and contaminating the second. To protect against the undesirable effects of contamination, each succeeding solution must be chemically bolstered to contain more of the active ingredient than might be needed strictly to react with components in the film or paper, if carryover did not occur.

Squeegees are important devices for minimizing carryout on the surfaces of photographic materials, and extra washes can be important means of reducing the carry over of unwanted contaminants. However, the fact remains that some unwanted material, even if it is water, will always be trapped within an emulsion and taken into the next processing tank. In addition, small amounts of some chemicals that were originally incorporated into the film or paper, including gelatin, will leach out into the solutions during processing. As previously stated, some chemicals produce byproducts by reacting with oxygen from the air. Finally, some portion of the processing chemicals will always have reacted with the emulsion, producing reaction byproducts (e.g., halides) that are released in one processing tank or another. The total effect of releasing chemicals from all of these sources into the solutions during processing, as well as partial oxidation by air, is known as "seasoning."

Therefore, some means of replenishing the lost components and removing the unwanted components or neutralizing their effects is necessary to operate a continuous process. This means that some waste will always be generated and needs to be treated in some fashion.

6.2 PHOTOPROCESSING EFFLUENT CHARACTERISTICS

6.2.1 Introduction

The photoprocessing industry is very diverse. It includes photofinishing laboratories, x-ray processing at medical and dental facilities and industrial sites, professional photographic operations, motion picture laboratories, processing systems for scientific uses such as astronomy and geology, aerial mapmaking and satellite photography, microfilm processors, graphic arts operations, and others.

Photographic effluents vary in composition because there are many different types of photographic processes and no two processing laboratories operate in the same manner. Processing laboratories vary greatly in size, wash water usage, daily operating time, volume of effluent, and the use of chemical recovery systems.

The actual effluent characteristics [2,7–10] for any photographic processing laboratory can best be determined by collecting a representative sample of the photoprocessing wastewater and having it analyzed by a certified analytical laboratory. However, although concentrations will vary, the effluent from most photoprocessing laboratories will generally be quite similar in chemical composition.

It is not within the scope of this section to provide the actual processing effluent characteristics of every photographic process. Tables 1 through 3 represent typical effluent concentration ranges for conventional color processes, plumbingless color processes, and black and white processes. The chemical concentrations (and other environmental parameters) of plumbingless processes (i.e., systems that do not utilize a conventional wash cycle) are usually quite high, but the loading of these ingredients in pounds or kilograms per day per unit of product processed will be similar to photographic processes using a conventional wash cycle. The plumbingless process is designed for use by relatively small processing operations; consequently, the total daily loading from these operations will likewise be small.

6.2.2 Environmental Parameters

The following sections discuss parameters that might be expected to occur in typical municipal sewer codes.

Table 1 Conventional Color Photoprocessing: Effluent Characteristics

	Typical concentration (mg/L or as noted)
Parameter	
Temperature	80–110°F
pH	6.5–9.0 units
BOD_5	200–3000
COD	400–5000
TDS	300–3000
TSS	<5 to 50
Phenolic compounds	None
Flammable; explosive	None
Detergents	Minimal
Oils and grease	None
NH_3-N	20–300
TKN	30–350
Thiosulfate	100–1000
Sulfate	50–250
Metals[a]	
Silver[b]	<5
Cadmium	<0.02
Chromium	<0.05
Iron	10–100
Lead	<0.02
Mercury	<0.0002
Nickel	<0.02
Zinc	<0.02–0.75

[a] Certain metals, including some of those listed and others, may be detected in trace amounts in photographic effluents even though they are not part of the photographic process, because they appear as contaminants coming from equipment, plumbing, and so on.
[b] After silver recovery (see Section 6.4.3). If iron replacement techniques are used for silver recovery, higher iron levels may be present.
BOD_5, five-day biochemical oxygen demand; COD, chemical oxygen demand; TDS, total dissolved solids; TSS, total suspended solids; TKN, total Kjeldahl nitrogen.

Temperature

The temperature of the most widely used photographic processes is in the 80–110°F (26.7–43.3°C) range. This temperature range should not present a problem to municipal sewer systems.

Five-Day Biochemical Oxygen Demand and Chemical Oxygen Demand

The five-day biochemical oxygen demand (BOD_5) test measures the quantity of oxygen that the effluent, chemical, or solution will consume over a five-day period through biological degradation. The BOD_5 concentration of effluent from a photographic processing laboratory will vary widely, depending on the amount of washwater used, the composition of the processing solutions, and the varying combinations of processing and nonprocessing waste.

Table 2 Plumbingless Color Photoprocessing: Effluent Characteristics

	Typical concentration (mg/L or as noted)
Parameter	
Temperature	<95°F
pH	6.5–9 units
BOD_5	5000–14,000
COD	30,000–36,000
TDS	60,000–90,000
TSS	10–50
Phenolic compounds	None
Flammable; explosive	None
Detergents	Minimal
Oils and grease	None
NH_3-N	6000–10,000
TKN	8000–13,000
Thiosulfate	20,000–25,000
Sulfate	3000–4000
Metals[a]	
Silver[b]	<5
Cadmium	<0.3
Chromium	<2
Iron	1400–2000
Lead	<1
Mercury	<0.0001
Nickel	<2
Zinc	<2

[a] See footnote of Table 1.
[b] After silver recovery. See footnote of Table 1.

In the chemical oxygen demand (COD) test, the chemical or sample in question is refluxed with potassium dichromate and concentrated sulfuric acid for 2 hours. The COD of photographic processing wastes is generally larger than the BOD_5, but the two analyses do not completely correlate on all samples because they do not measure the same oxygen-demanding chemicals. The COD test is much faster and more reproducible than the BOD_5 test.

Total Suspended Solids

Total suspended solids (TSS) are undissolved solid material carried in effluent. Photographic processing effluent is typically very low in suspended solids (less than 50 mg/L) and therefore should not present a problem to municipal treatment plants.

pH

This measurement is an indication of how acidic or alkaline (basic) the solution is, and almost every sewer code contains restrictions on the minimum and maximum pH of mixed effluent discharges. Most are in the range 6.0–10.0. The pH of individual processing solutions may range

Table 3 Black and White Photoprocessing: Effluent Characteristics

	Typical concentration (mg/L or as noted)
Parameter	
Temperature	<80–110°F
pH	6.5–9.0 units
BOD_5	300–5000
COD	2000–20,000
TDS	1500–30,000
TSS	<5–50
Phenolic compounds	<0.2
Flammable; explosive	None
Detergents	Minimal
Oils and grease	None
NH_3-N	350–4300
TKN	400–4500
Thiosulfate	1000–13,000
Sulfate	100–300
Metals[a]	
Silver[b]	<5
Cadmium	<0.02
Chromium	<0.05
Iron	<0.5
Lead	<0.5
Mercury	<0.0002
Nickel	<0.05
Zinc	<0.2

[a] See footnote of Table 1.
[b] After silver recovery. See footnote of Table 1.

from as low as 4 to as high as 12. However, the pH of overall photographic effluent usually does not present a problem to waste treatment systems since it is generally in the range 6.5–9.0.

Heavy Metals

Materials classed as heavy metals are commonly regulated by local sewer authorities. They are usually defined as those metals with a specific gravity greater than 5.0. This includes metals such as antimony, arsenic, cadmium, chromium, cobalt, copper, gold, iron, lead, manganese, mercury, molybdenum, nickel, silver, and zinc.

The concentration of some heavy metals in an effluent may be regulated because of the toxicity of these metals or their compounds. The toxicity can vary with the particular metal or compound and with the form in which the metal exists, for example, free ion, complex, or precipitate. Some metals may be quite toxic in one form while relatively nontoxic in another form; this property for metals to exist in various forms or species is known as "speciation." In the past, nearly all regulations were based on total metal concentration in an effluent; it has only been within the past few years that scientists have begun to realize the great importance of speciation, and that the species of a metal, plus its ability or inability to convert readily between species, should be the basis for regulation.

In addition to silver, some photosensitive photographic products do contain small amounts of metals that may appear in the effluent. Heavy metals frequently found in photographic processing effluent and that are commonly regulated include the following.

Silver. Silver compounds are the basic light-sensitive materials used in most of today's photographic films and papers. Neither elemental silver nor silver compounds are used as ingredients of packaged processing solutions.

During processing, particularly in the fixing bath, silver is removed from the film or paper and is carried out in the solution or wash overflow, usually in the form of a silver thiosulfate complex. Unlike free silver ion (Ag^+), which is toxic to microorganisms, the silver thiosulfate complex is relatively nontoxic and has no detrimental effect on the operation of a secondary waste treatment plant. Based on tests using fathead minnows, the silver thiosulfate complex has been shown to be at least 17,500 times less toxic than the free silver ion [11]. When this complex reaches a waste treatment plant, it is converted to insoluble silver sulfide (Ag_2S) through chemical or biological action and is collected as a solid sludge. Tests performed at Eastman Kodak Company laboratories have shown that a concentration of more than 300 mg/L of silver, present as silver sulfide in the activated sludge, does not interfere with the rate of normal biodegradation of photographic processing wastes [9]. This level of silver at which tests were conducted is much higher than would be expected in a typical municipal treatment plant [10].

Even though much of the silver will be removed in the secondary treatment plant and is not harmful to its operation, recovering the silver before discharge is a sound economic practice and therefore recommended. Not only does silver recovery have environmental benefits and conserve a valuable natural resource, but selling the recovered silver becomes a source of revenue to a photographic processing laboratory. When effective silver recovery practices are used, the residual amount of silver discharged in an effluent is not an environmental concern providing local sewer codes are met (see Section 6.4, "Silver Recovery"). In any event, extensive studies by many independent universities and institutions during the 1990s have shown that silver in a photographic effluent does not represent an environmental concern either from the standpoint of acute toxicity to aquatic organisms or effects on a treatment plant (see subsequent subsection, "Impact of Biotreatment on Silver") [12].

Iron. Iron compounds are commonly used in color photographic processing bleaches or bleach-fixes. Also, if silver recovery cartridges (steel wool metallic replacement cartridges) are used for silver recovery, the photographic effluent will contain iron. Iron is not a typical component of processing solutions for black and white processing.

The iron concentration in effluent is commonly regulated because it affects the appearance and taste of water and readily oxidizes to the reddish-colored ferric form that precipitates, causing rust stains. Iron in photographic effluent normally does not represent a toxicity problem because it is usually present only in the form of stable iron complexes.

Chromium. Certain bleaches used in black and white reversal processes, and historically some process systems cleaners, contain chromium compounds. Hexavalent chromium (Cr^{6+}) is regulated by federal and state agencies as a harmful compound. Some municipal sewer codes have specific limits for Cr^{6+}, Cr^{3+} (trivalent chromium), and total chromium. Cr^{6+} can be reduced by on-site treatment in the laboratory to Cr^{3+}, which is less hazardous (see subsequent subsection, "Chemical Recovery: Chromium").

Zinc. Zinc is present in the effluent of a few photographic color processes. It is a necessary nutrient for human and animal life. Since there is some concern that it may be toxic to fish at relatively low levels, and because it tends to impart an astringent taste to water, its concentration is usually regulated by local sewer authorities.

Cadmium. Cadmium is a known toxicant and therefore is regulated at low levels in many sewer codes. Cadmium has been removed from most photographic products in the past three decades and is present in only a very few specialized films. In these rare situations, small amounts of cadmium may leach out during processing and be detected in effluents. In other cases, if detected in the effluent it is usually due to a contaminant coming from plumbing, equipment, and so on. High concentrations of iron also can cause analytical interferences that may produce a false positive test for cadmium.

Other Trace Heavy Metals. Trace amounts of other heavy metals may sometimes be detected in photographic effluents. Although very small amounts are sometimes used in photographic emulsions, the more likely sources of these heavy metals may be the processing or mixing equipment, plumbing, impurities in processing chemicals, some metallic-replacement silver-recovery devices (MRCs) or in the incoming water supply. The detection of these heavy metals in photographic effluents may also be due to sampling and/or analytical interferences. In any case, their concentrations will usually occur at very low levels, if at all.

Phenolic Compounds

Phenol is not present in photographic effluents. Although some developing agents may be included by some administrative definitions in the broad classification of "phenolic compounds," they do not behave as phenol. They do not impart the undesirable taste and smell to water that phenolics do following chlorination. Developing agents can cause interference with the phenol wet chemical test method and yield a false positive response; however, they should not be considered phenol.

Cyano Complexes (Hexacyanoferrates)

Bleaches containing hexacyanoferrates (very stable complexes of iron and cyanide) are used in only a few photographic processes. These bleaches contain both ferri- and ferrocyanide, but the action of hypo and other chemicals in the effluent reduces most of the ferricyanide to ferrocyanide. A "total cyanide" analysis measures both of these ions, but they should not be confused with simple (free) cyanides. Unlike free cyanide, hexacyanoferrates have a low level of toxicity and are used in many common human applications, such as cosmetics (blue eyeshadow), paints, fabric pigments, and laundry bluing [5,6]. Cyano complexes are not typically found in black and white processing effluents (see subsequent section, "Ferrocyanide Precipitation").

It is possible that, during the chemical analysis for total cyanide, low levels of cyanide might be detected, due to a side reaction between any hydroxylamine and formaldehyde that might be present [13]. This reaction does not occur in the natural environment.

Thiocyanate

Thiocyanate is used in a few photographic processes. Thiocyanate should not be confused with cyanide. It is a different chemical substance with different properties, including much lower toxicity. Studies in the mid-1990s at the Georgia Institute of Technology [14] have shown that thiocyanate will biologically degrade in both aerobic secondary wastewater treatment plants [15] and in anaerobic systems [16] without producing cyanide.

Hydroquinone

Hydroquinone is commonly used in many black and white photographic developers as the reducing (developing) agent for silver. Hydroquinone, a chemical that occurs in nature, can be toxic to some organisms at relatively low concentrations. When discharged in mixed photographic processing effluents, it is present in low concentrations and readily biodegraded to innocuous products by biological treatment [17,18]. It also oxidizes quite readily with air alone, which is why a preservative like sodium sulfite is added to developer formulations. Ozone rapidly decomposes hydroquinone, with acetic acid as the end product [19].

Nitrogen

Relatively high concentrations of ammonia nitrogen may result from the use of ammonium fixers. Although some of the nitrogen-containing ions may be oxidized in a waste treatment plant, some of them may be carried through the plant and into the receiving body of water. If the waste has a pH greater than 8, some of the ammonium ion will be converted to free ammonia. Some stream standards and sewer codes do have limits on ammonia content. Ammonium ions, when present in a stream for long periods of time, will oxidize to nitrates through normal biological processes.

The TKN (total Kjeldahl nitrogen) test is an analytical method for measuring total (both inorganic and organic) nitrogen.

Phosphates and Nitrates

Metaphosphates are used as sequestering agents in some processing solutions to minimize sludging due to calcium in water. A few processing solutions contain trisodium phosphate or other phosphorus-containing compounds as buffers or calcium sequestrants.

Nitrates are present in only a few processing solutions and at levels that are not significant to treatment plants.

Detergents

Although detergents may be commonly used for cleaning purposes in processing laboratories, they are rarely used in processing solutions. However, a few photographic films contain components used to perfect the photographic image, which may leach into the photographic solutions and cause a positive test for detergents, depending on the analytical method chosen.

Color and Odor

Mixed processing wastes from conventional processing laboratories will generally have a very slight color and a scarcely detectable odor and, therefore, are not usually affected by effluent discharge regulations. Waste from plumbingless minilabs will have a darker color due mainly to the bleach or bleach-fix component, but the waste is emitted in very low volume compared to normal sewage.

Flammable and Explosive Materials

Conventional photographic processing solutions or effluent are not flammable or explosive.

Volatiles

Volatile compounds in industrial effluents are of concern to regulatory agencies because sufficient concentrations of vapors in an enclosed area can present a hazard, either as irritants to the skin, eyes, lungs, and other mucous membranes or (in the case of flammables) because they can be readily ignited. With the exception of small amounts of chemicals used in some film lubricants, soundtrack applicators, or film cleaners and lacquers, most solutions and chemicals used in photographic processing laboratories are not considered volatile.

Some solutions do contain ingredients such as acetic acid, formaldehyde, or ammonia, which are volatile. Formaldehyde emissions are regulated by the OSHA formaldehyde standard.

At a pH above 8.0, solutions containing ammonium compounds can give off ammonia. When chemicals such as these are used, adequate ventilation is needed.

States and localities may have codes that restrict the amount of volatile organic emissions that can be discharged to the air; exhausting process fumes to the outside air may require an air pollutant source emission permit. If large amounts of lacquers, lubricants, or cleaners are being used, the photoprocessing laboratory should check with the appropriate local or state agency to determine what limitations are imposed and what safeguards are required.

6.3 TREATING PHOTOPROCESSING WASTE

6.3.1 Introduction

Regardless of whether a waste issues from photoprocessing or any other source, all waste treatment technology can be broken down into one of three basic techniques: neutralization, oxidation/reduction, or separation.

Neutralization is the process of nullifying or dissipating the chemical effects or characteristics of a material while leaving it in place, without removing it from a chemical system, by adding one or more other ingredients. Examples would be pH adjustment with acid or alkali, solidification by mixing with a concretion additive, dilution with water, complexing of calcium or other metals by adding a sequestrant, blending of two waste streams to dilute and cancel out the effects of their individual ingredients, and so on.

Oxidation/reduction is the process of chemically changing the character of a material by supplying electrons to or removing electrons from the chemical structure, breaking down chemical bonds or forming new ones to create new compounds. Examples would include incineration, biological (secondary) treatment in which the waste acts as food for microorganisms, chlorination, electrolytic silver recovery, chemical oxidation with permanganate, peroxide, or ozone, chemical reduction of chromium 6+ to 3+ with bisulfite, burning of gasoline in an internal combustion engine, photolysis, corrosion of metals, natural decay of vegetation, and so on.

Separation is the process of physically separating one or more materials from a chemical system, generally without a chemical reaction but sometimes coincidental with it. Examples include filtration, activated carbon adsorption, reverse osmosis, ion exchange, settling, distillation and evaporation, freeze drying, solvent extraction, centrifuging, and numerous others.

Neutralization has limited use in waste treatment technology, except for the very critical use of pH adjustment. Oxidation/reduction and separation methods, either in individual processes or various sequential combinations, can be very effective and are in widespread use.

6.3.2 Secondary (Biological) Treatment

Biological waste treatment processes are a combination of oxidation/reduction and separation steps, assisted by appropriate pH adjustment. Microorganisms are intentionally encouraged to grow on the waste components, consuming some as food/fuel and assimilating others for cellular growth. Precipitation of materials occurs at various points of this cycle and at certain stages may be deliberately enhanced through the use of settling basins, skimmers, or filters to remove certain components as solids from the aqueous medium. Some waste components may tend to inhibit oxidation/reduction by being toxic to the microorganisms above a threshold concentration; others may form toxic or objectionable byproduct gases if not limited in concentration or chemically attenuated. The appropriate management of these parameters, including the use of sewer codes to limit input concentrations, analyses to detect the presence and/or concentration of components as well as end products, and equipment to conduct various process engineering operations as they are needed, allow the successful operation of such waste treatment processes.

Many different types and designs of secondary biological treatment systems are in use today, including activated sludge systems, sequencing batch reactors (SBRs), fixed film bioreactors, aerobic and facultative lagoons, trickling filters, oxidation ditches, and others.

Biodegradation Testing

Extensive studies have been conducted from the late 1960s to the present day at Eastman Kodak Company, and at many other outside institutions during this time, to evaluate the biodegradability of chemicals in typical photographic processing effluents [9,14,20–23]. The results indicated that these effluents could be treated biologically and that it is safe for processing laboratories to send their waste solutions to municipal secondary waste treatment plants, after practicing silver recovery and regeneration techniques that make sense economically and photographically. Of course, this recommendation depends on the concentration of the photographic effluent, size of the waste treatment plant, and nature and quantity of other wastes being simultaneously treated.

Some biological treatability studies were conducted using 5.5 and 55 L activated sludge units. The general results of the test were that, as long as not more than 10% by volume of conventional photographic processing effluent was treated in an activated sludge unit having a retention time of between 4 and 6 hours, no adverse effects would be seen on the activated sludge unit, and the photographic effluent would be adequately treated [9].

Similar tests were conducted by Hydroscience, Inc., in 1974 [18]. They undertook a study to determine the environmental effects of 45 selected photoprocessing chemicals and a typical photoprocessing effluent. This testing program was sponsored by the National Association of Photographic Manufacturers (NAPM, subsequently known as I3A, International Imaging Industry Association), which represented major segments of the photographic industry. The conclusions of these tests were the following:

1. Photoprocessing wastes do not present a hazard to biological systems encountered in conventional wastewater treatment schemes, nor to aquatic organisms indigenous to natural receiving waters. The assumption inherent in the above statement is that a photoprocessing waste is not directly discharged to a natural receiving water without adequate treatment before discharge.
2. The results indicate that photoprocessing wastes are amenable to biological treatment, with a removal efficiency approximately equivalent to that encountered with domestic wastewaters. The concentrations expected to occur in municipal or regional sewerage systems and in a receiving water upon treatment will not adversely affect natural biological systems.
3. The 45 photoprocessing chemicals investigated in this study have no significant impact on the activity or efficiency of the biomass of a conventional biological wastewater treatment system, at the concentrations realistically attainable in municipal or regional sewerage systems. For large laboratories having conventional processes that use water washes, the chemical concentrations expected to occur in the discharge may usually be estimated by applying a 100-fold dilution to the working solution concentrations. This mathematical estimate would not apply to minilabs, or to larger labs today that are installing plumbingless equipment to handle digital inputs.

Additional studies were also conducted by the J. B. Scientific Corporation [24] in the late 1970s. This investigation was also sponsored by NAPM. These studies reaffirmed that the use of biological treatment to handle photographic processing effluents is a preferred technique.

More recently, extensive rounds of testing have taken place over the past decade, starting with internal studies at Eastman Kodak Company in 1990–1991 [25]. This bench-scale work included various concentrations of the following processes in a simulated domestic wastewater: Kodak Flexicolor C-41, two versions of Kodak Ektacolor Process RA-4, Kodak X-Omat x-ray process, Kodak Ektachrome E-6, and the Kodak Ultratec graphic arts process. Other independent studies were completed at Clarkson University in 1990–1992 [21,26] and Georgia Institute of Technology in 1993–1996 [27–30]. These university studies focused mainly on the Kodak Flexicolor C-41 and Kodak Ektacolor Process RA-4 processes and their components.

A conference paper summarizing this and earlier work was presented in September 1997 at the 19th Brazilian Congress on Sanitary and Environmental Engineering [31]. In addition, more internal bench-scale work was carried out at Kodak in 1997–1998. These latter studies were specific to the fate and effects of minilab effluent and x-ray processing effluent on packaged wastewater treatment plants as they are found in shopping plazas and malls in Latin and South America.

All of the above studies have confirmed the previous findings. That is, in modest amounts (up to 10% of the hydraulic loading from conventional photoprocessing plants, equivalent to up to 1% hydraulic load from minilabs, and up to 5% v/v from x-ray processing operations), photoeffluents do not harm existing secondary waste treatment systems, and secondary waste treatment systems will satisfactorily treat and detoxify photoprocessing effluents [25,32]. Both aerobic and anaerobic small-scale systems have been tested [22,33], as has the fate and effects of silver on the anaerobic digestion process [34]. A large number of publications [14–16,20,22,23,27,32,34–39] resulted from independent university studies between 1990 and 1997.

Large-Scale Onsite Biotreatment

Conventional Activated Sludge. The use of biological treatment facilities operated by municipalities is a very effective way of handling photographic processing effluents. Tests by Eastman Kodak Company previously cited showed that it was feasible to use biological treatment to treat only photographic processing effluents [9]. These tests used a retention time of approximately 24 hours, indicating that extended aeration was necessary for treating sewage composed exclusively of photographic processing effluents without blending in domestic or other municipal waste. Further testing also showed that it was possible to pretreat photographic processing effluents using onsite treatment with biological techniques.

In addition to smaller bench-scale systems, an extended aeration activated sludge unit having a capacity of 76,000 L/day was operated by Kodak on photographic processing effluents [9]. Processing effluents were collected from six different color processes. The process solutions treated included effluents from Ektaprint R and Ektaprint 3 chemicals, and from the E-4, C-22, CRI-1, and K-12 processes. (These processes are now obsolete and have been superceded, although most of their chemical components are still used today in reformulated processes.) Once adequate mixed liquor suspended solids (MLSS) had built up in the aeration tank, the average weekly BOD_5 reductions ranged from 78 to 91%. During this time period, the MLSS ranged from slightly over 1100 to as high as 4600 mg/L. To obtain optimum efficiency, automatically backflushed sand filters were used to remove the very finely divided suspended solids that were produced in the system.

Although the use of activated sludge systems can provide an effective method for treating photoprocessing effluents, their operation and the subsequent handling of solids produced are both time-consuming and labor-intensive tasks. These can be difficult and uneconomical systems to operate onsite by a photoprocessing laboratory, unless it is very large and has a well-trained, permanent maintenance staff.

Rotating Biological Contactor. An alternative treatment technique is the rotating biological contactor. Small-scale testing has indicated that it is very effective and may be more economical than an activated sludge unit. Initial tests at Eastman Kodak Company indicated that with an effluent having a BOD_5 of approximately 600 mg/L, retention times of between 4 and 24 hours resulted in from 83 to 94% reductions in BOD_5 [40].

Additional testing with rotating biological contactors was conducted by Lytle [41]. In these experiments, photoprocessing effluents having BOD_5 concentrations between 440 and 1600 mg/L (with COD levels ranging from 1010 to 4120 mg/L) were effectively treated using retention times of 8 and 16 hours. In these tests, BOD_5 reductions of 72 to 95% and COD reductions between 55 and 72% were attained. Soluble silver levels were reduced from an initial 3–10 mg/L level to a final level of 0.15–0.75 mg/L. The precipitated silver was harvested with the sludge.

Further trials were conducted onsite in a production microfilm processing plant by Petche [42]. Black and white microfilm developer and fixer were treated in a 3200 L/day rotating biological contactor. The concentrations in the input solutions were as follows: BOD_5 292–1800 mg/L, COD 1090–8580 mg/L, and chlorine demand 80–2200 mg/L. Treatment produced BOD_5 reductions averaging 85%, COD reductions averaging 61%, and chlorine demand reductions of 98%.

Impact of Biotreatment on Silver

In addition to reducing the oxygen demand of photographic processing effluent, biological treatment also affects silver speciation. Depending on the size of the photoprocessing operation and complexity (and efficiency) of the silver recovery equipment, silver concentrations of between 0.5 and 10.0 mg/L may be found in photoprocessing effluents. The silver present in these effluents is in the form of a silver thiosulfate complex, which as previously noted is approximately 17,500 times less toxic than free silver ion [11]. Early studies by Dagon [9], Cooley *et al.* [10], J. B. Scientific Corp. [24], and Bard *et al.* [43] showed that in secondary waste treatment plants, the silver thiosulfate complex is converted by microorganisms into insoluble silver sulfide, with a small amount of metallic silver also formed. Both of these insoluble species are substantially removed from the secondary treatment plant effluent as insoluble sludge during the settling step.

In a 1974 study, Ericson and LaPerle [44] analyzed the effluent from the 76,000 L/day activated sludge unit operated at Kodak. Total silver concentration in the effluent after biological treatment was measured at 0.95 mg/L. This effluent was filtered through a 0.45 μ filter, then through an ultrafilter with a 1000 mol.wt. cutoff. The amount of insoluble silver was found to be 0.94 mg/L, primarily in the form of silver sulfide with some metallic silver also being present. The filtrate contained less than 0.01 mg/L soluble silver.

Subsequently, in 1979–1980, the Environmental Protection Agency (EPA) sponsored a study of publicly owned treatment works (POTWs) [45]. In this study, hundreds of influent, effluent, and sludge samples were analyzed for various chemicals, including silver. These analyses showed that almost all of the effluent silver was incorporated into the sludge.

In 1981, Lytle [46] carried out an analytical survey of six POTWs to evaluate speciation and fate of silver. In this study, the sources of silver were identified as municipal, photographic, and industrial. Samples of influent and effluent to and from the POTWs were analyzed for total silver and free silver ion. Regardless of the source of the silver, no significant amounts of free silver ion could be detected. Silver removal efficiencies were very high at all of the treatment plants studied. Those POTWs receiving photoprocessing effluents showed silver removal efficiencies equal to or greater than 90%. The concentration of free silver ion present in the effluents of the POTWs receiving photoprocessing waste was four parts per trillion. Lytle's study thus confirmed findings previously recorded by the EPA: regardless of its source, silver is rapidly and almost completely incorporated into the sludge during secondary waste treatment.

More recently, Pavlostathis and his students carried out a series of studies on photographic effluents over several years at two universities [14–16,21–30,32–39]. These proved conclusively that silver in photographic effluents did not adversely affect biodegradation processes under either aerobic or anaerobic conditions.

Over the past decade, a major breakthrough has been made in the understanding of silver behavior in the environment. Beginning in the late 1980s, it was noted by a growing number of researchers that the alleged concentrations of many metals that the U.S. Environmental Protection Agency (USEPA) and certain State agencies had established for numerous natural waters were not only in error by as much as several orders of magnitude, but in fact often could not be duplicated even by the persons who did the original sampling. The errors were traced to many sources of contamination during both field sampling and later analyses. "Clean" field sampling and laboratory analytical techniques were then established at the University of Wisconsin and elsewhere [47]. These proved that the previously reported concentrations for many metals including copper, zinc, and especially silver, may have been overestimated in certain bodies of water by as much as 1000 times or more. In contrast to earlier public fears, the actual amounts measured in lakes and streams by the new "clean" methods were often far below levels that might cause toxicity or environmental concerns. Similarly, it was realized that many previous and current laboratory toxicity studies with silver on various living organisms proved difficult to duplicate, suggesting that not all the proper parameters were being studied, measured or included during laboratory simulations.

At the same time, the Federal and many State governments in the United States and elsewhere were beginning to implement discharge regulations for metals based on the old assumptions. Industry became alarmed that without new factual data, very expensive corrective measures might be enforced, which would prove disastrous to U.S. industry without real justification and no significant environmental benefits.

The silver industry, and particularly the photographic industry, took the lead. Beginning in 1989, the Eastman Kodak Company issued the first contracts to the University of Wisconsin to study environmental transport, fate, and effects of silver. In 1992, it was agreed to set up a formal silver research program known as Argentum, where independent, unbiased research would be funded and results reported at an annual international conference. Research funding was expanded through the NAPM (now I3A). Besides Wisconsin–Madison, additional universities financially supported included McMaster, Texas A&M, Mississippi, SUNY Stony Brook, Wilfrid Laurier, Quebec, Clemson, Wisconsin–Superior, Delaware, Illinois Institute of Technology, Kentucky, Manhattan College, and others. The USEPA, certain States such as Wisconsin, Michigan, and Florida, the Canadian Government, other governmental and institutional sources, and most recently the Water Environment Research Foundation also supplied funding to U.S., Canadian, and European universities and private research institutions. In total more than 100 researchers participated in the investigations, which were reported in numerous individual research papers as well as the formal Argentum proceedings [48–53].

Significant breakthroughs were made, especially in the late 1990s, and reported in the final summary book that concluded the Argentum conference series [12]. (This book also contains hundreds of references to related individual research papers.) The most startling finding is the documentation that inorganic reduced sulfur, which ties up silver and many other metals, is ubiquitous in almost all natural waters throughout the world, regardless of whether fresh or saline, oxic or anoxic [54–58]. Other related findings were: (a) because silver is almost always complexed with sulfur species, resulting in extremely low toxicity, silver rarely if ever presents an acute threat to organisms in natural waters; (b) whether or not silver could be chronically toxic to certain organisms over a long period of time is not well established, because chronic toxicity would appear to occur by different mechanisms than acute toxicity; (c) a previous consideration by the USEPA that an acute-to-chronic mathematical ratio be used for regulatory purposes would not be a valid approach; and (d) laboratory toxicology studies on living organisms that had used only silver nitrate (free silver ion, or Ag^+) as the source of silver, without including sulfur or other complexing compounds such as dissolved organic matter (DOM) in the aquatic test medium, were artifacts that did not simulate real conditions in nature, were not valid except in very special cases, and the data reported should not be considered in formulating regulations [12]. Other studies showed that silver from photographic effluent was not toxic to humans [59].

In summary, clear scientific evidence has shown that silver in photoprocessing effluent is present as the thiosulfate complex, and this complex is satisfactorily removed during biological treatment such as in a municipal treatment plant. A very high percentage of the complex, essentially 100%, is converted to insoluble species, primarily silver sulfide and metallic silver, most of which is removed as sludge in the settling step. Practically no soluble forms of silver will be discharged in the liquid effluent from the treatment facility. Any fractional amount of particulate silver species that might be discharged would pose no acute threat to organisms in natural waters. Chronic effects are still unknown, although no deleterious chronic toxicity effects have thus far been observed, and silver exposure is shown from epidemiological studies not to pose a toxicity problem for humans.

Impact of Biotreatment on Specific Photoprocessing Chemicals

Recommendations from the photographic industry have always been that photographic processing effluents should never be discharged directly to a receiving body of water, but rather should be either treated on site, recovered for off site treatment, and/or discharged to a municipal secondary waste treatment plant as part of general building sewage. Certain constituents of photographic processing effluents can be relatively toxic if disposed directly into natural waters without treatment, as would many other materials, including human sewage.

Chemicals such as hydroquinone used in black and white developers and color developing agents used in color developers have high levels of toxicity associated with them, if they are not adequately treated [11]. Work conducted by Watson *et al.* [60] showed that while Kodak Color Developing Agent CD-3 was toxic at levels of approximately 1 mg/L to fathead minnows and *Daphnia* if untreated, significantly higher concentrations of this chemical could be safely discharged if it were first subjected to biological treatment. Levels of CD-3 as high as 30 mg/L were not toxic to aquatic species after such treatment. In addition, studies by Harbison and Belly [17] showed that hydroquinone is readily biodegradable and would not be of concern after treatment in a secondary waste treatment system.

Organic bleach components such as EDTA and PDTA have a relatively low rate of biodegradability although studies have shown they are biodegradable [61,62]. These compounds are considered nontoxic and are commonly encountered in domestic waste because they are used in other applications, mainly foods and agricultural applications. Inorganic bleach (ferricyanide, which is also used in a number of other domestic products) readily reacts with iron compounds in sewage, precipitates as a blue solid (ferric hexacyanoferrate) during secondary biological treatment, and is incorporated in the sludge.

Biotreatment is also highly effective in degrading thiosulfate and sulfite. Depending on the oxygen available during sewage treatment, end products are typically sulfate and sulfide. Both of these anions are useful in tying up trace metals that may enter the treatment plant from other sources. Acetic acid, gelatin, and a few other materials used in photography are common foodstuff components that biodegrade readily. None of these materials has an adverse impact on the operation of a secondary treatment plant.

6.3.3 Other Oxidation/Reduction Treatment Technologies

Other than biological treatment, there have been few instances in which oxidation/reduction technologies have been effectively used on any large, continuous scale to treat photoprocessing effluents for direct discharge to receiving bodies of water. Some have been used as pretreatment methods prior to discharge into sewer systems. Some of the more popular past methods are described below.

Ozone

Ozone is an extremely strong oxidizing gas that reacts with many organic compounds and some inorganic compounds to break them down to less environmentally objectionable materials. Experiments to evaluate the ability of ozone to decompose a number of photoprocessing chemicals were conducted between 1969 and 1973 by two of the authors [63].

As a result of this testing, the following conclusions were reached: (1) Ozone is particularly useful in breaking down many photographic chemicals, including developers and EDTA, to more easily biodegradable materials; (2) ozone is effective in lowering the BOD_5, COD, and chlorine demand of individual and combined photographic solutions; (3) ozone has no significant effect on acetic acid, acetate ion, or glycine and appears to degrade ferricyanide ion too slowly to be of practical importance; (4) thiosulfate, sulfite, formaldehyde, benzyl alcohol, hydroquinone, and ethylene glycol can all be degraded effectively to innocuous endproducts with ozone.

Four case histories that actually used ozone in production-scale photographic processing operations have been discussed by Hendrickson [64]. Although shown to be very effective in some applications, the use of ozone treatment alone to degrade photographic processing chemicals can be extremely expensive. Its use in combination with other techniques, such as biological treatment [65], may be much more economically desirable.

Experiments have also shown that the combination of ozone and ultraviolet light treatment can be effective in decomposing some processing chemicals. The economics and practical operating problems with this technique have not yet been established.

Chemical Oxidation/Reduction

Other oxidation techniques, such as the use of hydrogen peroxide [66–68], permanganate, persulfate, bromine, chlorine [69,70], perchlorate, hypochlorite, and numerous other chemical oxidants, have been investigated. Many of these have been used for the regeneration of photographic bleaches, either commercially or experimentally, and in the process were discovered to degrade trace components carried into the bleach solution. Concerns in modern times with these oxidizing agents, in addition to safety and cost of handling as well as general effectiveness, include the possibility of introducing yet another pollutant into the waste stream (e.g., manganese from permanganate), or forming byproducts that are even more toxic if the oxidation reaction is only partial (e.g., forming chloramines by the partial chlorination of organic amines). About the only chemical oxidants used on any significant commercial scale in photographic processing have been hydrogen peroxide [68] and "bromine sticks" (bromochlorodimethylhydantoin), the latter used in a unique application [71] to destroy residual thiosulfate while simultaneously curbing unwanted biological slime during a washwater recycling operation.

Chemical reduction has been commonly used in a number of operations in the past. As previously stated, dithionite and bisulfite have been used to reduce chromium 6+ to chromium 3+ in residues from systems cleaners or black and white reversal bleaches. Sodium borohydride has been used in certain special cases to precipitate silver as elemental silver metal from overflow fixer solutions, or other trace metals from a few processes (see Section 6.4.8, "Other Chemical Precipitants"). Sodium dithionite has been used in combination with ferrous sulfate to precipitate ferrocyanide from bleach wastes as insoluble ferrous ferrocyanide (see subsequent section "Ferrocyanide Precipitation") [72,73].

Electro-oxidation/Reduction

Electro-oxidation, the process of oxidizing dissolved compounds at the anode of an electrolytic cell, is a technology that many researchers have investigated for waste treatment applications, with varying results. Appropriate catalysts have been shown to enhance efficiency; however, there is difficulty in keeping the most effective of these from becoming poisoned, thus reducing their practical life. Heat build-up in the cell, cross-cell reactions whereby a component may be oxidized at one electrode and promptly reduced at the other for no net gain, gas formation, and undesired side reactions can all present practical operating problems.

In 1975, Kodak researchers publicly disclosed a method for generating sulfide ion upon demand by the deliberate, direct electro-oxidation (or alternatively, thermal treatment) of the waste silver-bearing fixer or bleach-fix [74]. This excess sulfide ion was then used to precipitate silver from the remainder of the solution in the same apparatus. The method had the advantage of not needing to prepare and store solutions of sodium sulfide, which can produce toxic as well as very odorous hydrogen sulfide (see Section 6.4.7, "Electrochemical Sulfide Precipitation").

In the early to mid-1990s, considerable research was carried out by Kodak to investigate new anode materials, electrolytic oxidation cell configurations including separate cathode and anode compartments, and doped anode materials for the treatment of photoeffluents [75]. The technology was proven feasible; it was possible to reduce oxygen demand by over 90%. However, electrolytic oxidation was deemed not practical at that time due to high energy costs and the high costs and very short life of electrode materials, which were observed to degrade over time. Electro-oxidation has also been tested in combination with other, less energy-intensive oxidation techniques such as biological treatment [65].

Currently such techniques are not being used as waste treatment technology with photoprocessing effluents. However, electrolytic cells have been used very successfully by the photographic industry for many years to reoxidize spent ferricyanide bleach (see subsequent subsection, "Bleach Regeneration").

Electrolytic reduction is most commonly used for recovering silver (and other metals) from solution, known as electroplating or electrowinning. This method is discussed extensively in the forthcoming section on silver. Electrochemical reduction has found no other widespread applications in the treatment of photographic wastes to date.

Incineration

Despite unfortunate and often misleading environmental publicity, and resulting adverse opinions expressed by the public, properly conducted incineration remains one of the most viable, universally applicable, and environmentally sound waste treatment methods known to mankind.

In 1989, Eastman Kodak Company contracted with the Mechanical Engineering Department of the University of Wisconsin–Madison to run batch incineration tests on waste solutions from the Kodak Flexicolor C-41 process, both combined and individual processing solutions [76,77].

The results represent only initial information from only a few solutions among a vast array of possible tests. However, the data were very encouraging: no significant residual organics were detected in the emissions when the solutions were incinerated with excess air at or above 800°C (1472°F), which was the lowest temperature tested. This is a reasonable operating condition for commercial incinerators. A second phase of these tests showed that emissions from incinerating photographic wastes were comparable to those generated by incinerating common foods.

Wet-Air Oxidation

Wet-air oxidation is the "pressure-cooking" of waste with air at high pressure and moderately high temperatures. Experiments were funded by Eastman Kodak Company in 1989 and again in 1991 for the Zimpro Passavant Corp. to conduct wet-air oxidation studies on Kodak Flexicolor C-41 process wastes at temperatures up to 280°C and pressures of 1500–2500 psi [78]. As in the case of the incineration tests above, the information obtained was only preliminary, and further follow-up work would be needed to prove its practicality and economic feasibility. However, the data did indicate that the treatment can achieve a high degree of destruction of certain developer components as well as EDTA, and can considerably increase the biodegradability of both the bleach or bleach-fix and the developer solutions.

Other High-Temperature Oxidation Techniques

A number of other state-of-the-art, high-temperature oxidation methods were investigated by the Environmental Sciences Section of Eastman Kodak Company in collaboration with universities and private research firms. The molten salt process (operating at up to 1700°F, developed by Rockwell International and the U.S. Bureau of Mines) showed particular promise from a technical standpoint. Extremely high-temperature (up to 25,000°C) plasma experiments at the University of Wisconsin combustion laboratory, and molten steel (up to 3300°F) evaluations with another manufacturer, presented practical implementation problems. In all cases practicality and economic considerations did not permit these approaches to be pursued beyond the investigatory stage.

6.3.4 Separation Methods

Evaporation

Evaporation experiments at atmospheric pressure to reduce the volume of waste photoprocessing solutions were first performed at Eastman Kodak Company in 1970 and have continued since, both at that company and others. Water was successfully removed, and sludges or slurries containing nearly all the residual chemicals occupied only 10–15% of the original volume. It was established that at higher temperatures near the boiling point of water, certain ingredients such as thiosulfate would decompose, liberating noxious gases such as sulfur dioxide or hydrogen sulfide, especially under acidic conditions. However, it was further discovered that an upper temperature limit of 160°F, along with a pH adjusted to near neutral, would minimize such gas formation, yet still allow reasonably rapid water removal. Suggestions made since that time, in response to inquiries from prospective manufacturers or users of evaporation equipment, were to limit the upper temperature to 140°F, thereby including a safety factor to cover cases of extreme solution aging or unforeseen equipment malfunctions.

A number of commercial evaporators for photographic wastes were produced by various firms between the late 1970s and the mid-1990s. Some manufacturers were able to overcome the thermal decomposition problem while maintaining a maximum evaporation rate by conducting the evaporation under vacuum, thereby lowering the boiling point of the solution to below 100°F. This, along with an appropriate pH adjustment, appears to virtually eliminate the formation of sulfurous gases [79].

A few other manufacturers proposed conducting atmospheric evaporation at higher temperatures, near the boiling point of water, and trapping the evolved gases in various absorptive or scrubbing media. However, in most cases this appeared to be a cumbersome and impractical answer, because the media would then need to be treated or discarded, and breakthrough of the odorous gases from the trap was generally unpredictable.

Over the past three decades, ammonium salts have become popular ingredients in fixers and bleaches because of their rapid penetration into gelatin emulsions, allowing shorter process times and less overall chemical consumption for significant environmental benefits. However, if the pH is too high, the evaporation of wastes containing ammonium salts can liberate ammonia gas. The best compromise, between the liberation of ammonia at alkaline pHs and sulfurous gas formation at low acidic pHs, appears to be to limit the pH for most combined photoprocessing waste solutions to a range between 6.0 and 6.8, aiming for about 6.2. This successfully ties up most of the ammonium salts in the residual sludge. In a few specific instances, such as high-ammonium x-ray fixers that had not aged significantly, pHs as low as 5.5 were successfully used without significant sulfurous gas formation.

Freezing and Freeze-Drying

Freeze and freeze-drying techniques were investigated to purify and separate components of photoprocessing waste, but no practical methods were proposed or are currently in commercial use.

Ion Exchange

Ion exchange is a process in which "ions held by electrostatic forces to charged functional groups on the surface of a solid are exchanged for ions of a similar charge in a solution in which the solid is immersed" [80]. Ion exchange is most commonly used to soften water by utilizing a cation-exchange resin to remove calcium and magnesium ions. In photoprocessing applications, anion-exchange resins were once widely used for removing halide ions for developer regeneration and recovering hexacyanoferrates from dilute solutions. Some large laboratories, especially motion picture labs, still use ion exchange to recover silver thiosulfate complexes from dilute solutions. Cation-exchange resins have been used to remove color developing agents from stop baths.

Precipitation

Precipitation, as described in this chapter, is the formation of insoluble solids through the reaction of two or more water-soluble species. Although most photographic chemicals are deliberately chosen to be soluble in water so they do not precipitate spontaneously during processing, in several instances a reagent can be found to solidify certain components for recovery (see Section 6.4, "Silver Recovery" and Section 6.5.2, "Chemical Recovery Methods") [72].

Settling and Decanting

After precipitation, methods must be found to separate the precipitate from the solution. Generally in photoprocessing applications, the solids are often either very finely divided or they form very colloidal, gelatinous suspensions. In either case, the materials are difficult to filter and often "blind" the fine-pore filters that would be required for adequate separation. The alternative is often to add a flocculating agent, allow the material to settle over a period of time, and then carefully decant the supernatant liquid to a point just over the sediment. By removing most of the water this way and allowing the mass to settle for a time, which sometimes promotes the growth of larger crystals or particles (known as "ripening"), direct separation by decantation may be possible or often the resulting mixture becomes easier to filter.

Filtration

Several different types of filters are used in photoprocessing applications. Woven, pleated paper, or particulate cartridge filters are often used on the processing machine itself to remove carried-in dirt or in regeneration operations. Sediments and precipitates that are formed either in waste mixtures or intentionally during recovery operations may need more elaborate filters. These should be oriented toward higher solids loading in production quantities and preferably be reusable. Filter types used in the past for recovering solids from photoprocessing chemical wastes have included filter presses, vacuum belt filters, pressure leaf filters, sand, carbon or other particulate filtration beds that can be back-washed, and others. To assist filtration, flocculants and filter aids have been included when necessary (see Section 6.4, "Silver Recovery" and Section 6.5.2, "Chemical Recovery Methods") [72].

A unique device that combined settling and decantation with filtration was patented by two of the present authors and colleagues [81] for use with the TMT silver recovery process (see Section 6.4.5, "The TMT Precipitation Method"). Owing to the fragile but adherent physical characteristics of the porous silver–TMT precipitate, it was possible to devise a filter vessel where freshly precipitated and flocculated "fines" were passed upward through a porous bed of the same slightly more dense, previously settled, silver–TMT material before they could encounter the pleated paper filter. The silver–TMT fines that normally would blind the pleated filter in a short time were thereby induced to adhere to the porous, already precipitated mass, which acted as its own filtration bed that grew in height with time. This extended the life of the backup paper-filter manifold and allowed a considerable weight of precipitate to collect in the vessel before it would be disconnected and shipped off for recovery.

Centrifuging

Centrifuging is the process of separating two or more materials of dissimilar specific gravity by a rotary spinning (centrifugal) action. Centrifuges are more expensive capital investments than filters, although they generally produce drier cakes of solids and may require less operating labor. In photographic processing, centrifuges have found limited use in some silver recovery applications, and in Kodachrome process operations for recovering color coupler (see subsequent subsection, "Coupler Recovery").

Reverse Osmosis

Reverse osmosis is the process of applying pressure to a salt solution on one side of a membrane, to force water through while retaining the salts on the original side (in effect "squeezing" fresh water out of saline water). The original stream is known as the "feed," the fresh water product as the "permeate," and the rejected concentrated solution as the "brine" or "concentrate." The process works best on dilute solutions such as wash water, where more than 90% of the original feed can be reclaimed before the salt buildup in the brine begins to hamper recovery. The advantages of the process are the very low energy needed to reclaim water as compared to distillation, and the lack of need for water treatment chemicals. Disadvantages include high capital costs, the size of equipment needed to obtain sufficient flow to keep up with a process, and frequent maintenance of the membrane system and pumps. Reverse osmosis has been tested for photoprocessing applications since the mid-1960s, but to date has not found widespread application. In the future, increased equipment reliability, better less fragile membranes, and changing energy cost considerations may widen its use.

Ultrafiltration and Nanofiltration

Ultrafiltration and nanofiltration appear similar to reverse osmosis superficially, but rather than relying on osmotic principles for removing dissolved materials, they are true filtration processes for removing finely divided particles using a membrane filter. Neither ultrafiltration nor nanofiltration have found widespread use in photoprocessing to date, because (1) simple wound cartridge filters are cheaper and more convenient to use for ordinary process dirt, and (2) situations where solids formation is deliberately induced for waste treatment reasons (see "Chemical Recovery Methods") need filters that do not blind easily and can handle large quantities of gelatinous solids. Possible applications continue to be explored.

In a specific motion picture process in 1995, the nanofiltration technique was shown to be very efficient in eliminating most of the pollutants from washes, and may enable the recycling of up to 80% of wash water. It is especially efficient in silver recovery. However, among the practical concerns, which have not yet been established through long-term studies, is membrane life. The stability of membranes could have significant impact on the final cost of using nanofiltration in film processing. In addition, the possibility of reusing recovered concentrate as a replenisher as well as recycling permeate water would bring a faster return on investment than recycling water alone [82].

In 1997 Eastman Kodak Company conducted studies on the use of nanofiltration and reverse osmosis to reuse water in the Ektachrome E-6, Ektachrome Reversal R-3, Flexicolor C-41, and Ektacolor RA-4 processes. Both technologies were shown to be feasible for reducing wash water usage by 85–90% [83].

6.4 SILVER RECOVERY

6.4.1 Introduction

As previously discussed, silver from photoprocessing operations is much less toxic than free silver ion. Silver is generally removed from photographic products during processing in the form of silver thiosulfate complex, $Ag(S_2O_3)_2^{3-}$. This complex has a dissociation constant of 5×10^{-14}; thus, it is virtually impossible for free silver ion (Ag^+) to be present at any significant concentration levels in photoprocessing effluents [10].

In black and white products, because the final image is metallic silver, the amount of silver removed during processing will depend on the amount of exposed image area. In color products, processing removes essentially all the silver from the emulsion. Although primarily found in fixers and bleach-fixes, small quantities of silver will also have seasoned the developer and bleach, and some will be carried over into solutions following the fixer or bleach-fix. (In modern minilabs, the stabilizers or final rinses often contain approximately 25% of the silver available for recovery.)

Even the very smallest processing operations have adequate economic and regulatory justification to recover silver from their exhausted fixers and bleach-fixes, low-flow washes, and stabilizers.

Whether the same justification applies to conventional washwater depends on the size of the operation, treatment alternatives, and applicable regulations. These washwaters usually contain about 2–5% of the potentially recoverable silver in the process. In conventional process machines that utilize multiple-stage, countercurrent, low-flow washes, the amount of silver in final washes may be considerably less.

The two most commonly used methods of silver recovery are electrolytic recovery and metallic replacement using steel wool cartridges [84]. Ion exchange has been used on a limited scale, generally by larger laboratories to treat dilute solutions such as wash waters. Precipitation with sulfide was used in the early part of the 20th century by many laboratories, but because of its handling hazards has largely been superceded by safer processes. It continues to be used in some situations today, particularly in developing countries, because of its low cost and ease of use despite the attendant personal hazards. A more recent development is the TMT (trimercapto-s-triazine) precipitation process, a safe process that can easily be automated, which is finding increasing use in larger U.S. and Canadian laboratories. Other precipitants have been tried by some commercial laboratories but are not widely used for various reasons.

Electrolytic silver recovery can be successfully applied to concentrated fixer and bleach-fix solutions in either an in-line or a terminal application. Other solutions that can be terminally desilvered include low-flow washes and plumbingless minilab stabilizer, but generally not high-flow washes. This method usually finds application in situations where higher concentrations of silver are encountered and must be removed rapidly, and/or where only a portion of the silver is to be removed continuously so the processing solution can be reused.

Metallic replacement is normally used with fixers and bleach-fixes in a terminal application, known as "tailing," to recover silver from solutions destined for the drain. In the past it was also used for in-line desilvering of bleach-fixes, where the addition of dissolved iron to the reclaimed solution was a benefit. It has also been successfully used to treat low-flow washes that have been combined with fixer and bleach-fix overflows. Metallic replacement has also been used on final washes alone, but with limited success.

Some large photoprocessing installations have used ion-exchange methods to recover silver from washwater or mixtures of washwater, fixers, and bleach-fixes. The destructive oxidation of bleach-fixes and fixers with chemicals such as hydrogen peroxide [68] and chlorine [69,70] also causes silver to precipitate as silver sulfide, which can be removed by settling or filtration.

If electrolytic silver recovery and/or metallic replacement are effectively used, a photoprocessing laboratory should be able to recover on site between 90 and 99% of the potentially recoverable silver. The effluent silver concentration from such a laboratory would be in the 1–5 mg/L range. Even higher concentrations will have no adverse impact on a secondary waste treatment plant or a receiving body of water [9,30,85] after the effluent is biologically treated. If ion exchange is used to treat the silver-bearing effluent, the treated stream would be expected to contain between 0.5 and 2 mg/L silver. If the fixer and bleach-fix overflows are first pretreated by electrolysis followed by ion-exchange treatment of the silver-bearing effluent, the final silver concentration can be reduced to the 0.1–0.5 mg/L range. The overall silver recovery efficiency can be 98–99+% of the potentially recoverable silver [86].

The EPA contracted with Versar, Inc., to provide a guidance document describing the control of water pollution in the photographic processing industry. This document, published in 1981, described the results of sampling effluents from 48 photoprocessing laboratories. The maximum silver concentration in the effluents averaged over a 30-day period was 1.1 mg/L, for laboratories using conventional silver recovery methods. Laboratories using conventional silver recovery plus the ion-exchange treatment of washwater averaged 0.4 mg/L over the period. Maximum single-day concentrations were 3.7 and 1.3 mg/L, respectively. The Versar report stated that more than 99% of the photoprocessing facilities it surveyed (over 1100 plants) discharged wastewater to POTWs [7].

In a 1980 study, total silver concentrations in effluents from two POTWs were reported as 0.2 and 0.004 mg/L, respectively, while both effluents contained free ionic silver concentrations of only 4×10^{-4} mg/L [46].

Subsequent studies in the 1990s by Shafer *et al.* [85,87], Adams and Kramer [88], Rozan and Hunter [89] and others [12], using "clean" analytical techniques (which had not been available previously), also proved conclusively that photographically derived silver (i.e., silver thiosulfate complex) could not survive in a treatment plant nor in a receiving stream, because en route it was always converted to silver sulfide. More than 100 researchers in North America and Europe addressed the subject of silver fate and effects in the environment in a designated silver research program (The Argentum Conferences) between 1989 and 2000 (see previous subsection, "Impact of Biotreatment on Silver") [48–53]. Moreover, a number of these scientists showed it was virtually impossible to measure any significant amount of free silver ion (Ag^+) in receiving bodies of water, regardless of the source of the silver and notwithstanding whether that water was aerobic or anaerobic, due to the ubiquitous presence of sulfide in practically all natural waters [54–58]. An entire final conference was devoted to this subject in 1999, with a follow-on meeting in 2000, and the aggregate conclusions of the decade's work published in 2002 as a guidance book for use by scientists and environmental regulatory agencies throughout the world [12]. These revolutionary findings not only eliminate silver as a metal of concern in terms of acute aquatic toxicity, but promise to change past conventional thinking on the toxicity of many other common metals in the natural environment as well.

6.4.2 Metallic Replacement

Metallic replacement has been an important means of recovering silver from fixers, bleach-fixes, and washwaters for many years. It can be used as a primary method of recovery or a secondary method following the electrolytic treatment of fixers or bleach-fixes.

Metallic replacement involves passing a silver-bearing solution through a vessel containing a more active metal in elemental form, usually iron. The reaction of dissolved silver with metallic iron is

$$Fe^0 + 2Ag(S_2O_3)_2{}^{3-} \longrightarrow Fe^{2+} + 2Ag^0 + 4S_2O_3{}^{2-} \qquad (8)$$

If conditions were ideal and no other reactions were involved, 1 g of steel wool would recover 3.86 g of silver. In actual practice, usually less than 1 g of silver is recovered per gram of steel wool. Most fixers, such as those used in x-ray and graphic arts processing, are acidic; therefore, a competing reaction that consumes much of the iron is the acidic dissolution of the steel wool. Spontaneous oxidation of iron with air (i.e., rusting) also occurs, particularly upon long-term standing under moist conditions. Insoluble iron hydroxide compounds are also formed when the solutions are at higher pH values, above 7.

A common size of steel wool cartridge (also originally known as a "chemical recovery cartridge" or "CRC," made by Kodak), contains about 8 lb of steel wool. However, many other manufacturers in the United States and overseas now produce such devices in various shapes, configurations, and sizes (categorically known as "metallic recovery cartridges" or "MRCs"), with various types and grades of steel wire filling.

Although a number of active metals (e.g., aluminum, manganese, zinc) can replace silver, steel wool has been shown to be the best choice from several technical, economic, safety, and environmental points of view. Certain metals higher than iron in the electromotive series, particularly aluminum, can react so violently with an acidic fixer that copious amounts of sulfurous gases, which are odorous and noxious, would be given off. Iron reacts more slowly and thus does not form such gases.

There are a number of advantages associated with metallic replacement, including low initial cost, simple nonelectrical installation, small size and low weight, little maintenance, and high efficiency of silver recovery if properly monitored. One disadvantage is the high shipping and refining costs after exhaustion, compared with those for silver flake; this may offset the lower initial cost of MRCs. Another disadvantage is the high iron concentration in the cartridge effluent, which has been measured to be as high as 3000 mg/L. This precludes the reuse of fixer after silver recovery with MRCs, produces a colored effluent, and could cause problems in meeting local sewer codes for color or iron content.

High silver-recovery efficiency can best be achieved if two CRCs are used in series and if they are well maintained. When the first cartridge shows exhaustion, it is removed. The cartridge in the second position is moved to the first position and a fresh cartridge placed in the second position.

Studies were conducted by Cooley [90] to determine how to optimize the use of CRCs. These showed that the optimum pH in actual laboratory practice is between 4.5 and 5.5. At lower pH values, the acidity will consume too much of the steel wool, whereas at higher values the reaction becomes slower as considerable amounts of iron hydroxide are formed that may obscure the surface of the steel wool. Silver deposited on the steel wool may also restrict the mass transfer of silver to the iron surface at higher pHs, while below pH 5.5 the iron may be continuously acid-etched to provide a fresh reaction surface.

6.4.3 Electrolytic Silver Recovery

Electrolysis, or more specifically electrowinning, is the most widely used and universally applicable method for silver recovery in the photoprocessing industry. An electrolytic silver-recovery cell consists of a cathode and an anode. Oxidation occurs at the anode (positive electrode) and reduction at the cathode (negative electrode). Silver deposits on the cathode during electrolysis when a direct current is passed through the silver-bearing photoprocessing solution. After sufficient silver has been plated, the cathode is removed from the system and the silver stripped off [91,92]. The primary reaction occurring at the cathode is

$$Ag(S_2O_3)_2^{3-} + e^- \longrightarrow Ag^0 + 2S_2O_3^{2-} \tag{9}$$

If the cathode voltage is allowed to become too high, thiosulfate could be reduced at the cathode as shown in the following equation:

$$S_2O_3^{2-} + 8e^- + 8H^+ \longrightarrow 2HS^- + 3H_2O \tag{10}$$

The production of sulfide is undesirable, because it will react with the silver complex to produce insoluble silver sulfide (known as "sulfiding"). Although from a recovery standpoint a small amount of silver sulfide can be tolerated, too high a level will result in a poor plate [93]. Additionally, if in-line fixer desilvering were being done, silver sulfide formation would contaminate the fixer and could damage the photographic product. Therefore, it is necessary to compromise on the voltage applied, to obtain optimum current efficiency while minimizing sulfide production.

Electrolytic silver recovery requires a larger capital expenditure than the use of MRCs and also necessitates an electrical connection. However, it has the advantage of yielding nearly pure silver, resulting in lower refining and shipping costs. A primary advantage from an environmental viewpoint is that it allows fixer reuse for many processes because it does not contaminate the fixer when properly controlled.

There are essentially two ways in which electrolytic silver recovery can be applied [92]. One involves its use in a terminal manner, and one concerns its application for the in-line desilvering of fixer. When used in a terminal manner, the silver-bearing solution is passed through the electrolytic cell to recover the silver and the desilvered solution is slowly discharged to the drain, perhaps through a secondary metallic replacement cartridge or using chemical precipitation for additional low-level silver recovery. An alternate terminal approach is to mix the electrolytically desilvered solution with silver-containing washwaters and pass the mixture through an ion-exchange system for further silver recovery.

Finally, part of the desilvered fixer or bleach-fix, if not mixed with other solutions or otherwise contaminated or altered during silver recovery, may be reused in making fresh replenisher, thus minimizing the environmental impact.

It is also possible to operate electrolytic equipment for the in-line desilvering of the fixer solution. The equipment is set to function so that the silver in the fixer tank is constantly maintained in the 0.5–1 g/L range. (This compares with typical silver concentrations in fixer tanks of 5–30 g/L when well seasoned.) Although careful control is essential to preclude the formation of silver sulfide, this method offers several environmental benefits. Depending on the process, the fixer replenishment rate can be reduced from 50 to 70% compared to the standard rate. Additionally, the lower silver level in the tank means that significantly less silver (only about 5–10% as much) will carry over to the washwaters, thus assuring that, overall, more silver is recovered and less is lost.

Several factors are involved in choosing and operating an electrolytic silver recovery unit [94,95]. The amount of current that a device delivers is important: low current density units can be used for desilvering fixers, but high current density is needed for bleach-fixes. Some method of agitation is required to keep the fresh silver-containing fixer in contact with the cathode, but too much turbulence that produces a vortex will whip air into the solution, consuming sulfite preservative and promoting sulfiding. A rotating cathode unit provides its own agitation, whereas a pump or impeller may be needed for a stationary cathode.

Some method for controlling the current is also important. Several methods are available, including timers, selective ion electrodes for online monitoring, and constant voltage operation including the more complicated use of potentiostatic control with IR (voltage) compensation [96,97]. The current density relative to the solution silver concentration should be high enough to desilver the solution in a reasonable time, yet low enough to prevent sulfiding. Well-designed controls step the current down in stages as silver is depleted from the solution.

In addition to the time, voltage, and current, pH is an important factor affecting electrolytic silver recovery. Tests have shown that the optimum pH for desilvering fixers is approximately 6.2, whereas the optimum for desilvering bleach-fixes is approximately 8–8.5. How high a pH can be used for desilvering bleach-fixes is limited by the evolution of ammonia. As the pH of the bleach-fix increases, a side reaction involving the reduction of iron is inhibited and the electrolytic silver-recovery efficiency increased [98].

6.4.4 Ion Exchange

If silver must be recovered from dilute solutions, ion exchange is the method of choice. This method can be applied to washwaters, mixtures of desilvered fixers and/or bleach-fixes with washwaters, and mixtures of silver-bearing fixers and/or bleach-fixes with washwaters. The lowest effluent silver concentrations can be attained when washwaters alone are treated, since input thiosulfate and silver levels are lowest. Treating a solution of desilvered fixer and/or bleach-fix mixed with washwater will produce the next lowest effluent silver levels.

The basic principle involves exchanging ions from solution with similarly charged ions bound to the resin, described as

$$R^+(S_2O_3)^{2-} + Ag(S_2O_3)_2^{3-} \longrightarrow 2S_2O_3^{2-} + R^+AgS_2O_3^- \qquad (11)$$

where R^+ represents the ion-exchange resin.

A weak-base anionic resin is generally used for silver recovery. Rohm & Haas (Philadelphia, PA) Amberlite IRA-68, a gel-type acrylic resin, has been shown to be an effective choice in various silver-recovery operations. Two approaches to recovering silver from dilute photoprocessing solutions are conventional ion exchange and *in situ* precipitation.

With conventional ion exchange, the silver-bearing solution is pumped through a vessel containing the resin [99]. Normally, two columns are used in series to minimize the loss of silver when the first column reaches exhaustion. Exchange occurs between the silver thiosulfate anion in the solution and the anion on the exchange site of the resin, which is usually thiosulfate. Silver will continue to be removed by the resin until the silver thiosulfate concentrations on the resin and in the solution reach equilibrium. The duty cycle is stopped when the effluent silver level reaches an undesirably high value, as the resin bed nears exhaustion. A concentrated thiosulfate solution is then pumped through the column as a resin regenerant. The silver thiosulfate on the exchange sites of the resin is replaced by the fresh, more concentrated thiosulfate; simultaneously, nearly all the silver previously on the resin has now been transferred to the thiosulfate regenerant, from which it can be removed electrolytically since it is now more concentrated. After a water rinse (which itself can be added to the washwater overflow stream for treatment, after being filtered), the system is then ready to recover additional silver. The bed can be reused for over 150 cycles, although eventually the resin deteriorates and must be replaced with fresh material. The electrolytically desilvered regenerant can also be reused for the next regeneration step after make-up chemicals have been added, to conserve chemical usage.

This technique has a very high degree of efficiency, about 98–99+%. Depending on the solutions treated, effluent silver levels in the 0.1–0.5 mg/L range can be attained, although levels in the 1–2 mg/L range are more common. The silver is recovered as high purity flake. Because thiosulfate is the only anion used, there is no contamination from other anions, which would otherwise need to be rinsed out and discarded. The regenerant solution and resin can be used for multiple cycles; therefore, operating costs are moderate. However, the initial capital cost for the ion-exchange and electrolytic silver-recovery equipment is high. Additionally, this technique, as described above, has been used successfully only by a few high-production laboratories with good maintenance capabilities. The most common operational problem is biological decomposition of the silver thiosulfate complex inside the column, which causes silver sulfide to form within the resin beads. Silver sulfide cannot be dissolved off the resin by the thiosulfate regenerant solution.

The *in situ* precipitation technique takes advantage of this natural tendency of the thiosulfate complex to decompose. It involves deliberately precipitating silver as silver sulfide inside the resin, with no attempt to regenerate [100,101]. Rohm & Haas Amberlite IRA-68 resin was also used with this method. After the silver thiosulfate complex has adsorbed on the resin, dilute sulfuric acid is then pumped through the column, intentionally converting the silver thiosulfate to silver sulfide. This acid is then neutralized and the column rinsed before the next cycle is run. The resin swells in size considerably to accommodate large amounts of silver sulfide, but it can be reused until its ultimate adsorption capacity (i.e., the capacity for swelling) is reached and the resin beads begin to fracture. The spent resin is finally removed from the column and incinerated to recover the silver. Effluent silver concentrations as low as 0.1 mg/L can be achieved using this method; however, levels of 1 mg/L are more common. The recovery efficiency attained is usually in the 98–99.5% range. This technique is generally recognized as the most efficient way of recovering silver from dilute photoprocessing solutions, since fewer steps, less labor, and simpler equipment are involved, and less silver is lost. However, the high cost of replacing the resin must be taken into account.

6.4.5 The TMT Precipitation Method

In the early 1990s, Eastman Kodak Company screened a large number of metal precipitating agents for recovery of silver from photoprocessing solutions, with the aim of improving silver recovery efficiency and simplifying operations. These efforts resulted in introducing the TMT (2,4,6-trimercapto-s-triazine) silver recovery process to the photoprocessing industry [102,103]. This technology offers low labor costs, easy process control, more efficient silver removal to lower levels, and less expensive refining costs compared to MRCs. The reagent, although relatively expensive, proved to be much less toxic and not hazardous to handle compared to sulfide precipitation or other metal precipitating agents.

Throughout the 1990s, Kodak investigated methods to improve the efficiency of the TMT process. Initial studies focused on using TMT as an alternative secondary recovery system, for removing residual silver from photoprocessing solutions that had first been electrolytically desilvered. Later a method was developed using TMT for primary silver recovery from mixed photographic solutions containing stabilizers [104]. Another process using TMT coupled with specific flocculating agents significantly improved the settling and filtration efficiency of the silver–TMT complex [105]. This invention proved very valuable when TMT is used for primary silver recovery. A high-pH rinse step was shown effectively to remove unwanted precipitates from reactor vessel walls [106]. Soon afterward, it was discovered that the unwanted precipitates could be minimized by raising the alkalinity of the TMT reagent solution itself during the continuous silver recovery process [107].

An automatic apparatus for continuous precipitation and flocculation was invented for various configurations and orientations [108], eliminating the need for manual mixing and reaction steps. A novel method and apparatus were also developed for collection and filtration of the silver–TMT precipitate [81], which negated the need for manual handling of the precipitate. These concepts were brought to market as equipment, the SR-2000 for small photofinishing operations such as minilabs, and the SR-2400 for large laboratories. At the same time a method was perfected for sequential addition of reagents to optimize particle agglomeration and growth [109].

Because the TMT reagent is expensive, an alternative chemical synthesis was conceived [110]; however, it proved not sufficiently cost effective to bring to production scale. In another attempt to lower costs, a method was invented to reduce the silver–TMT complex to metallic silver while liberating TMT reagent, which could in theory be reused [111]. Finally, a procedure was developed for efficient recovery of silver from solutions containing cations with higher valences (e.g., the alum hardening agent in x-ray fixer), by adding a chelating agent to complex the cations. The chelating agent can be added either before or during the silver recovery step, and the technique includes both batch and continuous silver recovery processes [112].

6.4.6 Sulfide Precipitation

Other precipitation methods have also been shown to be very effective for silver recovery [84]. In particular, precipitation with sodium sulfide has been used for desilvering both fixers and washwaters because silver sulfide is one of the most insoluble metallic species known, having a solubility product in water of 10^{-51}. This method involves first adjusting the pH of the normally acidic fixer with sodium hydroxide to an alkaline pH, to prevent the liberation of toxic hydrogen sulfide gas when the reactants are combined. Sodium sulfide is then added and the precipitated silver sulfide allowed to settle, after which it can be removed by decanting and/or filtration. This method has been used for many decades; however, although highly effective for removing silver, it is not very popular because of the "rotten egg" smell of sulfide and the potential hazard of forming hydrogen sulfide gas if personnel are not adequately trained. Silver levels as low as 0.01 mg/L have been attained in the laboratory.

As previously mentioned, sodium (or calcium) sulfide is cheap, readily available, quick-acting, and very simple to use. Despite its safety hazards it is currently being used extensively as an instant "curbside precipitant" by sidewalk entrepreneurs in some Third World countries, such as China, who purchase used fixers for their silver content.

In practice, the efficiency of sulfide precipitation is highly dependent on the filtration step. To avoid manual handling hazards and to ensure reliability, the technique was automated in the early 1970s by LaPerle [113]. This process uses automatic pH and specific ion electrodes to control the pH and add specific reagents as needed. An enclosed reaction tank and filtration system with automatic pumping cycles, plus an emergency override actuated by a hydrogen sulfide gas detector, virtually eliminated the safety hazard and much of the bad odor usually associated with handling sodium sulfide. Effluent silver could be reduced down to very low levels easily able to comply with environmental regulations. However, owing to subsequent developments with other methods, including less hazardous precipitants such as TMT, this automated method is not being used today.

6.4.7 Electrochemical Sulfide Precipitation

This technique, for generating sulfide ion directly from waste thiosulfate by electrolytic oxidation and using it to precipitate silver, was disclosed to the public by Kodak researchers in 1975, in lieu of seeking a patent [74]. In this method, an electrolytic cell is deliberately operated in such a manner as to produce excess sulfide ion directly from a small side stream of the silver-bearing fixer or bleach-fix solution containing thiosulfate. The presence of silver or iron helps somewhat to catalyze this reaction. Of course, a small amount of silver sulfide precipitate is produced within the side stream itself, but soluble sulfide ion is generated in great excess. This sulfide-laden side stream is then recombined with the parent solution to react with the remaining silver (silver thiosulfate complex), thus precipitating silver sulfide in the same manner as if the sulfide had been added as a separate solution. The silver sulfide is then precipitated and collected by settling and/or filtration with a flocculation step if needed. This method has the advantage of not having to mix or store odorous sulfide solution, but instead generates it only on demand from a portion of the waste-processing solution itself with the flip of an electric switch.

This electrochemical sulfiding reaction takes advantage of the well-known electrochemical phenomenon, sulfiding, long an operational headache for persons attempting to recover silver from fixer by electrolytic plating. However, very careful, continuous control is necessary, which may involve the use of expensive instrumentation. The pH must be kept neutral, because the same precautions concerning liberation of toxic hydrogen sulfide gas under acid conditions apply (see Section 6.4.6, "Sulfide Precipitation"), and ammonia can be generated at alkaline pHs if ammonium ion is present. In addition, the silver sulfide formed is a very fine precipitate that requires good filtration; the blinding of normal filters was routinely encountered during laboratory experiments. To the present authors' knowledge, no known commercial device currently uses this technology specifically for silver recovery.

6.4.8 Other Chemical Precipitants

Sodium borohydride can also be used to precipitate silver according to the reaction

$$BH_4^- + 2H_2O + 8Ag^+ \longrightarrow 8Ag^0 + 8H^+ + BO_2^- \qquad (12)$$

This method was used on occasion in the past for bleach-fixes, fixers, and washwaters [84]. Significantly more than the stoichiometric amount of borohydride is required to complete the reaction, and it must be performed under alkaline conditions. The silver recovered has a purity in the 90–95% range. Borohydride can be quite dangerous to handle, because it can explosively liberate hydrogen gas under acid conditions; therefore, this technique should only be used by chemically trained personnel.

Other chemical reducing agents have been tested experimentally for the precipitation of silver but are not in commercial use; therefore, they will not be discussed here.

6.5 REGENERATION, CHEMICAL RECOVERY, CONSERVATION, AND MANAGEMENT

6.5.1 Solution Regeneration Techniques

Introduction

The collection of processing tank overflows and subsequent regeneration of solutions for reuse can reduce the quantity of chemicals discharged by between 40 and 90%, depending on the solution. In addition to process and product modifications that have resulted in the use of fewer processing solutions, lower concentrations of chemicals in the solutions, and reduced replenishment rates [114], regeneration and reuse have significantly decreased the quantity of chemicals discarded from large photographic processing operations. The following section discusses the considerations that impact regeneration and reuse techniques for color developer solutions, desilvered fixers and bleach-fixes, and bleaches.

Whether or not a specific processing solution can be regenerated and/or reused depends on a number of factors. First, reuse requires that the reclaimed solution is or can be made photographically acceptable. That is, sensitometric measurements, generated from photographic test materials processed with the reclaimed solution, must meet specific quality standards.

Secondly, the practicality of regeneration techniques will depend on the size of the photographic processing operation and the consumption rate of the solutions. If sophisticated equipment and techniques are required, a small photoprocessing operation will most likely not have the technical expertise or be able to afford the capital expenditure needed to regenerate certain solutions.

In most photographic processes, fresh replenisher is added to each solution tank at a predetermined, fixed rate while the exposed product moves through the process, to maintain a certain minimal concentration of each required chemical in the processing solution. However, individual chemical constituents may be depleted at varying rates due to differences in exposure, size, or photographic characteristics of the photographic films or papers, differences in the rate they are fed through the process, effects of oxidation or carry-in of other solutions, and so on. While the replenisher formula attempts to correct for this as well as possible to maintain a chemical balance, the need to discard unwanted byproducts means that overflows from processing tanks will still contain large quantities of good chemicals that can potentially be reused. The exact techniques chosen for regeneration depend on the balance of unwanted vs. wanted components and their chemical nature.

Efforts by Eastman Kodak Company and other manufacturers have resulted in alternate options for specific photoprocessing solutions that use low replenishment rates. These produce much of the environmental benefit that can be realized by regeneration. Medium-to-large photoprocessing facilities are urged to explore the costs vs. benefits of these options for their own particular situation.

Color Developer Regeneration

Color developer solutions become exhausted through the oxidation of developing agent and the increased concentration of reaction products, which may significantly reduce the activity of the solution. In most cases, the limiting factor for reuse is the increased halide concentration. Therefore, to reuse color developers, the halide concentrations must be controlled. Two approaches are employed to control the halide level.

First, the undesirable reaction products such as halides can be removed. A strong-base anion-exchange resin such as Rohm & Haas IRA-400 or its successor is usually chosen to remove bromide and chloride ions from the solution. If proper techniques are used, this treatment may also remove other constituents such as color developing agents and oxidized color developing agents, but only to a very small extent. This regeneration method was applied to color developers from the color paper, reversal color paper, conventional color film, Kodak Ektachrome reversal film (in limited applications), Kodachrome reversal film, and Eastman Color motion picture processes [115–121]. After the halides have been removed, the purified solution is then collected and analyzed, and required make-up chemicals and water are added to bring it to replenisher strength. Previously, these regeneration techniques required considerable analytical capabilities. This limited the use of developer regeneration to only large laboratories having such facilities.

Secondly, some color developers can be reclaimed using the reconstitution technique. The overflow is collected and specific amounts of chemicals added to bring the solution back to replenishment strength. This method was used in some color processes including Eastman Color motion picture films and subsequently for color paper processing. The benefit of this approach over ion exchange is that minimal equipment is needed, desired chemicals such as color developing agent are not wasted by being adsorbed on the ion-exchange resin, and there is no need to discard regenerant solutions having high salt concentrations. During the past several years, developer regeneration kits have become available that minimize analytical demands, usually requiring only pH measurement and therefore making the technique available to smaller laboratories.

Besides significantly reducing the concentration of color developing agents discharged to sewer systems, the reuse of color developer solutions can produce substantial COD reductions. Use of developer regeneration for color paper presents minimal risks because the paper can always be reprinted if not satisfactory. However, application to color negative or, in particular, color reversal films represents a much greater risk because a customer's film may be ruined if the reused developer is out of specification.

Bleach Regeneration

As previously described, bleaches contain oxidizing agents that oxidize metallic silver to silver ion. Today most photographic processing bleaches use the selective oxidizing ability of ferric iron in a chelated form, often as an iron EDTA complex [122]. Other bleaching agents include iron PDTA, ferricyanide, and persulfate. Iron PDTA has replaced iron EDTA in some bleach formulations as a more environmentally acceptable bleaching agent because it is more active;

therefore, less is needed to obtain efficient conversion from metallic silver to silver halide in films [123,124].

Iron EDTA and iron PDTA bleaches used in color negative film and some color paper processes are usually regenerated in large photo labs. In most instances, the bleach overflow is collected and the ferrous ion oxidized to ferric by simple aeration, then make-up chemicals are added to bring the solution back to replenisher strength [125].

In the case of persulfate bleach, the overflow is simply restored to replenisher strength by running chemical analyses followed by specific chemical additions. Many persulfate bleaches require an activator or accelerator bath preceding the bleach. Often, the accelerator bath itself may also be reused by collecting the overflow solution and bringing it back to replenisher strength, through specific chemical additions.

Regeneration is attractive to many photofinishing laboratories as a cost-saving measure, and in most cases it will significantly reduce the amount of iron, chelating agent, and COD discharged to the environment.

Ferricyanide Bleach Management. Although most modern processes use alternative bleaches, ferricyanide is still the bleaching agent of choice for a few older processes such as the Kodachrome and Ektachrome Aerial film processes, and in some instances as an option in the Eastman Color motion picture processes. Ferricyanide bleaches underwent intensive study over the years for the development of regeneration and management methods.

In ferricyanide bleaches, ferricyanide ion is the oxidizing agent, which becomes reduced to ferrocyanide upon use. Together, these ions and related forms of the iron cyanide complex are known as hexacyanoferrates. The concentrations of hexacyanoferrates in an effluent can be minimized by a combination of reliable regeneration methods for the bleach overflow and recovery techniques for the fixer and wash waters.

Ferricyanide Bleach Regeneration. Ferricyanide bleach regeneration requires collecting the overflow and treating it with a strong oxidizing agent. Options include persulfate, peroxide, bromine, ozone, and electrolysis [126]. When persulfate (once the oxidant of choice) is used [127], the specific gravity of the solution may build up due to the formation of sulfate byproduct. Eventually, after several regeneration cycles, the sulfate concentration can grow high enough to reduce bleaching activity. This is usually remedied by discarding between 5 and 10% of the overflow. In lieu of hauling away all of the excess bleach, a precipitation technique can be used to prevent the wasted material from entering the sewer (see subsequent section, "Ferrocyanide Precipitation and Recovery"). Note that this same chemical recovery technique can be used to remove ferrocyanide from a fixing bath, when a fixer instead of a water wash step follows the bleach.

An alternative bleach oxidant to persulfate is ozone [128,129]. The use of ozone requires a fairly significant capital investment in equipment, and safeguards to minimize risk to personnel because ozone is a toxic and unstable gas. However, the specific gravity build-up problem attributable to persulfate is eliminated (see previous subsection, "Ozone").

Another technique having many of the advantages of ozone without the risk of a toxic gas is electrolytic bleach regeneration [130]. Ferrocyanide is oxidized to ferricyanide at the anode of an electrolytic cell. Because of the reduction reaction that occurs simultaneously at the cathode, the cell must be divided by some type of semipermeable membrane. Also, since some hydrogen is produced at the cathode, reliable exhaust ventilation is required. Commercial units are available (see previous subsection, "Electro-oxidation"). This is the most widely used method today.

Recovery of Hexacyanoferrate from Washwater. In some cases, it may be necessary to remove hexacyanoferrates from washwaters following a bleach or fixer. Because the complexes will be very diluted, it is not feasible to use precipitation techniques. Two options have been tried to concentrate these salts and allow them to be recovered:

1. Reverse osmosis can be used to concentrate the bleach components in washwater [126]. By using high pressure (300–600 psi), it is possible to produce a permeate stream containing 90% of the volume but only a small quantity of hexacyanoferrate. The smaller brine stream, although only about 10% of the flow, will contain almost all the hexacyanoferrate complex. Although this technique has been demonstrated repeatedly on laboratory and pilot scales, operational problems have limited its use in commercial practice (see previous section, "Reverse Osmosis").
2. Ion exchange has been proven in practice for removing hexocyanoferrate from washwaters [131]. Rohm & Haas Amberlite IRA-68 resin was used successfully. Experiments showed that 50–60 g of hexacyanoferrate could be collected on 1 L of resin before the effluent exceeded 1 mg/L hexacyanoferrate. Following ion-exchange treatment, the resin can be regenerated with a sodium hydroxide solution, producing a solution containing more than 25 g/L hexacyanoferrate (see previous subsection, "Ion Exchange").

Destruction of Hexacyanoferrate. Ferricyanide solutions were also treated by breaking down the hexacyanoferrate to innocuous products through severe oxidation methods. Hendrickson and Daignault [132] discussed the destruction of ferricyanide of hexacyanoferrate solution by chlorination and ozonation. The latter was shown to be a very slow process [63]. In one production system trace levels of hexacyanoferrate in water were eliminated by adding a halogenated compound, bromochlorohydantoin. Finally, although the chemical destruction of hexacyanoferrate solution by oxidation is technically possible, it is generally not economical on a practical basis.

Fixer and Bleach-Fix Reuse

The primary factors limiting the direct recycling of fixers and bleach-fixes are the build-up of silver, halide ions, sulfate, and, in some cases, oxidized developer products that can stain the product. However, with appropriate chemical treatment most fixers or bleach-fixes can be reused, although the degree of reuse possible varies from solution to solution.

When silver is recovered from the fixer or bleach-fix by plating at the cathode, the sodium sulfite or other preservative is consumed at the anode, causing a pH decrease. In addition, the fixer or bleach-fix will have been diluted by water or carried-in products from the preceding tanks, usually including developer oxidation products. To counter these effects, the unwanted components must be at least partially removed and make-up chemicals added to rejuvenate the desilvered overflow to replenisher strength.

The first step in regenerating either a fixer or bleach-fix involves collecting the overflow and desilvering the solution by electrolysis to a silver level of between 0.5 and 1.0 g/L. Lower levels are usually not attempted because of the possibility of sulfiding, which would irreversibly contaminate the solution.

A certain percentage of the desilvered overflow is saved for reuse. This can vary from a low of approximately 50% for the Kodacolor C-41 to as high as 75% with Kodachrome fixer. Make-up chemicals are added, and the fixer or bleach-fix can then be reused. The portion that cannot be reused should be further desilvered to a low silver level prior to disposal.

A more common technique used by many processing laboratories is in-line desilvering, as previously discussed. With this method, the electrolytic desilvering cell is plumbed directly to the fixer tank. The tank solution is continuously desilvered to silver levels in the range 0.5–1.0 g/L. This permits lowering of the replenisher rate and effectively reduces the amount of fixer used per unit of film or paper processed. Again, reductions in fixer use of 50–75% can be obtained, depending on the process [92].

In some black and white processes, a considerable amount of developer is carried into the fixer by the film. During electrolytic desilvering the developer is oxidized at the anode and, if allowed to build up, could form objectionable stain on the product. In at least one x-ray film process, a technique has been devised to pass the fixer through an ion-exchange cartridge after desilvering. The ion-exchange resin effectively removes the staining material so that at least 50% of the fixer can be reused. A similar method has been used in the past to remove excess bromide or iodide.

The regeneration method for desilvering and reusing bleach-fix varies slightly from the method for reusing fixer. In some color paper bleach-fixes, the overflow is collected and the pH adjusted to over 7 to facilitate electrolytic silver recovery; for other formulations such pH adjustment is unnecessary. After desilvering to between 0.5 and 1.0 g/L silver, a certain portion is saved for reuse. Aeration may be performed to reoxidize the ferrous salt back to ferric (although in many cases, the solution spontaneously reaerates itself simply upon standing or when being pumped to and mixed in the processing tank). Make-up chemicals are added together with water to bring the solution back to replenisher levels. A significant percentage of the bleach-fix can be reused.

Reuse of fixer and bleach-fix effectively reduces the ammonium ion, thiosulfate, BOD_5, and COD concentrations in the effluent [98]. Alternatively, use of low-replenishment rate, nonregenerated bleach-fix formulations will also significantly reduce the above-stated effluent parameters. Thus, bleach-fix regeneration is an effective silver management tool.

6.5.2 Chemical Recovery Methods

Introduction

Certain chemicals can be recovered individually from waste solutions even when the overall solution can no longer be salvaged. Often these are primary ingredients of a spent processing bath; in other situations, they may be foreign ingredients carried in from a preceding tank. Occasionally, it may be technically possible and economical to purify them in the laboratory and reuse them directly in a process. Other times the recovery may be done simply to extract a waste material that cannot be discharged to a sewer and isolate it for separate disposal.

Recovery of Color Developing Agents from Stop Baths

If color developing agents are carried over into bleaches or bleach-fixes, they are irreversibly oxidized and can no longer be reused. However, when a stop bath follows the color developer solution, the color developing agents remain essentially in their original form and can be recovered. A recovery method frequently used in the motion picture industry is to collect the stop bath and pass it through a column containing an ion-exchange resin such as the Rohm & Haas XAD class of resins [133]. The color developing agents will be adsorbed on the resin. A method developed by Linkopia in Sweden [134] then uses alkaline color developer solution to strip the color developing agents off the resin for reuse. This technique is useful when a processing laboratory must meet very stringent discharge limitations for color developing agents.

Coupler Recovery

In the Kodachrome film process, cyan, magenta, and yellow couplers are present in the three separate color developer solutions. (In all other processes, they are incorporated in individual layers within the color film or paper.) The Kodachrome couplers are all soluble in an alkaline solution but will precipitate at neutral or acid pH. Therefore, recovery becomes simple. The developer overflows are individually collected, pH is lowered (usually with sodium bisulfate, dilute sulfuric or acetic acid or carbon dioxide gas), and the precipitated coupler is then removed by centrifuging or filtration and dried for storage. When needed, the dried coupler can be resolubilized in alkali (usually preceded by a grinding or homogenizing step to make mixing easier) and reused. Analytical facilities are required for this technique [135].

Ferrocyanide Precipitation and Recovery

Although ferricyanide and ferrocyanide (together hexacyanoferrates) have relatively low toxicities, to the degree that they are even used as ingredients in foods [5,6], they can be of concern because they have the capability of being degraded very slowly (often a matter of months) through photolysis, especially by strong sunlight, to form low concentrations of free cyanide [11]. Therefore, the discharge of hexacyanoferrate from a photoprocessing operation must be controlled. Any overflow concentrates that cannot be regenerated and reused, and sometimes washes or fixers containing hexacyanoferrate, should be treated. A very effective technique has been developed for removing hexacyanoferrate complexes from an effluent by precipitation [72]. The ferricyanide in solution is reduced to ferrocyanide by the addition of dithionite or a similar reducing agent. The ferrocyanide is then precipitated by adding ferrous sulfate, to form insoluble ferrous ferrocyanide; this technique is generally used to recover ferrocyanide from either fixers or washes following bleaches. The precipitate is removed by filtration or centrifuging.

After washing off by-product salts, the precipitate can sometimes be reused in new bleach solutions by redissolving in alkali and reoxidizing it, although this latter method is usually economically feasible only for a very large production operation having good analytical capabilities [73].

Chromium Precipitation

As previously stated, chromium is a relatively rare metal in today's photographic processing laboratories, although it is used in a few black and white reversal bleaches as well as some process system cleaners. When a dichromate bleach is mixed with other processing solutions that are alkaline, and with solutions containing reducing agents such as thiosulfate and sulfite, the Cr^{6+} is reduced to Cr^{3+} and precipitated as chromium hydroxide or other insoluble particulate matter. The chromium 3+ precipitate is then removed during primary or secondary clarification at the wastewater treatment plant as a component of the sludge.

If necessary, because of sewer code discharge limits, chromium can be recovered from solution in the processing laboratory by precipitation. This is done by collecting the chromium-bearing overflow in tanks and reducing hexavalent chromium to the trivalent state by adding bisulfite or dithionite. The chromium is then precipitated as an amphoteric chromium hydroxide solid by adjusting the solution to pH 8 with an alkaline material such as dilute sodium hydroxide. This sludge can then be removed by settling and decantation, filtration, or similar techniques. The chromium hydroxide sludge will need to be managed according to applicable Federal or State waste management regulations.

Phosphate Recovery

During the 1970s there was great public pressure to remove phosphates from many discharge sources, primarily household detergents but including industrial wastes, because of the general absence of adequate municipal treatment facilities and the resulting eutrophication of streams. At that time, great effort was put into reformulating photographic solutions to exclude phosphates. Only small quantities remain today, generally as sequestrants to prevent calcium in hard water from sludging or crystallizing in process solutions and damaging emulsions, or for buffering high-pH developers. These levels are generally so low that no pretreatment would be required by the average laboratory that discharges to a POTW.

However, should phosphate removal be required because of particularly stringent discharge limitations, it can readily be precipitated by various agents, particularly lime (calcium hydroxide) or alum (potassium aluminum sulfate). Unfortunately, the resulting precipitate is a sticky mass resembling toothpaste or milk of magnesia, from which it is difficult to extract water. Successful ways of handling and dewatering this sludge include evaporation, filtration on a vacuum belt filter or in a filter press, or drying of the solution [72].

6.5.3 Conservation

A number of techniques can be applied in a photoprocessing laboratory to help reduce chemical discharges to the sewer. These methods include the use of squeegees, careful maintenance of replenishment rates, use of holding tanks and floating lids, and washwater conservation.

Squeegees

Squeegees are used to reduce the carry-over of processing solutions and washes. Both environmental and economic benefits result from their use. When properly installed and maintained, they can reduce carryover by 75% or more. This reduction can mean, in turn, that replenishment rates will be lowered significantly and regeneration efficiencies optimized. Squeegees are usually installed after all washes in a processing machine, and before and after all solutions being regenerated. Specifications for recommended placement of squeegees should be checked for each particular process to reduce carryover, minimize water spots, and prevent emulsion damage.

Squeegees exist as many different types, including wiper blades, vacuum squeegees, wringer-slinger, air squeegees, and so on [136–139]. The type to be used will depend on the photographic product being processed, the particular solution tank (e.g., some of the most flexible synthetic rubber squeegees used on downstream tanks could not survive the relatively harsh alkaline environment of the developer tank), the kind of processing machine, and the level of technical expertise available in the laboratory for installation and maintenance.

Replenishment Rates

The replenishment rate required for adequate processing of a photographic product will depend on many factors including the following: the nature of the photographic product, speed at which it is being transported through the machine, temperature of the solution (especially for developers), concentration of chemicals in the replenisher solution, rate of build-up of seasoning products in the solution, design of the machine (certain configurations introduce more air into the solution during operation than others), and the overall utilization (i.e., running vs. standby time) of the processing machine. Photographic product manufacturers will provide specifications for replenishment rates required for a particular product under varying processing conditions. An adequate safety factor is built into these recommendations. It is important that a processing laboratory use the recommended rates. Although too low a rate could cause adverse photographic results, too high a rate will result in a waste of chemicals and money. Therefore, all processing laboratories should periodically check and adjust processing solution replenishment rates.

Floating Lids

The use of floating lids on solution storage tanks will help reduce the aerial oxidation of easily oxidized chemicals, as well as evaporation. As the solution level drops, the lid stays in contact with the solution surface to eliminate any air space. It also safeguards against airborne dust and dirt and accidental contamination by other laboratory chemicals and materials. It helps protect the quality of the processed materials, helps extend solution life, and thus lessens the dumping of stored solutions gone "bad." All these factors play important roles in the environmental as well as economic health of a laboratory.

Washwater Conservation

Washwater conservation can be important for many reasons, including water shortages in some locations, energy and fee savings associated with reduced water consumption, and hydraulic limitations and/or discharge fees imposed on effluent discharged to municipal sewer systems [140]. A laboratory can conserve water in several ways. These include simple steps such as checking water supply equipment for leaks, using recommended wash rates, and running washwater only during actual processing.

Several methods can also be applied to reuse washwaters. Reverse osmosis and other membrane separation techniques have been chosen in a few instances, but their use is not widespread because of cost, maintenance, membrane life, and operational problems that to date remain largely unresolved.

Perhaps the most commonly practiced methods of reducing water use involve ion exchange. A process such as the former PACEX water recycling system incorporated ion-exchange technology together with filtration and bacterial control to remove contaminants [71]. In such a process, an anion-exchange resin removes silver thiosulfate complex from solution. An oxidizing agent subsequently oxidizes residual thiosulfate to sulfate and a biocide minimizes biological growth. A filter removes unwanted sediment. Depending on the process, reduction in washwater usage from 50–80% may be possible. This method not only conserves water and recovers silver but also provides energy savings by minimizing the quantity of fresh water that needs to be heated. Other manufacturers have announced similar water reclamation systems in the past.

Washwater reuse raises several technical concerns. In each case where such a system may be considered, the reputation and track record of the manufacturer should be checked. Some "water savers" simply filter and pump water back to the processing machine; however, the same result may be achievable merely by lowering the water input rate. Some more sophisticated water recirculation devices are available (especially for x-ray or graphic arts processes) that keep track of the amount of photographic material processed and sound an alarm after a predetermined amount is reached, indicating that the water should be changed.

On a number of occasions during the past century, "magnetic" or other mechanical water treatment devices that do not consume either energy or chemicals have been offered to the public, allegedly to purify water or stop the growth of microorganisms, but reportedly without any true scientific basis for making such claims [141–145]. We are not aware of any instance where these have been proven effective in a photographic laboratory.

The photoprocessing laboratory has the responsibility of ensuring that any recycling device operates properly and is constantly monitored for chemical build-up, such that the quality and storage life of processed films and papers do not suffer. Improper operation may damage the customer's photographic product. Also, washwater reuse will cause the effluents discharged from the photoprocessing laboratory to be more concentrated. Although the total chemical loading will not be higher, higher concentrations could result in exceeding the sewer code.

In 1981, the EPA published a "Guidance Document for the Control of Water Pollution in the Photographic Processing Industry." In its conclusions, the report states, "the agency does not recommend the use of concentration-based limitations for controlling pollutants at facilities in the industry" [7]. The report goes on to state that limits based on concentration tend to discourage water use reductions, whereas the agency encourages the reduction of wastewater quantities by various water-saving controls. Unfortunately, this recommendation has been largely ignored, and because discharge monitoring is more difficult by other methods, most municipalities continue to use concentration-based limitations. Therefore, before installing washwater conservation or reuse equipment, it is important to check its potential effect on the laboratory's continued ability to use the municipal sewer system.

6.5.4 Management of Laboratory Wastes

Good Housekeeping

Careful operating procedures in a photoprocessing laboratory can have a significant impact on the laboratory's typical effluent characteristics as measured over a period of time. Careful mixing of processing solutions will minimize dumping of concentrated processing solutions because of an error. Frequent maintenance of squeegees, tanks, and rollers, and readjustment of pH, temperatures, and replenishment rates will also aid in decreasing the waste of processing solutions and chemicals in daily operations, as well as enhancing quality of the product.

Holding Tanks

One way to lessen the potential impact of having to dump contaminated solution tanks to the sewer in batches is to use holding tanks. A holding tank should have enough capacity to contain the largest sudden processing solution dump that a laboratory might anticipate. The solution batch is pumped to the holding tank and then slowly released over an extended time interval through a valve to the drain, where it is combined with other normally generated processing and nonprocessing effluents. This approach will ensure that a sudden dump will not adversely affect the microorganisms in a secondary waste treatment plant.

Alternative Offsite Disposal and Waste Concentration Options

Some situations occur in which it is not possible to discharge processing solution overflows to a sewer. This may be due to stringent sewer codes that cannot be met or because no sewer system is available. Under these circumstances, it may be necessary to have the solutions hauled offsite by a waste disposal company. Such services will then treat the solutions at a centralized waste treatment facility. It is important that the company chosen be reputable and comply with all pertinent regulations; if not, the photoprocessing laboratory could be held liable for illegal disposal of the waste. The use of offsite disposal may require the processing laboratory to comply with its State's waste management requirements, including storage, safety factors, and labeling provisions.

Before signing a contract with such a waste disposal company, the laboratory manager and his or her chief technical assistant should personally visit the disposal facility. Facility personnel should be asked to describe or demonstrate their methods for handling, storing, and disposing of waste solutions. They should be able to produce permits and licenses to operate, as well as inspection records. Copies of such documents should be furnished to the prospective customer laboratory for its permanent files. The ultimate disposal of wastes from the treatment facility should be clearly stated, and the service company should be willing to furnish a signed statement to the laboratory, after each chemical pickup, that the laboratory no longer has title to the waste. The officers and principal stockholders of the company should be known, and their names checked with the state regulatory agency to be certain that they have no past record of waste-handling violations. A Dun and Bradstreet report or equivalent should be requested on the company to verify its financial stability, to ensure that it has sufficient funds to properly treat all wastes that may be in current storage at its treatment sites. The above requirements are subject to regulatory change; an attorney should be consulted before proceeding.

A laboratory that needs to use an off-site waste disposal company will often find it very advantageous to minimize waste volume. On-site evaporation of water from the waste is the most widely used technique. Properly designed evaporators are simple to operate, either manually or automatically, and can generally remove from 85 to 95% of the water from a photoprocessing waste, depending on the original composition. Up to 99% may be removed if only washwaters are evaporated, but this may be too costly an approach for these very dilute solutions because of the energy consumed. A better approach might be to use a water purification method such as ion-exchange to clean up the wash water, then simply concentrate the ion-exchange regeneration chemicals in the evaporator. Preadjusting the pH of the solution before evaporation may be necessary to prevent the formation of unwanted gases at high or very low pHs.

Before purchasing an evaporator, local regulatory codes should be checked to determine whether operating permits will be required. It is also important to make sure that the waste disposal firm will agree to handle the more concentrated waste. In addition, an economic check should be made to ensure that the smaller volume of more concentrated effluent will not cost as much to dispose of as the higher volume of more dilute, unevaporated effluent, when both waste-hauling and disposal charges and the purchase and operating costs of the evaporator are included (see previous section, "Evaporation").

In some branches of the photographic industry (particularly medical x-ray facilities and some minilabs), solution-service companies may contract to pick up silver-bearing waste from a photoprocessing customer, including overflow fixer, silver flake from electrolytic units, exhausted metallic replacement cartridges, and even scrap film and paper. The materials are processed and a credit for silver, minus the processing and refining charges, is returned to the laboratory. Sometimes, these companies will perform maintenance work on the processing machines and recovery devices, and may act as dealers to supply new chemicals and film or paper to the customer during their visit. With the advent of new waste regulations, these companies may be properly licensed to also haul away other wastes from the laboratory on routine pickup cycles. This option, although sometimes costly, frees the laboratory from having to worry about waste discharge and allows it to concentrate on processing photographic film and paper. The silver credit can help defray a significant part of the cost for such a service.

Summary

As a final comment, the management of any chemical waste discharges, including photographic processing wastes, is a constantly shifting target, both because of evolving regulations as well as advances in photography and waste treatment technology. Consultation with manufacturers of photographic products and reliable engineering and legal sources should be sought by any laboratory considering the appropriate, up-to-date management of its particular waste stream.

REFERENCES

1. Duffin, G.F. *Photographic Emulsion Chemistry*; Focal Press, Ltd.: London, 1966.
2. James, T.H. (Ed.) *The Theory of the Photographic Process*, 4th ed.; Macmillan: New York, 1977.
3. Mason, L.F.A. *Photographic Processing Chemistry*; Focal Press, Ltd.: London, 1985.
4. Proudfoot, C.N. (Ed.) *SPSE Handbook of Photographic Science and Engineering*, 2nd Ed.; The Society for Imaging Science and Technology: Springfield, VA, 1997; Sections 11, 12, 15.
5. Merck & Co., Inc. *The Merck Index*, 11th Ed.; Merck & Co.: Rahway, NJ, 1989; Item 3484, p. 550; Items 8562 and 8563, p. 1361; Item 3694, p. 631.
6. American Cyanamid Co. *The Chemistry of the Ferrocyanides*, Vol. VII. American Cyanamid Co.: New York, 1953.
7. Versar, Inc. *Guidance Document for the Control of Water Pollution in the Photographic Processing Industry*. EPA Rep. 440/1-81/082-9, April, Washington, DC, PB82-177643, pp. 111-1–111-62; I-3, I-5, II-1, V-1–V-25, 1981.
8. Petschke, D. (Ed.) Disposal and treatment of photographic effluent. In *Support of Clean Water*, Publ. J-55; Eastman Kodak Company: Rochester, NY, 1988.
9. Dagon, T.J. Biological treatment of photoprocessing effluents. J. Water Poll. Control Fed., **1973**, *45*, 2123–2135.
10. Cooley, A.C.; Dagon, T.J.; Jenkins, P.W.; Robillard, K.A. Silver and the environment. J. Imaging Technol. **1988**, *14* (*6*), 183–189.
11. Terhaar, C.J.; Ewell, W.S.; Dziuba, S.P.; Fassett, D.W. Toxicity of photographic processing chemicals to fish. Photo. Sci. and Eng. **1972**, *16* (*5*), 370–377.
12. Andren, A.W.; Bober, T.W. (Eds.) *Silver in the Environment: Transport, Fate and Effects. Research Findings of the Argentum International Conference Series, 1993–2000*. Society of Environmental Toxicology and Chemistry (SETAC): Pensacola, FL, 2002.
13. Owerbach, D. The analysis of cyanide in photographic processing effluents. In: *Analysis, Treatment and Disposal of Ferricyanide in Photographic Effluents–A Comendium,* Publication J-54; Eastman Kodak Company: Rochester, NY, 1990; 53–59.
14. Hung, C.-H. *Fate of Thiocyanate in Biological Treatment Processes*, MS thesis. Georgia Institute of Technology: Atlanta, GA, 1995.
15. Hung, C.-H.; Pavlostathis, S.G. Aerobic biodegradation of thiocyanate. Water Res. **1997**, *31*, 2761–2770.
16. Hung, C.-H.; Pavlostathis, S.G. Fate and transformation of thiocyanate and cyanate under methanogenic conditions. Appl. Microbiol. & Biotechnol. **1998**, *49*, 112–116.
17. Harbison, K.G.; Belly, R.T. The biodegradation of hydroquinone. Environ. Toxicol. Chem. **1982**, *1*, 9–15.
18. Hydroscience, Inc. *Environmental Effect of Photoprocessing Chemicals*, Vol. 1. National Association of Photographic Manufacturers: Harrison, NY, 1974.
19. Eisenhauer, H.R. The ozonation of phenolic wastes. J. Water Poll. Control Fed. **1968**, *40*, 1887.
20. Pavlostathis, S.G.; Sridhar, K. Treatment of photoprocessing effluents by the activated sludge process. J. Imaging Sci. & Technol. **1992**, *36* (*4*), 405–411.
21. Sridhar, K. *The Treatability of Photoprocessing Effluents by the Activated Sludge Process*, MS thesis. Clarkson University: Potsdam, NY, 1991.
22. Pavlostathis, S.G.; Morrison, D. Aerobic biodegradation potential of photoprocessing wastewater. Water Environ. Res. **1994**, *66*, 211–219.
23. Pavlostathis, S.G.; Jungee, S.A. Biological treatment of photoprocessing wastewaters. Water Sci. & Technol. **1994**, *29*, 89–98.
24. J. B. Scientific Corp. *Pathways of Photoprocessing Chemicals in Publicly Owned Treatment Works*. National Association of Photographic Manufacturers: Harrison, NY, 1977.
25. Unpublished data (1990–1991) Eastman Kodak Company: Rochester, NY, 1990–1991.
26. Morrison, D. *Fate and Effect of Photoprocessing Effluents on the Activated Sludge Process*, MS thesis. Clarkson University: Potsdam, NY, 1992.
27. Jungee, S.A. *Fate and Effect of Photoprocessing Wastewaters on the Anaerobic Digestion Process*, MS thesis. Georgia Institute of Technology: Atlanta, GA, 1993.

28. Schreiber, D. *Biological Oxidation of Thiosulfate*, MS thesis. Georgia Institute of Technology: Atlanta, GA, 1994.
29. Hung, C.-H. *Fate of Thiocyanate in Biological Treatment Processes*, MS thesis. Georgia Institute of Technology: Atlanta, GA, 1995.
30. Maeng, S.K. *Effect of a Silver-Bearing Photoprocessing Wastewater and Silver Complexes on Biological Treatment Processes*, MS thesis. Georgia Institute of Technology: Atlanta, GA, 1998.
31. Goettems, E.M.P.; Vacco, D. Tratamento e destino de efluentes fotográficos aspectos ambientais. *ABES – Associaço Brasileira de Engenharia Sanitária e Ambiental, I - 134, 19º Congresso Brasileiro de Engenharia Sanitaria e Ambiental*, Setembro, Foz Do Iguacu, Brasil.
32. Pavlostathis, S.G.; Morrison, D. Response of continuous-flow activated sludge reactors to photoprocessing wastewaters. Water Res. **1994**, *28*, 269–276.
33. Pavlostathis, S.G.; Sridhar, K. Anaerobic biodegradation potential of photoprocessing wastewater. Water Environ. Res. **1994**, *66*, 220–229.
34. Pavlostathis, S.G.; Maeng, S.K. Fate and effect of silver in the anaerobic digestion process. Water Res. **2000**, *34*, 3957–3966.
35. Pavlostathis, S.G.; Sridhar, K.; Morrison, D. Aerobic treatment of photoprocessing effluents. In: *Hazardous and Industrial Wastes*; Neufeld, R.D., Casson, L.W., Eds.; Technomic Publishing Co.: Lancaster, PA, 1991; 175–187.
36. Pavlostathis, S.G.; Sridhar, K. Anaerobic biodegradation potential of photoprocessing wastewaters. Water Environ. Res. **1994**, *66*, 220–229.
37. Schreiber, D.; Pavlostathis, S.G. Biological oxidation of thiosulfate. Water Res. **1998**, *32*, 1363–1372.
38. Pavlostathis, S.G.; Maeng, S.K. Aerobic biodegradation of a silver-bearing photoprocessing wastewater. Environ. Toxicol. & Chem. **1998**, *17*, 617–624.
39. Hung, C.-H.; Pavlostathis, S.G. Kinetics and modeling of autotrophic thiocyanate biodegradation. Biotechnol. & Bioeng. **1999**, *62*, 1–11.
40. Dagon, T.J. Photographic processing effluent control. J. Appl. Photo. Eng. **1978**, *2*, 62–71.
41. Lytle, P.E. Treatment of photofinishing effluents using rotating biological contactors (RBC's). J. Imaging Technol. **1984**, *10*, 221–226.
42. Petche, K.S. Meeting municipal sanitary sewer discharge permit limits for production operations at University Microfilms, International. Presented at *SPSE Ann. Conf.*, May 19, Boston, MA, 1989.
43. Bard, C.C.; Murphy, J.J.; Stone, D.J.; Terhaar, C.J. Silver in photoprocessing effluents. J. Water Pollut. Control Fed. **1976**, *8*, 389–394.
44. Ericson, F.A.; LaPerle, R.L. *The State of Silver Photographic Effluents After Secondary Treatment; The Effectiveness of Extended-Aeration Treatment in Removing Silver from Photographic Processing Effluents*; Eastman Kodak Company: Rochester, NY, 1974; Photographic Technol. Tech. Memo. TMB 74-36.
45. Feiler, H.D.; Storch, D.J.; Shattuck, A. *Treatment and Removal of Priority Industrial Pollutants at Publicly Owned Treatment Works*; Burns and Roe Industrial Services Corp.: Washington, DC, 1981; EPA Rep. 400/1-79-300, April, PB83-142414.
46. Lytle, P.E. Fate and speciation of silver in publicly owned treatment works. Environ. Toxicol. Chem. **1985**, *3*, 21–30.
47. Shafer, M.M. Sampling and analytical techniques for silver in natural waters. In: *Proceedings of the 3rd International Conference on Transport, Fate and Effects of Silver in the Environment (Argentum III)*, Washington, DC, 6–9 August, 1995; Andren, A.W., Bober, T.W., Eds.; University of Wisconsin Sea Grant Institute: Madison, WI, 1995; 99–111.
48. Andren, A.W.; Bober, T.W. (Eds.) *Proceedings of the 1st International Conference on Transport, Fate and Effects of Silver in the Environment (Argentum I)*, Madison, WI, USA, 8–10 August; University of Wisconsin Sea Grant Institute: Madison, WI, 1993.
49. Andren, A.W.; Bober, T.W. (Eds.) *Proceedings of the 2nd International Conference on Transport, Fate and Effects of Silver in the Environment (Argentum II)*, Madison, WI, 11–14 September; University of Wisconsin Sea Grant Institute: Madison, WI, 1994.

50. Andren, A.W.; Bober, T.W. (Eds.) *Proceedings of the 3rd International Conference on Transport, Fate and Effects of Silver in the Environment (Argentum III)*, Washington, DC, 6–9 August; University of Wisconsin Sea Grant Institute: Madison, WI, 1995.
51. Andren, A.W.; Bober, T.W. (Eds.) *Proceedings of the 4th International Conference on Transport, Fate and Effects of Silver in the Environment (Argentum IV)*, Madison, WI, USA, 25–28 August; University of Wisconsin Sea Grant Institute: Madison, WI, 1996.
52. Andren, A.W.; Bober, T.W. (Eds.) *Proceedings of the 5th International Conference on Transport, Fate and Effects of Silver in the Environment (Argentum V)*, Hamilton, Ontario, Canada, 28 Sept.–1 Oct.; University of Wisconsin Sea Grant Institute: Madison, WI, 1997.
53. Andren, A.W.; Bober, T.W. (Eds.) *Proceedings of the 6th International Conference on Transport, Fate and Effects of Silver in the Environment (Argentum VI)*, Madison, WI, USA, 21–25 August; University of Wisconsin Sea Grant Institute: Madison, WI, 1999.
54. Luther, G.W. III; Tsamakis, E. Concentration and form of dissolved sulfide in the oxic water column of the ocean. Marine Chem. **1989**, *27*, 165–177.
55. Radford-Knoery, J.; Cutter, G.A. Biogeochemistry of dissolved hydrogen sulfide species and carbonyl sulfide in the western North Atlantic Ocean. Geochim. Cosmochim. Acta **1994**, *58*, 5421–5431.
56. Luther, G.W., III; Theberge, S.M.; Rickard, D.S. Evidence for aqueous clusters as intermediates during zinc sulfide formation. Geochim. Cosmochim. Acta **1999**, *63*, 3159–3169.
57. Manolopoulos, H. *Metal Sulfides in Oxidizing Fresh Water Systems*, PhD thesis. McMaster University: Hamilton, Ontario, Canada, 2000; 179–184.
58. Rozan, T.F.; Lassman, M.E.; Ridge, D.P.; Luther, G.W., III Evidence for Fe, Cu, and Zn complexation as multinuclear sulfide clusters in oxic river waters. Nature **2000**, *406*, 879–882.
59. Juberg, D.A. A review of toxicity and epidemiological data for silver in animals and humans. *Proceedings of the 3rd International Conference on Transport, Fate and Effects of Silver in the Environment (Argentum III)*, Washington, DC, 6–9 August; Andren, A.W., Bober, T.W., Eds.; University of Wisconsin Sea Grant Institute: Madison, WI, 1995; 129–132.
60. Watson, H.M.; Boatman, R.; Ewell, W.S. *Simulated Secondary Waste Treatment of Kodak Color Developer CD-3*, Health and Environmental Laboratories Tech. Rep. ETS-TR-84-30; Eastman Kodak Company: Rochester, NY, 1984.
61. van Ginkel, C.G.; Vandenbroucke, K.L.; Stroo, C.A. Biological removal of EDTA in conventional activated-sludge plants operated under alkaline conditions. Biores. Technol. **1997**, *59*, 151–155.
62. Nortemann, B. Total degradation of EDTA by mixed cultures and a bacterial isolate. Appl. & Environ. Microbiol. **1992**, *58*(2), 671–676.
63. Bober, T.W.; Dagon, T.J. Treating photographic processing solutions and chemicals with ozone. J. Water Poll. Control Fed. **1975**, *47*, 2114–2129.
64. Hendrickson, T.N. *International Symposium on Ozone for Waste and Wastewater Treatment*; Industrial Ozone Institute, 1975; 578–586.
65. Fuji Film Co., Ltd. Treatment of Waste Photographic Processing Solutions. Japanese Patent No. JP04244299, January 9, 1992.
66. Knorre, H.; Maennig, D.; Stidetzel, K. Chemical treatment of effluent from photofinishing plants. J. Imaging Technol. **1988**, *4*, 154–156.
67. Medici, F. *Wastes from Photographic Processes*, ICP; Università di Roma la Sapienza: Rome, Italy, 1986; 63–66.
68. Hahn, F.; Meier, F. On the treatment of waste water from film developing plants. Chemiker-Zeitung **1971**, *95*, 467–471 (Translated from German by the Ralph McElroy Co., Custom Division, 504 West 24th, Austin, TX, USA).
69. Eastman Kodak Company. Alkaline Chlorination of Waste Photographic Processing Solutions Containing Silver. U.S. Patent No. 3,594,157, July 20, 1971.
70. Eastman Kodak Company. Destruction of EDTA by Alkaline Chlorination. U.S. Patent No. 3,767,572, Oct. 23, 1973.
71. Kreiman, R.T. Photo wash water recycling system utilizes ion exchange technology. J. Imaging Technol. **1984**, *10*, 244–246.

72. Bober, T.W.; Cooley, A.C. The filter press for the filtration of insoluble photographic processing wastes. J. Photo. Sci. and Eng. **1972**, *16* (*2*), 131–135.
73. Kleppe, J.W.; Vacco, D. *Settle and Decant Process for Ferrocyanide Removal from Fixer*; Eastman Kodak Company, Publ. J-54, 1979.
74. Bober, T.W.; Leon, R.B. Recovering metals from waste photographic processing solutions. Res. Disclosure **1975**, *37* (*5*), Item 13702.
75. Eastman Kodak Company. Method of Electrolysis Employing a Doped Diamond Anode to Oxidize Solutes in Wastewater. U.S. Patent No. 5,399,247, March 21, 1995.
76. Holm, C.E. (1990) Laboratory investigation of incineration of spent photofinishing liquids. MS thesis, Jan. Univ. of Wis., Madison, WI.
77. Ragland, K.W.; Holm, C.E.; Andren, A.W. Laboratory investigation of incineration of spent photofinishing liquids. Presented at *SPSE 6th International Symposium on Photofinishing Technology*, Las Vegas, NV, February 20, 1990.
78. Zimpro Passavant. Wet-air oxidation of three photographic developing solutions from Eastman Kodak, Unpublished report. November 9, Rothschild, WI, 1989.
79. Hoffman, F.; Vacco, D.; Yeaw, D. Photochemical waste reduction using vacuum distillation. Presented at the *8th International Symposium on Photofinishing Technologies*, Atlanta, GA, February 13–16, 1994.
80. Weber, W.J. Jr. *Physicochemical Processes for Water Quality Control*; Wiley Interscience: New York, 1972.
81. Eastman Kodak Company. *Apparatus for Removing a Component from Solution*, U.S. Patent 5,549,820. Washington, DC, August 27, 1996.
82. Unpublished data. Eastman Kodak Company: Rochester, NY, 1995.
83. Unpublished data. Eastman Kodak Company: Rochester: NY, 1997.
84. Technical publication. *Recovering Silver from Photographic Materials*, Publication J-10; Eastman Kodak Company: Rochester, NY, 1979.
85. Shafer, M.M.; Overdier, J.T.; Armstrong, D.E. Removal, partitioning, and fate of silver and other metals in wastewater treatment plants and effluent-receiving streams. Environ. Toxicolog. Chem. **1998**, *17*, 630–641.
86. Quinones, P.R. Optimizing silver recovery in photofinishing operations. J. Imaging Technol. **1985**, *11*, 43–50.
87. Shafer, M.M.; Armstrong, D.E.; Overdier, J.T. Silver levels and partitioning in effluent-receiving streams and a preliminary mass balance for silver in the Lake Michigan basin. *Proceedings of the 3rd International Conference on Transport, Fate and Effects of Silver in the Environment (Argentum III)*, Washington, DC, 6–9 August; Andren, A.W., Bober, T.W., Eds.; University of Wisconsin Sea Grant Institute: Madison, WI, 1995; 169–180.
88. Adams, N.W.H.; Kramer, J.R. Silver speciation in wastewater effluent, surface waters and porewaters. Environ. Toxicol. Chem. **1999**, *18*, 2667–2673.
89. Rozan, T.F.; Hunter, K.S. Effects of discharge on silver loading and transport in the Quinnipiac River, Connecticut. Sci. Total Environ. **2001**, *279*, 195–205.
90. Cooley, A.C. Silver recovery using steel wool metallic replacement cartridges. J. Imaging Technol. **1988**, *14*, 167–173.
91. Hickman, K.C.D. *et al.* Electrolysis of silver-bearing thiosulfate solutions. Indust. Eng. and Chem. **1933**, *25*, 202–212.
92. Cooley, A C.; Dagon, T.J. Current silver recovery practices in the photographic processing industry. J. Appl. Photo. Eng. **1976**, *2*, 36–41.
93. Cooley, A. C. The effect of the chemical components of fixer on electrolytic silver recovery. J. Imaging Technol. **1986**, *12*, 316–322.
94. Cooley, A. C. An engineering approach to electrolytic silver recovery systems design. J. Appl. Photo. Eng. **1982**, *8* (*4*), 171–180.
95. Cooley, A.C. A study of the major parameters for designing rotating cathode electrolytic silver recovery cells. J. Imaging Technol. **1984**, *10*, 226–232.
96. Cooley, A.C. Three-electrode control procedure for electrolytic silver recovery. J. Imaging Technol. **1984**, *10*, 233–238.

97. Branch, D.A. Silver recovery methods for photoprocessing solutions. J. Imaging Technol. **1988**, *14*, 160–166.
98. Krauss, S.J. Factors affecting the desilvering and reuse of bleach-fix. Presented at *SPSE Symposium on Environmental Issues in Photofinishing*, Los Angeles, CA, September 15–17, 1987.
99. Mina, R. Silver recovery from photographic effluents by ion-exchange methods. *J. Appl. Photo. Eng.* **1980**, *6*, 120–125.
100. Quinones, P.R. In-situ precipitation as the regeneration step in ion exchange for silver recovery. J. SMPTE **1984**, *93*, 800–807.
101. Lorenzo, G.A. In-situ ion exchange silver recovery for pollution control. J. Imaging Technol. **1988**, *14*, 174–177.
102. Spears, N.; Sentell, R. Silver recovery from photographic waste processing solutions by using the trisodium salt of 2,4,6-trimercapto-s-triazine. Presented at *7th International Symposium on Photofinishing Technology*, Las Vegas, NV, February 3–5; The Society for Imaging Science and Technology: Springfield, VA, 1992.
103. Kodak Environmental Information. *Silver Recovery Using KODAK Silver Recovery Agent for Large Photofinishing Labs*, Publ. CIS-155; September; Eastman Kodak Company: Rochester, NY, 2000.
104. Eastman Kodak Company. Process for Recovering Silver from Photographic Solutions. U.S. Patent No. 5,288,728. February 22 1994.
105. Eastman Kodak Company. Process of Recovering Silver from Photographic Solutions. U.S. Patent No. 5,437,792. August 1, 1995.
106. Eastman Kodak Company. Method for Flushing an Undesired Precipitate from Apparatus Surfaces. U.S. Patent No. 5,476,593, December 19, 1995.
107. Eastman Kodak Company. Process of Recovering Silver from Photographic Solutions and Inhibiting Formation of Undesirable Precipitate. U.S. Patent No. 5,496,474, March 5, 1996.
108. Eastman Kodak Company. Apparatus for Removing Silver from Spent Photoprocessing Solution. U.S. Patent No. 5,688,401, November 8, 1997.
109. Eastman Kodak Company. Methods for Removing Silver from Spent Photoprocessing Solution. U.S. Patent No. 5,695,645, December 9, 1997.
110. Eastman Kodak Company. Method of Making Trialkali and Triammonium Salts of TMT. U.S. Patent No. 5,563,267, October 8, 1996.
111. Eastman Kodak Company. Process for Recovering Mercapto-s-triazines from Silver Precipitate. U.S. Patent No. 5,759,410, June 2, 1998.
112. Eastman Kodak Company. Process for Recovery of Silver from Hardening Photoprocessing Solutions. U.S. Patent No. 5,961,939, October 5, 1999.
113. LaPerle, R.L. The removal of metals from photographic effluent by sodium sulfide precipitation. J. SMPTE **1976**, *85*, 206–216.
114. Cribbs, T.P.; Dagon, T.J. A Review of Waste Reduction Programs in the Photoprocessing Industry, CMA Waste Minimization Workshop Notebooks; Vol. I, C-5, Nov. 12; Chemical Manufacturer's Association: Washington, DC, 1987.
115. Daignault, L.G. Pollution control in the photoprocessing industry through regeneration and reuse. J. Appl. Photo. Eng. **1977**, *3*, 93–96.
116. Allen, L.E. Ion-exchange recovery techniques for the reuse of color developers. J. SMPTE **1979**, *88*, 165–167.
117. Kleppe, J.W. The application of an ion exchange method for color developer reuse. J. Appl. Photo. Eng. **1979**, *5*, 132–135.
118. Kleppe, J.W. Practical application of an ion-exchange method for color developer reuse. J. SMPTE **1979**, *88*, 168–171.
119. Meckl, H. Recycling of color paper developer. J. Appl. Photo. Eng. **1979**, *5*, 216–219.
120. Bard, C.C. Recovery and reuse of color developing agents. J. SMPTE **1980**, *89*, 225–228.
121. Burger, J.L.; Mina, R. An alternative ion-exchange regeneration system for recovery of Kodak Ektaprint 2 developer. J. Appl. Photo. Eng. **1983**, *9*, 71–75.
122. Dagon, T.J. Processing chemistry of bleaches and secondary processing solution and applicable regeneration techniques. J. Appl. Photo. Eng. **1976**, *2*, 42–45.

123. Foster, D.G.; Stephen, K.H. The latest in Process C-41 bleaching. Presented at *6th Annual Symposium on Photofinishing Technology*, Las Vegas, NV, February 19, 1990; The Society for Imaging Science and Technology: Springfield, VA, 1990.
124. Baughman, A.; Cribbs, T.A. Improved separate bleach and fixer for Process RA-4. Presented at *6th Annual Symposium on Photofinishing Technology*, Las Vegas, NV, February 19, 1990; The Society for Imaging Science and Technology: Springfield, VA, 1990.
125. Manual for KODAK Chemicals. *Using KODAK FLEXICOLOR Chemicals*, Kodak Processing Publication Z-131; Eastman Kodak Company: Rochester, NY, 1991.
126. Cooley, A.C. Regeneration and disposal of photographic processing solution containing hexacyanoferrate. J. Appl. Photo. Eng. **1976**, *2* (*2*), 61–64.
127. Hutchins, B.A.; West, L.A. The preparation or regeneration of a silver bleach by oxidizing ferrocyanide with persulfate. J. SMPTE **1957**, *66*, 764–768.
128. Bober, T.W.; Dagon, T.J. The regeneration of ferricyanide bleach using ozone – Part 1. Image Technol. **1972**, *14* (*4*), 13–16, 24, 25.
129. Bober, T.W.; Dagon, T.J. The regeneration of ferricyanide bleach using ozone – Part 2. Image Technol. **1972**, *14* (*5*), 19–24.
130. Kleppe, J.W.; Nash, C.R. A simplified electrolytic method for ferricyanide bleach regeneration. J. SMPTE **1978**, *87*, 4.
131. Brugger, D.A. Removal of hexacyanoferrate from selected photographic process effluents by ion exchange. J. SMPTE **1979**, *88*, 237–243.
132. Hendrickson, T.N.; Daignault, L.G. Treatment of photographic ferrocyanide-type bleach solution for reuse and disposal. J. SMPTE **1973**, *82*, 727–732.
133. Burger, J.L.; Fowler, H.E.; McPhee, B.A.; Yeager, J.E. Recovery of Kodak color developing agent CD-2 from process ECP-2A color developer and stop bath. J. SMPTE **1985**, *94*, 648–653.
134. Technical literature. Linkopia CD/Plus, Linkopia Technical Division: S-581 84 Linkoping, Sweden, 1993.
135. Technical manual. Recovery of KODAK coupling agent, C-16 and KODAK coupling agent, M-38 from KODACHROME film process solutions, K14-601, recovery of KODAK coupling agent, Y-55, K14-605. In: *Manual for Processing KODACHROME Film Process K-14*, Volume VI, Supplementary Information; Eastman Kodak Company: Rochester, NY, 1974.
136. Ott, H.F.; Dunn, J.E. The rotary-buffer squeegee and its use in a motion-picture film lubricator. J. SMPTE **1968**, *77*, 121–124.
137. Perkins, P.E. A review of the effects of squeegees in continuous-processing machines. J. SMPTE **1970**, *79*, 121–123.
138. Edgcomb, L.I.; Zankowski, J.S. Molded squeegee blades for photographic processing. J. SMPTE **1970**, *79*, 123–126.
139. Boutet, J.C. Spring-loaded wiper-blade squeegees. J. SMPTE **1972**, *1*, 792–796.
140. Fields, A.E. Reducing wash water consumption in photographic processing. J. Appl. Photo. Eng. **1976**, *2*, 128–133.
141. Hendricks, A.M. Water conditioning gadgets, fact or fancy? Presented at *NACE South Central Regional Meeting*, Oklahoma City, OK, October 1–4, 1967; National Association of Corrosion Engineers: Houston, TX, 1967.
142. Godard, H. (Ed.) Editorial. "Watch Out for Wondrous Water Treatment Witchcraft", Materials Performance; NACE Press, National Association of Corrosion Engineers: Houston, TX, 1974.
143. Newton, J. Cleanliness, not magnets, will wash away bugs, SPFE Newsletter, Jul./Aug. Society of Photo Finishing Engineers, Photo Marketing Assoc.: Jackson, MI, 1983.
144. Puckorius, P. How effective are mechanical devices for water treatment? In: *8th Annual Industrial Plant Energy Systems Guidebook*; McGraw-Hill, Inc, 1982.
145. Eliason, R.; Skrinde, R.T. Experimental evaluation of water conditioning performance. J. Am. Water Works Assoc. **1967**, *49*, 1179–1180.

7

Treatment of Soap and Detergent Industry Wastes

Constantine Yapijakis
The Cooper Union, New York, New York, U.S.A.

Lawrence K. Wang
Zorex Corporation, Newtonville, New York, U.S.A., and
Lenox Institute of Water Technology, Lenox, Massachusetts, U.S.A.

7.1 INTRODUCTION

Natural soap was one of the earliest chemicals produced by man. Historically, its first use as a cleaning compound dates back to Ancient Egypt [1–4]. In modern times, the soap and detergent industry, although a major one, produces relatively small volumes of liquid wastes directly. However, it causes great public concern when its products are discharged after use in homes, service establishments, and factories [5–22].

A number of soap substitutes were developed for the first time during World War I, but the large-scale production of synthetic surface-active agents (surfactants) became commercially feasible only after World War II. Since the early 1950s, surfactants have replaced soap in cleaning and laundry formulations in virtually all countries with an industrialized society. Over the past 40 years, the total world production of synthetic detergents increased about 50-fold, but this expansion in use has not been paralleled by a significant increase in the detectable amounts of surfactants in soils or natural water bodies to which waste surfactants have been discharged [4]. This is due to the fact that the biological degradation of these compounds has primarily been taking place in the environment or in treatment plants.

Water pollution resulting from the production or use of detergents represents a typical case of the problems that followed the very rapid evolution of industrialization that contributed to the improvement of quality of life after World War II. Prior to that time, this problem did not exist. The continuing increase in consumption of detergents (in particular, their domestic use) and the tremendous increase in production of surfactants are the origin of a type of pollution whose most significant impact is the formation of toxic or nuisance foams in rivers, lakes, and treatment plants.

7.1.1 Classification of Surfactants

Soaps and detergents are formulated products designed to meet various cost and performance standards. The formulated products contain many components, such as surfactants to tie up unwanted materials (commercial detergents usually contain only 10–30% surfactants), builders or polyphosphate salts to improve surfactant processes and remove calcium and magnesium ions, and bleaches to increase reflectance of visible light. They also contain various additives designed to remove stains (enzymes), prevent soil re-deposition, regulate foam, reduce washing machine corrosion, brighten colors, give an agreeable odor, prevent caking, and help processing of the formulated detergent [18].

The classification of surfactants in common usage depends on their electrolytic dissociation, which allows the determination of the nature of the hydrophilic polar group, for example, anionic, cationic, nonionic, and amphoteric. As reported by Greek [18], the total 1988 U.S. production of surfactants consisted of 62% anionic, 10% cationic, 27% nonionic, and 1% amphoteric.

Anionic Surfactants

Anionic surfactants produce a negatively charged surfactant ion in aqueous solution, usually derived from a sulfate, carboxylate, or sulfonate grouping. The usual types of these compounds are carboxylic acids and derivatives (largely based on natural oils), sulfonic acid derivatives (alkylbenzene sulfonates LAS or ABS and other sulfonates), and sulfuric acid esters and salts (largely sulfated alcohols and ethers). Alkyl sulfates are readily biodegradable, often disappearing within 24 hours in river water or sewage plants [23]. Because of their instability in acidic conditions, they were to a considerable extent replaced by ABS and LAS, which have been the most widely used of the surfactants because of their excellent cleaning properties, chemical stability, and low cost. Their biodegradation has been the subject of numerous investigations [24].

Cationic Surfactants

Cationic surfactants produce a positively charged surfactant ion in solution and are mainly quaternary nitrogen compounds such as amines and derivatives and quaternary ammonium salts. Owing to their poor cleaning properties, they are little used as detergents; rather their use is a result of their bacteriocidal qualities. Relatively little is known about the mechanisms of biodegradation of these compounds.

Nonionic Surfactants

Nonionic surfactants are mainly carboxylic acid amides and esters and their derivatives, and ethers (alkoxylated alcohols), and they have been gradually replacing ABS in detergent formulations (especially as an increasingly popular active ingredient of automatic washing machine formulations) since the 1960s. Therefore, their removal in wastewater treatment is of great significance, but although it is known that they readily biodegrade, many facts about their metabolism are unclear [25]. In nonionic surfactants, both the hydrophilic and hydrophobic groups are organic, so the cumulative effect of the multiple weak organic hydrophils is the cause of their surface-active qualities. These products are effective in hard water and are very low foamers.

Amphoteric Surfactants

As previously mentioned, amphoteric surfactants presently represent a minor fraction of the total surfactants production with only specialty uses. They are compounds with both anionic and cationic properties in aqueous solutions, depending on the pH of the system in which they work. The main types of these compounds are essentially analogs of linear alkane sulfonates, which provide numerous points for the initiation of biodegradation, and pyridinium compounds that also have a positively charged N-atom (but in the ring) and they are very resistant to biodegradation [26].

7.1.2 Sources of Detergents in Waters and Wastewaters

The concentrations of detergent that actually find their way into wastewaters and surface water bodies have quite diverse origins: (a) Soaps and detergents, as well as their component compounds, are introduced into wastewaters and water bodies at the point of their manufacture, at storage facilities and distribution warehouses, and at points of accidental spills on their routes of transportation (the origin of pollution is dealt with in this chapter). (b) The additional industrial origin of detergent pollution notably results from the use of surfactants in various industries, such as textiles, cosmetics, leather tanning and products, paper, metals, dyes and paints, production of domestic soaps and detergents, and from the use of detergents in commercial/industrial laundries and dry cleaners. (c) The contribution from agricultural activities is due to the surface runoff transporting of surfactants that are included in the formulation of insecticides and fungicides [27]. (d) The origin with the most rapid growth since the 1950s comprises the wastewaters from urban areas and it is due to the increased domestic usage of detergents and, equally important, their use in cleaning public spaces, sidewalks, and street surfaces.

7.1.3 Problem and Biodegradation

Notable improvements in washing and cleaning resulted from the introduction and increasing use of synthetic detergents. However, this also caused difficulties in sewage treatment and led to a new form of pollution, the main visible effect of which was the formation of objectionable quantities of foam on rivers. Although biodegradation of surfactants in soils and natural waters was inferred by the observation that they did not accumulate in the environment, there was widespread concern that their much higher concentrations in the effluents from large industrial areas would have significant local impacts. In agreement with public authorities, the manufacturers fairly quickly introduced products of a different type.

The surface-active agents in these new products are biodegradable (called "soft" in contrast to the former "hard" ones). They are to a great extent eliminated by normal sewage treatment, and the self-purification occurring in water courses also has some beneficial effects [28]. However, the introduction of biodegradable products has not solved all the problems connected to surfactants (i.e., sludge digestion, toxicity, and interference with oxygen transfer), but it has made a significant improvement. Studies of surfactant biodegradation have shown that the molecular architecture of the surfactant largely determines its biological characteristics [4]. Nevertheless, one of the later most pressing environmental problems was not the effects of the surfactants themselves, but the eutrophication of natural water bodies by the polyphosphate builders that go into detergent formulations. This led many local authorities to enact restrictions in or even prohibition of the use of phosphate detergents.

7.2 IMPACTS OF DETERGENT PRODUCTION AND USE

Surfactants retain their foaming properties in natural waters in concentrations as low as 1 mg/L, and although such concentrations are nontoxic to humans [24], the presence of surfactants in drinking water is esthetically undesirable. More important, however, is the generation of large volumes of foam in activated sludge plants and below weirs and dams on rivers.

7.2.1 Impacts in Rivers

The principal factors that influence the formation and stability of foams in rivers [27] are the presence of ABS-type detergents, the concentration of more or less degraded proteins and colloidal particles, the presence and concentration of mineral salts, the temperature and pH of the water. Additional very important factors are the biochemical oxygen demand (BOD) of the water, which under given conditions represents the quantity of biodegradable material, the time of travel and the conditions influencing the reactions of the compounds presumed responsible for foaming, between the point of discharge and the location of foam appearance, and last but not least, the concentration of calcium ion that is the main constituent of hardness in most natural waters and merits particular attention with regard to foam development.

The minimum concentrations of ABS or other detergents above which foam formation occurs vary considerably, depending on the water medium, that is, river or sewage, and its level of pollution (mineral or organic). Therefore, it is not merely the concentration of detergents that controls foam formation, but rather their combined action with other substances present in the waters. Various studies have shown [27] that the concentration of detergents measured in the foams is quite significantly higher, up to three orders of magnitude, than that measured at the same time in solution in the river waters.

The formation of foam also constitutes trouble and worries for river navigation. For instance, in the areas of dams and river locks, the turbulence caused by the intensive traffic of barges and by the incessant opening and closing of the lock gates results in foam formation that may cover entire boats and leave a sticky deposit on the decks of barges and piers. This renders them extremely slippery and may be the cause of injuries. Also, when winds are strong, masses of foam are detached and transported to great distances in the neighboring areas, causing problems in automobile traffic by deposition on car windshields and by rendering the road surfaces slippery. Finally, masses of foam floating on river waters represent an esthetically objectionable nuisance and a problem for the tourism industry.

7.2.2 Impacts on Public Health

For a long time, detergents were utilized in laboratories for the isolation, through concentration in the foam, of mycobacteria such as the bacillus of Koch (tuberculosis), as reported in the annals of the Pasteur Institute [27]. This phenomenon of extraction by foam points to the danger existing in river waters where numerous such microorganisms may be present due to sewage pollution. The foam transported by wind could possibly serve as the source of a disease epidemic. In fact, this problem limits itself to the mycobacteria and viruses (such as those of hepatitis and polio), which are the only microorganisms able to resist the disinfecting power of detergents. Therefore, waterborne epidemics could also be spread through airborne detergent foams.

7.2.3 Impacts on Biodegradation of Organics

Surfactant concentrations in polluted natural water bodies interfere with the self-purification process in several ways. First, certain detergents such as ABS are refractory or difficult to biodegrade and even toxic or inhibitory to microorganisms, and influence the BOD exhibited by organic pollution in surface waters. On the other hand, readily biodegradable detergents could impose an extreme short-term burden on the self-purification capacity of a water course, possibly introducing anaerobic conditions.

Surfactant concentrations also exert a negative influence on the bio-oxidation of certain substances, as evidenced in studies with even readily biodegradable substances [7]. It should be noted that this protection of substances from bio-oxidation is only temporary and it slowly reduces until its virtual disappearance in about a week for most substances. This phenomenon serves to retard the self-purification process in organically polluted rivers, even in the presence of high concentrations of dissolved oxygen.

An additional way in which detergent concentrations interfere with the self-purification process in polluted rivers consists of their negative action on the oxygen rate of transfer and dissolution into waters. According to Gameson [16], the presence of surfactants in a water course could reduce its re-aeration capacity by as much as 40%, depending on other parameters such as turbulence. In relatively calm waters such as estuaries, under certain conditions, the reduction of re-aeration could be as much as 70%. It is the anionic surfactants, especially the ABS, that have the overall greatest negative impact on the natural self-purification mechanisms of rivers.

7.2.4 Impacts on Wastewater Treatment Processes

Despite the initial apprehension over the possible extent of impacts of surfactants on the physicochemical or biological treatment processes of municipal and industrial wastewaters, it soon became evident that no major interference occurred. As mentioned previously, the greatest problem proved to be the layers of foam that not only hindered normal sewage plant operation, but when wind-blown into urban areas, also aided the probable transmission of fecal pathogens present in sewage.

The first unit process in a sewage treatment plant is primary sedimentation, which depends on simple settling of solids partially assisted by flocculation of the finer particles. The stability, nonflocculating property, of a fine particle dispersion could be influenced by the surface tension of the liquid or by the solid/liquid interface tension – hence, by the presence of surfactants. Depending on the conditions, primarily the size of the particles in suspension, a given concentration of detergents could either decrease (finer particles) or increase (larger particles) the rate of sedimentation [23]. The synergistic or antagonistic action of certain inorganic salts, which are included in the formulation of commercial detergent products, is also influential.

The effect of surfactants on wastewater oils and greases depends on the nature of the latter, as well as on the structure of the lipophilic group of the detergent that assists solubilization. As is the case, emulsification could be more or less complete. This results in a more or less significant impact on the efficiency of physical treatment designed for their removal. On the other hand, the emulsifying surfactants play a role in protecting the oil and grease molecules from attacking bacteria in a biological unit process.

In water treatment plants, the coagulation/flocculation process was found early to be affected by the presence of surfactants in the raw water supply. In general, the anionic detergents stabilize colloidal particle suspensions or turbidity solids, which, in most cases, are negatively charged. Langelier [29] reported problems with water clarification due to surfactants, although according to Nichols and Koepp [30] and Todd [31] concentrations of surfactants on the order of 4–5 ppm interfered with flocculation. The floc, instead of settling to the bottom, floats to the surface of sedimentation tanks. Other studies, such as those conducted by Smith *et al.* [32] and Cohen [10], indicated that this interference could be not so much due to the surfactants themselves, but to the additives included in their formulation, that is, phosphate complexes. Such interference was observed both for alum and ferric sulfate coagulant, but the use of certain organic polymer flocculants was shown to overcome this problem.

Concentrations of detergents, such as those generally found in municipal wastewaters, have been shown to insignificantly impact on the treatment efficiency of biological sewage treatment plants [33]. Studies indicated that significant impacts on efficiency can be observed only for considerable concentrations of detergents, such as those that could possibly be found in undiluted industrial wastewaters, on the order of 30 ppm and above. As previously mentioned, it is through their influence of water aeration that the surfactants impact the organics' biodegradation process. As little as 0.1 mg/L of surfactant reduces to nearly half the oxygen absorption rate in a river, but in sewage aeration units the system could be easily designed to compensate. This is achieved through the use of the alpha and beta factors in the design equation of an aeration system.

Surfactants are only partially biodegraded in a sewage treatment plant, so that a considerable proportion may be discharged into surface water bodies with the final effluent. The shorter the overall detention time of the treatment plant, the higher the surfactant concentration in the discharged effluent. By the early 1960s, the concentration of surfactants in the final effluents from sewage treatment plants was in the 5–10 ppm range, and while dilution occurs at the site of discharge, the resulting values of concentration were well above the threshold for foaming. In more recent times, with the advent of more readily biodegradable surfactants, foaming within treatment plants and in natural water bodies is a much more rare and limited phenomenon.

Finally, according to Prat and Giraud [27], the process of anaerobic sludge digestion, commonly used to further stabilize biological sludge prior to disposal and to produce methane gas, is not affected by concentrations of surfactants in the treated sludge up to 500 ppm or when it does not contain too high an amount of phosphates. These levels of concentration are not found in municipal or industrial effluents, but within the biological treatment processes a large part of the detergents is passed to the sludge solids. By this, it could presumably build up to concentrations (especially of ABS surfactants) that may affect somewhat the sludge digestion process, that is, methane gas production. Also, it seems that anaerobic digestion [34] does not decompose surfactants and, therefore, their accumulation could pose problems with the use of the final sludge product as a fertilizer.

The phenomena related to surface tension in groundwater interfere with the mechanisms of water flow in the soil. The presence of detergents in wastewaters discharged on soil for groundwater recharge or filtered through sand beds would cause an increase in headloss and leave a deposit of surfactant film on the filter media, thereby affecting permeability. Surfactants, especially those resistant to biodegradation, constitute a pollutant that tends to accumulate in groundwater and has been found to remain in the soil for a few years without appreciable decomposition. Because surfactants modify the permeability of soil, their presence could possibly facilitate the penetration of other pollutants, that is, chemicals or microorganisms, to depths where they would not have reached due to the filtering action of the soil, thereby increasing groundwater pollution [35].

7.2.5 Impacts on Drinking Water

From all the aforementioned, it is obvious that detergents find their way into drinking water supplies in various ways. As far as imparting odor to drinking water, only heavy doses of anionic surfactants yield an unpleasant odor [36], and someone has to have a very sensitive nose to smell detergent doses of 50 mg/L or less. On the other hand, it seems that the impact of detergent doses on the sense of taste of various individuals varies considerably. As reported by Cohen [10], the U.S. Public Health Service conducted a series of taste tests which showed that although 50% of the people in the test group detected a concentration of 60 mg/L of ABS in drinking water, only 5% of them detected a concentration of 16 mg/L. Because tests like this have been conducted using commercial detergent formulations, most probably the observed taste is not due to the surfactants but rather to the additives or perfumes added to the products. However, the actual limit for detergents in drinking water in the United States is a concentration of only 0.5 mg/L, less than even the most sensitive palates can discern.

7.2.6 Toxicity of Detergents

There is an upper limit of surfactant concentration in natural waters above which the existence of aquatic life, particularly higher animal life, is endangered. Trout are particularly sensitive to concentrations as low as 1 ppm and show symptoms similar to asphyxia [4]. On the other hand, numerous studies, which extended over a period of months and required test animals to drink significantly high doses of surfactants, showed absolutely no apparent ill effects due to digested detergents. Also, there are no instances in which the trace amounts of detergents present in drinking water were directly connected to adverse effects on human health.

River pollution from anionic surfactants, the primarily toxic ones, is of two types: (a) acute toxic pollution due to, for example, an accidental spill from a container of full-strength surfactant products, and (b) chronic pollution due to the daily discharges of municipal and industrial wastewaters. The international literature contains the result of numerous studies that have established dosages for both types of pollutional toxicity due to detergents, for most types of aquatic life such as species of fish.

7.3 CURRENT PERSPECTIVE AND FUTURE OUTLOOK

This section summarizes the main points of a recent product report [18], which presented the new products of the detergent industry and its proposed direction in the foreseeable future.

If recent product innovations sell successfully in test markets in the United States and other countries, rapid growth could begin again for the entire soap and detergent industry and especially for individual sectors of that industry. Among these new products are formulations that combine bleaching materials and other components, and detergents and fabric softeners sold in concentrated forms. These concentrated materials, so well accepted in Japan, are now becoming commercially significant in Western Europe. Their more widespread use will allow the industry to store and transport significantly smaller volumes of detergents, with the consequent reduction of environmental risks from housecleaning and spills. Some components of detergents such as enzymes will very likely grow in use, although the use of phosphates employed as builders will continue to drop for environmental reasons. Consumers shift to liquid formulations in areas where phosphate materials are banned from detergents, because they perceive that the liquid detergents perform better than powdered ones without phosphates.

In fuel markets, detergent formulations such as gasoline additives that limit the buildup of deposits in car engines and fuel injectors will very likely grow fast from a small base, with the likelihood of an increase in spills and discharges from this industrial source. Soap, on the other hand, has now become a small part (17%) of the total output of surfactants, whereas the anionic forms (which include soaps) accounted for 62% of total U.S. production in 1988. Liquid detergents (many of the LAS type), which are generally higher in surfactant concentrations than powdered ones, will continue to increase in production volume, therefore creating greater surfactant pollution problems due to housecleaning and spills. (Also, a powdered detergent spill creates less of a problem, as it is easier to just scoop up or vacuum.)

Changes in the use of builders resulting from environmental concerns have been pushing surfactant production demand. Outright legal bans or consumer pressures on the use of inorganic phosphates and other materials as builders generally have led formulators to raise the contents of surfactants in detergents. Builders provide several functions, most important of which are to aid the detergency action and to tie up and remove calcium and magnesium from the wash water, dirt, and the fabric or other material being cleaned. Besides sodium and potassium phosphates, other builders that may be used in various detergent formulations are citric acid and derivatives, zeolites, and other alkalis. Citric acid causes caking and is not used in powdered detergents, but it finds considerable use in liquid detergents. In some detergent formulations, larger and larger amounts of soda ash (sodium carbonate) are replacing inert ingredients due to its functionality as a builder, an agglomerating aid, a carrier for surfactants, and a source of alkalinity.

Incorporating bleaching agents into detergent formulations for home laundry has accelerated, because its performance allows users to curtail the need to store as well as add (as a second step) bleaching material. Because U.S. home laundry requires shorter wash times and lower temperatures than European home laundry, chlorine bleaches (mainly sodium hypochlorite) have long dominated the U.S. market. Institutional and industrial laundry bleaching, when done, has also favored chlorine bleaches (often chlorinated isocyanurates) because of their rapid action. Other kinds of bleaching agents used in the detergent markets are largely sodium perborates and percarbonates other than hydrogen peroxide itself.

The peroxygen bleaches are forecast to grow rapidly, for both environmental and technical reasons, as regulatory pressures drive the institutional and industrial market away from chlorine bleaches and toward the peroxygen ones. The Clean Water Act amendments are requiring lower levels of trihalomethanes (products of reaction of organics and chlorine) in wastewaters. Expensive systems may be needed to clean up effluents, or the industrial users of chlorine bleaches will have to pay higher and higher surcharges to municipalities for handling chlorine-containing wastewaters that are put into sewers. Current and expected changes in bleaching materials for various segments of the detergent industry are but part of sweeping changes to come due to environmental concerns and responses to efforts to improve the world environment.

Both detergent manufacturers and their suppliers will make greater efforts to develop more "environmentally friendly" products. BASF, for example, has developed a new biodegradable stabilizer for perborate bleach, which is now being evaluated for use in detergents. The existing detergent material, such as LAS and its precursor linear alkylbenzene, known to be nontoxic and environmentally safe as well as effective, will continue to be widely used. It will be difficult, however, to gain approval for new materials to be used in detergent formulations until their environmental performance has been shown to meet existing guidelines. Some countries, for example, tend to favor a formal regulation or law (i.e., the EEC countries) prohibiting the manufacture, importation, or use of detergents that are not satisfactorily biodegradable [28].

7.4 INDUSTRIAL OPERATION AND WASTEWATER

The soap and detergent industry is a basic chemical manufacturing industry in which essentially both the mixing and chemical reactions of raw materials are involved in production. Also, short- and long-term chemicals storage and warehousing, as well as loading/unloading and transportation of chemicals, are involved in the operation.

7.4.1 Manufacture and Formulation

This industry produces liquid and solid cleaning agents for domestic and industrial use, including laundry, dishwashing, bar soaps, specialty cleaners, and industrial cleaning products. It can be broadly divided (Fig. 1) into two categories: (a) soap manufacture that is based on the processing of natural fat; and (b) detergent manufacture that is based on the processing of

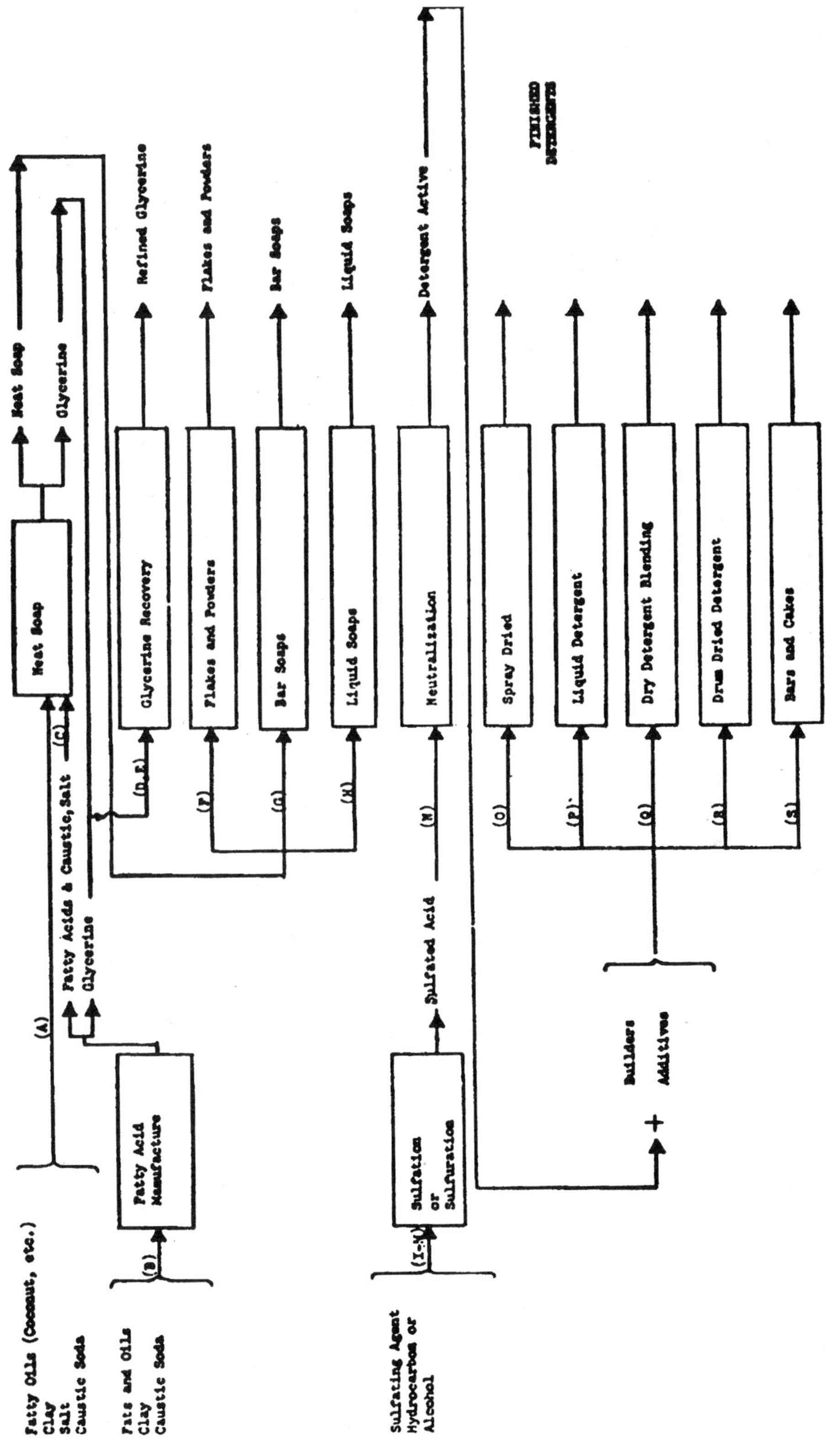

Figure 1 Flow diagram of soap and detergent manufacture (from Ref. 13).

petrochemicals. The information presented here includes establishments primarily involved in the production of soap, synthetic organic detergents, inorganic alkaline detergents, or any combinations of these, and plants producing crude and refined glycerine from vegetable and animal fats and oils. Types of facilities not discussed here include plants primarily involved in the production of shampoo or shaving creams/soaps, whether from soap or surfactants, and of synthetic glycerine as well as specialty cleaners, polishing and sanitation preparations.

Numerous processing steps exist between basic raw materials for surfactants and other components that are used to improve performance and desirability, and the finished marketable products of the soap and detergent industry. Inorganic and organic compounds such as ethylene, propylene, benzene, natural fatty oils, ammonia, phosphate rock, trona, chlorine, peroxides, and silicates are among the various basic raw materials being used by the industry. The final formulation of the industry's numerous marketable products involves both simple mixing of and chemical reactions among compounds such as the above.

The categorization system of the various main production streams and their descriptions is taken from federal guidelines [13] pertaining to state and local industrial pretreatment programs. It will be used in the discussion that ensues to identify process flows and to characterize the resulting raw waste. Figure 1 shows a flow diagram for the production streams of the entire industry. Manufacturing of soap consists of two major operations: the production of neat soap (65–70% hot soap solution) and the preparation and packaging of finished products into flakes and powders (F), bar soaps (G), and liquid soaps (H). Many neat soap manufacturers also recover glycerine as a byproduct for subsequent concentration (D) and distillation (E). Neat soap is generally produced in either of two processes: the batch kettle process (A) or the fatty acid neutralization process, which is preceded by the fat splitting process (B, C). (Note, letters in parentheses represent the processes described in the following sections.)

Batch Kettle Process (A)

This process consists of the following operations: (a) receiving and storage of raw materials, (b) fat refining and bleaching, and (c) soap boiling. The major wastewater sources, as shown in the process flow diagram (Fig. 2), are the washouts of both the storage and refining tanks, as well as from leaks and spills of fats and oils around these tanks. These streams are usually skimmed for fat recovery prior to discharge to the sewer.

The fat refining and bleaching operation is carried out to remove impurities that would cause color and odor in the finished soap. The wastewater from this source has a high soap concentration, treatment chemicals, fatty impurities, emulsified fats, and sulfuric acid solutions of fatty acids. Where steam is used for heating, the condensate may contain low-molecular-weight fatty acids, which are highly odorous, partially soluble materials.

The soap boiling process produces two concentrated waste streams: sewer lyes that result from the reclaiming of scrap soap and the brine from Nigre processing. Both of these wastes are low volume, high pH, with BOD values up to 45,000 mg/L.

Soap manufacture by the neutralization process is a two-step process:

fat + water → fatty acid + glycerine (*fat splitting*) (B)

fatty acid + caustic → soap (*fatty acid neutralization*) (C)

Fat Splitting (B)

The manufacture of fatty acid from fat is called fat splitting (B), and the process flow diagram is shown in Fig. 3. Washouts from the storage, transfer, and pretreatment stages are the same as those for process (A). Process condensate and barometric condensate from fat splitting will be contaminated with fatty acids and glycerine streams, which are settled and skimmed to recover

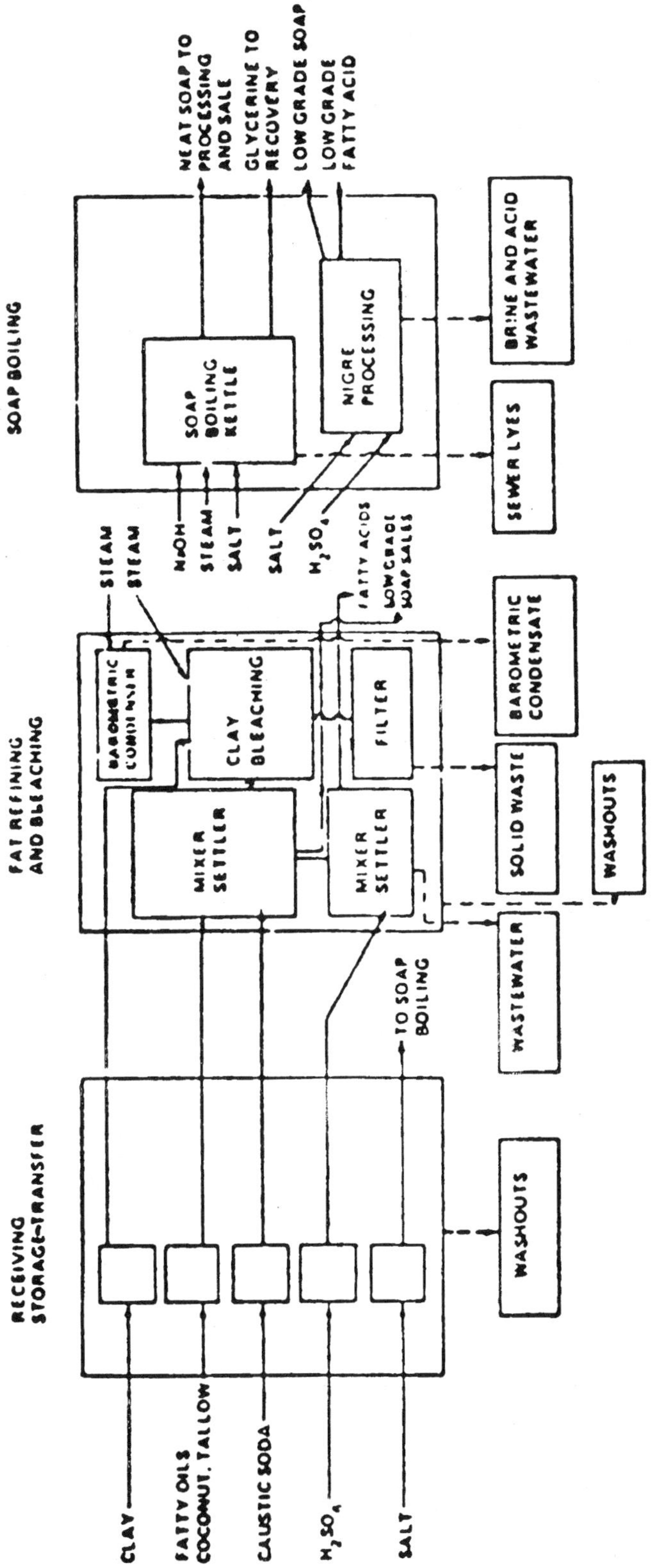

Figure 2 Soap manufacture by batch kettle (A) (from Ref. 13).

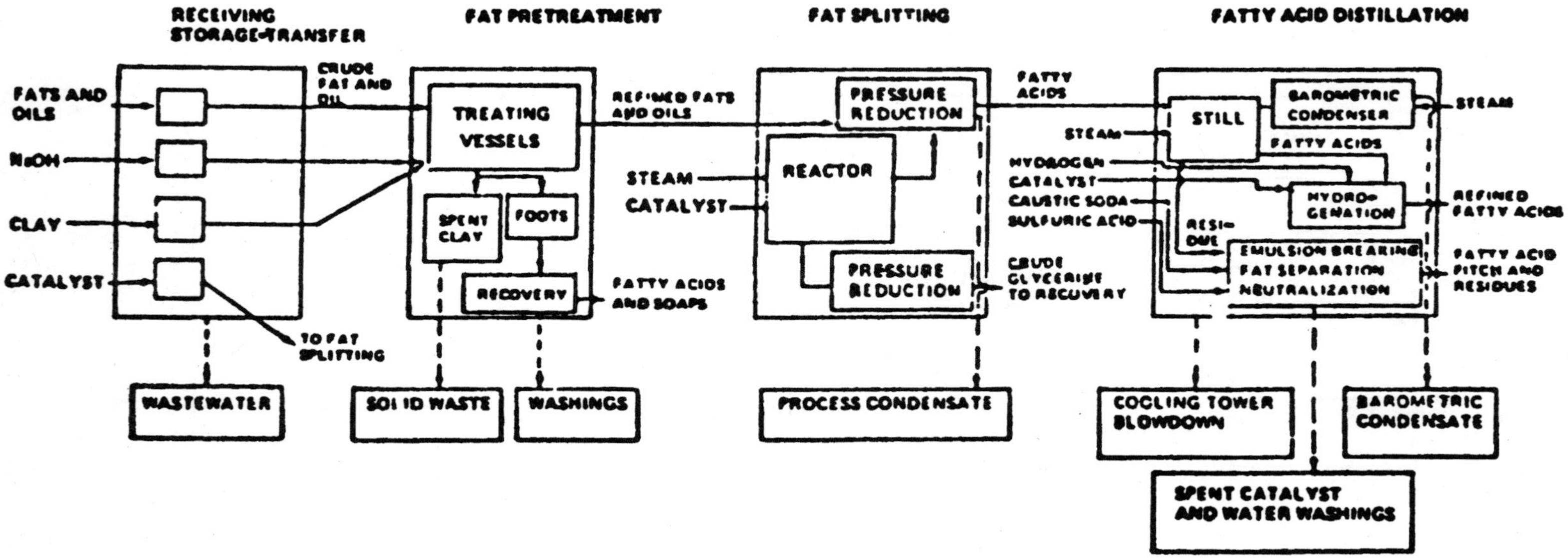

Figure 3 Fatty acid manufacture by fat splitting (B) (from Ref. 13).

the insoluble fatty acids that are processed for sale. The water will typically circulate through a cooling tower and be reused. Occasional purges of part of this stream to the sewer release high concentrations of BOD and some grease and oil.

In the fatty acid distillation process, wastewater is generated as a result of an acidification process, which breaks the emulsion. This wastewater is neutralized and sent to the sewer. It will contain salt from the neutralization, zinc and alkaline earth metal salts from the fat splitting catalyst, and emulsified fatty acids and fatty acid polymers.

Fatty Acid Neutralization (C)

Soap making by this method is a faster process than the kettle boil process and generates less wastewater effluent (Fig. 4). Because it is faster, simpler, and cleaner than the kettle boil process, it is the preferred process among larger as well as small manufacturers.

Often, sodium carbonate is used in place of caustic. When liquid soaps (at room temperature) are desired, the more soluble potassium soaps are made by substituting potassium hydroxide for the sodium hydroxide (lye). This process is relatively simple and high-purity raw materials are converted to soap with essentially no byproducts. Leaks, spills, storm runoff, and washouts are absent. There is only one wastewater of consequence: the sewer lyes from reclaiming of scrap. The sewer lyes contain the excess caustic soda and salt added to grain out the soap. Also, they contain some dirt and paper not removed in the strainer.

Glycerine Recovery Process (D, E)

A process flow diagram for the glycerine recovery process uses the glycerine byproducts from kettle boiling (A) and fat splitting (B). The process consists of three steps (Fig. 5): (a) pretreatment to remove impurities, (b) concentration of glycerine by evaporation, and (c) distillation to a finished product of 98% purity.

There are three wastewaters of consequence from this process: two barometric condensates, one from evaporation and one from distillation, plus the glycerine foots or still bottoms. Contaminants from the condensates are essentially glycerine with a little entrained salt. In the distillation process, the glycerine foots or still bottoms leave a glassy dark brown amorphous solid rich in salt that is disposed of in the wastewater stream. It contains glycerine, glycerine polymers, and salt. The organics will contribute to BOD, COD (chemical oxygen demand), and dissolved solids. The sodium chloride will also contribute to dissolved solids. Little or no suspended solids, oil, and grease or pH effect should be seen.

Glycerine can also be purified by the use of ion-exchange resins to remove sodium chloride salt, followed by evaporation of the water. This process puts additional salts into the wastewater but results in less organic contamination.

7.4.3 Production of Finished Soaps and Process Wastes

The production of finished soaps utilizes the neat soap produced in processes A and C to prepare and package finished soap. These finished products are soap flakes and powders (F), bar soaps (G), and liquid soap (H). See Figures 6, 7, and 8 for their respective flow diagrams.

Flakes and Powders (F)

Neat soap may or may not be blended with other products before flaking or powdering. Neat soap is sometimes filtered to remove gel particles and run into a reactor (crutcher) for mixing with builders. After thorough mixing, the finished formulation is run through various mechanical operations to produce flakes and powders. Because all of the evaporated moisture goes to the atmosphere, there is no wastewater effluent.

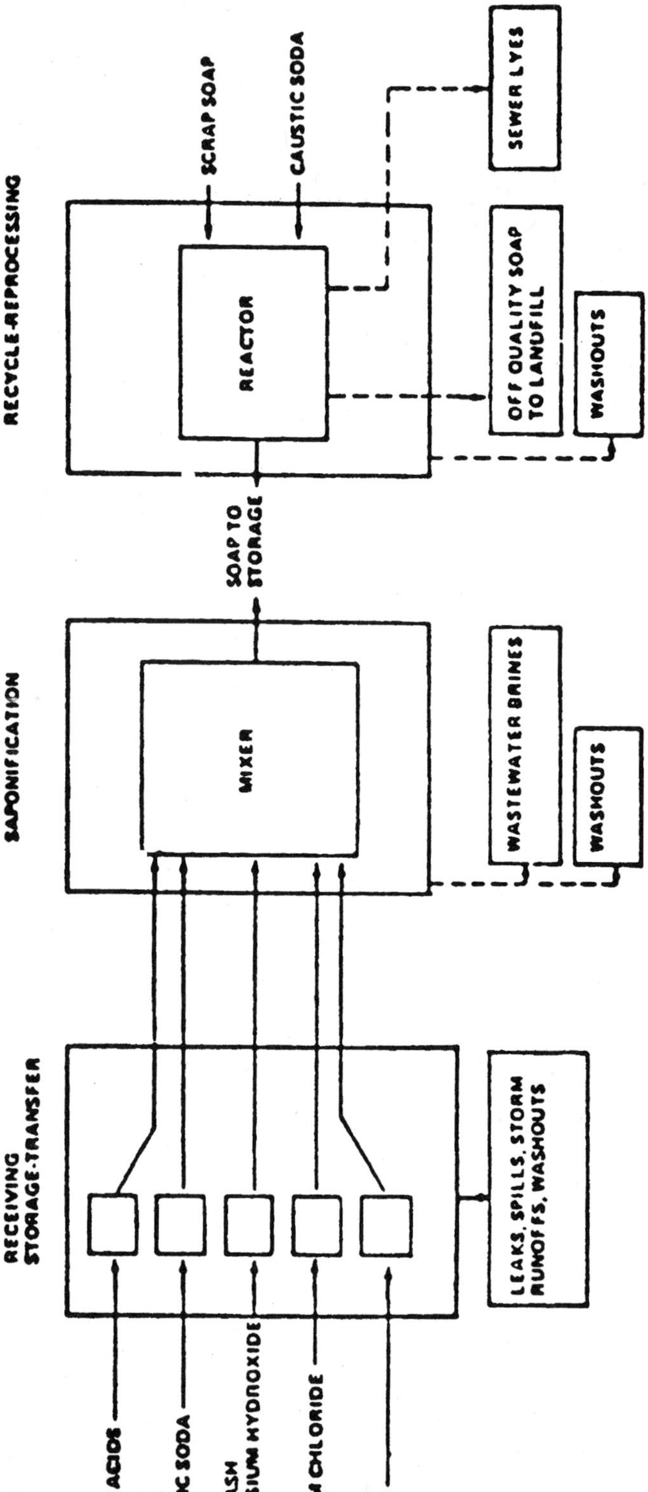

Figure 4 Soap from fatty acid neutralization (C) (from Ref. 13).

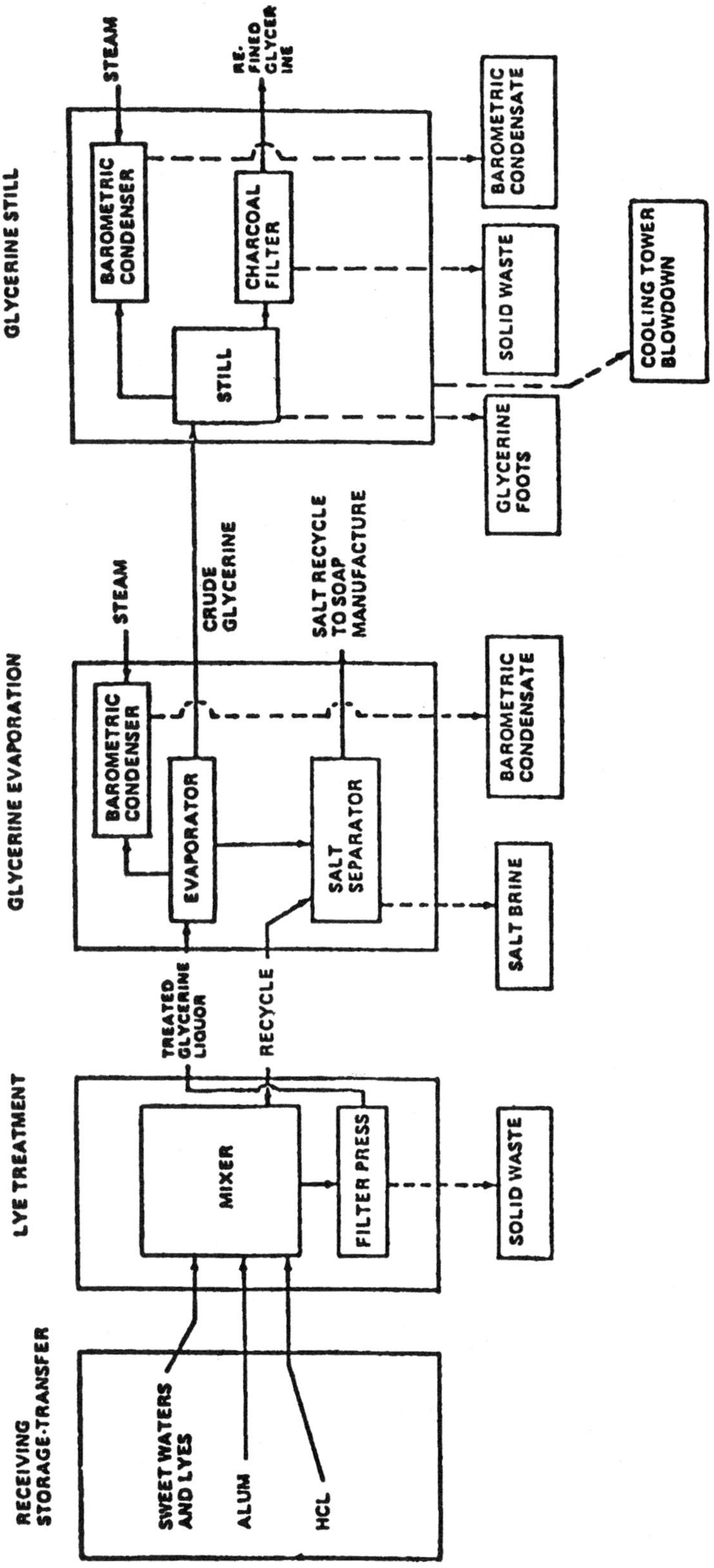

Figure 5 Glycerine recovery process flow diagram (D, E) (from Ref. 13).

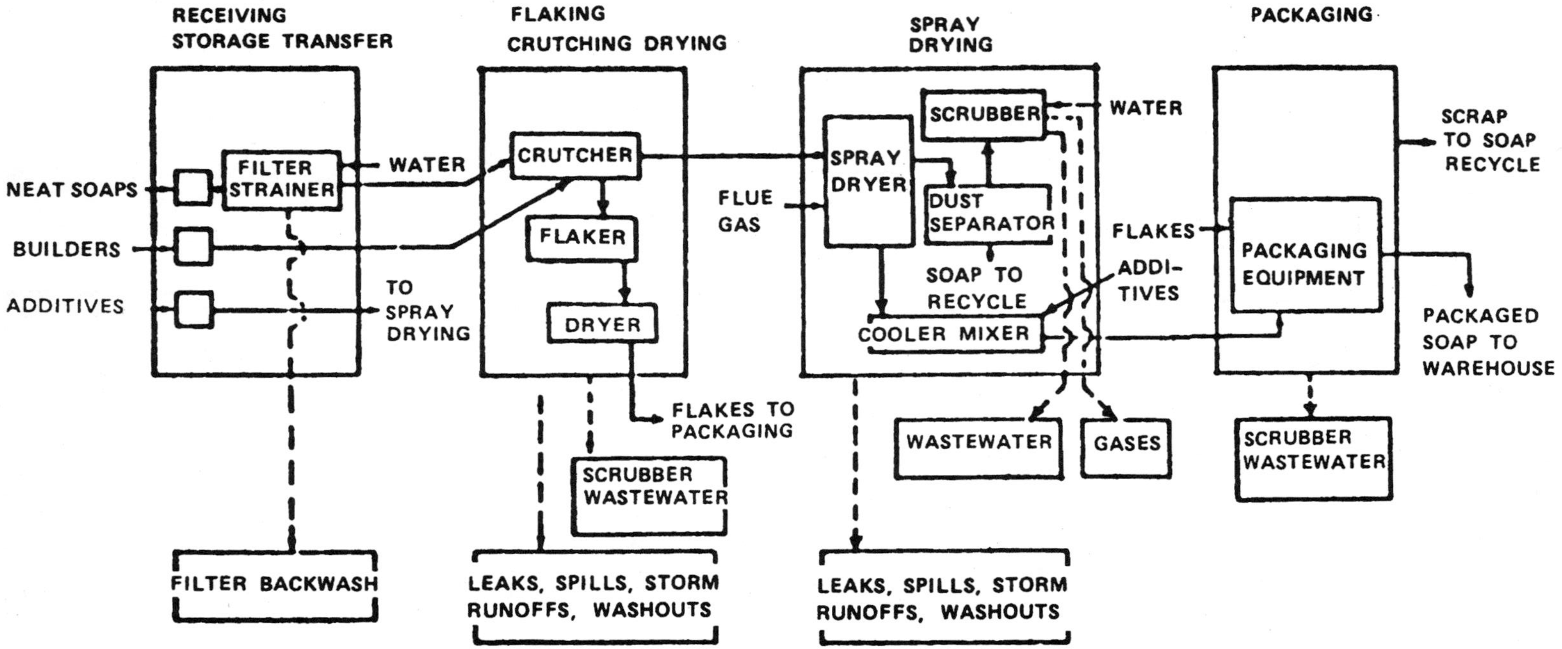

Figure 6 Soap flake and powder manufacture (F) (from Ref. 13).

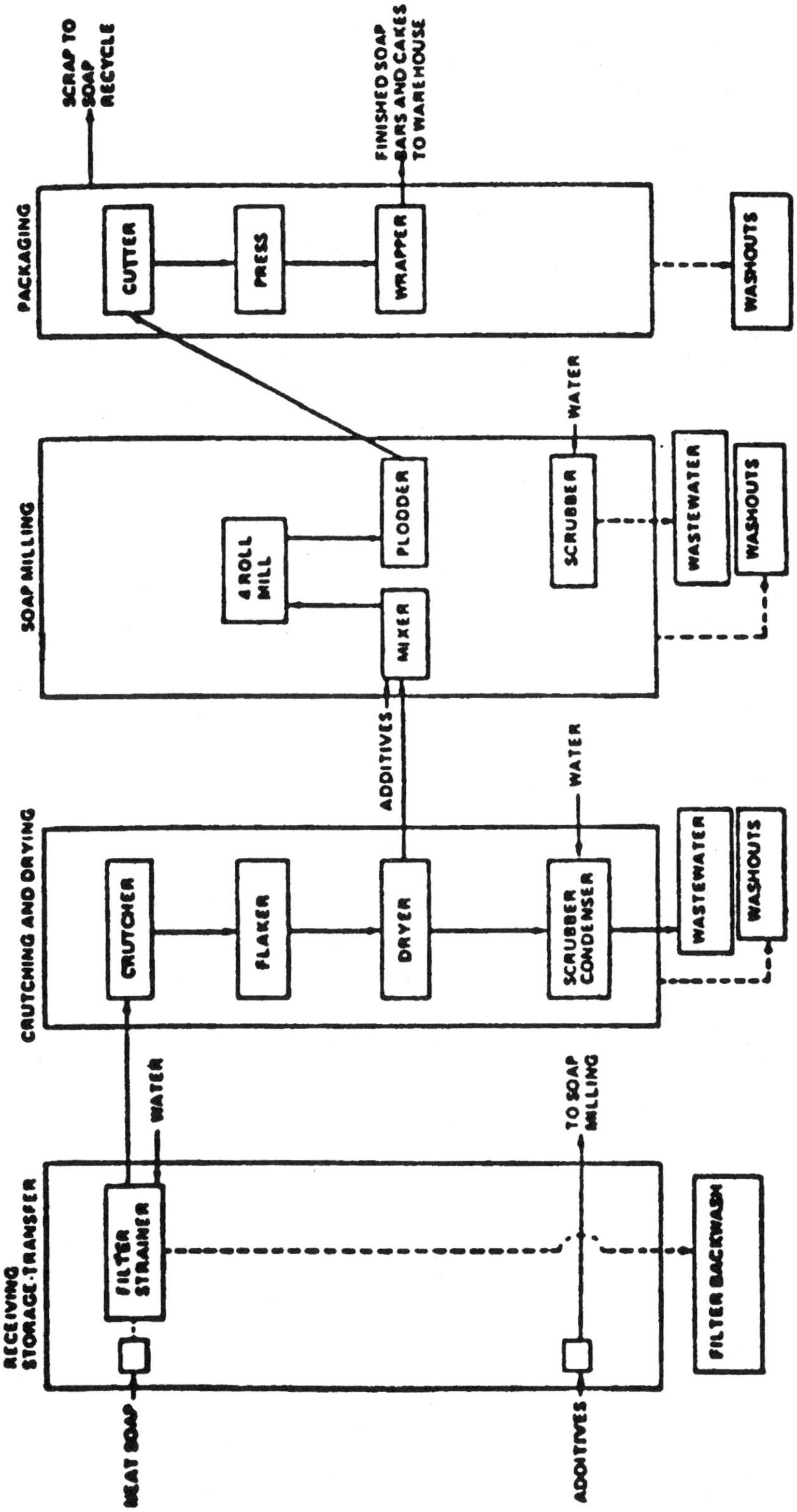

Figure 7 Bar soap manufacture (G) (from Ref. 13).

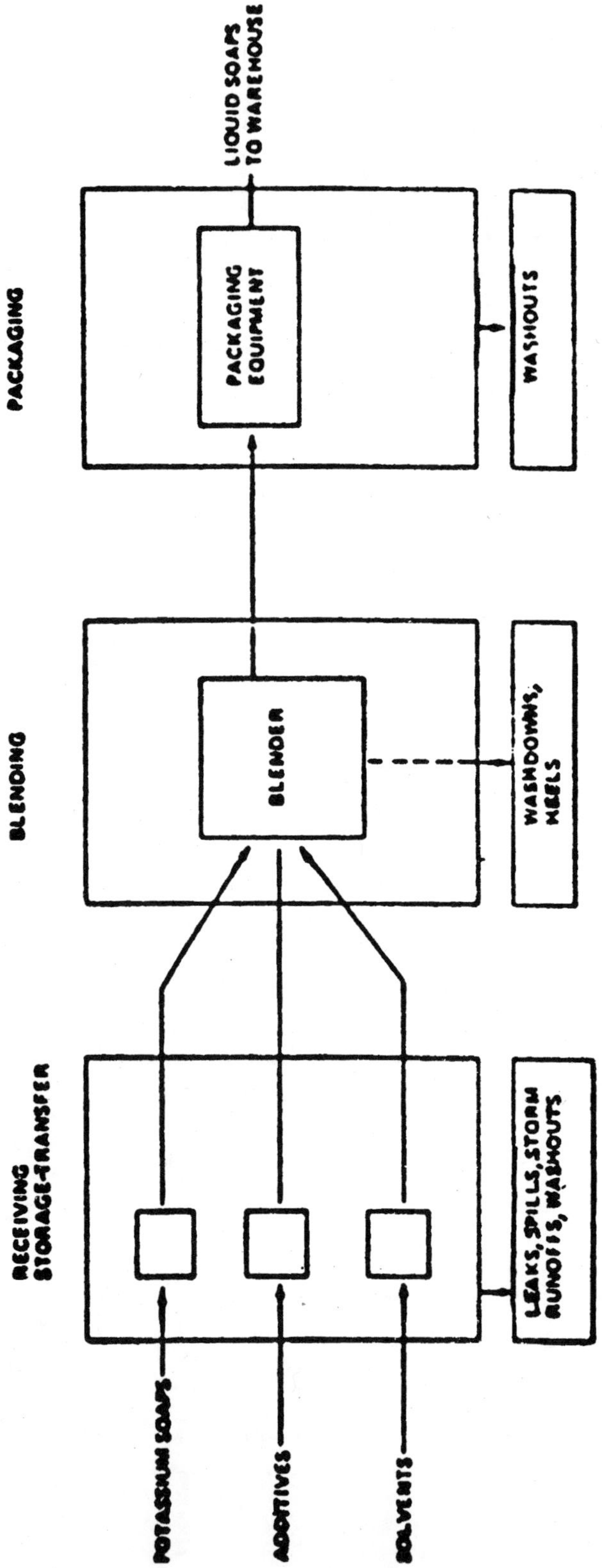

Figure 8 Liquid soap processing (H) (from Ref. 13).

Some operations will include a scrap soap reboil to recover reclaimed soap. The soap reboil is salted out for soap recovery and the salt water is recycled. After frequent recycling, the salt water becomes so contaminated that it must be discharged to the sewer. Occasional washdown of the crutcher may be needed. The tower is usually cleaned down dry. There is also some gland water that flows over the pump shaft, picking up any minor leaks. This will contribute a very small, but finite, effluent loading.

There are a number of possible effluents shown on the flow diagram for process F (Fig. 6). However, a survey of the industry showed that most operating plants either recycled any wastewater to extinction or used dry clean-up processes. Occasionally, water will be used for clean-up.

Bar Soaps (G)

The procedure for bar soap manufacture (O) will vary significantly from plant to plant, depending on the particular clientele served. A typical flow diagram for process O is shown in Figure 7. The amount of water used in bar soap manufacture varies greatly. In many cases, the entire bar soap processing operation is carried out without generating a single wastewater stream. The equipment is all cleaned dry, without any washups. In other cases, due to housekeeping requirements associated with the particular bar soap processes, there are one or more wastewater streams from air scrubbers.

The major waste streams in bar soap manufacture are the filter backwash, scrubber waters, or condensate from a vacuum drier, and water from equipment washdown. The main contaminant of all these streams is soap that will contribute primarily BOD and COD to the wastewater.

Liquid Soap (H)

In the making of liquid soap, neat soap (often the potassium soap of fatty acids) is blended in a mixing tank with other ingredients such as alcohols or glycols to produce a finished product, or the pine oil and kerosene for a product with greater solvency and versatility (Fig. 8). The final blended product may be, and often is, filtered to achieve a sparkling clarity before being drummed. In making liquid soap, water is used to wash out the filter press and other equipment. According to manufacturers, there are very few effluent leaks. Spills can be recycled or handled dry. Washout between batches is usually unnecessary or can be recycled to extinction.

7.4.4 Detergent Manufacture and Waste Streams

Detergents, as mentioned previously, can be formulated with a variety of organic and inorganic chemicals, depending on the cleaning characteristics desired. A finished, packaged detergent customarily consists of two main components: the active ingredient or surfactant, and the builder. The processes discussed in the following will include the manufacture and processing of the surfactant as well as the preparation of the finished, marketable detergent. The production of the surfactant (Fig. 1) is generally a two-step process: (a) sulfation or sulfonation, and (b) neutralization.

7.4.5 Surfactant Manufacture and Waste Streams

Oleum Sulfonation/Sulfation (I)

One of the most important active ingredients of detergents is the sulfate or sulfonate compounds made via the oleum route. A process flow diagram is shown in Figure 9. In most cases, the sulfonation/sulfation is carried out continuously in a reactor where the oleum (a solution of sulfur trioxide in sulfuric acid) is brought into contact with the hydrocarbon or alcohol and a

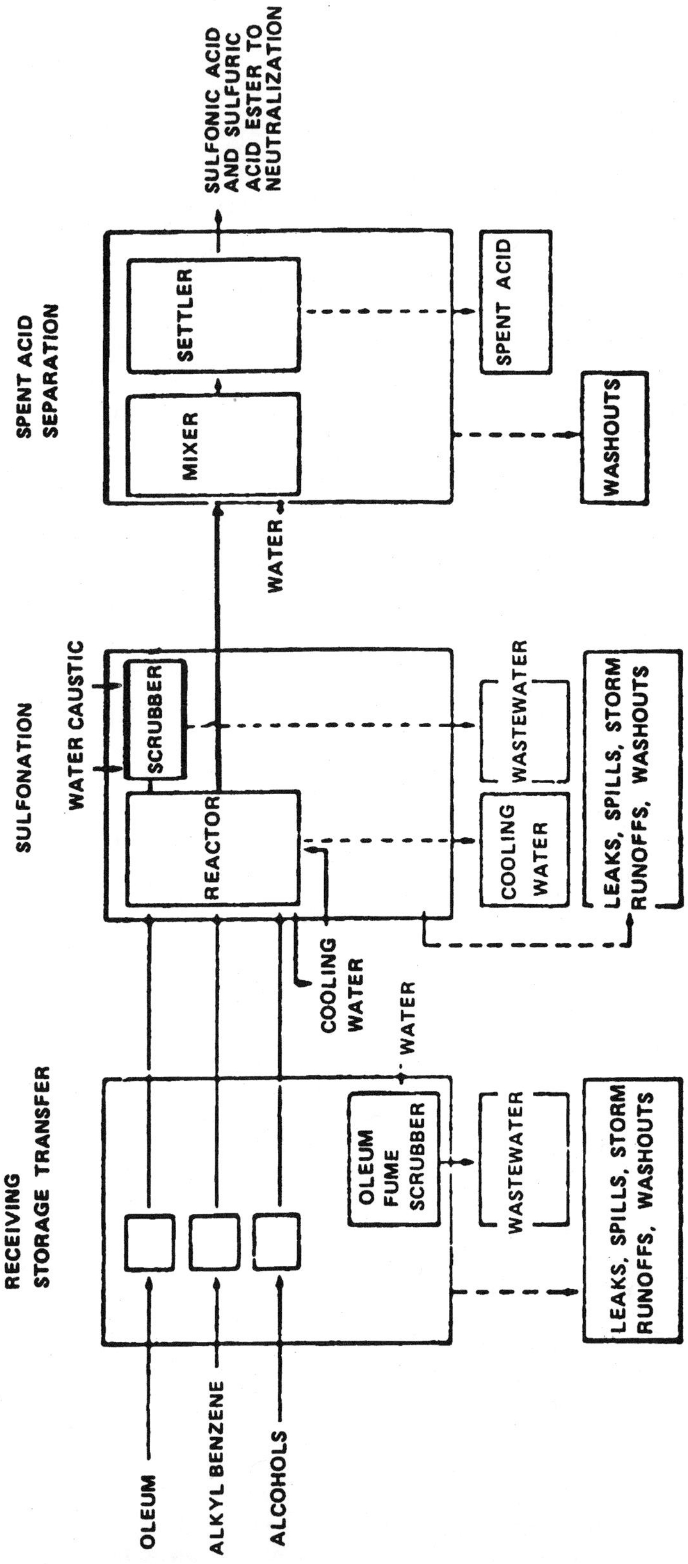

Figure 9 Oleum sulfation and sulfonation (batch and continuous) (I) (from Ref. 13).

rapid reaction ensues. The stream is then mixed with water, where the surfactant separates and is then sent to a settler. The spent acid is drawn off and usually forwarded for reprocessing, and the sulfonated/sulfated materials are sent to be neutralized.

This process is normally operated continuously and performs indefinitely without need of periodic cleanout. A stream of water is generally played over pump shafts to pick up leaks as well as to cool the pumps. Wastewater flow from this source is quite modest, but continual.

Air–SO_3 Sulfation/Sulfonation (J)

This process for surfactant manufacture has many advantages and is used extensively. With SO_3 sulfation, no water is generated in the reaction. A process flow diagram is shown in Figure 10. SO_3 can be generated at the plant by burning sulfur or sulfur dioxide with air instead of obtaining it as a liquid. Because of this reaction's particular tendency to char the product, the reactor system must be cleaned thoroughly on a regular basis. In addition, there are usually several airborne sulfonic acid streams that must be scrubbed, with the wastewater going to the sewer during sulfation.

SO_3 Solvent and Vacuum Sulfonation (K)

Undiluted SO_3 and organic reactant are fed into the vacuum reactor through a mixing nozzle. A process flow diagram is shown in Figure 11. This system produces a high-quality product, but offsetting this is the high operating cost of maintaining the vacuum. Other than occasional washout, the process is essentially free of wastewater generation.

Sulfamic Acid Sulfation (L)

Sulfamic acid is a mild sulfating agent and is used only in very specialized quality areas because of the high reagent price. A process flow diagram is shown in Figure 12. Washouts are the only wastewater effluents from this process as well.

Chlorosulfonic Acid Sulfation (M)

For products requiring high-quality sulfates, chlorosulfonic acid is an excellent corrosive agent that generates hydrochloric acid as a byproduct. A process flow diagram is shown in Figure 13. The effluent washouts are minimal.

Neutralization of Sulfuric Acid Esters and Sulfonic Acids (N)

This step is essential in the manufacture of detergent active ingredients as it converts the sulfonic acids or sulfuric acid esters (products produced by processes I–M) into neutral surfactants. It is a potential source of some oil and grease, but occasional leaks and spills around the pump and valves are the only expected source of wastewater contamination. A process flow diagram is shown in Figure 14.

7.4.6 Detergent Formulation and Process Wastes

Spray-Dried Detergents (O)

In this segment of the processing, the neutralized sulfonates and/or sulfates are first blended with builders and additives in the crutcher. The slurry is then pumped to the top of a spray tower of about 4.5–6.1 m (15–20 ft) in diameter by 45–61 m (150–200 ft) in height, where nozzles spray out detergent slurry. A large volume of hot air enters the bottom of the tower and rises to

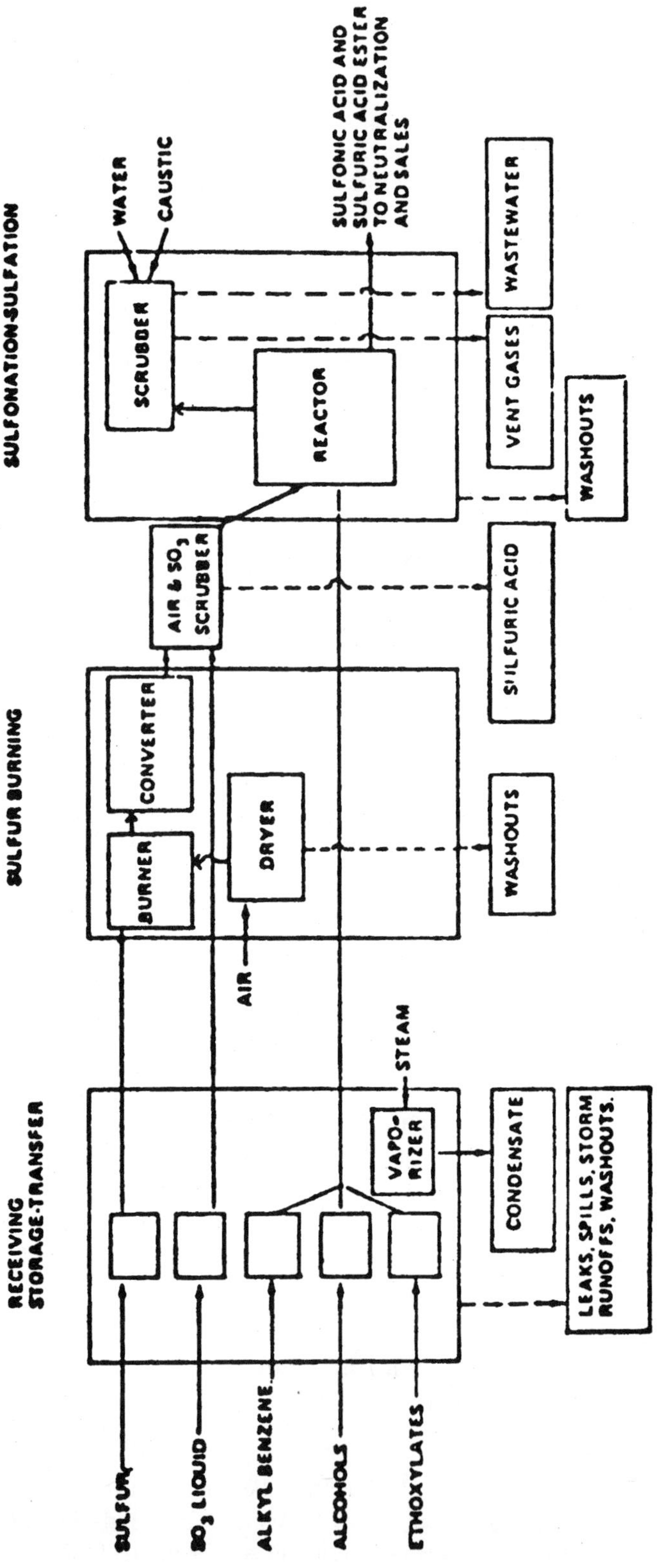

Figure 10 Air–SO_3 sulfation and sulfonation (batch and continuous) (J) (from Ref. 13).

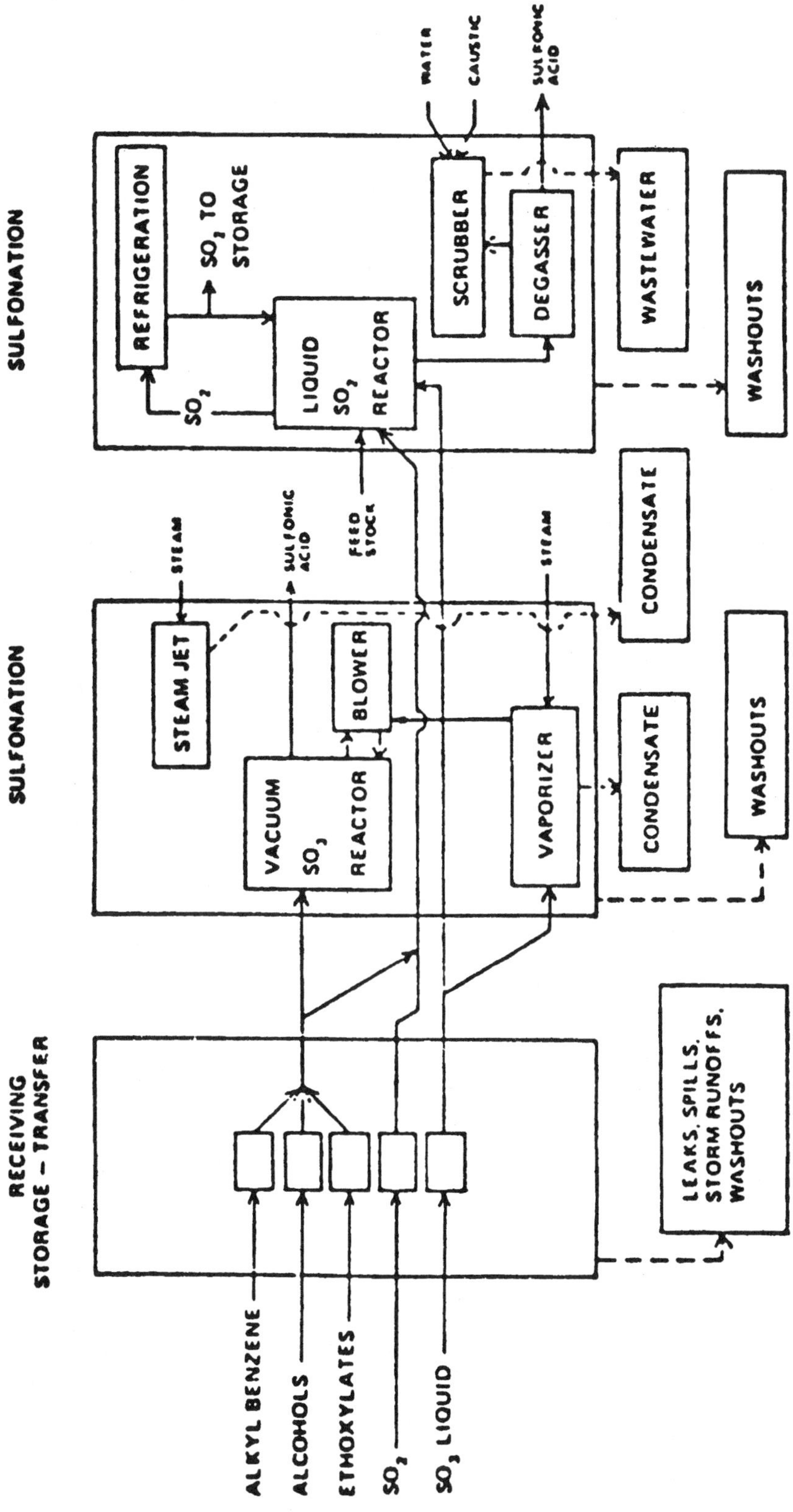

Figure 11 SO_3 solvent and vacuum sulfonation (K) (from Ref. 13).

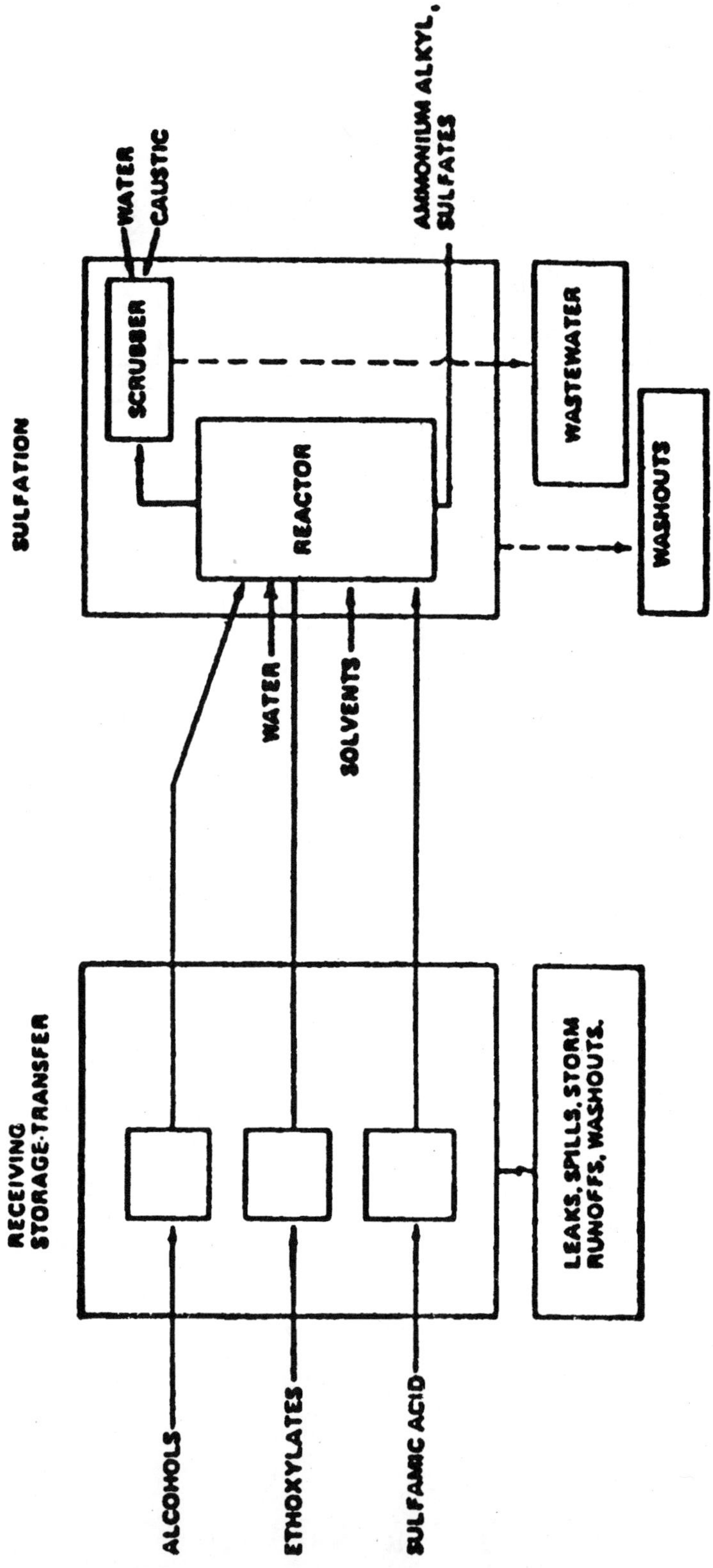

Figure 12 Sulfamic acid sulfation (L) (from Ref. 13).

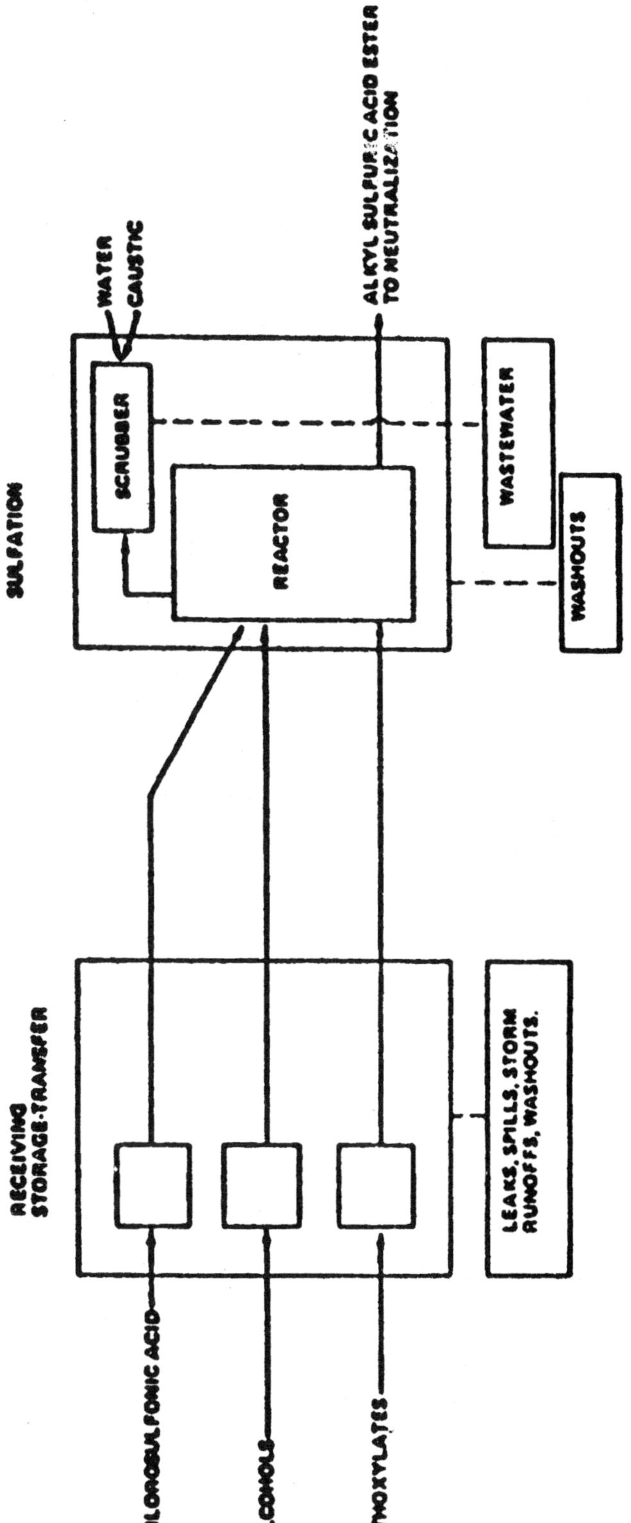

Figure 13 Chlorosulfonic acid sulfation (M) (from Ref. 13).

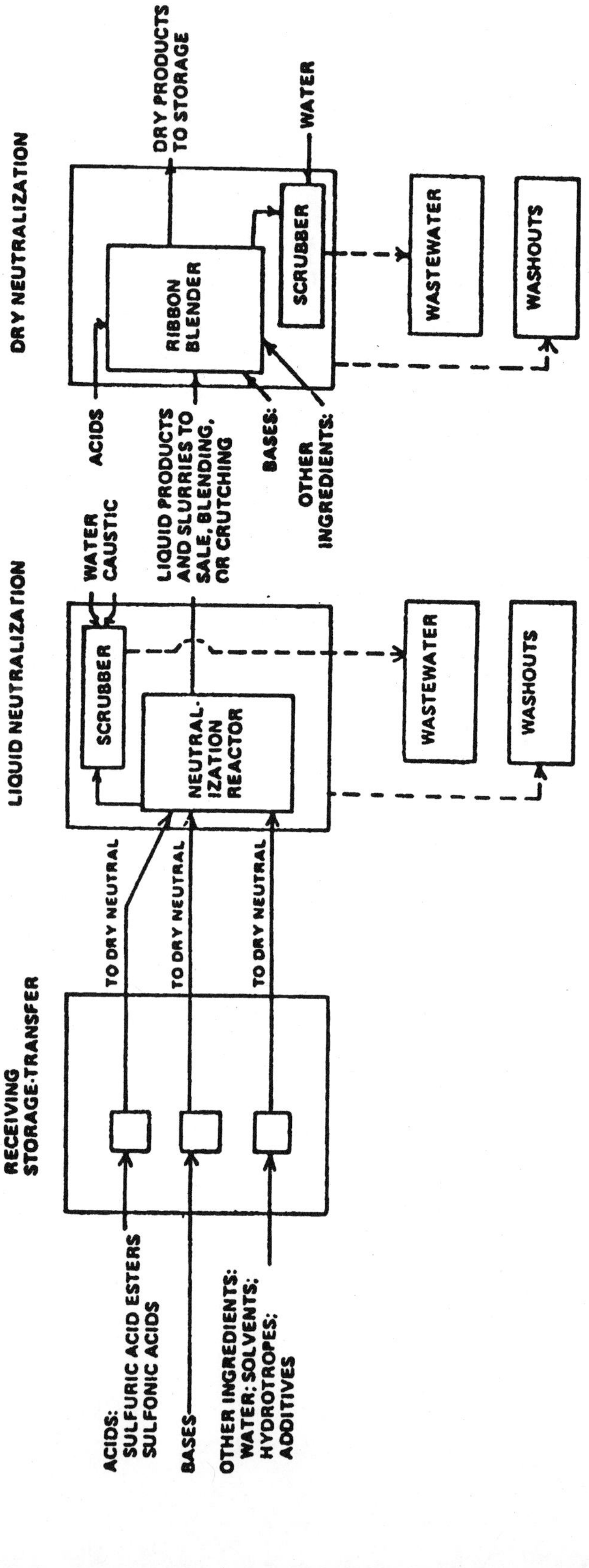

Figure 14 Neutralization of sulfuric acid esters and sulfonic acids (N) (from Ref. 13).

meet the falling detergent. The design preparation of this step will determine the detergent particle's shape, size, and density, which in turn determine its solubility rate in the washing process.

The air coming from the tower will be carrying dust particles that must be scrubbed, thus generating a wastewater stream. The spray towers are periodically shut down and cleaned. The tower walls are scraped and thoroughly washed down. The final step is mandatory because the manufacturers must be careful to avoid contamination to the subsequent formulation.

Wastewater streams are rather numerous, as seen in the flow diagram of Figure 15. They include many washouts of equipment from the crutchers to the spray tower itself. One wastewater flow that has high loadings is that of the air scrubber, which cleans and cools the hot gases exiting from this tower. All the plants recycle some of the wastewater generated, while some of the plants recycle all the flow generated. Owing to increasingly stringent air quality requirements, it can be expected that fewer plants will be able to maintain a complete recycle system of all water flows in the spray tower area. After the powder comes from the spray tower, it is further blended and then packaged.

Liquid Detergents (P)

Detergent actives are pumped into mixing tanks where they are blended with numerous ingredients, ranging from perfumes to dyes. A process flow diagram is shown in Figure 16. From here, the fully formulated liquid detergent is run down to the filling line for filling, capping, labeling, and so on. Whenever the filling line is to change to a different product, the filling system must be thoroughly cleaned out to avoid cross contamination.

Dry Detergent Blending (Q)

Fully dried surfactant materials are blended with additives in dry mixers. Normal operation will see many succeeding batches of detergent mixed in the same equipment without anything but dry cleaning. However, when a change in formulation occurs, the equipment must be completely washed down and a modest amount of wastewater is generated. A process flow diagram is shown in Figure 17.

Drum-Dried Detergent (R)

This process is one method of converting liquid slurry to a powder and should be essentially free of the generation of wastewater discharge other than occasional washdown. A process flow diagram is shown in Figure 18.

Detergent Bars and Cakes (S)

Detergent bars are either 100% synthetic detergent or a blend of detergent and soap. They are blended in essentially the same manner as conventional soap. Fairly frequent cleanups generate a wastewater stream. A process flow diagram is shown in Figure 19.

7.4.7 Wastewater Characteristics

Wastewaters from the manufacturing, processing, and formulation of organic chemicals such as soaps and detergents cannot be exactly characterized. The wastewater streams are usually expected to contain trace or larger concentrations of all raw materials used in the plant, all intermediate compounds produced during manufacture, all final products, coproducts, and byproducts, and the auxiliary or processing chemicals employed. It is desirable, from the

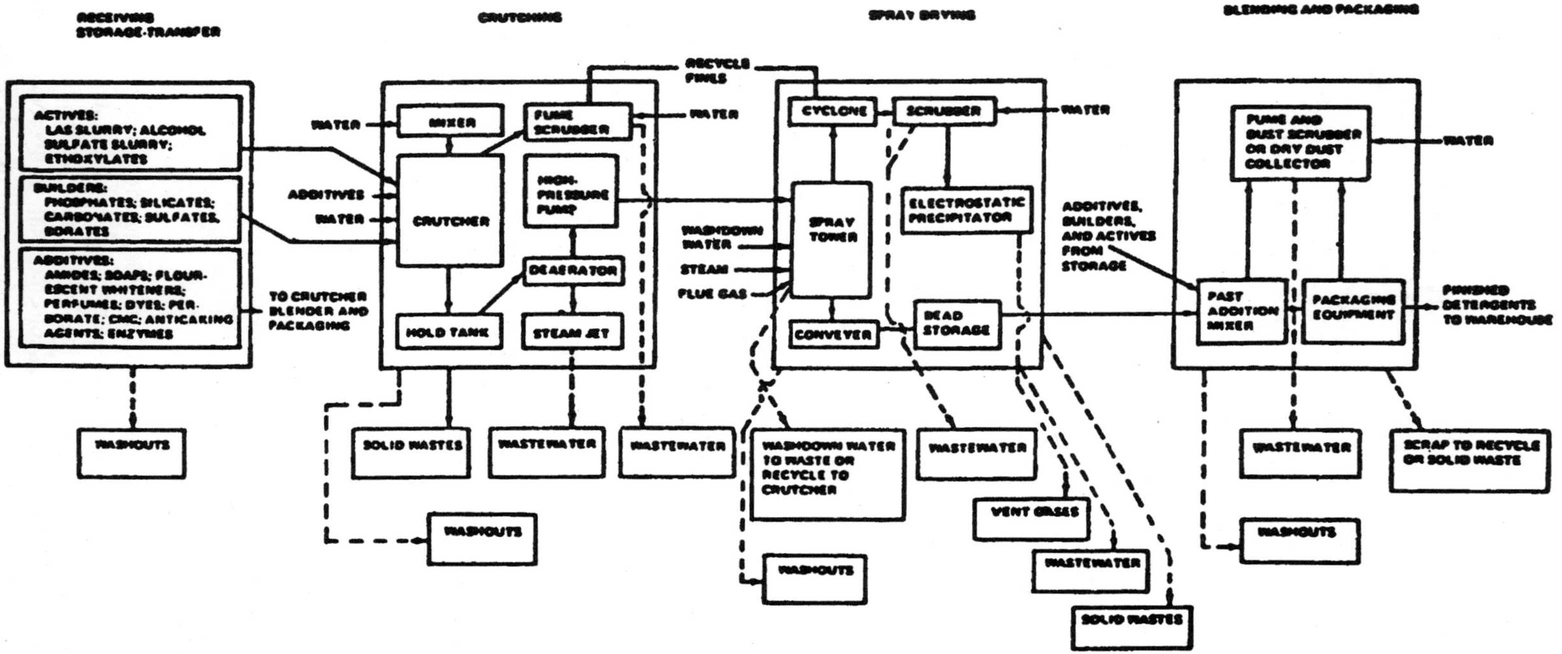

Figure 15 Spray-dried detergent production (O) (from Ref. 13).

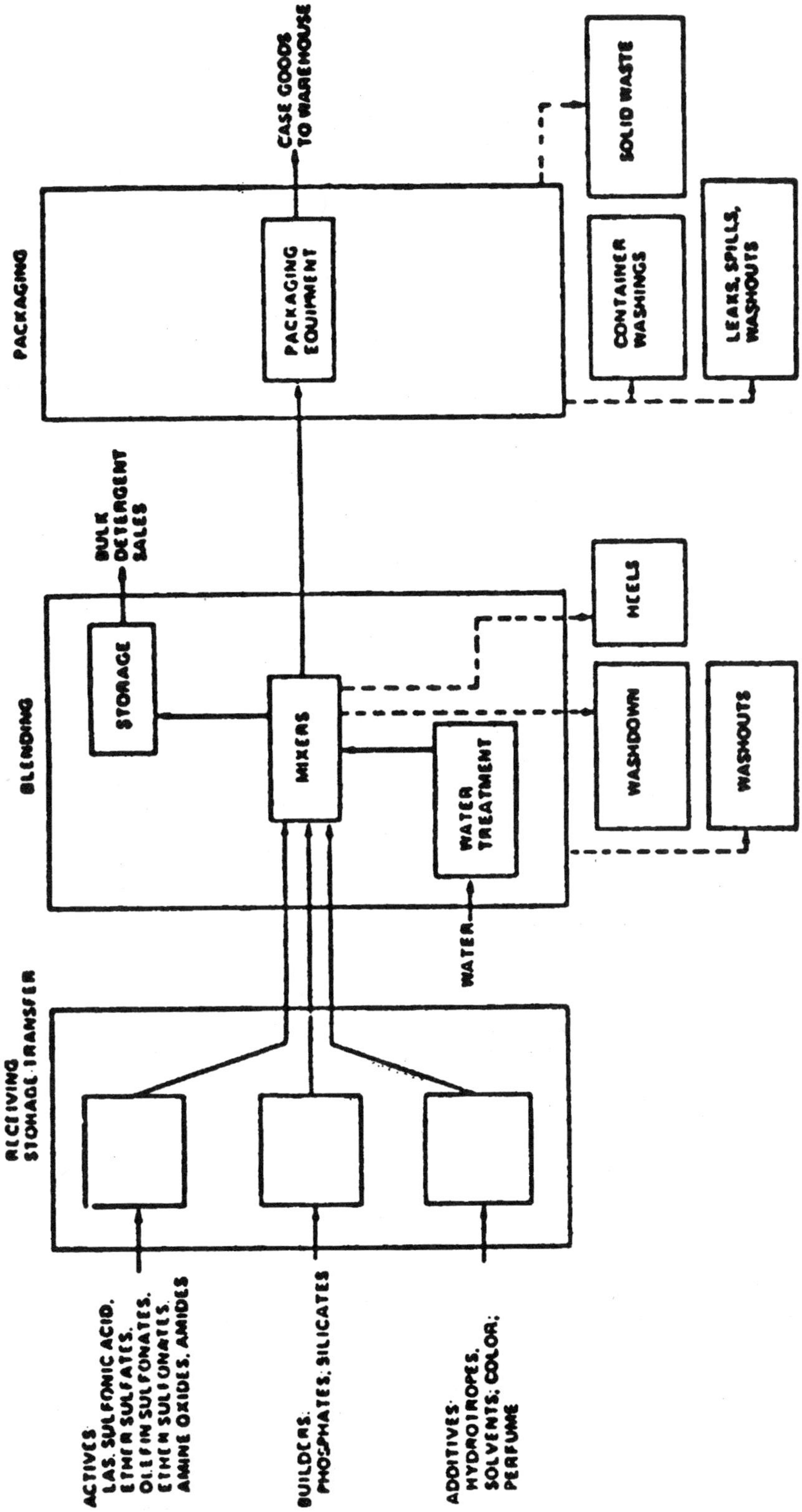

Figure 16 Liquid detergent manufacture (P) (from Ref. 13).

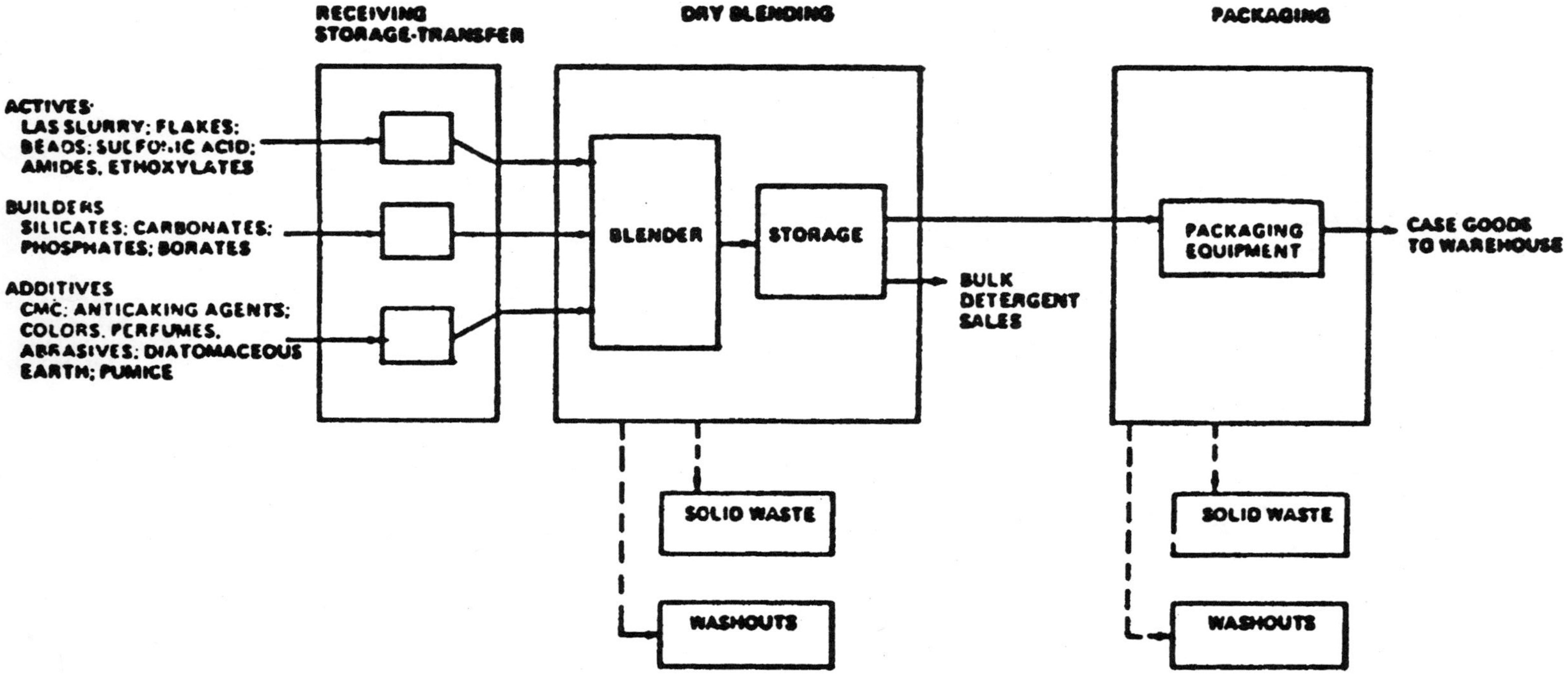

Figure 17 Detergent manufacture by dry blending (Q) (from Ref. 13).

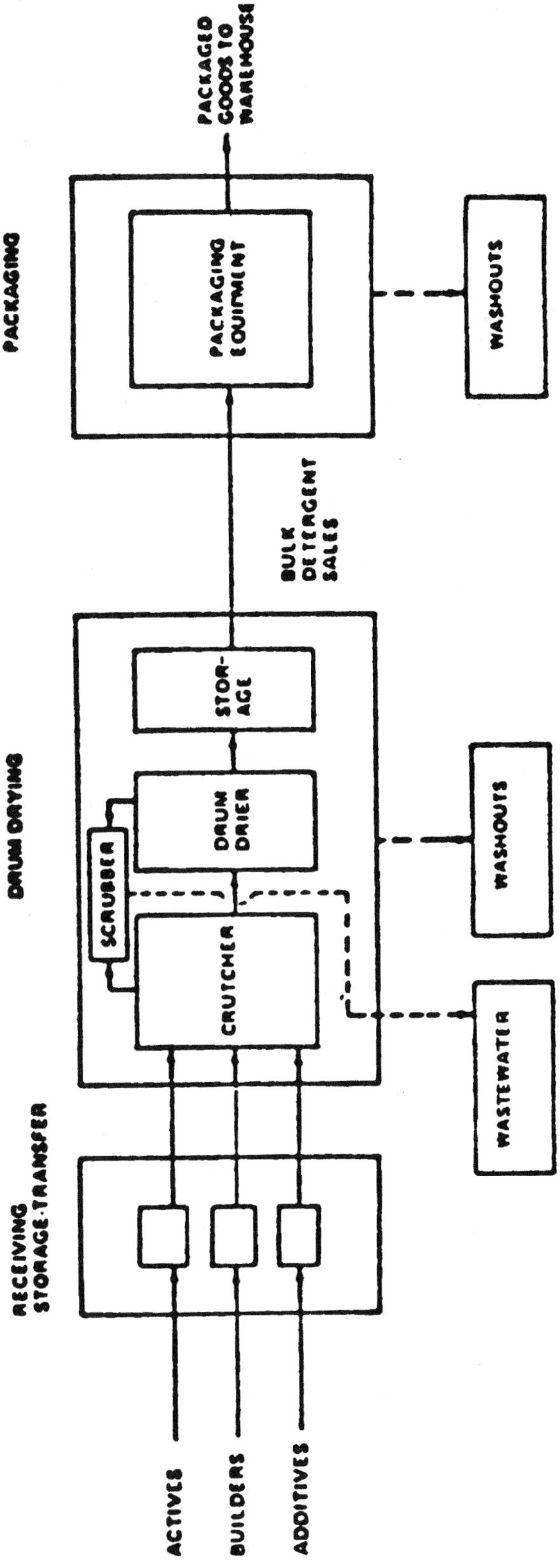

Figure 18 Drum-dried detergent manufacture (R) (from Ref. 13).

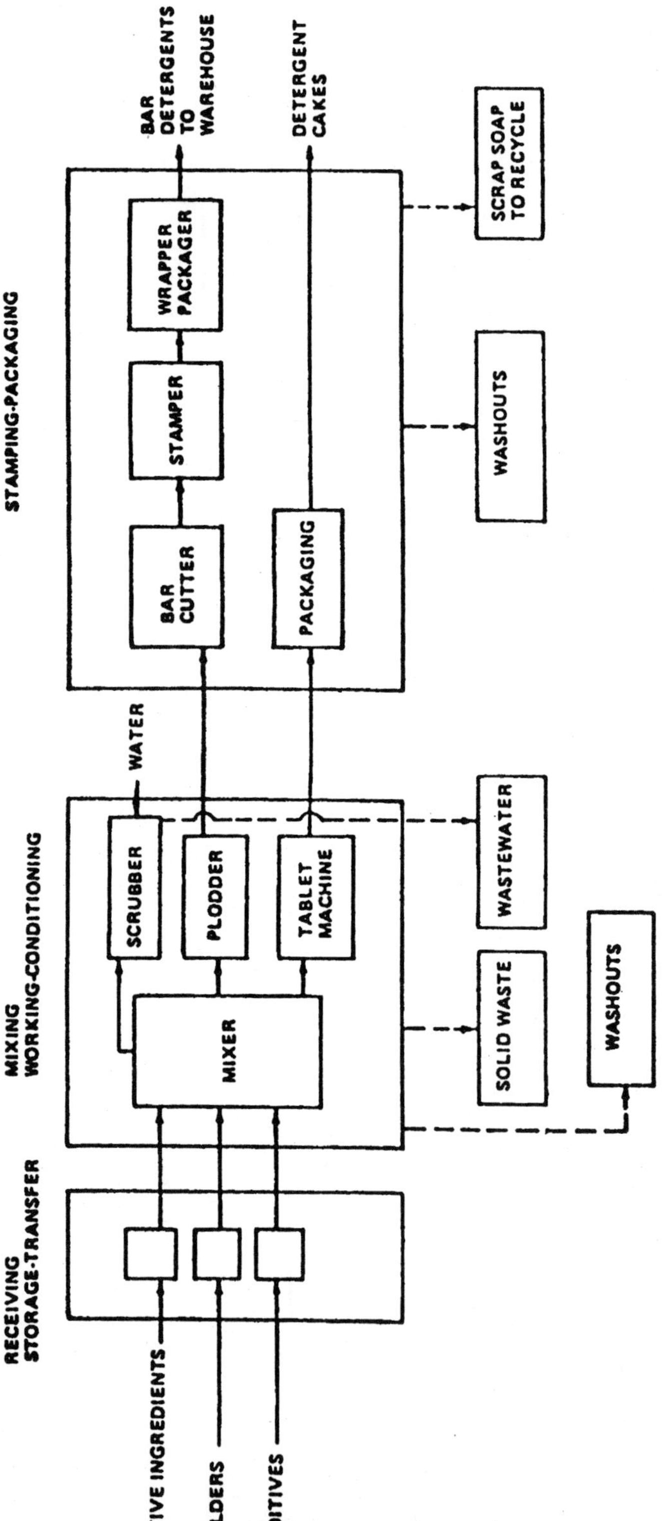

Figure 19 Detergent bar and cake manufacture (S) (from Ref. 13).

viewpoint of economics, that these substances not be lost, but some losses and spills appear unavoidable and some intentional dumping does take place during housecleaning and vessel emptying and preparation operations.

According to a study by the USEPA [12], which presents estimates of industrial wastewater generation as well as related pollution parameter concentrations, the wastewater volume discharged from soap and detergent manufacturing facilities per unit of production ranges from 0.3 to 2.8 gal/lb (2.5–23.4 L/kg) of product. The reported ranges of concentration (mg/L) for BOD, suspended solids, COD, and grease were 500–1200, 400–2100, 400–1800, and about 300, respectively. These data were based on a study of the literature and the field experience of governmental and private organizations. The values represent plant operating experience for several plants consisting of 24 hour composite samples taken at frequent intervals. The ranges for flow and other parameters generally represent variations in the level of plant technology or variations in flow and quality parameters from different subprocesses. In particular, the more advanced and modern the level of production technology, the smaller the volume of wastewater discharged per unit of product. The large variability (up to one order of magnitude) in the ranges is generally due to the heterogeneity of products and processes in the soap and detergent industry.

The federal guidelines [13] for state and local pretreatment programs reported the raw wastewater characteristics (Table 1) in mg/L concentration and the flows and water quality parameters (Table 2) based on the production or 1 ton of product manufactured for the subcategories of the industry. Most soap and detergent manufacturing plants contain two or more of the subcategories shown in Table 3, and their wastewaters are a composite of these individual unit processes.

7.5 U.S. CODE OF FEDERAL REGULATIONS

The information presented in this section has been taken from the U.S. Code of Federal Regulations (40 CFR), containing documents related to the protection of the environment [14], in particular, the regulations contained in Part 417, Soap and Detergent Manufacturing Point Source Category, pertaining to effluent limitations guidelines and pretreatment or performance standards for each of the 19 subcategories shown in Table 3.

The effluent guideline regulations and standards of 40 CFR, Part 417, were promulgated on February 11, 1975. According to the most recent notice in the Federal Register [15] regarding industrial categories and regulations, no review is under way or planned and no revision is proposed for the soap and detergent industry. The effluent guidelines and standards applicable to this industrial category include: (a) the best practicable control technology currently available (BPT); (b) the best available technology economically achievable (BAT); (c) pretreatment standards for existing sources (PSES); (d) standards of performance for new sources (NSPS); and (e) pretreatment standards for new sources (PSNS).

For all 19 subcategories of the soap and detergent manufacture industry, there are no pretreatment standards establishing the quantity and quality of pollutants or pollutant properties that may be discharged to a publicly owned treatment works (POTW) by an existing or new point source. If the major contributing industry is an existing point source discharging pollutants to navigable waters, it will be subject to Section 301 of the Federal Water Pollution Control Act and to the provisions of 40 CFR, Part 128. However, practically all the soap and detergent manufacturing plants in the United States discharge their wastewaters into municipal sewer systems. The effluent limitations guidelines for certain subcategories regarding BPT, BAT, and NSPS are presented in Tables 4–10.

Table 1 Soap and Detergent Industry Raw Wastewater Characteristics

Parameter	Batch kettle (A)	Fat splitting (B)	Fatty acid neutralization (C)	Glycerine concentration (D)	Glycerine distillation (E)	Flakes and powders (F)	Bar soap (G)	Liquid soap (H)
BOD (mg/L)	3600[a]	60–3600[a]	400				1600–3000[a]	
COD (mg/L)	4267[a]	115–6000[a]	1000					
TSS (mg/L)	1600–6420	115–6000	775					
Oil and grease (mg/L)	250[a]	13–760[a]	200[a]					
pH	5–13.5	High	High	Neutral	Neutral	Neutral	Neutral	Neutral
Chlorides (mg/L)	20–47 m[a]							
Zinc (mg/L)		Present						
Nickel (mg/L)		Present						

Parameter	Oleum sulfation and sulfonation (I)	Air sulfation and sulfonation (J)	SO_3 solvent and vacuum (K)	Sulfamic acid sulfation (L)	Chloro-sulfonic (M)	Neutral sulfuric (N)	Spray-dried (O)	Liquid detergent (P)	Dry blend (Q)	Drum-dried (R)	Bars and cakes (S)
BOD (mg/L)	75–2000[a]	380–520				8.5–6 m[a]	48–19 m[a]	65–3400[a]	Neg.		
COD (mg/L)	220–6000[a]	920–1589[a]				245–21 m[a]	150–60 m[a]	640–11 m[a]			
TSS (mg/L)	100–3000										
Oil and grease (mg/L)	100–3000[a]										
pH	1–2[a]	2–7[a]	Low	Low	Low	Low					
Surfactant (mg/L)	250–7000							60–2 m			
Boron (mg/L)	Present	Present	Present	Present	Present	Present	Present	Present	Present	Present	Present

[a]In high levels these parameters may be inhibitory to biological systems; m = thousands; BOD, biochemical oxygen demand; COD, chemical oxygen demand; TSS, total suspended solids.
Source: Ref. 10.

Table 2 Raw Wastewater Characteristics Based on Production

Parameter	Batch kettle (A)	Fat splitting (B)	Fatty acid neutralization (C)	Glycerine concentration (D)	Glycerine distillation (E)	Flakes and powders (F)	Bar soap (G)	Liquid soap (H)
Flow range (L/kkg)[a]	623/2500	3.3M/1924M	258			Neg.		Neg.
Flow type	B	B	B	B	B	B	B	B
BOD (kg/kkg)[b]	6	12	0.1	15	5	0.1	3.4	0.1
COD (kg/kkg)	10	22	0.25	30	10	0.3	5.7	0.3
TSS (kg/kkg)	4	22	0.2	2	2	0.1	5.8	0.1
Oil and grease (kg/kkg)	0.9	2.5	0.05	1	1	0.1	0.4	0.1

Parameter	Oleum sulfation and sulfonation (I)	SO_3 sulfation and sulfonation (J)	SO_3 solvent and vacuum sulfonation (K)	Sulfamic acid sulfation (L)	Chloro-sulfonic (M)	Neutral sulfuric acid esters (N)	Spray-dried (O)	Liquid detergent (P)	Dry blend (Q)	Drum-dried (R)	Bars and cakes (S)
Flow range (L/kkg)[a]	100/2740	249				10/4170	41/2084	625/6250			
Flow type	C	C	B	B	B	B&C	B	B	B	B	B
BOD (kg/kkg)[b]	0.2	3	3	3	3	0.10	0.1–0.8	2–5	0.1	0.1	7
COD (kg/kkg)	0.6	9	9	9	9	0.3	0.3–25	4–7	0.5	0.3	22
TSS (kg/kkg)	0.3	0.3	0.3	0.3	0.3	0.3	0.1–1.0		0.1	0.1	2
Oil and grease (kg/kkg)	0.3	0.5	0.5	0.5	0.5	0.1	Nil–0.3			0.1	0.2
Chloride (kg/kkg)					5						
Surfactant (kg/kkg)	0.7	3	3	3	3	0.2	0.2–1.5	1.3–3.3		0.1	5

[a] L/kkg, L/1000 kg product produced (lower limit/upper limit).
[b] kg/kkg, kg/1000 kg product produced.
B = Batch; C = Continuous; Neg. = Negligible; M = Thousand.
Source: Ref. 13.

Table 3 Soap and Detergent Categorization *Source*: Ref. 10.

Category	Subcategory	Code
Soap manufacture	Batch kettle and continuous	A
	Fatty acid manufacture by fat splitting	B
	Soap from fatty acid neutralization	C
	Glycerine recovery	
	Glycerine concentration	D
	Glycerine distillation	E
	Soap flakes and powders	F
	Bar soaps	G
	Liquid soap	H
Detergent manufacture	Oleum sulfonation and sulfation (batch and continuous)	I
	Air–SO_3 sulfation and sulfonation (batch and continuous)	J
	SO_3 solvent and vacuum sulfonation	K
	Sulfamic acid sulfation	L
	Chlorosulfonic acid sulfation	M
	Neutralization of sulfuric acid esters and sulfonic acids	N
	Spray-dried detergents	O
	Liquid detergent manufacture	P
	Detergent manufacture by dry blending	Q
	Drum-dried detergents	R
	Detergent bars and cakes	S

Source: Ref. 10.

Table 4 Effluent Limitations for Subpart A, Batch Kettle

	Effluent limitations [metric units (kg/1000 kg of anhydrous product)]	
Effluent characteristic	Maximum for any 1 day	Average of daily values for 30 consecutive days shall not exceed
(a) BPT		
BOD_5	1.80	0.60
COD	4.50	1.50
TSS	1.20	0.40
Oil and grease	0.30	0.10
pH	[a]	a
(b) BAT and NSPS		
BOD_5	0.80	0.40
COD	2.10	1.05
TSS	0.80	0.40
Oil and grease	0.10	0.05
pH	[a]	[a]

[a] Within the range 6.0–9.0.

BAT, best available technology economically achievable; NSPS, standards of performance for new sources.

Source: Ref. 14.

Table 5 Effluent Limitations for Subpart C, Soap by Fatty Acid

Effluent characteristic	Effluent limitations [metric units (kg/1000 kg of anhydrous product)] Maximum for any 1 day	Average of daily values for 30 consecutive days shall not exceed
(a) BPT		
BOD_5	0.03	0.01
COD	0.15	0.05
TSS	0.06	0.02
Oil and grease	0.03	0.01
pH	a	a
(b) BAT		
BOD_5	0.02	0.01
COD	0.10	0.05
TSS	0.04	0.02
Oil and grease	0.02	0.01
pH	a	a
(c) NSPS		
BOD_5	0.02	0.01
COD	0.10	0.05
TSS	0.04	0.02
Oil and grease	0.02	0.01
pH	a	a

[a] Within the range 6.0–9.0.
Source: Ref. 14.

7.6 WASTEWATER CONTROL AND TREATMENT

The sources and characteristics of wastewater streams from the various subcategories in soap and detergent manufacturing, as well as some of the possibilities for recycling and treatment, have been discussed in Section 7.4. The pollution control and treatment methods and unit processes used are discussed in more detail in the following sections. The details of the process design criteria for these unit treatment processes can be found in any design handbooks.

7.6.1 In-Plant Control and Recycle

Significant in-plant control of both waste quantity and quality is possible, particularly in the soap manufacturing subcategories where maximum flows may be 100 times the minimum. Considerably less in-plant water conservation and recycle are possible in the detergent industry, where flows per unit of product are smaller.

The largest in-plant modification that can be made is the changing or replacement of the barometric condensers (subcategories A, B, D, and E). The wastewater quantity discharged from these processes can be significantly reduced by recycling the barometric cooling water through fat skimmers, from which valuable fats and oils can be recovered, and then through the cooling towers. The only waste with this type of cooling would be the continuous small blowdown from

Table 6 Effluent Limitations for Subpart D, Glycerine Concentration

	Effluent limitations [metric units (kg/1000 kg of anhydrous product)]	
Effluent characteristic	Maximum for any 1 day	Average of daily values for 30 consecutive days shall not exceed
(a) BPT		
BOD_5	4.50	1.50
COD	13.50	4.50
TSS	0.60	0.20
Oil and grease	0.30	0.10
pH	a	a
(b) BAT		
BOD_5	0.80	0.40
COD	2.40	1.20
TSS	0.20	0.10
Oil and grease	0.08	0.04
pH	a	a
(c) NSPS		
BOD_5	0.80	0.40
COD	2.40	1.20
TSS	0.20	0.10
Oil and grease	0.08	0.04
pH	a	a

[a] Within the range 6.0–9.0.
Source: Ref. 14.

the skimmer. Replacement with surface condensers has been used in several plants to reduce both the waste flow and quantity of organics wasted.

Significant reduction of water usage is possible in the manufacture of liquid detergents (P) by the installation of water recycle piping and tankage and by the use of air rather than water to blowdown filling lines. In the production of bar soaps (G), the volume of discharge and the level of contamination can be reduced materially by installation of an atmospheric flash evaporator ahead of the vacuum drier. Finally, pollutant carryover from distillation columns such as those used in glycerine concentration (D) or fatty acid separation (B) can be reduced by the use of two additional special trays.

In another document [37] presenting techniques adopted by the French for pollution prevention, a new process of detergent manufacturing effluent recycle is described. As shown in Figure 20, the washout effluents from reaction and/or mixing vessels and washwater leaks from the paste preparation and pulverization pump operations are collected and recycled for use in the paste preparation process. The claim has been that pollution generation at such a plant is significantly reduced and, although the savings on water and raw materials are small, the capital and operating costs are less than those for building a wastewater treatment facility.

Besselievre [2] has reported in a review of water reuse and recycling by the industry that soap and detergent manufacturing facilities have shown an average ratio of reused and recycled water to total wastewater effluent of about 2:1. That is, over two-thirds of the generated wastewater stream in an average plant has been reused and recycled. Of this volume, about 66% has been used as cooling water and the remaining 34% for the process or other purposes.

Table 7 Effluent Limitations for Subpart G, Bar Soaps

	Effluent limitations [metric units (kg/1000 kg of anhydrous product)]	
Effluent characteristic	Maximum for any 1 day	Average of daily values for 30 consecutive days shall not exceed
(a) BPT		
BOD_5	1.02	0.34
COD	2.55	0.85
TSS	1.74	0.58
Oil and grease	0.12	0.04
pH	a	a
(b) BAT		
BOD_5	0.40	0.20
COD	1.20	0.60
TSS	0.68	0.34
Oil and grease	0.06	0.03
pH	a	a
(c) NSPS		
BOD_5	0.40	0.20
COD	1.20	0.60
TSS	0.68	0.34
Oil and grease	0.06	0.03
pH	a	a

[a]Within the range 6.0–9.0.
Source: Ref. 14.

7.6.2 Wastewater Treatment Methods

The soap and detergent manufacturing industry makes routine use of various physicochemical and biological pretreatment methods to control the quality of its discharges. A survey of these treatment processes is presented in Table 11 [13], which also shows the usual removal efficiencies of each unit process on the various pollutants of concern. According to Nemerow [38] and Wang and Krofta [39], the origin of major wastes is in washing and purifying soaps and detergents and the resulting major pollutants are high BOD and certain soaps (oily and greasy, alkali, and high-temperature wastes), which are removed primarily through air flotation and skimming, and precipitation with the use of $CaCl_2$ as a coagulant.

Figure 21 presents a composite flow diagram describing a complete treatment train of the unit processes that may be used in a large soap and detergent manufacturing plant to treat its wastes. As a minimum requirement, flow equalization to smooth out peak discharges should be utilized even at a production facility that has a small-volume batch operation. Larger plants with integrated product lines may require additional treatment of their wastewaters for both suspended solids and organic materials' reduction. Coagulation and sedimentation are used by the industry for removing the greater portion of the large solid particles in its waste. On the other hand, sand or mixed-bed filters used after biological treatment can be utilized to eliminate fine particles. One of the biological treatment processes or, alternatively, granular or powdered activated carbon is the usual method employed for the removal of particulate or soluble organics from the waste streams. Finally, as a tertiary step for removing particular ionized pollutants or

Table 8 Effluent Limitations for Subpart H, Liquid Soaps

Effluent characteristic	Effluent limitations [metric units (kg/1000 kg of anhydrous product)]: Maximum for any 1 day	Effluent limitations [metric units (kg/1000 kg of anhydrous product)]: Average of daily values for 30 consecutive days shall not exceed
(a) BPT		
BOD_5	0.03	0.01
COD	0.15	0.05
TSS	0.03	0.01
Oil and grease	0.03	0.01
pH	[a]	[a]
(b) BAT		
BOD_5	0.02	0.01
COD	0.10	0.05
TSS	0.02	0.01
Oil and grease	0.02	0.01
pH	[a]	[a]
(c) NSPS		
BOD_5	0.02	0.01
COD	0.10	0.05
TSS	0.02	0.01
Oil and grease	0.02	0.01
pH	[a]	[a]

[a] Within the range 6.0–9.0.
Source: Ref. 14.

total dissolved solids (TDS), a few manufacturing facilities have employed either ion exchange or the reverse osmosis process.

Flotation or Foam Separation

One of the principal applications of vacuum and pressure (air) flotation is in commercial installations with colloidal wastes from soap and detergent factories [20,40–42]. Wastewaters from soap production are collected in traps on skimming tanks, with subsequent recovery floating of fatty acids.

Foam separation or fractionation [40,41,43–45] can be used to extra advantage: not only do surfactants congregate at the air/liquid interfaces, but other colloidal materials and ionized compounds that form a complex with the surfactants tend to also be concentrated by this method. An incidental, but often important, advantage of air flotation processes is the aerobic condition developed, which tends to stabilize the sludge and skimmings so that they are less likely to turn septic. However, disposal means for the foamate can be a serious problem in the use of this procedure [46]. It has been reported that foam separation has been able to remove 70–80% of synthetic detergents, at a wide range of costs [2]. Gibbs [17] reported the successful use of fine bubble flotation and 40 mm detention in treating soap manufacture wastes, where the skimmed sludge was periodically returned to the soap factory for reprocessing. According to Wang [47–49], the dissolved air flotation process is both technically and economically feasible for the removal of detergents and soaps (i.e., surfactants) from water.

Table 9 Effluent Limitations for Subpart I, Oleum Sulfonation

Effluent characteristic	Effluent limitations [metric units (kg/1000 kg of anhydrous product)]	
	Maximum for any 1 day	Average of daily values for 30 consecutive days shall not exceed
(a) BPT		
BOD_5	0.09	0.02
COD	0.40	0.09
TSS	0.15	0.03
Surfactants	0.15	0.03
Oil and grease	0.25	0.07
pH	[a]	[a]
(b) BAT		
BOD_5	0.07	0.02
COD	0.27	0.09
TSS	0.09	0.03
Surfactants	0.09	0.03
Oil and grease	0.21	0.07
pH	[a]	[a]
(c) NSPS		
BOD_5	0.03	0.01
COD	0.09	0.03
TSS	0.06	0.02
Surfactants	0.03	0.01
Oil and grease	0.12	0.04
pH	[a]	[a]

[a] Within the range 6.0–9.0.
Source: Ref. 14.

Activated Carbon Adsorption

Colloidal and soluble organic materials can be removed from solution through adsorption onto granular or powdered activated carbon, such as the particularly troublesome hard surfactants. Refractory substances resistant to biodegradation, such as ABS, are difficult or impossible to remove by conventional biological treatment, and so they are frequently removed by activated carbon adsorption [11]. The activated carbon application is made either in mixed-batch contact tanks with subsequent settling or filtration, or in flow-through GAC columns or contact beds. Obviously, because it is an expensive process, adsorption is being used as a polishing step of pretreated waste effluents. Nevertheless, according to Koziorowski and Kucharski [22] much better results of surfactant removal have been achieved with adsorption than coagulation/settling. Wang [50–52] used both powdered activated carbon (PAC) and coagulation/settling/DAF for successful removal of surfactants.

Coagulation/Flocculation/Settling/Flotation

As mentioned previously in Section 7.2.4, the coagulation/flocculation process was found to be affected by the presence of surfactants in the raw water or wastewater. Such interference was observed for both alum and ferric sulfate coagulant, but the use of certain organic polymer

Table 10 Effluent Limitations for Subpart P, Liquid Detergents

	Effluent limitations [metric units (kg/1000 kg of anhydrous product)]	
Effluent characteristic	Maximum for any 1 day	Average of daily values for 30 consecutive days shall not exceed
(a) BPT[a]		
BOD_5	0.60	0.20
COD	1.80	0.60
TSS	0.015	0.005
Surfactants	0.39	0.13
Oil and grease	0.015	0.005
pH	[c]	[c]
(b) BPT[b]		
BOD_5	0.05	
COD	0.15	
TSS	0.002	
Surfactants	0.04	
Oil and grease	0.002	
pH	[c]	
(c) BAT[a]		
BOD_5	0.10	0.05
COD	0.44	0.22
TSS	0.01	0.005
Surfactants	0.10	0.05
Oil and grease	0.01	0.005
pH	[c]	[c]
(d) BAT[b]		
BOD_5	0.02	
COD	0.07	
TSS	0.002	
Surfactants	0.02	
Oil and grease	0.002	
pH	[c]	
(e) NSPS[a]		
BOD_5	0.10	0.05
COD	0.44	0.22
TSS	0.01	0.005
Surfactants	0.10	0.05
Oil and grease	0.01	0.005
pH	[c]	[c]
(f) NSPS[b]		
BOD_5	0.02	
COD	0.07	
TSS	0.002	
Surfactants	0.02	
Oil and grease	0.002	
pH	[c]	

[a]For normal liquid detergent operations.
[b]For fast turnaround operation of automated fill lines.
[c]Within the range 6.0–9.0.
Source: Ref. 14.

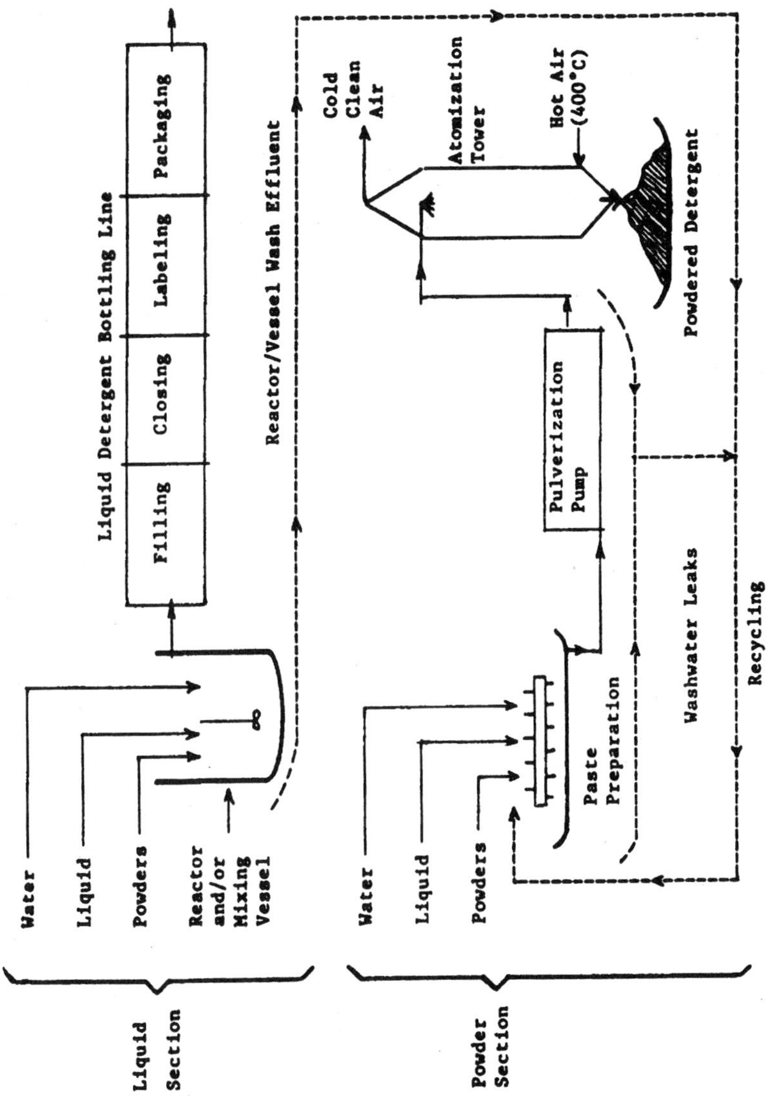

Figure 20 Process modification for wastewater recycling in detergent manufacture (from Ref. 37).

Table 11 Treatment Methods in the Soap and Detergent Industry

Pollutant and method	Efficiency (percentage of pollutant removed)
Oil and grease	
API-type separation	Up to 90% of free oils and greases. Variable on emulsified oil.
Carbon adsorption	Up to 95% of both free and emulsified oils.
Flotation	Without the addition of solid phase, alum, or iron, 70–80% of both free and emulsified oil. With the addition of chemicals, 90%.
Mixed-media filtration	Up to 95% of free oils. Efficiency in removing emulsified oils unknown.
Coagulation/sedimentation with iron, alum, or solid phase (bentonite, etc.)	Up to 95% of free oil. Up to 90% of emulsified oil.
Suspended solids	
Mixed-media filtration	70–80%
Coagulation/sedimentation	50–80%
BOD and COD	
Bioconversions (with final clarifier)	60–95% or more
Carbon adsorption	Up to 90%
Residual suspended solids	
Sand or mixed-media filtration	50–95%
Dissolved solids	
Ion exchange or reverse osmosis	Up to 90%

Source: Ref. 13.

flocculants was shown to overcome this problem. However, chemical coagulation and flocculation for settling may not prove to be very efficient for such wastewaters. Wastes containing emulsified oils can be clarified by coagulation, if the emulsion is broken through the addition of salts such as $CaCl_2$, the coagulant of choice for soap and detergent manufacture wastewaters [11]. Also, lime or other calcium chemicals have been used in the treatment of such wastes whose soapy constituents are precipitated as insoluble calcium soaps of fairly satisfactory flocculating ("hardness" scales) and settling properties. Treatment with $CaCl_2$ can be used to remove practically all grease and suspended solids and a major part of the suspended BOD [19]. Using carbon dioxide (carbonation) as an auxiliary precipitant reduces the amount of calcium chloride required and improves treatment efficiency. The sludge from $CaCl_2$ treatment can be removed either by sedimentation or by dissolved air flotation [39,53–56]. For monitoring and control of chemical coagulation, flocculation, sedimentation and flotation processes, many analytical procedures and testing procedures have been developed [57–64].

Ion Exchange and Exclusion

The ion-exchange process has been used effectively in the field of waste disposal. The use of continuous ion exchange and resin regeneration systems has further improved the economic feasibility of the applications over the fixed-bed systems. One of the reported [1] special

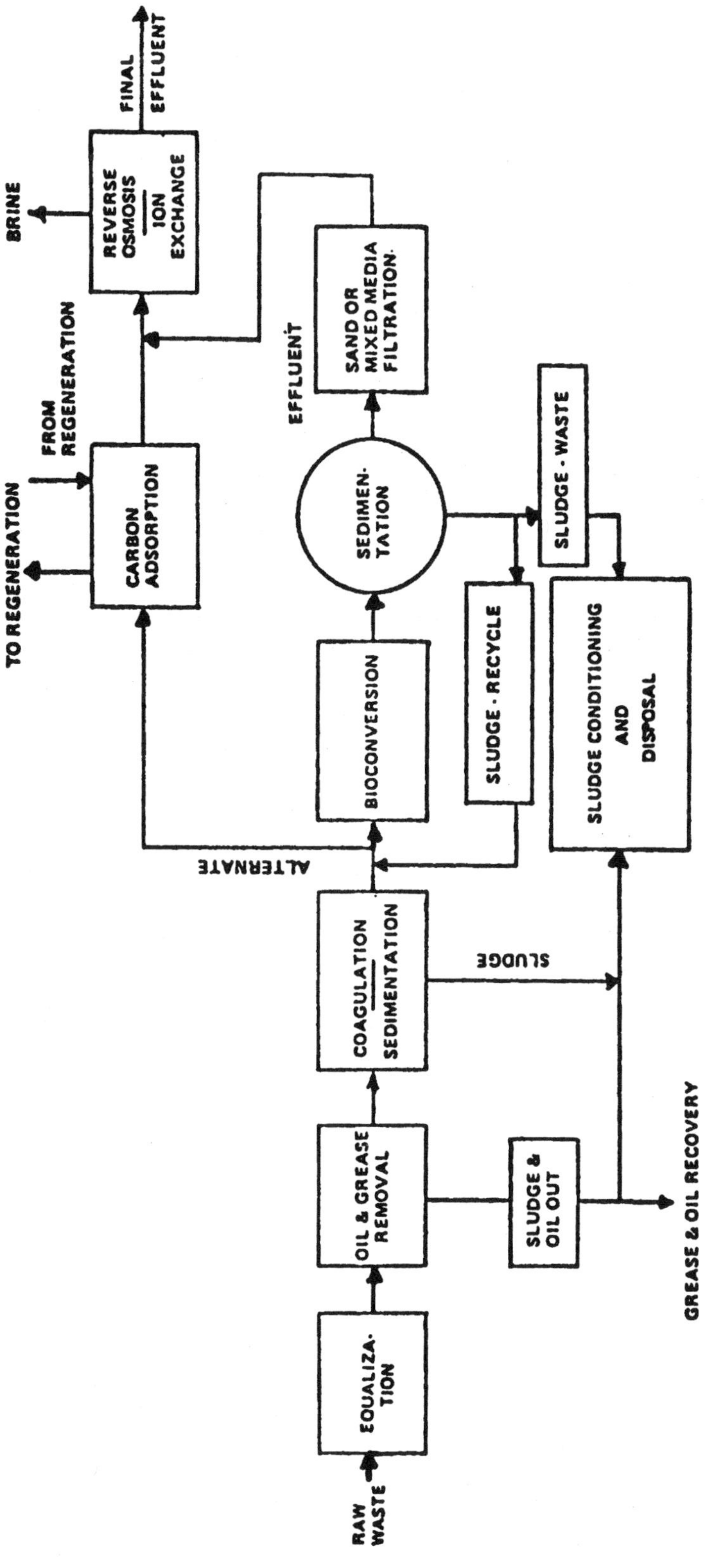

Figure 21 Composite flowsheet of waste treatment in soap and detergent industry (from Ref. 13).

applications of the ion-exchange resins has been the removal of ABS by the use of a Type II porous anion exchanger that is a strong base and depends on a chloride cycle. This resin system is regenerated by removing a great part of the ABS absorbed on the resin beads with the help of a mixture of hydrocarbons (HC) and acetone. Other organic pollutants can also be removed by ion-exchange resins, and the main problem is whether the organic material can be eluted from the resin using normal regeneration or whether it is economically advisable to simply discard the used resin. Wang and Wood [65] and Wang [51,52,66] successfully used the ion-exchange process for the removal of cationic surfactant from water.

The separation of ionic from nonionic substances can be effected by the use of ion exclusion [46]. Ion exchange can be used to purify glycerine for the final product of chemically pure glycerine and reduce losses to waste, but the concentration of dissolved ionizable solids or salts (ash) largely impacts on the overall operating costs. Economically, when the crude or sweet water contains under 1.5% ash, straight ion exchange using a cation and anion mixed bed can be used, whereas for higher percentages of dissolved solids, it is economically feasible to follow the ion exchange with an ion-exclusion system. For instance, waste streams containing 0.2–0.5% ash and 3–5% glycerine may be economically treated by straight ion exchange, while waste streams containing 5–10% ash and 3–5% glycerine have to be treated by the combined ion-exchange and ion-exclusion processes.

Biological Treatment

Regarding biological destruction, as mentioned previously, surfactants are known to cause a great deal of trouble due to foaming and toxicity [103] in municipal treatment plants. The behavior of these substances depends on their type [22], that is, anionic and nonionic detergents increase the amount of activated sludge, whereas cationic detergents reduce it, and also the various compounds decompose to a different degree. The activated sludge process is feasible for the treatment of soap and detergent industry wastes but, in general, not as satisfactory as trickling filters. The turbulence in the aeration tank induces frothing to occur, and also the presence of soaps and detergents reduces the absorption efficiency from air bubbles to liquid aeration by increasing the resistance of the liquid film.

On the other hand, detergent production wastewaters have been treated with appreciable success on fixed-film process units such as trickling filters [2]. Also, processes such as lagoons, oxidation or stabilization ponds, and aerated lagoons have all been used successfully in treating soap and detergent manufacturing wastewaters. Finally, Vath [102] demonstrated that both linear anionic and nonionic ethoxylated surfactants underwent degradation, as shown by a loss of surfactant properties, under anaerobic treatment.

Wang *et al.* [42,67,68] have developed innovative biological process and sequencing batch reactors (SBR) specifically for removal of volatile organic compounds (VOCs) and surfactants. Related analytical procedures [57–64,71–91] available for process monitoring and control are available in the literature.

7.7 CASE STUDIES OF TREATMENT FACILITIES

Soap and detergent manufacture and formulation plants are situated in many areas in the United States and other countries. At most, if not all of these locations, the wastewaters from production and cleanup activities are discharged to municipal sewer systems and treated together with domestic, commercial, institutional, and other industrial wastewaters. Following the precipitous reduction in production and use of "hard" surfactants such as ABS, no discernible problems in operation and treatment efficiency due to the combined treatment of surfactant manufacture wastes at these municipal sewage treatment plants (most of which employ biological processes) have been reported. In fact, there is a significantly larger portion of surfactants and related compounds being discharged to the municipal facilities from user sources. In most cases, the industrial discharge is simply surcharged due to its high-strength BOD concentration.

7.7.1 Colgate–Palmolive Plant

Possibly the most representative treatment facility that handles wastewaters from the production of soaps, detergents, glycerines, and personal care products is Colgate–Palmolive Company's plant at Jeffersonville, IN [3]. The production wastes had received treatment since 1968 [21] in a completely mixed activated sludge plant with a 0.6 MGD design flow and consisting of a 0.5 MG mixed equalization and storage basin, aeration basin, and final clarifier. The treated effluent was discharged to the Ohio River, combined with rain drainage and cooling waters. During operation, it was observed that waste overloads to the plant caused a deterioration of effluent quality and that the system recovered very slowly, particularly from surfactant short-term peaks. In addition, the fact that ABS had been eliminated and more LAS and nonionic surfactants were being produced, as well as the changes in product formulation, may have been the reasons for the Colgate treatment plant's generally less than acceptable effluent quality. (Note that 1 MG = 3785 m^3, 1 MGD = 3785 m^3/day.)

Owing to the fact that the company considered the treatment efficiency in need of more dependable results, in 1972–1973 several chemical pretreatment and biological treatment studies were undertaken in order to modify and improve the existing system. As a result, a modified treatment plant was designed, constructed, and placed in operation. A new 1.5 MG mixed flow and pollutant load equalization basin is provided prior to chemical pretreatment, and a flash mixer with lime addition precedes a flocculator/clarifier unit. Ahead of the pre-existing equalization and aeration basins, the capability for pH adjustment and nutrient supplementation was added. Chemical sludge is wasted to two lagoons where thickening and dewatering (normally 15–30% solids) take place.

The intermediate storage basin helps equalize upsets in the chemical pretreatment system, provides neutralization contact time, and allows for storage of pretreated wastewater to supply to the biological treatment unit whenever a prolonged shutdown of the chemical pretreatment occurs. Such shutdowns are planned for part of the weekend and whenever manufacturing stoppage occurs in order to cut down on costs. According to Brownell [3], waste loads to the pretreatment plant diminish during plantwide vacations and production shutdowns, and bypassing the chemical pretreatment allows for a more constant loading of the aeration basins at those times. In this way, the previously encountered problems in the start-up of the biological treatment unit after shutdowns were reduced.

The pollutant removal efficiency of this plant is normally quite high, with overall MBAS (methylene blue active substances) removals at 98–99% and monthly average overall BOD_5 removals ranging from 88 to 98% (most months averaging about 95%). The reported MBAS removals achieved in the chemical pretreatment units normally averaged 60–80%. Occasional high MBAS concentrations in the effluent from the chemical pretreatment system were controlled through the addition of $FeCl_2$ and an organic polymer that supplemented the regular dose of lime and increased suspended solids' capture. Also, high oil and grease concentrations were occasionally observed after spills of fatty acid, mineral oil, olefin, and tallow, and historically this caused problems with the biological system. In the chemical pretreatment units, adequate oil and grease removals were obtained through the addition of $FeCl_2$. Finally, COD removals in the chemical system were quite consistent and averaged about 50% (COD was about twice the BOD_5).

In the biological step of treatment, removal efficiency for BOD_5 was very good, often averaging over 90%. During normal operating periods, the activated sludge system appeared incapable of treating MBAS levels of over 100 lb/day (45.4 kg/day) without significant undesirable foaming. The BOD_5 loading was normally kept at 0.15–0.18 g/day/g (or lb/day/lb) MLVSS, but it had to be reduced whenever increased foaming occurred. Finally, suspended solids concentrations in the secondary clarifier effluent were occasionally quite high, although the overflow rate averaged only 510 gal/day/ft^2 and as low as 320 gpd/ft^2 (13–20.8 m^3/day/m^2). The use of polymer flocculants considerably improved the effluent turbidity, reducing it by 50–75%, and because higher effluent solids contribute to high effluent BOD_5, it was reduced as well. Therefore, although the Colgate–Palmolive waste treatment plant occasionally experiences operating problems, it generally achieves high levels of pollutant removal efficiencies.

Many analytical procedures have been developed for determination of MBAS [73,75] and COD/DO [61,89–91] concentrations in water and wastewater, in turn, for monitoring the efficiency of treatment processes.

7.7.2 Combined Treatment of Industrial and Municipal Wastes

Most soap and detergent manufacturing facilities, as mentioned previously, discharge their untreated or pretreated wastes into municipal systems. The compositions of these wastewaters vary widely, with some being readily biodegradable and others inhibitory to normal biological treatment processes. In order to allow and surcharge such an effluent to a municipal treatment plant, an evaluation of its treatability is required. Such a detailed assessment of the wastewaters discharged from a factory manufacturing detergents and cleaning materials in the vicinity of Pinxton, England, was reported by Shapland [92]. The average weekly effluent discharged from a small collection and equalization tank was 119 m^3/day (21.8 gpm), which contributes about 4% of the flow to the Pinxton sewage treatment plant.

Monitoring of the diurnal variation in wastewater pollutant strengths on different days showed that no regular diurnal pattern exists and the discharged wastewaters are changeable. In particular, the pH value was observed to vary rapidly over a wide range and, therefore, pH correction in the equalization tank would be a minimum required pretreatment prior to discharge into the sewers in such cases. The increase in organic loading contributed to the Pinxton plant by the detergent factory is much higher than the hydraulic loading, representing an average of 32% BOD increase in the raw influent and 60% BOD increase in the primary settled effluent, but it does not present a problem because the plant is biologically and hydraulically underloaded.

The treatability investigation of combined factory and municipal wastewaters involved laboratory-scale activated sludge plants and rolling tubes (fixed-film) units. The influent feed to these units was settled industrial effluent (with its pH adjusted to 10) mixed in various proportions with settled municipal effluent. The variation of hydraulic loading enabled the rotating tubes to be operated at similar biological loadings. In the activated sludge units, the mixed liquor suspended solids (MLSS) were maintained at about 3000 mg/L, a difficult task since frothing and floe break-up caused solids loss. The overall results showed that more consistent removals were obtained with the fixed-film system, probably due to the loss of solids from the aeration units [93].

At 3 and 6% by vol. industrial waste combination, slight to no biological inhibition was caused either to the fixed-film or activated sludge system. The results of sample analysis from the inhibitory runs showed that in two of the three cases, the possible cause of inhibition was the presence of chloroxylenes and brominated compounds. The third case represented only temporary inhibition, since the rolling tubes provided adequate treatment after a period of acclimation. Finally, the general conclusion reached in the investigation was that the detergent factory effluent may be accepted at 3% by vol. equalized flow to the municipal fixed-film treatment plant, that is, up to 200 m^3/day (36.7 gpm), without any noticeable efficiency reduction.

7.7.3 Treatability of Oily Wastes from Soap Manufacture

McCarty [94] addressed the subject of the treatability of animal and vegetable oils and fats in municipal treatment systems. In general, certain reported treatment difficulties in biological systems are attributed to the presence of fats, oils, and other "grease" components in wastewaters. However, as opposed to mineral-type oils, animal and vegetable oils and fats such as those discharged by soap manufacture plants are readily biodegradable and generally nontoxic, although differences exist as to the difficulties caused depending on the form (floatable or emulsified) and type (hydrocarbons, fatty acids, glycerides, sterols, etc.). In general, shorter-chain-length fatty acids, unsaturated acids, and soluble acids are more readily degraded than longer-chain, saturated, and insoluble ones. The more insoluble and larger fatty acid particles have been found to require greater time for degradation than those with opposite characteristics. It has also been reported that animal and vegetable oils, fats, and fatty acids are metabolized quickly in anaerobic systems and generate the major portion of methane in regular anaerobic sludge digestion.

McCarty [94] also reported on the results of laboratory investigations in the treatability of selected industrial oily wastes from soap manufacturing and food processing by the Procter & Gamble Co. in Cincinnati, OH, when combined with municipal sewage or sludge. The grease content of the industrial wastes was high in all cases, ranging from 13 to 32% of the waste COD, and it was about 2.9 g of COD per gram of grease. It was found that it is possible to treat about equal COD mixtures of the industrial wastes with municipal sewage using the activated sludge process and achieve removal efficiencies similar to those for municipal sewage alone.

The grease components of the industrial wastes were readily degraded by anaerobic treatment, with removal efficiencies ranging from 82 to 92%. Sludges from the anaerobic digestion of an industrial/municipal mixture could be dewatered with generally high doses of chemical conditioning ($FeCl_2$), but these stringent requirements seemed a result of the hard-to-dewater municipal waste sludge. In conclusion, the Procter & Gamble Co. industrial wastes were readily treated when mixed with municipal sewage without significant adverse impacts, given sufficient plant design capacity to handle the combined wastes hydraulically and biologically. Also, there was no problem with the anaerobic digestion of combined wastes, if adequate mixing facilities are provided to prevent the formation of scum layers.

For treatment process control, Wang [85–87] has developed rapid methods for determination of oil and grease and dissolved proteins in the wastewaters.

7.7.4 Removal of Nonionic Surfactants by Adsorption

Nonionic surfactants, as mentioned previously, have been widely adopted due to their characteristics and properties and, in particular, because they do not require the presence of undesirable phosphate or caustic builders in detergent formulation. However, the relatively lesser degree of biodegradability is an important disadvantage of the nonionic surfactants compared to the ionic ones. Adsorption on activated carbon and various types of clay particles is, therefore, one of the processes that has been effective in removing heterodisperse nonionic surfactants – those that utilize a polyhydroxyl alcohol as a lipophilic phase – from wastewaters [6]. In another study by Carberry and Geyer [5], the adsorptive capacity kinetics of polydisperse nonionic surfactants – those that utilize a hydrocarbon species as a lipophilic base – removal by granular activated carbon and clay were investigated. Both clay particulates of different types and various activated carbons were tested and proven efficient in adsorbing nonionic surfactants. Of all the clays and carbons studied, Bentolite-L appeared to be the superior adsorbent (9.95% mol/kg vs. 0.53 mol/kg for Hydrodarco 400), but reaction rate constants for all adsorbents tested appeared to be strikingly similar.

7.7.5 Removal of Anionic Detergents with Inorganic Gels

Inorganic gels exhibiting ion-exchange and sorption characteristics are more stable than synthetic organic resins, which have also been used for the removal of detergents from wastewaters [95]. The sorption efficiency and number of cycles for which inorganic gels can be used without much loss in sorption capacity would compensate the cost involved in their preparation. Zinc and copper ferrocyanide have been shown to possess promising sorption characteristics for cationic and anionic surfactants. Of the two, copper ferrocyanide is a better scavenger for anionic detergents, which have a relatively small rate and degree of biodegradation and their presence in raw water causes problems in coagulation and sedimentation.

The cation-exchange capacity of the copper ferrocyanide gel used was found to be about 2.60 meq/g and its anion-exchange capacity about 0.21 meq/g. In all cases of various doses of gel used and types of anionic surfactants being removed, the tests indicated that a batch contact time of about 12 hours was sufficient for achieving maximum removals. Trials with various fractions of particle size demonstrated that both uptake and desorption (important in material regeneration) were most convenient and maximized on 170–200 BSS mesh size particles. Also, the adsorption of anionic surfactants was found to be maximum at pH 4 and decreased with an increase in pH.

The presence of NaCl and $CaCl_2$ salts (mono and bivalent cations) in solution was shown to increase the adsorption of anionic surfactants in the pH range 4–7, whereas the presence of $AlCl_3$ salt (trivalent cation) caused a greater increase in adsorption in the same pH range. However, at salt concentrations greater than about 0.6 M, the adsorption of the studied anionic surfactants started decreasing. On the other hand, almost complete desorption could be obtained by the use of K_2SO_4 or a mixture of H_2SO_4 and alcohol, both of which were found to be equally effective. In conclusion, although in these studies the sorption capacity of the adsorbent gel was not fully exploited, the anionic detergent uptake on copper ferrocyanide was found to be comparable to fly ash and activated carbon.

7.7.6 Removal of Cationic Surfactants

There are few demonstrated methods for the removal of cationic surfactants from wastewater, as mentioned previously, and ion exchange and ultrafiltration are two of them. Chiang and Etzel [8] developed a procedure for selecting from these the optimum removal process for cationic surfactants from wastewaters. Preliminary batch-test investigations led to the selection of one resin (Rohm & Haas "Amberlite," Amb-200) with the best characteristics possible (i.e., high exchange capacity with a rapid reaction rate, not very fine mesh resin that would cause an excessive pressure drop and other operational problems, macroporous resin that has advantages over the gel structure resins for the exchange of large organic molecules) to be used in optimizing removal factors in the column studies vs. the performance of ultrafiltration membranes (Sepa-97 CA RO/UF selected). The cyclic operation of the ion-exchange (H^+) column consisted of the following steps: backwash, regeneration, rinse, and exhaustion (service).

The ion-exchange tests indicated that the breakthrough capacity or total amount adsorbed by the resin column was greater for low-molecular-weight rather than high-molecular-weight surfactants. Furthermore, the breakthrough capacity for each cationic surfactant was significantly influenced (capacity decreases as the influent concentration increases) by the corresponding relationship of the influent concentration to the surfactant critical micelle concentration (CMC). A NaCl/ethanol/water (10% NaCl plus 50% ethanol) solution was found to be optimum in regenerating the exhausted resin.

In the separation tests with the use of a UF membrane, the rejection efficiency for the C_{16} cationic surfactants was found to be in the range 90–99%, whereas for the C_{12} surfactants it ranged from 72 to 86%, when the feed concentration of each surfactant was greater than its corresponding CMC value. Therefore, UF rejection efficiency seems to be dependent on the respective hydrated micelle diameter and CMC value. In conclusion, the study showed that for cationic surfactants removal, if the feed concentration of a surfactant is higher than its CMC value, then the UF membrane process is found to be the best. However, if the feed concentration of a surfactant is less than its CMC value, then ion exchange is the best process for its removal.

Initial and residual cationic surfactant concentrations in a water or wastewater treatment system can be determined by titration methods, colorimetric methods, or UV method [69–71, 77–79,81]. Additional references for cationic surfactant removal are available elsewhere [44,45,51,65,66].

7.7.7 Adsorption of Anionic Surfactant by Rubber

Removal of anionic surfactants has been studied or reported by many investigators [96–101]. It has been reported [101] that the efficiency of rubber granules, a low-cost adsorbent material, is efficient for the removal of sodium dodecyl sulfate (SDS), which is a representative member of anionic surfactants (AS). Previous studies on the absorption of AS on various adsorbents such as alumina and activated carbon showed 80–90% removals, while the sodium form of type A Zeolite did not have a good efficiency; however, these adsorbing materials are not cost-effective. In this study, a very low-cost scrap rubber in the form of granules (the waste product of tires locally purchased for US$0.20 per kg) was used to remove AS from the water environment. Tires contain 25–30% by weight carbon black as reinforcing filler and hydroxyl and/or carboxyl groups; both the carbon black and carboxyl group are responsible for the high degree of adsorption. In addition to the abundance and low cost of the waste tire rubber, the advantage is the possibility of reusing the exhausted rubber granules as an additive to asphalt as road material.

Earlier, Shalaby and El-Feky [98] had reported successful adsorption of nonionic surfactant from its aqueous solution onto commercial rubber. The average size of sieved adsorbent granules used was 75, 150, and 425 m. It was observed that within 1 hour, with all three sizes, the removal of AS was the same, about 78%. But after 5 hours, the removal was found to be 90% for the 75 m average size, while it was only about 85% for the other two larger sizes (adsorption is a surface phenomenon and as the size decreases, the surface area increases). Tests performed with initial adsorbate (SDS) concentrations of 2, 4, and 6 mg/L and doses of adsorbent varying between 5 and 15 g/L showed a removal efficiency in all cases of 65–75% within 1 hour, which only increased to about 80% after 7 hours. The effect of solution pH on adsorption of AS by rubber granules was also studied over a pH range of 3–13 using an initial AS concentration and an adsorbent dose of 3 mg/L and 10 g/L, respectively. Over a 6 hour contact time, with increase of pH, the removal of AS decreased practically linearly from 86 to 72%, probably due to interference of OH^- ion, which has similar charge to that of AS.

The effect of Ca^{2+} ion, which is very common in waters, was investigated over a range of 0–170 ppm calcium and it was shown that about 80–89% removal of AS occurred throughout this range. Similarly high levels of AS removal (87–93%) were observed for iron concentrations from 20 to 207 ppm, possibly due to formation of insoluble salt with the anionic part of the surfactant causing increased removal. On the other hand, the ionic strength of the solution in the form of NO_3^- concentration ranging from 150 to 1500 ppm was shown to reduce SDS removal efficiency to 71–77%, while the effect of chloride concentration (in the range 15–1200 mg/L) on AS removal by rubber granules was found to be adverse, down to 34–48% of SDS, which might be due to competition for adsorbing sites.

For treatment process control, initial and residual anionic surfactant concentrations in a water treatment system can be determined by titration methods or colorimetric methods [75,76,80,84,90]. The most recent technical information on management and treatment of the soap and detergent industry waste is available from the state of New York [104].

REFERENCES

1. Abrams, I.M.; Lewon, S.M. J. Am. Water Works Assoc. **1962**, *54* (*5*).
2. Besselievre, E.B. *The Treatment of Industrial Wastes*; McGraw-Hill: New York, NY, 1969.
3. Brownell, R.P. Chemical-biological treatment of surfactant wastewater. *Proceedings of the 30th Industrial Waste Conference*, Purdue University, Lafayette, IN, 1975, Vol. 30, 1085.
4. Callely, A.G. *Treatment of Industrial Effluents*; Haisted Press: New York, NY, 1976.
5. Carberry, J.B.; Geyer, A.T. Adsorption of non-ionic surfactants by activated carbon and clay. *Proceedings of the 32nd Industrial Waste Conference*, Purdue University, Lafayette, IN, 1977, Vol. 32, 867.
6. Carberry, J.B. Clay adsorption treatment of non-ionic surfactants in wastewater. J. Water Poll. Control Fed. **1977**, *49*, 452.
7. Chambon, M.; Giraud, A. *Bull. Ac. Nt. Medecine (France)* **1960**, *144*, 623–628.
8. Chiang, P.C.; Etzel, J.E. Procedure for selecting the optimum removal process for cationic surfactants. In *Toxic and Hazardous Waste*; LaGrega, Hendrian, Eds.; Butterworth: Boston, MA, 1983.
9. Cohen, J.M. *Taste and Odor of ABS*; US Dept. of Health, Education and Welfare Dept.: Cincinnati, OH, 1962.
10. Cohen, J.M. J. Am. Water Works Assoc. **1959**, *51*, 1255–1266.
11. Eckenfelder, W.W. *Industrial Water Pollution Control*; McGraw-Hill: New York, NY, 1989.
12. USEPA. *Development Document on Guidelines for Soap and Detergent Manufacturing*, EPA-440/1-74-018a; US Government Printing Office: Washington, DC, Construction Grants Program, 1974.
13. USEPA. *Federal Guidelines on State and Local Pretreatment Programs*, EPA-430/9-76-017c; US Government Printing Office: Washington, DC, Construction Grants Program, 1977; 8-13-1–8-13-25.
14. Federal Register. *Code of Federal Regulations*; US Government Printing Office: Washington, DC, 1987; CFR 40, Part 417, 362–412.
15. Federal Register. *Notices, Appendix A, Master Chart of Industrial Categories and Regulations*; US Government Printing Office: Washington, DC, 1990; Jan. 2, *55* (*1*), 102–103.
16. Gameson, A.L.H. J. Inst. Water Engrs (UK) **1955**, *9*, 571.
17. Gibbs, F.S. The removal of fatty acids and soaps from soap manufacturing wastewaters. *Proceedings of the 5th Industrial Waste Conference*, Purdue University, Lafayette, IN, 1949, Vol. 5, p. 400.
18. Greek, B.F. Detergent industry ponders products for new decade. Chem. & Eng. News **1990**, *Jan. 29*, 37–60.
19. Gumham, C.F. *Principles of Industrial Waste Treatment*; Wiley: New York, NY, 1955.
20. Gurnham, C.F. (Ed.) *Industrial Wastewater Control*; Academic Press: New York, NY, 1965.

21. Herin, J.L. Development and operation of an aeration waste treatment plant. *Proceedings of the 25th Industrial Waste Conference*, Purdue University, Lafayette, IN, 1970, Vol. 25, p. 420.
22. Koziorowski, B.; Kucharski, J. *Industrial Waste Disposal*; Pergamon Press: Oxford, UK, 1972.
23. Payne, W.J. Pure culture studies of the degradation of detergent compounds. Biotech. Bioeng. **1963**, *5*, 355.
24. Swisher, R.D. *Surfactant Biodegradation*; Marcel Dekker: New York, NY, 1970.
25. Osburn, O.W.; Benedict, III H. Polyethoxylated alkyl phenols: relationship of structure to biodegradation mechanism. J. Am. Oil Chem. Soc. **1966**, *43*, 141.
26. Wright, K.A.; Cain, R.B. Microbial metabolism of pyridinium compounds. Biochem. J. **1972**, *128*, 543.
27. Prat; Giraud, A. *La Pollution des Eaux par les Detergents*, Report 16602; Scientific Committee of OECD: Paris, France, 1964.
28. OECD. *Pollution par les Detergents*, Report by Expert Group on Biodegradability of Surfactants: Paris, France, 1971.
29. Langelier, W.F. Proc. Am. Soc. Civil Engrs. **1952**, *78* (*118*), February.
30. Nichols, M.S.; Koepp, E. J. Am. Water Works Assoc. **1961**, *53*, 303.
31. Todd, R. Water Sewage Works **1954**, *101*, 80.
32. Smith, R.S. *et al.* J. Am. Water Works Assoc. **1956**, *48*, 55.
33. McGauhey, P.H.; Klein, S.A. *Sewage and Indust. Wastes*, **1959**, *31* (*8*), 877–899.
34. Lawson, R. *Sewage and Indust. Wastes*, **1959**, *31* (*8*), 877–899.
35. Robeck, G. *et al.* J. Am. Water Works Assoc. **1962**, *54*, 75.
36. Renn, E.; Barada, M. J. Am. Water Works Assoc. **1961**, *53*, 129–134.
37. Overcash, M.R. *Techniques for Industrial Pollution Prevention*; Lewis Publishers: MI, 1986.
38. Nemerow, N.L. *Industrial Water Pollution*; Addison-Wesley: Reading, MA, 1978.
39. Wang, L.K.; Krofta, M. *Flotation Engineering*, 3rd Ed.; Lenox Institute of Water Technology: Lenox, MA, 2002; 252p.
40. Wang, L.K.; Kurylko, L.; Hrycyk, O. Removal of Volatile Compounds and Surfactants from Liquid. US Patent No. 5,122,166, June 1992.
41. Hrycyk, O.; Kurylko, L.; Wang, L.K. Removal of Volatile Compounds and Surfactants from Liquid. US Patent No. 5,122,165, June 1992.
42. Wang, L.K.; Kurylko, L.; Wang, M.H.S. *Combined Coarse and Fine Bubbles Separation System.* US Patent No. 5,275,732, January 1994.
43. Wang, L.K. *Continuous Bubble Fractionation Process.* Ph.D. Dissertation; Rutgers University, NJ, USA, 1972; 171 p.
44. Wang, M.H.S.; Granstrom, M.; Wang, L.K. Lignin separation by continuous ion flotation: investigation of physical operational parameters. Water Res. Bull. **1974**, *10* (*2*), 283–294.
45. Wang, M.H.S.; Granstrom, M.; Wang, L.K. Removal of lignin from water by precipitate flotation. J. Environ. Eng. Div., Proc. of ASCE **1974**, *100* (*EE3*), 629–640.
46. Ross, R.D. (Ed.) *Industrial Waste Disposal*; Reinhold: New York, NY, 1968.
47. Wang, L.K. Treatment of various industrial wastewaters by dissolved air flotation. *Proceedings of the N.Y.–N.J. Environmental Expo.* Secaucus, NJ, October, 1990.
48. Wang, L.K. Potable water treatment by dissolved air flotation and filtration. J. AWWA **1982**, *74* (*6*), 304–310.
49. Wang, L.K. *The State-of-the-Art Technologies for Water Treatment and Management*, UNIDO Manual No. 8-8-95; United Nations Industrial Development Organization: Vienna, Austria, 1995; 145 p.
50. Wang, L.K. The adsorption of dissolved organics from industrial effluents onto activated carbon. J. Appl. Chem. Biotechnol. **1975**, *25* (*7*), 491–503.
51. Wang, L.K. Water treatment with multiphase flow reactor and cationic surfactants. J. AWWA **1978**, *70*, 522–528.
52. Wang, L.K. Application and determination of anionic surfactants. Indust. Engng Chem. **1978**, *17* (*3*), 186–195.
53. Krofta, M.; Wang, L.K. *Design of Dissolved Air Flotation Systems for Industrial Pretreatment and Municipal Wastewater Treatment – Design and Energy Considerations*; AIChE National Conference, Houston, TX, NTIS-PB83-232868, 1983, 30 p.

54. Krofta, M.; Wang, L.K. *Design of Dissolved Air Flotation Systems for Industrial Pretreatment and Municipal Wastewater Treatment – Case History of Practical Applications*; AIChE National Conference, Houston, TX, NTIS-PB83-232850, 1983, 25 p.
55. Krofta, M.; Wang, L.K. Development of a total closed water system for a deinking plant. *AWWA Water Reuse Symposium III* **1984**, *2*, 881–898.
56. Krofta, M.; Wang, L.K. *Flotation and Related Adsorptive Bubble Separation Processes*, 4th Ed.; Lenox Institute of Water Technology: Lenox, MA, 2001; 185 p.
57. Wang, L.K. Polyelectrolyte determination at low concentration. Indust. Engng Chem., Prod. Res. Devel. **1975**, *14* (*4*), 312–314.
58. Wang, L.K. *A Modified Standard Method for the Determination of Ozone Residual Concentration by Spectrophotometer*, PB84-204684; US Department of Commerce, National Technical Information Service: Springfield, VA, 1983; 16 p.
59. Wang, L.K. *Process Control Using Zeta Potential and Colloid Titration Techniques*, PB87-179099/AS; US Department of Commerce, National Technical Information Service: Springfield, VA, 1984, 126 p.
60. Wang, L.K. *Determination of Polyelectrolytes and Colloidal Charges*, PB86-169307; US Department of Commerce, National Technical Information Service: Springfield, VA, 1984; 47 p.
61. Wang, L.K. *Alternative COD Method for Reduction of Hazardous Waste Production*, PB86-169323; US Department of Commerce, National Technical Information Service: Springfield, VA, 1985; 7 p.
62. Wang, L.K. *Determination of Solids and Water Content of Highly Concentrated Sludge Slurries and Cakes*, PB85-182624/AS; US Department of Commerce, National Technical Information Service: Springfield, VA, 1985, 9 p.
63. Wang, L.K. *Laboratory Simulation and Optimization of Water Treatment Processes*, PB88-168414/AS; US Department of Commerce, National Technical Information Service: Springfield, VA, 1986; 54 p.
64. Chao, L.; Wang, L.K.; Wang, M.H.S. *Use of the Ames Mutagenicity Bioassay as a Water Quality Monitoring Method*, PB88-168422/AS; US Department of Commerce, National Technical Information Service: Springfield, VA, 1986; 25 p.
65. Wang, L.K.; Wood, G.W. *Water Treatment by Disinfection, Flotation and Ion Exchange Process System*, PB82-213349; US Department of Commerce, Nat. Tech. Information Service: Springfield, VA, 1982.
66. Wang, L.K. *Water Treatment by Disinfection, Flotation and Ion Exchange Process System*, PB82-213349; US Department of Commerce, National Technical Information Service: Springfield, VA, 1982; 115 p.
67. Wang, L.K.; Kurylko, L.; Wang, M.H.S. Sequencing Batch Liquid Treatment. US Patent No. 5,354,458, October 1994.
68. Wang, L.K.; Kurylko, L.; Hrycyk, O. Biological Process for Groundwater and Wastewater Treatment. US Patent No. 5,451,320, September 1995.
69. Wang, L.K. Neutralization effect of anionic and cationic surfactants. J. New Engl. Water Works Assoc. **1976**, *90* (*4*), 354–359.
70. Wang, L.K. *Cationic Surfactant Determination Using Alternate Organic Solvent*, PB86-194164/AS; US Department of Commerce, National Technical Information Service: Springfield, VA, 1983; 12 p.
71. Wang, L.K. *The Effects of Cationic Surfactant Concentration on Bubble Dynamics in a Bubble Fractionation Column*, PB86-197845/AS; US Department of Commerce, National Technical Information Service: Springfield, VA, 1984; 47 p.
72. Wang, L.K. A proposed method for the analysis of anionic surfactants. J. Am. Water Works Assoc. **1975**, *67* (*1*), 6–8.
73. Wang, L.K. Modified methylene blue method for estimating the MBAS concentration. J. Am. Water Works Assoc. **1975**, *67* (*1*), 19–21.
74. Wang, L.K. Analysis of LAS, ABS and commercial detergents by two-phase titration. Water Res. Bull. **1975**, *11* (*2*), 267–277.

75. Wang, L.K. Evaluation of two methylene blue methods for analyzing MBAS concentrations in aqueous solutions. J. Am. Water Works Assoc. **1975**, *67* (*4*), 182–184.
76. Wang, L.K. Determination of anionic surfactants with Azure A and quaternary ammonium salt. Analy. Chem. **1975**, *47* (*8*), 1472–1475.
77. Wang, L.K. Determining cationic surfactant concentration. Indust. Engng. Chem. Prod. Res. Devel. **1975**, *14* (*3*), 210–212.
78. Wang, L.K. A test method for analyzing either anionic or cationic surfactants in industrial water. J. Am. Oil Chemists Soc. **1975**, *52* (*9*), 340–346.
79. Wang, L.K. Rapid colorimetric analysis of cationic and anionic surfactants. J. New Engl. Water Works Assoc. **1975**, *89* (*4*), 301–314.
80. Wang, L.K. Direct two-phase titration method for analyzing anionic nonsoap surfactants in fresh and saline waters. J. Environ. Health **1975**, *38*, 159–163.
81. Wang, L.K. Analyzing cetyldimethylbenzylammonium chloride by using ultraviolet absorbance. Indus. Engng Chem. Prod. Res. Devel. **1976**, *15* (*1*), 68–70.
82. Wang, L.K. Role of polyelectrolytes in the filtration of colloidal particles from water and wastewater. Separ. Purif. Meth. **1977**, *6* (*1*), 153–187.
83. Wang, L.K. Application and determination of organic polymers. Water, Air Soil Poll. **1978**, *9*, 337–348.
84. Wang, L.K. Application and determination of anionic surfactants. Indus. Engng Chem. **1978**, *17* (*3*), 186–195.
85. Wang, L.K. *Selected Topics on Water Quality Analysis*, PB87-174066; US Department of Commerce, National Technical Information Service: Springfield, VA, 1982; 189 p.
86. Wang, L.K. *Rapid and Accurate Determination of Oil and Grease by Spectrophotometric Methods*, PB83-180760; US Department of Commerce, National Technical Information Service: Springfield, VA, 1983; 31 p.
87. Wang, L.K. *A New Spectrophotometric Method for Determination of Dissolved Proteins in Low Concentration Range*, PB84-204692; US Department of Commerce, National Technical Information Service: Springfield, VA, 1983; 12 p.
88. Wang, L.K.; DeMichele, E.; Wang, M.H.S. *Simplified Laboratory Procedures for DO Determination*, PB88-168067/AS; US Department of Commerce, National Technical Information Service: Springfield, VA, 1985; 13 p.
89. Wang, L.K.; DeMichele, E.; Wang, M.H.S. *Simplified Laboratory Procedures for COD Determination Using Dichromate Reflux Method*, PB86-193885/AS; US Department of Commerce, National Technical Information Service: Springfield, VA, 1986; 8 p.
90. AWWA, WEF, APHA. *Standard Methods for the Examination of Water and Wastewater*, AWWA/WEF/APHA, 2000.
91. Wang, L.K. *Recent Advances in Water Quality Analysis*, PB88-168406/AS; US Department of Commerce, National Technical Information Service: Springfield, VA, 1986; 100 p.
92. Shapland, K. Industrial effluent treatability. J. Water Poll. Control (UK), **1986**, p. 75.
93. Yapijakis, C. Treatment of soap and detergent industry wastes. In *Handbook of Industrial Waste Treatment*, Chap. 5; Wang, L.K., Wang, M.H.S., Eds.; Marcel Dekker, Inc.: New York, NY, 1992; 229–292.
94. McCarty, P.L. Treatability of oily wastewaters from food processing and soap manufacture. *Proceedings of the 27th Industrial Waste Conference*; Purdue University, Lafayette, IN, 1972; Vol. 27, p. 867.
95. Srivastava, S.K. *et al.* Use of inorganic gels for the removal of anionic detergents. *Proceedings of the 36th Industrial Waste Conference*, Purdue University, Lafayette, IN, 1981; Vol. 36, Sec. 20, p. 1162.
96. Bevia, F.R.; Prats, D.; Rico, C. Elimination of LAS during sewage treatment, drying and compostage of sludge and soil amending processes. In *Organic Contaminants in Wastewater, Sludge and Sediment*; Quaghebeur, D., Temmerman, I., Agelitti, G., Eds.; Elservier Applied Science: London, 1989.
97. Mathru, A.K.; Gupta, B.N. Detergents and cosmetics technology. Indian J. Environ Protection **1998**, *18* (*2*), 90–94.

98. Shalaby, M.N.; El-Feky, A.A. Adsorption of nonionic surfactant from its aqueous solution onto commercial rubber. J. Dispersion Sci. Technol. **1999**, *20* (*5*), 1389–1406.
99. Sing, B.P. Separation of hazardous organic pollutants from fluids by selective adsorption. Indian J. Environ. Protection **1994**, *14* (*10*), 748–752.
100. Wayt, H.J.; Wilson, D.J. Soil clean-up by in-situ surfactant flushing II: theory of miscellar solubilization. Separ. Sci. Technol. **1989**, *24*, 905–907.
101. Wang, L.K.; Yapijakis, C.; Li, Y.; Hung, Y.; Lo, H.H. Wastewater treatment in soap and detergent industry. OCEESA J. **2003**, *20* (*2*), 63–66.
102. Vath, C.A. Soap and Chem. Specif. **1964**, *March*.
103. Swisher, R.D. Exposure levels and oral toxicity of surfactants. Arch. Environ. Health **1968**, *17*, 232.
104. NYSDEC. Soap and Detergent Manufacturing Point Source Catergory. 4. CFR. Protection of Environment. Part 417. New York State Department of Environmental Conservation. Albany, New York. Feb. 2004.

8

Treatment of Textile Wastes

Thomas Bechtold and Eduard Burtscher
Leopold-Franzens-University, Innsbruck, Austria

Yung-Tse Hung
Cleveland State University, Cleveland, Ohio, U.S.A.

8.1. IDENTIFICATION AND CLASSIFICATION OF TEXTILE WASTES

8.1.1 Textile Processes

The production of textiles represents one of the big consumers of high water quality. As a result of various processes, considerable amounts of polluted water are released. Representative magnitudes for water consumption are 100–200 L of water per kilogram of textile product. Considering an annual production of 40 million tons of textile fibers, the release of wasted water can be estimated to exceed 4–8 billion cubic metres per year.

The production of a textile requires several stages of mechanical processing such as spinning, weaving, knitting, and garment production, which seem to be insulated from the wet treatment processes like pretreatment, dyeing, printing, and finishing operations, but there is a strong interrelation between treatment processes in the dry state and consecutive wet treatments.

For a long time the toxicity of released wastewater was mainly determined by the detection of biological effects from pollution, high bulks of foam, or intensively colored rivers near textile plants. Times have changed and the identification and classification of wastewater currently are fixed by communal regulations [1,2].

General regulations define the most important substances to be observed critically by the applicant, and propose general strategies to be applied for minimization of the release of hazardous substances. The proposed set of actions has to be integrated into processes and production steps [3]. Figure 1 gives a general overview of a textile plant and also indicates strategic positions for actions to minimize ecological impact. In this figure, the textile plant is defined as a structure that changes the properties of a textile raw material to obtain a desired product pattern. The activities to treat hazardous wastes can range from legal prohibition to cost-saving recycling of chemicals. Depending on the type of product and treatment, these steps can show extreme variability.

Normally the legal regulations are interpreted as a set of wastewater limits that have to be kept, but in fact the situation is more complex and at present a complex structure of actions has been defined and has described useful strategies to improve an actual situation.

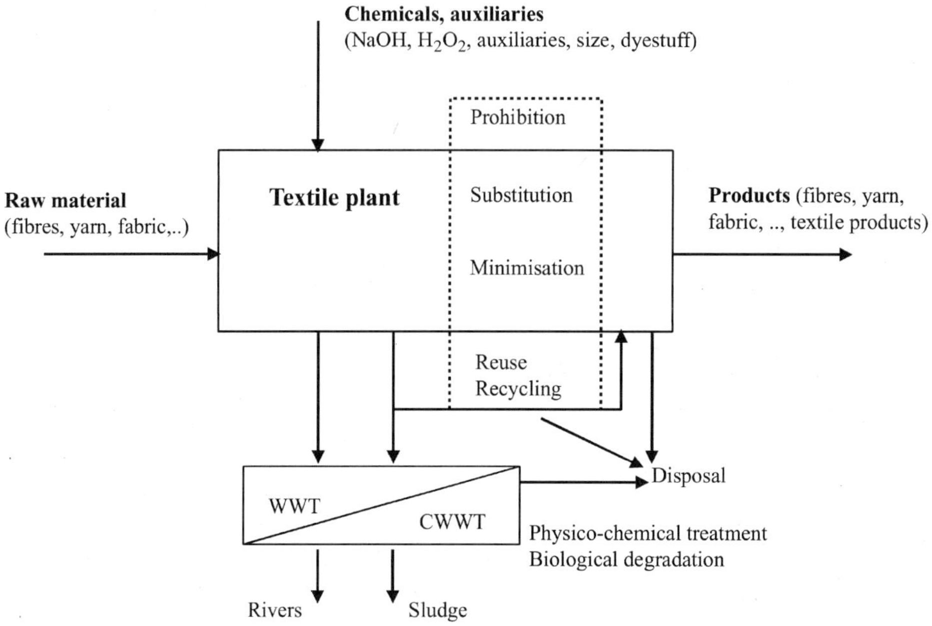

Figure 1 Flow structure of a textile plant (from Refs 2 and 3).

8.1.2 Strategies to Reach Existing Requirements

Figure 2 shows a general action path recommended to minimize a present problem in the wastewater released from a textile plant [3,4].

Replacement and Minimization

As a first step substances that are known to cause problems in the wastewater have to be replaced by less hazardous chemicals or the process itself should be reconsidered; for example,

- use of high-temperature dyeing (HT-dyeing) processes for polyester fibers (PES) instead of carrier processes;
- replacement of chloro-organic carriers;
- replacement of preservatives containing As, Hg, or Sn organic compounds;
- replacement of alkylphenolethoxylates (APEO) in surfactants [5];
- substitution of "chlorine" bleach for natural fibers by peroxide bleach processes;
- substitution of sizes with poor biodegradability, e.g., carboxymethylcellulose (CMC);
- replacement of "hard" complexing agents like ethylene-diamine-tetra-acetic acid (EDTA), phosphonates.

The implementation of these steps into a dyehouse reduces the chemical load of the released wastewater considerably. In particular the replacement of substances that exhibit high toxicity or very low biodegradability will facilitate the following efficient treatment of the wastewater.

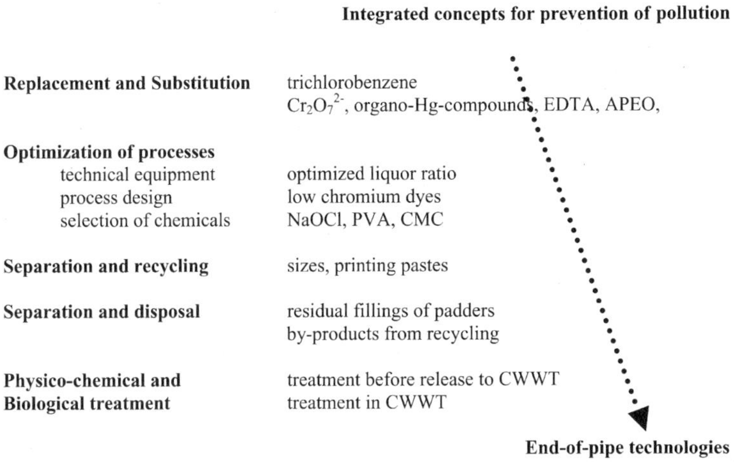

Figure 2 Action path for consideration and improvement of an existing situation (from Refs 1–3 and 9).

Optimization of Processes

The second general step recommended to improve an existing situation is the optimization of treatment steps with regard to a lowering of the released amounts of hazardous substances [6,7]. In many cases this strategy is more intelligent and less expensive than a concentration of activities on the final treatment of released effluents. Typical examples for possible optimization are:

- reconsideration of dyestuffs and machinery chosen in exhaust dyeing (degree of exhaustion, fixation, liquor ratio);
- optimization of dyes and reducing agent in sulfur dyeing;
- optimization of residual volumes of padders and printing machines;
- optimization of water consumption.

Separation and Recycling

Besides the replacement of substances, the improvement of processes on an optimization of the handling of rather concentrated liquors, for example, used in sizing, caustic treatment like mercerization, dyeing, finishing processes, or in textile printing processes is the next step. As a desired goal, a recycling of a main part of the substances should be attempted. Examples that can be mentioned include the recovery and regeneration of sizes and caustic soda solutions, and the recovery of lanolin from wool washing.

Separation and Treatment for Disposal or Drain

If regeneration is impossible, a separate collection of a certain type of waste and an optimized treatment of the concentrates is more efficient and cheaper than a treatment of the full waste stream. Such treatments will concentrate on a minimization of costs for disposal (e.g., disposal of sludge, printing pastes, chemical products) or reaching existing limits defined for various parameters analyzed in the wastewater, for example, pH value, content of heavy metals, chemical oxygen demand (COD), adsorbable halogenated organic compounds (AOX) [8].

General Wastewater Treatment

In any case the wastewater will finally be fed into rivers, lakes, or the sea; thus some wastewater treatments have to be performed before the textile effluents are released either to the communal wastewater treatment plant (CWWT) or into the rivers, lakes, and so on. Normally physical and (bio-) chemical treatments (e.g., adjustment of pH, temperature, sedimentation, flocculation) are performed in the textile plant, while the following biological treatment (aerobic, anaerobic degradation) is performed either in the textile plant or in a CWWT. The site of the biological treatment is dependent on the location of the textile plant; however, a biological treatment of textile effluents preceding release into surface water is state of the art.

8.1.3 Definitions and Limits

For a long time the treatment of textile effluents has concentrated mainly on two aspects: regeneration of concentrated effluents with regard to savings of chemicals and lowering of chemical costs and treatment of effluents with high toxicity.

Over the last decade the situation has changed and limits for a considerable number of compounds and parameters have been defined to avoid problems with regard to the following:

- biotoxicity (e.g., disturbance of biodegradation processes);
- heavy metal content (accumulation in sludge of CWWT);
- corrosion problems (e.g., sulfate can cause corrosion of concrete tubes);
- total COD/BOD load in the released effluents (capacity of the CWWT).

Table 1 gives an extract of important parameters for wasted water from textile plants, as defined by the Austrian Government [1]. The table contains limits defined for both direct release into surface water (rivers) and for release into a CWWT.

Table 1 can be used as a guide to define "hazardous" wastes from textile plants. Besides the direct toxicity of substances like chlorinated hydrocarbons, organo-Hg compounds or concentrated alkaline solutions, other parameters have been defined with regard to problems during biodegradation or accumulation in the sludge from CWWT. A particular situation is found with colored effluents, where limits for spectral absorption have been defined. While the toxicity of textile dyes is comparably low, these limits were derived from the visual aspect of the water released from a textile plant because they look "unhealthy."

As a result of these regulations, textile companies have to apply a strategic concept to lower both the daily load released into the wastewater stream and the concentrations of hazardous substances therein. On the basis of the action plan given in Figure 2, a stepwise improvement of the present situation of a plant has to be undertaken.

Owing to the extreme diversity of the textile processes and products, it is impossible to develop a realistic concept for an efficient wastewater treatment without detailed analysis of the particular situation of a textile plant. The more intelligently the applied technical concept has been designed, the lower will be the expected costs for installation and working of the equipment.

In the following sections techniques and technical solutions are given as examples that can be adapted to a certain problem.

To facilitate an overview and to consider the specific differences of textile fibers during pretreatment, dyeing, and finishing, the sections have been focused on the most important types of fibers: wool, cotton, and synthetic fibers. Mixtures of fibers can be seen as systems combining problems of the single fiber types. In Section 8.3 end-of-pipe technologies have been summarized.

Table 1 Representative Limits Defined for Release of Textile Waste Water

Limits for emission	Release into river	Release into CWWT
General parameters		
Temperature (°C)	30	40
Toxicity	<2	No hindrance of biodegradation
Filter residue (mg/L)	30	500
Sediments (mL/L)	<0.3	—
pH	6.5–8.5	6.5–9.5
Color, spectral coefficient of extinction:		
436 nm (yellow) (m^{-1})	7.0	28.0
525 nm (red) (m^{-1})	5.0	24.0
620 nm (blue) (m^{-1})	3.0	20.0
Inorganic parameters (mg/L)		
Aluminum	3	Limited by filter residue
Lead	0.5	0.5
Cadmium	0.1	0.1
Chromium total	0.5	1
Chromium-VI	0.1	0.1
Iron	2	Limited by filter residue
Cobalt	0.5	0.5
Copper	0.5	0.5
Zinc	2	2
Tin	1	1
Free chlorine (as Cl_2)	0.2	0.5
Chlorine total (as Cl_2)	0.4	1
Ammonium (as N)	5	—
Total phosphor (as P)	1	No problems in P elimination
Sulfate (as SO_4)	—	200
Organic parameters (mg/L)		
TOC (total organic carbon as C)	50	>70% biodegradation
COD (chemical oxygen demand as O_2)	150	>70% biodegradation
BOD_5 (biological oxygen demand as O_2)	20	—
AOX (adsorbable organic halogen as Cl)	0.5	0.5
Total hydrocarbon	5	15
VOX (volatile organic halogen)	0.1	0.2
Phenol index calculated as phenol	0.1	10
Total anionic and nonionic surfactants	1	No problems in sewer and CWWT

Source: Ref. 1.

8.1.4 IPPC Directive of the European Community

In the legislation of different national governments, some limits were defined especially for wastewater and air. The activities in Europe are covered by the Council Directive 96/61/EC concerning Integrated Pollution Prevention and Control (IPPC) [9]. This means that all environmental media (water, air, energy, ground) and a comprehensive description of the production have to be considered. In addition a broad harmonization of requirements for the approval of industrial plants can be reached.

The classification of a company as an IPPC plant is based on the definition of the work concerning plants for the pretreatment (operations such as washing, bleaching, mercerization) or dyeing of fibers or textiles where the treatment capacity exceeds 10 tons per day. As a firm basis of reference the capacity will be calculated as the potential output a company could have in 24 hours. Capacity means what a plant is designed for and not what is really achieved (actual production). The treatment of fibers and textiles covers fibers, yarns, and fabric in the wider sense of the word, that is, including knitted and woven materials and carpets. As most textiles are treated with continuous working machines with a very high theoretical maximum capacity, a lot of companies have to fulfill the directions for IPPC plants.

To reach the aim of the directive an efficient and progressive state of development is defined by the best available techniques (BAT). In practice, this means precaution against environmental pollution by the use of these techniques, special equipment and better way of production, and an efficient use of energy for prevention of accidents and provisions for a shutdown of a production plant. The term *best available techniques* is defined as the most effective and advanced stage in the development of activities and their methods of operation that indicate the practical suitability of particular techniques for providing in principle the basis for emission limit values designed to prevent and, where it is not practicable, generally to reduce emissions and the impact on the environment as a whole. These available techniques are developed on a scale that allows implementation under economically and technically viable conditions, taking into consideration the costs and advantages when the techniques are used.

In the best available technology reference document (BREF), particular attention is given to the processes of fiber preparation, pretreatment, dyeing, printing and finishing, but it also includes upstream processes that may have a significant influence on the environmental impact of textile processing. The treatment of all main fiber types as natural fibers (cotton, linen, wool, and silk), man-made fibers derived from natural polymers, such as viscose and celluloseacetate, as well as from synthetic polymers (such as polyester, polyamide, polyacrylnitrile, polyurethane, polypropylene) are described, including blends of these textile substrates. Beside general information about the industrial sector and the industrial processes, the situation in the plants is described by data about current emission and consumption.

A catalogue of emission reduction or other environmentally beneficial techniques that are considered to be most relevant in the determination of BAT (both generally and in specific cases) is given as a pool of possible techniques including both process integrated and end-of-pipe techniques, thus covering pollution prevention and pollution control measures. Techniques presented may apply to the improvement of existing installations, or to new installations, or a combination of both, considering various cost/benefit situations including both lower and higher cost techniques. To obtain a limitation of emission impact, different techniques are proposed corresponding to the basic possibilities for pollution prevention:

- handling of concentrates from various processes such as textile pretreatment, residual dye liquors from semicontinuous and continuous dyeing, residual printing pastes, residual finishing liquors, residues of prepared but not applied dyestuffs, textile auxiliaries, and so on;
- recovery of chemicals such as NaOH, sizing agents, indigo;
- assessment of textile auxiliaries aiming at a reduction of emissions of refractory and toxic compounds to water by substituting harmful substances with less harmful alternatives;

- reduction of releases to air from thermal treatment installations like stenter frames;
- reduction of releases to water by applying process-integrated measures and considering the available options for wastewater treatment; wastewater treatment including pretreatment onsite before discharge to the sewer as well as treatment of effluent onsite in case of discharge to rivers; efficiency of treatment of textile wastewater together with municipal wastewater;
- options for handling and treatment of residues and waste from different sources;
- minimizing of energy consumption used in energy-intensive processes such as pretreatment, fixation of dyes, finishing operation, and drying.

8.2. FIBER-SPECIFIC PROCESSES

The activities described in this section intend to minimize or avoid the release of chemicals into the stream wastewater by substitution, optimization, reuse, and recycling. Besides a lowering of the costs for following up general wastewater treatment, benefits due to minimization of chemical consumption are intended. As there are various specific problems arising from the particular treatment steps applied for different fibers, this section concentrates on the most important problems. Table 2 gives an overview of the annual production of textile fibers [10].

8.2.1 Protein Fibers: Wool

General

The annual production of wool is approximately 1.2 million tons, which corresponds to a share of 2% of the total production of textile fibers. A simplified route for the preparation, dyeing, and finishing of woolen textiles is shown in Figure 3.

Table 2 Annual Production of Textile Fibers 2001

Type of fiber	Mt/year
Man-made fibers	
Synthetics	31.6
Polyester	19.2
Polypropylene	5.8
Polyamide	3.7
Acrylics	2.6
Others	0.3
Cellulosics	2.7
Natural fibers	
Cotton	19.8
Jute	3.1
Ramie	0.2
Linen	0.6
Wool	1.2
Silk	0.1
Total	59.2

Mt, million tons.
Source: Ref. 10.

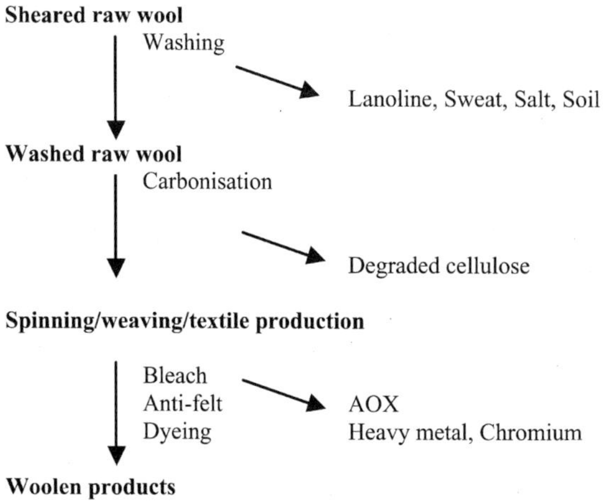

Figure 3 General processing route of woolen textiles (from Ref. 3).

Besides more general strategies of process optimization, three representative steps will be discussed in more detail because of their particular importance with regard to wastewater. The main problem resulting from these steps is given in parentheses:

- washing of raw wool (COD);
- antifelt treatment of wool (AOX);
- dyeing processes (chromium).

Washing of Raw Wool

The high content of impurities in raw wool has to be removed before further processing, for example, in carbonization, spinning, and weaving. As a considerable part of the raw material (approx. 30%) is removed and released into the wastewater, washing of raw wool can cause heavy pollution problems. These difficulties are not due to the toxicity of the released components, but result from the high concentrations and the load of organic material released in the form of dispersed and dissolved substances. Figure 4 gives an overview of a general set of techniques that can be applied to lower the initial COD in the effluent from approximately 80,000 mg/L to a final value of 12,000 mg/L [11,12].

The lanolin extracted from the wool is purified further for use in cosmetics, hand cream, boot-polish, and so on. Part of the permeate from the ultrafiltration is recycled to save fresh water. A particular advantage arises from the fact that the dissolved sweat components exhibit

Table 3 Average Composition of Raw Wool

Component	%
Fiber, protein	58
Wool-fat, lanolin, waxes	14
Soil, plant material (cellulose)	13
Sweat/salt, water soluble	5
Humidity	10

Source: Refs 3,11,12.

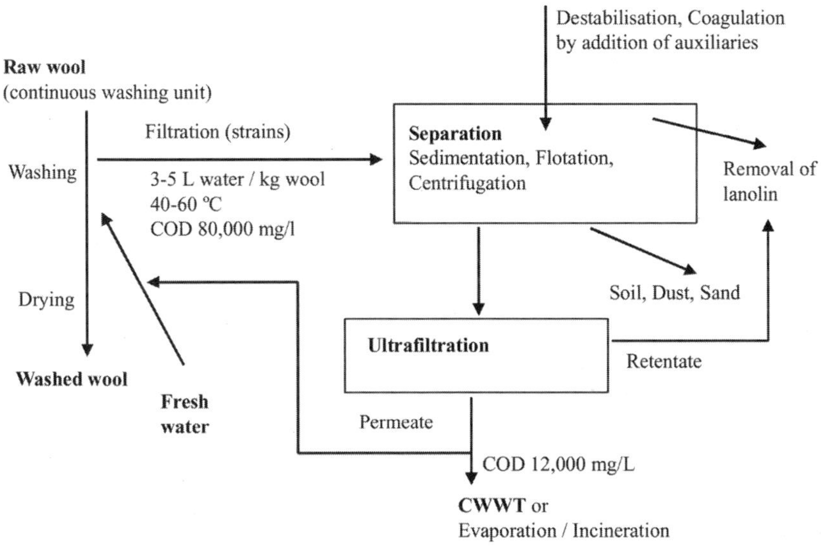

Figure 4 General scheme for the treatment of effluents from wool washing (from Refs 11–13).

distinct washing properties for raw wool and thus a certain content of dissolved sweat is favorable to improve the washing effect.

Various treatment concepts have been presented in the literature [11–13]. Besides the release of the pre-treated wastewater into the CWWT and aerobic biodegradation, in some cases evaporation of the wastewater and incineration of the residue are performed.

Antifelt Finishing of Wool

The surface of a wool hair is covered by keratin sheds, which cause a distinct tendency to shrinkage and formation of felts. This behavior is usually undesirable and thus an antifelt finishing is the most important treatment during the processing of woolen textiles. One of the most important standard procedures, the Hercosett finish, is based on the oxidative treatment of wool by application of compounds that release chlorine. Examples for applied chemicals are NaOCl, Cl_2 gas, and dichloroisocyanuric acid (DCCA) [14].

Such processes lead to the formation of adsorbable halogenated organic compounds (AOX) in high concentrations. Typical concentrations found in a continuous antifelt treatment are shown in Table 4. The high dissolved organic carbon (DOC) determined in the baths is one of the sources for the formation of high concentrations of chlorinated compounds. The formation of chlorinated products is the result of chemical reactions directly with the fiber, with organic compounds released from the fibers, and with added auxiliaries.

An average size of continuous treatment plant for antifelt treatment of wool releases approximately 140 g/hour AOX. As an optimization of the process is possible only within certain limits, alternative processes for an antifelt treatment have to be chosen to substitute the chlorination process, for example, enzymatic processes, oxidative processes ($KMnO_4$, persulfate), corona or plasma treatment. In many cases combinations with resin treatments are proposed.

Table 4 Concentrations for AOX Determined in the Chlorination Bath of the Chlorine-Hercosett Process

Parameter	Concentration
AOX	20 mg/L
$CHCl_3$	160–1200 μg/L
CCl_4	25–50 μg/L
DOC	1110 mg/L

Source: Refs 3,14.

Chromium in Wool Dyeing

A considerable part of the wool dyes contain Cr complexes. The average consumption of dyes used in 1992 is shown in Table 5. At this time approximately 70% of all dyes used contain chromium.

As shown in Table 1 the wastewater limit for chromium is 0.5–1 mg/L and Cr^{VI} is 0.1 mg/L. While conventional 1 : 2 and 1 : 1 dyes permit chromium concentrations in the dyebath at the end of the dyeing process of 3.0–13.0 mg/L Cr, the application of modern dyestuffs and optimized processes permits final concentrations to approximately 1 ppm. By general optimization of the process (e.g., dosage of acid), use of dyes with a high degree of exhaustion, and minimal concentration of free chromium [15], final bath concentrations below 4 ppm can be reached, even for black shades. By application of such procedures the exhaustion of the chromium should reach values of better than 95% of the initial value.

Owing to the low limits for concentrations of chromium the proposed processes for wastewater treatment concentrate on the removal, for example, by flocculation and precipitation, but as a result chromium-containing sludge/precipitate or concentrates are obtained that need further treatment.

8.2.2 Cellulose Fibers: Cotton

General

Cellulose fibers (Co, CV, CMD, CLY) represent the main group of textile fibers used [10]. In this section cotton will be considered as a representative type of fiber because the treatments for other cellulose fibers are similar in many cases, and often milder conditions are applied for other cellulose fibers.

Table 5 Dyestuff Consumption in Wool Dyeing

Dyestuff	%
1 : 2 Metal-complex[a]	35
Chromium dyes[a]	30
Acid dyes	28
1 : 1 Metal complex dyes[a]	4
Reactive dyes	3

[a] Contain Cr or Cr-salts are added.
Source: Refs 3,15.

Sources for textile effluents that need further treatment are found in all steps of processing. Table 6 shows a list of important parameters and wastes that require further treatment.

Sizing–Desizing

Before weaving, the warp is covered with a layer of polymer to withstand the mechanical stress (abrasion, tension) during weaving. These polymer coatings are so-called sizes. Normally native starch, modified starch like carboxymethyl-starch (CMS), carboxymethyl-cellulose (CMC), polyvinylalcohols (PVA), polyacrylates, and proteins can be used. The amount of added polymer for staple yarns like Co is between 8 and 20% of the weight of the warp. As a result, in many cases the final amount of polymer to be removed in the desizing step is approximately 5–10% of the weight of the fabric.

Sizing is not necessary in the case of knitted material, and much lower amounts are required for filament yarn (2–10% of the weight of the warp). The main problem resulting from the desizing step is the high load in COD found in the polymer-containing effluent. Table 7 summarizes the COD and biological oxygen demand (BOD) values determined for various sizes.

To estimate the COD/BOD load released from a desizing step, Eqs (1) and (2) can be used:

$$L_{\mathrm{COD}} = Cpm \times 10^{-3} \tag{1}$$

$$L_{\mathrm{BOD}} = Bpm \times 10^{-3} \tag{2}$$

Table 6 Processing of Cotton: Process Steps and Selected Parameters

Process step	Critical parameter	Component
Desizing	COD $\neq$ BOD	Starch, modified starch, PVA, polyacrylates
Scouring	COD $\neq$ BOD	Organic load released from cotton and added auxiliaries
	Complexing agents	EDTA, phosphonates
	pH	NaOH
Bleach		
Hypochlorite	AOX	Chlorinated compounds
Peroxide	Complexing agents	EDTA, phosphonates
Mercerization	pH	NaOH
Dyeing		
Direct	Salt	NaCl, Na_2SO_4
Reactive	Color	Hydrolyzed dyes
	Salt	NaCl, Na_2SO_4
	pH	NaOH
Vat	pH	NaOH
	Sulfate	Na_2SO_4, Na_2SO_3
Indigo	Color	Indigo
	Salt	Na_2SO_4
Printing	Printing pastes	Concentrated chemical load
	Washwater (COD, BOD, color)	Thickener, dyestuff
Finishing	Filling of padder	Concentrated chemical load

Source: Ref. 3.

Table 7 COD and BOD per Mass of Size Released

Type of size	COD C (mg/g)	BOD B (mg/g)
Starch	900–1000	500–600
CMC	800–1000	50–90
PVA	1700	30–80
Polyacrylate	1350–1650	<50
Galactomannane	1000–1150	400
PES-dispersion	1600–1700	<50
Protein	1200	700–800

Source: Ref. 3.

Desizing of $m = 1000$ kg of goods, which contain 5% of weight starch size ($p = 0.05$) cause a load $L_{COD} = 50$ kg and $L_{BOD} = 30$ kg. Using 10 L of water for desizing of 1 kg of fabric, a total volume of 10,000 L will be required and the load $L_{COD} = 50$ kg will be diluted in this volume. As a result, a COD value of 5000 mg/L can be calculated for the effluent.

Two different paths can be followed to describe the behavior of sizes released in effluents:

- Biodegradation, which refers to the complete biodegradation of sizes like starch. Here high values of COD are coupled to high BOD.
- Bioelimination is detected by BOD, which is rather low BOD, compared to the COD. In such cases the polymer is removed from the waste stream in the WWT/CWWT by flocculation, adsorption, hydrolysis, and, to a certain degree, by biodegradation. Representatives are PVA, CMC, and acrylate sizes [16,17].

The strategies to handle size-containing wastes are dependent on the type of size and particularly on the technique of desizing (Fig. 5). In the case of starch, the desizing step is usually performed by enzymatic degradation, and in some cases oxidative degradation is used. However, the starch is degraded and a reuse is not possible in such cases. The disadvantage of a high COD caused by the released partially degraded starch is accompanied by easy biodegradation, thus the effluents can be treated in a WWT/CWWT with sufficient capacity for biodegradation with no further problems.

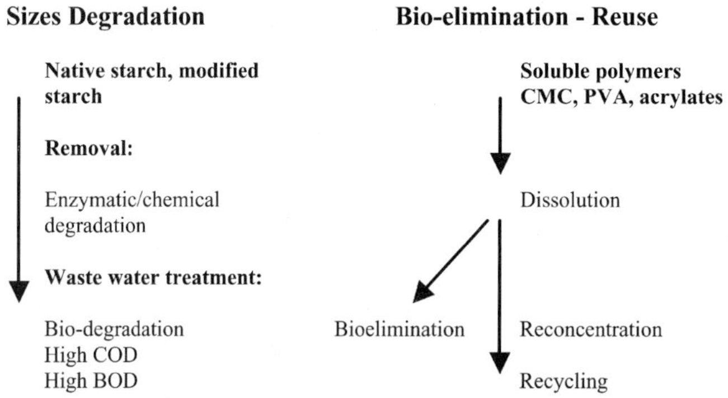

Figure 5 Desizing and treatment of size-containing wastes (from Refs 18–24).

Water-soluble sizes permit a recycling of the polymer for further weaving processes. Various techniques have been proposed to regenerate sizes released from the fabric. General requirements that have to be considered as fundamentals for possible reuse of sizes are summarized as the following:

- easy and short distance transportation of recovered size to sizing/weaving plant;
- known composition of sizes;
- development of standardized recipes;
- stable composition of recovered size/no degradation.

In practice, a recycling of sizes is hindered for a number of reasons. In many cases various qualities of fabric containing different sizes are treated in a dyehouse and the type of size is often not known. The selection of sizes with regard to easy biodegradation/bioelimination is necessary. When a regeneration is intended a direct interaction between the selection of size, desizing procedure, recycling processes, and the sizing/weaving process have to be considered.

Two general technological strategies have been developed and proposed:

- removal of water soluble sizes by washing;
- reconcentration in the washing machine or by UF/evaporation. Figure 6 gives an overview of these two techniques.

Washing techniques have been proposed for PVA and acrylate sizes [18]. When applying washing techniques the volume of concentrated washwater for each size is limited by the volume actually spent in the following up sizing process (e.g., 900 L in Fig. 6) [19–21]. The use of higher amounts of water would increase the mass of recovered size, but the dilution of the regenerate is too much and hinders a reuse without reconcentration. A typical balance for a full process for acrylate sizes is shown in Figure 7 [22].

The advantage of UF techniques is the higher rate of size recovery, because a reduction of volume is possible. In some cases an evaporation step is used as final concentration step because the viscosity of the sizes increases and the permeate flow is reduced substantially. Problems can result from a change in the composition of the size due to changes in the molecular weight distribution as a result of the cutoff of the UF membrane. Attention has to be paid to avoid biodegradation of the recovered sizes, which changes the properties of the polymer and causes intensive odor of the regenerates.

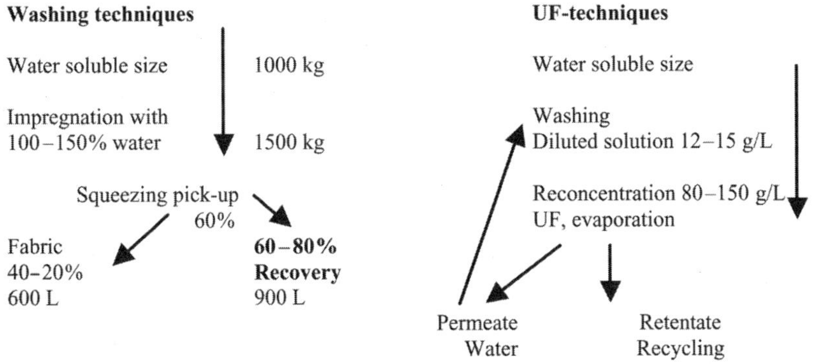

Figure 6 Recycling of sizes (from Refs 18–24).

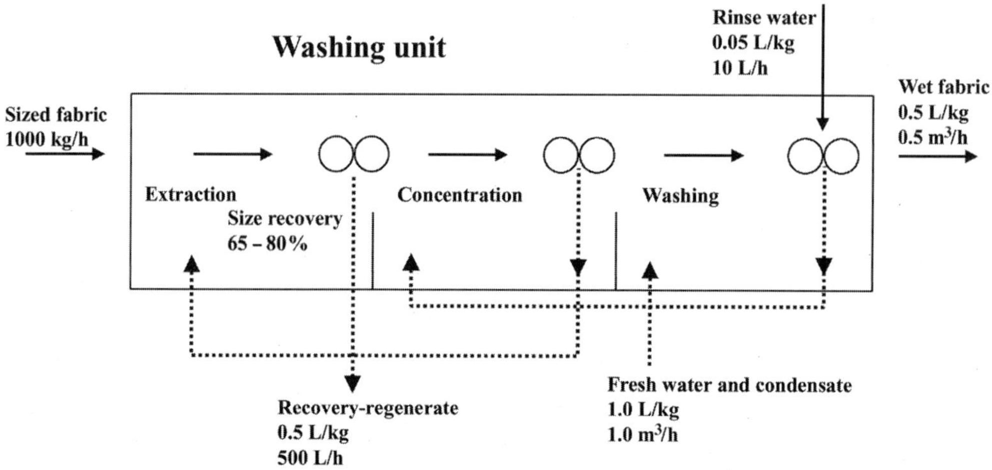

Figure 7 Recovery of sizes by washing techniques (from Refs 3 and 22).

In general, for a recovery of sizes, the following points have to be defined:

- establishment of continuous and time-stable conditions in sizing/desizing/regeneration/reuse;
- low amount of impurities in the regenerates due to dyes, colored fibers, dust from singing;
- establishment of an organizational structure that is able to handle the recovered products.

In many cases savings due to lowered costs for size and COD in the wastewater exceed the expenses for investment and running costs; thus acceptable data for ROI of less than two years are given in the literature [23,24].

Scouring, Alkaline Pretreatment, and Peroxide Bleach

A central step of pretreatment of natural cellulose fibers like cotton or linen for dyeing and printing is the alkaline scouring and bleach of the fibers. Figure 8 gives an overview for the pretreatment of cotton. Besides the destruction of the natural yellow-gray color of the fibers by the bleach chemicals, a considerable part of the organic compounds is removed from the fibers during the alkaline scouring step [3]. Average values of the compounds present in raw cotton are given in Table 8.

Assuming an average COD for the released compounds of 200 mgO_2/g, a total COD of 20 gO_2 per 1 kg of cotton is transported into the wastewater. In a batch treatment applying a liquor ratio of 1 : 10, 1 kg of cotton is extracted with 10 L of water, thus a COD of 2000 mg O_2/L can be estimated without consideration of the COD resulting from added auxiliaries or complexing agents. At present auxiliaries are usually in use that are easily biodegradable; thus after neutralization no problems should appear during the treatment in a CWWT. The main problem arising from alkaline scouring is therefore due to the considerable load in COD.

A typical recipe for alkaline scouring processes (liquor ratio 1 : 10) is as follows:

2–8 g/L NaOH;
0.3–3 g/L complexing agent (polyphosphate, carbohydrates, polyacrylate, phosphonate, nitrilo-tri-acetic acid (NTA);
0.5–3 g/L surfactant.

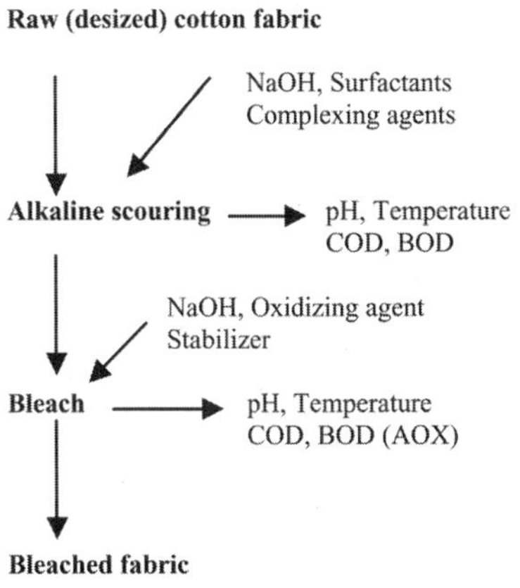

Figure 8 General scheme for the pretreatment of cotton (from Refs 3 and 27).

The total water consumption of the treatment including the rinsing step is approximately up to 50 L/kg. When the composition of an auxiliary is known, an estimation of the COD can be made by calculation of the oxygen demand for total oxidation. Examples are given below for Na–polyacrylate ($\text{—}CH_2\text{—}CHCO_2Na\text{—}$) and for Na–gluconate. Basing on Eqs (3) and (4), the oxygen demand for 1 g of compound can be calculated.

$$C_6H_{11}O_7Na + 5.5O_2 \rightleftarrows NaOH + 6CO_2 + 6H_2O \tag{3}$$

$$(\text{—}CH_2\text{—}CHCO_2Na\text{—}) + 3O_2 \rightleftarrows NaOH + 3CO_2 + 2H_2O \tag{4}$$

For the oxidation of 1 g of Na–gluconate, 810 mg of O_2 are required, and the oxidation of 1 g of Na–polyacrylate 1020 mgO_2 will be necessary. Technical products are mainly liquid formulations and the actual composition is given very rarely, but on the basis of the content of active compounds and an assumption of the chemical structure, an estimation of the contribution of the auxiliaries to the COD can be made.

The COD contribution of a recipe using 2 g/L of an auxiliary that contains 50% polyacrylate to the total COD in the wastewater will be approximately $\text{COD} = 2 \times 0.50 \times 1020 = 1020$ mgO_2/L.

Generally the treatment of waste water from alkaline scouring/bleaching (peroxide) processes will require an adjustment of pH and temperature, which is normally made by mixing with wastewater from other treatment steps. When surfactants, complexing agents, and so on,

Table 8 Average Composition of Raw Cotton

Component	%
Cellulose	80–90
Hemicellulose, pectin	4–6
Waxes, fat	0.5–1
Proteins	1.5
Minerals (Ca, Mg, K, Na, P)	1–2
Other components	0.5–1
Humidity	6–8

Source: Ref. 3.

with good biodegradability/bioelimination have been selected, the COD load is removed in CWWT without problems. The main load of the COD is due to the substances released from the fibers and added auxiliaries, thus an optimization of the load of COD released is limited to the auxiliaries only; however, these components will represent only a minor part of the total COD.

The application of chlorine bleach on the basis of hypochlorite/chlorite for the preparation of cotton/linen results in considerable formation of AOX in the effluents. Such processes should be replaced by bleach processes on the basis of peroxide. To obtain a sufficient degree of whiteness during the bleach, a two-step bleach (peracetic acid/peroxide) process has been proposed in the literature [25–27]. Such processes avoid the formation of chlorinated organic compounds (AOX).

Mercerization

Depending on conditions applied, the treatment of cotton textiles in concentrated alkaline solutions, for example, 300 g/L NaOH, leads to increased luster, improved dimensional stability, high uptake of dyes, and changes in strength and hand. Usually a continuous treatment process is applied. As a result enormous amounts of concentrated caustic soda solution have to be removed during the washing step. As a typical value approximately 300 g of NaOH are transported per 1 kg of cotton into the following up stabilization/washing baths. In the stabilization step the caustic soda is rapidly removed by washing with diluted caustic soda solutions. The effluents from the stabilization step contain approximately 40–60 g/L NaOH. Figure 9 gives an overview of the steps during mercerization of cotton.

The high costs for the consumed NaOH and the costs for neutralization of the NaOH in wastewater favor the recycling of NaOH by reconcentration procedures. Normally a re-concentration is made up to at least 400 g/L NaOH. Starting from a diluted NaOH containing 50 g/L NaOH, 7.8 L of water has to be removed to obtain 1 L NaOH with 440 g/L. The reconcentration is usually made by reboiling. For this purpose evaporation plants with several evaporation stages are in use. The use of several stages (normally at least three stages) is of

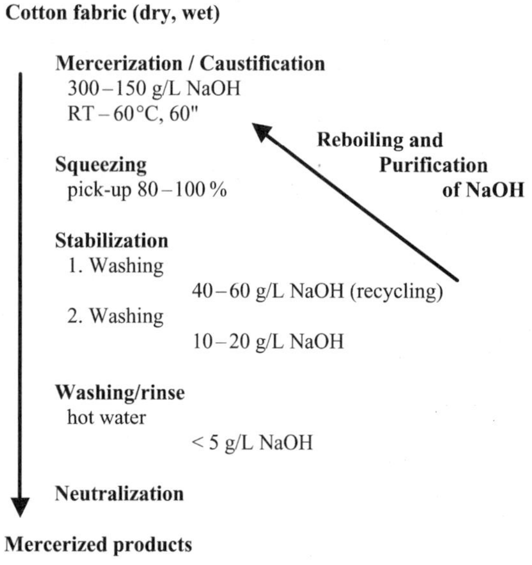

Figure 9 Mercerization of cotton (from Refs 28–31).

importance to keep the energy consumption of the process within reasonable dimensions. Typical values for energy consumption are 0.2–0.3 kWh/kg of evaporated water. Large amounts of waste energy are released from the condensation of the evaporated water and have to be used in the form of warm water. Care has to be taken to achieve a reuse of the warm water because the degree of heat recovery is essential to obtain an acceptable return on investment (ROI) of the unit [28].

Purification of the reboiled caustic soda is important to remove sizes (raw-mercerization), dyes (mercerization of dyed materials), fibers, and impurities released from the fibers. Important techniques are filtration, centrifugation, flotation processes, and oxidative processes [29–31]. The application of membrane processes for reconcentration is limited to low concentrations of NaOH because of the insufficient chemical stability of the membranes.

The reuse of the diluted caustic soda from the first stabilization compartment in other processes, for example, alkaline scouring, has been recommended. Problems can arise from variations in concentration and impurities present in the reused lye, so the recycling of the diluted NaOH for other treatment processes is not used widely. As the amount of caustic soda that can be reused for other processes is low compared to the amount of NaOH released from the mercerization step, regeneration by evaporation is normally the favored process.

Dyeing of Cellulose Fibers

Dyeing of cellulose textiles can be performed at all stages of textile processing, for example, fibers, yarn, fabric, or garment dyeing. Depending on the desired final properties of the dyed material, various classes of dyes are used, which are collected in gamuts of common application. Important classes of dyes are direct dyes, reactive dyes, and vat dyes, including indigo and sulfur dyes.

Wastewater problems mainly arise from three different sources:

- dyestuff: colored effluents, AOX, heavy metal content (Cu, Ni) [32];
- dispersing agents in dyestuff formulation: COD, poor biodegradability;
- auxiliaries, chemicals added: salt content (NaCl, Na_2SO_4), sulfide, pH value (NaOH, soda, silicates), COD (glucose, hydroxyacetone), N-content (urea).

Direct Dyes. For direct dyes a degree of fixation in the range 70–90% is given in the literature [33–35]. When optimized dyes and processes with a high degree of fixation are implemented into a dyehouse, problems of colored wastewater can be minimized. As heavy metal ions are mainly present in complexed form in the dyestuff, a lowering of the Cu and Ni content in the wastewater goes in parallel with an increase in dyestuff fixation. A similar situation is found with AOX values, which result from the halogen bound in the dyestuff molecules. In dyehouses where chlorine bleach has been substituted by other bleach chemicals, halogens bound in dyes can cause a main contribution to the AOX value found in the wastewater.

Reactive Dyes. The situation with regard to heavy metals (e.g., Cu, Ni from phthalocyanine dyes) and AOX from covalently bound halogen is comparable with direct dyes. Selection of processes with a high fixation of dyestuff yields a considerable decrease in Cu/Ni concentrations and AOX. For the fixation process certain amounts of alkaline are added to the dyebath. As the total amount of alkali used is low compared to the consumption of alkali during mercerization, scouring, and bleach, high pH due to the alkali from reactive dyeing is of minor relevance. Two main problems have to be mentioned in connection with reactive dyeing [36]:

- High load of soluble salt (NaCl, Na_2SO_4). For acceptable exhaustion of dyes, considerable concentrations of salt (up to 50 g/L) are required in exhaust dyeing processes. The release of the used dyebath transports a rather high load of salt into the wastewater stream. When a liquor ratio of 1 : 10 is applied, 10 L of dyebath are used for dyeing of 1 kg of goods, thus at a salt concentration of 50 g/L an amount of 0.5 kg salt is released for dyeing of 1 kg of goods.
- Colored wastewater. The problem of relatively high dyestuff concentrations in wastewater particularly arises when dyestuff exhaustion and fixation proceed only to a limited degree, typically only 70–80%, so that between 30 and 20% of the dye is released with the spent dyebath and the washing baths that follow. Such a situation is observed particularly with reactive dyeing processes where a covalent reaction of the dye with the fiber takes place but some of the reactive groups become hydrolyzed during dyeing and thus some dye remains unfixed in the dyebath. Depending on the general method of dyeing, two different qualities of colored wastewater can be identified (Fig. 10).

Particularly in the case of dyes with a limited degree of fixation the dyestuff content in the wasted water leads to intensively colored wastewater. As the reactive group of the unfixed dyestuff is hydrolyzed into an inactive form, a reuse is not possible. On the basis of an exhaust dyeing with 5% color depth, a liquor ratio of 1 : 10, and a degree of dyestuff fixation of 70–80% corresponding to 3.5–4 g/L of dye are fixed on the goods and 1.5–1 g/L of hydrolyzed dyes are released with the dyebath.

For exhaust dyeing processes a reduction of the liquor ratio leads to significant improvements. When the dyestuff fixation is known for a certain liquor ratio, the lowering of the amount of unfixed dye released into the wasted water can be estimated as a function of the liquor ratio (LR). The amount of dyestuff on the fiber, m_{DF}, can be calculated using Eq. (5), and the total amount of dyestuff in the dyebath, m_D, can be calculated using Eq. (6).

$$m_{DF} = m_F p_F \tag{5}$$

$$m_D = m_F c_D LR \tag{6}$$

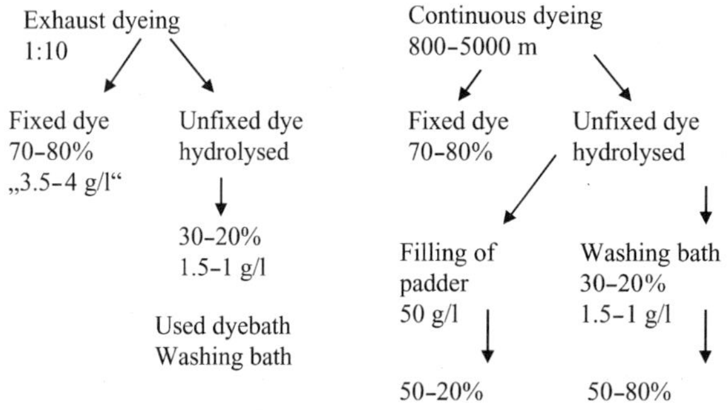

Figure 10 Sources for colored wastes from textile dyeing operations (from Ref. 55).

On the basis of Eqs (5) and (6), the part of dyestuff released as hydrolyzed dye can be estimated using Eq. (7).

$$L = \frac{m_F c_D LR}{m_F p_F + m_F c_D LR} = \frac{c_D LR}{p_F + c_D LR} \tag{7}$$

When a color depth of 5% (50 g dyestuff per 1 kg of goods) is used as basis for a calculation and a dyestuff fixation of 80% is observed at a liquor ratio of 1 : 10 (10 L of dyebath for 1 kg of goods) then a mass of 40 g dyestuff is fixed on the textile while 10 g remain in the dyebath as hydrolyzed dye. The dyestuff concentration c_D in the used bath is then 1 g/L ($p_F = 0.05$, $LR = 10$, $c_D = 1$ g/L).

While at *LR* 1 : 10 a fixation of 80% is observed, a reduction of *LR* to 1 : 5 lowers the losses of dyestuff to approximately 11% and a degree of fixation of 89% is expected. These results clearly indicate the importance of a low liquor ratio to optimize the degree of dyestuff fixation.

Another source of highly colored dyebaths is found in continuous dyeing processes where the last filling of the padder required to complete the process at well-defined conditions has to be withdrawn at the end of the padding process. Dyestuff concentrations of 50 g L^{-3} technical dyestuff are quite usual for such dye liquors.

For a dyestuff fixation of 70–80% and a color depth of 5% a concentration of 1.5–1 g/L hydrolyzed dye is expected in the wastewater, when 10 L of washing water is applied per 1 kg of goods. The emission of colored wastewater here can be divided into two different sources, the wastewater from the washing of the dyed material and the residual filling of the padder.

Depending on the length of the dyed piece (800–5000 m) the contribution of the filling of the padder to the total dyestuff concentration in the wasted water is estimated between 50 and 20%.

In general there are two different qualities of colored wasted water:

- The fillings of the padder. High dyestuff concentrations of approximately 50 g/L, high concentration of alkali;
- Spent dyebaths and washing baths. Low concentration of dyestuff, approximately 1 g/L, low concentration of alkali.

Besides an optimization of the dyestuff and the dyeing processes with regard to improved dyebath, exhaustion, the problem of colored wastewater released from dyehouses, has led to numerous technical developments proposed to overcome it.

A large number of techniques have been described in the literature, for example, dyestuff adsorption, oxidative and reductive treatments, electrochemical oxidation or reduction methods, electrochemical treatment with flocculation, membrane separation processes, and biological methods [37–55]. Each of these techniques offers special advantages, but they can also be understood as a source of coupled problems, for example, consumption of chemicals, increased COD, AOX, increased chemical load in the wastewater, and formation of sludge that has to be disposed.

The techniques for decolorization of dye-containing solutions can be applied at different stages:

- Treatment of concentrated dyestuff solutions (e.g., filling of padder), which is an efficient way to handle such concentrates, but as shown in Figure 10 usually only part of the released dyestuff is decolorized by treatment of such baths.

- Treatment of separately collected and reconcentrated baths that initially contain dyestuff concentrations of approximately 1 g/L and are reconcentrated to approximately 10–20 g/L dyestuff by membrane filtration. Such techniques yield considerable amounts of recyclable water, but care has to be taken to avoid any disturbing effect during reuse caused by salt and alkali content in the regenerate. The concentrated dyestuff solution can be treated with similar methods as concentrated dye solutions from fillings of padder.
- Treatment of the total wastewater: this technique will be discussed in Section 8.3, "End-of-pipe Technologies." The general scheme of such treatments is shown in Figure 11.

Vat Dyes. Vat dyes are normally present in their insoluble oxidized form. During their application in the dyeing process the dyestuffs are reduced in alkaline solution by addition of reducing agents, for example, dithionite, hydroxyacetone, formaldehydsulfoxylates. Vat dyes normally exhibit an excellent degree of fixation; thus, the problem of colored wastewater is of minor relevance. In addition, vat dyes are readily reoxidized in the wastewater into the insoluble oxidized form that precipitates and thus shows lower absorbance. The main problem in the wastewater released form reducing agents which cause certain load in the effluents (XX1). In the case of dithionite, sulfate is formed that can cause corrosion of concrete tubes, and in the case of hydroxyacetone, the COD is increased considerably. A substitution of the nonregenerable reducing agents by electrochemical reduction has been proposed in the literature [56].

Sulfur Dyes. Similar to the vat dyes, sulfur dyes are applied in reduced form. Owing to the lower redox potential of the dyes, reducing agents such as sulfide, polysulfide, glucose, hydroxyacetone, or mixtures of glucose with dithionite are in use. Sulfides should be replaced by other organic reducing agents mentioned above; in such cases the COD is increased but the products are easily biodegradable. In comparison to the vat dyes the degree of fixation is lower with sulfur dyes. As such, dyes are mainly used for dark shades and colored effluents have to be treated with methods similar to the processes mentioned with reactive dyes.

Indigo. Dyeing with indigo for the Denim market (jeans) is unique. Here a nonuniform dyeing through the cross-section of the yarn is the desired type of quality. There is only one dye in use, indigo. For this type of textile the warp is dyed before the weaving process and special techniques are applied on unique dyeing machines specialized to produce indigo-dyed warp yarn [57]. Figure 12 presents a scheme of the dyeing process. After the warp yarn has been wetted and squeezed, it is immersed into the dyebath, which contains the reduced indigo dye (from 1 to 5 g/L) for a few seconds. After mangling to 80–90% expression, the reduced dyestuff on the material is oxidized completely during an air passage that lasts for 60–120 s. The immersion/squeezing/

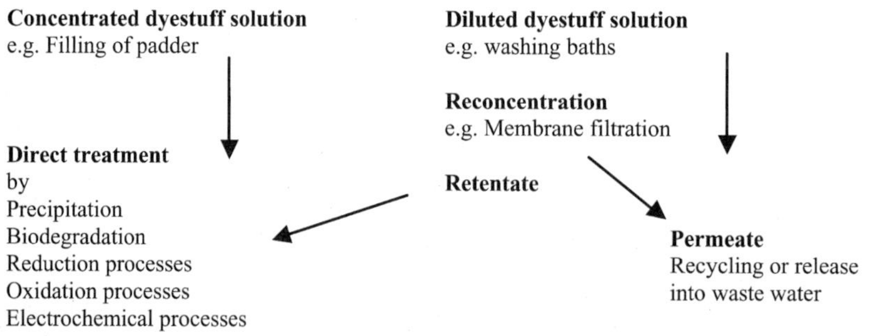

Figure 11 Treatment scheme for colored wasted water (from Ref. 54).

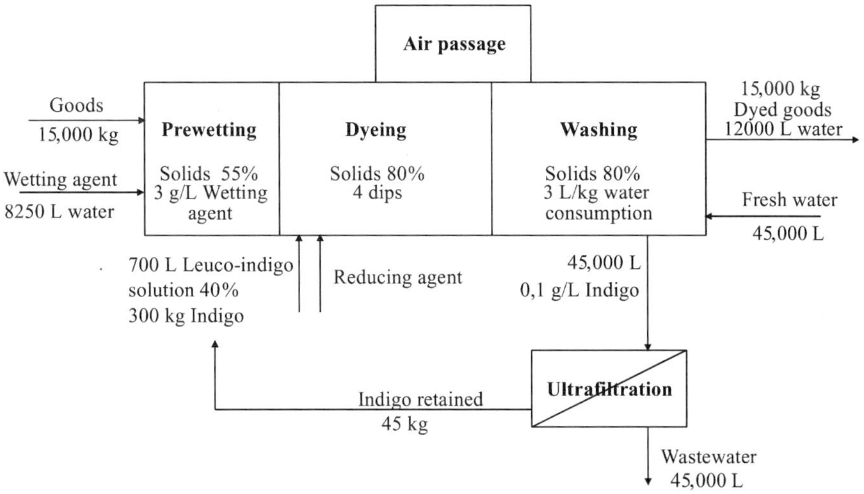

Figure 12 Flow scheme for indigo recovery in continuous yarn dyeing for denim (from Ref. 57).

oxidation cycle is repeated several times and the dyestuff is applied layer by layer. After the last oxidation passage the dyed material is washed and dried. Table 9 presents the typical data describing the production scale of such a dyeing unit. Two main difficulties exist at present:

- the indigo content in the wasted water, which causes colored wasted water;
- the sulfate or COD content in the washing water due to the use of dithionite or hydroxyacetone as reducing agents.

Table 9 Working Conditions and Production Data for a Full-Scale Indigo Dyeing Range

Production rate	
Cotton yarn	15,000 kg/day
	11.9 kg/min
Hours of operation	21 h/day
Warp speed	35 m/min
Depth of shade	2% indigo
Consumption of chemicals	
Reducing agent $Na_2S_2O_4$	50–126 kg/day
	40–100 g/min
Water	3–5 L/kg
	45–74 m^3/day
Composition of dyebath	
Wetting agent	0.5 g/L
Pre-reduced indigo	1–4 g/L
NaOH to maintain pH to	11.5–12.0
Temperature	20–30°C
Redox potential	<−700 mV

Source: Refs 57–59.

A considerable improvement of the situation could be obtained by the use of prereduced indigo instead of the reduction of the dyestuff in a stock vat [58]. By use of prereduced indigo the sulfate concentration in the wasted water can be lowered to approximately 50% of the initial value.

A recovery of the dispersed indigo from the wastewater can be obtained by use of UF. Owing to the low price of indigo the cost savings due to dyestuff recovery are poor compared to the investment. The problem of sulfate load can only be solved by the use of more expensive organic reducing agents, which can be degraded by anaerobic digestion [59]. Additional improvements are expected from the use of electrochemical methods for the reduction of dyestuff instead of nonregenerable reducing agents [57].

Figure 12 shows a flow scheme of a complete installation including the recycling of the diluted dyebath by ultrafiltration (UF) with regard to the dispersed oxidized indigo. The permeate is used as washing water or released, and after reduction of the dyestuff in a stock vat, the indigo-containing permeate is reused for dyeing processes.

The reuse of purified wastewater from dyeing processes for pretreatment processes has also been studied in detail [60].

8.2.3 Dyeing of Synthetic Fibers

Polyester PES

Polyester fibers represent the most important group of man-made fibers. With an annual production volume of 19.2 Mt, polyester fibers hold second position in world production of textile fibers [10]. Polyester is usually dyed with disperse dyes.

Three techniques are in use at present:

- High temperature (HT) processes. To exceed the glass-transition temperature processes and to achieve sufficient rate of dyeing and leveling, the temperature of the dyebath is elevated to 110–115°C in high-temperature dyeing apparatus. Normally such processes are limited to batch processes and specialized equipment has to be used to stand the high pressure.
- Dyeing with use of carriers. The addition of organic compounds of low molecular weight permits the temperature to be lowered below 100°C for polyester dyeing; thus dyeings can also be performed in normal pressure equipment. The chloro-organic compounds widely used in the 1970s have now been replaced by chlorine-free carriers such as aromatic esters, substituted phenols.
- Thermosol dyeing. The characteristics of low-molecular-weight polyester dyes can be utilized in thermosol dyeing processes. In this continuous dyeing process the material is impregnated with the dispersed dye, dried and heated to a temperature of approximately 200–210°C. The dyestuff is fixed by sublimation into the fiber.

Generally only low amounts of chemicals are added to the dyebaths and the degree of dyestuff fixation is high, so except for the application of carriers, which has to be considered carefully, and dispersing agents with limited biodegradability, the dyeing of polyester fibers causes minor problems with regard to the release of hazardous wastes [61].

An important innovative technique to replace water as the solvent in dyeing processes is the use of supercritical fluids, for example, supercritical CO_2 for dyeing processes. Successful trials have been conducted in various scales with different fibers and full-scale production has been performed in the case of PES dyeing [62,63]. Besides the handling of high pressure equipment, the development of special dyestuff formulations is required.

Elastomer Fibers: Elastan, Lycra^{TM}

An increasing percentage of textiles is now designed with elastic properties, which are obtained by the introduction of elastic fibers into them. The pretreatment of elastomer-containing fibers can be regarded as representative for the pretreatment of other man-made fibers. To improve the behavior of these fibers during spinning, winding, weaving, and knitting, considerable amounts of auxiliaries are added. Typical examples for such compounds are:

- fatty amines;
- polyethylene glycols;
- hydrocarbons;
- silicone compounds.

In particular in the case of elastomer fibers, such compounds (in many cases silicone compounds) add up to 2.5–8% of the weight of the fibers. Besides problems in removing these oily components during pretreatment, for example, washing of the textiles, the compounds are then detected in the wastewater in considerable amounts. As the addition of such auxiliaries is required for technical purposes, an optimization of the situation has to be achieved by direct cooperation between the fiber/yarn/fabric producer and the textile dyehouses.

8.2.4 Dyeing on Standing Dyebath

A method to lower the release of chemicals, auxiliaries, and residual dyestuff in exhaust processes is dyeing on a standing dyebath. In such a technique the exhausted dyebath, which contains the auxiliaries, chemicals (salt), and dyestuff is reused for the next dyeing after a replenishment of the exhausted dyestuff and lost chemicals. In fact, such techniques are not as widely in use as might be expected because a set of requirements has to be fulfilled to introduce them:

- no accumulation of chemicals (e.g., spent reducing agents in vat dyeing will lead to increasing salt concentration);
- no formation of dyestuff byproducts (hydrolyzed dye in reactive dyeing);
- the run of the dyeing process has to be suited for dyeing on a standing bath (no dosing of chemicals);
- the size of a batch that has to be dyed at the same conditions has to be significant.

Examples for such techniques are found in sulfur dyeing for black shades and in a special form in indigo dyeing for denim, where a continuous replenishment of the dyebath is performed for a long period of production.

8.2.5 Textile Printing Operations

Numerous variations of textile printing processes are found in textile production depending on the type of fiber, applied dyes, desired effect, and fashion.

At present, flat screen printing and rotary screen printing are the main techniques used. Here the dyestuff is dissolved/dispersed in a printing paste containing thickener and chemicals. With every change of color, the filling of the dosing unit and of the screen has to be withdrawn. As such, changes frequently involve considerable amounts of used printing pastes having to be handled. In addition, the equipment (screen, pumps, and containers) have to be cleaned, so a distinct load is released into the wastewater. This amount increases with shorter lengths of printed batch. Table 10 gives two examples for the composition of printing pastes.

Table 10 Composition of 1000 g Printing Pastes for Pigment Printing and Two-Phase Reactive Printing

Pigment printing	Mass (g)	Two-phase reactive printing	Mass (g)
Pigment	5–80	Dyestuff	1–100
Thickener (e.g., polyacrylate)	15–45	Urea	50
Emulsifier (e.g., fattyalcohol-polyglycolethers)	5–10	Alginate thickener	400
Binder (e.g., copolymers from butylacrylate, acrylonitrile, styrol)	60–80	m-Nitro-benzene-sulfonic acid Na-salt	15
Fixation agent (melamine formaldehyde condensation prod.)	5–10	Buffer (e.g., NaH_2PO_4)	2–3
Catalysator (e.g., $MgCl_2$)	0–2		
Softener (fatty acid ester)	5–10		
Anti-foam agent	0–3		
Water	ad 1000	Water	ad 1000

Source: Ref. 3.

A COD of reactive printing pastes of 150,000–200,000 mgO_2/kg for pigment paste values of up to 350,000 O_2/kg are realistic. Additional problems arise from the AOX content (chlorine containing dyestuff) and from heavy metal content resulting from metal ions complexed in the dyes (e.g., Co, Cu, Ni). Attention also has to be given to the use of antimicrobial agents in the printing pastes, which are added to block the microbial growth that results in degradation of the thickener and lowering of the viscosity of the printing paste.

Generally, any release of printing pastes into the wastewater should be avoided, and in many countries such action is forbidden. Figure 13 gives an overview of the possible proceedings to minimize chemical load in the wasted water from the release of printing pastes [64,65].

First the consumption of printing pastes has to be minimized by:

- Minimization of the required volumes to fill the equipment, e.g., printing screen, tubes, pumps, and container. By optimization, a filling of up to 8 kg can be reduced to a consumption less than 2 kg per filling.
- Exact calculation and metering of the consumption of printing paste to avoid excess of pastes.

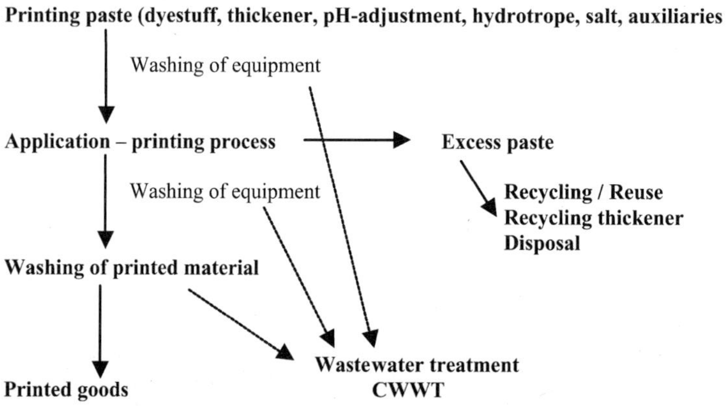

Figure 13 Minimization of chemical load from textile printing (from Ref. 57).

The minimization of the filling of the equipment is of particular importance for the production of short lengths, for example, during sample printing. In particular for the production of very short lengths (e.g., 120 m), a considerable portion of the printing paste is required for the filling of the printing machine. Depending on the coverage factor of a pattern, approximately 55–80% of the paste is used for printing, while 45–20% is spent for the filling of the printing machine, which is considerable with a mass of 5 kg in this example. When a length of 1000 m is produced the portion of paste spent for the filling reduces to 10–3% of the total mass of printing paste [66]. The high consumption of printing pastes for the production of short length samples causes high costs for the production of a collection of new patterns and thus at present digital printing techniques are recommended to substitute for the expensive full-scale production of design samples.

The high content of dissolved compounds and the broad variations in the concentration of dyes and auxiliaries make a direct recycling of pastes difficult. Supported by calculation programs, a certain portion of printing pastes can be added for the preparation of new pastes [67]. In the most simple case, the preparation of pastes for the printing of black color is carried out.

If disposal is necessary, various techniques can be used: drying and incineration, binding in concrete, and anaerobic degradation [64,65].

A recent technique to achieve a reuse of the thickener is the precipitation of the thickener by addition of organic solvent (e.g., methanol). After removal of the dyes and chemicals the thickener can be reused for the preparation of new pastes. The removed chemicals and dyes are collected and discarded [68]. By this method a considerable part of the COD-forming compounds can be recycled and the AOX and heavy metal content in the wastewater from textile printing can be reduced.

The replacement of classical textile printing techniques by digital printing techniques (ink-jet and bubble jet) is in full progress. Present limitations result from the availability of appropriate formulations of inks/dyes and fixation techniques. The comparable low production speed and limitations with regard to the quality of the textile material can be expected to be overcome within the next 5–10 years.

8.2.6 Finishing Processes

A great part of the variation in the final properties of a textile is adjusted for by finishing procedures, for example, wrinkle resistance, soil repellence, hydrophobic properties, flame retardance and antimicrobial properties [69]. In many cases chemicals are added by padding/squeezing followed by drying/fixation, for example, in a stenter.

Representative groups of chemicals used are:

- urea-formaldehyde resins for crosslinking of cellulose textiles, e.g., dimethylol-dihydroxyethylene-urea (DMDHEU);
- dispersions of polymers (polyacrylesters, polyethylene, silicones);
- fluorocarbon compounds.

The applied products are fixed on the textile by drying/curing, but similar to the pad batch dyeing procedures, the last filling of the padding unit needs additional attention. A release of such concentrated finishing baths can introduce a COD of up to 200,000 mgO_2/L of liquor [70].

In a first attempt the volumes of residual baths have to be optimized and a reorganization of the recipes with regard to feed of residual excess volumes of a finishing bath into similar finishing recipes is recommended [71]. If reuse is not possible, a careful check of recipes with regard to easy biodegradation/bioelimination is necessary.

8.3. END-OF-PIPE TECHNIQUES

8.3.1 First Steps

The application of end-of-pipe technologies as general procedures for the treatment of wastewater has changed from simple procedures to sophisticated concepts, applying a consecutive set of methods that has been adapted to the particular situation of a textile plant [72]. As already discussed in the previous sections, the separation of concentrated wastes and the treatment of small volumes of concentrates are much more efficient compared to a global treatment of mixed wastes.

Numerous techniques and types of equipment have been developed and tested in laboratory tests, on a pilot scale, or in full technical application. The introduction of a technique is always coupled to a general wastewater treatment concept and has to consider the individual situation of a textile producer [73–75].

As a first step, a separation of different types of wastewater into the following groups is recommended:

- Concentrated liquids: fillings of padders (dyeing, finishing), printing pastes, used dyebaths;
- Medium polluted wastes (e.g., washing, rinsing baths);
- Low to zero polluted wastes (e.g., cooling water).

Basic general procedures applied are:

- Collection and mixing of released baths to level pH and temperature maxima in the final wastewater stream;
- Adjustment of pH by neutralization. Cellulose dyeing and finishing companies mainly release alkaline baths, which can be neutralized by introduction of CO_2-containing waste gas from the power/steam generation plant [76].

8.3.2 Overview

According to Schönberger and Kaps [3], the various methods for the treatment of wastewater from textile plants can be divided into the groups given in Table 11.

Table 11 Techniques for Waste Water Treatment

Separation, concentration	Decompositon, degradation	Exchange processes
Membrane techniques: Microfiltration, ultrafiltration (UF), nanofiltration (NF), reverse osmosis (RO)	Oxidation: Aerobic, wet oxidation, ozonation, peroxides (incl. Fenton's reagents), electrochemical oxidation	Ion-exchange
Mechanical Processes: Sedimentation, filtration	Incineration	
Evaporation	Reduction: chemical, electrochemical	
Precipitation, flocculation		
Flotation		
Adsorption		
Formation of inclusion complexes		
Extraction processes		
Stripping		

Source: Ref. 3.

The application of a certain technology for wastewater treatment is dependent on the type of wastewater, thus different technologies have been proposed and are applied at present. Normally a combination of procedures and equipment are applied and a big variety of concepts have been realized. To facilitate an overview of the different techniques, the most important processes are discussed in this section. Full concepts that are specialized to a distinct situation are given in the references [77–82]. Some of the techniques have already been discussed in Section 8.2.

8.3.3 Desizing, Pretreatment

The anaerobic biodegradation of sizes is favorable because the aerobic degradation of size-containing waste water requires approximately 1 kWh/kg of BOD, while the anaerobic degradation yields 0.5–1.5 kWh/kg of BOD and in addition releases a lower volume of sludge. A general problem for biological treatment steps can be identified with the demand for a rather constant feed of load into the biological system to obtain constant conditions in microbial growth.

Theoretically, polymer-containing wastewater from desizing can be purified for water recycling by removal and reconcentration of the polymer by ultrafiltration or evaporation, but the high costs of investment and additional expenses for the disposal of the concentrate hinder the introduction of such techniques as a general treatment process.

For the degradation of polymers like PVA and carboxy-methyl-cellulose (CMC), low-pressure wet oxidation (5–20 bar, $<200^{\circ}C$) has been proposed [83]. In this process oxygen and a catalyst are used to destroy the organic material by oxidation.

The application of evaporation processes for purification and recycling of wastewater has been used in various concepts. The main problems that have to be considered are:

- energy consumption and heat recovery;
- incrustation and cleaning;
- corrosion;
- treatment of concentrated residues (e.g., incineration, disposal).

In many countries the disposal of the concentrated residues formed is rather complicated because this material has to be handled as hazardous waste.

The removal of fiber/yarn preparation during the pretreatment of knitted material can be identified as an important source of oil, grease, and silicones in wastewater. A general treatment can be performed by means of precipitation, flocculation, membrane filtration, and evaporation.

The removal of these components is required because these components are not biodegraded in the CWWT, but mainly adsorb on the sludge. When the sludge from the CWWT is used as fertilizer for farming, these components are transported to farmland and thus get released there. The reuse of bleach baths after catalase treatment has also been proposed in the literature [84].

8.3.4 Treatment of Wastewater from Dyeing Processes

The wastewater from dyeing processes contains a lot of components in various concentrations, for example, dyestuff, alkali, acid, salt, and auxiliaries [85]. In a first basic step, a separation of the wastewater stream according to the degree of chemical load should be performed.

A treatment of wastewater with low pollution for reuse can be achieved by the combination of:

- adjustment of pH and temperature;
- sedimentation, precipitation [86];

- flocculation ($Fe^{2+/3+}$, Al^{3+}, polyelectrolyte) [87];
- filtration (e.g., sand filter);
- adsorption (e.g., activated carbon) [79,88–93];
- ozone treatment.

In many cases the removal/destruction of the intensive color is the main goal to be achieved. Important techniques are given in the literature [66,94,95]:

Oxidative processes can be based on ozonation, UV treatment, hydrogen peroxide, and Fenton's reagent for the destruction of the chromopore [96–107].

Aerobic biodegradation processes often show unsatisfying results because a number of azo dyes are resistant to aerobic microbiological attack. The main process for removal of dyes in the aerobic part of a CWWT is based on an adsorption of the dyes on the biomass. Further problems in the destruction of chromophores result during the treatment of phthalocyanine dyes, anthraquinoid dyes, and vat and sulfur dyes, which contain rather persistent chromophores.

Reductive Processes. A reductive cleavage of the azo groups can be achieved by direct introduction of the dyes into the anaerobic step of a CWWT, but this method is restricted for heavy-metal-containing dyes, for example, phthalocyanine dyes, because of contamination of the sludge. In many cases the reductive destruction of colored dye baths is performed by the addition of reducing chemicals such as $Na_2S_2O_4$ and Fe^{2+} salts. As such processes generally lead to an increased load in the wastewater, such treatments should be replaced. The formation of aromatic amines as a result of the application of reducing conditions has to be considered in detail for every application.

Precipitation/Flocculation. Various chemicals can be added to textile wastewater to obtain precipitation/flocculation of colored substances:

- Addition of iron salt/$Ca(OH)_2$ is a rather simple and cheap method to form sludge, but the costs for separation and disposal of the sludge must be considered [108].
- Destabilization of the dissolved compounds by addition of iron or aluminum salts and addition of polyelectrolytes to support agglomeration and formation of larger size precipitation.

The removal of precipitate can be achieved by sedimentation, flotation, and filtration. If a recycling of water is intended, additional purification, for example, by adsorption methods, is needed to remove any added metal ions and flocculation auxiliaries.

At present these methods, which are based on the formation of a large amount of sludge containing substances of low/limited biodegradability should only be used after careful optimization of the process conditions.

Membrane Processes [109,110]. Depending on the desired application, membrane techniques can be divided into:

- micro-, ultrafiltration (e.g., polymers, pressure $p = 1–10$ bar);
- nanofiltration (e.g., organic molecules, $p = 10–40$ bar);
- reverse osmosis (e.g., salt, $p = 10–80$ bar).

In the case of purification of water the permeate is the cleaned water and the removed components are collected in the concentrate [111–113]. Various modules can be used, such as plate-modules, tubes, and capillary modules. For water purification and recycling processes the following aspects have to be considered:

- high permeate flow;
- selectivity;
- stability and life-time of membrane and equipment;

- cleaning of membrane;
- tendency for membrane fouling;
- costs.

Today numerous membrane filtration units for removal of dissolved dyes such as reactive dyes are in full-scale operation. The treatment of the remaining concentrates still remains difficult. At present the following have been proposed in the literature and tested in full-scale operation:

- evaporation, incineration;
- anaerobic degradation;
- electrochemical reduction.

Electrochemical Processes. The reductive cleavage of azo-group-containing dyes has been applied on a full scale for the decolorization of concentrates from batch dyeing. Depending on the color, decolorization of up to 80% of the initial absorbance can be obtained. Mixed processes consist of combinations of electrochemical treatment and precipitation by use of dissolving electrodes [43,49]. Such techniques have been described in the literature and have, in part, also been tested on a full scale. Anodic processes that form chlorine from oxidation of chloride have also been proposed to destroy dyes, but care has to be taken with regard to the chlorine and chlorinated products (AOX) formed [114,115].

A special technique proposed in the literature for the removal of dyes is the inclusion of dye into cave molecules such as crown-ethers/cucurbituril, but developments with regard to regeneration and disposal of the crown ether have to be performed to permit introduction into full-scale application [116].

Adsorption processes and ion-pair extraction processes can also be used to remove color from wastewater [117–119]. The main problem to be solved in adsorption processes is the further treatment of the loaded adsorbents (regeneration, disposal). A similar situation is found in ion-pair extraction, where a concentrated organic phase results from the process and further treatment of this product is required.

Evaporation can also be used to purify wastewater, particularly in the case of heavy-metal-containing wastewater where a removal of the heavy metal ions is achieved, but again the problem of further treatment or disposal of the formed concentrated residue has to be solved [80].

In many cases combinations of the techniques are applied to obtain an optimized process fitting on the individual situation of the textile dyehouse, for example:

- nanofiltraton–oxidation processes;
- nanofiltration–evaporation–oxidation;
- evaporation–oxidation.

Another full-scale process combines catalytic oxidation including biodegradation, adsorption, precipitation/flocculation, and reverse osmosis [120].

8.3.5 Wastewater from Printing and Finishing Processes

The main difference in the wastes from dyeing processes is identified in the presence of thickeners and, in some cases, additional difficulties can arise from the added auxiliaries and hydrotropes (e.g., urea).

As a result, a high COD is found in the effluents and end-of-pipe technologies that form sludge have to face a high amount of precipitate.

In pigment printing the dyestuff pigments are bound to the textile by means of a polymer binder system and no additional washing is performed; however, wastewater is released from the cleaning of the equipment and machinery.

Printing pastes should be recycled whenever possible. Disposal is possible by incineration and biological degradation. Problems can arise in biodegradation from preservatives added to the pastes to avoid microbial growth and in cases of high formaldehyde and heavy metal content.

As a high number of different chemicals is applied in finishing processes, reuse is difficult in many cases. A high number of the used compounds show low biodegradability, so disposal is recommended in many cases. Techniques proposed in the literature include incineration, low-pressure wet oxidation [H_2O_2, Fe salt, NaOH, $Ca(OH)_2$], and precipitation by addition of high concentrations of Na_2SO_4 [121].

8.3.6 General Treatment Procedures

For the treatment of already mixed wastewater, various methods have been proposed and tested in full-scale application; examples are:

- Oxidation processes: oxidation in the presence of carbon particles and coupled precipitation [$FeSO_4$, $Ca(OH)_2$, polyelectrolyte] [37];
- Biological oxidation/degradation including sedimentation;
- Coupling of physical processes (flotation, sedimentation) [82,122];
- Aerobic/anaerobic biological degradation [123–133].

In some cases (particularly reactive dyes) dyes can pass the aerobic, anaerobic degradation step and colored water is observed at the end of the treatment. In such cases a special treatment of the colored wastewater (reduction, adsorption, precipitation) has to be introduced [105,134–137]. In the presence of low concentrations of organic compounds, ozonation can be used as a final "polishing" step.

NOMENCLATURE

AOX	adsorbable halogenated compounds
APEO	alkylphenol-ethoxylates, surfactants
B	factors for BOD from Table 7 (mg/g)
BOD	biological oxygen demand (mg/L)
C	factors for COD from Table 7 (mg/g)
c_D	concentration of hydrolyzed dyestuff in spent dyebath (kg/L)
CLY	lyocell fiber
CMC	carboxymethyl cellulose (size, thickener for printing)
CMD	modal fiber
Co	cotton fiber
COD	chemical oxygen demand (mg/L)
CV	viscose fiber
CWWT	communal wastewater treatment plant (e.g., combination of sedimentation, aerobic treatment, anaerobic treatment, nitrification, and elimination of phosphor)
DOC	dissolved organic carbon (mg/L)
EDTA	ethylene-diamine-tetra-acetic-acid (complexing agent)
LR	liquor ratio as volume of dyebath per mass of goods (L/kg)
L	losses, part of dyestuff released into the wastewater stream (dimensionless)

L_{BOD}	released load in BOD (kgO_2)
L_{COD}	released load in COD (kgO_2)
m	mass of desized fabric (kg)
m_{DF}	mass of dyestuff fixed on the fiber (kg)
m_F	mass of goods (kg)
m_D	mass of dyestuff in spent dyebath (kg)
NF	nanofiltration
NTA	nitrilotriacetic acid (complexing agent)
p	mass of size in fabric (kg/kg)
p_F	fixation of dyestuff in dyed material (kg/kg)
PVA	polyvinyl alcohol (type of size)
RO	reverse osmosis
UF	ultrafiltration

REFERENCES

1. Anon. *Verordnung über die Begrenzung von Abwasseremissionen aus Textilbe-trieben 2/14.*, BGBL 1992/612, *Regulation of the Ministry of Agriculture and Forestry*: Austria, 1992.
2. Müller, K.; and Schönberger, H. Der Anhang 38 zur Abwasserverordnung. *Melliand Textil.* **2002**, *83*, 256–261.
3. Schönberger, H.; Kaps, U. *Reduktion der Abwasserbelastung in der Textilindustrie*, Umweltbundesamt; Berlin, 1994.
4. Schramm, W.; Jantschgi, J. Comparative assessment of textile dyeing technologies from a preventive environmental protection point of view. *J. Soc. Dyers Colour* **1999**, *115*, 130–135.
5. Naylor, C.G. Environmental fate and safety of nonylphenol ethoxylates. *Text. Chem. Color.* **1995**, *27*, 29–33.
6. Glober, B.; Lorrain H. Waste minimisation in the dyehouse. *Text. Chem. Color.* **1993**, *25*, 15–20.
7. Shah, H.A.; Sharma, M.A.; Doshi, S.M.; Pillay, G.R. Practical approach towards energy conservation, economy and effluent control in a process-house. *Colourage* **1989**, *15*, 20c.
8. Müller, B. Adsorbable organic halogens in textile effluents. *Rev. Prog. Coloration* **1992**, *22*, 14–21.
9. Anon. *Richtline 96/61/EG des Rates vom 24. Sept. 1996 über die integrierte Vermeidung und Verminderung der Umweltverschmutzung*, Amtsblatt der Europ. Gemeinschaften, L 257/26, 10.10.1996, 26–40.
10. Anon. World production: 59 million tons textile fibres. In *Man-Made Fibre Year Book*; IBP, International Business Press Publisher: Frankfurt a.M., Germany, August 2002; 21.
11. Gibson, M.D.M.; Morgan, W.V.; Robinson, B. Aspekte der Wollwäsche und Abwasseraufbereitung. *Textil Praxis Int.*, **1979**, **34**, 437–444, 701–708.
12. Hoffmann, R.; Timmer, G. Abwasserreinigung der BWK – Ergebnisse nach Optimierung des Gesamtsystems. *Melliand Textil.* **1994**, *75*, 831–837.
13. Hoffmann, R.; Timmer, G. Abwasserreinigung der BWK – eine Lösung für Wollwäschereien und kämmereien. *Melliand Textil.* **1991**, *72*, 562–566.
14. Heiz, H. Chlor/Hercosett-Ausrüstung von Wolle. *Textilveredlung*, **1981**, *16*, 43–53.
15. Duffield, P.A.; Holt, R.R.D.; Smith, J.R. Färben mit geringem Restchromgehalt im Abwasser. *Melliand Textil.* **1991**, *72*, 938–942.
16. Doser, C.; Zschocke, P.; Biedermann, J.; Süssmuth, R.; Trauter, J. Mikrobieller Abbau von Polyacrylsäureschlichten. *Textilveredlung* **1997**, *32*, 245–249.
17. Klein, M.; Zschocke, P.; Süssmuth, R.; Trauter, J. Über den anaeroben biologischen Abbau der Polyvinylalkohole. *Textilveredlung* **1997**, *32*, 241–245.
18. Rüttiger, W.; Schenk, W.; Würz, A. Recycling-Verfahren für Schlichtemittel. *Chem-Ing-Tech.* **1983**, *33*, 490–494.

19. Deschler, O. Schlichte-Rückgewinnung und Wiederverwendung – Resumée nach 5 Jahren Praxis. *Melliand Textil.* **1983**, *64*, 716–720.
20. Hoechst AG. German Patent Application, DAS 2808920, 02 March 1978.
21. BASF AG. German Patent Application, DAS 2543815, 01 October 1975.
22. Rüttiger, W. Abwasserentlastung durch Rückgewinnung und Aufkonzentrierung von Acrylat-Schlichten. *Textil Praxis Int.*, **1983**, *38*, 975–977, 1117–1121.
23. Byazeed, A.; Trauter, J. Untersuchungen von Veränderungen der physikalischen und technologischen Eigenschaften wasserlöslicher Schlichtemittel bei der Ultrafiltration. *Textil Praxis Int.*, **1992**, *47*, 220–229.
24. Trauter, J. Anwendung der Ultrafiltration für das Schlichtemittel- und Indigo-Recycling. *Melliand Textil.* **1993**, *74*, 559–563.
25. Olip, V. Peressigsäure eine Bleichchemikalie. *Melliand Textil.* **1992**, *73*, 819–822.
26. Wurster, P. Die Peressigsäurebleiche – eine Alternative zu Bleichverfahren mit halogenhaltigen Oxidationsmitteln. *Textil Praxis Int.* **1992**, *47*, 960–965.
27. Cheng, K.M. An improvement in effluent disposal with emphasis on cotton pretreatment processes. *Text. Chem. Color.* **1998**, *30* (3), 15–21.
28. Bechtold, T.; Gmeiner, D.; Burtscher, E.; Bösch, I.; Bobleter, O. Flotation of particles suspended in lye by the decomposition of hydrogen peroxide. *Separ. Sci. Technol.* **1989**, *24*, 441–451.
29. Galda, K. Rückgewinnung von Mercerisierlauge. *Melliand Textil.* **1998**, *79*, 38–39.
30. Son, E.J.; Choe, E.K. Nanofiltration membrane technology for caustic soda recovery. *Text. Chem. Color. Am. D.* **2000**, *32*, 46–52.
31. Bechtold, T.; Burtscher, E.; Sejkora, G.; Bobleter, O. Modern methods of lye recovery. *Int. Textil Bulletin* **1985**, *31*, 5–26.
32. Baughman, G.L. Fate of copper in copperized dyes during biological waste treatment I: Direct dyes. *Textile Chem. Color. Am. D.* **2000**, *32* (1), 51–55.
33. Anon. ATV-Arbeitsgruppe 7.2.23, Abwässer in der Textilindustrie – Arbeitsbericht der ATV-Arbeitsgruppe 7.2.23 "Textilveredlungsindustrie". *Korrespondenz Abwasser* **1989**, *36*, 1074–1084.
34. *Ecological and toxicological association of the dyestuff manufacturing industry (ETAD). E 3022 Environmental hazard and risk assessment in the context of EC directive 79/831/* from 31 May 1991.
35. Beckmann, W.; Sewekow, U. Farbige Abwasser aus der Reaktiv-färberei: Probleme und Wege zur Lösung. *Textil Praxis Int.* **1991**, *46*, 346–348.
36. Kwok, W.Y.; Xin, J.H.; Sin, K.M. Quantitative prediction of the degree of pollution of effluent from reactive dye mixtures. *Color. Technol.* **2002**, *118*, 174–180.
37. Marmagne, O.; Coste, C. Color removal from textile plant effluents. *Am. Dyest. Rep.* **1966**, *85*, 15–21.
38. Tsui, L.S.; Roy, W.R.; Cole M.A. Removal of dissolved textile dyes from wastewater by a compost sorbent. *J. Soc. Dyers Colour.* **2003**, *119*, 14–18.
39. Tokuda, J.; Ohura, R.; Iwasaki, T.; Takeuchi, Y.; Kashiwada, A.; Nango, M. Decoloration of azo dyes by hydrogen catalyzed by water-soluble manganese porphyrins. *Textile Res. J.* **1999**, *69*, 956–960.
40. Gregor, K.H. Aufbereitung und Wiederverwertung von Färbereiabwässern. *Melliand Textil.* **1998**, *79*, 643–646.
41. Perkins, W.S.; Walsh, W.K.; Reed, I.E.; Namboodri, C.G. A demonstration of reuse of spent dyebath water following color removal with ozone. *Text. Chem Color.* **1995**, *28*, 31–37.
42. Yoshida, Y.; Ogata, S.; Nakamatsu, S.; Shimamune, T.; Kikawa, K.; Inoue, H.; Iwakura, C. Decoloration of azo dye using atomic hydrogen permeating through a Pt-modified palladized Pd sheet electrode. *Electrochim. Acta* **1999**, *45*, 409–414.
43. Laschinger, M. Implementing dyehouse wastewater treatment systems. *Am. Dyest. Rep.* **1996**, *85*, 23–27.
44. Kolb, M.; Korger, P.; Funke, B. Entfärbung von textilem Abwasser mit Dithionit. *Melliand Textil.* **1988**, *69*, 286–287.

45. McClung, S.M.; Lemley, A.T. Electrochemical treatment and HPLC analysis of wastewater containing acid dyes. *Text. Chem. Color.* **1994**, *26*, 17–22.
46. Elgal, G.M. Recycling and disposing of dyebath solutions. *Text. Chem. Color.* **1986**, *18*, 15–20.
47. Burtscher, E.; Bechtold, T.; Amann, A.; Turcanu, A.; Schramm, C.; Bobleter, O. Aspekte der Teilstrombehandlung Textiler Abwässer unter besonderer Berücksichtigung der Farbigkeit. *Melliand Textil.* **1993**, *74*, 903–907.
48. Marzinkowski, J.M.; van Clewe, B. Wasserkreislaufführung durch Membranfiltration der farbigen Abwässer. *Melliand Textil.* **1998**, *79*, 174–177.
49. Wilcock, A.; Brewster, M.; Tincher, W. Using electrochemical technology to treat textile wastewater: three case studies. *Am. Dyest. Rep.* **1992**, *81*, 15–22.
50. Diaper, C.; Correia, V.M.; Judd, S.J. The use of membranes for the recycling of water and chemicals from dyehouse effluents: an economic assessment. *J. Soc. Dyers Colour.* **1996**, *112*, 273–281.
51. Knapp, J.S.; Zhang, F.; Tapely, K.N. Decolourisation of Orange II by a wood-rotting fungus. *J. Chem. Technol. Biot.* **1997**, *69*, 289–296.
52. Oxspring, A.A.; McMullan, G.; Smyth, W.F.; Marchant, R. Decolourisation and metabolism of the reactive textile dye, Remazol Black 5, by an immobilized microbial consortium. *Biotechnol. Lett.* **1996**, *18*, 527–530.
53. Nigam, P.; McMullan, G.; Banat, I.M.; Marchant, R. Decolourisation of effluent from the textile industry by a microbial consortium. *Biotechnol. Lett.* **1997**, *18*, 117–120.
54. Bechtold, T.; Burtscher, E.; Turcanu, A. Cathodic decolourisation of textile waste water containing reactive dyes using a multi-cathode electrolyser. *J. Chem Technol. Biot.* **2001**, *76*, 300–311.
55. Bechtold, T.; Mader, J.; Mader, C. Entfärbung von Reaktivfarbstoffen durch kathodische Reduktion. *Melliand Textil.* **2002**, *83*, 361–364.
56. Bechtold, T.; Burtscher, E.; Bobleter, O. Application of electrochemical processes and electroanalytical methods in textile chemistry. *Curr. Topics Electrochem.* **1998**, *6*, 97–110.
57. Bechtold, T.; Burtscher, E.; Kühnel, G.; Bobleter, O. Electrochemical processes in indigo dyeing. *J. Soc. Dyers Colour.* **1997**, *113*, 135–144.
58. DyStar/BASF Technical information Bulletin, TI/T 245e, 1994.
59. Rümmele, W. Lösung des Abwasserproblems in einer Indigofärberei. *Textilveredlung* **1989**, *24*, 96–97.
60. Denter, U.; Schollmeyer, E. Einsatz von mit Alkali- und Erdalkalisalzen belastetem Prozeßwasser in der Vorbehandlung von Textilien. *Textil Praxis Int.* **1991**, *46*, 644–646, 1343–1348.
61. Richter, P. Möglichkeiten zur Abwasserentlastung beim Färben von Polyesterfasern. *Melliand Textil.* **1993**, *74*, 872–875.
62. Bach, E.; Cleve, E.; Schüttken, J.; Schollmeyer, E.; Rucker, J.W. Correlation of solubility data of azo disperse dyes with the dye uptake of poly(ethylene terephthalate) fibres in supercritical carbon dioxide. *Color. Technol.* **2001**, *117*, 13–18.
63. Schmidt, A.; Bach, E.; Schollmeyer, E. Supercritical fluid dyeing of cotton modified with 2,4,6-trichloro-1,3,5-triazine. *J. Soc. Dyers Colour.* **2003**, *119*, 31–36.
64. Schönberger, H. Reduzierung der Abwasserbelastung in Textildruckereien durch produktionsintegrierte Massnahmen. Textilveredlung **1994**, *29*, 128–133.
65. Provost, J.R. Effluent improvement by source reduction of chemicals used in textile printing. *J. Soc. Dyers Colour.* **1992**, *108*, 260–264.
66. Pierce, J. Colour in textile effluents – the origins of the problem. *J. Soc. Dyers Colour.* **1994**, *110*, 131–133.
67. Brocks, J. Minimierung der Abwasserbelastung in der Druckerei. *Textil Praxis Int.* **1992**, *47*, 550–554.
68. Marte, W.; Meyer, U. Verdicker-Recyclierung im Textildruck. *Textil-veredlung* **1995**, *30*, 64–68.
69. Roberts, D.L.; Hall, M.E.; Horrocks, A.R. Environmental aspects of flame-retardant textiles – an overview. *Rev. Prog. Color.* **1992**, *22*, 48–57.
70. Puk, R.; Sedlak, D. Ein Konzept zur Minimierung von Restflotten. *Textil Praxis Int.* **1992**, *47*, 238–245.

71. Teichmann, R. Methodik zum Wiedereinsatz von konzentrierten Ausrüstungsmittel-Restflotten. *Textilveredlung* **1997**, *32*, 131–135.
72. Coia-Ahlman, S.; Groff, K.A. Textile wastes. *Res. J. Water Pollut. C.* **1990**, *62*, 473–478.
73. Höhn, W.; Reinigungmöglichkeiten für textile Abwässer im Überblick. *Melliand Textil.* **1998**, *79*, 647–649.
74. Park, J.; Shore, J. Water for the dyehouse: Supply, consumption, recovery and disposal. *J. Soc. Dyers Colour.* **1984**, *100*, 383–399.
75. Athanasopoulos, N. Abwässer bei der Veredlung von Baumwolle und Baumwollmischungen – von der Versuchsanlage zur Betriebsanlage. *Melliand Textil.* **1990**, *71*, 619–628.
76. Anon. Neutralisation alkalischer Abwässer mit Rauchgas. *Chemie-fasern/Textilindustrie* **1976**, *78*, 924.
77. Wilking, A.; Frahne, D. Textilabwasser – Behandlungsverfahren der 90er Jahre. *Melliand Textil.* **1993**, *74*, 897–900.
78. Sharma, M.A. Treatment of cotton textile mills effluent – A case study. *Colourage* **1989**, *36*, 15–21.
79. Schulze-Rettmer, R. Treatment of textile dyeing wastewater by adsorption/bio-oxidation process. *Text. Chem. Color.* **1998**, *30*, 19–23.
80. Richarts, F. Integrales Konzept der Ver- und Entsorgung von Textilveredlungsbetrieben. *Textil Praxis Int.* **1991**, *46*, 567–572.
81. Mukherjee, A.K.; Bhuvanesh, G.; Chowdhury, S.M.S. Separation dyes from cotton dyeing effluent using cationic polyelectrolytes. *Am. Dyest. Rep.* **1999**, *88*, 25–28.
82. Reid, R. On-site colour removal at Courtaulds Textiles. *J. Soc. Dyers Colour.* **1996**, *112*, 140–141.
83. Horak, O. Katalyische Naßoxidation von biologisch schwer abbaubaren Abwasserinhaltsstoffen unter milden Reaktionsbedingungen. *Chem. Ing. Tech.* **1990**, *62*, 555–557.
84. Tzanov, T.; Costa, S.; Guebitz, G.; Cavaco-Paulo, A. Effect of temperature and bath composition on the dyeing of cotton with catalase treated bleaching effluent. *J. Soc. Dyers Colour.* **2001**, *117*, 116–170.
85. Sewekow, U. Färbereiabwässer – behördliche Anforderungen und Problemlösungen. *Melliand Textil.* **1989**, *70*, 589–596.
86. Fiola, R.; Luce, R. Wastewater solutions for the blue jeans processing industry. *Am. Dyest. Rep.* **1998**, *87*, 54–55.
87. Papic, S.; Koprivanac, N.; Bozic, A.L. Removal of reactive dyes from wastewater using Fe(III) coagulant. *J. Soc. Dyers Colour.* **2000**, *116*, 352–359.
88. Netpradit, S.; Thiravetyan, P.; Towprayoon S. Application of "waste" metalhydroxide sludge for adsorption of azo reactive dyes. *Water Res.* **2003**, *37*, 763–772.
89. Kadirvelu, K.; Kavipriya, M.; Karthika, C.; Radhika, M.; Vennilamani N.; Pattabhi S. Utilization of various agricultural wastes for activated carbon preparation and application for the removal of dyes and metal ions from aqueous solutions. *Biores. Technol.* **2003**, *87*, 129–132.
90. Sun, Q.; Yang, L. The adsorption of basic dyes from aqueous solution on modified peat-resin particle. *Water Res.* **2003**, *37*, 1535–1544.
91. Puranik, S.A.; Rathi, A.K.A. Treatment of wastewater pollutant from direct dyes. *Am. Dyest. Rep.* **1999**, *88*, 42–50.
92. Gärtner, R.; Müller, W.; Schulz, G.; Lehr, T. Neue Sorptionsmaterialien auf Basis speziell aufbereiteter Polyamidabfälle als Adsorptivreiniger für Färbereiabwässer. *Melliand Textil.* **1996**, *77*, 67–72.
93. Laszo, J.A. Preparing an ion exchange resin from sugarcane bagasse to remove reactive dye from wastewater. *Text. Chem. Color.* **1996**, *85*, 13–17.
94. Beckmann, W.; Sewekow, U. Farbige Abwasser aus Reaktivfärberei: Probleme und Wege zur Lösung. *Textil Praxis Int.* **1991**, *46*, 445–449.
95. Cooper, P. Removing colour from dyehouse waste waters – a critical review of technology available. *J. Soc. Dyers Colour.* **1993**, *109*, 97–100.
96. Hickman, W.S. Environmental aspects of textile processing. *J. Soc. Dyers Colour.* **1993**, *109*, 32–37.
97. Huang, C.R.; Lin, Y.K.; Shu, H.Y. Wastewater decolorization and toc-reduction by sequential treatment. *Am. Dyest. Rep.* **1994**, *83*, 15–18.

98. Arslan-Alaton, I.; Balcioglu, I.A. Heterogenous photocatalytic treatment of dyebath wastewater in a TFFB reactor. *AATCC Review* **2002**, *2*, 33–36.
99. Hassan, M.M.; Hawkyard, C.J. Reuse of spent dyebath following decolorisation with ozone. *J. Soc. Dyers Colour.* **2002**, *118*, 104–111.
100. Fung, P.C.; Sin, K.M.; Tsui, S.M. Decolorisation and degradation kinetics of reactive dye wastewater by a UV/ultrasonic/peroxide system. *J. Soc. Dyers Colour.* **2000**, *116*, 170–173.
101. Tokuda, J.; Oura, R.; Iwasaki, T.; Takeuchi, Y.; Kashiwada, A.; Nango, M. Decolorisation of azo dyes with hydrogen peroxide catalyzed by manganese protoporphyrins. *J. Soc. Dyers Colour.* **2000**, *116*, 42–47.
102. Perkins, W.S. Oxidative decolorisation of dyes in aqueous medium. *Text. Chem. Color. Am. D.* **1999**, *1*, 33–37.
103. Ferrero, F. Oxidative degradation of dyes and surfactant in the Fenton and photo-Fenton treatment of dyehouse effluents. *J. Soc. Dyers Colour.* **2000**, *116*, 148–153.
104. Yang, Y.; Wyat, D.T.; Bahorsky, M. Decolorisation of dyes using UV/H_2O_2 photochemical oxidation. *Text. Chem. Color.* **1998**, *30*, 27–35.
105. Strickland, A.F.; Perkins, W.S. Decolorisation of continuous dyeing wastewater by ozonation. *Text. Chem. Color.* **1995**, *27*, 11–15.
106. Uygur, A. An overview of oxidative and photooxidative decolorisation treatments of textile waste waters. *J. Soc. Dyers Colour.* **1997**, *113*, 211–217.
107. Namboodri, C.G.; Walsh, W.K. Ultraviolet light/hydrogen peroxide system for decolorizing spent reactive waste water. *Am. Dyest. Rep.* **1996**, *85*, 15–25.
108. Kolb, M.; Funke, B.; Gerber, H.-P.; Peschen, N. Entfärbung von Abwasser aus Textilbetrieben mit Fe(II)/$Ca(OH)_2$. *Korrespondenz Abwasser* **1987**, *34*, 238–241.
109. Wehlmann, U. Reinigen von Abwasser aus der Textilveredlung mit Membranverfahren. *Melliand Textil.* **1997**, *78*, 249–252.
110. Majewska-Nowak, K.; Winnicki, T.; Wisniewski, J. Effect of flow conditions on ultrafiltration efficiency of dye solutions and textile effluents. *Desalination* **1989**, *71*, 127–135.
111. Marzinkowski, J.M.; van Clewe, B. Wasserkreislaufführung durch Membranfiltration der farbigen Abwässer. *Melliand Textil.* **1998**, *79*, 174–177.
112. Diaper, C.; Correia, V.M.; Judd, S.J. The use of membranes for the recycling of water and chemicals from dyehouse effluents: an economic assessment. *J. Soc. Dyers Colour.* **1996**, *112*, 270–280.
113. Schäfer, T.; Trauter, J.; Janitza, J. Aufarbeitung von Färbereiabwässern durch Nanofiltration. *Textilveredlung* **1997**, *32*, 79–83.
114. Vilaseca, M.M.; Gutierrez, M.C.; Crespi, M. Biologische Abbaubarkeit von Abwässern nach elektrochemischer Behandlung. *Melliand Textil.* **2002**, *83*, 558–560.
115. Brincell INC. 2109 West 2300 South, Salt Lake City, Utah.
116. Buschmann, H.-J.; Schollmeyer, E. Selektive Abtrennung von Schwermetallkationen aus Färbereiabwässern. *Melliand Textil.* **1991**, *72*, 543–544.
117. Buschmann, H.-J.; Schollmeyer, E. Die Entfärbung von textilem Abwasser durch Bildung von Farbstoffeinschlußverbindungen. *Textilveredlung* **1998**, *33*, 44–47.
118. Smith, B.; Konce, T.; Hudson, S. Decolorizing dye wastewater using chitosan. *Am. Dyest. Rep.* **1993**, *82*, 18–36.
119. Steenken-Richter, I.; Kermer, W.D. Decolorising textile effluents. *J. Soc. Dyers Colour.* **1992**, *108*, 182–186.
120. Anon. Reinigung und Teil-Wiederverwendung von Maschenwaren-Abwässern mit Hilfe von Kokskohle-Adsorbens. *Textil Praxis Int.* **1986**, *41*, 943–949.
121. Janitza, J.; Koscielski, S. Reinigung und Wiederverwertung von Druckereiabwässern. *Int. Text. Bull.* **1996**, *4/96*, 28–32.
122. Glöckler, R. Optimiertes Textilabwasserreinigungsverfahren. *Melliand Textil.* **1995**, *76*, 1020–1021.
123. Yang, Q.; Yang, M.; Pritsch, K.; Yediler, A.; Hagn, A.; Schloter, M.; Kettrup, A. Decolorization of synthetic dyes and production of manganese-dependent peroxidase by new fungal isolates. *Biotechnol. Lett.* **2003**, *25*, 709–713.

124. Abadulla, E.; Robra, K.; Gübitz, G.M.; Silva, L.M.; Cavaco-Paulo, A. Enzymatic decolorisation of textile dyeing effluents. *Textile Res. J.* **2000**, *70*, 409–414.
125. Churchley, J.H.; Greaves, A.J.; Hutchings, M.G.; Phillips, D.A.S.; Taylor, J.A. A chemometric approach to understanding the bioelimination of anionic, water-soluble dyes by a biomass – Part 3: Direct dyes. *J. Soc. Dyers Colour.* **2000**, *116*, 279–284.
126. Goncalves, I.M.C.; Gomes, A.; Bras, R.; Ferra, M.I.A.; Amorim, M.T.P.; Porter, R.S. Biological treatment of effluent containing textile dyes. *J. Soc. Dyers Colour.* **2000**, *116*, 393–397.
127. Metosh-Dickey, C.; Davis, T.M.; McEntire, C.A.; Christopher, J.; DeLoach, H.; Portier, R.J. COD, color, and sludge reduction using immobilized bioreactor technology. *Textile Chem. Color. Am. D.* **2000**, *32*, 28–31.
128. Moreira, M.T.; Mieglo, I.; Feijoo, G.; Lema, J.M. Evaluation of different fungal strains in the decolourisation of synthetic dyes. *Biotechnol. Lett.* **2000**, *22*, 1499–1503.
129. Willmott, N.; Guthrie, J.; Nelson, G. The biotechnology approach to colour removal from textile effluent. *J. Soc. Dyers Colour.* **1998**, *114*, 38–41.
130. Cao, H.; Hardin, I.R.; Akin D.E.; Optimization of conditions for microbial decolorisation of textile wastewater: starch as a carbon source. *AATCC Rev.*, **2001**, *7*, 37–42.
131. Beckert, M.; Ohmann, U.; Platzer, B.; Bäuerle, U.; Burkert, G. Langzeiterfahrungen beim Reinigen von Abwässern nach dem SB-Verfahren. *Melliand Textil.* **2000**, *81*, 64–68.
132. Feitkenhauer, H.; Meer, U.; Marte, W.; Integration der anaeroben Abwasservorbehandlung in das Wassermanagementkonzept. *Melliand Textil.* **1999**, *80*, 303–306.
133. Gähr, F.; Lehr, T. Verbesserung der anaeroben Abbaubarkeit von Teilströmen aus der Textilveredlung durch Ozonbehandlung. *Textilveredlung* **1997**, *32*, 70–73.
134. Mock, B.; Hamouda, H. Ozone application to color destruction of industrial wastewater – Part I: Experimental. *Am. Dyest. Rep.* **1998**, *87*, 18–22.
135. Streibelt, H.P. Abwasser-Reinigung und -Wiederverwertung in der Textilveredlung. *Chemiefasern/Textilindustrie.* **1986**, *36/88*, 401–402.
136. Küßner, J.; Janitza, J.; Koscielski, S. Entfärbung, Reinigung und Wiederverwertung von farbigen Abwässern der Textilveredlungsindustrie und umweltgerechte Entsorgung der Reinigungsnebenprodukte. *Textil Praxis Int.* **1992**, *47*, 736–741.
137. Wragg, P. Waste water recycling – a case study. *J. Soc. Dyers Colour.* **1993**, *109*, 280–282.

9
Treatment of Phosphate Industry Wastes

Constantine Yapijakis
The Cooper Union, New York, New York, U.S.A.

Lawrence K. Wang
Zorex Corporation, Newtonville, New York, U.S.A., and
Lenox Institute of Water Technology, Lenox, Massachusetts, U.S.A.

9.1 INTRODUCTION

The phosphate manufacturing and phosphate fertilizer industry includes the production of elemental phosphorus, various phosphorus-derived chemicals, phosphate fertilizer chemicals, and other nonfertilizer phosphate chemicals [1–30]. Chemicals that are derived from phosphorus include phosphoric acid (dry process), phosphorus pentoxide, phosphorus pentasulfide, phosphorus trichloride, phosphorus oxychloride, sodium tripolyphosphate, and calcium phosphates [8]. The nonfertilizer phosphate production part of the industry includes defluorinated phosphate rock, defluorinated phosphoric acid, and sodium phosphate salts. The phosphate fertilizer segment of the industry produces the primary phosphorus nutrient source for the agricultural industry and for other applications of chemical fertilization. Many of these fertilizer products are toxic to aquatic life at certain levels of concentration, and many are also hazardous to human life and health when contact is made in a concentrated form.

9.1.1 Sources of Raw Materials

The basic raw materials used by the phosphorus chemicals, phosphates, and phosphate fertilizer manufacturing industry are mined phosphate rock and phosphoric acid produced by the wet process.

Ten to 15 millions of years ago, many species of marine life withdrew minute forms of phosphorus dissolved in the oceans, combined with such substances as calcium, limestone, and quartz sand, in order to construct their shells and bodies [30]. When these multitudes of marine organisms died, their shells and bodies (along with sea-life excretions and inorganic precipitates) settled to the ocean bottom where thick layers of such deposits – containing phosphorus among other things – were eventually formed. Land areas that formerly were at the ocean bottom millions of years ago and where such large deposits have been discovered are now being commercially mined for phosphate rock. About 70% of the world supply of phosphate rock comes from such an area around Bartow in central Florida, which was part of the Atlantic Ocean 10 million years ago [1]. Other significant phosphate rock mining and processing operations can be found in Jordan, Algeria, and Morocco [28].

9.1.2 Characteristics of Phosphate Rock Deposits

According to a literature survey conducted by Shahalam [28], the contents of various chemicals found in the natural mined phosphate rocks vary widely, depending on location, as shown in Table 1. For instance, the mineralogical and chemical analyses of low-grade hard phosphate from the different mined beds of phosphate rock in the Rusaifa area of Jordan indicate that the phosphates are of three main types: carbonate, siliceous, and silicate-carbonate. Phosphate deposits in this area exist in four distinct layers, of which the two deepest – first and second (the thickness of bed is about 3 and 3.5 m, respectively, and depth varies from about 20 to 30 m) – appear to be suitable for a currently cost-effective mining operation. A summary of the data from chemical analyses of the ores is shown in Table 2 [28].

Screen tests of the size fraction obtained from rocks mined from these beds, which were crushed through normal crushers of the phosphate processing plant in the area, indicated that the best recovery of phosphate in the first (deepest) bed is obtained from phosphate gains recovered at grain sizes of mesh 10–20 (standard). The high dust (particles of less than 200 mesh) portion of 11.60% by wt. of the ores remains as a potential air pollution source; however, the chemical analyses of these ores showed that crushing to smaller grain sizes tends to increase phosphate recovery. The highest percentage of phosphate from the second bed (next deepest) is also recovered from grain sizes of 10–20 mesh; however, substantial amounts of phosphate are also found in sizes of 40–100 mesh. Currently, the crushing operation usually maintains a maximum grain size between 15 and 30 mesh.

The phosphate rock deposits in the Florida region are in the form of small pebbles embedded in a matrix of phosphatic sands and clays [31]. These deposits are overlain with lime

Table 1 Range of Concentrations of Various Chemicals in Phosphate Ores

Chemical	Range
Fluorine	2.8–5.6%[a]
Sulphur (SO_3)	0.8–7.52%[a]
Carbon (CO_2)	2.07–10.7%[a]
Strontium	180–1683 ppm[b]
Manganese	0.001–0.004%[a]
Barium	0.044–0.40%[a]
Chlorine	0.20–1.42%[a]
Zinc	59–765 ppm[b]
Nickel	7–244 ppm[b]
Cobalt	31–34 ppm[b]
Chromium	12–895 ppm[b]
Copper	18–46 ppm[b]
Vanadium	0.03–0.08%[a]
Cadmium	0.038–1.5 ppm[b]
Uranium	4–8 ppm[b]
P_2O_5	40–55%[c]
Silica	3–34%[c]
Carbon (C)	14–48%[c]

[a] % by wt.
[b] Parts per million.
[c] Kusaifa Rocks only (% by wt.).
Source: Ref. 28.

Table 2 Chemical Analysis of Different Size Fractions of Phosphate in Mining Beds at Rusaifa

Size fractions in mesh	First bed (average chemical composition)			Second bed (average chemical composition)			Fourth bed (average chemical composition)		
	P_2O_5%	$CaCO_3$%	Insoluble%	P_2O_5%	$CaCO_3$%	Insoluble%	P_2O_5%	$CaCO_3$%	Insoluble%
<10	21.35	10.01	36.46	15.73	35.64	23.35	17.99	47.95	2.08
10–20	21.03	10.67	36.94	18.04	32.60	21.26	21.33	36.58	2.25
20–30	21.03	10.46	37.30	19.63	29.23	19.82	26.95	30.81	2.48
30–40	21.82	10.56	36.21	21.80	25.24	17.72	29.38	28.51	2.00
40–60	22.04	10.65	33.70	25.96	22.40	14.78	32.26	20.66	2.49
60–100	24.12	11.27	29.64	26.65	18.13	11.75	30.88	26.76	1.23
100–150	24.32	10.90	28.64	26.76	21.71	13.17	26.55	33.51	1.41
150–200	25.92	11.95	24.30	24.46	23.18	14.45	23.18	40.25	2.52
>200	25.28	12.50	23.42	23.37	28.06	12.99	21.29	40.25	4.24
Average total sample	Chemical					25.19	34.21	2.32	
Mineralogical	47.0	31.0[a]	18.0[b]	27.0[a]	50.0	15.0[b]	55.0	35.0[a]	3.5[b]

[a] Carbonaceous materials.
[b] Silica.
Source: Ref. 28.

rock and nonphosphate sands and can be found at depths varying from a few feet to hundreds of feet, although the current economical mining operations seldom reach beyond 18.3 m (60 ft) of depth.

9.1.3 Mining and Phosphate Rock Processing

Mechanized open-cut mining is used to first strip off the overburden and then to excavate in strips the exposed phosphate rock bed matrix. In the Rusaifa area of Jordan, the stripping ratio of overburden to phosphate rock is about 7 : 1 by wt. [28]. Following crushing and screening of the mined rocks in which the dust (less than 200 mesh) is rejected, they go through "beneficiation" processing. The unit processes involved in this wet treatment of the crushed rocks for the purpose of removing the mud and sand from the phosphate grains include slurrification, wet screening, agitation and hydrocycloning in a two-stage operation, followed by rotating filtration and thickening, with a final step of drying the phosphate rocks and separating the dusts. The beneficiation plant makes use of about 85% of the total volume of process water used in phosphate rock production.

Phosphate rocks from crushing and screening, which contain about 60% tricalcium phosphate, are fed into the beneficiation plant for upgrading by rejection of the larger than 4 mm over-size particles. Two stages of agitation follow the hydrocycloning, the underflow of which (over 270 mesh particles) is fed to rotary filters from which phosphatic cakes results (with 16–18% moisture). The hydrocyclone overflow contains undesirable slimes of silica carbonates and clay materials and is fed to gravity thickeners. The thickener underflow consisting of wastewater and slimes is directly discharged, along with wastewater from dust-removing cyclones in the drying operation, into the nearby river.

In a typical mining operation in Florida, the excavated phosphate rockbed matrix is dumped into a pit where it is slurrified by mixing it with water and subsequently carried to a washer plant [31]. In this operation, the larger particles are separated by the use of screens, shaker tables, and size-separation hydrocyclone units. The next step involves recovery of all particles larger than what is considered dust, that is, 200 mesh, through the use of both clarifiers for hydraulic sizing and a flotation process in which selective coating (using materials such as caustic soda, fuel oil, and a mixture of fatty acids and resins from the manufacture of chemical wood pulp known as tall oil, or resin oil from the flotation clarifier) of phosphate particles takes place after pH adjustment with NaOH.

The phosphate concentration in the tailings is upgraded to a level adequate for commercial exploitation through removal of the nonphosphate sand particles by flotation [32], in which the silica solids are selectively coated with an amine and floated off following a slurry dewatering and sulfuric acid treatment step. The commercial quality, kiln-dried phosphate rock product is sold directly as fertilizer, processed to normal superphosphate or triple superphosphate, or burned in electric furnaces to produce elemental phosphorus or phosphoric acid, as described in Section 9.2.

9.2 INDUSTRIAL OPERATIONS AND WASTEWATERS

The phosphate manufacturing and phosphate fertilizer industry is a basic chemical manufacturing industry, in which essentially both the mixing and chemical reactions of raw materials are involved in production. Also, short- and long-term chemical storage and warehousing, as well as loading/unloading and transportation of chemicals, are involved in the operation. In the case of fertilizer production, only the manufacturing of phosphate fertilizers and mixed and blend fertilizers containing phosphate along with nitrogen and/or potassium is presented here.

Regarding wastewater generation, volumes resulting from the production of phosphorus are several orders of magnitude greater than the wastewaters generated in any of the other product categories. Elemental phosphorus is an important wastewater contaminant common to all segments of the phosphate manufacturing industry, if the phossy water (water containing colloidal phosphorus) is not recycled to the phosphorus production facility for reuse.

9.2.1 Categorization in Phosphate Production

As previously mentioned, the phosphate manufacturing industry is broadly subdivided into two main categories: phosphorus-derived chemicals and other nonfertilizer phosphate chemicals. For the purposes of raw waste characterization and delineation of pretreatment information, the industry is further subdivided into six subcategories. The following categorization system (Table 3) of the various main production streams and their descriptions are taken from the federal guidelines [8] pertaining to state and local industrial pretreatment programs. It will be used in the following discussion to identify process flows and characterize the resulting raw waste. Figure 1 shows a flow diagram for the production streams of the entire phosphate manufacturing industry.

The manufacture of phosphorus-derived chemicals is almost entirely based on the production of elemental phosphorus from mined phosphate rock. Ferrophosphorus, widely used in the metallurgical industries, is a direct byproduct of the phosphorus production process. In the United States, over 85% of elemental phosphorus production is used to manufacture high-grade phosphoric acid by the furnace or dry process as opposed to the wet process that converts phosphate rock directly into low-grade phosphoric acid. The remainder of the elemental phosphorus is either marketed directly or converted into phosphorus chemicals. The furnace-grade phosphoric acid is marketed directly, mostly to the food and fertilizer industries. Finally, phosphoric acid is employed to manufacture sodium tripolyphosphate, which is used in detergents and for water treatment, and calcium phosphate, which is used in foods and animal feeds.

On the other hand, defluorinated phosphate rock is utilized as an animal feed ingredient. Defluorinated phosphoric acid is mainly used in the production of animal foodstuffs and liquid fertilizers. Finally, sodium phosphates, produced from wet process acid as the raw material, are used as intermediates in the production of cleaning compounds.

Table 3 Categorization System in Phosphorous-Derived and Nonfertilizer Phosphate Chemicals Production

Main category	Subcategory	Code
1. Phosphorus-derived chemicals	Phosphorus production	A
	Phosphorus-consuming	B
	Phosphate	C
2. Other nonfertilizer phosphate chemicals	Defluorinated phosphate rock	D
	Defluorinated phosphoric acid	E
	Sodium phosphates	F

Source: Ref. 8.

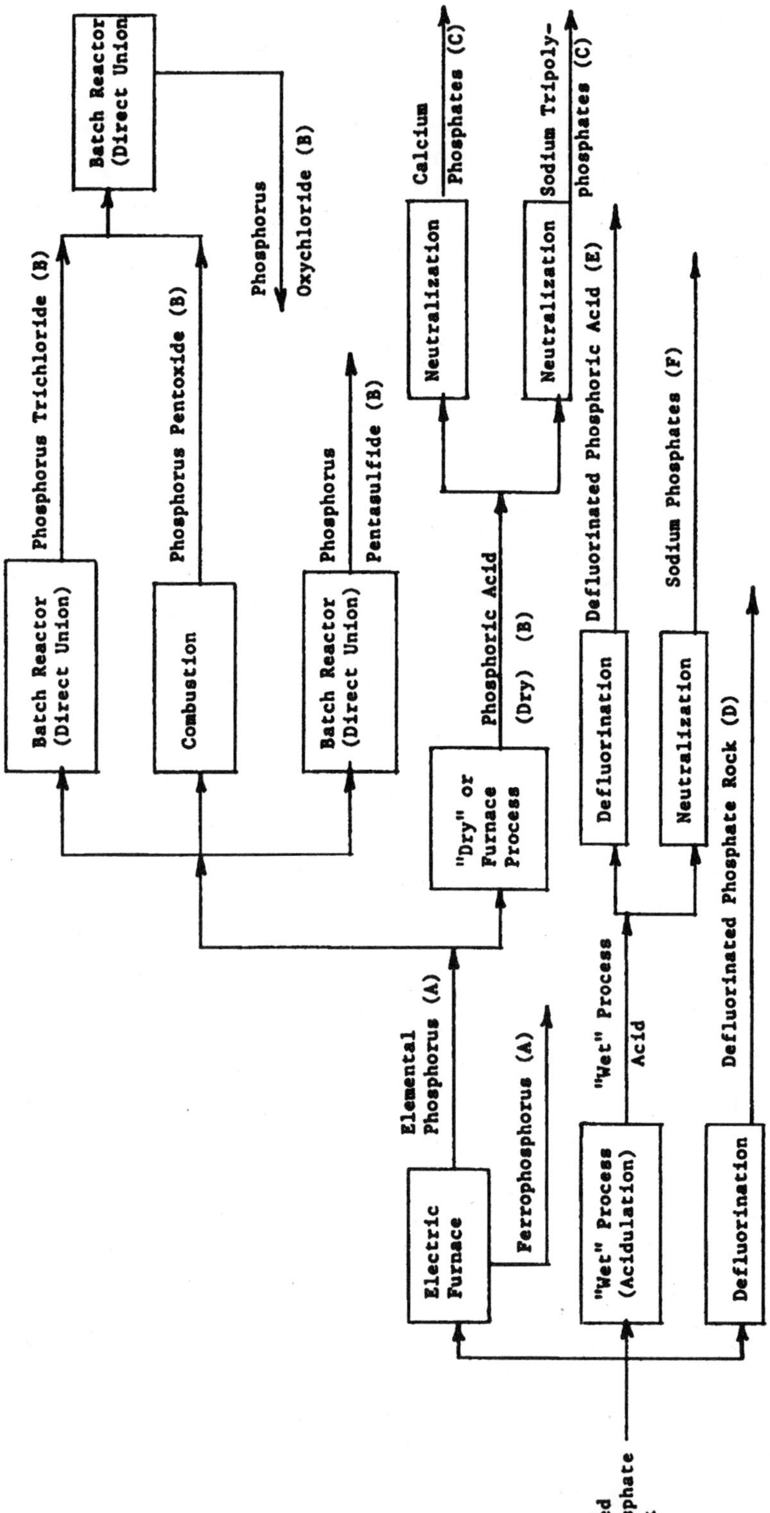

Figure 1 Phosphate manufacturing industry flow diagram (from Ref. 8).

9.2.2 Phosphorus and Phosphate Compounds

Phosphorus Production

Phosphorus is manufactured by the reduction of commercial-quality phosphate rock by coke in an electric furnace, with silica used as a flux. Slag, ferrophosphorus (from iron contained in the phosphate rock), and carbon monoxide are reaction byproducts. The standard process, as shown in Figure 2, consists of three basic parts: phosphate rock preparation, smelting in an electric furnace, and recovery of the resulting phosphorus. Phosphate rock ores are first blended so that the furnace feed is of uniform composition and then pretreated by heat drying, sizing or agglomerating the particles, and heat treatment.

The burden of treated rock, coke, and sand is fed to the furnace (which is extensively water-cooled) by incrementally adding weighed quantities of each material to a common conveyor belt. Slag and ferrophosphorus are tapped periodically, whereas the hot furnace gases (90% CO and 10% phosphorus) pass through an electrostatic precipitator that removes the dust before phosphorus condensation. The phosphorus is condensed by direct impingement of a hot water spray, sometimes enhanced by heat transfer through water-cooled condenser walls. Liquid phosphorus drains into a water sump, where the water maintains a seal from the atmosphere. Liquid phosphorus is stored in steam-heated tanks under a water blanket and transferred into tank cars by pumping or hot water displacement. The tank cars have protective blankets of water and are equipped with steam coils for remelting at the destination.

There are numerous sources of fumes from the furnace operation, such as dust from the raw materials feeding and fumes emitted from electrode penetrations and tapping. These fumes, which consist of dust, phosphorus vapor (immediately oxidized to phosphorus pentaoxide), and carbon monoxide, are collected and scrubbed. Principal wastewater streams consist of calciner scrubber liquor, phosphorus condenser and other phossy water, and slag-quenching water.

Phosphorus Consuming

This subcategory involves phosphoric acid (dry process), phosphorus pentoxide, phosphorus pentasulfide, phosphorus trichloride, and phosphorus oxychloride. In the standard dry process for phosphoric acid production, liquid phosphorus is burned in the air, the resulting gaseous phosphorus pentaoxide is absorbed and hydrated in a water spray, and the mist is collected with an electrostatic precipitator. Regardless of the process variation, phosphoric acid is made with the consumption of water and no aqueous wastes are generated by the process.

Solid anhydrous phosphorus pentaoxide is manufactured by burning liquid phosphorus in an excess of dried air in a combustion chamber and condensing the vapor in a roomlike structure. Condensed phosphorus pentaoxide is mechanically scraped from the walls using moving chains and is discharged from the bottom of the barn with a screw conveyor. Phosphorus pentasulfide is manufactured by directly reacting phosphorus and sulfur, both in liquid form, in a highly exothermic batch operation. Because the reactants and products are highly flammable at the reaction temperature, the reactor is continuously purged with nitrogen and a water seal is used in the vent line.

Phosphorus trichloride is manufactured by loading liquid phosphorus into a jacketed batch reactor. Chlorine is bubbled through the liquid, and phosphorus trichloride is refluxed until all the phosphorus is consumed. Cooling water is used in the reactor jacket and care is taken to avoid an excess of chlorine and the resulting formation of phosphorus pentachloride. Phosphorus oxychloride is manufactured by the reaction of phosphorus trichloride, chlorine, and solid phosphorus pentaoxide in a batch operation. Liquid phosphorus trichloride is loaded to the reactor, solid phosphorus pentoxide added, and chlorine bubbled through the mixture. Steam is

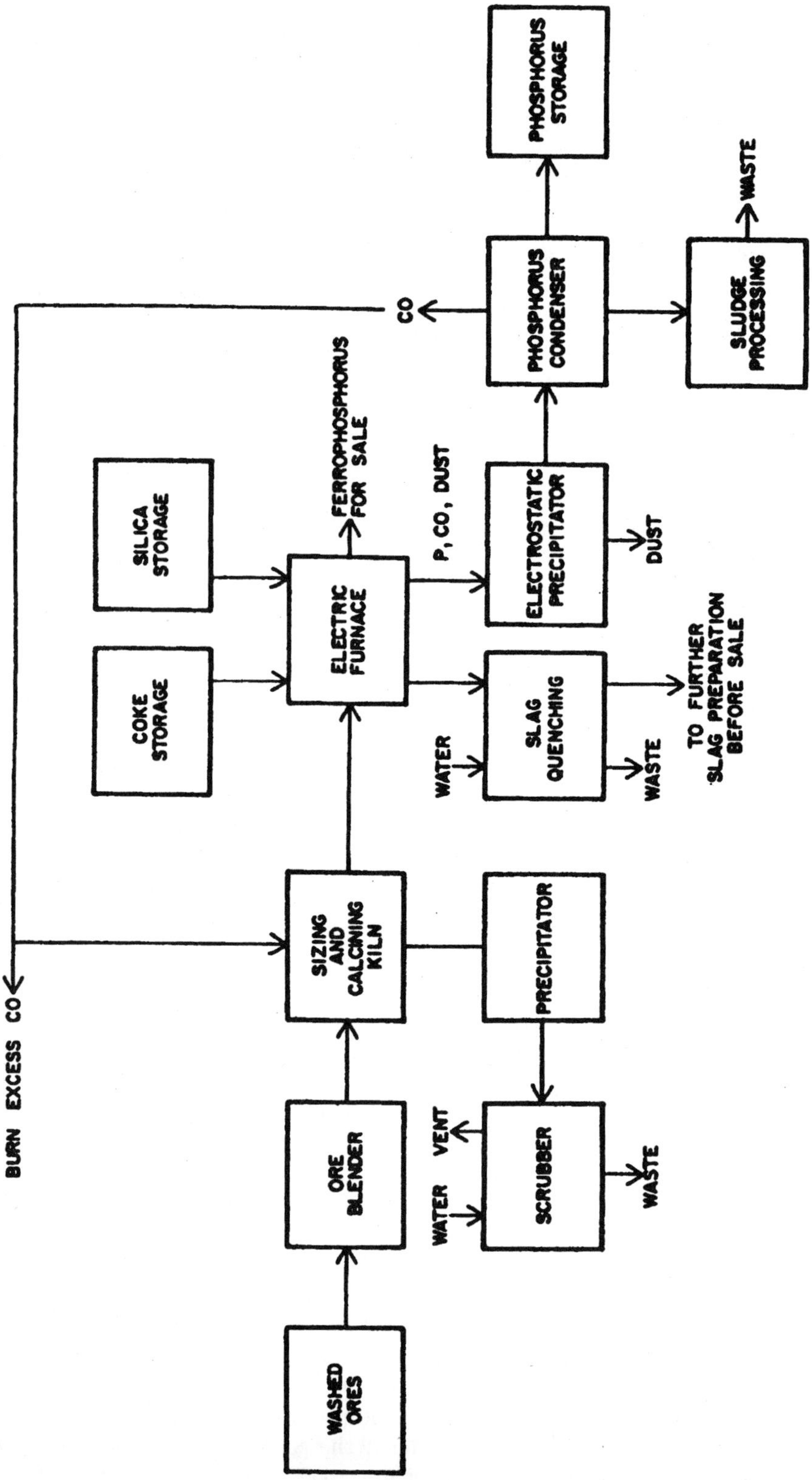

Figure 2 Standard phosphorus process flow diagram (from Ref. 8).

supplied to the reactor jacket, water to the reflux condenser is shut off, and the product is distilled over and collected.

Because phosphorus is transported and stored under a water blanket, phossy water is a raw waste material at phosphorus-consuming plants. Another source of phossy wastewater results when reactor contents (containing phosphorus) are dumped into a sewer line due to operator error, emergency conditions, or inadvertent leaks and spills.

Phosphate

This subcategory involves sodium tripolyphosphate and calcium phosphates. Sodium tripolyphosphate is manufactured by the neutralization of phosphoric acid by soda ash or caustic soda and soda ash, with the subsequent calcining of the dried mono- and disodium phosphate crystals. This product is then slowly cooled or tempered to produce the condensed form of the phosphates.

The nonfertilizer calcium phosphates are manufactured by the neutralization of phosphoric acid with lime. The processes for different calcium phosphates differ substantially in the amount and type of lime and amount of process water used. Relatively pure, food-grade monocalcium phosphate (MCP), dicalcium phosphate (DCP), and tricalcium phosphate (TCP) are manufactured in a stirred batch reactor from furnace-grade acid and lime slurry, as shown in the process flow diagram of Figure 3. Dicalcium phosphate is also manufactured for livestock feed supplement use, with much lower specifications on product purity.

Sodium tripolyphosphate manufacture generates no process wastes. Wastewaters from the manufacture of calcium phosphates are generated from a dewatering of the phosphate slurry and wet scrubbing of the airborne solids during product operations.

Defluorinated Phosphate Rock

The primary raw material for the defluorination process is fluorapatite phosphate rock. Other raw materials used in much smaller amounts, but critical to the process, are sodium-containing reagents, wet process phosphoric acid, and silica. These are fed into either a rotary kiln or a fluidized bed reactor that requires a modular and predried charge. Reaction temperatures are maintained in the 1205–1366°C range, whereas the retention time varies from 30 to 90 min. From the kiln or fluidized bed reactor, the defluorinated product is quickly quenched with air or water, followed by crushing and sizing for storage and shipment. A typical flow diagram for the fluidized bed process is shown in Figure 4.

Wastewaters are generated in the process of scrubbing contaminants from gaseous effluent streams. This water requirement is of significant volume and process conditions normally permit the use of recirculated contaminated water for this service, thereby effectively reducing the discharged wastewater volume. Leaks and spills are routinely collected as part of process efficiency and housekeeping and, in any case, their quantity is minor and normally periodic.

Defluorinated Phosphoric Acid

One method used in order to defluorinate wet process phosphoric acid is vacuum evaporation. The concentration of 54% P_2O_5 acid to a 68–72% P_2O_5 strength is performed in vessels that use high-pressure (30.6–37.4 atm or 450–550 psig) steam or an externally heated Dowtherm solution as the heat energy source for evaporation of water from the acid. Fluorine removal from the acid occurs concurrently with the water vapor loss. A typical process flow diagram for vacuum-type evaporation is shown in Figure 5.

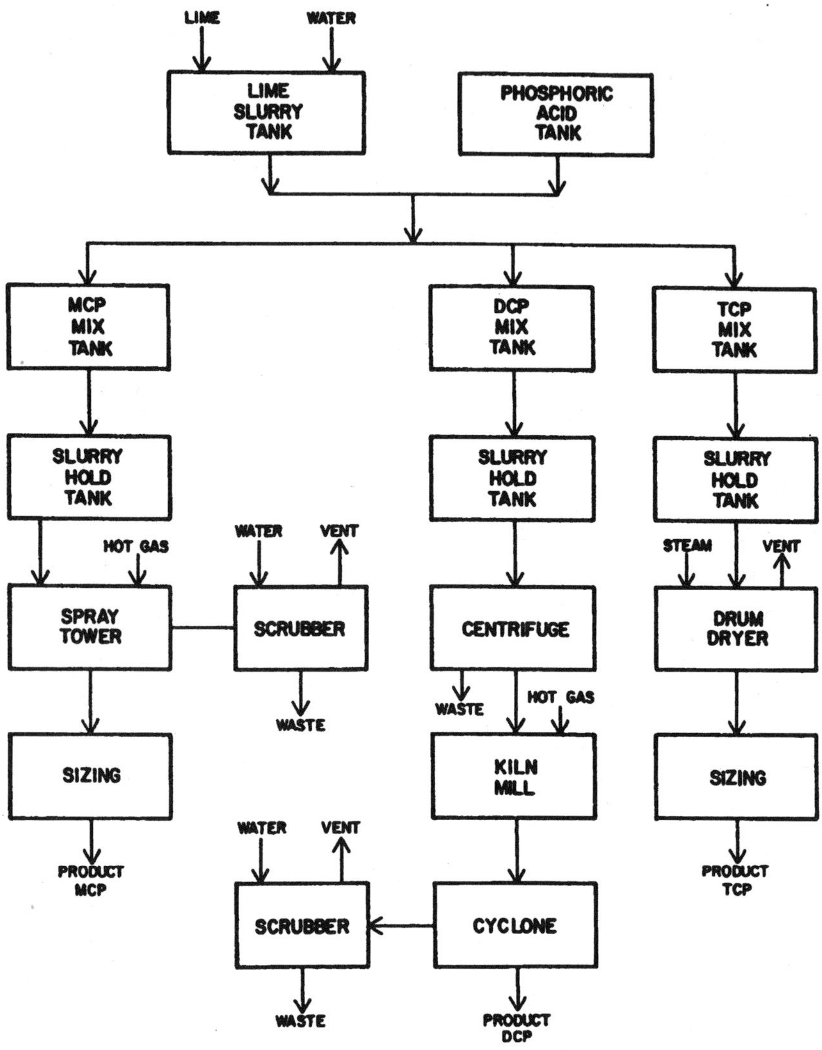

Figure 3 Standard process for food-grade calcium phosphates (from Ref. 8).

A second method of phosphoric acid defluorination entails the direct contact of hot combustion gases (from fuel oil or gas burners) with the acid by bubbling them through the acid. Evaporated and defluorinated product acid is sent to an acid cooler, while the gaseous effluents from the evaporation chamber flow to a series of gas scrubbing and absorption units. Finally, aeration can also be used for defluorinating phosphoric acid. In this process, diatomaceous silica or spray-dried silica gel is mixed with commercial 54% P_2O_5 phosphoric acid. Hydrogen fluoride in the impure phosphoric acid is converted to fluosilicic acid, which in turn breaks down to SiF_4 and is stripped from the heated mixture by simple aeration.

The major wastewater source in the defluorination processes is the wet scrubbing of contaminants from the gaseous effluent streams. However, process conditions normally permit the use of recirculated contaminated water for this service, thereby effectively reducing the discharged wastewater volume.

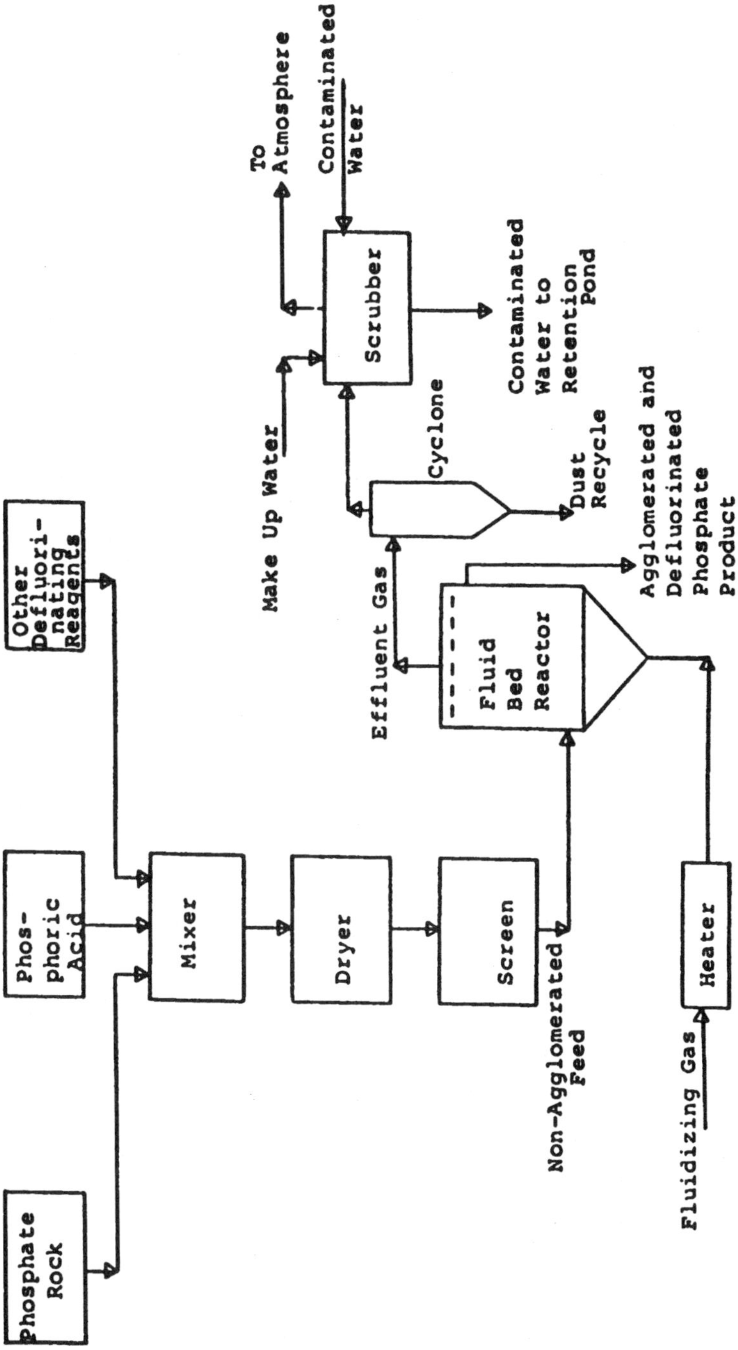

Figure 4 Defluorinated phosphate rock fluid bed process (from Ref. 8).

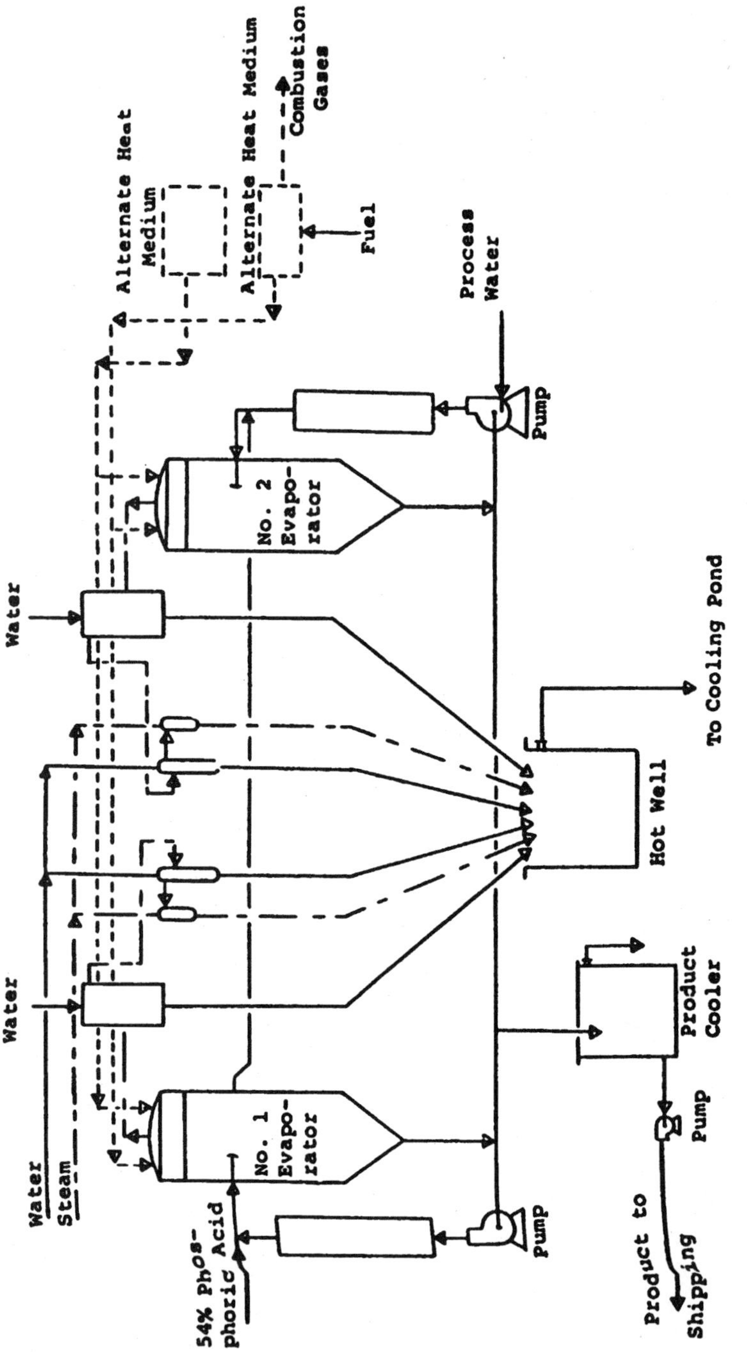

Figure 5 Defluorinated phosphoric acid vacuum process (from Ref. 8).

Sodium Phosphates

In the manufacture of sodium phosphates, the removal of contaminants from the wet process acid takes place in a series of separate neutralization steps. The first step involves the removal of fluosilicates with recycled sodium phosphate liquor. The next step precipitates the minor quantities of arsenic present by adding sodium sulfide to the solution, while barium carbonate is added to remove the excess sulfate. The partially neutralized acid still contains iron and aluminum phosphates, and some residual fluorine.

A second neutralization is carried out with soda ash to a pH level of about 4. Special heating, agitation, and retention are next employed to adequately condition the slurry so that filtration separation of the contaminants can be accomplished. The remaining solution is sufficiently pure for the production of monosodium phosphate, which can be further converted into other compounds such as sodium metaphosphate, disodium phosphate, and trisodium phosphate. A typical process flow diagram is shown in Figure 6. Wastewater effluents from these processes originate from leaks and spills, filtration backwashes, and gas scrubber wastewaters.

9.2.3 Categorization in Phosphate Fertilizer Production

The fertilizer industry comprises nitrogen-based, phosphate-based, and potassium-based fertilizer manufacturing, as well as combinations of these nutrients in mixed and blend fertilizer formulations. Only the phosphate-based fertilizer industry is discussed here and, therefore, the categorization mainly involves two broad divisions: (a) the phosphate fertilizer industry (A) and (b) the mixed and blend fertilizer industry (G) in which one of the components is a phosphate compound. The following categorization system of the various separate processes and their production streams and descriptions is taken from the federal guidelines [8] pertaining to state and local industrial pretreatment programs. It will be used in the discussion that ensues to identify process flows and characterize the resulting raw waste. Figure 7 shows a flow diagram for the production streams of the entire phosphate and nitrogen fertilizer manufacturing industry.

9.2.4 Phosphate and Mixed and Blend Fertilizer Manufacture

Phosphate Fertilizer (A)

The phosphate fertilizer industry is defined as eight separate processes: phosphate rock grinding, wet process phosphoric acid, phosphoric acid concentration, phosphoric acid clarification, normal superphosphate, triple superphosphate, ammonium phosphate, and sulfuric acid. Practically all phosphate manufacturers combine the various effluents into a large recycle water system. It is only when the quantity of recycle water increases beyond the capacity to contain it that effluent treatment is necessary.

Phosphate Rock Grinding. Phosphate rock is mined and mechanically ground to provide the optimum particle size required for phosphoric acid production. There are no liquid waste effluents.

Wet Process Phosphoric Acid. A production process flow diagram is shown in Figure 8. Insoluble phosphate rock is changed to water-soluble phosphoric acid by solubilizing the phosphate rock with an acid, generally sulfuric or nitric. The phosphoric acid produced from the nitric acid process is blended with other ingredients to produce a fertilizer, whereas the phosphoric acid produced from the sulfuric acid process must be concentrated before further use. Minor quantities of fluorine, iron, aluminum, silica, and uranium are usually the most serious waste effluent problems.

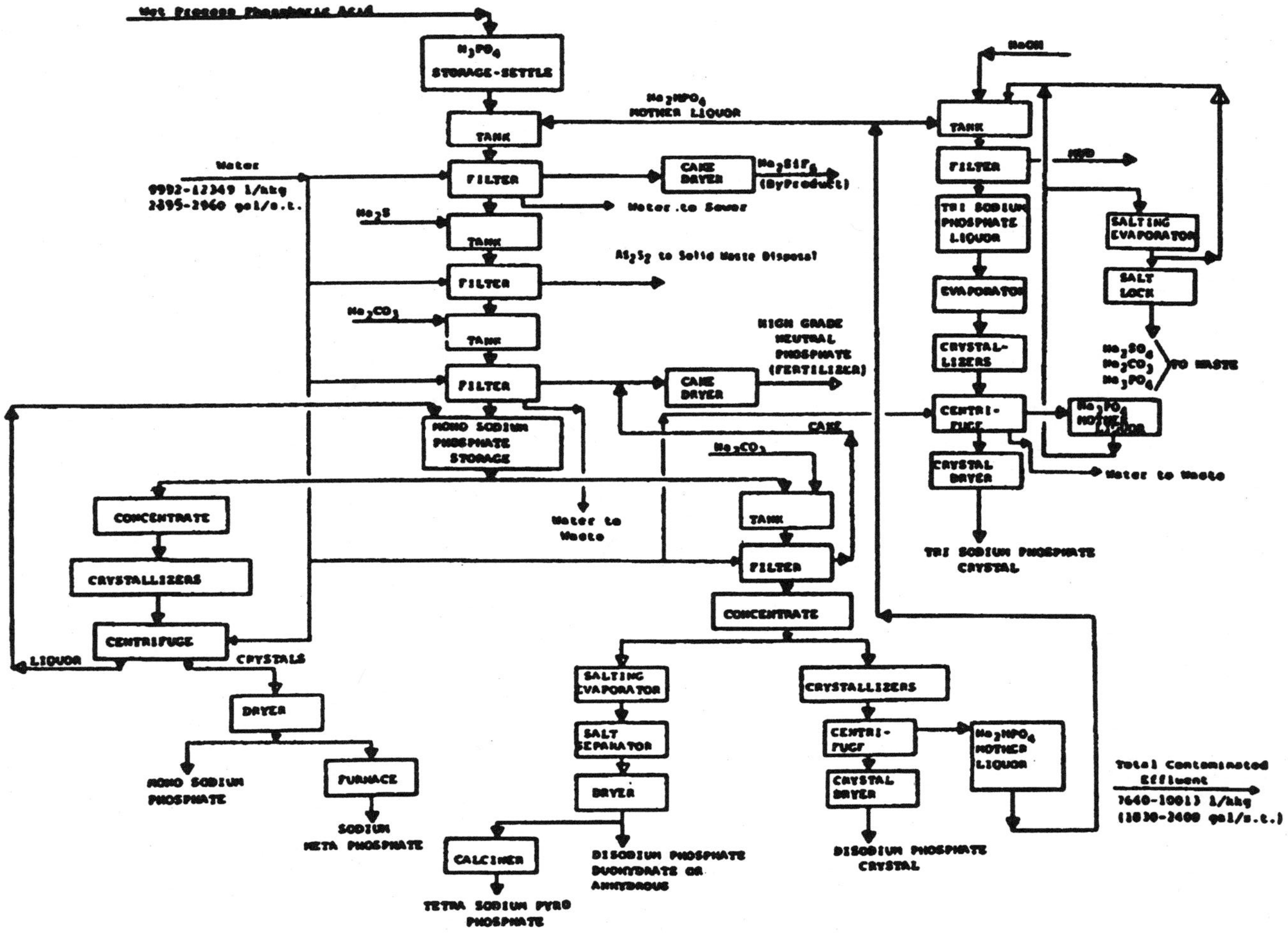

Figure 6 Sodium phosphate process from wet process (from Ref. 8).

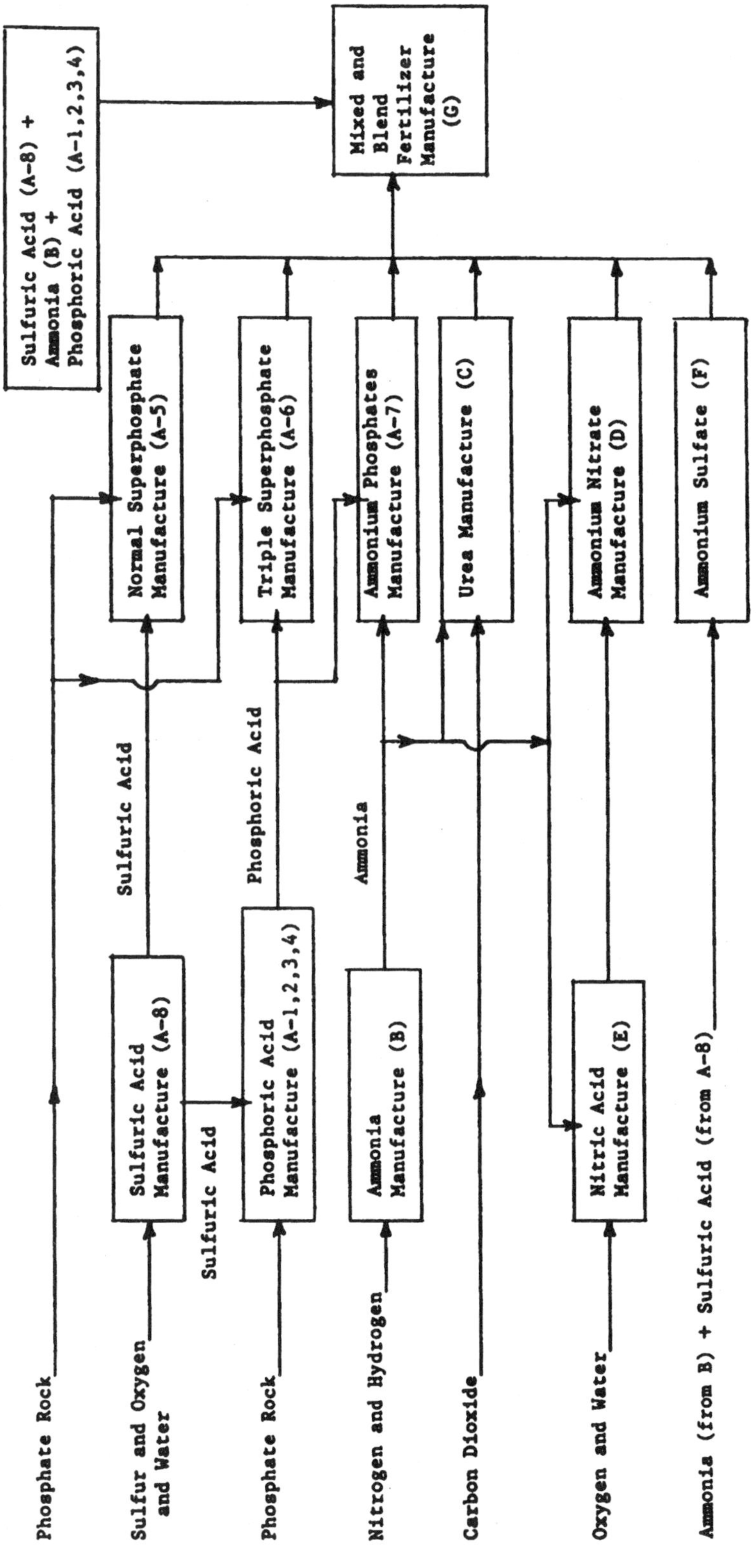

Figure 7 Flow diagram of fertilizer products manufacturing (from Ref. 8).

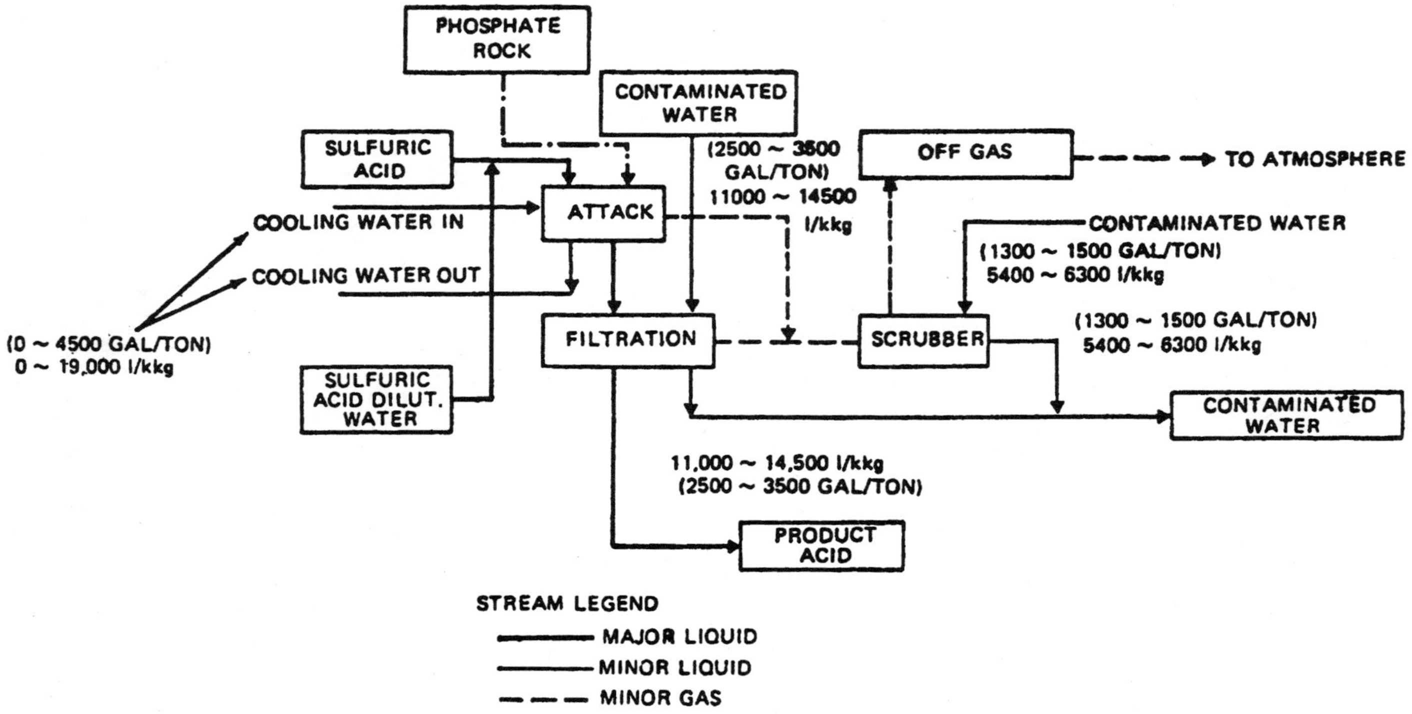

Figure 8 Wet process phosphoric acid (H_2SO_4) acidulation (from Ref. 8).

Phosphoric Acid Concentration. Phosphoric acid produced with sulfuric acid cannot be used for processing due to its very low concentration. It is therefore concentrated to 50–54% by evaporation. Waste streams contain fluorine and phosphoric acid.

Phosphoric Acid Clarification. When the phosphoric acid has been concentrated, iron and aluminum phosphates, gypsum and fluorosilicates become insoluble and can pose problems during acid storage. They are therefore removed by clarification and/or centrifugation.

Normal Superphosphate. Normal superphosphate is produced by the reaction between ground phosphate rock and sulfuric acid, followed by three to eight weeks of curing time. Obnoxious gases are generated by this process.

Triple Superphosphate (TSP). Triple superphosphate is produced by the reaction between ground phosphate rock and phosphoric acid by one of two processes. One utilizes concentrated phosphoric acid and generates obnoxious gases. The dilute phosphoric acid process permits the ready collection of dusts and obnoxious gases generated.

Ammonium Phosphate. Ammonium phosphate, a concentrated water-soluble plant food, is produced by reacting ammonia and phosphoric acid. The resultant slurry is dried, stored, and shipped to marketing.

Sulfuric Acid. Essentially all sulfuric acid manufactured in this industry is produced by the "contact" process, in which SO_2 and oxygen contact each other on the surface of a catalyst (vanadium pentaoxide) to form SO_3 gas. Sulfur trioxide gas is added to water to form sulfuric acid. The sulfur dioxide used in the process is produced by burning elemental sulfur in a furnace.

In addition, the process is designed to capture a high percentage of the energy released by the exothermic reactions occurring in the oxidation of sulfur to sulfur trioxide. This energy is used to produce steam, which is then utilized for other plant unit operations or converted into electrical energy. It is the raw water treatment necessary to condition water for this steam production that generates essentially all the wastewater effluent from this process.

Mixed and Blend Fertilizer (G)

Mixed Fertilizer. The raw materials used to produce mixed fertilizers include inorganic acids, solutions, double nutrient fertilizers, and all types of straight fertilizers. The choice of raw materials depends on the specific nitrogen, phosphate, potassium (N-P-K) formulation to be produced and on the cost of the different materials from which they can be made.

The mixed fertilizer process involves the controlled addition of both dry and liquid raw materials to a granulator, which is normally a rotary drum, but pug mills are also used. Raw materials, plus some recycled product material, are mixed to form an essentially homogeneous granular product. Wet granules from the granulator are discharged into a rotary drier, where the excess water is evaporated and dried granules from the drier are then sized on vibrating screens. Over- and undersized granules are separated for use as recycle material in the granulator. Commercial-product-size granules are cooled and then conveyed to storage or shipping.

Blend Fertilizer. Raw materials used to produce blend fertilizers are a combination of granular dry straight and mixed fertilizer materials with an essentially identical particle size. Although many materials can be utilized, the five most commonly used in this process are ammonium nitrate, urea, triple superphosphate, diammonium phosphate, and potash. These raw materials are stored in a multicompartmented bin and withdrawn in the precise quantities needed to produce the nitrogen-phosphorus-potassium (N-P-K) formulation desired. Raw material addition is normally done by batch weighing, and the combination of batch-weighed and granular raw materials is then conveyed to a mechanical blender for mixing. From the blender, the product is conveyed to storage or shipping.

9.2.5 Wastewater Characteristics and Sources

Wastewaters from the manufacturing, processing, and formulation of inorganic chemicals such as phosphorus compounds, phosphates, and phosphate fertilizers cannot be exactly characterized. The wastewater streams are usually expected to contain trace or large concentrations of all raw materials used in the plant; all intermediate compounds produced during manufacture; all final products, coproducts, and byproducts; and the auxiliary or processing chemicals employed. It is desirable from the viewpoint of economics that these substances not be lost, but some losses and spills appear unavoidable and some intentional dumping does take place during housecleaning, vessel emptying, and preparation operations.

The federal guidelines [8] for state and local pretreatment programs reported the raw wastewater characteristics (Table 4) in mg/L concentration, and flows and quality parameters (Table 5) based on the production of 1 ton of the product manufactured, for each of the six subcategories of the phosphate manufacturing industry. Few fertilizer plants discharge wastewaters to municipal treatment systems. Most use ponds for the collection and storage of wastewaters, pH control, chemical treatment, and settling of suspended solids. Whenever available retention pond capacities in the phosphate fertilizer industry are exceeded, the wastewater overflows are treated and discharged to nearby surface water bodies. The federal guidelines [8] reported the range of wastewater characteristics (Table 6) in mg/L concentrations for typical retention ponds used by the phosphate fertilizer industry.

The specific types of wastewater sources in the phosphate fertilizer industry are (a) water treatment plant wastes from raw water filtration, clarification, softening and deionization, which principally consist of only the impurities removed from the raw water (such as carbonates, hydroxides, bicarbonates, and silica) plus minor quantities of treatment chemicals; (b) closed-loop cooling tower blowdown, the quality of which varies with the makeup of water impurities and inhibitor chemicals used (note: the only cooling water contamination from process liquids is through mechanical leaks in heat exchanger equipment, and Table 7 shows the normal range of contaminants that may be found in cooling water blowdown systems [26]); (c) boiler blowdown, which is similar to cooling tower blowdown but the quality differs as shown in Table 8 [26]; (d) contaminated water or gypsum pond water, which is the impounded and reused water that accumulates sizable concentrations of many cations and anions, but mainly fluorine and phosphorus concentrations of 8500 mg/L F and in excess of 5000 mg/L P are not unusual; concentrations of radium 226 in recycled gypsum pond water are 60–100 picocuries/L, and its acidity reaches extremely high levels (pH 1–2); (e) wastewater from spills and leaks that, when possible, is reintroduced directly to the process or into the contaminated water system; and (f) nonpoint-source discharges that originate from the dry fertilizer dust covering the general plant area and then dissolve in rainwater and snowmelt, which become contaminated.

In the specific case of wastewater generated from the condenser water bleedoff in the production of elemental phosphorus from phosphate rock in an electric furnace, Yapijakis [33] reported that the flow varies from 10 to 100 gpm (2.3–23 m^3/hour), depending on the particular installation. The most important contaminants in this waste are elemental phosphorus, which is colloidally dispersed and may ignite if allowed to dry out, and fluorine which is also present in the furnace gases. The general characteristics of this type of wastewater (if no soda ash or ammonia were added to the condenser water) are given in Table 9.

As previously mentioned, fertilizer manufacturing may create problems within all environmental media, that is, air pollution, water pollution, and solid wastes disposal difficulties. In particular, the liquid waste effluents generated from phosphate and mixed and blend fertilizer production streams originate from a variety of sources and may be summarized [17,27] as follows: (a) ammonia-bearing wastes from ammonia production; (b) ammonium salts such as

Table 4 Phosphate Manufacturing Industry Raw Waste Characteristics

Parameter (mg/L except pH)	Phosphorus production (A)	Phosphorus-consuming (B)	Phosphate (C)	Defluorinated phosphate rock (D)	Defluorinated phosphoric acid (E)	Sodium phosphate (F)
Flow type	C	B	B	B	B	B
BOD_5			–	3	15	31
SS	100		24,000–54,000	16	30	416
						460
TDS			1900–7000[a]	2,250[a]	28780[a]	1640[a]
COD				48	306	55
pH (unit)				1.65[a]	1.29[a]	7.8
Phosphorus	21					
Phosphate	59		7000[a]			
Sulfate	260			350	4770	240
F	126			1930	967	15
HCl		0–800				
H_2SO_3		0–34				
$H_3PO_3 + H_3PO_4$		17–500				
HF, H_2SiF_6, H_2SiO_3			1900[a]			
Chloride				101	65	90
Calcium				40	1700[a]	95
Magnesium				12	106	
Aluminum				58	260	
Iron				8[a]	180[a]	
Arsenic				0.38[a]	0.83[a]	
Zinc				5.2[a]	5.3[a]	
Total acidity	128					
Total phosphorus				600	5590[a]	250

[a] In high levels, these parameters may be inhibitory to biological systems.
B = batch process.
C = continuous process.
BOD, biochemical oxygen demand; SS, suspended solids; TDS, total dissolved solids; COD, chemical oxygen demand.
Source: Ref. 8.

Table 5 Phosphate Manufacturing Industry Raw Waste Characteristics Based on Production

Parameter (kg/kkg except pH and flow)	Phosphorus production (A)	Phosphorus-consuming (B)	Phosphate (C)	Defluorinated phosphate rock (D)	Defluorinated phosphoric acid (E)	Sodium phosphate (F)
Flow range (L/kkg)	425,000	38,000	10,920	45,890	18,020–70,510	7640–10,020
Flow type	C	B	B	B	B	B
BOD_5					0.27–1.06	0.2–0.3
SS	42.5		22.5–50	0.73	0.54–2.11	3.5–4.6
TDS			4.0–14.6	103	519–2031	12.5–16.40
COD				2.2	5.5–21.5	0.4–0.52
pH (unit)				1.65	1.29	7.8
Phosphorus	9					
Phosphate	25		15			
Sulfate	111			16	86–336	1.8–2.36
F	53.5			88	17.4–68.1	0.1–0.13
HCl		0–3				
H_2SO_3		0–1.0				
$H_3PO_3 + H_3PO_4$		0.5–2.5				
HF, H_2SiF_6, H_2SiO_3			12			
Chloride				4.6	1.17–4.58	0.68–0.90
Calcium				1.8	30.6–120	0.72–0.94
Magnesium				0.6	1.9–7.43	
Aluminum				2.7	4.7–18.39	
Iron				0.37	3.2–12.52	
Arsenic				0.02	0.02–0.08	
Zinc				0.24	0.09–0.35	
Total acidity	54.5					
Total phosphorus				27.5	101–395	1.91–2.51

BOD, biochemical oxygen demand; SS, suspended solids; TDS, total dissolved solids; COD, chemical oxygen demand.
Source: Ref. 8.

Table 6 Raw Wastewater Characteristics of Phosphate Fertilizer Industry Retention Ponds

Quality parameter	Phosphate (A)
Suspended solids (mg/L)	800–1200
pH (unit)	1–2
Ammonia (mg/L)	450–500
Sulfate (mg/L)	4000
Chloride (mg/L)	58
Total phosphate (mg/L)	3–5M
Fluoride (mg/L)	6–8.5M
Aluminum (mg/L)	110
Iron (mg/L)	85
Radium 226 (picocuries/L)	60–100

M = thousand.
Source: Ref. 8.

ammonium phosphate; (c) phosphates and fluoride wastes from phosphate and superphosphate production; (d) acidic spillages from sulfuric acid and phosphoric acid production; (e) spent solutions from the regeneration of ion-exchange units; (f) phosphate, chromate, copper sulfate, and zinc wastes from cooling tower blowdown; (g) salts of metals such as iron, copper, manganese, molybdenum, and cobalt; (h) sludge discharged from clarifiers and backwash water from sand filters; and (i) scrubber wastes from gas purification processes.

Considerable variation, therefore, is observed in quantities and wastewater characteristics at different plants. According to a UNIDO report [34], the most important factors that contribute to excessive in-plant materials losses and, therefore, probable subsequent pollution are the age of the facilities (low efficiency, poor process control), the state of maintenance and repair (especially of control equipment), variations in feedstock and difficulties in adjusting processes to cope, and an operational management philosophy such as consideration for pollution control and prevention of materials loss. Because of process cooling requirements, fertilizer manufacturing facilities may have an overall large water demand, with the wastewater effluent discharge largely dependent on the extent of in-plant recirculation [17]. Facilities designed on

Table 7 Range of Concentrations of Contaminants in Cooling Water

Cooling water contaminant	Concentration (mg/L)
Chromate	0–250
Sulfate	500–3000
Chloride	35–160
Phosphate	10–50
Zinc	0–30
TDS	500–10,000
TSS	0–50
Biocides	0–100

TDS, total dissolved solids; TSS, total suspended solids.
Source: Ref. 26.

Table 8 Range of Concentrations of Contaminants in Boiler Blowdown Waste

Boiler blowdown contaminant	Concentration (mg/L)
Phosphate	5–50
Sulfite	0–100
TDS	500–3500
Zinc	0–10
Alkalinity	50–700
Hardness	50–500
Silica (SiO_2)	25–80

TDS, total dissolved solids.
Source: Ref. 26.

a once-through process cooling stream generally discharge from 1000 to over 10,000 m^3/hour wastewater effluents that are primarily cooling water.

9.3 IMPACTS OF PHOSPHATE INDUSTRY POLLUTION

The possibility of the phosphate industry adversely affecting streams did not arise until 1927, when the flotation process was perfected for increasing the recovery of fine-grain pebble phosphate [12]. A modern phosphate mining and processing facility typically has a 30,000 gpm (1892 L/s) water supply demand and requires large areas for clear water reservoirs, slime settling basins, and tailings sand storage. With the help of such facilities, the discharge of wastes into nearby surface water bodies is largely prevented, unless heavy rainfall inputs generate volumes that exceed available storage capacity.

According to research results reported by Fuller [12], the removal of semicolloidal matter in settling areas or ponds seems to be one of the primary problems concerning water pollution control. The results of DO and BOD surveys indicated that receiving streams were actually

Table 9 Range of Concentrations of Contaminants in Condenser Waste from Electric Furnace Production of Phosphorous

Quality parameter	Concentration or value
pH	1.5–2.0
Temperature	120–150°F
Elemental phosphorus	400–2500 Mg/L
Total suspended solids	1000–5000 Mg/L
Fluorine	500–2000 Mg/L
Silica	300–700 Mg/L
P_2O_5	600–900 Mg/L
Reducing substances (as I_2)	40–50 Mg/L
Ionic charge of particles	Predominantly positive (+)

Source: Ref. 15.

improved in this respect by the effluents from phosphate operations. On the other hand, no detrimental effects on fish were found, but there is the possibility of destruction of fish food (aquatic microorganisms and plankton) under certain conditions.

The wastewater characteristics vary from one production facility to the next, and even the particular flow magnitude and location of discharge will significantly influence its aquatic environmental impact. The degree to which a receiving surface water body dilutes a wastewater effluent at the point of discharge is important, as are the minor contaminants that may occasionally have significant impacts. Fertilizer manufacturing wastes, in general, affect water quality primarily through the contribution of nitrogen and phosphorus, whose impacts have been extensively documented in the literature. Significant levels of phosphates assist in inducing eutrophication, and in many receiving waters they may be more important (growth-limiting agent) than nitrogenous compounds. Under such circumstances, programs to control eutrophication have generally attempted to reduce phosphate concentrations in order to prevent excessive algal and macrophyte growth [14].

In addition to the above major contaminants, pollution from the discharge of fertilizer manufacturing wastes may be caused by such secondary pollutants as oil and grease, hexavalent chromium, arsenic, and fluoride. As reported by Beg et al. [3], in certain cases, the presence of one or more of these pollutants may have adverse impacts on the quality of a receiving water, due primarily to toxic properties, or can be inhibitory to the nitrification process. Finally, oil and grease concentrations may have a significant detrimental effect on the oxygen transfer characteristics of the receiving surface water body.

The manufacture of phosphate fertilizers also generates great volumes of solid wastes known as phosphogypsum, which creates serious difficulties, especially in large production facilities [18]. The disposal of phosphogypsum wastes requires large areas impervious to the infiltration of effluents, because they usually contain fluorine and phosphorus compounds that would have a harmful impact on the quality of a receiving water. Dumping of phosphogypsum in the sea would be acceptable only at coastal areas of deep oceans with strong currents that guarantee thorough mixing and high dilution.

9.4 U.S. CODE OF FEDERAL REGULATIONS

The information presented here has been taken from the U.S. Code of Federal Regulations, 40 CFR, containing documents related to the protection of the environment [9]. In particular, the regulations contained in Part 418, Fertilizer Manufacturing Point Source Category (Subpart A, Phosphate Subcategory, and Subpart O, Mixed and Blend Fertilizer Production Subcategory), and Part 422, Phosphate Manufacturing Point Source Category, pertain to effluent limitations guidelines and pretreatment or performance standards for each of the six subcategories shown in Table 3.

9.4.1 Phosphate Fertilizer Manufacture

The effluent guideline regulations and standards of 40 CFR, Part 418, were promulgated on July 29, 1987. According to the most recent notice in the Federal Register [10] regarding industrial categories and regulations, no review is under way or planned and no revision proposed for the fertilizer manufacturing industry. The effluent guidelines and standards applicable to this industrial category are (a) the best practicable control technology currently available (BPT); (b) the best available technology economically achievable (BAT); (c) the best conventional pollutant control technology (BCT); (d) new source performance standards (NSPS); and (e) new source pretreatment standards for new sources (NSPS).

The provisions of 40 CFR, Part 418, Subpart A, Phosphate Subcategory, are applicable to discharges resulting from the manufacture of sulfuric acid by sulfur burning, wet process phosphoric acid, normal superphosphate, triple superphosphate, and ammonium phosphate. The limitations applied to process wastewater, which establish the quantity of pollutants or pollutant properties that may be discharged by a point source into a surface water body after the application of various types of control technologies, are shown in Table 10. The total suspended solids limitation is waived for process wastewater from a calcium sulfate (phosphogypsum) storage pile runoff facility, operated separately or in combination with a water recirculation system, which is chemically treated and then clarified or settled to meet the other pollutant limitations. The concentrations of pollutants discharged in contaminated nonprocess wastewater, that is, any water including precipitation runoff that comes into incidental contact with any raw material, intermediate or finished product, byproduct, or waste product by means of precipitation, accidental spills, or leaks and other nonprocess discharges, should not exceed the values given in Table 10.

The provisions of Subpart G, Mixed and Blend Fertilizer Production Subcategory, are applicable to discharges resulting from the production of mixed fertilizer and blend fertilizer (or compound fertilizers), such as nitrogen/phosphorus (NP) or nitrogen/phosphorus/potassium (NPK) balanced fertilizers of a range of formulations. The plant processes involved in fertilizer compounding comprise mainly blending and granulation plants, with in-built flexibility to

Table 10 Effluent Limitations (mg/L) for Subpart A, Phosphate Fertilizer

Effluent characteristic	Maximum for any 1 day	Average of daily values for 30 consecutive days shall not exceed
(a) BPT		
Total phosphorus (as P)	105	35
Fluoride	75	25
TSS	150	50
(b) BAT		
Total phosphorus (as P)	105	35
Fluoride	75	25
(c) BCT		
TSS	150	50
(d) NSPS		
Total phosphorus (as P)	105	35
Fluoride	75	25
TSS	150	50
(e) Contaminated nonprocess wastewater		
Total phosphorus (as P)	105	35
Fluoride	75	25

BPT, best practicable control technology currently available; BAT, best available technology economically available; BCT, best conventional pollutant control technology; NSPS, standards of performance for new sources.
Source: Ref. 9.

produce NPK grades in varying proportions [22]. According to Subpart O, "mixed fertilizer" means a mixture of wet and/or dry straight fertilizer materials, mixed fertilizer materials, fillers, and additives prepared through chemical reaction to a given formulation, whereas "blend fertilizer" means a mixture of dry, straight, and mixed fertilizer materials. The effluent limitations guidelines for BPT, BCT, and BAT, and the standards of performance for new sources, allow no discharge of process wastewater pollutants to navigable waters. Finally, the pretreatment standards establishing the quantity of pollutants that may be discharged to publicly owned treatment works (POTW) by a new source are given in Table 11.

9.4.2 Phosphate Manufacturing

The effluent guideline regulations and standards of 40 CFR, Part 422, were promulgated on July 9, 1986. According to the most recent notice in the Federal Register [10] regarding industrial categories and regulations, no review is under way or planned and no revision proposed for the phosphate manufacturing industry. The effluent guidelines and standards applicable to this industrial category are (a) the best practicable control technology currently available (BPT); (b) the best conventional pollutant control technology (BCT); (c) the best available technology economically achievable (BAT); and (d) new source performance standards (NSPS).

The provisions of 40 CFR, Part 422, Phosphate Manufacturing, are applicable to discharges of pollutants resulting from the production of the chemicals described by the six subcategories shown in Table 3. The effluent limitations guidelines for Subpart D, Defluorinated Phosphate Rock Subcategory, are shown in Table 12, and the limitations for contaminated nonprocess wastewater do not include a value for total suspended solids (TSS). Tables 13 and 14 show the effluent limitations guidelines for Subpart E, Defluorinated Phosphoric Acid, and Subpart F, Sodium Phosphate Subcategories, respectively, and again the limitations for contaminated nonprocess wastewater do not include a value for TSS. As can be seen, only for Subpart F are the effluent limitations given as kilograms of pollutant per ton of product (or lb/1000 lb).

9.4.3 Effluent Standards in Other Countries

The control of wastewater discharges from the phosphate and phosphate fertilizer industry in various countries differs significantly, as is the case with effluents from other industries. The discharges may be regulated on the basis of the receiving medium, that is, whether the disposal is

Table 11 Effluent Limitations (mg/L except for pH) for Subpart G, Mixed and Blend Fertilizer

Effluent characteristic	Average of daily values for 30 consecutive days shall not exceed
BOD_5	–
TSS	–
pH	–
NH_3—N	30
NO_3—N	30
Total P	35

BOD, biochemical oxygen demand; TSS, total suspended solids.
Source: Ref. 9.

Table 12 Effluent Limitations (mg/L except for pH) for Subpart D, Defluorinated Phosphate Rock

Effluent characteristic	Maximum for any 1 day	Average of daily values for 30 consecutive days shall not exceed
(a) BPT and NSPS		
Total phosphorus (as P)	105	35
Fluoride (as F)	75	25
TSS	150	50
pH	[a]	[a]
(b) BPT and BAT for nonprocess wastewater, and BAT for process wastewater		
Total phosphorus (as P)	105	35
Fluoride (as F)	75	25
pH	[a]	[a]
(c) BCT		
TSS	150	50
pH	[a]	[a]

[a] Within the range 6.0–9.5.
BPT, best practicable control technology currently available; BAT, best available technology economically available; BCT, best conventional pollutant control technology; NSPS, standards of performance for new sources.
Source: Ref. 9.

to land, municipal sewer system, inland surface water bodies, or coastal areas. Consideration may be given to environmental, socio-economic, and water-quality requirements and objectives, as well as to an assessment of the nature and impacts of the specific industrial effluents, which leads to an approach of either specific industry subcategories or classification of waters, or on a case-by-case basis. To a more limited extent than in the United States, the Indian Central Board for Prevention and Control of Water Pollution established a fertilizer industry subcommittee that adopted suitable effluent standards, proposed effective pollution control measures, and established subcategories for the fertilizer industry [11].

Pollution control legislation and standards in many countries are based on the adoption of systems of water classification. This approach of environmental management can make use of either a broad system of classifications with a limited number of subcategories or, as in Japan, a detailed system of subcategories such as river groups for various uses, lakes, and coastal waters. Within such a framework, specific cases of discharge standards could also be considered under circumstances of serious localized environmental impacts. Other countries, such as the United Kingdom and Finland, have a more flexible approach and review discharge standards for fertilizer plants on a case-by-case basis, with no established uniform guidelines [17]. The assessment of each case is based on the nature and volume of the discharge, the characteristics of receiving waters, and the available pollution control technology.

9.5 WASTEWATER CONTROL AND TREATMENT

The sources and characteristics of wastewater streams from the various subcategories in phosphate and phosphate fertilizer manufacturing, as well as some of the possibilities for

Table 13 Effluent Limitations (mg/L except for pH) for Subpart E, Defluorinated Phosphoric Acid

Effluent characteristic	Maximum for any 1 day	Average of daily values for 30 consecutive days shall not exceed
(a) BPT and NSPS		
Total phosphorus (as P)	105	35
Fluoride (as F)	75	25
TSS	150	50
pH	[a]	[a]
(b) BAT for process and nonprocess wastewater, and BPT for nonprocess wastewater		
Total phosphorus (as P)	105	35
Fluoride (as F)	75	25
(c) BCT		
TSS	150	50
pH	[a]	[a]

[a] Within the range 6.0–9.5.
BPT, best practicable control technology currently available; BAT, best available technology economically available; BCT, best conventional pollutant control technology; NSPS, new source performance standards.
Source: Ref. 9.

recycling and treatment, were discussed in Section 9.2. The pollution control and treatment methods and unit processes used are discussed in more detail in the following. The details of the process design criteria for these unit treatment processes can be found in any design handbook.

9.5.1 In-Plant Control, Recycle, and Process Modification

The primary consideration for in-plant control of pollutants that enter waste streams through random accidental occurrences, such as leaks, spills, and process upsets, is establishing loss prevention and recovery systems. In the case of fertilizer manufacture, a significant portion of contaminants may be separated at the source from process wastes by dedicated recovery systems, improved plant operations, retention of spilled liquids, and the installation of localized interceptors of leaks such as oil drip trays for pumps and compressors [17]. Also, certain treatment systems installed (i.e., ion-exchange, oil recovery, and hydrolyzer-stripper systems) may, in effect, be recovery systems for direct or indirect reuse of effluent constituents. Finally, the use of effluent gas scrubbers to improve in-plant operations by preventing gaseous product losses may also prevent the airborne deposition of various pollutants within the general plant area, from where they end up as surface drainage runoff contaminants.

Cooling Water

Cooling water constitutes a major portion of the total in-plant wastes in fertilizer manufacturing and it includes water coming into direct contact with the gases processed (largest percentage) and water that has no such contact. The latter stream can be readily used in a closed-cycle system, but sometimes the direct contact cooling water is also recycled (after treatment to

Table 14 Effluent Limitations (mg/L except for pH) for Subpart F, Sodium Phosphates

Effluent characteristic	Maximum for any 1 day	Average of daily values for 30 consecutive days shall not exceed
(a) BPT		
TSS	0.50	0.25
Total phosphorus (as P)	0.80	0.40
Fluoride (as F)	0.30	0.15
pH	[a]	[a]
(b) BAT		
Total phosphorus (as P)	0.56	0.28
Fluoride (as F)	0.21	0.11
(c) NSPS		
TSS	0.35	0.18
Total phosphorus (as P)	0.56	0.28
Fluoride (as F)	0.21	0.11
pH	[a]	[a]
(d) BCT		
TSS	0.35	0.18
pH	[a]	[a]

[a] Within the range 6.0–9.5.
BPT, best practicable control technology currently available; BAT, best available technology economically available; BCT, best conventional pollutant control technology; NSPS, new source performance standards.
Source: Ref. 9.

remove dissolved gases and other contaminants, and clarification). By recycling, the amount of these wastewaters can be reduced by 80–90%, with a corresponding reduction in gas content and suspended solids in the wastes discharged to sewers or surface water [18,35,36]. Wang [35,36], Caswell [37], and Hallett [38] introduce new technologies and methodologies for cooling water treatment and recycle.

Phosphate Manufacturing

Significant in-plant control of both waste quantity and quality is possible for most subcategories of the phosphate manufacturing industry. Important control measures include stringent in-process abatement, good housekeeping practices, containment provisions, and segregation practices [8]. In the phosphorus chemicals industry (subcategories A, B, and C in Table 3), plant effluent can be segregated into noncontact cooling water, process water, and auxiliary streams comprising ion-exchange regenerants, cooling tower blowdowns, boiler blowdowns, leaks, and washings. Many plants have accomplished the desired segregation of these streams, often by a painstaking rerouting of the sewer lines. The use of once-through scrubber waste should be discouraged; however, there are plants that recycle the scrubber water from a sump, thus satisfying the scrubber water flow rate demands on the basis of mass transfer considerations while retaining control of water usage.

The containment of phossy water from phosphorus transfer and storage operations is an important control measure in the phosphorus-consuming sub-category B. Although displaced phossy water is normally shipped back to the phosphorus-producing facility, the usual practice in phosphorus storage tanks is to maintain a water blanket over the phosphorus for safety reasons.

This practice is undesirable because the addition of makeup water often results in the discharge of phossy water, unless an auxiliary tank collects phossy water overflows from the storage tanks, thus ensuring zero discharge. A closed-loop system is then possible if the phossy water from the auxiliary tank is reused as makeup for the main phosphorus tank.

Another special problem in phosphorus-consuming subcategory B is the inadvertent spills of elemental phosphorus into the plant sewer pipes. Provision should be made for collecting, segregating, and bypassing such spills, and a recommended control measure is the installation of a trap of sufficient volume just downstream of reaction vessels. In the phosphates subcategory C, an important area of concern is the pickup by stormwater of dust originating from the handling, storing, conveying, sizing, packaging, and shipping of finely divided solid products. Airborne dust can be minimized through air pollution abatement practices, and stormwater pickup could be further controlled through strict dust cleanup programs.

In the defluorinated phosphate rock (D) and defluorinated phosphoric acid (F) subcategories, water used in scrubbing contaminants from gaseous effluent streams constitutes a significant part of the process water requirements. In both subcategories, process conditions do permit the use of contaminated water for this service. Some special precautions are essential at a plant producing sodium phosphates (F), where all meta-, tetra-, pyro-, and polyphosphate wastewater in spills should be diverted to the reuse pond. These phosphates do not precipitate satisfactorily in the lime treatment process and interfere with the removal of fluoride and suspended solids. Because unlined ponds are the most common treatment facility in the phosphate manufacturing industry, the prevention of pond failure is vitally important. Failures of these ponds sometimes occur because they are unlined and may be improperly designed for containment in times of heavy rainfall. Design criteria for ponds and dikes should be based on anticipated maximum rainfall and drainage requirements. Failure to put in toe drainage in dikes is a major problem, and massive contamination from dike failure is a major concern for industries utilizing ponds.

The following are possible process modifications and plant arrangements [7] that could help reduce wastewater volumes, contaminant quantities, and treatment costs:

1. In ammonium phosphate production and mixed and blend fertilizer manufacturing, one possibility is the integration of an ammonia process condensate steam stripping column into the condensate-boiler feedwater systems of an ammonia plant, with or without further stripper bottoms treatment depending on the boiler quality makeup needed.
2. Contaminated wastewater collection systems designed so that common contaminant streams can be segregated and treated in minor quantities for improved efficiencies and reduced treatment costs.
3. In ammonium phosphate and mixed and blend fertilizer (G) production, another possibility is to design for a lower-pressure steam level (i.e., 42–62 atm) in the ammonia plant to make process condensate recovery easier and less costly.
4. When possible, the installation of air-cooled vapor condensers and heat exchangers would minimize cooling water circulation and subsequent blowdown.

In a recent document [25] presenting techniques adopted by the French for pollution prevention, a new process modification for steam segregation and recycle in phosphoric acid production is described. As shown in Figure 9, raw water from the sludge/fluorine separation system is recycled to the heat-exchange system of the sulfuric acid dilution unit and the wastewater used in plaster manufacture. Furthermore, decanted supernatant from the phosphogypsum deposit pond is recycled for treatment in the water filtration unit. The claim was that this process modification permits an important reduction in pollution by

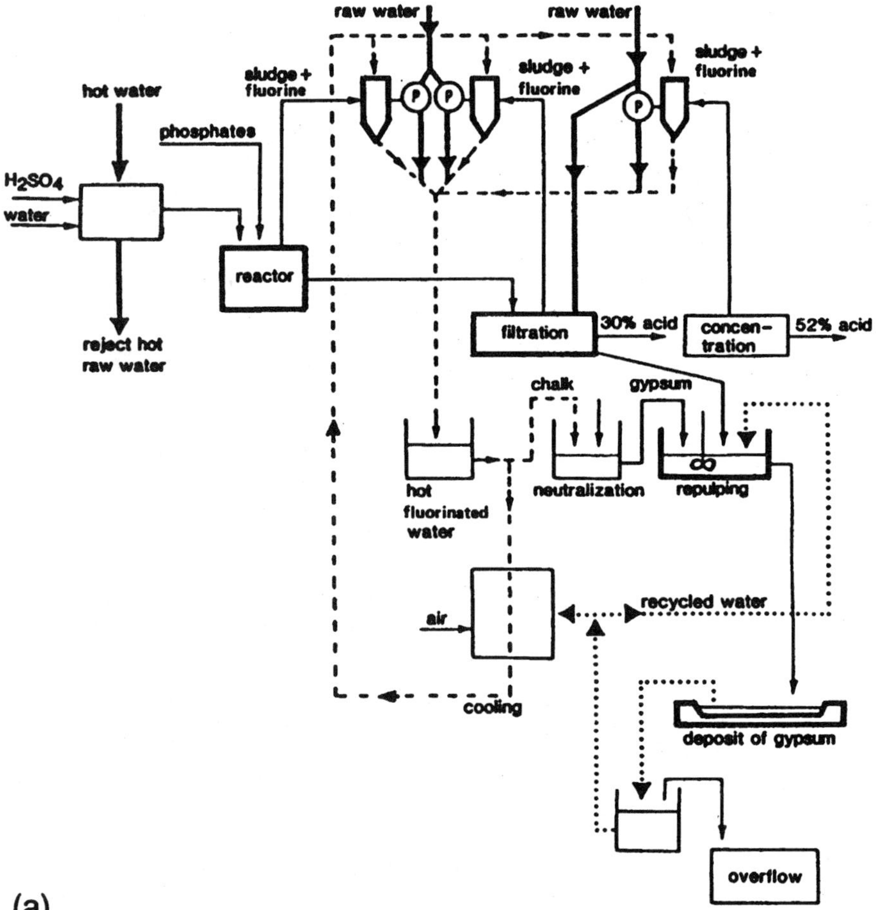

(a)

Figure 9 (a) The old process flow diagram and (b) the new process modification for steam segregation and recycle in phosphoric acid production (from Ref. 25).

fluorine, and that it makes the treatment of effluents easier and in some cases allows specific recycling. Finally, the new process produced a small reduction in water consumption, either by recycle or discharging a small volume of polluted process water downstream, and required no particular equipment and very few alterations in the mainstream lines of the old process.

9.5.2 Wastewater Treatment Methods

Phosphate Manufacturing

Nemerow [23] summarized the major characteristics of wastes from phosphate and phosphorus compounds production (i.e., clays, slimes and tall oils, low pH, high suspended solids, phosphorus, silica, and fluoride) and suggested the major treatment and disposal methods such as lagooning, mechanical clarification, coagulation, and settling of refined wastewaters. The

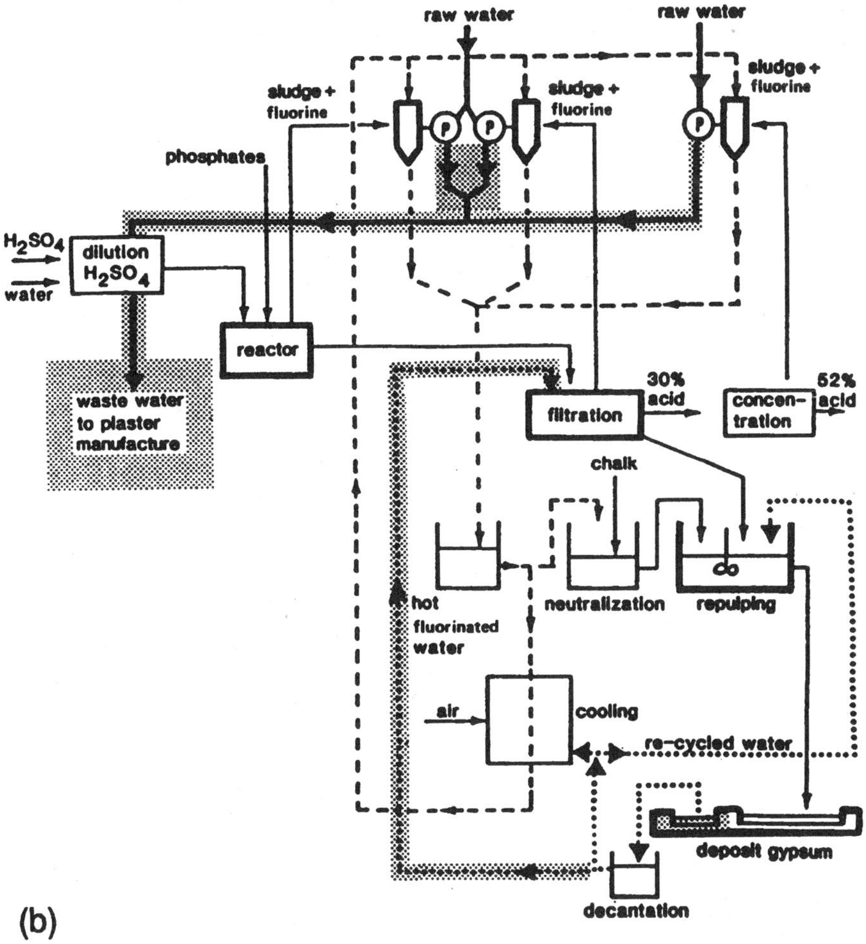

(b)

Figure 9 Continued.

various wastewater treatment practices for each of the six subcategories (Table 3) of the phosphate manufacturing industry were summarized by the USEPA [8] as shown in Table 15. The percent removal efficiencies indicated in this table pertain to the raw waste loads of process effluents from each of these six subcategories. As can be seen from Table 15, the predominant method of removal of primary pollutants such as TDS, TSS, total phosphate, phosphorus, fluoride, sulfate, and for pH adjustment or neutralization is lime treatment followed by sedimentation.

Phosphate Fertilizer Production

Contaminated water from the phosphate fertilizer subcategory A is collected in gypsum ponds and treated for pH adjustment and control of phosphorus and fluorides. Treatment is achieved by "double liming" or a two-stage neutralization procedure, in which phosphates and fluorides precipitate out [7]. The first treatment stage provides sufficient neutralization to raise the pH from 1 to 2 to a pH level of at least 8. The resultant effectiveness of the treatment depends on the

Table 15 Phosphate Manufacturing Industry Wastewater Treatment Practises and Unit Removal Efficiencies (%) and Effluent pH (unit)

Pollutant and method	Phosphorus production (A)	Phosphorus-consuming (B)	Phosphate (C)	Defluorinated phosphate rock (D)	Defluorinated phosphoric acid (E)	Sodium phosphate (F)
TDS						
Lime treatment and sedimentation[a]	99		99			
TSS						
Lime treatment and sedimentation[a]	99		99			
Flocculation, clarification, and dewatering		92				
Total phosphate						
Lime treatment and sedimentation[a]	97	73–97	97			
Phosphorus						
Lime treatment and sedimentation[a]				90	99	88
Flocculation, clarification, and dewatering		92				
Sulfate						
Lime treatment and sedimentation[a]	98		98			
Fluoride						
Lime treatment and sedimentation[a]	99		99	98	96	0
pH (effluent level)						
Lime treatment and sedimentation[a] (neutralization)				6–8[b]	6–8[b]	6–8[b]

[a] Preceded by recycle of phossy water and evaporation of some process water in subcategories A, B, and C.
TDS, total dissolved solids; TSS, total suspended solids.
[b] pH values in units.
Source: Ref. 8.

point of mixing of lime addition and on the constancy of pH control. Fluosilisic acid reacts with lime and precipitates calcium fluoride in this step of the treatment.

The wastewater is again treated with a second lime addition to raise the pH level from 8 to at least 9 (where phosphate removal rates of 95% may be achieved), although two-stage dosing to pH 11 may be employed. Concentrations of phosphorus and fluoride with a magnitude of 6500 and 9000 mg/L, respectively, can be reduced to 5–500 mg/L P and 30–60 mg/L F. Soluble orthophosphate and lime react to form an insoluble precipitate, calcium hydroxy apatite [17]. Sludges formed by lime addition to phosphate wastes from phosphate manufacturing or fertilizer production are generally compact and possess good settling and dewatering characteristics, and removal rates of 80–90% for both phosphate and fluoride may be readily achieved [13].

The seepage collection of contaminated water from phosphogypsum ponds and re-impoundment is accomplished by the construction of a seepage collection ditch around the perimeter of the diked storage area and the erection of a secondary dike surrounding the first [8]. The base of these dikes is usually natural soil from the immediate area, and these combined earth/gypsum dikes tend to have continuous seepage through them (Fig. 10). The seepage collection ditch between the two dikes needs to be of sufficient depth and size to not only collect contaminated water seepage, but also to permit collection of seeping surface runoff from the immediate outer perimeter of the seepage ditch. This is accomplished by the erection of the small secondary dike, which also serves as a backup or reserve dike in the event of a failure of the primary major dike.

The sulfuric acid plant has boiler blowdown and cooling tower blowdown waste streams, which are uncontaminated. However, accidental spills of acid can and do occur, and when they do, the spills contaminate the blowdown streams. Therefore, neutralization facilities should be supplied for the blowdown waste streams (Table 15), which involves the installation of a reliable pH or conductivity continuous-monitoring unit on the plant effluent stream. The second part of the system is a retaining area through which non-contaminated effluent normally flows. The detection and alarm system, when activated, causes a plant shutdown that allows location of the failure and initiation of necessary repairs. Such a system, therefore, provides the continuous protection of natural drainage waters, as well as the means to correct a process disruption.

Mixed fertilizer (subcategory G) treatment technology consists of a closed-loop contaminated water system, which includes a retention pond to settle suspended solids. The water is then recycled back to the system. There are no liquid waste streams associated with the blend fertilizer (subcategory G) process, except when liquid air scrubbers are used to prevent air pollution. Dry removals of air pollutants prevent a wastewater stream from being formed.

Phosphate and Fluoride Removal

Phosphates may be removed from wastewaters by the use of chemical precipitation as insoluble calcium phosphate, aluminum phosphate, and iron phosphate [5]. The liming process has been discussed previously, lime being typically added as a slurry, and the system used is designed as either a single- or two-stage one. Polyelectrolytes have been employed in some plants to improve overall settling. Clarifier/flocculators or sludge-blanket clarifiers are used in a number of facilities [11]. Alternatively, the dissolved air flotation (DAF) process is both technically and economically feasible for phosphate and fluoride removal, according to Wang *et al.* [39–43] and Krofta and Wang [44–48]. Both conventional biological sequencing batch reactors (SBR) and innovative physicochemical sequencing batch reactors (PC-SBR) have been proven to be highly efficient for phosphate and fluoride precipitation and removal [32,43].

A number of aluminum compounds, such as alum and sodium aluminate, have also been used by Layer and Wang [20] as phosphate precipitants at an optimum pH range of 5.5–6.5, as

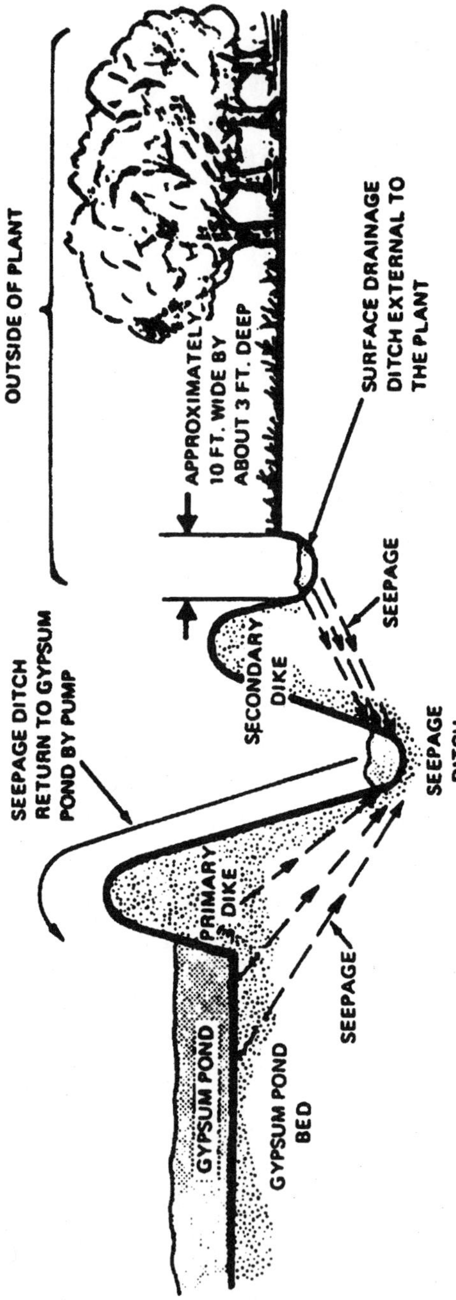

Figure 10 Phosphogypsum pond water seepage control (from Ref. 7).

have iron compounds such as ferrous sulfate, ferric sulfate, ferric chloride, and spent pickle liquor [17]. The optimum pH range for the ferric salts is 4.5–5, and for the ferrous salts it is 7–8, although both aluminum and iron salts have a tendency to form hydroxyl and phosphate complexes. As reported by Ghokas [13], sludge solids produced by aluminum and iron salts precipitation of phosphates are generally less settleable and more voluminous than those produced by lime treatment.

Removal of Other Contaminants

Compressor houses, tank farm areas, cooling water, and loading or unloading bays may be sources of oil in wastewater discharges from phosphate and fertilizer manufacturing. Oil concentrations may range from 100–900 mg/L and can be removed by such units as coke filters or recovered by the use of separators (usually operating in series to recover high oil levels). Chromates and dichromates are present in cooling tower blowdown, at levels of about 10 mg/L, because they are used in cooling water for corrosion inhibition. They may be removed (over 90%) from cooling tower discharges through chemical reduction (i.e., the use of sulfuric acid for lowering pH to less than 4 and addition of a reducing agent) and precipitation as the hydroxide by the addition of lime or NaOH. Chromates may also be recovered and reused from cooling tower blowdown by the use of ion exchange, at recovery levels in excess of 99%, employing a special weak base anion resin that is regenerated with caustic soda.

Phosphoric Acid Production

The use of the electric furnace process (Fig. 11) and acidulation of phosphate-bearing rock is made commercially to produce phosphoric acid. In the first method, elemental phosphorus is first produced from phosphate ore, coke, and silica in an electric furnace, and then the phosphorus is burned with air to form P_2O_5, which is cooled and reacted with water to form orthophosphoric acid (Section 9.2.2). Extremely high acid mist loadings from the acid plant are common, and there are five types of mist-collection equipment generally used: packed towers, electrostatic precipitators, venturi scrubbers, fiber mist eliminators, and wire mesh contactors [21]. Choosing one of these control equipments depends on the required contaminant removal efficiency, the required materials of construction, the pressure loss allowed through the device, and capital and operating costs of the installation (with very high removal efficiencies being the primary factor). The venturi scrubber is widely used for mist collection and is particularly applicable to acid

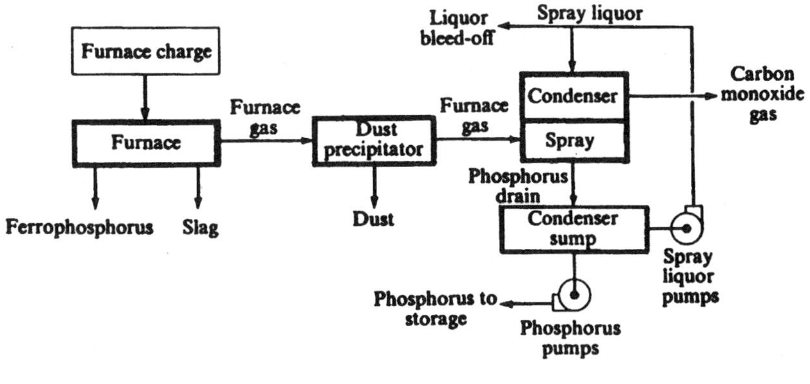

Figure 11 Flow diagram of electric furnace process for phosphorous production (from Ref. 15).

plants burning sludge. The sludge burned is an emulsion of phosphorus, water, and solids carried out in the gas stream from the phosphorus electric furnace as dust or volatilized materials. Impurities vary from 15 to 20% and the venturi scrubber can efficiently collect the acid mist and fine dust discharged in the exhaust from the hydrator.

Wet process phosphoric acid is made by reacting pulverized, beneficiated phosphate ore with sulfuric acid to form calcium sulfate (gypsum) and dilute phosphoric acid (see Section 9.2.4). The insoluble calcium sulfate and other solids are removed by filtration, and the weak (32%) phosphoric is then concentrated in evaporators to acid containing about 55% P_2O_5. Mist and gaseous emissions from the gypsum filter, the phosphoric acid concentrator, and the acidulation off-gas are controlled with scrubbers or other equipment. The preparation of the phosphate ore generates dust from drying and grinding operations, and this is generally controlled with a combination of dry cyclones and wet scrubbers [21]. The material collected by the cyclones is recycled, and the scrubber water discharged to the waste phosphogypsum ponds. Most frequently, simple towers and wet cyclonic scrubbers are used, but at some plants the dry cyclone is followed by an electrostatic precipitator.

9.6 CASE STUDIES OF TREATMENT FACILITIES

Phosphate production and phosphate fertilizer manufacturing facilities are situated in many areas in the United States (primarily in Florida and California) and in other countries such as Algeria, Jordan, and Morocco, as previously mentioned. The wastewaters from production and cleanup activities and surface runoff in most of these locations are stored, treated, and recycled, and the excess overflows are discharged into natural water systems. In those facilities where wastewater from production and cleanup activities and drainage are discharged into municipal water systems and treated together with domestic, commercial, institutional, and other industrial wastewaters, a degree of pretreatment is required to meet federal guidelines or local ordinances such as those presented in Section 9.4. For instance, according to the USEPA [6], the pretreatment unit operations required for the phosphate fertilizer industry comprise solids separation and neutralization, and it may be achieved by either a suspended biological process, a fixed-film biological process, or an independent physicochemical system.

9.6.1 Pebble Phosphate Mining Industry

In one of the earlier reports on the phosphate mining and manufacturing industry in Florida and its water pollution control efforts, Wakefield [31] gave the following generalized account. Because of the huge volumes of water being used for washing, hydraulic sizing, flotation, and concentration of phosphate ores (i.e., one of the main mines of a larger company requires about 60 MGD or 2.63 m^3/s), and since makeup water is not readily available and excess wastewater constitutes a major disposal problem, the recovery and reuse of water have always been of great importance. Waste products from the mining and processing operation consist of large quantities of nonphosphating sands and clays, together with unrecovered phosphatic materials less than 300 mesh in size, and they are pumped into huge lagoons. Easily settled sands fill the near-end, leaving the rest to be gradually filled with "slimes" (a semicolloidal water suspension), while a thin layer of virtually clear water at the surface of the lagoon flows over spillways and is returned to the washers for reuse.

The above ideal wastewater management, however, is infeasible during wet weather (rains of 3 in./day or 7.6 cm/day are frequent in the tropical climate of Florida), because more water goes into the lagoons than can be used by the washers, and the excess volume must be discharged into nearby surface waters. In particular in small streams, this results in highly turbid waters due to the fact that the larger of the slowly consolidating slimes is near the surface of the lagoons in most cases. Furthermore, there are no core walls (for cost reasons) in the large earthen dams forming the settling lagoons; rather, it is usual to depend on the slimes to seal their inner face and prevent excessive seepage. The entire operation, therefore, involves a delicate balance of slime input and weir discharge to accomplish the objective of a maximum of water reuse with a minimum of danger of dam failure and a minimum of turbid discharge to the stream. Such dams have failed very often and the pollution effect of the volume of, for example, a 100 acre (0.4 km^2) pond with 25–30 ft (7.6–9.1 m) of consolidated slime being discharged into a stream with a mean flow of possibly 300 cfs (8.5 m^3/s) has been devastating.

There are, however, other much more frequent situations when effluents of higher or lower turbidity are discharged to streams to protect dam structures or as excess flow when it rains heavily. These discharges of turbid wastewater volumes may be due to underdesigning the settling lagoons or because the wastewater slimes are of a more completely colloidal nature and do not clarify too well. They usually continue over extended periods and cause noticeable stream turbidities, although nothing approaching those encountered after a lagoon dam failure. When streams contain appreciable turbidity due to phosphate industry effluents, local sports fishermen claim that it ruins fishing (but mostly they just prefer fishing in clear waters), while industry managers have demonstrated that fish are not affected by stocking mined-out pits and settling ponds with bass and other species. Finally, a positive effect of discharging moderate levels of turbidity noted in a water treatment plant located downstream is a decrease in chemical costs, undoubtedly due to greater turbidity in the raw water that aids coagulation of color and other impurities.

9.6.2 Phosphate Industry Waste Disposal

In one of the earliest extensive studies and reports on the disposal of wastes from the phosphate mining and processing industry in Florida, Specht [29] reviewed the waste treatment and disposal practices in the various phases of phosphate and phosphorus manufacturing. Regarding waste disposal from mining and beneficiation operations, he reported the use of specially constructed settling areas for the clay and quartz sand separated in the washing and flotation processes and also for the clarification of water to be reused in the process or discharged into streams. As mining processes, the mined-out areas are then used as supplementary settling lagoons, with the wastewater circulating through them using specially made cuts, similar to the slow movement through settling areas that are frequently divided into compartments. During the dry season, as mentioned previously, very little (if any) water is wasted to the streams, but sometimes an estimated maximum of 10% of the total amount of water used is wasted at some facilities during the rainy season. This may represent a significant volume, given the large quantities of water needed in phosphate mining (2000–8000 gpm or 7.6–30 m^3/min) and at the recovery plants (4000–50,000 gpm or 15–190 m^3/min), depending on the size of the plant and its method of treatment.

Occasionally, the phosphate "slime" is difficult to settle in the lagoons because of its true colloidal nature, and the use of calcium sulfate or other electrolytes can promote coagulation, agglomeration, and settling of the particles. Usually an addition of calcium sulfate is unnecessary, because it is present in the wastewater from the sand-flotation process. Generally, it has been shown [33] that the clear effluent from the phosphate mining and beneficiation operation is not deleterious to fish life, but the occurrence of a dam break may result in adverse effects [19].

In superphosphate production, fluoride vapors are removed from the mixing vessel, den or barn, and elevators under negative pressure and passed through water sprays or suitable scrubbers. A multiple-step scrubber is required to remove all the fluorides from the gases and vapors, and the scrubbing water containing the recovered fluorosilisic acid and insoluble silicon hydroxide is recycled to concentrate the acid to 18–25%. The hydrated silica is removed from the acid by filtration and is washed with fresh water and then deposited in settling areas or dumps. In triple superphosphate (also known as double, treble, multiple, or concentrated superphosphate) manufacturing, the calcium sulfate cake from the phosphoric acid production is transferred into settling areas after being washed, where the solid material is retained. The clarified water that contains dissolved calcium sulfate, dilute phosphoric acid, and some fluorosilisic acid is either recycled for use in the plant or treated in a two-step process to remove the soluble fluorides, as described in Section 9.5.2. Water from plant washing and the evaporators may also be added to the wastes sent to the calcium sulfate settling area.

According to Specht [30], in the two-step process to remove fluorides and phosphoric acid, water entering the first step may contain about 1700 mg/L F and 5000 mg/L P_2O_5, and it is treated with lime slurry or ground limestone to a pH of 3.2–3.8. Insoluble calcium fluorides settle out and the fluoride concentration is lowered to about 50 mg/L F, whereas the P_2O_5 content is reduced only slightly. The clarified supernatant is transferred to another collection area where lime slurry is added to bring the solution to pH 7, and the resultant precipitate of P is removed by settling. The final clear water, which contains only 3–5 mg/L F and practically no P_2O_5, is either returned to the plant for reuse or discharged to surface waters. The two-step process is required to reduce fluorides in the water below 25 mg/L F, because a single-step treatment to pH 7 lowers the fluoride content only to 25–40 mg/L F. In the process where the triple phosphate is to be granulated or nodulized, the material is transferred directly from the reaction mixer to a rotary dryer, and the fluorides in the dryer gases are scrubbed with water.

In making defluorinated phosphate by heating phosphate rock, one method of fluoride recovery consists of absorption in a tower of lump limestone at temperatures above the dew point of the stack gas, where the reaction product separates from the limestone lumps in the form of fines. A second method of recovery consists of passing the gases through a series of water sprays in three separate spray chambers, of which the first one is used primarily as a cooling chamber for the hot exit gases of the furnace. In the second chamber, the acidic water is recycled to bring its concentration to about 5% equivalence of hydrofluoric acid in the effluent, by withdrawing acid and adding fresh water to the system. In the final chamber, scrubbing is supplemented by adding finely ground limestone blown into the chamber with the entering gases. Hydrochloric acid is sometimes formed as a byproduct from the fluoride recovery in the spray chambers and this is neutralized with NaOH and lime slurry before being transferred to settling areas.

9.6.3 Ammonium Phosphate Fertilizer and Phosphoric Acid Plant

The fertilizer industry is plagued with a tremendous problem concerning waste disposal and dust because of the very nature of production that involves large volumes of dusty material. Jones and Olmsted [16] described the waste disposal problems and pollution control efforts at such a plant, Northwest Cooperative Mills, in St. Paul, Minnesota. Two types of problems are associated with waste from the manufacture of ammonium phosphate: wastes from combining ammonia and phosphoric acid and the subsequent drying and cooling of the products, and wastes from the handling of the finished product arising primarily from the bagging of the product prior to shipping. Because the ammonia process has to be "forced" by introducing excess amounts of ammonia than the phosphoric acid is capable of absorbing, there is high ammonia content in the exhaust air stream from the ammoniator. Because it is neither economically sound nor environmentally acceptable to exhaust this to the atmosphere, an acid scrubber is employed to recover the ammonia without condensing with it the steam that nearly saturates the exhausted air stream.

Drying and cooling the products of ammonium phosphate production are conventionally achieved in a rotary drum, and a means must be provided to remove the dust particles from the air streams to be exhausted to the atmosphere. At the Minnesota plant, a high-efficiency dry cyclone recovery system followed by a wet scrubber was designed. In this way, material recovered from the dry collector (and recycled to the process) pays for the dry system and minimizes the load and disposal problem in the wet scrubber, because it eliminates the need for a system to recover the wet waste material that is discharged to the gypsum disposal pond for settling.

The remaining problem of removing dust from discharges to the ambient air originating from the bagging and shipping operations is the one most neglected in the fertilizer industry, causing complaints from neighbors. Jones and Olmsted [16] reported the installation of an elaborate, relatively expensive, system of suction pickups at each transfer point of the products in the entire bagging system and shipping platform areas. The collected dust streams are passed over a positive cloth media collector before discharge to the atmosphere, and the system recovers sufficient products to cover only operating expenses.

The filtered gypsum cake from phosphoric acid production, slurried with water to about 30%, is pumped to the settling lagoons, from where the clarified water is recycled to process. To provide a startup area, approximately two acres (or 8100 m^2) of the disposal area were black-topped to seal the soil surface against seepage, and the gypsum collected in this area was worked outward to provide a seal for enlarging the settling area. A dike-separated section of the disposal area was designated as a collection basin for all drainage waters at the plant site. From this basin, after the settling of suspended solid impurities, these waters are discharged to surface waters under supervision from a continuous monitoring and alarm system that guards against accidental contamination from any other source.

Air streams from the digestion system, vacuum cooler, concentrator, and other areas where fluorine is evolved are connected to a highly efficient absorption system, providing extremely high volumes of water relative to the stream. The effluent from this absorption system forms part of the recycled water and is eventually discharged as part of the product used for fertilizer manufacture. The Minnesota plant requires a constant recirculating water load in excess of 3000 gpm (11.4 m^3/min), but multiple use and recycle reduce makeup requirements to less than 400 gpm (1.5 m^3/min) or a mere 13% of total water use.

9.6.4 Rusaifa Phosphate Mining and Processing Plant

As reported by Shahalam *et al.* [28], Jordan stands third in the region following Morocco and Algeria with respect to the mining of phosphate rock and production of phosphates, and the Jordanian phosphate industry is bound to grow, with time creating additional environmental problems. The paper presented the results of a study assessing the phosphate deposits and pollution resulting from a phosphate processing operation in Rusaifa, Jordan. The beneficiation plant uses about 85% of the total process water, and the overflow from the hydrocyclones (rejected as slimes of silica carbonates and clay materials) is taken to a gravity thickener, the underflow (about 0.93 m^3/min or 245 gpm) from which is directly discharged to a nearby holding pond as wastewater and slimes (sludge containing about 25% solids by wt.). As shown in Table 16, the sludge contains significant amounts of fluorine, sulfate, P_2O_5, and organics, and it is unsuitable for direct or indirect discharge into natural waters, in accordance with U.S. 40 CFR

Table 16 Chemical Analysis of the Characteristics of Underflow Sludge from the Gravity Thickener

Parameter	Contents
pH	7.75
Solids	About 10% by sludge volume
P_2O_5	24.4% of dry solid
Total orthophosphate	0.30 mg/L
SiO_2 (inorganic)	14.16% of dry solid
Total SiO_2	16.83% of dry solid
Organics	1.00% of dry solid
Total chlorine	0.09% of dry solid
CaO	40.43% of dry solid
Fluorine	2.96% of dry solid
$CaSO_4$	0.32% of dry solid

Source: Ref. 28.

(see Section 9.4). The water portion of this thick sludge partially evaporates and the remainder percolates into the ground through the bottom of the holding pond.

The second major wastewater discharge (about 1.2 m^3/min or 318 gpm) is from the bottom of the scrubbers used after a dry dust collector cyclone to reduce the dust concentration in the effluent air stream from the phosphate dryer. The underflow of the scrubber contains a high concentration of dust and mud, laden with tiny phosphate organic and inorganic silica and clay particles, and is disposed of in a nearby stream. The main pollutant in the flow to the stream is P_2O_5 at a concentration of nearly 1200 mg/L and it remains mostly as suspended particles.

Finally, most of the P_2O_5 content in the solid wastes generated from Rusaifa Mining is mainly from overburdens and exists as solid particles. Because it does not dissolve in water readily, the secondary pollution potential with respect to P_2O_5 is practically nonexistent. However, loose overburdens in piles, when carried off by rainwater, may create problems for nearby stream(s) due to the high suspended solids concentration resulting in stormwater runoff. Also, loose overburdens blown by strong winds may cause airborne dust problems in neighboring areas; therefore, a planned land reclamation of the mined-out areas is the best approach to minimizing potential pollution from the solid wastes.

9.6.5 Furnace Wastes from Phosphorus Manufacture

The electric furnace process (Fig. 11) for the conversion of phosphate rock into phosphorus was described by Horton *et al.* [15] in a paper that also presented the results of a pilot plant study of treating the wastes produced. The process, as well as the handling of the various waste streams for pollution control, are discussed in Section 9.5.2. In processing the phosphate, the major source of wastewater is the condenser water bleedoff from the reduction furnace, the flow of which varies from 10 to 100 gpm (2.3–22.7 m^3/hour) and its quality characteristics are presented in Table 7.

Phossy water, a waste product in the production of elemental phosphorus by the electric furnace process, contains from 1000 to 5000 mg/L suspended solids that include 400–2500 mg/L of elemental phosphorus, distributed as liquid colloidal particles. These particles are usually positively charged, although this varies depending on the operation of the electrostatic precipitators. Furthermore, the chemical equilibrium between the fluoride and fluosilicate ions introduces an important source of variation in suspended solids that is a pH function. Commonly used coagulants such as alum or ferric chloride were unsatisfactory for wastewater clarification because of the positively charged particles, as were inorganic polyelectrolytes despite their improved performance. However, high-molecular-weight protein molecules at a suitable pH level (which varies for each protein) produced excellent coagulation and were highly successful in clarifying phossy water.

Horton *et al.* [15] investigated or attempted such potential treatment and disposal methods as lagooning, oxidizing, settling (with or without prior chemical coagulation), filtering, and centrifuging and concluded that the best solution appears to be coagulation and settling. The pilot units installed to evaluate this optimum system are shown in Figure 12, together with a summary of the experimental results. The proper pH for optimum coagulation with proteins alone was obscured at higher pH levels by the formation of silica, which tended to encrust the pipelines. It was found that the addition of clay, as a weighting agent with the coagulant, eliminated the scale problem without decreasing the settling rates. Finally, it was concluded that in the pilot plant it was possible to obtain a 40-fold concentration of suspended solids (or 25% solids) by a simple coagulation and sedimentation process.

9.6.6 Phosphate Fertilizer Industry in Eastern Europe

Koziorowski and Kucharski [18] presented a survey of fertilizer industry experience in Eastern European countries and compared it with the United States and Western European equivalents. For instance, they stated that HF and silicofluoric acid are evolved during the process of

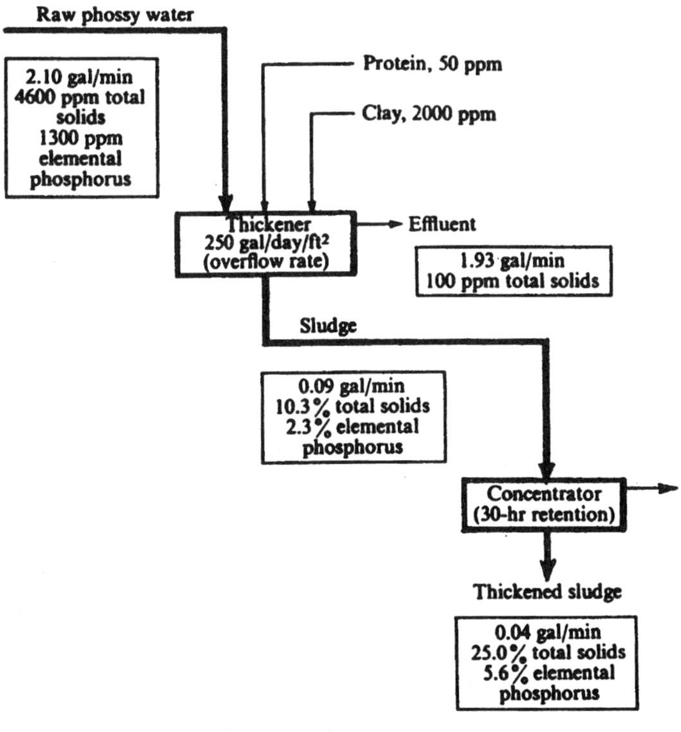

Figure 12 Summary of materials balance in a pilot plant for recovery of phosphorus from phossy water (from Ref. 15). 1 ppm = 1 mg/L.

dissolving phosphorite in the manufacture of normal superphosphate. These constituents are removed from the acidic and highly toxic gases by washing with water or brine in closed condenser equipment, and the wastes from this process are usually slightly acidic, clear, and colorless.

The production of superphosphate is often combined with the manufacture of sodium fluorosilicate and then the amount of wastes is larger.

Czechoslovakian experiments have shown that for every 1000 tons of superphosphate (20% P_2O_5 content) produced, 133 m^3 (4700 ft^3) of postcrystallization liquor from the crystallization of sodium fluorosilicate and 67 m^3 (2370 ft^3) of washings are discharged. The liquor contains 20–25 g/L NaCl, 25–35 g/L HCl, 10–15 g/L silisic acid, and 8–10 g/L sodium silicofluoride. For this waste, the most significant aspect of treatment is recovering the sodium silicofluoride from the brine used to absorb fluorine compounds from the gaseous waste streams, and this is relatively easy to accomplish since it settles nearly ten times as fast as silicic acid gel and, therefore, it is separated by sedimentation. The silicofluoride recovered is a valuable byproduct that, following filtration, washing with water, and drying, is used as a flux in enamel shops, glass works, and other applications.

In Czechoslovakian phosphate fertilizer plants, the superphosphate production wastewaters are further treated by neutralization on crushed limestone beds contained in special tanks that are followed by settling tanks for clarification of the wastewater. The beds have from three to five layers (with a minimum bed height of 0.35–1.60 m), treat a range of acidity of wastes from 438 to 890 meq/L, and are designed for a hydraulic load ranging from 0.13 to 0.52 cm^3/cm^2 s (1.9–7.7 gpm/ft^2) at operating temperatures of 20–28°C. This experience agrees with results reported from Polish plants, the limestone used contains 56% CaO, and it was found in practice that coarse particles of 3–5 mm give better results because less material is carried away. In the former USSR, superphosphate wastes were neutralized with powdered limestone or milk of lime.

The neutralized wastes leaving the settling tank contain primarily dissolved sodium and calcium chlorides. As previously mentioned, the manufacture of phosphate fertilizers also yields large quantities of phosphogypsum, which often contains significant amounts of fluorine and phosphorus compounds and requires large areas for dumping. It has been estimated that for each ton of phosphorite processed, a wet process phosphoric acid plant yields 1.4–1.6 tons of gypsum containing about 30% water and 66% calcium sulfate. This waste material has been used for the production of building materials such as plasterboard, ammonium sulfate, but primarily sulfuric acid and cement.

9.6.7 Phosphoric Acid and N-P-K Fertilizer Plant

According to the literature [3,17,33], the heterogeneous nature of fertilizer production plants precludes the possibility of presenting a "typical" case study of such a facility. Nevertheless, the wastewater flows, the characteristics, and the treatment systems for a phosphoric acid and N-P-K fertilizer plant were parts of a large fertilizer manufacturing facility. The full facility additionally included an ammonia plant, a urea plant, a sulfuric acid plant, and a nitric acid plant. The typical effluent flows were 183 m^3/hour (806 gpm) from the phosphoric plant and 4.4 m^3/hour (20 gpm) from the water treatment plant associated with it, whereas in the N-P-K plant they were 420 m^3/hour (1850 gpm) from the barometric condenser and 108 m^3/hour (476 gpm) from other effluent sources.

These wastewater effluents had quality characteristics that could be described as follows:

1. In the phosphoric acid plant, the contributing sources of effluent are the cooling tower bleedoff and the scrubber liquor solution that contains concentrations ranging for phosphate from 160 to 200 mg/L and for fluoride from 225 to 7000 mg/L.

2. In the water treatment plant, the wastewater effluent is slightly acidic in nature.
3. In the N-P-K plant, the barometric condenser effluent has a pH range of 5.5–8, and concentrations of ammonia-nitrogen of about 250 mg/L, fluoride of about 10 mg/L, and trace levels of phosphate.
4. The N-P-K plant's other effluents contain concentrations of ammonia-nitrogen of about 2000 mg/L, fluoride of about 350 mg/L, and phosphate of about 3000 mg/L.

The wastewater treatment systems utilized for the phosphoric acid and N-P-K plant effluents are shown in Figure 13. As can be seen, the cooling tower bleedoff and scrubber liquor from the phosphoric acid plant are treated together with N-P-K plant effluents by a two-stage lime slurry addition to precipitate out the phosphates and fluorides, reducing them to levels of less than 10 mg/L. The treated effluent pH is adjusted to 5.5–7 using sulfuric acid, and it is discharged to a river, while the precipitated slurry containing the phosphates and fluorides is disposed of in lagoons. As can be seen in the right-hand side of Figure 13, the effluent of the barometric condenser has its pH adjusted to 11 by adding lime to remove residual ammonia-nitrogen, and subsequently waste steam is introduced to remove free ammonia, and the final effluent is mixed with the water treatment plant effluent prior to discharge in a river.

9.6.8 Environmentally Balanced Industrial Complexes

Unlike common industrial parks where factories are selected simply on the basis of their willingness to share the real estate, environmentally balanced industrial complexes (EBIC) are a selective collection of compatible industrial plants located together in a complex so as to minimize environmental impacts and industrial production costs [24,33]. These objectives are accomplished by utilizing the waste materials of one plant as the raw materials for another with a minimum of transportation, storage, and raw materials preparation costs. It is obvious that when an industry neither needs to treat its wastes, nor is required to import, store, and pretreat its raw materials, its overall production costs must be reduced significantly. Additionally, any material reuse costs in an EBIC will be difficult to identify and more easily absorbed into reasonable production costs.

Such EBICs are especially appropriate for large, water-consuming, and waste-producing industries whose wastes are usually detrimental to the environment, if discharged, but they are also amenable to reuse by close association with satellite industrial plants using wastes from and

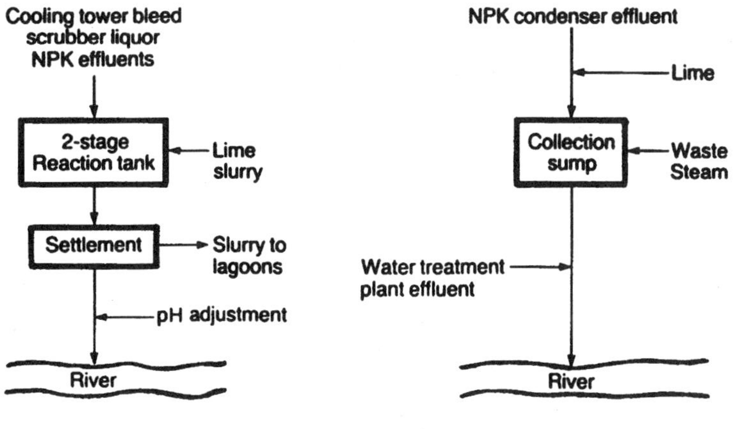

Figure 13 Phosphoric acid and N-P-K fertilizer production waste treatment (from Ref. 17).

producing raw materials for others within the complex. Examples of such major industries that can serve as the focus industry of an EBIC are fertilizer plants, steel mills, pulp and paper mills, and tanneries. Nemerow and Dasgupta [24] presented the example of a steel mill complex with a phosphate fertilizer and a building materials plant as the likely candidates for auxiliary or satellite industries (Fig. 14).

A second example presented was an EBIC centered phosphate fertilizer plant, with a cement production plant, a sulfuric acid plant, and a municipal solid wastes composting plant (its product to be mixed with phosphate fertilizer and sold as a combined product to the agricultural industry) as the satellite industries (Fig. 15). As previously mentioned, in the usual starting process of producing phosphoric acid and ammonium phosphate fertilizer by dissolving the phosphate rocks with sulfuric acid, a gypsumlike sludge is generated as a byproduct and some sulfur dioxide and fluorine are in the waste gases emitted at the high reaction temperatures. The large, relatively impure, quantities of phosphogypsum (5 vol. to 1 P_2O_5 fertilizer produced) are difficult to treat, and the fluorine present in the gas as hydrofluoric acid (concentrations from 1–10%, which is very low for commercial use) requires further costly and extensive treatment. Using such a fertilizer production facility as the focus industry of an EBIC would be a feasible solution to the environmental problems if combined, for example, with such satellite industries as: (a) a sulfuric acid plant to feed its products to the phosphoric acid plant and to use some of the hot water effluent from the cement plant and the effluent from the SO_2 scrubber of the phosphate fertilizer plant; (b) a municipal solid wastes composting plant utilizing hot water from the cement plant, serving as disposal facility for the garbage of a city, and producing composted organic solids to be used as fertilizer; (c) a cement and plasterboard production plant utilizing the phosphogypsum waste sludge in the manufacture of products for the construction industry, producing hot water effluent to be used as mentioned above, and waste dust collected by a dust filter and used as a filler for the soil fertilizer produced by mixing composted garbage and phosphate fertilizer.

Development of an environmentally balanced industrial complex has been discussed extensively [49] by the United Nations Industrial Development Organization (UNIDO), Vienna, Austria.

9.6.9 Fluoride and Phosphorus Removal from a Fertilizer Complex Wastewater

A laboratory-scale treatability study was conducted for the Mississippi Chemical Corporation to develop a physicochemical wastewater treatment process for a fertilizer complex wastewater to control nitrogen, phosphorus, and fluoride and to recover ammonia [2,33]. The removal technique investigated consisted of precipitation of fluorides, phosphorus, and silica by lime addition, a second stage required for the precipitation of ammonia by the use of phosphoric acid and magnesium, and a third stage for further polishing of the wastewater necessary to remove residual phosphate. The wastewater quality parameters included the following concentrations: fluoride 2000 mg/L, ammonia 600 mg/L, phosphorus as P_2O_5 (P) 145 mg/L [50], and an acidic pH level of 3.5. Ammonia removals of 96% were achieved and the insoluble struvite complex produced by the ammonia removal stage is a potentially commercial-grade fertilizer product, whereas the fluoride and phosphorus in the effluent fell below 25 and 2 mg/L, respectively.

The multistage treatment approach was to first remove the fluorides by lime precipitation, with the optimum removal (over 99%) occurring with a two-step pH adjustment to about 10.4 (removal was more a function of lime dosage, rather than a pH solubility controlled phenomenon). Each step was followed by clarification, and the required lime equivalent dosage was 180% of the calcium required stoichiometrically. The effluent from the first stage had an average fluoride content of 135 mg/L (93% removal) and a phosphorus content of less than 5 mg/L, whereas the hydraulic design parameters were a 15 mm per stage reaction time

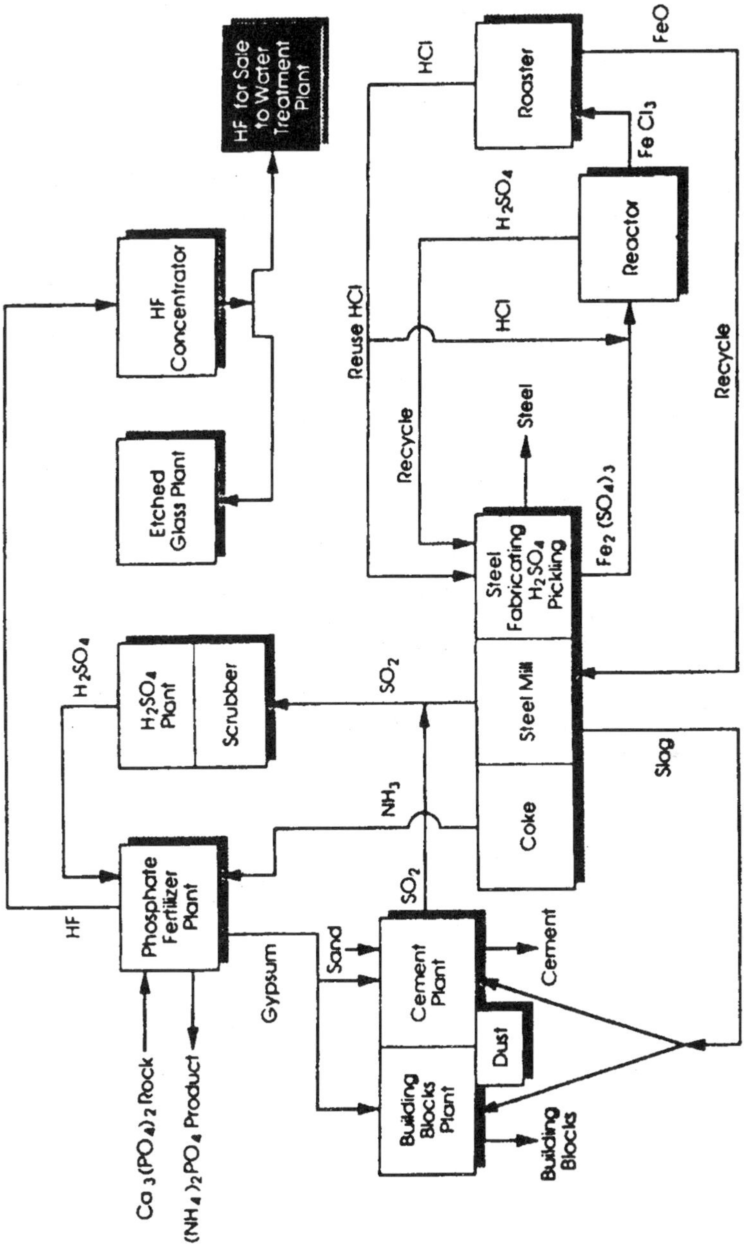

Figure 14 Example of environmentally balanced industrial complex (EBIC) centered about a steel mill plant (from Ref. 24).

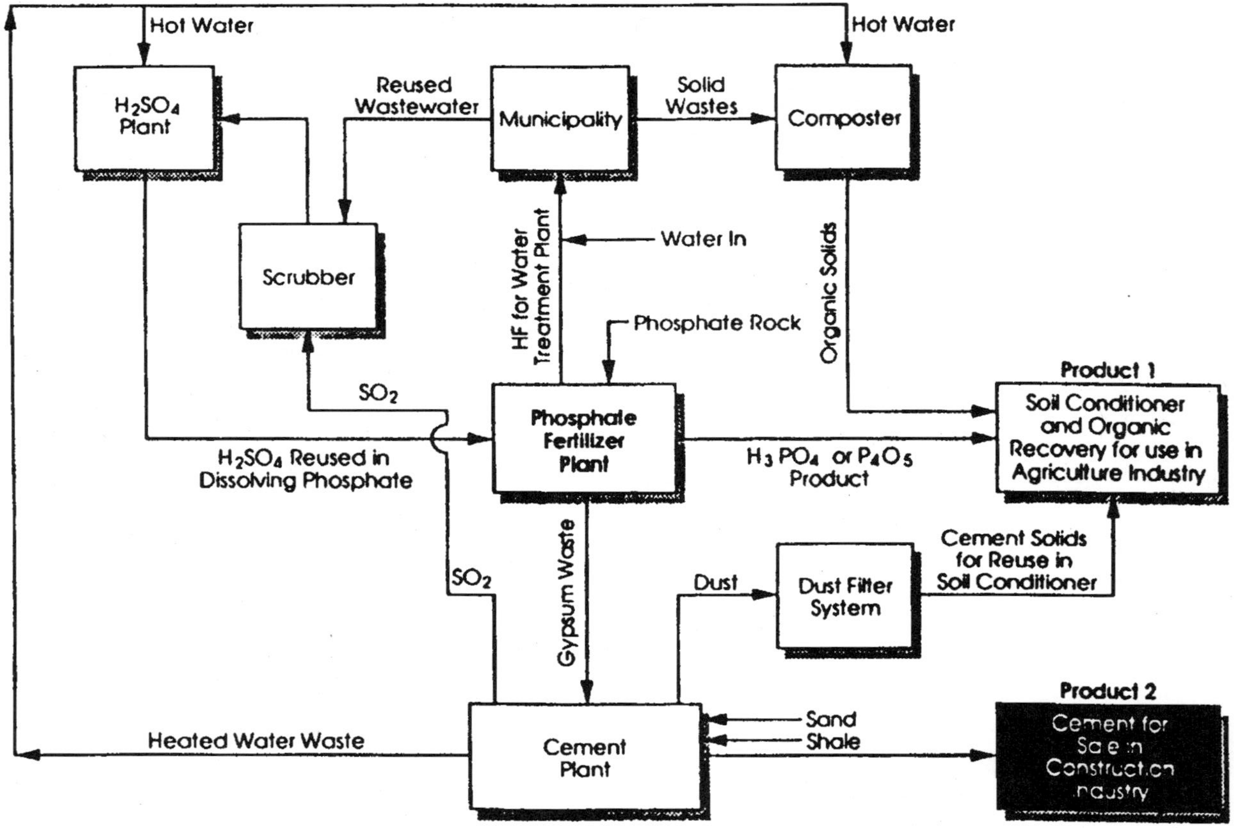

Figure 15 Example of an environmentally balanced industrial complex with a phosphate fertilizer plant as the focus industry (from Ref. 24).

and a 990 gpd/ft^2 (40.3 m^3/m^2/day) clarifier overflow rate. The resulting precipitated solids underflow concentration was 7.7% by weight.

The second-stage (ammonia removal) effluent contained unacceptable levels of F and P and had to be subjected to third-stage lime treatment. This raises the pH from 8.5 to 11.4 and produces an effluent with concentrations of F and P equal to 25 and 2 mg/L, respectively. The hydraulic design parameters were a 15 min reaction time and a 265 gpd/ft^2 (10.8 m^3/m^2/day) clarifier overflow rate. The resulting precipitated solids underflow concentration was 0.6% by wt. In all three stages, an anionic polymer was used to aid coagulation, solids settling, and effluent clarification.

The solids resulting from the first- and third-stage treatment consisted of calcium fluoride, calcium phosphate, and fluorapatite-type compounds. Typically, in the fertilizer industry, such sludges are disposed of by lagooning and subsequent landfilling. Other studies have investigated the recovery of the fluoride compounds, such as hydrofluoric and fluorosilisic acids, for use in the glass industry and in the fluoridation of drinking water supplies.

9.6.10 Phosphate Recovery by Crystallization Process

In municipal water applications and many industrial unit operations, phosphate-polluted wastewater is generated and, in general, conventional phosphate removal techniques are applied for treatment. These conventional processes are based on phosphate precipitation as calcium or iron salt or fixation in activated sludge. These processes, however, generate huge amounts of a water-rich sludge, which has to be disposed of at increasing costs. Typically, even after dewatering the sludge, the water content still is 60–85% and a relatively large part of the disposal costs comes from the expensive disposal of water. Owing to the high water content and the low quality of the sludge, reuse of phosphate is not an economically attractive option. Also, the area requirements for conventional phosphate precipitation is relatively high because the four process steps (coagulation, flocculation, sludge/water separation, dewatering) are performed serially.

An advanced alternative is to apply crystallization instead of precipitation [51,52]. The Crystalactor®, a fluid-bed type of crystallizer, has been developed by DHV (a multinational group headquartered in the Netherlands), to generate high-purity phosphate crystal pellets that can be reused in many applications and is more and more important because it is a sustainable solution to the environmental problems related to the mining and processing of natural phosphate resources. The chemistry of the process is comparable to conventional precipitation. By dosing of calcium or magnesium salt to the water (e.g., lime, calcium chloride, magnesium hydroxide, magnesium chloride), the solubility of calcium phosphate, magnesium phosphate, magnesium ammonium phosphate, or potassium magnesium phosphate is exceeded and subsequently phosphate is transformed from the aqueous solution into solid crystal material. The primary difference with conventional precipitation is that in the crystallization process the transformation is controlled accurately and that pellets with a typical size of approximately 1 mm are produced instead of fine dispersed, microscopic sludge particles. After atmospheric drying, readily handled and virtually water-free highly pure phosphate pellets are obtained. Several reuse options for the pellets are: (a) raw material for the production of phosphoric acid in either the wet or thermo production processes; (b) intermediate products for fertilizer formulation; (c) raw material for kettle food; and (d) direct use as (slow-release) fertilizer.

In the food industry, wastewaters with a high organic load are released. The wastewater of a Dutch potato processing plant is treated in an anaerobic biological reactor because of the low sludge production, the low energy consumption, and the biogas production. The effluent is polished in an aerobic biological treatment plant. Cost-effective phosphate removal by struvite crystallization in the Crystalactor was implemented by dosing magnesium chloride and sodium hydroxide solutions into a part of the anaerobic stage effluent [52]. A flow of a maximum of 150 m^3/hour with 120 mg/L P was successfully treated in a reactor with a diameter of 1.8 m to an effluent containing about 10 mg/L P at a pH of 8–8.5. The struvite (magnesium ammonium phosphate) was accepted by a fertilizer producer to be used directly in the granulation process for fertilizers.

For a chemical industry, a pilot plant was operated on a waste stream that contained very high P and K concentrations (several grams per liter). It seemed that the production of potassium magnesium phosphate was not successful for this specific wastewater; however, magnesium phosphate was easily produced in order to reduce P to the desired concentration. In a pilot test for a food industry on wastewater from a biological anaerobic pretreatment unit, containing around 150 mg/L P and 10 mol/L of inorganic carbon (high concentration of carbonates), magnesium phosphate was successfully produced without disturbance by the presence of inorganic carbon. The effluent soluble P concentration was about 10 mg/L. Finally, this technology has also found applications in water softening, fluoride removal, and heavy metal recovery [53].

9.6.11 Phosphate Acid Plant Wastewater Treatment

The effluent wastewater from a chemical plant manufacturing technical grade phosphoric acid, due to low BOD and COD ratio and very low pH, was not amenable to biological treatment. So, treatability studies were carried out using the Fenton reaction and physicochemical (coagulation-settling) treatment using lime, alum, Fe salts and polyaluminum chloride (PAC). The treatability studies indicated [54] that it is possible to remove 75–80% COD using Fenton's reagent at optimum doses of 1.0 g/L ferrous sulfate and 2 mL of 30% hydrogen peroxide. Simultaneously, significant quantities of suspended solids (SS), phosphate and fluoride were also removed. Polyaluminum chloride was found to be more effective towards SS, COD, phosphate, and fluoride removal, in comparison with other coagulants used in the many studies [50,54,55]. The addition of an anionic polyelectrolyte (Magnafloc 156) and PAC improved the performance further. A treatment scheme that consisted of neutralization (pH 4) + Fenton's reagent + neutralization (pH 7.5) + PAC/Magnafloc 156 was found to be effective in treating phosphoric acid plant wastewater to meet marine discharge standards [54].

9.6.12 Improvement and Phosphorus Removal in Wetlands and Sand Filters

Removal of phosphorus and nitrogen by a wetland process has been discussed by Hung *et al.* [56]. The efficiency of P removal in ecologically engineered wastewater treatment facilities, such as artificial wetlands and sand filters, was found to be enhanced by using a reactive sorbent. The sorbent must have a high P-sorption capacity and an adequate hydraulic conductivity. Several filter materials were tested in this study with regard to their P-sorption capacity [57]. The P-sorption experiments showed that, of the materials tested, the coarse (0.25–4 mm) blast furnace crystalline slag had the highest P-sorption capacity followed by the fine (0–0.125 mm) blast furnace crystalline slag, the fine amorphous slag, and the B horizon of a forest soil. The isotherm tests showed that the B horizon was the most efficient P retainer, followed in order by the coarse and the fine crystalline slag. It was also concluded that coarse crystalline slag possessed the highest hydraulic conductivity, suggesting that this material is the most suitable for use in ecological wastewater treatment systems.

9.6.13 Removal of Phosphate Using Feldspar

Feldspar, among many natural substances such as termite mount-clay, saw dust, kaolinite, and dolomite, offers significant removal ability for phosphate, sulfate, and color colloids. Optimization laboratory tests of parameters such as solution pH and flow rate, resulted in a maximum efficiency for removal of phosphate (42%), sulfate (52%), and color colloids (73%), x-ray diffraction, adsorption isotherms test and recovery studies suggest that the removal process of anions occurs via ion exchange in conjunction with surface adsorption. Furthermore, reaction rate studies indicated that the removal of these pollutants by feldspar follows first-order kinetics. Percent removal efficiencies, even under optimized conditions, will be expected to be somewhat less for industrial effluents in actual operations due to the effects of interfering substances [58].

9.6.14 Phosphate Biological Uptake at Acid pH

Some microorganisms store phosphate as polyphosphate, thereby removing it from solution, but current methods to induce this are unreliable. However, it was recently discovered that polyphosphate production is increased by acid shock [59]. These studies have led to the identification of a significant, yet previously recognized, microbial stress response at acidic pH levels, which may be a novel strategy for the "one-step" removal of phosphate from wastewater effluents. It was possible to increase the level of phosphate removal by the microflora of a conventional activated sludge plant – under fully aerobic conditions – by more than 50% if the operational pH was adjusted within the range 5.5–6.5, as opposed to within 7.2–7.7, which are typical in current practice. Similar results were obtained in four other activated sludge plants of varying influent characteristics; enhancement of phosphate removal at pH 5.5 varied between 56 and 142% and involved a considerable fraction of the microbial population – bacterial, yeast, and fungal. Further research to assess the economic viability of a low-pH phosphate removal system is being carried out in Northern Ireland and Britain.

9.6.15 Emerging Phosphorus Removal Technologies

More recent research on treatment or recovery of phosphates and phosphorus from wastewater can be found from the literature [57,58,60–69,73].

The use of combined biological and physicochemical treatment processes for phosphorus removal was originally conceived by Beer and Wang at Coxsackie Sewage Treatment Plant, NY [58], and by Krofta and Wang at the Lenox Institute of Water Technology, MA [65,66]. They successfully used ferric chloride, lime, and alum for precipitation of phosphate from the activated sludge aeration basin effluent. Wang and Aulenbach [67] have discussed the theory and principles of biological phosphorus uptake under aerobic conditions, biological phosphorus release under anoxic/anaerobic conditions, and physicochemical precipitation of released phosphorus (in phosphate form) by an innovative A/O process. Wang [68] has adopted a commercially available dissolved air flotation (DAF) clarifier [70] in a combined biological physicochemical process system for high-rate phosphorus removal.

Essentially, Wang's innovative technology [68] is a combined biological-chemical precipitation process involving the use of the following process steps:

1. Treatment of incoming wastewater (usually primary effluent) in an aeration basin to transfer phosphorus from the incoming wastewater to the "P-stripped activated sludge microorganisms" under aerobic conditions.
2. Separation of the spent activated sludge microorganisms from the aeration basin effluent by either a conventional sedimentation clarifier (with 2 hours definition time

or DT), or by a high-rate dissolved air flotation (DAF) clarifier (with 15 minutes DT) [65,66,68].

3. Discharge of the almost P-free clarifier effluent into a receiving water, while discharge of the P-concentrated clarifier sludge (either settled sedimentation clarifier sludge, or the floated DAF clarifier sludge) to a phosphate stripper (i.e., a thickener-type holding tank with 5–15 hours DT) where the clarifier sludge is subjected to anoxic/anaerobic conditions to induce phosphorus release from the clarifier sludge into the aqueous phase [68].
4. Chemical precipitation of the newly released phosphate in the phosphate stripper effluent (i.e., a P-rich, low-volume sidestream containing 40–80 mg/L P; amounting to about 10–15% of the total wastewater flow) using lime, ferric chloride, or alum, and subsequently flotation of the P-rich precipitated chemical sludge for reuse as a fertilizer using a high-rate DAF clarifier (15 min DT).
5. Collection of the phosphate-stripped activated sludge, which has extremely high phosphorus uptake capacity from the stripper.
6. Return of the phosphate-stripped activated sludge to the aeration basin for reuse in a new cycle, where the phosphate-stripped activated sludge microorganisms are again induced to take up dissolved phosphorus in excess of the amount required for growth under aerobic conditions.

The above P-removal process system reduces the volume of the wastewater to be treated (10–15% of total wastewater flow), thereby reducing the chemical dosage required, the amount of chemical sludge produced, and associated costs. Lime can be used to remove phosphorus from the stripper supernatant at lower pH levels (8.5–9.0) than normally required, although alum and ferric chloride are equally effective. The cycling of sludge through an anoxic phase may also assist in the control of hulking by the destruction of filamentous organisms to which hulking is generally attributed. The process is capable of reducing the total phosphorus concentration of typical municipal wastewaters to 1 mg/L or less. Adoption of a DAF clarifier instead of sedimentation clarifier for both secondary clarification and P-rich precipitated chemical sludge separation significantly reduces process time, and, in turn, saves overall capital and O&M costs [68].

Wang [68] has also successfully used conventional biological sequencing batch reactors (SBR) instead of conventional activated sludge aeration for steps 1 and 6 above, and has used physicochemical sequencing batch reactors (PC-SBR) instead of the stripper and DAF for steps 3 and 4 above. The readers are encouraged to further improve upon this emerging phosphorus removal technology. Descriptions and discussions of PC-SBR can be found from the literature [3,49,69,71].

REFERENCES

1. Anon. Phosphate, the servant of mankind. *Oil Power*, **1951**, *26* (*3*), 61–62.
2. Arnold, D.W.; Wolfram, W.E. Ammonia removal and recovery from fertilizer complex wastewaters. In *Proceedings of 30th Industrial Waste Conference*, Purdue University, Lafayette, IN, 1975; Vol. 30, 760–767.
3. Beg, S.A. *et al.* Effect of toxicants on biological nitrification for treatment of fertilizer industry wastewater. In *Proceedings of 35th Industrial Waste Conference*, Purdue University, Lafayette, IN, 1980; Vol. 35, 826–834.
4. Beg, S.A. *et al.* Inhibition of nitrification by arsenic, chromium and fluoride. *J. Water Poll. Control Fed.*, **1982**, *54*, 482–488.

5. USEPA. *Process Design Manual for Phosphorus Removal;* Office of Technology Transfer, U.S. Environmental Protection Agency: Washington, DC, 1971.
6. USEPA. *Pretreatment of Pollutants Introduced into Publicly Owned Treatment Works;* Federal Guidelines, Office of Water Program Operations, U.S. Environmental Protection Agency: Washington, DC, 1973.
7. USEPA. *Basic Fertilizer Chemicals*, EPA-440/1-74-Olla; Effluent Guidelines Division, U.S. Environmental Protection Agency: Washington, DC, 1974.
8. USEPA. *Federal Guidelines on State and Local Pretreatment Programs*, EPA-430/9-76-017c; Construction Grants Program, U.S. Environmental Protection Agency: Washington, DC, 1977.
9. Federal Register. *Code of Federal Regulations*, CFR 40; U.S. Government Printing Office: Washington, DC, 1987; 412–430, 729–739.
10. Federal Register. *Notices, Appendix A, Master Chart of Industrial Categories and Regulations;* U.S. Government Printing Office: Washington, DC, 1990; Vol. 55, No. 1, Jan. 2, 102, 103.
11. Fertilizer Association of India. Liquid effluents. In *Pollution Control in Fertilizer Industry*, Tech. Rep. 4, Part I, 1979.
12. Fuller, R.B. The position of the pebble phosphate industry in stream sanitation. *Sewage Wks. J.* **1949**, *9* (*5*), 944.
13. Ghokas, S.I. *Treatment of Effluents from a Fertilizer Complex*, M.S. thesis; University of Manchester, UK, 1983.
14. Griffith, E.J. Modem mankind's influence on the natural cycles of phosphorus. In *Phosphorus and the Environment;* Ciba Foundation, New Series, 57, 1978.
15. Horton, J.P. *et al.* Processing of phosphorus furnace wastes. *J. Water Poll. Control Fed.*, **1956**, *28 (1), 70–77.*
16. Jones, W.E.; Olmsted, R.L. Waste disposal at a phosphoric acid and ammonium phosphate fertilizer plant. In *Proceedings of the 17th Industrial Waste Conference*, Purdue University, Lafayette, IN, 1962; Vol. 17, 198–202.
17. Kiff, R.I. Water pollution control in the fertilizer manufacturing industry. In *Manufacturing and Chemical Industries*; Barnes, D. *et al.*, Eds.; Longman Scientific & Technical: Essex, UK, 1987.
18. Koziorowski, B.; Kucharski, J. *Industrial Waste Disposal;* Pergamon Press: Oxford, UK, 1972; 142–151.
19. Lanquist, E. *Peace and Alafia River Stream Sanitation Studies;* Florida State Board of Health, June, Suppl. II to Vol. II, 1955.
20. Layer, W.; Wang, L.K. *Water Purification and Wastewater Treatment with Sodium Aluminate*, Report PB 85-214-492/AS; U.S. Department of Commerce, National Technical Information Service: Springfield, VA, 1984.
21. Lund, H.F. (Ed.) *Industrial Pollution Control Handbook;* McGraw-Hill: New York, 1971; 14-6–14-8.
22. Markham, J.H. *Effluent Control on a Fertilizer Manufacturing Site;* Fertilizer Soc. of London, London, 1983; Proc. 213.
23. Nemerow, N.L. *Industrial Water Pollution;* Addison-Wesley: Reading, MA, 1978; 583–588.
24. Nemerow, N.L.; Dasgupta, A. Environmentally balanced industrial complexes. In *Proceedings of the 36th Industrial Waste Conference*, Purdue University, Lafayette, IN, 1981; Vol. 36, 982–989.
25. Overcash, M.R. *Techniques for Industrial Pollution Prevention;* Lewis Publishers: Michigan, 1986; 87–89.
26. Robasky, J.G.; Koraido, D.L. Gauging and sampling industrial wastewater. *J. Chem. Engrs.* **1973**, *80* (*1*), 111–120.
27. Search, W.J. *et al. Source Assessment, Nitrogen Fertilizer Industry Water Effluents*, Report PB 292 937; U.S. Department of Commerce, National Technical Information Service: Springfield, VA, 1979.
28. Shahalam, A.B.M. *et al.* Wastes from processing of phosphate industry. In *Proceedings of 40th Industrial Waste Conference*, Purdue University, Lafayette, IN, 1985; Vol. 40, 99–110.
29. Specht, R.C. *Phosphate Waste Studies*, Bull. 42; Florida Eng. and Ind. Exp. Sta., University of Florida: Gainesville, 1950.
30. Specht, R.C. Disposal of wastes from the phosphate industry. *J. Water Poll. Control Fed.* **1960**, *32* (*9*), 963–974.

31. Wakefield, J.W. Semi-tropical industrial waste problems. In *Proceedings of the 7th Industrial Waste Conference*, Purdue University, Lafayette, IN, 1952; Vol. 7, 495–508.
32. Krofta, M.; Wang, L.K. *Flotation and Related Adsorptive Bubble Separation Processes*, 4th Ed; Lenox Institute of Water Technology: Lenox, MA 2001; 185 p.
33. Yapijakis, C. Treatment of phosphate industry wastes. In *Handbook of Industrial Waste Treatment*, Wang, L.K., Wang, M.H.S. Eds.; Marcel Dekker: New York, NY, 1992; Chapter 8, 323–383.
34. UNIDO. *Minimizing Pollution from Fertilizer Plants*, Report, Expert Group Meeting, Helsinki, ID/WG 175/19, 1974.
35. Wang, L.K. *Pretreatment and Ozonation of Cooling Tower Water, Part I*, PB84-192053; U.S. Department of Commerce, National Technical Information Service: Springfield, VA, 1983; 34 p.
36. Wang, L.K. *Pretreatment and Ozonation of Cooling Tower Water, Part II*, PB84-192046; U.S. Department of Commerce, National Technical Information Service: Springfield, VA, 1983; 29 p.
37. Caswell, C.A. The fallacy of the closed-cycle cooling concept. *Ind. Water Engng* **1990**, *17* (*3*), 27–31.
38. Hallett, G.F. Area-view of proposed methods for evaluating cooling equipment. *Ind. Water Engng* **1990**, *17* (*3*), 12–26.
39. Wang, L.K.; Wang, M.H.S. Decontamination of groundwater and hazardous industrial effluents by high-rate air flotation process. Presented at Proc. Great Lakes 90 Conf., Hazardous Materials Control Res. Inst., Silver Springs, MD, September 1990.
40. Wang, L.K. *Preliminary Design Report of a 10-MGD Deep Shaft-Flotation Plant for the City of Bangor*, PB88-200597/AS; U.S. Department of Commerce, National Technical Information Service: Springfield, VA, 1987; 42 p.
41. Wang, L.K.; Wang, M.H.S. Treatment of storm run-off by oil–water separation, flotation, filtration and adsorption, part A: wastewater treatment. In *Proceedings of the 44th Industrial Waste Conference*, Purdue University, Lafayette, IN, 1990; 655–666.
42. Wang, L.K. *et al.* Treatment of storm run-off by oil-water separation, flotation, filtration and adsorption, part B: waste sludge management. In *Proceedings of the 44th Industrial Waste Conference*, Purdue University, Lafayette, IN, 1990; 667–673.
43. Wang, L.K.; Wang, P.; Clesceri, N. Groundwater decontamination using sequencing batch processes. *Water Treatment*, **1995**, *10*, 121–134.
44. Krofta, M.; Wang L.K. Development of innovative Sandfloat systems for water purification and pollution control. *ASPE Journal of Eng. Plumbing*, **1984**, *0* (*1*), 1–16.
45. Krofta, M.; Wang, L.K. Tertiary treatment of secondary effluent by dissolved air flotation and filtration. *Civil Engng for Practicing and Design Engrs*, **1984**, *3*, 253–272.
46. Krofta, M.; Wang, L.K. *Development of an Innovative and Cost-Effective Municipal-Industrial Waste Treatment System*, PB88-168109/AS; U.S. Department of Commerce, National Technical Information Service: Springfield, VA, 1985; 27 p.
47. Krofta, M.; Wang, L.K. Wastewater treatment by biological-physicochemical two-stage process system. *Proceedings of the 41st Industrial Waste Conference*; Lewis Publishers, Inc.: Chelsea, MI, 1987; 67–72.
48. Krofta, M.; Wang, L.K. Development of low-cost flotation technology and systems for wastewater treatment. In *Proceedings of the 42nd Industrial Waste Conference*, Purdue University, Lafayette, IN, 1988; p. 185.
49. Wang, L.K.; Krouzek, J.V.; Kounitson, U. *Case Studies of Cleaner Production and Site Remediation*; United Nations Industrial Development Organization (UNIDO): Vienna, Austria, 1995, Training Manual No. DTT-5-4-95, 136.
50. Krofta, M.; Wang, L.K. Winter operation of nation's first two potable flotation plants. In *Proceedings of 1987 Joint Conference of AWWA and WPCF*, Cheyenne, Wyoming, USA, Sept. 20–23, 1987.
51. Piekema, P.G.; Gaastra, S.B. Upgrading of a wastewater treatment plant: Several nutrients removal. *Eur. Water Pollut. Control* **1993**, *3* (*3*), 21–26.
52. Piekema, P.G.; Giesen, A. *Phosphate Recovery by the Crystallization Process: Experience and Developments*; DHV Water BV: P.O. Box 484, 3800 Al Amersfoort, The Netherlands, 2001.
53. Bouwman, J.G.M.A; Luisman, A.H.C. Aiming at zero discharge and total reuse. *Eur. Semicond.* **2000**, *Sept*, 25–27.

54. Nawghare, P.; Rao, N.N.; Bejankiwar, R.; Szyprkowicz. Treatment of phosphoric acid plant wastewater using Fenton's reagent and coagulants. *J. Environ. Sci. and Health* **2001**, *A 36 (10)*, 2011.
55. Wang, L.K. *Poly Iron Chloride and Poly Aluminum Chloride*, Technical Report LIWT/1-2002/252; Lenox Institute of Water Technology: Lenox, MA, 2002, 26 p.
56. Hung, Y.; Gubba, S.; Lo, H.H.; Wang, L.K.; Yapijakis, C.; Shammas, N.K. Application of wetland for wastewater treatment. *OCEESA J.* **2003**, *20 (1)*, 41–46.
57. Johansson, L. Industrial by-products and natural substrata as phosphorus sorbents. *Environ. Technol.* **1999**, *20 (3)*, 309.
58. Beer, C.; Wang, L.K. Full-scale operations of plug flow activated sludge systems. *J. New Engl. Water Poll. Control Assoc.* **1975**, *9(2)*, 145–173.
59. McGrath, J.W.; Quinn, J.P. *Phosphate Removal: a Novel Approach*; School of Biology and Biochemistry, Queens' University Belfast, Ireland, **2002**; http://www.qub.ac/uk/envres/EarthAir-water/phosphate-removal.htm
60. Akin, B.S.; Ugurlu, A. Enhanced phosphorus removal by glucose fed sequencing batch reactor. *J. Environ. Sci. and Health* **2001**, *A 36 (9)*, p 1757.
61. CEEP Phosphates, a sustainable future in recycling. Centre Europeen d' Polyphosphates, 1999.
62. Giesen, A. Eliminate sludge. *Ind. Wastewater* **1998**, *6*.
63. Martin, D.F.; Dooris, P.M.; Sumpter, D. Environmental impacts of phosphogypsum vs. borrow pits in roadfill construction. *J. Environ. Sci. and Health* **2001**, *A 36 (10)*, 1975.
64. Priyantha, N.; Pereira, S. Removal of phosphate, sulfate, and colored substances in wastewater effluents using Feldspar. *Water Res. Mgnt* **2000**, *14 (6)*, 417.
65. Krofta, M.; Wang, L.K. Improved biological treatment with a secondary flotation clarifier. *Civil Engng for Practicing and Design Engrs*, **1983**, *2*, 307–324.
66. Krofta, M.; Wang, L.K. Wastewater treatment by biological-physicochemical two-stage process systems. In *Proceedings of the 41st Industrial Waste Conference*, May, 1987; 67–72.
67. Wang, L.K.; Aulenbach, D. *BOD and Nutrient Removal by Biological A/O Process Systems*, PB88-168430/AS; U.S. Department of Commerce, National Technical Information Service: Springfield, VA, 1986; 12 p.
68. Wang, L.K. *An Emerging Technology for Phosphorus Removal from Wastewaters*, Technical Report LIWT/1-2002/253; Lenox Institute of Water Technology: Lenox, MA, 2002, 18 p.
69. Wang, L.K. Laboratory simulation of water and wastewater treatment processes. *Water Treatment*, **1995**, *10*, 261–282.
70. Anonymous. Flotation equipment. *Environ. Protection* **2003**, *14* (2), 137–138.
71. Wang, L.K.; Kurylko, L.; Wang, M.H.S. *Sequencing Batch Liquid Treatment*, U.S. Patent No. 5,354,458, October 11, 1994.
72. Wang, L.K. *Preliminary Design Report of a 10-MGD Deep Shaft-Flotation Plant for the City of Bangor, Appendix*, PB88-200605/AS; U.S. Department of Commerce, National Technical Information Service: Springfield, VA, 1987; 171 p.
73. State of Florida. Industrial Wastewater Program—Phosphate Industry. State of Florida, Dept. of Environmental Protection, Miami, FL, Feb. 2004. www.dep.state.fl.us.

10

Treatment of Pulp and Paper Mill Wastes

Suresh Sumathi
Indian Institute of Technology, Bombay, India

Yung-Tse Hung
Cleveland State University, Cleveland, Ohio, U.S.A.

10.1 POLLUTION PROBLEMS OF PULP AND PAPER INDUSTRIES

Pulp and paper mills are a major source of industrial pollution worldwide. The pulping and bleaching steps generate most of the liquid, solid, and gaseous wastes (Table 1) [1]. Pulping is a process in which the raw material is treated mechanically or chemically to remove lignin in order to facilitate cellulose and hemicellulose fiber separation and to improve the papermaking properties of fibers. Bleaching is a multistage process to whiten and brighten the pulp through removal of residual lignin. Pulping and bleaching operations are energy intensive and typically consume huge volumes of fresh water and large quantities of chemicals such as sodium hydroxide, sodium carbonate, sodium sulfide, bisulfites, elemental chlorine or chlorine dioxide, calcium oxide, hydrochloric acid, and so on. A partial list of the various types of compounds found in spent liquors generated from pulping and bleaching steps is shown in Table 2 [2–4]. The effluents generated by the mills are associated with the following major problems:

- Dark brown coloration of the receiving water bodies result in reduced penetration of light, thereby affecting benthic growth and habitat. The color responsible for causing aesthetic problems is attributable to lignin and its degradation products.
- High content of organic matter, which contributes to the biological oxygen demand (BOD) and depletion of dissolved oxygen in the receiving ecosystems.
- Presence of persistent, bio-accumulative, and toxic pollutants.
- Contribution to adsorbable organic halide (AOX) load in the receiving ecosystems.
- Measurable long-distance transport (>100 km) of organic halides (such as chloroguaiacols), thereby contaminating remote parts of seas and lakes [5].
- Cross-media pollutant transfer through volatilization of compounds and absorption of chlorinated organics to wastewater particulates and sludge.

Significant solid wastes from pulp and paper mills include bark, reject fibers, wastewater treatment plant sludge, scrubber sludge, lime mud, green liquor dregs, boiler and furnace ash. The bulk of the solid wastes are generated during wastewater treatment. Sludge disposal is a serious environmental problem due to the partitioning of chlorinated organics from effluents to

Table 1 Types of Pollutants Generated During Chemical (Kraft) Pulping and Bleaching Steps

Pollution generating step	Pollution output phase	Nature of pollution
Wood debarking and chipping, chip washing	Solid	Bark, wood processing residues
	Water	SS, BOD, color, resin acids
Chemical (Kraft) pulping, black liquor evaporation and chemical recovery steps	Air	Total reduced sulfur (hydrogen sulfide, methyl mercaptan, dimethyl sulfide, dimethyl disulfide), VOC
Wood chip digestion, spent pulping liquor evaporator condensates	Water	High BOD, color, may contain reduced sulfur compounds, resin acids
Pulp screening, thickening, and cleaning operations	Water	Large volume of waters with SS, BOD, color
Smelt dissolution, clarification to generate green liquor	Solid	Green liquor dregs
Recausticizing of green liquor, clarification to generate white liquor	Solid	Lime slaker grits
Chlorine bleaching of pulp	Water	BOD, color, chlorinated organics, resin acids
	Air	VOC
Wastewater treatment	Solid	Primary and secondary sludge, chemical sludge
	Air	VOC
Scrubbing for flue gases	Solid	Scrubber sludge
Recovery furnaces and boilers	Air	Fine and coarse particulates, nitrogen oxides, SO_2
	Solid	Ash

SS, suspended solids; VOC, volatile organics; BOD, biochemical oxygen demand.
Source: Ref. 1.

solids. The major air emissions are fine and coarse particulates from recovery furnaces and burners, sulfur oxides (SOx) from sulfite mills, reduced sulfur gases and associated odor problems from Kraft pulping and chemical recovery operations, volatile organic compounds (VOC) from wood chip digestion, spent liquor evaporation and bleaching, nitrogen oxides (NOx) and SOx from combustion processes. Volatile organics include carbon disulfide, methanol, methyl ethyl ketone, phenols, terpenes, acetone, alcohols, chloroform, chloromethane, and trichloroethane [1].

The extent of pollution and toxicity depends upon the raw material used, pulping method, and pulp bleaching process adapted by the pulp and paper mills. For example, the pollution load from hardwood is lower than softwood. On the other hand, the spent liquor generated from pulping of nonwood fiber has a high silica content. Volumes of wastewater discharged may vary from near zero to 400 m^3 per ton of pulp depending on the raw material used, manufacturing process, and size of the mill [6]. Thus, the variability of effluent characteristics and volume from one mill to another emphasizes the requirement for a variety of pollution prevention and treatment technologies, tailored for a specific industry.

Table 2 Low-Molecular-Weight Organic Compounds Found in the Spent Liquors from Pulping and Bleaching Processes

Class of compounds				
Acidic				
Wood extractives	Lignin/carbohydrate derived	Phenolic	Neutral	Miscellaneous
Category: Fatty acid	*Category: Hydroxy*	*Category: Phenolic*	Hemicelluloses	*Category: Dioxins*
Formic acid (S)	Glyceric acid	Monochlorophenols	Methanol	2,3,7,8-tetrachloro-dibenzodioxin (2,3,7,8-TCDD)
Acetic acid (S)	*Category: Dibasic*	Dichlorophenols	Chlorinated acetones	2,3,7,8-tetrachloro-dibenzofuran (2,3,7,8-TCDF)
Palmitic acid (S)	Oxalic acid	Trichlorophenols	Chloroform	*Wood derivatives*
Heptadecanoic acid (S)	Malonic acid	Tetrachlorophenol	Dichloromethane	Monoterpenes
Stearic acid (S)	Succinic acid	Pentachlorophenol	Trichloroethene	Sesquiterpenes
Arachidic acid (S)	Malic acid	*Category: Guaiacolic*	Chloropropenal	Diterpenes: Pimarol
Tricosanoic acid (S)	*Category: Phenolic acid*	Dichloroguaiacols	Chlorofuranone	Abienol
Lignoceric (S)	Monohydroxy benzoic acid	Trichloroguaiacols	1,1-dichloro-methylsulfone	*Juvabiones*
Oleic (US)	Dihydroxy benzoic acid	Tetrachloroguaiacol	Aldehydes	Juvabiol
Linolenic acid (US)	Guaiacolic acid	*Category: Catecholic*	Ketones	Juvabione
Behenic acid (S)	Syringic acid	Dichlorocatechols	Chlorinated sulfur	*Lignin derivatives*
Category: Resin acid		Trichlorocatechols	Reduced sulfur compounds	Eugenol
Abietic acid		*Category: Syringic*		Isoeugenol
Dehydroabietic acid		Trichlorosyringol		Stilbene
Mono and dichloro dehydrabietic acids		Chlorosyringaldehyde		Tannins (monomeric, condensed and hydrolysable)
Hydroxylated-dehydroabietic acid				Flavonoids
Levopimaric acid				
Pimaric acid				
Sandracopimaric acid				

S, saturated; US, unsaturated
Source: Refs 2–4.

The focus of this chapter is to trace the origin and nature of the major pollution (especially water) problems within the pulp and paper industries and to present an overview of the pollution mitigation strategies and technologies that are currently in practice or being developed (emerging technologies).

10.2 NATURE AND COMPOSITION OF RAW MATERIALS USED BY PULP AND PAPER INDUSTRIES

The pulp and paper industries use three types of raw materials, namely, hard wood, soft wood, and nonwood fiber sources (straw, bagasse, bamboo, kenaf, and so on). Hard woods (oaks, maples, and birches) are derived from deciduous trees. Soft woods (spruces, firs, hemlocks, pines, cedar) are obtained from evergreen coniferous trees.

10.2.1 Composition of Wood and Nonwood Fibers

Soft and hard woods contain cellulose (40–45%), hemicellulose (20–30%), lignin (20–30%), and extractives (2–5%) [7]. Cellulose is a linear polymer composed of β-*D*-glucose units linked by 1–4 glucosidic bonds. Hemicelluloses are branched and varying types of this polymer are found in soft and hard woods and nonwood species. In soft woods, galactoglucomannans (15–20% by weight) arabinoglucurono-xylan, (5–10% by weight), and arabinogalactan (2–3% by weight) are the common hemicelluloses, while in hard woods, glucuronoxylan (20–30% by weight) and glucomannan (1–5% by weight) are found [2,3]. Lignin is a complex heterogeneous phenylpropanoid biopolymer containing a diverse array of stable carbon–carbon bonds with aryl/alkyl ether linkages and may be cross-linked to hemicelluloses [8]. Lignins are amorphous, stereo irregular, water-insoluble, nonhydrolyzable, and highly resistant to degradation by most organisms and must be so in order to impart resistance to plants against many physical and environmental stresses. This recalcitrant biopolymer is formed in plant cell walls by the enzyme-catalyzed coupling of p-hydroxycinnamyl alcohols, namely, p-coumaryl, coniferyl, and sinapyl alcohols that make up significant proportion of the biomass in terrestrial higher plants. In hardwoods, lignin is composed of coniferyl and sinapyl alcohols and in softwoods is largely a polymer of coniferyl alcohol. The solvent extractable compounds of wood termed as "extractives" include aliphatics such as fats, waxes, and phenolics that include tannins, flavonoids, stilbenes, and terpenoids. Extractives comprise 1–5% of wood depending upon the species and age of the tree. Terpenoids that include resin acids are found only in softwood and are derived from the "pitch" component of wood. Compared to wood, the structures of nonwood species are not well studied. Grasses usually contain higher amounts of hemicelluloses, proteins, silica, and waxes [9]. On the other hand, grasses contain lower lignin content compared to wood and the bonding of lignin to cellulose is weaker and therefore easier to access.

10.3 PULPING PROCESSES

The steps involved in pulping are debarking, wood chipping, chip washing, chip crushing/digestion, pulp screening, thickening, and washing (Fig. 1). The two major pulping processes that are in operation worldwide are mechanical and chemical processes. Mechanical pulping methods use mechanical pressure, disc refiners, heating, and mild chemical treatment to yield pulps. Chemical pulping involves cooking of wood chips in pulping liquors containing chemicals under high temperature and pressure. Other pulping operations combine thermal, mechanical, and/or chemical methods. Characteristic features of various pulping processes are summarized in Table 3 and are further described shortly in the following subsections [3,10–12].

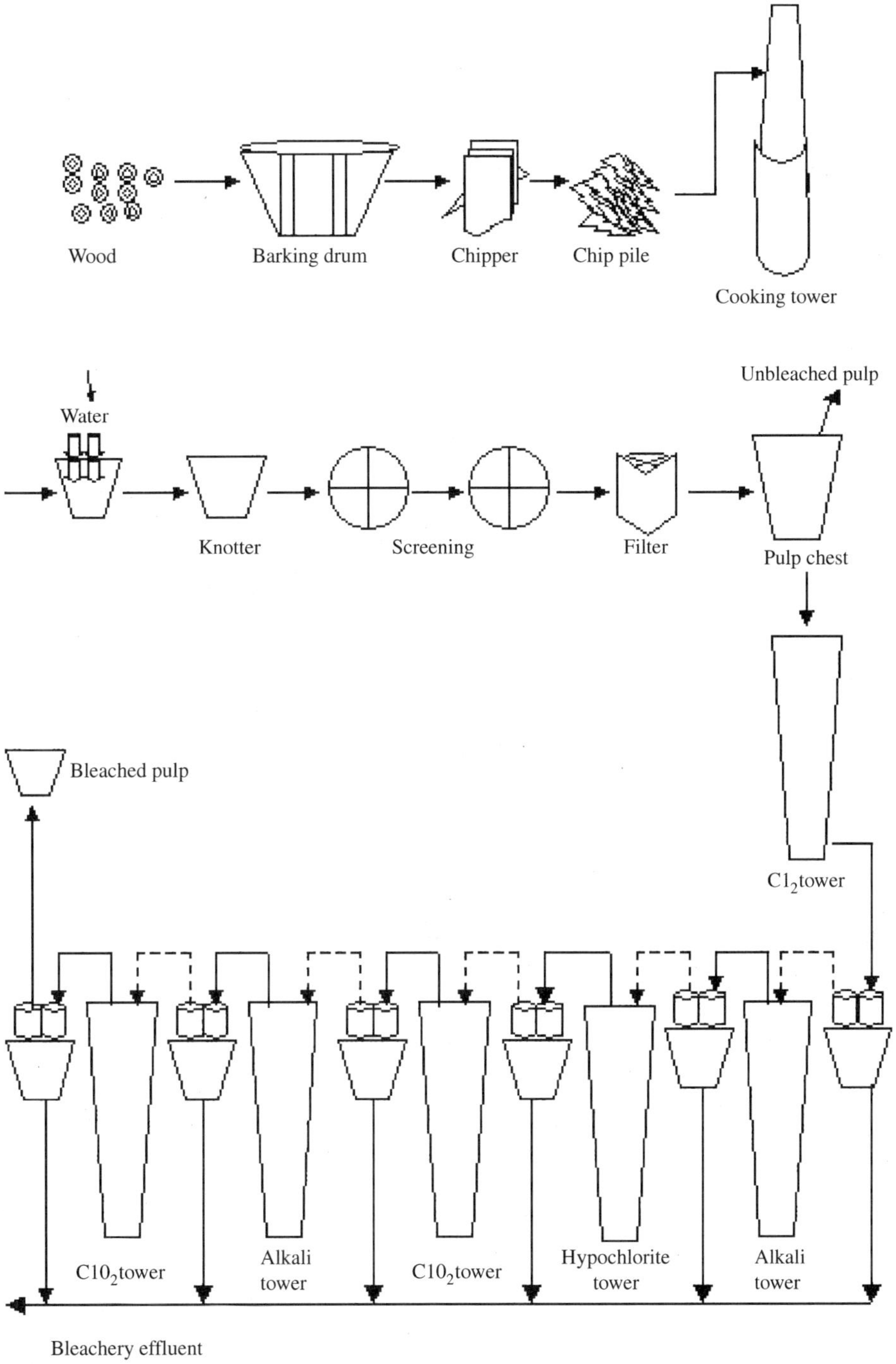

Figure 1 Steps involved in the pulping and pulp bleaching processes (from Ref. 2).

Table 3 Comparison of Various Pulping Processes

	Name of the pulping process				
Process features	Mechanical	CTMP	NSSC	Kraft	Sulfite
Pulping mechanism	Grinding stone, double disc refiners, steaming, followed by refining in TMP process	Chemical treatment using NaOH or $NaHSO_3$ + steaming followed by mechanical refining	Continuous digestion in $Na_2SO_3 + Na_2CO_3$ liquor using steam followed by mechanical refining	Cooking at 340–350°F, 100–135 psi for 2–5 hours in NaOH, Na_2S, and Na_2CO3; efficient recovery of chemicals	Sulfonation at 255–350°F, 90–110 psi for 6–12 hours in H_2SO_3 and Ca, Na, NH_4, $Mg(HSO_3)_2$
Cellulosic raw material	Hard woods like poplar and soft woods like balsam, fir, hemlock	Hard and soft woods	Hard woods like aspen, oak, alder, birch, and soft wood sawdust and chips	Any type of hard and soft wood, nonwood fiber sources	Any hard wood and nonresinous soft woods
Pulp properties	Low-strength soft pulp, low brightness	Moderate strength	Good stiffness and moldability	High-strength brown pulps, difficult to bleach	Dull white-light brown pulp, easily bleached, lower strength than Kraft pulp
Typical yields of pulp	92–96%	88–95%	70–80%	65–70% for brown pulps, 47–50% for bleachable pulps, 43–45% after bleaching	48–51% for bleachable pulp, 46–48% after bleaching
Paper products	Newspaper, magazines, inexpensive writing papers, molded products	Newspaper, magazines, inexpensive writing papers, molded products	Corrugating medium	Bags, wrappings, gumming paper, white papers from bleached Kraft pulp, cartons, containers, corrugated board	Fine paper, sanitary tissue, wraps, glassine strength reinforcement in newsprint

TMP, thermomechanical pump; CTMP, chemi-thermomechanical pump; NSSC, neutral sulfite semichemical pulp.
Source: Refs 3, 10, and 12.

Nonconventional pulping methods such as solvent pulping, acid pulping, and biopulping are discussed in subsection 10.9.1.

10.3.1 Mechanical Pulps

Stone-Ground Wood Pulp

Wood logs are pushed under the revolving grindstone and crushed by mechanical pressure to yield low-grade pulps. Lignin is not removed during this process and therefore imparts a dark color to the pulp and paper product.

Refiner Mechanical Pulp

Wood chips are passed through a narrow gap of a double-disc steel refiner consisting of stationary and rotating plates having serrated surfaces. This process results in the mechanical separation of fibers that are subsequently frayed for bonding. The strength of the refiner pulp is better than that of ground-wood pulps.

Thermomechanical Pulp (TMP)

Wood chips are preheated in steam before passage through disc refiners. Heating is meant for softening the lignin portion of wood and to promote fiber separation. This pulp is stronger than that produced by the ground-wood process.

10.3.2 Semichemical Pulp

Wood chips are processed in mild chemical liquor and subjected to mechanical refining using disc refiners. Semichemical pulping liquors have variable composition ranging from sodium hydroxide alone, alkaline sulfite (sodium sulfite + sodium carbonate), mixtures of sodium hydroxide and sodium carbonate, to Kraft green or white liquors [3]. Sodium sulfite/ sodium carbonate liquor is most commonly used and the pulp product obtained thereafter is referred to as neutral sulfite semichemical (NSSC) pulp.

10.3.3 Chemithermo Mechanical Pulp (CTMP)

This process involves a mild chemical treatment of wood chips in sodium hydroxide or sodium bisulfite before or during steaming. Chemically treated chips are passed through mechanical disc refiners.

10.3.4 Chemical Pulps

Chemical pulping of wood is commonly carried out according to the Kraft (sulfate) or sulfite processes [13]. These methods are described in the following subsections.

Kraft Pulping

Kraft pulping involves the cooking of wood chips at 340–350°F and 100–135 psi in liquor that contains sodium hydroxide, sodium sulfide, and sodium carbonate. This process promotes cleavage of the various ether bonds in lignin and the degradative products so formed dissolve in alkaline pulping liquor. The Kraft process normally incorporates several steps to recover chemicals from the spent black liquor [3].

Sulfite Pulping

The sulfite process solubilizes lignin through sulfonation at 255–350°F under 90–110 psi. The pulping liquors are composed of mixture of sulfurous acid (H_2SO_3) and bisulfites (HSO_3^{2-}) of ammonium, sodium, magnesium, or calcium, and lignin is separated from the cellulose as lignosulfonates [3]. Bisulfite pulping is performed in the pH range of 3–5 while acid sulfite pulping is carried out with free sulfurous acid at pH 1–2. Sulfite pulping mills frequently adapt methods for the recovery of SO_2, magnesium, sodium, or ammonium base liquors [3].

10.4 COMPOSITION OF SPENT PULPING LIQUORS

10.4.1 Kraft Pulping Liquors (Black Liquors)

During Kraft pulping, about 90–95% of the reactive biopolymer, namely lignin, becomes solubilized to form a mixture of lignin oligomers that contribute to the dark brown color and pollution load of pulping liquors. Lignin oligomers that are released into the spent liquors undergo cleavage to low-molecular-weight phenylpropanoic acids, methoxylated and/or hydroxylated aromatic acids. In addition, cellulose and hemicelluloses that are sensitive to alkali also dissolve during the pulping processes [13]. Black liquors generated from the Kraft pulping process are known to have an adverse impact on biological treatment facilities and aquatic life. Emissions of total reduced sulfur (TRS) and hazardous air pollutants (HAP) are also generated. Black liquors typically consist of the following four categories of compounds derived from dissolution of wood [3]:

- ligninolytic compounds that are polyaromatic in nature;
- saccharic acids derived from the degradation of carbohydrates;

Table 4 Components of Kraft Black Liquor and Characteristics of Kraft Evaporator Condensate

Kraft black liquor characteristics	
Component	Weight %, dry solids basis
Lignin	30–45
Hemicellulose and sugars	1
Hydroxy acids	25–35
Extractives	3–5
Acetic acid	2–5
Formic acid	3–5
Methanol	1
Sulfur	3–5
Sodium	17–20
Kraft liquor evaporator condensate characteristics	
COD	1000–33,600 mg/L
Major organic component	Methanol, 60–90% of COD
Anaerobic degradability	80–90% of COD
Compounds that inhibit anaerobic metabolism	Reduced sulfur, resin acids, fatty acids, volatile terpenes

COD, chemical oxygen demand.
Source: Refs 3 and 6.

- solvent extractives that include fatty acids and resin acids;
- low-molecular-weight organic acids.

Table 4 shows the typical ranges of black liquor constituents and characteristics of Kraft evaporator condensates. The composition of liquors may vary significantly, depending upon the type of raw material used. Inorganic constituents in black liquor are sodium hydroxide, sodium sulfate, sodium thiosulfate, sodium sulfide, sodium carbonate, and sodium chloride [11].

10.4.2 Sulfite Pulping Liquors (Red Liquors)

Table 5 summarizes the composition of ammonia, sodium, magnesium, and calcium base sulfite pulping liquors. In general, spent ammonia base liquors have higher BOD_5, COD and dissolved organics and exhibit more toxicity as compared to sodium, calcium, or magnesium base liquors. Higher toxicity is attributed to ammoniacal compounds in the spent liquors. The sulfite-spent liquors contain COD values typically ranging from 120–220 g/L and 50–60% of these are lignosulfonates [6]. The sulfite-spent liquor evaporator condensates have COD values in the range of 7500–50,000 mg/L. The major organic components in the condensates are acetic acid (30–60% of COD) and methanol (10–25% of COD). Anaerobic biodegradability of the condensates is typically 50–90% of COD and sulfur compounds are the major inhibitors of methanogenic activity [6].

Table 5 Composition of Ammonia, Sodium, Magnesium, and Calcium Base Sulfite Pulping Liquors

Parameter	Ammonia base mill[a]	Sodium base mill[b]	Magnesium base mill[c]	Calcium base mill[d]
Pulp liquor volume (m^3/ODT)	9.46	7.10	6.08	9.28
pH range	1.5–3.3	2.1–4.8	~3.4	5.3
BOD (kg/ODT)	413	235	222	357
COD (kg/ODT)	1728	938	975	1533
Dissolved organics (kg/ODT)	1223	595	782	1043
Dissolved inorganics (kg/ODT)	12.5	226	126	250
Lignin as determined by UV absorption (kg/ODT)	892	410	501	800
Total sugars (kg/ODT)	288	137	129	264
Reduced sugars (kg/ODT)	212	74	106	238
Toxicity emission factor[e] (TEF)	3663	714	–	422

[a]Average data based on 4 mills; [b]Average data based on 12 mills; [c] Average data based on 2 mills; [d]Composition of one mill; [e]Toxicity emission factors are based on static 96 hour bioassays and factored to the volume of liquor production. ODT = Oven dried ton of pulp.
Source: Refs 3 and 10.

10.4.3 Thermomechanical Pulp (TMP), CTMP, and Semichemical Pulping Liquors

Thermomechanical pulp (TMP) and CTMP pulping liquors exhibit COD values in the ranges of 1000–5600 mg/L and 2500–13,000 mg/L, respectively [6]. Lignin derivatives can constitute anywhere from 15 to 50% of the soluble COD values in these spent liquors. The composition of spent NSSC pulping liquors and evaporator condensates are shown in Table 6. In general, anaerobic biodegradability of semichemical pulping and CTMP effluents are low as well as inhibitory to methanogenic metabolism [6].

10.4.4 Spent Liquors from Agro-Residue Based Mills

Agro-residue mills typically employ a soda or alkaline sulfite pulping process [14]. Typical compositions of the spent liquors generated from the small-scale, agro-residue utilizing pulp and paper mills are shown in Table 7. It is evident from the table that 45–50% of the total solids is represented by lignin. Most of the lignin present in the black liquor is the high-molecular-weight fraction, a key factor contributing to low BOD/COD ratio.

10.5 TOXICITY OF PULPING LIQUORS

A number of studies have evaluated the toxicity of pulping liquors, in particular the black liquors generated from Kraft mills. Table 8 shows a partial representation of toxicity data compiled by the NCASI (National Council of the Paper Industry for Air and Stream Improvement) and McKee and Wolf for Kraft mill pulping wastewaters [15,16]. The table indicates that hydrogen sulfide, methyl mercaptan, crude sulfate soap, salts of fatty and resin acids are particularly toxic

Table 6 Composition of Spent NSSC Pulping Liquor

Spent NSSC pulping liquor characteristics	
Parameter	Average value
Total solids (%)	12
Volatile solids (% of total solids)	48
COD (mg/L)	40,000
BOD_5 (mg/L)	25,000
Wood sugars (mg/L)	7000
Lignin (mg/L)	45,000
Acetate (mg/L)	18,000
pH range	6.5–8.5
Anaerobic degradability	NR
Compounds that have the potential to inhibit anaerobic process	Tannins, sulfur compounds
NSSC pulping liquor condensate characteristics	
COD	7000 mg/L
Major organic component	Acetic acid, 70% of COD
Anaerobic degradability	NR
Inhibitors of anaerobic degradation process	Sulfur compounds

NR, not reported; COD, chemical oxygen demand; BOD, biochemical oxygen demand.
Source: Refs 3 and 6.

Table 7 Characteristics of Agro-Residue Based Spent Black Liquors

Parameter	Mill 1 (bagasse, wheat straw, and lake reed used as raw material)	Mill 2 (wheat straw used as the raw material)	Mill 3 (rice straw used as the raw material)
pH	9.7	10.2	8.8
Total solids (g/L)	44	42	38
Silica % (w/w) as SiO_2	2.4	3.2	12.0
Total organics % (w/w)	74.4	74.0	76.7
Lignin (g/L)	16.0	13.2	14.4
COD (mg/L)	48,700	45,600	40,000
BOD (mg/L)	15,500	13,800	16,500
COD/BOD	3.4	3.3	2.4

BOD, biochemical oxygen demand; COD, chemical oxygen demand.
Courtesy of MNES and UNDP India websites, Ref. 14.

to Daphnia and fish populations. Among the toxic pollutants, compounds such as sodium hydroxide, hydrogen sulfide, and methyl mercaptan fall under the EPA's list of hazardous substances. Extractive compounds such as resin acids are known to contribute up to 70% of the total toxicity of effluents generated from chemical and mechanical pulping processes [17]. The concentrations of resin acids in the pulp mill discharges are two to four times higher than their LC_{50} values (0.5–1.7 mg/L) [17]. Some reports suggest that the transformation products of resin acids such as retene, dehydroabietin, and tetrahydroretene induce mixed function monooxygenases (MFO) in fish populations [17,18]. Hickey and Martin in 1995 found a correlation between the extent of resin acid contamination in sediments and behavior modification in benthic invertebrate species [19]. Johnsen et al. in 1995 reported that TMP mill effluents containing resin acids were lethal to rainbow trout following 2–4 weeks exposure at 200-fold dilution [20]. McCarthy et al. in 1990 demonstrated that resin acids are toxic to methanogens, thereby inhibiting the performance of these bacteria in anaerobic reactors [21].

Table 8 Toxicity of the Components of Kraft Pulp Mill Wastewaters

	Minimum lethal dose (ppm)	
Compound	Daphnia	Fish
Sodium hydroxide	100	100
Sodium sulfide	10	3.0
Methyl mercaptan	1.0	0.5
Hydrogen sulfide	1.0	1.0
Crude sulfate soap	5–10	5.0
Sodium salts of fatty acids	1.0	5.0
Sodium salts of resin acids	3.0	1.0
Sodium oleates	–	5.0
Sodium linoleate	–	10.0
Sodium salts of abietic acids	–	3.0

Source: Refs 15 and 16.

10.6 PULP BLEACHING PROCESSES

About 5–10% of the original lignin cannot be removed from the pulp without substantial damage to the cellulosic fraction. Removal of the residual lignin, which is responsible for imparting dark color to the pulp, and the production of white pulp, requires a series of steps employing bleach chemicals (Fig. 1). Pulp bleaching is normally accomplished by sequential treatments with elemental chlorine (C_1), alkali (E_1), chlorine dioxide (D_1), alkali (E_2), and chlorine dioxide (D_2). The C stage consists of charging a slurry of the pulp (at 3–4% consistency) with elemental chlorine (60–70 kg/ton of pulp) at 15–30°C at pH 1.5–2.0 [2]. The chlorinated pulp slurry (at 10% consistency) is treated with alkali (35–40 kg/ton of pulp) at 55–70°C and pH 10–11. An optional hypochlorite (H) stage is introduced between the E_1 and D_1 stages for increasing the brightness of pulp. During the conventional bleaching, approximately 70 kg of each ton of pulp is expected to dissolve into the bleaching liquors [2]. The largest quantity of pulp is dissolved during the C_1 and E_1 stages. Alternate pulp bleaching techniques such as the elemental chlorine free (ECF), total chlorine free (TCF), and bio bleaching are described in subsection 10.9.2.

10.6.1 Compounds Formed during Chlorine Bleaching Process

During pulp bleaching, lignin is extensively modified by chlorination (C stage) and dissolved by alkali (E stage) into the bleaching liquor. The E stage is intended for dissolving the fragmented chloro-lignin compounds and removal of noncellulosic carbohydrates. The most important reactions are oxidation and substitution by chlorine, which lead to the formation of chlorinated organic compounds or the AOX (Table 2). Chlorine bleaching liquors exhibit COD values ranging from 900–2000 mg/L and 65–75% of this is from chlorinated lignin polymers [6]. The types of chlorinated compounds found in the spent bleach liquors and their concentrations depend upon the quantity of residual lignin (Kappa number) in the pulp, nature of lignin, bleaching conditions such as chlorine dosage, pH, temperatures, and pulp consistencies. The spent liquors generated from the conventional pulping and bleaching processes contain approximately 80% of the organically bound chlorine as high-molecular-mass material (MW above 1000) and 20% as the low-molecular-mass (MW of less than 1000) fraction [22].

The high-molecular-mass compounds, referred to as chlorolignins, cannot be transported across the cell membranes of living organisms and are likely to be biologically inactive. Nevertheless, these compounds are of environmental importance because they carry chromophoric structures that impart light-absorbing qualities to receiving waters. Long-term and low rates of biodegradation may generate low-molecular-weight compounds, causing detrimental effects on biological systems.

Efforts have been made to characterize the nature and content of individual components that are present in the low-molecular-mass fraction of the total mill effluents, which include the spent chlorination and alkali extraction stage liquors [2,4]. Approximately 456 types of compounds have been detected in the conventional bleach effluents, of which 330 are chlorinated organic compounds [22]. The compounds may be lumped into three main groups, namely, acidic, phenolic, and neutral (Table 2). Acidic compounds are further divided into the five categories of acids: fatty, resin, hydroxy, dibasic, and aromatic acids. The most important fatty acids are formic and acetic acids. The dominant resin acids are abietic and dehydroabietic acids. Among the hydroxy acids identified, glyceric acid predominates. Dibasic acids such as oxalic, malonic, succinic, and malic acids are derived from the lignin and carbohydrate fraction of wood and are present in significant amounts in the mill bleach effluents. Aromatic acids are formed from residual lignin through the oxidation of phenylpropanoid units and comprise four major categories: monohydroxy (phenolic), ortho-dihydroxy (catecholic), methoxy-hydroxy (guaiacolic), and dimethoxy-hydroxy (syringic) acids. The principal phenolics are chlorinated phenols, chlorinated catechols, chlorinated guaiacols, and chlorinated vanillin, derived from the chlorination and oxidative cleavage of lignin. The major neutral compounds are methanol, hemicellulose, and trace concentrations of aldehydes, ketones, chlorinated acetones, dichloromethane, trichloroethene, chloropropenal, chlorofuranone, chloroform, chlorinated sulfur derivatives, and 1,1-dichloromethylsulfone. In addition to the abovementioned compounds, the spent bleaching liquors have been reported to contain about 210 different chlorinated dioxins that belong to the two families: polychlorinated dibenzodioxins (PCDDs) and polychlorinated dibenzofurans (PCDFs) [22].

10.7 TOXICITY OF SPENT BLEACH LIQUORS

Compounds responsible for imparting toxicity to the spent bleach effluents originate during the chlorination (C) stage and caustic extraction (E) stages. The major classes of toxic compounds are resin acids, fatty acids, and AOX. Fatty and resin acids in bleach liquors often originate from the washing of unbleached pulps. They are recalcitrant to biodegradation as well as inhibitory to the anaerobic process. Adsorbable organic halides are the products of lignin degradation formed exclusively during the C stage of pulp bleaching and dissolved into the bleaching liquors during the E stage. About 1–3% of the AOX fraction is extractable into nonpolar organic solvents and is referred to as extractable organic halide (EOX). This extractable fraction poses greater environmental risks than the remaining 99% of the AOX and comprises compounds that are lipophilic with the ability to penetrate cell membranes and potential to bioaccumulate in the fatty tissues of higher organisms. Dioxins, in particular 2,3,7,8-tetrachlorodibenzodioxin (2,3,7,8-TCDD) and 2,3,7,8-tetrachlorodibenzofuran (2,3,7,8-TCDF) are highly toxic, bioaccumulable, carcinogenic, and cause an adverse impact on almost all types of tested species [2,22,23]. Additionally, the abovementioned dioxins and the other unidentified components of bleach liquors are also endocrine disrupting chemicals (EDC) that decrease the levels and activity of the estrogen hormone, thereby reducing reproductive efficiency in higher organisms [24]. However, limited information is available regarding these undesirable, genetically active, and endocrine-disrupting pollutants in receiving waters; further research is essential in this direction. Table 9 summarizes some findings related to the toxicity and impact of bleach mill discharges on selected aquatic organisms [25–31].

10.8 STRATEGIES FOR POLLUTION CONTROL IN PULP AND PAPER INDUSTRIES

Traditionally, discharge limits have been set for lumped environmental parameters such as BOD_5, COD, TSS, and so on. However, on account of the adverse biological effects of chlorinated organics coupled to the introduction of stricter environmental legislation, pulp and paper mills are faced with the challenges of not only reducing the BOD and suspended solids, but also controlling the total color as well as AOX in the effluents prior to discharge. In recent years,

Table 9 Summary of Selected Toxicology Studies to Assess the Ecological Impacts of Bleach Mill Effluents

Bleach process adopted by the mill	Organism studied	Physiological/biochemical effect(s)/ levels of toxicity	Research group
Kraft mill using 100% chlorine dioxide	Coastal fish community	High levels of mortality and low embryo quality	Sandstrom, 1994 [25]
New and old wood pulp bleaching employing various bleach sequences	Mesocosm and fish biomarker tests	Elemental chlorine containing bleach sequence, CEHDED was the most toxic	Tana et al., 1994 [26]
Kraft mill using 100% chlorine dioxide	Baltic sea amphipod (crustacean)	Reduced swimming activity	Kankaanpaa et al., 1995 [27]
Kraft bleach mill effluent produced by oxygen delignification or 100% chlorine dioxide	Freshwater fish	Induction of mixed function oxidase (MFO) enzymes following exposure to 4% and 12% effluent in artificial streams	Bankey et al., 1995 [28]
Kraft mill using 100% ClO_2	Aquatic organisms	Overall toxicity pattern of effluents in the bioassay was: Untreated ECF > untreated TCF > Secondary treated ECF > secondary treated TCF	Kovacs et al., 1995 [29]
Kraft mill using 100% ClO_2	Fish populations	Changes in the reproductive development – reduction in gonad size, depression of sex hormones following exposure to bleach effluents subjected to secondary treatment	Munkittrick et al., 1997 [30]
Kraft mill that had used elemental chlorine historically	Microbial community and diatom species in lake sediments sampled from 2–8 cm depths	Drop in the ATP content, depressed butyrate-esterase activity indicating toxicity to microorganisms, and reduction in diatom species richness	Mika et al., 1999 [31]

the pulp and paper industry has taken great strides forward in recognizing and solving many of the environmental problems by adopting two strategies:

1. Pollution reduction measures within plants that include minimization of spills and modifications in the process through adaptation of cleaner technologies as alternatives to conventional technologies.
2. End-of-pipe pollution treatment technologies, which are essential either as a supplement or as backup measures to pollution reduction techniques in order to meet the effluent regulation standards.

These two approaches are equally important in meeting environmental regulations and are addressed in separate sections.

10.9 POLLUTION REDUCTION THROUGH PLANT PROCESS MODIFICATIONS

10.9.1 Nonconventional Pulping Technologies

Industries have developed alternate pulping techniques that do not use the conventional cooking methods. Some of these techniques are described briefly in the following subsections. Readers may note that some of these processes have not yet reached commercial stages.

Organic Solvent Pulping

Organic solvents such as methanol, ethanol, and other alcohols are used for pulping. This process is economical for small- to medium-scale mills with significant recovery of chemicals for reuse. However, pulping must be conducted in enclosed containers to prevent the loss of volatile solvents and for workers' safety. Additionally, some of these processes are more energy intensive than traditional methods. Major benefits include the elimination of odorous sulfur-containing compounds in the effluents and air.

Acid Pulping

Wood chips are treated with acetic acid at pressures that are significantly lower than those used for Kraft pulping. Drawbacks include loss of acid, although recovery is possible through the energy-intensive distillation process.

Biopulping

This method utilizes whole cells of microorganisms and microbial enzymes such as xylanases, pectinases, cellulases, hemicellulases, and ligninases, or their combinations, for pulping herbaceous fibers and improving the properties of pulp derived from wood [32]. Pretreatment of wood chips with lipases is known to reduce the problematic oily exudates during the pulping process as well as improving the texture of paper through the specific degradative action of these enzymes on pitch-derived extractives such as fatty acids and waxes. The innovative approach of using microorganisms or microbial enzymes to reduce the consumption of chemicals in the pulp and paper industry is known as biopulping. Biopulping has generated much interest among the pulp and paper industries because of the following advantages:

- Reduction in the chemical and energy requirements per unit of pulp produced. Thus, the process is expected to be cost effective and more affordable for medium- and small-scale mills.
- Reduction in the pollution load due to reduced application of chemicals.
- The yield and strength properties of the pulp are comparable (sometimes even better) to those obtained through conventional pulping techniques.

Nonwood fibers are more responsive to the action of pulping enzymes compared to wood, presumably due to lower lignin content and weaker hemicellulose–lignin bonds. This is clearly advantageous for developing countries, which are faced with the problem of shrinking forest wood resources. However, further research is required to optimize the conditions required for enzymatic pulping of herbaceous fibers and commercialization of the process.

10.9.2 Cleaner Pulp Bleaching Technologies

The use of elemental chlorine in pulp bleaching has been gradually discontinued in several countries to prevent the toxic effects of chlorinated organics in receiving waters and to meet regulatory requirements. Most nations have imposed stringent regulatory limits on AOX, ranging from 0.3 to 2.0 kg/ton of pulp [22]. Cleaner bleaching methods have been developed by industries based on elemental chlorine free (ECF), total chlorine free (TCF), microbial systems (bio-bleaching), extended delignification, and methods for monitoring and improved control of bleaching operations. Each of these approaches is discussed in the following subsections.

Elemental Chlorine Free (ECF) and Total Chlorine Free (TCF) Bleaching

Elemental chlorine has been replaced by chlorine dioxide and hypochlorite in the ECF bleach sequence, while oxygen, ozone, caustic soda, and hydrogen peroxide have been advocated for TCF bleaching of softwood and hardwood Kraft pulps. Benefits include significant reduction in the formation of chlorinated organics or their elimination and lower ecological impacts. Two Finnish mills eliminated elemental chlorine from the bleach sequence and substituted chlorine dioxide, thereby sharply reducing the concentration of chlorinated cymenes [33]. In another Finnish example, levels of chlorinated polyaromatic hydrocarbons in mill wastes were substantially reduced during production of bleached birch Kraft pulp without the use of elemental chlorine as compared to pine pulp bleached with elemental chlorine [34,35]. Research has also been conducted on the optimal usage of agents such as ozone and hydrogen peroxide [36,37]. However, alternatives such as ozonation, oxygenation, and peroxidation are not economically viable for medium- or small-capacity mills due to higher capital investments and plant operation costs.

Biobleaching

Biobleaching processes based on the pretreatment of pulp with microbial whole cells or enzymes have emerged as viable options. A number of studies examined the direct application of white rot fungi such as *Phanerochaete chrysosporium* and *Coriolus versicolor* for biobleaching of softwood and hardwood Kraft pulps [38–44]. It has been found that fungal treatment reduced the chemical dosage significantly as compared to the conventional chemical bleach sequence and enhanced the brightness of the pulp. Specific features of the fungal-mediated biobleaching processes are:

- Action through delignification that commences at the onset of the secondary metabolic (nitrogen starvation) phase in most fungi.
- Delignification is an enzymatic process mediated through the action of extracellular enzymes.
- The growth phase of the fungus has an obligate requirement for a primary substrate such as glucose.

The major drawbacks of the fungal bleaching process are that it is extremely slow for industrial application and requires expensive substrates for growth. To overcome these problems, enzyme preparations derived from selected strains of bacteria or fungi are recommended. The enzymatic method of pulp bleaching is being increasingly preferred by a number of pulp and paper industries, especially in the West, because it is a cost-effective and environmentally sound technology [32]. The distinct advantages of enzyme-mediated pulp bleaching are:

- minimal energy input;
- specificity in reactivity, unlike that of chemicals;

- reduced dosage of bleach chemicals in the downstream steps;
- improved quality of pulp through bleach boosting;
- reduced load of AOX in the effluents.

Two categories of enzymes, namely xylanases and peroxidases (lignin degrading), have been identified in the pulp bleaching processes. Of the two classes, the use of xylanases has achieved enormous success in aiding pulp bleaching [45]. Xylanase enzymes apparently cause hydrolytic breakdown of xylan chains (hemicellulose) as well as the cleavage of the lignin–carbohydrate bonds, thereby exposing lignin to the action of subsequent chemical bleaching steps [46]. However, most biobleaching studies using xylanases have been carried out with either hardwood or softwood, while nonwood resources are being increasingly used as the chief agricultural raw material for pulp production. Therefore, further research with regard to enzyme applications for nonwood pulp bleaching is warranted.

It is unlikely that xylanase treatment alone will completely replace the existing chemical bleaching technology, because this enzyme does not act directly on lignin, a crucial color-imparting polymer of the pulp. Nonspecific oxido-reductive enzymes such as lignin peroxidase, manganese peroxidase and in particular, laccases, which are lignolytic, are likely to be more effective in biobleaching [47]. The abovementioned enzymes can also act on a wide variety of substrates and therefore have significant potential for applications to pulp and paper effluent treatment [48,49]. The applicability of the laccase mediator system for lignolytic bleaching of pulps derived from hard wood, soft wood, and bagasse has been reviewed and compared by Call and Mucke [47]. The major advantage of enzymatic bleaching is that the process may be employed by the mills over and beyond the existing technologies with limited investment. Furthermore, there is ample scope for the improvement of the process in terms of cost and performance.

Extended Delignification

The key focus of this process is on the enhanced removal of lignin before subjecting the pulp to bleaching steps [50,51]. Such internal process measures also imply cost savings during the subsequent chemical bleaching steps and have a positive impact on the bleach effluent quality parameters such as COD, BOD, color, and AOX. Extended delignification may be achieved through:

- *Extended cooking*. This can be done by enhancing cooking time or temperature or by multiple dosing of the cooking liquors.
- *Oxygenation*. The pulp is mixed with elemental oxygen, sodium, and magnesium hydroxides under high pressure. An example is the PRENOX process [50]. According to Reeve, about 50% of the world capacity for Kraft pulp production incorporated oxygen delignification by the year 1994 [52].
- *Ozonation*. Ozone and sulfuric acid are mixed with the pulp in a pressurized reactor.
- *Addition of chemical catalysts*. Compounds such as anthraquinone or polysulfide or a mixture of the two are introduced into the Kraft cooking liquor.

Improved Control of Bleaching Operations

Installation of online monitoring systems at appropriate locations and controlled dosing of bleach chemicals can aid in the reduction of chlorinated organics in effluents.

10.10 TREATMENT OF PULP AND PAPER WASTEWATERS

Plant process modifications and cleaner technologies have the potential to reduce the pollution load in effluents. However, this approach cannot eliminate waste generation. End-of-pipe pollution treatment technologies are essential for meeting the prescribed limits for discharged pollutant concentrations such as color, AOX, BOD, COD, and so on. Assessment of the water quality of receiving ecosystems and periodic ecological risk assessments are required to validate the effectiveness of various treatment methods [53]. The most common unit processes employed by the pulp and paper mills during preliminary, primary, secondary, and tertiary (optional) stages of effluent treatment are listed in the flow sheet shown in Figure 2. Process technologies that are currently applied can be broadly classified as the physico-chemical and biological treatment methods. These technologies are discussed in the following subsections.

10.10.1 Physico-Chemical Processes

Several physico-chemical methods are available for the treatment of pulping and pulp bleaching effluents. The most prominent methods are membrane separation, chemical coagulation, and precipitation using metal salts and advanced oxidation processes.

Membrane Separation Techniques

Membrane processes operate on the basis of the following mechanisms:

- pressure driven, which includes reverse osmosis (RO), ultrafiltration (UF), and nanofiltration (NF);
- concentration driven, which includes diffusion dialysis, vapor permeation, and gas separation;
- electrically driven, which includes electrodialysis;
- temperature difference driven, including membrane distillation.

Membrane filtration (UF, RO, and NF) is a potential technology for simultaneously removing color, COD, AOX, salts, heavy metals, and total dissolved solids (TDS) from pulp mill effluents, resulting in the generation of high-quality effluent for water recycling and final discharges. The possibility of obtaining solid free effluents is a very attractive feature of this process. Ultrafiltration was used by Jonsson et al. [54] for the treatment of bleach plant effluents.

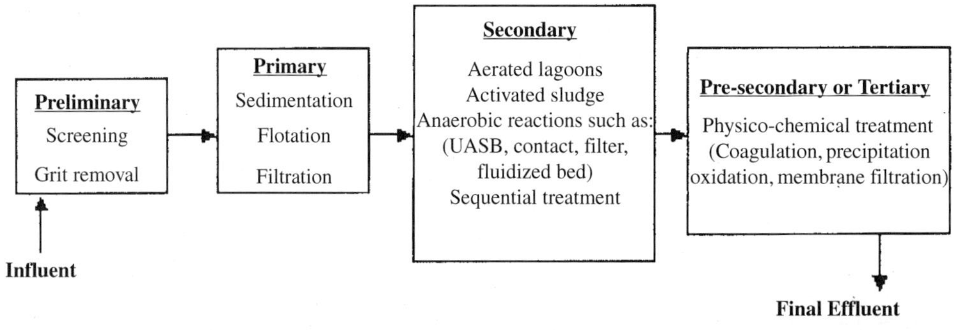

Figure 2 Flow sheet showing the unit processes employed by pulp and paper mills for effluent treatment.

Sierka et al. [55] described a study that compared the efficiencies of UF alone and UF in combination with RO for the removal of color and total organic carbon (TOC) in the D_o (acid stage) wastewaters discharged from the Weyerhaeuser Grande Priare pulp mill, which produces 300,000 tons of bleached Kraft bleached pulp per year. The bleach plant of this Kraft mill typically employs five stages ($ECD_oE_{op}DED$ sequence) for the production of tissue and specialty grade paper. D_o stage wastewaters were sterilized using 0.45 μm filters and subsequently passed through Amicon-stirred UF cells fitted with membranes having cutoff values of 500 Daltons (D) (YCO5), 1000 D (YM1), 3000 D (YM3) or 10,000 D (YM10). Table 10 summarizes the characteristics of the permeate and concentrates obtained following the ultrafiltration of D_o wastewater using various membranes. Based on these results, Sierka et al. concluded that most of the color (50%) is due to organic compounds with molecular size above 3000 D. Table 11 presents the results of additional studies conducted on D_o stage effluents that involved pretreatment by UF followed by RO. Clearly, the combination of the UF and RO steps gave excellent results by removing 99% of the color and more than 80% of the TOC from the D_o stage effluent.

Koyuncu et al. [56] presented pilot-scale studies on the treatment of pulp and paper mill effluents using two-stage membrane filtrations, ultrafiltration and reverse osmosis [56]. The combination of UF and RO resulted in very high removals of COD, color, and conductivity from the effluents. At the end of a single pass with seawater membrane, the initial COD, color and conductivity values were reduced to 10–20 mg/L, 0–100 PCCU (platinum cobalt color units) and 200–300 μs/cm, respectively. Nearly complete color removals were achieved in the RO experiments with seawater membranes.

A distinct advantage of the membrane technology is that it can be utilized at the primary, secondary, or tertiary phases of water treatment. Some membranes can withstand high concentrations of suspended solids, which presents a possible direct application for separating mixed liquor suspended solids (MLSS) in an activated sludge plant (membrane bioreactor) to replace the conventional sedimentation tank. Key variable parameters of membrane technology include variation in membrane pore size, transmembrane pressure, cross-flow velocity, temperature, and back flushing. The major disadvantages are high capital and maintenance costs, accumulation of reject solutes and decrease in the membrane performance, membrane fouling, and requirement for the pretreatment of discharges.

Table 10 Ultrafiltration Characteristics of D_o Wastewater Using Different Membranes

	Membrane			
	YCO5	YM1	YM3	YM10
UF input/output	500 D	1000 D	3000 D	10,000 D
TOC (mg/L)				
Feed	792.5	792.5	792.5	792.5
Permeate	282	465	546	634
Concentrate	1739	1548	1158	1033
Color (PCCU[a])				
Feed	1700	1700	1700	1700
Permeate	107	334	835	1145
Concentrate	3066	2972	2221	1876

[a] Platinum cobalt color units.
Source: Ref. 55.

Table 11 Characteristics of D_o Wastewater Subjected to Ultrafiltration and Reverse Osmosis

Input/output of UF/RO unit	pH	TOC (mg/L)	Color (PCCU[a])
Feedwater to UF unit	7.00	825	1750
Composite permeate from UF unit	–	555	1231
Feedwater to RO unit	–	555	1231
Composite permeate from RO unit	6.44	70.68	13.79
Concentrate from RO unit	6.47	3219	2741

[a] Platinum cobalt color units.
Experimental conditions: Pressure = 1104 kPa; temperature = 40°C; batch volume = 4 L; cutoff value of the UF membrane = 8000 D.
Source: Ref. 55.

Chemical Coagulation and Precipitation

This method relies on the addition of metal salts to cause agglomeration of small particles into larger flocs that can be easily removed by settling. The effectiveness of this process is dependent upon the nature of coagulating agent, coagulant dosage, pH, ionic strength, and the nature and concentration of compounds present in wastewaters. The not-so-easily biodegradable fraction of pulping and bleaching effluents consists of polar and hydrophobic compounds, notably resin acids, long-chain fatty acids, aromatic acids and phenols, lignin, and terpenes. Almost all of these toxic compounds can be effectively removed through coagulation using chloride and sulfate salts of Fe^{3+} and Al^{3+}. Typically, these trivalent cations remain in solution at acidic pH and form metal hydroxides that aggregate rapidly at higher pH conditions. Hydrogen bonding, electrostatic and surface interactions (adsorption) between the metal hydroxides and organic anions (containing hydroxyl and carboxyl groups) lead to the formation of metal hydroxide–organic compound precipitates [57,58]. Dissolved organics are also removed by physical adsorption to flocs.

Chemical precipitation of mill effluents from CTMP, BKME (bleached Kraft mill effluent), NSSC, E & C bleach discharges have been extensively studied by Stephenson and Duff [59] using alum, lime, ferric chloride, ferrous sulfate, magnesium hydroxide, polyimine, polymers, and alum in combination with lime. They observed removal of 88% of total carbon and 90–98% of color and turbidity from mechanical pulping effluents using Fe^{3+}/Al^{3+} salts. In another publication, Stephenson and Duff reported significant reduction in the toxicity of wastewaters subsequent to the chemical coagulation process [60]. Ganjidoust et al. [61] compared the effects of a natural polymer, chitosan, and synthetic polymers, namely hexamethylene diamine epichlorohydrin polycondensate (HE), polyethyleneimine (PEI), polyacrylamide (PAM), and a chemical alum coagulant, alum on the removal of lignin (black liquor color and total organic carbon) from alkaline pulp and paper industrial wastewater. They observed that PAM, a nonionic polymer, had a poor effect, whereas HE and PEI, which are cationic polymers, coagulated 80% of the color and 30% of the total organic carbon from alkaline black liquor wastewater by gravity settling in 30 min. Alum precipitation removed 80% of the color and 40% of total organic carbon. By comparison, the natural coagulant chitosan was the most effective; it eliminated up to 90% of the color and 70% of the total organic carbon, respectively.

The major disadvantages of coagulation and precipitation are the generation of chemical sludge and the need for subsequent treatment of the sludge to eliminate the adsorbed toxic pollutants prior to disposal.

Advanced Oxidation Processes

Destruction of chromophoric and nonchromophoric pollutants in pulp and paper effluents may be achieved by advanced oxidation methods such as photocatalysis, photo-oxidation using hydrogen peroxide (H_2O_2)/UV or ozone (O_3)/UV systems, Fenton-type reactions, wet oxidation, and by employing strong oxidants such as ozone.

Photocatalysis has gained attention for its application to aqueous phase and wastewaters for near total oxidation and elimination of organic compounds [62]. The process involves mixing wastewater with aeration in a reactor at 20–25°C and the introduction of titanium dioxide (TiO_2) followed by irradiation using a UV lamp. Irradiation by UV light generates an electron hole on TiO_2 surface, which reacts with the adsorbed organic compounds or water molecules. TiO_2 can be provided as a suspension or as covered supports (immobilized on beads, inside tubes of glass/teflon, fiberglass, woven fibers, etc.). Various research groups have shown that photocatalysis is nonselective and that there is a nearly parallel reduction in the color, lignosulfonic acids, and other organic compounds in the treated pulp and paper mill effluents. Balcioglu et al. [63] observed enhanced biodegradability (increase in BOD_5/COD ratio) of raw Kraft pulp bleaching effluents and improved quality of the biologically pretreated effluents following TiO_2 photocatalytic oxidation. Yeber et al. [64] described the photocatalytic (TiO_2 and ZnO) treatment of bleaching effluents from two pulp mills. Photocatalysis resulted in the enhanced biodegradability of effluents with concomitant reduction in the toxicity.

Photo-oxidation systems using H_2O_2/UV or O_3/UV combinations generate hydroxyl radicals that are short lived but extremely powerful oxidizing organics through hydrogen abstraction. The result is the onsite total destruction of refractory organics without generation of sludges or residues. Wastewater is injected with H_2O_2 or saturated with O_3 and irradiated with UV light at 254 nm in a suitable reactor with no additional requirement for chemicals. The rate of oxidative degradation is generally much higher than systems employing UV or O_3 alone. Legrini et al. [62] have extensively reviewed the experimental conditions used by various researchers for conducting the photo-oxidation process as well as their application for removal of various types of organic compounds.

Fenton's reactions involving hydrogen peroxide (H_2O_2) and ferrous ion as the solution catalyst are an effective option for effluent treatment. Fenton's reaction as described by Winterbourn [65] requires a slightly acidic pH and results in the formation of highly reactive hydroxyl radicals ($^{\cdot}OH$), which are capable of degrading many organic pollutants. Rodriguez et al. [66] evaluated Fenton-type reactions facilitated by catecholic compounds such as 2,3-dihydroxybenzoic acid, 3,4-dihydroxybenzoic acid, and 1,2-dihydroxybenzene for treating pulp bleaching effluent. Their research indicated that 2,3-dihydrobenzoic acid was the most effective compound in enhancing hydroxyl radical formation in the iron–hydrogen peroxide reaction system at pH 4.0 with the concomitant reduction in the AOX concentration and toxicity of the bleach effluents.

Wet oxidation is a process where organic contaminants in liquids or soils are extracted into an aqueous phase and reacted with an oxidant at high temperature (220–290°C) and pressures (100–250 bar) to promote rapid destruction. Laari et al. [67] evaluated the efficiency of wet oxidation for the treatment of TMP processing waters. The major objective of this research was to reduce the concentration of lipophilic wood extractives (LWE) and to treat concentrated residues from evaporation and membrane filtration by low-pressure catalytic wet oxidation.

The wet oxidation of membrane and evaporation concentrates was effective in reducing 50% of the COD at 150°C and enhancing the biodegradability of wastewater.

Oxidants such as chlorine, oxygen, ozone, and peroxide have been proposed for the treatment of pulp bleach effluents. Ozonation has been reported to reduce the toxicity of CTMP and bleached Kraft mill (BKM) effluents at low dosages [22]. Hostachy et al. [68] reported detoxification and an increase in the biodegradability of bleach effluents by ozonation at low dosages [0.5–1 kg/ADMT (air-dried metric ton)] of pulp. The researchers observed significant elimination of the residual COD by catalyzed ozone treatment of hardwood and softwood pulp and paper mill final discharges. Such a treatment method may allow for reutilization of treated process waters and reduce consumption of freshwater during pulping steps. Helbe et al. [69] described a tertiary treatment process involving ozonation in combination with a fixed-bed biofilm reactor for the reuse of treated effluent in a pulp and paper industry. Sequential ozonation and bioreactor treatment gave maximum elimination of COD, color, and AOX from biologically treated effluent with minimum dosage of ozone. Further, the authors suggested that two-stage ozonation with intermediate biodegradation is more effective in terms of achieving higher removal of persistent COD.

The advantages of the various oxidation processes include nonselective and rapid destruction of pollutants, absence of residues, and improved biodegradability of the effluents. Some of the disadvantages are extremely short half-life of the oxidants and high expense of their generation.

10.10.2 Biological Processes

The most commonly used biological treatment systems for the pulp and paper mill discharges are activated sludge plants, aerated lagoons, and anaerobic reactors. Sequential aerobic-anaerobic systems (and vice versa) are a recent trend for handling complex wastewaters of pulp and paper mills that contain a multitude of pollutants. The application of various types of biological reactor systems for treating pulp and paper mill effluents are discussed in the following subsections.

Activated Sludge Process

This conventional aerobic biological treatment train consists of an aeration tank with complete mixing (for industrial discharges) followed by a secondary clarifier and has been typically used for the reduction of COD, BOD, TSS, and AOX in pulp and paper mill waste effluents. Oxygen is provided to the aerobic microorganisms through aeration or by using pure oxygen as in the deep shaft systems. Bajpai [22] has reviewed the efficiencies of activated sludge plants and reported that the overall removal of AOX can range from 15 to 65%, while the extents of removal of individual chlorinated organics such as chlorinated phenols, guaiacols, catechols, and vanillins can vary from 20 to 100%. Biotransformation and biodegradation seem to be the important mechanisms for reduction in the overall AOX concentrations and hydraulic retention time (HRT) is the key operating parameter.

There are a number of full-scale activated sludge plants that are in operation in countries such as the United States, Canada, and Finland, which treat effluents from Kraft, sulfite, TMP, CTMP, and newsprint mills [22]. Schnell et al. [70] reported the effectiveness of a conventional activated sludge process operating at an alkaline-peroxide mechanical pulping (APMP) plant at Malette Quebec, Canada. The full-scale plant achieved 74% reduction in filterable COD and nearly complete elimination of BOD_5, resin acids, and fatty acids in the whole mill effluent. The treated effluent tested nontoxic as measured by a Microtox assay. Saunamaki [71] reported excellent performances of activated sludge plants in Finland that were designed according to the low loading or extended aeration principle. Control of nutrients, aeration, low loading rates, introduction of equalization and buffer basins seemed to be the key process control parameters for successful treatment. BOD_7 and COD removal averaged 94 and 82%, respectively, at paper mills while at pulp mills, the values were 82 and 60%, respectively. All paper mill activated sludge plants required dosing of nitrogen and phosphorus. Narbaitz et al. [72] evaluated the impacts of adding powdered activated carbon to a bench-scale activated sludge process ($PACT^{TM}$) fed with low-strength Kraft pulp mill wastewaters. Enhanced removal of AOX and marginal improvement in the levels of COD and toxicity reduction as compared to the conventional activated sludge process was observed.

Two common operational problems encountered during the treatment of pulp and paper wastewaters in activated sludge plants are:

- Limiting concentrations of nitrogen and phosphorus (N and P) that are vital for maintenance of active microbial population in an activated sludge plant.
- Growth of filamentous organisms or formation of pinpoint flocs that negatively impact the sludge settling rates, thereby reducing the effluent quality.

The problem of nutrient deficiency is frequently overcome through the external addition of nutrients with optimization of their dosage. A major drawback of supplementation is the requirement for extensive monitoring of treated effluents for N and P prior to discharge to avoid adverse environmental impacts such as the eutrophication of receiving waters. Alternate approaches have been investigated, such as the selection and incorporation of bacteria capable of fixing atmospheric nitrogen (nitrogen fixers) in the biological reactors or addition of solid N and P sources with low solubility to prevent excess loadings in the final effluents. As an example, Rantala and Wirola [73] have demonstrated the success of using a solid source of phosphorus with low solubility in activated sludge plants fed with CTMP mill wastewater. They observed that the total phosphorus concentration in the effluents was more than 2 mg/L in the activated sludge reactor fed with liquid phosphoric acid and less than 0.5 mg/L if fed with solids such as apatite or raw phosphate. Based on a full-scale trial study at a CTMP mill, the authors concluded that the addition of nutrient in the form of apatite is a viable alternative for reducing phosphorus load in the treated effluents.

Conventional practices for controlling sludge bulking are through chlorination or peroxidation of sludge or addition of talc powder. Clauss et al. [74] discussed two case studies on the application of fine Aquatal (product designed by Luzenac–Europe), a mineral talc-based powder, to activated sludge plants for counteracting the floc settlability problems. In the first case, Aquatal was added to aeration tanks to control sludge volume index (SVI) and reduce the concentration of suspended solids in the effluents. In a second case study, Aquatal was introduced to prevent sludge blanket bulking. In both cases, the mineral powder additive resulted in the formation of compact, well-structured heavier flocs that displayed increased settling velocities, and good thickening and dewatering properties. However, a major drawback of this method is that it addresses the symptoms of the problem rather than the root cause. A permanent solution based on the comparison of physiology, substrate requirement and degradation kinetics of floc forming and filamentous bacteria is needed.

A number of case studies have reported on the improvement of existing activated sludge plants in the pulp and paper industries through modifications. Two case studies are presented below.

(a) A Case Study on the Up-Gradation of an Activated Sludge Plant in Poland. Hansen et al. [75] described the up-gradation of an existing activated sludge plant of 400,000 ADMT pulp capacity mill in Poland that produces unbleached Kraft pulp. The discharge limits as set by

the Polish authorities were 150 mg/L of COD, 15 mg/L of BOD, 60 mg/L of SS and 88,000 m^3/day of water to the receiving river from the year 2000. To meet the new demands, two additional FlooBed reactors with a total volume of 50% of the existing activated sludge plant were installed. The three biological reactors were operated in series, with the activated sludge plant as the third stage, as shown in Figure 3. FlooBed is an activated sludge tank with microorganisms supported as thin films over floating carrier materials. The carrier is made of polyethylene with an area of 200 m^2/m^3 for biofilm growth. All three stages of the biological reactor were amended with the required concentrations of urea and phosphate. The up-graded plant operation was commissioned in 1998 and the efficiency of COD reduction was reported to have increased from 51 to 90%. The first-stage FlooBed reactor removed most of the easily biodegradable fraction, while the second FlooBed reactor mainly degraded the not-so-easily biodegradable fraction with continuing action on the easily biodegradable compounds. The third bioreactor (existing activated sludge plant) acted as a polisher and handled the residual biodegradable contaminants. The color of the untreated mill discharge was dark brown, while the effluent from the third-stage bioreactor was reported to be clear. According to the authors, the prescribed discharge limits were successfully met by the up-graded activated sludge plant. Since February 1999, the discharge from the plant is reported to have stabilized at 4 kg COD/ton of pulp produced.

(b) A Case Study on the Up-Gradation of an Activated Sludge Plant in Denmark. Andreasen et al. [76] presented a case study on the successful up-gradation of a Danish pulp industry activated sludge plant with an anoxic selector to reduce bulking sludge problems (Fig. 4). The wastewater of the pulp mill contained large amounts of biodegradable organics with insufficient concentrations of N and P. This condition led to excessive growth of filamentous microbes and poor settling properties of the sludge. The DSVI (sludge volume index) often exceeded 400 mL/g of suspended solids and, as a result, the sludge escaped from the settlers and caused a 70% reduction in the plant capacity. A selector dosed with nitrate was installed ahead of the activated sludge plant to remove a large fraction of easily degradable COD under denitrification conditions. Installation of this anoxic selector significantly improved the DSVI to less than 50 mL/g and enhanced the performance of settlers. Sludge loading of 20–30 kg COD/kg VSS corresponding to a removal rate of 16 kg filterable COD/kg NO_3-N and retention time of 17–22 min were chosen for the optimal performance of the selector. The dosing of nitrate was maintained above 1 mg/L in the selector to avoid anaerobic conditions. Phosphorus was not added due to stringent effluent discharge standards.

Aerated Lagoons (Stabilization) Basins

Aerated lagoons are simple, low-cost biological treatment systems that have been explored in laboratory-scale, pilot-scale, and full-scale studies for the treatment of pulp and paper industrial effluents. Distinct advantages of stabilization basins are lower energy requirement for operation and production of lower quantities of prestabilized sludge. In developed countries like Canada and the United States, the earliest secondary treatment plants for the treatment of pulp and paper effluents were aerated stabilization basins, while in developing countries such as India and China these simple, easy to operate, systems continue to be the most popular choice. Aerated lagoons have masonry or earthen basins that are typically 2.0–6.0 m deep with sloping sidewalls and use mechanical or diffused aeration (rather than algal photosynthesis) for the supply of oxygen [77]. Mixing of biomass suspension and lower hydraulic retention time (HRT) values prevent the growth of algae. Aerated lagoons are classified on the basis of extent of mixing. A completely mixed lagoon (also known as aerated stabilization basin, ASB) is similar to an activated sludge process where efficient mixing is provided to supply adequate concentrations of oxygen and to

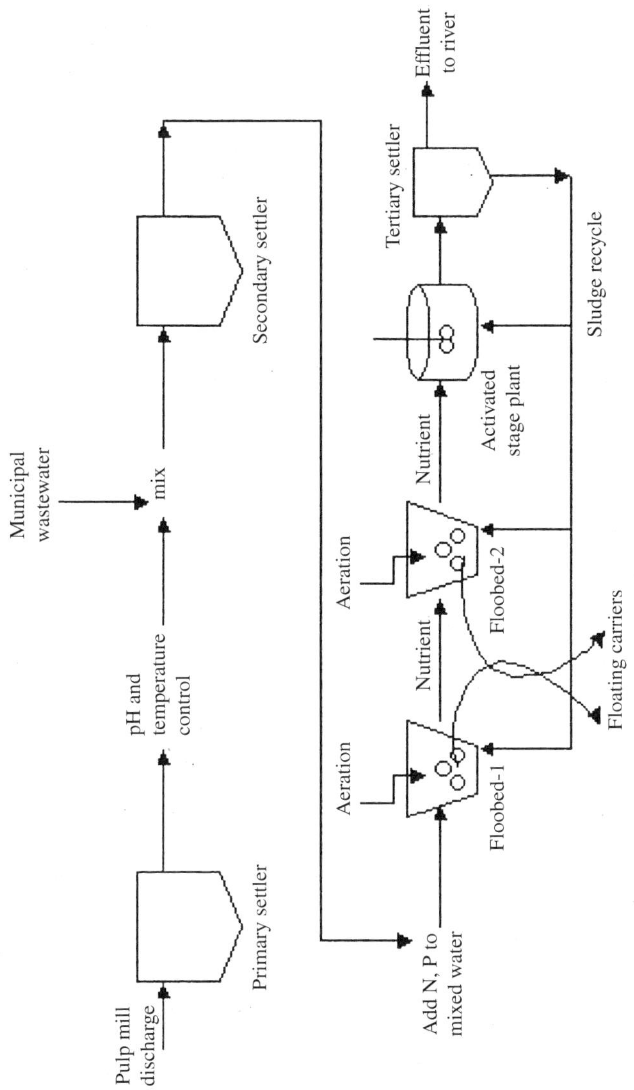

Figure 3 Up-gradation of an existing activated sludge plant in Poland by installation of FlooBed reactors (from Ref. 75).

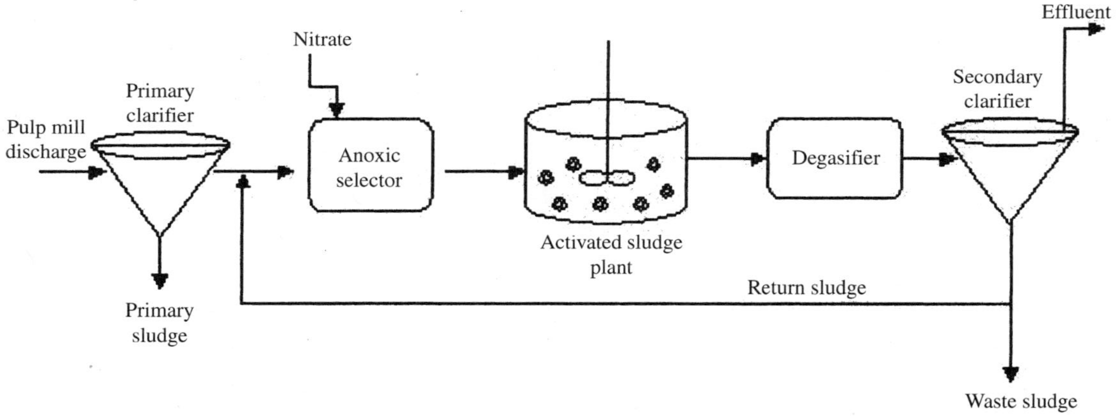

Figure 4 Up-gradation of a Danish pulp and paper mill activated sludge plant through installation of an anoxic selector (from Ref. 76).

keep all of the biomass in suspension (Fig. 5a). However, the system does not include a mechanism for recycling biomass or solids; consequently, the HRT approaches the SRT value. Aerobic bacteria oxidize a portion of the biodegradable organics into carbon dioxide and water, and the rest is utilized to generate biomass components. Several completely mixed aerated lagoons may be linked in series to increase the HRT/SRT value, thereby facilitating further stabilization of synthesized biomass and organic solids under aerobic conditions. In a partially mixed aerated lagoon (also known as facultative stabilization basin, FSB), the power input adequately satisfies the system's oxygen requirements but is insufficient for keeping the solids in suspension. This allows for settlement of biosolids by gravity sedimentation and subsequent benthal stabilization through anaerobic processes. Thus, the biological activity in facultative lagoons is partially aerobic and anaerobic. Partial mix lagoons are generally designed to include two to three cells in series (Fig. 5b). Typically the first cell is completely mixed with intense aeration while the final cell may have very low mixing in which the biomass is allowed to settle down to form benthal deposit. Growth of algae in the settling lagoon is prevented by minimum aeration and limiting HRT value of the overlying clear water zone.

Aerated lagoons have been employed as full-scale treatment systems or as polishing units in Kraft, TMP, and CTMP mills for the removal of BOD, low-molecular-weight AOX, resin and fatty acids [22]. Typical HRT values range from 5 to 10 days. Bajpai [22] compared the reduction efficiencies of individual chlorophenols across aerated lagoons and noted that the values ranged from 30 to 90%. Overall reduction of AOX in bleached kraft mill effluents typically vary from 15 to 60%. Removal of resin and fatty acids in CTMP effluents occur through aerobic oxygenation and degradation with efficiencies exceeding 95%. Welander et al. [78] observed significant improvement in the efficiency of aerated lagoons by installing a support matrix for microbial growth in 20 m^3 pilot-scale plants at two Swedish pulp and paper mills. The two plants were operated for nearly a year and exhibited 60–70% reduction in COD and phosphorus levels. However, efficiencies were much lower for full-scale plants. Kantardjieff and Jones [79] conducted pilot-scale studies on a Canadian integrated sulfite pulp and paper mill effluent using an aerobic biofilter (1 m^2, 3 m depth) as the main unit and aerated stabilization basins (3 m^3) as the polishing stage. The biofilter unit treated the most concentrated sulfite mill effluent and the resulting effluent was mixed with remaining mill wastewaters to be treated in the polishing ASB unit. Characteristics of the raw wastewater, biofilter, and ASB treated mill discharges are summarized in Table 12. In the final design, the ASB had two sections and was operated as a completely mixed system with a total HRT of 2.5 days. The final effluents met the prescribed discharge permit limits and were reported to be nontoxic.

Laboratory-scale treatability studies were conducted by Hall and Randle [80] to monitor and compare the performance of an activated sludge system, ASB, and FSB, operated in parallel to treat Kraft mill wastewaters. Results indicated that FSB and ASB achieved higher removal efficiencies of total and filterable AOX as compared to the activated sludge process under varying temperatures and SRT values. Higher removal rates of chlorinated organics were observed in FSB when the SRT value was increased from 15 to 30 days. The principal removal mechanism seemed to be sorption of AOX to biomass, settling, and anaerobic benthic dechlorination and degradation of the sorbed AOX. Slade et al. [81] evaluated three aerated stabilization basins in New Zealand, which treated elemental chlorine free (ECF) integrated bleached Kraft mill effluents. All three treatment systems achieved 90% removal of BOD without nutrient supplementation. Aerated basin receiving wastewater with a higher BOD : N ratio (100 : 0.8) exhibited nitrogen fixation capability. For phosphorus limited or lower BOD : N (100 : 2) ratio waste streams, benthic recycling seemed to be a crucial mechanism for nutrient supply in aerated basins. Bailey and Young [82] conducted toxicity tests using *Ceriodaphnia*

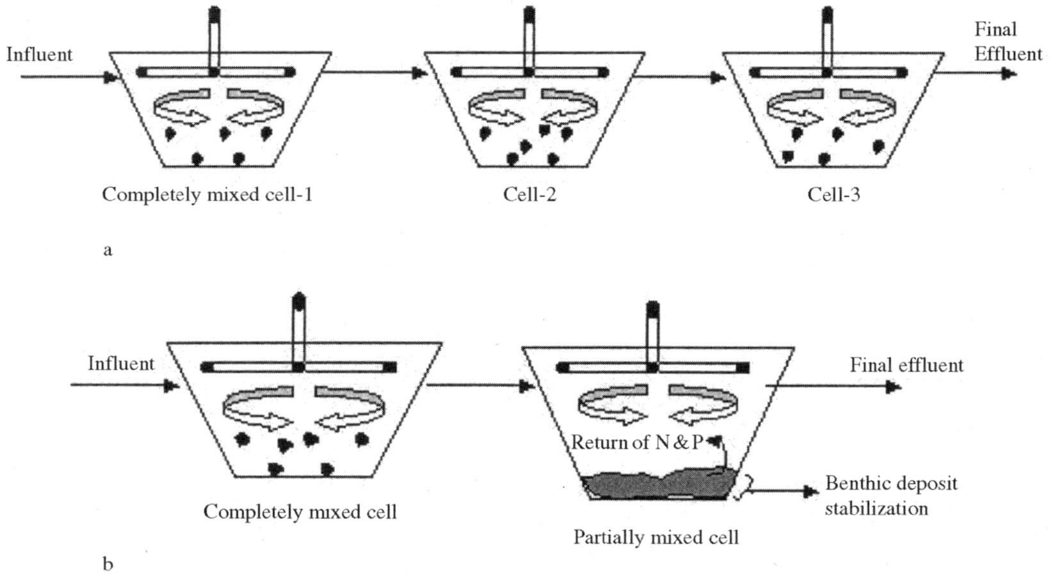

Figure 5 Types of aerated lagoons: (a) Biotransformation of organics and stabilization of biomass under aerobic conditions; (b) Biotransformation of organics under aerobic conditions followed by benthal stabilization of biosolids under anaerobic conditions (from Ref. 77).

Table 12 Characteristics of the Untreated and Biologically Treated Sulfite Mill Effluent

Parameters	Influent (average value) to aerobic filter	Aerobic filter effluent (average value)	Influent (average value) to aerated lagoon[a]	Aerated lagoon effluent (average value)
COD_{total} (mg/L)	2920	1493	1508	795
$COD_{soluble}$ (mg/L)	2737	1383	1305	721
BOD_{total} (mg/L)	795	144	320	50
SS (mg/L)	56	53	102	30
VSS (mg/L)	49	43	96	28
COD load ($kg/m^3/d$)	12	–	–	–
BOD load ($kg/m^3/d$)	34	–	–	–
COD removal (%)	–	49	–	44
BOD removal (%)	–	82	–	85

[a] Influent composed of a portion of aerobic filter effluent mixed with the remaining mill effluents.
Source: Ref. 79

dubia, *Selenastrum capricornutum*, and rainbow trout, and suggested that mill effluents treated by ASBs exhibit less toxicity than other treatment methods.

Anaerobic Treatment Processes

Anaerobic processes have been employed to stabilize sewage sludge for more than a century. The application of this process for high-strength industrial wastewater treatment began with the development of high rate anaerobic reactors [83,84]. A spectrum of innovative reactors ranging from suspended to attached growth systems or a combination of both (hybrid) operate currently with a range of HRT and SRT values. Retention of biomass is accomplished through the sedimentation of microbial flocs or granules, use of reactor configuration that retains sludge, or immobilization on fixed surfaces or carrier materials [77]. High-rate reactors typically achieve 80–90% reduction in BOD_5, with biogas and methane production of 0.5 m^3/kg COD and 0.35 m^3/kg COD, respectively [83]. Biomass generation ranges from 0.05–0.1 kg VSS/kg COD removed. The various types of bioreactor configurations that are employed to treat industrial wastewaters include: (a) upflow anaerobic sludge blanket (UASB), (b) anaerobic contact (AC), (c) anaerobic filter (AF), (d) hybrid UASB with filter (UASB/AF), (e) expanded granular sludge blanket (EGSB), (f) fluidized bed (FB), (g) down-flow stationary fixed film (DSFF), and (h) anaerobic lagoons. Hulshoff Pol et al. [85] reported that in 1997, about 61% of the full-scale industrial anaerobic plants were designed as UASB-type reactors, while the rest employed contact processes (12%), lagoons (7%), filters (6%), hybrid reactors (4%), EGSB reactors (3%), fluidized bed reactors (2%), and fixed film reactors (2%).

Application of Anaerobic Bioreactors in the Pulp and Paper Industries. A number of factors govern the choice of a treatment process and reactor. The preferred choice for treatment of pulp and paper mill effluents is anaerobic degradation because these industries typically generate high-strength wastewaters with the potential to recover energy in the form of biogas. Moreover, anaerobic microorganisms are reported to be more efficient in dehalogenating and degrading chlorinated organics compared to aerobic microorganisms. Additional factors such as lower capital investment and limitation of land area often translate into a reactor that can accommodate high organic and hydraulic loadings with the least maintenance and operation problems. However, assessing the suitability of an anaerobic process and systematic evaluation of reactor configurations are essential prior to the full-scale implementation, in view of the heterogeneous nature of pulp and paper mill effluents. Laboratory-scale and pilot-scale studies on specific mill effluents must address the following key issues:

- Toxicity of the wastewater, especially to the methanogenic population. In general, wastewaters from chemical, NSSC pulping spent liquor condensates, and TMP are nontoxic. On the other hand, unstable anaerobic operations have been noticed with untreated NSSC spent liquors, effluents from debarking, CTMP, and chemical bleaching. Resin acids, fatty acids, terpenes, condensed and hydrolyzable tannins, sulfate, sulfite, reduced sulfur compounds, alkylguaiacols, and chlorinated phenols have been reported to be highly inhibitory to the methanogenic population [86]. Inhibitory waste streams must be diluted or treated by methods such as precipitation, aerobic biodegradation, autooxidation, and polymerization for the selective removal of toxic compounds before anaerobic treatment.
- Anaerobic biodegradability of the components in various effluents (lignin derivatives, resin and fatty acids are known to be highly resistant to anaerobic degradation).
- Maximum loading capacity and reliability of the process under fluctuating loads and shock loading conditions.
- Ease of start-up following interruption of the process.
- Cost of construction, operation and maintenance of reactors.
- Recovery of chemicals and energy.

Anaerobic reactor configurations that have found application in the treatment of pulp and paper mill effluent include anaerobic contact, UASB, anaerobic filter, UASB/AF hybrid, and fluidized bed reactors. Specific features of these reactors are described in the following sections.

Anaerobic Contact Reactor. The anaerobic contact system as illustrated in Figure 6 consists of a completely mixed anaerobic reactor with suspended growth of biomass, a degasifier unit, and a sedimentation unit intended for the separation of clarified effluent from biosolids. Part of the biomass is recycled to the bioreactor through a recycle line. The purpose of the degasifier is to remove gases such as carbon dioxide and methane and to facilitate efficient settling of solids. The volumetric loading rate (VLR) varies from 0.5 to 10 kg COD/m^3 day with HRT and SRT values in the range 0.5–5 days and 10–20 days, respectively [77]. Volatile suspended solids (VSS) in the bioreactor typically range from 4–6 g/L to 25–30 g/L. The process is applicable to wastewaters containing high concentrations (>40%) of suspended solids. Major disadvantages of contact systems are poor settleability of sludge and susceptibility to shock loadings and toxicity.

Up-Flow Anaerobic Sludge Blanket (UASB) Reactor. The UASB reactor is a suspended growth reactor in which the microorganisms are encouraged to develop into dense, compact, and readily settling granules (Fig. 7). Granulation is dependent upon the environmental conditions of the reactor and facilitates the maintenance of high concentrations (20–30 g/L) of VSS in the reactor. The flow of influent starts at the bottom of the reactor, passes through the blanket of dense granules in the bottom half portion of the reactor and reaches into the gas–liquid–solid separator located in the top portion of the reactor. The gas–liquid–solid separating device consists of a gas collection hood and a settler section. Most of the treatment occurs within the blanket of granules. The gas collected in the hood area of the bioreactor is continually removed while the liquid flows into the settler section for liquid–solid separation and settlement of solids back into the reactor. The combined effects of wastewater flow (upflow liquid velocity of 1 m/hour) and biogas production facilitate mixing and contact between the wastewater and microorganisms in the granules. The volumetric loading rates vary from 10 to 30 kg COD/m^3 day

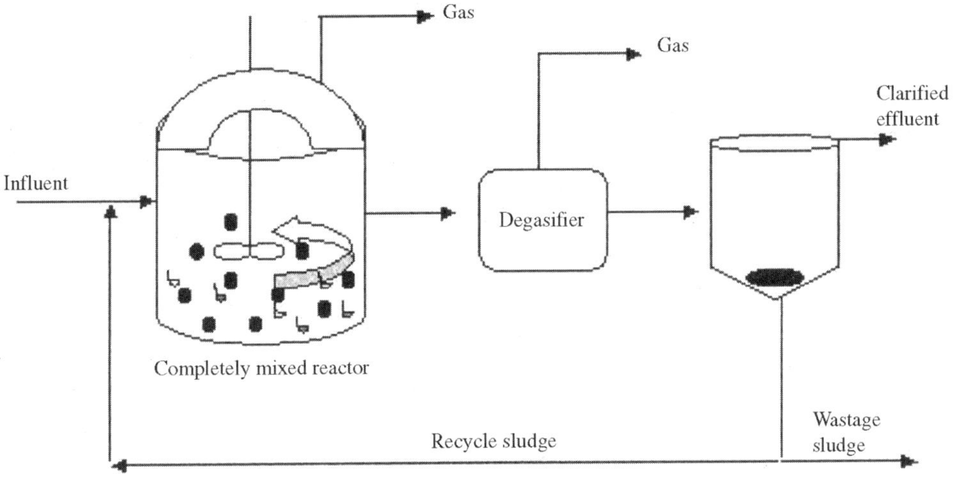

Figure 6 Diagrammatic representation of anaerobic contact process (from Ref. 77).

and typical HRT values range from 4 to 12 hours. The UASB reactors can handle effluents with a high content of solids. However, the quality of granules and hence the performance of the reactor is highly dependent upon the toxicity and other characteristics of wastewater. A modified version of the UASB reactor is the expanded granular sludge blanket (EGSB) reactor, which is

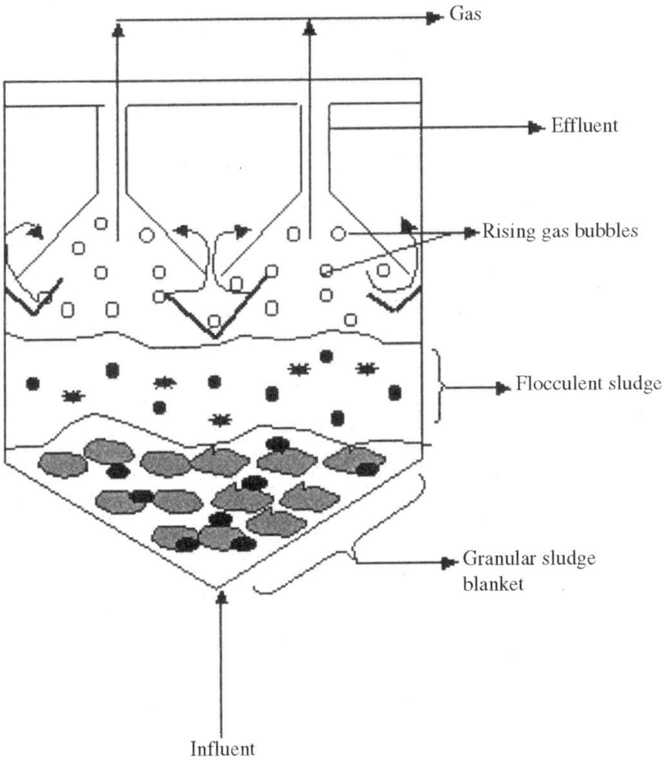

Figure 7 Diagrammatic representation of upflow anaerobic sludge blanket reactor process (from Ref. 77).

designed for higher up-flow velocity (3–10 m/hour) of liquid. Higher flow velocity is achieved by using tall reactors or recycling of treated effluent or both. The VLR in EGSB reactors ranges from 20 to 40 kg COD/m^3 day. Another modification to the UASB is an internal circulation reactor (IC) that has two UASB compartments on top of each other with biogas separation in each stage [6].

Anaerobic Filter. An anaerobic filter consists of packed support media that traps biomass as well as facilitates attached growth of biomass as a biofilm (Fig. 8). Such a reactor configuration helps in the retention of suspended biomass as well as gas–liquid–solid separation. The flow of liquid can be upward or downward, and treatment occurs due to attached and suspended biomass. Treated effluent is collected at the bottom or top of the reactor for discharge and recycling. Gas produced in the media is collected underneath the bioreactor cover and transported for storage or use. Volumetric loading rates vary from 5 to 20 kg COD/m^3 day with HRT values of 0.5–4 days.

Hybrid UASB/Filter Reactor. This hybrid reactor is a suspended growth reactor primarily designed as a UASB reactor at the bottom with packing media (anaerobic filter) on the top of the reactor. The influent is uniformly distributed at the bottom of the reactor and flows upwards sequentially through the granular sludge blanket and filter media where gas–liquid–solid separation takes place. Treated effluent is collected at the top for discharge or recycling. Gas collected under the cover of the bioreactor is withdrawn for use or storage. Process loadings for this system are similar to those of UASB.

Fluidized Bed Reactor. Fluidized bed systems are upflow attached growth systems in which biomass is immobilized as a thin biofilm on light carrier particles such as sand (Fig. 9). A high specific surface area of carrier particles facilitates accumulation of VSS concentrations ranging from 15 to 35 g/L in the bioreactor. The upflow velocities are much higher compared to UASB, AF, or hybrid reactors, preventing the growth of suspended biomass. Carrier particles with biomass are fluidized to an extent of 25–300% of the resting bed volume by the upward flowing influent and the recirculating effluent. Such a system allows for high mass transfer rates with minimal clogging and reduces the risk of toxic effects of the incoming wastewater. HRT values ranging from 0.2 to 2 days and VLR above 20 kg COD/m^3 day are common [77].

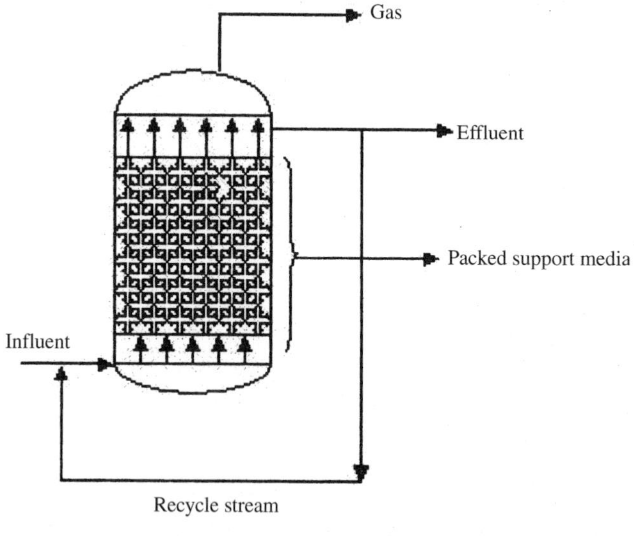

Figure 8 Diagrammatic representation of anaerobic filter process (from Ref. 77).

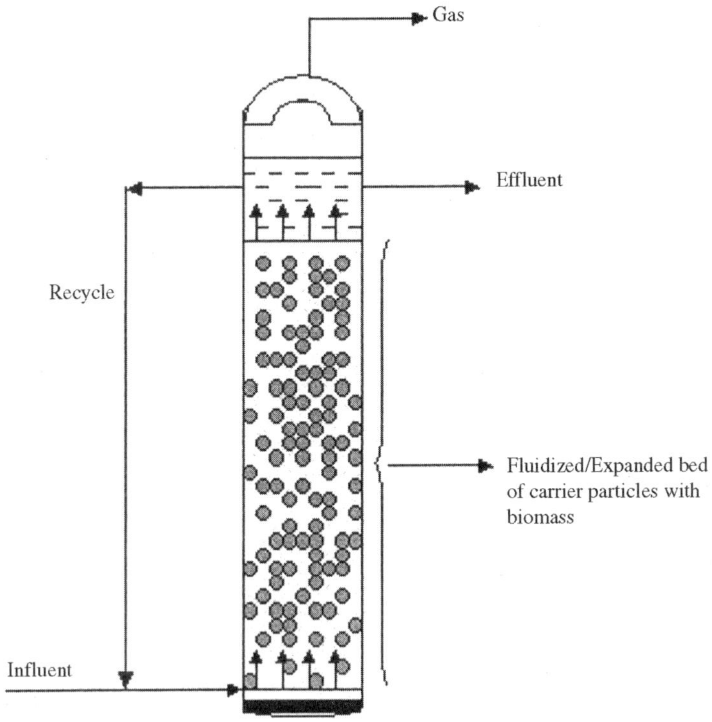

Figure 9 Diagrammatic representation of fluidized and expanded bed reactor process (from Ref. 77).

Anaerobic Technologies Suppliers and Anaerobic Plants in the Pulp and Paper Industry. Table 13 shows the major suppliers of full-scale anaerobic treatment plants for industrial wastewater and the corresponding technologies. It may be noted from this table that a significant number of installations for pulp and paper mill wastewater treatment are performed by Paques BV, The Netherlands, and Bioethane Systems International, Canada. In total, 89 installations were performed in the year 1998, of which 69 UASB, 15 anaerobic contact, and 3 fluidized bed reactors were chosen [6]. Table 14 summarizes the performance of selected full-scale applications of anaerobic technologies in the pulp and paper industry. The UASB reactor configuration is distinctively the best choice, followed by the anaerobic contact process. There are limited applications of the other anaerobic reactor configurations such as fluidized bed and anaerobic filter. Agro-pulping black liquors have a high content of nonbiodegradable lignin that contributes to 50% of the COD; subsequently, there is a need to develop viable solutions to handle such wastewaters.

Thermophilic Anaerobic Reactor Applications. Pulp and paper industries typically discharge warm (50°C) effluents, and conventional reactors operating under mesophilic conditions require cooling of such wastewaters. Attempts have been made periodically by various groups to investigate the possibility of applying thermophilic anaerobic processes to pulp and paper discharges, but to date there is no conclusive evidence to prove the superior performances of thermophilic reactors as compared to their mesophilic counterparts.

Lepisto and Rintala [87] used four different types of thermophilic (55°C) anaerobic processes, namely an upflow anaerobic sludge blanket (UASB) digester, UASB enriched with sulfate, UASB with recirculation, and fixed-bed digester with recirculation for investigating the

Table 13 Commercial Suppliers of Full-Scale Anaerobic Treatment Plants for Industrial Wastewaters

Process configuration	Technology supplier/Trade name. Number of plants: pulp and paper/total number of industries				
	ADI[a]	Biothane[b]	Degremont[c]	Paques[d]	Purac[e]
UASB	–	Biothane UASB 17/189	Anapulse 3/44	Biopaq 49/266	–
Contact process	ADI-BVF 1/67	Biobulk 0/4	Analift 1/22	–	Anamet 13/63
IC reactor	–	–	–	Biopaq-IC 1/40	
Anaerobic filter	–	–	Anafiz 1/8	–	–
Fluidized bed reactor	–	–	Anaflux 3/13	–	–

[a]ADI Systems Inc., USA; [b]Biothane Systems International, Canada; [c]Degremont, France; [d]Paques BV, The Netherlands; [e]Purac AB, Sweden.
Courtesy of UNDP India website, Ref. 6.

removal of chlorinated phenolics from soft wood bleaching effluents. All four processes eliminated/reduced chlorinated phenols, catechols, guiacols, and hydroquinones from bleach mill effluents. The ranges of COD and AOX removal were 30–70% and 25–41%, respectively, in the four reactors. Jahren et al. [88] treated TMP whitewaters using three different types of thermophilic anaerobic reactor configurations. The anaerobic hybrid reactor consisted of a UASB and a filter that degraded 10 kg COD/m^3/day. The anaerobic multistage reactor, composed of granular sludge and carrier elements, gave a degradation rate of 9 kg COD/m^3/day at loading rates of 15–16 kg COD/m^3/day and HRT of 2.6 hours. The anaerobic moving bed biofilm digester handled loading rates of up to 1.4 kg COD/m^3/day.

Sequential Anaerobic-Aerobic Treatment Systems

Biological reactors employing combination anaerobic and aerobic environments can be more effective for the detoxification of pulp and paper mill discharges through the following processes:

- Reduction of biodegradable organics under anaerobic and aerobic conditions;
- Transformation and degradation of chlorinated compounds presumably via reductive dehalogenation and subsequent aerobic metabolism;
- Aerobic metabolism of extractable compounds such as resin acids via hydroxylation reactions.

Haggblom and Salkinoja-Salonen [89] treated Kraft pulp bleaching in an anaerobic fluidized bed reactor, followed by an aerobic trickling filter. The sequential treatment process reduced 65% of AOX and 75% of the chlorinated phenolic compounds. The anaerobic reactor was efficient in dechlorination, thereby eliminating most of the toxicity and improving biodegradability of the subsequent aerobic reactor at shorter retention times. The researchers identified two species of *Rhodococcus* bacteria that were capable of degrading polychlorinated phenols, guaiacols, and syringols in the bleaching effluents. Wang et al. [90] examined continuous-flow sequential reactors operated in anaerobic-aerobic and aerobic-aerobic modes. The objective of this research was to enhance reductive dehalogenation and degradation of

Table 14 Commercial Suppliers of Full-Scale Anaerobic Treatment Plants for Industrial Wastewaters

	Paper mill	Influent type and characteristics	Supplier	Reactor details	Start-up year	Plant performance
1	Macmillan Bloedel Ltd., Canada	Corrugated Cardboard/NSSC Flow: 6300 m^3/day COD load: 107 T/day	Biothane	UASB V = 7000 m^3 (2 × 3500 m^3 R) VLR = 15.4 kg	1989	Gas production: 1140 m^3/hour COD reduction: 55% BOD reduction: 85%
2	Stone Container, Canada	CTMP/NSSC recycle Flow: 656 m^3/day COD load: 7.7 T/day	Biothane	UASB V = 15,600 m^3 (2 × 7800 m^3 R) VLR = 12 kg	1988	BOD reduction: 85%
3	Industriewater, The Netherlands	Total Flow from 3 mills = 12,400 m^3/day COD load: 22 T/day	Paques BV	UASB V = 2184 m^3 VLR = 6–7 kg	1990	Gas production: 2000–3600 m^3/hour COD reduction: 70% BOD reduction: 80%
4	Satia Paper Mills Ltd., India	Chemical pulping using agricultural residues Flow: 4500 m^3/day COD load: 53 T/day	Paques BV	UASB V = 5200 m^3 (2 × 2600 m^3 R) VLR = 10 kg	1997	Gas production: 10,000–12,000 m^3/hour COD reduction: 50–60% BOD reduction: 60–70%

(continues)

Table 14 Continued

	Paper mill	Influent type and characteristics	Supplier	Reactor details	Start-up year	Plant performance
5	Papetries Lecoursonnois Paper Mill, France	Corrugated medium and coated paper Flow: 600 m^3/day COD load: 3.6 T/day	Degremont	Fluidized bed capacity of the reactor = 20–30 m^3/hour	1994	COD reduction: 75% BOD reduction: 85%
6	Modo Paper AB, Sweden	Sulfite condensate Flow: 6000 m^3/day COD load: 72 T/day	Purac AB	V = 30,000 m^3 (2 × 15,000 m^3 R) Reactor loading rate = 125 T COD/day	1984	Gas production: 23,000 m^3/hour COD reduction: 81% BOD reduction: 98%
7	Pudumjee Pulp and Paper Mills, India	40 tpd bagasse bleached pulp Flow: 2200 m^3/day COD load: 20–22 T/day	Sulzer Brothers Ltd., Switzerland	Continuous stirred tank (An-OPUR) V = 13,000 m^3 (2 × 6500 m^3 R) VLR = 3–5 kg	1989	Gas production: 5000–6000 m^3/hour COD reduction: 60–70% BOD reduction: 85–90%

R = Reactor; VLR (volumetric loading rate) expressed as COD/m^3/day; T = Ton.
Courtesy of UNDP India website, Ref. 6.

organics. The researchers noted that the anaerobic-aerobic reactor improved the biodegradability of bleach wastewaters and performed better compared to the aerobic-aerobic reactor. Rintala and Lepisto [91] performed experiments on a mixture comprised of 20% Kraft mill chlorination (C) stage effluent, 30% alkaline extraction stage (E) effluent, and 50% tap water in anaerobic-aerobic and aerobic post-treatment reactors (250 mL volume) operated at 55°C and partially packed with polyurethane. They observed that the second stage of the anaerobic-aerobic reactor removed negligible COD if the anaerobic stage performed well. However, the aerobic reactor exhibited excellent removal efficiencies during periods of anaerobic reactor upsets. Lee et al. [92] evaluated a continuous-flow sequential anaerobic-aerobic lagoon process for the removal of AOX from Kraft effluents on bench and pilot scales. Bench-scale studies demonstrated AOX removal efficiencies of approximately 70% at HRT values of 2, 5, and 10 days. In comparison, aerobic lagoons removed only 20, 35, and 36% of AOX at the above HRT values.

Mycotic Systems for the Removal of Color and AOX

Fungal systems, particularly white-rot fungi such as *Phanerochaete chrysosporium*, *Ganoderma lacidum*, and *Coriolus versicolor* have been investigated for their abilities to decolorize and degrade Kraft pulp bleaching effluents [93–95]. For instance, Huynh et al. [96] used the MyCoR (mycelial color removal) process for the treatment of E_1 stage effluents. The MyCoR process utilizes a fixed film reactor such as a rotating biological contactor (RBC) for the surface growth of *Phanerochaete chrysosporium*. A related process known as MyCoPOR uses polyurethane foam cubes (surface area of 1 cm^2) as porous carrier material in trickling filter reactors for supporting the growth of *Phanerochaete chrysosporium*. The immobilized fungus is capable of removing color and AOX, and has the capacity to degrade several chlorinated phenols such as 2,4,6-trichlorophenol, polychlorinated guaiacols, and polychlorinated vanillins, all of which are toxic compounds that are present in the pulp bleaching effluents. Prasad and Joyce [97] employed *Phanerochaete chrysosporium* in a rotating biological contactor for treatment of E_1 stage effluent from a Kraft mill bleach plant effluent containing approximately 190 mg/L of AOX. The extent of reduction in AOX, color, COD, and BOD levels were 42, 65, 45, and 55%, respectively, in the aerobic fungal reactor system through biotransformation mechanisms. Efficiency of the treatment process remained constant for up to 20 days with replacement of effluent every 2 days. A subsequent anaerobic treatment removed additional 40% AOX, 45% of soluble COD, and 65% of total BOD at a loading rate of 0.16 kg COD/m^3 day. Thus, the fungal-anaerobic system proved to be more effective than the fungal process alone. Taseli and Gokcay [98] observed 70% removal of AOX by a fungal culture immobilized on glass wool packed in an upflow column reactor. Optimum dechlorination by the fungus was observed at pH 5.5 and 25°C with HRT of 7–8 hour and required very low levels of carbon and dissolved oxygen. *Coriolus versicolor* has shown promising results in terms of color removal, COD reduction, and degradation of chlorolignins in bleach effluents and Kraft mill liquors [22]. Various researchers have applied the fungus in the form of mycelial pellets or as mycelium, entrapped within calcium alginate beads. The fungus has an obligate requirement for simple growth substrates such as glucose, sucrose, and starch for effective performance.

Mechanisms of White-Rot Fungal Mediated Degradation and Decolorization Processes. The ligninolytic enzyme systems of white-rot fungi exhibit catalytic activities that are beneficial for the transformation and mineralization of a wide range of organopollutants (including halogenated organics) with structural similarities to lignin [48,49]. Currently, three types of extracellular lignin modifying enzymes (LMEs), namely lignin peroxidase (LiP), Mn dependent peroxidase (MnP), and laccase (Lac) are known to aid in the catabolism of lignin [99].

LiP catalyzes the oxidation of a low-molecular-weight redox mediator, veratryl alcohol, which in turn mediates one-electron oxidation of lignin to generate aryl cation radicals [100]. The radicals facilitate a wide variety of reactions such as carbon–carbon cleavage, hydroxylation, demethylation, and so on. Dezotti et al. [101] reported enzymatic removal of color from extraction stage effluents using lignin and horseradish peroxidases immobilized on an activated silica gel support.

MnP catalyzes hydrogen-peroxide-dependent oxidation of Mn^{2+} to Mn^{3+}, which in turn oxidizes phenolic components of lignin [102]. Oxidative demethylation, dechlorination, and decolorization of bleach plant effluent by the MnP of *Phanerochaete chrysosporium* has been demonstrated [103,104].

Laccase generates free radicals from low-molecular-weight redox mediators such as 3-hydroxanthranilate and initiates condensation of phenolic compounds [105]. The Lac-catalyzed polymerization reaction has the potential for enhancing the subsequent lime-induced precipitation of high-molecular-weight chromophoric pollutants. Archibald et al. [106] and Limura et al. [107] demonstrated efficient decolorization and up to 85% removal of chloroguaiacol using laccase elaborated by *Trametes versicolor*. Davis and Burns [108] demonstrated decolorization of pulp mill bleach effluents by employing laccase that was covalently immobilized on activated carbon. They reported color removal at the rate of 115 units/unit enzyme/hour.

Treatment of bleach mill effluents using the white-rot fungi is promising and offers the option to expand the range of pollutants that cannot be biodegraded by the prokaryotes (bacteria). White-rot fungal remediation may be particularly suited for those recalcitrant compounds for which bioavailability and toxicity are the key issues.

10.11 TREATMENT OF WASTE GAS EMISSIONS FROM PULP AND PAPER INDUSTRIES

A spectrum of gaseous phase pollutants is emitted from pulp and paper mills, which include VOCs, total reduced sulfur compounds (TRS), NOx, and SOx. Air pollution control, particularly nuisance odor abatement, has gained importance in recent years. VOCs formed during pulping, bleaching, and liquor evaporation are conventionally treated by physico-chemical methods such as adsorption to activated coal filters, absorption, thermal oxidation (incineration), catalytic oxidation, and condensation [109]. However, the major limitations of these air pollution control technologies include energy costs, transfer of pollutants from one phase to another, or generation of secondary pollutants. A more recent trend is the development of a low-cost and effective biological treatment of air through usage of biofilters and bioscrubbers that can remove trace concentrations of pollutants. The biological vapor phase remediation process involves three steps: the transfer of pollutants from gaseous phase to liquid phase; transfer from liquid to microbial surface and uptake of pollutants by the microorganisms; and finally transformation and/or degradation of the pollutants by microbial enzymes. Most of the biodegradative enzymes concerned are intracellular. A brief description of the biofilters and bioscrubbers is provided in the following subsections.

10.11.1 Biofilters

A typical biofilter setup consists of a blower, humidification chamber and a biofilter unit, and an additional polishing unit (optional) such as granular activated carbon backup. The biofilter is composed of microbial communities supported on a packing surface material such as wet peat, wood chips, charcoal, compost, diatomaceous earth pellets, and so on. The microorganisms derive all their nutrients from the surrounding liquid film; therefore, an additional device for direct sprinkling of water on the packing surface is usually included. The operational steps in biofiltration involve contacting the waste air with a stream of water in a humidifier and passing the moisture loaded air at a slow rate through microbial film supported on packing surfaces. The superficial gas flow varies from 1 to 15 cm/s for a bed height of 1–3 m. Removal efficiencies of 90% can be achieved at volumetric loading rates of 0.1–0.25 kg organics/m^3/day. The major drawbacks of biofilters are the lack of control over pH and space requirements.

Characteristics of Contaminants Treated by Biofilters and Their Removal Mechanisms

Biofilter units, which have been used conventionally for odor control in wastewater treatment plants, have gained acceptance over the last decade for control of VOCs in industries such as the pulp and paper mills. The technology is best suited for contaminated air streams that have trace concentrations of readily soluble and biodegradable VOCs. Eweis et al. [109] reviewed research findings related to the application of biofiltration for removal of VOCs ranging in concentration from 0.01 to 3000 ppm. Types of contaminants that can be removed through biofiltration include reduced sulfur compounds, ammonia, NOx, and chlorinated organic compounds. Most full-scale biofilter applications have been for odor control through removal of sulfide ranging in concentrations from less than a ppm to several hundred ppm. H_2S is highly soluble in water and oxidized by the aerobic, chemoautotophic bacteria to H_2SO_4, which results in highly acidic conditions during biofiltration. Although sulfur-oxidizing bacteria function well at acidic pH, there is a need to control pH effectively for maintaining the optimal performance of other groups of bacteria. The most effective method for removal of NOx from air streams is based on denitrification, which requires an electron donor. Chlorinated compounds may be dechlorinated to produce HCl with the concomitant drop in pH.

10.11.2 Bioscrubbers

The bioscrubber is a modification of biofilter technology. Contaminated air is drawn into a chamber and sprayed with a fine mist of liquid stream containing a suspension of microorganisms. The liquid is continually circulated between the spray chamber and an activated sludge process unit where biodegradation occurs.

10.11.3 Biotreatment of Flue Gases

Nitrogen and sulfur oxides released in flue gases are removed by scrubbing, which involves dissolution of these gases in a solution of $NaHCO_3$ and Fe(II)-EDTA. A new system based on the regeneration of scrubbing solution via biological denitrification and desulfurization is shown in the following flow chart [110]:

Influent gas (SOx and NOx) $\longrightarrow$ Scrubber $\longrightarrow$ Denitrification tank

$\longrightarrow$ UASB reactor ($S^{4+} \longrightarrow S^{2-}$) $\longrightarrow$ Aerobic reactor ($S^{2-} \longrightarrow S^0$)

$\longrightarrow$ Filter $\longrightarrow$ Recovery of elemental sulfur

10.12 CONCLUSIONS

Over the last decade, much interest has been generated in monitoring environmental problems and associated risks of wastes, in particular, wastewaters generated by the pulp and paper industries. A major goal is to reassess the target pollutant levels and consider the use of risk-based discharge permit values rather than the absolute endpoint values. This risk-based approach requires analytical tools that can quantify the ecotoxic characteristics of discharges rather than the absolute concentration of specific pollutants or the values of lumped pollution parameters such as BOD, COD, and so on.

The development of pollution treatment strategies, technologies, and their implementation in pulp and paper industries requires an integrated holistic approach that requires a detailed understanding of the manufacturing processes as well as the physical, chemical, and biological properties of the multitude of pollutants generated. The nature and extent of pollution varies significantly from one mill to another. Therefore, the selection of treatment technology(ies) and optimization of the process calls for laboratory- and pilot-scale studies to be conducted with actual wastewater and processes under consideration. Finally, the success of a specific process such as biological treatment is highly dependent upon the preceding operations such as segregation of mill process streams, primary treatment such as filtration, chemical precipitation, and oxidation. Physico-chemical treatment methods such as the oxidation processes can effectively increase the biodegradability of recalcitrant pollutants in the subsequent bioreactor. Thus, choosing the right combination and sequence of treatment methods is the key to the successful handling of pollution problems in the pulp and paper industries.

REFERENCES

1. Smook, G.A. (Ed.) *Handbook for Pulp and Paper Technologists*, 2nd Ed.; Angus Wilde: Vancouver, Canada, 1992.
2. Kringstad, K.P.; Lindstrom, K. Spent liquors from pulp bleaching (critical review). *Environ. Sci. Technol.* **1984**, *18*, 236A–247A.
3. U.S. EPA. *Technical support document for best management practises for spent liquor management, spill prevention and control*; USEPA: Washington, DC, 1997.
4. Sharma, C.; Mohanty, S.; Kumar, S.; Rao, N.J. Gas chromatographic determination of pollutants in Kraft bleachery effluent from Eucalyptus pulp. *Anal. Sci.* **1999**, *15*, 1115–1121.
5. Grimvall, A.; Boren, H.; Jonsson, S.; Karlsson, S.; Savenhed, R. Organohalogens of natural and industrial origin in large recipients of bleach-plant effluents. *Water. Sci. Technol.* **1991**, *24 (3/4)*, 373–383.
6. Rintala, J.A.; Jain, V.K.; Kettunen, R.H. Comparative status of the world-wide commercially available anaerobic technologies adopted for biomethanation of pulp and paper mill effluents. *4th International Exhibition and Conference on Pulp and Paper industry, PAPEREX-99*, New Delhi, India, 14–16 December, 1999.
7. Sjostrom, E. (Ed.) *Wood Chemistry, Fundamentals and Applications*; Academic Press: New York, 1981.
8. Sjostrom, E. (Ed.) *Wood Chemistry, Fundamentals and Applications*, 2nd Ed.; Academic Press: San Diego, CA, 1993.
9. Folke, J. Environmental aspects of bagasse and cereal straw for bleached pulp and paper. *Conference on Environmental Aspects of Pulping Operations and their Wastewater Implications*, Edmonton, Canada, 27–28 July 1989.
10. Kocurek, M.J.; Ingruber, O.V.; Wong, A. (Eds.) *Pulp and Paper Manufacture, Volume 4, Sulfite Science and Technology*; Joint Textbook Committee of the Paper Industry: Tappi, Atlanta, GA, 1985.

11. Green, R.P.; Hough, G. (Eds.) *Chemical Recovery in the Alkaline Pulping Process*, Revised Edition; Tappi, Atlanta, GA, 1992.
12. Biermann, C.J. (Ed.) *Essentials of Pulping and Paper Making*; Academic Press: New York, 1993.
13. Rydholm, S.A. (Ed.) *Pulping Processes*; Interscience: New York, 1965.
14. Jain, R.K.; Gupta, A.; Dixit, A.K.; Mathur, R.M.; Kulkarni, A.G. Enhanced biomethanation efficiency of black liquor through lignin removal process in small agro based paper mills. *Bioenergy News* **2001**, *5* (*3*), 2–6.
15. National Council of the Paper Industry for Air and Stream Improvement, Inc. (NCASI). *The toxicity of Kraft pulping wastes to typical fish food organisms*, Technical Bulletin No. 10; NCASI: New York, 1947.
16. Mckee, J.E.; Wolf, H.W. (Eds.) *Water Quality Criteria*, 2nd Ed., Publication 3-A; State Water Resources Control Board, The Resources Agency of California: Pasadena, CA, 1963.
17. Liss, S.N.; Bicho, P.A.; McFarlane, P.N.; Saddler, J.N. Microbiology and degradation of resin acids in pulp mill effluents: a mini review. *Can. J. Microbiol.* **1997**, *75*, 599–611.
18. Lepp, H.J.T.; Oikari, A.O.J. The occurrence and bioavailability of retene and resin acids in sediments of a lake receiving BKME (bleached Kraft mill effluent). *Water. Sci. Technol.* **1999**, *40* (*11–12*), 131–138.
19. Hickey, C.W.; Martin, M.L. Relative sensitivity of five benthic invertebrate species to reference toxicants and resin acid contaminated sediments. *Environ. Toxicol. Chem.* **1995**, *14*, 1401–1409.
20. Johnsen, K.; Mattsson, K.; Tana, J.; Stuthridge, T.R.; Hemming, J.; Lehtinen, K.-J. Uptake and elimination of resin acids and physiological responses in rainbow trout exposed to total mill effluent from an integrated newsprint mill. *Environ. Toxicol. Chem.* **1995**, *14*, 1561–1568.
21. McCarthy et al. Role of resin acids in the anaerobic toxicity of chemi-thermomechanical pulp wastewaters. *Water. Res.* **1990**, *24*, 1401–1405.
22. Bajpai, P. Microbial degradation of pollutants in pulp mill effluents. *Adv. Appl. Microbiol.* **2001**, *48*, 79–134.
23. Kringstad, K.P.; Ljungquist, P.O.; de Sousa, F.; Stromberg, L.M. Identification and mutagenic properties of some chlorinated aliphatic compounds in the spent liquor from Kraft pulp chlorination. *Environ. Sci. Technol.* **1981**, *15*, 562–566.
24. McMaster, M.E.; Van Der Kraak, G.J.; Munkittrick, K.R. An evaluation of the biochemical basis for steroidal hormonal depression in fish exposed to industrial wastes. *J. Great Lakes Res.* **1996**, *22*, 153–171.
25. Sandstrom, O. Incomplete recovery in a fish coastal community exposed to effluent from a Swedish bleached Kraft mill. *Can. J. Fish. Aqua. Sci.* **1994**, *51*, 2195–2202.
26. Tana, J.; Rosemarin, A.; Lehtinee, K.-J.; Hardig, J.; Grahn, O.; Landner, L. Assessing impacts on Baltic coastal ecosystems with mesocosm and fish biomarker tests: a comparison of new and old pulp bleaching technologies. *Sci. Total. Environ.* **1994**, *145*, 213–214.
27. Kankaanpaa, H.; Lauren, M.; Mattson, M.; Lindstrom, M. Effects of bleached Kraft mill effluents on the swimming activity of Monoporeia affinis (Crustacea, Amphipoda). *Chemosphere* **1995**, *31*, 4455–4473.
28. Bankey, L.A.; Van Vels, P.A.; Borton, D.L.; LaFleur, L.; Stegeman, J.J. Responses of cytochrome P4501A1 in freshwater fish exposed to bleached Kraft mill effluent in experimental stream channels. *Can. J. Fish. Aqua. Sci.* **1995**, *52*, 439–447.
29. Kovacs, T.G.; Gibbons, J.S.; Trembaly, L.A.; O'Connor, B.I.; Martel, P.H.; Vos, R.H. The effects of a secondary treated bleached Kraft mill effluent on aquatic organisms as assessed by short term and long term laboratory tests. *Ecotox. Environ. Safe.* **1995**, *31*, 7–22.
30. Munkittrick, K.R.; Servos, M.R.; Carey, J.H.; Van Der Kraak, G.J. Environmental impacts of pulp and paper wastewater: evidence for a reduction in environmental effect at north American pulp mills since 1992. *Water. Sci. Technol.* **1997**, *35* (*2–3*), 329–338.
31. Mika, A.K.; Liukkonen, M.; Wittmann, C.; Suominen, K.P.; Salkinoja-Salonen, M.S. Integrative assessment of sediment quality history in pulp mill recipient area in Finland. *Water. Sci. Technol.* **1999**, *40* (*11–12*), 139–146.

32. Kirk, T.K.; Jeffries, T.W. Roles for microbial enzymes in pulp and paper processing. In *Enzymes for Pulp and Paper Processing*; Jeffries, T.W.; Viikari, L.; Eds.; American Chemical Society Symposium, **1996**, Series 655, 2–14.
33. Rantio, T. Chlorinated cymenes in the effluents of two Finnish pulp mills in 1990–1993. *Chemosphere* **1995**, *31*, 3413–3423.
34. Koistinen, J.; Paasivirta, J.; Nevalainen, T.; Lahtipera, M. Chlorophenanthrenes, alkylchlorophenanthrenes and alkylchloronapthalenes in Kraft mill products and discharges. *Chemosphere* **1994**, *28*, 1261–1277.
35. Koistinen, J.; Paasivirta, J.; Nevalainen, T.; Lahtipera, M. Chlorinated fluorenes and alkylfluorenes in bleached Kraft pulp and pulp mill discharges. *Chemosphere* **1994**, *28*, 2139–2150
36. Mielisch, H.J.; Odermatt, J.; Kordsachia, O.; Patt, R. TCF bleaching of Kraft pulp: Investigation of the mixing conditions in an MC ozone stage. *Holzforshung* **1995**, *49*, 445–452.
37. Axegard, P.; Bergnor, E.; Elk, M.; Ekholm, U. Bleaching of softwood Kraft pulps with H_2O_2, O_3 and ClO_2. *Tappi J.* **1996**, *79*, 113–119.
38. Mehta, V.; Gupta, J.K. Biobleaching eucalyptus Kraft pulp with *Phanerochaete chrysosporium* and its effect on paper properties. *Tappi J.* **1992**, *75*, 151–152.
39. Reid, I.D.; Paice, M.G.; Ho, C.; Jurasek, L. Biological bleaching of softwood Kraft pulp with the fungus, *Trametes (Coriolus) versicolor. Tappi J.* **1990**, *73*, 149–153.
40. Kirkpatrick, N.; Reid, I.D.; Ziomek, F.; Paice, M.G. Biological bleaching of hardwood Kraft pulp using *Trametes (Coriolus) versicolor* immobilized in polyurethane foam. *Appl. Microbiol. Biotechnol.* **1990**, *33*, 105–108.
41. Paice, M.G.; Jurasek, L.; Ho, C.; Bourbonnais, R.; Archibald, F. Direct biological bleaching of hardwood kraft pulp with the fungus, *Coriolus versicolor. Tappi J.* **1989**, *72*, 217–221.
42. Tran, A.V.; Chambers, R.P. Delignification of an unbleached hardwood Kraft pulp by *Phanerochaete chrysosporium. Appl. Microbiol. Technol.* **1987**, *25*, 484–490.
43. Ziomek, E.; Kirkpatrick, N.; Reid, I.D. Effect of polymethylsioloxane oxygen carriers on the biological bleaching of hardwood Kraft pulp by *Trametes versicolor. Appl. Microbiol. Biotechnol.* **1991**, *25*, 669–673.
44. Murata, S.; Kondo, R.; Sakai, K.; Kashino, Y.; Nishida, T.; Takahra, Y. Chlorine free bleaching process of Kraft pulp using treatment with the fungus *IZU-154. Tappi J.* **1992**, *75*, 91–94.
45. Viikari, L.; Pauna, M.; Kantelinen, A.; Sandquist, J.; Linko, M. Bleaching with enzymes. In *Proceedings of the Third International Conference on Biotechnology in the Pulp and Paper Industry*, Stockholm, **1986**, 67–69.
46. Daneault, C.; Leduce, C.; Valade, J.L. The use of xylanases in Kraft pulp bleaching: a review. *Tappi J.* **1994**, *77*, 125–131.
47. Call, H.P.; Mucke, I. History, overview and applications of mediated lignolytic systems, especially laccase-mediator systems (Lignozym-process). *J. Biotechnol.* **1997**, *53*, 163–202.
48. Hammel, K.E. Organopollutant degradation by ligninolytic fungi. *Enzyme Microb. Technol.* **1989**, *11*, 776–777.
49. Hammel, K.E. Organopollutant degradation by ligninolytic fungi. In *Microbial Transformation and Degradation of Toxic Organic Chemicals*; Young, L.Y.; Cerniglia, C.E., Eds.; Wiley-Liss: New York, 1995; 331–346.
50. Gullichsen, J. Process internal measures to reduce pulp mill pollution load. *Water Sci. Technol.* **1991**, *24* (3/4), 45–53.
51. McDonough, T. Bleaching agents (pulp and paper). In *Kirk-Othmer Encyclopedia of Chemical Technology*; Grayson, M., Ed., 4th Ed.; John Wiley and Sons: New York, 1992; 301–311.
52. Reeve, D.W. Introduction to the principles and practice of pulp bleaching. In *Pulp Bleaching: Principles and Practice*; Dence, C.W., Reeve, D.W. Eds., Tappi,; Atlanta, GA, 1996; 2–24.
53. Sprague, J.B. Environmentally desirable approaches for regulating effluents from pulp mills. *Water Sci. Technol.* **1991**, *24 (3/4)*, 361–371.
54. Jonsson, A.-S.; Jonsson, C.; Teppler, M.; Tomani, P.; Wannstrom, S. Treatment of paper coating color effluents by membrane filtration. *Desalination* **1996**, *105*, 263–276.

55. Sierka, R.A.; Cooper, S.P.; Pagoria, P.S. Ultrafiltration and reverse osmosis treatment of an acid stage wastewater. *Water Sci. Technol.* **1997**, *35 (2–3)*, 155–161.
56. Koyuncu, I.; Yalcin, F.; Ozturk, I. Color removal of high strength paper and fermentation industry effluents with membrane technology. *Water Sci. Technol.* **1999**, *40 (11–12)*, 241–248.
57. Randtke, S.J. Organic contaminant removal by coagulation and related process combination. *J. Am. Wat. Wks. Assoc.* **1988**, *80*, 40–56.
58. Licsko, I. Dissolved organics removal by solid–liquid phase separation (adsorption and coagulation). *Water Sci. Technol.* **1993**, *27* (11), 245–248.
59. Stephenson, R.J.; Duff, S.J.B. Coagulation and precipitation of a mechanical pulping effluent–I. Removal of carbon, color and turbidity. *Water Res.* **1996**, *30*, 781–792.
60. Stephenson, R.J.; Duff, S.J.B. Coagulation and precipitation of a mechanical pulping effluent–II. Toxicity removal and metal recovery. *Water Res.* **1996**, *30*, 793–798.
61. Ganjidoust, H.; Tatsumi, K.; Yamagishi, T.; Gholian, R.N. Effect of synthetic and natural coagulant on lignin removal from pulp and paper wastewater. *Water Sci. Technol.* **1997**, *35 (2–3)*, 291–296.
62. Legrini, O.; Oliverous, E.; Braun, E.M. Photochemical process for water treatment. *Chem. Rev.* **1993**, *93*, 671–698.
63. Balcloglu, A.I.; Cecen, F. T. reatability of Kraft pulp bleaching wastewater by biochemical and photocatalytic oxidation. *Water Sci. Technol.* **1999**, *40* (1), 281–288.
64. Yeber, M.C.; Rodriguez, J.; Baeza, J.; Freer, J.; Zaror, C.; Duran, N.; Mansilla, H.D. Toxicity abatment and biodegradability enhancement of pulp mill bleaching effluent by advanced chemical oxidation. *Water Sci. Technol.* **1999**, *40 (11–12)*, 337–342.
65. Winterbourn, C. Toxicity of iron and hydrogen peroxide: The Fenton reaction. *Toxic. Lett.* **1995**, *82/83*, 969–974.
66. Rodriguez, J.; Contreras, D.; Parra, C.; Freer, J.; Baeza, J.; Duran, N. Pulp mill effluent treatment by Fenton-type reaction catalyzed by iron complexes. *Water Sci. Technol.* **1999**, *40 (11–12)*, 351–355.
67. Laari, A.; Korhonen, S.; Tuhkanen, T.; Verenich, S.; Kallas, J. Ozonation and wet oxidation in the treatment of thermomechanical pulp (TMP) circulation waters. *Water Sci. Technol.* **1999**, *40 (11–12)*, 51–58.
68. Hostachy, J.-C.; Lenon, G.; Pisicchio, J.-L.; Coste, C.; Legay, C. Reduction of pulp and paper mill pollution by ozone treatment. *Water Sci. Technol.* **1997**, *35 (2–3)*, 261–268.
69. Helbe, A.; Schlayer, W.; Liechti, P.-A.; Jenny, R.; Mobius, C.H. Advanced effluent treatment in the pulp and paper industries with a combined process of ozonation and fixed bed biofilm reactors. *Water Sci. Technol.* **1999**, *40 (11–12)*, 345–350.
70. Schnell, A.; Sabourin, M.J.; Skog, S.; Garvie, M. Chemical characterization and biotreatability of effluents from an integrated alkaline-peroxide mechanical pulping/machine finish coated (APMP/MFC) paper mill. *Water Sci. Technol.* **1997**, *35 (2–3)*, 7–14.
71. Saunamaki, R. Activated sludge plants in Finland. *Water Sci. Technol.* **1997**, *35 (2–3)*, 235–243.
72. Narbaitz, R.M.; Droste, R.L.; Fernandes, L.; Kennedy, K.J.; Ball, D. PACT^{TM} process for treatment of Kraft mill effluent. *Water Sci. Technol.* **1997**, *35 (2–3)*, 283–290.
73. Rantala, P.-R.; Wirola, H. Solid, slightly soluble phosphorus compounds as nutrient source in activated sludge treatment of forest industry wastewaters. *Water Sci. Technol.* **1997**, *35 (2–3)*, 131–138.
74. Clauss, F.; Balavoine, C.; Helaine, D.; Martin, G. Controlling the settling of activated sludge in pulp and paper wastewater treatment plants. *Water Sci. Technol.* **1999**, *40 (11–12)*, 223–229.
75. Hansen, E.; Zadura, L.; Frankowski, S.; Wachoevicz, M. Upgrading of an activated sludge plant with floating biofilm carrier at frantschach Swwiecie S.A. to meet the new demands of year 2000. *Water Sci. Technol.* **1999**, *40 (11–12)*, 207–214.
76. Andreasen, K.; Agertved, J.; Petersen, J.-O.; Skaarup, H. Improvement of sludge settleability in activated sludge plants treating effluent from pulp and paper industries. *Water Sci. Technol.* **1999**, *40 (11–12)*, 215–221.
77. Grady, Jr. C.P.L.; Daigger, G.T.; Lim, H.C. (Eds.) *Biological Wastewater Treatment*; Marcel Dekker: New York, 1999.

78. Welander, T.; Lofqvist, A.; Selmer, A. Upgrading aerated lagoons at pulp and paper mills. *Water Sci. Technol.* **1997**, *35* (*2–3*), 117–122.
79. Kantardjieff, A.; Jones, J.P. Practical experiences with aerobic biofilters in TMP (thermomechanical pulping) sulfite and fine paper mills in Canada. *Water Sci. Technol.* **1997**, *35* (*2–3*), 227–234.
80. Hall, E.R.; Randle, W.G. AOX removal from bleached Kraft mill wastewater. A comparison of three biological treatment processes. *Water Sci. Technol.* **1992**, *26* (*1–2*), 387–396.
81. Slade, A.H.; Nicol, C.M.; Grigsby, J. Nutrients within integrated bleached Kraft mills: sources and behaviour in aerated stabilization basins. *Water Sci. Technol.* **1999**, *40* (*11–12*), 77–84.
82. Bailey, H.C.; Young, L. A comparison of the results of freshwater aquatic toxicity testing of pulp and paper mill effluents. *Water Sci. Technol.* **1997**, *35* (*2–3*), 305–313.
83. Hall, E.R. Anaerobic treatment of wastewaters in suspended growth and fixed film processes. In *Design of Anaerobic Processes for the Treatment of Industrial and Municipal Wastes*; Malin, J.F.; Pohland, F.G., Eds.; Technomics: Lancaster, PA, 1992, 41–118.
84. Iza, J.; Colleran, E.; Paris, J.M.; Wu, W.M. International workshop on anaerobic treatment technology for municipal and industrial wastewaters – summary paper. *Water Sci. Technol.* **1991**, *24* (*8*), 1–16.
85. Hulshoff Pol, L.; Euler, H.; Eitner, A.; Grohganz, D. GTZ sectoral project "Promotion of anaerobic technology for the treatment of municipal and industrial sewage and wastes". *Proceedings of the 8th International Conference on Anaerobic Digestion*, Sendai, Japan, May 25–29, 1997; Vol II, 285–292.
86. Sierra-Alvarez, R.; Field, J.A.; Kortekaas, S.; Lettinga, G. Overview of the anaerobic toxicity caused by organic forest industry wastewater pollutants. *Water Sci. Technol.* **1994**, *29* (*5–6*), 353–363.
87. Lepisto, R.; Rintala, J.A. The removal of chlorinated phenolic compounds from chlorine bleaching effluents using thermophilic anaerobic processes. *Water Sci. Technol.* **1994**, *29* (*5–6*), 373–380.
88. Jahren, S.J.; Rintala, J.A.; Odegaard, H. Anaerobic thermophilic (55°C) treatment of TMP whitewater in reactors based on biomass attachment and entrapment. *Water Sci. Technol.* **1999**; *40* (*11–12*), 67–76.
89. Haggblom, M.; Salkinoja-Salonen, M. Biodegradability of chlorinated organic compounds in pulp bleaching effluents. *Water Sci. Technol.* **1991**, *24* (*3/4*), 161–170.
90. Wang, X.; Mize, T.H.; Saunders, F.M.; Baker, S.A. Biotreatability test of bleach wastewaters from pulp and paper mills. *Water Sci. Technol.* **1997**, *35* (*2–3*), 101–108.
91. Rintala, J.A.; Lepisto, R. Thermophilic anaerobic-aerobic and aerobic treatment of Kraft bleaching effluents. *Water Sci. Technol.* **1993**, *28* (*2*), 11–16.
92. Lee, E.G.-H.; Crowe, M.F.; Stutz, H. Anaerobic-aerobic lagoon treatment of Kraft mill effluent for enhanced removal of AOX. *Water Pollut. Res. J. Can.* **1993**, *28*, 549–569.
93. Eaton, D.; Chang, H.-M., Kirk, T.K. Fungal decolorization of Kraft bleach effluents. *Tappi J.* **1980**, *63*, 103–106.
94. Wang, S.-H.; Ferguson, J.F.; McCarthy, J.L. The decolorization and dechlorination of Kraft bleach plant effluent solutes by use of three fungi: *Ganderma lacidum, Coriolus versicolor* and *Hericium erinaceum. Holzforschung*, **1992**, *46*, 219–233.
95. Livernoche, D.; Jurasek L.; Desrochers, M.; Dorica J.; Veliky, I.A. Removal of color from Kraft mill wastewaters with the cultures of white-rot fungi and with immobilized mycelium of *Coriolus versicolor. Biotechnol. Bioengg.* **1983**, *25*, 2055–2065.
96. Huynh, V.-B.; Chang, H.-m.; Joyce, T.W.; Kirk, T.K. Dechlorination of chloro-organics by a white-rot fungus. *Tappi J.* **1985**, *68*, 98–102.
97. Prasad, D.Y.; Joyce, T.W. Sequential treatment of E_1 stage Kraft bleach plant effluent. *Biores. Technol.* **1993**, *44*, 141–147.
98. Taseli, B.K.; Gokcay, C.F. Biological treatment of paper pulping effluents by using a fungal reactor. *Water Sci. Technol.* **1999**, *40* (*11–12*), 93–100.
99. Pointing, S.B. Feasibility of bioremediation by white-rot fungi. *Appl. Microbiol. Biotechnol.* **2001**, *57*, 20–33.
100. Reddy, C.A.; D'Souza, T.M. Physiology and molecular biology of the lignin peroxidases of *Phanerochaete chrysosporium. FEMS Microbiol. Rev.* **1994**, *13*, 137–152.

101. Dezotti, M.; Innocentini-Mei, L.H.; Duran, N. Silica immobilized enzyme catalyzed removal of chlorolignins from eucalyptus Kraft effluent. *J. Biotechnol.* **1995**, *43*, 161–167.
102. Wariishi, H.; Valli, K.; Gold, M.M. H. Manganese (II) oxidation by manganese peroxidase from the basidiomycete *Phanerochaete chrysosporium*. *J. Biol. Chem.* **1992**, *267*, 23688–23695.
103. Michel, F.C.J.; Dass, S.B.; Grulke, E.A.; Reddy, C.A. Role of manganese peroxidases (MNP) and lignin peroxidases (LiP) of *Phanerochaete chrysosporium* in the decolorization of Kraft bleach plant effluent. *Appl. Environ. Microbiol.* **1991**, *57*, 2368–2375.
104. Jaspers, C.J.; Jiminez, G.; Penninck, M.J. Evidence for a role of manganese peroxidase in the decolorization of Kraft pulp bleach plant effluent *Phanerochaete chrysosporium*: effects of initial culture conditions on enzyme production. *J. Biotechnol.* **1994**, *37*, 229–234.
105. Bourbonnais, R.; Paice, M.G.; Freiermuth, B.; Bodie, E.; Borneman, S. Reactivities of various mediators and laccases with Kraft pulp and lignin model compounds. *Appl. Environ. Microbiol.* **1997**, *63*, 4627–4632.
106. Archibald, F.S.; Paice, M.G.; Jurasek, L. Decolorization of Kraft bleachery effluent chromophores by *Coriolus (Trametes) versicolor*. *Enzyme Microb. Technol.* **1990**, *12*, 846–853.
107. Limura, Y.; Hartkainen, P.; Tatsumi. K. Dechlorination of tetrachloroguaiacol by laccase of white-rot basidiomycete Coriolus versicolor. *Appl. Microbiol. Biotechnol.* **1996**, *45*, 434–439.
108. Davis, S.; Burns, R.G. Covalent immobilization of laccase on activated carbon for phenolic treatment. *Appl. Microbiol. Biotechnol.* **1992**, *37*, 474–479.
109. Eweis, J.B.; Ergas, S.J.; Chang, D.P.Y.; Schroeder, E.D. (Eds.) *Bioremediation Principles*; McGraw-Hill, Singapore, 1998.
110. Vandevivere, P.; Verstraete, W. Environmental applications, In Basic Biotechnology; Ratledge, C., Kristiansen, B., Eds., 2nd Ed.; Cambridge University Press, Cambridge, UK, 2001; 531–557.

11

In-Plant Management and Disposal of Industrial Hazardous Substances

Lawrence K. Wang
Zorex Corporation, Newtonville, New York, U.S.A., and
Lenox Institute of Water Technology, Lenox, Massachusetts, U.S.A.

11.1 INTRODUCTION

If the hazardous substances at industrial, commercial, and agricultural sites can be properly handled, stored, transported, and/or disposed of, there will be no environmental pollution, and no need to embark on any site remediation. With this concept in mind, the goal of in-plant hazardous waste management is to achieve pollution prevention and human-health protection at the sources where there are hazardous substances. This chapter begins with hazardous waste terminologies and characteristics. Special emphasis is placed on the manifest system, hazardous substances storage requirements, underground storage tanks, above-ground storage tanks, hazardous substances transportation, hazardous waste handling, and disposal.

11.1.1 General Introduction and Objectives

Most hazardous wastes are produced in the manufacturing of products for domestic consumption, or various industrial applications. Rapid development and improvement of industrial technologies, products, and practises frequently increase the generation rate of hazardous substances (including both useful materials and waste materials). These hazardous substances, which can be in the form of gas, liquid, or solid, must be properly handled in order to protect the plant personnel, the general public, and the environment.

The term "hazardous substance" refers to any raw materials, intermediate products, final products, spent wastes, accidental spills, leakages, and so on, that are hazardous to human health and the environment. Technically speaking, all ignitable, corrosive, reactive (explosive), toxic, infections, carcinogenic, and radioactive substances are hazardous [1–3].

Legally radioactive substances (including radioactive wastes) are regulated by the Nuclear Regulatory Commission (NRC), while all other hazardous substances (excluding radioactive substances) are mainly regulated by the U.S. Environmental Protection Agency (USEPA), the Occupational Safety and Health Administration (OSHA), and the state environmental protection agencies [4–22]. Guidelines and recommendations by the National Institute for Occupational Safety and Health (NIOSH), the American Conference of Governmental Industrial Hygienists (ACGIH), American Water Works Association (AWWA), American Public Health Association (APHA), Water Environmental Federation (WEF), American Institute of Chemical

Engineers (AIChE), and the American Society of Civil Engineers (ASCE) are seriously considered by practicing environmental engineers and scientists (including chemical/civil/mechanical engineers, biologists, geologists, industrial hygienists, chemist, etc.) in their decision-making process when managing, handling, and/or treating hazardous substances.

In the past 25 years, industry, government, and the general public in the industrially developed as well as developing countries have become increasingly aware of the need to respond to the industrial hazardous substance problems.

Some hazardous wastes, or mixture of hazardous wastes (such as cyanides, hydrogen sulfide, and parathion) are extremely or acutely hazardous because of their high acute toxicity. These extremely hazardous wastes, if human exposure should occur, may result in disabling personal injury, illness, or even death.

Dioxin-contaminated sites, which pose a human health threat, have been the subject of recent analyses by the Centers for Disease Control (CDC) in Atlanta, GA. It has been determined by CDC that 1 ppb of dioxin is detrimental to public health and that people should be dissociated from the hazard. A level of 1 ppb of dioxin (2,3,7,8-TCDD) in soil is recommended as an action level. In cases where soil concentrations exceed 1 ppb, it is recommended by CDC that potential human exposure to the contamination be examined further. If there is human exposure to 1 ppb or higher on a regular basis, cleanup is indicated. A substance that may be more toxic and hazardous than dioxin is expected to be discovered in the near future.

Although the properties of hazardous substances may sound alarming, the managerial skills and technologies used to handle, store, or treat hazardous substances are available. Modern technology exists to build and maintain environmentally sound industrial facilities that effectively produce useful products and, at the same time, render hazardous waste inert. Environmental laws, rules, regulations, and guidelines also exist to ensure that the modern technology will be adopted by owners or plant managers of industrial facilities for environmental protection.

This chapter is intended for the plant owner, the plant engineer/manager, their contractors, their consulting engineers, and the general public. This chapter may be used:

1. As a management and planning tool by industrial and technical personnel; and
2. As a reference document and an educational tool by any individuals who want to review important aspects of in-plant air quality, water quality, safety, and health protection at industrial sites having hazardous substances.

This chapter is not a comprehensive information source on occupational safety and health. It provides a general guideline for industrial and technical personnel at industrial sites to understand or familiarize themselves with:

- hazardous substance classification;
- environmental hazards and their management;
- hazardous air quality management;
- hazardous water quality management;
- hazardous solid waste (including asbestos) management;
- monitoring and analysis of hazardous samples;
- measuring instruments for environmental protection;
- hazardous waste generator status, and the regulatory requirements;
- hazardous waste and waste oil documentation requirements;
- hazardous waste and waste oil storage and shipping requirements;
- emergency preparation and response procedures;
- responsibilities and management strategies of very small quantity generator (VSQG), small quantity generator (SQG), and large quantity generator (LQG) of hazardous wastes;
- an example for managing hazardous wastes generated at medical offices;

- an example for managing hazardous wastes generated at graphic artists, printers, and photographers; and
- two case histories for disposing of photographic wastes by a very small quantity generator (VSQG) and a large quantity generator (LQG).

11.1.2 Hazardous Waste Classification

The first step of site management is to determine whether or not the waste generated or an accidental release (i.e., spill of leaks of chemical/biological substances) occurring on an industrial site is hazardous.

Common hazardous wastes include: (a) waste oil, (b) solvents and thinners, (c) acids and bases/alkalines, (d) toxic or flammable paint wastes, (e) nitrates, perchlorates, and peroxides, (f) abandoned or used pesticides, and (g) some wastewater treatment sludges. Special hazardous wastes include: (a) industrial wastes containing the USEPA priority pollutants, (b) infectious medical wastes, (c) explosive military wastes, and (d) radioactive wastes or releases.

In general, there are two ways a waste or a substance may be identified as hazardous – it may be listed in the Federal and/or the State regulations or it may be defined by its hazardous characteristics.

Hazardous waste may be a listed discarded chemical, an off-specification product, an accidental release, or a liquid or solid residue from an operation process, which has one or more of the characteristics below:

- ignitable (easily catches fire, flash point below 140°F);
- corrosive (easily corrodes materials or human tissue, very acidic or alkaline, pH of <2 or >12.5);
- reactive (explosive, produces toxic gases when mixed with water or acid);
- toxic (can leach toxic chemicals as determined by a special laboratory test); and
- radioactive.

The hazardous waste identification regulations that define the characteristics of toxicity, ignitability, corrosivity, reactivity, and the tests for these characteristics, differ from state to state. In addition, concentration limits may be set out by a state for selected persistent and bioaccumulative toxic substances that commonly occur in hazardous substances. For example, the California Hazardous Waste Control Act requires the California State Department of Health Services (CDHS) to develop and adopt by regulation criteria and guidelines for the identification of hazardous wastes and extremely hazardous wastes.

In the State of California, a waste or a material is defined as hazardous because of its toxicity if it meets any of the following conditions: (a) acute oral LD_{50} of less than 5000 mg/kg; (lethal oral dose for 50% of an exposed population); (b) acute dermal LD_{50} of less than 4300 mg/kg; (c) acute 8 hour inhalation LC_{50} of less than 10,000 ppm; (d) acute aquatic 96 hour LC_{50} of less than 500 mg/L measured in waste with specified conditions and species; (e) contains 0.001% by weight, or 10 ppm, of any of 16 specified carcinogenic organic chemicals; (f) poses a hazard to human health or the environment because of its carcinogenicity, acute toxicity, chronic toxicity, bioaccumulative properties, or persistence in the environment; (g) contains a soluble or extractable persistent or bioaccumulative toxic substance at a concentration exceeding the established Soluble Threshold Limit Concentration (STLC); (h) contains a persistent or bioaccumulative toxic substance at a total concentration exceeding its Total Threshold Limit Concentration (TTLC); (i) is a listed hazardous waste (California list consistent with the Federal RCRA list) designated as toxic; and (j) contains one or more materials with an 8-hour LC_{50} or LCLo of less than 10,000 ppm and the LC_{50} or LCLo is exceeded in the head space vapor (lethal inhalation concentration for 50% of an exposed population).

A waste or a material is designated as "extremely hazardous" in the State of California if it meets any of the following criteria: (a) acute oral LD_{50} of less than or equal to 50 mg/kg; (b) acute dermal LD_{50} of less than or equal to 50 mg/kg; (c) acute inhalation LC_{50} of less than or equal to 100 ppm; (d) contains 0.1% by weight of any of 16 specified carcinogenic organic chemicals; (e) has been shown through experience or testing to pose an extreme hazard to the public health because of its carcinogenicity, bioaccumulative properties, or persistence in the environment; (f) contains a persistent or bioaccumulative toxic substance at a total concentration exceeding its TTLC as specified for extremely hazardous waste; and (g) is water-reactive (i.e., has the capability to react violently in the presence of water and to disperse toxic, corrosive, or ignitable material into the surroundings).

The carcinogenic substances specified in the California criteria for hazardous and extremely hazardous materials have been designated potential carcinogens by OSHA. Under the California criteria, these substances cause a material to be designated as hazardous if they are present at a concentration of 0.001% by weight (10 ppm). A material containing 0.1% of these substances is designated extremely hazardous. The carcinogenic chemicals are the following: 2-acetylaminofluorence, acrylonitrile, 4-aminodiphenyl, benzidine and its salts, bis(chloromethyl) ether (CMME), 1,2-dibromo-3-chloropropane (DBCP), 3,3-dichlorobenzidine and its salts (DCB), 4-dimethylaminoazobenzene (DAB), ethyleneimine (EL), alpha-naphthylamine (1-NA), beta-naphthylamine (2-NA), 4-nitrobiphenyl (4-NBP), n-nitrosodimethylamine (DMN), beta-propiolactone (BPL), and vinyl chloride (VCM).

California criteria for defining hazardous wastes that are ignitable and reactive are identical to Federal criteria for hazardous wastes under RCRA defined at 40 CFR, Part 261. The California corrosivity criteria differ from the Federal criteria only in the addition of a pH test for nonaqueous wastes.

Because each state has its own criteria for defining hazardous wastes, the plant manager of an industrial site having hazardous substances should contact the local state environmental protection agency for the details.

In the State of Massachusetts, the waste generated on the site is considered "acutely hazardous" (equivalent to "extremely hazardous" as defined by the State of California) if it is on the list of "acutely hazardous wastes" published by the State of Massachusetts and/or Federal governments. These acutely hazardous wastes are extremely toxic or reactive and are regulated more strictly than other hazardous wastes. In order to find out if the waste on the site is hazardous, or even acutely hazardous, a plant manager may also check with: (a) the supplier of the product (request a hazardous material safety data sheet); (b) laboratories; (c) trade associations; and/or (d) environmental consulting engineers and scientists. In addition, self-reviewing the State and/or Federal hazardous waste regulations for the purpose of verification is always required.

Radioactive wastes are, indeed, hazardous, but are only briefly covered in this chapter. The readers are referred elsewhere [23–25] for detailed technical information on management of radioactive wastes.

Noise hazard at an industrial site should also be properly controlled. The readers are referred to another source [26] for detailed noise control technologies.

11.2 MANAGEMENT OF ENVIRONMENTAL HAZARDS AT INDUSTRIAL SITES

Environmental hazards are a function of the nature of the industrial site as well as a consequence of the work being performed there. They include (a) chemical exposure hazards, (b) fire and explosion hazards, (c) oxygen deficiency hazards, (d) ionizing radiation hazards, (e) biological hazards, (f) safety hazards, (g) electrical hazards, (h) heat stress hazards, (i) cold exposure hazards, and (j) noise hazards. Both the hazards and the solutions are briefly described in this section [21].

11.2.1 Chemical Exposure Hazards

Preventing exposure to hazardous industrial chemicals is a primary concern at industrial sites. Most sites contain a variety of chemical substances in gaseous, liquid, or solid form. These substances can enter the unprotected body by inhalation, skin absorption, ingestion, or through a puncture wound (injection). A contaminant can cause damage at the point of contact or can act systemically, causing a toxic effect at a part of the body distant from the point of initial contact.

Chemical exposure hazards are generally divided into two categories: acute and chronic. Symptoms resulting from acute exposures usually occur during or shortly after exposure to a sufficiently high concentration of a hazardous contaminant. The concentration required to produce such effects varies widely from chemical to chemical. The term "chronic exposure" generally refers to exposures to "low" concentrations of a contaminant over a long period of time. The "low" concentrations required to produce symptoms of chronic exposure depend upon the chemical, the duration of each exposure, and the number of exposures. For either chronic or acute exposure, the toxic effect may be temporary and reversible, or may be permanent (disability or death). Some hazardous chemicals may cause obvious symptoms such as burning, coughing, nausea, tearing eyes, or rashes. Other hazardous chemicals may cause health damage without any such warning signs (this is a particular concern for chronic exposures to low concentrations). Health effects such as cancer or respiratory disease may not become manifest for several years or decades after exposure. In addition, some hazardous chemicals may be colorless and/or odorless, may dull the sense of smell, or may not produce any immediate or obvious physiological sensations. Thus, a worker's senses or feelings cannot be relied upon in all cases to warn of potential toxic exposure to hazardous chemicals.

Many guidelines for safe use of chemicals are available in the literature [27,28].

11.2.2 Explosion and Fire Hazards

There are many potential causes of explosions and fires at industrial sites handling hazardous substances: (a) chemical reactions that produce explosion, fire, or heat; (b) ignition of explosive or flammable chemicals; (c) ignition of materials due to oxygen enrichment; (d) agitation of shock- or friction-sensitive compounds; and (e) sudden release of materials under pressure [21,29].

Explosions and fires may arise spontaneously. However, more commonly, they result from site activities, such as moving drums, accidentally mixing incompatible chemicals, or introducing an ignition source (such as a spark from equipment) into an explosive or flammable environment. At industrial sites, explosions and fires not only pose the obvious hazards of intense heat, open flame, smoke inhalation, and flying objects, but may also cause the release of hazardous chemicals into the environment. Such releases can threaten both plant personnel on site and members of the general public living or working nearby.

To protect against the explosion and fire hazard, a plant manager should (a) have qualified plant personnel field monitor for explosive atmospheres and flammable vapors, (b) keep all potential ignition sources away from an explosive or flammable environment, (c) use non-sparking, explosion-proof equipment, and (d) follow safe practises when performing any task that might result in the agitation or release of chemicals.

11.2.3 Oxygen Deficiency Hazards

The oxygen content of normal air at sea level is approximately 21%. Physiological effects of oxygen deficiency in humans are readily apparent when the oxygen concentration in the air decreases to 16%. These effects include impaired attention, judgment, and coordination, and increased breathing and heart rate. Oxygen concentrations lower than 16% can result in nausea and vomiting, brain damage, heat damage, unconsciousness, and death. To take into account individual physiological responses and errors in measurement, concentrations of 19.5% oxygen or lower are considered to be indicative of oxygen deficiency.

Oxygen deficiency may result from the displacement of oxygen by another gas, or the consumption of oxygen by a chemical reaction. Confined spaces or low-lying areas are particularly vulnerable to oxygen deficiency and should always be monitored prior to entry. Qualified plant personnel should always monitor oxygen levels and should use atmosphere-supplying respiratory equipment [21].

11.2.4 Ionizing Radiation Hazards

Radioactive materials emit one or more of three types of harmful radiation: alpha, beta, and gamma. Alpha radiation has limited penetration ability and is usually stopped by clothing and the outer layers of the skin. Alpha radiation poses little threat outside the body, but can be hazardous if materials that emit alpha radiation are inhaled or ingested. Beta radiation can cause harmful "beta burns" to the skin and damage the subsurface blood system. Beta radiation is also hazardous if materials that emit beta radiation are inhaled or ingested. Use of protective clothing, coupled with scrupulous personal hygiene and decontamination, affords good protection against alpha and beta radiation.

Gamma radiation, however, easily passes through clothing and human tissue and can also cause serious permanent damage to the body. Chemical-protective clothing affords no protection against gamma radiation itself; however, use of respiratory and other protective equipment can help keep radiation-emitting materials from entering the body by inhalation, ingestion, infection, or skin absorption.

If levels of radiation above natural background are discovered, a plant manager should consult a health physicist. At levels greater than 2 mrem/hour, all industrial site activities should cease until the site has been assessed by an industrial health scientist or licenced environmental engineers.

11.2.5 Biological Hazards

Wastes from industrial facilities, such as a biotechnology firms, hospitals, and laboratories, may contain disease-causing organisms that could infect site personnel. Like chemical hazards, etiologic agents may be dispersed in the environment via water and wind. Other biological hazards that may be present at an industrial site handling hazardous substances include poisonous plants, insects, animals, and indigenous pathogens. Protective clothing and respiratory equipment can help reduce the chances of exposure. Thorough washing of any exposed body parts and equipment will help protect against infection [30,31].

11.2.6 Safety Hazards

Industrial sites handling hazardous substances may contain numerous safety hazards, such as (a) holes or ditches, (b) precariously positioned objects, such as drums or boards that may fall,

(c) sharp objects, such as nails, metal shards, and broken glass, (d) slippery surfaces, (e) steep grades, (f) uneven terrain, and (g) unstable surfaces, such as walls that may cave in or flooring that may give way.

Some safety hazards are a function of the work itself. For example, heavy equipment creates an additional hazard for workers in the vicinity of the operating equipment. Protective equipment can impair a worker's ability, hearing, and vision, which can result in an increased risk of an accident.

Accidents involving physical hazards can directly injure workers and can create additional hazards, for example, increased chemical exposure due to damaged protective equipment, or danger of explosion caused by the mixing of chemicals. Site personnel should constantly look out for potential safety hazards, and should immediately inform their supervisors of any new hazards so that proper action can be taken [1,21,31].

11.2.7 Electrical Hazards

Overhead power lines, downed electrical wires, and buried cables all pose a danger of shock or electrocution if workers contact or sever then during site operations. Electrical equipment used on site may also pose a hazard to workers. To help minimize this hazard, low-voltage equipment with ground-fault interrupters, and water-tight, corrosion-resistant connecting cables should be used on site. In addition, lightning is a hazard during outdoor operations, particularly for workers handling metal containers or equipment. To eliminate this hazard, weather conditions should be monitored and work should be suspended during electrical storms. An additional electrical hazard involves capacitors that may retain a charge. All such items should be properly grounded before handling. OSHA's standard 29 CFR, Part 1910.137, describes clothing and equipment for protection against electrical hazards.

11.2.8 Heat Stress Hazards

Heat stress is a major hazard, especially for workers wearing protective clothing. The same protective materials that shield the body from chemical exposure also limit the dissipation of body heat and moisture. Personal protective clothing can therefore create a hazardous condition. Depending on the ambient conditions and the work being performed, heat stress can occur within as little as 15 minutes. It can pose as great a danger to worker health as chemical exposure. In its early stages, heat stress can cause rashes, cramps, discomfort, and drowsiness, resulting in impaired functional ability that threatens the safety of both the individual and coworkers.

Continued heat stress can lead to stroke and death. Careful training and frequent monitoring of personnel who wear protective clothing, judicious scheduling of work and rest periods, and frequent replacement of fluids can protect against this hazard [21].

11.2.9 Cold Exposure Hazards

Cold injury (frostbite and hypothermia) and impaired ability to work are dangers at low temperatures and when the wind-chill factor is low. To guard against them, the personnel at an industrial site should (a) wear appropriate clothing, (b) have warm shelter readily available, and (c) carefully schedule work and rest periods, and monitor workers' physical conditions.

11.2.10 Noise Hazards

Work around large equipment often creates excessive noise. The effects of noise can include (a) workers being startled, annoyed, or distracted, (b) physical damage to the ear, pain, and temporary and/or permanent hearing loss, and (c) communication interference that may increase potential hazards due to the inability to warn of danger and the proper safety precautions to be taken.

If plant workers are subjected to noise exceeding an 8 hour, time-weighted average sound level of 90 dBA (decibels on the A-weighted scale), feasible administrative or engineering controls must be utilized. In addition, whenever employee noise exposure equals or exceeds an 8 hour, time-weighted average sound level of 85 dBA, workers must administer a continuing, effective hearing conservation program as described in OSHA regulation 29 CFR, Part 1910.95, [1,21,26].

11.3 MANAGEMENT OF AIR QUALITY AT INDUSTRIAL SITES

11.3.1 Airborne Contaminants

The U.S. Environmental Protection Agency (USEPA) has estimated that about 30% of commercial and industrial buildings cause "sick building syndrome." Alternatively the health problems associated with such buildings can also be called "building syndrome," "building-related illness," or "tight building syndrome." As a rule of thumb, to be considered as causing "sick building syndrome" a commercial/industrial building must have at least 20% of its occupants' complaints last for more than two weeks, with symptom relief when the occupants leave the sick building.

At an industrial site, occupants complain when they experience respiratory problems, headache, fatigue, or mucous membrane irritation of their eyes, noses, mouths, and throats.

The following contaminants in air are caused by the building materials [1,32,33,61]:

- Formaldehyde: from particle board, pressed wood, urea-formaldehyde foam insulation, plywood resins, hardwood paneling, carpeting, upholstery;
- Asbestos: from draperies, filters, stove mats, floor tiles, spackling compounds, older furnaces, roofing, gaskets, insulation, acoustical material, pipes, etc.;
- Organic vapors: from carpet adhesives, wool finishes, etc.;
- Radon: from brick, stone, soil, concrete, etc.;
- Synthetic mineral fibers: from fiberglass insulation, mineral wood insulation, etc.; and
- Lead: from older paints.

The following contaminants in air are caused by the use of various building equipments [33–36,66,70–75,79–81]:

- Ammonia: from reproduction, microfilm, and engineering drawing machines;
- Ozone: from electrical equipment and electrostatic air cleaners;
- Carbon monoxide, carbon dioxide, sulfur dioxide, hydrogen cyanide, particulates, nitrogen dioxide, benzoapryene, etc.: from combustion sources including gas ranges, dryers, water heaters, kerosene heaters, fireplaces, wood stoves, garage, etc.;
- Aminos: from humidification equipment;
- Carbon, powder, methyl alcohol, trinitrofluorene, trinitrofluorenone: from photocopying machines;
- Methacrylates: from signature machines;
- Methyl alcohol: from spirit duplication machines;
- Dusts: from various industrial equipments; and

- Microorganisms including bacteria, protozoa, virus, nematodes, and fungi: from stagnant water in central air humidifier, microbial slime in heating, ventilation, and air conditioning (HVAC) systems, fecal material of pigeons in HVAC units, etc.

Certain common contaminants in air are caused by the building inhabitants and hazardous substance releases:

- Formaldehyde: from smoking, waxed paper, shampoo, cosmetics, and medicine products, etc.;
- Acetone, butyric acid, ethyl alcohol, methyl alcohol, ammonia, odors: from biological effluents;
- Asbestos: from talcum powder, hot mittens;
- Nicotine, acrolein, carbon monoxide: from smoking;
- Vapors and dusts: from personal care products, cleaning products, fire retardants, insecticides, fertilizers, adhesives, carbonless paper products, industrial hazardous substance releases, etc.;
- Vinyl chloride: from aerosol spray; and
- Lead: from lead-containing gasoline.

Any real property, the expansion, redevelopment, or reuse of which may be complicated by the presence of one or more of the above hazardous substances is termed "brownfield" [37,38,70,84].

11.3.2 Health Effects

Various airborne contaminant sources and the health effects of each specific pollutant are described below in detail.

Carbon Monoxide

Carbon monoxide (CO) is a common colorless and odorless pollutant resulting from incomplete combustion. One of the major sources of CO emission in the atmosphere is the gasoline-powered internal combustion engine. The chemical can be a fatal poison. It can be traced to many sources, including incomplete incineration, unvented gas appliances and heaters, malfunctioning heating systems, kerosene heaters, and underground or connected garages. Environmental tobacco smokes is another major source of CO. The gas ties up hemoglobin from binding oxygen and may cause asphyxiation. Fatigue, headache, and chest pain are the result of repeated exposure to low concentrations. Impaired vision and coordination, dizziness, confusion, and death may develop at the high concentration exposure levels [32,33].

Carbon Dioxide

Carbon dioxide (CO_2) is a colorless and odorless gas. It is an asphyxiant-causing agent. A concentration of 10% can cause unconsciousness and death from oxygen deficiency. The gas can be released from industrial studies [39], automobile exhaust, environmental tobacco smoke (ETS), and inadequately vented fuel heating systems. It is heavy and accumulates at low levels in depressions and along the floor.

Nitrogen Oxides

Nitrogen oxides, which are mainly released from industrial stacks, include nitrous oxide (N_2O), nitric oxide (NO), nitrogen dioxide (NO_2), nitrogen trioxide (N_2O_3), nitrogen tetraoxide (N_2O_4), nitrogen pentoxide (N_2O_5), nitric acid (HNO_5), and nitrous acid (HNO_2). Nitrogen dioxide is the most significant pollutant. The nature of the combustive process varies with the concentration of nitrogen oxides. Inhalation of nitrogen oxides may cause irritation of the eyes and mucous membranes. Prolonged low-level exposure may stain skin and teeth yellowish and brownish. Chronic exposure may cause respiratory dysfunction. Nitrogen oxides partially cause acid rains.

Sulfur Dioxide

Sulfur dioxide (SO_2) is a colorless gas with a strong odor and is the major substance causing acid rains. The major emission source of the gas is fuel or rubber tire combustion from industry [40]. Excess exposure may occur in industrial processes such as ore smelting, coal and fuel oil combustion, paper manufacturing, and petroleum refining. The chemical has not been identified as a carcinogen or co-carcinogen by the data, but short-term acute exposures to a high concentration of sulfur dioxide suggest adverse effects on pulmonary function [33].

Ozone

Ozone (O_3) is a powerful oxidizing agent. It is found naturally in the atmosphere by the action of electrical storms. The major indoor source of ozone is from electrical equipment and electrostatic air cleaners. The indoor ozone concentration is determined by ventilation. It depends on the room volume, the number of air changes in the room, room temperature, materials, and the nature of surfaces in the room. Ozone is irritating to the eyes and all mucous membranes. Pulmonary edema may occur after exposure has ceased [32,33].

Radon

Radon is a naturally occurring radioactive decay product of uranium. A great deal of attention centers around radon^{222}, which is the first decay product of radium^{228}. Radon and radon daughters have been found to contribute to lung cancer; USEPA estimates that radon may cause 5000 to 20,000 lung cancer deaths per year in the United States. The released energy from radon decay may damage lung tissue and lead to lung cancer. Smokers also may have a higher risk of developing lung cancer induced by radon.

Radon is present in the air and soil. It can leak into the indoor environment through dirt floors, cracks in walls and floors, drains, joints, and water seeping through walls. Radon can be measured by using charcoal containers, alpha-track detectors, and electronic monitors. Results of the measurement of radon decay products and the concentration of radon gas are reported as "working levels (WL)" and "picocuries per liter" (pCi/L), respectively. The continuous exposure level of 4 pCi/L or 0.02 WL has been used by USEPA and CDC as a guidance level for further testing and remedial action [33].

Once identified, the risk of radon can be minimized through engineering controls and practical living methods. The treatment techniques include sealing cracks and other openings in basement floors, and installation of sub-slab ventilation. Crawl spaces should also be well ventilated. Radon-contaminated groundwater can be treated by aerating [41–43] or filtering through granulated activated carbon [43,44].

Asbestos

Asbestos is a naturally occurring mineral and was widely used as an insulation material in building construction [35]. Asbestos possesses a number of good physical characteristics that make it useful as thermal insulation and fire-retardant material. It is electrically nonconductive, durable, chemical resistant, and sound absorbent. However, lung cancer and mesothelioma have been found to be associated with environmental asbestos exposure. USEPA has listed asbestos as a hazardous air pollutant since 1971. The major route of exposure is the respiratory system. Adverse health effects include asbestosis, lung cancer, mesothelioma, and other diseases. The latency period for asbestos diseases varies from 10 to 30 years [33].

Formaldehyde

Formaldehyde (HCHO) is a colorless gas with a pungent odor. Formaldehyde has found wide industrial usage as a fungicide and germicide, and in disinfectants and embalming fluids. The serious sources of indoor airborne formaldehyde are furniture, floor underlayment insulation, and environmental tobacco smoke. Urea formaldehyde (UF) is mixed with adhesives to bond veneers, particles, and fibers. It has been identified as a potential hazardous source.

Formaldehyde gas may cause severe irritation to the mucous membranes of the respiratory tract and eyes. Repeated exposure to formaldehyde may cause dermatitis either from irritation or allergy. The gas can be removed from the air by an absorptive filter of potassium permanganate-impregnated alumina pellets or fumigation using ammonia. Exposure to formaldehyde may be reduced by using exterior grade pressed wood products that contain phenol resins. Maintaining moderate temperature and low humidity can reduce emissions from formaldehyde-containing material. The chemical is intensely irritating to mucous membranes of the upper respiratory tract, the eyes, and skin. Repeated exposure may cause dermatitis and skin sensitization. This substance has been listed as a carcinogen.

Pesticides

Pesticides are used to kill household insets, rats, cockroaches, and other pests. Pesticides can be classified based on their chemical nature or use as organophosphates, carbonates, chlorinated hydrocarbons, bipyridyls, coumarins and indandiones, redenticides, fungicides, herbicides, fumigants, and miscellaneous insecticides. The common adverse effects are irritation of the skin, eyes, and upper respiratory tract. Prolonged exposure to some chemicals may cause damage to the central nervous system and kidneys [32,33].

Volatile Organic Compounds

The sources of volatile organic compounds (VOCs) include building materials, maintenance materials, building inhabitants, and gasoline spills/leaks. Building materials include carpet adhesives and wool finishes. Maintenance materials include varnishes, paints, polishes, and cleaners. Volatile organic compounds may pose problems for mucous surfaces in the nose, eyes, and throat. Chemicals that have been recognized as a cancer-causing agent include, at least, perchloroethylene used in dry cleaning, chloroform from laboratories, gasoline from gas stations, etc. [33,42].

Lead

Lead has been widely used in the storage battery industry, the petroleum industry, pigment manufacturing, insecticide production, the ceramics industry, and the metal products industry. Most of the airborne lead that has been identified comes from combustion of gasoline [33,79] and removal of lead paint [34].

Respirable Particles

Respirable particles are 10 or less micrometers in aerodynamic diameter. The sources of respirable particles include kerosene heaters, paint pigments, insecticide dusts, radon, and asbestos. The particles may irritate the eyes, nose, and throat and may contribute to respiratory infections, bronchitis, and lung cancer.

Tobacco Smoke

Environmental tobacco smoke (ETS) is a major indoor pollutant. Both the National Research Council (NRC) and USEPA have indicated that passive smoking significantly increases the risk of lung cancer in adults and respiratory illness in children. It is composed of irritating gases and carcinogenic tar particles. Nonsmokers breathing ETS are called "involuntary smokers," "passive smokers," or "second-hand smokers." There are more than 4700 chemical compounds in cigarette combustion products, such as carbon monoxide, carcinogenic/tars, hydrogen cyanide, formaldehyde, and arsenic. Of the chemicals, 43 have been recognized as carcinogens.

Environmental tobacco smoke (ETS) is a suspected source of many pollutants causing impaired health. A plant manager should either ban indoor smoking, or assign smoking areas at an industrial site. The most common impact in children from ETS is the development of wheezing, coughing, and sputum. According to 1986 reports by NRC, the risk of lung cancer is about 30% higher for nonsmoking spouses of smokers than for nonsmoking spouses of nonsmokers. Some studies also showed that ETS has been associated with an increased risk of heart disease [33].

PCB (Polychlorinated Biphenyl)

Polychlorinated biphenyls (PCBs) are a family of compounds that were used extensively in electrical equipment, such as transformers, because of their insulating and heat transferring qualities. They are suspected human carcinogens and have been linked to liver, kidney, and other health problems. It is known that PCBs can be transported by air, and this is thought to be one of the major ways in which they circulate around the world, explaining why they are found in the Arctic and Antarctic. Indian women dwelling on Cornwall Island located in the Canadian portion of the reservation have elevated levels of toxic PCBs in their breast milk. The PCB contamination does not appear to come from fish, but from air the women breathe every day [45].

Chlorofluorocarbon (CFC) and Freon

Freon is a commercial trademark for a series of fluorocarbon products used in refrigeration and air-conditioning equipment, as aerosol propellants, blowing agents, fire extinguishing agents, and cleaning fluids and solvents. Many types contain chlorine as well as fluorine, and should be called chlorofluorocarbons (CFCs) [85,86].

According to USEPA, roughly 28% of the ozone depletion attributed to chlorofluorocarbon (CFC) is caused by coolants in refrigerators and mobile air-conditioners. This being the case, it is necessary to analyze such issues as the refrigerants themselves used in air-conditioners, the types of air-conditioning resulting in CFC emissions, and the environmental fate, human toxicity, and legislation applying to these refrigerants.

The two most common CFC refrigerants in use today for air-conditioning purposes are Refrigerant 12 (CCl_2F_2) and Refrigerant 22 ($CHClF_2$). Refrigerant 12 was the first fluorocarbon-type refrigerant developed and used commercially. Its high desirability in air-conditioning applications arises from its extremely low human toxicity, good solubility, lack of effect on elastomers and other plastics, and reasonable compression ratio. Refrigerant 22, another commonly used air-conditioning coolant, although much safer to stratospheric ozone (because of the hydrogen molecule contained), tends to enlarge elastomers and weaken them, thus causing leakage wherever there is a rubber seal [46]. Of the CFC-12 used for refrigeration in the United States, 41% is used by vehicle air-conditioners. However, because vehicle air-conditioners are particularly prone to leaks and need frequent replacements of refrigerant, they use 75% of the country's replacement CFC-12.

The acute health effects of Refrigerant 12 are (a) irritation of mouth, nose, and throat; (b) irregular heart beat; and (c) dizziness and light headiness. Chronic health effects are not known at this time. The acute health effects of Refrigerant 22 are (a) heart palpitations; (b) tightness in the chest; (c) difficulty in breathing. Chronic health effects include irregular heat rhythms and skipped beats, and possible damage to the liver, kidneys, and blood.

Dioxins

Dioxins form a family of aromatic compounds known chemically as di-benzo-p-dioxins. Each of these compounds has a nucleus triple ring structure consisting of two benzene rings interconnected to each other through a pair of oxygen atoms. Dioxin compound generally exists as colorless crystalline solid at room temperatures, and is only slightly soluble in water and most organic liquids. They are usually formed through combustion processes involving precursor compounds. Once formed, the dioxin molecule is quite stable.

Dioxins are not decomposed by heat or oxidation in a 700°C incinerator, but pure compounds are largely decomposed at 800°C. Chlorinated dioxins lose chlorine atoms on exposure to sunlight and to some types of gamma radiation, but the basic dioxin structure is largely unaffected. The biological degradation rate of chlorinated dioxins is slow, although measured rates differ widely.

Incineration has been well organized as one of the best demonstrated and available technologies for waste destruction by direct heat, thus the volume and toxicity of the remaining residuals can be reduced.

Most interest has been directed toward the isomer 2,3,7,8-TCDD, which is among the most toxic compounds known. Experimental animals are exceedingly sensitive to TCDD. The LD_{50}, the dose that kills half of a test group, for 2,3,7,8-TCDD is 0.6 μ/kg of body weight for male guinea pigs. Humans exhibit symptoms effecting on enzyme and nervous systems, and muscle and joint pains [46].

Dioxin can enter a person through (a) dermal contact, absorption through skin; (b) inhalation, breathing of contaminated air; and (c) ingestion, eating contaminated materials such as soil, food, or drinking water contaminated by dioxin. In assessing these three routes, control of the physical and chemical properties of TCDD in the environment are containment, capping, and monitoring.

Under existing USEPA regulations, dioxin-bearing wastes may be stored in tanks, placed in surface impoundments and waste piles, and placed in landfills. However, in addition to meeting the Resource Conservation and Recovery Act (RCRA) requirements for these storage and disposal processes, the operators of these processes must operate in accordance with a management plan for those wastes that is approved by USEPA. Factors to be considered include: (a) volume, physical, and chemical characteristics of wastes, including their potential to migrate through soil or to volatilize or escape into the atmosphere; (b) the alternative properties of underlying and surrounding soils or other materials; (c) the mobilizing properties of other materials codisposed with these wastes; and (d) the effectiveness of additional treatment, design, or monitoring techniques.

Additional design, operating, and monitoring requirements may be necessary for facilities managing dioxin wastes in order to reduce the possibility of migration of these wastes to groundwater, surface water, or air so as to protect human health and the environment.

11.3.3 Air Emission Control

Air emission control technologies reduce levels of particulate emission and/or gaseous emission. Some air emission control equipment, such as dry injection units, fabric filters, cyclones, and electrostatic precipitators, are mainly designed to control particulate emissions. Others, such as dry scrubbers, thermal oxidizers, granular activated carbon, adsorption filters, and coalescing filters, control mainly gaseous pollutants including oily vapor. Air emission control equipment such as wet scrubbers and cartridge filters can control both particulate and gaseous emissions. Any gaseous effluent discharge at an industrial site that handles hazardous substances will normally require a discharge permit from one or more regular agencies.

For indoor air quality control, in addition to the air emission control technologies identified above, ventilation and air conditioning are frequently adopted by plant managers [36,85,86].

11.4 MANAGEMENT OF WATER QUALITY AT INDUSTRIAL SITES

11.4.1 Waterborne Contaminants and Health Effects

All point source and nonpoint source wastewaters at an industrial site must be properly managed for source separation, waste minimization, volume reduction, collection, pretreatment, and/or complete end-of-pipe treatment [39,47]. When industrial waste is not disposed of properly, hazardous substances may contaminate a nearby surface water (river, lake, sea, or ocean) and/or groundwater. Any hazardous substance release, either intentionally or unintentionally, increases the risk of water supply contamination and human disease. Major waterborne contaminants and their health effects are listed below.

Arsenic (As)

Arsenic occurs naturally and is also used in insecticides. It is found in tobacco, shellfish, drinking water, and in the air in some locations. The standard allows for 0.05 mg of arsenic per liter of water. If persons drink water that continuously exceeds the standard by a substantial amount over a lifetime, they may suffer from fatigue and loss of energy. Extremely high levels can cause poisoning.

Barium

Although not as widespread as arsenic, barium also occurs naturally in the environment in some areas. It can also enter water supplies through hazardous industrial waste discharges or releases. Small doses of barium are not harmful. However, it is quite dangerous when consumed in large quantities. The maximum amount of barium allowed in drinking water by the standard is 1.0 mg/L of water.

Cadmium

Only minute amounts of cadmium are found in natural waters in the United States. Hazardous waste discharges from the electroplating, photography, insecticide, and metallurgy industries can increase cadmium levels. Another common source of cadmium in drinking water is from galvanized pipes and fixtures if the pH of a water supply is not properly controlled. The sources of cadmium exposure are the foods we eat and cigarette smoking. The maximum amount of cadmium allowed in drinking water by the standard is 0.01 mg/L of water.

Chromium

Chromium is commonly released to the environment from the electroplating industry and is extremely hazardous. Some studies suggest that in minute amounts, chromium may be essential to human beings, but this has not been proven. The standard for chromium is 0.05 mg/L of water [76].

Lead

Lead sources include lead and galvanized pipes, auto exhausts, and hazardous waste releases. The maximum amount of lead permitted in drinking water by the standards is 0.05 mg/L of water. Excessive amounts well above this standard may result in nervous system disorders or brain or kidney damage [69].

Mercury

Large increases in mercury levels in water can be caused by industrial and agricultural use and waste releases. The health risk from mercury is greater from mercury in fish than simply from water-borne mercury. Mercury poisoning may be acute, in large doses, or chronic, from lower doses taken over an extended time period. The maximum amount of mercury allowed in drinking water by the standard is 0.002 mg/L of water. That level is 13% of the total allowable daily dietary intake of mercury.

Selenium

Selenium is found in meat and other foods due to water pollution. Although it is believed to be essential in the diet, there are indications that excessive amounts of selenium may be toxic. Studies are under way to determine the amount required for good nutrition and the amount that may be harmful. The standard for selenium is 0.01 mg/L of water. If selenium came only from drinking water, it would take an amount many times greater than the standard to produce any ill effects.

Silver

Silver is some times released to the environment by the photographic industry, and is considered to be toxic at high concentration. Because of the evidence that silver, once absorbed, is held indefinitely in tissues, particularly the skin, without evident loss through usual channels of elimination or reduction by transmigration to other body sites, and because of other factors, the maximum amount of silver allowed in drinking water by the standard is 0.05 mg/L of water.

Fluoride

High levels of fluoride in drinking water can cause brown spots on the teeth, or mottling, in children up to 12 years of age. Adults can tolerate ten times more than children. In the proper amounts, however, fluoride in drinking water prevents cavities during formative years. This is why many communities add fluoride in controlled amounts to their water supply. The maximum amount of fluoride allowed in drinking water by the standard ranges from 0.4 to 2.4 mg/L depending on one average maximum daily air temperature. The hotter the climate, the lower the amount allowed, for people tend to drink more in hot climates. In this hot area, the maximum contaminant level for fluoride is 2.0 mg/L of water.

Nitrate

Nitrate in drinking water above the standard poses an immediate threat to children under three months of age. In some infants, excessive levels of nitrate have been known to react with the hemoglobin in the blood to produce an anemic condition commonly known as "blue baby." If the drinking water contains an excessive amount of nitrate, it should not be given to infants under three months of age and should not to be used to prepare formula. The standard allows for 10.0 mg of nitrate (as N) per liter of water. Nitrate can be removed from water by ion exchange, RO, or distillation [48].

Pesticides

Millions of pounds (1 lb = 0.454 k) of pesticides are used on croplands, forests, lawns, and gardens in the United States each year. A large quantity of hazardous pesticides is also released by the pesticide industry to the environment. These hazardous pesticides drain off into surface waters or seep into underground water supplies. Many pesticides pose health problems if they get into drinking water and the water is not properly treated. The maximum limits for pesticides in drinking water are: (a) endrin, 0.0002 mg/L; (b) lindane, 0.004 mg/L; (c) methoxychlor, 0.1 mg/L; (d) toxaphene, 0.005 mg/L; (e) 2,4-D, 0.1 mg/L; and (f) 2,4,5-TP silvex, 0.01 mg/L.

Priority Pollutants

Many toxic organic substances, known as the USEPA priority pollutants, are cancer-causing substances and, in turn, are hazardous substances. Both the U.S. Drinking Water Standards and the Massachusetts Drinking Water Standards give maximum contaminant levels (MCL) for benzene, carbon tetrachloride, p-dichlorobenzene, 1,2-dichloroethane, 1,2-dichloroethylene, 1,1,1-trichloroethane, trichloroethylene (TEC), vinyl chloride, and total trihalomethanes (TTHM) in drinking water. In Massachusetts, monitoring for 51 unregulated VOCs is also required. In addition, the State of Massachusetts has announced the Massachusetts Drinking Water Guidelines, giving the lowest practical quantization limit (PQL) for 40 contaminants that have no regulated MCLs, but are evaluated on a case-by-case, on-going basis. More toxic priority pollutants may be incorporated into this list for enforcement by the State. Plant managers and consulting engineers should contact the home state for specific state regulations.

Microorganisms

Pathogenic microorganisms from the biotechnology industry, agricultural industry, hospitals, and so on may cause waterborne diseases, such as typhoid, cholera, infectious hepatitis, dysentery, etc. Coliform bacteria regulated by both the Federal and the State governments are only an indicator showing whether or not the water has been properly disinfected. For a disinfected water, a zero count on coliform bacteria indicates that the water is properly disinfected, and other microorganisms are assumed to be sterilized.

Radionuclides

Gross alpha particle activity, gross beta particle activity, and total radium 226 and 228 are found from radioactive wastes, uranium deposits, and certain geological formations, and are a cancer-causing energy. The MCLs for gross alpha particle activity, gross beta particle activity, and total radium 226 and 228 are set by the USEPA at 15 pCi/L, 4 mrem/year, and 5 pCi/L, respectively. Again the Massachusetts Drinking Water Guidelines are more stringent, and include additional photon activity, tritium, strontium-90, radon-222, and uranium for State enforcement. Radon in groundwater can be effectively removed by granular activated carbon [44]. In a recent decision having potentially broad implications, a U.S. federal Court of Appeals has upheld USEPA regulations establishing standards for radionuclides in public water systems [49].

PCBs, CFCs, and Dioxin

Polychlorinated biphenyls (PCBs), CFCs, petroleum products, and dioxin are major toxic contaminants in air (Section 11.3.2), soil (Section 11.5.3), and also in water. The readers are referred to Sections 11.3.2 and 11.5.3 for details about PCB characteristics, health effects, treatment technologies, and so on. For water quality management, they have been included in the list of the USEPA priority pollutants [86].

Asbestos

Asbestos is an airborne contaminant (Section 11.3.2), a hazardous solid waste (Section 11.5), and also a waterborne contaminant, regulated by many states. The health effect of asbestos in water, however, is not totally known.

11.4.2 Water Pollution Prevention and Control

Depending on the State where the industrial plant is located, an aqueous effluent from a pretreatment facility or a complete end-of-pipe treatment facility can be discharged into a river, a lake, or an ocean, only if it meets the pretreatment standards and the effluent discharge standards established by the regulatory agencies, in accordance with the National Pollutant Discharge Elimination System (NPDES) or the State Pollutant Discharge Elimination System (SPDES). The standards can be industry-specific, chemical-specific, or site-specific, or all three. The readers are referred to other chapters of this handbook series for the details.

The plant manager of an industrial site having hazardous substances must establish an in-plant hazardous substance management program to ensure that the plant's hazardous substances will not be released by accident, or by neglect, to the plant's soil and groundwater.

Once a groundwater or a surface water is contaminated, the cleanup cost is very high. In general, a contaminated groundwater or surface water must be decontaminated to meet the Federal and the State drinking water standards and the State Guidelines if the groundwater or surface water source is also a potable water supply source. Even if a receiving water (either a surface water or a groundwater) is not intended to be used as a water supply source, the cleanup cost and the loss of revenue can be as high as hundreds of millions of dollars. Pollution prevention before contamination occurs is always better and more economical than pollution control after contamination occurs.

11.4.3 A Case History of Water Pollution by PCB Release

Polychlorinated biphenyls (PCBs) are colorless toxic organic substances that cause cancer and birth defects. There are more than 200 different types of PCBs, ranging in consistency from heavy, oily liquids to waxy solids, and each type further varying in the number and location of chlorine atoms attached to its molecular carbon rings. They are fire resistant and do not conduct heat or electricity well. Accordingly they have numerous commercial applications as insulation in electrical systems, for example, for transformers.

Owing to a lack of environmental knowledge and governmental guidance, General Electric Company released about 500,000 lb of hazardous PCBs into Hudson River in New York State between 1947 and 1976 from its plants in Fort Edward and Hudson Falls. Hudson River is one of North America's great mountain streams, cruising through gorges, crashing over boulders, churning into a white-water delight, and eventually reaching the great Atlantic Ocean. For centuries, the great Hudson has been a reliable water resource for navigation, fishing, boating, swimming, winter sports, water supply, and natural purification. Around Glens Falls, the Hudson runs into civilization, into industry, and, in turn, into an industrial disaster: the pollution of more than 185 miles of the river with over half a million pounds of hazardous and poisonous PCBs.

In 1977, PCB production was banned in the United States, and its release to the Hudson was stopped. Since 1976, the State of New York has banned all fishing on the river between Bakers Fall in the Village of Hudson Fall and the Federal dam at Troy. Most affected has been the commercial striped bass fishery, which once earned New Yorkers $40 million a year. Now the river is no longer suitable for swimming or any water contact sports, and of course, definitely not suitable for domestic water supply. The loss of its recreation and water supply revenues is simply too high to be priced. In 1983, the USEPA declared the Hudson River, from Hudson Falls to New York City, one of the Nation's largest and most complicated Superfund toxic-waste sites.

Now the New York State Department of Environmental Conservation and some environmental groups have advocated dredging the PCB-contaminated river bottom and transferring the PCB-containing sediment to a landfill site. Even though the cleanup costs, now estimated to run as high as $300 million U.S. dollars, are acceptable to U.S. tax payers, a landfill site to receive the PCB-contaminated sediment still cannot be found because of public resistance [50].

This is a typical environmental disaster that the industry must not forget and must not repeat. For more information on PCB pollution and management, the readers are referred to the literature [46,51].

11.5 MANAGEMENT OF HAZARDOUS SOLID WASTES AT INDUSTRIAL SITES

11.5.1 Disposal of a Large Quantity of Hazardous Solid Wastes

When disposed of improperly, hazardous solid wastes may contaminate air, soil, and/or groundwater, and increase the risk of human disease and environmental contamination.

Inevitably, some hazardous solid wastes generated at an industrial site must be discarded. Rusted, old containers or equipment might be targets for plantwide cleaning. Some industrial materials or products, such as half-used cans of paint or chemical might be discarded. Or the owner or plant manager might want to dispose of some products that are too old to be sold, or some building material (such as asbestos) that is too hazardous for everyday use.

A large quantity of any hazardous solid wastes can only be properly transported or disposed of by licenced or certified environmental professionals. Small quantity of hazardous wastes, however, can be handled by a plant manager.

11.5.2 Disposal of a Small Quantity of Hazardous Solid Wastes

Right now there is no easy way to dispose of very small quantity of hazardous household products, such as pesticides, batteries, outdated medicines, paint, paint removals, used motor oil, wool preservatives, acids, caustics, and so on. There are no places that accept such small quantities of wastes as generated by a small industrial/commercial site. For now, the best disposal techniques are listed in Table 1, which is recommended by the Massachusetts Department of Environmental Management, Bureau of Solid Waste Disposal.

Table 1 Methods for Disposal of Small Quantities of Common Hazardous Wastes

Product	Take to a hazardous waste collection site (or store until available)	Wrap in plastic bag, put in trash, and alert the collector	Wash down drain with lots of water	Take to a special recycling center (not paper recycling)	Give to a friend to use, with careful instructions	Return to the manufacturer or to the retailer
Acids (strong)	Best	Never	Never	Unavailable	Impractical	Impractical
Acids (weak)	Best	4th best	3rd best	Unavailable	2nd best	Impractical
Banned pesticides	2nd best	Never	Never	Never	Never	Best
Batteries	3rd best	Never	Impractical	Best	Never	2nd best
Caustics	Best	3rd best	4th best	Unavailable	2nd best	Impractical
Pesticide containers	Best	2nd best	Impractical	Unavailable	Impractical	Impractical
Flammables	Best	3rd best	Never	Unavailable	2nd best	Impractical
Outdated medicines	Best	3rd best	2nd best	Never	Never	Impractical
Paint	2nd best	3rd best	Never	Unavailable	Best	Impractical
Paint remover	Best	Never	Never	Unavailable	2nd best	Impractical
Pesticides	Best	3rd best	Never	Unavailable	2nd best	Impractical
Used motor oil	3rd best	Never	Never	Best	Never	2nd best
Wood preservatives	Best	2nd best	Never	Unavailable	3rd best	Impractical

Note: Strong acids include battery acid, murintic acid, and hydrochloric acid. Weak acids include acetic acid, toilet bowl cleaner, and lactic acid. Banned pesticides include Silvex, Mirex, Aldrin, Chlordane, DDT, and Heptachlor. Caustics include oven cleaner and drain cleaner. Flammables include alcohol, acetone, turpentine, lacquer, and paint thinner. Pesticides include rodent poisons, insecticides, weed killer, and other herbicides and fungicides. Pesticide containers should be triple-rinsed, and the contents sprayed on crops or yard, before discarding.

Small quantities of hazardous solid wastes (such as potassium dichromate, lead nitrate, silver nitrate, asbestos, etc.), liquid chemicals (such as chloroform, PCB, methylene chloride, etc.), petrochemicals (such as gasoline, No. 2 fuel oil, etc.), or pure metals (such as mercury, sodium, etc.), which are stored in bottles or cans, however, are not considered to be hazardous "household products." Accordingly these nonhousehold hazardous solid wastes, even in small quantities, can only be properly disposed of by licenced or certified environmental professionals.

11.5.3 Hazardous and Infectious Solid Wastes

A few selected hazardous solid wastes, and hazardous liquid wastes stored in drums/tanks, are described below for reference.

Infectious and Hazardous Medical Wastes

In a 1987 *Federal Register* notice, USEPA first defined the three waste categories (pathological waste, laboratory waste, isolation waste) below, which should be treated as infectious:

1. Pathological waste: Surgical or operating room specimens (like body parts) and other potentially contaminated waste from outpatient areas and emergency rooms.
2. Laboratory waste: Pathological specimens (all tissues, blood specimens, excreta, and secretions obtained from patients or laboratory animals) and other potentially contaminated wastes.
3. Isolation waste: Disposable equipment and utensils (like syringes and swabbing) from rooms of patients suspected to have a communicable disease.
4. General hospital waste: Cafeteria garbage, disposal gowns, drapes, packaging, etc., representing about 85% of total hospital waste.
5. Hazardous waste: Dental clinics, chemotherapy wastes (some) listed as hazardous by USEPA, and low-level radioactive waste.

Incineration has been common practise in hospitals for decades. It is quick, easy, and especially handy for rendering the more repulsive wastes unrecognizable. It also reduces waste volume by up to 90%, leaving mostly ashes behind, for landfilling. Because of their comparatively small size, hospital incinerators have until recently been exempted from federal rules that control air emissions of larger incinerators, like mass-burn facilities. According to the November 1987 USEPA report, there were 6200 hospital incinerators around the United States. Only 1200 are "controlled-air" incinerators, a relatively new design that limits the air in the burn chamber, ensuring more complete incineration. However, even the 1200 controlled-air models do not necessarily have stacks equipped with scrubbers to prevent acid gas and dioxin emissions [46,52].

In many states, regulations only require that hospital incinerators not create a public nuisance usually recognized as odors and smoke opacity. Disposal costs for these medical wastes are becoming stiffer, just as surely as they are for infectious and other hazardous/toxic wastes. This adds another incentive to incinerate. It may be possible that a good deal of hospital waste could be separated, reduced, and recycled. While infectious waste is obviously not recyclable, the amount of waste designated infectious can be greatly reduced by separating materials to avoid excess contamination [74].

Health officials are increasingly concerned about disposal of infectious, radioactive, and toxic medical wastes that have become major components in the treatment and diagnosis of many diseases. Legal complications in handling medical wastes are another issue. There are, for example, no federal regulations for disposal of medical waste. State and local regulations are widely divergent.

Petroleum Contaminated Soil

Petroleum (crude oil) is a highly complex mixture of paraffinic, cycloparaffinic (naphthenic), and aromatic hydrocarbons, containing low percentages of sulfur and trace amounts of nitrogen and oxygen compounds. The most important petroleum fractions, obtained by cracking or distillation, are various hydrocarbon gases (butane, ethane, propane), naphtha of several grades, gasoline, kerosene, fuel oils, gas oil, lubricating oils, paraffin wax, and asphalt. From the hydrocarbon gases, ethylene, butylene, and propylene are obtained. About 5% of the petroleum (crude oil) consumed in the United States is used as feedstocks by the chemical industries. The rest is consumed for production of various products, such as gasoline, fuel oils, and so on, introduced above. The crude oil, when spilled or leaked, will contaminate the soil because it is flammable, and moderately toxic by ingestion. One of the major components of petroleum product is benzene, which is a known human carcinogen.

Gasoline, fuel oils, and lubricating oils are three major pollutants among the petroleum family members, and are therefore introduced in more detail.

Gasoline is a mixture of volatile hydrocarbons suitable for use in a spark-ignited internal combustion engine and having an octane number of at least 60. The major components are branched-chain paraffins, cycloparaffins, and aromatics. The present source of gasoline is petroleum, but it may also be produced from shale oil and Athabasca tar sands, as well as by hydrogenation or gasification of coal. There are many different kinds of gasolines:

- Antiknock gasoline: a gasoline to which a low percentage of tetra-ethyl-lead, or similar compound, has been added to increase octane number and eliminate knocking. Such gasolines have an octane number of 100 or more and are now used chiefly as aviation fuel.
- Casinghead gasoline: see natural gasoline (below).
- Cracked gasoline: gasolines produced by the catalytic decomposition of high-boiling components of petroleum, and having higher octane ratings (80–100) than gasoline produced by fractional distillation. The difference is due to the prevalence of unsaturated, aromatic, and branched-chain hydrocarbons in the cracked gasoline.
- High-octane gasoline: a gasoline with an octane number of about 100.
- Lead-free gasoline: an automotive fuel containing no more than 0.05 g of lead per gallon, designed for use in engines equipped with catalytic converters.
- Natural gasoline: a gasoline obtained by recovering the butane, pentane, and hexane hydrocarbons present in small proportions in certain natural gases. Used in blending to produce a finished gasoline with adjusted volatility, but low octane number. Do not confuse with natural gas (q.v.).
- White gasoline: an unleaded gasoline especially designed for use in motorboats; it is uncracked and strongly inhibited against oxidation to avoid gum formation, and is usually not colored to distinguish it from other grades. It also serves as a fuel for camp lanterns and portable stoves.
- Polymer gasoline: a gasoline produced by polymerization of low-molecular-weight hydrocarbons such as ethylene, propane, and butanes. It is used in small amounts for blending with other gasoline to improve its octane number.
- Pyrolysis gasoline: gasoline produced by thermal cracking as a byproduct of ethylene manufacture. It is used as a source of benzene by the hydrodealkylation process.
- Reformed gasoline: a high-octane gasoline obtained from low-octane gasoline by heating the vapors to a high temperature or by passing the vapors through a suitable catalyst.
- Straight-run gasoline: gasoline produced from petroleum by distillation, without use of cracking or other chemical conversion processes. Its octane number is low.

Fuel oil is any liquid petroleum product that is burned in a furnace for the generation of heat, or used in an engine for the generation of power, except oils having a flash point below 100°F and oil burned in cotton or wool burners. The oil may be a distillated fraction of petroleum, a residuum from refinery operations, a crude petroleum, or a blend of two or more of these.

ASTM has developed specifications for six grades of fuel oil. No. 1 is a straight-run distillate, a little heavier than kerosene, used almost exclusively for domestic heating. No. 2 (diesel oil) is a straight-run or cracked distillate used as a general purpose domestic or commercial fuel in atomizing-type burners. No. 4 is made up of heavier straight-run or cracked distillates and is used in commercial or industrial burner installations not equipped with preheating facilities. The viscous residuum fuel oils, Nos. 5 and 6, sometimes referred to as bunker fuels, usually must be preheated before being burned. ASTM specifications list two grades of No. 5 oil, one of which is lighter and under some climatic conditions may be handled and burned without preheating. These fuels are used in furnaces and boilers of utility power plants, ships, locomotives, metallurgical operations, and industrial power plants.

Lubrication oil is a selected fraction of refined mineral oil used for lubrication of moving surfaces, usually metallic, and ranging from small precision machinery (watches) to the heaviest equipment. Lubricating oils usually have small amounts of additives to impart special properties such as viscosity index and detergency. They range in consistency from thin liquids to greaselike substances. In contract to lubricating greases, lube oils do not contain solid or fibrous minerals.

The major petroleum release sources are bulk gasoline terminals, bulk gasoline plants, service stations, and delivery tank trucks. USEPA estimates there are approximately 1500 bulk terminals, 15,000 bulk plants, and 390,000 gasoline service stations in the United States, of which some 180,000 are retail outlets [46]. Fuel oil release is mainly caused by underground storage tank leakage. Lubricating oil release, however, is mainly caused by neglect or intentional dump.

Release of gasoline, lubricating oil, and fuel oils to the soil occurs from spills, leaks, loading and unloading operations. Disposal of petroleum-contaminated soil is now one of the major environmental tasks.

Dioxin

Dioxin (2,3,7,8-tetrachlorodibenzo-p-dioxin; TCDD) is among the most toxic compounds known today. It is an airborne contaminant from an incineration process, which has been described in Section 11.3.2. Dioxin also frequently occurs as an impurity in the herbicide 2,4,5-T. Accordingly, when the herbicide 2,4,5-T is applied to crops, dioxin is also released to the soil. Any spills of dioxin also cause soil contamination. It may be removed by extraction with coconut-activated carbon. Its half-life in soil is about one year.

PCBs

Polychlorinated Diphenyl (PCB) is an airborne contaminant (Section 11.3.2), a waterborne contaminant (Section 11.4.1) and also a contaminant in soil due to PCB releases, such as spills, leakages, and landfills. Before the United States banned manufacture of PCBs in 1979, Monsanto had produced more than 1 billion pounds. Practises one thought acceptable and hazard-free in the past have led to PCB releases into the environment. Such practises were conducted by industries using PCBs in processes and products and discharging the PCB-containing waste into rivers and streams. Other PCB-containing waste was disposed of in landfills. When used in transformers and electrical capacitors, PCB compartments are sealed and in place for the life of the equipment. Occasionally seals will leak or external structures are damaged, resulting in leakage. The following are applications in which PCBs have been found and hence are potential sources: (a) cooling and insulating fluids for transformers; (b) dielectric impregnating for capacitors; (c) flame retardants for resins and plastics in the electrical industry; (d) formulations in paints and printing inks; (e) water-repellent additives; (f) dye carrier for pressure-sensitive copy paper; (g) incombustible hydraulic fluids; and (h) dust control agents for road construction.

Other Organic and Inorganic Contaminants

In addition to gasoline, CFC, and so on, various other organic and inorganic compounds such as heavy metals, sulfides, and cyanides on the USEPA Priority Pollutants List, and subject to various water quality criteria, guidelines, etc., when released can also contaminate the soil. The contaminated soil then becomes a hazardous solid waste which must be properly disposed of [63–86].

11.5.4 Disposal of Hazardous and Infectious Wastes

Incineration has been used extensively in hospitals for disposal of hospital wastes containing infectious and/or hazardous substances. Most hospital incinerators (over 80%), however, are outdated or poorly designed. Modern incineration technology, however, is available for complete destruction of organic hazardous and infectious wastes. In addition, adequate air pollution control facilities, such as scrubbers, secondary combustion chambers, stacks, and so on, are needed to prevent acid gas, dioxin, and metals from being discharged from the incinerators.

The same modern incinerators equipped with scrubbers, bag-filters, electro-precipitators, secondary combustion chambers, stacks, etc., are equally efficient for disposal of hazardous PCBs, dioxin, USEPA priority pollutants, and so on, if they are properly designed, installed, and managed. Incineration technology is definitely feasible, and should not be overlooked. The only residues left in the incinerators are small amount of ashes containing metals. The metal-containing ashes may be solidified and then disposed of on a landfill site.

Environmentalists and ecologists, however, oppose construction of any new incinerators and landfill facilities. They would like to close all existing incineration and landfill facilities, if possible. They are wrong. Unless human civilization is to go backward, there will always be hazardous and infectious wastes produced by industry. These wastes must go somewhere. A solution must be found.

It is suggested that waste minimization, spill prevention, leakage prevention, volume reduction, waste recycle, energy conversion, and conservation be practiced by the industry as well as the community. Innovative technology must be developed, and good managerial methods must be established for this practise. With all these improvements, modern incinerators and landfill facilities may still be needed, but their numbers and sizes will be significantly reduced.

Section 11.15 introduces a case history showing how an organic hazardous waste can be reused as a waste fuel in the cement industry. A cement plant is a manufacturing plant needed by our civilization. With special managerial arrangements and process modification, a cement kiln can be operated for production of cement as well as for incineration of hazardous waste. Because hazardous waste can replace up to 15% of fuel for this operation, the industry not only saves 15% of energy cost, but also solves a hazardous waste disposal problem. It should be noted that modern incineration and air purification technologies are still required. In this case the cement kiln acts like an incinerator. It is not necessary for the community or the waste-producing industry to build an incinerator solely for waste disposal.

Section 11.14 presents two case histories: (a) disposal of photographic wastes by a large quantity generator; and (b) disposal of photographic wastes by a small quantity generator. In general, it is economically feasible for a large quantity generator to pretreat its wastes, aiming at regulatory compliance. A small quantity generator with in-house engineering support may also pretreat its wastes, and discharge the pretreated effluent to a receiving water or a POTW. Without in-house engineering support, it would be more cost-effective for the small quantity generator to hire an outside engineering consultant and/or an outside general contractor for proper onsite storage of its hazardous/infectious wastes, subsequent transportation of its wastes by a licenced transporter, and final offsite disposal of its wastes by a licenced facility.

Section 11.13 presents an example showing how a medical office manages its hazardous wastes and what the regulatory requirements are.

Friable asbestos is hazardous, and should be properly disposed of following governmental requirements and guidelines presented in Section 11.6.

11.6 DISPOSAL OF HAZARDOUS ASBESTOS

11.6.1 Asbestos, Its Existence and Releases

The term "asbestos" describes six naturally occurring fibrous minerals found in certain types of rock formations. Of that general group, the minerals chrysolite, amosite, and crocidolite have been most commonly used in building products. Under the Clean Air Act of 1970, the USEPA has been regulating many asbestos-containing materials (ACM), which, by USEPA definition, are materials with more than 1% asbestos. "Friable asbestos" includes any materials that contain greater than 1% asbestos, and that can be crumbled, pulverized, or reduced to powder by hand pressure. This asbestos may also include previously nonfriable material that becomes broken or damaged by mechanical force. The Occupational Safety and Health Administration's (OSHA) asbestos construction standard in Section K, "Communication of Hazards to Employees," specifies labeling many materials containing 0.1% or more asbestos [20,22,53].

Asbestos became a popular commercial product because it is strong, will not burn, resists corrosion, and insulates well. When mined and processed, asbestos is typically separated into very thin fibers. When these fibers are present in the air, they are normally invisible to the naked eye. Asbestos fibers are commonly mixed during processing with material that binds them together so that they can be used in many different products. Because these fibers are so small and light, they remain in the air for many hours if they are released from ACM in a building. When fibers are released into the air they may be inhaled by people in the building.

In July 1989, USEPA promulgated the Asbestos Ban and Phase-down Rule. The rule applies to new product manufacture, importation, and processing, and essentially bans almost all asbestos-containing products in the United States by 1997. This rule does not require removal of ACM currently in place in buildings. In fact, undisturbed materials generally do not pose a health risk; they may become hazardous when damaged, disturbed, or deteriorate over time and release fibers into building air. Controlling fiber release from ACM in a building or removing it entirely is termed "asbestos abatement," aiming at mainly friable asbestos.

Asbestos has been mainly used as building construction materials for many years. Their applications and releases include the following situations.

Vinyl Floor Tiles and Vinyl Sheet Flooring

Asbestos has been added to some vinyl floor tiles to strengthen the product materials, and also to decorate the exposed surfaces. Asbestos is also present in the backing in some vinyl sheet flooring. The asbestos is often bound in the tiles and backing with vinyl or some type of binder. Asbestos fibers can be released if the tiles are sanded or seriously damaged, or if the backing on the sheet flooring is dry-scraped or sanded, or if the tiles are severely worn or cut to fit into place.

Pipe Insulation

Hot water and steam pipes in some older homes may be covered with an asbestos-containing material, primarily as thermal insulation to reduce heat loss, and to protect nearby surfaces from the hot pipes. Pipes may also be wrapped in an asbestos "blanket" or asbestos paper tape.

Asbestos-containing insulation has also been used on furnace ducts. Most asbestos pipe insulation in homes is preformed to fit around various diameter pipes. This type of asbestos-containing insulation was manufactured from 1920 to 1972. Renovation and home improvements may expose and disturb the asbestos-containing materials.

Wall and Ceiling Insulation

Buildings constructed between 1930 and 1950 may contain insulation made with asbestos. Wall and ceiling insulation that contains asbestos is generally found inside the wall or ceiling ("sandwiched" behind plaster walls). The asbestos is used as material for thermal insulation, acoustical insulation, and fire protection. Renovation and home improvements may expose and disturb the materials.

Appliances

Some appliances, such as toasters, popcorn poppers, broilers, dishwashers, refrigerators, ovens, ranges, clothes dryers, and electric blankets are, or have been, manufactured with asbestos-containing parts or components for thermal insulation. As a typical example, hair dryers with asbestos-containing heat shields were only recalled in 1979. Laboratory tests of most hair dryers showed that asbestos fibers were released during use.

Roofing, Shingles, and Siding

Some roofing shingles, siding shingles, and sheets have been manufactured with asbestos using Portland cement as a binding agent. The purposes for the addition of asbestos are strength enhancement, thermal insulation, acoustical insulation, and fire protection. Because these products are already in place and outdoors, there is likely to be little risk to human health. However, if the siding is worn or damaged, asbestos may be released.

Ceilings and Walls with Patching Compounds and Textured Paints

Some large buildings built or remodeled between 1978 and 1987 may contain a crumbly, asbestos-containing material that has been sprayed onto the ceiling or walls. Some wall and ceiling joints may be patched with asbestos-containing material manufactured before 1977. Some textured paint sold before 1978 contained asbestos. Sanding or cutting a surface with the building materials that may contain asbestos will release asbestos to the air, and thus should be avoided.

Stoves, Furnaces, and Door Gaskets

Asbestos-containing cement sheets, millboard, and paper have been used frequently in buildings when wood-burning stoves have been installed. These asbestos-containing materials were used as thermal insulation to protect the floor and walls around the stoves. On cement sheets, the label may tell the plant manager if they contains asbestos. The cement sheet material will probably not release asbestos fibers unless scraped. This sheet material may be coated with a high temperature paint, which will help seal any asbestos into the material. Asbestos paper or millboard were also used for this type of thermal insulation. If these materials were placed where they are subjected to wear, there is an increased possibility that asbestos fibers may be released. Damage or misuse of the insulating material by sanding, drilling, or sawing will also release asbestos fibers.

Oil, coal, or wood furnaces with asbestos-containing insulation and cement are generally found in some older buildings. Updating the system to oil or gas can result in removal or damage to the old insulation. If the insulation on or around the furnaces is in good condition, it is best to leave it alone. If the insulation is in poor condition, or pieces are breaking off, there will be an asbestos release.

Some door gaskets in furnaces, ovens, and wood and coal stoves may contain asbestos. The asbestos-containing door gaskets on wood and coal-burning stoves are subject to wear and can release asbestos fibers under normal use conditions. Handle the asbestos-containing material as little as possible.

11.6.2 Health Risk of Asbestos

Asbestos has been shown to cause cancer of the lung and stomach according to studies of workers and others exposed to asbestos. There is no level of exposure to asbestos fibers that experts can assume is completely safe.

Some asbestos materials can break into small fibers that can float in the air, and these fibers can be inhaled. These tiny fibers are small, cannot be seen, and can pass through the filters of normal vacuum cleaners and get back into the air. Once inhaled, asbestos fibers can become lodged in tissue for a long time. After many years, cancer or other sickness can develop. In order to be a health risk, asbestos fibers must be released from the material and be present in the air for people to breathe. A health risk exists only when asbestos fibers are released from the material or product. Soft, easily crumbled asbestos-containing material, previously defined as "friable asbestos," has the greatest potential for asbestos release and therefore has the greatest potential to create health risks.

Asbestos fibers, in particular in friable asbestos, can cause serious health problems. If inhaled, they can cause diseases that disrupt the normal functioning of the lungs. Three specific diseases – asbestoses (a fibrous scarring of lungs), lung cancer, and mesothelioma (a cancer of the lining of the chest or abdominal cavity) – have been linked to asbestos exposure. These diseases do not develop immediately after inhalation of asbestos fibers; it may be 20 years or more before symptoms appear. In general, as with cigarette smoking and the inhalation of tobacco smoke, the more asbestos fibers a person inhales, the greater the risk of developing an asbestos-related disease.

11.6.3 Identification of Asbestos

Plumbers, building contractors, or heating contractors are often able to make a reasonable judgment about whether or not a product contains asbestos, based on a visual inspection. In some cases, the plant manager may want to have the material analyzed. Such analysis may be desirable if the industrial plant has a large area of damaged material or if the plant manager is preparing a major renovation that will expose material contained behind a wall or other barrier.

A list of 221 laboratories receiving initial accreditation to perform bulk asbestos analysis during the second quarter of 1989 has been released by the National Institute of Standards and

Technology, Gaithersburg, MD. There are two types of air sampling techniques:

1. Personal air sampling (required by OSHA) is designed to measure an individual worker's exposure to fibers while the worker is conducting tasks that may disturb ACM. The sampling device is worn by the worker and positioned so that it samples air in the worker's breathing zone.
2. Area (or ambient) air sampling is conducted to get an estimate of the numbers of airborne asbestos fibers present in a building. It is used as an assessment tool in evaluating the potential hazard posed by asbestos to all building occupants.

11.6.4 Operation and Maintenance (O&M) Program

The principal objective of an O&M program is to minimize exposure of all building occupants to asbestos fibers. To accomplish this objective, an O&M program includes work practises to (a) maintain ACM in good condition, (b) ensure proper cleanup of asbestos fibers previously released, (c) prevent further release of asbestos fibers, and (d) monitor the condition of ACM.

The methods for monitoring/correcting the condition of ACM include: (a) "surfacing ACM" (asbestos-containing material that is sprayed on or otherwise applied to surfaces, such as acoustical plaster on ceilings and fireproofing materials on structural members, or other materials on surfaces for acoustical, fireproofing, or other purposes); (b) "thermal system insulation" (TSI) (asbestos-containing material applied to pipes, fittings, boiler, breaching, tanks, ducts, or other interior structural components to prevent heat loss or gain or water condensation); and (c) "miscellaneous ACM" (interior asbestos-containing building material on structural components, structural members or fixtures, such as floor and ceiling tiles; does not include surfacing material or thermal system insulation).

The O&M program can be divided into three types of projects: (a) those that are unlikely to involve any direct contact with ACM; (b) those that may cause accidental disturbance of ACM; and (c) those that involve relatively small disturbances of ACM.

First, a person who may be the plant manager, a principal member of staff, or an outside asbestos consultant should be installed as the Asbestos Program Manager in order to establish and implement an O&M program. The appointed Asbestos Program Manager shall have overall responsibility for the asbestos control program. He/she may develop and implement the O&M program, establish training and experience requirements for contractors' workers, supervise and enforce work practises with assistance of work crew supervisors, and conduct periodic reinspections and be responsible for record keeping. This Asbestos Program Manager should be properly trained in O&M program development and implementation. An asbestos contractor may be hired to provide services for ACM abatement and for building decontamination following a fiber release episode. In addition to the abovementioned Asbestos Program Manager, the plant manager, asbestos consultant, asbestos contractor, a communications person, a record-keeping person, a lawyer, and the federal, state, and local government advisors may also get involved in the O&M program. Secondly, a physical and visual inspection of the building is to be conducted and bulk samples of such materials are to be taken to determine if ACM is present. Then an ACM inventory can be established, and the ACM's condition and potential for disturbance can be assessed.

An official O&M program is to be developed based on the inspection and assessment data, as soon as possible if ACM is located. Either the Asbestos Program Manager or a qualified consultant should develop the O&M program. The written O&M program should state clearly the O&M policies and procedures for that building, identify and describe the administrative line of authority for that building, and should clearly define the responsibilities of key participants, such as the Asbestos Program Manager and custodial and maintenance supervisors and staff. The written O&M program should be available and understood by all participants involved in the management and operations of the building.

In general the O&M program developed for a particular building should include the following O&M program elements:

- Notification: a program to tell workers, tenants, and building occupants where ACM is located, and how and why to avoid disturbing the ACM. All persons affected should be properly informed.
- Surveillance: regular ACM surveillance to note, assess, and document any changes in the ACM's condition by trained workers or properly trained inspectors. Air monitoring to detect airborne asbestos fibers in the building may provide useful supplemental information when conducted along with a comprehensive visual and physical ACM inspection/reinspection program. Air samples are most accurately analyzed using transmission electron microscopy (TEM).
- Controls: work control/permit system to control activities that might disturb ACM. This system requires the person requesting work to submit a job request form to the Asbestos Program Manager before any work is begun.
- Work practices: O&M work practises to avoid or minimize fiber release during activities affecting ACM.
- Record keeping: to document O&M activities. OSHA and USEPA have specific requirement for workers exposed to asbestos.
- Worker protection: medical and respiratory protection programs, as applicable.
- Training: the Asbestos Program Manager, and custodial and maintenance staff training. The building owner should make sure that the O&M program developed is site-specific and tailored for the building. The O&M program should take into account use, function, and design characteristics of a particular building.

The O&M program once established shall be implemented and managed conscientiously and reviewed periodically. Alternatives on control options that may be implemented under an O&M program include: (a) repair, (b) encapsulation, (c) enclosure, (d) encasement, and (e) minor removal. The abatement actions other than O&M can also be selected when necessary. For instance, removal of ACM before renovations may be necessary in some instances.

11.6.5 O&M Training Program

Properly trained custodial and maintenance workers are critical to a successful A&M program. The following items are highlighted training requirements:

1. OSHA and USEPA require a worker training program for all employees exposed to fiber levels at or above the action level (0.1 f/cc, 30 min time-weighted average or TWA).
2. Some states and municipalities may have specific work training requirements.
3. At least three levels of maintenance worker training can be identified: (a) Level 1 Awareness training for workers involved in activities where ACM may be accidentally disturbed (may range from 2 to 8 hours); (b) Level 2 Special O&M training for maintenance workers involved in general maintenance and incidental ACM repair tasks (at least 16 hours); (c) Level 3 Abatement worker training for workers who may conduct asbestos abatement. This work involves direct, intentional contact with ACM. "Abatement worker" training courses that involve 24 to 32 hours of training fulfill this level of training.

11.6.6 General Guidelines for Handling Asbestos-Containing Materials

If the plant manager thinks that a material contains asbestos, and the material must be banned, rubbed, handled, or taken apart, he/she should hire a trained, asbestos-removal contractor before taking any risky action. In order to determine the experience and skill of a prospective asbestos-removal contractor, the contractor should be asked these questions:

1. Is the contractor certified? (Ask to see the certificate).
2. Have the contractor and the contractor's workers been trained?
3. Does the contractor have experience of removing asbestos from buildings?
4. Will the contractor provide a list of references from people for whom he/she has worked with asbestos?
5. Will the contractor provide a list of places where he/she has worked with asbestos?
6. Will the contractor use the "wet method" (water and detergent)?
7. Will the contractor use polyethylene plastic barriers to contain dust?
8. Will the contractor use a HEPA (high efficiency particulate air) filter vacuum cleaner?
9. Will the contractor's workers wear approved respirators?
10. Will the contractor properly dispose of the asbestos and leave the site free of asbestos dust and debris?
11. Will the contractor provide a written contract specifying these procedures?

The plant manager or the owner of an industrial site must make sure to hire a certified, trained, and experienced asbestos contractor who follows the following General Guidelines for Handling Products Containing Asbestos established by the U.S. Consumer Product Safety Commission and the U.S. Environmental Production Agency [22]:

1. The contractor should seal off the work area from the rest of the residence and close off the heating/air conditioning system. Plastic sheeting and duct tape may be used, which can be carefully sealed with tape when work is complete. The contractor should take great care not to track asbestos dust into other areas of the residence.
2. The work site should be clearly marked as a hazard area. Only workers wearing disposable protective clothing should have access. Household members and their pets should not enter the area until work is completed and inspected.
3. During the removal of asbestos-containing material, workers should wear approved respirators appropriate for the specific asbestos activity. Workers should also wear gloves, hats, and other protective clothing. The contractor should properly dispose of all of this equipment (along with the asbestos material) immediately after using it.
4. The contractor should wet the asbestos-containing material with a hand sprayer. The sprayer should provide a fine mist, and the material should be thoroughly dampened, but not dripping wet. Wet fibers do not float in the air as readily as dry fibers and will be easier to clean up. The contractor should add a small amount of a low sudsing dish or laundry detergent to improve the penetration of the water into the material and reduce the amount of water needed.
5. The contractor should assure that if asbestos-containing material must be drilled or cut, it is done outside or in a special containment room, with the material wetted first.
6. The contractor should assure that, if the material must be removed, it is not broken into small pieces, as asbestos fibers are more likely to be released. Pipe insulation is usually installed in preformed blocks and should be removed in complete pieces.

7. The contractor should place any material that is removed and any debris from the work in sealed, leak-proof, properly labeled, plastic bags (6 mm thick) and should dispose of them in a proper land-fill. The contractor should comply with Health Department instructions about how to dispose of asbestos-containing material.
8. The contractor should assure that after removal of the asbestos-containing material, the area is thoroughly cleaned with wet mops, wet rags, or sponges. The cleaning procedure should be repeated a second time. Wetting will help reduce the chance that the fibers are spread around. No asbestos material should be tracked into other areas. The contractor should dispose of the mop heads, rags, and sponges in the sealed plastic bags with the removed materials.
9. Plant personnel, if trained but not certified, can perform minor repairs (approximately the size of a hand), taking special precautions regarding dust, sweep, or vacuum particles suspected of containing asbestos. The fibers are so small that they cannot be seen and can pass through normal vacuum cleaner filters and get back into the air. The dust should be removed by a wet-mopping procedure or by specially designed "HEPA" vacuum cleaners used by trained asbestos contractors.

11.6.7 Environmental Regulations on ACM Mandatory Requirements

Regulations

There are several important OSHA and USEPA regulations that are designed to protect workers. They are summarized here, as guidance. OSHA has specific requirements concerning worker protection and procedures used to control ACM. These include the OSHA construction industry standard for asbestos (29 CFR1926.58), which applies to O&M work, and the general industry asbestos standard (29 CFR1910.1001). State-delegated OSHA plans, as well as local jurisdictions, may impose additional requirements.

The OSHA standards generally cover private sector workers and public sector employees in states that have an OSHA state plan. Public sector employees, or certain school employees, who are not already subject to a state OSHA plan are covered by the USEPA "Worker Protection Rule" (Federal Register: February 25, 1987; 40 CFR 763, Subpart G, Abatement Projects; Worker Protection, Final Rule).

The OSHA standards and the USEPA Worker Protection Rule require employers to address a number of items, which are triggered by exposure of employees to asbestos fibers. Exposure is discussed in terms of fibers per cubic centimeter (cc) of air. A cc is a volume approximately equivalent to that of a sugar cube.

Two main provisions of the regulations fall into the federal category of "Permissible Exposure Limits" (PELs) to airborne asbestos fibers. They are:

1. An 8 hour time-weighted average limit (TWA) of 0.2 fiber per cubic centimeter (f/cc) of air based on an 8 hour time-weighted average (TWA) sampling period. This is the maximum level of airborne asbestos, on average, that any employee may be exposed to over an 8 hour period (normal work shift).
2. Excursion limit (El): 1.0 f/cc as averaged over a sampling period of 30 minutes.

These levels trigger mandatory requirements, which include the use of respirators and protective clothing, the establishment of "regulated areas," the posting of danger signs, as well as the use of engineering controls and specific work practices [20,53].

OSHA regulations also establish an "action level": 0.1 f/cc for an 8 hour TWA. Employee training is required once an action level of 0.1 f/cc and/or the "excursion limit" is reached.

This training must include topics specified by the OSHA rules. If an employee is exposed at or above the action level for a period of 30 days or more in a calendar year, medical surveillance is required according to the OSHA construction industry asbestos standard.

Medical Examination and Medical Surveillance

OSHA also requires medical examinations under its "General Industry Standard" for any employee exposed to fiber levels in the air at or above the OSHA "action level" (0.1 f/cc) and/or the "excursion limit" (1.0 f/cc). In both cases – the action level and excursion limit – the OSHA medical examination requirement applies if the exposure occurs for at least one day per year.

Medical surveillance is defined as "a periodic comprehensive review of a worker's health status." The required elements of an acceptable medical surveillance program are listed in the OSHA standards for asbestos. According to those regulations, participation in a medical surveillance program is required for any employee who is required to wear a negative pressure, air-purifying respirator. Replacement, annual, and termination physical exams are also required for these employees. However, a termination exam is only necessary under the construction industry standard (which applies to custodial and maintenance employees) if a physician recommends it. While not mandatory, USEPA and NIOSH recommend physical examinations, including cardiac and pulmonary tests, for any employee required to wear a respirator by the building owner. These tests determine whether workers will be unduly stressed or uncomfortable when using a respirator [20].

11.6.8 Notification Requirements

USEPA or the State [if the State has been delegated authority under National Emission Standards for Hazardous Air Pollutants (NESHAP)] must be notified before a building is demolished or renovated. The following information is required on the NESHAP notice: (a) name and address of the building owner or manager; (b) description and location of the building; (c) estimate of the approximate amount of friable ACM present in the facility; (d) scheduled starting and completion dates of ACM removal; (e) nature of planned demolition or renovation and method(s) to be used; (f) procedures to be used to comply with the requirements of the regulation; and (g) name, address, and location of the disposal site where the friable asbestos waste material will be deposited.

The notification requirements do not apply if a building owner plans renovation projects that will disturb less than the NESHAP limits of 160 square feet of friable ACM on facility components or 260 linear feet of friable ACM on pipes (quantities involved over a one-year period). For renovation operations in which the amount of ACM equals or exceeds the NESHAP limits, notification is required as soon as possible.

11.6.9 Emissions Control, Waste Transportation, and Waste Disposal

The NESHAP asbestos rule prohibits visible emissions to the outside air by requiring emission control procedures and appropriate work practises during collection, packaging, transportation, or disposal of friable ACM waste. All ACM must be kept wet until sealed in a leak-tight container that includes the appropriate label. The following table provides a simplified reference for building owners regarding the key existing NESHAP requirements.

Under the expanded authority of RCRA, a few states have classified asbestos-containing waste as a hazardous waste, and require stringent handling, manifesting, and disposal procedures. In those cases, the state hazardous waste agency should be contacted before disposing of asbestos for approved disposal methods and record-keeping requirements, and for a list of approved disposal sites.

Friable asbestos is also included as a hazardous substance under USEPA's CERCLA regulations. The owner or manager of a facility (e.g., building, installation, vessel, landfill) may have some reporting requirements, for example, the U.S. Department of Transportation (USDOT) requirements for asbestos transport activities under the Hazardous Materials Transportation Act of 1975 (HMTA). The HMTA regulatory program applies to anyone who transports hazardous materials, or arranges for their transportation or shipment, and to anyone who manufactures, reconditions, repairs, tests, or marks packages or containers for use in the transportation of hazardous materials [49 USC Sec. 1804(a)].

USDOT has designated asbestos as a hazardous material for the purposes of transportation, and has issued requirements for shipping papers, packaging, marking, labeling, and transport vehicles applicable to shipment and transportation of asbestos materials (49CFR173.101). Commercial asbestos must be transported in rigid, leak-tight packages: in bags or other nonrigid packaging in close freight containers, motor vehicles or rail cars loaded by the consignor and unloaded by the consignee exclusively, or bags or other nonrigid packages that are dust- and sift-proof in strong fiber board or wooden boxes (49CFR173.1090).

Specific regulations exist for the transport of asbestos materials by highway [53]. Asbestos must be loaded, handled, and unloaded using procedures that minimize occupational exposure to airborne asbestos particles released in association with transportation. Any asbestos contamination of transport vehicles also must be removed using such procedures (49CFR177.844). Additional motor carrier's safely regulations apply to common, contract, and private carriers of property by motor vehicle, as defined under these regulations (49CFR Parts 390–397).

11.7 MONITORING AND ANALYSIS OF AIR, WATER, AND CONTAMINATED MATERIALS

11.7.1 General Approach

Because airborne and volatile contaminants can present a significant threat to industrial workers' health and safety, identification and quantification of these airborne and volatile contaminants through air/soil monitoring is an essential component of a health and safety program at an industrial site having hazardous substances. The purpose of air and soil monitoring is to identify and quantify airborne and volatile hazardous contaminants in order to determine the level of plant worker's protection needed.

In general, there are two principal approaches available for identifying and/or quantifying airborne contaminants as well as volatile contaminants in soil:

1. The first approach: onsite use of direct-reading instruments as initial qualitative identification or screening (note: the airborne/volatile contaminant, or the class to which it belongs, is demonstrated to be present but quantitative determination of its exact concentration must await subsequent testing); and
2. The second approach: laboratory analysis of air and/or soil samples (note: the air sample can be obtained by gas sampling bag, filter, sorbent, and wet-contaminant collection methods).

Care must be taken in sampling of contaminated air, soil, water, or materials in order to obtain representative samples, and, in turn, to gain meaningful results. In general, the onsite use of direct-reading instruments for qualitative analysis and the onsite sampling of contaminated air, soil, water, or materials are performed by a licenced engineer, a licenced geologist, or a certified technician. The subsequent quantitative laboratory analysis, if required, can be performed by either a certified laboratory or a licenced engineering firm, depending on the environmental quality parameters.

For instance, air samples and the building material samples contaminated by formaldehyde and lead are routinely sampled by an engineering technician under the supervision of a licenced engineer. The samples are shipped to a certified laboratory for quantitative analysis by the licenced engineer.

In another common case, soil that may be contaminated by volatile gasoline is routinely qualitatively tested with a direct-reading instrument and sampled by an engineer/scientist under the supervision of either a licenced engineer or geologist. The contaminated soil is qualitatively identified and/or documented and shipped by the engineer/scientist, quantitatively analyzed by a certified laboratory, and its quantitative data interpreted by the licenced engineer/geologist.

In New York and Massachusetts where PCB contamination is always a possibility, the laboratory tests required by the state environmental protection agencies for analysis of a petroleum-contaminated soil are as follows: (a) flash point; (b) total petroleum hydrocarbon (TPH); (c) PCB screening; (d) total organic halides (TOH); (e) reactivity of cyanide and sulfide; (f) BTEX or equivalent; (g) eight metals under TCLP (Toxicity Characteristics Leaching Procedure) for USTs; and (h) full range of tests under TCLP for ASTs and spills.

In still another case, airborne asbestos is frequently qualitatively identified and/or sampled by either a licenced engineer or a certified asbestos contractors, and quantitatively analyzed by a certified laboratory. The building material, such as the insulation for the plumbing system, however, can only be removed by State-certified asbestos contractor. The readers are referred to Section 11.6.3 for air sampling and identification of asbestos-containing materials.

A continuous contaminant source monitor can provide both industrial plants and regulatory agencies with numerous benefits. A properly installed and operated continuous monitoring system can yield a large amount of data on source air emissions or source effluent discharges. This information is beneficial, because it establishes a reliable foundation upon which important decisions can be made.

11.7.2 Measuring Instruments

Reliable measurements of airborne volatile or hazardous substances in the field using onsite instruments are useful for: (a) selecting personal protective equipment at an industrial site; (b) delineating areas where protection is needed; (c) assessing the potential health effects of hazardous exposure; (d) determining the need for specific medical monitoring; and (e) providing an early warning for personnel evacuation due to contamination, when necessary.

The National Pollutant Discharge Elimination System (NPDES) reporting requirements for effluent testing allow alternate methods of analysis to be substituted for the prescribed methods if prior approval has been obtained from the U.S. Environmental Protection Agency (USEPA) regional administrator having jurisdiction where the discharge occurs.

Steps an individual permit holder must take to use an alternate test procedure for regulatory reporting of specific discharges follow. An alternate test procedure differs from those published in the Federal Register for NPDES-certification purposes (Source: Federal Register, Title 40, Chapter 1, Subchapter D, Part 136: Vol. 38, No. 199, Oct. 16, 1973; Vol. 41, No. 232, Dec. 1, 1976). Many Hach methods (Hach Company, Loveland, CO, USA) are identical to these published methods and thus are approved by USEPA and highly recommended by the authors for rapid field testing of effluent samples.

Direct-reading instruments have been developed as early warning devices for use at various industrial sites, where a leak or an accident could release a high concentration or high dose of a known chemical or known radiation into the environment. They provide information on flammable, or explosive atmospheres, oxygen deficiency, certain gases and vapors, or ionizing radiation, at the time of measuring, enabling rapid decision making by the plant managers. Direct-reading instruments, which can be either batch monitoring systems or continuous monitoring systems, are the primary tools of initial site characterization. The readers are referred to Chapter 1 entitled "Onsite Monitoring and Analyses of Industrial Pollutants" for more information on several common direct-reading field instruments and their conditions and/or hazardous substances they measure.

As a minimum, the flame ionization detector (FID) or the photo-ionization detector (PID) must be available at industrial sites handling hazardous substances.

11.8 HAZARDOUS WASTE GENERATOR STATUS AND REGULATORY REQUIREMENTS

11.8.1 Hazardous Waste Generators

Regulations

In general, two activities determine the generator category of an industrial plant: the rate at which the plant generates and how much the plant stores (accumulates). Under new, more flexible regulations, the amount and length of time an industrial plant can accumulate wastes may vary according to the type of waste. In the State of Massachusetts, there are three generator statuses, which are introduced below as a typical example.

1. Large Quantity Generator (LQG): generates more than 1000 kg (2200 lb) of hazardous waste in a month; once the first 1000 kg has been accumulated, the waste must be shipped within 90 days; there is no limit to the amount that can be accumulated.
2. Small Quantity Generator (SQG): generates less than 1000 kg of hazardous waste in a month, and/or less than 1 kg of acutely hazardous waste (acutely hazardous waste is listed in the State regulations).
3. Very Small Quantity Generator (VSQG): generates less than 100 kg of hazardous waste in a month, and generates no acutely hazardous waste.

Other State governments in the United States have similar regulatory requirements. The maximum monthly volume of waste oil and maximum monthly volume of all other hazardous waste generated at an industrial plant site can be estimated and regulated according to the State of Massachusetts "Guide to Determining Status and Regulatory Requirements" (Table 2).

An Example in Massachusetts

An industrial plant in Massachusetts generates 60 gallons of spent solvent and 550 gallons (2081.75 L) of waste oil in a month. According to the Guide (Table 2); the plant is a Small Quantity Generator (SQG) of hazardous waste because it produces more than 100 kg but less than 1000 kg, and the plant is also a Large Quantity Generator (LQG) of waste oil because the plant produces more than 1000 kg. The plant's regulatory status is found in Table 2, under line 5 (SQG for HW; LQG for WO).

Reading across the columns, on line 5, the plant may accumulate its solvent for as long as 180 days, or until the plant has reached a volume of 2000 kg (500 gallons; 1892.5 L) in

Table 2 Guide to Determining Status and Regulatory Requirements for Hazardous Waste Management

Regulatory status of co.						
Hazardous waste (HW)[c]	Waste oil (WO)[c]	Accumulation time, HW (days)	Accumulation HW volume in tanks (kg)	Accumulation HW volume in containers (kg)	Manifest usage requirement	Permission for self-transport HW
LQG	LQG	90	No limit	No limit	Yes	No
LQG	SQG	90	No limit	No limit	Yes	No
LQG	VSQG	90	No limit	No limit	Yes[a]	No
LQG	None	90	No limit	No limit	Yes	No
SQG	LQG	180	6000[b]	2000	Yes	No
SQG	SQG	180	6000[b]	2000	Yes	No
SQG	VSQG	180	6000[b]	2000	Yes[a]	No
SQG	None	180	6000[b]	2000	Yes	No
VSQG	LQG	No limit	600	600	Yes[a]	Yes
None	LQG	N/A	N/A	N/A	Yes	No
VSQG	SQG	No limit	600	600	Yes[a]	Yes
VSQG	VSQG	No limit	600	600	Yes[a]	Yes
VSQG	None	No limit	600	600	Yes[a]	Yes
None	SQG	N/A	N/A	N/A	Yes	No
None	VSQG	N/A	N/A	N/A	Yes[a]	Yes

Regulatory status of co.						
Hazardous waste (HW)[c]	Waste oil (WO)[c]	Accumulation time, WO (days)	Accumulation WO volume in tanks (kg)	Accumulation WO volume in containers (kg)	Manifest usage requirement	Permission for self-transport WO
LQG	LQG	90	No limit	No limit	Yes	No
LQG	SQG	180	6000[b]	2000	Yes	No
LQG	VSQG	No limit	600	600	Yes[a]	Yes
LQG	None	N/A	N/A	N/A	Yes	No
SQG	LQG	90	No limit	No limit	Yes	No
SQG	SQG	180	6000[b]	2000	Yes	No
SQG	VSQG	No limit	600	600	Yes[a]	Yes
SQG	None	N/A	N/A	N/A	Yes	No
VSQG	LQG	90	No limit	No limit	Yes[a]	No
None	LQG	90	No limit	No limit	Yes	No
VSQG	SQG	180	6000[b]	2000	Yes[a]	No
VSQG	VSQG	No limit	600	600	Yes[a]	Yes
VSQG	None	N/A	N/A	N/A	Yes[a]	Yes
None	SQG	180	6000[b]	2000	Yes	No
None	VSQG	No limit	600	600	Yes[a]	Yes

Note: This matrix guide does not reflect **acutely** hazardous wastes.
[a]A manifest must be used for the VSQG category unless self-transported.
[b]When accumulating in both tanks and containers, the total accumulation cannot exceed 6000 kg and the container accumulation cannot exceed 2000 kg.
[c]LQG = 1000 or more kg per month of waste generation; SQG = 100–999 kg per month of waste generation; VSQG = less than 100 kg per month of waste generation.

containers (Table 2), whichever happens first (column 3). The plant must ship its waste oil every 90 days regardless of the volume. The plant manager must obtain an USEPA Identification Number and use a manifest for both wastes. The plant manager must manage his/her waste according to the accumulation area standards and must fulfill the emergency preparation and response requirements listed in subsequent sections. The plant manager, however, is not required to file an annual report or a contingency plan or provide full personnel training, which is necessary for larger generators.

11.8.2 Hazardous Waste and Waste Oil Documentation Using a Manifest

As a generator, an industrial plant always retains responsibility for hazardous waste. If the plant's waste is dumped or disposed of improperly, the plant manager and the owner will be held responsible. It is therefore important that the plant manager or the owner knows where the plant's waste is going and whether or not it is handled properly and safely [73].

U.S. Federal law (the Recourse Conservation and Recovery Act of 1976, known as RCRA) requires a national "cradle to grave" tracking system for hazardous waste. In the State of Massachusetts, for instance, every shipment of hazardous waste by a large or small generator must be transported by a licenced hauler and sent to a licenced treatment, storage, or disposal facility (TSD) or a permitted recycling facility, and it must be accompanied by a multipart shipping document, called the Uniform Hazardous Waste Manifest.

In the State of Massachusetts, the plant manager or a designated consulting engineer must use the Massachusetts Manifest form unless the plant is sending its waste to a facility out of state, in which case the plant manager should contact the other state to find out which form to use. The plant manager or the plant's consulting engineer will be responsible for completing the generator portion of the manifest. Directions for the distribution of the copies are printed on the manifest. A copy will be returned to the industrial plant when the disposal facility or the recycling facility has accepted its shipment.

If the industrial plant's manager or consulting engineer does not receive a copy of the manifest from the receiving facility (i.e., the disposal facility and/or the recycling facility) within 35 days of the date when the plant's waste was shipped, the transporter or the operator of the facility must be contacted to determine the status of the waste. If the plant has still not received the manifest within 45 days, an Exception Report, explaining the efforts the plant has taken, must be filed with the State's Division of Hazardous Waste and with the State where the designated facility is located.

For all generators, copies of all manifests and any records of tests and analyses carried out on the hazardous waste must be kept for at least three years, and for the duration of any enforcement action.

The most common problems in completing the manifest are clerical. For clarity, because this is a multiple carbonless copy form of about eight pages, typing is strongly recommended. The generator should check for legibility of all copies before transferring the manifest to the transporter at the time of shipment. The generator must ensure that all information is complete and accurate by reviewing the following summary when completing the manifest.

1. The plant's federal Identification (ID) Number must be correctly stated. The plant's specific location must have an ID number to use the manifest.
2. The identification number of the transporter and the receiving facility and their valid hazardous waste licenses must be double checked with the State's regulatory agency.
3. If there is a second transporter, the generator has the responsibility to select this second transporter and both the generator and the second transporter must complete certain portions of the manifest.
4. The generator shall have a program to reduce the volume and toxicity of waste generated, which is a national requirement of all generators and is intended to encourage good management practise. Large quantity generators are required to report how they are reducing waste in their annual report.

5. The contents of the shipment must be fully and accurately described, packed, marked, and labeled.
6. Any special handling instructions must be clearly given. The generator can list an alternative receiving facility and must list, in the case of an international shipment, the city and State at which the shipment leaves the United States.
7. If more than four wastes are included in a single shipment, a second prenumbered manifest must be used. When more than two transporters are used for one shipment, the State requirements must be reviewed. In the State of Massachusetts an eight-part Massachusetts Continuation Sheet, numbered to match the first manifest, should be used.
8. Instructions regarding the use and distribution of the manifest copies that are stated on the manifest must be reviewed. The generator retains certain copies at the time of shipment. One copy should be mailed to the manifest office of the State in which the destination facility is located. One copy is returned to the generator by the receiving facility when the shipment arrives. The generator copies must be kept in the file for at least three years. If a signed manifest copy from the destination facility is not received within 35 days, the generator must investigate and file an Exception Report with the State Enforcement Section within 45 days of shipment if the signed copy has still not been received.

When a small or a very small quantity generator is to ship only waste oil or a very small quantity generator is to spill other waste, a transporter's log instead of a manifest may be used for that shipment. However, the generator must register on a prescribed form with the State of Massachusetts.

11.8.3 The USEPA Identification Number (USEPA-ID)

In order to have an industrial hazardous waste accepted by a licenced hauler or treatment/ storage facility, the industrial plant (i.e., the generator) must be assigned a number, with a special prefix for the plant location. This number will be entered on each manifest.

In order to get an USEPA-ID, the plant manager shall call or contact the State government for an application for an USEPA Identification Number. The completed application should be mailed to the state office listed in the instruction. While a plant is waiting for a permanent USEPA-ID number, the plant can obtain a temporary USEPA-ID number over the telephone.

The USEPA-ID number is site-specific. The State Division of Hazardous Waste must be notified in writing, or on a specified form, of any change in the generator's address, contact person, or generator status.

11.8.4 Shipping Hazardous Waste

All hazardous waste must be transported in containers [24,54] that are labeled with the words HAZARDOUS WASTE, the name of the waste, type of hazard (e.g., toxic, flammable), and generator's name, address, and USEPA-ID number.

A list of licenced transporters and facilities for treatment, storage, or disposal is always available from the State government. Many transporters are authorized to assist the plant manager in preparing the plant's hazardous waste for shipment.

A summary of recommended procedures for shipping hazardous wastes from an industrial plant to another location is now given below:

1. Select a licenced transporter and a hazardous waste facility that will receive the plant's waste;

2. Identify the waste based on a licenced engineer's testing or a certified laboratory testing prior to shipping the waste;
3. Obtain a federal identification (USEPA-ID) number by requesting a required form (such as Notification of Hazardous Waste Activity Form) from a State regulatory agency (note: the identification number is specific to the location, not the hazardous waste);
4. Obtain a manifest for a shipment of waste destined for disposal in a State (note: this specific State's manifest form with a preprinted State document number is required); and
5. Ship the plant's waste in accordance with federal transportation regulations (CFR Title 49, Part 100–177).

11.8.5 Hazardous Waste Storage Standards for an Accumulation Area

The accumulation or storage area of an industrial plant (i.e., a generator) must meet the following conditions for both containers and tanks in accordance with the home State regulations. The Massachusetts hazardous waste regulations (310 CMR 30.000) are listed below as a reference:

1. Above-ground tanks and containers must be on a surface that does not have any cracks or gaps and is impervious to the hazardous wastes being stored;
2. The area must be secured against unauthorized entry;
3. The area must be clearly marked (e.g., by a visible line or tape, or by a fence) and be separate from any points of generation;
4. The area must be posted with a sign. "HAZARDOUS WASTE" in capital letters at least one inch high (1 in. = 2.54 cm);
5. An outdoor area must have secondary containment, such as a dike, which will hold any spill or leaks at (a) 10% of the total volume of the containers, or (b) 110% of the volume of the largest container, whichever is larger; and
6. Any spillage must be promptly removed: in general, if the hazardous waste being stored has no free liquids, no pad is required, provided that the accumulation area is sloped, or the containers are elevated.

11.8.6 Standards for Waste Containers and Tanks

General Massachusetts standards (310 CMR 30.680–30.690) for waste containers and tanks in accordance with the same Massachussetts hazardous waste regulations (310 CMR 30.000) are given below as a reference:

1. Each container and tank must be clearly and visibly labeled throughout the period of accumulation with the following:
   (a) the words "HAZARDOUS WASTE,"
   (b) the name of the waste (e.g., waste oil, acetone),
   (c) the type of hazard(s) (e.g., ignitable, toxic, dangerous when wet, corrosive), and
   (d) the date on which the accumulation begins;
2. Each container must be in good condition;
3. Wastes of different types must be segregated; for example, this includes not mixing waste oil or used fuel oil with other wastes; be careful not to put incompatible wastes in the same container or put wastes in unwashed containers that previously stored incompatible wastes;
4. Separate containers of incompatible wastes by a dike or similar structure;

5. Each container holding hazardous wastes must be tightly closed throughout the period of accumulation, except when the waste is being added or removed;
6. Containers holding ignitable or reactive wastes must be at least 15 m (50 ft) away from the property line; if this is not possible or practical, the plant manager representing the generator must store such containers in compliance with all applicable local ordinances and bylaws; and
7. Inspect the accumulation area at least once a week for any leaking or deterioration of all containers; there must be enough aisle space between the containers to allow for inspections.

11.8.7 Criteria for Accumulation Time Limits

If an industrial plant is classified as a small quantity generator (SQG), the plant manager may accumulate up to 2000 kg or 4400 lb in containers, or up to 6000 kg (approximately 1650 gal or 6245 L) in tanks for as long as 180 days according to Massachusetts regulations 310 CMR 30.351. If both tanks and containers are used to store hazardous waste and/or waste oil, the total waste that can be accumulated at any one time may not be determined by adding the two limits. The 180 day clock may be started when a total of 100 kg, (approximately 25 gal or 94.63 L) is accumulated, if the containers are redated at that time.

11.8.8 Criteria for Satellite Accumulation

Additional flexibility is offered by allowing an industrial plant to accumulate up to 55 gal (or 208.18 L) of hazardous waste, or one quart (or 1 L) of acutely hazardous waste, at each point where the plant generates its waste if the plant meets the following conditions:

1. The waste must be generated from a process at the location of the satellite accumulation;
2. Each satellite accumulation area can have only one container for each waste stream in use at a time;
3. Each satellite accumulation area must be managed by a person who is directly responsible for the process producing the waste; and
4. The waste must be moved to the main designated accumulation area within three days after the container is full.

11.8.9 Criteria for Accumulation of Waste Oil in Underground Storage Tanks

The Massachusetts criteria (310 CMR 30.690) for accumulating waste oil in underground storage tanks (USTs), including those resting directly on the ground, are generalized below:

1. For leak detection in old tanks containing waste oil that were installed before October 15, 1983 under a grandfather clause, a dipstick test must be conducted every 30 days; a more than $\frac{1}{2}$ in. (1.27 cm) difference in level within a 24 hour period must be reported to the State government; underground tanks containing other hazardous wastes must undergo a tightness test, and must be monitored on a daily basis;
2. Tanks installed after the effective date (October 15, 1983) of a new Massachusetts law regarding underground storage tanks must have secondary containment and a monitoring system or be constructed of a corrosion-resistant material; and
3. A log must be kept of all test results for at least three years.

11.9 STORAGE TANK INSPECTION AND LEAK DETECTION

11.9.1 Requirements for Underground Storage Tanks

The State of New York [11–14,55] has promulgated rules and regulations for the early detection of leaks or potential leaks of petroleum bulk storage by plant owners and operators. In the State of New York [14], underground tanks shall be checked for leakage using one or more of the following:

1. Inventory monitoring may be used if it detects a leak of one percent (1%) of flowthrough plus 130 gal on a monthly basis and is coupled with an annual tightness test. Inventory monitoring must be done.
2. Weekly monitoring of the interstitial space of a double-walled tank may be practiced using pressure monitoring, vacuum monitoring, electronic monitoring, or manual sampling.
3. Vapor wells for monitoring soils in the excavation zone may be used. Vapor monitoring systems must be designed and installed by a qualified engineer or technician in accordance with generally accepted practises. Wells must be protected from traffic, permanently labeled as a "monitoring well" or "test well – no fill" and equipped with a locking cap, which must be locked when not in use so as to prevent unauthorized access and tampering. Vapor monitoring may be used only under the following conditions: (a) soils in the excavation zone must be sufficiently porous to allow for the movement of the vapors from the tank to the vapor sensor; gravel, coarse and crushed rocks are examples of porous soils; (b) the stored substance or a tracer compound placed in the tank must be sufficiently volatile so as to be detectable by the vapor sensor; (c) vapor monitoring must not be hindered by groundwater, rainfall, or soil moisture such that a release could go undetected for more than 30 days; (d) background contamination must not mask or interfere with the detection of a release; (e) the system must be designed and operated to detect increases in vapors above background levels; monitoring must be carried out at least once per week; and (f) the number and positioning of vapor monitoring wells must be sufficient to ensure detection of releases from any portion of the tank and must be based on a scientific study; wells must be at least four inches in diameter.
4. Groundwater monitoring wells designed and installed by a qualified engineer or technician may be used. Wells must be protected from traffic, permanently labeled as a "monitoring well" or "test well – no fill" and equipped with a locking cap that must be locked when not in use to prevent unauthorized access and tampering. Groundwater monitoring may be used only under the following conditions: (a) the substance stored must be immiscible in water and have a specific gravity of less than one; (b) the groundwater table must be less than 20 ft from the ground surface; the hydraulic conductivity of the soil between the tank and well must not be less than one hundredth (0.01) cm/s; gravel and coarse to medium sand are examples of such soil; (c) the slotted portion of the well casing must be designed to prevent migration of soil into the well and must allow entry of the hazardous substances into the well under both high and low groundwater conditions; (d) wells must be at least four inches in diameter and be sealed from the ground surface to the top of the filter pack to prevent surface water from entering the well; (e) wells must be located within the excavation zone or as close to it as technically feasible; (f) the method of monitoring must be able to detect at least one-eighth ($\frac{1}{8}$) of an inch of free product on top of the groundwater; monitoring must be carried out once per week; and (g) the number and positioning of

the groundwater monitoring well(s) must be sufficient to ensure detection of releases from any portion of the tank and must be based on a scientific study.

5. Automatic tank gauging equipment may be used if it can detect a leak of two-tenths (0.2) of a gallon per hour or larger with a probability of detection of 95% and probability of false alarm of 5% or less. Monitoring must be carried out once per week; or
6. Other equivalent methods as approved by the Department if the method can detect a leak of two-tenths (0.2) of a gallon per hour with a 95% probability of detection and probability of false alarm of 5%.

In the State of New York, underground and on-ground piping shall also be checked for leakage by the owner or the plant manager according to the general guidelines established by the Department of Environmental Conservation [14].

11.9.2 Requirements for Aboveground Storage Tanks

While leak detection is not emphasized for aboveground storage tanks (ASTs), daily inspections, monthly inspections, annual inspections, and five-year inspections are legally required by the State of New York for AST owners or operators [14].

Daily Inspection

The owner or operator must visually inspect the aboveground storage equipment for spills and leaks each operating day. In addition, the owner or operator must check to ensure that drain valves are closed if not in use and there are no unpermitted discharges of contaminated water or hazardous substances.

Monthly Inspections

The owner or operator must conduct comprehensive monthly inspections of aboveground storage equipment. This inspection includes: (a) identifying cracks, area of wear, corrosion, poor maintenance and operating practises, excessive settlement of structures, separation or swelling of tank insulation, malfunctioning equipment, safety interlocks, safety trips, automatic shutoffs, leak detection, and monitoring, warning, or gauging equipment that may not be operating properly; (b) visually inspecting dikes and other secondary containment systems for erosion, cracks, evidence of releases, excessive settlement, and structural weakness; (c) checking on the adequacy of exterior coatings, corrosion protection systems, exterior welds and rivets, foundations, spill control equipment, emergency response equipment, and fire extinguishing equipment; (d) visual checking of equipment, structure, and foundations for excessive wear or damage; (e) reviewing the State compliance; and (f) performing monthly release detection, which meets the performance standards established by the State.

Annual Inspections

The structure-to-electrolyte potential of corrosion protection systems used to protect aboveground tank bottoms and connecting underground pipes must be inspected annually.

Five-Year Inspections

The owner or operator must inspect aboveground piping systems and all aboveground tanks; the inspection must be consistent with a consensus code, standard, or practise and be developed by a nationally recognized association or independent testing laboratory and meet the specifications of this subdivision; based on the inspection, an assessment and evaluation must be made of system tightness, structural soundness, corrosion, wear and operability; reinspection is required no later than every five years from the date of the initial inspection or regulatory deadline, whichever occurs first, except as follows. If thinning of 1 mL per year or greater occurs on the pipe or tank walls, or the expected remaining useful life as determined by the above inspections is less than ten years, then reinspection must be performed on the tank or pipe at one-half of the remaining useful life.

11.9.3 Tank and Pipeline Leak Tests

Tracer tank and pipeline leak tests developed by Tracer Research Corporation do not require that tanks or pipelines be taken out of service during any testing procedures.

The leak tests have demonstrated the capability for unambiguously detecting, quantifying, and locating leaks as small as 0.05 gal/hour in underground and aboveground storage tanks and pipelines. Storage tanks containing fuels, lubricants, heating oils, solvents, wastewater, volatile or nonvolatile chemicals, and hazardous wastes are easily tested regardless of size or type.

Leak Testing for Underground Storage Tanks

This section introduces a five-step procedure developed by Tracer Research Corporation for conducting the leak testing for an underground storage tank.

1. Step 1. Leak testing is performed by adding a small amount of a special volatile chemical tracer to the contents of a tank or pipeline; these chemicals are selected for their compatibility with tank and pipeline systems, as well as the lack of their presence in the environment around the tank; the tracer is added at a concentration of only a few PPM, and thus has no impact on the physical properties of tank and pipeline contents.
2. Step 2. Tracer mixes evenly in tanks, pipelines, and the vapor space inside a tank, by diffusion and product use.
3. Step 3. If a tank or pipeline leaks, the tracer is released into the surrounding soil where it rapidly volatizes; after the tracer has had time to disperse and migrate through the soil away from the leak (usually about two weeks), soil gas samples are collected from the probes surrounding the tanks and pipelines.
4. Step 4. Samples are analyzed for tracer and hydrocarbon vapors by means of a very sensitive gas chromatograph; the presence of tracer vapors, which can be detected in the low parts-per-trillion, provides unambiguous information about the occurrence of leakage and its location.
5. Step 5. Because information about site contamination is important, the plant manager is provided with a hydrocarbon site survey at the same time; hydrocarbon vapor maps serve to show the magnitude of leakage and the extent of the contamination if leaks are detected; if no leaks are detected, the absence of hydrocarbons confirms this finding.

Leak Testing for Pipelines

The tracer pipeline leak testing, which is similar to the tracer tank leak testing, is effective for locating leaks in all types of pipeline installations, including pipe buried under pavement, airline runways, buildings, or underwater. Where leaks are known to exist, the tracer leak test is effective in determining their location without expensive excavation.

The testing method can be retrofitted to existing underground piping. Where the pipeline runs under soil cover, a special leak detection hose that is permeable to the tracer is buried approximately 0.61 m (2 ft) deep in a ditch running above the pipeline. One sample from the hose can provide monitoring coverage of up to 152.4 m (500 ft) of pipeline. At new installations, the leak detection hose is installed adjacent to the pipe at the time of burial. This installation is very low cost and provides unique sensitivity. When a pipeline runs under concrete or pavement, it is monitored by a series of probes placed 7.62 m (25 ft) apart, installed through the pavement.

Leak Testing for Aboveground Storage Tanks

Aboveground tank testing is performed by inserting vapor sampling probes under the tank bottom. To ensure detection of leakage from any point on the tank floor, evacuation probes are placed under the perimeter of the tank and one or more air injection probes are placed beneath the center of the tank. A program of air injection and/or evacuation is initiated to collect samples from under the tank. These samples are analyzed for the presence of tracer.

In the case of a facility that has multiple tanks in close proximity to each other, different tracer compounds can be used so that sample analysis will rapidly identify a specific leaking tank.

Leak Testing for Tank Farms

The tracer leak tests are also economical means for testing aboveground storage tanks at large tank installations, such as jet fuel systems at military bases, large airport hydrant fuel systems, terminals and refineries. Important benefits result from the fact that the testing is implemented by placing tracer in the receiving tanks where incoming product is stored. The product is released to other parts of the system and the same tracer is used to test all the portions of the system that contain or transport the product.

11.10 EMERGENCY PREPARATION AND RESPONSE

11.10.1 Emergency Equipment

To minimize the risk of fire, explosion, or release of hazardous wastes that may contaminate the environment, an industrial plant classified as a generator is required to have the following on site, and immediately accessible to its hazardous waste handling area:

- an alarm or communication system that can provide emergency instruction to employees;
- a telephone, two-way radio, or other device that can summon police, fire, or emergency response teams;
- portable fire extinguishers and/or fire control equipment (e.g., foam, inert gas), spill control equipment, and decontamination equipment; and
- adequate supply and pressure of water, automatic sprinklers or water sprays, or foam-producing equipment.

All equipments identified above are required unless the hazards posed by the plant's wastes do not require one of them. In such a case, an approval from the regulatory agency is required. The equipment, when provided, must be periodically tested and properly maintained so it will work during an emergency.

11.10.2 Emergency Preparation

An industrial plant classified as a generator must thoroughly familiarize each of its employees with all the waste handling and emergency procedures that may be needed for each of their jobs. An employee must have immediate access to alarm or communication devices, either directly or through another employee, whenever hazardous waste is being handled. If the plant's operation is at any time being handled by a single employee, that person must have immediate access to a telephone or two-way radio.

For easy movement of employees and emergency equipment, the plant manager must mark all exits clearly and maintain adequate aisle space in the area of hazardous waste handling.

11.10.3 Liaison With Local Authorities

A generator and its designated consulting engineer must make every reasonable attempt to carry out the following arrangements, with regard to the waste produced by the generator:

- Familiarizing the plant's local police department, fire department, local boards of health, and any emergency response teams with the hazardous nature of the plant's waste; the layout of the plant site, including entrances and evacuation routes, and the location where the plant's employees usually work;
- Familiarizing local hospitals with the hazards of the plant's waste and the types of injuries that could result from any accidents;
- Obtaining agreements with emergency response teams and contractors, and local boards of health; and
- Making an agreement with the regulatory agency and service agency that will have primary emergency authority, and specifying others as support, if more than one police and/or fire department might respond to an emergency.

If such arrangements cannot be made, a copy of a signed and dated letter from the plant, the generator, to the State or local entity, which demonstrates an effort to make these arrangements must be considered sufficient, if an approval from the State or local entity can be obtained.

11.10.4 Emergency Coordinator

The industrial plant, the generator, must designate at least one employee to be on call (or on the premises) at all times. This person is the emergency coordinator and is responsible for coordinating all emergency response measures. Alternatively, a licenced consulting engineer can also be retained by the generator to be its emergency coordinator.

11.10.5 Emergency Response

It is generally required by the State regulations that the generator have posted next to each telephone near the plant's waste generation area the following:

- Name(s) and telephone number(s) of the plant's emergency coordinator(s);
- Location(s) of the fire control equipment and any fire alarms;
- Telephone numbers of the National Response Center, the fire department, the police department, and the ambulance department, or if there is a direct alarm system, instructions on how to use it; and
- Evacuation routes, where applicable.

If any of the following emergencies occur, the plant manager or the assigned emergency coordinator should immediately perform the following:

- Fire. Attempt to extinguish the fire and/or calling the fire department;
- Hazardous chemical/oil spill or leak. Contain the flow as quickly as possible and as soon as possible clean up the waste and any soil or other materials that may have become contaminated with waste;
- A hazardous chemical/oil release (spill or leak) or threat of release, fire or explosion of hazardous waste that may threaten human health or the environment. (a) Call the appropriate State environmental protection agency's regional office, or (b) Call the State police if the incident occurs after 5 p.m., or on a day that the State environmental protection agency is closed, and (c) Call the National Response Center, which usually has a 24-hour toll-free number.

11.11 MANAGEMENT OF AN INDUSTRIAL SITE CLASSIFIED AS A VERY SMALL QUANTITY GENERATOR

11.11.1 Registration

If an industrial plant in Massachusetts generates less than 100 kg a month of hazardous waste, and no acutely hazardous waster, the plant is eligible to register as a very small quantity generator. To qualify as a very small quantity generator (VSQG), the plant manager must register a waste management plan with the appropriate State environmental protection agency. If the plant does not register as a VSQG, it will be subject to the more stringent SQG regulations.

11.11.2 Treatment/Disposal Options

As a registered VSQG, an industrial plant has the following options for handling the waste:

1. The plant may recycle or treat its waste, provided the process described in the plant's registration is acceptable to the appropriate State environmental protection agency;
2. The plant may transport its waste to another generator who is in compliance with the regulations and who will count the plant's waste as part of their generation; or
3. The plant may transport its waste in the plant's own vehicle to a licenced treatment, storage or disposal facility, or permitted recycling facility, or use a licenced transporter and a manifest form, which requires an USEPA-ID number; or
4. The plant may use a licenced transporter and a manifest form, which requires an USEPA-ID number.

11.11.3 Self-Transport Option

As a registered VSQG, an industrial plant may transport its own hazardous waste under certain conditions in accordance with the appropriate State regulations. The following are the Massachusetts regulations (310 CMR 30.353), which are presented as a typical example:

1. The plant transports only the waste that the plant generated on its premises.
2. The plant does not transport more than 200 kg at one time.
3. The plant's waste must be in containers that are:
   (a) no larger than 55 gal or 208.18 L in volume;
   (b) compatible with the waste;

  (c) tightly sealed;
  (d) labeled as "HAZARDOUS WASTE";
  (e) labeled with the name of the waste and the type of hazard (i.e., ignitable); and
  (f) tightly secured to the vehicle.
4. The plant does not transport incompatible wastes in the same shipment.
5. In the event of a spill or leak of hazardous waste that may threaten human health or the environment, the plant or its designated consulting engineer should notify the appropriate State environmental protection agency, the State police, the local fire department, and the National Response Center, as described previously.
6. The plant must have a copy of its registration with the State in the vehicle.
7. The plant must be in compliance with the federal Department of Transportation and State Department of Public Safety requirements, if any.

11.11.4 Record-Keeping

If an industrial plant in Massachusetts, for instance, is not using a licenced transporter but is transporting its own wastes, this plant does not need an USEPA-ID number or manifest form. The plant must, however, keep a record of the type and quantity, as well as the date, method of transport, and treatment/disposal of its waste(s). The plant manager needs proof of the receipt of the waste by the facility and/or generator.

All generators must keep receipts or manifests of waste shipped, and records of waste analysis for at least three years, or for the duration of any enforcement action by the appropriate State environmental protection agency.

11.11.5 Accumulation Limits

The plant as a very small quantity generator (VSQG) in Massachusetts may accumulate up to 600 kg (approximately 165 gal or three 55 gal drums) of hazardous waste in containers that meet the standards introduced previously, with no time limit.

11.12 MANAGEMENT OF AN INDUSTRIAL SITE CLASSIFIED AS A LARGE QUANTITY GENERATOR OR A SMALL QUANTITY GENERATOR

11.12.1 Registration

The amount and length of time a large industrial plant accumulates its wastes may vary according to the type of waste. The Massachusetts Guide to Determining Status and Regulatory Requirements (Table 2) or equivalent should be used as a guide to determine the plant's generator category (Regulatory Status) for hazardous waste and waste oil [6].

For example, a plant in Massachusetts must be registered as a Large Quantity Generator (LQG) if it produces more than 2200 lb (1000 kg) of hazardous waste, not including waste oil, or one quart (1 kg) or more of acutely hazardous waste, as defined in the December 1992 Massachusetts regulations [4–10], in a month's time. There is no limit to the amount that can be accumulated by the plant, but the waste must be shipped within 90 days. A generator not in Massachusetts must contact the local State agency in order to obtain the most recent regulations for its home state.

If a Massachusetts plant produces less than this amount each month, the plant is classified as a small (SQG) or very small (VSQG) quantity generator and is subject to less stringent requirements, as discussed previously.

If a Massachusetts plant produces more than 1000 kg (approximately 265 gal) of waste oil in a month, the plant's waste oil must be shipped within 90 days but the plant is not subject to certain written plans and reports under Massachusetts Management Requirements. The plant may, however, be classed as a small quantity generator (SQG) or very small quantity generator (VSQG) of other hazardous wastes.

As a large (LQG) or small (SQG) quantity generator of hazardous waste in the State of Massachusetts, the plant is required to:

1. Notify the US Environmental Protection Agency (USEPA), and obtain an USEPA Identification Number for the industrial site;
2. Identify and segregate the plant's hazardous wastes;
3. Label the plant's waste as hazardous waste, describing the waste and the hazards associated with it and the date when accumulation began in each container;
4. Store the plant's waste by type in separate containers that are tightly sealed, and provide appropriate aisle space to meet fire codes;
5. Use a licenced hazardous waste transporter and/or a licenced treatment, storage, or disposal facility, under the condition that the plant, as a generator, has ultimate legal responsibility for the plant's hazardous wastes;
6. Use a uniform hazardous waste manifest as a shipping document for all plant wastes, including waste oil; and
7. Keep records of waste analyses, reports, and manifests for at least three years.

In Massachusetts, there are additional requirements for Large Quantity Generators (LQG) of hazardous waste (but not waste oil):

1. Each manifest must contain a certification that the plant has a program in place to reduce the volume and toxicity of waste generated, as much as is economically practicable;
2. A Biennial Report summarizing the plant's manifest shipments for the previous years must be submitted to the State environmental protection agency;
3. A training program is required for all personnel involved in managing hazardous waste. A written plan is required to specify how the plant's personnel will be familiarized with procedures for using and repairing emergency and monitoring equipment, how the plant's personnel will respond to fire or explosions, potential groundwater or surface water contamination, how to shutdown operations, what the job title and description of each position will be related to hazardous waste management with the requisite qualifications and duties, what training will be provided, and what the qualifications of the relevant training personnel will be; and
4. A written contingency plan is prepared based on a Spill Prevention Control and Countermeasures (SPCC) Plan, or similar emergency plan, which describes the layout of the plant, emergency equipment and handling procedures, places where the plant personnel would normally be working, entrances and exits and evacuation routes; a list including the names, addresses, and telephone numbers of the emergency coordinator(s) must be distributed to local fire and police departments, the mayor, board of health and emergency response teams.

11.13 MANAGEMENT OF HAZARDOUS INDUSTRIAL WASTE FROM MEDICAL OFFICES: AN EXAMPLE

11.13.1 Hazardous and Infectious Wastes from Medical Offices

Federal and State laws define waste as "hazardous" if it is ignitable, corrosive, reactive, or toxic. Other wastes are listed by name. These may differ from lists of hazardous materials, which are regulated by OSHA and Right-to-Know. The Standard Industrial Classification (SIC) of a physician's medical office and a dentist's office are 8011 and 8021, respectively.

If a medical office has photoprocessing waste, typically from x-ray processes, which leaches silver in a concentration of 5 mg/L or more, or has a dental waste which leaches mercury in a concentration of 0.2 mg/L or more, this medical office is a "generator" of hazardous waste, of which concentrations are determined by an extraction procedure toxicity test.

Syringes, sharps, blood products, and the like from hospitals are considered infectious waste and are regulated by the US Department of Public Health. It is recommended that any infectious waste from a medical office be placed in rigid containers and steam-sterilized or autoclaved. A method and a facility for disinfecting and compacting infectious wastes, such as disposal diapers, animal beddings, and so on, have recently been developed by Wang and Wang [59].

11.13.2 Waste Disposal

If the amount of hazardous waste a medical office produces in a month is less than 25 gal (95 L), this medical office qualifies as a very small quantity generator (VSQG) in Massachusetts. As a VSQG, the medical office is required to register with the State regulatory agency, label its wastes as hazardous, and ship it with a licenced hazardous waste hauler or precious metal transporter to a licenced treatment or disposal facility.

The disposal options of this medical office as a VSQG are listed below [5]:

1. The generator may want to reclaim silver from the x-ray waste itself. If its silver recovery equipment is hard-piped and connected to the photoprocessor, this generator is currently exempt from recycling permits.
2. The generator may ship its silver waste to a reclaimer. Be aware that it is the generator's responsibility as the generator of the waste to know where its waste is going and how it is handled.
3. If the generator is a registered VSQG, it may transport its waste to another generator or a receiving facility as long as it carries its VSQG registration in its vehicle, does not transport more than 55 gal (208.18 L) at a time, obtains a receipt for its waste, and retain the receipts for at least three years.
4. Some liquid residues can be discharged to the sewer if they are not classified as hazardous waste. The generator should call its local sewer authority for information. If the generator discharges the waste to a septic tank or dry well, it needs a groundwater discharge permit from the State regulatory agency.

11.14 MANAGEMENT OF HAZARDOUS INDUSTRIAL WASTES FROM GRAPHIC ARTS, PRINTERS, AND PHOTOGRAPHERS: AN EXAMPLE

11.14.1 Requirements

Each State has its own requirements and regulations for management of hazardous wastes at industrial sites. This section presents the Massachusetts requirements for graphic artists, printers, and photographers as a typical example.

Massachusetts Law requires industrial plants that produce hazardous waste to: (a) identify their wastes; (b) count their wastes to determine monthly quantities; (c) manage their wastes properly, based on the State requirements on monthly quantities of hazardous wastes that can be stored; (d) apply for a federal USEPA-ID if the industrial plant is a small quantity generator (SQG) or very small quantity generator (VSQG); or (e) register with the state division of hazardous waste if the plant qualifies as a very small quantity generator (VSQG).

11.14.2 Hazardous Waste Identification

The hazardous wastes generated from graphic artists, printers, and photographers can be identified by their specific wastes number, hazard condition and SIC as shown in Tables 3 and 4.

To identify other hazardous wastes in shop, the three types of the material safety data sheet (MSDS) provided by the supplier of the product should be reviewed. A plant manager can also find out the hazardous ingredients in the processing chemical and refer to the State Hazardous Waste Regulations or call the State Division of Hazardous Waste.

It should be noted that ink and paint wastes may contain certain metals that make the waste "FP Toxic." For more information, the readers are referred to the MSDS, to talk to the manufacturer or an environmental consultant, or have a certified laboratory conduct an Extraction Procedure Toxicity Test on the waste in question.

The following summarizes the current Massachusetts regulations for disposal of photographic wastes containing silver (waste number D011 shown in Table 3).

If less than 25 gal (95 L) of spent fixer is generated each month (assuming there are no other wastes, or if there are other wastes, the total quantity, excluding waste oil, does not exceed 25 gal or 95 L), a generator will face the following situations:

1. The generator may register as a very small quantity generator, and/or obtain a USEPA-ID;
2. No recycling permit is required for the generator; registration is sufficient;
3. The generator may treat spent fixer for reclaiming the silver at the site of generation, or the generator may ship the spent fixer waste offsite with a licenced hazardous waste transporter or a State-approved precious metal transporter and recycling facility, or self-transport up to 55 gal at a time to another generator or a receiving facility;

Table 3 Identification of Hazardous Wastes

Typical waste	Waste number	Hazard
A. Spent solvents		
Ethyl alcohol, isopropanol	D001	Ignitable
Methylene chloride,		
Trichloroethylene	F001	Toxic
Ethyl benzene	F003	Toxic
B. Ink/paint wastes	D001	Ignitable
C. Ink/paint wastes containing metals such as:		
Chromium	D007	EP Toxic
Lead	D008	EP Toxic
D. Etch and acid baths	D002	Corrosive
E. Spent photographic wastes		
Containing silver	D011	Toxic

Table 4 Standard Industrial Classification (SIC)

Type of business	SIC
Graphic arts, photographic labs	7333
Advertising/art	7311
Commercial printing	2751
Miscellaneous publishing	2741

4. If the generator transports to another generator or an authorized facility, the generator must obtain a receipt for the generator's waste and retain records for a minimum of three years.

If more than 25 gal (95 L) of hazardous waste, including spent fixer, is generated in a month, the generator will face different situations:

1. The generator needs an USEPA-ID number and must use a manifest if it ships its waste offsite and may use a licenced hazardous waste transporter or a precious metal transporter and recycling facility;
2. The generator can use a recovery device directly connected by pipe to the film processor at the site of generation (no recycling permit is required: operation can begin within 10 days of the receipt of the application if applicant does not hear from the State); and
3. The generator must meet the concentration limits of the local sewer authority if discharging the waste to a sewer system is intended, or obtain a groundwater discharge permit from the State Division of Water Pollution Control if the generator discharges to a septic system or other groundwater disposal. If the waste or its pretreated effluent meets silver concentration limits of less than 5 mg/L of silver, the waste or the effluent is not classified as a hazardous waste.

11.14.3 A Case History for Disposal of Photographic Wastes by a Large Quantity Generator (LQG)

A graphic arts company in Farmingdale, New York, produced four wastewater streams at Outfall Nos. 001, 002, 003, and 004, as shown in Figure 1.

Outfall	Wastewater flow (gal/day)
001	2000 (average)
002	30,000 (average)
003	5000 (average)
004	5000 (average)

Note: 1 gal/day = 3.785 L/day.

The four wastewater streams discharged at Outfall Nos. 001, 002, 003, and 004 were photographic process wastewater, cooling water (noncontact), sanitary waste A, and sanitary waste B, respectively. A State Pollutant Discharge Elimination System (SPDES) discharge permit was issued to the company in compliance with the Environmental Conservation Law of

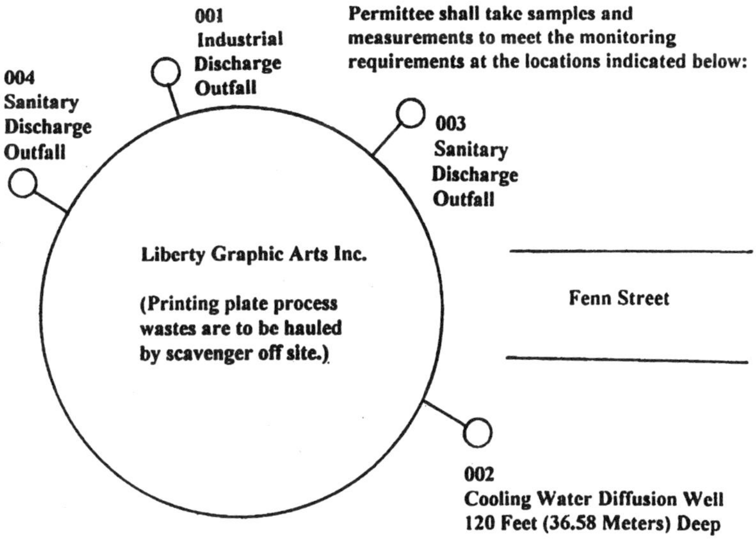

Figure 1 Monitoring locations at Liberty Graphic Arts, Inc.

New York State [13] and the Federal Clean Water Act, as amended. Specifically, the company was authorized to discharge its treated effluents from the company's facility to a nearby receiving water. Table 5 indicates the final effluent limitations and monitoring requirements specified by the SPDES discharge permit.

The company was required to take samples and measurements to meet the monitoring requirements at the outfall location Nos. 001, 002, 003 and 004, as indicated in Figure 1.

The photographic wastewater from the company was collected for treatment in accordance with the technologies described by Bober *et al.* [57] before being discharged to a State-approved receiving water.

A small quantity of wastewater containing extremely toxic pollutants, however, was held and hauled by an approved scavenger.

11.14.4 A Case History for Disposal of Photographic Wastes by a Very Small Quantity Generator (VSQG)

Environmental Situations

A small printing company in Lenox, Massachusetts, United States, produced 24 gal/month (91 L/month) of industrial wastewater mainly consisting of the following two spent chemicals:

- Spent Kodak ultratek fixer and replenisher were accumulated (3 parts water per 1 part fixer dilution; about 100 gal or 378.5 L accumulated); and
- Spent Kodak ultratec developer and replenisher were accumulated (1 part water per 5 parts developer; about 100 gal or 378.5 L accumulated).

The MSDS of both diluted spent chemicals were obtained from the chemical supplier for review by the company's consulting engineer. The following are the chemical descriptions and disposal methods from the MSDS:

- Kodak ultratek fixer and replenisher. This chemical formulation has a high biological oxygen demand, and it is expected to cause significant oxygen depletion in aquatic systems. It is expected to have a low potential to affect aquatic organics. It is expected

Table 5 Final Effluent Limitations and Monitoring Requirements Specified by the New York State SPDES Discharge Permit

During the period April 1, 1981, to April 1, 1986, the discharges from the permitted facility shall be limited and monitored by the permittee as specified below:

Outfall number	Effluent parameter	Discharge limitations other units (specify)		Monitoring requirements measurement sample	
		Daily avg.	Daily max.	Frequency	Type
001	Process flow, photographic waste		–	Continuous	Recorded
	Nitrogen, total – monitor only		10 mg/L	Monthly	Composite
	Cadmium, total – monitor only		–	Monthly	Composite
	Dissolved solids, total		1000 mg/L	Monthly	Composite
	Chemical oxygen demand		150 mg/L	Monthly	Composite
	Color, units – monitor only		–	Monthly	Composite
	Iron, total		0.6 mg/L	Monthly	Composite
	Phenols		0.002 mg/L	Monthly	Composite
	Zinc, total		5.0 mg/L	Monthly	Composite
	pH Units – Range	6.5–8.5		Daily	Grab
002	Cooling water, noncontact – no chemical treatment allowed				
003	Sanitary wastes only				
004	Sanitary wastes only				

Notes:
1. All wastewater discharge from the printing plate process are held and hauled by an approved scavenger.
2. Approximate flows are as follows:

Outfall	Flow (gal/day)
001	2000
002	30,000
003	5000
004	5000

to have a moderate potential to affect secondary waste treatment microorganisms. It is expected to have a moderate to high potential to affect the germination and growth of some plants. The components of this chemical formulation are biodegradable and are not likely to bioconcentrate. If diluted with a large amount of water, this chemical formulation released directly or indirectly into the environment is not expected to have a significant impact.

- Kodak ultratec developer and replenisher. This formulation is a strongly alkaline aqueous solution, and this property may cause adverse environmental effects. It has a low biological oxygen demand and is expected to cause little oxygen depletion in aquatic systems. It is expected to have a high potential to affect aquatic organisms and a moderate potential to affect secondary waste treatment microorganisms and the germination and growth of some plants. The organic components of this chemical formulation are readily biodegradable and are not likely to bioconcentrate. The direct instantaneous discharge to a receiving body of water of an amount of this chemical formulation that will rapidly produce, by dilution, a final concentration of 0.1 mg/L or less is not expected to cause an adverse environmental effect. After dilution with a large amount of water, followed by a secondary waste treatment, the chemicals in this formulation are not expected to have any adverse environmental impact.

Both spent chemicals were analyzed by a certified laboratory. The analytical data of the two spent chemicals were:

- Spent Kodak ultratek fixer and replenisher (i.e., spent fixer): COD = 161,000 mg/L, silver = 1384 mg/L, total solid = 6%; and
- Spent Kodak ultratec developer and replenisher (i.e., spent developer): COD = 103,000 mg/L, silver = 0 mg/L.

The company produced less than 100 kg (220 lb or approximately 25 gal) of hazardous waste a month. Accordingly, it was eligible to be registered as a very small quantity generator (VSQG).

To qualify as a VSQG, the company owner notified the Massachusetts Department of Environmental Protection (DEP) on a two-part registration form, which listed the types of hazardous waste generated, the amount of each in gallons per month, and the proposed disposal, treatment, storage, and/or recycling destination of the waste.

The owner's registration was effective as soon as it was received by the State. Renewal would occur after one year.

As a VSQG, the company had never accumulated more than 165 gal (624.5 L) at any one time. As a registered VSQG, the company's owner also tried four different treatment and disposal methods for his hazardous wastes.

Options for Recycling, Treatment, and POTW Discharge

Initially the owner tried to recycle or treat his wastes, because the process described in his registration was acceptable to the Massachusetts DEP. It appeared that the silver in the spent fixer had to be removed, and the remaining pollutants were mainly biodegradable organics. Four silver removal methods were considered by the owner [58].

Chemical Recovery Cartridge Metallic Replacement method

In this method, a metal (usually iron) in a chemical recovery cartridge (CRC) reacts with the silver thiosulfate in the spent fixer and goes into solution. The less active metal (silver) settles out as a solid. To bring the silver into contact with the iron, the spent fixer is passed through the CRC container, which is filled with steel wool. The steel wool provides the source of iron to replace the silver. The main advantages of this CRC method are the very low initial cost (cartridges cost about US$60) and the simplicity of installation; only a few simple plumbing connections (shown in Fig. 2) are required.

The main disadvantages, compared to the electrolytic method, are that the silver is recovered as a sludge, making it more difficult to determine the exact amount recovered. The recovered sludge containing silver requires more refining processes than the plate silver obtained

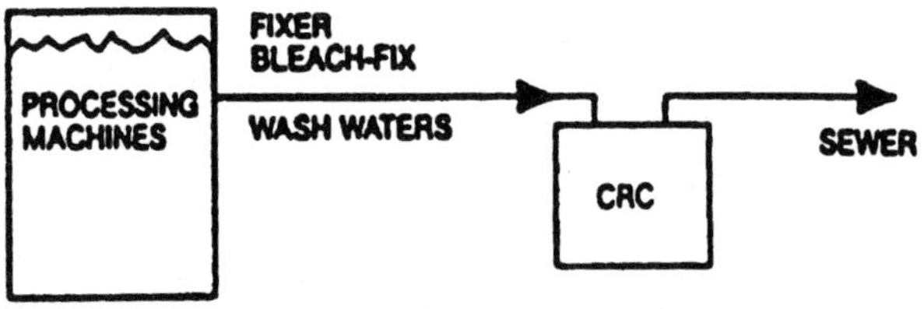

Figure 2 Chemical recovery cartridges (CRCs). (Courtesy of Eastman Kodak Co., NY.)

from electrolytic methods, if silver recovery and silver refining are both intended. Also, CRCs cannot be reused. They must be replaced when they are exhausted.

In summation, the silver chemical recovery cartridge method (Fig. 2) can achieve silver recovery efficiencies of greater than 90%. However, it is difficult to achieve this level of recovery consistently, making it an unreliable choice if the operator needs to meet low silver discharge limits. Another problem with the chemical recovery cartridge method is that as silver is recovered, the steel wool becomes soluble, producing iron levels in the effluent as high as 3000 mg/L. Iron is regulated to levels well below those concentrations by many sewer codes.

Electrolytic Silver Recovery Method

In this method, the silver-bearing solution is passed between two electrodes through which a controlled direct electric current flows as shown in Figure 3. Silver plates out on the cathodes as almost pure metal.

The advantages of the electrolytic method is that silver is recovered in an almost pure form, making it easier to handle and less costly to refine. With careful monitoring, it also permits fixer reuse for some processes. It also avoids the need to store and replace cartridges, as with the

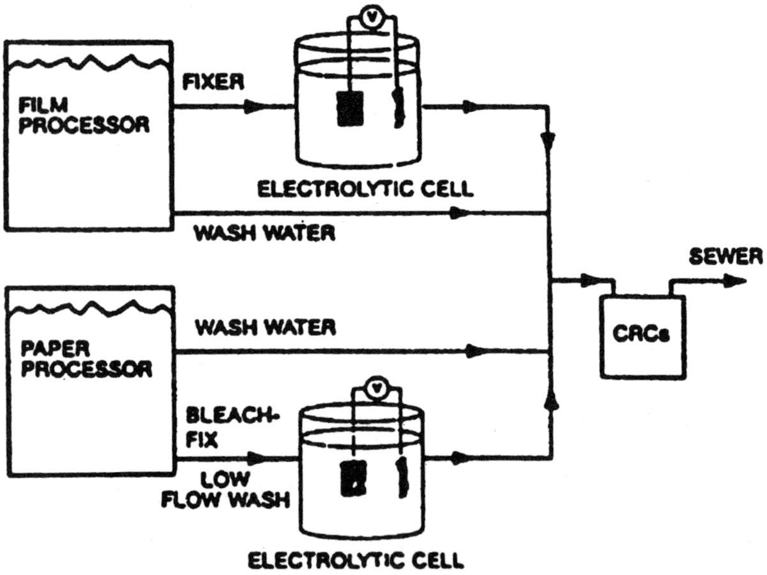

Figure 3 Electrolytic cells plus CRCs. (Courtesy of Eastman Kodak Co., NY.)

metallic replacement method. Recovery efficiency is typically 93–97%, and by maintaining the correct mix of processing effluent, can be as high as 99%.

The disadvantages of electrolytic methods are the difficulty in reducing silver in the effluent to very low levels, and the careful monitoring required to avoid silver sulfide formation. Initial capital investment is high. None of these disadvantages is a serious deterrent. The concentration that can be achieved depends on how low the current density can be set with the unit. As the silver concentration gets lower, the current density can be set lower to prevent silver sulfide from forming. With low current densities, a large cathode area is needed to achieve the necessary silver recovery rate. In order to reduce the residual silver concentration in the electrolytic cell effluent further, at least one CRC is used for finally polishing the electrolytic cell effluent (Fig. 3).

Conventional Ion-Exchange Method

There are two ion-exchange methods that have been used in photoprocessing laboratories to recover silver from dilute solutions: conventional ion exchange (Fig. 4) and in situ ion exchange (Fig. 5). With both of these ion-exchange methods, the silver is removed by pumping it through a column of anion-exchange resin. The difference between the two ion-exchange methods is the regeneration step.

In the conventional ion-exchange method (Fig. 4), the silver is removed from the resin by regenerating it with thiosulfate solution. The silver is then removed from the regenerant by running it through an electrolytic cell. The greatest advantage of using the conventional ion-exchange method for silver recovery is that the operator can reduce the silver in the processing effluent to very low levels (0.1–2 mg/L). In areas that strictly regulate the discharge of silver, it may be the only recovery method that is satisfactory.

The conventional ion-exchange method also has some major disadvantages, such as the high capital investment (both an ion-exchange unit and an electrolytic unit are needed), and the increased complexity of operation (only a few high-volume laboratories have used this method successfully). However, it remains an option for those laboratories that must meet strict limits on

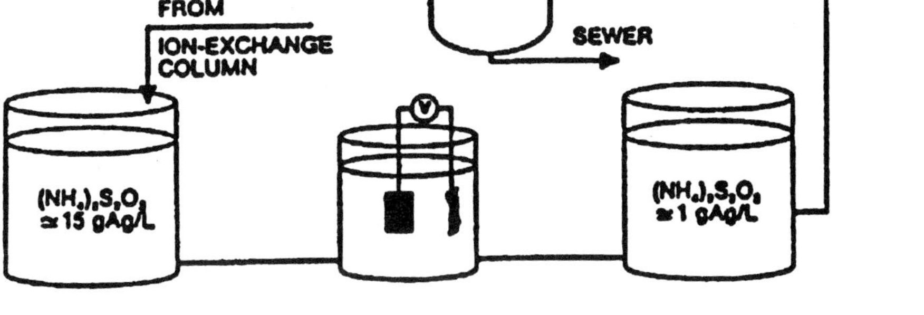

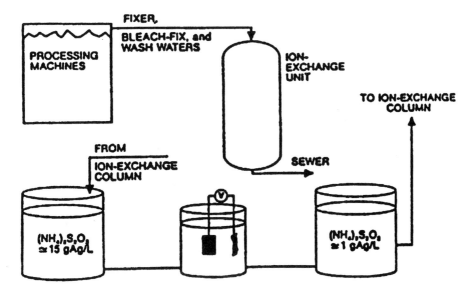

Figure 4 Conventional ion exchange. (Courtesy of Eastman Kodak Co., NY.)

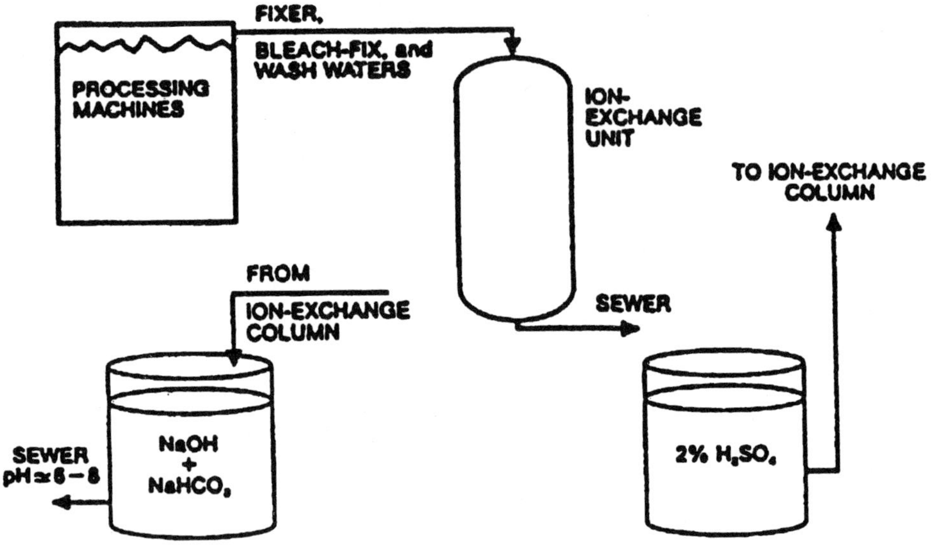

Figure 5 In situ precipitation. (Courtesy of Eastman Kodak Co., NY.)

the amount of silver discharged. It is also critical that the operator dilutes the concentrate with the proper amount of wash water prior to ion-exchange treatment; too high a thiosulfate concentration in the solution being treated will cause silver to leak through the column.

In Situ Ion Exchange Method

With the in situ ion-exchange method (Fig. 5), dilute sulfuric acid is used to precipitate the silver in the resin beads as silver sulfide instead of removing it with regenerant. The resin that is inside the ion exchange unit is used for many cycles without a loss in capacity. When the resin eventually loses its capacity to recover silver, or when there is sufficient silver to make recovery worthwhile, it is sent to a silver refiner who incinerates it to remove the silver. This may occur after between six months and a year.

The advantages of using the in situ ion exchange method for silver recovery is similar to that of using the conventional ion exchange method. The disadvantages of the in situ method are that it requires a greater capital investment, and more chemical handling than with either chemical recovery cartridges or electrolytic cells. Also, the pH of the spent regenerant must be adjusted as it is discharged from the columns to prevent the formation of sulfur dioxide, and to be sure the discharge meets the local sewer codes.

Ion-exchange methods are not recommended to be used by VSQG to recover silver from spent fixers or bleach-fixes. They are suitable only for recovering silver from dilute solutions, like washwater, or a combination of fixer, bleach-fix, and washwater.

Electrolytic Cell Plus In Situ Ion-Exchange Method

Figure 6 illustrates a combined system involving the use of both the electrolytic cell and the *in situ* ion-exchange unit. The combined system (Fig. 6) produces an excellent effluent with lower residual silver in comparison with the chemical recovery cartridge method (Fig. 2), electrolytic silver recovery method (Fig. 3), the conventional ion-exchange method (Fig. 4), and

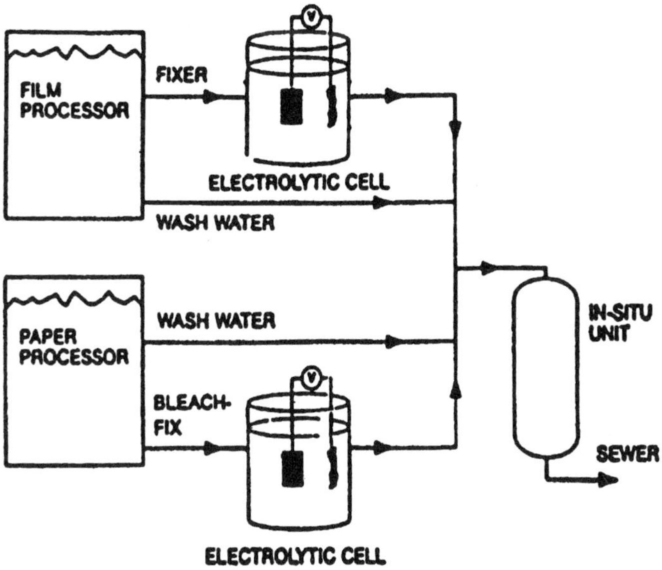

Figure 6 Electrolytic plus in situ precipitation. (Courtesy of Eastman Kodak Co., NY.)

the in situ ion exchange (Fig. 5). The disadvantage of the combined system is its high capital and operating costs.

Hydroxide Precipitation Method

In this method, sodium hydroxide, potassium hydroxide, calcium hydroxide, magnesium hydroxide, or sodium aluminate can be fed to the spent fixer for precipitation of silver ions as insoluble silver hydroxide precipitates. Figure 7 indicates that the residual silver concentration in the hydroxide precipitation treated effluent can be about 1 mg/L at pH 12 [19].

The advantage of this method is its low cost. Its disadvantages are: (a) the residual silver concentration in the treated effluent (about 1 mg/L at pH 12) may exceed the local regulatory agency's effluent limit on silver; (b) the hydroxide sludges produced in the hydroxide precipitation method require further thickening and dewatering treatment and final disposal; and (c) refining silver from the precipitated silver hydroxide sludges is difficult [87].

Sulfide Precipitation Method

In this method, sodium sulfide, potassium sulfide, and/or ferrous sulfide can be dosed to the spent fixer during mixing at an alkaline pH range, for precipitation of silver ions as insoluble silver sulfide precipitates [87]. Figure 7 [19,59] indicates that the residual silver concentration in the sulfide precipitation treated effluent can be below 10^{-9} mg/L in the entire alkaline range, and can be as low as 10^{-12} mg/L at pH 10.5.

There are two advantages for this method: (a) the capital and operating costs are low; and (b) silver removal efficiency is extremely high. There are a few disadvantages for the method: (a) the sulfide sludges produced in the sulfide precipitation method require further thickening and dewatering treatment, and final disposal; (b) refining of silver from the precipitated silver sulfide sludges is not easy; and (c) hydrogen sulfide toxic gas may be produced from the sulfide precipitation process system if the pH of the spent fixer is controlled in an acid range by accident.

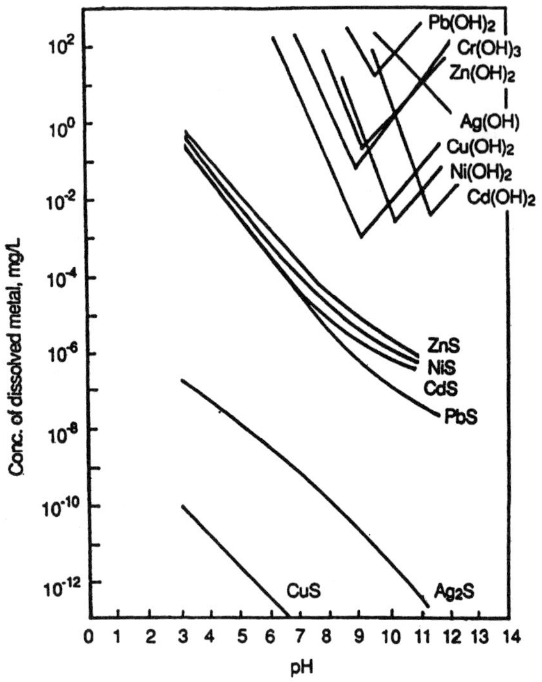

Figure 7 Solubilities of metal hydroxides and metal sulfides. (Courtesy of USEPA.)

Process Comparison and Selection

Selection of a suitable method for silver removal depends on many factors; what processes the company uses, what volume of wastes the company produces, what kind of training and technical knowledge the company's personnel has, whether the company wants to reuse the company's fixer or bleach-fix, how much the company wants to spend for recovery equipment, and what the environmental concerns are, such as how strict the effluent discharge limits are. Just considering these factors makes choosing a silver recovery method very much an individual decision for each company.

Table 6 summarizes the silver removal efficiencies of the various process methods identified above. The silver concentration that can be discharged to a treatment plant or to a receiving body of water is often regulated even though silver in photographic effluent is in a form

Table 6 Typical Silver Concentrations in Effluent After Recovery

Recovery method	Silver concentration (mg/L)
Chemical recovery cartridges (CRCs)	10–20
Electrolytic (with tailing CRCs)	1–5
In situ ion exchange	0.1–2
Conventional ion exchange	0.5–2
Electrolytic (with tailing in situ ion exchange)	<0.1–2
Hydroxide precipitation	1–5
Sulfide precipitation	<0.1

that is not harmful. Therefore, cost is only one of the primary considerations in choosing the company's silver-recovery method.

The company had considered all silver recovery options. The main factor was the company's processing volume. Other economic considerations were the price of silver and operating and refining costs. If the company's processing volume were to be high, the company would probably want to make frequent cartridge replacements or set aside a large amount of storage area for spent and replacement cartridges. Although initial capital investment with electrolytic recovery cells was higher than with chemical recovery cartridges, there would not be the recurring cost of equipment replacement. If the company were to use an electrolytic cell, the company's refining costs for the recovered silver would be much lower than with other methods because the silver-plated out would usually be more than 95% pure.

If the company were a large-volume operation, the in situ ion exchange would also be an option. The company could use this method for primary treatment. The company could also use it to tail an electrolytic unit if the company were first to dilute the discharge from the electrolytic cell with washwater. Using this method would enable the company to recover the maximum amount of silver and to minimize the amount of silver discharged. It would require a greater capital investment and more chemical handling than with electrolytic cells or chemical recovery cartridges in accordance with the information from Kodak Company [58].

However, the company's wastewater volume was actually very low, and chemical recovery cartridges, hydroxide precipitation tanks, and sulfide precipitation tanks become reasonable choices for silver recovery. Chemical recovery cartridges and the two types of precipitation tanks were all very simple to install. The costs for purchasing, installing, operating, and monitoring this equipment are very low compared with other methods.

In comparison with the silver recovery/removal efficiencies of the chemical recovery cartridge (CRC) method, the hydroxide precipitation method, and the sulfide precipitation method shown in Table 6, the two precipitation methods appeared to be a better choice than the CRC method.

Considering the silver removal efficiencies of the various process methods in Table 6 (the lower the residual silver concentration in effluent after treatment, the higher the silver removal efficiency), the company's local code limits for silver, the ease of process operation, the safety, the costs, the volume of waste production, and the silver content in the spent fixer, the company finally selected the hydroxide precipitation method as the first-stage treatment, and the sulfide precipitation method as the second-stage treatment for silver removal from the spent fixer. It should be noted, however, that sulfide precipitation alone would have been sufficient. After silver was significantly removed from the spent fixer, both the treated spent fixer and the untreated spent developer were mixed together, forming a pretreated combined wastewater for possible discharge to a POTW. The analytical data of the pretreated combined wastewater are: COD = 132,000 mg/L; silver = <0.1 mg/L; and pH = 9.5. At this stage, the pretreated combined wastewater was no longer considered to be hazardous because it contained only a high concentration of biodegradable organics in terms of 132,000 mg/L of COD.

It is important to note that if the precipitation tanks were hard-piped and connected to the company's processing units, the pretreated combined wastewater would not be considered to be a hazardous waste legally, and would be allowed to be discharged into the POTW without any legal problems. The precipitation tanks of the company, however, were not hard-piped to the company's processing units. An application for a Permit for Sewer System Extension or Connection was then officially filed at the local town of Lenox, Massachusetts, by the consulting engineer, on behalf of the company. It was proposed that a permit be issued by the town and the State for the company to discharge a design flow ranging from 5 gpd (average) to 20 gpd (maximum) of the aforementioned pretreated combined wastewater into an existing Lenox

POTW system (1 gpd = 3.785 L/day). It should be noted that actual company's wastewater flow was only 24 gal/month (90.84 L/month). The average sewage flow of the Lenox POTW was 0.4 MGD (1514 m^3/day), and the BOD/COD ratio of the waste was determined to be 0.65. If a permit were issued to the company for discharging the pretreated wastewater to the Lenox POTW, an increase in silver concentration would be negligible, and an increase in BOD in the Lenox POTW would only be by about 1 mg/L during discharging of the pretreated wastewater at an instantaneous flow as high as 5 gpm (18.93 L/min). Besides, the organics in the pretreated wastewater were biodegradable in accordance with the MSDS. Under normal situation, a sewer discharge permit could have been granted because the company's pretreated wastewater would not adversely affect the normal operation of biological wastewater treatment at the Lenox POTW. The town of Lenox was too small to have a licenced engineer to handle the legal case. The company was advised by the town to haul their small quantity of pretreated wastewater to the nearby city of Pittsfield's POTW for disposal because there was an agreement between the town and the city.

The transportation of the pretreated wastewater from the company to the city, which was only 6 miles away, had to comply with all government rules and regulations because the pretreated wastewater was legally considered to be a hazardous waste, although technically it was not. The company faced a transportation problem because of its high cost.

Option of Transporting Wastes to Another Generator

The company could transport its untreated wastewaters or pretreated wastewater(s) to another generator who is in compliance with the regulations and who will count the company's waste as part of another generator's waste. Another generator, J.F. Co., Inc., was found in Springfield, Massachusetts, which was about 80 miles away from the company. J.F. Co., Inc., agreed to accept the spent Kodak Ultratec Fixer and Replenisher (i.e., the spent fixer) containing 1400 mg/L silver for silver recovery at a cost to the Lenox Company of US$2.00 per gallon, delivered to J.F. Co., Inc., in Springfield. While the cost quoted for disposal of the spent fixer was reasonable, the company in Lenox faced two other problems: (a) the spent developer containing no silver but a high concentration of dissolved organic carbon (DOC) still needed to be disposed of; and (b) the Lenox company, which was the original generator, would have to take full responsibility for whatever the actions were to be taken by the other generator, which was not well-known in the field.

The company was discouraged by the option.

Option of Transporting Wastes to a Licenced Facility

The company in Lenox also had an option to transport its wastes to a licenced treatment, storage or disposal facility, or permitted recycling facility, with the facility's permission. There were many licenced facilities in Massachusetts that were willing to accept the company's spent fixer and spent developer for final disposal.

As a registered VSQG, the company might transport its own hazardous waste under the following conditions: (a) the company transports only the waste that is generated on its premises (no problem); (b) the company does not transport more than 55 gal or 208.2 L at one time (no big problem but time consuming); (c) the company does not transport incompatible wastes in the same shipment (no problem); (d) the company's waste is in containers that are tightly sealed, labeled as "HAZARDOUS WASTE," with the name of the waste and the type of hazard, and are tightly secured to the vehicle (no problem); (e) the company keeps a copy of its registration as a VSQG in the vehicle while transporting its waste (no problem); (f) the company is in compliance with all USDOT and Massachusetts Department of Public Safety requirements (a problem to

VSQG); and (g) in the event of a spill or leak of hazardous waste that may threaten humans or the environment, the company shall notify the Massachusetts DEP or the State Police (no problem).

The company's official Massachusetts DEP assigned SIC number was 2751. The two wastewater classifications and waste numbers were as follows: (a) spent fixer: waste number is D011, classified as "toxic"; and (b) spent developer: no waste number, classified as "corrosive," neutralization is recommended but not required before transportation.

If the company decided not to use a licenced transporter but would be transporting its own wastes, the company did not need an USEPA-ID number or manifest form. The company must, however, keep a record of the type and quantity, as well as the date, method of transport, and treatment/disposal of its waste(s). The company would need proof of the receipt of the waste by the treatment/disposal facility. The company or its consulting engineer must keep receipts or manifests of waste shipped and records of waste analysis for at least three years, or for the duration of any enforcement action by the Massachusetts DEP.

Apparently this option was technically and economically feasible for the small company in Lenox. Unfortunately small (SQG) or very small (VSQG) generators similar to this Lenox company simply cannot find out the latest USDOT requirements, the Massachusetts Department of Public Safety requirements, the Massachusetts Department of Environmental Protection requirements by themselves without hiring a consulting engineer. The State regulatory agency(s) should provide more technical assistance to VSQG and SQG, whenever possible.

Option of Using a Licenced Transporter and Facility

The company's owner finally decided to use a licenced transporter and a licenced facility for transportation, treatment, and disposal of its untreated wastes, even though the option of recycling, treatment, POTW discharge, and the option of self-transportation to a licenced facility were equally feasible. The licenced transporter and the licenced facility can be owned by two different firms or by one firm. In this particular case, the company in Lenox, which was a VSQG, selected an environmental service company in Albany, New York, which was a licenced transporter as well as a licenced facility. The costs for picking up, transporting, and disposing of six 55 gal drums of photographic developer and fixer solution in 1991 are documented in Table 7.

The prices given in Table 7 were based upon the following conditions:

- Free and easy access for the transporter/facility personnel to work site;
- Applicable taxes and state regulatory fees are not included in quoted process;
- A fuel usage surcharge of half of 1% will be added to the invoice total to cover rising fuel cost;

Table 7 Costs for Picking Up, Transporting and Disposing of Chemicals

Method	Waste material	Container	Charge
A. Picking up for disposal	Spent Kodak ultratec developer and replenisher	55 gal drum	$145/drum
B. Picking up for disposal	Spent Kodak ultratek fixer and replenisher	55 gal drum	$130/drum
C. Transportation of drum to the Massachusetts disposal facility (maximum 6 drums per load)			US$300.00/load

- Waste material conforms to waste profile sheets;
- All drums are centrally located and DOT approved;
- Drums are in shippable condition; and
- Transportation rate allows one hour for loading time; additional time required will be billed at US$75.00 per hour.

11.15 RECYCLING OF HAZARDOUS INDUSTRIAL WASTES AS WASTE-DERIVED FUELS

11.15.1 Introduction and Objective

Southdown, Inc., Houston, TX, engages in the cement, ready-mixed concrete, concrete products, construction aggregates, and hazardous waste management industries throughout the United States. According to Southdown, they are making a significant contribution to both the environment and energy conservation through the utilization of waste-derived fuels as a supplemental fuel source. Cement kiln energy recovery is an ideal process for managing certain organic hazardous wastes. The burning of wastes or hazardous wastes as supplemental fuel in the cement and other industries is their engineering approach.

By substituting only 15% of its fossil fuel needs with solid hazardous waste fuel, a modern dry-process cement plant with an annual production capacity of 650,000 tons of clinker can save the energy equivalent of 50,000 barrels of oil (or 12,500 tons of coal) a year. Southdown typically replaces 10–20% of the fossil fuels it needs to make cement with hazardous waste fuels.

By using hazardous waste fuels, the nation's hazardous waste (including infectious waste) problem can be at least partially solved economically.

11.15.2 Cement Kiln Energy Recovery System

The cement kiln is a long, inclined cylinder that can be hundreds of feet in length and up to 15 ft in diameter. Raw materials, such as limestone, clay, sand, and a small amount of iron-containing substances enter at one end and cement clinker, the product, exits at the other. Material temperatures required to make cement clinker must be maintained at a minimum of 2450°F while gas temperatures inside the kiln can reach 3500°F. During operation, the kiln slowly rotates to ensure a thorough blending and "cooking" of the raw materials. These raw materials are heated using fossil fuels (about 85%) along with hazardous waste fuels (about 15%) in the huge kiln at such high temperature until they chemically combine to become marble-sized nodules called "clinker." The clinker is then mixed with gypsum and ground to a fine powder to make cement. Cement, in turn, is a key ingredient in concrete, which is a vital component of the world's roads, buildings, houses, and offices.

Cement kilns manage destruction of organics in hazardous waste through a high-temperature combustion. This involves heating the waste to a sufficient temperature, keeping it in the kiln for enough time, and providing the waste with sufficient oxygen. Because this method destroys organic chemical wastes, such as paint thinners, printing inks, and industrial cleaning solvents, combustion has become the preferred method of managing them and utilizing their BTU value. The conditions in the kiln ensure, and U.S. Environmental Protection Agency (USEPA) regulations require, that 99.99% or more of organic hazardous wastes are destroyed, that is, converted to carbon dioxide and water vapor.

Exhaust gases leaving the kiln pass through highly efficient air pollution control devices such as baghouse filters or electrostatic precipitators. The high temperatures required to make cement destroy 99.99% or more of the organic hazardous wastes. The content of hydrocarbons and carbon monoxide in stack emissions is monitored to ensure that the combustion process is optimized. When combustion is efficient, emission of carbon monoxide is minimized and hydrocarbons disappear. In this way, operators are assured that a destruction efficiency of 99.99% or more is always maintained and that they stay within the stringent limits on emissions set by USEPA, which has identified ten metals that must be controlled. The list includes antimony, arsenic, barium, beryllium, cadmium, chromium, lead, mercury, silver, and thallium. All cement kilns that want to recycle hazardous waste as fuels will have to meet stringent limits on emissions of these metals.

11.15.3 Cement Kiln Monitoring and Control

Under the newly adopted federal regulations for facilities using hazardous waste fuels, cement kilns must comply with stringent testing and permitting requirements before they can recycle the wastes. These procedures ensure that cement companies wanting to recycle hazardous wastes as fuel will do so safely. Those facilities unable to meet the rigorous RCRA standards will not be allowed to burn hazardous waste fuel. Under the USEPA BIF rule ("Burning Hazardous Waste in Boilers and Industrial Furnaces"), cement kilns recycling hazardous waste as fuel are now perhaps the most regulated form of thermal treatment. Major components of the regulatory approach include monitoring and control – allowing operators to detect problems in the process and control the system on a continuing basis. This ensures the process always stays within a safe window of operating conditions – and that emissions always remain within the strict limits prescribed by USEPA.

Under USEPA's BIF rule, manufacturers are required to closely monitor numerous conditions in the kiln and to observe limits on the following aspects of the process: (a) the maximum feed rate of hazardous waste fuel; (b) the maximum feed rate of metals from both raw materials and fuels; (c) the maximum feed rate of chlorine from raw materials and fuels; (d) the maximum feed rate of raw materials; (e) the maximum temperature at the inlet to the air pollution control devices; (f) the maximum concentration of carbon monoxide and total hydrocarbons in the flue gas; (g) the maximum temperature in the combustion zone or minimum temperature at the kiln inlet; and (h) any decrease of pressure at the baghouses or any decline in the strength of the electric field of electrostatic precipitators (both are types of air pollution control devices).

Cement manufacturers use a number of quality control measures. Key among these is careful selection of fuels for recycling. Fuels that contain metals above specified levels, for instance, will be rejected. For that reason, each shipment of fuel is carefully analyzed to determine its ingredients. If the fuel fails to meet predetermined specifications, it will not be used.

There are two primary sources of controls on recycling hazardous waste fuels in cement kilns. First, cement kiln operations are tightly regulated on both the Federal and State level. These regulations cover everything from the transportation of the fuel to the conditions that must be maintained in the kiln. Using USEPA's new, highly sensitive Toxicity Characteristics Leaching Procedure (TCLP), scientists have confirmed that the chemical reactions that must take place in order to make cement prevent unacceptable concentrations of metals from being released from cement or the concrete. Second, because the chemistry of cement-making is both sensitive and precise, manufacturers cannot afford to put anything into their kilns that could produce variations in the clinker. If they did, the cement would not meet the rigorous, industry-wide product quality standards set by the American Society of Testing and Materials (ASTM). ASTM specifies tests and test methods [60] to ensure uniform controls on cement producers nationwide. Before the product can be called Portland cement, tests must show it has the required chemical composition. It also must pass tests measuring physical qualities, such as strength and particle fineness. In this way, product quality is assured regardless of what raw materials or fuels are used.

11.15.4 Permit System for Process Operation, Waste Transportation, and In-Plant Waste Handling

All cement kilns burning hazardous waste fuels will have to obtain a permit from USEPA and local regulatory agencies. Because the permitting process can be lengthy, cement kilns already burning hazardous wastes will be subject to regulation almost immediately under what is known as "interim status." Interim status is a standard regulatory approach used when new regulations are approved under RCRA to bring existing facilities under the new regulations without delay.

Trucks transporting hazardous wastes to cement kilns are regulated by Federal and State transportation agencies. This means they are controlled every step of the way. All trucks must meet U.S. Department of Transportation standards, which require all hazardous wastes to be transported under strict conditions in specially designed containers. State transportation agencies test and licence truck drivers to ensure they understand the precautions required with these fuels.

Both Federal and State regulations under the RCRA specify storage and handling requirements designed to ensure safe operations. For example, the facilities for unloading, storing, and transporting hazardous wastes to the kiln are built with government-approved systems designed to prevent sparks and accidental fires. Such areas are also designed to meet or exceed Federal and State standards for environmental safety, including secondary containment in the unlikely event of a spill.

11.15.5 Health Effects and Risks

Because cement kilns effectively destroy more than 99.99% of organic chemical wastes and emissions are tightly controlled by the BIF rule and other regulations, only minute amounts of organic compounds are emitted and testing has indicated that these emissions are independent of fuel type. In fact, organic emissions are sometimes reduced through the use of waste fuels. The quantity is so small that it does not present a perceptible increase in risks to public health or the environment. Cement kiln exhaust gases typically contain less than one-tenth of the hydrocarbons present in automobile exhaust gases.

Because cement kilns are so good at destroying organic chemical wastes, emissions of dioxins – or any other type of products of incomplete combustion (PIC) – are so low they pose no danger to the environment. In the case where some of the hazardous waste fuels used contain toxic dioxin, the cement kiln temperatures of 1650°F will destroy dioxins in less than one second. Because cement kilns operate at much higher temperatures (at least 2450°F), and because the burning wastes have an average residence time in the kiln of at least two seconds, any dioxins are destroyed. However, dioxin waste is never accepted by Southdown for use in its cement kilns.

Cement made with hazardous waste fuels contains essentially the same amount of metals as cement made using traditional fossil fuels, such as coal, coke, or oil. Also, tests show cement made with hazardous waste fuels has essentially identical leaching characteristics as those of cement produced solely with traditional fuels. This means the metals are no more likely to leach out of the cement made using hazardous waste fuels than if it were made using coal, coke, or oil.

The TCLP tests are performed by subjecting samples to a much harsher environment than would be encountered in natural surroundings. The samples of concrete are pulverized to maximize exposure to the acid used. Next, a particularly harsh solution of acetic acid is employed. Acid can leach out much higher concentrations of metals than liquids to which concrete is normally exposed, such as rain or groundwater. Finally, the amount of acid solution used is very large in comparison to the amount of concrete.

Health risks to residents near cement kilns may actually decrease when hazardous waste fuels are used. This is because the permit needed to recycle hazardous waste fuels requires more stringent emissions controls than those for cement kilns using only fossil fuels. Also, fossil fuels contain natural impurities that are reduced or no longer emitted when some types of hazardous waste fuels are used.

11.15.6 Southdown Experience in Waste Fuel Selection

Production of high-quality cement and compliance with environmental regulations are Southdown's top priorities. Therefore, great care is taken to ensure that only those wastes that can be safely recycled and that are compatible with the cement manufacturing process are used. Cement production requires fuels with a high energy value. Waste materials that provide enough heat include such familiar items as paint thinners, printing inks, paint residues, and industrial cleaning solvents. Cement kilns can also help alleviate one of the most difficult solid waste problems – scrap tires, which take up valuable landfill space. Tires (also known as "tire-derived fuel") can be used as an efficient fuel in cement kilns.

Before Southdown accepts any waste materials for recycling as fuel, a chemical analysis must be performed to identify their chemical composition. Wastes that cannot be blended to meet standards for content, heat value, and compatibility with cement production are not accepted. For instance, cement cannot be made with fuels that have a high chlorine content.

Both fossil fuels and hazardous waste fuels used in Southdown cement kilns contain metals. The raw materials (limestone, clay, sand) used to make cement clinker also contain metals. In fact, certain metals, such as iron and aluminum, are essential components of the final product. While metals cannot be destroyed, the Southdown cement kiln process effectively manages them in the following ways: (a) cement kiln operators limit emissions by carefully restricting the metals content in wastes accepted for recycling; (b) dust particles containing metals are returned to the kiln through closed-loop mechanisms, where metals are chemically bonded into the cement clinker; (c) particles not returned to the kiln are captured in state-of-the-art pollution control devices; and (d) small amounts are emitted from the stack in quantities strictly limited by USEPA's BIF rule.

Electrostatic precipitators and baghouses are used to catch dust particles containing metals. Electrostatic precipitators use an electrical field to remove the particles. Baghouses use fiberglass filters, similar to vacuum cleaner bags, to catch them. The majority of theses particles, called cement kiln dust (CKD), are trapped by this equipment and returned to the kiln for incorporation into the cement clinker. Under USEPA's BIF rule, Southdown tests its cement kiln dust to judge whether it is hazardous. If the CKD does not meet the standards set under the BIF rule, it must be disposed of in accordance with USEPA's strict hazardous waste regulations. For that reason, Southdown does not accept fuels that would cause the waste CKD to fail this test.

11.15.7 Southdown Experience in Product Quality Monitoring

A concrete made from Southdown cement is called a "Southdown concrete." Even under the TCLP testing extreme conditions, the amount of metals that leached out of the Southdown concrete were many orders of magnitude below the standards set by USEPA. In many cases the levels were, in fact, below the limits of detection for the test. One historical use of Southdown concrete has been for pipes used to transport drinking water. Drinking water is routinely tested to show that it meets Federal standards for a wide variety of contaminants, including metals. If metals leaching from the concrete pipes were a concern after many years of use, either USEPA or another recognized scientific organization would have sounded a warning.

The water distribution system in the city of Dayton, OH, uses Southdown concrete water mains to deliver water to its citizens. Routine sampling and testing of Dayton's water supply by the city's Department of Water consistently shows that the levels of metals are well below the Ohio EOA Community Drinking Water Standards, and that these levels have remained constant throughout a nine-year testing period from 1982 to 1990. Because metal leaching has not occurred, there is no reason for concern over the safety of Southdown concrete pipes to transport drinking water.

REFERENCES

1. Wang, L.K.; Wang, M.H.S.; Wang, P. *Management of Industrial Hazardous Substances at Industrial Sites*; United Nations Industrial Development Organization (UNIDO): Vienna, Austria, 1995; Training Manual no. 4-4-95, 104 p.
2. Wang, L.K. *Case Studies of Cleaner Production and Site Remediation.* United Nations Industrial Development Organization (UNIDO): Vienna, Austria, April 1995; Training Manual no. 5-4-95, 136 p.
3. WPCF. *Hazardous Waste Treatment Processes*; Water Pollution Control Federation: Alexandria, VA, 1990.
4. Massachusetts DEP. *Massachusetts Hazardous Waste Regulations, 310CMR30.000*; Massachusetts Department of Environmental Protection: Boston, MA, 2000.
5. Massachusetts DEP. *Massachusetts Hazardous Waste Information for Medical Offices*; Massachusetts Department of Environmental Protection: Boston, MA, 1988.
6. Massachusetts DEP. *Large Quantity Generator Fact Sheet*; Massachusetts Department of Environmental Protection: Boston, MA, 2000.
7. Massachusetts DEP. *Small Quantity Generator Fact Sheet*; Massachusetts Department of Environmental Protection: Boston, MA, 2000.
8. Massachusetts DEP. *Application for an USEPA Identification Number*; Massachusetts Department of Environmental Protection: Boston, MA, 1992.
9. Massachusetts DEP. *Guide for Determining Status and Regulatory Requirements*; Massachusetts Department of Environmental Protection: Boston, MA, 1992.
10. Massachusetts DEP. *How Many of Your Common Household Products are Hazardous?* Massachusetts Department of Environmental Protection: Boston, MA, January 1983.
11. New York DEC. *Petroleum Bulk Storage*, Parts 612, 613, and 614; NYS Department of Environmental Conservation: Albany, NY, December 27, 1985.
12. New York DEC. *Supporting Documents for Chemical Bulk Storage Regulations*, Parts 595, 596, and 597; NYS Department of Environmental Conservation: Albany, NY, October 1987.
13. New York DEC. *Water Pollution Control and Enforcement Laws, and Environmental Conservation Law of the State of New York*; NYS Department of Environmental Conservation: Albany, NY, 1992.
14. New York DEC. *Chemical Bulk Storage*, Parts 595, 596, 597, 598, and 599; NYS Department of Environmental Conservation: Albany, NY, May 1993.
15. USEPA. *Scientific and Technical Assessment Report on Cadmium*, USEPA-600/6-75-003; U.S. Environmental Protection Agency: Washington, DC, 1975.
16. USEPA. *Federal Register*; U.S. Environmental Protection Agency: Washington, DC, November 28, 1980.
17. USEPA. *Field Standard Operating Procedures for the Decontamination of Response Personnel*, Publication No. FSOP-7; U.S. Environmental Protection Agency: Washington, DC, January 1985.
18. USEPA. *Reclamation and Redevelopment of Contaminated Land: US Case Studies*, Report No. USEPA/600/2-86/066; U.S. Environmental Protection Agency: Washington, DC, August 1986.

19. USEPA. *Solid Waste and Emergency Response*, Report No. USEPA/625/6-87-015; U.S. Environmental Protection Agency: Washington, DC, 1987.
20. USEPA. *Managing Asbestos in Place – A Building Owner's Guide to Operations and Maintenance Programs for Asbestos-Containing Materials*, Report No. TS-799; U.S. Environmental Protection Agency: Washington, DC, July 1990.
21. USGPO. *Occupational Safety and Health Guidance Manual for Hazardous Waste Site Activities*, Publication No. DHHS-NIOSH-85-115; U.S. Government Printing Office: Washington, DC, October 1985.
22. USGPO. *Asbestos in the Home*; U.S. Government Printing Office: Washington, DC, August 1989.
23. Aulenbach, D.B.; Ryan, R.M. Management of radioactive wastes. In *Handbook of Environmental Engineering, Volume 4, Water Resources and Natural Control Processes*; Wang, L.K., Pereira, N.C. Eds.; The Humana Press: Clifton, NJ, 1986; 283–372.
24. Centofanti, L.F. Halting the legacy. Environ. Protection **2002**, *13* (*5*), 30.
25. Wang, L.K. *Biological Process for Groundwater and Wastewater Treatment*. U.S. Patent No. 5,451,320, September 19, 1995.
26. Jensen, P. Noise control. In *Handbook of Environmental Engineering, Volume 1, Air and Noise Pollution Control*; Wang, L.K., Pereira, N.C., Eds.; Clifton, NJ: 1979, 411–474.
27. Gallery, A.G. Disinfect with sodium hypochlorite – safety guidelines. Chem. Engng Prog. **2003**, *99* (*3*), 42–47.
28. USEPA. *Chemical Aids Manual for Wastewater Treatment Facilities*; U.S. Environmental Protection Agency: Washington, DC, 1979; 430/9-79-018.
29. Swichtenberg, B. Firefighting. Water Engng and Mgnt **2003**, *150* (*3*), 8–9.
30. NRC. *Arsenic*; Committee on Medical and Biological Effects of Environmental Pollutants, National Research Council, National Academy of Sciences: Washington, DC, 1977; ISBN 0-709-02604-0.
31. Wirth, N. Hazardous chemical safety. Operations Forum, Water Environment Federation, Alexandria, VA, **1998**, *10* (*8*).
32. Cook, C.; McDaniel, D. IAQ problem solving. Environ. Protection **2002**, *13* (*5*), 24.
33. Ferrante, L.M. Indoor/in-plant air quality. Natl Environ. J. **1993**, *March–April*, 36–40.
34. Wang, L.K.; Wu, B.C.; Zepka, J. *An Investigation of Lead Content in Paints and PCB content in Water Supply for Eagleton School*; U.S. Department of Commerce, National Technical Information Service: Springfield, VA, 1984; Report No. PB86-169315, 23 p.
35. Wang, L.K.; Zepka, J. *An Investigaiton of Asbestos Content in Air for Eagleton School*; U.S. Department Commerce, National Technical Information Service: Springfield, VA, 1984; Report No. PB86-194172/AS, 17 p.
36. Wang, M.H.S.; Wang, L.K. Ventilation and air conditioning. In *Handbook of Environmental Engineering, Volume 1, Air and Noise Pollution Control*; Wang, L.K., Pereira, N.C., Eds.; The Humana Press: Clifton, NJ, 1979; 271–353.
37. Nevius, J.G. Brownfields and green insurance. Environ. Protection **2003**, *14* (*2*), 27–27.
38. Addlestone, S.I. Waste makes haste: top issues in waste management in 2003. Environ. Protection **2003**, *14* (*1*), 34–37.
39. Wang, L.K.; Wang, M.H.S. *Control of Hazardous Wastes in Petroleum Refining Industry, Symposium on Environmental Technology and Managements*; U.S. Department of Commerce, National Technical Information Service: Springfield, VA, 1982; Report No. PB82-185273, 60–77.
40. Wang, L.K.; McGinnis, W.C.; Wang, M.H.S. *Analysis and Formulation of Combustible Components in Wasted Rubber Tires*; U.S. Department of Commerce, National Technical Information Service: Springfield, VA, 1985;. Report no. PB86-169281/AS, 1985; 39 p.
41. Hrycyk, O.; Kurylko, L.; Wang, L.K. *Removal of Volatile Compounds and Surfactants from Liquid*. U.S. Patent No. 5,122,166, June 1992.
42. Wang, L.K.; Hrycyk, O.; Kurylko, L. *Removal of Volatile Compounds and Surfactants from Liquid*. U.S. Patent No. 5,122,165, June 1992.
43. Wang, L.K.; Kurylko, L.; Wang, M.H.S. *Contamination Removal System Employing Filtration and Plural Ultraviolet and Chemical Treatment Steps and Treatment Mode Controller*. U.S. Patent No. 5,190,659, March 1993.

44. Wang, L.K.; Kurylko, L.; Wang, *M.H.S. Method and Apparatus for Filtration with Plural Ultraviolet Treatment Stages*. U.S. Patent No. 5,236,595, August 17, 1993.
45. Clabby, C. PCBs threaten a way of life. Times Union **1993**, July 15, A-1, A-12.
46. Cheremisinoff, P.N. Focus on high hazard pollutants. Pollut. Engng **1990**, *22* (*2*), 67–79.
47. Krofta, M.;Wang, L.K. *Hazardous Waste Management in Institutions and Colleges*; U.S. Department of Commerce, National Technical Information Service: Springfield, VA, 1985; Report No. PB86-194180/AS, June 1985.
48. Desilva, F.J. Nitrate removal by ion exchange. Water Qual. Prod. **2003**, *8* (*4*), 9–30.
49. Kucera, D. Court upholds U.S. EPA's radionuclide rules. Water Engng Mgnt **2003**, *150* (*4*), 7.
50. Environmental Control Library. PCB. *Current Awareness* **1993**, *93* (*13*), 5–7.
51. Martin, W.H. Risk management of PCB transformers. Pollut. Engng **1990**, *March*, 74–77.
52. Cheremisinoff, P.N. Spill and leak containment and emergency response. Pollut. Engng **1989**, *21* (*13*), 42–51.
53. Steinway, D.M. Scope and numbers of regulations for asbestos-containing materials, abatement continue to grow. Hazmat World **1990**, *April*, 32–58.
54. Hannak, W.R. Hazardous waste shipping containers. Environ. Protection **2002**, *13* (*1*), 34.
55. Newton, J.J. Are you an SQG? Pollut. Engng **1989**, *21* (*13*), 64–66.
56. Wang, L.K.; Cheryan, M. Application of membrane technology in food industry for cleaner production. Water Treatment **1995**, *10* (*4*), 283–298.
57. Bober, T.W.; Dagon, T.J.; Fowler, H.E. Treatment of photographic processing wastes. In *Handbook of Industrial Waste Treatment*; Wang, L.K., Wang, M.H.S., Eds.; Marcel Dekker, Inc.: New York, NY, 1992; 173–227.
58. Kodak Company. *Choices: Choosing the Right Silver Recovery Method for Your Need–Environment*; Kodak Company: Rochester, NY,1987.
59. Wang, L.K.; Wang, M.H.S. *Method and Apparatus for Purifying and Compacting Solid Wastes*. U.S. Patent No. 5,232,584, August 7, 1983.
60. ASTM. *ASTM Standards on Environmental Site Characterization*; ASTM, 2002; 1827 p.
61. DeChacon, J.R.;Van Houten, N.J. Indoor air quality. Natl Environ. J. **1991**, *16–18 Nov*.
62. Drill, S.; Konz, J.H. Mahar; Morse, M. *The Environmental Lead Problem: An Assessment of Lead in Drinking Water from a Multi-Media Perspective*; U.S. Environmental Protection Agency, Criteria and Standards Division, PB-296 566, May 1979.
63. Hall, S.K. Oil spills at sea. Pollut. Engng, **1989**, *21* (*13*), 59–63.
64. Heinold, D.; Smith, D. Finding the weak links. Environ. Protection **2003**, *14* (*2*), 56–65.
65. Munshower, F.F. Microelements and their role in surface mine planning. In *Coal Development: Collected Papers, Volume II*; Coal Development Workshops in Grand Junction, Colorado and Casper, Wyoming, Bureau of Land Management, July 1983.
66. Nagl, G.J. Air: controlling hydrogen sulfide emissions. Environ. Technol. **1999**, *9* (*7*), 18–23.
67. NRC. *Nickel*; Committee on Medical and Biological Effects of Environmental Pollutants, Division of Medical Sciences, National Research Council, National Academy of Sciences: Washington, DC, 1975; ISBN 0-309-02314-9.
68. NRC. *Zinc*; Committee on Medical and Biological Effects of Environmental Pollutants, National Research Council, National Academy of Sciences, prepared for the U.S. Environmental Protection Agency: Washington, DC, 1978; USEPA-600/1-78-034.
69. NRC. *Lead in the Human Environment*; Committee on Lead in the Human Environment, National Research Council, National Academy of Sciences: Washington, DC, 1980.
70. Qudir, R.M. The brownfield challenge. Environ. Protection **2002**, *13* (*3*), 34–39.
71. Ryan, J.A. Factors affecting plant uptake of heavy metals from land application of residuals. In *Proceedings of the National Conference on Disposal of Residues on Land*, St. Louis, Missouri, September 13–15, 1976.
72. Spicer, S. Accident releases lethal gas. Operations Forum, Water Environment Federation, Alexandria, VA, **2001**, *13* (*8*).
73. Turner, P.L. Preparing hazardous waste transport manifests. Environ. Protection **1987**, *3* (*10*), 12–16.

74. VanDenBos, A.; Izadpanah, A. A new partnership for handling medical waste. Environ. Protection **2002**, *13* (*5*), 26.
75. Wang, L.K. *Prevention of Airborn Legionnaires' Disease by Formation of a New Cooling Water for Use in Central Air Conditioning Systems*; U.S. Department of Commerce, National Technical Information Service, Report No. PB85-215317/AS, 1984; 92 p.
76. Wang, L.K. *Design of Innovative Flotation – Filtration Wastewater Treatment Systems for a Nickel–Chromium Plating Plant*; U.S. Department of Commerce, National Technical Information Service, Report No. PB88-200522/AS, 1984; 50 p.
77. Wang, L.K.; Pressman, M.; Shuster, W.W.; Shade, R.W.; Bilgen, F.; Lynch, T. Separation of nitrocellulose fine particles from industrial effluent with organic polymers. Can. J. Chem. Eng. **1982**, *60*, 116–122.
78. Wang, L.K.; Wu, B.C. Treatment of groundwater by dissolved air flotation systems using sodium aluminate and lime as flotation aids. OCEESA J. **1984**, *1* (*3*), 5–18; NTIS Report No. PB85-167229/AS.
79. Wang, M.H.S.; Wang, L.K.; Simmons, T.; Bergenthal, J. Computer-aided air quality management. J. Environ. Mgnt **1979**, *9*, 61–87.
80. Wang, L.K. *Identification, Transfer, Acquisition and Implementation of Environmental Technologies Suitable for Small and Medium Size Enterprises*; United Nations Industrial Development Organization (UNIDO), Vienna, Austria, 1995; Technical paper no. 9-9-95, 5 p.
81. Wang, L.K. *Liquid Treatment System with Air Emission Control.* U.S. Patent No. 5,399,267, March 21, 1995.
82. Wastewater Engineers, Inc. Case histories in wastewater recycling. Environ. Technol. **1999**, *9* (*7*), 24–26.
83. Wang, L.K.; Wang, P.; Celesceri, N.L. Groundwater decontamination using sequencing batch processes. Water Treatment **1995**, *10* (*2*), 121–134.
84. Wang, L.K. *Site Remediation Technology*. U.S. Patent No. 5,552,051, September 3, 1996.
85. Wang, L.K.; Pereira, N.C.; Hung, Y. *Air Pollution Control Engineering*. Humana Press, Totowa, NJ. 2004.
86. Wang, L.K.; Pereira, N.C.; Hung, Y. *Advanced Air and Noise Pollution Control.* Humana Press, Totowa, N.J. 2004.
87. Wang, L.K.; Hung, Y.; Shammas, N.K. *Physicochemical Treatment Processes*. Humana Press, Totowa, NJ. 2004.

12

Application of Biotechnology for Industrial Waste Treatment

Joo-Hwa Tay, Stephen Tiong-Lee Tay, and Volodymyr Ivanov
Nanyang Technological University, Singapore

Yung-Tse Hung
Cleveland State University, Cleveland, Ohio, U.S.A.

12.1 BIOTREATABILITY OF INDUSTRIAL HAZARDOUS WASTES

Environmental biotechnology concerns the science and practical knowledge relating to the use of microorganisms and their products. Biotechnology combines fundamental knowledge in microbiology, biochemistry, genetics, and molecular biology, and engineering knowledge of the specific processes and equipment. The main applications of biotechnology in industrial hazardous waste treatment are: prevention of environmental pollution through waste treatment, remediation of polluted environments, and biomonitoring of environment and treatment processes. The common biotechnological process in the treatment of hazardous waste is the biotransformation or biodegradation of hazardous substances by microbial communities.

Bioagents for hazardous waste treatment are biotechnological agents that can be applied to hazardous waste treatment including bacteria, fungi, algae, and protozoa. Bacteria are microorganisms with prokaryotic cells and typically range from 1 to 5 μm in size. Bacteria are most active in the biodegradation of organic matter and are used in the wastewater treatment and solid waste or soil bioremediation. Fungi are eukaryotic microorganisms that assimilate organic substances and typically range from 5 to 20 μm in size. Fungi are important degraders of biopolymers and are used in solid waste treatment, especially in composting, or in soil bioremediation for the biodegradation of hazardous organic substances. Fungal biomass is also used as an adsorbent of heavy metals or radionuclides. Algae are saprophytic eukaryotic microorganisms that assimilate light energy. Algal cells typically range from 5 to 20 μm in size. Algae are used in environmental biotechnology for the removal of organic matter in waste lagoons. Protozoa are unicellular animals that absorb organic food and digest it intracellularly. Typical cell size is from 10 to 50 μm. Protozoa play an important role in the treatment of industrial hazardous solid, liquid, and gas wastes by grazing on bacterial cells, thus maintaining adequate bacterial biomass levels in the treatment systems and helping to reduce cell concentrations in the waste effluents.

Microbial aggregates used in hazardous waste treatment. Microorganisms are key biotechnology agents because of their diverse biodegradation and biotransformation abilities and their small size. They have high ratios of biomass surface to biomass volume, which ensure high rates of metabolism. Microorganisms used in biotechnology typically range from 1 to 100 μm in size. However, in addition to individual cells, cell aggregates in the form of flocs, biofilms, granules, and mats with dimensions that typically range from 0.1 to 100 mm may also be used in biotechnology. These aggregates may be suspended in liquid or attached to solid surfaces. Microbial aggregates that can accumulate in the water–gas interface are also useful in biotechnology applications in hazardous waste treatment.

Microbial communities for hazardous waste treatment. It is extremely unusual for biological treatment to rely solely on a single microbial strain. More commonly, communities of naturally selected strains or artificially combined strains of microorganisms are employed. Positive or negative interactions may exist among the species within each community. Positive interactions, such as commensalism, mutualism, and symbiosis, are more common in microbial aggregates. Negative interactions, such as amensalism, antibiosis, parasitism, and predation, are more common in natural or engineering systems with low densities of microbial biomass, for example, in aquatic or soil ecosystems.

12.1.1 Industrial Hazardous Solid, Liquid, and Gas Wastes

Hazardous Waste

Industrial wastes are identified as hazardous wastes by the waste generator or by the national environmental agency either because the waste component is listed in the List of Hazardous Inorganic and Organic Constituents approved by the national agency or because the waste exhibits general features of hazardous waste, such as harming human health or vital activity of plants and animals (acute and chronic toxicity, carcinogenicity, teratogenicity, pathogenicity, etc.), reducing biodiversity of ecosystems, flammability, corrosive activity, ability to explode, and so on. The United States annually produces over 50 million metric tonnes of federally regulated hazardous wastes [1].

Hazardous Substances

It is estimated that approximately 100,000 chemical compounds have been produced industrially [2,3] and many of them are harmful to human health and to the environment. However, only 7% of the largest-volume chemicals require toxicity screening [2]. In the United States, the Agency for Toxic Substances and Disease Registry (ATSDR) and the Environmental Protection Agency (EPA) maintain a list, in order of priority, of substances that are determined to pose the most significant potential threat to human health due to their known or suspected toxicity. This Comprehensive Environmental Response, Compensation, and Liability Act (CERCLA) Priority List of Hazardous Substances was first issued in 1999 and includes 275 substances (www.atsdr.cdc.gov/clist.html).

Application of Biotechnology in the Treatment of Hazardous Substances From the CERCLA Priority List

The CERCLA Priority List of Hazardous Substances has been annotated with information on the types of wastes and the possible biotechnological treatment methods, as shown in Table 1. The remarks on biotreatability of these hazardous substances are based on data from numerous papers, reviews and books on this topic [4–8]. Databases are available on the biodegradation of hazardous substances. For example, the Biodegradative Strain Database [9] (bsd.cme.msu.edu) can be used to select suitable microbial strains for biodegradation applications, while the

Table 1 Major Hazardous Environmental Pollutants and Applicability of Biotechnology For Their Treatment

1999 Rank	Substance name	Type of waste (S = solid, L = liquid, G = gas)	Biotechnological treatment with formation of nonhazardous or less hazardous products
1	Arsenic	S, L	Bioreduction/biooxidation following immobilization or dissolution
2	Lead	S,L	Bioimmobilization, biosorption, bioaccumulation
3	Mercury	S,L,G	Bioimmobilization, biovolatilization, biosorption
4	Vinyl chloride	L,G	Biooxidation by cometabolization with methane or ammonium
5	Benzene	L,G	Biooxidation
6	Polychlorinated biphenyls	S,L	Biooxidation after reductive or oxidative biodechlorination
7	Cadmium	S,L	Biosorption, bioaccumulation
8	Benzo(A)pyrene	S,L	Biooxidation and cleavage of the rings
9	Polycyclic aromatic hydrocarbons	S,L,G	Biooxidation and cleavage of the rings
10	Benzo(B)fluoranthene	S,L	Biooxidation and cleavage of the rings
11	Chloroform	L,G	Biooxidation by cometabolization with methane or ammonium
12	DDT, P,P′-	S,L	Biooxidation after reductive or oxidative biodechlorination
13	Aroclor 1260	S,L	Biooxidation after reductive or oxidative biodechlorination
14	Aroclor 1254	S,L	Biooxidation after reductive or oxidative biodechlorination
15	Trichloroethylene	L,G	Biooxidation by cometabolization with methane or ammonium
16	Chromium, hexavalent	S,L	Bioreduction/bioimmobilization, biosorption
17	Dibenzo(A,H)anthracene	S,L	Biooxidation and cleavage of the rings
18	Dieldrin	S,L	Biooxidation after reductive or oxidative biodechlorination
19	Hexachlorobutadiene	L,G	Biooxidation after reductive or oxidative biodechlorination
20	DDDE, P,P′-	S,L	Biooxidation after reductive or oxidative biodechlorination
21	Creosote	S,L	Biooxidation and cleavage of the rings
22	Chlordane	S,L	Biooxidation after reductive or oxidative biodechlorination
23	Benzidine	L,G	Biooxidation and cleavage of the rings
24	Aldrin	S,L	Biooxidation
25	Aroclor 1248	S,L	Biooxidation after reductive or oxidative biodechlorination
26	Cyanide	S,L,G	Removal by ferrous ions produced by bacterial reduction of Fe(III)

(continues)

Table 1 Continued

1999 Rank	Substance name	Type of waste (S = solid, L = liquid, G = gas)	Biotechnological treatment with formation of nonhazardous or less hazardous products
27	DDD, P,P′-	S,L	Biooxidation after reductive or oxidative biodechlorination
28	Aroclor 1242	S,L	Biooxidation after reductive or oxidative biodechlorination
29	Phosphorus, white	S,L,G	
30	Heptachlor	L,G	
31	Tetrachloroethylene	L,G	Biooxidation by cometabolization with methane or ammonium
32	Toxaphene	S,L	Reductive (anaerobic) dechlorination
33	Hexachlorocyclohexane, gamma-	S,L,G	Biooxidation by white-rot fungi
34	Hexachlorocyclohexane, beta-	S,L,G	Biooxidation by white-rot fungi
35	Benzo(A)Anthracene	S,L	Biooxidation and cleavage of the rings
36	1,2-Dibromoethane	L,G	Biooxidation by cometabolization with methane or ammonium
37	Disulfoton	S,L	Biooxidation
38	Endrin	S,L	Biooxidation
39	Beryllium	S,L	Biosorption
40	Hexachlorocyclohexane, delta-	S,L,G	Biooxidation by white-rot fungi; biooxidation after reductive or oxidative biodechlorination
41	Aroclor 1221	S,L	Biooxidation after reductive or oxidative biodechlorination
42	Di-*N*-Butyl phthalate	L,G	Biooxidation
43	1,2-Dibromo-3-chloropropane	L,G	Biooxidation after reductive or oxidative biodechlorination
44	Pentachlorophenol	L,G	Biooxidation after reductive or oxidative biodechlorination
45	Aroclor 1016	S,L	Biooxidation after reductive or oxidative biodechlorination
46	Carbon tetrachloride	L,G	Biodechlorination and biodegradation
47	Heptachlor epoxide	L,G	
48	Xylenes, total	S,L,G	Biooxidation
49	Cobalt	S,L	Biosorption
50	Endosulfan sulfate	S,L	Biosorption
51	DDT, O,P′-	S,L	Biooxidation by white-rot fungi
52	Nickel	S,L	Biosorption
53	3,3′-Dichlorobenzidine	L,G	Biooxidation after reductive or oxidative biodechlorination

54	Dibromochloropropane	L,G	Biooxidation after reductive or oxidative biodechlorination
55	Endosulfan, alpha	S,L	Biooxidation by fungi or bacteria
56	Endosulfan	S,L	Biooxidation by fungi or bacteria
57	Benzo(K)fluoranthene	S,L	Biooxidation and cleavage of the rings
58	Aroclor	S,L	Biooxidation after reductive or oxidative biodechlorination
59	Endrin ketone	S,L	
60	Cis-Chlordane	S,L	Biooxidation after reductive or oxidative biodechlorination
61	2-Hexanone	L,G	
62	Toluene	L,G	Biooxidation and cleavage of the ring
63	Aroclor 1232	S,L	Biooxidation after reductive or oxidative biodechlorination
64	Endosulfan, beta	S,L	Biooxidation by fungi and bacteria
65	Methane	G	Biooxidation by methanotrophic bacteria
66	Trans-Chlordane	S,L,G	
67	2,3,7,8-Tetrachlorodibenzo-p-dioxin	S,L	Biooxidation after reductive or oxidative biodechlorination
68	Benzofluoranthene	S,L	Biooxidation and cleavage of the rings
69	Endrin aldehyde	S,L	
70	Zinc	S,L	Microbial immobilization/solubilization
71	Dimethylarsinic acid	S,L	
72	Di(2-ethylhexyl)phthalate	S,L	Biooxidation and cleavage of the rings
73	Chromium	S,L	Microbial reduction/oxidation followed immobilization or solubilization
74	Methylene chloride	L,G	Biooxidation by cometabolization with methane or ammonium
75	Naphthalene	S,L,G	Biooxidation and cleavage of the rings
76	Methoxychlor	S,L	Biooxidation after reductive or oxidative biodechlorination
77	1,1-Dichloroethene	L,G	Biooxidation by cometabolization with methane or ammonium
78	Aroclor 1240	S,L	Biooxidation after reductive or oxidative biodechlorination
79	*Bis*(2-chloroethyl) ether	L,G	
80	1,2-Dichloroethane	L,G	Biooxidation by cometabolization with methane or ammonium
81	2,4-Dinitrophenol	S,L,G	Biooxidation
82	2,4,6-Trinitrotoluene	S, L,G	Biooxidation
83	2,4,6-Trichlorophenol	S,L,G	Biooxidation
84	Chlorine	L,G	Removal by ferrous or manganese ions produced by bacterial reduction of Fe(III) and Mn(IV)
85	Cyclotrimethylenetrinitramine (Rdx)	S,L	

(continues)

Table 1 Continued

1999 Rank	Substance name	Type of waste (S = solid, L = liquid, G = gas)	Biotechnological treatment with formation of nonhazardous or less hazardous products
86	1,1,1-Trichloroethane	L,G	Biooxidation by cometabolization with methane or ammonium
87	Ethylbenzene	L,G	Biooxidation and cleavage of the rings
88	1,1,2,2-Tetrachloroethane	L,G	Biooxidation by cometabolization with methane or ammonium
89	Thiocyanate	S,L	Removal by ferrous or manganese ions produced by bacterial reduction of Fe(III) and Mn(IV)
90	Asbestos	S,G	
91	4,6-Dinitro-o-cresol	S,L	Biooxidation
92	Uranium	S,L	Bioleaching of uranium from minerals
93	Radium	S,L	
94	Radium-226	S,L	
95	Hexachlorobenzene	L,G	
96	Ethion	S,L	
97	Thorium	S,L	
98	Chlorobenzene	S,L,G	Biooxidation after reductive or oxidative biodechlorination
99	Barium	S,L	Biosorption
100	2,4-Dinitrotoluene	S,L	Biooxidation
101	Fluoranthene	S,L	Biooxidation and cleavage of the rings
102	Radon	G	
103	Radium-228	S,L	
104	Thorium-230	S,L	
105	Diazinon	S,L	
106	Bromine	G	Binding with Fe or Mn reduced by bacteria
107	1,3,5-Trinitrobenzene	S,L,G	Biodegradation
108	Uranium-235	S,L	Biosorption/bioleaching and oxidation/reduction mediated by other elements oxidized or reduced by microorganisms
109	Tritium	S,L	
110	Uranium-234	S,L	Biosorption/bioleaching and oxidation/reduction mediated by other elements oxidized or reduced by microorganisms

111	Thorium-228	S,L	
112	*N*-Nitrosodi-*N*-propylamine	S,L,G	
113	Cesium-137	S,L	Bioimmobilization/biosorption
114	Hexachlorocyclohexane, alpha-	S,L	Biooxidation after reductive or oxidative biodechlorination
115	Chrysene	S,L	Biooxidation and cleavage of the rings
116	Radon-222	G	
117	Polonium-210	S,L	
118	Chrysotile asbestos	S,G	
119	Thorium-227	S,L	
120	Potassium-40	S,L	Bioaccumulation
121	Coal tars	S,L	Biooxidation
122	Plutonium-238	S,L	Biosorption
123	Thoron (Radon-220)	G	
124	Copper	S,L	Biosorption
125	Strontium-90	S,L	Bioimmobilization/solubilization
126	Cobalt-60	S,L	Biosorption
127	Methylmercury	L,G	Biodegradation
128	Chlorpyrifos	S,L	
129	Lead-210	S,L	Biosorption
130	Plutonium-239	S,L	Biosorption
131	Plutonium	S,L	Biosorption
132	Americium-241	S,L	
133	Iodine-131	S,L	
134	Amosite asbestos	S,G	
134	Guthion	S,L	
136	Bismuth-214	S,L	
136	Lead-214	S,L	Biosorption
138	Chlordecone	S,L	
138	Plutonium-240	S,L	Biosorption
138	Tributyltin	S,L	Biodetoxication
141	Manganese	S,L	Microbial reduction/oxidation
142	S,S,S-Tributyl phosphorotrithioate	S,L,G	
143	Selenium	S,L	Microbial reduction/oxidation

(*continues*)

Table 1 Continued

1999 Rank	Substance name	Type of waste (S = solid, L = liquid, G = gas)	Biotechnological treatment with formation of nonhazardous or less hazardous products
144	Polybrominated biphenyls	S,L	Biooxidation after reductive or oxidative biodechlorination
145	Dicofol	S,L	
146	Parathion	S,L	Biodegradation by enzymes of genetically engineered strains
147	Hexachlorocyclohexane, technical	S,L	Biooxidation after reductive or oxidative biodechlorination
148	Pentachlorobenzene	L,G	Biooxidation after reductive or oxidative biodechlorination
149	Trichlorofluoroethane	L,G	Biooxidation by cometabolization with methane or ammonium
150	Treflan (Trifluralin)	S,L	
151	4,4′-Methylenebis(2-chloroaniline)	S,L	
152	1,1-Dichloroethane	L,G	Biooxidation by cometabolization with methane or ammonium
153	𝒟𝒟𝒟, O,P′-	S,L	Biooxidation after reductive or oxidative biodechlorination
154	Hexachlorodibenzo-p-dioxin	S,L	Biooxidation after reductive or oxidative biodechlorination
155	Heptachlorodibenzo-p-dioxin	S,L	Biooxidation after reductive or oxidative biodechlorination
156	2-Methylnaphthalene	S,L	Biooxidation and cleavage of the rings
157	1,1,2-Trichloroethane	L,G	Biooxidation by cometabolization with methane or ammonium
158	Ammonia	L,G	Biooxidation (nitrification) followed denitrification; bioremoval by combined IRB/IOB biotechnology
159	Acenaphthene	S,L	
160	1,2,3,4,6,7,8,9-Octachlorodibenzofuran	S,L	Biooxidation after reductive or oxidative biodechlorination
161	Phenol	L,G	Biooxidation and cleavage of the rings; anaerobic biodegradation
162	Trichloroethane	L,G	Biooxidation by cometabolization with methane or ammonium
163	Chromium(Vi) trioxide	S,L	
164	1,2-Dichloroethene, trans-	L,G	Biooxidation by cometabolization with methane or ammonium
165	Heptachlorodibenzofuran	S,L	Biooxidation after reductive or oxidative biodechlorination
166	Hexachlorocyclopentadiene	L,G	Biooxidation after reductive or oxidative biodechlorination
167	1,4-Dichlorobenzene	L,G	Biooxidation after reductive or oxidative biodechlorination
168	1,2-Diphenylhydrazine	L,G	
169	Cresol, para-	S,L,G	
170	1,2-Dichlorobenzene	L,G	Biooxidation after reductive or oxidative biodechlorination

171	Lead-212	S,L	
172	Oxychlordane	S,L	Biooxidation after reductive or oxidative biodechlorination
173	2,3,4,7,8-Pentachlorodibenzofuran	S,L	Biooxidation after reductive or oxidative biodechlorination
174	Radium-224	G	
175	Acetone	L,G	
176	Hexachlorodibenzofuran	S,L	Biooxidation after reductive or oxidative biodechlorination
177	Benzopyrene	S,L	Biooxidation and cleavage of the rings
177	Bismuth-212	S,L	
179	Americium	S,L	
179	Cesium-134	S,L	Biosorption
179	Chromium-51	S,L	Bioreduction/biooxidation
182	Tetrachlorophenol	L,G	Biooxidation after reductive or oxidative biodechlorination
183	Carbon disulfide	L,G	
184	Chloroethane	L,G	Biooxidation by cometabolization with methane or ammonium
185	Indeno(1,2,3-Cd)pyrene	S,L	Biooxidation and cleavage of the rings
186	Dibenzofuran	S,L	Biooxidation and cleavage of the rings
187	p-Xylene	L,G	Biooxidation and cleavage of the rings
188	2,4-Dimethylphenol	L,G	Biooxidation and cleavage of the rings
189	Aroclor 1268	S,L	Biooxidation after reductive or oxidative biodechlorination
190	1,2,3-Trichlorobenzene	L,G	Biooxidation after reductive or oxidative biodechlorination
191	Pentachlorodibenzofuran	S,L	Biooxidation after reductive or oxidative biodechlorination
192	Hydrogen sulfide	L,G	Biooxidation by aerobic or microaerophilic bacteria; binding with ferrous ions produced by iron-reducing bacteria; biooxidation by phototrophic bacteria
193	Aluminum	S,L	
194	Tetrachloroethane	L,G	Biooxidation by cometabolization with methane or ammonium
195	Cresol, Ortho-	L,G	Biooxidation and cleavage of the rings
196	1,2,4-Trichlorobenzene	L,G	Biooxidation after reductive or oxidative biodechlorination
197	Hexachloroethane	L,G	Biooxidation after reductive or oxidative biodechlorination
198	Butyl benzyl phthalate	S,L	Biooxidation and cleavage of the rings
199	Chloromethane	L,G	Biooxidation by cometabolization with methane or ammonium
200	Vanadium	S,L	Biosorption
201	1,3-Dichlorobenzene	L,G	Biooxidation after reductive or oxidative biodechlorination
202	Tetrachlorodibenzo-p-dioxin	S,L	Biooxidation after reductive or oxidative biodechlorination

(continues)

Table 1 Continued

1999 Rank	Substance name	Type of waste (S = solid, L = liquid, G = gas)	Biotechnological treatment with formation of nonhazardous or less hazardous products
203	2-Butanone	G	Biooxidation
204	*N*-Nitrosodiphenylamine	S,L	
205	Pentachlorodibenzo-p-dioxin	S,L	Biooxidation after reductive or oxidative biodechlorination
206	2,3,7,8-Tetrachlorodibenzofuran	S,L	Biooxidation after reductive or oxidative biodechlorination
207	Silver	S,L	Biosorption
208	2,4-Dichlorophenol	L,G	Biooxidation after reductive or oxidative biodechlorination
209	1,2-Dichloroethylene	L,G	Biooxidation after reductive or oxidative biodechlorination
210	Bromoform	L,G	Biooxidation by cometabolization with methane or ammonium
211	Acrolein	L,G	
212	Chromic acid	S,L	
213	2,4,5-Trichlorophenol	L,G	Biooxidation after reductive or oxidative biodechlorination
214	Nonachlor, trans-	S,L	
215	Coal tar pitch	S,L	Biooxidation and cleavage of the rings
216	Phenanthrene	S,L	Biooxidation and cleavage of the rings
217	Nitrate	S,L	Microbial denitrification
218	Arsenic trioxide	S,L	
219	Nonachlor, *cis-*	S,L	
220	Hydrazine	L,G	
221	Technetium-99	S,L	Biosorption
222	Nitrite	S,L	Microbial denitrification
223	Arsenic acid	S,L	Bioreduction
224	Phorate	S,L	
225	Bromodichloroethane	L,G	Biooxidation by cometabolization with methane or ammonium
225	Dimethoate	S,L	
227	Strobane	S,L	
228	Naled	S,L	
229	Arsine	S,L	Biooxidation
230	4-Aminobiphenyl	S,L	

230	Pyrethrum	S,L	
230	Tetrachlorobiphenyl	S,L	Biooxidation after reductive or oxidative biodechlorination
233	Dibenzofurans, chlorinated	S,L	Biooxidation after reductive or oxidative biodechlorination
233	Ethoprop	S,L	
233	Nitrogen dioxide	G	Bioreduction
236	Carbophenothion	S,L	
236	Thorium-234	S,L	
238	Dichlorvos	S,L	
238	Ozone	G	
238	Palladium	S,L	
241	Calcium arsenate	S,L	Bioreduction; bioaccumulation
241	Carbon-14	S,L,G	
241	Europium-154	S,L	
241	Krypton-85	G	
241	Mercuric chloride	S,L	Bioimmobilization; biomethylation
241	sodium-22	S,L	Bioaccumulation
241	Strontium-89	S,L	Biosorption
241	Sulfur-35	S,L,G	Biooxidation/bioreduction
241	Uranium-233	S,L	Bioaccumulation/biosorption or bioleaching
250	2,4-D Acid	S,L	
251	Antimony	S,L	Biooxidation/bioreduction
252	Cresols	L,G	Biooxidation and cleavage of the rings
253	Pyrene	S,L	Biooxidation and cleavage of the rings
254	2-Chlorophenol	L,G	Biooxidation after reductive or oxidative biodechlorination
255	Dichlorobenzene	S,L,G	Biooxidation after reductive or oxidative biodechlorination
256	Formaldehyde	L,G	
257	*N*-Nitrosodimethylamine	S,L	
258	Chlorodibromomethane	L,G	Biooxidation by cometabolization with methane or ammonium
259	Sutan	S,L	
260	Dichloroethane	S,L,G	Biooxidation by cometabolization with methane or ammonium
261	1,3-Dinitrobenzene	S,L,G	Biodegradation
262	Dimethyl formamide	S,L	Biodegradation
263	1,3-Dichloropropene, *cis*-	S,L,G	

(continues)

Table 1 Continued

1999 Rank	Substance name	Type of waste (S = solid, L = liquid, G = gas)	Biotechnological treatment with formation of nonhazardous or less hazardous products
264	Ethyl ether	L,G	
265	4-Nitrophenol	L,G	Biodegradation
266	1,3-Dichloropropene, *trans-*	L,G	
267	Trichlorobenzene	L,G	Biooxidation after reductive or oxidative biodechlorination
268	Fluoride	S,L	
269	1,2-Dichloropropane	L,G	Biooxidation after reductive or oxidative biodechlorination
270	2,6-Dinitrotoluene	L,G	Biodegradation
271	Methyl parathion	S,L	
272	Methyl isobutyl ketone	L,G	
273	Octachlorodibenzo-p-dioxin	S,L	Biooxidation after reductive or oxidative biodechlorination
274	Styrene	S,L	Biooxidation and cleavage of the rings
275	Fluorene	S,L	Biooxidation and cleavage of the rings

University of Minnesota Biocatalysis/Biodegradation Database (umbbd.ahc.umn.edu) can be used to predict biodegradation pathways and biodegradation metabolites. Approximately two-thirds of the hazardous substances mentioned in the CERCLA Priority List of Hazardous Substances can be treated by different biotechnological methods.

Production of Hazardous Wastes

The toxic substances appear mostly in: (a) the waste streams of manufacturing processes of commercial products; (b) the wastes produced during the use of these products, or (c) the post-manufacturing wastes related to the storage of these commercial products. Some toxic substances appear as constituents of commercial products that are disposed of once their useful lives are over [2]. If these products are disposed of in a landfill, product deterioration will eventually lead to release of toxic chemicals into the environment. The annual world production of hazardous wastes is estimated to range from 20×10^6 to 50×10^6 metric tonnes. These hazardous wastes include oil-polluted soil and sludges, hydroxide sludges, acidic and alkaline solutions, sulfur-containing wastes, paint sludges, halogenated organic solvents, nonhalogenated organic solvents, galvanic wastes, salt sludges, pesticide-containing wastes, explosives, and wastewaters and gas emissions containing hazardous substances [3].

Secondary Hazardous Wastes

Secondary wastes are generated from the collection, treatment, incineration, or disposal of hazardous wastes, such as sludges, sediments, effluents, leachates, and air emissions. These secondary wastes may also contain hazardous substances and must be treated or disposed of properly to prevent secondary pollution of underground water, surface water, soil or air.

Oil and Petrochemical Industries as Sources of Hazardous Organic Wastes

The petrochemical industry is a major source of hazardous organic wastes, produced during the manufacture or use of hazardous substances. The recovery, transportation, and storage of raw oil or petrochemicals are major sources of hazardous wastes, often produced as the consequence of technological accidents. Seawater and fresh water pollution due to oil and oil-product spills, underground or soil pollution due to land spills or leakage from pipelines or tanks, and air pollution due to incineration of oil or oil sludge, are major cases of environmental pollution. Gasoline is the main product in the petrochemical industry and consists of approximately 70% aliphatic linear and branched hydrocarbons, and 30% aromatic hydrocarbons, including xylenes, toluene, di- and tri-methylbenzenes, ethylbenzenes, benzene and others. Other pure bulk chemicals used for chemical synthesis include formaldehyde, methanol, acetic acid, ethylene and polyethylenes, ethylene glycol and polyethylene glycols, propylene, propylene glycol and polypropylene glycols, and such aromatic hydrocarbons as benzene, toluene, xylenes, styrene, aniline, phthalates, naphthalene, and others.

Hazardous Wastes of Other Chemical Industries

The hazardous substances contained in solid, liquid, or gaseous wastes may include products from the pesticide and pharmaceutical industries. The paint and textile industries produce hazardous solid, liquid, and gaseous wastes that contain diverse organic solvents, paint and fiber preservatives, organic and mineral pigments, and reagents for textile finishing [3]. The pulp industry generates wastewater that contains chlorinated phenolic compounds produced in the chlorine bleaching of pulp. Widely used wood preservatives are usually chlorinated or unchlorinated monocyclic and polycyclic aromatic hydrocarbons. The explosives industry generates wastes containing recalcitrant chemicals with nitrogroups [3].

Xenobiotics and Their Biodegradability

Organic substances, synthesized in the chemical industry, are often hardly biodegradable. The substances that are not produced in nature and are slowly/partially biodegradable are called xenobiotics. Vinylchloride (a monomer for the plastic industry), chloromethanes and chloroethylenes (solvents), polychlorinated aromatic hydrocarbons (pesticides, fungicides, dielectrics, wood preservatives), organophosphate- and nitro-compounds are examples of xenobiotics. The biodegradability of xenobiotics can be characterized by biodegradability tests such as: rate of CO_2 formation (mineralization rate), rate of oxygen consumption (respirometry test), ratio of BOD to COD (oxygen used for biological or chemical oxidation), and by the spectrum of intermediate products of biodegradation.

Hazardous Wastes of Nonchemical Industries

The coal industry, mining industry, hydrometallurgy, and metal industry are sources of solid and liquid wastes that may contain heavy metals, sulfides, sulfuric and other acids, and some toxic reagents used in industrial processes. The electronics and mechanical production industries are sources of hazardous wastes containing organic solvents, surfactants, and heavy metals. Nuclear facilities produce solid and liquid wastes containing radionuclides. Large-scale accidents on nuclear facilities serve as potential sources of radioactive pollution of air and soil, and the polluted areas can be as large as the combined areas of several states.

12.1.2 Suitability of Biotechnological Treatment for Hazardous Wastes

Comparison of Different Treatments of Hazardous Wastes

Usually, the hazardous substance can be removed or treated by physical, chemical, physicochemical, or biological methods. Advantages and disadvantages of these methods are shown in Table 2. The advantages of biotechnological treatment of hazardous wastes are biodegradation or detoxication of a wide spectrum of hazardous substances by natural microorganisms and availability of a wide range of biotechnological methods for complete destruction of hazardous wastes without production of secondary hazardous wastes. However, to intensify the biotreatment, nutrients and electron acceptors must be added, and optimal conditions must be maintained. On the other hand, there may be unexpected or negative effects mediated by microorganisms, such as emission of odors or toxic gases during the biotreatment, and it may be difficult to manage the biotreatment system because of the complexity and high sensitivity of the biological processes.

Cases When Biotechnology is Most Applicable for the Treatment of Hazardous Wastes

The main considerations for application of biotechnology in hazardous waste treatment are as follows:

1. Reasonable rate of biodegradability or detoxication of hazardous substance during biotechnological treatment; such rates are derived from a knowledge of the optimal treatment duration;
2. Necessity to have low volume or absence of secondary hazardous substances produced during biotechnological treatment;

Table 2 Advantages and Disadvantages of Different Treatments of Hazardous Wastes

Method of treatment	Advantages	Disadvantages
Physical treatment (sedimentation, volatilization, fixation, evaporation, heat treatment, radiation, etc.)	• Required time is from some minutes to some hours.	• High expenses for energy and equipment.
Chemical treatment (oxidation, incineration, reduction, immobilization, chelating, transformation)	• Required time is from some minutes to some hours.	• High expenses for reagents, energy, and equipment; air pollution due to incineration.
Physico-chemical treatment (adsorption, absorption)	• Required time is from some minutes to some hours.	• High expenses for adsorbents; formation of secondary hazardous waste.
Biotechnological aerobic treatment (oxidation, transformation, degradation)	• Low volume or absence of secondary hazardous wastes. • Process can be initiated by natural microorganisms or small quantity of added microbial biomass. • Wide spectrum of degradable substances and diverse methods of biodegradability.	• Some expenses for aeration, nutrients, and maintenance of optimal conditions. • Required time is from some hours to days. • Unexpected or negative effects of microorganisms-destructors. • Low predictability of the system because of complexity and high sensitivity of biological systems.
Biotechnological anaerobic treatment (reduction, degradation)	• Low volume or absence of secondary hazardous wastes • Process can be initiated by natural microorganisms or small quantity of added microbial biomass. • Wide spectrum of degradable substances and diverse methods of biodegradability.	• Required time is from some days to months • Emission of bad smelling or toxic gases. • Unexpected or negative effects of microorganisms-destructors. • Low predictability of the system because of complexity and high sensitivity of biological systems.
Landfilling (as a combination of physical and biological treatment)	• Low expenses for landfilling.	• Harmful air emissions; leaching; expensive land use. • Required time is some years.

3. Biotechnological treatment is more cost-effective than other methods; the low cost of biotechnological treatment is largely attributed to the small quantities or total absence of added reagents and microbial biomass to start up the biotreatment process;
4. Public acceptance of biotechnological treatment is better than for chemical or physical treatment.

However, the efficiency of actual biotechnological application depends on its design, process optimization, and cost minimization. Many failures have been reported on the way from bench laboratory-scale to field full-scale biotechnological treatment because of variability, instability, diversity, and heterogeneity of both microbial properties and conditions in the treatment system [10].

Treatment Combinations

In many cases, a combination of physical, chemical, physico-chemical, and biotechnological treatments may be more efficient than one type of treatment (Table 3). Efficient pretreatment schemes, used prior to biotechnological treatment, include homogenization of solid wastes in water, chemical oxidation of hydrocarbons by H_2O_2, ozone, or Fenton's reagent, photochemical oxidation, and preliminary washing of wastes by surfactants.

Roles of Biotechnology in Hazardous Waste Management

Biotechnology can be applied in different fields of hazardous waste management (Table 4): hazardous waste identification by biotechnological tests of toxicity and pathogenicity; prevention of hazardous waste production using biotechnological analogs of products; hazardous

Table 3 Examples of Combinations of Different Treatments

Combination of treatments	Example of combination
Physical and biotechnological treatment	• Thermal pretreatment of waste can enhance the biodegradability of hazardous substance. • Homogenization/suspension of solid wastes or nondissolved sludges in water will increase the surface of the waste particles and, as result of this, the rate of biodegradation will also be increased.
Chemical and biotechnological aerobic treatment	• Preliminary chemical oxidation of aromatic hydrocarbons by H_2O_2 or ozone will improve the biodegradability of these hazardous substances because of the cleavage of aromatic rings.
Biotechnological and chemical treatment	• Reduction of Fe(III) from nondissolved iron hydroxides will produce dissolved Fe(II) ions, which can be used for the precipitation of organic acids or cyanides.
Physico-chemical and biotechnological treatment	• Preliminary washing of wastes polluted by hydrophobic substances by water or solution of surfactants will remove these molecules from the waste; thus, the hydrophobic substances of suspension will be degraded faster than if attached to the particles of hazardous waste.
Biotechnological anaerobic and aerobic treatment	• Anaerobic treatment will perform anaerobic dechlorination of hazardous substances; it will enhance following aerobic treatment.

Table 4 Roles of Biotechnological Applications in Hazardous Waste Management

Type of waste management	Examples of biotechnological application
Hazardous waste identification	• Detection of toxicity, mutagenicity, or pathogenicity by conventional methods or by fast biotechnological tests.
Prevention of hazardous waste production	• Production, trade, or use of specific products containing nonbiodegradable hazardous substances may be banned based on biotechnological tests of biodegradability and toxicity. • Selection of environmentally preferred products based on biotechnological tests of biodegradability and toxicity. • Replacement of chemical pesticides, herbicides, rodenticides, termiticides, fungicides, and fertilizers by biodegradable and nonpersistent in the environment biotechnological analogs.
Hazardous waste collection	• Production and use of biodegradable containers. • Biotechnological formation of chemical substances (H_2S, Fe^{2+}) used for the collection of hazardous substances.
Hazardous waste reduction	• Biotreatment and biodegradation of hazardous waste. • Immobilization of hazardous substances from the streams. • Solubilization of hazardous substances from waste.
Hazardous waste toxicity reduction	• Biodegradation of hazardous substances. • Immobilization/solubilization of hazardous substances. • Biotransformation and detoxication of hazardous substances.
Hazardous waste recycling	• Solubilization/precipitation and recycling of heavy metals from waste. • Bioassimilation, precipitation, and recycling of ammonia, nitrate, and nitrite.
Hazardous waste incineration	• Sorption of hazardous products of combustion and their biodegradation.
Hazardous waste landfilling	• Inoculation of landfill for faster biodegradation. • Biotreatment of landfill leachate.

waste collection in biodegradable containers; hazardous waste toxicity reduction by biotreatment/biodegradation/bioimmobilization of hazardous substances; and hazardous waste recycling by recycling of nutrients during hazardous waste treatment.

12.1.3 Biosensors of Hazardous Substances

An important application of biotechnology in hazardous waste management is the biomonitoring of hazardous substances. This includes monitoring of biodegradability, toxicity, mutagenicity, concentration of hazardous substances, and monitoring of concentration and pathogenicity of microorganisms in untreated wastes, treated wastes, and in the environment [11,12].

Whole-Cell Biosensors

Simple or automated offline or online biodegradability tests can be performed by measuring CO_2 or CH_4 gas production or O_2 consumption [13]. Biosensors may utilize either whole bacterial cells or enzymes to detect specific molecules of hazardous substances. Toxicity can be monitored specifically by whole-cell sensors whose bioluminescence may be inhibited by the presence of hazardous substance. The most popular approach uses cells with an introduced luminescent reporter gene to determine changes in the metabolic status of the cells following intoxication [14]. Nitrifying bacteria have multiple-folded cell membranes that are sensitive to all membrane-disintegrating substances. Therefore, respirometric sensors that measure the respiration rates of these bacteria can be used for toxicity monitoring in wastewater treatment [15]. Another approach involves amperometric measurements of oxidized or reduced chemical mediators as an indicator of the metabolic status of bacterial or eukaryotic cells [14]. Biosensors measuring concentrations of hazardous substances are often based on the measurement of bioluminescence [16]. This toxicity sensor is a bioluminescent toxicity bioreporter for hazardous wastewater treatment. It is constructed by incorporating bioluminescence genes into a microorganism. These whole-cell toxicity sensors are very sensitive and may be used online to monitor and optimize the biodegradation of hazardous soluble substances. Similar sensors can be used for the measurement of the concentration of specific pollutants. A gene for bioluminescence has been fused to the bacterial genes coding for enzymes that metabolize the pollutant. When this pollutant is degraded, the bacterial cells will produce light. The intensity of biodegradation and bioluminescence depend on the concentration of pollutant and can be quantified using fiber-optics online. Combinations of biosensors in array can be used to measure concentration or toxicity of a set of hazardous substances.

Microbial Test of Mutagenicity

The mutagenic activity of chemicals is usually correlated with their carcinogenic properties. Mutant bacterial strains have been used to determine the potential mutagenicity of manufactured or natural chemicals. The most common test, proposed by Ames in 1971, utilizes back-mutation in auxotrophic bacterial strains that are incapable of synthesizing certain nutrients. When auxotrophic cells are spread on a medium that lacks the essential nutrients (minimal medium), no growth will occur. However, cells that are treated with a tested chemical that causes a reversion mutation can grow in minimal medium. The frequency of mutation detected in the test is proportional to the potential mutagenicity and carcinogenicity of the tested chemical. Microbial mutagenicity tests are used widely in modern research [17–19].

Molecular Sensors

Cell components or metabolites capable of recognizing individual and specific molecules can be used as the sensory elements in molecular sensors [11]. The sensors may be enzymes, sequences of nucleic acids (RNA or DNA), antibodies, polysaccharides, or other "reporter" molecules. Antibodies, specific for a microorganism used in the biotreatment, can be coupled to fluorochromes to increase sensitivity of detection. Such antibodies are useful in monitoring the fate of bacteria released into the environment for the treatment of a polluted site. Fluorescent or enzyme-linked immunoassays have been derived and can be used for a variety of contaminants, including pesticides and chlorinated polycyclic hydrocarbons. Enzymes specific for pollutants and attached to matrices detecting interactions between enzyme and pollutant are used in online biosensors of water and gas biotreatment [20,21]

Detection of Bacterial DNA Sequences by Oligonucleotide Probe or Array

A useful approach to monitor microbial populations in the biotreatment of hazardous wastes involves the detection of specific sequences of nucleic acids by hybridization with complementary oligonucleotide probes. Radioactive labels, fluorescent labels, and other kinds of labels are attached to the probes to increase sensitivity and simplicity of the hybridization detection. Nucleic acids that are detectable by the probes include chromosomal DNA, extra-chromosomal DNA such as plasmids, synthetic recombinant DNA such as cloning vectors, phage or virus DNA, rRNA, tRNA, and mRNA transcribed from chromosomal or extra-chromosomal DNA. These molecular approaches may involve hybridization of whole intact cells, or extraction and treatment of targeted nucleic acids prior to probe hybridization [22–24]. Microarrays for simultaneous semiquantitative detection of different microorganisms or specific genes in the environmental sample have also been developed [25–27].

12.2 AEROBIC, ANAEROBIC, AND COMBINED ANAEROBIC/AEROBIC BIOTECHNOLOGICAL TREATMENT

Relation of Microorganisms to Oxygen. The evolution from an anaerobic atmosphere to an aerobic one resulted in the creation of anaerobic (living without oxygen), facultative anaerobic (living under anaerobic or aerobic conditions), microaerophilic (preferring to live under low concentrations of dissolved oxygen), and obligate aerobic (living only in the presence of oxygen) microorganisms. Some anaerobic microorganisms, called tolerant anaerobes, have mechanisms protecting them from exposure to oxygen. Others, called obligate anarobes, have no such mechanisms and may be killed after some seconds of exposure to aerobic conditions. Obligate anaerobes produce energy from: (a) fermentation (destruction of organic substances without external acceptor of electrons); (b) anaerobic respiration using electron acceptors such as CO_2, NO_3^-, NO_2^-, Fe^{3+}, SO_4^{2-}; (c) anoxygenic ($H_2S \rightarrow S$) or oxygenic ($H_2O \rightarrow O_2$) photosynthesis. Facultative anaerobes can produce energy from these reactions or from the aerobic oxidation of organic matter. The following sequence arranges respiratory processes according to increasing energetic efficiency of biodegradation (per mole of transferred electrons): fermentation $\rightarrow$ CO_2 respiration ("methanogenic fermentation") $\rightarrow$ dissimilative sulfate reduction $\rightarrow$ dissimilative iron reduction ("iron respiration") $\rightarrow$ nitrate respiration ("denitrification") $\rightarrow$ aerobic respiration.

12.2.1 Aerobic Microorganisms and Aerobic Treatment of Solid Wastes

Such xenobiotics as aliphatic hydrocarbons and derivatives, chlorinated aliphatic compounds (methyl, ethyl, methylene, and ethylene chlorides), aromatic hydrocarbons and derivatives (benzene, toluene, phthalate, ethylbenzene, xylenes, and phenol), polycyclic aromatic hydrocarbons, halogenated aromatic compounds (chlorophenols, polychlorinated biphenyls, dioxins and relatives, DDT and relatives), AZO dyes, compounds with nitrogroups (explosive-contaminated waste and herbicides), and organophosphate wastes can be treated effectively by aerobic microorganisms.

Conventional Composting of Organic Wastes

Technologically, composting is the simplest way to treat solid waste containing hazardous substances. Composting converts biologically unstable organic matter into a more stable humuslike product that can be used as a soil conditioner or organic fertilizer. Additional benefits of composting of organic wastes include prevention of odors from rotting wastes, destruction of pathogens and parasites (especially in thermophilic composting), and retention of nutrients in the endproducts. There are three main types of composting technology: the windrow system, the static pile system, and the in-vessel system.

Windrow System

Composting in windrow systems involves mixing an organic waste with inexpensive bulking agents (wood chips, leaves, corncobs, bark, peanut and rice husks) to create a structurally rigid matrix, to diminish heat transfer from the matrix to the ambient environment, to increase the treatment temperature and to increase the oxygen transfer rate. The mixed matter is stacked in 1–2 m high rows called windrows. The mixtures are turned over periodically (2 to 3 times per week) by mechanical means to expose the organic matter to ambient oxygen. Aerobic and partially anaerobic microorganisms, which are present in the waste or were added from previously produced compost, will grow in the organic waste. Owing to the biooxidation and release of energy, the temperature in the pile will rise. This is accompanied by successional changes in the dominant microbial communities, from less thermoresistant to more thermophilic ones. This composting process ranges from 30 to 60 days in duration.

Static Pile System

The static pile system is an intensive biotreatment because the pile of organic waste and bulking agent is intensively aerated using blowers and air diffusers. The pile is usually covered with compost to remove odor and to maintain high internal temperatures. The aerated static pile process typically takes 21 days, after which the compost is cured for another 30 days, dried and screened to recycle the bulking agent.

In-Vessel Composting

In-vessel composting results in the most intensive biotransformation of organic wastes. In-vessel composting is performed in partially or completely enclosed containers in which moisture content, temperature, and oxygen content in the gas can be controlled. This process requires little space and takes some days for treatment, but its cost is higher than that of open systems.

Composting of Hazardous Organic Wastes

Hazardous wastes can be treated in all the systems mentioned above, but long durations are usually needed to reach permitted levels of pollution. The choice of the system depends on the required time and possible cost of the treatment. Time of the treatment decreases, but the costs increase in the following sequence: windrow system → static pile system → in-vessel system. To intensify the composting of hazardous solid waste, the following pretreatments can be used: mechanical disintegration and separation or screening to improve bioavailability of hazardous substances, thermal treatment, washing out of hazardous substances from waste by water or surfactants to diminish their content in waste, or application of H_2O_2, ozone, or Fenton's reagent as a chemical pretreatment to oxidize and cleave aromatic rings of hydrocarbons. There are many reports of successful applications of all types of composting for the treatment of crude-oil-impacted soil, petrochemicals-polluted soil, and explosives-polluted soil.

Application of Biotechnology in/on the Sites of Postaccidental Wastes

This direction of environmental biotechnology is known as soil bioremediation. There are many options in the process design described in the literature [7,28,29].

The main options tested in the field are as follows:

- Engineered in situ bioremediation (in-place treatment of a contaminated site);
- Engineered onsite bioremediation (the treatment of a percolating liquid or eliminated gas in reactors placed on the surface of the contaminated site). The reactors used for

this treatment are suspended biomass stirred-tank bioreactors, plug-flow bioreactors, rotating-disc contactors, packed-bed fixed-biofilm reactors (biofilter), fluidized bed reactors, diffused aeration tanks, airlift bioreactors, jet bioreactors, membrane bioreactors, and upflow bed reactors [30].

- Engineered ex situ bioremediation (the treatment of contaminated soil or water that is removed from a contaminated site).

The first option is used when the pollution is not strong, time required for the treatment is not a limiting factor, and there is no pollution of groundwater. The second option is usually used when the level of pollution is high and there is secondary pollution of groundwater. The third option is usually used when the level of pollution is so high that it diminishes the biodegradation rate due to toxicity of substances or low mass transfer rate. Another reason for using this option might be that the conditions insite or onsite (pH, salinity, dense texture or high permeability of soil, high toxicity of substance, and safe distance from public place) are not favorable for biodegradation.

Artificial Formation of Geochemical Barrier

One aim of using biotechnology is to prevent the dispersion of hazardous substances from the accident site into the environment. This can be achieved by creating physical barriers on the migration pathway with microorganisms capable of biotransforming the intercepted hazardous substances, for example, in polysaccharide (slime) viscous barriers in the contaminated subsurface. Another approach, which can be used to immobilize heavy metals in soil after pollution accidents, is the creation of biogeochemical barriers. These geochemical barriers could comprise gradients of H_2S, H_2, or Fe^{2+} concentrations, created by anaerobic sulfate-reducing bacteria (in the absence of oxygen and the presence of sulfate and organic matter), fermenting bacteria (after addition of organic matter and in the absence of oxygen), or iron-reducing bacteria [in the presence of Fe(III) and organic matter], respectively. Other bacteria can form a geochemical barrier for the migration of heavy metals at the boundary between aerobic and anaerobic zones. For example, iron-oxidizing bacteria will oxidize Fe^{2+} in this barrier and produce iron hydroxides that can diminish the penetration of ammonia, phosphate, organic acids, cyanides, phenols, heavy metals, and radionuclides through the barrier.

12.2.2 Aerobic Biotechnological Treatment of Wastewater

Treatment in Aerobic Reactors

Industrial hazardous wastewater can be treated aerobically in suspended biomass stirred-tank bioreactors, plug-flow bioreactors, rotating-disc contactors, packed-bed fixed-biofilm reactors (or biofilters), fluidized bed reactors, diffused aeration tanks, airlift bioreactors, jet bioreactors, membrane bioreactors, and upflow bed reactors [28,30].

One difference between these systems and the biological treatment of nonhazardous wastewater is that the exhaust air may contain volatile hazardous substances or intermediate biodegradation products. Therefore, the air must be treated as secondary hazardous wastes by physical, chemical, physico-chemical, or biological methods. Other secondary hazardous wastes may include the biomass of microorganisms that may accumulate volatile hazardous substances or intermediate products of their biodegradation. This hazardous liquid or semisolid waste must be properly treated, incinerated, or disposed.

Treatment of Wastewater with Low Concentration of Hazardous Substance

Wastewater with low concentrations of hazardous substances may reasonably be treated using biotechnologies such as granular activated carbon (GAC) fluidized-bed reactors or co-metabolism. Granulated activated carbon or other adsorbents ensure sorption of hydrophobic hazardous substances on the surface of GAC or other adsorbent particles. Microbial biofilms can also be concentrated on the surface of these particles and can biodegrade hazardous substances with higher rates compared to situations when both substrate and microbial biomass are suspended in the wastewater. Cometabolism refers to the simultaneous biodegradation of hazardous organic substances (which are not used as a source of energy) and stereochemically similar substrates, which serve as a source of carbon and energy for microbial cells. Biooxidation of the hazardous substance is performed by the microbial enzymes due to stereochemical similarity between the hazardous substance and the substrate. The best-known applications of cometabolism are the biodegradation/detoxication of chloromethanes, chloroethanes, chloromethylene, and chloroethylenes by enzyme systems of bacteria for oxidization of methane or ammonia as the main source of energy. In practise, the bioremediation is achieved by adding methane or ammonia, oxygen (air), and biomass of methanotrophic or nitrifying bacteria to soil and groundwater polluted by toxic chlorinated substances.

Combinations of Aerobic Treatment with Other Treatments

To intensify the biotreatment of hazardous liquid waste, the following pretreatments can be used: mechanical disintegration/suspension of hazardous hydrophobic substances to improve the reacting surface in the suspension and increase the rate of biodegradation; removal from wastewater or concentration of hazardous substances by sedimentation, centrifugation, filtration, flotation, adsorption, extraction, ion exchange, evaporation, distillation, freezing separation; preliminary oxidation by H_2O_2, ozone, or Fenton's reagent to produce active oxygen radicals; preliminary photo-oxidation by UV and electrochemical oxidation of hazardous substances.

Application of Microaerophilic Microorganisms in Biotechnological Treatment

Some aerobic microorganisms prefer low concentrations of dissolved oxygen in the medium for growth, for example, concentrations below 1 mg/L. These microorganisms include filamentous hydrogen sulfide-oxidizing bacteria (e.g., *Beggiatoa* spp.), pathogenic bacteria (e.g., *Campylobacter* spp., *Streptococcus* spp., and *Vibrio* spp.), microaerophilic spirilla (e.g., *Magnetospirillum* spp.) and neutrophilic iron-oxidizing bacteria. Iron-oxidizing bacteria can produce sheaths or stalks that act as organic matrices upon which the deposition of ferric hydrooxides can occur [31,32]. Some microaerophiles are active biodegraders of organic pollutants in postaccident sites [33], while other microaerophiles form H_2O_2 to oxidize xenobiotics. Sheaths of neutrophilic iron-oxidizing bacteria can adsorb heavy metals and radionuclides from hazardous streams.

12.2.3 Aerobic Biotechnological Treatment of Hazardous Waste Gas

Biodegradable Hazardous Gases

The CERCLA Priority List of Hazardous Substances contains many substances released in industry as gaseous hazards and which can be treated biotechnologically (Table 1), including the following: chloroform, trichloroethylene, 1,2-dibromoethane, 1,2-dibromo-3-chloropropane, carbontetrachloride, xylenes, dibromochloropropane, toluene, methane, methylene chloride, 1,1-dichloroethene, *bis*(2-chloroethyl) ether, 1,2-dichloroethane, chlorine, 1,1,2-trichloroethane, ethylbenzene, 1,1,2,2-tetrachloroethane, bromine, methylmercury, trichlorofluoroethane, 1,1-dichloroethane, 1,1,2-trichloroethane, ammonia, trichloroethane, 1,2-dichloroethene, carbon disulfide, chloroethane, p-xylene, hydrogen sulfide, chloromethane, 2-butanone, bromoform, acrolein, bromodichloroethane, nitrogen dioxide, ozone, formaldehyde, chlorodibromomethane, ethyl ether, and 1,2-dichloropropane.

Reactors

The common way to remove vaporous or gaseous pollutants from gas or air streams is to pass contaminated gases through bioscrubbers containing suspensions of biodegrading microorganisms or through a biofilter packed with porous carriers covered by biofilms of degrading microorganisms. Depending on the nature and volume of polluted gas, the biofilm carriers may be cheap porous substrates, such as peat, wood chips, compost, or regular artificial carriers such as plastic or metal rings, porous cylinders and spheres, fibers, and fiber nets. The bioscrubber contents must be stirred to ensure high mass transfer between the gas and microbial suspension. The liquid that has interacted with the polluted gas is collected at the bottom of the biofilter and recycled to the top part of the biofilter to ensure adequate contact of polluted gas and liquid and optimal humidity of biofilter. Addition of nutrients and fresh water to the bioscrubber or biofilter must be made regularly or continuously. Fresh water can be used to replace water that has evaporated in the bioreactor. If the mass transfer rate is higher than the biodegradation rate, the absorbed pollutants must be biodegraded in an additional suspended bioreactor or biofilter connected in series to the bioscrubber or absorbing biofilter.

Applications

The main application of biotechnology for the treatment of hazardous waste gases is the bioremoval of biodegradable organic solvents. Other important applications include the biodegradation of odors and toxic gases such as hydrogen sulfide and other sulfur-containing gases from the exhaust ventilation air in industry and farming. Industrial ventilation air containing formaldehyde, ammonia, and other low-molecular-weight substances can also be effectively treated in the bioscrubber or biofilter.

12.2.4 Anaerobic Microorganisms and Anaerobic Biotechnological Treatment of Hazardous Wastes

Fermentation and Anaerobic Respiration

The main energy-producing pathways in anaerobic treatment are fermentation (intramolecular oxidation/reduction without external electron acceptor) or anaerobic respiration (oxidation by electron acceptor other than oxygen). The advantage of anaerobic treatment is that there is no need to supply oxygen in the treatment system. This is useful in cases such as bioremediation of clay soil or high-strength organic waste. However, anaerobic treatment may be slower than aerobic treatment, and there may be significant outputs of dissolved organic products of fermentation or anaerobic respiration.

Biotreatment by Facultative Anaerobic Microorganisms

Facultative anaerobic microorganisms may be useful when integrated together with aerobic and anaerobic microorganisms in microbial aggregates. However, this function is still not well studied. One interesting and useful feature in this physiological group is the ability in some representatives (e.g., *Escherichia coli*) to produce an active oxidant, hydrogen peroxide, during normal aerobic metabolism [34].

Biotreatment by Anaerobically Respiring Bacteria

Aerobic respiration is more effective in terms of output of energy per mole of transferred electrons than fermentation. Anaerobic respiration can be performed by different groups of prokaryotes with such electron acceptors as NO_3^-, NO_2^-, Fe^{3+}, SO_4^{2-}, and CO_2. Therefore, if the concentration of one such acceptor in the hazardous waste is sufficient for the anaerobic respiration and oxidation of the pollutants, the activity of the related bacterial group can be used for the treatment. CO_2-respiring prokaryotes (methanogens) are used for methanogenic biodegradation of organic hazardous wastes in anaerobic reactors or in landfills. Sulfate-reducing bacteria can be used for anaerobic biodegradation of organic matter or for the precipitation/immobilization of heavy metals of sulfate-containing hazardous wastes. Iron-reducing bacteria can produce dissolved Fe^{2+} ions from insoluble Fe(III) minerals. Anaerobic biodegradation of organic matter and detoxication of hazardous wastes can be significantly enhanced as a result of precipitation of toxic organics, acids, phenols, or cyanide by Fe(II). Nitrate-respiring bacteria can be used in denitrification, that is, reduction of nitrate to gaseous N_2. Nitrate can be added to the hazardous waste to initiate the biodegradation of different types of organic substances, for example, polycyclic aromatic hydrocarbons [35]. Nitrogroups of hazardous substances can be reduced by similar pathway to related amines.

Biotreatment of Hazardous Waste by Anaerobic Fermenting Bacteria

Anaerobic fermenting bacteria (e.g., from genus *Clostridium*) perform two important functions in the biodegradation of hazardous organics: they hydrolyze different natural polymers and ferment monomers with production of alcohols, organic acids, and CO_2. Many hazardous substances, for example, chlorinated solvents, phthalates, phenols, ethyleneglycol, and polyethylene glycols can be degraded by anaerobic microorganisms [28,36–38]. Fermenting bacteria perform anaerobic dechlorination, thus enhancing further biodegradation of chlorinated organics. There are different biotechnological systems to perform anaerobic biotreatment of wastewater: biotreatment by suspended microorganisms, anaerobic biofiltration, and biotreatment in upflow anaerobic sludge blanket (UASB) reactors [7,28].

Application of Biotechnology in Landfilling of Hazardous Solid Wastes

Landfilled organic and inorganic wastes are slowly transformed by indigenous microorganisms in the wastes [39]. Organic matter is hydrolyzed by bacteria and fungi. Amino acids are degraded via ammonification with formation of toxic organic amines and ammonia. Amino acids, nucleotides, and carbohydrates are fermented or anaerobically oxidized with formation of organic acids, CO_2, and CH_4. Xenobiotics and heavy metals may be reduced, and subsequently dissolved or immobilized. These bioprocesses may result in the formation of toxic landfill leachate, which can be detoxicated by aerobic biotechnological treatment to oxidize organic hazards and to immobilize dissolved heavy metals.

12.2.5 Combined Anaerobic/Aerobic Biotreatment of Wastes

A combined anaerobic/aerobic biotreatment can be more effective than aerobic or anaerobic treatment alone. The simplest approach for this type of treatment is the use of aerated stabilization ponds, aerated and nonaerated lagoons, and natural and artificial wetland systems, whereby aerobic treatment occurs in the upper part of these systems and anaerobic treatment occurs at the bottom end. A typical organic loading is 0.01 kg BOD/m^3 day and the retention time varies from a few days to 100 days [30]. A more intensive form of biodegradation can be achieved by combining aerobic and anaerobic reactors with controlled conditions, or by integrating anaerobic and aerobic zones within a single bioreactor. Combinations or even alterations of anaerobic and aerobic treatments are useful in the following situations: (a) biodegradation of chlorinated aromatic hydrocarbons including anaerobic dechlorination and aerobic ring cleavage; (b) sequential nitrogen removal including aerobic nitrification and anaerobic denitrification; (c) anaerobic reduction of Fe(III) and microacrophilic oxidation of Fe(II) with production of fine particles of iron hydroxide for adsorption of organic acids, phenols, ammonium, cyanide, radionuclides, and heavy metals.

12.2.6 Biotechnological Treatment of Heavy Metals-Containing Waste and Radionuclides-Containing Waste

Liquid and solid wastes containing heavy metals may be successfully treated by biotechnological methods. The effects of microorganisms on metals are described below.

Direct Reduction/Oxidation, or Reduction/Oxidation Mediated by Other Metals or Microbial Metabolites

Some metals such as iron are reduced or oxidized by specific enzymes of microorganisms. Microbial metabolism generates products such as hydrogen, oxygen, H_2O_2, and reduced or oxidized iron that can be used for oxidation/reduction of metals. Reduction or oxidation of metals is usually accompanied by metal solubilization or precipitation.

Effect of Microbial Metabolites

Solubilization or precipitation of metals may be mediated by microbial metabolites. Microbial production of organic acids in fermentation or inorganic acids (nitric and sulfuric acids) in aerobic oxidation will promote formation of dissolved chelates of metals. Microbial production of phosphate, H_2S, and CO_2 will stimulate precipitation of nondissolved phosphates, carbonates, and sulfides of heavy metals, for example arsenic, cadmium, chromium, copper, lead, mercury, nickel; production of H_2S by sulfate-reducing bacteria is especially useful to remove heavy metals and radionuclides from sulfate-containing mining drainage waters, liquid waste of nuclear facilities, drainage from tailing ponds of hydrometallurgical plants; wood straw or saw dust. Organic acids, produced during the anaerobic fermentation of cellulose, may be preferred as a source of reduced carbon for sulfate reduction and further precipitation of metals.

Biosorption

The surface of microbial cells is covered by negatively charged carboxylic and phosphate groups, and positively charged amino groups. Therefore, depending on pH, there may be significant adsorption of heavy metals onto the microbial surface [7]. Biosorption, for example by fungal fermentation residues, is used to accumulate uranium and other radionuclides from waste streams.

Degradation of Minerals

Metal-containing minerals, for example sulfides, can be oxidized and metals can be solubilized. This approach is used for the bioleaching of heavy metals from sewage sludge [40,41] before landfilling or biotransformation.

Volatilization

Some metals, arsenic and mercury for example, may be volatilized by methylation due to activity of anaerobic microorganisms. Arsenic can be methylated by methanogenic *Archaea* and fungi to volatile toxic dimethylarsine and trimethylarsine or can be converted to less toxic nonvolatile methanearsonic and dimethylarsinic acids by algae [42].

Combination of Methods

In some cases the methods may be combined. Examples would include the biotechnological precipitation of chromium from Cr(VI)-containing wastes from electroplating factories by sulfate reduction to precipitate chromium sulfide. Sulfate reduction can use fatty acids as organic substrates with no accumulation of sulfide. In the absence of fatty acids but with straw as organic substrate, the direct reduction of chromium has been observed without sulfate reduction [43].

Biodegradation of Organometals

Hydrophobic organotins are toxic to organisms because of their solubility in cell membranes. However, many microorganisms are resistant to organotins and can detoxicate them by degrading the organic part of them [5].

12.3 ENHANCED BIOTECHNOLOGICAL TREATMENT OF INDUSTRIAL HAZARDOUS WASTES

Several key factors are critical for the successful application of biotechnology for the treatment of hazardous wastes: (a) environmental factors, such as pH, temperature, and dissolved oxygen concentration, must be optimized; (b) contaminants and nutrients must be available for action or assimilation by microorganisms; (c) content and activity of essential microorganisms in the treated waste must be sufficient for the treatment.

12.3.1 Enhancement of Biotreatment by Abiotic Factors

Optimal Temperature

Psychrophilic microorganism have optimal growth temperatures below 15°C. These organisms may be killed by exposure to temperatures above 30°C. Mesophilic microorganisms have optimal growth temperatures in the range between 20 and 40°C. Thermophiles grow best above 50°C. Some bacteria can grow up to temperatures where water boils; those with optimal growth temperatures above 75°C are categorized as extreme thermophiles. Therefore, the biotreatment temperature must be maintained at optimal growth temperatures for effective biotreatment by certain physiological groups of microorganisms. The heating of the treated waste can come from microbial oxidation or fermentation activities provided there is sufficient heat generation and good thermoisolation of treated waste from the cooler surroundings. The bulking agent added to solid wastes may also be used as a thermoisolator.

Optimum pH

The pH of natural microbial biotopes vary from 1 to 11: volcanic soil and mine drainage have pH values between 1 and 3; plant juices and acid soils have pH values between 3 and 5; fresh and sea water have pH values between 7 and 8; alkaline soils and lakes, solutions of ammonia, and rotten organics have pH values between 9 and 11. Most microbes grow most efficiently within the pH range 5–9. They are called neutrophiles. Species that have adapted to grow at pH values lower than 4 are called acidophiles. Species that have adapted to grow at pH values higher than 9 are called alkalophiles. Therefore, pH of the treatment medium must be maintained at optimal values for effective biotreatment by certain physiological groups of microorganisms. The optimum pH may be maintained physiologically, by addition of pH buffer or pH regulator as follows: (a) control of organic acid formation in fermentation; (b) prevention of formation of inorganic acids in aerobic oxidation of ammonium, elemental sulfur, hydrogen sulfide, or metal sulfides; (c) assimilation of ammonium, nitrate, or ammonium nitrate, leading to decreased pH, increased pH, or neutral pH, respectively; (d) pH buffers such as $CaCO_3$ or $Fe(OH)_3$ can be used in large-scale waste treatment; (e) solutions of KOH, NaOH, NH_4OH, $Ca(OH)_2$, HCl, or H_2SO_4 can be added automatically to maintain the pH of liquid in a stirred reactor. Maintenance of optimum pH in treated solid waste or bioremediated soil may be especially important if there is a high content of sulfides in the waste or acidification/alkalization of soil in the bioremediation process.

Enhancement of Biodegradation by Nutrients and Growth Factors

The major elements that are found in microbial cells are C, H, O, N, S, and P. An approximate elemental composition corresponds to the formula $CH_{1.8}O_{0.5}N_{0.2}$. Therefore, nutrient amendment may be required if the waste does not contain sufficient amounts of these macroelements. The waste can be enriched with carbon (depending on the nature of the pollutant that is treated), nitrogen (ammonium is the best source), phosphorus (phosphate is the best source), and/or sulfur (sulfate is the best source). Other macronutrients (K, Mg, Na, Ca, and Fe) and micronutrients (Cr, Co, Cu, Mn, Mo, Ni, Se, W, V, and Zn) are also essential for microbial growth and enzymatic activities and must be added into the treatment systems if present in low concentrations in the waste. The best sources of essential metals are their dissolved salts or chelates with organic acids. The source of metals for the bioremediation of oil spills may be lipophilic compounds of iron and other essential nutrients that can accumulate at the water–air interface where hydrocarbons and hydrocarbon-degrading microorganisms can also occur [44]. In some biotreatment cases, growth factors must also be added into the treated waste. Growth factors are organic compounds such as vitamins, amino acids, and nucleosides that are required in very small amounts and only by some strains of microorganisms called auxotrophic strains. Usually those microorganisms that are commensals or parasites of plant and animals require growth factors. However, sometimes these microorganisms may have the unique ability to degrade some xenobiotics.

Increase of Bioavailability of Contaminants

Hazardous substances may be protected from microbial attack by physical or chemical envelopes. These protective barriers must be destroyed mechanically or chemically to produce fine particles or waste suspensions to increase the surface area for microbial attachment and subsequent biodegradation. Another way to increase the bioavailability of hydrophobic substances is washing of waste or soil by water or a solution of surface-active substances (surfactants). The disadvantage of this technology is the production of secondary hazardous waste because chemically produced surfactants are usually resistant to biodegradation. Therefore, only easily biodegradable or biotechnologically produced surfactants can be used for pretreatment of hydrophobic hazardous substances.

Enhancement of Biodegradation by Enzymes

Extracellular enzymes produced by microorganisms are usually expensive for large-scale biotreatment of organic wastes. However, enzyme applications may be cost-effective in certain situations. Toxic organophosphate waste can be treated using the enzyme parathion hydrolase produced and excreted by a recombinant strain of *Streptomyces lividans*. The cell-free culture fluid contains enzymes that can hydrolyze organophosphate compounds [4]. Future applications may be related with cytochrome-P450-dependent oxygenase enzymes that are capable of oxidizing different xenobiotics [45].

Enhancement of Biodegradation by Aeration and Oxygen Supply

Concentrations of dissolved oxygen may be very low (7–8 mg/L) and can be rapidly depleted during waste biotreatment with oxygen consumption rates ranging from 10 to 2000 g O_2/L hour. Therefore, oxygen must be supplied continuously in the system. Supply of air in liquid waste treatment systems is achieved by aeration and mechanical agitation. Different techniques are employed to supply sufficient quantities of oxygen in fixed biofilm reactors, in viscous solid wastes, in underground layers of soil, or in aquifers polluted by hazardous substances. Very often the supply of oxygen is the critical factor in the successful scaling up of bioremediation technologies from laboratory experiments to full-scale applications [10]. Air sparging in situ is a commonly used bioremediation technology, which volatilizes and enhances aerobic biodegradation of contamination in groundwater and saturated soils. Successful case studies include a 6–12 month bioremediation project that targeted both sandy and silty soils polluted by petroleum products and chlorinated hydrocarbons [46]. Application of pure oxygen can increase the oxygen transfer rate by up to five times, and this can be used in situations with strong acute toxicity of hazardous wastes and low oxygen transfer rates, to ensure sufficient oxygen transfer rate in polluted waste.

Enhancement of Biodegradation by Oxygen Radicals

In some cases, hydrogen peroxide has been used as an oxygen source because of the limited concentrations of oxygen that can be transferred into the groundwater using aboveground aeration followed by reinjection of the oxygenated groundwater into the aquifer or subsurface air sparging of the aquifer. However, because of several potential interactions of H_2O_2 with various aquifer material constituents, its decomposition may be too rapid, making effective introduction of H_2O_2 into targeted treatment zones extremely difficult and costly [47]. Pretreatment of wastewater by ozone, H_2O, by TiO_2-catalyzed UV-photooxidation, and electrochemical oxidation can significantly enhance biodegradation of halogenated organics, textile dyes, pulp mill effluents, tannery wastewater, olive-oil mills, surfactant-polluted wastewater, and pharmaceutical wastes, and diminish the toxicity of municipal landfill leachates. In some cases, oxygen radicals generated by Fenton's reagent ($Fe^{2+} + H_2O_2$ at low pH), and iron peroxides [Fe(VI) and Fe(V)] can be used as the oxidants in the treatment of hazardous wastes.

Enhancement of Biodegradation by Biologically Produced Oxygen Radicals

Many microorganisms can produce and release to the environment such toxic metabolites of oxygen as hydrogen peroxide (H_2O_2), superoxide radical ($O_2^{\bullet}$), and hydroxyl radical ($OH^{\bullet}$).

Lignin-oxidizing "white-rot" fungi can degrade lignin and all other chemical substances due to intensive generation of oxygen radicals, which oxidize the organic matter by random incorporation of oxygen into the molecule. Not much is known about the biodegradation ability of H_2O_2-generating microaerophilic bacteria.

Enhancement of Biodegradation by Electron Acceptors Other Than Oxygen

Dissolved acceptors of electrons such as NO_3^-, NO_2^-, Fe^{3+}, SO_4^{2-}, and HCO_3^- can be used in the treatment system when oxygen transfer rates are low. The choice of the acceptor is determined by economical and environmental factors. Nitrate is often proposed for bioremediation [35] because it can be used by many microorganisms as an electron acceptor. However, it is relatively expensive and its supply to the treatment system must be thoroughly controlled because it can also pollute the environment. Fe^{3+} is an environmentally friendly electron acceptor. It is naturally abundant in clay minerals, magnetite, limonite, goethite, and iron ores, but its compounds are usually insoluble and it diminishes the rate of oxidation in comparison with dissolved electron acceptors. Sulfate and carbonate can be applied as electron acceptors in strictly anaerobic environments only. Other disadvantage of these acceptors is that these anoxic oxidations generate toxic and foul smelling H_2S or "greenhouse" gas CH_4.

12.3.2 Enhancement of Biotreatment by Biotic Factors

Reasons for Bioenhancement of the Treatment

Addition of microorganisms (inoculum) to start up or to accelerate the biotreatment process is a reasonable strategy under the following conditions:

1. If microorganisms that are necessary for hazardous waste treatment are absent or their concentration is low in the waste;
2. If the rate of bioremediation performed by indigenous microorganisms is not sufficient to achieve the treatment goal within the prescribed duration;
3. If the acclimation period is too long;
4. To direct the biodegradation/biotreatment to the best pathway from many possible pathways;
5. To prevent growth and dispersion in the waste treatment system of unwanted or nondetermined microbial strains, which may be pathogenic or opportunistic.

Application of Acclimated Microorganisms

A simple way to produce a suitable microbial inoculum is the production of an enrichment culture, which is a microbial community containing one or more dominant strains naturally formed during cultivation in a growth medium modeling the hazardous waste under defined conditions. If the cultivation conditions change, the dominant strains in the enrichment culture may also change. Another approach involves the use of part of the treated waste containing active microorganisms as inoculum to start up the process. Application of acclimated microorganisms in an enrichment culture or in biologically treated waste may significantly decrease the start-up period for biotechnological treatment. In cases involving treatment of toxic substances and high death rates of microorganisms during treatment, regular additions of active microbial cultures may be useful to maintain constant rates of biodegradation.

Selection and Use of Pure Culture

Notwithstanding the common environmental engineering practise of using part of the treated waste as inoculum, applications of defined pure starter cultures have the following theoretical advantages: (a) greater control over desirable processes; (b) lower risk of release of pathogenic or opportunistic microorganisms during biotechnological treatment; (c) lower risk of accumulation of harmful microorganisms in the final biotreatment product. Pure cultures that are most active in biodegrading specific hazardous substances can be isolated by conventional microbiological methods, quickly identified by molecular–biological methods, and tested for pathogenicity and biodegradation properties. The biomass of pure culture can be produced in a large scale in commercial fermenters, then concentrated and dried for storage before field application. Therefore, it is not only the biodegradation abilities of pure cultures, but also the suitability for industrial production of dry biomass that must be taken into account in the selection of pure culture for the biotechnological treatment of industrial hazardous waste. Generally, Gram-positive bacteria are more viable after drying and storage than Gram-negative bacteria. Spores of Gram-positive bacteria can form superstable inocula.

Construction of Microbial Community

A pure culture is usually active in the biodegradation of one type of hazardous substance. Wastes containing a variety of hazardous substances must be treated by a microbial consortium comprising a collection of pure cultures most active in the degradation of the different types of substances. However, even in cases involving a single hazardous substance, degradation rates may be higher for a collection of pure cultures acting mutually (symbiotically) than for single pure cultures. Mutualistic relationships between pure cultures in an artificially constructed or a naturally selected microbial community may be based on the sequential degradation of xenobiotic, mutual exchange of growth factors or nutrients between these cultures, mutual creation of optimal conditions (pH, redox potential) and gradients of concentrations. Mutualistic relationships between the microbial strains are more clearly demonstrated in dense microbial aggregates such as biofilms, flocs, and granules used for biotechnological treatment of hazardous wastes.

Construction of Genetically Engineered Microorganisms

Microorganisms suitable for the biotreatment of hazardous substances can be isolated from the natural environment. However, their ability for biodegradation can be modified and amplified by artificial alterations of the genetic (inherited) properties of these microorganisms. The description of the methods is given in many books on environmental microbiology and biotechnology [7,28]. Natural genetic recombination of the genes (units of genetic information) occurs during DNA replication and cell reproduction, and includes the breakage and rejoining of chromosomal DNA molecules (separately replicated sets of genes) and plasmids (self-replicating minichromosomes containing several genes). Recombinant DNA techniques or genetic engineering can create new, artificial combinations of genes, and increase the number of desired genes in the cell. Genetic engineering of recombinant microbial strains suitable for the biotreatment may involve the following steps: (a) DNA is extracted from the cell and cut into small sequences by specific enzymes; (b) the small sequences of DNA can be introduced into DNA vectors; (c) the vector (virus or plasmid) is transferred into the cell and self-replicated to produce multiple copies of the introduced genes; (d) the cells with newly acquired genes are selected based on activity (e.g., production of defined enzymes, biodegradation capability) and stability of acquired genes. Genetic engineering of microbial strains can create (transfer) the ability to biodegrade xenobiotics or amplify this ability through the amplification of related genes. Another approach is the construction of hybrid metabolic pathways to increase the range of biodegraded xenobiotics and the rate of biodegradation [48]. The desired genes for biodegradation of different xenobiotics can be isolated and then cloned into plasmids. Some plasmids have been constructed containing multiple genes for the biodegradation of several xenobiotics simultaneously. The strains containing such plasmids can be used for the bioremediation of sites heavily polluted by a variety of xenobiotics. The main problem in these applications is maintaining the stability of the plasmids in these strains. Other technological and public concerns include the risk of application and release of genetically modified microorganisms in the environment.

Application of Microbial Aggregates and Immobilized Microorganisms

Self-aggregated microbial cells of biofilms, flocs, and granules, and artificially aggregated cells immobilized on solid particles are often used in the biotreatment of hazardous wastes. Advantages of microbial aggregates in hazardous waste treatment are as follows: (a) upper layers and matrix of aggregates protect cells from toxic pollutants due to adsorption or detoxication; therefore microbial aggregates or immobilized cells are more resistant to toxic xenobiotics than suspended microbial cells; (b) different or alternative physiological groups of microorganisms (aerobes/anaerobes, heterotrophs/nitrifiers, sulfate-reducers/sulfur-oxidizers) may coexist in aggregates and increase the diversity of types of biotreatments, leading to higher treatment efficiencies in one reactor; (c) microbial aggregates may be easily and quickly separated from treated water. Microbial cells immobilized on carrier surfaces such as GAC that can adsorb xenobiotics will degrade xenobiotics more effectively than suspended cells [49]. However, dense microbial aggregates may encounter problems associated with diffusion limitation, such as slow diffusion both of the nutrients into and the metabolites out of the aggregate. For example, dissolved oxygen levels can drop to zero at some depth below the surface of microbial aggregates. This distance clearly depends on factors such as the specific rate of oxygen consumption and the density of biomass in the microbial aggregate. When the environmental conditions within the aggregate become unfavorable, cell death may occur in zones that do not receive sufficient nutrition or that contain inhibitory metabolites. Channels and pores in aggregate can facilitate transport of oxygen, nutrients, and metabolites. Channels in microbial spherical granules have been shown to penetrate to depths of 900 μm [50] and a layer of obligate anaerobic bacteria was detected below the channelled layer [51]. This demonstrates that there is some optimal size or thickness of microbial aggregates appropriate for application in the treatment of hazardous wastes.

REFERENCES

1. Levin, M.A.; Gealt, M.A. Overview of biotreatment practices and promises. In *Biotreatment of Industrial and Hazardous Wastes*; Levin, M.A., Gealt, M.A., Eds.; McGraw-Hill, Inc.: NY, 1993; 1–18.
2. Geizer, K. Source reduction: quantity and toxicity. Part 6B. Toxicity reduction. In *Handbook of Solid Waste Management*, 2nd Ed.; Kreith, F., Tchobanoglous, G., Eds.; McGraw-Hill: NY, 2002; 6.27–6.41.
3. Swoboda-Goldberg, N.G. Chemical contamination of the environment: sources, types, and fate of synthetic organic chemicals. In *Microbial Transformation and Degradation of Toxic Organic Chemicals*; Young, L.Y., Cerniglia, C., Eds.; Wiley-Liss: NY, 1995; 27–74.

4. Coppella, S.J.; DelaCruz, N.; Payne, G.F.; Pogell, B.M.; Speedie, M.K.; Karns, J.S.; Sybert, E.M.; Connor, M.A. Genetic engineering approach to toxic waste management: case study for organophosphate waste treatment. Biotechnol. Prog. **1990**, *6*, 76–81.
5. Gadd, G.M. Microbial interactions with tributyltin compounds: detoxification, accumulation, and environmental fate. Sci. Total Environ. **2000**, *258*, 119–227.
6. Gealt, M.A.; Levin, M.A.; Shields, M. Use of altered microorganisms for field biodegradation of hazardous materials. In *Biotreatment of Industrial and Hazardous Wastes*; Levin, M.A., Gealt, M.A., Eds.; McGraw-Hill, Inc: NY, 1993; 197–206.
7. Moo-Young, M.; Anderson, W.A.; Chakrabarty, A.M. Eds. *Environmental Biotechnology: Principles and Applications*; Kluwer Academic Publishers: Dordrecht, 1996.
8. Sayles, G.D.; Suidan, M.T. Biological treatment of industrial and hazardous wastewater. In *Biotreatment of Industrial and Hazardous Wastes*; Levin, M.A., Gealt, M.A., Eds.; McGraw-Hill, Inc.: NY, 1993; 245–267.
9. Urbance, J.W.; Cole, J.; Saxman, P.; Tiedje, J.M. BSD: the biodegradative strain database. Nucl. Acids Res. **2003**, *31*, 152–155.
10. Talley, J.W.; Sleeper, P.M. Roadblocks to the implementation of biotreatment strategies. Ann. N. Y. Acad. Sci. **1997**, *829*, 16–29.
11. Burlage, R.S. Emerging technologies: bioreporters, biosensors, and microprobes. In *Manual of Environmental Microbiology*; Hurst, C.J., Crawford, R.L., McInerney, M.J., Eds.; ASM Press: Washington, DC, 1997; 115–123.
12. Wood, K.V.; Gruber, M.G. Transduction in microbial biosensors using multiplexed bioluminescence. Biosens. Bioelectron. **1996**, *11*, 207–214.
13. Reuschenbach, P.; Pagga, U.; Strotmann, U. A critical comparison of respirometric biodegradation tests based on OECD 301 and related test methods. Water Res. **2003**, *37*, 1571–1582.
14. Bentley, A.; Atkinson, A.; Jezek, J.; Rawson, D.M. Whole cell biosensors – electrochemical and optical approaches to ecotoxicity testing. Toxicol in Vitro **2001**, *15*, 469–475.
15. Inui, T.; Tanaka, Y.; Okayas, Y.; Tanaka, H. Application of toxicity monitor using nitrifying bacteria biosensor to sewerage systems. Water Sci. Technol. **2002**, *45*, 271–278.
16. Lajoie, C.A.; Lin, S.C.; Nguyen, H.; Kelly, C.J. A toxicity testing protocol using a bioluminescent reporter bacterium from activated sludge. J. Microbiol. Methods **2002**, *50*, 273–282.
17. Czyz, A.; Jasiecki, J.; Bogdan, A.; Szpilewska, H.; Wegrzyn, G. Genetically modified *Vibrio harveyi* strains as potential bioindicators of mutagenic pollution of marine environments. Appl. Environ. Microbiol. **2000**, *66*, 599–605.
18. Hwang, H.M.; Shi, X.; Ero, I.; Jayasinghe, A.; Dong, S.; Yu, H. Microbial ecotoxicity and mutagenicity of 1-hydroxypyrene and its photoproducts. Chemosphere **2001**, *45*, 445–451.
19. Yamamoto, A.; Kohyama, Y.; Hanawa, T. Mutagenicity evaluation of forty-one metal salts by the umu test. Biomed. Mater. Res. **2002**, *59*, 176–183.
20. Dewettinck, T.; Van Hege, K.; Verstraete, W. The electronic nose as a rapid sensor for volatile compounds in treated domestic wastewater. Water Res. **2001**, *35*, 2475–2483.
21. Nielsen, M.; Revsbech, N.P.; Larsen, L.H.; Lynggaard-Jensen, A. On-line determination of nitrite in wastewater treatment by use of a biosensor. Water Sci. Technol. **2002**, *45*, 69–76.
22. Hatsu, M.; Ohta, J.; Takamizawa, K. Monitoring of *Bacillus thermodenitrificans* OHT-1 in compost by whole cell hybridization. Can. J. Microbiol. **2002**, *48*, 848–852.
23. Nogueira, R.; Melo, L.F.; Purkhold, U.; Wuertz, S.; Wagner, M. Nitrifying and heterotrophic population dynamics in biofilm reactors: effects of hydraulic retention time and the presence of organic carbon. Water Res. **2002**, *36*, 469–481.
24. Sekiguchi, Y.; Kamagata, Y.; Ohashi, A.; Harada, H. Molecular and conventional analyses of microbial diversity in mesophilic and thermophilic upflow anaerobic sludge blanket granular sludges. Water Sci. Technol. **2002**, *45*, 19–25.
25. Fredrickson, H.L.; Perkins, E.J.; Bridges, T.S.; Tonucci, R.J.; Fleming, J.K.; Nagel, A.; Diedrich, K.; Mendez-Tenorio, A.; Doktycz, M.J.; Beattie, K.L. Towards environmental toxicogenomics – development of a flow-through, high-density DNA hybridization array and its application to ecotoxicity assessment. Sci. Total. Environ. **2001**, *274*, 137–149.

26. Koizumi, Y.; Kelly, J.J.; Nakagawa, T.; Urakawa, H.; El-Fantroussi, S.; Al-Muzaini, S.; Fukui, M.; Urushigawa, Y.; Stahl, D.A. Parallel characterization of anaerobic toluene- and ethylbenzene-degrading microbial consortia by PCR-denaturing gradient gel electrophoresis, RNA–DNA membrane hybridization, and DNA microarray technology. Appl. Environ. Microbiol. **2002**, *68*, 3215–3225.
27. Loy, A.; Lehner, A.; Lee, N.; Adamczyk, J.; Meier, H.; Ernst, J.; Schleifer, K.H.; Wagner, M. Oligonucleotide microarray for 16S rRNA gene-based detection of all recognized lineages of sulfate-reducing prokaryotes in the environment. Appl. Environ. Microbiol. **2002**, *68*, 5064–5081.
28. Evans, G.M.; Furlong, J.C. *Environmental Biotechnology: Theory and Applications*; John Wiley & Sons, Ltd.: Chichester, 2003.
29. Rittman, B.; McCarty, P.L. *Environmental Biotechnology: Principles and Applications*; McGraw-Hill: Boston, 2002.
30. Armenante, P.M. Bioreactors. In *Biotreatment of Industrial and Hazardous Wastes*; Levin M.A., Gealt, M.A., Eds.; McGraw-Hill, Inc.: New York, 1993; 65–112.
31. Emerson, D. Microbial oxidation of Fe(II) at circumneutral pH. In *Environmental Microbe–Metal Interactions*; Lovley, D.R., Ed.; ASM Press: Washington, DC, 2000; 31–52.
32. Emerson, D.; Moyer, C.L. Neutrophilic Fe-oxidizing bacteria are abundant at the Loihi Seamount hydrothermal vents and play a major role in Fe oxide deposition. Appl. Environ. Microbiol. **2002**, *68*, 3085–3093.
33. Holden, P.A.; Hersman, L.E.; Firestone, M.K. Water content mediated microaerophilic toluene biodegradation in arid vadose zone materials. Microb. Ecol. **2001**, *42*, 256–266.
34. Gonzalez-Flecha, B.; Demple, B. Homeostatic regulation of intracellular hydrogen peroxide concentration in aerobically growing *Escherichia coli*. J. Bacteriol. **1997**, *179*, 382–388.
35. Eriksson, M.; Sodersten, E.; Yu, Z.; Dalhammar, G.; Mohn, W.W. Degradation of polycyclic aromatic hydrocarbons at low temperature under aerobic and nitrate-reducing conditions in enrichment cultures from northern soils. Appl. Environ. Microbiol. **2003**, *69*, 275–284.
36. Marttinen, S.K.; Kettunen, R.H.; Sormunen, K.M.; Rintala, J.A. Removal of bis(2-ethylhexyl)phthalate at a sewage treatment plant. Water Res. **2003**, *37*, 1385–1393.
37. Otal, E.; Lebrato, J. Anaerobic treatment of polyethylene glycol of different molecular weights. Environ. Technol. **2002**, *23*, 1405–1414.
38. Borch, T.; Ambus, P.; Laturnus, F.; Svensmark, B.; Gron, C. Biodegradation of chlorinated solvents in a water unsaturated topsoil. Chemosphere **2003**, *51*, 143–152.
39. Tchobanoglous, G.; Theisen, H.; Vigil, S.A. *Integrated Solid Waste Management: Engineering Principles and Management Issues*; McGraw-Hill Book Co.: Singapore, 1993.
40. Ito, A.; Takachi, T.; Aizawa, J.; Umita, T. Chemical and biological removal of arsenic from sewage sludge. Water Sci. Technol. **2001**, *44*, 59–64.
41. Xiang, L.; Chan, L.C.; Wong, J.W. Removal of heavy metals from anaerobically digested sewage sludge by isolated indigenous iron-oxidizing bacteria. Chemosphere **2000**, *41*, 283–287.
42. Tamaki, S.; Frankenberger, W.T. Jr. Environmental biochemistry of arsenic. Rev. Environ. Contam. Toxicol. **1992**, *124*, 79–110.
43. Vainshtein, M.; Kuschk, P.; Mattusch, J.; Vatsourina, A.; Wiessner, A. Model experiments on the microbial removal of chromium from contaminated groundwater. Water Res. **2003**, *37*, 1401–1405.
44. Atlas, R.M. Bioaugmentation to enhance microbial bioremediation. In *Biotreatment of Industrial and Hazardous Wastes*; Levin, M.A., Gealt, M.A., Eds.; McGraw-Hill, Inc.: New York, 1993; 19–37.
45. De Mot, R.; Parret, A.H. A novel class of self-sufficient cytochrome P450 monooxygenases in prokaryotes. Trends Microbiol. **2002**, *10*, 502–508.
46. Bass, D.H.; Hastings, N.A.; Brown, R.A. Performance of air sparging systems: a review of case studies. J. Hazard. Mater. **2000**, *72*, 101–119.
47. Zappi, M.; White, K.; Hwang, H.M.; Bajpai, R.; Qasim. The fate of hydrogen peroxide as an oxygen source for bioremediation activities within saturated aquifer systems. J. Air. Waste Manag. Assoc. **2000**, *50*, 1818–1830.
48. Ensley, B.D. Designing pathways for environmental purposes. Curr. Opin. Biotechnol. **1994**, *5*, 249–252.

49. Vasilyeva, G.K.; Kreslavski, V.D.; Oh, B.T.; Shea, P.J. Potential of activated carbon to decrease 2,4,6-trinitrotoluene toxicity and accelerate soil decontamination. Environ. Toxicol. Chem. **2001**, *20*, 965–971.
50. Tay, J.-H.; Ivanov, V.; Pan, S.; Tay, S.T.-L. Specific layers in aerobically grown microbial granules. Lett. Appl. Microbiol. **2002**, *34*, 254–258.
51. Tay, S.T.-L.; Ivanov, V.; Yi, S.; Zhuang, W.-Q.; Tay, J.-H. Presence of anaerobic Bacteroides in aerobically grown microbial granules. Microbial Ecol. **2002**, *44*, 278–285.

13

Treatment of Dairy Processing Wastewaters

Trevor J. Britz and Corné van Schalkwyk
University of Stellenbosch, Matieland, South Africa

Yung-Tse Hung
Cleveland State University, Cleveland, Ohio, U.S.A.

13.1 INTRODUCTION

The dairy industry is generally considered to be the largest source of food processing wastewater in many countries. As awareness of the importance of improved standards of wastewater treatment grows, process requirements have become increasingly stringent. Although the dairy industry is not commonly associated with severe environmental problems, it must continually consider its environmental impact – particularly as dairy pollutants are mainly of organic origin. For dairy companies with good effluent management systems in place [1], treatment is not a major problem, but when accidents happen, the resulting publicity can be embarrassing and very costly.

All steps in the dairy chain, including production, processing, packaging, transportation, storage, distribution, and marketing, impact the environment [2]. Owing to the highly diversified nature of this industry, various product processing, handling, and packaging operations create wastes of different quality and quantity, which, if not treated, could lead to increased disposal and severe pollution problems. In general, wastes from the dairy processing industry contain high concentrations of organic material such as proteins, carbohydrates, and lipids, high concentrations of suspended solids, high biological oxygen demand (BOD) and chemical oxygen demand (COD), high nitrogen concentrations, high suspended oil and/or grease contents, and large variations in pH, which necessitates "specialty" treatment so as to prevent or minimize environmental problems. The dairy waste streams are also characterized by wide fluctuations in flow rates, which are related to discontinuity in the production cycles of the different products. All these aspects work to increase the complexity of wastewater treatment.

The problem for most dairy plants is that waste treatment is perceived to be a necessary evil [3]; it ties up valuable capital, which could be better utilized for core business activity. Dairy wastewater disposal usually results in one of three problems: (a) high treatment levies being charged by local authorities for industrial wastewater; (b) pollution might be caused when untreated wastewater is either discharged into the environment or used directly as irrigation water; and (c) dairy plants that have already installed an aerobic biological system are faced with the problem of sludge disposal. To enable the dairy industry to contribute to water conservation, an efficient and cost-effective wastewater treatment technology is critical.

Presently, plant mangers may choose from a wide variety of technologies to treat their wastes. More stringent environmental legislation as well as escalating costs for the purchase of fresh water and effluent treatment has increased the impetus on improving waste control. The level of treatment is normally dictated by environmental regulations applicable to the specific area. While most larger dairy factories have installed treatment plants or, if available, dispose of their wastewater into municipal sewers, cases of wastewater disposal into the sea or disposal by means of land irrigation do occur. In contrast, most smaller dairy factories dispose of their wastewater by irrigation onto lands or pastures. Surface and groundwater pollution is therefore a potential threat posed by these practises.

Because the dairy industry is a major user and generator of water, it is a candidate for wastewater reuse. Even if the purified wastewater is initially not reused, the dairy industry will still benefit from in-house wastewater treatment management, because reducing waste at the source can only help in reducing costs or improving the performance of any downstream treatment facility.

13.2 DAIRY PROCESSES AND COMPOSITION OF DAIRY PRODUCTS

Before the methods of treatment of dairy processing wastewater can be appreciated, it is important to be acquainted with the various production processes involved in dairy product manufacturing and the pollution potential of different dairy products (Table 1). A brief summary of the most common processes [8] is presented below.

13.2.1 Pasteurized Milk

The main steps include raw milk reception (the first step of any dairy manufacturing process), pasteurization, standardization, deaeration, homogenization and cooling, and filling of a variety of different containers. The product from this point should be stored and transported at 4°C.

13.2.2 Milk and Whey Powders

This is basically a two-step process whereby 87% of the water in pasteurized milk is removed by evaporation under vacuum and the remaining water is removed by spray drying. Whey powder can be produced in the same way. The condensate produced during evaporation may be collected and used for boiler feedwater.

13.2.3 Cheese

Because there are a large variety of different cheeses available, only the main processes common to all types will be discussed. The first process is curd manufacturing, where pasteurized milk is mixed with rennet and a suitable starter culture. After coagulum formation and heat and mechanical treatment, whey separates from the curd and is drained. The finished curd is then salted, pressed, and cured, after which the cheese is coated and wrapped. During this process two types of wastewaters may arise: whey, which can either be disposed of or used in the production of whey powder, and wastewater, which can result from a cheese rinse step used during the manufacturing of certain cheeses.

Table 1 Reported BOD and COD Values for Typical Dairy Products and Domestic Sewage

Product	BOD_5 (mg/L)	COD (mg/L)	Reference
Whole milk	114,000	183,000	4
	110,000	190,000	5
	120,000		6
	104,000		7
Skim milk	90,000	147,000	4
	85,000	120,000	5
	70,000		6
	67,000		7
Buttermilk	61,000	134,000	4
	75,000	110,000	5
	68,000		7
Cream	400,000	750,000	4
	400,000	860,000	5
	400,000		6
	399,000		7
Evaporated milk	271,000	378,000	4
	208,000		7
Whey	42,000	65,000	4
	45,000	80,000	5
	40,000		6
	34,000		7
Ice cream	292,000		7
Domestic sewage	300	500	4, 5

BOD, biochemical oxygen demand; COD, chemical oxygen demand.
Source: Refs 4–7.

13.2.4 Butter

Cream is the main raw material for manufacturing butter. During the churning process it separates into butter and buttermilk. The drained buttermilk can be powdered, cooled, and packed for distribution, or discharged as wastewater.

13.2.5 Evaporated Milk

The milk is first standardized in terms of fat and dry solids content after which it is pasteurized, concentrated in an evaporator, and homogenized, then packaged, sterilized, and cooled for storage. In the production of sweetened condensed milk, sugar is added in the evaporation stage and the product is cooled.

13.2.6 Ice Cream

Raw materials such as water, cream, butter, milk, and whey powders are mixed, homogenized, pasteurized, and transferred to a vat for ageing, after which flavorings, colorings, and fruit are added prior to freezing. During primary freezing the mixture is partially frozen and air is incorporated to obtain the required texture. Containers are then filled and frozen.

13.2.7 Yogurt

Milk used for yogurt production is standardized in terms of fat content and fortified with milk solids. Sugar and stabilizers are added and the mixture is then heated to 60°C, homogenized, and heated again to about 95°C for 3–5 minutes [9]. It is then cooled to 30–45°C and inoculated with a starter culture. For set yogurts, the milk base is packed directly and the retail containers are incubated for the desired period, after which they are cooled and dispatched. For stirred yogurts, the milk base is incubated in bulk after which it is cooled and packaged, and then distributed.

13.2.8 Wastewater from Associated Processes

Most of the water consumed in a dairy processing plant is used in associated processes such as the cleaning and washing of floors, bottles, crates, and vehicles, and the cleaning-in-place (CIP) of factory equipment and tanks as well as the inside of tankers. Most CIP systems consist of three steps: a prerinse step to remove any loose raw material or product remains, a hot caustic wash to clean equipment surfaces, and a cold final rinse to remove any remaining traces of caustic.

13.3 CHARACTERISTICS AND SOURCES OF WASTEWATER

The volume, concentration, and composition of the effluents arising in a dairy plant are dependent on the type of product being processed, the production program, operating methods, design of the processing plant, the degree of water management being applied, and subsequently the amount of water being conserved. Dairy wastewater may be divided into three major categories:

1. Processing waters, which include water used in the cooling and heating processes. These effluents are normally free of pollutants and can with minimum treatment be reused or just discharged into the storm water system generally used for rain runoff water.
2. Cleaning wastewaters emanate mainly from the cleaning of equipment that has been in contact with milk or milk products, spillage of milk and milk products, whey, pressings and brines, CIP cleaning options, and waters resulting from equipment malfunctions and even operational errors. This wastewater stream may contain anything from milk, cheese, whey, cream, separator and clarifier dairy waters [10], to dilute yogurt, starter culture, and dilute fruit and stabilizing compounds [9].
3. Sanitary wastewater, which is normally piped directly to a sewage works.

Dairy cleaning waters may also contain a variety of sterilizing agents and various acid and alkaline detergents. Thus, the pH of the wastewaters can vary significantly depending on the cleaning strategy employed. The most commonly used CIP chemicals are caustic soda, nitric acid, phosphoric acid, and sodium hypochloride [10]; these all have a significant impact on wastewater pH. Other concerns related to CIP and sanitizing strategies include the biochemical oxygen demand (BOD) and chemical oxygen demand (COD) contributions (normally $<10\%$ of total BOD concentration in plant wastewater), phosphorus contribution resulting from the use of phosphoric acid and other phosphorus-containing detergents, high water volume usage for cleaning and sanitizing (as high as 30% of total water discharge), as well as general concerns regarding the impact of detergent biodegradability and toxicity on the specific waste treatment facility and the environment in general [11].

Dairy industry wastewaters are generally produced in an intermittent way; thus the flow and characteristics of effluents could differ between factories depending on the kind of products produced and the methods of operation [12]. This also influences the choice of the wastewater treatment option, as specific biological systems have difficulties dealing with wastewater of varying organic loads.

Published information on the chemical composition of dairy wastewater is scarce [10]. Some of the more recent information available is summarized in Tables 2 and 3. Milk has a BOD content 250 times greater than that of sewage [23]. It can therefore be expected that dairy wastewaters will have relatively high organic loads, with the main contributors being lactose, fats, and proteins (mainly casein), as well as high levels of nitrogen and phosphorus that are largely associated with milk proteins [12,17]. The COD and BOD for whey have, for instance, been established to be between 35,000–68,000 mg/L and 30,000–60,000 mg/L, respectively, with lactose being responsible for 90% of the COD and BOD contribution [24].

13.4 TREATMENT OPTIONS

The highly variable nature of dairy wastewaters in terms of volumes and flow rates (which is dependent on the factory size and operation shifts) and in terms of pH and suspended solid (SS) content (mainly the result of the choice of cleaning strategy employed) makes the choice of an effective wastewater treatment regime difficult. Because dairy wastewaters are highly biodegradable, they can be effectively treated with biological wastewater treatment systems, but can also pose a potential environmental hazard if not treated properly [23]. The three main options for the dairy industry are: (a) discharge to and subsequent treatment of factory wastewater at a nearby sewage treatment plant; (b) removal of semisolid and special wastes from the site by waste disposal contractors; or (c) the treatment of factory wastewater in an onsite wastewater treatment plant [25,26]. According to Robinson [25], the first two options are continuously impacted by increasing costs, while the control of allowable levels of SS, BOD, and COD in discharged wastewaters are also becoming more stringent. As a result, an increasing number of dairy industries must consider the third option of treating industrial waste onsite. It should be remembered, however, that the treatment chosen should meet the required demands and reduce costs associated with long-term industrial wastewater discharge.

13.4.1 Direct Discharge to a Sewage Treatment Works

Municipal sewage treatment facilities are capable of treating a certain quantity of organic substances and should be able to deal with certain peak loads. However, certain components found in dairy waste streams may present problems. One such substance is fat, which adheres to the walls of the main system and causes sedimentation problems in the sedimentation tanks. Some form of onsite pretreatment is therefore advisable to minimize the fat content of the industrial wastewater that can be mixed with the sanitary wastewater going to the sewage treatment facility [6].

Dairy industries are usually subjected to discharge regulations, but these regulations differ significantly depending on discharge practises and capacities of municipal sewage treatment facilities. Sewer charges are based on wastewater flow rate, BOD_5 mass, SS, and total P discharged per day [10]. Some municipal treatment facilities may demand treatment of high-strength industrial effluents to dilute the BOD load of the water so that it is comparable to that of domestic sewage [7].

Table 2 Chemical Characteristics of Different Dairy Plant Wastewaters

Industry	BOD_5 (mg/L)	COD (mg/L)	pH	FOG (g/L)	TS (mg/L)	TSS (mg/L)	Alkalinity (mg/L as $CaCO_3$)	Reference
Cheese								
14 Cheese/whey plants	565–5722	785–7619	6.2–11.3	–	1837–14,205	326–3560	225–1550	10
Cheese/whey plant	377–2214	189–6219	5.2	–	–	188–2330	–	13
Cheese factory	–	5340	5.22	–	4210	–	335	14
Cheese factory	–	2830	4.99	–	–	–	–	15
Cheese processing industry	–	63,300	3.38	2.6	53,200	12,500	–	16
Cheese/casein product plant	–	5380	6.5	0.32	–	–	–	15
Cheese/casein product plant	8000	–	4.5–6.0	0.4	–	–	–	17
Milk								
Milk processing plant	–	713–1410	7.1–8.1	–	900–1470	360–920	–	18
Milk/yogurt plant	–	4656	6.92	–	2750	–	546	14
Milk/cream bottling plant	1200–4000	2000–6000	8–11	3–5	–	350–1000	150–300	19, 20
Butter/milk powder								
Butter/milk powder plant	–	1908	5.8	–	1720	–	532	14
Butter/milk powder plant	1500	–	10–11	0.4	–	–	–	17
Butter/Comte´cheese plant	1250	2520	5–7	–	–	–	–	21
Whey								
Whey wastewater	35,000	–	4.6	0.8	–	–	–	17
Raw cheese whey	–	68,814	–	–	3190	1300	–	22

BOD, biological oxygen demand; COD, chemical oxygen demand; TS, total solids; TSS, total suspended solids; FOG, fats, oil and grease.

Table 3 Concentrations of Selected Elements in Different Dairy Wastewaters

Industry	Total P (mg/L)	PO_4-P (mg/L)	TKN (mg/L)	NH_4-N (mg/L)	Na^+ (mg/L)	K^+ (mg/L)	Ca^{2+} (mg/L)	Mg^{2+} (mg/L)	Reference
Cheese									
14 Cheese/whey plants	29–181	6–35	14–140	1–34	263–1265	8.6–155.5	1.4–58.5	6.5–46.3	16
Cheese/whey plant	0.2–48.0	0.2–7.9	13–172	0.7–28.5	–	–	–	–	13
Cheese factory	45	–	102	–	550	140	30	35	15
Cheese/casein product plant	85	–	140	–	410	125	70	12	15
Cheese/casein product plant	100	–	200	–	380	160	95	14	17
Milk									
Milk/cream bottling plant	–	20–50	50–60	–	170–200	35–40	35–40	5–8	19, 20
Butter/milk powder									
Butter/milk powder plant	35	–	70	–	560	13	8	1	17
Butter/Comté cheese plant	50	–	66	–	–	–	–	–	21
Whey									
Whey wastewater	640	–	1400	–	430	1500	1250	100	17
Raw cheese whey	379	327	1462	64.3	–	–	–	–	22

In a recent survey conducted by Danalewich et al. [10] at 14 milk processing plants in Minnesota, Wisconsin, and South Dakota, it was reported that four facilities directed both their mixed sanitary and industrial wastewater directly to a municipal treatment system, while the rest employed some form of wastewater treatment. Five of the plants that treated their wastewater onsite did not separate their sanitary wastewater from their processing wastewater, which presents a major concern when it comes to the final disposal of the generated sludge after the wastewater treatment, since the sludge may contain pathogenic microorganisms [10]. It would thus be advisable for factories that employ onsite treatment to separate the sanitary and processing wastewaters, and dispose of the sanitary wastewater by piping directly to a sewage treatment facility.

13.4.2 Onsite Pretreatment Options

Physical Screening

The main purpose of screens in wastewater treatment is to remove large particles or debris that may cause damage to pumps and downstream clogging [27]. It is also recommended that the physical screening of dairy wastewater should be carried out as quickly as possible to prevent a further increase in the COD concentration as a result of the solid solubilization [28]. Wendorff [7] recommended the use of a wire screen and grit chamber with a screen aperture size of 9.5 mm, while Hemming [28] recommended the use of even finer spaced mechanically brushed or inclined screens of 40 mesh (about 0.39 mm) for solids reduction. According to Droste [27], certain precautionary measures should be taken to prevent the settling of coarse matter in the wastewater before it is screened. These requirements include the ratio of depth to width of the approach channel to the screen, which should be 1 : 2, as well as the velocity of the water, which should not be less than 0.6 m/s. Screens can be cleaned either manually or mechanically and the screened material disposed of at a landfill site.

pH Control

As shown in Table 2, large variations exist in wastewater pH from different dairy factories. This may be directly attributed to the different cleaning strategies employed. Alkaline detergents generally used for the saponification of lipids and the effective removal of proteinacous substances would typically have a pH of 10–14, while a pH of 1.5–6.0 can be encountered with acidic cleaners used for the removal of mineral deposits and acid-based sanitizers [11,29]. The optimum pH range for biological treatment plants is between 6.5 and 8.5 [30,31]. Extreme pH values can be highly detrimental to any biological treatment facility, not only for the negative effect that it will have on the microbial community, but also due to the increased corrosion of pipes that will occur at pH values below 6.5 and above 10 [6]. Therefore, some form of pH adjustment as a pretreatment step is strongly advised before wastewater containing cleaning agents is discharged to the drain or further treated onsite. In most cases, flow balancing and pH adjustment are performed in the same balancing tank. According to the International Dairy Federation (IDF) [30], a near-neutral pH is usually obtained when water used in different production processes is combined. If pH correction needs to be carried out in the balancing tank, the most commonly used chemicals are H_2SO_4, HNO_3, NaOH, CO_2, or lime [30].

Flow and Composition Balancing

Because discharged dairy wastewaters can vary greatly with respect to volume, strength, temperature, pH, and nutrient levels, flow and composition balancing is a prime requirement for any subsequent biological process to operate efficiently [28]. pH adjustment and flow balancing can be achieved by keeping effluent in an equalization or balancing tank for at least 6–12 hours [7]. During this time, residual oxidants can react completely with solid particles, neutralizing cleaning solutions. The stabilized effluent can then be treated using a variety of different options.

According to the IDF [30], balance tanks should be adequately mixed to obtain proper blending of the contents and to prevent solids from settling. This is usually achieved by the use of mechanical aerators. Another critical factor is the size of the balance tank. This should be accurately determined so that it can effectively handle a dairy factory's daily flow pattern at peak season. It is also recommended that a balancing tank should be large enough to allow a few hours extra capacity to handle unforeseen peak loads and not discharge shock loads to public sewers or onsite biological treatment plants [30].

Fats, Oil, and Grease Removal

The presence of fats, oil, and grease (FOG) in dairy processing wastewater can cause all kinds of problems in biological wastewater treatment systems onsite and in public sewage treatment facilities. It is therefore essential to reduce, if not remove FOG completely, prior to further treatment. According to the IDF [32], factories processing whole milk, such as milk separation plants as well as cheese and butter plants, whey separation factories, and milk bottling plants, experience the most severe problems with FOG. The processing of skim milk seldom presents problems in this respect.

As previously mentioned, flow balancing is recommended for dairy processing plants. An important issue, however, is whether the FOG treatment unit should be positioned before or after the balancing tank [32]. If the balancing tank is placed before the FOG unit, large fat globules can accumulate in the tank as the discharged effluent cools down and suspended fats aggregate during the retention period. If the balancing tank is placed after the FOG removal unit, the unit should be large enough to accommodate the maximum anticipated flow from the factory. According to the IDF [32], it is generally accepted that flow balancing should precede FOG removal. General FOG removal systems include the following.

Gravity Traps. In this extremely effective, self-operating, and easily constructed system, wastewater flows through a series of cells, and the FOG mass, which usually floats on top, is removed by retention within the cells. Drawbacks include frequent monitoring and cleaning to prevent FOG buildup, and decreased removal efficiency at pH values above 8 [32].

Air Flotation and Dissolved Air Flotation. Mechanical removal of FOG with dissolved air flotation (DAF) involves aerating a fraction of recycled wastewater at a pressure of about 400–600 kPa in a pressure chamber, then introducing it into a flotation tank containing untreated dairy processing wastewater. The dissolved air is converted to minute air bubbles under the normal atmospheric pressure in the tank [6,32]. Heavy solids form sediment while the air bubbles attach to the fat particles and the remaining suspended matter as they are passed through the effluent [6,9,25]. The resulting scum is removed and will become odorous if stored in an open tank. It is an unstable waste material that should preferably not be mixed with sludge from biological and chemical treatment processes since it is very difficult to dewater. FOG waste should be removed and disposed of according to approved methods [32]. DAF components require regular maintenance and the running costs are usually fairly high.

Air flotation is a more economical variation of DAF. Air bubbles are introduced directly into the flotation tank containing the untreated wastewater, by means of a cavitation aerator coupled to a revolving impeller [32]. A variety of different patented air flotation systems are available on the market and have been reviewed by the IDF [32]. These include the "Hydrofloat," the "Robosep," vacuum flotation, electroflotation, and the "Zeda" systems.

The main drawback of the DAF [25], is that only SS and free FOG can be removed. Thus, to increase the separation efficiency of the process, dissolved material and emulsified FOG solutions must undergo a physico-chemical treatment during which free water is removed and waste molecules are coagulated to form larger, easily removable masses. This is achieved by recirculating wastewater prior to DAF treatment in the presence of different chemical solutions such as ferric chloride, aluminum sulfate, and polyelectrolytes that can act as coalescents and coagulants. pH correction might also be necessary prior to the flotation treatment, because a pH of around 6.5 is required for efficient FOG removal [32].

Enzymatic Hydrolysis of FOG. Cammarota et al. [33] and Leal et al. [34] utilized enzymatic preparations of fermented babassu cake containing lipases produced by a *Penicillium restrictum* strain for FOG hydrolysis in dairy processing wastewaters prior to anaerobic digestion. High COD removal efficiencies as well as effluents of better quality were reported for a laboratory-scale UASB reactor treating hydrolyzed dairy processing wastewater, and compared to the results of a UASB reactor treating the same wastewater without prior enzymatic hydrolysis treatment.

13.4.3 Treatment Methods

Biological Treatment

Biological degradation is one of the most promising options for the removal of organic material from dairy wastewaters. However, sludge formed, especially during the aerobic biodegradation processes, may lead to serious and costly disposal problems. This can be aggravated by the ability of sludge to adsorb specific organic compounds and even toxic heavy metals. However, biological systems have the advantage of microbial transformations of complex organics and possible adsorption of heavy metals by suitable microbes. Biological processes are still fairly unsophisticated and have great potential for combining various types of biological schemes for selective component removal.

Aerobic Biological Systems. Aerobic biological treatment methods depend on microorganisms grown in an oxygen-rich environment to oxidize organics to carbon dioxide, water, and cellular material. Considerable information on laboratory- and field-scale aerobic treatments has shown aerobic treatment to be reliable and cost-effective in producing a high-quality effluent. Start-up usually requires an acclimation period to allow the development of a competitive microbial community. Ammonia-nitrogen can successfully be removed, in order to prevent disposal problems. Problems normally associated with aerobic processes include foaming and poor solid–liquid separation.

The conventional *activated sludge process* (ASP) is defined [35] as a continuous treatment that uses a consortium of microbes suspended in the wastewater in an aeration tank to absorb, adsorb, and biodegrade the organic pollutants (Fig. 1). Part of the organic composition will be completely oxidized to harmless endproducts and other inorganic substances to provide energy to sustain the microbial growth and the formation of biomass (flocs). The flocs are kept in suspension either by air blown into the bottom of the tank (diffused air system) or by mechanical aeration. The dissolved oxygen level in the aeration tank is critical and should preferably be 1–2 mg/L and the tank must always be designed in terms of the aeration period and cell residence time. The mixture flows from the aeration tank to a sedimentation tank where the activated sludge flocs form larger particles that settles as sludge. The biological aerobic metabolism mode is extremely efficient in terms of energy recovery, but results in large quantities of sludge being produced (0.6 kg dry sludge per kg of BOD_5 removed). Some of the sludge is returned to the aeration tank but the rest must be processed and disposed of in an environmentally acceptable

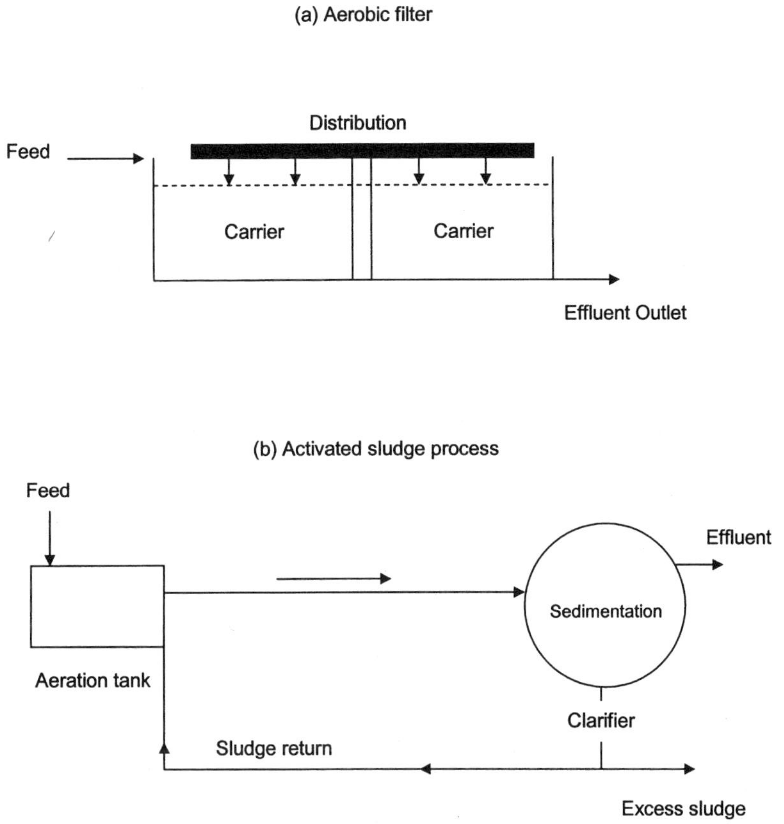

Figure 1 Simplified illustrations of aerobic wastewater treatment processes: (a) aerobic filter, (b) activated sludge process (from Refs. 31, and 35–37).

manner, which is a major operating expense. Many variations of the ASP exist, but in all cases, the oxygen supplied during aeration is the major energy-consuming operation. With ASPs, problems generally encountered are bulking [17], foam production, precipitation of iron and carbonates, excessive sludge production, and a decrease in efficiency during winter periods.

Many reports show that ASP has been used successfully to treat dairy industry wastes. Donkin and Russell [36] found that reliable COD removals of over 90% and 65% reductions in total nitrogen could be obtained with a milk powder/butter wastewater. Phosphorus removals were less reliable and appeared to be sensitive to environmental changes.

Aerobic filters such as conventional trickling or percolating filters (Fig. 1) are among the oldest biological treatment methods for producing high-quality final effluents [35]. The carrier media (20–100 mm diameter) may consist of pumice, rock, gravel, or plastic pieces, which is populated by a very diverse microbial consortium. Wastewater from a storage tank is normally dosed over the medium and then trickles downward through a 2 m medium bed. The slimy microbial mass growing on the carrier medium absorbs the organic constituents of the wastewater and decomposes them aerobically. Sludge deposits require removal from time to time. Aerobic conditions are facilitated by the downward flow and natural convection currents resulting from temperature differences between the air and the added wastewater. Forced ventilation may be employed to enhance the decomposition, but the air must be deodorized by passing through clarifying tanks. Conventional filters, with aerobic microbes growing on rock or gravel, are limited in depth to about 2 m, as deeper filters enhance anaerobic growth with subsequent odor problems. In contrast, filters with synthetic media can be fully aerobic up to about 8 m [37]. The final effluent flows to a sedimentation or clarifying tank to remove sludge and solids from the carrier medium.

It is generally recommended that organic loading for dairy wastewaters not exceed 0.28–0.30 kg BOD/m^3 and that recirculation be employed [38]. A 92% BOD removal of a dairy wastewater was reported by Kessler [4], but since the BOD of the final effluent was still too high, it was further treated in an oxidation pond.

An inherent problem is that trickling filters can be blocked by precipitated ferric hydroxide and carbonates, with concomitant reduction of microbial activity. In the case of overloading with dairy wastewater, the medium becomes blocked with heavy biological and fat films. Maris et al. [39] reported that biological filters are not appropriate for the treatment of high-strength wastewaters, as filter blinding by organic deposition on the filter medium is generally found.

The *rotating biological contactors* (RBC) design contains circular discs (Fig. 2) made of high-density plastic or other lightweight material [35]. The discs, rotating at 1–3 rpm, are placed on a horizontal shaft so that about 40–60% of the disc surface protrudes out of the tank; this allows oxygen to be transferred from the atmosphere to the exposed films. A biofilm develops on the disc surface, which facilitates the oxidation of the organic components of the wastewater. When the biofilm sludge becomes too thick, it is torn off and removed in a sedimentation tank. Operation efficiency is based on the g BOD per m^2 of disc surface per day [35]. Rusten and his coworkers [40] reported 85% COD removal efficiency with an organic loading rate (OLR) of 500 g COD/m^3 hour while treating dairy wastewater.

The RBC process offers several advantages over the activated sludge process for use in dairy wastewater treatment. The primary advantages are the low power input required, relative ease of operation and low maintenance. Furthermore, pumping, aeration, and wasting/recycle of solids are not required, leading to less operator attention. Operation for nitrogen removal is also relatively simple and routine maintenance involves only inspection and lubrication.

The *sequencing batch reactor* (SBR) is a single-tank fill-and-draw unit that utilizes the same tank (Fig. 2) to aerate, settle, withdraw effluent, and recycle solids [35]. After the tank is filled, the wastewater is mixed without aeration to allow metabolism of the fermentable compounds. This is followed by the aeration step, which enhances the oxidation and biomass formation. Sludge is then settled and the treated effluent is removed to complete the cycle. The SBR relies heavily on the site operator to adjust the duration of each phase to reflect fluctuations in the wastewater composition [41]. The SBR is seen as a good option with low-flow applications and allows for wider wastewater strength variations. Eroglu et al. [42] and Samkutty et al. [43] reported the SBR to be a cost-effective primary and secondary treatment option to handle dairy plant wastewater with COD removals of 91–97%. Torrijos et al. [21] also demonstrated the efficiency of the SBR system for the treatment of wastewater from small cheese-making dairies with treatment levels of >97% being obtained at a loading rate of 0.50 kg COD/m^3 day. In another study, Li and Zhang [44] successfully operated an SBR at a hydraulic retention time (HRT) of 24 hours to treat dairy waste with a COD of 10 g/L. Removal efficiencies of 80% in COD, 63% in total solids, 66% in volatile solids, 75% Kjeldahl nitrogen, and 38% in total nitrogen, were obtained.

In areas where land is available, *lagoons/ponds/reed beds* (Fig. 2) constitute one of the least expensive methods of biological degradation. With the exception of aerated ponds, no mechanical devices are used and flow normally occurs by gravity. As result of their simplicity and absence of a sludge recycle facility, lagoons are a favored method for effective wastewater treatment. However, the lack of a controlled environment slows the reaction times, resulting in

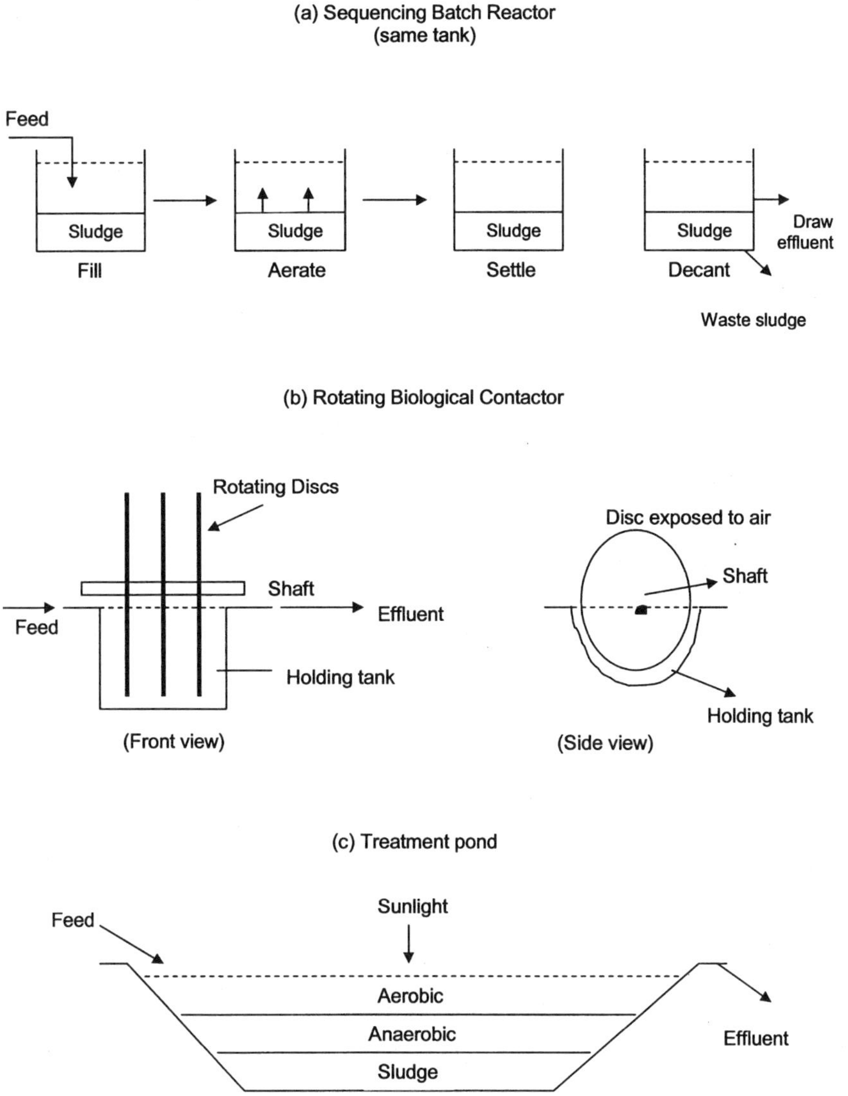

Figure 2 Simplified illustrations of aerobic wastewater treatment processes: (a) sequencing batch reactor, (b) rotating biological contactor, (c) treatment pond (from Refs. 35, 40, 42, 45, 47–49).

long retention times of up to 60 days. Operators of sites in warmer climates may find the use of lagoons a more suitable and economical wastewater treatment option. However, the potential does exist for surface and groundwater pollution, bad odors, and insects that may become a nuisance.

Aerated ponds are generally 0.5–4.0 m deep [45]. Evacuation on the site plus lining is a simple method of lagoon construction and requires relatively unskilled attention. Floating aerators may be used to allow oxygen and sunlight penetration. According to Bitton [46], aeration for 5 days at 20°C in a pond normally gives a BOD removal of 85% of milk wastes. Facultative ponds are also commonly used for high-strength dairy wastes [47]. Although ponds/lagoons are simple to operate, they are the most complex of all biologically engineered degradation systems [48]. In these systems, both aerobic and anaerobic metabolisms occur in addition to photosynthesis and sedimentation. Although most of the organic carbon is converted to microbial biomass, some is lost as CO_2 or CH_4. It is thus essential to remove sludge regularly to prevent buildup and clogging. The HRT in facultative ponds can vary between 5 and 50 days depending on climatic conditions.

Reed-bed or wetland systems have also found widespread application [49]. A design manual and operating guidelines were produced in 1990 [49,50]. Reed beds are designed to treat wastewaters by passing the latter through rhizomes of the common reed in a shallow bed of soil or gravel. The reeds introduce oxygen and as the wastewater percolates through it, aerobic microbial communities establish among the roots and degrade the contaminants. Nitrogen and phosphorus are thus removed directly by the reeds. However, reed beds are poor at removing ammonia, and with high concentrations of ammonia being toxic, this may be a limiting factor. The precipitation of large quantities of iron, manganese, and calcium within the reed beds will also affect rhizome growth and, in time, reduce the permeability of the bed. According to Robinson et al. [49], field studies in the UK have shown that reed beds have enormous potential and in combination with aerobic systems, provide high effluent quality at reasonable cost.

Anaerobic Biological Systems. Anaerobic digestion (AD) is a biological process performed by an active microbial consortium in the absence of exogenous electron acceptors. Up to 95% of the organic load in a waste stream can be converted to biogas (methane and carbon dioxide) and the remainder is utilized for cell growth and maintenance [51,52]. Anaerobic systems are generally seen as more economical for the biological stabilization of dairy wastes [14], as they do not have the high-energy requirements associated with aeration in aerobic systems. Anaerobic digestion also yields methane, which can be utilized as a heat or power source. Furthermore, less sludge is generated, thereby reducing problems associated with sludge disposal. Nutrient requirements (N and P) are much lower than for aerobic systems [37], pathogenic organisms are usually destroyed, and the final sludge has a high soil conditioning value if the concentration of heavy metals is low. The possibility of treating high COD dairy wastes without previous dilution, as required by aerobic systems, reduces space requirements and the associated costs [53]. Bad odors are generally absent if the system is operated efficiently [51,54].

The disadvantages associated with anaerobic systems are the high capital cost, long start-up periods, strict control of operating conditions, greater sensitivity to variable loads and organic shocks, as well as toxic compounds [55]. The operational temperature must be maintained at about 33–37°C for efficient kinetics, because it is important to keep the pH at a value around 7, as a result of the sensitivity of the methanogenic population to low values [48]. As ammonia-nitrogen is not removed in an anaerobic system, it is consequently discharged with the digester effluent, creating an oxygen demand in the receiving water. Complementary treatment to achieve acceptable discharge standards is also required.

The *anaerobic lagoon* (anaerobic pond) (Fig. 3) is the simplest type of anaerobic digester. It consists of a pond, which is normally covered to exclude air and to prevent methane loss to the atmosphere. Lagoons are far easier to construct than vertical digester types, but the biggest drawback is the large surface area required.

In New Zealand, dairy wastewater [51] was treated at 35°C in a lagoon (26,000 m^3) covered with butyl rubber at an organic load of 40,000 kg COD per day, pH of 6.8–7.2, and HRT of 1–2 days. The organic loading rate (OLR) of 1.5 kg COD/m^3 day was on the low side. The pond's effluent was clarified and the settled biomass recycled through the substrate feed. The clarified effluent was then treated in an 18,000 m^3 aerated lagoon. The efficiency of the total system reached a 99% reduction in COD.

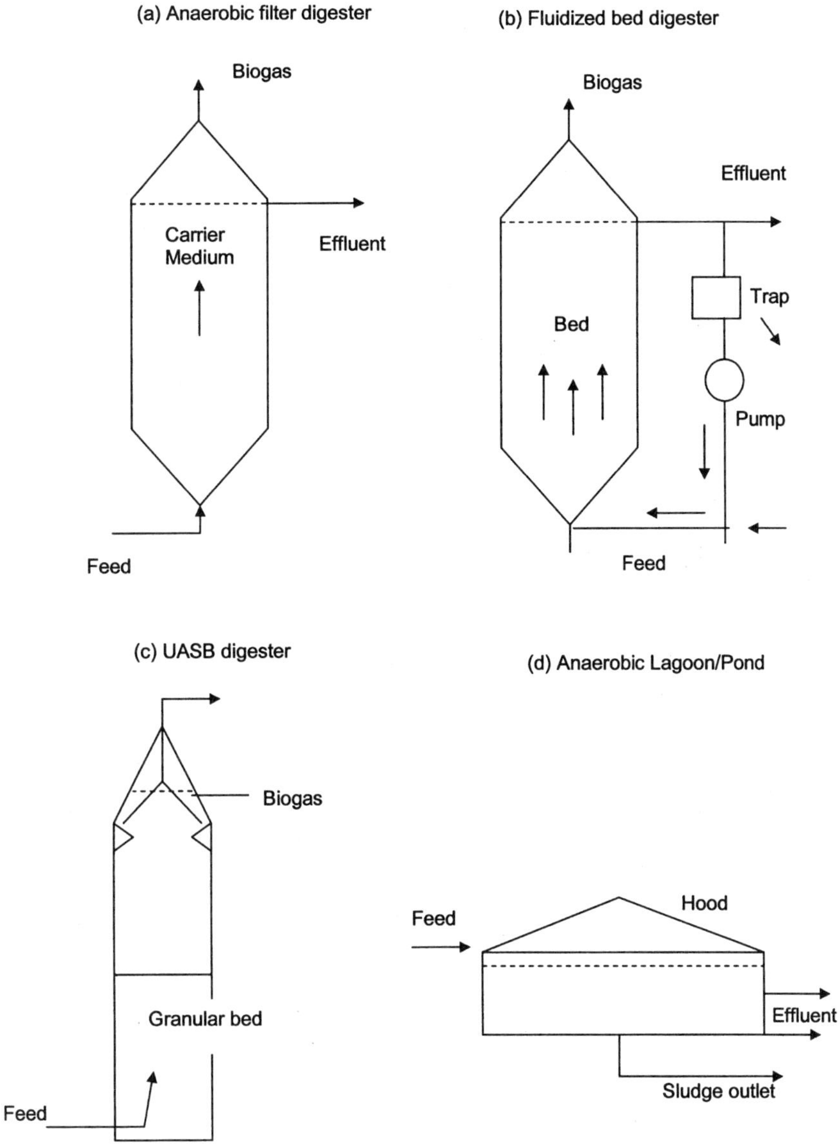

Figure 3 Simplified illustrations of anaerobic wastewater treatment processes: (a) anaerobic filter digester, (b) fluidized-bed digester, (c) UASB digester, (d) anaerobic lagoon/pond (from Refs. 31, 35, 51, 58, 70).

Completely stirred tank reactors (CSTR) [56] are, next to lagoons, the simplest type of anaerobic digester (Fig. 4). According to Sahm [57], the OLR rate ranges from 1 to 4 kg organic dry matter m^{-3} day^{-1} and the digesters usually have capacities between 500 and 700 m^3. These reactors are normally used for concentrated wastes, especially those where the polluting matter is present mainly as suspended solids and has COD values of higher than 30,000 mg/L. In the CSTR, there is no biomass retention; consequently, the HRT and sludge retention time (SRT) are not separated, necessitating long retention times that are dependent on the growth rate of the

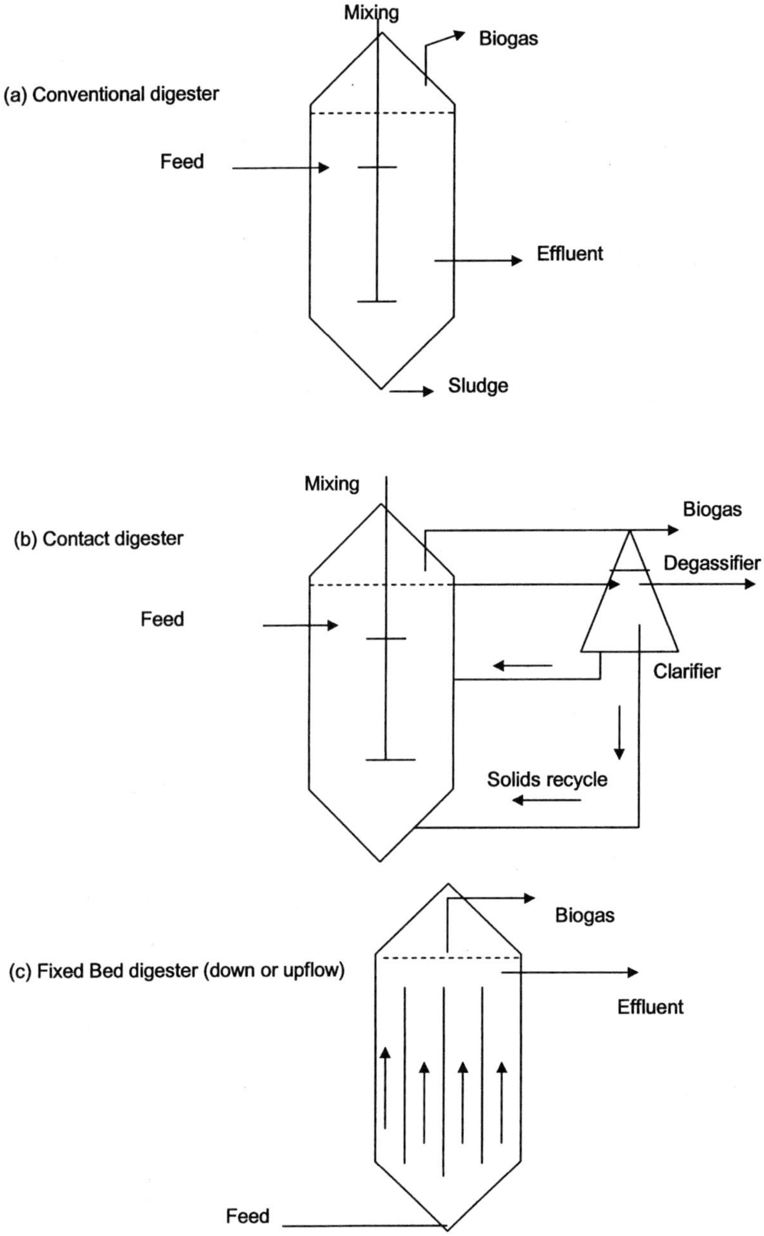

Figure 4 Simplified illustrations of anaerobic wastewater treatment processes: (a) conventional digester, (b) Contact digester, (c) fixed-bed digester (from Refs. 31, 57, 58, 60, 64, 66, 79).

slowest-growing bacteria involved in the digestion process. Ross [58] found that the HRT of the conventional digesters is equal to the SRT, which can range from 15 to 20 days.

This type of digester has in the past been used by Lebrato et al. [59] to treat cheese factory wastewater. While 90% COD removal was achieved, the digester could only be operated at a minimum HRT of 9.0 days, most probably due to biomass washout. The wastewater, consisting of 80% washing water and 20% whey, had a COD of 17,000 mg/L. While the CSTR is very useful for laboratory studies, it is hardly a practical option for full-scale treatment due to the HRT limitation.

The anaerobic *contact process* (Fig. 4) was developed in 1955 [60]. It is essentially an anaerobic activated sludge process that consists of a completely mixed anaerobic reactor followed by some form of biomass separator. The separated biomass is recycled to the reactor, thus reducing the retention time from the conventional 20–30 days to <1.0 days. Because the bacteria are retained and recycled, this type of plant can treat medium-strength wastewater (200–20,000 mg/L COD) very efficiently at high OLRs [57]. The organic loading rate can vary from 1 to 6 kg/m^3 day COD with COD removal efficiencies of 80–95%. The treatment temperature ranges from 30–40°C. A major difficulty encountered with this process is the poor settling properties of the anaerobic biomass from the digester effluent. Dissolved air flotation [61] and dissolved biogas flotation techniques [62] have been attempted as alternative sludge separation techniques. Even though the contact digester is considered to be obsolete there are still many small dairies all over the world that use the system [63].

The upflow *anaerobic filter* (Fig. 3) was developed by Young and McCarty in 1969 [64] and is similar to the aerobic trickling filter process. The reactor is filled with inert support material such as gravel, rocks, coke, or plastic media and thus there is no need for biomass separation and sludge recycling. The anaerobic filter reactor can be operated either as a downflow or an upflow filter reactor with OLR ranging from 1 to 15 kg/m^3 day COD and COD removal efficiencies of 75–95%. The treatment temperature ranges from 20 to 35°C with HRTs in the order of 0.2–3 days. The main drawback of the upflow anaerobic filter is the potential risk of clogging by undegraded suspended solids, mineral precipitates or the bacterial biomass. Furthermore, their use is restricted to wastewaters with COD between 1000 and 10,000 mg/L [58]. Bonastre and Paris [65] listed 51 anaerobic filter applications of which five were used for pilot plants and three for full-scale dairy wastewater treatment. These filters were operated at HRTs between 12 and 48 hours, while COD removal ranged between 60 and 98%. The OLR varied between 1.7 and 20.0 kg COD/m^3 day.

The *expanded bed* and/or *fluidized-bed digesters* (Fig. 3) are designed so that wastewaters pass upwards through a bed of suspended media, to which the bacteria attach [66]. The carrier medium is constantly kept in suspension by powerful recirculation of the liquid phase. The carrier media include plastic granules, sand particles, glass beads, clay particles, and activated charcoal fragments. Factors that contribute to the effectiveness of the fluidized-bed process include: (a) maximum contact between the liquid and the fine particles carrying the bacteria; (b) problems of channeling, plugging, and gas hold-up commonly encountered in packed-beds are avoided; and (c) the ability to control and optimize the biological film thickness [57]. OLRs of 1–20 kg/m^3 day COD can be achieved with COD removal efficiencies of 80–87% at treatment temperatures from 20 to 35°C.

Toldrá et al. [67] used the process to treat dairy wastewater with a COD of only 200–500 mg/L at an HRT of 8.0 hours with COD removal of 80%. Bearing in mind the wide variations found between different dairy effluents, it can be deduced that this particular dairy effluent is at the bottom end of the scale in terms of its COD concentration and organic load. The dairy wastewater was probably produced by a dairy with very good product-loss control and rather high water use [68].

The *upflow anaerobic sludge blanket* (UASB) reactor was developed for commercial purposes by Lettinga and coworkers at the Agricultural University in Wageningen, The Netherlands. It was first used to treat maize-starch wastewaters in South Africa [69], but its full potential was only realized after an impressive development program by Lettinga in the late 1970s [70,71]. The rather simple design of the UASB bioreactor (Fig. 3) is based on the superior settling properties of a granular sludge. The growth and development of granules is the key to the success of the UASB digester. It must be noted that the presence of granules in the UASB system ultimately serves to separate the HRT from the solids retention time (SRT). Thus, good granulation is essential to achieve a short HRT without inducing biomass washout. The wastewater is fed from below and leaves at the top via an internal baffle system for separation of the gas, sludge, and liquid phases. With this device, the granular sludge and biogas are separated. Under optimal conditions, a COD loading of 30 kg/m^3 day can be treated with a COD removal efficiency of 85–95%. The methane content of the biogas is between 80 and 90% (v/v). HRTs of as low as 4 hours are feasible, with excellent settling sludge and SRT of more than 100 days. The treatment temperature ranges from 7 to 40°C, with the optimum being at 35°C.

Goodwin et al. [72] treated a synthetic ice cream wastewater using the UASB process at HRTs of 18.4 hours and an organic carbon removal of 86% was achieved. The maximum OLR was 3.06 kg total organic carbon (TOC) per m^3 day. Cheese effluent has also been treated in the UASB digester at a cheese factory in Wisconsin, USA [73]. The UASB was operated at an HRT of 16.0 hours and an OLR of 49.5 kg COD/m^3 day with a plant wastewater COD of 33,000 mg/L and a COD removal of 86% was achieved. The UASB digester was, however, only a part of a complete full-scale treatment plant. The effluent from the UASB was recycled to a mixing tank, which also received the incoming effluent. Although the system is described as an UASB system, it could also pass as a separated or two-phase system, since some degree of pre-acidification is presumably attained in the mixing tank. Furthermore, the pH in the mixing tank was controlled by means of lime dosing when necessary. The effluent emerging from the mixing tank was treated in an aerobic system, serving as a final polishing step, to provide an overall COD removal of 99%.

One full-scale UASB treatment plant [51] in Finland at the Mikkeli Cooperative Dairy, produces Edam type cheese, butter, pasteurized and sterilized milk, and has a wastewater volume of 165 million liters per year. The digester has an operational volume of 650 m^3, which includes a balancing tank of 300 m^3 [74,75]. The COD value was reduced by 70–90% and 400 m^3 biogas is produced daily with a methane content of 70%, which is used to heat process water in the plant.

One of the most successful full-scale 2000 m^3 UASB described in the literature was in the UK at South Caernarvon Creameries to treat whey and other wastewaters [76]. The whey alone reached volumes of up to 110 kiloliters (kL) per day. In the system, which included a combined UASB and aerobic denitrification system, COD was reduced by 95% and sufficient biogas was produced to meet the total energy need of the whole plant. The final effluent passed to a sedimentation tank, which removed suspended matter. From there, it flowed to aerobic tanks where the BOD was reduced to 20.0 mg/L and the NH_3-nitrogen reduced to 10.0 mg/L. The effluent was finally disposed of into a nearby river. The whey disposal costs, which originally amounted to £30,000 per year, were reduced to zero; the biogas also replaced heavy fuel oil costs. On full output, the biogas had a value of up to £109,000 per year as an oil replacement and a value of about £60,000 as an electricity replacement. These values were, however, calculated in terms of the oil and electricity prices of 1984, but this illustrates the economic potential of the anaerobic digestion process.

The *fixed-bed digester* (Fig. 4) contains permanent porous carrier materials and by means of extracellular polysaccharides, bacteria can attach to the surface of the packing material and still remain in close contact with the passing wastewater. The wastewater is added either at the bottom or at the top to create upflow or downflow configurations.

A downflow fixed-film digester was used by Cánovas-Diaz and Howell [77] to treat deproteinized cheese whey with an average COD of 59,000 mg/L. At an OLR of 12.5 kg COD/m^3 day, the digester achieved a COD reduction of 90–95% at an HRT of 2.0–2.5 days. The deproteinized cheese whey had an average pH of 2.9, while the digester pH was consistently above pH 7.0 [78].

A laboratory-scale fixed-bed digester with an inert polyethylene bacterial carrier was also used by De Haast et al. [79] to treat cheese whey. The best results were obtained at an HRT of 3.5 days, with 85–87% COD removal. The OLR was 3.8 kg COD/m^3 day and biogas yield amounted to 0.42 $m^3/kg\ COD_{added}$ per day. The biogas had a methane content of between 55 and 60%, and 63.7% of the calorific value of the substrate was conserved in the methane.

In a *membrane anaerobic reactor system* (MARS), the digester effluent is filtrated by means of a filtration membrane. The use of membranes enhances biomass retention and immediately separates the HRT from the SRT [68].

Li and Corrado (80) evaluated the MARS (completely mixed digester with operating volume of 37,850 L combined with a microfiltration membrane system) on cheese whey with a COD of up to 62,000 mg/L. The digester effluent was filtrated through the membrane and the permeate discharged, while the retentate, containing biomass and suspended solids, was returned to the digester. The COD removal was 99.5% at an HRT of 7.5 days. The most important conclusion the authors made was that the process control parameters obtained in the pilot plant could effectively be applied to their full-scale demonstration plant.

A similar membrane system, the anaerobic digestion ultrafiltration system (ADUF) has successfully been used in bench- and pilot-scale studies on dairy wastewaters [81]. The ADUF system does not use microfiltration, but rather an ultrafiltration membrane; therefore, far greater biomass retention efficiency is possible.

Separated phase digesters are designed to spatially separate the acid-forming bacteria and the acid-consuming bacteria. These digesters are useful for the treatment of wastes either with unbalanced carbon to nitrogen (C : N) ratios, such as wastes with high protein levels, or wastes such as dairy wastewaters that acidify quickly [51,68]. High OLRs and short HRTs are claimed to be the major advantages of the separated phase digester.

Burgess [82] described two cases where dairy wastewaters were treated using a separated phase full-scale process. One dairy had a wastewater with a COD of 50,000 mg/L and a pH of 4.5. Both digester phases were operated at 35°C, while the acidogenic reactor was operated at an HRT of 24 hours and the methanogenic reactor at an HRT of 3.3 days. In the acidification tank, 50% of the COD was converted to organic acids while only 12% of the COD was removed. The OLR for the acidification reactor was 50.0 kg COD/m^3 day, and for the methane reactor, 9.0 kg COD/m^3 day. An overall COD reduction of 72% was achieved. The biogas had a methane content of 62%, and from the data supplied, it was calculated that a methane yield ($Y_{CH4}/COD_{removed}$) of 0.327 $m^3/kg\ COD_{removed}$ was obtained.

Lo and Liao [83,84] also used separated phase digesters to treat cheese whey. The digesters were described as anaerobic rotating biological contact reactors (AnRBC), but can really be described as tubular fixed-film digesters orientated horizontally, with internally rotating baffles. In the methane reactor, these baffles were made from cedar wood, as the authors contend that the desired bacterial biofilms develop very quickly on wood. The acidogenic reactor was mixed by means of the recirculation of the biogas. However, it achieved a COD reduction of only 4%. More importantly, the total volatile fatty acids concentration was increased from 168 to 1892 mg/L. This was then used as substrate for the second phase where a COD reduction of up to 87% was achieved. The original COD of the whey was 6720 mg/L, which indicates that the whey was diluted approximately ten-fold.

Many other examples of two-phase digesters are found in the literature. It was the opinion of Kisaalita et al. [85] that two-phase processes may be more successful in the treatment of lactose-containing wastes. The researchers studied the acidogenic fermentation of lactose, determined the kinetics of the process [86], and also found that the presence of whey protein had little influence on the kinetics of lactose acidogenesis [87]. Venkataraman et al. [88] also used a two-phase packed-bed anaerobic filter system to treat dairy wastewater. Their main goals were to determine the kinetic constants for biomass and biogas production rates and substrate utilization rates in this configuration.

Land Treatment

Dairy wastewater, along with a wide variety of other food processing wastewaters, has been successfully applied to land in the past [31]. Interest in the land application of wastes is also increasing as a direct result of the general move of regulatory authorities to restrict waste disposal into rivers, lakes, and the ocean, but also because of the high costs of incineration and landfilling [89]. Nutrients such as N and P that are contained in biodegradable processing wastewaters make these wastes attractive as organic fertilizers, especially since research has shown that inorganic fertilizers might not be enough to stem soil degradation and erosion in certain parts of the world [89,90]. Land application of these effluents may, however, be limited by the presence of toxic substances, high salt concentrations, or extreme pH values [89]. It might be, according to Wendorff [7], the most economical option for dairy industries located in rural areas.

Irrigation

The distribution of dairy wastewaters by irrigation can be achieved through spray nozzles over flat terrain, or through a ridge and furrow system [7]. The nature of the soil, topography of the land and the waste characteristics influence the specific choice of irrigation method. In general, loamy well-drained soils, with a minimum depth to groundwater of 1.5 m, are the most suitable for irrigation. Some form of crop cover is also desirable to maintain upper soil layer porosity [31]. Wastewater would typically percolate through the soil, during which time organic substances are degraded by the heterotrophic microbial population naturally present in the soil [7]. An application period followed by a rest period (in a 1 : 4 ratio) is generally recommended.

Eckenfelder [31] reviewed two specific dairy factory irrigation regimes. The first factory produced cream, butter, cheese, and powdered milk, and irrigated their processing wastewaters after pretreatment by activated sludge onto coarse and fine sediments covered with reed and canary grass in a 1 : 3 application/rest ratio. The second factory, a Cheddar cheese producer, employed only screening as a pretreatment method and irrigated onto Chenango gravel with the same crop cover as the first factory, in a 1 : 6 application/rest ratio.

Specific wastewater characteristics can have an adverse effect on a spray irrigation system that should also be considered. Suspended solids, for instance, may clog spray nozzles and render the soil surface impermeable, while wastewater with an extreme pH or high salinity might be detrimental to crop cover. Highly saline wastewater might further cause soil dispersion, and a subsequent decrease in drainage and aeration, as a result of ion exchange with sodium replacing magnesium and calcium in the soil [31]. The land application of dairy factory wastewater, which typically contains high concentrations of sodium ions, might thus be restricted [89]. And although milk proteins and lactose are readily degradable by anaerobic bacteria naturally present in the soil, FOG tends to be more resistant to degradation and will accumulate under anaerobic conditions [7].

According to Sparling et al. [15] there is little published information relating the effect that long-term irrigation of dairy factory effluent may have on soil properties. Based on the irrigation data Degens et al. [91] and Sparling et al. [15] investigated the effect that long-term dairy wastewater irrigation can have on the storage and distribution of nutrients such as C, N, and P, and the differences existing between key soil properties of a long-term irrigation site (22 years) and a short-term irrigation site (2 years). Degens et al. [91] reported that irrigation had no effect on total soil C in the 0–0.75 m layer, although redistribution of C from the top 0–0.1 m soil had occurred, either as a result of leaching caused by the irrigation of highly alkaline effluents, or as a result of increased earthworm activity. The latter were probably promoted by an increased microbial biomass in the soil, which were mostly lactose and glucose degraders. It was also reported that about 81% of the applied P were stored in the 0–0.25 m layer compared to only 8% of the total applied N. High nitrate concentrations were measured in the groundwater below the site, and reduced nitrogen loadings were recommended in order to limit nitrogen leaching to the environment [91]. In contrast to the results reported by Degens et al. (2000) for a long-term irrigated site, Sparling et al. [15] found no redistribution of topsoil C in short-term irrigated soils, which was probably the result of a lower effluent loading. Generally, it was found that hydraulic conductivity, microbial content, and N-cycling processes all increased substantially in long-term irrigated soils. Since increases in infiltration as well as biochemical processing were noted in all the irrigated soils, most of the changes in soil properties were considered to be beneficial. A decrease in N-loading was, however, also recommended [15].

13.4.4 Sludge Disposal

Different types of sludge arise from the treatment of dairy wastewaters. These include: (a) sludge produced during primary sedimentation of raw effluents (the amounts of which are usually low); (b) sludge produced during the precipitation of suspended solids after chemical treatment of raw wastewaters; (c) stabilized sludge resulting from the biological treatment processes, which can be either aerobic or anaerobic; and (d) sludge generated during tertiary treatment of wastewater for final suspended solid or nutrient removal after biological treatment [92]. Primary sedimentation of dairy wastewater for BOD reduction is not usually an efficient process, so in most cases the settleable solids reach the next stage in the treatment process directly. An important advantage of anaerobic processes is that the sludge generated is considerably less than the amount produced by aerobic processes, and it is easier to dewater. Final wastewater polishing after biological treatment usually involves chemical treatment of the wastewater with calcium, iron, or aluminum salts to remove dissolved nutrients such as nitrogen and phosphorus. The removal of dissolved phosphorus can have a considerable impact on the amount of sludge produced during this stage of treatment [92].

The application of dairy sludge as fertilizer has certain advantages when compared to municipal sludge. It is a valuable source of nitrogen and phosphorous, although some addition of potassium might be required to provide a good balance of nutrients. Sludge from different factories will also contain different levels of nutrients depending on the specific products manufactured. Dairy sludge seldom contains the same pathogenic bacterial load as domestic sludge, and also has considerably lower heavy metal concentrations. The recognition of dairy sludge as a fertilizer does, however, depend on local regulations. Some countries have limited the amount of sludge that can be applied as fertilizer to prevent nitrates from leaching into groundwater sources [92].

According to the IDF [92], dairy sludge disposal must be reliable, legally acceptable, economically viable, and easy to conduct. Dairy wastewater treatment facilities are usually small compared to sewage treatment works, which means that thermal processes such as drying and incineration can be cost-prohibitive for smaller operations. It is generally agreed that disposal of sludge by land spraying or as fertilizer is the least expensive method. If the transport and disposal of liquid sludge cannot be done within reasonable costs, other treatment options such as sludge thickening, dewatering, drying, or incineration must be considered. Gravity thickeners are most commonly used for sludge thickening, while the types of dewatering machines most commonly applied are rotary drum vacuum filters, filter presses, belt presses, and decanter centrifuges [92].

13.5 POLLUTION PREVENTION

Reduction of wastewater pollution levels may be achieved by more efficiently controlling water and product wastage in dairy processing plants. Comparisons of daily water consumption records vs. the amount of milk processed will give an early indication of hidden water losses that could result from defective subfloor and underground piping. An important principle is to prevent wastage of product rather than flush it away afterwards. Spilled solid material such as curd from the cheese production area, and spilled dry product from the milk powder production areas should be collected and treated as solid waste rather than flushing them down the drain [6].

Small changes could also be made to dairy manufacturing processes to reduce wastewater pollution loads, as reviewed by Tetrapak [6]. In the cheese production area, milk spillage can be restricted by not filling open cheese vats all the way to the rim. Whey could also be collected sparingly and used in commercial applications instead of discharging it as waste.

Manual scraping of all accessible areas after a butter production run and before cleaning starts would greatly reduce the amount of residual cream and butter that would enter the wastewater stream. In the milk powder production area, the condensate formed could be reused as cooling water (after circulation through the cooling tower), or as feedwater to the boiler. Returned product could be emptied into containers and used as animal feed [6]. Milk and product spillage can further be restricted by regular maintenance of fittings, valves, and seals, and by equipping fillers with drip and spill savers. Pollution levels could also be limited by allowing pipes, tanks, and transport tankers adequate time to drain before being rinsed with water [8].

13.6 CASE STUDIES

13.6.1 Case Study 1

A summary of a case study as reported by Rusten et al. [93] is presented for the upgrading of a cheese factory additionally producing casein granules.

Background

The authors described how a wastewater treatment process of a Norwegian cheese factory, producing casein granules as a byproduct, was upgraded to meet the wastewater treatment demands set by large increases in production and stricter environmental regulations. The design criteria were based on the assumption that the plant produced an average amount of 150 m^3/day of wastewater, which had an average organic load of 200 kg BOD/day with an average total phosphorous (TP) load of 3.5 kg TP/day and a pH range between 2 and 12.

Requirements

It was required that the treatment plant be able to remove more than 95% of the total BOD (>95% total COD). The specific amount of phosphorous that could be allowed in the discharged wastewater was still being negotiated with the authorities. The aim however, was to remove as much phosphorous as possible. The pH of the final effluent had to be between 6.5 and 8.0.

The Final Process

A flow diagram of the final process is summarized in Figure 5.

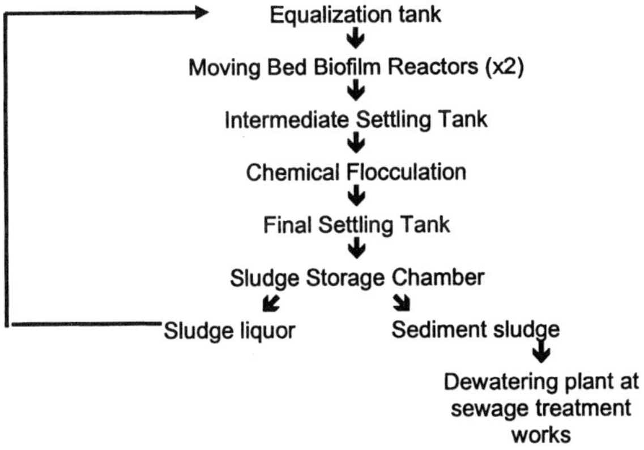

Figure 5 Flow diagram of the final process of Case Study 1.

Process Efficiency

After modifications, the average organic load was 347 kg COD/day with average removal efficiency of 98% for both the total COD and the total phosphorous content. Extreme pH values in the incoming wastewater were also efficiently neutralized in the equalization tank, resulting in a 7.0–8.0 pH range in the reactors.

13.6.2 Case Study 2

A summary of a case study reported by Monroy et al. [94] is presented.

Background

As with the first case study, the authors reported on how an existing wastewater treatment system of a cheese manufacturing industry in Mexico, which was operating below the consents, could be upgraded so that the treated wastewater could meet the discharge limits imposed by local environmental authorities. The factory produced an average wastewater volume of 500 m^3/day with an average composition (mg/L) of 4430 COD, 3000 BOD_5, 1110 TSS, and 754 FOG.

Requirements

Environmental regulations required the treated wastewater to have less than 100 mg/L BOD, 300 mg/L COD, 100 mg/L TSS, and 15 mg/L FOG. The pH of the discharged effluent had to be between 6.0 and 9.0. The old treatment system was not effective enough to reduce the BOD, COD, TSS, and FOG to acceptable levels, although the final pH of 7.5 was within the recommended range. The factory was looking for a more effective treatment system that could utilize preexisting installations, thereby reducing initial investment costs, and also have low operation costs.

The Final Process

A flow diagram of the final process is summarized in Figure 6.

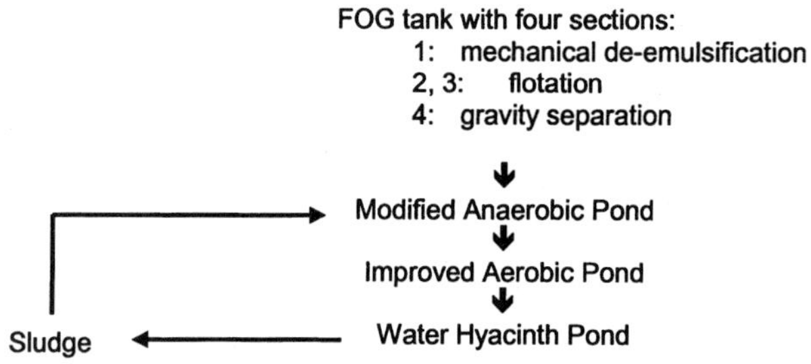

Figure 6 Flow diagram of the final process of Case Study 2.

Process Efficiency

Pollution levels in the raw wastewater were first reduced by initiating an "in-factory" wastewater management program, which resulted in greater pH stability and lower phosphorous levels (by recycling certain cleaning chemicals and substituting others) as well as reduced levels of salt (by concentrating and drying brine). The modified wastewater treatment process resulted in an overall removal efficiency of 98% BOD (final concentration = 105 mg/L), 96% COD (final concentration = 225 mg/L), 98% TSS (final concentration = 24 mg/L), and 99.8% FOG (final concentration = 1.7 mg/L). The modifications ultimately resulted in a total operating cost increase of 0.4% at the factory.

13.6.3 General Conclusions: Case Studies

All wastewater treatment systems are unique. Before a treatment strategy is chosen, careful consideration should be given to proper wastewater sampling and composition analysis as well as a process survey. This would help prevent an expensive and unnecessary or overdesigned treatment system [95]. A variety of different local and international environmental engineering firms are able to assist in conducting surveys. These firms can also be employed to install effective patented industrial-scale installations for dairy processing wastewater treatment.

13.7 CONCLUSIONS

As management of dairy wastes becomes an ever-increasing concern, treatment strategies will need to be based on state and local regulations. Because the dairy industry is a major water user and wastewater generator, it is a potential candidate for wastewater reuse. Purified wastewater can be utilized in boilers and cooling systems as well as for washing plants, and so on. Even if the purified wastewater is initially not reused, the dairy industry will still benefit directly from in-house wastewater treatment, since levies charged for wastewater reception will be significantly reduced. In the United Kingdom, 70% of the total savings that have already been achieved with anaerobic digestion are due to reduced discharge costs [96]. The industry will also benefit where effluents are currently used for irrigation of pastures, albeit in a more indirect way. All these facts underline the need for efficient dairy wastewater management.

Before selecting any treatment method, a complete process evaluation should be undertaken along with economic analysis. This should include the wastewater composition, concentrations, volumes generated, and treatment susceptibility, as well as the environmental impact of the solution to be adopted. All options are expensive, but an economic analysis may indicate that slightly higher maintenance costs may be less than increased operating costs. What is appropriate for one site may be unsuitable for another.

The most useful processes are those that can be operated with a minimum of supervision and are inexpensive to construct or even mobile enough to be moved from site to site. The changing quantity and quality of dairy wastewater must also be included in the design and operational procedures. From the literature it appears as if biological methods are the most cost-effective for the removal of organics, with aerobic methods being easier to control, but anaerobic methods having lower energy requirements and lower sludge production rates. Since no single process for treatment of dairy wastewater is by itself capable of complying with the minimum effluent discharge requirements, it is necessary to choose a combined process especially designed to treat a specific dairy wastewater.

REFERENCES

1. Russell, P. Effluent and waste water treatment. Milk Ind. Int. **1998**, *100 (10)*, 36–39.
2. Strydom, J.P.; Mostert, J.F.; Britz, T.J. Effluent production and disposal in the South African dairy industry–a postal survey. Water SA **1993**, *19* (*3*), 253–258.
3. Robinson, T. The real value of dairy waste. Dairy Ind. Int. **1997**, *62* (*3*), 21–23.
4. Kessler, HG. (Ed.) *Food Engineering and Dairy Technology*; Verlag: Freisburg, Germany, 1981.
5. Odlum, C.A. Reducing the BOD level from a dairy processing plant. *Proc. 23rd Int. Dairy Cong.*, Montreal, Canada, October 1990.
6. Tetrapak. *TetraPak Dairy Processing Handbook*; TetraPak Printers: London, UK, 1995.
7. Wendorff, W.L. Treatment of dairy wastes. In *Applied Dairy Microbiology*, 2nd Ed.; Marth, E.H., Steele, J.L., Eds.; Marcel Dekker Inc: New York, 2001; 681–704.
8. Steffen, Robertson, Kirsten Inc. *Water and Waste-water Management in the Dairy Industry*, WRC Project No. 145 TT38/89. Water Research Commission: Pretoria, South Africa, 1989.
9. Tamime, A.Y.; Robinson, R.K. (Eds.) *Yoghurt Science and Technology*; Woodhead Publishing Ltd: Cambridge, England, 1999.
10. Danalewich, J.R.; Papagiannis, T.G.; Belyea, R.L.; Tumbleson, M.E.; Raskin, L. Characterization of dairy waste streams, current treatment practices, and potential for biological nutrient removal. Water Res. **1998**, *32* (*12*), 3555–3568.
11. Bakka, R.L. Wastewater issues associated with cleaning and sanitizing chemicals. Dairy Food Environ. Sanit. **1992**, *12* (*5*), 274–276.
12. Vidal, G.; Carvalho, A.; Méndez, R.; Lema, J.M. Influence of the content in fats and proteins on the anaerobic biodegradability of dairy wastewaters. Biores. Technol. **2000**, *74, 231–239.*
13. Andreottola, G.; Foladori, P.; Ragazzi, M.; Villa, R. Dairy wastewater treatment in a moving bed biofilm reactor. Wat. Sci. Technol. **2002**, *45* (*12*), 321–328.
14. Strydom, J.P.; Britz, T.J.; Mostert, J.F. Two-phase anaerobic digestion of three different dairy effluents using a hybrid bioreactor. Water SA **1997**, *23*, 151–156.
15. Sparling, G.P.; Schipper, L.A.; Russell, J.M. Changes in soil properties after application of dairy factory effluent to New Zealand volcanic ash and pumice soils. Aust. J. Soil. Res. **2001**, *39*, 505–518.
16. Hwang, S.; Hansen, C.L. Characterization of and bioproduction of short-chain organic acids from mixed dairy-processing wastewater. Trans. ASAE **1998**, *41* (*3*), 795–802.
17. Donkin, J. Bulking in aerobic biological systems treating dairy processing wastewaters. Int. J. Dairy Tech. **1997**, *50*, 67–72.
18. Samkutty, P.J.; Gough, R.H. Filtration treatment of dairy processing wastewater. J. Environ. Sci. Health. **2002**, *A37* (*2*), 195–199.
19. Ince, O. Performance of a two-phase anaerobic digestion system when treating dairy wastewater. Wat. Res. **1998**, *32* (*9*), 2707–2713.
20. Ince, O. Potential energy production from anaerobic digestion of dairy wastewater. J. Environ. Sci. Health. **1998**, *A33* (*6*), 1219–1228.

21. Torrijos, M.; Vuitton, V.; Moletta, R. The SBR process: an efficient and economic solution for the treatment of wastewater at small cheese making dairies in the Jura Mountains. Wat. Sci. Technol. **2001**, *43*; 373–380.
22. Malaspina, F.; Cellamare, C.M.; Stante, L.; Tilche, A. Anaerobic treatment of cheese whey with a downflow-upflow hybrid reactor. Biores. Technol. **1996**, *55*, 131–139.
23. Burton, C. FOG clearance. Dairy Ind. Int. **1997**, *62* (*12*), 41–42.
24. Berruga, M.I.; Jaspe, A.; San-Jose, C. Selection of yeast strains for lactose hydrolysis in dairy effluents. Int. Biodeter. Biodeg. **1997**, *40* (*2–4*), 119–123.
25. Robinson, T. How to be affluent with effluent. The Milk Ind. **1994**, *96* (*4*), 20–21.
26. Gough, R.H.; McGrew, P. Preliminary treatment of dairy plant waste water. J. Environ. Sci. Health. **1993**, *A28* (*1*), 11–19.
27. Droste, R.L. (Ed.) *Theory and Practice of Water and Wastewater Treatment*; John Wiley & Sons Inc: New York, USA, 1997.
28. Hemming, M.L. The treatment of dairy wastes. In *Food Industry Wastes: disposal and Recovery*; Herzka, A., Booth, R.G., Eds.; Applied Science Publishers Ltd: Essex, 1981.
29. Graz, C.J.M.; McComb, D.G. Dairy CIP–A South African review. Dairy, Food Environ. Sanit. **1999**, *19* (*7*), 470–476.
30. IDF. Balance tanks for dairy effluent treatment plants. Bull. Inter. Dairy Fed. **1984**, Doc. No. 174.
31. Eckenfelder, W.W. (Ed.) *Industrial Water Pollution Control*; McGraw-Hill Inc: New York, USA, 1989.
32. IDF. Removal of fats, oils and grease in the pretreatment of dairy wastewaters. Bull. Inter. Dairy Fed. **1997**, Doc. No. 327.
33. Cammarota, M.C.; Teixeira, G.A.; Freire, D.M.G. Enzymatic pre-hydrolysis and anaerobic degradation of wastewaters with high fat contents. Biotech. Lett. **2001**, *23*, 1591–1595.
34. Leal, M.C.M.R.; Cammarota, M.C.; Freire, D.M.G.; Sant'Anna Jr, G.L. Hydrolytic enzymes as coadjuvants in the anaerobic treatment of dairy wastewaters. Brazilian J. Chem. Eng. **2002**, *19* (*2*), 175–180.
35. Smith, P.G.; Scott J.S. (Eds.) *Dictionary of Water and Waste Management*; IWA Publishing. Butterworth Heinemann: Oxford, UK, 2002.
36. Donkin, J.; Russell, J.M. Treatment of a milk powder/butter wastewater using the AAO activated sludge configuration. Water Sci. Tech. **1997**, *36*, 79–86.
37. Thirumurthi, D. Biodegradation of sanitary landfill leachate. In *Biological Degradation of Wastes*, A.M. Martin, Ed. Elsevier Appl. Sci.; London, UK, 1991; 208.
38. Herzka, A.; Booth, R.G. (Eds.) *Food Industry Wastes: Disposal and Recovery*; Applied Science Publishers Ltd: Essex, UK, 1981.
39. Maris, P.J.; Harrington, D.W.; Biol, A.I.; Chismon, G.L. Leachate treatment with particular reference to aerated lagoons. Water Poll. Cont. **1984**, *83*, 521–531.
40. Rusten, B.; Odegaard, H.; Lundar, A. Treatment of dairy wastewater in a novel moving-bed biofilm reactor. Water Sci. Tech. **1992**, *26* (*3/4*), 703–709.
41. Gough, R.H.; Samkutty, P.J.; McGrew, P.; Arauz, A.; Adkinson, A. Prediction of effluent biochemical oxygen demand in a dairy plant SBR waste water system. J. Environ. Sci. Health. **2000**, *A35*, 169–175.
42. Eroglu, V.; Ozturk, I.; Demir, I.; Akca, A. Sequencing batch and hybrid reactor treatment of dairy wastes. In *Proc 46th Purdue Ind. Wast. Conf.*, West Lafayette, IN, 1992; 413–420.
43. Samkutty, P.J.; Gough, R.H.; McGrew, P. Biological treatment of dairy plant wastewater. J. Environ. Sci. Health **1996**, *A31*, 2143–2153.
44. Li, X.; Zhang, R. Aerobic treatment of dairy wastewater with sequencing batch reactor systems. Bioproc. Biosys. Eng. **2002**, *25*, 103–109.
45. Tanaka, T. Use of aerated lagoons for dairy effluent treatment. *Sym. Dairy Effl. Treat*, Kollenbolle, Denmark, May 1973.
46. Bitton, G. (Ed.) *Wastewater Microbiology*; Wiley Press: New York, 1994.
47. Sterritt, R.M.; Lester, J.N. (Eds.) *Microbiology for Environmental and Public Health Engineers*; E & FN Spon., London, UK, 1988.

48. Thirumurthi, D. Biodegradation in waste stabilization ponds (facultative lagoons). In *Biological Degradation of Wastes*; Martin, A.M., Ed. Elsevier Applied Sci.; New York, 1991; 231–235.
49. Robinson, H.D.; Barr, M.J.; Formby, B.W.; Formby, B.W.; Moag, A. The treatment of landfill leachates using reed bed systems. *IWEM Annual Training Day*, October 1992.
50. European Water Pollution Control Association (EWPCA). *European Design and Operations Guidelines for Reed-Bed Treatment Systems*, Report to EC/EWPCA Treatment Group, P.F. Cooper, Ed.; August 1990.
51. IDF. Anaerobic treatment of dairy effluents – The present stage of development. Bull. Inter. Dairy Fed., Doc. **1990**, *252*.
52. Weber, H.; Kulbe, K.D.; Chmiel, H.; Trösch, W. Microbial acetate conversion to methane: kinetics, yields and pathways in a two-step digestion process. Appl. Microb. Biotech. **1984**, *19*, 224–228.
53. Lema, J.M.; Mendez, R.; and Blazquez, R. Characteristics of landfill leachates and alternatives for their treatment: a review. Water Air Soil Pollut. **1988**, *40*, 223–227.
54. Strydom, J.P.; Mostert, J.F.; Britz, T.J. Anaerobic treatment of a synthetic dairy effluent using a hybrid digester. Water SA **1995**, *21* (*2*), 125–130.
55. Britz, T.J.; Van Der Merwe, M.; Riedel, K.-H.J. Influence of phenol additions on the efficiency of an anaerobic hybrid digester treating landfill leachate. Biotech. Lett. **1992**, *14*, 323–327.
56. Feilden, N.E.H. The theory and practice of anaerobic reactor design. Proc. Biochem. **1983**, *18*, 34–37.
57. Sahm, H. Anaerobic wastewater treatment. Adv. Biochem. Eng. Biotech. **1984**, *29*, 83–115.
58. Ross, W.R. *Anaerobic Digestion of Industrial Effluents With Emphasis on Solids-Liquid Separation and Biomass Retention*, Ph.D. Thesis, University of the Orange Free State Press, South Africa, 1991.
59. Lebrato, J.; Perez-Rodriguez, J.L.; Maqueda, C.; Morillo, E. Cheese factory wastewater treatment by anaerobic semicontinuous digestion. Res. Cons. Recyc. **1990**, *3*, 193–199.
60. Schroepfer, G.J.; Fuller, W.J.; Johnson, A.S.; Ziemke, N.R.; Anderson, J.J. The anaerobic contact process as applied to packinghouse wastes. Sew. Ind. Was. **1955**, *27*, 460–486.
61. Speece, R.E. Advances in anaerobic biotechnology for industrial waste water treatment. *Proc. 2nd Int. Conf. Anaerobic Treat. Ind. Wast. Wat.*, Chicago, Il, USA, 1986; 6–17.
62. Ross, W.R.; De Villiers, H.A.; Le Roux, J.; Barnard, J.P. Sludge separation techniques in the anaerobic digestion of wine distillery waste. *Proc. 5th Int. Symp. Anaerobic Digestion*, Bologna, Italy, May 1988, 571–574.
63. Ross, W.R. Anaerobic treatment of industrial effluents in South Africa. Water SA **1989**, *15*, 231–246.
64. Young, J.C.; McCarty, P.L. The anaerobic filter for waste treatment. J. Wat. Poll. Cont. Fed. **1969**, *41*, 160–173.
65. Bonastre, N.; Paris, J.M. Survey of laboratory, pilot and industrial anaerobic filter installations. Proc. Biochem. **1989**, *24*, 15–20.
66. Switzenbaum, M.S.; Jewell, W.J. Anaerobic attached-film expanded-bed reactor treatment. J. Wat. Poll. Cont. Fed. **1980**, *52*, 1953–1965.
67. Toldrá, F.; Flors, A.; Lequerica, J.L.; Vall S.S. Fluidized bed anaerobic biodegradation of food industry wastewaters. Biol. Wast. **1987**, *21*, 55–61.
68. Strydom, J.P.; Mostert, J.F.; Britz, T.J. Anaerobic digestion of dairy factory effluents. WRC Report No 455/1/01. ISBN 1868457249; Water Research Commission: Pretoria, South Africa, 2001.
69. Hemens, J.; Meiring, P.G.; Stander, G.J. Full-scale anaerobic digestion of effluents from the production of maize-starch. Wat. Wast. Treat. **1962**, *9* (*1*), 16–35.
70. Lettinga, G.; Van Velsen, A.F.M.; Hobma, S.W.; De Zeeuw, W.; Klapwijk, A. Use of the upflow sludge blanket (USB) reactor concept for biological wastewater treatment especially for anaerobic treatment. Biotech. Bioeng. **1980**, *22*, 699–734.
71. Lettinga, G.; Hulshoff-Pol, L.W. UASB-process design for various types of wastewaters. Water Sci. Tech. **1991**, *24*, 87–107.
72. Goodwin, J.A.S.; Wase, D.A.J.; Forster, C.F. Anaerobic digestion ice-cream wastewaters using the UASB process. Biol. Wast. **1990**, *32*, 125–144.

73. De Man, G.; De Bekker, P.H.A.M.J. New technology in dairy wastewater treatment. Dairy Ind. Int. **1986**, *51* (*5*), 21–25.
74. Carballo-Caabeira, J. Depuracion de augas residuales de centrales lecheras. Rev. Española de Lech. **1990**, *13* (*12*), 13–16.
75. Ikonen, M.; Latola, P.; Pankakoski, M.; Pelkonen, J. Anaerobic treatment of waste water in a Finnish dairy. Nord. Mejeriind. **1985**,*12* (*8*), 81–82.
76. Anon. South Caernarvon Creameries converts whey into energy. Dairy Ind. Int. **1984**, *49* (*10*), 16–17.
77. Cánovas-Diaz, M.; Howell, J.A. Down-flow anaerobic filter stability studies with different reactor working volumes. Proc. Biochem. **1987**, *22*, 181–184.
78. Cánovas-Diaz, M.; Howell, J.A. Stratified ecology techniques in the start-up of an anaerobic down-flow fixed film percolating reactor. Biotech. Bioeng. **1987**, *10*, 289–296.
79. De Haast, J.; Britz, T.J.; Novello, J.C. Effect of different neutralizing treatments on the efficiency of an anaerobic digester fed with deproteinated cheese whey. J. Dairy Res. **1986**, *53*, 467–476.
80. Li, A.Y.; Corrado J.J. Scale up of the membrane anaerobic reactor system. *Proc. 40th Ann. Purdue Ind. Wast. Conf.*, West Lafayette, IN, 1985; 399–404.
81. Ross, W.R.; Barnard, J.P.; De Villiers, H.A. The current status of ADUF technology in South Africa. In *Proc. 2nd Anaerobic Digestion Symp*, University of the Orange Free State Press: Bloemfontein, South Africa, 1989; 65–69.
82. Burgess, S. Anaerobic treatment of Irish creamery effluents. Proc. Biochem. **1985**, *20*, 6–7.
83. Lo, K.V.; Liao, P.H. Digestion of cheese whey with anaerobic rotating biological contact reactor. Biomass **1986**, *10*, 243–252.
84. Lo, K.V.; Liao, P.H. Laboratory scale studies on the mesophilic anaerobic digestion of cheese whey in different digester configurations. J. Agric. Eng. Res. **1988**, *39*, 99–105.
85. Kisaalita, W.S.; Pinder, K.L.; Lo, K.V. Acidogenic fermentation of lactose. Biotech. Bioeng. **1987**, *30*, 88–95.
86. Kissalita, W.S.; Lo, K.V.; Pinder, K.L. Kinetics of whey-lactose acidogenesis. Biotech. Bioeng. **1989**, *33*, 623–630.
87. Kisaalita, W.S.; Lo, K.V.; Pinder, K.L. Influence of whey protein on continuous acidogenic degradation of lactose. Biotech. Bioeng. **1990**, *36*, 642–646.
88. Venkataraman, J.; Kaul, S.N.; Satyanarayan, S. Determination of kinetic constants for a two-stage anaerobic up-flow packed bed reactor for dairy wastewater. Biores. Technol. **1992**, *40*, 253–261.
89. Cameron, K.C.; Di, H.J.; McLaren, R.G. Is soil an appropriate dumping ground for our wastes? Aust. J. Soil Res. **1997**, *35*, 995–1035.
90. Obi, M.E.; Ebo, P.O. The effects of organic and inorganic amendments on soil physical properties and maize production in a severely degraded sandy soil in Southern Nigeria. Biores. Technol. **1995**, *51*, 117–123.
91. Degens, B.P.; Schipper, L.A.; Claydon, J.J.; Russell, J.M.; Yeates, G.W. Irrigation of an allophonic soil with dairy factory effluent for 22 years: responses of nutrient storage and soil biota. Aust. J. Soil. Res. **2000**, *38*, 25–35.
92. IDF. Sludge from dairy effluent treatment plants – 1998 survey. International Dairy Federation Draft paper: IDF-group B 18/19, 1999.
93. Rusten, B.; Siljudalen, J.G.; Strand, H. Upgrading of a biological-chemical treatment plant for cheese factory wastewater. Wat. Sci. Tech. **1996**, *43* (*11*), 41–49.
94. Monroy H.O.; Vázquez M.F.; Derramadero, J.C.; Guyot, J.P. Anaerobic-aerobic treatment of cheese wastewater with national technology in Mexico: the case of "El Sauz". Wat. Sci. Tech. **1995**, *32* (*12*), 149–156.
95. Ardundel, J. (Ed.) *Sewage and Industrial Effluent Treatment*; Blackwell Science Ltd: Oxford, England, 1995.
96. Senior, E. Wealth from Waste. In *Proc. 1st Anaerobic Digestion Symp*; University of the Orange Free State Press: Bloemfontein, South Africa, 1986; pp. 19–30.

14

Seafood Processing Wastewater Treatment

Joo-Hwa Tay and Kuan-Yeow Show
Nanyang Technological University, Singapore

Yung-Tse Hung
Cleveland State University, Cleveland, Ohio, U.S.A.

14.1 INTRODUCTION

The seafood industry consists primarily of many small processing plants, with a number of larger plants located near industry and population centers. Numerous types of seafood are processed, such as mollusks (oysters, clams, scallops), crustaceans (crabs and lobsters), saltwater fishes, and freshwater fishes. As in most processing industries, seafood-processing operations produce wastewater containing substantial contaminants in soluble, colloidal, and particulate forms. The degree of the contamination depends on the particular operation; it may be small (e.g., washing operations), mild (e.g., fish filleting), or heavy (e.g., blood water drained from fish storage tanks).

Wastewater from seafood-processing operations can be very high in biochemical oxygen demand (BOD), fat, oil and grease (FOG), and nitrogen content. Literature data for seafood processing operations showed a BOD production of 1–72.5 kg of BOD per tonne of product [1]. White fish filleting processes typically produce 12.5–37.5 kg of BOD for every tonne of product. BOD is derived mainly from the butchering process and general cleaning, and nitrogen originates predominantly from blood in the wastewater stream [1].

It is difficult to generalize the magnitude of the problem created by these wastewater streams, as the impact depends on the strength of the effluent, the rate of discharge, and the assimilatory capacity of the receiving water body. Nevertheless, key pollution parameters must be taken into account when determining the characteristics of a wastewater and evaluating the efficiency of a wastewater treatment system. Section 2 discusses the parameters involved in the characterization of the seafood processing wastewater.

Pretreatment and primary treatment for seafood processing wastewater are presented in Section 14.3. These are the simplest operations to reduce contaminant load and remove oil and grease from an effluent of seafood processing wastewater. Common pretreatments for seafood-processing wastewater include screening, settling, equalization, and dissolved air flotation.

Section 14.4 focuses on biological treatments for seafood processing wastewater, namely aerobic and anaerobic treatments. The most common operations of biological treatments are also described in this section.

Section 14.5 discusses the physico-chemical treatments for seafood processing wastewater. These operations include coagulation, flocculation, and disinfection. Direct disposal of seafood processing wastewaters is discussed in Section 14.6. Potential problems in land application are highlighted. General seafood processing plant schemes are presented in Section 14.7. Economic considerations are always the most important factors that influence the final decision for selecting processes for wastewater treatment. The economic issues related to wastewater treatment process are discussed in Section 14.8.

14.2 SEAFOOD-PROCESSING WASTEWATER CHARACTERIZATION

Seafood-processing wastewater characteristics that raise concern include pollutant parameters, sources of process waste, and types of wastes. In general, the wastewater of seafood-processing wastewater can be characterized by its physico-chemical parameters, organics, nitrogen, and phosphorus contents. Important pollutant parameters of the wastewater are five-day biochemical oxygen demand (BOD_5), chemical oxygen demand (COD), total suspended solids (TSS), fats, oil and grease (FOG), and water usage [2]. As in most industrial wastewaters, the contaminants present in seafood-processing wastewaters are an undefined mixture of substances, mostly organic in nature. It is useless or practically impossible to have a detailed analysis for each component present; therefore, an overall measurement of the degree of contamination is satisfactory.

14.2.1 Physico-Chemical Parameters

pH

pH serves as one of the important parameters because it may reveal contamination of a wastewater or indicate the need for pH adjustment for biological treatment of the wastewater. Effluent pH from seafood processing plants is usually close to neutral. For example, a study found that the average pH of effluents from blue crab processing industries was 7.63, with a standard deviation of 0.54; for non-Alaska bottom fish, it was about 6.89 with a standard deviation of 0.69 [2]. The pH levels generally reflect the decomposition of proteinaceous matter and emission of ammonia compounds.

Solids Content

Solids content in a wastewater can be divided into dissolved solids and suspended solids. However, suspended solids are the primary concern since they are objectionable on several grounds. Settleable solids may cause reduction of the wastewater duct capacity; when the solids settle in the receiving water body, they may affect the bottom-dwelling flora and the food chain. When they float, they may affect the aquatic life by reducing the amount of light that enters the water.

Soluble solids are generally not inspected even though they are significant in effluents with a low degree of contamination. They depend not only on the degree of contamination but also on the quality of the supply water used for the treatment. In one analysis of fish filleting wastewater, it was found that 65% of the total solids present in the effluent were already in the supply water [3].

Odor

In seafood-processing industries, odor is caused by the decomposition of the organic matter, which emits volatile amines, diamines, and sometimes ammonia. In wastewater that has become septic, the characteristic odor of hydrogen sulfide may also develop. Odor is a very important issue in relation to public perception and acceptance of any wastewater treatment plant. Although relatively harmless, it may affect general public life by inducing stress and sickness.

Temperature

To avoid affecting the quality of aquatic life, the temperature of the receiving water body must be controlled. The ambient temperature of the receiving water body must not be increased by more than 2 or 3°C, or else it may reduce the dissolved oxygen level. Except for wastewaters from cooking and sterilization processes in canning factories, fisheries do not discharge wastewaters above ambient temperatures. Therefore, wastewaters from canning operations should be cooled if the receiving water body is not large enough to restrict the change in temperature to 3°C [4].

14.2.2 Organic Content

The major types of wastes found in seafood-processing wastewaters are blood, offal products, viscera, fins, fish heads, shells, skins, and meat "fines." These wastes contribute significantly to the suspended solids concentration of the waste stream. However, most of the solids can be removed from the wastewater and collected for animal food applications. A summary of the raw wastewater characteristics for the canned and preserved seafood processing industry is presented in Table 1.

Wastewaters from the production of fish meal, solubles, and oil from herring, menhaden, and alewives can be divided into two categories: high-volume, low-strength wastes and low-volume, high-strength wastes [5].

High-volume, low-strength wastes consist of the water used for unloading, fluming, transporting, and handling the fish plus the washdown water. In one study, the fluming flow was estimated to be 834 L per tonne of fish with a suspended solids loading of 5000 mg/L. The solids consisted of blood, flesh, oil, and fat [2]. The above figures vary widely. Other estimates listed herring pump water flows of 16 L/s with total solids concentrations of 30,000 mg/L and oil concentrations of 4000 mg/L. The boat's bilge water was estimated to be 1669 L per ton of fish with a suspended solids level of 10,000 mg/L [2].

Stickwaters comprise the strongest wastewater flows. The average BOD_5 value for stickwater has been listed as ranging from 56,000 to 112,000 mg/L, with average solids concentrations, mainly proteinaceous, ranging up to 6%. The fish-processing industry has found the recovery of fish solubles from stickwater to be at least marginally profitable. In most instances, stickwater is now evaporated to produce condensed fish solubles. Volumes have been estimated to be about 500 L per ton of fish processed [2].

The degree of pollution of a wastewater depends on several parameters. The most important factors are the types of operation being carried out and the type of seafood being processed. Carawan [2] reported on an EPA survey with BOD_5, COD, TSS, and fat, oil and grease (FOG) parameters. Bottom fish was found to have a BOD_5 of 200–1000 mg/L, COD of 400–2000 mg/L, TSS of 100–800 mg/L, and FOG of 40–300 mg/L. Fish meal plants were reported to have a BOD_5 of 100–24,000 mg/L, COD of 150–42,000 mg/L, TSS of 70–20,000 mg/L, and FOG of 20–5000 mg/L. The higher numbers were representative of bailwater only. Tuna plants were reported to have a BOD_5 of 700 mg/L, COD of 1600 mg/L,

Table 1 Raw Wastewater Characteristics of the Canned and Preserved Seafood-Processing Industries

Effluent	Flow (L/day)	BOD_5 (mg/L)	COD (mg/L)	TSS (mg/L)	FOG (mg/L)
Farm-raised catfish	79.5K–170K	340	700	400	200
Conventional blue crab	2650	4400	6300	420	220
Mechanized blue crab	75.7K–276K	600	1000	330	150
West coast shrimp	340K–606K	2000	3300	900	700
Southern nonbreaded shrimp	680K–908K	1000	2300	800	250
Breaded shrimp	568K–757K	720	1200	800	–
Tuna processing	246K–13.6M	700	1600	500	250
Fish meal	348K–378.5K[a]	100–24M[a]	150–42K[a]	70–20K[a]	20K–5K[a]
All salmon	220K–1892.5K	253–2600	300–5500	120–1400	20–550
Bottom and finfish (all)	22.71K–1514K	200–1000	400–2000	100–800	40–300
All herring	110K	1200–6000	3000–10,000	500–5000	600–5000
Hand shucked clams	325.5K–643.5K	800–2500	1000–4000	600–6000	16–50
Mechanical clams	1135.5K–11.4M	500–1200	700–1500	200–400	20–25
All oysters	53K–1211K	250–800	500–2000	200–2000	10–30
All scallops	3.785K–435K	200K–10M	300–11,000	27–4000	15–25
Abalone	37.85K–53K	430–580	800–1000	200–300	22–30

BOD_5, five day biochemical oxygen demand; COD, chemical oxygen demand; TSS, total suspended solids; FOG, fat, oil, and grease.

[a] Higher range is for bailwater only; K = 1000; M = 1,000,000.

Source: Ref. 2.

TSS of 500 mg/L, and FOG of 250 mg/L. Seafood-processing wastewater was noted to sometimes contain high concentrations of chlorides from processing water and brine solutions, and organic nitrogen of up to 300 mg/L from processing water.

Several methods are used to estimate the organic content of the wastewater. The two most common methods are biochemical oxygen demand (BOD) and chemical oxygen demand (COD).

Biochemical Oxygen Demand

Biochemical oxygen demand (BOD) estimates the degree of contamination by measuring the oxygen required for oxidation of organic matter by aerobic metabolism of the microbial flora. In seafood-processing wastewaters, this oxygen demand originates mainly from two sources. One is the carbonaceous compounds that are used as substrate by the aerobic microorganisms; the other source is the nitrogen-containing compounds that are normally present in seafood-processing wastewaters, such as proteins, peptides, and volatile amines. Standard BOD tests are conducted at five-day incubation for determination of BOD_5 concentrations.

Wastewaters from seafood-processing operations can be very high in BOD_5. Literature data for seafood processing operations show a BOD_5 production of one to 72.5 kg of BOD_5 per ton of product [1]. White fish filleting processes typically produce 12.5–37.5 kg BOD_5 for every ton of product. The BOD is generated primarily from the butchering process and from general cleaning, while nitrogen originates predominantly from blood in the wastewater stream [1].

Chemical Oxygen Demand

Another alternative for measuring the organic content of wastewater is the chemical oxygen demand (COD), an important pollutant parameter for the seafood industry. This method is more convenient than BOD_5 since it needs only about three hours for determination compared with five days for BOD_5 determination. The COD analysis, by the dichromate method, is more commonly used to control and continuously monitor wastewater treatment systems. Because the number of compounds that can be chemically oxidized is greater than those that can be degraded biologically, the COD of an effluent is usually higher than the BOD_5. Hence, it is common practise to correlate BOD_5 vs. COD and then use the analysis of COD as a rapid means of estimating the BOD_5 of a wastewater.

Depending on the types of seafood processing, the COD of the wastewater can range from 150 to about 42,000 mg/L. One study examined a tuna-canning and byproduct rendering plant for five days and observed that the average daily COD ranged from 1300 to 3250 mg/L [2].

Total Organic Carbon

Another alternative for estimating the organic content is the total organic carbon (TOC) method, which is based on the combustion of organic matter to carbon dioxide and water in a TOC analyzer. After separation of water, the combustion gases are passed through an infrared analyzer and the response is recorded. The TOC analyzer is gaining acceptance in some specific applications as the test can be completed within a few minutes, provided that a correlation with the BOD_5 or COD contents has been established. An added advantage of the TOC test is that the analyzer can be mounted in the plant for online process control. Owing to the relatively high cost of the apparatus, this method is not widely used.

Fats, Oil, and Grease

Fats, oil, and grease (FOG) is another important parameter of seafood-processing wastewater. The presence of FOG in an effluent is mainly due to the processing operations such as canning, and the seafood being processed. The FOG should be removed from wastewater because it usually floats on the water's surface and affects the oxygen transfer to the water; it is also objectionable from an aesthetic point of view. The FOG may also cling to wastewater ducts and reduce their capacity in the long term. The FOG of a seafood-processing wastewater varies from zero to about 17,000 mg/L, depending on the seafood being processed and the operation being carried out.

14.2.3 Nitrogen and Phosphorus

Nitrogen and phosphorus are nutrients that are of environmental concern. They may cause proliferation of algae and affect the aquatic life in a water body if they are present in excess. However, their concentration in the seafood-processing wastewater is minimal in most cases. It is recommended that a ratio of N to P of 5 : 1 be achieved for proper growth of the biomass in the biological treatment [6,7].

Sometime the concentration of nitrogen may also be high in seafood-processing wastewaters. One study shows that high nitrogen levels are likely due to the high protein content (15–20% of wet weight) of fish and marine invertebrates [8]. Phosphorus also partly originates from the seafood, but can also be introduced with processing and cleaning agents.

14.2.4 Sampling

Of equal importance is the problem of obtaining a truly representative sample of the stream effluent. The samples may be required not only for the 24 hour effluent loads, but also to determine the peak load concentrations, the duration of peak loads, and the occurrence of variation throughout the day. The location of sampling is usually made at or near the point of discharge to the receiving water body, but in the analysis prior to the design of a wastewater treatment, facility samples will be needed from each operation in the seafood-processing facility. In addition, samples should be taken more frequently when there is a large variation in flow rate, although wide variations may also occur at constant flow rate.

The particular sampling procedure may vary, depending on the parameter being monitored. Samples should be analyzed as soon as possible after sampling because preservatives often interfere with the test. In seafood-processing wastewaters, there is no single method of sample preservation that yields satisfactory results for all cases, and all of them may be inadequate with effluents containing suspended matter. Because samples contain an amount of settleable solids in almost all cases, care should be taken in blending the samples just prior to analysis. A case in which the use of preservatives is not recommended is that of BOD_5 storage at low temperatures (4°C), which may be used with caution for very short periods, and chilled samples should be warmed to 20°C before analysis. For COD determination, the samples should be collected in clean glass bottles, and can be preserved by acidification to a pH of 2 with concentrated sulfuric acid. Similar preservation can also be done for organic nitrogen determination. For FOG determination, a separate sample should be collected in a wide-mouth glass bottle that is well rinsed to remove any trace of detergent. For solids determination, an inspection should be done to ensure that no suspended matter adheres to the walls and that the solids are refrigerated at 4°C to prevent decomposition of biological solids. For the analysis of phosphorus, samples should be preserved by adding 40 mg/L of mercuric chloride and storing in well-rinsed glass bottles at −10°C [4].

14.3 PRIMARY TREATMENT

In the treatment of seafood-processing wastewater, one should be cognizant of the important constituents in the waste stream. This wastewater contains considerable amounts of insoluble suspended matter, which can be removed from the waste stream by chemical and physical means. For optimum waste removal, primary treatment is recommended prior to a biological treatment process or land application. A major consideration in the design of a treatment system is that the solids should be removed as quickly as possible. It has been found that the longer the detention time between waste generation and solids removal, the greater the soluble BOD_5 and COD with corresponding reduction in byproduct recovery. For seafood-processing wastewater, the primary treatment processes are screening, sedimentation, flow equalization, and dissolved air flotation. These unit operations will generally remove up to 85% of the total suspended solids, and 65% of the BOD_5 and COD present in the wastewater.

14.3.1 Screening

The removal of relatively large solids (0.7 mm or larger) can be achieved by screening. This is one of the most popular treatment systems used by food-processing plants because it can reduce the amount of solids being discharged quickly. Usually, the simplest configuration is that of flow-through static screens, which have openings of about 1 mm. Sometimes a scrapping mechanism may be required to minimize the clogging problem in this process.

Generally, tangential screening and rotary drum screening are the two types of screening methods used for seafood-processing wastewaters. Tangential screens are static but less prone to clogging due to their flow characteristics (Fig. 1), because the wastewater flow tends to avoid clogging. The solids removal rates may vary from 40 to 75% [4]. Rotary drum screens are mechanically more complex. They consist of a drum that rotates along its axis, and the effluent enters through an opening at one end. Screened wastewater flows outside the drum and the retained solids are washed out from the screen into a collector in the upper part of the drum by a spray of the wastewater.

Fish solids dissolve in water with time; therefore immediate screening of the waste streams is highly recommended. Likewise, high-intensity agitation of waste streams should be minimized before screening or even settling, because they may cause breakdown of solids rendering them more difficult to separate. In small-scale fish-processing plants, screening is often used with simple settling tanks.

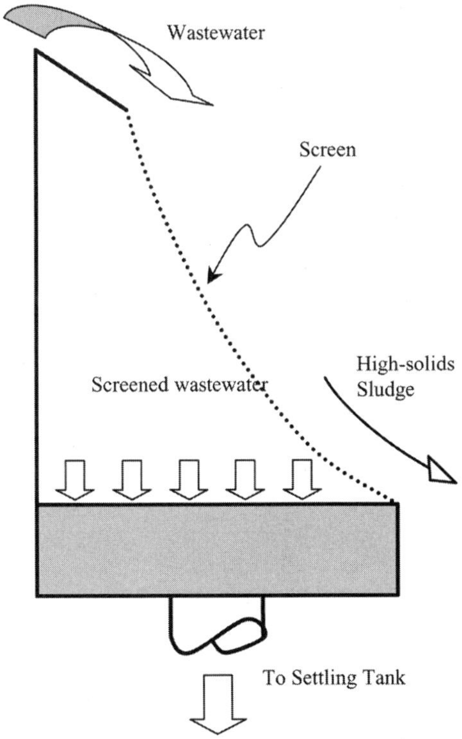

Figure 1 Diagram of an inclined or tangential screen.

14.3.2 Sedimentation

Sedimentation separates solids from water using gravity settling of the heavier solid particles [9]. In the simplest form of sedimentation, particles that are heavier than water settle to the bottom of a tank or basin. Sedimentation basins are used extensively in the wastewater treatment industry and are commonly found in many flow-through aquatic animal production facilities. This operation is conducted not only as part of the primary treatment, but also in the secondary treatment for separation of solids generated in biological treatments such as activated sludge or trickling filters. Depending on the properties of solids present in the wastewater, sedimentation can proceed as discrete settling, flocculent settling, or zone settling. Each case has different characteristics, which will be outlined.

Discrete settling occurs when the wastewater is relatively dilute and the particles do not interact. A schematic diagram of discrete settling is shown in Fig. 2.

Calculations can be made on the settling velocity of individual particles. In a sedimentation tank, settling occurs when the horizontal velocity of a particle entering the basin is less than the vertical velocity in the tank. The length of the sedimentation basin and the detention time can be calculated so that particles with a particular settling velocity (V_c) will settle to the bottom of the basin [9]. The relationship of the settling velocity to the detention time and basin depth is:

$$V_c = \frac{\text{depth}}{\text{detention time}} \tag{1}$$

For flocculent suspension, the formation of larger particles due to coalescence depends on several factors, such as the nature of the particles and the rate of coalescence. A theoretical analysis is not feasible due to the interaction of particles, which depends, among other factors, on the overflow rate, the concentration of particles, and the depth of the tank.

Zone settling occurs when the particles do not settle independently. In this case, an effluent is initially uniform in solids concentration and settles in zones. The clarified effluent and compaction zones will increase in size while the other intermediate zones will eventually disappear.

The primary advantages of using sedimentation basins to remove suspended solids from effluents from seafood-processing plants are: the relative low cost of designing, constructing, and operating sedimentation basins; the low technology requirements for the operators; and the demonstrated effectiveness of their use in treating similar effluents. Therefore proper design,

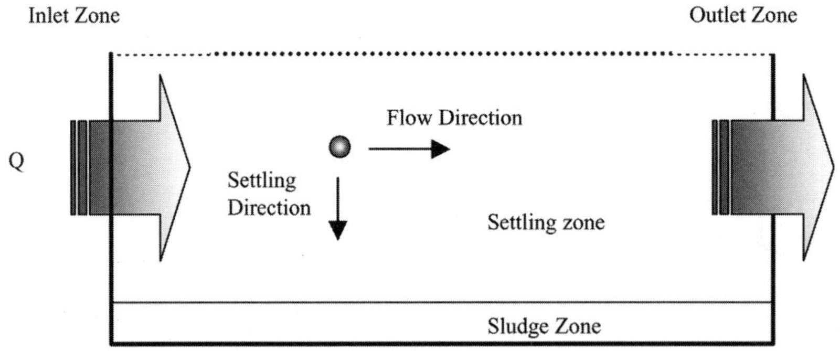

Figure 2 Schematics of discrete settling.

construction, and operation of the sedimentation basin are essential for the efficient removal of solids. Solids must be removed at proper intervals to ensure the designed removal efficiencies of the sedimentation basin.

Rectangular settling tanks (Fig. 3) are generally used when several tanks are required and there is space constraint, because they occupy less space than several circular tanks. Usually there is a series of chain-driven scrapers used for removal of solids. The sludge is collected in a hopper at the end of the tank, where it may be removed by screw conveyors or pumped out.

Circular tanks are reported to be more effective than rectangular ones. The effluent in a circular tank circulates radially, with the water introduced at the periphery or from the center. The configuration is shown in Fig. 4. Solids are generally removed from near the center, and the sludge is forced to the outlet by two or four arms provided with scrapers, which span the radius of the tank. For both types of flows, a means of distributing the flow in all directions is provided. An even distribution of inlet and outlet flows is important to avoid short-circuiting in the tank, which would reduce the separation efficiency.

Generally, selection of a circular tank size is based on the surface-loading rate of the tank. It is defined as the average daily overflow divided by the surface area of the tank and is expressed as volume of wastewater per unit time and unit area of settler (m^3/m^2 day), as shown in Eq. (2). This loading rate depends on the characteristics of the effluent and the solids content. The retention time in the settlers is generally one to two hours, but the capacity of the tanks must be determined by taking into account the peak flow rates so that acceptable separation is obtained in these cases. Formation of scum is almost unavoidable in seafood-processing wastes, so some settling tanks are provided with a mechanism for scum removal.

Selection of the surface loading rate depends on the type of suspensions to be removed. The design overflow rates must be low enough to ensure satisfactory performance at peak rates of flow, which may vary from two to three times the average flow.

$$V_o = \frac{Q}{A} \tag{2}$$

where V_o = overflow rate (surface-loading rate) (m^3/m^2 day), Q = average daily flow (m^3/day), and A = total surface area of basin (m^2).

The area A is calculated by using inside tank dimensions, disregarding the central stilling well or inboard well troughs. The quantity of overflow from a primary clarifier Q is equal to the wastewater influent, and since the volume of the tank is established, the detention period in the tank is governed by water depth. The side water depth of the tank is

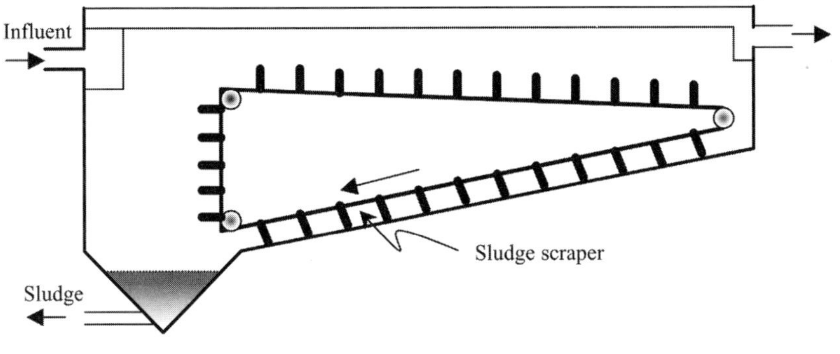

Figure 3 Diagram of a rectangular clarifier.

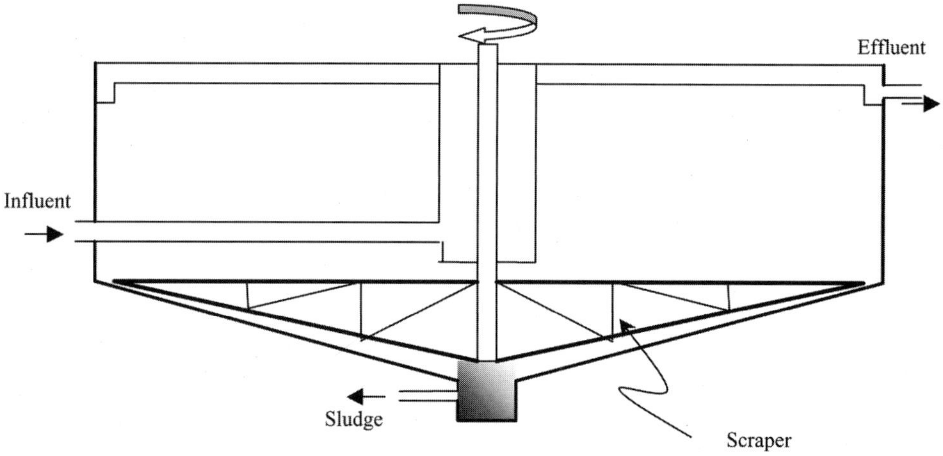

Figure 4 Diagram of radial flow sedimentation tank.

generally between 2.5 and 5 m. Detention time is computed by dividing the tank volume by influent flow uniform rate equivalent to the design average daily flow. A detention time of between 1.5 and 2.5 hours is normally provided based on the average rate of wastewater flow. Effluent weir loading is equal to the average daily quantity of overflow divided by the total weir length expressed in m^3/m day.

$$T = \frac{24V}{Q} \tag{3}$$

where $T =$ detention time (hour), $Q =$ average daily flow (m^3/day), and $V =$ basin volume (m^3).

Temperature effects are normally not an important consideration in the design. However, in cold climates, the increase in water viscosity at lower temperatures retards particles settling and reduces clarifier performance.

In cases of small or elementary settling basins, the sludge can be removed using an arrangement of perforated piping placed at the bottom of the settling tank [10]. The pipes must be regularly spaced, as shown in Figure 5, to be of a diameter wide enough to be cleaned easily in case of clogging. The flow velocities should also be high enough to prevent sedimentation. Flow in individual pipes may be regulated by valves. This configuration is best used after screening and is also found in biological treatment tanks for sludge removal.

Inclined tube separators are an alternative to the above configurations for settling [11]. These separators consist of tilted tubes, which are usually inclined at 45–60°. When a settling particle reaches the wall of the tube or the lower plate, it coalesces with another particle and forms a larger mass, which causes a higher settling rate. A typical configuration for inclined media separators is shown in Figure 6.

14.3.3 Flow Equalization

A flow equalization step follows the screening and sedimentation processes and precedes the dissolved air flotation (DAF) unit. Flow equalization is important in reducing hydraulic loading in the waste stream. Equalization facilities consist of a holding tank and pumping equipment designed to reduce the fluctuations of the waste streams. The equalizing tank will store excessive

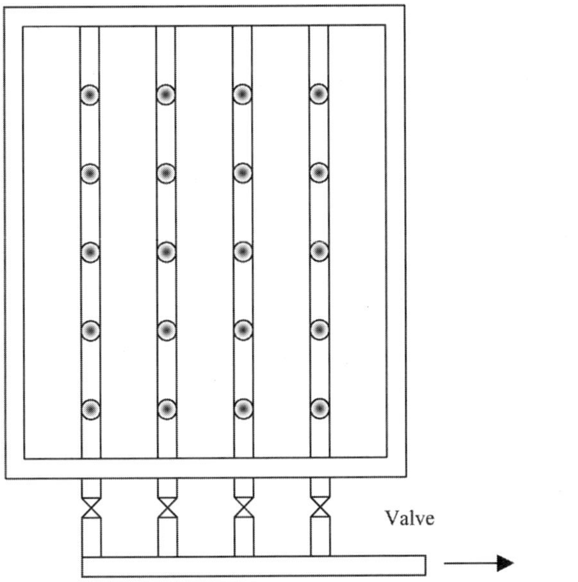

Figure 5 Pipe arrangement for sludge removal from settling tanks.

hydraulic flow surges and stabilize the flow rate to a uniform discharge rate over a 24 hour day. The tank is characterized by a varying flow into the tank and a constant flow out.

14.3.4 Separation of Oil and Grease

Seafood-processing wastewaters contain variable amounts of oil and grease, which depend on the process used, the species processed, and the operational procedure. Gravitational separation may be used to remove oil and grease, provided that the oil particles are large enough to float towards the surface and are not emulsified; otherwise, the emulsion must be first broken by pH adjustment. Heat may also be used for breaking the emulsion but it may not be economical unless there is excess steam available. The configurations of gravity separators of oil–water are similar to the inclined tubes separators discussed in the previous section.

14.3.5 Flotation

Flotation is one of the most effective removal systems for suspensions that contain oil and grease. The most common procedure is that of dissolved air flotation (DAF), which is a waste-treatment process in which oil, grease, and other suspended matter are removed from a waste stream. This treatment process has been in use for many years and has been most successful in removing oil from waste streams. Essentially, DAF is a process that uses minute air bubbles to remove the suspended matter from the wastewater stream. The air bubbles attach themselves to a discrete particle, thus effecting a reduction in the specific gravity of the aggregate particle to less than that of water. Reduction of the specific gravity for the aggregate particle causes separation from the carrying liquid in an upward direction. Attachment of the air bubble to the particle induces a vertical rate of rise. The mechanism of operation involves a clarification vessel where

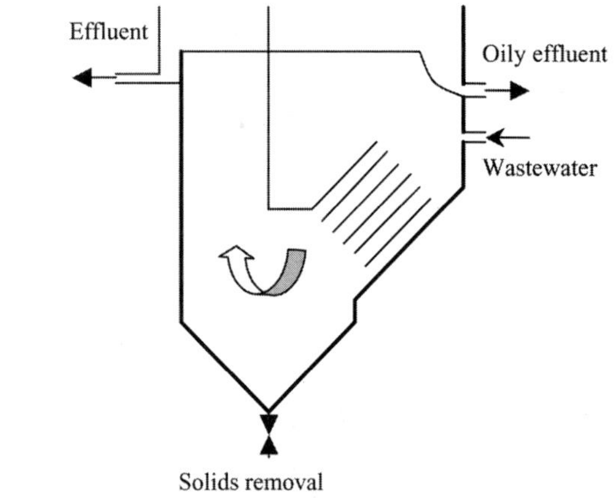

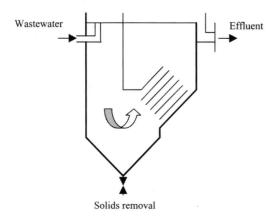

Figure 6 Typical configurations for inclined media separators.

the particles are floated to the surface and removed by a skimming device to a collection trough for removal from the system. The raw wastewater is brought in contact with a recycled, clarified effluent that has been pressurized through air injection in a pressure tank. The combined flow stream enters the clarification vessel and the release of pressure causes tiny air bubbles to form and ascend to the surface of the water, carrying the suspended particles with their vertical rise. A schematic diagram of the DAF system is shown in Figure 7.

Key factors in the successful operation of DAF units are the maintenance of proper pH (usually between 4.5 and 6, with 5 being most common to minimize protein solubility and break up emulsions), proper flow rates, and the continuous presence of trained operators.

In one case, oil removal was reported to be 90% [12]. In tuna processing wastewaters, the DAF removed 80% of oil and grease and 74.8% of suspended solids in one case, and a second case showed removal efficiencies of 64.3% for oil and grease and 48.2% of suspended solids. The main difference between these last two effluents was the usually lower solids content of the second [13]. However, although DAF systems are considered very effective, they are probably not suitable for small-scale seafood-processing facilities due to the relatively high cost. It was reported that the estimated operating cost for a DAF system was about US$250,000 in 1977 [14].

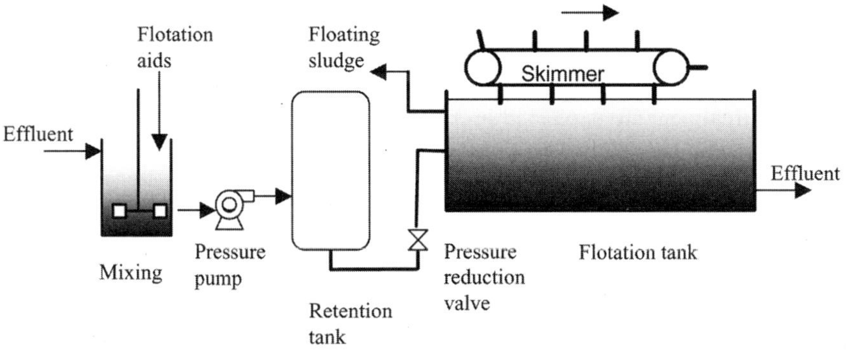

Figure 7 Diagram of a dissolved air flotation (DAF) system.

14.4 BIOLOGICAL TREATMENT

To complete the treatment of the seafood-processing wastewaters, the waste stream must be further processed by biological treatment. Biological treatment involves the use of microorganisms to remove dissolved nutrients from a discharge [15]. Organic and nitrogenous compounds in the discharge can serve as nutrients for rapid microbial growth under aerobic, anaerobic, or facultative conditions. The three conditions differ in the way they use oxygen. Aerobic microorganisms require oxygen for their metabolism, whereas anaerobic microorganisms grow in absence of oxygen; the facultative microorganism can proliferate either in absence or presence of oxygen although using different metabolic processes. Most of the microorganisms present in wastewater treatment systems use the organic content of the wastewater as an energy source to grow, and are thus classified as heterotrophes from a nutritional point of view. The population active in a biological wastewater treatment is mixed, complex, and interrelated. In a single aerobic system, members of the genera *Pseudomonas*, *Nocardia*, *Flavobacterium*, *Achromobacter*, and *Zooglea* may be present, together with filamentous organisms. In a well-functioning system, protozoas and rotifers are usually present and are useful in consuming dispersed bacteria or nonsettling particles.

Biological treatment systems can convert approximately one-third of the colloidal and dissolved organic matter into stable endproducts and convert the remaining two-thirds into microbial cells that can be removed through gravity separation. The organic load present is incorporated in part as biomass by the microbial populations, and almost all the rest is liberated gas. Carbon dioxide (CO_2) is produced in aerobic treatments, whereas anaerobic treatments produce both carbon dioxide and methane (CH_4). In seafood-processing wastewaters, the nonbiodegradable portion is very low.

The biological treatment processes used for wastewater treatment are broadly classified as aerobic and anaerobic treatments. Aerobic and facultative microorganisms predominate in aerobic treatments, while only anaerobic microorganisms are used for the anaerobic treatments.

If microorganisms are suspended in the wastewater during biological operation, this is known as a "suspended growth process," whereas the microorganisms that are attached to a surface over which they grow are said to undergo an "attached growth process."

Biological treatment systems are most effective when operating continuously 24 hours a day and 365 days a year. Systems that are not operated continuously have reduced efficiency because of changes in nutrient loads to the microbial biomass. Biological treatment systems also generate a consolidated waste stream consisting of excess microbial biomass, which must be properly disposed. Operation and maintenance costs vary with the process used.

The principles and main characteristics of the most common processes used in seafood-processing wastewater treatment are explained in this section.

14.4.1 Aerobic Process

In seafood processing wastewaters, the need for adding nutrients (the most common being nitrogen and phosphorus) seldom occurs, but an adequate provision of oxygen is essential for successful operation. The most common aerobic processes are activated sludge systems, lagoons, trickling filters and rotating disc contactors. The reactions occurring during the aerobic process can be summarized as follows:

$$\text{Organic} + O_2 \longrightarrow \text{cells} + CO_2 + H_2O$$

Apart from economic considerations, several factors influence the choice of a particular aerobic treatment system. The major considerations are: the area availability; the ability to operate intermittently is critical for several seafood industries that do not operate in a continuous fashion or work only seasonally; the skill needed for operation of a particular treatment cannot be neglected; and finally the operating and capital costs are also sometimes decisive. Table 2 summarizes these factors when applied to aerobic treatment processes.

The considerations for rotating biological contactors (RBC) systems are similar to those of trickling filters.

Activated Sludge Systems

In an activated sludge treatment system, an acclimatized, mixed, biological growth of microorganisms (sludge) interacts with organic materials in the wastewater in the presence of excess dissolved oxygen and nutrients (nitrogen and phosphorus). The microorganisms convert the soluble organic compounds to carbon dioxide and cellular materials. Oxygen is obtained from applied air, which also maintains adequate mixing. The effluent is settled to separate

Table 2 Factors Affecting the Choice of Aerobic Processes

(A) Operating characteristics			
System	Resistance to shock loads of organics or toxics	Sensitivity to intermittent operations	Degree of skill needed
Lagoons	Maximum	Minimum	Minimum
Trickling filters	Moderate	Moderate	Moderate
Activated	Minimum	Maximum	Maximum
(B) Cost considerations			
System	Land needed	Initial costs	Operating costs
Lagoons	Maximum	Minimum	Minimum
Trickling filters	Moderate	Moderate	Moderate
Activated	Minimum	Maximum	Maximum

Source: Ref. 10.

biological solids and a portion of the sludge is recycled; the excess is wasted for further treatment such as dewatering. These systems originated in England in the early 1900s. The layout of a typical activated sludge system is shown in Figure 8.

Most of the activated sludge systems utilized in the seafood-processing industry are of the extended aeration types: that is, they combine long aeration times with low applied organic loadings. The detention times are 1 to 2 days. The suspended solids concentrations are maintained at moderate levels to facilitate treatment of the low-strength wastes, which usually have a BOD_5 of less than 800 mg/L.

It is usually necessary to provide primary treatment and flow equalization prior to the activated sludge process, to ensure optimum operation. A BOD_5 and suspended solids removals in the range of 95–98% can be achieved. However, pilot- or laboratory-scale studies are required to determine organic loadings, oxygen requirements, sludge yields, sludge settling rates, and so on, for these high-strength wastes.

In contrast to other food-processing wastewaters, seafood wastes appear to require higher oxygen availability to stabilize them. Whereas dairy, fruit, and vegetable wastes require approximately 1.3 kg of oxygen per kg of BOD_5, seafood wastes may demand as much as 3 kg of oxygen per kg of BOD_5 applied to the extended aeration system [2].

The most common types of activated sludge process are the conventional and the continuous flow stiffed tanks, as shown in Figure 8, in which the contents are fully mixed. In the conventional process, the wastewater is circulated along the aeration tank, with the flow being arranged by baffles in plug flow mode. This arrangement demands a maximum amount of oxygen and organic load concentration at the inlet. A typical conventional activated sludge process is shown in Figure 9. Unlike the conventional activated sludge process, the inflow streams in the completely mixed process are usually introduced at several points to facilitate the homogeneity of the mixing such that the properties are constant throughout the reactor if the mixing is completed. This configuration is inherently more stable in terms of perturbations because mixing causes dilution of the incoming stream into the tank. In seafood-processing wastewaters the perturbations that may appear are peaks of concentration of organic load or flow peaks. Flow peaks can be damped in the primary treatment tanks. The conventional configurations would require less reactor volume if smooth plug flow could be assured, which usually does not occur.

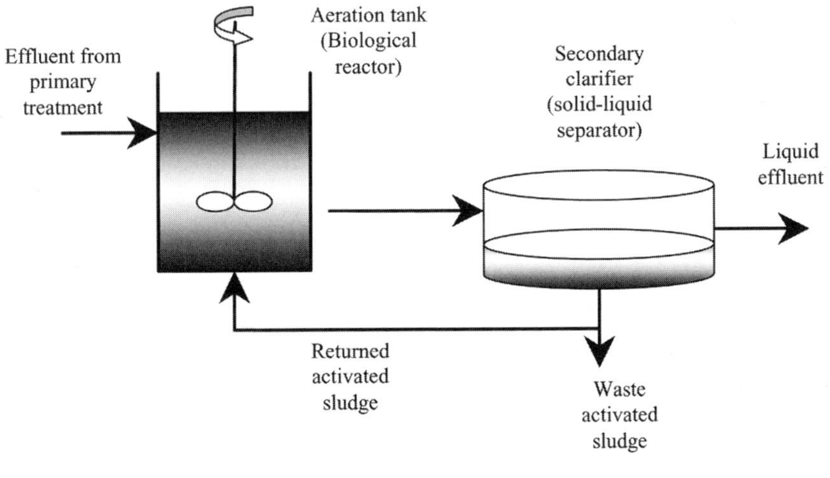

Figure 8 Diagram of a simple activated sludge system.

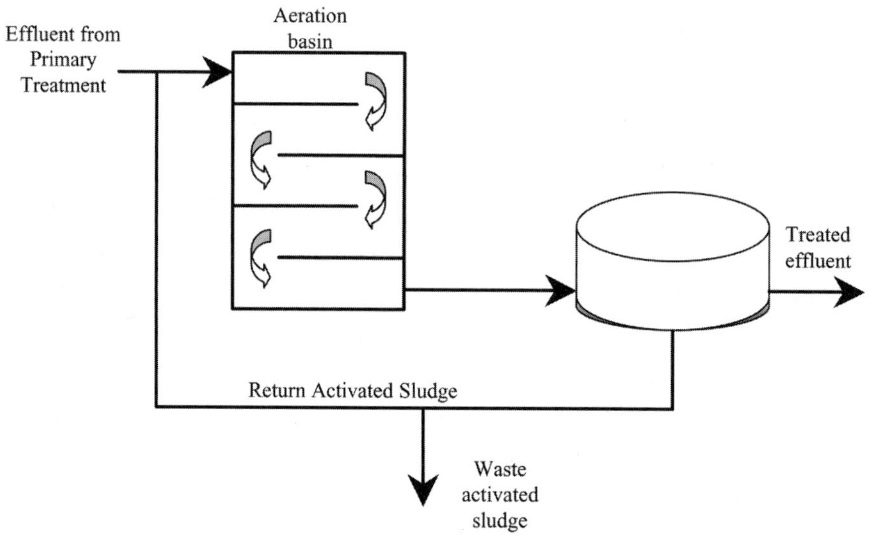

Figure 9 Diagram of a conventional activated sludge process.

In activated sludge systems, the cells are separated from the liquid and partially returned to the system; the relatively high concentration of cells then degrades the organic load in a relatively short time. Therefore there are two different resident times that characterize the systems: one is the hydraulic residence time (θ_H) given by the ratio of reactor volume (V_R) to flow of wastewater (Q_R):

$$\theta_H = \frac{V_R}{Q_R} \tag{4}$$

The other is the cell residence time (θ_C), which is given by the ratio of cells present in the reactor to the mass of cells wasted per day. Typical θ_H values are in the order of 3 to 6 hours, while θ_C fluctuates between 3 and 15 days.

To ensure the optimum operation of the activated sludge process, it is generally necessary to provide primary treatment and flow equalization prior to the activated sludge process. Pilot- or laboratory-scale studies are required to determine organic loadings, oxygen requirements, sludge yields, and sludge settling rates for these high-strength wastes. There are several pieces of information required to design an activated sludge system through the bench-scale or pilot-scale studies:

- BOD_5 removal rate;
- oxygen requirements for the degradation of organic material and the degradation of dead cellular material (endogenous respiration);
- sludge yield, determined from the conservation of soluble organics to cellular material and the influx of inorganic solids in the raw waste;
- solid/liquid separation rate: the final clarifier would be designed to achieve rapid sedimentation of solids, which could be recycled or further treated. A maximum surface settling rate of 16.5 m^3/m^2 day has been suggested for seafood-processing wastes [2].

Typically, 85–95% of organic load removals can be achieved in activated sludge systems. Although used by some large seafood-processing industries that operate on a year-round basis, activated sludge may not be economically justified for small, seasonal seafood processors because of the requirement of a fairly constant supply of wastewater to maintain the microorganisms.

Aerated Lagoons

Aerated lagoons are used where sufficient land is not available for seasonal retention or land application and economics do not justify an activated sludge system. Efficient biological treatment can be achieved by the use of the aerated lagoon system. It was reported to have removal efficiency of 90–95% of BOD_5 in seafood-processing wastewater treatment [2]. The major difference with respect to activated sludge systems is that the aerated lagoons are basins, normally excavated in earth and operated without solids recycling into the system. The ponds are between 2.4 and 4.6 m deep, with 2 to 10 days retention and achieve 55–90% reduction in BOD_5. Two types of aerated lagoons are commonly used in seafood-processing wastewater treatment: completely mixed lagoons and facultative lagoons. In the completely mixed lagoon, the concentrations of solids and dissolved oxygen are uniformly maintained and neither the incoming solids nor the biomass of microorganisms settle, whereas in the facultative lagoons, the power input is reduced, causing accumulation of solids in the bottom that undergo anaerobic decomposition, while the upper portions are maintained in an aerobic state (Fig. 10).

The major operational difference between these lagoons is the power input, which is in the order of 2.5–6 W/m^3 for aerobic lagoons while the requirement for facultative lagoons is of the order 0.8–1 W/m^3. Reduction in biological activity can occur when the lagoons are exposed to low temperatures and eventually ice formation. This problem can be partially alleviated by increasing the depth of the basin.

If excavated basins are used for settling, care should be taken to provide a residence time long enough for the solids to settle, and provision should also be made for the accumulation of sludge. There is a very high possibility of offensive odor development due to the decomposition of the settled sludge, and algae might develop in the upper layers causing an increased content of suspended solids in the effluent. Odors can be minimized by using minimum depths of up to 2 m, whereas algae production can be reduced with a hydraulic retention time of fewer than 2 days.

Solids will also accumulate all along the aeration basins in the facultative lagoons and even at corners, or between aeration units in the completely mixed lagoon. These accumulated

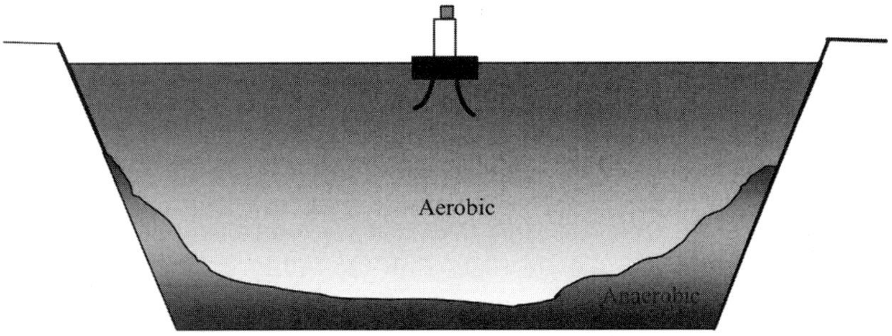

Figure 10 Diagram of facultative aerated lagoon.

solids will, on the whole, decompose at the bottom, but since there is always a nonbiodegradable fraction, a permanent deposit will build up. Therefore, periodic removal of these accumulated solids is necessary.

Stabilization/Polishing Ponds

A stabilization/polishing ponds system is commonly used to improve the effluent treated in the aerated lagoon. This system depends on the action of aerobic bacteria on the soluble organics contained in the waste stream. The organic carbon is converted to carbon dioxide and bacterial cells. Algal growth is stimulated by incident sunlight that penetrates to a depth of 1–1.5 m. Photosynthesis produces excess oxygen, which is available for aerobic bacteria; additional oxygen is provided by mass transfer at the air–water interface.

Aerobic stabilization ponds are 0.18–0.9 m deep to optimize algal activity and are usually saturated with dissolved oxygen throughout the depth during daylight hours. The ponds are designed to provide a detention time of 2–20 days, with surface loadings of 5.5–22 g BOD_5/day/m^2 [2]. To eliminate the possibility of shortcircuiting and to permit sedimentation of dead algal and bacterial cells, the ponds usually consist of multiple cell units operated in series. The ponds are constructed with inlet and outlet structures located in positions to minimize shortcircuiting due to wind-induced currents; the dimensions and geometry are designed to maximize mixing. These systems have been reported achieving 80–95% removal of BOD_5 and approximately 80% removal of suspended solids, with most of the effluent solids discharged as algal cells [2].

During winter, the degree of treatment decreases markedly as the temperature decreases and ice cover eliminates algal growth. In regions where ice cover occurs, the lagoons may be equipped with variable depth overflow structures so that processing wastewater flows can be stored during the winter. An alternative method is to provide long retention storage ponds; the wastes can then be treated aerobically during the summer prior to discharge.

Aerobic stabilization ponds are utilized where land is readily available. In regions where soils are permeable, it is often necessary to use plastic, asphaltic, or clay liners to prevent contamination of adjacent groundwater.

Trickling Filters

The trickling filter is one of the most common attached cell (biofilm) processes. Unlike the activated sludge and aerated lagoons processes, which have biomass in suspension, most of the biomass in trickling filters are attached to certain support media over which they grow (Fig. 11).

Typical microorganisms present in trickling filters are *Zoogloea*, *Pseudomonas*, *Alcaligenes*, *Flavobacterium*, *Streptomyces*, *Nocardia*, fungi, and protozoa. The crux of the process is that the organic contents of the effluents are degraded by these attached growth populations, which absorb the organic contents from the surrounding water film. Oxygen from the air diffuses through this liquid film and enters the biomass. As the organic matter grows, the biomass layer thickens and some of its inner portions become deprived of oxygen or nutrients and separate from the support media, over which a new layer will start to grow. The separation of biomass occurs in relatively large flocs that settle relatively quickly in the supporting material. Media that can be used are rocks (low-rate filter) or plastic structures (high-rate filter). Denitrification can occur in low-rate filters, while nitrification occurs under high-rate filtration conditions; therefore effluent recycle may be necessary in high-rate filters.

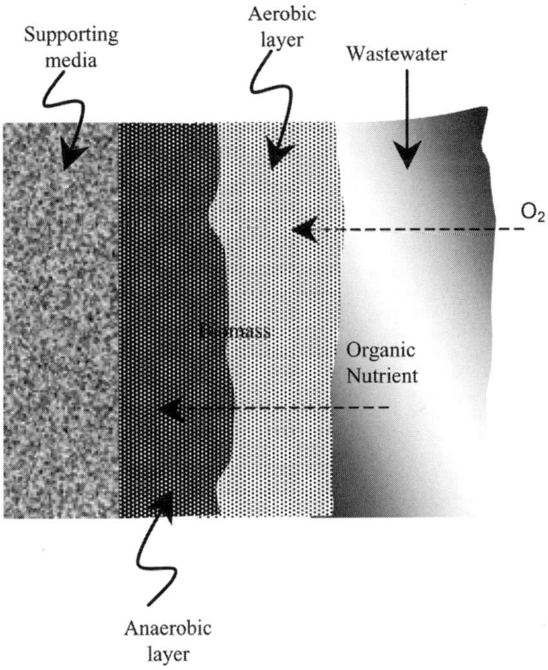

Figure 11 Cross-section of an attached growth biomass film.

In order to achieve optimum operation, several design criteria for trickling filters must be followed:

- roughing filters may be loaded at a rate of 4.8 kg BOD_5/day/m^3 filter media and achieve BOD_5 reductions of 40–50%;
- high-rate filters achieve BOD_5 reductions of 40–70% at organic loadings of 0.4–4.8 kg/BOD_5/day/m^3; and
- standard rate filters are loaded at 0.08–0.4 kg/BOD_5/day/m^3 and achieve BOD_5 removals greater than 70% [2].

The trickling filter consists of a circular tank filled with the packing media in depths varying from 1 to 2.5 m, or 10 m if synthetic packing is used. The bottom of the tank must be constructed rigidly enough to support the packing and designed to collect the treated wastewater, which is either sprayed by regularly spaced nozzles or by rotating distribution arms. The liquid percolates through the packing and the organic load is absorbed and degraded by the biomass while the liquid drains to the bottom to be collected.

With regard to the packing over which the biomass grows, the void fraction and the specific surface area are important features; the first is necessary to ensure a good circulation of air and the second is to accommodate as much biomass as possible to degrade the organic load of the wastewaters. Although more costly initially, synthetic packings have a larger void space, larger specific area, and are lighter than other packing media. Usually, the air circulates naturally, but forced ventilation is used with some high-strength wastewaters. The latter may be used with or without recirculation of the liquid after the settling tank. The need for recirculation is dictated by the strength of the wastewater and the rate of oxygen transfer to the biomass. Typically, recirculation is used when the BOD_5 of the seafood-processing wastewater to be treated exceeds 500 mg/L. The BOD_5 removal efficiency varies with the organic load imposed but usually fluctuates between 45 and 70% for a single-stage filter. Removal efficiencies of up to 90% can be achieved in two stages [4]. A typical unit of a trickling filter is shown in Figure 12.

Rotating Biological Contactors (RBC)

Increasingly stringent requirements for the removal of organic and inorganic substances from wastewater have necessitated the development of innovative, cost-effective wastewater treatment alternatives in recent years. The aerobic rotating biological contactor (RBC) is one of the biological processes for the treatment of organic wastewater. It is another type of attached growth process that combines advantages of biological fixed-film (short hydraulic retention time, high biomass concentration, low energy cost, easy operation, and insensitivity to toxic substance shock loads), and partial stir. Therefore the aerobic RBC reactor is widely employed to treat both domestic and industrial wastewater [16–18]. A schematic diagram of the rotating biological contactor (RBC) unit is shown in Fig. 13; it consists of closely spaced discs mounted on a common horizontal shaft, partially submerged in a semicircular tank receiving wastewater. When water containing organic waste and nutrients flows though the reactor, microorganisms consume the substrata and grow attached to the discs' surfaces to about 1–4 mm in thickness; excess is torn off the discs by shearing forces and is separated from the liquid in the secondary settling tank. A small portion of the biomass remains suspended in the liquid within the basin and is also responsible in minor part for the organic load removal.

Aeration of the culture is accomplished by two mechanisms. First, when a point on the discs rises above the liquid surface, a thin film of liquid remains attached to it and oxygen is transferred to the film as it passes through air; some amount of air is entrained by the bulk of liquid due to turbulence caused by rotation of discs. Rotation speeds of more than 3 rpm are seldom used because this increases electric power consumption while not sufficiently increasing oxygen transfer. The ratio of surface area of discs to liquid volume is typically 5 L/m^2. For high-strength wastewaters, more than one unit in series (staging) is used.

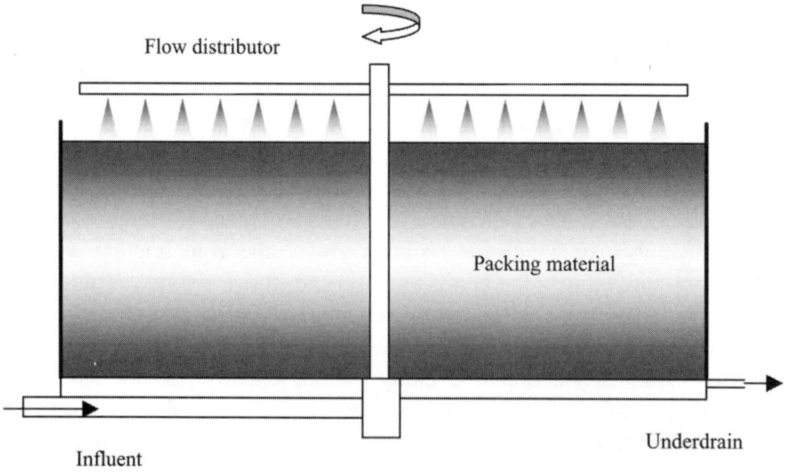

Figure 12 Sketch of a trickling filter unit.

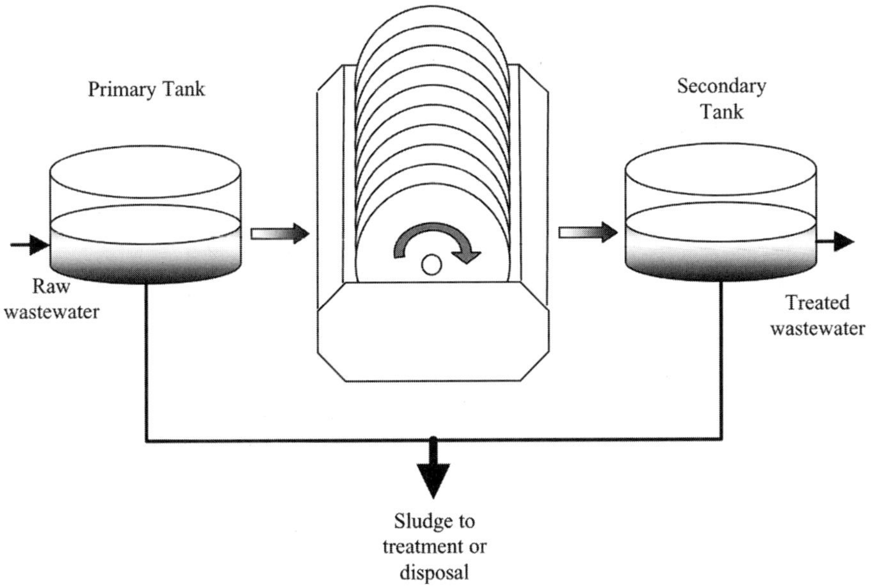

Figure 13 Diagram of a rotating biological contactor (RBC) unit.

14.4.2 Anaerobic Treatment

Anaerobic biological treatment has been applied to high BOD or COD waste solutions in a variety of ways. Treatment proceeds with degradation of the organic matter, in suspension or in a solution of continuous flow of gaseous products, mainly methane and carbon dioxide, which constitute most of the reaction products and biomass. Its efficient performance makes it a valuable mechanism for achieving compliance with regulations for contamination of recreational and seafood-producing wastes. Anaerobic treatment is the result of several reactions: the organic load present in the wastewater is first converted to soluble organic material, which in turn is consumed by acid-producing bacteria to produce volatile fatty acids, plus carbon dioxide and hydrogen. The methane-producing bacteria consume these products to produce methane and carbon dioxide. Typical microorganisms used in this methanogenic process are *Metanobacterium*, *Methanobacillus*, *Metanococcus*, and *Methanosarcina*. These processes are reported to be better applied to high-strength wastewaters, for example, blood water or stickwater. The scheme of reactions during anaerobic treatment is summarized in Fig. 14.

Digestion Systems

Anaerobic digestion facilities have been used for the management of animal slurries for many years, they can treat most easily biodegradable waste products, including everything of organic or vegetable origin. Recent developments in anaerobic digestion technology have allowed the expansion of feedstocks to include municipal solid wastes, biosolids, and organic industrial waste (e.g., seafood-processing wastes). Lawn and garden, or "green" residues, may also be included, but care should be taken to avoid woody materials with high lignin content that requires a much longer decomposition time [19]. The digestion system seems to work best with a feedstock mixture of 15–25% solids. This may necessitate the addition of some liquid,

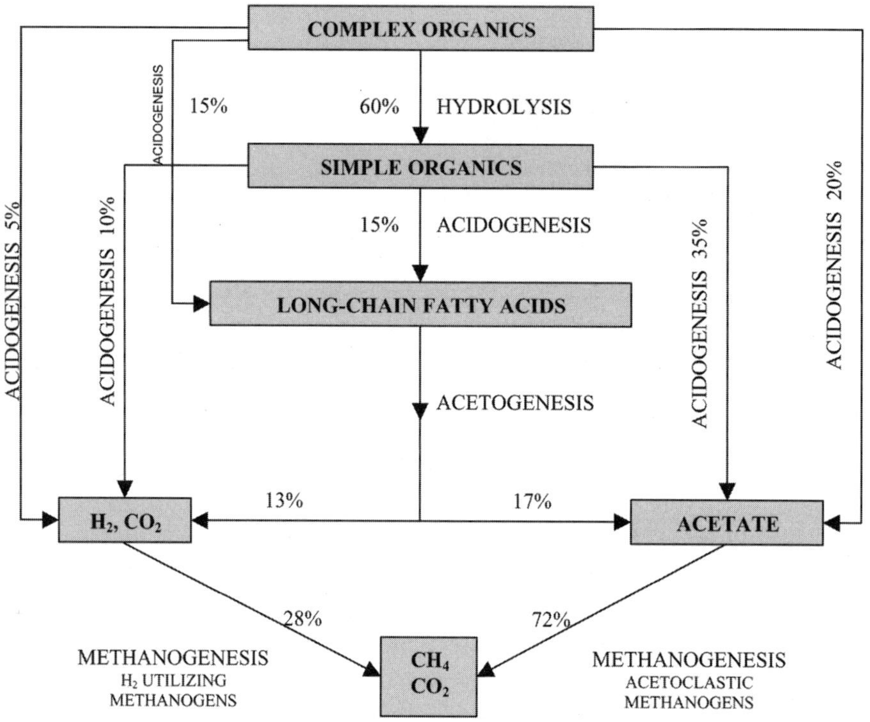

Figure 14 Scheme of reactions produced during anaerobic treatment.

providing an opportunity for the treatment of wastewater with high concentrations of organic contaminants. A typical anaerobic system diagram is shown in Fig. 15.

The flow of anaerobic digestion resembles that of an activated sludge process except that it occurs in the absence of oxygen. Therefore, it is essential to have a good sealing of the digestion tanks since oxygen kills some of the anaerobic bacteria present and presence of air may easily disrupt the process. From the anaerobic digester the effluent proceeds to a degasifier and to a settler from which the wastewater is discharged and the solids are recycled. The need for recycling is attributed to the fact that anaerobic digestion proceeds at a much slower rate than aerobic processes, thereby requiring more time and more biomass to achieve high removal efficiencies. The amount of time required for anaerobic digestion depends upon its composition and the temperature maintained in the digester because anaerobic processes are also sensitive to temperature. Mesophilic digestion occurs at approximately 35°C, and requires 12–30 days for processing. Thermophilic processes make use of higher temperatures (55°C) to speed up the reaction time to 6–14 days. Mixing the contents is not always necessary, but is generally preferred, as it leads to more efficient digestion by providing uniform conditions in the vessel and speeds up the biological reactions.

Anaerobic processes have been applied in seafood-processing wastewaters, obtaining high removal efficiencies (75–80%) with loads of 3 or 4 kg of COD/m^3 day [20,21].

In total, 60–70% of the gas produced by a balanced and well-functioning system consists of methane, with the rest being mostly carbon dioxide and minor amounts of nitrogen and hydrogen. This biogas is an ideal source of fuel, resulting in low-cost electricity and providing steam for use in the stirring and heating of digestion tanks.

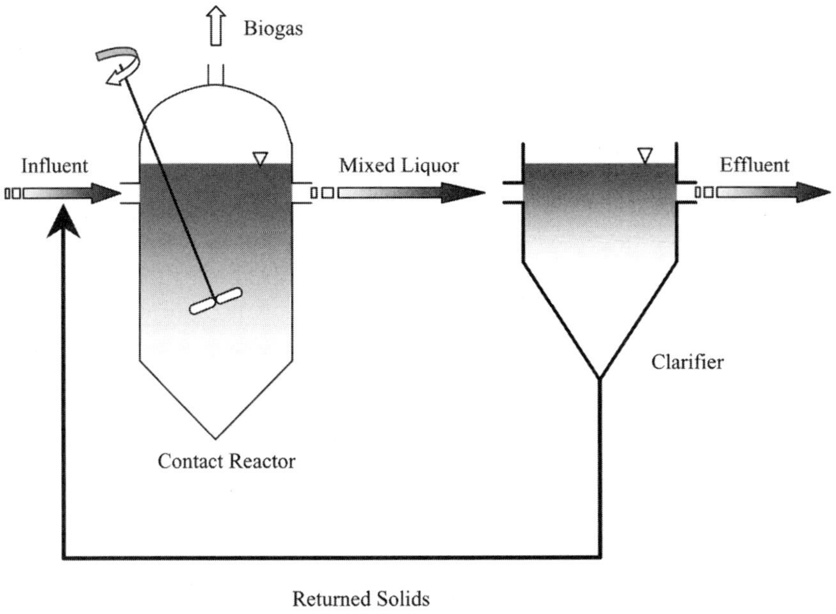

Figure 15 Diagram of an anaerobic digestion process.

Imhoff Tanks

The Imhoff tank is a relatively simple anaerobic system that was used to treat wastewater before heated digesters were developed. It is still used for plants of small capacity. The system consists of a two-chamber rectangular tank, usually built partially underground (Fig. 16).

Wastewater enters into the upper compartment, which acts as a settling basin while the settled solids are stabilized anaerobically at the lower part. Shortcircuiting of the wastewater can

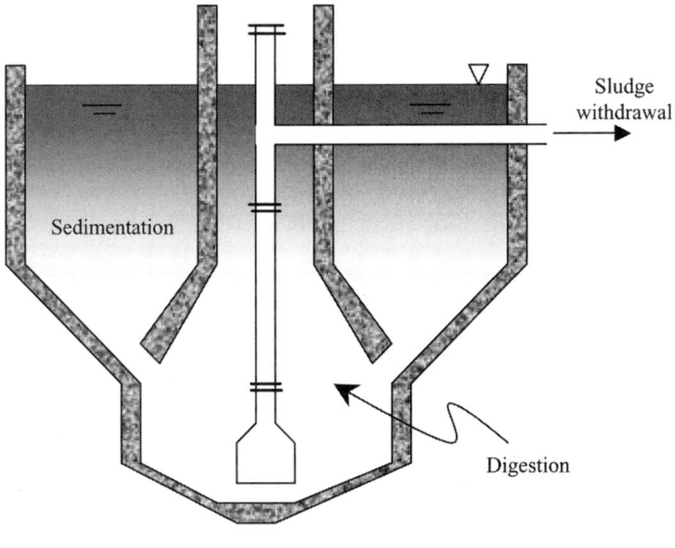

Figure 16 An Imhoff tank.

be prevented by using a baffle at the entrance with more than one port for discharge. The lower compartment is generally unheated. The stabilized sludge is removed from the bottom, generally twice a year, to provide ample time for the sludge to stabilize, although the removal frequency is sometimes dictated by the convenience of sludge disposal. In some cases, these tanks are designed with inlets and outlets at both ends, and the wastewater flow is reversed periodically so that the sludge at the bottom accumulates evenly. Although they are simple installations, Imhoff tanks are not without inconveniences; foaming, odor, and scum can form. These typically result when the temperature falls below 15°C and causes a process imbalance in which the bacteria that produce volatile acids predominate and methane production is reduced. This is why in some cases immersed heaters are used during cold weather. Scum forms because the gases that originate during anaerobic digestion are entrapped by the solids, causing the latter to float. This is usually overcome by increasing the depth in the lower chamber. At lower depths, bubbles form at a higher pressure, expand more when rising, and are more likely to escape from the solids. Odor problem is minimal when the two stages of the process of acid formation and gas formation are balanced.

14.5 PHYSICO-CHEMICAL TREATMENTS

14.5.1 Coagulation/Flocculation

Coagulation or flocculation tanks are used to improve the treatability of wastewater and to remove grease and scum from wastewater [9]. In coagulation operations, a chemical substance is added to an organic colloidal suspension to destabilize it by reducing forces that keep them apart, that is, to reduce the surface charges responsible for particle repulsions. This reduction in charges is essential for flocculation, which has the purpose of clustering fine matter to facilitate its removal. Particles of larger size are then settled and clarified effluent is obtained. Fig. 17 illustrates the coagulation/flocculation and settling of a seafood-processing wastewater.

In seafood processing wastewaters, the colloids present are of an organic nature and are stabilized by layers of ions that result in particles with the same surface charge, thereby increasing their mutual repulsion and stabilization of the colloidal suspension. This kind of wastewater may contain appreciable amounts of proteins and microorganisms, which become charged due to the ionization of carboxyl and amino groups or their constituent amino acids.

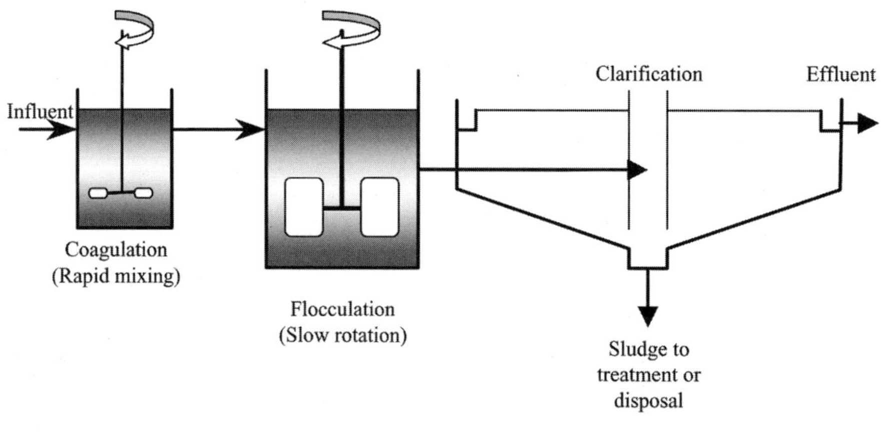

Figure 17 Chemical coagulation process.

The oil and grease particles, normally neutral in charge, become charged due to preferential absorption of anions, which are mainly hydroxyl ions.

Several steps are involved in the coagulation process. First, coagulant is added to the effluent, and mixing proceeds rapidly and with high intensity. The purpose is to obtain intimate mixing of the coagulant with the wastewater, thereby increasing the effectiveness of destabilization of particles and initiating coagulation. A second stage follows in which flocculation occurs for a period of up to 30 minutes. In the latter case, the suspension is stirred slowly to increase the possibility of contact between coagulating particles and to facilitate the development of large flocs. These flocs are then transferred to a clarification basin in which they settle and are removed from the bottom while the clarified effluent overflows.

Several substances may be used as coagulants. The pH of several wastewaters of the proteinaceous nature can be adjusted by adding acid or alkali. The addition of acid is more common, resulting in coagulation of the proteins by denaturing them, changing their structural conformation due to the change in their surface charge distribution. Thermal denaturation of proteins can also be used, but due to its high energy demand, it is only advisable if excess steam is available. In fact, the "cooking" of the blood–water in fishmeal plants is basically a thermal coagulation process.

Another commonly used coagulant is polyelectrolyte, which may be further categorized as cationic and anionic coagulants. Cationic polyelectrolytes act as a coagulant by lowering the charge of the wastewater particles because wastewater particles are negatively charged. Anionic or neutral polyelectrolyte are used as bridges between the already formed particles that interact during the flocculation process, resulting in an increase of floc size.

Since the recovered sludges from coagulation/flocculation processes may sometimes be added to animal feeds, it is advisable to ensure that the coagulant or flocculant used is not toxic.

In seafood-processing wastewaters there are several reports on the use (at both pilot plant and working scale) of inorganic coagulants such as aluminum sulfate, ferric chloride, ferric sulfate, or organic coagulants [22–25].

On the other hand, fish scales are reported to be used effectively as an organic wastewater coagulant [26]. These are dried and ground before being added as coagulant in powder form. Another marine byproduct that can be used as coagulant is a natural polymer derived from chitin, a main constituent of the exoskeletons of crustaceans, which is also known as chitosan.

14.5.2 Electrocoagulation

Electrocoagulation (EC) has also been investigated as a possible means to reduce soluble BOD. It has been demonstrated to reduce organic levels in various food- and fish-processing waste streams [27]. During testing, an electric charge was passed through a spent solution in order to destabilize and coagulate contaminants for easy separation. Initial test results were quickly clarified with a small EC test cell – contaminants coagulated and floated to the top. Analytical test results showed some reduction in BOD_5, but not as much as originally anticipated when the pilot test was conducted. Additional testing was carried out on site on a series of grab samples; however, these runs did not appear to be as effective as originally anticipated. The pH was varied in an attempt to optimize the process, but BOD_5 reductions of only 21–33% were observed. Also, since metal electrodes (aluminum) were used in the process, the presence of metal in the spent solution and separated solids posed a concern for byproduct recovery. Initial capital outlays and anticipated operating costs were not unreasonable (US$140,000 and US$40,000, respectively), but satisfactory BOD_5 reductions could not be achieved easily. It was determined that long retention times would be needed in order to make EC work effectively.

14.5.3 Disinfection

Disinfection of seafood-processing wastewater is a process by which disease-causing organisms are destroyed or rendered inactive. Most disinfection systems work in one of the following four ways: (i) damage to the cell wall, (ii) alteration of cell permeability, (iii) alteration of the colloidal nature of protoplasm, or (iv) inhibition of enzyme activity [9,15].

Disinfection is often accomplished using bactericidal agents. The most common agents are chlorine, ozone (O_3), and ultraviolet (UV) radiation, which are discussed in the following sections.

Chlorination

Chlorination is a process commonly used in both industrial and domestic wastewaters for various reasons. In fisheries effluents, however, its primary purpose is to destroy bacteria or algae, or to inhibit their growth. Usually the effluents are chlorinated just before their final discharge to the receiving water bodies. For this process either chlorine gas or hypochlorite solutions may be used, the latter being easier to handle. In waste solutions, chlorine forms hypochlorous acid, which in turn forms hypochlorite.

$$Cl_2 + H_2O \longrightarrow HOCl + H^+ + Cl^-$$

$$HOCl \longrightarrow H^+ + OCl^-$$

A problem that may occur during chlorination of fisheries effluents is the formation of chloramines. These wastewaters may contain appreciable amounts of ammonia and volatile amines, which react with chlorine to give chloramines, resulting in an increased demand for chlorine to achieve a desired degree of disinfection. The proportions of these products depend on the pH and concentration of ammonia and the organic amines present. Chlorination also runs the risk of developing trihalomethanes, which are known carcinogens. Subsequently, the contact chamber must be cleaned regularly.

The degree of disinfection is attributed to the residual chlorine present in water. A typical plot of the breakpoint chlorination curve with detailed explanation is shown in Fig. 18.

Initially, the presence of reducing agents reduce an amount of chlorine to chloride and makes the residual chlorine negligible (segment A–B). Further addition of chlorine may result in the formation of chloramines. These appear as residual chlorine but in the form of combined chlorine residual (segment B–C). Once all the ammonia and organic amines have reacted with the added chlorine, additional amounts of chlorine result in the destruction of the chloramines by oxidation, with a decrease in the chlorine residual as a consequence (segment C–D). Once this oxidation is completed, further addition of chlorine results in the appearance of free available chlorine. Point D on the curve is also known as "breakpoint chlorination." The goal in obtaining some free chlorine residual is to achieve disinfection purpose.

Chlorination units consist of a chlorination vessel in which the wastewater and the chlorine are brought into contact. In order to provide sufficient mixing, chlorine systems must have a chlorine contact time of 15–30 minutes, after which it must be dechlorinated prior to discharge. A schematic diagram of the systems is presented in Fig. 19.

The channels in this contact basin are usually narrow in order to increase the water velocity and hence reduce accumulation of solids by settling. However, the space between the channels should allow for easy cleaning. The levels of available chlorine after the breakpoint should comply with the local regulations, which usually vary between 0.2 and 1 mg/L. This value strongly depends on the location of wastewater to be discharged, because residual chlorine in treated wastewater effluents was identified, in some cases, as the main toxicant suppressing

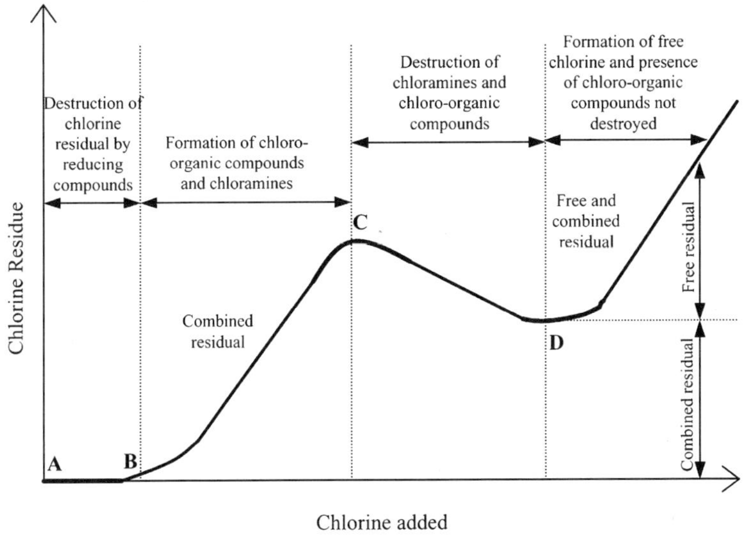

Figure 18 Breakpoint chlorinating curve (from Ref. 9).

the diversity, size, and quantity of fish in receiving streams [28]. Additionally, the chlorine dosage needed to achieve the residual effect required varies with the wastewater considered: 2–8 mg/L is common for an effluent from an activated sludge plant, and can be about 40 mg/L in the case of septic wastewater [6,7].

Ozonation

Ozone (O_3) is a strong oxidizing agent that has been used for disinfection due to its bactericidal properties and its potential for removal of viruses. It is produced by discharging air or oxygen across a narrow gap with application of a high voltage. An ozonation system is presented in Fig. 20.

Ozonation has been used to treat a variety of wastewater streams and appears to be most effective when treating more dilute types of wastes [29]. It is a desirable application as a

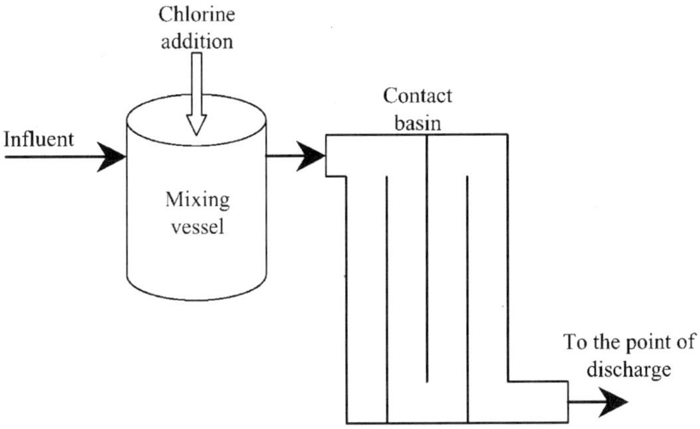

Figure 19 Schematics of a chlorination system.

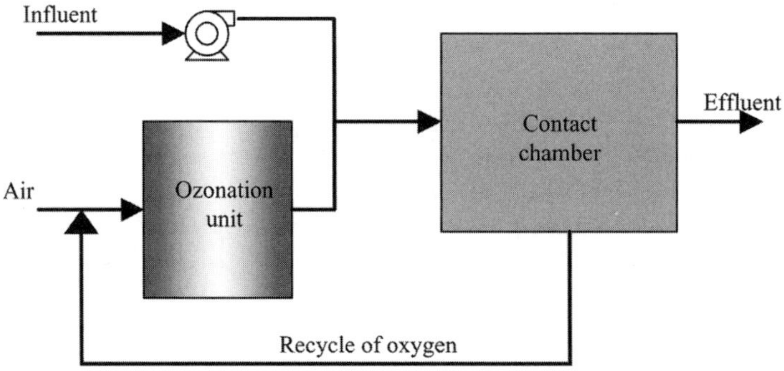

Figure 20 Simplified diagram of an ozonation system.

polishing step for some seafood-processing wastewaters, such as from squid-processing operations, which is fairly concentrated [30].

Ozone reverts to oxygen when it has been added and reacted, thus increasing somewhat the dissolved oxygen level of the effluent to be discharged, which is beneficial to the receiving water stream. Contact tanks are usually closed to recirculate the oxygen-enriched air to the ozonation unit. Advantages of ozonation over chlorination are that it does not produce dissolved solids and is affected neither by ammonia compounds present nor by the pH value of the effluent. On the other hand, ozonation has been used to oxidize ammonia and nitrites presented in fish culture facilities [31].

Ozonation also has limitations. Because ozone's volatility does not allow it to be transported, this system requires ozone to be generated onsite, which requires expensive equipment. Although much less used than chlorination in fisheries wastewaters, ozonation systems have been installed in particular in discharges to sensitive water bodies [4,32,33].

Ultraviolet (UV) Radiation

Disinfection can also be accomplished by using ultraviolet (UV) radiation as a disinfection agent. UV radiation disinfects by penetrating the cell wall of pathogens with UV light and completely destroying the cell and/or rendering it unable to reproduce.

However, UV radiation system might have only limited value to seafood-processing wastewater without adequate TSS removal since the effectiveness decreases when solids in the discharge block the light. This system also requires expensive equipment with high maintenance [34]. Nevertheless, UV radiation and other nontraditional disinfection processes are gaining acceptance due to stricter regulations on the amount of residual chlorine levels in discharged wastewaters.

14.6 LAND DISPOSAL OF WASTEWATER

Land application of wastewater is a low capital and operating cost method for treating seafood-processing wastes, provided that sufficient land with suitable characteristics is available. The ultimate disposal of wastewater applied to land is by one of the following methods:

- percolation to groundwater;
- overland runoff to surface streams;
- evaporation and evapo-transpiration to the atmosphere.

Generally, several methods are used for land application, including irrigation, surface ponding, groundwater recharge by injection wells, and subsurface percolation. Although each of these methods may be used in particular circumstances for specific seafood-processing waste streams, the irrigation method is most frequently used. Irrigation processes may be further divided into four subcategories according to the rates of application and ultimate disposal of liquid. These are overland flow, normal irrigation, high-rate irrigation, and infiltration–percolation.

Two types of land application techniques seem to be most efficient, namely infiltration and overland flow. As these land application techniques are used, the processor must be cognizant of potential harmful effects of the pollutants on the vegetation, soil, surface and groundwaters. On the other hand, in selecting a land application technique one must be aware of several factors such as wastewater quality, climate, soil, geography, topography, land availability, and return flow quality.

The treatability of seafood-processing wastewater by land application has been shown to be excellent for both infiltration and overland flow systems [2]. With respect to organic carbon removal, both systems have achieved pollutant removal efficiencies of approximately 98 and 84%, respectively. The advantage of higher efficiency obtained with the infiltration system is offset somewhat by the more expensive and complicated distribution system involved. Moreover, the overland flow system is less likely to pollute potable water supplies.

Nitrogen removal is found to be slightly more effective with infiltration land application when compared to overland flow application. However, the infiltration type of application has been shown to be quite effective for phosphorus and grease removal, and thus offers a definite advantage over the overland flow if phosphorus and grease removal are the prime factors. One factor that may negate this advantage is soil conditions are not favorable for phosphorus and grease removal and chemical treatment is required.

Irrigation is a treatment process that consists of a number of segments:

- aerobic bacterial degradation of the deposited suspended materials and evaporation of water and concentration of soluble salts;
- filtration of small particles through the soil cover, and biological degradation of entrapped organics in the soil by aerobic and anaerobic bacteria;
- adsorption of organics on soil particles and uptake of nitrogen and phosphorus by plants and soil microorganisms;
- uptake of liquid wastes and transpiration by plants;
- percolation of water to groundwater.

The importance of these processes depends on the rate of application of waste, the characteristics of the waste, the characteristics of soil and substrata, and the type of cover crop grown on the land.

14.6.1 Loading Rates

Application rates should be determined by pilot plant testing for each particular location. The rate depends on whether irrigation techniques are to be used for roughing treatment or as an ultimate disposal method.

This method has both hydraulic and organic loading constraints for the ultimate disposal of effluent. If the maximum recommended hydraulic loading is exceeded, the surface runoff would increase. Should the specified organic loading be exceeded, anaerobic conditions could develop with resulting decrease in BOD_5 removal and the development of odor problem. The average applied loadings of organic suspended solids is approximately 8 g/m^2; however, loadings up to

22 g/m^2 have also been applied successfully [2]. A resting period between applications is important to ensure survival of the aerobic bacteria. The spray field is usually laid out in sections such that resting periods of 4–10 days can be achieved.

14.6.2 Potential Problems in Land Application with Seafood-Processing Wastewater

Two potential problems may be encountered with land application of seafood-processing wastewaters: the presence of disease-producing bacteria and unfavorable sodium absorption ratios of the soil. A key to minimizing the risk of spreading disease-producing bacteria can be accomplished by using low-pressure wastewater distribution systems to reduce the aerosol drift of the water spray. With respect to unfavorable sodium absorption ratios associated with the soil type, the seafood processor should be aware that clay-containing soils will cause the most serious sodium absorption problem. Sandy soils do not appear to be affected by unfavorable sodium absorption ratios and seem to be the best suited for accepting the high sodium chloride content found in most meat packing plant wastewaters.

As seafood-processing plant wastewaters are applied to land, certain types of grasses have been found to be compatible with these wastewaters. These are Bermuda NK-32, Kentucky-31 Tall Fescue, Jose Wheatgrass, and Blue Panicum [2]. In addition, it was reported that the southwestern coast of the United States, with its arid climate, mild winters, and vast available land areas, presents ideal conditions for land application treatment systems.

In some cases, the use of land application systems by today's seafood processors is feasible. However, in many cases, land disposal of seafood-processing wastes must be ruled out as a treatment alternative. Coastal topographic and soil characteristics, along with high costs of coastal property are the two major factors limiting the use of land application systems for treating seafood-processing wastes.

14.7 GENERAL SEAFOOD-PROCESSING PLANT SCHEMES

Seafood processing involves the capture and preparation of fish, shellfish, marine plants and animals, as well as byproducts such as fish meal and fish oil. The processes used in the seafood industry generally include harvesting, storing, receiving, eviscerating, precooking, picking or cleaning, preserving, and packaging [2]. Figure 21 shows a general process flow diagram for seafood processing. It is a summary of the major processes common to most seafood processing operations; however, the actual process will vary depending on the product and the species being processed.

There are several sources that produce wastewater, including:

- fish storage and transport;
- fish cleaning;
- fish freezing and thawing;
- preparation of brines;
- equipment sprays;
- offal transport;
- cooling water;
- steam generation;
- equipment and floor cleaning.

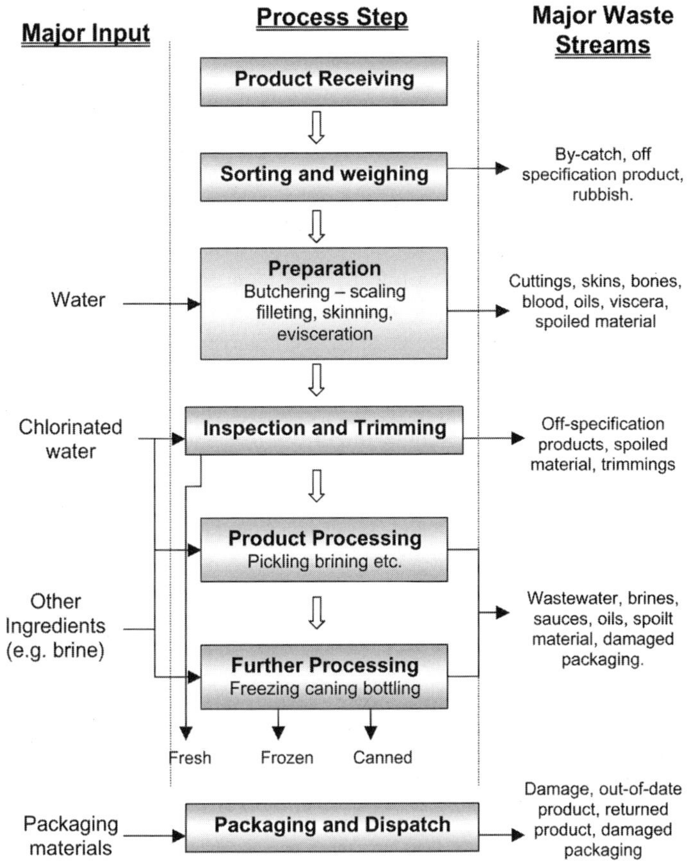

Figure 21 General process flow diagram for seafood processing operations.

Organic material in the wastewater is produced in the majority of these processes. However most of it originates from the butchering process, which generally produces organic material such as blood and gut materials. The volume and quality of wastewater in each area is highly dependent on the products or species being processed and the production processes used.

Most seafood processors have a high baseline water use for cleaning plant and equipment. Therefore, water use per unit product decreases rapidly as production volume increases. Reducing wastewater volumes tends to have a significant impact on reducing organic loads as these strategies are typically associated with reduced product contact and better segregation of high-strength streams.

Water consumption in seafood-processing operations has traditionally been high to achieve effective sanitation. Industry literature indicates that water use varies widely throughout the sector, from 5 to 30 L/kg of product. Several factors affect water use, including the type of product processed, the scale of the operation, the process used, and the level of water minimization in place [1]. General cleaning contributes significantly to total water demand so smaller scale sites tend to have significantly higher water use per unit of production. Thawing operations can also account for up to 50% of the wastewater generated. A figure for water use of around 5–10 L/kg is typical of large operations with dedicated, automated, or semi-automated equipment that have implemented water minimization practises.

14.8 ECONOMIC CONSIDERATIONS OF SEAFOOD-PROCESSING WASTEWATER TREATMENT

Economic considerations are always the most important parameters that influence the final decision as to which process should be chosen for wastewater treatment. In order to estimate cost, data from the wastewater characterization should be available together with the design parameters for alternative processes and the associated costs. Costs related to these alternative processes and information on the quality of effluent should also be obtained prior to cost estimation in compliance with local regulations.

During the design phase of a wastewater treatment plant, different process alternatives and operating strategies could be evaluated by several methods. This cost evaluation can be achieved by calculating a cost index using commercially available software packages [36,37]. Nevertheless, actual cost indices are often restrictive, since only investment or specific operating costs are considered. Moreover, time-varying wastewater characteristics are not directly taken into account but rather through the application of large safety factors. Finally, the implementation of adequate control strategies such as a real-time control is rarely investigated despite the potential benefits [38,39]. In order to avoid these problems, a concept of MoSS-CC (Model-based Simulation System for Cost Calculation) was introduced by Gillot et al. [40], which is a modeling and simulation tool aimed at integrating the calculation of investment and fixed and variable operating costs of a wastewater treatment plant. This tool helps produce a holistic economic evaluation of a wastewater treatment plant over its life cycles.

14.8.1 Preliminary Costs of a Wastewater Treatment Plant

Several methods may be used to assess the preliminary costs of a wastewater treatment plant to facilitate a choice between different alternatives in the early phase of a process design. One method is cost functions [41–45]. Examples of different investment and operating cost functions are presented in Tables 3, 4, and 5. These cost functions were developed for the MoSS-CC modeling tool.

Another method was developed by EPA to estimate the construction costs for the most common unitary processes of wastewater treatment, as presented in Table 6. This was developed for municipal sewage treatment and may not be entirely applicable for small wastewater treatment plants, but it is useful for preliminary estimation and comparison among alternatives [4].

14.8.2 Cost of Operation and Maintenance

Several main factors influence the costs of operation and maintenance, including energy costs, labor costs, material costs, chemical costs, and cost of transportation of sludges for final disposal and discharge of treated wastewater. The relative importance of these items varies significantly depending on the location, the quality of the effluent discharged, and on the specific characteristics of the wastewater being treated [4].

The total operating cost of a wastewater treatment plant may be related to global plant parameters (e.g., average flow rate, population equivalent), generally through power laws [46–48]. However, such relationships apply to the average performance of plants and often suffer from a high uncertainty, unless very similar plant configurations are considered [40]. In terms of

Table 3 Examples of Investment Cost Functions

Unit	Item	Cost function	Parameter	Parameter range	Reference	Cost unit
Influent pumping station	Concrete	$2334Q^{0.637}$	Q = flow rate (m^3/hour)	250–4000	45	Euro of 1998
	Screws	$2123Q^{0.540}$				
	Screening	$3090Q^{0.349}$				
Any unit	Excavation	$2.9(\pi/4D^2H)$	D = diameter (m)	Not defined	44	Can$ of 1995
	Compaction	$24.1 \times 0.4(\pi/4D^2)$	H = height (m)			
	Concrete base	$713.9 \times 0.5(\pi/4D^2)$				
	Concrete wall	$933.6 \times 0.5\pi DH$				
Oxidation ditch	Concrete	$10304V^{0.477}$	V = volume (m^3)	1100–7700	45	Euro of 1998
	Electromech.[a]	$8590OC^{0.433}$	OC = oxygen capacity (kgO_2/hour)			
Settler	Concrete	$2630A^{0.678}$	A = area (m^2)	175–1250	45	Euro of 1998
	Electromech.[a]	$6338A^{0.325}$				
	Concrete	$150(A/400)^{0.56}$	A	60–400	41, 42	Can$*1000 of 1990
		$150(A/400)^{1.45}$		400–800		
	Electromech.[a]	$60(A/220)^{0.62}$		60–7000		
Sludge pump	Electromech.[a]	$9870IQ^{0.53}$	Q, I = Engin. Index[b]	Not defined	52	US$ of 1971
	Electromech.[a]	$5038Q^{0.304}$	Q	35–2340	45	Euro of 1998

[a]Electromech. = electromechanical equipment; [b]Engineering News Record Index = index used to update costs in United States.
Source: Ref. 40

Table 4 Examples of Fixed Operating Cost Functions

Cost item	Cost function			Reference
	Formula	Symbols	Units	
Normal O&M	$L = U_c PE$	L = labor U_c = unit cost PE = population equivalent	man-hour/year man-hour/year/PE —	53
Clarifier mechanism	$P = \theta A^b$	P = power θ, b = constant A = area	kW — m^2	44
Mixers	$P = P_s V$	P = power P_s = specific power V = volume	kW kW/m^3 m^3	53
Small equipment (supplies, spare parts...)	$C = U_c PE$	C = cost U_c = unit cost PE = population equivalent	Euro/year Euro/year/PE —	5
Analyses	$C = U_c PE$	C = cost U_c = unit cost PE = population equivalent	Euro/year Euro/year/PE —	

Source: Ref. 40

cost functions evaluations, some possible models in generic form for the fixed and variable operation costs are illustrated in Tables 4 and 5, respectively.

Capital Costs

These comprise mainly the unit construction costs, the land costs, the cost of the treatment units, and the cost of engineering, administration, and contingencies. The location should be carefully evaluated in each case because it affects the capital costs more than the operating costs [4]. When comparing different alternatives, special attention should be paid to the time and space scales chosen [38], since it may influence the choice of the implemented cost functions [49]. At best, an overall plant evaluation over the life span of the plant should be conducted [40].

Estimation of Total Costs

The total cost of a plant is normally determined by using the present worth method [50]. All annual operating costs for each process are converted into their corresponding present value and added to the investment cost of each process to yield the net present value (NPV). The net present value of a plant over a period of n years can be determined as:

$$\text{NPV} = \sum_{k=1}^{N} IC_k + \left[\frac{1-(1+i)^{-n}}{i}\right] \sum_{k=1}^{N} OC_k \tag{5}$$

Table 5 Example of Variable Operating Cost Functions

Cost item	Cost function			Reference
	Formula	Symbols	Units	
Pumping power	$P = Qwh/\eta$	Q = flow rate P = power w = specific liquid weight h = dynamic head η = pump efficiency	m^3/s kW N/m^3 m^3/s —	54
Aeration power (fine bubble aeration)	$q_{air} = f(K_L a_f)$ $P = f(q_{air})$	q_{air} = air flow rate P = power $K_L a_f$ = oxygen transfer coefficient in field conditions	Nm^3/hour kW 1/hour	53, 55
Sludge thickening dewatering and disposal	$C = U_c TSS$	C = cost U_c = unit cost TSS = excess sludge	Euro/year Euro/t TSS t	5
Chemicals consumption	$C = U_c C_n$	C = cost U_c = unit cost C_n = consumption	Euro/year Euro/kg kg	40
Effluent taxes (organic matter and nutrient)	$L = U_c^* (k_{org} \cdot N_{org} + k_{nut} \cdot N_{nut})$	U_c = unit cost $N_{org} = f(Q, \text{BOD, TSS, COD})$ $N_{nut} = f(Q, N, P)$	Euro/unit	38

Source: Ref. 40.

where IC_k represents the investment cost of a unit k, and OC_k the operating cost, i is the interest rate, and N is the number of units. The results could also be expressed as equivalent annual worth (AW):

$$\text{AW} = \frac{i(1+i)^n}{(1+i)^n - 1} \sum_{k=1}^{N} IC_k + \sum_{k=1}^{N} OC_k \tag{6}$$

For small wastewater treatments plants, an initial estimate of the total cost can be obtained from the cost of a similar plant with a different capacity, a relationship derived from costs relationships in chemical industries. The cost of plants of different sizes is related to the ratio of their capacity raised to the 0.6 power [4]:

$$\text{Capital}_2 = \text{Capital}_1 \times \left(\frac{\text{Capacity}_2}{\text{Capacity}_1}\right)^{0.6} \tag{7}$$

where $\text{Capital}_{1,2}$ = capital costs of plants 1 and 2, and $\text{Capacity}_{1,2}$ = capacity of plants 1 and 2.

The operation and maintenance costs can be estimated by a similar formula:

$$\text{OM}_2 = \text{OM}_1 \times \left(\frac{\text{Capacity}_2}{\text{Capacity}_1}\right)^{0.85} \tag{8}$$

where $\text{OM}_{1,2}$ = operation and maintenance costs of plants 1 and 2, $\text{Capacity}_{1,2}$ = capacity of plants 1 and 2.

Table 6 Construction Costs for Selected Unitary Operations of Wastewater Treatment

Liquid stream	Correlation
Preliminary treatment	$C = 5.79 \times 10^4 \times Q^{1.17}$
Flow equalization	$C = 1.09 \times 10^5 \times Q^{0.49}$
Primary sedimentation	$C = 1.09 \times 10^5 \times Q^{1.04}$
Activated sludge	$C = 2.27 \times 10^5 \times Q^{0.17}$
Rotating biological contactor	$C = 3.19 \times 10^5 \times Q^{0.92}$
Chemical addition	$C = 2.36 \times 10^4 \times Q^{1.68}$
Stabilization pond	$C = 9.05 \times 10^5 \times Q^{1.27}$
Aerated lagoon	$C = 3.35 \times 10^5 \times Q^{1.13}$
Chlorination	$C = 5.27 \times 10^4 \times Q^{0.97}$
Solids stream	**Correlation**
Sludge handling	$C = 4.26 \times 10^4 \times Q^{1.36}$
Aerobic digestion	$C = 1.47 \times 10^5 \times Q^{1.14}$
Anaerobic digestion	$C = 1.12 \times 10^5 \times Q^{1.12}$
Incineration	$C = 8.77 \times 10^4 \times Q^{1.33}$

C represents the cost in USD and Q represents the flow rate of the wastewater to be treated.
Source: EPA, 1978.

An alternative procedure for developing cost models for wastewater treatment systems includes the preparation of kinetic models for the possible treatment alternatives, in terms of area and flow rates at various treatment efficiencies, followed by the computation of mechanical and electrical equipment, as well as the operation and maintenance costs as a function of the flow rates [51].

ACKNOWLEDGMENTS

The assistance provided by Mr Lam Weh Yee is gratefully acknowledged.

REFERENCES

1. Environment Canada. *Canadian Biodiversity Strategy: Canadian response to the Convention on Biological Diversity*, Report of the Federal Provincial Territorial Biodiversity Working Group; Environmant Canada: Ottawa, 1994.
2. Carawan, R.E.; Chambers, J.V.; Zall, R.R. *Seafood Water and Wastewater Management*, 1979. The North Carolina, Agricultural Extension Service. U.S.A.
3. Gonzalez, J.F.; Civit, E.M.; Lupin, H.M. Composition of fish filleting wastewater. Environ. Technol. Lett. **1983**, *7*, 269–272.
4. Gonzalez, J.F. *Wastewater Treatment in the Fishery Industry*, FAO fisheries Technical Paper, 1996; 355.
5. Alexandre, O.; Grand d'Esnon, A. Le cout des services d'assinissement ruraux. Evaluation des couts d'investissement et d'exploitation. TSM, 7/8, 1998; 19–31. (In French.)
6. Metcalf and Eddy, Inc. *Wastewater Engineering: Treatment, Disposal, Reuse*. McGraw-Hill Book Co.: New York, 1979.
7. Eckenfelder, W.W. *Principles of Water Quality Management*; CBI Publishing Co.: Boston, 1980.

8. Sikorski, Z. *Seafood Resources: Nutritional Composition and Preservation*; CRC Press, Inc.: Boca Raton, FL, 1990.
9. Metcalf and Eddy, Inc. *Wastewater Engineering: Treatment and Disposal*, 3rd Ed.; revised by Tchobanoglous, G., Burton, F.; McGraw-Hill, Inc.: New York, 1991.
10. Rich, L.G. *Low Maintenance, Mechanically Simple Wastewater Treatment Systems*; McGraw-Hill Book Co.: New York, 1980.
11. Hansen, S.P.; Culp, G.L. Applying shallow depth sedimentation theory. J. Am. Water Works Assoc. **1967,** *59*; 1134–1148.
12. Illet, K.J. Dissolved air flotation and hydrocyclones for wastewater treatment and by-product recovery in the food process industries. Water Services **1980**, *84*; 26–27.
13. Ertz, D.B.; Atwell, J.S.; Forsht, E.H. Dissolved air flotation treatment of seafood processing wastes – an assessment. In *Proceedings of the Eighth National Symposium on Food Processing Wastes*, EPA-600/Z-77-184, August 1977; p. 98.
14. Anon. *Environmental Assessment and Management of the Fish Processing Industry*, Sectoral studies series No. 28; UNIDO: Vienna, Austria, 1986.
15. Henry, J.G.; Heinke, G.W. *Environmental Science and Engineering*, 2nd Ed.; Prentice-Hall, Inc.: Upper Saddle River, NJ, 1996; 445–447.
16. Tokus, R.Y. Biodegradation and removal of phenols in rotating biological contactors. Water Sci. Technol. **1989**, *21*, 1751.
17. Gujer, W.; Boller, M. A mathematical model for rotating biological contactors. Water Sci. Technol. **1990**, *22*, 53–73.
18. Ahn, K.H.; Chang, J.S. Performance evaluation of compact RBC-settling tank system. Water Sci. Technol. **1991**, *23*, 1467–1476.
19. WRF (World Resource Foundation). Preserving Resources Through Integrated Sustainable Management of Waste; WRF, 1997.
20. Balslev-Olesen, P.; Lyngaard, A.; Neckelsen, C. Pilot-scale experiments on anaerobic treatment of wastewater from a fish processing plant. Water Sci. Technol. **1990**, *22*, 463–474.
21. Mendez, R.; Omil, F.; Soto, M.; Lema, J.M. Pilot plant studies on the anaerobic treatment of different wastewaters from a fish canning factory. Water Sci. Technol. **1992**, *25*, 37–44.
22. Johnson, R.A.; Gallager, S.M. Use of coagulants to treat seafood processing wastewaters. J. Water Pollut. Control Feder. **1984**, *56*, 970–976.
23. Nishide, E. Coagulation of fishery wastewater with inorganic coagulants. Bull. College of Agriculture and Veterinary Medicine: Nihon University, Japan, **1976**, *33*, 468–475.
24. Nishide, E. Coagulation of fishery wastewater with inorganic coagulants. Bull. College of Agriculture and Veterinary Medicine: Nihon University, Japan, **1977**, *34*, 291–294.
25. Ziminska, H. Protein recovery from fish wastewaters. In *Proceedings of the Fifth International Symposium on Agricultural Wastes*, American Society of Agriculture Engineering: St. Joseph, MI, 1985; 379.
26. Hood, L.F.; Zall, R.R. Recovery, utilization and treatment of seafood processing wastes. In *Advances in Fish Science and Technology*. Conell, J.J., Ed.; Fishing News Books, Ltd.: Surrey, England, 1980.
27. Beck, E.C.; Giannini, A.P.; Ramirez, E.R. Electrocoagulation clarifiers food wastewater. Food Technol. **1974**, *28* (*2*), 18–22.
28. Paller, M.H.; Lewis, W.M.; Heidinger, R.C.; Wawronowicz, J.L. Effects of ammonia and chlorine on fish in streams receiving secondary discharges. J. Water Pollut. Control Feder. **1983**, *55*, 1087–1097.
29. Ismond, A. End of pipe treatment options. Presented at *Wastewater Technology Conference and Exhibition*, Vancouver, BC, 1994.
30. Park, E.; Enander, R.; Brnett, S.M.; Lee, C. Pollution prevention and biochemical oxygen demand reduction in a squid processing facility. J. of Cleaner Production **2000**, 9 (*200*) 341–349.
31. Monroe, D.W.; Key, W.P. The feasibility of ozone for purification of hatchery waters. Ozone Sci. Engng. **1980**, *2*, 203–224.
32. Rosenthal, H.; Kruner, G. Efficiency of an improved ozonation unit applied to fish culture situations. Ozone, Sci. Engng. **1985**, *7*, 179–190.
33. Stover, E.L.; Jover, R.N. High level ozone disinfection of wastewater for shellfish discharges. Ozone Sci. Engng. **1980**, *1*, 335–346.

34. Whiteman, C.T.; Mehan, G.T.; Grubbs, G.H. et al. *Development Document for Proposed Effluent Limitations Guidelines and Standard for the Concentrated Aquatic Production Industry Point Source Category*, USEPA 2002; Chapter 7.
35. UNEP. 1998.
36. McGhee, T.J.; Mojgani, P.; Viicidomina, F. Use of EPA's CAPDET program for evaluation of wastewater treatment alternatives. J. Water Pollut. Control Fed. **1983**, *55* (*1*), 35–43.
37. Spearing, B.W. Sewage treatment optimization model – STOM – the sewage works in a personal computer. Proc. Instn. Civ. Engrs. Part 1 **1987**, *82*, 1145–1164.
38. Vanrolleghem, P.A.; Jeppsson, U.; Cartensen, J.; Carlsson, B.; Olsson, G. Integration of wastewater treatment plant design and operation – a systematic approach using cost functions. Water Sci. Technol. **1996**, 34 (*3–4*), 159–171.
39. Ekster, A. Automatic waste control. Water Environ. Technol. **1998**, *10* (*8*), 63–64.
40. Gillot, S.; Vermeire, P.; Grootaerd, H.; Derycke, D.; Simoens, F.; Vanrolleghem, P.A. *Integration of Wastewater Treatment Plant Investment and Operating Costs for Scenario Analysis Using Simulation*. In: Proceedings 13th Forum Applied Biotechnology. Med. Fac. Landbouww. Univ. Gent, Belgium, 64/5a, (1999), 13–20.
41. Wright, D.G.; Woods, D.R. Evaluation of capital cost data. Part 7: Liquid waste disposal with emphasis on physical treatment. Can. J. Chem. Eng. **1993**, *71*, 575–590.
42. Wright, D.G.; Woods, D.R. Evaluation of capital cost data. Part 8: Liquid waste disposal with emphasis on biological treatment. Can. J. Chem. Eng. **1993**, *72*, 342–351.
43. Agences de l'eau, Ministere de l'Environment. Approche technico-economique des couts d'investissement des stations d'epuration. Cahier Technique, 1995; 48 p. (In French.)
44. Fels, M.; Pinter, J.; Lycon, D.S. Optimized design of wastewater treatment systems: Application to the mechanical pulp and paper industry: I. Design and cost relationships. Can. J. Chem. Eng. **1997**, *75*, 437–451.
45. Vermeire, P. *Economishe optimalisatie van waterzuiveringsstations. Ontwikkeling van investeringskostenfunties voor Vlaanderen* (in Dutch). Engineers Thesis. Faculty of Agricultural and Applied Biological Sciences. Univ. Gent, Belgium, 1999, pp. 101.
46. Smeers, Y.; Tyteca, D. A geometric programming model for optimal design of wastewater treatment plants. Opn. Res. **1984**, *32* (*2*), 314–342.
47. Balmer, P.; Mattson, B. Wastewater treatment plant operation costs. Water Sci. Technol. **1994**, *30* (*4*), 7–15.
48. Water Environment Research Federation (WERF). Benchmarking wastewater operations – collection, treatment, and biosolids management – Final report. Project 96-CTS-5, 1997.
49. Rivas, A.; Ayesa, E. Optimum design of activated sludge plants using the simulator DAISY 2.0. In *Measurement and Modelling in Environmental Pollution*; San Jose, R., Brebbia, C.A., Ed.; Computational Mechanics Publications: Southampton, Boston, 1997.
50. White, J.A.; Agee, M.H.; Case, K.E. *Principles in Engineering Economic Analysis*; John Wiley & Sons, 1989.
51. Uluatam, S.S. Cost models for small wastewater treatment plants. Int. J. Environ. Studies **1991**, *37*, 171–181.
52. Tyteca, D. Mathematical models for cost effective biological wastewater treatment. In *Mathematical Models in Biological Wastewater Treatment*; Jorgensen and Gromiec: Amsterdam, 1985.
53. Jacquet, P. Een globale kostenfuctie voor tuning en evaluatie van op respirometrie gabaseerde controle algoritmen voor actiefsliprocessen. Engineers Thesis. Faculty of Agricultural and Applied Biological Sciences, University Gent: Belgium, 1999; 122. (In Dutch.)
54. ASCE. *ASCE Standard Measurement of Oxygen Transfer in Clean Water*; American Society of Civil Engineers, 1992.
55. Gillot, S.; De Clercq, B.; Defour, D.; Simoens, F.; Gernaey, K.; Vanrolleghem, P.A. Optimization of wastewater treatment plant design and operation using simulation and cost analysis. 72nd Annual Conference WEFTEC 1999, New Orleans, USA, 9–13 October, 1999.
56. Environmental Protection Agency (EPA). *Construction Costs for Municipal Wastewater Treatment Plants: 1973–1977*. Technical Report MCD-37, USEPA, Washington, D.C., USA, 1978

15
Treatment of Meat Wastes

Charles J. Banks and Zhengjian Wang
University of Southampton, Southampton, England

15.1 THE MEAT INDUSTRY

The meat industry is one of the largest producers of organic waste in the food processing sector and forms the interface between livestock production and a hygienically safe product for use in both human and animal food preparation. This chapter looks at this interface, drawing its boundaries at the point of delivery of livestock to the slaughterhouse and the point at which packaged meat is shipped to its point of use. The chapter deals with "meat" in accordance with the understanding of the term by the United States Environmental Protection Agency (USEPA) [1] as all animal products from cattle, calves, hogs, sheep and lambs, and from any meat that is not listed under the definition of poultry. USEPA uses the term "meat" as synonymous with the term "red meat." The definition also includes consumer products (e.g., cooked, seasoned, or smoked products, such as luncheon meat or hams). These specialty products, however, are outside the scope of the current text. The size of the meat industry worldwide, as defined above, can thus be judged by meat production (Table 1), which globally is around 140 million tons (143 million tonnes) for major species, with about one-third of production shared between the United States and the European Union. The single largest meat producer is China, which accounts for 36% of world production.

The first stages in meat processing occur in the slaughterhouse (abattoir) where a number of common operations take place, irrespective of the species. These include holding of animals for slaughter, stunning, killing, bleeding, hide or hair removal, evisceration, offal removal, carcass washing, trimming, and carcass dressing. Further secondary operations may also occur on the same premises and include cutting, deboning, grinding, and processing into consumer products.

There is no minimum or maximum size for a slaughterhouse, although the tendency in Europe is towards larger scale operations because EU regulations on the design and operation of abattoirs [2] have forced many smaller operators to cease work. In the United States there are approximately 1400 slaughterhouses employing 142,000 people, yet 3% of these provide 43% of the industry employment and 46% of the value of shipments [1]. In Europe slaughterhouses tend to process a mixed kill of animals; whereas in the United States larger operations specialize in processing one type of animal and, if a single facility does slaughter different types of meat animals, separate lines or even separate buildings are used [3].

Table 1 Meat Production Figures (×1000) and Percentage of Global Production by the United States and European Union (EU)

	Global tons/year (tonnes/year)	USA tons/year (tonnes/year)	%	EU tons/year (tonnes/year)	%
Beef[a]	49,427 (50,220)	12,138 (12,333)	24.6	7136 (7250)	14.4
Lamb[b]	6872 (6982)	111 (113)	1.6	1080 (1097)	15.7
Pork[a]	84,115 (85,465)	8831 (8973)	10.5	17,519 (17,800)	20.8
Total	140,414 (142,667)	21,081 (21,419)	15.0	25,734 (26,147)	18.3

Figures derived from a wide range of statistics provided by the U.S. Department of Agriculture Foreign Agricultural Service.
[a] Provisional figures for 2002.
[b] Figures for 1997.

15.2 PROCESSING FACILITIES AND WASTES GENERATED

As a direct result of its operation, a slaughterhouse generates waste comprised of the animal parts that have no perceived value to the slaughterhouse operator. It also generates wastewater as a result of washing carcasses, processing offal, and from cleaning equipment and the fabric of the building. The operations taking place within a slaughterhouse and the types of waste and products generated are summarized in Fig. 1. Policies on the use of blood, gut contents, and meat and bone meal vary between different countries. Products that may be acceptable as a saleable product or for use in agriculture as a soil addition in one country may not be acceptable in another. Additionally, wastes and wastewaters are also generated from the stockyards, any rendering process, cooling facilities for refrigeration, compressors and pumps, vehicle wash facilities, wash rooms, canteen, and possibly laundry facilities.

15.2.1 Waste Characteristics and Quantities Generated

In general the characteristics of the solid wastes generated reflect the type of animal being killed, but the composition within a particular type of operation is similar regardless of the size of the plant. The reason for this is that the nature of the waste is determined by the animal itself and the quantity is simply a multiplication of the live weight of material processed. For example, the slaughter of a commercial steer would yield the products and byproducts shown in Table 2.

As can be seen the noncommercial sale material represents a little over 50% of the live weight of the animal, with about 25% requiring rendering or special disposal. The other 25% has a negative value and, because of its high water content, is not ideally suited to the rendering process. For this reason alternative treatment and disposal options have been sought for nonedible offal, gut fill, and blood, either separately or combined together, and in some cases combined with wastewater solids. The quantity of waste from sheep is again about 50% of the live weight, while pigs have only about 25% waste associated with slaughter.

Other solid waste requiring treatment or disposal arises mainly in the animal receiving and holding area, where regulations may demand that bedding is provided. In the European Union the volume of waste generated by farm animals kept indoors has been estimated by multiplying the number of animals by a coefficient depending on types of animals, function, sex, and age. Examples of coefficients that can be used for such calculations are given in Table 3 [5]. These

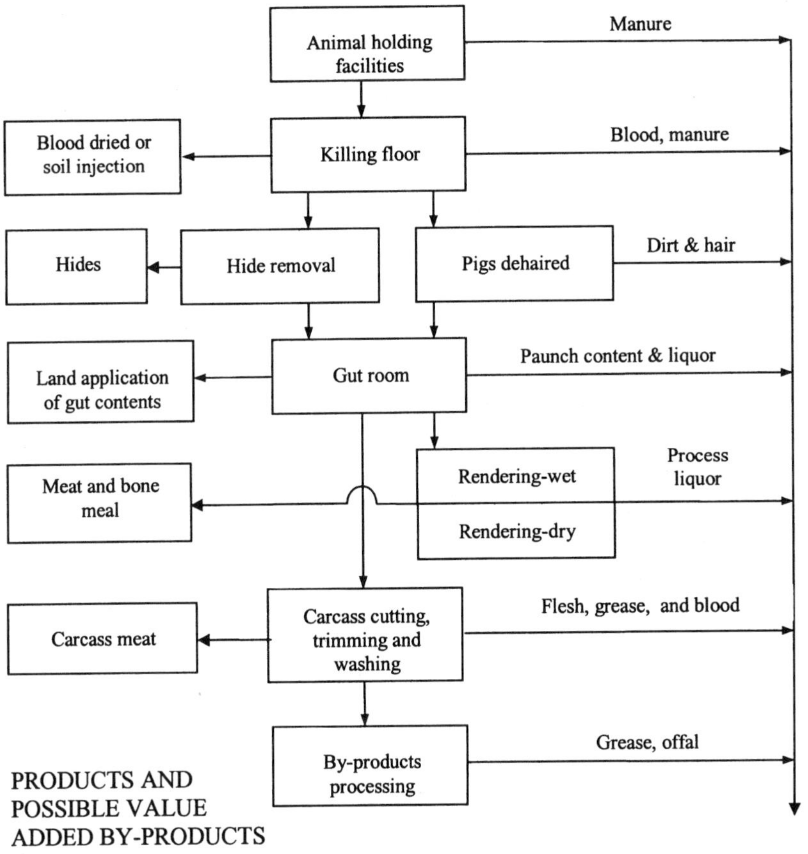

Figure 1 Flow diagram indicating the products and sources of wastes from a slaughterhouse.

figures are for normal farm conditions and may vary for temporary holding accommodation depending on feeding and watering regimes.

For the purposes of waste treatment, volume is not as useful as knowing the pollution load. Denmead [6] estimated that 8.8 lb (4 kg) dry organic solids/cattle and 1.65 lb (0.75 kg) dry organic solids/sheep or lamb would be produced during an overnight stock of animals in the holding pens of a slaughterhouse.

Table 2 Raw Materials Segregated from a Commercial Steer (990 lb or 450 kg Live Weight

Edible meat	Edible offals	Hide	High-grade fat	Bone and meat trim	Nonedible offal and gut fill	Blood	BSE suspect material
350 lb	35 lb	70 lb	100 lb	110 lb	245 lb	35 lb	45 lb
160 kg	15 kg	32 kg	45 kg	50 kg	112 kg	16 kg	20 kg
	Commercial sale		Byproducts for rendering		Waste		Special disposal

Source: Ref. 4.

Table 3 Waste Generated for Cattle and Pigs of Different Ages and Sexes (*Source:* Ref. 5)

Animal category	Quantity (L/day)
Cattle	
Less than 1 year	11.4
Between 1 and 2 years	20
More than 2 years	40
Pigs	
Less than 44 lb (20 kg)	2.1
Fattening pigs more than 44 lb (20 kg)	4.3
Breeding pigs	8.6
Covered sows	14.3

Once on the slaughter line, the quantity of waste generated depends on the number of animals slaughtered and the type of animal. Considering the total annual tonnage of animals going to slaughter there is surprisingly little information in the scientific literature on the quantities of individual waste fractions destined for disposal. The average weight of wet solid material produced by cutting and emptying of the stomachs of ruminants was estimated by Fernando [7] as 60 lb (27 kg) for cattle, 6 lb (2.7 kg) for sheep and 3.7 lb (1.7 kg) for lambs. Pollack [8] gave a much higher estimate for the stomach contents of cattle at 154 lb (70 kg) per head, and 2.2 lb (1 kg) per animal for pigs. There is a more consistent estimate of the quantity of blood produced: Brolls and Broughton [9] reported average weight of wet blood produced is around 32 lb per 1000 lb of beef animal (14.5 kg per 454 kg); Grady and Lim [10] likewise reported 32.5 lb of blood produced per 1000 lb (14.7 kg per 453 kg) of live weight; and Banks [4] indicated 35 lb of blood produced per 990 lb (16 kg per 450 kg) of live weight.

Wastewater Flow

Water is used in the slaughterhouse for carcass washing after hide removal from cattle, calves, and sheep and after hair removal from hogs. It is also used to clean the inside of the carcass after evisceration, and for cleaning and sanitizing equipment and facilities both during and after the killing operation. Associated facilities such as stockyards, animal pens, the steam plant, refrigeration equipment, compressed air, boiler rooms, and vacuum equipment will also produce some wastewater, as will sanitary and service facilities for staff employed on site: these may include toilets, shower rooms, cafeteria kitchens, and laboratory facilities. The proportions of water used for each purpose can be variable, but as a useful guide the typical percentages of water used in a slaughterhouse killing hogs is shown in Fig. 2 [11].

Johnson [12] classified meat plant wastewater into four major categories, defined as manure-laden; manure-free, high grease; manure-free, low grease; and clear water (Table 4).

The quantity of wastewater will depend very much on the slaughterhouse design, operational practise, and the cleaning methods employed. Wastewater generation rates are usually expressed as a volume per unit of product or per animal slaughtered and there is a reasonable degree of consistency between some of the values reported from reliable sources for different animal types (Table 5). These values relate to slaughterhouses in the United States and

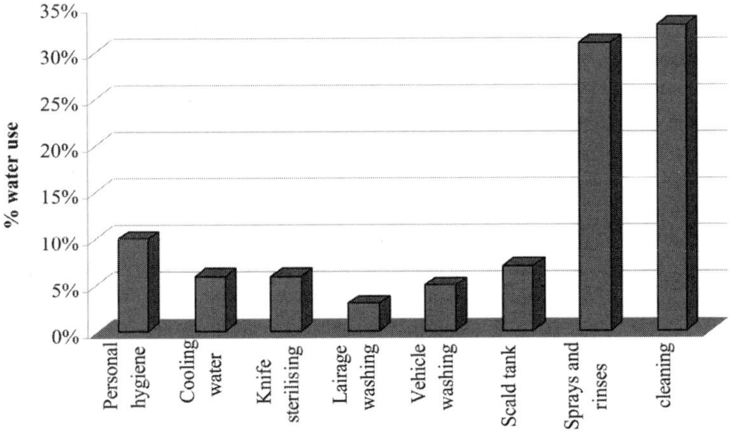

Figure 2 Percentage water use between different operations in a typical slaughterhouse killing hogs (from Ref. 11).

Europe, but the magnitude of variation across the world is probably better reflected in the values given by the World Bank [13], which quotes figures between 2.5 and 40 m^3/ton or tonne for cattle and 1.5–10 m^3/ton or tonne for hogs.

The rate of water use and wastewater generation varies with both the time of day and the day of the week. To comply with federal requirements for complete cleaning and sanitation of equipment after each processing shift [1], typical practise in the United States is that a daily processing shift, usually lasting 8–10 hours, is followed by a 6–8 hours cleanup shift. Although the timing of the processing and cleanup stages may vary, the pattern is consistent across most

Table 4 Examples of Wastewater Types and Arisings from Slaughtering and Processing

Wastewater category	Examples
Manure-laden	Holding pens, gut room washwaters, scald tanks, dehairing and hair washing, hide preparation, bleed area cleanup, laundry, casing preparation, catch basins
Manure-free, high grease water	Drainage and washwater from slaughter floor area (except bleeding and dehairing), carcass washers, rendering operations
Manure-free, low grease water (slaughterhouse)	Washwater from nonproduction areas, finished product chill showers, coolers and freezers, edible and inedible grease, settling and storage tank area, casing stripper water (catch basin effluent), chitterling washwater (catch basin effluent), tripe washers, tripe and tongue scalders
Manure-free, low grease water (cutting rooms, processing and packing)	Washwater from nonproduction areas, green meat boning areas, finished product packaging, sausage manufacture, can filling area, loaf cook water, spice preparation area
Clear water	Storm water, roof drains, cooling water (from compressors, vacuum pumps, air conditioning) steam condenser water (if cooling tower is not used or condensate not returned to boiler feed), ice manufacture, canned product chill water

Source: Ref. 12.

Table 5 Wastewater Generation Rate from Meat Processing

Meat type	Slaughterhouse	Packinghouse	Reference
Cattle	• 312–601 gal/10^3 lb LWK (2604–5015 L/tonne)		14
	• 395 gal/animal (1495 L/animal)	• 2189 gal/animal (8286 L/animal)	15
	• 345–390 gal/10^3 lb LWK (2879–3255 L/tonne)	• 835 gal/10^3 lb LWK (6968 L/tonne)	1
	• 185–264 gal/animal (700–1000 L/animal)		11
	• 256 gal/10^3 lb LWK (2136 L/tonne)		16
	• 185–265 gal/animal (700–1003 L/animal)		17
	• 300–4794 gal/10^3 lb (2500–40,000 L/ tonne)	• 240–7190 gal/10^3 lb (2000–60,000 L/ tonne)	13
Hog	• 243–613 gal/10^3 lb LWK (2028–5115 L/tonne)	• 1143 gal/10^3 lb LWK (9539 L/tonne)	1
	• 155 gal/10^3 lb LWK (1294 L/tonne)	• 435–455 gal/10^3 lb LWK (3630–3797 L/tonne)	18
	• 143 gal/animal (541 L/animal)	• 552 gal/animal (1976 L/animal)	15
	• 60–100 gal/animal (227–379 L/animal)		17
	• 42–61 gal/animal 160–230 L/animal)		11
	• 269 gal/10^3 lb LWK (2245 L/tonne)		19
	• 180–1198 gal/10^3 lb (1500–10,000 L/ tonne)		13
Sheep	• 26–40 gal/animal (100–150 L/animal)		11
Mixed	• 359 gal/animal (1359 L/animal)	• 996 gal/animal (3770 L/animal)	15
		• 38–80 gal/animal (144–189 L/animal)	18
		• 1500 gal/10^3 lb LWK (12,518 L/animal)	12
	• 606–6717 L/10^3 lb LWK (1336–14,808 L/tonne)		20
	• 152–1810 gal/animal (575–6852 L/animal)		21
	• 599–1798 gal/10^3 lb (5000–15,000 L/tonne)		9

LWK, live weight kill.

slaughterhouses worldwide; hence the nature of the wastewater and its temperature will show a marked differentiation between the two stages. During the processing stage water use and wastewater generation are relatively constant and at a low temperature compared to the cleanup period. Water use and wastewater generation essentially cease after the cleanup period until processing begins next day.

Wastewater Characteristics

Effluents from slaughterhouses and packing houses are usually heavily loaded with solids, floatable matter (fat), blood, manure, and a variety of organic compounds originating from proteins. As already stated the composition of effluents depends very much on the type of production and facilities. The main sources of water contamination are from lairage, slaughtering, hide or hair removal, paunch handling, carcass washing, rendering, trimming, and cleanup operations. These contain a variety of readily biodegradable organic compounds, primarily fats and proteins, present in both particulate and dissolved forms. The wastewater has a high strength, in terms of biochemical oxygen demand (BOD), chemical oxygen demand (COD), suspended solids (SS), nitrogen and phosphorus, compared to domestic wastewaters. The actual concentration will depend on in-plant control of water use, byproducts recovery, waste separation source and plant management. In general, blood and intestinal contents arising from the killing floor and the gut room, together with manure from stockyard and holding pens, are separated, as best as possible, from the aqueous stream and treated as solid wastes. This can never be 100% successful, however, and these components are the major contributors to the organic load in the wastewater, together with solubilized fat and meat trimmings.

The aqueous pollution load of a slaughterhouse can be expressed in a number of ways. Within the literature reports can be found giving the concentration in wastewater of parameters such as BOD, COD, and SS. These, however, are only useful if the corresponding wastewater flow rates are also given. Even then it is often difficult to relate these to a meaningful figure for general design, as the unit of productivity is often omitted or unclear. These reports do, however, give some indication as to the strength of wastewaters typically encountered, and some of their particular characteristics, which can be useful in making a preliminary assessment of the type of treatment process most applicable. Some of the reported values for typical wastewater characterization parameters are listed along with the source reference in Table 6. These values could be averaged, but the value of such an exercise would be limited as the variability between the wastewaters, for the reasons previously mentioned, is considerable. At best it can be concluded that slaughterhouse wastewaters have a pH around neutral, an intermediate strength in terms of COD and BOD, are heavily loaded with solids, and are nutrient-rich.

It is therefore clear that for the purposes of design of a treatment facility a much better method of assessing the pollution load is required. For this purpose the typical pollution load resulting from the slaughter of a particular animal could be used, but as animals vary in weight depending upon their age and condition at the time of slaughter, it is better to use the live weight at slaughter as the unit of productivity rather than just animal numbers. Some typical pollution loads per unit of productivity are given in Table 7 along with the source references for different types of slaughtering operations.

Very little information is available on where this pollution load arises within the slaughterhouse, as waste audits on individual process streams are not commonly reported. Nemerow and Agardy [15] describe the content of individual process wastes from a slaughterhouse (Table 8). It can be seen that the two most contaminated process streams are related to blood and paunch contents. Blood and meat proteins are the most significant sources of nitrogen in the wastewater and rapidly give rise to ammonical nitrogen as breakdown occurs.

The wastewater contains a high density of total coliform, fecal coliform, and fecal streptococcus groups of bacteria due to the presence of manure material and gut contents. Numbers are usually in the range of several million colony forming units (CFU) per 100 mL. It is also likely that the wastewater will contain bacterial pathogens of enteric origin such as *Salmonella* sp. and *Campylobacter jejuni*, gastrointestinal parasites including *Ascaris* sp., *Giardia lamblia*, and *Cryptosporidium parvum*, and enteric viruses [1]. It is therefore essential

Table 6 Reported Chemical Compositions of Meat Processing Wastewater

	Type of meat			
Item	Hog	Cattle	Mixed	Reference
pH			7.1–7.4	12
			6.5–8.4	9
			7.0	22
			6.3–10.5	23
		6.7–9.3		24
			6.5–7.2	25
	7.3			26
		6.0–7.5		27
			6.7	28
			7.3–8.0	29
COD (mg/L)			960–8290	9
			1200–3000	30
			583	22
		3000–12,873		24
	3015			26
		2100–3190		27
			5100	28
			12,160–18,768	29
BOD (mg/L)	2220	7237		1
			900–2500	12
			600–2720	9
	1030–1045	448–996	635–2240	15
			700–1800	30
			404	22
			950–3490	23
		900–4620		24
			944–2992	25
	1950			26
		975–3330		27
			3100	28
			8833–11,244	29
Suspended solids (SS) (mg/L)	3677	3574		1
			900–3200	12
			300–4200	15
	633–717	467–820	457–929	30
			200–1000	22
			1375	23
			381–3869	24
		865–6090		26
	283		310	28
			10,588–18,768	29
Nitrogen (mg/L)	253	378		1
			22–510	9
	122	154	113–324	15

(continues)

Table 6 Continued

Item	Type of meat			Reference
	Hog	Cattle	Mixed	
			70–300	30
			152	22
			89–493	23
		93–148		24
			235–309	25
	14.3			26
			405	28
			448–773	29
Phosphorus (mg/L)	154	79		1
		26		24
	5.2			26
			30	28

that slaughterhouse design ensures the complete segregation of process washwater and strict hygiene procedures to prevent cross-contamination. The mineral chemistry of the wastewater is influenced by the chemical composition of the slaughterhouse's treated water supply, waste additions such as blood and manure, which can contribute to the heavy metal load in the form of copper, iron, manganese, arsenic, and zinc, and process plant and pipework, which can contribute to the load of copper, chromium, molybdenum, nickel, titanium, and vanadium.

15.3 WASTEWATER MINIMIZATION

As indicated previously, the overall waste load arising from a slaughterhouse is determined principally by the type and number of animals slaughtered. The partitioning of this load between the solid and aqueous phases will depend very much upon the operational practises adopted, however, and there are measures that can be taken to minimize wastewater generation and the aqueous pollution load.

Minimization can start in the holding pens by reducing the time that the animals remain in these areas through scheduling of delivery times. The incorporation of slatted concrete floors laid to falls of 1 in 60 with drainage to a slurry tank below the floor in the design of the holding pens can also reduce the amount of washdown water required. Alternatively, it is good practise to remove manure and lairage from the holding pens or stockyard in solid form before washing down. In the slaughterhouse itself, cleaning and carcass washing typically account for over 80% of total water use and effluent volumes in the first processing stages. One of the major contributors to organic load is blood, which has a COD of about 400,000 mg/L, and washing down of dispersed blood can be a major cause of high effluent strength. Minimization can be achieved by having efficient blood collection troughs allowing collection from the carcass over several minutes. Likewise the trough should be designed to allow separate drainage to a collection tank of the blood and the first flush of washwater. Only residual blood should enter a second drain for collection of the main portion of the washwater. An efficient blood recovery

Table 7 Pollutant Generation per Unit of Production for Meat Processing Wastewater

	Type of meat			
Parameter	Hog	Cattle	Mixed	Reference
BOD	16.7 lb/10^3 lb or kg/tonne LWK	38.4 lb/10^3 lb or kg/tonne LWK		1
	6.5–9.0 lb/10^3 lb or kg/tonne		1.9–27.6 lb/10^3 lb or kg/tonne	12
			1.1–1.2 lb/hog-unit 2.4–2.6 Kg/hog-unit	18
			8.6–18.0 lb/10^3 lb or kg/tonne	31
Suspended solids	13.3 lb/10^3 lb or kg/tonne	11.1 lb/10^3 lb or kg/tonne		1
			1.2–53.8 lb/10^3 lb or kg/tonne	12
			5.5–15.1 lb/10^3 lb or kg/tonne	31
Total volatile solids (VS)			3.1–56.4 lb/10^3 lb or kg/tonne	12
Grease			0.2–10.2 lb/10^3 lb or kg/tonne	31
Hexane extractables	3.7 lb/10^3 lb or kg/tonne	6.2 lb/10^3 lb or kg/tonne		1
Total Kjeldahl nitrogen	1.3 lb/10^3 lb or kg/tonne	1.2 lb/10^3 lb or kg/tonne		1
Total phosphorus	0.8 lb/10^3 lb or kg/tonne	0.2 lb/10^3 lb or kg/tonne		1
Fecal coliform bacterial	6.2×10^{10} CFU/10^3 lb	2.9×10^{10} CFU/10^3 lb		1
	13.6×10^{10} CFU/tonne	6.4×10^{10} CFU/tonne		

LWK, live weight kill; CFU, colony forming unit.

Table 8 Typical Wastewater Properties for a Mixed Kill Slaughterhouse

Source	SS (mg/L)	Organic-N (mg/L)	BOD (mg/L)	pH
Killing floor	220	134	825	6.6
Blood and tank water	3690	5400	32,000	9.0
Scald tank	8360	1290	4600	9.0
Meat cutting	610	33	520	7.4
Gut washer	15,120	643	13,200	6.0
Byproducts	1380	186	2200	6.7

Original data from US Public Health Service and subsequently reported in Refs. 15 and 33.
SS, suspended solids; BOD, biochemical oxygen demand.

system could reduce the aqueous pollution load by as much as 40% compared to a plant of similar size that allows the blood to flow to waste [18].

The second area where high organic loads into the wastewater system can arise is in the gut room. Most cattle and sheep abattoirs clean the paunch (rumen), manyplies (omasum), and reed (abomasum) for tripe production. A common method of preparation is to flush out the gut manure from the punctured organs over a mechanical screen, and allow water to transport the gut manure to the effluent treatment system.

Typically the gut manure has a COD of over 100,000 mg/L, of which 80% dissolves in the washwater. Significant reductions in wastewater strength can be made by adopting a "dry" system for removing and transporting these gut manures. The paunch manure in its undiluted state has enough water present to allow pneumatic transport to a "dry" storage area where a compactor can be used to reduce the volume further if required. The tripe material requires washing before further processing, but with a much reduced volume of water and resulting pollution load.

The small and large intestines are usually squeezed and washed for use in casings. To reduce water, washing can be carried out in two stages: a primary wash in a water bath with continuous water filtration and recirculation, followed by a final rinse in clean potable water. Other measures that can be taken in the gut room to minimize water use and organic loadings to the aqueous stream include ensuring that mechanical equipment, such as the hasher machine, are in good order and maintained regularly.

Within the slaughtering area and cutting rooms, measures should be adopted to minimize meat scraps and fatty tissue entering the floor drains. Once in the drains these break down due to turbulence, pumping, or other mechanical actions (e.g., on screens), leading to an increase in effluent COD. These measures include using fine mesh covers to drains, encouraging operators to use collection receptacles for trimmings, and using well-designed equipment with catch trays. Importantly, a "dry" cleaning of the area to remove solid material, for example using cyclonic vacuum cleaners, should take place before any washdown.

Other methods can also be employed to minimize water usage. These will not in themselves reduce the organic load entering the wastewater treatment system, but will reduce the volume requiring treatment, and possibly influence the choice of treatment system to be employed. For example, high-strength, low-volume wastewaters may be more suited to anaerobic rather than aerobic biological treatment methods. Water use minimization methods include:

- the use of directional spray nozzles in carcass washing, which can reduce water consumption by as much as 20%;
- use of steam condensation systems in place of scald tanks for hair and nail removal;
- fitting washdown hoses with trigger grips;
- appropriate choice of cleaning agents;
- reuse of clear water (e.g., chiller water) for the primary washdown of holding pens.

15.4 WASTEWATER TREATMENT PROCESSES

The degree of wastewater treatment required will depend on the proposed type of discharge. Wastewaters received into the sewer system are likely to need less treatment than those having direct discharge into a watercourse. In the European Union, direct discharges have to comply with the Urban Waste Water Treatment Directive [32] and other water quality directives. In the United States the EPA is proposing effluent limitations guidelines and standards (ELGs) for the

Meat and Poultry Products industries with direct discharge [1]. These proposed ELGs will apply to existing and new meat and poultry products (MPP) facilities and are based on the well-tested concepts of "best practicable control technology currently available" (BPT), the "best conventional pollutant control technology" (BCT), the "best available technology economically achievable" (BAT), and the "best available demonstrated control technology for new source performance standards" (NSPS). In summary, the technologies proposed to meet these requirements use, in the main, a system based on a treatment series comprising flow equalization, dissolved air flotation, and secondary biological treatment for all slaughterhouses; and require nitrification for small installations and additional denitrification for complex slaughterhouses. These regulations will apply to around 6% of an estimated 6770 MPP facilities.

There is some potential, however, for segregation of wastewaters allowing specific individual pretreatments to be undertaken or, in some cases, bypass of less contaminated streams. Depending on local conditions and regulations, water from boiler houses and refrigerating systems may be segregated and discharged directly or used for outside cleaning operations.

15.4.1 Primary and Secondary Treatment

Primary Treatment

Grease removal is a common first stage in slaughterhouse wastewater treatment, with grease traps in some situations being an integral part of the drainage system from the processing areas. Where the option is taken to have a single point of removal, this can be accomplished in one of two ways: by using a baffled tank, or by dissolved air flotation (DAF). A typical grease trap has a minimum detention period of about 30 minutes, but the period need not to be greater than 1 hour [33]. Within the tank, coagulation of fats is brought about by cooling, followed by separation of solid material in baffled chambers through natural flotation of the less dense material, which is then removed by skimming.

In the DAF process, part of the treated water is recycled from a point downstream of the DAF. The recycled flow is retained in a pressure vessel for a few minutes for mixing and air saturation to take place. The recycle stream is then added to the DAF unit where it mixes with the incoming untreated water. As the pressure drops, the air comes out of solution, forming fine bubbles. The fine bubbles attach to globules of fat and oil, causing them to rise to the surface where they collect as a surface layer.

The flotation process is dependent upon the release of sufficient air from the pressurized fluid when the pressure is reduced to atmospheric. The nature of the release is also important, in that the bubbles must be of reasonably constant dimensions (not greater than 130 microns), and in sufficient numbers to provide blanket coverage of the retaining vessel. In practise, the bubble size and uniform coverage give the appearance of white water. The efficiency of the process depends upon bubble size, the concentration of fats and grease to be separated, their specific gravity, the quantity of the pressurized gas, and the geometry of the reaction vessel.

Figure 3 shows a schematic diagram of a typical DAF unit. The DAF unit can also be used to remove solids after screening, and in this case it usually incorporates chemical dosing to bring about coagulation and flocculation of the solids. When used for this purpose, the DAF unit will remove the need for a separate sedimentation tank.

Dissolved air flotation has become a well-established unit operation in the treatment of abattoir wastes, primarily as it is effective at removing fats from the aqueous stream within a short retention time (20–30 minutes), thus preventing the development of acidity [18]. Since the 1970s, DAF has been widely used for treating abattoir and meat-processing wastes. Some early

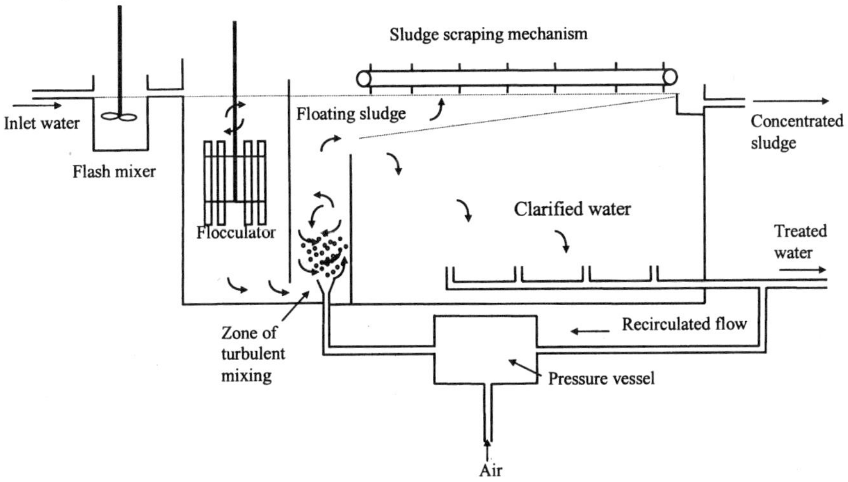

Figure 3 Schematic diagram of typical DAF unit.

texts mention the possibility of fat and protein recovery using DAF separation [9,34]. Johns [14] reported, however, that such systems had considerable operating problems, including long retention times and low surface overflow rates, which led to solids settling, large volumes of putrefactive and bulky sludge with difficult dewatering properties, and sensitivity to flow variations.

DAF units are still extensively used within the industry, but primarily now as a treatment option rather than for product recovery. The effectiveness of these units depends on a number of factors and on their position within the series of operations. The efficiency of the process for fat removal can be reduced if the temperature of the water is too hot (>100°F or 38°C); the increase in fat recovery from reducing the wastewater temperature from 104 to 86°F (40 to 30°C) is estimated to be up to 50% [35]. Temperature reduction can be achieved by wastewater segregation or by holding the wastewater stream in a buffer or flow equalization tank. Operated efficiently in this manner the DAF unit can remove 15–30% COD/BOD, 30–60% SS, and 60–90% of the oil and grease without chemical addition. Annual operating costs for DAF treatment remain high, however, indicating that the situation has not altered significantly since Camin [36] concluded from a survey of over 200 meat packing plants in the United States that air flotation was the least efficient treatment in terms of dollars per weight of BOD removed.

Chemical treatment can improve the pollution removal efficiency of a DAF unit, and typically ferric chloride is used to precipitate proteins and polymers used to aid coagulation. The adjustment of pH using sulfuric acid is also reported to be used in some slaughterhouses to aid the precipitation of protein [37]. Travers and Lovett [38] reported enhanced removal of fats when a DAF unit was operated at pH 4.0–4.5 without any further chemical additions. Such a process would require substantial acid addition, however.

A case study in a Swiss slaughterhouse describes the use of a DAF plant to treat wastewater that is previously screened at 0.5 mm (approx 1/50 inch) and pumped to a stirred equalization tank with five times the volumetric capacity of the hourly DAF unit flow rate [39,40]. The wastewater, including press water returns, is chemically conditioned with iron(III) for blood coagulation, and neutralized to pH 6.5 with soda lime to produce an iron hydroxide floc, which is then stabilized by polymer addition. This approach is claimed to give an average of

80% COD removal, between 40 and 60% reduction in total nitrogen, a flotation sludge with 7% dry solids with a volume of 2.5% of the wastewater flow. The flotation sludge can then be dewatered further with other waste fractions such as slurry from vehicle washing and bristles from pig slaughter to give a fraction with around 33% dry solids.

It must be borne in mind that although chemical treatment can be used successfully to reduce pollution load, especially of soluble proteinaceous material, it results in much larger quantities of readily putrescible sludge. It will, however, significantly reduce the nutrient load onto subsequent biological processes.

In many existing plants a conventional train of unit operations is used, in which solids are removed from the wastewater using a combination of screens and settlement. Screening is usually carried out on a fine-mesh screen (1/8 to 1/4 inch aperture, or 0.3–0.6 cm), which can be of a vibrating, rotating, or mechanically cleaned type. The screen is designed to catch coarse materials such as hair, flesh, paunch manure, and floating solids. Removals of 9% of the suspended solids on a 20-mesh screen and 19% on a 30-mesh screen have been reported [15]. The coarser 20-mesh screen gives fewer problems of clogging, but even so the screen must be provided with some type of mechanism to clean it. In practise mechanically cleaned screens using a brush type of cleaner give the best results. Finer settleable solids are removed in a sedimentation tank, which can be of either a rectangular or circular type. The size and design of sedimentation tanks varies widely, but Imhoff tanks with retentions of 1–3 hours have been used in the past in the United States and are reported to remove about 65% of the suspended solids and 35% of BOD [18]. The use of a deep tank can lead to high head loss, or to the need for excavation works to avoid this. For this reason, longitudinal or radial flow sedimentation tanks are now preferred for new installations in Europe. The usual design criteria for these when dealing with slaughterhouse wastewaters is that the surface loading rate should not exceed 1000 gal/ft^2 day (41 m^3/m^2 day).

As discussed above, the nature of operations within a slaughterhouse means that the wastewater characteristics vary considerably throughout the course of a working day or shift. It is therefore usually necessary to include a balancing tank to make efficient use of any treatment plant and to avoid operational problems. The balancing tank should be large enough to even out the flow of wastewater over a 24 hour period. To be able to design the smallest, and therefore most economical, balancing tank requires a full knowledge of variations in flow and strength throughout the day. This information is often not available, however, and in this case it is usual to provide a balancing tank with a capacity of about two-thirds of the daily flow.

Secondary Treatment

Secondary treatment aims to reduce the BOD of the wastewater by removing the organic matter that remains after primary treatment. This is primarily in a soluble form. Secondary treatment can utilize physical and chemical unit processes, but for the treatment of meat wastes biological treatment is usually favored [41].

Physico-Chemical Secondary Treatment

Chemical treatment of meat-plant wastes is not a common practise due to the high chemical costs involved and difficulties in disposing of the large volumes of sludge produced. There are, however, instances where it has been used successfully. Nemerow and Agardy [15] report a treatment facility that used $FeCl_3$ to reduce the BOD from 1448 to 188 mg/L (87% reduction) and the suspended solids from 2975 to 167 mg/L (94% reduction), with an operation cost of US$68 per million gallons. Using chlorine and alum in sufficient quantities could also significantly reduce the BOD and color of the wastes, but once again the chemical costs are high.

With this approach the BOD of raw wastewaters ranging from 1500 to 3800 mg/L can be reduced to between 400 and 600 mg/L. Dart [18] reported a 64% reduction in BOD using alumina-ferric as a coagulant with a dosing rate equivalent to 17 mg/L of aluminum. Chemical treatment has also been used to remove phosphates from slaughterhouse wastewater. Aguilar et al. (2002) used $Fe_2(SO_4)_3$, $Al_2(SO_4)_3$, and poly-aluminum chloride (PAC) as coagulants with some inorganic products and synthetic polyelectrolytes to remove approximately 100% orthophosphate and between 98.93 and 99.90% total phosphorus. Ammonia nitrogen removal was very low, however, despite an appreciable removal of albuminoidal nitrogen (73.9–88.77%).

The chemical processes described rely on a physical separation stage such as sedimentation, as illustrated in Fig. 4, or by using a DAF unit (see "Primary Treatment" section and Fig. 3). Using this approach coupled with sludge dewatering equipment it is possible to achieve a good effluent quality and sludge cake with a low water content.

Biological Secondary Treatment

Using biological treatment, more than 90% efficiency can be achieved in pollutant removal from slaughterhouse wastes. Commonly used systems include lagoons (aerobic and anaerobic), conventional activated sludge, extended aeration, oxidation ditches, sequencing batch reactors, and anaerobic digestion. A series of anaerobic biological processes followed by aerobic biological processes is often useful for sequential reduction of the BOD load in the most economic manner, although either process can be used separately. As noted above, slaughterhouse wastewaters vary in strength considerably depending on a number of factors. For a given type of animal, however, this variation is primarily due to the quantity of water used within the abattoir, as the pollution load (as expressed as BOD) is relatively constant on the basis of live weight slaughtered. Hence the more economical an abattoir is in its use of water, the stronger the effluent will be, and vice versa. The strength of the organic degradable matter in the wastewater is an important consideration in the choice of treatment system. To remove BOD using an aerobic biological process involves supplying oxygen (usually as a component in air) in proportion to the quantity of BOD that has to be removed, an increasingly expensive process as

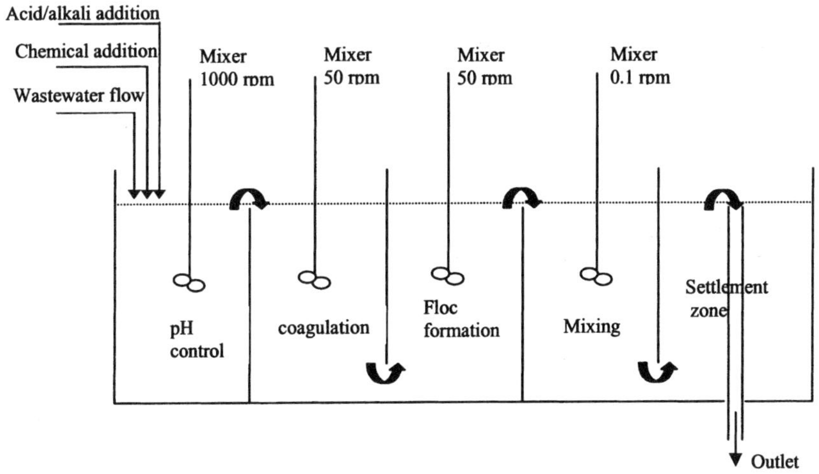

Figure 4 Typical chemical treatment and conditioning system.

the BOD increases. On the other hand an anaerobic process does not require oxygen in order to remove BOD as the biodegradable fraction is fermented and then transformed to gaseous endproducts in the form of carbon dioxide (CO_2) and methane (CH_4).

15.4.2 Anaerobic Treatment

Anaerobic digestion is a popular method for treating meat industry wastes. Anaerobic processes operate in the absence of oxygen and the final products are mixed gases of methane and carbon dioxide and a stabilized sludge. Anaerobic digestion of organic materials to methane and carbon dioxide is a complicated biological and chemical process that involves three stages: hydrolysis, acetogenesis, and finally methanogenesis. During the first stage, complex compounds are hydrolyzed to smaller chain intermediates. In the second stage acetogenic bacteria convert these intermediates to organic acids and then ultimately to methane and carbon dioxide via the methanogenesis phase (Fig. 5).

In the United States anaerobic systems using simple lagoons are by far the most common method of treating abattoir wastewater. These are not particularly suitable for use in the heavily populated regions of western Europe due to the land area required and also because of the difficulties of controlling odors in the urban areas where abattoirs are usually located. The extensive use of anaerobic lagoons demonstrates the amenability of abattoir wastewaters to anaerobic stabilization, however, with significant reductions in the BOD at a minimal cost.

The anaerobic lagoon consists of an excavation in the ground, giving a water depth of between 10 and 17 ft (3–5 m), with a retention time of 5–15 days. Common practise is to provide two ponds in series or parallel and sometimes linking these to a third aerobic pond. The pond has no mechanical equipment installed and is unmixed except for some natural mixing brought about by internal gas generation and surface agitation; the latter is minimized where possible to prevent odor formation and re-aeration. Influent wastewater enters near the bottom of the pond and exits near the surface to minimize the chance of short-circuiting. Anaerobic ponds can provide an economic alternative for purification. The BOD reductions vary widely, although

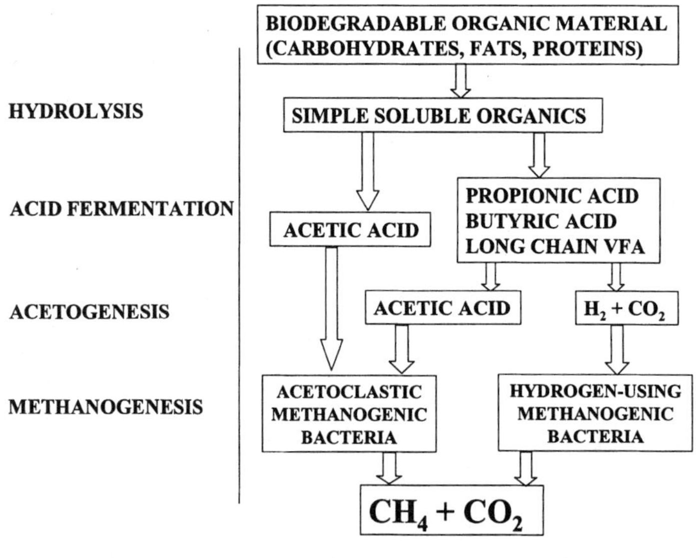

Figure 5 The microbial phases of anaerobic digestion.

excellent performance has been reported in some cases, with reductions of up to 97% in BOD, up to 95% in SS, and up to 96% in COD from the influent values [14,20,42]. Table 9 summarizes some of the literature data on the performance of anaerobic lagoons for the treatment of slaughterhouse wastes. The use of anaerobic lagoons in New Zealand is reported by Cooper et al. [30].

Anaerobic lagoons are not without potential problems, relating to both their gaseous and aqueous emissions. As a result of breakdown of the wastewater, methane and carbon dioxide are both produced. These escape to the atmosphere, thus contributing to greenhouse gas emissions, with methane being 25 times more potent than carbon dioxide in this respect. Gaseous emissions also include the odoriferous gases, hydrogen sulfide and ammonia. The lagoons generally operate with a layer of grease and scum on the top, which restricts the transfer of oxygen through the liquid surface, retains some of the heat, and helps prevent the emission of odor. Reliance on this should be avoided wherever possible, however, since it is far from a secure means of preventing problems as the oil and grease cap can readily be broken up, for example, under storm water flow conditions. Odor problems due to anaerobic ponds have a long history: even in the 1960s when environmental awareness was lower and public threshold tolerances to pollution were higher, as many as nine out of ten anaerobic lagoons in the United States were reported as giving rise to odor nuisance [43]. A more satisfactory and environmentally sound solution is the use of membrane covers that prevent odor release, while at the same time allowing collection of the biogas that can be used as fuel source within the slaughterhouse. This sort of innovation moves the lagoon one step closer to something that can be recognized as a purpose-built treatment system, and provides the opportunity to reduce plant size and improve performance.

The use of fabricated anaerobic reactors for abattoir wastewater treatment is also well established. To work efficiently these are designed to operate either at mesophilic (around 95°F or 35°C) or thermophilic (around 130°F or 55°C) temperatures. Black et al. [47] reported that the practicality of using anaerobic digestion for abattoir wastewater treatment was established in the 1930s. Their own work concerned the commissioning and monitoring of an anaerobic contact process installed at the Leeds abattoir in the UK. The plant operated with a 24 hour retention time at a loading of 29.3 lb BOD/10^3 gal (3.5 kg BOD/m^3) and showed an 88–93% reduction in BOD, giving a final effluent concentration of around 220 mg/L. Bohm [48] conducted trials using a 106 ft^3 (3 m^3) anaerobic contact process at a loading of 21.7 lb BOD/10^3 gal day (2.6 kg BOD/m^3 day), with a removal efficiency of 80%. An economic evaluation of the process showed savings on effluent disposal charges. The review by Cillie et al. [49] refers to work by Hemens and Shurben [50] showing a 95% BOD reduction from an influent BOD of 2000 mg/L.

Table 9 Treatment of Meat Industry Wastes by Anaerobic Lagoon.

Loading rate [lb/10^3 gal day (kg BOD/m^3 day)]	Retention time (days)	Depth [feet (m)]	BOD removal (%)	Reference
–	16	6.9 (2.1)	80	43
1.1 (0.13)	7–8	15.1 (4.6)	60	31
1.6 (0.19)	5	14.1 (4.3)	80	31
1.7 (0.20)	–	10.5 (3.2)	86	31
3.4 (0.41)	3.5	15.1 (4.6)	87	27
1.8 (0.21)	1.2	15.1 (4.6)	58	44
1.3 (0.15)	11	8.9 (2.7)	92	45
1.3 (0.16)	–	15.1 (4.6)	65	46

Gas production was only just sufficient to maintain the digester temperature of 91°F (33°C), however. The Albert Lee plant in Minnesota, Unites States, is also mentioned, in which an anaerobic contact digester with vacuum degassing operating at a retention time of 30 hours achieved a 90% reduction in BOD. Work is also described at the Lloyd Maunder Ltd abattoir in Devon, UK, again using an anaerobic contact digester. This achieved 90% BOD removal, but only a low gas production. In the conclusion of their review Cillie et al. [49] state that the most successful anaerobic plants for industrial waste liquids seem to be those dealing with slaughterhouse and meat-packing wastes.

Kostyshyn et al. [24] used both mesophilic and thermophilic anaerobic contact processes as an alternative to physico-chemical treatment over an eight-month trial period. At a loading rate of 22.9 lb COD/10^3 gal day (2.75 kg COD/m^3 day) and a retention time of 2.5 days they achieved an average of 93.1% BOD removal and 74.9% COD removal. The process appears to be able to operate successfully at loadings of up 20.9 lb COD/10^3 gal day (2.5 kg COD/m^3 day). This is possible because the anaerobic contact process maintains a high biomass density and long solids retention time (SRT) in the reactor by recirculation of sludge from a separation stage, which usually involves sedimentation. The high biomass density, long SRT, and elevated temperature enable a short hydraulic retention time. As with most anaerobic reactor systems, however, they are expensive to install and require close technical supervision.

Anaerobic filters have also been applied to the treatment of slaughterhouse wastewaters. These maintain a long SRT by providing the microorganisms with a medium that they can colonize as a biofilm. Unlike conventional aerobic filters, the anaerobic filter is operated with the support medium submerged in an upflow mode of operation. Because anaerobic filters contain a support medium, there is potential for the interstitial spaces within the medium to become blocked, and effective pretreatment is essential to remove suspended solids as well as solidifiable oils, fats, and grease.

Andersen and Schmid [51] used an anaerobic filter for treating slaughterhouse wastewater, and encountered problems with grease in the startup period. The problem was solved by introducing dissolved air flotation as a pretreatment for the removal of grease. The filter showed between 62 and 93% removal of COD over a trial period of 22 weeks, but the authors concluded that the process required close supervision and emphasized the need for good pretreatment. Arora and Routh [29] also used an anaerobic filter with a 24 hour retention time and loads of up to 58.4 lb COD/10^3 gal day (7.0 kg COD/m^3 day). Treatment efficiency was up to 90% at loadings up to 45.9 lb COD/10^3 gal day (5.5 kg COD/m^3 day). Festino and Aubart [52,53] used an anaerobic filter for wastewaters containing less than 1% solids, but the main focus of their work was on the high solids fraction of abattoir wastes in complete mix reactors. Generally speaking, a safe operational loading range for a mesophilic anaerobic filter appears to be between 16.7 and 25.0 lb COD/10^3 gal day (2–3 kg COD/m^3 day), and at this loading a COD reduction of between 80 and 85% might conservatively be expected.

The third type of high-rate anaerobic system that can be applied to slaughterhouse wastewaters is the upflow anaerobic sludge blanket reactor (UASB). This is basically an expanded-bed reactor in which the bed comprises anaerobic microorganisms, including methanogens, which have formed dense granules. The mechanisms by which these granules form are still poorly understood, but they are intrinsic to the proper operation of the process. The influent wastewater flows upward through a sludge blanket of these granules, which remain within the reactor as their settling velocity is greater than the upflow velocity of the wastewater. The reactor therefore exhibits a long sludge retention time, high biomass density per unit reactor, and can operate at a short HRT.

UASB reactors overcome the limitations of anaerobic contact plant and anaerobic filters, yet their application to slaughterhouse wastewater appears limited to laboratory- and pilot-scale reactors. The reason for this is the difficulties in trying to form stable granules when dealing with slaughterhouse wastewater, and this may be due to the high fat concentrations [54].

Although anaerobic processes have generally shown good results in the treatment of abattoir wastewaters, some problems have also been reported. Nell and Krige [55] comment in their paper on aerobic composting systems that in the anaerobic process the high organic content leads to a resistance to fermentation and there is a tendency towards scum formation. The work carried out at thc Lloyd Maunder Ltd. Plant [49] reports the buildup of scum in the digestion process. Grease was also shown to be a problem in the digester operated by Andersen [51]. Cooper et al. [30], in the paper on abattoir waste treatment in New Zealand, state that the use of anaerobic contact and anaerobic filters is not economic as the energy content in the fat is adsorbed and not really broken down in the anaerobic process. This demonstrates the need for proper pretreatment and for an energy balance as part of the design work.

There is a substantial amount of evidence at laboratory, pilot, and full scale that anaerobic systems are suitable for the treatment of abattoir wastewaters. There is also evidence that with the weaker abattoir wastewaters with BODs around 2000 mg/L, gas production is only just sufficient to maintain reactor temperature as might be predicted from thermodynamics. Table 10 summarizes some results achieved using anaerobic reactors of different types applied to slaughterhouse wastewaters.

15.4.3 Aerobic Treatment

Aerobic biological treatment for the treatment of biodegradable wastes has been established for over a hundred years and is accepted as producing a good-quality effluent, reliably reducing influent BOD by 95% or more. Aerobic processes can roughly be divided into two basic types: those that maintain the biomass in suspension (activated sludge and its variants), and those that retain the biomass on a support medium (biological filters and its variants). There is no doubt that either basic type is suitable for the treatment of slaughterhouse wastewater, and their use is well documented in works such as Brolls and Broughton [9], Dart [31], and Kaul [68], where aerobic processes are compared with anaerobic ones. In selecting an aerobic process a number of factors need to be taken into account. These include the land area available, the head of water available, known difficulties associated with certain wastewater types (such as bulking and stable foam formation), energy efficiency, and excess biomass production. It is important to realize that the energy costs of conventional aerobic biological treatment can be substantial due to the requirement to supply air to the process. It is therefore usual to only treat to the standard required, as treatment to a higher standard will incur additional cost. For example, in order to convert ammonia to nitrate requires 4.5 moles of oxygen for every mole of ammonia converted. In effect this means that a 1 mg/L concentration of ammonia has an equivalent BOD of 4.5 mg/L. It is therefore only usual to aim for the conversion of ammonia to nitrate when this is required.

The most common aerobic biological processes used for the treatment of meat industry wastes are biological filtration, activated sludge plants, waste stabilization ponds, and aerated lagoons.

Waste Stabilization Ponds

A waste stabilization pond (WSP) is the simplest method of aerobic biological treatment and can be regarded as bringing about the natural purification processes occurring in a river in a more restricted time and space. They are often used in countries where plenty of land is available and weather conditions are favorable. In the United States, WSPs with depths of between 1.5 and 9 ft (0.5–2.7 m; typical value 4 ft or 1.2 m) have been used. A typical BOD loading of

Table 10 Anaerobic Treatment of Abattoir Wastes

Reactor type	Loading rate [lb COD/ft^2 day (kg COD/m^3 d)]	Retention time	Temperature (°C)	Removal (%)	Gas production	Reference
Lagoon	0.1–0.6 (0.016–0.068)	10–12 days	Ambient	82.6 (BOD)	–	30
Contact	10.0–18.4 (1.2–2.2)	1–1.7 days	35	–	–	56
AF[a]	16.7 (2.0)	–	–	85.0 (COD)	–	6
AF[a]	45.9 (5.5)	1 day	37	90.5 (COD)	–	29
Two stage	–	1 day	30–40	–	0.2–0.3 m^3 CH_4/kg COD removed	57
AF[a]	6.7–30.0 (0.8–3.6)	1.4 day	32	62–92 (COD)	–	51
AF[a]	35.9–50.1 (4.3–6.0)	0.71 day	35	49–57 (COD)	0.8–2.2 mL CH_4/g COD removed	58
CSTR[b]	7.7 (0.92)	23 days	35	56.6 (COD)	0.2 m^3 CH_4/kg COD removed	59
CSTR[b]	24.3–73.0 lb VS/10^3 gal day (2.9–8.75 g VS/L-day)	12 days	35–55	45–65 (COD)	0.30–0.43 m^3 CH_4/kg COD removed	60
Contact	22.9 (2.75)	2.5 days	35	84.5 (COD)	0.28 m^3 CH_4/kg COD removed	24
UASB[c]	20.9–162.7 (2.5–19.5)	1.7–9 hours	30	53–67	0.82–5.2	61
	25–100 (3.0–12.0)	5–10 hours	20	40–62 (COD)	1.22–3.2 kg CH_4 – COD/m^3 d	
UASB[c]	4.2–167 (0.5–20)	0.5–1.7 days	30	68.4–82.3 (COD)	–	62
Contact	8.3 (1.0)	3.3 days	22	70.0 (COD)	–	63
Contact	133.5 lb TS/10^3 gal day (16 kg TS/m^3 day)	10 days	55	27.0 (TS)	0.08 m^3 CH_4/kg TS added	64
AF[a]	16.7–154.4 (2–18.5)	5–0.5 days	–	27–85 (COD)	–	65
ABR[d]	5.6–39.5 (0.67–4.73)	0.1–1.1 days	25–35	75–90 (COD)	0.07–0.15 m^3 CH_4/kg COD removed	66
Two stage UASB[c]	125.2 (15)	5.5 hours	18	90.0 (COD)	–	67

[a]AF, anaerobic filter
[b]CSTR, classic continuous stirred tank reactor
[c]UASB, upflow anaerobic sludge blanket
[d]ABR, anaerobic baffle reactor.
VS, volatile solids; TS, total solids

20–30 lb BOD/day acre (22–34 kg BOD/ha day) with a typical retention time of 30 to 120 days has been reported [18]. Such ponds are often used in series and can incorporate an anaerobic pond as the first stage (see Section 15.4.2), followed by a facultative pond and maturation ponds. By using a long total retention and low overall BOD loading a good-quality effluent can be achieved. As a stand-alone system the facultative pond may be expected to give between 60 and 90% BOD/COD reduction and between 10 and 20% reduction in total nitrogen. When coupled with maturation lagoons a further 40–70% reduction in BOD/COD can be achieved, primarily as a result of the settlement and breakdown of biomass generated in the facultative pond. This will result in an overall suspended solids reduction of up to 80% [35].

In both the facultative and maturation ponds the oxygen required for the growth of the aerobic organisms is provided partly by transfer across the air/water interface and partly by algae as a result of photosynthesis. This leads to a very low operating cost as there is no requirement for mechanically induced aeration. Conditions in WSPs are not easily controlled due to the lack of mixing, and organic material can settle out near the inlet of the pond causing anaerobic conditions and offensive smells, especially when treating meat industry wastes that contain grease and fat materials. It is therefore not uncommon to find that the facultative pond may also be fitted with a floating surface aerator to aid oxygen transfer and to promote mixing. There is a point, however, when the oxygen input by mechanical means exceeds that naturally occurring by surface diffusion and photosynthesis: at this point the facultative lagoon is best described as an aerated lagoon. The design of a WSP system depends on a number of climatic and other factors: excellent guidance can be found in the USEPA design manual and the work of Mara and Pearson [69,70].

Biological Filters

Biological filters can also be used for treating meat industry wastes. In this process the aerobic microorganisms grow as a slime or film that is supported on the surface of the filter medium. The wastewater is applied to the surface and trickles down while air percolates upwards through the medium and supplies the oxygen required for purification (Fig. 6). The treated water along with

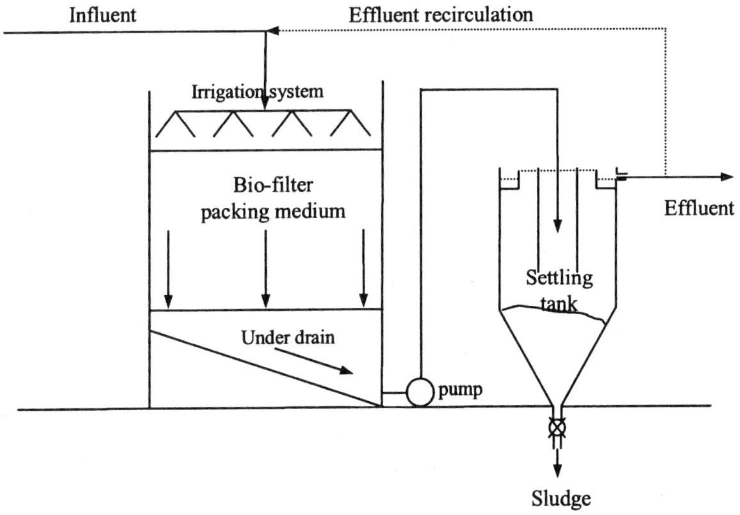

Figure 6 Typical biological filtration treatment system.

any microbial film that breaks away from the support medium collects in an under-drain and passes to a secondary sedimentation tank where the biological solids are separated. Trickling filters require primary treatment for removal of settleable solids and oil and grease to reduce the organic load and prevent the system blocking. Rock or blast furnace slag have traditionally been used as filter media for low-rate and intermediate-rate trickling filters, while high-rate filters tend to use specially fabricated plastic media, either as a loose fill or as a corrugated prefabricated module. The advantage of trickling filters is their low energy requirement, but the disadvantage is the low loading compared to activated sludge, making the plant larger with a consequent higher capital cost. Hydraulic loading rates range from 0.02–0.06 gal/ft^2 day (0.001–0.002 m^3/m^2 day) for low-rate filters to 0.8–3.2 gal/ft^2 day (0.03–0.13 m^3/m^2 day) for high-rate filters. Organic loading rates range from 5–25 lb BOD/10^3 ft^2 day to 100–500 lb BOD/10^3 ft^2 day (0.02–0.12 kg/m^2 day to 0.49–2.44 kg/m^2 day). The overall BOD removal efficiency can be as great as 95%, but this is dependent on the loading applied and the mode of operation. A typical performance envelope for biological filters operating with a plastic support medium is given in Fig. 7.

Because of the relatively high strength of slaughterhouse wastewater, biological filters are more suited to operation with effluent recirculation, which effectively increases surface hydraulic loading without increasing the organic loading. This gives greater control over microbial film thickness. In the United States, high-rate single-stage percolating filters with high recirculation ratios have been used. An overall BOD removal of 92–98% was reported using a high-rate filter with a BOD loading of 2.6–3.8 lb BOD/10^3 gal media day (0.31–0.45 kg BOD/m^3 media day) and a recirculation ratio of about 5 : 1 for treating preliminary treated slaughterhouse wastes [71]. Dart [18] reported that a high-quality effluent with 11 mg/L of BOD and 25 mg/L SS could be obtained using alternating double filtration (ADF) at a loading rate of 2.8 lb/10^3 gal day (0.34 kg BOD/m^3 day) for treating screened and settled abattoir waste; the influent was diluted 1 : 1 with recirculated effluent. Higher loadings with a BOD of between 17 and 33 lb/10^3 gal (2–4 kg BOD/m^3) and a surface hydraulic loading of 884 gal/ft^2 day (1.5 m^3/m^2 hour) and recirculation ratios of 3–4 are given as a typical French design guideline aimed at providing a roughing treatment in reactors 13.1 ft (4 m) high [14]. Such a design is likely to give a BOD removal of less than 75% (Fig. 7) and not to provide any nitrification.

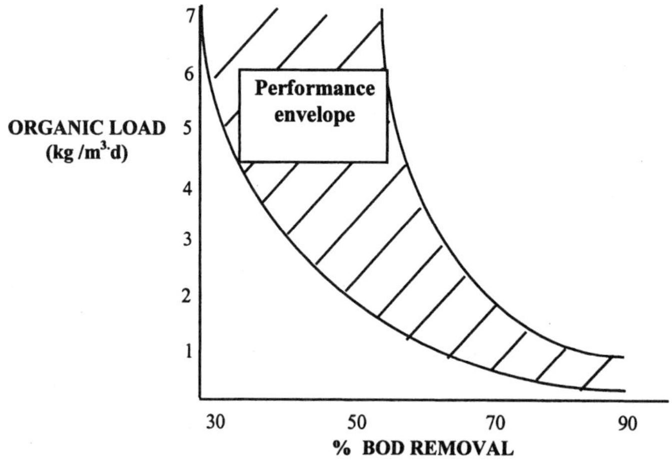

Figure 7 Performance envelope for high rate biological filtration.

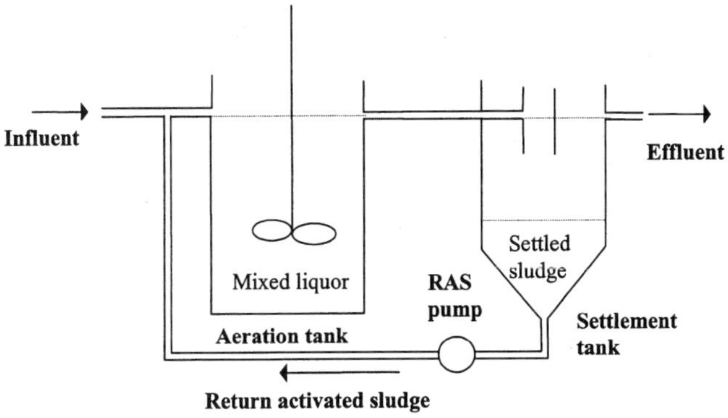

Figure 8 Schematic for a completely mixed continuous flow activated sludge plant.

Dart [31] summarized the performance of some high-rate filtration plants treating meat industry wastes (Table 11).

Biological filters have not been widely adopted for the treatment of slaughterhouse wastewaters despite the lower operating costs compared with activated sludge systems. Obtaining an effluent with a low BOD and ammonia in a single-reactor system can provide conditions suitable for the proliferation of secondary grazing macro-invertebrate species such as fly larvae, and this may be unacceptable in the vicinity of a slaughterhouse. There is also the need for very good fat removal from the influent wastewater flow, as this will otherwise tend to coat the surface of the biofilm support medium. The use of traditional biological filtration for abattoir wastewater treatment is discussed by Philips [72], and further reviewed by Parker et al. [73].

Rotating Biological Contactors

Rotating biological contactors (RBCs) are also fixed biofilm reactors, which consist of a series of closely spaced circular discs mounted on a longitudinal shaft. The discs are rotated, exposing the attached microbial mass alternately to air and to the wastewater being treated, and allowing the adsorption of organic matter, nutrients, and oxygen. Typical design values for hydraulic and

Table 11 Treatment of Meat Industry Wastewaters by High-Rate Biological Filtration

	BOD load		
Medium	(lb/10^3 gal day)	(kg/m^3 day)	BOD removal (%)
Cloisonyle	67.6	8.1	75
Flocor	14.2	1.7	72
Flocor	15.0	1.8	85
Flocor	20.0	2.4	66
Flocor	25.0	3.0	50
Flocor	25.9	3.1	60
Flocor	26.7	3.2	60
Rock	12.5	1.5	61
Unspecified PVC	10.0	1.2	74

Source: Ref. 31.

organic loading rates for secondary treatment are 2–4 gal/ft^2 day (0.08–0.16 m^3/m^2 day) and 2.0–3.5 lb total BOD/10^3 ft^2 day (0.01–0.017 kg BOD/m^2 day) respectively, with effluent BOD concentrations ranging from 15 to 30 mg/L. For secondary treatment combined with nitrification, typical hydraulic and organic loading rate design values are 0.75–2 gal/ft^2 day and 1.5 to 3.0 lb total BOD/10^3 ft^2 day respectively (0.03–0.08 m^3/m^2 day and 0.007–0.014 kg BOD/m^2 day), producing effluent BOD concentrations between 7 and 15 mg/L and NH_3 concentrations of less than 2 mg/L [74]. The above performance figures are typical of this type of unit, but are not necessarily accurate when applied to the treatment of slaughterhouse wastewaters. Bull et al. [75] and Blanc et al. [76] reported that the performance of RBCs appeared inadequate when compared to activated sludge or high-rate biological filtration. Another report of RBC use in slaughterhouse wastewater treatment is given by Bilstad [77], who describes the upgrading of a plant using one of these systems.

Aerated Filters

These comprise of an open tank containing a submerged biofilm support medium, which can be either static or moving. The tank is supplied with air to satisfy the requirements of the bio-oxidation process. There are a number of proprietary designs on the market, but each works on the principle of retaining a high concentration of immobilized biomass within the aerobic reaction tank, thus minimizing the need for secondary sedimentation and sludge recycle. The major differences between the processes are the type of biomass support medium, the mechanism of biofilm control, and whether or not the support medium is fixed or acts as an expanded or moving bed. As an example of the use of such a process, a Wisconsin slaughterhouse installed a moving-bed biofilm reactor (MBBR) to treat a wastewater flow of 168,000 USgal/day, with surge capabilities to 280,000 gal/day (636 and 1060 m^3/day). Average influent soluble BOD and soluble COD concentrations were 1367 mg/L, and 1989 mg/L, respectively. The Waterlink, Inc., process selected used a small polyethylene support element that occupied 50% of the 9357 ft^3 (265 m^3) volume provided by two reactors in series to give 10 hours hydraulic retention time at average flows and six hours at peak hydraulic flow [78]. Effluent from the second MBBR was sent to a dissolved air flotation unit, which removed 70–90% of the solids generated. The average effluent soluble BOD and COD were 59 mg/L and 226 mg/L, respectively.

Activated Sludge

The activated sludge process has been successfully used for the treatment of wastewaters from the meat industry for many decades. It generally has a lower capital cost than standard-rate percolating filters and occupies substantially less space than lagoon or pond systems. In the activated sludge process the wastewaters are mixed with a suspension of aerobic microorganisms (activated sludge) and aerated. After aeration, the mixed liquor passes to a settlement tank where the activated sludge settles and is returned to the plant inlet to treat the incoming waste. The supernatant liquid in the settlement tank is discharged as plant effluent. Air can be supplied to the plant by a variety of means, including blowing air into the mixed liquor through diffusers; mechanical surface aeration; and floor-mounted sparge pipes. All the methods are satisfactory provided that they are properly designed to meet the required concentration of dissolved oxygen in the mixed liquor (greater than 0.5 mg/L) and to maintain the sludge in suspension; for nitrification to occur it may be necessary to maintain dissolved oxygen concentrations above 2.0 mg/L.

The activated sludge process can be designed to meet a number of different requirements, including the available land area, the technical expertise of the operator, the availability of sludge disposal routes, and capital available for construction. Excellent descriptions of the process can be found in many texts: Metcalf & Eddy provides many good examples [74]. The first step in the design of an activated sludge system is to select the loading rate, which is usually defined as the mass ratio of substrate inflow to the mass of activated sludge (on a dry weight basis); this is commonly referred to as the food to microorganism (F:M) ratio and is usually reported as lb BOD/lb MLSS day (kg BOD/kg MLSS day). For conventional operation the range is 0.2–0.6; the use of higher values tends to produce a dispersed or nonflocculent sludge and lower values require additional oxygen input due to high endogenous respiration rates. Systems with F:M ratios above 0.6 are sometimes referred to as high rate, while those below 0.2 are known as extended aeration systems (Table 12). The latter, despite their higher capital and operating costs are commonly chosen for small installations because of their stability, low sludge production, and reliable nitrification. Because of the stoichiometric relationship between F:M ratio and mean cell residence time (MCRT), high-rate plants will have an MCRT of less than 4 days and extended aeration plants of greater than 13 days. Because of the low growth rates of the nitrifying bacteria, which are also influenced markedly by temperature, the oxidation of ammonia to nitrates (nitrification) will only occur at F:M ratios less than 0.1. It is also sometimes useful to consider the nitrogen loading rate, which for effective nitrification should be in the range 0.03–0.08 lb N/lb MLSS-day (kg N/kg MLSS day).

Conventional plants can be used where nitrification is not critical, for example as a pretreatment before sewer discharge. One of the main drawbacks of the conventional activated sludge process, however, is its poor buffering capability when dealing with shock loads. This problem can be overcome by the installation of an equalization tank upstream of the process, or by using an extended aeration activated sludge system. In the extended aeration process, the aeration basin provides a 24–30 hour (or even longer) retention time with complete mixing of tank contents by mechanical or diffused aeration. The large volume combined with a high air input results in a stable process that can accept intermittent loadings. A further disadvantage of using a conventional activated sludge process is the generation of a considerable amount of surplus sludge, which usually requires further treatment before disposal. Some early work suggested the possible recovery of the biomass as a source of protein [30,79], but concerns over the possible transmission of exotic animal diseases would make this unacceptable in Europe [80]. The use of extended aeration activated sludge or aerated lagoons minimizes biosolids production because of the endogenous nature of the reactions. The size of the plant and the additional aeration required for sludge stabilization does, however, lead to increased capital and operating costs. Considering the high concentrations of nitrogen present in slaughterhouse wastewater, ammonia removal is often regarded as essential from a regulatory standpoint for direct discharge, and increasingly there is a requirement for nutrient removal. It is therefore not surprising that most modern day designs are of an extended aeration type so as to promote

Table 12 Classification of Activated Sludge Types Based on the F:M Ratio Showing Appropriate Retention Times and Anticipated Sludge Yields

Mode of operation	F:M ratio	Retention time (hours)	Sludge yield [lb SS/ lb BOD (kg SS/kg BOD)]	BOD removal efficiency (%)
High rate	0.6–0.35	1	1.0	60–70
Conventional	0.2–0.6	6–10	0.5	90–95
Extended aeration	0.03–0.2	24+	0.2	90–95

Typical values derived from a wide range of sources.

reliable nitrification as well as to minimize sludge production. Efficient designs will also attempt to recover the chemically bound oxygen in nitrate through the process of denitrification, thus reducing treatment costs and lowering nitrate concentrations in the effluent.

Design criteria and loadings for activated sludge treatment have been widely reported and reliable data can be found in a number of reports [9,34,81–83].

In recent years a great deal of interest has been shown in the use of sequencing batch reactors (SBRs) for food-processing wastewaters, as these provide a minimum guaranteed retention time and produce a high-quality effluent. A batch process also often fits well with the intermittent discharge of an industrial process working on one or two shifts. Advantages are an ideal plug flow that maximizes reaction rates, ideal quiescent sedimentation, and flow equalization inherent in the design. Decanting can be achieved using floating outlets and adjustable weirs, floating aerators are commonly employed, and an anoxic fill overcomes problems of effluent turbidity [84] as well as providing ideal conditions for denitrification reactions.

Hadjinicolaou [85] described using a pilot-scale SBR activated sludge system for the treatment of cattle slaughterhouse wastewaters. The system was operated on a 24 hour cycle and 97.8% of COD removal was achieved with an influent COD concentration of 3512 mg/L. A case study showing the use of an SBR in conjunction with an anaerobic lagoon has shown the potential of the system for both organic load reduction and nutrient removal [86]. The main effluent stream from the slaughterhouse containing some blood, fat, and protein enters a save-all for separation of fat and settleable solids. The flow is then equally split, one portion going to the anaerobic lagoon, which also receives clarified effluent from settling pits that are used to collect manure and paunch material, and the other to the SBR; the effluent from the lagoon subsequently also enters the SBR. The average ratio of BOD : total N entering the SBR is 3 : 1, which provides sufficient carbon to achieve complete nitrification and denitrification. The SBR has a cycle during which filling takes place over an 11 hour period corresponding to the daily operation and cleaning cycle of the slaughterhouse. The aeration period is 12 hours, settlement 1 hour and decanting to a storage lagoon over 3 hours. The total volume of the plant is 0.33 Mgal (1250 m^3), with a daily inflow of 66,000 gal (250 m^3) at a BOD of 600–800 mg/L, total N of 200 mg/L, and total P of 40 mg/L. The plant is reported to achieve a final effluent with values less than 2 mg/L NH_3-N, 10 mg/L NO_3-N, total P 20 mg/L, BOD 20 mg/L, and SS 20 mg/L. Additionally, all surplus activated sludge from the SBR is returned to the anaerobic lagoon.

15.5 SOLID WASTES

If good operational practise is followed in the slaughterhouse, the solids and organic loading entering the aqueous phase can be minimized. The separated solids still require treatment prior to disposal, however, and traditional rendering of some of these fractions is uneconomic because of the high water and low fat content. These fractions are the gut manures, the manure and bedding material from holding pens, material from the wastewater screens and traps on surface drains, sedimentation or DAF sludge, and possibly hair where no market exists for this material. Other high-protein and fat-containing residues such as trimmings, nonedible offal, and skeletal material can be rendered to extract tallow and then dried to produce meat and bone meal. The traditional rendering process is not within the scope of the present chapter, but consideration is given to the disposal of the other fractions as these may appear in the form of a wastewater sludge, although in an efficient slaughterhouse they would be “dry” separated.

Manures from stockyards and holding pens are likely to be similar in composition to the animal slurries that are generated on a farm. Typical characteristics of these are reported by Gendebien et al. [5] and are shown in Table 13. It is important that as much of the material as

Table 13 Comparison of Farm Collected Animal Slurries to Manure Washed From a Stockyard Cleaning Operation

	Concentration (g/L)		
Characteristics	Stockyard[a]	Cattle slurry[b]	Pig slurry[b]
Total suspended solids	0.173	10–180	10–180
Volatile suspended solids	0.132	10–107	34–70
Organic nitrogen	0.011	2–18	2–16
Ammonia nitrogen	0.08	0.6–2.2	2.1–3.6
BOD	0.64	27[c]	37[c]

[a]Derived from Nemerow and Agardy [15].
Source: Refs. 5, 15, 87
[b]Derived from Gendebien et al. [5].
[c]Derived from American Society of Agricultural Engineers [87].
Source: Refs. 5, 15, 87

possible is removed for further processing before the holding pen area is washed down as otherwise this will result in a high-volume, high water content waste flow that can only be handled in the wastewater treatment system. For example, results from a survey of Chicago stockyards (Table 13) by Nemerow and Agardy [15] showed the effluent to be weak in strength compared to animal slurries collected on farms for land spreading.

Gut manures that can be dry separated for separate processing also have a very high pollutant load and solids content, as indicated in Table 14.

15.5.1 Land Disposal

Land disposal of abattoir solid wastes, either by land spreading or landfill, has been a common practise for many years, but concern over the transmission of exotic animal diseases has already seen a decline in this practise in Europe over the past several years. The EU Animal By-products Regulations [80] now prohibit land disposal of all animal wastes with the exception of manures and digestive tract contents, and these only when "the competent authority does not consider them to present a risk of spreading any serious transmissible disease." The only restriction on digestive tract contents is that spreading is limited to nonpasture land. In the EU operators will

Table 14 Analysis of Paunch Contents of Ruminant Animals and Mixed Gut Material

Type of animal	Total solids (%)	COD [lb/10^3 lb (g/kg)]	BOD (mg/L)	Reference
Cattle			50,000	7
	14.2	134		88
	12.7	134		88
	14			8
Sheep			30,000	7
Lamb	9.3	165		88
Pig	26			8
Mixed kill	4.7–9.7			7
	22		50,000	9
	2.4–21		6000–41,000	5

also no longer be able to spread untreated blood from abattoirs onto land or dispose of it down wastewater drainage systems for treatment by water companies. The blood will need to be treated in an approved rendering, biogas, or composting plant before it can be land-spread. The regulations will vary from country to country, but it should be noted that land-spreading of any abattoir waste is liable to cause public nuisance through odor and environmental concerns, and has potential for disease transmission. It is therefore beneficial to treat the waste by a stabilization process before land application, and where this is not possible it is imperative that land application is undertaken with great care. The rate of application of the waste should be based on the level of plant nutrients present.

Where there are no country-specific regulations, as a general rule all abattoir wastes should be injected into the soil to reduce odor and avoid any potential pathogen transmission, and should not be surface-spread on pasture land or forage crops. If these materials are surface-spread on arable land, they should be incorporated immediately by plowing. Injection into grassland should be followed by a minimum interval of three weeks before the grass is used for grazing or conservation. Storage time for the materials should be kept to a minimum to avoid further development of odors [5].

The regulations currently in force in the United States for the land application of slaughterhouse-derived biosolids are given in the USEPA's Guide to Field Storage of Biosolids [89].

15.5.2 Composting

Composting can be defined as the biological degradation of organic materials under aerobic conditions into relatively stable products, brought about by the action of a variety of microorganisms such as fungi, bacteria, and protozoa. The process of composting may be divided into two main stages: stabilization and maturation. During stabilization, three phases may be distinguished: first a phase of rising temperature, secondly the thermophilic phase where high temperature is maintained, and thirdly the mesophilic phase where the temperature gradually decreases to ambient. During the first phase, a vigorous multiplication of mesophilic bacteria is observed, and a transformation of easily oxidized carbon compounds, rich in energy, to compounds of lower molecular structure takes place. Excess energy results in a rapid rise in temperature and leads to the thermophilic phase when less easily degraded organic compounds are attacked. When the energy source is depleted the temperature decreases gradually to ambient. Actinomycetes and fungi become active in the mesophilic phase, during which biological degradation of the remaining organic compounds is slowly continued. At this stage the composting material is stabilized but not yet matured. During maturation, mineralization of organic matter continues at a relatively slow rate until a carbon : nitrogen (C : N) ratio of 10 : 1 is reached and the volatile matter content falls below 50%. Only then is the compost completely matured.

Composting of dewatered sedimentation tank solids from a slaughterhouse in mechanically turned open windrows was described by Supapong [90]. The material was kept in windrows for 40 days, and the temperature fluctuated between 149 and 158°F (65–70°C) for up to three weeks. The resulting product was a friable, odorless, and microbiologically satisfactory material whose bulk was only one-fifth of the original volume. It contained 0.5 and 3% by weight of phosphorus and nitrogen, respectively, and was an acceptable soil conditioner. Nell and Krige [55] conducted an in-vessel composting process for treating solid wastes mainly consisting of paunch and pen manure. The reactor was an insulated rotating stainless steel drum of 10 ft^3 (280 L) capacity. After four days retention in the reactor, the waste reached the stabilization stage, and after a further 50 days the composting was completed. The final product had the following characteristics: pH 8.6, 65.1% moisture content, 55.3% of volatile matter, 2.1% of nitrogen, and 17.9% of carbon. The Australian Environmental Protection Authority [91] suggests paunch contents can be efficiently and economically disposed of by composting as long as offensive odors are not generated, and state that the most suitable composting techniques are turned windrows and aerated static piles.

15.5.3 Anaerobic Digestion

Anaerobic digestion of abattoir solid wastes is not common in the United States, UK, or elsewhere, despite the potential for stabilization of the solid residues with the added bonus of fuel gas production. Cooper et al. [30] looked at the potential in New Zealand for production of methane from both the solid and liquid fraction of abattoir wastes. Based on tests carried out by Buswell and Hatfield in 1939, they concluded that paunch contents and fecal matter would not give an economic return. In these very early tests it was reported that a retention time of 38–40 days might be required and that the expected gas yield would be 2500 ft^3/lb solids added (156 m^3/kg). In the UK the first of a new generation of well-mixed digestion plants to treat slaughterhouse wastes was installed in 1984 to treat all the paunch wastes, blood, and settlement tank solids produced by a small abattoir in Shropshire. The operation and performance of a 3531 ft^3 (100 m^3) demonstration-scale anaerobic digester treating cattle and lamb paunch contents, blood, and process wastewaters from a slaughterhouse was described by Banks [4]. Anaerobic digestion of the solid fraction of abattoir wastes suffers from low methane production and solid reduction as well as requiring a longer retention time compared to sewage and food processing wastes [30]. Steiner et al. [60] reported the failure of a digester when treating a mixture of abattoir wastes. The mixture contained 13% of rumen and intestine contents, 25% of manure from animal buildings, 44% of surplus sludge from an aerobic sewage treatment plant, and 19% fat derived from the fat separator, and exhibited a COD of 165 g/L, a BOD of 112 g/L, a dry weight of 120 g/L, and a volatile solids concentration of 105 g/L consisting of 25% fat and 23% protein. The experiment was carried out in a cylindrical completely mixed reactor with a capacity of 0.07 ft^3 (2 L). When the organic loading rate was raised to more than 73 lb VS/10^3 gal day (8.75 g VS/L day), digestion failure occurred and was caused by enrichment of volatile acids in the digester. In his paper, Banks [4] also mentioned serious problems associated with the accumulation of ammonia concentration in the process. Several other authors also indicate that where blood and fat form a significant proportion of the feedstock it is found to be digestible in only limited quantities due to an inhibitory effect on methanogenesis, thought to be caused by accumulation of toxic intermediates produced by the hydrolysis/acidification stages [57,92,93].

Using a two-stage anaerobic process, Banks and Wang [94] successfully overcame the toxicity problems associated with the accumulation of ammonia and volatile fatty acids when treating a mixture of cattle paunch contents and cattle blood. The first-stage reactor was operated in a hydraulic flush mode to maintain a significantly shorter liquid retention time than the solids retention time of the fibrous components in the feedstock. The first-stage reactor was run in this mode using solids retention times of 5, 10, 15, 20, and 30 days with liquid retention of between 2 and 5 days. Up to 87% solid reductions were achieved compared to a maximum of 50% when the control reactor was operated in single-pass mode with solids and liquid retentions of equal duration. The liquid effluent from the first stage hydrolysis reactor was treated by a second-stage completely-mixed immobilized-cell digester. Operated at a retention time of between 2 and 10 days with loading rates in the range of 36–437 lb/10^3 ft^3 day (0.58–7.0 kg COD/m^3 day), the second stage reactor achieved a COD removal of 65–78% with a methane conversion efficiency between 2 and 4 ft^3 CH_4/lb COD removed (0.12–0.25 m^3 CH_4/kg COD removed).

Other than these few reports there has been little research on the anaerobic digestion of the solid waste fraction and it is clear that certain conditions and waste types lead to operational instability. Early work questions the economic viability of the digestion process when used only for the treatment of paunch content and intestinal fecal material and it may be necessary to look at the codigestion of slaughterhouse waste fractions with other waste materials. One successful operation is the Kristianstad biogas plant in Sweden, which coprocesses organic household waste, animal manure, gastrointestinal waste from two slaughterhouses, biosludge from a distillery, and some vegetable processing waste [95]. The slaughterhouse waste fraction is 24,600 tonnes per annum of a total throughput of 71,200 tonnes which is treated in the 1.2 Mgal (4500 m^3) digester. The plant biogas production was equivalent to 20,000 MWh and the digester residue is returned to the land as a fertilizer. The plant represents an environmentally friendly method of waste treatment and appears to have overcome the problems of trying to digest slaughterhouse solid wastes in isolation.

REFERENCES

1. USEPA. *Development document for the proposed effluent limitations guidelines and standards* for the meat and poultry products industry. Office of Water (4303T), United States Environmental Protection Agency (USEPA), EPA-821-B-01-007, 2002.
2. Council of the European Communities. Council Directive on health problems affecting the production and marketing of meat products and certain other products of animal origin. *Official J. Eur. Comm.* **1977**, *L 026* (31 October 1977), 0085–0100.
3. Warris, P.D. *Meat Science: An Introductory Text*; CABI Publishing: New York, 2000.
4. Banks, C.J. Anaerobic digestion of solid and high nitrogen content fractions of slaughterhouse wastes. In *Environmentally Responsible Food Processing*; Niranjan, K., Okos, M.R., Rankowitz, M., Eds.; Vol. AIChE Symposium Series. American Institute of Chemical Engineers: New York, 1994; 103–109.
5. European Commission. *Survey of wastes spread on land – Final report*; Gendebien, A., Ferguson, R., Brink, J., Horth, H., Sullivan, M., Davis, R., Brunet, H., Dalimier, F., Landrea, B., Krack, D., Perot, J., and Orsi, C; Report No. CO 4953–2; Directorate-General for Environment, 2001.
6. Denmead, C.F. *Methane production from meat industry wastes and other potential methods for their utilization*, Publication no. 602; Meat Industry Research Institute; Hamilton, New Zealand, 1977.
7. Fernando, T. Utilization of Paunch Content Material by ultrafiltration. *Process Biochem.* **1980**, *15*, 7–9.
8. Pollack, H. Biological waste disposal from slaughterhouses. In *Anaerobic Digestion and Carbohydrate Hydrolysis of Wastes*; Ferrero, G.L., Ferranti, M.P. Naveau, H., Eds.; Elsevier Applied Science Publishers: London and New York, 1984: 323–330.
9. Brolls, E.K.; Broughton, M. The treatment of effluents arising from the animal by-products industry. In *Food Industry Wastes: Disposal and Recovery*; Herzka, A., Booth, R.G., Eds.; Applied Science Publishers: London and New Jersey, 1981; 184–203.
10. Grady, C.P.L.; Lim, H.C. *Biological Wastewater Treatment: Theory and Applications*; Marcel Dekker Inc: New York, 1980.
11. AEA Technology. *Environmental Technology Best Practice Programme: Reducing Water and Effluent Costs in Red Meat Abattoirs*, Report no. GG234; AEA Technology: Harwell, UK, 2000.
12. Johnson, A.S. Meat. In *Industrial Wastewater Control*; Gurnham, C.F., Ed.; Academic Press: New York and London, 1965.
13. World Bank. Meat processing and rendering. In *Pollution Prevention and Abatement Handbook*; World Bank: Washington DC, 1988: 336–340.
14. Johns, M.R. Developments in wastewater treatment in the meat processing industry: a review. *Biores. Technol.* **1995**, *54*, 203–216.

15. Nemerow, N.L.; Agardy, F.J. Origin and characteristics of meat-packing wastes. In *Strategies of Industrial and Hazardous Waste Management*; Agardy, F.J., Ed., Van Nostrand Reinhold: New York, 1998; 427–432.
16. Carawan, R.E.; Pilkington, D.H. Reduction in load from a meat-processing plant – beef. Randolph Packing Company/North Carolina Agricultural Extension Service, 1986.
17. UNEP. Cleaner production in meat processing. COWI Consulting Engineers/UNEP/Danish Environmental Protection Agency, 2000.
18. Dart, M.C. Treatment of meat trade effluents. In *Practical Waste Treatment and Disposal*; Dickinson, D., Ed.; Applied Science Publishers Ltd: London, **1974**; 75–86.
19. Denker, D.O.; Grothman, D.L.; Berthouex, P.M.; Scully, L.J.P.; Kerrigan, J.E.O. *Characterization and potential methods for reducing wastewater from in-plant hog slaughtering operations*; Interim Report to Mayer and Company/University of Wisconsin, 1973.
20. USEPA. *Development document for effluent limitation guidelines and new source performance standards for red meat segment of the meat product and rendering processing point source category*; Effluent Guidelines Division, Office of Air and Water Programs, USEPA, EPA-440/1–74–012a, 1974.
21. Macon, J.A.; Cote, D.N. *Study of meat packing wastes in North Carolina*; Industrial Extension Service, School of Engineering, North Carolina State College, 1961.
22. Millamena, S.M. Ozone treatment of slaughterhouse and laboratory wastewaters. *Aquacult. Eng.* **1992**, *11*, 23–31.
23. Lloyd, R.; Ware, G.C. Anaerobic digestion of waste waters from slaughterhouses. *Food Manuf.* **1956**, *31*, 511–515.
24. Kostyshyn, C.R.; Bonkoski, W.A.; Sointio, J.E. Anaerobic treatment of a beef processing plant wastewater: a case history. *Proceedings of 42nd Industrial Waste Conference*, Purdue University, Lafayette, IN, 1987, 673–692.
25. Jayangoudar, I.; Thanekar, A.; Krishnamoorthi, K.P.; Satyanarayana, S. Growth potentials of algae in anaerobically treated slaughterhouse waste. *Indian J. Environ. Health* **1983**, *25*, 209–213.
26. Gariepy, S.; Tyagi, R.D.; Couillard, D.; Tran, F. Thermophilic process for protein recovery as an alternative to slaughterhouse wastewater treatment. *Biol. Waste.* **1989**, *29*, 93–105.
27. Enders, K.E.; Hammer, M.; Weber, C.L.; Anaerobic lagoon treatment of slaughterhouse waste. *Water Sewage Works* **1968**; *115*, 283–288.
28. Borja, R.; Banks, C.J.; Wang, Z. Performance and kinetics of an Upflow Anaerobic Sludge Blanket (UASB) reactor treating slaughterhouse wastewater. *J. Environ. Sci. Heal. A* **1994**, *29*, 2063–2085.
29. Arora, H.C.; Routh, T. Treatments of slaughterhouse effluents by anaerobic contact filter. *Indian Assoc. Water Pollut. Control* **1980**, *16*, 67–78.
30. Cooper, R.N.; Heddle, J.F.; Russell, J.M. Characteristics and treatment of slaughterhouse effluents in New Zealand. *Prog. Wat. Treat.* **1979**, *11*, 55–68.
31. Dart, M.C. Treatment of waste waters from the meat industry. *Process Biochem.* **1974**, *9*, 11–14.
32. Council of the European Communities. Urban Waste Water Treatment Directive 91/271/EEC. *Off. J. Eur. Comm.*, **1991**, *L135/40-52*, (30 May 1991).
33. Eldridge, E.F. Meat-packing and slaughterhouse wastes. In *Industrial Waste Treatment Practice*; McGraw-Hill, London and New York, 1942.
34. Hopwood, D. Effluent treatment in meat and poultry processing industries. *Process Biochem.* **1977**, *12*, 5–8.
35. Meat and Livestock Australia Ltd. *Eco-Efficiency Manual for Meat Processing*, ABN 39 081 678 364 (MLA), 2002.
36. Camin, K.Q. Cost of waste treatment in the meat packing industry. In *Proceedings of 25th Purdue Industrial Waste Conference*, Purdue University Lafayette, IN, 1970; 193–202.
37. Masse, D. I.; Masse, L. Characterization of wastewater from hog slaughterhouses in Eastern Canada and evaluation of their in-plant wastewater treatment systems. *Can. Agr. Eng.* **2000**, *42*, 131–137.
38. Travers, S.M.; Lovett, D.A. Pressure flotation of slaughterhouse wastewaters using carbon dioxide. *Water Res.* **1985**, *19*, 1479–1482.

39. Hans Huber. *Wastewater treatment in slaughterhouses and meat processing factories*; Technical brochure; Hans Huber AG, Maschinen-und-Andagenbau: Berching, Germany, 2002.
40. Hans Huber. *Slaughterhouse wastewater treatment in combination with flotate sludge dewatering*, practice report/application info; Hans Huber AG, Maschinen-und-Andagenbau: Berching, Germany, 2002.
41. Peavy, H.S.; Rowe, D.R.; Tchobanoglous, G. *Environmental Engineering*; McGraw-Hill: New York, 1986.
42. USEPA. *Development document for effluent limitation guidelines and new source performance standards for the poultry segment of the meat product and rendering processing point source category*; USEPA, EPA-440/1-75-031b, 1975.
43. Steffen, A.J. Stabilisation ponds for meat packing wastes. *J. Wat. Pollut. Control Fed.* **1963**; *35*, 440–444.
44. Rollag, D.A.; Dornbush, J.N. Design and performance evaluation of an anaerobic stabilization pond system for meat-processing wastes. *J. Wat. Pollut. Control Fed.* **1966**, *38*, 1805–1812.
45. Witherow, J.L. Small meat-packers waste treatment systems. In *Proceedings of 28th Industrial Waste Conference*, Purdue University, Lafayette, IN, **1973**; 994–1009.
46. Wymore, A.H.; White, J.E. Treatment of slaughterhouse waste using anaerobic and aerated lagoons. *Water Sewage Works* **1986**; *115*, 492–498.
47. Black, M.G.; Brown, J.M.; Kaye, E. Operational experiences with an abattoir waste digester plant at Leeds. *Water Pollut. Control*, **1974**, (*73*), 532–537.
48. Bohm, J.L. Digestion anaerobie des effluents d'abattoirs dans une unite pilote de 3000 litres epuration et production d'energie. *Entropie* **1986**, 130–131, 83–87.
49. Cillie, G.G.; Henzen, M.R.; Stander, G.J.; Baillie, R.D. Anaerobic digestion IV. The application of the process in waste purification. *Water Res.* **1969**, *3*, 623–643.
50. Hemens, J.; Shurben, D.G. Anaerobic digestion of wastewaters from a slaughterhouse. *Food Trade Rev.* **1959**, *29*, 2–7.
51. Andersen, D.R.; Schmid, L.A. Pilot plant study of an anaerobic filter for treating wastes from a complex slaughterhouse. In *Proceedings of 40th Purdue Industrial Waste Conference*, Purdue University, Lafayette, IN, 1985; 87–98.
52. Festino, C.; Aubart, C. Optimisation of anaerobic digestion of slaughterhouse wastes and mixtures of animal wastes with sewage sludges and slaughterhouse wastes. *Entropie* **1986**, 130–131, 20.
53. Narbonne, C.; Fromantin-Souli. Epuration d'effluents liquides et valorisation energetique de dechets solides d'abattoir par voie anaerobie. *Entropie* **1986**, 130–131, 57–60.
54. Rajeshwari, K.V.; Balakrishnan, M.; Kansal, A.; Lata, K.; Kishore, V.V.N. State-of-the-art of anaerobic digestion technology for industrial wastewater treatment. *Renew. Sust. Energy. Rev.* **2000**, *4*, 135–156.
55. Nell, J.H.; Krige, P.R. The disposal of solid abattoir waste by composting. *Water Res.* **2000**, *5*, 1177–1189.
56. Meat Industry Research Institute. The anaerobic treatment of effluent from meat processing operations; Denmead, C.F. Publication no. 405, 1974.
57. Vollmer, H.; Scholz, W. Recycle international *In 4th International Recycling Congress*; Thome-Kozmiensky, K.J., Ed. Berlin, **1984**; 667–671.
58. Wheatley, A.D.; Cassell, L. Effluent treatment by anaerobic biofiltration. *Water Pollut. Control* **1985**, *84*, 10–22.
59. Campagna, R.; Del Medico, G.; Pieroni, M. Methane from biological anaerobic treatment of industrial organic wastes. In *3rd EC Conference on Energy from Biomass*; Elsevier Applied Science Publishers: London & New York, 1985.
60. Steiner, A.E.; Wildenauer, F.X.; Kandler, O. Anaerobic digestion and methane production from slaughterhouse wastes. In *3rd Energy Conservation Conference*; Elsevier Applied Science Publishers: London & New York, 1985.
61. Sayed, S.; van Campen, L.; Lettinga, G. Anaerobic treatment of slaughterhouse waste using a granular sludge UASB reactor. *Biol. Waste.* **1987**; *21*, 11–28.

62. Sayed, S.; De Zeeuw, W. The performance of a continuously operated flocculent sludge UASB reactor with slaughterhouse wastewater. *Biol. Waste.* **1988**, *24*, 213–226.
63. Ross, W.R. Anaerobic treatment of industrial effluents in South Africa. *Water SA* **1989**, *15*, 231–246.
64. Marchaim, U.; Levanon, D.; Danai, O.; Musaphy, S. A suggested solution for slaughterhouse wastes: uses of the residual materials after anaerobic digestion. *Biores. Technol.* **1991**, *37*, 127–134.
65. Tritt, W.P. The anaerobic treatment of slaughterhouse wastewater in fixed-bed reactors. *Biores. Technol.* **1992**, *41*, 201–207.
66. Polprasert, C.; Kemmadamrong, P.; Tran, F.T. Anaerobic Baffle Reactor (ABR) process for treating a slaughterhouse wastewater. *Environ. Technol.* **1992**; *13*, 857–865.
67. Sayed, S.K.I.; van der Spoel, H.; Truijen, G.J.P. A complete treatment of slaughterhouse wastewater combined with sludge stabilization using two stage high rate UASB process. *Water Sci. Technol.* **1993**, *27*, 83–90.
68. Kaul, S.N. Biogas from industrial wastewaters. *J. IPHE India* **1986**, *3*, 5–17.
69. Mara, D.D.; Pearson, H.W. *Design Manual for Waste Stabilisation Ponds in Mediterranean Countries*; Lagoon Technology International: Leeds, 1998.
70. USEPA. *Design Manual: Municipal Wastewater Stabilization Ponds*; USEPA, EPA 625/1-83-016, 1983.
71. US Department of Health Education and Welfare. *An Industrial Waste Guide to the Meat Industry*; USDHEW. US Public Health Service Publication No. 386, 1958.
72. Philips, S.A. Wastewater treatment handles cattle killing waste. *Water Sewage Works* **1975** *122*, 50–51.
73. Parker, D.S.; Lutz, M.P.; Pratt, A.M. New trickling filter applications in the USA. *Water Sci. Technol.* **1990**, *22*, 215–226.
74. Tchobanoglous, G.; Burton, F.L. *Wastewater Engineering: Metcalf & Eddy*; McGraw-Hill: New York, 1991.
75. Bull, M.A.; Sterritt, R.M.; Lester, J.N. The treatment of wastewaters from the meat industry: a review. *Environ. Tech. Lett.* **1982**, *3*, 117–126
76. Blanc, F.C.; O'Shaughnessy, J.C.; Corr, S.H. Treatment of soft drink bottling wastewater from bench-scale treatability to full-scale operation. In *Proceedings of 38th Purdue Industrial Waste Conference*; Purdue University, Lafayette, In, 1984; 243–256.
77. Bilstad, T. Upgrading slaughterhouse effluent with rotating biological contactors. In *Proceedings of the 1st International Conference on Fixed-Film Biological Processes*; Wu, Y.C., Smith, E.D. Miller, R.D., Eds. Kings Island: Ohio, USA, 1982; 892–912.
78. Bibby, J. *Innovative Technology Reduces Costs for a Meat Processing Plant*; Waterlink, Inc. 2002 (www.waterlink.com).
79. Gariepy, S.; Jyagi, R.D.; Couillard, D.; Tran, F. Thermophylic process for protein recovery as an alternative to slaughterhouse wastewater treatment. *Biol. Waste.* **1989**, *29*, 93–105.
80. Council of the European Communities. Regulation (EC) no. 1774/2002 of the European Parliament and of the Council of 3 October 2002. Laying down health rules concerning animal by-products not intended for human consumption. *Offi. J. Eur. Comm.*, **2002**. *L 273/1-95* (10 October 2002).
81. Heddle, J.H. Activated sludge treatment of slaughterhouse wastes with protein recovery. *Water Res.* **1979**, *13*, 581–584.
82. Lovett, D.A.; Travers, S.M.; Davey, K.R. Activated sludge treatment of slaughterhouse wastewater – I. Influence of sludge age and feeding pattern. *Water Res.* **1984**, *18*, 429–434.
83. Travers, S.M.; Lovett, D.A. Activated sludge treatment of slaughterhouse wastewater – II. Influence of dissolved oxygen concentration. *Water Res.* **1984**, *18*, 435–439.
84. Irvine, R.L.; Miler, G.; Bhamrah, A.S. Sequencing batch treatment of wastewater in rural areas. *J. Wat. Pollut. Control Fed.* **1979**, *51*, 244.
85. Hadjinicolaou, J. Evaluation of a controlled condition in a sequencing batch reactor pilot plant operation for treatment of slaughterhouse wastewaters. *Can. Agr. Eng.* **1989**, *31*, 249–264.
86. EIDN. *Nutrient Removal System for Treatment of Abattoir Wastewater. Environmental Case Studies Directory*; Environmental Industries Development Network, www.eidn.co.au/Poowon.htm, 2002.

87. American Society of Agricultural Engineers. *Manure Production and Characteristics*; Standard ASAE D384.1, 1999.
88. Wang, Z. *Evaluation of a Two Stage Anaerobic Digestion System for the Treatment of Mixed Abattoir Wastes*. PhD Thesis, University of Manchester Institute of Science and Technology (UMIST), UK, 1996.
89. USEPA. *Guide to Field Storage of Biosolids*; Office of Wastewater, EPA/832-B-00-007, United States Environmental Protection Agency: Washington, DC, 2000.
90. Meat Industry Research Institute. *Stabilization of Save-All Bottom Solids Including Paunch Material, by Composting*; Supapong, B. Publication no. 336, 1973.
91. Australian Environment Protection Authority. Authorised officers manual, EPA 95/89, 1995.
92. Rooke, C.D. *Pilot Scale Studies on the Anaerobic Digestion of Abattoir Waste*; BSc Industrial Biology Dissertation, Southbank Polytechnic: Borough Rd, London, UK, 1988.
93. Cox, D.J.; Banks, C.J.; Hamilton, I.D.; Rooke, C.D. *The Anaerobic Digestion of Abattoir Waste at Bishops Castle Meat Co. Part I and II*. Meat and Livestock Commission Report, MLC: Milton Keynes, UK, 1987.
94. Banks, C.J.; Wang, Z. Development of a two phase anaerobic digester for the treatment of mixed abattoir wastes. *Water Sci. Technol.* **1999**, *40*, 67–76.
95. Centre for Renewable Energy, Environmental Technology Support Unit (ETSU). *Co-Digestion of Manure with Industrial and Household Waste*, Technical Brochure No. 118. AEA Technology: Harwell, UK, 2000.

16
Treatment of Palm Oil Wastewaters

Mohd Ali Hassan and Shahrakbah Yacob
University Putra Malaysia, Serdang, Malaysia

Yoshihito Shirai
Kyushu Institute of Technology, Kitakyushu, Japan

Yung-Tse Hung
Cleveland State University, Cleveland, Ohio, U.S.A.

16.1 INTRODUCTION

This chapter discusses the palm oil extraction process, wastewater treatment systems, and future technologies and applications for the palm oil industry. Crude palm oil (CPO) is extracted from the mesocarp of the fruitlets while palm kernel oil is obtained from the kernel (Fig. 1). The oil contents originating from mesocarp and kernel are 20 and 4%, respectively. Palm oil is a semisolid oil, rich in vitamins and several major fatty acids: oleic, palmitic, and linoleic. To produce palm oil, a considerable amount of water is needed, which in turn generates a large volume of wastewater. Palm oil mills and palm oil refineries are two main sources of palm oil wastewater; however, the first is the larger source of pollution and effluent known as palm oil mill effluent (POME). An estimated 30 million tons of palm oil mill effluent (POME) are produced annually from more than 300 palm oil mills in Malaysia. Owing to the high pollution load and environmental significance of POME, this chapter shall place emphasis on its treatment system.

16.1.1 Production of Crude Palm Oil (CPO)

It is important to note that no chemicals are added in the extraction of oil from the oil palm fruits, therefore making all generated wastes nontoxic to the environment. The extraction of crude palm oil involves mainly mechanical and heating processes, and is illustrated in several steps below (Fig. 2).

Sterilization

To ensure the quality and the productivity of palm oil mill, the fresh fruit bunches (FFB) must be processed within 24 hours of harvesting. Thus, most of the palm oil mills are located in close

Figure 1 (A) Cross-section cutting of oil palm fruit showing shell, mesocarp, and kernel sections; (B) manual harvesting of fresh fruit bunches using sickle. (Courtesy of Malaysian Palm Oil Board.)

proximity to the oil palm plantation. During sterilization, the FFB is subjected to three cycles of pressures (30, 35, and 40 psi) for a total holding time of 90 minutes. There are four objectives of the FFB sterilization: (a) to remove external impurities, (b) to soften and loosen the fruitlets from the bunches, (c) to detach the kernels from the shells, and, most importantly, (d) to deactivate the enzymes responsible for the buildup of free fatty acids. The sterilization process acts as the first contributor to the accumulation of POME in the form of sterilizer condensate.

Bunch Stripping

Upon completion of the sterilization, the "cooked" FFB will be subjected to mechanical threshing to detach the fruitlets from the bunch. At this stage the loose fruitlets are transferred to the next process while the empty fruit bunches (EFB) can be recycled to the plantation for mulching or as organic fertilizer.

Digestion and Pressing

The digester consists of a cylindrical vessel equipped with stirrer and expeller arms mainly to digest and press the fruitlets. Steam is introduced to facilitate the oil extraction from the digested mesocarp. At the end of the process, oil and pressed cake comprising nuts and fiber are produced. The extracted oil will then be purified and clarified in the next stage. At the same time the fiber and nuts are separated in the depericarper column. The waste fiber is then burnt for energy generation inside the boiler.

Oil Clarification and Purification

As the name of this process implies, the extracted oil is clarified and purified to produce CPO. Dirt and other impurities are removed from the oil by centrifugation. Before the CPO is transferred to the storage tank, it is subjected to high temperatures to reduce the moisture content in the CPO. This is to control the rate of oil deterioration during storage prior to processing at the palm oil refinery. The sludge, which is the byproduct of clarification and purification procedures, is the main source of POME in terms of pollution strength and quantity.

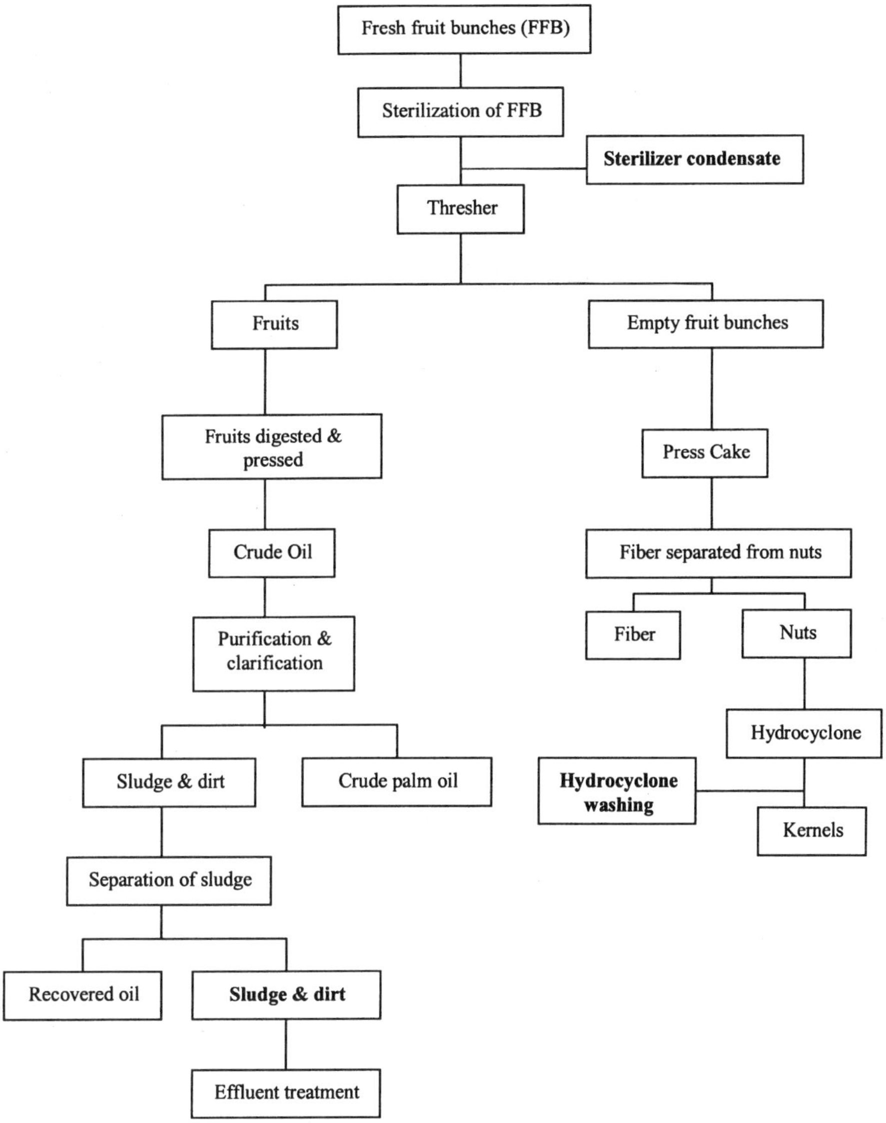

Figure 2 Flow diagram of crude palm oil extraction processes and sources of POME.

Nut Cracking

At this point, the nuts from the digestion and pressing processes are polished (to remove remnants of fiber) before being sent to the nut-cracking machine or ripple mill. The cracked mixture of kernels and shells is then separated in a winnowing column using upwards suction (hydrocyclone) and a clay bath. The third source of POME is the washing water of the hydrocyclone. The kernel produced is then stored before being transferred to palm kernel mill for oil extraction. Shell wastes will join the fiber at the boiler for steam and power generation.

16.1.2 Production of Refined Bleached Deodorized Palm Oil (RBDPO)

The refining of CPO employs physical/steam refining in which steam distillation is used to separate free fatty acids under high temperature and vacuum (Fig. 3). It consists of two main processes as follows.

Pretreatment

Before the actual refining process is carried out, the CPO is pretreated with phosphoric acid to eliminate impurities such as gums and trace metals. A bleaching technique is then used to remove phosphoric acid and its content under vacuum, followed by a filtration method. Solid waste in the form of sludge is disposed and buried in a landfill.

Deodorization

At this stage, steam is introduced under a vacuum condition to strip the pretreated oil of volatile free fatty acids, odoriferous compounds, and unstable pigments. The distillate for the deodorization process will form the main source of palm oil refinery effluent (PORE). The distillate has a free fatty acid content of approximately 80–90%. After the refining process, the oil is known as refined, bleached and deodorized palm oil (RBDPO). Further process such as fractionation of RBDPO will separate palm olein and stearin based on the different melting points of each component.

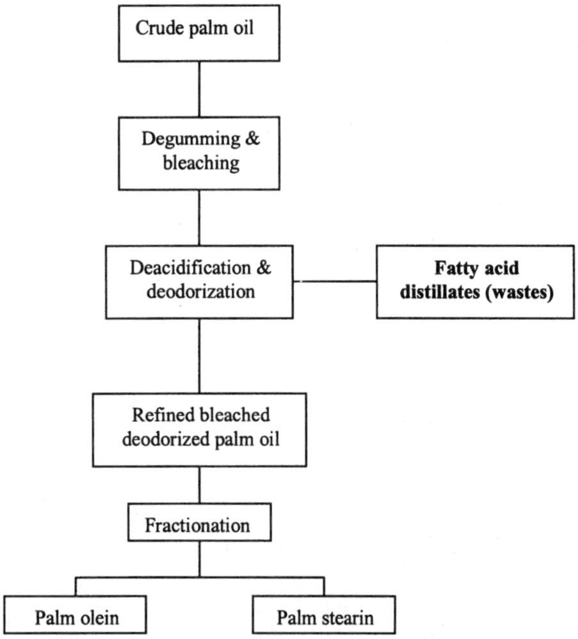

Figure 3 Flow diagram of physical refining process of crude palm oil and source of PORE.

16.2 PALM OIL MILL EFFLUENT (POME)

Palm oil mill effluent originates from two main processes: sterilization and clarification stages, as the condensate and clarification sludge, respectively (Fig. 2). The clarification sludge shows higher level of solid residues compared to the sterilizer condensate. Both contain some level of unrecovered oils and fats. The final POME would of course include hydrocyclone washing and cleaning up processes in the mill [1]. Approximately 1–1.5 tons of water are required to process 1 ton of FFB.

16.2.1 Properties of POME

Based on the process of oil extraction and the properties of FFB, POME is made up of about 95–96% water, 0.6–0.7% oil, and 4–5% total solid, including 2–4% suspended solids, which are mainly debris from palm mesocarp [2]. No chemicals are added during the production of palm oil; thus it is a nontoxic waste. Upon discharge from the mill, POME is in the form of a highly concentrated dark brown colloidal slurry of water, oil, and fine cellulosic materials. Owing to the introduction of heat (from the sterilization stage) and vigorous mechanical processes, the discharge temperature of POME is approximately 80–90°C. The chemical properties of POME vary widely and depend on the operation and quality control of individual mills [3]. The general properties of POME are indicated in Table 1.

Apart from the organic composition, POME is also rich in mineral content, particularly phosphorus (18 mg/L), potassium (2270 mg/L), magnesium (615 mg/L) and calcium (439 mg/L) [2]. Thus most of the dewatered POME dried sludge (the solid endproduct of the POME treatment system) can be recycled or returned to the plantation as fertilizer.

16.2.2 Biological Treatment

Owing to its chemical properties, POME can be easily treated using a biological approach. With high organic and mineral content, POME is a suitable environment in which microorganisms can thrive. Hence it could harbor a consortium of microorganisms that will consume or break down the wastes or pollutants, turning them into harmless byproducts. In some cases, these byproducts have high economic value and can be used as potential renewable sources or energy. In order to achieve such a goal, a suitable mixed population of microorganisms must be introduced and the

Table 1 Chemical Properties of Palm Oil Mill Effluent (POME)

Chemical property	Average	Range
pH	4.2	3.4–5.2
BOD (mg/L)	25,000	10,250–43,750
COD (mg/L)	50,000	15,000–100,000
Oil and grease (mg/L)	6000	150–18,000
Ammoniacal nitrogen (mg/L)	35	4–80
Total nitrogen (mg/L)	750	180–1400
Suspended solid (mg/L)	18,000	5000–54000
Total solid (mg/L)	40,000	11,500–78,000

Source: Refs. 3, 4.

process should be optimized. Three biological processes are currently employed by the industry as a series of anaerobic, facultative anaerobic, and aerobic treatments. However, the major reduction of POME polluting strength – up to 95% of its original BOD – occurs in the first stage, that is, during the anaerobic treatment [4].

The anaerobic process involves three main stages; hydrolytic, acidogenic, and methanogenic. In the first stage, hydrolytic microorganisms secrete extracellular enzymes to hydrolyze the complex organic complexes into simpler compounds such as triglycerides, fatty acids, amino acids, and sugars. These compounds are then subjected to fermentative microorganisms that are responsible for their conversion into short-chain volatile fatty acids – mostly acetic, propionic, butyric acids, and alcohols. In the final stage, there are two separate biological transformations: first, the conversion of acetic acid into methane and carbon dioxide by methanogens; secondly, the conversion of propionic and butyric acids into acetic acid and hydrogen gas before being consumed by the methanogens. The endproducts of the anaerobic degradation are methane and carbon dioxide. Traces of hydrogen sulfide are also detected as the result of the activity of sulfate-reducing bacteria in the anaerobic treatment. The biochemical oxygen demand (BOD) at the first two stages remains at the same level as when it entered the anaerobic treatment, because only the breakdown of the complex compounds to a simpler mixture of organic materials has occurred. Only after the methanogenic stage will the BOD be reduced significantly.

16.2.3 Wastewater Treatment Systems for POME

The choice of POME wastewater treatment systems is largely influenced by the cost of operation and maintenance, availability of land, and location of the mill. The first factor plays a bigger role in the selection of the treatment systems. In Malaysia, the final discharge of the treated POME must follow the standards set by the Department of Environment (DOE) of Malaysia, which is 100 mg/L of BOD or less (Table 2) regardless of which treatment system is being utilized.

Pretreatment

Prior to the primary treatment, the mixed raw effluent (MRE, a mixture of wastewater from sterilization, clarification, and other sources) will undergo a pretreatment process that includes the removal of oil and grease, followed by a stabilization process. The excess oil and grease is extracted from the oil recovery pit using an oil skimmer. In this process, steam is continuously

Table 2 Environmental Regulations for Watercourse Discharge for Palm Oil Mill Effluent (POME)

Parameters	Level
BOD (mg/L)	100
Suspended solids (mg/L)	400
Oil and grease (mg/L)	50
Ammoniacal nitrogen (mg/L)	150
Total nitrogen (mg/L)	200
pH	5–9

Source: Ref. 5.

supplied to the MRE to aid the separation between oil and liquid sludge. The recovered oil is then reintroduced to the purification stage. The process will prevent excessive scum formation during the primary treatment and increase oil production. The MRE is then pumped into the cooling and mixing ponds for stabilization before primary treatment. No biological treatment occurs in these ponds. However, sedimentation of abrasive particles such as sand will ensure that all the pumping equipment is protected. The retention time of MRE in the cooling and mixing ponds is between 1 and 2 days.

Ponding System

The ponding system is comprised of a series of anaerobic, facultative, and algae (aerobic) ponds. These systems require less energy due to the absence of mechanical mixing, operation control, or monitoring. Mixing is very limited and achieved through the bubbling of gases; generally this is confined to anaerobic ponds and partly facultative ponds. On the other hand, the ponding system requires a vast area to accommodate a series of ponds in order to achieve the desired characteristics for discharge. For example, in the Serting Hilir Palm Oil Mill, the total length of the wastewater treatment system is about 2 km, with each pond about the size of a soccer field (Fig. 4). Only a clay lining of the ponds is needed, and they are constructed by excavating the earth. Hence, the ponding system is widely favored by the palm oil industry due to its marginal cost.

In constructing the ponds, the depth is crucial for determining the type of biological process. The length and width differ based on the availability of land. For anaerobic ponds, the optimum depth ranges from 5 to 7 m, while facultative anaerobic ponds are 1–1.5 m deep. The effective hydraulic retention time (HRT) of anaerobic and facultative anaerobic systems is 45 and 20 days, respectively. A shallower depth of approximately 0.5–1 m is required for aerobic ponds, with an HRT of 14 days. The POME is pumped at a very low rate of 0.2 to 0.35 kg $BOD/m^3 \cdot day$ of organic loading. In between the different stages of the ponding system, no pumping is required as the treated POME will flow using gravity or a sideways tee-type subsurface draw-off system. Under these optimum conditions, the system is able to meet the requirement of DOE. The number of ponds will depend on the production capacity of each palm oil mill.

One problem faced by pond operators is the formation of scum, which occurs as the bubbles rise to the surface, taking with them fine suspended solids (Fig. 5). This results from the presence of oil and grease in the POME, which are not effectively removed during the pretreatment stage. Another disadvantage of the ponding system is the accumulation of solid sludge at the bottom of the ponds (Fig. 6). Eventually the sludge and scum will clump together inside the pond, lowering the effectiveness of the pond by reducing the volumetric capacity and HRT. When this happens, the sludge may be removed by either using submersible pumps or excavators. The removed sludge is dewatered and dried before being used as fertilizer. The cleanup is normally carried out every 5 years or when the capacity of the pond is significantly reduced.

Open Digester and Ponding Systems

This system is a combination of an open digester tank and a series of ponding systems (Fig. 7). The anaerobic digestion is carried out in the digester, then in the facultative anaerobic and algae ponds. It has been shown that by using an open digester, a better reduction of BOD can be achieved in a shorter time. Digesters are constructed of mild steel at various volumetric capacities ranging from 600 up to 3600 m^3. The treatment of treated POME from the digester

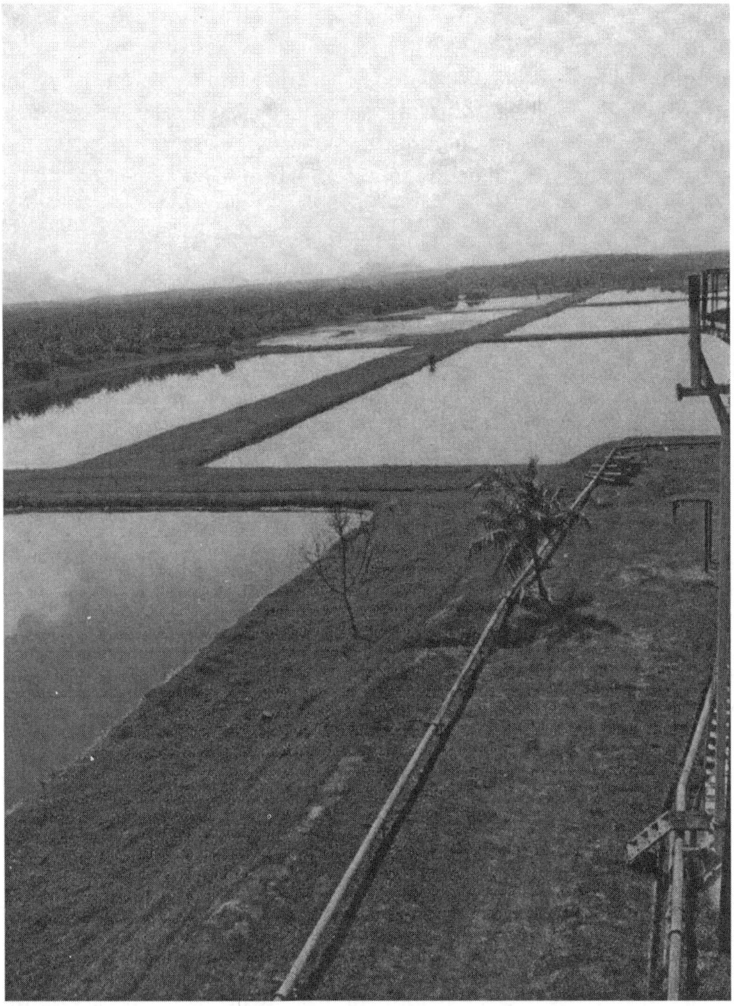

Figure 4 A series of ponds for POME treatment occupying a large land area. (Courtesy of Felda Palm Industries.)

will start at the facultative ponds, followed by the algae ponds. A description of the ponding systems is outlined in the previous section "Pretreatment."

The HRT of the digester is only 20–25 days and has a higher organic loading of 0.8–1.0 BOD $kg/m^3 \cdot day$ compared to anaerobic ponds. With minimal financial input from the operators, no mechanical mixing equipment is installed in the digesters. Using the same principle as anaerobic ponds, mixing of POME is achieved via bubbling of biogas. Occasionally, the mixing is also achieved when the digester is being recharged with fresh POME. The treated POME is then overflowed into the ponding system for further treatment.

Although the digester system has been proven to be superior to anaerobic ponds, it also has similar problems of scum formation and solid sludge accumulation. Another serious problem is

Figure 5 Active bubbling of gases leading to the formation of scum.

the corrosion of the steel structures due to long exposure to hydrogen sulfide. Incidents such as burst and collapsed digesters have been recorded. Accumulated solids could be easily removed using the sludge pipe located at the bottom of the digester. The dewatered and dried sludge can then be disposed for land application.

Extended Aeration

To complement the previous systems, mechanical surface aerators can be introduced at the aerobic ponds (Fig. 8). This effectively reduces the BOD through aerobic processes. The aerators are normally installed at the end of the ponding system before discharge. However, this happens only where land area is a constraint and does not permit extensive wastewater treatment. Otherwise, aerators must be provided to meet DOE regulations.

16.3 PALM OIL REFINERY EFFLUENT (PORE)

Following the production of CPO from the palm oil mill, the CPO is then subjected to further refining before it can be categorized as edible oil. Even after the clarification and purification processes, there are still large amounts of impurities such as gums, pigments, trace of metals, and soluble fats that cause unpleasant taste, odor, and color. There are three common types of operation in the palm oil refineries: (a) physical refining and dry fractionation, (b) physical refining and detergent fractionation, and (c) physical and chemical refining with dry/detergent fractionation [6].

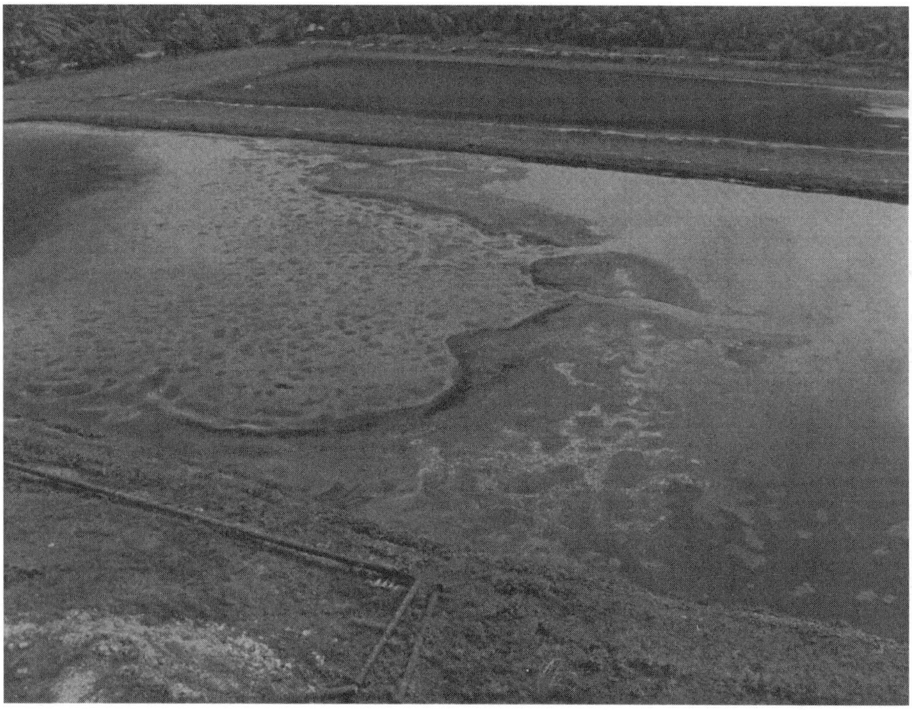

Figure 6 Formation of islets of sludge in the middle of the pond. (Courtesy of Felda Palm Industries.)

16.3.1 Chemical Properties of PORE

The main sources of PORE are water from the deodorization process and cleaning operations within the mill (Fig. 3). The characteristics of PORE are very much dependent on the type of process employed. The main chemical properties of PORE are as described in Table 3.

Figure 7 A series of 3600 m^3 open digesters for POME treatment. (Courtesy of Felda Palm Industries.)

Figure 8 An aerator system installed to accelerate BOD reduction at the aerobic pond. (Courtesy of Malaysian Palm Oil Board.)

In comparison with POME, PORE is less polluting. This is largely due to the absence of oil and grease, and its low organic load. From Table 3, it is obvious why most of the palm oil refineries in Malaysia have adopted physical refining and dry fractionation to produce edible oil. Not only does the system reduce the effluent problem, but higher yield and oil purity with lower operating costs are obtained.

Table 3 Chemical Properties of Palm Oil Refinery Effluent (PORE) Based on Different Operations

	Type of refinery processes		
Chemical properties	Physical refining and dry fractionation	Physical refining and detergent fractionation	Physical and chemical refining with dry/detergent fractionation
Temperature (°C)	35	42	57
pH	5.3	4.9	3.0
BOD (mg/L)	530	2640	4180
COD (mg/L)	890	5730	7700
Total solids (mg/L)	330	1170	2070
Suspended solids (mg/L)	50	12	6
Phosphorus (mg/L)	4	1	12
Total fatty matter (mg/L)	220	1580	3550

Source: Ref. 3.

16.3.2 Wastewater Treatment Systems for PORE

Unlike POME wastewater treatment systems, the PORE system is more systematic and predictable. Most PORE systems involve biological processes, with some using physical and chemical methods such as sedimentation, dissolved air flotation after coagulation and flocculation using lime, alum and polyelectrolytes [8].

Pretreatment of PORE

The first step in ensuring satisfactory performance of a PORE treatment plant is to remove oil and fat from the MRE. The separation is carried out using several methods such as fat traps, tilted-plate separators, and dissolved flotation units. Beside physical separations, the addition of chemical flocculants and coagulants also helps in reducing the total fatty matter and other suspended solids. Before the commencement of the biological treatment, the pH of the PORE is adjusted to the desired level as pH plays an important role in the optimum biodegradation of PORE.

Activated Sludge System

Many palm oil refineries use activated sludge systems to treat PORE. This is because of land constraints (for ponding systems) and effective removal of BOD in a short HRT [6]. This system can be very effective if the level of total fatty matter is kept low after the pretreatment stage. The presence of fatty matter in the activated sludge systems will contribute not only higher BOD but the formation of scum. This leads to poor performance of the system.

The treatment is carried out by bringing PORE into contact with a mixed population of aerobic microorganisms in the controlled environment of the activated sludge system. In this process, oxygen is supplied via aeration or vigorous agitation for the oxidation of organic wastes to carbon dioxide. After the treatment, the suspended biomass is separated in the clarifier. The biomass is reintroduced back into the activated sludge systems as "return activated sludge." This is to ensure the density of microorganisms is maintained at an optimum level for maximum removal of BOD. The supernatant from the clarifier can then be safely discharged into the watercourse. The HRT of PORE and cell residence time are 1–2 days and 3–10 days, respectively. Using this system, a final BOD and suspended solids of 20 and 30 mg/L, respectively, can be obtained with 1500–2000 mg/L of mixed liquor suspended solids. Apart from the energy requirement to operate the treatment facilities, additional nutrients are normally added to the effluent. This is because the effluent from the palm oil refinery is low in nitrogen content, which is essential for the growth of aerobic micro-organisms. A ratio of BOD : N of 100 : 5 is kept constant throughout the process.

16.4 POTENTIAL TECHNOLOGIES AND COMMERCIAL APPLICATIONS OF PALM OIL WASTES

16.4.1 POME Treatment

Evaporation Technology

In one study, a 200 L single-effect evaporator was constructed to test the evaporation technique in POME treatment [8]. It used the principle of rapid heating to vaporize water at 600 mm Hg and 80°C using a plate heat exchanger. Staggered feeding of fresh POME was introduced into the evaporator when the liquor dropped by half of the initial volume. The feeding was carried out until the accumulated solid sludge reached the pre-set level of 30%. The solid was then discharged before the new cycle began. The single-effect evaporator was able to recover 85% of water from POME with a good quality distillate of 20 mg/L BOD. The distillate could be recycled as process water or feedwater for the boiler in the mill. Even though the system promises a significant reduction of liquid waste (and thus less dependence on vast land area for ponding systems), the energy required for heating may impose financial constraints for the mill operator. Moreover, the mill may have to make a big investment in equipment, skilled operators, and maintenance. Further studies are being carried out to produce cost-effective systems such as utilizing excess organic biomass from the mill as an energy source.

High-Technology Bioreactor Design

There have been numerous studies to optimize the anaerobic treatment of POME using various designs of bioreactor. Laboratory-scale studies have been carried out to evaluate the effectiveness of anaerobic filters (AF) and a fluidized-bed reactor (FBR) in treating POME [9]. About 90% of the fed COD was effectively removed by both reactor systems. However, when the COD loading was increased, a significant reduction in terms of COD removal was recorded in the FBR system, while clogging of the filter was evident in the AF reactor. A higher COD removal efficiency was reported [10] when using a modified anaerobic baffled reactor (MABR). The system also demonstrated a short retention time of three days. Despite the good potential of the bioreactor systems for POME treatment, none has been implemented at a larger scale.

Power Generation: Closed Digester

The composition of biogas emitted from an open digester tank and the lagoon was lower than that reported for laboratory studies [2]. The biogas composition was 40% methane and 60% carbon dioxide for the open digester tank, and 55% methane and 45% carbon dioxide in anaerobic lagoons. In terms of energy value, it is comparable to commercially available gas fuels as shown in Table 4. The potential energy that could be generated from 1 m^3 of biogas is 1.8 kWh [11].

A closed digesting system was tested to improve the anaerobic digestion of POME, leading to the production of biogas. Using the same design of open digester, a fixed or floating cover is included, equipped with the other facilities such as gas collector, safety valves, and monitoring facilities.

Compost

Based on our research, dewatered POME sludge can be composted with domestic wastes and bulking agents such as shredded wood and sawdust. A modified composter from a cement mixer with insulated drum was used as a reactor to run the composting process. Experimental

Table 4 Comparisons Between Methane Derived from Anaerobic Digestion of POME and Other Gas Fuels

Chemical properties	Methane	Natural gas	Propane
Gross calorific value (kcal/kg)	4740–6150	907	24,000
Specific gravity	0.847–1.002	0.584	1.5
Ignition temperature (°C)	650–750	650–750	450–500
Inflammable limits (%)	7.5–21.0	5–15	2–10
Combustion air required (m^3/m^3)	9.6	9.6	13.8

Source: Ref. 11.

parameters such as aeration, pH, temperature, C/N ratio, and moisture content were controlled and monitored during the fermentation phase of the composting process. It took about 40 days to completely convert the POME sludge into compost via the solid substrate fermentation process with mixed microbial inoculum. The carbon content decreased towards the end of the composting process, which resulted in a decrease of the C/N ratio from 30 to 20. The low C/N ratio of the final compost product was very important as an indicator of maturity. The characteristics of the final compost products for POME sludge were similar to commercial composts and complied with US Environmental Protection Agency (EPA) standards, especially in heavy metal content and total coliforms. Planting out tests with leafy vegetables showed satisfactory performance [12].

Organic Acids

Two-stage fermentation was carried out in a study where POME was used as substrate for volatile fatty acids (VFA) production by continuous anaerobic treatment using a locally fabricated 50 L continuous stirred tank reactor (CSTR). The highest VFA obtained was at 15 g/L at pH 6.5, 30°C, 100 rpm, sludge to POME ratio 1 : 1, HRT 4 days, without sludge recycle. The highest BOD removal corresponded with the high production of organic acids. The organic acids produced from POME were then recovered and purified using acidification and evaporation techniques. A clarified concentrated VFA comprised of 45 g/L acetic, 20 g/L propionic and 22 g/L butyric acids were obtained with a recovery yield of 76% [13].

Production of Polyhydroxyalkanoates

The organic acids from treated POME can be used to biologically synthesize polyhydroxyalkanoates (PHA), a bacterial bioplastic. The concentrated organic acids obtained were used in a fed-batch culture of *Alcaligenes eutrophus* for the production of PHA. About 45% PHA content in the dry cells could be obtained, corresponding to a yield of 0.32 from acetic acid. The overall volumetric productivity of PHA is estimated at 0.09 g PHA/L hour. This indicates that the application of a high-density cell culture to produce bioplastic from POME can be achieved [14].

Biological Hydrogen

Another potential application of POME as a renewable resource of energy is the production of biological hydrogen via a fermentation process. The main purpose of producing biological hydrogen is to offer an alternative source of energy to fossil fuels. The major advantage of biological hydrogen is the lack of polluting emission since the utilization of hydrogen, either via combustion or fuel cells, results in pure water [15]. Currently, two proposed systems produce biological hydrogen using photoheterotrophic and heterotrophic bacteria. However, the latter is most suitable for POME due to limited light penetration caused by the sludge particles as experienced during the production of PHA by phototrophic *Rhodobacter sphaeroides* [16]. Moreover, it would be costly to construct and maintain a photobioreactor at a commercial-scale operation.

In the anaerobic degradation of POME, complex organic matter is converted into a mixture of methane and carbon dioxide in a network of syntrophic bacteria. Prior to this, fermentative and acetogenic bacteria first convert organic matter into a mixture of VFA and hydrogen before being consumed by methanogenic bacteria. Based on the metabolic activities of these microorganisms in POME degradation, a system combining the organic acids and biological hydrogen production is suggested. However, the utilization of biological hydrogen from POME is still at the planning stage. Major development in terms of selection of suitable microorganisms and optimization of process conditions is required for cost-effective production of hydrogen. Nevertheless, this technology promises a means to conserve the environment by generating clean energy.

16.4.2 PORE Treatment

Sequential Batch Reactor System

A new technology using the sequential batch reactor (SBR) technique has been shown to provide an effective treatment of PORE [7] as shown in Fig. 9. Among the advantages of SBR over the conventional activated sludge are an automated control system, more versatility, stability, and the ability to handle high fluctuations in organic loading. A consistent output of BOD below 50 mg/L was observed. With this system, the hydraulic retention time and solid sludge content could be controlled, thus eliminating the need for clarifier and sludge recycling facilities.

16.5 FUTURE TRENDS

From the preceding section, several potential and emerging technologies for POME wastewater treatment system can be integrated into the palm oil mill operation (Fig. 10). The strategy is to combine the existing wastewater treatment system with the production of appropriate bioproducts, towards zero discharge for the palm oil industry [17]. In anaerobic treatment, methanogenic activity will be suppressed or inhibited in order to extract the organic acids produced. This in turn shall lower the greenhouse gases (methane and carbon dioxide) emissions from the anaerobic digestion, thus reducing the effects of global warming. Further separation and purification processes are needed before organic acids can be utilized as a substrate for

Figure 9 A pilot plant sequential bioreactor system tested for POME treatment. (Courtesy of Malaysian Palm Oil Board.)

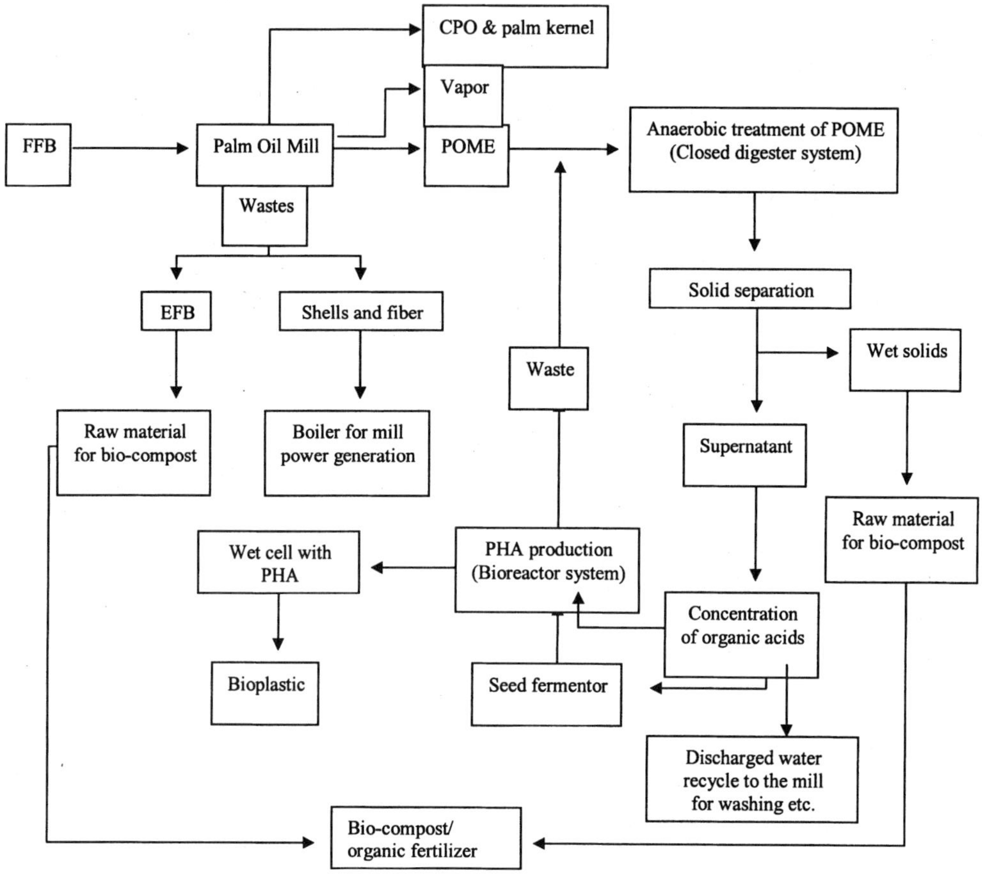

Figure 10 Proposed integrated palm oil production and POME wastewater treatment system (from Ref. 17).

PHA-producing microorganisms. The solid wastes (sludges) generated from the wastewater treatment system will be used as a mixture with EFB to form biocompost.

Wastes generated from the palm oil mill contain a high percentage of degradable organic material and can be converted into value-added products and chemicals. It is expected that changes in the technologies in POME treatment could lead to a substantial reduction in terms of waste discharged. On the other hand, the palm oil industry will experience a sustainable growth by addressing the excessive pollution issue through development of biowastes as alternative sources of renewable energy and valued chemicals. This in turn shall generate additional revenue for the industry. Finally, better-integrated waste management is associated with other environmental benefits such as reduction of surface waterbody and groundwater contamination, less waste of land and resources, lower air pollution, and a reduction of accelerating climate changes.

REFERENCES

1. Agamuthu, P. Palm oil mill effluent – treatment and utilization. In *Waste Treatment Plant*; Sastry, C.A., Hashim, M.A., Agamuthu, P., Eds.; Narosa Publishing House: New Delhi, India, 1995; 338–360.

2. Ma, A.N. Treatment of palm oil mill effluent. In *Oil Palm and the Environment – A Malaysian Perspective*; Singh, G., Lim, K.H., Leng, T., David, L.K., Eds.; Malaysian Oil Palm Growers Council: Selangor, Malaysia, 1999; 113–126.
3. Basiron, Y.; Darus, A. The oil palm industry – from pollution to zero waste. *The Planter* **1995**, *72 (840)*, 141–165.
4. Ma, A.N. Environment management for the palm oil industry. *Palm Oil Develop.* **1999**, *30*, 1–10.
5. Md. Noor, M. Environmental legislation: Environmental quality act, 1974. In *Oil Palm and the Environment – A Malaysian Perspective*; Singh, G., Lim, K.H., Leng, T., David, L.K., Eds.; Malaysian Oil Palm Growers Council: Selangor, Malaysia, 1999; 261–264.
6. Ma, A.N. Treatment of palm oil refinery effluent. In *Oil Palm and the Environment – A Malaysian Perspective*; Singh, G., Lim, K.H., Leng, T., David, L.K., Eds.; Malaysian Oil Palm Growers Council: Selangor, Malaysia, 1999; 127–136.
7. Sastry, C.A. Waste treatment case studies. In *Waste Treatment Plant*; Sastry, C.A., Hashim, M.A., Agamuthu, P., Eds.; Narosa Publishing House: New Delhi, 1995; 361–379.
8. Ma, A.N. Evaporation technology for pollution abatement in palm oil mills. In *Proceedings of the National Seminar on Palm Oil Milling, Refining Technology and Quality*; Chang, A.K.C., Ed.; PORIM: Selangor, Malaysia, 1997; 167–171.
9. Borja, R.; Banks, C.J. Comparison of an anaerobic filter and an anaerobic fluidized bed reactor treating palm oil mill effluent. *Process Biochem.* **1995**, *30 (6)*, 511–521.
10. Faisal, M; Unno, H. Kinetic analysis of palm oil mill wastewater treatment by a modified anaerobic baffled reactor. *Biochem. Eng. J.* **2001**, *9*, 25–31.
11. Ma, A.N.; Toh, T.S.; Chua, N.S. Renewable energy from oil palm industry. In *Oil Palm and the Environment – A Malaysian Perspective*; Singh, G., Lim, K.H., Leng, T., David, L.K., Eds.; Malaysian Oil Palm Growers Council: Selangor, Malaysia, 1999; 253–260.
12. Abdul Rahman, A.R.; Baharum, Z.; Hassan, M.A.; Idris, A. Bioreactor composting of selected organic sludges. *Proceedings of the 13th National Biotechnology Seminar*, Penang, Malaysia, 2001.
13. Noraini, A.R.; Hassan, M.A.; Shirai, Y.; Karim, M.I.A. Production of organic acids from palm oil mill effluent during continuous anaerobic treatment. *Asia-Pac. J. Mol. Biol.* **1999**, *7 (2)*, 179–184.
14. Hassan, M.A.; Shirai, Y.; Umeki, H.; Yamazumi, H.; Jin, S.; Yamamoto, S.; Abdul Karim, M.I.; Nakanishi, K.; Hashimoto, K. Acetic acid separation from anaerobically treated palm oil mill effluent by ion exchange resin for the production of polyhydroxyalkanoate by *Alcaligenes eutrophus. Biosci. Biotech. Biochem.* **1997**, *61 (9)*, 1465–1468.
15. Claassen, P.A.M; van Lier, J.B.; Lopez Contreras, A.M.; van Niel, E.W.J.; Sijtsma, L.; Stams, A.J.M.; de Vries, S.S.; Weusthuis, R.A. Utilization of biomass for the supply of energy carrier. *Appl. Microbiol. Biot.* **1999**, *52*, 741–755.
16. Hassan, M.A.; Shirai, Y.; Kusubayashi, N.; Abdul Karim, M.I.; Nakanishi, K.; Hashimoto, K. The production of polyhydroxyalkanoate from anaerobically treated palm oil mill effluent by *Rhodobacter sphaeroides. J. Ferment. Bioeng.* **1997**, *83 (3)*, 485–488.
17. Hassan, M.A.; Nawata, O.; Shirai, Y.; Noraini, A.R.; Yee, P.L.; Karim, M.I.A. A proposal for zero emission from palm oil industry incorporating the production of polyhydroxyalkanoates from palm oil mill effluent. *J. Chem. Eng. Jpn.* **2002**, *35 (1)*, 9–14.
18. Malaysia Palm Oil Board. The processing sector, In *Oil Palm Industry in Malaysia*; Salleh, M., Ed.; Malaysia Palm Oil Board: Selangor, Malaysia, 2000; 10–17.

17
Olive Oil Waste Treatment

Adel Awad and Hana Salman
Tishreen University, Lattakia, Syria

Yung-Tse Hung
Cleveland State University, Cleveland, Ohio, U.S.A.

17.1 INTRODUCTION

The extraction and use of olive oil has been linked to Mediterranean culture and history since 4000 BC. Several terms used today are reminders of this ancient heritage. For example, the Latin words *olea* (oil) and *olivum* (olive) were derived from the Greek word *elaia.* As a dietary note, olive oil is high in nutrition, and appears to have positive effects in the prevention and reduction of vascular problems, high blood pressure, arteriosclerosis, thrombosis, and even some types of cancer [1].

The social and economic importance of the olive production sector may be observed by considering some representative data. In the European Union (EU), there are about 2 million companies related to olives and olive oil. Worldwide olive oil production is about 2.6 million tons per year, 78% (about 2.03 million tons) of which are produced in the EU (main producers: Spain, Greece, and Italy). Other main producers are Turkey (190,000 tons), Tunisia (170,000 tons), Syria (110,000 tons), and Morocco (70,000 tons). More than 95% of the world's olives are harvested in the Mediterranean region. In Spain alone, more than 200 million olive trees out of the total world number of 800 million are cultivated on an area of approximately 8.5 million ha. Within Spain, 130 million olive trees are found in Andalusia, where about 15% of the total arable land is used for olive cultivation [2].

According to the FAOSTAT database [3], the total waste generated by olive oil production worldwide in 1998 was 7.3 million tons, 80% of which was generated in the EU and 20% generated in other countries. In Spain, the top olive oil producer, the generated waste in 1998 alone was 2.6 million tons, or about 36% of the waste generated worldwide.

Approximately 20 million tons of fresh water are required for olive oil production in the Mediterranean area, resulting in up to 30 million tons of solid–liquid waste (*orujo* and *alpeorujo*) per year. By comparison, the annual amount of sewage sludge in Germany is 55 million m^3, with 5% dry solid matter content [4].

17.2 OLIVE OIL MILL TECHNOLOGY

The olive oil extraction industry is principally located around the Mediterranean, Aegean, and Marmara seas, and employs a very simple technology (Fig. 1). First, the olives are washed to remove physical impurities such as leaves, pieces of wood, as well as any pesticides. Afterwards, the olives are ground and mixed into paste. Although a large variety of extracting systems are available, two methods are generally employed: traditional pressing and modern centrifuging. Pressing is a method that has evolved since ancient times, while centrifuging is a relatively new technology. Figures 2 and 3 are schematic drawings of the two systems. Figure 2 represents the traditional discontinuous press of olive oil mills, while Fig. 3 represents more recent continuous solid/liquid decanting system (three-phase decanting mills). Both systems (traditional and three-phase decanter) generate one stream of olive oil and two streams of wastes, an aqueous waste called *alpechin* (black water) and a wet solid called *orujo.* A new method of two-phase decanting, extensively adopted in Spain and growing in popularity in Italy and Greece, produces one stream of olive oil and a single stream of waste formed of a very wet solid called *alpeorujo.*

Looking at milling systems employed worldwide, a greater percentage of centrifuge systems are being used compared to pressing systems. Because of the higher productivity of the more modern centrifuge systems, they are capable of processing olives in less time, which is a requisite for a final quality product [5].

Furthermore, in contrast to the three-phase decanter process, the two-phase decanter does not require the addition of water to the ground olives. The three-phase decanter requires up to 50 kg water for 100 kg olive pulp in order to separate the latter into three phases: oil, water, and solid suspension [6]. This is necessary, since a layer of water must be formed with no bonds to the oil and solid phase inside the decanter. Thus, up to 60 kg of alpechin may be produced from 100 kg olives. Alpechin is a wastewater rich in polyphenols, color, and soluble stuffs such as sugar and salt [7].

In the two-phase decanter, there must be no traces of water inside the decanter to prevent water flowing out with the oil and reducing the paste viscosity, which leads to improved oil extraction [8]. The two-phase decanter process is considered more ecological, not only because it reduces pollution in terms of the alpechin, but since it requires less water for processing [9]. Depending on the preparation steps (ripeness, milling, malaxing time, temperature, using enzymes or talcum, etc.), the oil yield using the two-phase decanter may be higher than that using the three-phase decanter [10]. The oil quality is also different in each process. In the case

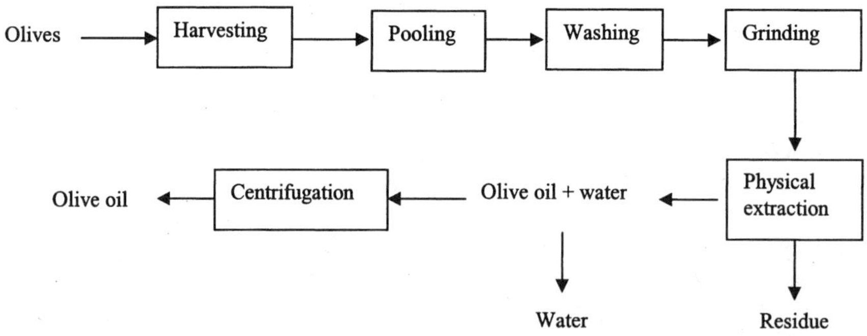

Figure 1 Technology generally used to produce olive oil (from Ref. 5).

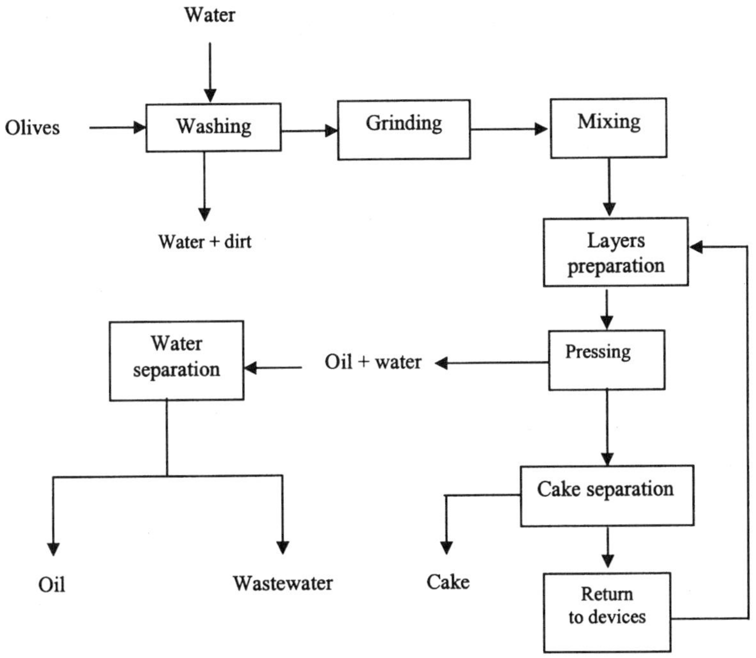

Figure 2 Traditional pressing for olive oil production (from Ref. 5).

of the three-phase decanter, the main part of the polyphenols will be washed out in the alpechin phase. These chemicals, which also provide antioxidation protection, are sustained in the oil phase using the two-phase decanter; the results are better conditions for a long oil shelf life as well as a more typical fruit taste [11].

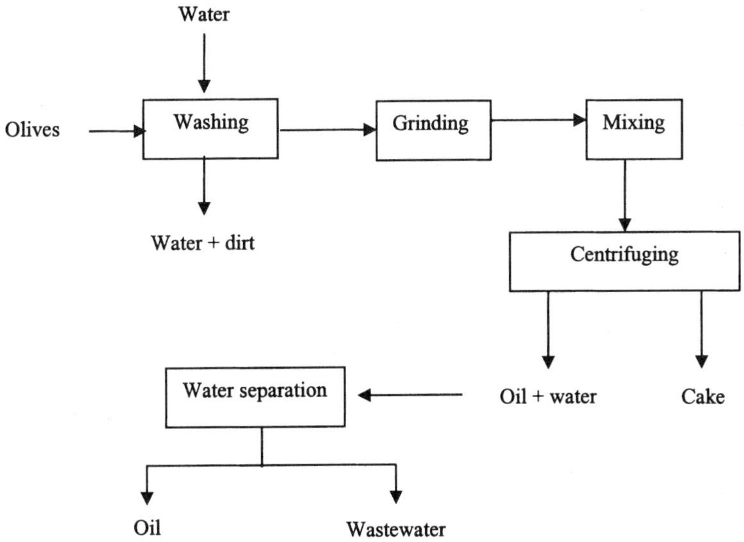

Figure 3 Modern centrifuging for olive oil production (three-phase decanter) (from Ref. 5).

The alpeorujo (solid/liquid waste) has a moisture content of 60–65% at the decanter output while the moisture content of the solid waste using the three-phase decanter is about 50%, and by traditional pressing is about 25%. One drawback is that two-phase alpeorujo is more difficult to store due to its humidity. Comparing the three different solids (orujo press cake, three-phase decanter orujo, and two-phase decanter alpeorujo), the two-phase decanter alpeorujo is the best residue to be reprocessed for oil [9].

17.3 OLIVE OIL WASTEWATER CHARACTERISTICS

The olive consists of flesh (75–85% by weight), stone (13–23% by weight) and seed (2–3% by weight) [12].The chemical composition of the olive is shown in Table 1. The quantities and composition of olive mill waste (OMW) vary considerably, owing to geographical and climatic conditions, tree age, olive type, extraction technology used, use of pesticides and fertilizers, harvest time, and stage of maturity.

In waste generated by olive oil mills, the only constituents found are produced either from the olive or its vegetation water, or from the production process itself. Auxiliary agents, which are hardly used in production, may be influenced and controlled by process management. Therefore, they are not important to the composition of wastewater. However, the composition of the olive and its vegetation wastewater cannot be influenced; thus the constituents of vegetation wastewater are decisive for the expected pollution load. Table 2 summarizes some literature data concerning the constituents of olive oil wastewater [13–25]. The variations of maximum and minimum concentrations of olive oil wastewater resulting from both methods (traditional presses and decanter centrifuge) are also presented, according to the International Olive Oil Council (IOOC) in Madrid [26], in Table 3.

Wastewater from olive oil production is characterized by the following special features and components [27]:

- color ranging from intensive violet–dark brown to black;
- strong olive oil odor;
- high degree of organic pollution (COD values up to 220 g/L, and in some cases reaching 400 g/L) at a COD/BOD_5 ratio between 1.4 and 2.5 and sometimes reaching 5 (difficult to be degraded);

Table 1 Composition of Olives

Constituents	Pulp	Stone	Seed
Water	50–60	9.3	30
Oil	15–30	0.7	27.3
Constituents containing nitrogen	2–5	3.4	10.2
Sugar	3–7.5	41	26.6
Cellulose	3–6	38	1.9
Minerals	1–2	4.1	1.5
Polyphenol (aromatic substances)	2–2.25	0.1	0.5–1
Others	–	3.4	2.4

Note: Values in percent by weight (%).
Source: Ref. 12.

Table 2 Summary of the Constituents of Olive Oil Wastewater (Alpechin) According to Different Literature Data

Parameter	Pompei[13] (1974)	Fiestas[14] (1981)	Garcia[18] (1989)[a]	Steegmans[15] (1992)	Hamdi[16] (1993)	Borja[25] (1995)	Beccari[23] (1996)[f]	Ubay[22] (1997)[e]	Zouari[24] (1998)[c]	Andreozzi[17] (1998)	Beltran-Heredia[21] (2000)[d]	Kissi[20] (2001)[b]	Rivas[19] (2001)[a]
pH	–	4.7	–	5.3	3–5.9	5.2	5.06	4.7	–	5.09	13.6	4.2	12.9
Chemical oxygen demand, COD (g/L)	195	–	15–40	108.6	40–220	60	90 (filtered 63)	115–120	225	121.8	6.7	50	24.45
Biochemical oxygen demand in 5 days, BOD_5 (g/L)	38.44	–	9–20	41.3	23–100	–	–	–	58	–	4.3	–	14.8
Total solids, TS (g/L)	–	1–3	–	19.2	1–20	48.6	51.5	8.5–9 (SS)	–	102.5	22.9	4 (SS)	–
Organic total solids (g/L)	–	–	–	16.7	–	41.9	37.2	–	190	81.6	4.6	–	–
Fats (g/L)	–	–	–	2.33	1–23	–	–	7.7	–	9.8	–	–	–
Polyphenols (g/L)	17.5	3–8	0.5	0.002	5–80	0.3	3.3	–	–	6.2	0.12	12	0.833
Volatile organic acids (g/L)	–	5–10	–	0.78	0.8–10	0.64	15.25	–	–	0.96	–	–	–
Total nitrogen (g/L)	0.81	0.3–0.6	–	0.6	0.3–1.2	0.16 ($N\text{-}NH_4$)	0.84	0.18	1.2	0.95	–	–	–

[a]Wastewater generated in the table olive processing industries during different stages including washing of fruits, debittering of green olives (addition of sodium hydroxide), fermentation and packing.
[b]Other parameters were measured such as: color (A_{395}) = 16; Cl^- = 11.9 g/L; K^+ = 2.5 g/L; NH_4^+ = 0.15 g/L.
[c]Since the dark color of olive oil mill effluent was difficult to determine quantitavely, the optical value (OD) at 390 nm was measured; this value was 8.5.
[d]Represents wastewater generated in table olive processing plant (black olives). Aromatic compounds (A) = 17 were determined by measuring the absorbance of the samples at 250 nm (the maximum absorbance wavelength of these organic compounds.
[e]Represents concentrated black water from a traditional olive oil mill plant. Other parameters were measured such as SS = 8.5–9 g/L, Total P = 1.2 g/L.
[f]Other parameters were measured such as TC = 25.5 g/L, Total P = 0.58 g/L, Lipids = 8.6 g/L.
Source: Refs. 13–25.

Table 3 Maximum and Minimum Concentration Values of Olive Oil Wastewater According to Applied Type of Technology

	Technology type	
Parameters	Centrifuge	Traditional presses
pH	4.55–5.89	4.73–5.73
Dry matter (g/L)	9.5–161.2	15.5–266
Specific weight	1.007–1.046	1.02–1.09
Oil (g/L)	0.41–29.8	0.12–11.5
Reducing sugars (g/L)	1.6–34.7	9.7–67.1
Total polyphenols (g/L)	0.4–7.1	1.4–14.3
O-diphenols (g/L)	0.3–6	0.9–13.3
Hydroxytyrosol (mg/L)	43–426	71–937
Ash (g/L)	0.4–12.5	4–42.6
COD (g/L)	15.2–199.2	42.1–389.5
Organic nitrogen (mg/L)	140–966	154–1106
Total phosphorus (mg/L)	42–495	157–915
Sodium (mg/L)	18–124	38–285
Potassium (mg/L)	630–2500	1500–5000
Calcium (mg/L)	47–200	58–408
Magnesium (mg/L)	60–180	90–337
Iron (mg/L)	8.8–31.5	16.4–86.4
Copper (mg/L)	1.16–3.42	1.10–4.75
Zinc (mg/L)	1.42–4.48	1.6–6.50
Manganese (mg/L)	0.87–5.20	2.16–8.90
Nickel (mg/L)	0.29–1.44	0.44–1.58
Cobalt (mg/L)	0.12–0.48	0.18–0.96
Lead (mg/L)	0.35–0.72	0.40–1.85

Source: Ref. 26.

- pH between 3 and 5.9 (slightly acid);
- high content of polyphenols, up to 80 g/L; other references up to 10 g/L [28];
- high content of solid matter (total solids up to 102.5 g/L);
- high content of oil (up to 30 g/L).

Table 4 compares the composition values of olive oil mill wastewater (A and B) with those of municipal wastewater (C). While the ratio COD/BOD_5 in both types of wastewater is rather close (between 1.5 and 2.5), there is a big difference between the two for the ratio (BOD : N : P); olive oil wastewater (100 : 1 : 0.35) highly deviates from that in municipal wastewater (100 : 20 : 5).

Based on Tables 2 and 3, the phenols and the organic substances responsible for the high COD value must be considered as problematic for treatment of this wastewater, and the presence of inhibitory or toxic substances may seriously affect the overall treatment system. Therefore, the chemical oxygen demand (COD), the total aromatic content (A), and the total phenolic content (TPh) are mostly selected as representative parameters to follow the overall purification process [19,21,29].

The terms and definitions for the waste resulting from the different oil extraction processes are neither standardized nor country specific [30]. Table 5 shows the nominations found in the Mediterranean countries, while Table 6 shows the most common terminology used in these countries with descriptions.

Table 4 Comparison of Composition Values of Olive Oil Wastewater from a Small Mill (A) and a Big Mill (B) with Municipal Wastewater (C)

	Source of liquid waste		
Parameter	A	B	C
pH	4.5–5.3	5.3–5.7	7–8
BOD_5 (g/L)	15–65	17–41	0.1–0.4
COD (g/L)	37–150	30–80	0.15–1
Total solids (g/L)	24–115	19–75	0.35–1.2
Volatile solids (g/L)	20–97	17–68	0.18–0.6
Suspended solids (g/L)	5.7–14	0.7–26	0.1–0.35
Fats and oils (g/L)	0.046–0.76	0.1–8.2	0.05–0.1
Total nitrogen (g/L)	0.27–0.51	0.3–0.48	0.02–0.08
Total phosphorus (g/L)	0.1–0.19	0.075–0.12	0.006–0.02
COD/BOD_5	2.3–2.5	1.8–2	1.5–2.5
BOD_5 : N : P	100 : 0.98 : 0.37	100 : 1.3 : 0.34	100 : 20 : 5

Between 400 and 600 L of liquid waste are generated per ton of processed olives from the traditional presses used for olive oil extraction, which are operated discontinuously. Depending on its size, the capacity of such an olive oil mill is about 10 to 20 tons of olives per day. With a capacity of 20 tons of olives per day and a process-specific wastewater volume of 0.5 m^3 per ton of olives, the daily wastewater can range up to 10 m^3 per day.

Compared to the traditional presses, twice the quantity of wastewater (from 750 to 1200 L per ton of olives) is produced with the three-phase decanting method. Depending on their size, the capacities of the olive oil mills are also between 10 and 20 tons of olives per day. With a capacity of 20 ton of olives per day and a process specific wastewater volume of about 1 m^3 per ton of olives, the daily wastewater volume from a continuous process is up to 20 m^3 per day. The concentration of the constituents in wastewater from traditional presses is therefore twice as high as in the wastewater resulting from three-phase decanting. In general, the organic pollution

Table 5 Nominations of Waste Resulting from Different Oil Extraction Processes as Found in the Mediterranean Area

	Pressing	Three-phase decanting	Two-phase decanting
Solid	Orujo (Sp) Pirina (Gr, Tk) Hask (It, Tu) Grignons (Fr)	Orujo (Sp) Grignons (Fr) Pirina (Gr, Tk) Hask (It, Tu) Orujillo (Sp) after de-oiling of solid waste	Alpeorujo (in two-phase decanting mainly alpeorujo is produced)
Wastewater	Alpechin (Sp) Margine (Gr) Jamila (It)	Alpechin (Sp) Margine (Gr) Jamila (It)	Alpechin
Oil (from de-oiling of solid waste)	–	Orujooil	Orujooil

Note: Sp, Spanish; Gr, Greek; It, Italian; Tu, Tunisian; Tk, Turkish; Fr, French.
Source: Ref. 30.

Table 6 Terminology of the Olive Oil Sector Related with Waste

Name	Description
Flesh, pulp (En)	Soft, fleshy part of the olive fruit
Pit, husk, stone (En)	Nut, hard part of the olive
Kernel, seed (En)	Softer, inner part of the olive
Alpeorujo, orujo de dos fases, alperujo (Sp)	Very wet solid waste from the two-phase decanters
Orujo, orujo de tres fases (Sp)	
Pirina (Gr/Tk)	
Pomace (It)	Wet solid waste from the three-phase decanters and presses
Grignons (Fr)	
Husks (It/Tu)	
Orujillo (Sp)	De-oiled orujo, de-oiled alpeorujo
Alpechin (Sp)	Liquid waste from the three-phase decanters and presses
Margine (Gr)	
Jamila (It)	
Alpechin-2 (Sp)	
Margine-2 (Gr)	Liquid fraction from secondary alpeorujo treatment (second decanting, repaso, etc.)
Jamila-2 (It)	

Note: En, English; Sp, Spain; Gr, Greek; It, Italian; Tu, Tunisian; Tk, Turkish; Fr, French.
Source: Ref. 1.

load in wastewater from olive oil extraction processes is practically independent of the processing method and amounts to 45–55 kg BOD_5 per ton of olives [31].

The input–output analysis of material and energy flows of the three production processes (press, two-phase, and three-phase decanting) is shown in Table 7. The basis of reference is one metric ton of processed olives.

17.3.1 Design Example 1

What is the population equivalent (pop. equ.) of the effluents discharged from a medium-sized oil mill processing about 15 tons (33,000 lb) of olives per day by using the two systems of traditional pressing or continuous centrifuging?

Solution

Traditional pressing of olives results in a wastewater volume of approximately 600 L (159 gal) per ton of olives; thus wastewater flow rate $= 15\ \text{T} \times 0.6\ \text{m}^3/\text{T} = 9\ \text{m}^3/\text{day}$ (2378 gal/day). Assuming a BOD_5 concentration of 40 g/L (0.34 lb/gal), the resulting total BOD_5 discharged per day $= 9\ \text{m}^3/\text{day} \times 40\ \text{kg/m}^3 = 360$ kg BOD_5/day (792 lb/day).

$$BOD_5 \text{ per person} = 54 - 60\ \text{g/p.day}\ (0.119 - 0.137\ \text{lb/p.day})$$

then

$$\text{Pop. equ.} = \frac{360}{0.06} = 6000 \text{ persons}$$

Continuous centrifuging (three-phase decanting) of olives results in a wastewater volume of approximately 1000 L (264.2 gal) per ton of olives, thus wastewater flow rate =

Table 7 An Input–Output Analysis of Material and Energy Flows of the Production Processes Related to One Ton of Processed Olives

Production process	Input	Amount of input	Output	Amount of output
Traditional pressing process	Olives	1000 kg	Oil	200 kg
	Washing water	0.1–0.12 m^3	Solid waste (25% water + 6% oil)	400 kg
	Energy	40–63 kWh	Wastewater (88% water)	600 L[a]
Three-phase decanters	Olives	1000 kg	Oil	200 kg
	Washing water	0.1–0.12 m^3	Solid waste (50% water + 4% oil)	500–600 kg
	Fresh water for decanter	0.5–1 m^3	Wastewater (94% water +1% oil)	1000–1200 L[b]
	Water to polish the impure oil	10 kg		
	Energy	90–117 kWh		
Two-phase decanter	Olives	1000 kg	Oil	200 kg
	Washing water	0.1–0.12 m^3	Solid waste (60% water +3% oil)	800–950 kg
	Energy	<90–117 kWh		

[a]According to International Olive Oil Council: (400–550 L/Ton processed olives)
[b]According to International Olive Oil Council: (850–1200 L/Ton processed olives)
Source: Ref. 1.

15 T × 1 m^3/T = 15 m^3/day (3963 gal/day). Assuming a BOD_5 concentration of about 23 g BOD_5/L (0.192 lb/gal), the resulting total BOD_5 discharged per day is:

$$15\,\text{m}^3/\text{day} \times 23\,\text{kg/m}^3 = 345\,\text{kg/day}\ (759\,\text{lb/day})$$

then

$$\text{Pop. equ.} = \frac{345}{0.06} = 5750\ \text{persons}$$

17.4 ENVIRONMENTAL RISKS

Olive oil mill wastewaters (OMW) are a major environmental problem, in particular in Mediterranean countries, which are the main manufacturers of olive oil, green and black table olives. In these countries, the extraction and manufacture of olive oil are carried out in numerous small plants that operate seasonally and generate more than 30 million tons of liquid effluents (black water) [16], called "olive oil mill wastewaters" (OMW) each year. These effluents can cause considerable pollution if they are dumped into the environment because of their high organic load, which includes sugar, tannins, polyphenols, polyalcohols, pectins, lipids, and so on. Seasonal operation, which requires storage, is often impossible in small plants [32]. In fact, 2.5 L of waste are released per liter of oil produced [28].

Olive oil mill wastewaters contain large concentrations of highly toxic phenol compounds (can exceed 10 g/L) [33]. Much of the color of OMW is due to the aromatic compounds present, which have phytotoxic and antibacterial effects [34,35].

Despite existing laws and regulations, disposal of untreated liquid waste into the environment is uncontrolled in most cases. When it is treated, the most frequent method used is to retain the effluent in evaporation ponds. However, this procedure causes bad odors and risks polluting surface waters and aquifers. Therefore, this process presents an important environmental problem. Table 8 displays the risks that arise from direct disposal of olive oil mill wastewater (OMW) in the environment (soil, rivers, ground water). Examples of the risks [2] are described in the following sections.

17.4.1 Discoloring of Natural Waters

This is one of the most visible effects of the pollution. Tannins that come from the olive skin remain in the wastewater from the olive oil mill. Although tannins are not harmful to people, animals, or plants, they dye the water coming into contact with them dark black-brown. This undesired effect can be clearly observed in the Mediterranean countries [2].

17.4.2 Degradability of Carbon Compounds

For the degradation of the carbon compounds (BOD_5), the bacteria mainly need nitrogen and phosphorus besides some trace elements. The $BOD_5:N:P$ ratio should be 100 : 5 : 1. The optimal ratio is not always given and thus an excess of phosphorus may occur [36].

17.4.3 Threat to Aquatic Life

Wastewater has a considerable content of reduced sugar, which, if discharged directly into natural waters, would increase the number of microorganisms that would use this as a source of

Table 8 The Environmental Risks Resulting from the Direct Disposal of the Olive Oil Mill Liquid Water Without Treatment

Pollutants	Medium/environment	Effects
Acids	Soil	Destroys the cation exchange capacity of soil
Oil		Reduction of soil fertility
Suspended solids		Bad odors
Organics	Water	Consumption of dissolved oxygen
Oil		Eutrophication phenomena
Suspended solids		Impenetrable film
		Aesthetic damage
Acids	Municipal wastewater sewerage	Corrosion of concrete and metal canals/pipes
Suspended solids		Flow hindrance
		Anaerobic fermentation
Acids	Municipal wastewater treatment plants	Corrosion of concrete and metal canals/pipes
Oil		Sudden and long shocks to activated sludge and trickling filter systems
Organics		
Nutrient imbalance		Shock to sludge digester

Source: Refs 2 and 15.

substrate. The effect of this is reduction of the amount of oxygen available for other living organisms, which may cause an imbalance of the whole ecosystem.

Another similar process can result from the high phosphorus content. Phosphorus encourages and accelerates the growth of algae and increases the chances of eutrophication, destroying the ecological balance in natural waters. In contrast to nitrogen and carbon compounds, which escape as carbon dioxide and atmospheric nitrogen after degradation, phosphorus cannot be degraded but only deposited. This means that phosphorus is taken up only to a small extent via the food chain: plant $\rightarrow$ invertebrates $\rightarrow$ fish $\rightarrow$ prehensile birds.

The presence of such a large quantity of nutrients in the wastewater provides a perfect medium for pathogens to multiply and infect waters. This can have severe effects on the local aquatic life and humans that may come into contact with the water [2].

17.4.4 Impenetrable Film

The lipids in the wastewater may form an impenetrable film on the surface of rivers, their banks, and surrounding farmlands. This film blocks out sunlight and oxygen to microorganisms in the water, leading to reduced plant growth in the soils and river banks and in turn erosion [2].

17.4.5 Soil Quality

The waste contains many acids, minerals, and organics that could destroy the cation exchange capacity of the soil. This would lead to destruction of microorganisms, the soil–air and the air–water balance, and therefore a reduction of the soil fertility [15].

17.4.6 Phytotoxicity

Phenolic compounds and organic acid can cause phytotoxic effects on olive trees. This is of dire importance since wastewater can come into contact with crops due to possible flooding during the winter. The phenols, organic, and inorganic compounds can hinder the natural disinfection process in rivers and creeks [2].

17.4.7 Odors

Anaerobic fermentation of the wastewater causes methane and other gases (hydrogen sulfide, etc.) to emanate from natural waters and pond evaporation plants. This leads to considerable pollution by odors even at great distances [2].

Other risks could be referred to in this respect, such as agricultural-specific problems arising from pesticides and other chemicals, although their effect in olive cultivation is less pronounced than other fields of agriculture. The main problem is soil erosion caused by rainwater, which results in steeper slopes and increases difficulty in ploughing. Soil quality and structure also influence erosion caused by rain. At present, protective measures such as planting of soil-covering species or abstention from ploughing are hardly used.

17.5 LIQUID WASTE TREATMENT METHODS

Disposal and management of highly contaminated wastewater constitute a serious environmental problem due to the biorecalcitrant nature of these types of effluents, in most cases. Generally, biological treatment (mainly aerobic) is the preferred option for dealing with urban and industrial effluents because of its relative cost-effectiveness and applicability for treating a wide variety of hazardous substances [19]. Nevertheless, some drawbacks may be found when applying this technology. For instance, some chemical structures, when present at high concentrations, are difficult to biodegrade because of their refractory nature or even toxicity toward microorganisms. Thus, several substances have been found to present inhibitory effects when undergoing biological oxidation. Among them, phenolic compounds constitute one of the most important groups of pollutants present in numerous industrial effluents [37]. Owing to the increasing restrictions in quality control of public river courses, development of suitable technologies and procedures are needed to reduce the pollutant load of discharges, increase the biodegradability of effluent, and minimize the environmental impact to the biota.

Industries that generate nonbiodegradable wastewater showing high concentrations of refractory substances (chiefly phenol-type compounds) include the pharmaceutical industry, refineries, coal-processing plants, and food-stuff manufacturing. The olive oil industry (a common activity in Mediterranean countries), in particular, generates highly contaminated effluents during different stages of mill olive oil production (washing and vegetation waters).

Therefore, most treatment processes used for high-strength industrial wastewaters have been applied to olive oil mill effluents (OME). Yet, OME treatment difficulties are mainly associated with: (a) high organic load (OME are among the strongest industrial effluents, with COD up to 220 g/L and sometimes reaching 400 g/L); (b) seasonal operation, which requires storage (often impossible in small mills); (c) high territorial scattering; and (d) presence of organic compounds that are difficult to degrade by microorganisms (long-chain fatty acids and phenolic compounds of the C-7 and C-9 phenylpropanoic family) [23].

Furthermore, a great variety of components found in liquid waste (alpachin) and solid waste (orujo and alpeorujo) require different technologies to eliminate those with harmful effects on the environment. Most used methods for the treatment of liquid waste from olive oil production are presented in Table 9. They correspond to the current state-of-art-technologies and are economically feasible. These methods are designed to eliminate organic components and to reduce the mass. In some cases, substances belonging to other categories are also partly removed. In practise, these processes are often combined since their effects differ widely [1]. Therefore, methods should be used in combination with each other.

The following key treatment methods are mainly applied to liquid waste. Some of these methods can also be used in the treatment of liquid–solid waste (alpeorujo), for example, treatment by fungi, evaporation/drying, composting, and livestock feeding. However, those methods tested at laboratory scale must be critically examined before applying them at industrial or full-scale, in order to meet the local environmental and economical conditions.

Regarding the olive oil industry, it should always be considered that complicated treatment methods that lack profitable use of the final product are not useful, and all methods should have a control system for the material flows [38].

17.5.1 Low-Cost Primitive Methods

These methods are mostly applied in the developing countries producing olive, due to their simplicity and low costs. Of these methods, the most important are:

- Drainage of olive oil mill liquid waste in some types of soils, with rates up to 50 m^3/ha-year (in the case of traditional mills) and up to 80 m^3/ha-year (in the case of decanting-based methods), or to apply the olive oil mill liquid wastes to the irrigation water for a rate of less than 3%. These processes are risky because they decrease the fertility of the soil. This calls for greater care and scientific research into these methods prior to agronomic application.

Table 9 Treatment Methods for the Liquid and Solid Waste from Olive Oil Production

Treatment method of (alpechin)
Low-cost primitive methods
• Drainage in soil
• Simple disposal in evaporation ponds
• Mixing with solid waste in sanitary landfills
Aerobic treatment
Anaerobic treatment
Combined biological treatment methods
Wet air oxidation and ozonation
Fungal treatment
Decolorization
Precipitation/flocculation
Adsorption
Filtration (biofiltration, ultrafiltration)
Evaporation/drying
Electrolysis
Bioremedation and composting[a]
Livestock feeding[a]
Submarine outfall

[a] These recycling methods can be used for liquid as well as solid waste from olive oil production.

Products resulting from treatment may be reused, for instance, as fertilizer or fodder in agriculture. For all methods, waste that is not suited for reuse can be disposed at landfills.

- Simple disposal and retention in evaporation ponds (large surface and small depth ponds), preferably in distant regions, to be dried by solar radiation and other climatic factors. This method does not require energy or highly trained personnel. Drawbacks are associated with the evaporation process, which generates odors and additional risks for the aquatic system of the area (filtration phenomena, surface water contamination, etc.). In addition, the disadvantages include: the need for large areas for drying in selected regions with impermeable (clay) soil distant from populated areas; the requirement, in most cases, for taking necessary precautions to prevent pollutants reaching the groundwater through placement of impermeable layers in the ground and walls of ponds; ineffective in higher rainfall regions; emergence of air pollutants caused by decomposition of organic substances (ammonia-hydrocarbon volatile compounds). This method is being applied in many countries of the Mediterranean area. In Spain alone, there are about 1000 evaporation ponds, which improve the water quality, but the ponds themselves caused serious negative environmental impacts. Dried sludge from corporation ponds can be used as fertilizer, either directly or composted with other agricultural byproducts (e.g., grape seed residues, cotton wastes, bean straw) [39].
- Mixing the olive oil mill liquid wastes with municipal solid wastes in sanitary landfills leads to increased organic load on site. Consideration should be made regarding the pollutants that may reach the groundwater, in addition to the risks of combustion due to generation of combustible hydrocarbon gases. These factors should be taken into account in designing and establishing landfills, not forgetting the necessity to collect

and treat the drainage wastewater resulted from applying this method. This method is cost-effective and is suitable for final disposal of the wastes, with the property of obtaining energy from the generated gases. Nevertheless, there are drawbacks such as the air pollution caused by the decomposition, the need for advanced treatment for the highly polluted collected drainage wastewater, and the need for using large areas of land and particular specifications.

17.5.2 Aerobic Treatment

When biodegradable organic pollutants in olive oil mill wastewater (alpechin) are eliminated by oxygen-consuming microorganisms in water to produce energy, the oxygen concentration decreases and the natural balance in the water body is disturbed. To counteract an overloading of the oxygen balance, the largest part of these oxygen-consuming substances (defined as BOD_5) must be removed before being discharged into the water body. Wastewater treatment processes have therefore been developed with the aim of reducing the BOD_5 concentration as well as eliminating eutrophying inorganic salts, that is, phosphorus and nitrogen compounds, ammonium compounds, nonbiodegradable compounds that are analyzed as part of the COD, and organic and inorganic suspended solids [38].

In aerobic biological wastewater treatment plants, the natural purification processes taking place in rivers are simulated under optimized technical conditions. Bacteria and monocellular organisms (microorganisms) degrade the organic substances dissolved in water and transform them into carbonic acid, water, and cell mass. The microorganisms that are best suited for the purification of a certain wastewater develop in the wastewater independently of external influences and adapt to the respective substrate composition (enzymatic adaptation). Owing to the oxidative degradation processes, oxygen is required for wastewater treatment. The oxygen demand corresponds to the load of the wastewater.

Two types of microorganisms live in waters: suspended organisms, floating in the water, and sessile organisms, which often settle on the surface of stones and form biofilms. Biofilm processes such as fixed-bed or trickling filter processes are examples of the technical application of these natural processes [38].

Treatment of Olive Oil Mill Wastewaters in Municipal Plants

Municipal wastewater is unique in that a major portion of the organics are present in suspended or colloidal form. Typically, the BOD in municipal sewage consists of 50% suspended, 10% colloidal, and 40% soluble parts. By contrast, most industrial wastewater are almost 100% soluble. In an activated sludge plant treating municipal wastewater, the suspended organics are rapidly enmeshed in the flocs, the colloids are adsorbed on the flocs, and a portion of the soluble organics are absorbed. These reactions occur in the first few minutes of aeration contact. By contrast, for readily degradable wastewaters, that is, food processing, a portion of the BOD is rapidly sorbed and the remainder removed as a function of time and biological solids concentration. Very little sorption occurs in refractory wastewaters. The kinetics of the activated sludge process will therefore vary depending on the percentage and type of industrial wastewater discharged to the municipal plant and must be considered in the design calculations [40].

The percentage of biological solids in the aeration basin will also vary with the amount and nature of the industrial wastewater. Increasing the sludge age increases the biomass percentage as volatile suspended solids undergo degradation and synthesis. Soluble industrial wastewater will increase the biomass percentage in the activated sludge.

A number of factors should be considered when discharging industrial wastewaters, including olive oil mill effluents, into municipal plants [40]:

- *Effect on effluent quality*. Soluble industrial wastewaters will affect the reaction rate K. Refractory wastewaters such as olive oil mills, tannery, and chemical will reduce K, while readily degradable wastewaters such as food processing and brewery will increase K.
- *Effect on sludge quality*. Readily degradable wastewaters will stimulate filamentous bulking, depending on basin configuration, while refractory wastewaters will suppress filamentous bulking.
- *Effect of temperature*. An increased industrial wastewater input, that is, soluble organics, will increase the temperature coefficient θ, thereby decreasing efficiency at reduced operating temperatures.
- *Sludge handling*. An increase in soluble organics will increase the percentage of biological sludge in the waste sludge mixture. This will generally decrease dewaterability, decrease cake solids, and increase conditioning chemical requirements. One exception is pulp and paper-mill wastewaters in which pulp and fiber serve as a sludge conditioner and enhances dewatering rates.

It is worth pointing out that certain threshold concentrations for inhibiting agent and toxic substances must not be exceeded. Moreover, it should be noted that most industrial wastewaters are nutrient deficient, that is, they lack nitrogen and phosphorus. Municipal wastewater with a surplus of these nutrients will provide the required nutrient balance.

The objective of the activated sludge process is to remove soluble and insoluble organics from a wastewater stream and to convert this material into a flocculent microbial suspension that is readily settleable and permits the use of gravitational solids liquid separation techniques. A number of different modifications or variants of the activated sludge process have been developed since the original experiments of Arden and Lockett in 1914 [40]. These variants, to a large extent, have been developed out of necessity or to suit particular circumstances that have arisen. For the treatment of industrial wastewater, the common generic flow sheet is shown in Fig. 4.

The activated sludge process is a biological wastewater treatment technique in which a mixture of wastewater and biological sludge (microorganisms) is agitated and aerated. The biological solids are subsequently separated from the treated wastewater and returned to the aeration process as needed. The activated sludge process derives its name from the biological mass formed when air is continuously injected into the wastewater. Under such conditions, microorganisms are mixed thoroughly with the organics under conditions that stimulate their growth through use of the organics as food. As the microorganisms grow and are mixed by the agitation of the air, the individual organisms clump together (flocculate) to form an active mass of microbes (biologic floc) called activated sludge [41].

In practise, wastewater flows continuously into an aeration tank where air is injected to mix the activated sludge with the wastewater and to supply the oxygen needed for the organisms to break down the organics. The mixture of activated sludge and wastewater in the aeration tank is called mixed liquor. The mixed liquor flows from the aeration tank to a secondary clarifier where the activated sludge is settled out. Most of the settled sludge is returned to the aeration tank (return sludge) to maintain a high population of microbes to permit rapid breakdown of the organics. Because more activated sludge is produced than is desirable in the process, some of the return sludge is diverted or wasted to the sludge handling system for treatment and disposal.

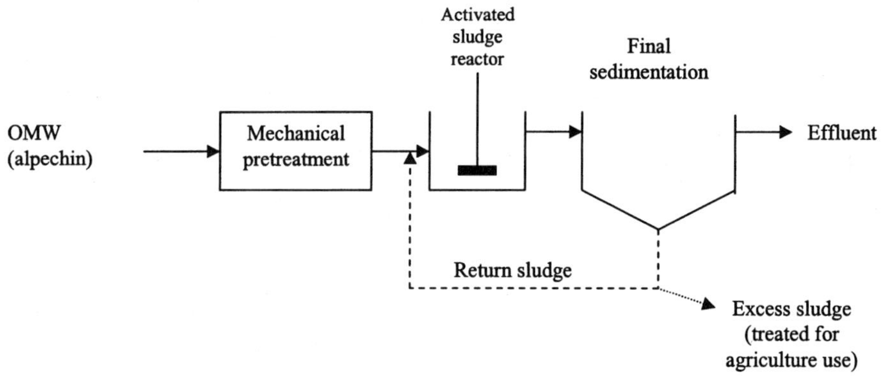

Figure 4 Aerobic treatment (activated sludge plant).

Biofilm processes are used when the goal is very far-reaching retention and concentration of the biomass in a system. This is especially the case with slowly reproducing microorganisms in aerobic or anaerobic environments. The growth of sessile microorganisms on a carrier is called biofilm. The filling material (e.g., in a trickling filter stones, lava slag, or plastic bodies) or the filter material (e.g., in a biofilter) serve as carrier. The diffusion processes in biofilm plants are more important than in activated sludge plants because unlike activated sludge flocs the biofilms are shaped approximately two-dimensionally. On the one hand, diffusion is necessary to supply the biofilm with substrate and oxygen; on the other hand, the final metabolic products (e.g., CO_2 and nitrate) must be removed from the biofilm.

For treatment of industrial wastewater, trickling filters are often used. A trickling filter is a container filled completely with filling material, such as stones, slats, or plastic materials (media), over which wastewater is applied. Trickling filters are a popular biological treatment process [42]. The most widely used design for many years was simply a bed of stones, 1–3 m deep, through which the wastewater passed. The wastewater is typically distributed over the surface of the rocks by a rotating arm. Rock filter diameters may range up to 60 m. As wastewater trickles through the bed, a microbial growth establishes itself on the surface of the stone or packing in a fixed film. The wastewater passes over the stationary microbial population, providing contact between the microorganisms and the organics. The biomass is supplied with oxygen using outside air, most of the time without additional technical measures. If the wastewater is not free of solid matter (as in the case of alpechin), it should be prescreened to reduce the risk of obstructions.

Excess growths of microorganisms wash from the rock media and would cause undesirably high levels of suspended solids in the plant effluent if not removed. Thus, the flow from the filter is passed through a sedimentation basin to allow these solids to settle out. This sedimentation basin is referred to as a secondary clarifier, or final clarifier, to differentiate it from the sedimentation basin used for primary settling. An important element in trickling filter design is the provision for return of a portion of the effluent (recirculation) to flow through the filter. Owing to seasonal production of wastewater and to the rather slow growth rates of the microorganisms, these processes are less suited for the treatment of alpechin, compared to the activated sludge process.

Another worthwhile aerobic treatment method developed by Balis and his colleagues [38] is the bioremediation process, based on the intrinsic property of an *Azotobacter vinelandii* strain (strain A) to proliferate on limed olive oil mill wastewater. More specifically, the olive mill wastewater is pretreated with lime to pH 7–8 and then is fed into an aerobic bioreactor equipped with a rotating wheel-type air conductor. The reactor is operated in a repeated fed batch culture fashion with a cycle time of 3 days. During each cycle, the Azotobacter population proliferates and fixes molecular nitrogen. It concomitantly produces copious amounts of slime and plant growth promoting substances. The endproduct is a thick, yellow-brown liquid. It has a pH of about 7.5–8.0, it is nonphytotoxic, soluble in water, and can be used as liquid fertilizer over a wide range of cultivated plants (olives, grapes, citrus, vegetables, and ornamentals). Moreover, there is good evidence that the biofertilizer induces soil suppressiveness against root pathogenic fungi, and improves soil structure. A medium-scale pilot plant of 25 m^3 capacity has been constructed in Greece by the Olive Cooperative of Peta near Arta with the financial support of the General Secretariat of Science and Technology of Greece. The plant has been operating since 1997. The local farmers use the liquid biofertilizer that is produced to treat their olive and citrus groves.

In short, it has been demonstrated that free-living N_2-fixing bacteria of Azotobacter grow well in olive mill wastewater and transform the wastes into a useful organic fertilizer and soil conditioner. For further details in this regard, refer to Section 17.5.17 (Bioremediation and Composting).

The following case study explains the influence of aerobic treatments for already fermented olive oil mill wastewater (OMW), on the anaerobic digestion of this waste.

Case Study

This kinetic study [25] allows intercomparison of the effects of different aerobic pretreatments on the anaerobic digestion of OMW, previously fermented with three microorganisms (*Geotrichum condidum*, *Azotobacter chroococcum*, and *Aspergillus terreus*). The OMW used was obtained from a continuous olive-processing operation. The bioreactor used was batch fed and contained sepiolite as support for the mediating bacteria. The results of the microtox toxicity test expressed as toxic units (TU) for both pretreated and untreated OMW are as follows:

- prior to inoculation (untreated OMW): TU = 156;
- after fermentation with Geotrichum: TU = 64;
- after fermentation with Azotobacter: TU = 32;
- after fermentation with Aspergillus: TU = 20.

The influence of the different aerobic pretreatments on the percentages of elimination of COD and total phenol contents are indicated in Table 10.

Table 10 Influence of Different Aerobic Pretreatments on the Percentages of Elimination of COD and Total Phenol Contents

Pretreatment	Elimination COD %	Elimination phenols %
Geotrichum	63.3	65.6
Azotobacter	74.5	90.0
Aspergillus	74.0	94.3

Source: Ref. 25.

A kinetic model was developed for the estimation of methane production (G) against time (t), represented in the following equation:

$$G = G_M\left[1 - \exp\left(-\frac{AXt}{S_0}\right)\right], \quad \text{over the COD range studied } (3.9 - 14.5\,\text{g/L})$$

where G_M is the maximum methane volume obtained at the end of digestion time, S_0 is the initial substrate concentration, X is the microorganism concentration, and A is the kinetic constant of the process, which was calculated using a nonlinear regression. This kinetic parameter was found to be influenced by the pretreatment carried out, and was 4.6, 4.1, and 2.3 times higher for Aspergillus-, Azotobacter-, and Geotrichum-pretreated OMWs than that obtained in the anaerobic digestion of untreated OMW. The kinetic constant increased as the phenolic compound content and biotoxicity of the pretreated OMWs decreased.

The final conclusion that can be drawn from this work is that aerobic pretreatment of the OMW with different microorganisms (Geotrichum, Azotobacter, and Aspergillus) considerably reduces the COD and the total phenolic compound concentration of waste that is responsible for its biotoxicity. This fact is shown through enhancement of the kinetic constant for the anaerobic digestion process, and a simultaneous increase in the yield coefficient of methane production.

Case studies regarding the role and importance of the aerobic treatment process combined with chemical oxidation such as wet air oxidation (WAO) are found in Section 17.5.9.

17.5.3 Design Example 2

An olive oil mill is to treat its wastewater in an extended aeration activated sludge plant. The final effluent should have a maximum soluble BOD_5 of 20 mg/L during the olive mill operation season. This plant is to be designed under the following conditions: $Q = 60\ m^3/day$ (15,850 gal/day); S_0 (diluted) = 800 mg/L; S_e = 20 mg/L; X_v = 3000 mg/L; $a = 0.50$; $a' = 0.6$; $b = 0.10$ at 20°C; $\theta = 1.065$; $K = 6.0$/day at 20°C; and $b' = 0.12$/day.

Solution

$$t = \frac{S_0(S_0 - S_e)}{KS_eX_v}$$

$$t = \frac{800(800 - 20)}{6(20)(3000)} = 1.73 \text{ days}$$

$$\frac{F}{M} = \frac{S_0}{X_vt} = \frac{800}{3000 \times 1.73} = 0.154$$

The degradable fraction is determined by:

$$X_d = \frac{0.8}{1 + 0.26\theta c}$$

Assuming $\theta c = 25$ day (SRT)

$$X_d = \frac{0.8}{1 + 0.2 \times 0.1 \times 25} = 0.53$$

The aeration basin volume is: $60\ m^3/day \times 1.73\ day = 104\ m^3$ (27,421 gal). The sludge yield can be computed as:

$$\Delta X_v = aS_r - bX_d X_v t$$
$$\Delta X_v = 0.5 \times 780\,mg/L - 0.10 \times 0.53 \times 3000\,mg/L \times 1.73$$
$$\Delta X_v = 115\,mg/L$$
$$\Delta X_v = 115\,mg/L \times 60\,m^3/day \times 10^{-3}$$
$$= 7.0\,kg/day (15.4\,lb/day)$$

Check the sludge age:

$$\theta_c = \frac{\forall X_v}{\Delta X_v} = \frac{104 \times 3000}{7 \times 1000} = 45\,day$$

or

$$\theta_c = \frac{27{,}421\,gal \times 8.34 \times 10^{-6} \times 3000}{15.4} = 45\,day$$

Compute the oxygen required:

$$O_2/day = a' S_r Q + b' X_d X_v \forall$$
$$O_2/day = (0.6 \times 780 \times 60 + 0.12 \times 0.53 \times 3000 \times 104)10^{-3}$$
$$O_2/day = 48\,kg/day = 2\,kg/hour\ (4.4\,lb/hour)$$

The oxygen needed can also be calculated directly from the approximate relation:

$$2.0 - 2.5\,kg\ O_2/kg\ BOD_5$$
$$O_2/day = 60\,m^3/day \times 800\,g\ BOD_5/m^3 \times 10^{-3} \times 2\,kg\ O_2/kg\ BOD_5$$
$$O_2/day = 96\,kg\ O_2/day\ (4\,kg/hour)\ (8.8\,lb/hour)$$

Compute the effluent quality at 15°C:

$$K_{15^\circ} = 6 \times 1.065^{(15-20)} = 4.38/day$$

$$S_e = \frac{S_0{}^2}{KX_v t + S_0} = \frac{800^2}{4.38 \times 3000 \times 1.73 + 800}$$

$$S_e = 27\,mg/L$$

The effluent quality at 10°C:

$$K_{10^\circ} = 6 \times 1.065^{(10-20)} = 3.19/day$$

$$S_e = \frac{(800)^2}{3.19 \times 3000 \times 1.73 + 800}$$

$$S_e = 37\,mg/L$$

17.5.4 Anaerobic Treatment

Anaerobic processes are increasingly used for the treatment of industrial wastewaters. They have distinct advantages including energy and chemical efficiency and low biological sludge yield, in addition to the possibility of treating organically high-loaded wastewater (COD $>$ 1500 mg/L), with the requirement of only a small reactor volume.

Anaerobic processes can break down a variety of aromatic compounds. It is known that anaerobic breakdown of the benzene nucleus can occur by two different pathways, namely, photometabolism and methanogenic fermentation. It has been shown that benzoate, phenylacetate, phenylpropionate, and annamate were completely degraded to CO_2 and CH_4. While long acclimation periods were required to initiate gas production, the time required could be reduced by adapting the bacteria to an acetic acid and substrate before adapting them to the aromatic.

Chmielowski et al. [43] showed that phenol, p-cresol, and resorcinol yielded complete conversion to CH_4 and CO_2.

Principle of Anaerobic Fermentation

In anaerobic fermentation, roughly four groups of microorganisms sequentially degrade organic matter. Hydrolytic microorganisms degrade polymer-type material such as polysaccharides and proteins to monomers. This reduction results in no reduction of COD. The monomers are then converted into fatty acids (VFA) with a small amount of H_2. The principal organic acids are acetic, propionic, and butyric with small quantities of valeric. In the acidification stage, there is minimal reduction of COD. Should a large amount of H_2 occur, some COD reduction will result, seldom exceeding 10%. All formed acids are converted into acetate and H_2 by acetogenic microorganisms. The breakdown of organic acids to CH_4 and CO_2 is shown in Fig. 5. Acetic acid and H_2 are converted to CH_4 by methanogenic organisms [40].

The specific biomass loading of typical anaerobic processes treating soluble industrial wastewaters is approximately 1 kg COD utilized/(kg biomass-day). There are two classes of methanogenes that convert acetate to methane, namely *Methanothrix* and *Methanosarcina*. Methanothrix has a low specific activity that allows it to predominate in systems with a low steady-state acetate concentration. In highly loaded systems, Methanosarcina will predominate with a higher specific activity (3 to 5 times as high as Methanothrix) if trace nutrients are

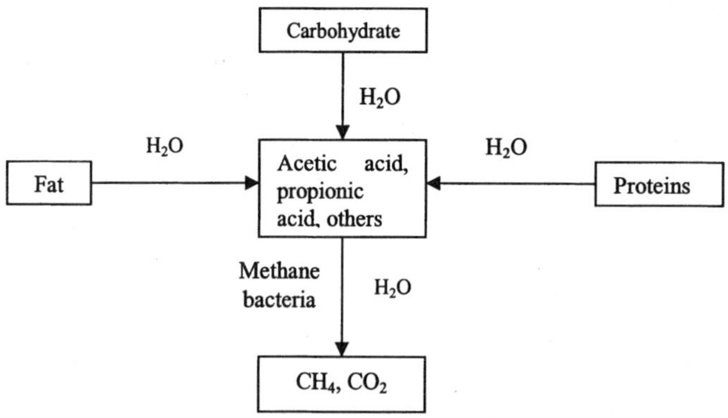

Figure 5 Anaerobic degradation of organics (from Ref. 46).

available. At standard temperature and pressure, 1 kg of COD or ultimate BOD removed in the process will yield 0.35 m^3 of methane [40].

The quantity of cells produced during methane fermentation will depend on the strength and character of the waste, and the retention of the cells in the system.

In comparing anaerobic processes and aerobic processes, which require high energy and high capital cost and produce large amounts of secondary biological sludge, the quantity of excess sludge produced is 20 times lower in anaerobic processes. This can be explained by the fact that with the same organic load under oxygen exchange about 20 times less metabolic energy is available for the microorganisms. Anaerobic wastewater treatment methods are mainly used for rather high-loaded wastewaters with a COD of 5000 up to 40,000 mg/L from the food and chemical industry [2]. Unfortunately, these methods are normally employed strictly as pretreatment measures. Aerobic follow-up treatment, for example, in a downstream-arranged activated sludge plant, is possible and recommended (Fig. 6).

Factors Affecting Anaerobic Process Operation

The anaerobic process functions effectively over two temperature ranges: the mesophilic range of 85–100°F (29–38°C) and the thermophilic range of 120–135°F (49–57°C). Although the rates of reaction are much greater in the thermophilic range, the maintenance of higher temperatures is usually not economically justifiable.

Methane organisms function over a pH range of 6.6–7.6 with an optimum near pH 7.0. When the rate of acid formation exceeds the rate of breakdown to methane, a process unbalance results in which the pH decreases, gas production falls off, and the CO_2 content increases [40]. pH control is therefore essential to ensure a high rate of methane production. According to German literature, the tolerable pH range for anaerobic microorganisms is between 6.8 and 7.5. This means that the anaerobic biocenosis is very pH-specific [38].

With regard to the influence of initial concentration on anaerobic degradation, preliminary laboratory and pilot-scale experimentation on diluted olive oil mill effluents (OME) [44] showed that the anaerobic contact process was able to provide high organic removal efficiency (80–85%) at 35°C and at an organic load lower than 4 kg COD/m^3-day; however, in particular at high feed concentration, the process proved unstable due to the inhibitory effects of substances

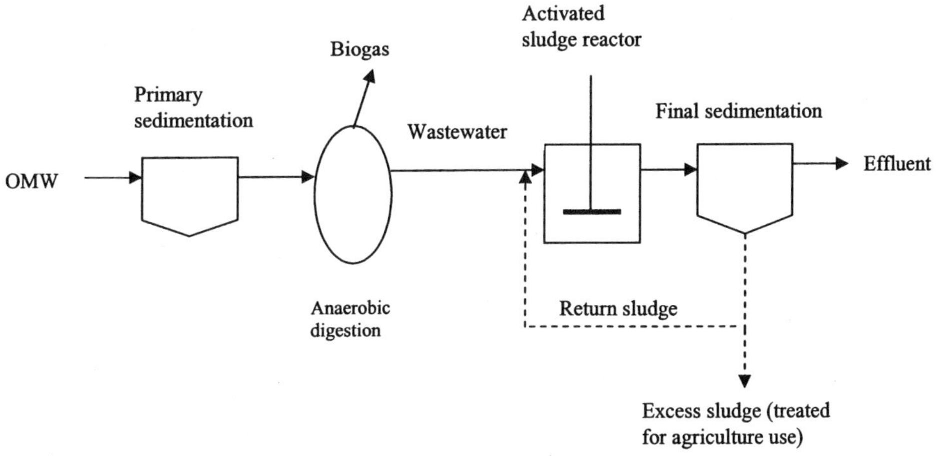

Figure 6 Anaerobic–aerobic treatment method.

such as polyphenols. Moreover, additions of alkalinity to neutralize acidity and ammonia to furnish nitrogen for cellular biosynthesis were required.

To overcome these difficulties and improve process efficiency and stability, there are basically two methods that may be adopted [23]: (a) the treatment of combined OME and sewage sludge in contact bioreactors; and (b) operation with more diluted OME in high-rate bioreactors (such as UASB reactors and fixed-bed filters).

In the first method, conventional digesters can be overloaded with concentrated soluble wastes such as OME, and still operate satisfactorily. Moreover, nutrients such as ammonia and buffers are provided by degradation of proteineous substances from sludge. On this basis, laboratory-scale experimentation [45] has shown that removal efficiencies of 65 and 37% in terms of COD and VSS, respectively, were obtained at 35°C and at an organic load of 4.2 kg COD/m^3-day (66% from sewage sludge, 34% from OME). Higher OME additions led to process imbalance due to the inhibitory effects of polyphenols. This method, based on anaerobic contact digestion of combined OME and sewage sludge, seems to be suitable only for those locations where the polluting load due to the OME is lower than the domestic wastewater load. In this regard it is worth considering that during the olive oil milling season, OME pollution largely exceeds that from domestic wastewater [23].

With regard to the second method, based on the use of high-rate bioreactors, experimentation on UASB reactors [46,47] showed that COD removal efficiencies of about 70–75% were obtained at 37°C and at an organic load in the range 12–18 kg COD/m^3-day by adopting a dilution ratio in the range of 1 : 8 to 1 : 5 (OME: tap water; diluted OME initial concentration in the range 11–19 g COD/L). Slightly less satisfactory results were obtained by using anaerobic filters filled with macroreticulated polyurethane foam [45].

It is important to note that immobilization of methanogenic bacteria may decrease the toxicity of phenolic compounds. Another pilot-scale anaerobic–aerobic treatment of OME mixed with settled domestic wastewater [48] produced a final COD concentration of about 160 mg/L, provided that a dilution ratio of 1 : 60 to 1 : 100 was adopted, corresponding to a COD load ratio equal to 3 : 1 for OME and domestic wastewater, respectively. This ratio is typical for those locations with a high density of olive oil mills. However, in addition to the high value required for the dilution ratio, the final effluent did not comply with legal requirements in terms of color and nitrogen [23].

The aforementioned data clearly show that in the treatment of OME, even when carried out with the use of most appropriate technology, that is, anaerobic digestion, it was difficult to reach the treatment efficiencies required by national regulations throughout the Mediterranean area. In particular, methanogenesis, which represents the limiting step in the anaerobic digestion of soluble compounds, is severely hindered by the inhibition caused by the buildup of volatile fatty acids (VFAs) and/or the presence of a high concentration of phenolic compounds and/or oleic acid in the OME. As for phenol, 1.25 g/L leads to 50% activity reduction of acetate-utilizing methanogens [49]. As for oleic acid, it is reported that 5 mM is toxic to methanogenic bacteria [50].

The reader may refer to the following Case Study V to better understand the mechanism of biodegradation of the main compounds contained in the OME in relation to pH, temperature, and initial concentration of effluents, and in particular the mutual coherence of the two successive partial stages occurring in anaerobic digestion of OME, acidogenesis, and methanogenesis.

Anaerobic Treatment Systems of Wastewater

Seasonal operation of olive oil mills is not a disadvantage for anaerobic treatment systems because anaerobic digesters can be easily restarted after several months of mill shutdown [51].

At present there are no large-scale plants. However, the anaerobic contact reactors and upflow sludge-blanket reactors have been mainly studied using several pilot tests (Fig. 7), besides other tested reactors such as anaerobic filters and fluidized-bed reactors.

Sludge retention is decisive for the load capacity and thus the field of application of an anaerobic reactor. In the UASB reactor, favorable sludge retention is realized in a simple way. Wastewater flows into the active space of the reactor, passing from the bottom to the top of the reactor. Owing to the favorable flocculation characteristics of the anaerobic-activated sludge, which in higher-loaded reactors normally leads to the development of activated sludge grains and to its favorable sedimentation capacity, a sludge bed is formed at the reactor bottom with a sludge blanket developing above it. To avoid sludge removal from the reactor and to collect the biogas, a gas-sludge separator (also called a three-phase separator) is fitted into the upper part of the reactor. Through openings in the bottom of this sedimentation unit, the separated sludge returns into the active space of the reactor. Because of this special construction, the UASB reactor has a very high load capacity. In contrast to the contact sludge process, no additional sedimentation tank is necessary, which would require return sludge flow for the anaerobic activated sludge, resulting in a reduction of the effective reactor volume. Several studies on anaerobic treatment of olive oil wastewaters have been carried out, and data from different publications are listed in Table 11.

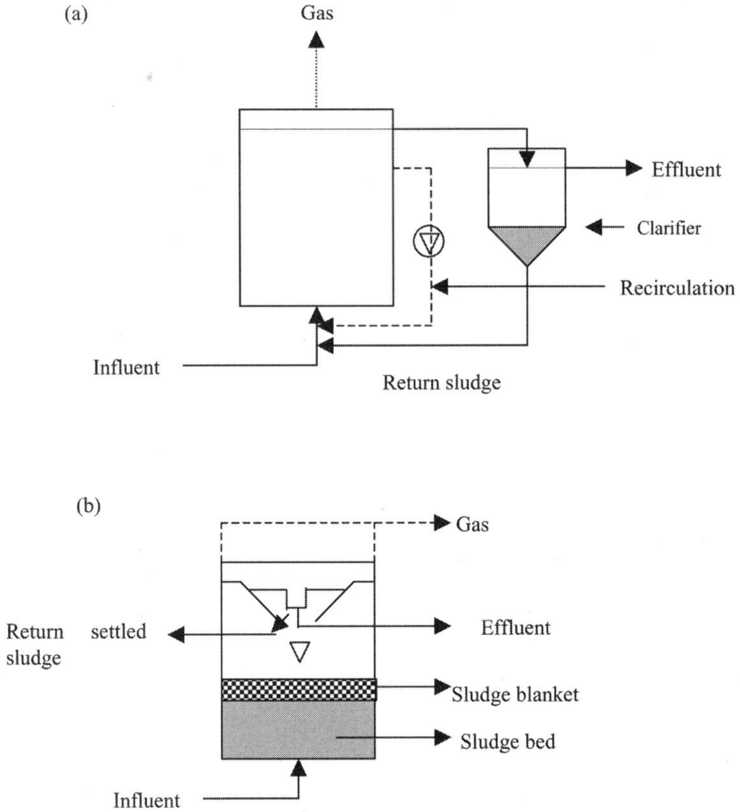

Figure 7 Anaerobic treatment processes: (a) Contact sludge reactor; (b) UASB reactor.

Table 11 Summary of the Data from Different Publications Related to Anaerobic Treatment of Olive Oil Wastewater

	Fiestas (1981)[14]	FIW[38]	Aveni (1984)[44 a]	FIW[38]	FIW[38]	FIW[38]	Steegmans (1992)[15]	Ubay (1997)[22]
Treatment process	Contact process	UASB reactor	Contact process	Conventional reactor	UASB reactor	Packed-bed reactor	UASB reactor	UASB reactor
Influent	33–42 g BOD_5/L	4–6 g COD/L	–	20–65 g COD/L	5–15 g COD/L	45–50 g COD/L	26.7 g COD/L	5–22.6 g COD/L
Volumetric loading	1.2–1.5 kg BOD/ (m^3*day)	15–20 kg COD/ (m^3*day)	4 kg COD/ (m^3*day)	20–65 kg COD/ (m^3*day)	5–21 kg COD/ (m^3*day)	–	1.59 kg COD/ (m^3*day)	5–18 kg COD/ (m^3*day)
Purification efficiency	80–85% BOD	70% COD	80–85% COD	80–85% COD	70–80% COD	45–55% COD	55.9% COD	70–75% COD
Gas production	700 L/kg BOD_{elim}	–	–	550 L/kg COD_{elim}	8000 L/ (m_r^3*day)	300–600 L/kg COD_{elim}	50–100 L CH_4/ kg COD_{elim}	350 L CH_4/kg COD_{elim}
Methane content	70%	–	–	50–70 %	70–80 %	84 %	70%	–

[a]Based on laboratory and pilot experimentation on diluted olive oil mill effluents.
Source: Refs. 14, 15, 22, 38, 44.

Case Studies

Many anaerobic pilot plants have been applied successfully in treating OMW in various parts of the world. The following describe some of these pilot plants and tests.

Case Study I. The search for an economic treatment process for wastewater from an olive oil extraction plant in Kandano (region of Chania, Crete) led to the concept of a pilot plant. The goal was to study the efficiency of separate anaerobic treatment of the settled sludge and of the sludge liquor from the settling tank (Fig. 8) [38].
Description of the plant:

- delivery, storage container;
- settling tank with a capacity of 650 m^3;
- anaerobic digester (volume: 16 m^3) for the sludge;
- UASB (upflow anaerobic sludge blanket) reactor (volume: 18 m^3) for the sludge liquor.

The plant can receive one-sixth of the total wastewater volume produced. The daily influent is 30 m^3. The wastewater is collected in a storage container where its quality and quantity are analyzed. The raw wastewater is then retained for ten days in the settling tank where the particular substances settle.

Two separate zones are formed:

- the supernatant zone;
- the thickening and scraping zone.

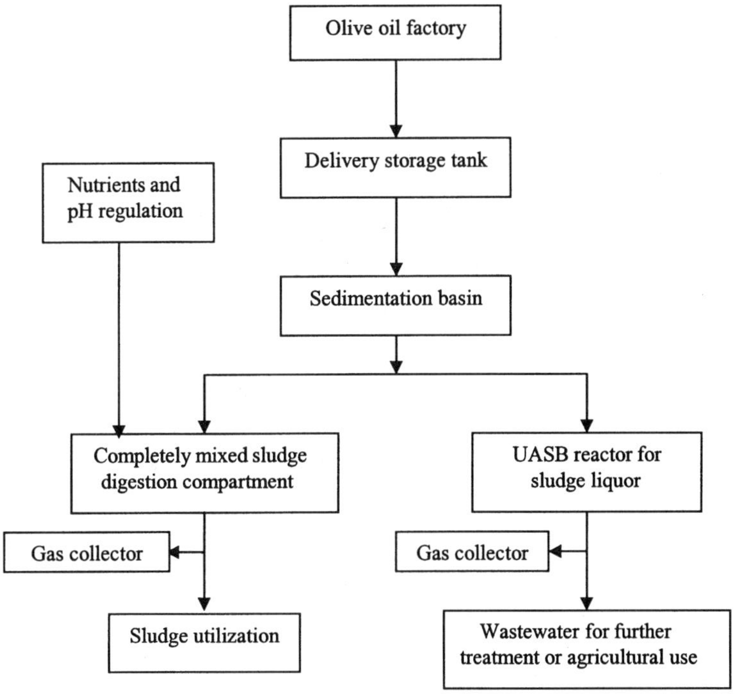

Figure 8 Pilot plant for treatment of wastewater from olive oil extraction in Kandano (a region of Chania, Crete) (from Ref. 38).

Both the preclarified sludge liquor and the primary sludge withdrawn are anaerobically treated in parallel. There is the risk of scum layer formation in the settling tank, which may lead to strong odors. This problem can be solved by covering the tank or using a scraper bridge.

The preclarified sludge liquor is preheated and fed into the UASB reactor. The biogas obtained is withdrawn from the upper part of the reactor and conducted to the gas storage room. The liquid phase is submitted to sedimentation, then stored in a container.

After the addition of nutrients and pH regulation, the primary sludge, showing a high water content (65–80%), is fed into a completely mixed digester. The biogas is again withdrawn from the upper part of the digester and conducted to the gas storage room. The treated liquid phase is conducted to the settling tank and then to the collecting container. At this point, the biogas is incinerated.

To build a plant that treats 30 m^3 per day, a surface of at least 1 ha is necessary, at the cost of about 150,000 Euro. This sum does not include the construction costs for a soil filter or an irrigation system because these strongly depend on the location of the plant. At least 50% of the staff should be skilled workers, including a chemical engineer who is in charge of plant operation. Because of its high realization costs, this method is suited for industrial-scale oil mills, or as a central treatment facility for several oil mills.

The biogas may be used by the plant itself, or it may be fed into the public supply grid. The liquid phase, designated to be spread on agricultural land, is stored in an open pit. After drying, the solids can be sold as soil-improving material or as humus after having been mixed with vegetable residues. There are no odor nuisances from escaping liquids from the digesters, and maintenance costs are moderate. If the treated wastewater is additionally submitted to soil filtration and then used for irrigation or as fertilizer, the water cycle is closed, thus solving the problem of olive oil waste.

Case Study II. A pilot plant was operated between January 1993 and April 1994 to treat the wastewater from an oil mill in the region of Kalyvia/Attica (Fig. 9) [38].

Description of the plant:

- delivery, storage tank with a volume of 20 m^3 for the total quantity of margine produced;
- settling tank with a volume of 4 m^3;
- UASB reactor with a working volume of 2 m^3, additionally equipped with a high-performance heat exchanger to maintain the temperature during the mesophile phase;
- fixed-bed reactor with a working volume of 2 m^3, a high-performance heat exchanger, and recirculation system;
- gas storage room;
- seven tests (mesophile phase) have been carried out under varying operational conditions.

The organic load was degraded by 88–89%. During the fourth test, the phenol content was reduced by 74–75%, while the biogas production was 21–23 L gas per liter of bioreactor volume.

Foregoing the addition of CaO and expensive processing equipment facilitates the treatment for wastewater from oil mills. Plant investments can be quickly amortized by methane production.

Case Study III. A pilot test has been carried out in Tunisia with a sludge-bed reactor and an anaerobic contact reactor, followed by a two-stage aerobic treatment [15,38]. To compare the two different anaerobic processes, the semitechnical pilot plant was designed with parallel streams. The goal was not only to determine parameters and values for design and operation of optimal anaerobic–aerobic treatment, dependent on the achievable purification

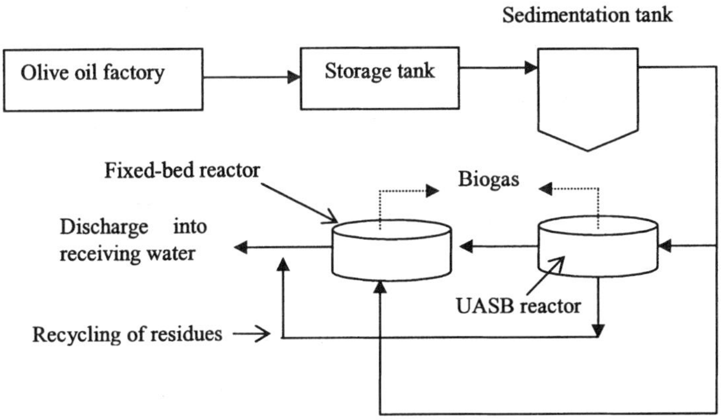

Figure 9 Pilot plant for treatment of wastewater from an olive oil mill in the region of Kalyvia, Attica (from Ref. 38).

capacity, but also to examine, modify, and further develop the process technology with regard to optimizing the purification capacity of the single stages, the total purification capacity, and process stability.

The tests determined that both anaerobic–aerobic procedures proved successful in the treatment of liquid waste from olive oil production. Comparing the anaerobic contact process with the bed process, neither is clearly favored. Both procedures lead to nearly the same results with regard to pretreatment of liquid waste from olive oil production.

Case Study IV. The anaerobic treatability of olive mill effluent was investigated using a laboratory-scale UASB reactor (with active volume of 10.35 L) operating for about 6 months. The black water collected from a traditional olive oil extraction plant in Gemlik village (Turkey) was used as the feed [22].

Active anaerobic sludge retained in the UASB reactor after a previous study was used as the seed. During the startup, pH was maintained in the range 6.8–8.0 and the average temperature was kept at mesophilic operating conditions (34°C) in the reactor. NaOH solution was added directly to the reactor to maintain the required pH levels when it was necessary. Urea was added to the feed to provide COD : N : P ratio of 350 : 5 : 1 in the system due to N deficiency of the feed.

In the first part of this study, the reactor was operated with feed COD concentrations from 5000 to 19,000 mg/L and a retention time of 1 day, giving organic loading rates (OLR) of 5–18 kg COD/m^3-day. Soluble COD removal was around 75% under these conditions. In the second part of the study, feed COD was varied from 15,000 to 22,600 mg/L while retention times ranged from 0.83 to 2 days; soluble COD removal was around 70%. A methane conversion rate of 0.35 m^3 per kg COD removed was achieved during the study. The average volatile solids or biomass (VS) concentration in the reactor had increased from 12.75 g/L to 60 g/L by the end of the study. Sludge volume index (SVI) determinations performed to evaluate the settling characteristics of the anaerobic sludge in the reactor indicating excellent settleability with SVI values of generally less than 20 mL/g. Active sludge granules ranging from 3 to 8 mm in diameter were produced in the reactor.

In short, it may be concluded that anaerobic treatment may be a very feasible alternative for olive mill effluents, but additional posttreatment, such as aerobic treatment, would be needed to satisfy discharge standards required for receiving waters (river, lake).

Case Study V. This experiment aimed at gaining better insight into the degradation of the main compounds contained in the OME, in particular, the interaction between the two successive stages occurring in the anaerobic digestion: acidogenesis and methanogenesis [23].

Fresh OME was obtained from the olive oil continuous centrifuge processing plant of Montelibretti (Rome). The tests were carried out in 500 mL glass bottles with perforated screw tops with latex underneath, which served to ensure that the bottles were airtight. These bottles were filled with OME diluted in distilled water to obtain the required concentration (in the range of 10–60 g COD/L). The inoculum was obtained from a sludge anaerobic digester at the East Rome wastewater treatment plant. The main results that can be drawn from this study are as follows.

Under the most favorable conditions (pH 8.5, 35°C, initial concentration 10 g COD/L, acclimatized inoculum) the OME were degraded with a high conversion yield (70–80%), both in acidogenic and methanogenic tests. Most of the lipids were degraded both in acidogenesis and methanogenesis tests. On the other hand, polyphenol-like substances were not degraded at all in acidogenic conditions, whereas they were partially removed in methanogenic conditions. Such a difference has been observed both in OME and synthetic solutions. A little methanogenic activity, established in acidogenic conditions because of the partial degradation of the chemical inhibitor, seems to be the key factor determining lipids degradation, even in acidogenesis tests.

It was also experimentally reported that polyphenol degradation is directly related to the presence of an intense methanogenic activity. In addition, bioconversion yields of OME in acidogenesis are remarkably less sensitive to the effect of pH and substrate concentrations than in methanogenesis. This result might lead to adoption of two-phase anaerobic digestion of OME as a suitable process for optimizing its performance. It is our recommendation that further research be conducted in this scope.

17.5.5 Design Example 3

The design of an anaerobic contact reactor to achieve 90% removal of COD from a wastewater flow 180 m^3/day (47,600 gal/day) resulted from a group of neighboring olive mills. The following conditions apply: total influent COD = 13,000 mg/L; nonremovable COD = 2500 mg/L; removable COD (COD_R) = 10,500 mg/L; and COD to be removed = 90%. The process parameters are: sludge age (SRT) = 15 days (minimum); temperature = 35°C; $a = 0.14$ mg VSS/mg COD_R; $b = 0.02$ mg VSS/mg VSS-day; $K' = 0.0005$ L/mg-day; $X_v = 5000$ mg/L.

Solution

(a) The digester volume from the kinetic relationship:

$$\text{Detention time, } t = \frac{S_r}{X_v K' S} = \frac{(10{,}500)(0.9)}{(5000)(0.0005)(1050)} = 3.6\,\text{day}$$

The digester volume is therefore:

$$\forall = (3.6\,\text{day})(180\,\text{m}^3/\text{day}) = 648\,\text{m}^3\ (0.1712\,\text{MG})$$

Check SRT from the equation:

$$\mathrm{SRT} = \frac{X_v t}{\Delta X_v} = \frac{X_v t}{aS_r - bX_v t}$$

$$= \frac{(5000)(3.6)}{(0.14)(9450) - (0.02)(5000)(3.6)} = 18.7\,\text{day}$$

This is in excess of the recommended SRT of 15 days to ensure the growth of methane formers. (b) The sludge yield from the process is:

$$\Delta X_v = aS_r - bX_v t$$
$$= (0.14)(9450) - (0.02)(5000)(3.6) = 963\,\text{mg/L}$$

$$\Delta X_v = 963\,\text{mg/L} \times 180\,\text{m}^3/\text{day}$$
$$= 173.34\,\text{kg/day}\ (381.35\,\text{lb/day})$$

(c) Gas production:

$$G = 0.351(S_r - 1.42\Delta X_v),$$

where $G = \text{m}^3$ of CH_4 produced/day

$$G = 0.351[(9.450)(180) - (1.42)(173.34)]$$

$$= 0.351(1701 - 246.14) = 511\,\text{m}^3\ \text{CH}_4/\text{day}$$

or

$$G = 5.62(S_r - 1.42\Delta X_v),$$

where $G = \text{ft}^3$ of CH_4 produced/day

$$G = 5.62[(9450)(0.0476\ \text{MG/day})(8.34) - (1.42)(381.35)]$$

$$= 18{,}040\,\text{ft}^3/\text{day}\ (511\,\text{m}^3/\text{day})$$

Gas production can be also determined by using the approximate estimation, which is 1 kg COD_{elim} yields about 0.3–0.5 m^3 of methane. Therefore, total gas production:

$$G = 9.45\,\text{kg COD/m}^3 \times 180\,\text{m}^3/\text{day} \times 0.3\,\text{m}^3\ \text{CH}_4/\text{kg COD}$$

$$= 510\,\text{m}^3\ \text{CH}_4/\text{day}$$

(d) Heat required can be estimated by calculating the energy required to raise the influent wastewater temperature to 35°C (95°F) and allowing 1°F (0.56°C) heat loss per day of detention time. Average wastewater temperature = 24°C (75.2°F) and heat transfer efficiency = 50%.

$$\text{Btu}_{req} = \frac{W(T_i - T_e)}{E} \times (\text{specific heat})$$

$$= \frac{(47{,}600\,\text{gal/day})(8.34\,\text{lb/gal})(95°\text{F} + 3.6°\text{F} - 75.2°\text{F})}{0.5} \times \left(\frac{1\,\text{Btu}}{1\,\text{lb°F}}\right)$$

$$= 18{,}600{,}000\,\text{Btu}\ (19{,}625{,}000\,\text{kJ})$$

The heat available from gas production is:

$$\begin{aligned} \text{Btu}_{\text{available}} &= (18{,}040\,\text{ft}^3\ \text{CH}_4/\text{day})(960\,\text{Btu}/\text{ft}^3\ \text{CH}_4) \\ &= 17{,}320{,}000\,\text{Btu}/\text{day}\ (18{,}300{,}000\,\text{kJ}/\text{day}) \end{aligned}$$

$$\text{External heat of } 18{,}600{,}000 - 17{,}320{,}000 = 1{,}280{,}000\,\text{Btu}/\text{day}$$

1,325,000 kJ/day should be supplied to maintain the reactor at 35°C (95°F).
(e) Nutrient required as nitrogen is:

$$\begin{aligned} N &= 0.12\Delta X_v = 0.12 \times 173.34\,\text{kg}/\text{day} \\ &= 20.80\,\text{kg}/\text{day}\ (45.8\,\text{lb}/\text{day}) \end{aligned}$$

The phosphorus required is:

$$\begin{aligned} P &= 0.025\Delta X_v = 0.025 \times 173.34\,\text{kg}/\text{day} \\ &= 4.33\,\text{kg}/\text{day}\ (9.534\,\text{lb}/\text{day}) \end{aligned}$$

Remarks

1. The effluent from the anaerobic plant does not achieve the national quality criteria of the water resources because of the high values of residual COD_R (10% = 1050 mg/L) and nonremovable COD (2500 mg/L). Therefore, we recommend that an aerobic treatment process (such as activated sludge) follow the anaerobic process to produce an effluent meeting the quality limits.
2. Another suggestion is to apply wet air oxidation (WAO) as a pretreatment step to remove biorecalcitrant compounds, which leads to the reduction of anaerobic reactor volume and also to the reduction of energy consumption. This combined WAO–anaerobic process achieves an overall performance to meet the national regulations of Mediterranean countries.

17.5.6 Combined Biological Treatment Processes

The following models are suggested for combined biological treatment processes of OMW. It has been referred to as the combined treatment in order to realize the following: partial treatment by high organic load in the first phase and full treatment by low organic load in the second phase.

Treatment on Site

Before discharge to a nearby water recourse, OMW could be subjected to either of the two hereafter proposed complete treatment systems.

Anaerobic–Aerobic Treatment. The combined model "anaerobic–aerobic treatment" (Fig. 10) may be considered quite practical, both environmentally and economically. This method can be applied without serious emissions into air, water, and soil, keeping to the key objectives of environmental policy adopted worldwide.

Anaerobic processes are especially suited for the treatment of high-load wastewater with a COD concentration of thousands (mg/L) in industry. Moreover, the climatic conditions in the olive-growing and production countries are optimal for anaerobic processes.

Combining anaerobic and aerobic processes lessens the disadvantages resulting from separate applications. The first step includes the advantages of the anaerobic process concerning degradation efficiency, energy self-sufficiency, and minimal excess sludge production. The

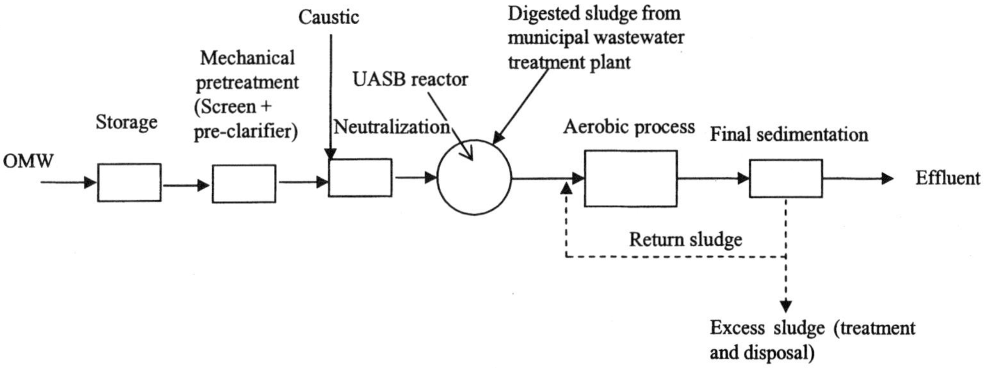

Figure 10 Combined anaerobic–aerobic treatment model (on site).

disadvantages of aerobic treatment are nearly compensated by the anaerobic preliminary stage. The high quantity of excess sludge that normally results is strongly reduced. At the same time, the aeration energy needed for the aerobic process is also considerably minimized. With regard to treatment efficiency, plant reliability, and costs, the anaerobic–aerobic model well suits the treatment of olive oil mill wastewater (alpechin) from both ecological and economical aspects [38].

Two-Stage Aerobic Treatment. This is a combined treatment model of two-stage aerobic treatment based on an activated sludge process, as illustrated in Fig. 11.

Treatment in Combination with Municipal Wastewater. In the case where full treatment onsite is not possible, OMW after pretreatment should be drained to a municipal wastewater treatment plant in the vicinity. Figure 12 illustrates clearly the combined treatment of OMW with municipal wastewater, where two streams (a and b) are suggested.

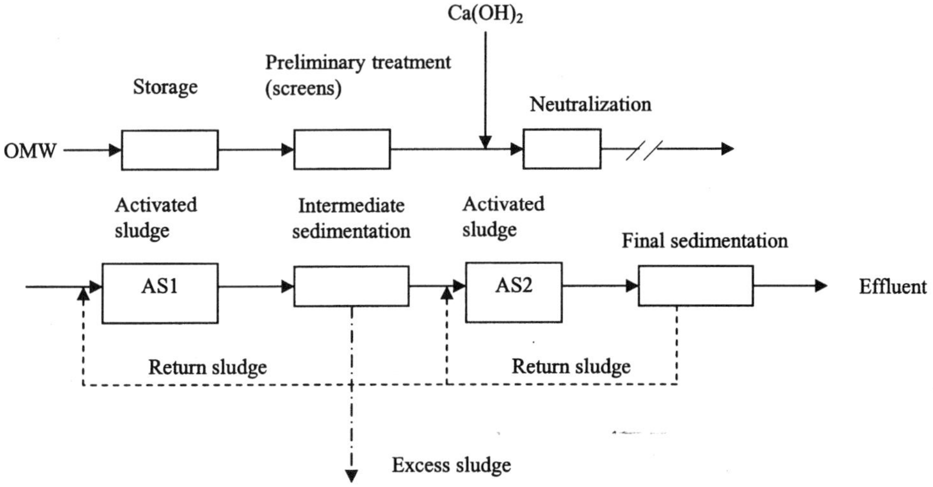

Figure 11 Combined treatment model of two-stage activated sludge process (on site). (*Note*: In dispensing with the primary sedimentation tank, it is recommended here to recirculate the return sludge from the final sedimentation to both the AS1 and AS2. Consequently, excess sludge will be discharged only from the intermediate sedimentation tank.)

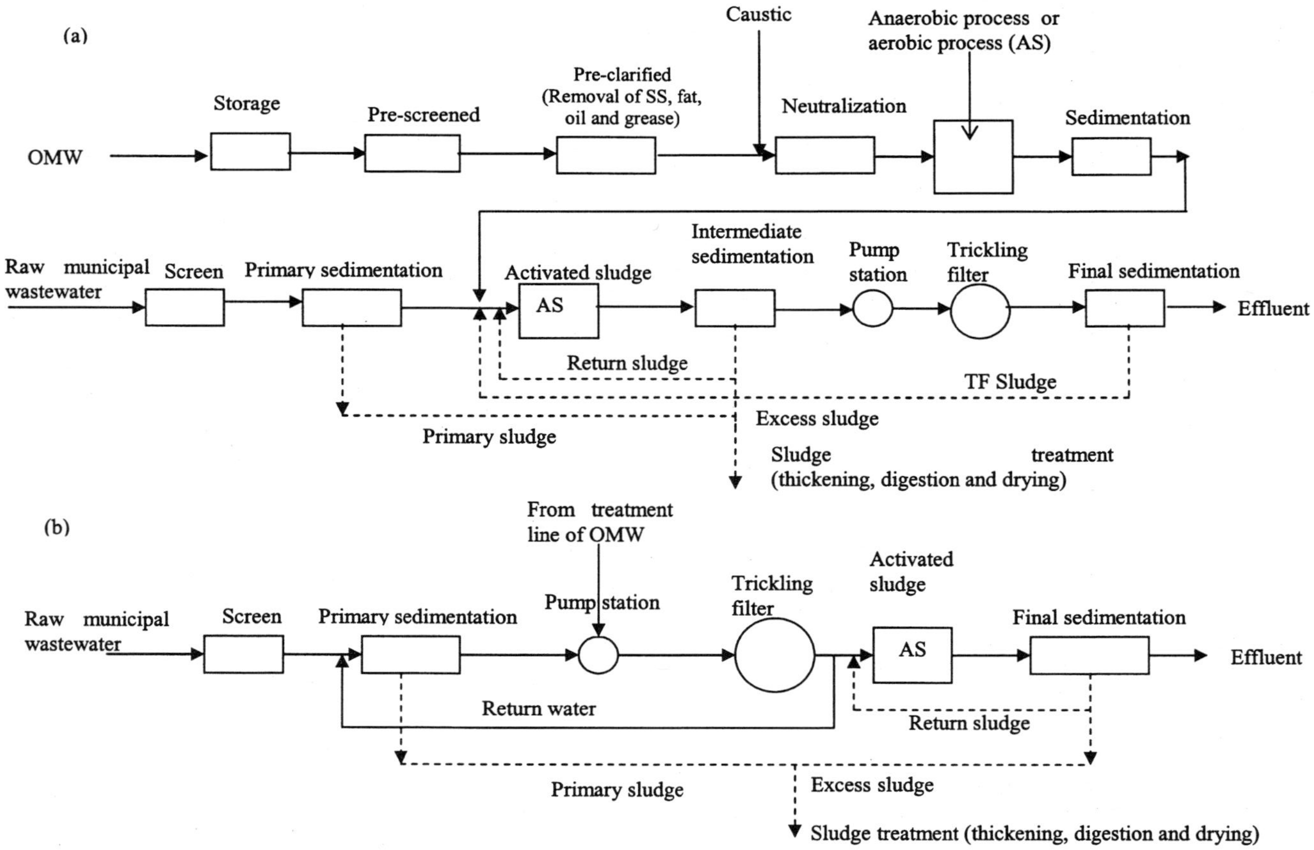

Figure 12 Combined treatment of OMW with municipal wastewater. (*Note*: Aerobic process may need addition of nutrients in order to maintain the ratio COD : N : P at 100 : 5 : 1, this ratio being commonly satisfactory for microorganism growth and activity.) (a) Where the activated sludge process is before a trickling filter process is preferable to line (b) in general, with the consideration that line (b) (trickling filter–activated sludge combined model) dispenses with the intermediate sedimentation basin.

The aforementioned combined models suggested for treatment of OMW realize different degrees of efficiency depending on the wastewater characteristics, discharge regulations, organic load in each phase, type and number of phases within the treatment line or plant. In this respect it is necessary that the treated wastewater meet the quality criteria of the water resources (drinking, irrigation, recreation, etc.), where it is supposed to be discharged. In the event the treated wastewater is intended to be used directly for irrigation, it should meet local criteria adopted in that country or those adopted by the Food and Agriculture Organization (FAO).

17.5.7 Design Example 4

To continue Example 3, assuming that an air-activated sludge plant follows the anaerobic process, a design for this plant is required under the following conditions to produce an effluent with a COD of 30 mg/L. The aerobic process parameters are: $T = 20^\circ C$; $a = 0.5$; $F/M = 0.3\,\text{day}^{-1}$; $a' = 0.55$; $X_v = 2500$ mg/L; $b = 0.15/\text{day}$ at 20°C; power = 1.5 lb O_2/(hp-hour) (0.91 kg O_2/kW).

Solution

$$t = \frac{S_0}{X_v(F/M)} = \frac{1050}{2500 \times 0.3} = 1.4\,\text{day}$$

$$S_r = S_0 - S_e = 1050 - 30 = 1020\,\text{mg/L}$$

$$K = \frac{(S_0 S_r)}{t S_e X_v} = \frac{1050 \times 1020}{1.4 \times 30 \times 2500} = 10.2/\text{day}$$

The aeration tank volume is:

$$\forall = Q \cdot t = 180 \times 1.4 = 252\,\text{m}^3\ (66{,}640\,\text{gal})$$

Calculate the degradable fraction X_d using the following equation:

$$\begin{aligned} X_d &= \frac{aS_r + bX_v t - [(aS_r + bX_v t)^2 - 4bX_v t \times 0.8aS_r]^{1/2}}{2bX_v t} \\ &= \frac{(0.5 \times 1020) + (0.15 \times 2500 \times 1.4) - [\ldots\ldots\ldots]^{1/2}}{2 \times 0.15 \times 2500 \times 1.4} \\ &= \frac{(510 + 525) - [(510 + 525)^2 - (4 \times 525 \times 0.8 \times 510)]^{1/2}}{2 \times 525} \\ &= \frac{1035 - 463}{1050} = 0.545 \end{aligned}$$

The oxygen required is:

$$\begin{aligned} O_2/\text{day} &= (a'S_r + 1.4bX_d X_v t)Q \\ &= [(0.55 \times 1020) + (1.4 \times 0.15 \times 0.545 \times 2500 \times 1.4)] \times 47{,}600\,\text{gal} \times 8.34 \times 10^{-6} \\ &= 382\,\text{lb/day} = 16\,\text{lb/hour}\ (7.3\,\text{kg/hour}) \end{aligned}$$

The power required is:

$$\text{hp} = \frac{\text{O}_2/\text{hour}}{[1.5\,\text{lb O}_2/(\text{hp-hour})]} = \frac{16}{1.5}$$

$$= 10.7\,\text{hp}\ (8\,\text{kW})$$

Other olive oil mills wishing to economize their operations would like to join the abovementioned combined anaerobic–aerobic plant for the treatment of their wastewater (45 m^3/day), without affecting the plant's efficiency.

- Compute the new effluent from the anaerobic process assuming (X_v) remains the same; what will the new gas production be?
- What modifications to the aerobic process must be made to maintain the same effluent quality? Assume the sludge settling characteristics are the same as originally and the volatile content of the sludge is 75%.

Solution

The load to the plant is increased to 225 m^3/day (59,400 gal/day).

(a) *Anaerobic process*. New effluent concentration; from example 3: $SRT_{min} = 15$ days; $T = 35°C$; $a = 0.14$; $b = 0.02$; $k' = 0.0005$ L/(mg-day); $X_v = 5000$ mg/L; $COD_R = 10{,}500$ mg/L; and volume = 648 m^3 (0.1712 MG).

The new detention time is:

$$t' = \frac{\forall}{Q} = \frac{648}{225} = 2.9\,\text{day}$$

The COD effluent from the anaerobic process can be estimated by:

$$\text{COD}_E = \frac{\text{COD}_{removed}}{X_v K' t'} = \frac{(\text{COD}_R - \text{COD}_E)}{X_v K' t'}$$

$$= \frac{\text{COD}_R}{(1 + X_v K' t')}$$

$$= \frac{10{,}500}{(1 + 5000 \times 0.0005 \times 2.9)}$$

$$= 1273\,\text{mg/L}$$

The COD removed is:

$$\text{COD}_{removed} = \text{COD}_R - \text{COD}_E$$

$$= 10{,}500 - 1273$$

$$= 9227\,\text{mg/L}$$

Check SRT using the equation:

$$\text{SRT} = \frac{X_v t'}{\Delta X_v} = \frac{X_v t'}{a\text{COD}_{\text{removed}} - bX_v t'}$$

$$= \frac{5000 \times 2.9}{(0.14 \times 9227) - (0.02 \times 5000 \times 2.9)}$$

$$= 14.5\,\text{day} \approx 15\,\text{day} \qquad \text{OK}$$

New gas production. The sludge yield is:

$$\Delta X_v = (a\text{COD}_{\text{removed}} - bX_v t')Q$$
$$= (0.14 \times 9227 - 0.02 \times 5000 \times 2.9)\,\text{mg/L} \times 59{,}400\,\text{gal/day}$$
$$\times\, 8.34 \times 10^{-6}\,(\text{lb/MG})/\text{mg/L}$$
$$= 496.4\,\text{lb/day}\ (225.36\,\text{kg/day})$$

The mass of COD removed per day is:

$$S_r = \text{COD}_{\text{removed}} \times Q$$
$$= 9227\,\text{mg/L} \times 59{,}400\,\text{gal/day} \times 8.34 \times 10^{-6}$$
$$= 4571\,\text{lb/day}\ (2076\,\text{kg/day})$$

or

$$S_r = 9227\,\text{mg/L} \times 225\,\text{m}^3/\text{day} \times 10^{-3} = 2076\,\text{kg/day}$$

The methane production can be estimated from:

$$G = 5.62(S_r - 1.42\Delta X_v)$$

where G is given in ft^3 of CH_4/day

$$G = 5.62(4571 - 1.42 \times 496.4)$$
$$= 21{,}727\,\text{ft}^3/\text{day}\ (615\,\text{m}^3/\text{day})$$

(b) *Aerobic process.* The new detention time is:

$$t' = \frac{252\,\text{m}^3}{225\,\text{m}^3/\text{day}} = 1.12\,\text{day}$$

The new COD removed:

$$S_r' = S_0' - S_e = 1273 - 30 = 1243\,\text{mg/L}$$

From the equation:

$$\frac{S_0 - S_e}{X_v t} = K\frac{S_e}{S_0}$$

By rearrangement, the new MLVSS are obtained as

$$X_v' = (S_0' S_r')/(t' S_e K)$$
$$= (1273) \times (1243)/(1.12 \times 30 \times 10.2)$$
$$= 4617\,\text{mg VSS/L}$$

and the MLSS are:

$$\text{MLSS} = 4617/0.75 = 6156\,\text{mg/L}$$

The new F/M is:

$$(F/M)' = S_0'/(X_v' t')$$
$$= 1273/(4617 \times 1.12) = 0.25/\text{day}$$

Power increase, the new degradable factor is:

$$X_d' = 0.50$$

The new oxygen required is:

$$\text{O}_2/\text{day} = (a' S_r' + 1.4 b X_d' X_v' t')Q$$
$$= (0.55 \times 1243 + 1.4 \times 0.15 \times 0.5 \times 4617 \times 1.12) \times 59{,}400\,\text{gal} \times 8.34 \times 10^{-6}$$
$$= 608\,\text{lb/day} = 25.3\,\text{lb/hour}\ (11.5\,\text{kg/hour})$$

The new power required is:

$$h'p = \frac{25.3\,\text{lb/hour}}{1.5} = 16.9\,\text{hp}\ (12.6\,\text{kW})$$

The power increase is:

$$\text{hp}_{\text{inc.}} = 16.9 - 10.7 = 6.2\,\text{hp}\ (4.6\,\text{kW})$$

17.5.8 Design Example 5

A 7500 m^3/day (2.0 million gal/day) municipal activated sludge plant operates at an F/M of 0.3 day^{-1}. A group of olive oil mills needs to discharge 450 m^3/day (0.12 million gal/day) of wastewater with a BOD of 8000 mg/L to the plant. What pretreatment is requested of the mills to reduce the BOD in their wastewater, in order to win the plant's approval?

Solution

(a) Municipal sewage: flow = 7500 m^3/day (2.0 million gal/day); S_0 (BOD) = 300 mg/L; Soluble BOD = 100; $F/M = 0.3$; X_v = 2500 mg/L; S_e (soluble) = 10 mg/L; $K = 8$/day at 20°C. (b) Olive mill wastewater: flow = 450 m^3/day (0.12 MG/day); S_0 (BOD) = 8000 mg/L; $K = 2.6$/day at 20°C; estimated MLVSS = 3500 mg/L.

Detention time is:

$$\frac{F}{M} = \frac{S_0}{X_v t}$$

$$t = \frac{300}{2500 \times 0.3} = 0.4\,\text{day}$$

Average reaction rate K will be:

$$\frac{7500(8) + 450(2.6)}{7950} = 7.7/\text{day}$$

The new detention time is $0.4 \times 7500/7950 = 0.38$. The influent to the plant to meet the permit can be calculated:

$$\frac{S_0 - S_e}{X_v t} = K\frac{S_e}{S_0}$$

$$S_0^2 - S_e S_0 - S_e K X_v t = 0$$

$$S_0 = \frac{S_e + \sqrt{{S_e}^2 + 4S_e K X_v t}}{2} = \frac{10 + \sqrt{100 + (4 \times 10 \times 7.7 \times 3500 \times 0.38)}}{2}$$

$$= 325\,\text{mg/L of soluble BOD}$$

The concentration of BOD in the pretreated mill wastewater can then be calculated by a material balance:

$$Q_s(S_{0,s}) + Q_I(S_{0,I}) = (Q_s + Q_I)S_{0,s+I}$$
$$7500(100) + 450(S_{0,I}) = 7950(325)$$

or

$$2.0(100) + 0.12(S_{0,I}) = 2.12(325)$$
$$S_{0,I} = 4075\,\text{mg/L}$$

Pretreatment is required to reduce about 50% of the BOD in the mill wastewater.
(c) *Temperature effects*: Determine the change in MLVSS that will be required when the temperature coefficient θ increases from 1.015 to 1.04 due to an increase in soluble mill wastewater BOD:

$$\frac{K20}{K10} = (1.015)^{10} = 1.16\,\text{sewage}$$

$$\frac{K20}{K10} = (1.04)^{10} = 1.48\,\text{sewage–mill–wastewater}$$

The increase in MLVSS can be calculated as:

$$\frac{1.48}{1.16} \times 2500\,\text{mg/L} = 3190\,\text{mg/L}$$

Remarks

1. To achieve the BOD reduction of about 50% in the olive oil mill effluents, the anaerobic process should be recommended as pretreatment.
2. The municipal activated sludge plant could not achieve the quality limits or criteria of the water resources because of the high value of BOD in the mill wastewater (4075 mg/L). In such a case, an additional aerobic degradation stage is needed, such as activated sludge or trickling filter as illustrated in Figure 12.

17.5.9 Wet Air Oxidation and Ozonation

The clear advantages of the anaerobic process make it the process of choice for treating olive oil effluents [52]. However, many problems concerning the high toxicity and inhibition of biodegradation of these wastes have been encountered during anaerobic treatments, because some bacteria, such as methanogens, are particularly sensitive to the organic contaminants present in the OME. The biorecalcitrant and/or inhibiting substances, essentially phenolic compounds (aromatics), severely limit the possibility of using conventional wastewater anaerobic digestions [53] or lead to difficulties in the anaerobic treatment of OME [23].

Moreover, it was proved that the anaerobic sludge digestion of OME in UASB-like reactors was unstable after a relatively short period of activity [54]. Consequently, anaerobic biological treatment as a unique process showed limited efficiency in the removal of aromatics. Therefore, other treatments such as chemical oxidation have been investigated for olive oil mill wastewater and for table olive wastewater purification, with encouraging results.

This chemical oxidation proved to be very effective in treating wastewaters that contain large quantities of aromatics [55,56]. Recently, integrated chemico-physical and biological technologies have been developed as efficient processes to achieve high purification levels in wastewaters characterized by difficult biotreatability [57].

The effectiveness of the combination of chemical oxidation and biological degradation relies on the transformation of nonbiodegradable substances into biogenic compounds readily assimilated by microorganisms [57].

Principle of Wet Air Oxidation (WAO)

The type of chemical preoxidation used in integrated processes is highly dependent on the characteristics and nature of the wastewater to be treated. Thus, in the case of effluents with a high content of phenol-type substances, oxidizing systems based on the use of oxygen or ozone at high temperatures and pressures have been shown to readily degrade phenolic structures [58]. Wet air oxidation (WAO) is an oxidation process, conducted in the liquid phase by means of elevated temperatures (400–600 K) and pressures (0.5–20 MPa). The oxidant source is an oxygen-containing gas (usually air).

As pressure increases, the temperature rises, which leads to an increasing degree of oxidation. With far-reaching material conversion, only the inorganic final stages of CO_2 and water (and possibly other oxides) are left. With incomplete degradation, the original components (which often are nondegradable) are decomposed to biodegradable fragments. Therefore, it is useful to install a biological treatment stage downstream of the wet oxidation stage (Fig. 13) (Case Study I).

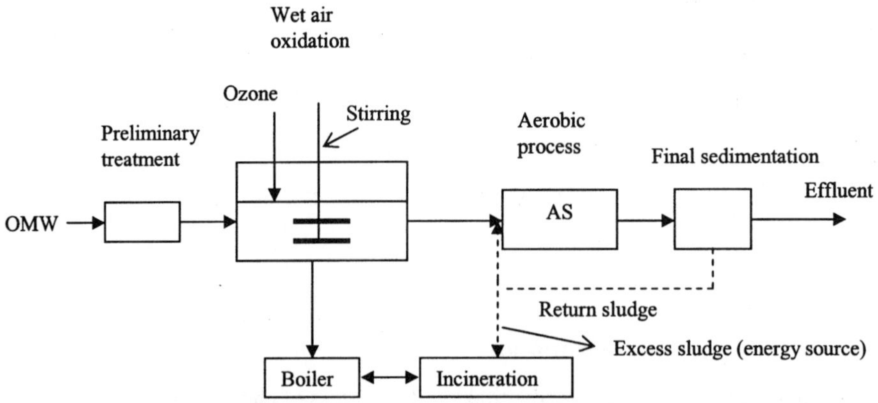

Figure 13 Wet air oxidation–aerobic process.

On other hand, Beltran-Heredia et al. [21] applied an opposite arrangement, that is, aerobic degradation followed by ozonation, in normal conditions where the temperature and the pH values were varied (Case Study II). Oxidizing chemicals are also used instead of oxygen so that even hardly degradable constituents of liquid waste from olive oil production can be destroyed or attacked. Possible oxidizing agents are ozone (O_3) or hydrogen peroxide (H_2O_2) [59].

The utilization of H_2O_2 has turned out to be environmentally friendly because this oxidizing agent has no negative effects. However, since H_2O_2 quickly undergoes decomposition, its ability to be stored is limited. The OH radicals formed during H_2O_2 decomposition have oxidative effects. Using suitable agents [e.g., titanium dioxide (TiO_2)] or UV radiation, the development of radicals can be considerably forced [38].

In oxidation systems, ozone in particular has many of the oxidizing properties desirable for use in water and wastewater treatment; it is a powerful oxidant capable of oxidative degradation of many organic compounds, is readily available, soluble in water, and leaves no byproducts that need to be removed. In addition, it may also be used to destroy bacteria, odors, taste, and coloring substances.

It has been reported in the literature that anions of phenolic compounds are more reactive towards oxidative processes than the noncharged species [58,60].

Case Studies

Case Study I. A considerable amount of work has been devoted to the integrated wet air oxidation–aerobic biodegradation process (Fig. 14) in treating olive-processing wastewater in the province of Badajoz, Spain [19]. The most representative parameters are the COD and BOD_5, with values of 24.45 and 14.8 g O_2/L respectively, and phenolic content 833 mg phenol/L. Chemical oxygen demand (COD) conversion in the range 30–60% (6 hours of treatment) was achieved by WAO using relatively mild conditions (443–483 K and 3.0–7.0 MPa of total pressure using air). Also noticed was a significant removal of phenolic content at the end of WAO process with conversion values 95%. Use of the homogeneous catalysts such as radical promoters (hydrogen peroxide) resulted in a higher efficiency of the process (between 16 and 33% COD removal improvement, depending on operating conditions). Biodegradability tests conducted after the oxidation pretreatment showed the positive effect of the WAO pretreatment on the aerobic biological oxidation of wastewater. Acclimation of microorganisms to oxygenated species formed in a chemical preoxidation step enhanced the efficiency of the biodegradation.

In conclusion, if WAO is used as a pretreatment step, the advantages associated with the use of the previous oxidation are based on the higher biodegradation rate and better properties of the activated sludge used in the biodegradation process to remove biorecalcitrant compounds.

As inferred and reported from this work [19], the following conclusions may be drawn:

- The WAO process may become thermally self-sustaining, because the COD of the influent is well above 15 g/L. In this case, the wastewater stream would not be diluted and more severe conditions should be applied.
- The seasonal character of these activities (fruit and vegetable related processes) may allow for the use of WAO mobile units, capable of processing up to a maximum of 400–500 L/hour of wastewater (more than needed for these types of industries). As a result, a permanent location is not needed, with subsequent savings in fixed capital costs.
- Use of *in situ* WAO shows additional advantages regarding necessary barreling and hauling to appropriate wastewater plants.

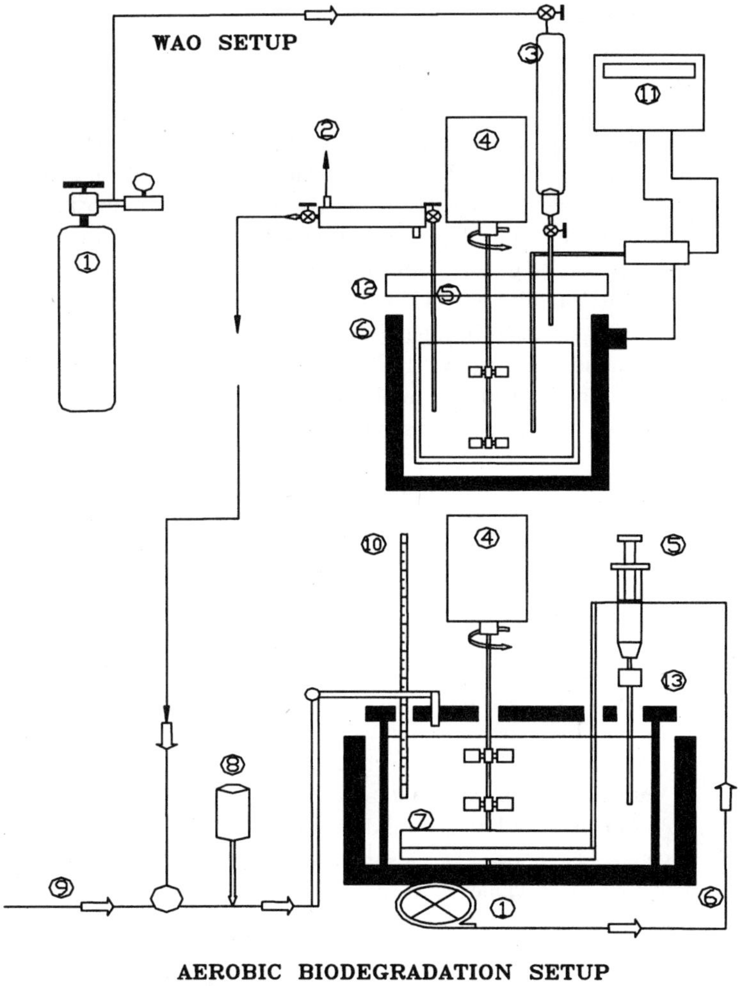

Figure 14 Experimental setup of WAO–aerobic processes (from Ref. 19). 1 = Air Cylinder; 2 = Cooling Water; 3 = Injection Port; 4 = Stirring System; 5 = Sampling Port; 6 = Thermostatic Bath; 7 = Porous Plate; 8 = pH Controller; 9 = Dilution Waterline; 10 = Thermometer; 11 = Temperature Controller; 12 = High-Pressure Reaction Vessel; 13 = Glass Bioreactor.

- The consequences of WAO pretreatment may also affect the operability of aerobic biological treatment itself. Thus the benefits are as follows. (a) The biodegradation rate was observed to increase from a nonpretreated effluent to a WAO pretreatment wastewater, which would imply a lower total volume of biological reactor and lower energy consumption (requirements for mixing and aeration) to achieve an overall performance to meet the limits of the environmental legislation. (b) The sludge volume index (SVI) decreased if the WAO pretreatment was applied. An average 20% decrease was observed for biological experiments using pretreated wastewater. This would help to prevent operational problems usually found in activated sludge plants, such as bulking sludge, rising sludge, and nocardia foam, and would allow a wider food-to-microorganisms (F/M) ratio for operation in the aeration tank and lower total volume of the secondary clarifier. (c) An excess of generated sludge as a result of biological oxidation could be recycled as an energy source by combustion or anaerobic treatment to use in the wastewater treatment plant, or it could even be treated by the same WAO system.

Case Study II. The original black-olive wastewater was obtained from a table olive processing plant in the Extremadura community (Spain). The treatment was carried out by ozonation, aerobic biological degradation, and the combination of two successive steps: an aerobic biological process followed by ozonation. For this purpose, the chemical oxygen demand (COD), the total aromatic content (A), and the total phenolic content (Tph), were selected as representative parameters to follow the overall purification process.

The experimental results [21] given for ozonation, where the temperature (10, 20, and 30°C) and the pH (7.9 and 13.6) were varied, are as follows: the COD conversions ranged between 42 and 55% depending on the operating conditions; the conversions of the total phenolic and aromatic compounds are around 75 and 67%, respectively.

A direct influence of temperature and pH on the COD and the phenolic compounds degradation was also observed. Thus, it may be concluded that ozone is an excellent oxidizing agent in the specific destruction of phenolic and aromatic compounds.

The experimental results from the aerobic biological treatment were as follows: the COD conversions ranged between 76 and 90%; the conversions of aromatic compounds ranged between 16 and 35%; and conversions ranged between 53 and 80% for total phenolics.

The combined process of an aerobic degradation followed by an ozonation produced a higher COD, phenolic and aromatic removal efficiency. This combined process reached a degradation level that cannot be obtained by any chemical or biological process individually under the same operating conditions.

There was a clear improvement in the second stage relative to ozonation, and biological pretreatment also led to an increase in the kinetic parameters. This implied that the aerobic pretreatment enhanced the later ozone oxidation by removing most of the biodegradable organic matter, while the ozonation step degraded some of the nonbiodegradable organic matter plus most of the phenolic compounds not removed previously.

Case Study III. This research focuses on the degradation of the pollutant organic matter present in wastewater obtained from an olive oil production plant located at the Extremadura Community (Spain), by combining two successive steps: (a) ozonation followed by aerobic degradation, and (b) aerobic degradation followed by ozonation. For this purpose, the chemical oxygen demand (COD), the total aromatic content (A) and the total phenolic content (Tph), were selected as criteria to monitor the overall degradation process [32]. The combined OMW degradation processes were studied with the goal of evaluating the influence of each respective pretreatment on the second stage. The first combined process (C-1) comprised ozone oxidation pretreatment followed by aerobic biodegradation. Table 12 summarizes the operating conditions, the initial and final COD concentrations, and the conversion values obtained (X_{COD}) in each stage individually considered, as well as the conversion achieved by the overall process. The total conversion obtained by the successive stage (C-1) was 84.6%, a higher value than achieved by either single process under the same operating conditions. This suggests that ozone pretreatment enhances the subsequent aerobic process, probably by removing some phenolic compounds capable of inhibiting biological oxidation. Similar to combination (C-1), the overall process achieved, by the second combined process (C-2), 81.8% degradation, which was greater than that obtained by the individual chemical or biological processes under the same operating conditions (Table 12). This suggests that aerobic pretreatment enhanced the subsequent ozone oxidation by removing most of the biodegradable organic matter. The ozonation step then

Table 12 Treatment of Olive Mill Wastewaters by Ozonation, Aerobic Degradation, and the Combination of Both Treatment Methods

C-1 Ozonation followed by aerobic degradation	C-2 Aerobic degradation followed by ozonation
C-1-A Ozonation stage Operating conditions: T = 20°C; Po_3 = 1.73 kPa; pH = 7; CODo = 34.05 g dm^{-3} Substrate removal obtained: COD_l = 29.9 g dm^{-3}; X_{COD} = 12.2%	C-2-A Aerobic degradation stage Operating conditions: X = 0.53 g dm^{-3}; CODo = 41.95 g dm^{-3} Substrate removal obtained: COD_l = 11.07 g dm^{-3}; X_{COD} = 73.6%
C-1-B Aerobic degradation stage Operating conditions: X = 0.59 g dm^{-3}; CODo = 29.85 g dm^{-3} Substrate removal obtained: COD_l = 5.22 g dm^{-3}; X_{COD} = 82.5%	C-2-B Ozonation stage Operating conditions: T = 20°C; Po_3 = 1.69 kPa; pH = 7; CODo = 10.95 g dm^{-3} Substrate removal obtained: COD_l = 7.63 g dm^{-3}; X_{COD} = 30.3%
Total removal in process C-1: X_{COD} = 84.6%	Total removal in process C-2: X_{COD} = 81.8%

Source: Ref. 32.

degraded some of the nonbiodegradable organic matter and much of the residual phenolic compounds.

In conclusion, the study shows that ozonation of OMW achieves a moderate reduction in the COD, and significant removal of aromatic and total phenolic compounds. The microbial aerobic treatment achieves significant removal of COD and phenolics but with less elimination of aromatic substances. The two processes combined, as presented in this case study, achieve higher COD removal efficiency than treatment by either stage separately under the same operating conditions. Together, the two processes may be used to treat OMW to meet discharge criteria or norms and reach treatment efficiencies required by national regulations, particularly in Mediterranean countries.

17.5.10 Fungal Treatment

Several types of industrial wastes contain phenols. Many of these compounds are extremely harmful as they are highly toxic both towards microorganisms and vertebrates [61]. Enzymatic approaches to removing phenols have been tried for some years as they have several advantages compared with the conventional methods (solvent extraction, chemical oxidation, absorbance on active carbons, etc.) [62].

Recently, results have been obtained for the removal of phenols using phenol oxidizers, which catalyze oxidative coupling reactions of phenol compounds and do not require hydrogen peroxide (H_2O_2) [63]. Olive oil mill wastewaters (OMW) contain large concentrations of phenol compounds, which are highly toxic. The structure of the aromatic compounds present in OMW can be assimilated to many of the components of lignin [64].

However, some microorganisms actively degrade lignin, among which the "white-rot" fungi are particularly efficient. These organisms utilize mainly peroxidazes and phenol oxidizers [65]. Potential applications of white-rot fungi and their enzymes are gaining increasing importance in the detoxification of industrial wastewaters, reducing the toxicity of many aromatic compounds (pesticides, disinfectants, phenols) in several types of polluted environments.

Case Studies

Case Study I. This study investigates the application of "white-rot" basidiomycete *Pleurotus ostreatus* and the phenol oxidizers it produces, for reducing the phenol content and the toxicity of the olive wastewater at an olive oil factory in Abruzzo, Italy [61]. It was found that up to 90% of the phenols present in OMW could be removed by treatment with phenol oxidizer from a mixture containing aromatic compounds extracted from OMW, although no concomitant decrease of toxicity was observed.

Results show that *P. ostreatus* removed phenols and detoxified OMW diluted to 10% in the absence of any external added nutrient; the diluted wastewaters were also clarified from this treatment in a relatively short time (100 hours). The detoxifying activity of *P. ostreatus* was concomitant with a progressively increasing phenol oxidase expression. It was noticed that after 100 hours incubation with *P. ostreatus*, the concentration of phenol compounds decreased by 90% and the toxicity towards *Bacillus cereus* was reduced seven-fold compared with that of untreated waste.

Case Study II. This study focused on the ability of white-rot fungi isolated from Moroccan OMW (classified as *Phanerochaete chrysosporium* Burdsall M_1 to modify the polluting properties of diluted OMW in comparison with that of *P. ostreatus*. Olive oil mill wastewater (OMW) was collected from an olive oil factory in Marrakech, Morocco [20].

In order to study the effects of fungal treatment on OMW, two different white-rot fungi were tested in batch cultures of diluted OMW (20%). The maximum reduction of phenol content and COD was 62 and 52% for *P. ostreatus*, whilst it was 82 and 77% for *Ph. chrysosporium* after 15 days of treatment. The time course of absorbance decrease is similar to that of phenol content and COD reduction for both fungi, suggesting the existence of a correlation between these parameters and the colored components present in OMW. The results obtained indicate that *Ph. chrysosporium* is able to decolorize OMW and to degrade its phenolic component more efficiently than *P. ostreatus* can.

Toxicity tests performed on *B. cereus* revealed that fungal treatment of the waste (20 or 50%) causes the complete loss of OMW toxicity after 15 days of treatment. The optimal decolorization temperature for *Ph. chrysosporium* Busdsall M_1 was 28°C. Furthermore, the optimal pH for *Ph. chrysosporium* OMW treatment was in the 4.0–5.0 range. Since the pH of diluted OMW was between 4.0 and 5.0, the process did not require any pH alteration of the effluent.

Degradation of 20 or 50% OMW, expressed as color, phenol, and COD removal, was almost the same after 15 days of fungal growth. Hence, not only is this fungus able to grow in 50% OMW as the sole carbon source, but the degradation rate of the effluent increases in these cultural conditions. This proves that the isolated *Ph. chrysosporium* strain, which is able to grow using diluted OMW, and to notably reduce color, phenol content, and COD, would be a good candidate for the effective treatment of this wastewater.

17.5.11 Decolorization

Investigation of the effect of oxidative coloration on the methangenetic toxicity and anaerobic biodegradability of aromatics showed that their oxidized solutions were less biodegradable in proportion to their color [66]. In contrast, the aerobic processes can have substantial aromatic removal efficiency, but these processes require sizeable energy expenditures in oxygen transfer and sludge handling [67].

An important step in the degradation of olive oil wastewater is the breakdown of colored polymeric phenolics (decolorization) to monomers, which may subsequently be mineralized.

A significant correlation has been demonstrated between sewage decolorization and reduction of total organic carbon and phenolic content. However, decolorization of wastewaters appears to be associated only with a partial depolymerization. A decrease in the content of the lower molecular mass components and an increase in the proportion of components of intermediary molecular mass have also been demonstrated.

Crude oil wastewater and solutions of its brownish pigment change in both color and solubility as the result of pH modification. It appears that sewage decolorization may be produced simply by a process of adsorption or by adsorption associated with subsequent chemical modification of chromophores.

The effluent is acidified as a consequence of fungal growth. A considerable decrease in pH and an elevated adsorption of lignin-derived products onto the biological matrix suggested that the decolorization process was an indirect effect of culture acidification. The sewage decolorization eventually stops with time, suggesting that the putative enzymes responsible for decolorization have a defined lifetime.

Many recalcitrant compounds from olive oil mill wastewater are present in the colored fraction. Optimum culture conditions will be identified for the decolorization of that sewage by *Phanerochaete flavido-alba* for subsequent use in bioremediation assays. Of several media tested, nitrogen-limited *P. flavido-alba* cultures containing 40 μg/mL Mn(II) were the most efficient at decolorizing oil wastewater. Decolorization was accompanied by a 90% decrease in the phenolic content of the wastewater. Concentrated extracellular fluids alone (showing manganese peroxidaze, but not lignin peroxidaze activity) did not decolorize the major olive oil wastewater, suggesting that mycelium binding forms part of the decolorization process [38].

In batch cultures, or when immobilized on polyurethane, *Ph. chrysosporium* is able to degrade the macromolecular chromophores of oil wastewater and decrease the amount of phenolic compounds with low molecular weight. *Pleurotus ostreatus* and *Lentinus edodes* also decrease the total phenolic content and reduce the color of cultures containing oil wastewater.

Decolorization of juices and wastewaters by Duolite XAD 761 resin is widely used on an industrial scale and is particularly useful for the removal of color, odor, and taste from various organic solutions in the food and pharmaceutical industries. It removes color, protein, iron complexes, tannins, hydroxymethyl furfural and other ingredients responsible for off-flavors, according to the Duolite Company. The degree of adsorption tends to increase with molecular weight in a given homologous series and has more affinity for aromatic than aliphatic compounds. Recovery of coloring compounds and pigments from agroindustrial products is a common practise [24].

The following case study offers detailed information about the efficiency of resin application in decolorization of olive mill effluents.

Case Study

Chemical and physical treatments of olive oil mill effluent (OME) were performed in this study [24]. The goal was to evaluate the efficiency of aromatic removal from undiluted OME through precipitation by iron sulfate and lime, adsorption on a specific resin, and chemical oxidation by hydrogen peroxide prior to anaerobic digestion as the final treatment method, in order to reduce the toxic effect of OME on bacterial growth and to reduce the coloring compounds in undiluted OME. Olive oil mill effluent was obtained from a local olive oil mill in Tunis and stored at $-20^{\circ}C$. The main findings from this case study are as follows:

1. With regard to the decolorization of OME by iron as a complexing agent, it was noticed that many of the organic and inorganic OME components are susceptible to precipitation by iron. The decrease in the color of OME resulted in a decrease in COD.

The maximum amount of COD and OD removal that could be attained was close to 70% by using 30 g/L of ammonium iron(III) sulfate. Moreover, it seems that the removal of OME color corresponded to the same degree of COD removal. This means that COD is mostly due to the aromatic compounds that are responsible for the color. The complexing effect of iron was complete after 3 hours.

2. As for decolorization of OME by lime treatment and pure calcium hydroxide, the removal efficiency increased with increasing lime concentration. In total, 55% of COD and 70% of color (OD_{390nm}) removal were reached. However, for economic and biological considerations, treatment with 10 g/L calcium hydroxide was sufficient. The effect of lime was complete after 12 hours. It may be concluded that using only 10 g/L of iron and lime as complexing agents was sufficient to precipitate more than 50% of the initial COD and remove 50% of the initial color within a short contact time.
3. With regard to decolorization of OME by resin treatment, the Duolite XAD 761 resin as aromatic adsorbent was used in a column (28 cm long, 1.5 cm in diameter, and with a total volume of 50 cm^3). The results obtained after treating one, two, or three bed volumes of OME, were as follows: COD removal varied between 63 and 75%, and color decrease varied between 52 and 66% for OD_{280nm} and between 51 and 64% for OD_{390nm}. It was also shown that the coloring components in OME are the compounds most responsible for its pollution potential (COD). It may be concluded that the aromatic adsorbent resin retained more than 50% of the coloring compounds (chromophores) corresponding to removal of more than 60% of the initial COD after treating three bed volumes of crude OME. The efficiency depended on the volume treated.
4. As for oxidation of OME by hydrogen peroxide, it has already been shown before (Section 17.5.9) that chemical oxidation is very effective in treating wastewaters that contain large quantities of aromatics. The study was limited to the use of hydrogen peroxide (H_2O_2) concentrations of up to 3%. The effect of H_2O_2 on OME is clear: H_2O_2 removed the substituents of the aromatic rings, which resulted in a decrease in length of the coloring compounds in OME. However, they were not completely degraded, leading to shorter wavelength absorption. This chemical treatment was efficient in color removal but only 19% COD removal was possible. In all cases, simple aromatics were reduced, as determined by GPC analysis.
5. With regard to anaerobic digestion of pretreated OME, the anaerobic digestion of crude and treated OME was elucidated in order to evaluate the efficiency of the physical and chemical pretreatments of OME (Fig. 15). In general, it may be concluded that each pretreatment was efficient in removing the toxic effect in OME. The anaerobic digestibility of OME was improved, with iron and lime, and no inhibition was observed on methanogenic activity. Oxidation of coloring compounds in OME by H_2O_2 removed their toxic effect and did not generate new toxic chemicals to bacterial growth. Separation of aromatics by resin treatment seemed to be the most effective in removing the inhibitory effect of OME prior to anaerobic digestion. Nevertheless, the choice from these different alternatives must be based on economic considerations.

The following process was proposed for reducing environmental pollution by aromatic compounds: Physico-chemical reduction of most toxic compounds of OME, followed by anaerobic microbial decomposition of the main pollutants up to an insignificant amount (see Section 17.5.10 for case studies about the role of fungal treatment in decolorization of OME).

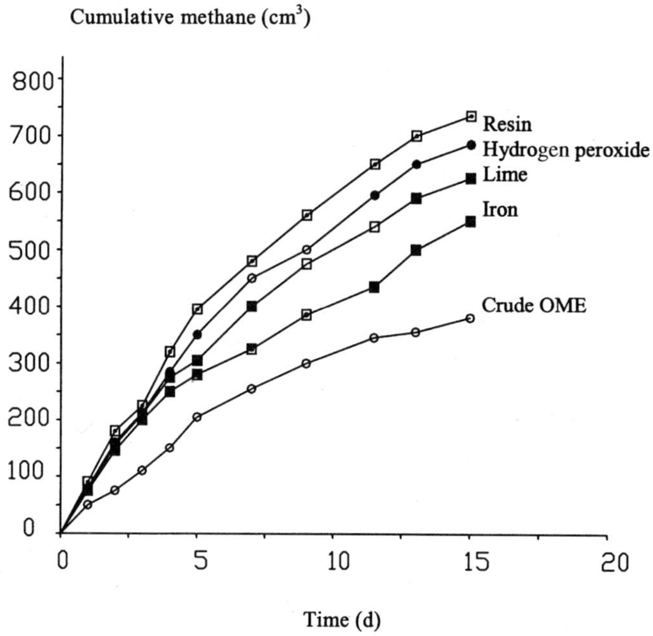

Figure 15 Methanogenic activity in relation to different treatments of OME (role of different treatments in decolorization of OME) (from Ref. 24).

17.5.12 Precipitation/Flocculation

Precipitation involves transforming a water-soluble substance into its insoluble particular form by means of a chemical reaction. Certain chemicals cause precipitation when they react with dissolved and suspended organic compounds. By adding flocculants and coagulation aids, the finest suspended compounds or those dissolved in colloidal form are then transformed into a separable form. This means that, in contrast to precipitation, flocculation is not a phase-transition process [38]. The wastewater may be further treated by activated carbon, ultrafiltration, or reverse osmosis. Figure 16 gives a general concept of the precipitation–flocculation process.

Iron sulfate and aluminum sulfate are commonly used as efficient chelating agents of complex organic compounds in certain wastewaters [68]. Their adsorption capacity is complex and depends on the composition of the precipitated molecule. Lime stabilization is a recognized means of treating municipal sludge prior to land application [69]. The addition of lime temporarily halts biological activity. Moreover, lime renders organic molecules more accessible to microorganisms [70].

In wastewater from olive oil mills (OMW), a purification efficiency of almost 70% of the organic and inorganic components could be removed or complexed by lime (calcium hydroxide) [24]. Disadvantages include the high consumption of chemicals and the large quantities of sludge formed in the process (about 20% of treated alpechin) [38]. For more information about the efficiency of lime and iron as complexing agents in removing COD and color from OMW, refer to the case study presented in Section 17.5.11 (Decolorization).

A proposed plant in Madrid for combined precipitation/flocculation treatment of OMW is presented as a good example of a complete treatment system [38]. This system consists of four phases. In the first phase, a flocculent is added, followed by discharge, filtration, or

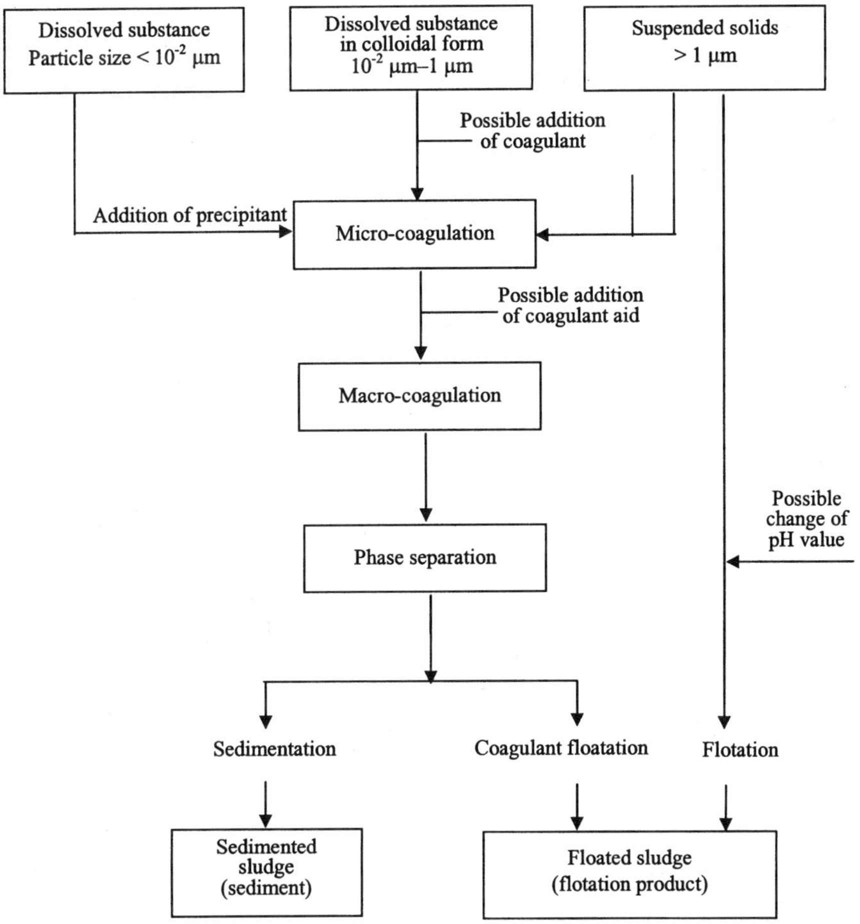

Figure 16 Precipitation–flocculation processes in general (from Ref. 38).

centrifugation. The resulting liquid has a dark red color, and its BOD_5 is about 10,000 mg/L. In the second phase, another flocculation occurs where the smaller size of the flocs are separated through filtration, and its BOD_5 reaches 8000 mg/L. The sludge from these two stages combined is 12% of the original alpechin. The third phase is biological and occurs in three or four stages in purification towers with a separation device for the solids (biomass) and biomass recirculation. The resulting wastewater has a BOD_5 of 2000 mg/L. The fourth phase consists of the filtration of the wastewater, ultrafiltration, and reverse osmosis. The concentrated and thickened sludge from the previous phase is then dried by means of band filters for further use as fertilizer.

17.5.13 Adsorption

Currently, the most commonly used methodologies for the treatment of aromatic-bearing wastewaters include solvent extraction, physical adsorption separation, and chemical oxidation [67]. The adsorption method, which refers to bonding of dissolved compounds (adsorbate) at the surface of solid matter (adsorbent), for example, activated carbon and bentonite, is used for adsorption of dissolved organic pollutants in water. In the field of olive oil wastewater, these are coloring substances (mainly tannic acid), hardly or nonbiodegradable pollutants, bactericidal or inhibiting compounds, which have to be removed. Adsorption not only takes place at the visible surface of the solid, but also in its pores. Activated carbon is especially suited because of its large inner surface (500–1500 m^2/g) and its high adsorptive capacity, but unfortunately it cannot be reused. However, the calorific value is very high so it can be incinerated without problems [38]. Activated carbons are the most common adsorbent, and they are made from different plants, animal residues, and bituminous coal [71,72]. Depending on the composition of the industrial wastewater, one type of carbon may be superior to another [73]. Between 60 and 80% of the organic constituents from alpechin can be adsorbed by activated carbon.

Strong contamination has negative effects on the workability of the plant; thus the alpechin should be pretreated, for example in an activated sludge tank (Fig. 17) [38].

The use of bentonite as an adsorbent for cleaning vegetable oils suggests its applications to reduce lipid inhibition on thermophilic anaerobic digestion [74]; bentonite was added to a synthetic substrate (glyceride trioleate, GTO) and turned out to stimulate methane production by binding the substrate on its surface and thus lowering GTO concentration in the liquid phase.

Laboratory-scale experiments were carried out on fresh OME obtained from an olive oil continuous centrifuge processing plant located in Itri, Lazio, Italy, in order to identify pretreatment type and condition capable of optimizing OME anaerobic digestion in terms of both kinetics and methane yield [75]. In this regard, a set of tests was carried out to evaluate the effect of adding bentonite to OME, both untreated and pretreated with $Ca(OH)_2$. Significant results were obtained by adding $Ca(OH)_2$ (up to pH 6.5) and 15 g/L of bentonite, and then feeding the mixture to the anaerobic biological treatment without providing an intermediate phase separation. Indeed, the biodegradable matter adsorbed on the surface of bentonite was gradually released during the biotreatability test, thus allowing the same methane yield (referred to the total COD contained in untreated OME) both in scarcely diluted (1 : 1.5) pretreated OMW and in very diluted (1 : 12) untreated OME.

These results suggest the application of a continuous process combining pretreatment [with $Ca(OH)_2$ and bentonite] and anaerobic digestion without intermediate phase separation [75]. Specific resin is an economic adsorbent alternative for separating complex organic compounds from wastewater. The Duolite XAD 761 resin is used industrially for the adsorption of mono- and polyaromatic compounds. A considerable number of experiments have focused on removal of coloring compounds in OME by resin treatment [24]. Crude OME was passed through a resin (Duolite XAD 761) column (28 cm long, 1.5 cm in diameter, and with a total volume of 50 cm^3) according to the suggested operating conditions reported by the Duolite

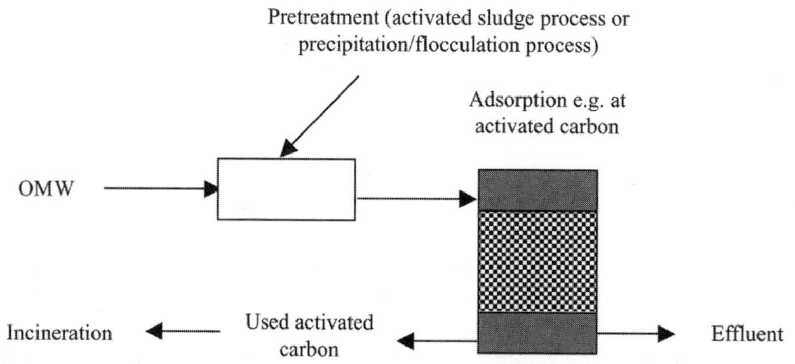

Figure 17 Adsorption process for treatment of olive oil mill wastewater (from Ref. 38).

Company. The pH of the resin was almost 4, and the pH of OME was corrected to 4 using 2 mol/L HCl. The OME was passed through the resin bed at a rate of 50 cm^3/hour. Table 13 shows the results obtained after treating one, two, and three bed volumes of OME. With such treatment, it is clear that the removal of COD up to 75% and decrease in color (OD_{280nm} and OD_{390nm}) up to 66.3 and 63.5%, respectively, could be achieved. Efficiency of the resin treatment decreased with OME volume, due to the saturation of the resin. Moreover, the ratio OD_{280nm}/OD_{390nm} remained constant (almost 5) in crude and treated OME, which meant that adsorption of organic compounds on the resin occurred with the same degree of affinity. On the other hand, the decrease in OME color corresponded to the same degree of COD removal. (For more information about this process, refer to Section 17.5.11).

17.5.14 Biofiltration and Ultrafiltration

Physical processes including filtration, centrifugation, sedimentation, and ultrafiltration are highly efficient methods for phase separation. Filtration processes are used to remove solid material as far as possible from the wastewater. Particles and liquid are separated as a result of pressure difference between both sides of the filter, which enables the transport of water through the filter. During the filtering process, the solids accumulate in the filter and reduce the pore volume, resulting in a change of resistance to filtration and of the filtrate quality. As soon as the admissible resistance to filtration is reached, the filter must be backwashed by forcing clean water backwards through the filter bed. The washwater is a waste stream that must be treated [76].

Compounds that are already dissolved cannot be treated, except by biofiltration. In this case, the filter serves also as nutrient for bacteria so that dissolved organic substance can be aerobically degraded. The purification capacity of biofiltration plants is between 70 and 80%. Up to 100% of the solids can be reduced.

A prerequisite for biofiltration is sufficient oxygen supply. If the alpechin is insufficiently treated, the filter will be quickly clogged. The material kept back in the filter can be used in agriculture (Fig. 18).

A promising alternative method is based on a chemico-physical pretreatment that removes lipids and polyphenols as selectively as possible before biological treatment. In this regard, the potential of filtration applied with other techniques for removal of COD, lipids, and polyphenols from OME has been studied in the following example [75].

A laboratory-scale experiment was carried out in order to choose the pretreatment operating conditions capable of optimizing the anaerobic digestion of OME in terms both of

Table 13 Treatment of OME Through Duolite XAD 761 Resin

OME	OD (280 nm)	OD (280 nm) removal (%)	OD (390 nm)	OD (390 nm) removal (%)	OD (280 nm)/OD (390 nm) ratio	COD (g/dm^3)	COD removal (%)
Crude OME	45.1	–	8.5	–	5.3	147	–
[V(o)/V(r)] = 1	15.2	66.3	3.1	63.5	4.9	37	75
[V(o)/V(r)] = 2	18.7	58.5	3.6	57.6	5.2	43.4	70.1
[V(o)/V(r)] = 3	21.7	51.8	4.2	50.6	5.1	54	63.2

Note: OD: optical density measures qualitatively the color darkness of OME. The OD values were measured at 390 nm and 280 nm.
Source: Ref. 24

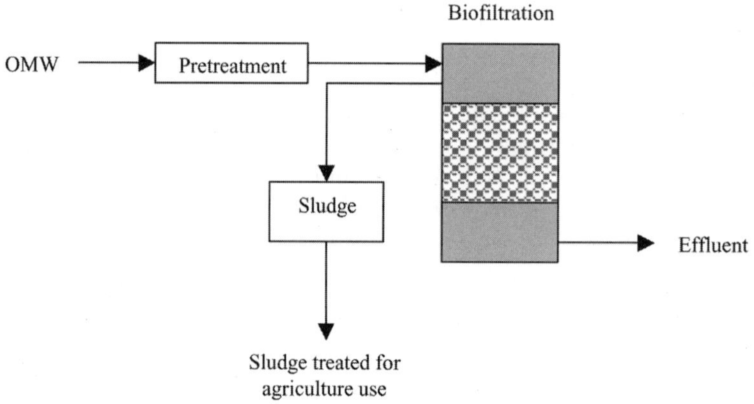

Figure 18 Biofiltration process for treatment of olive oil mill wastewater (from Ref. 38).

kinetics and biomethane yield. Fresh OME was obtained from an olive oil continuous centrifuge processing plant located in Itri, Italy. The OME (pH 4.4, total COD = 92.6 g/L) contained 5.1 g/L of polyphenols, 3.1 g/L of oleic acid, and 11.1 g/L of lipids. The first set of pretreatment tests was carried out by using only physical methods of phase separation: sedimentation, centrifugation, filtration, and ultrafiltration. In the sedimentation phase, after two hours of magnetic stirring, 50 mL of OME were left undisturbed for 24 hours. Afterwards, the OME were centrifuged at 4600 rpm for 15 minutes. The resulting intermediate phase was filtered under vacuum on filter at several pore sizes (25, 11, 6, and 0.45 μm). After filtration on 0.45 μm filters, 20 mL of OME were ultrafiltrated on membranes at 1000 and 10,000 D cutoff threshold (a micron ultrafiltration cell; operating pressure, 4 bar by nitrogen gas).

Table 14 shows the results obtained. The highest removals of oleic acid (99.9%) and polyphenols (60.2%) were obtained through ultrafiltration (at 1000 D). However, COD removed by this technique (65.1%) was much higher than COD associated to lipids and polyphenols removal. While very efficient as a separation technique, ultrafiltration subtracts too much biodegradable COD from the pretreated OME, thus lowering the potential for methane production.

Table 14 Removal of COD, Oleic Acid, and Polyphenols from OME by Means of Physical Methods of Separation

Method of separation	Removal of COD (%)	Removal of oleic acid (%)	Removal of polyphenols (%)
Sedimentation	38.4	96.1	0
Cenrifugation	38.6	95.4	10.2
Filtration [pore size (μm)]			
25	36.7	96.6	12.2
11	37.6	97.6	13.4
6	38.9	98.1	13.4
0.45	40.3	99.0	13.1
Ultrafiltration [cutoff (D)]			
10,000	51.5	99.8	37.2
1000	65.1	99.9	60.2

Source: Ref. 75.

Therefore, ultrafiltration is considered here as a separation technique with poor selectivity. Moreover, the application of ultrafiltration to OME pretreatment might encounter serious problems of membrane fouling as well as of treatment of the concentrated stream. Among the other separation techniques, centrifugation demonstrated the important advantages of producing smaller volumes of separated phases. Further details about this and other sets of pretreatment tests in connection with anaerobic biotreatability may be found in Ref. 75.

17.5.15 Evaporation/Drying

Evaporation is a method used to concentrate non-steam-volatile wastewater components. The evaporation plant contains a vapor condenser by which vapor and steam-volatile compounds are separated from the concentrate. While the concentrate is then recycled into the evaporator, the exhaust steam can be used for indirect heating of other evaporator stages (Fig. 19).

The degree of concentration of the wastewater components depends on different factors, for example [38]:

- reuse of the concentrate (e.g., reuse in production, use as fodder, recovery of recyclable material);
- type of disposal of the concentrate (e.g., incineration, landfill)
- properties of the concentrate (e.g., viscosity, propensity to form incrustation, chemical stability).

Advantages of this method include:

- the residue (dried oil wastes) can be reused as fodder and fertilizer;
- only a small area is needed;
- exhaust steam can be reused as energy;
- considered state of the art in the food industry [38].

Disadvantages are:

- the exhaust steam from evaporation is organically polluted and needs treatment;
- rather high operation and maintenance costs;

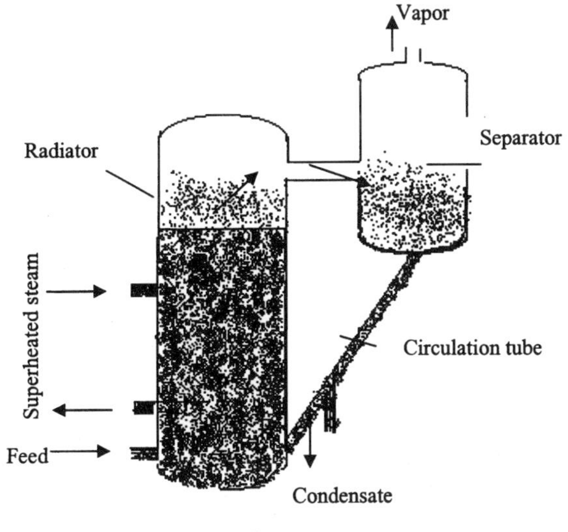

Figure 19 Evaporation/drying processes for treatment of olive oil mill wastewater (from Ref. 38).

- requires high energy;
- requires trained personnel.

Details about drying processes, including case studies for the treatment of olive oil mill wastes, are discussed in Section 17.6.2.

17.5.16 Electrolysis

There are methods still in the experimental stage for treatment of olive oil mill wastewater, one of which is electrolysis. This method is based on electrolytic oxidation of margine constituents, using titanium/platinum for the anode and stainless steel for the cathode. The following data are drawn from laboratory experience (Fig. 20) [38].

The process has the following components:

- electrolytical cell;
- recirculation reactor;
- margine input;
- pH control;
- cooling system.

The performance of the electrolytic cell was tested with a 4% NaCl density in the margine (alpechin) at 42°C, with the temperature remaining constant during the course of the experiment. Four tests lasting 10 hours each were carried out under the same conditions. After 10 hours of electrolysis, the organic load was reduced by 93% in COD and by 80.4% in TOC (total organic carbon). The greatest disadvantage of this method is its high energy consumption (12.5 kW per kg of margine). Therefore it should be applied only as part of the biological pretreatment of the wastewater. Energy consumption then reaches 4.73 kW/kg within the first three hours [38].

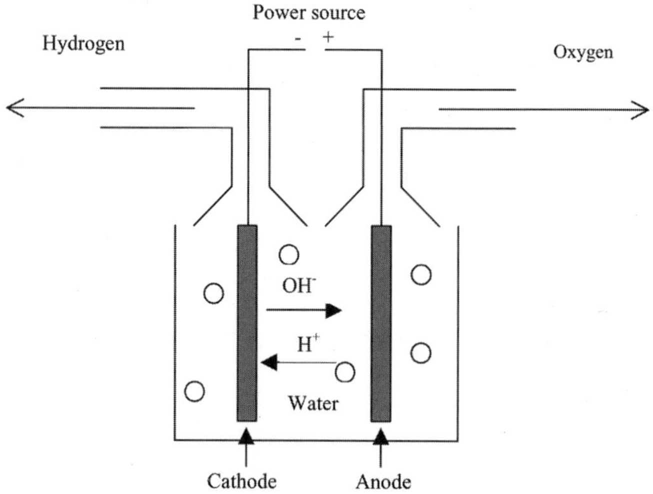

Figure 20 Experimental setup of electrolysis for olive oil wastewater treatment (from Ref. 38).

17.5.17 Bioremediation and Composting

The aim of bioremediation is to repurpose the liquid waste (alpechin) or the liquid fraction of alpeorujo (aqueous fraction that can be separated from fresh alpeorujo by percolation or soft pressing) by diverse aerobic fermentation. The composting of the solid waste (orujo) or the solid fraction of alpeorujo produces a useful material for plant growth.

Several years' research work at the Laboratory for Microbiology of the Athens University has shown that margine (alpechin) is a good substrate for certain microbial fauna. It is especially useful for producing fertilizer for agricultural purposes. Under aerobic conditions, the margine content aids the qualitative breeding of nitrogen-consuming bacteria, especially of acetobacter. This feature was taken into consideration when developing a treatment method for the wastewaters from olive oil production with high organic load. Using this method, a substrate for soil improvement with high nutrient content is obtained from the wastewaters.

Case Studies

Case study I. A pilot plant was put into operation in an oil mill of the Romano-Pylias region. The first big treatment unit was built in 1997 within the framework of the LIFE program for a total of six oil mills in the region of Kalamata (Peloponnes). In addition, a second plant with lower performance was built to treat wastewater from the oil mill in the Arta district. The method consists of two phases [38]. In Phase 1, the margine is neutralized by adding CaO at a pH between 7 and 8. The substrate is mixed in a reactor, which is equipped with a mechanical stirring device. The undiluted residues from the decanter are fed into the stirring reactor. In Phase 2, the contents of the stirring reactor are fed into the bioreactor where sessile microorganisms (especially *Acetobacter vinelandii*) degrade the substances with phytotoxic effect. These bacteria consume nitrogen and take in oxygen from atmospheric air, which is provided by a turntable air distribution system. This leads to increased nitrogen consumption of the bacteria, degradation of the phytotoxic substances, formation of polymers, and secretion of reproduction factors like auxines, cytocynines, which support plant growth.

Retention time in the reactor is 3 days (repeated fed batch culture). The advantages of this method lie in the possibility of applying it directly to olive oil mill wastewater without oil separation, and the high removal efficiency of COD and decolorization.

We propose the possibility of replacing the bioreactor (Phase 2) with the process of natural composting, where the content of Phase 1 is to be mixed, in a well studied way, with municipal solid waste. On the other hand, the main disadvantages here are the long duration (one month or more) needed for aerobic degradation and the need for a large area to conduct the aerobic process.

The final product from the bioreactor or from the natural waste composting plant has a pH of 7.5–8, and, mixed with any quantity of water, can be used to improve soil. Moreover, it has the following characteristics:

- It shows a high content of organic nitrogen (by consumption of atmospheric nitrogen), and substances like auxines support plant growth.
- All nutrients and trace elements present in the olive can be found again in the substrate improved soils.
- The product is able to improve the soil structure and to increase its water retention capacity, due to the biopolymers contained therein.

Case study II. A study was carried out on isolating bacteria from the alpeorujo composting system at Kalamata, Greece [77]. The main results were:

- Identifying bacterial diversity using biochemical techniques of lipid analysis and the molecular biological techniques.
- Demonstration of detoxification of compost by indigenous bacteria.
- Possibility of using a combination of traditional microbiological and modern molecular biological approaches, to follow the changes in microbial flora within the composting material in a qualitative manner.

Strain A of *Azotobacter vinelandii* was used as an agent in the bioremediation process, which was studied in an aerobic, biowheel-type bioreactor, under nonsterile conditions. Before inoculation, the pH of the liquid function of alpeorujo was adjusted to 8.5 by adding CaO. The inoculation was then added at a rate of 10^5 cells/cm^3. The main experimental findings were:

- The alpeorujo liquid fraction (ALF) is very phytotoxic, and inhibitory to the growth of pleurotus and other fungi and many bacteria.
- When ALF is diluted with water (10-fold or more) it can be used as substrate for Azotobacter, Fusarium, Pleurostus and some yeasts (Candida).
- *A. vinelandii* (strain A), while it can degrade and utilize phenolic compounds, grows slowly during the first three days because of the antimicrobial properties of OMW.

Standard bioremediation conditions are of major importance, since (a) the OMW quality is largely dependent on the olive mill machinery and storage facilities and on the quality of the raw material (olives); and (b) bioremediation cycles are performed during wintertime in plants that are exposed to variable environmental conditions.

A continuous composting process was followed. It was observed that alpeorujo, unlike the extracted press cake of the three-phase decanters, is highly unsuitable and cannot be used as a pleurotus substrate. This is due to its high concentration of phenolics. This toxicity is more acute in the pulp fraction of alpeorujo. The wet olive pulp represents 60% of alpeorujo. It is acidic (pH 4.6–4.8), almost black in color mass with moisture content of 65–67% (wet basis), having a smooth doughlike structure. It is also rich in organic and inorganic constituents, especially potassium. Nevertheless its chemical composition is not compatible with the composting process, and so the olive pulp poses quite a serious obstacle to waste treatment and hinders alpeorujo recycling efforts.

In the course of this case study, the possibility of composting both alpeorujo and pulp was also investigated. The major experimental findings were:

- Composting of alpeorujo is feasible when it is mixed with bulky material at a proportion of 3 : 1.
- The mature alpeorujo compost or compost taken from the end of the thermophilic phase offers an ideal microbial consortium to act as starter.
- For alpeorujo and deoiled alpeorujo a self-sustainable composting process was elaborated. Bulky material is only required for the initiation of the process.

In addition, a novel thermophilic process of composting based on the use of hydrogen peroxide (H_2O_2) was developed, due to the fact that hydrogen peroxide exerts a triggering effect on the composting process. The key points include:

- The long-term rise of temperature reflects intensification of microbiological activity in the catabolic processes.

- The formation of glucose from cellulose yields hydrogen peroxide, hydroxyl, and superoxide radicals that are needed to initiate in a snowball reaction the breakdown of the lignin skeleton.
- Similar evidence has bean reported in the case of the brown rot fungus *Gloephyllum trabeum*.

These findings have led to the establishment of a new method for assessing compost stability [77].

With regard to positive effect on plant growth and control of soil fungal pathogens, it was noticed that *A. vinelandii* possesses the ability to inducing soil suppressiveness against some notorious soil-borne root pathogens such as *Pythium*, *Phytophthora*, and *Rhizoctonia* species through its intrinsic ability to produce siderophores.

At the end of this project, the compost produced satisfied farmers, who expressed commercial interest in its use. The compost extract gave similar or even better control against potato blight when compared with commercial organic preparations. Therefore, composting and subsequent utilization in agriculture appears to be the most suitable procedure for treatment of (solid–liquid) waste (alpeorujo). However, large-scale application and more intensive investigation must follow before these procedures may be introduced to the market.

17.5.18 Livestock Feeding

Several methods may be used to enrich OMW with fungi and yeasts so that it becomes suitable for animal feed. The following is a summary of successful experiments performed in Greece as part of the Improlive project, an "International Project to Improve Environmental Compatibility in Olive Oil Production" (during the period 1997–1999) within the European FAIR Programme "Quality of Life and Management of Living Resources."

Case Study

Research [78] was conducted by the University of Athens (1997–1999) with the objective of enriching the two-phase system waste "alpeorujo" with fungal or yeast protein through microbial fermentation and subsequent amino acid production. To give a clear picture of the microorganisms (such as fungi, yeasts, and bacteria) present in the alpeorujo, various techniques and methodologies were applied: serial dilution and selective culture media, application of different inoculation techniques and enrichment of cultures and subcultures, as well as variation in growth temperature and anaerobic conditions. The isolated microorganisms were analyzed for their morphological and biochemical features, then classified into 27 bacteria strains, nine yeasts and three more fungal strains. In order to study the fermentation of bacteria and yeasts, the microcosm system was selected, while a solid-state fermentation bioreactor was used for the fungal strain of *Paecilomyces variotii*. In the microcosm system, and as for as the bacteria concerned, their population declined immediately after inoculation and showed no survival after 72 hours. Total sugars and tannins of the fermented products decreased shortly after each growth cycle of the inoculums. Total lipid content increased after fermentation in all cases.

The microcosm system was followed by solid-state fermentation experiments, which were used to study the growth and activity of selected strains of yeasts and fungi and relevant control conditions, leading to findings such as (a) protein content increased after fermenting the substrate (alpeorujo) with *P. variotii*; (b) the best growth temperature is 35°C for *P. variotii*; (c) long-term experiments are suitable for the best fermentation of alpeorujo substrate. Another step performed was the enrichment of alpeorujo with molasses, which is an inexpensive, renewable industrial byproduct with a very high sugar concentration.

The following conclusions may be drawn from the case study:

- The main constituents of alpeorujo are tannins, lipids, proteins, sugars, and lignocellulosic materials. The chemical profile of alpeorujo makes it adequate for supporting microbial growth by providing plenty of carbon, nitrogen, and energy sources. The results confirm this assumption: alpeorujo is a suitable substrate for the growth of fungi and yeasts and metabolite production.
- Apart from the aerobic bacteria growing at 30°C, several thermophilic bacteria have been isolated and identified, in addition to yeasts (for example, *Candida* genus) and fungi such as *Rhizopus* and *Penicillium.*
- The enrichment of alpeorujo with molasses produced satisfactory results. The increase in the final protein content is around 45%. This increase is a very positive result for the use of the waste material.
- The industrial application of *P. variotii* as a means of increasing the protein content seem feasible, giving the excellent ability to grow in a variety of high-polluted industrial effluents, such as molasses, wood hydrolysates, and spent sulfite liquor. This fungus has an optimum growth at 35°C, while the optimum pH was 4.

The enrichment of alpeorujo with molasses could be a good solution to increase the final protein content and for the optimization of waste materials to be used as animal feed or food additives.

The final conclusion is that *P. variotii* is a fungus that can better utilize the substrate and grows well in it. The resulting increase in the final protein content allows for the possibility of using it as an animal feed or as a feed additive. In addition, not only the fresh but also dried (solid/liquid) waste can be used for fermentation experiments. It is more convenient, however, to use the latter since it is easily handled as a substrate. Further experiments are needed to test the nutrition value of the derived products and their safety for animal consumption.

17.5.19 Ocean Outfalls

The authors proposed for study and application the following method for disposal of olive oil mill wastewaters through submarine outfalls. This section will introduce this method and present its advantages, defects, success conditions, quality limits of sea water, design criteria of marine outfalls, and the required specific pretreatment.

Significance of Submarine Outfalls

Discharge of sewage to the sea through sea outfalls was introduced more than 50 years ago. Outfalls can range in length from a few hundred meters up to more than 15 km; diameters typically vary from 0.5 m up to 8 m and the number of diffuser ports can range from one to several hundred. Sea conditions vary significantly from protected estuaries to open coasts with strong currents and breaking waves [79].

The discharge of industrial and domestic wastewater through submarine outfalls and diffuser systems is one of the most economic solutions for the final disposal process in coastal areas. This disposal system represents a viable alternative for the many population and industrial centers of the world located on sea coasts, particularly for developing countries where financial resources are limited. The capital costs of constructing inland treatment works are often similar to those for an equivalent marine treatment scheme. However, the operational cost of inland treatment is much greater.

Diffusion of industrial and domestic wastewater into marine receiving water, after the degree of treatment deemed necessary for a location, from a properly designed and sited marine outfall system is one of the most environmentally safe options for populations near open coast areas. Such systems can make maximum utilization of the natural assimilating capacity of the sea water environment, which serves as a treatment and disposal facility, and when properly planned, will not produce an undesirable impact upon marine water.

Specific Pretreatments and Quality Limits

Marine treatment via a sea outfall must be considered as a part of the wastewater treatment in conjunction with land treatment, and is one of the most efficient processes to treat effluents with high contaminations. However, since wastewater discharged from inadequately designed or poorly maintained sea outfalls can be a major source of pollution in many coastal areas, the EPA and the EEC have developed some restrictive legislation regarding this issue [80].

In some cases, sea outfalls are used to discharge toxic effluents without proper pretreatment and consequently are responsible for some ecological damage. However, it is widely accepted by scientists and engineers that the use of long sea outfall with an adequate control of the discharged effluent quality is an environmentally safe, waste disposal option.

Materials diffused through marine outfalls may or may not affect the ecology of the receiving water area. Consequently, the oceanography, biology, and ecology of receiving water areas were studied to determine sensitivities to contaminants and design allowing diffusion below sensitivity levels. By satisfying these requirements, marine outfalls could have a positive impact on the coastal water including the presence of fertilizers, such as nitrogen, phosphorus, and carbon in wastewater that maintain life productivity [81].

Sea discharge of industrial and municipal effluents should meet the quality limits of coastal waters used for fisheries, swimming, and recreational purposes, and meet the design criteria given at national level. If a coastal country has no such limits or standards, it may benefit from other countries' experience in this respect. Turkey is a good example in the Mediterranean area (Tables 15 and 16).

Table 15 Required Characteristics of Industrial Wastewater for Sea Discharge in Turkey

Parameter	Value	Remarks
pH	6–9	
Temperature (°C)	35	
SS (mg/L)	350	
Oil and grease (mg/L)	10	
Floating matter	None	
BOD_5 (mg/L)	250	
COD (mg/L)	400	
Total N (mg/L)	40	
Total P (mg/L)	10	
Surface active agents (mg/L)	10	
Other parameters	–	Special care for hazardous wastes

Source: Ref. 82.

Table 16 Design Criteria for Marine Outfalls Systems in Turkey

Parameter	Limits
Temperature	2°C (max) increase after initial dilution
Total coliform (fecal coliform bacteria/100 mL)	1000 in 90% of samples
Initial dilution (D_x)	40 (min)
Discharge depth (m)	20 (min)
Discharge length (m)	1300 m (min) for discharge depth less than 20 m

Source: Ref. 82.

If the receiving water body and/or wastewater characteristics are not deemed acceptable, then marine outfall is not permitted [82]. Table 17 shows the necessity for pretreating some polluting constituents such as particle, oil, grease, and floatables prior to sea discharge through submarine outfalls, with special concentration on refractory substances and heavy metals that require specific treatment at source in conformity with the quality limits of the sea water.

Table 18 presents the removal of significant constituents by pretreatments (milliscreens or rotary screens and by primary sedimentation) [83]. It is noted that the main differences in effluent characteristics relate to the removal of settleable solids and suspended solids and, to a lesser extent, to removal of grease. However, milliscreens remove floatables and particulate fat, which is the material of significance regarding aesthetic impact on the marine environment. The only adverse impact of the discharge of grease relates to slick formation, but when initial dilution is sufficient, the concentration of such material in the mixed effluent/sea water plume is very low and this problem is eliminated [84].

In addition, the data show that screens with openings of less than 1.0 mm require extensive maintenance for cleaning whereas those of 1.0 mm do not.

Disposal of OME Through Submarine Outfalls

With regard to olive industry wastewater, which is mainly characterized by a high content of polyphenols, fats, COD, and solid matters, Table 17 shows that sea water can play a role in the treatment and disposal of biodegradable organics. Refractory organics should be subjected to proper treatment at the source (mill). Fats, floatables, settleable and suspended solids should be pretreated by rotary screens or milliscreens and primary sedimentation. It is possible to treat polyphenols by the decolorization process, which has demonstrated significant correlation

Table 17 The Role of Sea Water in Removal of Wastewater Constituents and the Required Pretreatment Process Prior to Sea Discharge Through Submarine Outfalls

Constituent	Pretreatment	The required process
Particle	Partly needed	Mechanical pretreatment (preliminary treatment + primary sedimentation)
Oil, fats, and floatables	Needed	
Biodegradable organics	Not needed	–
Nutrients	Not needed	–
Pathogenic bacteria	Not needed	–
Refractory organics	Needed	Proper treatment at source
Heavy metals	Needed	Proper treatment at source

Table 18 Removal of Wastewater Constituents by Milliscreens and Primary Treatment

	Percentage removal		
	Milliscreens		
Constituent	0.5 mm apertures	1.0 mm apertures	Primary treatment
Settleable solids	43	23	95–100
Suspended solids	15	10	50
Oil and grease	43	30	50–55
Floatable solids	99	96	95–100

Source: Ref. 82.

between the sewage decolorization and reduction of total organic carbon and phenolic content. It is also advisable to conduct intensive research about sea water's role in reducing these compounds. In cases where pH is less than or equal to 5, it is necessary to apply neutralization within the pretreatment. The criteria given in Table 16 can be referred to for planning and designing the submarine outfalls. Other references provide further details about design criteria and modeling [85]. For economic reasons, it is recommended that several neighboring mills associate in one submarine outfall.

The possible impact of effluents on public health and the environment (aesthetic) should be assessed through monitoring stations for effluent discharge and bathing water (Fig. 21) to achieve national or international standards (fats, COD, and polyphenols).

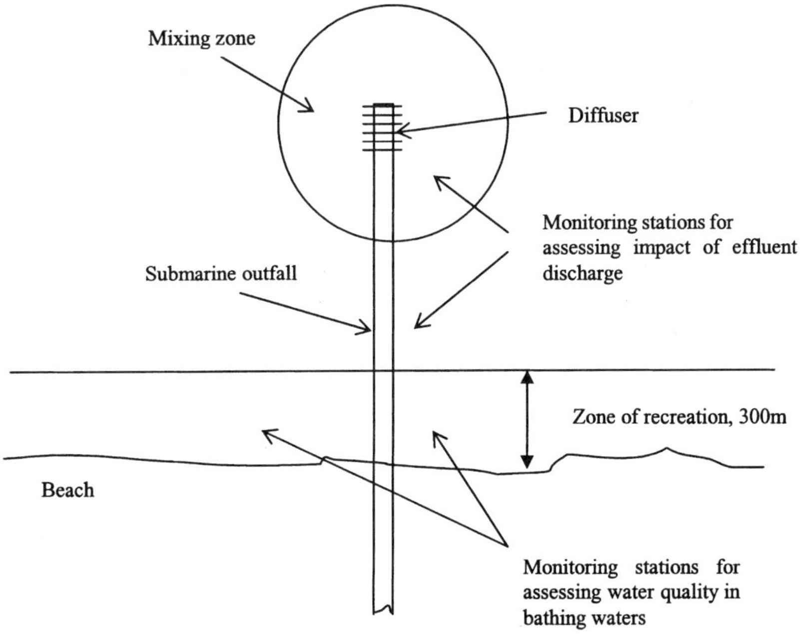

Figure 21 Monitoring stations location for olive oil mill wastewater discharge through submarine outfalls (from Ref. 48).

17.6 SOLID WASTE TREATMENT METHODS

Many of the abovementioned treatment methods for liquid waste are suitable for the treatment of solid/liquid waste arising from the two-phase decanter (alpeorujo). Some of these methods are also appropriate for the treatment of solid waste (orujo), such as recycling methods (composting and livestock feeding). In this respect, a distinction should be made between aerobic treatment systems for liquid waste (such as activated sludge, trickling filter, bioremediation) and aerobic treatment systems for solid waste (such as composting).

Based on the various experiments and published research for waste, especially solid waste and liquid–solid waste, we can propose suitable methods for treating waste from olive oil mills (Table 19). These treatments are classified into three groups: physical, biotechnological, and chemical processes [1]. At the same time, it should be realized that no specific treatment or solution can be generalized. Each case must be studied and evaluated according to local circumstances.

17.6.1 Biotechnological Processes

Biotechnological processes mainly include aerobic (composting), anaerobic (mixed fermentation), solid fermentation, and fungal treatments. A detailed description of methodologies, results, and case studies related to these processes was discussed in Section 17.5.

Other points of considerable importance can be added in this respect [1]:

- Because olive oil mills are operated over limited periods, that is, about three months only per year, an ideal treatment method would be one that could be shifted for treating other types of waste after the end of the olive oil production season.
- The composting method for solid waste treatment is preferable to other methods. This process takes place without serious emissions into air, water, or soil and therefore conforms to the key objectives of Mediterranean environmental policy. Since operational and personnel costs are rather low, this process might also be accepted by plant operators.

Table 19 Treatment Methods for the Solid Waste from Olive Oil Production

Treatment method of orujo and alpeorujo
Physical processes
• Drying
• Evaporation
• Thermal treatment
Biotechnological treatment
• Aerobic (composting)
• Solid fermentation
• Anaerobic/mixed fermentation
• Fungal treatment
Chemical processes
• Incineration
• Combustion
• Pyrolysis
• Gasification

Adapted from Ref. 1.

- The costs of a composting plant strongly depend on the sales potential for the final product in the individual countries. In Greece, for example, higher receipts from compost selling are possible than in Spain. As a result the total costs of a plant also change [2].
- The start-up time of the compost process is only two weeks. It runs in a cycle, which means that additional structuring material is required only in the beginning, and the compost itself is used later as structuring material. The final product is of a high quality and well suited to be used as fertilizer in agriculture.
- Anaerobic treatment by itself is not suitable for solid waste because of its low water content. Problems with mixing and clogging may arise during treatment. Moreover, anaerobic treatment requires further treatment measures, causing additional costs. Another problem is the long start-up time of the process after a longer shutdown period. These problems were behind the breakdown of anaerobic plants in Greece. In the meantime, these plants have been shut down. An economically reasonable solution is to combine this treatment for existing fermentation plants. For this purpose, however, the local situation must be suitable, that is, the fermentation plant should have free capacity and be situated near the olive oil production to avoid high transportation costs and start of digestion of the solid waste.

17.6.2 Physical Processes

Evaporation/drying processes and their advantages and disadvantages in liquid waste treatment have already been discussed. In solid waste treatment, these processes can be discussed in detail as follows. Two of the most important problems related with the treatment of solid waste or solid/liquid waste (alpeorujo) are the optimization of drying and oil recovery by physical means (to get, as much as possible, olive oil instead of orujo oil).

The following case studies discuss new driers based on the combination of fluidized and moving beds, in addition to different pilot-plant treatments of pit separation, drying in a ring drier, and deoiling solid waste in oil mills.

Principle of Fluidized/Moving Beds (Flumov)

The fluidized/moving bed (flumov) combines a fluidized bed with a top section in the form of a fixed/moving bed. The main problem that must be dealt with is the control of the circulation of solids to obtain almost-perfect mixing flow of the solids through the fluidized bed and a plug-flow of the solids in the moving bed (Fig. 22).

The drying of solid waste (or alpeorujo) is required before this waste may be used to recover orujo oil by extraction with hexane and for other processes such as the production of compost, activated coal, biopolymers, and so on. The classical driers, for example, rotary kilns (trommels) and trays, have a low thermal efficiency due to the poor air–solid contact and can present several problems because of the high moisture and sugar contents of the alpeorujo. The presence of the moving zone in flumov allows the fresh product feed to have a higher degree of moisture. Moreover, it favors the solid transport to the fluidized bed contactor, since part of the water is eliminated in the moving zone and the solid enters into the fluidized zone with a relatively low level of moisture [86].

We were particularly interested in confirming the filtering action of the moving bed zone. The filter effectiveness would improve the performance of conventional filtering units usually required for eliminating the suspended solids in the outgoing gas, and even eliminate the necessity of using these units. The stability of the vault, which forms between both beds, requires the input of secondary air into the conical zone to regulate the flow rate of solids from moving

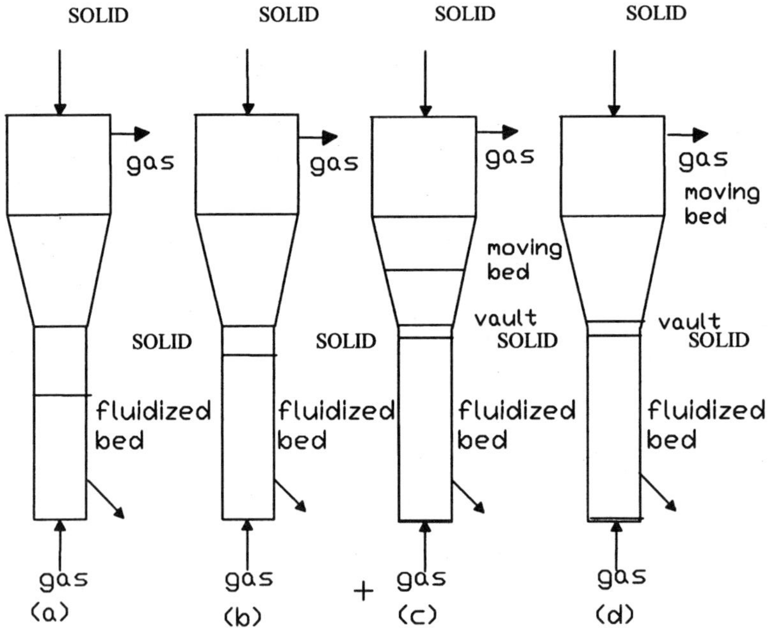

Figure 22 Concept of the flumov state: (a) fluidized bed; (b) expanded fluidized bed; (c) formation of vault; (d) fluidized moving bed (from Ref. 1).

bed to the fluidized bed. The experimental results of residence time distribution of the solid agree with combined models of flow and illustrate the almost plug-flow in the moving bed and the perfect mixing in the fluidized zone of the flumov. The filtering effectiveness of the moving zone is very high and the fines in the output air are mostly eliminated.

Case Studies

Case Study I: Flumov Drier. A fluidized/moving bed drier was constructed and operated [87,88]. It consisted of a cylinder 5.4 cm (inner diameter) and 40 cm height (fluidized bed zone) jointed by a conical device to an upper cylinder 19.2 cm (inner diameter) and 30 cm height (moving bed zone). The feed and removal of solids is made with the aid of J-valves especially designed for this work [89]. The system is a small pilot plant capable of treating up to 5 kg/hour of solid or solid–liquid waste (alpeorujo) (Fig. 23). The drying of waste was studied in batch, semibatch, and continuous operation. Several runs were made in both a conventional fluidized bed drier and a flumov drier with input air between 70 and 200°C and temperature inside the beds between 50 and 150°C. Fresh alpeorujo contained 50–60% moisture (wet basis) and the dried alpeorujo obtained was rather homogeneous. The extracted oil had the same quality as the oil obtained from dried alpeorujo obtained by other drying methods. The filtering effectiveness of the moving bed was very high. In order to solve the operative problems derived from the high moisture content of alpeorujo and the high viscosity of the semidried one, two solutions were found: mixing dry and wet alpeorujo and using pulses of a secondary air injection into the conical zone. Using these two conditions, the dry/wet mixture circulated more effectively along the whole system than the fresh wet alpeorujo. The feeding from the moving bed to the fluidized zone was also well controlled, the air–solid contact improved and the flumov drier was able to operate at a low temperature, about 60°C, inside the fluidized zone (implying a better thermal efficiency balance and allowed for improvement in the dry solid characteristics).

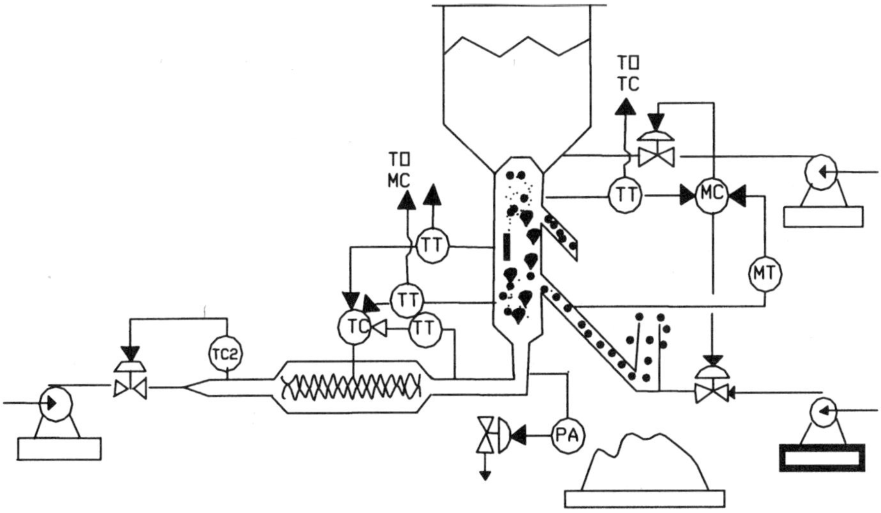

Figure 23 Drier with the implemented advanced control prototype (from Ref. 1). Temperature (TC) and Moisture (MC) control system; Pressure alarm (PA) and temperature transmitters (TT). Control prototype designed by Cognito Quam Electrotechnologies Ltd.

The energy consumption of the flumov drier was between 0.71 and 1.11 kWh/kg water. The mechanical power consumption was similar to other industrial driers, 0.05 kWh/kg water. From the results obtained in the small pilot plant, the flumov drier is a feasible and competitive solution for drying waste. The possibility of drying at low temperatures resulted in a better thermal efficiency balance, lower operating and energetic costs, and improved solid characteristics in use of subsequent solid treatments (high quality of the orujo oil extracted). The main advantages of the system are: reduced total volume, filtering capacity, and ability of using low temperature sources to recover heat from several systems, for example, combined cycle systems and exhaust gases. The details about the control system and prototype, and moisture sensor are in the reference materials [1,89].

Case Study II: Ring Drier. (a) Deoiling of the waste. In southern Spain, Westfalio Separator A.G. installed a batch pilot plant with a capacity of approximately 1 m^3 per batch (Fig. 24) [1]. This plant allowed for an efficient pretreatment of solid/liquid waste (alpeorujo), the separation of the phases as well as a subsequent drying. Owing to product variation, the actual daily quality of the waste was determined as a basis for the planning of the tests. Thus, for each sample a standard test was carried out and several runs were carried out under different process combinations in order to reach a better deoiling of the fresh waste. For this aim, the pits were partially separated, different malaxing times were tested, enzymes or talcum were added to the malaxing process, small quantities of water were added, or other measures were tested for an improvement of the oil yield.

All these measures changed the characteristics of the raw material and consequently contributed in improving the drying process of the deoiled waste. After the deoiling, different intermediate products were generated, that is, partially deoiled orujo and partially depitted orujo. The following parameters were adjusted or the following aids were used [1].

- Enzymes: combination of pectinase and cellulase;
- Talcum: type "talco" 2%;

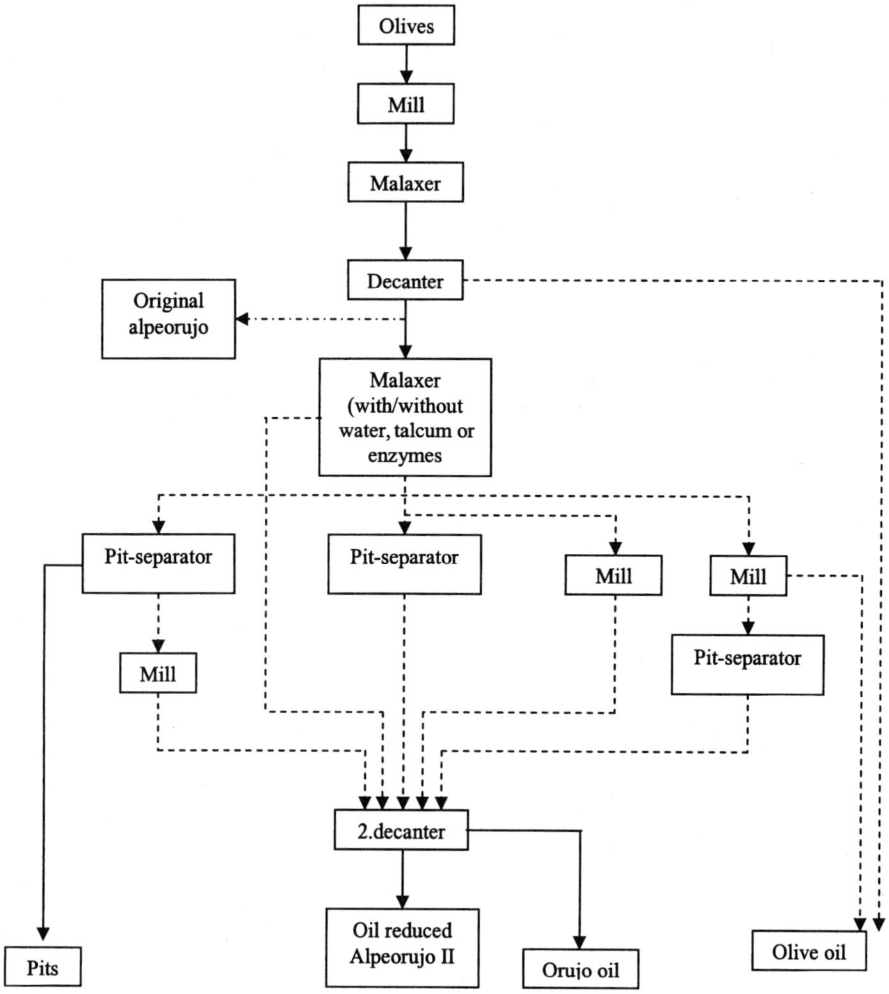

Figure 24 Flow sheet of deoiling pilot plants (from Ref. 1).

- Mill: 2.5, 3, and 4 mm screen;
- Pit separator: 3 mm and 4 mm screen;
- Storage tank: 2 m^3;
- Malaxer: 1 m^3;
- Feed rate: up to 500 kg/h (waste).

(b) Drying of the waste. For the drying, a ring drier was installed to dry different alpeorujos. The intermediate products generated by the deoiling pilot plant were stored and dried. This drier was fueled by propane gas, and hot air was produced with this gas heater. The temperature of this hot air can be varied between 160 and 400°C. With the help of the horizontal screw, one part of the dried waste was mixed with the raw stuff. Both pit-reduced waste and simple deoiled waste were dried as a result. By using the ring drier, the humidity of the waste (alpeorujo) was reduced to approximately 10–15%. The dried material is a powder, the fractions of which are: pit fragments, skin, fruit flesh particles, or agglomerates. The thermal energy requirement for the drier is 1.13 kWh/kg evaporated water. After drying, the oil content vs. drying substance (DS) is sometimes higher than the original material. Another conclusion can be drawn here from deoiling and drying of waste in ring drier. Pit separation before processing is a good solution, in order to produce pit that can be used as a fuel directly in the oil mill, and can raise the throughput. On the other hand, the oil yield is a little bit lower than in the basic version. It is worth noting that drying of solid or solid/liquid waste (alpeorujo) is supposed to precede composting or combustion, and is even indispensable for the latter.

17.6.3 Chemical Processes

Incineration plants are widely known as the conventional means for municipal solid waste treatment for many decades up to the present day. This method, which consists of oxidation of organic substances in high temperatures, has its advantages and disadvantages. Pyrolysis, in-contrast to incineration, is a thermic-decomposed reaction of materials containing a high percentage of carbon (without oxygen) in high temperatures. Thus, pyrolysis is a reduction process that might trigger degasification. It is possible to introduce gasification when there is a partial reaction of coke and water with oxygen. These substances react to carbon oxide and hydrogen. The heat obtained in this process helps to crack heavy molecules. Although the pyrolysis can be used to recycle solid residuals and produce heat, it has not become widespread for technical and/or economical reasons [90]. Additionally, there are no known successful applications, even at pilot plant scale of either incineration or pyrolysis in treatment of olive oil mill waste.

We will discuss a new technique that applies combustion and gasification together in a pilot plant, and has tested successfully in treatment of olive oil mill solid waste concentration [1]. This technique depends on a fluidized moving system, which is a good concept of the gasifier because of the special configuration of the reactor zones. In the bottom part of the gasifier, the fluidized bed permits the required combustion, which represents exothermic reactions, necessary to maintain the thermal balance inside the whole reactor. In the upper part, the moving bed zone does not allow the combustion process to occur but only the endothermic gasification processes. This is due to the fact that the rising gas that reaches the moving bed contains a very low concentration of oxygen and has a high temperature (800–850°C). So only the gasification process can be performed in the moving bed.

Case Study

A fluidized/moving bed reactor was designed to serve as combustor and gasifier. The pilot plant was capable of processing 1–5 kg/hour of solids. The control system in the reactor could regulate the mass flow of air, temperature, and level in the fluidized bed and solid feed. The gasifier is a flumov system, a rather new concept of reactor, and was based on a combination of (a) fluidized bed in the bottom part, where mainly combustion processes take part, and (b) moving bed in the upper part, where the solids are preheated and gasified (Fig. 25) [86]. A special characteristic of the flumov system is that the moving bed filters the flue gases.

The solid used for gasification was orijullo (deoiled orujo and deoiled alpeorujo) of mean particle size 1.4 mm, and pits (ground stone) of mean particle size 2.57 mm. The fluidized bed was filled with sand of mean particle size 0.21 mm, or in some runs, with dolomite with a mean particle size 0.35 mm.

The ultimate analysis of orujillo and stone showed that both have the same composition (dry ash free analysis: 47% C, 6% H, 1% N, 46% O, and <0.01% S). The content of ash is about 3.2% by weight. One of the main elements in ash is potassium (8–30% in K_2O), the ingredient

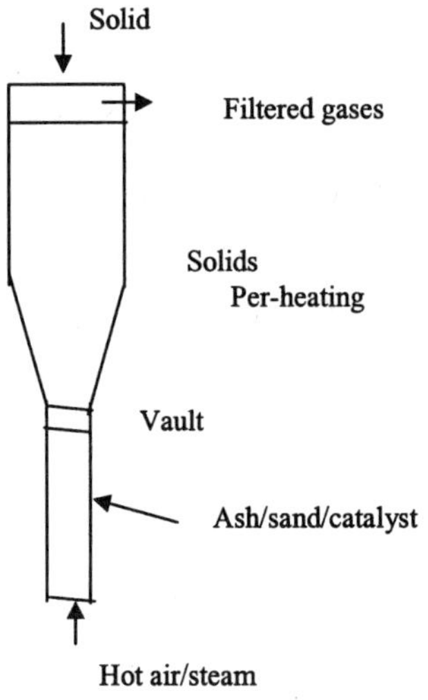

Figure 25 Flumov gasifier (from Ref. 86).

that makes the ashes useful as fertilizer additives. The main process operation variables were temperature, air/water ratio, and equivalent ratio (ER). The presence of sand and dolomite in the fluidized bed had no positive effect on the tar production in the moving bed nor on the flue gas composition (10% H_2, 2% CH_4, 8% CO). Many runs were carried out to find out the best operating conditions, both in combustion and in gasification to obtain the best thermal efficiency. The optimal operating conditions for obtaining the best flue gas were:

- Equivalent ratio (ER = actual air/stoichiometric): 0.20–0.30;
- Temperature in the moving bed: 750–800°C;
- Temperature in the fluidized bed: 800–824°C;
- Throughput: 400–500 kg solid/hm^2 fluidized bed;
- Airflow rate: 1.3 Nm^3/hour;
- Water/air ratio: 0.2 kg water/kg air.

An assessment of the energetic validation by combustion and gasification of orujillo and pits was made. The gas produced in the fluidized/moving bed gasifier supported the expected composition of gasification flue gases and could be suitable for applications in the electrical power production by means of classical explosion motors.

17.6.4 Examples of Technologies and Treatments

After reviewing various case studies applied in different regions, we can conclude that the most appropriate treatment depends not only on intrinsic factors but also on the capacity and system of production of the plants (olive mills and extraction plants and other industries or activities) [1].

As an example, the present practise in Greece and Italy is decanting in three-phase conditions (Fig. 26) with generation of alpechin and treatment of orujo in extraction plants that use hexane to extract the orujo oil. Part of the deoiled orujo (orujillo) is used to dry wet orujo in its own extraction plant. The excess orujillo is sold as solid fuel (ceramic manufacture furnaces, cement kilns, domestic heating), or used as raw material for composting and as additive for animal feed.

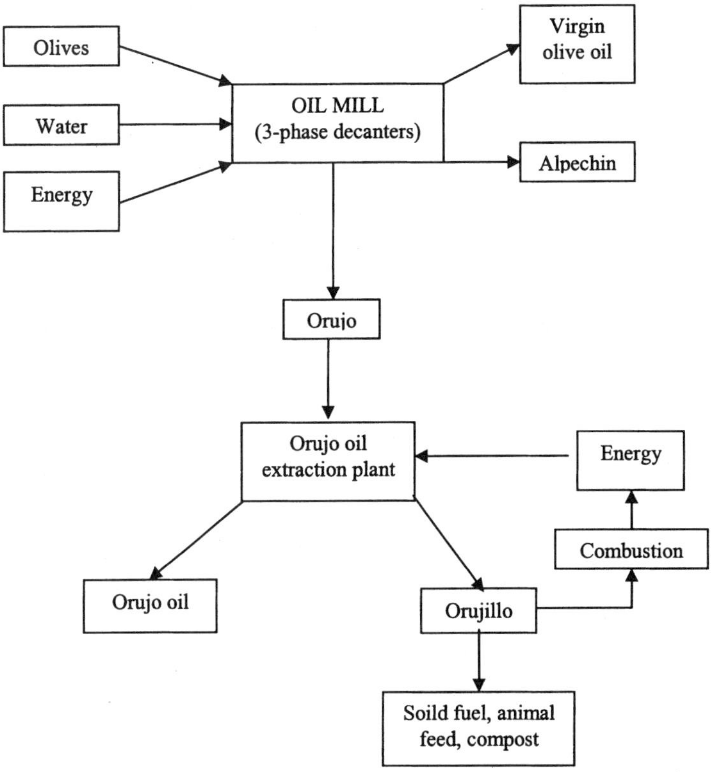

Figure 26 Common integration of treatments for orujo from three-phase decanting method (from Ref. 1).

Spain is a different case, especially in the southern regions, where production is carried out almost exclusively by medium and big cooperatives, and where the two-phase decanting method has been adopted by more than 95% of producers (Fig. 27). The main waste is alpeorujo. Nowadays the "repaso" or second decanting of alpeorujo in the same oil mill is producing a new kind of wastewater, not equal to alpachin but nevertheless representing a growing environmental problem. The orujo oil can still be extracted by extraction plants, but the oil content decreases over time due to the deoiling of alpeorujo made in the oil mills. This means that some producers have decided to burn deoiled alpeorujo to produce electricity. Recent normative, with assured advantages for producers of energy from biomass, has also contributed to the use of orujillo as fuel in small electrical power plants (15 MW). Other new applications such as the production of active coal are also emerging [1].

Currently there is a tendency in some countries to move from the traditional pressing system to the three-phase system and from three-phase to the two-phase system, so the use of different models is constantly changing. Since there are no general unified solutions, every case should be studied according to the local conditions.

As we have seen in the previous section, in the case of waste resulting from the two-phase decanting process, separation into pulp, alpeorujo liquid fraction (ALF), and pits allows for the application of selective treatments and techniques such as composting, bioremediation, and gasification. Another valuable point is worth mentioning here: mixing alpeorujo with other wastes such as molasses improves the production of animal feed with a high protein content.

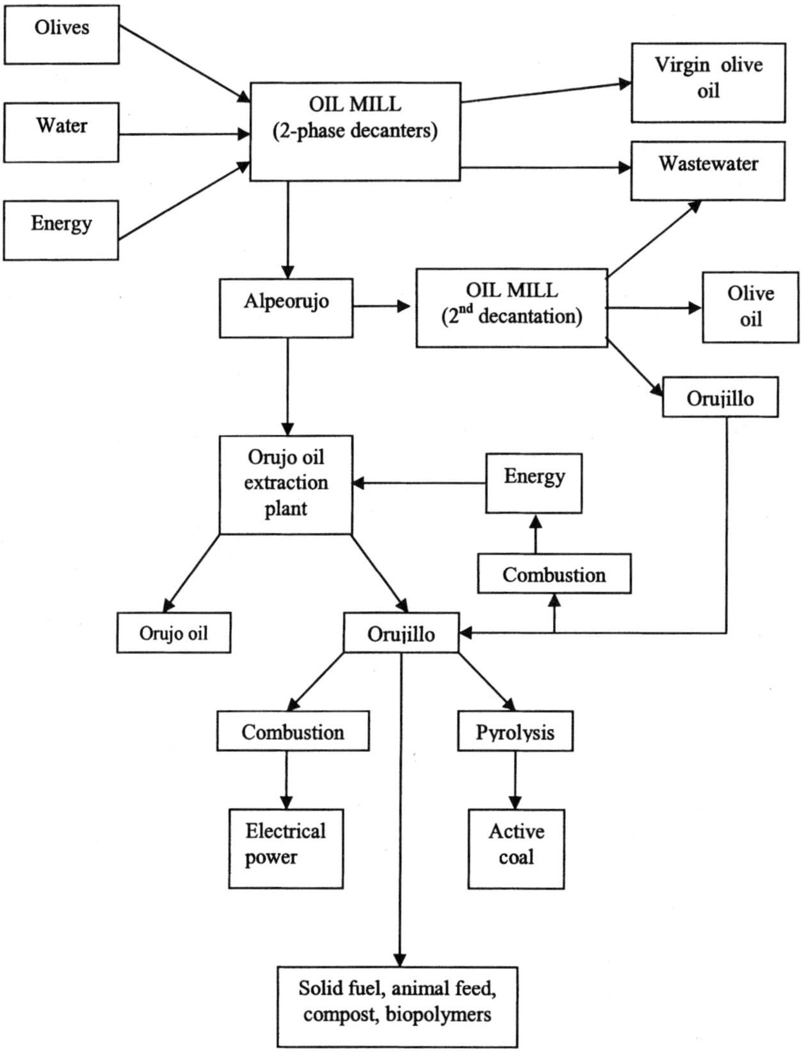

Figure 27 Common integration of treatments for orujo from two-phase decanting method (from Ref. 1).

With regard to the energy value of wastes, it is important to consider that the integration of energy cycles will optimize costs and environmental impacts, for example, by burning pits to dry, or predry the waste or alpeorujo, and combustion/gasification of it to recover energy and combustible gases to obtain and use electrical energy [1].

Furthermore, there should be always specific training programs for operators and supervisors in oil mills and related waste treatment units.

17.7 ECONOMY OF TREATMENT PROCESSES

Many food-processing-related industries, including the manufacture of olive oil and table olives, are of a seasonal nature, and consequently waste is not generated throughout the entire year.

Capital and operating costs of an *in situ* complete treatment (physical-chemical and biological processes) of these waste streams are inevitably high [91]. Thus, if a factory is located in an urban area, the most common practise for dealing with these kinds of effluents is to deliver the industrial effluent to the nearest municipal wastewater treatment plant and to pay the appropriate fee. However, the presence of inhibitory or toxic substances may have a serious effect on the overall treatment system, particularly the biological treatment process, from an operational and economical viewpoint. Thus, in the activated sludge process, phenol-type compounds in concentrations of >200 mg/L and >10 mg/L are known to inhibit carbonaceous removal and nitrification respectively [92]. As a result, some action must be taken before discharging these industrial effluents into municipal sewers and treatment facilities.

As discussed before, several anaerobic processes or techniques have been applied only to the treatments of diluted OME, such as an upflow anaerobic sludge blanket (UASB) reactor, a combined sludge blanket reactor with fixed-bed filter, anaerobic contact reactors, and anaerobic filters. In these biological treatments, OME has to be diluted prior to biological digestion, otherwise the bioreactors need high volumes due to the relatively low loading rates that could be applied and the high pollution potential of OME. At the same time, physical and chemical methods are widespread and applied for treatment of OME. These methods, as discussed before (treatment sections), are considered partial treatments, for example, precipitation by iron and lime, adsorption on a specific resin, and chemical oxidations by hydrogen peroxide and ozone. It was noticed that each pretreatment was efficient in removing the toxic effect of OME. Furthermore, the aerobic pretreatment of OME with different microorganisms (such as *Azotobacter* and *Aspergillus*) reduces considerably the COD and the total phenolic compounds concentration of the waste, which is responsible for its biotoxicity.

It is important to consider that any of these alternatives (physical, chemical, biological) must depend on economic factors, taking into account the possible combination of two or more alternatives. The physical or chemical pretreatment of OME could resolve the problems of time-variable composition and of pollution potential [24]. As a result, dilution for further biological treatment could be reduced, which is an important factor in the evaluation of its economy. The precise evaluation of the cost and feasibility of each of these treatment alternatives depends on several factors, such as capacity of production, waste amount, waste state (liquid or solid), site requirements, specific training of the workers, noise and odor emissions, industrial and agriculture–ecological surroundings, local laws [93].

As reported in the literature, wet air oxidation (WAO) is an economically acceptable technology used to treat aqueous wastes containing oxidizable pollutants at concentrations too high or too toxic for aerobic biological treatment [94]. An exhaustive economic evaluation of WAO is a rather difficult task, given the high number of parameters involved in the process. Thus for a continuous process, there are several operating variables (influent flow rate, temperature, pressure, contamination level, cooling and steam process water temperature, effluent temperature, final contamination level, biodegrability, etc.). Obviously, kinetic and thermodynamic data of the wastewater to be processed must also be considered (specific heat, heat of reaction, rate constants, etc.). These parameters will determine the residence time of the wastewater in the reactor and the energy needed and released in the process [19,95].

An economic assessment compared WAO and incineration processes for treatment of industrial liquid waste with a high content of phenol-type substances. The outcome was that incineration resulted in roughly four times the expense of WAO [96,97].

Another example focuses on solid waste treatment by gasifier/combustion flumov system to produce the optimal flue gas. Economic and industrial estimations were made of the gasifier's industrial design. The size and cost of a gasifier for treating 15 T/hour of solids capacity was estimated at 3.6 million euro (fluidized bed 2.6 × 8 m, moving bed 8 × 8 m) [1].

As previously discussed, it is important from an economic perspective to develop profitable uses for the final waste product, such as organic fertilizer, soil conditioner, and livestock feed. In this regard, it is worth pointing out that an opportunity exists to obtain a new type of renewable and low-cost activated carbon (J-carbon) from the processed solid residue of olive mill products. This is due to the fact that olive mills generate a huge amount of waste, which can be suitable as a raw material with economic value, and as a supportive means for pollutant removal from wastewater [98]. A study was performed to compare the capability of J-carbon with commercial activated carbon to remove ammonia (NH_3), total organic carbon (TOC), and some special organics from Flexsy's (Rubber) wastewater treatment plant as tertiary treatment [99].

In this regard the final result was that the J-carbon has almost similar behavior and efficiency as the commercial activated carbons (powder activated carbon and granular activated carbon). Therefore, it was concluded that the J-carbon, as well as other commercial activated carbons, could be used in the treatment of industrial wastewater to improve efficiency of the treatment plant. The exhausted carbon would be settled by gravity and disposed of with the sludge as a carbon–sludge mixture. Thus, there would be no need for regeneration since the J-carbon is a renewable and very low-cost adsorbent.

17.8 SUMMARY

This chapter is based around the fact that the olive oil industry is in continuous growth due to its nutritious and economic importance, particularly for Mediterranean countries. This is accompanied by vast waste generation from different olive oil technologies (traditional and pressing decanting processes). The wastewater is mainly characterized by a high degree of organic pollution, polyphones, and aromatics forming inhibitor or toxic substances, which constitute a serious environmental problem for soil, rivers, and groundwater.

The great variety of components found in liquid waste and solid waste requires different appropriate technologies to eliminate those that have harmful effects on the environment. From an economic perspective it is important to develop profitable uses for the final waste product such as organic fertilizer, soil conditioner, and livestock feed.

The optimal disposal and management of olive oil mill waste should be viewed within a multidisciplinary integrated frame that comprises specific procedures such as extraction by decanter centrifuge, liquid/solid waste treatments, aerobic bioremediation and composting, enrichment of waste with fungal/yeast protein, drying and gasification in fluidized moving beds, recovery of orujo oil, and recovery of energy and combustible gases.

Prospective researches should take into consideration the new advances in biotechnology, treatment reactors, control, new products and processes, composting from different wastes mixtures, all for the service of minimizing the impact on the environment, and reducing the use of valuable natural and living resources within the course of sustainable development.

REFERENCES

1. Aragón, J.M. *Improvement of Treatments and Validation of the Liquid–Solid Waste from the Two-Phase Olive Oil Extraction*, Final Report, (Annex A2) Project IMPROLIVE (FAIR. CT96-1420); Department of Chemical Engineering, Univesidad Complutense de Madrid: Madrid, Spain, 2000.

2. Stölting, B.; Bolle, W.F. Treatment processes for liquid and solid waste from olive oil production. In *Proceedings of Workshop "IMPROLIVE-2000"*, (Annex A1), Seville, Spain, April 13–14, 2000; 29–35.
3. FAOSTAT data base. http://appsfao.org/default.htm (2000).
4. IMPROLIVE Project. http://fiw.rwth-aachen.de/info/improlive (2002).
5. Fedeli, E. Olive oil technology. SSOG, Milano, *Toronto Special, OLIVE*, **1993**, *45*, 20–23.
6. Hruschka, S.; Geissen, K. Global contemplation of the olive oil production with regard to the economy, quality and ecology. In *Proceedings of Symposium "Olive Waste,"* Culture Center of Kalamata, Greece, November 5–8, 1997.
7. Kiritsakis, A.K. Olive oil – from the tree to the table. In *Food and Nutrition*; Press Inc.: Trumbull, CT, USA, 1998.
8. Baccioni, L. Producing first-rate extra virgin oil. *Oil and Fats Intern.*, **1999**, *15* (*1*), 30–31.
9. Gasparrini, R. Treatment of olive oil residues. *Oil and Fats Intern.*, **1999**, *15* (*1*), 32–33.
10. Boskou, D. Olive oil. In *Chemistry and Technology*; AOCS-Press: New York, 1996.
11. Kiritsakis, A.K. The olive oil quality. In *8th Congress Seccion Latinoamericana AOCS*, Santiago de Chile, October 24–27, 1999.
12. Durán, R.M. Relationship between the composition and ripening of olive and quality of the oil. In *Proceedings of I Intl, Sympos, on Olive Growing*, Rallo, L. Universidad Autonoma de Cordoba, Acta Horticulturae 286, 1989.
13. Pompei, C.; Codovilli, F. Risultari preliminari sul trattamento di deparazione delle acque di vegetazione delle olive per osmosi inversa. Scienza e Tecnologia Degli Alimenti **1974**, 363–364.
14. Fiestas, R. et al. The anaerobic digestion of wastewater from olive oil extraction. In *Anaerobic Digestion*; Traue-muende: Poster, Germany, 1981.
15. Steegmans, S.R. *Optimierung der anaeroben Verfahrenstechnik zur Reinigung von organischen hochverschmutzten Abwässern aus der Oliven Ölgewinnung*; Forschungsinstitut fuer Wassertechnologie an der RWTH Aachen (Hrsg.), Forschungsbericht AZ 101/81 der Oswald-Schulze-Stiftung: Aachen, 1992.
16. Hamdi, M. Toxicity and biogradability of olive mill wastewaters in batch anaerobic digestion. Bioprocess Engineering, Heft **1993**, *8*, 79.
17. Andreozzi, R. et al. Integrated treatment of olive oil mill effluents (OME): Study of ozonation coupled with anaerobic digestion. J. Wat. Res. **1998**, *32* (*8*), 2357–2364.
18. Garcia, G.P.; Garrido, A.; Chakman, A.; Lemonier, J.P.; Overrend, R.P.; Chornet E. Purifiction de aguas residuales ricas en poliven-oles: Aplication de la oxidactión húmeda a los effluentes acuosos derivados de las industrias olivarerns. Grasos y Aceites **1989**, *40* (*4–5*), 291–295.
19. Rivas, F.J.; Beltran, F.J., Gimeno, O.; Alvares, P. Chemical-biological treatment of table olive manufacturing wastewater. J. Environ. Engng ASCE, **2001**, *127* (*7*), 611–619.
20. Kissi, M.; Mountador, M.; Assobhei, O.; Gargivlo, E.; Palmieri, G.; Giardira, P.; Sannia, G. Roles of two white-rot basidiomycete fungi in decolorisation and ditoxification of olive mill wastewater. J. Appl. Microbiol. Biotech. **2001**, *57*, 221–226.
21. Beltran-Heredia, J.; Torregrosa, J.; Dominguez, J.R.; Garcia, J. Treatment of black-olive wastewaters by ozonation and aerobic biological degradation. J. Wat. Res. **2000**, *34* (*14*), 3515–3522.
22. Ubay, G.; Özturk, I. Anaerobic treatment of olive mill effluents. J. Wat. Sci. and Technol. **1997**, *36* (*2*), 287–294.
23. Beccari, M.; Bonemazzi, M.; Majone, M.; Riccardi, C. Interaction between acidogenesis and methanogenesis in the anaerobic treatment of olive oil mill effluents. J. Wat. Res. **1996**, *30* (*1*), 183–189.
24. Zouari, N. Decolorization of olive oil mill effluent by physical and chemical treatment prior to anaerobic digestion. J. Chem. Technol. Biotechnol. **1998**, *73*, 297–303.
25. Borja, R.; Martin, A.; Alonso, V.; Garcia, I.; Banks, C.J. Influence of different aerobic pretreatments on the kinetics of anaerobic digestion of olive mill wastewater. J. Wat. Res. **1995**, *29* (*2*), 489–495.
26. IOOC. *Olive Oil Quality Improvement*, Technical Report. International Olive Oil Council: Madrid.
27. Lopez, R. et al. Land treatment of liquid wastes from the olive oil industry (Alpechin). Fresenivs Envir. Bull. *1*, **1992**, 129–134.

28. Borja, R.; Martin, A.; Maestro, R.; Alba, J.; Fiestas, J.A. Enhancement of the anaerobic disgestion of olive mill wastewaters by removal of phenolic inhibitors. J. Processes Biochem. **1992**, *27*, 231–237.
29. Annesini, M.C. et al. Treatment of olive oil wastes by distillation. J. Effluent and Water Treatment **1983**, *June*.
30. Geissen, K. *Aufbereitung von Alpeorujo aus der Olivenoelproduction in Spanien*; Westfalia Separator AG (HRSG), Interne Firmenschrift: Oelda, Germany, 1995.
31. Özturk, I.; Ubay, G.; Sakar, S. *Anaerobic Treatability of Olive Mill Effluents and Bioenergy Recovery*, Research Report; The Scientific and Technical Research Council of Turkey, Project No. DEBCAG-56, Turkey, 1990.
32. Benitez, F.J.; Beltran-Heredia, J.; Torregrosa, J.; Acero, J.L. Treatment of olive mill wastewaters by ozonation, aerobic degradation and the combination of both treatments. J. Chem. Technol. Biotechnol. **1999**, *74*, 639–646.
33. Wu, J.; Taylor, K.E.; Biswas, N. Optimization of the reaction conditions for enzymatic removal of phenol from wastewaters in the presence of polyethlene glycerol. J. Wat. Res. **1993**, *27*, 1701–1706.
34. Perez, D.J.; Esteban, E.; Gomez, M.; Dolaraf, G. Effect of wastewater from olive processing on seed germination and early plant growth of different vegetable species. J. Environ. Sci. Health. **1986**, *49*, 349–357.
35. Yesilada, O.; Sam, M. Toxic effects of biodegraded and detoxified olive oil mill wastewaters on the growth of Pseudomonas aeruginosa. Toxic. Environ. Chem. **1998**, *65*, 87–94.
36. Bahlo, K.; Wach, G. *Naturnahe Abwasserreinigung*, 3. Aufl.; Oekobuch Verlag: Staufen bei Freiburg, Germany, 1995.
37. Kumaran, P.; Paruchuri, L. Kinetics of phenol biotransformation. J. Wat. Res. **1997**, *31* (*1*), 11–22.
38. FIW (Forshungsinstitut für Wasser and Abfallwirtschaft, RWTH Aachen, Germany)–IMPROLIVE web site. www.fiw.rwth-aachen.de/improlive/improlive.htm.
39. J.G. Press Inc., Seville. Recycling olive mill residuals in Andalusia. Bio-cycle **1995**, *36* (*6*), 24.
40. Eckenfelder, W.W. *Industrial Water Pollution Control*, 2nd Ed.; McGraw-Hill International, 1989.
41. Davis, M.; Cornwell, D. *Introduction to Environmental Engineering*, 2nd Ed.; McGraw-Hill International, 1991.
42. US-EPA. *Environmental Pollution Control Alternatives, Municipal Wastewater*; Environmental Protection Agency: Washington DC, 9–12.
43. Chmielowski, J. et al. Zesz. Nauk. Politech Slaska Inz. (Polish) **1965**, *8*, 97.
44. Aveni, A. Biogas recovery from olive oil mill wastewater by anaerobic digestion. In *Anaerobic Digestion and Carbohydrate Hydrolysis of Waste*; Ferrero, G.L., Ferranti, M.P., Neveau, H. Eds.; Elsevier, 1984; 489–490.
45. Carrieri, C.; Balice, V.; Rozzi, R. Comparison of three anaerobic treatment processes on olive mills effluents. In *Proceeding. Int. Conf. on Environment Protection*, Ischia, Italy, October 5–7, 1988.
46. Boari, G.; Brunetti, A.; Passin, R.; Rozzi, A. Anaerobic digestion of olive oil mill wastewaters. J. Agric. Wastes **1984**, *10*, 161–175.
47. Özturk, I.; Sakar, S.; Ubay, G.; Eroglu, V. Anaerobic treatment of olive mill effluents. In *Proceedings of the 46th Industrial Waste Conf.*, Purdue University, West Lafayette, IN, May 14–16, 1991.
48. Rozzi, A.; Boari, G.; Liberti, L.; Santori, M.; Limoni, N.; Menegatti, S.; Longobardi, C. Anaerobic-aerobic combined treatment of urban and olive oil mill effluents – pilot scale experimentation. J. Ing. Sanit. **1989**, *37* (*4*), 44–54.
49. Wang, Y.T.; Gabbard, H.D.; Pai, P.C. Inhibition of acetate methanogenesis by phenols. Environ. Engng **1991**, *117*, 487–500.
50. Koster, I.W.; Gramer, A. Inhibition of methanogenesis from acetate in granular sludge by long-chain fatty acids. J. Appl. Environ. Microbiol. **1980**, *53*, 403–409.
51. Craveiro, A.M.; Rocha, B.M. Anaerobic digestion of vinasse in high-rate reactors. In *proceedings of NWA-EWPCA Conference (Aquatech 86) on Anaerobic Treatment*, Amsterdam, September 15–19, 1986; 307–320.
52. Borja, R.; Martin, A.; Garrido, A. Anaerobic digestion of black-olive wastewater. J. Biores. Technol. **1993**, *45*, 27–32.

53. Capasso, R.; Cristinzio, G.; Evidente, A.; Scognamiglio, F. Isolation, spectroscopy and selective phyto-toxic effects of polyphenols from vegetable wastewaters. J. Phytochemistry **1992**, *31*, 4125–4128.
54. Zouari, N.; Ellouz, R. Toxic effect of colored olive compounds on the anaerobic digestion of olive oil mill effluent in UASB like reactors. J. Chem. Technol. Biotechnol. **1996**, *16*, 414–420.
55. Eisenhauser, H.R. Ozonation of phenolic wastes. J. Wat. Pollut. Control Fed. **1986**, *40*, 188–193.
56. Gould, J.P.; Weber, W.J. Jr. Oxidation of phenols by ozone. J. Wat. Pollut. Control Fed. **1976**, *48*, 47–52.
57. Scott, J.P.; Ollis, D.F. Integration of chemical and biological oxidation processes for water treatment: review and recommendation. Environ. Prog. **1995**, *14* (*2*), 88–103.
58. Kolaczkowski, S.T.; Beltran, F.J.; McLurgh, D.B.; Rivas, F.J. Wet air oxidation of phenol: Factors that may influence global kinetics. *Trans. Instn. Chem. Engs, Part B, Process, Safety and Environ. Protection* **1997**, *75*, 257–265.
59. Debellefontaine, H.; Chakchouk, M.; Foussard, J.N.; Tissot, A.; Striolo, P. Treatment of organic aqueous wastes: Wet air oxidation and wet peroxide oxidation. J. Environ. Pollut. **1996**, *92* (*2*), 155–164.
60. Beltran, F.J.; Kolaczkowski, S.T.; Crittenden, B.D.; Rivas, F.J. Degradation of o-chlorophenol with ozone in water. *Trans. Instn Chem. Engrs, Part B, Process, Safety and Environ. Protection* **1993**, *71*, 57–65.
61. Martirani, L.; Giardina, P.; Marzullo, L.; Sannia, J. Reduction of phenol content and toxicity in olive oil mill wastewaters with the ligninolytic fungus Pleurotus ostreatus. J. Wat. Res. **1996**, *30* (*8*), 1914–1918.
62. Nannipieri, P.; Ballag, J.M. Use of enzymes to detoxify pesticide-contaminated soils and waters. J. Environ. Qual. **1991**, *20*, 510–517.
63. Atlow, S.C.; Bonadonna, M.; Apero, L.C.; Klibanov, A.M. Dephenolization of industrial wastewaters catalyzed by polyphenol oxidase. J. Biotechnol. Bioeng. **1984**, *26*, 599–603.
64. Sanjust, E.; Pompei, R.; Rescigno, A.; Rinaldi, A.; Ballero, M. Olive milling wastewaters as a medium for growth of four pleurotus species. J. Appl. Biochem. Biotech. **1991**, *31*, 223–235.
65. Eriksson, K.E.L.; Blanchette, R.A.; Ander, P. Biodegradation of lignin. In *Microbial and Enzymatic Degradation of Wood and Wood Components*; Timmel, T.E., Eds.; Springer-Verlag: Berlin, 1990; 215–232.
66. Field, J.A.; Lettinga, G. The effect of oxidative coloration on the methanogenic toxicity and anaerobic biodegradability of phenols. J. Biol. Wastes **1989**, *29*, 161–179.
67. Wang, Y.T.; Suidan, M.T.; Rittman, B.E. Anaerobic treatment of phenol by an expanded-bed reactor. J. Wat. Pollut. Control Fed. **1986**, *58*, 227–232.
68. Jackson-Moss, C.A.; Duncan, J.R. The effect of aluminum on anaerobic digestion. Biotech. Lett. **1991**, *13*, 143–148.
69. Christy, R.W. Sludge disposal using lime. J. Wat. Environ. Technol. **1990**, *4*, 56–61.
70. Bevins, R.E.; Longmaid, F.M. Stabilization of sewage-sludge cake by addition of lime and other materials. J. Wat. Pollut. Control **1984**, 9–12.
71. Pollard, S.J.T., Thompson, F.E.; McConnachie, G.L. Microporous carbons from moringa oleifera husks for water purification in less developed countries. J. Wat. Res. **1995**, *29*, 337–374.
72. Wigmans, T. Industrial aspects of production and use of activated carbons. J. Carbon. **1989**, *27*, 13–22.
73. Lankford, P.W.; Eckenfeldes, W.W. *Toxicity Reduction in Industrial Effluents*; VAN Nostrand Reinhold: New York, 1990.
74. Angelidaki, I.; Peterson, S.P.; Ahring, B.K. Effect of lipids on thermophilic anaerobic digestion and reduction of lipid inhibition upon addition of bentonite. J. Appl. Microbiol. Biotechnol. **1990**, *33*, 469–472.
75. Beccari, M.; Majone, M.; Riccardi, C.; Savarese, F.; Torrisi, L. Integrated treatment of olive oil mill effluents: effect of chemical and physical pretreatment on anaerobic treatability. Wat. Sci. & Technol. **1999**, *40* (*1*), 347–355.
76. Davis, M.L.; and Cornwell, D.A. *Introduction to Environmental Engineering*, 2nd Ed.; McGraw-Hill Int., 1991.

77. Balis, C.; Antonokou, M. Composting and bioremedation. In *Proceedings of Workshop "IMPROLIVE-2000"*, (Annex A1), Seville, Spain, April 13–14, 2000; 13–18.
78. Giannoutsou, E.P.; Karagouni, A.D. Olive oil waste: could microbial fermentation be the solution. In *Proceedings of Workshop "IMPROLIVE-2000"*, (Annex A1), Seville, Spain, April 13–14, 2000; 23–28.
79. Larsen, T.; Burrows, R.; Engedahl, L. Unsteady flow and saline intrusion in long sea outfalls. J. Wat. Sci. & Technol. **1992**, *25* (*9*), 225.
80. Monteiro, A.J.; Neves, J.R.; Sousa, R.E. Modeling transport and dispersion of effluent outfalls. J. Wat. Sci. & Technol. **1992**, *25* (*9*), 143.
81. Garber, F.G.; Neves, J.R.; Roberts, P. Marine disposal systems. J. Wat. Sci. & Technol. **1992**, *25* (*9*), IX.
82. Özturk, I.; Eroglu, V.; Akkoyunlu, A. Marine outfall applications on the Turkish coast of the Black Sea. J. Wat. Sci. Technol. **1992**, *25* (*9*), 204.
83. Bannatyne, A.N.; Speir, J. Milli-screening – a pretreatment option for marine disposal. In *Proceedings of the IAWPRC Marine Disposal Seminar*, Rio de Janeiro, 1987, *18*, 11.
84. Ludwig, G.R. *Environmental Impact Assessment – Sitting and Design of Submarine Outfalls*; The Monitoring and Assessment Research Center (MARC) and WHO, 1988, Report 43, 5–6.
85. Awad, A. *Submarine Outfalls of Wastewater*, 1st Ed.; Kuwait Foundation for the Advancement of Science, Department of Authorship, Translation and Publications: Kuwait, 1998.
86. Aragón, J.M.; Palancar, M.C.; Serrano, M.; Torrecilla, J.S. Fluidised/moving beds: applications to driers and gasifiers. In *Proceedings of Workshop "IMPROLIVE-2000"*, (Annex A1), Seville, Spain, April 13–14, 2000; 19–22.
87. Aragón, J.M.; Palancar, M.C.; Torrecilla, J.S.; Aparicio, J.T. Modeling fluidized bed dryers by artificial neural network. In *Chemical and Engineering Congress*, Chisa 98, 1998.
88. Aragón, J.M.; Palancar, M.C.; Torrecilla, J.S.; Serrano, M. Drying a high viscosity solid–liquid waste in fluidized-moving bed. In *Chemical and Engineering Congress*, Chisa 98, 1998.
89. Aragón, J.M.; Palancar, M.C.; Serrano, M.; Torrecilla, J.S. Design and fluid-dynamic behaviour of J-valve systems for feeding and discharging solid in fluidized beds. In *Chemical and Engineering Congress*, Chisa 98, 1998.
90. Awad, A. Pyrolysis for disposing municipal wastes. In *Proceedings of the Atmospheric Pollution Symposium*, Damascus, August 1985; 219–225.
91. Kim, J.S.; Kim, B.G.; Lee, C.H.; Kim, S.W.; Jee Koh, J.H.; Fan, A.G. Development of clean technology in alcohol fermentation industry. J. Cleaner Prod. **1997**, *5* (*4*), 263–267.
92. Metcalf & Eddy (Eds.) *Wastewater Engineering*, 3rd Ed.; McGraw-Hill, Int., 1991.
93. European Commission. *Improlive 2000*, Final Report (Accompanying Measures), Program (Quality of Living Sources), No. QLK1-1995-30011, Madrid, June 2000.
94. Baillod, C.; Faith, B. *Wet Oxidation and Ozonation of Specific Organic Pollutants*, EPA-600-2-83-060; US. Environmental Protection Agency: Washington DC, 1983.
95. Baillod, C.R.; Lamparter, R.A.; Barra, B.A. Wet oxidation for industrial waste treatment. J. Chem. Engng Prog. **1985**, *March*, 52–56.
96. Zimpro Environmental Inc. *Commercial-Scale Wet Air Oxidation Process*; Rothschild, WI, 1990.
97. Zimpro Environmental Inc. Wet air oxidation cleans up black wastewater. J. Chem. Engng **1993**, *Sept.*, 175–176.
98. Gharaibeh, S.H.; Abu El Sha'r, W.Y.; Al Kofahi, M.M. Removal of selected heavy metals from aqueous solutions using processed solid residue of olive mill products. J. Wat. Res. **1998**, *32* (*2*), 498–502.
99. Gharaibeh, S.H.; Moore, S.V.; Buck, A. Effluent treatment of industrial wastewater using processed solid residue of olive mill products and commercial activated carbon. J. Chem. Technol. Biotechnol. **1998**, *71*, 291–298.

18
Potato Wastewater Treatment

Yung-Tse Hung and Howard H. Lo
Cleveland State University, Cleveland, Ohio, U.S.A.

Adel Awad and Hana Salman
Tishreen University, Lattakia, Syria

18.1 INTRODUCTION

In the past two decades, the potato industry has experienced rapid growth worldwide, accompanied by a staggering increase in the amount of water produced. It is estimated that the US potato industry alone generates about 1.3×10^9 kg of wastes each year [1]. Large volumes of wastewater and organic wastes are generated in potato processing as result of the water used in washing, peeling, and additional processing operations.

The potato industry is well known for the vast quantities of organic wastes it generates. Treatment of industrial effluents to remove organic materials, however, often changes many other harmful waste characteristics. Proper treatment of potato processing wastewaters is necessary to minimize their undesirable impact on the environment.

Currently, there is an increasing demand for quality improvement of water resources in parallel with the demand for better finished products. These requirements have obliged the potato industry to develop methods for providing effective removal of settleable and dissolved solids from potato processing wastewater, in order to meet national water quality limits. In addition, improvement and research have been devoted to the reduction of wastes and utilization of recovered wastes as byproducts.

This chapter discusses (a) the various potato processing types and steps including their sources of wastewaters; (b) characteristics of these wastewaters; (c) treatment methods in detail with relevant case studies and some design examples; and (d) byproduct usage.

18.2 POTATO PROCESSING AND SOURCES OF WASTEWATER

High-quality raw potatoes are important to potato processing. Potato quality affects the final product and the amount of waste produced. Generally, potatoes with high solid content, low reducing sugar content, thin peel, and of uniform shape and size are desirable for processing. Potatoes contain approximately 18% starch, 1% cellulose, and 81% water, which contains dissolved organic compounds such as protein and carbohydrate [2]. Harvesting is an important operation for maintaining a low level of injury to the tubers. Improved harvesting machinery reduces losses and waste load.

The type of processing unit depends upon the product selection, for example, potato chips, frozen French fries and other frozen food, dehydrated mashed potatoes, dehydrated diced potatoes, potato flake, potato starch, potato flour, canned white potatoes, prepeeled potatoes, and so on. The major processes in all products are storage, washing, peeling, trimming, slicing, blanching, cooking, drying, etc.

18.2.1 Major Processing Steps

Storage

Storage is needed to provide a constant supply of tubers to the processing lines during the operating season. Potato quality may deteriorate in storage, unless adequate conditions are maintained. The major problems associated with storage are sprout growth, reducing sugar accumulation, and rotting. Reduction in starch content, specific gravity, and weight may also occur. Handling and storage of the raw potatoes prior to processing are major factors in maintaining high-quality potatoes and reducing losses and waste loads during processing.

Washing

Raw potatoes must be washed thoroughly to remove sand and dirt prior to processing. Sand and dirt carried over into the peeling operation can damage or greatly reduce the service life of the peeling equipment. Water consumption for fluming and washing varies considerably from plant to plant. Flow rates vary from 1300 to 2100 gallons per ton of potatoes. Depending upon the amount of dirt on the incoming potatoes, wastewater may contain 100–400 lb of solids per ton of potatoes. For the most part, organic degradable substances are in dissolved or finely dispersed form, and amount to 2–6 lb of BOD_5 (biological oxygen demand) per ton of potatoes [3].

Peeling

Peeling of potatoes contributes the major portion of the organic load in potato processing waste. Three different peeling methods are used: abrasion peeling, steam peeling, and lye peeling. Small plants generally favor batch-type operation due to its greater flexibility. Large plants use continuous peelers, which are more efficient than batch-type peelers, but have high capital costs [4].

Abrasion peeling is used in particular in potato chip plants where complete removal of the skin is not essential. High peeling losses, possibly as high as 25–30% may be necessary to produce a satisfactory product.

Steam peeling yields thoroughly clean potatoes. The entire surface of the tuber is treated, and size and shape are not important factors as in abrasion peeling. The potatoes are subjected to high-pressure steam for a short period of time in a pressure vessel. Pressure generally varies from 3 to 8 atmospheres and the exposure time is between 30 and 90 seconds. While the potatoes are under pressure, the surface tissue is hydrated and cooked so that the peel is softened and loosened from the underlying tissue. After the tubers are discharged from the pressure vessel, the softened tissue is removed by brushers and water sprays [4]. Screens usually remove the peelings and solids before the wastewater is treated.

Lye peeling appears to be the most popular peeling method used today. The combined effect of chemical attack and thermal shock softens and loosens the skin, blemishes, and eyes so that they can be removed by brushes and water sprays. Lye peeling wastewater, however, is the most troublesome potato waste. Because of the lye, the wastewater pH is very high, usually between 11 and 12. Most of the solids are colloidal, and the organic content is generally higher than for the other methods. The temperature, usually from 50 to 55°C, results in a high dissolved starch content, and the wastewater has a tendency to foam.

The quality of the peeling waste varies according to the kind of potato processing product, peeling requirements, and methods. Table 1 represents the difference in waste quality among the peeling methods in potato processing plants.

18.2.2 Types of Processed Potatoes

Potato Chips

The processing of potatoes to potato chips essentially involves the slicing of peeled potatoes, washing the slices in cool water, rinsing, partially drying, and frying them in fat or oil. White-skinned potatoes with high specific gravity and low reducing sugar content are desirable for high-quality chips. A flow sheet of the process is shown in Fig. 1 [3].

Frozen French Fries

For frozen French fries and other frozen potato production, large potatoes of high specific gravity and low reducing sugar content are most desirable. After washing, the potatoes are peeled by the steam or lye method. Peeling and trimming losses vary with potato quality and are in the range 15–40%. After cutting and sorting, the strips are usually water blanched. Because the blanching water is relatively warm, its leaching effect may result in high dissolved starch content in the wastewater. Surface moisture from the blanching step is removed by hot air

Table 1 Wastewater Quality in the Different Applied Peeling Methods in Potato Processing Plants

	Potato peeling method		
Parameters	Abrasion[a]	Steam[b]	Lye[c]
Flow (gal/ton, raw potato)	600	625	715
BOD	20 lb/ton (4000 ppm)	32.6 lb/ton (6260 ppm)	40 lb/ton (6730 ppm)
COD	–	52.2 lb/ton (10,000 ppm)	65.7 lb/ton (11,000 ppm)
Total solids	–	53.2 lb/ton (10,200 ppm)	118.7 lb/ton (20,000 ppm)
Volatile solids	–	46.8 lb/ton (9000ppm)	56.4 lb/ton (9500 ppm)
Suspended solids	90 lb/ton (18,000 ppm)	26.8 lb/ton (5150 ppm)	49.7 lb/ton (8350 ppm)
pH	–	5.3	12.6

[a] Waste quality in a dehydration plant [5].
[b] Waste quality in a potato flour plant [6].
[c] Waste quality in a potato flake plant [6].
Source: Refs 5 and 6.

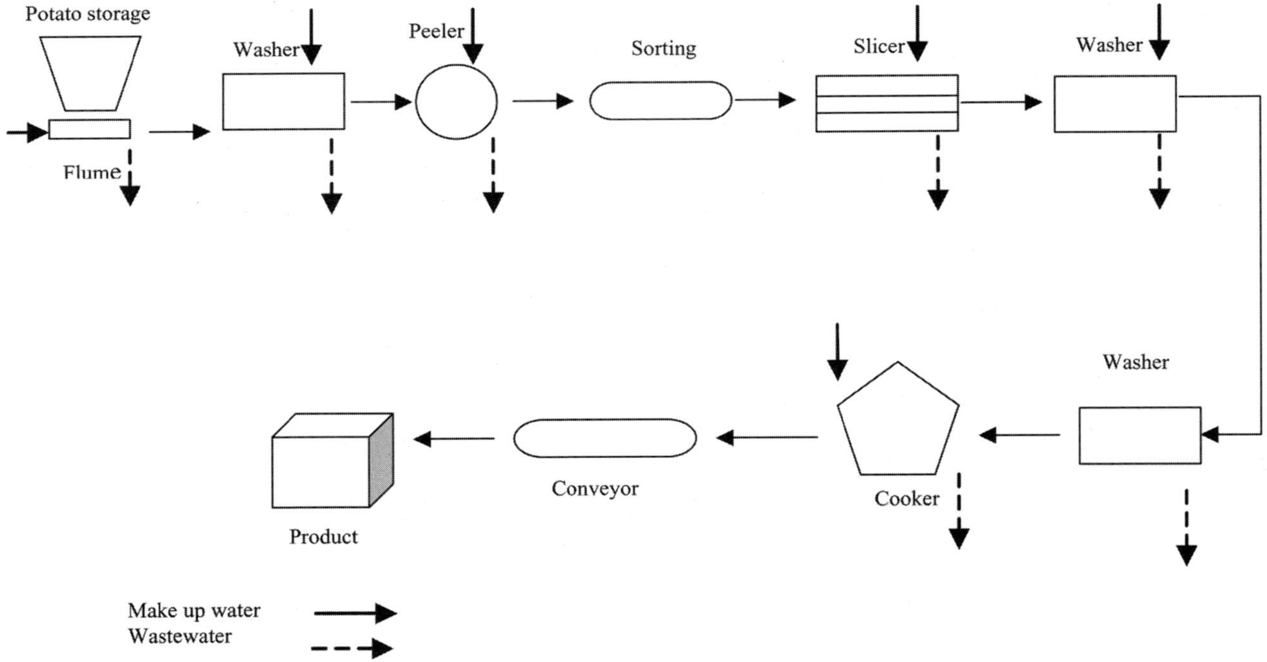

Figure 1 Typical potato chip plant (from Ref. 3).

prior to frying. After frying, the free fat is removed on a shaker screen and by hot air stream. The fries are then frozen and packed. Figure 2 is a flow diagram of the French fry process [3].

Dehydrated Diced Potato

Potatoes with white flesh color and low reducing sugar content are desirable for dice production. After washing and preliminary sorting, the potatoes are peeled by the steam or lye method. Minimum losses amount to 10%. One important factor during trimming is minimizing the exposure time. The tubers are cut into different sized pieces. After cutting and washing, the dice are blanched with water or steamed at 200–212°F. Following blanching, a carefully applied rinsing spray removes surface gelatinized starch to prevent sticking during dehydration. Sulfite is usually applied at this point as a spray solution of sodium sulfite, sodium bisulfite, or sodium metabisulfite. Calcium chloride is often added concurrently with sodium bisulfite or sodium metabisulfite. Following drying, the diced potatoes are screened to remove small pieces and bring the product within size specification limits. Finally, the potatoes are packed in cans or bags [3].

Dehydrated Mashed Potatoes: Potato Granules

Potato granules are dehydrated single cells or aggregated cells of the potato tuber that are dried to about 6–7% moisture content. A flow diagram of the potato granules is shown in Fig. 3. After peeling and trimming, the potatoes are sliced to obtain more uniform cooking. The slices are cooked in steam at atmospheric pressure for about 30–40 minutes. After cooking is completed, the slices are mixed with the dry add-back granules and mashed to produce a moist mix. This mix is cooled and conditioned by holding for about one hour before further mixing and then dried to about 12–13% moisture content [3,4].

Potato Flakes

Potato flakes are a form of dehydrated mashed potatoes that have been dried on a steam-heated roll as a thin sheet and then broken into small pieces for packaging. Potatoes for flake processing have the same characteristics as those for potato granule processing. A flow diagram of the process is shown in Fig. 4. After prewashing, the potatoes are lye or steam peeled. Following trimming, the tubers are sliced into 0.25–0.50 in. slices and washed prior to precooking in water at 160–170°F for about 20 minutes [6]. After cooking, the potatoes are mashed and then dried on a single drum drier in the form of a sheet. The sheet is broken into flakes of a convenient size for packaging.

Potato Starch

Potato starch is a superior product for most of the applications for which starch is used. Figure 5 shows a flow diagram of a typical starch plant. After fluming and washing, the potatoes are fed to a grinder or hammer mill and disintegrated to slurry, which is passed over a screen to separate the freed starch from the pulp. The pulp is passed to a second grinder and screened for further recovery of starch. The starch slurry, which is passed through the screen, is fed to a continuous centrifuge to remove protein water, which contains soluble parts extracted from the potato. Process water is added to the starch, and the slurry is passed over another screen for further removal of pulp. Settling vats in series are used to remove remaining fine fibers. The pure starch settles to the bottom while a layer of impurities (brown starch) forms at the top. The latter is removed to the starch table consisting of a number of settling troughs for final removal of white starch. The white starch from the settling tanks and the starch table is dried by filtration or centrifugation to a moisture content of about 40%. Drying is completed in a series of cyclone driers using hot air [3].

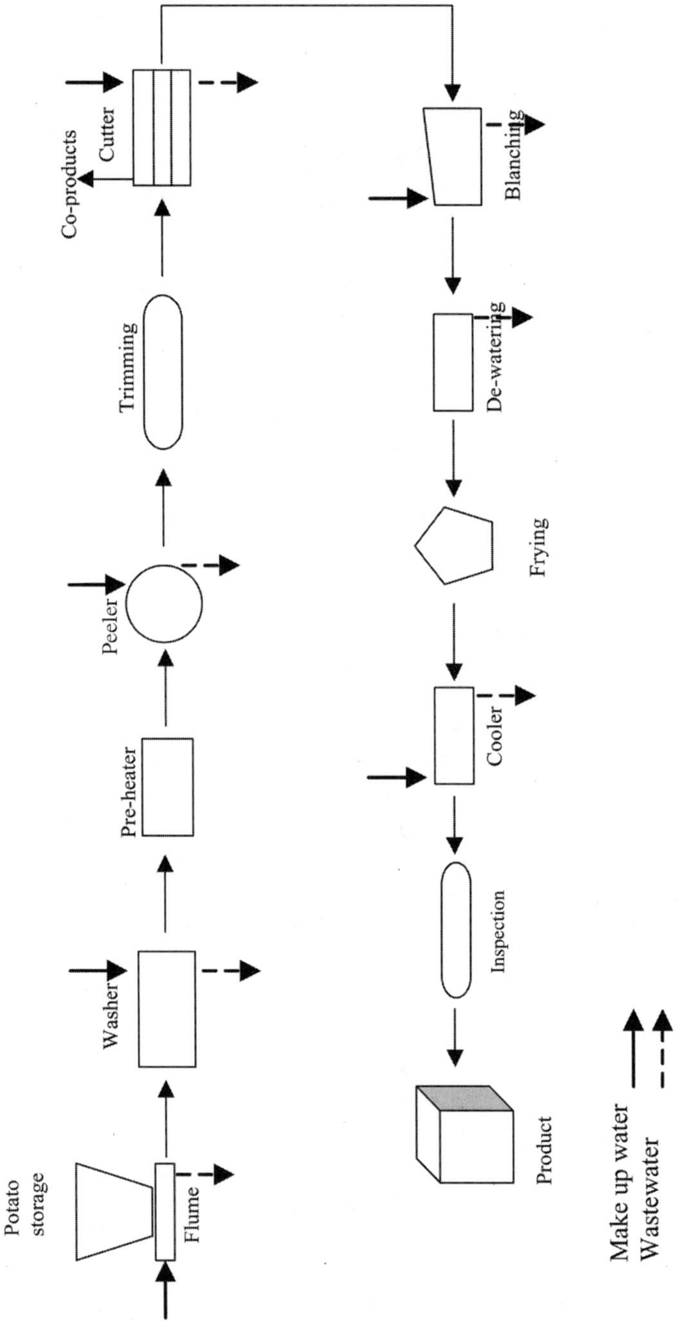

Figure 2 Typical French fry plant (from Ref. 3).

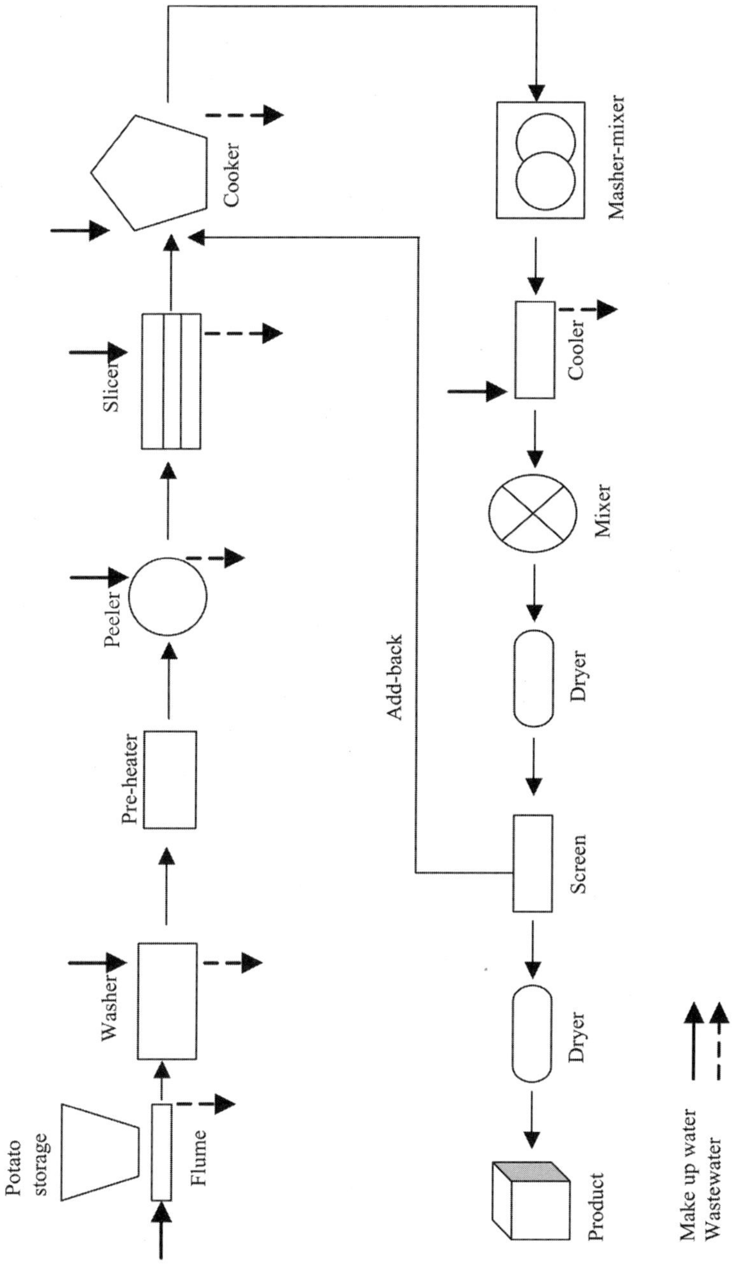

Figure 3 Typical potato granule plant (from Ref. 3).

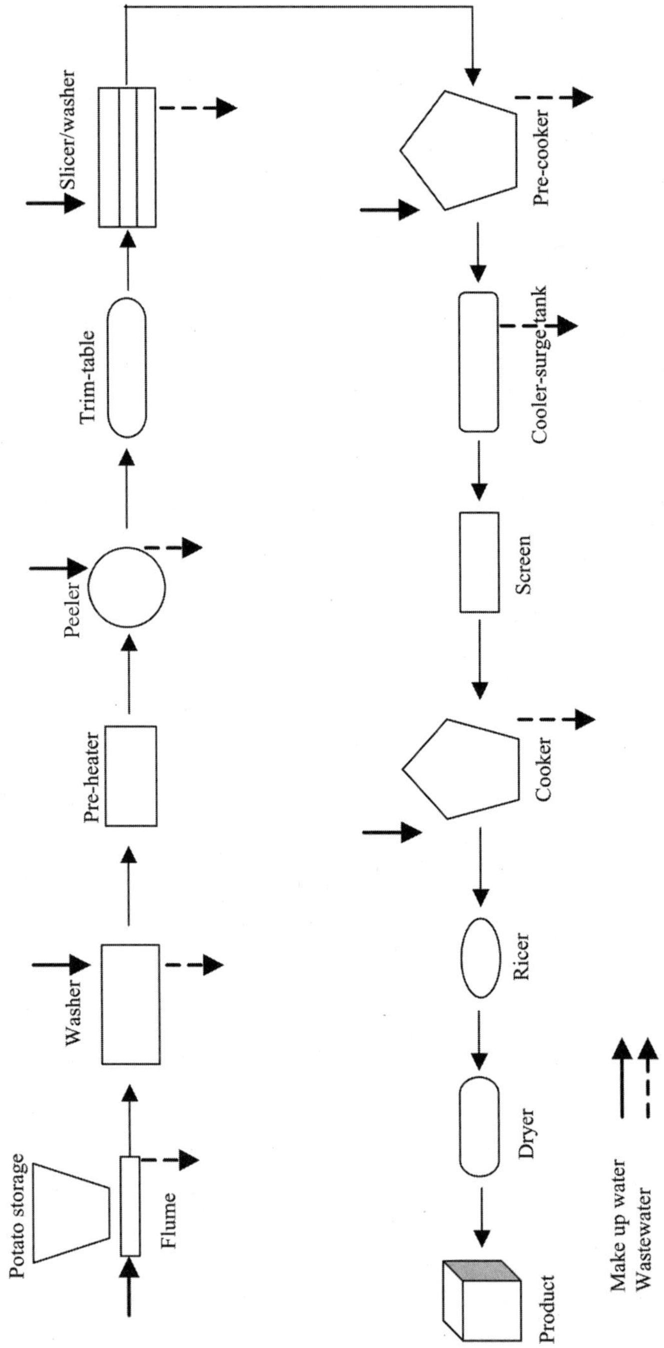

Figure 4 Typical potato flake plant (from Ref. 3).

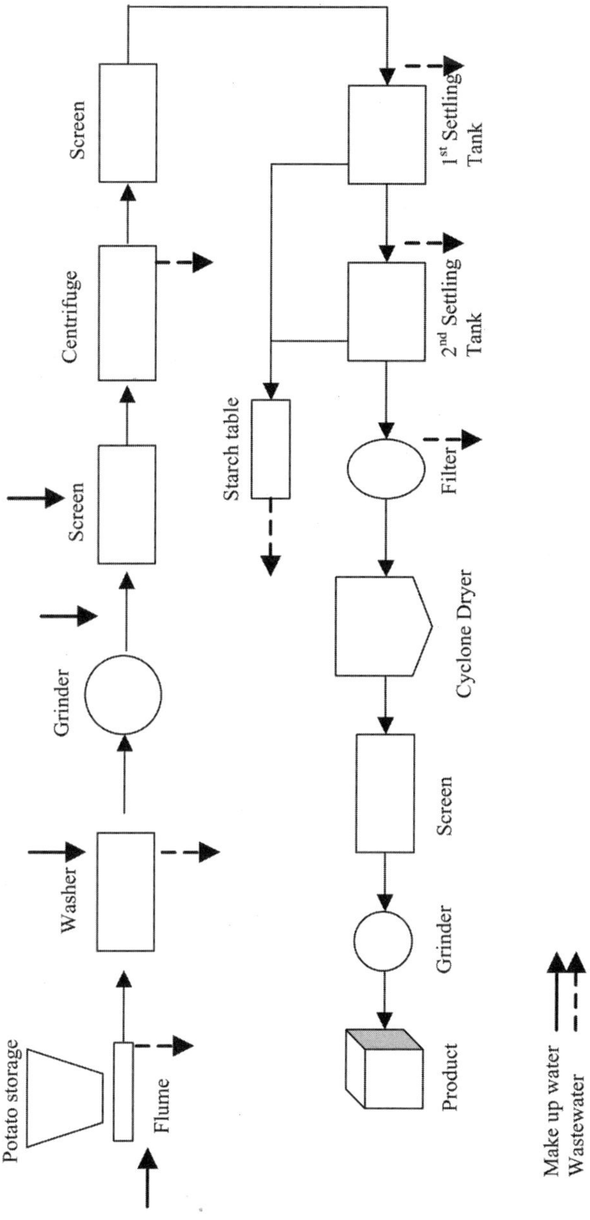

Figure 5 Typical potato starch plant (from Ref. 3).

Potato Flour

Potato flour is the oldest commercial processed potato product. Although widely used in the baking industry, production growth rates have not kept pace with most other potato products. A flow diagram of the process is shown in Fig. 6. After the prewash, the potatoes are peeled, usually with steam. Trimming requirements are not as high as for most potato products. The flaking operation requires well-cooked potatoes; the tubers are conveyed directly from the cooker to the dryer, where 4–5 applicator rolls along one side of the drum contribute a thin layer of potato mesh. The mesh is rapidly dried and scraped off the drum at the opposite side by a doctor knife. The dried sheets are passed to the milling system where they are crushed by a beater or hammer mill and then screened to separate granular and fine flour [3].

Besides the above products, other types include canned potatoes, prepeeled potatoes, and even alcohol. The quantities and qualities of the wastewaters resulting from the mentioned potato processing plants are discussed in the next section.

18.3 CHARACTERISTICS OF POTATO PROCESSING WASTEWATER

18.3.1 Overview

Because potato processing wastewater contains high concentrations of biodegradable components such as starch and proteins [7,8], in addition to high concentrations of chemical oxygen demand (COD), total suspended solids (TSS) and total kjeldahl nitrogen (TKN) [9], the potato processing industry presents potentially serious water pollution problems. An average-sized potato processing plant producing French fries and dehydrated potatoes can create a waste load equivalent to that of a city of 200,000 people. About 230 million liters of water are required to process 13,600 tons of potatoes. This equals about 17 L of waste for every kilogram of potatoes produced. Raw potato processing wastewaters can contain up to 10,000 mg/L COD. Total suspended solids and volatile suspended solids can also reach 9700 and 9500 mg/L, respectively [10]. Wastewater composition from potato processing plant depends on the processing method, to a large extent. In general, the following steps are applied in potato processing: washing the raw potatoes; peeling, which includes washing to remove softened tissue; trimming to remove defective portions; shaping, washing, and separation; heat treatment (optional); final processing or preservation; and packaging.

The potato composition used in potato processing operations determines the components of the resultant waste stream. Foreign components that may accompany the potato include dirt, caustic, fat, cleaning and preserving chemicals. A typical analysis of potato waste solids from a plant employing steam or abrasive peeling is shown in Table 2. Generally, the various waste streams are discharged from the potato plant after being combined as effluent. It is difficult to generalize the quantities of wastewater produced by specific operations, due to the variation in process methods. Many references and studies in this respect show wide variations in water usage, peeling losses, and methods of reporting the waste flow. Several publications on the characteristics of wastewaters resulting from various types of potato processing are summarized in Table 3 for French fries [11,12], Table 4 for starch plants [12], and Table 5 for the other types of potato processing plants (chips, flakes, flour, mashed) [13–18].

Processing involving several heat treatment steps such as blanching, cooking, caustic, and steam peeling, produces an effluent containing gelatinized starch and coagulated proteins. In contrast, potato chip processing and starch processing produce effluents that have unheated components [11].

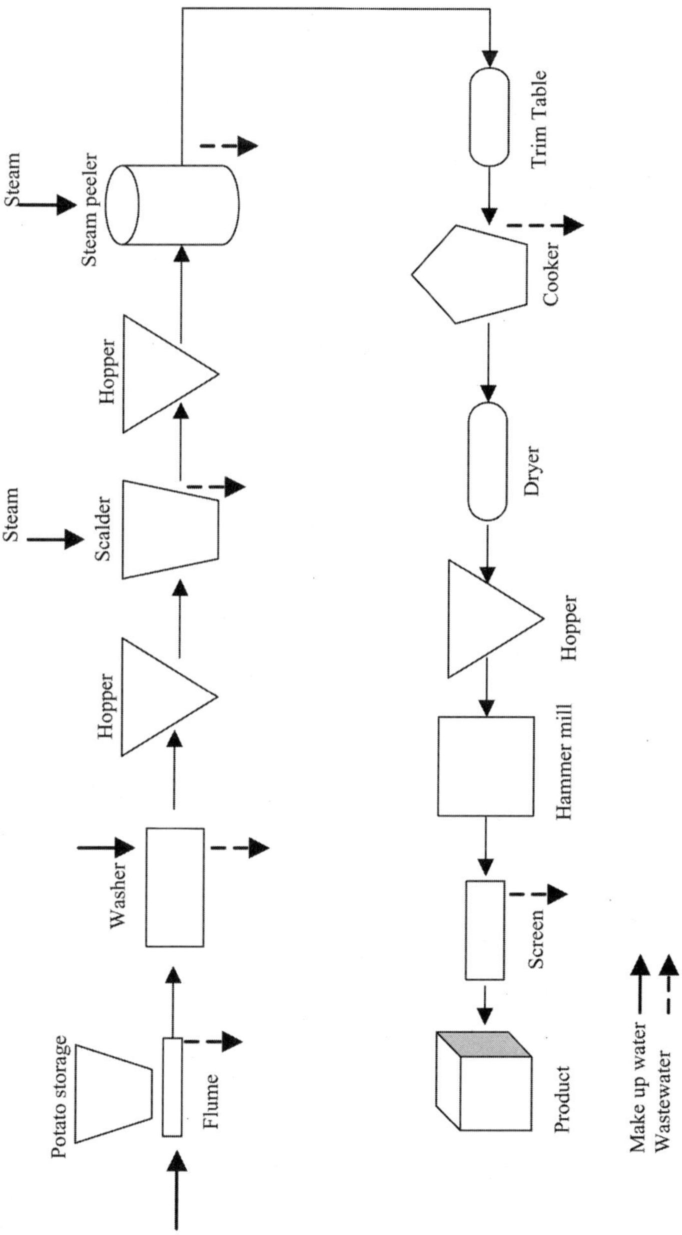

Figure 6 Typical potato flour plant (from Ref. 3).

Table 2 Composition Percentage of Potato Waste Solids

Component	Amount (%)
Total organic nitrogen as N	1.002
Carbon as C	42.200
Total phosphorus as P	0.038
Total sulfur as S	0.082
Volatile solid	95.2

Source: Ref. 11.

As for the starch plant effluent, the resulting protein water and pulp form about 95% of the total organic load in the effluent. Table 4 represents the composition of waste streams of starch plants and summarizes a survey of five starch plants in Idaho/United States, with and without pulp.

It is evident that if the pulp is kept and not wasted, the organic load is significantly reduced. Potato pulp has been proven to be a valuable feed for livestock when mixed with other ingredients and thus represents a valuable by-product [19]. Protein water is difficult to treat because of the high content of soluble organic water [3].

In plants of joint production of starch and alcohol found in some countries, the pulp and protein water from the starch production is used for alcohol fermentation. As for the wastewater streams in French fries plants, it can be noted from Table 3 that the spray washer forms the main organic load (BOD and COD) in comparison to other waste streams. The large variations in wastewater composition can be observed in the potato processing plants as presented in Table 5, particularly in COD and TSS concentrations and pH values.

Depending on the abovementioned characteristics of potato processing wastewater, the following should be highly considered:

- Potential methods for reducing the load of waste production including in-plant measures for water conservation, byproduct recovery, and water recycling.
- Choosing the wastewater treatment systems that take into account the wide variations of wastewater compositions, due to wide variation in potato processing steps and methods, in order to reduce the wastewater contaminants for meeting in-plant reuse or the more stringent effluent quality standards required in the potato processing industry.

18.3.2 Case Study [20]

J.R. Simplot Company, an international agribusiness company, operated a potato processing plant in Grand Forks, North Dakota, United States. The company's frozen potato product line, which was produced locally in Grand Forks, consists of more than 120 varieties of French fries and formed products. In all, J.R. Simplot produced more than 2 billion pounds of French fries annually, making it one of the largest processors of frozen potatoes. Its local plant in Grand Forks employed nearly 500 people.

Sources of Wastewater [20]

The main sources of wastewaters consist of silt water and process wastewater. The silt waste resulted from raw potato washing and fluming operations. It contained a large amount of soil removed from the raw potatoes. Process wastewater results from potato processing operations including peeling, cutting, blanching, and packing. The process wastewater included caustic

Table 3 Characteristics of Wastewater from French Fries Plants

Parameters	French fries						French fries and starch plant					
	Spray washer	Trimming	Cutting	Inspection	Blanch	Plant composite	Caustic peel	Wash water	Peel waste	Trim table	Blanch waste	Plant effluent
COD (mg/L)	2830	45	150	32	1470	1790	–	100–250	10,000–12,000	150–200	600–700	6450
BOD (mg/L)	1950	30	77	5	1020	1150	4300	–	–	–	–	4100
Total solids (mg/L)	14,900	270	880	260	2283	8100	11,550	700	10,000–15,000	600	1600	7794
Suspended solids (mg/L)	2470	7	16	15	60	1310	–	–	–	–	–	4050
Settleable solids (mg/L)	–	–	–	–	–	–	–	2.0–5.5	200–400	0.6	2–3	–
Total nitrogen (mg/L)	60	–	–	–	–	20	–	–	–	–	–	224
Total phosphorus (mg/L)	81	27	29	14	160	80	–	–	–	–	–	23
pH	11.5	6.9	7.2	6.9	4.7	11.1	–	7.0	–	6.2	5.1	10.7

Source: Refs. 11 and 12.

Table 4 Characteristics of Wastewater from Starch Plants

Type of waste	Plant capacity (tons/day)	Flow rate (gal/ton)	BOD		COD		Solid content (%wt)	Protein in solid (%wt)
			mg/L	lb/ton	mg/L	lb/ton		
Waste stream								
Flume water	–	1740[a]	100	0.4	260	1.5	–	–
Protein water	–	670	5400	30.1	7090	40.3	1.7	38.5
First starch washwater	–	155	1680	2.2	2920	3.3	0.46	31.1
Second starch washwater	–	135	360	0.4	670	0.8	–	–
Brown starch water	–	30	640	0.2	1520	0.4	0.81	–
Starch water	–	25	150	0.0	290	0.0	–	–
Pulp (dry basis)[b]	–	–	–	24.8	–	56.8	–	–
Total organic load without pulp								
Plant I	200	–	–	45.3	–	–	–	–
Plant II	250	–	–	27.7	–	–	–	–
Plant III	150	–	–	26.2	–	–	–	–
Plant IV	62.5	–	–	31.7	–	–	–	–
Plant V	180	–	–	35.0	–	–	–	–
Average				33.3				
Total organic load with pulp								
Plant I	200	–	–	70.1	–	–	–	–
Plant II	250	–	–	52.5	–	–	–	–
Plant III	150	–	–	51.0	–	–	–	–
Plant IV	62.5	–	–	56.5	–	–	–	–
Plant V	180	–	–	59.8	–	–	–	–
Average				58.1				

[a]No recirculation.
[b]An average of 55.5 lb of pulp (on dry basis) were produced per ton of potatoes processed.
Source: Ref. 12.

Table 5 Characteristics of Wastewater from Different Potato Processing Plants

Parameters	Wastewater after settling (Austerman-Haun, *et al.* 1999)[13]	Wastewater after screening and presettlement (Zoutberg and Eker, 1999)[14]			Wastewater from potato chips plant (Hadjivassilis, et al. 1997)[8]	(Kadlec, et al. 1997)[15]	Wastewater influent (Hung, 1989)[16]	
		Smith food	Peka Kroef	Uzay Gida			Wastewater from potato juice	Wastewater from mashed potato
Total daily flow (m^3/day)	1700	912	1600	890	115	–	–	–
Hourly peak flow (m^3/hour)	–	(38 av.)	90 (67 av.)	(37 av.)	15	–	–	–
COD (mg/L)	4000	5000	7500	4500	7293	1100–3100	2546	1626
BOD (mg/L)	–	–	–	–	5450	–	–	–
Total suspended solids (mg/L)	–	–	–	–	1300	280–420	18,107	33,930
VSS	–	–	–	–	–	–	–	–
Total TKN (mg/L)	120	286 (max. 400)	50–200	20–70	–	95–145	–	–
Total P (mg/L)	60	–	10–50 (PO_4-P)	2–10 (PO_4-P)	–	10–15	–	–
pH	6.6 (adjusted)	4.5–7.5	4.5 (after buffering)	5–9	4–10	–	7.6	7.3

(continues)

Table 5 Continued

Parameters	Wastewater from potato starch	Primary settling tank effluents (Hung, 1984)[17]	Potato chips (slicing and washing) (Cooley et al. 1964)	Potato flakes (slicing, washing, precooking and cooling) (Cooley et al. 1964)[6]	Potato flour (raw screened waste) (Cooley et al., 1964)[6]	Potato flour (raw screened waste) (Olson et al., 1965)[18]
Total daily flow (m^3/day)	–	–	1140 gal/ton (4.3 m^3/t)	1540 gal/ton (5.8 m^3/t)	–	–
Hourly peak flow (m^3/hour)	–	–	–	–	–	–
COD (mg/L)	1270	2500	7953	4373	12,582	8314
BOD (mg/L)	–	–	2307	2988	7420	3314
Total suspended solids (mg/L)	62,444	500	5655	1276	6862	4398
VSS	–	450	6685	4147	6480	3019
Total TKN (mg/L)	–	–	–	–	–	–
Total P (mg/L)	–	–	–	–	–	–
pH	7.8	6.7	7.4	5.2	4.2	6.9

Source: Refs. 6, 8, 13–18.

potato peeler and barrel washer discharges, as well as all other liquid wastes from the processing operations, including cleanup water.

Characteristics of Wastewater [20]

The characteristics of the potato processing wastewater were influenced by potato processing operations. Potato peeling was the first stage of potato processing. Caustic soda was used to soften the potato skin so that it can be removed by the scrubbing and spraying action of the polisher. The liquid effluent from the polisher, which contained a majority of the contaminants of wastewater, accounted for about 75% of the alkalinity of the wastewater from the plant. It was also high in COD and BOD, with values of about 2000 and 1000 mg/L, respectively. The TDS (total dissolved solids) and TSS (total suspended solids) were about 29,000 and 4100 mg/L, respectively.

Polished potatoes were then conveyed to the cutter. The degree of size reduction depended upon the requirements of the final product. Here the surface of the potato and the amount of water used for washing determine the quantity of soluble constituent in the waste stream. The pH of the stream was about 7. The COD and BOD values were about 50% of those of the effluent from the polisher. The TDS and TSS were approximately 1390 and 460 mg/L, respectively. The blanching process removed reducing sugar, inorganic salts, gelatinized starch, and smaller amounts of protein and amino acids. The effluent stream from this operation had pH 6.2, total dissolved solids 1500 mg/L, phenols 8.2 mg/L, COD 1000 mg/L, and BOD 800 mg/L, respectively.

The wastewater treatment processes used in the plant included shaker, primary settling tank, aerated lagoon, and final settling tank. The effluent from the final settling tank was discharged to the municipal sewer and was transported to Grand Forks Municipal Wastewater Treatment Plant, Grand Forks, North Dakota, for treatment. A portion of the final settling tank effluent was treated by tertiary sand filter. The filtered water was reused inside the plant.

During the period of September 1978 to March 1979, primary effluent had an average concentration of 4250 mg/L COD and 3000 mg/L TSS. After primary settling tank treatment, the effluent had an average concentration of 2500 mg/L COD and 500 mg/L TSS. After the aerated lagoon and final settling tank treatment, the effluent had an average concentration of 410 mg/L COD and 350 mg/L TSS and pH 7.55. The aerated lagoon had 4900 mg/L MLSS (mixed liquor suspended solids) and 4100 mg/L MLVSS (mixed liquor volatile suspended solids). The onsite treatment plant removed 90.35% COD and 88.33% TSS.

18.4 TREATMENT METHODS

Wastewater from fruit and vegetable processing plants contains mainly carbohydrates such as starches, sugars, pectin, as well as vitamins and other components of the cell wall. About 75% of the total organic matter is soluble; therefore, it cannot be removed by mechanical or physical means. Thus, biological and chemical oxidations are the preferred means for wastewater treatment [21,22].

In the United States, there are three geographical areas of major potato processing activity: (a) Idaho, eastern Oregon, and eastern Washington; (b) North Dakota and Minnesota; and (c) Maine. Most plants are located in sparsely populated areas where the waste load from the plants is extremely large compared to the domestic sewage load [11]. By contrast, potato chips and prepeeled potato plants, while expanding in number and size, are largely located near metropolitan areas, where the waste effluent is more easily handled by municipal facilities. In general, these plants are much smaller than French fry or dehydrated potato plants and produce less waste load.

18.4.1 Waste Treatment Processes

An integrated waste treatment system usually consists of three phases: primary treatment, secondary treatment, and advanced treatment. Primary treatment involves the removal of suspended and settleable solids by screening, flotation, and sedimentation. Secondary treatment involves the biological decomposition of the organic matter, largely dissolved, that remains in the flow stream after treatment by primary processes. Biological treatment can be accomplished by mechanical processes or by natural processes.

The flow from the biological units is then passed through secondary sedimentation units so that the biological solids formed in the oxidation unit may be removed prior to the final discharge of the treated effluent to a stream. When irrigation is used as the secondary treatment system, bacteria in the topsoil stabilize the organic compounds. In addition, the soil may accomplish removal of some ions by adsorption or ion exchange, although ion exchange in some soils may fail. In all cases, great importance should be given to the steps that contribute to reducing the waste load in the plant itself. As for the industrial wastewaters, most of them require equalization (buffering) and neutralization prior to biological treatment, according to the characteristics of the resultant effluents.

In many parts of the world, potato processing wastewater treatment systems employed primary treatment from 1950 until 1970 to 1980. Thereafter, potato processing plants involved either secondary treatment or spray irrigation systems. Currently the most commonly used treatment methods, particularly in the United States, depend on screening, primary treatment, and settling of silt water in earthen ponds before discharging to municipal sewers or separate secondary treatment systems.

Many countries that have potato processing industries have determined current national minimum discharge limits following secondary treatment or in-land disposal. For example, the US Environmental Protection Agency (EPA) has proposed nationwide such limits for potato processing effluents [12].

To meet national effluent limits or standards, advanced waste treatment is needed in many cases to remove pollutants that are not removed by conventional secondary treatment. Advanced treatment can include removal of nutrients, suspended solids, and organic and inorganic materials. The unit processes for treating potato processing effluent are shown in sequence in Table 6.

Figure 7 illustrates a general treatment concept typical for the treatment of potato processing effluent: advanced treatment is added as a result of the growing environmental requirements. Currently, different treatment units are combined as a highly effective system for the secondary (biological) treatment that covers both anaerobic and aerobic processes. Note that it is quite acceptable and applicable that wastewater after preclarification (screening and primary treatment) can be discharged into the public sewer system to be treated together with sewage water in the municipal treatment plants.

The following describes in detail the current wastewater treatment units and subsystems.

In-Plant Treatment

Minimizing waste disposal problems requires reduction of solids discharged into the waste stream and reduction of water used in processing and clean-up. To reduce the solids carried to waste streams, the following steps should be undertaken [11]:

- improvement of peeling operation to produce cleaner potatoes with less solids loss;
- reduction of floor spillage;

Table 6 Treatment Units, Unit Operation, Unit Processes, and Systems for Potato Processing Wastewater

Treatment unit or subsystem	Unit operation/unit process/ treatment system	Remarks
In-plant	• Conservation and reuse of water • Process revisions • Process control • New products	• Reduction of waste flow and load
Pretreatment	• Screening (mesh size: 20 to 40 per inch)	• 10–25% BOD_5 removal
Primary treatment	• Sedimentation • Flotation • Earthen ponds	• 30–60% BOD_5 removal • 20–60% COD removal
Equalization	• Balancing tank/buffer tank	• Constant flow and concentration
Neutralization	• Conditioning tank	• pH and temperature corrections
Secondary treatment		• 80–90% BOD_5 removal
1. Aerobic processes	• Natural systems – Irrigation land treatment – Stabilization ponds and aerated lagoons – Wetland systems • Activated sludge • Rotating biological contactors • Trickling filters	• 70–80% COD removal
2. Anaerobic processes	• Upflow anaerobic sludge blanket (UASB) reactors • Expended granular sludge bed (EGSB) reactors • Anaerobic contact reactors • Anaerobic filters and fluidized-bed reactors	• 80–90% BOD_5 removal • 70–80% COD removal
Advanced treatment	• Microstraining • Granular media filtration • Chemical coagulation/ sedimentation • Nitrification–denitrification • Air stripping and ion exchanging • Membrane technology (reverse osmosis, ultrafiltration)	• 90–95% BOD_5 removal • 90–95% COD removal (Sometimes >95%)

Notes: BOD_5 and COD removal percentage depended on experience of the German and other developed countries. There are other advanced treatment methods (not mentioned in this table) used for various industrial wastewater such as activated carbon adsorption, deep well injection, and chlorination that are not expected to be highly used in potato processing wastewater treatment.

- collection of floor waste in receptacles instead of washing them down the drains;
- removal of potato solids in wastewater to prevent solubilization of solids.

Water volume can be reduced by reusing process water, with several advantages. First, the size of wastewater treatment facilities can be decreased accordingly. Secondly, with

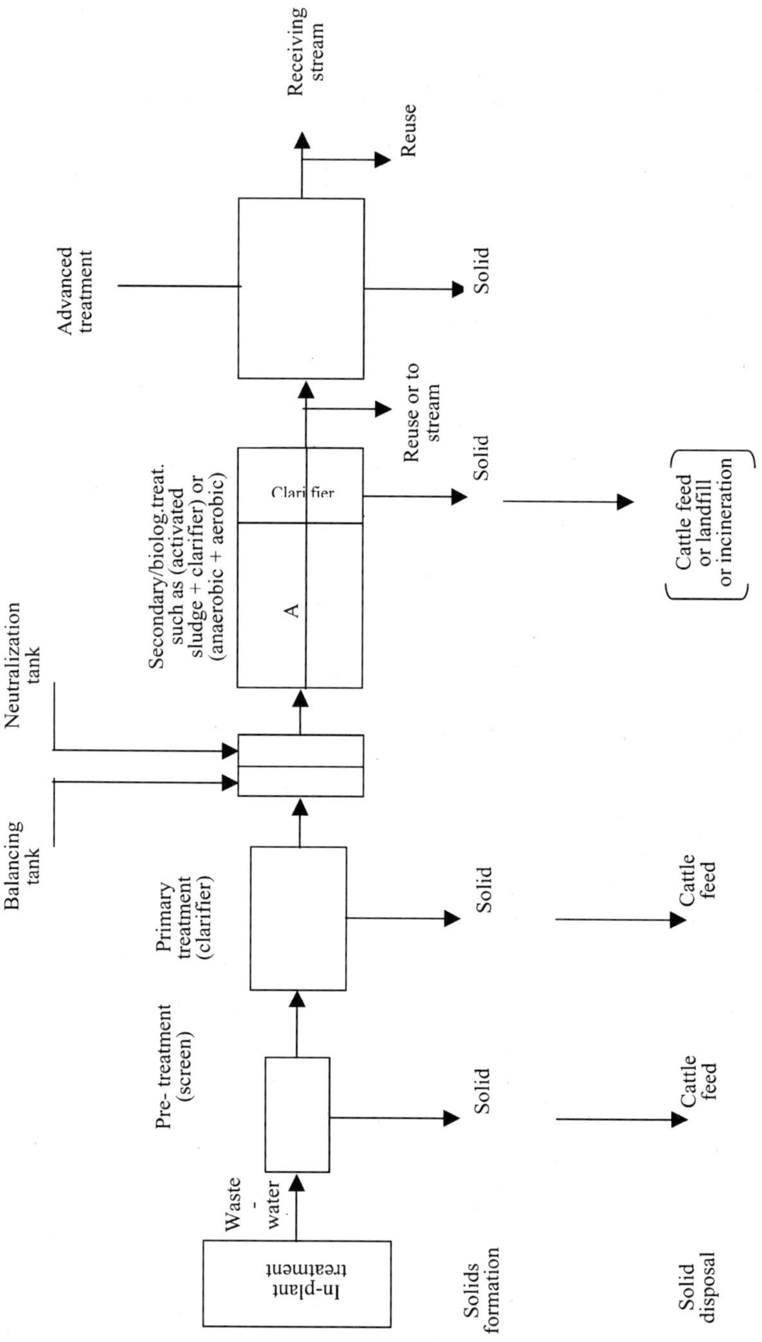

Figure 7 General treatment scheme for potato processing effluent.

concentration of the waste, the efficiency of a primary settling tank is increased. In the final processing stages, chlorinated water should be utilized to prevent bacterial contamination of the product. Other steps to reduce wastewater volume include alternate conveying methods of transporting potatoes other than water fluming, improved cleaning facilities for equipment and floors (high-pressure nozzles, shut-off nozzles for hoses), collecting clean waste streams, and discharge to natural drainage or storm water systems.

Pretreatment (Screening)

Typically, the screen is the first device encountered by wastewater entering the treatment plant. Screening is often used to remove large pieces of waste so that the water can be reused within the processing plant. Three types of screens are commonly used: stationary gravity screens, rotary screens, and vibratory screens. These units are similar to screens used in dewatering products during processing. Coarse solids are normally removed in a fine screen with a mesh size of 1 mm. The simplest type of stationary screen consists of a number of bars eventually spaced across the wastewater channel (bar rack). In modern wastewater treatment plants, the racks are cleaned mechanically. Rotary screens are used to a large extent and a variety of types are available. The most common type is the drum screen, which consists of a revolving mesh where wastewater is fed into the middle of the drum, and solids are retained on the peripheral mesh as the water flows outward. Another type of rotary screen is the disc screen, which is a perforated plate of wire mesh disc set at right angles to the waste stream. The retained solids are removed at the top of the disc by brushes or water jets. Vibratory screens may have reciprocating orbital or rocking motion, or a combination of both. The wastewater is fed into the horizontal surface of the screen, and the water passing through the retained solids is bounced across the screen to a discharge point.

The waste screen should be carefully located and elevated. Plant wastewaters can be collected in a sump pit below the floor level of the plant, from which they are pumped to the screen. The screen is elevated so that the solid wastes may fall by gravity into a suitable hopper. Then, the water flows down into the primary treatment equipment or to the sewer. With suitable elevations, the screen can be located below the level of the plant drains. After screening, the solid waste is conveyed up to the waste hopper and the water pumped into the clarifier, or other disposal system.

Primary Treatment

Sedimentation. Sedimentation is employed for the removal of suspended solids from wastewater. After screening, wastewater still carries light organic suspended solids, some of which can be removed from the wastewater by gravity in sedimentation tanks called clarifiers. These tanks/clarifiers can be round or rectangular, are usually about 3.5 m deep, and hold the wastewater for periods of two to three hours [23]. The required geometry, inlet conditions, and outlet conditions for successful operation of such units are already known. The mass of settled solids is called raw sludge, which is removed from the clarifiers by mechanical scrapers and pumps. Floating materials such as oil and grease rise to the surface of the clarifier, where they are collected by a surface skimming system and removed from the tank for further processing.

Figures 8 and 9 show cross-sections of typical rectangular and circular clarifiers. Construction materials and methods vary according to local conditions and costs.

In the primary treatment of potato wastes (Fig. 10), the clarifier is typically designed for an overflow rate of 800–1000 gal/(ft^2-day) (33–41 m^3/m^2-day) and a depth of 10–12 ft (3–3.6 m). Most of the settleable solids are removed from the effluent in the clarifier. The COD

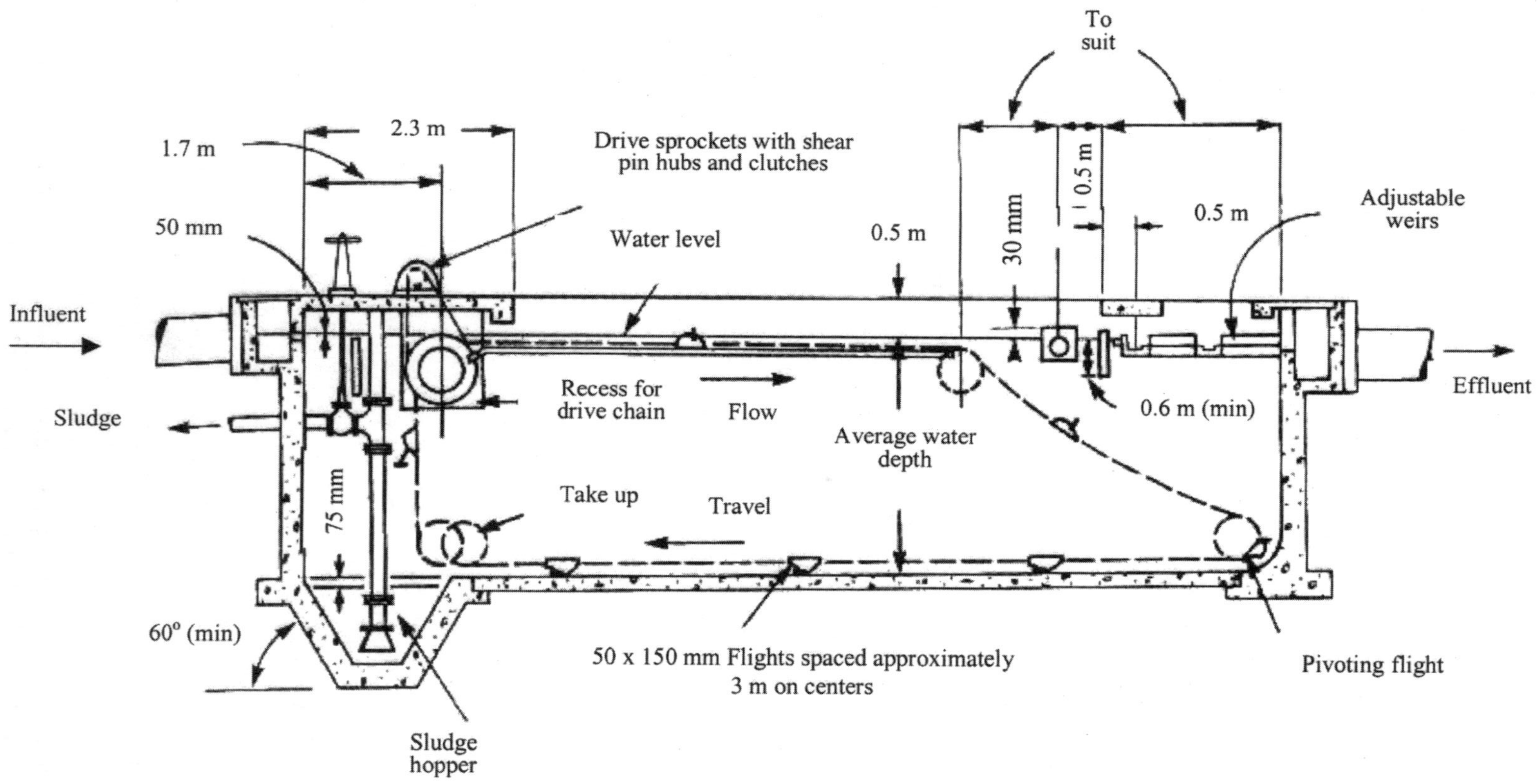

Figure 8 Rectangular primary clarifier.

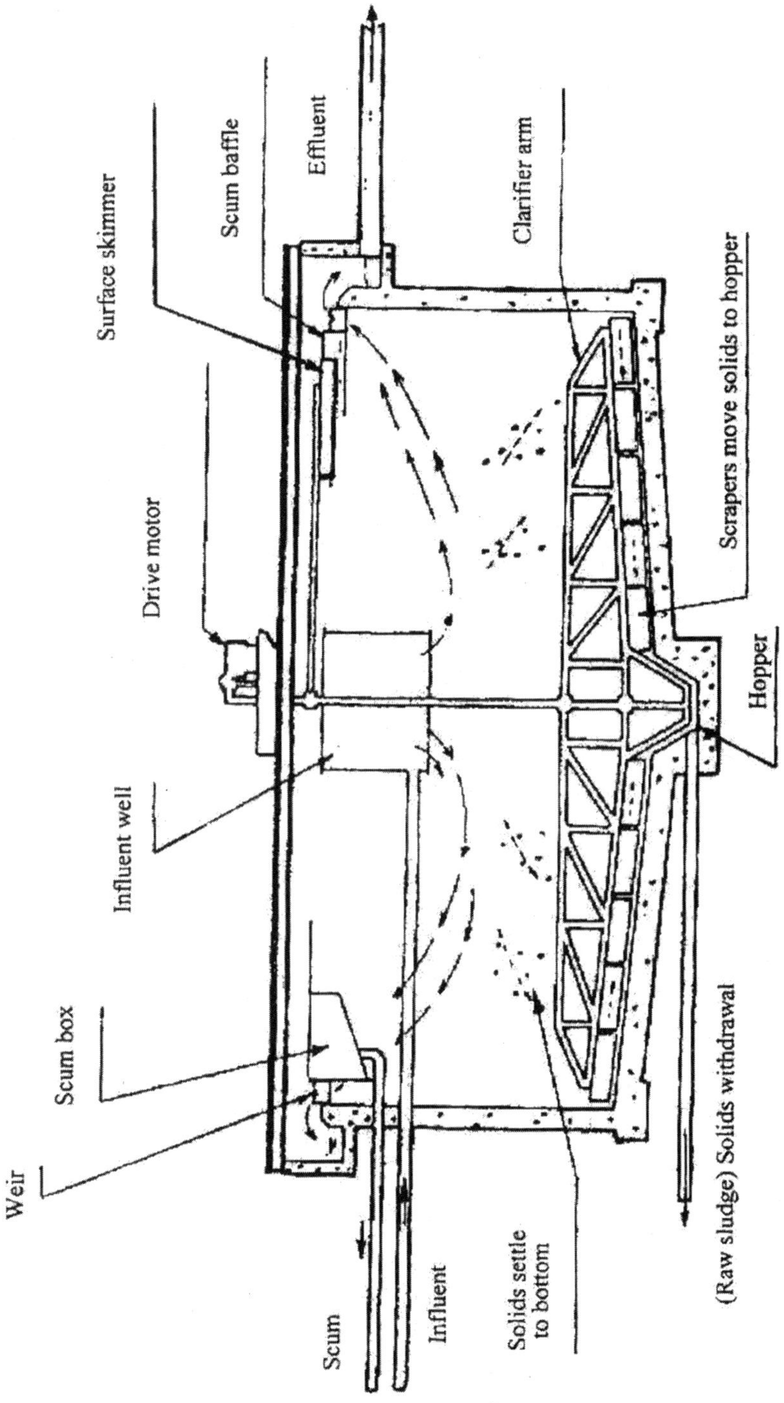

Figure 9 Circular primary clarifier.

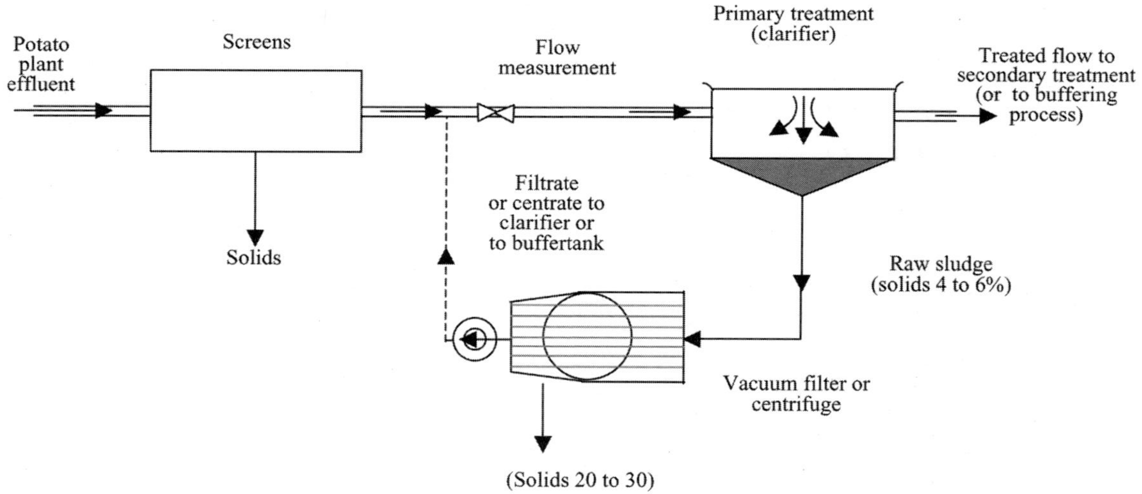

Figure 10 Schematic representation of primary treatment for potato wastes (from Ref. 11).

removal in this primary treatment is generally between 40–70% [11]. In comparison with cornstarch wastes, it was reported that BOD removals of 86.9% were obtained from settling this kind of waste [24].

To reduce the volume of the settled waste, which contains 4–6% solids, vacuum filters or centrifuges are used.

Withdrawal of the underflow from the bottom of the clarifier is accomplished by pumping. The resulting solids from caustic peeling have a high pH. The optimum pH level for best vacuum filtration of solids differs from plant to plant. However, when the underflow withdrawal is adjusted to hold the solids in the clarifier for several hours, biological decomposition begins and the pH of the solids falls greatly. At a pH of between 5 and 7, these solids will dewater on a vacuum filter without the addition of coagulating chemicals.

As for the solids resulting from steam or abrasive peeling operations, these will also undergo biological degradation in a few hours. With a longer duration, however, dewatering of solids becomes more difficult.

Flotation. Flotation is another method used for the removal of suspended solids and oil and grease from wastewater. The pretreated waste flow is pressurized to 50–70 lb/in^2 (345–483 kPa or 3.4–4.8 atm) in the presence of sufficient air to approach saturation [24]. When this pressurized air–liquid mixture is released to atmospheric pressure in the flotation unit, minute air bubbles are released from the solution. The suspended solids or oil globules are floated by these minute air bubbles, which become enmeshed in the floc particles. The air–solids mixture rises to the surface, where it is skimmed off by mechanical collectors. The clarified liquid is removed from the bottom of the flotation unit. A portion of the effluent may be recycled back to the pressure chamber.

The performance of a flotation system depends upon having sufficient air bubbles present to float substantially all of the suspended solids. This performance in terms of effluent quality and solids concentration in the float, is related to an air/solids ratio that is usually defined as mass of air released per mass of solids in the influent waste.

Pressure, recycle ratio, feed solid concentration, and retention period are the basic variables for flotation design. The effluent's suspended solids decrease and the concentration of solids in the float increase with increasing retention period. When the flotation process is used for primary clarification, a detention period of 20–30 min is adequate for separation and concentration. Rise velocity rates of 1.5–4.0 gal/(min-ft^2) [0.061–0.163 m^3/(min-m^2)] are commonly applied [24].

Major components of a flotation system include a pressurizing pump, air-injection facilities, a retention tank, a backpressure regulating device, and a flotation unit, as shown in Fig. 11. The pressurizing pump creates an elevated pressure to increase the solubility of air. Air is usually added through an injector on the suction side of the pump or directly to the retention tank. The air and liquid are mixed under pressure in a retention tank with a detention time of 1 to 3 min. A backpressure regulating device maintains a constant head on the pressurizing pump.

Equalization

Equalization is aimed at minimizing or controlling fluctuations in wastewater characteristics for the purpose of providing optimum conditions for subsequent treatment processes. The size and type of the equalization basin/tank used varies with the quantity of waste and the variability of the wastewater stream. In the case of potato processing wastewater, the mechanically pretreated or preclarified wastewater flows into a balancing tank (buffer tank). Equalization serves two purposes: physical homogenization (flow, temperature) and chemical homogenization (pH,

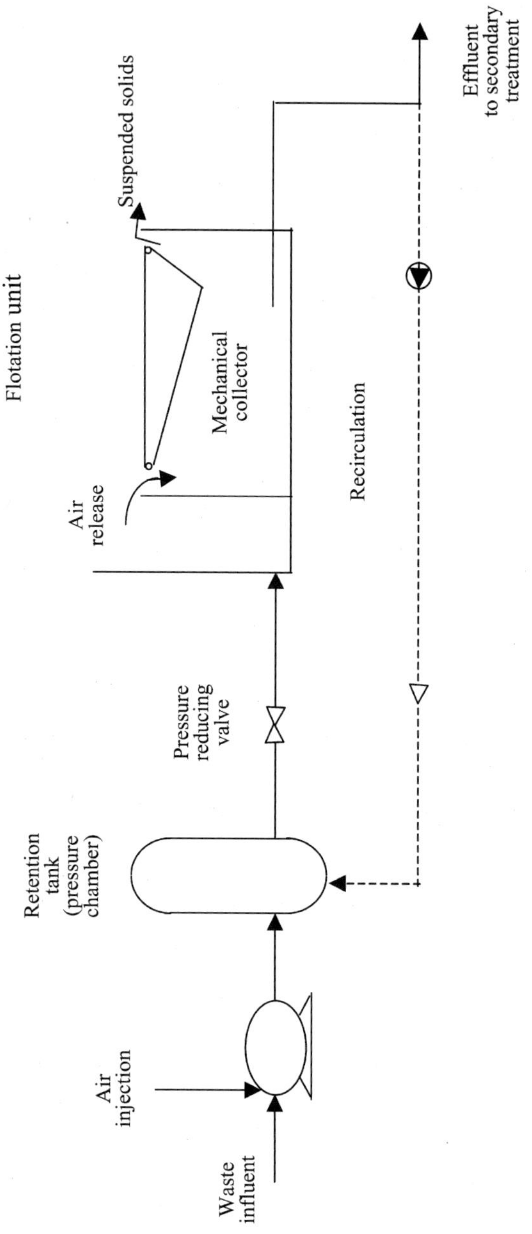

Figure 11 Schematic diagram of flotation system (from Ref. 24).

nutrients, organic matter, toxicant dilution). For proper homogenization and insurance of adequate equalization of the tank content, mixing is usually provided, such as turbine mixing, mechanical aeration, and diffused air aeration. The most common method is to use submerged mixers.

Neutralization

Industrial wastewaters that contain acidic or alkaline materials should be subjected to neutralization prior to biological treatment or prior to discharge to receiving wastes. For biological treatment, a pH in the biological system should be maintained between 6.5 and 8.5 to ensure optimum biological activity. The biological process itself provides neutralization and a buffer capacity as a result of the production of CO_2, which reacts with caustic and acidic materials. Therefore, the degree of the required preneutralization depends on the ratio of BOD removed and the causticity or acidity present in the waste [24].

As for potato processing wastewater in general, the water from the balancing tank (buffer tank) is pumped into a conditioning tank where the pH and temperature of the wastewater are controlled or corrected. Continuous monitoring of the pH of the influent is required by dosing a caustic or acidic reagent, according to the nature of resulting wastewater. The required caustic or acidic reagent for dosing in the neutralization process is strongly related to the different peeling methods used in the potato processing plant, since peeling of potatoes forms the major portion of the organic load in potato processing waste. Three different peeling methods are used extensively today: abrasion peeling, steam peeling, and lye peeling. Between lye and steam peeling wastes, the biggest difference is the pH of the two wastes. While steam peeling wastes are usually almost neutral (pH values vary between 5.3 and 7.1), lye peeling wastes have pH values from 11 to 12 and higher [3].

Secondary Treatment

Secondary treatment is the biological degradation of soluble organic compounds from input levels of 50–1000 mg/L BOD or more to effluent levels typically under 15–20 mg/L. In all cases, the secondary treatment units must provide an environment suitable for the growth of biological organisms that carry out waste treatment. This is usually done aerobically, in an open aerated tank or lagoon. Also, wastewaters may be pretreated anaerobically, in a pond or a closed tank. After biotreatment, the microorganisms and other carried-over solids are allowed to settle. A fraction of this sludge is recycled in certain processes. However, the excess sludge, along with the sedimented solids, must be disposed of after treatment.

As for potato waste, the most full-scale secondary treatment systems have been applied since 1968, although considerable research works of a pilot-plant scale have been conducted prior to that date. The description or characteristic data of these pilot-scale secondary treatment designs have been presented in detail [11]. Among the different known aerobic processes for secondary treatment of wastewater, we concentrate here on the most common treatment processes for potato processing wastewater with relevant case studies.

Natural Treatment Systems: Irrigation Land Treatment. Land treatment of food-processing wastewater resulting from meat, poultry, dairy, brewery, and winery processes has proved successful mainly through spray irrigation, applied as various types and methods in many areas. By 1979, there were an estimated 1200 private industrial land-treatment systems [24]. Potato processing wastewater can be utilized as irrigation water to increase the crop yield, because they are not polluted biologically. Irrigation systems include ones in which loading rates are about 2–4 in./week (5–10 cm/week).

Factors such as the crops grown, soil type, groundwater, and weather determine the required land area for irrigation. Some potato processors choose land disposal systems (spray or flood irrigation) because other treatment systems, while they give a higher efficiency rate, are exposed to operational problems.

Loamy, well-drained soil is most suitable for irrigation systems. However, soil types from clays to sands are acceptable. A minimum depth to groundwater of 5 ft (1.5 m) is preferred to prevent saturation of the root zone [24]. If a 5 ft depth is not available due to higher groundwater, underdrained systems can be applied without problems. As for potential odors issued from spray irrigation, they can be controlled by maintaining the wastewater in a fresh condition in order not to become anaerobic.

Water-tolerant grasses have proved to be the most common and successful crops for irrigation disposal, due to their role in maintaining porosity in the upper soil layers. The popular cover crop is reed canary grass (*Phalaris arundinacea*), which develops extensive roots that are tolerant to adverse conditions. In addition, water-tolerant perennial grasses have been widely used because they are able to absorb large quantities of nitrogen, require little maintenance, and maintain high soil filtration rates.

In some cases, wastewaters have been sprayed into woodland areas. Trees develop a high-porosity soil cover and yield high transpiration rates. Irrigation systems normally consist of an in-plant collection system, screens, low-head pump station, pressure line, pumping reservoir, high-head irrigation pumps, distribution piping, spray nozzles, and irrigation land. It is preferable in this respect to preclarify the potato processing wastewater by using a primary settling tank with a minimum 1.5 hour detention time to decrease the suspended solids content, in order to prevent closing of spray nozzles and soil. If the effluent has excess acid or alkali, it should be neutralized prior to discharging to land so that cover crops may be protected. Groundwater contamination from irrigation can be a serious problem and must be addressed during the predesign phase of a project, with the consideration that continuous monitoring of groundwater is necessary at all times in the irrigated area.

Design Example 1. A potato processing industry plans to treat its resultant wastewater by a land irrigation system. Determine the area required under the specific conditions: flow = 0.2 MG/day (756 m^3/day), BOD concentration = 2600 mg BOD/L, N concentration = 100 mg N/L. The regulation limits are: loading rates are 2 in./week (5 cm/week) and 535 lb BOD/acre-day (0.06 kg/m^2-day), nitrogen loading rate for crop's need of grass is 250 lb N/acre (0.028 kg/m^2) (the spraying period for the grass is 16 weeks).

Solution: Prescreened wastewater: assuming that 20% BOD is removed by using fine screen with mesh size 1 mm. Residual BOD: 2600 × 0.8 = 2080 mg/L.

$$\frac{Qm}{A} = \frac{r}{258}$$

where Qm is in million gallons per day, A is in acres, and r is the average wastewater application rate (inches per week).

$$\frac{0.2}{A} = \frac{2}{258}$$

and $A = 26$ acres (10.5 ha = 105,000 m^2).

$$\text{Daily loading of BOD} = \frac{2080\,\text{mg BOD}}{\text{L}} \times 0.2\,\text{MG/day} \times 8.34\frac{\text{lb/MG}}{\text{mg/L}}$$

$$= 3469.4\,\text{lb/day}\ (1575\,\text{kg/day})$$

$$A = \frac{3469.4\,\text{lb/day}}{535\,\text{lb/acre.day}} = 6.5\,\text{acres}\ (2.6\,\text{ha} = 26{,}000\,\text{m}^2)$$

$$\text{Total loading of N} = \frac{100\,\text{mg N}}{\text{L}} \times 0.2\,\text{MG/day} \times 8.34\frac{\text{lb/MG}}{\text{mg/L}}$$

$$\times\ 16\ \text{weeks} \times 7\ \text{days/week}$$

$$= 18{,}682\,\text{lb N}\ (8482\,\text{kg})$$

$$A = \frac{18{,}682\,\text{lb N}}{250\,\text{lb N/acre}} = 75\,\text{acres}\ (30.4\,\text{ha} = 304{,}000\,\text{m}^2)$$

or

$$\frac{Qm}{A} = \frac{NC}{58.4\,nT}$$

where NC is nitrogen removal by the growing crop (lb/acre), n is nitrogen concentration of the wastewater (mg/L), and T is the number of weeks of the irrigation season.

$$\frac{0.2}{A} = \frac{250}{58.4 \times 100 \times 16}$$

and $A = 75$ acres (30.4 ha = 304,000 m^2) or, in metric units:

$$\frac{Qm}{A} = \frac{143NC}{nT}$$

$$\frac{756\,\text{m}^3/\text{day}}{A} = \frac{143\ (0.028\,\text{kg/m}^2)}{100 \times 16}$$

where $A = 304{,}000$ m^2 = 30.4 ha (75 acres).

The area required is 75 acres (30.4 ha).

Natural Treatment Systems: Stabilization Ponds and Aerated Lagoons. A wastewater pond, sometimes called a stabilization pond, oxidation pond, or sewage lagoon, consists of a large, shallow earthen basin in which wastewater is retained long enough for natural processes of treatment to occur. Oxygen necessary for biological action is obtained mainly from photosynthetic algae, although some is provided by diffusion from the air. Lagoons differ from ponds in that oxygen for lagoons is provided by artificial aeration.

Depending on the degree of treatment desired, waste stabilization ponds may be designed to operate in various ways, including series and parallel operations. In some cases such as industrial wastewater treatment, they are referred to as tertiary ponds (polishing or maturation ponds), in order to remove residual pollutants and algae prior to effluent discharges.

The majority of ponds and lagoons serving municipalities and industries are of the facultative type, where the wastewater is discharged to large ponds or lagoons. Usually the ponds vary from 3 to 6 ft (0.9 to 1.8 m) deep, for a period of 3 weeks and longer, while lagoons vary from 6 to 15 ft (1.8 to 4.6 m), for a period of 2 weeks and longer.

Climatic conditions play an important role in the design and operation of both ponds and lagoons. Air temperature has a great effect on the success of this type of treatment. Within naturally occurring temperature ranges, biological reactions roughly double for each 10°C increment in water temperature. This fact encourages countries with warmer climates to utilize ponds and lagoons for wastewater treatment, particularly where land is abundant, thus providing considerable savings in both capital and operating costs.

The use of a stabilization pond in treating combined wastewaters of potato processing wastewaters and domestic wastewaters has been examined [25]. Extensive treatment loading rates for stabilization ponds were recommended in the range 5.6–6.7 kg BOD/1000 m^3-day.

High-strength wastewaters require long detention times, increasing heat loss, and decreasing efficiency in cold climates. Additionally, highly colored wastewaters cannot be treated effectively by facultative ponds, where oxygen generation is supplied mainly by photosynthesis, which depends on light penetration. Therefore, it is necessary to use aerated lagoons in which the required oxygen is supplied by diffused or mechanical aeration units. The biological life in such lagoons contains a limited number of algae and is similar to that found in an activated sludge system. In addition, aerated lagoons prevent the completion of anaerobic conditions with their attendant odor problems.

There are two types of aerated lagoons: aerobic and facultative lagoons. They are primarily differentiated by the power level employed. In aerobic lagoons, the power level is sufficiently high to maintain all solids in suspension and may vary from 14 to 20 hp/MG (2.8–3.9 W/m^3) of lagoon volume, depending on the nature of the suspended solids in the influent wastewater [24].

In facultative lagoons or aerobic–anaerobic lagoons, the power level employed is only sufficient to maintain a portion of the suspended solids in suspension, where the oxygen is maintained in the upper liquid layers of the lagoon. The employed power level in such lagoons for treating industrial wastewater is normally lower than 1 W/m^3.

As for the design of facultative ponds and aerated lagoons, several concepts and equations have been employed, and they can be found in many publications. The following is a design example for the treatment plant of potato processing wastewater.

Design Example 2. A potato processing wastewater flow of 1150 gal/ton of raw potatoes (4.35 m^3/ton) has a BOD of 2400 mg/L and a VSS content of 450 mg/L (nondegradable). It is to be pretreated in an aerobic lagoon with a retention period of one day. The k is 36/day; the raw potatoes processed are 150 tons/day. Estimate the following: the effluent soluble BOD concentration; the effluent VSS concentration; the oxygen required in mass/day; where $a = 0.5$, $a' = 0.55$, $b = 0.15$/day.

Solution: Effluent soluble BOD (Se), by rearranging the equation:

$$\frac{S_e}{S_o} = \frac{1 + bt}{akt}$$

$$S_e = \frac{S_o(1 + bt)}{akt} = \frac{2400\,\text{mg/L}(1 + 0.15/\text{day} \times 1\,\text{day})}{0.5 \times 36 \times 1\,\text{day}}$$

$$S_e = 153\,\text{mg/L}$$

Effluent volatile suspended solids ($VSS_{effl.}$): the mixed liquor volatile suspended solids can be predicted from the equation:

$$X_v = \frac{aS_r}{1 + bt} + X_i$$

where X_i = influent volatile suspended solids not degraded in the lagoon.

$$\begin{aligned} X_v &= \frac{0.5(2400 - 153)\,\text{mg/L}}{1 + 0.15/\text{day} \times 1.0\,\text{day}} + 450\,\text{mg/L} \\ &= 977 + 450 \\ &= 1427\,\text{mg/L} \end{aligned}$$

Oxygen required, using equation:

$$\begin{aligned} O_R &= [a'(S_o - S_e) + 1.4bX_v t]Q \\ &= [0.55(2400 - 153)\,\text{mg/L} + 1.4 \times 0.15/\text{day} \times 977\,\text{mg/L} \times 1\,\text{day}] \\ &\quad \times 4.35\,\text{m}^3/\text{ton} \times 150\,\text{ton/day} \\ &= (1235.85 + 205.17)\ 652.5 \times 10^{-3} \\ &= 940.27\,\text{kg/day}\ (2069\,\text{lb/day}) \end{aligned}$$

Remark: The pretreated wastewater in an aerobic lagoon can be discharged to a municipal treatment system, or to facultative ponds followed the aerobic lagoon.

Natural Treatment System: Wetland Systems. Wetland treatment technology of wastewater dates back to 1952 in Germany, starting with the work of Seidel on the use of bulrushes to treat industrial wastewaters. In 1956, Seidel tested the treatment of dairy wastewater with bulrushes, which may be regarded as the first reported application of wetland plants in food processing industries [26].

Throughout the last five decades, thousands of wetland treatment systems have been placed in operation worldwide. Most of these systems treat municipal wastewater, but a growing number of them involve industrial wastewaters. Frequently targeted pollutants are BOD, COD, TSS, nitrogen, phosphorus, and metals.

The design and description of treatment wetlands involves two principal features, hydraulics and pollutant removal [9], while the operational principles include biodegradation, gasification, and storage. Food-processing wastes are prime candidates for biodegradation. The attractive features of wetland systems are moderate capital cost, very low operating cost, and environmental friendliness. The disadvantage is the need for large amounts of land.

Reed beds in both horizontal and vertical flows have been successfully used in treating wastewater of the potato starch industry [27]. Several types of meat processing waters have been successfully treated using wetland systems [28–30]. The vertical flow of the integrated system has been used with favorable results in several domestic wastewater treatment applications [31–33].

Engineered natural systems have been used successfully to treat high-strength water from potato processing. Such integrated natural systems consist in general of free water surface and vertical flow wetlands, and a facultative storage lagoon (Fig. 12) [34]. (For a detailed description of wetland components with regard to their operational results and performance refer to case studies.)

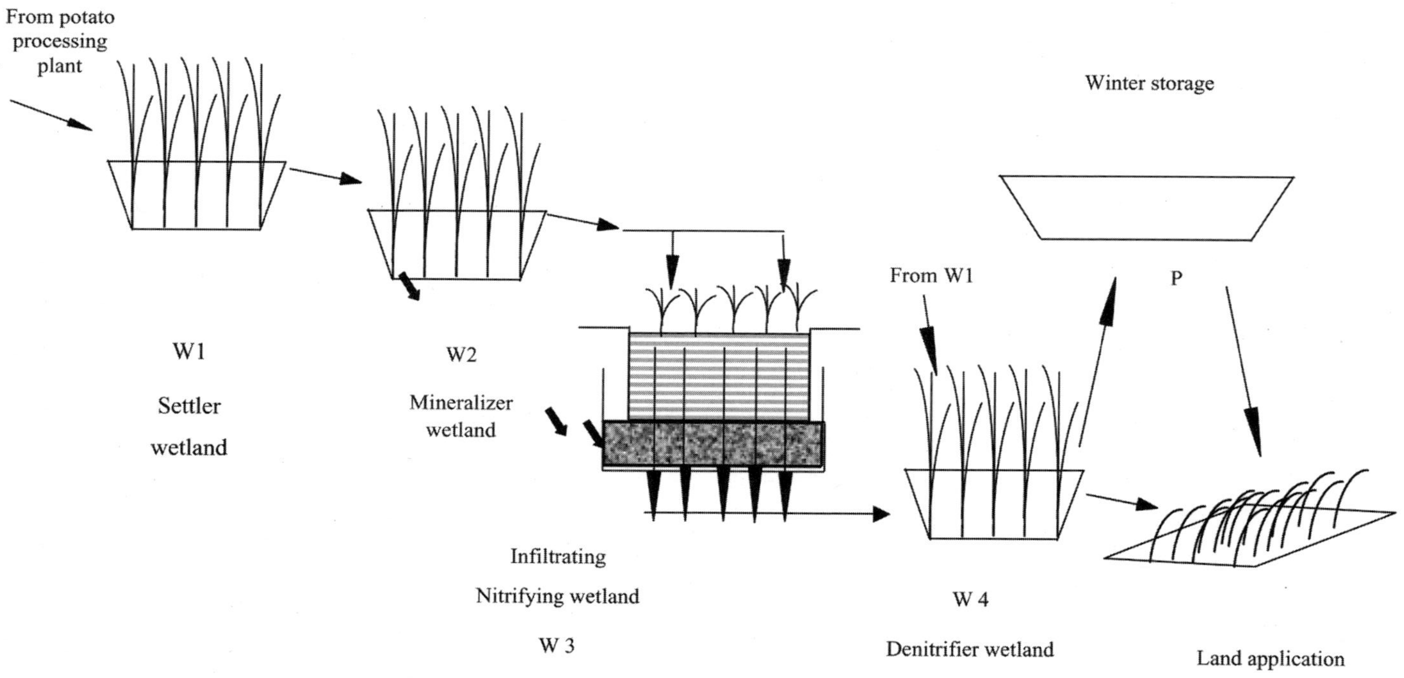

Figure 12 Schematic layout of an integrated natural system (wetland) for treatment of potato processing wastewater (from Ref. 15).

Case Studies

Case Study I. A full-scale integrated natural system has been used to treat high-strength potato processing water for 2 years [34]. The integrated natural system consists of free water surface and vertical flow wetland, and a facultative storage lagoon. Wetland components were designed for sequential treatment of the wastewater. Wastewater is pumped from a primary clarifier to ten hectares of free water surface wetlands constructed for sedimentation and mineralization of wastewater (W1/W2). The process water from the W1/W2 wetlands is sprayed onto 4 hectares of vertical flow wetland (W3) for oxidation of carbon and nitrogen. These wetlands were filled with 0.9 m of a local sand (D50 = 2.6 mm) excavated on site. These vertical flow wetlands were operated as intermittent sand filters with duty cycles of 6–72 hours. They were not planted with *Phragmites australis* due to poor growth when sprayed with the wastewater [15]. Water flows by gravity from the W3 into 2 hectares of denitrifying free water surface wetlands (W4). Raw process water is supplemented to augment denitrification in the wetlands. Treated process water flows into a 0.48 million m^3 lagoon (126 million gallon), which provides facultative treatment and storage prior to land application (Fig. 12).

The wetlands were constructed in stages throughout 1994 and 1995 in Connell, Washington. Connell is located centrally in the Colombia Basin, which is an arid agricultural area sustained by irrigation water from the Colombia River. All wetlands were lined with 1.0 mm (40 mil) HDPF liner impregnated with carbon black for UV resistance. All free water surface wetlands had 20–30 cm (8–12 in.) of native soil placed on the liners as soil for *Typha* sp. and 2 spaces of *Scirpus* sp.

The wetlands system is designed to treat an annual average flow of 1.4 mgd (approx. 5300 m^3/day) of wastewater with an annual average concentration of 3150 mg/L COD, 575 mg/L TSS, 149 mg/L TKN, and 30 ml/L NH_4-N. The winter design temperature was 1°C, with the consideration that the flow to the engineered natural system was lower in the winter season, due to operational difficulties in the water supply system.

Regarding the operational results of the integrated natural system, there were excellent reductions of TSS and COD, while organic nitrogen was effectively mineralized. TKN was reduced by about two-thirds, which is the requisite amount for balancing irrigation and nitrogen supply to the crop [15].

The net COD removal through the system was greater than 90% all year round. The W1/W2 wetlands removed about 85–90% of the COD, and 80–90% of the TSS. The average COD loading to the W1/W2 was 0.5 kg/m^3-day (31 lb/1000 ft^3-day) and 0.3 kg/m^3-day (18 lb/1000 ft^3-day) for the summer and winter, respectively. This loading rate is similar to the low rate covered anaerobic lagoons used for COD reduction in food processing. The effluent concentrations from the wetlands are lower in COD and TSS than from equivalently loaded covered anaerobic lagoons [35,36].

The effluent TSS from W1/W2 wetlands is consistently less than 75 mg/L. The W1/W2 wetland plants have proven to be very effective in solids removal. The TSS concentration increases in the lagoon due to algae growth.

In terms of nitrogen removal, the treatment objective of the system is a 53% reduction in total nitrogen (TN). The wastewater application permit requires an annual nitrogen load of 500 kg/ha-year on 213 hectares of land used to grow alfalfa and other fodder crops. The results related to TN removal indicate that the wetlands operate better than design expectation.

With regard to organic carbon, the potato water mineralizes very rapidly so that >60% of the organic carbon was mineralized to NH_4-N prior to entering the wetlands. This mineralization continued in the W1/W2 wetlands so that <15 mg/L organic nitrogen remained.

More than 60% of the TN entering the W1/W2 wetlands was in the form of NH_4-N, and 10–20% of the NH_4 was removed from the W1/W2. The pH in the W1/W2 was always >7.0 and may have contributed to volatilization of NH_4-N. The NH_4-N removal through the vertical flow wetlands averages 85% during the summer and 30–50% during the winter.

Removal of nitrate and nitrite is critical for compliance with TN removal goals in order to minimize the amount of oxidized N applied in land. Reduction of COD or BOD is often viewed as a prerequisite to establishment of nitrifying conditions [37]. Dissolved oxygen is slightly higher in the winter, but most of the system is anoxic except for the vertical flow component. Alkalinity is sufficient to support nitrification (ca. 1000 mg/L) [15]. The majority of the denitrification occurred in the W4 wetlands. Endogenous carbon in the W4 wetland was inadequate to support significant denitrification. Addition of raw potato water allows >90% denitrification, but also resulted in increased effluent NH_4-N concentrations. Approximately 5–7 NO_3-N were removed for each NH_4-N added.

Regarding the problem of odor, which generates from the decomposition of potato products, the strongest odors arose from the death of a large population of purple sulfur bacteria in the W1/W2 wetlands and the resulting sulfides >40 mg/L.

The integrated natural system is effective in reducing sulfate concentrations, from about 40 mg/L to 10 mg/L, in wetland W1. Because W1 is devoid of oxygen, sulfate has been reduced to sulfides or sulfur, including the possibility of hydrogen sulfide formation. The effluent of the treatment system has no serious odors. The final product is high-quality water with available nutrients and no odor problem during land application.

In comparing this integrated natural system with other treatment wetlands for treating food processing wastewaters, such as meat processing waters, it may be concluded that potato processing water is comparable to meat processing effluents in treatability [15]. Furthermore, it has been demonstrated that the use of this full-scale engineered natural system is a cost-effective treatment alternative for high-strength industrial wastewater. Continued research and development in operations and design of the full-scale system have resulted in better performance than that of the original design.

Activated Sludge Processes. In these processes, the preclarified wastewater is discharged into aeration basins/tanks, where atmospheric oxygen is diffused by releasing compressed air into the wastewater or by mechanical surface aerators. Soluble and insoluble organics are then removed from the wastewater stream and converted into a flocculent microbial suspension, which is readily settleable in sedimentation basins, thus providing highly treated effluent.

There is a number of different variants of activated sludge processes such as plug-flow, complete mixing, step aeration, extended aeration, contact stabilization, and aerobic sequential reactors. However, all operate essentially in the same way. These variants are the result of unit arrangement and methods of introducing air and waste into the aeration basin and they have, to a large extent, been modified or developed according to particular circumstances.

For the treatment of food and vegetable industrial wastewater, the common activated sludge methods are shown in Fig. 13.

With regard to potato wastewater treatment, the first full-scale activated sludge system was applied in the United States toward the end of the 1970s, by the R.T. French Company for treating their potato division wastewaters in Shelley, Idaho. Thereafter, many other potato processors installed biological treatment systems, most of which were activated sludge processes (Table 7).

Hung and his collaborators have conducted extensive research in various treatment processes for potato wastewater [10,16,17,20,38–41]. These included activated sludge processes with and without addition of powdered activated carbon, a two-stage treatment system of an activated sludge process followed by biological activated carbon columns, a two-stage

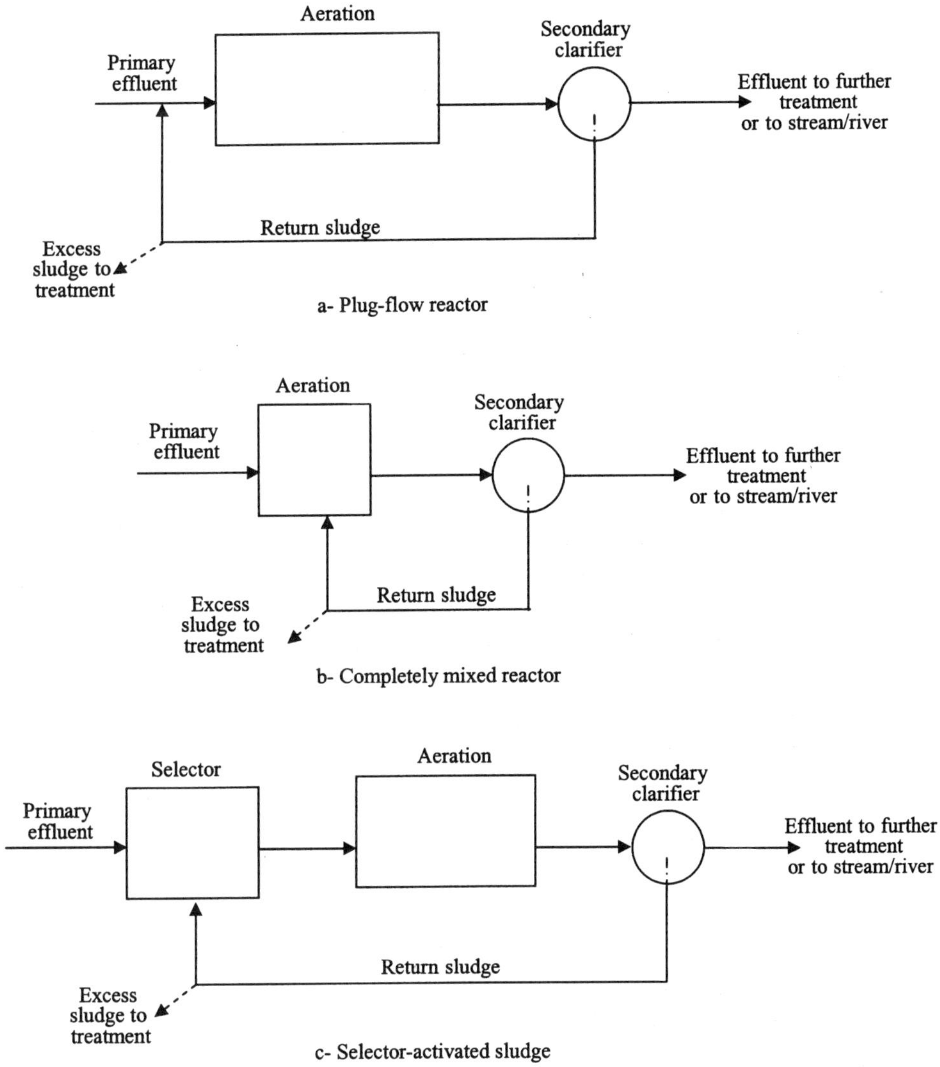

Figure 13 Flow sheets of activated sludge processes.

treatment system of an anaerobic filter followed by an activated sludge process, anaerobic digestion, and bioaugmentation process in which bacterial culture products were added to the activated sludge and anaerobic filter processes, and activated carbon adsorption process. In a laboratory study activated sludge treatment removed 86–96% of COD from primary settled potato wastewaters with 2500 mg/L COD and 500 mg/L TSS. Activated sludge followed by activated carbon adsorption removed 97% COD from primary settled wastewaters with a final effluent COD of 24 mg/L [17]. The hydraulic detention time in the aeration tank was 6.34 hours and in the sludge was 20 days.

A comparison study for potato wastewater treatment was conducted for a single-stage treatment system activated sludge reactor with and without addition of powdered activated carbon (PAC) and a two-stage treatment system using activated sludge followed by the

Table 7 Data of Various Full-Scale Secondary Treatment Designs (*Source*: Refs 11 and 12)

Treatment process and process modification	Type of process water	Volumetric organic loading	Detention time	BOD removal (%)	Remarks
Complete mixing activated sludge	Dry caustic peel	32–39 lb/ (1000 ft^3.day)	2 days	73	Sludge bulking
Complete mixing activated sludge	Lye peel	28–84 lb/ (1000 ft^3.day)	1–2 days	70–90	Removal varies with sludge bulking
Complete mixing activated sludge	Lye peel	60–180 lb/ (1000 ft^3.day)	14 hours	87	Sludge bulking will reduce removal
Multiple aerated lagoons	Lye peel	3–6 lb/ (1000 ft^3.day) in aerated lagoons	16–20 days in aerated lagoons 105 days in aerobic lagoons	98	Algal blooms will reduce removal
Anaerobic pond and lye peel activated sludge	Lye peel	25–80 lb/ (1000 ft^3.day) to activated sludge	1 day	95	Sludge bulking will reduce removal
Activated sludge and lye peel aerated lagoons	Lye peel	60–150 lb/ (1000 ft^3.day) in aeration basin	14 hours in aerated basin	99	Sludge bulking and algal blooms will reduce removal
		55 lb/ac in aerated lagoons 8.5 lb/ac in aerobic lagoons	52 days in aerated lagoon 60 days in aerobic lagoon		

Note: lb/(1000 ft^3-day) = 0.016 kg/(m^3-day). Excess sludge: 0.2–0.5 lb/lb COD removed at about 2.0% solid concentration.

biological activated carbon (BAC) column [10,41]. The primary settled wastewater contained 2668–3309 mg/L COD. Results indicated that 92% of COD was removed in the non-PAC activated sludge reactors, while 96% COD was removed in the PAC activated sludge reactors. For the non-PAC activated sludge process, increasing hydraulic detention time in the aeration tank from 8 to 32 hours reduced effluent COD from 304 to 132 mg/L. With the addition of powdered activated carbon in the activated sludge tank, effluent COD was further improved to 78 mg/L at a hydraulic detention time of 32 hours. The BAC column removed 85% from activated sludge reactor effluents with a final effluent COD of 34 mg/L.

Bioaugmentation processes with addition of bacterial culture product have been used to improve the removal efficiency of organic pollutants and to reduce the amount of sludge in municipal wastewater treatment systems, particularly in activated sludge treatment processes. Three different systems, namely extended aeration, aerated lagoon, and oxidation ditch have been used. In all three cases, bioaugmentation improved sludge settleability and BOD and COD removal efficiency [42].

Bioaugmentation with addition of bacterial culture product LLMO (live liquid microorganisms) to the activated sludge reactor was investigated for treatment of potato wastewater [38]. Influent with 2381 mg/L COD was decreased to 200 mg/L in the bioaugmented activated sludge reactor and to 236 mg/L in the nonbioaugmented activated sludge reactor. The bioaugmented reactor can operate at a higher F/M ratio and a lower MLVSS level than the nonbioaugmented reactor and achieves a better COD removal efficiency. Effect of types of bacterial culture product addition to the activated sludge reactors on reactor performance have been studied [39]. Types of LLMO used included S1, G1, E1, N1, and New 1 LLMO. S1 LLMO was found to be the most effective, and removed 98% TOC (total organic carbon) and reduced 67% VSS (volatile suspended solids). The effect of bioaugmentation on the treatment performance of a two-stage treatment system using an anaerobic filter followed by an activated sludge process for treating combined potato and sugar wastewater was investigated [40]. The combined wastewater had 435 mg/L TOC. The bioaugmented two-stage treatment system had a better TOC removal efficiency and at a shorter hydraulic detention time of the aeration tank than the nonbioaugmented treatment system. The final effluent TOC was 75 mg/L and 89 mg/L at a hydraulic detention time of aeration tank of 12 hours and 24 hours for the bioaugmented and nonbioaugmented treatment systems, respectively.

Research on the treatment of potato processing wastewater showed that the major disadvantages of full-scale aerobic treatment are high power consumption, the large amount of sludge needing handling, and maintenance, in addition to the costs of sludge dewatering and sludge disposal (dumping and incineration), increasing substantially over the years. As a result, most potato processing companies have turned to the use of anaerobic treatment with various type of reactors followed by aerobic treatment.

Design Example 3. Continuing design example 2, a municipal extended aerobic activated sludge plant receives potato processing wastewater and has a combined BOD_5 of 450 mg/L. The return sludge has a concentration of 7000 mg/L from the secondary clarifier. Determine the required recycle ratio to the activated sludge reactor with an organic loading of 0.10 g BOD/g VSS, in order to produce an effluent meeting national discharge limits.

Solution: The organic loading (OL) can be expressed by:

$$OL = \frac{QS_o}{Q_R X_{vr}} = 0.10\,\text{g BOD/g VSS} \tag{1}$$

where Q is the flow, S_o the influent BOD, Q_R the recycle flow, and X_{vr} the volatile suspended solids concentration in the recirculation line expressed in g VSS/L.

Assuming 85% VSS for the recirculation, $X_{vr} = 0.85$, $X_r = 0.85 \times 7000 = 5950$ mg VSS/L = 5.95 g VSS/L. The required recycle ratio can be calculated from Eq. (1).

$$Q_R = \frac{QS_o}{OL \cdot X_{vr}} = \frac{450\,\text{mg BOD/L} \times Q}{100\,\text{mg BOD/g VSS} \times 5.95\,\text{g VSS/L}}$$

$$= 0.756Q$$

Design Example 4. A municipal conventional activated sludge treatment plant is planning to receive the potato processing wastewater given in design example 2, without pretreatment (in an aerobic lagoon). Determine what changes need to be made in the processing conditions of the plant to avoid filamentous bulking. Assume: $T = 20°C$, $a' = 0.55$, $b' = 0.15/$day, $X = 0.6$, $N_b = 1.5$ lb O_2/(hp.hour).

For the potato processing wastewater (example 2): BOD concentration = 2400 mg/L, Flow = 1150 gal/ton × 150 t/day = 172,500 gal/day or = 4.35 m^3/ton × 150 t/day = 652.5 m^3/day.

Solution: The municipal activated sludge treatment plant before potato processing discharge has the following characteristics: $Q_{bef.} = 2.5$ MG/day (9450 m^3/day), $S_{inf.} = 300$ mg/L, $S_e = 10$ mg/L, $S_{r,b} = 300 - 10 = 290$ mg/L, $t_b = 6$ hours = 0.25 day, $X_{v,b} = 3000$ mg/L, (F/M) = 0.3/day.

The dissolved oxygen required can be taken from reference (International water pollution control, Figs. 6–15): $DO_b = 1.7$ mg/L. The oxygen needed can be calculated by equation:

$$
\begin{aligned}
O_{R,b} &= (a'S_{r,b} + b'XX_{v,b}t_b)Q_b \\
&= (0.55 \times 290 + 0.15 \times 0.60 \times 3000 \times 0.25)\text{mg/L} \\
&\quad \times 2.5\,\text{MGD} \times 8.34(\text{lb/MG})/(\text{mg/L}) \\
&= 2733\,\text{lb/day}\ (1241\,\text{kg/day}) \\
&= 113.9\,\text{lb/hour}\ (51.71\,\text{kg/hour})
\end{aligned}
$$

The power requirement is:

$$HP_b = O_{R,b}/N_b = \frac{113.9\,\text{lb/hour}}{1.5\,\text{lb/(hp.hour)}} = 76\,\text{HP}\ (57\,\text{kW})$$

After the potato industry discharge in the municipal activated sludge plant, the following will apply. Assume for the MLVSS, the value $X_{v,a} = 4000$ mg/L.

$$
\begin{aligned}
Q_{after} &= Q_{before} + Q_{ind} = 2.5 + 0.1725 = 2.6725\,\text{MG/day (m}^3\text{/day)} \\
S_{inf.a} &= \frac{Q_b S_{inf.b} + Q_{ind} S_{ind}}{Q_a} \\
&= \frac{(2.5 \times 300) + (0.1725 \times 2400)}{2.6725} = 43{,}505\,\text{mg/L}
\end{aligned}
$$

The BOD removed will be:

$$S_{r,a} = 435.5 - 10 = 425.5\,\text{mg/L}$$

The new retention time will be:

$$t_a = t_b \frac{Q_b}{Q_a} = 0.25\,\text{day} \frac{2.5}{2.6725} = 0.234\,\text{day}$$

The new F/M ratio can be computed using the equation:

$$(\text{F/M})_a = \frac{S_{inf.a}}{X_{v,a} \cdot t_a} = \frac{435.5}{4000 \times 0.234} = 0.465\,\text{day}$$

From the reference mentioned above, the dissolved oxygen required is: $DO_a = 3.6$ mg/L. Assuming the same values for a', b' and X, the oxygen required can be computed:

$$
\begin{aligned}
O_{R,a} &= (0.55 \times 425.5 + 0.15 \times 0.60 \times 4000 \times 0.234)\,\text{mg/L} \\
&\quad \times 2.6725\,\text{MGD} \times 8.34\ (\text{lb/MG})/(\text{mg/L}) \\
&= 7093.7\,\text{lb/day}\ (3220.5\,\text{kg/day}) \\
&= 295.6\,\text{lb/hour}\ (134.2\,\text{kg/hour})
\end{aligned}
$$

The oxygen saturation at 20°C is: $C_s = 9.2$ mg/L. The new N_a:

$$N_a = N_b \frac{(C_s - DO_a)}{(C_s - DO_b)} = \frac{1.5\,\text{lbO}_2}{(\text{hp.hour})} \times \frac{9.2 - 3.6}{9.2 - 1.7}$$

$$= 1.12\,\text{lb/(hp.hour)}\ (0.68\,\text{kg/kW.hour})$$

The power required is:

$$HP_a = O_{R,a}/N_a = \frac{295.6\,\text{lb/hour}}{1.12\,\text{lb/(hp.hour)}} = 264\,\text{HP}\ (197\,\text{kW})$$

The additional power required is:

$$HP_{add} = HP_a - HP_b = 264 - 76 = 188\,\text{HP}\ (140\,\text{kW})$$

Remark: To avoid the filamentous bulking in the conventional activated sludge plant, the following modifications are needed:

- increasing the MLVSS from 3000 to 4000 mg/L;
- increasing the power required from 76 HP (57 kW) to 264 HP (197 kW), in addition to the necessity to control the bulking.

Rotating Biological Contactors. The rotating biological contactor (RBC) is an aerobic fixed-film biological treatment process. Media in the form of large, flat discs mounted on a horizontal shaft are rotated through specially contoured tanks in which wastewater flows on a continuous basis. The media consist of plastic sheets ranging from 2 to 4 m in diameter and up to 10 mm thick. Spacing between the flat discs is approximately 30–40 mm. Each shaft, full of medium, along with its tanks and rotating device, forms a reactor module. Several modules may be arranged in parallel and/or in series to meet the flow and treatment requirements (Fig. 14). The contactor or disc is slowly rotated by power supplied to the shaft, with about 40% of the surface area submerged in wastewater in the reactor.

A layer of 1–4 mm of slime biomass is developed on the media (equivalent to 2500–10,000 mg/L in a mixed system) [24], according to the wastewater strength and the rotational speed of the disc. The discs, which develop a slime layer over the entire wetted surface, rotate through the wastewater and contact the biomass with the organic matter in the waste stream and then with the atmosphere for absorption of oxygen. Excess biomass on the media is stripped off by rotational shear forces, and the stripped solids are held in suspension with the wastewater by the mixing action of the discs. The sloughed solids (excess biomass) are carried with the effluent to a clarifier, where they are settled and separated from the treated wastewater.

The RBC system is a relatively new process for wastewater treatment; thus full-scale applications are not widespread. This process appears to be well suited to both the treatment of industrial and municipal wastewater. In the treatment of industrial wastewaters with high BOD levels or low reactivity, more than four stages may be desirable. For high-strength wastewaters, the first stage can be enlarged to maintain aerobic conditions. An intermediate clarifier may be

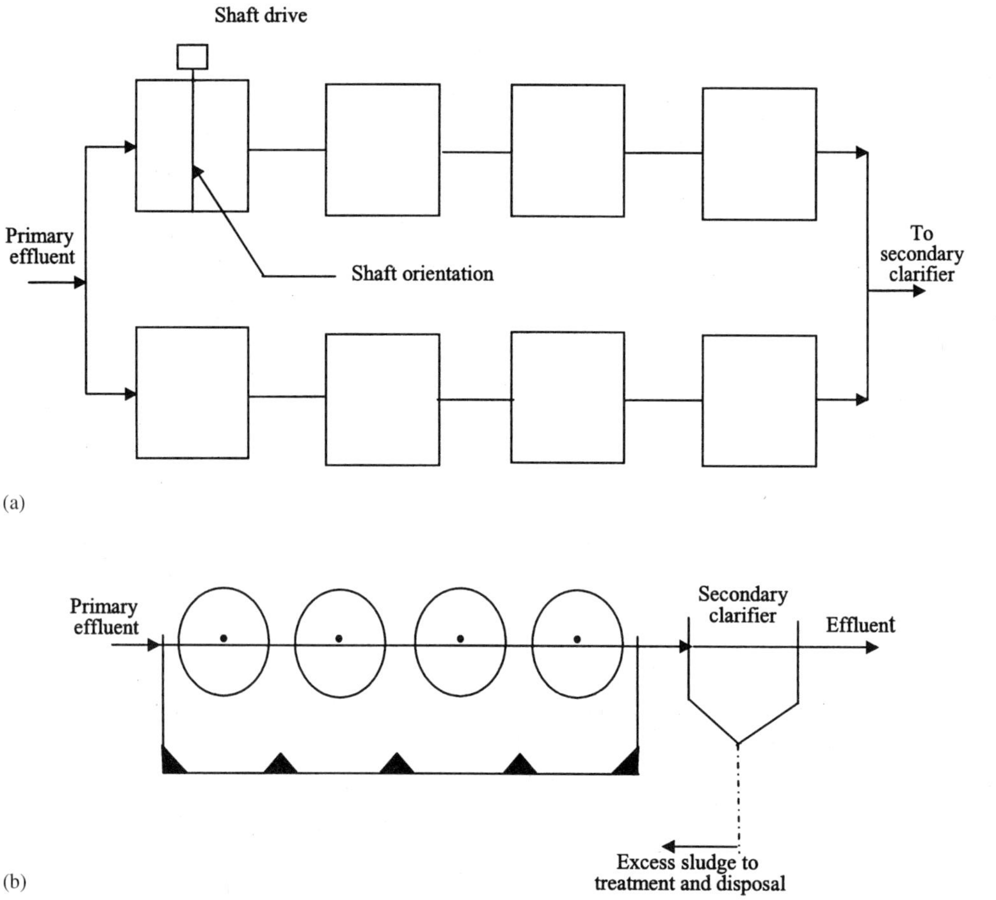

Figure 14 Rotating biological contactor system. (a) Flow-sheet of typical staged rotating biological contactors (RBCs). (b) Schematic diagram of the RBCs.

employed where high solids are generated to avoid anaerobic conditions in the contactor basins. Currently used media consist of high-density polyethylene with a specific surface of 37 ft^2/ft^3 (121 m^2/m^3). One module or unit, 17 ft (3.7 m) in diameter by 25 ft (7.6 m) long, contains approximately 10,000 m^2 of surface area for biofilm growth. This large amount of biomass permits a short contact time, maintains a stable system under variable loading, and should produce an effluent meeting secondary-treatment limits or standards.

Recirculating effluent through the reactor is not necessary. The sloughed solids (biomass) are relatively dense and settle well in the secondary clarifier. Low power requirement and simple operating procedure are additional advantages. A 40-kW motor is sufficient to turn the 3.7×7.6 m unit previously described [43]. Therefore, it can be clearly realized that the RBC can be applied successfully for treatment of potato processing effluents, in particular for values of BOD_5 and COD concentrations not exceeding, in the main, 5000 to 6000 mg/L in the wastewater stream. Depending on these properties, the data taken from case studies for treating contaminated wastewater with BOD_5 and COD concentrations close to those found in wastewater from potato processing, can be of much benefit. These data are based on the experience published by USEPA [44]. Table 8 summarizes the experience represented in design criteria and performance of the applied RBC for treating landfill leachate, which can be successfully applied to the potato processing industry within the range of pollutant concentrations mentioned above. However, an optimum design can be achieved by a pilot-plant study of the RBC.

Design Example 5. Design a rotating biological contactor (RBC). Determine the surface area required for an RBC system to treat preclarified potato processing wastewater with a flow of 150,000 gal/day (567 m^3/day) and BOD concentration of 4000 mg/L, with a maximum system effluent of 20 mg BOD/L. Minimum temperature is expected to be 32°C (90°F). The selected plastic medium is manufactured in 8 m shaft lengths, with each shaft containing 1.2×10^4 m^3 of surface area.

Solution: RBC performance:

$$\frac{4000 - 20}{4000} \times 100 = 99.5\%$$

No temperature correction in loading is needed, because the wastewater temperature is >55°F (13°C). Based on the hydraulic surface loading, the selected design value of Table 8 is: Hydraulic loading rate = 1.2 gal/ft^2-day (49 L/m^2-day).

Table 8 Design Criteria and Performance of Rotating Biological Contactors [44]

Parameter	Range	
(a) Design criteria		
MLSS (mg/L)	3000–4000	
MLVSS (mg/L)	1500–3000	
F/M (lb BOD/lb MLVSS/day)	0.05–0.3	
Maximum BOD volumetric loading (lb BOD/1000 ft^3/day)	15–60	
Maximum BOD surface loading (lb BOD/1000 ft^2/day)	0.05–0.7 (4–8 g BOD_5/m^2.day according to German experience)	
Number of stages per train	1–4	
Hydraulic surface loading (gal/day/ft^2)	0.3–1.5	
HRT (days)	1.5–10	
Compound	**Influent (mg/L)**	**Removal (%)**
(b) Performance		
SCOD	800–5200	55–99
$SBOD_5$	100–2700	95–99
$TBOD_5$	3000	99+
TOC	2100	99
DOC	300–2000	63–99
NH_4-N	100	80–99

Remark: These design and performance data are based on results of different references including EPA publications that handle landfill leachate treatment.

Disc area is calculated directly in a simple form:

$$A_{\rm d} = \frac{150{,}000\,{\rm gal/day}}{1.2\,{\rm gal/ft^2\text{-}day}} = 125{,}000\,{\rm ft}^2$$

$$= \frac{567\,{\rm m^3/day}}{0.049\,{\rm m^3/m^2\text{-}day}} = 11{,}600\,{\rm m}^2 = 1.16 \times 10^4\,{\rm m}^2$$

Based on the organic surface loading, normally adopted in Germany, the selected design value of Table 8 is: Organic loading rate = 4 g BOD/m^2-day.

$$\text{Influent BOD loading} = \frac{567\,{\rm m^3/day} \times 4000\,{\rm mg/L}}{1000} = 2268\,{\rm kg/day}$$

Disc area is:

$$A'_{\rm d} = \frac{2268\,{\rm kg\ BOD/day}}{4\,{\rm g/m^2\text{-}day}} \times \frac{1000\,{\rm g}}{1\,{\rm kg}} = 567{,}000\,{\rm m}^2 = 5.67 \times 10^5\,{\rm m}^2$$

In comparing $A_{\rm d}$ and $A'_{\rm d}$, it is clear that the required disc area will be:

$$A'_{\rm d} = 5.67 \times 10^5$$

$$\text{Modules number} = \frac{5.67 \times 10^5\,{\rm m}^2}{1.2 \times 10^4\,{\rm m^2/Module}} = 47\,\text{Modules}$$

On average, 50 modules are required for the first stage of wastewater treatment.

For potato industrial wastewater, a minimum of four stages (200 modules) in series will be required. These can be placed in two lines, each line to contain four stages.

Anaerobic Treatment Systems. With more than 1800 plants worldwide using different applications (food processing, chemical industry, pulp and paper industry), anaerobic treatment has gained widespread use as a reliable and efficient means for reduction of COD [45]. Of all anaerobic processes, those technologies based on high-rate, compact, granular biomass technology, such as upflow anaerobic sludge blanket (UASB) and expended granular sludge bed (EGSB), have a leading position (more than 750 plants) [14].

A large number of analyses have been carried out since 1958, when the first full-scale anaerobic wastewater treatment plants were introduced. In Germany alone there are currently 125 methane reactors treating industrial wastewater. Forty-three plants are working with a contact process, 38 plants run sludge blanket reactors, and 33 plants work with fixed-film methane reactors. The other 11 plants have completely stirred tank reactors (CSTR), self-made contribution, hybrid reactors, or other unnamed reactor types [13].

Table 9 gives an overview of the typical problems and solutions in various food and beverage industries, including potato processing and potato starch industries, for all kinds of anaerobic reactor systems. This experience gathered by German researchers reveals that each industry has its own specific problems. Therefore, specific investigations should be undertaken to find the relevant solutions. Furthermore, these data show that it is possible to treat several different industrial wastewaters together in one plant, which is particularly beneficial for small factories, especially in the food industry [13].

Batch mesophilic anaerobic digestion processes for potato wastewater treatment have been conducted [16]. After 33 days of anaerobic digestion at a reactor pH of 6.5–7.3 and at a temperature of 22°C the batch treatment process removed 84, 82, and 90% COD from potato

Table 9 Several Food and Beverage Industries With Their Special Problems and Solutions (*Source*: Ref. 13)

Industry	Special problem	Solution
Potato processing industry	Solids	Sieve, acidification tank, EGSB methane reactor
Potato and wheat starch industry	Precipitation of MAP (magnesium ammonium phosphate)	pH regulation
Beet sugar factories	Lime precipitation	Cyclone
	pH lower than 5 in the pond system	Lowering the pH in the circuit system
Pectin factories	High nitrate concentrations over 1000 mg NO_3-N/L	Denitrification stage before methane reactor
Breweries	Considerable pH variations	Equalizing tanks, pH regulation
	Kieselguhr contents	Treatment together with municipal sludge
	Aluminum precipitation in the acidification stage	Settling tank
Distilleries (alcohol production from molasses slops)	Discontinuous production	Equalizing tanks and pH regulation
Anaerobic pretreatment of wastewater from different industries in one plant	Different small factories with high loaded wastewater and campaign processing	Anaerobic pretreatment of the wastewater mixture of a brewery, two vegetable, and one fish processing factory at the municipal sewage treatment plant
Anaerobic/aerobic treatment	Carbon : nitrogen relation bulking sludge	Bypassing the anaerobic stage, pretreatment

Source: Ref. 13.

juice, mashed potato, and potato starch wastewater, respectively. Hydrolysis played an important role in the anaerobic digestion process by converting the particulate substrate in the mashed potato and potato starch wastewaters to soluble substrate, which was subsequently utilized by anaerobes for production of organic acids and methane production.

Based on the wastewater composition (average data of settled samples: COD 4000 mg/L; total N 120 mg/L; total P 60 mg/L), wastewater from the potato processing industry is very well suited for anaerobic treatment. Accordingly, there are over 50 anaerobic plants in this sector of the industry worldwide, the majority of which consist of UASB reactors. More recently, the EGSB process (high-performance UASB), developed from the UASB, has been implemented. In the potato processing industry, several UASB plants have been built by Biothane Systems Inc. and its worldwide partners for customers such as McCain Foods (French fries) and Pepsico (potato crisps). Recently, other Biothane UASB plants have joined the Pepsico network, such as Greece (Tasty Foods, Athens), Turkey (Ozay Gida, Istanbul) and Poland (E. Wedel, Warsaw) [14].

An important prerequisite is that the influent to the UASB reactor must be virtually free of suspended solids, since the solids would displace the active pellet sludge in the system. The newly developed EGSB reactors are operated with a higher upflow velocity, which causes a partial washout of the suspended solids [14]. EGSB technology is capable of handling wastewater of fairly low temperatures and considerable fluctuations in COD composition and load throughout the year.

A description of the first large-scale EGSB (Biobed reactor) in Germany will be presented in case studies to follow.

Comparison Between Biothane UASB Reactors and Biobed EGSB Reactors [14]. The UASB technology (Fig. 15) and the EGSB technology (Fig. 16) both make use of granular anaerobic biomass. The processes have the same operation principles, but differ in terms of geometry, process parameters, and construction materials.

In both processes, wastewater is fed into the bottom of the reactor through a specially designed influent distribution system. The water flows through a sludge bed consisting of anaerobic bacteria, which develop into a granular form. The excellent settleability (60–80 m/hour) of these anaerobic granules enables high concentrations of biomass in a small reactor volume. The granules do not contain an organic carrier material, such as sand or basalt.

In the sludge bed, the conversion from COD to biogas takes place. In both reactor types, the mixture of sludge, biogas, and water is separated into three phases by means of a specially designed three-phase, separator (or settler) at the top of the reactor. The purified effluent leaves the reactor via effluent laundries, biogas is collected at the top, and sludge settles back into the active volume of the reactor.

One of the most important design parameters for both types of reactors is the maximum allowable superficial upflow liquid velocity in the settler. Upflow velocities in excess of this maximum design value result in granular sludge being washed out of the reactor. The Biobed EGSB settler allows a substantially higher upstream velocity (10 m/hour) than the Biothane UASB settler (1.0 m/hour).

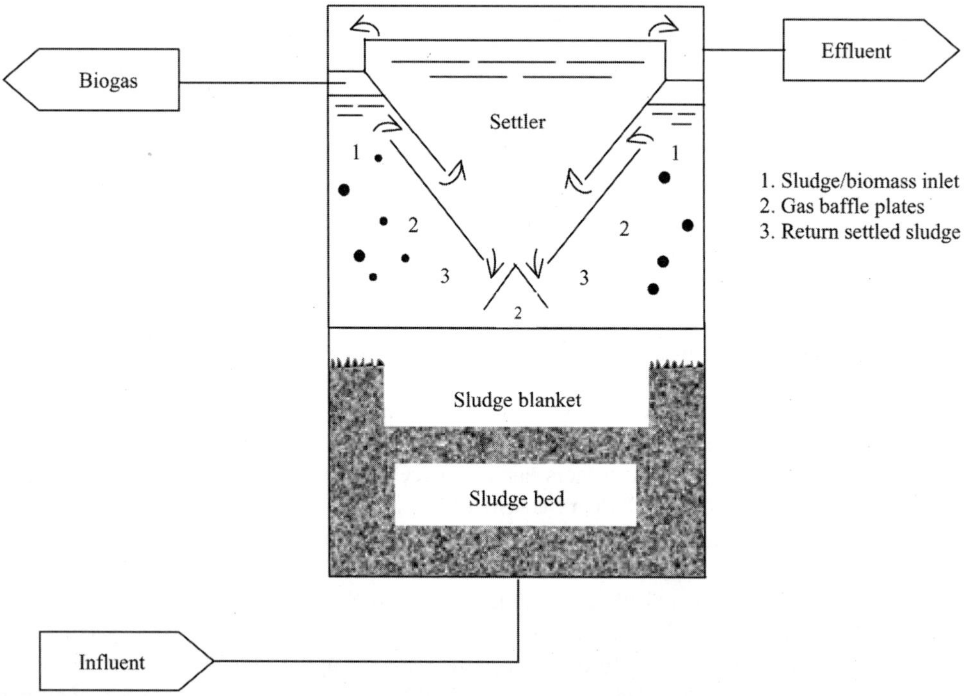

Figure 15 A cross-section of the Biothane UASB reactor (from Ref. 14).

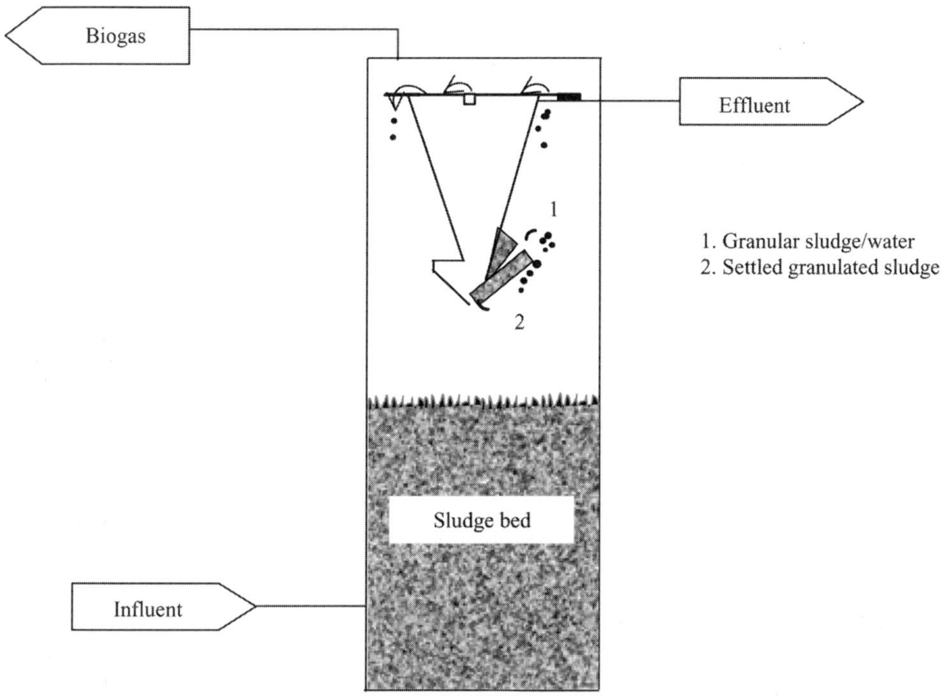

Figure 16 A cross-section of the Biogas EGSB reactor (from Ref. 14).

Another important design parameter is the maximum COD load allowed. The Biobed EGSB process operates under substantial higher COD loads (30 kg/m^3-day) than the Biothane UASB process (10 kg/m^3-day). The result of this is that for a given COD load, the Biobed EGSB reactor volume is smaller than for a Biothane UASB reactor. Biothane UASB reactors are typically rectangular or square, with an average height of 6.0 m and are usually constructed of concrete. Biobed EGSB reactors have a substantially smaller footprint. These high and narrow tanks are built in FRP (fiber glass reinforced plastic) or stainless steel and have a typical height of 12–18 m. The height of the granular sludge bed in the Biothane UASB reactor varies between 1 and 2 m and in the Biobed EGSB between 7 and 14 m. A Biobed EGSB reactor is normally built as a completely closed reactor resulting in a system with zero odor emission. Additionally, a Biobed EGSB reactor can be operated under overpressure, thereby making any use of gasholders and biogas compressors redundant. The general differences between the processes are shown in Table 10.

Wastewater in the potato processing industry contains substantial amounts of suspended solids. The Biothane UASB process is characterized by longer hydraulic retention times than the Biobed EGSB process. As a consequence, use of the Biothane UASB process results in a greater removal of suspended solids and therefore higher overall COD removal efficiencies. The Biobed EGSB process has been designed mainly for removal of soluble COD. Therefore, the use of Biobed EGSB in the potato processing industry is emphasized for those applications where the anaerobic effluent will be discharged to a sewer or to a final aerobic post-treatment.

Thermophilic UASB Reactors. In general, hot wastewater streams discharge from food industries including vegetable processing. These streams are generated from high temperature unit operations and are highly concentrated due to enhanced dissolution of organic material at

Table 10 Comparison of the Main Characteristic Parameters of Biothane UASB and Biobed EGSB (*Source*: Ref. 14)

Parameter	Unit	Biothane UASB	Biobed EGSB
Load	kg COD/m^3.day	10	30
Height	m	5.5–6.5	12–18
Toxic Components		+/−	++
V_{liquid} settler	m/hour	1.0	10
V_{liquid} reactor	m/hour	<1.0	<6.0
V_{gas} reactor	m/hour	<1.0	<7.0

Source: Ref. 14.

elevated temperatures. Anaerobic treatment, especially the thermophilic process, offers an attractive alternative for the treatment of high-strength, hot wastewater streams [46].

In the thermophilic process, the most obvious benefits compared with the mesophilic anaerobic process involve increased loading rate and the elimination of cooling before treatment. Furthermore, the heat of the wastewater could be exploited for post-treatment, which, for example, if realized and mixed with sewage water could assist in obtaining nitrification with a normally low sewage temperature (less than 10°C) [46].

Loading rates of up to 80 kg COD/m^3-day and more have been reached in laboratory-scale thermophilic reactors treating volatile fatty acids (VFA) and glucose [47,48], acetate and sucrose [49,50] and thermomechanical pulping white water [51].

As mentioned before, during the past half century, anaerobic treatment of food processing wastewaters has been widely studied and applied using mesophilic processes. In many cases, compared with single aerobic treatment, anaerobic treatment of food industry wastewaters is economical due to decreased excess sludge generation, decreased aeration requirement, compact installation, and methane energy generation. Thermophilic anaerobic treatment of food industry wastewaters, such as vinasse [52] and beer brewing [53] wastewaters, has been studied on laboratory and pilot scales.

The removal efficiencies of pollutants in these thermophilic reactors have been found to be very satisfactory. For example, in UASB reactors treating brewery wastewater and volatile fatty acids (VFA) at 55°C with loading rates of 20–40 kg COD/m^3-day, the COD removals reached over 80% in 50–60 days.

Thermophilic anaerobic processes have been used for the treatment of high solids content in vegetable waste (slop) from distillery [24–29 kg total solids (TS)/m^3] [54] and potato sludge [42 kg suspended solids (SS)/m^3] [55]. This technology has also been applied on a laboratory scale for the treatment of vegetable processing wastewaters in UASB reactors at 55°C, where the wastewater streams result from steam peeling and blanching of different processed vegetables (carrot, potato, and swede) [46]. For further information about this application, refer to the case studies.

Case Studies

Case Study I. This study examines the first EGSB operating in a German potato processing factory [13]. A wastewater flow of 1700 m^3/day passed through a screen and a fat separator into a 3518 m^3 balancing tank (weekly balance 30% constant retention) that also served as an acidification tank. Owing to the high retention time, it may be assumed that a nearly complete acidification took place, between 40 and 50% related to filtered COD. The methane reactor had a height of 14 m with a water volume of 750 m^3. The feeding of the reactor occurred at a constant rate from a conditioning tank (pump storage reservoir), where the recirculation flow mixed with the influent and the pH was adjusted to 6.6, using sodium hydroxide. The effluent from the methane reactor passed through a lamella separator for the removal of solids, which could also be placed between the acidification and methane reactor. The anaerobically treated wastewater was fed into the municipal wastewater treatment plant.

With an average filtered COD of 3500 mg/L in the influent, the efficiency of the anaerobic treatment was 70–85%, resulting in a biogas production with about 80% methane content. The concentration of filterable solids in the influent fluctuated between 500 and 2500 mg/L. According to operational experience in this anaerobic system, these values have not caused any considerable deterioration of the pellet sludge structure during operation.

Case Study II. This study addresses the anaerobic treatment of wastewater from the potato processing industry. A Biothane UASB reactor and Biobed EGSB reactor were installed at two different potato processing facilities in the Netherlands [14]. The first example is Smiths Food, which produces potato chips. They chose the Biothane UASB anaerobic treatment process for bulk COD removal from their wastewater and aerobic final treatment to meet the discharge limits. Figure 17 shows the flow scheme of this process. Coarse solids are removed in a parabolic screen (mesh size 1 mm). After this screen, the water enters a preclarifier designed at a surface load of 1 m/hour for removal of suspended solids and residual fat, oil, and grease. The settled solids are dewatered in a decanter and the water flows by gravity into a buffer tank of 400 m^3. From the buffer tank, the water is pumped to a conditioning tank for pH and temperature correction. Conversion of COD takes place in the Biothane UASB reactor. The total anaerobic plant has a COD removal efficiency of approximately 80%. The remaining COD and kjeldahl nitrogen is removed in the aerobic post-treatment.

The final COD concentration is less than 100 mg/L and the K_j-N concentration is less than 10 m/L. The final effluent is discharged to the municipal sewer. The performance of the combined UASB anaerobic-carrousel aerobic wastewater treatment plant of Smiths Food is specified in Table 11.

The second example is Peka Kroef, which produces potato and vegetable-based half products for the salad industry in Europe. Owing to the specific characteristics of the resulting wastewater (low temperature, COD load fluctuations, COD composition fluctuations, high suspended solids concentration) an alternative for the conventional UASB, the EGSB technology, was tested. Extensive laboratory research showed good results with this type of anaerobic treatment at temperatures of 20–25°C.

Figure 18 shows the flow scheme of the EGSB process at Peka Kroef. The wastewaters from the potato and the vegetable processing plants follow similar but separate treatment lines. Coarse solids are removed in parabolic screens and most of the suspended solids in a preclarifier. The settled solids are dewatered in a decanter and the overflow is fed into a buffer tank of 1000 m^3. The anaerobic plant consists of two identical streets, giving Peka Kroef a high degree of operational flexibility. From the buffer tank the water is pumped to the conditioning tanks where the pH of the wastewater is controlled. Wastewater is then pumped to the Biobed EGSB reactors where the COD conversion takes place. The conditioning tanks and the anaerobic reactors operate under 100 mbar pressure and are made from FRP. It is possible to operate without a gasholder or a compressor. In addition, the EGSB reactor guarantees operating under a "zero odor emission" and supports the aerobic post-treatment in order to increase nitrogen and phosphorus removal for final discharge to the sewer. Initial results of this Biobed reactor in the potato processing industry are very promising.

Case Study III. In this study, vegetable processing wastewaters were subjected to thermophilic treatment in UASB reactors at 55°C [46]. The high-strength wastewater streams, coming from steam peeling and balancing of carrot, potato, and swede were used. The

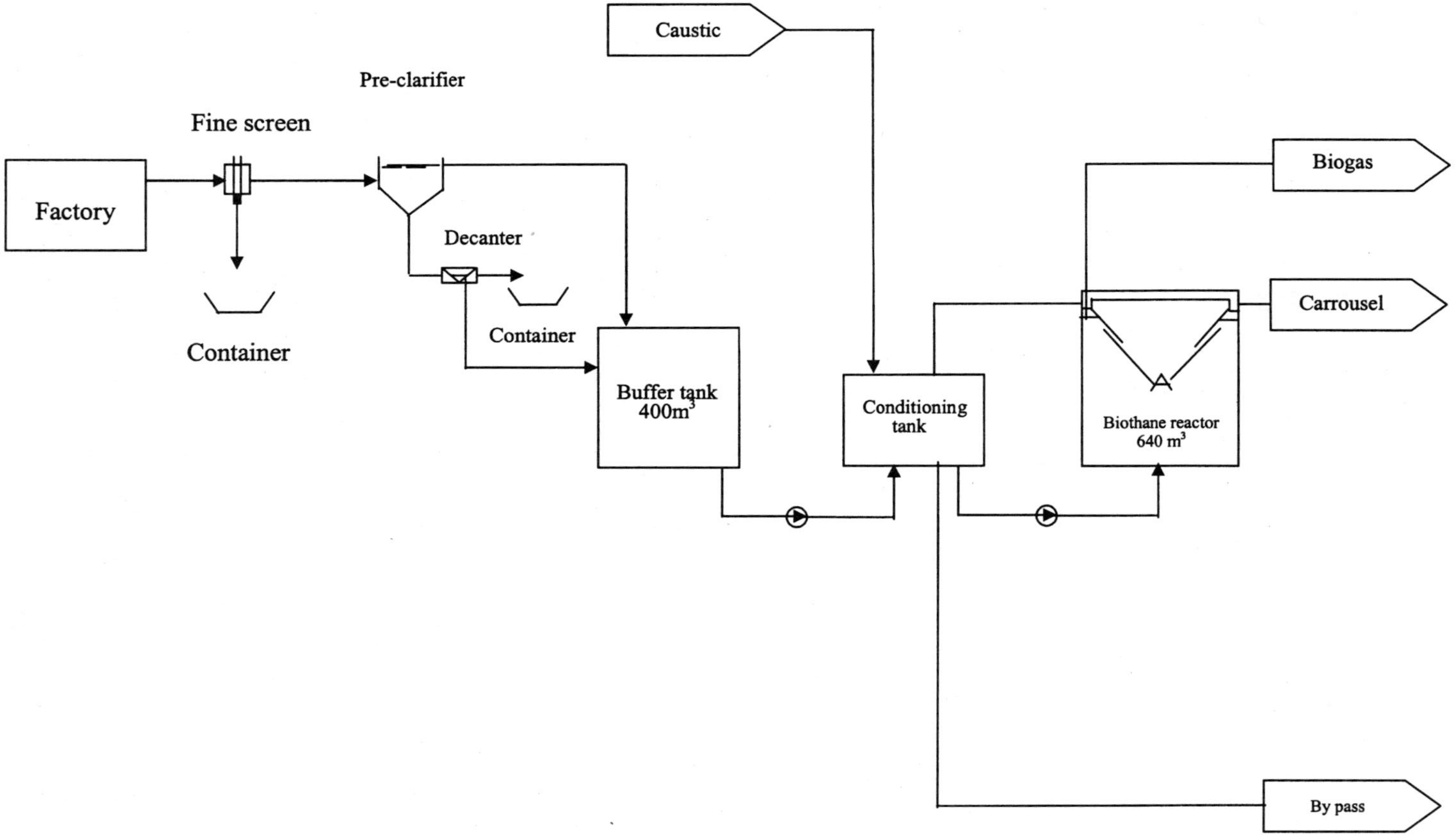

Figure 17 Schematic representation of the pretreatment stage and anaerobic treatment stage at Smiths Food (from Ref. 14).

Table 11 Performance Data of Wastewater Treatment Plant at Smiths Food (*Source*: Ref. 14)

Parameter	Unit	Value	Efficiency
Influent (data after primary clarifier)			
Flow	m^3/day	517	
t-COD	mg/L	4566	
s-COD	mg/L	2770	
SS	mg/L	890	
Anaerobic effluent			
t-COD	mg/L	926	80%
s-COD	mg/L	266	90%
SS	mg/L	600	
TKN	mg/L	196	
Aerobic (final) effluent			
t-COD	mg/L	165	96%
s-COD	mg/L	60	98%
BOD	mg/L	17	
SS	mg/L	82	
TKN	mg/L	4	

Source: Ref. 14.

wastewater characteristics are summarized in Table 12. Carbohydrates contributed 50–60% of the COD in different wastewaters.

The reactors were inoculated with mesophilic granular sludge. Stable thermophilic methanogenesis with about 60% COD removal was reached within 28 days. During the 134 day study period, the loading rate was increased up to 24 kg COD/m^3·day. High treatment efficiency of more than 90% COD removal and concomitant methane production of 7.3 m^3 CH_4/m^3·day were achieved.

The highest specific methanogenic activity (SMA) reported in this study was 1.5 g CH_4-COD/g VSS-day, while SMA_s of 2.0 and 2.1 g COD/g VSS-day have been reported with sludge from 55°C UASB reactors treating other food industry wastewaters [52,53].

Key points of interest that can be drawn from this case study are as follows:

- The results support the previous finding that 55°C UASB reactors can be started with mesophilic granular sludge as inoculum.
- The anaerobic process performance was not affected by the changes in the wastewater due to the different processing vegetables.
- The achieved loading rates and COD removals demonstrated that the thermophilic high-rate anaerobic process is a feasible method to treat hot and concentrated wastewaters from vegetable processing.

Design Example 6. Design an anaerobic process reactor to achieve 85% removal of COD from a preclarified wastewater flow 360 m^3/day (95,100 gal/day) resulting from a potato factory, depending on the steam peeling method, where total influent COD = 5000 mg/L, COD to be removed = 85%, pH = 6.2, and temperature = 30°C. The anaerobic process parameters are: sludge age (SRT) = 20 days (minimum), temperature = 35°C, $a = 0.14$ mg VSS/mg COD, $b = 0.021$ mg VSS/(mg VSS-day), $K = 0.0006$ L/(mg VSS-day), $X_v = 5500$ mg/L.

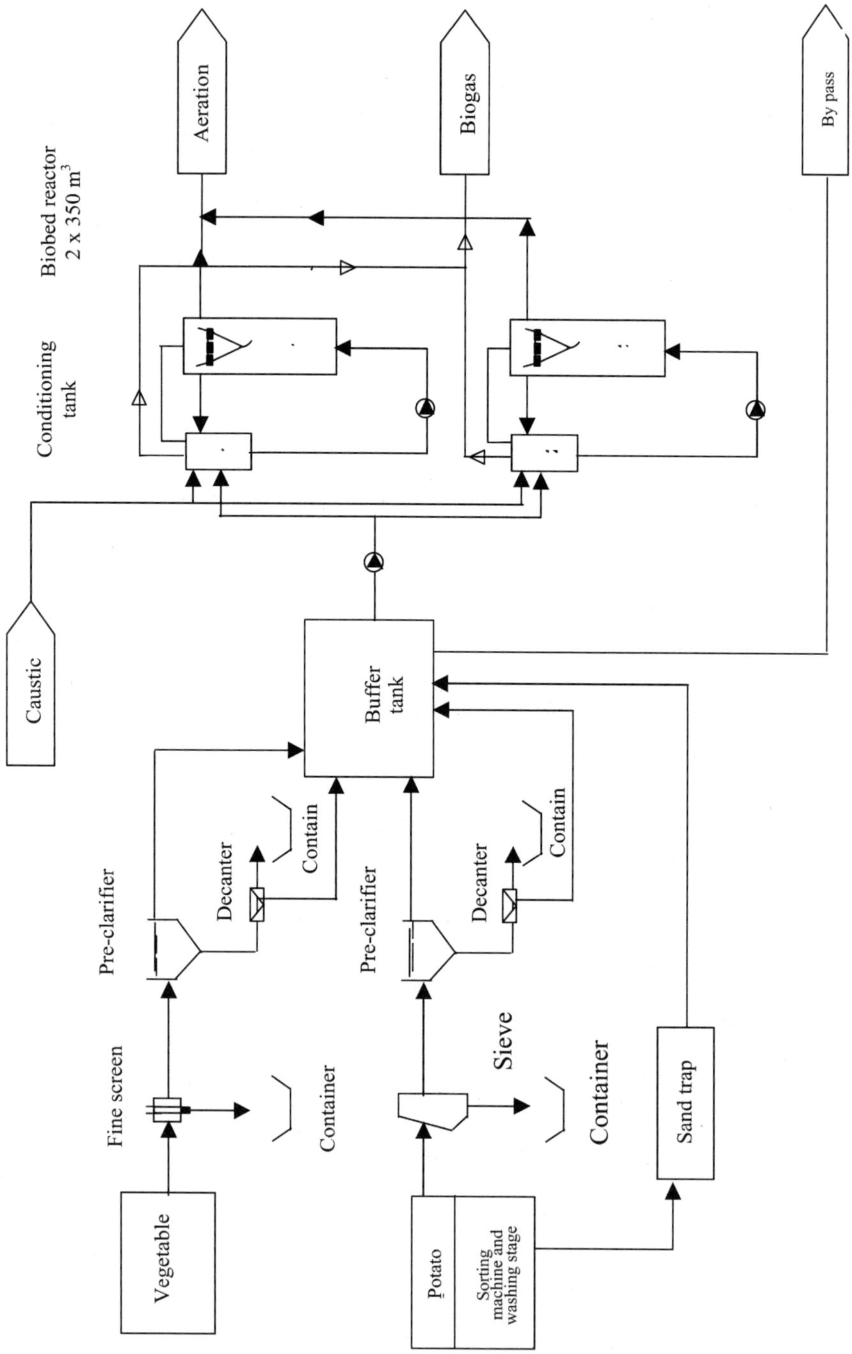

Figure 18 Schematic representation of the pretreatment stage and anaerobic treatment stage at Peka Kroef (from Ref. 14).

Table 12 Characteristics of Vegetable Processing Wastewaters After Removing Solids Through Settling and Drum

Unit	Raw material	Total COD (g/L)		Soluble COD (g/L)	
		Average	Range	Average	Range
Steam peeling	Carrot	19.4	17.4–23.6	17.8	15.1–22.6
	Potato	27.4	13.7–32.6	14.2	11.7–17.5
Blanching	Carrot	45.0	26.3–71.4	37.6	22.1–45.8
	Potato	39.6	17.0–79.1	31.3	10.9–60.6
	Swede	49.8	40.5–59.1	49.4	40.5–58.3

Source: Ref. 46.

Solution: Prior to anaerobic treatment of potato processing wastewater, it is important to provide favorable conditions for the anaerobic process through equalization and neutralization of the influent. Because the preclarified wastewater is almost neutral, there is no need for neutralization, and accordingly no need for correction of pH and temperature. Buffering of the wastewater is necessary here, to guarantee constant or near-constant flow. Total daily flow (average) = 360 m^3/day. Flow (average after buffering) = 15 m^3/hour, assuming that retention time is approximately 1 day in the buffer tank (balancing tank), with volume = 350 m^3. Influent COD (average) = 5000 mg/L. (Exact calculation of the buffer tank requires data plotted as the summation of inflow vs. time of day.)

Digester volume from the kinetic relationship:

$$\text{Detention time: } t = \frac{S_r}{X_v \cdot K \cdot S} = \frac{5000 \times 0.85}{5500 \times 0.0006 \times 750} = 1.72\,\text{day}$$

The digester volume is therefore:

$$\forall = (1.72\,\text{day})(360\,\text{m}^3/\text{day}) = 620\,\text{m}^3 \;(0.1638\,\text{MG})$$

Check SRT from the equation:

$$\text{SRT} = \frac{X_v t}{\Delta X_v} = \frac{X_v t}{aS_r - bX_v t}$$

$$= \frac{5500 \times 1.72}{0.14 \times 4250 - 0.021 \times 5500 \times 1.72} = 24\,\text{day}$$

This is in excess of the recommended SRT of 20 days to ensure the growth of methane formers.

$$\text{Daily COD load} = 5000\,\text{mg/L} \times 360\,\text{m}^3/\text{day} \times \frac{1}{1000} = 1800\,\text{kg COD/day}$$

$$\text{Design volumetric loading} = \frac{1800\,\text{kg/day}}{620\,\text{m}^3} = 3.0\,\text{kg/m}^3 \cdot \text{day}$$

This value is acceptable for a conventional anaerobic contact process. In the case of a UASB reactor, the organic loading can be easily increased to 10 $kg/m^3 \cdot day$, that is, it is sufficient to have only one-third or less of the calculated volume (about 200 m^3), to achieve the same performance.

In the case of the expanded granular sludge bed (EGSB) reactor, the organic loading can be increased up to 30 $kg/m^3 \cdot day$, where the required volume becomes only:

$$\frac{1800\,kg/day}{30\,kg/m^3 \cdot day} = 60\,m^3$$

The sludge yield from the process is:

$$\begin{aligned}\Delta X_v &= aS_r - bX_v t \\ &= (0.14)(4250) - (0.021)(5500)(1.72) = 396.34\,mg/L\end{aligned}$$

$$\begin{aligned}\Delta X_v &= 396.34\,mg/L \times 360\,m^3/day \times \frac{1}{1000} \\ &= 142.7\,kg/day\ (314\,lb/day)\end{aligned}$$

Gas production

$$G = 0.351(S_r - 1.42\Delta X_v)$$

where $G = m^3$ of CH_4 produced/day

$$\begin{aligned}G &= 0.351[(4250)(360) - (1.42)(142.7)] \\ &= 0.351\ (1530 - 202.63) = 465\,m^3CH_4/day\end{aligned}$$

or

$$G = 5.62(S_r - 1.42\Delta X_v)$$

where $G = ft^3$ of CH_4 produced/day

$$\begin{aligned}G &= 5.62[(4250)(0.0951\,MG/day)(8.34) - (1.42)(314)] \\ &= 16{,}433.5\,ft^3/day\ (465\,m^3/day)\end{aligned}$$

Heat required can be estimated by calculating the energy required to raise the influent wastewater temperature to 35°C (95°F) and allowing 1°F (0.56°C) heat loss per day of detention time. Average wastewater temperature = 30°C (86°F) and heat transfer efficiency = 50%.

$$\begin{aligned}BTU_{req.} &= \frac{W(T_i - T_e)}{E} \times (\text{specific heat}) \\ &= \frac{(95{,}100\,gal/day)(8.34\,lb/gal)(95° + 1.72°F - 86°)}{0.5} \times \left(\frac{1\,B_{tu}}{1\,lb.°F}\right) \\ &= 17{,}004{,}792\,BTU\ (17{,}940{,}055\,KJ)\end{aligned}$$

The heat available from gas production is $BTU_{avail.}$ = (16,433.5 ft^3 CH_4/day) (960 BTU ft^3 CH_4) = 15,776,160 BTU/day (16,643,850 kJ/day). External heat of 17,004,792 − 15,776,160 = 1,228,832 BTU/day (1,296,207 kJ/day) should be supplied to maintain the reactor at 35°C (95°F).

Nutrients required: the nitrogen required is:

$$N = 0.12\,\Delta X_v = 0.12 \times 142.7\,kg/day = 17.124\,kg/day\ (37.673\,lb/day)$$

The phosphorus required is:

$$P = 0.025\Delta X_v = 0.025 \times 142.7\,\text{kg/day} = 3.568\,\text{kg/day}\ (7.85\,\text{lb/day})$$

Remarks:

1. The effluent from the anaerobic plant alone does not meet the national minimum discharge limits because of the high values of residual COD (15% = 750 mg/L). Therefore it is recommended here to handle the anaerobic process effluent in an aerobic post-treatment (such as activated sludge). The final effluent of this combination of anaerobic and aerobic treatment processes can certainly be discharged to the central sewerage system or reused within the factory.
2. The equalization (buffering) was indicated in this example to dampen the fluctuations in potato processing wastewater flow that occur on a daily or longer term basis. It must be noted that optimum equalization of both flow and concentration are not achievable in a single process. To equalize flows, the buffer tank at certain times should be empty. To equalize concentration, the tank should always be full. Nevertheless, a tank that equalizes flows will also produce some reduction in peak concentration. Optimally, the organic loading to the anaerobic process reactor is constant over a 24 hour period. Equalization of flow was intended to be considered and simplified in this design example.

Advanced Treatment

Advanced wastewater treatment comprises a large number of individual treatment processes that can be utilized to remove organic and inorganic pollutants from secondary treated wastewater. The following treatment processes presented can be used to meet the effluent discharge requirements for potato processing plants. These may include suspended solids, BOD, nutrients, and COD.

Microstraining. Microstrainers consist of motor-driven drums that rotate about a horizontal axis in a basin, which collects the filtrate. The drum surface is covered by a fine screen with openings ranging from 23 to 60 μm. It has been reported that effluent suspended solids and BOD from microstrainers following an activated sludge plant have a ranges of 6–8 mg/L and 3.5–5 mg/L, respectively [56].

The head loss of the drum is less than 12–18 in (30–46 cm) of water. Peripheral drum speeds vary up to 100 ft/min (30.5 m/min) with typical hydraulic loadings of 0.06–0.44 m/min (1.5–10 gal/ft^2-min) on the submerged area; the backwash flow is normally constant and ranges up to 5% of the product water [57]. Periodic cleaning of the drum is required for slime control.

Granular Media Filtration. Granular filtration employing mixed media or moving bed filters plays an important role in improving the secondary effluent quality, where most of the BOD is found in bacterial solids. Therefore, removal of the suspended solids greatly improves the effluent quality. Granular filtration is generally preferred to microstraining, which is associated with greater operational problems and lower solids removal efficiencies.

Effective filter media sizes are generally greater than 1 mm. Filtration rates range from 0.06 to 0.5 m/min (1.5 to 12 gal/ft^2-min) with effluent suspended solids from 1 to 10 mg/L. This represents a reduction of 20 to 95% from the concentration in the filter influent [57,58]. Secondary effluent should contain less than 250 mg/L of suspended solids in order to make filtration more suitable [11]. In the case of higher concentrations of suspended solids, the secondary effluent should be first led to polishing ponds (maturation ponds) or subjected to chemical coagulation and sedimentation.

Chemical Coagulation Followed by Sedimentation. Phosphorus is a nutrient of microscopic and macroscopic plants, and thus can contribute to the eutrophication of surface waters. Phosphorus may be removed biologically or chemically. In some cases, chemicals may be added to biological reactors instead of being used in separate processes while in others, biologically concentrated phosphorus may be chemically precipitated. Chemical phosphorus removal involves precipitation with lime, iron salts, or alum. Lime should be considered for this purpose if ammonia removal is also required for pH adjustment. For low effluent phosphorus concentrations, effluent filtration may be required due to the high phosphorus content of the effluent suspended solids.

Whatever coagulant is employed, a large quantity of sludge is produced. Sludge lagoons can be considered as an economical solution to sludge disposal, although this treatment requires considerable land area.

Improved removal of phosphorus without any chemical addition can be obtained by a biological process that employs an anoxic or anaerobic zone prior to the aeration zone. When this process is used to maximize phosphate removal (sometimes called a sequencing batch reactor), it is possible to reduce the phosphorus content to a level of about 1 mg/L, with no chemical addition.

The principle of bio-P removal is the exposure of the organisms to alternating anaerobic and aerobic conditions. This can be applied with or without nitrogen removal. The alternating exposure to anaerobic and to aerobic conditions can be arranged by recirculation of the biomass through anaerobic and aerobic stages, and an anoxic stage if nitrogen removal is also required. General flowsheets of these processes are shown in Fig. 19.

As for potato processing wastewater, which often contains high concentrations of nutrients (N and P compounds), it is recommended here to apply biological phosphorus removal including an anoxic stage for the advanced treatment.

The abovementioned role of chemical coagulation may be followed by sedimentation in the reduction of nutrients. This method can also be applied to treat potato processing wastes in general [59]. Coagulating and flocculating agents were added to wastewater from abrasive-peeled, lye-peeled, and steam-peeled potato processing. Total suspended solid and COD concentrations were significantly reduced with chemical and polymer combination treatments, at adjusted pH levels.

Nitrification–Denitrification. Based on water quality standards and point of discharge, municipal treatment plants may be: (a) free from any limits on nitrogen discharges, (b) subject to limits on ammonia and/or TKN, (c) subject to limits on total nitrogen. Nitrogen can be removed and/or altered in form by both biological and chemical techniques. A number of methods that have been successfully applied can be found in many publications. Biological removal techniques include assimilation and nitrification–denitrification. Occasionally, nitrification is adequate to meet some water quality limitations where the nitrogenous oxygen demand (NOD) is satisfied and the ammonia (which might be toxic) is converted to nitrate. According to USEPA publications, the optimum pH range for nitrification has been identified as between 7.2 and 8.0. Regarding the effect of temperature, it has been noted that nitrification is more affected by low temperature than in the case of BOD removal [60].

Nitrification can be achieved in separate processes after secondary treatment or in combined processes in which both BOD and NOD are removed. In combined processes the ratio of BOD to TKN is greater than 5, while in separate processes the ratio in the second stage is less than 3 [57].

Denitrification is a biological process that can be applied to nitrified wastewater in order to convert nitrate to nitrogen. The process is anoxic, with the nitrate serving as the electron acceptor for the oxidation of organic material.

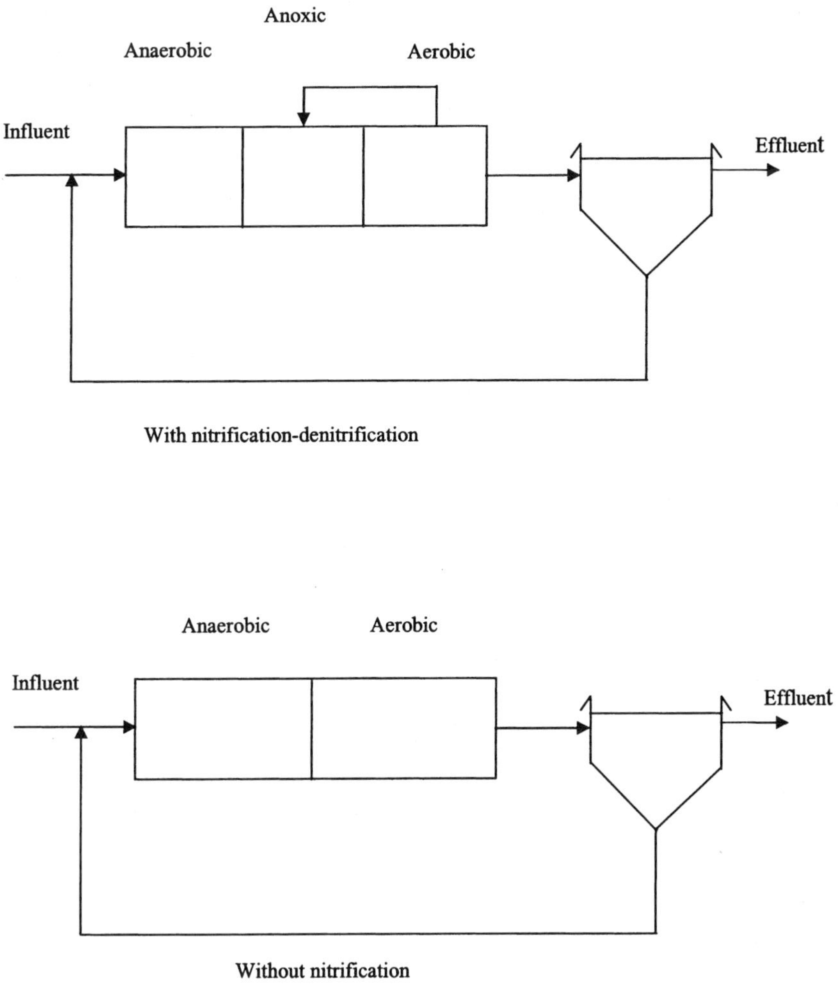

Figure 19 General flow sheets of biological phosphorus removal with and without nitrification–denitrification (from Ref. 24).

There is a variety of alternatives for the denitrification process such as suspended growth and attached growth systems with and without using methanol as a carbon source. Chemical nitrogen-removal processes generally involve converting the nitrogen to a gaseous form (N_2) and ammonia (NH_3). The processes of major interest include break-point chlorination, ion exchange, and air stripping. Natural zeolitic tuffs play an important role as ion exchange media for ammonium and phosphate removal through columns or batch reactors [61], where the total volume treated between generation cycles depends on the ammonium concentration in the wastewater and the allowed concentration in the effluent. The wastewater itself can be stripped of ammonia if it is at the requisite pH (10.5–11.5) and adequate air is provided. The feasibility of stripping the wastewater itself depends on whether the necessary pH can be achieved at moderate cost. The air stream carries with it the stripped ammonia to be released to the atmosphere. When the ammonia is dissolved in the solution, it forms the ammonium salt of the acid, which has an economic value as a fertilizer to the soil.

Regarding land-application systems for treatment of potato processing wastewaters, they may be satisfactory regarding nitrogen removal with no need for additional biological or chemical treatment.

Membrane Technology. Membrane technology encompasses a wide range of separation processes from filtration and ultrafiltration to reverse osmosis. Generally, these processes produce a very high quality effluent defined as membrane filtration and refer to systems in which discrete holes or pores exit the filter media, generally in the order of 10^2 to 10^4 nm or larger. The difference in size between the pore and the particle to be removed determines the extent of filtration efficiency. The various filtration processes in relation to molecular size can be found in Ref. 24.

The criteria for membrane technology performance are related to the degree of impermeability (the extent of membrane's detention of the solute flow) or the degree of permeability (the extent of membrane's allowance of the solute flow). The design and operating parameters for a reverse osmosis system are presented in detail in Ref. 62.

Regarding potato processing wastewaters, reverse osmosis and ultrafiltration have been used for treating wastewater for the recovery of sweet potato starch [63]. They may also be successful for application within in-plant treatment and recycling systems. Other advanced treatment methods used for various industrial wastewaters such as activated carbon adsorption, deep well injection, and chlorination, are not suitable for potato processing wastewater treatment due to their high costs of application.

It is worth mentioning that important research has been carried out regarding the treatment of potato processing wastewaters by the activated carbon adsorption process used as an advanced treatment method. It was reported that activated carbon adsorption treatment following complete mix activated sludge treatment removed 97% COD from primary settled potato processing wastewaters with an effluent COD of 24 mg/L [17]. In addition, it was concluded that powdered activated carbon was more effective than granular activated carbon in removing COD from activated sludge treated effluents.

18.4.2 Bases of Potato Processing Effluent Treatment

For an existing plant, it is necessary to measure the flow of all waste streams and determine the quantity and character of the pollutants found in these flows. The reduction of wastewater discharge into the final plant effluent and the reduction of water flow throughout the plant is of major importance. For a proposed new plant for which the waste treatment units must be designed, information may be found in the literature for a similar installation. In most cases, however, a reasonable estimate of the waste flow may be determined from the estimated capacity of the plant, the recovery of product expected, and the type of screening and clarification equipment to be installed. It is necessary to have accurate estimates of water usage and methods of reuse in application. For preliminary estimates, it can be assumed that a lb (or 1 kg) of dry potato solids exerts a BOD of 0.65 lb (or 0.65 kg) and a COD of 1.1 lb (or 1.1 kg) [11].

18.5 BYPRODUCT USAGE

18.5.1 In-Plant Usage of Potato Scraps

Plants processing French fries have developed additional product lines to utilize small potatoes (chopped or sliced), cutter scraps, slivers, and nubbins. These are processed similarly to French fries and include potato patties, mashed or whipped potatoes, diced potatoes, potato puffs, and hash browns [64].

18.5.2 Potato Peels

Approximately two million tons per year of potato peels are produced from potato processing as byproducts [65]. Potato peels provide a good source of dietary fiber, particularly when processed by a lye-peeling technique [66]. Potato peels contain 40 g dietary fiber/100 g dry matter, depending on the variety of potato processed and the method of peeling [67]. Application of extruded and unextruded potato peels as a source of dietary fiber in baked goods has been evaluated [1]. Acceptable muffins were made with a 25% replacement potato peel for wheat flour. Potato peels were also found to prolong muffin shelf-life by controlling lipid oxidation [65]. Extrusion cooking of potato peels affects the color of baked goods, and some physical and chemical properties of the peels [67]. Potato peels have also been used in limited quantities in a commercial snack food potato skin type product.

18.5.3 Potato Processing Wastes as Soil Conditioner

Potato processing solid wastes are often applied to agricultural land as a disposal medium. Research supports this method [68]. Solid potato processing wastes containing nitrogen are obtained by filtering or centrifuging the settled solids from the primary clarifiers. Wastes are applied to land and used for crops, which utilize the applied nitrogen. The soil does not accumulate the nitrogen or other organic waste and becomes increasingly fertile with continued wastewater application. Additionally, potato processing wastewater was found to be effective in promoting corn growth as effectively as commercial ammonium nitrate fertilizers, when applied at optimum nitrogen levels [69]. Applying wastewater and solid wastes from potato processing provides an effective method of applying reusable nutrients that would be otherwise wasted, and thus reduces pollution levels in municipal waterways.

18.5.4 Potato Wastes as Substrate for Organic Material Production

Potato wastes have also been evaluated as a potential source from which to produce acetone, butanol, and ethanol by fermentation techniques [70]. This application of biotechnology in membrane extraction resulted in a procedure to extract a biofuel that utilizes potato wastes as a renewable resource.

18.5.5 Cattle Feed

Filter cakes and dry potato peels are used as an excellent carbohydrate source in cattle feed. Using potato wastes instead of corn in cattle feed does not affect the metabolic state or milk status of the cattle [71]. Typically, potato wastes are fed in a dry, dewatered form. The use of wet potato wastes in cattle feed has been investigated to reduce drying expenditures. Wet potato processing wastes can be introduced into cattle feed up at to 20% without negative results.

The issue of dry vs. wet application of potato processing wastes was also explored. Again, dry potato wastes are expensive due to the drying processes used to stabilize the wastes. Wet wastes must be used quickly and within a close proximity to the potato processing wastes site due to microbial and enzymatic spoilage of the waste. Barley straw has been investigated as silage material to be mixed with wet potato wastes to absorb excess moisture [72]. Problems encountered with this procedure are due to elevated pH levels being attained following five weeks of storage. Elevated pH levels can permit growth of toxigenic bacteria.

Carbohydrate-rich potato wastes can also be converted to protein for additional nutrients for animal feed [1]. Research indicates that starchy substances such as potato wastes can be converted to "microbial biomass protein" by digestion with a amylolytic, acidophilic, thermophilic fungus. The fungus hydrolyzes starch, under specific high-temperature/low-pH conditions. Utilizing nitrogen in the potato wastes, the fungus produces protein which is filtered, and has been shown to be nutritionally effective in animal feeding trials if supplemented with methionine. Limitations of this process include the short time that wastes are viable for this treatment. Wastes can become toxic to fungus during storage. Potato and corn single-cell protein was also used in place of soybean meal as a source of supplemental protein in cattle feed. Results indicate the substitution can be made, if in conjunction with soybean meal protein for growing steers [73].

18.5.6 Potato Pulp Use

Processing potato starch results in potato pulp as a major byproduct, particularly in Europe. Research indicates that potato pulp can be fractionated to produce several commercially viable resources. Pectin and starch can be isolated, as well as cellulase enzyme preparation [74]. It was hypothesized that ethanol production would be feasible, but low sugar concentration prevented this. Potato pulp may also have applications for reuse in the following industries: replacement of wood fiber in paper making, and as a substrate for yeast production and B_{12} production [74]. Potato pulp isolated from potato starch production can be isolated and sold as pomace [75]. Protein can also be isolated from the starch processing wastewater and sold as fractionated constituents [74].

In summary, new technologies have served to minimize potato processing wastes and appropriate means of utilizing the rich byproducts are still under research. The vast quantities of wastes will continue to be minimized and byproducts have found new applications as renewable resources and potential energy sources. All of these goals will continue to be realized as research leads to the development of unique technologies to treat wastes, minimize the impact on the environment, reduce use of valuable natural resources, and reduce the impact of waste effluent.

REFERENCES

1. Stevens, C.A.; Gregory, K.F. Production of microbial biomass protein from potato processing wastes by cephalosporium eichhorniae. Appl. Environ. Microbiol. **1987**, *53*, 284–291.
2. Vegt, A.; Vereijken. M. Eight year full-scale experience with anaerobic treatment of potato processing effluent, In *Proceedings of the 46th Industrial Waste Conference*, Purdue University, West Lafayette, IN, 1992; 395–404.
3. Guttormsen, K.G.; Carlson, D.A. *Current Practice in Potato Processing Waste Treatment*, Water Pollution Research Series, Report No. DAST-14; Federal Water Pollution Control Federation, U.S. Department of the Interior: Washington, DC, 1969.
4. Talburt, W.F.; Smith, O. *Potato Processing*; Van Nostrand Reinhold Company: New York, 1967.
5. Gray, H.F.; Ludwig, H.F. Characteristics and treatment of potato dehydration wastes. Sewage Works, **1943**, *15*, 1.
6. Cooley, A.M.; Wahl, E.D.; Fossum, G.O. Characteristics and amounts of potato wastes from various process stream. In *Proceedings of the 19th Industrial Waste Conference*, Purdue University, West Lafayette, IN, 1964; 379–390.
7. Abeling, U.; Seyfried, C.F. Anaerobic-aerobic treatment of potato-starch wastewater. Water Sci. Technol. **1993**, *28* (*2*), 165–176.
8. Hadjivassilis, I.; Gajdos, S.; Vanco, D.; Nicolaou, M. Treatment of wastewater from the potato chips and snacks manufacturing industry. Water Sci. Technol. **1997**, *36* (*2–3*), 329–335.

9. Kadlec, R.H. Deterministic and stochastic aspecting constructed wetland performance and design. Water Sci. Technol. **1997**, *35* (*5*), 149–156.
10. Hung, Y.T. Tertiary treatment of potato processing waste by biological activated carbon process. Am. Potato J. **1983**, *60* (*7*), 543–555.
11. Pailthorp, R.E.; Filbert, J.W.; Richter, G.A. Treatment and disposal of potato wastes. In *Potato Processing*; Talburt, W.F., Smith, O., Eds.; Van Nostrand Reinhold Co.: New York, 1987; 747–788.
12. USEPA. *Development Document for Proposed Effluent Limitation Guidelines and New Source Performance Standards for the Citrus, Apple and Potato Segment of the Canned and Preserved Fruits and Vegetables Processing Plant Source Category*, EPA-440/1–73/027; U.S. Environmental Protection Agency: Washington, DC, 1973.
13. Austerman-Haun, U.; Mayer, H.; Seyfried, C.F.; Rosenwinkel, K.H. Full scale experiences with anaerobic/aerobic treatment plants in the food and beverage industry. Water Sci. Technol. **1999**, *40*(*1*), 305–325.
14. Zoutberg, G.R.; Eker, Z. Anaerobic treatment of potato processing wastewater. Water Sci. Technol. **1999**, *40* (*1*), 297–304.
15. Kadlec, R.H.; Burgoon, P.S.; Henderson, M.E. Integrated natural systems for treating potato processing wastewater. Water Sci. Technol. **1997**, *35* (*5*), 263–270.
16. Hung, Y.T. Batch mesophilic anaerobic digestions of potato wastewaters. *Am. Potato J.* **1989**, *66* (*7*), 437–447.
17. Hung, Y.T. Treatment of potato processing wastewaters by activated carbon adsorption process. Am. Potato J. **1984**, *61* (*1*), 9–22.
18. Olson, O.O.; Van Heuvelen, W.; Vennes, J.W. Experimental treatment of potato wastes in North Dakota. In *Proceeding of the International Symposium, Utilization and Disposal of Potato Wastes*; New Brunswick Research and Productivity Council: New Brunswick, Canada, 1965; 316–344.
19. Dickey, H.C.; Brugman, H.H.; Highlands, M.E.; Plummer, B.E. The use of by-products from potato starch and potato processing. In *Proceeding of the International Symposium, Utilization and Disposal of Potato Wastes*; New Brunswick Research and Productivity Council: New Brunswick, Canada, 1965; 106–121.
20. Hung, Y.T.; Priebe, B.D. *Biological Activated Carbon Process for Treatment of Potato Processing Wastewater for In-Plant Reuse*, Report No. 81-10-EES-01; Engineering Experimental Station, University of North Dakota: Grand Forks, North Dakota, 1981.
21. Loehr, R.C. Biological processes. In *Agricultural Wastes Management Problems, Processes and Approaches*; Academic Press: New York, 1974; 129–182.
22. Bertola, N.; Palladino, L.; Bevilacqua, A.; Zaritzky, N. Optimisation of the design parameters in an activated sludge system for the wastewater treatment of a potato processing plant. Food Eng. **1999**, *40*, 27–33.
23. Davis, M.L.; Cornwell, D.A. *Introduction to Environmental Engineering*, 2nd Ed.; McGraw-Hill, International: New York, 1991.
24. Eckenfelder, W.W. *Industrial Water Pollution Control*, 2nd Edition; McGraw-Hill, International: New York, 1989.
25. Fossum, G.O.; Cooley, A.M.; Wahl, G.D. Stabilization ponds receiving potato wastes with domestic sewage. In *Proceedings of the 19th Industrial Waste Conference*, Purdue University, West Lafayette, IN, 1964; 96–111.
26. Bastian, R.K.; Hammer, D.A. The use of constructed wetlands for wastewater treatment and recycling. In *Constructed Wetlands for Water Quality Improvement*; Moshiri, G.A; Ed.; Lewis Publishers: Boca Raton, FL 1993; 59–68.
27. de Zeeuv, W.; Heijnen, G.; de Vries, J. Reed bed treatment as a wastewater (post) treatment alternative in the potato starch industry. In *Constructed Wetlands in Water Pollution Control (Adv. Water Pollut, Control No. 11)*; Cooper, P.F., Findlate, B.C., Eds.; Pergamon Press: Oxford, UK 1990; 551–554.
28. Van Oostrom, A.J. Nitrogen removal in constructed wetlands treating nitrified meat processing effluent. Water Sci. Technol. **1995**, *33* (*3*), 137–148.

29. Van Oostrom, A.J.; Russel, J.M. Denitrification in constructed wastewater wetlands receiving high concentrations of nitrate. Water Sci. Technol. **1992**, *29* (*4*), 7–14.
30. Van Oostrom, A.J.; Cooper, R.N. Meat processing effluent treatment in surface-flow and gravel-bed constructed wastewater wetlands. In *Constructed Wetlands in Water Pollution Control (Adv. Water Pollut. Control No. 11)*; Cooper, P.F., and Findlater, B.C., Eds.; Pergamon Press: Oxford, UK, 1990; 321–332.
31. Haberl, R.; Partler, R.; Mayer, H. Constructed wetlands in Europe. Water Sci. Technol. **1995**, *33* (*3*), 305–315.
32. Burka, U.; Lawrence, P.C. A new community approach to wastewater treatment with higher plants. In *Constructed Wetlands for Water Pollution Control (Adv. Water Pollut, Control No. 11)*; Cooper, P.F., Findlater, B.C., Eds.; Pergamon Press: Oxford, UK, 1990; 359–371.
33. Bahlo, K.E.; Wach, F.C. Purification of domestic sewage with and without faeces by vertical intermittent filtration in reed and rush beds. In *Constructed Wetlands for Water Pollution Control (Adv. Water Pollut, Control No. 11)*; Cooper, P.F., Findlater, B.C., Eds.; Pergamon Press: Oxford, UK, 1990; 215–221.
34. Burgoon, P.S.; Kadlec, R.H.; Henderson, M. Treatment of potato processing wastewater with engineered natural systems. Water Sci. Technol. **1999**, *40* (*3*), 211–215.
35. Cocci, A.A.; Page, I.C.; Grant, S.R.; Landine, R.C. Low-rate anaerobic treatment of high-strength industrial wastewater: ADI-BVF case histories. In *Seminar of Anaerobic Treatment for Industrial Wastes*; East Syracuse, New York, 1997.
36. Malina, J.F.; Pohland, F.C. *Design of Anaerobic Processes for the Treatment of Industrial and Municipal Wastes*; Technomic Publishing Company, Inc.: Lancaster, PA, 1992.
37. Metcalf and Eddy, Inc. *Wastewater Engineering*; McGraw-Hill: New York, 1991.
38. Liyah, R.Y.; Hung, Y.T. Bio-augmented activated sludge treatment of potato wastewaters. Acta Hydrochim Hydrobiol. **1988**, *16* (*2*), 223–230.
39. Hung, Y.T.; Howard, H.L.; Javaid, A.M. Effect of bio-augmentation on activated sludge treatment of potato wastewater. Environ. Stud. **1994**, *45*, 98–100.
40. Hung, Y.T.; Jen, P.C. Anaerobic filter followed by activated sludge process with bio-augmentation for combined potato and sugar wastewater treatment. In *Proceedings of 1987 Food Processing Waste Conference*, Atlanta, Georgia, September 1–2, 1987.
41. Shih, J.K.C.; Hung, Y.T. Biological treatment of potato processing wastewaters. Am. Potato J. **1987**, *64* (*9*), 493–506.
42. Chambers, D.A. Improving removal performance reliability of a wastewater treatment system through bio-augmentation. In *Proceedings of the 36th Industrial Waste Conference*, Purdue University, West Lafayette, IN, 1981; 631.
43. Peavy, H.S.; Rowe, D.R.; Tchobanoglous, C. *Environmental Engineering* 1st Ed.; McGraw-Hill, International: New York, 1985.
44. USEPA. *Ground-Water and Leachate Treatment Systems*, EPA/625/R-94/005; Environmental Protection Agency: Washington, DC, 1995.
45. Hulshoffpol, L.; Hartlieb, E.; Eitner, A.; Grohganz, D. GTZ sectorial project promotion of anaerobic technology for the treatment of municipal and industrial sewage and wastes. In *Proceedings of the 8th International Conference on Anaerobic Digestions*; Sendai, Japan, **1997**, *2*, 285–292.
46. Lepisto, S.S.; Rintala, J.A. Start-up and operation of laboratory-scale thermophilic upflow anaerobic sludge blanket reactors treating vegetable processing wastewaters. Chem. Technol. Biotechnol. **1997**, *68*, 331–339.
47. Wiegant, W.M.; de Man, A.W.A. Granulation of biomass in the thermophilic upflow anaerobic sludge blanket reactor treating acidified wastewaters. Biotechnol. Bioeng. **1986**, *28*, 718–727.
48. Wiegant, W.M.; Lettinga, G. Thermophilic anaerobic digestion of sugars in upflow anaerobic sludge blanket reactors. Biotechnol. Bioeng. **1985**, **27**, 1603–1607.
49. Van Lier, J.B.; and Lettinga, G. Limitations of thermophilic anaerobic wastewater treatment and the consequences for process design. In *Proceedings of International Meeting on Anaerobic Processes for Bioenergy and Environment*, Copenhagen, Denmark, January 25–27, 1995, 1995; Section 16.

50. Uemura, S.; Harada, H. Microbial characteristic of methanogenic sludge consortia developed in thermophilic UASB Reactors. Appl. Microbiol. Biotechnol. **1995**, *39*, 654–660.
51. Rintala, J.; Lepistö, S. Anaerobic treatment of thermomechanical pulping whitewater at 35–70°C. Water Res. 1992, *26*, 1297–1305.
52. Souza, M.E.; Fuzaro, G.; Polegato, A.R. Thermophilic anaerobic digestion of vinasse in pilot plant UASB reactor. Water Sci. Technol. **1992**, *25* (*7*), 213–222.
53. Ohtsuki, T.; Tominaga, S.; Morita, T.; Yoda, M. Thermophilic UASB system start-up and management-change in sludge characteristics in the start-up procedure using mesophilic granular sludge. In *Proceedings of Seventh International Symposium on Anaerobic Digestion*, Cape Town, South Africa, January 23–27, 1994; 348–357.
54. Garavini, B.; Mercuriali, L.; Tilche, A.; Xiushan, Y. Performance characteristics of a thermophilic full scale hybrid reactor treating distillery slops. In *Poster-Papers of the Fifth International Symposium on Anaerobic Digestion,* Bologna, Italy, 22–26 May, 1988; Tilche, A; Rozzi, A; Eds.; 509–515.
55. Trösch, W.; Chmiel, H. Two-stage thermophilic anaerobic digestion of potato wastewater-experience with laboratory, pilot scale and full-scale plants. In *Poster-Papers of the Fifth International Symposium on Anaerobic Digestion,* Bologna, Italy, 22–26 May, 1988; Tilche, A; Rozzi, A; Eds.; 599–602.
56. Lynam, B.; Ettelt, G.; McAloon, T. Tertiary treatment at Metro Chicago by means of rapid sand filtration and microstrainers. Water Pollut. Control Feder. **1969**, *41*, 247.
57. McGhee, T.J. *Water Supply and Sewerage*, 6th Ed.; McGraw-Hill, International: New York, 1991.
58. Ripley, P.G.; Lamb, G. Filtration of effluent from a biological-chemical system. Water Sewage Works. **1973**, *12* (*2*), 67.
59. Karim, M.I.A.; Sistrunk, W.A. Treatment of potato processing wastewater with coagulating and polymeric flocculating agents. Food Sci. **1985**, *50*, 1657–1661.
60. Sutton, P.M. et al. Efficacy of biological nitrification. Water Pollut. Control Feder. **1975**, *47*, 2665.
61. Awad, A.; Garaibeh, S. Nutrients removal of biological treated effluent through natural zeolite. In *Proceedings of the Second Syrian-Egyptian Conference in Chemical Engineering*; Al-Baath University: Homs, Syria, 20–22 May 1997; 616–640.
62. Agardi, F.J. Membrane processes. In *Process Design in Water Quality Engineering*; Tackston, E.L.; and Eckenfelder, W.W., Eds.; Jenkins Publishing Co.: Austin, Texas, 1972.
63. Chiang, B.H.; Pan, W.D. Ultrafiltration and reverse osmosis of the wastewater from sweet potato starch process. Food Sci. **1986**, *51* (*4*), 971–974.
64. Talburt, W.F.; Weaver, M.L.; Renee, R.M.; Kueneman, R.W. Frozen french fries and other frozen potato products. In *Potato Processing*; Talburt, W.F, Smith, O; Eds.; Van Nostrand Reinhold Co.: New York, 1987; 491–534.
65. Arora, A.; Camire, M.E. Performance of potato peels in muffins and cookies. Food Res. Inter. **1994**, *27*, 15–22.
66. Smith, O. Potato Chips. In *Potato Processing*; Talburt, W.F, and Smith, O; Eds.; Van Nostrand Reinhold Co.: New York, 1987; 371–474.
67. Arora, A.; Jianxin, Z.; Camire, M.E. Extruded potato peel functional properties affected by extrusion conditions. Food Sci. **1993**, *58* (*2*), 335–337.
68. Smith, J.H. Decomposition of potato processing wastes in soil. *Environ. Qual.* **1986**, *15*(*1*), 13–15.
69. Smith, J.H.; Hayden, C.W. Nitrogen availability from potato processing wastewater for growing corn. Environ. Qual. **1984**, *13* (*1*), 151–158.
70. Grobben, N.G.; Egglink, G.; Cuperus, F.P.; Huizing H.J. Production of acetone, butanol and ethanol (ABE) from potato wastes: fermentation with integrated membrane extraction. Appl. Microbiol. Biotechnol. **1993**, *39*, 494–498.
71. Onwubuwmell, C.; Huber, J.T.; King, K.J.; Johnson, C.O.L.E. Nutritive value of potato processing wastes. Dairy Sci. **1985**, *68* (*5*), 1207–1214.
72. Sauter, E.A.; Hinman, D.D.; Parkinson, J.F. The lactic acid and volatile fatty acid content and in vitro organic matter digestibility of silages made from potato processing residues and barley. Anim. Sci. **1985**, *60* (*5*), 1087–1094.

73. Hsu, J.C.; Perry, T.W.; Mohler, M.T. Utilization of potato-corn biosolids single-cell protein and potato-corn primary waste by beef cattle. Anim Sci. **1984**, *58* (*5*), 1292–1299.
74. Kingspohn, U.; Bader, J.; Kruse, B.; Kishore, P.V.; Schugerl, K.; Kracke-Helm, H.A.; Likidis, Z. Utilization of potato pulp from potato starch processing. Proc. Biochem. **1993**, *28*, 91–98.
75. Treadway, R.H. Potato Starch. In *Potato Processing*; Talburt, W.F, Smith, O. Eds.; Van Nostrand Reinhold Co.: New York, New York, 1987; 647–666.

19
Stormwater Management and Treatment

Constantine Yapijakis
The Cooper Union, New York, New York, U.S.A.

Robert Leo Trotta
Sullivan County Division of Public Works, Monticello, New York, U.S.A.

Chein-Chi Chang
District of Columbia Water and Sewer Authority, Washington, D.C., U.S.A.

Lawrence K. Wang
Zorex Corporation, Newtonville, New York, U.S.A., and
Lenox Institute of Water Technology, Lenox, Massachusetts, U.S.A.

19.1 CONSIDERATIONS FOR STORMWATER MANAGEMENT AND TREATMENT

19.1.1 Pollution Aspects and Considerations

The pollution aspects of stormwater are related to the substances that become entrained in it from its point of origin to its point of discharge into a water body. Stormwater originates from the clouds and its first contamination is from pollution sources contained within the air we breathe. Most notable and well known is the pollution related to acid rain. Acid rain is generally stormwater that has absorbed airborne contaminants propagated by the burning of sulfur-bearing fuels used for heating and power generation. The oxidation of the sulfur and subsequent reaction with atmospheric water vapor produces sulfuric acid. This is but one example of a mechanism that contributes to the contamination of stormwater. Further details with respect to acid rain and other pathways involving the entrapment of pollutants in stormwater are discussed in Section 19.3.

In industry there are many compounds that in the presence of water and other substances could lead to the development of acidic, caustic, or poisonous characteristics in stormwater. Of particular interest in this regard is the possible entrainment of nutrients, organics, inorganics, heavy metals, pesticides, volatile organics, oils, greases, and other pollutants. The contaminants can enter into stormwater in the form of liquids, floatables, grit, settleable solids, suspended solids, soluble substances, and dissolved gases. These substances in significant concentrations can have an adverse impact on fish and plant life contained within the water body receiving the stormwater discharge, as well as wildlife that utilizes the water resources. Furthermore, when such water bodies are either tributary to or directly used as drinking water supplies, the contaminated stormwater could contribute to the destruction of the surface water supply.

Similarly, groundwater drinking water supplies can be polluted by contaminated stormwater. The stormwater enters groundwater supplies through points of recharge from surface waters and through percolation into the soil.

19.1.2 Federal Stormwater Regulations

In the United States, federal laws have dictated the course of measures implemented on federal, state, and local levels to control discharges into the nation's surface waters. In the past, the laws focused on control of wastewater discharges; however, more recent considerations have been with respect to combined sewer overflows and industrial stormwater discharges. Combined sewer overflow is the discharge to water bodies from combined sewers that occurs as a result of a storm event (they normally convey only sanitary flows during dry weather). Currently, industries that are connected to such systems are regulated through pretreatment regulations administered on a local level in accordance with federal and state program requirements.

Federal regulation of stormwater originated with the 1987 Clean Water Act amendments, which mandated the establishment of a permit system for point sources of stormwater discharges into waters of the United States [1]. The permit requirements initially developed by the US Environmental Protection Agency (USEPA) mandated the issuance of State National Pollutant Discharge Elimination System (NPDES) permits for five categories of stormwater discharges based on the Code of Federal Regulations (40 CFR 126.26), only three of which have a primary impact on the industrial and business sector. Two stormwater rules followed in 1990 and 1992: the "stormwater application rule" and the "stormwater implementation rule." The stormwater application rule of November 1990 identified the types of facilities subject to permitting under the NPDES program, and the stormwater implementation rule of April 1992 described the requirements of NPDES permits [2]. Phase I of the stormwater application rule applied to heavy industrial discharges, as well as large and medium municipal separate storm sewers and operators of large construction sites. The Phase II rule expanded the Phase I authority to include small municipal separate storm sewers and small construction sites.

Industrial facilities are required to comply with stormwater rules if they meet the following criteria. The facilities fall within one of the following categories if they discharge stormwater via one or more point sources into US waters:

- either engaged in industrial activities;
- already covered under an NPDES permit;
- identified by the USEPA as contributing to a water quality violation.

Note that the stormwater rules are not applicable in the following situations.

- nonpoint source discharges of stormwater;
- discharges of stormwater to municipal sewer systems that are combined stormwater and sanitary sewers;
- discharges of stormwater to groundwater.

The Multi-Sector General permit and the Individual permit are the two types of stormwater discharge permits currently issued to industrial dischargers by the NPDES permitting authority.

Multi-Sector General Permit

The Multi-Sector General Permit (MSGP) is the simplest form of NPDES permit coverage that industrial facilities can obtain, although there are circumstances that would cause a facility to be ineligible for MSGP coverage. Industrial facilities that have activities covered under one or more of the industrial sectors in the MSGP are eligible for coverage. To obtain MSGP coverage, the facilities must submit a Notice of Intent (NOI) for coverage, and prepare and implement a Storm Water Pollution Prevention Plan (SP3). The MSGP contains industrial-specific requirements for stormwater monitoring, reporting, and best management practices (BMPs) to minimize contamination of runoff.

Individual Permit

The Individual Permit requires the preparation and submittal of NPDES forms 1 and 2F, which request specific information about the facility, the industrial operations, and the results of stormwater sampling, analysis, and flow measurement. A facility-specific Individual Permit is issued by the NPDES permitting authority and typically contains discharge limits, monitoring, reporting requirements, and may require implementation of BMPs or pollution prevention measures.

Construction General Permit

The Construction General Permit is applicable to construction projects at industrial facilities that disturb one or more acres of land area. The permitting process is the same as for the MSGP: submittal of an NOI for coverage and implementation of an SP3 that focuses on BMPs during construction.

Stormwater Pollution Prevention Plan (SP3)

Among the important requirements of MSGP is the development and implementation of an SP3. The goal of SP3 is to reduce or eliminate the amount of pollutants in stormwater discharges from an industrial site. The SP3 must be developed with input from a designated Pollution Prevention Team. The SP3 must identify all potential pollutant sources and include descriptions of control measures to eliminate or minimize contamination of stormwater. The SP3 must contain the following [3]:

- a map of the industrial facility identifying the areas that drain to each stormwater discharge point;
- identification of the manufacturing or other activities that takes places within each area;
- identification of the potential sources of pollutants within each area;
- an inventory of materials that can be exposed to stormwater;
- an estimate of the quantity and type of pollutants likely to be contained in the stormwater runoff;
- a history of spills or leaks of toxic or otherwise hazardous material for the past three years.

Best Management Practices

Best Management Practices (BMPs) must be identified that should include good housekeeping practices, structural control measures, a preventive maintenance program for stormwater control measures, and procedures for spill prevention and response. As needed, traditional stormwater management controls, such as oil/water separators and retention/equalization devices must also be included [4].

For facilities that are subject to the Emergency Planning and Community Right-to-Know Act (EPCRA 313) reporting, the SP3 must address those areas where the listed Section 313 "toxic water priority chemicals" are stored, processed, or handled. These areas typically require stricter BMPs in the form of structural control measures.

A certification of nonstormwater dischargers. The facility must have piping diagrams that confirm no nonstormwater connections to the storm sewer. Otherwise, all outfalls must be tested to insure that there are no connections of sewers that carry other than stormwater.

A record-keeping system must be developed and maintained, as well as an effective program for training employees in matters of controls and procedures for pollution prevention.

19.2 QUANTITY AND QUALITY

19.2.1 Hydrologic Considerations

Meteorologists collect data on, report on, and work with the total depths of rainfall events of various durations. Engineers, on the other hand, use the average rainfall intensity (ratio of total depth and duration of an event) as the primary parameter for their work, implicitly assuming that the intensity of a rainfall event is constant during its occurrence. Extensive presentations of the following concepts may be found in any book on hydrology for engineers [5,6].

Rainfall Depth, Duration, and Frequency

Many different empirical formulas have been proposed by researchers to describe the presumed relationships between rainfall intensity and the duration frequency of an event or between rainfall depth and duration frequency. Such relationships are derived from statistical analysis either of point rainfall data, that is, precipitation events as measured by a single rain-gage station, or of data from networks of rain gages. The point data and their evaluation results are statistically adequate to define the main temporal variations of the characteristics of storm events. One observation is that as the duration of a storm event decreases, the average rainfall intensity increases given a specific frequency of return. Another observation useful in design is that, as the frequency of the return increases, the average rainfall intensity decreases given a specific duration. Data from networks of rain gages and their evaluation results are statistically sufficient to define the main spatial variation characteristic of storm events. The observation is that the more limited the area over which a storm event is occurring, the higher the value of the average rainfall intensity as compared to the maximum observed point rainfall intensity within the event area. For design purposes, the ratio of the spatial average to the point temporal average rainfall intensity (corresponding to identical frequencies of return) is required in order to adjust a design storm event point depth to account for spatial variation.

Probable Maximum Rainfall

Certain critical storm events are used in estimating flood flow peak design values by US water resources agencies such as the Corps of Engineers. As reported by Riedel et al. [7], one such critical storm event is the probable maximum precipitation. This is defined as the critical depth–duration–area rainfall relationship for a specific area during the seasons of the year, resulting from a storm event of the most critical meteorological conditions. The probable maximum rainfall is based on the most effective combination of factors that control rainfall intensity. Annual probable maximums may be less important than seasonal maximums, in flooding situations that may occur in combination with snowmelt runoff.

Evapotranspiration and Interception of Rainfall

Evaporation is the process by which precipitated water is lost to the runoff process by transference from land and water masses of the earth to the atmosphere, in the form of vapor.

Transpiration is water loss to the atmosphere through the action of plants that absorb it with their roots and let it escape through pores in their leaves. From the practical viewpoint of water resources engineers, only total evapotranspiration (i.e., combined evaporation and transpiration) is of interest. Various investigators have proposed theoretical, analytical, or empirical methods for estimating evapotranspiration losses, but no system has been found acceptable under all encountered conditions. An additional part of the precipitation volume from a storm event is intercepted by the vegetation cover of a drainage area until it evaporates and, therefore, it is lost to the runoff process. The volume of intercepted water depends on the storm event character, the species and density of plants and trees, and the season.

Depression Storage and Infiltration Losses

Precipitation that is also lost to the surface runoff process may infiltrate into the ground or become trapped in the many ground depressions from where it can only escape through evaporation or infiltration. Owing to the fact that there is extreme variability in the characteristics of land depressions and insufficient measurements, no generalized relationships with enough specified parameters for all situations are possible. Nevertheless, a few rational models and values of the range of depression storage losses have been reported in the literature. Infiltration losses are a very significant parameter in the distribution of the water volume from a storm event. As accurate as possible estimates of infiltrating volumes must, therefore, be made since they affect the timing, distribution, and magnitude of precipitation surface runoff. The type and extent of the vegetal cover, the condition and properties of the surface crust and the soil, and the rainfall intensity are among the factors that may influence the rate of infiltration f. No satisfactory general relationship exists. Instead, hydrograph analyses and infiltrometer studies are methods used for infiltration capacity estimates. For small urban areas that respond rapidly to storm inputs, more precise values of infiltration rates are sometimes needed, whereas on large watersheds where long-duration storm events generate the peak flow conditions, average or representative values may suffice.

19.2.2 Surface Runoff

Runoff Flows and Hydrographs

When considering stormwater management, surface runoff is the main concern. However, the relationship between precipitation and runoff is most complex and influenced by such storm event characteristics as pattern, antecedent events, and watershed parameters. Many approximate formulas, therefore, have been developed and empirical methods such as the rational formula or site-specific equations can estimate the peak runoff rate, in cases where it is sufficient for the analysis and design of simple stormwater systems. Calculations of runoff volumes using sound rational equations based on physical principles and hydrographs are necessary in cases where a more detailed analysis of the system hydrology and hydraulics is needed. A hydrograph is a continuous graph showing the magnitude and time distribution of the main parameters, stage and discharge, of surface runoff or stream flow. It can, therefore, be a stage hydrograph or a discharge hydrograph (more common) and it is influenced by the physical and hydrological characteristics of the drainage basin. The discharge shown by a hydrograph at any time is the additive result of the direct surface runoff, interflow, groundwater or base flow, and channel precipitation. A typical hydrograph is shown in Fig. 1.

Drainage Basin Characteristics

The shape of the flood hydrograph from a catchment area is a function of the hydrologic input to that region and of the catchment characteristics, such as area, shape, channel, and overland

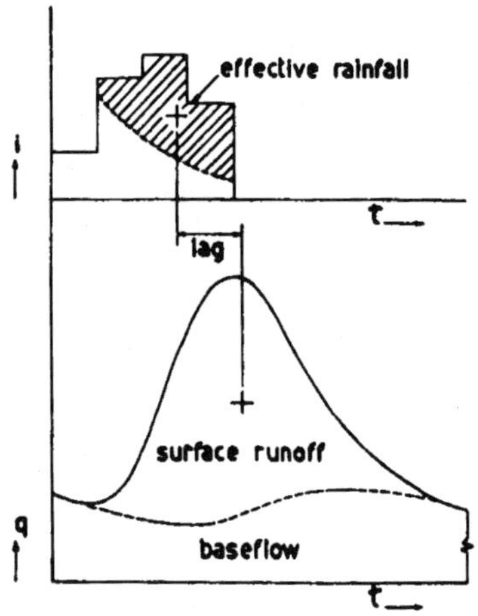

Figure 1 Rainfall/runoff relationship.

slopes, soil types and their distribution, type and extent of vegetative cover, and other geological and geomorphological watershed features. One of the primary measures of the relative timing of hydrologic events is basin lag t_l. Basin lag is defined as the time between the center of mass of the rainfall excess producing surface runoff and the peak of the hydrograph produced. The lag time is influenced by such parameters as the shape and average slopes of the drainage area, the slope of the main channel, channel geometry, and the storm event pattern. Various investigators have proposed relationships predictive of basin lag, but Snyder's equation [8], based on the data from large natural watersheds, is the most widely used and adapted by others

$$t_l = Ct(Lca\,L)^{0.3} \tag{1}$$

where t_l = basin lag (hour), Ct = coefficient depending on basin properties, Lca = distance (miles) along the main stream from the base gage to a point opposite the basin centroid, and L = maximum travel distance (miles) along the main stream (1 mile = 1609 m). The Soil Conservation Service [9] defines t_l as

$$t_l = 0.6\,t_c \tag{2}$$

where t_c (hour) is the time of concentration, another primary measure of the relative timing of hydrologic events. The time of concentration is usually defined as the sum of the overland travel time from the furthest basin point and the channel travel time to the outlet of concern.

Runoff and Snowmelt Runoff Determination

Water resource engineers are involved in estimating stream flows using one of two approaches. The first, an indirect approach in which runoff is estimated based on observed or expected precipitation, will be discussed in Section 19.2.3. The second method is based on the direct analyses of recorded runoff data without consideration of corresponding rainfall data. These types of analyses are usually frequency studies to evaluate the probability of occurrence of a specific runoff event, to determine the risk associated with a design or operation alternative. Such frequency analyses usually determine maximums or floods and minimums or droughts. However, when existing runoff records are short-term or incomplete, the frequency analyses cannot be very reliable. In certain cases, sequential generating techniques or time-series analyses are used to develop synthetic records of runoff for any desired length of time. In many areas, such as mountainous watersheds, snowmelt runoff is the dominant source of stream flows. For instance, Goodell [10] has reported that as much as 90% of the annual water supply volume in the high-elevation watersheds of the Colorado Rockies may originate in snowfall accumulations. Some of the greatest flood flows may be caused by a combination of very large rainstorms and simultaneous snowmelt. Adequate knowledge of the extent and other characteristics of snow packs within a watershed, therefore, is very important in stream flow forecasting. Investigators have followed various approaches to runoff determination from snowmelt, which range from simple correlation analyses that ignore the physical snowmelt process to sophisticated methods using physical equations. The US Army Corps of Engineers [11] conducted extensive studies that produced several general equations for snowmelt (in./day) during rainfree periods and periods of rain, both for open or partly covered areas and for heavily forested watersheds (Note 1 in./day = 2.54 cm/day).

Overland Flow Routing

Watershed overland flow simulation, as well as flood forecasting and reservoir design, generally uses some type of flow-routing methodology. Routing may be employed to predict the temporal and spatial variations of the outflow hydrograph from a watershed receiving a known volume of precipitation. There are two types of routing: hydrologic, which employs the continuity equation with a relationship between storage and discharge within the system, and hydraulic, which uses both the continuity and momentum equation. The latter better describes the flow dynamics through use of the partial differential equations for unsteady flow in open channels. In hydrologic routing, watershed runoff is considered modified by two kinds of storage, channel and reservoir, and the watershed can be considered [12] as reservoirs in series with an individual relationship between storage and outflow. The assumption is that each reservoir is instantaneously full and discharges into the one following, and so on. The Muskingum method or the concept of routing a time–area histogram can also be used to derive an outflow hydrograph from a watershed [5]. In hydraulic routing, the two routed flow components (the overland and channel flow) are considered and the watershed is described mathematically by defining the various phases of flow of the effective rainfall through its boundaries. The resulting computer programs are very complex and, therefore, most applications use simplifications in overland flow routing. Empirical equations are usually used to estimate the lag or overland flow travel time t_o. For instance, the Federal Aviation Agency [13] uses the following equation for airfield drainage problems, but it has also been used frequently for overland flow in urban basins:

$$t_o = \frac{1.8(1.1 - C)L^{0.50}}{(S^{0.35})} \tag{3}$$

where t_o = overland travel time (min), C = rational formula runoff coefficient, L = length of overland flow path (ft), and S = average surface slope (%). Another equation applied to surface runoff from developed areas and proposed by Morgali and Linsley [14] is:

$$t_o = \frac{0.94(Ln)^{0.6}}{(i^{0.4})(S^{0.03})} \tag{4}$$

where n = Manning's roughness coefficient (Table 1) [15], i = effective precipitation intensity (in./hour), and S = average overland slope (ft/ft).

The above equation needs to be solved by iteration since both i and t_o are unknown. Table 1 presents the values of Manning's roughness coefficient recommended by Kibler et al. [15] for small urban or developing watersheds. An equation recommended by the Soil Conservation

Table 1 Manning Roughness Coefficients

Type of surface	Manning's n
Dense grass or forest	0.40
Pasture or average grass cover	0.20
Poor grass, moderately bare surface	0.10
Smooth, bare, packed soil, free of stones	0.05
Smooth impervious surface	0.035

Source: Ref. 15.

Service [16] for agricultural and rural watersheds is:

$$t_o = \frac{60(L^{0.8})[(1000/CN) - 9]^{0.7}}{1900\, S^{0.5}} \tag{5}$$

where L = hydraulic length of the longest flow path (ft), CN = SCS runoff curve number (Table 2) [16], and S = average watershed slope (%).

In mixed areas the formula overestimates t_o, and the SCS [16] recommends the use of factors to correct for channel improvement (Fig. 2) and impervious areas (Fig. 3). McCuen et al. [17] presented revised lag factors, found to yield more accurate estimates, in place of the SCS ones (Fig. 4). Finally, for channel flow in a catchment area, the well-known Manning's formula may be used to estimate velocities and channel travel time.

Land Use Effects

Drainage basin characteristics, such as slope, size of impervious portion, soil and rock type, and vegetal cover, affect the magnitude and distribution variation of runoff. Therefore, any modifications of these due to human actions and land use changes will have varying impacts on both the quantity and quality of runoff. Land use changes, in particular, that alter both the form of the drainage network and the watershed surface characteristics [18] may increase or decrease the runoff volume from a given site, as well as the peak and overland travel time of a flood. Activities that impact on the infiltration rate and surface storage of a catchment area are most important considering their effect on flow volume, peak rate, and overland lag. Industrial operations that may cause such impacts on stormwater management can include wildscape clearing and grading for buildings and parking lots, felling of forests and drainage of swamps to open up land, and stormwater drainage infrastructure built where there was once a rural area. In such cases, the natural drainage systems are altered and supplemented by manmade stormwater drainage and flood alleviation schemes such as channels, storm drains, flood embankments, and flood storage or infiltration ponds. In general, land use practices that decrease flow volume also decrease the peak rate of flow, and vice versa [5]. On the other hand, reductions in the time lag or concentration time of a drainage basin affect the frequency or reduce the return period of a certain flow.

19.2.3 Design Considerations

Industrial parks and individual industrial sites (including agricultural industry activities) comprise either urbanized drainage areas or small rural watersheds. Methods that have been found appropriate for stormwater management in these cases include peak flow formulas, urban runoff models, and small watershed simulation procedures. Some of these are described in the following subsections.

Table 2 SCS Runoff Curve Number

	Cover		Hydrologic soil group			
Land use	Treatment or practice	Hydrologic condition	A	B	C	D
Fallow	Straight row	–	77	86	91	94
Row crops	Straight row	Poor	72	81	88	91
	Straight row	Good	67	78	85	89
	Contoured	Poor	70	79	84	88
	Contoured	Good	65	75	82	86
	Contoured	Poor	66	74	80	82
	Contoured and terraced	Good	62	71	78	81
	Contoured and terraced					
Small grain	Straight row	Poor	65	76	84	88
	Contoured	Good	63	75	83	87
	Contoured and terraced	Poor	63	74	82	85
		Good	61	73	81	84
		Poor	61	72	79	82
		Good	59	70	78	81
Close-seeded	Straight row	Poor	66	77	85	88
legumes[a] or	Straight row	Good	58	72	81	87
rotation	Contoured	Poor	64	75	83	85
meadow	Contoured	Good	55	69	78	84
	Contoured and terraced	Poor	63	73	80	82
	Contoured and terraced	Good	51	67	76	81
Pasteur or range	Contoured	Poor	68	79	86	89
	Contoured	Fair	49	69	79	84
	Contoured	Good	39	61	74	80
		Poor	47	67	81	88
		Fair	25	59	75	83
		Good	6	35	70	79
Meadow		Good	30	58	71	78
Woods		Poor	45	66	77	83
		Fair	36	60	73	79
		Good	25	55	70	77
Farmsteads		–	59	74	82	86
Roads						
Dirt[b]		–	72	82	87	89
Hard surface[b]		–	74	84	90	92

[a] Close-drilled or broadcast.
[b] Including right of way.
Source: Ref. 16.

Rational Formula

The most common empirical procedure used for designing small drainage systems is the rational formula

$$Q = CIA \tag{6}$$

where Q = peak runoff flow (cfs), C = runoff coefficient (Table 3) [16], ratio of runoff/rainfall, I = average effective rainfall intensity (in./hour) with a duration equal to the time of concentration, and A = drainage area (acres).

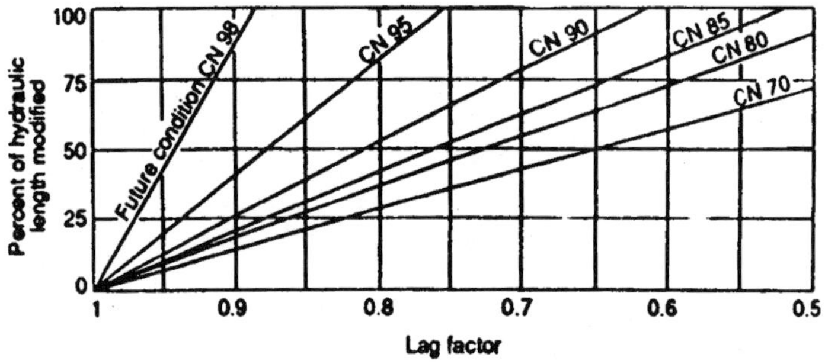

Figure 2 Lag adjustment factors for Eq. (5) when the main channel has been hydraulically improved.

Assumptions made for the application of the rational formula include:

- return periods for rainfall and runoff are considered to be equal;
- runoff coefficient selected is considered constant for the entire design storm and also from storm to storm;
- design rainfall intensity is read from a locally derived intensity/duration/frequency curve;
- rainfall intensity is considered constant over the entire watershed and design storm event; and
- in practice, a composite weighted average C is estimated for the various surface types of the study area.

SCS TR-55 Method

As mentioned previously, the Soil Conservation Service [16] report on *Urban Hydrology for Small Watersheds*, known as Technical Release No. 55, provides a simple rainfall/runoff method for peak flow estimates based on the 24 hour net rain depth and the time of concentration t_o. This is a graphical approach assuming homogeneous watersheds where the land use and soil type are represented by a single parameter, the runoff curve number (CN). The SCS peak discharge graph shown in Fig. 5 [16] is applied only when the peak flow is designed for 24 hour, type II storm distributions (typical of thunderstorms experienced in all US states except the Pacific Coast ones).

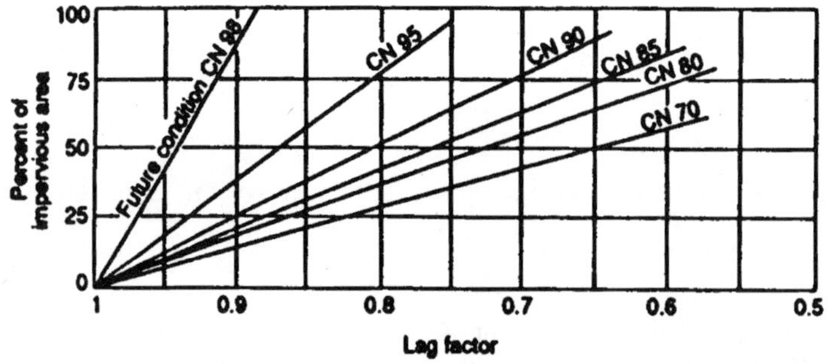

Figure 3 Lag adjustment factors for Eq. (5) when impervious areas occur in the watershed.

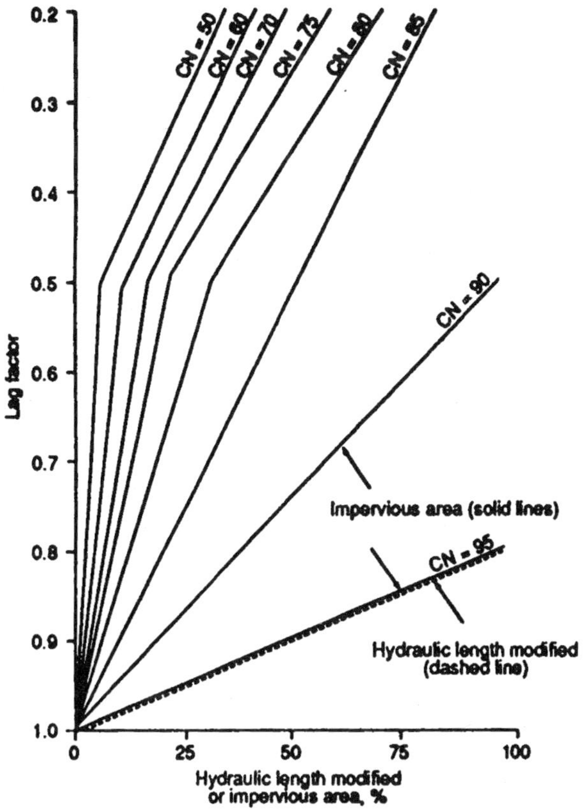

Figure 4 Proposed lag factor vs. hydraulic length modified or impervious area (from Ref. 17).

Unit Hydrograph

One method that has been used extensively to predict flood peak flows and flow hydrographs from storm events is the unit hydrograph method. The unit graph for a watershed is defined as the hydrograph showing the runoff rates of a given one-day storm event producing a 1 in. depth of runoff over the watershed. Time periods other than one day are also used for the derivation of unit hydrographs. The assumption made, generally not true, in the application of this method is that the rainfall is distributed in the same spatial and temporal pattern for all storm events. The development of unit hydrographs for other than the derived duration is accomplished by the use of a method called the S hydrograph, which employs a unit hydrograph to form an S hydrograph resulting from a continuous rainfall.

Synthetic Unit Hydrograph Formulas

Peak flows draining from urban and small catchment areas can be estimated from critical or design storm information using the synthetic unit hydrograph technique. One technique employed by the US Corps of Engineers [19] uses Snyder's method of synthesizing unit hydrographs, according to which a storm with a duration given by

$$t_r = \frac{t_0}{5.5} \tag{7}$$

Table 3 Runoff Coefficients, C

Description of area of surface	C factor
Business	–
Downtown	0.70–0.95
Neighborhood	0.50–0.70
Residential	
Single-family	0.30–0.50
Multiunits, detached	0.40–0.60
Multiunits, attached	0.60–0.75
Residential (suburban)	0.50–0.70
Apartment	0.50–0.70
Industrial	
Light	0.50–0.80
Heavy	0.60–0.90
Parks, cemeteries	0.10–0.25
Playgrounds	0.20–0.35
Railroad yard	0.20–0.35
Unimproved	0.10–0.30
Pavement	
Asphaltic and concrete	0.70–0.95
Brick	0.70–0.85
Roofs	0.75–0.95
Lawns, sandy soil	
Flat, 2%	0.05–0.10
Average, 2–7%	0.10–0.15
Steep, 7%	0.15–0.20
Lawns, heavy soil	
Flat, 2%	0.13–0.17
Average, 2–7%	0.18–0.22
Steep, 7%	0.25–0.35

and a lag time given by

$$t_0 = Ct(L\,Lca)^{0.3} \tag{8}$$

produces a peak flow for 1 in. of excess rain given by

$$Qp = \frac{640\,CpA}{(t_0)} \tag{9}$$

where t_r = duration of the unit rainfall excess (hour), t_0 = lag time (hour) from the centroid of unit rainfall excess to the peak of the synthetic unit hydrograph, Ct = coefficient representing variations of watershed slopes and storage, L = length (miles) of the main stream channel from the outlet to the farthest point (1 mile = 1609 m), Lca = length (miles) along the main channel to the point opposite the watershed centroid, Qp = peak flow of synthetic unit hydrograph (cfs) (1 cfs = 0.02832 m^3/s), Cp = coefficient of peak flow accounting for watershed retention or storage capacity, and A = watershed area ($mile^2$) (1 $mile^2$ = 2.59 km^2).

According to another similar method developed by the Soil Conservation Service [20] for constructing synthetic unit hydrographs, the peak flow produced from a storm event with a

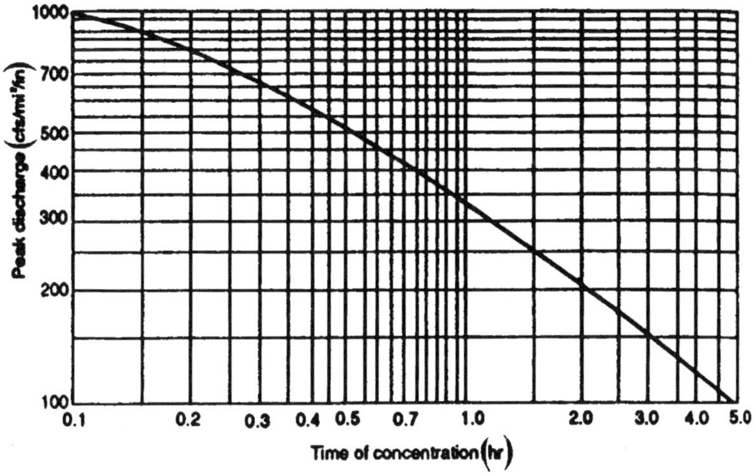

Figure 5 Peak discharge vs. time of concentration t_0 for 24 hour Type II storm distribution (from Ref. 16).

duration D is equal to

$$Qp = \frac{484A}{t_p} \tag{10}$$

where t_p = time (hour) from the beginning of the effective rain to the time of the peak runoff flow, which by definition is $t_p = t_0 + D/2$, where D = duration (hour) of rainfall, equal to $0.133t_c$, where t_c is the time of concentration.

Both of the above empirical equations apply for a certain duration D of 1 in. net rain. The Qp from either formula can be multiplied by the actual net p (peak flow) for other storm events with equal D but different depths. Peak flows for storm events with longer or shorter durations D have to be estimated using unit hydrograph methods.

Urban Runoff Models

Urban runoff computer modeling attempts to quantify all relevant phenomena from rainfall to resulting runoff. This requires the determination of a design storm minus the losses to arrive at a net rainfall rate, the use of overland flow equations to find and route the gutter flow, the routing of flow in stormwater drains, and the determination of the outflow hydrograph. Most urban runoff models deal only with single storm events and, if the errors made are small and noncumulative, the predicted runoff is valid. More recently, the trend has been a continuous-time simulation of many storm and dry periods. The following urban watershed models are representative of many others used.

The Road Research Laboratory and Illinois Simulator is an urban runoff model that uses the time–area runoff routing method; it was developed in England. Another very widely accepted and used storm runoff model is the EPA Stormwater and Management Model (SWMM), which is designed to simulate the runoff of a catchment area for any predescribed storm event pattern. It can determine, for short-duration rainfalls, the locations and magnitudes of local flood flows and also the quantity and quality of runoff at several locations both in the system and in receiving water bodies. Finally, the University of Cincinnati Urban Runoff Model is similar to the SWMM and divides a drainage area into subcatchments with closely matched characteristics whose flows are routed overland into gutters and sewers.

Design Storms

Most designs of major structures involving hydrologic analyses utilize a flood magnitude that is considered critical. When possible, stream flow records are analyzed, but usually design flood hydrographs have to be synthesized from available storm records using rainfall/runoff procedures. Typical storm depths required for major structure design are the probable maximum precipitation or PMP (discussed in Section 19.2.1) and the standard project storm (SPS). The latter is crucial in the design of large dams and its value is usually obtained from the record of large storm events in the neighborhood of the drainage area of study. The SPS is patterned after a storm event that has caused the most critical rainfall depth/area/duration relationship and also includes the effect of snowmelt. In general, the SPS rainfall is approximately 50% of the PMP. Finally, frequency curves can be plotted and used in major and minor structure design in cases where extensive records are available.

Customarily, frequency-based floods are not a part of the design criteria for major structures, but they are commonly used in minor structure design.

Stormwater Retention Basins

As previously discussed, land use changes impact on stormwater runoff. Volumes and peak flows will increase following urbanization, that is, industrial park development, when previously natural pervious land is covered by such structures as buildings, roadways, and parking lots, and when natural storage areas or depressions have been eliminated and the vegetal cover removed. On the other hand, it may be desirable at an industrial site to collect contaminated runoff and hold it either for treatment or slower release to a water body. Detention ponds are constructed to alleviate some of the above problems and serve as holding and treatment facilities. Their primary objective is to reduce the peak flows of surface runoff and the peak loads of pollutants they carry into receiving water bodies. The design of detention basins is carried out on the basis of hydraulic and hydrologic principles, but it is usually not known which design storm would result in the largest retention storage volume. Determination of the required size must, therefore, be accomplished by designing for several critical storm events with various intensities and durations and for several antecedent soil moisture conditions and flow release rates. If the storm event runoff can be approximated by a triangular hydrograph (when the duration of an event is equal to or less than the time of concentration t_c), the required storage [21] in the retention basin is

$$S = 0.083(T_s + t_c)(Qp - Q_0) \tag{11}$$

where $S =$ storage volume required (acre-ft), $T_s =$ duration (hour) of rainfall event, $t_c =$ time of concentration (hour), $Qp =$ peak inflow rate (cfs) (1 cfs $= 0.2832\ m^3/s$), and $Q_0 =$ peak outflow rate (cfs).

If the storm event runoff can be approximated by a trapezoidal hydrograph (when duration is longer than the time of concentration), the required storage [21] in the retention basin is

$$S = 0.083[Qp\ T_s - Q_0(T_s + t_c) + Q_0^2 t_c/Qp] \tag{12}$$

All parameters and units are the same as above. Of course, more elaborate routing techniques and other methods could also be used, but the above approximate, simple method has been found to give very good results.

19.2.4 Quality Considerations

The traditional target of water pollution control, in general, and industrial pollution control, in particular, has been discharges from point sources because they were relatively easy to monitor and treat. However, for the past decade or so, more and more attention has been paid to contributions of pollutants from nonpoint sources, including industrial activities, due to the stormwater runoff process. Runoff erodes, washes off, and carries all sorts of pollutants from large surface areas, pervious and impervious, and eventually discharges them into receiving water bodies.

Point and Nonpoint Source Pollution

Conventional pollution control is geared toward point sources that provide relatively concentrated pollutants and steady flows that do not fluctuate excessively. Nonpoint sources, on the other hand, are much more diffuse and governed by randomly occurring and intermittent storm events. Point sources of pollution discharge into surface water bodies at specific points, and they can be measured and their pollutional load directly estimated. Such typical point sources include industrial and sewage treatment plant discharges, combined sewer overflow, and collected and piped leachate from sanitary landfills.

Diffuse or nonpoint sources typically include urban runoff, construction activities, mining operations, agriculture and animal farming, atmospheric deposition, erosion of virgin lands and forests, and transportation; of these, the major problems stem from urban runoff and agricultural activities [22]. Pollution from urban runoff contains such quality parameters as organic pollutants, heavy toxic metals, coliforms and pathogens, suspended solids, oil and grease, nitrogen and phosphorus, and toxic priority pollutants. Obviously, industrial zones and highly dense urban areas produce higher pollution loads and contribute a greater variety of pollutants. Agricultural activities primarily contribute nitrogen and phosphorus, pesticides and insecticides, organic substances, pathogens, and soil erosion products. Their concentrations depend on the application of fertilizers and pesticides and various tillage activities. Finally, atmospheric deposition has also been designated [22] as a major nonpoint source of lead, phosphorus, PCBs, and acidity (in the form of acid precipitation).

Estimates of Stormwater Pollutant Loadings

Pollutant loadings resulting from nonpoint sources are commonly expressed in terms of mass/area/time (i.e., tons/mile^2/day) as compared to mass/time (i.e., lb/day) for point sources. Following the identification and listing of all the contributing diffuse sources in the study area, their loads have to be characterized and quantified. Wu and Ahlert [23] defined three levels of detail at which estimates of pollutant loads contributed by stormwater runoff might be required: mean yearly loads, assumed to be spread uniformly over both wet and dry periods; interevent loads, which take into account the variations that occur from storm to storm; and intraevent loads, which consider the transient water quality state during an individual storm event. Concentration values from nonpoint source sampling and runoff flows are usually used to estimate a pollutant load per individual storm. An average or expected mean concentration for the storm event is multiplied by the volume of runoff, and these load per storm values from many representative storm events are used for the calculation of a yearly pollutant discharge from the study area [24].

Another technique for estimating yearly pollutant loads has been presented by Smith and Stewart [25] among others, and it employs simple modeling by regression analysis. Plots of log-load vs. log-runoff volume from a group of storm events are used for regression analysis that reflects the log-normal distribution of the data and may be employed to predict a pollutant load based on a runoff volume. As mentioned at the beginning of this section, yearly loads are commonly normalized on the basis of the catchment areas, but other methods are also used to express loads as a function of curb length, population density, drainage density, and a specific land use [24]. The latter is quite adequate as a first approximation, but it may often lead to results that will deviate significantly from measured values [26]. Typical pollutant loads from various nonpoint sources and land uses are presented in a later section.

Water Quality Models

Wu and Ahlert [23] categorized the models used for estimating stormwater pollutant loadings into four types: zero-order or empirical methods, direct methods, statistical methods, and descriptive methods. A variety of diffuse source-estimating models are available in each of the four categories, and a few representative ones are mentioned here. The choice and use of a specific model is a function of the resources available to a project and the particular characteristics of the nonpoint sources involved. The empirical methods constitute the simplest approach and involve the application of unit load rates for the particular water quality parameter to the catchment areas with different land uses within a watershed. These rates are taken, if possible, from local water quality monitoring reports or from the literature on similar (hydrologically and based on land use) watersheds. The direct methods calculate the average pollutant loads as the product of the average flow and concentration measured from the study area, if we assume that the flow and concentration are independent variables. The mean concentration can be found from the literature or derived on the basis of flow-weighting from grab samples or estimated with the help of a statistical method.

Classical statistical methodologies, such as multiple linear regression analysis or discriminant analysis, can be applied to available water quality data for the prediction of pollutant loads from nonpoint sources. In studies where multiple linear regression analysis has been used, the dependent variables were total storm event loads or mean concentrations, and the independent variables were parameters such as climate, land use, topography, and season [18]. Finally, the descriptive methods fall into two types: first, loading functions based on the universal soil loss equation (which will be discussed in a later section) that predict sediment loads on which potency factors can be applied to estimate other pollutant loadings; and second, simulation methods allowing for transient effects during storm events, on the basis of the modeling of dust and dirt accumulation and their removal rates by both rainfall and street-sweeping practices. Regarding the prediction of stormwater runoff loadings in an urbanized drainage area, among the more widely known and applied water quality models are the Stormwater Management Model (SWMM) of Lager et al. [27] and the Storage, Treatment and Overflow Runoff Model (STORM) of the US Army Corps of Engineers [19]. Although the water quality components of these two models are practically the same, STORM provides continuous simulations of storm events over an extended time frame, as opposed to SWMM that handles isolated events. However, STORM has by far the more simple hydrologic component.

Major Quality Parameters of Concern

A detailed overview of pollution parameters stemming from various diffuse sources discharging stormwater runoff contaminated by industrial activities is presented in Section 19.3. The type of pollution and the extent and diversity of contamination, as well as the measures to be undertaken for prevention and treatment, depend of course on the particular type of activity and the chemicals utilized, manufactured, stored, or transported. In each case, care should be taken to identify and subtract background mineral and organic natural pollution concentrations in order to account for the actual pollutant loading contribution by the industrial activities themselves. Major water quality parameters of concern will undoubtedly include BOD, suspended solids, pathogens, mineral oil and grease, heavy metals, nutrients such as N and P, insecticides and pesticides, but also all sorts of trace man-made substances from the EPA's priority pollutants list. Of greater concern should be the pollutants that may bioaccumulate and concentrate in deposits, soils, or sludges, to eventually impart long-term and difficult to clean up toxicity in surface and ground freshwater bodies.

19.2.5 Erosion, Scouring, and Sedimentation

Erosion Process and Controlling Factors

The land surface loses soil particles continually and they are transported downstream with overland runoff and in stream flow until deposition occurs in lakes, estuaries, or coastal areas. As reported by Novotny and Chesters [28], soil particles by themselves comprise a major pollutant, in the form of suspended solids, turbidity, and sedimentation of waterways. Additionally, eroded soil (and especially its fines) is a primary carrier of many other pollutants from contaminated industrial sites: heavy metals, nutrients, insecticides and pesticides, PCBs, and other organic and inorganic toxic substances. Also, large amounts of particulate sediment result in the runoff from pervious and impervious urban areas, and it contains pollutants emanating from traffic, combustion processes, other air pollution sources, and all kinds of urban litter and spills. The rate of erosion and, therefore, its capacity as a carrier of other pollutants are affected and controlled by many factors. Some of the more important include the rainfall regime, vegetal cover of the watershed, soil type and its infiltration capacity, and land slope. On the other hand, industrial activities that may cause excessive erosion will include various agricultural practices, intensive cultivation close to a stream, residential or commercial construction, unstable road banks, surface mining, and animal feedlots close to a stream.

Suspended Sediment Transport

Eroded soil particles and the pollutants they carry move as suspended sediment or washload in the flowing water in overland flow or streams and as bed load that slides and rolls along the channel bottom. The processes are not independent because suspended material at one river stretch may turn into bed load at another. Measurements of sediment movement in lowland, largely agricultural areas indicate that washload may account for 90–95% of the total sediment load [28]. The transport of washload (the generally accepted limit of particles is 0.06 mm, i.e., clay and silt fractions) depends more on the availability of such sediments from upstream sources and not flow characteristics. However, suspended sediments (including silts and clays) will settle in low-velocity zones in slow-moving streams or lakes and reservoirs, where the transport and settling rates of fine sediments are controlled by flow conditions. Williams and Berndt [29] proposed an empirical formula for channel sediment delivery based on studies of Texas watersheds and assuming that sediment deposition depends on particle settling velocity

$$DR = \exp[BT_i(D_i)^{0.5}] \tag{13}$$

where DR = delivery (close to unity where transport is controlled by sediment availability, and <1 where sediment transport capacity of stream is exceeded due to flow conditions), B = routing coefficient determined from field data (range is 4.9–6.3, with an average value of 5.3), T_i = travel time (hour) from source of sediment i to watershed outlet, D_i = the median particle diameter of the sediment (mm) for the source (subwatershed) i. The DR factor is applied to the soil loss estimate that can be obtained from the Universal Soil Loss Equation (discussed in a later section). On the other hand, bed-load transport estimates have been based on the equation proposed by duBoys [6]

$$G_i = \frac{\gamma T_o}{w(T_o - T_c)} \tag{14}$$

where Gi = rate of bed-load transport per unit width of stream (lb/ft-s) (1 lb/ft-s = 1.488 kg/m-s), γ = empirical coefficient depending on the size and shape of sediment particles (Table 4), w = specific weight of water (64 lb/ft^3), T_o = shear at stream bed (lb/ft^2), equal to wHS, where H (ft) = stream depth and S (ft/ft) = energy slope (1 lb/ft^2 = 4.8824 kg/m^2), and T_c = shear value at which transport begins (Table 4). A DR–watershed size relationship has been proposed by Roehl [30] and used as a first-step guide (Fig. 6). Finally, a correlation of DR with channel density and soil texture, proposed by McElroy et al. [31], is shown in Fig. 7.

Drainage Basin Sediment Production

Average annual sediment production values from a catchment area depend on factors such as soil type, land uses, topography, and the existence of lakes and reservoirs. Stream flow sampling can yield relationships of suspended sediment discharge and water flow, such as the typical sediment-rating curve shown in Fig. 8. With the long-term sediment/flow relationship established, it can be combined with a long-term flow–frequency curve to obtain average annual production values. Data from over 250 drainage areas from around the world were analyzed by Fleming [32], who proposed the following relationship:

$$Q_s = aQ^n \tag{15}$$

where Q_s = mean annual suspended sediment load (tons) (1 ton = 907.2 kg = 2000 lb), a, n = coefficients based on various vegetal covers (Table 5), and Q = mean annual stream flow (cfs) (1 cfs = 28.32 L/s).

Errors of up to ±50% may be expected from such relationships and, although they offer an estimate of the order of magnitude of sediment yields, their results should be compared with sediment data on similar watersheds in the same region, if possible [6].

Universal Soil Loss Equation

The Universal Soil Loss Equation (USLE) [33] has been traditionally used in erosion modeling to predict sediment loads resulting from upland (sheet) soil erosion for a certain time period:

$$A = RK(LS)CP \tag{16}$$

where A = soil loss (ton/ha) for a storm event, R = rainfall erosion factor, K = soil erodibility factor, LS = slope–length factor, C = vegetal cover factor, and P = erosion control practices factor. As discussed previously, this soil loss estimate has to be adjusted by the sediment delivery ratio (DR). An extended discussion of the use of the USLE and estimates of the various factors may be found in Novotny and Chesters [28] or Wanielista [34].

19.3 PATHWAYS FOR CONTAMINATION

Contaminated stormwater runoff or nonpoint pollution accounts for more than 50% of the total water quality problem [28]. In many areas, diffuse pollution sources such as runoff from agricultural activities, strip mining, urban stormwater, and runoff from construction sites are becoming major water quality problems. Nonpoint pollution involves not only the usual

Table 4 γ, Empirical Coefficient and τ_c Shear Values

	γ		τ_c	
Particle diameter (mm)	ft^6/lb^2-s	m^6/kg^2-s	lb/ft^2	kg/m^2
1/8	0.81	0.0032	0.016	0.078
1/4	0.48	0.0019	0.017	0.083
1/2	0.29	0.0011	0.022	0.107
1	0.17	0.0007	0.032	0.156
2	0.10	0.0004	0.051	0.249
4	0.06	0.0002	0.090	0.439

pollution parameters, but also serious problem contaminants such as PCBs, acid rain, and pesticides, which do not have a parallel in the traditional point source environmental pollution control. The following discussion presents most of the major nonpoint contaminant sources, with an emphasis on industrial activities, the kinds of problems caused, and some considerations for their control. Figure 9 illustrates generally classified pathways for stormwater contaminants at an industrial site.

19.3.1 Atmospheric Impacts

Atmospheric contaminants, in dissolved, gaseous, or particulate form, enter stormwater runoff through either the process of precipitation or as dustfall; gases also enter by direct absorption at the earth's surface. The deposition rates of particulate atmospheric contaminants in US urban areas vary from 3.5 to over 35 tons/km^2/month [26], and the higher rates are found in congested industrial areas and business districts. In addition to particulate matter, many other contaminants are contained in, transported by, or deposited from atmospheric fallout, such as nitrogen and phosphorus, sulfur dioxide, toxic heavy and trace metals, pesticides and insecticides, fungi and pollen, methane and mercaptans, fly ash, and soil particles. Dustfall rates vary significantly from area to area and are largest in the central United States, with the geometric means of dustfall

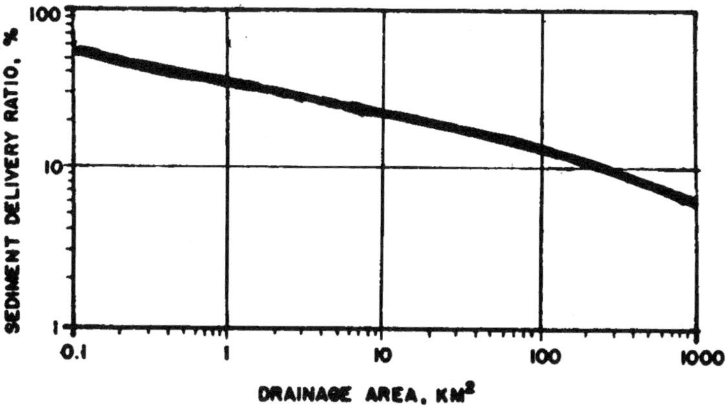

Figure 6 Sediment delivery ratio factor vs. the watershed area (from Ref. 30).

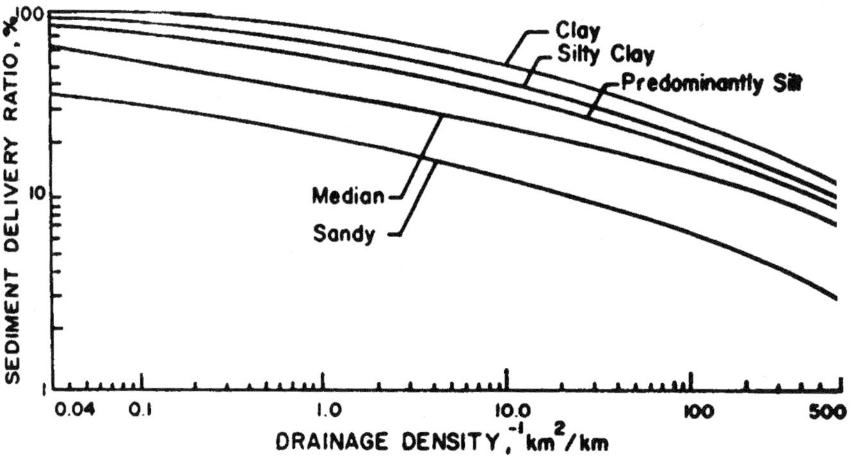

Figure 7 Sediment delivery factor vs. drainage density and soil texture (from Ref. 31).

values ranging from 2.8 to 144 tons/mi^2/month [34]. Although contaminant concentrations in rural dustfall are related closely to soil conditions, urban dustfall is related more to local air pollution problems. Fly ash from industrial coal-burning activities and disintegration of urban litter is one more important source of atmospheric contaminant contribution, especially in the vicinity of industrial and urban centers [28].

19.3.2 Acid Precipitation Impacts

One particularly important contamination due to atmospheric precipitation results from the combustion of organic fuels (especially coal) containing sulfur and nitrogen, which appear in gaseous endproducts (SOx and NOx); these gases react with H^+ in atmospheric moisture to produce acid rain (pH of less than about 5.6, which is the pH of "normal" rainwater in equilibrium with dissolved CO_2). Although the major part of acidity in precipitation is attributed

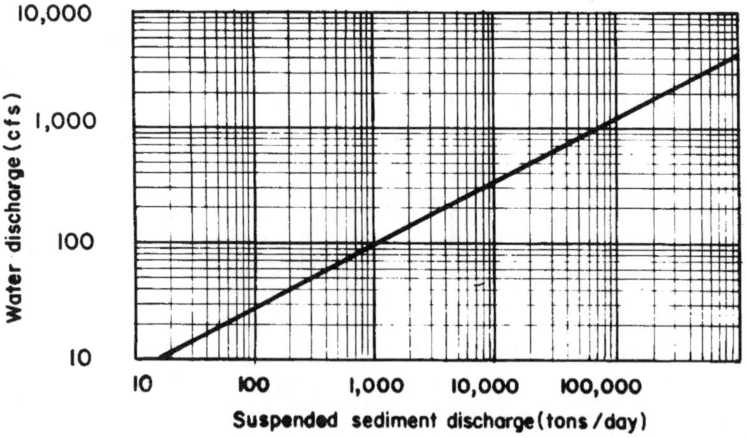

Figure 8 Typical sediment–rating curve.

Table 5 Coefficients Based on Vegetal Cover

		a	
Vegetal cover	*n*	For Q_s (tons)	For Q_s (metric tons)
Mixed broadleaf and coniferous	1.02	117	106
Coniferous forest and tall grassland	0.82	3523	3196
Short grassland and scrub	0.65	19,260	17,472
Desert and scrub	0.72	37,730	34,228

to energy production (power plants), nitrates and therefore NOx are attributed mostly to agricultural and traffic sources. For instance, while in New York State and parts of New England, the data showed that 60–70% of acidity is due to sulfuric acid and 30–40% to nitric acid, these proportions are reversed in areas of heavy traffic, such as southern California [35]. The lowest pH values of precipitation are usually measured near large coal-burning power plants or smelting operations and in heavy-traffic corridors. Since the 1950s, the trend to meet local air pollution standards by building taller stacks has worsened the acidity of rain (longer travel times in the atmosphere, longer reaction times for pollutants) and turned the local problems into regional ones. Acidification of lakes in watersheds with low-buffer capacity soils (those lacking $CaCO_3$) has been occurring in the northeastern United States, southeastern Canada, and Sweden. The resulting low pH has caused severe decreases in or elimination of fish populations and has adversely affected the biota of streams and lakes and the terrestrial ecosystems of watersheds.

19.3.3 Housecleaning and Site Drainage

During normal housecleaning operations inside an industrial plant or at the site, there are intentional or accidental releases of pollutants that may find their way into either surface runoff or the stormwater drainage system of the industrial site. Regular operations in and around a plant involve cleaning up spills, washing vessels and all sorts of containers, and washing floors in the production buildings and warehouses. In many cases, the drains may be connected to the stormwater drainage system of the site, thereby causing direct contamination. On the other hand, accidental spills in parking lots, unloading areas, driveways, and roads within the site and intentional discharges of waste storage and disposal areas provide a variety of pollutants that contaminate the surface runoff originating at the site.

19.3.4 Raw Material Stockpiles

Bulk materials are often stockpiled outdoors at industrial sites, mining locations, or transportation facilities, for example, coal stockpiled in coal terminals or power plants. In such cases, several impacts occur from discharges of untreated leachate generated by precipitation and/or contaminated stormwater runoff into surface or subsurface water bodies. Contaminants found in these discharges depend on the nature, purity, and time of exposure of the stockpiled bulky raw materials. There are two pathways, producing two types of contaminated wastewaters. First, when precipitation runoff occurs, it causes particulates to wash off and be carried away from the stockpile surface. Secondly, rainwater or snowmelt slowly percolates through the stockpile, dissolving some of the chemicals in addition to concentrating pile-bound particulates and

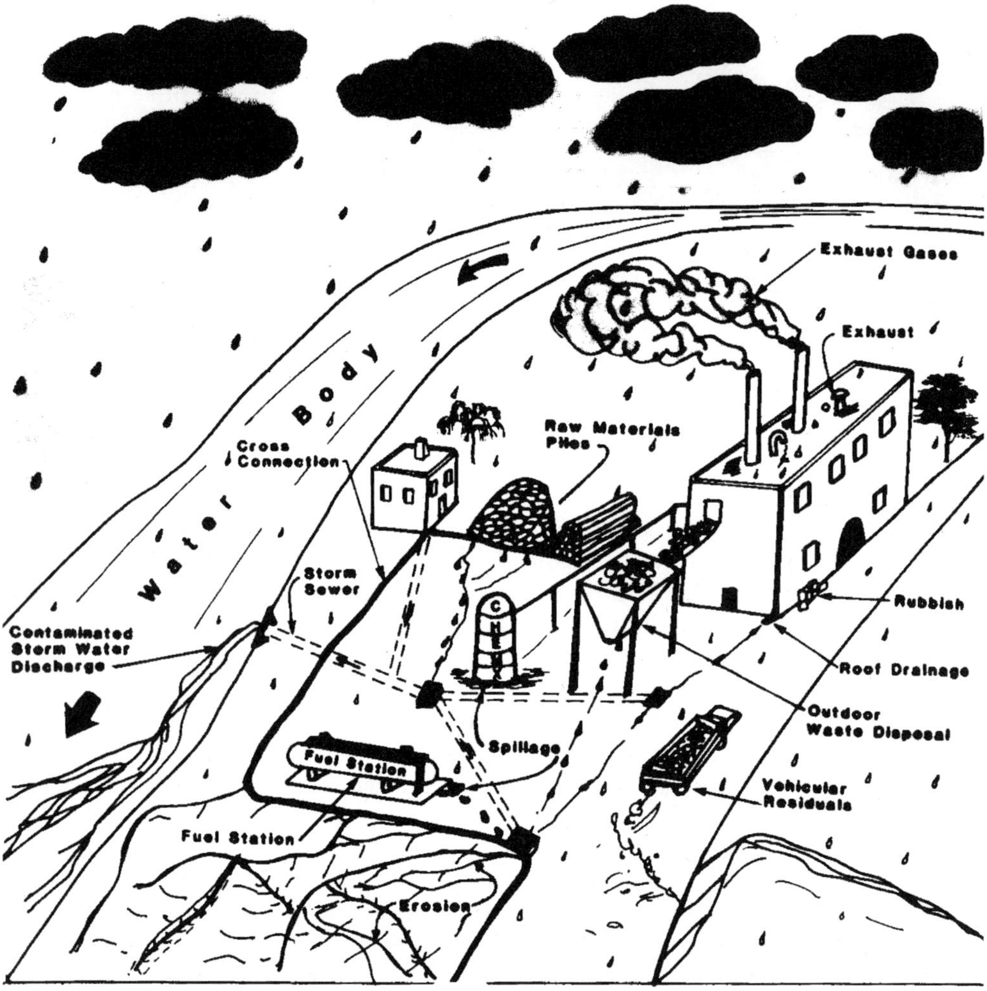

Figure 9 Typical pathways for stormwater contamination. (Courtesy of O'Brien & Gere Engineers, Inc.).

appearing as leachate. There is a complex relationship between runoff and leachate; that is, in the outer crust of the stockpile, it is possible that the two terms become synonymous as runoff percolates below the surface and re-emerges to join the main surface streams. The quality of contaminated runoff from a stockpiling area or a bulk material transport terminal, in general, would not meet the federal or state criteria for discharges into a surface water body without treatment. The first approach to the problem should be prevention, collection, reduction of runoff and leachate volume, and the second, treatment of the resulting contaminated stormwater prior to leaving the industrial facility. Sheets of plastic and other material or permanent cover structures should be used for reduction of the volume and degree of contaminant concentration in the stormwater drainage or leachate. Lining of the stockpile areas and installation of collection and containment piping, ditches, berms, and other structures could alleviate problems and aid in the subsequent treatment of the stormwater, which will depend on the nature of contamination.

19.3.5 Spent Material Stockpiles

Many industries have designated locations outdoors where they store, stockpile, or dispose of their wastes. These may include an area adjacent to the buildings where drums with waste are stored or an open pit for liquid wastes, an area within the industrial site where spent solid wastes or powder are stockpiled, and/or an infiltration pond or aerated lagoon where liquid wastes are discharged. In such cases, several pathways of contamination can exist due to precipitation carrying away pollutants through surface runoff or leachate reaching surface or subsurface water bodies. Contaminants in the stormwater runoff would depend on the nature, mixture, and time of exposure of the spent materials and waste disposal systems. The mechanisms of contamination are the same as discussed in Section 19.3.4. Similarly, the first approach to the problem should be prevention, proper storage and disposal of the industrial wastes, collection and reduction of runoff and leachate volume, and the second, treatment of the resulting polluted stormwater.

19.3.6 Roadway Drainage

There is significant awareness of highway and roadway stormwater runoff as a nonpoint source threatening the quality of water resources. The impact from this source of contamination can generally be separated into changes in the quantity of runoff due to the creation of large impervious areas and changes in the quality of runoff due to a change in the character of the catchment surface involving depositions from vehicular traffic and accidental spills of chemicals. The accumulation on highways and roadways, railroad tracks and yards, and urban streets of materials that can be removed by stormwater runoff, such as high amounts of heavy metals attributed to emissions and to the breakdown of road surface materials and vehicle parts, asbestos from clutch plates and brake linings, motor oil and fuel and gasoline spills, tire and wheel abrasion particles, spills of various chemicals due to traffic accidents, and deicing salts are all contributors of pollution. In general, contaminants found in highway and roadway runoff are similar in type and quantity to those found in stormwater runoff from urban areas and, in particular, industrial zones [36].

19.3.7 Land Use Impacts

The land use effects on stormwater runoff quality depend on the prevailing activities taking place within the area of concern, their intensity, and the resulting nonpoint sources of contamination. The problem of land use and its effect on water quality is primarily associated with urban, industrial, and agricultural developments. In rural areas, animal barnyards and feedlots may cause severe contamination of water bodies, as can overfertilization and intensive pesticide application in farmland without adequate erosion controls. Land uses usually requiring more intensive control measures typically include industrial areas, mining operations, animal farms, and construction areas. Nevertheless, areas classified into a single land use category can have diverse characteristics, such as topography, soil types, and slopes, and they therefore can generate wide-ranging volumes of flow and quantities of contaminants [28]. Several of these land use categories and their impact with regard to stormwater runoff pollution are discussed in the following sections.

19.3.8 Mining Drainage

Stormwater runoff generated from some mining operations may pose certain serious problems to the quality of water resources, but mining cannot be viewed as a homogenous source of nonpoint contamination [37]. Many different minerals are mined, coal and metal ores among them, each causing its own type of pollution problems. Mining nonpoint sources include runoff and leachate from abandoned mines, as well as from inactive access roadways and old tailings and spoil piles, resulting in sediment, salts, metals, and acid drainage discharges. Even though active mining operations also cause similar pollution problems, they are considered to be point source problems and are regulated under state and federal National Pollutant Discharge Elimination System (NPDES) permits [37]. In addition, the Surface Mining Control and Reclamation Act (SMCRA) of 1977 includes requirements for the collection and treatment of active coal mine runoff to meet point source discharge criteria. In strip-mining operations, enormous, bare surface land areas are exposed and the consequence is huge soil erosion yields. On the other hand, acid mine drainage has lethal and sublethal effects on the biota within surface water bodies. Methods for controlling pollution from active mines are available and required by SMCRA for all new mines, the key to prevention being proper site planning. Also, methods are available for solving many contamination problems related to surface mining, such as those regarding land areas and adding topsoil for revegetation in abandoned mines to control the excessive erosion and runoff of pollutants from the area. Additional best management practices would include [26,37] the sealing of abandoned mines and/or diverting the surface runoff to reduce drainage contamination; mixing of fine and coarse materials to stabilize mill tailings; equalizing the flow of and treating acid mine drainage by neutralization; compounding of hazardous materials using asphalt or concrete or capping with clay to assure permanent storage and leachate reduction; and containment of leached materials by use of ditches, dikes, and impoundments.

19.3.9 Construction Site Runoff

Nonpoint source pollution resulting from construction activities has very high localized impacts on water quality. Sediment is the main construction site contaminant, but the stormwater runoff may contain other pollutants such as fertilizers and nutrients, pesticides and insecticides (used at construction sites), petroleum products and construction chemicals (cleaning solvents, paints, asphalt, acids, etc.), and debris. Erosion rates from construction sites may be 10 to 20 times and runoff flow rates can be up to 100 times those from agricultural lands [37]. Some of the pollution control methods that could be used are protection of disturbed areas from rainfall and flowing runoff water, dissipation of the energy of runoff, trapping of transported sediment, and good housekeeping practices to prevent the other pollutants mentioned above from being transported by stormwater runoff. Finally, each construction project should be planned and managed by considering drainage problems and contamination, avoiding critical areas on and adjacent to the construction area, and attempting to minimize impacts on natural drainage systems.

19.3.10 Agricultural Industry

The nature and extent of agricultural nonpoint source pollution are directly related to the way and intensity with which the land is used. For instance, raw cropping usually involves not only a great deal of land disruption, but also the application of fertilizers and pesticides. According to the USEPA [37], therefore, agricultural activities constitute the most pervasive cause of water pollution from nonpoint sources. Actually, pollution from agriculture has various sources, each with different associated impacts, which may be categorized as follows: nonirrigated croplands, both row (i.e., corn and soybeans) and field (i.e., wheat); irrigated croplands; animal production on rangeland and pastureland; and livestock facilities. The latter two activities will be discussed in Section 19.3.11. The discharged contaminants from agricultural croplands include eroded sediments and washed out fertilizers, nutrients and organics from manure applications, traces of pesticides and herbicides, and leached plant residues. Pesticides and other organic chemicals become airborne, especially when sprayed by plane or helicopter, and travel long distances before they precipitate on the earth or directly on surface water bodies. Excessive applications of fertilizers or manure to cropland result in nitrogen, phosphorus, and potassium contributions to runoff that accelerate the eutrophication of lakes and reservoirs and cause high nitrate levels in groundwater. On the other hand, irrigation return flows would carry salts and other minerals to surface water bodies or aquifers by percolation, and manure application would contribute bacterial contamination. Since the majority of pollutants from agricultural activities are carried by stormwater runoff, the usual soil conservation practices such as contour farming, strip cropping, terracing, and crop rotation are the most effective ways of controlling pollution. Also, the use of grassed waterways, runoff diversions and retention basins, crop management practices, timing of pesticide and fertilizer applications, and management of quantity and timing of irrigation water are additional pollution control techniques. Finally, manure should be incorporated immediately into the soil and not spread on frozen ground to reduce runoff losses.

19.3.11 Animal Farming

Rangeland and pastureland used in livestock farming can contribute significant amounts of sediment and nutrients to stormwater runoff, especially when overgrazing is allowed. Rangeland and pastureland erosion is a fundamental problem and, therefore, management practices that maintain or improve the condition of such lands can significantly reduce their erodibility. Manure from livestock contributes nitrogen, phosphorus, potassium, ammonia, fecal bacteria, and other contaminants that pollute water resources. Runoff from more contained livestock areas, such as feedlots and barnyards, contributes much higher concentrations of pollutants, but the NPDES permit program regulates only the large operations [37]. Wherever livestock has access to highly erodible areas and water courses, pastures may become a local pollution hazard, unless the animals are kept away by fencing. Finally, good feedlot waste management would include runoff diversions, ponds, and scraping that control liquid and solid animal waste.

19.3.12 Silvicultural Nonpoint Sources

Virgin or very lightly developed forests and woodlands are the best protection from the sediments and contaminants carried by stormwater runoff. Such areas often generate very small amounts of runoff because their soil can absorb large quantities of water due to high permeability, they can retard and retain surface runoff, and due to tree canopy and groundcover, they can also significantly reduce soil erosion losses. Forest and rangelands contribute contamination that includes sediments, wildlife and vegetation decay endproducts, as well as pesticides and nutrients depending on forest management practices. Silvicultural activities are actually comprised of a number of different operations (i.e., road building, pesticide and herbicide application, harvesting and logging, removal of trees from the harvesting site, and preparing the site for revegetation), each of which has a different potential for nonpoint source pollution [37]. Clear cutting or uncontrolled logging operations may greatly reduce the resistance of a forest to soil erosion. Observations and records show that almost all sediment contributed to surface water bodies from forests originates in clear cut areas and logging roads [38], especially roads that disrupt or infringe on natural drainage channels.

19.3.13 Industrial Urban Areas

Industrial activities in urban areas range from workshops and light manufacturing contributing relatively small amounts of contaminants to heavy and wet industries such as steel mills, cement manufacturing, meat packing, textiles, and beer production. Their discharges of pollutants to water bodies result in minor to major problems and originate from point and nonpoint sources. The latter have been presented in detail in previous sections. The main source of nonpoint industrial pollution is air pollutant release and subsequent deposition within an industrial area, but in general, assessing the contribution of nonpoint pollution from manufacturing activities is quite complex. The extent of soil contamination in certain industrial sites is of such magnitude that it necessitates the collection and treatment of the entire volume of stormwater runoff. Pollutants emanating from industrial areas include heavy toxic metals (such as copper, cadmium, lead, zinc), particulates, inorganic and organic priority pollutants, nutrients, petroleum products and other liquid contaminants from intentional dumping and accidental spills and leaks, either at transfer and manufacturing areas or at storage and waste disposal sites.

Observation and sampling data indicate that particulate and other pollutant loads in industrial areas are, in general, higher than those for residential or commercial areas. For instance, the sampling program of New York City's Industrial Pretreatment Program [39] showed that during the one to two months of summer vacation (shutdown of most operations), there was a dramatic reduction (anywhere from 70–90%) in the concentrations of all heavy metals in samples of stormwater runoff from industrial areas. This result signifies that a considerable amount of contaminants attributed to stormwater runoff is still industrial in origin, indirectly contributed through spills and intentional discharges. Similar results were also reported for heavy metals (Cr, Cu, Ni, Pb, and Zn) in urban runoff by Wilber and Hunter [40], who found significant increases in concentration when land use percentages increased for industrial use from 5 to 12% and for commercial use from 42 to 58% (residential land use decreased from 53 to 30%).

19.3.14 Landfill Runoff and Leachate

The usual method of industrial solid waste disposal involves landfilling either in sanitary or secure landfills. Nevertheless, prior to the 1970s when landfill leachate was recognized as a potential source of groundwater and surface water pollution, only about 25% of the landfills in the United States were classified as sanitary, and none as secure landfills [28]. As previously discussed in Sections 19.3.4 and 19.3.5 with regard to stockpiles of raw and spent materials, the generation of leachate results from the infiltration and percolation of rainfall, stormwater runoff, and groundwater through an operational or closed solid waste landfill. In general, it is not possible to exactly predict leachate characteristics because of variable landfill constituents; however, leachate contamination may severely pollute neighboring ground and surface waters. Solid waste disposal sites produce a highly mineralized leachate containing pollutants such as heavy toxic metals, chlorides, nutrients, volatile organics, cyanides, chlorinated hydrocarbons, and other priority organic pollutants. Techniques of leachate pollution control would include minimization of leachate formation by reducing water infiltration; collection and treatment; detoxification or immobilization of hazardous industrial wastes prior to landfilling; development of landfill sites on uplands, not on floodplains, and in soils of low permeability.

19.4 MANAGEMENT

No program for management would be adequate without first considering the strategy by which to avoid the initial problem of handling contaminated stormwater. That is, first, plan for the construction and operation of facilities that seek to avoid the contamination of stormwater through inadvertent contact with (a) raw materials, (b) manufacturing processes, (c) finished products, and (d) waste materials present at the industrial site. Secondly, when stormwater is likely to contact contaminants at a site, measures for minimizing such contact to the best practical means are recommended to reduce the quantity of contaminated stormwater that may be generated. The following sections describe the elements of planning and implementing a strategy for contaminated stormwater management. When treatment becomes a necessity, Sections 19.5 and 19.6 of this chapter should be reviewed, as well as the many chapters on treatment processes that comprise this handbook.

19.4.1 Site Planning and Practices

The goal of the planning process is to reduce the incidence of contamination of stormwater. In instances where it is not practical to eliminate the contact of stormwater with contaminating substances, the secondary goal of minimizing the exposure and containment of contaminated stormwater should be reached. In this case, the quantity of contaminated stormwater would be minimized.

Volume/Rate Reduction

The volume of stormwater to be exposed to contaminants can be reduced through the minimization of impervious areas. Open areas where rainfall does not contact with potential sources for contamination could be graded to minimize erosion. In addition to grading, ground cover should be utilized, which would impede the velocities of sheet flow as well as aid in the retention and subsequent evaporation process. Through increased retention by ground cover, flow reduction would be furthered through the increase in the quantity of percolation. Nonpoint drainage should be channeled to discharge into natural water courses, wetlands, or areas for seepage. Through the use of porous pavements in yard areas requiring hard surfaces, it may be possible to maximize the natural drainage of uncontaminated stormwaters. In areas where the underlying soils might be impermeable, an underdrain system may be necessary or conventional collection systems could be utilized. More important for paved or otherwise impervious areas is to identify areas that are not exposed to sources of contamination and to route the stormwater collected from such areas separately from areas that may have contaminated stormwater. In so doing, the quantity of contaminated stormwater to be later handled will be reduced. Another means for reducing the rates of flows with subsequent volume reduction is through the use of storage or retention areas. This method is elaborated in the following subsection.

Storage/Retention

The storage/retention of stormwater can be utilized for (a) flow reduction, (b) flow equalization, (c) source control, and (d) increasing the effectiveness of natural drainage. Storage/retention can be provided economically when flat rooftops are available, where drainage can be routed to natural or man-made depressions (basins) or through the use of dikes or earthen berms. Limited storage can be provided by taking advantage of the volume within stormwater sewers/pipelines when available [41].

Flow equalization and flow reduction can be utilized to reduce the possibility of erosion, which would otherwise occur through the discharge of stormwater during peak rainfall intensities. Essentially, the storage capacity provided would act as a shock absorber to decrease the peaking effects through dampening. The discharge of stormwater from storage/retention areas could be controlled hydraulically through the use of weirs, orifices, or tipping devices. The discharge of stored flows at controlled rates on open lands could facilitate the percolation process, as well as deliver stormwater to the points of discharge at a lesser rate. Either way, the impact on the discharge water course would be reduced, as opposed to receiving the stormwater at high rates. In the event that stormwater treatment is to be provided, storage/retention would help decrease the size of facilities by decreasing the peak hydraulic requirements.

Finally, storage/retention of stormwater is useful in the segregation of flows from contaminated and uncontaminated areas. Stormwater could be captured and conveyed separately from areas that may contain sources of contamination. Many site-specific issues have to be examined prior to the use of storage/retention. The actual case-by-case intent and goals would govern the exact methodology to be implemented.

Housekeeping Practices

Good housekeeping practices can have a positive effect on the minimization or elimination of sources of stormwater contamination. In general, a program to keep yard areas and stormwater collection and conveyance items clean and free of refuse, raw materials, products, solids, or other materials that may contaminate stormwater should be maintained.

The routine cleaning of parking areas, storage yards, stockyards, streets, and sidewalks to remove debris and potentially harmful materials should be performed. Catch basin and storm sewer cleaning should be performed as needed to remove deposits that can contaminate stormwater and/or decrease the capacity of the conveyance system. When contaminating materials come into contact with areas subjected to stormwater, these materials should be cleaned up prior to a storm. Equipment and cleaning materials should be kept on site where hazardous materials may be utilized. Workers should be trained in the procedures necessary to provide a proper cleanup or decontamination of areas where contaminating materials have been inadvertently placed.

19.4.2 Cross Connections

Cross connections include the connection of sanitary drains, process/production area drains, or floor drains from material/product area drains, or floor drains from material/product areas to separate storm sewers. These types of connections can be found in existing older facilities, as well as those that have been renovated or expanded. Such connections are generally illegal and cause the contamination of stormwater. These connections may also result in the discharge of contaminants during dry weather.

When planning a stormwater management program, a check for cross connections can be made by examining storm sewer lines during dry weather. If any flow is present in a storm sewer during dry weather, it very likely may be due to cross connection(s) or infiltration sources. Infiltration is the flow of groundwater into a storm sewer through cracks in manholes or piping, when the groundwater table is above the storm sewer/structures. In order to ascertain whether dry weather storm sewer discharge is likely to be from a cross connection or groundwater source, an analytical testing program would need to be implemented. Upon the determination that cross connection(s) are present, the identification of points connected would need to be made. Dye or smoke testing could be used to determine the points of connection. Actual protocol for locating cross connections would vary case by case.

19.4.3 Spill Prevention Program

Spill prevention programs are most utilized in the hazardous waste field, where specific measures have been implemented for the prevention/containment of hazardous material spillage. Spills can occur due to (a) the rupturing of a storage vessel, (b) overfilling a vessel, (c) breaks in conveyance pipelines or equipment, and (d) inadvertent discharge via an open drain connection. The majority of spill prevention protocols developed to date have been in the area of liquid spill prevention; however, many of the same concepts apply to solid material spills, where such solids are also stored in vessels (hoppers, silos, etc.). When potentially contaminating liquids or materials are stored either in vessels (liquids/solids) or stockyard piles (solids) subjected to stormwater, spill prevention and containment should be implemented. The measures to be implemented should include both the storage area and areas designated for the delivery of liquids/materials (transfer or unloading areas).

Areas where delivery vehicles will be parked during the transfer of liquids/materials should be within a diked area or otherwise enclosed area where drainage can be isolated in the event of a spill. Drains from these areas should normally be closed and routinely opened to allow uncontaminated stormwater to be released and spillage or contaminated stormwater to be collected. In the event this area becomes contaminated, the extent of contamination will be limited to the delivery area from where stormwater can be pumped out and decontaminated.

The storage vessels subjected to stormwater should be contained within a diked area or containment wall/tank suitable to contain the contents of the vessel should a rupture of the vessel or interconnecting piping occur. Similarly, this would limit the quantity of stormwater that may be contaminated and facilitate the collection and disposal of contaminated water. As an example, aboveground oil storage facilities have been regulated by state legislature, local regulations, and fire codes. These provide an excellent example of preventive measures and generally offer protection for stormwater.

19.5 ONSITE VS. OFFSITE TREATMENT

Depending on the regulatory requirements governing the need for treatment of contaminated stormwater, along with the possible availability of offsite treatment, a decision as to whether to provide onsite or offsite treatment will need to be made. Therefore, prior to proceeding, an industry should (a) investigate whether or not offsite treatment is available, (b) identify if offsite facilities provide the level of treatment required, (c) ascertain tipping fees or costs to be incurred, and (d) determine the reliability of the offsite facility. Once options have been identified, the direction to be taken will be a function of reliability, environmental soundness, regulatory acceptability, economics, and company policy. Offsite treatment facilities include (a) municipal wastewater treatment plants, (b) private residential wastewater treatment plants, and (c) private commercial wastewater treatment facilities. The ability of any of these facilities to provide industrial stormwater treatment would be a function of its available treatment capacity and regulatory permit requirements.

Small industrial sites may be able to truck collected contaminated stormwater to a treatment facility, whereas larger sites would need to construct conveyance pipelines. Depending upon the proximity of the offsite facilities, conveyance requirements and associated costs would need to be considered. In certain instances where several industries requiring treatment are in close proximity to one another, a central offsite treatment facility should be considered. In such cases, privatization of such a facility may offer an advantage to the utilizing industries.

The above discussion briefly presented a few of the various considerations to be weighed prior to making a decision on offsite vs. onsite treatment. In fact, many other factors need to be considered on a case-by-case basis. In particular, the quantity of stormwater to be treated and the needs for bypassing flows are to be considered as well. In order to properly evaluate the requirements and measures to be implemented, a detailed study of the needs for treatment may be a necessity.

19.6 CONTAMINATED STORMWATER TREATMENT

Treatment of contaminated stormwater varies with the type and degree of contamination found to be present. As a general guide to contaminated stormwater treatment, the numerous books and reports describing treatment processes and specific contaminant handling should be referenced. Books describing unit operations and processes that may be applicable in the handling of specific types of pollutants in industrial wastewater might be especially helpful. Hence, this section will not attempt to reiterate the wealth of information contained in the volumes of this handbook, but rather to describe basic processes and concepts applicable to contaminated stormwater treatment.

19.6.1 Batch vs. Online Treatment Capability

The variability of stormwater flow due to varying storm intensity (in./hour) and magnitude (total cumulative inches of precipitation) makes the sizing of treatment units a more complicated matter as compared to sizing such units for sanitary wastewater flows. One consideration in designing a system is with respect to handling the contaminated stormwater (CSW) on a batch or online basis. Batch treatment is a method of treating contaminated water in treatment units on a cyclic operational basis. A specific volume of CSW would enter the units at a designated rate and reside in the treatment units a defined length of time while undergoing treatment. Following the treatment of the first volume (batch) of CSW, subsequent volumes would be processed until the total volume has been treated. Storage is inherent to this type of treatment scheme since the entire volume of CSW generated by a storm may not likely be treated by a single batch run. Hence, storage may be necessary to contain CSW until it can be processed.

Online treatment capacity refers to a treatment system that is designed on a rate per unit time basis. Unit processes would then need to be designed to operate at a rate equal to the rate at which CSW is generated or through the use of flow equalization some lesser throughput rate. Flow equalization in this case would be provided by a storage basin designed to contain the CSW sent to the plant that would exceed the plant's capacity. In most cases, a simple overflow weir arrangement could be used to route flows in excess of the design capacity of the online treatment system to the flow equalization storage basin.

Whether batch treatment, complete online capacity, or online capacity plus flow equalization is to be utilized is a function of (a) the quantity of CSW to be handled over time, (b) space requirements for each system, (c) reliability, (d) ease of operation, and (e) overall capital and operating costs. Since requirements will be different for each application based on design storm criteria, treatment stipulations, and contaminated stormwater characteristics, judgment as to which methodology should be utilized will vary on a case-by-case basis.

19.6.2 Treatment Processes

Although depending on the characteristics of contaminated stormwater a variety of treatment methods may be applicable, for the purposes of this section the eight most promising processes are discussed broadly. These include (a) skimming, (b) screening, (c) sedimentation, (d) concentration, (e) filtration/straining, (f) dissolved air flotation, (g) biological, and (h) chemical treatment. The following subsections provide a brief description of these treatment methods.

Skimming

Skimming can remove a wide variety of floatable materials including wood, paper, plastics, styrofoam, grease, oil, and other buoyant materials. In order to accomplish skimming, a basin of adequate size (10–60 mm retention time) is necessary to allow for floatable material to surface. Normally a barrier wall or partition is helpful at the surface and extending into the water columns to help trap the floating material. The stormwater entering the tank would flow beneath the wall/partition while floatables would collect upstream. Through the use of an adjustable weir or a slotted pipe (skim trough) located on the upstream side, the floatables can be removed from the surface of the basin. Where sedimentation is being utilized for solids removal, skimming can be incorporated into the same tankage.

Skimmings can be diverted to an overflow structure where grease and oils can be removed, as well as solid floatable materials. Following the removal of floatable solids through mechanical separation utilizing a rake, basket, or conveyor or other means to remove the solids from the liquid, the remaining floating greases and oils can be drained by gravity or pumped from the surface of the stormwater. The remaining stormwater can be returned to the head end of the skimming tank. When floatable solids are not as present, a commercially available oil/water separator can be utilized. Approximately 80–90% of the grease and oils should be removed as well as 80–95% of the floatable solids. In certain cases, capture may be enhanced by injecting air at a low rate to the bottom of the skimming basin.

Screening

Various types of screening equipment are available to remove debris that may be too large for skimming or material that cannot be skimmed. In their simplest form, round and rectangular bars arranged in a parallel 30–60° incline framework within a channel (bar screens) can provide for screening. Debris will deposit on these bar screens and can be removed manually with a rake. Mechanically cleaned screens are available that can either be run for continuous cleaning or put on a timer for discontinuous screening. Bar spacings vary between 10–75 mm, depending on application. In practice, several sets of screens in series have been used successfully. Normally, coarse screens, with 50–75 mm spacing, preceding medium screens, 10–25 mm, can provide a significant degree of screening.

Sedimentation

Grit, settleable, and suspended solids, can be removed from contaminated stormwater using the sedimentation treatment process. This process relies on gravity to remove particles and materials. Depending on treatment and solids handling requirements, it may be desirable to remove grit separately from suspended and settleable solids. Normally, grit may require mechanical means of conveyance, while suspended and settleable solids can be pumped to a dewatering device or truck.

A typical sedimentation tank for suspended/settleable solids removal may be rectangular or circular in configuration, with a water depth of 10–12 ft and surface areas varying on the basis of particle settling characteristics to achieve from 300 to 1200 gpd/ft^2 surface loading rates. In certain cases, chemical flocculants can be added to enhance the settling of suspended solids. Removal efficiencies expected for settleable solids range from 30 to 90%. Suspended solids removal efficiencies range from 20 to 60% without chemical addition and 60 to 70% with chemical addition. Alum, ferrous sulfate, lime, and polymers are examples of chemical additions that may enhance settling.

In addition to the removal of settleable and suspended solids, sedimentation will remove a portion of the biological oxygen demand (BOD) associated with organic solids removal. Hence, 25–40% BOD removal may be achieved without chemical addition, whereas a slightly higher amount may result with chemical addition. Other compounds and contaminants such as metals and pesticides will be removed in proportion to their presence in the solids removed.

Sedimentation basins/tanks can double as storage tanks. This would be possible through the dewatering of the tank once flows to it have ceased. During the start of a storm, the tank would begin to fill up. Overflows from the tank would therefore be treated through sedimentation. As previously mentioned, the removal of floatables could be incorporated into the design. Figure 10 [42] illustrates a conventional system for sedimentation.

Concentration

Various types of pollutant concentrators are known as methods for treating contaminated water. These devices through circular, spiral, or helical motion impart secondary fluid motion through long-path geometric flow patterns, thereby enabling their separating solids from the contaminated water. The secondary flow stream containing the solid particles entrained is collected and conveyed separately from the main water throughput. Hence, the solids can be conveyed for further treatment or degritting/dewatering. Three such devices currently applicable to stormwater treatment include the vortex regulator, swirl concentrator, and spiral flow (helical bend) regulator. Each of these three devices is illustrated in Fig. 11 [43].

The vortex regulator was developed in England and basically consists of a circular channel. The flow enters at the periphery and exits at the center of the device. The circular flow pattern causes the solids to move toward the center where they are collected and removed. Similarly, with the swirl concentrator the flow enters at the periphery and exits toward the center. Solids following the secondary flow stream are collected separately. A scum ring or baffle at the surface of the swirl concentrator can be used to remove floatable materials. The spiral flow regulator, otherwise known as a helical bend device, is also illustrated in Figure 11 [43]. Through the use of an arc, this device causes the development of secondary fluid motion, thereby separating the stream of solids from the main flow stream. The bend angle is normally between 60 and 90°.

Of these three devices, the vortex regulator and swirl concentrator appear to have better application for the removal of solids from contaminated stormwater flow found at industrial facilities. Although all three devices provide treatment, the helical bend device is more suited to combined sewer overflow (CSO) handling rather than solely contaminated stormwater. Hence, it should be considered when combined sewer systems are utilized. The vortex regulator and swirl concentrator are applicable outside of a combined sewer system. However, when they are used solely for contaminated stormwater treatment, consideration should be given to the discharge of the solids flow stream to a sanitary or combined sewer if allowable. In other cases, the secondary solids stream will have to receive additional treatment or degritting/dewatering. Reportedly [43], up to 98% suspended solids removal for vortex regulators may be possible. Solids removal efficiencies for swirl concentrators [34] range from 40 to 60% for suspended solids and from 50 to 90% for settleable solids. Generally, these devices tend to offer an economic advantage over sedimentation or straining.

Filtration/Straining

Various types of filtration/straining equipment are available. Some of these include gravity filters, pressure filters, microstrainers, drum screens, and disc screens. Filtration generally uses a filter media or fine fabric to separate solids from contaminated water. Straining generally utilizes

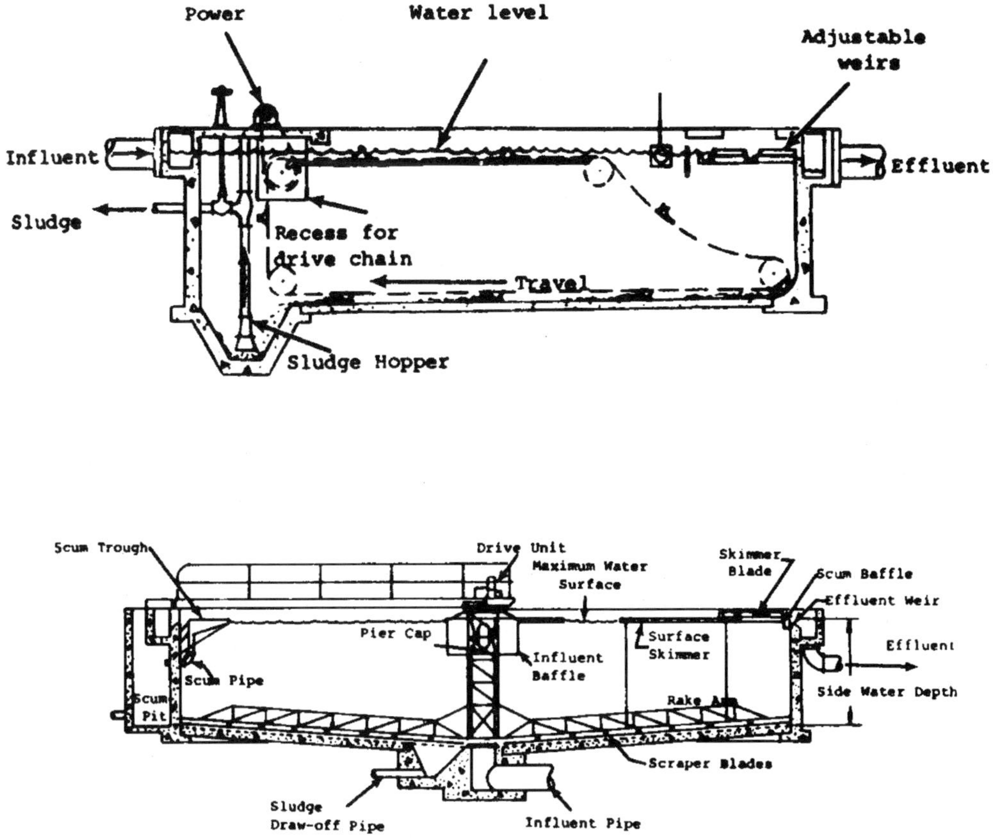

Figure 10 Rectangular and circular sedimentation basins (from Ref. 42).

a wire mesh or coarse fabric to provide a lesser degree of removal efficiency. Strainers provide a removal efficiency ranging from 10 to 55% for suspended solids and 60 to 95% for settleable solids depending on screen sizing. Filtration equipment including microstrainers (depicted in Fig. 12) [42] can provide 50–95% suspended solids removal. Surface loading rates for filtration and fine media microstrainer units range from 2 to 10 gal/min/ft^2. Rotary drum strainers and high-rate coarse screen microstrainers have surface loading rates ranging from 20 to 100 gal/min/ft^2. Specific performance details and sizing requirements can be obtained from manufacturers.

Dissolved Air Flotation

The flotation process is useful for removing solids, oils, and grease from contaminated water. Dissolved air is released into the influent contaminated water that enters the flotation tank. The dissolved air bubbles attach to solid particles, oil droplets, or grease, thereby decreasing their specific gravity. Through buoyancy, these materials float to the surface of the tank where they can be skimmed. Figure 13 [43] illustrates a typical dissolved air flotation unit.

Removal of suspended solids using the dissolved air flotation process ranges from 40 to 65% without the use of chemicals and 80 to 93% with chemicals such as alum, ferric chloride, or polymers. Oil and grease removal ranges from 60 to 80% without chemical addition and 85 to

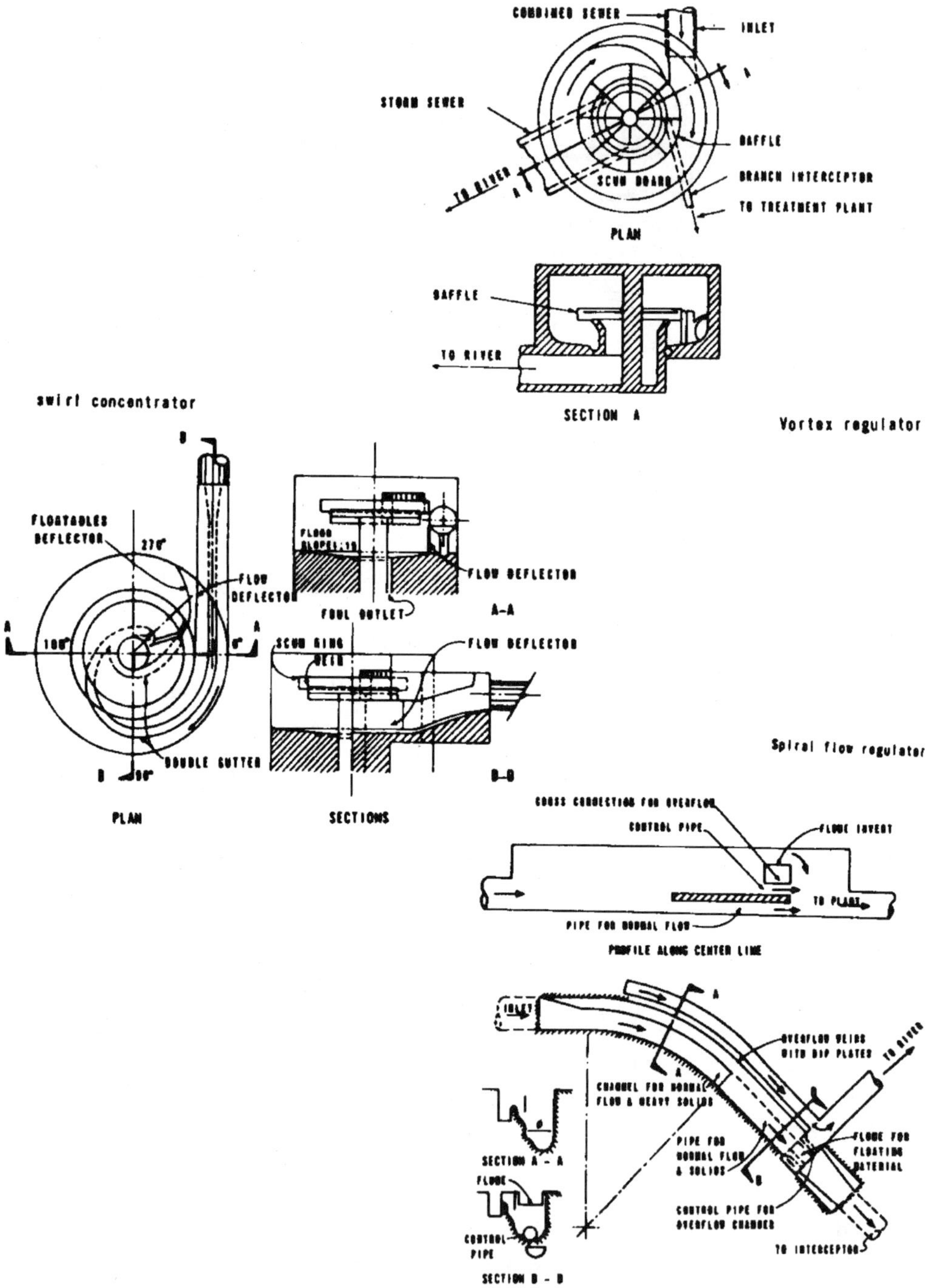

Figure 11 Vortex regulator, swirl concentrator, and spiral flow (helical bend) regulator (from Ref. 43).

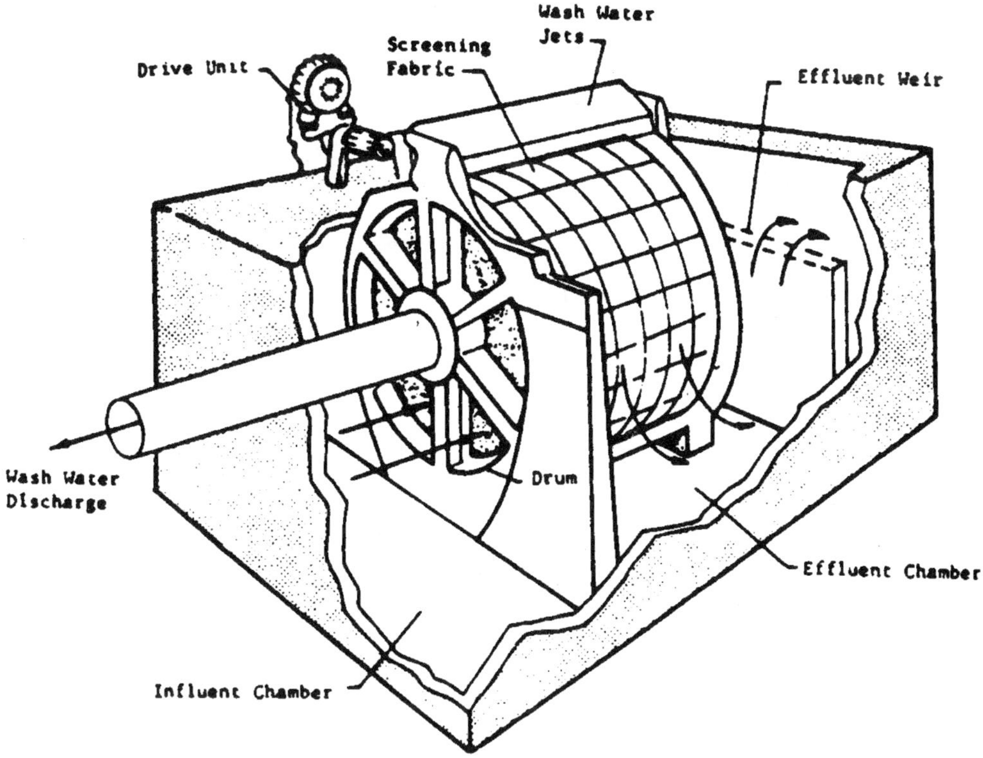

Figure 12 Microstrainer (from Ref. 42).

95% with chemical addition. Common surface loading rates range from 500 to 8000 gal/day/ft^2. Imperial Oil Company, Inc. presented a case history for its treatment of storm runoff by dissolved air flotation [44,45].

Biological Treatment

Generally, the colloidal and soluble substances found in contaminated stormwater can contain biological oxygen demand (BOD) that cannot be removed through sedimentation or filtration/straining. In order to remove these substances biologically, the stormwater would need to contact with microorganisms that through metabolic processes remove the BOD. Unlike wastewater treatment, in which a continuous supply of waste is available to be treated, the discontinuous supply of contaminated stormwater presents a unique problem that needs to be addressed in order to make biological treatment applicable. The biological treatment process relies on the supply of contaminants within the wastewater as a food source to keep the microorganisms alive and thriving. Hence, the lack thereof during dry weather could cause the failure of such a treatment system.

Several methods of maintaining a biological treatment system during dry weather include (a) during dry weather using the system to treat sanitary flows, (b) using a combination of storage and treatment to maintain the system, and (c) if another biological treatment system is on site, using the waste biological solids to maintain the level of microorganisms. In the first case, the substitution of sanitary flow for storm flow, if adequate sanitary flow is present, would maintain

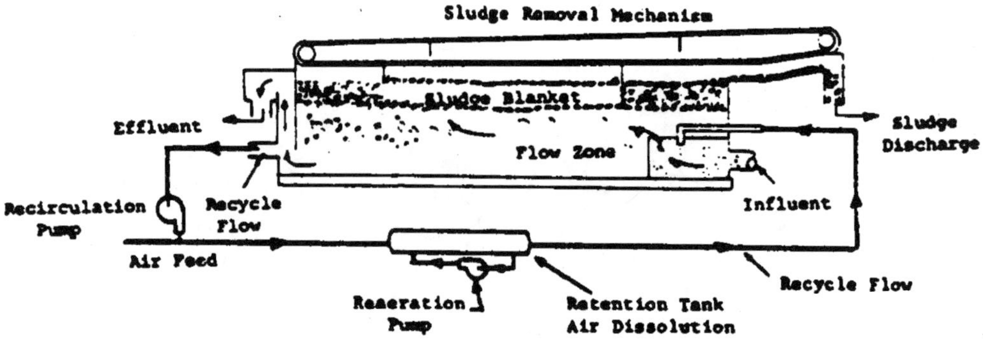

Figure 13 Air flotation unit (from Ref. 43).

the biology of the system. During dry weather, the treated sanitary discharge could be made to the existing sanitary municipal system if available.

In the second case, storage is used along with a biological treatment system sized to work in a manner that keeps it running during dry periods. This treatment system would operate during dry weather using the contaminated stormwater captured during wet weather within the storage tank. Naturally, the system would operate at average to peak capacity during a storm.

In the third case in which a sanitary treatment system is on site or nearby, the level of active organisms would be maintained by importing organisms from the other facility. This method would likely be subject to the operation of the facility supply, the microorganisms, and may not be as reliable as the other two strategies. Biological systems that may have better application for contaminated stormwater treatment include trickling filters, rotating biological contactors, and aerobic/facultative lagoons. These are illustrated in Fig. 14 [46], Fig. 15 [43], and Fig. 16 [43]. In certain cases, land/wetland application may also be applicable. All of these systems have the common element of long residence time for solids and ability to maintain biomass. Expected treatment efficiencies range from 75–95% BOD_5 reduction. Details for the performance of these systems are widely available in EPA publications and other chapters in this handbook.

Chemical Treatment

Chemical treatment includes (a) the use of flocculants and coagulants to augment the sedimentation or flotation process; (b) carbon adsorption; (c) ion exchange; and (d) chlorination for disinfection. The following is a brief description of each of these processes.

Flocculants are chemicals that facilitate the massing together of particles into a clump known as a floc. Flocs tend to settle faster than the individual particles, thereby increasing sedimentation efficiency. Flocculants can be either organic or inorganic in origin, and they include activated silica, certain clays, fine sands, and trade products such as Magnafloc, Purifloc, and Superfloc.

Coagulation is the process by which colloidal particles are destabilized. Normally, colloidal particles are charged and tend to repel each other, thereby maintaining themselves in suspension. Through the use of a coagulant, the electrical charge of the particles can be neutralized. Once neutralized, the particles will no longer repel one another and begin to coagulate/flocculate and settle (precipitate). Coagulates include such compounds as aluminum sulfate, ferric chloride, ferric sulfate, and copper sulfate.

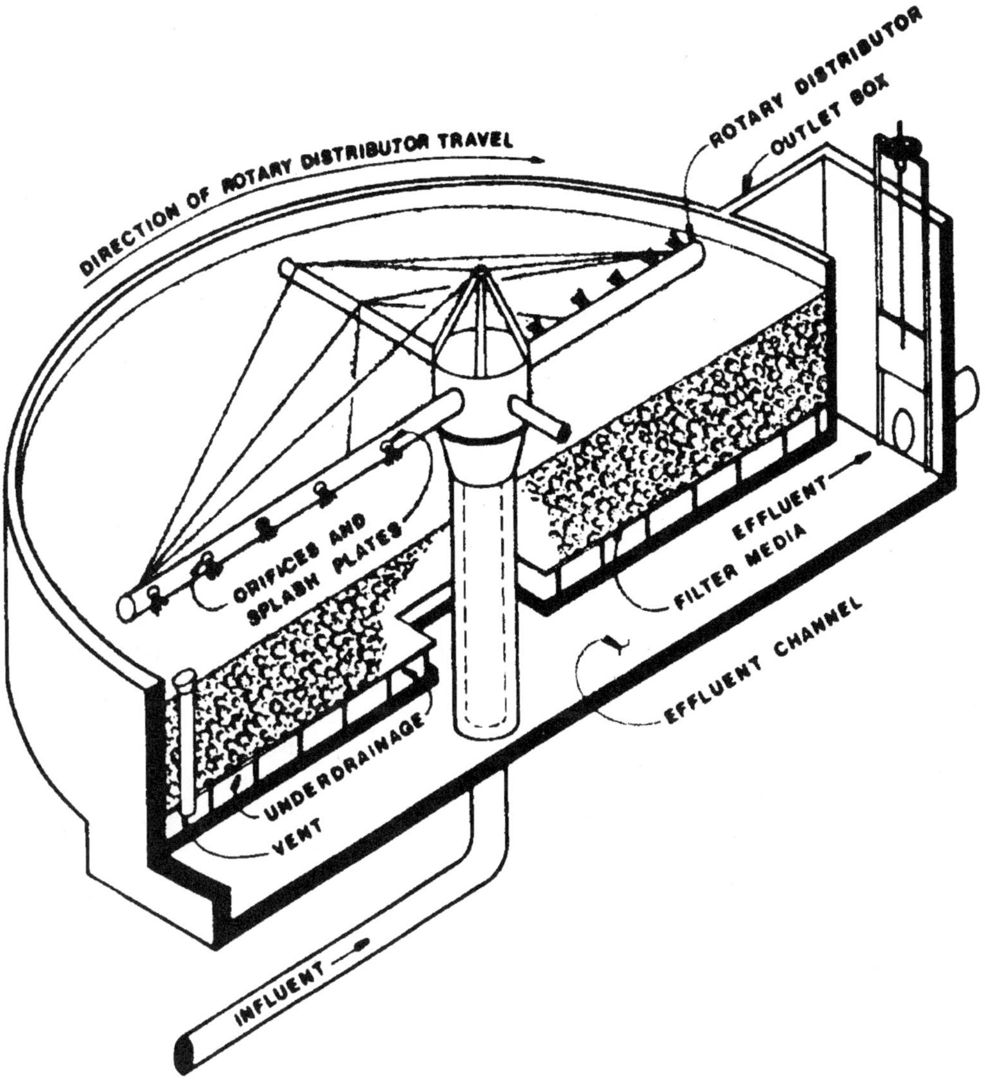

Figure 14 Trickling filter (from Ref. 46).

Activated carbon has the ability to adsorb contaminants from the stormwater. This adsorptive property lends itself to the removal of soluble BOD, organics, and toxic substances. Carbon can be used in powdered or granular form. In the powdered or granular form, it may be added to the contaminated stormwater directly and in a mixed contacting environment remove impurities. The granular form can also be used in packed bed or as a filtration media, whereby the contaminated stormwater may be passed through and impurities adsorbed. Activated carbon is also somewhat effective in removing various inorganic contaminants, such as copper, lead, cyanide, and chromium. Ion exchange is only suitable for use after all particulate matter has been removed from the contaminated stormwater. This is due to the subsequent plugging of the ion exchange media that would result if this matter is not removed. Ion exchange media come in two types: cationic and anionic. Cationic media are typically used to remove positively charged ions,

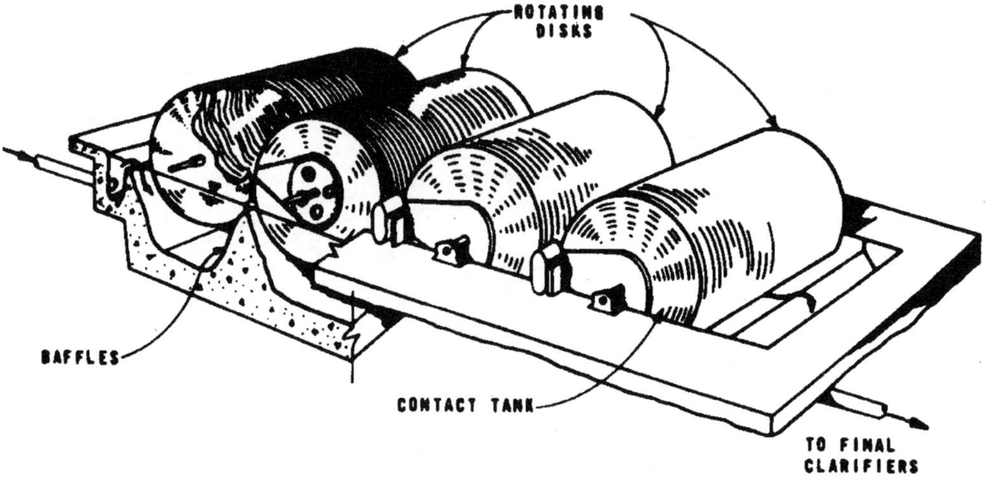

Figure 15 Rotating biological contactors (from Ref. 43).

whereas anionic media remove negatively charged ionic substances from the water. Generally, contaminants from electroplating and metal-finishing industries may be effectively removed using the ion exchange process (e.g., cadmium, chromium, and nickel).

Chlorination is but one means for disinfecting contaminated stormwater. The object of the disinfecting process is basically to kill pathogenic bacteria. This disinfecting method has been known to be the most common and economical method of choice in the wastewater treatment field. However, methods should be investigated that may, on a case-by-case basis, offer some advantage in certain instances. Conventionally, a 15 mm contact time has been used to facilitate the required kill; however, the contact time can be decreased by increasing the intensity of mixing as well as the dosage of chlorine.

Common varieties of chlorine used for chlorination are chlorine gas, chlorine dioxide, and sodium or calcium hypochlorite. Chlorine gas, though a better disinfectant, is hazardous and, therefore, not the first choice. Both chlorine dioxide and sodium hypochlorite have found wider use in chlorination.

Sand Filter Structure

Sand filter structures, similar to those used in potable water and industrial treatment, have recently been introduced in urban runoff management. They differ from those described under infiltration practices by being installed in a structural box and having a surface effluent, instead of being a soil amendment with underdrain system. The structures have been used intensively in the urban environment due to the lack of spaces for large basin treatments. The interested reader can access several web sites, such as *http://www.epa.gov/owm/mtb/sandfltr.pdf* and *http://p2library.nfesc.navy.mil/P2_Opportunity_Handbook/10-1.html* for more detailed information.

Bioretention Systems

Bioretention systems are utilizing natural vegetal systems to remove pollutants from stormwater runoff. The bioretention system, which may consist of a grass buffer strip, sand bed, ponding area, vegetal layers or plants, and soil is designed to treat stormwater sheet flow. Recently, low impact development (LID) has been used as an alternative design for traditional stormwater

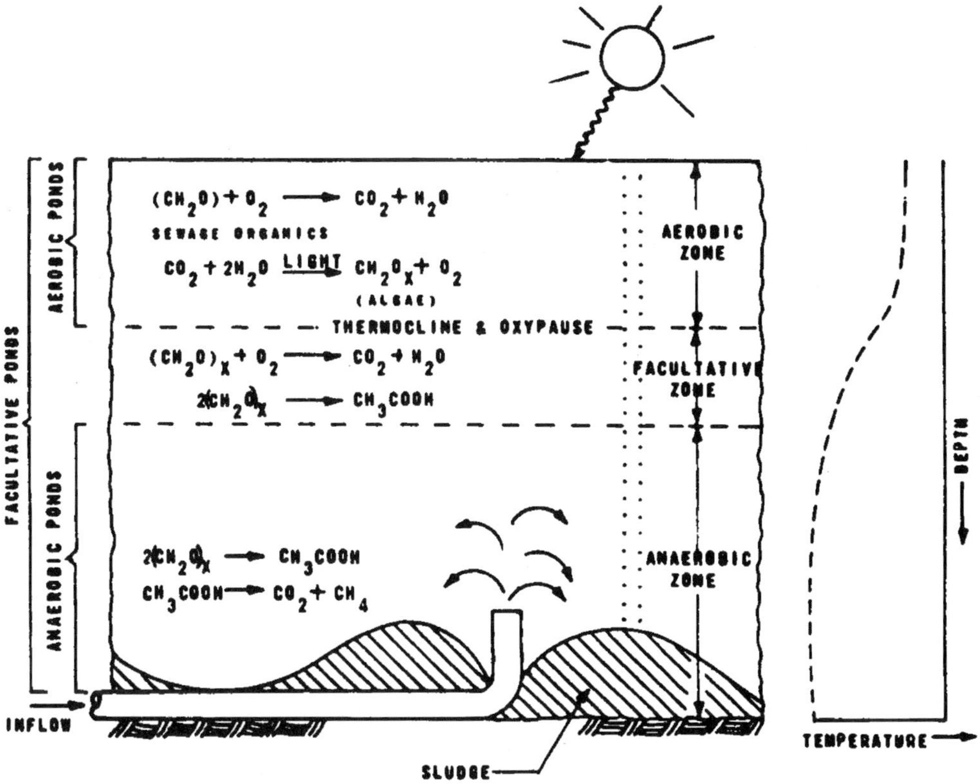

Figure 16 Facultative lagoon (from Ref. 43).

management design. The LID is trying to achieve stormwater management controls by fundamentally changing conventional site design to create a friendly environmental design that mimics natural watershed conditions. Bioretention systems can be considered as a LID design. Readers are referred to several web sites, such as *http://www.epa.gov/owm/mtb/biortn.pdf* and *http://www.state.nj.us/dep/watershedmgt/DOCS/BMP_DOCS/bmp2003pdfs/Stormwater_9.1.pdf* for more information on bioretention systems.

19.7 CASE STUDIES

In the following, information and conclusions from several papers relevant to stormwater pollution from industrial activities and facilities are summarized as typical case studies of what has already been discussed in this chapter.

19.7.1 Hazardous Material Spill Control

Mandel [47] reported that no convenient method exists to neutralize and solidify spills of acids and bases and to render these materials as noncorrosive solids. Also, there is no convenient method for adsorbing common fuel and organic solvent spills and for elevating their flashpoint to above 60°C (140°F). Furthermore, to date there is still room for improvement of the methods for application of cleanup chemicals at a safe distance from a spill in order to minimize the risk to emergency personnel. Present techniques of spill cleanup utilize mostly manual application methods that, in most cases, only absorb the spilled materials or employ time-consuming applications of neutralizing nonsolidifying materials. In most instances, therefore, the end result of the cleanup effort yields a waste material that is still considered hazardous and has to be handled and disposed of as such.

In a market survey of persons employed in US industries utilizing or producing hazardous materials, with responsibilities in environmental quality, environmental engineering or management, and industrial hygiene and safety, the following information was collected from the 1800 respondents. Some of the major industries surveyed were fertilizers and agricultural chemicals, chemicals and petrochemicals, explosives, drugs and cosmetics, food and beverages, fats and oils, plastics, paints, petroleum refineries, fibers and textiles, soaps and detergents, and semiconductors. The respondents listed the major areas in which spills occurred at their facilities and the percentage of the number of spills in each area as follows: shipping 9%, receiving 8%, warehouse 11%, manufacturing 30%, process piping 14%, tank storage 17%, laboratories 10%, and offsite transport 4%. The average spill size in these typical industrial facilities was determined to be 11.2 gal, with a spill frequency per plant being eight spills per year. Interestingly, the number of spills reported above 55 gal (a standard drum) was statistically insignificant.

The respondents indicated that, in their view, the spill incidents were unpredictable and posed a threat to the environment and safety and caused business interruption and profit loss. They further reported that in the surveyed plants the hazardous materials mainly fell into three broad chemical classes: acids, bases, and organics. Therefore, the bulk belonged to the corrosive and ignitable classes of hazardous materials. According to the information provided in the questionnaires, current methods of spill cleanup included containment, soak-up or absorbing, neutralizing, increasing viscosity, or diverting to a sump. Most of these methods were considered time-consuming and dangerous.

19.7.2 Overland Flow Treatment of Feedlot Runoff

The results of a six-month pilot-scale study in Oklahoma [48], which was followed by a six-month field test, indicated that the overland flow method of wastewater treatment has potential for treatment of runoff from beef cattle feedlots in warm, subtropical, humid regions. Confinement feeding of beef cattle is used extensively in the plains states of the United States, and rainfall runoff from feedlots has been identified as a major cause of fish kills in this region. Therefore, this runoff needs to be retained and treated before it can be discharged to surface water bodies without serious impacts, because it contains high concentrations of suspended solids, organics that create BOD, and nutrients that accelerate eutrophication. However, retention and treatment of the runoff at cattle feedlots located in this subhumid area of the plains states are major problems.

The application of wastewater to the land for reuse or treatment can be an efficient and economical management approach when land is readily available, as it is in the vicinity of many feedlot operations. One method of land-based wastewater management that has been used successfully is the overland flow treatment under intermittent or continuous use. The system can be managed to achieve the efficient removal of suspended solids, BOD, and nutrients from concentrated wastewaters.

The pilot-scale study was conducted on plots 9 m (30 ft) long with a 4.5% slope, where the native vegetation of mixed grasses was left undisturbed. The study showed that liquified loadings of 5–7.5 cm (2–3 in.)/week applied at instantaneous loading rates of less than 2.5 mm

(0.10 in.)/hour appeared suitable for field testing with dosing frequencies in the range of daily to three times per week. The field study was conducted at a 12,000-head capacity feedlot and utilized a four-component train for runoff collection and treatment. The treatment train included, as can be seen in Fig. 17 [48], collection lagoons, a storage reservoir, the overland flow areas, and a final polishing pond. Data from the six-month operation are summarized in Tables 6 and 7 [48], and corroborated the results of the pilot-scale study, and indicated that inclusion of the final polishing pond substantially improved the overall performance of the treatment system [62].

19.7.3 Petroleum Refinery Runoff Management

Stalzer and McArdle [49] reported on a SOHIO Toledo refinery, 2 km^2 (500 acre) area, stormwater runoff management study in which a nonsegregated oily sewer system collected both process wastewater and stormwater runoff. The handling of runoff at such a large petroleum refinery site requires adequate diversion and impounding capacity to store excessive storm flows for treatment after the runoff subsides. The paper presented the case history of upgrading the refinery's wastewater treatment system to efficiently manage the runoff while maximizing overall treatment efficiency. One of the primary objectives of the study was to estimate the peak flow rate and total impounding volume required for design storm events using the Illinois Urban Drainage Area Simulator (ILLUDAS) for stormwater runoff computer modeling. By establishing a hydraulic profile of the oily sewer at the wastewater treatment system, the diversion and impounding system, and primary wastewater treatment units, it was determined that the influent weir to the oil/water separator controlled the water level in the sewer for a considerable distance upstream. A water level sensor at a lift station downstream of the oil/water separator was used to monitor changes in flows and determine the sequence for opening

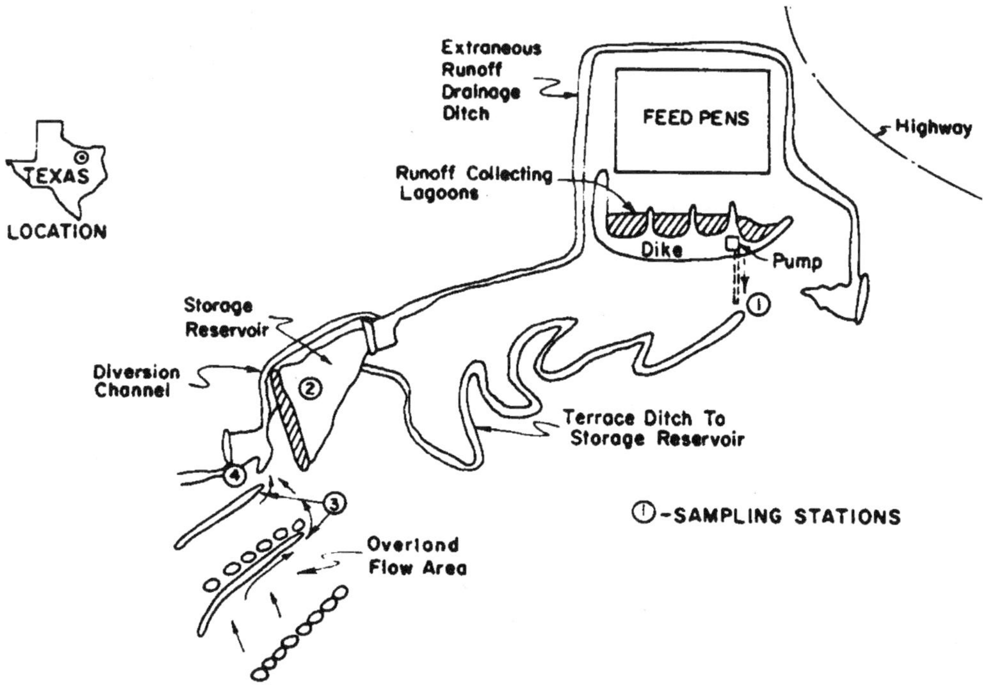

Figure 17 Feedlot and treatment train (from Ref. 48).

Table 6 Water Quality Data for the Runoff Collecting Lagoons

Date	Parameter concentration (mg/L)				
	TDS	TSS	COD	BOD	T-P
10-1-70	1316	174	620	165	21.3
11-4-70	1122	107	314	15	17.0
12-2-70	1059	216	386	23	11.1
1-5-71	959	267	396	–	11.3
2-4-71	1128	174	403	65	13.6
2-24-71	1169	136	298	31	12.6
3-25-71	1046	292	569	80	7.9
Mean	1114	195	426	63	13.5

Source: Ref. 48.

the diversion valves. This was based on the lift pump having a maximum capacity of 0.5 m^3/s (8000 gpm). For use as a design basis, the 25 year, 1 hour and 10 year, 24 hour storm events were chosen to establish the design peak flow rate (4.3 m^3/s or 152 cfs) and total impounding volume (based on a peak rate of 1.93 m^3/s or 68 cfs), respectively. To determine the required impounding capacity from the 10 year storm hydrograph, the amount of the runoff that can be treated during the storm was determined based on the maximum allowable wastewater treatment system flow rate and the dry weather flow rate. This flow rate was then subtracted from the hydrograph of the storm, with the area under the resultant hydrograph equal to the required impounding volume.

Because of the variability in dry flow rates, the base flow used to calculate the required impounding volume was not the average, but a 90% maximum dry weather flow based on four years of data to ensure some conservatism in the design. Even though the chance of a design storm actually occurring at the same time as the 90-percentile dry flow rate may have low probability, the additional cost for extra impounding volume must be weighed against the liabilities of not having sufficient volume. On the other hand, to enhance the conservative nature of the overall design, peak flows used in designing for diversion into the impounding basin and overflow from the basin under flood conditions were based on the 100 year storm. As flows

Table 7 Water Quality Data for the Farm Pond Discharge

Date	Days of operation	Parameter concentration (mg/L)					
		TDS	TSS	COD	BOD	T-P	T-N
10-1-70	0	342	10	78	2	0.4	4.4
11-4-70	35	437	19	71	2	0.2	3.5
12-2-70	63	477	11	92	2	0.2	3.6
1-5-71	97	848	24	149	5	1.0	6.7
2-4-71	107	832	8	166	5	0.5	6.3
2-24-71	127	780	6	134	6	0.4	5.1
3-25-71	158	874	6	183	12	1.0	8.0
Mean	–	656	12	125	5	0.5	5.4

Source: Ref. 48.

increase in the oily sewer due to stormwater runoff, a dam in the sewer with a flow meter and valve combination allows a preset maximum flow to the wastewater treatment system. As a result, the water level upstream of the dam increases until it begins spilling over a diversion weir into the impounding basin, thereby maintaining a constant optimum flow rate through the treatment system.

19.7.4 Pesticide Loads in Surface Runoff

Li et al. [50] developed two simple models or loading functions for estimating mean annual pesticide mass loads in surface runoff from agricultural areas. The loading functions were regression equations derived from 100 year simulation runs of a daily pesticide runoff model. Simulation runs were made for 1920 different cases, including 12 locations in the eastern United States and 80 combinations of pesticide half-lives (3–150 days) and adsorption partition coefficients (0.1–200 cm^3/g). Published research data indicated that surface runoff, soil erosion, chemical persistence, and strength of adsorption to soil particles are major determinants of pesticide runoff, whereas runoff and soil erosion are, in turn, determined by soil, ground cover, and weather patterns. The hydrology component was based on the US Soil Conservation Service (SCS) curve number (CN) runoff equation and a daily version of the Universal Soil Loss Equation (USLE).

Two different regression equations explained the highest fractions of observed variation in pesticide runoff. Regression Model A, $P = a_0 + a_1X$, explains 71–94% of pesticide runoff variation (P = mean annual pesticide runoff, percent of application) and is a particularly simple function, requiring only mean annual soil erosion (X in mg/ha) to obtain an estimate of pesticide runoff. Model B, $P = b_0 + b_1X + b_2Qm$, explains 85–96% of pesticide runoff variations and requires both mean annual soil erosion (X) and surface runoff volume (Qm = mean surface runoff during month of pesticide application, mm). The regression parameters a_0, a_1, b_0, b_1, and b_2 are provided in tables calculated by Li et al. [50]. Except for the more persistent and strongly adsorbed pesticides, standard errors of estimate are substantially smaller for Model B, indicating that it will generally provide a more accurate estimate of pesticide runoff. However, Model A is easier to use as soil erosion is routinely calculated by the USLE.

19.7.5 Treatment of Base Metal Mining Drainage

Huck et al. [51] reported on operational experience with a base metal mine drainage treatment pilot plant, established at the site of a northeastern New Brunswick, Canada, mining and smelting mill. The objective was to develop and demonstrate new and existing technology for the removal of heavy metals and the neutralization of such effluents. Three minewaters, characterized as strong, weak, and moderately strong, were evaluated, as indicated in Table 8 [51]. The use of a polymeric flocculant improved settling, yielded greater reliability of operation, and appeared to reduce metal concentrations in all cases. The optimum treatment configuration for the three minewaters is shown in Fig. 18 [51] and consists of a once-through operation using polymer and two-stage neutralization. The following conclusions were drawn from the data obtained in the study:

1. Using conventional neutralization, precipitation (optimum precipitation pH was 9.5), and sedimentation, effluent metal concentrations of less than 0.3 mg/L lead, 0.6 mg/L zinc, 0.1 mg/L copper, and 0.6 mg/L iron were achieved with all three base metal mine drainage wastewaters.
2. The initial minewater metals concentration strength had little effect on the effluent metal concentrations.

Table 8 Minewater Characterization[a]

Drainage	Strong minewater		Weak minewater		Moderately strong minewater	
	Mean	Range	Mean	Range	Mean	Range
Constituent[b]						
pH	–	2.4–3.2	–	2.8–3.3	–	2.3–2.9
Sulfate	7090	1860–14,300	1020	810–1790	4530	2350–7290
Acidity	6030	4550–9650	800	510–1530	5290	2600–15000
Lead	4.3	0.9–9	1.3	0.1–5	1.2	0.03–3
Zinc	1160	735–1,590	108	22–175	540	390–723
Copper	10	15–17	20	12–52	50	24–76
Iron	1580	815–3210	68	24–230	720	350–1380
Suspended solids	–	75–260[c]	–	5–90	–	5–190
Ionic ratios[d]						
Zn : Fe	0.7		1.6		0.8	
Cu : Fe	0.006		0.3		0.07	
Fe : SO_4	0.2		0.07		0.2	

[a] Based on daily means. All constituents except pH reported in mg/L.
[b] For the first run. The second run was similar, although the iron was somewhat lower and the range was narrower.
[c] Based on data for the month of June only.
[d] By weight.
Source: Ref. 51.

3. The improvement of effluent metal concentrations with a low ratio ($\leq$2 : 1) sludge recycle appeared to be greatest for the weakest minewater. The improvement was similar to that achieved with polymer addition. High recycle reduced dissolved zinc but increased total metals.
4. Reactor contact times and the number of neutralization stages did not significantly affect results. However, as precipitation pH (8.9–10.5 gave approximately equal effluent qualities) control under neutralization load variation is better with staging, two-stage treatment was therefore chosen as the optimum process configuration.
5. Calcium sulfate deposition presented difficult but surmountable problems in the operation of the treatment process, and it could be controlled partly through sludge recycling. With minewater sulfate loadings of less than 2500 mg/L, calcium sulfate scaling on reactor surfaces was minimal and a daily cleaning of the pH probes was adequate. With concentrations in excess of this, probes required cleaning every 8 hour and light coatings formed on pipes and tanks within a few weeks.
6. Effluent polishing techniques such as sand filtration, coprecipitation, lagooning, or the use of an alternative reagent demonstrated that better effluent quality may be attained.

19.7.6 Arsenic Removal from Combined Runoff and Wastewater

The feasibility of removing hazardous soluble arsenic (+5) and other conventional pollutants from combined storm runoff and process wastewater by oil/water separation, dissolved air flotation (DAF), filtration, and granular activated carbon (GAC) adsorption was fully demonstrated for an oil-blending company in New Jersey [44,45,52]. The oil separated from the raw combined wastewater by the American Petroleum Institute (API) oil/water separators was

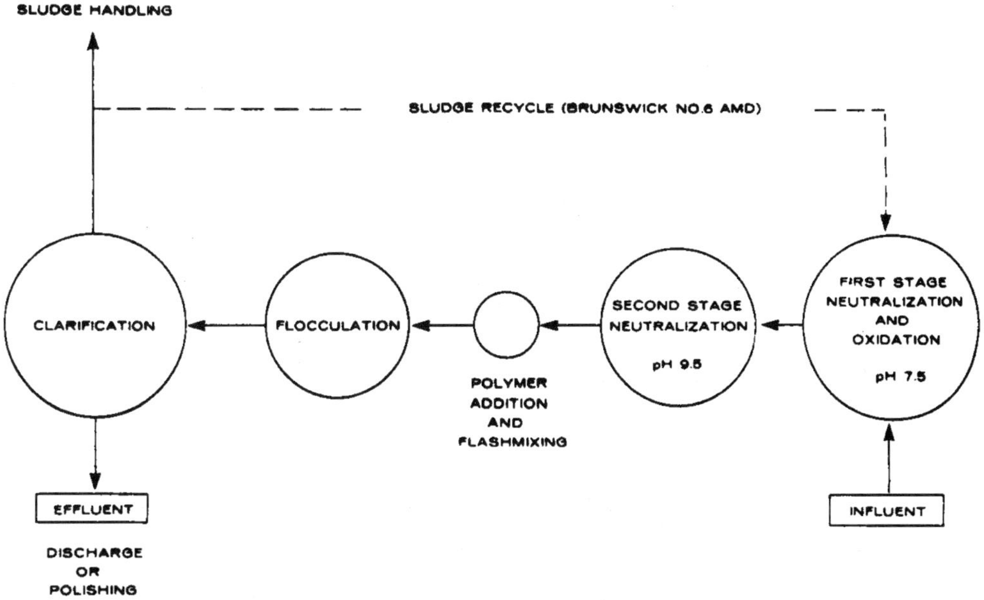

Figure 18 Optimum process configuration (from Ref. 51).

virgin and therefore skimmed off, dried, and reused. The oil/water separator effluent containing 1.01 mg/L of arsenic, 3 NTU of turbidity, 50 units of color, 28.5 mg/L of oil and grease (O&G), and 83 mg/L of chemical oxygen demand (COD) was fed to a DAF clarifier for removal of arsenic by 90.1%, turbidity by 30%, color by 43%, O&G by 43.2%, and COD by 32.5%. Either ferric chloride or ferric sulfate was an effective coagulant for arsenic removal.

The same oil/water separator effluent was also successfully treated by a DAF-filtration clarifier. Reductions of arsenic, turbidity, color, O&G, and COD were 90.6, 93.3, 98, 74.7, and 51.8%, respectively. Although DAF and DAF-filtration both proved to be excellent pretreatment processes, granular activated carbon (GAC) post-treatment removed 100% of soluble arsenic. Table 9 [44,45,52] presents some important operational data generated at the oil-blending company.

19.7.7 Recent Advances in Stormwater Management and Treatment

There are many new developments in the area of stormwater management and treatment [44,45,52–61]. A useful guide for dealing with the stormwater management permit program has been developed by Barron [53]. Coppes and LeMire [54] suggest many solutions to stormwater management by combining both technology and regulations. Treatment of stormwater at airport terminals is reported by the editor of Public Works [55], while treatment of combined stormwater and wastewater at oil company is reported by Wang [52,60].

It is reported by Bolender and Shuler [56] that municipalities are shifting perspective in their stormwater management design. Aboveground infiltration facilities, although land-intensive, can alleviate concerns about underground construction and maintenance. While involved in a long-term sewer separation project, Nashua, New Hampshire, United States, has realized the benefits of proactively addressing its stormwater treatment problems [57]. The city has used the inline stormwater treatment facility to protect its area surface water successfully.

Table 9 Treatment of Storm Runoff by Dissolved Air Flotation and Sand Filtration

Treatment efficiency	Test 1	Test 2	Test 3
Influent arsenic (mg/L)	0.0015	0.518	1.010
Removal by flotation (%)	33.3	79.5	90.1
Removal by flotation/filtration (%)	100.0	90.3	90.6
Influent turbidity (NTU)	3.4–3.6	3.5	3.0
Removal by flotation (%)	35.7	45.7	30.0
Removal by flotation/filtration (%)	93.0	93.7	93.3
Influent color (unit)	48–49	49.0	50.0
Removal by flotation (%)	50.5	40.8	43.0
Removal by flotation/filtration (%)	98.9	97.9	98.0
Influent O&G (mg/L)	29.2	29.3	28.5
Removal by flotation (%)	52.7	51.5	43.2
Removal by flotation/filtration (%)	60.2	61.1	74.7
Influent COD (mg/L)	85.0	86.0	83.0
Removal by flotation (%)	37.6	46.5	32.5
Removal by flotation/filtration (%)	58.8	53.5	51.8
Influent pH (unit)	7.0–7.1	7.1–7.1	7.2–7.2
After flotation (unit)	6.9–7.0	7.1–7.2	6.9–7.1
After flotation/filtration (unit)	6.9–7.0	7.0–7.1	7.0–7.1

Chemical flotation treatment: Sodium aluminate, 15 mg/L as Al_2O_3, and either ferric chloride or ferric sulfate, 15 mg/L as Fe.
Sand specification: ES = 0.85 mmn, UC = 1.55, depth = 11 in.
Source: Ref. 52.

Contech Construction Products Inc. [58] has built an underground stormwater detention for the University of Illinois at Urbana-Champaign (UIUC), in Illinois. Boneyard Creek regularly flooded parts of the UIUC campus. As part of the design-supply team to bring Boneyard Creek under control, Contech supplied nearly 9300 ft of 96 in. diameter Aluminized Steel Type 2 corrugated pipe for a subsurface drainage system and stream enclosure. Now, a multi-use corridor has been reclaimed from the Boneyard that includes a park with bikie trials.

Loudon County, Virginia, United States, is taking a different approach to its quality assurance program [59], and is developing a storm sewer inventory GIS from a GPS field survey. Its quality assurance program will ensure an accurate stormwater GIS.

Klahn examined the top issues in water quality in 2003 [61]. The US Environmental Protection Agency (USEPA) has proposed modifications to its Sanitary Sewer Overflows (SSO) rule, addressing the issue of capacity and notification in the event of an overflow, as well as limitations on enforcement. The USEPA has estimated that there are over 40,000 overflows of sanitary sewers each year. Without a permit, these types of discharges are prohibited under the Clean Water Act. Municipal sanitary sewer systems holding SSO permits are often subject to enforcement actions for overflows that exceed the scope of their permits. The EPA has acknowledged that many municipalities have vastly undersized sanitary sewer systems. The new proposed rule maintains the requirements for permits that include technology-based effluent limitations and addresses the issues of sewer system capacity, operation, and maintenance. One USEPA estimate suggests that municipalities will need to spend between \$93.5 million and \$126.5 million each year to accommodate the proposed rule changes. Funding acquisition for this is undetermined and this funding gap has been the subject of congressional hearings and analyses from both inside and outside industry. The USEPA has also announced that the rule will address local conditions and allow each region to set up an SSO enforcement structure so that municipalities can avoid enforcement actions from overflows beyond their control. The EPA expected to publish the rule in the Federal Register during the fall of 2002 and begin the public comment period. Although the rule is specific to SSOs, the agency is also seeking comment on the application of the SSO rules to Combined Sewer Overflows (CSOs).

The issue of Concentrated Animal Feeding Operation (CAFO) has also been reported by Klahn [61]. The USEPA has also taken up modification to CAFO regulations. In January 2001, the USEPA proposed two rules, one that addressed CAFO operations regarding manure and other process waters, and another that established CAFO technology-based treatment requirements for feedlots. In the first regulation, the USEPA proposed changes to the definition of CAFOs, changes to the means by which CAFOs could land-apply manure and other process waters, changes requiring processors to be co-permittees, and changes to ensure public access to information regarding CAFO permits. In the second regulation, the USEPA proposed certain minimum types of practices, such as containment pond construction and setback requirements, to avoid contamination of groundwater with CAFO wastes. USEPA is also considering the merits of a suggestion that livestock producers who adopt voluntary performance standards, equivalent to best available technology, could opt out of regulation altogether.

As comments and supporting data were received, the USEPA amended its analytical approach and thus modified its economic and financial analysis of the impact to industry from the regulation. The agency issued a final rule in early 2003. Several states have substantively addressed CAFOs with more restrictive regulations than those proposed by USEPA [62].

REFERENCES

1. Weiss, K. Controlling pollutants in runoff from industrial facilities, *National Conference On Urban Runoff Management*, EPA Seminar, Center for Envir. Res. Info.; U.S. Environmental Protection Agency, Cincinnati, OH, 1995; EPA/625/R-95/003.
2. Yapijakis, C.; Llorente, M.T. Industrial sites stormwater management: regulations and treatment, *Proceedings of International Conference on Sustainable Land Use and Management*, Canakkale, Turkey, Tubitak, June 10–13, 2002; 312–317.
3. Woodar, F. *Industrial Waste Treatment Handbook*; Butterworth and Heinemann Publishers: Boston, 2001.
4. Yapijakis, C. Stormwater runoff pollution and prevention at industrial sites, *Proceedings of International Conference on Protection and Restoration of the Environment VI*, University of Thessaly, Volos, Greece, 2002.
5. Viessman, W., Jr.; Lewis, G.L. *Introduction to Hydrology*; Prentice Hall: New Jersey, 2003.
6. Linsley, R.K.; Kohler, M.A.; Paulhus, J.L. *Hydrology for Engineers*; McGraw-Hill: New York, 1982.
7. Reidel, J.T.; Appleby, J.F.; Schloemer, R.W. *Seasonal Variation of the Probable Maximum Precipitation East of the 105th Meridian for Areas from 10 to 1000 Square Miles and Durations of 6, 12, 24, and 48 Hours*, Hydrometeorological Rep. 33; U.S. Weather Bureau: Washington, DC, 1967.
8. Snyder, F.F. Synthetic unit graphs. Trans. Amer. Geophys. Union, **1938**, *19*, 447–454.
9. Soil Conservation Service (SCS). *Hydrology*, National Engineering Handbooks, Sec. 4; U.S. Department of Agriculture: Washington, DC, 1972.
10. Goodell, B.C. Snowpack management for optimum water benefits, *Proceedings of ASCE Water Resources Engineering Conference*, Denver, Colorado, 1966.
11. U.S. Army Corps of Engineers. *Snow Hydrology*; North Pacific Div.: Portland, Oregon, 1956.
12. Nash, J.E. The form of the instantaneous unit hydrograph. Intern. Assoc. Sci. Hydro. **1957**, *3*, 14–121, Publ. 45.
13. Federal Aviation Agency (FAA). *Department of Transportation Advisory Circular on Airport Drainage*, Rept. A/C 150-5320-5B; FAA: Washington, DC, 1970.

14. Morgali, J.R.; Linsley, R.K. Computer simulation of overland flow. J. Hydraul Div. ASCE **1965**, *91*.
15. Kibler, D.F. *Recommended Hydrologic Procedures for Computing Urban Runoff from Small Developing Watersheds*; Pennsylvania Institute for Research on Land and Water Resources: University Park, PA, 1982.
16. SCS. Urban Hydrology for Small Watersheds, SCS TR-55; SCS: Washington, DC, 1975.
17. McCuen, R.H.; Rawls, W.J.; Wong, S.L. *Evaluation of the SCS Urban Flood Frequency Procedures*; U.S. SCS: Beltsville, MD, 1981.
18. Hall, M.J. *Urban Hydrology*; Elsevier: London, 1984.
19. U.S. Army Corps of Engineers. *Storage, Treatment, Overflow, Runoff Model "STORM". Generalized Computer Program Users' Manual*; USACE Hydrologic Eng. Ctr.: Davis, CA, 1974.
20. Mockus, V. *Use of Storm and Watershed Characteristics in Synthetic Hydrograph Analysis and Application*; U.S. Dept. of Agriculture: Washington, DC, 1957.
21. Boyd, M.J. *Urban Stormwater Quality, Management and Planning*; Yen, B. C., Ed.; Water Res. Publ.: Littleton, CO, 1982; p. 370.
22. Council on Environmental Quality (CEO). *Environmental Quality–1979*, 10th Ann. Rep.; CEO, U.S. Government Printing Office: Washington, DC, 1980.
23. Wu, J.S.; Ahlert, R.C. Assessment of methods for computing storm runoff loads. Wat. Res. Bull. 1978, 14, 429–439.
24. Whipple, W.; Grigg, N.C.; Grizzard, T.; Randall, C.W.; Shubinski, R.P.; Tucker, L.C. *Urbanizing Areas*; Prentice-Hall: Englewood Cliffs, NJ, 1983.
25. Smith, K.V.; Stewart, D.A. Statistical models of river loadings of nitrogen and phosphorus in the Lough Nengh system. Water Res. **1977**, *11*, 631.
26. Krenkel, P.A.; Novotny, V. *Water Quality Management*; Academic Press: New York, 1980.
27. Lager, J.A.; Shubinski, R.P.; Russell, L.W. Development of a simulation model for stormwater management. J. Water Poll. Control Fed. **1971**, *43*, 2424.
28. Novotny, V.; Chesters, G. *Handbook of Nonpoint Pollution – Sources and Management*; Van Nostrand Reinhold: New York, 1981.
29. Williams, J.R.; Berndt, H.D. Sediment yield computed with universal equation. J. Hydraul. Div. ASCE **1972**, *98* (*Hy 12*), 2087.
30. Roehl, J.W. *Sediment Source Areas, Delivery Ratios and Influencing Morphological Factors*, Publ. 59: IASH Comm. on Land Erosion, 1962; 202–213.
31. McElroy, A.D.; Chic, S.Y.; Nebgen, J.W.; Aleti, A.; Bennett, F.W. *Loading Functions for Assessment of Water Pollution from Nonpoint Sources*, EPA/600/2-76/151; U.S. Environmental Protection Agency: Washington, DC, 1976.
32. Fleming, G. Design curves for suspended load estimation. Proc. Inst. Civil Eng. **1969**, *43*, 1–9.
33. Weschmeier, W.H.; Smith, D.D. *Predicting Rainfall-Erosion Losses from Cropland Costs of the Rocky Mountains*, Agricultural Handbook 282; U.S. Dept. of Agriculture: Washington, DC, 1965.
34. Wanielista, M.P. *Storm Water Management – Quantity and Quality;* Ann Arbor Science Publishers: Ann Arbor, MI, 1978.
35. Glass, N.R.; Glass, G.E.; Rennie, P.J. Effects of acid precipitation, *Proceedings 4th Annual Energy Research and Development Conference*; ACS: Washington, DC, 1979.
36. Sylvester, R.O.; DeWalle, F.B. *Character and Significance of Highway Runoff Waters*, Res. Rept. 7.1: Wash. State Highway Dept. Res. Program: Olympia, Washington, 1972.
37. USEPA. *Report to Congress: Nonpoint Source Pollution in the U.S. Office of Water Program Operations*; Water Planning Div., U.S. Environmental Protection Agency: Washington, DC, 1984.
38. Beschta, K.L. Long term patterns of sediment prediction following road construction and logging in Oregon coast range. Water Res. Res. **1978**, *14*, 1011.
39. Metcalf & Eddy, Inc. *New York City Industrial Pretreatment Program Report*, Task 7 – Sampling Program; Metcalf & Eddy: New York, 1984.
40. Wilber, W.G.; Hunter, J.V. Contributions of metal resulting from stormwater runoff and precipitation in Lodi, New Jersey. In *Urbanization and Water Quality Control*; American Water Resources Assoc., June 1975; pp. 45–54.
41. American Society of Civil Engineers (ASCE) and Water Pollution Control Federation (WPCF). *Design and Construction of Storm and Sanitary Sewers*; ASCE: New York, 1969.

42. USEPA. *Innovative and Alternative Technology Assessment Manual*, EPA 430/9-78-009; U.S. Environmental Protection Agency: Cincinnati, OH, 1980.
43. USEPA. *Urban Stormwater Management and Technology: An Assessment*, EPA 670/2-74-040; U.S. Environmental Protection Agency: Cincinnati, OH, 1974.
44. Wang, L.K.; Mahoney, W.J. Treatment of storm runoff by oil–water separation, flotation, filtration and adsorption. Part A: Wastewater treatment, *Proceedings of the 44th Industrial Waste Conference*, Purdue University, Lafayette, IN, 1989; Vol. 44, pp. 655–666.
45. Wang, L.K.; Wang, M.H.S.; Mahoney, W.J. Treatment of storm runoff by oil–water separation, flotation, filtration and adsorption. Part B: Waste sludge management, *Proceedings of the 44th Industrial Waste Conference*, Purdue University, Lafayette, IN, 1989; Vol. 44, pp. 655–666.
46. USEPA. Process Control Manual for Aerobic Biological Wastewater Treatment Facilities, EPA 430/9-77-006; U.S. Environmental Protection Agency: Washington, DC, 1977.
47. Mandel, F.S. Novel hazardous material spill control agents and methods of application, *Proceedings HAZMACON '87*, Santa Clara, CA, 1987; p. 442.
48. USEPA. *Feasibility of Overland-Flow Treatment of Feedlot Runoff*, Office of Res. and Devel. Rept., EPA-660/2-74-062; U.S. Environmental Protection Agency: Washington, DC, Dec. 1974.
49. Stalzer, R.B.; McArdle, G.W. A petroleum refinery stormwater run-off study, *Proceedings of the 38th Industrial Waste Conference*, Purdue University, Lafayette, IN, 1983; pp. 893–902.
50. Li, W.; Merrill, D.E.; Haith, D.A. Loading functions for pesticide runoff. Res. J. WPCF, **1990**, *62*, 16–26.
51. Huck, P.M.; Le Clair, B.P.; Shibley, P.W. *Operational Experience with a Base Metal Mine Drainage Pilot Plant*, Tech. Dev. Rep. EPS 4-WP-74-8; Water Pollution Control Directorate for Environment: Canada, Sept. 1974.
52. Wang, L.K.; Wang, M.H.S. Decontamination of groundwater and hazardous industrial effluents by high rate air flotation processes, *Proceedings of GreatLakes '90 Conference*, Hazardous Materials Control Res. Inst., Silver Springs, MD, 1990.
53. Barron, J. The scoop on NPDES permits. Environ. Protect. **2002**, *13* (*10*), 32.
54. Coppes, B.A.; LeMire, R. Stormwater management solutions. Environ. Protect. **2002**, *13* (*11*), 33.
55. Fumess, R. In the field: stormwater treatment at a new continental airport terminal. Water Eng. Mnt. **2003**, *150* (*3*), 30–36.
56. Bolender, B.N.; Shuler, J.A. Municipalities' shifting perspective in stormwater management design. Public Works **2003**, *134* (*2*), 26–29.
57. Editor. In-line stormwater treatment protects area surface water. Public Works **2003**, *134*, (*2*), 30–33.
58. Contech Construction Products Inc. Underground storm water detention. Public Works **2003**, *134* (*2*), 3.
59. Cobb, J. Quality assurance procedures ensure an accurate stormwater GIS. Public Works **2003**, *134* (*3*), 22–25.
60. Wang, L.K. *Stormwater Treatment and Management*, LIWT/Jan-10-2001/375; Lenox Institute of Water Technology: Lenox, MA, 2001; 255p.
61. Klahn, S. Plumbing the murky depths: examining the top issues in water quality in 2003. Environ. Protect. **2003**, *14* (*1*), 29–33.
62. Chen, J.P.; Zou, S.; Hung, Y.T.; Wang, L.K. Livestock waste treatment. In: Handbook of Industrial and Hazardous Wastes Treatment. (Wang, L.K.; Hung, Y.T.; Lo, H.H.; Yapijakis, C. eds.). Marcel Dekker Inc., NY. 2004; pp. 1051–1076.

20
Site Remediation and Groundwater Decontamination

Lawrence K. Wang
Zorex Corporation, Newtonville, New York, U.S.A., and
Lenox Institute of Water Technology, Lenox, Massachusetts, U.S.A.

20.1 INTRODUCTION

20.1.1 Summary

Hazardous waste pollution, hazardous waste terminologies, various onsite, offsite, *in situ*, and *ex situ* environmental remediation technologies, and case histories are presented in this chapter. The topics of soil remediation technologies covered here include excavation, stabilization, solidification, vapor stripping, vacuum extraction, thermal desorption, incineration, starved air combustion, pyrolysis, hot air enhanced stripping, steam enhanced stripping, thermal extraction, subsurface volatilization and ventilation, vitrification, soil surfactant flushing, soil washing, soil bioremediation, bioventing, slurry bioreactor, chemical treatment, KPEG treatment, and natural attenuation. The topics of groundwater decontamination technologies covered here include air stripping, ultraviolet radiation, oxidation, carbon adsorption, groundwater bioremediation, sewer discharge, liquid/liquid (oil/water) separation, free product recovery, in situ flushing, trenching, containerizing, and dissolved air flotation.

20.1.2 Site Remediation and Groundwater Decontamination: a Joint UN–USEPA Effort

At the end of 1993, the United Nations Industrial Development Organization (UNIDO), the World Bank, and the United Nations Environment Programme Industry and Environment Programme Activity Centre (UNEP/IEPAC) started issuing new Industrial Pollution Prevention and Abatement Guidelines. In later years, pollution prevention, waste minimization, and manufacturing process integration together have been referred to as "cleaner production" by the international community in order to build awareness of sustainable industrial development, sustainable agricultural development, and environmental protection. The objectives of all these international efforts are to disseminate information on pollution prevention options, end-of-pipe treatments, and cleaner production technologies. The emphasis of the international efforts has been on pollution prevention at source, treatment at the end of pipe, and manufacturing process integration through cleaner production, because there is increasing evidence of the economic and environmental benefits to be realized by preventing or reducing pollution, rather than by managing hazardous wastes after they have been produced, and the environment has been polluted.

Until recently, industry has not been overly concerned with cleaner production, hazardous waste management, and environmental protection, so there have been many direct and indirect damages caused to the environment by mishandling of hazardous wastes. This chapter will discuss various in situ, ex situ, onsite, and offsite technologies for site remediation and groundwater decontamination, assuming that the worse situation has happened – the environment has already been polluted by the hazardous wastes.

Site remediation and groundwater decontamination are pressing issues in all industrial and developing countries, especially for European countries due to limited availability of land. As a result, much progress is being made in the development of various technologies for effectively remediating contaminated industrial, agricultural, and commercial sites. These site remediation technologies, developed by Holland, Germany, and Belgium, include vacuum extraction of volatile organic compounds from contaminated soils, in situ washing of cadmium-polluted soil, high-temperature slagging incineration of low-level radioactive wastes, *in situ* steam stripping, and a number of bioremediation and soil washing operations. The United Nations (UN) and the US Environmental Protection Agency (USEPA) have played the leadership roles in information dissemination, technology promotion, in-depth R&D, and commercialization of most of the site remediation technologies for the benefit of entire world [1–25].

20.1.3 Terminologies

The hazardous substances at contaminated sites cannot be properly managed without knowing the correct terminologies. According to the 1978 Resource Conservation and Recovery Act (RCRA) of the United States, a waste is considered hazardous when it poses a treat to human health or the environment. The US Comprehensive Environmental Response, Compensation and Liability Act (CERCLA; otherwise known as Superfund) was established in 1980 [1]. Under the 1984 reauthorization of the RCRA, the USEPA land disposal restrictions (LDRs) (also known as land bans) of 1985–1990 were imposed. Using the toxicity characteristic leaching procedure (TCLP), a concentration of any listed constituent in the leachate at or above these levels designates the wastes as hazardous. The waste remains hazardous until treated to reduce its leachability below the TC levels. The heavy metal levels apply not only to the definition of a hazardous waste, but to the LDR maximum leaching levels for disposal of "characteristic waste" at an RCRA treatment, storage, and disposal facility (TSDF), otherwise known as a secure landfill.

At an industrial, commercial, or agricultural site that has been contaminated by hazardous wastes, both the environmental samples (such as contaminated soil, air, or groundwater), and hazardous wastes (such as PCB-containing transformers, waste oil, waste gasoline, old chemicals, spent activated carbons, precipitated heavy metals, etc.) must be handled with care in accordance with government rules and regulations and standard engineering practices.

Characterization of hazardous wastes and environmental samples [26] is a critical step in determining how a hazardous waste or sample should be handled. The first step in waste and sample characterization is to determine the phase of the wastes or samples. Nonaqueous-phase liquids (NAPLs) are organic liquids that are relatively insoluble in water. There are two classifications of nonaqueous-phase liquids:

1. Light nonaqueous-phase liquids (LNAPLs), such as jet fuel, kerosene, gasoline, and nonchlorinated industrial solvents (benzene, toluene, etc.), which have densities smaller than water, and will tend to float vertically through aquifers.
2. Dense nonaqueous-phase liquids (DNAPLs), such as chlorinated industrial solvents (methylene chloride, trichloroethylene, trichloroethane, dichlorobenzene, *trans*-1,2-dichloroethylene, etc.), which have densities greater than water, and will tend to sink vertically through aquifers.

The next step is to determine whether or not the hazardous wastes or samples can be handled separately, together in bulk or in packaged form. Only the qualified environmental engineers can wisely decide how the hazardous wastes or samples should be properly handled. Mixing small quantities of hazardous substances with other nonhazardous substances, water, or soil, may generate larger quantities of hazardous wastes, creating more environmental troubles, or even danger. There are two kinds of hazardous wastes to be handled:

1. *Designated hazardous waste*: a waste that is specifically listed by the national government (such as USEPA) as hazardous (such as hydrogen cyanide).
2. *Characteristic hazardous waste*: a waste that exhibits any one of the characteristics of ignitability, corrosiveness, reactivity, or extractive procedure (EP) toxicity [21].

Furthermore, an ignitable waste is defined as any liquid with a flash point of less than 60°C, any nonliquid that can cause a fire under certain conditions, or any waste classified by the national government (such as the US Department of Transportation in the United States) as a compressed ignitable gas or oxidizer. A corrosive waste is defined as any aqueous material that has a pH less than or equal to 2, a pH greater than or equal to 12.5, or any material that corrodes SAE 1020 steel at a rate greater than 0.25 in. per year. (1 in. = 2.54 cm). A reactive waste is defined as one that is unstable, changes form violently, is explosive, reacts violently with water, forms an explosive mixture with water, or generates toxic gases in dangerous concentrations. An extractive procedure toxicity (EP Toxicity) waste is one whose extract contains concentrations of certain constituents in excess of those stipulated by the national government's drinking water standards (such as the USEPA Safe Drinking Water Act).

The third step is to determine whether or not the hazardous wastes or samples should be treated or handled *in situ* or *ex situ*, which are defined as follows [22–28]:

1. *In situ treatment*: the hazardous wastes or environmental samples are not removed from the storage or disposal area to be processed. In general, treatment is accomplished by mixing the reagent into the waste storage zone by some mechanical means such as auger, backhoe, rotary tilling device, etc. Site remediation by "*in situ* solidification" is a typical example.
2. *Ex situ treatment*: the hazardous wastes or environmental samples are removed from the storage or disposal area to be processed elsewhere through a mechanical system. Soil remediation by excavation and incineration is a typical example. Another example is application of the "pump-and-treat" technology for groundwater decontamination.

Another step is to decide whether or not the ex situ treatment should be carried out onsite or offsite, which are defined as follows [22–25]:

1. *Onsite treatment*: the hazardous wastes or environmental samples are not removed from the contaminated site to be processed. Any kind of *in situ* treatment is onsite treatment. Application of the pump-and-treat technology for groundwater decontamination at the contaminated site is an *ex situ* treatment as well as an onsite treatment. Onsite treatment systems consist mainly of mobile or transportable equipment, installation, labor, and support services.
2. *Offsite treatment*: the hazardous wastes or environmental samples are removed from the contaminated site to be processed. If the contaminated soil must be excavated from the site, and transported to another location for incineration, it is an *ex situ* treatment as well as an offsite treatment. Offsite treatment systems involve mainly fixed operations using nonmobile or nontransportable equipment.

Mobile operations are generally taken to mean that the process equipment is on wheels and that the entire site remediation operation can be rapidly moved, set up, and ready for operation at a new contaminated site, within a few days. Transportable operations mean that the process equipment may be broken down into a number of segments that must be transported separately and are assembled at the operational site, often within a few weeks or months.

Once an industrial, agricultural, or commercial site is seriously contaminated by the hazardous waste, the government will list the site as a hazardous waste contaminated site, or a Superfund site. Delisting is an amendment to the lists of hazardous wastes or hazardous waste sites, granted by the national government when it is shown that a specific waste stream or waste site no longer has the hazardous characteristics for which it was originally listed.

Restoration of any industrial, agricultural, commercial or even residential sites that have been seriously contaminated by hazardous wastes is termed "site remediation." A contaminated site may involve contaminated soil and/or groundwater. Purification of any groundwater by either *in situ* or *ex situ* means is called groundwater decontamination. Site remediation is a broader term that includes groundwater decontamination.

Where water penetrates, some of the hazardous wastes dissolve; there is no such thing as a completely insoluble material. Accordingly, when a hazardous waste, treated or not, is exposed to water, a rate of dissolution can be measured. This process is termed "leaching." The water with which we start is the "leachant," and the contaminated water that has passed through the waste is the "leachate." The capacity of hazardous waste material to leach is called its "leachability."

A test can be conducted in situ, ex situ, onsite, or offsite, either using an actual waste sample, or a simulated synthetic waste sample, to determine whether or not a particular process method or equipment can be used to treat the waste sample. Such a test is called a treatability test or treatability study. Ambient air monitoring in the field can provide immediate data about contaminants and speed up cleanups [27].

Hyperspectral imaging has been employed by Howard and Pacifici [28] in environmental site assessments to detect and identify contaminated areas. Groundwater monitoring is also advancing due to a new technology for sampling and installing monitoring wells [29]. Parish and Fournier [30] offer a method for comparing horizontal wells with vertical wells for subsurface remediation.

20.2 SITE REMEDIATION MANAGEMENT

Analytical methods for determination of the concentrations of pollutants in solid wastes and hazardous wastes can be found from governmental agencies [3,21]. Because most site remediation projects involve the use of onsite treatment systems, it is necessary to define the required onsite service as follows, in normal chronological sequence: (a) obtaining samples of the hazardous waste, (b) preliminary laboratory treatability test, (c) preliminary quote, (d) meeting with customer, field sampling, and preliminary meetings with the regulatory agency, (e) final laboratory treatability tests, (f) firm quotation to customer, (g) regulatory approval, (h) mobilization, (i) setup at job site, or the contaminated site, (j) site remediation, treatment of the wastes and environmental samples, (k) close-down and cleanup at job site and return to home base, (l) final laboratory leaching and physical tests on solid and/or groundwater produced in job to satisfy contract requirements and protect warranty, (m) completion of a final project report, and (n) possible follow-up sampling and laboratory testing of waste samples at various times if required by contract or desired by contractor for information or warranty protection.

When groundwater is contaminated by hazardous wastes, the groundwater can either be treated in place using *in situ* technologies, or be pumped from subsurface to the ground surface for *ex situ* treatment. The later *ex situ* groundwater decontamination technology is also called the pump-and-treat technology.

The best demonstrated available technologies (BDAT) recommended by USEPA and many industrial nations are presented in the following sections [2–20,22–25,31–34]. The BDAT particularly recommended by the industrial nations and international communities for site remediation considerations are: incineration, soil washing, chemical treatment, low-temperature thermal desorption, and solidification.

Butcher and Dresser [35] offer tips for handling public meetings concerning releases of contaminants to industrial or commercial sites.

20.2.1 Soil Decontamination

According to the frequency of applications or popularity, the most popular soil decontamination technologies are the following, in decreasing order:

1. Excavation;
2. In situ subsurface volatilization and ventilation/aeration;
3. Bioremediation;
4. Thermal destruction or incineration;
5. Soil vapor stripping or soil vacuum extraction;
6. Soil washing or soil scrubbing;
7. Stabilization and solidification;
8. Natural attenuation; and
9. Chemical treatment (pH adjustment).

20.2.2 Groundwater Decontamination

The most popular groundwater decontamination technologies are the following, in decreasing order:

1. Air stripping;
2. Carbon adsorption;
3. Bioremediation;
4. Sewer discharge;
5. Liquid/liquid (oil/water) separation;
6. *In situ* flushing;
7. Trenching; and
8. Containerizing.

20.3 EXCAVATION

Contaminated soil may be excavated by mechanical means for treatment and/or disposal, or treated *in situ*. Excavation can be completed in a few days or take several months depending on site-specific complexities. Excavation is the unit operation used most commonly to remove the contaminated soil. However, its applicability so far is limited to small volumes of contaminated soil and shallow excavations.

20.4 IN SITU STABILIZATION AND SOLIDIFICATION OF CONTAMINATED SOILS

The process terms of chemical fixation, immobilization, stabilization, and solidification have been used interchangeably. The following are the common terminologies.

20.4.1 Stabilization

"Stabilization" refers to those techniques that reduce the hazard potential of a waste by converting the contaminants into their least soluble, mobile, or toxic form. The physical nature and handling characteristics of the waste are not necessarily changed by stabilization.

20.4.2 Solidification

"Solidification" refers to techniques that encapsulate the waste in a monolithic solid of high structural integrity. The encapsulation may be of fine waste particles (micro-encapsulation) or of a large block or container of wastes (macro-encapsulation). Solidification does not necessarily involve a chemical interaction between the wastes and the solidifying reagents, but may mechanically bind the waste into the monolith. Contaminant migration is restricted by vastly decreasing the surface area exposed to leaching, and/or by isolating the wastes within an impervious capsule.

20.4.3 Process Description

Solidification and stabilization are nevertheless used interchangeably in the field [2,4]. In actual site remediation operation, the process immobilizes contaminants in soils and sludges by binding them in a concretelike, leach-resistant matrix. Contaminated hazardous waste materials are collected, screened to remove oversized material, and introduced to a batch mixer. The hazardous waste material is then mixed with water; a chemical reagent; some selected additives; and fly ash, kiln dust, or cement. After it is thoroughly mixed, the treated waste is discharged from the mixer. Treated waste is a solidified mass with significant unconfined compressive strength (UCS), high stability, and a rigid texture similar to that of concrete.

This process treats soils and sludges contaminated with toxic organic compounds, hazardous metals, inorganic compounds, and oil and grease. Batch mixers of various capacities can treat different volumes of hazardous waste.

The solidification and stabilization process (Fig. 1) was once demonstrated in December 1988 at the Imperial Oil Company, Champion Chemical Company Superfund site, in Morganville,

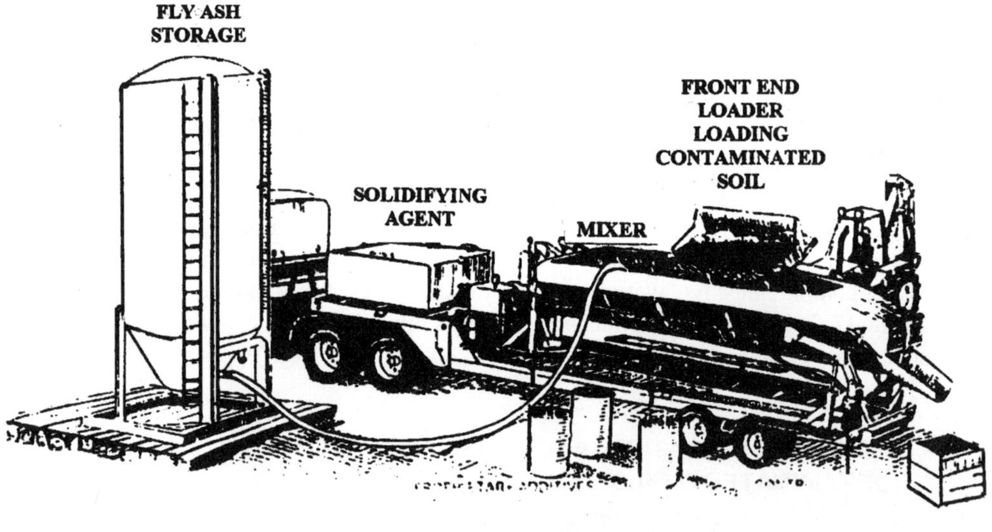

Figure 1 Solidification process equipment. (Courtesy of USEPA.)

New Jersey. This location formerly contained both chemical processing facilities and oil reclamation facilities. Soils, filter cake, and oily wastes from an old storage tank were treated during the demonstration. These wastes were contaminated with petroleum hydrocarbons, polychlorinated biphenyls (PCB), other organic chemicals, and hazardous heavy metals.

A Technology Evaluation Report [5], an Applications Analysis Report [6], and a Demonstration Bulletin [7] are available from the USEPA, Washington, DC, United States. Long-term chemical and physical monitoring and mineralogic analyses have also been conducted by USEPA.

Key findings from the solidification and stabilization process demonstration are summarized below:

1. Extract and leachate analyses showed that heavy metals in the untreated waste were immobilized.
2. The process solidified both solid and liquid wastes with high organic content (up to 17%), as well as oil and grease.
3. Volatile organic compounds in the original waste were not detected in the treated waste.
4. Physical test results of the solidified waste showed: (a) UCS ranging from 390 to 860 pounds per square inch (psi); (b) very little weight loss after 12 cycles of wet and dry and freeze and thaw durability tests; (c) low permeability of the treated waste; and (d) increased density after treatment.
5. The solidified waste increased in volume by an average of 22%. Because of solidification, the bulk density of the waste material increased by about 35%.
6. Trace amounts of semivolatile organic compounds were detected in the treated waste and the toxicity characteristic leaching procedure (TCLP) extracts from the treated waste, but not in the untreated waste or its TCLP extracts. The presence of these compounds is believed to result from chemical reactions in the waste treatment mixture.
7. The oil and grease content of the untreated waste ranged from 2.8 to 17.3% (28,000 to 173,000 ppm). The oil and grease content of the TCLP extracts (USEPA, 1980) from the solidified waste ranged from 2.4 to 12 ppm.
8. The pH of the solidified waste ranged from 11.7 to 12.0. The pH of the untreated waste ranged from 3.4 to 7.9.
9. No PCBs were detected in any extracts or leachates from the treated waste.
10. Visual observation of solidified waste revealed dark inclusions about 1 mm in diameter. Ongoing microstructural studies are expected to confirm that these inclusions are encapsulated wastes.

The USEPA Risk Reduction Engineering Laboratory, Cincinnati, OH, United States, may be contacted for further information on this stabilization and solidification process.

20.5 IN SITU SOIL VAPOR STRIPPING OR SOIL VACUUM EXTRACTION

Soil vapor stripping (SVS), soil vapor extraction (SVE), soil venting (SV), vacuum extraction (VE), and soil vacuum extraction (SVE) are the terms used interchangeably to describe a process that removes volatile organic compounds (VOC) from the vadose, or unsaturated soil zone, by vacuum stripping. These compounds can often be removed from the vadose zone before they contaminate groundwater. The extraction process uses readily available equipment, including extraction and monitoring wells, manifold piping, a vapor and liquid separator, a vacuum pump, and an emission control device, such as an activated carbon adsorption filter. After the contaminated area is completely defined, extraction wells are installed and connected by piping to the vacuum extraction and treatment system.

First, a vacuum pump draws the subsurface contaminants from the extraction wells to the liquid/gas separator. The vapor-phase contaminants are then treated with an activated carbon adsorption filter or a catalytic oxidizer before the gases are discharged to the atmosphere. Subsurface vacuum and soil vapor concentrations are monitored with vadose zone monitoring wells.

The technology is effective in most hydrogeological settings, and can reduce soil contaminant levels from saturated conditions to a nondetectable level. The process even works in less permeable soils (clays) with sufficient porosity. Dual vacuum extraction of groundwater and vapor quickly restores groundwater quality to drinking water standards. In addition, the technology is less expensive than other remediation methods, such as incineration. Figure 2 illustrates the SVS or VE process. Typical contaminant recovery rates range from 20 to 2500 lb/day (1 lb = 454 g), depending on the degree of site contamination and the VOCs to be removed.

The VE or SVS technology effectively treats soils containing virtually any VOCs and has successfully removed over 40 types of chemicals from soils and groundwater, including toxic organic solvents and gasoline- and diesel-range hydrocarbons. Nevertheless, the range of applicability of VE or SVS processes is bounded by the following constraints [34]:

1. The hazardous substances to be removed must be volatile or at least semivolatile (a vapor pressure of 0.5 torr or greater);
2. The hazardous substances to be removed must have relatively low water solubility or the soil moisture content must be quite low;
3. The hazardous substances to be removed must be in the vadose zone (above the groundwater table) or, in the case of LNAPLs, floating on it;
4. The soil must be sufficiently permeable to permit the vapor extraction wells to draw air through all of the contaminated domains at a reasonable rate.

The SVS or VE process cannot remove heavy metals, most pesticides, water-soluble solvents (acetone, alcohols, etc.), and PCBs because their vapor pressures in moist soils are too low.

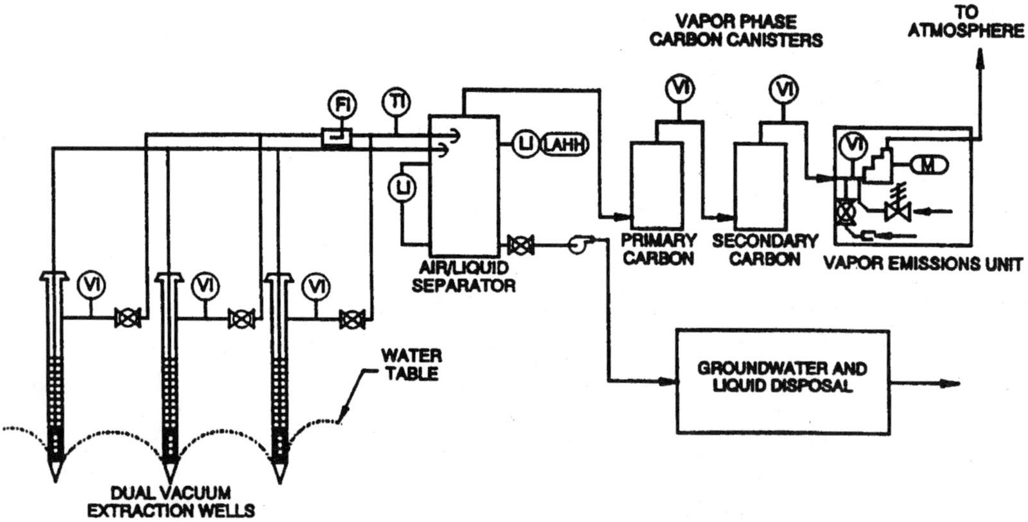

Figure 2 In situ vacuum extraction process diagram. (Courtesy of USEPA.)

The technology is relatively cheap and rapid, has a comparatively low environmental impact, and results in elimination of the contaminated hazardous substances or its concentration into a small volume of highly concentrated, easily handled waste that may be disposed of by incineration or recycled for reuse.

The SVS or VE process was first demonstrated at a Superfund site in Puerto Rico. Terra Vac has since applied the technology at 15 additional Superfund sites and at more than 400 other waste sites throughout the United States, Europe, and Japan.

The process (Fig. 2) was demonstrated under USEPA supervision at the Groveland Wells Superfund site in Groveland, MA, United States, in 1987–1988. The technology successfully remediated soils contaminated by trichloroethene (TCE). The USEPA Technology Evaluation Report [8] and the USEPA Applications Analysis Report [7] have been published. During the Groveland Wells demonstration, four extraction wells pumped contaminants to the process system. During a 56-day operational period, 1300 lb (1 lb = 454 g) of VOCs, mainly TCE, were extracted from both highly permeable strata and less permeable clays. The vacuum extraction process achieved nondetectable VOC levels at some locations, and reduced the VOC concentration in soil gas by 95%. Average reductions were 92% for sandy soils and 90% for clays. Field evaluations have yielded the following conclusions:

1. VOCs can be reduced to nondetectable levels; however, some residual VOC concentrations usually remained in the treated soils.
2. Volatility of the contaminants and site soils is a major consideration when applying this technology.
3. Pilot demonstrations are necessary at sites with complex geology or contaminant distributions.
4. Treatment costs are typically \$40 per ton of soil, but can range from \$10 to \$150 per ton of soil, depending on requirements for gas effluent or wastewater treatment (1989 costs).
5. Contaminants should have a Henry's constant of 0.001 or higher.

20.6 EX SITU AND IN SITU LOW-TEMPERATURE THERMAL DESORPTION

There are three types of thermal treatment for site mediation: (a) incineration; (b) *in situ* thermal extraction process; and (c) thermal desorption. Only thermal desorption is introduced here. In a thermal desorption reactor, the moisture, volatile organic compounds (VOCs), semivolatile organic compounds (SVOCs), and volatile inorganics in the contaminated soil or hazardous wastes are reduced by the elevated high temperature, without combusting the solid materials. For this reason, the thermal desorption process is also called a pyrolysis process. For economic reasons, the moisture content of the contaminated soil or hazardous wastes must be reduced as much as possible through mechanical means prior to thermal desorption [34].

The following are the basic types of process equipment that have been developed and commercially available for the thermal desorption of hazardous organic and inorganic chemicals from contaminated soils and solids.

20.6.1 Ex Situ Rotary Thermal Desorption Dryer

This consists of a cylinder that is slightly inclined from the horizontal and revolves at about five to eight revolutions per minute. The inside of the dryer is usually equipped with flights or baffles throughout its length to break up the contaminated soils or solids. Wet cake is mixed with previously heat-dried soils or solids in a pug mill. The system may include cyclones for soils/solids and gas separation, dust collection scrubbers, and a gas incineration step.

20.6.2 Ex Situ Hot Oil Heated Screws (Conveyors)

Multiple screws, or augers, are used to heat, mix, and convey the soil inside enclosed shells or troughs. Contaminated soil is fed into one end of the process reactor, which has a hot oil heat transfer fluid circulating inside the screw shaft, the screw flights, and the outer vessel's shell. Heat is conducted to the soil from the hot oil, and the VOCs, SVOCs, inorganic volatile compounds, and water are vaporized. Vapors are ducted to a gas treatment system. Using commercial heat transfer fluids, it can routinely heat the soil to about 275°C, and it is effective for the decontamination of light solvents, fuel products, and some SVOCs.

20.6.3 Ex Situ Molten Salt Heated Screws (Conveyors)

The design of a molten salt heated screw is similar to the hot oil heated screws, except that a molten salt heat transfer system is used instead of a hot oil heat transfer system in order to reach higher operating temperatures, up to 450°C. Soil temperatures of up to 400°C have been achieved when using molten salts.

20.6.4 Ex Situ Electric Resistance Heated Screws (Conveyors)

The design of electric resistance heated screws is similar to the hot oil heated screws, except that electric resistance elements are attached to the outer wall of the screw conveyors for heating. The soil is heated up to 1100°C by a combination of conduction and radiation from the heated outer wall. Several such heated screws are manifolded together to make a unit of commercial capacity. PCBs, and other VOCs, and SVOCs can be effectively removed using this high-temperature thermal desorption system. The desorbed gases from the heated screws can be collected and treated in either condensation or afterburner gas systems.

20.6.5 Ex Situ Steam or Hot Air Heated Screw Dryer

This design is similar to that of the hot oil heated screws, except that steam or hot air will be used for heating and thermal desorption. This type of dryer is still in the developmental stage.

20.6.6 Ex Situ Fluidized Bed Dryer

This consists of a vertically oriented reactor through which hot gases are circulated from bottom to top. The contaminated soils and hazardous wastes are fed downward into the reactor, where they are suspended by the upward flowing gas stream. The gas flow rate can be adjusted until the drag force on the soil particles from the flowing gas compensates for the force of gravity, allowing the solid particles to be suspended in a fluidized bed in the center of the dryer reactor. High heat transfer efficiency can be reached with this kind of thermal desorption reactor. This type of process equipment has been fully commercialized.

20.6.7 Ex Situ Microwave or Radio-Frequency Thermal Desorption

This process reactor is similar to a household microwave oven. The microwave dryer consists of a chamber that is connected to a microwave generator by wave guides. The contaminated

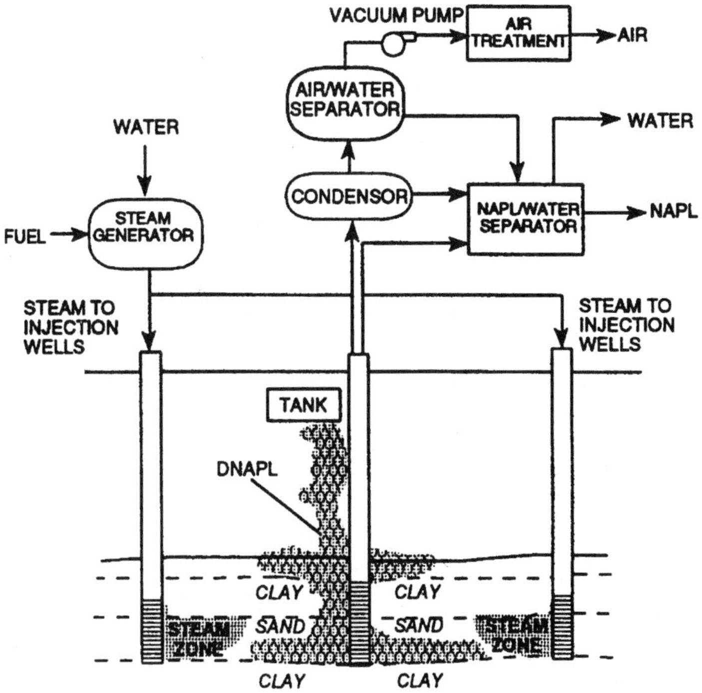

Figure 3 In situ thermal extraction process diagram. (Courtesy of USEPA.)

soil or hazardous wastes are placed into the chamber, and the radio frequency radiation is focused on them by the wave guides. By using microwaves, the heating energy is focused inside the soil particles, achieving better thermal desorption efficiency. Also, the microwave generator can be remotely controlled without exposing workers to a contaminated environment. This type of thermal desorption unit is in the developmental stage because scaling up is not cheap [19,25].

20.6.8 In Situ Radio-Frequency Radiation

This is used for the in-place thermal desorption of contaminated soil. Radio-frequency source electrodes are placed either in or on the ground in the contaminated area, and energy is transmitted to the contaminated soil mass. A fume hood is erected over the contaminated area, and the vaporized VOCs, SVOCs, and inorganic volatile compounds are collected and treated in either condensation or afterburner gas systems.

20.6.9 In Situ Stream or Hot Air Heated Mixing Augers

This system is applied to contaminated soil directly in the field by the use of large, vertical stream or air-heated soil mixing augers common to the construction industry for boring holes. Steam or hot air is injected into the contaminated soil, in place, through the auger as a hole is bored. The VOCs, SVOCs, and inorganic volatile compounds are collected in a hood and treated in a condensation or afterburner gas treatment system.

All the *in situ* or *ex situ* thermal desorption systems described above have generally been configured with two types of gas treatment systems attached to the primary soil desorption unit: (a) a condensation gas treatment system, and (b) an afterburner treatment system.

A condensation gas treatment system (Fig. 3) recovers the bulk of the organic substances (as a concentrated liquid or sludge) using a condenser, a cyclone separator, baghouse, or filter for particulate removal, a granular activated carbon (GAC) adsorber for vapor reduction, an afterburner, or catalytic oxidizer for residual VOCs and SVOCs emission control.

An afterburner gas treatment system employs mainly a combustion chamber to destroy the separated VOCs and SVOCs. The system also needs supplemental process equipment, such as a cyclone separator, baghouse or filter for particulate removal, and/or acid gas control, depending on the gaseous waste streams.

20.7 INCINERATION, THERMAL DESTRUCTION, STARVED AIR COMBUSTION AND HIGH-TEMPERATURE PYROLYSIS

There are three operational modes of high-temperature thermal treatment reactor: (a) incineration or thermal destruction; (b) starved air combustion or thermal gasification; and (c) high-temperature pyrolysis.

High-temperature incineration is one of the five promising site remediation technologies being used and is continuously studied by industrial nations [4,25,36]. Incineration or thermal destruction is a two-step process involving drying and combustion after preliminary drying. A typical feed soil is composed of the soils contaminated by volatile organic compounds (VOCs) or semivolatile organic compounds (SVOCs). At very high temperatures (over 1000°C) and in the presence of oxygen, the organic contents in the contaminated soil in an incinerator are burned and converted to carbon dioxide gas, water steam, and small amounts of organic vapors, which are then collected and treated in an afterburner gas treatment system. The soil after incineration is clean, disinfected, and ready to be returned to the site.

The starved air combustion (SAC) or thermal gasification process utilizes equipment and process flows similar to incineration except that less than the theoretical amount of air for complete combustion is supplied. Auxiliary fuel may be required, depending on the volatiles in the contaminated soil. The high temperature decomposes or vaporizes the hazardous organic matters. The gas-phase reactions are pyrolytic or oxidative, depending on the concentration of oxygen remaining in the gaseous stream. The dried soil or solid residue is dark in color. The SAC has a higher thermal efficiency than incineration due to the lower quantity of air required for the process. In addition, capital economies can be realized due to the smaller gas handling requirements. Again, an afterburner gas treatment system will be required for air stream purification. The soil after SAC is clean, disinfected, and ready to be returned to the site.

The high-temperature pyrolysis process utilizes equipment and process flow diagrams similar to incineration except that it is operated in the absence of oxygen, but at a high temperature. It should be noted that while the low-temperature pyrolysis process, also known as thermal desorption, is one of the best demonstrated available technologies (BDAT) for site remediation, high-temperature pyrolysis is still in the research stage.

The above are three different operational modes for high-temperature thermal treatment reactors. Theoretically, each type of high-temperature thermal treatment reactor can be operated in three different operational modes, depending on the amount of oxygen to be supplied to the reactor.

There are different types of thermal treatment reactors or incineration reactors: (a) the rotary kiln furnace; (b) the multiple hearth furnace; and (c) the fluidized bed furnace.

The rotary kiln furnace, or rotary kiln incinerator, is unique in that it is designed to allow a portion of its hazardous waste load to be charged in batch rather than continuous mode. In this batch mode of operation, solid contaminated soils, solid wastes, and "containerized" liquid wastes are introduced through entrance chutes, typically concurrent with the gas flow. Kiln angle and rotation speed continuously expose fresh surface for oxidation, determine the residence times of noncombustible materials, and provide for continuous ash removal. Upon entry into the incinerator, the liquid waste container, typically cardboard, plastic, or steel drums, ruptures or burns, exposing the contents to the hot kiln environment. The hazardous liquid then rapidly vaporizes and reacts with the excess oxygen present in the combustion gases from the continuous primary flame. An afterburner and other supplemental air treatment equipment will be required for purification of the produced gaseous streams.

The fluidized bed furnace (FBF) is a vertically oriented, cylindrically shaped, refractory lined, steel shell, which contains a sand bed and fluidizing air distributor. The FBF is normally available in diameters of 9–25 ft and heights of 20–60 ft (1 ft = 0.3048 m). The sand bed is approximately 2.4 ft thick and rests on a refractory lined air distribution grid through which air is injected at a pressure of 3–5 psi to fluidize the bed. Bed expansion is approximately 80–90%. The temperature of the bed is controlled at between 1400 and 1500°F. Ash is carried out of the top of the furnace and is removed by air pollution control devices. It is effective for incineration of "containerized" liquid hazardous wastes.

20.8 IN SITU HOT AIR/STEAM ENHANCED STRIPPING AND IN SITU THERMAL EXTRACTION

This process (Fig. 3) is a modification to the soil vapor stripping (SVS) or soil vacuum extraction (SVE) process presented earlier. Again there are many terminologies that are being used interchangeably for description of this same process, because the terminology has not been standardized [22–25,34]:

- in situ thermal extraction (ISTE);
- thermally enhanced vapor stripping (TEVS);
- vacuum-assisted steam stripping (VASS);
- steam/hot air stripping (SHAS);
- in situ steam extraction (ISSE);
- in situ steam enhanced extraction (ISSEE);
- steam injection/vapor extraction (SIVE);
- in situ steam/hot air extraction (ISSHAE), etc.

This process can be operated under two environmental conditions, as follows.

20.8.1 Operation Above the Water Table, or in the Vadose Zone

Hot air and/or steam is first injected into the soil and them removed, possibly under vacuum, together with the desorbed volatile organic compounds (VOCs) and semivolatile organic compounds (SVOCs). Then, gas steam should undergo treatment for air purification. The condensed steam should be pumped from the ground and treated. The required supplemental process equipment includes: demisters, scrubbers, condensers, chillers, heaters, and so on, which are all well-established technologies. The system should be properly operated so that the vadose zone does not become saturated with water and exhibit reduced or no permeability for the gases and vapors targeted for removal. General site requirements include: adequate soil permeability, penetrable soils for insertion of augers, wells, minimal subsurface obstacles, and appropriate ambient temperatures in the range 20–100°F [15].

20.8.2 Operation Both Above and Below the Water Table

Steam is introduced to the soil through injection wells screened in contaminated zones both above and below the groundwater table. The steam flow sweeps contaminants to extraction wells. Groundwater and liquid contaminants are pumped from the extraction wells; steam, air, and vaporized contaminants are then extracted under vacuum. After the soil is heated by steam injection, the injection wells can introduce additional agents to facilitate the cleanup. Recovered vapors pass through a condenser. The resulting condensate is combined with pumped liquids for processing in separation equipment. This in-situ thermal extraction (ISTE) process to be operated both above and below the water table will enhance the soil vapor extraction (SVE) and pump-and-treat processes used to treat VOCs and SVOCs. Heating the soil with steam injection is an effective and relatively inexpensive technique to raise a target soil volume to a nearly uniform temperature.

In general, the separated nonaqueous-phase liquids (NAPL) from either of the above two operations can be recycled or disposed of, and the water treated prior to discharge. The noncondensable gases are directed to a vapor treatment system that consists of: (a) oxidation equipment, (b) activated carbon filters, or (c) treatment onsite in a catalytic destruction process.

In general, the process to be operated either above or below the water table uses conventional injection, extraction, and monitoring wells, off-the-shelf piping, steam generators, condensers, heat exchangers, separation equipment, vacuum pumps, and vapor emission control equipment.

Specifically, the in situ thermal extraction (ISTE) process to be operated both above and below the water table removes VOCs and SVOCs from contaminated soils and groundwater. The process primarily treats chlorinated solvents such as trichloroethene (TCE), perchloroethene (PCE), and dichlorobenzene; hydrocarbons such as gasoline, diesel, and jet fuel; and mixtures of these compounds. The process can be applied to rapid cleanup of source areas such as dense NAPL pools below the water table surface, light NAPL pools floating on the water table surface, and NAPL contamination remaining after conventional pumping techniques. Subsurface conditions are amenable to biodegradation of residual contaminants, if necessary, after application of the thermal process. A cap must exist to implement the process near the surface. For dense NAPL compounds in high concentrations, a barrier must be present or created to prevent downward percolation of the NAPL. The process is applicable in less permeable soils using novel delivery systems such as horizontal wells. For more information about this technology, the reader is referred to USEPA, Risk Reduction Engineering Laboratory, 26 West Martin Luther King Drive, Cincinnati, OH 45268, United States.

20.9 IN SITU SUBSURFACE VOLATILIZATION AND VENTILATION (COMBINED SATURATED ZONE SPARGING AND IN SITU VADOSE ZONE VAPOR STRIPPING)

The contaminated soil and groundwater in the saturated zone can be remediated for VOCs removal through sparging. The technology involves the use of combined saturated zone sparging and in situ vadose zone vapor stripping [34]. It is also called subsurface volatilization and ventilation [22–25], in situ sparging, in situ air stripping, in situ aeration, and aeration curtain.

There are two broad approaches to the process, which involves sparging volatile organics compounds (VOCs) from the saturated zone using compressed air:

1. *Throughout the contaminated zone.* Individual sparging wells are placed with a combination of saturated zone sparging and in situ vadose zone vapor stripping throughout the contaminated zone to remove VOCs and SVOCs from across a wide area. Wells are screened over a narrow interval located at the bottom of an aquifer or below the

deepest contamination within the aquifer. Compressed air is forced from the well screen and flows radically outward and upward through the contaminated soils. As the air bubbles move upward through the contaminated groundwater and soils, VOCs and SVOCs dissolved in the groundwater and absorbed to the soil particles' surface are volatilized and swept to the unsaturated zone with the air bubbles. The extracted air is then collected by vacuum through the screened vacuum extraction well, and further purified by air purification means (such as dryer, activated carbon, or equivalent) before its release to the ambient air. Biodegradation may occur within the remediation system, thus reducing the need for off-gas treatment.

2. *Combination of saturated zone sparging and in situ vadose zone vapor stripping to form aeration curtains oriented at right angles to the flow of the groundwater plume.* Aeration curtains can be created in trenches backfilled with porous media. The trenches have a horizontal slotted pipe (air injection well, or air distribution pipe) near the bottom of the trench to supply compressed air. As the groundwater flows through the trench, the rising air bubbles strip the VOCs and SVOCs to the top of the trench, reaching the unsaturated zone with the air bubbles. The extracted air is then collected by vacuum through the screened vapor recovery pipe (or vacuum extraction well) and further purified by air purification means (such as dryer, activated carbon, or equivalent) before its release to the ambient air. Biodegradation may occur within the remediation system, thus reducing the need for off-gas treatment.

A well-established subsurface volatilization and ventilation system (SVVS) is presented below as a case study. The SVVS (Fig. 4) was developed by Billings and Associates, Inc. (BAI), Albuquerque, NM, United States, and operated by several other firms under a licencing agreement, uses a network of injection and extraction wells (collectively, a reactor nest) to treat subsurface

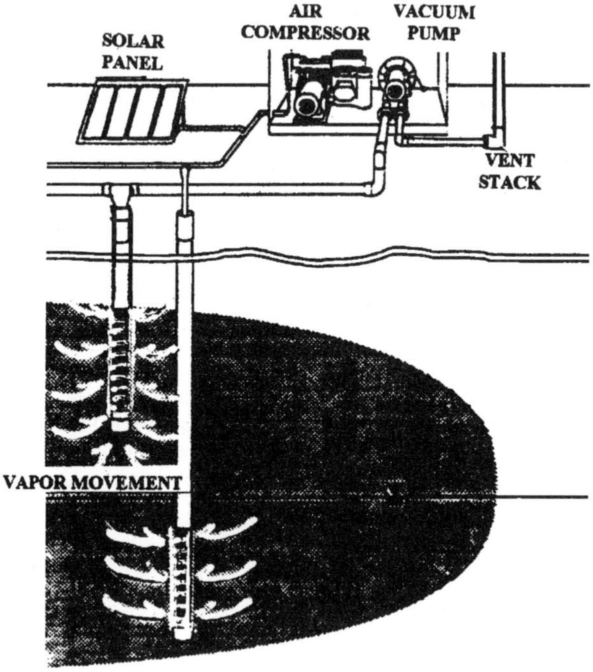

Figure 4 Subsurface volatilization and ventilation system (SVVS). (Courtesy of USEPA.)

VOCs and SVOCs contamination through *in situ* biodegradation using compressed air below the water table combined with soil vacuum extraction in the vadose zone (above the water table). Each system is custom designed to meet site-specific conditions. A series of compressed air injection wells and vacuum extraction wells is installed at a site. One or more vacuum pumps create negative pressure to extract contaminant vapors, while an air compressor simultaneously creates positive pressure, sparging air through the subsurface treatment area. This placement allows the groundwater to be used as a diffusion device. Control is maintained at a vapor control unit that houses pumps, control valves, gages, and other process control hardware.

The number and spacing of the wells depends on the modeling results of applying a design parameter matrix, as well as the physical, chemical, and biological characteristics of the site. The exact depth of the injection wells and screened intervals are additional design considerations.

To enhance vaporization, solar panels are occasionally used to heat the injected compressed air. Additional valves for limiting or increasing air flow and pressure are placed on individual reactor nest lines (radials) or, at some sites, on individual well points. Depending on groundwater depths and fluctuations, horizontal vacuum screens, "stubbed" screens, or multiple-depth completions can be applied. The system is dynamic: positive and negative air flow can be shifted to different locations at the site to place the most remediation stress on the areas requiring it. Negative pressure is maintained at a suitable level to prevent the escape of vapors.

Because it provides oxygen to the subsurface, the SVVS, or equivalent, can enhance *in situ* bioremediation at a site. Thus, it can decrease site remediation time significantly. These processes are normally monitored by measuring dissolved oxygen levels in the aquifer, recording carbon dioxide levels in transmission lines and at the emission point, and periodically sampling microbial populations. If air quality permits require, VOC emissions can be treated by a biological treatment process unit that uses indigenous microbes from the site.

The SVVS, or equivalent, is applicable to sites with leaks or spills of gasoline, diesel fuels, and other hydrocarbons, including halogenated compounds. The system is very effective on methyl tertiary-butyl ether (MTBE), benzene, toluene, ethylbenzene, and xylene (BTEX) decontamination. It can also contain contaminant plumes through its unique vacuum and air injection techniques.

The technology should be effective in treating soils contaminated with virtually any material that has some volatility or is biodegradable. The technology can be applied to contaminated soil, sludges, free-phase hydrocarbon product, and groundwater. By changing the injected gases to induce anaerobic conditions and by properly supporting the microbial population, the SVVS can remove nitrates from groundwater. The aerobic SVVS or equivalent raises the redox potential of groundwater to precipitate and remove heavy metals.

20.10 EX SITU AND IN SITU VITRIFICATION

Vitrification is a process of melting contaminated soil, buried hazardous wastes, or toxic sludges at a temperature as high as 1600–2000°C, in an electric furnace or in place at a contaminated site, to render the materials nonhazardous. The final nonhazardous product is a glassy and/or crystalline solid matrix that is resistant to leaching and more durable than natural granite or marble. If the vitrification process is carried out in an electric furnace, it is called ex situ vitrification (ESV). If it is carried in place at a contaminated site, it is called in situ vitrification (ISV).

The technology is based on the concept of joule heating to melt the contaminated soil or sludges electrically in order to destroy toxic organic and inorganic contaminants by pyrolysis. It was initially developed by the US Department of Energy (USDOE) to provide enhanced isolation of previously disposed radioactive wastes. Today over 160 bench-scale (10 kW, 5–10 kg), engineering-scale (30 kW, 0.05–1 ton), pilot-scale (500 kW, 10–50 ton), and large-scale (3755 kW, 400–1000 tons) vitrification tests have been conducted and have demonstrated the general feasibility and its widespread applications in treating or containing hazardous wastes: contaminated soil sites, burial grounds, and storage tanks that contain hazardous materials in the form of either sludge or salt cakes, process sludges, and many others.

A case history of ex situ vitrification using electric furnace vitrification is presented first. The ex situ vitrification technology uses an electric furnace to convert contaminated soils, sediments, and sludges into oxide glasses at over 1500°C, chemically rendering them nontoxic and suitable for landfilling as nonhazardous materials. Successful vitrification of soils, sediments, and sludges requires: (a) development of glass compositions tailored to a specific waste, and (b) glass melting technology that can convert the waste and additives into a stable glass without producing toxic emissions. There are two types of melters:

- *Electric melter.* In an electric melter, glass, which is an ionic conductor of relatively high electrical resistivity, stays molten with joule heating. Such melters process waste under a relatively thick blanket of feed material, which forms a counterflow scrubber that limits volatile emissions. Commercial electric melters have significantly reduced the loss of inorganic volatile constituents such as boric anhydride (B_2O_3) or lead oxide (PbO). Because of its low emission rate and small volume of exhaust gases, electric melting is a promising technology for incorporating waste into a stable glass.
- *Fossil fuel melter.* In contrast, fossil fuel melters have large, exposed molten glass surface areas from which hazardous constituents can volatilize. Because of its high toxic emission rate, a fossil fuel melter may not be more beneficial than an electric melter for vitrifying toxic wastes.

Ex situ vitrification using an electric melter and furnace (Fig. 5) stabilizes inorganic components found in hazardous waste. In addition, the high temperature involved in glass production (over 1500°C) decomposes anthracene, *bis*(2-ethylhexyl phthalate), and

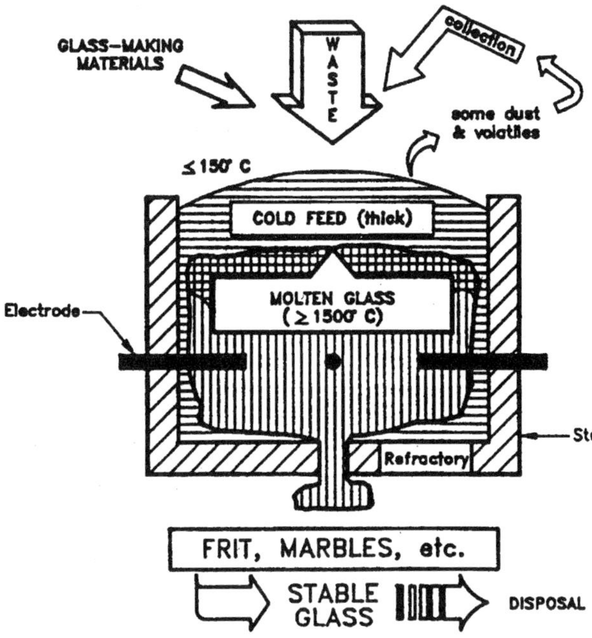

Figure 5 Electric furnace vitrification system. (Courtesy of USEPA.)

pentachlorophenol in the waste. The decomposition products can easily be removed from the low volume of melter off-gas. Several glass compositions suitable for processing synthetic soil matrix have been developed and subjected to toxicity characteristic leaching procedure testing (TCLP). Ten independent replicates of the preferred composition produced the results in Table 1 for the ex situ vitrification through electric melting.

The mean analyte concentrations were less than 10% of the remediation limit at a statistical confidence of 95%.

The readers are referred to Ferro Corporation, Independence, OH, United States, and Geosafe Corporation, Richland, Washington, United States, for detailed information regarding the vitrification process [20].

Many large-scale in situ vitrification (ISV) processes have been developed. To accomplish ISV, four electrodes are inserted into the contaminated soil to the desired treatment depth. In addition to the four electrodes, the supplemental components include: the off-gas hood to cover the contaminated area and the electrodes, an offer-gas trailer with off-gas purification units (quench, scrubber, dewatering unit, heater, filter, and adsorber), a support trailer for holding cooler, instrumentation, and support transformer, an electrical trailer holding a main transformer, and a backup generator.

There are three operational sequences of the ISV process:

1. *Initiation of vitrification.* To initiate the ISV process, a conductive mixture of flaked graphite and glass frit is placed among the electrodes to act as the starter path for the electric circuit. An electric current passed between the electrodes through the graphite and frit path initiates the vitrification melting process. Eventually the graphite starter path is consumed by oxidation and the electric current is transferred to the surrounding molten soil, which is then electrically conductive.
2. *Subsidence during vitrification.* As the melt grows downward and outward, the nonvolatile elements become part of the melt matrix and the organic compounds are destroyed by pyrolysis. The pyrolyzed byproducts migrate to the surface of the vitrified zone, where they combust in the presence of air. Inorganic materials are dissolved into or are encapsulated in the vitrified mass. Convective currents within the melt uniformly mix materials that are present in the soil.
3. *Vitrification completion and backfill.* When the desired melt depth and volume have been achieved, the electric current is discontinued and the molten volume is allowed to cool and solidify. During the process, a hood is placed over the affected area to

Table 1 Remediation Limits Vs. TCLP Analyte Concentration of Glass Replicates From Vitrification Process (Courtesy of USEPA)

	TCLP analyte concentration (parts per million)	
Metal	Remediation limit	Mean of glass replicates
As	5	<0.100
Cd	1	<0.010
Cr	5	0.019
Cu	5	0.355
Pb	5	0.130
Ni	5	<0.010
Zn	5	0.293

collect any combustion gases and entrained particles for off-gas treatment. A backfill with clean top soil on the top of the vitrified monolith will complete the ISV process.

20.11 IN SITU SOIL SURFACTANT FLUSHING AND EX SITU SOIL WASHING

The soil surfactant flushing is defined as a process for in situ treatment of the contaminated soil or other matrix with surfactant solution, while soil surfactant washing is defined as a process for soil excavation, slurry preparation, and subsequent ex situ treatment aboveground with surfactant solution. So soil flushing is an in situ treatment process, and soil washing is an ex situ treatment process, both of which involve the use of surfactant solutions.

Surfactants are amphipathic molecules or ions. One portion of the surfactant molecule is hydrophilic (water loving), while another portion is hydrophobic (water hating). Hydrophilic portions are ionic or polar heads. Hydrophobic portions are tails containing 12 or more carbon atoms as hydrocarbon chains.

In the presence of water and air, the surfactants tend to concentrate at solid/water interfaces and air/water interfaces of water mixtures. By concentrating at the air/water and solid/water interfaces of the water mixture, the surfactant species are able to reduce the surface tension of the contaminated soil particles, thereby enhancing the chances for separation of contaminants from the soil particles.

Many basic and applied engineering projects have been conducted by researchers [34,37]. The readers are referred to an excellent book by Wilson and Clarke [34] for the theory and principles of flushing and washing. The in situ soil flushing process is still in experimental stage. A typical large-scale ex situ soil washing process is described below [16,18,25].

An ex situ soil washing process system (Fig. 6) is a water-based volume reduction process used to treat excavated soil. The system may be applied to contaminants concentrated in the fine-size soil fraction (silt, clay, and soil organic matter) or contamination associated with the coarse (sand and gravel) soil fraction.

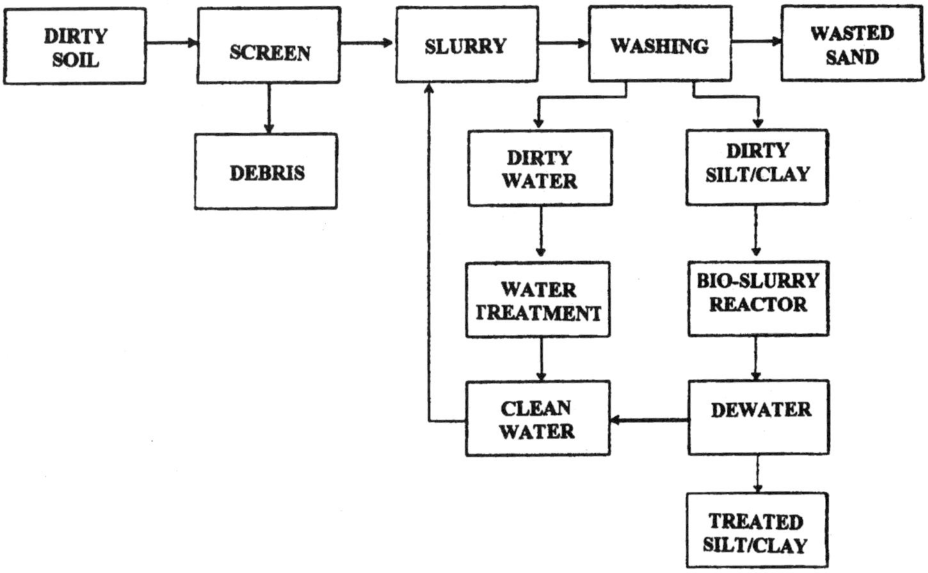

Figure 6 Soil washing system process diagram. (Courtesy of USEPA.)

As a part of the soil washing process, debris is removed from the soil, and the soil is mixed with water and subjected to various unit operations common to the mineral processing industry. These operations can include mixing trammels, pug mills, vibrating screens, froth flotation cells, attrition scrubbing machines, hydrocyclones, screw classifiers, and various dewatering operations.

The core of the soil washing process is a multistage, countercurrent, intensive scrubbing circuit with interstage classification. The scrubbing action disintegrates soil aggregates, freeing contaminated fine particles from the coarser material. In addition, surface contamination is removed from the coarse fraction by the abrasive scouring action of the particles themselves. Contaminants may also be solubilized, as dictated by solubility characteristics or partition coefficients. Contaminated residual products can be treated by other methods. Process water is normally recycled after biological or physical treatment. Contaminated fines may be disposed of off site, incinerated, stabilized, and biologically treated.

This ex situ soil washing system was initially developed Bio Trol, Inc., Eden Prairie, MN, United States, to clean soils contaminated with hazardous wood preserving wastes, such as polynuclear aromatic hydrocarbons (PAH) and pentachlorophenol (PCP). The system may also be applied to soils contaminated with petroleum hydrocarbons, pesticides. PCBs, various industrial chemicals, and hazardous metals.

The soil washing system was demonstrated under the SITE Program in 1989 at the MacGillis and Gibbs Superfund site in New Brighton, Minnesota, United States [16,18,25]. A pilot-scale unit with a treatment capacity of 500 lb/hour operated 24 hours/day during the demonstration. Feed for the first phase of the demonstration (2 days) consisted of soil contaminated with 130 ppm PCP and 247 ppm total PAHs; feed for the second phase (7 days) consisted of soil containing 680 ppm PCP and 404 ppm total PAHs. Contaminated soil washing process water was treated biologically in a fixed-film reactor and recycled. A portion of the contaminated soil washing fins was treated biologically in a three-stage, pilot-scale EIMCO Biolift reactor system supplied by the EIMCO Process Equipment Company. Key findings from the BioTrol demonstration are summarized below.

1. Feed soil (dry weight basis) was successfully separated into 83% washed soil, 10% woody residues, and 7% fines. The washed soil retained about 10% of the feed soil contamination; 90% of this contamination was contained within the woody residues, fines, and process wastes.
2. The soil washer removed up to 89% PCP and 88% total PAHs, based on the difference between concentration levels in the contaminated (wet) feed soil and the washed soil.
3. The system degraded up to 94% PCP in the process water during soil washing. PAH removal could not be determined because of low influent concentrations.
4. Cost of a commercial-scale soil washing system, assuming use of all three technologies, was estimated to be $168 per ton. Incineration of woody material accounts for 76% of the cost (1989 costs).

20.12 BIOREMEDIATION FOR SOIL AND/OR GROUNDWATER DECONTAMINATION

20.12.1 Bioremediation

Many terminologies are being used in the field of environmental biotechnology. They are briefly defined as follows for the purpose of clarification and comparison:

- *Biological treatment.* Any kind of water treatment, waste treatment, or even air treatment involving mainly the use of living organisms, especially microorganisms for breaking down organic substances in the influent under aerobic, anaerobic, or anoxic conditions. The influent can be wastewater, sludge, solid waste, hazardous waste, contaminated soil, ground water, river water, lake water, storm runoff water, landfill leachate, or a contaminated air stream.

- *Biological waste treatment.* Biological treatment stated above to be used only for treatment of mainly wastewaters or hazardous wastes.
- *Biodegradation.* An action or reaction for breaking of organic compounds by living organisms, especially microorganisms, resulting in the formation of simpler intermediate compounds, carbon dioxide, water, and other gases. Alternatively, the disappearance of environmentally undesirable properties of a substances.
- *Mineralization.* Complete breaking down of organic compounds by living organisms, especially microorganisms, resulting in the formation of carbon dioxide, water, and other minerals or gases.
- *Biotransformation.* Biological conversion of some characteristic property (i.e., altering the toxicity, form, and mobility) of the original compounds with no decrease in molecular complexity.
- *Biostimulation.* Addition of nutrients, change of pH or temperatures, or optimization of soil or groundwater environmental conditions (such as humidity, porosity of soil) in order to enhance the efficiency of biological treatment.
- *Bioaugmentation.* Addition of microorganisms to a process system or a contaminated site to degrade specific contaminants readily.
- *Biofiltration, air biofilter or vapor phase bioreactor.* The terms loosely used by practising engineers for a biological filter (with microorganisms attached on the filter media) for purification of air streams, aiming at removal of toxic organics and odors [67].
- *Bioventing.* The use of soil venting, or soil vacuum extraction, to promote aerobic biodegradation in soils is termed bioventing. Soil aeration and not vapor extraction is the primary purpose of the bioventing process.
- *Bioremediation, bioreclamation, enhanced biodegradation, or enhanced bioremediation.* Site remediation, groundwater decontamination, or environmental restoration through alteration, or optimization of environmental factors to enhance biological treatment.
- *Bioremediation process for soil decontamination.* This relies mainly on the soil microorganisms, soil nutrients, and oxygen (enhanced by aeration), and may be assisted by adding genetically engineered microorganisms to the contaminated soil.
- *Bioremediation process for groundwater decontamination.* This may be accomplished by: (a) adding nutrients and/or oxygen, or hydrogen peroxide, to the aquifer to enhance the growth and activity of indiginous microorganisms; (b) injecting the aquifer with genetically engineered microorganisms.

The theory, principles, and applications of all biological treatment and reactions are alike, and can be found elsewhere [17,25,38,59–60,66–67]. Vandenbergh and Saul [39] report a special bioremediation process that accelerates natural degradation of groundwater and soil contaminants.

Only the bioremediation processes that are suitable to remove hazardous substances from contaminated soil and groundwater will be introduced here. There are four in situ bioremediation technologies: (a) enhanced bioremediation; (b) bioventing; (c) anaerobic–aerobic sequential processes; and (d) sequencing batch reactor. Each is separately introduced below.

20.12.2 Enhanced Bioremediation System

An enhanced bioremediation system has been used for removal of petroleum hydrocarbon from contaminated soil and groundwater. Contamination has been caused by a leaky underground storage tank (UST) upgradient of groundwater flow. Direction of the subsurface plume has been known based on available geological data for the region. It appears that the river nearby has not been polluted. To remove the hazardous contaminants by the enhanced bioremediation system will include the following engineering tasks:

- Install monitoring wells to confirm or determine the degree of contamination, and the direction of subsurface plume.
- Install Bentonite slurry cutoff along the river bank that intersects the subsurface plume to prevent river water contamination.
- Install the fuel oil collection trench, and extract free NAPL contaminants from the wells or trenches that intersect the subsurface plume. LNAPL lies on the water table, while DNAPL concentrates on impervious soil layers beneath the water table.
- Install groundwater pumping wells, treatment facilities (such as bioreactors, spray irrigation) and injection wells for pumping, treating, and reinjecting the groundwater to the subsurface, respectively.
- Add nutrients (inorganic soluble nitrogen and phosphorus compounds) and an oxygen source (sparged air or oxygen, hydrogen peroxide, or nitrate) to the groundwater, either above or below ground, and/or to soil for biostimulation.
- Add selected microorganisms to the contaminated soil for bioaugmentation.

To operate an enhanced bioremediation system will include all of the above. Accordingly, the enhanced bioremediation is defined as a complete technology system involving monitoring, pollution prevention, free contaminants removal, groundwater decontamination by biological treatment (either above or below ground), groundwater reinjection, biostimulation (adding nutrients, oxygen source) to contaminated groundwater and/or soil, and bioaugmentation to the contaminated soil.

20.12.3 Bioventing System

A typical bioventing system for soil decontamination is now introduced. Bioventing is defined as an in situ biotechnology for aerobic biodegradation of organic contaminants in soils using soil venting. Soil aeration and not vapor extraction is the primary purpose of bioventing, potentially making the technology effective in removing, through in situ biodegradation, organic contaminants having both high and low volatilities and water solubilities. Bioventing can remediate soils with low water permeabilities, such as silty and clay soils, as long as some air flow paths exist. Costs for bioventing should be comparable to or slightly lower than those for soil venting, excluding aboveground vapor treatment costs. Smaller pumps and decreased pumping rates are required to maintain minimal oxygen levels for aerobic respiration than are required by conventional soil venting.

20.12.4 Anaerobic/Aerobic Sequential Bioremediation System

An anaerobic/aerobic sequential bioremediation system (Fig. 7) for removal of PCE is now introduced. It has been demonstrated that sequential anaerobic/aerobic biodegradation of PCE is feasible if the proper conditions can be established. The anaerobic process can potentially completely dechlorinate PCE. However, conversion of vinyl chloride (VC) to ethylene is the slowest step in this process. Of the chlorinated ethenes, VC is the most amenable to treatment by aerobic methanotrophic processes. Therefore, a two-step process is thought to be the most efficient. The first step is anaerobic, which rapidly dechlorinates PCE and trichloroethylene (TCE) to break down products 1,2-dichloroethylene (DCE) and VC. Since the anaerobic dechlorination of DCE and VC to ethylene can be quite slow, a second aerobic step is

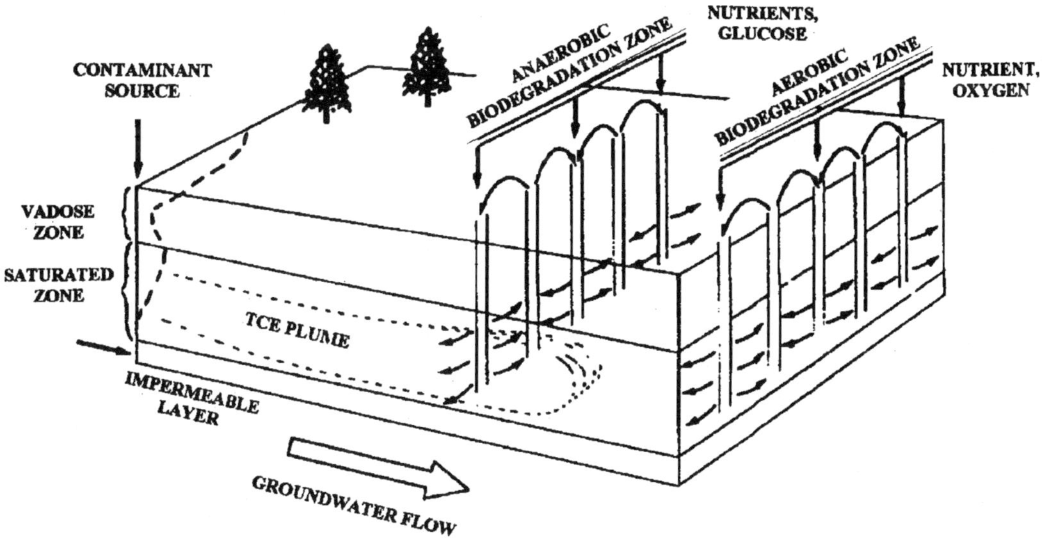

Figure 7 Anaerobic/aerobic sequential bioremediation system. (Courtesy of USEPA.)

implemented that can more quickly complete the remediation process. The schematic diagram in Fig. 7 illustrates this technology.

In practical operation of an anaerobic/aerobic sequential bioremediation system, care must be taken to create and maintain the proper in situ conditions for chlorinated ethene degradation in an aquifer. Carbon and mineral nutrients should be injected and delivered into an aquifer contaminated with PCE or TCE. Groundwater chemical conditions should be monitored within and downgradient of the anaerobic treatment zone to gage the efficiency of the anaerobic process. If volatile organic compound analyses show that the resulting downgradient breakdown products include TCE, DCE, or VC, oxygen and methane will be added to the groundwater to stimulate aerobic degradation by indigenous methanotrophic bacteria. It has been demonstrated that this anaerobic/aerobic sequential bioremediation technology removes PCE, TCE, DCE, and VC from groundwater. The readers are referred to ABB Environmental Services, Inc., Wakefield, MA, United States, for the details of this commercially available process.

20.12.5 Sequencing Batch Reactor

A sequencing batch reactor (SBR) process has been successfully demonstrated for both soil decontamination [40] and groundwater decontamination [25,37,41]. An SBR system is very similar to a continuous complete mix activated sludge process system, except that SBR is operated as a batch unit. An SBR has the smallest footprint, and it is mobile and easy for field operation. A modern SBR process system can also be a physiochemical process, or a combined physicochemical and biological process [45,68].

20.12.6 Combined Sequencing Batch Reactor and Membrane Bioreactor (SBR–MBR)

A membrane bioreactor (MBR) process consists of two principal components: (a) a biological reactor tank, and (b) an ultrafiltration (UF), or microfiltration (MF) membrane to retain biological solids within this biological reactor tank [42]. The membrane may either be internal or external to the reactor tank. The membrane provides a barrier to keep microorganisms in the reactor and renders a clarified, solids-free effluent stream for discharge. The microorganisms are essentially the same as those found within activated sludge systems. A combined sequencing batch reactor and membrane bioreactor (SBR–MBR) is technically and economically feasible for groundwater decontamination applications [23,25], because the SBR–MBR process equipment is commercially available [43] and mobile, and can be quickly erected on site for a remediation project.

20.13 SLURRY BIOREACTOR SYSTEM FOR SOIL DECONTAMINATION

The first ex situ bioremediation process to be introduced here is the Eimco Biolift Slurry Reactor, developed by Eimco Process Equipment Company, Salt Lake City, Utah, United States. The process (Fig. 8) has successfully removed toxic polynuclear aromatic hydrocarbons (PAH) in soil. Traditional biological treatments, such as landfarming and in situ bioremediation, may not reduce PAHs in soil to target levels in a timely manner. Slurry reactors are more efficient for bioremediation and more economical than thermal desorption and incineration.

In a typical onsite bioremediation project, a mixing bioreactor, a first slurry bioreactor, a second fermentation slurry bioreactor in semicontinuous plug-flow mode, and a third slurry bioreactor may be operated in the following manner for contaminants removal:

1. The mixing bioreactor receives and mixes the contaminated soil, makeup process water and supplements of salicylate and succinate as nutrients. Salicylate induces the naphthalene degradation operon on PAN plasmids. This system has been shown to degrade phenanthrene and anthracene. The naphthalene pathway may also play a role in carcinogenic PAH (CPAH) metabolism. Succinate is a byproduct of naphthalene metabolism and serves as a general carbon source. The first reactor in series will remove easily degradable carbon and increase biological activity against more recalcitrant PAHs (i.e., three-ring compounds and higher).

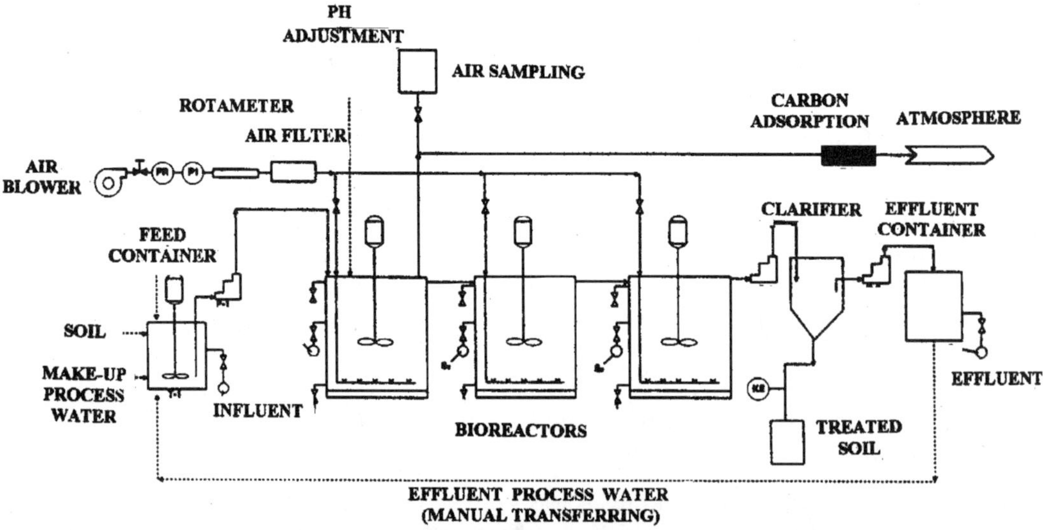

Figure 8 Biological slurry reactor system for soil decontamination. (Courtesy of USEPA.)

2. The fresh contaminated soil slurry and nutrients in the mixing bioreactor are pumped to the first slurry bioreactor (60 L) for treatment.
3. The effluent from the first slurry bioreactor overflows to the second fermentation slurry bioreactor (10 L) in series, where Fenton's reagent is added to accelerate oxidation for four- to six-ring PAHs. Fenton's reagent (hydrogen peroxide in the presence of iron salts) produces a free radical that can effectively oxidize multiring aromatic hydrocarbons.
4. The third slurry bioreactor (60 L) in series is used as a polishing reactor to remove any partially oxidized contaminants remaining after Fenton's reagent is added.
5. Slurry is removed from this third soil slurry bioreactor and clarified using gravity settling techniques The treated or reclaimed soil is settled at the bottom of the clarifier, ready to be returned to the site.
6. The effluent from the settling clarifier is pumped to a final effluent container, from where the effluent can be partially discharged and partially returned to the very first mixing bioreactor.

Operation of the slurry bioreactor system for soil decontamination will increase the rate and extent of PAH biodegradation, making bioslurry treatment of impacted soils and sludges a more effective and economically attractive remediation option. This technology is applicable to PAH-contaminated soils and sludges that can be readily excavated for slurry reactor treatment. Soils from coal gasification sites, wood-treating facilities, petrochemical facilities, and coke plants are typically contaminated with PAHs.

20.14 ANAEROBIC–AEROBIC FIXED FILM BIOLOGICAL SYSTEM FOR GROUNDWATER DECONTAMINATION

Another ex situ bioremediation process introduced here is an anaerobic–aerobic fixed film biological groundwater treatment system (Fig. 9) commercially known as the immobilized cell bioreactor biotreatment system. It is designed to remove organic contaminants (including nitrogen-containing compounds and toxic chlorinated solvents) from contaminated groundwater, and other heavily polluted aqueous streams. This groundwater decontamination system offers improved treatment efficiency by using (a) a unique reactor medium that maximizes biological activity in the reactor, and (b) a bioreactor design that maximizes contact between the biofilm and the contaminants.

This anaerobic–aerobic fixed film biological groundwater treatment system has a completely enclosed headspace, eliminating the possibility of air stripping of volatile organics or intermediates. These features result in quick, complete degradation of target contaminants to carbon dioxide, water, and biomass. Additional advantages include (a) high treatment capacity, (b) compact and mobile system design suitable for site remediation, and (c) reduced operations and maintenance costs resulting from simplified operation and low sludge production.

Basic system components of the anaerobic–aerobic fixed film biological groundwater treatment system include the bioreactors, media, mixing tanks and pumps, feed pump, recirculation pump, and a blower to provide air to the aerobic bioreactor. Depending on the specifics of the influent groundwater streams, some standard pretreatments, such as pH adjustment or oil and water separation, may be required. Effluent clarification is not required for the system to operate, but may be required to meet specific discharge requirements. The system is designed to treat 10 million gallons per day of contaminated groundwater streams, and has been successfully applied to groundwater containing contaminants including polynuclear aromatic hydrocarbons (PAH), phenols, gasoline, chlorinated solvents, diesel fuel, and chlorobenzene.

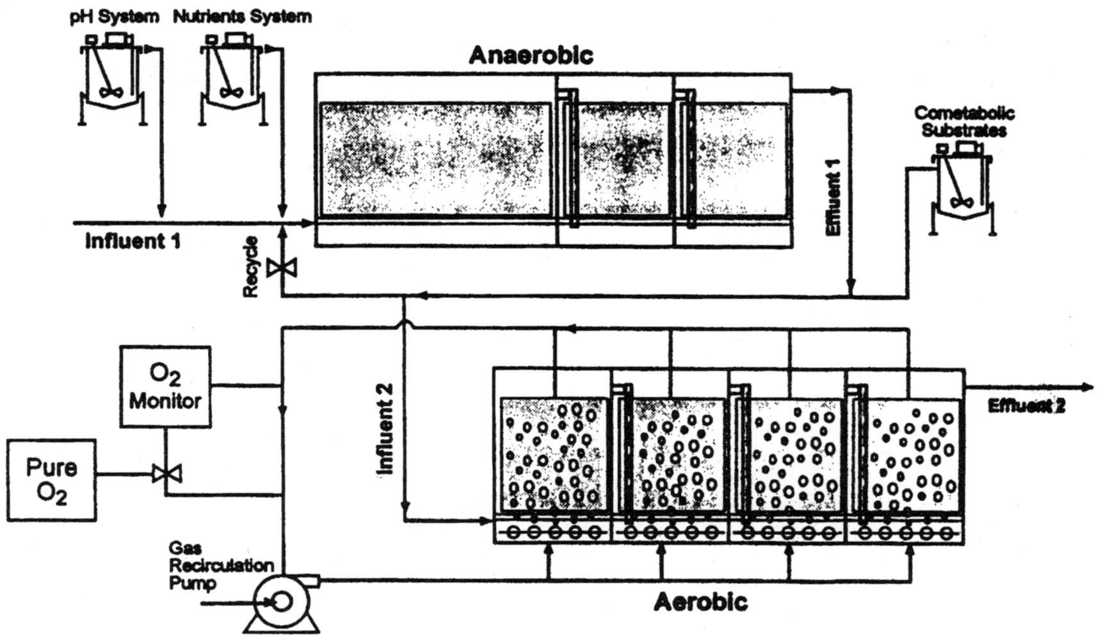

Figure 9 Dual anaerobic/aerobic immobilized cell bioreactor system diagram. (Courtesy of USEPA.)

The readers are referred to the USEPA Risk Reduction Engineering Laboratory, Cincinnati, OH, for more detailed information regarding this process system.

20.15 CHEMICAL TREATMENT (pH ADJUSTMENT, KPEG TREATMENT)

Theoretically chemical treatment involves the use of any kind of chemical(s) that can react with the hazardous wastes or soil, and convert them to either nonhazardous or less hazardous compounds. Chemical treatment is important when the soil or groundwater is contaminated by acidic chemicals, alkaline chemicals, toxic heavy metals [44], or toxic organics that cannot be treated by biological processes [69].

The most common chemical treatment processes for site remediation include:

- *pH adjustment.* Acidic chemicals are used to neutralize basic hazardous wastes or soils, while alkaline chemicals are used to neutralize acidic hazardous wastes or soils.
- *KPEG chemical treatment.* Chemical reagents prepared from polyethylene glycols and potassium hydroxide (KPEG) have been demonstrated under mild conditions (25–140°C) to dehalogenate PCBs, PCDDs, and PCDFs with destruction efficiencies exceeding 99.9999%. The reaction mechanism is nucleophilic substitution at an aromatic carbon.
- *Chemical precipitation and stabilization.* Chemical precipitation agents (such as hydroxides and sulfides) are used to precipitate, fix, and separate heavy metals, in turn, to purify the soil and groundwater [44].

The pH adjustment is one of the most common site remediation technologies, while KPEG chemical treatment of one of the five best demonstrated available technologies (BDAT).

In the KPEG reagent preparation, KOH reacts with HO-PEG (poly ethylene glycol; with a molecular weight approximately = 400) to form KO-PEG (alkoxide) and water. The KO-PEG (alkoxide) in turn reacts initially with one or more of the chlorine atoms on the aryl ring (aryl-Cl) to produce aryl-O-PEG (ether) and KCl (potassium chloride) salt.

In 1986, a 2700 gallon KPEG reactor was used in Montana, on a wood-preserving site, and in Washington, on a waste disposal site, to successfully detoxify PCDDs and PCDFs (120 ppb to 200 ppm) in 17,000 gallons of liquid waste to nondetectable levels. A reactor designed to treat both liquids and solids was tested by US Department of Defense sites. These field studies validated conditions for destruction of PCBs, PCDDs, and PCDFs, to acceptable levels required by the regulations [4].

Wang *et al.* [23–25,41,45,68,69] have developed a physicochemical sequencing batch reactor (PC-SBR) process, which is identical to a conventional biological sequencing batch reactor (SBR), except that the PC-SBR is a 100% physicochemical process. The PC-SBR has been adopted in full scale for recovering toxic chromium from tannery wastewater for reuse at Germanakos SA Tannery near Athens in Greece [23]. Naturally, PC-SBR can be adopted for chemical treatment of contaminated soil slurry or contaminated groundwater. Since the process equipment of conventional biological SBR can be adopted for the PC-SBR process operation, the PC-SBR process equipment is considered to be commercially available [46].

Figure 10 shows a typical chemical treatment system for removal of toxic chromium from a contaminated groundwater. The process is developed by Geochem of Terra Vac Co., Lakewood, CO. In operation, the contaminated groundwater is brought to the surface and treated using conventional treatment systems, such as ferrous ion (Fig. 10). Next, a reductant is added to the treated water, which is reinjected around the plume margin. Here it reacts with and reduces residual levels of chromium, forming a precipitate. Such reinjection creates a "barrier" of elevated water levels around the plume, enhancing the gradient and associated hydraulic control.

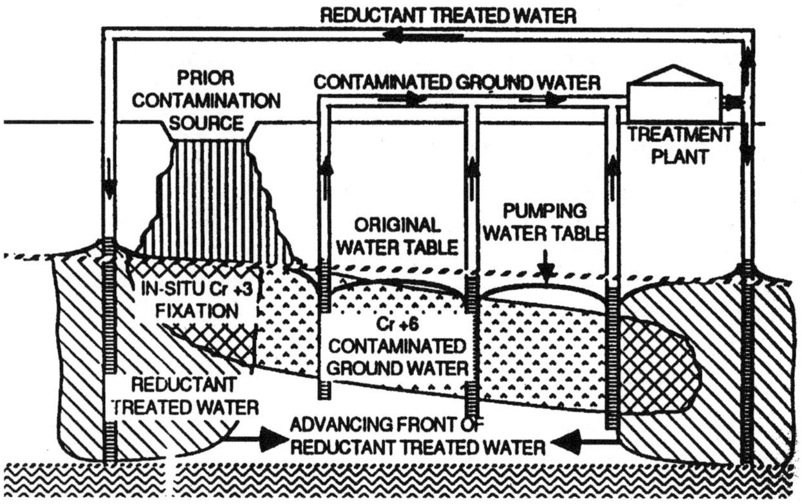

Figure 10 Chemical treatment of contaminated groundwater for chromium removal. (Courtesy of USEPA.)

The reinjection also allows for in situ reduction and subsequent fixation of residual chromium. The process is capable of treating dissolved hexavalent chromium in groundwater at concentrations ranging from the detection limit to several hundred ppm.

20.16 ULTRAVIOLET RADIATION AND OXIDATION FOR GROUNDWATER DECONTAMINATION

This ultraviolet (UV) radiation and oxidation process uses UV radiation, ozone, and hydrogen peroxide to destroy toxic organic compounds, particularly chlorinated hydrocarbons, in contaminated groundwater. The process oxidizes compounds that are toxic or refractory (resistant to biological oxidation) to parts per million (ppm) or parts per billion (ppb) levels.

The UV radiation and oxidation system (Fig. 11) consists of a treatment tank module, an air compressor and ozone generator module, and a hydrogen peroxide feed system. The system can be skid-mounted and portable, and may permit onsite treatment of a wide variety of liquid wastes, such as industrial wastewater, groundwater, and leachate. Treatment tank size is determined by the expected wastewater flow rate and the necessary hydraulic retention time needed to treat the contaminated water. The approximate UV intensity, and ozone and hydrogen peroxide doses, are determined by pilot-scale studies [61,62].

Treatment tank influent is simultaneously exposed to UV radiation, ozone, and hydrogen peroxide to oxidize the organic compounds. Off-gas from the treatment tank passes through an ozone destruction unit, which reduces ozone levels before air venting. The ozone destruction unit also destroys VOC stripped off in the treatment tank. Effluent from the treatment tank is tested and analyzed before disposal.

The UV radiation and oxidation system treats contaminated groundwater, industrial wastewaters, and leachates containing halogenated solvents, phenol, penta-chlorophenol, pesticides, polychlorinated biphenyls, explosives, benzene, toluene, ethyl-benzene, xylene, methyl tertiary butyl ether, and other organic compounds. The system also removes low-level total organic compounds, chemical oxygen demand, and biochemical oxygen demand.

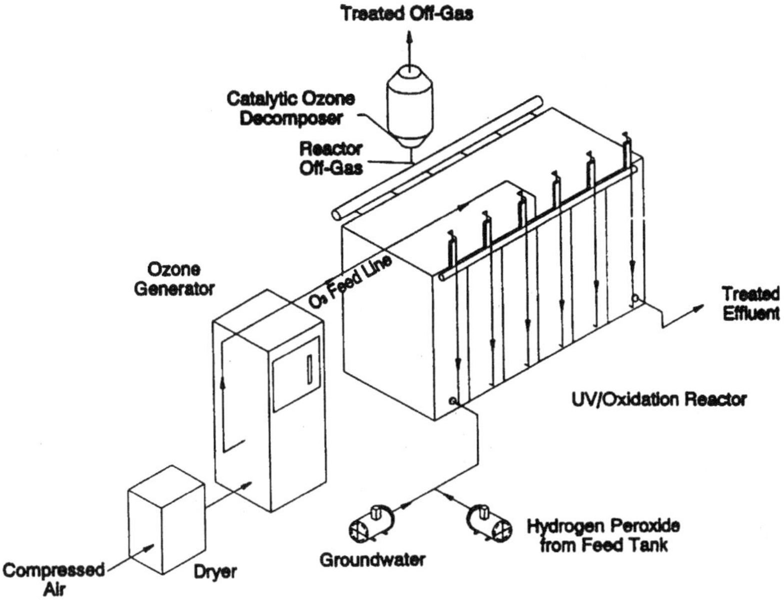

Figure 11 Ultraviolet radiation and chemical oxidation system. (Courtesy of USEPA.)

A field-scale demonstration was completed in March 1989 at the Lorentz Barrel and Drum Company site in San Jose, CA, United States, under the supervision of the USEPA. The test program was designed to evaluate system performance for several combinations of five operating parameters: (a) influent pH, (b) retention time, (c) ozone dose, (d) hydrogen peroxide dose, and (e) UV radiation intensity [11,12].

The UV radiation and oxidation technology is fully commercial, with over 30 systems installed. Flow rates ranging from 5 gallons per minute (gpm) to 1050 gpm are in use at various industries and site remediations, including aerospace, US Department of Energy, US Department of Defense, petroleum, pharmaceutical, automotive, wood-treating, and municipal facilities.

Such UV oxidation technology has been included in Records of Decision for several Superfund sites where groundwater pump-and-treat remediation methods will be used. Contaminated groundwater treated by the system met regulatory standards at the appropriate parameter levels. Out of 44 VOCs in the wastewater, trichloroethene, 1,1- dichloroethane, and 1,1,1-trichloroethane were chosen as indicator parameters. All three are relatively refractory to conventional oxidation.

The ozone destruction unit reduced ozone to less than 0.1 ppm, with efficiencies greater then 99.99%. The VOCs present in the air within the treatment system were not detected after passing through the ozone destruction unit. The UV/oxidation reactor system produced no harmful air emissions. Total organic carbon removal was low, implying partial oxidation of organics without complete conversion to carbon dioxide and water. The UV itself is also an effective chemical reduction process (e.g., dechlorination process) in case the groundwater is contaminated by chlorine or similar oxidizing chemicals [47].

Energy and Environmental Engineering, Inc, East Cambridge, MA, has applied the principles of chemical oxidation, UV radiation, and photocatalysts, and developed a commercial process known as the PhotoCat process. Table 2 shows some representative

Table 2 Removal of Organic Contaminants From Groundwater Using a Peroxide Oxidation–UV Radiation Process

Contaminant species	Molecular weight	Feed concn. (ppm)	Equivalence ratio	Resistance time (seconds)	Product conc C/C_0
Chlorobenzene	112	50	3.3	3.7	0.04
Benzene/toluene	78/93	100/100	3.5	3.8	0.4/0.4
Yellow 106	1374	110	4.2	7.7	0.08
Yellow 49	438	110	1.0	23.0	0.007
Blue 41	463	115	0.6	11.5	0.001
Red 83	1025	140	1.0	7.7	0.008

ppm, parts per million by weight.
Source: Courtesy of USEPA.

results from groundwater decontamination projects using the UV lamp and hydrogen peroxide system.

20.17 AIR STRIPPING FOR GROUNDWATER DECONTAMINATION

There are two types of pump-and-treat air stripping processes for decontamination of groundwater: (a) the countercurrent air stripping tower; and (b) air sparging system.

The countercurrent air stripping tower is very similar to a cooling tower, except that it is not for cooling. The contaminated groundwater is pumped from the underground, and is fed into the countercurrent air stripping tower from the top. Through a distribution means, the influent groundwater becomes water drops falling downward from the reactor tower, while the low-pressure air stream flows upward from the reactor bottom, powered by an air blower. The upward air stream strips out the VOCs and SVOCs from the water drops falling downward in the countercurrent air stripping tower. The purified water effluent is discharged from the tower bottom and may either be further treated by a granular activated carbon (GAC) process for final polishing, or reinjected into the underground. The air effluent collected from the tower top can be purified by either a condensation gas treatment system, or an afterburner gas treatment system before its discharge into the ambient air environment.

The air sparging system is very similar to the conventional aeration tank used in the activated sludge process, except that no microorganisms are seeded in the air sparging reactor, and the reactor depth is usually shallow and the reactor top is enclosed. Diffused air bubbles are generated at the reactor bottom to strip out the VOCs and SVOCs from the bulk of the groundwater in the reactor. The water effluent from the air sparging reactor may either be further treated by a GAC process for final polishing, or reinjected into the underground. The air effluent collected from the top by the reactor's enclosure can be purified by either a condensation gas treatment system, or an afterburner gas treatment system.

One of the advantages of using the air sparging system is that the reactor can be seeded with microorganisms and nutrients for biological treatment of the contaminated groundwater [25,37,48–50]. So the reactor is a combined air stripping and biological treatment unit.

Both of the above processes are fully developed and widely adopted for decontamination of groundwater. Although corrective actions for groundwater may entail treatment, containment, or dilution technologies, the environmental engineers in the field prefer to adopt treatment technologies, especially by air stripping treatment. Air stripping has been applied in 51% of all corrective actions requiring removal of VOCs from groundwater in the United States [4].

20.18 GRANULAR ACTIVATED CARBON ADSORPTION FOR GROUNDWATER DECONTAMINATION

Adsorption is a process by which a solute, a liquid organic pollutant in this case, accumulates or concentrates on the internal and external surface of a solid, such as granular activated carbon.

Among all groundwater decontamination technologies available and feasible, granular activated carbon (GAC) adsorption is the second most popular process adopted by practicing environmental engineers. In the United States, GAC has been applied in 27% of the corrective actions requiring VOCs removal from contaminated groundwater [4]. GAC adsorbers are usually packed in similar manner to sand filters, except that GAC firms the media instead of sand.

In process operation, there are two types of applications. When the GAC adsorbers are used for groundwater treatment, they are "liquid phase GAC adsorbers." When the GAC adsorbers are used for air stream purification , they are "gas phase GAC adsorbers" [37,48–50].

The GAC adsorbers are very effective for VOCs and SVOCs reduction. Because GAC costs are high, GAC adsorbers are usually used for final polishing of either the water stream or air stream. Periodically, the spent GAC must be either disposed of properly, or regenerated for reuse. Regeneration of GAC for desorption of VOCs and SVOCs can be carried out either in situ, ex situ, onsite, or offsite. Without regeneration, the spent GAC, which contains VOCs and SVOCs, may be classified as hazardous waste.

The gas phase GAC adsorber's adsorption efficiency can be significantly increased if the humidity of the contaminated air stream can be reduced.

The most cost-effective and popular GAC regeneration process the is low-temperature thermal desorption process discussed earlier. The most efficient GAC regeneration process is high-temperature pyrolysis (in the absence of oxygen), discussed earlier. A complete onsite GAC regeneration process has been developed by Wangs *et al.* [37].

Purus, Inc., San Jose, CA, has developed a similar air stripping–adsorption system (Fig. 12) for groundwater decontamination. The difference between Wang's system [37] and the Purus system (Fig. 12) is that the former uses GAC, while the latter uses polymeric adsorbent. A detailed operational diagram of a polymeric adsorption system is shown in Fig. 13.

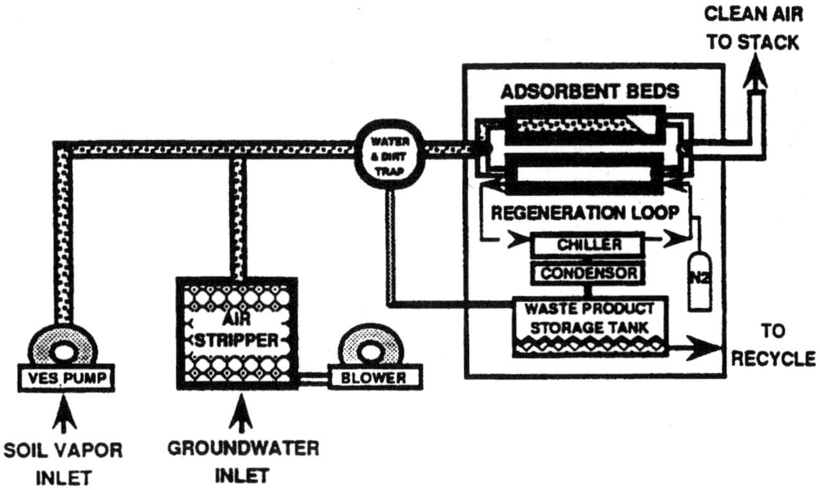

Figure 12 Combined air stripping and polymeric adsorption system for groundwater decontamination. (Courtesy of USEPA.)

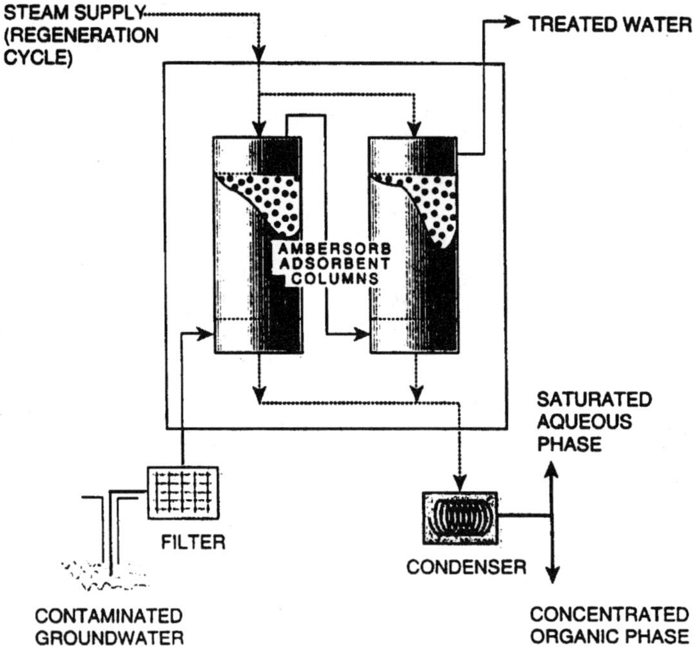

Figure 13 Polymeric adsorption system for groundwater decontamination. (Courtesy of USEPA.)

20.19 SEWER DISCHARGE FOR GROUNDWATER TREATMENT

Sewer discharge is the third most popular method for groundwater decontamination, as popular as bioremediation. It has been applied in 13% of the corrective actions requiring VOCs removal from a contaminated groundwater source. In the case where there is a publicly owned treatment works (POTW) near the contaminated site, which has spare capacity to treat the groundwater, sewer discharge can be very attractive. The ease of operation, high removal efficiency, and relatively low cost account for its popularity and simplicity.

In operation, the contaminated groundwater is pumped from the underground, to the POTW for proper treatment. The treated effluent can be discharged either into a nearby river, or reinjected into the underground to balance the water table. The POTW can use biological and/or physical-chemical process systems for groundwater treatment.

20.20 LIQUID/LIQUID SEPARATION AND FREE PRODUCT RECOVERY FOR GROUNDWATER DECONTAMINATION

Many liquid/liquid separation (such as oil/water separation) reactors are commercially available and technically feasible for separation of either LNAPL (such as oil) or DNAPL (such as chloroform) from a contaminated groundwater source. There are three types of oil and water separation processes:

20.20.1 Gravity Separation Reactor

One liquid separates from another liquid based on their difference in densities. Groundwater density is one. For LNAPL (such as oil), where density is less than one, if will therefore float on the top of groundwater if sufficient separation detention time is given. For DNAPL (such as chloroform), density is greater than one, thus it will settle near or at the bottom of the groundwater, if sufficient detention time is provided.

20.20.2 Absorption Reactor

Absorption is a process by which a solute, a liquid organic pollutant in this case, penetrates into the bulk phase of a solid (such as a sponge). When it is not clear whether adsorption or absorption is involved, the term sorption is sometimes used. Absorption can be either an in situ process, or an ex situ process, depending on how and when this process is to be applied.

20.20.3 Atomizing or Nebulization Reactor

This is a modification of gravity separation in which the mixture of the two liquids to be separated are forced to form extremely fine liquid drops at a temperature just over boiling point of the first liquid (water), but below the boiling point of the second liquid (oil) in order to enhance the efficiency of liquid/liquid separation. The first liquid drops (water) with lower boiling point (such as water at 100°C) are vaporized at the controlled reactor temperature (such as 105°C, which is higher than the first liquid's boiling point), while the second liquid drops (oil) with much higher boiling point resettle and regroup to be the liquid again because the controlled reactor temperature is well below the boiling point of the second liquid drops (oil). The means used for atomizing or nebulization is called an atomizer or a nebulizer. This advanced liquid–liquid separation method is used for separation of small amount of free hazardous liquid from another liquid (such as water) [33].

The liquid/liquid separation process (such as oil–water separation) has been applied in 13% of the corrective actions requiring VOCs reduction from a contaminated groundwater source. The same process equipment can be used for separation of water–LNAPL, water–DNAPL, or even LNAPL–DNAPL.

Free product recovery usually involves groundwater pumping, which controls groundwater flow within a sphere of influence, and the depressed groundwater table around the pumping zone eventually accentuates free product (free hazardous liquid) accumulation and subsequent removal Free hazardous liquid is generally recovered from various subsurface settings in large enough quantities that it can be removed by mechanical means. The free hazardous liquid may collect on impermeable strata, collect in small subsurface basins, or enter an aquifer where the free hazardous liquid will either float (such as LNAPL) on the water, or sink (such as DNAPL) to the bottom, depending on the density of the free hazardous liquids. The pump-and-treat method is used for recycling or disposal of the free hazardous liquids.

20.21 NATURAL ATTENUATION, NATURAL FLUSHING, TRENCH, AND CONTAINERIZING

It is believed that "mother nature" has her own self-purification capacity.

Natural attenuation (or natural dilution) has been applied in 2% of the corrective actions requiring soil remediation, in which LNAPL contamination of soil is simply diluted.

Natural flushing involves flushing the contaminated aquifer with large amount of clean water (such as river water, lake water, or another source of clean groundwater). The term is specifically applied to groundwater decontamination relying on dilution with clean water. About 3.5% of groundwater decontamination projects have adopted natural flushing for dilution of LNAPL, and refilling of an aquifer with clean water.

Trench technology has been used in about 3.5% of groundwater decontamination projects. It is frequently used in conjunction with a big bioremediation project. For instance, a Bentonite slurry trench cutoff wall can be built to intersect the subsurface plume to prevent the aquifer from further contamination at downgradient. Then a free product recovery trench can be built before the Bentonite slurry trench cutoff wall to directly intersect the subsurface plume to recover either LNAPL near the water table, or recover DNAPL near the aquifer bottom, or the impervious layer.

Containerizing involves collection of heavily contaminated groundwater in containers for treatment, disposal, or storage onsite or offsite. This is an undesirable method, but it has been encountered 1.8% of contaminated sites conducting groundwater purification.

20.22 DISSOLVED AIR FLOTATION FOR GROUNDWATER DECONTAMINATION

Dissolved air flotation (DAF) is a new process for groundwater decontamination; however, it is the most feasible and most cost-effective process to be used on a contaminated site due to its high treatment efficiency, air stripping effect, extremely short detention time, small footprint, high mobility, simplicity, and low cost.

The DAF process is used to remove LNAPL and suspended solids from contaminated groundwater by rising microbubbles by decreasing their apparent density. Dissolved air flotation consists of saturating a portion or all of the groundwater feed, or a portion of recycled effluent with air at a pressure of 25–90 psi. The pressurized groundwater is held at this pressure for about 30 seconds in a retention tank and then released to atmospheric pressure to the DAF chamber. The sudden reduction in pressure results in the release of microscopic air bubbles, which attach themselves to LNAPL and suspended particles in the groundwater in the DAF chamber. This results in agglomeration which, due to the entrained air, have greatly increased vertical rise rates of about 0.5–2 ft/min (1 ft/min = 0.375 m/min). The floated materials rise to the surface to form a scum layer (called "float"). Specially designed sludge scoops or flight scrapers on top continuously remove the float. The retention time in the DAF chambers is usually about 5–20 minutes. The effectiveness of DAF depends upon both the attachment of bubbles to the NAPL and other contaminants, and the air stripping effect of bubbles, resulting in cost-effective removal of the contaminants in groundwater.

The contaminants that can be removed by DAF include VOCs, SVOCs, heavy metals, surfactants (used in soil washing or flushing), phenols, and many other toxic organic and inorganic hazardous substances.

Because the detention time of DAF is only about 5–20 minutes (in comparison with conventional biological treatment requiring 7–9 hours detention time), and DAF has the air stripping effect and the highest mobility, it is the most cost-effective process to treat groundwater at a contaminated industrial site.

There are two types of DAF process equipment available for groundwater treatment:

1. Continuous DAF: the groundwater is continuously fed to DAF for treatment, and the effluent is reinjected into the aquifer for recharge. Continuous DAF process equipment is commercially available [51,52], and can all be on wheels ready for site remediation action [22–25,39,49,53–56].
2. Sequencing batch reactor (SBR–DAF): the reactor is similar to conventional SBR, except that DAF is used instead of sedimentation. It can be: (a) a 100% biological SBR–DAF process if microorganisms and nutrients are seeded, (b) a 100% physical-

chemical process if no microorganisms are seeded, or (c) a combined physical-chemical-biological process, if microorganisms and nutrients are seeded, and chemical is added for enhancing the treatment. Since the process reactor is simple, it can be quickly erected on site for groundwater decontamination applications. This SBR–DAF process was developed by Dr. Lawrence K. Wang and his colleagues [25,41,45,56], but its process equipment is still not commercially available.

A small volume of the recovered float (i.e., floated scum) having high concentrations of hazardous substances (NAPL, or heavy metals, or surfactants) must be properly disposed. The treated groundwater can be reinjected into the aquifer. The released air stream can be collected for treatment by either the condensation gas treatment system, or the afterburner gas treatment system, which have been discussed previously.

Ultraviolet, hydrogen peroxide, GAC, and so on, can all be easily be adopted as a part of pre- or post-treatment to DAF, if a high quality of effluent meeting drinking water standards is desired.

20.23 PERVAPORATION PROCESS FOR GROUNDWATER DECONTAMINATION

Section 20.12.6 discussed the application of a combined Sequencing Batch Reactor and Membrane Bioreactor (SBR–MBR) system to groundwater decontamination. Figure 14 shows the flow diagram of the pervaporation process (one of the membrane processes), which is a feasible process for removing VOCs from contaminated water. Permeable membranes that preferentially adsorb VOCs are used to partition VOCs from the contaminated water. The VOCs diffuse from the membrane–water interface through the membrane by vacuum. Upstream of the system's vacuum vent, a condenser traps and contains the permeating vapors, condensing the vapor to liquid while

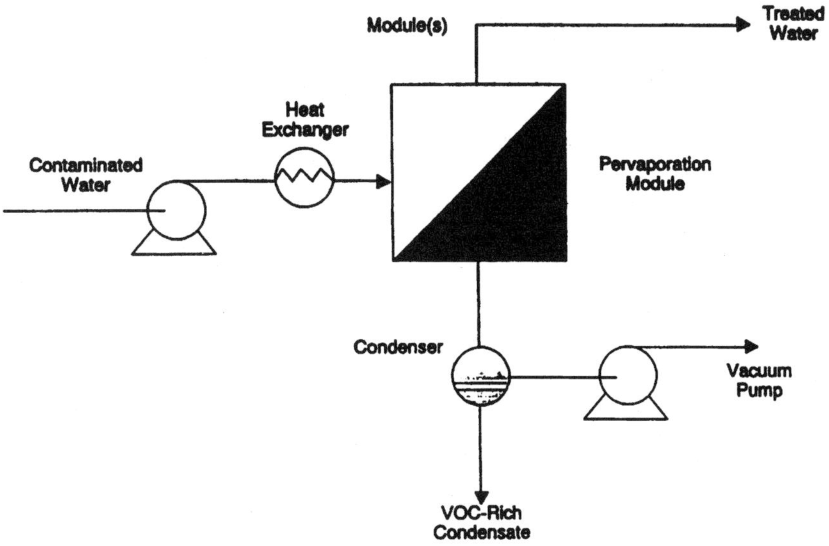

Figure 14 Cross-flow pervaporation system for groundwater decontamination. (Courtesy of USEPA.)

alleviating fugitive emissions. The condensed organic materials represent only a fraction of the initial wastewater volume and may be subsequently disposed of at significant cost savings. The membrane modules consist of hollow fibers with well-defined alignment that results in high mass transfer efficiencies, minimal pressure drip, and low operating costs. The process is commercially available from Zenon Environmental Inc., Ontario, Canada.

20.24 GLOSSARY [22–25,57–60,66–69]

Adsorption. A process whereby a solute (such as a liquid organic pollutant) accumulates or concentrates on the surface of a solid, such as granular activated carbon, polymeric adsorbent, activated alumina, etc.

Aerobic. Involving conditions in which oxygen is available.

Air biofilter (vapor phase bioreactor). Terms used to describe a biological filter (with microorganisms attached to the filter media) used for purification of air streams that aims to remove toxic organics or odors.

Air stripping. A process in which volatile organic compounds, semi-volatile organic compounds, and volatile inorganic compounds are stripped out from a contaminated soil, groundwater, or industrial effluent by air.

Anaerobic. Involving conditions in which oxygen is not available.

Asbestos. Six naturally occurring fibrous minerals most commonly used in building products, but also popular in other commercial products because they are strong, do not burn, resist corrosion, and insulate well. Because asbestos fibers are small and light, they remain in the air for many hours if they are released and may be inhaled by people in the building.

Bioaugmentation. The addition of microorganisms to a waste treatment process unit, an ex situ hazardous material processing system, or a contaminated site to degrade specific contaminants readily and biologically.

Biodegradation. The process by which living organisms, especially microorganisms, break down organic compounds, resulting in the formation of simpler intermediate compounds, carbon dioxide, water and other gases; or resulting in the disappearance of environmentally undesirable properties of substances.

Biological treatment. Any environmental treatment mainly involving the use of living organisms, especially microorganisms, for breaking down organic substances.

Biological waste treatment. Biological treatment of wastewater, solid waste, or hazardous waste.

Bioremediation (bioreclamation). Site remediation, groundwater decontamination, or environmental restoration through alteration or optimization of environmental factors to enhance biological treatment.

Biostimulation. The use of additional nutrients, a change in pH or temperature, or the optimization of soil or groundwater environmental conditions (i.e., humidity or porosity of soil) in order to enhance the efficiency of biological treatment.

Biotransformation. Biological conversion of some characteristic property (i.e., toxicity, form, or mobility) of the original compounds with no decrease in molecular complexity.

Bioventing. The in situ use of soil venting or soil vacuum extraction to promote aerobic biodegradation in soils; soil aeration, not vapor extraction, is the primary purpose of the bioventing process.

BTEX. Benzene, toluene, ethylbenzene, and xylene, which are a group of common toxic compounds.

Carbon adsorption. Adsorption (i.e., of hazardous substances or impurities) that occurs on either powdered, granular, or fiber carbon molecules.

Characteristic waste. A waste that exhibits any one of the characteristics of ignitability, corrosiveness, reactivity, or toxicity.

Chemical treatment. The use of any chemical(s) that can react with an influent material (water, effluent, hazardous waste, or contaminated soil) to render the influent material nonhazardous, less hazardous, or cleaner.

Cleaner production. Industrial production that creates fewer hazardous and other polluting byproducts.

Containerizing. The collection of heavily contaminated groundwater into containers for onsite or offsite treatment, disposal, or storage.

Corrosive waste. Any aqueous material that has a pH less than or equal to 2, a pH greater than or equal to 12.5, or any material that corrodes SAE 1020 steel at a rate greater than 0.25 in/year (1 in. = 2.54 cm)

Decarbonation. A process in which carbon dioxide or carbon monoxide is removed from a liquid medium.

Dematerialization. In an ideal industrial ecological system, material for industrial development is reduced for environmental conservation.

Dense nonaqueous-phase liquids (DNAPLs). Liquids that have densities greater than that of water, and will tend to sink vertically through aquifers, for example, chlorinated industrial solvents (methylene chloride, trichloroethylene, trichloroethane, dichlorobenzene, *trans*-1,2-dichloroethylene, etc.).

Design for environment. A profitable industrial system designed for sustainable environmental conservation.

Designated waste (listed hazardous waste). A waste specifically listed in the United States by the USEPA as hazardous.

Disposal site. Legally, any structure, well, pit, pond, lagoon, impoundment, ditch, landfill, or other place or area, excluding ambient air or surface water, where uncontrolled oil or hazardous material has come to be located as a result of any spilling, leaking, pouring, abandoning, emitting, emptying, discharging, injecting, escaping, leaching, dumping, discarding, or otherwise disposing of such oil or hazardous material. The term shall not include any site containing only oil or hazardous materials that are lead-based paint residues emanating from a point of original application of such paint, result from emissions from the exhaust of an engine, are building materials still serving their original intended use or emanating from such use, or result from release of source, byproduct, or special nuclear material from a nuclear incident, as those terms are defined elsewhere.

Dissolved air flotation. A cost-effective liquid treatment process in which pressurization is applied to dissolve air in water (under high pressure), then depressurization is applied to release and generate extremely fine air bubbles (under normal 1 atmospheric pressure) for separation of impurities, pollutants, light nonaqueous-phase liquids (LNAPLs) and other suspended particles from a contaminated water.

DNAPLs. Dense nonaqueous-phase liquids.

Emergency response action. A response action that is taken in response to a sudden release or a threat of release of oil or hazardous material.

Emission. Discharge of a gas into atmospheric circulation.

End-of-pipe treatments. Decontamination technologies that remove hazardous pollutants before industrial effluents are released into the surrounding environment. This is distinguished from technologies that prevent such hazardous substances from being created in the first place.

Enhanced biodegradation (enhanced bioremediation). A complete technology system involving monitoring of hazardous substances, pollution prevention, removal of free contaminants, groundwater decontamination by biological treatment (either above or below ground), groundwater reinjection, biostimulation (i.e., adding nutrients, bioventing) to contaminated groundwater and/or soil, and bioaugmentation of contaminated soil.

Environment. Waters, land, surface, or subsurface strata, or ambient air of the planet of Earth.

Environmental restoration. The restoration of environmental quality by removing hazardous environmental pollutants from an area.

Excavation. An engineering process in which soil is removed from the ground for direct application, or for decontamination and reuse.

Exposure. Any contact with or ingestion, inhalation, or assimilation of oil or hazardous materials, including irradiation.

Ex situ *treatment.* Site remediation treatment in which the hazardous wastes or environmental samples are removed from the storage or disposal area to be processed elsewhere through a mechanical system. Soil remediation by excavation and incineration is a typical example. Another example is application of the "pump-and-treat" technology for groundwater decontamination.

Ex situ *vitrification.* A vitrification process that is carried out in an electric furnace (see *Vitrification*).

Facility. An entire contiguous plant or installation, including the contiguous grounds of such plant or installation, which is under common ownership.

Free contaminant. An environmental contaminant that is in pure form almost free from impurities, water, or soil.

Free product recovery. Recovery of an environmental contaminant, which is almost free from impurities, water, or soil, and can be reused as an industrial product.

Friable asbestos. Soft, easily crumbled asbestos-containing material that has the greatest potential for asbestos release and therefore has the greatest potential to create health risks. This asbestos may also include previously nonfriable material that becomes broken or damaged by mechanical force.

Granular activated carbon (GAC). An insoluble, granular, and porous carbon particle that can adsorb impurities or pollutants onto the surface of the carbon particle's micropores (see *Adsorption* and *Carbon adsorption*).

Greenhouse gas. A gas that accelerates global warming if it is released into the environment (i.e., carbon dioxide, methane).

Groundwater decontamination. The purification of groundwater that has been contaminated by hazardous wastes.

Halogen. One of the elements of Group VIIA of the Periodic Table (i.e., fluorine, chlorine, bromine, iodine, and astatine).

Hazardous material. Legally, material including, but not limited to, any material in whatever form that, because of its quantity, concentration, chemical, corrosive, flammable, reactive, toxic, infectious, or radioactive characteristics, either separately or in combination with any substance or substances, constitutes a present or potential threat to human health, safety, welfare, or to the environment, when improperly stored, treated, transported, disposed of, used, or otherwise managed. The term shall not include oil, but shall include waste oil and all those substances that are included by the government.

Hazardous substance. Any raw material, intermediate product, final product, spent waste, accidental spill, leakage, etc., that can be in the form of gas, liquid, or solid, and is ignitable, corrosive, reactive (explosive), toxic, infectious, carcinogenic, and/or radioactive, and is harmful to living beings and the environment.

Hazardous waste. A waste that can be in the form of gas, liquid, or solid, and is ignitable, corrosive, reactive (explosive), toxic, infectious, carcinogenic, and/or radioactive, and is harmful to living beings and the environment.

Heavy metals. Dense metals, specifically cadmium, mercury, lead, copper, silver, zinc, and chromium, which may be found in the waste stream in soluble or insoluble toxic forms.

Hot air (enhanced) stripping. A soil remediation process in which hot air without steam is used for enhanced stripping removal of volatile pollutants from the contaminated soil.

Ignitable waste. Any liquid with a flash point of less than 60°C (140°F), any nonliquid that can cause a fire under certain conditions, or any waste classified by the US Department of Transportation (USDOT) as a compressed ignitable gas or oxidizer.

Imminent hazard. A hazard that would pose a significant or otherwise unacceptable risk of harm to health, safety, public welfare, or the environment if it were present for even a short period of time.

Incineration (thermal destruction). A two-step hazardous waste treatment process involving drying and combustion of hazardous materials.

Industrial ecology (IE). A branch of systems science of sustainability or a framework for designing and operating industrial systems as sustainable living systems interdependent with natural systems. It seeks to balance environmental and economic performance within the emerging understanding of local and global ecological constraints.

Infectious waste. A waste that potentially carries a communicable disease; legal definitions can also include general, radioactive, and toxic chemical wastes from hospitals.

In situ *flushing.* See *Natural attenuation.*

In situ *subsurface ventilation/aeration.* A soil and groundwater remediation process in which air diffuses through spargers in the saturated zone of contaminated groundwater, and air flows through the contaminated soil in the vadose zone, both for stripping removal of volatile organic compounds and semivolatile organic compounds from the groundwater and the soil, respectively, under in situ conditions.

In situ *thermal extraction.* See *Thermal extraction.*

In-situ *treatment.* Site remediation treatment in which the hazardous wastes or environmental samples are not removed from the storage or disposal area to be processed. In general, treatment is accomplished by mixing a reagent into the waste storage zone by some mechanical means such as auger, backhoe, rotary tilling device, etc. Site remediation by "in situ solidification" is a typical example.

In situ *vitrification.* A vitrification process that is carried out in place at a contaminated site (see *Vitrification*).

In situ *volatilization.* See *In situ subsurface ventilation/aeration.*

KPEG treatment. A chemical treatment process in which chemical reagents prepared from polyethylene glycols and potassium hydroxide (KPEG) are used under mild temperatures (25–140°C) to dehalogenate the pollutants (such as PCBs, PCDDs, and PCDFs) containing halogens.

Leachability. The capacity of hazardous waste material to leach into passing water.

Leachant. The passing water into which hazardous wastes leach.

Leachate. Contaminated water that has passed through solid waste or hazardous waste.

Leaching. The process by which hazardous wastes dissolve into water that passes through the waste and thus contaminates the surrounding soils and groundwater.

Light nonaqueous-phase liquids (LNAPLs). Liquids that have densities smaller than that of water and will tend to float vertically through aquifers, such as jet fuel, kerosene, gasoline, and nonchlorinated industrial solvents (benzene, toluene, etc.).

LNAPLs. Light nonaqueous-phase liquids.

Manifest system. A government hazardous waste documentation system that ensures proper handling when waste is transported to another site for disposal.

Material balance. An accounting of the weights of material entering and leaving a process or system, usually made on a time-related basis.

Material substitution. An industrial ecological process in which a different new material is used for substitution of an old material in order to improve production, conserve energy, and/or protect the environment.

Migration pathway. A pathway by which an oil or hazardous material is transported at or from a disposal site.

Mineralization. The complete breaking down of organic compounds by living organisms, especially microorganisms, resulting in the formation of carbon dioxide, water, and other minerals or gases.

MTBE. Methyl tertiary butyl ether, which is a gasoline (or similar fuel) additive as well as a groundwater contaminant.

NAPLs. Nonaqueous-phase liquids.

Natural attenuation (natural flushing, natural dilution). A groundwater remediation technique that dilutes hazardous substances in a contaminated aquifer by flushing it with a large amount of clean water (i.e., river water, lake water).

Nebulization. A process or act of nebulizing, atomization. A process of causing a liquid to form small droplets.

Nonaqueous-phase liquids (NAPLs). Organic liquids that are relatively insoluble in water. There are two classifications of nonaqueous-phase liquids: light nonaqueous-phase liquids (LNAPLs) and dense nonaqueous-phase liquids (DNAPLs).

Offsite treatment. Site remediation treatment in which the hazardous wastes or environmental samples are removed from the contaminated site to be processed. If the contaminated soil must be excavated from the site and transported to another location for incineration, it is an ex situ treatment as well as an offsite treatment. Offsite treatment systems involve mainly fixed operations using nonmobile or nontransportable equipment.

Oil. Insoluble or partially soluble oils of any kind or origin or in any form, including, without limitation, crude or fuel oils, lube oil or sludge, asphalt, insoluble or partially insoluble derivatives of mineral, animal, or vegetable oils.

Onsite treatment. Site remediation treatment in which the hazardous wastes or environmental samples are not removed from the contaminated site to be processed. Any kind of in situ treatment is onsite treatment. Application of the pump-and-treat technology for groundwater decontamination at the contaminated site is an ex situ treatment as well as an onsite treatment. Onsite treatment systems consist mainly of mobile or transportable equipment, installation, labor, and support services.

Oxidation. A process reaction that increases the proportion of oxygen, acid-forming element, or radical in a compound. Generally, any process reaction that involves the loss of electrons from an atom.

Oxidizer. Any material that yields oxygen readily to stimulate the combustion or oxidation of organic matter.

PCB. Polychlorinated biphenyls.

pH adjustment. A chemical treatment (see *Chemical treatment*) involving the use of acids or bases for precipitation of pollutants, or for neutralization.

Process substitution. An industrial ecological methodology in which a different new process is used for substitution of an old process in order to improve production, conserve energy, and/or protect the environment.

Pump-and-treat technology. A groundwater decontamination technology in which contaminated groundwater is pumped from below the surface to the ground surface for ex situ treatment.

Pyrolysis (thermal desorption). The process by which moisture, volatile organic compounds (VOCs), semivolatile organic compounds (SVOCs), and volatile inorganics in contaminated soil or hazardous wastes are reduced by elevating the temperature without combusting the solid materials. This is a process similar to incineration except that it is operated in the absence of oxygen but at high temperatures.

Reactive waste. A waste that is unstable, changes form violently, is explosive, reacts violently with water, forms an explosive mixture with water, or generates toxic gases in dangerous concentrations.

Recycling. The process by which materials otherwise destined for disposal are collected, reprocessed or remanufactured, and reused.

Refuse derived fuel (RDF). Boiler fuel made by shredding and screening solid waste into a material of relatively uniform handling and combustion properties. Often, recyclables can be recovered from the RDF process.

Release. Any spilling, leaking, pumping, pouring, emitting, emptying, discharging, injecting, escaping, leaching, dumping, or disposing into the environment.

Resource recovery. A term describing the extraction and utilization of materials and energy from the waste stream. The term is sometimes used synonymously with energy recovery.

Route of exposure. A mechanism including, but not limited to, ingestion, inhalation, dermal absorption, and transpiration by which an oil or hazardous material comes into contact with a human or environmental receptor.

SBR. Sequencing batch reactor, which is a batch biological process involving the use of activated sludge process principles.

Sewer discharge. Discharge of a treated or untreated effluent, or contaminated groundwater to a sewerage system for subsequent treatment by a publicly-owned treatment works (POTW).

Sheen. Appearance of any oil or waste oil on the surface of any river, stream, lake, pond, spring, impoundment, estuary, or coastal water that is caused by the release of such oil or waste oil.

Site. Any building, structure, installation, ground, equipment, pipe or pipeline including any pipe discharging into a sewer or publicly-owned treatment works, well, pit, pond, lagoon, impoundment, ditch, landfill, storage container, motor vehicle, rolling stock, or aircraft, or any other place or area where oil or hazardous material has been deposited, stored, disposed of, or placed, or otherwise come to be located. The term shall not include any consumer product in consumer use or any vessel.

Site remediation. The restoration of any industrial, agricultural, commercial or residential sites whose soil or groundwater have been seriously contaminated by hazardous wastes.

Slurry bioreactor. An ex situ biological soil remediation technology that is more efficient than other bioremediation technologies and more economical than thermal desorption and incineration. It is also more efficient than other biological treatments in reducing PAHs in soil to target levels in a timely manner.

Soil surfactant flushing. An in situ site remediation process for contaminated soil that uses surfactant solution.

Soil surfactant washing. An ex situ site remediation process for soil excavation, slurry preparation, and subsequent ex situ treatment aboveground with surfactant solution.

Soil vacuum extraction (SVE). One of a number of terms used interchangeably to describe a process that removes volatile organic compounds (VOC) from the vadose, or unsaturated soil zone by vacuum stripping.

Soil vapor extraction (SVE). One of a number of terms used interchangeably to describe a process that removes volatile organic compounds (VOC) from the vadose, or unsaturated soil zone by vacuum stripping.

Soil vapor stripping (SVS). One of a number of terms used interchangeably to describe a process that removes volatile organic compounds (VOC) from the vadose, or unsaturated soil zone by vacuum stripping.

Soil venting (SV). One of a number of terms used interchangeably to describe a process that removes volatile organic compounds (VOC) from the vadose, or unsaturated soil zone by vacuum stripping.

Soil washing (soil scrubbing). A soil remediation water-based volume reduction process in which debris is removed from the soil, and the soil is mixed with water and subjected to various operations common to the mineral processing industry in order to break down contaminants.

Solidification. Hazardous waste treatment techniques that encapsulate the waste in a monolithic solid of high structural integrity.

Source reduction. The design, manufacture, acquisition, and reuse of materials so as to minimize the quantity and/or toxicity of waste produced. Source reduction prevents waste either by redesigning products or by otherwise changing societal patterns of consumption, use, and waste generation.

Stabilization techniques. Hazardous waste treatment techniques that reduce the hazard potential of a waste by converting the contaminants into their least soluble, least mobile, or least toxic form. The physical nature and handling characteristics of the waste are not necessarily changed by stabilization.

Starved air combustion (SAC) (thermal gasification). A hazardous waste treatment process similar to incineration except that less than the theoretical amount of air necessary for complete combustion is supplied.

Steam (enhanced) stripping. A soil remediation process in which hot air with steam is used for enhanced stripping removal of volatile pollutants from the contaminated soil (see *Hot air* (enhanced) *stripping* for comparison).

Storage. The interim containment of solid waste, hazardous waste, contaminated soil or groundwater, in an approved manner, after generation and prior to ultimate disposal or reuse.

Substantial hazard. A hazard that would pose a significant or otherwise unacceptable risk of harm to health, safety, public welfare, or the environment if it continued to be present for several years.

Subsurface volatilization. See in situ *subsurface ventilation/aeration.*

Subsurface ventilation. See in situ *subsurface ventilation/aeration.*

Surfactant. Any compound that reduces surface tension when dissolved in a solution, or that reduces the tension at the interface of two liquids or a liquid and a solid. There are three basic classes of surfactants: detergents, wetting agents, and emulsifiers.

Surfactant solution. A solution containing surfactants, or surface active molecules or ions. One portion of the surfactant molecule is hydrophilic (water loving), while the other part is hydrophobic (water hating).

Sustainable. Something (i.e., agriculture, industry, development) that can continue at its present (or a comparable) level in the future and does not interfere with future well-being by degrading or overusing natural resources.

Sustainability. The degree of capability for something (i.e., agriculture, industry, development) that can continue at its present (or a comparable) level in the future and does not interfere with future well-being by degrading or overusing natural resources.

SVOC. Semivolatile organic compounds (see *Volatile compounds/solids* and *VOC*).

System. A set of elements that are interrelated in a structured way; the elements are perceived as a whole with a purpose.

TCE. Trichloroethene, which is a common toxic organic solvent.

TCLP. Toxicity characteristic leaching procedure.

Thermal desorption (pyrolysis). The process by which moisture, volatile organic compounds (VOCs), semivolatile organic compounds (SVOCs), and volatile inorganics in contaminated soil or hazardous wastes are reduced by elevating the temperature without combusting the solid materials.

Thermal destruction (incineration). A two-step hazardous waste treatment process involving drying and combustion of hazardous materials.

Thermal gasification. A hazardous waste treatment process similar to incineration except that less than the theoretical amount of air necessary for complete combustion is supplied [see *Starved air combustion* (*SAC*)].

Threat of release. A substantial likelihood of a release that requires action to prevent or mitigate damage to the environment that may result from such release. Circumstances that represent a threat of release include, but are not limited to, sites or vessels containing or conducting an amount of oil or hazardous material in excess of the reportable quantity for that oil or hazardous material where no release has occurred but where (a) corrosion, damage, malfunction or other conditions are visible, known to exist or should be known to exist; and (b) where these conditions are likely to result in a release.

Treatability test (treatability study). A test using an actual or simulated waste sample to determine whether or not a particular process method or equipment can be used to treat the waste in question.

Trench. (a) An excavation made for installing pipes, masonry walls, Bentonite slurry wall, and for other construction or environmental purposes. A trench is distinguished from a ditch in that the opening is temporary and is eventually backfilled; (b) a relatively long but narrow structural or erosional feature of the Earth's subsurface.

Trenching (trench technology). A common remediation process for decontamination of soil and groundwater involving the construction of trenches filled with Bentonite slurry or other materials (see *Trench*). For instance, a Bentonite slurry trench cutoff wall can be built to intersect the subsurface plume to prevent the aquifer from further contamination downgradient. Then, a free product recovery trench can be built before the Bentonite slurry trench cutoff wall to directly intersect the subsurface plume to recover either LNAPL near the water table or DNAPL near the aquifer bottom or impervious layer.

Ultraviolet irradiation. Very short wavelengths of light not visible to the human eye, sometimes used for disinfection.

UNIDO. United Nations Industrial Development Organization.

USEPA. United States Environmental Protection Agency.

Vapor. The gaseous phase of matter that normally exists in a liquid or solid state.

Vaporization. The process by which a substance such as water changes from the liquid or solid phase to the gaseous phase.

Vapor pressure. According to kinetic theory, liquids as well as gases are in constant agitation, and molecules are constantly flying from the surface of the liquid into the atmosphere above. In open systems most of these particles never return, and the liquid is said to be undergoing evaporation. In a closed system, however, particles return to the liquid phase in proportion to their concentration in the gaseous phase. Eventually the rate of return equals the rate of flight, and a condition of equilibrium is established. The vapor is then said to be saturated. The pressure exerted by the vapor under these conditions is known as the vapor pressure. The vapor pressure of all liquids increases with temperature.

Vacuum stripping. See *Vacuum extraction.*

Vacuum extraction (VE). One of a number of terms used interchangeably to describe a process that removes volatile organic compounds (VOCs) from the vadose, or unsaturated soil zone by vacuum stripping.

Vadose zone. Unsaturated soil zone above the water table.

Vapor stripping. See *Soil vapor stripping.*

Vessel. Every description of watercraft, aircraft, or other artificial contrivance used, or capable of being used, as a means of transportation on water or air.

Vitrification. A process of melting contaminated soil, buried hazardous wastes, or toxic sludges at a temperature as high as 1600–2000°C, in an electric furnace or in place at a contaminated site to render the materials nonhazardous. The final nonhazardous product is a glassy and/or crystalline solid matrix that is resistant to leaching and more durable than natural granite or marble.

VOC. Volatile organic compounds, referring to volatile organic solvents and emissions (see *Volatile compounds/solids*).

Volatile compounds/solids. Solids, dissolved or suspended, that are primarily organic, and can be lost on ignition of the total solids.

Volume reduction. The processing of waste materials so as to decrease the amount of space the materials occupy, usually by compacting or shredding (mechanical), incineration (thermal), or composting (biological).

Waste derived fuel (WDF). See *Refuse derived fuel.*

Waste exchange. An organization that assists in the transfer of wastes from a generating company to another company for use as a feedstock. There are primarily two types of waste exchanges: active and passive. An active exchange intervenes between the generator and the potential user (e.g., by taking possession of the waste or locating wastes for "clients"). A passive exchange involves the exchange of information only. Frequently a passive exchange will maintain a catalog or newsletter identifying generators who can provide wastes, and consumers who can use wastes as feedstock. Generally the type, quantity, and description of the waste available/desired is provided. Interested parties are then responsible for contacting each other directly and the waste exchange plays no additional role.

Waste minimization. The reduction, to the extent feasible, of industrial/hazardous waste that is generated or subsequently, treated, stored, or disposed of. It includes any source reduction or recycling activity undertaken by a generator that results in either (a) the reduction of total volume or quantity of hazardous waste, or (b) the reduction of toxicity of hazardous waste, or both, so long as the reduction is consistent with the goal of minimizing present and future threats to human health and the environment.

Waste oil. Used and/or reprocessed, but not subsequently re-refined, oil that has served its original intended purpose. Such oil includes, but is not limited to, used and/or reprocessed fuel oil, engine oil, gear oil, cutting oil, and transmission fluid.

Waste reduction. Reduction in the volume and toxicity of waste. As used in this manual, it includes all source reduction and recycling practices performed on gaseous, aqueous, or solid hazardous or toxic wastes. Treating or emitting, discharging, or disposing of wastes after they have been generated is not waste reduction. Reducing the federal government regulated hazardous wastes by proportionally increasing water discharges of toxic substances is also not an acceptable form of waste reduction.

Water table. The depth below which the ground is saturated with groundwater.

Zero discharge. No discharge of a selected effluent or combination of several effluents (gaseous effluent, liquid effluent, and/or solid effluent) from a site or installation into the environment. For instance, an industrial plant may have successfully accomplished zero discharge of industrial wastewater effluent, but may still violate air emission standards.

Zero emission. No discharge of any gas from a site or installation into atmospheric circulation.

ACKNOWLEDGMENT

This chapter is a revised, updated, and condensed version of four technical reports [22–25] that were all completed under the financial sponsorship of the United Nations Industrial Development Organization (UNIDO), Vienna, Austria, and when the author was a UNIDO Senior Advisor. The scientific views expressed in this chapter are solely of the author and not necessarily those of the sponsoring organization or the UN member countries. Requests for further information dissemination and technical assistance shall be directed to the author.

REFERENCES

1. US Congress. Comprehensive Environmental Response, Compensation, and Liability Act, PL96-510, 1980.
2. USEPA. *Guide to the Disposal of Chemically Stabilized and Solidified Waste*, SW872; US Environmental Protection Agency: Washington, DC, 1982.
3. USEPA. *Test Methods for Evaluating Solid Waste*, SW-846; US Environmental Protection Agency: Washington, DC, 1986.
4. USEPA. *The 14th Annual Research Symposium: Land Disposal, Remedial Action, Incineration and Treatment of Hazardous Waste*; USEPA Hazardous Waste Environmental Research Laboratory: Cincinnati, OH, USA, 1988.
5. USEPA. *Technology Evaluation Report*, EPA/540/5-89/005a; US Environmental Protection Agency: Washington, DC, USA, 1989.
6. USEPA. *Application Analysis Report*, EPA/540/A5-89/005; US Environmental Protection Agency: Washington, DC, USA, 1989.
7. USEPA. *Demonstration Bulletin*, EPA/540/M5-89/005; US Environmental Protection Agency: Washington, DC, USA, 1989.
8. USEPA. *Technology Evaluation Report*, EPA/540/5-89/003a; US Environmental Protection Agency: Washington, DC, USA, 1989.
9. USEPA. *Application Analysis Report*, EPA/540/A5-89/003; US Environmental Protection Agency: Washington, DC, USA, 1989.
10. USEPA. *Assessment of International Technologies for Superfund Application: Technology Identification and Selection*, EPA/600/2-89/017; US Environmental Protection Agency: Washington, DC, USA, 1989.
11. USEPA. *Technology Evaluation Report*, EPA/540/5-89/012; US Environmental Protection Agency: Washington, DC, USA, 1989.
12. USEPA. *Application Analysis Report*, EPA/540/A5-89/012; US Environmental Protection Agency: Washington, DC, USA, 1989.
13. USEPA. *Handbook on In Situ Treatment of Hazardous Waste-Contaminated Soils*, EPA/540/2-90/002; US Environmental Protection Agency: Washington, DC, USA, 1990.
14. USEPA. *Toxic Treatments In Situ Stream/Hot-Air Stripping Technology*, EPA/540/A5-90/800; US Environmental Protection Agency: Washington, DC, USA, 1991.
15. USEPA. *Engineering Bulletin: In Situ Steam Extraction*, EPA/540/2-91/005; US Environmental Protection Agency: Washington, DC, USA, 1991.

16. USEPA. *Application Analysis Report*, EPA/540/A5-91/003; US Environmental Protection Agency: Washington, DC, USA, 1991.
17. USEPA. *Innovative Treatment Technologies: Semi-Annual Status Report*, EPA/540/2-91/001; US Environmental Protection Agency: Washington, DC, USA, 1991.
18. USEPA. *Technology Evaluation Reports*, PB92-115310, PB92-115328, & PB92-115336; US Environmental Protection Agency: Washington, DC, USA, 1992.
19. USEPA. *Radio-Frequency Heating, KAI Technologies, Inc.*, EPA/540/R-94/528; US Environmental Protection Agency: Washington, DC, USA, 1995.
20. USEPA. *Geosafe Corporation In Situ Vitrification*, EPA/540/R-94/520; US Environmental Protection Agency: Washington, DC, USA, 1995.
21. US Fed. Reg. *EP Toxicity Test Procedure*, 40CFR Part 261.24, Appendix II, May 19, 1980; Washington, DC, USA, 1980.
22. Wang, L.K.; Wang, M.H.S.; Wang, P. *Management of Industrial Hazardous Substances at Industrial Sites*; United Nations Industrial Development Organization (UNIDO): Vienna, Austria, 1995, Training Manual No. 4-4-95, 104 p.
23. Wang, L.K. *Case Studies of Cleaner Production and Site Remediation*; United Nations Industrial Development Organization (UNIDO), Vienna, Austria, 1995; Training Manual No. 5-4-95, 136 p.
24. Wang, L.K. *Identification, Transfer, Acquisition and Implementation of Environmental Technologies Suitable for Small and Medium Size Enterprises*; United Nations Industrial Development Organization (UNIDO), Vienna, Austria, 1995; Technical paper No. 9-9-95, 5 p.
25. Wang, L.K. *Site Remediation and Groundwater Decontamination Technologies*; United Nations Industrial Development Organization (UNIDO), Vienna, Austria, 1996; Training Manual No. 1-15-96, 46 p.
26. ASTM. *ASTM Standards on Environmental Site Characterization*; ASTM, 2002; 1827pp.
27. Ebersold, P.; Young, W. Here and now. Environ. Protect. **2002**, *13* (*5*), 28.
28. Howard, H.T.; Pacifici, K.H. The evolution of remote sensing. Environ. Protect. **2002**, *13* (*4*), 28.
29. McCall, W. Getting a direct push. Environ. Protect. **2002**, *13* (*9*), 49.
30. Parish, B.C.; Fournier, L.B. A new slant on drilling. Environ. Protect. **2002**, *13* (*5*), 32.
31. Lesage, S.; Jackson R.E. (Eds.) Groundwater Contamination and Analysis at Hazardous Waste Sites; Marcel Dekker: NYC, NY, 1992; 545pp.
32. Wang, L.K.; Wang, M.H.S. (Eds.) *Handbook of Industrial Waste Treatment*; Marcel Dekker, Inc.: NY, USA, 1992; 392p.
33. Weber, R.E.; Wang, L.K.; Pavlovich, J.J. Separation of Liquids with Different Boiling Points with Nebulizing Chamber. US Patent No. 5,156,747, October 1992.
34. Wilson, D.J.; Clarke, A.N. (Eds). *Hazardous Waste Site Soil Remediation*; Marcel Dekker: NYC, NY, USA, 1994; 567 pp.
35. Butcher, S.W.; Dresser, T.H. Setting the stage. Environ. Protect. **2002**, *13* (*6*), 64.
36. Wang, L.K.; Pereira, N.C. *Handbook of Environmental Engineering: Solid Waste Processing and Resource Recovery*, Volume 2; Humana Press: Clifton, NJ, USA, 1980; 151–225.
37. Wang, L.K.; Kurylko, L.; Hrycyk, O. Site Remediation Technology. US Patent No. 5,552,051, September 3, 1996.
38. Wang, L.K.; Pereira, N.C. *Handbook of Environmental Engineering: Biological Treatment Processes*, Volume 3; Humana Press: Clifton, NJ, USA, 1986; 498pp.
39. Vandenbergh, P.A.; Saul, M. Rushing remediation. Environ. Protect. **2002**, *13* (*6*), 43.
40. Cassidy, D.P. Biological surfactant production in a biological slurry reactor treating diesel fuel contaminated soil. Water Environ. Res. **2001**, *73* (*1*), 87–94.
41. Wang, L.K.; Wang, P.; Clesceri, N.L. Groundwater decontamination using sequencing batch processes. Water Treatment **1995**, *10* (*2*), 121–134.
42. Togna, P.; Enegess, D. Wastewater treatment goes MBR. Environ. Protect. **2003**, *14* (*3*), 43–47.
43. Editor. Membrane microfiltration and ultrafiltration. Water Qual. Prod. **2002**, *7* (*8*), 64–65.
44. Swartzbaugh, J.; Sturgill, J.; Williams, H.D.; Cormier, B. Remediating sites contaminated with heavy metals. Hazardous Mater. Control **1992**, *5* (*6*), 36–46.

45. Wang, L.K.; Kurylko, L.; Wang, M.H.S. Sequencing Batch Liquid Treatment. US Patent No. 5,354,458, October 1994.
46. Editor. Sequencing batch reactors. Pollut. Engng **2000**, *32* (*12*), 75.
47. Shipe, B. The case for UV in dechlorination applications. Water Conditioning and Purification **2003**, *45* (*1*), 34–36.
48. Hrycyk, O.; Kurylko, L.; Wang, L.K. Removal of Volatile Compounds and Surfactants from Liquid. US Patent No. 5,122,166, June 1992.
49. Wang, L.K.; Kurylko, L.; Hrycyk, O. Removal of Volatile Compounds and Surfactants from Liquid. US Patent No. 5,122,165, June 1992.
50. Wang, L.K.; Kurylko, L.; Hrycyk, O. Biological Process for Groundwater and Wastewater Treatment. US Patent No. 5,451,320, September 19, 1995.
51. Editor. Flotation equipment. Environ. Protect. **2003**, *14* (*2*), 137–138.
52. Editor. Buyer's guide: flotation equipment. Water Environ. Fed. **2002/2003**, *14*, 89.
53. Wang, L.K.; Wu, B.C. Treatment of groundwater by dissolved air flotation systems using sodium aluminate and lime as flotation aids. OCEESA J. **1984**, *1* (*3*), 15–18. (NTIS Report No. PB85-167229/AS.)
54. Wang, L.K. *Design of Innovative Flotation–Filtration Wastewater Treatment Systems for a Nickel-Chromium Plating Plant*; US Dept. of Commerce, National Technical Information Service, Report No. PB88-200522/AS, 1984; 50 p.
55. Wang, L.K. Decontamination of Groundwater and Hazardous Industrial Effluents by High-Rate Air Flotation Processes. In *Great Lakes '90 Conference Proceedings*, *Hazardous Materials Control Research Institute*, Silver Spring, Maryland, USA, September 1990.
56. Wang, L.K. Modern Technologies for Prevention and Control of Groundwater Contamination. In *Proceedings of New York & New Jersey Environmental Exposition*, *NYNJEE*, Belmont, MA, USA, October 1990.
57. Wang, L.K. *Environmental Engineering Glossary*; Arvin Corporation (formerly Calspan Corporation): Buffalo, NY, 1972; 424p.
58. Krofta, M.; Wang, L.K. Hazardous Waste Management in Institutions and Colleges; US Department of Commerce, National Technical Information Service: Springfield, VA, 1985; Report No. PB86-194180/AS.
59. Nevius, J.G. Brownfields and green insurance. Environ. Protect. **2003**, *14* (*2*), 27–27.
60. Qadir, R.M. The brownfield challenge. Environ. Protect. **2002**, *13* (*3*), 34–39.
61. Wang, L.K.; Kurylko, L.; Wang, M.H.S. Contamination Removal System Employing Filtration and Plural Ultraviolet and Chemical Treatment Steps and Treatment Mode Controller. US Patent No. 5,190,659, March 1993.
62. Wang, L.K.; Kurylko, L.; Wang, M.H.S. Method and Apparatus for Filtration with Plural Ultraviolet Treatment Stages. US Patent No. 5,236,595, August 17, 1993.
63. Wang, L.K.; Kurylko, L. Liquid Treatment System with Air Emission Control. US Patent No. 5,399,267, March 21, 1995.
64. Wang, L.K.; Kurylko, L.; Wang, M.H.S.. Water and Wastewater Treatment System. US Patent No. 5,240,600, August 1993.
65. Wang, L.K.; Kurylko, L.; Wang, M.H.S. Combined Coarse and Fine Bubble Separation System. US Patent No. 5,275,732, January 1994.
66. Addlestone, S.I. Waste makes haste: top issues in waste management in 2003. Environ. Protect. **2003**, *14* (*1*), 34–37.
67. Adler, S.F. Biofiltration – a primer. Chem. Engng. Prog. **2001**, *97* (*4*), 33–41.
68. Wang, L.K.; Pereira, N.C.; Hung, Y.T. Biological Treatment Processes. Humana Press, Inc., Totowa, NJ. 2004
69. Wang, L.K.; Pereira, N.C.; Hung, Y.T. Physicochemical Treatment Processes. Humana Press, Inc., Totowa, NJ. 2004.

21
Pollution Prevention

J. Paul Chen
National University of Singapore, Singapore

Thomas T. Shen
Independent Environmental Advisor, Delmar, New York, U.S.A

Yung-Tse Hung
Cleveland State University
Cleveland, Ohio, U.S.A.

Lawrence K. Wang
Zorex Corporation, Newtonville, New York, U.S.A. and
Lenox Institute of Water Technology, Lenox, Massachusetts, U.S.A.

21.1 INTRODUCTION

We are witnessing the evolution of a fully industrialized world, with global industrial production, global markets, global telecommunication, global transportation, and global prosperity. This prospect brings with it the realization that current patterns of industrialization will not be adequate to sustain environmentally safe growth and therefore needs drastic improvement. What is urgently needed is a total management systems approach to modern civilization by focusing on pollution prevention activities as the first step.

In the past, pollution control by media-specific control technologies has improved environmental quality to a certain extent. Generally, however, it not only fails to eliminate pollutants, but waste treatment processes have produced a large amount of sludge and residue that require further treatment prior to disposal so that they will not create secondary pollution. Waste treatment systems require investment in design, installation, operation, and maintenance, but these systems contribute no financial benefit to the industrial production. Pollution control technologies may also transfer pollutants from one environmental medium (air, water, or land) to another, causing potential secondary pollution problems that require further treatment and disposal. Pollution control technologies addressed only short-term problems, rather than eliminate pollutants. Costs of pollution control, cleanup, and liability have risen every year, as have the costs of resource inputs, energy, and raw materials. Through many years of research, we are beginning to understand the complexities of pollution management problems [1–10].

Some professionals still believe that pollution control via end-of-pipe strategies, such as a wastewater treatment plant, flue gas cleaning system, land disposal, or incineration can solve pollution problems. This is because such equipment or systems limit the release of harmful pollutants compared to uncontrolled discharge into the environment. As with pain-relieving medication, pollution control methods attempt, although imperfectly, to minimize the effect of releasing pollutants into the environment. Some releases and effects are curtailed, but the original toxic or environmentally harmful pollutants and products remain behind or are transformed into different hazardous substances to some degree.

It is important to realize that pollution prevention applies beyond industrial sectors to a variety of economic sectors and institutional settings. Many organizations and institutions can apply pollution prevention concepts, which not only reduce generation of pollutants and wastes, but also minimize use of certain environmentally harmful products and services. In practice, pollution prevention approaches can be applied to all pollution-generating activities, including energy production and consumption, transportation, agriculture, construction, land use, city planning, government activities, and consumer behavior.

Economics plays an increasingly important role in environmental decision making. The resolution of virtually every environmental issue involves an economic component. Traditionally, most industry and business decision makers who invested in control technologies such as waste treatment and disposal facilities considered these nonproductive, because such added costs to production would be hard to recover. Product prices could be increased to offset these costs, but this was not an option in a competitive market. Such perspectives seemed valid in the past because decision makers did not know the various benefits of pollution prevention that will be described later in this chapter.

Accepting the primacy of pollution management strategy and preventive technologies does not mean abandoning traditional waste management strategy and pollution control technologies or the government regulatory and legal systems designed to ensure their implementation. In fact, not all waste and pollution can be eliminated or prevented, either immediately or in the long run. The remaining waste that cannot be prevented needs to be adequately treated and disposed of. What is absolutely crucial, however, is to recognize the importance of pollution prevention in the hierarchy of environmental options [1].

This chapter highlights the concept and applications of pollution prevention, focusing on the expanding environmental problems from municipal and industrial wastes to toxic chemicals, hazardous products and services, as well as the pollution management challenges to search for new cost-effective technologies such as pollution prevention (P2). Discussions include P2 laws and regulations, project feasibility analyses, implementation, as well as systematic examination of industrial P2. The purpose is to provide readers with a better understanding of pollution sources and pollution prevention. While subtopics may not necessarily be covered in depth, references can provide additional P2 knowledge and information.

21.2 TRADITIONAL MANAGEMENT OF POLLUTANTS AND WASTEWATER

The demand for fresh water rises continuously as the world's population grows. From 1940 to 1990, withdrawals of fresh water from rivers, lakes, reservoirs, and other water sources increased about fourfold. Water is used for various purposes. In the United States, irrigation, electric power generation, and other utilities respectively consume 39, 39, and 12% of water; industry and mining uses 7%, and the rest is used for agricultural livestock and commercial purposes. Table 1 gives a list of major parameters that have great environmental impacts.

Wastewater contains mainly human sewage, industrial wastewater, and agricultural chemicals such as fertilizers and pesticides. According to the US Environmental Protection Agency (USEPA), some 37% of lakes and estuaries, and 36% of rivers are too polluted for basic uses such as fishing or swimming during all or part of the year [2]. In developing nations, more

Table 1 Typical Parameters in Wastewater

Item	Name
Physical parameters	Color; conductivity; settleable solids; suspended solids; temperature; turbidity
Chemical parameters	pH; alkalinity or acidity; arsenic; hardness; biochemical oxygen demand (BOD); chemical oxygen demand (COD); total organic carbon (TOC); aluminum; cadmium; calcium; hexavalent chromium; total chromium; copper; iron; lead; magnesium; manganese; mercury; nickel; zinc; total phosphate; ammonium nitrate; total nitrogen; cyanide; oil and grease pesticides; fluoride; sulfate; phenol; surfactants; chlorinated hydrocarbons
Biological parameters	Coliform bacteria; fecal streptococci bacteria

than 95% of untreated urban sewage is discharged into rivers and bays, which can result in a serious human health hazard. For example, in China, the fastest developing country in the last 20 years, overall municipal wastewater treatment is still less than 5%.

Industrial processes of all types almost invariably produce wastes having numerous sources, forms, and names. For example, wastewater in a petroleum refinery is generated by units when water is contacted with process materials in desalting, stream stripping, and washing operations throughout the refinery processes. In addition, wastewater can be generated by utility systems, from boiler feedwater treatment processes, boiler blowdown, and cooling tower blowdown. The strength and quantity of the wastewater is dependent on the design and operation of the processes.

Until the middle of the 20th century, industrial wastes were considered only a casual nuisance and were handled as such by generators. Industrial plants of the time disposed of most wastes by burial in landfills, discharge into seepage basins, or by pumping directly to a body of water or into a deep well. Refinements were added over the years; for example, much waste was drummed and the containerized waste sent for offsite disposal. However, little if any thought was given to the fact that these wastes, once generated, ultimately ended up being released to the environment unless they were destroyed by treatment.

Industrial waste generators have been made increasingly aware of the nature of their wastes and the problems that waste disposal imposes on our environment. Spurred by mandates from the USEPA as well as by their own sense of corporate responsibility, industries addressed air pollution emissions, wastewater discharges, industrial hygiene/worker safety, and a variety of related issues. With rare exception, however, the actual generation of wastes was never questioned.

New information regarding industrial wastes was developed and complementary federal regulations required industries to reexamine the overall concept of waste generation. First, it was determined that many chemicals present in industrial wastes exerted a permanent deleterious effect on human health. In fact, exposure to some chemicals can alter human genetic material so that the effects of exposure are passed on to future generations.

Secondly, industrial wastes that are not properly treated and disposed of will ultimately release that constituents to the environment. For example, wastes disposed of in landfills may release constituents to subsurface aquifers that serve as drinking water supplies.

Thirdly, testing methods have been developed to evaluate whether an industrial waste contains any constituents of concern to human health or the environment. Furthermore, the tests determine whether and at what rate a waste will release constituents into the environment.

Wastes that contain any of an extensive list of hazardous constituents or that exhibit a hazard characteristic or that are generated by certain industrial processes are referred to as hazardous wastes under the Resource Conservation and Recovery Act (RCRA). About 400 million metric tons of hazardous wastes are generated each year. The United States alone produces about 250 million metric tons; 70% comes from the chemical industry. The treatment, storage, and disposal of these wastes are now governed by strict regulations.

Wastewater treatment and disposal for industrial residues have assumed growing importance. In particular, those wastes defined as RCRA-hazardous require meticulous attention to treatment and ultimate disposal. During the late 1980s, federal regulations were enacted eliminating any form of land disposal for a variety of hazardous wastes, thereby making imperative the treatment of these wastes to render them nonhazardous.

21.3 POLLUTION PREVENTION: MOTIVATION AND CONCEPT

21.3.1 Motivation

According to Webster's Dictionary, the environment is the complex of climatic, economic, and biotic factors that act upon an organism or an ecological community and ultimately determine its form and survival. It is the aggregate of social and cultural conditions that influence the life of an individual or human behavior such as production and consumption.

Environmental pollution is formed as a result of inefficiencies in manufacturing processes, both operational practices and improperly designed and utilized equipment. Pollutants can be unused raw materials, on byproducts resulting from production processes. Pollution represents a loss in profits during manufacturing. It also can be a result of careless human activities in social developments. Releasing pollutants and wastes into the environment creates pollution. Environmental pollution from human activities is never avoidable.

End-of-pipe measures include wastewater treatment, hazardous waste incineration, landfills, and monitoring equipment. They have been used in environmental protection for many years and act as an important component in the P2 in environmental protection. In the last 20 years, however, many environmental accidents, complaints, and concerns have pressured industries to shift from the traditional end-of-pipe approaches to sound pollution prevention strategies.

Public concern about the environment continues to grow. Public education through various media, such as school, television, and the Internet, has become powerful tools for spreading information about the environment and its impact on human health. Protection of the environment increasingly becomes a social responsibility. With increasing understanding of pollutants and their long-term consequences in the environment, some pollutants that were considered less harmful become more important. Dioxin is a good example. Pollution is no longer a site-specific problem; it has become a global issue. For example, mercury has been detected in deep-sea animals (e.g., salmon), which are not supposed to be exposed to polluted environments. The mercury accumulation in the animals is a result of its transport in seawater.

Pollution means loss of raw materials and production of wastes (which are also byproducts). These activities can definitely cause a loss in profits. In addition, pollution created in the workplace can pose either high or low risks to workers, and faces the public most of the time. For example, an improperly operated swine farm can cause water pollution as well as unpleasant odors. The property value in industrial estates can be depreciated and the image of companies deteriorate.

21.3.2 Principles of Environmental Pollution

Socioeconomic development is necessary for meeting people's basic needs of food, clothing, transportation, and shelter, and also to improve living standards; however, such development must be sustainable. That means development should be balanced with the environment. Environmental laws and regulations have focused on media-specific, end-of-pipe, and commend-and-control of pollutants and wastes. Such pollution control technologies have reduced pollution to a certain extent, but are not cost-effective and need to be upgraded to pollution prevention whenever possible. With that recognition, Shen [3] has addressed environmental pollution from a practical point of view by outlining three principles of environmental pollution, which are comparable to some of the thermodynamic laws familiar to most engineers and scientists. Table 2 gives these three important principles of environmental pollution.

21.3.3 Concept

Environmental practitioners in various organizations define P2 based on their own understanding and applications, resulting in somewhat different interpretations. Essentially, it means to prevent or reduce the sources of pollution before problems occur [1]. It is generally contrasted with the media-specific and end-of-pipe control approaches. The difference between pollution prevention and pollution control can be illustrated by the following instances. Vaccines prevent illnesses, while antibiotics control illnesses; seat belts prevent injury, while casts and crutches help cure injury from car accidents. The P2 concept and practices find broad applications such as waste minimization, clean production, green chemistry, green product, waste utilization, ISO 14000, and a number of other related terminologies.

Table 2 Shen's Three Principles of Environmental Pollution

Principle	Description
Pollution from human activities is unavoidable	Pollution is created by releasing pollutants and wastes into the environment as well as by producing certain environmentally harmful products and services as a result of careless human activities related to social and economic development.
Prevent pollution whenever possible	As a result of the first law, pollution needs to be cost-effectively managed. Pollution can be prevented or minimized, but may not be completely eliminated. The remaining residual pollution from human activities must be properly treated and disposed in order to protect human health and the environment.
Minimal pollution is acceptable	Ecosystems can safely handle and assimilate certain amounts of pollution. If pollution is within the environmental quality standards, its impacts to human health and the environment can be acceptable. We must work within the confines of the natural laws to prevent pollution problems in a new planned and economically feasible fashion.

Note. Human activities cover production, distribution, transport, storage, mining, urban development, construction, consumption, and services. The word *products* can be industrial, agricultural, mineral, structural, commercial, and others. The word *services* can be conceptual, technical, and physical such as professional and nonprofessional, government and nongovernment services, including design, plan, operation, construction, transportation systems, repair, maintenance, education and training, management, and others.
Source: Ref. 1.

According to the Pollution Prevention Act of 1990 and other related regulations, the United States defines pollution prevention as follows [4]:

- Reduction of the amount of any hazardous substance, pollutant, or contaminant reentering any waste stream or otherwise released into the environment prior to recycling, treatment, and disposal.
- Reduction of the hazards to public health and the environment associated with the release of such substances, pollutants, or contaminants.
- Reduction or elimination of the creation of pollutants through (a) increased efficiency in the use of raw materials; or (b) protection of natural resources by conservation.

The Canadian Ministry of Environment defines pollution prevention as any action that reduces or eliminates the creation of pollutants or wastes at the source, achieved through activities that promote, encourage, or require changes in the basic behavioral patterns of industrial, commercial, and institutional generators or individuals.

Traditionally, pollution prevention was defined more narrowly as waste reduction or toxic material cutback at sources, focused on waste releases from existing manufacturing operations. Releases of waste from production operations, including those from stacks, vents, and outfalls (called point sources) and those from leaks, open vats, paint areas, and other nonconfined sources (called fugitive emissions) are often the major sources of pollution. Certain products, while leaving the manufacturing plant for distribution through transport, storage, consumption, as well as used-product disposal can cause serious environmental pollution problems, such as hazardous waste treatment, disposal, and remedial sites.

The definition of P2 needs to be updated as our knowledge increases. It should mean a broader sense of minimizing or eliminating the sources of the pollution from every place where they are created in industry, agriculture, commercial establishments, government and nongovernment organizations, and homes. It seeks not only to eliminate or reduce pollutants and wastes, but also certain harmful products and services. It optimizes total materials cycle from virgin material, to finished material, to components, to product, to obsolete product, to ultimate disposal, and also to various technical and nontechnical services. Pollution prevention includes practices that reduce or eliminate the creation of pollutants through increased efficiency in the use of raw materials, energy, water, or other resources, or protection of natural resources by conservation. In practice, pollution prevention approaches can be applied not only to industrial sectors, but all sectors of our society, including energy production and consumption, construction, transport, land use, city planning, government activities, and consumer behavior [5].

21.3.4 Industrial Pollution Prevention

Industrial operations traditionally have adopted a variety of media-specific waste management techniques to control releases of pollutants and wastes. Most environmental legislation in the past had little economic incentive for industries to properly manage their wastes and manufacture green products. P2 is a relatively new pollution management strategy that involves prevention of pollutant and waste as well as promotion of environmentally friendly products and services. As mentioned in Section 21.3.2, pollution should be prevented whenever possible – from product design, production, distribution, and consumption activities. In the event that waste may not be completely prevented, the remaining residual waste from the manufacturing facilities should then be properly treated and disposed in a safe way.

Pollution prevention is the logical extension of pollution control. Environmental management strategies are gradually being transformed as more professionals adopt the pollution prevention concept. It should be emphasized that there are many sources of environmental pollution. Industry is only one sector of pollution sources and surely it is the major one because of waste quantity and toxicity. Other pollution sectors include agriculture, commerce, mining, transport, energy, construction, and consumption. P2 technology in the energy sector, for example, can reduce environmental damages from extraction, processing, transport, and combustion of fuels. Its activities include: (a) increasing efficiency in energy use; (b) substituting fossil fuels by renewable energies; and (c) design changes that reduce the demand for energy [2]. More detailed P2 methods and technologies used in the industrial sector are described in Sections 21.4 and 21.7.

21.4 P2 TECHNOLOGIES AND BENEFITS

21.4.1 P2 Technologies

Today's rapidly changing technologies, industrial processes, and products may generate pollutants that, if improperly managed, could threaten public health and the environment. Many pollutants, when mixed, can produce hazards through heat generation, fire, explosion, or release of toxic substances. To prevent these hazards, pollution generators must be required to describe and characterize their pollutants accurately, by including information as to the type and the nature of the pollutants, chemical compositions, hazardous properties, and special handling instructions.

In practice, preventive technologies not only reduce the generation of waste materials, but also encourage environmentally friendly products and services. It can be applied also to all pollution-generating activities, including energy production and consumption, transportation, agriculture, construction, land use, city planning, government activities, and consumer behavior. In the energy sector, for example, pollution management can reduce environmental damages from extraction, processing, transport, and combustion of fuels. Major preventive technologies applied in industrial processes are described in Section 21.8.

Pollution prevention is receiving widespread emphasis internationally within multinational organizations and within individual countries. The driving force behind the emphasis is the concept of sustainable development and the hold that this concept has over planning strategies and long-term solutions to global limits and north–south economic issues. Examples of some pollution prevention technologies are:

- Raw material substitution – eliminating or reducing a hazardous constituent used either in the product or during manufacture of the product.
- End product substitution – producing a different product that accomplishes the same function with less pollution than the original product.
- Process modification – changing the process design to reduce waste generation.
- Equipment redesign – changing the physical design of equipment to reduce waste generation.
- Direct recycling – reusing materials directly in the manufacturing process without prior treatment. These materials would otherwise become wastes.
- Good housekeeping – instituting new procedures, such as preventive maintenance, to reduce waste generation.
- Inventory control – minimizing the quantities of raw materials or manufactured product in stock, to eliminate surplus that could become waste when the product is changed or discontinued [2].

In the energy sector, for example, pollution prevention can reduce environmental damage from extraction, processing, transport, and combustion of fuels. Pollution prevention technologies include: (a) increasing efficiency in energy use; (b) substituting fossil fuels by renewable energies; and (c) design changes that reduce the demand for energy. During the past few years, considerable progress and success have been achieved in attaining pollution prevention in various sectors of our society.

21.4.2 P2 Benefits

The most important benefit of P2 is that it can achieve national environmental goals while coinciding with the industry's interests [6]. Businesses will have strong economic incentives to reduce the toxicity and volume of the waste they generate. Some reported benefits of P2 practices are that it can:

- avoid inadvertent transfer of pollutants across media by treatment and disposal systems;
- reduce the risks from any release of pollutants and wastes into the environment;
- protect natural resources for future generations;
- save money by preventing excessive use of raw material, energy, and natural resources;
- reduce costs of regulatory violations and costs for waste treatment and disposal;
- avoid long-term potential liabilities associated with releases of wastes and disposal sites;
- increase production efficiency and company reputation;
- improve product quality for world trade market competition.

With P2, some wastes can be reused as raw materials. Waste reduction means increasing production efficiency and generating more profits. Reducing wastes also provides upstream benefits because it reduces ecological damage from raw material extraction and pollutant release during the production process as well as waste recycling, treatment, and disposal operations. A company with effective, ongoing P2 plans may well be the lowest-cost producer and enjoy significant benefits in a competitive world market as a result. Costs per unit produced will drop as P2 measures reduce liability risks and operating costs. Cost savings from prevention come not only from avoiding environmental costs such as hazardous waste disposal fees, but also from avoiding costs that are often more challenging to count, such as those resulting from injuries to workers and ensuing losses in productivity. In that sense, prevention is not only an environmental activity, but also a tool to promote workers' health and safety. Furthermore, P2 activities may enhance the company's public image, public health, and public relations. Among all the benefits, the economic benefits of P2 have proven to be the most compelling argument for industry and business to undertake prevention projects [7].

Many successful P2 cases are available in the literature (Table 3) [8–13]. The P2 program in the USEPA website (www.epa.gov) provides a series of good examples. A study was carried out by Bendavid-Val *et al.* [9] to compare the cost saving from the adoption of the P2. It can be seen from Table 3 that the nine randomly selected plants had different savings ranging from 0 to 100%. Among them, four plants had a saving of 100%. Another example is the dramatic reduction of the wastes from the pulp and paper processing industry [10]. It was reported that, through the implementation of P2 programs, the industry has witnessed the reduction of its biochemical oxygen demand (BOD) and total suspended solids (TSS) by 75 and 45%, respectively, from 1975 to 1988. The US paper recovery rate increased from 22.4% in 1970 to 45.2% in 1997. Other items, such as emissions of total reduced sulfur, SO_2, and ClO_2, were also reduced significantly.

Table 3 List of Industrial Case Studies in P2

Case	Industrial operation/product	P2 action	Main benefits
1	Cardboard box manufacturer and printer	"Good housekeeping" to reduce ink wastes	90% savings in waste disposal and reduction of costs for raw materials
2	Manufacturer of sliding rear windows of automotive industry	Installation of an in-line computer monitoring system and replacement of a pump	90% reduction of hazardous wastes and improvement of safety
3	Brewing	Use waste as fertilizer	Reduction of waste disposal
4	Furniture manufacturing	Hazardous waster reuse	Reduction of waste disposal
5	Textile printing	Solvent recovery	100% cost saving
6	Manufacturer of automatic fluid controls	Replacement of trichloroethylene (TCE) with a waterbased, nontoxic detergent cleaner	Elimination of TCE emissions and related wastes and improvement of safety

The benefits from the P2 exercises based on long-term evaluation are obvious; however, their short-term advantages may not be significant. Sometimes implementation of P2 may even cause a negative impact on industrial performances. Sarkis and Cordeiro [11] carried out an empirical evaluation of environmental efficiencies (by end-of-pipe or P2 approaches) and US corporate financial performance. Interestingly, they found that there was a negative correlation between the above two performances. The negative relationship became more obvious when P2 was implemented. The corporate greening could cause depreciation of stock values. However, higher pollution levels can negatively affect a firm's market values. Therefore, a sound balance must be carefully maintained.

21.5. P2 LAWS AND REGULATIONS

21.5.1 Federal Regulations and Laws

In the United States, Congress enacted the Clean Water Act (CWA) of 1972 to achieve a goal of "fishable and swimmable" surface waters. It covers regulations of wastewater discharges [12]. Most industries must meet discharge standards for various pollutants. Specific methods of control such as pollution prevention are not specified. Many facilities use pollution prevention as a means of reducing the cost of compliance with federal regulations. State and local authorities also have responsibilities to implement the provisions of the CWA. These authorities must enforce the federal guidelines at a minimum, but may choose to enforce more stringent requirements. Some localities include pollution prevention planning requirements into discharge permits [12].

The Emergency Planning and Community Right-to-Know Act (EPCRA, also known as SARA Title III) requires certain companies to submit an annual report of the amount of listed "toxic chemicals" entering the environments. Source reduction and waste management information must be provided for the listed toxic chemicals.

The Resource Conservation and Recovery Act (RCRA), and Hazardous and Solid Waste Amendments (HSWA) to RCRA require that the reduction or elimination of hazardous waste generation at the source should take priority over other management methods such as treatment and disposal. Hazardous waste generators are required to certify on their hazardous waste manifests that they have programs in place to reduce the volume or quantity and toxicity of hazardous waste generated to the extent economically practicable. Materials that are recycled may be exempt from RCRA regulations if certain conditions are met.

The Water Quality Act of 1987 further strengthened the CWA, and amendments to the Safe Drinking Water Act required numerous treatment facility upgrades. Although all these acts are dramatic in their protection of US citizens against waterborne diseases and the improvement of water quality, they placed little emphasis on source reduction or elimination of the root cause of pollution.

To address this issue of the need of regulation upgrading, the Pollution Prevention Act of 1990 was passed. It formalized a national policy and commitment to waste reduction, functioning primarily to promote the consideration of pollution prevention measures at the federal government level. This act crosses media boundaries by establishing a national policy on pollution prevention, including programs that emphasize source reduction, reuse, recycling, and training. All these areas are key to the successful implementation of a P2 industrial wastewater management program [1]. According to the act, the USEPA should review existing and proposed programs and new regulations to determine their effect on source reduction [12]. Source reduction activities among the USEPA programs and other federal agencies are coordinated. It provides public access to environmental data and fosters the exchange of source reduction information. It establishes pollution prevention training programs for Federal and State environmental officials. Finally, the USEPA is required to facilitate adoption of source reduction by businesses, as well as identify and make recommendations to Congress to eliminate barriers to source reduction.

Since 1990, the USEPA has implemented a diverse set of programs and initiatives to meet their obligations defined by the law. A series of achievements has been reported, including 33/50, Climate Wise, Green Lights, Energy Star, WAVE, the Pesticide Environmental Stewardship Program, Indoor Air, Indoor Radon, Design for the Environment, the Environmental Leadership Program, and the Common Sense Initiative [12]. For example, reduction of a series of key pollutants was achieved through the 33/50 programs [13]. The Program targeted 17 priority chemicals (e.g., benzene, tetrachloroethylene, and toluene) and set as its goal a 33% reduction in releases and transfers of these chemicals by 1992 and a 50% reduction by 1995, measured against a 1988 baseline. Its primary purpose was to demonstrate whether voluntary partnerships could augment the USEPA's traditional command-and-control approach by bringing about targeted reductions more quickly than would regulations alone. The program sought to foster a pollution prevention ethic, encouraging companies to consider and apply pollution prevention approaches to reducing their environmental releases rather than traditional end-of-pipe methods for treating and disposing of chemicals in waste. The 33/50 Program achieved its goal in 1994, one year ahead of schedule, primarily through program participants' efforts. Facilities also reduced releases and transfers of the other 33/50 chemicals by 50% from 1988 to 1995 [13].

Traditionally, environmental laws and regulations have controlled the releases of pollutants and wastes. Only in recent years have laws and regulations gradually covered the production of certain environmentally unfriendly products and services that also caused environmental pollution. For example, DDT, CFCs, asbestos, leaded gasoline, certain kinds of plastics, medicines, cosmetics, fertilizers, pesticides, and herbicides have been restricted in production. Similarly, consulting services in designing products and process, in equipment manufacturing and supply, and in education and training reduce significantly adverse impacts of the environmental quality.

Effective pollution management requires cost-effective regulations and standards, followed by a combination of incentives and partnership approaches, and monitoring activities to enforce the standards. Some targeting will be required toward the most polluting subsectors or the most polluted regions. If there is sufficient institutional capacity to implement industry-specific or other-specific programs, government agencies may also provide information and other incentives to encourage the adoption of new and emerging preventive technologies for various pollution sources to protect the environmental and natural resources.

21.5.2 State Pollution Prevention Laws

Many US states (e.g., Arizona, California, Minnesota, and Texas) have passed laws to incorporate aspects of P2 into RCRA and EPCRA requirements. The laws require manufacturers that produce wastes to develop a source reduction and waste minimization plan, including an implementation schedule, and to track and report waste reduction progress. A number of states have implemented voluntary pollution prevention programs. The foundation of these programs is generally educational outreach and technical assistance mechanisms. Tables 4 and 5 give lists of state mandatory and voluntary P2 programs [12]. The following are some examples of provisions from state laws.

The Arizona law applies only to facilities that must file the annual Toxic Chemical Release Inventory Form R required by EPCRA Section 313 or during the preceding 12 months generated an average of 1 kg per month of an acutely hazardous waste. The California law only applies to facilities that generate more than 12,000 kg of hazardous waste or 12 kg of extremely hazardous waste in a calendar year. The programs require facilities to perform P2 planning that identifies waste sources and specific technical steps that can be taken to eliminate or reduce the generation of hazardous wastes. The facilities are required to submit progress reports with the length of time between reports ranging from one to two years [12].

Table 4 List of State Mandatory Pollution Prevention Programs

Mandatory P2 programs	Statute
Arizona	AZ Rev. Stat. Ann. 49-961 to -73
California	CA Health & Safety Code 25244.12 to .24
Georgia	GA Code Ann. 12-8-60 to -83
Louisiana	LA Rev. Stat. Ann. 30.2291 to .2295
Maine	ME Rev. Stat. Ann., tit. 38, 2301 to 2312
Massachusetts	MA Ann. Laws ch. 211,1 to 23
Minnesota	MN Stat. Ann. 115D.01 to .12
Mississippi	MS Code Ann. 49-31-1 to -27
New Jersey	NJ Stat. Ann. 13: 1D-35 to -50
New York	NY Envtl Conserv. Law 27-0900 to -0925
Oregon	OR Rev. Stat. 465.003 to .037
Tennessee	TN Code Ann. 68-212-301 to -312
Texas	TX Title 30, Ch 335
Washington	WA Rev. Code 70.95C.010 to .240

Source: Ref. 12.

Table 5 List of State Voluntary Pollution Prevention Programs

Voluntary P2 programs	Statute
Alaska	AK Stat. 46.06.021 to .041
Colorado	CO Rev. Stat. Ann. 25-16.5-101 to -110
Connecticut	CT Gen. Stat. Ann Appendix Pamphlet, P.A. 9 1-376
Delaware	7 DE Code Ann. 7801 to 7805
Florida	FL Stat. Ann. 403 .072 to .074
Illinois	IL Ann. Stat. Ch. 111 ,7951 to 7957
Indiana	IN Code Annl 3-9-1 to-7
Iowa	IA Code Ann. 455B.516 to .518
Kentucky	KY Rev Stat. Ann. 224.46-3 10 to -325
Ohio	HB 147, HB 592
Rhode Island	RI Gen. Laws 37-15.1-1 to .11
South Carolina	SC Code Ann. 68-46-301 to -312
Wisconsin	WI Stat. Ann. 144.955

Source: Ref. 12.

21.5.3 Local Pollution Prevention Requirements

According to the CWA, qualified local publicly owned treatment works (POTWs) are given the authority to administer pretreatment programs (e.g., regulation of industrial dischargers). The POTWs can have the authority to implement regulations that are more stringent than federal guidelines (e.g., 40 CFR 433). A number of local agencies therefore use this authority to reduce the impact of industrial discharges on the operation of the POTW, reduce the concentration of toxic pollutants in POTW sludge, and/or to reduce the mass of pollutants discharged by the POTW. This can be accomplished by lowering the permissible concentration limits of industrial discharges below the federal standards.

One local program administered by the Palo Alto Regional Water Quality Control Plant (RWQCP) incorporates P2 requirements into pretreatment discharge permits. This is one of the first examples of the use of such requirements in place of more traditional pollutant concentration limitations. In response to its own stringent copper discharge limit, the RWQCP had to reduce the copper content of the influent at the plant. This effort focused upon all sources of copper mainly targets on computer parts manufacturers, who are given the choice between mass-based discharge limits or concentration limits in the P2 program. There were a total of 13 facilities in the RWQCP service area in 1995. One made an unrelated decision to move out of the service area; eight facilities chose the concentration-based limits and installation of Reasonable Control Measures (RCMs) [12].

21.6 POLLUTION PREVENTION FEASIBILITY ANALYSES

The level of required analysis depends on the complexity of the considered pollution prevention project. A simple, low-capital cost improvement such as preventive maintenance would not need much analysis to determine whether it is technically, environmentally, and economically feasible. On the other hand, input material substitution could affect a product specification, while a major modification in process equipment could require large capital expenditures. Such changes could also alter process waste quantities and compositions, thus requiring more systematic evaluation.

Various options of pollution prevention projects may be evaluated, depending on the resources currently available. It may be necessary to postpone feasibility analyses for some options; however, all options should be evaluated eventually. This section describes how to screen and narrow identified options to a few that will be evaluated in greater detail. Detailed analysis includes evaluation of technical, environmental, economical, and institutional feasibilities. It is important to note that many of the issues and concerns during pollution prevention feasibility analyses are interrelated.

21.6.1 Technical Feasibility Analysis

Technical feasibility analysis requires comprehensive knowledge of pollution prevention techniques, vendors, relevant manufacturing processes, and the resources and limitations of the facility. The analysis can involve inspection of similar installations, obtaining information from vendors and industry contacts, and using rented test units for bench-scale experiments when necessary. Some vendors will install equipment on a trial basis and payment after a prescribed time, if the user is satisfied.

Technical analysis should determine which technical alternative is the most appropriate for the specific pollution prevention project in question. Such analysis considers a number of factors and asks very detailed questions to ensure that the pollution prevention technique will work as intended. Examples of facility-related questions to be considered include:

- Will it reduce waste?
- Is space available?
- Are utilities available or must these be installed?
- Is the new equipment or technique compatible with current operating procedures, workflow, and production rates?
- Will product quality be maintained?
- How soon can the system be installed?
- How long will production be stopped in order to install the system?
- Is special expertise required to operate or maintain the new system? Will the vendor provide acceptable service?
- Will the system create other environmental problems?
- Is the system safe?
- Are there any regulatory barriers?

All affected groups in the facility should contribute to and review the results of the technical analysis. Prior consultation and review with the affected groups (e.g., production, maintenance, purchasing) will ensure the viability and acceptance of an option. If a change in production methods is necessary, the project's effects on the quality of the final product must be determined.

21.6.2 Environmental Feasibility Analysis

The environmental feasibility analysis weighs the advantages and disadvantages of each option with regard to the environment. Most housekeeping and direct efficiency improvements have obvious advantages. Some options require a thorough environmental evaluation, especially if they involve product or process changes or the substitution of raw materials. The environmental option of pollution prevention is rated relative to the technical and economical options with respect to the criteria that are most important to the specific facility. The criteria may include:

- reduction in waste quantity and toxicity;
- risk of transfer to other media;

- reduction in waste treatment or disposal requirements;
- reduction in raw material and energy consumption;
- impact of alternative input materials and processes;
- previous successful use within the company or in other industry;
- low operating and maintenance costs;
- short implementation period and ease of implementation;
- regulatory requirements.

The environmental evaluation is not always so clear-cut. Some options require a thorough environmental evaluation, especially if they involve product or process changes or the substitution of raw materials. To make a sound evaluation, information should be gathered on the environmental aspects of the relevant product, raw material, or constituent part of the process. This information would consider the environmental effects not only of the production phase and product life cycle but also of extracting and transporting the alternative raw materials and of treating any unavoidable waste. Energy consumption should also be considered. To make a sound choice, the evaluation should consider the entire life cycle of both the product and the production process.

21.6.3 Economic Feasibility Analysis

Economic feasibility analysis is a relatively complex topic, which is only briefly discussed here. Economic analysis deals with the allocation of scarce, limited resources to various pollution prevention modifications, and compares various investments to help determine which investments will contribute most to the company.

A benefit is usually defined as anything that contributes to the objectives of the pollution prevention project; costs are defined as anything that detracts from the achievement of a project's objectives. Normally, benefits and costs are evaluated from the perspective of whether they contribute to (or detract from) the maximization of a company's income. Economic cost–benefit analysis uses a number of measures of profitability such as net present value, internal rate of return, and benefit–cost ratio.

When measuring savings, it is important to look at not only the direct savings but also the indirect savings of pollution prevention. In addition, there are intangible benefits that are difficult to quantify in financial terms; nevertheless, they are an important aspect of any pollution prevention project, and should be factored into the decision-making process. The economic feasibility analysis of pollution prevention alternatives examines the incremental costs and savings that will result from each pollution prevention option. Typically, pollution prevention measures require some investment on the part of the operator, whether in capital or operating costs. The purpose of economic feasibility analysis is to compare those additional costs to the savings (or benefits) of pollution prevention.

For most capital investments, the direct cost factors are the only ones considered when project costs are being estimated. For pollution prevention projects, direct cost factors may only be a net cost, even though a number of the components of the calculation will represent savings. Therefore, confining the cost analysis to direct costs may lead to the incorrect conclusion that pollution prevention is not a sound business investment. In performing the economic analysis, various costs must be considered. As with any project, the direct costs should be broken down as:

- *Capital expenditures* – for purchasing process equipment, additional equipment, materials, site preparation, designing, purchasing, installation, utility connections, training costs, start-up cost, permitting costs, initial charge of catalysts and chemicals, working capital, and financing charges.

- *Operating costs* – typically associated with costs of raw materials, water and energy, maintenance, supplies, labor, waste treatment, transportation, handling, storage, and disposal, and other fees. Revenues may partially offset operating costs from increased production or from the sale or reuse of byproducts or wastes.

Unlike more familiar capital investments, indirect costs for P2 are likely to represent a significant net savings. Indirect costs are hidden in the sense that they are either allocated to overhead rather than to their source (production process or product), or altogether omitted from the project financial analysis. A necessary first step in including indirect costs in an economic analysis is to estimate and allocate them to their source. Indirect costs may include:

- administrative costs;
- regulatory compliance costs such as permitting, record keeping and reporting, monitoring, manifesting;
- insurance costs;
- workman's compensation; and
- onsite waste management and control equipment operation costs.

Estimating and allocating future liability costs involves much uncertainty. It may be difficult to estimate liabilities from actions beyond our control, such as an accidental spill by a waste hauler. It is also difficult to estimate future penalties and fines for noncompliance with regulatory standards that do not exist yet. Similarly, it is difficult to estimate personal injury and property damage claims that result from consumer misuse, disposal of waste later classified as hazardous, or claims of accidental release of hazardous waste after disposal. Allocation of future liabilities to the products or production processes also presents practical difficulties in a cost assessment.

A pollution prevention project can benefit from water, energy, and material savings as well as from waste reduction, recycling, and reuse. It may also deliver substantial benefits from an improved product and company image or from improved employee health. These benefits remain largely unexamined in environmental investment decisions. Although they are often difficult to measure, they should be incorporated into the assessment whenever feasible. At the very least, they should be highlighted for managers after presenting costs that can be the more easily quantified and allocated. Intangible benefits may include:

- increased sales due to improved product quality, enhanced company image, and consumer trust in products;
- improved supplier–customer relationship;
- reduced health maintenance costs;
- increased productivity due to improved employee relations; and
- improved relationships with regulators.

21.6.4 Institutional Feasibility Analysis

Institutional analysis is concerned with evaluating the strengths and weaknesses of the company's involvement in the implementation and the operation of investment in pollution prevention projects. It includes, for example:

- staffing profiles;
- task analysis and definitions of responsibility;
- skill levels;
- processes and procedures;

- information systems and flows for decision making; and
- policy positions on pollution prevention priorities.

The analysis should cover managerial practices, financial processes and procedures, personnel practices, staffing patterns, and training requirements. Issues of accountability need to be addressed. Proper incentives, in terms of money and career advancements, will encourage employees to achieve pollution prevention goals.

21.7 P2 PROJECT IMPLEMENTATION AND REVISION

After a pollution prevention project or plan of program is established and its technical, environmental, economic, and institutional feasibilities are analyzed, team members will be able to more easily encourage management to implement chosen projects. All members of the company may not embrace a pollution prevention project immediately, especially if they do not fully understand the benefits and the cost savings of pollution prevention. To implement a pollution prevention plan or program most effectively, the true cost of waste generation and management must be constantly emphasized. The true cost includes all environmental compliance costs such as manifesting, training, reporting, accident preparedness; future liability costs; and intangible costs such as product acceptance, labor relations, and public image.

This section describes the essential elements and methods of (a) understanding processes and wastes, (b) selecting a pollution prevention project, (c) obtaining funding, (d) implementing projects through various engineering steps, (e) reviewing and revising projects, and (f) project progress monitoring and revising.

21.7.1 Understanding Processes and Wastes

Understanding processes is important as it can provide useful information on both quantity and quality of waste. It includes the following aspects: (a) gathering background information, defining production units, (b) characterization of general process, (c) understanding unit processes, and (d) performing materials balance. A detailed discussion is given in Section 21.9.

21.7.2 Selecting Projects

Final selection of a project from among the various proposed projects for implementation depends primarily upon the pollution prevention feasibility analyses. The selection should generally rely on the hierarchy for waste reduction, which emphasizes more source reduction; results of the waste reduction assessment; availability of specific clean technologies or procedural applications; qualitative assessment of technical and economic feasibility; institutional feasibility; and other considerations. The next step is to develop a schedule for implementation. The selected pollution prevention projects should be flexible enough to accommodate possible alternatives or modifications. The pollution prevention team should be willing to do background and support work, and anticipate potential problems in implementing projects.

21.7.3 Obtaining Funding

The pollution prevention team must seek funding for those selected projects that will require expenditures. Within a company, there are probably other projects such as expanding production capacity or moving into new product lines that compete with the pollution prevention projects for funding. If the team is part of the overall budget decision-making process, it can make an informed decision on whether the selected pollution project should be implemented right away or whether it can await the next capital budgeting period. The team needs to ensure that the pollution prevention projects will be reconsidered at that time.

Some companies will have difficulty raising funds internally for capital investment, especially companies in developing countries. External funds are available to implement pollution prevention projects. Private sector financing includes bank loans and other conventional sources of financing. Financial institutions and international organizations (e.g., Asian Development Bank) are becoming more cognizant of the sound business aspects of pollution prevention and cleaner production [15]. Government financing should be available in some cases to help small- and medium-sized plants.

A strong engineering approach helps ensure proper implementation of the selected projects. Outside process engineering support may be required if company personnel do not have the time to implement tasks. Many pollution prevention projects may require changes in operating procedures, purchasing methods, materials inventory control, equipment modification, or new equipment. Such changes may affect a company's policies and procedures. However, the implementation phases resemble those of most other company projects.

Personnel who will be directly affected by the project (line workers and engineers) should participate from the start. Those personnel indirectly affected (e.g., controllers, purchasing agents) should also participate as project implementation proceeds. Any additional training requirements should be identified and arrangements made for instruction. All employees should be periodically informed of the project status and should be educated as to the benefits of the projects to them and to the company. Encouraging employee feedback and ideas may ease the natural resistance to change.

21.7.4 Project Implementation

Implementation of a pollution prevention project will generally follow the procedures established by the company for implementing any new procedure, process modification, or equipment change. Implementing a major pollution prevention project typically involves several steps:

- *Preparing a detailed design.* It is helpful to discuss the project with representatives from production, maintenance, safety, and other departments who may be affected by the change or who may have suggestions regarding equipment manufacturers, layout, scheduling, or other aspects of implementation.
- *Preparing a construction bid package and equipment specifications.* If construction is required, details of the necessary construction will need to be assembled into a construction bid package. Depending on the established procedures in the company, specifications for new equipment or particular manufacturers and models may be necessary.
- *Selecting construction staff and purchasing materials.* Construction may be performed by an in-house or outside company, depending on cost and availability.
- *Installing new equipment.* Construction will generally involve installation of the necessary equipment. Timing and scheduling of installation may be critical in some operations.

- *Training personnel*. Personnel from maintenance as well as production may need training. Proper operation and maintenance are critical for effective, safe, and trouble-free operation. Training is sometimes available from equipment vendors.
- *Starting operation*. Extra care is necessary during the initial stages of operation to ensure that all proceeds smoothly; first impressions of the effectiveness of the system are often lasting.
- *Monitoring and evaluating performance*. Monitoring is typically an integral part of operations, and is performed by the production staff. Depending on the complexity of the chosen pollution prevention project, only a few of these steps may be necessary.

21.7.5 Reviewing and Revising Projects

The pollution prevention process does not end with implementation. After the pollution prevention plan is implemented, we need to track its effectiveness and compare it to previous technical and economic assessments. Options that do not meet the original performance expectations may require rework or modification. This can be done through the knowledge gained by continuing to evaluate and fine-tune the pollution prevention projects. The success of the initial pollution prevention project may be only the first step along the road to establishing pollution prevention as either a stand-alone program or as an important criterion for consideration in other ongoing plant programs within the company. Developing the pollution prevention ethic is usually a step-by-step evolutionary process. It starts slowly and gradually builds momentum. Each successful pollution prevention experience provides even more incentive for management to support and diversify pollution prevention within the company.

To ensure that the pollution prevention momentum is preserved from the end of one pollution prevention project to the beginning of the next, it is important to provide an opportunity for feedback and evaluation. This opportunity should be used not only to critique the previous effort, but also to promote it and raise its visibility; that is, to "wave the banner" for pollution prevention and demonstrate that pollution prevention has even broader applications in the company. In this way we can use each pollution prevention experience as a promotional device for the next project, so that the role of pollution prevention within the company will continue to grow. With time, pollution prevention will become a natural part of the company's infrastructure and operating practices.

21.7.6 Project Progress Monitoring

In order to track the facilities' achievements and the overall effectiveness of the regulatory approach, pollution prevention related data must be available. Industry also needs independently verifiable data, which effectively measure pollution prevention progress both to demonstrate compliance with newly evolving environmental laws and to promote an environmentally sensitive image.

A successful pollution prevention measurement needs to identify the specific objective, resource availability, and proper approach. If the industrial plant has the appropriate resources, the pollution prevention measurement can help develop a rapid understanding of the relationship between wastes and the manufacturing process.

The needs cover pollution prevention data acquisition, data analysis, and methods of measuring pollution prevention progress with the emphasis on source reduction in various industrial processes. It also discusses toxic release inventories, material accounting data, and expansion of the Community-Right-to-Know program used by the US regulatory agencies to measure the progress of pollution prevention and project revisions.

Measuring pollution prevention progress is to evaluate the progress against the established goals, which can be different in various settings; it includes data acquisition and analysis. Different data collection approaches might be required in different types of industrial processes and we must select a quantity in terms of waste volume or waste toxicity. In the analysis, the most cost-effective methods must be selected for specific situations. Semiquantitative data are easier to obtain but have less utility.

Methods of measuring pollution prevention progress consist of toxic release inventory, materials accounting data and evaluation. Experience indicates that changes in reporting requirements, and in respondents' understanding of them, can introduce errors in analyses.

21.8 P2 APPLICATIONS IN INDUSTRIAL PLANTS

Industry involves thousands of products and production processes, resulting in a decentralized enterprise system. Technical progress toward pollution prevention is likewise decentralized, since it is driven by economic consideration, and frequently specific to one of thousands of different industrial processes. Current pollution prevention technology generally employs conventional engineering approaches. Even as priority shifts from waste treatment and control to prevention, engineers will still employ available technology at first to achieve their objectives. In the future, however, process modifications and friendly product designs will become more innovative for the environment.

Pollution prevention techniques for industrial manufacturing facilities such as waste minimization and source reduction can be understood by observing the path of material as it passes through an industrial site. Even before materials arrive at the site, we could avoid toxic materials when less toxic substitutes exist. Pollution prevention technologies for industries can be generalized into five groups: improved plant operations, in-process recycling, process modification, materials and product substitutions, and materials separations.

21.8.1 Improved Plant Operations

Manufacturers could implement a variety of improved management or "housekeeping" procedures that would aid pollution reduction; they could conduct environmental audits, establish regular preventive maintenance, specify proper material handling procedures, implement employee training, as well as record and report data.

Environmental Audits

Environmental audits may be conducted in many different settings by individuals with varied backgrounds and skills, but each audit tends to contain certain common elements. It is better to identify and correct problems associated with plant operation to minimize waste generation. One aspect of improved plant operation is cost saving. Production costs and disposal costs can be cut simultaneously by improving plant operation. The practice of environmental auditing also examines critically the operations on a site and, if necessary, identifying areas for improvement to assist the management to meet requirements. The essential steps are as follows.

(1) Collecting information and facts,
(2) Evaluating that information and facts,
(3) Drawing conclusions concerning the status of the programs audited with respect to specific criteria,

(4) Identifying aspects that need improvement, and
(5) Reporting the conclusions to appropriate management.

Audits enable manufacturers to inventory and trace input chemicals and to identify how much waste is generated through specific processes. Consequently, they can effectively target the areas where waste can be reduced and formulate additional strategies to achieve reductions. The audit consists of a careful review of the plant's operations and waste streams and the selection of specific streams and/or operations to assess. It is an extremely useful tool in diagnosing how a facility can reduce or eliminate hazardous and nonhazardous wastes. It focuses on regulatory compliance and environmental protection.

Regular Preventive Maintenance

Preventive maintenance involves regular inspection and maintenance of plant equipment, including lubrication, testing, measuring, replacement of worn or broken part, and operational conveyance systems. Equipment such as seals and gaskets should be replaced periodically to prevent leaks. The benefits of preventive maintenance are increased efficiency and longevity of equipment, fewer shutdowns and slowdowns due to equipment failure, and less waste from rejected, off-specification products. Maintenance can directly affect and reduce the likelihood of spills, leaks, and fires. An effective maintenance program includes identification of equipment for inspection, periodic inspection, appropriate and timely equipment repairs or replacement, and maintenance of inspection records.

Corrective maintenance is needed when the design levels of a process change and adjustments to indirect factors are required. This type of maintenance includes recognizing the signs of equipment failure and anticipating what repairs or adjustments need to be made to fix the problem or improve the overall efficiency of the machinery. Visual inspection ensures that all of the elements of the process system are working properly. However, routine inspections are not a substitute for the more thorough annual compliance inspections. After each visual inspection, it is important to document the results and evaluate the effectiveness of corrective previous actions. Any necessary future corrective action should also be identified.

In reducing fugitive emissions, conscientious leak detection and repair programs have proven to be extremely effective at a fraction of the cost of replacing conventional equipment with leafless technology components. Besides being expensive, changing to leafless technology is not always feasible, and reduces emissions over well-maintained, high-quality conventional equipment only marginally.

Material Handling and Storage

Material handling and storage operations can cause two types of fugitive emissions: (a) low-level leaks from process equipment, and (b) episodic fugitive emissions, where an event such as equipment failure results in a sudden large release. Often, methods for reducing low-level equipment leaks result in fewer episodes, and vice versa. Methods for reducing or eliminating both types of fugitive emissions can be divided into two groups: (a) leak detection and repair and (b) equipment modification. Such emissions can be prevented by good practices. Proper materials handling and storage ensures that raw materials reach a process without spills, leaks, or other types of losses that could result in waste generation. Some basic guidelines for good operation practices are suggested to reduce wastes by:

- spacing containers to facilitate inspection;
- labeling all containers with material identification, health hazards, and first aid recommendations;

- stacking containers according to manufacturers' instructions to prevent cracking and tearing from improper weight distribution;
- separating different hazardous substances to prevent cross-contamination and facilitate inventory control; and
- raising containers off the floor to inhibit corrosion from "sweating" concrete.

Spills and leaks are major sources of pollutants in industrial processes and material handling. When material arrives at a facility, it is handled and stored prior to use; material may also be stored during stages of the production process. It is important to prevent spillage, evaporation, leakage from containers or conduits, and shelf-life expirations. Standard operating procedures to eliminate and minimize spills and leaks should take place regularly. Better technology might consist of tighter inventory practices, seal-less pumps, welded rather than flanged joints, bellows seal valves, floating roofs on storage tanks, and rolling covers vs. hinged covers on openings. While these techniques are not novel, they could still lead to large replacement costs if a company has many locations where leakage can occur. Conversely, they could provide large economic benefits by reducing the loss of valuable materials.

Waste from storage vessels takes many forms, from emissions due to vapor displacement during loading and unloading of storage tanks, to wastes formed during storage, to the storage containers themselves if they are discarded. Reducing waste from storage vessels therefore consists of a variety of activities. Storage tanks for storing organic liquids are found at petroleum refineries, organic chemical manufacturing facilities, bulk storage and transport facilities, and other facilities handling organic liquids. They are used to dampen fluctuation in input and output flow. Storage tanks can be disastrous sources of waste when weakly active undesired reactions leak out, and it is important to monitor temperatures where this can occur and to design tanks so that heat dissipation effects dominate. Inadequate heat dissipation is of particular concern in the storage of bulk solids and viscous liquids. Other aspects of pollution prevention for storage tanks involving tank bottoms, standing and breathing losses, and emissions due to the loading and unloading of storage tanks.

Vapors that are displaced in loading and off-loading operations can be a significant source of VOC emissions from storage containers. Vapor recovery devices that trap and condense displaced gases reduce losses due to loading and unloading of fixed-roof storage tanks by 90–98%. Vapor balance, where vapors from the container being filled are fed to the container being emptied, is another technique that can be applied in some cases to reduce emissions. Spills due to overfilling of storage containers are another source of emissions that occur during loading and unloading operations. These spills can be prevented through the use of appropriate overflow control equipment and/or overflow alarms.

Employee Training

Employee training is paramount to successful implementation of any industrial pollution prevention program. All the plant operations staff should be trained according to the objectives and the elements of the program. Training should address, among other things, spill prevention, response, and reporting procedures; good housekeeping practices; material management practices; and proper fueling and storage procedures. Properly trained employees can more effectively prevent spills and reduce emission of pollutants.

Well-informed employees are also better able to make valuable waste reduction suggestions. Plant personnel should comprehend fully the costs and liabilities incurred in generating wastes. They should have a basic idea of why and where waste is produced and whether the waste is planned or unplanned.

Employee training can take place in three stages: prior to job assignment, during job training, and ongoing throughout employment. Before beginning an assignment, employees should become familiar with toxic properties and health risks associated with exposures to all hazardous substances that they will be handling. In addition, employees should learn the consequences of fire and explosion involving these substances. Finally, they should learn what protective clothing or gear is required and how to use it. During job training, employees should learn how to operate the equipment safely, the methods and signs of material releases, and what procedures to follow when a spill or leak occurs. Ongoing education includes regular drill, updates on operating and cleanup practices, and safety meetings with other personnel.

The challenge of education and training today is how to integrate air–water–land pollution management through integrated waste prevention prior to the application of waste treatment and disposal technologies. A multimedia plan would help to implement pollution prevention. Cross-disciplinary education and training will enable trainees to understand the importance of multimedia pollution prevention principles and strategies so that they can carry out such principles and strategies for pollution prevention.

Operating Manual and Record Keeping

Over the past decade, governments and industry trade organizations have developed guides and handbooks for reducing wastes. These materials are useful for analyses of individual facilities, although some guides attempt to be more general. Good facility documentation can have many benefits for the plant, including waste reduction. Facility documentation of process procedures, control parameters, operator responsibilities, and hazards in a manual or set of guidelines will contribute to safe and efficient operation. It also promotes consistency, thereby lessening the likelihood of producing an unacceptable product, which must be discarded. A facility operating manual will assist the operators in monitoring waste generation and identifying unplanned waste releases and will also assist operators in responding to equipment failure.

Diligent record keeping with regard to waste generation, waste handling and disposal costs, and spills and leaks helps to identify areas where operating practices might be improved and later will help in assessing the results of those improved practices. Record keeping can be instrumental in assuring compliance with environmental regulations and is a sign of concern and good faith on the part of the company. An industrial pollution prevention program should document spills or other discharges, the quality and quantity of accidental releases, site inspections, maintenance activities, and any other information that would enhance the effectiveness of the program. In addition, all records should be retained for at least three years.

21.8.2 Process Modification

Pollution can be prevented in many ways specific to particular processes. Many industrial plants have prevented pollution successfully by modifying production processes. Such modifications include adopting more advanced technology through process variable controls, changing cleaning processes, chemical catalysts, segregating and separating wastes as follows.

Process Variable Controls

Temperature and pressure applications are critical variables as materials are reacted and handled in industrial processes. They can significantly alter the formation of toxins. Improvements include better control mechanisms to meter materials into mixtures; better sensors to measure reactions; more precise methods, such as lasers, to apply heat; and computer assists to automate the activity.

Changing Cleaning Processes

The cleaning of parts, equipment, and storage containers is a significant source of contamination. Toxic deposits are common on equipment walls. The use of solvents to remove such contamination creates two problems: disposal of the contaminants and emissions from the cleaning process itself. Some changes include the use of water-based cleansers vs. toxic solvents, nonstick liners on equipment walls, nitrogen blankets to inhibit oxidation-induced corrosion, and such solvent-minimizing techniques as high-pressure nozzles for water rinse-out.

Chemical Catalysts

Because catalysts facilitate chemical reactions, they welcome pollution prevention research. Better catalysts and better ways to replenish or recycle them would induce more complete reactions and less waste. Substitution of feedstock materials that interact better with existing catalysts can accomplish the same objective.

Coating and Painting

The paints and coatings industry will have to change technologies to accommodate environmental preventive goals. Manufacturers of architectural coatings under increasing environmental regulations will reduce the volatile organic compounds contained in their coatings by displacing oil-based products with water-based coatings. In particular, the paint industry will center its research upon reformulations and increasing the efficiency of coating applications via water-based paints, powder coatings, high-solids enamels, reactive diluents, and radiation curable coatings. For the common source of toxic waste, technical improvements include better spray equipment, such as electrostatic systems and robots, and alternatives to solvents, such as bead blasting.

Segregating and Separating Wastes

A drop of pollutant in a pure solution creates a container of pollution. Segregating wastes and nonwastes reduces the quantity of waste that must be handled. Various technical changes and modifications provide more precise and reliable separation of materials unavoidably mixed together in a waste stream by taking advantage of different characteristics of materials, such as boiling or freezing points, density, and solubility. Separation techniques such as distillation, supercritical extraction, membranes, reverse osmosis, ultrafiltration, electrodialysis, adsorption, separate pollutants or mixed wastes back to their constituent parts (Table 6). Although simple in principle, these processes become high-tech in the precision with which they are applied to facilitate other options in the hierarchy such as recycling, treatment, and disposal.

Support Activities

Garages, motor pools, powerhouses, boilers, and laboratories – all can produce wastes that must be addressed. Their sources of pollution may be significantly reduced through improvement of operation practices.

21.8.3 In-Process Recycling

Materials are processed frequently in the presence of heat, pressure, and/or catalysts, to form products. As materials are reacted, combined, shaped, painted, plated, and polished, excess materials not required for subsequent stages become waste, frequently in combination with toxic

Table 6 Applicable Treatment Technologies for Wastewater Reuse

Treatment technology	Applications
Reverse osmosis	Remove BOD, COD, TSS, TDS, nitrogen, and phosphorus
Electrodialysis	Remove TDS and recover metal salts
Micro/Ultrafiltration	Remove TSS, turbidity, and oil
Ion exchange	Remove TDS and toxic metal ions; reduce hardness
Activated carbon adsorption	Remove many organic and inorganic compounds
Sedimentation	Remove solids that are more dense than water
Filtration	Dewater sludges and slurries; remove TSS from liquid
Evaporation	Treat hazardous wastes, solvent wastes with nonvolatile constituents; separate dissolved and suspended solids
Dewatering	Reduce the moisture content of sludges

solvents used to cleanse the excess from the product. The industry disposes of these wastes either by recycling them into productive reuse or by discharging them as wastes into the air, water, or land. Costly treatment is often required to reduce the toxicity and pollutants in the waste discharge before final disposal. These liquid, solid, or gaseous wastes at each stage of the production process are the source of pollution problems. Onsite recycling of process waste back into the production process will often allow manufacturers to reduce pollution and save costs for less waste treatment and disposal.

For example, solvents are being recycled in many industrial processes. The current goal of solvent recycling is to recover and refine its purity similar to virgin solvent for reuse in the same process, or of sufficient purity to be used in another process application. Recycling activities may be performed either onsite or offsite. Onsite recycling activities include: (a) direct use or reuse of the waste material in a process. It differs from closed-loop recycling, in that wastes are allowed to accumulate before reuse; and (b) reclamation is achieved by recovering secondary materials for a separate end-use or by removing impurities so that the waste may be reused.

Advantages of onsite recycling include:

- less waste leaving the facility;
- control of reclaimed solvent's purity;
- reduced liability and cost of transporting waste offsite;
- reduced reporting (manifesting);
- possible lower unit cost of reclaimed solvent.

Disadvantages of onsite recycling must also be considered, including:

- capital outlay for recycling equipment;
- liabilities for worker health, fires, explosions, leaks, spills, and other risks as a result of improper equipment operation;
- possible need for operator training; and additional operating costs.

Offsite commercial recycling services are well suited for small quantity generators of waste since they do not have sufficient volume of waste solvent to justify onsite recycling. Commercial recycling facilities are privately owned companies that offer a variety of services ranging from operating a waste recycling unit on the generator's property to accepting and recycling batches of solvent waste at a central facility.

21.8.4 Materials and Product Substitutions

The issues involving materials and product substitutions are complex and include economic and consumer preferences as well as technological considerations. Obviously, the use of less toxic materials in production can effectively prevent pollution in a decentralized society. Scientists and engineers are actively evaluating and measuring material toxicity and developing safer materials. Likewise, the life-cycle approach requires that products be designed with an awareness of implication from the raw material stage through final disposal stage. Examples of the product life-cycle applications include substitutes for fast-food packaging, disposal of diapers, plastic containers, and certain drugs and pesticides.

Materials Substitution

Industrial plants could use less hazardous materials and/or more efficient inputs to decrease pollution. Input substitution has been especially successful in material coating processes, with many companies substituting water-based for solvent-based coatings. Water-based coatings decrease volatile organic compound emissions, while conserving energy. Substitutes, however, may take a more exotic form, such as oil derived from the seed of a native African plant, *Vernonia galamensis*, to substitute for traditional solvents in alkyd resin paints.

Product Substitution

Manufacturers could also reduce pollution by redesigning or reformulating endproducts to be less hazardous. For example, chemical products could be produced as pellets instead of powder, decreasing the amount of waste dust lost during packaging. Unbleached paper products could replace bleached alternatives. With uncertain consumer acceptance, redesigning products could be one of the most challenging avenues for preventing pollution in the industrial sector. Moreover, product redesign may require substantial alterations in production technology and inputs, but refined market research and consumer education strategies, such as product labeling, will encourage consumer support.

Changes in endproducts could involve reformulation and a rearrangement of the products' requirements to incorporate environmental considerations. For example, the endproduct could be made from renewable resources, have an energy-efficient manufacturing process, have a long life, and be nontoxic as well as easy to reuse or recycle. In the design of a new product, these environmental considerations could become an integral part of the program of requirements.

In both the redesign of existing products and the design of new products, additional environmental requirements will affect the methods applied and procedures followed. These new environmental criteria will be added to the list of traditional criteria. Environmental criteria for product design include:

- using renewable natural resource materials;
- using recycled materials;
- using fewer toxic solvents or replacing solvents with less toxic replacements;
- reusing scrap and excess material;
- using water-based inks instead of solvent-based ones;
- reducing packaging requirements;
- producing more replaceable component parts;
- minimizing product filter;
- producing more durable products;
- producing goods and packaging that can be reused by consumers; and
- manufacturing recyclable final products.

21.8.5 Materials Separation

In the chemical process industry, separation processes account for a significant portion of investments and energy consumption. For example, distillation of liquids is the dominant separation process in the chemical industry. Pollution preventive technology aims to find methods that provide a sharper separation than distillation, thus reducing the amounts of waste, improving the use of raw materials, and yielding better energy economy. The relationship of various separation technologies to particle size is given in Figure 1 [1]. The figure describes the physical selection parameter with respect to particle size for various separation techniques.

In examining separation equipment for waste reduction, three levels of analysis can be considered. One level of analysis involves minimizing the wastes and emissions that are routinely generated in the operation of the equipment. A second level of analysis seeks to control excursions in operating conditions. The third level of analysis seeks to improve the design efficiency of the separation units. Waste reduction opportunities derived from each of these levels of analysis are presented below.

Distillation columns produce wastes by inefficiently separating materials, through off-normal operation, and by generating sludge in heating equipment. The following solutions to these waste problems have been proposed:

- Increase the reflux ratio, add a section to the column, retray/repack the column, or improve feed distribution to increase column efficiency.
- Insulate or preheat the column feed to reduce the load on the reboiler. A higher reboiler load results in higher temperatures and more sludge generation.
- Reduce the pressure drop in the column, which lowers the load on the reboiler.
- In addition, vacuum distillation reduces reboiler requirements, which reduces sludge formation.
- Changes in tray configurations or tower packing may prevent pollution from distillation processes.
- Another method for preventing pollution from distillation columns involves reboiler redesign.

Supercritical Extraction

Supercritical extraction is essentially a liquid extraction process employing compressed gases instead of solvents under supercritical conditions. The extraction characteristics are based on the solvent properties of the compressed gases or mixtures. Researchers have known about the solvent power of supercritical gases or liquids for more than 100 years, but the first industrial application did not begin until the late 1970s.

From an environmental point of view, the choice of extraction gas is critical, and to date, only the use of carbon dioxide would qualify as an environmentally benign solution. From a chemical engineering point of view, the advantage offered by supercritical extraction is that it combines the positive properties of both gases and liquids, that is, low viscosity with high density, which results in good transport properties and high solvent capacity. In addition, under supercritical conditions, solvent characteristics can be varied over a wide range by means of pressure and temperature changes.

Membranes

Membrane technology offers other new techniques for combining reaction and separation activities when the product molecules are smaller than the reactant molecules. Removal of

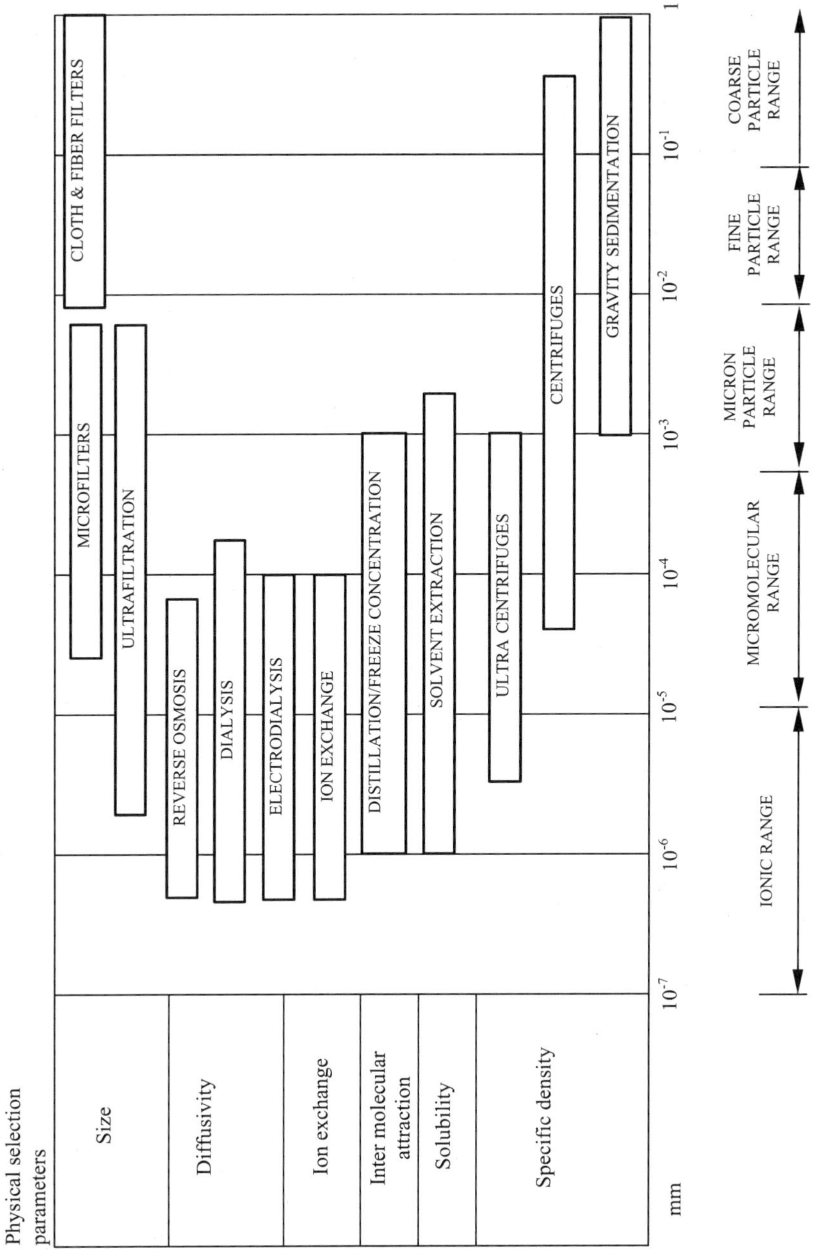

Figure 1 Separation technologies for particles of various sizes.

product also makes membrane reactors advantageous if the product can react with a reactant to form a waste. Membranes are important in modern separation processes, because they work on continuous flows, are easily automated, and can be adapted to work on several physical parameters, such as molecular size, ionic character of compounds, polarity, and hydrophilic/hydrophobic character of components.

Microfiltration, ultrafiltration, and reverse osmosis differ mainly in the size of the molecules and particles that can be separated by the membrane (Table 6). Liquid membrane technology offers a novel membrane separation method in that separation is affected by the solubility of the component to separate into a liquid membrane rather than by its permeation through pores, as is the case in conventional membrane processes, such as ultrafiltration and reverse osmosis. The component to be separated is extracted from the continuous phase to the surface of the liquid membrane, through which it diffuses into the interior liquid phase. Promising results have been reported for a variety of applications, and it is claimed to offer distinct advantages over alternative methods, but liquid membrane extraction is not yet widely available.

Ultrafiltration

Ultrafiltration separates two components of different molecular mass. The size of the membrane pores constitutes the sieve mesh covering a range on the order of 0.002–0.05 microns. The permeability of the membrane to the solvent is generally quite high, which may cause an accumulation of the molecular phase close to the surface of the membrane, resulting in increased filtration resistance, that is, membrane polarization and back diffusion. However, the application of transmembrane feed flow is being used effectively to reduce membrane polarization.

Reverse Osmosis

Reverse osmosis is generally based on the use of membranes that are permeable only to the solvent component, which in most applications is water. The osmotic pressure due to the concentration gradient between the solutions on both sides of the membrane must be counteracted by an external pressure applied on the side of the concentrate in order to create a solvent flux through the membrane. Desalting of water is one area where reverse osmosis is already an established technique. The major field for future work will be increasing the membrane flux and lowering the operating pressure currently required in demineralization and desalination by reverse osmosis.

Electrodialysis

Electrodialysis is used to separate ionic components in an electric field in the presence of semipermeable membranes, permeable only to anions or cations. Applications are demineralization and desalination of brackish water or recuperation of ionic components such as hydrofluoric acid.

Adsorption

Adsorption involves physical and/or chemical interactions between the molecules in gas or liquid and a solid surface. It can be used to remove a pollutant from a gas or liquid stream. Gas adsorption processes, for example, can be used to separate a wide range of materials from process gas streams. Normally adsorption processes are considered for use when the pollutant is fairly dilute in the gas stream. The magnitude of adsorption force, which determines the efficiency, depends on the molecular properties of the solid surface and the surrounding conditions.

Adsorbents may be polar or nonpolar; however, polar sorbents will have a high affinity for water vapor and will be ineffective in gas streams that have any appreciable humidity. Gas streams associated with industrial processes will be humid or even saturated with water vapor. Activated carbon, a non-polar adsorbent, is effective at removing most volatile organic compounds (Table 6). Examples of adsorbents for separating gaseous pollutants include activated carbon, activated alumina, silica gel, molecular sieves, charcoal, and Zeolite.

21.8.6 Solvent Alternative Technologies

The following alternative technologies are being investigated and implemented at the Douglas Aircraft Company (DAC) [1].

High Solids Topcoats

This is an alternative technology to painting aircraft with conventional topcoats, which contain high levels of solvents for sprayability and drying, while it was recently acceptable to simply substitute exempt solvents, such as 1,1,1-trichloroethane, in order to comply with regulations.

Chromium Elimination

This covers a variety of processes, such as painting, sealing, plating, and chemical processing. Chromium has long been a main ingredient of many airframe processes because of its excellent corrosion and wear-resistance properties. In 1990 DAC expanded the use of a thin film sulfuric acid anodize process to include some commercial work, thus reducing the use of the popular chromic acid anodize. Chrome-free aircraft sealers, alternative plating technologies, and nonchromated deoxidizers are also being researched by DAC engineers.

Alkaline/Aqueous Degreasing Technologies

There are cleaning processes that uses solvent vapors alone to effectively remove a variety of contaminants from the workpiece. It is a relatively simple, one step process that provides a clean, dry part ready for subsequent processing. Tests are presently being conducted on various immersion-type cleaners to replace solvent vapor degreasing. Some of the candidates are aqueous cleaners, terpene-based cleaners, and the use of ultrasonic technology with immersion cleaners.

Aqueous cleaners are typically alkaline in nature, their pH being in the range 9–11. Many chemical suppliers already provide alkaline cleaners on the open market. Alkaline/aqueous cleaning is presently the leading contender to replace vapor degreasing, but the implementation of this will require some change in process and equipment, which will require some operator training and/or familiarity.

Alternative Handwipe Solvents Cleaners

These are used extensively for cleanup and repair during the manufacture and assembly of transport aircraft. For example, the solvents used for vapor degreasing may be ozone depletors and/or carcinogens. The Douglas Aircraft Company's engineers are working with their suppliers to develop cleaners that will work effectively at ambient temperatures to remove the common aircraft industry contaminants.

CFC Elimination

Chlorofluorocarbons (CFCs) are used as cleaners of small electronic parts such as printed circuit boards (PCBs), in wire assembly areas, and in many maintenance tasks. They are also used in air conditioners and machine tool chillers, and as propellants in aerosol can applications. There are several lubricants and mold release compounds at DAC that are comprised of CFCs. These are not used in high volume, but do provide an opportunity for reduction of CFC emissions. DAC has begun the process of substituting the propellants used in some of aerosol lubricants.

Resource Recovery and Waste Minimization

This is another broad category that encompasses many technologies: chemical processing, waste disposal, recycling, and housekeeping are just a few. There are many opportunities for improvement under this category as environmental technology continues to advance.

Waste minimization was highlighted during DAC's recent implementation of the Total Quality Management philosophy. This effort enlightened and encouraged every employee to consider his/her impact on the environment and the workplace. An effective example of waste minimization was accomplished by simply reducing the size of their vendor-provided wipe rags. An onsite survey conducted to evaluate the usage of wipe rags discovered that the three-foot square rags were too large for convenient wipe operations. The supplier agreed to provide smaller rags at less cost to DAC, thus reducing both the volume and weight of rag-generated wastes. Because of the wide variety of uses for wipe rags, they are liable to become contaminated with many products including hazardous substances, requiring the disposal of these rags as hazardous waste.

Each of these projects contributes to eliminate the negative impact of manufacturing processes upon the environment. At DAC they are working diligently with their suppliers and subcontractors to develop, test, and implement new alternative technologies. Alternative technologies are becoming increasingly necessary to meet the ever-tightening demands of an aware public when it comes to environmental legislation. It is noteworthy that environmental professionals tend to share technological developments and breakthroughs to lead society to be come a better, cleaner, and healthier place to live.

Chlorinated hydrocarbon solvents and chlorofluorocarbons are used extensively in cleaning operations in the Department of Energy (DOE) defense program, the nuclear weapons complex, the Department of Defense (DOD) weapons refurbishment facilities, and in industry. A *Solvent Utilization Handbook* has been published by their joint task force to provide guidelines for the selection of nontoxic environmentally safe substitute solvents for these operations. The information contained will include cleaning performance, corrosion testing, treatability operations, recycle/recovery techniques, volatile organic compound emissions and control techniques, as well as other information. The *Handbook* will be updated on an annual basis with information on new solvent substitutes that appear in the marketplace. The handbook database is under revision. Toxicological information, handling and disposal, and economics of solvent usage will also be included in the updated handbook.

A series of databases has been developed for implementation of P2. For example, Krewer *et al.* [16] developed software called PoProf for selection of solvent and waste minimization. It was based on two principles: (a) the selected solvents for a given process exhibit good environmental behavior in addition to good performance; and (b) the waste from the process can be minimized.

21.8.7 Pollution Prevention Incentives

Economic, regulatory, and institutional incentives are needed for adopting pollution prevention plans and programs, including:

- Effectiveness of incentive regulatory policy, economic, and legislative efforts is critical in controlling human behavior.
- Assessment of tools for identifying contradictory regulations.
- Comparative evaluation of levels of benefit/cost derived from pollution prevention by industries and societies.
- Internalization and externalization.
- Evaluation of waste/cost accounting systems for small and large industries.
- Comparative economics of disposal, treatment, and pollution prevention with regard to products and processes.
- Effectiveness of taxes and tax credits in affecting pollution generation.
- Use of regulatory approaches to active pollution prevention objectives.
- Effectiveness of technical assistance programs.
- Research stimulation in the private and public sectors.
- Usefulness of grant programs.
- Economic modeling.

21.9 SYSTEMATIC ANALYSIS OF POLLUTION GENERATION AND PREVENTION

A generalized industrial manufacturing plant is illustrated in Figure 2. As shown, mass and energy flow enter the manufacturing processes, which include raw materials, energy (e.g., heat and electricity), water used for manufacture and/or cooling purpose, and air. In order to enhance chemical reactions, catalysts, which may be very expensive (e.g., rhodium, gold, and silver), are added into the reactors. The inflow can include the above substances and energy. Through the manufacture, profitable products are produced, together with byproducts and wastes. Byproducts have their own values only when they are used for adjustable applications; otherwise, they can become wastes.

Wastes can be categorized as harmless or harmful. The former essentially does not have an environmental impact, while the latter is important. Identification of harmful wastes, design of new manufacturing processes, and retrofits of existing plants can be conducted with help of knowledge-based approaches and/or numerical optimization approaches. Conceptual tools have also been used in the development stages of a design. A hierarchical decision procedure described by Douglas is a good example [17].

The knowledge-based system, sometimes called an expert system, is a system of rules based on an area of expert proven knowledge. It also can be used for hierarchical design and review procedures. Computer programs based on the system can simulate human thought processes and can therefore be used to design *cleaner* manufacturing facilities to produce less polluted (or *greener*) products. This system is essentially dependent upon a long-term accumulation of experts' knowledge. It can be used for new plant design as well as retrofit of an old plant. More recently, Halim and Srinivasan [18] developed an intelligent system for qualitative waste minimization analysis. A knowledge-based expert system, called ENVOP Expert was used to identify practical and cost-effective P2 programs. A case study of the hydrodealkylation process was tested with satisfactory results.

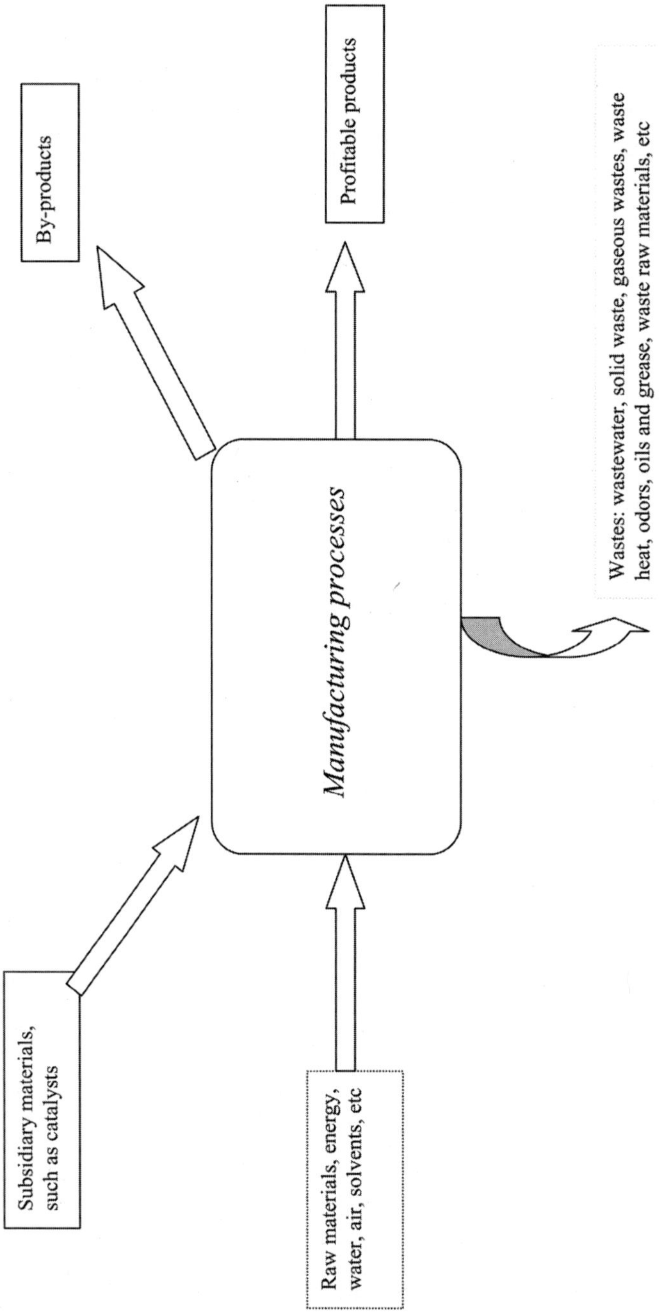

Figure 2 Illustration of manufacture production and subsequent waste generation.

Numerical optimization approaches are based on several considerations, such as energy consumption and mass transfer. Economic analysis together with consumption of both energy and mass have been incorporated by some researchers.

The well-cited pinch analysis (or pinch technology) originally developed based on fundamental thermodynamics has been used to analyze heat flows through industrial processes. It can be used for reduction of energy consumption. It also can be used to minimize wastewater in process industries [19]. Through water reuse and/or other the internal rearrangements in the manufacturing facility, the emission of waste to the environment can therefore be minimized. This approach was used for P2 in a citrus plant [20]. An initial diagnosis indicated that the maximum theoretical freshwater consumption or wastewater generation was reduced by 31%.

Single- and multi-objective optimization approaches have been used in the analysis of pollution prevention. The integration approach has been used in pollution prevention/ wastewater minimization programs [21–24]. The fundamentals of the approach are optimization/minimization of capital and operating costs with minimum waste production and energy consumption. A series of case studies is available in the literature. For example, Parthasarathy and Krishnagopalan [25] used mass integration for the systematic reallocation of aqueous resources in a Kraft pulp mill. An optimal allocation of chloride in different streams throughout the plant was achieved, which led to a built-up concentration below undesirable levels. More importantly, the freshwater requirement was reduced by 57%. For more technical information on P2 and case histories see Chapter 1.

REFERENCES

1. Shen, T.T. *Industrial Pollution Prevention Book*, 2nd Ed.; Springer-Verlag: Germany, 1999.
2. USEPA. *Pollution Prevention 1997 – A National Progress Report*, EPA-742-R-97-000; USEPA: Washington, DC, 1997.
3. Shen, T.T. *New Directions for Environmental Protection*; The Chinese–American Academic and Professional Society Annual Meeting, New York, August, 2002.
4. Hagler Bailly Consulting, Inc. *Introduction to Pollution Prevention*, Training Manual, EPA-742-B-95-003; Hagler Bailly Consulting, Inc.: Arlington, VA, July, 1995.
5. Shen, T.T. *Sustainable Development: Strategy and Technology*. Keynote speech at the Sustainable Development and Emerging Technology Forum sponsored by the United Nations and hosted by the Chinese Ministry of Science and Technology, Beijing, April 2002.
6. USEPA. *Facility Pollution Prevention Guide*, EPA/600/R-92/088; USEPA: Washington, DC, 1992.
7. Ling, J. *Industrial Waste Management*. A speech published in the Vital Speeches of the Day, Vol. LXIV, No. 9, February 18, 1998, 284–288.
8. Overcash, M. The evolution of US pollution prevention, 1976–2001: a unique chemical engineering contribution to the environment – a review. *J. Chem. Technol. Biotechnol.* **2002**, *77*, 1197.
9. Bendavid-Val, A.; Overcash, M.; Kramer, J.; Ganguli, S. *EP3 – Environmental Pollution Prevention Project Paper*; US Agency for International Development, project No 936-5559, Washington, DC, 1992; 71.
10. Das, L.K.; Jain, A.K. Pollution prevention advances in pulp and paper processing, *Environ. Prog.* **2001**, *20*(2), 87.
11. Sarkis, J.; Cordeiro, J.J. An empirical evaluation of environmental efficiencies and firm performance: pollution prevention versus end-of-pipe practice. *Eur. J. Opl Res.* **2001**, *135*, 102.
12. USEPA. *Printed Wiring Board Pollution Prevention and Control Technology: Analysis of Updated Survey Results*, EPA 744-R-95-006; USEPA: Washington, DC, 1995.
13. USEPA. *33/50 Program, The Final Record*, Office of Pollution Prevention and Toxics, EPA-745-R-99-004; USEPA: Washington, DC, March, 1999.
14. Metcalf and Eddy, Inc. *Wastewater Engineering Treatment, Disposal, and Reuse*, 4th Ed. McGraw-Hill, New York, 2002.

15. Evans, J.W.; Hamner, B. Cleaner production at the Asian Development Bank. *J. Cleaner Prod.* **2003**, *11*, 639.
16. Krewer, U.; Liauw, M.A.; Ramakrishna, M.; Babu, M.H.; Raghavan, K.V. Pollution prevention through solvent selection and waste minimization. *Indust. Engng Chem. Res.* **2002**, *41*, 4534.
17. Douglas, J.M. Process synthesis for waste minimization. *Indust. Engng Chem. Res.* **1992**, *31*(1), 238.
18. Halim, I.; Srinivasan, R. Integrated decision support system for waste minimization analysis in chemical processes. *Environ. Sci. Technol.* **2002**, *36*, 1640.
19. Wang, Y.P.; Smith, R. Wastewater minimization. *Chem. Engng Sci.* **1994**, *49*(7), 981.
20. Thevendiraraj, S.; Klemes, J.; Paz, D.; Aso, G.; Cardenas, G.J. Water and wastewater minimization study of a citrus plant *Res. Conserv. Recycling* **2003**, *37*, 227.
21. El-Halwagi, M.M. *Pollution Prevention Through Process Integration: Systematic Design Tools*; Academic Press: San Diego, 1997.
22. Alva-Argaez, A.; Kokossis, A.C.; Smith, R. Wastewater minimization of industrial systems using an integrated approach. *Comput. Chem. Engng* **1998**, *22*, 741.
23. Savelski, M.J.; and Bagajewicz, M.J. On the optimality conditions of water utilization systems in process plants with single contaminants. *Chem. Engng Sci.* **2000**, *55*, 5035.
24. Bagajewicz, M.; Rodera, H.; Savelski, M. Energy efficient water utilization systems in process plants. *Comput. Chem. Engng* **2002**, *26*, 59.
25. Parthasarathy, G.; Krishnagopalan, G. Systematic reallocation of aqueous resources using mass integration in a typical pulp mill. *Adv. Environ. Res.* **2001**, *5*, 61.

22

Treatment of Pesticide Industry Wastes

Joseph M. Wong
Black & Veatch, Concord, California, U.S.A.

22.1 INTRODUCTION

Pesticides are chemical or biological substances intended to control weeds, insects, fungi, rodents, bacteria, and other pests. They protect food crops and livestock, control household pests, promote agricultural productivity, and protect public health. The importance of pesticides to modern society can be summarized by a statement made by Norman E. Borlaug, the 1970 Nobel Peace Prize winner: "Let's get our priorities in perspective. We must feed ourselves and protect ourselves against the health hazards of the world. To do that, we must have agricultural chemicals. Without them, the world population will starve" [1].

However, the widespread use of pesticides has also caused significant environmental pollution problems. Examples of these include the biological concentration of persistent pesticides (e.g., DDT) in food chains and contamination of surface and groundwater used for drinking sources. Because they can affect living organisms, pesticides are highly regulated in the United States to ensure that their use will be safe for humans and the environment. Recently, the National Research Council's Committee on the Future Role of Pesticides in US Agriculture conducted a comprehensive study and concluded that although they can cause environmental problems, chemical pesticides will continue to play a role in pest management for the foreseeable future. In many situations, the benefits of pesticide use are high relative to risks or there are no practical alternatives [2].

This chapter deals with the characterization, environmental regulations, and treatment and disposal of liquid wastes generated from the pesticide industry.

22.2 THE PESTICIDE INDUSTRY

The pesticide industry is an important part of the economy. Worldwide and US pesticide sales in 1990 were expected to reach more than \$20 billion and \$6 billion, respectively (*Chemical Week*, January 3, 1990). Usually the highest usage of pesticides is in agriculture, accounting for about 80% of production [3]. Agricultural pesticide use in the United States averaged 1.2 billion pounds of ingredient in 1997, and was associated with expenditures exceeding \$11.9 billion. This use involved over 20,700 products and more than 890 active ingredients [2]. Household and garden pesticide uses are other significant markets. The United States constituted about 40% of the world market for household pesticides, with annual sales exceeding $1 billion in 2002 [4]. China is the second largest national market with over $580 million of household insecticides purchased each year [5]. The United States also dominates the world market for garden pesticides with sales of over $1.5 billion per year. The United Kingdom is a distant second with sales of $155 million [5].

Pesticides are classified according to the pests they control. Table 1 lists the various pesticides and other classes of chemical compounds not commonly considered pesticides but included among the pesticides as defined by US federal and state laws [1]. The four most widely used types of pesticides are: (a) insecticides, (b) herbicides, (c) fungicides, and (d) rodenticides [6].

The major components of the pesticide industry include manufacturing and formulation/packaging [7]. During manufacture, specific technical grade chemicals are made. Formulating/packaging plants blend these chemicals with other active or inactive ingredients to achieve the endproducts' desired effects, and then package the finished pesticides into marketable containers. A brief overview of these sectors of the industry follows.

Table 1 Pesticide Classes and Their Uses

Pesticide class	Function
Acaricide	Kills mites
Algicide	Kills algae
Avicide	Kills or repels birds
Bactericide	Kills bacteria
Fungicide	Kills fungi
Herbicide	Kills weeds
Insecticide	Kills insects
Larvicide	Kills larvae (usually mosquito)
Miticide	Kills mites
Molluscicide	Kills snails and slugs (may include oysters, clams, mussels)
Nematicide	Kills nematodes
Ovicide	Destroys eggs
Pediculicide	Kills lice (head, body, crab)
Piscicide	Kills fish
Predicide	Kills predators (coyotes, usually)
Rodenticide	Kills rodents
Silvicide	Kills trees and brush
Slimicide	Kills slimes
Termiticide	Kills termites
Chemicals classed as pesticides not bearing the -cide suffix:	
Attractant	Attracts insects
Chemosterilant	Sterilizes insects or pest vertebrates (birds, rodents)
Defoliant	Removes leaves
Desiccant	Speeds drying of plants
Disinfectant	Destroys or inactivates harmful microorganisms
Growth regulator	Stimulates or retards growth of plants or insects
Pheromone	Attracts insects or vertebrates
Repellent	Repels insects, mites and ticks, or pest vertebrates (dogs, rabbits, deer, birds)

Source: Ref. 1

22.2.1 Pesticide Manufacturing

There are more than 100 major pesticide manufacturing plants in the United States. Figure 1 presents the geographical locations of these plants [7]. Specific pesticide manufacturing operations are usually unique and are characteristic only of a given facility.

Almost all pesticides are organic compounds that contain active ingredients for specific applications. Based on 500 individual pesticides of commercial importance and perhaps as many as 34,000 distinct major formulated products, pesticide products can be divided into six major groups [8]:

1. Halogenated organic.
2. Organophosphorus.
3. Organonitrogen.
4. Metallo-organic.
5. Botanical and microbiological.
6. Miscellaneous (not covered in the preceding groups).

Plants that manufacture pesticides with active ingredients use diverse manufacturing processes, including synthesis, separation, recovery, purification, and product finishing such as drying [9].

Chemical synthesis can include chlorination, alkylation, nitration, and many other substitution reactions. Separation processes include filtration, decantation, extraction, and centrifugation. Recovery and purification are used to reclaim solvents or excess reactants as well as to purify intermediates and final products. Evaporation and distillation are common recovery and purification processes. Product finishing may involve blending, dilution, pelletizing, packaging, and canning. Examples of production facilities for three groups of pesticides follow.

Halogenated Aliphatic Acids

Figure 2 shows a simplified process flow diagram for halogenated aliphatic acid production facilities [8]. Halogenated aliphatic acids include chlorinated aliphatic acids and their salts, for example, TCA, Dalapon, and Fenac herbicides. Chlorinated aliphatic acids can be prepared by nitric acid oxidation of chloral (TCA) or by direct chlorination of the acid. The acids can be sold as mono- or dichloro acids, or neutralized to an aqueous solution with caustic soda. The neutralized solution is generally fed to a dryer from which the powdered product is packaged.

As shown on Figure 2, wastewaters potentially produced during the manufacture of halogenated aliphatic acids include the following:

- vent gas scrubber water from the caustic soda scrubber;
- wastewater from the chlorinator (reactor);
- excess mother liquor from the centrifuges;
- process area cleanup wastes;
- scrubber water from dryer units;
- washwater from equipment cleanout.

Nitro Compounds

This family of organonitrogen pesticides includes the nitrophenols and their salts, for example, Dinoseb and the substituted dinitroanilines, trifluralin, and nitralin. Figure 3 shows a typical commercial process for the production of a dinitroaniline herbicide [8]. In this example, a chloroaromatic is charged to a nitrator with cyclic acid and fuming nitric acid. The crude product is then cooled to settle out spent acid, which can be recovered and recycled. Oxides of nitrogen

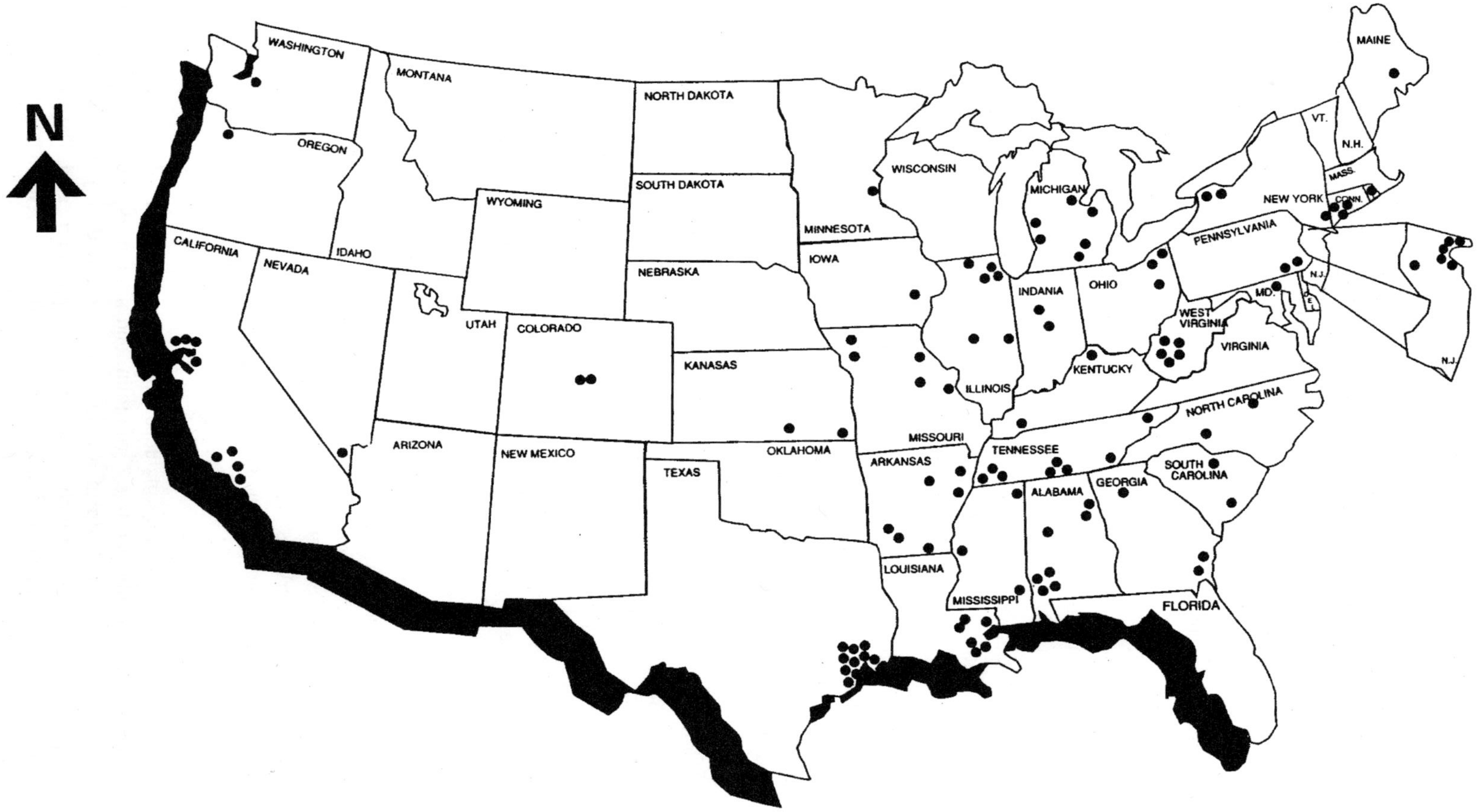

Figure 1 Geographical distribution of major pesticide manufacturers in the United States. Most of the plants are located in the eastern half of the continent (from Ref. 7).

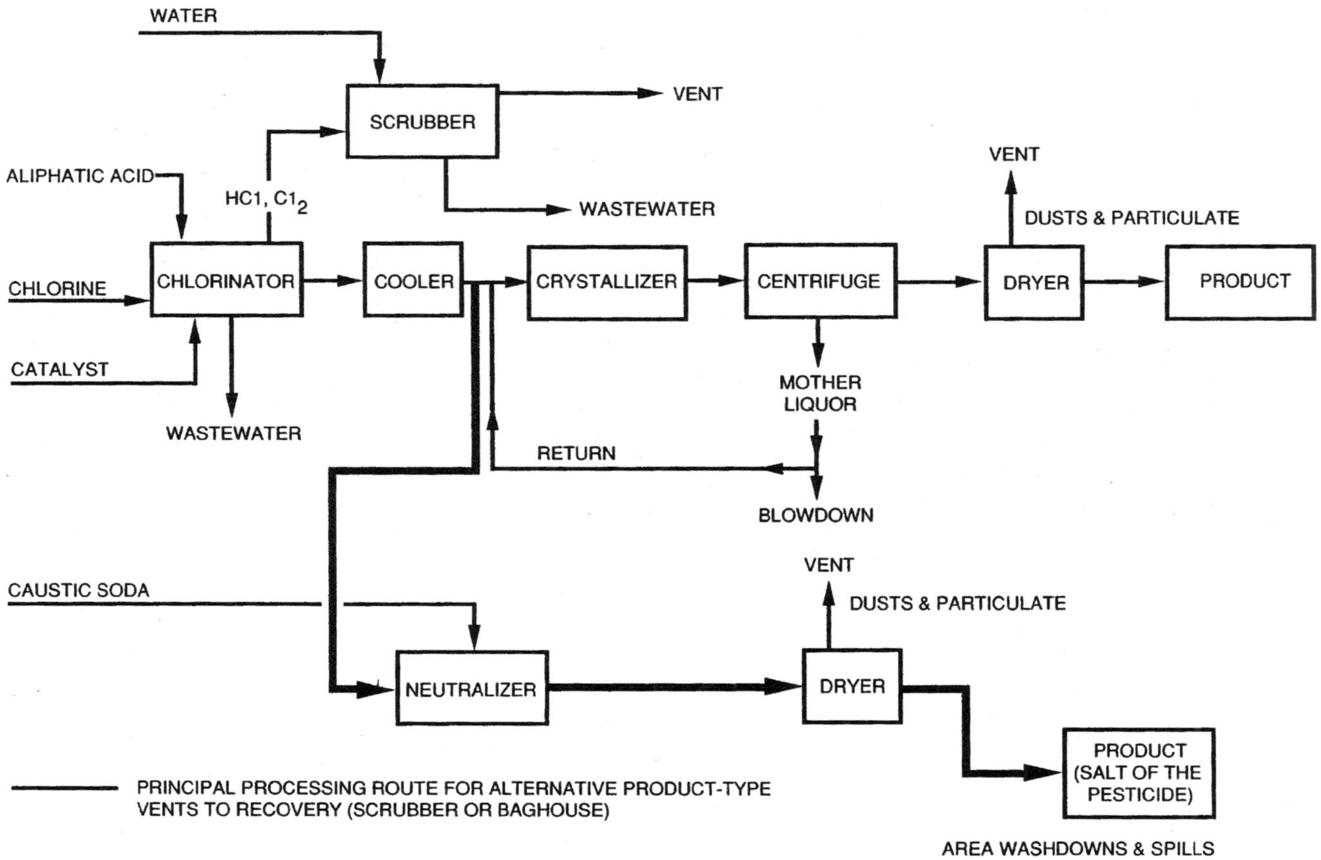

Figure 2 General process flow diagram for halogerated aliphatic acid production facilities. Major processes for pesticide production, including chlorination, cooling, crystallization, centrifying, and drying. The salt of the pesticide is produced by another route (from Ref. 8).

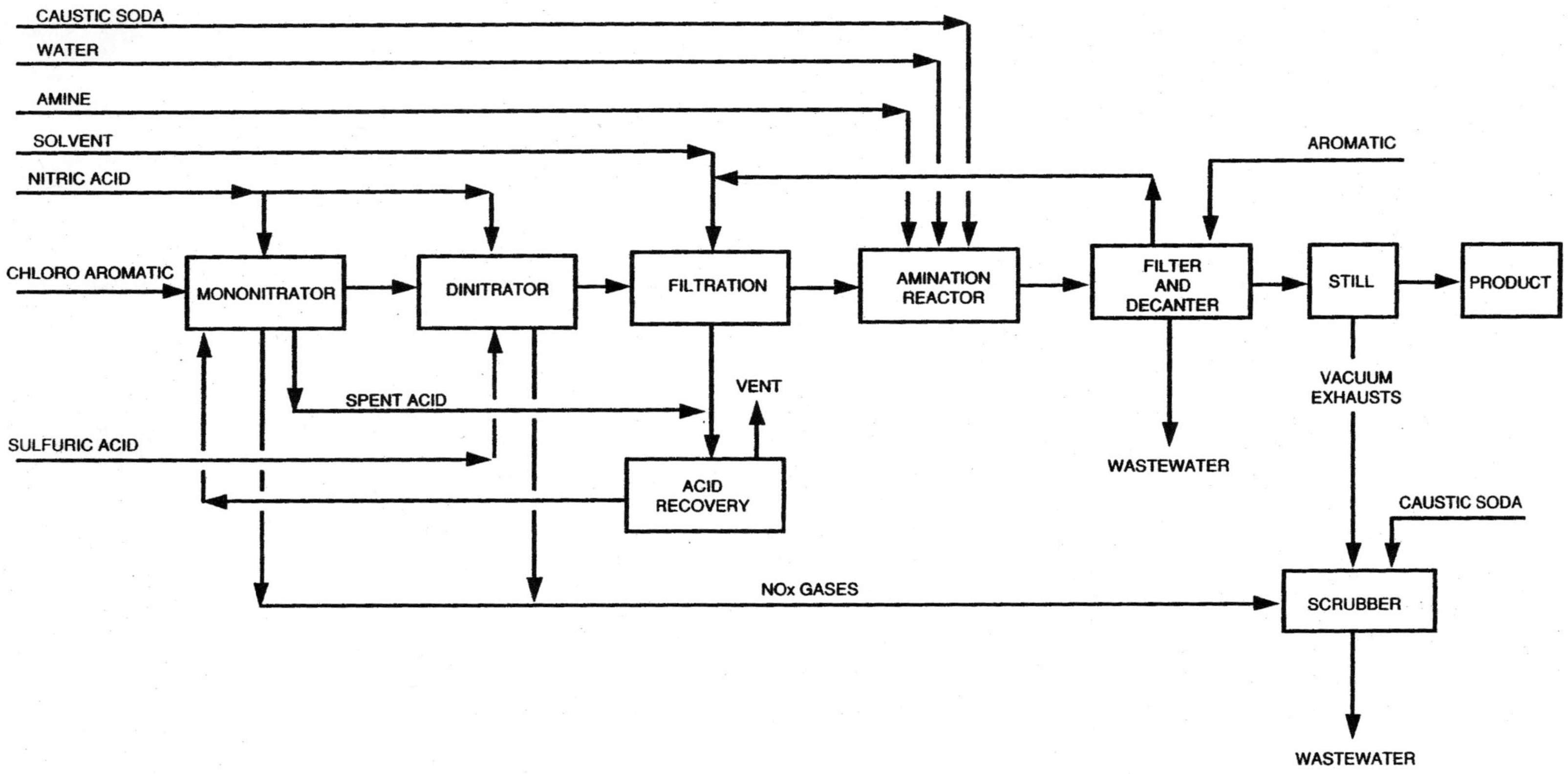

Figure 3 General process flow diagram for nitro-type pesticides. Major processes for pesticide production are mononitration, dinitration, filtration, amination, filtering, and vacuum distillation (from Ref. 8).

are vented and caustic scrubbed. The mononitrated product is then charged continuously to another nitrator containing 100% sulfuric acid and fuming nitric acid at an elevated temperature.

The dinitro product is then cooled and filtered (the spent acid liquor is recoverable), the cake is washed with water, and the resulting washwater is sent to the wastewater treatment plant. The dinitro compound is then dissolved in an appropriate solvent and added to the amination reactor with water and soda ash. An amine is then reacted with the dinitro compound. The crude product is passed through a filter press and decanter and finally vacuum distilled. The saltwater layer from the decanter is discharged for treatment. The solvent fraction can be recycled to the reactor, and vacuum exhausts are caustic scrubbed. Still bottoms are generally incinerated.

Wastewaters potentially generated during the manufacture of the nitro family of pesticides include the following:

- aqueous wastes from the filter and the decanter;
- distillation vacuum exhaust scrubber wastes;
- caustic scrubber wastewaters;
- periodic kettle cleanout wastes;
- production area washdowns.

Metallo-Organic Pesticides

Metallo-organic active ingredients mean organic active ingredients containing one or more metallic atoms, such as arsenic, mercury, copper, and cadmium, in the structure. Figure 4 shows a general process flow diagram for arsenic-type metallo-organic pesticide production [8]. Monosodium acid methanearsenate (MSMA) is the most widely produced organoarsenic herbicide in this group.

The first step of the process is performed in a separate, dedicated building. The drums of arsenic trioxide are opened in an air-evacuated chamber and automatically dumped into 50% caustic soda. A dust collection system is used. The drums are carefully washed with water, the washwater is added to the reaction mixture, and the drums are crushed and sold as scrap metal. The intermediate sodium arsenite is obtained as a 25% solution and is stored in large tanks prior to further reaction. In the next step, the 25% sodium arsenite is treated with methyl chloride to produce the disodium salt DSMA (disodium methanearsenate, hexahydrate). This DSMA can be sold as a herbicide; however, it is more generally converted to MSMA, which has more favorable application properties [8].

To obtain MSMA, the DSMA solution is partially acidified with sulfuric acid and the resulting solution concentrated by evaporation. As the aqueous solution is being concentrated, a mixture of sodium sulfate and sodium chloride precipitates out (about 0.5 kg per 100 kg of active ingredient). These salts are a troublesome disposal problem because they are contaminated with arsenic. The salts are removed by centrifugation, washed in a multistage, countercurrent washing cycle, and then disposed of in an approved landfill.

Methanol, a side product of methyl chloride hydrolysis, can be recovered and reused. In addition, recovered water is recycled. The products are formulated on site as solutions and are shipped in 1 to 30 gallon containers.

Wastewaters that can be generated from the production of these pesticides include the following:

- spillage from drum washing operations;
- washwater from product purification steps;
- scrub water from the vent gas scrubber unit;
- process wastewater;

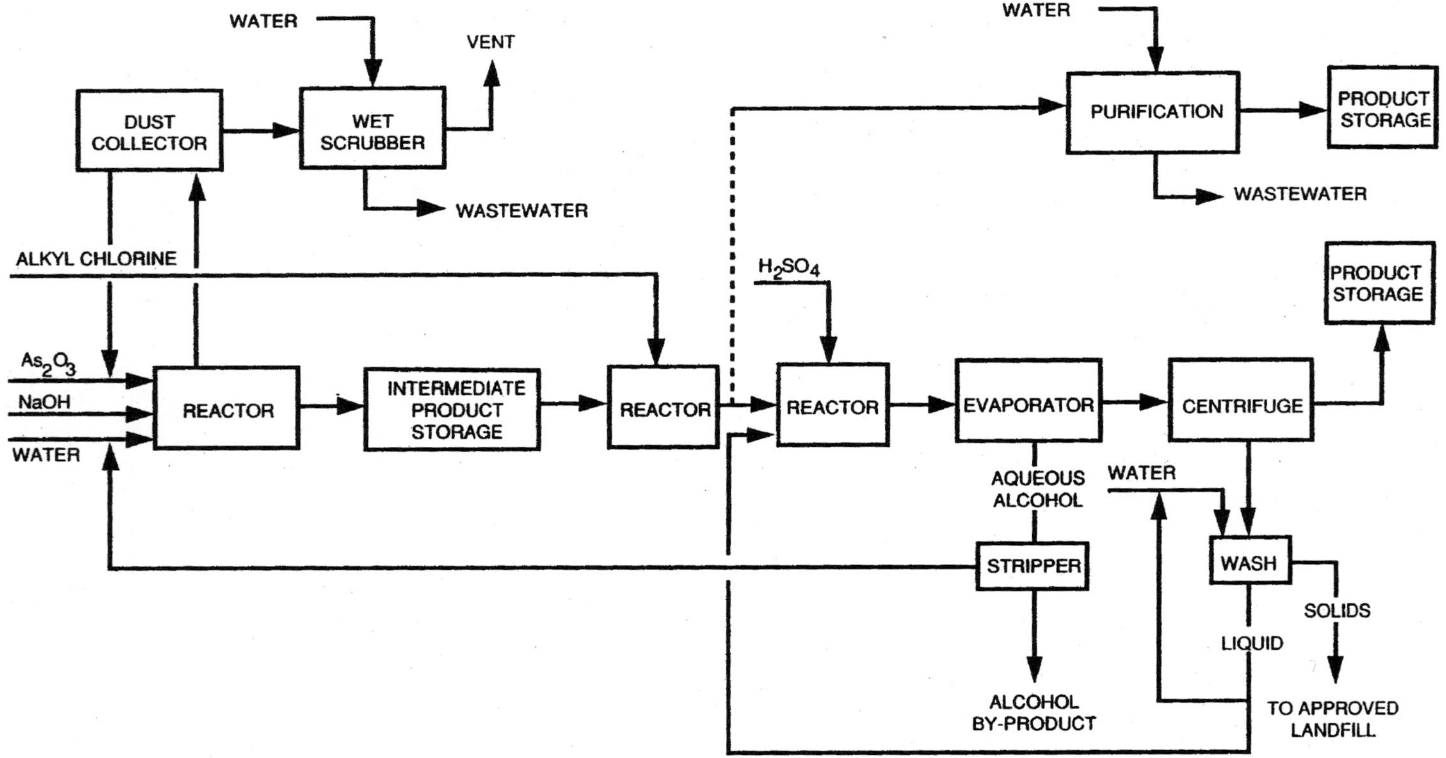

Figure 4 General process flow diagram for arsenic-type metallo-organic production. Sodium arsenate is formed in the first reactor, disodium methanearsenate (DSMA) in the second reactor; DSMA is purified as a product or further changed to monosodium methanearsenate (MSMA) by acidification and purified (from Ref. 8).

- area washdowns;
- equipment cleanout wastes.

22.2.2 Pesticide Formulating/Packaging

After a pesticide is manufactured in its relatively pure form (the technical grade material) the next step is formulation – processing a pesticide compound into liquids, granules, dusts, and powders to improve its properties of storage, handling, application, effectiveness, or safety [9]. The technical grade material may be formulated by its manufacturer or sold to a formulator/packager.

In the United States, there are more than a thousand pesticide formulating/packaging plants covering a broad range of formulations [7]. Many small firms have only one product registration, and produce only a few hundred pounds of formulated pesticides each year. However, USEPA [7] identified one plant operating in the range of 100 million pounds of formulated product per year. The approximate production distribution of formulators/packagers is presented in Table 2 [7].

The most important unit operations involved in formulation are dry mixing and grinding of solids, dissolving solids, and blending [8]. Formulation systems are virtually all batch-mixing operations. The units may be completely enclosed within a building or may be in the open, depending primarily on the geographical location of the plant. Production units representative of the liquid and solid formulation/packaging equipment in use as well as wastewater generation are described in the following.

Liquid Formulation Units

A typical liquid formulation unit is depicted in Fig. 5 [8]. Until it is needed, technical grade pesticide is usually stored in its original shipping container in the warehouse section of the plant. When this material is received in bulk, however, it is transferred to holding tanks for storage.

Batch-mixing tanks are frequently open-top vessels with a standard agitator and may or may not be equipped with a heating/cooling system. When solid technical grade material is used, a melt tank is used before this solid material is added to the mix tank. Solvents are normally stored in bulk tanks and are either metered into the mix tank or are determined by measuring the tank level. Necessary blending agents (emulsifiers, synergists, etc.) are added directly. From the mix tank, the formulated material is frequently pumped to a holding tank before being put into containers for shipment. Before packaging, many liquid formulations must be filtered by conventional cartridge filters or equivalent polishing filters.

Air pollution control equipment used on liquid formulation units typically involves exhaust systems at all potential sources of emission. Storage and holding tanks, mix tanks, and

Table 2 Formulator/Packager Production Distribution

Production (million lb/year)	Formulator/ Packagers (%)
<0.5	24
>0.5 to <5.0	41
>5.0 to <50	35
Total	100

Source: Ref. 7.

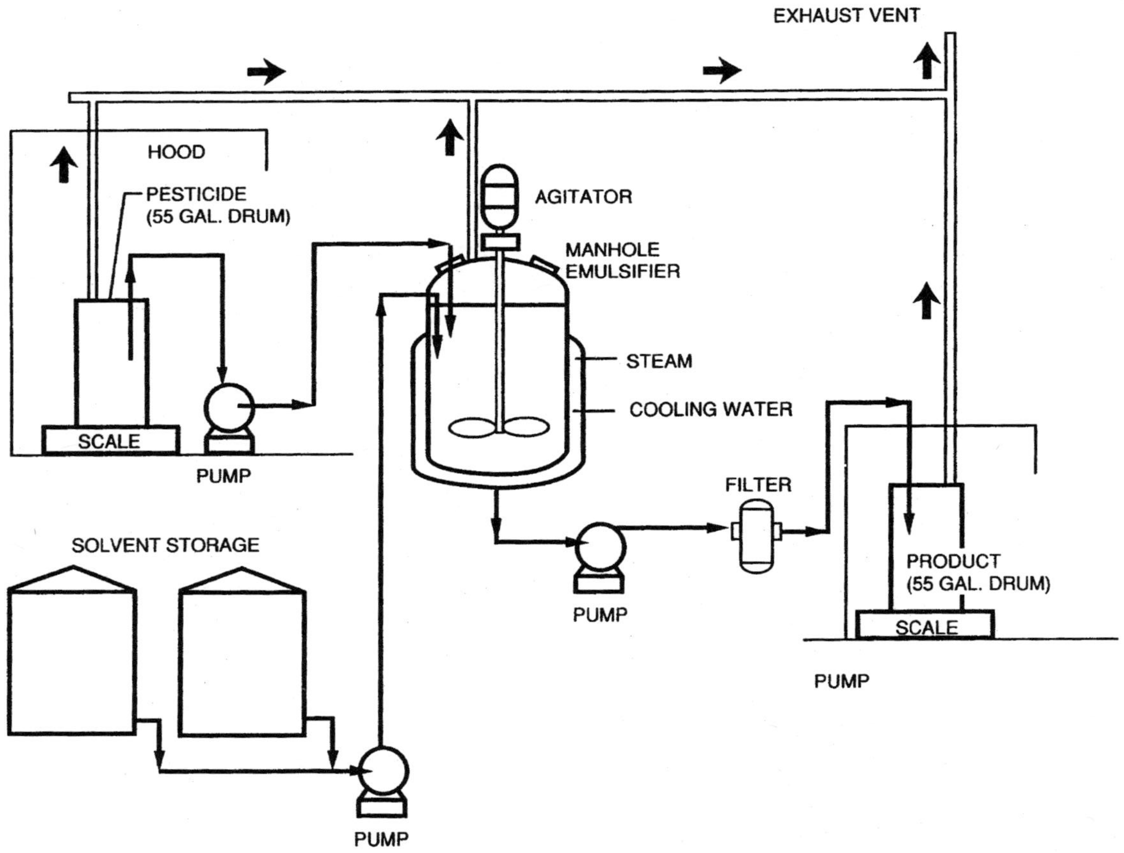

Figure 5 Liquid formulation unit. Technical grade pesticide products are blended with solvents and emulsifiers or other agents in a mix tank. Formulated products are filtered before packaging (from Ref. 8).

container-filling lines are normally provided with an exhaust connection or hood to remove any vapors. The exhaust from the system normally discharges to a scrubber system or to the atmosphere.

Dry Formulation Units

Dry products can include dusts, powders, and granules. Dusts and powders are manufactured by mixing technical grade material with the appropriate inert carrier and grinding the mixture to obtain the correct particle size. Several rotary or ribbon blender-type mixers mix the product. Figure 6 shows a typical dry formulation unit for pesticides [8].

Baghouse systems efficiently control particulate emissions from grinding and blending processes. Vents from feed hoppers, crushers, pulverizers, blenders, mills, and cyclones are typically routed to baghouses for product recovery. This method is preferable to using wet scrubbers. However, even scrubber effluent can be largely eliminated by recirculation.

Granules are formulated in systems similar to the mixing sections of dust plants. The active ingredient is adsorbed onto a sized, granular carrier such as clay or a botanical material. This is accomplished in various capacity mixers that generally resemble cement mixers. If the technical grade material is a liquid, it can be sprayed directly onto the granules. Solid material is usually melted or dissolved in a solvent to provide adequate dispersion on the granules. Screening to remove fines is the last step prior to intermediate storage before packaging.

Packaging and Storage

Packaging the finished pesticide into a marketable container is the last operation conducted at a formulation plant. This operation is usually carried out in conventional filling and packaging units. By moving from one unit to another, the same liquid filling line is frequently used to fill products from several formulation units. Packages of almost every size and type are used, including 1, 2, and 5 gallon cans, 30 and 55 gallon drums, glass bottles, bags, cartons, and plastic jugs.

Aerosol products (for home use) undergo leak testing in a heated water bath to comply with US Department of Transportation regulations. This water bath also serves as a quality control checkpoint for leaks. Bath water must be kept clean for inspection.

Generally, onsite storage is minimized. The storage facility is often a building completely separate from the formulation and filling operation or is at least located in the same building but separate from the formulation units to avoid contamination and other problems. Technical grade material, except for bulk shipments, is usually stored in a special section of the product storage area.

Wastewater Sources

In pesticide formulating/packaging plants, wastewaters can be generated at several sources, including the following [8]:

- formulation equipment cleanup;
- spill washdown;
- drum washing;
- air pollution control devices;
- area runoff;
- laboratory drains.

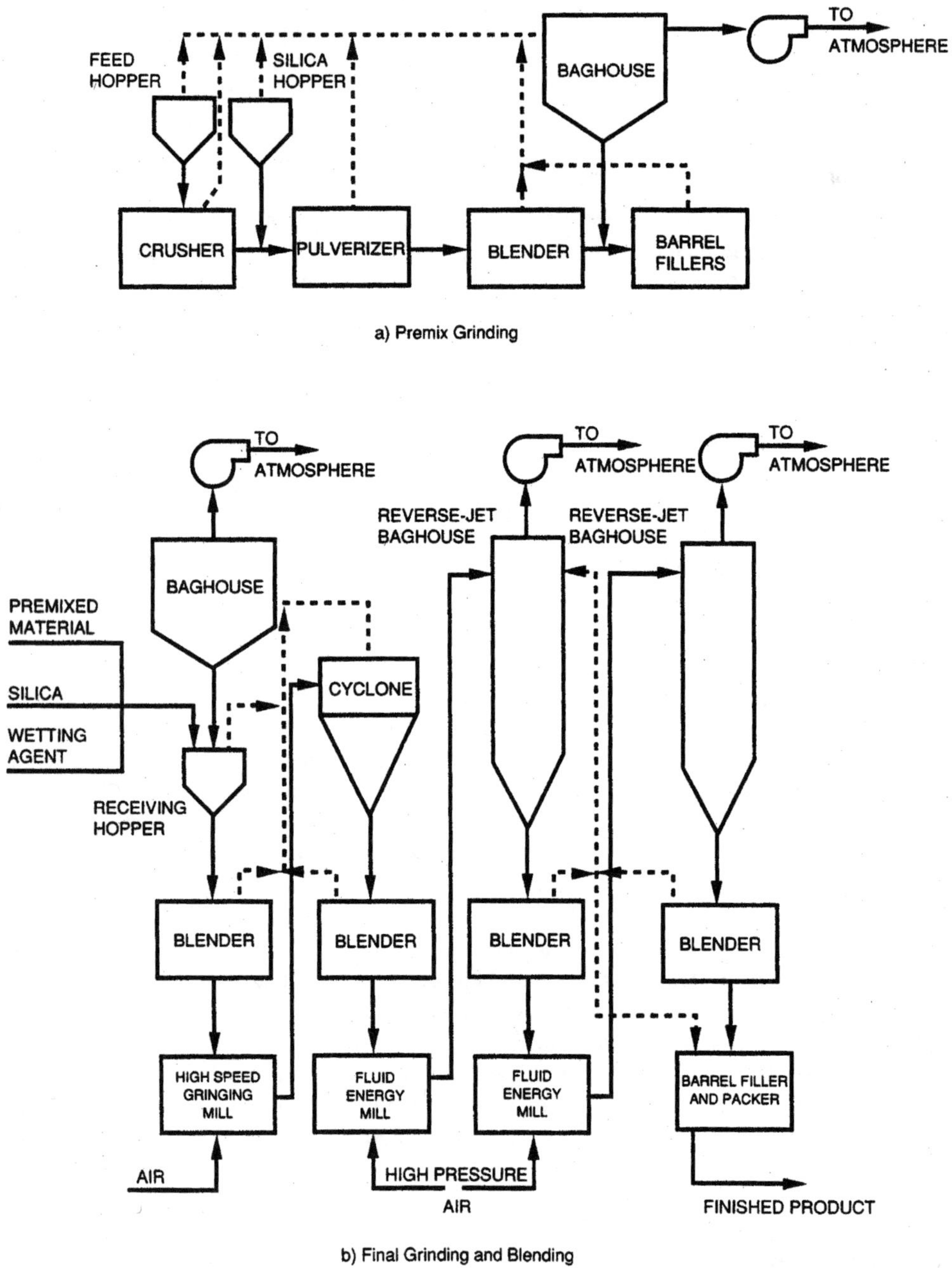

Figure 6 Dry formulation unit. Technical grade products are ground and mixed with appropriate inert materials; the premixed material is further blended with more inert materials and wetting agents in several steps to obtain the correct particle size (from Ref. 8).

The major source of contaminated wastewater from formulation plants is equipment cleanup. Formulation lines, including filling equipment, must be cleaned periodically to prevent cross-contamination of product. Sometimes equipment is washed with formula solvent and rinsed with water. Hence, this waste may contain pesticide ingredients as well as solvents.

For housekeeping purposes, most formulators clean buildings that house formulation units on a routine basis. Prior to washdown, as much dust, dirt, and so on as possible is swept and vacuumed. The washdown wastewater, which generally contains pesticide ingredients, is normally contained within the building and is disposed of in whatever manner is used for other contaminated wastewater.

A few formulation plants process used pesticide drums so they can be sold to a drum reconditioner or reused by the formulator for appropriate products, or simply to decontaminate the drums before they are disposed of. Drum-washing procedures range from a single rinse with a small volume of caustic solution or water to complete decontamination and reconditioning processes. Hence, drum-washing wastewater usually contains caustic solution as well as washed pesticide ingredients in the drums.

Water-scrubbing devices are often used to control emission to the air. Most of these devices generate wastewater streams that are potentially contaminated with pesticide ingredients. Although the quantity of water in the system is high – about 20 gallons per 1000 cfm – water consumption is kept low by a recycle–sludge removal system.

Natural runoff at formulating/packaging plants, if not properly handled, can become a major factor in the operation of wastewater systems simply because of the relatively high flow and because normal plant wastewater volumes are generally extremely low. Isolation of runoff from contaminated process areas or wastewaters, however, eliminates its potential for becoming significantly contaminated with pesticide ingredients. Hence, the content of area runoff depends on the degree of weather protection and area isolation. Modern stormwater pollution prevention regulations in the United States have virtually eliminated this pollution source.

Most of the larger formulation plants have some type of control laboratory on site. Wastewater from the control laboratories relative to the production operations can range from an insignificantly small, slightly contaminated stream to a rather concentrated source of contamination. In many cases, this stream can be discharged into the sanitary sewer. Larger, more highly contaminated streams, however, must be treated along with other contaminated wastewaters.

22.3 WASTE CHARACTERISTICS

Wastewater sources from pesticide manufacturing and formulating/packaging facilities have been described in the previous section. This section discusses wastewater quality and quantity.

22.3.1 Pesticide Manufacturing

Because of the variety and uniqueness of pesticide manufacturing processes and operations, the flow and characteristics of wastewater generated from production plants vary broadly. In 1978, 1979, 1980, 1982, and 1984, the USEPA conducted surveys to obtain basic data concerning manufacturing, disposal, and treatment as well as to identify potential sources of priority pollutants in pesticide manufacturers [7]. The results of these surveys and USEPA's interpretations and evaluations are summarized in the following.

Wastewater Flows

Based on survey results from individual plants, USEPA determined the amount of flow per unit of pesticide production (gal/1000 lb) and the amount of flow (million gallons per day, or MGD) at these plants.

Figure 7 presents a probability plot of the flow ratio (gal/1000 lb) for 269 of the 327 pesticide process areas for which data were available [7]. Significant information in this figure shows that 11% of all pesticide processes have no flow, 50% of all pesticide processes have flows equal to or less than 1000 gal/1000 lb, and 84% have flows equal to or less than 4500 gal/1000 lb.

Figure 8 presents a probability plot of pesticide wastewater flows (MGD) at individual plants [7]. This figure shows that 50% of all plants have flows less than 0.01 MGD, and that virtually all plants (98%) have flows less than 1.0 MGD.

Wastewater Constituents

Because of the nature of pesticides and their components, wastewaters generated from manufacturing plants usually contain toxic (e.g., toxic priority pollutants as defined by USEPA) and conventional pollutants. Based on the results of the surveys and process evaluations, USEPA determined the pollutants or groups of pollutants likely to be present in raw wastewater from these facilities. The agency also selected raw waste loads for these pollutants in order to design treatment and control technologies. The approach taken was to design for the removal of maximum priority pollutant raw waste concentrations as reported in the surveys. Table 3 presents the summary of these raw waste load design levels [7].

The pollutants or groups of pollutants likely to be present in raw wastewater include volatile aromatics, halomethanes, cyanides, haloethers, phenols, polynuclear aromatics, heavy

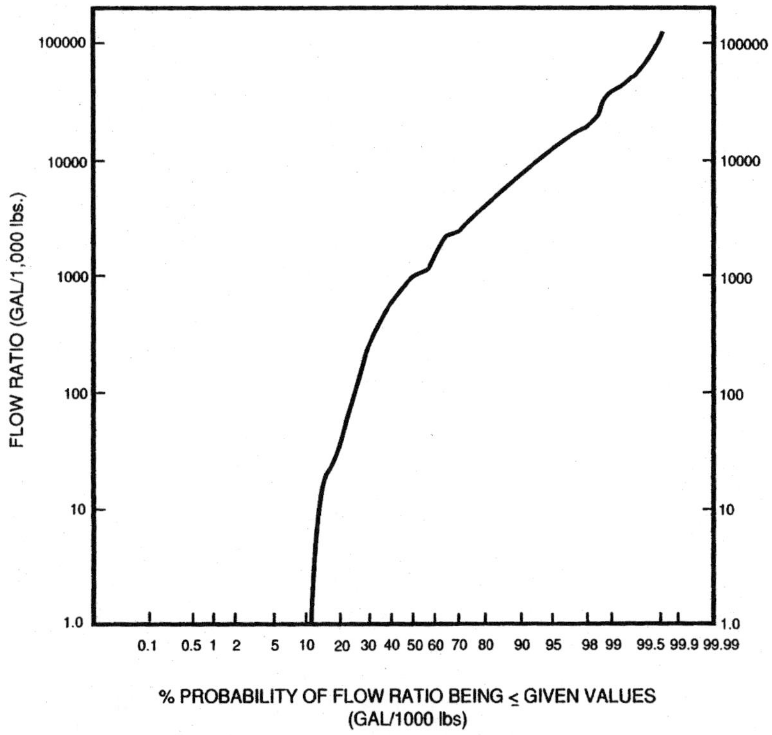

Figure 7 Probability plot of pesticide product flow ratios. Of pesticide production processes, 11% have no flow, 50% have flows less than 1000 gal/1000 lb; 84% have flows less than 4500 gal/1000 lb (from Ref. 7).

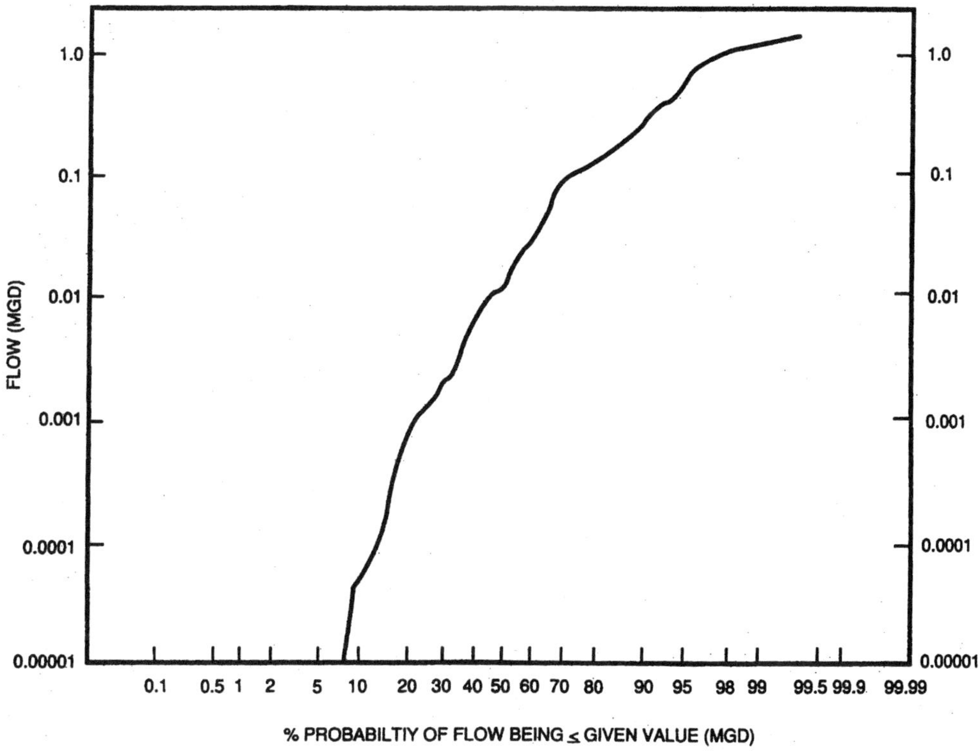

Figure 8 Probability plot of pesticide product wastewater flows. Of pesticide manufacturing plants, 50% have flows less than 0.01 MGD; 98% have flows less than 1.0 MGD (from Ref. 7).

metals, chlorinated ethanes and ethylenes, nitrosamines, phthalates, dichloropropane and dichloropropene, pesticides, dienes, TCDD, and other common constituents such as BOD, COD, and TSS. The sources and significance of these pollutants are briefly discussed [7].

Volatile Aromatics

Benzene and its derivatives are used widely throughout the chemical industry as solvents and raw materials. Mono-, di-, and trichlorobenzenes are used directly as pesticides for their insecticidal and fungicidal properties. Benzene, toluene, and chlorobenzene are used as raw materials in the synthesis of at least 15 pesticides, although their main use is as a carrier solvent in 76 processes. Additional priority pollutant aromatics and chlorinated aromatics exist as impurities or as reaction byproducts because of the reactions of the basic raw materials and solvent compounds.

Halomethanes

Halomethanes, including methylene chloride, chloroform, and carbon tetrachloride (di-, tri-, and tetrachloromethane, respectively), are used mainly as raw materials and solvents in approximately 28 pesticide processes. Bromomethanes can be expected in at least five pesticides as raw materials, byproducts, or impurities and in the case of methyl bromide, can function as a fumigant.

Table 3 Summary of Raw Waste Load Design Levels.

Pollutant group	Design level (mg/L)	Detected pesticide wastewaters at design level[a] (%)
Volatile aromatics	127–293,000	24
Halomethanes	122–2,600	23
Cyanides	5,503	6.0
Haloethers	0.582	17
Phenols	100–42,000	45
Nitro-substituted aromatics	ND[b]	100
Polynuclear aromatics	1.06–1.2	25
Metals		
Copper	4,500	17
Zinc	247	100
Chlorinated ethanes and ethylenes	98–10,000	18
Nitrosamines	1.96	100
Phthalates	ND	100
Dichloropropane and dichloropropene	ND	100
Pesticides	10–11,200	45
Dienes	2,500–15,000	50
TCDD	0.022	100
Miscellaneous	N/A[c]	N/A
PCBs	N/A	N/A
Benzidine	N/A	N/A
BOD	1,470	33
COD	3,886	45
TSS	266	14

[a]Remainder of known pesticide wastewaters are below design level prior to biological oxidation.
[b]ND = not detected.
[c]NA = not applicable.
Source: Ref. 7.

Cyanide

Cyanide is a known or suspected pollutant in approximately 24 pesticide processes. The primary raw materials that favor the generation of cyanides as either byproducts or impurities are cyanamides, cyanates, thiocyanates, and cyanuric chloride. Cyanuric chloride is used exclusively in the manufacture of triazine pesticides.

Haloethers

Five compounds classified as priority pollutants contain an ether moiety and halogen atoms attached to the aryl and alkyl groups. Five pesticides are suspected to contain at least one compound from this class. *Bis*(2-chloroethyl) ether (BCEE) is used as a raw material in two pesticides; BCEE itself functions as a fungicide or bactericide in certain applications. In the other three pesticides, the ethers are suspected to be present as raw material impurities.

Phenols

Phenols are compounds having the hydroxyl (OH) group attached directly to an aromatic ring. Phenols commonly found in pesticide wastewaters include chlorophenols, nitrophenols, and methylphenols (cresols). These compounds may be found throughout the pesticide industry as raw materials, impurities in raw materials, or as byproducts of reactions using related compounds such as chlorobenzenes. The presence of nitrated phenols is expected in six pesticides. Methylated phenols are not expected to be significant because they are not used as raw materials, but they may appear as impurities of reaction from one pesticide because of using 4-methylthio-m-cresol as a raw material.

Polynuclear Aromatics

Seventeen priority pollutant compounds can be classified as polynuclear aromatics (PNA). These compounds consist of two or more benzene rings that share a pair of carbon atoms. They are all derived from coal tar, with naphthalene being the largest constituent. Naphthalene derivatives such as alpha-naphthylamine and alpha-naphthol are used in some pesticide processes; therefore, naphthalene is by far the most prevalent PNA priority pollutant in the industry. Acenaphthene, anthracene, fluorene, fluoranthene, and phenathrene are found as raw material impurities. Acenaphthene is found in one pesticide process as a raw material. The remaining ten PNAs are not suspected to be present in pesticide processes.

Heavy Metals

In the pesticide industry, metals are used principally as catalysts or as raw materials that are incorporated into the active ingredients, for example, metallo-organic pesticides. Priority pollutant metals commonly incorporated into metallo-organic pesticides include arsenic, cadmium, copper, and mercury. For metals not incorporated into the active ingredients, copper is found or suspected in wastewaters from at least eight pesticides, where it is used as a raw material or catalyst; zinc becomes part of the technical grade pesticide in seven processes; and mercury is used as a catalyst in one pesticide process. Nonpriority pollutant metals such as manganese and tin are also used in pesticide processes.

Chlorinated Ethanes and Ethylenes

The chlorinated ethanes and ethylenes are used as solvents, cleaning agents, and intermediates. Vinyl chloride (chloroethylene) is used in the production of plastic polyvinyl chloride (PVC). In the pesticide industry, approximately 23 products are suspected to contain a member of this group of priority pollutants. The main pollutants include 1,2-dichloroethane, which is used as a solvent in seven pesticides and tetrachloroethylene, which is used as a solvent in two pesticides.

Nitrosamines

N-nitrosamines are a group of compounds characterized by a nitroso group (N=O) attached to the nitrogen of an aromatic or aliphatic secondary amine. N-nitrosodi-N-propylamine is a suspected reaction byproduct from the nitrosation of di-N-propylamine. Two pesticides are suspected to contain some form of nitrosamines.

Phthalates

Phthalate esters are used widely as plasticizers in commercial polymers and plastic endproducts such as PVC. One phthalate classified as a priority pollutant is suspected to be present in three pesticide processes. Dimethyl phthalate is known to be a raw material in two products.

Dichloropropane and Dichloropropene

1,3-Dichloropropene is a raw material in one pesticide. 1,3-Dichloropropene and the combined pollutants 1,2-dichloropropane-1, 3-dichloropropene are pesticide products as well as priority pollutants and function as insecticidal fumigants.

Priority Pollutant Pesticides

There are only 18 priority pollutants commonly classified as pesticides. Only two are still in production: heptachlor and chlordane. Aldrin, dieldrin, and endrin aldehyde are suspected as reaction byproducts in the endrin process. Heptachlor epoxide occurs as a reaction byproduct in both chlordane and heptachlor manufacturing. DDD, DDE, and DDT can occur as a reaction byproduct in the manufacture of endosulfan.

Dienes

Four manufactured pesticides and two pesticides currently not manufactured use a priority pollutant diene as a raw material. The basic material for all six pesticides is hexachlorocyclopentadiene (HCCPD). The priority pollutant hexachlorobutadiene is suspected to be present in the pesticide wastewater because it is a byproduct of HCCPD synthesis and is used as a solvent in manufacturing mirex.

TCDD

2,3,7,8-Tetrachlorodibenzo-p-dioxin (TCDD) is believed to be a byproduct in chemical processing generated by a halophenol or chlorobenzene starting material. An intermediate reaction will occur at an elevated temperature (equal to or greater than 160°C), an alkaline condition, or in the presence of a free halogen. The end reaction results in either direct dioxin, intermediate dioxin, or predioxin formation that will ultimately form dibenzo-p-dioxins [10]. TCDD is suspected in wastewaters from pesticide manufacture that uses such raw materials as 2,4,5-trichlorophenol (2,4,5-T) and 1,2,4,5-tetra-chlorobenzene, which are characteristic of TCDD precursors. A TCDD level as high as 111 mg/L has been found in drums of waste from the production of the pesticide 2,4,5-T.

Other Pollutants

The pesticide industry routinely monitors conventional and nonconventional pollutants in manufacturing wastewaters. According to the USEPA surveys [7], chemical oxygen demand (COD) concentrations ranged from 14.0 mg/L to 1,220,000 mg/L; Total organic carbon (TOC) ranged from 53.2 mg/L to 79,800 mg/L; biochemical oxygen demand (BOD) ranged from nondetected to 60,000 mg/L; and total suspended solids (TSS) ranged from 2.0 mg/L to 4090 mg/L. Many other pollutants can be present in pesticide wastewaters that are not unique to this industry, including pollutants such as ammonia, oil and grease, fluoride, and inorganic salts. Nonpriority pollutant pesticides would naturally occur in their manufacturing wastewaters due to imperfect separations.

22.3.2 Pesticide Formulating/Packaging

Washing and cleaning operations provide the principal sources of wastewater in formulating and packaging operations. Because these primary sources are associated with cleanup of spills, leaks, area washdowns, and stormwater runoff, there is apparently no basis from which to correlate the pollutants generated to the product made.

According to USEPA's survey [8] of 71 pesticide formulating/packaging plants, 59 reported no generation of wastewater. For the plants that generated wastewater, neither the rate of production nor the type of product formulated had a direct bearing on the quality or quantity of wastewater generated. The three largest plants of a major pesticide formulator each generated less than 5800 gal/day. Other plants generated from 5 to 1000 gal/day. The average flows generated in formulating/packaging plants were between 50 and 1000 gal/day [11].

The pollutants contained in the wastewaters are expected to be similar to those from manufacturing facilities. Pesticides and solvents are the principal pollutants of concern. Although their volumes are small, the wastewaters from pesticide formulating/packaging plants could be highly contaminated and toxic.

22.4 ENVIRONMENTAL REGULATIONS

Many federal and state regulations govern the registration, manufacture, transportation, sale, use, and disposal of pesticides in the United States. Pesticides are regulated by the USEPA primarily under the Federal Insecticide, Fungicide, and Rodenticide Act (FIFRA) and the Federal Food, Drug, and Cosmetic Act (FDCA). The FIFRA requires pesticides to be registered by USEPA and authorizes the agency to prescribe conditions for their use. The FDCA requires the agency to establish maximum acceptable levels of pesticide residues in foods. The transportation of hazardous pesticides is regulated by the Hazardous Materials Transportation Act (HMTA). In addition, certain states such as California and Florida aggressively enforce their own pesticide laws.

The disposal of pesticides and pesticide wastes is regulated by the Clean Air Act (CAA), the Clean Water Act (CWA), the Resource Conservation and Recovery Act (RCRA), and the Comprehensive Environmental Response, Compensation, and Liability Act (CERCLA). This section deals with the regulations for liquid waste disposal, which is mainly under the CWA. However, when the waste is disposed of as a hazardous waste, it is regulated by the RCRA.

22.4.1 Clean Water Act

The US Congress enacted the Federal Water Pollution Control Act (FWPCA) in 1972. The act was significantly amended in 1977 and has since become known as the CWA. It was again amended by the Water Quality Act of 1987. The CWA applies to all industries that generate wastewater discharges. Some of its provisions are particularly applicable to the pesticide industry.

Effluent Guidelines for Pesticides

Under Section 304 of the CWA, USEPA was required to establish "effluent guidelines" for a number of different industrial categories by specifying the effluent limits that must be met by dischargers in each category. Two types of standards were required for each industry: (a) effluent limitations that require the application of the best practicable control technology (BPT) currently available, and (b) effluent limitations that require application of the best available technology (BAT).

Effluent limitations reflecting BPT currently available for the pesticide manufacturing and formulating industrial category were promulgated by USEPA on April 25, 1978 (43 Federal

Regulation 17,785, 1978). The pesticide industry was divided into three subcategories under the BPT regulations: (a) organic pesticide chemicals manufacturing, (b) metallo-organic pesticide chemical manufacturing, and (c) pesticide chemicals formulating and packaging.

For the first subcategory, the rules limit the number of pounds or kilograms of COD, BOD, TSS, and pesticide chemicals that a plant may discharge during any 1 day or any 30 consecutive days. Table 4 presents the BPT effluent limitations for the organic pesticide chemicals manufacturing subcategory (40 CFR pt. 455). For the second and third subcategories, the regulations permit "no discharges of process wastewater pollutants into navigable waters." The BPT regulations are based on pesticide removal by hydrolysis or adsorption followed by biological treatment [3].

The USEPA issued BAT regulations for the pesticide industry in October 1985 (50 Federal Regulation 40701, 1985). However, four chemical companies and three chemical trade organizations challenged these regulations in *Chemical Specialties Manufacturers Association vs. EPA*, No. 86-8024 (11th Cir. July 25, 1986), modified (11th Cir. August 29, 1986). As a result, the agency voluntarily withdrew its regulations and, on remand by the Eleventh Circuit, agreed to initiate a new round of rule making on the pesticide industry standards (51 Federal Regulation 44,911, 1986). The new regulations were later proposed by USEPA in 1992 [12] and finalized in 1996 (61 FR 57551, No. 6, 1996). All of the updated effluent guidelines and standards for the pesticide manufacturing and formulation industries are included in 40 CFR Part 455 – Pesticide Chemicals.

Pretreatment Standards for Pesticides

Section 306(b) of the CWA requires USEPA to promulgate pretreatment standards applicable to the introduction of wastes from industry and other nondomestic sources into publicly owned treatment works (POTWs). USEPA issued the General Pretreatment Regulations on June 26, 1978, and amended these regulations several times in the following years (40 CFR pt. 403).

The pretreatment standards for existing and new sources for the organic pesticide chemicals manufacturing subcategory were promulgated on September 28, 1993 (58 Federal

Table 4 BPT[b] Effluent Limitations for Organic Pesticide Chemicals Manufacturing Subcategory

Effluent characteristics	Maximum for any 1 day	Average of daily values for 30 consecutive days shall not exceed
COD	13.000	9.0000
BOD_5	7.400	1.6000
TSS	6.100	1.8000
Organic pesticide chemicals	0.010	0.0018
pH	(′)[a]	(′)

Source: 40 CFR 455.22.

[a] (′) Within the range 6.0–9.0.

[b]BPT, best practicable control technology currently available.

Note: For COD, BOD_5, and TSS, metric units: Kilogram/1,000 kg of total organic active ingredients. English units: Pound/1,000 lb of total organic active ingredients. For organic pesticide – metric units: Kilogram/1,000 kg of organic pesticide chemicals. English units: Pound/1,000 lb of organic pesticide chemicals.

Register 50690). The main concern for this subcategory is the discharge of priority pollutants into POTWs. Table 6 in 40 CFR Part 455 listed 24 priority pollutants with maximum daily and maximum monthly discharge limitations. With the exception of cyanide and lead, all the priority pollutants are organic compounds. Presently there are no pretreatment standards for the metallo-organic pesticide chemicals manufacturing subcategory. The pretreatment standard for the pesticide chemicals formulating and packaging subcategory is no discharge of process wastewater pollutants to POTWs (40 CFR Part 455.46).

The general pretreatment regulations prohibit an industry or nondomestic source from introducing pollutants that will pass through or interfere with the operation or performance of POTWs [40 CFR Section 403.5(a)]. In addition, the CWA requires USEPA to establish "categorical" pretreatment standards, which apply to existing or new industrial users in specific categories (40 CFR Section 403.6). The discharge of wastewater from the pesticide industry to POTWs will also be subject to the general discharge prohibitions against "pass through" and "interference" with the POTWs. These pretreatment requirements are usually enforced by POTWs, with approved pretreatment programs. As an example, Table 5 shows the general

Table 5 Industrial Waste Pretreatment Limits for a Publicly Owned Treatment Works

Toxic substance	Maximum allowable concentration (mg/L)
Aldehyde	5.0
Antimony	5.0
Arsenic	1.0
Barium	5.0
Beryllium	1.0
Boron	1.0
Cadmium	0.7
Chlorinated hydrocarbons including, but not limited to, pesticides, herbicides, algaecides	Trace
Chromium, total	1.0
Copper	2.7
Cyanides	1.0
Fluorides	10.0
Formaldehydes	5.0
Lead	0.4
Manganese	0.5
Mercury	0.010
Methyl ethyl ketone and other water-insoluble ketones	5.0
Nickel	2.6
Phenol and derivatives	30.0
Selenium	2.0
Silver	0.7
Sulfides	1.0
Toluene	5.0
Xylene	5.0
Zinc	2.6
pH, su	5.0 to 10.5

Source: City of San Jose (California) Municipal Code, 1988.

industrial effluent limits established by the City of San Jose, CA (San Jose Municipal Code, 1988).

Toxic Pollutant Effluent Standards

Section 307 of the CWA requires USEPA to maintain and publish a list of toxic (priority) pollutants, to establish effluent limitations for the BAT economically achievable for control of such pollutants, and to designate the category or categories of sources to which the effluent standards shall apply [3]. Effluent standards have been promulgated for the following toxic pollutants: aldrin/dieldrin; DDT, DDD, and DDE; endrin; toxaphene; benzidine; and PCBs (40 CFR 129.4). These standards, which may be incorporated into National Pollutant Discharge Elimination System (NPDES) permits, limit or prohibit the discharge of process wastes or other discharge from manufacturing processes into navigable waters. For example, any discharge of aldrin or dieldrin is prohibited for all manufacturers (40 CFR 129.100(b)(3)).

Water Quality-Based Limitations

In the United States, as control of conventional pollutants has been significantly achieved, increased emphasis is being placed on reduction of toxic pollutants. The USEPA has developed a water quality-based approach to achieve water quality where treatment control-based discharge limits have proved to be insufficient [13].

The procedures for establishing effluent limitations for point sources discharging to a water quality-based segment generally involves the use of some type of mathematical model or allocation procedure to apportion the allowable loading of a particular toxicant to each discharge in the segment. These allocations are generally made by the state regulatory agency and reviewed, revised, and approved by the USEPA in accordance with Section 303 of the CWA.

To control the discharge of toxic pollutants in accordance with Section 304(1) of the CWA, state and regional regulatory agencies may also establish general effluent limitations for a particular water body. For example, Table 6 shows the discharge limits for toxic pollutants

Table 6 Effluent Limitations for Selected Toxic Pollutants for Discharge to Surface Waters (All Values in μg/L)

	Daily average	
	Shallow water	Deep water
Arsenic	20	200
Cadmium	10	30
Chromium(VI)	11	110
Copper	20	200
Cyanide	25	25
Lead	5.6	56
Mercury	1	1
Nickel	7.1	71
Silver	2.3	23
Zinc	58	580
Phenols	500	500
PAHs	15	150

Source: Water Quality Control Plan, San Francisco Bay Basin, 1986.

established by the San Francisco Bay Regional Water Quality Control Board in 1986. This regional agency has also adopted biomonitoring and toxicity requirements for municipal and industrial dischargers. Biomonitoring, or whole-effluent toxicity testing, has become a requirement for most discharges in the United States. As of 1988, more than 6000 discharge permits incorporated toxicity limits to protect against acute and chronic toxicity [13] and practically all discharge permits in the United States have toxicity limits as of 2003.

When a discharge exceeds the toxicity limits, the discharger must conduct a toxicity identification/reduction evaluation (TI/RE). A TI/RE is a site-specific investigation of the effluent to identify the causative toxicants that may be eliminated or reduced, or treatment methods that can reduce effluent toxicity.

22.4.2 Resource Conservation and Recovery Act

The Resource Conservation and Recovery Act (RCRA) was enacted in 1976 and was revised substantially by the Hazardous and Solid Waste Amendment (HSWA) of 1984 (40 CFR pts. 260–280). The RCRA regulates the management of "solid wastes" that are "hazardous." The definition of "solid wastes" in these regulations generally encompasses all "discarded" materials (including solid, liquid, semisolid, and contained gaseous materials) and many "secondary materials" (e.g., spent solvents, byproducts) that are recycled or reused rather than discarded [3]. Products such as commercial pesticides are not ordinarily "solid wastes," but they become solid wastes if and when they are discarded or stored, treated, or transported prior to such disposal.

The "solid wastes" that are RCRA hazardous wastes are those either listed in 40 CFR pt. 261, or exhibit one of the four "characteristics" [ignitability, corrosivity, reactivity, and "extraction procedure" (EP) toxicity] identified in Part 261 [a more stringent Toxicity Characteristic Leaching Procedure (TCLP) replaced EP in 1986 (51 Federal Regulation 21,648 1986)]. Both the characteristics and the lists sweep many pesticides and pesticide wastes into the RCRA regulatory program.

The USEPA has developed extensive lists of waste streams (40 CFR Sections 261.31, 261.32) and chemical products (40 CFR Section 261.33) that are considered hazardous wastes if and when disposed of or intended for disposal. The waste streams listed in Sections 261.31 and 261.32 include numerous pesticide manufacturing and formulating process wastes. The lists of commercial chemical products in Section 261.33 include two sublists; both include numerous insecticides, herbicides, and other pesticides. The E List (Table 7) identifies pesticides and other commercial chemicals regulated as "acutely hazardous wastes" when discarded. The F List (Table 8) identifies pesticides that are regulated as toxic (hazardous) wastes when discarded.

Listed pesticides (formulated, manufacturing-use, and off-specification) are regulated as hazardous wastes under the RCRA if they are discarded rather than used for their intended purposes. State listings are often more extensive. Both onsite and offsite disposal options are regulated under the RCRA. Onsite facilities that generate more than 1 kg/month of acutely hazardous wastes in the RCRA E List or 1000 kg/month of any waste as defined in 40 CFR 261.31, 261.32, or 261.33 will require an RCRA hazardous waste permit for treatment or for storage for more than 90 days. Offsite disposal must be handled by an RCRA-permitted facility.

22.5 CONTROL AND TREATMENT FOR PESTICIDE MANUFACTURING WASTES

The management of wastes from pesticide manufacturing plants includes source control, in-plant control/treatment, end-of-pipe treatment, and other control methods for concentrated

Table 7 Pesticide Active Ingredients That Appear on the RCRA Acutely Hazardous Commercial Products List (RCRA E List)

Acrolein	Endrin
Aldicarb	Famphur
Aldrin	Fluoroacetamide
Allyl alcohol	Heptachlor
Aluminum phosphide	Hydrocyanic acid
4-Aminopyridine	Hydrogen cyanide
Arsenic acid	Methomyl
Arsenic pentoxide	Alpha-naphthylthiourea (ANTU)
Arsenic trioxide	Nicotine and salts
Calcium cyanide	Octamethylpyrophosphoramide (OMPA, schradan)
Carbon disulfide	Parathion
p-Chloroaniline	Phenylmercuric acetate (PMA)
Cyanides (soluble cyanide salts)	Phorate
Cyanogen	Potassium cyanide
2-Cyclohexyl-4,6-dinitrophenol	Propargyl alcohol
Dieldrin	Sodium azide
0,0-Diethyl S-[2-ethylthio)ethyl] phosphorodithioate (disulfoton, Di-Syston®)	Sodium cyanide
0,0-Diethyl 0-pyrazinyl phosphorothioate (Zinophos®)	Sodium fluoroacetate
Dimethoate	Strychnine and salts
0,0-Dimethyl 0-p-nitrophenyl phosphorothioate (Methyl parathion)	0,0,0,0-tetraethyl dithiopyrophosphate (sulfotepp)
4,6-Dinitro-o-cresol and salts	Tetraethyl pyrophosphate
4,6-Dinitro-o-cyclohexylphenol	Thallium sulfate
2,4-Dinitrophenol	Thiofanox
Dinoseb	Toxaphene
Endosulfan	Warfarin
Endothall	Zinc phosphide

Note: There are currently no inert pesticide ingredients on the RCRA E List.
Source: 40 CFR 261.33(e).

wastes such as incineration. Source control can reduce the overall pollutant load that must be treated in an end-of-pipe treatment system. In-plant control/treatment reduces or eliminates a particular pollutant before it is diluted in the main wastewater stream, and may provide an opportunity for material recovery. End-of-pipe treatment is the final stage for meeting regulatory discharge requirements and protection of stream water quality. These and other control techniques are discussed in more detail in the following sections.

22.5.1 Source Control

Source control and waste minimization can be extremely effective in reducing the costs for in-plant controls and end-of-pipe treatment, and in some cases can eliminate the need for some treatment units entirely. The first step is to prepare an inventory of the waste sources and continuously monitor those sources for flow rates and contaminants. The next step is to develop in-plant operating and equipment changes to reduce the amount of wastes. The following are some of the techniques available for the pesticides manufacturing facilities.

Table 8 Pesticides and Inert Pesticide Ingredients Contained on the RCRA Toxic Commercial Products List (RCRA F List)

Active ingredients

Acetone
Acrylonitrile
Amitrole
Benzene
Bis(2-ethylhexyl) phthalate
Cacodylic acid
Carbon tetrachloride
Chloral (hydrate)
Chlorodane, technical
Chlorobenzene
4-Chloro-m-cresol
Chloroform
o-Chlorophenol
4-Chloro-o-toluidine hydrachloride
Creosote
Cresylic acid (cresols)
Cyclohexane
Cyclohexanone
Decachlorooctahydro-1,3,4-metheno-2H-cyclobuta[c,d]-pentalen-2-one (Kepone, chlordecone)
1,2-dibromo-3-chloropropane (DBCP)
Dimbutyl phthalate
S-2,3-(Dichloroallyl diisopropylthiocarbamate) (diallate, Avadex)
o-Dichlorobenzene
p-Dichlorobenzene
Dichlorodifluoromethane (Freon 12®)
3,5-Dichloro-N-(1,1-dimethyl-2-propynyl) benzamide (pronamide, Kerb®)
Dichloro diphenyl dichloroethane (DDD)
Dichloro diphenyl trichloroethane (DDT)
Dichloroethyl ether
2,4-Dichlorophenoxyacetic, salts and esters (2,4-D)
1,2-Dichloropropane
1,3-Dichloropropene (Telone)
Diethyl phthalate
Epichlorohydrin (1-chloro-2,3-epoxypropane)
Ethyl acetate
Ethyl 4,4′-dichlorobenzilate (chlorobenzilate)
Ethylene dibromide (EDB)
Ethylene dichloride
Ethylene oxide
Formaldehyde
Furfural
Hexachlorobenzene
Hexachlorocyclopentadiene
Hydrofluoric acid
Isobutyl alcohol
Lead acetate
Lindane
Maleic hydrazide
Mercury
Methyl alcohol (methanol)
Methyl bromide
Methyl chloride
2,2′-Methylenebis (3,4,6-trichlorophenol) (hexachlorophene)
Methylene chloride
Methyl ethyl ketone
4-Methyl-2-pentanone (methyl isobutyl ketone)
Naphthalene
Nitrobenzene
p-Nitrophenol
Pentachloronitrobenzene (PCNB)
Pentachlorophenol
Phenol
Phosphorodithionic acid, 0,0-diethyl, methyl ester
Propylene dichloride
Pyridine
Resorcinol
Safrole
Selenium disulfide
1,2,4,5-Tetrachlorobenzene
1,1,2,2-Tetrachloroethane
2,3,4,6-Tetrachlorophenol
Thiram
Toluene
1,1,1-Trichloroethane
Trichloroethylene
Trichloromonofluoromethane (Freon 11®)
2,4,5-Trichlorophenol
2,4,6-Trichlorophenol
2,4,5-Trichlorophenoxyacetic acid (2,4,5-T)
2,4,5-Trichlorophenoxypropionic acid (Silvex)
Xylene

(*continues*)

Table 8 Continued

Inert ingredients	
Acetone	Formaldehyde
Acetonitrile	Formic acid
Acetophenone	Isobutyl alcohol
Acrylic acid	Maleic anahydride
Aniline	Methyl alcohol (methanol)
Benzene	Methyl ethyl ketone
Chlorobenzene	Methyl methacrylate
Chloroform	Naphthalene
Cyclohexane	Saccharin and salts
Cyclohexanone	Thiourea
Dichlorodifluoromethane (Freon 12®)	Toluene
Diethyl phthalate	1,1,1-Trichloroethane
Dimethylamine	1,1,2-Trichloroethane
Dimethyl phthalate	Trichloromonofluoromethane (Freon 11^R)
1,4-Dioxane	Vinyl chloride
Ethylene oxide	Xylene

Source: 40 CFR 261.33(f).

Waste segregation is an important step in waste reduction. Process wastewaters containing specific pollutants can often be isolated and disposed of or treated separately in a more technically efficient and economical manner. Highly acidic and caustic wastewaters are usually more effectively adjusted for pH prior to being mixed with other wastes. Separate equalization for streams of highly variable characteristics is used by many plants to improve overall treatment efficiency [7].

Wastewater generation can be reduced by general good housekeeping procedures such as substituting dry cleanup methods for water washdowns of equipment and floors. This is especially applicable for situations where liquid or solid materials have been spilled. Flow measuring devices and pH sensors with automatic alarms to detect process upsets are two of many ways to effect reductions in water use. Prompt repair and replacement of faulty equipment can also reduce wastewater losses.

Barometric condenser systems can be a major source of contamination in plant effluents and can cause a particularly difficult problem by producing a high-volume, dilute waste stream [8]. Water reduction can be achieved by replacing barometric condensers with surface condensers. Vacuum pumps can replace steam jet eductors. Reboilers can be used instead of live steam; reactor and floor washwater, surface runoff, scrubber effluents, and vacuum seal water can be reused.

In some cases, wastewater can be substantially reduced by substituting an organic solvent for water in the synthesis and separation steps of the production process, with subsequent solvent recovery. Specific pollutants can be eliminated by requesting specification changes from raw material suppliers in cases when impurities are present and known to be discharged in process wastewaters [7].

Raw material recovery can be achieved through solvent extraction, steam-stripping, and distillation operations. Dilute streams can be concentrated in evaporators and then recovered. Recently, with the advent of membrane technology, reverse osmosis (RO) and ultrafiltration (UF) can be used to recover and concentrate active ingredients [14].

22.5.2 In-Plant Control/Treatment

There are six primary in-plant control methods for removal of priority pollutants and pesticides in pesticide manufacturing plants. These methods include steam-stripping, activated carbon adsorption, chemical oxidation, resin adsorption, hydrolysis, and heavy metals separation. Steam-stripping can remove volatile organic compounds (VOCs); activated carbon can remove semivolatile organic compounds and many pesticides; and resin adsorption, chemical oxidation, and hydrolysis can treat selected pesticides [7]. Heavy metals separation can reduce toxicity to downstream biological treatment systems. Discussion of each of these methods follows.

Steam-Stripping

Steam-stripping is similar to distillation. Steam contacts the wastewater to remove the soluble or sparingly soluble VOCs by driving them into the vapor phase. The steam, which behaves both as a heating medium and a carrier gas, can be supplied as live or reboiled steam. As shown in Fig. 9 [11], a steam-stripping system generally includes an influent storage drum, feed/bottom heat exchangers, pumps, a stripping column (packed column or tray tower), an overhead condenser, an effluent storage drum, and sometimes a reflux drum. Reflux is used to enrich or concentrate the VOCs in the condensate. Enrichment of the condensate could provide higher energy content so that it can be burned for energy recovery [9].

In the pesticide industry, steam-stripping has proven effective for removing groups of priority pollutants such as volatile aromatics, halomethanes, and chloroethanes as well as a variety of nonpriority pollutant compounds such as xylene, hexane, methanol, ethylamine, and ammonia [11]. Thus, this process is used to reduce or remove organic solvents from waste

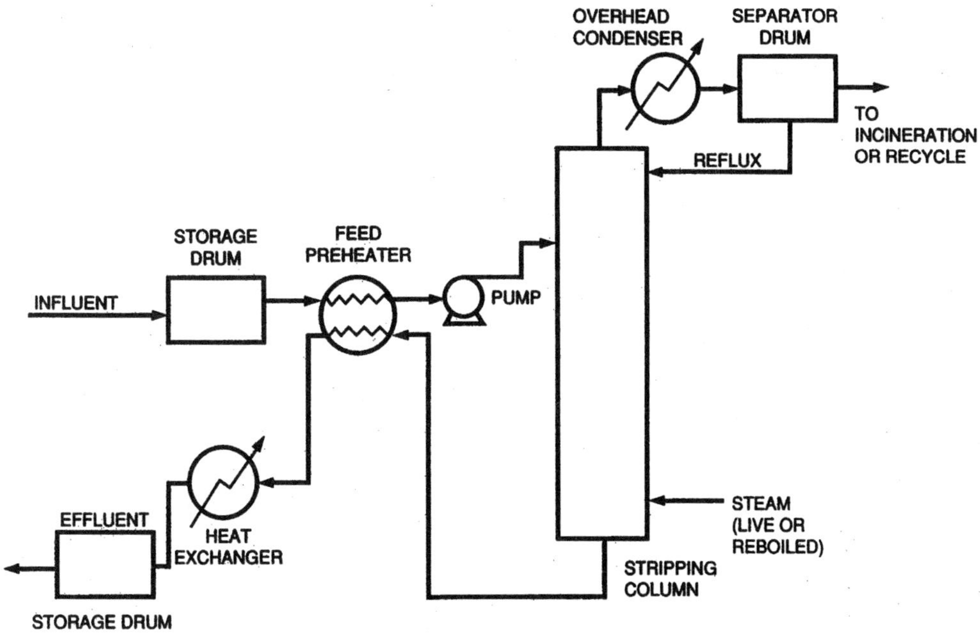

Figure 9 Steam-stripping flow diagram. The influent is heated by the stripper effluent before entering the stripping column near the top; the liquid stream flows downward through the packing, and steam flows upward, carrying volatile compounds; the overhead is condensed and liquid returned to the column; volatile compounds are either recycled or incinerated (from Ref. 11).

streams. A comprehensive study on steam-stripping of organic priority pollutants indicated that effluent concentrations of these pollutants can be reduced to as low as 0.05 mg/L from influent concentrations at their solubility [15]. Pesticides usually have high molecular weights and low volatility and are not effectively removed by steam-stripping.

One variation of steam-stripping is vacuum-stripping, which uses vacuum to create the driving force for pollutant separation. Vacuum strippers normally operate at an absolute pressure of 2 in. of mercury. At least eight pesticide manufacturing plants in the United States use steam-stripping or vacuum-stripping for VOCs removal [7]. The flow rates vary from 0.01 to 0.09 MGD. For example, one pesticide plant uses a steam-stripper to remove methylene chloride from a segregated stream with a flow rate of 0.0165 MGD. The stripper contains 15 ft of packing consisting of 1 in. polypropylene saddles. The steam feed rate is about 1860 lb/hour. Stripped compounds are recycled to the process, thus realizing a net economic savings.

Activated Carbon Adsorption

Activated carbon adsorption is a well-established process for adsorption of organics in wastewater, water, and air streams. Granular activated carbon (GAC) packed in a filter bed or of powdered activated carbon (PAC) added to clarifiers or aeration basins is used for wastewater treatment. In the pesticide industry, GAC is much more widely used than PAC. Figure 10 shows the process flow diagram of a GAC system with two columns in series, which is common in the pesticide industry [11].

Activated carbon studies on widely used herbicides and pesticides have shown that it is successful in reducing the concentration of these toxic compounds to very low levels in wastewater [16]. Some examples of these include BHC, DDT, 2,4-D, toxaphene, dieldrin, aldrin, chlordane, malathion, and parathion. Adsorption is affected by many factors, including

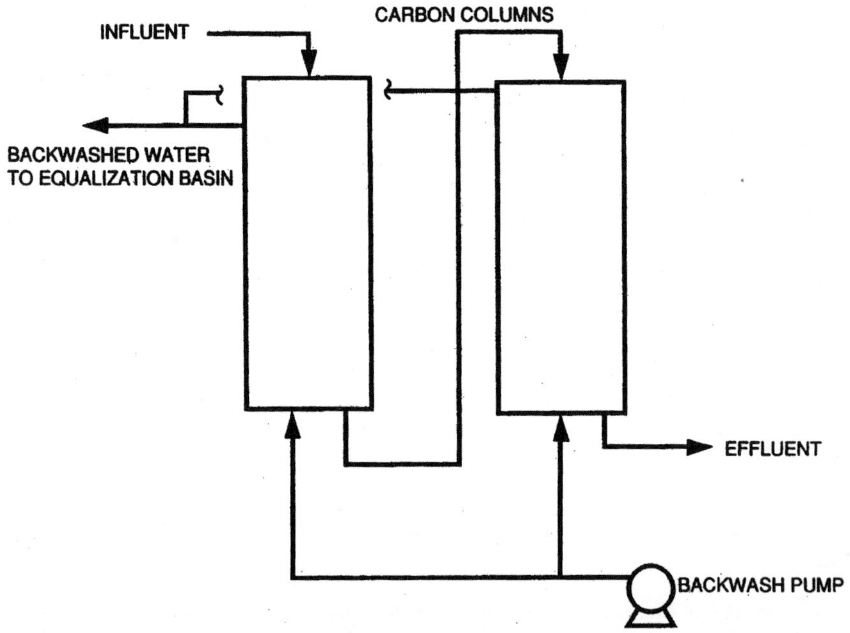

Figure 10 Carbon adsorption flow diagram. The carbon columns are operated in series; backwash water is provided by a pump (from Ref. 11).

molecular size of the adsorbate, solubility of the adsorbate, and pore structure of the carbon. A summary of the characteristics of activated carbon treatment that apply to the pesticide industry follows [11]:

1. Increasing molecular weight is conducive to better adsorption.
2. The degree of adsorption increases as adsorbate solubility decreases.
3. Aromatic compounds tend to be more readily absorbed than aliphatics.
4. Adsorption is pH-dependent; dissolved organics are generally adsorbed more readily at a pH that imparts the least polarity to the molecule.

According to the USEPA surveys, at least 17 pesticide plants in the United States use GAC treatment [7]. Flow rates vary from a low of 0.0004 MGD to a high of 1.26 MGD (combined pesticide flow). Empty bed contact times of the GAC systems vary from a low of 18 minutes to a high of 1000 minutes. The majority of these plants use long contact times and high carbon usage rate systems that are applied as a pretreatment for removing organics from concentrated waste streams. Three plants operate tertiary GAC systems that use shorter contact times and have lower carbon usage rates. Most of the full-scale operating data from the GAC plants indicate a 99% removal of pesticides from the waste streams. The common surface loading rate for primary treatment is 0.5 gallon per minute per square foot (gpm/ft^2) and for tertiary treatment, 4 gpm/ft^2.

Activated carbon adsorption is mainly a waste concentration method. The exhausted carbon must be regenerated or disposed of as hazardous waste. For GAC consumptions larger than 2000 lb/day, onsite regeneration may be economically justified [7]. Thermal regeneration is the most common method for GAC reactivation, although other methods such as washing the exhausted GAC with acid, alkaline, solvent, or steam are sometimes practised for specific applications [17].

Figure 11 shows a typical flow diagram for a thermal regeneration system [11]. Thermal regeneration is conventionally carried out in a multiple hearth furnace or a rotary kiln at

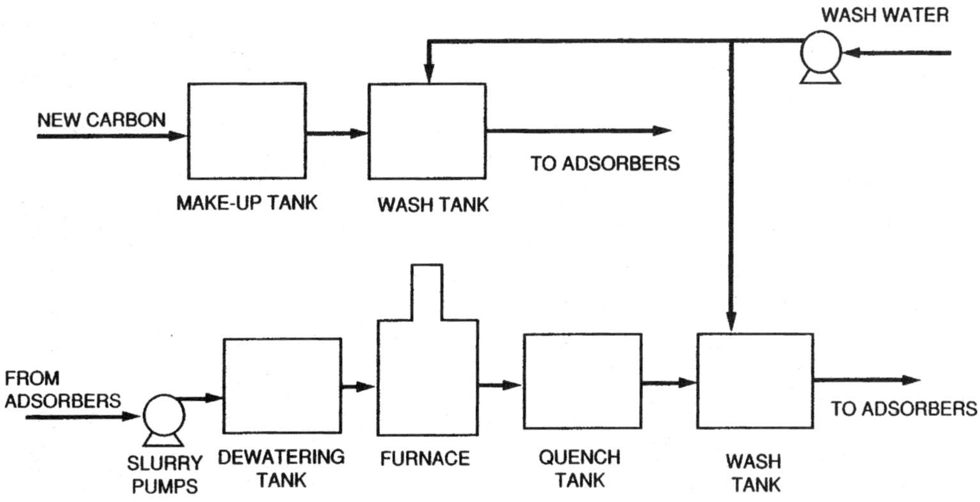

Figure 11 Carbon regeneration flow diagram. Exhausted carbon is sluiced from adsorbers, dewatered, and regenerated in a thermal furnace (multiple hearth, rotary kiln, infrared, or fluidized bed); the regenerated carbon is quenched and washed before returning to the adsorbers; new carbon is washed and added to make up for the loss during regeneration (from Ref. 11).

temperatures from 870 to 980°C. The infrared furnace is a newer type and was installed in a pesticide plant for a GAC system treating mainly aqueous discharge from vacuum filtration of the mother liquor [7]. Infrared furnace manufacturers have claimed ease of operation with quick startup and shutdown capabilities [18]. Another newer type of reactivation process is the fluidized bed process where the GAC progresses downward through the reactivator counterflow to rising hot gases, which carry off volatiles as they dry the spent GAC and pyrolyze the adsorbate. Both the infrared furnace and fluidized bed reactivation processes have been pilot-tested by USEPA in drinking water treatment plants [18].

Other adsorbing materials besides GAC have also been investigated for treating pesticide-containing wastewaters [19]. Kuo and Regan [20] investigated the feasibility of using spent mushroom compost as an adsorption medium for the removal of pesticides including carbaryl, carbofuran, and aldicarb from rinsate. The adsorption of carbamate pesticides on the sorbent exhibited nonlinear behavior that could be characterized by the Freundlich isotherm. Competitive adsorption was observed for pesticide mixtures with adsorption in the order: carbaryl > carbofuran > aldicarb. In another study, Celis and coworkers [21] studied montmorillonites and hydrotalcite as sorbent materials for the ionizable pesticide imazamox. At the pH of the sorbent [6–7], the calcined product of hydrotalcite was found to be the best sorbent for imazamox anion. Sudhakar and Dikshit [22] found that wood charcoal removed up to 95% of endosulfan, an organochlorine insecticide, from water. The sorption followed second-order kinetics with an equilibrium time of 5 hours. In a separate study, pine bark, a wood industry byproduct, was evaluated as an economical adsorbent for tertiary treatment of water contaminated with various organochlorine pesticides [23].

Chemical Oxidation

Oxidizing agents have been shown to be extremely effective for removing many complex organics from wastewater, including phenols, cyanide, selected pesticides such as ureas and uracils, COD, and organo-metallic complexes [11]. Many oxidants can be used in wastewater treatment. Table 9 shows the oxidation potentials for common oxidants [24]. The most widely used oxidants in the

Table 9 Oxidation Potential of Oxidants

Relative oxidation power ($Cl_2 = 1.0$)	Species	Oxidative potential (V)
2.23	Fluorine	3.03
2.06	Hydroxyl radical	2.80
1.78	Atomic oxygen (singlet)	2.42
1.52	Ozone	2.07
1.31	Hydrogen peroxide	1.78
1.25	Perhydroxyl radical	1.70
1.24	Permanganate	1.68
1.17	Hypobromous acid	1.59
1.15	Chlorine dioxide	1.57
1.10	Hypochlorous acid	1.49
1.07	Hypoiodous acid	1.45
1.00	Chlorine	1.36
0.80	Bromine	1.09
0.39	Iodine	0.54

Source: Ref. 24.

pesticide industry are chlorine and hydrogen peroxide (H_2O_2). However, the use of chlorine may create objectionable chlororganics such as chloromethanes and chlorophenols in the wastewater. When organic pollutant concentrations are very high, the use of chemical oxidation may be too expensive because of the high chemical dosages and long retention time required.

At least nine United States pesticide manufacturers use chemical oxidation to treat wastewater [7]. In these systems, more than 98% of cyanide, phenol, and pesticides are removed; COD and other organics are reduced considerably. Some plants use chemical oxidation to reduce toxic compounds from the wastewater to make the streams more suitable for subsequent biological treatment.

Reynolds *et al.* [25] conducted a comprehensive review of aqueous ozonation of five groups of pesticides: chlorinated hydrocarbons, organophosphorus compounds, phenoxyalkyl acid derivatives, organonitrogen compounds, and phenolic compounds. Generally, chlorinated compounds were more resistant to ozonation than the other groups. With the exception of a few pesticides, most of the compounds in the four other groups could achieve complete destruction upon ozonation. The presence of bicarbonate ions could decrease reaction rates by acting as free radical scavengers. Contact times and pH were important parameters. Atrazine destruction by ozonation was evaluated in a bench-scale study in the presence of manganese [26]. Mn-catalyzed ozonation was enhanced in the presence of a small amounts of humic substances (1 mg/L as DOC).

A newer development in chemical oxidation is the combination of ultraviolet (UV) irradiation with H_2O_2 and/or ozone (O_3) oxidation. This combination generates hydroxyl radical, which is a stronger oxidant than ozone or H_2O_2. The UV light also increases the reactivities of the compounds to be oxidized by exciting the electrons of the molecules to higher energy levels [27]. As a result, lower chemical dosages and much higher reaction rates than other oxidation methods can be realized. When adequate chemical dosages and reaction times are provided, pesticides and other organic compounds can be oxidized to carbon dioxide, inorganic salts, and water [28]. Beltran *et al.* [29] evaluated atrazine removal in bubble reactors by treating three surface waters with ozone, ozone in combination with H_2O_2 or UV radiation. Surface water with low alkalinity and high pH resulted in the highest atrazine removal, and ozonation combined with H_2O_2 or UV radiation led to higher atrazine removal and higher intermediates formation as compared to single ozonation or UV radiation.

The UV/O_3 process has been shown to be effective in destroying many pesticides in water [30]. Pilot tests conducted in California on synthetic pesticide wastewaters demonstrated that 15 mg/L each of organic phosphorous, organic chlorine, and carbamate pesticides can be UV-oxidized to nondetectable concentrations [31]. Figure 12 shows a UV/oxidation process flow diagram with the option of feeding both O_3 and H_2O_2. The combination of O_3 and H_2O_2 without UV can also generate the powerful hydroxyl radicals and can result in catalyzed oxidation of organics [32].

The UV/O_3 process was investigated as a pretreatment step to biological treatment by measuring biodegradability (BOD_5/COD), toxicity (ED_{50}), and mineralization efficiency of treated pesticide-containing wastewater [33]. The investigator found that after treatment of an industrial pesticide wastewater by the UV/O_3 process for one hour, COD was reduced by only 6.2% and TOC by merely 2.4%. However, the value of BOD_5/COD increased significantly so that the wastewater was easily biodegradable ($BOD_5/COD > 0.4$) and the toxicity obviously declined (EC_{50} reduction $> 50\%$). The UV light intensity used was 3.0 mW/cm^2 and O_3 supply rate was 400 g/m^3/hour. The investigator concluded that using UV/O_3 as pretreatment for a biological unit is an economical approach to treating industrial wastewaters containing xenobiotic organics as most part of the mineralization work is done by the biological unit rather than photolytic ozonation.

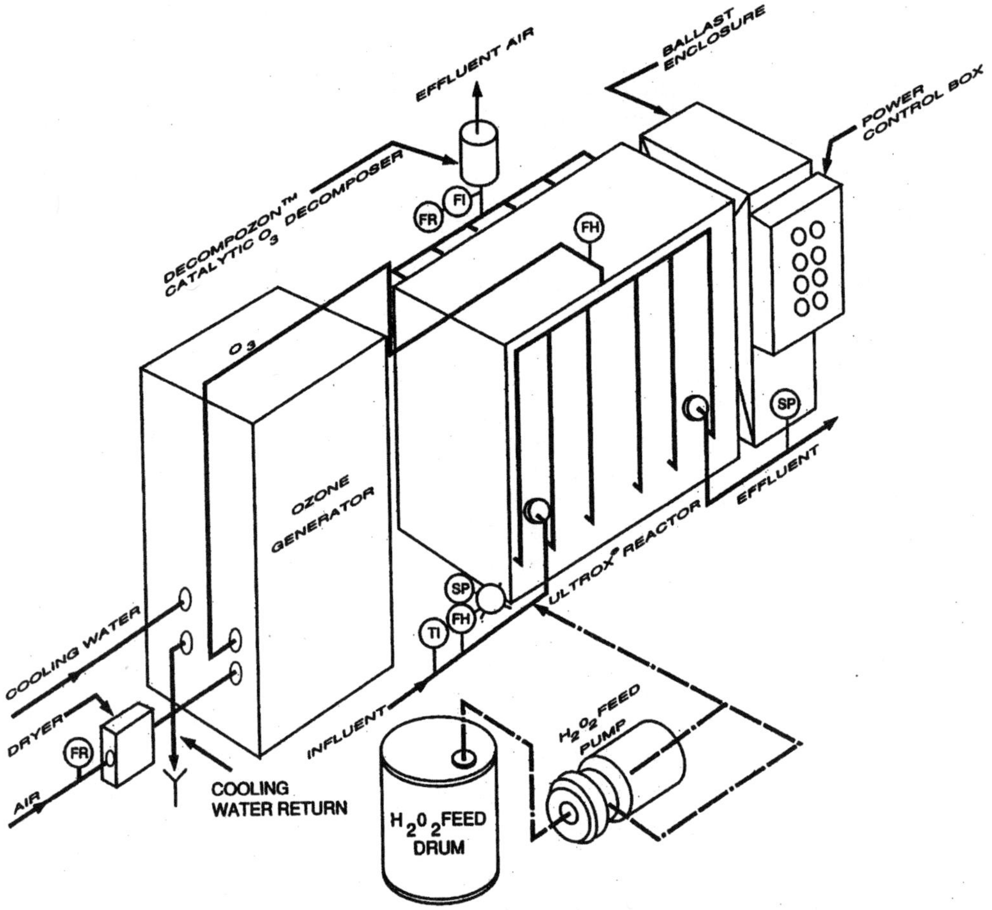

Figure 12 Ultrox® ultraviolet/oxidation process flow schematic. Equipment includes an O_3 generation and feed system and an oxidation reactor mounted with UV lamps inside; H_2O_2 feed is optional. (Courtesy of Ultrox International.)

Balmer and Sulzberger [34] found that the kinetics of atrazine degradation by hydroxyl radicals in photo-Fenton systems were controlled by iron speciation, which further depended upon pH and oxalate concentration. Nguyen and Zahir [35] found that the photodecomposition of the herbicide methyl viologen with UV light was a hemolytic process leading to the formation of methyl pyridinium radicals, which then underwent photolysis at a much faster rate, producing environmentally benign byproducts. In a separate study, Lu [36] investigated the photocatalytic oxidation of the insecticide propoxur, in the presence of TiO_2 supported on activated carbon. Photodegradation of the insecticide followed a pseudo-first-order kinetics described by the Langmuir–Hinshelwood equation. Photocatalytic oxidation of the fungicide metalaxyl in aqueous suspensions containing TiO_2 was explained in terms of the Langmuir–Hinshelwood kinetic model [37].

Resin Adsorption

Adsorption by synthetic polymeric resins is an effective means for removing and recovering specific chemical compounds from wastewater. The operation is similar to that of GAC adsorption. Polymeric adsorption can remove phenols, amines, caprolactam, benzene, chlorobenzenes, and chlorinated pesticides [11]. The adsorption capacity depends on the type and concentration of specific organics in the wastewater as well as pH, temperature, viscosity, polarity, surface tension, and background concentration of other organics and salts. For example, a high salt background will enhance phenol adsorption; increasing the pH will cause the adsorptive capacity to change sharply because the phenolic molecule goes from a neutral, poorly dissociated form at low and neutral pH to an anionic charged dissociated form at high pH [7].

The binding energies of the resin are normally lower than those of activated carbon for the same organic molecules, which permits solvent and chemical regeneration and recovery. Regeneration can be conducted with caustic or formaldehyde or in solvents such as methanol, isopropanol, and acetone. Batch distillation of regenerant solutions can be used to separate and return products to the process.

The USEPA surveys identified four resin adsorption systems in the pesticide industry [7]. Phenol, pesticide, and diene compounds are all effectively removed by these systems. At least one system realized a significant product recovery via regeneration and distillation. The design surface loading rates vary from 1.0 to 4.0 gpm/ft^2 with empty bed contact times of 7.5 to 30 minutes.

Amberlite XAD-4 resin, a synthetic, polymeric adsorbant, was used in one pesticide plant to treat an influent with 1000 mg/L of para-nitrophenol (PNP). With an effluent PNP concentration of 1.0 mg/L, the capacity of the resin was 3.3 lb PNP/cu ft of resin. Kennedy [38] conducted a study regarding the treatment of effluent from a manufacturer of chlorinated pesticides with Amberlite XAD-4 and GAC. Results indicated that the leakage of unadsorbed pesticides from the XAD-4 column was significantly lower than that from the GAC column. An economic analysis indicated that pesticide waste treatment via XAD-4 resin and chemical regeneration would be more economical than GAC adsorption using external thermal regeneration. Chemical regeneration becomes more advantageous because of its feasibility for regenerant recovery and reuse and recycle of adsorbed materials.

Hydrolysis

Hydrolysis is mainly an organic detoxification process. In hydrolysis, a hydroxyl or hydrogen ion attaches itself to some part of the pesticide chemical molecule, either displacing part of the group or breaking a bond, thus forming two or more new compounds. The agents for acid hydrolysis most commonly used are hydrochloric acid and sulfuric acid [11]. Alkaline hydrolysis uses sodium hydroxide most frequently, but the alkaline carbonates are also used. Sometimes high temperature and pressure or catalytic enzymes are required to attain a reasonable reaction time.

Hydrolysis can detoxify a wide range of aliphatic and aromatic organics such as esters, ethers, carbohydrates, sulfonic acids, halogen compounds, phosphates, and nitriles. It can be conducted in simple equipment (in batches in open tanks) or in more complicated equipment (continuous flow in large towers). However, a potential disadvantage is the possibility of forming undesirable reaction products. This possibility must be evaluated in bench- and pilot-scale tests before hydrolysis is implemented.

The primary design parameter to be considered in hydrolysis is the half-life of the original molecule, which is the time required to react 50% of the original compound. The half-life is generally a function of the type of molecule hydrolyzed and the temperature and pH of the reaction. Figure 13 shows the effect of pH and temperature for the degradation of malathion by hydrolysis [11].

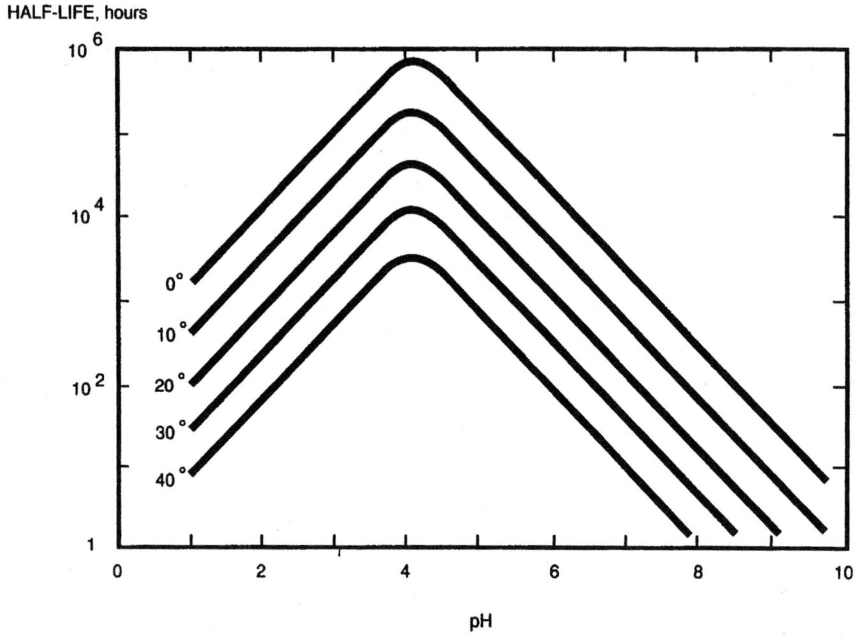

Figure 13 Effect of pH and temperature on malathion degradation by hydrolysis (temperature in degrees C); degradation is faster at higher temperatures and pH values further away from 4.0 to 4.2 (from Ref. 11).

In a study the insecticide carbofuran was hydrolyzed to carbofuran-phenol and monomethyl amine in an anaerobic system. Carbofuran-phenol was resistant to further degradation, while monomethyl amine was further mineralized in the methanogenic culture. Huang and Stone [39] found that the hydrolysis of the secondary amide naptalam, which has a carboxylate side group, was inhibited by dissolved metal ions such as Cu^{2+} and Zn^{2+} and by Al_2O_3 and FeOOH surfaces. In contrast, the hydrolysis of secondary amide propanil and tertiary amide furalaxyl, which lack carboxylate side groups, was unaffected by the presence of Cu^{2+} during the 45-day reaction period. In a separate study, Skadberg *et al.* [40] investigated the stimulation of 2,6-DCP transformation using electric current under varying pH, current, and Cu concentrations. Formation of H_2 at the cathode was found to induce dechlorination with simultaneous removal of Cu.

The USEPA surveys identified nine pesticide plants using full-scale hydrolysis treatment systems [7]. In the industry, a detention time of up to 10 days is used to reduce pesticide levels by more than 99.8%, resulting in typical effluent less than 1 mg/L. The effluents are treated further in biological treatment systems, GAC systems, or chemical oxidation systems, or are discharged to POTWs, if permitted.

Heavy Metals Separation

Metallic ions in soluble form are commonly removed from wastewater by conversion to an insoluble form followed by separation processes such as flocculation, sedimentation, and filtration. Chemicals such as lime, caustic soda, sulfides, and ferrous or ferric compounds have been used for metals separation. Polymer is usually added to aid in flocculation and sedimentation.

For removing low levels of priority metal pollutants from wastewater, using ferric chloride has been shown to be an effective and economical method [41]. The ferric salt forms iron oxyhydroxide, an amorphous precipitate in the wastewater. Pollutants are adsorbed onto and trapped within this precipitate, which is then settled out, leaving a clear effluent. The equipment is identical to that for metal hydroxide precipitation. Trace elements such as arsenic, selenium, chromium, cadmium, and lead can be removed by this method at varying pH values. Alternative methods of metals removal include ion exchange, oxidation or reduction, reverse osmosis, and activated carbon.

At least three pesticide plants use priority pollutant metals separation systems in the United States [7]. One plant uses hydrogen sulfide precipitation to remove copper from its pesticide wastewater. The operating system consists of an agitated precipitator to which the H_2S is added, a soak vessel to which sulfur dioxide is added, a neutralization step using ammonia, and a gravity separation and centrifuging process. Copper is removed from an influent level of 4500 mg/L to 2.2 mg/L.

A second plant uses sodium sulfide for the precipitation of copper from pesticide wastewater. Effluent copper concentration can be lowered to 23 μg/L in this wastewater.

A third plant uses a chemical precipitation step for removing arsenic and zinc from contaminated surface water runoff. Ferric sulfate and lime are alternately added while the wastewater is vacuum-filtered and sludge is contract-hauled. The entire treatment system consists of dual-media filtration, carbon adsorption, ion exchange, chemical precipitation, and vacuum filtration. Sampling results across the entire treatment system indicated that arsenic was reduced from 6.9 to 0.2 mg/L and zinc from 0.34 to 0.11 mg/L.

One caution about metals removal for wastewater with complex organics is that precipitation may be hindered by the formation of soluble metal complexes. Bench- and pilot-scale tests are required for new applications of technology on a particular wastewater stream. Porras-Rodriguez and Talens-Alesson [42] found that flocs resulting from the adsorption of Al^{3+} to lauryl sulfate micelles possessed pollutant-sequestering properties. In studies conducted by these researchers, the pesticide 2,4-D appeared to associate with the micelle-bound Al^{3+} following a Guoy–Chapman–Stern isotherm.

22.5.3 End-of-Pipe Treatment Methods

End-of-pipe treatment methods commonly used in the pesticide industry include equalization, neutralization, biological treatment, and filtration. These methods are discussed as follows.

Equalization

Equalization consists of a wastewater holding vessel or a pond large enough to dampen flow and/or pollutant concentration variation that provides a nearly constant discharge rate and wastewater quality. Capacity is determined by wastewater volume and composition variability. The equalization basin may be agitated or may use a baffle system to prevent short circuiting. Aeration is sometimes needed to prevent septicity. Equalization is used prior to wastewater treatment processes that are sensitive to fluctuation in waste composition or flow, such as biological treatment processes. The recommended detention time for equalization in the pesticide industry is 12 hours prior to pretreatment and 24 hours prior to biological treatment [7].

Neutralization

Neutralization is practised in the pesticide industry to raise or lower the pH of a wastewater stream to meet discharge requirements or to facilitate downstream treatment. Alkaline wastewater may be neutralized with hydrochloric acid, carbon dioxide, sulfur dioxide, and most commonly, sulfuric acid. Acidic wastewater may be neutralized with limestone or lime slurries, soda ash, caustic soda, or anhydrous ammonia. Often a suitable pH can be achieved through mixing acidic and alkaline process wastewaters. Selection of neutralizing agents is based on cost, availability, safety, ease of use, reaction byproducts, reaction rates, and quantities of sludge formed.

In the pesticide industry, neutralization is provided prior to GAC and resin adsorption, pesticide hydrolysis, and biological treatment. The neutralization basin is sized on the basis of an average retention time of 6 minutes and 70 horsepower per million gallons for mixing requirements [7].

Biological Treatment

Biological treatment processes are widely used throughout the pesticide industry to remove organic pollutants measured by parameters such as BOD and COD. Biological treatment involves using microorganisms (bacteria) under controlled conditions to consume organic matter in the wastewater as their food, making this a useful process for removing certain organic materials from the wastewater. Because the process deals with living organisms, every factor influencing the growth and health of the culture must be considered, including an adequate food supply (organic materials), the availability of proper nutrients (phosphorus and nitrogen), a temperate climate, and a nontoxic and relatively uniform environment free of temperature shocks and similar disturbances.

Biological treatment processes are probably the most cost-effective techniques for treating aqueous waste streams containing organic contaminants [43]. Because of the potential presence of toxic materials in pesticide wastewater that may inhibit biological treatment, physical/chemical processes are usually used before biological treatment. Some materials that may inhibit or interfere with biological treatment are heavy metals, cyanides, chlorinated organic compounds, and high salt content ($>$20,000 to 30,000 mg/L).

Many pesticides are complex compounds that may not be easily biodegraded. Some factors that affect biodegradability include [44]:

- *Solubility and availability.* Compounds in emulsified or chelated forms are not readily available to microorganisms and are removed slowly. A prime example is DDT and many of its isomers, which are extremely insoluble in water.
- *Molecular size.* The physical size of complex molecules often limits the approach of enzymes and reduces the rate at which organisms can break down the compound. Many pesticide compounds and their isomers are of large and complex structure, making them resistant to degradation. Examples are some carbamates and carboxylic acid-based compounds.
- *Molecular structure.* In general, aliphatic (straight and cyclic) compounds are more degradable than aromatic compounds. Thus some pesticide compounds and parts of some molecules can be degraded easily while other parts cannot. In some cases, partial degradation will occur, but the pesticidal activity of the waste stream may not be reduced significantly if the toxic components of the compound are bioresistant.
- *Substitutions.* The substitution of elements other than carbon in the molecular chain often make the compound more resistant. Esters and epoxides, salts, and so on are more resistant than the base pesticidal compound.
- *Functional groups.* Halogen substitution to an aromatic compound renders it less degradable. The number of substitutions and the location are important. Chlorophenols are an excellent example of increasing resistance with increasing substitution. Amino and hydroxyl substitutions often increase degradability.

Under proper conditions, biological treatment effectively removes priority pollutants, nonconventional pollutants (TOC), and conventional pollutants (e.g., BOD). The mechanism of pollutant removal may be one or more of the following: (a) biological degradation of the pollutant, (b) adsorption of the pollutant onto sludge, which is separately disposed, or (c) volatilization of the pollutant into the air.

The USEPA [7] surveys identified 31 pesticide plants using biological treatment processes to treat wastewater, including: (a) 14 aerated lagoon systems with detention times ranging from approximately 2 days to 95 days, (b) 13 activated sludge systems with detention times from 7.15 hours to 79 hours, and (c) four trickling filter systems. Biochemical oxygen demand (BOD) removals ranging from 87.4 to 98.8% were achieved at major industry biological treatment systems. Chemical oxygen demand (COD) removals at these plants ranged from 60.5 to 89.7%.

Removal of priority pollutant phenols reached more than 90% when influent levels were high (i.e., 60–1000 mg/L). However, if influent levels were at approximately 1 mg/L or less, the removal ranged from 4.5% or less to 97.6%. Cyanide removals reached more than 50% when influent levels were more than 1 mg/L, and less than 50% for raw wasteloads less than 1 mg/L. Volatile priority pollutants were removed from more than approximately 1 mg/L down to their detection limits of 0.005 to 0.01 mg/L.

Approximately 50% of priority pollutant metals, including copper and zinc, were removed at influent concentrations of 1 mg/L or less. These metals were adsorbed onto sludge because they were not volatile or biodegradable. Priority pollutant dienes were not expected to be biodegraded or volatilized due to their relatively low solubility, but like metals, will adsorb on sludge. Pesticides were removed in biological systems to varying degrees, based on the characteristics of the individual compound. Some plants were achieving removals in excess of 50% at influent concentrations of approximately 1 mg/L.

One concern about biological treatment is the potential toxicity of pesticides, which could inhibit microorganism growth. Results from bench-scale treatability studies performed by one pesticide plant showed that a pesticide in concentrations up to 3000 mg/L did not inhibit aerobic degradation of sewage at typical aerator food-to-microorganism ratios [7].

Chemagro Corporation used an activated sludge system to degrade wastewater with a mixture of organophosphates such as guthion, meta-systox, coumaphos, and fenthion [45]. The system used a first-stage activated sludge process to absorb some of the shock this waste had on the microbes and a second-stage activated sludge process for the ultimate degradation. Atkins [44] reported that using a trickling filter ahead of an activated sludge was successful for the treatment of 2,4-D waste. The trickling filter can handle large pulse loads of toxics because the contact time is short, and thus it is a good pretreatment device for biologically degradable toxic wastes.

A sequential batch reactor (SBR) is a variation of activated sludge using the sequential steps of fill, aerate, settle, and withdrawal. It has become more popular recently because of the advances in instrumentation and automatic control technology. Mangat and Elefsiniotis [46] examined the biodegradation of herbicide 2,4-D by SBRs. After four months of acclimation, they obtained more than 99% removal of 2,4-D at steady-state operation. In addition, they demonstrated that the removal rates of 2,4-D were influenced by the types of supplemental substrates (phenol and dextrose). A sequential batch biofilm reactor (SBBR) incoculated with *Agrobacterium radiobacter* strain J14a was used to treat formulated atrazine rinsate from an agricultural chemical formulation facility [47]. The SBBR reduced 30 mg/L of atrazine to less than 1 mg/L within 12 hours at 22°C. Galluzo and co-workers [48] found that aerobic cometabolic biodegradation of atrazine was responsible for the removal of 30–35% of the pesticide in continuous flow packed columns with humic acid and peptone–tryptone–yeast extract–glucose as the primary substrate.

The use of PAC in combination with activated sludge (PACT®) is a biophysical process that has high potential to improve biological treatment for pesticide wastewater due to the adsorption of toxic compounds by activated carbon. Although the PACT® process has not been widely used on a full-scale basis in the pesticide industry, this trend is expected to change when regulatory agencies adopt more stringent discharge requirements. One particular requirement is whole effluent toxicity reduction, which has become a criterion for many direct discharges in the United States since the late 1980s [13].

Wong and Maroney [49] reported on a pilot plant comparison of PACT® and extended aeration (activated sludge) for treating petroleum refinery wastewater. Results indicated that although both processes performed similarly in COD removal, only the PACT® system yielded an effluent meeting the discharge requirements for whole effluent toxicity reduction. Similar results in toxicity reduction have been reported for wastewaters from other industries [50].

Another important advance in biological treatment is the use of membrane bioreactors (MBR). The MBR uses a UF membrane system immersed in an aeration tank to accomplish both biochemical oxidation and excellent solid/liquid separation, without the need for a separate clarification step [51]. The MBR can offer reliability, compactness, and excellent treated water quality as the MBR effluent is of UF permeate quality. It is particularly attractive in situations when long solids retention times are required, such as for biological degradation of pesticide ingredients. Successful treatment of herbicide-contaminated wastewater in an MBR by sulfate-reducing consortia was reported by Gonzalez-Gonzalez *et al.* [52].

Filtration

Multimedia filtration has been used in the pesticide industry for two purposes: (a) suspended solids removal prior to activated carbon or resin adsorption applied as a pesticide removal pretreatment, and (b) tertiary polishing after biological oxidation and before tertiary activated carbon treatment. With more stringent discharge requirements adopted by regulatory agencies, the increasing use of filtration for tertiary polishing is widespread in every industry, including the pesticide industry. The common design criteria for dual- or multimedia filtration for this industry include a surface loading rate of 4 gpm/ft^2 with a run length of 12 hours [7]. With the advent of membrane technologies, low-pressure membrane processes such as UF and microfiltration (MF) are replacing granular media in certain applications, especially when water reuse is practised. The effluent turbidity from UF/MF system is usually below 0.1 nephelometric turbidity unit (NTU) as these membranes exclude particles larger than 0.01–0.1 micron [53].

22.5.4 Control Methods for Concentrated Wastes

The pesticide industry generates many concentrated wastes that are considered hazardous wastes. These wastes must be detoxified, pretreated, or disposed of safely in approved facilities. Incineration is a common waste destruction method. Deep well injection is a common disposal method. Other technologies such as wet air oxidation, solvent extraction, molten-salt combustion, and microwave plasma destruction have been investigated for pesticide waste applications.

Incineration

Incineration is an established process for virtually complete destruction of organic compounds. It can oxidize solid, liquid, or gaseous combustible wastes to carbon dioxide, water, and ash. In the pesticide industry, thermal incinerators are used to destroy wastes containing compounds such as hydrocarbons (e.g., toluene), chlorinated hydrocarbons (e.g., carbon tetrachloride), sulfonated solvents (e.g., carbon disulfide), and pesticides [7]. More than 99.9% pesticide removal, as well as more than 95% BOD, COD, and TOC removal, can be achieved if sufficient temperature, time, and turbulence are used.

Sulfur- and nitrogen-containing compounds will produce their corresponding oxides and should not be incinerated without considering their effects on air quality. Halogenated hydrocarbons not only may affect air quality but also may corrode the incinerator. Also, organometallic compounds containing cadmium, mercury, and so on, are not recommended for incineration because of the potential for air and solid waste contamination.

Many types of incinerators may be used for thermal destruction of hazardous wastes, including the following basic types [54]:

- multiple hearth;
- fluidized bed;
- liquid injection;
- fume;
- rotary kiln;
- multiple chamber;
- cyclonic;
- auger combustor;
- ship-mounted.

Each of these incinerators has advantages and disadvantages that must be evaluated before final process selection. Figure 14 shows a typical flow diagram of an incineration system incorporating any of these incinerators [11]. Residence times and operating temperature ranges for the various types of incinerators are listed in Table 10 [54]. A matrix matching waste types against incineration equipment is presented in Table 11 [54]. This matrix offers a general guideline for using different types of incinerators for different wastes (e.g., solid, liquid, and fume).

In addition to using the proper type of incinerator and operating conditions to destroy the pesticide wastes, the incineration system must be equipped with the proper emission controls to ensure that toxic gases and particulates do not escape into the environment [55]. The ash (which may contain hazardous substances) must be properly disposed. Many wet collection systems (scrubbers) can be used for removing gaseous pollutants. The various types of scrubbers available include venturi, plate, packed tower, fiber bed, spray tower, centrifugal, moving bed, wet cyclone, self-induced spray, and jet. Dry collection equipment is available for the removal of particulate pollutants and includes settling chambers, baffle chambers, skimming chambers, dry cyclones, impingement collectors, electrostatic precipitators, and fabric filters. The incinerator ash, scrubber water, and particulate collection can then be landfilled, chemically treated, or otherwise processed for disposal.

The USEPA surveys identified at least 14 pesticide plants using incineration for flows ranging up to 39,000 gal/day and heat capacities up to 77 million Btu/hour [7]. Many incinerators are devoted entirely for the destruction of pesticide wastes, but in some cases, only a small part of the capacity is devoted for this purpose.

As an example of incinerator use in the pesticide industry, one plant operates two incinerators to dispose of wastewater from six pesticide products [7]. They are rated at heat release capacities of 35 and 70 million Btu/hour and were designed to dispose of two different wastes. The first primary feed stream consists of approximately 95% organics and 5% water. The second stream consists of approximately 5% organics and 95% water. The energy generated in burning the primary stream is anticipated to vaporize all water in the secondary stream and to oxidize all the organics present. Wastes from two of the six pesticide processes use 0.55% and 4.68% of the incinerator capacity, respectively. The volume of the combined pesticide

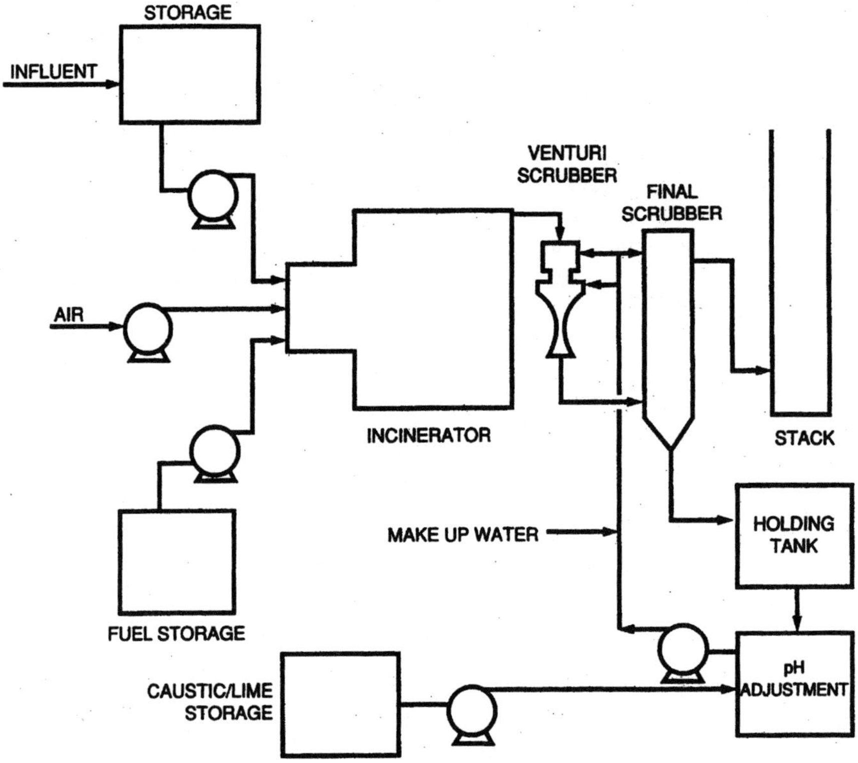

Figure 14 Incineration system flow diagram. Waste is incinerated in the presence of air and supplemental fuel; the incinerator can be multiple hearth, fluidized bed, liquid injection, rotary kiln, or other types; caustic or lime scrubbers are used to remove gaseous pollutants from exhaust gases (from Ref. 11).

wastewater incinerated is 0.0074 MGD. The scrubber effluent is discharged from the tertiary treatment system at a rate of 0.992 MGD.

Since 1974, the USEPA has conducted many incineration tests for pesticide destruction. Most pesticides tested were capable of being destroyed to an efficiency of more than 99.99%. The only exception was Mirex, with 98–99% destruction. However, investigators felt that destruction could be improved to the 99.99% level with a somewhat more effective incinerator design. Incineration has become very controversial in recent years because of the potential to generate dioxin under high temperature conditions.

Deep Well Disposal

Deep well disposal involves injecting liquid wastes into a porous subsurface stratum that contains noncommercial brines [57]. The wastewaters are stored in sealed subsurface strata isolated from groundwater or mineral resources. Disposal wells may vary in depth from a few hundred feet (100 m) to 15,000 ft (4570 m), with capacities ranging from less than 10 to more than 2000 gpm. The disposal system consists of the well with high-pressure injection pumps and pretreatment equipment necessary to prepare the waste for suitable disposal into the well.

Table 10 Operating Parameters for Incinerators

Incinerator type	Temperature range (°F)	Residence time
Multiple hearth	1400–1800	0.25–1.5 hour
Fluidized bed	1400–1800	Seconds–hours
Liquid injection	1800–3000	0.1–2 s
Fume	1400–3000	0.1–2 s
Rotary kiln	1500–3000	Liquids and gases: seconds Solids: hours
Multiple chamber	1000–1800	Liquids and gases: seconds Solids: minutes
Cyclonic	1800–3000	0.1–2 s
Auger combustor	1400–1800	Seconds–hours
Ship-mounted	1800–3000	0.1–2 s

Source: Ref. 54.

In the United States, injection wells are classified into three categories: Class 1 wells are used to inject hazardous wastes; Class 2 wells are used to inject fluids brought to the surface in connection with the production of oil and gas or for disposal of salt water; and Class 3 covers solution mining wells [58]. Class 1 wells are heavily regulated by the USEPA and state agencies because of the potential for groundwater contamination.

The USEPA surveys identified 17 pesticide plants using deep well injection for the disposal of wastewater [7]. One plant used incinerators to remove pesticides as well as benzene and toluene from the wastewater before disposal by deep well injection. Using deep well injection to dispose of hazardous wastes is expected to decrease in the future because of more stringent regulatory requirements and increased concerns about the long-term fate of these wastes in the injection zone.

Other Technologies

Other potential technologies that can be applied to the treatment of concentrated wastes from pesticide manufacturing include wet air oxidation, solvent extraction, molten salt combustion, and microwave plasma destruction.

Table 11 Matrix of Incinerator Application

	Waste					
Incinerator	Solid	Sludge	Slurry	Liquid	Fume	Containerized
Multiple hearth	✓	✓	✓	✓		
Fluidized bed	✓	✓	✓	✓	✓	
Liquid injection			✓	✓	✓	
Fume					✓	
Rotary kiln	✓	✓	✓	✓	✓	✓
Multiple chamber	✓	✓	✓	✓	✓	✓
Cyclonic		✓	✓	✓	✓	
Auger combustor	✓	✓	✓	✓	✓	
Ship-mounted		✓	✓	✓		

Source: Ref. 54.

Wet air oxidation (WAO) is a liquid phase oxidation and/or hydrolysis process at elevated temperature (175–345°C) and pressure (300–3000 psig) in the presence of oxygen. The WAO process can be used as a pretreatment step to destroy toxics or substantially reduce organics before using other conventional treatment processes. When raw waste loads reach a level of 20,000–30,000 mg/L COD, the process becomes thermally self-sustaining [7]. Phenols, cyanides, nitrosoamines, dienes, and pesticides have been shown to be effectively removed by WAO. Zimpro [59] reported that in pilot-plant tests using WAO, a wastewater composite of about 40 pesticides showed a 99+% pesticide destruction and 85% COD reduction. Another investigation indicated that the pesticide Amiben was degraded by 88% to 99.5% and atrazine by 100% [60].

The use of solvent extraction as a unit process operation is common in the pesticide industry; however, it is not widely practised for removing pollutants from waste effluents. Solvent extraction is most effectively applied to segregated process streams as a roughing treatment for removing priority pollutants such as phenols, cyanide, and volatile aromatics [7]. One pesticide plant used a full-scale solvent extraction process for removing 2,4-D from pesticide process wastewaters. As a result, 2,4-D was reduced by 98.9%, from 6710 mg/L to 74.3 mg/L.

Molten-salt combustion is a process by which hazardous materials can be oxidized below the surface of a salt or salt mixture in the molten state [27]. Molten sodium carbonate and a molten mixture of sodium carbonate and sodium sulfate (90 : 10, w/w) have been used. Operating temperatures range from 800 to 1000°C. Hazardous materials and air are fed into the combustion chamber below the surface of the melt. Generally, the heat produced during oxidation is adequate for maintaining the salt or salts in the molten state. Oxidation products include carbon dioxide, steam, and elemental gases such as nitrogen. Hydrochloric acid and sulfur dioxide, which form during the oxidation of chlorine- and sulfur-bearing compounds, react with sodium carbonate. In bench-scale tests, pesticides such as DDT, chlordane, and 2,4-D had more than 99.9% destructions [60].

In microwave plasma destruction, organic material is channeled through a plasma detector tube where destruction is initiated by microwave radiation-producing electrons. The electrons react with the organic molecules to form free radicals and final simple reaction products such as SO_2, CO_2, CO, H_2O, HPO_3, $COCl_2$, and Br_2 [60]. In bench-scale tests, the plasma method resulted in extensive detoxification (>99% destruction) for several pesticides, including malathion, phenylmercuric acetate (PMA), and Kepone® [55].

22.6 CONTROL AND TREATMENT FOR PESTICIDE FORMULATING WASTES

Management of wastes in pesticide formulating/packaging plants is simpler than in manufacturing plants because the volumes are much smaller. In the past, evaporation was the predominant disposal technique for wastewater generated in formulating plants [8]. However, due to concerns over air pollution and other nonwater quality environmental impacts, evaporation is currently not favored. Contract hauling for incineration is the recommended disposal method for small volumes of concentrated wastes, which are handled and transported as hazardous wastes in compliance with RCRA regulations.

For high-flow plants, treatment and recycle/reuse of the wastewater is recommended. The physical/chemical treatment processes used in pesticide manufacturing plants can be applied to formulating plants. The USEPA evaluated treatment and recycle technology for four plants that discharge high volumes of formulating/packaging wastewater [7]. These plants confirmed that treatment and recycle technology are feasible in their facilities but identified selected production processes that are not amenable to reuse. Such processes demand high purity source water to guarantee product integrity. The water volume requirements are low; therefore, wastewater from these processes is contract hauled and incinerated.

One of the four plants presently treats and reuses 75% of its waste stream as vent scrubber washwater. A second plant incinerates formulating/packaging process waste and discharges incinerator blowdown that contains levels of pesticides measured as not detected.

Waste reduction/minimization have also been evaluated and practised by the pesticide formulating/packaging industry [61]. Some techniques include using high-pressure spray nozzles to wash tanks and to clean production floors and other equipment, which can reduce wastewater volumes by at least 50%. Some plants use storage tanks to hold wash liquids (water or solvents) to be used for makeup purposes when the same product is formulated again [62]. This procedure reduces the total quantity of washwater discharged and minimizes product loss. Other techniques used in the pesticide manufacturing plants can also be applied in formulating/packaging plants and are not repeated here.

REFERENCES

1. Ware, G.W. *Pesticides: Theory and Application*; W. H. Freeman and Company; San Francisco, 1983.
2. National Research Council. The *Future Role of Pesticides in U.S. Agriculture*; The National Academies Press: Washington, DC, 2000.
3. McKenna, Conner & Cuneo. *Pesticide Regulation Handbook*, Revised Edition; Executive Enterprises Publications Co., Inc.: New York, 1987.
4. Worldwatch Institute. *Vital Signs 2002*; Worldwatch Institute: Washington, DC, 2003.
5. Agrow Reports. *World Non-Agricultural Pesticide Market*; PJB Publications: London, 2000.
6. Coble, H.D. *Pesticide, The World Book Encyclopedia*; World Book, Inc.: Chicago, 1989; 317–318.
7. USEPA. *Development Document for Effluent Limitations Guidelines, and Standards for the Pesticide Point Source Category*, Report EPA 440/1-85-079; USEPA, 1985.
8. USEPA. *Development Document for Interim Final Effluent Limitations Guidelines for the Pesticide Chemical Manufacturing Point Source Category*, Report EPA 440/1-75/060-d; USEPA, 1976.
9. Wong, J.M. Pesticides Wastewater Management. Paper presented at the *3rd Annual Hazardous Materials Management Conference and Exhibition/West*, Long Beach, California, 1987.
10. Dryden, F.E. *et al. Assessment of Dioxin-Forming Chemical Processes*, prepared for USEPA IERL, Cincinnati, OH; Walk, Haydel and Associates, Inc.: New Orleans, Louisiana, 1979.
11. USEPA. *Development Document for Effluent Limitations Guidelines and Standards for the Pesticides Point Source Category*, Report EPA 440/1-82/079-b; USEPA, 1982.
12. USEPA. *Development Document for Best Available Technology and New Source Performance for the Pesticide Chemical Industry, Proposed*, Report EPA/821-R-92.005; USEPA, 1992.
13. Thomas, N.A. Use of biomonitoring to control toxics in the United States. Wat. Sci. Technol. **1988**, *20*, 10.
14. World Bank Group. *Pollution Prevention and Abatement Handbook – Pesticides Manufacturing*; World Bank Group, July 1998.
15. Hwang, S.T.; Fahrenthold, P. Treatability of the organic priority pollutants by steam stripping. Am. Inst. Chem. Engrs. Symposium Series, Water–1979 **1980**, *197* (*76*): 37–60.
16. Becker, D.L.; Wilson, S.C. The use of activated carbon for the treatment of pesticides and pesticide waste. In *Carbon Adsorption Handbook*; Cheremisinoff, P.N., Ellenworth, F., Eds., Ann Arbor Science: Ann Arbor, Michigan, 1978.
17. Lyman, W.J. Applicability of carbon adsorption to the treatment of hazardous industrial wastes. In *Carbon Adsorption Handbook*; Cheremisinoff, P.N., Ellenworth, F., Eds.; Ann Arbor Science: Ann Arbor, Michigan, 1978.
18. Clark, R.M.; Lykins, B.W., Jr. *Granular Activated Carbon Design, Operation and Cost*; Lewis Publishers: Chelsea, MI, 1989.

19. Bhandari, A.; Xia, K.; Starrett, S.K. Pesticides and herbicides. Wat. Environ. Res. Literature Review, **2000**, *72*, 5.
20. Kuo, W.S.; Regan, R.W., Sr. Removal of pesticides from rinsate by adsorption using agricultural residuals as medium. J. Environ. Sci. Health, **1999**, *B34*, 431.
21. Celis, R.; Koskinen, W.C.; Cecchi, A.M.; Bresnahan, G.A.; Carrisoza, M.J.; Ulibarri, M.A.; Pavlovic, I.; Hermosin, M.C. Sorption of the ionizable pesticide Imazamox by organo-clays and organohydrotalcites. J. Environ. Sci. Health, **1999**, *B34*, 929.
22. Sudhakar, Y.; Dikshit, A.K. Kinetics of endosulfan sorption on two wood charcoal. J. Environ. Sci. Health, **1999**, *B34*, 587.
23. Bras, I.P.; Santos, L.; Alves, A. Organochlorine pesticides removal by pinus bark sorption. Environ. Sci. Technol., **1999**, *33*, 631.
24. Hager, D.G.; Smith, C.E. The UV–hydrogen peroxide process: An emerging technology for groundwater treatment. Paper presented at *HazMat West 85*, Long Beach, California, 1985.
25. Reynolds, G.; Graham, N.; Perry, R.; Rice, R.B. Aqueous ozonation of pesticides: A review. Ozone Sci. Engg. **1989**, *11* (*4*), 339–382.
26. Ma, J.; Graham, N.J.D. Degradation of atrazine by manganese-catalyzed ozonation: Influence of humic substances. Wat. Res. **1999**, *33*, 785.
27. Tucker, S.P.; Carson, G.A. Deactivation of hazardous chemical wastes. Environ. Sci. Technol. **1985**, *19* (*3*), 215–220.
28. Zeff, J.D. New developments in equipment for detoxifying halogenated hydrocarbons in water and air. Paper presented at the *Halogenated Solvents Alliance Meeting*, San Francisco, September, 1985.
29. Beltran, F.J.; Rivas, J.; Acedo, B. Atrazine removal by ozonation processes in surface waters. J. Environ. Sci. Health, **1999**, *B34*, 229.
30. Mauk, C.E.; Prengle, H.W.; Payne, N.E. *Oxidation of Pesticides by Ozone and Ultraviolet Light*; Houston Research Inc.: Houston, TX, 1976.
31. Ultrox International. UV/ozone treatment of pesticide and groundwater, prepared for Department of Health Services, California, Grant No. 85-00169, 1988.
32. Glaze, W.H.; Kang, J.W.; Chapin, D.H. The chemistry of water treatment processes involving ozone, hydrogen peroxide and ultraviolet radiation. Ozone Sci. Engrg. **1987**, *9* (*4*), 335.
33. Kuo, W.S. Effects of photolytic ozonation on biodegradability and toxicity of industrial wastewater. Environ. Sci. Health **1999**, *A34* (*4*), 919–933.
34. Balmer, M.E.; Sulzberger, B. Atrazine degradation in irradiated iron/oxalate systems: Effect of pH and oxalate. Environ. Sci. Technol. **1999**, *33*, 2418.
35. Nguyen, C.; Zahir, K.O. UV induced degradation of herbicide methyl viologen: kinetics and mechanism and effect of ionic media on degradation rates. J. Environ. Sci. Health, **1999**, *B34*, 1.
36. Lu, M. Photocatalytic oxidation of Propoxur insecticide with titanium dioxide supported on activated carbon. J. Environ. Sci. Health **1999**, *B34*, 207.
37. Topalov, A.; Molnar-Gabor, D.; Csnadi, J. Photocatalytic oxidation of the fungicide Metalaxyl dissolved in water over TiO_2. Wat. Res. **1999**, *33*, 1372.
38. Kennedy, D.C. Treatment of effluent from manufacture of chlorinated pesticides with a synthetic, polymeric adsorbent, Amberlite XAD-4. Environ. Sci. Technol. **1973**, *7* (*2*), 138–141.
39. Huang, C.; Stone, A.T. Hydrolysis of naptalam and structurally related amides: inhibition by dissolved metal ions and metal hydroxide surfaces. J. Agric. Food Chem. **1999**, *47*, 4425.
40. Skadberg, B.; Geoly-Horn, S.L.; Sangamalli, V.; Flora, J.R.V. Influence of pH, current, and copper on the biological dechlorination of 2,6-dichlorophenol in an electrochemical cell. Wat. Res. **1999**, *33*, 1997.
41. Merrill, D.T.; Manzione, M.A.; Peterson, J.J.; Parker, D.S.; Chow, W.; Hobbs, A.D. Field evaluation of arsenic and selenium removal by iron coprecipitation. J. WPCF **1986**, *58*, 18–26.
42. Porras-Rodriguez, M.; Talens-Alesson, F.I. Removal of 2,4-dichlorophenoxyacetic acid from water from adsorptive micellar flocculation. Environ. Sci. Technol. **1999**, *33*, 3206.
43. Kiang, Y.H.; Metry, A.A. *Hazardous Waste Processing Technology*; Ann Arbor Science: Ann Arbor, MI, 1982.

44. Atkins, P.R. *The Pesticide Manufacturing Industry – Current Waste Treatment and Disposal Practices*, USEPA Project No. 12020 FYE, 1972.
45. Lue-Hing, C.; Brady, S.D. Biological treatment of organic phosphorus pesticide wastewaters. Purdue Univ. Eng. Ext. Servo 132 (pts. 1/2), 1968;1166–1177.
46. Mangat, S.S.; Elefsiniotis, P. Biodegradation of the herbicide 2,4-dichlorophenoxyacetic acid (2,4-D) in sequencing batch reactor. Wat. Res. **1999**, *33*, 861.
47. Portzman, R.S.; Lee, P.H.; Ong, S.K.; Moorman, T.B. Treatment of formulated Atrazine rinsate by agrobacterium radiobactor strain J14a in a sequencing batch biofilm reactor. Wat. Res. **1999**, *33*, 1399.
48. Galluzo, M.J.; Banrji, S.K.; Bajpai, R.; Surampalli, R.Y. Atrazine removal through biofiltration. Practice Periodical of Hazardous, Toxic, and Radioactive Waste Manage. **1999**, *3*, 163.
49. Wong, J.M.; Maroney, P.M. Pilot plant comparison of extended aeration and PACT® for toxicity reduction in refinery wastewater. *Proceedings, 44th Purdue Industrial Waste Conference*, West Lafayette, IN, 1989.
50. Zimpro, Inc. CIBA-GEIGY Meeting Tough Bioassay Test. Reactor, June 1986; 13–14.
51. Stephenson, T.; Judd, S.; Jeferson, B.; Brindle, K. *Membrane Bioreactors for Wastewater Treatment*; IWA Publishing: London, 2000.
52. Gonzalez-Gonzalez, L.R.; Buenrostro-Zagal, J.F.; Luna-Martinez, A.D.; Sandoval-Gomez, Y.G.; Schettino-Bermudez, B.S. Treatment of an herbicide-contaminated wastewater in a membrane bioreactor by sulfate-reducing consortia. *Proceedings, 7th International Symposium on In Situ and On-Site* Bioremediation, Orlando, Florida, June 2–5, 2003.
53. Wong, J.M. Technologies and case histories for industrial wastewater recovery and reuse. *Proceedings, Seminar on Application of Biotechnology in Industrial Wastewater Treatment and Reuse*, Kaohsiung, Taiwan, ROC, October 30, 2002.
54. Kiang, Y.H.; Metry, A.A. *Hazardous Waste Processing Technology*; Ann Arbor Science: Ann Arbor, MI, 1982.
55. Dillon, A.P. *Pesticide Disposal and Detoxification Processes and Techniques*; Noyes Data Corporation: Park Ridge, NJ, 1981.
56. Oberacker, D.A. Incineration options for disposal of waste pesticides. In *Pesticide Waste Disposal Technology*; Bridges, J.S., Dempsey, C.R., Eds.; Noyes Data Corporation: Park Ridge, NJ, 1988.
57. Eckenfelder, W.W., Jr. *Industrial Water Pollution Control*, 2nd ed.; McGraw-Hill: New York, 1989.
58. McNally, R. Tougher rules challenge future for injection wells. Petrol. Eng. Int. **1987**, *July*, 28–30.
59. Zimpro, Inc. *Report on Wet Air Oxidation for Pesticide Chemical Manufacturing Wastes*, prepared for G. M. Jett, USEPA; Rothchild: Wisconsin, 1980.
60. Honeycutt, R.; Paulson, D.; LeBaron, H.; Rolofson, G. Chemical treatment options for pesticide wastes disposal. In *Pesticide Waste Disposal Technology*; Bridges, J.S., Dempsey, C.R., Eds.; Noyes Data Corporation: Park Ridge, NJ, 1988.
61. Lewis, D.A. Waste minimization in the pesticide formulation industry. J. APCA **1988**, *38*; (*10*), 1293–1296.
62. World Bank Group. *Pollution Prevention and Abatement Handbook – Pesticides Formulation*; World Bank Group, July 1998.

23
Livestock Waste Treatment

J. Paul Chen and Shuaiwen Zou
National University of Singapore, Singapore

Yung-Tse Hung
Cleveland State University
Cleveland, Ohio, U.S.A.

Lawrence K. Wang
Zorex Corporation, Newtonville, New York, U.S.A., and
Lenox Institute of Water Technology, Lenox, Massachusetts, U.S.A.

23.1 INTRODUCTION

Livestock and poultry production involves the conversion of raw feeds such as grass into valuable products such as meat, milk, and eggs to meet daily human consumption. In the United States, the total amount of chickens consumed increased by 27.5% from 1992 to 2001. The chicken consumption per capita is projected to grow from 78 pounds in 1992 to 92 pounds in 2004 (Fig. 1). The US Department of Agriculture forecasted US agricultural exports for fiscal year 2003 at $57.5 billion, a $4 billion increase over the expected $53.5 billion for fiscal year 2002. Export sales at this level would be the highest since 1997, only $2.3 billion below the all-time record of $59.9 billion in 1996.

In the past half century, livestock and poultry agriculture has experienced a rapid transition from small flocks and herds to large-scale intensive (concentrated) productions in a few locations due to global industrialization. This has served to reduce operational costs as well as improve profitability and increased coordination between animal feeding operations and processing firms. Operations have tended to cluster near slaughtering and manufacturing plants as well as near end-consumer markets, which has encouraged safety standards in the food supply and convenience of food delivery [1].

The consolidation and intensification of production, while perceived as sound from economic and management perspectives, often do not give full consideration to the potential environmental impact. One of the obvious problems is considerable odorous emissions produced from livestock and poultry production sites. In recent years, attention has shifted to their impact on water, soil, and air quality. One new challenge is the increase of methane in the atmosphere, a potent greenhouse gas, which can contribute to global climate change. The gas is produced by ruminant animals, such as cattle, sheep, buffalo, and goats. Because of their special digestive systems, they can produce methane. Globally, ruminant livestock produce about 80 million tons of methane annually, accounting for about 22% of global methane emissions from human-related activities. Livestock production systems can also emit other greenhouse gases

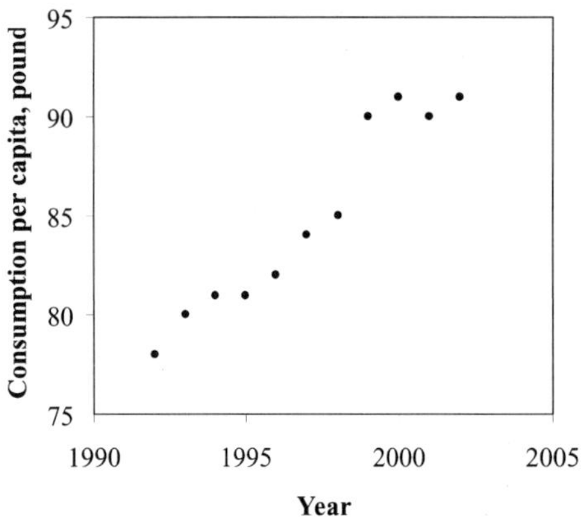

Figure 1 Chicken consumption in United States from 1992 to 2004.

such as nitrous oxide and carbon dioxide. The most promising approach for reducing methane emissions from the US livestock is to improve the productivity and efficiency of livestock production [2–6].

Unavoidable and less desirable livestock and poultry waste is excreted in solid, liquid, and gaseous forms. In past years, especially before industrialization, the livestock and poultry excreta was predominantly used as fertilizer and soil conditioner, which was termed as manure. The geographic concentrations of livestock production has led to a challenge in land availability. The volume of the excreta or manure generated today could become a major obstacle to future development of the livestock and poultry industries if the environmental impact is not properly managed. For example, North Carolina is one of the leading states in livestock and poultry production. In 1999, there were approximately 5.5 million animal units (AU) in the state. Animal production operation produced about 33 million tons of fresh manure, which consisted of 240,000 tons of nitrogen (N), 182,000 tons of phosphorus (P_2O_5), 169,000 tons of potash (K_2O), and other nutrients [2–6].

Potential pollutants are mainly nitrogen and phosphorus; others are solids, pathogens, and odorous compounds. Manure can also be a source of salts and various trace elements (e.g., zinc and copper), as well as pesticides, antibiotics, and hormones. It is well known that manure has nutrients (i.e., N, P, and K) that have potential value as fertilizers [7]. However, the nutrients are of relatively low concentrations, which makes them more costly to store than apply as fertilizer. Fresh swine manure, for example, contains 0.6% N, 0.45% P, and 0.45% K. The pollutants can be released into the water environment through discharge or runoff if the manure and wastewater are not properly handled and managed. As a result, laws and regulations have been applied to restrict agricultural practices and impose penalties for exceeding land application limits in an effort to control agricultural impacts. In the future, sound nutrient management practices will be mandated to manage both feed and manure to minimize the environmental impact of livestock and poultry agriculture (USEPA, 2003) [3–6].

23.2 CHARACTERISTICS AND IMPACT OF WASTES

23.2.1 Livestock and Poultry Wastes

Livestock and poultry wastes are mainly composed of excreta and associated losses, bedding, washwaters, sprinkling waters from livestock cooling, precipitation polluted by falling on or

Table 1 Manure Characteristics of Dairy Cattle

	Weight of dairy cattle (lbs)[a]				
	150	250	500	1000	1400
Daily production (lbs/day)	12	20	41	82	115
Daily production (gallons/day)	1.5	2.4	5.0	9.9	13.9
Total solids (lbs/day)	1.6	2.6	5.2	10.4	14.6
Volatile solids (lbs/day)	1.3	2.1	4.3	8.6	1.7
BOD_5 (lbs/day)	0.26	0.43	0.86	1.70	2.38
Nitrogen (lbs/day)	0.06	0.10	0.20	0.41	0.57
Phosphate (lbs/day)	0.023	0.045	0.082	0.166	0.23
Potash (lbs/day)	0.048	0.084	0.169	0.325	0.45

[a]1 lb = 453.6 g.

flowing onto an animal feeding operation, and other materials polluted by livestock [1,3–5]. Livestock and poultry wastes can be categorized as solid, semisolid, and liquid wastes. Solid waste contains more than 20% solids, such as bedding and waste feed. Waste is deemed semisolid if it has 10–20% solids. Liquid waste has less than 10% solids. For systems designed to transfer wastes by pumping, the optimum liquid waste contains less than 4% solids.

Both quantity and quality of livestock waste can vary significantly from farm to farm; they are also greatly dependent on animal species, age, ration, and production systems. Tables 1 and 2 show the characteristics for dairy cattle, sheep, horse, and poultry species. Typical waste parameters of dairy, swine, and chicken are listed in Table 3. The impact of age, species, ration, and production system are as follows:

- *Animal age.* Manure from mature animals is not as biologically stable as that from younger ones. The larger an animal, the more manure it produces (Table 1). A rough estimate of the amount of manure produced per day is 8% of the animal's body weight.
- *Animal species.* Livestock manure characteristics can differ greatly among animal species. For example, since nonruminant animals cannot digest high-cellulose feed, they can produce relatively more manure than ruminant animals.

Table 2 Manure Characteristics of Sheep, Horse and Poultry.

			Poultry	
Animal species	Sheep	Horses	Layers	Broilers
Size of animal (lbs)	100	1000	4	2
Daily production (lbs/day)	4.0	45	0.21	0.14
Daily production (gallons/day)	0.46	5.63	0.027	0.018
Total solids (lbs/day)	1.00	9.4	0.053	0.036
Volatile solids (lbs/day)	0.85	7.5	0.037	0.025
BOD_5 (lbs/day)	0.09	—	0.014	0.0023
Water Content (%)	75	79.5	74.8	74.8
Nitrogen (lbs/day)	0.045	0.27	0.0029	0.0024
Phosphate (lbs/day)	0.015	0.105	0.0025	0.00123
Potash (lbs/day)	0.039	0.205	0.0014	0.0009

Table 3 Typical Wastewater Parameters of Dairy, Cattle, Swine, and Chicken

Parameter	Dairy manure	Swine manure	Chicken manure
BOD_5 (mg/L)	14,000	28,000	36,000
DM (%)	10	10	18
TS (mg/L)	100,000	9,800	180,000
Total nitrogen (mg/L)	3,800	4,600	1,090
Total phosphorus (mg/L)	800	1,600	940

Note: DM, dry matter.

- *Ration.* Diet affects the characteristics of the manure. Digestibility, protein, and fiber content of rations are important factors. For instance, cattle fed high-concentrate rations do not produce as much manure as cattle fed high-roughage rations.
- *Production system.* The production system affects whether the manure is solid, semisolid, or liquid. For example, if large amounts of bedding are used, the manure will be more solid. On the other hand, if a flush-type handling system is used, the manure will be more liquid due to the manure being diluted by the flush water.

Manufacturing plants producing different products will in turn produce different amounts and types of wastes. For instance, the average waste volume coefficients for the dairy industry in general, and cheese producers in particular, were 2.43 and 3.14 m^3 wastewater/ton milk processed, respectively [8].

Three main crop nutrients – nitrogen, phosphorus, and potassium – and the dry matter (DM) content are the most important in livestock manure due to their great environmental impact on rivers and estuaries [9].

Nitrogen is a major concern in livestock and poultry manure. It exists in the forms of organic nitrogen, ammonia, nitrite, and nitrate. Total Kjeldahl nitrogen (TKN) includes organic nitrogen and ammonia.

Manure degradation is a source of nitrogen oxides (NO_x), contributing to accumulation of greenhouse gases. Volatilization of ammonia (NH_3) can causes "acid rain," which acidifies soils and woodlands. It was reported that animal sources in Western Europe are responsible for 50% of the acid precipitation. Emissions of nitrous oxide (N_2O) during the nitrification–denitrification cycle can contribute to ozone depletion [1,5,10].

Application of manure as a fertilizer can result in excessive quantities of nitrogen, which may contaminate ground or surface waters through surface runoff and leaching. Ammonia can be oxidized to nitrite and then to nitrate, which depletes dissolved oxygen (DO) in the water and thus causes toxicity to fish and other living organisms in surface waters [2–5].

Nitrate leaching is considered another major concern to livestock farms today. The transformation of nitrogen is illustrated in Figure 2. Nitrogen enters to the waters by runoff, erosion, leaching, and volatilization (e.g., NH_3). The amount may be significant. Normally, the nitrogen loss is 31–50%, 60–70%, and 75% for poultry, cattle, and swine, respectively. In all manure land application operations, about 13% available manure nitrogen exceeds crop system need. Nitrogen conversion in manure can also be a source of odors. Ammonia may become a health hazard affecting the performance, morbidity, and mortality of the animals and poultry. In drinking water, nitrate concentrations above 10 ppm are a health threat to humans, particularly infants [2–5,11].

Phosphorus is another key element in livestock and poultry waste. The typical phosphorus lost is about 15%. In all manure land application operations, about 25% available manure phosphorus exceeds crop system need. In other words, phosphorus is over-applied. Unlike nitrogen, excess phosphorus in manure does not leach through the soil into the groundwater. It is

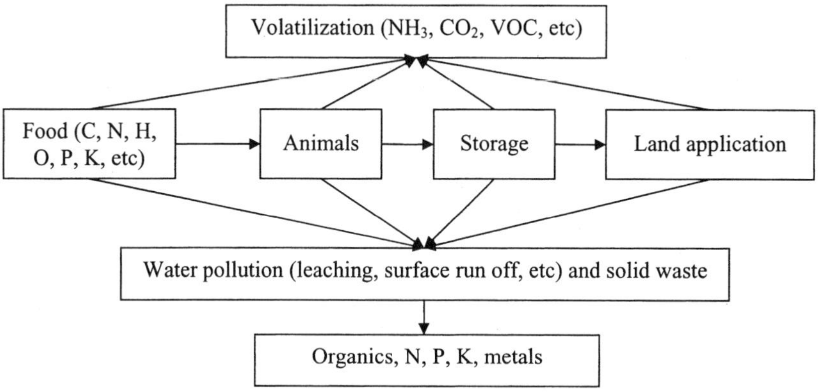

Figure 2 Transformation of pollutants released to the environment.

not toxic and has minimal environmental impact. However, its presence could cause additional consumption of alum in drinking water treatment if it precipitates with aluminum ions (Al^{3+}). Phosphorus is a nutrient that limits biological activity in most of the clear water lakes, reservoirs, and streams. Hydrogen, oxygen, carbon dioxide (CO_2), and nitrogen of sufficient quantities are naturally present in the water environment to support the growth of algae and plants. Insufficient phosphorus in most inland water bodies keeps the clear water lakes and streams from being congested with plant growth. If concentration of phosphorus exceeds a critical value, acceleration of eutrophication will be observed. As a result, the DO can be depleted, which can cause the death of aquatic species and release of toxins from the blue-green algae. These consequences in turn can increase the treatment costs of drinking water. Advanced eutrophication can also reduce aquatic wildlife populations and species diversity by lowering DO and increasing the 5-day biochemical oxygen demand (BOD).

Potassium is a crop nutrient that is found in livestock manure in significant amounts. Potassium is the main cation within cells in plants. It plays an important role in guard cell operation and functions as a cofactor for some enzymes. Potassium is a limiting plant nutrient as plants have a higher demand for it. Potassium deficiency is particularly common on heavily cropped land. As a result, addition of potassium fertilizers is necessary for the maintenance of soil fertility. Potassium also plays a key role in control of the balance of electrical potentials across cell membranes in human beings and is involved in nerve impulses [11]. It can contribute to the salinity of manure, which in turn may change the salinity in both surface and subsurface waters contaminated by the manure. Fertility of the soil may be reduced if potassium content is too high.

In manure waste, solids, organic matter, pathogens, salts, trace elements, odorous compounds, antibiotics, as well as pesticides and hormones, play important roles in water pollution.

The sources of solids include animal manure, spilled feed, bedding and litter materials, hair, feathers, and corpses. Solids measured as dry matter (DM), total suspended solids (TSS) or total solids (TS), are the mass of solids, as a percentage of the overall mass of diluted livestock manure. Solids may be measured as total dissolved solids (TDS). The presence of the solids can result in high turbidity of waters; they can also act as adsorption sites for metals and other contaminates.

Manure contains a high organic content. Its wastewater normally has a BOD_5 of 10,000–50,000 mg/L. Such high contamination levels can significantly reduce the DO in the waters due to biochemical reactions, which can be harmful to fish and other aquatic organisms. The reactions can also change the natural acidity of water, also harmful to aquatic life [1,6].

Salts from livestock waste, such as calcium, magnesium, sodium, potassium, chloride, sulfate, carbonate, bicarbonate, and nitrite, can be released to the water environment. They result from undigested feed. Higher accumulation of the salts can deteriorate soil structure, lower permeability, contaminate surface and ground waters, and reduce crop yields. Trace elements in the salts, such as arsenic, copper, lead, boron, molybdenum, mercury, and nickel are also important because of their higher toxicity [2–4].

Livestock manure contains a significant amount of pathogens including bacteria, viruses, protozoa, and parasites. They enter the waters through various pathways. The obvious species are the protozoa *Cryptosporidium parvum* and *Giardia* species, and bacteria *Escherichia coli* and *Salmonella* species. They are normally associated with health-related problems, such as food-borne disease. They may be able to survive and remain infectious for a long period of time. These negative effects can render waters unsuitable for consumption or recreational purposes [2–5,12].

Antibiotics, pesticides, and hormones are used in livestock production, and are eventually released to the aqueous environment. Use of antibiotics can cause their presence in waters as well as the development of antibiotic-resistant pathogens. Pesticide and hormones can result in various long-term environmental problems, especially through the food chains [6,11].

23.2.2 Mechanisms for Transporting Contaminate in Waters

Contaminants from livestock and poultry waste (e.g., N, P, and organics) reach surface and groundwaters by various mechanisms. The pollutant entrance to the waters can be via runoff, erosion, leaching, and many other paths.

Runoff occurs when the water flows across ground surfaces. It becomes a major path when rainfall takes place soon after the livestock manure is applied, over-applied, or misapplied. In an area where the soil surface is worn away by the action of wind and water, pollutant surface discharge can easily occur. Erosion is a main effect for the transportation of phosphorus in waters. Phosphorus is strongly adsorbed onto soils.

Owing to operational failures in manure land application (e.g., pump failures and pipe breakage), surface discharges via spills can occur. Direct contact between confined animals and waters (e.g., rivers and streams) can also cause transport of contaminants. This is mainly due to historical practices. Farms were normally located near rivers and streams because of the convenience for water access for livestock production as well as waste discharges. This is still a practice in some developing countries.

The above are the mechanisms for the contamination of surface waters. Additionally, the soluble pollutants can penetrate the aquifer and cause groundwater pollution. The transport is significantly dependent on the properties of soils, water table, weather, and other factors. Many field observations show that more than half of earthen lagoons in the United States have groundwater problems. Volatilization of manure pollutants (e.g., NH_3) and subsequent deposition can also contribute to water pollution.

23.3 US EPA REGULATIONS

The US livestock industry has undergone dramatic changes in the last 20 years [5]. The livestock and poultry industry tends to have fewer, but larger and more intensive specialized operations, which results in more concentrated manure and other animal wastes. This can lead to a significant increase in the contaminants in the watercourses in industrial operational areas. Wastes may be more concentrated in nonagricultural areas where there is inadequate land to accommodate the application of the manure (e.g., as a fertilizer).

23.3.1 Current Regulations

The National Effluent Limitations Guidelines (NELG) and standards for feedlots and the National Pollutant Discharge Elimination System (NPDES) Permit Program for Concentrated Animal Feeding Operations (CAFOs) were issued on February 14, 1974, and March 18, 1976, respectively [2–6].

Under the NPDES CAFOs regulations, CAFOs were defined as facilities [or Animal Feeding Operations (AFOs)] with 1000 or more animal units (AU). An AFO with 300 to 1000 AU was defined as a CAFO if it discharged pollutants through a manmade device or if the pollutants were discharged to waters that ran through the facility or otherwise came into contact with the confined animals. The AFOs were not termed as CAFOs if they discharged only during a 25 year, 24 hour storm. Any AFO with less than 300 AU could be designated as a CAFO if it did not meet the discharge criteria specified and was determined to be a significant contributor of pollution [3–6].

The NELG issued in 1974 were applicable to those AFOs in specified sectors with as many as or more than 1000 AU that were to be issued an NPDES permit. It did not allow discharges of pollutants from CAFOs into the nation's waters except when a chronic or catastrophic storm caused an overflow from a AFO that had been designed, constructed, and operated to contain manure, process wastewater, and runoff resulting from a 25 year, 24 hour storm [3–5].

The goals for the above regulations were for owners and operators of AFOs to take action so that pollution could be minimized in the areas where the livestock productions were operated. All AFOs should develop and implement technically sound, economically feasible, and site-specific comprehensive nutrient management plans (CNMPs) in order to accomplish the objective. The actions included development and implementation of CNMPs, promotion of voluntary and incentive-based programs, improvement of the existing regulatory program, and development of technologically sound technologies.

23.3.2 New Developments in US Regulations

The USEPA published regulations for the NELG and the NPDES Permit Regulations for CAFOs in January 2003. The final rule focuses on the largest operations and those having the greatest environmental risk. It strengthens the current regulations for reduction of pollution from AFOs and promotes innovation [3]. Any AFO, according to the new regulations, must provide storage for manure and wastewater. The new rule ensures that land application of manure by CAFOs use nutrients for agricultural activities. Changes to the existing regulations include:

- All CAFOs, all large chicken operations, large swine nurseries, and heifer operations must apply for a NPDES permit;
- AFOs may elect to use innovative technologies and alternative management practices;
- CAFOs must implement nutrient management plans that include appropriate best management practices to protect waters;
- CAFOs must report key information about their operations annually.

23.4 LIVESTOCK WASTE STORAGE TECHNOLOGIES AND MANAGEMENT

Livestock and poultry wastes are normally stored in animal shelters of animal feeding operations until they are removed for treatment or utilization. An animal shelter can be an anaerobic lagoon, a storage pond, a manure pit, a below/aboveground storage tank or other structure used to house animals in production agriculture. These systems should be properly designed, constructed, and managed to ensure that the waste has no adverse effects on the surrounding environments (ground and surface waters and odors).

In general, the service life and durability, foundation, structure loading, and structural design must be in accordance with the national, state, and local government standards and rules. Odors from livestock production facilities (e.g., H_2S due to anaerobic digestion of manure) must be prevented from releasing to the environment. Hence, adequate separation distance between the production plants and residential areas, and the appropriate orientation of the plants must be ensured.

Anaerobic lagoons have multiple functions of storage and decomposition of manure. Biogas of methane can be produced during the anaerobic biological process. The volume of the lagoon consists of minimum treatment volume, sludge volume, livestock manure and wastewater volume, net precipitation volume (i.e. precipitation minus evaporation), volume due to a 25 year, 24 hour storm event, runoff volume, and additional safety volume (freeboard, normally height of >1 ft) [1]. The minimum treatment volume can be estimated by the volatile solids loading rate ranging from 3 to 7 pounds per 1000 ft^3 per day; the manure and wastewater volume is based on a storage period of 90–365 days. A typical depth of 8–20 ft is normally used. The sides of the lagoon should be sloped with a horizontal to vertical ratio of 2 : 1 to 3 : 1. Lagoons with two stages are usually used in the storage. A cover is used so that odors can be avoided and methane is collected. An optional solid separation unit can be used before the waste enters lagoons. Impervious liners are normally used for prevention of possible groundwater pollution. An extremely high hydraulic retention time (HRT) of more than 200 days is normal in the lagoon operation. The averaged removal efficiencies of COD, TSS, total nitrogen (TN), phosphorus, and potassium are 90, 94, 75, 87, and 45%, respectively [2].

A *storage pond* is similar to the anaerobic lagoon, except that the minimum treatment volume is not added. Thus, its function is mainly to store the manure, but not to actively treat it. The organic carbon, phosphorus, and nitrogen can be removed mainly because of physical actions, rather than chemical. The removal of sludge can lead to reduction of these three elements. Some nitrogen, in the form of ammonia, is reduced as a result of volatilization.

Solid wastes can be stored in *manure pits* that are located inside or outside of buildings. A typical storage period of 5 to 12 months is used. Semisolid wastes are often pumped to pits (Fig. 3), where solids are separated from liquids. A pit should be lined with an impermeable soil

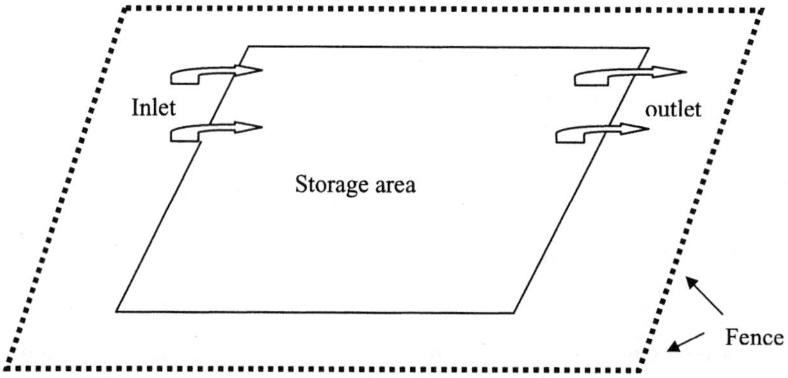

Figure 3 A pit to contain semisolid or liquid wastes.

(such as certain clays), concrete, and a heavy plastic liner in order to prevent groundwater contamination. The design life of earthen pits is approximately 10 years. Manure stored in earthen pits can form a nearly impermeable seal of organic matter and microorganisms on the bottom and sides.

The function of *below/above ground storage tanks* is the same as storage pits and earthen lagoons. The tanks are commonly constructed of moderate-cost concrete, higher-cost glass-lined steel, or others materials. Both tanks are suitable for storage of both semisolid (slurry) and liquid manure. Owing to the high cost of storage volume, tanks are not usually used to contain large volumes of lot runoff. It is important to minimize the runoff area draining into the tank to increase ease of management and reduce tank size. Storage tanks can be located above grade, below, or partially below grade. Below-grade tanks are easy to fill by scraping, whereas above-grade tanks may require pumps for filling (Fig. 4). For open storage structures, it is important to minimize odor and sight nuisances. Tanks should be located at least 300 ft from water wells. Open tanks should be fenced, as necessary, to exclude animals and children. Tanks filled by scraping should have guardrails or grates to prevent machinery as well as animals and people from entering the tank.

Plumbing for the facility must be properly designed with safeguards against ruptures and leaks. Excess water into an animal shelter can result in possible environmental consequences. Dry waste can become wet waste and be harder to handle, while a wet waste system can have more liquid than the system is designed to handle. Ruptures of waste-handling pipes can result in direct discharge to the environment.

Proper ventilation design within the animal shelter optimizes animal health and minimizes odor emissions. Ventilation design should be in accordance with national standards. An erosion and sediment control plan should be developed for the facility to minimize erosion during construction. The finished surfaces should be shaped to provide positive drainage from the animal shelter. All the disturbed area for the construction of the animal shelters should be established to permanent vegetation according to national standards.

Animal manure storage systems should be regularly inspected to ensure proper functioning of all design features that protect the environment. Pipes that convey waste should be regularly inspected to repair any leaks. Concrete cracks or plumbing leaks that could release waste should also be immediately repaired. Dust within the shelter is known to contribute to odors and should be controlled or removed [12,13].

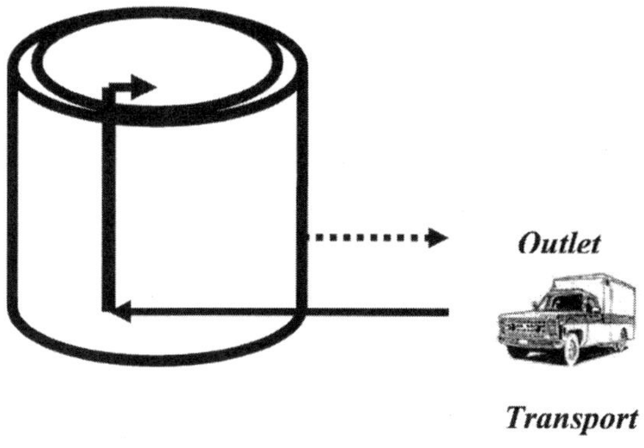

Figure 4 An aboveground storage tank for livestock manure.

23.5 GENERAL INTRODUCTION OF LIVEȘTOCK WASTE TREATMENT

Solid livestock waste can be used as a fertilizer due to its nutritional values (e.g., N, P, and K) and converted to useful energy (as biogas of methane) because of its high organic carbon content. These benefits can be obtained only when the waste is carefully treated; additional facilities usually must be built and operational costs budgeted.

In most cases, however, the waste is usually discharged into nearby waters. Owing to the negative environmental impact, livestock and poultry waste must be treated before it is released. Depending on the goals and characteristics of the waste, solid–liquid separation, biological treatment, chemical treatment, conversion (to useful materials), and composting may be employed. In biological treatment, both aerobic and anaerobic processes may be used. The process configuration varies from lagoon, ponds, oxidation ditches, to plug flow reactor [2]. Air pollution control, which is not a topic in this series, may also require consideration due to its greenhouse effect.

23.6 SOLID–LIQUID SEPARATION

Solid–liquid separation technologies are used to separate the partial organic and inorganic solids from liquid manure. The goal is to reduce the solids content for subsequent handling and treatment, and to recover solids for further usage as fertilizer and/or other applications. Effective separation can lead to many benefits, such as reduction of organic matter in the liquid fraction, concentration of nutrients in the solid fraction, ease in transport and handling, reduction of odor emissions, reduction of size of the lagoon or storage pond, and flexibility for ultimate disposal and use of livestock waste.

The separated solids can be used for composting, soil amendments, animal feed supplements, or for generating biogas (methane). Composted material can have some applications, such as bedding in barns. The separated liquid fraction could be recycled as flush water or stored and land applied. A basic arrangement for a mechanical solid–liquid separation system is illustrated in Fig. 5.

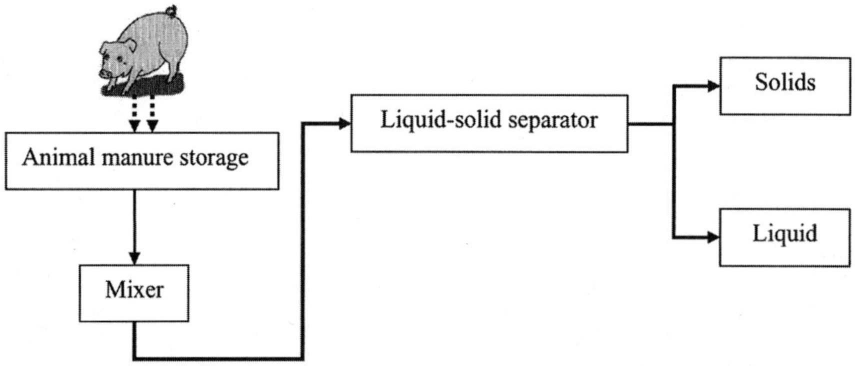

Figure 5 Basic arrangement for a mechanical solid–liquid separation system.

Separation of solids from the liquid portion is normally achieved by using gravity or a mechanical device. Mechanical separation can be achieved via. screen, press, and centrifuge. The advantage of the mechanical separation is its higher efficiency, while disadvantages include higher energy consumption and difficulty in management. In addition, it may not be cost-effective if used for small operations.

There are two characteristics of solids in liquids that are used as the basis for separation. One is based on the density difference between the solids and the water solution, while the other is based on the physical size and shape of the solids. A basic requirement for efficient separation is the continuous agitation of the manure. Otherwise, the relatively fast sedimentation processes that occur during storage can reduce its efficiencies [5,6].

Gravity settling is effective in a manure stream when it contains less than 1% total solids. Settling is typically accomplished through a series of sedimentation ponds or tanks. Y- and V-shaped gutters under slatted floors are useful for swine operations. The former is more difficult to build and clean than the latter, even though the operations are similar. The sideslope of the gutters is 1 : 1 and 0.75 : 1 for farrowing and nurseries operations, respectively [1,3–6].

In the screening process, solids in the liquid manure are collected on the screen while the liquids pass through. The separation efficiency is dependent on the size of the solids, physical properties of the screen (e.g., opening) and operating conditions (e.g., flow rate). Stationary screens, rotating drum screens, and vibrating screens are commonly used.

In addition to the above two types of separators, other separators use smaller screen sizes and pressure to "squeeze" the manure through screw press, belt press, or centrifuge. These processes can handle manure with higher solid content and are more efficient than typical screen separators; however, they often require more capital investment and operational costs.

23.7 BIOLOGICAL TREATMENT

Biological treatment is the most common and extensively used process for livestock and poultry manure wastes. Manure stored in earthen basins, pits, or tanks or spread on land undergoes biological reactions, which typically are not carefully controlled and take a long time. Therefore, well-designed and operated biological treatment systems become important for enhancing treatment efficiencies. The objectives of the treatment are stabilization of manure, removal of odor, removal of organic matter, nitrification, removal of nutrients, and recovery of energy (e.g., methane) [22,23].

23.7.1 Anaerobic Treatment

Anaerobic decomposition is one of the most common natural processes and has been extensively used in waste treatment. In the absence of free oxygen, anaerobic microorganisms can decompose the complex organic compounds. If livestock manure is stored in a container at a temperature ranging from 0 to 150°C, the waste solids and dissolved organic compounds can be decomposed, yielding carbon dioxide, methane, ammonia, hydrogen sulfide, and a series of simple organic compounds. pH, temperature, presence of toxicity, and alkalinity are important to anaerobic processes. pH must be maintained above 6.5.

Manure is a complex mixture of carbohydrates, proteins, and fats, while anaerobic decomposition is a complex process involving numerous individual reactions and a variety of enzyme-excreting bacteria. The latter can essentially be considered as a breakdown process that converts complex manure constituents into simpler, more stable, and more easily used endproducts. The microorganisms responsible for the process experience survival, growth, and reproduction. The biodegradation can often be simplified and described as a three-stage process. The first stage is hydrolysis, which converts the complex organic compounds to fatty acids, monosaccharides, amino acids, purines and pyrimidines, and other simple compounds. The second stage is acid formation, in which compounds from the hydrolysis are converted to a series of simple organic acids (e.g., acetic acid). The third stage is the acid recovery stage, which involves a more sensitive group of bacteria, the methane formers, whose role is to metabolize the organic acids, converting them to carbon dioxide and methane.

Anaerobic Lagoon

The anaerobic lagoon is most commonly used to treat livestock waste. A livestock lagoon contains manure diluted with building washwater, rainfall, water wastage, and surface runoff. In the earthen and pondlike lagoon, the waste becomes partially liquefied and stabilized by the biological reactions. Anaerobic bacteria can decompose more complex organic matters per unit lagoon volume than aerobic bacteria. The process is predominantly used for treatment of concentrated wastes. Because the anaerobic lagoon does not require DO, it can be much deeper and permit a small surface area for a given volume. Anaerobic decomposition of livestock waste can lead to emission of a series of odorous gases, which are mainly carbon dioxide, methane, hydrogen sulfide, and ammonia. The advantages of lagoon systems for treatment of livestock waste include storage and disposal flexibility, less land requirement, liquid recycling for pit waste removal, land application by simple irrigation, and lower labor and operational costs. However, it has disadvantages such as loss of nutrient value, offensive odors (H_2S and NH_3), and possible groundwater pollution.

Slightly different from the conventional anaerobic processes, biological treatment in a lagoon follows four steps: sedimentation of manure solids at the bottom of the lagoon; biological conversion of the settled solids into organic acids and other byproducts; conversion of organic acids to methane and carbon dioxide; and emission of unpleased odorous byproducts, such as hydrogen sulfide, ammonia, and volatile organic compounds (VOC).

A lagoon performs these functions in five zones as shown in Figure 6. In the operation, liquid levels are not allowed to drop below the minimum drawdown to maintain treatment and sludge storage functions. In addition, liquids must be maintained below the maximum operating level to prevent overflow and to protect the embankment from waves. The volume sandwiched between the minimum drawdown and maximum operation levels is used as effluent storage.

Anaerobic lagoon design is based on the following considerations: loadings, volume, operating levels, shape, site investigation, land application, irrigation equipment, sludge removal, and solids separation [12]. A typical volatile solids loading rate [VSLR, gram of volatile solids (VS) fed per liter reactor volume per day] for a nonheated covered anaerobic lagoon is 0.24 g L^{-1} day^{-1}; hydraulic retention time is 65 days [14].

Two-stage lagoons provide certain advantages over single primary lagoons as demonstrated in Figure 7. The main advantage of the secondary treatment is that it can reduce odors and the possibility that disease may be transmitted when the lagoon water is used for flushing gutters. Multistage lagoons work well for livestock manure treatment, especially when the treated manure is used for irrigation or recirculation in a flush-type handling system [1,6].

Secondary lagoons provide temporary storage prior to land application. A second stage also allows a maximum liquid volume to be maintained in primary anaerobic lagoons for stabilizing incoming wastes. It also provides some insurance against disease organisms being returned from the primary lagoon before a reasonable die-off period. Pumping from a secondary

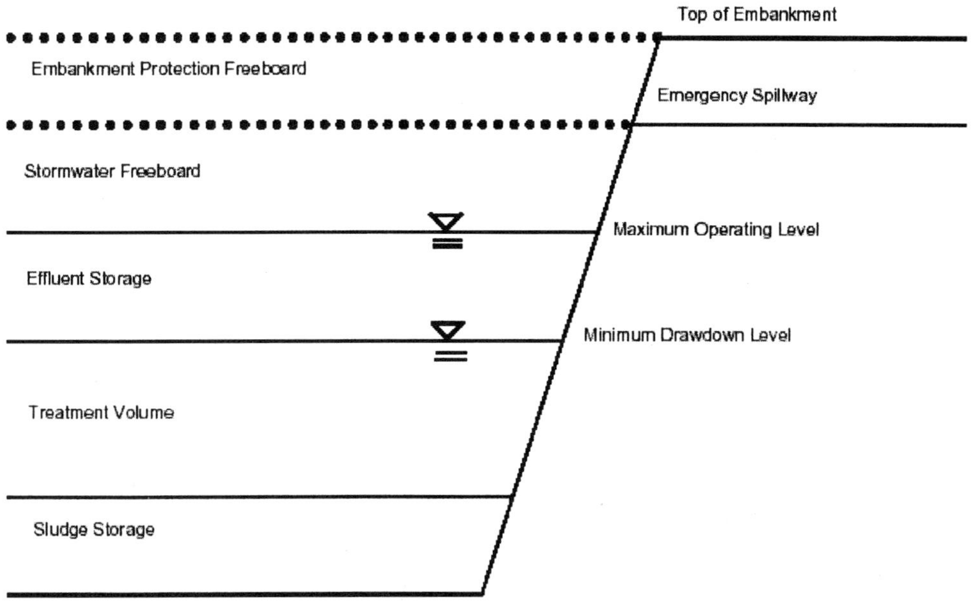

Figure 6 Schematic illustration of anaerobic lagoon.

lagoon reduces the solids pickup common in primary lagoons due to seasonal water turnovers and biological mixing [5,6].

Anaerobic digestion in lagoons is operated at normal temperatures ranging from 3 to 35°C (sometimes termed as psychrophilic). The retention time ranges from 30 to 90 days. Higher temperatures in the operational areas would cause lower retention time. Removal efficiencies for COD, TSS, TN, phosphorus, and potassium are 70–90, 75–95, 25–35, 50–80, and 30–50%, respectively. Higher removal of pathogen from the animal manure may also be achieved [4–6].

Covered anaerobic lagoons are used for manure digestion as well as for collection of biogas. Normally, methane is dominant in the biogas. Its percentage is 70–85%, depending on the chemical compositions of the manure and operational conditions. Carbon dioxide is the second dominant species (20–30%); H_2S and NH_3 are present in lower levels. When applied, air pollution can be virtually eliminated.

Anaerobic Digester

Like an anaerobic lagoon, the anaerobic digester is used to break down complex organic materials by a microbial population in the absence of DO. Anaerobic digesters can be designed and managed to optimize the bacterial decomposition of organic matter under more controlled conditions than those of anaerobic lagoons. Completely stirred tank reactors (CSTR) and plug-flow (PF) reactors are commercially available. Covers floated on the surface of the manure are used to collect methane as well as to minimize odors. Approximately 60–80% of biogas is methane.

Temperature is crucial to the production of methane. As in other anaerobic processes, two optimal temperature peaks are 32–40°C (mesophilic) and 57–68°C (thermophilic). The mesophilic and thermophilic digesters are operated with retention times of 12–20 days and 6–12 days, respectively. Since higher temperatures can shorten the retention time, the digesters can be heated.

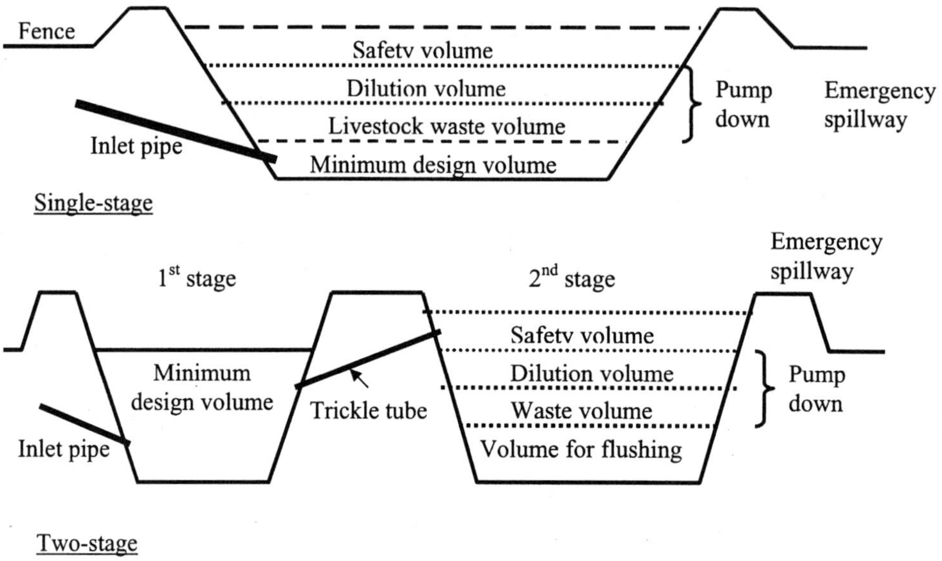

Figure 7 Single- and two-stage anaerobic lagoons.

A complete anaerobic PF digestion system used on livestock manure is shown in Figure 8. It is suitable for animal manures that contain a solids concentration of 11–13%. The raw manure slurry enters one end of a rectangular tank and decomposes as it moves through the tank. New material added to the tank pushes older material to the opposite end. Coarse solids in manure form a viscous material as they are digested, limiting solids separation in the digester tank [1,4–6].

A PF digester requires normal maintenance. When digesters are operated at high temperatures (>120°F), they provide better treatment and biogas production. However, there

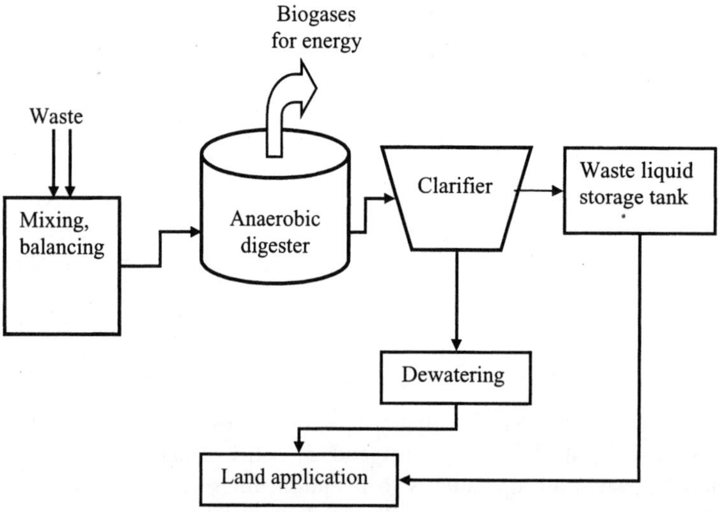

Figure 8 A typical anaerobic digestion system for livestock waste treatment.

have also been some successful applications in 60–75°F, with lower treatment efficiencies offset by higher retention times. The typical hydraulic retention time (HRT) and the removal efficiencies for COD and TSS are 18–20 days, 35–70%, and 20–45%, respectively. A 99% removal for pathogen can be achieved.

Inside the digester, suspended heating pipes allow hot water to circulate and keep the slurry at 77–104°F, a temperature range suitable for methane-producing bacteria. The hot water can come from recovered waste heat from an engine generator fueled with digester gas or from burning digester gas directly in a boiler.

The CSTR digesters can handle manures with TSS of 3–10%, and in large volumes. The reactor is a large, vertical, poured concrete or steel circular container. Manure is collected in a mixing pit by either a gravity-flow or pump system and deliberately mixed within the digester reactor. The mixing process creates a homogeneous substrate that prevents the formation of a surface crust and keeps solids in suspension. Mixing and heating improve digester efficiency. A CSTR digester shown in Figure 9 can be operated at either mesophilic or thermophilic temperature ranges with an HRT of 12–20 days. A VSLR of CSTR is 1.3 g L^{-1} day^{-1} [14]. The removal efficiencies for COD and TSS are 35–70, and 25–50%, respectively. A 99% reduction for pathogen can be achieved. However, CSTR digesters are more expensive to construct, operate, and maintain than PF digesters.

Anaerobic Contact Digester

The contact digester is a high-rate process that retains bacterial biomass by separating and concentrating the solids in a separate reactor and returning the solids to the influent. More of the degradable waste can be converted to gas since a substantial portion of the bacterial mass is conserved. The contact digester can be either CSTR or PF and can be operated in the thermophilic or mesophilic range. The contact reactor can treat waste of intermediate strength (COD of 2000–10,000 mg/L). The advantages are relatively low construction cost and good treatment capacity for higher TSS waste, while the disadvantages are the limited loading capacity and the poor settleability of the biomass [15].

Anaerobic Sequencing Batch Reactor

The anaerobic sequencing batch reactor (ASBR) is an anaerobic activated sludge system in which all processes occur in the same reactor. Similar to other sequencing batch reactors, an

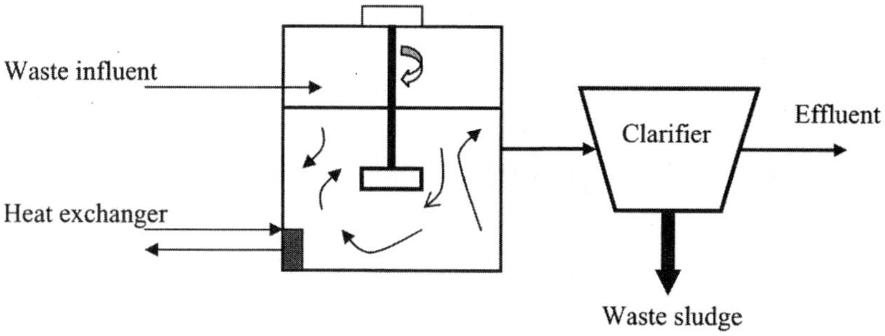

Figure 9 A CSTR digester for organic biodegradation.

ASBR is operated in sequences: fill phase, reaction phase, settle phase, draw phase, and idle phase. Biological degradation of manure occurs in the reaction phase; while the separation of activated sludge from the treated water takes place in the settle phase. This reactor can provide good treatment for animal manure waste. For example, the removal efficiencies for COD and ammonia are 95 and 98%, respectively. While this technology provides a compact, land-safe treatment system, it requires significant process control and is equipment intensive. An ASBR with a VSLR of 1.7 g L^{-1} day^{-1} and HRT of 15 days was successfully used in the full-scale treatment of swine manure [14].

Anaerobic Filters

An anaerobic filter is a column filled with various types of solid media, where manure flows either up or down the column, contacting the media on which anaerobic bacteria grow and are retained. Because the bacteria are retained on the media and not washed off in the effluent, long solids retention time can be achieved with reasonably short HRT. The anaerobic filter can be much smaller than other types of digesters with equivalent treatment efficiencies. The flow capacity of anaerobic filters ranges from 200 to 25,000 gallons per day. The filters can be operated in parallel when higher flow rate is applied. Higher COD and SS removal can be achieved. A VSLR of 4 g L^{-1} day^{-1} and an HRT of 3–9 days can be used in the treatment of swine manure [16].

Other Anaerobic Systems

There are many other anaerobic treatment systems for livestock manure, including upflow anaerobic sludge bed reactors (UASB), anaerobic fluidized and attached-film expanded-bed reactors (AAFEB), anaerobic rotating biological contactors (ARBC), and anaerobic baffled reactors (ABR) [19]. Results from pilot studies have shown various advantages. Among them, ABR demonstrated a better treatment efficiency than the other processes. The HRT is 14 day, while the VSLR is 4–8 g L^{-1} day^{-1} and gas production is 0.72–1.46 L/g VS. However, very few of these technologies have found full-scale industrial applications.

23.7.2 Aerobic Treatment

Aerobic treatment is a biological degradation and purification process commonly used for domestic wastewater. The theories, design procedures, and operations are well documented. Treatment results are strongly dependent on the oxygen supply and solid retention time (SRT). The end products of the biodegradation are water, carbon dioxide, nitrates, sulfates, and other simple molecules. Figure 10 shows a diagram of a typical aerobic treatment system.

Aerobic treatment can eliminate manure odor and is usually suitable for separated slurry or dilute waste. Solids in manure increase the amount of oxygen demand and the energy required for mixing. Slurry aeration allows microorganisms to metabolize dissolved components such as organic acids and phenols, which are responsible for most offensive odor emissions. Because complete stabilization of livestock manure by aerobic treatment is normally not economically justifiable, lower levels of aeration are recommended for partial odor control.

Aerobic Lagoon

Aerated lagoons use mechanical aeration to provide dissolved oxygen for the biological treatment of wastewater. Oxygen is supplied either from the atmosphere by means of mechanical aeration or from algae as a result of the photosynthetic process. The main advantages of aerobic lagoons are that bacterial degradation tends to be more complete than anaerobic digestion, with relatively lower odor.

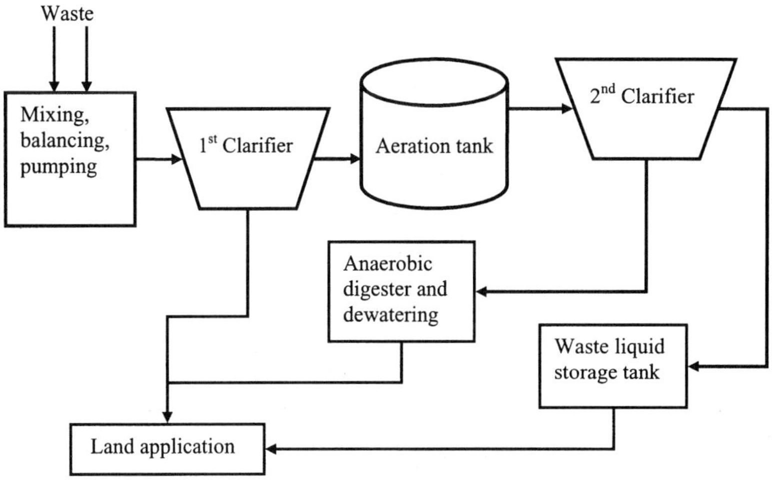

Figure 10 A typical aerobic treatment system.

In naturally aerobic lagoons, oxygen diffusion occurs across the water surface. Algae also generate oxygen through photosynthesis, which takes place when sunlight can penetrate the water depths. Water depths range from 3 to 5 ft. Because of the need for oxygen transfer, naturally aerobic lagoons are designed on the basis of surface area rather than volume. The

Table 4 Livestock Aerobic Lagoon Design Criteria

Animal type	Animal unit[a]	Average animal live weight (lbs)	Naturally aerobic lagoon surface area[b] (ft^2)	Mechanically aerated lagoon: Surface area[c] (ft^2)	Mechanically aerated lagoon: Aeration horsepower[d] (hp)
Dairy	Per head	1,400	2,030	104	0.10
Beef	Per head	800	1,150	44	0.044
Veal	Per head	200	180	6.3	0.0063
Swine					
Weanling-to-feeder	Per head	30	80	1.7	0.0017
Feeder-to-finish	Per head	135	350	7.8	0.0078
Farrow-to-meanling	Per sow	433	745	17	0.017
Farrow-to-feeder	Per sow	522	900	20	0.020
Farrow-to-finish	Per sow	1,417	3,660	82	0.082
Poultry					
Layer	Per bird	4.0	11.5	0.32	0.00032
Pullet	Per bird	1.5	4.3	0.12	0.00012

[a]One-time animal or bird capacity.
[b]Loading rate 50 lbs BOD_5/surface acre/day; mean liquid depth, 4 ft.
[c]1000 ft^2/hp of aeration and a minimum liquid depth = 10 ft.
[d]50% satisfaction of waste COD and oxygen transfer rate of 3 lbs/hp hour.

USDA Soil Conservation Service recommends a maximum daily loading rate of 50 pounds of BODs per acre of lagoon surface. According to these design criteria, Table 4 gives the amount of surface area required to maintain naturally aerobic lagoon conditions. Vast amounts of land are required for naturally aerobic lagoons – as much as 25 times more surface area and 10 times more volume than an anaerobic lagoon 10 ft deep. Thus, naturally aerobic lagoons are impractical for primary oxidation and are generally not recommended for treatment of livestock production wastes [11,12].

Aerobic Digester

Aerobic digestion is based on the biological principle that microorganisms metabolize their own cellular mass under aerobic conditions when the available food supply in the surrounding wastewater is inadequate. This phenomenon is called endogenous respiration. Cell tissue is oxidized aerobically to carbon dioxide, water, and ammonia.

Some advantages of aerobic digestion systems over anaerobic processes are: lower BOD concentrations in the supernatant, production of less odorous and humuslike biologically stable material, and volatile solids reduction. The disadvantages include higher operational cost and temperature impact. Activated sludge process in both continuously operated and batch reactors can be used.

Autoheated Aerobic Digester

This digester takes advantage of the heat produced during the oxidation of waste to raise the system temperature above ambient levels. The design and operation of conventional aerobic biological treatment processes can be applied to this digester system. Also, the removal of COD, TSS, and TN with similar efficiencies can be achieved for this system.

In the autoheated aerobic digester, higher oxygen transfer efficiency can be achieved. Mesophilic (30°C) and thermophilic (55–60°C) temperatures can be attained for different regions. Under these temperatures, ammonia removal due to the air stripping mechanism and nitrification–denitrification can be achieved. A TSS of 3–5% influent is important for maintaining thermophilic temperature. One of the system's appealing characteristics is the significantly high and rapid reduction in pathogen densities.

23.8 PHYSICOCHEMICAL TREATMENT

Physicochemical technologies can be used to treat livestock manure. Coagulation, sedimentation, filtration, and other processes separate the solids from the liquid, thus removing COD, BOD_5, TSS, nitrogen, and phosphorus [24]. Few pilot- or full-scale studies are available.

Miner *et al.* [9] used a chemical process to treat effluent from a manure anaerobic lagoon by alum and polymer. Results showed that the BOD_5 and COD were reduced by 40 and 60%, respectively; the effluent had low TSS. Westerman and Bicudo [17] used a tangential flow separation system to treat swine manure waste. The system consisted of a screening, a lime slurry tank, two chemical dosage tanks ($FeCl_3$ and polymer coagulation reactions), a tangential flow separator, and finally a thickening tank. The waste had a flow rate of 110 m^3/day (20 gpm), TSS of 8014 mg/L, COD of 7780 mg/L, TKN of 1368 mg/L, and total phosphorus (TP) of 354 mg/L. The dosage amounts of lime, $FeCl_3$, and polymer (PERCOL) were 1650–3500, 100–400, and 10–40 mg/L, respectively. Treatment results showed that the reduction efficiencies of COD, TSS, TKN, and TP were 49, 82, 22, and 90%, respectively. In addition, copper and zinc were reduced by 87%. The removal mechanisms were coagulation, precipitation, and sedimentation.

23.9 CONSTRUCTED WETLAND

Constructed wetlands (CWs) is an artificial wastewater treatment system involving many mechanisms such as aerobic and anaerobic biodegradation, sedimentation, filtration, adsorption/biosorption, plant and microbe uptake, reduction, oxidation, and volatilization [18]. A typical treatment system for manure treatment consists of: (a) a manure treatment system (e.g., anaerobic lagoon and digester); (b) CWs; and (c) storage pond.

Good effluent from the CWs is normally achieved and can be applied on agricultural lands or discharged to nearby water courses [19]. The operational cost for CWs is very low; the disadvantages include high land requirement and emission of methane, a greenhouse gas that potentially causes global warming. It was estimated that the biological activities in CWs contribute some 22% of global methane emissions [20].

Two principal types of CWs are used: free water surface systems (FWS) and subsurface flow systems (SFS) (Fig. 11). The FWS is also called a surface flow (SF), while SFS is termed subsurface flow (SF). The FWS are basins or channels that are carefully graded to ensure uniform flow, planted with emergent vegetation. Water flows over the ground at relatively shallow depths of 30 cm; the shallow water is exposed to the air. The FWS resembles natural marshes, providing wildlife habitat and aesthetic benefits and water treatment. The near surface layer is aerobic, while the deeper layer is anaerobic. Both constructional and operational costs for the FWS are low; however, the land requirement is high.

Based on the USEPA wetland design manual, the design equation for the FWS is [18]

$$\frac{C_e}{C_o} = A e^{-0.7 K_T A_v^{1.75} LWdn/Q} \tag{1}$$

where C_e is effluent BOD_5 (mg/L); C_o is influent BOD_5 (mg/L); A is fraction of BOD_5 not removed as settleable solids near headworks of the system; A_v is specific surface area for microbial activity (m^2/m^3); L is length of system (parallel to water flow) (m); W is width of

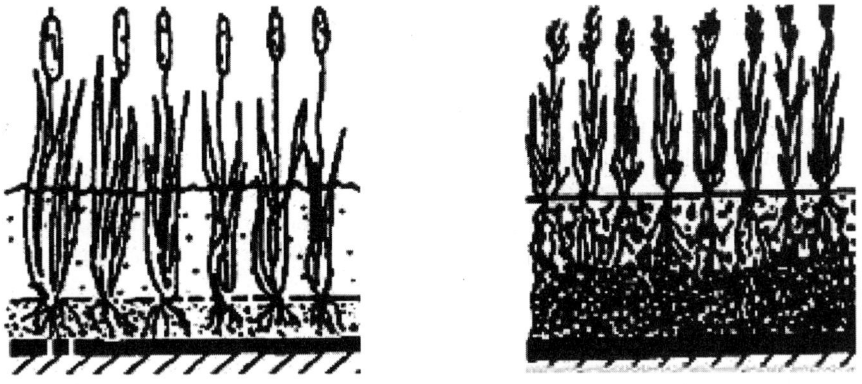

Figure 11 Illustration of constructed wetland systems.

system (m); d is design depth of system (m); n is porosity of system; and Q is average hydraulic loading, m^3/day.

In SFS, water level is below ground; water flows through a sand or gravel bed. It functions like a series of horizontal trickling filters. Emergent plants with extensive root systems are grown within the media. In the United States, flow path is normally horizontal, while some European applications use vertical flow paths. Because of its nature, SFS is suitable for waste that has less TSS and a low flow rate. It tolerates cold temperature, can be operated with low odor emission, and has a lower land requirement than FWS. However, both constructional and operational costs are higher than FWS. Based on the USEPA wetland design manual, the design equation for the SFS is [18]:

$$C_e = e^{\frac{-A_S K_T d n}{Q}} \tag{2}$$

where C_e is effluent BOD_5 (mg/L); C_o is influent BOD_5 (mg/L); K_T is temperature-dependent first-order reaction rate constant (d^{-1}); t is hydraulic residence time (day); Q is average flowrate through the system (m^3/day); d is depth of submergence (m); n is porosity of the bed; and A_S is surface area of the system (m^2).

The Gulf of Mexico Program (GMP) conducted a survey of treatment performances of 68 constructed wetland sites with a total of 135 pilot- and full-scale operations, and developed the Livestock Wastewater Treatment Wetland Database. It showed that the average concentration reduction efficiencies were 65, 53, 48, 42, and 42%, for BOD_5, TSS, NH_3-N, total nitrogen, and TP, respectively [21]. The reduction of fecal coliforms was 97%. Treatment performance of cattle feeding, dairy, poultry, and swine waste is illustrated in Table 5. As shown, good effluent can be produced for most cases. The survey showed that the hydraulic loading rate ranged from 3.8 to 5.5 cm/day.

23.10 COMPOSTING

Composting is a biological process in which organic matter is degraded to a relatively stable humuslike material. The process must take place under aerobic conditions to minimize odor problems. When finished, compost becomes dark brown or black, has a slight earthy or musty odor, and a loose texture. The process can stabilize organic matter manure and reduce the volume of waste. The compost also has some fertilizer values; weed seeds and pathogens can be destroyed.

Composting requires air, moisture, and a specific proportion of carbon to nitrogen. The rate of composting can be controlled by adjusting air, moisture, as well as carbon and nitrogen content. In addition, particle size of the manure and temperature play important roles. Smaller particle size can lead to faster reaction kinetics. Optimum moisture content for composting is 50–60%. At high moisture content, voids fill with liquids, and aeration is hindered. At low moisture content, microbial activity is hindered. Typically solid livestock manure has 25% solids; therefore, additional drying is required [2–6].

Aeration allows manure with higher moisture contents to be composted. Fresh manure can be directly composted if adequate aeration is provided. Over-aeration should be avoided as it can reduce the decomposition rate due to air cooling of the compost. Compost is generally piled into 6–8 ft windrows, which are turned at 1–60 day intervals. Compost may be aerated by forcing air through a huge rotating drum in which the compost rolls for 2–10 days.

The optimal composting temperature is 135°C. If this temperature is sustained long enough, pathogens, weed seeds, and insect eggs and larvae are destroyed. Usually, the inside of

Table 5 Average Treatment Performance of Constructed Wetlands for Livestock Waste

	BOD_5			TSS			NH_3-N			Total nitrogen		
Waste	Influent (mg/L)	Effluent (mg/L)	Removal (%)	Influent (mg/L)	Effluent (mg/L)	Removal (%)	Influent (mg/L)	Effluent (mg/L)	Removal (%)	Influent (mg/L)	Effluent (mg/L)	Removal (%)
Cattle feeding	137	24	83	291	55	81	5.1	2.2	57	—	—	—
Dairy	442	141	68	1,111	592	47	105	42	60	103	51	51
Poultry	153	115	25	—	—	—	74	59	20	89	70	22
Swine	104	44	58	128	62	52	366	221	40	407	248	39

Source: Ref. 21.

the pile reaches the optimal temperature, but the outside of the pile does not. Turning the pile can help to ensure that all of the material will be composted adequately.

The ideal carbon–nitrogen ratio is about 30 : 1. Carbon and nitrogen serve as energy sources and nutrients for microorganisms, respectively. Livestock manure also has a high nitrogen content for efficient composting. The composting process works best when animal livestock and plant products are mixed together. Nitrogen escapes as ammonia during composting if the C/N ratio is less than 20 : 1, although too much carbon can reduce the decomposition rate. Chopped straw, paper, and sawdust can be added to increase the ratio for success of manure composting.

There are three principal processing systems of composting: windrow, static pile, and in-vessel. In windrow composting, the manure is placed in long, narrow piles or windrows and is frequently agitated or turned to facilitate biological oxidation. Oxygen is supplied primarily by natural air movement instead of the artificial aeration used in conventional wastewater treatment. In addition, stirring the windrows can add oxygen and speed drying as well as eliminate odors. The manure is heated (120 and 160°F) and turned several times to mix in oxygen. The windrow, which has a width of 10–20 ft and a height of <3 ft, is suitable for large quantities of manure. Complete composting by the system can be achieved in a few weeks [2,5,6].

In static pile composting, the mixed manure is stacked in perforated plastic pipes. Air is supplied to the system through suction or positive pressure. In in-vessel composting, the manure is mixed with a bulking agent and then placed in containers or vessels. Controlled oxygen is supplied with a specific detention time [2,4–6].

Manure composting requires careful control and constant attention. Composting windrows must be sheltered from rain and snow to prevent leaching of nitrogen into the soil, and runoff should be collected. Composting can take place in winter, but should be sheltered to control moisture.

23.11 DEWATERING OF MANURE

Dewatering or drying may be used to further remove the water content from livestock and poultry manure. Fresh poultry manure has a dry matter content of over 20%, which makes it more suitable for drying. The drying can be carried out in two ways [2,6]:

1. *High-temperature drying.* This is carried out at elevated temperatures, which leads to dehydration of the manure. The dryers consist of oil burners supplying hot air to remove moisture from a continuous stream of waste material that is conveyed through the drying chamber on perforated drying trays. The initial temperature range of the drying air is 500–600°C.
2. *Low-temperature drying.* This process is carried out at temperatures equal to or very slightly above ambient temperature. Moisture is removed at a slower rate than that of high-temperature drying.

23.12 CASE STUDIES

23.12.1 Swine Waste Treatment

A swine farm discharged wastewater with supernatant COD of 4200–6000 mg/L and NH_3-N of 450–850 mg/L. The following three processes were used to treat the waste: (a) aerobic treatment; (b) anaerobic treatment followed by aerobic treatment; and (c) coagulation, air stripping for nitrogen removal, and aerobic treatment.

The characteristics of the influent and effluent, together with the operational parameters, are demonstrated in Table 6. As shown, the third process provides the best treatment results.

Table 6 Water Quality of Effluent by the Three Treatment Processes for Swine Waste Treatment

	Loading (kg/m2-day)		Influent (mg/L)		Effluent (mg/L)		
Process	COD	NH_3-N	COD	NH_3-N	COD	BODs	NH_3-N
1	0.6–0.7	0.08–0.10	4,000	570	700–800	21–25	300–500
2	1.5–2.0*	0.23–0.31*	5,600	850	700–800	23–25	900–1,000
	0.8–0.9**	0.5–0.55**					
3	2.0–3.0	0.16–0.18	3,600	180	<400	18–21	5–28

Note: 1, Aerobic treatment; 2, Anaerobic + aerobic treatment (*anaerobic treatment, **aerobic treatment); 3, Coagulation + air stripping + aerobic treatment.

Through the coagulation, most of the suspended solids could be removed, which resulted in some COD removal. The solution pH, at the same time, was increased, which provided a favorable operating condition for NH_3-N removal by the air stripping process. This nitrogen removal reduced the operating cost for the subsequent biological nitrogen treatment.

In the coagulation process, lime [$Ca(OH)_2$] was used. When dissolved in a water solution, it forms a floc, which can remove the suspended solids by various mechanisms, such as adsorption. A linear relationship was found between the COD removal efficiency and the dosage of lime:

$$\text{COD removal }\% = 0.01 \times C \tag{3}$$

where C is the concentration of lime (mg/L). The concentration of lime ranges from 2000 to 4000 mg/L.

It was found that the solution pH increased from 8.5 to 10.5 when the lime dosage was increased from 500 to 6000 mg/L. A higher pH can significantly enhance the percentage of NH_3 in the solution, which can be more easily removed by the air stripping. At pH 8, the NH_3-N removal was almost zero; however, it increased to 90% when the pH was increased to 10.5. The operating condition in the coagulation process was then set at lime of 3000–6000 mg/L (dependent on the temperature of the wastewater). In the air stripping process, the aeration time was set at 2–3 hours, pH was at 10–10.5, and the air to water ratio was 10 : 1.

After coagulation and air stripping treatment, the wastewater had a COD of 2600–3500 mg/L and NH_3-N of 140–180 mg/L. It was then treated by a sequencing batch reactor, whose mixed liquid suspended solids was 3500–4500 mg/L. The sludge loading was 0.5–0.7 kg COD/m^2 day. As shown in Table 6, the effluent was quite acceptable.

23.12.2 Chicken Waste Treatment

ABC Farm is the first fully automated farm in City of XYZ. It occupies an area of 10.4 hectares. It houses 500,000 chickens in 16 layers houses (where the eggs are laid and collected) and 6 grower houses (where the chicks are reared). The daily production of eggs is about 290,000.

The mechanical washing operation and transportation heavily contributes to the pollutant load due to egg breakage. It is estimated that about 5–10% of the eggs are broken, spilling contents into the washwater. The total volume of egg washwater is small, only about 5–10% of the plant's total water usage; however, it contains a high concentration of pollutants.

The waste treatment process is illustrated in Figure 12. The first unit operation is the screening, a mechanized sloping stationary screen solid–liquid separation device shown in

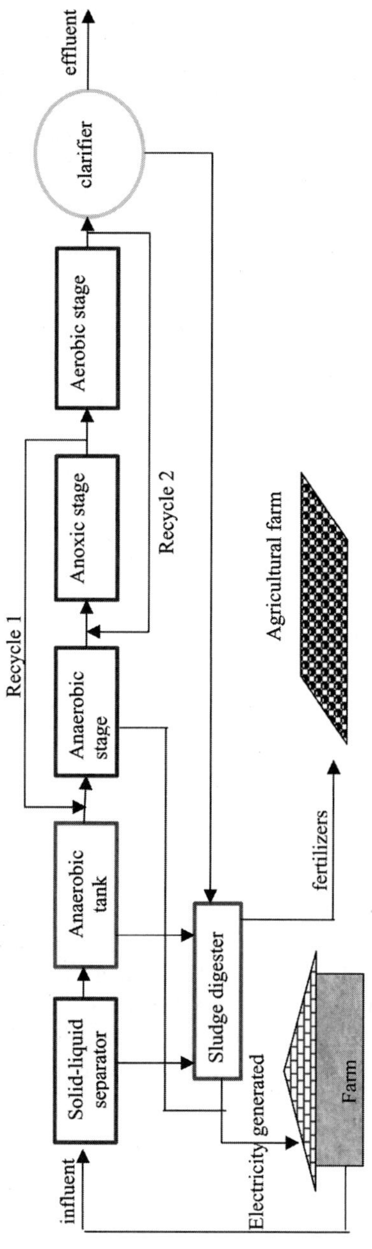

Figure 12 Illustration of treatment process for chicken waste.

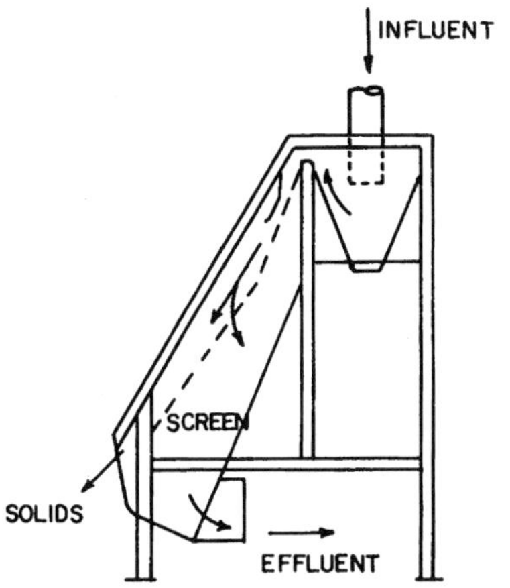

Figure 13 Mechanized sloping stationary screen solid-liquid separation device.

Figure 13. It is used to retain solid waste such as poultry feathers, broken eggs, wasted feed, and manure. Solids separation may exclude as much as 50% of layer manure solids. The screen is a wedge-wire screen and the screen surface is medium, ranging from 0.01 to 0.1 in. The device is specially designed to be in a triangular shape as this helps the solid waste to "slide" down the screen (through gravity) to the lower edge of the device where the solids are collected. The separated solids, due to the high nutritious value (broken eggs, manure, and feed), are then used as fertilizers. The effluent is transported to the anaerobic tank for further treatment. A completely mixed anaerobic activated sludge system is used. The temperature of the reactor is maintained between 30 and 38°C. Anaerobic processes can achieve high levels of organic removal, with an efficiency of up to 99%. The sludge produced in the anaerobic reactor is directed to the sludge digester, where the sludge is stabilized. Methane produced is collected and used to generate electricity to run the farm. The unremoved nitrogen and phosphorus is then directed to the VIP process for further treatment.

Odor from the lagoons is particularly severe when temperature increases to the point that biological activities become more intense. One method to reduce the odor is to reduce the loading rates, or to increase the treatment volume relative to organic loading. Lagoons in the warmer regions can be sized at higher loading rates because warmer water temperatures promote more rapid biological breakdown. Chemical and biological additives, masking agents, and other proprietary products can be used to improve lagoon performance, liquefy the accumulated solids and reduce odors. Another method is to add a lightly loaded, second-stage lagoon to an adequately sized primary anaerobic lagoon. The purpose of this second-stage lagoon is to further treat and store the effluent so as to lower the odor potential of the effluent. In this way, effluent can be used in land application or recycled as flushwater. For more detailed information on various biological treatment processes see Ref. 22–23.

REFERENCES

1. Barker J.C. *Lagoon Design and Management for Livestock Waste Treatment and Storage*; North Carolina Cooperative Extension Service: 1996.
2. USEPA. *Development document for the proposed revisions to the national pollutant discharge elimination system regulation and the effluent guidelines for concentrated animal feeding operations*, EPA821-R-01-003; USEPA: Washington, DC, 2001.

3. USEPA, *NPDES permit regulation and effluent limitations guidelines for concentrated animal feeding operations*, EPA821-F-03-003, Washington, DC, 2003.
4. USEPA, National pollutant discharge elimination system permit regulation and effluent limitation guidelines and standards for concentrated animal feeding operations (CAFOs); Final rule. *Federal Register*, **2003**, *68(29)*.
5. USEPA, *Proposed regulations to address water pollution from concentrated animal feeding operations*, EPA833-F-00-016; USEPA: Washington, DC, 2003.
6. USEPA, *National Pollutant Discharge Elimination System Permit Regulation and Effluent Limitation Guidelines and Standards for Concentrated Animal Feeding Operations (CAFOs)*, FRL2040-AD19; USEPA, 2003.
7. Araji, A.A.; Abdo, Z.O.; Joyce, P. Efficient use of animal manure on cropland – economic analysis. *Biores. Technol.*, **2001**, *79*, 179–191.
8. Danalewich, J.R.; Papagiannis, T.G.; Belyea, R.L.; Tumbleson, M.E.; Raskin, L. Characterization of dairy waste streams, current treatment practices, and potential for biological nutrient removal. *Water Res.*, **1998**, *32(12)*, 3555–3568.
9. Miner, J.R.; Goh, A.C.; Taiganides, E.P. Chemical treatment of anaerobic swine manure lagoon effluent and contents. *Trans. ASAE*, **1983**, *26(4)*, 1177–1180.
10. Miner, J.R.; Humenik, F.J.; Overcash, M.R. *Managing Livestock Waste to Preserve Environmental Quality*; Iowa State University Press, 2000.
11. Jackson, J; Jackson, J. *Environmental Science, The Natural Environment and Human Impact*; Longman: Singapore, 1996.
12. Department of Agriculture Natural Resources Conservations Service (USDA-NRCS). Conservation Practice Standard, *Waste Treatment Lagoon*, No Code 359; USDA-NRCS: Mississippi, 2000.
13. US Department of Agriculture Natural Resources Conservations Service (USDA-NRCS). *Environmental Considerations for Design and Construction of Poultry and Livestock Shelters*, Alabama Guide Sheet No. AL 313C; USDA-NRCS: Mississippi, 2001.
14. Angenent, L.T.; Sung, S.; Raskin, L. Methanogenic population dynamics during startup of a full-scale anaerobic sequencing batch reactor treating swine waste. *Water Res.*, **2002**, *36*, 4648–4654.
15. Sharma, V.K.; Testa, C.; Castelluccio, G. Anaerobic treatment of semi-solid organic waste. *Energy Convers. Mgnt.* **1999**; *40*, 369–384.
16. Boopathy, R. Biological treatment of swine waste using anaerobic baffled reactors, *Biores. Technol.* **1998**, *64*, 1–6.
17. Westerman, P.W.; Bicudo, J.R. Tangential flow separation and chemical enhancement to recover swine manure solids, nutrients and metals. *Biores. Technol.*, **2000**, *73*, 1–11.
18. USEPA, *Design Manual, Constructed Wetlands and Aquatic Plant Systems or Municipal Wastewater Treatment*, EPA/625/1-88/022, USEPA, Office of Research and Development; Cincinnati, OH, September 1988.
19. Hammer, D.A. Designing constructed wetlands systems to treat agricultural nonpoint source pollution. *Ecol. Enging.*, **1992**, *1*, 49–82.
20. El-Fadel, M.; Massoud, M. Methane emission from wastewater management. *Environ. Pollut.*, **2001**, *114*, 177–185.
21. Knight, R.L.; Payne, V.W.E.J.; Borer, R.E.; Clarke, R.A.J.; Pries, J.H. Constructed wetlands for livestock wastewater management. *Ecol. Engng.*, **2000**, *15*, 41–55.
22. Wang, L.K.; Pereira, N.C.; Hung, Y.T. (eds.). *Biological Treatment Processes.* Humana Press: Totowa NJ, 2004.
23. Wang, L.K.; Hung, Y.T.; Shammas, N.K. (eds.). *Advanced Biological Treatment Processes.* Humana Press: Totowa NJ, 2004.
24. Wang, L.K.; Hung, Y.T.; Shammas, N.K. (eds.). *Physicochemical Treatment Processes.* Humana Press: Totowa NJ, 2004.

24
Soft Drink Waste Treatment

J. Paul Chen and Swee-Song Seng
National University of Singapore, Singapore

Yung-Tse Hung
Cleveland State University, Cleveland, Ohio, U.S.A.

24.1 INTRODUCTION

The history of carbonated soft drinks dates back to the late 1700s, when seltzer, soda, and other waters were first commercially produced. The early carbonated drinks were believed to be effective against certain illnesses such as putrid fevers, dysentery, and bilious vomiting. In particular, quinine tonic water was used in the 1850s to protect British forces abroad from malaria.

The biggest breakthrough was with Coca-Cola, which was shipped to American forces wherever they were posted during World War II. The habit of drinking Coca-Cola stayed with them even after they returned home. Ingredients for the beverage included coca extracted from the leaves of the Bolivian Coca shrub and cola from the nuts and leaves of the African cola tree. The first Coca-Cola drink was concocted in 1886. Since then, the soft drink industry has seen its significant growth.

Table 1 lists the top 10 countries by market size for carbonated drinks, with the United States leading the pack with the largest market share. In 1988 the average American's

Table 1 Top Ten World Market Size in Carbonated Soft Drinks, 1988

Rank	Country	1000 million liters
1	United States	42.7
2	Mexico	8.4
3	China	7.0
4	Brazil	5.1
5	West Germany	4.6
6	United Kingdom	3.5
7	Italy	2.6
8	Japan	2.5
9	Canada	2.4
10	Spain	2.3

Source: Ref. 1.

consumption of soft drinks was 174 L/year; this figure has increased to approximately 200 L/year in recent years. In 2001, the retail sales of soft drinks in the United States totaled over $61 billion. The US soft drink industry features nearly 450 different products, employs more than 183,000 nationwide and pays more than $18 billion annually in state and local taxes.

The soft drink industry uses more than 12 billion gallons of water during production every year. Therefore, the treatment technologies for the wastewater resulting from the manufacturing process cannot be discounted. This chapter reviews the technologies that are typically used to treat soft drink wastewater.

24.1.1 Composition of Soft Drinks

The ingredients of soft drinks can vary widely, due to different consumer tastes and preferences. Major components include primarily water, followed by carbon dioxide, caffeine, sweeteners, acids, aromatic substances, and many other substances present in much smaller amounts. Table 2 lists calories and components of major types of soft drinks.

Water

The main component of soft drinks is water. Regular soft drinks contain 90% water, while diet soft drinks contain up to 99% water. The requirement for water in soft drink manufacturing is that it must be pure and tasteless. For this reason, some form of pretreatment is required if the tap water used has any kind of taste. The pretreatment can include coagulation–flocculation, filtration, ion exchange, and adsorption.

Carbon Dioxide

The gas present in soft drinks is carbon dioxide. It is a colorless gas with a slightly pungent odor. When carbon dioxide dissolves in water, it imparts an acidic and biting taste, which gives the drink a refreshing quality by stimulating the mouth's mucous membranes. Carbon dioxide is delivered to soft drink factories in liquid form and stored in high-pressure metal cylinders.

Carbonation can be defined as the impregnation of a liquid with carbon dioxide gas. When applied to soft drinks, carbonation makes the drinks sparkle and foam as they are dispensed and consumed. The escape of the carbon dioxide gas during consumption also enhances the aroma since the carbon dioxide bubbles drag the aromatic components as they move up to the surface of the soft drinks. The amount of the carbon dioxide gas producing the carbonation effects is specified in volumes, which is defined as the total volume of gas in the liquid divided by the volume of the liquid. Carbonation levels usually vary from one to a few volumes of carbon dioxide. Figure 1 shows the typical carbonation levels for a range of well-known drinks [1].

In addition, the presence of carbon dioxide in water inhibits microbiological growth. It has been reported that many bacteria die in a shorter time period in carbonated water than in noncarbonated water.

Caffeine

Caffeine is a natural aromatic substance that can be extracted from more than 60 different plants including cacao beans, tea leaves, coffee beans, and kola nuts. Caffeine has a classic bitter taste that enhances other flavors and is used in small quantities.

Table 2 List of Energy and Chemical Content per Fluid Ounce

Flavor types	Calories	Carbohydrates (g)	Total sugars (g)	Sodium (mg)	Potassium (mg)	Phosphorus (mg)	Caffeine (mg)	Aspartame (mg)
Regular								
Cola or Pepper	12–14	3.1–3.6	3.1–3.6	0–2.3	0–1.5	3.3–6.2	2.5–4.0	0
Caffeine-free cola or Pepper	12–15	3.1–3.7	3.1–3.7	0–2.3	0–1.5	3.3–6.2	0	0
Cherry cola	12–15	3.0–3.7	3.0–3.7	0–1.2	0–1.0	3.9–4.5	1.0–3.8	0
Lemon-lime (clear)	12–14	3.0–3.5	3.0–3.5	0–4.6	0–0.3	0–0.1	0	0
Orange	14–17	3.4–4.3	3.4–4.3	1.1–3.5	0–1.4	0–5.0	0	0
Other citrus	10–16	2.5–4.1	2.5–4.1	0.8–4.1	0–10.0	0–0.1	0–5.3	0
Root beer	12–16	3.1–4.1	3.1–4.1	0.3–5.1	0–1.6	0–1.6	0	0
Ginger ale	10–13	2.6–3.2	2.6–3.2	0–2.3	0–0.3	0–trace	0	0
Tonic water	10–12	2.6–2.9	2.6–2.9	0–0.8	0–0.3	0–trace	0	0
Other regular	12–18	3.0–4.5	3.0–4.5	0–3.5	0–2.0	0–7.8	0–3.6	0
Juice added	12–17	3.0–4.2	3.0–4.2	0–1.8	2.5–10.0	0–6.2	0	0
Diet								
Diet cola or pepper	<1	0–0.1	0	0–5.2	0–5.0	2.1–4.7	0–4.9	0–16.0
Caffeine-free diet cola, pepper	<1	0–0.1	0	0–6.0	0–10.0	2.1–4.7	0	0–16.0
Diet cherry cola	<1	0–<0.04	0–trace	0–0.6	1.5–5.0	2.3–3.4	0–3.8	15.0–15.6
Diet lemon-lime	<1	0–0.1	0	0–7.9	0–6.9	0–trace	0	0–16.0
Diet root beer	<2	0–0.4	0	3.3–8.5	0–3.0	0–1.6	0	0–17.5
Other diets	<6	0–1.5	0–1.5	0–8.0	0.3–10.1	0–trace	0–5.8	0–17.0
Club soda, Seltzer, sparkling water	0	0	0	0–8.1	0–0.5	0–0.1	0	0
Diet juice added	<3	0.1–0.5	0.1–0.5	0–1.8	0–9.0	0–5.0	0	11.4–16.0

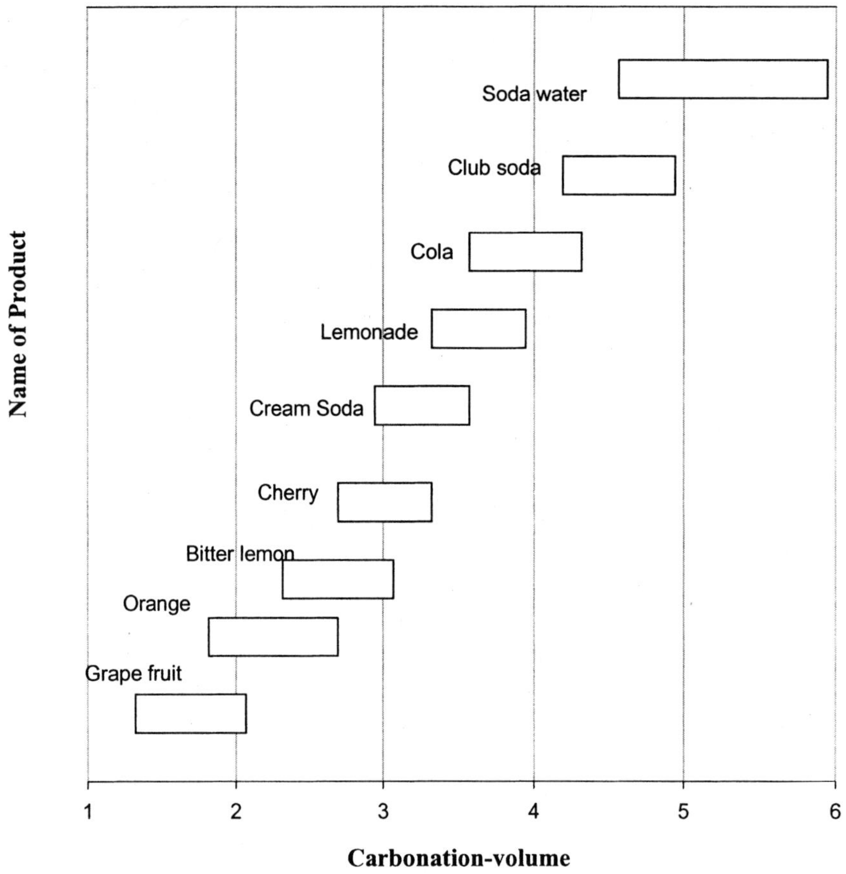

Figure 1 Carbonation levels of various popular soft drinks.

Sweeteners

Nondiet and diet soft drinks use different types of sweeteners. In nondiet soft drinks, sweeteners such as glucose and fructose are used. Regular (nondiet) soft drinks contain about 7–14% sweeteners, the same as fruit juices such as pineapple and orange. Most nondiet soft drinks are sweetened with high fructose corn syrup, sugar, or a combination of both. Fructose is 50% sweeter than glucose and is used to reduce the number of calories present in soft drinks.

In diet soft drinks, "diet" or "low calorie" sweeteners such as aspartame, saccharin, suralose, and acesulfame K are approved for use in soft drinks. Many diet soft drinks are sweetened with aspartame, an intense sweetener that provides less than one calorie in a 12 ounce can. Sweeteners remain an active area in food research because of the increasing demand in consumer's tastes and preferences.

Acids

Citric acid, phosphoric acid, and malic acid are the common acids found in soft drinks. The function of introducing acidity into soft drinks is to balance the sweetness and also to act as a preservative. Its importance lies in making the soft drink fresh and thirst-quenching. Citric acid is naturally found in citrus fruits, blackcurrants, strawberries, and raspberries. Malic acid is found in apples, cherries, plums, and peaches.

Other Additives

Other ingredients are used to enhance the taste, color, and shelf-life of soft drinks. These include aromatic substances, colorants, preservatives, antioxidants, emulsifying agents, and stabilizing agents.

24.1.2 Manufacturing and Bottling Process of Soft Drinks

The manufacturing and bottling process for soft drinks varies by region and by endproducts. Generally, the process consists of four main steps: syrup preparation; mixing of carbonic acid, syrup and water; bottling of the soft drink; and inspection.

Syrup Preparation

The purpose of this step is to prepare a concentrated sugar solution. The types of sugar used in the soft drinks industry include beet sugar and glucose. For the production of "light" drinks, sweeteners or a combination of sugar and sweeteners is used instead. After the preliminary quality control, other minor ingredients such as fruit juice, flavorings, extracts, and additives may be added to enhance the desired taste.

Mixing of Carbonic Acid, Syrup, and Water

In this second step, the finished syrup, carbonic acid, and water of a fixed composition are mixed together in a computer-controlled blender. This is carried out on a continuous basis. After the completion of the mixing step, the mixed solution is conveyed to the bottling machine via stainless steel piping. A typical schematic diagram of a computer-controlled blender is shown in Fig. 2.

Bottling of Soft Drinks

Empty bottles or cans enter the soft drinks factory in palletized crates. A fully automated unpacking machine removes the bottles from the crates and transfers them to a conveyer belt. The unpacking machines remove the caps from the bottles, then cleaning machines wash the bottles repeatedly until they are thoroughly clean. The cleaned bottles are examined by an inspection machine for any physical damage and residual contamination.

Inspection

This step is required for refillable plastic bottles. A machine that can effectively extract a portion of the air from each plastic bottle is employed to detect the presence of any residual foreign substances. Bottles failing this test are removed from the manufacturing process and destroyed.

A typical bottling machine resembles a carousel-like turret. The speed at which the bottles or cans are filled varies, but generally the filling speed is in excess of tens of thousands per hour. A sealing machine then screws the caps onto the bottles and is checked by a pressure tester machine to see if the bottle or can is properly filled. Finally, the bottles or cans are labeled, positioned into crates, and put on palettes, ready to be shipped out of the factory.

Before, during, and after the bottling process, extensive testing is performed on the soft drinks or their components in the laboratories of the bottling plants. After the soft drinks leave the manufacturing factory, they may be subjected to further testing by external authorities.

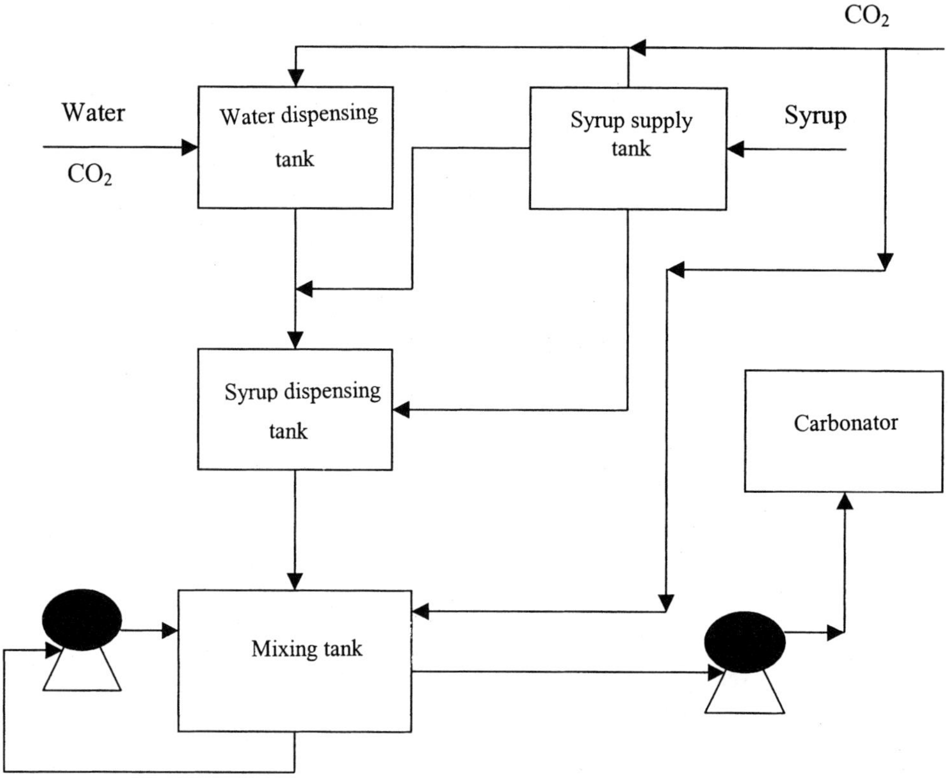

Figure 2 Schematic diagram of a computer-controlled blender.

24.2 CHARACTERISTICS OF SOFT DRINK WASTEWATER

Soft drink wastewater consists of wasted soft drinks and syrup, water from the washing of bottles and cans, which contains detergents and caustics, and finally lubricants used in the machinery. Therefore, the significant associated wastewater pollutants will include total suspended solids (TSS), five-day biochemical oxygen demand (BOD_5), chemical oxygen demand (COD), nitrates, phosphates, sodium, and potassium (Table 2). Table 3 gives a list of typical wastewater parameters. As shown, higher organic contents indicate that anaerobic treatment is a feasible process.

24.3 BIOLOGICAL TREATMENT FOR SOFT DRINK WASTEWATER

Biological treatment is the most common method used for treatment of soft drink wastewater because of the latter's organic content (Table 3). Since BOD_5 and COD levels in soft drink wastewaters are moderate, it is generally accepted that anaerobic treatment offers several advantages compared to aerobic alternatives. Anaerobic treatment can reduce BOD_5 and COD from a few thousands to a few hundreds mg/L; it is advisable to apply aerobic treatment for further treatment of the wastewater so that the effluent can meet regulations. High-strength wastewater normally has low flow and can be treated using the anaerobic process; low-strength wastewater together with the effluent from the anaerobic treatment can be treated by an aerobic process.

Table 3 Soft Drink Wastewater Characteristics

Item	Value (mg/L)
COD	1200–8000
BOD_5	600–4500
Alkalinity	1000–3500
TSS	0–60
VSS	0–50
NH_3-N	150–300
PO_4-P	20–40
SO_4	7–20
K	20–70
Fe	10–20
Na	1500–2500
Ni	1.2–2.5
Mo	3–8
Zn	1–5
Co	3–8

A complete biological treatment includes optional screening, neutralization/equalization, anaerobic and aerobic treatment or aerobic treatment, sludge separation (e.g., sedimentation or dissolved air flotation), and sludge disposal. Chemical and physical treatment processes (e.g., coagulation and sedimentation/flotation) are occasionally used to reduce the organic content before the wastewater enters the biological treatment process. Since the wastewater has high sugar content, it can promote the growth of filamentous bacteria with lower density. Thus, dissolved air flotation may be used instead of the more commonly used sedimentation.

24.4 AEROBIC WASTEWATER TREATMENT

Owing to the high organic content, soft drink wastewater is normally treated biologically; aerobic treatment is seldom applied. If the waste stream does not have high organic content, aerobic treatment can still be used because of its ease in operation. The removal of BOD and COD can be accomplished in a number of aerobic suspended or attached (fixed film) growth treatment processes. Sufficient contact time between the wastewater and microorganisms as well as certain levels of dissolved oxygen and nutrients are important for achieving good treatment results. An aerobic membrane bioreactor (MBR) for organic removal as well as separation of biosolids can be used in the wastewater treatment.

24.4.1 Aerobic Suspended Growth Treatment Process

Aerobic suspended growth treatment processes include activated sludge processes, sequencing batch reactors (SBR), and aerated lagoons. Owing to the characteristics of the wastewater, the contact time between the organic wastes and the microorganisms must be higher than that for domestic wastewater. Processes with higher hydraulic retention time (HRT) and solids retention time (SRT), such as extended aeration and aerated lagoon, are recommended to be used.

O'Shaughnessy et al. [2] reported that two aerobic lagoons with volume of 267,800 gallons each were used to treat a wastewater from a Coca Cola bottling company. Detention time was 30 days; the design flow was 20,000 gpd. A series of operational problems occurred in the early phase, including a caustic spill incident, continuous clogging of air diffusers, and bad effluent quality due to shock loading (e.g., liquid sugar spill). Failure to meet effluent standards was a serious problem in the treatment plant. It was observed that the effluent BOD_5 and COD were above 100 and 500 mg/L, respectively. This problem, however, was solved by addition of potassium; the effluent BOD_5 decreased to 60 mg/L.

Tebai and Hadjivassilis [3] used an aerobic process to treat soft drink wastewater with a daily flow of 560 m^3/day, BOD_5 of 564 mg/L, and TSS of 580 mg/L. Before beginning biological treatment, the wastewater was first treated by physical and chemical treatment processes. The physical treatment included screening and influent equalization; in the chemical treatment, pH adjustment was performed followed by the traditional coagulation/flocculation process. A BOD_5 and COD removal of 43.2 and 52.4%, respectively, was achieved in the physical and chemical treatment processes. In the biological treatment, the BOD_5 loading rate and the sludge loading rate were 1.64 kg BOD_5/day m^3 and 0.42 kg BOD_5/kg MLSS day; the BOD_5 and COD removal efficiencies were 64 and 70%, respectively. The biological treatment was operated at a high-rate mode, which was the main cause for the lower removal efficiencies of BOD_5 and COD.

24.4.2 Attached (Fixed Film) Growth Treatment Processes

Aerobic attached growth treatment processes include a trickling filter and rotating biological contactor (RBC). In the processes, the microorganisms are attached to an inert material and form a biofilm. When air is applied, oxidation of organic wastes occurs, which results in removal of BOD_5 and COD.

In a trickling filter, packing materials include rock, gravel, slag, sand, redwood, and a wide range of plastic and other synthetic materials [4]. Biodegradation of organic waste occurs as it flows over the attached biofilm. Air through air diffusers is provided to the process for proper growth of aerobic microorganisms.

An RBC consists of a series of closely placed circular discs of polystyrene or polyvinyl chloride submerged in wastewater; the discs are rotated through the wastewater. Biodegradation thus can take place during the rotation.

A trickling filter packed with ceramic tiles was used to treat sugar wastewater. The influent BOD_5 and COD were 142–203 mg/L and 270–340 mg/L; the organic loading was from 5 to 120 g BOD_5/m^2 day. Removal efficiencies of BOD_5 of 88.5–98% and COD of 67.8–73.6% were achieved. The process was able to cope effectively with organic shock loading up to 200 g COD/L [5].

An RBC was recommended for treatment of soft drink bottling wastewater in the Cott Corporation. The average wastewater flow rate was 60,000 gpd; its BOD_5 was 3500 mg/L; and TSS was of the order of 100 mg/L. Through a laboratory study and pilot-plant study, it was found that RBC demonstrated the capability of 94% BOD_5 removal at average loading rate of 5.3 lb BOD_5 applied per 1000 square feet of media surface [6].

24.5 ANAEROBIC WASTEWATER TREATMENT

The anaerobic process is applicable to both wastewater treatment and sludge digestion. It is an effective biological method that is capable of treating a variety of organic wastes. Because the anaerobic process is not limited by the efficiency of the oxygen transfer in an aerobic process, it is more suitable for treating high organic strength wastewaters ($\geq$5 g COD/L). Disadvantages of the process include slow startup, longer retention time, undesirable odors from production of hydrogen sulfite and mercaptans, and a high degree of difficulty in operating as compared to aerobic processes. The microbiology of the anaerobic process involves facultative and anaerobic microorganisms, which in the absence of oxygen convert organic materials into mainly gaseous carbon dioxide and methane.

Two distinct stages of acid fermentation and methane formation are involved in anaerobic treatment. The acid fermentation stage is responsible for conversion of complex organic waste (proteins, lipids, carbohydrates) to small soluble product (triglycerides, fatty acids, amino acids, sugars, etc.) by extracellular enzymes of a group of heterogeneous and anaerobic bacteria. These small soluble products are further subjected to fermentation, β-oxidations, and other metabolic processes that lead to the formation of simple organic compounds such as short-chain (volatile) acids and alcohols. There is no BOD_5 or COD reduction since this stage merely converts complex organic molecules to simpler molecules, which still exert an oxygen demand. In the second stage (methane formation), short-chain fatty acids are converted to acetate, hydrogen gas, and carbon dioxide in a process known as acetogenesis. This is followed by methanogenesis, in which hydrogen produces methane from acetate and carbon dioxide reduction by several species of strictly anaerobic bacteria.

The facultative and anaerobic bacteria in the acid fermentation stage are tolerant to pH and temperature changes and have a higher growth rate than the methanogenic bacteria from the second stage. The control of pH is critical for the anaerobic process as the rate of methane fermentation remains constant over pH 6.0–8.5. Outside this range, the rate drops drastically. Therefore, maintaining optimal operating conditions is the key to success in the anaerobic process [7]. Sodium bicarbonate and calcium bicarbonate can be added to provide sufficient buffer capacity to maintain pH in the above range; ammonium chloride, ammonium nitrate, potassium phosphate, sodium phosphate, and sodium tripolyphosphate can be added to meet nitrogen and phosphorus requirements.

A number of different bioreactors are used in anaerobic treatment. The microorganisms can be in suspended, attached or immobilized forms. All have their advantages and disadvantages. For example, immobilization is reported to provide a higher growth rate of methanogens since their loss in the effluent can be diminished; however, it could incur additional material costs. Typically, there are three types of anaerobic treatment processes. The first one is anaerobic suspended growth processes, including complete mixed processes, anaerobic contactors, anaerobic sequencing bath reactors; the second is anaerobic sludge blanket processes, including upflow anaerobic sludge blanket (UASB) reactor processes, anaerobic baffled reactor (ABR) processes, anaerobic migrating blanket reactor (AMBR) processes; and the last one is attached growth anaerobic processes with the typical processes of upflow packed-bed attached growth reactors, upflow attached growth anaerobic expanded-bed reactors, attached growth anaerobic fluidized-bed reactors, downflow attached growth processes. A few processes are also used, such as covered anaerobic lagoon processes and membrane separation anaerobic treatment processes [4].

It is impossible to describe every system here; therefore, only a select few that are often used in treating soft drink wastewater are discussed in this chapter. Figure 3 shows the schematic diagram of various anaerobic reactors, and the operating conditions of the corresponding reactors are given in Table 4.

24.5.1 Upflow Anaerobic Sludge Blanket Reactor

The upflow anaerobic sludge blanket reactor, which was developed by Lettinga, van Velsen, and Hobma in 1979, is most commonly used among anaerobic bioreactors with over 500 installations

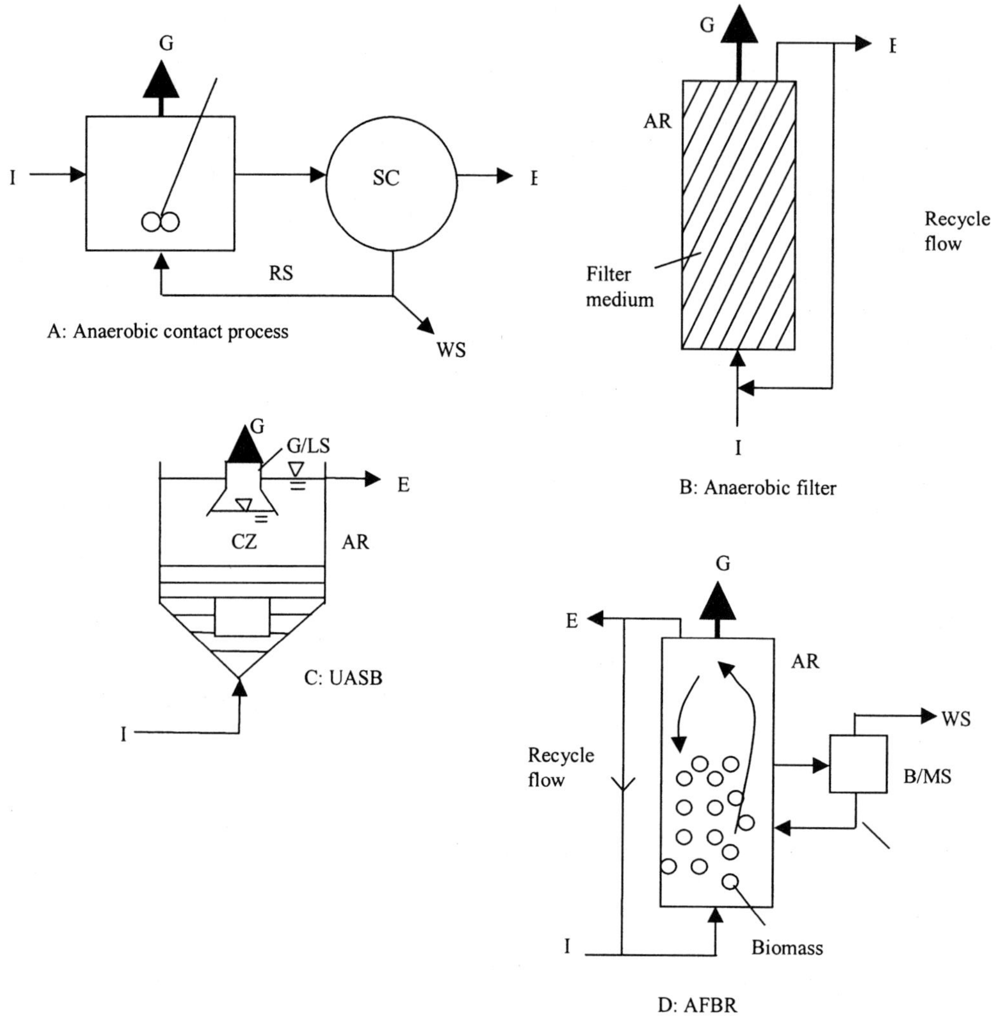

Figure 3 Schematic diagram of various anaerobic wastewater treatment reactors. AR: anaerobic reactor; B/MS: biofilm/media separator; CZ: clarification zone; E: effluent; G: biogas; G/LS: gas-liquid separator; I: influent; RS: return sludge; SC: secondary clarifier; SZ: sludge zone; WS: waste slude.

treating a wide range of industrial wastewaters [4]. The UASB is essentially a suspended-growth reactor with the fixed biomass process incorporated. Wastewater is directed to the bottom of the reactor where it is in contact with the active anaerobic sludge solids distributed over the sludge blanket. Conversion of organics into methane and carbon dioxide gas takes place in the sludge blanket. The sludge solids concentration in the sludge bed can be as high as 100,000 mg/L. A gas–liquid separator is usually incorporated to separate biogas, sludge, and liquid. The success of UASB is dependent on the ability of the gas–liquid separator to retain sludge solids in the system. Bad effluent quality occurs when the sludge flocs do not form granules or form granules that float.

The UASB can be used solely or as part of the soft drink wastewater treatment process. Soft drink wastewater containing COD of 1.1–30.7 g/L, TSS of 0.8–23.1 g/L, alkalinity of

Table 4 Operating Conditions of Common Anaerobic Reactors

Reactor type	AC	UASB	AF	AFBR
Organic loading (kg COD/ m^3-day)	0.48–2.40	4.00–12.01	0.96–4.81	4.81–9.61
COD removal (%)	75–90	75–85	75–85	80–85
HRT (hour)	2–10	4–12	24–48	5–10
Optimal temperature (°C)		30–35 (mesophilic) 49–55 (thermophilic)		
Optimal pH		6.8–7.4		
Optimal total alkalinity ($mgCaCO_3/L$)		2000–3000		
Optimal volatile acids (mg/L as acetic acid)		50–500		

AC, anaerobic contactor; UASB, upflow anaerobic sludge bed; AF, anaerobic filter; AFBR, anaerobic fluidized bed reactor.
Source: Ref. 7.

1.25–1.93 g $CaCO_3$/L, nitrogen of 0–0.05 g N/L and phosphate of 0.01–0.07 gP/L was treated by a 1.8 L UASB reactor [8]. The pH of wastewater was 4.3–13.0 and temperature was between 20 and 32°C. The highest organic loading reported was 16.5 kg COD m^{-3} day^{-1}. A treatment efficiency of 82% was achieved.

The "Biothane" reactor is a patented UASB system developed by the Bioethane Corporation in the United States. Its industrial application in wastewater treatment systems was described by Zoutberg and Housley [9]. The wastewater mainly consists of waste sugar solution, product spillage, and wastewater from the production lines. The flow rate averages about 900 m^3/day with an average BOD and COD load of 2340 and 3510 kg/day, respectively. The soft drink factory was then producing 650 × 10^6 L of product annually, with three canning lines each capable of producing 2000 cans/min and three bottling lines each capable of filling 300 bottles/min. A flow diagram of the "Biothane" wastewater treatment plant is shown in Fig. 4. Monitoring of the plant could be performed on or off site. A supervisory control and data acquisition system (SCADA) was responsible for providing continuous monitoring of the process and onsite equipment. In normal operation, COD removal of 75–85% was reported with 0.35 m^3 of biogas produced per kg COD.

24.5.2 Anaerobic Filters

The anaerobic filter was developed by Yong and McCarty in the late 1960s. It is typically operated like a fixed-bed reactor [10], where growth-supporting media in the anaerobic filter contacts wastewater. Anaerobic microorganisms grow on the supporting media surfaces and void spaces in the media particles. There are two variations of the anaerobic filters: upflow and downflow modes. The media entraps SS present in wastewater coming from either the top (downflow filter) or the bottom (upflow filter). Part of the effluent is recycled and the magnitude of the recycle stream determines whether the reactor is plug-flow or completely mixed. To prevent bed clogging and high head loss problems, backwashing of the filter must be periodically performed to remove biological and inert solids trapped in the media [7]. Turbulent fluid motion that accompanies the rapid rise of the gas bubbles through the reactor can be helpful to remove solids in the media [10].

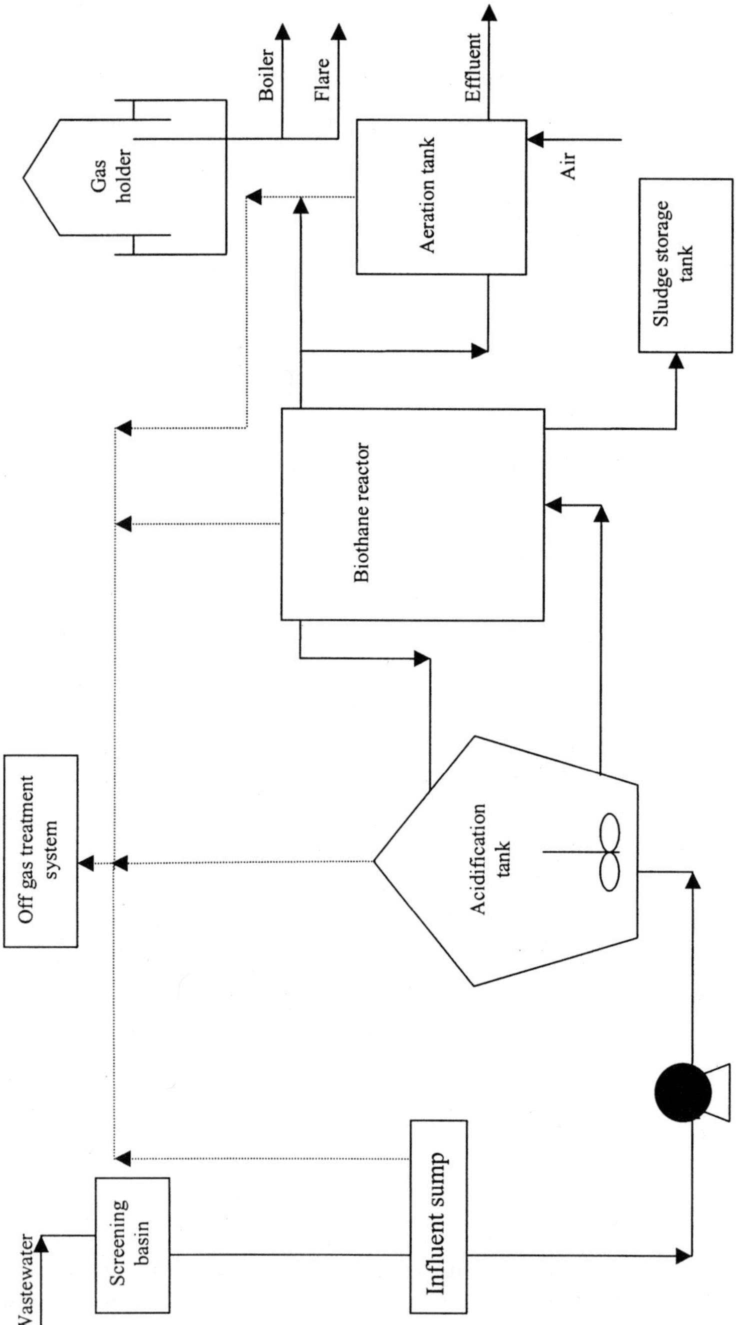

Figure 4 Flow diagram of the "Biothane" wastewater treatment plant.

Siino *et al.* [11] used an anaerobic filter to treat soluble carbohydrate waste (soft drink wastewater). At an HRT of 1.7 days, organic loading of 44–210 lb. COD/1000 ft^3/day, and SRT of 137 days, removal of 85–90% of COD ranging from 1200 to 6000 mg/L can be achieved. The percentage of methane ranged from 60 to 80%; its product was 0.13–0.68 ft^3/day. COD removal efficiency (E %) can be estimated by the following equation:

$$E = 93(1 - 1.99/\text{HRT}) \tag{1}$$

24.5.3 Anaerobic Fluidized Bed Reactor

Soft drink wastewater can also be treated by an anaerobic fluidized bed reactor (AFBR), which is similar in design to the upflow expanded-bed reactor. Influent wastewater enters the reactor from the bottom. Biomass grows as a biolayer around heavy small media particles. At a certain upflow velocity, the weight of the media particles equals the drag force exerted by the wastewater. The particles then become fluidized and the height of the fluidized bed is stabilized.

Packing size of 0.3–0.8 mm and upflow liquid velocities of 10–30 m/hour can be used in order to provide 100% bed expansion. The high flow velocity around the media particles provides good mass transfer of the dissolved organic matter from the bulk liquid to the particle surface. The bed depth normally ranges from 4 to 6 m. Sand, diatomaceous earth, anion and cation exchange resins, and activated carbon can be used as packing materials [4]. The overall density of media particles decreases as the biomass growth accumulates on the surface areas. This can cause the biomass attached media particles to rise in the reactor and eventually wash out together with the effluent. To prevent this from occurring, a portion of the biomass attached particles is wasted and sent to a mechanical device where the biomass is separated from the media particles. The cleaned particles are then returned to the reactor, while the separated biomass is wasted as sludge [7,12]. Owing to the high turbulence and thin biofilms developed in an AFBR, biomass capture is relatively weak; therefore, an AFBR is better suited for wastewater with mainly soluble COD.

Borja and Banks [13] reported that bentonite, saponite, and polyurethane were respectively used as the suspended support materials for three AFBRs. The composition and parameters of the soft drink wastewater were: total solids (TS) of 3.7 g/L; TSS of 2.9 g/L; volatile suspended solids (VSS) of 2.0 g/L; COD of 4.95 g/L; volatile acidity (acetic acid) of 0.12 g/L; alkalinity of 0.14 g $CaCO_3$/L; ammonium of 5 mg/L; phosphorus of 12 mg/L; pH of 4.8. The average COD removal efficiencies for the three reactors were 89.9% for bentonite, 93.3% for saponite, 91.9% for polyurethane. The amount of biogas produced decreases with increasing HRT. The percentages of methane were 66.0% (bentonite), 72.0% (saponite), and 69.0% (polyurethane).

Borja and Banks [14] used zeolite and sepiolite as packing materials in AFBRs to treat soft drink wastewater. On average, the COD removal of 77.8% and yield coefficient of methane was 0.325 L CH_4/g COD destroyed. The effluent pH was around 7.0–7.3 in all reactors. The content of methane in the biogas ranges from 63 to 70%.

Hickey and Owens [15] conducted a pilot-plant study on the treatment of soft drink bottling wastewater using an AFBR. Diluted soda syrup was used as the substrate, and nitrogen and phosphorus were added with a COD : N : P ratio of 100 : 3 : 0.5. An organic loading rate of 4.0–18.5 kg COD/m^3 day results in BOD_5 and COD removal of 61–95% and 66–89%, respectively. Within this organic loading range, the solids production varies from 0.029 to 0.083 kg TSS/kg COD removed. Methane gas was produced at a rate of 0.41 L/g COD destroyed. The composition of the biogas consists of 60% methane and 40% CO_2.

24.5.4 Combined Anaerobic Treatment Process

A combination of different anaerobic reactors has been used to treat soft drink wastewater. It has been reported that treatment efficiency and liability for combined reactors are better than those of a single type of reactor. Several examples are given below.

Stronach *et al.* [16] reported that a combination of upflow anaerobic sludge blanket reactor, anaerobic fluidized-bed reactor, and anaerobic filter was used to treat fruit processing and soft drink wastewater with TSS, COD, and pH of 160–360 mg/L, 9–15 g/L, and 3.7–6.7, respectively. The organic loadings were 0.75–3.00 kg COD m^{-3} day^{-1} for all three different reactors. COD removal efficiency >79% was achieved. The AFBR performed better than the UASB and the AF in terms of COD removal efficiency and pH stability; however, the methane production was the greatest in the UASB.

Vicenta *et al.* [17] reported that a 68 L semipilot scale AF installed in series with a UASB was used to treat bottling wastes (bottling washing water and spent syrup wastewater). At an organic loading of 0.59 and 0.88 kg COD m^{-3} day^{-1} for the AF and UASB respectively, an overall COD removal of 75% was achieved. The hydraulic retention time (HRT) for the AF and UASB was maintained at 3.4 and 2.2 days, respectively. An average gas yield of 0.83 L per L of influent was produced.

Silverio *et al.* [18] used a series of UASB and upflow AF and trickling filter to treat bottling wastewater with pH of 7.6, COD of 7500 mg/L, TSS of 760 mg/L, and alkalinity of 370 mg $CaCO_3$/L, respectively. The total capacity of the reactors in series is 239 L. An organic loading of 2.78 kg COD m^{-3} day^{-1} and HRT of 2.5 days achieved COD removal of 73% and gas yield of 1 L per L of wastewater in the UASB. The COD level of the effluent from the AF after the UASB further dropped to 550 mg/L and corresponded to a removal efficiency of 87%. The HRT and organic loading in the AF were 2.2 days and 0.88 kg COD m^{-3} day^{-1}, respectively. Incorporation of the trickling filter further reduced the COD level of the effluent to 100 mg/L [18]. All biological treatment processes are discussed in detail in Wang et al. [19] and Wang et al. [20].

REFERENCES

1. Mitchell, A.J. *Formulation and Production of Carbonated Soft Drinks*; Blackie: Glasgow and London, 1990.
2. O'Shaughnessy, J.C.; Blanc, F.C.; Corr, S.H.; Toro, A. Enhanced treatment of high strength soft drink bottling wastewaters, *42nd Annual Purdue Industrial Waste Conference*, 1987; 607–618.
3. Tebai, L.; Hadjivassilis, I. Soft drinks industry wastewater treatment. *Water Sci. Technol.*, **1992**, *25*, 45–51.
4. Metcalf and Eddy. *Wastewater Engineering: Treatment Disposal Reuse*, 4th ed.; McGraw-Hill, 2003.
5. Hamoda, M.F.; Al-Sharekh, H.A. Sugar wastewater treatment with aerated fixed-film biological systems. *Water Sci. Technol*, **1999**, *40*, 313–321.
6. Blanc, F.C.; O'Shaughnessy, J.C.; Miller, C.H. Treatment of bottling plant wastewater with rotating biological contactors, *33rd Annual Purdue Industrial Waste Conference*, 1978; 614–623.
7. Liu, H.F. Wastewater treatment. In *Environmental Engineers's Handbook*, 2nd ed.; Lewis Publishers: Boca Raton, New York, 1997; 714–720.
8. Kalyuzhnyi, S.V.; Saucedo, J.V.; Martinez, J.R. The anaerobic treatment of soft drink wastewater in UASB and hybrid reactors. *Appl. Biochem. Biotech.*, **1997**; *66*, 291–301.
9. Housley, J.N.; Zoutberg, G.R. Application of the "Biothane" wastewater treatment system in the soft drinks industry. *J. Inst. Water. Env. Man.* **1994**; *8*, 239–245.
10. Rittmann, B.E.; and McCarty, P.L. Anaerobic treatment by methanogenesis. In *Environmental Biotechnology: Principles and Applications*; McGraw Hill: New York, 2001; 573–579.

11. Siino, F.J.; Blanc, F.C.; O'Shaughnessy, J.C. Performance of an anaerobic filter treating soluble carbohydrate waste, *40th Annual Purdue Industrial Waste Conference*, 1985; 785–793.
12. Heijnen, J.J.; Mulder, A.; Enger, W.; Hoeks, F. Review on the application of anaerobic fluidized bed reactors in wastewater treatment. *Chem. Eng. J. & Biochem. Eng. J.*, **1989**; *41*, B37–50.
13. Borja, R.; Banks, C.J. Semicontinuous anaerobic digestion of soft drink wastewater in immobilized cell bioreactors. *Biotechnol. Lett.*, **1993**; *15*, 767–772.
14. Borja, R.; Banks, C.J. Kinetics of anaerobic digestion of soft drink wastewater in immobilized cell bioreactors. *J. Chem. Technol. Biot.*, **1994**; *60*, 327–334.
15. Hickey, R.F.; Owens, R.W. Methane generation from high-strength industrial wastes with the anaerobic biological fluidized bed. *Biotechnol. Bioeng. Symp.*, **1981**; *11*, 399–413.
16. Stronach, S.M.; Rudd, T.; Lester, J.N. Start-up of anaerobic bioreactors on high strength industrial wastes. *Biomass*, **1987**; *13*, 173–197.
17. Vicenta, M.; Pacheco, G.; Anglo, P.G. Anaerobic treatment of distillery slops, coconut water, and bottling waste using an upflow anaerobic filter reactor. In *Alternative Energy Sources 8: Solar Energy Fundamentals & Applications*, Vol 1, Hemisphere Publication: 1989; 865–875.
18. Silverio, C.M.; Anglo, P.G.; Luis, Jr, V.S.; Avacena, V.P. Anaerobic treatment of bottling wastewater using the upflow anaerobic reactor system. In *Alternative Energy Sources 8: Solar Energy Fundamentals & Applications* Vol 1; Hemisphere Publication: 1989; 843–853.
19. Wang, L.K.; Pereira, N.C.; Hung, Y.T. (eds.) *Biological Treatment Processes*. Humana Press: Totowa NJ, 2004.
20. Wang, L.K.; Hung, Y.T.; Shammas, N.K. (eds.) *Advanced Biological Treatment Processes* Humana Press: Totowa NJ, 2004..

25
Bakery Waste Treatment

J. Paul Chen, Lei Yang, and Renbi Bai
National University of Singapore, Singapore

Yung-Tse Hung
Cleveland State University, Cleveland, Ohio, U.S.A.

25.1 INTRODUCTION

The bakery industry is one of the world's major food industries and varies widely in terms of production scale and process. Traditionally, bakery products may be categorized as bread and bread roll products, pastry products (e.g., pies and pasties), and specialty products (e.g., cake, biscuits, donuts, and specialty breads). In March 2003, there were more than 7000 bakery operations in the United States (Table 1) with more than 220,000 employees. More than 50% of bakery businesses are small, having fewer than 100 employees [1].

The bakery industry has had a relatively low growth rate. Annual industry sales were $14.7 billion, $16.6 billion, and $17.7 billion in 1998, 2000, and 2002, respectively; the average weekly unit sales were $9,890, $10,040, and $10,859 during the same periods. Industry sales increased 6.5%, only 1.6% ahead of the compounded rate of inflation, according to *www.bakery-net.com*. Production by large plant bakers contributes more than 80% of the market's supply, while master bakers sell less than 5% [1].

The principles of baking bread have been established for several thousand years. A typical bakery process is illustrated in Fig. 1. The major equipment includes miller, mixer/kneading machine, bun and bread former, fermentor, bake ovens, cold stage, and boilers [2–4]. The main processes are milling, mixing, fermentation, baking, and storage. Fermentation and baking are normally operated at 40°C and 160–260°C, respectively. Depending on logistics and the market, the products can be stored at 4–20°C.

Flour, yeast, salt, water, and oil/fat are the basic ingredients, while bread improver (flour treatment agents), usually vitamin C (ascorbic acid), and preservatives are included in the commercial bakery production process.

Flour made from wheat (e.g., hard wheats in the United States and Canada) contains a higher protein and gluten content. Yeast is used to introduce anaerobic fermentation, which produces carbon dioxide. Adding a small amount of salt gives the bread flavor, and can help the fermentation process produce bread with better volume as well as texture. A very small quantity of vegetable oil keeps the products soft and makes the dough easier to pass through the

Table 1 Bakery Industry Market in the United States

Number of employees	Number of businesses	Percentage of businesses	Total employees	Total sales	Average employees/businesses
Unknown	1,638	23.65	N/A	N/A	N/A
1	644	9.30	644	487	1
2–4	1,281	18.50	3,583	505.5	3
5–9	942	13.60	6,138	753	7
10–24	1,117	16.13	16,186	1,208.1	14
25–49	501	7.23	17,103	1,578.7	34
50–99	287	4.14	18,872	23,51.7	66
100–249	305	4.40	45,432	10,820.5	149
250–499	130	1.88	43,251	6,909.1	333
500–999	70	1.01	45,184	3,255	645
1,000–2,499	7	0.10	8,820	N/A	1,260
2,500–4,999	2	0.03	7,295	760.2	3,648
10,000–14,999	1	0.01	11,077	N/A	11,077
Total/Average	6,925	100.00	223,585	28,628.8	32

Note: data include bread, cake, and related products (US industry code 2051); cookies and crackers (US industry code 2052); frozen bakery products, except bread (US industry code 2053); sales are in $US.
Source: Ref. 1.

manufacturing processes. Another important component in production is water, which is used to produce the dough. Good bread should have a certain good percentage of water. Vitamin C, a bread improver, strengthens the dough and helps it rise. Preservatives such as acetic acid are used to ensure the freshness of products and prevent staling. The ratio of flour to water is normally 10 : 6; while others are of very small amounts [3–6].

During the manufacturing process, 40–50°C hot water mixed with detergents is used to wash the baking plates, molds, and trays. Baking is normally operated on a single eight-hour shift and the production is in the early morning hours.

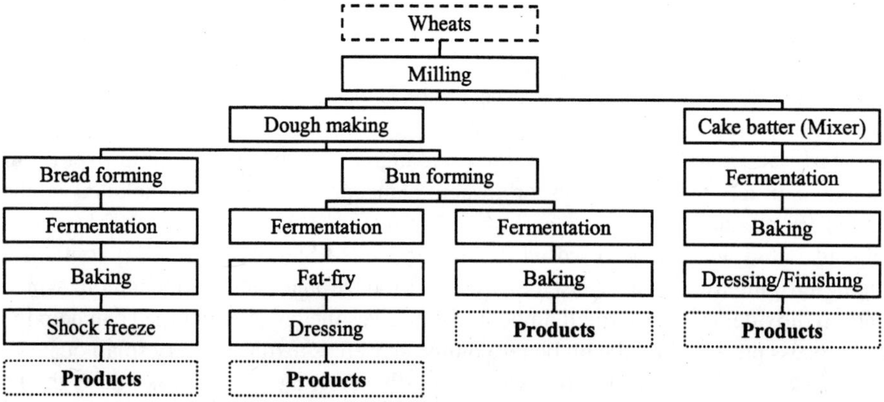

Figure 1 General production process diagram of bakery industry.

25.2 BAKERY INDUSTRY WASTE SOURCES

The bakery industry is one of the largest water users in Europe and the United States. The daily water consumption in the bakery industry ranges from 10,000 to 300,000 gal/day. More than half of the water is discharged as wastewater. Facing increasing stringent wastewater discharge regulations and cost of pretreatment, more bakery manufacturers have turned to water conservation, clean technology, and pollution prevention in their production processes.

As shown in Figure 1, almost every operation unit can produce wastes and wastewaters. In addition, other types of pollution resulting from production are noise pollution and air pollution.

25.2.1 Noise

Noise usually comes from the compressed air and the running machines. It not only disturbs nearby residents, but can harm bakery workers' hearing. It is reported that sound more than 5 dB(A) above background can be offensive to people. A survey of bakery workers' exposure showed that the average range is 78–85 dB(A), with an average value of 82 dB(A). Ear plugs can help to effectively reduce the suffering. Other noise control measures include the reduction of source noise, use of noise enclosures, reduction of reverberation, and reduction of exposure time [2,7].

25.2.2 Air Pollution

The air pollution is due to emission of volatile organic compounds (VOC), odor, milling dust, and refrigerant agent. The VOC can be released in many operational processes including yeast fermentation, drying processes, combustion processes, waste treatment systems, and packaging manufacture. The milling dust comes from the leakage of flour powder. The refrigerant comes from the emissions leakage of the cooling or refrigeration systems. All of these can cause serious environmental problems. The controlling methods may include treatment of VOC and odor, avoidance of using the refrigerants forbidden by laws, and cyclic use of the refrigerants.

25.2.3 Wastewater

Wastewater in bakeries is primarily generated from cleaning operations including equipment cleaning and floor washing. It can be characterized as high loading, fluctuating flow and contains rich oil and grease. Flour, sugar, oil, grease, and yeast are the major components in the waste.

The ratio of water consumed to products is about 10 in common food industry, much higher than that of 5 in the chemical industry and 2 in the paper and textiles industry [3,6]. Normally, half of the water is used in the process, while the remainder is used for washing purposes (e.g., of equipment, floor, and containers).

Typical values for wastewater production are summarized in Tables 2–4 [3,8,9]. Different products can lead to different amounts of wastewater produced. As shown in Table 2, pastry production can result in much more wastewater than the others. The values of each item can vary significantly as demonstrated in Table 3. The wastewater from cake plants has higher strength than that from bread plants. The pH is in acidic to neutral ranges, while the 5-day biochemical oxygen demand (BOD_5) is from a few hundred to a few thousand mg/L, which is much higher than that from the domestic wastewater. The suspended solids (SS) from cake plants is very high. Grease from the bakery industry is generally high, which results from the production operations. The waste strength and flow rate are very much dependent on the operations, the size of the plants, and the number of workers. Generally speaking, in the plants with products of bread, bun, and roll, which are termed as dry baking, production equipment (e.g., mixing vats and baking pans) are cleaned dry and floors are swept before washing down. The wastewater from cleanup

Table 2 Summary of Waste Production from the Bakery Industry

Manufacturer	Products	Wastewater production (L/tonne-production)	COD (kg/tonne-production)	Contribution to total COD loading (%)
Bread and bread roll	Bread and bread roll	230	1.5	63
Pastry	Pies and sausage rolls	6000	18	29
Specialty	Cake, biscuits, donuts, and Persian breads	74	–	–

Source: Ref. 3.

has low strength and mainly contains flour and grease (Table 3). On the other hand, cake production generates higher strength waste, which contains grease, sugar, flour, filling ingredients, and detergents.

Due to the nature of the operation, the wastewater strength changes at different operational times. As demonstrated in Table 3, higher BOD_5, SS, total solids (TS), and grease are observed from 1 to 3 am, which results from lower wastewater flow rate after midnight.

Bakery wastewater lacks nutrients; the low nutrient value gives $BOD_5:N:P$ of 284 : 1 : 2 [8,9]. This indicates that to obtain better biological treatment results, extra nutrients must be added to the system. The existence of oil and grease also retards the mass transfer of oxygen. The toxicity of excess detergent used in cleaning operations can decrease the biological treatment efficiency. Therefore, the pretreatment of wastewater is always needed.

25.2.4 Solid Waste

Solid wastes generated from bakery industries are principally waste dough and out-of-specified products and package waste. Solid waste is the loss of raw materials, which may be recovered by cooking waste dough to produce breadcrumbs and by passing cooked product onto pig farmers for fodder.

25.3 BAKERY WASTE TREATMENT

Generally, bakery industry waste is nontoxic. It can be divided into liquid waste, solid waste, and gaseous waste. In the liquid phase, there are high contents of organic pollutants including chemical oxygen demand (COD), BOD_5, as well as fats, oils, and greases (FOG), and SS. Wastewater is normally treated by physical and chemical, biological processes.

Table 3 Wastewater Characteristics in the Bakery Industry

Type of bakery	pH	BOD_5 (mg/L)	SS (mg/L)	TS (mg/L)	Grease (mg/L)
Bread plant	6.9–7.8	155–620	130–150	708	60–68
Cake plant	4.7–8.4	2,240–8,500	963–5,700	4,238–5,700	400–1,200
Variety plant	5.6	1,600	1,700	–	630
Unspecified	4.7–5.1	1,160–8,200	650–13,430	–	1,070–4,490

Source: Refs. 8 and 9.

Table 4 Average Waste Characteristics at Specified Time Interval in a Cake Plant

Time interval	pH	BOD_5 (mg/L)	SS (mg/L)	TS (mg/L)	Grease (mg/L)
3 am–8 am	7.9	1480	834	3610	428
9 am–12 am	8.6	2710	1080	5310	457
1 pm–6 pm	8.1	2520	795	4970	486
7 pm–12 pm	8.6	2020	953	3920	739
1 am–3 am	8.9	2520	1170	4520	991

Source: Ref. 9.

25.4 PRETREATMENT SYSTEMS

Pretreatment or primary treatment is a series of physical and chemical operations, which precondition the wastewater as well as remove some of the wastes. The treatment is normally arranged in the following order: screening, flow equalization and neutralization, optional FOG separation, optional acidification, coagulation–sedimentation, and dissolved air flotation. The pretreatment of bakery wastewater is presented in Fig. 2.

In the bakery industry, pretreatment is always required because the waste contains high SS and floatable FOG. Pretreatment can reduce the pollutant loading in the subsequent biological and/or chemical treatment processes; it can also protect process equipment. In addition, pretreatment is economically preferable in the total process view as compared to biological and chemical treatment.

25.4.1 Flow Equalization and Neutralization

In bakery plants, the wastewater flow rate and loading vary significantly with the time as illustrated in Table 4 [8,9]. It is usually economical to use a flow equalization tank to meet the peak discharge demand. However, too long a retention time may result in an anaerobic environment. A decrease in pH and bad odors are common problems during the operations.

25.4.2 Screening

Screening is used to remove coarse particles in the influent. There are different screen openings ranging from a few μm (termed as microscreen) to more than 100 mm (termed as coarse screen). Coarse screen openings range from 6 to 150 mm; fine screen openings are less than 6 mm. Smaller opening can have a better removal efficiency; however, operational problems such as clogging and higher head lost are always observed.

Fine screens made of stainless material are often used. The main design parameters include velocity, selection of screen openings, and head loss through the screens. Clean operations and waste disposal must be considered. Design capacity of fine screens can be as high as 0.13 m^3/s; the head loss ranges from 0.8 to 1.4 m. Depending on the design and operation, BOD_5 and SS removal efficiencies are 5–50% and 5–45%, respectively [8,9].

25.4.3 FOG Separation

As wastewater may contain high amount of FOG, a FOG separator is thus recommended for installation. Figure 3 gives an example of FOG separation and recovery systems [4]. The FOG

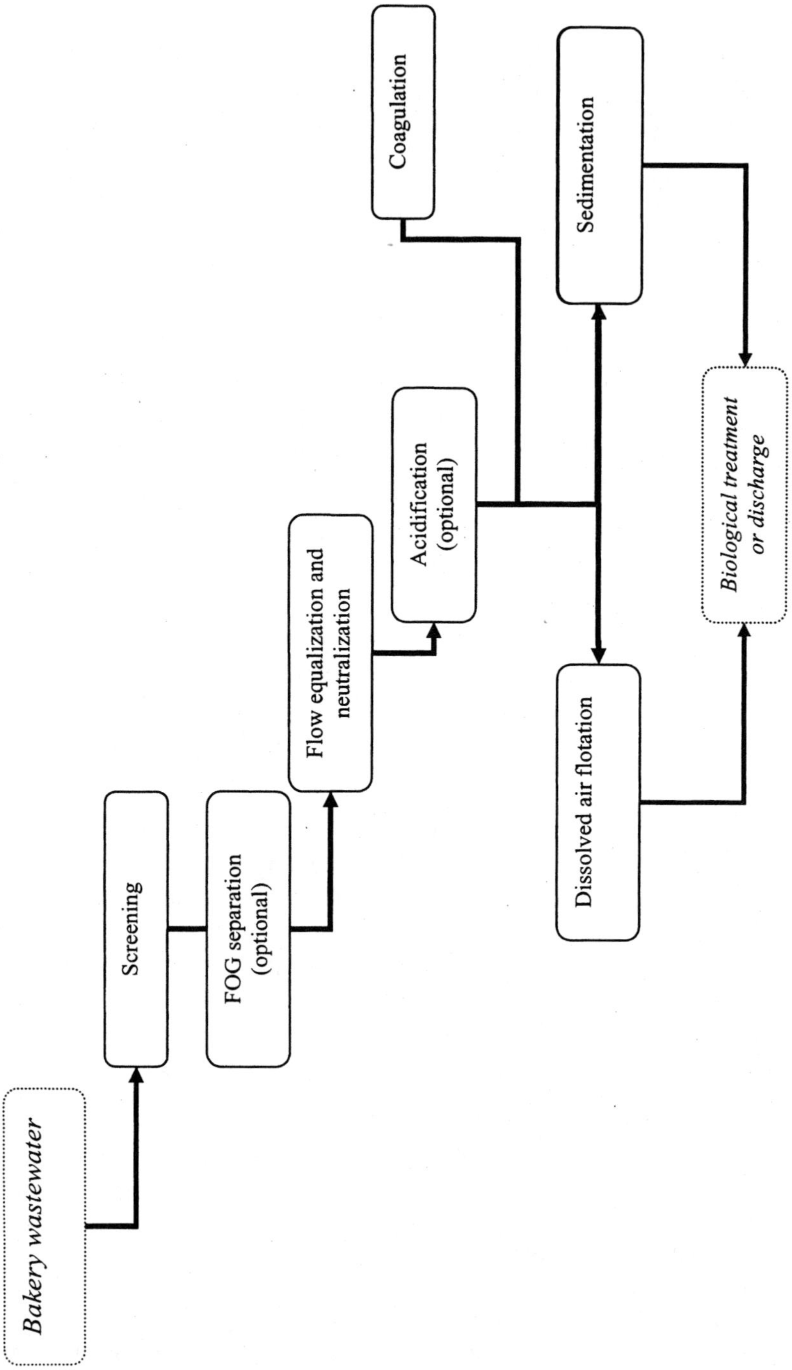

Figure 2 Bakery wastewater pretreatment system process flow diagram.

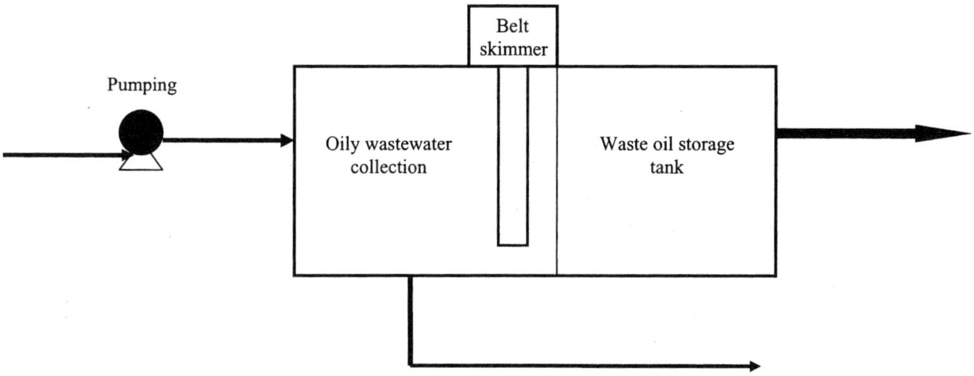

Figure 3 Fats, oils, and grease (FOG) separation unit.

can be separated and recovered for possible reuse, as well as reduce difficulties in the subsequent biological treatment.

25.4.4 Acidification

Acidification is optional, depending on the characteristics of the waste. Owing to the presence of FOG, acid (e.g., concentrated H_2SO_4) is added into the acidification tank; hydrolysis of organics can occur, which enhances the biotreatability. Grove et al. [10] designed a treatment system using nitric acid to break the grease emulsions followed by an activated sludge process. A BOD_5 reduction of 99% and an effluent BOD_5 of less than 12 mg/L were obtained at a loading of 40 lb $BOD_5/1000\ ft^3$ and detention time of 87 hour. The nitric acid also furnished nitrogen for proper nutrient balance for the biodegradation.

25.4.5 Coagulation–Flocculation

Coagulation is used to destabilize the stable fine SS, while flocculation is used to grow the destabilized SS, so that the SS become heavier and larger enough to settle down. The Coagulation–flocculation process can be used to remove fine SS from bakery wastewater. It normally acts as a preconditioning process for sedimentation and/or dissolved air flotation.

The wastewater is preconditioned by coagulants such as alum. The pH and coagulant dosage are important in the treatment results. Liu and Lien [11] reported that 90–100 mg/L of alum and ferric chloride were used to treat wastewater from a bakery that produced bread, cake, and other desserts. The wastewater had pH of 4.5, SS of 240 mg/L, and COD of 1307 mg/L. Values of 55% and 95–100% for removal of COD and SS, respectively, were achieved. The optimum pH for removal of SS was 6.0, while that for removal of COD was 6.0–8.0. It was also found that $FeCl_3$ was relatively more effective than alum. Yim *et al.* [8] used coagulation–flocculation to treat a wastewater with much higher waste strength. Table 5 gives the treatment results. Owing to the higher organic content, SS, and FOG, coagulants with high dosage of 1300 mg/L were applied [8,9]. The optimal pH was 8.0. As shown, removal for the above three items was fairly high, suggesting that the process can also be used for high-strength bakery waste. However, the balance between the cost of chemical dosage and treatment efficiency should be justified.

25.4.6 Sedimentation

Sedimentation, also called clarification, has a working mechanism based on the density difference between SS and the water, allowing SS with larger particle sizes to more easily settle

Table 5 Comparison of Different Bakery Waste Pretreatment Methods

	BOD_5		SS		FOG	
Coagulant	Influent (mg/L)	Removal (%)	Influent (mg/L)	Removal (%)	Influent (mg/L)	Removal (%)
Ferric sulfate	2780	71	2310	94	1450	93
Alum	2780	69	2310	97	1450	96

Source: Ref. 9.

down. Rectangular tanks, circular tanks, combination flocculator–clarifiers, and stacked multilevel clarifiers can be used[6].

25.4.7 Dissolved Air Flotation (DAF)

Dissolved air flotation (DAF) is usually implemented by pumping compressed air bubbles to remove fine SS and FOG in the bakery wastewater. The wastewater is first stored in an air pressured, closed tank. Through the pressure-reduction valves, it enters the flotation tank. Owing to the sudden reduction in pressure, air bubbles form and rise to the surface in the tank. The SS and FOG adhere to the fine air bubbles and are carried upwards. Dosages of coagulant and control of pH are important in the removal of BOD_5, COD, FOG, and SS. Other influential factors include the solids content and air/solids ratio. Optimal operation conditions should be determined through the pilot-scale experiments. Liu and Lien [11] used a DAF to treat a wastewater from a large-scale bakery. The wastewater was preconditioned by alum and ferric chloride. With the DAF treatment, 48.6% of COD and 69.8% of SS were removed in 10 min at a pressure of 4 kg/cm^2, and pH 6.0. Mulligan [12] used DAF as a pretreatment approach for bakery waste. At operating pressures of 40–60 psi, grease reductions of 90–97% were achieved. The BOD_5 and SS removal efficiencies were 33–62% and 59–90%, respectively.

25.5 BIOLOGICAL TREATMENT

The objective of biological treatment is to remove the dissolved and particulate biodegradable components in the wastewater. It is a core part of the secondary biological treatment system. Microorganisms are used to decompose the organic wastes [6,8–15].

With regard to different growth types, biological systems can be classified as suspended growth or attached growth systems. Biological treatment can also be classified by oxygen utilization: aerobic, anaerobic, and facultative. In an aerobic system, the organic matter is decomposed to carbon dioxide, water, and a series of simple compounds. If the system is anaerobic, the final products are carbon dioxide and methane.

Compared to anaerobic treatment, the aerobic biological process has better quality effluent, easies operation, shorter solid retention time, but higher cost for aeration and more excess sludge. When treating high-load influent (COD > 4000 mg/L), the aerobic biological treatment becomes less economic than the anaerobic system. To maintain good system performance, the anaerobic biological system requires more complex operations. In most cases, the anaerobic system is used as a pretreatment process.

Suspended growth systems (e.g., activated sludge process) and attached growth systems (e.g., trickling filter) are two of the main biological wastewater treatment processes. The activated sludge process is most commonly used in treatment of wastewater. The trickling filter is easy to control, and has less excess sludge. It has higher resistance loading and low energy cost. However, high operational cost is its major disadvantage. In addition, it is more sensitive to temperature and has odor problems. Comprehensive considerations must be taken into account when selecting a suitable system.

25.6 AEROBIC TREATMENT

25.6.1 Activated Sludge Process

In the activated sludge process, suspended growth microorganisms are employed. A typical activated sludge process consists of a pretreatment process (mainly screening and clarification), aeration tank (bioreactor), final sedimentation, and excess sludge treatment (anaerobic treatment and dewatering process). The final sedimentation separates microorganisms from the water solution. In order to enhance the performance result, most of the sludge from the sedimentation is recycled back to the aeration tank(s), while the remaining is sent to anaerobic sludge treatment. A recommended complete activated sludge process is given in Fig. 4.

The activated sludge process can be a plug-flow reactor (PFR), completely stirred tank reactor (CSTR), or sequencing batch reactor (SBR). For a typical PFR, length–width ration should be above 10 to ensure the plug flow. The CSTR has higher buffer capacity due to its nature of complete mixing, which is a critical benefit when treating toxic influent from industries. Compared to the CSTR, the PFR needs a smaller volume to gain the same quality of effluent. Most large activated sludge sewage treatment plants use a few CSTRs operated in series. Such configurations can have the advantages of both CSTR and PFR.

The SBR is suitable for treating noncontinuous and small-flow wastewater. It can save space, because all five primary steps of fill, react, settle, draw, and idle are completed in one tank. Its operation is more complex than the CSTR and PFR; in most cases, auto operation is adopted.

The performance of activated sludge processes is affected by influent characteristics, bioreactor configuration, and operational parameters. The influent characteristics are wastewater flow rate, organic concentration (BOD_5 and COD), nutrient compositions (nitrogen and phosphorus), FOG, alkalinity, heavy metals, toxins, pH, and temperature. Configurations of the bioreactor include PFR, CSTR, SBR, membrane bioreactor (MBR), and so on. Operational parameters in the treatment are biomass concentration [mixed liquor volatile suspended solids concentration (MLVSS) and volatile suspended solids (VSS)], organic load, food to microorganisms (F/M), dissolved oxygen (DO), sludge retention time (SRT), hydraulic retention time (HRT), sludge return ratio, and surface hydraulic flow load. Among them, SRT and DO are the most important control parameters and can significantly affect the treatment results. A suitable SRT can be achieved by judicious sludge wasting from the final clarifier. The DO in the aeration tank should be maintained at a level slightly above 2 mg/L. The typical design parameters and operational results are listed in Table 6.

Owing to the high organic content, it is not recommended that bakery wastewater be directly treated by aerobic treatment processes. However, there are a few cases of this reported in the literature, including a study from Keebler Company [4]. The company produces crackers and cookies in Macon, Georgia. The FOG and pH of the wastewater from the manufacturing facility were observed as higher than the regulated values. Wastewater was treated by an aerobic activated sludge process, which included a bar screen, nutrient feed system, aeration tank, clarifier, and sludge storage tank. Because of the large quantities of oil in the water (Table 7), two FOG separators as shown in Figure 3 (discussed previously) were installed in the oleo/lard

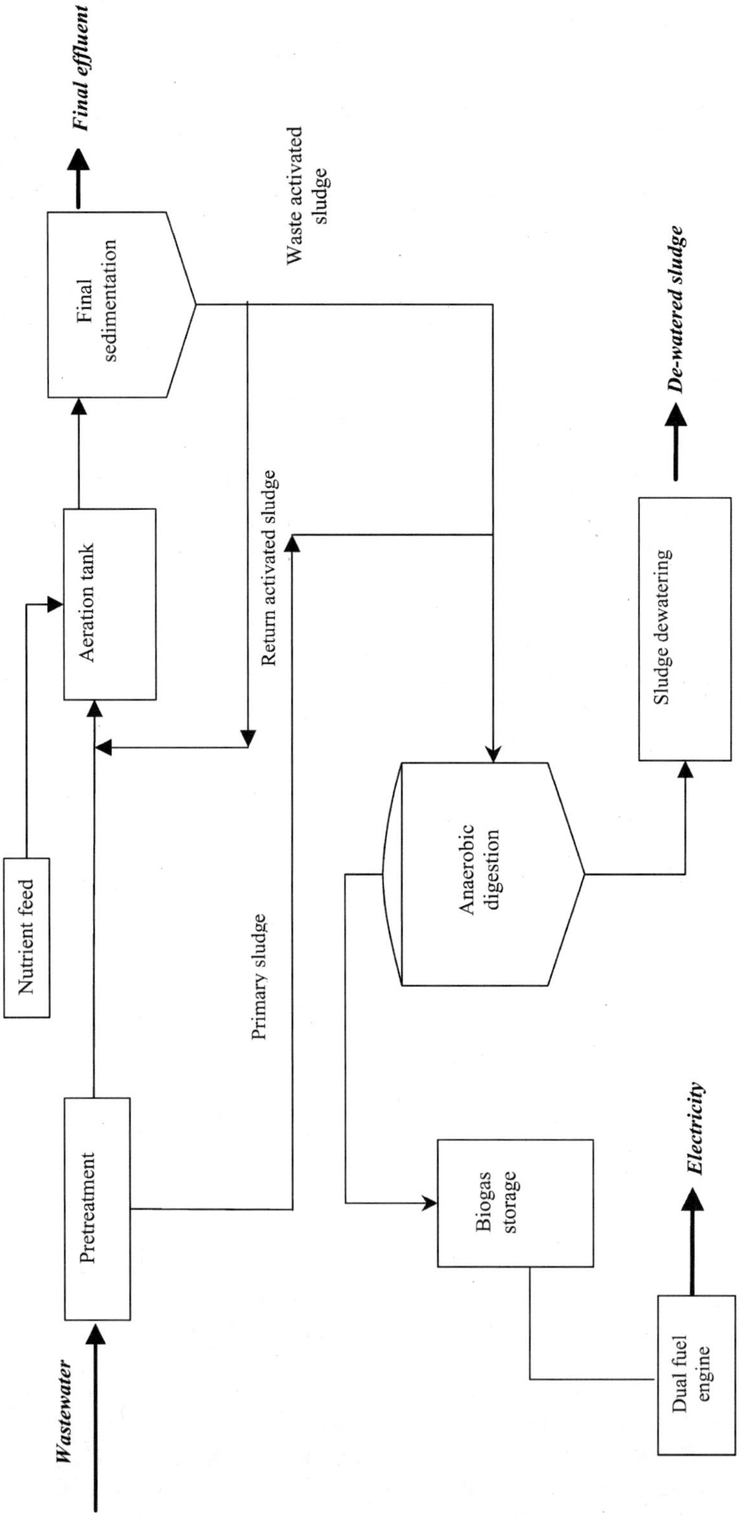

Figure 4 Process flow diagram of activated sludge treatment of bakery wastewater.

Table 6 Design and Performance of Activated Sludge Processes

Activated sludge processes	Extended	Conventional	High rate
F/M (kg BOD_5/kg MLSS · day)	0.06–0.2	0.3–0.6	0.5–1.9
MLSS (g/L)	4–7.5	1.9–4	5–12
HRT (hour)	18–36	4–10	2–4
SRT (day)	20–30	5–15	3–8
BOD_5 removal (%)	> 95	95	70–75
VLR (kg BOD_5/m^3 · day)	0.2–0.4	0.4–1.0	2–16

Note: F/M, food to microorganisms ratio; MLSS, mixed liquid suspended solids; SRT, sludge retention time; HRT, hydraulic retention time; BOD_5, five-day biochemical oxygen demand; VLR, volumetric loading rate.

storage area and the coconut oil spray machines. Characteristics of influent and effluent as well as design parameters are given in Table 7. As shown, the company had favorable treatment results; the effluent was good enough for direct discharge to a nearby watercourse. Owing to the poor nutrient content in the influent, nutrient was fed directly into the aeration tank. Not all the added nitrogen was consumed in the treatment, thus the total Kjedahl nitrogen (TKN) concentration in the effluent was higher than that in the influent. The high HRT in Table 7 shows that the process was not in fact economical. The bakery wastewater treatment can be more cost-effective if the waste is first treated by an anaerobic process and then an aerobic process.

25.6.2 Trickling Filter Process

Aerobic attached-growth processes include tricking filters (biotower) and rotating biological contactors (RBC). In these processes, microorganisms are attached onto solid media and form a layer of biofilm. The organic pollutants are first adsorbed to the biofilm surface, oxidation reactions then occur, which break the complex organics into a group of simple compounds, such as water, carbon dioxide, and nitrate. In addition, the energy released from the oxidation together with the organics in the waste is used for maintenance of microorganisms as well as synthesis of new microorganisms.

Table 7 Summary of Wastewater Treatment in the Keebler Company

Parameter	Influent: Design basis[a]	Influent: Operation[b]	Effluent[b]
Flow rate (gpd)	51,200	37,000	–
PH	5.6	6.0	6.8
TCOD (mg/L)	1620	830	65
SCOD (mg/L)	–	290	40
$TBOD_5$ (mg/L)	891	500	39
$SBOD_5$ (mg/L)	–	175	24
TS (mg/L)	756	–	11[b]
FOG (mg/L)	285	–	3[b]
TKN (mg/L)	–	2	5
PO_4-P (mg/L)	–	3	3

[a]Based on historical pretreatment program monitoring data.[b]Based on operation in August 1988. Operational parameters: HRT = 2.8 day; MLSS = 3300 mg/L; VSS = 2600 mg/L; DO = 2.2 mg/L; F/M = 0.07 1b BOD/1b VSS · day. Yield = 0.32; clarifier overflow rate = 118 gpd/ft^2; clarifier solids loading rate = 5 1b/ft^2 · day.
Source: Ref. 4.

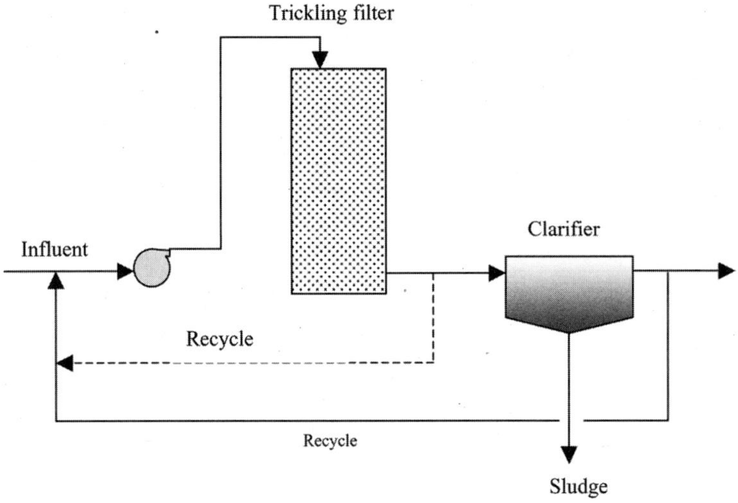

Figure 5 Flow diagram of trickling filter for bakery wastewater treatment.

The tricking filter can be used to treat bakery wastewater. Solid media such as crushed rock and stone, wood, and chemical-resistant plastic media are randomly packed in the reactor. Figure 5 shows a typical trickling filter, which can be used for the bakery wastewater treatment. Surface area and porosity are two important parameters of filter media. A large surface area can cause accumulation of a large amount of biomass and result in high treatment efficiency; large porosity would lead to higher oxygen transfer rate and less blockage. A common problem in trickling filter systems is the excess growth of microorganisms, which can cause serious blockage in the medium and reduce the porosity. Typical design parameters and performance data for aerobic trickling filters are listed in Table 8. Keenan and Sabelnikov [14] demonstrated that a biological system containing a mixing-aeration tank and biological filter (trickling filter) was able to eliminate grease and oil in bakery waste. A dramatic reduction of FOG content from 1500 mg/L to less than 30 mg/L was achieved. This system was fairly stable during 20 months of continuous operation.

25.7 ANAEROBIC BIOLOGICAL TREATMENT

Bakery waste contains high levels of organics, FOG, and SS, which are treated using the preferred method of anaerobic treatment processes. There are different types of anaerobic

Table 8 Design and Performance of Trickling Filter

Type of filter	BOD_5 loading ($kg/m^3 \cdot day$)	Hydraulic loading ($m^3/m^2 \cdot day$)	Depth (m)	BOD_5 removal (%)	Medium
Low rate	0.07–0.4	1–3	1.8–2.4	95	Rock, slag
Mid-range rate	0.2–0.45	3–7	1.8–2.4	–	Rock, slag
High rate	0.5–1	6–20	1–1.8	50–70	Rock

processes available on the market, such as CSTR, AF, UASB, AFBR, AC, and ABR. The most obvious operational parameters are high SRT, HRT, and biomass concentration. Anaerobic processes have been widely used in treatment of a variety of food processing and other wastes since they were first developed in the early 1950s. Figure 6 illustrates a typical anaerobic treatment process for bakery wastewater.

In addition to accommodating organic waste treatment, anaerobic treatment can produce methane, which can be used for production of electricity (Fig. 6). The disadvantages, however, include complexity in operation, sensitivity to temperature and toxicity, time-consuming in startup, and susceptibility to process upset. Table 9 gives a summary of design and performance of typical anaerobic treatment processes.

Anaerobic processes are suitable for a variety of bakery wastewater. For example, an anaerobic contactor was successfully used to treat wastewater from a production facility of snack cake items [13]. The waste strength was extremely high as demonstrated in Table 10. The BOD_5 to COD ratio of the raw wastewater was 0.44. An anaerobic contact reactor was used, similar to that in Figure 6, except that two bioreactors were operated in series. As shown in Table 10, the system provides good treatment results. The removal efficiencies for BOD_5, COD, TSS, and FOD were above 96%. The treated stream can be directly discharged to the domestic sewage systems. Alternatively, a subsequent aerobic treatment can be used to further reduce the waste strength and the effluent can then be discharged to a watercourse.

25.8 AIR POLLUTION CONTROL

While air pollution in the bakery industry may be not serious, it can become a concern if not properly handled. Dust, VOC, and refrigerant are three main types of air pollutants.

25.8.1 Dust

Flour production workers are usually harmed by dust pollution. Lengthy exposure time at a high exposure level can cause serious skin and respiration diseases. The control approaches include prevention of the leakage of flour power, provision of labor protection instruments, and post treatment. Filters and scrubbers are commonly used.

25.8.2 Refrigerant

In the chilling, freezing storage or transport of bakery products, a large amount of refrigerant is used. Chlorofluorocarbons (CFCs) and hydrochlorofluorocarbons (HCFCs) are the common refrigerants and can damage the ozone layer. They can be retained in the air for approximately 100 years. Owing to the significantly negative environmental effects, replacement chemicals such as hydrofluorocarbons (HFC) have been developed and used. Another measure is the prevention of the refrigerant leakage.

25.8.3 VOC

Several measures can be used to control VOC pollution, including biological filters and scrubbers.

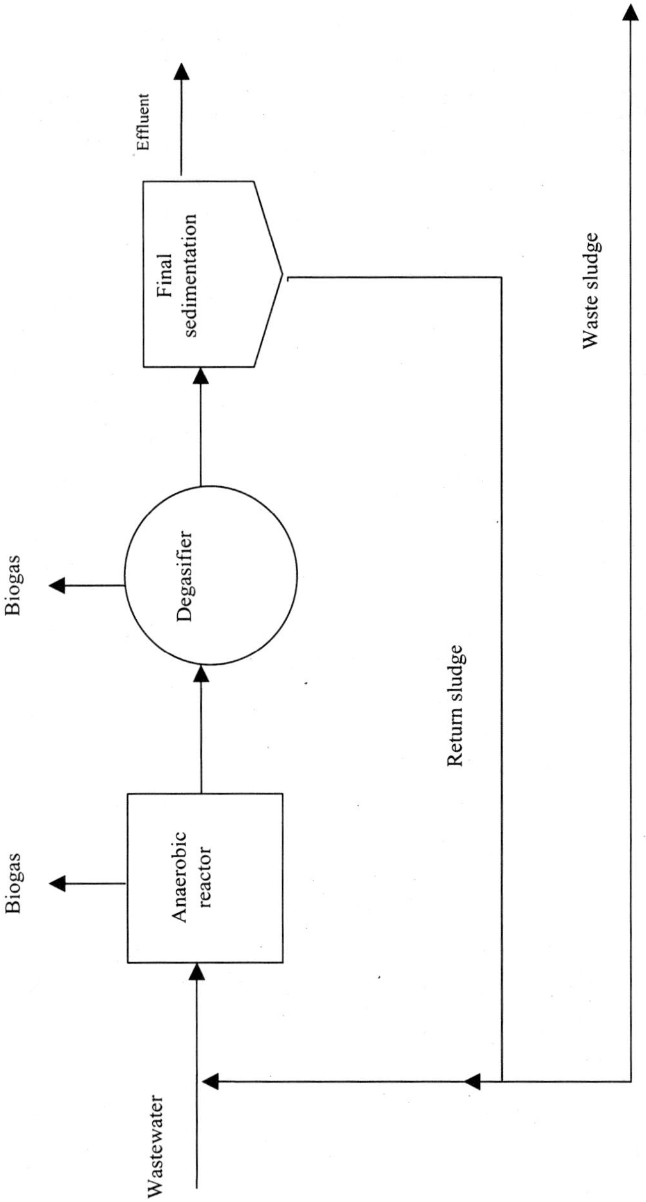

Figure 6 Schematic of anaerobic contact process.

Table 9 Design and Performance of Anaerobic Treatment Processes

Reactor	Influent COD (g/L)	HRT (day)	VLR (kg COD/m^3 · day)	Removal (%)
AF	3–40	0.5–13	4–15	60–90
AC	3–10	1–5	1–3	40–90
AFBR	1–20	0.5–2	8–20	80–99
UASB	5–15	2–3	4–14	85–92

25.9 SOLID WASTE MANAGEMENT

Bakery solid waste includes stale bakery products, dropped raw materials (e.g., dough), and packages. The most simple and common way is to directly transport these to landfill or incineration. Landfill can cause the waste to decompose, which eventually leads to production of methane (a greenhouse gas) and groundwater pollution (organic compounds and heavy metals). Incineration of bakery waste can also release nitrogen oxide gases.

Reclamation of the bakery waste can play an important role in its management. The waste consists primarily of stale bread, bread rolls, and cookies – all of which contain high energy and can be fed directly to animals, such as swine and cattle. Another application is to use the waste for production of valuable products. For example, Oda *et al.* [15] successfully used bakery waste to produce lactic acid with a good conversion efficiency of 47.2%.

25.10 CLEANER PRODUCTION IN THE BAKERY INDUSTRY

25.10.1 Concepts

The production of bakery products involves many operation units that may cause a variety of wastes. Most bakery industries are of small or medium size, and are often located in densely populated areas, which makes environmental problems more critical. Nevertheless, the conventional "end-of-pipe" treatment philosophy has its restrictions in dealing with these problems. It only addresses the result of inefficient and wasteful production processes, and should be considered only as a final option.

Table 10 Performance of Anaerobic Contact Process

Parameter	Raw water (mg/L) Range	Raw water (mg/L) Average	Clarifier effluent (mg/L) Range	Clarifier effluent (mg/L) Average	Average removal (%)
BOD_5	906–24,000	9,873	65–267	145	98.5
COD	2,910–50,400	23,730	315–1,340	642	97.3
TS	848–36,700	15,127	267–1,260	502	96.7
FOG	429–10,000	5,778	9–113	41	99.3

Operational parameters: Bioreactor: HRT = 7.8 day; SRT = 50 day; volumetric BOD_5 loading = 1.3 kg BOD_5/m^3 · day, volumetric COD loading = 3.0 kg COD/m^3 · day. Clarifier: Overflow rate = 3.7 m^3/m^2 · day; HRT = 16 hour, solids loading = 20.5 m^3/m^2 · day, clarification efficiency = 91%.
Source: Ref. 13.

Manufacturing will always cause direct or indirect pollution of the environment. It is hard to realize “zero discharge,” and waste treatment is always expensive. Cleaner production (CP) has two key components: maximization of waste reduction and minimization of raw material usage and energy consumption. The United Nations Environment Program (UNEP) defines CP as [7]:

> The continuous application of an integrated preventive environmental strategy to processes, products, and services to increase overall efficiency, and reduce risks to humans and the environment. Cleaner Production can be applied to the processes used in any industry, to products themselves and to various services provided in society.

Cleaner production results from one or a combination of conserving raw materials, water, and energy; eliminating toxic and dangerous raw materials; and reducing the quantity and toxicity of all emissions and wastes at source during the production process. It aims to reduce the environmental, health, and safety impacts of products over their entire life-cycles, from raw materials extraction, through manufacturing and use, to the “ultimate” disposal of the product. It implies the incorporation of environmental concerns into designing and delivering services [3,7].

In the CP process, raw materials, water, and energy should be conserved, their emission or wastage should be reduced, and application of toxic raw materials must be avoided. It is also important to reduce the negative impacts during the whole production life-cycle, from the design of the production to the final waste disposal. The main steps of a CP assessment are outlined in Fig. 7. The CP can be illustrated by the following example.

25.10.2 A Case Study in Country Bake Pty. Ltd.

Country Bake Pty. Ltd. [3] is a well-known bakery in Queensland, Australia, that produces mainly bread and bread rolls, as well as pastry products and cakes. Production is highly automated, and CP was carried out at the bakery to improve its operational efficiency.

Staff Awareness and Management Expectation

An initial brief survey showed that general awareness of CP at the manufacturing facility was fairly low before its implementation. The staff felt that changes were most likely to be in the areas of general housekeeping and minor process improvements. However, workers were keen on voluntary improvements to their operations as CP could lead to reduction of environmental and health risk liability, less operating costs through better waste and energy management, and reduction of environmental impact. In addition, both management and labor believed that higher business profitability as well as improvement of the company’s public image could be achieved through exercising CP.

Assessment of Waste

Areas of waste generation were identified and characterized. It was found that water usage was 719,000 L/week, with about 59% used in production, while the remainder was ultimately discharged as wastewater from cleaning and other ancillary uses. The pastry area and bread and bread rolls area contributed 35 and 36% of wastewater volume, respectively. Other wastewater arose from the boiler, the crate wash, and the staff amenities. In terms of COD loading, the pastry area, bread and bread rolls area, and night cleaning contributed 29, 25, and 38%, respectively. The characterization of wastes can be found in Table 2.

Approximately 1.7 tons of dough per week was lost in the waste stream, leading to a loss of 0.5% of the total mass of ingredients (or a loss of $4000/month). Pancoat oil and white oil

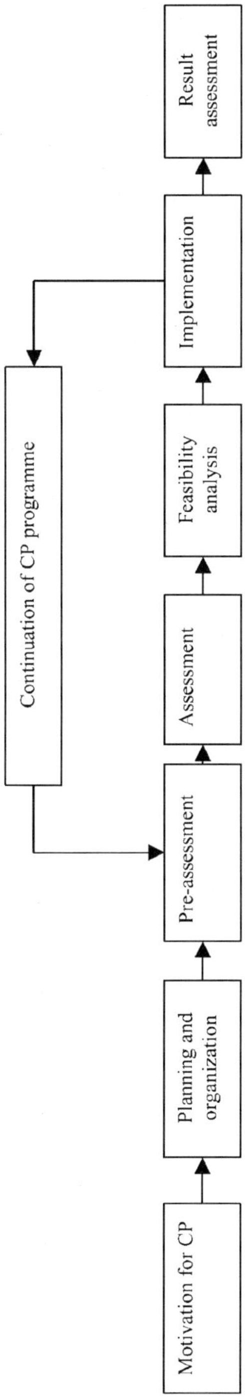

Figure 7 Outline of CP assessment process.

were used in production, most of which were lost and became the main contributors to the FOG in the waste stream. Monthly cost for their purchase was $13,140. Prevention of oil loss therefore could lead to significant savings for the bakery.

CP Strategies

Three CP strategies were proposed. The first was to reduce the COD load of wastewater discharged from the bread/bread roll area. Some dough material still fell on the floor and ultimately found its way to the drains. The following approaches were used for reclaiming and recycling the material: relocation of drains for easier collection of dough and installation of screens at drain points to capture fallen dough. A second strategy was to reduce the volumes of wastewater discharged from the pastry area by modification of cleaning practices, elimination or reuse of water discharges from the vacuum pump, and reuse of water discharges from the blast chiller. The last strategy was to reduce the loss of oil by modifications of equipment.

Staff Involvement

Cleaner production cannot be implemented well without great enthusiasm and commitment of the staff to CP, as they are the first to fulfill the CP. The company developed 12 work teams made up of individuals from the major functional work areas. These teams met regularly to discuss issues relevant to their specific work areas. These teams assumed responsibility for driving CP in the workplaces. Team leaders who were trained by the UNEP Working Group conducted a series of training programs for the remaining staff. Finally, the staff was rewarded for their implementation of CP.

Cost-Saving Benefits

Through implementation of CP in production, it was estimated that a total monthly saving of $27,700 could be achieved.

REFERENCES

1. D&B Sales & Marketing Solutions. *Poultry Slaughtering and Processing Report.* http://www.zapdata.com/, 2003.
2. Kannan, P; Boie, W. Energy management practices in SME – case study of a bakery in Germany. Energy Conv. Mgnt 2003, *44*, 945–959.
3. Gainer, D.; Pullar, S.; Lake, M.; Pagan, R. The Country Bake Story – How a modern bakery is achieving productivity and efficiency gains through cleaner production. *Sustainable Energy and Environmental Technology – Challenges and Opportunities, Proceedings*, Gold Coast, 14–17, June, 1998. 573–578.
4. Givens, S.; Cable, J. Case study – A tale of two industries, pretreatment of confectionary and bakery wastewaters. *1988 Food Processing Waste Conference*, presented by the Georgia Tech Research Institute, Atlanta, Georgia, October 31 – November 2, 1988.
5. Dalzell, J.M. *Food Industry and the Environment in the European Union – Practical Issues and Cost Implications*, 2nd Ed., Aspen Publishers, Inc.: Gaithersburg, Maryland, 2000.
6. Metcalf and Eddy. *Wastewater Engineering: Treatment Disposal Reuse*, 4th Ed.; McGraw-Hill, 2002.
7. Nations Environment Programme (UNEP). (http://www.uneptie.org/pc/cp/home.htm). 2003.
8. Yim, B.; Young, R.H.F.; Burbank, N.C. Dugan, G.L. Bakery waste: its characteristics, Part I. Indust. Wastes **1975** *March/April*, 24–25.
9. Yim, B.; Young, R.H.F.; Burbank, N.C.; Dugan, G.L. Bakery waste: its characteristics and treatability, Part II. Indust. Wastes **1975** *September/October*, 41–44.

10. Grove, C.S. Jr.; Emerson, D.B.; Dul, E.F.; Schlesiger, H.; Brown, W. Design of a treatment plant for bakery wastes. *24th Purdue Industrial Waste Conference (PIWC)*, Lafayetle, IN; 1969; 155–178.
11. Liu, J.C.; Lien, C.S. Pretreatment of bakery wastewater by coagulation–flocculation and dissolved air flotation. Water Sci. Technol. **2001**, *43*, 131–137.
12. Mulligan, T. Bakery sewage disposal. *Proceedings of the 1967 Meeting of the American Society of Bakery Engineers*, 1967; 254–263.
13. Shin, B.S.; Eklund, C.W.; Lensmeyer, K.V. Bakery waste treatment by an anaerobic contact process. Res. J. Water Pollut. Control **1990**, *62* (*7*), 920–925.
14. Keenan, D.; Sabelnikov, A. Biological augmentation eliminates grease and oil in bakery wastewater. Water Environ. Res. **2000**, *72*(*2*), 141–146.
15. Oda, Y.; Park, B.S.; Moon, K.H.; Tonomura, K. Recycling of bakery wastes using an amylolytic lactic acid bacterium. Biores. Technol. **1997**, *60*, 101–106.

26
Explosive Waste Treatment

J. Paul Chen, Shuaiwen Zou, and Simo Olavi Pehkonen
National University of Singapore, Singapore

Yung-Tse Hung
Cleveland State University, Cleveland, Ohio, U.S.A.

Lawrence K. Wang
Zorex Corporation, Newtonville, New York, U.S.A., and
Lenox Institute of Water Technology, Lenox, Massachusetts, U.S.A.

26.1 INTRODUCTION

In the early 20th century, more than 60 highly explosive compounds were developed. Testing and use of high explosives has extensively contaminated soil, sediments, and water with toxic explosive residues at a large number of government installations. The end of the Cold War resulted in a significant surplus of both conventional and nuclear weapons. The United States and the former Soviet Union, together with their allies and the People's Republic of China are destroying large quantities of weapons. As a result, a great many high explosives are being released directly and indirectly to the environment. These are highly toxic and mutagenic and classified as environmental hazards and priority pollutants by the US Environmental Protection Agency (USEPA) [1–3].

Large amounts of highly explosive wastes (HEWs) need to be treated. The United States, Europe, and the former Soviet Union have produced more than 360,000 tons of HEW per year. It is estimated that 2,4,6-trinitrotoluene (TNT) alone is produced in amounts close to 1,000,000 kg a year. It has been well documented that the toxicity of these energetic chemicals can be accumulated in bodies of aquatic organisms and terrestrial species such as earthworms and mammals. Under the Intermediate-Range Nuclear Forces Treaty (INF) and Strategic Arms Reduction Treaties, the US Department of Defense demilitarization program destroys a significant amount of weapons, which generates large amounts of high explosive contaminated wastewaters that needs to be treated [4–6]. In this chapter, we will discuss the generation of HEWs, and the respective treatment technologies.

26.2 CHARACTERIZATION OF EXPLOSIVE WASTEWATER

26.2.1 General Aspect

"Explosive waste" normally refers to propellants, explosives, and pyrotechnics (PEP), which belong to the more general category of energetic materials. These materials are susceptible to initiation, or self-sustained energy release, when they are present in sufficient quantities and exposed to stimuli (e.g., heat, shock, and friction). Each reacts differently; all will burn, but explosives and propellants can detonate under certain conditions (e.g., confinement). Figure 1 outlines the various categories of energetic materials.

Wastewaters containing explosives are produced at military installations carrying out manufacturing, and loading, assembly, and packing (LAP) of munitions, as well as washout or deactivation/demilitarization operations. The LAP generates wastewater from cleanup and washdown operations. Deactivation is accomplished by washing out or steaming out the explosives from bombs and shells. Explosives contaminated waters are normally subdivided into two categories based on the color of the wastewater:

- Red water, which comes strictly from the manufacture of 2,4,6-trinitrotoluene (TNT); and
- Pink water, which includes any washwater associated with LAP operations or with the deactivation of munitions involving contact with finished TNT.

A list of explosive compounds is given in Table 1. The explosives-associated compounds (XACs) enter the natural subsurfaces from several types of sources:

- production facilities, e.g., wastewater lagoons and filtration pits;
- solid waste destruction facilities, e.g., burn pits and incineration wastes;
- packing or warehouse facilities; and
- dispersed, unexploded ordnance, e.g., from firing ranges.

The wastewater from manufacturing and packing operations has posed the greatest threat to groundwater. In the United States, royal demolition explosives (RDX) is presently the most important military high explosive because TNT is no longer manufactured, although it is a

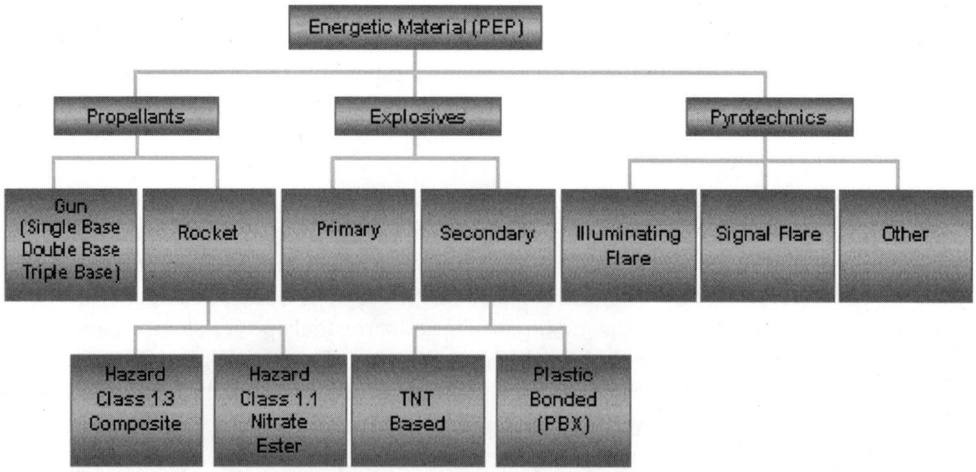

Figure 1 Categories of energetic materials.

Table 1 List of High Explosives

TNT (2,4,6-Trinitrotoluene)	Picrates
RDX (Cyclo-1,3,5-trimethylene-2,4,6-trinitramine)	TNB (Trinitrobenzenes)
Tetryl (*N*-Methyl-*N*,2,4,6-tetranitrobenzeneamine)	DNB (Dintrobenzenes)
2,4-DNT (2,4-Dinitrotoluene)	Nitroglycerine
2,6-DNT (2,6-Dinitrotoluene)	Nitrocellulose
HMX (1,3,5,7-Tetranitro-1,3,5,7-tetraazocyclooctane)	AP (Ammonium perchlorate)
Nitroaromatics	Nitroglycerine

widely used military explosive. Energetic materials are classified into propellants, explosives, and PEP as shown in Figure 1. Characterization and environmental impact of PEP are summarized below.

26.2.2 Explosives

Explosives are commonly classified as primary and secondary based on their susceptibility to initiation. Primary explosives such as lead azide and lead styphenate are highly susceptible to initiation; as they are used to ignite secondary explosives, they are referred to as initiating explosives.

Secondary explosives include TNT, cyclotrimethylenetrinitroamine or cyclo-1,3,5-trimethylene-2,4,6-trinitramine or royal demolition explosives (RDX or cyclonite), cyclotetramethylenetetranitramine (cyclo-1,3,5,7-tetramethylene-2,4,6,8-tetranitramine or high melting explosives (HMX), and tetryl. They are much more prevalent at military sites than primary explosives. As they are formulated to detonate only under specific circumstances, they are used as main charge or bolstering explosives. Secondary explosives can be loosely categorized into melt–pour explosives, which are based on TNT, and plastic-bonded explosives (PBX), which are based on a binder and crystalline explosive such as RDX. They can also be classified according to their chemical structure as nitroaromatics such as TNT, and nitramines such as RDX.

Trinitrotoluene can have three structural isomers: 2,3,5-TNT, 2,4,6-TNT and 2,4,5-TNT. The symmetrical 2,4,6-trinitrotoluene is the most commonly found form and thus in most publications that discuss munitions compounds, the simple name TNT refers to this isomer. TNT is still the most widely used military explosive due to its low melting point of 80.1°C, stability, low sensitivity to impact, and relatively safe methods of manufacture. RDX, or cyclonite as it is commonly referred to, is an explosive used extensively as a propellant for propelling artillery shells and in projectiles. It is used in the above applications because it offers higher energy, higher density, and lower flame temperatures. It is often used in binary mixtures with TNT. It finds its way into the munition wastewaters during manufacturing and blending operations [3]. The general chemical and physical properties of TNT, RDX, and HMX are illustrated in Table 2.

The nitro-organics found in the waste streams from army ammunition plants (AAPs), are resistant to aerobic biodegradation and are toxic in nature in addition to being suspected carcinogens. These waste streams, if released untreated, would cause serious contamination in natural surface and subsurface systems, thereby affecting the health of humans and may have a detrimental effect on the environment. For example, HMX has adverse effects on the central nervous system (CNS) in mammals and has also been classified as a class D carcinogen by the USEPA. Therefore, army installations are prohibited from discharging wastewater into the environment prior to meet a low interim discharge limit of 1 ppm of total nitrobodies. The USEPA has ambient criteria of 0.06 mg/L for TNT and 0.3 mg/L for RDX/HMX, and drinking water criteria of 0.049 mg/L for TNT and 0.035 mg/L for RDX/HMX [4–7].

Table 2 Properties of Important Explosive Wastes

Munitions	TNT	RDX	HMX
Chemical name	2,4,6-Trinitrotoluene	Hexahydro-1,3,5-trinitro-s-triazine, cyclotrimethylenetrinitramine, 1,3,5-trinitro-1,3,5-triazocyclohexane, Hexogen, cyclonite, and T4	Octahydro-1,3,5,7-tetranitro-1,3, 5,7-tetrazocine, cyclotetramethylene-tetramine, and octogen
Chemical formula	$C_7H_5N_3O_6$	$C_3H_6N_6O_6$	$C_4H_8N_8O_8$
Molecular weight	227.15	222.26	296.16
Physical property	Light yellow or buff crystalline solid; soluble in alcohol, ether and hot water; detonates at around 240°C	Colorless polycrystalline material; soluble in acetone; insoluble in water, alcohol, carbon tetrachloride, and carbon disulfide; slightly soluble in methanol and ether	Colorless polycrystalline material; similar to replaced RDX but has a higher density and much higher melting point; highly soluble in dimethylsulfoxide
Usage	TNT is the most common military explosive because of its ease of manufacture and its suitability for melt loading, either as the pure explosive or as binary mixtures with RDX or HMX	RDX is used extensively as the base charge in detonators. Its most common uses are as an ingredient in castable TNT-based binary explosives and as the primary ingredient in plastic-bonded explosives. Mixtures are used as the explosives fill in almost all types of munitions	Because of its high density, HMX has replaced RDX in explosive applications for which energy and volume are important. It is used in castable TNT-based binary explosives, as the main ingredient in high-performance plastic-bonded explosives, and in high-performance solid propellants
Hazard	Flammable, dangerous fire risk, moderate explosion risk. Toxic by ingestion, inhalation, and skin absorption	High explosive, easily initiated by mercury fulminate. Toxic by inhalation and skin contact. 1.5 times as powerful as TNT	High explosive; toxic inhalation and skin contact

26.2.3 Propellants

Propellants include both rocket and gun propellants. Most rocket propellants are either based on: (a) a rubber binder, ammonium perchlorate (AP) oxidizer, and a powdered aluminum (Al) fuel; or (b) a nitrate ester, usually nitroglycerine (NG), nitrocellulose (NC), HMX, AP, or polymer-bound low NC. If a binder is used, it is normally an isocyanate-cured polyester or polyether. Some propellants contain combustion modifiers, such as lead oxide.

Gun propellants are usually single base (NC), double base (NC and NG), or triple base [NC, NG, and nitroguanidine (NQ)]. Some of the newer, lower vulnerability gun propellants contain binders and crystalline explosives and are similar to PBX.

26.2.4 Pyrotechnics

Pyrotechnics include illuminating flares, signaling flares, colored and white smoke generators, tracers, incendiary delays, fuses, and photo-flash compounds. Pyrotechnics are composed of an inorganic oxidizer and metal powder in a binder. Illuminating flares contain sodium nitrate, magnesium, and a binder. Signaling flares contain barium, strontium, or other metal nitrates.

26.2.5 Safety Aspects

Owing to the nature of explosive wastes, strict safety precautions must be taken at sites contaminated with explosive wastes to avoid initiation. The US Army Environmental Center (USAEC) has developed protocols for identifying sites that require explosives safety precautions and for handling explosives wastes at these sites.

With the protocol, people can determine quickly and inexpensively whether materials are susceptible to initiation and propagation by analyzing the composition of samples from the site. According to the deflagration-to-detonation test, soils containing more than 12% secondary explosives by weight are susceptible to initiation by flame; according to the shock gap test, soils containing more than 15% secondary explosives by weight are susceptible to initiation by shock. As a conservative limit, USAEC considers all soils containing more than 10% secondary explosives by weight to be susceptible to initiation and propagation and exercises a number of safety precautions when sampling and treating these soils. Sampling and treatment precautions are exercised when handling soils that contain even minute quantities of primary explosives.

Work, sampling, and health and safety plans for explosives waste sites should incorporate safety provisions that normally would not be included in work and sampling plans for other sites. The most important safety precaution is to minimize exposure, which involves minimizing the number of workers exposed to hazardous situations, the duration of exposure, and the degree of hazard.

26.3 OVERVIEW OF TREATMENT TECHNOLOGIES

26.3.1 Basic Mechanisms of Remediation

Based on treatment location, there are *in situ* and *ex situ* treatments. Each treatment includes physical, chemical, and biological processes, which are discussed in Sections 26.4 and 26.5. There are many technologies available in the market for remediation of contaminated sites. We can classify them into three categories. They can be used separately and in most of cases in conjunction to remediate explosive waste contaminated sites. They are: (a) alteration of contaminants; (b) separation of contaminants from environmental media; and (c) immobilization of contaminants. As no single technology can remediate an entire site, several treatment technologies must usually be combined for cost-effective treatment.

Alteration of contaminant is usually referred to as destruction of the contaminant in most environmental engineering text books. Strictly speaking, the term *alteration* is more suitable, because treatment technologies in most cases are not able to eliminate the contaminants. There is always a question of treatment efficiency. In addition, a series of byproducts may be produced during the treatment. Destruction of contaminant can be fulfilled by altering the chemical structures of the contaminants by physical, biological, and chemical treatment approaches. These destruction technologies can be applied in situ or ex situ to contaminated media.

Separation of contaminants from environmental media includes physical and chemical processes. The separated pollutants are eventually transported to treatment stations for their final destruction. This process is applied to contaminated sites. Typical examples are soil washing and pump-and-treat technologies. The technologies and the possible integration must be carefully selected based on the natural characteristics of water, soil, and sediment in order to achieve the most cost-effective treatment. For example, more air than water can be moved through soil. Therefore, for a volatile contaminant in soil that is relatively insoluble in water, soil vapor extraction would be more efficient than soil flushing or washing.

Immobilization of contaminants includes stabilization, solidification, and containment technologies, such as placement in a secure landfill or construction of slurry walls. As no immobilization technology is permanently effective, monitoring and maintenance are important to prevent secondary pollution, such as leaching.

26.3.2 In Situ Treatment

In *in situ* treatment or disposal, contaminated surface water, groundwater, soil, and sediment are treated directly in contaminated sites. This method does not require movement of contaminated materials. The main advantage is significant cost savings; however, it generally requires longer time periods. In addition, there is less certainty about the uniformity of treatment because of the variability in water and soil characteristics and because the efficacy of the process is more difficult to verify. Like treatment technologies for other contaminants, physical, physicochemical, and biological processes are available.

In Situ Physical and Physicochemical Treatment

Physical and physicochemical treatments use the physical and chemical properties of the contaminants or the contaminated medium to destroy and separate the contamination. For contaminated surface and subsurface waters, and leachate, the following technologies can be used: adsorption, membrane separation, air sparging, bioslurping, chemical oxidation, directional wells, dual phase extraction, thermal treatment, hydrofracturing enhancements, in-well air stripping, and passive/reactive treatment walls. For contaminated soil, sediment, bedrock, and sludge, the available technologies include: chemical oxidation, electrokinetic separation, fracturing, soil flushing, soil vapor extraction, solidification/stabilization, and thermal treatment [3,8–13]. Physical and physicochemical treatment technologies are normally costly even though removal of contaminants can be completed in short time periods in comparison with biological treatment. Residuals after treatment will require further disposal, which will add to the total project costs.

In Situ Biological Treatment

In in situ bioremediation processes, microorganisms are stimulated to grow and use the organic contaminants (e.g., TNT) as a food and energy source. In order to facilitate the processes, a favorable environment for the microorganisms must be provided and maintained. This can be done by providing sufficient oxygen, nutrients, and moisture, and controlling the temperature and pH. Sometimes, microorganisms adapted for degradation of the specific contaminants are added to enhance the process.

Bioventing, enhanced bioremediation, and phytoremediation are commonly used to treat contaminants in soil, sediment, bedrock, and sludge, while enhanced bioremediation, monitored natural attenuation, and phytoremediation are employed to remediate contaminated surface and subsurface waters [3]. These bioremediation techniques have been successfully used in the remediation of various contaminated sites; they have very limited effect on inorganic contaminants.

Compared to physicochemical treatments, the operating costs for biological treatments are typically lower. The residuals after biological treatments are normally less harmful and thus no further treatment or disposal is required. However, in a few cases, the biological residuals could be more toxic and may be mobilized in natural subsurface waters if no careful control is performed. If it occurs, in situ bioremediation must be performed above a low permeability soil layer in order to avoid further groundwater pollution. In addition, a good groundwater monitoring system must be installed in the treatment site.

26.3.3 Ex Situ Treatment

In ex situ treatment or disposal, contaminated surface water, groundwater, soil, and sediment are first transported from the contaminated sites and subsequently treated by various processes. Unlike in situ treatment, ex situ treatment requires shorter time periods. Ex situ remediation can be easily monitored and controlled, which is much better than in situ treatment. However, it requires transportation of contaminated substances, such as pumping of groundwater, which can lead to significantly higher operating costs. Like treatment technologies for other contaminants, physical, physico-chemical, and biological processes are available.

Ex Situ Physical and Physicochemical Treatment

Ex situ physical and physicochemical treatment processes destroy, separate, or concentrate contaminants based on the physical properties of the contaminants or the contaminated medium. For contaminated soil, sediment, bedrock, and sludge, chemical extraction, chemical reduction/oxidation, dehalogenation, soil washing, solidification/stabilization, hot-gas decontamination, incineration, open burn/open detonation, pyrolysis, thermal desorption, and landfill cap can be used. Contaminated groundwater, surface water, and leachate can be remediated by adsorption, advanced oxidation processes, air stripping, coagulation, and flocculation. Among the above technologies, advanced oxidation processes (e.g., UV oxidation) and thermal treatment processes (e.g., incineration) are destruction technologies; all other technologies are separation technologies, which either separate or concentrate pollutants.

Ex situ physical and physicochemical remediation can be completed in shorter time periods than biological treatment. Treatment residuals from separation techniques will require treatment or disposal, which will add to the total project costs and may require permits. For example, adsorption can remove TNT species from aqueous solutions. The water after treatment becomes less contaminated; however, further disposal of TNT adsorbed chemically onto activated carbons is problematic.

Ex Situ Biological Treatment

In ex situ biological treatment, wastewater, soil, and sediment are first moved to treatment stations and treated biologically. For soil, sediment, bedrock, and sludge, biopiles, composting, landfarming, slurry phase biological treatment can be used; natural subsurface and surface waters can be remediated by bioreactors and constructed wetlands. A series of bioreactors have been widely studied and applied [12]. If based on oxygen environments, there are aerobic and anaerobic treatments; if based on reactor configuration, there are continuously stirred tank reactors (CSTR), plug-flow reactors (PFR), fixed-bed reactors (FBR), and fluidized-bed reactors.

Ex situ biological remediation generally requires shorter time periods than in situ remediation. There is more certainty about the uniformity of treatment because of the ability to monitor and manage the treatment. However, ex situ treatment requires transportation of contaminated water, soil, and sediment (e.g., pumping of groundwater), leading to increased costs and inconvenience in management. Other properties of ex situ biological remediation are similar to those of in situ treatment as discussed previously.

26.4 PHYSICAL AND PHYSICOCHEMICAL TREATMENT

The most common method for transformation of nitro-aromatics at present is incineration. However, while incineration has been demonstrated to be an effective technology, issues such as safety, noise, air emissions, costs, regulatory requirements, etc., have motivated research in physical and physico-chemical treatment technologies. Several commonly used technologies are summarized below.

26.4.1 Advanced Oxidation Treatment

Oxidative processes have been used for the treatment of wastewater contaminated with organic compounds. Direct oxidation of aqueous solutions containing organic contaminants can be carried out under a variety of conditions. Catalytic oxidation of aromatics contaminated wastewater has shown that the ring cleavage step is very fast at moderate pressures and temperatures. Oxidation may also be carried out by enervation of hydroxyl radicals in sufficient quantities by ultraviolet radiation alone or in the presence of an oxidant chemical such as ozone (O_3) and hydrogen peroxide (H_2O_2). These types of oxidation processes are referred as advanced oxidation processes (AOP).

Hydrogen peroxide is a relatively inexpensive and readily available chemical. Photolysis of H_2O_2 is the most direct method for the generation of hydroxyl radicals ($HO^{\bullet}$) according to the following reaction:

$$H_2O_2 + h\nu \rightarrow 2HO^{\bullet} \tag{1}$$

The excited hydrogen peroxide molecule is cleaved into two hydroxyl radicals, which initiate the chain decomposition:

$$HO^{\bullet} + H_2O_2 \rightarrow {}^{\bullet}OOH + H_2O_2 \tag{2}$$

$${}^{\bullet}OOH + H_2O_2 \rightarrow HO^{\bullet} + H_2O + O_2 \tag{3}$$

$$2{}^{\bullet}OOH \rightarrow H_2O + O_2 \tag{4}$$

At the same time, these radicals initiate the degradation process by abstracting a hydrogen atom from a high explosives molecule:

$$RH + HO^{\bullet} \rightarrow R^{\bullet} + H_2O \tag{5}$$

Direct ultraviolet photolysis and a combination of ultraviolet photolysis with hydrogen peroxide oxidation have been used to treat TNT, RDX, and HMX wastewater [13]. With a low-pressure mercury UV lamp, more than 70% of RDX/HMX can be oxidized in about 10 min. Complete oxidation is achieved in 25 min. When ultraviolet oxidation is applied, 70% degradation of TNT is achieved; however, the process takes a rather long period due to the complicated structure of TNT.

Hydrogen peroxide in combination with UV radiation does not significantly enhance degradation of RDX/HMX. However, it greatly improves the photolytic degradation of TNT. When the H_2O_2/TNT ratio is increased to a higher level, hydrogen peroxide has a negative effect on the UV photolytic process [13].

26.4.2 Adsorption by Activated Carbon

The adsorption of TNT and other explosive wastes on activated carbon has been one of the most common treatment technologies used by military ammunition plants. This technology is effective at removing a wide variety of explosive contaminants from water, but is nondestructive and expensive to operate. Moreover, after the carbon is exhausted, it has to be incinerated or disposed of into a hazardous waste disposal site. The disposal of used carbon is very costly.

Batch reactors and fixed-bed reactors can be used for the remediation. The batch reactor is flexible to operate; the fixed-bed reactor is more convenient to use. The adsorption capacity and treatment efficiency are dependent on physical and chemical properties of carbon, solution chemistry, as well as operating conditions [3,8,9,12,14,15].

A successful example was illustrated in McAlester Army Ammunition Plant (McAAP) [7]. A process for pink water treatment includes storage tanks, coagulation/flocculation, settling tanks, sand filters, and activated carbon adsorption columns. When the concentration approaches the pretreatment discharge limit of 1 mg/L TNT, the carbon columns are replaced [7].

Heilmann and coworkers [5] used a combination of adsorption and alkaline hydrolysis to treat HMX and other explosive waste. The waste first enters the adsorption reactor, where the contaminants are adsorbed. The spent carbons are then sent to alkaline hydrolysis reactors for desorption. This system provides a better treatment result and can provide a higher degree of flexibility in operation.

26.5 BIOLOGICAL TREATMENT

26.5.1 Mechanisms

Biological remediation of explosive wastes mainly occurs via compound oxidation and reduction. Oxidation takes place when oxygen is the reactant and oxygenase or peroxidase enzymes act as catalysts to cleave the aromatic ring. Reduction is the more common mechanism for nitroaromatics and occurs when the nitroaromatic compound is reduced to arylamines via hydrolytic deamination, acetylation, reductive deamination, and finally cyclization.

It was found that the *Pseudomonas* species degraded both DNT and TNT aerobically with supplemental glucose as a carbon source [16,17]. Reduction of the nitro groups took place only at the para position and proceeded through hydroxylamino-nitrotoluene to aminonitrotoluene. Haidour and Ramos [18] observed 2-hydroxylamino-4,6-dinitrotoluene,4-hydroxylamino-2,6-dinitrotoluene, 4-amino-2,6-dinitrotoluene, 2-amino-4,6-dinitrotoluene, and 2,4-diamino-nitro-toluene as the products of TNT degradation with *Pseudomonas* sp. Boopathy *et al.* [19] reported the anaerobic degradation of TNT under different electron accepting conditions by a soil bacterial consortium. Hughes et al. [20] demonstrated the ability of *Clostridium acetobutylicum* to reduce TNT to 2,4-dihydroxylamino-6-nitrotoluene and then to phenol products via the Bamberger rearrangement. The transformation pathway is shown in Fig. 2 [20].

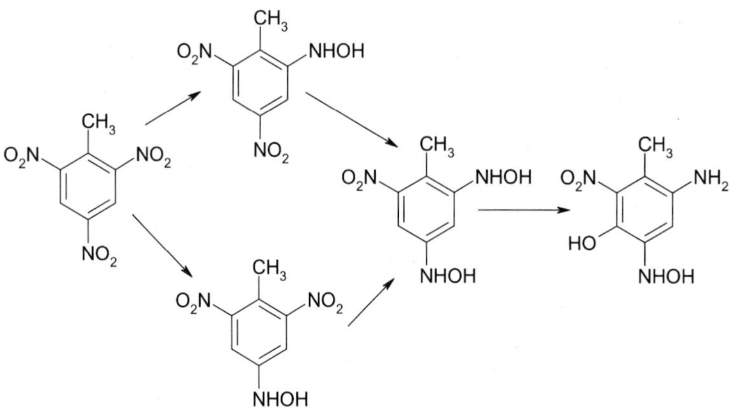

Figure 2 Pathway of TNT transformation observed in *Clostridium acetobutylicum* crude cell extracts.

Biotransformation of RDX has been observed by a number of researchers. Young *et al.* [21] found a bacterial consortium in horse manure capable of degrading RDX at the rate of 0.022 L/g cells per hour. Most of the research in the biological degradation of RDX has been carried out under anaerobic conditions.

Kitts et al. [22] isolated three different genera of bacteria, which were able to degrade RDX. The most effective degrader of these three isolates was identified as *Morganella morganii.* One pathway for the biotransformation of RDX is a stepwise reduction of each of three nitro groups in RDX to form nitroso groups, as shown in Fig. 3 [22,23]. Detail of site remediation and ground water decontamination technologies are presented in another chapter by Wang [27].

26.5.2 Operation Conditions

As with other organics, biodegradation of explosive wastes is influenced by temperature, oxygen supply, nutrient supply, pH, the availability of the contaminant to the microorganism, the concentration of the wastes, and the presence of substances toxic to the microorganisms (e.g.,

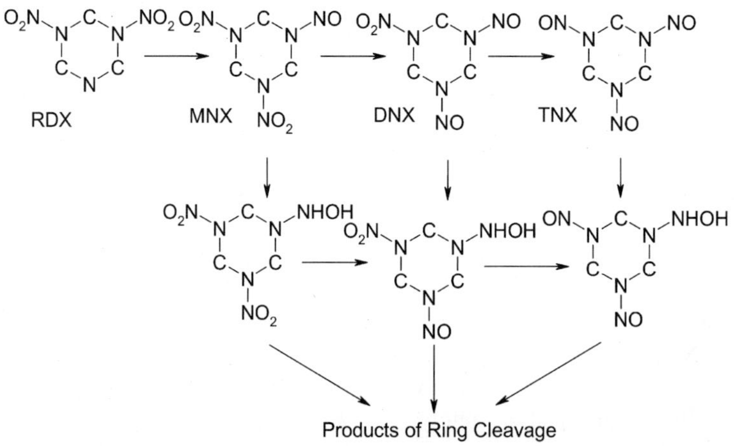

Figure 3 Stepwise reduction of RDX through reduction of nitroso groups.

mercury) [3,12]. Therefore, operation conditions must be carefully controlled in order to reach optimal treatment results for explosive wastes.

Nutrients required for cell growth are nitrogen, phosphorus, potassium, sulfur, magnesium, calcium, manganese, iron, zinc, and copper. If nutrients are not available in sufficient amounts, microbial activity will stop. This is common in explosive wastes. Phosphates are suspected to cause soil plugging as a result of their reaction with minerals, such as iron and calcium. They form stable precipitates that fill the pores in the soil and aquifer. Nitrogen and phosphorus are the nutrients most likely to be deficient in the contaminated environment and thus are usually added to the bioremediation system in a useable form (e.g., as ammonium for nitrogen and as phosphate for phosphorus) [3]. pH affects the solution chemistry, for example, solubility of compounds, which can affect biological activity. Many metals potentially toxic to microorganisms are insoluble at elevated pH; therefore, elevating the pH of the treatment system can reduce the risk of poisoning the microorganisms. Oxygen level is very important in the biological remediation of explosive wastes.

Temperature affects microbial activity in the environment. The biodegradation rate would slow down with decreasing temperature; thus, in northern climates bioremediation may be ineffective during part of the year unless it is carried out in a climate-controlled facility. The microorganisms remain viable at temperatures below freezing and will resume activity when the temperature rises. Provisions for heating the bioremediation site, such as use of warm air injection, may speed up the remediation process. Too high a temperature, however, can be detrimental to some microorganisms, essentially sterilizing the aquifer. It can also affect nonbiological losses of contaminants, mainly through the evaporation of contaminants at high temperatures. The solubility of contaminants typically increases with increasing temperature; however, some hydrocarbons are more soluble at low temperatures than at high temperatures. Additionally, oxygen solubility decreases with increasing temperature.

26.5.3 Bioreactors

There are many bioreactors available in the market for bioremediation of explosive wastes [3,12]. Successful examples include: (a) granular activated carbon–fluidized bed reactor (GAC–FBR) [24], (b) fungal bioremediation [25], (c) membrane bioreactor [6], and (d) constructed wetland [3,26]. All these technologies are able to treat the contaminants below regulated levels.

REFERENCES

1. Drucker, M.P. Pollution control and waste minimization in military facilities. In *Professional Engineering for Pollution Control and Waste Minimization*; Wise, D.L., Trantolo D., Ed.; Marcel Dekker, New York, 1994.
2. Noyes, R. *Chemical Weapons Destruction and Explosive Waste/Unexploded Ordnance Remediation*; Noyes Publications: New Jersey, 1996.
3. US Army Environmental Center. *FY 2000–Annual Report*, SFIM-AEC-ET-TR-200116; Pollution Prevention & Environmental Technology Division, U.S. Army Environmental Center, 2000.
4. Maloney, S.W.; Meenakshisundaram, D.; Mehta, M.; Pehkonen, S.O. *Electrochemical reduction of nitro-aromatic compounds, product studies and mathematical modeling*, technical report; U.S. Army, Corps of Engineers, CERL: Champaign, IL; 1999, Report Number 99/85, ADANumber 371059, 01 Oct 1999.

5. Heilmann, H.M.; Stenstrom, M.K.; Hesselmann, R.P.X.; Wiesmann, U. Kinetics of the aqueous alkaline homogenous hydrolysis of high explosive 1,3,5,7-tetraaza-1,3,5,7-tetranitrocyclooctane (HMX). Water Sci. Technol. **1994**, *30*, 53–61.
6. Zoh, K.D.; Stenstrom, M.K. Application of a membrane bioreactor for treating explosives process wastewater. Water Res. **2002**, *36*, 1018–1024.
7. US Department of Defense, Environmental Security Technology Certification Program. Mineralization of TNT, RDX and by-products in an anaerobic granular activated carbon-fluidized bed reactor; U.S. Department of Defense, 2003; CP-0004.
8. Chen, J.P.; Yiacoumi, S. Transport Modeling of depleted uranium (DU) in subsurface systems. Water, Air, Soil Pollut. **2002** *140* (*1–4*), 173–201.
9. Chen, J.P.; Wang, L. Characterization of metal adsorption kinetic properties in batch and fixed-bed reactors., Chemosphere **2004**, *54* (*3*), 397–404.
10. Chen, J.P.; Lie, D.; Wang, L.; Wu, S.N.; Zhang, B.P. Dried waste activated sludge as biosorbents for metal removal: adsorptive characterization and prevention of organic leaching. J. Chem. Technol. Biotechnol. **2002**, *77* (*6*), 657–662.
11. Chen, J.P.; Lim, L.L. Key factors in chemical reduction by hydrazine for recovery of precious metals. Chemosphere **2002**, *49* (*4*), 363–370.
12. Metcalf and Eddy. *Wastewater Engineering: Treatment Disposal Reuse*, 4th ed.; McGraw-Hill, 2003.
13. Alnaizy, R.; Akgerman, A. Oxidative treatment of high explosives contaminated wastewater. Water Res. **1999**, (*33*), 2021–2030.
14. Chen, J.P.; Lin, M.S. Equilibrium and kinetics of metal ion adsorption onto a commercial H-type granular activated carbon: experimental and modeling studies. Water Res. **2001**, *35* (*10*), 2385–2394.
15. Rajagopal, C.; Kapoor, J.C. Development of adsorptive removal process for treatment of explosives contaminated wastewater using activated carbon. J. Hazardous Mater. **2001**, *B87*, 73–98.
16. Parrish, F.W. Fungal transformation of 2,4-dinitrotoluene and 2,4,6-trinitrotoluene. Appl. Environ. Microbiol. **1997**, *47*, 1295–1298.
17. Schackmann, A.; Muller, R. Reduction of nitro-aromatic compounds by different species under aerobic conditions. Appl. Environ. Microbiol. **1991**, *34*, 809–813.
18. Haidour, A.; Ramos, J.L. Identification of products resulting from the biological reduction of TNT and DNT by *Pseudomonas* sp. Environ. Sci. Technol. **1996**, *30*, 2365–2370.
19. Boopathy, R.; Wilson, M.; Kulpa, C.F. Anaerobic removal of TNT under different electron accepting conditions: lab study. Water Environ. Res. **1993**, *65*, 271–275.
20. Hughes, J.B.; Wang, C.; Yesland, K.; Richardson, A.; Bhadra, R.; Bennett, G.; Rudolph, F. Bamberger rearrangement during TNT metabolism by *C. acetobutylicum*. Environ. Sci. Technol. **1998**, *32*, 494–500.
21. Young, D.M.; Kitts, C.L.; Unkefer, P.J.; Ogden, K.L. Biological breakdown of RDX in slurry reactors. Biotechnol. Bioeng. **1997**, *56*, 258–267.
22. Kitts, C.L.; Cunningham, D.P.; Unkefer, J.P. Isolation of three hexahydro-1,3,5-trinitro-1,3,5-triazine-degrading species of the family *Enterobacteriaceae* from nitramine explosive contaminated soil. Appl. Environ. Microbiol. **1994**, *60*, 4608–4611.
23. McCormick, N.G.; Cornell, J.H.; Kaplan, A.M. Biodegradation of hexahydro-1,3,5-trinitro-1,3,5-triazine. Appl. Environ. Microbiol. **1981**, *42*, 817–823.
24. Maloney, S.W.; Adrian, N.R.; Hickey, R.F.; Heine, R.L. Anaerobic treatment of pinkwater in a fluidized bed reactor containing GAC. J. Hazardous Mater. **2002**, *92*, 77–88.
25. Bennett, J.W. Prospects for fungal bioremediation of TNT munition waste. Int. Biodeter. Biodegrad. **1994**, *34* (*1*), 21–34.
26. Lorin, R. Constructed wetlands: passive system for wastewater treatment, Technology status report prepared for the US EPA Technology Innovation Office under a national network of environmental management studies fellowship, 2001.
27. Wang, L.K. *Site remediation and groundwater decontamination.* In: Handbook of Industrial and Hazardous Wastes Treatment. Wang, L.K.; Hung, Y.T.; Lo, H.H.; Yapijakis. C. (eds.) Marcel Dekker: New York, 2004, 923–969.

27
Food Waste Treatment

Masao Ukita and Tsuyoshi Imai
Yamaguchi University, Yamaguchi, Japan

Yung-Tse Hung
Cleveland State University, Cleveland, Ohio, U.S.A.

27.1 INTRODUCTION

Food processing industries occupy an important position economically and generate large volumes of mostly biodegradable wastes. However, hazardous wastes are also occasionally generated depending on situations such as contamination by pesticides or herbicides, and pathogens. Even simply unbalanced localization may induce unsuitable accumulation or putrefaction of organic wastes. Discarded gourmet foods might also generate hazardous wastes.

Wastes derived from food industries are categorized into three groups: (a) manufacturing losses, (b) food products thrown away as municipal solid waste (MSW), and (c) discarded wrappers and containers. These groups may be further divided into liquid and solid wastes. This chapter will focus on the background and issues surrounding food wastes from a structural point of view, liquid wastes and wastewater treatment systems, and solid wastes and hazardous wastes of the US and Japanese food industries. Although both countries are in developed stages, they offer contrasting pictures of food waste treatment.

Additionally, several topics will be introduced regarding recent technologies relating to food wastes. These are (a) examples of fermentation factories, (b) cassava starch industries, (c) resource recovery by UASB (up-flow anaerobic sludge bed reactor) or EGSB (extended granular sludge bed reactor), (d) reduction and reuse of wastewater, (e) zero-emission of beer breweries, and (f) technology for garbage recycling.

27.2 STRUCTURAL POINT OF VIEW

The recycling of food wastes should be considered as part of the long-term sustainability of agriculture. As Japan is a typical island state, the undesirable influence of oversea-dependent food production has become obvious. Although free trade systems are commonly accepted in the world today, reconsideration of them may be necessary concerning food and feed from environmental aspects. Figure 1 shows the food and feed cycle in Japan in 1998 on the basis of nitrogen (N), in 10^3 tons/year.

From this figure, it is obvious that the nitrogen cycles originally closed, are very open, because of the consumption of chemical fertilizer and large amount of imported food and feed.

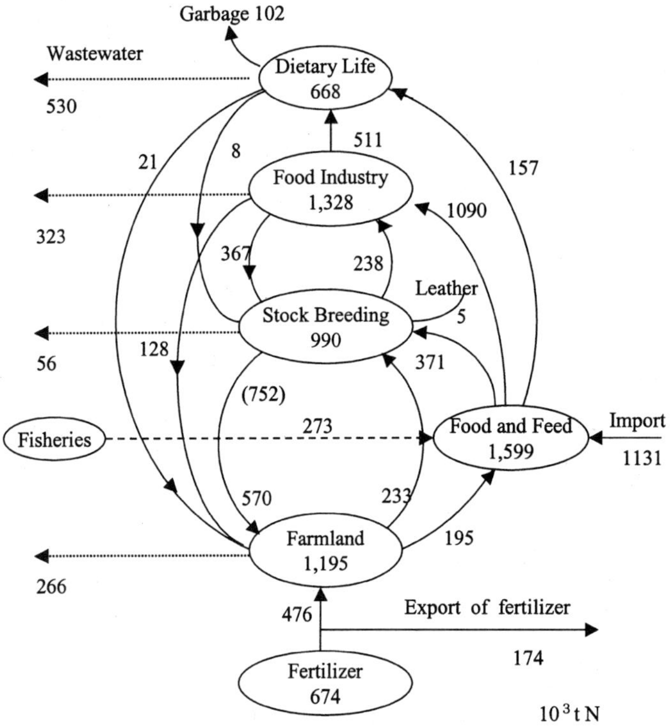

Figure 1 Nitrogen cycle relating to food in Japan.

The rate of Japan's self supply of domestic food was 41% in 1970, 32% in 1990, and 29% in 1998 for N, and 33% (1970), 29% (1990), and 28% (1998) for Phosphorus, excluding grass feed [1]. These facts make the recycling of food wastes difficult in various phases. We have not enough farmlands for food wastes to be recycled. The supply of composts to paddy field for rice plantation decreased from 5.07 ton/ha/year in 1965 to 1.25 ton/ha/year in 1997 [2]. Figure 2 shows a comparison of food balance between Japan and the United States.

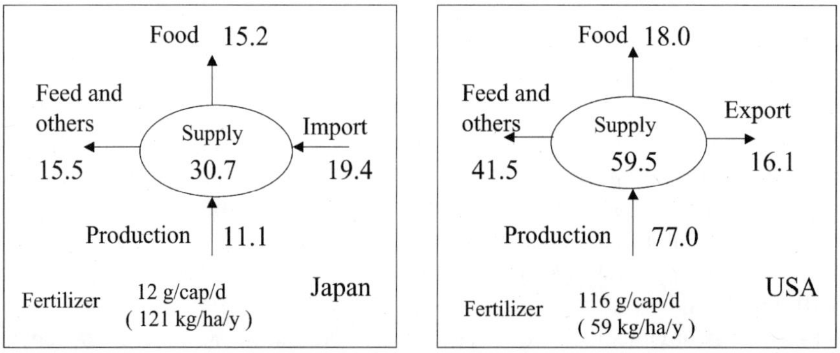

Figure 2 Nitrogen balance of food and feed in Japan and the United States (1992–1994).

Based on international statistics on agriculture, forestry, and fisheries [3], the United States exports food and feed of 4.2 g N/capita/day. However Japan imports food and feed at a rate as high as 19.4 g N/capita/day. The supply of food is 15.2 g/capita/day in Japan, and 18.0 g/capita/day in the United States. The ratio of the amount of food to be recycled to farmland vs. chemical fertilizer consumption is $15.5:12 = 1.3$ in Japan, $41.5:116 = 0.36$ in the United States. The consumption of chemical fertilizer on farmland is 121 kg N/ha and 59 kg N/ha for Japan and for the United States, respectively. Considering these situations, it easy to understand the difficulty of food waste recycling in Japan, which uses more than twice as much foreign farmland overseas as domestic farmland.

These realities also profoundly affect issues of eutrophication in water, not only shortage of the demand for recycled food wastes. The principle that organic wastes should return to the land needs to be enforced.

27.3 LIQUID WASTES FROM FOOD INDUSTRIES

27.3.1 Wastewater Treatment Systems for Food Processing

Different sources contribute to the generation of wastewater in food processing industries, including meat processing, dairy products, seafood and fish processing, fruits and vegetable processing, starch and gluten products, confectionery, sugar processing, alcoholic/nonalcoholic beverages, and bean products.

Wastewaters released from these industries are turbid, with high concentrations of biochemical oxygen demand (BOD), Fats, oils and grease (FOG), suspended solids (SS), and usually nitrogen and phosphorus. Hazardous chemical content is generally low. Other characteristics of food processing wastewater are (a) large seasonal variation, (b) large hourly variation and concentration in daytime, (c) factories are often of small scale, (d) sometimes unbalanced ratio of BOD : N : P that induces the bulking of sludge, and (e) colored effluent.

Usually it is desirable to group wastewater as high concentration, medium concentration, and low concentration. High-concentration wastewater may sometimes be concentrated further, treated, and recycled or disposed as solid wastes. Medium-concentration wastewater may be treated on site or discharged into public sewers. Low-concentration wastewater such as indirect cooling water may be discharged without any treatment.

Decreasing the pollutant load of wastewater requires the reduction of both water consumption and pollutant, as well as reduction of the opportunity for interaction between the two. This may be accomplished by the following measures [4]: (a) reducing the amount of washing water for raw materials and reusing the water [5]; (b) mechanical separation and obtaining concentrated wastewater by saving processing water; (c) minimization of over spills during bottling processes [6] and; (d) reducing the amount of water used to wash tanks and containers after operations.

Typically, the wastewater is subjected to pH adjustment and chemical/physical processes that cause the pollutants to form flocs for subsequent removal. Activated sludge processes are generally employed in food industries. Sometimes advanced treatment systems are used such as coagulation and filtration or other innovated technologies. Recent wastewater treatment technologies are listed in Table 1 [7–12].

Sequencing batch reactors (SBR) are often employed in small food processing factories and have been observed to improve the activated sludge process for the removal efficiency of nitrogen and phosphorus [13]. The conversion of aeration tanks to include anaerobic mixing capabilities increases the removal efficiency of phosphorus and is also effective in preventing bulking. One way to improve the nitrification or the removal efficiency of refractory organics is

Table 1 Recent Progress in Biological Treatment

Type of reactor	Characteristic point	Advantage
Aerobic treatment		
Sequencing batch reactor	Automatic sequential control	Space saving, removal of nutrients
Anerobic–aerobic method	One sludge method	Removal of P, resistant to bulking
Membrane separating method	0.04–0.1 μUF hollow fiber membrane	Space saving, removal of microorganisims
Moving-bed reactor	PP, PU, PE, Activated carbon, etc., are used	Enhance nitrification
Entrapped media method	PEG, PEG-PPG, Calcium arginate, etc., are used	Enhance nitrification
Anaerobic treatment		
UASB	Self-Granulation	High efficiency
EGSB	Higher velocity of upflow	Higher efficiency for medium concentration

Source: Refs. 7–12.

to use the membrane separating method of activated sludge, MF/UF bioreactors, moving-bed or fluidized-bed bioreactors, and entrapped media bioreactors. However, this is not frequently used due to its high costs. Accompanying the development of membrane technology, MF/UF bioreactors may become popular.

Anaerobic treatment systems have been salvaged by adopting UASB [11] or EGSB [12] processes for saving energy. It should also be noted that the primary processing of agro-industries has shifted back to the site of production of raw materials, often in developing countries. These include sugar cane mills and sugar refineries, cassava starch factories, and alcohol fermentation using molasses. Developed countries then import processed raw materials or crude products for further refining or applications. The wastewater can be recycled after treatment, usually through oxidation ponds or stabilization ponds for irrigation on farms; biomass wastes may also be used as fuel for factory operations.

Labor-intensive industries such as sea-food processing also tend to shift to developing countries where cheap labor is available.

27.3.2 Effluent Guidelines and Standards for Food Processing

Table 2 shows the water quality of untreated wastewater from food processing industries in Japan 20 years ago [14]. The flow rate shown represents the values for standard size factories. It should be noted that the oxidation efficiency of COD_{Mn} (JIS K102) may be about one-third.

Table 3 shows the present state of effluent quality in food processing in Japan [15]. The regulation of BOD or COD_{Mn} has already been enacted for a long period, therefore the values for them are converged near the range of standards without large variance. On the other hand, the regulation for TN and TP began only recently, and then only for the specified enclosed water areas such as important lakes, inner sea areas like Seto Inland Sea, Tokyo Bay, and Ise Bay.

Table 4 shows the effluent guidelines by USEPA [16] together with Japanese guidelines [17], which were recently revised for the specified sea areas. The Japanese guidelines are set for the specified wastewater as mentioned above. The USEPA values for effluent limits are noticeably larger than the Japanese values. As a further reference, Table 5 shows the effluent standards for food processing industries of the Tokyo Metropolitan Government [18]. For a flow

Table 2 Characterization of Raw Wastewater from Food Processing in Japan

	Flow rate (m^3/day)	BOD_5 (mg/L)	COD_{Mn} (mg/L)	SS (mg/L)	TN (mg/L)	TP (mg/L)
Meat processing	830	600	400	300	50	15
Dairy products	820	250	170	200	35	5
Seafood cans	530	2700	1700	450	210	75
Fish paste products	650	800	600	500	150	50
Vegetable pickles	440	2300	2500	1000	100	30
Animal feeds	1490	600	300	100	50	10
Starch and gluten	2160	2300	1300	800–1300	130	30–40
Bread and cookies	540	1300	800	900	30	15
Frozen cooked products	400	440	170	200	30	5
Beet sugar processing	4600	450	300	100	25	5
Beer	8500	1000	500	300	40	10
Spirit	1170	500	200	300	20	10
Seasoning chemicals	6500	1000	680	300	460	50
Soy source and amino acids	1090	1000	300	200	100	15

Note: After the investigation by Japanese EPA in 1979–1980, before treatment; flowrate: factories of standard scale. BOD_5, five-day biochemical oxygen demand; COD, chemical oxygen demand; SS, suspended solids; TN, total nitrogen; TP, total phosphorus.
Source: Ref. 14.

rate of more than 500 m^3/day, the criteria of BOD, SS, TN, and TP are set to be 20, 40, 20, and 2–3 mg/L, respectively.

27.4 SOLID WASTES FROM FOOD PROCESSING

Two groups of solid wastes are generated in food industries. One group is organic residual wastes such as sludge from wastewater treatment and food wastes or garbage accompanied with consumption. Another group is solid wastes such as vessels, containers, and wrappers. Among the wastes of this group, plastic wastes should be noted in particular.

27.4.1 Organic Residual Wastes

Table 6 shows the estimated amount of bio-organic wastes in Japan [19]. Wastes from food processing industries amount to only 62,000 tons of N. This is smaller than the value shown in Figure 1 because the statistics of food wastes are not sufficiently arranged.

Table 3 Present State of Effluent Quality in Food Processing in Japan

Categories of specified plants	Number of samples	Flow rate (m^3/day)	Relative varience	BOD_5 (mg/L)	Relative varience	COD_{Mn} (mg/L)	SS (mg/L)	n-hex. extract (mg/L)	TN (mg/L)	TP (mg/L)
Meat and dairy products	528	649	1.1	12	0.2	16	13	2.0	10.9	2.77
Fishery products	443	317	2.1	19	0.3	25	28	3.8	21.7	4.01
Vegetables and fruits products	404	451	1.5	29	3.1	27	15	1.3	7.4	1.41
Soy source and amino acids	133	2,144	5.5	10	0.1	23	14	2.0	11.2	2.42
Sugar processing	36	21,850	0.8	28	0.4	19	15	0.0	4.2	0.19
Bread and cookies	53	465	2.5	7	0.1	14	11	0.9	8.3	2.43
Rice cake and kouji	46	312	0.9	17	0.2	21	21	1.8	5.2	5.26
Beverage	399	1,068	1.9	12	0.2	14	11	0.9	5.2	1.45
Animal feeds and organic fertilizer	61	952	1.6	18	0.2	20	17	1.3	36.5	0.91
Animal oil processing	56	3,852	2.3	51	2.2	20	17	7.1	11.4	2.29
Yeast	3	6,762	1.1	16	0.2	3	14	–	1.4	0.01
Starch and gluten	43	4,785	1.4	49	0.4	47	53	0.6	6.4	2.65
Glucose and maltose	12	7,239	1.3	22	0.2	16	16	0.6	3.3	1.21
Noodle products	98	305	0.9	5	0.1	9	8	1.3	4.7	1.31
Bean curd and processing	187	429	1.9	10	0.1	15	12	1.3	12.3	2.24
Instant coffee	4	2,170	0.5	7	0.0	12	3	0.8	5.8	0.64
Frozen cooked products	94	275	0.9	11	0.1	16	12	1.8	8.4	2.00

Note: Number of samples for some items is less than that for flow rate.
Source: Ref. 15.

Table 4 Effluent Guidelines for Food Processing in the United States and Japan

Category of industries	Japanese guidelines for effluent[a]			USEPA guidelines (BPT)[b]		USEPA guidelines (SPN)[c]	
	COD_{Mn}	TN	TP	BOD_5	TSS	BOD_5	TSS
Meat processing	30–80	10–60	1–16	370–740	450–900	370–740	450–900
Seafood products							
Tuna processing	30–120	10–55	3–12		3300–8300		
Fish meal processing				3900–7000	1500–3700	3800–6700	1500–3700
Dairy Products							
Fluid products	20–50	10–30	1–16	1350–3375	2025–5506	370–740	463–925
Dry milk				650–1625	975–2438	18–36	225–450
Canned vegetables and fruits	30–80	10–30	3–12	140–420	80–240		
Cereal processing							
Flour and other grain mill products	30–70	10–30	1.5–7.5				
Animal feeds	20–90	10–30	1–3.5				
Starch and gluten	40–80	10–30	1.5–10	2000–6000	2000–6000	100–3000	1000–3000
Bread and cookies	40–70	10–30	1.5–7.5				
Sugar processing							
Beet sugar	30–80	10–30	1.5–7.5	2200–3300			
Raw cane sugar processing				630–1140	470–1410		
Cane sugar refining	30–80	10–30	1.5–7.5	430–1190	90–270	90–180	35–110
Oil mill and processing	30–80	10–30	1–7.5				
Wines and beverage							
Beer	30–70	10–30	1.5–4				
Spirit	20–50	10–30	1.5–4				
Yeast	90–130	10–30	1.5–7.5				
Seasoning	20–100	10–145	1.5–9				

[a]Effluent limitations guidelines for specified wastewater from specified plants. Local governments select a value in the range from lower values and higher values.
[b]Effluent limitations guidelines for existing point sources attainable by the application of the best practicable control technology currently available. Lower values: maximum for any 1 day; higher values: average for concecutive 30 days.
[c]Standards of performance for new sources of effluent discharged into navigable waters.
Source: Refs 16 and 17.

For organic food wastes, the options of feedstuff use, composting, biogas then composting, and heat recovery are adopted for treatment and recycling.

Feedstuff Use

Industrial wastes from food processing are still recycled as feed or organic fertilizer to a fair extent. As shown in Table 7, 5.5 million tons of food processing byproducts are used for feeding

Table 5 Effluent Standards for Food Processing Industries by Tokyo Metropolitan Government

	Flow rate (m^3/day)	BOD (mg/L)	COD_{Mn} (mg/L)	SS (mg/L)	TN (mg/L)	TP (mg/L)	Odor intensity
Existing	50–500	25	25	50	30	6	4
	>500	20	20	40	20	3	4
New sources	50–500	20	20	40	25	3	4
	>500	20	20	40	20	2	4

Source: Ref. 18.

Table 6 Amount of Bio-organic Wastes in Japan

	Generation	Dry matter		Contents		Generation of N, P	
	(10^4 tons)	(parts)	(10^4 tons)	Nitrogen	P_2O_5	N	P
Agriculture							
Rice stems	1094			0.60	0.20	6.6	0.96
Straw	78			0.40	0.20	0.3	0.07
Rice husks	232			0.60	0.20	1.4	0.20
Stockbreeding							
Manure	9430			0.79	0.13	74.9	11.96
Residues	167			5.01	7.13	8.4	5.20
Forestry							
Bark	95			0.53	0.08	0.5	0.03
Sawdust	50			0.15	0.03	0.1	0.01
Wood waste	402			0.15	0.03	0.6	0.05
Food processing							
Animal/plant residues	248	0.28	69	1.41	0.53	1.0	0.16
Sludge	1504	0.05	75	7.01	4.02	5.3	1.32
Construction waste							
Wood waste	632			0.15	0.03	0.9	0.08
Municipal solid waste							
Garbage	2028	0.29	588	1.41	0.53	8.3	1.36
Wood and bamboo	247			0.76	0.19	1.9	0.20
Others							
Sewage sludge	8550	0.02	171	5.18	5.37	8.9	4.01
Nightsoil sludge	1995			0.6	0.10	12.0	0.87
Joukasou septage	1359	0.02	27	5.18	5.37	1.4	0.63
Farm sewage sludge	32	0.02	0.6	5.18	5.37	0.0	0.01
Total	28,143					132.3	27.13

Source: Ref. 19.

Table 7 Feedstuff Use of Byproducts from Food Processing

Byproducts of:	Generation, *a* (10^3 t/year)	Use for feed, *b* (10^3 t/year)	*b*/*a* (%)
Fruit juice	116	98	84
Vegetable can	80	72	90
Wine	3030	2707	89
Starch processing	1162	80	7
Bean processing	795	411	52
Sugar	1858	1350	73
Fish processing	58	24	41
Bread and malt	29	16	55
Total	7128	5487	77

Source: Ref. 20.

[20]. The rate of use is 77%. Other than that, rice bran, wheat bran, and plant oil residues and BMP, and others are used for general feedstuff products. Feeding use of residual food for pigs has drastically decreased from 206 kg/head/year in 1965 to 6 kg/head/year in 1997 in Japan [2]. Although this system is a good option for recycling, it is a general tendency that pig farms has gradually shifted far from residential areas and have changed to modernized farms.

Composting and Biogas Production

Composting is a traditional, reliable method of recycling food wastes. This will be discussed later together with biogas production.

Incineration With Energy Recovery

Utilization of biomass as it relates to CO_2 reduction against global warming has been focused on recently. For this purpose, woody biomass is more suitable, and organic wastes including various minerals may not be appropriate for incineration.

27.4.2 Vessels, Containers, and Wrapping Wastes

Another type of waste relating to food industries is the waste originating from containers, vessels, bottles, and wrapping materials. These wastes occupy a large portion of municipal solid waste (MSW). Among these wastes, plastic wastes in particular should be focused on from an environmental standpoint.

One company, FP Co. Ltd., has developed a good recycling system for polystylene paper (PSP) trays in Japan. They employ a whole network of transportation systems from factories through markets and back to factories again. The recycling is a tray-to-tray system. However, the rate of recycling is restricted one-third, because the efficiency of transportation for wastes of PSP tray is one-third of that for the products that can be packed compactly.

Polyethylene terephtalate (PET) bottles are the most suitable wastes for material recycling. They can be recycled as polyester fiber products through PET flake and to raw chemicals through chemical recycling. However, the amount of incinerated PET bottles has still been increasing because the rate of consumption continues to exceed the recycling effort. Chemical recycling to obtain the monomer of dimethyl terephtalate (DMT) has been conducted successfully using recycled PET bottles collected by municipalities [21]. The company Teijin Co. Ltd. plans to transport PET bottles amounting to 60,000 t/year to its factory in the Yamaguchi Prefecture.

Recently, other plastic wastes used for wrapping and vessels have been recycled by means of gasification, liquifaction to oil, or heat recovery in the blast furnaces of steel industries (substituting cokes) and cement kilns (substituting coal). However, there are complicated arguments as to whether the direct incineration for heat recovery is more environmentally friendly than options through gas and oil as mentioned above.

27.5 HAZARDOUS WASTES FROM FOOD PROCESSING

27.5.1 Management of Chemicals Based on EPCRA or PRTR

Chemicals commonly encountered in food processing are listed in Table 8, relating to the Emergency Planning and Community Right-to-Know Act (EPCRA) in the United States [22]. The Pollutant Release and Transfer Register (PRTR) system was also enacted in Japan from the 2001 fiscal year. Figure 3 describes the chemicals used in food processing [22]. Similarly, it is applicable for other materials included in the food itself.

Food that has been accidentally contaminated by pesticides, herbicides, or fumigants may also be treated as hazardous waste. Chlorine is frequently used for sanitary cleaning in food processing at the end of daily operations. Therefore chlorinated organic compounds should be noted in the wastewater treatment plants of food industries. It is very possible that wastewater contains certain levels of trihalomethane and related compounds.

27.5.2 Accidentally Contaminated Food Wastes

Food products contaminated with pathogenic microbes or food poisoning sometimes result in hazardous wastes. The incident of Kanemi rice oil contaminated by PCB in 1968 is still discussed with regards to dioxins (DXNs) as the possible cause of the Kanemi Yusho disease. Two recent examples discussed below include the treatment of contaminated milk products and the issues relating to the issues of BSE.

Table 8 Chemicals Commonly Encountered in Food Processing

Purposes	Chemicals used
Water treatment	Chlorine, chlorine dioxide
Refrigerant uses	Ammonia, ethylene glycol, freon gas
Food ingredients	Phosphoric acid, various food dyes, various metals, peracetic acid
Reactants	Ammonia, benzoyl peroxide, Cl_2, ClO_2, ethylene oxide, propylene oxide, phosphoric acid
Catalysts	Nickel and nickel compounds
Extraction/carrier solvents	n-Butyl alcohol, dichloromethane, n-hexane, phosphoric acid, cyclohexane, tert-butyl alcohol
Cleaning/disinfectant uses	Chlorine, chlorine dioxide, formaldehyde, nitric acid, phosphoric acid, 1,1,1-trichloroethane
Wastewater treatment	Ammonia, hydrochloric acid, sulfuric acid
Fumigants	Bromomethane, ethylene oxide, propylene oxide, bromine
Pesticides/herbicides	Various pesticides and herbicides
Byproducts	Ammonia, chloroform, methanol, hydrogen fluoride, nitrate compounds
Can making/coating	Various ink and coating solvents, various listed metals, various metal pigment compounds

Source: Ref. 22.

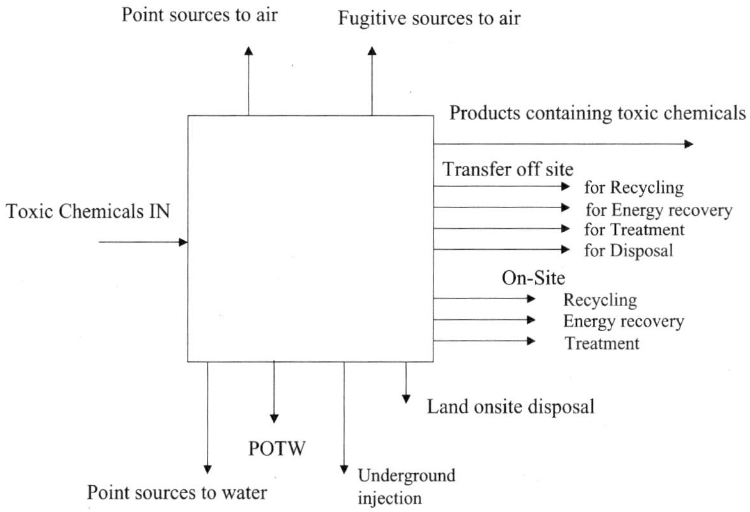

Figure 3 Possible release and other waste management types for EPCRA.

Contaminated Milk with Enterotoxin A

In June 2000, skim milk contaminated by *Staphylococcus aureus* led to a huge outbreak of food poisoning in Japan. The milk was contaminated in April because of an electricity outage that lasted several hours in the factory of Yukijirushi Co. Ltd. Afterwards, contaminated milk products were widely distributed and 14,780 persons exhibited food poisoning symptoms [23]. The milk products produced by the company were removed from market displays, and most of them were incinerated as hazardous wastes.

Treatment of Bone and Meat Powder Suspicious of BSE

The Japanese government has prohibited the import of bone powder from England since March 1996 and also from EU countries since January 2001 to prevent the introduction of Bovine Spongiform Encephalopathy (BSE), known as "mad cow" disease. After three infected cows were reported in Japan, bone and meat powder (BMP) rendered in domestic bone-boiling factories has been prohibited as use as feedstuff and also as fertilizer for a certain period. Although BMP is a good fertilizer, especially for fruit trees, consumers have avoided it because of the risk of BSE. Therefore BMP has been labeled a hazardous waste that is incinerated or treated in cement factories. In total, 8.5% of cow's body is bone, corresponding to 20% of meat. It is said that the amount that should be incinerated is near 1 thousand tons daily [24].

The BMP is transported to cement factories, packed in plastic bags of 15–20 kg each. The bags are thrown down into the end of the kiln at 1000°C. The contents of raw material for cement are shown in Table 9. For 1 ton of cement clinker 2 kg of BMP are used. The percentage of BMP among raw materials is 0.13% by dry base. The receiving toll is around 40,000 yen/t of BMP. Since prion is a kind of protein, it must be decomposed at very high temperatures in the kiln.

27.5.3 Incineration Ash of Food Wastes

Incineration is the most popular method of disposing of combustible solid wastes in Japan, especially for MSW from restaurants, hotels, and supermarkets. For a long time incineration was perceived as a progressive method and the percentage of incineration of MSW was near 80%.

Table 9 Percentages of Raw Materials of Various Kinds of Cement

	Portland cement original	Portland cement recent	Eco-cement
Lime stone	78	76.5	52
Clay	16	1	
Silicate	4	7.5	
Iron source	2	2	1
Pretreated ash of MSW		1	39
Others (coal ash, sludge, plastics)		12	8
Total	100	100	100

However, the evaluation of this method has changed with the discovery of dioxins (DXNs). After incineration, ash containing DXNs is inevitably generated, resulting in another type of hazardous wastes derived from food products. Recently, it has become popular in Japan to recycle the ash to cement raw material. Table 8 illustrates two methods for this type of recycling: (a) eco-cement which uses MSW ash at 39% of total amount of raw material, (b) ordinary cement, which uses pretreated ash at 0.5–0.7% of raw material, as shown in Fig. 4. In Yamaguchi Prefecture 50,000 tons of MSW ash can be recycled by this second method.

This method is considered to be tentative as the final means of recycling. Essentially the goal should be to recycle back to farmland not only major nutrients like N and P but also various minor nutrients like Ca, Mg, Fe, Mn, Cu, Zn, Mo, and B. This means mineral resources derived from food are contaminated by heavy metals through incinaration of garbage together with various other wastes.

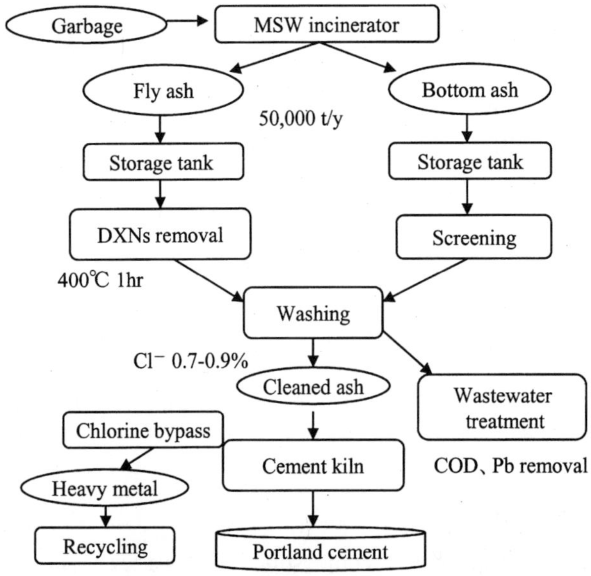

Figure 4 Ash recycling for Portland cement feedstock.

27.6 RECENT TECHNOLOGIES ON FOOD WASTES TREATMENT

27.6.1 Waste Management in Fermentation Industries

Fermentation industries cover a wide range of food processing from the traditional industries of breweries, soy source, miso, pickles, to yeast, alcohol, amino acids, nucleic acids, antibiotics, enzyme, and other bio-active fine chemicals. Usually, the harvest rate of these products is not high except for traditional fermentation, which typically has a large pollutant load. However, the possibility of resources recovery is also high because hazardous chemicals are rarely used.

The wastewater comes from (a) the mother liquid after harvesting the products, (b) cleaning water of cells or reactors, (c) condensates from the evaporator, (d) spent eluting solution in purifying processes, and (e) ammonium sulfate, used in salt crystallization of enzymes, and others.

The following information is cited mainly from a thesis [25] published in 1983. While details may have changed since then, the fundamental structures remain largely the same.

Alcohol

Molasses and sweet potatoes are used as the raw materials for alcohol fermentation. In Japan, to improve wastewater quality, the trend of importing crude alcohol and refining it increased in the 1970s. Here the case of alcohol production using sugar cane molasses as raw material is introduced. It is also replaced for example by acetic acid.

The raw materials necessary to produce 1 kL of 95% alcohol are 3.12 tons of molasses, 1.2 kg of urea, and 1.2 kg of ammonium sulfate. The input of N and P is 14.2 kg/kL and 2.18 kg/kL, respectively. The main part of wastewater is the evaporation residues of 10 kL/kL. The water quality of this is 30 g/L for COD_{Mn}, 1.1 g/L for TN, and 0.2 g/L for TP. Therefore the unit loading factor is 300 kg/kL for COD_{Mn}, 11 kg/kL for TN, and 2 kg/kL for TP. The loss of N likely occurs in the process. Methane fermentation was employed before the 1970s; however, it has been replaced by ocean dumping and by drying for reuse to organic fertilizer. Dried matter of distillery waste from the maize alcohol process is valuable and popular as distillery feeds in the United States. On the other hand, distillery wastes from processes using sugar cane molasses have high color and ash content, were recycled to organic fertilizer through concentration, humination with acid treatment to reduce viscosity, adding calcium and magnesium silicate, drying, and palletizing. Consequently, the cleaning water of reactors and the condensates from the evaporation process of distillery wastes are generated as the wastewater.

Bread Yeast

Bread yeast is still produced by fermentation using molasses. As shown in Fig. 5, this process includes cultivation, separation, pressurized filtration, and the addition of baking powder. Wastewater derives mainly from centrifugal supernatant and the cell-washing process.

The estimated mass balance of this process is shown in Table 10 assuming that the yeast content is 2.4% of N and 0.36% of P, and the harvest vs. molasses is 75% and sugar content of molasses is 55%. The rate of loss to wastewater was estimated to be 8% for N and 3% for P. This factory in Hyogo Prefecture had sophisticated treatment systems for its wastewater. The fraction of high concentration was further concentrated and dried to make organic fertilizer. The removal rates of COD_{Mn}, TN, and TP are 99, 72, and 44%, respectively. The unit loading factor vs. product was around 3.3 kg of COD_{Mn}/t of yeast, 0.58 kg TN/t and 0.058 kg TP/t respectively.

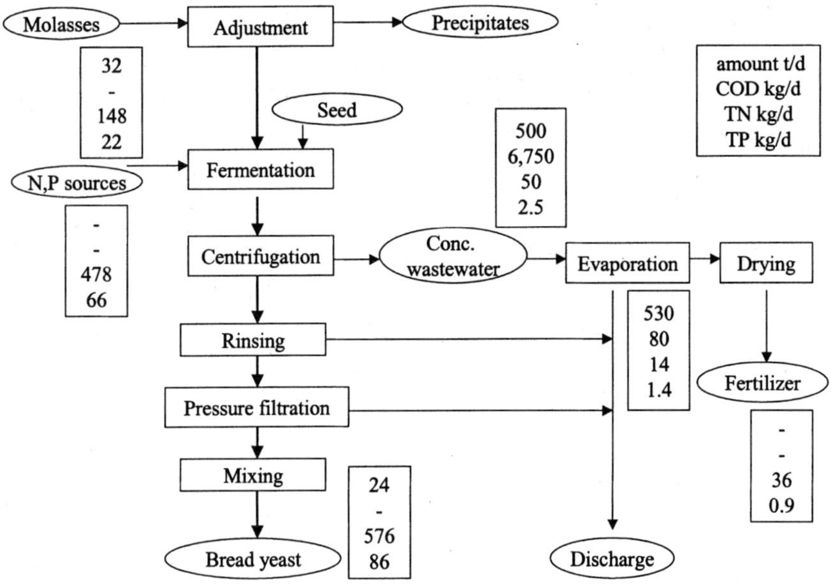

Figure 5 Estimated flow of N and P in a bread yeast production factory.

Mono-Sodium Glutamate

Mono-sodium glutamate (MSG) is a product that originates from Japan, and is produced mostly by fermentation. Glucose, acetic acid, or molasses are used as the raw material for the source of carbon, while ammonia and urea may be used as the source of nitrogen. Figure 6 shows the flow sheet for one factory in Japan. The main part of the wastewater comes from the crystal separating process.

This factory employs concentration and wet combustion processes. Through this treatment, COD was reduced from 2.4 t/day to 0.05 t/day with a removal rate of 98%. The unit loading factor of the discharging stage was estimated to be 15 kg COD/t of MSG, similarly 30 kg/t for TN and 1.5 kg/t for TP. Assuming the input of N and P of 4.6 t/day and 0.88 t/day, respectively, the rate of loss was 22% for N and 57% for P, as shown in Table 11.

Nucleotides

Through hydrolysis of RNA, nucleotides such as inosinic acid and guanyl acid are formed and used for seasonings. Ribonucleotides are produced by the combination of extraction,

Table 10 Wastewater Load of Bread Yeast Production

	Amount of use, *a* (kg/day)	Generated load, *b* (kg/day)*	Loss rate, *b*/*a* (%)	Discharged load, *c* (t/day)	Reduction rate, 1-*c*/*b* (%)	Unit loading factor (kg/t yeast)**
COD_{Mn}	–	6750	–	80	99	33
TN	626	50	8	14	72	16
TP	88	3	3	1	44	0.06

*Excluding other nontreated wastewater; **Assuming moisture of yeast is 68%.

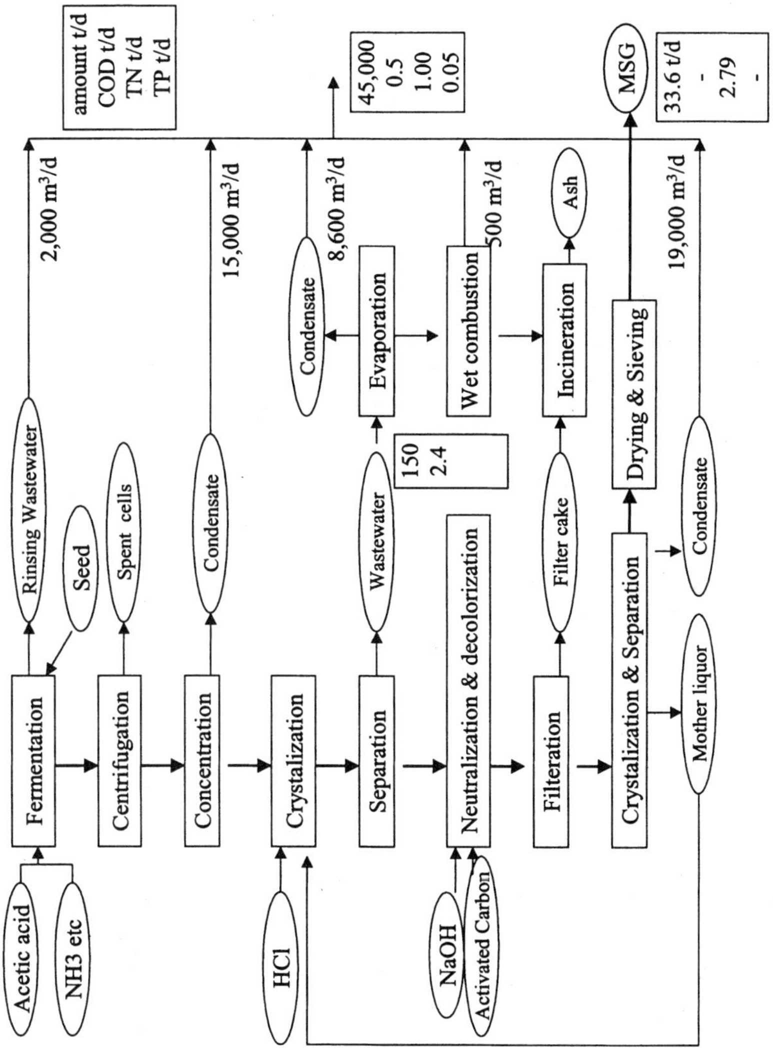

Figure 6 Estimated flow of wastewater in the production of MSG.

Table 11 Wastewater Load of MSG Production

	Amount of use, a (t/day)	Generated load, b (t/day)	Loss rate, b/a (%)	Discharged load, c (t/day)	Reduction rate, $1 - c/b$ (%)	Unit loading factor (t/t MSG)
COD_{Mn}	–	–	–	0.50	–	0.015
TN	4.6	1.8	39.1	1.00	44.4	0.030
TP	0.09	0.09	100.0	0.05	43.2	0.002

fermentation, and chemical synthesis. During fermentation, yeast capable of accumulating RNA is cultured and forms ribonucleotides through enzymic hydrolysis of extracted RNA from the yeast. Finally, ribonucleotides are purified through ion exchange resins, crystallized, and dried to become the products. The main part of wastewater is derived from the yeast separation and purifying processes as shown in Fig. 7.

As these products consist of finer materials, the loss rate may be larger than in the case of MSG. Therefore, various byproducts are recovered. The spent cell of the yeast is utilized for feedstuff and the concentrated part of the wastewater from the purifying process is used as liquid fertilizer after further concentration. A fraction of the wastewater of medium concentration is treated by the activated sludge process. After separating the yeast cells, the wastewater is treated by an activated sludge process of deep shaft aeration. Molasses had been used previously in this process, but acetic acid replaced it as a countermeasure for reducing wastewater load. Table 12 shows the balance of N and P as the result of estimation shown in Table 13. The rate of loss is very large at

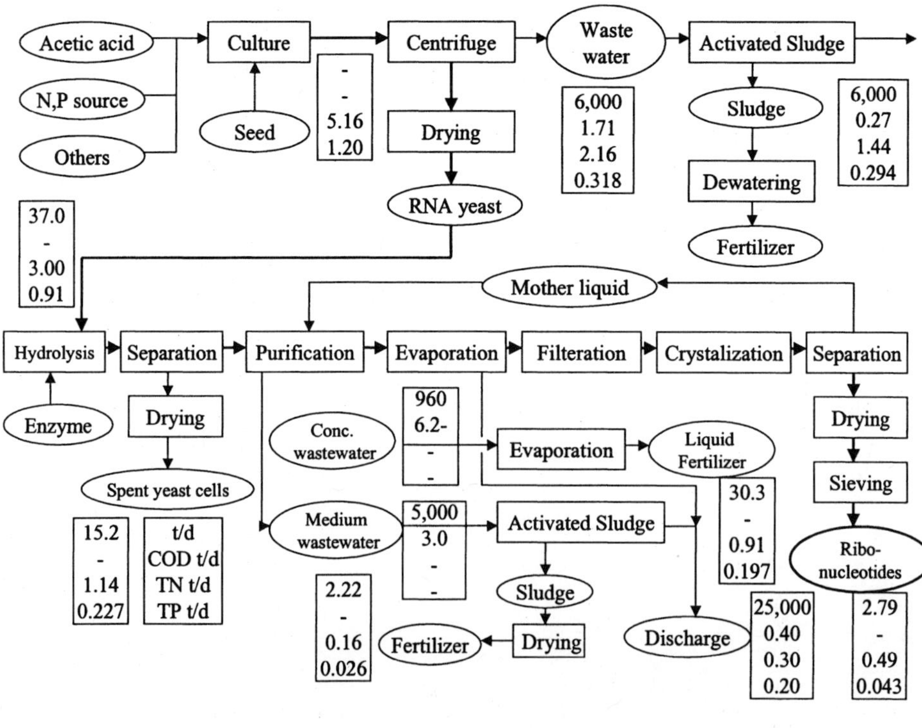

Figure 7 Estimated flow of N and P in the production of ribonucleotides.

Table 12 Wastewater Load of Ribonucleotides Production

	Amount of use, a (t/day)	Generated load, b (t/day)	Loss rate, b/a (%)	Discharged load, c (t/day)	Reduction rate, $1 - c/b$ (%)	Unit loading factor (t/t of nucleotide)
COD_{Mn}	–	11.01	–	0.67	94	0.24
TN	5.16	3.90	76	1.74	55	0.62
TP	1.24	0.74	62	0.49	33	0.17

76% for TN and 62% for TP. The unit loading factor of COD_{Mn}, TN, and TP after treatment was 240 kg/t, 620 kg/t, and 170 kg/t, respectively. The removal rate was 94% for COD_{Mn}, 56% for TN, and 33% for TP.

Example of Comprehensive Fermentation Factory

During the 1970s, a fermentation factory in Yamaguchi Prefecture produced alcohol, MSG, lysine, antibiotics, and organic combined fertilizer and used molasses to produce alcohol, MSG, and yeast. It is introduced as an example, although its present situation has changed.

Previously, the factory employed a combination of methane fermentation and biological aerobic treatment. With the advance of environmental criteria, the following countermeasures were conducted, in stages: (a) to make the substrate concentration higher, (b) reuse of wastewater to substrate solution, (c) conversion of raw material from molasses to acetic acid, and (d) improvement of extraction and purifying processes. For treatment processes, measures included, (a) classification of wastewater, and (b) possible resource recovery of valuable matters. Figure 8 shows the flow of wastewater treatment of this factory in 1976.

The values shown in Figure 8 are the flow rate, daily load of COD_{Mn}, TN, and TP. Table 14 summarizes the estimation process of the balance shown in Fig. 7. In fermentation

Table 13 Estimation of N and P Balance of Ribonucleotides Production Starting from RNA Yeast

	Amount (t/day)[c]	Content		Amount of N and P	
		N (%)	P (%)	N (t/day)	P (t/day)
Input					
RNA yeast	37.0	8.1	2.4	3.00	0.88
Output[a]					
Nucleotides seasonings[b]	2.8	11.8	5.8	0.33	0.16
Spent yeast cells	15.2	7.5	1.5	1.14	0.23
Liquid fertilizer	30.3	3.0	0.7	0.91	0.20
Sludge fertilizer	2.2	8.0	1.3	0.16	0.03
Wastewater	28,000			0.30	0.20
Total				2.84	0.81

[a]Ignoring the input of enzyme and the output of filter cake.
[b]Taking average of N, P contents for 5′-GMP 2Na $7H_2O$ and 5′-IMP 2Na 7-$8H_2O$.
[c]Assuming 27 days/month and 330 days/year.

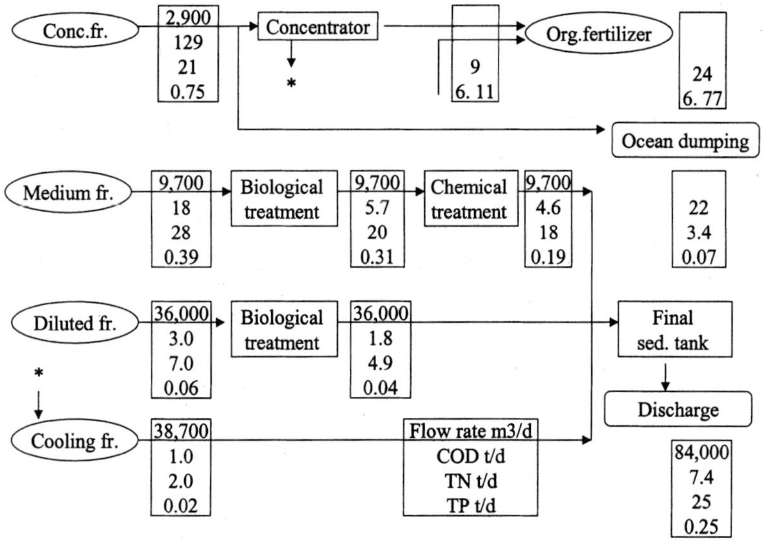

Figure 8 Estimated flow of wastewater in a comprehensive fermentation factory.

factories, batch processes are generally adopted, so that it is difficult to estimate the reliable balance. As shown in Table 15, 76 tons of N were used; 74% of it was generated as wastewater and 25 tons were discharged to the coastal water. As for P, 2.01 tons were used, excluding organic fertilizer production, and 1.2 tons, corresponding to 77% of use, were generated and

Table 14 Estimation of N and P Balance in a Comprehensive Fermentation Factory

	Amount	Content		Amount of N and P[a]	
	(t/year)	N (%)	P (%)	N (t/day)	P (t/day)
Input					
Molasses	282,000	0.43	0.07	3.67	0.598
Soy bean husk	3,000	3.57	0.67	0.32	0.061
Ammonia	28,800	82.25	0	71.78	0
Calcium perphosphate	6,000	0	7.42	0	1.349
Phosphate and others					5.68[b]
Subtotal				75.8	7.69
Output					
Mono-sodium glutamate	15	8.28	0	5.12	0
Amino acids	4,800	15	0	2.18	0[c]
Seasonings	7,080	7	1	1.5	0.215
Organic fertilizer	85,898	9.11	2.6	23.71	6.768
Yeast	3,012	7	1	0.64	0.091
Recycled cells	195	7	1	0.51	0.072
Ocean dumping				3.4	0.07
Removal				12.3	0.22
Discharge	84,000			24.7	0.25
Subtotal				74.1	7.69

[a]Assuming 330 working days/year; [b]estimated from the balance; [c]lysine, leucine, valine.

Table 15 N and P Balance in a Fermentation Factory

	Use, *a* (t/day)	Generated load, *b* (t/day)	Loss rate, *b/a* (%)	Discharged load, *c* (t/day)	Reduction rate, 1 − *c/b* (%)	Recovery, *d* (t/day)	Recovery rate, *d/b* (%)
COD_{Mn}	–	150	–	7.4	95	106	71
TN	76	56	74	25	55	16	29
TP	7.7	1.2	16	0.3	79	0.7	57
	2.01[a]		60				

[a]Excluding the input to fertilizer production.

0.25 tons were discharged to the coastal water. The COD_{Mn} generated was 150 t/day and discharged 7.4 t/day. The production of organic fertilizer had been increasing year by year at that time, and 71, 29, and 57% of COD_{Mn}, TN, and TP, respectively, were recovered as organic fertilizer out of generation load.

27.6.2 Agro-Industries in Tropical Countries

Cassava is a key food product in many tropical countries. In 1997, 165 million tons were produced worldwide for food and feed. Cassava chips, pellet, and starch are major exports for Thailand, which produces 18 million tons annually. Eight million tons were exported annually in 1992–94 to Europe, mainly for the feedstuff, but has decreased since then [26]. The processes for starch are peeling, washing, grating, starch extraction, settling, drying, milling, and sieving.

Solid waste from the extraction process of cassava is known as cassava pulp [27]. The quantity of discharged pulp (60–75% moisture content) is about 15–20% of the root weight (65–75% moisture content) being crushed. This is equivalent to about 1.5–2.0 million tons of pulp discharged each year from 10 million tons of root crushed. Discharged pulp still contains a high starch content (around 50% dry basis). This is due to the inextricable starch that is trapped inside the cells. Pulp is sun dried to reduce the moisture content and used as filler in animal feed. Environmental problems from the solid waste occur only if the storage of pulp is badly managed and it becomes exposed to rain. Utilization of pulp as a substrate for industrial fermentation has been investigated, but to date there has been no success. An attempt to extract the starch from the pulp by means of enzyme hydrolysis has been reported. Treatment of pulp with a mixture of cellulase and pectinase increased starch recovery by 40%.

Liquid waste is discharged from the factory at about 10–30 m^3/ton of starch produced. This waste has a high BOD and COD content. In Thailand, simple treatment of the wastewater is practised. The first stage is screening to remove insoluble debris such as peel and woody parts. Wastewater is treated in an open-type anaerobic pond followed by oxidation ponds or aerated lagoons and finally a polishing pond before final discharge. Balogoparan and Padmaja [29] reported cyanide concentration between 10.4 and 274 mg/L in the final effluent, and also a high concentration of cyanide in groundwater near the factory ranging between 1.2 and 1.6 mg/L [28].

Most factories prefer to build a "no discharge" system; this means more ponds are prepared for retaining the treated wastewater. A huge land area is required for wastewater because of the long retention times required. Some factories near towns have had to improve their wastewater treatment processes. These factories now employ closed-type anaerobic reactors or an activated sludge process. The composition of wastewater from five factories during the year 1997 is shown in Table 16. Regulation of the Ministry of Industry, Thailand, allows a BOD in discharge wastewater of 20 mg/L.

Table 16 Wastewater Characteristics From Representative Cassava Starch Factories

	Factories				
	1	2	3	4	5
Production, *a* (t/day)	70	90	155	300	141
Wastewater, *b* (m^3/day)	2350	2700	4100	2936	1770
b/*a* (m^3/t)	33	32	25	11	14
pH	4.8	5.0	5.7	8.3	5.3
COD (g/L)	13.0	16.3	15.0	19.3	19.2
BOD (g/L)	6.5	11.6	10.6	12.6	9.4
TS (g/L)	13.2	16.4	12.5	19.8	18.0
SS (g/L)	7.4	8.0	6.8	7.0	7.8

Source: Ref. 27.

Table 17 shows other examples of agro-industrial wastewater treatment in Thailand. Similarly to that mentioned above, the wastewaters are generally treated with lagoon systems. In the case of the palm oil mill, through 81 days of retention time, the effluent is utilized for irrigation of palm fields [1]. A type of closed system is realized in this case. However, if other systems can recover energy at low cost, they would be more preferable.

The color of the effluent has also become a serious problem of public interest. The colors of food processing wastewater are caused generally by melanoidines formed through the reaction of sugars with amino acids and polyphenols derived from lignin materials. It is not easy to decompose these materials biologically, although many researchers have tried to use effective

Table 17 Examples of Agro-Industry Wastewater in Thailand

	Raw material (t/day)	Treatment	Wastewater (m^3/day)	Water quality (mg/L)				
				BOD	COD	SS	TN	TP
Fishery products								
Influent	500	4 ponds (42,000 m^2)	3,000	1,020	1,950	368	46	12
Effluent				20	385	19	2	14
Criteria				20	120	30–150		
Palm oil mill								
Influent	600	22 ponds (10,120 m^2)	300	50,000				
Effluent				100				
Criteria				60				
Palm oil mill								
Influent				25,000	50,000	18,000	750	180
Effluent			3 m^3/t of crude oil					
Criteria				100		400	200	

Source: Ref. 1.

fungi. It should be noted that for the decolorization of alcohol distillery wastewater using *Mycelia sterilia*, even if melanoidines are removed, residual polyphenols cause color through oxidation [29]. Ozone, electrolysis, and coagulation with chitosan are effective [30]. However, these processes have not prevailed because of the high cost to date.

27.6.3 UASB and EGSB Treatment Systems

Anaerobic treatment, especially thermophilic treatment, offers an attractive alternative for the treatment of high-strength, hot wastewater. The thermophilic process, compared to the mesophilic anaerobic process, has the advantages of increased loading rate and the elimination of cooling before treatment. Furthermore, the heat content of the wastewater would be available for post-treatment. Loading rates up to 80 $kgCOD/m^3/day$ and more have been reached in laboratory-scale thermophilic reactors treating glucose, acetate, and sucrose and thermomechanical pulping whitewater. Table 18 shows the results of food wastewater treatment by thermophilic UASB at 55°C together with the examples for pulp mill wastewater [31–40]. For alcohol distillery wastewater at the loading rate of 100 $kgCOD/m^3/day$, successful removal efficiencies were reported. Rintala and Lipisto [42] reported 70°C thermopilic UASB experiment using pulp mill wastewater; however, the COD removal was not high at 56% at the loading rate of 41 $kgCOD/m^3/day$ [41].

Table 19 summarizes successful performances using EGSB in China. Biogas recovered at the rate of 0.4–0.6 m^3/day vs. 1 kL of 95% alcohol has been used as supplementary fuel for coal [42].

Among seven beet sugar processing factories in Hokkaido, Japan, three use UASB. Biogas is generated at the rate of 7000 N m^3/day under the condition of 15 $kgCOD/m^3/day$. It is used as fuel for boilers and dryers of beet pulp. Starch processing also generates wastewater of high BOD. Four factories in Japan use UASB reactors and generate biogas of 8000–9000 N m^3/day under the condition of 15 $kgCOD/m^3/day$.

Thermophilic anaerobic treatment of hot vegetable processing wastewaters deriving from steam peeling and blanching of carrot, potato, and swede, was studied in laboratory-scale UASB reactors at 55°C [43]. The reactors were inoculated with mesophilic granular sludge. Stable thermophilic methanogenesis with about 60% COD removal was reached within 28 days. During the 134 day study period the loading rate was increased up to 24 $kgCOD/m^3/day$. More than 90% COD removal and methane production of 7.3 m^3CH_4/day were achieved. The anaerobic process performance was not affected by the changes in the wastewater due to the different processed vegetables. The wastewater characteristics are summarized in Table 20, and the water qualities of influent and effluent in the experiments are shown in Table 21.

Several studies have also attempted to use membrane technology in combination with anaerobic packed bed reactors [44]. Three different methane fermentation processes were evaluated using soybean processing wastewater: (a) Process A – acidification (empty bed volume: 1 m^3) and methane reactors (empty bed volume: 2 m^3); (b) Process B – acidification and methane reactors followed by membrane (Polysulfone and PVA, MW cutoff, approx. 15,000); and (c) Process C – acidification reactor, membrane, and methane reactors. The characteristics of the wastewater are BOD 1000 mg/L, COD 1629 mg/L, VSS 693 mg/L, protein 544 mg/L, and lipid 23 mg/L.

Process B showed a COD removal of 77.7% by decreasing the free SS in the treated water. Higher acetic acid and propionic acid concentrations were found as residual in the treated water. The rate of methane conversion was 68.9%.

Process C showed a remarkable removal of COD by 92.4% and methane conversion of 83.4%. Process C gave noteworthy improvement in results compared with process A. It has been

Table 18 Results of Wastewater Treatment by Thermophilic UASB

Reactor volume (L)	Temperature (°C)	Inoculum sludge	Types of wastewater	Wastewater conc. (mgCOD/L)	Volumetric loading rate ($kgCOD/m^3/day$)	COD removal rate (%)	Reference
75,000	55–57	Methophilic digested sludge and cow dung	Alcohol wastewater	31,500	26.5	71.7	31
5.5	55	Thermophilic granular sludge	Alcohol wastewater	15,400	98.3	58	32
126	55	Thermophilic digested sludge	Alcohol wastewater	10,000	28	41.5	33
0.12	55	Methophilic granular sludge	Pulp wastewater	1,900–2,200	14.1	45	34
0.12	55	Thermophilic granular sludge	Pulp wastewater	1,900–2,200	17.7	56	34
0.12	55	Thermophilic granular sludge	Pulp wastewater	1,900–2,200	40.8	56	34
71.5	50–53	—	Alcohol wastewater	19,200–25,600	9.6	91.2	35
46.2	51–54	—	Acetic acid production	6,600–8,500	5.0–6.0	93	36
1,400	55	Methophilic granular sludge	Beer wastewater	2,000	50	82	36
2	55	Methophilic granular sludge	Coffee processing	1,000–4,000	4	>70.0	37
0.25	55	—	Pulp wastewater	1,900	2.6	26.0–42.0	38
8	55	Thermophilic granular sludge	Alcohol wastewater	3,000–10,000	100	85.0–90.0	39
11.9	55	Methophilic granular sludge	Alcohol wastewater	9,000	30	85	40

Source: Refs. 31–40.

Table 19 Performance of EGSB in Alcohol Wastewater in China

Place	Production (kL/year)	Alcohol (%)	Raw material	Wastewater (m^3/day)	COD before EGSB (mg/L)	COD after EGSB (mg/L)	Biogas generation (m^3/day)
Nin Bo	35,000	95	Cassava	1,500	25,000–30,000	1,000–1,500	20,000–23,000
Nan Hai	50,000	38	Rice	550	30,000–35,000	1,000–1,500	10,000
Nin He	60,000	95	Maize	1,500	25,000–30,000	1,000–1,500	25,000
Tan Chen	15,000	95	Potato	700	10,000–14,000	500–1,000	3,000

Source: Ref. 42.

concluded that process C was excellent and recommended when wastewater enriched with VSS is treated. The combination of membrane with UASB was also studied. However, the results are not feasible for practical use [45].

27.6.4 Zero-Emission in Beer Breweries

Waste recycling systems in beer breweries are very complete. Kirin Beer Co. Ltd. has achieved zero-emission for its industrial wastes since 1998. Table 22 shows the amounts for each of the wastes and their uses [46]. Moreover, the emission factor of wastes has itself also decreased from 0.205 kg/L in 1996 to 0.140 kg/L in 2001. Wastewater is treated by a UASB reactor and activated sludge method in 10 out of 12 factories in this company. In fact, 18,860 $\times$ 10^3 m^3 of wastewater generate 4800 tons of methane gas from UASB reactors, corresponding to 5200 kL of oil. The biogas is used for the fuel of boiler and cogeneration systems. In another big beer company in Japan, we can see similar situations. They treat wastewater in eight out of nine factories by UASB or EGSB reactors and activated sludge. They produce 8315 tons of methane from 14,652 $\times$ 10^3 m^3 of wastewater.

27.6.5 Recycling of Garbage

Composting

Composting has long been a traditional technology, but new composting technologies have also been developed. Generally there are mainly two types of technology, both dependent on using microorganisms. One type uses a thermophilic bacillus, which is effective in enhancing the initial decomposing phase. Many types of microbial additives have been developed by

Table 20 Characteristics of Vegetable Processing Wastewater after Removing Solids

Unit operation	Raw material	Total COD		Soluble COD			
		Average	Range	Average	Range		
Steam peeling	Carrot	19.4	17.4–23.6	17.8	15.1–22.6	Exp.1	1–73 days
	Potato	27.4	13.7–32.6	14.2	11.7–17.5	Exp.2	38–46 days
Blanching	Carrot	45	26.3–71.4	37.6	22.1–45.8	Exp.1	1–73 days
	Potato	39.6	17.0–79.1	31.3	10.9–60.6	Exp.2	38–46 days
	Swede	49.8	40.5–59.1	49.4	40.5–58.3	Exp.2	1–38 days

Source: Ref. 43.

Table 21 Quality of Influent and Effluent for Thermophilic UASB in Vegetable Food Processing

	Influent		Effluent	
	BOD	COD	BOD	COD
Exp. 1	6,900–19,900	9,500–27,600	660–1,400	1,200–2,100
Exp. 2	6,700–7,600	10,500–11,800	1,400–1,600	2,000–2,300

Source: Ref. 43.

companies. Some employ a heat-shock process at 75°C for 3 hours before adding the microbes. Figure 9 shows a flow of this type adopted in Komagane City [47].

Another type of composting uses mixed culture mainly constituted by lactic acid bacteria (EM) under anaerobic condition. Figure 10 shows an example of this process adopted in a small town in Japan. Usually the addition of a relatively large amount of rice husk may be needed for the production of valuable compost.

A combination of both types of composting has also been tested. In one case, relatively fresh fish refuse was preheated at 80°C for several hours with the addition of a thermophilic bacillus, and then EM was added. In this case EM was expected to work after packaging in plastic bags. This product may be used as chicken feed.

Usually, garbage includes salt and oily materials, which makes it difficult to use as a sole source of compost. Often after drying or primary fermentation through consuming electric

Table 22 Recycling of Wastes from Beer Breweries (Kirin Beer Co. Ltd. in 1988)

Wastes	Amount (tons/year)	Uses
Spent grain	359,919	Feedstuff, fertilizer, heat recovery
Refined dregs	2,618	Feedstuff, fertilizer
Excess yeast	7,168	Foodstuff, medicine
Used diatomaceous earth	17,556	Cement feedstock, soil conditioner
Wastewater sludge	35,319	Fertilizer
Paper waste		
Labels	3,051	Recycled paper
Danball	2,167	
Bags for raw materials	103	
Used paper	400	
Wood pallet	3,231	Fuel
Glass wastes	57,409	Raw material of glass
Caps	337	Raw material of steel
Aluminum cans	493	Raw material of aluminum
Steel cans	1,008	Raw material of steel
Plastic wastes		
Plastic box	6,795	
Plastics	553	Reductant for high furnace
Used oil	141	Fuel, recycled oil
Incineration ash	152	Cement feedstock, soil conditioner
Others	1,869	
	500,289	Recycling rate: 100%

Source: Ref. 46.

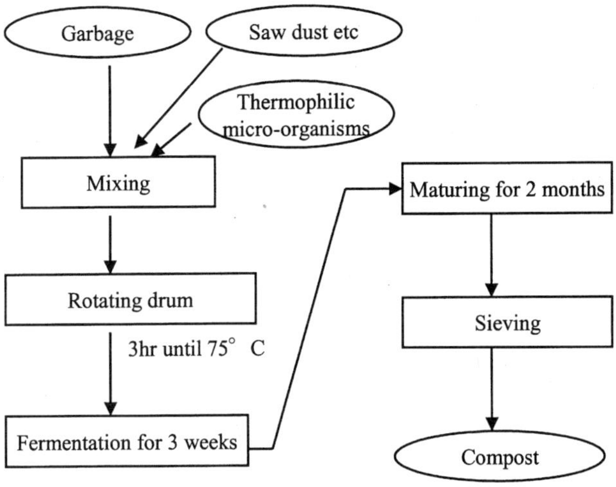

Figure 9 Composting process using thermotolerant microorganisms.

energy at home, restaurant, and supermarket, pretreated garbage is transported and mixed to livestock manure compost at a certain proportion, 20% for instance.

In Japan, many companies intend to produce compost using the garbage collected from commercial sectors. However, as mentioned previously, due to the difficult situation surrounding organic fertilizer, a high quality of compost is needed. Under these circumstances, a pilot study was conducted in Kitakyushu City to produce polylactic acid (PL) through glucosidation, lactic acid fermentation, butyl ester formation, hydrolysis, and polymerization [48]. The cost of

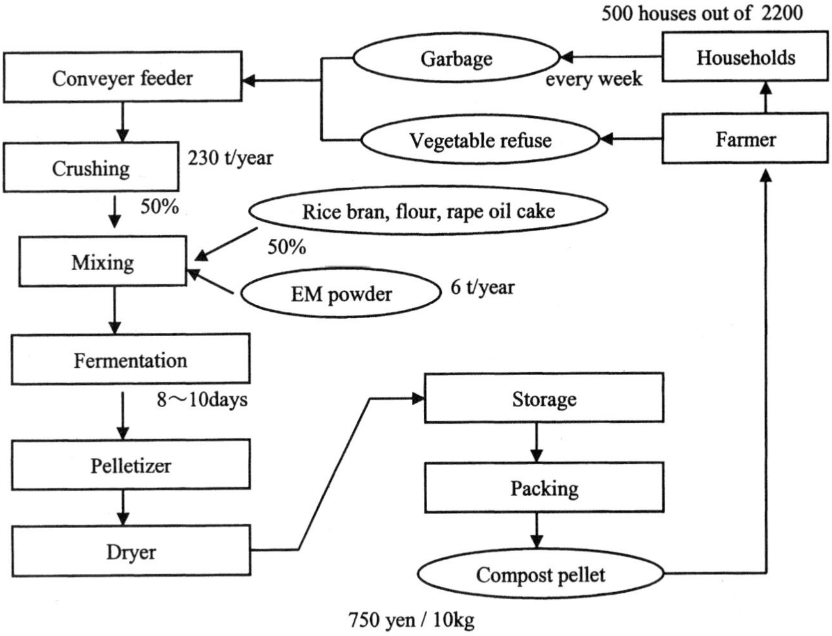

Figure 10 Composting process of garbage using EM in a small town in Japan.

3.5 $/kg PL was estimated for the plant of 100 tons of garbage/day. The harvest rate of PL is assumed to be 5% vs. garbage. If the garbage treatment cost is 30 yen/kg, then it corresponds to $600/125 = 4.5$ $/kg PL, and it is expected to be feasible. This project is challenging, however, and may encounter other difficulties.

Biogas Production

This system also seems feasible, especially since the capacities of night soil treatment plants have become excessive, accompanied by the spread of publicly owned sewerage systems. Therefore a part of these night soil plants have been reformed to treat night soil, sludge from septic tanks, and garbage together. Table 23 shows a list of such plants [49]. There are also several plants treating garbage solely by methane fermentation in Japan. It is said that the production of biogas corresponds to 100–200 N m^3/ton of garbage. Table 24 shows the results of comparing a biogas system to an incineration system with power generation. The energy efficiency of the biogas system is estimated to be better than that of an incineration system. If the heat value of garbage becomes higher, then incineration with power generation becomes advantageous [50].

27.7 CONCLUSION

This chapter discussed the structural point of view and the characterization of food processing relating to liquid, solid, and hazardous wastes. Focusing on case studies mainly in Japan the details of food waste generation and treatment have been presented, sometimes in comparison with cases in the United States. We can summarize the conclusions and recommendations on several points as follows.

As a main principle, the self-supply of food should be a goal for all countries. The local unbalance between food production and consumption makes food wastes hazardous and induces eutrophication of water bodies, nitrate pollution of groundwater, and ruins farmland soil because of limited recycling of minerals. Therefore, the free trade of primary products should be re-examined in light of environmental issues.

Table 23 Sludge Recycling Centers Using Anaerobic Treatment in Japan

Prefecture	Region	Construction period	Amount of wastes night soil and sludge (kL/day)	Garbage, etc. (t/day)	Energy use	Technology
Nara	Jouetsu	1999	240	8.0	a	Thermophilic
Nagano	Shimoina	1999	16	8.0	a	Thermophilic
Niigata	Higashi Kanbara	1999	22.3	3.5	b	Mesophilic
Nara	Ikoma	2000	80	1.3	a	Thermophilic
Miyazaki	Kusima	2000	35	0.9	b	Mesophilic
Nara	Nara	2001	90	3.0	b	Mesophilic
Hokkaido	Minami Souya	2002	15	16.0	a	Thermophilic
Hokkaido	Nisi Tenpoku	2002	20	8.0	b	Thermophilic
Miyagi	Mutsunokuni	2002	105	1.0	b	Thermophilic
Nagasaki	Kamigotou	2002	69	3.0	b	Thermophilic

a: power generation with cogeneration; b: heat use only.
Source: Ref. 49.

Table 24 Energy Efficiency of Biogas Production System and Incineration with Power Generation

	Unit	Incineration with power generation	Biogas production system
Biogas generated	Nm^3/t		100
Calorific value generated	Kcal/t	1,000,000[a]	5,000,000
Electric power generated	KWh/t	150	150
Efficiency assumed	%	13	26
Self-consumption of electricity	KWh/t	100	50[b]
Net gain of energy	KWh/t	50	100

[a]Assuming garbage of 1000 kcal/kg; [b]Not including wastewater treatment.
Source: Ref. 50.

Generally, food processing does not generate chemical hazards. However, attention should be given to chlorine used for cleaning and sanitation leading to chlorinated byproducts in wastewater. If contaminated by poisonous materials or pathogens, food may change to hazardous wastes. Treatment of BMP by cement kiln is a typical example of this.

As previously discussed, incineration of food wastes together with other miscellaneous wastes is not a suitable solution because of the generation of hazardous ash containing DXNs and heavy metals; doing so also threatens food recycling efforts. A recommended option would be composting followed by the combination of biogas production and composting of the sludge.

During the past 30 years, wastewater from food processing in Japan has been significantly improved due to better wastewater treatment systems and conversion of raw materials or processes. It is notable that primary processes generating much waste have been shifted to other countries where the raw materials are produced and sometimes to developing countries where cheap labor is supplied.

Anaerobic treatment systems lost their popularity in wastewater treatment fields for failing to meet strict environmental criteria. Recently, however, these systems have regained attention because of their ability to save energy and also reflecting the development of UASB or EGSB technology.

In relation to ISO 14000, some companies have targeted and attained zero emission in their industries. However, zero emission should also cover all products, including containers and wrappers.

ACKNOWLEDGMENTS

The authors thank Dr. H. Nakanishi, Emeritus Professor of Yamaguchi University, and Dr. I. Fukunaga, Professor of Osaka Human Science University, and other individuals for giving us valuable input.

REFERENCES

1. Ukita, M.; Prasertsan, P. Present state of food and feed cycle and accompanying issues around Japan. *Water Sci. Technol.*, **2002**, *45*(*12*), 13–21.
2. Ushikubo, A. Present state of food wastes and countermeasures for Food Recycling Act; http://mie.lin.go.jp/summary/recycle/recyle.htm (in Japanese), 2002.

3. Ministry of Agriculture, Forestry and Fisheries. *International Statistics on Agriculture, Forestry and Fisheries*; MAFF, 1998.
4. Fukunaga, I. Recent advances of the treatment and disposal of wastewater and solid waste in food industry. *Foods and Food Ingredients J. Japan*, **1995**, *165*, 21–30 (in Japanese).
5. Mavrov; Belieres. Reduction of water consumption and wastewater quantities in the food industry by water recycling using membrane processes. *Desalination* **2000**, *131*, 75–86.
6. Ridgway; Henthorn; Hull. Controlling of overfilling in food processing. *J. Mater. Process. Technol.*, **1999**, *93*, 360–367.
7. Norcross, K.L. Sequencing batch reactors – An overview. *Water Sci. Technol.* **1992**, *26 (9–11)*, 2523–2526.
8. Wentzel, M.C.; Ekama, G.A.; Marais, G.V.R. Kinetic of nitrification denitrification biological excess phosphorus removal systems: A review. *Water Sci. Technol.* **1990**, *17* (*11–12*), 57–71.
9. Zhou, H.; Smith, D.E. Advanced technologies in water and wastewater treatment. *J. Environ. Eng. Sci.* **2002**, *1*, 247–264.
10. Sumino, T. Immobilization of nitrifying bacteria by polyethylene glycol pre-polymer, *J. Ferment. Bioeng.* **1991**; *73*, 37–42.
11. Lettinga, G. *et al.* Use of upflow sludge blanket reactor concept for biological wastewater treatment, especially for anaerobic treatment. *Biotechnol. Bioeng.* **1980**, *22*, 699–734.
12. Kato, M.; Field, J.A.; Versteeg, P.; Lettinga, G. Feasibility of the expanded granular sludge bed (EGSB) reactors for the anaerobic treatment of low strength soluble wastewaters. *Biotechnol. Bioeng.* **1994**, *44*, 469–479.
13. US Environmental Protection Agency. *Wastewater Technology Fact Sheet Sequencing Batch Reactors*, EPA/832/F-99/073; US Environmental Protection Agency, Office of Water: Washington, DC, 1999.
14. Nippon Suido Consultant Co. Ltd. *Report of the Investigation on Instruction for Reducing Nitrogen and Phosphorus Load*(in Japanese); Nippon Suido Consultant Co. Ltd., 1982.
15. Ministry of Environment Japan. *Report on the Discharge of Water Pollutant in 2002* (in Japanese); Ministry of Environment Japan, 2003.
16. USEPA. *Electronic Code of Federal Regulations*, http://ecfr1.access.gpo.gov/otcgi/cfr/, 2003.
17. Central Council of Environment. *Draft on the Regulation of COD, Nitrogen and Phosphorus* (in Japanese), http://www.env.go.jp/press/press.php3?serial = 1317 (in Japanese), 2000.
18. Bureau of Environment Tokyo Metropolitan Government. *Criteria of wastewater* (in Japanese); Bureau of Environment: Tokyo, 2002. http//www.kankyo.metro.tokyo.jp/kaizen/kisei/mizu/kijun/np.htm.
19. The Research Group for Biological Organic Wastes. *Present State and Problems on Biological Organic Wastes* (in Japanese); Research Group for Biological Organic Works, 1999.
20. Japan Livestock Industry Association. *Report on Promoting Feed Use of Un-utilized Resources* (in Japanese); Japan Livestock Industry Association, 1996.
21. Teijin Ltd. *Technical Report on PET recycling*; Teijin Ltd., 2000.
22. USEPA. *EPCRA Section 313 Reporting Guidance for Food Processors*; USEPA: Washington, DC, 1998.
23. The final report of the investigation on the food poisoning accidents by Yukijirushi Milk Company; http://www.mhlw.go.jp/topics/001/tp1220-1.html (in Japanese).
24. Endo, K. The recode of the discussion in budget committee in Japanese Parliament, The 1st Mar. (in Japanese), 2002.
25. Ukita, M. *Fundamental Research on the Behavior of Nitrogen and Phosphorus and on the Mechanisms of Eutrophication in Japan*; Thesis of Kyoto University (in Japanese), 1987.
26. Titapiwatanakun, B. *Report of Strategy Agricultural Comodity Project*: Cassava; Department of Agricultural and Resource Economics, Faculty of Economics: Kasetsart University, 1997.
27. Klanorang, S.; Kuakoon, P.; Sittichoke, W.; Christopher, O. Cassava Starch Technology: The Thai Experience. *Trends Food Sci. Technol.*, **2000**, *52*, 439–449.
28. Balagopalan, C.; Padmaja, G. Cyanogen accumulation in environment during processing of casaba (*Manihot esculenta* Grantz) for starch and sago. *Water Air Solid Pollu.* **1998**, 102, 407–413.

29. Nagano, A. *Study on the Decolorization of Molasses Wastewater*; Thesis of Nagaoka University of Technology, Japan, 2000.
30. Kumar, M.N.V.R. *Chitin and Chitosan for Versatile Applications*, http://www.geocities.com/mnvrk/chitin.html, 2003.
31. Souza, M.E.; Fuzaro, G.; Polegato, A.R. Thermophilic anaerobic digestion of vinasse in pilot plant UASB reactor. *Wat. Sci. Technol.*, **1992**, *25* (*7*), 213.
32. Wiegant, W.M.; Claasseen, J.A.; Lettinga, G. Thermophilic anaerobic digestion of high strength wastewaters. *Biotechnol. Bioeng.*, **1986**, *27* (*9*), 1374.
33. Harada, H.; Uemura, S.; Chen, A.C.; Jayadevan, J. Anaerobic treatment of a recalcitrant distillery wastewater by a thermophilic UASB reactor. *Biores. Technol.*, **1996**, *55*, 215–221.
34. Rintala, J.; Martin, J.S.L.; Lettinga, G. Thermophilic anaerobic treatment of sulfate-rich pulp and paper integrate process water. *Wat. Sci. Technol.*, **1991**, *24* (*3/4*), 149.
35. Zuxuan, Z.; Zepeng, C.; Zeshu, Q. Status quo and prospects on the study of anaerobic digestion for industrial wastewater in China In *Proceedings of 4th International Symposium on Anaerobic Digestion*, 1985; 259.
36. Ohtsuki, T.; Tominaga, S.; Morita, T.; Yoda, M. Thermophilic UASB system start-up and management change in sludge characteristics in the start-up procedure using mesophilic granular sludge In *Proceedings of 7th International Symposium on Anaerobic Digestion*, 1994; 348.
37. Daoming, S.; Forster, C.F. An examination of the start-up of a thermophilic upflow sludge blanket reactor treating a synthetic coffee waste. *Environ. Technol.*, **1994**, *14*, 965.
38. Lepisto, S.S.; Rintala, J. The removal of chlorinated phenolic compounds from chlorine bleaching effluents using thermophilic anaerobic processes. *Water Sci. Technol.*, **1994**, *29* (*5/6*), 373.
39. Harada, H.; Syutsubo, K.; Ohashi, A.; Sekiguchi, Y.; Tagawa, T. Realization of super high-rate anaerobic wastewaters treatment by a novel multi-staged thermophilic UASB reactor. *Environ. Eng. Res.*, **1997**, *34*, 327–336.
40. Syutsubo, K.; Harada, H.; Ohashi, A.; Suzuki, H. An effective start-up of thermophilic UASB reactor by seeding mesophilically grown granular sludge. In *Proceedings 8th International Conference on Anaerobic Digestion*, 1997; Vol. 1, 388–396.
41. Rintala, J.; Lepisto, S.S. Anaerobic treatment of thermomechanical pulping wastewater at 35–70°C. *Water Res.*, **1992**, *26* (*10*), 1297.
42. Zhang J.; Shanghai Communication University. Personal communication, 2003.
43. Satu S. Lepisto. Start-up and operation of laboratory-scale thermophilic upflow anaerobic sludge blanket reactors treating vegetable processing wastewaters. *J. Chem. Technol. Biotechnol.*, **1997**, *68*, 331–339.
44. Yushina, Y.; Husegawa, J. Process performance comparison of membrane introduced anaerobic digestion using food industry wastewater. *Desalination*, **1998**, *27*, 413–421.
45. Inoue, Y.; Doi, K.; Kamiyama, K. Granule formation facilitation of up flow type sludge blanket reactor using UF membrane. In *Proceedings of 45th JSCE Annual Conference*, 1991; Vol. 45/II, 1084–1085 (in Japanese).
46. Hibana, K. Tackling with zero emission in beer brewery. In *Practices of the Introducing Zero-Emission in Industries*; NTS Publisher, 2001; 147–168 (in Japanese).
47. Sakai, S. Booklet "Let's make compost and fermented feed from garbage" (in Japanese), 2000.
48. Shirai, Y. *The introduction of research*, http://www.lsse.kyutech.ac.jp/lsse_j/kyokan_shirai.html, 2003.
49. Li, G. *The application of methane fermentation technology in sludge recycling center, New century of environmental engineering*; The Committee of Environmental Engineering JSCE, 2003; 73–74 (in Japanese).
50. Kawano, T. *Biogas production from garbage, new century of environmental engineering*; The Committee of Environmental Engineering JSCE, 2003; 73–74 (in Japanese).

28
Treatment of Landfill Leachate

Michal Bodzek and Joanna Surmacz-Gorska
Silesian University of Technology, Gliwice, Poland

Yung-Tse Hung
Cleveland State University, Cleveland, Ohio, U.S.A.

28.1 INTRODUCTION

Municipal waste landfills are one of the most popular methods for waste disposal. Over the years they have evolved significantly because of growing environmental protection demands. Initially, landfills were fenced in far from residential areas, but now they are highly engineered facilities that enable secure waste disposal and restrict environmental impact.

The major threats of landfills are related to leachate discharge into the environment. A badly designed or managed landfill can be the source of groundwater and soil pollution because of seeped leachate. Leachate arises from rainwater percolating through the waste deposit and washing out solubilized and suspended organic compounds as well as inorganic salts. The quality and amount of washed-out organic compounds depend on disposed wastes and their decomposition phase. Since landfill wastes are subjected to mainly anaerobic biological processes, their intermediate and final products are present in the leachate. Among them are substances especially harmful to the environment, such as AOX (adsorbable halogenated organics), PCB (polychlorinated biphenyls), heavy metals, and so on. High concentrations of ammonia nitrogen, chlorides, and sulfates also exist. Therefore, many countries regulate leachate as hazardous industrial waste. However, in practice, raw leachate is often treated with wastewater in municipal wastewater treatment plants (MWWTP) without any additional processes. This method is recognized as subpar [1–3] because it can disturb wastewater treatment plant performance in several ways; namely, it can provoke the following nitrification difficulties:

- Nitrites build-up or complete inhibition of both phases of nitrification;
- Lack of organic carbon in sufficient amount for nitrites and nitrates reduction;
- Heavy metals present in leachate and adsorbed in sludge can make impossible further sludge utilization in agriculture.

However, the presence of toxic and persistent organic compounds in leachate is the most important problem in its treatment in MWWTP. Organic compounds from the Priority Pollutants List are detected in leachate in trace amounts, but they are biologically nondegradable; therefore, in MWWTP they are diluted and disposed into the environment. The toxic organic compounds are carried through surface water or groundwater to the water reservoir and must be removed with very expensive processes before the water is introduced into water supply systems. Thus, appropriate methods guaranteeing proper environmental protection must be considered in light of the growing concern about water quality and environmental standards in worldwide leachate treatment.

28.2 LEACHATE CHARACTERIZATION

28.2.1 Leachate Generation

Organic and mineral compounds generated as products of waste mineralization within biological processes and accompanying physical and chemical processes are washed out by percolating rainwater through the deposit of wastes in landfill and form heavy polluted waters, or leachate. Leachate generation accompanies landfill during its exploitation and a long time after its closing and recultivation. The composition and amount of leachate depend on many factors, among others:

- quality of wastes and its crumbling;
- techniques of landfilling and degree of waste compaction;
- age of landfill;
- biochemical and physical processes of waste decomposition;
- moisture and absorption capacity of wastes;
- precipitation, humidity, and evapotranspiration rate;
- topography of landfill site;
- lining system;
- hydrogeology;
- vegetation [4].

Owing to the synergism of the abovementioned factors, it is difficult to predict the amount and quality of leachate at a given landfill. For instance, data from 15 German landfills showed a range of leachate amounts from 0.4 to 10.6 m^3/ha-day with an average value of 4.7 m^3/ha-day [4].

Precipitation and climate have the strongest influence on leachate generation, causing the amount to vary during the year. Absorption capacity of wastes is another factor affecting leachate production. Initial moisture of municipal wastes depends on type of waste, seasonal trends, and treatment after collection and amounts on average to 35% of dry weight [5]. Additionally, wastes can absorb liquid up to the moment when downward percolation begins. The absorption capacity is influenced by waste density and pathways of liquid infiltrating through the deposit of wastes. Generally, an increase of waste density decreases leachate production [4,5].

The amount of leachate generated in municipal landfill can be calculated with the following water balance equation [5]:

$$LP = P - (R + \Delta U + ET + \Delta U_w) \qquad (1)$$

where LP = leachate production, P = precipitation, R = surface runoff, ΔU = changes in soil moisture storage, ET = evaporation from soil/evapotranspiration from a vegetated surface, and ΔU_W = changes in moisture content in wastes.

The amount of leachate can also be calculated using the simplified equation [6]

$$Q = P - R - E \qquad [\text{mm/m}^2] \qquad (2)$$

where Q = amount of leachate, P = mean height of annual precipitation calculated on landfill surface unit, R = surface runoff from landfill surface unit, and E = evaporation as a part of rainfall calculated on landfill surface unit.

For water balance estimation, many proposed methods utilize lysimeter measurements as well as multiyear experiences and observations made at actual landfills.

Since leachate is considered harmful to environment wastewater, it is important to minimize its production. A strategy of leachate control comprises the following techniques [7]:

1. *Control of waste input.* This consists of reduction to a minimum amount of landfilled wastes, especially biodegradable and hazardous wastes, by separate collection, recycling, composting, incineration, etc. An optimal solution that is launched nowadays is landfilling only inert material that is nonbiodegradable and nontoxic.
2. *Control of water input.* In the case of nonbiodegradable wastes, hazardous potential water infiltration may be stopped by means of top sealing. If wastes are biodegradable, some water is necessary for their biostabilization. In this case, limiting water input is indispensable for restricting leachate production. This may be realized by:
   - landfill organizing possibly in places of low precipitation;
   - covering of landfill with soil suitable for vegetation;
   - vegetation of the topsoil with plants enhancing evapotranspiration;
   - top surface lining;
   - restriction in sludge disposal;
   - surface water collection and disposal outside of landfill;
   - high compaction of wastes;
   - unequal settlement prevention;
   - mobile roof above the waste deposit.
3. *Control of landfill processes.* Organic carbon from biodegradable wastes should be transferred into the gas phase rather than into the liquid phase. This strategy is based on methanogenic phase enhancement.
4. *Control of leachate disposal into the environment.* Law regulations in many countries resulted in development of environment protection techniques as landfill lining, drainage, and collection systems, which assist in safely removing leachate from landfill. For leachate treatment, a combination of different biological, chemical, and physical processes is required. Moreover, environmental monitoring is applied to control the effectiveness of landfill systems performance.

28.2.2 Leachate Composition

More than 200 organic compounds have been identified in leachate. They may be classified as cyclic hydrocarbons, bicyclic compounds, aromatic hydrocarbons, substituted benzenes, alcohols and ethers, cyclic ethers, ketones and ene-ones, acids and esters, phenols, phthalates, furans and nitrogen-, phosphorus-, sulfur- and silica-containing compounds, and others that remain unidentified. Among the abovementioned compounds are 35 substances recognized as priority pollutants. These are chloro- and dichlorobenzene, toluene, ethylbenzene, xylenes, styrene, naphthalene, methyl-, di-methyl-, tri-methylnaphthalenes, 1,1′-biphenyl, phenantrene, 9H-fluorene, fluoranthene, pyrene, methyl-, tri-chloro-, tetra-chloro-, and penta-chloro-phenols, nonyl phenol, phthalates, dibenzofuran, chloroaniline, phosphoric acid tributyl, and triphenyl ester [8,9]. For this reason, leachate is recognized in many countries as having a hazardous impact on the environment.

Many factors mentioned in previous chapters influence leachate production and composition, resulting in a different amount and quality of leachate produced in a particular landfill. Moreover, the composition of leachate is changed significantly by the anaerobic processes occurring in the deposit of wastes and age of landfill [4,10,11]. As an example, Table 1 shows the composition of leachate from different sources, which point at highly varying ranges of respective parameters.

Larger divergences are visible in the ranges of heavy metal concentrations in leachate (Table 2). The variation of data is affected on the one hand by different waste composition and landfilling technology and, on the other hand, by differences in protocols for sampling, filtration, and storage of leachate samples [13]. Jensen and Christensen [13] stressed the importance of colloidal matter contents in leachate because of the high affinity of colloids for heavy metals. Different contents of this fraction in leachate can result in the above-mentioned divergences in heavy metal concentrations. Moreover, heavy metal complexes with colloidal fraction are probably more resistant to treatment and more mobile than the dissolved fraction of heavy metals [13].

When landfill is less than 3–5 years old, there are many organic compounds in the leachate that result from the first acidogenic phase of anaerobic waste decomposition. In this case, both COD (chemical oxygen demand) and BOD_5 (five-day biochemical oxygen demand) reach very high concentrations and pH is low due to the considerable amounts of volatile fatty acids produced in this phase. The BOD_5/COD ratio achieves values possibly higher than 0.7, due to high biodegradability of organic compounds contained in such a leachate (Table 3). With the biodegradation process it is noted that the above mentioned parameters have undergone changes. The COD and BOD_5 values diminish, but not equally. In leachate from ageing landfill, both the concentration of easily biodegradable compounds and the BOD_5/COD ratio decrease. When disposed wastes are stabilized completely, the BOD_5/COD ratio reaches the value of 0.1. This occurs due to the decreasing amount of biodegradable products of waste stabilization and increasing amount of waste stabilization products resistant to biological decomposition. All organic substances detected in leachate may be ranked according to their biodegradability, from easily biodegradable ones such as volatile fatty acids, low-molecular aldehydes and amino acids, and carbohydrates, to less biodegradable ones such as hydrolyzing amino acids, humic and fulvic acids [16].

Growing amounts of humic and fulvic acids in leachate are responsible for a very low BOD_5/COD ratio because their structures undergo changes with the ageing of landfill and then become more difficult for bacteria to react. In leachate from young landfills, the aromatic ring in humic acid molecules is less condensed and the molecules are small in size. With landfill ageing, humic acid molecules are larger, with a more condensed aromatic ring. This means that the humification degree of leachate increases with the age of disposed wastes [17]. The presence of humic and fulvic acids in leachate is important not only because of its resistance to biodegradability but also because of its affinity for pollutants. They are responsible for transport and behavior of pollutants in the environment, such as heavy metals and hydrophobic pollutants [17].

Table 1 Landfill Leachate Composition From Three Different Sources

Parameter	Range	Range	Range
pH	4.5–9.0	5.8–7.5	5.3–8.5
COD (mg/L)	500–60,000	100–62,000	150–100,000
BOD_5 (mg/L)	20–40,000	2–38,000	100–90,000
Sulfate (mg SO_4/L)	10–1,750	60–460	10–1,200
Chloride (mg Cl/L)	100–5,000	100–3,000	30–4,000
Ammonia nitrogen (mg N-NH_4/L)	30–3,000	5–1,000	1–1,200

Note: COD, chemical oxygen demand; BOD_5, five-day biochemical oxygen demand.
Source: Refs 4,10–12.

Table 2 Heavy Metals Concentrations in Landfill Leachate

Parameter	German landfills Ref. 10	German landfills Ref. 14 cit. after 13	UK landfills Ref. 4	Danish landfills Ref. 13	Polish landfills Ref. 15
Cd (μg/L)	0.5–140	2–20	5–100	0.2–3.6	10–159
Pb (μg/L)	8–1020	50–200	50–600	0–16	160–740
Cu (μg/L)	4–1400	20–100	10–150	2–34	90–2670
Hg (μg/L)	0.2–50	–	–	–	–
Cr (μg/L)	30–1600	100–500	50–1000	0–188	–
Ni (μg/L)	–	100–400	50–1700	28–84	–
Zn (μg/L)	–	500–2000	50–1300	85–5310	860–3540

Source: Refs. 4, 10, 13, 15.

As landfill ages, the pH of leachate also undergoes changes. Owing to the stability of the second, methanogenic phase of anaerobic waste decomposition, the pH of leachate increases to 8.5–9.0 [18]. The degree to which landfill age and waste decomposition influence the BOD_5/COD ratio and pH of leachate is presented in Table 3.

Sometimes the leachate composition from a young landfill does not correspond with the data presented in this chapter and Table 3. There are young landfills where leachate pH reaches a high level of 8–8.5 and the BOD_5/COD ratio is lower than 0.7. The COD and BOD_5 values may be far lower than is described in literature; they do not exceed on average 3000 and 1500 mg/L, respectively [20]. This type of leachate composition is usually affected by several factors in landfilling techniques:

- disposal of new waste layers on the old, partially stabilized ones;
- recycling of untreated leachate into the landfill;
- disposal of anaerobic stabilized sewage sludge into the landfill;
- retention and mixing of leachate from old and new sectors of landfill in one tank.

The first three factors above are responsible for accelerating the methanogenic phase of waste biostabilization. Volatile fatty acids in leachate from fresh wastes, when passing through the old layers of stabilized waste, are further decomposed by methanogenic bacteria. Additionally, recycled raw leachate passing again through the waste deposit are treated, and biodegradable substances are removed in larger amounts when compared to that of one passage by landfill. Sewage sludge being disposed into the landfill after anaerobic stabilization is also a source of large amounts of methanogenic bacteria and causes a quicker methanogenic

Table 3 Landfill Age Influence on BOD_5/COD Ratio and pH of Leachate

Landfill age	Degree of waste decomposition	pH of leachate	BOD_5/COD ratio
Young (<5 years)	Fresh, not decomposed wastes	<6.5	0.7
Mature (ageing)	Partially decomposed	6.5–7.5	0.5 → 0.3
5–10 years	Partially stabilized wastes		
Old (>10 years)	Well-stabilized wastes	>7.5	0.1

Source: Refs. 18, 19.

phase of waste stabilization. The last factor is responsible only for the average leachate composition.

28.2.3 Toxicity of Leachate

Leachate consists of a mixture of organic and inorganic compounds, many of which have a hazardous impact on the environment. Assessment of their toxicity as separate substances is insufficient and risks underestimation. Therefore, the toxicity evaluation must examine leachate as a complete mixture and is performed by means of biological screening methods (biotests).

Toxicity tests are generally grouped as acute and chronic. Acute toxicity is measured at short-time exposure and it is expressed as mortality (for fish), immobility (for crustaceans), and reduced photosynthesis (algae), or reduced light emission (bacteria). Chronic toxicity evaluates long-time exposure at concentrations proven to be nonlethal in acute toxicity tests. For bacteria, algae, and small crustaceans, which have a short life-cycle, chronic tests may comprise several generations. For organisms with longer life-cycles such as fish, tests are focused on important life stages such as reproduction, embryo and larval growth, and survival [21]. The most popular biological screening methods are presented in Table 4.

Species used for toxicity examination display different levels of sensitivity. This is seen in Fig. 1, which presents data from leachate toxicity tests conducted at eight landfills in Sweden [21]. For these eight landfills, the most sensitive species were algae, followed by crustacean, fish, plant, and bacteria (Microtox). However, for other landfills, this "sensitivity chain" can be completely different [21].

Effect concentrations are given as percentage of leachate responsible (EC or LC). For Microtox tests, the percentage of inhibition is reported as 100% in the leachate [21].

Table 4 The Most Popular Biotests Applied for Leachate Toxicity Evaluation

Trophic level	Organism	Effect parameter
Fish	Zebra fish Guppy Rainbow trout Fathead minnow	Reduced survival of larvae or fingerlings
Crustacea	*Daphnia magna* *Ceriodaphnia dubia* *Mysidopsis*	Reduced survival of larvae
Plants	*Lemna minor* (Duckweed) Radish Sorghum	Inhibition of growth (chlorophyll *a* and weight) Reduced germination or seedling growth
Algae	*Selenastrum capricornutum* *Nitzschia palea* *Skeletonema costatum*	Inhibition of photosynthesis Inhibition of cell growth
Bacteria	*Photobacterium phosphoreum* (Microtox)	Inhibition of light emission
	Salmonella typhimurium (Ames test)	Revertants

Source: Ref. 21.

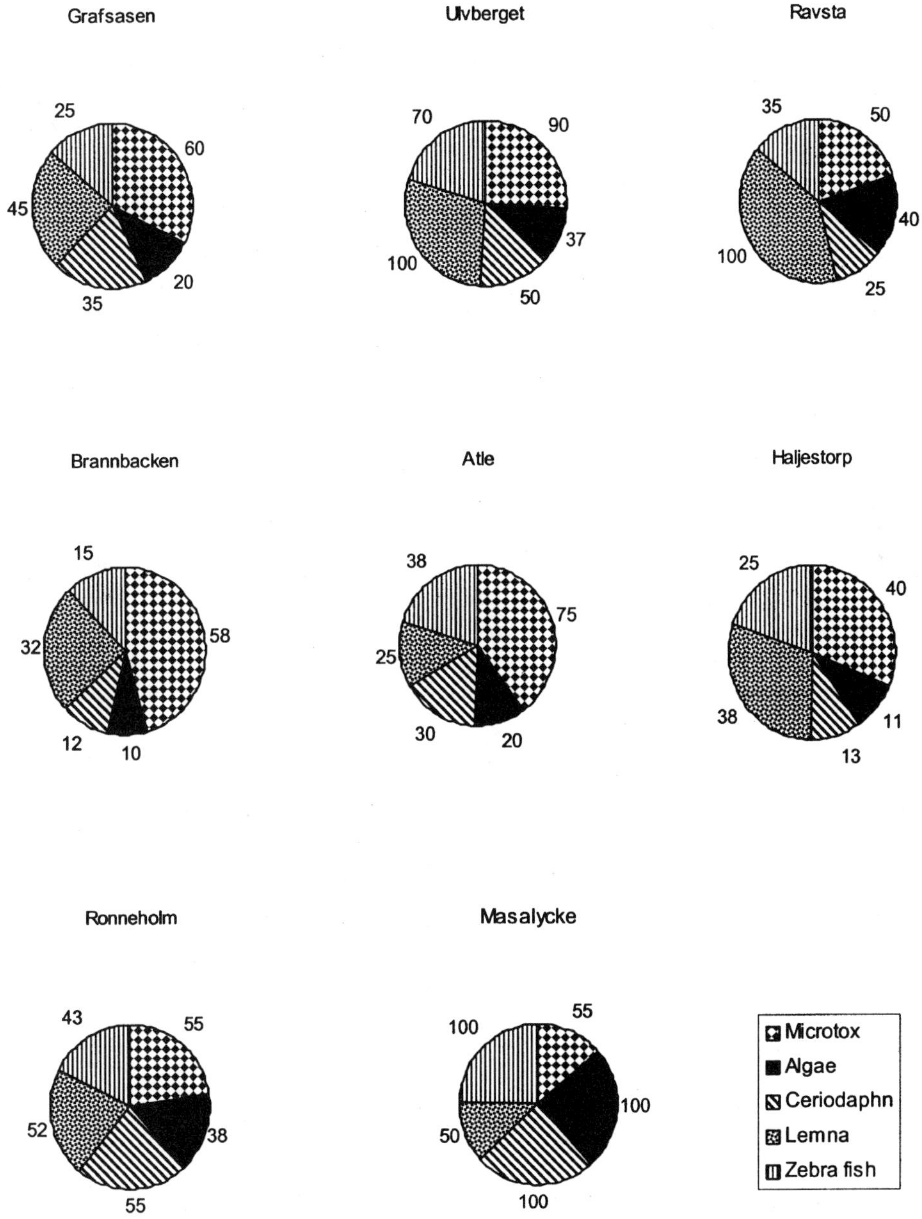

Figure 1 Leachate biotests of eight sanitary landfills in Sweden. Effect concentrations are given as percentage of leachate responsible for 50% effect (EC or LC). For Microtox tests percentage of inhibition in 100% leachate is reported (from Ref. 21).

As shown in Fig. 1, particular organisms display different levels of sensitivity to toxic substances in leachate. Investigations in leachate toxicity show that no organism or species would be completely resistant to the toxic effect of leachate components [21–24]. These results are not confirmed by the presence of biocenoses formed at biological treatment processes. Biocenoses active in biological treatment processes are very differentiated, mixed bacterial populations (apart from organisms from higher trophic levels, which are not as essential to treatment effectiveness). The normally demonstrate much higher resistance to toxic substances than test organisms used in ecotoxicology. On the other hand, biocenoses are less resistant to other factors such as organic compounds overloading, which can disturb homeostasis.

Sensitivity of activated sludge to leachate can be estimated by means of oxygen uptake rate and/or dehydrogenase activity measurements. Both parameters are good indicators of the whole metabolism of microorganisms because they reflect processes arising in the respiratory chain, where all metabolic pathways converge. Therefore, any disturbance in any metabolic pathway exerts its influence on processes in the respiratory chain. Sensitivity of activated sludge to toxic substances is measured as inhibition of oxygen uptake rate or dehydrogenase activity resulting from biomass contact with different concentrations of examined substances [20,25].

An example of lack of sensitivity of activated sludge to leachate is presented in Fig. 2 and 3, which show the examination results of leachate from three landfills in Poland's region of Upper Silesia [20]. An increase of oxygen uptake rate and dehydrogenase activity was observed. This was independent of the percentage and origin of leachate found in an activated sludge sample. Oxygen uptake rate grew from 3 to 48% in comparison with a control sample. The increase of oxygen uptake rate depended on amount and quality of introduced leachate and amount of biodegradable compounds in leachate, as well as on activity of activated sludge measured as oxygen uptake rate before leachate introduction.

Results of dehydrogenase activity measurements confirmed data obtained in experiments with oxygen uptake rate. None of the leachate doses introduced into the activated sludge samples affected the decrease of dehydrogenase activity. This confirmed the lack of sensitivity of acitvated sludge to components of examined leachate. Even the growth of activity was observed with the increase of leachate doses in the samples. All these results proved that concentrations of xenobiotics present in leachate did not inhibit metabolic activity of microorganisms, but simultaneously only small amounts of these components could be used in biological processes of oxidation by nonadapted biocenosis [20].

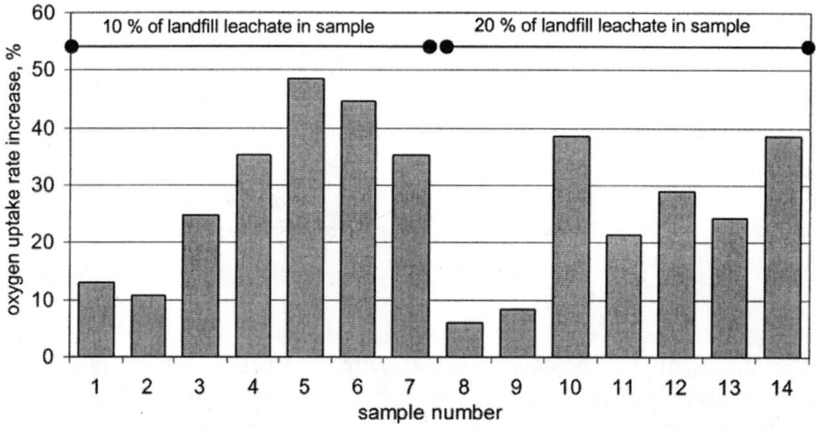

Figure 2 Activated sludge oxygen uptake rate increase influenced by different percentage of leachate from landfill in Dąbrowa Górnicza (Poland) (from Ref. 20).

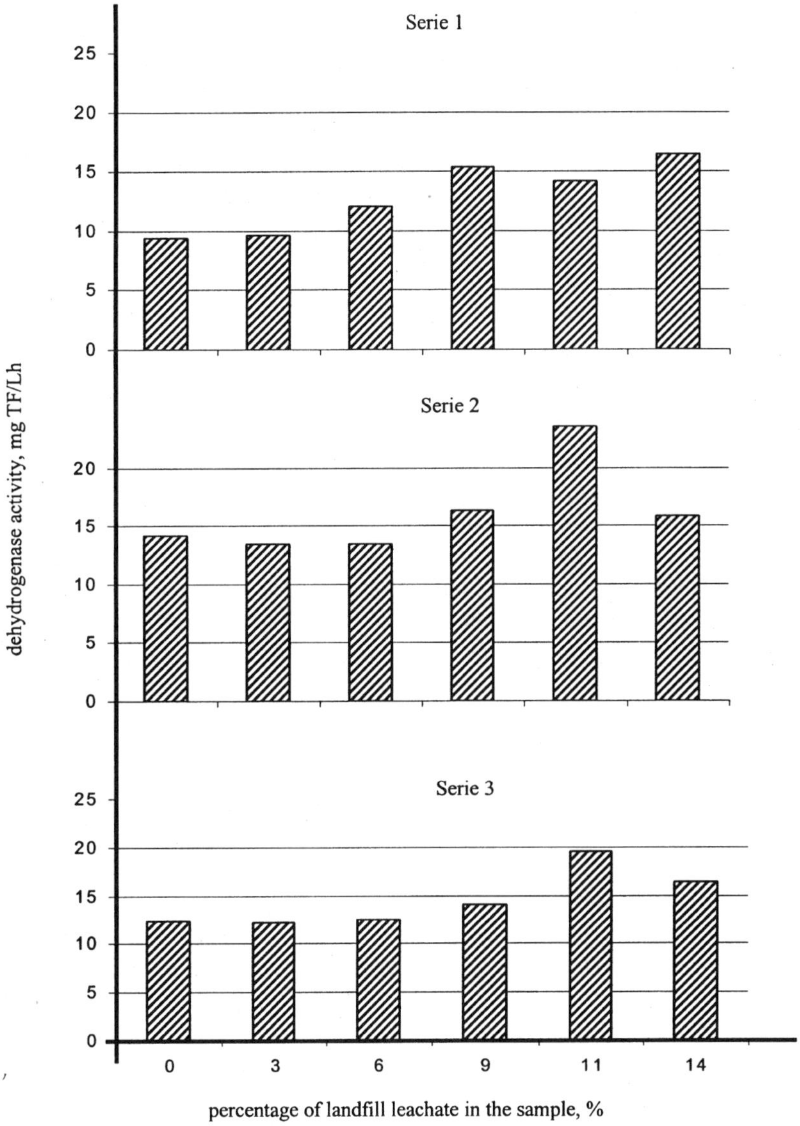

Figure 3 Activated sludge dehydrogenase activity increase influenced by different percentage of leachate from three landfills in Poland. Series 1, leachate from landfill in Dąbrowa Górnicza; series 2, leachate from landfill in Siemianowice; series 3, leachate from landfill in Świętochłowice (from Ref. 15).

28.3 BIOLOGICAL PROCESSES

28.3.1 Natural Systems

Treatment on Land

Treatment on land is based on the natural attenuation processes occurring in soils. Leachates are sprayed on grass vegetated lands, peat bogs, forest areas, or old, closed landfills. This technique is recommended for "weak" leachate. The typical hydraulic loading rate acceptable in practice is approximately 56 m^3/ha-day. Results achieved with the described method resulted in reduction of pollutants at the hydraulic loading rate of 45 m^3/ha-day by grass irrigation. The obtained effects are as follows: BOD, 95%; COD, 85%;, ammonia nitrogen, 67%; suspended solids, 64% [4]. However, it should be mentioned that ammonia nitrogen removal observed during treatment on land can be slower than in conventional wastewater treatment systems, according to Tyrrel and colleagues [26].

Other problems also accompany this method. Improperly selected hydraulic loading rate can destroy soil structure and disturb plant vegetation. Moreover, the method is restricted because heavy metals and refractory organic compounds are present in leachate, which can accumulate in soil and plants. Problems and limitations in this application technique are also generated in the form of odors as well as by the freezing of spraying devices in wintertime [4].

Lagoons and Wetlands

Lagoons and wetlands may be used for leachate treatment when low investment and operational cost, and simple and robust installation are required. This technique also does not demand a highly qualified staff [27]. In the United Kingdom lagoons and wetlands are very popular for leachate treatment. They operate as polishing stages after other biological processes [28].

The first automated plant in the United Kingdom designed to treat large quantities of strong leachate was built at Bryn Posteg Landfill Site in central Wales in the early 1980s. The plant consisted of a large aerated lagoon as a main element and was capable of treating up to 150 m^3 of leachate per day. Effluent was discharged to local wastewater treatment works. The long retention time of the leachate and high efficiency of floating aerators allowed high standards of effluent to be achieved. In the winter, at a mean temperature 4°C and mean hydraulic loading rate of 84 m^3/day, COD removal reached 98%. Both BOD_5 and ammonia nitrogen had a 99% removal efficiency. High effectiveness (above 90%) in Fe and Zn removal was also achieved in suspended solids [27].

Normally, leachate is treated in a complex of lagoons and wetlands where flexible anaerobic and aerobic conditions are assured. The leachate retention time reaches 10–20 days and hydraulic loading rate varies from several tens to several hundreds of m^3 per day [4,29,30]. Plants applied in constructed wetlands are terrestrial or aquatic species. Wetlands with reed (e.g., *Phragmites australis*) and willow (e.g., species *Salix iminalis*, *Salix dasyclados*, *Salix cinerea*, and a cultivar *Salix delamere*) beds are the most popular [29,30]. However, experiments with other macrophytes such as *Stenotaphrum secundatum* (tropical to subtropical terrestrial species), *Lemna minor*, *Eichhornia crassipes*, and *Myriophyllum verticellatum* (free-floating aquatic species) have also been performed [31].

Figure 4 presents the process scheme of a leachate treatment plant near Oslo, Norway, that treated leachate from an old landfill receiving domestic and commercial wastes as well as

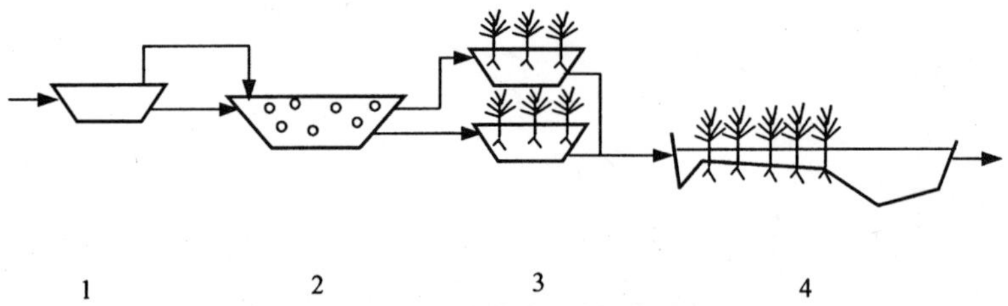

Figure 4 Landfill leachate treatment plant with anaerobic lagoon (1), aerobic lagoon (2), two parallel horizontal subsurface flow constructed wetlands (3), and one free water surface constructed wetland (4) (from Ref. 30).

sludge. The plant consists of an anaerobic lagoon, aerated lagoon, two parallel horizontal subsurface flow constructed wetlands, and one free water surface constructed wetland. The mean retention time of leachate is approximately 40 days at a mean hydraulic loading rate of 120 m^3/day. After 6 months of plant operation, the overall removal for COD, BOD, N, P, Fe, and pathogens (*E. coli*) varied from 65 to 95% (without data from wintertime) [30].

Satisfactory results are achieved in lagoons and wetlands, due to their ability to take large volumes of significantly diluted, treated leachate. Thus, the fluctuation of loads and hydraulic loading rates do not significantly influence treatment results. Also, the influence of toxicity and heavy metals on microorganisms and plants is limited. Disadvantages of lagoons and wetlands include the large area required for their construction and problems with maintaining effluent high standards during wintertime [4,30].

28.3.2 Conventional Systems

Anaerobic Systems

Anaerobic treatment of leachate is similar to the anaerobic processes occurring inside municipal waste landfills. Organic substances are first hydrolyzed to simpler forms. Amino acids are produced from proteins, monosaccharides from polysaccharides, fatty acids from lipids, purines and pyrimidines from nucleic acids. Then hydrolyzed products are subjected to a fermentation process, forming higher organic acids–nonmethanogenic substrates (~75% of organic matter) such as propionate and butyrate. Nonmethanogenic substrates as well as small quantities of input organic matter are transformed by acetogenesis and dehydrogenation and, as a result, methanogenic, substrates are generated. Among the methanogenic substrates are H_2, CO_2, formate, methanol, methylamines, and acetate, which is the main substrate. This entire phase is called acidogenesis because its main products are fatty acids, whose presence is responsible for the pH decrease. Next, the methanogenic phase occurs, and methane in combination with CO_2 is produced (Fig. 5). In each phase, another group of microorganisms is active. One group of bacteria is responsible for hydrolysis of organic polymers to their monomers. The second group-decomposes monomers into organic acids. There are facultative and obligate anaerobic bacteria called acidogenic bacteria. Methanogenic bacteria, which are anaerobes responsible for methane and CO_2 production from acetate as well as from hydrogen and carbon dioxide, are the last group of bacteria involved in anaerobic processing of organic matter decomposition [32].

Anaerobic treatment is applied mainly to young landfill leachate, which is characterized by very high values of BOD_5 from several to several tens of thousands of milligrams per liter. Moreover, it contains large quantities of easily biodegradable organic pollutants susceptible to fermentation such as volatile fatty acids, alcohols, and aldehydes [11,34,35]. Low BOD_5 values applied in anaerobic treatment vary from 1000 to 1500 mg/L [33]. Kettunen and Rintala [35] write, however, that that favorable effects can be achieved at leachate COD values higher than 800 mg/L and BOD/COD ratios higher than 0.3.

Anaerobic processes for landfill leachate treatment have been examined in typical systems applied in wastewater treatment and systems:

- UASB (upflow anaerobic sludge blanket) reactor [35];
- AF (anaerobic filters) [18];
- ASBR (anaerobic sequencing batch reactor) [36];
- AHBF (anaerobic hybrid bed filter – merged upflow sludge blanket at the bottom and anaerobic filter on the top of the filter) [36].

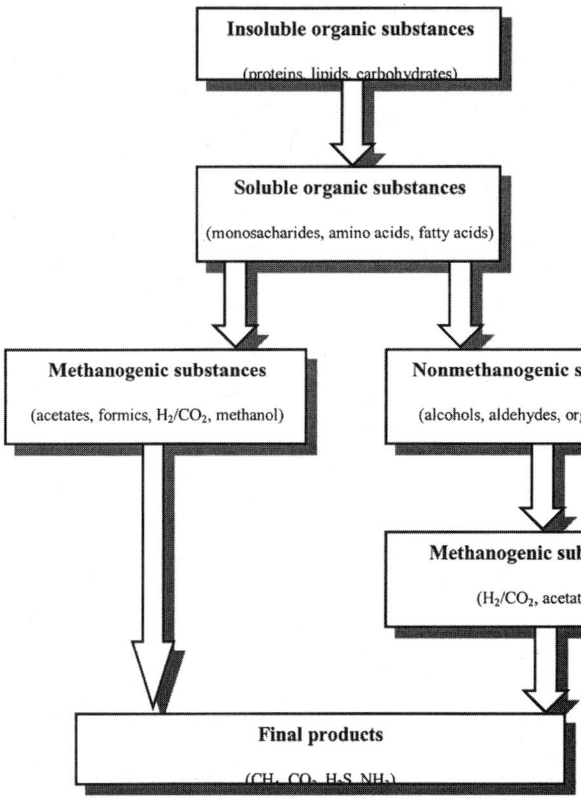

Figure 5 Diagram of organic matter decomposition within anaerobic process. (From Refs. 32, 33.)

Recent literature indicates that anaerobic processes allow for the complete removal of organic carbon from leachate of 70–80% and BODs removal beyond 90% [35,36]. Table 5 shows details concerning the results reached in different reactor types presented in literature during the 1980s and late 1990s. All results are comparable and differences are provoked probably more by leachate composition and organic loading rate than by type of reactor.

Some difficulties in anaerobic leachate treatment must also be mentioned. Disadvantages of anaerobic process that should be taken into account during leachate treatment are:

- Decrease of anaerobic treatment efficiency along with the development of methanogenic phase of waste decomposition in the landfill. Consequently, when the methanogenic phase is predominant, the change of the anaerobic process into an aerobic process is necessary.
- Possible lack of ammonia nitrogen removal. This component is one of the major pollutants in leachate.
- Considerable vulnerability to changes of environmental factors such as temperature, pH, as well as to toxic substances [11,41].

The influence of temperature on anaerobic process efficiency is illustrated in Table 5. In experiments carried out by Kettunen and Rintala [35], the decrease of temperature from 21–23°C to 13°C lowered the anaerobic process efficiency by 8–22%. Methanogenic phase inhibition caused the drop in COD removal, as evidenced by the increase of volatile acids (acetic

Table 5 Different Anaerobic Reactors Performance

Type of reactor	Volume of reactor (L)	Hydraulic retention time (days)	Temperature (°C)	COD removal (%)	Biogas production (biogas L/kg COD destroyed)	Methane contents in biogas (%)	Source Ref.
Anaerobic filter	24	5	25–34	70	400–500	80	[38, cit. after 37]
Packed bed anaerobic reactors	3	8	Ambient, 21–25	95	580	74	[18]
UASB reactor	12.7	1.9	29	83	499	75.5	[39, cit. after 37]
Full-scale AHBFs (anaerobic filter/UASB)	135,000	NM	35	95	500	72	[40, cit. after 37]
AHBFs (anaerobic filter/UASB)	2.75 (available volume)	2.4	35	37.5–76	NM	NM	[36]
ASBR	2	1.5–10	35	57–74	NM	NM	[36]
UASB reactor	40	0.96	23	57–63	NM	NM	[35]
		0.54	21	71–77			
		1.3	13	49–55			

Note: NM, not mentioned.
Source: Refs. 18, 35–37.

and propionic acids) concentrations in the effluent. The authors also described the significant quantities of inorganic precipitates accumulated in the reactor sludge. At longer operation times of the reactor, the problem of accumulated precipitates can arise, which authors suggested remedying by periodical replacement of the sludge [35].

Aerobic Systems

Aerobic decomposition of pollutants in leachate is based on processes with suspended growth microorganisms–activated sludge and/or attached growth microorganisms–with different types of fixed film biological reactors.

Activated sludge is one of the basic methods applied for municipal landfill leachate treatment. This process is used for biochemical oxidation of biodegradable organic compounds, but mainly for the biological removal of nitrogen [20,42–46].

Efficiency of organic compounds removal by means of activated sludge varies strongly and depends on leachate composition to a major extent, mainly on leachate BOD but also on applied process parameters. Klimiuk and Kulikowska [46] reached 83 and 71% of COD removal in leachate from young landfill in activated sludge SBR at a hydraulic retention time of 12 and 2 days, respectively, and a sludge age of 58 and 19 days, respectively. Similar results were also achieved by Bae and colleagues [42] in a three-stage system consisting of denitrifying filter and two steps of activated sludge with total hydraulic retention time of 14 days. In both abovementioned cases, the BOD_5/COD ratio in raw leachate was unknown.

In leachate treatment using biofilms, the leachate composition and carrier type are both important in affecting achieved results. For leachate with a BOD_5/COD ratio of 0.6 applied in biofilters, special open-cell plastic foam characterized by large surface, porosity, and absorption capacity, as well as easy access of leachate into inside parts of the carrier, gave the following treatment results:

- BOD_5 removal up to 96%;
- COD removal to 80%;
- TKN removal to 90%;
- total nitrogen (all organic and inorganic forms), 50% [47].

Effects of nitrification at leachate treatment usually reached over 90% of ammonia nitrogen oxidation and allowed for the achievement of ammonia nitrogen concentrations in effluents at the level of 10 mg N/L and even [20,42,45,48]. However, these results were reached in different systems of activated sludge, starting with a one-stage system with several phases up to complex multistage systems, where different hydraulic retention times, sludge ages, organic loading rates, and nitrogen loading rates in activated sludge are applied. Multistage biological systems were among the first plants used for sanitary landfill leachate treatment. For example, Figure 6 shows the first leachate treatment plant in Germany, which consisted of four stages of activated sludge [49].

The different composition of leachate is the reason why many systems are used for its treatment. For instance, different ammonia nitrogen concentrations exist in leachate from different landfills. They may differ even by several hundreds of milligrams per litre. Differences also exist in quantities of refractory organic compounds and toxic substances. In the case of high concentrations of ammonia nitrogen accompanied by large amounts of refractory and toxic substances, activated sludge is not sufficient and other solutions are required, such as extra application of granular activated carbon in the biological reactor. Horan and colleagues [44] characterized leachate by average concentrations of COD, 2450 mg/L; ammonia nitrogen, 744 mg/L; and low BOD_5/COD ratio, 0.08, and applied granular activated carbon to 125 g/L of

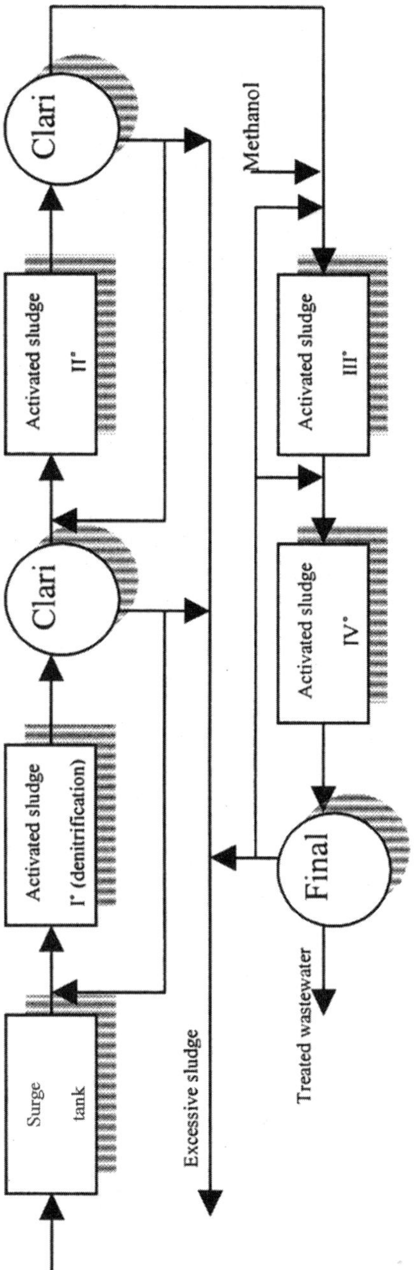

Figure 6 Multistage, biological leachate treatment by means of activated sludge (from Ref. 49).

mixed liquor. They obtained 90% of ammonia nitrogen oxidation, total BOD removal and 60% of COD removal. For leachate with higher concentrations of COD and ammonia nitrogen, in amounts of 5000 mg/L and 1800 mg/L, respectively, and low BOD_5/COD ratio (0.2), Loukidou and Zouboulis [50] applied 90 g/L granular activated carbon and received removal level of 90% of BOD, 81% of COD, and 85% of ammonia nitrogen. These results are comparable. In both cases, the function of granular activated carbon not only supported attached growth of bacteria but also the absorption of refractory organic compounds and toxic substances. The carbon allowed the bacteria to oxidize ammonia nitrogen and biodegradable substances. As a disadvantage of this process, Loukidou and Zouboulis [50] pointed to the large amount of residual suspended solids requiring separation and treatment as well as increased operational costs due to the addition of activated carbon.

In addition to high concentrations and toxicity of leachate compounds, there are problems with weak sedimentation properties of activated sludge, which leads to microorganisms washout from installations, and high sensitivity of microorganisms to low temperatures [3]. A decrease of temperature from 20 to 10°C within the activated sludge process can lower the nitrification rate by as much as 50% [51]. The influence of temperature on microorganisms of activated sludge was also examined by Ilies and Mavinic [52]. They controlled the activated sludge reaction at a temperature decrease from 17 to 10°C in the Bardenpho system, which consisted of two aerobic and two anoxic reactors. Leachate in this treatment process originated from the landfill in the methanogenic phase and was characterized by a low BOD value and ammonia nitrogen concentrations of up to 2200 mg N/L. The experiments showed that the temperature decrease from 17 to 14°C had no influence on the intrification process. The drop of nitrification efficiency between 10 and 30% started only after the temperature decreased to 10°C. The unexpected denitrification process appeared to be much more vulnerable to the temperature decrease. Its inhibition started at 14°C and at 10°C denitrification efficiency was lower than 5% of its potential. Figure 7 illustrates the influence of temperature on the activated sludge process.

Increased efficiency of leachate treatment by means of biological processes is achieved by biofilms application. An attached biomass on different moving and fixed carriers is often used

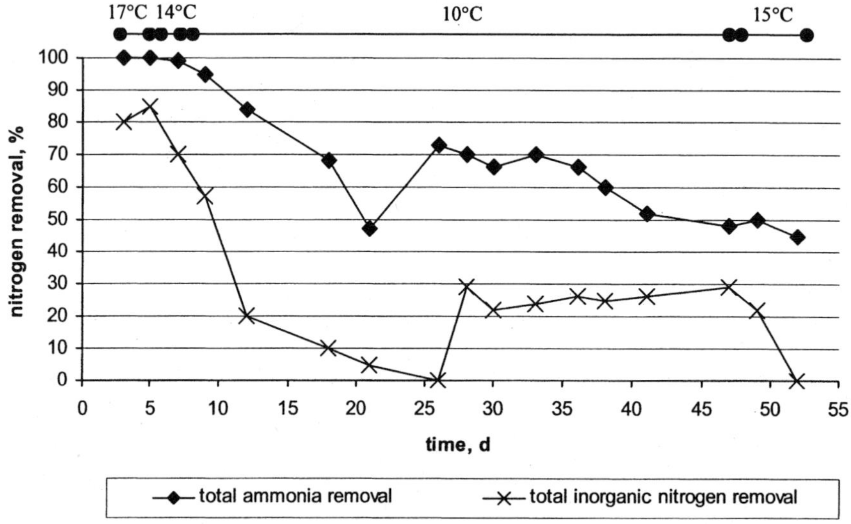

Figure 7 Temperature influence on nitrification and denitrification efficiency in Bardenpho system (from Ref. 52).

for leachate treatment [3,20,44,47,50,53–56]. Biofilm reactors are widely used because microorganisms change their properties when attached to biofllm. They are more resistant to changes of environmental parameters and toxic substances than microorganisms of activated sludge and are not washed out from reactors to as high a degree as activated sludge. The carrier application for microorganisms attachment is widely used for high concentration removal of ammonia nitrogen by means of biological nitrification. The bacteria responsible for this process (nitrifiers) are recognized as vulnerable to temperature and pH changes as well as the presence of toxic substances; moreover, they grow very slowly [51]. Benefits from attached-growth nitrifying bacteria application are well illustrated by their reaction to a decrease in temperature. A temperature decrease from 20 to 5°C resulted in a nitrification rate drop from 6.2 g $NH4^+$-N/m^3hour to 4.8 gNH_4^+-N/m^3hour, or about 22.6%. In the case of activated sludge nitrification, a smaller temperature decrease (from 20 to 10°C) resulted in a 50% nitrification efficiency decrease [51].

System with attached-growth microorganisms have also some disadvantages. The cost of the carrier, which can comprise a significant part of total installation costs, is one. Another is the need to maintain high concentration of dissolved oxygen in reactors with a moving bed in order to assure maximal nitrification rate [54].

Almost all types of filters are applied for landfill leachate treatment–moving-bed filters [3,44,50,54], rotating biological contactors [20,48,53,55], and trickling filters [47]. Extremely favorable results of organic compounds and nitrogen removal are achieved on rotating biological contactors. The thick biofilm formed on contactor discs facilitates quick oxidation processes (Fig. 8).

Figures 9 and 10 present an example of results achieved for a rotating biological contactor treating leachate characterized by a BOD_5/COD ratio equal 0.6, and COD and ammonia nitrogen average concentrations amounting to 1638 mg/L and 404 mg/L, respectively. The contactor treated leachate at organic and nitrogen loading rates of up to 27 g COD/m^2 day and

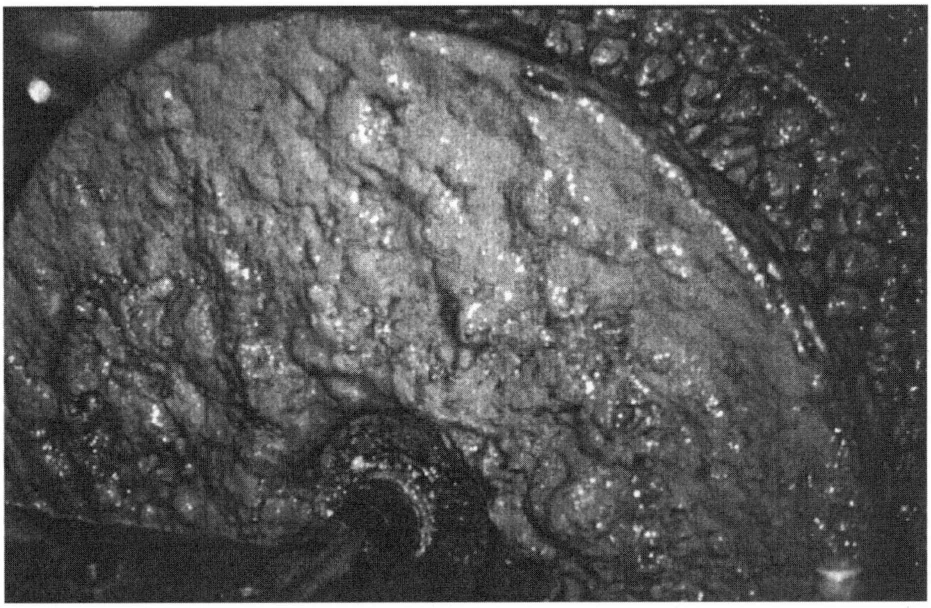

Figure 8 Biofilm on rotating biological contactor discs (from Ref. 20).

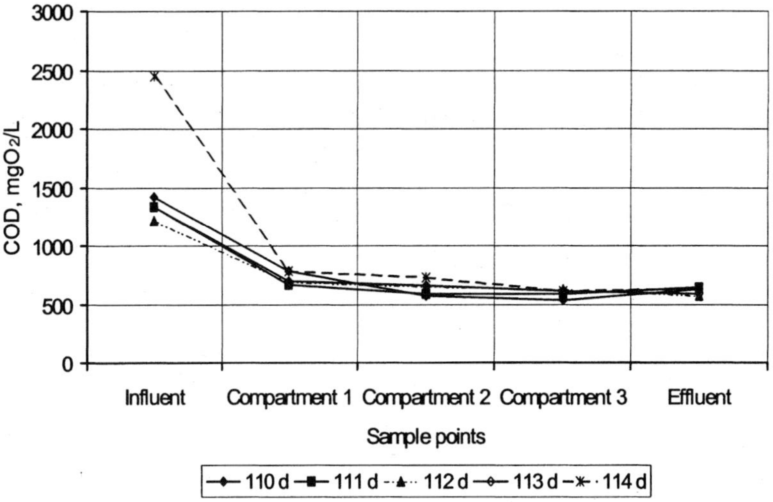

Figure 9 Chemical oxygen demand (COD) removal from leachate for a rotating biological contactor (from Ref. 20).

5.64 g NH_4^+-N/m^2day. Organic compounds and ammonia nitrogen were oxidized very quickly in the first compartment of the contactor. In the next two compartments, only nitrites oxidation to nitrates was observed [20].

A long retention time of biomass in rotating biological contactors and significant thickness of biofilm facilitate nitrogen removal in addition to classic nitrification (oxidation of ammonia nitrogen first to nitrites and then to nitrates) in a deaminification process. This phenomenon is observed in treatment of leachate without biodegradable organic matter. The efficiency of this process has reached up to 70% of total nitrogen removal [48,55].

In general, removal of nitrogen from landfill leachate requires an external source of organic carbon. This occurs in a given biological treatment system because the ratio of biodegradable organic carbon to nitrogen in leachate is too low. The lack of an external source of organic carbon is responsible for the presence of oxidized forms of nitrogen up to several hundreds of milligrams per liter in effluents from leachate treatment plants. The problem with denitrification because of insufficient amounts of organic carbon is observed even then, when nitrification concludes with nitrites as the main products and by requirement for organic carbon is limited [42,45]. Therefore, research on low-cost denitrification systems is carried out without the costs for external sources of organic carbon. Jokela and colleagues [56] carried out investigations on denitrification processes with landfill solid wastes as a source of organic carbon. They hypothesized that for small landfill technology, leachate would be introduced again after nitrification into the landfill body where oxidized nitrogen would be subjected to denitrification with organic carbon originating from solid wastes. Contrary to expectations, the experiments carried out on a laboratory scale with rates up to 3.8 g NO_x-N/t TS_{waste}-day showed a lack of denitrification influence on methanogenesis in a landfill. The authors attributed this to a much lower oxidized nitrogen loading rate used in their experiments in comparison with loading rates applied in experiments where a negative influence on methanogenesis was observed. They pointed to the need for further research to verify the denitrification sufficiency of a landfill body for long-term operation [56].

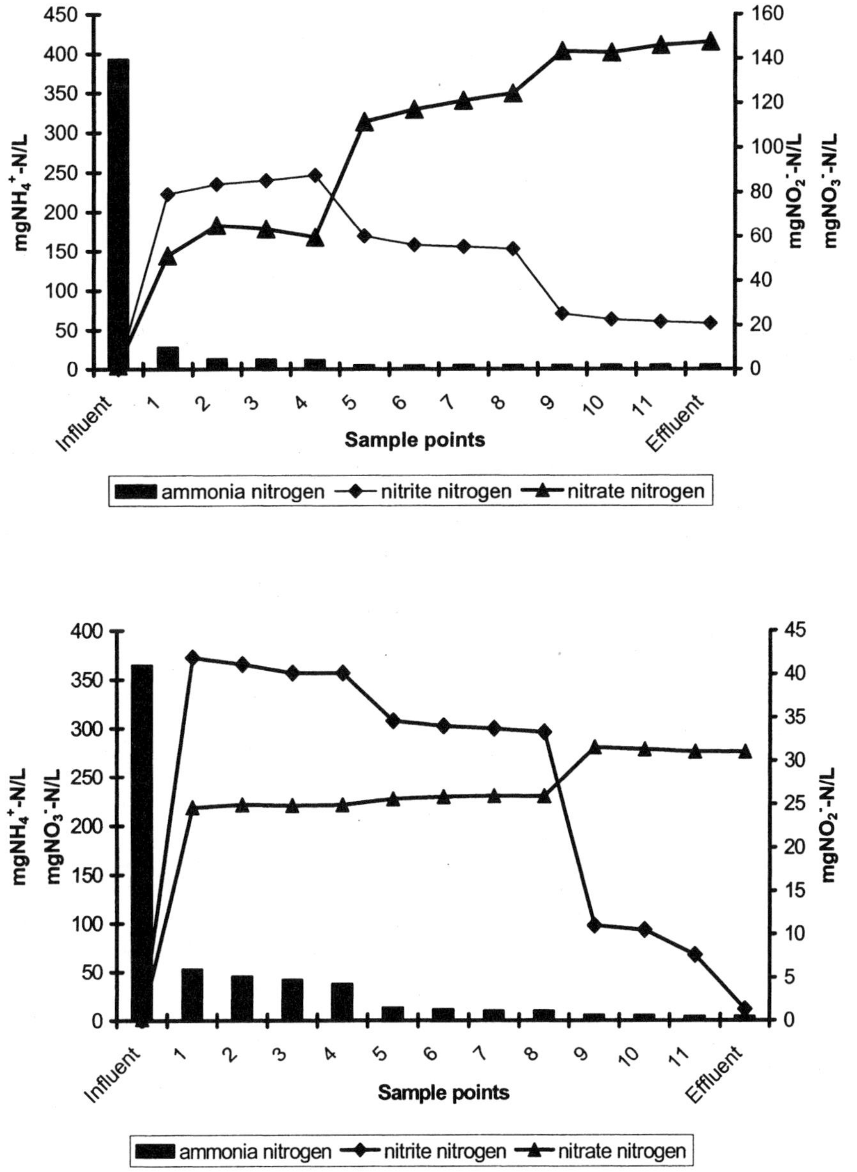

Figure 10 Nitrification performance in a rotating biological contractor (from Ref. 15).

28.4 CHEMICAL PROCESSES

Among the chemical processes applied for landfill leachate treatment, advanced oxidation processes play a key role. Their effectiveness results from an increased oxidation rate arising from hydroxyl radicals $OH^{\bullet}$. These free radicals are generated from added oxidants and are characterized by high reactivity, low selectivity towards organic compounds and high oxidation potential ($E^{\circ} = 2.80$ V) [57]. Generally, advanced oxidation processes may be divided into two groups: chemical oxidation (without photolysis) and photochemical (enhanced by photolysis).

Within the first group are the following processes (Fig. 11):

- ozone oxidation at high pH;
- ozone and H_2O_2; and
- oxidation with Fenton reagent.

In the second group, the processes used for leachate treatment are as follows (Fig. 11):

- ozone oxidation combined with UV light;
- H_2O_2 oxidation combined with UV light;
- oxidation with Fenton reagent combined with UV light: Photo-Fenton reaction;
- photocatalysis [57–59].

28.4.1 Oxidation With Fenton Reagent

Organic compounds oxidation may be effected by means of Fenton reagent assisted, or not assisted, by photolysis [57,60–62]. The Fenton reaction consists of a reaction of hydrogen peroxide with ferrous sulfate at a low pH [57,61]. As a result, free hydroxyl radicals, characterized by high reactivity, are produced according to the following reaction [3]:

$$Fe^{2+} + H_2O_2 \longrightarrow Fe^{3+}OH^- + OH^{\bullet\bullet} \quad (3)$$

Then, in chain reactions (4) and (5), organic compounds, marked as HRH, are oxidized, resulting in organic radicals $RH^{\bullet}$ production. Subsequently, oxygen is required. Unless the recombination

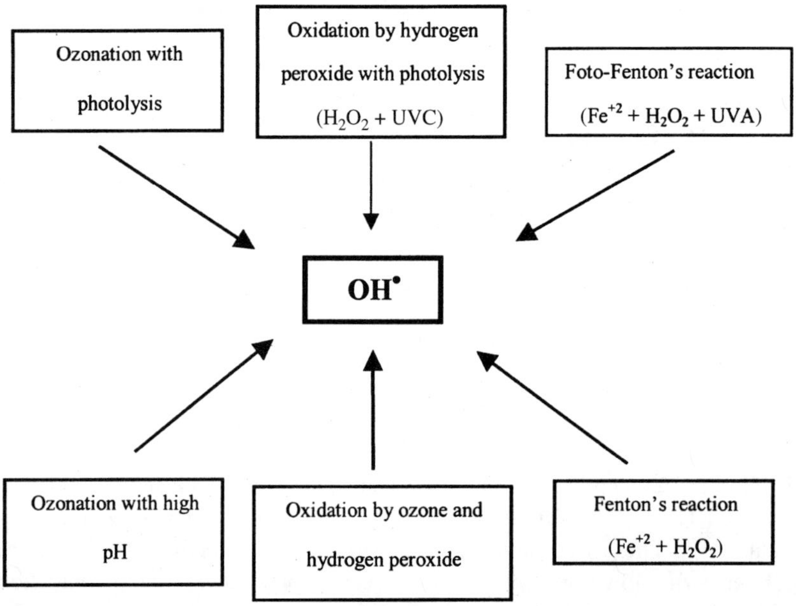

Figure 11 Chemical and photochemical processes used for landfill leachate treatment (From Refs 57, 58).

of organic radicals takes place [reaction (6)], the decomposition of organic compounds cannot proceed:

$$HRH + OH^{\bullet} \longrightarrow H_2O + RH^{\bullet} \quad (4)$$

$$RH^{\bullet} + O_2 \longrightarrow O_2RH^{\bullet} \quad (5)$$

$$RH^{\bullet} + RH^{\bullet} \longrightarrow HRRH \quad (6)$$

With the Fenton reagent, very different organic compounds present in industrial wastewater [63,64], rain water [65], as well as in landfill leachate are oxidized [20,60,66]. Therefore, the reaction is completed at different doses of ferrous sulfate and hydrogen peroxide. The reciprocal relation between ferrous sulfate and hydrogen peroxide is established experimentally. Although the molar ratio $[H_2O_2]:[Fe^{2+}]$ in Fenton reagent is applied usually within range from 5 : 1 to 10 : 1 [62], it is often accepted even at the level 1 : 1. However, a dose of hydrogen peroxide is varied with oxygen demand for oxidation of particular amounts of organic compounds, that is, with particular values of COD [61]. Normally, a low pH of 3–3.5 is recommended because a higher pH would foster coagulation domination of the oxidation process [20,63].

When landfill leachates are treated with a Fenton reagent, 60–80% of COD removal can be achieved [20,63]. The process efficiency depends, among other facts, on molecular weight of organic compounds subjected to oxidation. The higher the molecular weight reached by means of the Fenton reaction, the better it would be. Yoon and colleagues [60] showed that using a Fenton reagent removes 72–89% of organic compounds having molecular weight over 500 and no more than 43% of organic compounds characterized by molecular weight below 500.

The combination of a typical Fenton reaction with UV irradiation is also called the Photo-Fenton reaction. Additional UV irradiation at different wavelengths increases the organic compounds' oxidation rate. UV light of wavelengths lower than 300 nm generates more free hydroxyl radicals because of extra photolysis of hydrogen peroxide; it is also responsible for direct photolysis of organic compounds. However, UV light of wavelengths above 300 nm allows ferrous ions to regenerate [57,61]. The mechanism of the Photo-Fenton reaction is illustrated in Fig. 12.

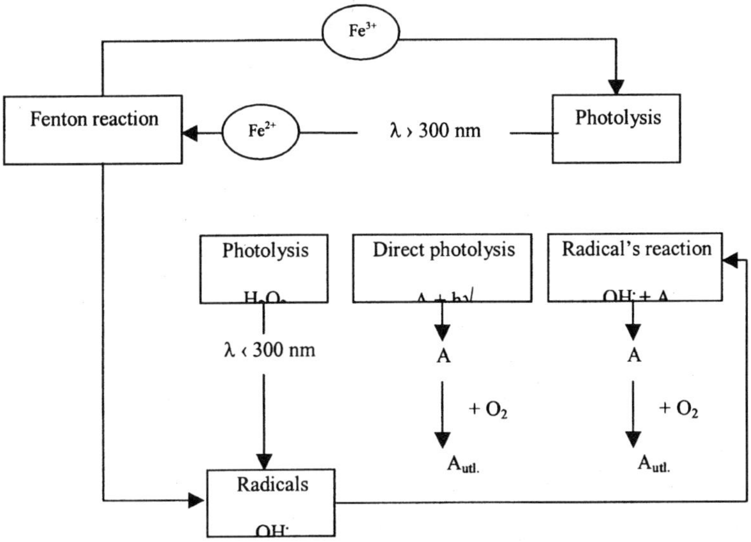

Figure 12 The mechanism of the Photo-Fenton reaction (From Refs 57, 61).

With the Photo-Fenton reagent, it is possible to increase the rate of organic compound oxidation tenfold in comparison with the traditional Fenton reagent. However, favorable results can be achieved if the content of carbonates in treated leachate is lowered. This is necessary since carbonates react with hydroxyl radicals making them inactive, according to reactions (7) and (8):

$$OH^{\bullet} + HCO_3^{-} \xrightarrow{\nu} H_2O + CO_3^{-} \tag{7}$$

$$OH^{\bullet} + CO_3^{2-} \xrightarrow{\nu} HO^{-} + CO_3^{-} \tag{8}$$

A similar effect is caused by excessive amounts of ferrous ions, which are also responsible for the turbidity of wastewater and obstruct the UV light beam's penetration inside the wastewater, decreasing the treatment process efficiency [61].

28.4.2 Oxidation with Ozone and Hydrogen Peroxide

During oxidation of organic compounds by means of ozone at high pH, some chain reactions occur that result in hydroxyl radicals, among others. Free radicals are responsible for oxidation of organic substances according to reactions (4) and (5), but the decomposition of organic matter can also be caused by ozone alone [58]. Ozone oxidation requires a high pH; carbonate ions will therefore disturb the process since they are scavangers of free radicals. This was described in the discussion of Fenton reagent. For landfill leachate, characterized by COD of 900–2100 mg/L, ozone doses are included in the range 1.8–3.5 g O_3/g COD. They allow the removal of 25–90% of COD and AOX [67].

Ozone oxidation enhanced by photolysis is also a very efficient method that produces favorable results. Chemical oxygen demand and AOX reduction may be achieved at the level of 80% with ozone dose of 1.6 g O_3/g COD. This method offers the technical possibility for COD values in biologically treated leachate to decrease to lower than 60 mg/L and for AOX to be less than 70 μg/L [68]. It is also possible to remove 100% of phenols, 23–96% of hydrocarbons and biphenyl, as well as 74% of dioxins and furans [69].

Combination of ozone and hydrogen peroxide also results in free hydroxyl radicals according to reactions (9) to (12) [70]:

$$H_2O_2 \longrightarrow HO_2^{-} + H^{+} \tag{9}$$

$$HO_2^{-} + O_3 \longrightarrow HO_2^{\bullet} + O_3^{\bullet -} \tag{10}$$

$$O_3^{\bullet} + H^{+} \longrightarrow HO_3^{\bullet} \tag{11}$$

$$HO_3^{\bullet} \longrightarrow OH^{\bullet} + O_2 \tag{12}$$

Summarily, it can be written as:

$$H_2O_2 + 2O_3 \longrightarrow 2OH^{\bullet} + 3O_2 \tag{13}$$

Free radicals can also be produced by means of hydrogen peroxide UV irradiation alone, as described in reaction (14):

$$H_2O_2 \xrightarrow{h\surd} 2OH^{\bullet} \tag{14}$$

as well as by ozone irradiation in a water environment according to reaction (15) [17]:

$$O_3 + H_2O \xrightarrow{h\surd} O_2 + H_2O_2 \xrightarrow{h\surd} 2OH^{\bullet} \tag{15}$$

Ozone irradiated with UV light decomposes to oxygen and hydrogen peroxide, which further decomposes to free radicals. Moreover, hydrogen peroxide reacts with ozone, according to summary reaction (13), producing another extra free radical.

Application of hydrogen peroxide oxidation with photolysis for biologically pretreated leachate allows removal of 60–95% of COD as well as 85–90% of AOX. Because oxidation results from free radical activity, an acidic pH of 2–4 is required. The decrease of pH results in low concentrations of carbonate ions HCO_3^- and CO_3^{2-}, which are scavengers of free radicals.

The efficiency of hydrogen peroxide oxidation enhanced by photolysis depends on the energy of emitted radiation. However, energy consumption depends on the kind of the lamps applied. With low-pressure lamps, energy consumption varies from 100 to 200 kWh/kg COD, and for high-pressure lamps, the amounts range from 400 to 700 kWh/kg $COD_{removed}$. Because only high-pressure lamps can be used on a technical scale for landfill leachate treatment, the high consumption of energy is a drawback of this application method [67,72].

Chemical methods are applied for landfill leachate treatment after biological pretreatment, which allows for removal of nitrogen and biodegradable organic compounds. Thus, the chemical treatment methods are focused on removal of refractory organic compounds as a main goal [72]. However, the advanced oxidation processes also have disadvantages, which are presented in Table 6. Exploitation difficulties can be classified as follows:

- The formation of scale on UV lamps as a result of precipitation. This weakens UV light penetration into the place of reaction.
- Limitation of the reaction rate because of free hydroxyl radicals inactivation as a result of reaction with carbonate ions.

These drawbacks are eliminated by upgrading the reactors, that is, separating the source of UV light from the reaction environment [69] or applying extra beds made of catalyst, accelerating mineralization of adsorbed organic compounds by radicals produced on the catalyst surface [72].

28.4.3 Photocatalysis

The mechanism of photocatalysis consists of electron ejection from the valence hand to the conduction band of the catalyst. The TiO_2 semiconductor is the most popular catalyst. The electron ejection to the conduction band and h^+ hole creation in the valence band results from catalyst irradiation with UV light having energy equal to or higher than the band gap (Fig. 13).

Table 6 Main Difficulties of Some Methods of Chemical Oxidation Application

Oxidation process	Comments
Ozonation	Good results are achieved at high pH and low alkalinity
Ozonation enhanced by hydrogen peroxide	Good results are reached at high pH
Ozonation enhanced by UV light	The formation of scale on UV lamps and inside the pipes due to precipitation of residues; insufficient UV light transmission into leachate; low concentration of carbonate ions required, recommended pH range 5–8
Hydrogen peroxide oxidation enhanced by UV light	High energy consumption, low transparency for UV light in leachate, recommended pH range 2–4
Oxidation with Fenton reagent	Sludge precipitation, increase of sulfates concentration, acidic pH required (~3)

Source: Ref. 73.

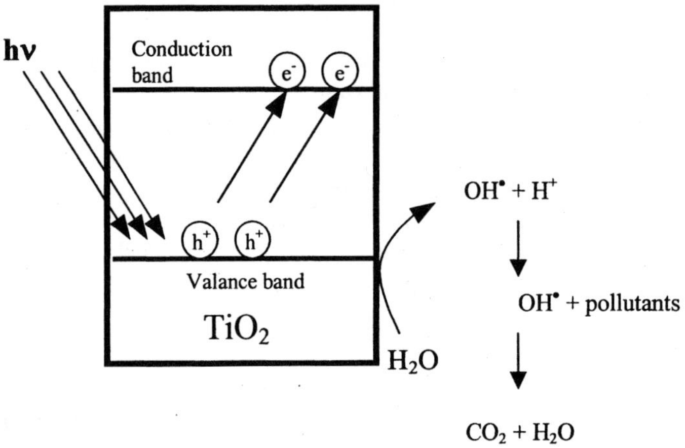

Figure 13 Scheme of photocatalysis with TiO_2 (from Ref. 74).

This energy, for TiO_2, is equal to 3.2 eV and corresponds with the energy of light of wavelength 400 nm. Therefore UV irradiation of wavelength $300\,nm < \lambda < 400\,nm$, being a part of sunlight, can also excite the electrons of the valence band [74].

The irradiation of energy higher than 3.2 eV creates a pair of electron – "hole":

$$TiO_2 + h\nu \longrightarrow e^- + h^+ \tag{16}$$

As a result, free hydroxyl radicals are produced on the surface of the catalyst and/or direct oxidation of organic compounds R takes place according to the following equations:

$$h^+ + H_2O \longrightarrow OH^{\bullet} + H^+ \tag{17}$$

$$h^+ + OH^- \longrightarrow OH^{\bullet} \tag{18}$$

$$h^+ + R \longrightarrow R^+ \tag{19}$$

However, the ejected electrons react with dissolved oxygen, and free radicals are also created:

$$e^- + O_2 \longrightarrow O_2^{\bullet -} \tag{20}$$

$$e^- + O_2 + 2H^+ \longrightarrow H_2O_2 \tag{21}$$

$$H_2O_2 + e^- \longrightarrow OH^{\bullet} + H^- \tag{22}$$

Further oxidation proceeds with the mechanisms described earlier.

The photocatalysis of leachate is a very promising method. The removal of COD can reach 50% [59] at optimal conditions, but there are also some obstacles. First, the dark color of leachate makes light penetration into the reactor difficult. Therefore, reactors of specific construction are required. The concentrations of inorganic ions and pH also influence photocatalysis efficiency. Cho and colleagues [59] pointed to the acidic pH ranging from 3 to 5 as optimal and explained that at low pH, the carbonates concentration lowered the protection of free radicals against inactivation and promoted organic matter sorption on the catalyst surface.

The presence of ions as chlorides, sulfates, and carbonates in very high concentrations (even a few grams per liter) makes the process slower and less effective. Anions such as chlorides or sulfates react with free radicals and form radicals of lower reactivity than hydroxyl radicals. This results in a decreased process rate. Carbonates are responsible for process inhibition because they are free radical scavengers [59].

28.5 CASE STUDIES OF LANDFILL LEACHATE TREATMENT

In practice, leachate treatment facilities very often consist of combined biological and chemical stages. As an example, the following systems, delivered by WEDECO company, combine a biological stage with ozonation enhanced by UV light, as presented below.

During the period 1991–2002, 40 different installations were constructed with the procedures presented in Table 7. They operate with a flow rate varying from 10,000 to 150,000 m^3/year. The COD exists in raw leachate at levels of 2000–4000 mg/L and AOX of 1.5–3.5 mg/L and are reduced to tolerable levels in effluents.

28.5.1 Singhofen Landfill Leachate Treatment Plant

This facility, operating in a procedure merging two biological stages with an ozonation step placed between them, treats 29,200 m^3/year and consists of [75,76]:

- a raw leachate basin (1000 m^3);
- a biological stage (activated sludge) composed of
  — 1 initial denitrification reactor (c. 390 m^3),
  — 3 nitrification reactors (c. 3×350 m^3),
  — a sedimentation tank and sand filter;
- an uozonation + UV stage composed of
  — intermittent reservoirs (3×30 m^3),
  — ozone generators (2×5 kg ozone/hour),
  — UV medium pressure lamps (6×10 kW),
  — reaction tanks (3×10 m^3),
  — a sand–gravel filter;
- a subsequent biological stage composed of
  — tanks (total volume 60 m^3),
  — aerated packed bed.

The required standards of effluent disposed into the environment for the facility are: COD, 200 mg/L; NH_4^+-N, 50 mg/L; AFS (filterable substances), 20 mg/L; and AOX, 0.5 mg/L.

Since January 1994 the plant operated only with the two first steps. Biodegradable organic compounds were oxidized in the biological stage. Nitrogen compounds were nitrified to nitrates and part of them reduced to gaseous nitrogen within denitrification. Then the biologically pretreated leachate was directed to an ozone/UV stage where refractory organic compounds, and especially AOX, were degraded.

The facility's performance showed two effects. The first proved that UV light was not necessary to fulfill required standards. But the second effect indicated a BOD_5 increase of up to

Table 7 Landfill Leachate Treatment Facilities

Procedure	Number of installations	Startup years
Biology + ozonation	8	1991–1995
Biology + ozonation + biology	10	1993–1996
BIOQUINT® (biology + ozone recycle)	17	1995–2002
Biology + activated carbon	5	1996–1998

Source: Refs. 75, 76.

100 mg/L after the ozonation step. Therefore, a second biological stage has been included in the system after ozonation since December 1995. This allowed the reduction of ozone mass requirement for oxidation because of the biodegradation potential of the second biological treatment step. A comparison of the results obtained in the systems is presented in Table 8.

The second biological stage applied after ozonation allowed for a reduction of ozone consumption by about 20%, whereas the biological removal of COD increased by about 10% (Fig. 14). At the same time, the effluent directed into the environment fulfilled the standard requirements; for instance, an average COD concentration lower than 200 mg/L, among others.

28.5.2 Asbach Landfill Leachate Treatment Plant

The facility in Asbach (Germany) has worked as the BIOQUINT® system since 1998 and treats up to 26,000 m^3/year. The installation consists of:

- a raw leachate basin (1000 m^3);
- a biological stage with nitrification fixed-bed biofilter (45 m^3) and two denitrification fixed-bed biofilters (2 × 10 m^3);
- ozonation (4 kg O_3/hour).

The required standards for effluent disposed into the environment are: pH6.5 − 9.0, COD < 200 mg/L, NH_4^+-N < 50 mg/L, NO_2^--N < 2 mg/L, N_{total} < 70 mg/L, and AOX < 0.5 mg/L.

A schematic of the facility is presented in Fig. 15. The system is supplied with methanol and phosphoric acid (if an external source of organic carbon for denitrification or phosphorus for growth of bacteria is needed). From the denitrification step, leachate is directed to the aerobic bioreactor for further COD removal in order to fulfill standard requirements.

Since March 1998, when a startup phase was finished, the facility treated leachate successfully and fulfilled all required limits. In February, methanol was added to the system and an increase of COD in the effluent was observed. However during March, because of adaptation, the quality of treated leachate was improved and from the end of March, the effluent met the standard requirements. Summarized average results of plant operation are presented in Table 9.

Table 8 Comparison of the Results achieved in Two Systems

Parameter	System: Biology + ozonation	System: Biology + ozonation + biology
Raw water flow rate (m^3/hour)	2.9	2.9
Influent COD (mg/L)	1800	1800
Effluent biology I COD (mg/L)	800	800
Effluent ozonation COD (mg/L)	150	300
Effluent biology II COD (mg/L)	–	150
Total COD removal (mg/L)	1650	1650
COD removal biology I + II (mg/L)	1000	1150 (1000 + 150)
COD removal ozonation (mg/L)	650	500
Ozone mass requirement (COD removal with ozonation × 2.5) (g/hour)	4.712	3.625
Factor (ozone mass/COD removed within ozonation + biology) (kg O_3/kg $COD_{removed}$)	2.5	1.9

Source: Ref. 76.

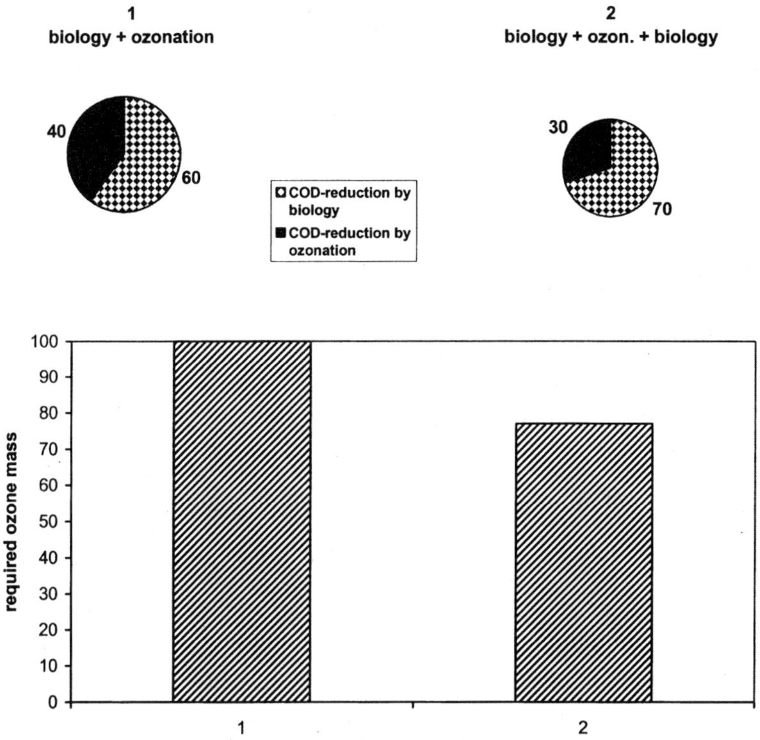

Figure 14 Chemical oxygen demand (COD) removal within biological and ozone steps of the leachate treatment plant in Singhofen (from Ref. 76).

The ozone dosage required for COD removal according to the limits was set in the range 0.9–1.2 kg O_3/kg $COD_{removed}$. This allowed for reduction of the COD, on average, by 71–82%, whereas the biological stage assures COD removal of 40%. At the same time, total nitrogen removal reached 98%.

The improved version of the BIOQUINT® system was applied at the landfill leachate treatment plant in Friedrichshafen, Germany, where the biological stage consists of nitrification and denitrification activated sludge reactors having volumes 100 and 25 m^3, respectively, followed by an ozonation/biological recycle step. The ozone step (1.5 kg O_3/hour) merged with a fixed-bed bioreactor (10 m^3) allowed for elimination of nonbiodegradable COD that had not been removed in the first biological stage. The achieved average COD reduction of 82% was

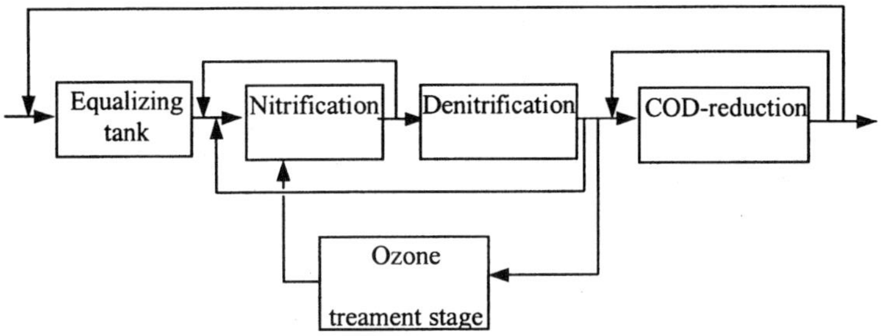

Figure 15 Schematic of the facility in Asbach (from Ref. 76).

Table 9 Table Results of Asbach Plant Operation

Parameter (mg/L)	Influent			Effluent		
	Max	Min	Average	Max	Min	Average
COD	735	422	548	193	82	150
NH_4^+-N	441	158	292	<0.1	<0.1	<0.1

Source: Ref. 76.

higher than that of the Asbach plant. Nitrogen elimination was similar, reaching 98%, and AOX elimination was 87% [75,76].

28.6 MEMBRANE TECHNIQUES IN LANDFILL LEACHATE TREATMENT

28.6.1 Introduction

Landfill leachate poses a considerable hazard to the environment, because of the toxic and hazardous compounds that can originate as a result of the soluble components of solid and liquid wastes being leached into surface and groundwater. Landfill leachate is a complex waste with considerable variations in both composition and volumetric flow [77] The volume of leachate produced is strongly related to the climate and other parameters that influence leachate production, and composition is strongly dependent on the landfill age, as well as the degree of compaction and the composition of the waste. Landfill leachates are therefore a mixture of organic and inorganic components, and their composition is determined by the composition of the waste material at the landfill and the age of the landfill [77].

Landfill leachate must be collected and treated in order to avoid contamination of natural waters. Their loading, frequently changing composition, and volume make the treatment of these wastewaters much more difficult than the treatment of municipal wastewaters [78]. Thus, landfill leachate is comparable to complex industrial waste streams, which contain both toxic organic and inorganic contaminants. Table 10 presents various alternative applicable methods for partial or total treatment or neutralization of dump effluents [78].

The assessment of data of leachate from more than 150 landfills [77] has shown that the level of components dissolved in leachate from different kinds of landfills covers the

Table 10 Treatment and Neutralization Methods Involving Landfill Leachate

Group of methods	Method
Biological	Aerobic biological treatment
	Anaerobic biological treatment
Physicochemical	Chemical precipitation
	Oxidizing
	Adsorption on activated carbon
	Reverse osmosis and other membrane methods
	Stripping (expelling)
Others	Joint treatment with municipal sewage
	Recirculation to the waste dump
	Lagooning and recirculation

Source: Ref. 78.

range 2–15 g/L. From this, the fraction of organic components is considered to cover the range 0.1–3 g/L. That is much smaller than the inorganic part, indicated to be in the range 1.6–14.3 g/L, including ammonia at values of 0.3–2 g/L [77].

The effectiveness of biological treatment and other physicochemical processes such as evaporation is much lower during the winter season, when the production of effluents is the highest. The variations in volume and composition hamper an efficient traditional treatment (aerobic or anaerobic biological methods; physicochemical treatment such as chemical precipitation, oxidation with ozone, hydrogen peroxide or UV, and activated carbon adsorption). Because such contaminants are not conducive to treatment with conventional biological processes, new regulations tend to limit the discharge of such complex wastes to municipal sewers. Even by combining biological treatment with adsorption onto active carbon, or with oxidation of a part of the dissolved organic material using ozone or other oxidizing agents, only partial destruction of contaminants will be achieved. It will not reach the purification level needed to fully reduce the negative impact of landfill leachate on the environment [78]. One aspect involves the "hard COD," which is not biodegradable and thus will remain in the water discharged after being treated with the mentioned processes. This causes future problems as a consequence of the effects of long-term accumulation.

The overall strategy for the treatment of landfill leachate is difficult to specify in general terms due to different composition, dumping site, and age of the wastes [77]. Therefore, more effective treatment methods of this material have had to be developed, and membrane technology offers the best solution [78–83]. The use of reverse osmosis either as a main step in the landfill leachate treatment chain or as a single step has shown to be a very successful means of achieving full purification.

The following process systems are currently suggested for the treatment of landfill leachate [78,84]:

- Biological pretreatment involving nitrification and denitrification – physicochemical treatment by means of flocculants and active carbon. The remaining part is dehydrated and removed as hazardous waste.
- Two-stage treatment with the use of reverse osmosis (1° – tubular modules; 11° – spiral modules); neutralization of retentate involving the evaporation process, drying, and disposal as hazardous waste.
- Methods affiliating biological treatment involving the nitrification and denitrification with reverse osmosis (tubular modules); neutralization of retentate involving the evaporation process, drying, and disposal as hazardous waste.

The membrane system that can be applied is a dual-step reverse osmosis (RO) [85], preceded only by a cartridge filtration or a biological treatment.

28.6.2 Application of Reverse Osmosis

Reverse osmosis is a pressure-driven membrane process. Owing to the ability of modern, high-rejection reverse osmosis membranes to retain both organic and inorganic contaminants dissolved in water with high efficiency, a wide variety of industrial wastewaters can be treated. Thus, reverse osmosis is useful as a main step in the purification of liquid waste such as landfill leachate and contributes to solving the growing water pollution problems. The application potential of this technique for the treatment of landfill leachate has been recently investigated [77–94].

The method of reverse osmosis does not lead to the neutralization of wastes; most frequently, the wastes are concentrated and returned to the dump, and the recovered and treated water may be released to surface waters or to the sewage system. Furthermore, such features as high efficiency, easy handling, low energy consumption, and optimized operation costs are expected by potential customers, who at the same time want to use membrane technology under difficult working conditions and as much as possible in the unattended operation mode.

To construct a full-scale reverse osmosis plant, the following data should be evaluated [78,84]:

- quantity of leachate;
- quality of leachate in connection with pretreatment of raw leachate;
- possible concentration factor to keep the required permeate quality;
- kind of modules, membrane materials, and cleaning procedures;
- process parameters (pressure, temperature, and cross-flow velocity) influencing the permeate flux and permeate quality);
- treatment of retentate (concentrate).

Pretreatment is recommended in the form of filtration on sand filters or drum-type filters to ensure the protection of pumps, membranes, and measurement equipment against destruction.

Apart from the required highly resistant membranes, the treatment of landfill leachate necessitates the use of open-channel module systems. In most applications, tubular and spiral modules are used [78]. In comparison with sea water desalination, there are more problems with scaling, fouling, and especially biofouling.

The disc-tube module (DT-module), which makes use of selected membranes and a special plant design, can meet further requirements expected by customers in this application including easy handling, low energy consumption, and optimized operation costs [81,82]. The DT-module is a fully modular design comprising the following components:

- membrane cushions;
- hydraulic carrier disc with internal O-ring gasket;
- end flanges with lip seals;
- tubular pressure vessels;
- central rod with tailpiece and nut.

This has been validated by more than 100 plants that have been in continuous operation – for many years and under rough working conditions – on landfill sites in Europe, North America, and Asia [81,82].

The quality of permeate should correspond with the respective requirements involving the release of wastewaters to the environment. It should be controlled by a respective selection of volumetric concentration factor C_{FV}. If the permeate is to be returned to the dump, the value of the C_{FV} parameter is 5, which is appropriate in light of economic reasons [84]. However, if the retentate has to be utilized by means of a thermal method, then the value C_{FV} can be higher than 5, which contributes to the reduction of costs involving this process [84]. If the reverse osmosis concentration process is limited by high concentrations of calcium or iron compounds as well as by high concentrations of ammonia nitrogen or organic chlorine compounds, the economics of reverse osmosis may be questioned for the treatment of raw effluents.

A strong relationship exists between other process parameters [84]. The size of plant, operating pressure, temperature, and recirculation rates determine the quality and power consumption of the plant. These parameters should be evaluated during pilot tests. The tests should also answer the question if advanced biological pretreatment is necessary or the purification of raw leachate is sufficient and economical.

The treatment of diluted leachate and raw leachate from the methaogenic phase does not always give satisfactory results, especially for ammonia [84]. The process should be carried out at pH 6.5 and requires the addition of sulfuric acid and precipitation of gypsum inhibitors. Two-stage reverse osmosis must be applied since treatment of the permeate from the first stage is necessary. In the first stage of reverse osmosis, the tubular membrane should always be used, and a spiral one in the second stage. For dilute leachate, it is possible to use a cellulose acetate (CA)-membrane in the first stage (higher permeate flux) and composite in the second stage. The disadvantages of such a solution include difficulty in reaching a low concentration of ammonia and the hydrolysis of membrane material.

For raw leachate from the methanogenic phase, it is possible to meet requirements by treating raw leachate with reverse osmosis. However, in such a situation, it is necessary to use the best quality composite membranes in the first and second stage reverse osmosis installation. The maximum concentration factors are 3.5 and 5 in the particular stages of the RO process, and are linked to the quality of the effluent.

Influence of Pressure and Chemical Oxygen Demand of Raw Leachate on Reverse Osmosis

In spite of much experimental work reported at laboratory and industrial scales, a systematic investigation of the performance of leachate disposal by a reverse osmosis unit under various sets of operating conditions is of great economical importance, especially using a wide range of operating pressures and of COD level of the leachate.

The effect of operating pressure on the permeate flux rate through the membrane has been investigated for values of COD within the range 0–1749 mg/L [95]. With a constant value of COD and salts content, the permeate flux increases linearly upon increasing the operating pressure, as expected. In fact, according to the solution–diffusion model [96] that excludes convective flow, the steady-state permeation flux of the solvent (water) through the membrane for a dilute solution (J_w) is given by:

$$J_w = \frac{P_w}{l}(\Delta p - \Delta \pi) \tag{23}$$

where Δp = transmembrane pressure, $\Delta\pi$ = osmotic pressure difference between the feed and permeate at the membrane surface, P_w/l = the pure water permeability constant, l = membrane thickness, and P_w = specific hydraulic permeability, which is given by the relationship [84]:

$$P_w = \frac{D_{w,m} c_{w,m} V_w}{RT} \tag{24}$$

where $D_{w,m}$ = water diffusion coefficient, $c_{w,m}$ = water concentration in the membrane, V_w = water partial molar volume, R = gas constant, and T = absolute temperature.

This means that the influence of leachate COD and sodium content on osmotic pressure is strong, while it is quite negligible on water permeability. The osmotic pressure due to COD is possible to evaluate based on the assumption that the contributions of inorganic salts and organic compounds in the leachate are additive with respect to the overall osmotic pressure. In order to determine the contribution of sodium salts, the sodium content is attributed to sodium chloride and sodium sulfate in the molar ratio reported in Ref. 95. The values of the osmotic pressure (π) due to each salt are calculated by the general equation for dilute solutions [95]:

$$\pi = g v c R T \tag{25}$$

where: c = solute concentration, v = number of ions from dissociation, and g = osmotic coefficient.

In general, g is a function of the solution composition. For example, g amounts to 0.937 and 0.762 for sodium chloride and for sodium sulfate, respectively [95]. Assuming that the concentration of solute in the permeate is negligible, the difference of osmotic pressure within the membrane coincides with the osmotic pressure in the concentrated solution.

The overall osmotic pressure appears to be mainly affected by COD. This means that the reduction of the permeate flux, observed upon increasing the leachate concentration, is essentially due to the presence of the organic compounds described by the COD parameter. The estimated values of the osmotic pressure of COD increase linearly with the COD content according to the following relationship [95]:

$$\pi = 0.0031\,\mathrm{COD} \tag{26}$$

A second important problem in leachate treatment is to evaluate the effect of the operating pressure on the COD rejection coefficient defined as:

$$R = 1 - \frac{c_p}{c_r} \tag{27}$$

where: c_r and c_p = solute concentrations, that is, the COD of the feed and the permeate streams, respectively.

It was stated that R is almost independent of the COD of the feed stream, but significantly dependent on the operating pressure. Investigations have shown that the COD rejection coefficient increases from 96 to 98% upon increasing the applied pressure from 2.0 to 5.3 MPa [84,97]. Equation (28) is usually used to express the relationship between the rejection coefficient and the applied pressure difference across a membrane [95]:

$$R = 0.991 \frac{\Delta p - \Delta\pi}{\Delta p - \Delta\pi + 0.469} \tag{28}$$

where $\Delta\pi$ is the experimental osmotic pressure for the considered COD.

It is well known that with increasing Δp, R tends toward an asymptotic value, lower than 1. Equation (28) can be also used to determine the minimum pressure required to achieve the maximum COD allowed in the permeate stream.

Equally important is the influence of the leachate's COD on the removal of heavy metals by reverse osmosis. Chian *et al.* [98] examined the operation yield with respect to the behavior of three heavy metals, Zn, Cu, and Cd, whose requirements for the discharge of the leachate are very different. The adopted concentration of these metals in the feed stream, equal to 20 mg/L, is higher than the usual ones in the leachate. The obtained results, reported in Table 11 [98], lead to the following observations:

- The removal of the three examined metals by RO in the absence of organic contaminants is always higher than 98%; This is in agreement with the data reported in the literature.

Table 11 Removal of Cu, Zn, and Cd From Aqueous Solutions at Two COD Contents

	Rejection coefficient (%)			
	2.0 MPa		5.3 MPa	
Metal	Without COD	With COD of 1265 mg/L	Without COD	With COD of 1265 mg/L
Cu	99.2	98.5	99.2	98.7
Zn	98.1	97.2	98.3	97.3
Cd	99.6	99.5	99.6	99.5

Source: Ref. 98.

- The COD of the leachate water lowers the removal of Cu and Zn, but to a different extent.

Examples of Reverse Osmosis Application to Landfill Leachate Treatment Plants

Figure 16 presents the diagram of a reverse osmosis system used in the treatment of landfill leachate from the dump at Wijster in Holland and at Rastorf in Plön, Germany [77,89]. Both plants consist of two stages: the first stage is equipped with tubular modules and the second with spiral-wound modules. The retentate from the first stage is returned to the landfill and the permeate is treated in the second stage. The retentate from the second stage is recycled to stage one and the permeate is discharged. The leachate is prefiltered and the pH is adjusted before the reverse osmosis membranes, but there is no other pretreatment. The operating parameters are presented in Table 12 [77].

The retention of pollutants is very good. For example, the retention of COD and NH_4^+-N is higher than 99% at the Rastorf plant. After a few years of exploitation, a four-stage evaporation plant has been commissioned with a steam-drying plant to treat the reverse osmosis retentate at the plant at Rastorf. The granulate from the steam-drying plant is now transported to a special waste depot. Table 13 provides data demonstrating the effectiveness of this system according to Rautenbach [91] and Table 14 does likewise, according to Mulder [89].

Using ROCHEM disc-tube module (DT-module) and the adequate membrane optimized results, different landfill leachate treatment plants were installed in Germany in the period 1988–1993 [79,81,82,99], that is:

- Schwabach (1988): two RO stages with DT-modules;
- Ihlenberg (former Schönberg) (1990): 36 m^3/hour feed capacity (reverse osmosis as main step of a landfill leachate purification process avoiding any wastewater, the most modern and largest multistage plant that has been built to date for this application);
- Kolenfeld (1990): plant with two stages with DT-module, ca. 30 m^3/day feed capacity;
- In 1991 plants started operation at the sites of Wiershop, Susel, Lochau, Höfer, and further systems in Schönberg.
- In 1992 systems were installed in Lüchow-Dannenberg and Burgdorf.

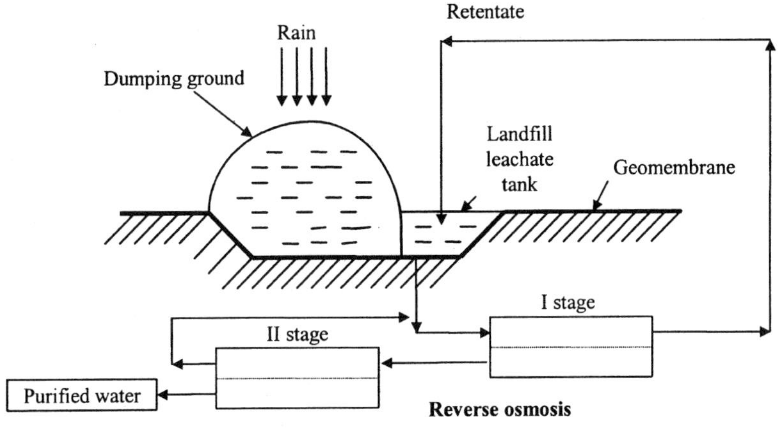

Figure 16 Diagram of landfill leachate treatment with the use of reverse osmosis. (From Refs 78–90.)

Table 12 Operating Parameters for Two Commercial Reverse Osmosis Plants Treating Landfill Leachate

	Rastorf		Wijster	
Treated leachate (m^3/day)	70		1056	
Total reduction of leachate volume (%)	71		54	
Conductivity of the feed (mS/m)	1500		2500	
	Stage one	Stage two	Stage one	Stage two
Membrane area (m^2)	210	120	1702	691
Temperature (°C)	30	30	25–30	25–30
Pressure (MPa)	3	2	4	5
Flux (L/m^2/hour)	15–18	30	18	34
Membrane lifetime (year)	1.5	2	2	3.5

Source: Refs 77, 89.

The data presented in Table 15 show some examples for the plant with two stages at Schwabach [81,82,99].

The successful operation of reverse osmosis in the plant of the municipal waste landfill of Ihlenberg is an example that has been realized for landfill leachate purification. With two-stage RO, the rejection for salts and organic contaminations is about 99% (Table 16) [81,95]. Depending on salt content in the feedwater and operating time between cleaning cycles, operating pressure ranges between 3.6 and 6.0 MPa with a specific permeate flux amounting to 15 L/m^2hour can be achieved.

Similar results are reported from other plants. Table 16 presents typical plant performance in leachate purification using DT-modules. These data covering long-term experience have also been confirmed by the results of more than 150 other systems in operation on different landfills worldwide [99].

Table 13 Treatment Effectiveness of Landfill Leachate With the Use of Reverse Osmosis

Parameter	Raw landfill leachate	Permeate	Retention coefficient (%)
pH	8.6	5.25	–
Conductivity (μS/cm)	17,700	314	98.23
COD (mg/L)	2,560	21	99.18
BOD (mg/L)	65	0.6	99.08
NH_4^+ (mg/L)	465	6	98.71
Cl^- (mg/L)	4,219	72	98.71
SO_4^{2-} (mg/L)	2,800	12	99.57
Ca^{2+} (mg/L)	160	2.1	98.69
Zn^{2+} (μg/L)	220	100	54.55
Cu^{2+} (μg/L)	640	28.8	95.50
Pb^{2+} (μg/L)	20	5	75.00
Cd^{2+} (μg/L)	<10	1	>90
Hg^{2+} (μg/L)	<1	0.6	>40

Source: Ref. 91.

Table 14 Effectiveness of the Two-Stage Osmosis System for the Treatment of Landfill Leachate at Wijster (Holland)

Parameter	Raw landfill leachate	Permeate from second stage of reverse osmosis	System effectiveness (%)
BOD (mg/L)	60	2	93.7
COD (mg/L)	1600	8	99.5
Nitrogen (mg/L)	480	10	97.4
Chloride (mg/L)	2250	45	97.9
pH	8.2	5.6	–
Capacity (m^3/hour)	35	27	–

Source: Refs 78, 89.

The results prove that reverse osmosis is a very effective instrument for the purification of landfill leachate, if all design criteria and requirements specific to landfill leachate have been taken into consideration and an adapted module system is used.

Treatment of Reverse Osmosis Retentate

The purification of landfill leachate with membrane techniques helps prevent contamination of natural water sources. Besides this ecological aspect, commercial feasibility must also be considered. In this regard, membrane filtration has proved to be a justifiable and economic solution in most cases, even when the overall costs for the purification are compared with other approaches. This evaluation includes the handling of the retentate (concentrate) produced in the reverse osmosis plant, because of high concentrations of organic and inorganic compounds that must be further treated.

In the past, treatment of reverse osmosis retentate was considered a very expensive treatment step. In many old plants in operation, this step consisted of evaporation and drying followed by deposition of the dry residues in a special landfill. A good example is the treatment system used for leachate of a landfill in Mechernich, Germany [85], schematically represented in

Table 15 Landfill Leachate Treated by the Reverse Osmosis Plant with DT-Module in Two Stages in Schwabach

Parameter	Feed water (mg/L)	Permeate, second stage (mg/L)	Retention coefficient (%)
Sulfate	22,100	4.80	99.9
Chloride	6,360	14.0	99.8
Ammonium	1,955	42.0	97.9
Nitrate	455	18.0	96.0
COD	912	15.0	98.4
TOC	289	4.00	98.6
Hydrocarbons	13.4	0.30	97.8
Nickel	2.78	0.10	96.4
Chromium	2.18	0.10	95.4
Copper	0.97	0.10	89.6
Vanadium	290	2.20	99.2

Source: Refs 81, 82, 99.

Table 16 Typical Plant Performance in Leachate Purification

Parameter	Raw leachate	Permeate I	Permeate II	Retention coefficient (%)
pH	7.7	6.8	6.6	–
Conductivity (μS/cm)	17,250	382	20	99.9
COD (mgO_2/L)	1,800	15	<15	>99.2
Ammonium (mg/L)	366	9.8	0.66	99.9
Chloride (mg/L)	2,830	48.4	1.9	99.9
Sodium (mg/L)	4,180	55.9	2.5	99.9
Heavy metals (mg/L)	0.25	<0.005	<0.005	>98

Source: Refs. 81, 95.

Fig. 17 [80]. The total capacity is 150 m^3/day; the first RO unit uses tubular membranes in order to avoid fouling; the second reverse osmosis unit uses spiral-wound modules. The concentrate from the second RO is brought back to the inlet of the first RO unit. The concentrate from the first reverse osmosis unit is evaporated in two stages, upon which the residue is dried in a fluidized bed. The dried material is landfilled; the distillate from the evaporation is brought back to the second RO unit.

This scheme can be simplified for relatively new landfills where the leachate is relatively biodegradable, and concentrate can then be directly reinjected into the landfill. During the recirculation, the quality of the concentrate will improve and equilibrium between leaching and biodegradation will be reached for the organic fraction [81,82]. However, this method is not applicable for older landfills and for leachate containing a large inorganic fraction. Evaporation and drying is then necessary and may be followed by a solidification, so that a material is obtained with low water permeability and with low leaching of heavy metals, which can be landfilled without any additional environmental risk.

Today, other possibilities that meet the best ecological and economical requirements are considered. These are [81,82]:

- Transport of the concentrate to an incineration plant equipped for the burning of liquid hazardous waste. This option allows energy to be extracted from the waste material. The incineration must be carried out in a rotating kiln furnace, which provides better incineration than a grate furnace and is thus more suitable for hazardous materials.
- The solidification of the concentrate with different materials, as fly ash [82] or sludge from wastewater treatment plants [82], and disposal of this kind of dry residue on the

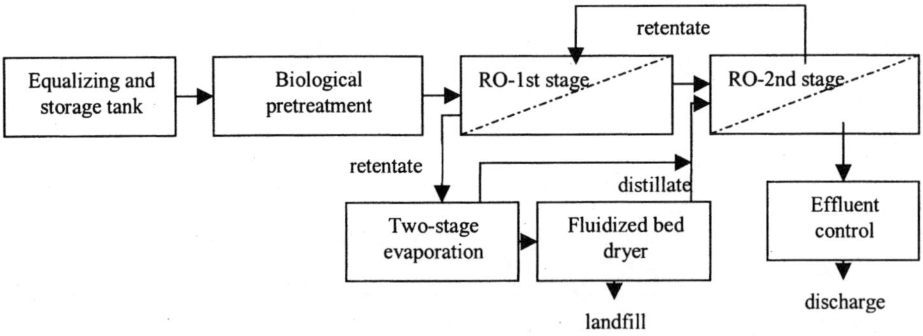

Figure 17 Schematic of a landfill leachate treatment in Mechernich, Germany (From Refs 80, 85).

landfill itself If these solids are produced under controlled conditions, no further pollution is to be expected.

- The abovementioned controlled reinjection of the concentrate into changing areas of the landfill in order to improve the biochemical degradation process in the waste itself and accelerate the immobilization of the organic matter.

Process Improvement With Reduction of Retentate Volume

Improvement of membrane technology can be carried out using the reverse osmosis process with high operating pressure close to 12 MPa instead of the standard 6.5 MPa. This is exemplified by the DT–reverse osmosis module [81,82,99]. In such a solution, recovery of permeate has been achieved by osmotic pressure and the concentration factor of dissolved compounds can be increased tenfold to 90%. This technique is frequently combined with a seeding technique, which allows for elimination of the membrane scaling problem and makes it possible to achieve permeate recovery of up to 95% (20-fold) [81]. The $CaSO_4$ crystals are separated from the retentate, supporting further concentration of the retentate.

Even higher permeate recovery can be reached by applying the hybrid membrane system containing reverse osmosis and nanofiltration techniques. Nanofiltration allows for separation of monovalent and divalent ions as well as organic substances of molar mass above 300–500 Da. Table 17 presents some examples for rejection coefficients of the substances in landfill leachate [81].

For this application, a very specific membrane and membrane module must be selected. These requirements are met by Rochem's DFT-module, in which open-channel construction is combined with narrow gap technique [81,82].

By combining nanofiltration with reverse osmosis, high-pressure reverse osmosis and crystallization of calcium salts, it is possible to reach permeate recovery amounting to 95–97.5% [81,82]. This should be an operating standard today. Figure 18 shows an example of such a system for the purification of landfill leachate [81].

28.6.3 Combined Activated Sludge and Reverse Osmosis Treatment Process

Principle of the Process

The process of reverse osmosis combined with biological treatment has proved to be very effective in the treatment of landfill leachate [84–86,89,90,93,100]. The biological process aims

Table 17 Rejection of Substances from Landfill Leachate by Means of Nanofiltration

Parameter	Raw leachate	Permeate	Retention coefficient (%)
Conductivity (μS/cm)	61	43	29.5
COD (mgO_2/L)	17,000	700	95.8
Ammonia (mg/L)	3,350	1,420	57.6
Sulfate (mg/L)	31,200	2,345	92.5
Chloride (mg/L)	12,800	17,700	−38.9
Calcium (mg/L)	2,670	187	93.0
Magnesium (mg/L)	1,030	72.7	92.9
Sodium (mg/L)	10,900	5,010	54.0

Source: Ref. 81.

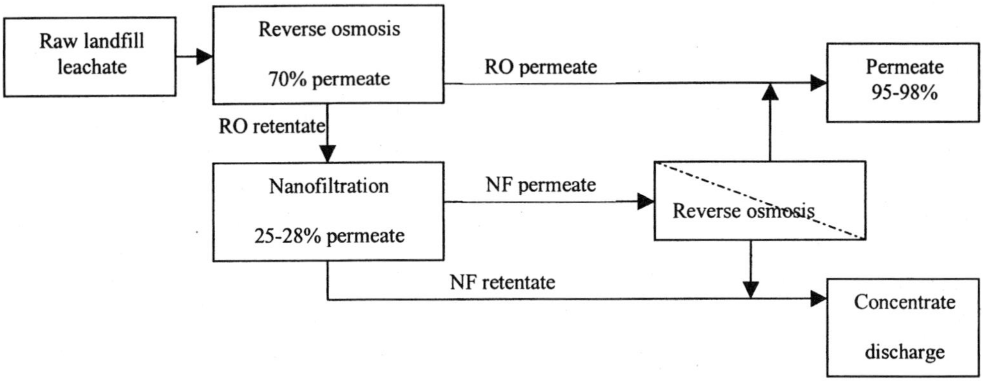

Figure 18 Hybrid membrane system for landfill leachate treatment with high permeate recovery (from Ref. 81).

at the removal of substances undergoing biodegradation, especially in the case of effluent heavily loaded with organic substances. As a result of reverse osmosis, refractive substances and inorganic salts are removed. The flux leaving the biological installation is subjected to further treatment with the use of reverse osmosis. With minimum requirements involving the quality of permeate, a single-stage reverse osmosis may be sufficient.

There are two pretreatment possibilities of landfill leachate before reverse osmosis with the biological method [84]. The first involves activated sludge with high sludge loading above 0.25 kg/kg·day. With this solution, it is possible to diminish only TOC and AOX. The advantages of such a solution are as follows:

- buffer for COD/BOD peaks;
- removal of scaling and fouling causing compounds; and
- enhancement of permeate flux and concentration factor in the reverse osmosis step.

The second pretreatment involves advanced biological pretreatment including nitrification and denitrification. Figure 19 presents the diagram of a pilot installation utilizing that concept for the treatment of outflows [84].

The task of fabric–drum filters is to enhance the nitrification process and to protect reverse osmosis membranes. The advantages of such a system include a fixed film of nitrifying bacteria and enhanced N-loading rate due to optimum BOD/N ratios of less than 0.1.

The application of an affiliate system of biological treatment and reverse osmosis is more economical than a system where only reverse osmosis is applied, since the permeate flux and/or obtainable concentration factor are considerably increased. Furthermore, the nitrification process lowers the pH of the pretreated stream, and the quantity of added acid is considerably lowered. The procedure of membrane cleaning is carried out with detergent solution and citric acid, with much lower frequency as compared to the previous method. Modules for reverse osmosis are always equipped with tubular membranes from cellulose acetate (retention coefficient of NaCl, 95%) and composite membranes of the retention coefficient equalling 99%. The utilization of retentate most frequently consists in the application of a single-stage evaporator with forced circulation of the liquid and drying the concentrate from the evaporator in the drying machine [114]. As a result, solid dry waste, suitable for storing, is obtained.

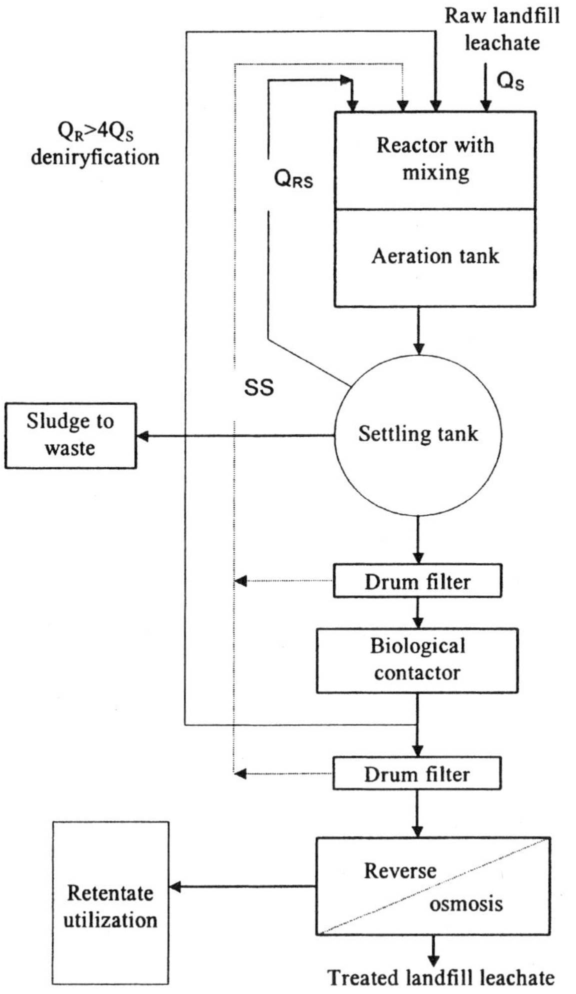

Figure 19 Diagram of the installation for multistage treatment of landfill leachate. (From Refs 78, 84.)

The introduction of a biological pretreatment to the system for the treatment of landfill leachate with the application of reverse osmosis results not only in the removal of substances undergoing biodegradation, but it has many other important advantages [84].

- only refractive substances remain in the landfill leachate;
- higher permeate fluxes and higher volumetric capacities of the permeate can be obtained in the process of reverse osmosis, because the biofouling of osmotic membranes is eliminated;
- the removal of calcium and iron compounds effected by aeration of the outflow is taking place, which prevents the formation of membrane scale;
- nitrogen compounds that cause problems in the course of concentrate evaporation after reverse osmosis are removed;
- treatment costs are lowered.

To sum up, biological pretreatment has a strong influence on the reverse osmosis feed. It diminishes not only organic loading factors but also conductivity of each leachate. This offers a possibility for evaluating the effect of primary purification on the RO process. Figure 20 shows the permeate flux vs. the volumetric concentration factor [84]. The more biological pretreatment diminishes the conductivity, the higher is the possible concentration factor. This phenomenon can be explained by the fact that the osmotic pressure of inorganic salts is higher than that of organic salts or ammonia.

Examples of Hybrid Process Biological Treatment and Reverse Osmosis

The first example of hybrid process biological treatment and reverse osmosis is operating on a technological scale in Leipzig-Seehausen and was constructed by the firm Keramchemie GmbH (Germany) [93]. It has a capacity of 136 m^3/day.

Another leachate treatment plant at Mechernich, near Köh, Germany [80,85] belongs to the first generation of plants of that type. The process consists of biological pretreatment with nitrification and denitrification and a following two-stage reverse osmosis plant with evaporation and granular drying of the concentrates. The process conception, developed by the author Seyfried [85] is shown in Fig. 21. In Table 18, the operating results for the leachate treatment plant from 1994 are presented [85].

Comparing the results with values for indirect discharge in accordance with the local effluent requirements as well as with the standards in Germany for direct discharge, it can be clearly seen that the permeate parameters are considerably better than the values required [85].

Modification of Performance by the Application of a Nanofiltration Process

Frequently, conditions of leachate treatment within the landfill change considerably in comparison with the specifications for the planning of the plant. Thus, the task is to utilize the potential of the plant optimally under these new conditions. The most important element of the enlargement conception is the membrane change in the reverse osmosis plant. One replacement of reverse osmosis by nanofiltration membranes. The plant could then be used with these membranes without the need to effect comprehensive conversion works. Similar to reverse osmosis, nanofiltration

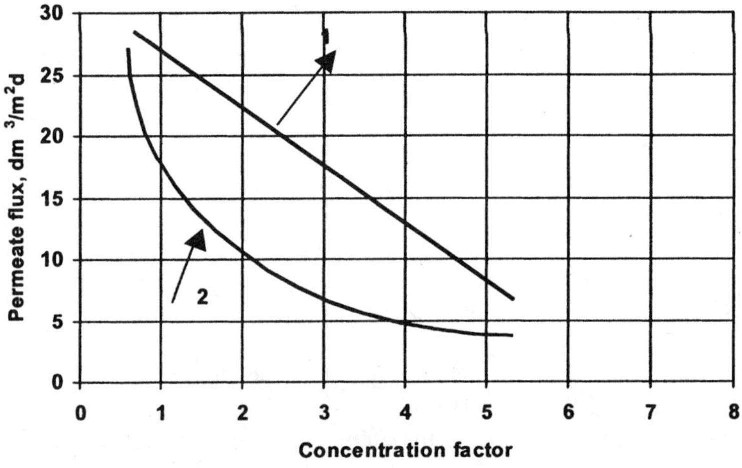

Figure 20 Permate flux in RO process as a function of concentration factor (1, leachate after biological pretreatment; 3, raw leachate) (From Refs 78–84).

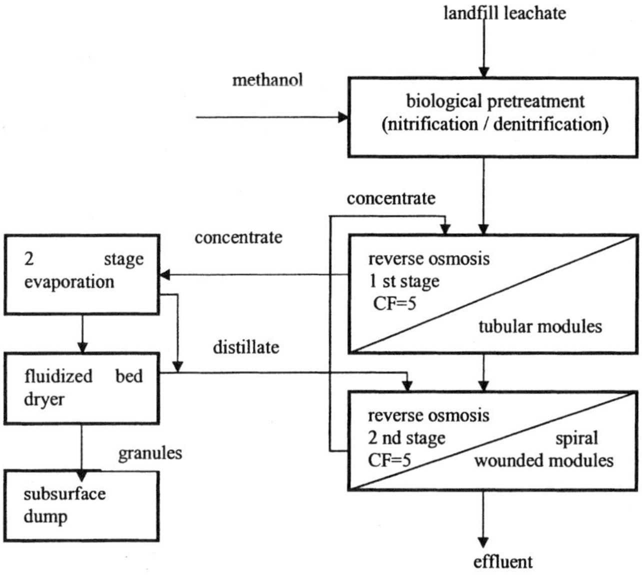

Figure 21 Schematic of the leachate treatment plant for the landfill at Mechernich (from Ref. 85).

membranes are characterized by the fact that they have a high retention of organic substances from molecular weight ca. 200 g/mol or from a size of approximately 1 nm. Furthermore, nanofiltration membranes are ion selective, which means that the salts with monovalent anions are able to pass through the membrane to a high degree, whereas the salts with polyvalent anions are retained [91,96,101]. Table 19 presents the retention coefficients of large-scale, single-stage reverse osmosis plants with composite membranes for biologically pretreated leachate in comparison with nanofiltration membranes as determined in medium-scale examinations.

By using nanofiltration membranes instead of the first-stage reverse osmosis plant, the organic components, heavy metals, and the bivalent organic salts would be retained to a great

Table 18 Effectiveness of the Leachate Treatment Plant at Mechernich (1994)

	Influent leachate		Biological treatment		Reverse osmosis permeate	
Parameter	Average	Max	Average	Max	Average	Max
COD (mg/L)	3176	6440	1301	1755	16.8	48
BOD_5 (mg/L)	1062	4540	23.7	64	2.11	5.5
Organic N (mg/L)	251	1449	82.1	125	5.71	9.8
NH_4-N (mg/L)	884	1153	1.9	92	0.48	4.2
NO_3-N (mg/L)	32	110	129	440	10.0	22.5
NO_2-N (mg/L)	0.1	0.6	2.4	75.2	0.12	0.79
N_{inorg} (mg/L)	916	1173	131.2	430	10.6	20.6
Pb (mg/L)	0.593	0.9	0.142	0.19	<0.001	<0.001
AOX (mg/L)	1261	2750	775	1360	<0.01	0.015
Cl^- (mg/L)	2172	5430	2010	2300	29.0	73

Source: Ref. 85.

Table 19 Reference Data for the Retention of Reverse Osmosis and Nanofiltration Membranes for Biologically Pretreated Leachate

Parameter	Retention coefficient of RO membranes (%)	Retention coefficient of NF membranes (%)
COD	92–97	85–95
AOX	90–96	80–91
NO_3-N	83–93	5–20
Cl^-	85–95	5–20
SO_4^{2-}	95–99	90–96
Heavy metals	88–97	85–96

Source: Refs 91, 96, 101.

extent. The monovalent ions as chloride and all the nitrogen ions would pass through the membrane. In such a situation, biological pretreatment would have to be designed in a way to fulfill the requirements on the discharge of the leachate concerning the inorganic nitrogen compounds. In the nanofiltration plant, a volumetric concentration factor amounting to 10 would be possible, so little volume of retentate should be sent to the evaporation plant, and the evaporation and drying plant would therefore be sufficiently dimensioned. Apart from this, recirculation of the retentate from nanofiltration to biological pretreatment in order to increase COD elimination [102] would only be possible to a limited degree, since without the extraction of sulfates and heavy metals, they would accumulate in the biological stage and soon result in considerable problems with calcium sulfate scaling. Figure 22 presents the concept of nanofiltration application for leachate treatment with biological pretreatment.

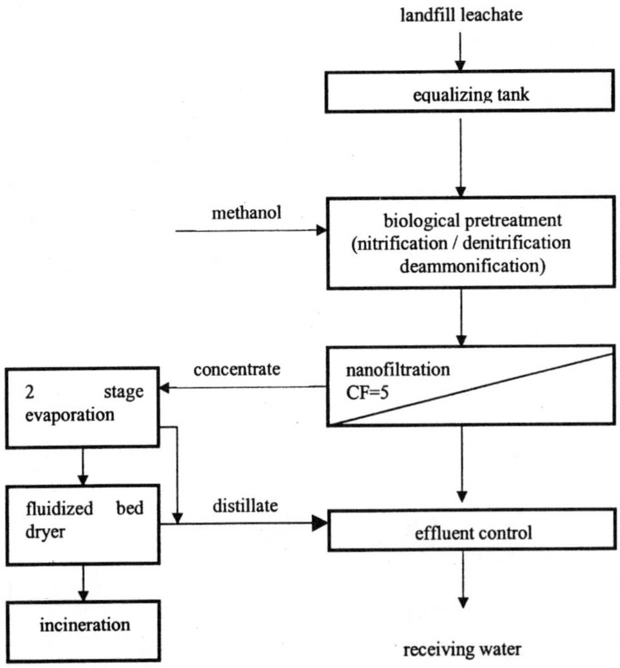

Figure 22 Process conception for capacity enlargement of the leachate treatment plant (From Refs 78, 82).

Application of Ultrafiltration Using Activated Sludge and Reverse Osmosis

As reported in recent literature, ultrafiltration is introduced into the technological cycle for total neutralization of landfill leachate. The ultrafiltration may be treated as a filter before the reverse osmosis stage, both in the system of biological treatment [90,93] and in two-stage reverse osmosis [94]. Figure 23 presents the solution concept [90] of the affiliated method consisting of the following stages:

- biological pretreatment and ultrafiltration;
- reverse osmosis method;
- evaporation of retentate.

In the course of biological treatment, some of the organic compounds undergo biodegradation, whereas the remaining ones, together with nitrogen compounds, are removed in effect by flocculation. The first stage of the process is carried out in the denitrification reactor, in the reactor with activated sludge, and in the ultrafiltration system. The purpose of the ultrafiltration process is to remove the activated sludge from the effluent. It also replaces a drum-type filter, biological reactor and settling tank used in the conventional system for the treatment of effluents by means of the affiliated biological method and reverse osmosis. Organic refractive compounds and inorganic compounds left after the first stage are removed using reverse osmosis. The volumetric permeate capacity in the reverse osmosis is 75–80%, and the permeate may be discharged to surface waters. The retentate undergoes recirculation to the dump or is further utilized using evaporation and drying.

Advanced Landfill Leachate Treatment Using an Integrated Membrane Process

A landfill leachate treatment plant must be frequently retrofitted to improve treatment efficiency by using an integrated membrane process composed of a membrane bioreactor (MBR) and reverse osmosis [100]. In general, proper treatment of landfill leachate is difficult because of its refractory contaminants and widely varying volume due to irregular rainfall. Therefore, various advanced physicochemical treatment processes have been developed. These processes are mainly composed

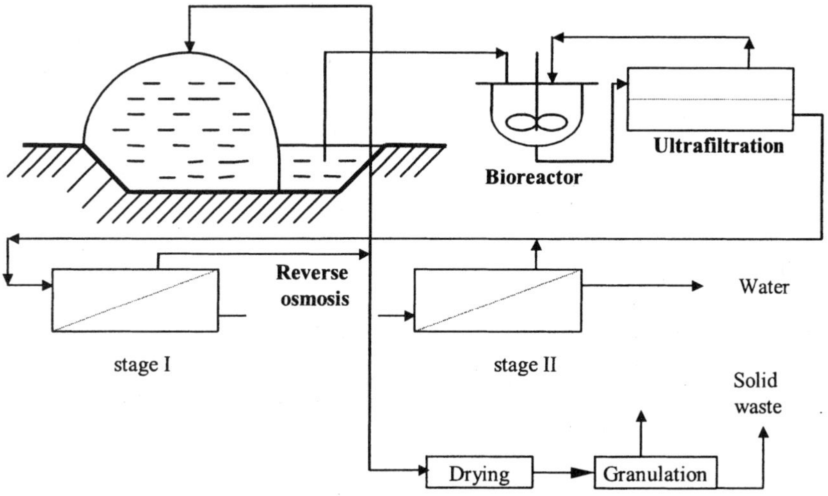

Figure 23 Diagram for the treatment of landfill leachate using the affiliated biological method and reverse osmosis with the use of ultrafiltration (From Refs 78, 90).

of biological and physicochemical processes and are very effective and economical in the early stage of landfill. However, as landfill ages, the characteristics of landfill leachate change to lower biodegradable organic matters and highly concentrated nitrogenous compounds. In this case, biological oxidation of organic compounds and denitrification of nitrogenous compounds barely occurs. To solve these problems, it is necessary to use the integrated membrane process, that is, MBR, to enhance the efficiency of biodegradable organic matters and nitrification rate and reverse osmosis in removing the remaining inorganic nitrogenous ions and nonbiodegradable matters. As a result, the effluent water quality can meet the enhanced regulation limits.

This process is exemplified by a landfill leachate treatment plant located in the Chung-Nam Province in Korea [100]. The novel process was developed as an integrated membrane system. It is composed of on MBR using a submerged membrane named KIMAS (Kolon Immersed Membrane Activated Sludge) and a reverse osmosis process using a spiral-wound membrane module. The MBR has an anaerobic (anoxic) and aerobic part using the submerged hollow fibre membrane. Then, reverse osmosis process was installed to remove nonbiodegradable compounds and inorganic nitrogenous ions. Both membrane specifications are summarized in Table 20 [100]. Figure 24 shows the diagram of the retrofitted landfill leachate treatment plant with MBR [100].

By adopting MBR, stable retention of biomass is enabled to improve the removal efficiency of BOD_5 and nitrification rate. The BOD removal is easy for the MBR and frequently amounts to 97% [100], but nitrogen removal is not as easy.

Nitrification is important in nitrogen removal by the RO membrane, as the negatively charged RO membrane removes the negatively charged nitrate and nitrite more effectively than the positively charged ammonium ions or neutral ammonia [104].

MBR also proves effective in pretreatment for the reverse osmosis membrane by removing suspended solids completely, which prevents channel clogging and enables the adoption of a spiral-wound RO module.

The subsequent RO membrane effectively removes the dissolved nonbiodegradable organic matters and negatively charged nitrified compounds. The average influent COD of 1017 mg/L may be lowered to 32 mg/L in order to effect the overall removal rate of 97% [100]. Table 21 presents the effectiveness of a landfill leachate treatment system consisting of an MBR–reverse osmosis system.

In summary, the integrated membrane process (MBR + RO) has showed a stable removal efficiency in spite of the unbalanced carbon to nitrogen ratio. From an economic point of view, lower operating costs may be achieved as compared to those of the conventional advanced treatment. The integrated membrane process greatly enhances treatment efficiency and does not remove biodegradable organic and nitrogenous matters of landfill leachate.

28.1 LANDFILL LEACHATE TREATMENT BY COAGULATION–FLOCCULATION

28.7.1 Coagulation–Flocculation

The application of a coagulation–flocculation process to landfill leachate treatment results in the removal of organic and inorganic substances in the form of colloids, suspended solids, and

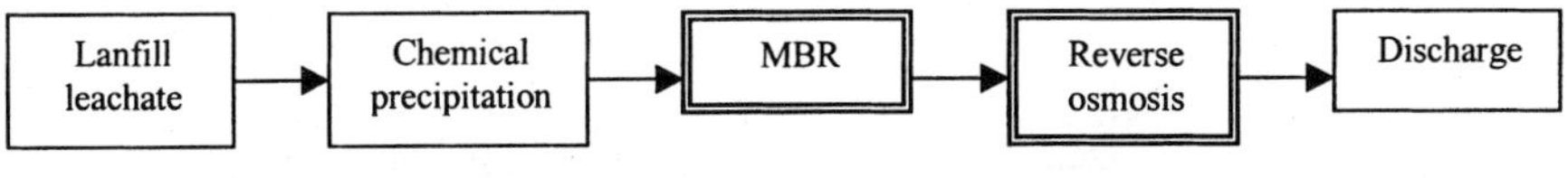

Figure 24 Process flow diagram of integrated membrane process for landfill leachate treatment (MBR, membrane bioreactor) (from Ref. 100).

Table 20 Membrane Specifications

	Microfiltration	Reverse osmosis
Manufacturer	SK Chemicals (Korea)	Filmtec (USA)
Model	KIMAS	SW-4040
Module type	Submerged	Spiral wound
Membrane material	Polysulfone	Polyamide
Pore size	0.1 μm	99% NaCl rejection
Membrane area	5 m^2	6.7 m^2
Operation	Low-pressure suction	Pressurization

Source: Ref. 100.

water-dissolved compounds. In this way, it is possible to effect the reduction not only of COD but also of color and turbidity [20]. First the organics of molar mass above 1000 g/mol are removed [105]. These are mainly humic acids. The process is most effective for landfill leachate with low BOD_5 (<300–500 mgO_2/L) and BOD_5/COD ratio <0.1, that is, biological pretreatment [105]. Coagulation is generally more effective for young leachate than for old [20]. The percentage removal of COD and TOC obtained by coagulation–flocculation is 10–25% with young leachate, but is highest (50–65%) with leachate of low BOD_5/COD ratio (stabilized leachate or leachate pretreated by biological processes) [20,105]. According to Yoon *et al.* [60], coagulation–flocculation removes more organics (59–73%) of molecular weight larger than 500 and only up to 18% of organics of molecular weight smaller than 500.

Aluminum sulfate (alum), ferrous sulfate, ferric chloride, and ferric chlorosulfate are commonly used coagulants, but ferrous salts are not as effective as other coagulants [20,105–109]. Moreover, iron salts seem to be more efficient than aluminum salts. Nonionic, cationic, or anionic polyelectrolytes may be used as coagulant aids to increase the floc settling rate, without really improving the removal efficiency of turbidity [60].

The most important parameters of the process are coagulant dose and pH of the process system. The coagulant dose should be individually determined for each leachate. For example, the dose of Fe^{3+} in one industrial installation was 0.25–0.50 kg/m^3 of leachate [105]. If ferric salts are used, pH of flocculation is 4.5–4.8, whereas for aluminum salts, it is 5–5.5 [105].

There have been some attempts [20,110] to apply electro-coagulation to the treatment of landfill leachate, using anode made of aluminum or iron as a source of coagulation ions. With this method, it is possible to obtain removal of COD amounting to 30–50%. Substances with high molecular weight (MW) are removed by coagulation and substances with low MW by electrolysis.

Table 21 Efficiencies of Hybrid Membrane Process for Landfill Leachate Treatment (Unit: mg/L)

Parameter	Influent	MBR permeate	RO permeate	Average retention coefficient (%)	Regulation criterion in Korea (1997)
BOD_5	100–500	4.3–29	1–7	97	<50
COD	400–1500	211–856	6–72	97	<600
SS	200–1000	1–5	1–1.6	99	<50
NH_4^+-N	200–1400	100–408	10–47	96	<100
NO_3^--N	28–251	34–378	7–23	93	<200(NO_x)

Note: SS, suspended solids.
Source: Ref. 100.

28.7.2 Coagulation–Flocculation as a Pretreatment Mode of Reverse Osmosis

For many landfills that are 10 years old or more in Europe and the United States, leachates are becoming increasingly stabilized. Processes such as biological degradation, physico-chemical treatment and activated carbon adsorption are rarely suitable for effectively treating these leachates. In contrast, reverse osmosis effects at least a 95% reduction in both COD and total dissolved solid (TDS) [19,111].

However, the feasibility of landfill leachate reverse osmosis is highly conditioned by the choice of feed pretreatment mode. Biological pretreatments often prove ineffective as reverse osmosis pretreatments [19,112]. In contrast, lime precipitation or coagulation–flocculation could be adapted [19,109,112]. In the same way, microfiltration and ultrafiltration have proved to be suitable, provided that they are preceded by physico-chemical processes such as coagulation–flocculation or lime precipitation [92,94].

In any case, coagulation–flocculation is a suitable pretreatment mode for the removal of colloidal particles, which are the principal reverse osmosis foulants of landfill leachate. However, this treatment process is often recommended for stabilized leachate COD removal [19,113].

It is important to optimize the coagulation–flocculation process parameters, first the optimal coagulant dose and pH, as well as to improve the flocculation by adding flocculants. The optimal coagulant dose is similar for aluminum sulfate and ferric chloride. However, ferric chloride is more effective than aluminum sulfate in both turbidity (94 or 87%) and COD removal (55 or 42%) [19]. With increased coagulant dose, the percentage sludge volume (mL of sludge per 100 mL of leachate) increases gradually, giving a maximum (45%) for coagulant dose near optimal coagulant dose and decreasing beyond [19]. The floc diameters are less then 0.1 mm [19], so the sedimentation time is not long [19].

Lime milk (50 g/L) and sodium hydroxide solution (8 mol/L) are frequently used for pH control during the coagulation process [19]. Using the optimal coagulant dose, experiment results indicate that turbidity and COD removal decrease with the increase of pH [19]. By considering optimal pH as being the pH at which maximum turbidity removal is obtained, it seems that optimal coagulant dose influences the optimal pH, that is, the optimal pH corresponds to pH achieved at optimal coagulant dose. Turbidity reduction is less efficient in the alkaline region than in the 4.4–5.5 pH region [19,113,114]. Again, it must be noted that organic colloids are favorably flocculated for pH between 4 and 4.5 with ferric salts and between 5 and 6 with aluminum salts [19,115,116].

If coagulation–flocculation is carried out in the alkaline region ($pH > 9$), $Fe(OH)_2$ precipitates more, and iron removal is more efficient (than that in pH 4.5–5.5. However, turbidity reduction is less considerable; the supernatant could foul a reverse osmosis membrane by its residual turbidity. Because of its alkaline pH, the supernatant would foul the reverse osmosis membrane by $CaCO_3$ scaling. In this case, the supernatant must be pretreated once again (before a subsequent reverse osmosis process) to remove residual turbidity and especially to remove $CaCO_3$ by, for example, adding CO_2 until pH 9 is reached (to precipitate $CaCO_3$) and acidification (by adding HCl) until a pH of 4.5–6.5 is achieved [19,92].

To improve flocculation quality, not only the optimal coagulant dose should be used, but also organic nonionic and ionic polymers to aid the sedimentation process.

The research performed to date [19] (Table 22) shows that the best results are obtained with aluminum sulfate/anionic polymer and ferric chloride/nonionic polymer combinations. According to the nature of the most effective flocculant (anionic with alum and nonionic with ferric chloride), the supernatant's zeta potential is probably negative or zero when alum or ferric

Table 22 Coagulation Performance Using Ferric Chloride and Alum and Flocculation With Polymers

Coagulant (dose used) (Polymer)	Aluminum sulfate (0.035 mol Al/L) (Anionic)	Ferric chloride (0.035 mL Fe/L) Nonionic
Doses tested (mg/L)	0–90	0–110
Optimal dose (mg/L)	40	40
Floc size (mm)	2–4	2–4
Sedimentation time (min)	30	30
Percentage COD reduction	35–45	45–55
Percentage turbidity reduction	90	95

Source: Ref. 19.

chloride are respectively used. Nevertheless, with a polymer dose of 40 mg/L and above, flocs have a size of 2–4 mm, and the required sedimentation time is reduced to 30 min [19].

28.8 ADSORPTION IN LANDFILL LEACHATE TREATMENT PROCESS

Adsorption is nonselective to dissolved organics, which is why the process can be used for landfill leachate treatment [20]; so during the process, biodegradable and nonbiodegradable substances as well as humic substances and organic halogens can be removed. The concentration of adsorbed organic halogens (AOX) can be lowered even below 1 mg/L [105]. After landfill leachate treatment, the adsorbent bed should be regenerated.

Granular or powdered activated carbon is the most commonly used adsorbent [19,105,116,117]. Carbon adsorption permits 50–70% removal of both COD and ammonia nitrogen. Other materials, tested as adsorbent, have given close treatment performances to those obtained with activated carbon. These are zeolite, vermiculite, illite, keolinite, activated alumina, and municipal waste incinerator bottom ash [19,20,105,112].

Experiments with pilot installations have shown that adsorption with powdered activated carbon is more effective then with granular activated carbon columns. The effectiveness of adsorption carried out in tanks with powered carbon can be improved by aeration [20,105]. One gram of powdered activated carbon allows, in optimal conditions, for the removal from leachate of 250–500 mg of COD and 500 mg of AOX [20,105].

Very frequently, two-stage adsorption is applied at an industrial scale. For example, in the landfill at Henhagen (Germany), powdered carbon is added to landfill leachate in a stirred tank in the first stage and in the second stage, leachate is flowed through the filter filled with activated carbon [105]. Such purified leachate is then treated biologically together with communal sewage.

In many cases, adsorption with activated carbon is combined with biological processes [20], so activated carbon occurs as a double-acting process, that is, as adsorbent and as a carrier of microorganisms. In such a situation it is possible to remove 50–70% of COD and ammonia nitrogen [20,44,115,117,118].

This is exemplified by the leachate generated at the landfills of Golstar and Göttingen in Germany, where a full-scale technology for the treatment consists of biological treatment followed by nitrification and ultrafiltration as the main treatment steps [119,120]. As the treated effluent still showed an unacceptable concentration of nonbiodegradable organics and AOX, the adsorption by activated carbon was used as the final treatment step in order to meet the specifications. The carbon column adsorption studies show that nonbiodegradable and color-causing components remaining after biological treatment can be removed by using carbon adsorption techniques. The carbon used must have high efficiency to adsorb nonbiodegradable compounds, and the AOX concentration was brought down to an acceptable level (0.5 mg/L) [119]. The physicochemical characteristics of the effluent after carbon adsorption are listed in Table 23.

On an industrial scale, adsorption is usually carried out in a continuous flow column. The kinetics of the column is determined by the diffusion of the solute from the bulk solution to the adsorbent and/or by the diffusion of the solute from the outer surface of the particle to the inner adsorption site [119]. The adsorption kinetics in a fixed bed of carbon is more complex than in batch reactors [121]. The contact time in packed beds of carbon is often very short; consequently the transfer of the adsorbate from the bulk liquid phase to the external and internal surfaces of the carbon must be accomplished quickly [122]. The overall rate of adsorption may be limited by film diffusion, fluid-phase pore diffusion, internal surface diffusion, or reaction with the solid surfaces of the adsorbent [119], and the effective rate of adsorption is controlled by the step that exerts the largest resistance to transfer. Many researchers have advanced mathematical models to calculate diffusion coefficients, mass-transfer coefficients, and so forth. In an effort to describe the adsorption kinetics, the Loebenstein model [123] has been used in work with landfill leachate and data obtained for COD during column experiments. The equation derived from this model is:

$$\log\left(\frac{C_0}{C-1}\right)=\frac{k_1XW}{Q}-\frac{k_1C_0V}{Q} \tag{29}$$

where C_0 = initial concentration of the influent, C = equilibrium concentration, k_1 = mass transfer coefficient, Q = flow rate, X = adsorption capacity at saturation, W = mass of carbon, and V = throughput volume.

Column studies also indicate physical adsorption since multimolecular layer formation was observed in the second column and both columns could be readily regenerated for carbon. The results demonstrate that in multimolecular adsorption, low molecular weight compounds are preferentially adsorbed, leaving less space for high molecular weight compounds. The

Table 23 Chemical Characteristics of the Effluent after Adsorption Treatment

Parameters	Concentration after biological treatment	Concentration after adsorption with activated carbon
pH	7.5	8.53–8.66
Colour	Chocolate color (yellowish red)	Colorless
Conductivity (μS/cm)	–	14,400
Total dissolved solids (mg/L)	7,940	7,940
Chemical oxygen demand (COD) (mg/L)	879–940	80.4
Suspended solids (mg/L)	Nondetectable	Nondetectable
Nitrate (mg/L)	1,082–1,159	1,020
Nitrite (mg/L)	4.9–6.4	4.2
Chloride (mg/L)	1,950	1,950
Sulfate (mg/L)	138–159	159
Phosphate (mg/L)	41.4	Nondetectable
Biochemical oxygen demand (BOD) (mg/L)	3.2–3.6	1.2
Adsorbed organic halogen (AOX) (mg/L)	2.6–2.0	0.47–0.41

Source: Refs 119, 120.

adsorption capacity of both columns was found to be similar. However, the higher transfer rate of the first column is an indication that adsorption-desorption phenomena takes place during the adsorption process.

Adsorption onto activated carbon may also be used in the final treatment of retentate, coming from reverse osmosis or nanofiltration of landfill leachate. Such a solution has been used in the treatment process of landfill leachate in Lüneburg, Germany [20]. Concentrated organics (also AOX) in retentate are retained onto the activated carbon bed, so there is no need to apply an expensive process such as evaporation and drying [20,124].

NOMENCLATURE

c	solute concentration (mg/L)
C	equilibrium concentration (mg/L)
C_0	initial concentration of the influent (mg/L)
c_r and c_p	solute concentrations (mg/L)
$c_{w,m}$	water concentration in the membrane (mg/L)
$D_{w,m}$	water diffusion coefficient (m^2/s)
E	evaporation as a part of rainfall calculated on landfill surface unit (mm/m^2 year)
ET	evaporation from soil/evapotranspiration from a vegetated surface (mm/m^2 year)
g	osmotic coefficient
k_1	mass transfer coefficient
l	membrane thickness (μm)
LP	leachate production (m^3/year)
P	mean height of annual precipitation calculated on landfill surface unit (mm/year)
P	precipitation (mm/year)
Q	amount of leachate (m^3/day)
Q	flow rate (m^3/day)
R	gas constant (J/mol K)
R	surface runoff from landfill surface unit (m^3/m^2 year)
R	surface runoff (m^3/m^2 year)
T	absolute temperature (K)
v	number of ions from dissociation
V	throughput volume (m^3)
V_w	water partial molar volume
W	mass of carbon
X	adsorption capacity at saturation
Δp	transmembrane pressure (MPa)
ΔU	changes in soil moisture storage (m^3/m^3)
ΔU_W	changes in moisture content in wastes (m^3/m^3)
$\Delta \pi$	osmotic pressure difference between the feed and permeate at the membrane surface (MPa)

REFERENCES

1. Peters, Th.; Stanford, P. L'osmose inverse et le disc-tube modulé dans le traitement des lixiviats. Tribune de l'eau **1993**, *566* (*6*), 67–72.

2. Keramchemie GmbH. Multi-stage leachate treatment plant for a landfill site in Germany. Waste Mgnt. Recycling Int. **1995**, 73–74.
3. Welander, U.; Henrysson, T.; Welander, T. Nitrification of landfill leachate using suspended-carrier biofilm technology. Water Res. **1997**, *31*, 2351–2355.
4. Harrington, D.W.; Maris, J.P. The treatment of leachate: a UK perspective. Water Pollut. Control **1986**, 45–56.
5. Blakey, N.C. Model prediction of landfill leachate production. In *Landfilling of Waste: Leachate*; Christensen, T.H., Cossu, R., Stegmann, R., Ed.; Elsevier Applied Science: London and New York, 1992; 17–34.
6. Piotrowska, H.; Kwiatkowski-Bluhm, J.; Litwin, B. *Municipal Waste Landfill Design and Control*; MGPiB, OBREM: Łódź (in Polish), 1993.
7. Christensen, T.H.; Cossu, R.; Stegmann, R. Landfill leachate: An introduction. In *Landfilling of Waste: Leachate*; Christensen, T.H., Cossu, R., Stegmann, R., Ed.; Elsevier Applied Science: London and New York 1992; 3–14.
8. Bauer, M.J.; Herrmann, R.; Martin, A.; Zellmann, H. Chemodynamics, transport behavior and treatment of phtalic acid esters in municipal landfill leachate. Water Sci. Technol. 1998, *38*, 185–192.
9. Paxéus, N. Organic compounds in municipal landfill leachate. Water Sci. Technol. 2000, *42*, 323–333.
10. Bretschneider, H.; Löcher, K. (Eds). *Taschbuch der Wasserwirtschaft*; Paul Parey: Hamburg/Berlin, 1993.
11. Edeline, F. L'épuration des lixiviats de décharge. Tribune de l'eau **1987**; *566* (*6*), 57–65.
12. Andreottola, G.; Cannas, P. Chemical and biological characteristics of landfill leachate. In *Landfilling of Waste: Leachate*; Christensen, T.H., Cossu, R., Stegmann, R., Eds.; Elsevier Applied Science: London and New York, 1992; 65–88.
13. Jensen, D.L.; Christensen, T.H. Colloidal and dissolved metals in leachate from four Danish landfills. Water Res. **1999**, *33*, 2139–2147.
14. Ehnig, H.-J. Quality and quantity of sanitary landfill leachate. Waste Mgmt. Res. **1983**, *1*, 53–68.
15. Surmacz-Gorska, J. *Organic Compounds and Ammonia Nitrogen Removal from Municipal Landfill Leachate*; Publishers of Silesian University of Technology: Gliwice (in Polish), 2000.
16. Szpadt, R. Municipal waste landfills, leachate characteristics. Civil Engng Municipal Mgmt. **1988**, *4*, 72–76 (in Polish).
17. Kang, K.-H.; Shin, H.S.; Park, H. Characterization of humic substances present in landfill leachate with different landfill ages and its implications. Water Res. **2002**, *36*, 4023–4032.
18. Henry, J.G.; Prasad, D.; Young, H. Removal of organics from leachate by anaerobic filter. Water Res. **1987**, *21*, 1395–1399.
19. Amokrane, A.; Comel, C.; Veron, J. Landfill leachate pretreatment by coagulation–flocculation. Water Res. **1997**, *31*, 2775–2782.
20. Surmacz-Gorska, J. *Degradation of Organic Compounds in Municipal Landfill Leachate*; Publishers of Environmental Engineering Committee of Polish Academy of Sciences: Lublin (in Polish), 2001.
21. Kristensen, P. Ecotoxicological characteristics of landfill leachate. In *Landfilling of Waste: Leachate*; Christensen, T.H., Cossu, R., Stegmann, R., Ed.; Elsevier Applied Science: London and New York, 1992; 89–105.
22. Brack, W.; Rottler, H.; Frank, H. Volatile fractions of landfill leachate and their effect on *Chlamydomonas Reinhardtii: in vivo* chlorophyll a fluorescence. Environ. Toxicol. Chem. **1998**, *17*, 1982–1991.
23. Diehl, K.; Hagendorf, U. (Ed.) *Datensammlung Bioteste – Erhebungen, Bewertung, Empfehlungen. Texte 9*; Umweltbundesamt: Berlin, 1998.
24. Schrab, G.E.; Brown, K.W.; Donnelly, K.C. Acute and genetic toxicity of municipal landfill leachate. Water Air Soil Pollut. **1993**, *69*, 99–112.
25. ISO 8192. Water quality – Test for inhibition of oxygen consumption by activated sludge, first edition; 1986- 07–15.
26. Tyrrel, S.F.; Leeds-Harrison, P.B.; Harrison, K.S. Removal of ammoniacal nitrogen from landfill leachate by irrigation onto vegetated treatment plannes. Water Res. **2002**, *36*, 291–299.
27. Robinson, H. Aerated lagoons. In *Landfilling of Waste: Leachate*; Christensen, T.H., Cossu, R., Stegmann, R., Ed.; Elsevier Applied Science: London and New York, 1992; 203–210.

28. Robinson, H. Exporting waste expertise. Water 21 **1999**, *Sept-Oct.*, 35–36.
29. Kowalik, P.; Slater, F.M.; Randerson, P.F. Constructed wetlands for landfill leachate treatment. In *Ecotechnics for a Sustainable Society*; Thofelt, L., Englund, A., Ed.; Londyn, 1995; 189–200.
30. Maehlum, T. Treatment of landfill leachate in on-site lagoons and constructed wetlands. Water Sci. Technol. **1995**, *32*, 129–135.
31. Cossu, R.; Haarstad, K.; Lavagnolo, M.C.; Littarru, P. Removal of municipal solid waste COD and NH_4-N by phyto-reduction: A laboratory-scale comparison of terrestrial and aquatic species at different organic loads. Ecol. Engng. **2001**, *16*, 459–470.
32. Metcalf and Eddy. *Wastewater Engineering: Treatment, Disposal and Reuse*; McGraw-Hill, Inc.: New York, 1991.
33. Hattmann, L. *Biological Wastewater Treatment*; Publishers Instalator Polski: Warszawa (Polish edition), 1996.
34. Nedwell, D.B.; Reynolds, P.J. Treatment of landfill leachate by methanogenic and sulphate-reducing digestion. Water Res. **1996**, *30*, 21–28.
35. Kettunen, R.H.; Rintala, J.A. Performance of an on-site UASB reactor treating leachate at low temperature. Water Res. **1998**, *32*, 537–546.
36. Timur, H.; Özturk, I. Anaerobic treatment of leachate using sequencing batch reactor and hybrid bed filter. Water Sci. Technol. **1997**, *36*, 501–508.
37. Vicevic, G.P.; Top, P.J.; Laughlin, R.G.W. Aerobic and anaerobic fixed film biological reactors. In *Landfilling of Waste: Leachate*; Christensen, T.H., Cossu, R., Stegmann, R., Ed.; Elsevier Applied Science: London and New York, 1992; 229–243.
38. Henry, J.G.; Prasad, D.; Scarcello, J.; Hilgerdenoar, M. Treatment of landfill leachate by anaerobic filter, Part 2: Pilot studies. Water Pollut. Res. J. Can. **1983**, *18*, 45–46.
39. Young, C.P.; Maris, P.J. Optimization of on-site treatment of leachate by a two phase biological process. In *Proceedings of Safe Waste 87 Conference*, Cambridge, UK. Industrial Seminars, Tunbridge Wells, UK, 1987; 334–339.
40. Wright, P.J.; Austin, T.P. Nova Scotia landfill leachate treatment facility is first of its kind in Canada. Environ. Sci. Eng. **1988**, *1*, 12–14.
41. Sung, M.S.; Chang, D.; Lee, H.Y. Performance improvement of an unstable anaerobic leachate treatment system in an industrial waste landfill. Water Sci. Technol. **1997**, *36*, 333–340.
42. Bae, J-H.; Kim, S-K.; Chang, H-S. Treatment of landfill leachate: ammonia removal via nitrification and denitrification and further COD reduction via Fenton's treatment followed by activated sludge. Water Sci. Technol. **1997**, *36*, 341–348.
43. Diamadopoulos, E.; Samaras, P.; Dabou, X.; Sakellaropoulos, G.P. Combined treatment of landfill leachate and domestic sewage in a sequencing batch reactor. Water Sci. Technol. **1997**, *36*, 61–68.
44. Horan, N.J.; Gohar, H.; Hill, B. Application of granular activated carbon–biological fluidised bed for the treatment of landfill leachate containing high concentrations of ammonia. Water Sci. Technol. **1997**, *36*, 369–375.
45. Martienssen, M.; Schöps, R. Biological treatment of leachate from solid waste landfill sites – alterations in the bacterial community during the denitrification process. Water Res. **1997**, *31*, 1164–1170.
46. Klimiuk, E.; Kulikowska, D. Biological treatment of landfill leachate by activated sludge in SBR reactors. Biotechnologia **1999**, *2*, 190–202 (in Polish).
47. Jowett, E.C. Bio-remediation of landfill leachate with remote monitoring and on-site disposal. In *Phytoremediation and Innovative Strategies for Specialized Remedial Applications*; Leeson, A., Alleman, B.C., Ed.; Battelle Press: Columbus, Richland, 1999; 259–264.
48. Hippen, A.; Helmer, C.; Kunst, S.; Rosenwinkel, K.-H.; Seyfried, C.F. Sludge liquor treatment with aerobic deamonification. In *Proceedings of 3rd International Conference "WASTEWATER '99"*, Teplice, Czech Republic, 1999; 97–104.
49. Doppert, M. Erste komunale Deponiesickerwasser – Aufbereitung nach der Turmbiologie. Wasser Luft und Boden **1988**, *10*, 12–17.
50. Loukidou, M.X.; Zouboulis, A.I. Comparison of two biological treatment processes using attached-growth biomass for sanitary landfill leachate treatment. Environ. Pollut. **2001**, *111*, 273–281.
51. Eckenfelder, W.W.; Grau, P. (Eds.) *Activated Sludge Process Design and Control: Theory and Practice*; Technomic: Lancaster, Basel, 1992.

52. Ilies, P.; Mavinic, D.S. The effect of decreased ambient temperature on the biological nitrification and denitrification of a high ammonia landfill leachate. Water Res. **2001**, *35*, 2065–2072.
53. Spengel, D.B.; Dzombak, D.A. Treatment of landfill leachate with rotating biological contactors: bench-scale experiments. Res. J. WPCF **1991**, *63*, 971–981.
54. Welander, U.; Henrysson, T.; Welander, T. Biological nitrogen removal from municipal landfill leachate in a pilot scale suspended carrier biofllm process. Water Res. **1998**, *32*, 1564–1570.
55. Siegrist, H.; Reithaar, S.; Laais, P. Nitrogen loss in a nitrifying rotating contactor treating ammonium rich leachate without organic carbon. Water Sci. Technol. **1998**, *37*, 589–591.
56. Jokela, J.P.Y.; Kettunen, R.H.; Sormunen, K.M.; Rintala, J.A. Biological nitrogen removal from municipal landfill leachate: low-cost nitrification in biofilters and laboratory scale in-situ denitrification. Water Res. **2002**, *36*, 4079–4087.
57. Kim, S.-M. *Untersuchungen zur Abwasserreinigung mit Hilfe des Photo-Fenton-Prozesses*; CutecSchriftenreihe-31: Clausthal, 1997.
58. Weichgrebe, D. *Beitrag zur Chemisch-Oxidativen Abwasserbehandlung*; Cutec-Schriftenreihe-12: Clausthal, 1994.
59. Cho, S.P.; Hong, S.C.; Hong, S.-I. Photocatalytic degradation of the landfill leachate containing refractory matters and nitrogen compounds. Appl. Catalysis B: Environ. **2002**, *39*, 125–133.
60. Yoon, J.; Cho, S.; Cho, Y.; Kim, S. The characteristics of coagulation of Fenton reaction in the removal of landfill leachate organics. Water Sci. Technol. **1998**, *38*, 209–214.
61. Kim, S.-M.; Geissen, S.-U.; Vogelpohl, A. Landfill leachate treatment by a photoassisted Fenton reaction. Water Sci. Technol. **1997**, *35*, 239–248.
62. Bachmann, I. *Zum Abbauverhalten Chlorierter Organischer Verbindungen in Wässriger Lösung bei der UV/Oxidationsmittelbehandlung*: GWA, 155, Aachen, 1996.
63. Kang, S.-F.; Chang, H.-M. Coagulation of textile secondary effluents with Fenton's reagent. Water Sci. Technol. **1997**, *36*, 215–222.
64. Kuo, W.G. Decolorizing dye wastewater with fenton's reagent. Water Res. **1992**, *26*, 881–886.
65. Pfister, S.; Ohse, A.; Winter, J. Weitergehende Mischwasserbehandlung durch Fentons Reagenz. Korrespondenz Abwasser **1997**, *1*, 67–79.
66. Bae, J.H.; Cho, K.W.; Lee, S.J.; Bum, B.S.; Yoon, B.H. Effects of leachate recycle and anaerobic digester sludge recycle on methane production from solid wastes. Water Sci. Technol. **1998**, *38*, 159–168.
67. Steensen, M. Removal of non-biodegradable organics from leachate by chemical oxidation. In *SARDINIA '93 Proceedings of 4th International Landfill Symposium*, CISA, Cagliari (I) **1993**; 945–958.
68. Leitzke, O. Chemische oxidation mit Ozon und UV-licht unter einem druck von 5 bar absolut und Temperaturen zwischen 10 und 60°C. Wasser-Abwasser gwf **1993**, *134*, 202–207.
69. Wenzel, A.; Gahr, A.; Niessner, R. TOC-removal and degradation of pollutants in leachate using a thin-film photoreactor. Water Res. **1999**, *33*, 937–946.
70. Doré, M.; Chramosta, N.; Karpel Vel Leitner, N.; De Laat, J. Overview of oxidation by ozone/ultraviolet lamp and H_2O_2. In *Proceedings of IWSA-AWWA Conf. Advanced Oxidation Processes*, Miami, Florida (USA), 1992; 1–14.
71. Guittonneau, S.; De Laat, J.; Duguet, J.P.; Bonnel, C.; Doré, M. Oxidation of parachloronitrobenzene in dilute aqueous solution by O_3 + UV and H_2O_2 + UV: a comparative study. Ozone Sci. Engng. **1990**, *12*, 73–94.
72. Steensen, M. Chemical oxidation for the treatment of leachate – process comparison and results from full-scale plants. Water Sci. Technol. **1997**, *35*, 249–256.
73. Collivignarelli, C.; Avezzu, F.; Bertanza, G. Wet oxidation and other oxidation processes in leachate treatment (not published).
74. Robert, D.; Gauthier, A. Prospects for a supported photocatalyst in the detoxification of drinking water. Water Qual. Int. **1998**, *November/December*, 27–28.
75. Ried, A. Unpublished data delivered by WEDECO Company.
76. Ried, A.; Mielcke, J. The state of development and operational experience gained with processing leachate with a combination of ozone and biological treatment. In *Proceedings of 14th Ozone World Congress*, 1999.

77. Linde, K.; Jönsson, A.; Wimmerstedt, R. Treatment of three types of landfill leachate with reverse osmosis. Desalination **1995**, *101*, 21–30.
78. Bodzek, M. Membrane techniques in wastewater treatment. In *Water Management Purification & Conservation in Arid Climate*; Goosen, M.F.A., Shayya, W.H., Ed.; Technomic Publishing Company: Lancaster-Basel 1999; 121–184.
79. Peters, T. Purification of landfill leachate with membrane technology. Water Qual. Int. **1996**, *9–10*, 18–22.
80. Van der Bruggen, B.; Lejon, L.; Vandecasteele, C. Reuse, treatment and discharge of the concentration of pressure driven membrane processes. Environ. Sci. Technol. **2003**, *37*, 3733–3738.
81. Peters, A. Purification of landfill leachate with membrane filtration. Purification & Separation **1998**, Jan/Feb, 33–36.
82. Peters, A. Purification of landfill leachate with reverse osmosis and nanofiltration. Desalination **1998**, *119*, 289–293.
83. Chiang, L.C.; Chang, J.E.; Wen, T.C. Indirect oxidation effect in electrochemical oxidation treatment of landfill leachate. Water Res. **1995**, *29*, 671–678.
84. Weber, B.; Holz, F. Landfill leachate treatment by reverse osmosis. In *Effective Industrial Processes: Benefits and Opportunities*; Tuner, M.K., Ed.; Elsevier Applied Science: London-New York, 1991, 143–154.
85. Baumgarten, G.; Seyfried, C.F. Experiences and new developments in biological pre-treatment and physical post treatment of landfill leachate. Water Sci. Technol. **1996**, *34*, 445–453.
86. Harris, J.M.; Gaspar, J.A. Management of leachate from sanitary landfills. AIChE Symp. Series **1988**, *84*, 171–184.
87. Lema, J.M. Mendez, R.; Blasquez, R. Characteristics of landfill leachate and alternatives for their treatment: A review. Water. Air Soil Pollut. **1988**, *40*, 223–250.
88. Linde, K.; Jönsson, A.-N.; Wimmerstedt, R. Treatment of three different types of landfill leachate with reverse osmosis. In *Preprints of 7th International Symposium, Synthetic Membranes in Science and Industry*; Töbingen, Germany, 1994; 488–491.
89. Mulder, M. The use of membrane processes in environmental problems. An introduction. In *Membrane Processes in Separation and Purification*; Crepso, J.G., Böddeker, K.W., Ed.; Kluwer Academic Publishers: Dordrecht-Boston-London, 1994; 229–262.
90. Noworyta, A. Opportunities and challenges in environmental problems with membrane, biotechnology and hybrid processes. In *Towards Hybrid Membrane and Biotechnology Solutions for Polish Environmental Problems*; Howell, J.A., Noworyta, A., Ed.; Wroclaw Technical University Press: Wroclaw, Poland, 1995; 7–27.
91. Rautenbach, R.; *Membrane Processes*; John Wiley and Sons: New York, 1989.
92. Slater, C.; Ahlert, R.; Uchrin, C. Treatment of landfill leachate by reverse osmosis. Environ. Progress **1983**, *2*, 251–256.
93. Friöhlich, S. Removal of ecological damages with physical, chemical and biological methods. Examples of technological schemes of realized installations for treatment of soils, ground waters and landfill leachate. Lecture on International Symposium *Industrial and domestic wastes – strategies and methods*; Międzyzdroje, Poland (not published), 1995.
94. Syzdek, A.C.; Ahlert, R.C. Separation of landfill leachate with polymeric ultrafiltration membranes. J. Hazardous Mater **1984**, *9*, 209–220.
95. Chinese, A.; Ranauro, R.; Verdone, N. Treatment of landfill leachate by reverse osmosis. Water Res. **1999**, *33*, 647–652.
96. Mulder, M. *Basic Principles of Membrane Technology*; Kluwer Academic Publishers: Dordrecht, 1991.
97. Kinman, R.N.; Nutini, D.L. Reverse osmosis treatment of landfill leachate. In *45th Purdue Industrial Waste Conference Proceedings*, Lewis Publishers, Michigan, USA, 1991; 617–622.
98. Chian, E.S.K.; Asce, M.; De Walle, F.B. Sanitary landfill leachate and their treatment. Am. Soc. of Civil Eng., Environ. Div. J. **1976**, *102*, 411–431.
99. Peters, A. Purification of landfill leachate with reverse osmosis and DT-module. In *Membrane Technology: Application to Industrial Wastewater Treatment*; Caetano, A. *et al.*, ed.; Kluwer Academic Publishers, 1995; 175–185.

100. Ahn, W.Y.; Kang, M.S.; Yim, S.K.; Choi, K.H. Advanced landfill leachate treatment using an integrated membrane processes. Desalination **2002**, *149*, 109–114.
101. Rautenbach, R.; Gröschl, A. Separation potential of nanofiltration membranes. Desalination **1990**, *77*, 73–84.
102. Rautenbach, R.; Mellis, R. Erhöhter biologischer Abbaugard durch Kombination einer biologischen Reinigungsstufe mit Membranverfahren. Korrespondenz Abwasser **1993**, *40*, 1138–1142.
103. Verstraete, W.; Vanstaen, H.; Voets, J.P. Adaptation to nitrification of activated sludge systems treating highly nitrogenous waters. J. WPCF **1997**, *July*, 1604–1608.
104. Elimelech, M.; Chen, W.H.; Waypa, J.J. Measuring the zeta (electrokinetics) potential of reverse osmosis membranes by streaming potential analyser. Desalination **1994**, *95*, 269–286.
105. Rosik-Dulewska, Cz. *Principle of Waste Management*; Ekoinzynieria Press: Lublin (in Polish), 1999.
106. Thorton, R.J.; Blanc, F.C. Leachate treatment by coagulation and precipitation. J. Environ. Eng. Div. **1973**, *99*, 535–544.
107. Cho, S.; Boyle, W.C.; Ham, R.K. Chemical treatment of leachate from sanitary landfills. J. Water Pollut. Fed. **1974**, *46*, 1776–1791.
108. Keenan, J.D.; Steiner, R.L.; Fungaroli, A.A. Chemical-physical leachate treatment. J. Environ. Eng. **1983**, *109*, 1371–1384.
109. Slater, C.S.; Ahlert, R.C.; Uchrin, G. Physico-chemical pretreatment of landfill leachate using coagulation. J. Environ. Sci. Hlth. **1983**, *18*, 125–134.
110. Tsai, C.T.; Lin, S.T.; Shue, Y.C.; Su, P.L. Electrolysis of soluble organic matter in leachate from landfills. Water Res. **1997**, *31*, 3073–3081.
111. Bilstad, T.; Madland, M.V. Leachate minimization by reverse osmosis. Water Sci. Technol. **1992**, *25*, 117–120.
112. Chian, E.S.K.; Dewalle, F.B. Sanitary landfill leachate and their treatment. J. Environ. Eng. Div. **1976**, *102*, 411–431.
113. Erig, H.J. Treatment of sanitary landfill leachate. Biological treatment. Waste Manage. Res. **1984**, *2*, 131–152.
114. Stegmann, R.; Ehrig, H.J. Operation and design of biological leachate treatment plant. Prog. Water Technol. **1980**, *12*, 919–947.
115. Babcock, D.B.; Singer, P.C. Chlorination and coagulation of humic and fulvic acids. J. Am. Water Works Assoc. **1979**, *71*, 149–152.
116. Semmens, M.L.; Field, T.K. Coagulation: experiences in organics removal. J. Am. Water Works Assoc. **1980**, *72*, 476–483.
117. Iwami, N.; Imai, A.; Inamori, Y.; Sudo, R. Treatment of landfill leachate containing refractory organics and ammonium nitrogen by the microorganisms attached activated carbon fluidised bed. Water Sci. Technol. **1992**, *26*, 1999–2002.
118. Suidan, M.T.; Schroeder, A.T.; Nath, R.; Krishnan, E.R.; Brenner, R.C. Treatment of carcela leachate by carbon-assisted anaerobic fluidised bed. Water Sci. Technol. **1993**, *27*, 273–282.
119. Morawe, B.; Ramteke, D.S.; Vogelpohl, A. Activated carbon column performance studies of biologically treated landfill leachate. Chem. Eng. Processing **1995**, *34*, 299–303.
120. Lübbecke, S.; Geissen, S.U.; Vogelpohl, A. Nitrifikation in einem Hochleistungssystem. Chem.-Ing. Tech. **1993**, *65*, 1341–1343.
121. Hand, D.W.; Crittenden, J.C.; Arora, H.; Miller, J.M.; Lykins, B.W. Designing fixed bed adsorbers to remove mixtures of organics. J. Am. Water Works Assoc. **1989**, *81*, 67–71.
122. Wilmanski, K.; Lipinski, K. Adsorption kinetics in GAC systems for water treatment. J. Environ. Eng. **1989**, *115*, 91–108.
123. Loebenstein, W.V. Batch adsorption from solution. J. Res. Natl. Bur. Stand. **1962**, *66A*, 503.
124. Schalk, I.; Timm, C.; Ringe, H. Weitergehende Sickerwassrreinigung mit Membranbiologie und Nanofiltration – Betriebsergebnisse der Zentraldeponie Lüneburg. Korrespondenz Abwasser **1999**, *46*, 1253–1258.

29

On-Site Monitoring and Analysis of Industrial Pollutants

Jerry R. Taricska
Hole Montes, Inc., Naples, Florida, U.S.A.

Yung-Tse Hung
Cleveland State University,
Cleveland, Ohio, U.S.A.

Kathleen Hung Li
NEC Business Network Solutions,
Irving, Texas, U.S.A.

29.1 MONITORING AND ANALYSIS OF INDUSTRIAL POLLUTANTS

The primary objective of monitoring and analysis of industrial pollutants is to determine the identity, concentrations, and extent of toxic chemical contamination. An essential element of any monitoring work is rapid sampling and analytical support, to match the quality objectives. Precision and accuracy are major analytical considerations that define the quality of the results.

Precision refers to the reproducibility of measurements and shows the variability of a group of measurements compared to the average value. It is generally expressed as standard deviation, or coefficient of variation, range, and relative range.

Accuracy refers to the lack of bias in measurements. Sources of bias include the sampling process, contamination, preservation, handling, sample matrix, sample preparation, and analytical techniques. Sampling accuracy can be determined by measuring the concentration of contaminations in blanks in the field and analytical accuracy can be determined by using known and unknown quality control samples and matrix spikes. Accuracy is generally reported as percent recovery, or percent bias. A 100% recovery means a completely accurate result. Percent bias measures the difference of the results from the true ones. A zero percent bias indicates a completely accurate result.

Field screening can be described as an investigation technique used at a site to rapidly determine (analytically) the presence of environmental contaminants and the approximate concentration of the target compound. A target compound list is a list created by the US Environmental Protection Agency (USEPA) that identifies those compounds of interest during Superfund site remediation.

The appropriate type of sampling and analysis at a given site depends primarily on the intended end use of the results and associated quality requirements. Data quality is defined by the level of analytical support appropriate to various data uses. According to the USEPA, five levels of analytical support are defined, as follows:

- *Level I.* Field screening or analyses using portable instruments. Results are often not compound specific and not quantitative, but results are available in real time. It is typically the least costly of the analytical options.
- *Level II.* Field analysis using more sophisticated field portable analytical instruments. In some cases, the instruments can be set up in a mobile laboratory on site. There is a wide range in the quality of data that can be generated, from qualitative to quantitative.
- *Level III.* Laboratory analyses using methods other than the Contract Laboratory Program (CLP)–Routine Analytical Services (RAS). This level is used primarily in support of engineering studies using standard USEPA-approved procedures. Some procedures may be equivalent to CLP–RAS, but without the CLP requirement for documentation.
- *Level IV.* CLP RAS. This level is characterized by rigorous QA/QC protocols and documentation and provides qualitative and quantitative analytical data.
- *Level V.* Nonstandard methods. Analyses that may require method modification and/or development.

Generally, the quality of data increases from Level I to V. However, with appropriate QA/QC procedures, the quality of Level II and III data can be equal to that of Level IV and V data. The quality of Level IV data may not be higher than those of other levels, but it provides sufficient documentation to allow qualified personnel to review and evaluate data quality. In other words, it provides data of known quality.

Level I and II analyses are defined as field screening and analytical methods, which are the onsite monitoring and analytical technologies to be discussed in this chapter.

Level I analyses are generally defined as field screening with the objectives of generating data to be used in refining sampling plans and determining the extent of contamination and conserving other analytical support resources. They provide data for on site real-time total vapor measurements, evaluation of existing conditions, sample location optimization, extent of contamination, and health and safety evaluations.

Level II analyses provide real-time data for ongoing field activities, or provide initial data to be used as the basis for determining laboratory support. Level II analyses can measure many organic compounds such as volatiles, base neutral acid, extractable organics, pesticides, and PCBs (polychlorinated biphenyls) from soil, water, air, and waste materials. By using field atomic adsorption and other instruments, inorganic analyses can also be conducted.

Level II analyses are used for onsite real-time data development, extent of contamination and cleanup operation, and the assessment of health and safety. Generally, gas chromatograph, mass spectrometry, and other instruments are used. The ability to determine accuracy and precision depends on the quality assurance and quality control (QA/QC) steps taken, which include documentation of blanks, and runs of standards. Blanks are samples designed to monitor the introduction of artifacts into the measurement process.

The universal blank for aqueous samples is reagent water, but no universal blank exists for soil samples. Several types of blanks are used for monitoring various processes:

- *Laboratory blank.* This blank sample is taken during sample preparation and tested to determine contamination in sample preparation and analyses.
- *Holding blank.* This blank sample is stored with the sample and tested to determine contamination in sample storage, preparation, and analyses.

- *Trip blank.* This blank sample (not open in the field) is shipped with the field sample to determine contamination from preservation, site conditions and transport, as well as sample storage, preparation, and analyses.
- *Field blank.* This blank sample is open in the field to determine contamination from the atmosphere as well as those activities listed under the *trip blank.*
- *Wash Blank.* This blank sample is a sample taken by pouring aqueous reagent over or through the sample collection devices and is tested to determine the cleanliness of sampling equipment as well as those activities listed under the *field blank.*

The accuracy and precision of Level I or II analyses depend on the matrix and contaminants being sampled, and most importantly, upon the skill of the analyst. Level II analyses based in a mobile laboratory using state-of-the-art instrumentation can allow mobile laboratory capabilities to approach the analytical range and performance achieved in offsite analytical facilities.

The mobility of field equipment utilized in Level I and II analyses may be categorized as portable, field friendly, or mobile. Portable equipment can be hand held, easily carried by one person, and requires no external power source. It includes photoionization detection, flame ionization detection, suitcase-sized gas chromatographs, X-ray fluorescence devices, and self-contained immunoassays. Field-friendly equipment is rugged, requires limited external power sources, and can be easily transported in a vehicle. Mobile equipment can be carried in a mobile laboratory. It includes most analytical instruments. Owing to power source requirements, many instruments cannot be used in mobile laboratories. Additional information on onsite monitoring and analysis of industrial pollutants can be found elsewhere[1–4].

29.2 FIELD ANALYTICAL SCREENING

The Field Analytical Screening Project (FASP) was established by the USEPA in response to the concerns raised by Congress and the public to improve the characterization of hazardous waste sites by enhancing onsite analytical capabilities for the investigation of these sites. FASP is used to generate quick turnaround screening data where rigorously qualified CLP data is not required. It is used mainly to help determine the extent of contamination and in choosing samples to document a contaminant release. A Base Support Facility is located in each region to serve as a staging area for all field activities. Instrumentation is stored at the facility and transferred to the support vehicle as needed.

In general, the FASP emphasizes the screening of target compounds known to be on site from previous sample analysis. This data is used to complement CLP data as required by the users and allows for customized screening of only the compounds of interest, thereby shortening the time required. The nationally coordinated program includes analytical methods, QA/QC protocols, safety procedures, personnel training, analytical instrumentation, and data quality objectives. All these aspects of the centrally managed project are essential in providing consistent, reliable data that can be used with confidence by the USEPA in all regions.

The FASP is intended to support site investigation activities in which target compounds are identified from prior analysis. Samples are screened specifically for these target compounds and the methods are optimized for them. The program, however, is not well suited for screening for a broad range of unknown contaminants. Uncharacterized sites and unknown contaminants require the usual CLP analysis. The program is best utilized on large sites that require screening of a small number of contaminants in many samples. Screening methods are customized for these target compounds, which allows for rapid screening at a lower cost.

Onsite feedback is one of most important benefits the FASP can provide to the inspection team at the site. Onsite feedback of FASP screening results can aid the field investigation teams in site characterization. This includes finding "hot spots" onsite and determining the extent of contamination. It is also used to direct ongoing work, redirect sampling efforts, and modify work plans. In removal operations, screening data can determine if enough contaminated material has been removed, while the removal equipment is still on site. When results from FASP indicate that enough hazardous material has been removed, a sample sent to the CLP can be used for confirmation.

Optimization of air monitoring programs is another benefit of onsite screening. The time between sampling and analysis of volatile organic compounds must be minimized to prevent artifact formation and breakthrough, which complicate and/or invalidate the results. It is best to perform air monitoring on site where holding times are minimized and sampling problems can be corrected while the samplers are still there.

Onsite screening can be used in conjunction with soil gas sampling for organic contaminants to help to achieve the best results. Soil–gas samples should be analyzed on site for the same reasons as for air samples.

The FASP is implemented through seven phases, as follows:

- *Phase 1.* Screening for volatile organics in water, soil, and grab air samples;
- *Phase 2.* Screening for PCBs in water and soil;
- *Phase 3.* Screening for chlorinated pesticides in water and soil;
- *Phase 4.* Screening for selected base/neutral organics in water and soil;
- *Phase 5.* Screening for selected acid-fraction organics in water and soil;
- *Phase 6.* Screening for selected metals in water and soil;
- *Phase 7.* Special screening, that is, composite air sampling, GC/MS, herbicides, purge, trap, etc.

This phased approach allows EPA regions to select analytical capabilities based on their particular requirements. Each phase establishes specific capabilities while building on the previous phase.

29.3 ONSITE MONITORING AND ANALYTICAL TECHNOLOGIES

The advantages of onsite monitoring and analysis may be summarized as follows. Onsite monitoring and analyses provide "real-time," onsite feedback, which can aid in identifying hot spots and the extent of contamination, redirecting sampling efforts, modifying work plans, making sure all contaminated materials have been removed in the removal and cleanup projects, while the equipment is still on the site. It actually possesses capabilities beyond that of CLP for air monitoring, soil gas sampling, and screening.

Onsite monitoring and analyses are best suited to perform customized analysis. In the later stages of cleanup operations, it is necessary to perform site-specific, contaminant-specific analyses, which can be accomplished by field-specific detectors, grab air sampling, or soil gas sampling.

Cost savings are an important advantage of onsite monitoring and analyses. By directing extremely contaminated samples through the CLP laboratory, onsite monitoring and analyses eliminate many negatives, minimize full analyses at the CLP laboratory, and make more effective use of those resources. Standby time for sampling and cleanup personnel is also minimized, which results in overall cost savings.

The onsite monitoring and analyses of metals by using x-ray fluorescence will next be discussed, followed by onsite monitoring and analyses of organics by using immunoassays. The other onsite techniques will also be evaluated and assessed.

29.4 X-RAY FLUORESCENCE

X-ray fluorescence (XRF), as used for metal analyses, incorporates a flux of high-energy x-rays to bombard metal samples, causing the metal elements in the sample to emit energy at various characteristic wavelengths. The wavelengths produced are then separated by instruments into a spectrum that contains energy peaks characteristic of the metal elements present. The intensity of energy produced for each metal element is directly proportional to the concentration of the element present.

This technique can be used to screen soil and sediment samples for concentrations of lead, zinc, copper, arsenic, iron, and chromium. Because XRF is a nondestructive technique, samples can be stored for future reference after analysis.

X-ray fluorescence lacks the sensitivity or precision of atomic absorption or other conventional methods. Cadmium, manganese, barium, and mercury can only be detected at high concentrations due to spectral overlap of other elements. Sample matrix effects may cause significant accuracy problems and can never be fully eliminated. Matrix effects include nonhomogeneity, surface conditions, and spectral interferences. Its detection limit is directly related to the total number of x-rays counted and the number of x-rays due to interference and background. The detection limits range from 15 mg/kg for arsenic to 140 mg/kg for iron. Accuracy of XRF, as determined by Student's *t*-test and Wilcoxin's test, shows agreement between this technique and CLP data at a 95% confidence level for arsenic, copper, lead, zinc, and iron with sample numbers ranging from 26 to 45. Precision of XRF is shown to have a coefficient of variation of $\pm 20\%$ at the detection limit and $\pm 5\%$ at higher values for all elements tested.

Sample preparation includes drying a 5 g sample and grinding it to a fine powder. Sample calibration includes measurement of pure element calibration standards, measurement of site-specific samples with known analytic concentrations, input of calibration standard concentrations, and calculation of calibration coefficients. Midpoint standards should be re-checked after five samples. If deviation of standards is greater than 3%, the instrument needs recalibration.

The XRF analyzer must be set up, calibrated, and programmed before samples are screened. Maintenance includes checking the probe for cleanliness and dryness and checking for source decay. Generally, it takes about 10 to 30 minutes for sample preparation, depending on the moisture content of the samples. The actual analysis time is about 5 minutes per sample.

Field-portable XRF instruments have been utilized in the minerals industry for about 12 years. Documented use of their application for hazardous waste began in 1985. The Smuggler Mountain Site near Aspen, CO, was the first published use of this instrument to determine the boundaries of criteria levels of 1000 mg/kg lead and 10 mg/kg cadmium in soils and tailing.

The potential application of field-portable XRF instruments can be almost universal in that virtually any media containing elements heavier in atomic weight than silicon can be qualitatively or quantitatively analyzed. The instruments are able to distinguish background areas from contaminated areas; differentiate uncontaminated soils from soils containing phototoxic concentration of metals, and identify the metals; distinguish homes painted with lead-based paint; distinguish and rank water samples containing as little as 5 mg/L zinc and copper; and determine volume of contaminated material for removal.

X-ray fluorescence screening analysis of low-level metal contamination has been proven to be valuable in the investigation of hazardous waste sites. By using minimal sample preparation and field sampling methods, the results were comparable to laboratory results using conventional methods such as atomic absorption. Multiple-elemental analysis was performed on soil samples with particular interest in lead, arsenic, chromium, and copper levels. Detection limits achieved for some elements were 10 ppm. The XRF inorganic results were used in mapping and contouring the extent of contamination of a hazardous waste site containing organic and inorganic contaminations. The cost savings compared to usual inorganic analytical services are quite significant. The small amount of sample necessary and minimal sample preparations reduce health and safety problems, and the amount of sample disposal. The results obtained from XRF screening analyses show good correlation with other types of inorganic analyses.

Based on results achieved, the field-portable x-ray fluorescence (FPXRF) has been shown in the field to be capable of providing the data necessary for screening and characterizing many inorganic contaminants, both rapidly and inexpensively. Its sensitivity and detection limits can be materially improved with high-resolution detection systems designed for field-portable instruments, and calibration constraints can be relaxed with incorporation of fundamental parameter techniques and better software systems.

A high-resolution portable x-ray fluorescence HgI spectrometer has been used for field screening of hazardous wastes. Its multi-element capabilities were illustrated with measurements on chemically analyzed samples representing materials collected from several hazardous waste sites containing different metallic pollutants in a variety of soil matrices. The range of the analyzer extends from Ca to U, and a typical configuration provides for about 20 elements that are simultaneously reported together with the analytical precision. Minimum detection limits for most elements are in the range of 50–200 mg/kg with a 200 s analyzing time. The solid-state detector is operated near ambient temperature and affords an energy resolution of better than 300 eV for the MnK x-rays. Intrinsic detector efficiency exceeds 60% for energies up to 100 keV. Internal storage provides for the retention of 30 spectra and 100 multi-element analytical reports. A "fundamental parameters" based analysis algorithm is used to compute elemental concentrations. This computational approach, together with the comprehensive element coverage, permits "standardless" measurements over a wide range of material compositions.

Energy-dispersive XRF spectrometry (EDXRF) has been shown to characterize low-concentration soil contaminants. Inorganic pollutants were determined in contaminated soils with EDXRF using a thermoelectrically cooled Si (Li) detector. A field mobile laboratory van or trailer accommodated the system since the electrically cooled detector did not require cryogenic cooling. Soil sample preparation for the analysis was minimal, thus short turnaround times were realized between sampling and reporting results. Detection limits below 20 ppm were obtained for the elements of environmental concern, such as Mn, Fe, Cu, Zn, Pb, As, and Cd.

29.5 IMMUNOASSAYS

Immunoassays for trace organic analysis for individual chemicals are in varying stages of development. Immunoassay development for pentachlorophenol is nearly complete.

The principle of immunoassay for trace organic analysis is based on a competitive inhibition enzyme immunoassay (EIA) that incorporates monoclonal antibodies produced via cell fusion (hybridoma) technology. The EIA steps include the addition of reagents such as antibodies and samples or standards to a microtiter plate, a three-hour incubation, and spectrophotometric analysis of microtiter plate. A colorimetric reaction indicates the concentration of target compound by reference to standard curve.

The detection limit for pentachlorophenol without sample concentration is 25 μg/L. The detection limit for other chemicals depends upon the binding strength of antibodies for the target compound and is compound specific. Each assay is highly selective for single compounds with minor or no cross-reactivity. Antibodies can also be developed and exhibit broad cross-reactivity with compounds of similar structure. The accuracy and precision of immunoassay are not available. However, preliminary results show +10 to 20% variability between samples. One limitation of the antibody method is that they cannot detect organic compounds of molecular weight less than 100 atomic units.

In calibration, a standard curve and controls, along with samples, are needed. The standard curve is developed using known concentrations of the target compound and occupies 18 wells on the plate.

Because antibodies recognize and bind to specific chemical structures, sample cleanup and purification may not be critical in many cases. Sample preparation for soil samples requires extraction or solvent exchange into a polar solvent prior to introduction into the immunoassay. Aqueous samples can rapidly and inexpensively analyze in high volume.

The analysis time for immunoassay is about 4 hours per plate for up to 24 samples per plate. Several plates may be run concurrently. Immunoassays are simple to perform. Highly trained operators are not required. Methods are being developed for analysis of dioxins, pesticides, PCBs, benzene, phenols and so on.

A self-contained immunoassay test system was reported to offer self-measuring of reagents, in addition to internal controls for validating performance of the assay. It was designed on the principle that an optimum system would be a zero technique immunoassay that integrates reagents and the delivery system, and that is self-contained and disposable.

The utility of the test system has been assessed by applying the technology to two assays. One assay developed required part per billion level sensitivity and organic solvent extraction to solubilize Aflatoxin B1, a product fungal contamination in feed commodities. The assay worked well in the presence of high organic solvent. Current procedures for Aflatoxin B1 include thin-layer chromatography or high-pressure liquid chromatography. This test system for Aflatoxin B1 correlates well with these procedures. Another application of the technology was developed to detect the antibiotic penicillin in milk samples. Existing methodologies for detection of penicillin include culturing with the organism *Bacillus stearothermophilus* or a binding assay using radioactively labeled penicillin. The correlation between the test system and these methodologies was excellent.

This self-contained immunoassay system can be used to conduct semiquantitative analysis of low molecular weight contaminants and residues by individuals less skilled than trained laboratory personnel.

A rapid, disposable, self-contained, sensitive enzyme immunoassay device was reported to allow untrained personnel to test for pesticides or other specific environmental contaminants in water or soil. The Pinch Test format can detect paraoxon at one micromolar in water, with positive results indicated by clearly visible color development within 10 minutes. It is operational in salt, brackish, and fresh raw water. This format is designed to have all dry components and to have an ambient shelf life of greater than one year. It has been claimed that the assay format can easily be adapted to measure low levels of environmental chemical hazards such as PCBs, pentachlorophenol, and other compounds.

Two prototype immunodetection systems suitable for environmental field monitoring were reported, one based upon a chromatographic approach and the other utilizing an optical fiber immunosensor design. The performance of these systems has been shown with reference to the immunochemical detection of low levels of bacteria in environmental samples. A wide range of analytes, including toxic chemicals and hazardous microorganisms, may be monitored successfully using one or another of the sensors being developed for use.

The application of monoclonal antibodies for the immunoassay to detect polychlorinated dibenzo-p-dioxins and dibenzo-p-furans (PCDD and PCDF) in a wide range of contaminated samples was reported. For 15 different samples, a direct comparison was made between the levels of PCDD and PCDF contamination determined by conventional GC/MS and those determined by immunoassay. Samples included fly ash, soil, technical grade chemicals, motor oils, PCB transformer oil, and silt bottom residues. These ranged in contamination from less than 1 ppb to several thousand ppb of PCDD and PCDF. Good correlation was demonstrated between the results of the immunoassay and those of conventional GC/MS analyses in spite of these differences in the exact sample matrix detected by the two different techniques. In general, the immunoassay requires substantially less sample cleanup than did GC/MS, therefore offering substantially reduced cost and time for sample analysis.

A laser/microbial bioassay technology was reported as being able to detect various classes of chemical toxicants present in the environment. The system reportedly is capable of identifying and quantifying toxicants in various matrices, which are likely to contain the contaminants even at low levels. The system is field portable, rapid, and sensitive, while still incorporating a high level of specificity for numerous toxic chemicals and chemical classes. The system is also capable of distinguishing between those substances that possess cytotoxicity only and those that have genotoxic properties.

The bioassay system consists of a laser photometer that makes 1200 measurements per second at 15 unique angles over a 180° arc and an isogenic set of *Bacillus subtilis* mutants. The laser detects toxicity to the bacteria by differential light scattering of a fine beam, which is measured quantitatively in terms of intensity at various angles. The parameters recorded are the number of bacteria, their size, shape, and distribution, and any increases or decreases in all of these parameters. Because the system also includes a metabolic activation capability in a unique "solvent" cocktail, carcinogenic chemicals such as benzo(a)pyrene amd dimethyl hydrazine are easily identified and quantified. Individual samples can be read and analyzed by an integrated computerized system so that 5 to 10 various concentrations of an aqueous sample can be processed in just 66 minutes.

An enzyme immunoassay device was reported to be rapid, sensitive, self-contained, disposable, and designed to allow unskilled personnel to perform in field situations. The device was developed for drugs detection in urine or for environmental contamination on surface soil or water. The analyte in the test sample competes with an enzyme–analyte conjugate for a limited number of immobilized antibody sites. It was claimed to detect analytes at 10 ppb in biological fluids, water, and soil, with positive results indicated by clearly visible color development within 10 minutes. It was designed to have all dry components and to have an ambient shelf life of greater than one year. It was claimed to be readily adaptable for use with numerous low molecular weight analytes.

Individual field- and laboratory-based immunoassays have been conducted in the field for the detection and measurement of pentachlorophenol (PCP) at hazardous waste sites. Although the immunoassay methods evaluated are less precise than traditional methods, they have attractive features such as rapid turnaround time for total analysis and low cost per sample. They are also field portable and require minimal training to use the immunoassay method. Detection limits and linear dynamic ranges are comparable to those of traditional methods. Although an immunoassay can be developed for a specific target analyte, it may be subject to certain interference effects as well as from matrix effects unless a more involved extraction is used. However, most cross-reacting compounds are structurally related to the compounds of interest, and due to their intrinsic toxicity, they are also frequently analytes of interest for regulatory monitoring. No evidence of false negative results was found, which is an important feature of a screening method. The majority of false positives occurred at values near the lower limit of detection for each method; this amount might be reduced by simple changes in the acceptance criteria for the method. While insufficient for initial site characterization, the information provided by immunoassay analysis is frequently useful for detailed characterization of previously identified compounds of interest. The methods were found to be able to provide appropriate information for rapid onsite field decision making. The high sample capacity of the plate immunoassay provides a low-cost screening alternative to higher cost laboratory analysis, the results of which are available only after a few weeks.

The real-time detection of biological aerosols has been investigated by using immunoassay. The rapid immunoassay technology and a real-time air sampling capability were integrated. A two-stage air sampler with impingement capability was developed and combined with an immunologically based biosensor to affect a real-time aerosol detection capability. The sampler concentrated and impinged 100 L of air into 100 μL of fluid. The impinged sample was then mixed with the immunoreagents, and the resulting immune complex was trapped onto a nitrocellulose filter, while unreacted materials were washed away. An enzymatic tag was used, which affects a pH change in the substrate buffer. The light-addressable potentiometric sensor was then used to determine the slope of the resultant pH change. This technology is still in the early stages of development, but shows great promise.

29.6 FIBER OPTIC CHEMICAL SENSOR

It is possible to use fiber-optic sensors for *in situ* environmental monitoring of various contaminants in groundwater, air, and soil. Additionally, fiber-optic sensors can determine contamination levels in μg/L. While the techniques are still developmental, the goal of fiber-optic sensors is to use optical fiber as a light tube in remote fiber spectroscopy mode (RFS), or as a sensor in fiber optical chemical sensing mode (FOCS). Remote fiber spectroscopy provides light transmission over large distances with low light loss; transmission depends on the core having a higher index of refraction than the cladding. Configurations include use of laser-induced fluorescence, absorption spectroscopy, Raman spectroscopy, and multivariate spectral analysis. A FOCS unit includes a chemical-specific sensor and can operate in at least three modes: fluorescence, absorption, and refraction. It can respond to a class of compounds or a single species.

The detection limit for RFS can be at low μg/L range for aromatic gasoline components using laser-induced fluorescence; and at μg/L level for organophosphorus pesticides and other compounds using Raman spectroscopy. The detection limit for FOCS can be at low μg/L range for chloroform.

The accuracy of fiber-optic sensors is potentially higher than conventional sampling techniques, because no sample storage is required. The precision has been found to be within 3% when analyses were repeated for chloroform from one well using an organic chloride FOCS.

RFS has a low developmental cost but high operational and equipment cost. Currently, there is a limited amount of field experience with most RFS and FOCS applications. However, research and development efforts are under way by several agencies and companies. Fiber-optic sensors can be a powerful tool in *in situ*, real-time monitoring of environmental contamination. It is particularly suited for inaccessible or hazardous situations such as groundwater, and spill cleanup monitoring.

A fiber-optic chemical sensor has been developed to measure gasoline as a liquid, a vapor dissolved in water, and as a gasoline–water emulsion. The sensor was based on a special coating, which has a high affinity to gasoline. The complete sensor system consists of a portable spectrometer and the gasoline sensor. The components of the spectrometer include a tungsten–halogen lamp as the light source, narrow band filters, a dichotic mirror, a photodiode detector, and associated electronics. A fiber-optic cable is utilized to direct light into and out of the instrument. The chemical sensing material is incorporated onto the side of a short fiber-optic core. One end of this probe is coupled to the long cable and other end is impregnated with a fluorescent dye. The fluorescence intensity of this dye is modulated by gasoline in the sensing region of the probe. The variation of this fluorescence intensity provides quantitative results.

This fiber-optic chemical sensor was found to be specific to gasoline in the presence of other mixtures such as kerosene and jet fuel. Sensitivity was as high as 100% at concentrations less than 10 μL/L. Substituted aromatic hydrocarbons were more responsive than the aliphatic.

A prototype fiber-optic chemical sensing probe based on Surface Enhanced Raman Spectroscopy was used on silver electrodes. The signal is obtained by the Raman spectroscopy from the scattering of a molecule absorbed onto a roughened metal substrate. The surface-enhanced Raman spectroscopy is chemically specific, and increases the selectivity of the adsorption of organic molecules on metal substrates and potential dependence of electro-sorption. The technique is capable of providing information on *in situ* organic contamination that is both sensitive and selective. Two major limitations of this or any other Raman-based technique are size and high power consumption, which must be overcome in order to apply these techniques in the field.

Efforts have been made to reduce the size and power requirement of the fiber-optic surface-enhanced Raman system. The compounds to be identified were adsorbed onto a metal microbase that is microlithographically produced with submicron resolution. Less than 1% of a monolayer of a Raman active target compound offers a high signal-to-noise ratio. By depositing the microbase on the exterior of a fiber-optic cable, convenient field screening or monitoring is permitted. By using highly effective microbases, it is possible to sufficiently reduce laser power requirements to allow an economical but complete system to be housed in a suitcase.

A novel porous fiber-optic sensor was developed at the Fiber Optic Materials Research Program, Rutgers University. The sensor is based on a porous optical fiber device. A small region of a fiber (0.5 cm) is made porous and a chemical reagent is immobilized in the pores. The porous section is an integral part of the fiber. Preliminary experimental results for detecting ammonia and moisture in the gaseous state and pH in the liquid state suggest excellent sensitivity, reproducibility, and a wide dynamic range. The results also suggest the feasibility of developing a wide range of sensors for online monitoring of groundwater contaminants using the porous glass fiber approach.

Porous glass fiber-optic sensors have been developed and demonstrated for pH and temperature measurements. Porous glass fiber-optic sensors are made by selective leaching of a jacketed borosilicate glass fiber. The degree of leaching can be controlled to provide a monolithic structure with a predetermined pore size, which can be varied to allow these sensors to be used for measurements in liquids, gases, or mixed matrices such as sludge. The monolithic structure also maintains a strong attachment between the sensor portion and the rest of the fiber, which acts as a light pipe. Since the sensor is an integral part of the fiber, losses between the sensor and light pipe regions are minimized. The end of the sensor is coated with a thin, porous layer of gold to reflect the incident and response radiation back into the light pipe for analysis.

A variety of analytes can be measured with porous glass fiber-optic sensors by changing the active species bonded to the large internal surface area of the sensor. The pH and temperature sensors use dyes as the active species. Other active species such as enzyme and other biochemicals can be attached to the internal glass surface. Sensitivity can be controlled by varying the length of the porous sensor region and by varying the concentration of the bonded species. The sensitivity of porous sensors is 200 times higher than that of a two-fiber sensor because the porous sensor is able to capture essentially all the fluorescence given off within the fiber. Only a fraction of the fluorescence is captured in the receiving fiber of the two-fiber sensors, and the rest is lost in the open space between the two fibers.

A family of novel fiber-optic sensors is being developed for online monitoring of chemical species in gases and liquids. The sensors utilize porous polymer or glass optical fibers in which selective chemical reagents have been immobilized. These reagents react with the analyte of interest, resulting in a change in the optical properties of the sensor (absorption, transmission, fluorescence). Using this approach, low parts per billion in detection level of the aromatic fuel vapors, benzene, toluene, and xylene, and hydrazines has been demonstrated, as have sensors for ethylene vapor. Also relevant to groundwater monitoring is the development of a pH Optrode System for the pH range 4–8, with additional Optrodes for lower pH ranges. Sensors utilizing optical waveguides offer many advantages for hazardous waste monitoring applications including size, near real-time response, and low manning and expertise requirements. Additionally, porous glass and polymer optical fibers offer significant advantages in these applications because their large interactive surface area dramatically improves sensitivity. They also provide a continuous optical path. This minimizes mechanical and optical coupling losses. Furthermore, sensor interfaces can be developed to allow operation of multisensors. These chemical Optrodes can be applied in a variety of environmental monitoring scenarios, as well as to developmental bioreactors, control of process streams, and industrial hygiene.

Fiber-optic chemical sensors have been used in groundwater and vadose zone monitoring. The principle of detection is a quantitative, irreversible chemical reaction that forms visible light-absorbing products. Modifications in the measurement scheme have lowered the detection limits significantly for several priority pollutants. The sensor has been evaluated against gas chromatographic standard measurements and has demonstrated accuracy and sensitivity sufficient for the environmental monitoring of trace levels of the contaminants trichloroethylene (TCE) and chloroform.

A new field screening method couples a fiber-optic-based chemical sensor system to a truck-mounted cone penetrometer. The system provides the capability for real-time, simultaneous measurement of chemical contaminants and soil type to depths of 50 m. Standard sampling rates yield a vertical spatial resolution of approximately 2 cm as the penetrometer probe is pushed into the ground at a rate of 1 m per minute. Laboratory and field tests indicate that the fluorometer system is quantitative for direct determination of fuel products (diesel fuel, marine fuel, and JP-5) in soil (sands) for concentrations in the range of 100–10,000 ppm. The greatest utility of the system is for rapid screening for petroleum–oil–lubricant contamination in order to more precisely locate contaminated zones, and thus significantly reduce the number of monitoring wells required for site characterization. Field experience indicates that the optical fiber system is rugged enough to withstand normal deployment procedures with the penetrometer system and that the sapphire viewing window appears to be self-cleaning, thereby avoiding memory effects.

29.7 SOIL GAS SAMPLING

Soil gas sampling and analysis is finding widespread acceptance as a preliminary screening method for determining subsurface contamination. The growing concern for leakage from underground storage tanks of fuels and chemicals has prompted a need for faster and better ways to assess those subsurface problems. Soil gas sampling followed by field monitors such as a portable gas chromatograph, is a power tool to locate and pinpoint subsurface contamination. The difficulties of this technology lie in the sampling. At the present time, there is no field method that is truly adequate.

The USEPA has compiled several available soil gas sampling technologies, which have been successfully utilized on a site-specific basis. These include industrial hygiene samplers, mini-barrel sampler, one-liter syringe, direct injection-stopper, perforated tube, Tenax tubes, downhole profiler, and direct injection-auger.

29.7.1 Industrial Hygiene Samplers

Passive soil gas sampling for organic vapor monitoring can be accomplished using industrial hygiene samplers, which utilize open, inverted one-quart metal cans containing an activated carbon, buried at a depth of 1 ft in the area to be sampled. The carbon is exposed for a time determined by the sampling rate for the target chemical and anticipated soil concentrations of the target chemical for a sampling period ranging from 8 hours to one month. After the sampling period is completed, solvent desorption from the activated carbon is completed in field laboratory and analysis accomplished by gas chromatographic (GC) detector. Soil gas concentration of the chemical by volume is calculated using the results of the analysis, sample exposure time, and sampling rate for the chemical of interest. This does not provide groundwater contamination concentrations in a direct manner.

The sample collection time varies depending on analytical sensitivity, expected concentration of contaminant, etc. Sample processing involves about a half hour of solvent desorption and 15 to 30 minutes of GC analysis per sample.

29.7.2 Mini-Barrel Sampler

This method collects soil gas samples when low concentration is expected. Soil aliquots are directly removed and prepared. Headspace gas is withdrawn and injected into the GC.

A coring tool is used along with an auger or backhoe to collect the sample. Samples are forced into screw-cap vials by a sampling extruder. Deionized or distilled water is added to the vial, which is then heated, agitated, and samples are withdrawn for GC analysis. This method is suited for low-concentration contaminants with low vapor pressures or where sample contaminants need to be concentrated in headspace to be within detection limits of the analytical instrument.

29.7.3 One-Liter Syringe

This method rapidly samples soil gas for onsite analysis to determine the contaminant plume. Repeated sampling is not possible once the auger is removed.

To obtain a sample, a hole is drilled using hollow-stem auger. A length of Teflon® tubing attached to a 1 L gas-tight syringe is lowered to the desired sampling depth. The Teflon line is conditioned by drawing a 500 mL air sample. The syringe is detached and evacuated, then the syringe is reattached and a 1 L sample is drawn, which is injected into a purge and trap unit for concentration and GC analysis.

This method has a high correlation with other sampling techniques. Correlation coefficients were 0.68 with the direct injection-auger, 0.88 with the direct injection-stopper, and 0.90 with the Tenax tube. Its sensitivity depends on GC used. The concentrating techniques used overcome some field GC limitations. This method requires about 15 to 30 minutes for drilling the borehole, 1 minute for sample collection, and 30 minutes for GC analysis. The syringes and tubing must be cleaned and flushed with nitrogen gas. The auger and syringes require decontamination before reuse.

This technique uses a simple sample collection and performs better than direct injection-auger sampling.

29.7.4 Direct Injection-Stopper

In a direct injection-stopper, a 12–17 ft (1 ft = 30.84 cm) deep borehole is drilled using a hollow-stem auger with 8 in. (1 in. = 2.54 cm) outside diameter. A sample probe is inserted and borehole walls are allowed to collapse as the auger is pulled. The hole is sealed with a bentonite slurry plug, and the probe is sealed with a stopper and screw cap. Samples are collected by syringe after a two-day equilibration period and analyzed by a GC.

The results from this method correlated well with those from other sampling methods at the same location. Correlation coefficients were 0.78 with direct injection-auger, 0.87 with one-liter method, 0.73 with headspace method, and 0.98 with Tenax.

This method requires 15 to 30 minutes to drill boreholes, 2 days for probe equilibration, and 1 minute to collect the sample.

29.7.5 Perforated Tube

In this method, a nonreactive rod or pipe with a perforated tip is driven into the ground to the desired sampling depth. A pump is used to withdraw soil gas through Teflon tubing, from which grab samples can be taken using a syringe, or composite samples obtained by adding Tenax or other adsorbent material in the collection line along with a flow meter to measure air volume sampled. A water trap is used in some applications.

Aerobic degradation of hydrocarbon may occur in some areas at shallow sampling depths. Sample leakage from the probe or syringe is possible. Flushing of sampling equipment with nitrogen gas is effective in preventing contamination. Carrier gas samples may be injected into a GC to check for contamination. Teflon may be subject to carryover from high liquid- or gas-phase concentrations.

29.7.6 Tenax Tubes

Stainless steel desorption tubes packed with Tenax are suspended inside a stopper sampling probe and connected to a pump. Three liters of soil gas are drawn through each tube, concentrating the contaminant on the adsorbing material. Contaminants are driven off thermally and analyzed using a GC.

A sampling probe must be installed and Tenax tubes preconditioned, because prepacked Tenax tubes are often contaminated. The sampling pumping may be quite lengthy. This method has high correlation with other sampling techniques: 0.80 with direct injection-auger, 0.84 with headspace sampler, 0.90 with one-liter technique, 0.98 with direct injection-stopper.

This method takes 15 to 30 minutes for drilling the borehole, 10 minutes for installing the sampling probe, 45 minutes for sample collection, and 25 minutes for GC analysis.

This technique allows long-term and time-averaged sampling. It has a comparatively easy sample collection. It may not show long-term changes in soil gas concentration unless probe contents are mixed to provide equilibrium along the probe length before sampling. It provides more information about widespread, low-level contamination than the direct injection-auger method but no more detailed plume map. It allows more complete desorption than activated carbon and is easier to clean for reuse. Samples may be refrigerated for future analysis.

29.7.7 Downhole Profiler

A Teflon collection hose with an attached chamber is lowered into a borehole, and the sample is withdrawn using a pump. A gas-tight syringe is used to take a sample from a septum fitting on the pump. Both grab and composite samples may be taken. The analysis time is about 40 minutes per sample.

This is one of several soil gas sampling methods used to determine the vertical contamination gradient, identify hot spots, predict emission rates, and assess migration pathways as determined by soil type and strata graphic. Soil temperature increase due to boring may influence results.

29.7.8 Direct Injection-Auger

A 12–17 ft (1 ft = 30.48 cm) deep borehole is drilled using a hollow-stem auger with an outside diameter of 8 in. When the desired sampling depth is reached, a 500 mL gas-tight, side port syringe is lowered and filled. Syringe contents are injected directly into a GC for analysis.

The results from this method correlated well with results from other sampling methods at the same location: 0.68 for one-liter method, 0.72 for headspace method, 0.78 for direct injection-stopper, and 0.80 for Tenax. It takes 15 to 30 minutes to drill a borehole, and 1 minute for sample collection.

29.7.9 Other Methods

A cone penetrometer with associated sensors and detectors has been used, along with a prototype soil gas sampling system referred to as TerraTrog, to quantitatively sample hazardous soil gases and vapors. The TerraTrog can be deployed by a cone penetrometer to depths of 100 ft far less expensively than drilling monitoring wells. It comprises two modules: an implant of small dimensions containing a gas-permeable membrane of high diffusion impedance, and a sampling and calibration interfaces with a pneumatic manifold. It relies only on soil gas diffusion for subsurface soil gas collection and a carrier gas stream flowing at a slight positive pressure for lifting the sample to the surface. Since the sampling is diffusion limited by a membrane of known impedance, the sampling rate and sample size are independent of soil permeability. Owing to its small diameter and ability to descend down standard well casings, it can also be used for routine groundwater monitoring.

29.8 HEADSPACE TECHNIQUES FOR VOLATILE ORGANIC COMPOUNDS

It has been demonstrated that volatile organic compounds (VOC) were detected more frequently than other types of priority pollutants in groundwater. This suggested that VOC scans might be used as a screening technique to establish the extent of organic contamination of hazardous waste sites. The headspace techniques that sample the vapor headspace above the solution have been widely used to measure VOC in a liquid or soil sample. The multiple head space extraction method can be applied in various techniques. The USEPA has compiled available head space techniques for VOC determinations.

29.8.1 Automated Headspace Sampler

A field laboratory is required for an automated headspace sampler interfaced to a GC to screen VOCs in soil and water. Identification and quantification are carried out by comparing standard peak retention times and peak areas with sample. The water method consists of transferring 1 mL of water sample to a 3 mL reaction vial. For soil, the method consists of transferring 1 g of sample to a 3 mL reaction vial. One mL of surrogate standard is added to each sample and the temperature is set at 60°C. The sampler is loaded and analyzed. Detection limits are 10 μg/L for most VOCs in water, 20 μg/L for some halogenated aromatics, and 10 μg/L for most VOCs in soil. At low concentrations of 10–50 μg/L, there is good correlation to GC/MS data. At higher levels, there is a discrepancy between data sources by a factor of two. Bias ranged from high to low with increasing concentrations.

Column and detector cleaning and reconditioning are generally not required because of the nature of headspace analysis. Samples must equilibrate in a sampler one hour before analysis. A three-point standard calibration curve is necessary to accurately quantify compounds detected by GC. Single-point calibration is adequate for semiquantitative data. The analysis time is 30 to 40 minutes per sample.

29.8.2 Ion Detector

An ion trap detector is used together with a gas-chromatographic/mass-spectrometer (GC/MS) system to screen water, soil, and sediment samples for VOCs. The GC separates the compounds and introduces them into the MS, which provides positive identification of peaks. Quantification is determined by peak areas on GC. The method involves collecting the desired sample in a 40 mL vial, preparing the soil sample, and sampling the vapor headspace above the sample. It is then analyzed on an ion trap detector. Air is screened directly by collecting the sample and injecting it into GC for analysis.

The detection limit is in μg/L for most VOCs. A six-point calibration curve yields a correlation coefficient of 0.964 and showed 20% relative standard deviation. It takes 10 to 20 minutes to analyze a sample.

29.8.3 Organic Vapor Analyzer

This method utilizes a portable GC equipped with a flame ionization detector. Samples are collected in a 40 mL vial. The sample is prepared if it is soil or sediment, and the vapor headspace above the aqueous solution is sampled. The vapor sample is analyzed by the organic vapor analyzer. Identification and quantification are determined by comparing the peaks of standards to those of the samples.

This analyzer measures volatile organics only. Highly volatile organics such as methane tend to skew analysis for total VOCs. Light VOCs such as vinyl chloride are rapidly eluted from the column and are difficult to detect. Response is given in methane equivalent.

The detection limit is 500 μg/L for most volatile organics. Calibration errors for benzene ranged from 25% of value at a detection limit at 0.74 mg/kg to 14% at higher concentrations at 165 mg/kg. It takes about one hour to analyze 20 samples.

The columns must be cleaned every three months. The interpretation of results requires a trained chemist. The column can easily be saturated by concentrated samples, resulting in an inoperable unit.

29.8.4 HNU

HNU is a portable photoionization detector that requires internal electronic calibration as well as calibration to a known standard. After a sample is collected, it can be analyzed by inserting the probe of the HNU into the headspace of the vial.

This detector measures total organic vapor concentration. Response to VOC varies with the probe used. While the detector is insensitive to methane, it may detect unsaturated hydrocarbons, chlorinated hydrocarbons, aromatics, nitrogen and sulfur compounds, aldehydes, ketones, alcohols, acids, and others. It is, however, unable to identify individual compounds. Total response is reported as benzene equivalent. High ambient humidity causes erratic responses.

The detection limit is 100 $\mu g/L$ for most volatile organics. Linear operating range for most compounds is 100–60,000 $\mu g/L$. Its useful range extends to 200,000 $\mu g/L$.

Three probes are available that vary in sensitivity to various organic compounds. A 9.5 electron volt (eV) probe detects aromatics and large molecules, while the 10.7 eV probe detects the aforementioned compounds plus vinyl chloride, methyl ethyl ketone (MEK), trichloroethylene (TCE), and other 2-4 carbon compounds. The 11.7 eV probe detects the above compounds plus halocarbons, methanol, and other single carbon compounds. The response is given in less than 5 s.

29.8.5 Mobile GC

A GC is equipped with a photoionization detector, flame ionization detector, or electron capture detector to screen water, air, soil, and sediment samples. When sampling a water sample, the desired sample is collected in a 40 mL vial, and the vapor head space above the aqueous solution analyzed. The air sample is screened directly by collecting the sample and injecting it into the GC for analysis. Identification and quantification is determined by comparing peak retention times and areas of a standard solution to those of the samples.

The mobile GC system is good for most VOCs at low concentrations. The instrument and sample need to be at the same temperature as required to free the organic vapor. Complex samples give co-eluting peaks. Identifications are considered tentative for soil.

The detection limit is 1 $\mu g/L$ for aromatics, and 40 $\mu g/L$ for some chlorinated hydrocarbons. Analysis time is 10 to 20 minutes per sample.

Field gas chromatography has been used for head space analysis. To analyze head space above drinking water samples, the samples can be rapidly screened by simply injecting 200 μL of headspace gas into the GC septum and observing the total response of the detector in the backflush configuration. When the backflush peak exceeds some low limit, the presence of parts per billion (ppb) levels of dissolved volatiles is indicated. At this point, a sample can be injected into the septum using the GC configuration and rapid analysis with rather good resolution can be achieved for the typical list of volatiles found in contaminated groundwater. Identification is performed by comparing peak retention time to known standard mixtures in the field. Quantification is achieved by comparing the unknown peaks to known standards with identical retention times. Where retention times are ambiguous, it is a simple matter to change the column and repeat the sample and standard to determine retention times again on a different packed column.

Efforts have been made to evaluate and improve head space methods for screening soils for VOC in the field. The research involved comparing several extraction procedures using soils from actual waste sites, and determining the agitation and mixing necessary to achieve equilibrium. Headspace was analyzed using a relatively simple portable gas chromatograph with a short column. The results were variable and showed that several procedures should be attempted and the results evaluated before selecting a screening procedure to be used onsite-specific contaminated soil.

Field headspace for field soil gas analysis and a standard method of analysis of volatile petroleum hydrocarbons in water and soil have been compared. Twelve sites were evaluated for hydrocarbon contamination associated with the use of existing underground storage facilities. Using probe-driven technology, soil gas, groundwater, and/or soil samples were taken from each sampling location. Each soil gas sample, groundwater sample, and soil sample was analyzed on location using a field-mobilized gas chromatograph. Each groundwater sample and soil sample was also analyzed by a contracted laboratory using the appropriate standard method. The correlation between field analytical results and the laboratory analytical results was 0.87 over four orders of magnitude for 25 samples. The correlation for toluene in soil gas samples vs. toluene in soil samples was 0.81 over five orders of magnitude. It can be concluded that soil gas investigations are a useful complement to soil and water sampling approaches to site evaluations. It is important to use appropriate action levels for soil gas measurements. For locations where soil gas measurements are inappropriate, soil or water samples can be collected. It should be noted that these results were obtained using laboratory-grade analytical equipment, which was mobilized for field use. Some soil gas investigations are conducted using hand-held instruments or portable gas chromatographs with little or no temperature control of the sample stream. Since these devices are not as reliable, caution should be used in applying these results to those approaches.

A manual, static headspace GC method is being developed and used in the field for the screening of gasoline-contaminated groundwater and soil samples. The method has focused primarily on the analyses of benzene (B), toluene (T), ethylbenzene (E), and the three xylene isomers (X) (collectively abbreviated as BTEX). It also allows for the determination of methyl-t-butyl ether (MTBE), trichloroethylene (TCE) and tetrachloroethylene in the head space above the aqueous layer as detected by a photoionization detector (PID) of a field-portable gas chromatograph. The test is performed from the same vial where the sample is collected, which reduces the chances of sample loss. Good agreement was found between the field static head space method, a laboratory-based manual static method, and a laboratory-based purge-and-trap method.

An aqueous headspace standard was compared with a SUMMA canister air standard for volatile organic compound field screening. According to both theoretical and experimental results, the aqueous head space standard is a suitable VOC standard for ambient air field screening analysis. The field screening headspace standard is easy to prepare with materials that are readily available in any environmental laboratory. Very little time is needed for preparation and the cost to prepare this type standard is minimal. Canister standards possess high accuracy for most of the VOCs and reflect real concentrations of the VOCs inside. They are easy to store and transport, reusable, and are stable in terms of temperature and time. However, compared with aqueous headspace standards, canister standards are relatively expensive, and are only recommended to those laboratories that have a canister analysis setup.

29.8.6 Multiple Headspace Extraction

A multiple headspace extraction technique was used in conjunction with a portable GC for quantitative analysis of volatile organics in water. This method, first proposed by McAuliffe, consists of a repeated analysis of vapor over liquid after replacing the analyzed equilibrated gas by an equal volume of pure air with its subsequent equilibrium.

More accurate data can be obtained by using the multiple headspace method as compared with the standard single extraction procedure. With the multiple headspace method, the influence of a sample matrix on the phase equilibrium is eliminated. This method simultaneously determines the analyte contents in the equalized gas in the headspace and measures the partition coefficient for the liquid sample under investigation. Therefore, this method provides a useful field ability to determine volatile organics in all types of aqueous samples. It requires no external aqueous calibration standard and raises no concerns about matrix effects, which are likely to occur in complex samples found at hazardous waste sites.

29.9 AIR MONITORING FOR VOLATILE ORGANICS

29.9.1 Programmed Thermal Desorber (PDT) and Gas Chromatography

This method uses a field sampling pump and programmed thermal desorber (PDT) to determine the organic contaminants in air samples as a time-weighted average. Identification and quantification is done by comparing peak retention times and heights of peaks with standards. Sample acquisition involves collecting an adequate sample (usually 10–30 L) of ambient air using a Tenax or activated carbon tube attached to a pump. The sample is thermally desorbed in the PTD. Two samples are withdrawn from the PTD and analyzed on the GC. The first sample is usually a small volume preliminary sample, and the second is a larger volume sample, which is based on results from the first.

The detection limits of this method are: 1.0 μg/L using the Photovac or Finnigan GC/MS for most volatile organics and 500 μg/L using Foxboro OVA for most volatile organics. It has good separation of peaks, and 99% recovery of 11 spiked vinyl chloride samples using freshly packed coconut charcoal tubes; lower recovery in other tests is possibly related to the spiking method. The standard deviation of spiked vinyl chloride samples is about 10%.

This method takes 4–8 hours for sample collection, 5 minutes for desorbing, and 5 minutes for analysis. The interpretation of results requires a trained chemist.

29.9.2 Other Methods

Portable gas chromatographs (PGC) capable of direct detection of ambient concentrations of toxic organic vapors in air were operated in field studies, while simultaneous data were taken for comparison by the Canister/TO-14 Method. Samples were obtained downwind of Superfund hazardous waste sites, highways, chemical plants, and in locations where there was concern about odors or nasal/respiratory irritation. Reasonable agreement between methods was found, even though sampling techniques were not equivalent.

A high-speed gas chromatography from optimizing column operating conditions has been shown to minimize the injection band width, thus shortening the separation time to a few seconds. The system was evaluated using common organics including alkanes, aromatics, alcohols, ketones, and chlorinated hydrocarbons. Quantitative trapping and reinjection was achieved for all tested compounds. Limits of detection (LOD) for many compounds, based on a 1 cm^3 gas sample, were less than 1 ppb. By using the cold trap inlet with a low dead volume detector and a high-speed electrometer, the efficiency available from commercial capillary columns can be better utilized and retention times for some routine separations may be reduced to a few seconds.

Two different types of direct sampling mass spectrometers were evaluated for use as rapid screening tools for volatile organics in a wide range of environmental matrices. They are a commercially available ion trap mass spectrometer (ITMS) and a specially designed tandem source glow discharge quadruple mass spectrometer. Both are equipped with versatile sampling interfaces, which enable direct monitoring of volatile organics at ppb levels in air, water, and soil samples. Direct sampling mass spectrometry does not utilize chromatographic or other separation steps prior to admission of samples into the analyzer. Instead, individual compounds are measured using one or more of the following methods: spectral subtraction, selective chemical ionization, and tandem mass spectrometry (MS/MS). For air monitoring applications, an active "sniffer" probe is used to achieve instantaneous response. Water and soil samples are analyzed by means of high-speed direct purge into the mass spectrometer. Both instruments provide a range of ionization options for added selectivity and the ITMS can also provide high-efficiency collision-induced dissociation MS/MS for target compound analysis.

A fully portable GC/MS system was based on the combination of an automated vapor sample inlet, a "transfer line" gas chromatography module system. The unit weighs approximately 70–75 lb (1 lb = 454 g) and uses 150–200 W of battery power. The MS and computer are carried in front of the operator by means of a shoulder harness whereas the battery pack, carrier gas supply, and roughing vacuum system are carried as a backpack. Air samples can be analyzed using a special automated air sampling inlet. The system is designed to be supported by a vehicle transportable docking station.

Newly designed instrumentation for multimedia environmental trace organic analysis was described for onsite application. The automated prototype units feature advanced sample processing with interfaces for online analyses with chromatographic and/or spectral detectors. Thermal sample processing is provided by an Environmental Pyroprobe Analyzer, including modules for purge and trap/thermal desorption, dynamic headspace, and pyrolysis. Nonthermal multisample processing is conducted with a Chemical Hazards Automated Multi Processor based on supercritical fluid extraction and specialty interface units. Analyses of low ppb levels of vapors, aerosols/particulates, gasoline, and soils illustrate the proven capabilities of the integrated modular systems.

Field GC/MS applications have utilized Bruker Instruments' mobile mass spectrometer. The MS, initially designed for NATO as a chemical warfare detector, was manufactured from the outset as a field instrument. The MS was transported from site to site in a midsized truck and was battery operated for about 8–10 hours at ambient conditions. Simple field methods have been developed based on analyte introduction by thermal desorption followed by fast GC separation and MS detection. The goal was to provide a practical GC/MS tool that can deliver the quality of data required for the study with minimal sample cleanup.

Thermal extraction was offered as a fast and safe alternative to classical, cumbersome solvent extractions for a wide range of semivolatile pollutants in GC/MS analysis. Samples are loaded into porous quartz crucibles with no preparation other than weighing required prior to analysis. Analytes are volatized into the helium carrier gas flow at controlled preprogrammable temperature profiles and subsequently cryo-condensed onto a conventional gas chromatographic column. The method was demonstrated by analyzing for a representative group of organic pollutants covering a wide range of polarity/volatility contained in natural soil matrices at concentrations as low as 0.5 ppm using a Pyran Thermal Chromatograph. Average correlation coefficients for calibration curves range from 0.938 to 0.997 for compounds less volatile than naphthalene. Naphthalene and more volatile compounds experienced variable losses during open-air sample loading. Diakylphthalates underwent partial decomposition during the thermal extraction process. Recoveries varied depending on soil types as well as on the physical and chemical nature of analytes, with generally the highest thermal extraction yields for river silt and the lowest yields for clay. Typical recoveries were 10–30% for polynuclear aromatic hydrocarbons, 60–70% for hexachlorobenzene, and nearly 100% for chloronaphthalenes. However, the pesticide aldrin showed recoveries of at most 19%. A majority of the analytical results are within an accepted range for quantitative analysis. With sample turnaround times of typically 30–60 minutes this instrument should greatly facilitate many onsite monitoring and analytical efforts.

Examples have been presented on the use of a mobile ITMS for onsite environmental screening and monitoring of vapors by GC/MS. The instrument is built around a miniaturized ITMS system, with a novel direct vapor sampling inlet and coupled to a high-speed transfer line GC column (short capillary column with fixed pressure drop). The column is temperature controlled inside the standard ion trap transfer line housing. This provides for high-speed analyses using an automated sampling system constructed with only inert materials in the sample path. The system demonstrated the following capabilities: detection limits of less than 10 ppb for a variety of volatile organic compounds; selective analysis of 21 compounds or more in a single one minute chromatogram with boiling point windows depending on column type and temperature; repetitive sampling as frequent as every 10 seconds for monitoring transient vapor concentrations; and direct variation of sample size with sample pulse time to readily optimize GC resolution vs. ultimate sensitivity. The instrument is rugged enough for most field screening and hazardous waste site investigations.

A second-generation transportable gas chromatograph/ion trap detector was developed for the *in situ* characterization of chemical waste sites. The instrument is extensively based on commercial instrumentation and can be used for field analysis of volatile organic compounds in soil and water. A purge and trap GC is used for sampling and separation of VOCs from the environmental matrix before their introduction to the ion trap detector for mass spectral analysis. A secondary microprocessor controls the sampling and GC hardware in parallel with the ion trap detector, which in turn is controlled by the host PC. It has been claimed that the ion trap detector provides many advantages as a mass analyzer, such as being simple to maintain and operate. The high sensitivity of the ion trap and the inherent universality of the modular mass spectrometer system are the most important features for a field analytical instrument. The instrument provides high specificity for compound identification due to the two-dimensional information provided by chromatographic retention time and mass spectral library identification. Mobile ion trap mass spectrometers operating in the MS/MS mode have been successfully applied for the direct, continuous, or near-continuous analysis of permanent gases and condensable vapors. Ion trap mass spectrometer systems have also been developed for rapid screening of volatile organics in environmental matrices by MS/MS techniques.

29.10 ANALYSIS OF POLYCHLORINATED BIPHENYL

The USEPA has compiled the following field methods for the analysis of polychlorinated biphenyl (PCB).

29.10.1 Hexane/Methanol/Water Extraction

As in other methods, hexane/methanol/water extraction requires a field laboratory with a GC and linearized electron capture detector for PCB analysis of soil samples. Identification is carried out by comparing peak retention time with external standards. Quantification is determined by comparing peak heights and volumes of the standard and sample. Sample preparation consists of mixing 0.8 g of soil with a 1:4:5 ratio of distilled water/methanol/hexane. An optional step is to dry and grind the sample before extraction. The sample is agitated and allowed to sit, allowing the hexane layer to separate. The hexane layer is transferred to a test tube containing sulfuric acid and mixed. This step is optional, as it is used to eliminate matrix interferences. The sample is then withdrawn from the hexane layer for GC analysis.

This method is most appropriate for Aroclors 1242, 1248, 1254, and 1260, and good for 1016, 1221, and 1232. If the concentrations are above 100,000 μg/kg, these contaminants will be underestimated by 60%. The results are approximations. The detection limit is 200 μkg. The recovery of Aroclor 1242 spike is 80–105%. Based on the results of 300 samples, the accuracy is equivalent to CLP when the concentration is below 100,000 μg/kg. Relative standard deviation of four samples is 10–12%.

The calibration is determined by peak heights and retention times of PCB standards. Standards and blanks should be run every ten samples. The analysis time is 5 to 10 samples per hour.

29.10.2 Hexane Extraction

Sample preparation for water consists of adding 1.5 mL hexane to 15 mL of water, mixing and separating the hexane layer. Sample preparation for soil consists of mixing 2 g of soil with 2 g sodium sulfate. Then 10 mL of hexane is added to the sample, mixed with an ultrasonic probe, and the hexane layer separated, which is ready for GC analysis.

The detection limits are 25 μg/L in water and 2500 μg/kg in soil. The analysis time is 20 to 25 minutes per sample.

29.10.3 Hexane/Methanol Extraction

Sample preparation for water consists of adding 1 mL hexane to 100 mL of water, mixing sample, and separating the hexane layer. This step is repeated once. Then 1 mL sulfuric acid is added to the hexane extract and mixed, which is ready for GC analysis. Sample preparation for soil consists of mixing 1–2 g soil, 2 mL methanol, and 10 mL hexane. The hexane layer is separated, and 1 mL sulfuric acid is added, and mixed, which is ready for GC analysis.

This method is most appropriate for Aroclor 1232, 1242, 1248, 1254, and 1260. The detection limits are 200 μg/kg in water, and 100 μg/kg in soil. The average percent recovery is 104%, the average relative standard deviation, 14%. The analysis time is 5 to 10 samples per hour.

29.10.4 Hexane/Acetone Extraction

Sample preparation consists of mixing 10–15 g of soil with a UV grade 1:1 hexane/acetone solution, followed by extraction. Florisil SepPak® is used to adsorb interferences from solution. The extract from SepPak is then screened by GC analysis. This method is best suited for Aroclor 1232, 1242, 1248, 1254, and 1260. The detection limit is 2000 μkg. The accuracy is about 5% of true value. The mean recovery of the soil spike is 97–98% with 11 soil spikes tested. One sample run in triplicate gave a standard deviation of 0.231. Two duplicate samples differed by 2–18%. Relative standard deviation of the soil spike was ± 14% with 11 samples tested. The analysis time is 30 minutes per sample.

29.10.5 Other Methods

Room temperature phosphorescence (RTP) has been used as a screening technique for PCBs. It is based on detecting the phosphorescence emitted from organic compounds adsorbed on solid substrates at ambient temperatures, which is unlike conventional low-temperature phosphorimetry. The technique is used to obtain a solution containing the materials to be analyzed using rapid extraction procedures (1–3 min). A few microliters of the sample solution are then spotted on a filter paper. The spot is dried for about three minutes with a heating lamp then transferred to the sample compartment of the spectrometer. Measurements are performed with a commercial spectrofluorimeter equipped with a phosphoroscope. The sensitivity and selectivity of RTP can be enhanced by mixing the sample or pretreating the filter paper with a heavy-atom salt solution. Salts such as thallium acetate or lead acetate are efficient in enhancing phosphorescence quantum yields for most PCBs. The RTP technique approach was claimed to offer several advantages including rapid analysis, simple setup, field applicability, and low per cost analysis. These features make RTP suitable for screening where a rapid estimation for specific PCBs is needed. Field screening analysis allows rapid decisions in a cleanup operation and reduces the need for either return visits to a site by a cleanup crew, or extensive and costly laboratory analyses of samples that contain no detectable levels of PCBs. Field screening techniques also reduce the cost of remedial actions by preventing unnecessary excavation of uncontaminated soil.

A competitive inhibition enzyme immunoassay (EIA) was developed for the determination of PCBs. The test was claimed to be capable of analyzing for PCBs in the field in 15 minutes from prepared sample, using no specialized equipment. The EIA specificity was restricted to PCBs with high sensitivity for Aroclors' 1016, 1232, 1242, 1248, 1254, and 1260, and moderate sensitivity for Aroclor 1221. Matrix and solvent interferences were minimal. It was capable of direct analysis of PCBs at low ppb levels in water. Unoily soils were analyzed using a solvent extraction and direct EIA of the diluted extract. Oily soils were analyzed using a solvent extraction and direct EIA of an increased volume of the diluted extract to correct for inefficient partitioning from the oil phase.

29.11 ANALYSIS OF PESTICIDES

Two methods for pesticides analysis have been used in the field. Both require GC and linearized electron capture detector.

29.11.1 Hexane/Methanol Extraction

Sample preparation for water consists of adding 1 mL hexane to 100 mL water, mixing, and separating the hexane layer. This step is repeated once. Sample preparation for soil consists of mixing 1–2 g of soil, 2 mL methanol, and 10 mL hexane, and separating the hexane layer. Then, 1 mL sulfuric acid is added to the extract and mixed. If the pesticide is sensitive to acidification, this step is omitted. The hexane extract is ready for GC analysis.

This is a simple and rapid determination of most pesticides except endrin ketone and methoxychlor. The detection limits are 100 μg/L in water and 20 μg/kg in soil. The recovery is 8–107% for three water matrix spikes of 13 compounds, and 26–200% for three soil matrix spikes of 13 compounds. The analysis time is 30 to 45 minutes per sample.

29.11.2 Hexane Extraction

Sample preparation for water consists of mixing 15 mL of water with 1.5 mL hexane and the separating hexane layer. Sample preparation for soil consists of mixing 2 g soil with 2 g sodium sulfate. To this, 10 mL hexane is added, mixed with an ultrasonic probe, and the hexane layer is separated, and ready for analysis.

The detection limits are 5 μg/L in water, and 500 μg/kg in soil. The analysis time is 20 to 25 minutes per sample.

29.11.3 Thermal Extraction

Thermal extraction techniques have been investigated as a rapid alternative to classical soil analyses by solvent extraction GC/MS. Samples were heated according to a preprogrammed temperature profile and evolving volatiles analyzed by in-line GC/MS methods. Tedious wet extraction procedures were completely eliminated. A comparison study of thermal and wet extraction procedures was conducted using pesticide containing soil standards with pesticide contents ranging from 1 to 1000 ppm. The system was shown to be capable of providing rapid analyses of different soils for most of the pesticides studied. Virtually no background signal from organic materials contained in the soils was observed, and clean total ion chromatograms were obtained. The results showed that no free exchange exists of adsorbed pesticides and their isotopic analogs spiked onto the soils before analysis. Calculated recoveries based on internal standard quantification generally decreased.

REFERENCES

1. SENSPOL European Union Thematic Network. *Response to New Pollution Challenges*; 2003.
2. SENSPOL European Union Thematic Network. *Monitoring Freshwater Sediments*; 2003. http://www.cranfield.ac.uk/biotech/senspol/senspol_sediment.pdf.
3. US Department of Defense. *Case Narrative: Al Jubayl, Saudi Arabia*; Department of Defense, 2001. http://www.gulflink.osd.mil/al_jub_ii/al_jub_ii_tabe.htm#TAB%20E%20-%20Saudi%20Environmental%20Monitoring.
4. Argonne National Laboratory. *Environmental Monitoring at Argonne National Laboratory*; 2001. http://www.anl.gov/OPA/env/EMfacts.html.
5. Tang, D.T.; Hung, Y.T. Personal communication, 2002.
6. Tang, D.T. Onsite monitoring and analyses of industrial pollutants, unpublished paper, 1999.

30
Treatment of Rubber Industry Wastes

Jerry R. Taricska
Hole Montes, Inc., Naples, Florida, U.S.A.

Lawrence K. Wang
Zorex Corporation, Newtonville, New York, U.S.A., and
Lenox Institute of Water Technology, Lenox, Massachusetts, U.S.A.

Yung-Tse Hung
Cleveland State University,
Cleveland, Ohio, U.S.A.

Joo-Hwa Tay
Nanyang Technological University, Singapore

Kathleen Hung Li
NEC Business Network Solutions, Irving, Texas, U.S.A.

30.1 INDUSTRY DESCRIPTION

30.1.1 General Description

The US rubber processing industry encompasses a wide variety of production activities ranging from polymerization reactions closely aligned with the chemical processing industry to the extrusion of automotive window sealing strips. The industry is regulated by seven Standard Industrial Classification (SIC) codes [1]:

- SIC 2822: Synthetic Rubber Manufacturing (vulcanizable elastomers);
- SIC 3011: Tire and Inner Tube Manufacturing;
- SIC 3021: Rubber Footwear;
- SIC 3031: Reclaimed Rubber;
- SIC 3041: Rubber Hose and Belting;
- SIC 3069: Fabricated Rubber Products, Not Elsewhere Classified; and
- SIC 3293: Rubber Gaskets, Packing, and Sealing Devices.

Approximately 1650 plants exist in the United States and have production ranges from 1.6×10^3 kkg/year (3.5×10^6 lb/year) to 3.7×10^8 Kkg/year (8.2×10^8 lb/year). Table 1 presents a summary of the rubber processing industry regarding the number of subcategories and the number and types of dischargers. Table 2 presents a subcategory profile of best practical control technology currently available (BPT) regulations (daily maximum and 30-day averages) [2]. The effluent limitations are shown as kilogram of pollutants per 1000 kg of raw material processed (kg/kkg).

Table 1 Industry Summary

Industry: Rubber processing
Total number of subcategories: 11
Number of subcategories studied: 3[a]
Number of dischargers in industry:
• Direct: 1054
• Indirect: 504
• Zero: 100

[a]Wet digestion, although not a paragraph 8 exclusion, was not studied because of the lack of plant-specific data. Emulsion and solution crumb rubber, although candidates for exclusion, were studied, because data were available
Source: USEPA.

The rubber processing industry is divided into 11 subcategories based on raw waste loads as a function of production levels, presence of the same or similar toxic pollutants resulting from similar manufacturing operations, the nature of the wastewater discharges, frequency and volume of discharges, and whether the discharge is composed of contact or noncontact wastewater. Other primary considerations are treatment facilities and plant size, age, and location. The 11 subcategories are listed below. A brief description of each subcategory follows.

- Subcategory 1: Tire and Inner Tube Manufacturing;
- Subcategory 2: Emulsion Crumb Rubber Production;
- Subcategory 3: Solution Crumb Rubber Production;
- Subcategory 4: Latex Rubber Production;
- Subcategory 5: Small-Sized General Molding, Extruding, and Fabricating Rubber Plants;
- Subcategory 6: Medium-Sized General Molding, Extruding, and Fabricating Rubber Plants;
- Subcategory 7: Large-Sized General Molding, Extruding, and Fabricating Rubber Plants;
- Subcategory 8: Wet Digestion Reclaimed Rubber;
- Subcategory 9: Pan, Dry Digestion, and Mechanical Reclaimed Rubber;
- Subcategory 10: Latex-Dipped, Latex-Extruded, and Latex Molded Goods;
- Subcategory 11: Latex Foam.

Subcategory 1. Tire and Inner Tube Manufacturing

The production of tires and inner tubes involves three general steps: mixing and preliminary forming of the raw materials, formation of individual parts of the product, and constructing and curing the final product. In total, 73 plants use these general steps to produce tires in the United States.

The initial step in tire construction is the preparation or compounding of the raw materials. The basic raw materials for the tire industry include synthetic and natural rubber, reinforcing agents, fillers, extenders, antitack agents, curing and accelerator agents, antioxidants, and pigments. The fillers, extenders, reinforcing agents, pigments, and antioxidant agents are added and mixed into the raw rubber stock. This stock is nonreactive and can be stored for later use. When curing and accelerator agents are added, the mixer becomes reactive, which means it has a short shelf-life and must be used immediately.

Table 2 BPT Limitations for Subcategories of Rubber Processing Industry (kg/kkg of raw material)

Pollutant	Tire and inner tube plants[b]		Emulsion crumb rubber		Solution crumb rubber		Latex rubber		Small GMEF[c]		Medium GMEF[c]	
	Daily max	30-day avg.[a]	Daily max	30-day avg.[a]	Daily max	30-day avg.[a]	Daily max	30-day avg.[a]	Daily max	30-day avg.[a]	Daily max	30-day avg.[a]
COD			12.0	8.0	5.9	3.9	10.0	6.8				
BOD_5			0.60	0.40	0.60	0.40	0.51	0.34				
TSS	0.096	0.064	0.98	0.65	0.98	0.65	0.82	0.55	1.3	0.64	0.80	0.40
Oil and grease	0.024	0.016	0.24	0.16	0.24	0.16	0.21	0.14	0.70	0.25	0.42	0.15
Lead									0.0017	0.0007	0.0017	0.0007
Zinc												
pH[d]												

Pollutant	Large GMEF[c]		Wet digestion reclaimed		Pan, dry digestion, mechanical reclaimed		LDEM[e]		Latex foam	
	Daily max	30-day avg.[a]	Daily max	30-day avg.[a]	Daily max	30-day avg.[a]	Daily max	30-day avg.[a]	Daily max	30-day avg.[a]
COD			15	6.1	6.2[f]	2.8				
BOD_5							3.7	2.2	2.4	1.4
TSS	0.50	0.25	1.0	0.52	0.38	0.19	7.0	2.9	2.3	0.94
Oil and grease	0.26	0.093	0.40	0.14	0.40	0.14	2.0	0.73		
Lead	0.00017	0.0007								
Zinc									0.058	0.024
Chromium							0.0086[g]	0.0036		

[a]Computed from average daily value taken over 30 consecutive days.
[b]Oil and grease limitations for nonprocess wastewater from plants placed in operation before 1959: daily max = 10 mg/L; 30-day avg. = 5 mg/L.
[c]General molded, extruded, and fabricated rubber.
[d]Limitation is 6–9 pH units for all subcategories.
[e]Latex-dipped, latex-extruded, and latex-molded goods.
[f]Allowable when the pan, dry digestion, mechanical reclaimed processes are integrated with a wet digestion reclaimed rubber process.
[g]Allowable when plants employ chromic acid for cleaning operations.
Source: USEPA.

After compounding, the stock is sheeted out in a roller mill and extruded into sheets or pelletized. This new rubber stock is tacky and must be coated with an antitack solution, usually a soapstone solution or clay slurry, to prevent the sheets or pellets from sticking together during storage.

The rubber stock, once compounded and mixed, must be molded or transformed into the form of one of the final parts of the tire. This consists of several parallel processes by which the sheeted rubber and other raw materials, such as cord and fabric, are made into the following basic tire components: tire beads, tire treads, tire cords, and the tire belts (fabric). Tire beads are coated wires inserted in the pneumatic tire at the point where the tire meets the wheel rim (on which it is mounted); they ensure a seal between the rim and the tire. The tire treads are the part of the tire that meets the road surface; their design and composition depend on the use of the tire. Tire cords are woven synthetic fabrics (rayon, nylon, polyester) impregnated with rubber; they are the body of the tire and supply it with most of its strength. Tire belts stabilize the tires and prevent the lateral scrubbing or wiping action that causes tread wear.

The processes used to produce the individual tire components usually involve similar steps. First, the raw stock is heated and subjected to a final mixing stage before going to a roller mill. The material is then peeled off rollers and continuously extruded into the final component shape. Tire beads are directly extruded onto the reinforcing wire used for the seal, and tire belt is produced by calendering rubber sheet onto the belt fabric.

The various components of the tire are fitted together in a mold to build green, or uncured, tires which are then cured in an automatic press. Curing times range from less than one hour for passenger car tires to 24 hours for large, off-the-road tires. After curing, the excess rubber on the tire is ground off (deflashed) to produce the final product.

This subcategory is often subdivided into two groups of plants: (a) those starting operations prior to 1959, (applies to 39 plants) and (b) those starting operations after 1959. This subdivision must be recognized in applying limitations on plant effluents of oil and grease because BPT limitations are different for the two groups of plants. For plants placed in operation after 1959, the 30-day average oil and grease limitation is 0.016 kg/kkg of product. For plants placed in operation prior to 1959, the limitation is the same (0.016 kg/kkg) but only for process wastewater. Process wastewater for these pre-1959 plants comes from soapstone solution applications, steam cleaning operations, air pollution control equipment, unroofed process oil unloading areas, mold cleaning operations, latex applications, and air compressor receivers. Water used only for tread cooling and discharges from other areas of such plants is classified as nonprocess wastewater, in which oil and grease levels are limited to 5 mg/L as a 30-day average and 10 mg/L as a daily maximum.

Emulsion polymerization, the traditional process for synthetic rubber production, is the bulk polymerization of droplets of monomers suspended in water. Emulsion polymerization is operated with sufficient emulsifier to maintain a stable emulsion and is usually initiated by agents that produce free radicals. This process is used because of the high conversion and the high molecular weights that are possible. Other advantages include a high rate of heat transfer through the aqueous phase, easy removal of unreacted monomers, and high fluidity at high concentrations of product polymer. Over 90% of styrene butadiene rubber (SBR) is produced by this method. Approximately 17 plants use the emulsion crumb rubber process.

Raw materials for this process include styrene, butadiene, catalyst, activator, modifier, and soap solution.

Polymerization proceeds stepwise through a train of reactors. This reactor system contributes significantly to the high degree of flexibility of the overall plant in producing different grades of rubber. The reactor train is capable of producing either “cold” (277–280 K, 103–206 kPa) or “hot” (323 K, 380–517 kPa) rubber. The cold SBR polymers, produced at the lower temperature and stopped at 60% conversion, have improved properties when compared to hot SBRs. The hot process is the older of the two. For cold polymerization, the monomer–additive emulsion is cooled prior to entering the reactors. Each reactor has its own set of cooling coils and is agitated by a mixer. The residence time in each reactor is approximately one hour. Any reactor in the train can be bypassed. The overall polymerization reaction is ordinarily carried to no greater than 60% conversion of monomer to rubber since the rate of reaction falls off beyond this point and product quality begins to deteriorate. The product rubber is formed in the milky white emulsion phase of the reaction mixture called latex. Short stop solution is added to the latex exiting the reactors to quench the polymerization at the desired conversion. The quench latex is held in blowdown tanks prior to the stripping operation.

The stripping operation removes the excess butadiene by vacuum stripping, and then removes the excess styrene and water in a perforated plate stripping column. The water and styrene from the styrene stripper are separated by decanting and the water is discharged to the treatment facility. The recovered monomers are recycled to the monomer feed stage. The latex is now stabilized and is precipitated by an electrolyte and a dilute acid. This coagulation imparts different physical characteristics to the rubber depending on the type of coagulants used. Carbon black and oil can be added during this coagulation/precipitation step to improve the properties of the rubber. This coagulated crumb is separated from the liquor, resuspended and washed with water, then dewatered, dried, and pressed into bales for shipment. The underflow from the washing is sent to the wastewater treatment facility.

Subcategory 3: Solution Crumb Rubber Production

Solution polymerization is bulk polymerization in which excess monomer serves as the solvent. Solution polymerization, used at approximately 13 plants, is a newer, less conventional process than emulsion polymerization for the commercial production of crumb rubber. Polymerization generally proceeds by ionic mechanisms. This system permits the use of stereospecific catalysts of the Ziegler–Natta or alkyl lithium types which make it possible to polymerize monomers into a *cis* structure characteristic that is very similar to that of natural rubber. This *cis* structure yields a rubbery product, as opposed to a *trans* structure which produces a rigid product similar to plastics.

The production of synthetic rubbers by solution polymerization processes is a stepwise operation very similar in many aspects to production by emulsion polymerization. There are distinct differences in the two technologies, however. For solution polymerization, the monomers must be extremely pure and the solvent should be completely anhydrous. In contrast to emulsion polymerization, where the monomer conversion is taken to approximately 60%, solution polymerization systems are polymerized to conversion levels typically in excess of 90%. The polymerization reaction is also more rapid, usually being completed in 1 to 2 hours.

Fresh monomers often have inhibitors added to them while in storage to prevent premature polymerization. These inhibitors and any water that is present in the raw materials must be removed by caustic scrubbers and fractionating drying columns to provide the solution process with the high purity and anhydrous materials needed. The purified solvent and monomers are then blended into what is termed the "mixed feed," which may be further dried in a desiccant column.

The dried mixed feed is now ready for the polymerization step, and catalysts can be added to the solution (solvent plus monomers) just prior to the polymerization stage or in the lead polymerization reactor.

The blend of solution and catalysts is polymerized in a series of reactors. The reaction is highly exothermic and heat is removed continuously by either an ammonia refrigerant or by chilled brine or glycol solutions. The reactors are similar in both design and operation to those used in emulsion polymerization. The mixture leaves the reactor train as a rubber cement, that is, polymeric rubber solids dissolved in solvent. A short stop solution is added to the cement after the desired conversion is reached.

The rubber cement is then sent to storage tanks where antioxidants and extenders are mixed in. The rubber cement is pumped from the storage tank to the coagulator where the rubber is precipitated with hot water under violent agitation. The solvent and unreacted monomer are first steam stripped overhead and then condensed, decanted, and recycled to the feed stage. The bottom water layer is discharged to the wastewater treatment facility.

The stripped crumb slurry is further washed with water, then dewatered, dried, and baled as final product. Part of the water from this final washing is recycled to the coagulation stage, and the remainder is discharged for treatment.

Subcategory 4: Latex Rubber Production

The emulsion polymerization process is used by 17 production facilities to produce latex rubber products as well as solid crumb rubber. Latex production follows the same processing steps as emulsion crumb rubber production up to the finishing process. Between 5 and 10% of emulsion polymerized SBR and nearly 30% of nitrile rubber production (NBR) are sold as latex. Latex rubber is used to manufacture dipped goods, paper coatings, paints, carpet backing, and many other commodities.

Monomer conversion efficiencies for latex production range from 60% for low-temperature polymerization to 98% for high-temperature conversion.

The monomers are piped from the tank farm to the caustic soda scrubbers where the inhibitors are removed. Soap solution, catalysts, and modifiers are added to produce a feed emulsion which is fed to the reactor train. Fewer reactors are normally used than the number required for a crumb product line. When polymerization is complete, the latex is sent to a holding tank where stabilizers are added.

A vacuum stripper removes any unwanted butadiene, and the steam stripper following it removes the excess styrene. Neither the styrene nor butadiene is recycled. Solids are removed from the latex by filters, and the latex may be concentrated to a higher solids level.

Subcategories 5, 6, 7: Small-, Medium-, and Large-Sized General Molding, Extruding, and Fabricating Plants

These three closely related subcategories are divided based on the volume of wastewater emanating from each. These subcategories include a variety of processes such as compression molding, transfer molding, injection molding, extrusion, and calendering. An estimated 1385 plants participate in these subcategories.

A common step for all of the above processes is the compounding and mixing of the elastomers and compounding ingredients. The mixing operation is required to obtain a thorough and uniform dispersion of the rubber and other ingredients. Wastewater sources from the mixing operation generally derive from leakage of oil and grease from the mixers.

Compression molding is one of the oldest and most commonly used manufacturing processes in the rubber fabrication industry. General steps for the processes include warming the raw materials, preforming the warm stock into the approximate shape, cooling and treating with antitack solution, molding by heat and pressure, and finally deflashing. Major products from this process include automotive parts, medical supplies, and rubber heels and soles.

Transfer molding involves the forced shifting of the uncured rubber stock from one part of the mold to another. The prepared rubber stock is placed in a transfer cavity where a ram forces the material into a heated mold. The applied force combined with the heat from the mold softens the rubber and allows it to flow freely into the entire mold. The molded item is cured, then removed and deflashed. Final products include V-belts, tool handles, and bushings with metal inserts.

Injection molding is a sophisticated, continuous, and essentially automatic process that uses molds mounted on a revolving turret. The turret moves the molds through a cyclic process that includes rubber injection, curing, release agent treatment, and removal. Deflashing occurs after the product has been removed. A wide range of products is made by this process, including automotive parts, diaphragms, hot-water bottles, and wheelbarrow tires.

The extrusion process takes unvulcanized rubber and forces it trough a die, which results in long lengths of rubber of a definite cross-section. There are two general subdivisions of this technique; one extrudes simple products and the other builds products by extruding the rubber onto metal or fabric reinforcement. Products from these techniques include tire tread, cable coating, and rubber hose.

Calendering involves passing unformed or extruded rubber through a set or sets of rolls to form sheets or rolls of rubber product. The thickness of the material is controlled by the space between the rolls. The calender may also produce patterns, double the product thickness by combining sheets, or add a sheet of rubber to a textile material. The temperature of the calender rolls is controlled by water and steam. Products produced by this process include hospital sheeting and sheet stock for other product fabrication.

This subcategory represents a process that is used to recover rubber from fiber-bearing scrap. Scrap rubber, water, reclaiming and defibering agents, and plasticizers are placed in a steam-jacketed, agitator-equipped autoclave. Reclaiming agents used to speed up depolymerization include petroleum and coal tar-base oils and resins as well as various chemical softeners such as phenol alkyl sulfides and disulfides, thiols, and amino acids. Defibering agents chemically do the work of the hammer mill by hydrolyzing the fiber; they include caustic soda, zinc chloride, and calcium chloride.

A scrap rubber batch is cooked for up to 24 hours and then discharged into a blowdown tank where water is added to facilitate subsequent washing operations. Digester liquor is removed by a series of screen washings. The washed rubber is dewatered by a press and then dried in an oven. Two major sources of wastewater are the digester liquor and the washwater from the screen washings.

Two rubber reclaiming plants use the wet digestion method for reclamation of rubber.

Subcategory 9: Pan, Dry Digestion, and Mechanical Reclaimed Rubber

This subcategory combines processes that involve scrap size reduction before continuing the reclaiming process. The pan digestion process involves scrap rubber size reduction on steel rolls, followed by the addition of reclaiming oils in an open mixer. The mixture is discharged into open pans, which are stacked on cars and rolled into a single-cell pressure vessel where live steam is used to heat the mixture. Depolymerization occurs in 2 to 18 hours. The pans are then discharged and the cakes of rubber are sent on for further processing. The steam condensate is highly contaminated and is not recycled.

The mechanical rubber reclaiming process, unlike pan digestion, is continuous and involves fiber-free scrap being fed into a horizontal cylinder containing a screw that works the scrap against the heated chamber wall. Reclaiming agents and catalysts are used for depolymerization. As the depolymerized rubber is extruded through an adjustable orifice, it is quenched. The quench vaporizes and is captured by air pollution control equipment. The captured liquid cannot be reused and is discharged for treatment.

Subcategory 10: Latex-Dipped, Latex-Extruded, and Latex-Molded Goods

These three processes involve the use of latex in its liquid form to manufacture products. Latex dipping consists of immersing an impervious male mold or article into the latex compound, withdrawing it, cleaning it, and allowing the adhering film to air dry. The straight dip process is replaced by a coagulant dip process when heavier films are desired. Fabric or other items may be dipped in latex to produce gloves and other articles. When it has the required coating, the mold is leached in pure water to improve physical and electrical properties. After air drying, the items are talc-dusted or treated with chlorine to reduce tackiness. Water is often used in several processes, for makeup, cooling, and stripping. Products from dipping include gloves, footwear, transparent goods, and unsupported mechanical goods.

Latex molding employs casts made of unglazed porcelain or plaster of paris. The molds are dusted with talc to prevent sticking. The latex compound is then poured into the mold and allowed to develop the required thickness. The mold is emptied of excess rubber and then oven dried. The mold is removed and the product is again dried in an oven. Casting is used to manufacture dolls, prosthetics, printing matrices, and relief maps.

Subcategory 11: Latex Foam

No latex foam facilities are known to be in operation at this time.

30.1.2 Wastewater Characterization

The raw wastewater emanating from rubber manufacturing plants contains toxic pollutants that are present due to impurities in the monomers, solvents, or the actual raw materials, or are associated with wastewater treatment steps. Both inorganic and organic pollutants are found in the raw wastewater, and classical pollutants may be present in significant concentrations. Wastewater from reclaimed rubber manufacturing had 16,800–63,400 mg/L total solids, 1000–24,000 mg/L suspended solids, 3500–12,500 mg/L BOD (biochemical oxygen demand), 130–2000 mg/L chlorides, pH of 10.9–12,2, wile wastewaters from synthetic rubber manufacturing had 1900–9600 mg/L total solid, 60–3700 mg/L suspended solids, 75–1600 mg/L BOD, and pH of 3.2–7.9 [3].

Table 3 presents an industry-wide profile of the concentration of toxic pollutants found at facilities in each subcategory (no data are available for Subcategories 9, 10, and 11). Table 4 gives a subcategory profile of the pollutant loadings (no data are available for Subcategories 8, 10, and 11). These tables were prepared from available screening and verification sampling data. The minimum detection limit for toxic pollutants is 10 μg/L and any value below 10 μg/L is presented in the following tables as BDL, below detection limit.

In-plant management practices may often control the volume and quality of the treatment system influent. Volume reduction can be attained by process wastewater segregation from noncontact water, by recycling or reuse of noncontact water, and by the modification of plant processes. Control of spills, leakage, washdown, and storm runoff can also reduce the treatment system load. Modifications may include the use of vacuum pumps instead of steam ejectors, recycling caustic soda solution rather than discharging it to the treatment system, and incorporation of a more efficient solvent recovery system.

30.1.3 Tire and Inner Tube Manufacturing

The tire and inner tube manufacturing industry has several potential areas for wastewater production, but water recycle is used extensively. The major area for water use is in processes

Table 3 Concentrations of Toxic Pollutants Found in the Rubber Processing Industry by Subcategory, Verification, and Screening

	Tire and inner tube manufacturing							
	Treatment influent				Treatment effluent			
Toxic pollutants (μg/L)	Number of samples	Average	Median	Maximum	Number of samples	Average	Median	Maximum
Metals and Inorganics								
Chromium	1	10			1	BDL		
Copper	1	BDL			0			
Lead	2	25		50	0			
Zinc	5	260	150	770	1	330		
Phenols								
2,4,6-Trichlorophenol	0				1	<14		
Aromatics								
Toluene	0				1	<10,000		
Halogenated aliphatics								
1,2-*Trans*-dichloroethylene	0				1	16		
Methylene chloride	0				2	<5,000		<10,000
Trichloroethylene	0				1[a]	40		
Pesticides and metabolites								
Isophorone	0				1	BDL		
			Emulsion crumb rubber manufacturing					
Metals and Inorganics								
Cadmium	2	46		90	1			BDL
Chromium	5	230	250	720	2	140		220
Copper	1			200	0			
Lead	1			390	0			

(*continues*)

Table 3 *Continued*

	Emulsion crumb rubber manufacturing							
	Treatment influent				Treatment effluent			
Toxic pollutants (μg/L)	Number of samples	Average	Median	Maximum	Number of samples	Average	Median	Maximum
Mercury	3	BDL	BDL	BDL	3	BDL	BDL	BDL
Nickel	2	380		590	1			400
Selenium	1			20	1			<24
Zinc	3	100	BDL	290	2	BDL		BDL
Phthalates								
Bis(2-ethylhexyl)phthalats	3	310	260	530	3	250	200	430
Dimethyl phthalate	1			11	2	BDL		14
Nitrogen compounds								
Acrylonitrile[b]	4	BDL		BDL	4			BDL
Phenols								
2-Nirophenol	1			BDL	1			BDL
Phenol	3	180	57	440	3	30	19	37
Aromatics								
Acenapthene[c]	1			BDL	1			BDL
Acenapthylene[c]	1			BDL	1			BDL
Benzene								
Benzopyrene[c]								
Ethylbenzene								
Napthalene[c]								
Toluene								
Halogenated aliphatics								
Dichlorobromoethane	1			>3,100	1			BDL

Carbon tetrachloride	1			BDL	1			BDL
Chloroform	3	130	100(c)	270	2	BDL		BDL
1,1-Dichloroethane	1			BDL	1			BDL
1,2-Dichloroethane	1			93	0			
1,2-*Trans*-dichloroethylene	1			16	0			
Methylene chloride	3	29	15	73	3	220	150	520
1,1,2,2-Tetrachloroethane	1			BDL	1			BDL
			Solution crumb rubber manufacturing					
Metals and Inorganics								
Metals and inorganics								
Cadmium	3	31	BDL	90	2	BDL		BDL
Chromium	4	350	310	720	3	170	67	410
Copper	3	72	BDL	200	2	BDL		14
Lead	1			390				
Mercury	3	BDL	BDL	BDL	2	BDL		BDL
Nickel	1	160			0			
Zinc	2	8,100		16,000	1			190,000
Phthalates								
Bis(2-ethylhexyl)phthalate	3	260	140	530	3	190	120	430
Dimethyl phthalate	1			BDL	1			BDL
Phenols								
Phenol	3	210	180	440	3	15	BDL	37
Aromatics								
Acenapthene	1			BDL	1			BDL
Acenapthylene	1			BDL	1			BDL
Benzopyrene	1			BDL	1			BDL
Benzene	3	1,200	50	3,400	3	BDL	BDL	10
Ethylbenzene	2	BDL		10	2	BDL		10
Tolunene	4	BDL	BDL	10	5	88	BDL	420

(continues)

Table 3 *Continued*

Toxic pollutants (µg/L)	Number of samples	Average	Median	Maximum	Number of samples	Average	Median	Maximum
	Solution crumb rubber manufacturing / Tire and inner tube manufacturing							
	Treatment influent				Treatment effluent			
Halogenated aliphatics								
Carbon tetrachloride	1	35			1			1,400
Chloromethane	1	4,900			1			2,200
Chloroform	2	BDL		BDL	2	BDL		BDL
1,2-*Trans*-dichloroethylene								
Methylene chloride	2	BDL		15	2	<260		520
1,1,2,2-Tetrachloroethane	1			BDL	1			BDL
1,1,2,2-Trichloroethane	1			BDL	1	BDL		BDL
Trichloroethylene	1			BDL	1			BDL
Pesticides and metabolites								
Acrolein	1			BDL	1			BDL
	General molding, extruding, and fabricating							
Metals and Inorganics								
Lead	1			20	0			
Mercury	0				1			BDL
Zinc	0				1			970
Phthalates								
Bis(2-ethylhexyl)phthalate	1			17	2	BDL		16
Di-*n*-butyl phthalate	0				1			36
Nitrogen compounds								
N-nitrosodiphenylamine	2	35		53	0			

Phenols						
Pentachlorophenol	1		BDL	0		12,000
Phenol	0			1		
Aromatics						
Benzene	0			1		BDL
Halogenated aliphatics						
Chloroform	1		25	2	BDL	10
1,1-Dichloroethane	0			1		110
1,2-Dichloroethane	0			1		BDL
1,2-*Trans*-dichloroethylene	0			1		290
1,1,2,2-Tetrachloroethane	0			1		BDL
1,1,1-Trichloroethane	0			1		7,100
1,1,2-Trichloroethane	0			1		BDL
Trichloroethylane	0			1		1,600
Wet digestion reclaimed rubber						
Metals and Inorganics						
Cadmium[d]	1		10	0		
Lead	1		50	0		
Zinc[d,e]	2	250	350	0		
Nitrogen compounds						
Phenol[e]	1		BDL	0		
Pesticides and metabolites						
Isophorone[e]	1		BDL	0		

Analytic methods: V.7.3.29, Data sets 1,2.
BDL, below detection limit.
[a]40 μg/L of trichloroethylene also measured in city water.
[b]Detection limit of acrylonitrile by direct aqueous injection was 2300 μg/L.
[c]This value believed to be a glassware contaminant.
[d]These pollutants appear to be attributed to tire operation.
[e]Wastewater is from both tire and reclaiming processes.
Source: USEPA.

Table 4 Industry Profile of Toxic and Classical Pollutant Loadings, Verification, and Screening Data (Toxic Pollutants Kg/kkg)

	Tire and inner tube manufacturing							
	Treatment influent				Treatment effluent			
Toxic Pollutants (Kg/Mg)	Number of samples	Average	Median	Maximum	Number of samples	Average	Median	Maximum
Toxic metals								
Chromium	0				1			0.000005
Copper	1			0.001	0			
Lead	1			0.001	0			
Zinc	3	0.003	0.004	0.006	1			0.0007
Toxic organics								
Phenol	1			BDL	0			
Methylene chloride	0				1			BDL
2,4,6-Trichlorophenol	0				1			BDL
Isophorone	1			BDL	1			BDL
Classical pollutants (kg/day)								
TSS	4	590	200	2,000	47	270	32	2,400
Oil and grease	8	17	7.2	120	35	7.3	2.1	42
pH, pH units	10	7.6	2.4	9.4	44	7.5	7.5	10.3
Toxic pollutants (kg/Mg)	Emulsion crumb rubber manufacturing							
Toxic metals								
Cadmium	2	0.0004		0.0006	1			0.00001
Chromium	5	2.6	0.003	13	4	3.0	0.0005	12
Copper	1			0.003	0			
Lead	1			0.006	0			
Mercury	2	0.00003		0.00003	2	0.00002		0.00002
Nickel	2	0.006		0.008	1			0.005
Selenium	1			<1.0	1			1.3
Zinc	3	0.002	BDL	0.005	2	BDL		BDL

Toxic organics								
Bis(2-ethylhexyl)phthalate	3	<2.4	0.0017	<7.3	3	<2.3	0.0016	<7.0
Dimethyl phthalate	2	0.0002		0.0002	2	0.0001		0.0002
Acrylonitrile	4	BDL	BDL	BDL	5	<240	BDL	<1,200
N-nitrosodiphenylamine	1			BDL	1			BDL
2-Nitrophenol	1			<0.5	1			0.26
Phenol	4	0.75	0.003	3.0	4	<0.25	0.0004	<0.98
Benzene	3	0.01	0.0007	0.01	2	0.0003		0.0005
Ethylbenzene	5	<0.01	BDL	<0.05	4	<0.001	BDL	<0.005
Nitrobenzene	1			<0.0004	1			<0.0004
Toluene	6	<0.009	0.002	<0.05	5	<0.001	0.000001	
Carbon tetrachloride	1			0.00001	1			<0.000002
Chloroform	3	0.14	0.0004	0.40	2	0.04		0.09
1,1-Dichloroethane	1			0.00002	1			<0.00002
1,1-*Trans*-dichloroethylene	1			ND	0			
1,2-Dichloroethane	1			ND	0			
Methylene chloride	3	<1.3	0.0002	<3.8	3	<1.9	0.00007	<5.7
1,1,2,2-Tetrachloroethane	1			0.00002	1			0.000001
Acenapthene	1			BDL	1			BDL
Acenapthylene	1			BDL	1			BDL
Napthalene	1			BDL	1			BDL
Benzo-pyrene	1			BDL	1			BDL
Dichlorobromomethane	1			<1.6	1			7.0
Acrolein	1			BDL	1			BDL
			Solution crumb rubber manufacturing					
Metals and Inorganics								
Toxic metals								
Cadmium	3	0.01	0.0007	0.04	2	0.5		0.09
Chromium	4	<4.3	0.0006	<17	3	<0.42	0.0004	<1.3
Copper	3	<0.09	0.00007	<0.28	2	<0.17		<0.34
Lead	1			0.006	0			
Mercury	1			0.00003	1			0.00001
Nickel	1			0.003	0			
Zinc	2	0.07		0.14	1			2.0

(*continues*)

Table 4 *Continued*

	Solution crumb rubber manufacturing							
	Treatment influent				Treatment effluent			
Toxic Pollutants (Kg/Mg)	Number of samples	Average	Median	Maximum	Number of samples	Average	Median	Maximum
Toxic organics								
Bis(2-ethylhexyl)phthalate	3	<1.8	0.007	<5.4	3	<2.7	0.006	<8.1
Dimethyl phthalate	1			0.0001	1			0.00008
Phenol	3	2.4		7.1	3	<0.25	0.0005	<0.76
Benzene	4	43	0.0004	130	4	<0.002	<0.0001	<0.007
Ethylbenzene	2	0.00005		0.00005	2	<0.00001		<0.00002
Toluene	4	0.001	0.00006	0.004	5	0.003	0.0001	0.007
Carbon tetrachloride	1			0.0003	1			0.0001
Chloroform	2	0.06		0.12	2	0.03		0.06
Methylene chloride	2	0.0001		0.0002	2	0.004		0.007
1,1,2,2-Tetrachloroethane	1			0.004	1			<0.007
1,1,2-Trichloroethane	1			<0.0000008	1			<0.000001
Trichloroethylene	1			<0.0000008	1			<0.000001
Acenapthene								
Acenapthylene								
Benzo-pyrene								
Chloromethane	1			0.04	2	0.02		0.02
Acrolein	1			BDL	1			BDL
Classical pollutants (kg/day)								
BOD_5	6	2,900	900	15,000	9	200	86	1,100
COD	4	2,300	2,500	4,400	8	310	320	1,200
TSS	4	540	920	1,200	8	270	85	1,100
Oil and grease	3	96	110	130	7	<25	11	<92
pH (pH units)	1			9.5	4	6.8	7.5	8.2

				Latex rubber manufacturing				
Metals and Inorganics								
Toxic metals								
Chromium	2	BDL		BDL	2	BDL		BDL
Zinc	2	BDL		BDL	2	BDL		BDL
Toxic organics								
Bis(2-ethylhexyl)phthalate	1			0.0004	1			<0.00004
Di-*n*-butyl phthalate	1			BDL	1			BDL
Acrylonitrile	2	BDL		BDL	2	BDL		BDL
Pentachlorophenol	1			0.0001	1			<0.00004
Phenol	1		0.0001		1			<0.00004
Benzene	1			BDL	1			BDL
Ethylbenzene	3	0.002	BDL	0.006	3	<0.000007	BDL	<0.00002
Toluene	2	BDL		BDL	2	BDL		BDL
Methylene chloride	1			BDL	1			BDL
Butylbenzyl phthalate	1			BDL	1			BDL
Napthalane	1			BDL	1			BDL
Classical pollutants (kg/day)								
BOD_5	0				5	86	15	340
COD	0				3	120	150	160
TSS	1			640	5	130	12	590
Oil and grease	0				3	2.8	3.2	4.0
pH (pH units)	0				3	10	10	10.0
Toxic pollutants (kg/Mg)				General molding, extruding, and fabricating rubber manufacturing				
Metals and Inorganics								
Toxic metals								
Lead	1			0.003	1			0.0001
Zinc	0				1			0.14
Toxic organics								
Bis(2-ethylhexyl)phthalate	1			0.002	0			0.005

(*continues*)

Table 4 *Continued*

	General molding, extruding, and fabricating rubber							
	Treatment influent				Treatment effluent			
Toxic Pollutants (Kg/Mg)	Number of samples	Average	Median	Maximum	Number of samples	Average	Median	Maximum
Di-*n*-butyl phthalate	0				1			
N-nitrosodiphenylamine	1			0.0007	0			
Pentachlorophenol	1			0.00003	0			
Phenol	0				1			1.7
Benzene	0				1			0.001
Chloroform	0				1			0.0003
1,1-Dichloroethane	0				1			0.02
1,1-*Trans*-dichloroethylene	0				1			0.04
1,2-Dichloroethane	0				1			0.0006
Tetrachloroethylene	0				1			0.0006
1,1,1-Trichloroethane	0				1			1.0
1,1,2-Trichloroethane	0				1			0.0002
Trichloroethylene	0				1			0.23

Analytic methods: V.7.3.29, Data sets 1,2.
BDL, below detection limit.
ND, not detected.
Source: USEPA.

requiring noncontact cooling. The general practice of the industry is to recirculate the majority of this water with a minimal blowdown to maintain acceptable concentrations of dissolved solids. Another water use area is contact water used in cooling tire components and in air pollution control devices. This water is also recirculated. Steam condensate and hot and cold water are used in the molding and curing areas. The majority of the water is recycled back to the boiler or hot water tank for use in the next recycle. Soapstone areas and plant and equipment cleanup are the final water use areas. Most facilities try to recycle soapstone solution because of its high solids content. Plant and equipment cleanup water is generally sent to the treatment system. Table 5 presents a summary of the potential wastewater sources for this subcategory.

Grease, oils, and suspended solids are the major pollutants within this industry. Organic pollutants, pH, and temperature may also require treatment. The organics present are due generally to poor housekeeping procedures.

30.1.4 Emulsion Crumb Rubber Production

In-process controls for the reduction of wastewater flows and loads for emulsion crumb rubber plants include recycling of finishing line wastewaters and steam stripping of heavy monomer decanter wastewater. Recycling of finishing line wastewater occurs at nearly all emulsion crumb plants with the percent recycle depending primarily upon the desired final properties of the crumb. Approximately 75% recycle is an achievable rate, with recycle for white masterbatch crumb below this level and that for black masterbatch crumb exceeding it.

Organic toxic pollutants found at emulsion crumb rubber plants come from the raw materials, impurities in the raw materials, and additives to noncontact cooling water. BOD, COD, and TSS levels may also reach high loadings.

Table 5 Summary of Potential Process-Associated Wastewater Sources from the Tire and Inner Tube Industry

Plant area	Source	Nature and origin of wastewater contaminants
Oil storage	Runoff	Oil
Compounding	Washdown, spills, leaks, discharges from wet air pollution equipment	Solids from soapstone dip tanks; oil from seals in roller mills; oil from solids from Banbury seals; solids from air pollution equipment discharge
Bead, tread, tube formation	Washdown, spills, leaks	Oil and solvent-based cements from the cementing operation; oil from seals in roller mills
Cord and belt formation	Washdown, spills, leaks	Organics and solids from dipping operation; oil from seals, in roller mills, calenders, etc.
Green tire painting	Washdown, spills, air pollution equipment	Organics and solids from spray-painting operation; soluble organics and solids from air pollution equipment discharge
Molding and curing	Washdown, leaks	Oil from hydraulic system; oil from presses
Tire finishing	Washdown, spills, air pollution equipment	Solids and soluble organics from painting operations; solids from air pollution equipment discharge

Source: USEPA.

Table 6 lists potential wastewater sources and general wastewater contaminants for the emulsion crumb rubber industry.

30.1.5 Solution Crumb Rubber Production

Solution crumb rubber production plants have lower raw wastewater loads than emulsion crumb plants because of the thorough steam stripping of product cement to remove solvent and permit effective coagulation. Recycling in this industry is comparable to that in the emulsion crumb industry, with about 75% of the wastewater being recirculated.

Toxic pollutants found in the wastewater streams are normally related to solvents and solvent impurities, product additives, and cooling water treatment chemicals. Table 7 presents a listing of the potential wastewater sources and the associated contaminants for this industry.

30.1.6 Latex Rubber Production

No in-process contact water is currently used by the latex rubber industry. No raw material recycling is practised because of poor control of monomer feeds and the buildup of impurities in the water.

Organic toxic pollutants and chromium are present in the raw wastewater and normally consist of raw materials, impurities, and metals used as cooling water corrosion inhibitors.

Table 8 presents potential wastewater sources and general contaminants for this industry.

30.1.7 General Molding, Extruding, and Fabricating Rubber Plants

Toxic pollutants resulting from production processes within this industry are generally the result of leaks, spills, and poor housekeeping procedures. Pollutants include organics associated with the raw materials and lead from the rubber curing process.

Table 6 Summary of Wastewater Sources From Emulsion Crumb Rubber Production Facilities

Processing unit	Source	Nature of wastewater contaminants
Caustic soda scrubber	Spent caustic solution	High pH, alkalinity, and color. Extremely low average flow rate
Monomer recovery	Decant water layer	Dissolved and separable organics. Source of high BOD and COD discharges
Coagulation	Coagulation liquor overflow	Acidity, dissolved organics, suspended and high dissolved solids, and color. High wastewater flow rates relative to other sources
Crumb dewatering	Crumb rinse water overflow	Dissolved organics, and suspended and dissolved solids. Source of highest wastewater volume from emulsion crumb rubber production
Monomer strippers	Stripper cleanout rinse water	Dissolved organics, and suspended and dissolved solids. High quantities of uncoagulated latex
Tanks and reactors	Cleanout rinse water	Dissolved organics, and suspended and dissolved solids. High quantities of uncoagulated latex
All plant areas	Area washdowns	Dissolved and separable organics, and suspended and dissolved solids

Source: USEPA.

Table 7 Summary of Wastewater Sources From Solution Crumb Rubber Production

Processing unit	Source	Nature of wastewater contaminants
Caustic soda scrubber	Spent caustic solution	High pH, alkalinity, and color. Extremely low average flow rate
Monomer and solvent drying columns	Water removed from monomers and solvent	Dissolved and separable organics. Very low flow
Solvent purification	Fractionator bottoms	Dissolved and separable organics.
Monomer recovery	Decant water layer	Dissolved and separable organics.
Crumb dewatering	Crumb rinse water overflow	Dissolved organics, and suspended and dissolved solids. Source of highest volume wastewater flow
All plant areas	Area washdowns	Dissolved and separable organics, and suspended and dissolved solids

Source: USEPA.

30.1.8 Rubber Reclamation

Wastewater effluents from this subcategory contain high levels of toxic organic and inorganic pollutants. These pollutants generally result from impurities in the tires and tubes used in the reclamation process. The wastewater from the pan process is of low volume [0.46 m^3/kkg (56 gal/1000 lb)], but is highly contaminated, requiring treatment before discharge. The mechanical reclaiming process uses water only to quench the reclaimed rubber, but it uses a much higher quantity (1.1 m^3/kkg). Steam generated from the quenching process is captured in a scrubber and sent to the treatment system. Wet digestion uses 5.1 m^3 of water per kkg (610 gal/1000 lb) of product in processing, of which 3.4 m^3/kkg (407 gal/1000 lb) of product is used in air pollution control.

30.1.9 Latex-Dipped, Latex-Extruded, and Latex-Molded Goods

Wastewater sources in this subcategory are the leaching process, makeup water, cooling water, and stripping water. Toxic pollutants are present at insignificant levels in the wastewater discharges.

Table 8 Summary of Wastewater Sources From Latex Rubber Production

Processing unit	Source	Nature of Wastewater contaminants
Caustic soda scrubber	Spent caustic solution	High pH, alkalinity, and color. Extremely low average flow rate
Excess monomer stripping	Dacent water layer	Dissolved and separable organics
Latex evaporators	Water removed during latex concentration	Dissolved organics, suspended and dissolved solids. Relatively high wastewater flow rates
Tanks, reactors, and strippers	Cleanout rinse water	Dissolved organics, suspended and dissolved solids. High quantities of uncoagulated latex
Tank cars and tank trucks	Cleanout rinse water	Dissolved organics, suspended and dissolved solids. High quantities of uncoagulated latex
All plant areas	Area washdowns	Dissolved and separable organics, and suspended and dissolved solids

Source: USEPA.

30.1.10 Latex Foam

No information is available on the wastewater characteristics of this subcategory.

30.2 PLANT-SPECIFIC DESCRIPTION

Only two subcategories of the rubber industry have not been recommended as Paragraph 8 exclusions of the NRDC Consent Decree: Wet Digestion Reclaimed Rubber, and Pan, Mechanical, and Dry Digestion Reclaimed Rubber. Of these two, plant specific data are available only for the latter. Of the nine remaining subcategories, plant-specific information is available only for Emulsion Crumb Rubber and Solution Crumb Rubber, and is presented below. Two plants in each subcategory are described. They were chosen as representative of their subcategories based on available data.

Plant 000012 produces 3.9×10^4 kkg/year (8.7×10^7 lb/year) of emulsion crumb rubber, primarily neoprene. The contact wastewater flow rate is approximately 8.45 m^3/day (2.25×10^3 gpd) and includes all air pollution control equipment, sanitary waste, maintenance and equipment cleanup, and direct contact wastewater. The treatment process consists of activated sludge, secondary clarification, sludge thickening, and aerobic sludge digestion. Noncontact wastewater, with a flow rate of approximately 1.31×10^5 m^3/day (3.46×10^7 gpd), is used on a once-through basis and is returned directly to the river source. Contact wastewater is also returned to the surface stream after treatment.

Plant 000033 produces three types of emulsion crumb rubber in varying quantities. Styrene butadiene rubber (SBR) forms the bulk of production, at nearly 3.7×10^5 kkg/year (8.2×10^8 lb/year), with nitrile butadiene rubber (NBR) and polybutadiene rubber (PBR) making up the remainder of production [4.5×10^3 kkg/year (1.0×10^7 lb/year) and 4.5×10^3 kkg/year, respectively]. Wastewater consists of direct contact process water, noncontact blowdown, and noncontact ancillary water. The total flow of contact water is approximately 1.27×10^4 m^3/day (3.355×10^6 gpd), and the total flow of noncontact water is 340.4 m^3/day (9×10^4 gpd). Treatment of the wastewater consists of coagulation, sedimentation, and biological treatment with extended aeration. Treated wastewater is discharged to a surface stream.

Tables 9 and 10 present plant-specific toxic pollutant data for the selected plants. Table 11 gives plant-specific classical pollutant data, including BPT regulations set for each specific plant.

30.2.1 Solution Crumb Rubber Production

Plant 000005 produces approximately 3.2×10^4 kkg/year (7.0×10^7 lb/year) of isobutene–isopropene rubber. Wastewater generally consists of direct processes and MEC water. Contact wastewater flow rate is approximately 1040 m^3/day (2.75×10^5 gpd), and noncontact water flows at about 327 m^3/day (8.64×10^4 gpd). Treatment consists of coagulation, flocculation, and dissolved air flotation, and the treated effluent becomes part of the noncontact cooling stream of the onsite refinery.

Plant 000027 produces polyisoprene crumb rubber [4.5×10^4 kkg/year (1.0×10^8 lb/year)] polybutadiene crumb rubber, and ethylene-propylene-diene-terpolymer rubber [EPDM; 4.5×10^4 kkg/year (1.0×10^8 lb/year)]. Wastewater consists of contact process water, MEC, cooling tower blowdown, boiler blowdown, and air pollution control. Wastewater is produced at about 12,100 m^3/day (3.2×10^6 gpd). Treatment consists of API separators, sedimentation, stabilization, and lagooning, followed by discharge to a surface stream.

Table 9 Plant-Specific Verification Data for Emulsion Production Plant 000012

	Local in process line				
Pollutant	Stripper decant	Spray wash water	Treatment influent	Treatment effluent	Raw intake water
Toxic pollutant (μg/L)					
Cadmium	<1	<1	<2	<1	<1.0
Mercury	1.5	1.7	2.5	1.6	1.5
Nickel	60	690	610	400	<10
Bis(2-ethylhexyl)phthalate	290	490	260	<230	260
Dimethyl phthalate	<14	<14	<14	<14	<16
N-nitrosodiphenylamine	1.5	<1.0	5.2	<2.0	<1.0
Phenol	19	29	41	19	<2
Nitrobenzene	<30	<30	<30	<30	<30
Toluene	70	<0.5	250	<0.5	<0.5
Carbon tetrachloride	41	0.1	4.7	<0.2	0.3
Chloroform	110	14	27	4.1	8.5
1,1-Dichloroethylene	51	<1.7	<1.7	<1.7	<1.7
Methylene chloride	4.8	1.0	<0.1	1.0	<0.1
Tetrachloroethylene	<0.1	<0.1	1.4	<0.1	<0.1
1,1,1-Trichloroethane	<1.6	0.3	<1.1	0.3	0.2

Analytic methods: V.7.3.29, Data set 2.
Flow rate (cu. m/day): contact = 8.45; noncontact = 131,000.
Source: USEPA.

Tables 12 and 13 show plant specific toxic pollutant data for the above plants. Classical pollutant data and BPT regulations are presented in Table 14.

30.2.2 Dry Digestion Reclaimed Rubber

A data summary for plant 000134 is given in Table 15. Production, wastewater flow, and treatment data are currently not available for a plant within this subcategory.

30.3 POLLUTANT REMOVABILITY

In this industry, numerous organic compounds, BOD, and COD are typically found in plant wastewater effluent. Industrywide flow and production data show that these pollutants can be reduced by biological treatment. In emulsion crumb and latex plants, uncoagulated latex contributes to high suspended solids. Suspended solids are produced by rubber crumb fines and include both organic and inorganic materials. Removal of such solids is possible using a combination of coagulation/flocculation and dissolved air flotation.

Solvents, extender oils, and insoluble monomers are used throughout the rubber industry. In addition, miscellaneous oils are used to lubricate machinery. Laboratory analysis indicates the presence of oil and grease in the raw wastewater of these plants. Oil and grease entering the wastewater streams are removed by chemical coagulation, dissolved air flotation, and, to some extent, biological oxidation.

Table 10 Plant-Specific Verification Data for Emulsion Crumb Rubber Production Plant 000033

	Location in process line							
Pollutant (μg/L)	SBR stripper	Finishing comp.	NBR finishing	Treatment influent	Treatment effluent	NBR decant	Raw intake, well	Raw intake, river
Cadmium	<1	80	<1	40	40	<2	<1	<1
Chromium	6	400	20	250	220	10	6	5
Copper	71	80	<1	1,400	410	<1	1	<1
Mercury	0.8	63	2.2	3.2	3.1		0.7	0.6
Selenium	<4	<30	<6	<20	<25	<6	<4	<4
Bis(2-ethylhexyl)phthalate	<350	<210	<170	<130	<130	<120	<110	<110
Acrylonitrile	<26	<23	94	32	<23	48,000	<23	<23
2-Nitrophenol	<4	<4	<17	<10	<5	<5	<4	<4
Phenol	41	67	32	61	<20	<16	10	<3
Ethylbenzene	<38	<0.1	<0.1	<0.1	<0.1	<23	<0.1	<0.1
Toluene	<0.1	<0.1	<0.1	<0.1	<0.1	<25	<0.1	<0.1
Chloroform	1.5	2.5	5.2	8.3	1.8	37	1.2	41
Dichlorobromomethane	<0.3	<0.1	<0.5	<0.3	<0.1	5.2	<0.1	6.2
Methylene chloride	<110	<0.1	<80	<67	<110	180	<2	<2

Analytic methods: V.7.3.29, Data set 2.
Flow rate (cu.m/day): SBR–contact = 10,200, noncontact = 190; NBR–contact = 1,250, noncontact = 75.7; PBR–contact = 1,250, noncontact = 75.7; Total–contact = 12,700, noncontact = 340.
Source: USEPA.

Table 11 Plant-Specific Classical Pollutant Data for Selected Emulsion Crumb Rubber Production Plants, Verification Data

	Waste load, plant 000012					
Parameter	Influent		Effluent		BPT regulation	
BOD_5	1,200	(2,600)	5.0	(11)	44	(97)
COD	2,100	(4,600)	130	(280)	880	(1,900)
TSS	8	(18)	35	(77)	71	(160)
Oil and grease	<8	(<18)	8	(18)	18	(39)
pH (pH units)					6 to 9	
Phenol	0.014	(0.03)	30	(67)		
	Waste load, plant 000033					
BOD_5	2,700	(5,900)	140	(320)	460	(1,000)
COD	8,600	(19,000)	2,700	(5,900)	9,200	(20,000)
TSS	2,100	(4,700)	240	(540)	750	(1,700)
Oil and grease	240	(530)	140	(310)	180	(410)
pH (pH units)					6 to 9	
Phenol	4.8	(10.5)	0.35	(0.75)		

Analytic methods: V.7.3.29, Data set 2.
Blanks indicate data not available.
Source: USEPA.

Wastewater sampling indicates that toxic pollutants found in the raw wastewater can be removed. Biological oxidation (activated sludge) adequately treats all of the organic toxic pollutants identified in rubber industry wastewater streams. Significant removal of metals was also observed across biological treatment. The metals are probably absorbed by the sludge mass and removed with the settled sludge. Treatment technologies currently in use are described in the following subcategory descriptions.

30.3.1 Emulsion Crumb Rubber Plants

There are a total of 17 plants in the United States producing emulsion-polymerized crumb rubber. Five of these plants discharge to POTWs; 10 discharge to surface streams; one plant discharges to an evaporation pond; and one plant employs land application with hauling of settled solids. Of the five plants discharging to POTWs, four pretreat using coagulation and primary treatment and one employs equalization with pH adjustment. All 10 of the plants discharging to surface streams employ biological waste treatment ranging from conventional activated sludge to nonaerated wastewater stabilization lagoons.

Organic pollutants are generally found to be reduced to insignificant levels ($<10\ \mu g/L$) by biological treatment. Most metals are also found to be reduced across biological treatment; they are generally at very low levels in the treated effluent. However, significant metal concentrations may be found in some treated effluent.

At emulsion crumb rubber facilities, a well-operated biological treatment facility permits compliance with BPT limitations and reduces organic toxic pollutant levels. Toxic metals that may not be reduced include chromium, cadmium, copper, selenium, and mercury. Tables 16 and 17 show pollutant removal efficiencies at two emulsion crumb plants.

Table 12 Plant-Specific Verification Data for Solution Crumb Rubber Production Plant 000005

	Location in process line		
Pollutant (μg/L)	Screen tank 1 and 2 comp.	Expeller 1 and 2 comp.	DAF influent
Cadmium	<1[a]	<1[a]	<1[a]
Chromium	3	6	75
Copper	6	7	9
Zinc	14,000	12,000	14,000
Bis(2-ethylhexyl)phthalate	<60	100	200
Phenol	9	<5	7
Benzene	<22	<13	<22
Ethylbenzene	<38	<2	<12
Toluene	<26	<3	<26
Carbon tetrachloride	0.06	0.06	35
Chloroform	0.90	0.88	2.2
Methyl chloride	14,000	2,600	4,900
Methylene chloride	<1	<1	<1
1,1,2-Trichloroethane	<0.1	<0.1	<0.1
Trichloroethylene	<0.1	<0.1	<0.1

	Location in process line			
	DAF effluent	Well water (a)	Boiler feedwater (a)	Boiler blowdown
Cadmium	<1[a]	<1	<1	
Chromium	410	3	3	5,700
Copper	14	3	7	
Zinc	13,000	30	13,000	3,700
Bis(2-ethylhexyl)phthalate	24	98	50	
Phenol	5	<2	6	
Benzene	<110	<43	<43	
Ethylbenzene	<38	<110	<110	
Toluene	<26	<72	<72	
Carbon tetrachloride	14	0.10	0.06	
Chloroform	1.3	1.0	0.98	
Methyl chloride	2,000	190	31	
Methylene chloride	<1	<1	35	
1,1,2-Trichloroethane	<0.1	<0.1	<0.1	
Tetrachloroethylene	<0.1	<0.1	<0.1	

Analytic methods: V.7.3.29, Data set 2.
[a]Screening data.
Flow rate (cu.m/day): contact = 1000; noncontact = 327.
Source: USEPA.

30.3.2 Solution Crumb Rubber Plants

There are 13 solution crumb rubber plants in the United States. Twelve of these plants discharge treated wastewater to surface streams; the other plant discharges its treated wastewater into a neighboring oil refinery's noncontact cooling water system.

Table 13 Plant-Specific Verification Data for Solution Crumb Rubber Production Plant 000027

Pollutant (μg/L)	SN/CB process	EPDM process	Treatment influent	Treatment effluent	Well water	Boiler blowdown
Cadmium	<1	<1	<1	<1	<1	<1
Chromium	450	820	440	19	2	2,600
Copper	4	<2	<7	<5	2	<1
Mercury	1.8	2.3	1.1	2.0	4.0	1.4
Bis-(2-ethylhexyl) phthalate	77	120	<140	<120	170	<46
Phenol	13	670	180	<12	<2	7
Benzene	<0.1	39,000	3,300	<0.1	<0.1	<0.1
Ethylbenzene	<0.1	<0.1	<0.1	<0.1	<0.1	<0.1
Toluene	<0.1	<43	<0.1	<0.1	<0.1	<0.1
Chloroform	1.0	22	3.2	0.9	1.1	1.0
1,1,2,2-Tetrachloroethene	<0.1	<0.1	<0.1	<0.1	<0.1	<0.1

Analytic methods: V.7.3.29, Data set 2.
Total flow rate: 12,100 cu. m/day.
Source: USEPA.

Ten of the plants discharging to surface streams employ some form of biological treatment for waste load reduction. Two of the plants discharging to surface streams use in-process controls, oil removal, and primary treatment prior to discharge. In-process control employed at one plant consists of steam stripping of wastewaters, while in-process control at the second plant was not disclosed. The plant discharging to the oil refinery noncontact cooling water system used coagulation, flocculation, and dissolved air flotation prior to discharge.

The results of the verification program showed that all organic toxic pollutants were reduced across biological treatment. Chloromethane, used as a solvent at plant 000005, was present at significant levels in treated effluent.

Tables 18 and 19 show pollutant removal efficiencies at two selected solution crumb rubber plants.

30.3.3 Latex Rubber Plants

There are 17 latex rubber production facilities in the United States. Of these, nine plants discharge to POTWs; seven discharge to surface streams; and one employs land application with contractor disposal of solids. All seven plants discharging to surface streams employ biological treatment before discharge. Pretreatment for the POTW dischargers consists of coagulation, flocculation, and primary treatment for seven of the nine dischargers, equalization for one discharger, and biological treatment for the other plant.

30.3.4 Tire and Inner Tube Manufacturing

There are a total of 73 tire and inner tube manufacturing facilities in the United States, of which 39 were placed in operation prior to 1959. Twenty-three of the pre-1959 plants do not treat their wastewaters, and six of these plants discharge to POTWs. A total of 17 plants placed in operation since 1959 provide no treatment of their wastewaters, and 10 of these plants discharge into POTWs.

Table 14 Plant-Specific Classical Pollutant Data for Solution Crumb Rubber Production Plant

	Waste load											
	Plant 00005						Plant 000027					
Parameter	Influent		Treated effluent		BPT regulation		Influent		Treated effluent		BPT regulation	
BOD_5	95	(210)	68	(150)	51	(110)	1,200	(2,700)	<90	(<200)	160	(360)
COD	250	(550)	140	(300)	500	(1,100)	2,700	(5,900)	450	(1,000)	1,600	(3,600)
TSS	19	(41)	11	(25)	83	(180)	1,300	(2,800)	11	(25)	270	(590)
Oil and grease	100	(230)	14	(30)	20	(45)	45	(100)	<90	(<200)	66	(150)
pH (pH units)					6 to 9						6 to 9	
Phenol	<0.01	(0.01)	<0.01	(0.01)			1.0	(2.3)	0.16	(0.35)		

Analytic methods: V.7.3.29, Data set 2.
Parameter data: kg/day (lb/day).
Source: USEPA.

Table 15 Plant-Specific Verification Data for Pan, Dry Rubber Digestion, and Mechanical Reclaiming Plant 000134

	Location in process line			
Pollutant (μg/L)	Treatment influent, automatic sampler	Treatment effluent, grab composite		Treatment effluent, automatic samples
Cadmium	<1	<1		3
Chromium	6	4		21
Copper	31	<1		12
Lead	70	290		670
Mercury		1.9		2.3
Zinc	100	2,700		2,500
Bis(2-ethylhexyl)phthalate	16,000	<80		4,300
Di-*n*-butyl phthalate				
2,4-Dimethylphenol	58,000	56,000		15,000
Phenol	26,000	21,000		4,900
Benzene	<60	<10		<0.1
Chlorobenzene				
Ethylbenzene	8,600	<0.1		<0.1
Toluene	2,700	<0.1		<0.1
Acenaphthylene	<33	<8		<8
Anthracene				
Phenanthrene	1,400[c]	<49		<300
Fluorene	2,000[c]	<40		<12
Naphthalene	100,000[c]	<12		<44
Pyrene	6,700[c]	<10		<14
Chloroform	1.9	1.3		1.4
Methylene chloride				
	Cooling tower blowdown, grab composite	Steam condensate, grab composite	Boiler blowdown	Intake water
Cadmium	<1	35	1	<1
Chromium	<2	33	2	<1

(continues)

Table 15 *Continued*

Pollutant (μg/L)	Location in process line			
	Treatment influent, automatic sampler	Treatment effluent, grab composite		Treatment effluent, automatic samples
Copper	3	20	6	<1
Lead	29	330	22	10
Mercury	0.5	1.0	0.9	0.8
Zinc	100		220	30
Bis(2-ethylhexyl)phthalate	120	2,800	940	1,300
Di-*n*-butyl phthalate		1,900		
2,4-Dimethylphenol	<510	730	11	<6
Phenol	<130	950	12	<4
Benzene	<0.4	27[b]	<0.1	<0.1
Chlorobenzene		25,000[b]		
Ethylbenzene	0.1	<0.1[b]	<0.1	<0.1
Toluene	<0.1	<0.1[b]	<0.1	<0.1
Acenaphthylene	<8	<16	<8	<8
Anthracene		<190		
Phenanthrene	<110	<140	340	<4
Fluorene	<12	<12	<12	<12
naphthalene	<17	<1,400[c]	<12	<12
Pyrene	<32	<29	<8	15
Chloroform	4.9	3.3[b]	1.0[a]	36[a]
Methylene chloride		1,300[b]		

Analytic methods: V.7.3.29, Data set 2.
Blanks indicate data not available.
[a]Based on first 24-hour composite sample.
[b]Based on second and third 24-hour composite samples.
[c]Interference may have caused this value to be too high.
Source: USEPA.

Table 16 Toxic Pollutant Removal Efficiency at Emulsion Crumb Rubber Plant 000012, Verification Data (Treatment Technology: Activated Sludge, Discharge Point: Surface Stream)

	Concentration (μg/L)		
Pollutant	Influent	Effluent	Percent removal
Cadmium	1	<1	NM
Mercury[a]	2.5	1.6	36
Nickel	610	400	34
Bis(2-ethylhexyl)phthalate[b]	260	220	15
Dimethyl phthalate	<14	<14	NM
N-nitrosodiphenylamine	5.2	1.6	69
Phenol	41	19	54
Nitrobenzene	<30	<30	NM
Toluene	250	<0.1	>99
Carbon tetrachloride	4.7	0.1	98
Chloroform	27	4.1	85
1,1-Dichloroethylene	<1.7	<1.7	NM
Methylene chloride	<0.1	0.9	NM
Tetrachloroethene	1.4	<0.1	>93
1,1,1-Trichloroethane	1.0	3.3	NM

Analytic methods: V.7.3.29, Data set 2.
NM, not meaningful.
[a]Intake measured at 1.5 μg/L, making plant's contribution minimal.
[b]Analytical methodology for phthalates is questionable. Therefore, significance of values reported is unknown.
Source: USEPA.

The toxic pollutants present in raw wastewaters from tire and inner tube manufacturing operations are volatile organic pollutants that are used as degreasing agents in tire production. These toxic pollutants (methylene chloride, toluene, trichloroethylene) were found to be reduced to insignificant levels across sedimentation ponds.

30.3.5 Rubber Reclamation Plants

There are nine rubber reclaiming plants in the United States. Two of these use wet digestion, and all nine use pan, mechanical, and dry digestion. Eight of the plants discharge to POTWs. The other plant employs cartridge filtration and activated carbon for oil removal, followed by activated sludge. Table 20 shows the pollutant removal efficiency at a dry digestion reclaiming plant.

30.3.6 Rubber Fabricating Operations

Rubber fabricating operations include latex-dipped, extruded, and molded goods (LDEM), and general molded, extruded, and fabricated rubber (GMEF). There are an estimated 1385 rubber fabricating plants in the United States.

No treatment method descriptions are currently available for this subcategory. Wastewater treatment technology consistent with equalization and sedimentation may permit compliance with BPT regulations.

Table 17 Toxic Pollutant Removal Efficiency at Emulsion Crumb Rubber Plant 000033, Verification Data

	Concentration (μg/L)		
Pollutant	Influent	Effluent	Percent removal
Cadmium[a]	40	40	0
Chromium[a]	250	220	12
Copper[a]	1,400	410	71
Mercury[a]	3.2	4.9	NM
Selenium[a]	<20	20	NM
Bis(2-ethylhexyl) phthalate[b] (range)	100 (65–140)	94 (59–130)	6
Acrylonitrile	32,000	<23,000	>28
2-Nitrophenol	9	3	67
Phenol	60	19	68
Ethylbenzene	<0.1	<0.1	NM
Toluene	<0.1	<0.1	NM
Chloroform	8.2	1.8	78
Dichlorobromomethane	0.3	0.1	67
Methylene chloride[c]	66	110	NM

Analytic methods: V.7.3.29, Data set 2.
NM, not meaningful.
[a]Found at potentially significant levels in treatment effluent although generally higher than during screening.
[b]Analytical methodology for phthalates is questionable; therefore, significance of values reported is unknown.
[c]Suspected contaminant from glassware cleaning procedures or analytical methods.
Treatment technology: Primary flocculation/separation, aerated lagoons.
Discharge point: Surface stream.
Source: USEPA.

30.4 TREATMENT METHODS

The treatment methods for ruber wastewaters consist of various biological processes, and physico-chemical processes including coagulation, ozonation, activated carbon adsorption, aeration, sulfonation, chlorination, and aeration, and biological nutrient removal processes. The purpose of the treatment is to meet USEPA effluent limitations [4].

30.4.1 Coagulation

Municipal wastewaters containing synthetic latexes have been treated with coagulants with 84% BOD removal efficiency using $Al_2(SO_4)_3$, $Fe_2(SO_4)_3$, and $FeCl_3$ [5].

30.4.2 Odor Control

For butadiene wastewaters with initial odor concentration of 1200, it was reduced to 100 and 250 after bubble aeration and spray aeration, respectively. For styrene wastewaters with initial odor concentration of 1000, it was reduced to 50 with bubble aeration and there was no odor reduction using spray aeration [6]. Sulfonation treatment for butadiene wastewater with initial odor

Table 18 Toxic Pollutant Removal Efficiency at Solution Crumb Rubber Plant 000005

	Concentration (μg/L)		
Pollutant	Influent	Effluent	Percent removal
Cadmium	<1	<1	NM
Copper	9	14	NM
Chromium	75	410	NM
Zinc	14,000	13,000	7
Bis(2-ethylhexyl)phthalate[a]	180	24	87
Phenol	7	5	29
Benzene	<22	110[b]	NM
Ethylbenzene	<46	<39	NM
Toluene	<26	<26	NM
Carbon tetrachloride	35	14	60
Chloroform	2.2	1.3	41
Methyl chloride	4,900	2,200[c]	55
Methylene chloride	<1.0	<1.0	NM
1,1,2-Trichloroethane	<0.1	<0.1	NM
Trichloroethylene	<0.1	<0.1	NM

Analytic methods: V.7.3.29, Data set 2; NM, not meaningful.
[a]Analytical methodology for phthalates is questionable; therefore, significance of values reported is unknown.
[b]Average of 320 μg/L, <11 μg/L.
[c]Found at significant levels in treatment effluent.
Treatment technology: Primary flocculation/clarification (DAF).
Discharge point: Treated effluent is discharged to a nearly oil refinery's cooling water system.
Source: USEPA.

Table 19 Toxic Pollutant Removal Efficiency at Solution Crumb Rubber Plant 000027, Verification Data

	Concentration (μg/L)		
Pollutant	Influent	Effluent	Percent removal
Cadmium	<1	1	NM
Chromium	440	20	95
Copper	7	5	29
Mercury[a]	1.1	2.0	NM
Bis(2-ethylhexyl)phthalate[b]	120	110	8
Phenol	180	11	94
Benzene	3,200	<0.1	>99
Ethylbenzene	<0.1	<0.1	NM
Toluene	<0.1	<0.1	NM
Chloroform	3.0	0.9	70
1,1,2,2-Tetrachloroethane	<0.1	<0.1	NM

Analytic methods: V.7.3.29, Data set 2; NM, not meaningful.
[a]Intake measured at 4 μg/L, making plant's contribution zero.
[b]Analytical methodology for phthalates is questionable; therefore, significance of values reported is unknown.
Treatment technology: Sedimentation, waste stabilization lagoons.
Discharge point: Surface stream.
Source: USEPA.

Table 20 Toxic Pollutant Removal Efficiency at Dry Digestion Reclaiming Plant 000134, Verification Data

	Concentration (μg/L)				
Pollutant (μg/L)	Influent	Effluent	Percent removal	Cooling tower blowdown (μg/L)[a]	Percent removal
Cadmium	1	3	NM	<1	>67
Chromium	6	21	NM	2	90
Copper	28	12	57	2	83
Lead[b]	70	670	NM	29	96
Mercury		2.3		0.5	78
Zinc	100	2,500	NM	100	96
Bis(2-ethylhexyl)phthalate[c]	16,000	4,200	74	100	98
2,4-Dimethylphenol	58,000	25,000	57	120	>99
Phenol	26,000	4,900	81	27	>99
Benzene	60	<0.1	>99	<0.1	NM
Ethylbenzene	8,600	<0.1	>99	<0.1	NM
Toluene	2,700	<0.1	>99	<0.1	NM
Acenaphthylene	<33	<8	NM	<8	NM
Fluorene	2,000	<12	>99	<12	NM
Naphthalene[d]	100,000	42	>99	13	69
Phenanthrene	1,300	300	77	<4	>99
Pyrene	6,800	11	>99	4	64
Chloroform	1.9	1.4	26	4.9	NM

Analytic methods: V.7.3.29, Data set 2.
NM, not meaningful.
[a]Effluent from treatment goes into cooling tower and is discharged with noncontact cooling water as cooling tower blowdown.
[b]Potentially significant levels observed in cooling tower blowdown.
[c]Analytical methodology for phthalates is questionable; therefore, significance of values raported is unknown.
[d]Significance of blowdown value questionable due to high detection limit, low values observed in the carbon column effluent (treatment influent), and the fact that compound is not a metabolic byproduct of activated sludge treatment.
Treatment technology: Cartridge filtration, activated carbon (oil removal), activated sludge sedimentation.
Discharge point: Noncontact cooling water system, blowdown of this system to surface stress.
Source: USEPA.

concentration of 4100 reduced odor to 250 and 500 using 100 ppm Na_2SO_3 and Na_2S respectively after 17 days of treatment. For styrene wastewaters of initial odor concentration of 128, it was reduced to 4 after 9 days treatment with 100 ppm Na_2SO_3 and for the same wastewaters with initial odor concentration of 65 ppm, it was reduced to 4 ppm using Na_2S using the same dosage and same duration of treatment [6].

30.4.3 Biological Treatment

A trickling filter has been used to treat neutralized rubber wastewaters with initial BOD of 445 mg/L. It removed 92.1% BOD with a 24-hour detention time [5]. The activated sludge process has removed 85% BOD from combined rubber and domestic wastewater wastewaters [7].

30.4.4 Nutrient Removal

Attached-growth waste stabilization ponds have been used to remove 65–70% TKN (total Kjeldahl nitrogen), and 70–83% NH_3-N from concentrated latex and rubber sheet plant wastewaters [8]. A combined algae and water hyacinth system has been used to remove 96.41% COD, 98.93% TKN, 99.28% NH_3-N, 100% NO_2-N, and 100% NO_3-N [9].

30.5 TREATMENT TECHNOLOGY COSTS

The investment cost, operating, and maintenance costs, and energy costs for the application of control technologies to the wastewaters of the rubber processing industry have been analyzed. These costs were developed to reflect the conventional use of technologies in this industry. Several unit operation/unit process configurations have been analyzed for the cost of application of technologies and to select BPT and BAT level of treatment. The applicable tretment technologies, cost methodology, and cost data are available in a detailed presentation [10].

REFERENCES

1. US Department of Labor – Occupational Safety and Health Administration. *SIC Division Structure*, http://www.osha.gov/cgi-bin/sic/sicser5, 2003.
2. USEPA. *Subchapter N – Effluent Guidelines and Standards in CFR Title 40*, Protection of Environment; http://www.epa.gov/docs/epacfr40/chapt-I.info/subch-N.htm, 2003.
3. Sechrist, W.D.; Chamberlain, N.S. Chlorination of phenol bearing rubber wastes. In *Proceedings of 6th Industrial Waste Conference*, Purdue University, Lafasette, IN, November 1951; 396.
4. USEPA. *Development Document for Effluent Limitation Guidelines and New Source Performance Standards for the Tire and Synthetic Segment of the Rubber Processing Point Source Category*, US Environmental Protection Agency: Washington, DC. http://www.epa.gov/cgi-bin/, 1974.
5. Morzycki, J. Effluents from sewage contaminated with latex. Chem. Abs. **1966**, *64*, 3192.
6. Black, O.R. Study of wastes from rubber industry. Sewage Works J. **1946**, *18*, 1169.
7. Mills, R.E. Progress report on the bio-oxidation of phenolic and 2,4-D waste waters. In *Proceedings of 4th Ontario Industrial Waste Conference*, June 1957; 30.

8. Rakkoed, A.; Danteravanich, S.; Puetpaiboon, U. Nitrogen removal in attached growth waste stabilization ponds of wastewater from a rubber factory. Water Sci. Technol. **1999**, *40(1)*, 45–52.
9. Bich, N.N.; Yaziz, M.I.; Bakti, N.A.K. Combination of *Chlorella vulgaris* and *Eichhornia crassipes* for wastewater nitrogen removal. Water Res. **1999**, *33* (*10*), 2357–2362.
10. USEPA and Envirodyne Engineers, Inc. *Review of the Best Available Technology for the Rubber Processing Point Source Category*, Technical Report No. 68-01-4673; US Environmental Protection Agency: Washington, DC, 1978.

31

Treatment of Timber Industry Wastes

Lawrence K. Wang
Zorex Corporation, Newtonville, New York, U.S.A., and
Lenox Institute of Water Technology, Lenox, Massachusetts, U.S.A.

31.1 INTRODUCTION

The timber products processing industry encompasses manufacturers and processors who use forest materials to produce their goods and merchandise. Fifteen distinct subcategories of manufacturers and/or processors are engaged in the utilization of timber. This chapter addresses three major subsections of the entire industry: (a) wood preserving, both steaming and Boulton processes; (b) insulation board manufacturing; and (c) both wet–wet (S1S or smooth one side) and wet–dry (S2S or smooth two side) hardboard manufacturing.

The number of dischargers in the timber products processing industry in the United States can be broken down as follows: (a) 19 direct dischargers; (b) 55 indirect dischargers; and (c) 172 zero dischargers.

31.1.1 Water Pollution

The timber industry treats timber and wood products with chemical preservatives to protect the wood from degradation due to various organisms including fungi, and insects such as borers and termites. This treatment extends the range of applications and the service life of the wood. By design, the chemicals used to protect wood must be toxic to the target organisms, but they may also affect nontarget organisms and the environment [1].

The following groups of preservatives are commonly used for wood preservation: (a) copper chrome arsenate (CCA); (b) copper-based alternatives to CCA [ammoniacal copper quaternary (ACQ) and copper azole]; (c) boron; (d) creosote; and (e) pyrethroid- and metal-based light organic solvent preservatives (LOSPs). Section 31.2.1 presents a complete list of toxic chemicals used in wood preservation in the United States.

Copper chrome arsenate (also known as CCA or chromated copper arsenate) consists of three metals: copper, chromium, and arsenic. All three metals pose a risk to the environment. Both hexavalent chromium and arsenic can cause cancer in humans. The CCA concentrate is diluted with water to create a working solution that is used in the pressure treatment of timber.

CCA-treated timber is commonly a greenish color, but this is also often the case with the other copper-based preservatives. CCA-treated timber is registered for use by the industrial countries under their laws. The registered uses include internal building uses and external uses such as decks, walkways, fences, playground equipment and retaining walls, and some marine water applications such as wharfs and jetty piles.

31.1.2 Air Pollution and Health Hazards

Published results of scientific studies indicate that copper, chromium, and arsenic slowly leach from CCA-treated timber products. All three metals pose a risk to human health and the environment, but also exist naturally in the environment in varying concentrations.

Based on currently available evidence, CCA-treated wood does not pose any significant risk to the public. However, as arsenic is a known human carcinogen, it would be prudent to avoid unnecessary exposure to it. Some common sense tips to minimize unnecessary exposure to CCA-treated timber are: (a) treated wood should never be burned in open fires, stoves, fireplaces, or residential boilers; (b) hands should always be washed thoroughly after contact with any treated wood, especially before eating and drinking; (c) food should not come into direct contact with any treated wood; and (d) precautions should be taken to wear protective gear when working with CCA-treated wood.

31.1.3 Solid Wastes Disposal

Small quantities of household CCA-treated timber waste (e.g., offcuts from a small job) could be placed in the owner's rubbish bin, with the rest of owner's household waste. CCA-treated timber waste from larger household building and demolition jobs is classified as inert waste, and can be disposed of to most suburban landfills.

CCA-treated timber waste from industrial sources should only be disposed of to certain landfills. CCA-treated timber waste must not be burned or used as such.

31.1.4 Global Trends

The timber industry is moving away from using pesticides such as CCA and creosote. The United States and Canada have moved to phase out the use of CCA to treat timbers intended for residential uses after December 2003 and Europe after June 2004. This trend appears to be driven by recent European risk assessments of arsenic, the application of the "precautionary principle", and by perceived consumer demand shifts in North America [1].

The United States Environmental Protection Agency (USEPA) granted the cancellation of CCA registration for most residential uses of treated timber following an application to do so from the wood preservation industry. From January 1, 2004, the USEPA will not allow CCA products to be used to treat timber intended for most residential uses. The USEPA is continuing an assessment of the risks posed to children by arsenic leaching from CCA-treated timber. The European Commission has announced a partial prohibition on the use of CCA-treated timber, to take effect after June 2004. In addition to the residential uses being restricted in the United States and Canada, CCA preservatives will also not be used in the EU for timber destined for marine and most agricultural uses.

The New Zealand Environmental Risk Management Authority (ERMANZ) has decided not to change the registration of CCA following a recent review of the potential public health risks arising from the continuing use of CCA-treated timber. ERMANZ found that the extent of any risk to public health arising from CCA remains unclear but is considering further investigation into the possible environmental and occupational health risks arising from CCA.

The Ministry of Environment and the Ministry of Health of Manatu Hauora [2] announced their Guidelines for Selected Timber Treatment Chemicals in 1997.

31.1.5 Cleaner Production and Economy

Cleaner production aims at avoiding the generation of waste and emissions, by making more efficient use of materials and energy, through modifications in the production processes, input materials, operating practices, and/or products and services [3,4]. Van Berkel [4] illustrates how the timber industry and other industries were able to implement cleaner production practices and technologies.

Aruna and Mercer [5] state the timber economy of the Mid-Atlantic Region in the United States. Specifically, the health and sustainability of ecosystems in an eight state region for the forest industries (including the timber industry) have been assessed. Wang and colleagues [6], meanwhile, present the costs of various wastewater treatment systems.

31.2 INDUSTRY SUBCATEGORY DESCRIPTION

General descriptions and process descriptions for the major subcategories of the timber products processing point source category are introduced in this section.

31.2.1 Wood Preserving

Creosote, pentachlorophenol (PCP), and formulations of water-soluble inorganic chemicals are the three most prevalent types of preservatives used in wood preserving. The most common of inorganic chemicals are the salts of copper, chromium, and arsenic. Fire retardants are formulations of salts, the principal ones being borates, phosphates, and ammonium compounds. Of plants in the United States, 80% use at least two of the three types of preservatives. Many plants treat with one or two preservatives plus a fire retardant.

There are two basic steps in the wood preserving process: (a) preconditioning the wood to reduce its natural moisture content and to increase permeability; and (b) impregnating the wood with the desired preservatives.

The preconditioning step may be performed by one of several methods including (a) seasoning or drying wood in large, open yards; (b) kiln drying; (c) steaming the wood at elevated pressure in a retort followed by application of a vacuum; (d) heating the stock in a preservative bath under reduced pressure in a retort (Boulton process); or (e) vapor drying, heating of the unseasoned wood in a solvent to prepare it for preservative treatment. All of these preconditioning methods have, as their objective, the reduction of the moisture content in the unseasoned stock to a point where the requisite amount of preservative can be retained in the wood.

Conventional steam conditioning (open steaming) is a process in which unseasoned or partially seasoned stock is subjected to direct steam impingement at an elevated pressure in a retort. The maximum permissible temperature is set by industry standards at 118°C and the duration of the steaming cycle is limited by these standards to no more than 20 hours. Steam condensate formed in the retort exits through traps. The condensate is discharged to oil–water separators for removal of free oils. Removal of emulsified oils requires further treatment.

In closed steaming, a widely used variation of conventional steam conditioning, the steam needed for conditioning is generated *in situ* by covering the coils in the retort with water from a reservoir and heating the water by passing process steam through the coils. The water is returned to the reservoir after oil separation and reused during the next steaming cycle. There is a slight increase in volume of water in the storage tank after each cycle due to water exuded from the wood. A small blowdown from the storage tank is necessary to remove this excess water and to control the level of wood sugars in the water.

Modified closed steaming is a steam conditioning process variation in which steam condensate is allowed to accumulate in the retort during the steaming operation until it covers the heating coils. At that point, direct steaming is discontinued and the remaining steam required for the cycle is generated within the retort by utilizing the heating coils. Upon completion of the steaming cycle, and after recovery of oils, the water in the cylinder is discarded.

Preconditioning is accomplished in the Boulton process by heating the stock in a preservative bath under reduced pressure in the retort. The preservative serves as a heat transfer medium. After the cylinder temperature has been raised to operating temperature, a vacuum is drawn, and water, which is removed in vapor form from the wood, passes through a condenser to an oil–water separator. At this point low-boiling fractions of the preservative are removed. The Boulton cycle may have a duration of 48 hours or longer for large poles and piling. This fact accounts for the lower production per retort day as compared to plants that steam condition.

The vapor-drying process consists of exposing wood in a closed vessel to vapors from any one of the many organic chemicals that are immiscible with water and have a narrow boiling range.

The following is a summary of toxic pollutants found in significant quantities in the wood preserving category [7,8]:

- Pentachlorophenol
- Phenol
- 2,4-Dimethylphenol
- 2,4-Dichlorophenol
- Copper
- Chromium
- 3,4-Benzofluoranthene
- Benzo(k)fluoranthene
- Pyrene
- Benzo(a)pyrene
- Indeno (1,2,3-cd)pyrene
- Benzo(ghi)perylene
- Arsenic
- Nickel
- Zinc
- Fluorene
- Phenol
- Fluoranthene
- Chrysene
- *Bis*(2-ethylhexyl)phthalate
- Naphthalene
- Acenaphthylene

31.2.2 Insulation Board Manufacturing

Insulation board is a form of fiberboard, which in turn is a broad generic term applied to sheet materials constructed from ligno-cellulosic fibers. Insulation board is a "noncompressed" fiberboard, which is differentiated from "compressed" fiberboards, such as hardboard, on the basis of density. Densities of insulation board range from about 0.15 to 0.50 g/cm^3 (9.5–31 lb/ft^3).

There are more than 16 insulation board plants in the United States with a combined annual production capacity of over 330 million square meters (3600 million square feet) on a 13 mm (0.5 in.) basis. Sixteen of the plants use wood as a raw material for some or all of their production. One plant uses bagasse exclusively, and one plant uses waste paper exclusively for raw material. Four plants use mineral wood, a nonwood-based product, as a raw material for part of their insulation board production. Five plants produce hardboard products as well as insulation board at the same facility.

Insulation board can be formed from a variety of raw materials including wood from softwood and hardwood species, mineral fiber, waste paper, bagasse, and other fibrous materials. In this section, only those processes employing wood as raw materials are considered. Plants utilizing wood may receive it as roundwood, fractionated wood, and/or whole tree chips. Fractionated wood can be in the form of chips, sawdust, or planer shavings.

The toxic pollutants found in significant quantities in insulation board manufacturing wastewater are:

- Copper
- Nickel
- Zinc
- Phenol
- Benzene
- Toluene

31.2.3 Hardboard Manufacturing

Hardboard is a form of fiberboard, which is a broad generic term applied to sheet materials constructed from ligno-cellulosic fibers. Hardboard is a "compressed" fiberboard, with a density greater than 0.50 g/cm^3 (31 lb/ft^3). The thickness of hardboard products ranges between 2 and 13 mm (nominal 1/12 to 7/16 in.).

Production of hardboard by the wet process method is usually accomplished by thermomechanical fiberization of the wood furnish. One plant produces wet–dry hardboard using primarily mechanical refining.

Dilution of the wood fiber with fresh or process water then allows the formation of a wet mat of a desired thickness on a forming machine. This wet mat is then pressed either wet or after drying. Chemical additives help the overall strength and uniformity of the product. The uses of manufactured products are many and varied, requiring different processes and control measures. The quality and type of board are important in the end use of the product.

Hardboard that is pressed wet immediately following forming of the wet-lap is called wet–wet or smooth-one-side (S1S) hardboard; that which is pressed after the wet-lap has been dried is called wet-dry or smooth-two-side (S2S) hardboard.

There are about 16 wet process hardboard plants in the United States, representing an annual production in excess of 1.5 million metric tons per year. Seven of the plants produce only S1S hardboard. Nine plants produce S2S hardboard. Of these nine, five plants also produce insulation board, while three plants also produce S1S hardboard.

The toxic pollutants found in significant quantities in the hardboard manufacturing wastewater are [8]:

- Copper
- Phenol
- Chromium
- Benzene

- Nickel
- Toluene
- Zinc

31.3 WASTEWATER STREAMS

The timber products processing industry was analysed in a screening program for the 129 USEPA priority pollutants. Those pollutants detected in screening were further analyzed in a verification sampling analysis. The minimum detection limit for toxic organics is 10 μg/L and for toxic metals, 2 μg/L. Any concentration below its detection limit is presented in the following tables as BDL, below detection limit.

31.3.1 Wood Preserving Wastewater

The quantity of wastewater generated by a wood preserving plant is a function of the method of conditioning used, the moisture content of the wood being treated, and the amount of rainwater draining toward the treating cylinder. Most wood preserving plants treat stock having a wide range of moisture contents. Although most plants use predominantly one of the major conditioning methods, many plants use a combination of several conditioning methods, and the actual quantity of wastewater generated by a specific plant may vary considerably. The average wastewater volume from 14 Boulton plants is reported to be 21,200 L/day (5600 gal/day) or 139 L/m^3 (1.03 gal/ft^3) of production. The average wastewater volume for eight closed loop steaming plants is 5200 L/day (1370 gal/day) or 60 L/m^3 (0.45 gal/ft^3). The average wastewater volume for 10 plants that treat significant amounts of dry stock is 13,300 L/day (3510 gal/day) or 121 L/m^3 (0.91 gal/ft^3). Additionally the average wastewater volume for 14 open steaming plants is 32,300 L/day (9250 gal/day) or 236 L/m^3 (1.87 gal/ft^3).

Tables 1A and B present the concentrations of toxic pollutants found in streaming subcategory raw wastewater and in Boulton subcategory raw wastewater, respectively. Table 1C presents the concentrations of toxic pollutants found in both streaming and Boulton subcategory treated effluent (Source: USEPA).

31.3.2 Insulation Board Manufacturing Wastewater

Insulation board plants responding to the data collection portfolio reported fresh water usage rates ranging from 95,000 to 5,700,000 L/day for process water (0.025–1.5 MGD). One insulation board plant that also produces hardboard in approximately equal amounts, uses over 15 million L/day (4 MGD) of fresh water for process water.

Water becomes contaminated during the production of insulation board primarily through contact with the wood during fiber preparation and forming operations. The vast majority of pollutants are fine wood fibers and soluble wood sugars and extractives. More specifically, potential sources of wastewater in an insulation board plant include: (a) chip washwater; (b) process whitewater generated during fiber preparation (refining and washing); (c) process whitewater generated during forming; and (d) wastewater generated during miscellaneous operations (dryer washing, finishing, housekeeping, etc.)

The average unit flow for an insulation board mechanical refining plant is 8.3 L/kg (2000 gal/ton), assuming the plant produces a full line of insulation board products and practises internal recycling to the extent practicable. Table 2 presents concentrations of toxic pollutants found in insulation board manufacturing raw wastewater [8].

Table 1A Concentrations of Toxic Pollutants Found in Streaming Subcategory Raw Wastewater

Toxic pollutant (μg/L)	Range	Median
Metals and inorganics		
Antimony	BDL–47	BDL
Arsenic	3–14,000	33
Beryllium	BDL–19	BDL
Cadmium	BDL–10	BDL
Chromium	1–98	23
Copper	8–850	120
Lead	BDL–91	14
Mercury	BDL–BDL	BDL
Nickel	3–150	28
Selenium	BDL–7	BDL
Silver	BDL–6	BDL
Thallium	BDL–10	BDL
Zinc	120–820	310
Phthalates		
Bis(2-ethylhexyl)phthalate	BDL–440	BDL
Phenols		
2-Chlorophenol	BDL–42	BDL
2,4-Dimethylphenol	BDL–6,600	130
Pentachlorophenol	1,200–160,000	16,000
Phenol	1,400–87,000	16,000
2,4,6-Trichlorophenol	BDL–530	BDL
Monocyclic aromatics		
Benzene	BDL–2,800	1,000
Ethylbenzene	37–2,100	380
Toluene	27–3,200	500
Polycyclic aromatic hydrocarbons		
Acenaphthene	1,100–55,000	1,500
Acenaphthylene	BDL–1,200	720
Anthracene/phenanthrene	2,000–39,000	6,500
Benzo(a)anthracene	BDL–7,700	160
Benzo(a)pyrene	BDL–2,700	BDL
Benzo(b)fluoranthene	BDL–1,700	BDL
Benzo(ghi)perylene	BDL–320	BDL
Benzo(k)fluoranthene	BDL–3,900	BDL
Chrysene	BDL–4,700	73
Dibenzo(ah)anthracene	BDL–430	BDL
Fluoranthene	630–35,000	1,600
Fluorene	820–48,000	1,500
Indeno(1,2,3-cd)pyrene	BDL–5,500	BDL
Naphthalene	380–45,000	2,200
Pyrene	360–22,000	810
Halogenated aliphatics		
Methyl chloride	BDL–700	77
Chloroform	BDL–20	BDL

BDL, below detection limit.
Source: USEPA.

Table 1B Concentrations of Toxic Pollutants Found in Boulton Subcategory Raw Wastewater

Toxic pollutant (μg/L)	Range	Median
Metals and inorganics		
Antimony	BDL–13	3
Arsenic	3–14	7
Beryllium	BDL–2	BDL
Cadmium	BDL–5	BDL
Chromium	4–3,900	9
Copper	80–1,600	110
Lead	BDL–14	5
Mercury	BDL–3.7	BDL
Nickel	20–210	94
Selenium	2–53	3
Silver	BDL–2	BDL
Thallium	BDL–2	BDL
Zinc	320–26,000	840
Phthalates		
Bis(2-ethylhexyl)phthalate	BDL–1,500	430
Phenols		
2-Chlorophenol	BDL	
2,4-Dimethylphenol	BDL	
Pentachlorophenol	27,000	
Phenol	71	
2,4,6-Trichlorophenol	BDL	
Monocyclic aromatics		
Benzene	BDL	
Ethylbenzene	BDL	
Toluene	BDL	
Polycyclic aromatic hydrocarbons		
Acenaphthene	BDL–2,800	BDL
Acenaphthylene	BDL–2,100	BDL
Anthracene/phenanthrene	BDL–1,500	920
Benzo(a)anthracene	BDL–34	BDL
Benzo(a)pyrene	BDL	
Benzo(b)fluoranthene	BDL	
Benzo(ghi)perylene	BDL	
Benzo(k)fluoranthene	BDL	
Chrysene	BDL–18	BDL
Dibenzo(ah)anthracene	BDL	
Fluoranthene	BDL–280	BDL
Fluorene	BDL–820	BDL
Indeno(1,2,3-cd)pyrene	BDL	
Naphthalene	BDL–3,100	BDL
Pyrene	BDL	
Halogenated aliphatics		
Methyl chloride	2,600	
Chloroform	BDL	

BDL, below detection limit.
Source: USEPA.

Table 1C Concentrations of Toxic Pollutants Found in Both Streaming and Boulton Subcategory Treated Effluent

Toxic pollutant (μg/L)	Range	Median
Metals and inorganics		
Antimony	BDL–14	BDL
Arsenic	2–7,000	35
Beryllium	BDL–13	BDL
Cadmium	BDL–7	BDL
Chromium	BDL–4,000	8
Copper	18–270	57
Lead	BDL–37	4
Mercury	BDL–2	BDL
Nickel	2–150	18
Selenium	BDL–39	2
Silver	BDL–4	BDL
Thallium	BDL–7	BDL
Zinc	47–41,000	200
Phthalates		
Bis(2-ethylhexyl)phthalate	BDL–300	BDL
Phenols		
2-Chlorophenol	BDL	
2,4-Dimethylphenol	BDL–140	BDL
Pentachlorophenol	32–8,300	2,700
Phenol	BDL–16,000	15
2,4,6-Trichlorophenol	BDL	
Monocyclic aromatics		
Benzene	BDL–33	10
Ethylbenzene	BDL–20	BDL
Toluene	BDL–140	23
Polycyclic aromatic hydrocarbons		
Acenaphthene	BDL–18,000	90
Acenaphthylene	BDL–190	BDL
Anthracene/phenanthrene	BDL–37,000	59
Benzo(a)anthracene	BDL–3,400	BDL
Benzo(a)pyrene	BDL–290	BDL
Benzo(b)fluoranthene	BDL–2,500	BDL
Benzo(ghi)perylene	BDL–63	BDL
Benzo(k)fluoranthene	BDL–210	BDL
Chrysene	BDL–19,000	BDL
Dibenzo(ah)anthracene	BDL	
Fluoranthene	BDL–17,000	110
Fluorene	BDL–16,000	36
Indeno(1,2,3-cd)pyrene	BDL–110	BDL
Naphthalene	BDL–36,000	33
Pyrene	BDL–9,400	77
Halogenated aliphatics		
Methyl chloride	13–1,900	140
Chloroform	BDL–23	BDL

BDL, below detection limit.
Source: USEPA.

Table 2 Concentrations of Toxic Pollutants Found in Insulation Board Subcategory Raw Wastewater, USEPA Verification Data

Toxic pollutant (μg/L)	Number of Samples	Range	Median
Metals and inorganics			
Antimony	4	0.67–3	1.5
Arsenic	4	1.6–3.3	2.5
Beryllium	4	0.5–0.83	0.5
Cadmium	4	0.5–1.0	0.66
Chromium	4	1.3–11	4.9
Copper	4	200–450	310
Lead	4	1.3–21	3.3
Mercury	4	1–7.5	5.8
Nickel	4	8.8–240	58
Selenium	4	3.3–5.0	4.5
Silver	4	0.5–0.6	0.5
Thallium	4	0.5–0.83	0.7
Zinc	4	250–720	530
Toxic organics			
Chloroform	3	BDL–20	BDL
Phenol	3	BDL–40	BDL
Benzene	3	BDL–70	50
Toluene	3	BDL–60	40

BDL, below detection limits.
Source: USEPA.

31.3.3 Hardboard Manufacturing Wastewater

Significant amounts of water are required for production of hardboard by wet process. Plants responding to the data collection portfolio reported fresh water usage rates for process water ranging from approximately 190,000 to 19 million L/day (0.05–5 MGD). One plant produces both hardboard and insulation board in approximately equal amounts, and reported fresh water use of over 15 million L/day (4 MGD).

Process water becomes contaminated during the production of hardboard primarily through contact with the wood raw material during the fiber preparation, forming, and, in the case of S1S hardboard, pressing operations. The vast majority of pollutants in the wastewater consist of fine wood fibers, soluble wood sugars, and extractives. Additives not retained in the board also add to the pollutant load.

Process whitewater is the water used to process and transport the wood from the fiber preparation stage through mat formation. Process whitewater produced by the dewatering of stock at any stage of the process is usually recycled to be used as stock dilution water. However, in order to avoid undesirable effects in the board when elevated concentrations of suspended solids and dissolved organic materials occur, excess process whitewater is discarded.

Potential wastewater sources in the production of wet process hardboard include: (a) chip washwater; (b) process whitewater generated during fiber preparation (refining and washing); (c) process whitewater generated during forming; (d) hot press squeezeout water; and (e) wastewater generated during miscellaneous operations (dryer washing, finishing, housekeeping, etc.).

A unit flow of 12 L/kg (2800 gal/ton) is considered to be representative of an S1S hardboard plant that produces a full line of hardboard products and that practises internal recycling to the extent practicable. A unit flow of 24.6 L/kg (5900 gal/ton) is considered to be representative of an S2S hardboard manufacturing plant that produces a full line of hardboard products and practises internal recycling to the extent possible.

Available data analyses list primarily metals and inorganics as toxic pollutants; no base/neutrals data are presented. Table 3 presents concentrations and pollutant loadings for toxic and classical pollutants found in hardboard manufacturing raw wastewater.

31.4 WASTEWATER TREATMENT

The following sections address the current level of in-place treatment technology and the raw and treated effluent loads and percent reduction for several pollutants and several plants. Information is organized with respect to the aforementioned subcategories.

31.4.1 Wood Preserving Wastewater Treatment

The following sections present the current level of in-place treatment technology for Boulton–no dischargers, Boulton-indirect dischargers, steaming–no dischargers, steaming–direct dischargers, and steaming–indirect dischargers, respectively.

Current level of in-place technology, Boulton, without any dischargers is summarized as follows:

- primary gravity oil–water separation: 83% of all plants;
- oil separation by dissolved air flotation (DAF): 4% of all plants;
- evaporation ponds: 63% of all plants;
- spray or soil irrigation: 4% of all plants;
- cooling tower evaporation: 17% of all plants;
- thermal evaporation: 4% of all plants;
- effluent recycle to boilers or condensers: 17% of all plants;
- no discharge: 8% of all plants.
- of course, the above plants may use more than one technology.

Current level of in-place technology, Boulton, with indirect dischargers is summarized below, with the understanding that the following plants may use more than one technology:

- primary gravity oil–water separation: 100% of all plants;
- chemical flocculation and/or oil absorptive media: 36% of all plants; and
- biological treatment: 18% of all plants.

Current level of in-place technology, steaming, without any dischargers is summarized below knowing that some plants use more than one technology:

- primary gravity oil–water separation: 80% of all plants;
- chemical flocculation and/or oil absorptive media: 9.2% of all plants;
- sand filtration: 14% of all plants;
- oxidation lagoon: 7.7% of all plants;
- aerated lagoon: 15% of all plants;
- spray irrigation: 14% of all plants;
- holding basin: 35% of all plants;
- thermal evaporation: 3.1% of all plants;
- solar evaporation pond: 38% of all plants;

Table 3 Concentrations and Loadings of Toxic and Classical Pollutants Found in Hardboard Manufacturing Subcategory Raw Wastewater, USEPA Verification Data

Pollutant	Number of samples	Concentration (μg/L)		Loading (mg/Mg)	
		Range	Median	Range	Median
Toxic pollutant					
Metals and Inorganic [a]					
Antimony	6	0.5–8	2.6	9–100	52
Arsenic	6	1–1.3	1.2	12–26	15
Beryllium	6	0.5–0.67	0.5	5–13	7.5
Cadmium	6	0.5–5	0.5	7–60	18
Chromium	6	1–420	52	17–5,500	240
Copper	6	33–530	350	440–14,000	3,600
Lead	6	2–55	4	20–800	51
Mercury	6	0.05–18	1.3	1.2–310	14
Nickel	6	3.3–270	7.5	60–2,400	100
Selenium	6	0.8–3.8	2.1	18–60	23
Silver	6	0.5–7	0.58	5–180	9
Thallium	6	0.5–1.5	0.58	5–13	11
Zinc	6	190–2,300	660	3,000–24,00	8,000
Phenols					
Phenol[b]	2	BDL–680			
Phenol[c]	3	BDL–300	BDL[c]		
Monocyclic aromatics					
Benzene[b]	2	BDL–80			
Benzene[c]	3	BDL–90	BDL[c]		
Ethylbenzene[b]	2	BDL–20			
Toluene[b]	2	15–70			
Toluene[c]	3	BDL–60	10		
Halogenated aliphatics					
Chloroform[b]	2	BDL–20			
Chloroform[c]	3	BDL–20	BDL		
1,1,2-Trichloroethane[c]	3	BDL–90	5		
Pesticides and metabolites[d]					
Aldrin			<0.001		
PCBs			0.015		
Chlordane			<0.001		
Meptachlor			<0.001		
Classical pollutant			<0.001		
BOD_5	4			1.9–120	37
Total phenols	4	BDL–8,900	340	0.003–0.04	0.009

BDL, below detection limits.
[a]S1S and S2S combined for metals – no observed difference.
[b]S1S type hardboard; no loading data.
[c]S2S type hardboard; no loading data.
[d]S1S and S2S processes combined; number of plants was not specified.
Source: USEPA.

- spray-assisted solar evaporation: 26% of all plants;
- effluent recycle to boiler or condenser: 17% of all plants.

Current level of in-place technology, steaming, with direct discharger (note, there is only one plant) is summarized as follows:

- primary gravity oil–water separation: 100% of all plants;
- chemical flocculation and/or oil absorptive media: 100% of all plants;
- aerated lagoon: 100% of all plants;
- holding basin: 100% of all plants;
- spray-assisted solar evaporation: 100% of all plants;
- effluent recycle to boiler or condenser: 100% of all plants.

Current level of in-place technology, steaming, with indirect dischargers is summarized below knowing that some plants use more than one technology:

- primary gravity oil–water separation: 97% of all plants;
- chemical flocculation and/or oil absorptive media: 26% of all plants;
- sand filtration: 13% of all plants;
- oxidation lagoon: 3.2% of all plants;
- aerated lagoon: 6.4% of all plants;
- holding basin: 55% of all plants;
- spray-assisted solar evaporation: 6.4% of all plants;
- effluent recycle to boiler or condenser: 6.4% of all plants.

Table 4 presents average raw and treated waste loads and percent removal for COD, phenols, oil and grease, and pentachlorophenol [8]. Table 5 presents average raw and treated waste loads and percent removals of methylene chloride, trichloromethylene, benzene, ethylbenzene, and toluene for plants with current BPT in place. Table 6 presents similar data for base/neutral toxic pollutants for current BPT in place [8]. Table 7 presents similar data for wood preserving phenols for plants with current pretreatment technology in place and current BPT in place [8]. In addition, Tables 8 and 9 present data for average raw and treated waste loads and percent removals of metals for plants with current BPT in place [8].

31.4.2 Insulation Board Manufacturing Wastewater Treatment

The following is a summary of the current level of in-place treatment technology for six plants:

1. Use of a clarifier and aerated lagoon system for wastewater treatment at a structure/decorative insulation board plant using thermomechanical process.
2. Use of oxygen-activated sludge system and clarifier for wastewater treatment at an insulation board/hardboard plant using thermomechanical process.
3. Use of an aerated lagoon, evaporaton pond, and self-contained discharger (irrigation) system for wastewater treatment at an insulation board plant using mechanical process.
4. Use of a clarifier and activated sludge system for wastewater treatment at a structure/decorative insulation board plant using mechanical process.
5. Use of a floc-clarifier, aerated lagoon, and discharge to POTW system for wastewater treatment at a structure/decorative insulation board plant using mechanical process.
6. Use of a settling pond, aerated lagoon, and oxidation pond system for wastewater treatment at an insulation board plant producing S2S hardboard and using thermomechanical process.

Table 4 Wood Preserving Classical Pollutant Data Averages for Plants with Current BPT In Place

Pollutant	Number of plants	Number of samples	Waste load (kg/1,000 cu.m) Raw	Waste load (kg/1,000 cu.m) Treated	Percent removal
COD	4	6	500	96	81
Phenols	4	6	38	0.16	99
Oil and grease	4	6	69	<13	>81
Pentachlorophenol	3	5	<4.3	0.16	<96

Source: USEPA.

Table 5 Wood Preserving Volatile Organic Analysis Data Averages for Plants with Current BPT In Place

Pollutant	Number of plants	Waste load (kg/10 cu.m) Raw	Waste load (kg/10 cu.m) Treated	Percent removal
Methylene chloride	3	78	69	12
Trichloromethylene	3	<0.16	<3.2	NM
Benzene	3	320	<4.8	>98
Ethylbenzene	3	1,600	<1.6	>99
Toluene	3	380	<14	>96

NM, not meaningful.
Source: USEPA.

Table 6 Wood Preserving Base/Neutrals Analysis Data Averages for Plants With Current BPT In Place (*Source*: USEPA)

Pollutant	Number of plants	Number of samples	Waste load (kg/10E6 cu.m) Raw	Waste load (kg/10E6 cu.m) Treated	Percent removal
Fluoranthene	3	4	850	140	83
Benzo(b)fluoranthene	3	4	<140	<22	NM
Benzo(k)fluoranthene	3	4	210	<24	>88
Pyrene	3	4	640	51	92
Benzo(a)pyrene	3	4	<160	<29	NM
Indeno(1,2,3-CD)pyrene	3	4	<120	<16	NM
Benzo(ghi)perylene	3	4	<24	<6.4	NM
Phenanthrene/anthracene	3	4	1,900	<110	>94
Benzo(a)anthracene	3	4	<210	<38	NM
Dibenzo(a,h)anthracene	3	4	<8	<1.6	NM
Naphthalene	3	4	>3,000	<6.4	>99
Acenaphthene	3	4	700	35	95
Acenaphthylene	3	4	78	<3.2	>96
Fluorene	3	4	540	<24	>96
Chrysene	3	4	<180	<24	NM
Bis(2-ethylhexyl)phthalate	3	4	<3.2	<1.6	NM

NM, not meaningful.
Source: USEPA.

Table 7 Wood Preserving Phenols Analysis Data Averages for Plants with Current BPT In Place

Pollutant	Number of plants	Waste load (kg/1000 cu.m) Raw	Treated	Percent removal
Phenols	3	5,600	<3.2	>99
2-Chlorophenol	3	<6.4	<1.6	NM
2,4-Dimethyl phenol	3	700	<16	>98
2,4,6-Trichlorophenol	3	<80	<1.6	NM
Pentachlorophenol	5	1,200	220	82

NM, not meaningful.
Source: USEPA.

Tables 10–14 present treated effluent characteristics and various average raw and treated waste characteristics and removals for the insulation board manufacturing subcategory [8].

31.4.3 Hardboard Manufacturing Wastewater Treatment

The following is a summary of the current level of in-place treatment technology for 13 hardboard manufacturing plants:

1. Use of an activated sludge and aerated lagoon system for wastewater treatment at a hardboard manufacturing plant producing S1S and S2S hardboards.
2. Use of a lime neutralization and discharge to POTW system for wastewater treatment at a hardboard manufacturing plant producing S1S hardboards.

Table 8 Wood Preserving Heavy Metals in Organic Preservatives Analysis Data Averages for Plants With Current BPT In Place

Pollutant	Number of plants	Waste load (kg/10E6 cu.m) Raw	Treated	Percent removal
Antimony	4	3.5	<1.3	>63
Arsenic	4	990	540	45
Beryllium	4	<0.16	<0.16	NM
Cadmium	4	<0.16	<0.32	NM
Chromium	4	1.9	1.6	16
Copper	4	7.7	5.6	27
Lead	4	6.9	<3.4	>51
Mercury	4	<0.16	<0.16	NM
Nickel	4	1.6	1.6	0
Selenium	4	0.32	0.32	0
Silver	4	<0.16	<0.16	NM
Thallium	4	<0.16	<0.16	NM
Zinc	4	26	15	42

NM, not meaningful.
Source: USEPA.

Table 9 Wood Preserving Heavy Metals in Both Organic and Inorganic Preservatives Analysis Data Averages for Plants with Current BPT In Place

Pollutant	Waste load (kg/10E6 cu.m) Raw	Treated	Percent removal
Antimony	<0.16	<0.16	NM
Arsenic	40	40	0
Beryllium	<0.16	<0.16	NM
Cadmium	0.32	1.6	NM
Chromium	7.2	15	NM
Copper	24	29	NM
Lead	5.0	4.8	4
Mercury	0.48	0.16	67
Nickel	30	5.4	82
Selenium	<0.16	<0.16	NM
Silver	<0.16	<0.16	NM
Thallium	<0.16	<0.16	NM
Zinc	37	50	NM

NM, not meaningful.
Source: USEPA.

Table 10 Insulation Board Thermomechanical Refining Treated Effluent Characteristics, Annual Average

Plant number	Production		Flow		BOD		TSS	
	Mg/day	tons/day	1000 L/Mg	1000 gal/ton	kg/Mg	lb/ton	kg/Mg	lb/ton
537	145	160	1.75	0.419	1.9	3.8	1.1	2.3
108[a]	605	665[b]	51.3	12.3	4.1	8.1	12	24
103	359	395[b]	21.9	5.26	2.2	4.3	0.94	1.9

[a]Data are taken before paper wastewater is added.
[b]Includes both insulation board and hardboard production.
Source: USEPA.

Table 11 Insulation Board Mechanical Refining Raw Wastewater and Treated Effluent Characteristics, Annual Average

Plant number	BOD (kg/Mg) Raw waste	Treated effluent	Percent reduction	TSS (kg/Mg) Raw waste	Treated effluent	Percent reduction
360	4.6	1.0	76	0.88	1.2	NM
36	21	0.28	99	31	1.5	95
889	1.3	0.07	95	0.46	0.16	65

NM, not meaningful.
Source: USEPA.

Table 12 Insulation Board Thermomechanical Refining Raw Wastewater and Treated Effluent Characteristics, Annual Average (*Source*: USEPA)

	BOD (kg/Mg)			TSS (kg/Mg)		
Plant number	Raw waste	Treated effluent	Percent reduction	Raw waste	Treated effluent	Percent reduction
537	24	2.2	92	39	1.3	97
108	30	4.1	86	29	12	58
1035	43	2.2	95		0.94	

Source: USEPA.

Table 13 Raw and Treated Effluent Loadings and Percent Reductions for Insulation Board Metals

	Plant 360			Plant 183		
	Waste load (mg/Mg)			Waste load (mg/Mg)		
Pollutant	Raw	Treated	Percent reduction	Raw	Treated	Percent reduction
Antimony	2.1	18	NM	25	21	16
Arsenic	13	6	54	27	13	52
Beryllium	4.2	2.1	50	7	12	NM
Cadmium	2.8	3.5	NM	8	13	NM
Chromium	6	22	NM	60	20	67
Copper	1,900	900	53	2,300	20	99
Lead	6	6	0	170	21	88
Mercury	2.1	0.4	81	41	13	68
Nickel	800	600	25	850	900	NM
Selenium	14	7	50	35	25	29
Silver	2.1	2.1	0	4.9	17	NM
Thallium	2.8	8	<0	4.1	4.1	0
Zinc	3,000	1,400	53	4,200	4,800	NM

NM, not meaningful.
Source: USEPA.

Table 14 Industrial Board Toxic Pollutant Data, Organics

	Average concentration (μg/L)				
	Raw wastewater			Treated effluent	
Pollutant	Plant 183	Plant 36	Plant 531	Plant 36	Plant 537
Chloroform	20	BDL	BDL	BDL	BDL
Benzene	70	40	BDL	BDL	BDL
Toluene	60	40	BDL	BDL	BDL
Phenol	BDL	40	BDL	BDL	BDL

BDL, below detection limits.
Source: USEPA.

3. Use of an activated sludge, humus ponds, aerated lagoons, and settling pond system for wastewater treatment at a hardboard manufacturing plant producing S1S and S2S hardboards.
4. Use of a settling ponds system for wastewater treatment at a hardboard manufacturing plant producing S1S hardboards.
5. Use of an air ponds, Infilco Aero Accelerators, aerated lagoons, and facultative lagoon system for wastewater treatment at a hardboard manufacturing plant producing S2S hardboards.
6. Use of a settling pond and aerated lagoon system for wastewater treatment at a hardboard manufacturing plant producing S1S hardboards.
7. Use of a nonspecified system for wastewater treatment at a hardboard manufacturing plant producing S2S hardboards.
8. Use of a settling ponds and aerated lagoon system for wastewater treatment at a hardboard manufacturing plant producing S1S and S2S hardboards.
9. Use of a settling ponds and aerated lagoon system for wastewater treatment at a hardboard manufacturing plant producing S1S hardboards.
10. Use of an activated sludge, settling ponds and aerated lagoon system for wastewater treatment at a hardboard manufacturing plant producing S1S hardboards.
11. Use of an activated sludge and aerated lagoon system for wastewater treatment at a hardboard manufacturing plant producing S1S and S2S hardboards.
12. Use of an aerated lagoons and settling ponds system for wastewater treatment at a hardboard manufacturing plant producing S1S hardboards.
13. Use of a clarifier, aerated lagoon and oxidation ponds system for wastewater treatment at a hardboard manufacturing plant producing S2S hardboards.

Tables 15 and 16 present treated effluent characteristics and various raw and treated waste characteristics and percent removals for the hardboard manufacturing subcategory [8].

31.5 TREATMENT TECHNOLOGY COSTS

The investment cost, operating and maintenance costs, and energy costs for the application of control technologies to the wastewaters of the timber products processing industry have been analyzed. These costs were developed to reflect the conventional use of technologies in this

Table 15 S1S Hardboard Subcategory Toxic Pollutant Data, Organics

	Average concentration (μg/L)			
	Raw wastewater		Treated effluent	
Pollutant	Plant 207	Plant 931	Plant 207	Plant 931
Chloroform	BDL	20	BDL	BDL
Benzene	BDL	80	10	80
Ethylbenzene	20	BDL	BDL	BDL
Toluene	15	70	BDL	70
Phenol	BDL	680	BDL	20

BDL, below detection limits.
Source: USEPA.

Table 16 S2S Hardboard Subcategory Toxic Pollutant Data, Organics

	Average concentration (μg/L)			
	Raw wastewater		Treated effluent	
Pollutant	Plant 1	Plant 943	Plant 1	Plant 943
Chloroform	20	BDL	BDL	BDL
1,1,2 Trichloroethane	BDL	90	BDL	BDL
Benzene	90	BDL	40	BDL
Toluene	60	10	30	BDL
Phenol	300	BDL	BDL	BDL

BDL, below detection limits.
Source: USEPA.

industry. Several unit operation and unit process configurations have been analyzed for the cost of application of technologies and to select BPT and BAT level of treatment. A detailed presentation of the applicable treatment technologies, cost methodology, and cost data are available in the literature [6].

REFERENCES

1. Australian EPA. *Questions and Answers on the Wood Preservation Industry*; Australian Environmental Protection Authority, June 19, 2003.
2. Ministries of Environment & Health. *Health and Environmental Guidelines for Selected Timber Treatment Chemicals*; Manatu Hauora, June 1997; pp. 37, http://www.mfe.govt.nz/publications/hazardous/timber-guide-jun97/chapter-1-jun97.pdf
3. Wang, L.K.; Krouzek, J.V.; Kounitson, U. *Case Studies of Cleaner Production and Site Remediation*; United Nations Industrial Development Organization (UNIDO): Vienna, Austria, 1995; Training Manual No. DTT-5-4-95, 136.
4. Van Berkel, C.W.M. Cleaner production: a profitable road for sustainable development of Australian industry. *Clean Air*, 1999, *33(4)*, 33–38.
5. Aruna, P.B.; Mercer, D.E. The timber economy of the Mid-Atlantic Region. In *Proceedings of the 1998 Southern Forest Economics Workshop*, March 25, 1998; 172–179.
6. Wang, J.C.; Aulenbach, D.B.; Wang, L.K. Energy models and cost models for water pollution controls. In *Clean Production*; Misra, K.B., Ed.; Springer-Verlag: Berlin, Germany, 685–720.
7. Famisan, G.; Grubbs, S.; Willey, T. *Sources of Soil and Ground Water Contaminants: Hazardous Wastes*; Civil Engineering Department, Virginia Tech.: VA, 1999; http://www.cee.vt.edu/program_areas/environmental/teach/gwprimer/group16/hazwaste.html
8. USEPA. *Development Document for Effluent Limitations Guidelines and Standards for the Timber Products Processing Point Source Category*, EPA-440/1-79/023-b; US Environmental Protection Agency: Washington, DC, 1979; 427p.

32

Treatment of Power Industry Wastes

Lawrence K. Wang
Zorex Corporation, Newtonville, New York, U.S.A. and
Lenox Institute of Water Technology, Lenox, Massachusetts, U.S.A.

32.1 INTRODUCTION

32.1.1 Steam Electric Power Generation Industry

The steam electric power generation industry is defined as those establishments primarily engaged in the steam generation of electrical energy for distribution and sale. Those establishments produce electricity primarily from a process utilizing fossil-type fuel (coal, oil, or gas) or nuclear fuel in conjunction with a thermal cycle employing the steam–water system as the thermodynamic medium. The industry does not include steam electric power plants in industrial, commercial, or other facilities. The industry in the United States falls under two Standard Industrial Classification (SIC) Codes: SIC 4911 and SIC 4931.

There are about 1000 steam electric power generating plants in operation in the United States. Of these plants, approximately 35% generate in excess of 500 megawatts (MW) and approximately 12% generate 25 MW or less. These steam electric power generating plants represent about 79% of the entire electric utility generating capacity, and they generate about 85% of electricity produced by the entire electric utility industry. Within the steam electric power generation industry, plants built after 1970 represent 44% of the total capacity, and plants built before 1960 represent 26% of capacity.

"Small units" are defined by the US Environmental Protection Agency (USEPA) as generating units of less than 25 MW capacity. "Old units" are defined as generating units of 500 MW or greater rated net generating capacity that were first placed into service on or before January 1, 1970, as well as any generating unit of less than 500 MW capacity first placed in service on or before January 1, 1974.

The term "10-year, 24-hour rainfall event" refers to a rainfall event with a probable recurrence interval of once in 10 years as defined by the National Weather Service.

32.1.2 Power Generation, Waste Production, and Effluent Discharge

In the operation of a power plant, combustion of fossil fuels – coal oil, or gas – supplies heat to produce steam, which is used to generate mechanical energy in a turbine. This energy is subsequently converted by a generator to electricity. Nuclear fuels, currently uranium, are used in a similar cycle except that the heat is supplied by nuclear fusion wastewater discharge. A number of different operations by steam electric power plants discharge chemical wastes. Many wastes are discharged more or less continuously as long as the plant is operating. These include wastewaters from the following sources: cooling water systems, ash handling systems, wet-scrubber air pollution control systems, and boiler blowdown. Some wastes are produced at regular intervals, as in water treatment operations, which include a cleaning or regenerative step as part of their cycle (ion exchange, filtration, clarification, evaporation). Other wastes are also produced intermittently but are generally associated with either the shutdown or startup of a boiler or generating unit, such as during boiler cleaning (water side), boiler cleaning (fire side), air preheater cleaning, cooling tower basin cleaning, and cleaning of miscellaneous small equipment.

The discharge frequency for these varies from plant to plant. Some or all of the various types of wastewater streams occur at almost all of the plant sites in the industry. However, most plants do not have distinct and separate discharge points for each source of wastewater; rather, they combine certain streams prior to final discharge.

Additional wastes exist that are essentially unrelated to production. These depend on meteorological or other factors. Rainfall runoff, for example, causes drainage from coal piles, ash piles, floor and yard drains, and from construction activity.

The summary for the steam electric power generating (utility) point source category in terms of the number of dischargers in industry is as follows:

- direct dischargers in industry: 1050;
- indirect dischargers in industry: 100;
- zero dischargers in industry: 10.

Current BPT regulations for the steam electric power industry for generating, small and old units can be found elsewhere [1].

32.2 INDUSTRY SUBCATEGORY AND SUBDIVISIONS

Subcategories for the steam electric utility point source category are developed according to chemical waste stream origin within a plant. This approach is a departure from the usual method of subcategorizing an industry according to different types of plants, products, or production processes. Categorization by waste source provides the best mechanism for evaluating and controlling waste loads since the steam electric power plant waste stream source has the strongest influence on the presence and quantity of various pollutants as well as on flow. The breakdown of the stream electric power generation industry into subcategories and subdivisions is based on similarities in wastewater characteristics throughout the industry. The eight broad subcategories and their subdivisions are presented below:

1. Once-through cooling water.
2. Recirculating cooling system blowdown.
3. Ash transport water:
   - fly ash transport;
   - bottom ash transport.
4. Low volume wastes:
   - clarifier blowdown;
   - makeup water filter backwash;
   - ion exchange softener regeneration;
   - evaporator blowdown;
   - lime softener blowdown;
   - reverse osmosis brine;

- demineralizer regenerant;
- powdered resin demineralizer;
- floor drains;
- laboratory drains;
- sanitary wastes; and
- diesel engine cooling system discharge.

5. Metal cleaning wastes:
   - boiler tube cleaning;
   - cleaning rinses;
   - fireside wash; and
   - air preheater wash.
6. Ash pile, chemical handling, and construction area runoff coal pile runoffs.
7. Coal pile runoff.
8. Wet flue gas cleaning blowdown.

32.2.1 Once-Through Cooling Water Subcategory

In a steam electric power plant, cooling water is utilized to absorb heat that is liberated from the steam when it is condensed to water in the condensers. The cooling water is withdrawn from a water source, passed through the system, and returned directly to the water source. Shock (intermittent) chlorination is employed in many cases to minimize the biofouling of heat transfer surfaces. Continuous chlorination is used only in special situations. Based on 308 data, approximately 65% of the existing steam electric power plants have once-through cooling water systems.

32.2.2 Recirculating Cooling Water Subcategory

In a recirculating cooling water system, the cooling water is withdrawn from the water source and passed through condensers several times before being discharged to the receiving water. After each pass through the condenser, heat is removed from the water through evaporation. Evaporation is carried out in cooling ponds or canals, in mechanical draft evaporative cooling towers, and in natural draft evaporative cooling towers. In order to maintain a sufficient quantity of water for cooling, additional makeup water must be withdrawn from the water source to replace the water that evaporates.

When water evaporates from the recirculating cooling water system, the dissolved solids content of the water remains in the system, and the dissolved solids concentration tends to increase over time. If left unattended, the formation of scale deposits will result. Scaling due to dissolved solids buildup is usually controlled through the use of a bleed system called cooling tower blowdown. A portion of the cooling water in the system is discharged via blowdown, and since the discharged water has a higher dissolved solids content than the intake water used to replace it, the dissolved solids content of the water in the system is reduced.

Chemicals such as sulfuric acid are used to control scaling in the system. Biofoulants such as chlorine and hypochlorite are widely used by the industry. These additives are discharged in the cooling tower blowdown.

32.2.3 Ash Transport Water Subcategory

Steam electric power plants using oil or coal as a fuel produce ash as a waste product of combustion. The total ash product is a combination of bottom ash and fly ash. Because the ash composition of oil is much less than that of coal, the presence of ash is an extremely important consideration in the design of a coal-fired boiler. Improper design leads to the accumulation of ash deposits on furnace walls and tubes, leading to reduced heat transfer, increased pressure drop, and corrosion. Accumulated ash deposits are removed and transported to a disposal system.

The method of transport may be either wet (sluicing) or dry (pneumatic). Dry handling systems are more common for fly ash than bottom ash. The dry ash is usually disposed of in a landfill, but the ash is also sold as an ingredient for other products. Wet ash handling systems produce wastewaters that are either discharged as blowdown from recycle systems or are discharged to ash ponds and then to receiving streams in recycle and once-through systems.

Ash From Oil-Fired Plants

Fly ash is a light material, which is carried out of the combustion chamber in the flue gas stream. The ash from fuel oil combustion is usually in the form of fly ash. The many elements that may appear in oil ash deposits include vanadium, sodium, and sulfur.

Ash From Coal-Fired Plants

More than 90% of the coal used by electric utilities is burned in pulverized coal boilers. In these boilers, 65–80% of the ash produced is in the form of fly ash. This fly ash is carried out of the combustion chamber in the flue gases and is separated from these gases by electrostatic precipitators and/or mechanical collectors. The remainder of the ash drops to the bottom of the furnace as bottom ash. While most of the fly ash is collected, a small quantity may pass through the collectors and be discharged to the atmosphere. The vapor is that part of the coal material that is volatilized during combustion. Some of these vapors are discharged into the atmosphere; others are condensed onto the surface of fly ash particles and may be collected in one of the fly ash collectors.

32.2.4 Low-Volume Wastes Subcategory

Low-Volume Blowdowns

Low-volume wastes include wastewaters from all sources except those for which specific limitations are otherwise established in 40 CFR 423. Waste sources include, but are not limited to, wastewaters from wet scrubber air pollution control systems [2], ion exchange water treatment systems, water treatment evaporator blowdown, laboratory and sampling streams, floor drainage, cooling tower basin cleaning wastes, and blowdown from recirculating house service water systems. Sanitary wastes and air conditioning wastes are specifically excluded from the low-volume waste subcategory.

Boiler Blowdown

Power plant boilers are either of the once-through or drum-type design. Once-through boilers operate under supercritical conditions and have no wastewater streams directly associated with their operation. Drum-type boilers operate under subcritical conditions where steam generated in the drum-type units is in equilibrium with the boiler water. Boiler water impurities are concentrated in the liquid phase. Boiler blowdown serves to maintain concentrations of dissolved and suspended solids at acceptable levels for boiler operation. The sources of impurities in the blowdown are the intake water, internal corrosion of the boiler, and chemicals added to the boiler. Phosphate is added to the boiler to control solids deposition.

In modern high-pressure systems, blowdown water is normally of better quality than the water supply. This is because plant intake water is treated using clarification, filtration, lime/lime soda softening, ion exchange, evaporation, and in a few cases reverse osmosis to produce makeup for the boiler feedwater. The high-quality blowdown water is often reused within the plant for cooling water makeup or it is recycled through the water treatment and used as boiler feedwater.

32.2.5 Metal Cleaning Wastes Subcategory

Metal cleaning wastes result from cleaning compounds, rinse waters, or any other waterborne residues derived from cleaning any metal process equipment, including, but not limited to, boiler tube cleaning, boiler fireside cleaning, and air preheater cleaning.

Boiler Tube Cleaning

Chemical cleaning is designed to remove scale and corrosion products that accumulate in the steam side of the boiler. Hydrochloric acid, which forms soluble chlorides with the scale and corrosion products in the boiler tubes, is the most frequently used boiler tube cleaning chemical. In boilers containing copper, a copper complexer is used with hydrochloric acid to prevent the replating of dissolved copper onto steel surfaces during chemical cleaning operations. If a complexer is not used, copper chlorides, formed during the cleaning reaction, react with boiler tube iron to form soluble iron chlorides while the copper is replated onto the tube surface.

Alkaline cleaning (flush/boil-out) is commonly employed prior to boiler cleaning to remove oil-based compounds from tube surfaces. These solutions are composed of trisodium phosphate and a surfactant and act to clear away the materials that may interfere with reactions between the boiler cleaning chemicals and deposits.

Citric acid cleaning solutions are used by a number of utilities in boiler cleaning operations. The acid is usually diluted and ammoniated to a pH of 3.5 and then used for cleaning in a two-stage process. The first stage involves the dissolution of iron oxides. In the second stage, anhydrous ammonia is added to raise the pH of the cleaning solution to between 9 and 10 and air is bubbled through the solution to dissolve copper deposits.

Ammoniated EDTA has been used in a wide variety of boiler cleaning operations. The cleaning involves a one-solution, two-stage process. During the first stage, the solution solubilizes iron deposits and chelates the iron solution. In the second stage, the solution is oxidized with air to induce iron chelates from ferric to ferrous and to oxidize copper deposits into solution where the copper is chelated. The most prominent use of this agent is in circulating boilers that contain copper alloys.

When large amounts of copper deposits in boiler tubes cannot be removed with hydrochloric acid due to the relative insolubility of copper, ammonia-based oxidizing compounds have been effective. Used in a single separate stage, the ammonia sodium bromate step includes the introduction into the boiler system of solutions containing ammonium bromate to rapidly oxidize and dissolve the copper.

The use of hydroxyacetic/formic acid in the chemical cleaning of utility boilers is common. It is used in boilers containing austenitic steels because its low chloride content prevents possible chloride stress corrosion cracking of the austenitic-type alloys. It has also found extensive use in the cleaning operations for once-through supercritical boilers. Hydroxyacetic/formic acid has chelation properties and a high iron pick-up capability; thus it is used on high iron content systems. It is not effective on hardness scales.

Sulfuric acid has found limited use in boiler cleaning operations. It is not feasible for removal of hardness scales due to the formation of highly insoluble calcium sulfate. It has found some use in cases where a high-strength, low-chloride solvent is necessary. Use of sulfuric acid requires high water usage in order to rinse the boiler sufficiently.

Boiler Fireside Washing

Boiler firesides are commonly washed by spraying high-pressure water against boiler tubes while they are still hot.

Air Preheater Washing

Air preheaters employed in power generating plants are either the tubular or regenerative types. Both are periodically washed to remove deposits that accumulate. The frequency of washing is typically once per month; however, frequency variations ranging from 5 to 180 washings per year are reported. Many preheaters are sectionalized so that heat transfer areas may be isolated and washed without shutdown of the entire unit.

32.2.6 Ash Pile, Chemical Handling, and Construction Area Runoff Subcategory

Fly ash and bottom ash stored in open piles, chemicals spilled in handling, and soil distributed by construction activities will be carried in the runoff caused by precipitation events.

32.2.7 Coal Pile Runoff Subcategory

In order to ensure a consistent supply of coal for steam generation, plants typically maintain an outdoor 90-day reserve supply. The piles are usually not enclosed, so the coal comes in contact with moisture and air, which can oxidize metal sulfides to sulfuric acid. Precipitation then results in coal pile runoff with minerals, metals, and low pH (occasionally) in the stream.

32.2.8 Wet Flue Gas Cleaning Blowdown Subcategory

Depending on the fossil fuel sulfur content, an SO_2-scrubber may be required to remove sulfur emissions in the flue gases. These scrubbing systems result in a variety of liquid waste streams depending on the type of process used. In all of the existing FGD (flue gas desulfurization) systems, the main task of absorbing SO_2 from the stack gases is accomplished by scrubbing the existing gases with an alkaline slurry. This may be preceded by partial removal of fly ash from the stack gases. Existing FGD processes may be divided into two categories: nonregenerable FGD processes include lime, limestone, and lime/limestone combination, and double alkali systems.

In the lime or limestone FGD process, SO_2 is removed from the flue gas by wet scrubbing with a slurry of calcium oxide or calcium carbonate [3]. The waste solid product is disposed by ponding or landfill. The clear liquid product can be recycled. Many of the lime or limestone systems discharge scrubber waters to control dissolved solids levels.

A number of processes can be considered double alkali processes, but most developmental work has emphasized sodium-based systems, which use lime for regeneration. This system pretreats the flue gas in a prescrubber to cool and humidify the gas and to reduce fly ash and chlorides. The gas passes through an absorption tower where SO_2 is removed into a scrubbing solution, which is subsequently regenerated with lime or limestone in a reaction tank.

The disadvantage of all nonregenerable systems is the production of large amounts of throwaway sludges. Onsite disposal is usually performed by sending the waste solids to a settling pond. The supernatant from the ponds may be recycled; however, according to 308 data, 82% of the plants with FGD systems discharged the supernatant into surface waters.

32.3 WASTEWATER

32.3.1 Characterization

Wastewater produced by a steam electric power plant can result from a number of operations at the site. Many wastewaters are discharged more or less continuously as long as the plant is operating. These include wastewaters from the following sources: cooling water systems, ash handling systems, wet-scrubber air pollution control systems, and boiler blowdown. Some wastes are produced at regular intervals, as in water treatment operations that include a cleaning or regenerative step as part of their cycle (ion exchange, filtration, clarification, evaporation). Other wastes are also produced intermittently but are generally associated with either the shutdown or startup of a boiler or generating unit such as during boiler cleaning (water side), boiler cleaning (fire side), air preheater cleaning, cooling tower basin cleaning, and cleaning of miscellaneous small equipment. Additional wastes exist that are essentially unrelated to production. These depend on meteorological or other factors. Rainfall runoff, for example, causes drainage from coal piles, ash piles, floor and yard drains, and from construction activity. A diagram indicating potential sources of wastewaters containing chemical pollutants in a coal-fueled steam electric power plant is shown in Figure 1.

Data on wastestream characteristics presented in this section are based on the results of screening sampling carried out at eight plants, verification sampling carried out at 18 plants, and periodic surveillance and analysis sampling carried out as part of compliance monitoring at eight plants. These data were stored on a computerized data file [1]. All waste streams discussed in this chapter were analyzed during the screening program, while the verification program focused on the following waste streams: once-through cooling water, cooling tower blowdown, and ash handling waters. The wastewater characteristics of the various waste streams are discussed in the following sections. Where they are available, only verification data are presented. Where verification data are limited or not available, screening and/or surveillance and analysis data are presented. The data source is clearly indicated in each table and in the text.

The following is a summary of all priority pollutants detected in any of the waste streams from steam electric power plants:

- Benzene
- Chlorobenzene
- 1,2-Dichloroethane
- 1,1,1-Trichloroethane
- 1,1,2-Trichloroethane
- 2-Chloronaphthalene
- Chloroform
- 2-Chlorophenol
- 1,2-Dichlorobenzene
- 1,4-Dichlorobenzene
- 1,1-Dichloroethylene
- 1,2-*trans*-Dichloroethylene
- 2,4-Dichiorophenol

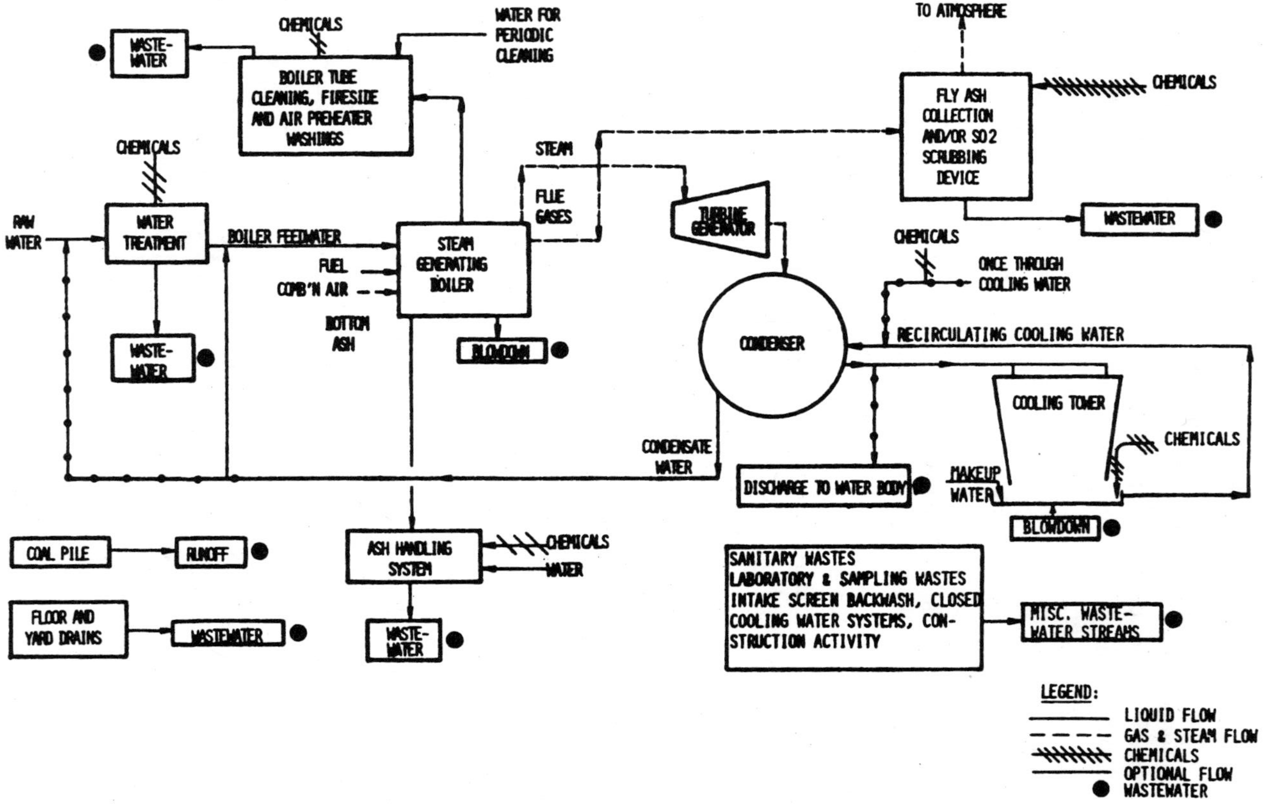

Figure 1 Potential sources of wastewater in a stream electric power generation plant. (Courtesy of USEPA.)

- Ethylbenzene
- Methylene chloride
- Bromoform
- Dichlorobromomethane
- Trichlorofluoromethane
- Chlorodibromomethane
- Nitrobenzene
- Pentachlorophenol
- Phenol
- *Bis*(2-ethylhexyl) phthalate
- Butyl benzyl phthalate
- Di-*n*-butyl phthalate
- Di-*n*-octyl phthalate
- Diethyl phthalate
- Dimethyl phthalate
- Tetrachloroethylene
- Toluene
- Trichloroethylene
- 4,4-DDD
- Antimony (total)
- Arsenic (total)
- Asbestos (total-fibers/L)
- Beryllium (total)
- Cadmium (total)
- Chromium (total)
- Copper (total)
- Cyanide (total)
- Lead (total)
- Mercury (total)
- Nickel (total)
- Selenium (total)
- Silver (total)
- Thallium (total)
- Zinc (total)

32.3.2 Cooling Water

In general, wastewater characteristics of once-through cooling water and recirculating cooling water systems are similar. Pollutants discharged from both systems are caused by the erosion or corrosion of construction materials plus the chemical additives used to control erosion, scaling, and biological growth (biofouling). The wastewater generated from a recirculating cooling water system also depends on the design limits for dissolved solids in the system.

Erosion

The fill material in natural draft cooling towers is frequently asbestos cement. Erosion of this fill material may result in the discharge of asbestos in cooling water blowdown. In a testing program for detection of asbestos fibers in the waters of 18 cooling systems, seven of the 18 sites contained detectable concentrations of chrysotile asbestos in the cooling tower waters at the time of sampling.

Corrosion

Corrosion is an electrochemical process that occurs when metal is immersed in water and a difference in electrical potential between different parts of the metal causes a current to pass through the metal between the region of lower potential (anode) and the region of higher potential (cathode). The migration of electrons from anode to cathode results in the oxidation of the metal at the anode and the dissolution of metal ions into the water.

Copper alloys are used extensively in power plant condensers, and as a result, copper can usually go into a corrosion product film or directly into solution as an ion or as a precipitate in the initial stages of condensation by tube corrosion. As corrosion products form and increase in thickness, the corrosion rate decreases until a steady state is achieved. Studies indicate that copper release is a function of flow rate more so than of the salt content of the makeup water.

Data on copper concentrations in both once-through cooling and recirculatory cooling systems indicate that corrosion products are more of a problem in cooling tower blowdown than in once-through systems discharge. The concentration of pollutants (via evaporation) in recirculating systems probably accounts for most of the difference in the level of metals observed between once-through discharge and cooling tower blowdown.

Chemical Treatment

Chemical additives are needed at some plants with recirculating cooling water systems in order to prevent corrosion and scaling. Chemical additives are also occasionally used at plants with once-through cooling water systems for corrosion controls.

Scaling occurs when the concentration of dissolved materials, usually calcium- and magnesium-containing species, exceeds their solubility levels. The addition of scaling control chemicals allows a higher dissolved solids concentration to be achieved before scaling occurs.

Therefore, the amount of blowdown required to control scaling can be reduced. Chemicals added to once-through cooling water to control corrosion or to recirculating cooling water to control corrosion and scaling is usually present in the discharges. Chromium and zinc are the active components of most of the popular corrosion inhibitors.

The solvent and carrier components that may be used in conjunction with scaling and corrosion control agents are as follows:

- Dimethyl formamide
- Methanol
- Ethylene glycol monomethyl ether
- Ethylene glycol monobutyl ether
- Methyl ethyl ketone
- Glycols to hexylene glycol
- Heavy aromatic naphthalene
- Cocoa diamine
- Sodium chloride
- Sodium sulfate
- Polyoxyethylene glycol
- Talc
- Sodium aluminate
- Monochlorotoulene
- Alkylene oxide–alchohol glycol ethers

Chlorine and hypochlorite are used to control biofouling in both once-through and recirculating cooling water systems. The addition of chlorine to the water causes the formation of toxic compounds and chlorinated organics, which may be priority pollutants.

Eleven plants with once-through cooling water systems were sampled as part of the verification program and the surveillance and analysis sampling efforts. Four of these plants have estuarine or salt water intakes, and the remaining seven plants have fresh water intakes. Sampling was carried out only during the period of chlorination. Samples were analyzed for all organic priority pollutants except the pesticides, and for total organic carbon and total residual chlorine (nine plants). Table 1 is a summary of the data collected in the verification and surveillance and analysis sampling efforts. Only the priority pollutants that were detected are shown.

The following is a summary of once-through cooling system flow rates based on responses to 308 USEPA questionnaires.

1. The once-through cooling system flow rate range of 239 steam electric power generation plants using coal is 0.189–6,280,000 m^3/plant (mean 1,130,000 m^3/plant), which is equal to 0.001–209,000 m^3/MW (mean 4,310,000 m^3/MW).
2. The once-through cooling system flow rate range of 104 steam electric power generation plants using gas is 0.29–7,230,000 m^3/plant (mean 783,000 m^3/plant), which is equal to 0.006–13,800,000 m^3/MW (mean 2,410,000 m^3/MW).
3. The once-through cooling system flow rate range of 138 steam electric power generation plants using oil is 0.007–26,700,000 m^3/plant (mean 1,490,000 m^3/plant), which is equal to 0.00004–219,000 m^3/MW (mean 5260 m^3/MW).

The fuel designations in the above surveyed data were determined by the fuel that contributes the most Btu for power generation during the survey.

The surveyed data also indicate that there were net increases in all of the following compounds: total dissolved solids, total suspended solids, total organic carbon, total residual chlorine, free available chlorine 2,4-dichlorophenol, 1,2-dichlorobenzene, phenolics, chromium, lead, copper, mercury, silver, iron, arsenic, zinc, barium, calcium, manganese, sodium, methyl chloride, aluminum, boron, and titanium.

Eight power plants with cooling towers were sampled at intake and discharge points during the verification sampling program. The results of the verification sampling program for cooling tower blowdown (recirculating cooling water) are presented in Table 2A and 2B for intake information and discharge information, respectively. Only the priority pollutants that were detected are shown.

The following is a summary of cooling tower blowdown flow rates based on responses to 308 questionnaires:

1. The cooling tower blowdown flow rate range of 82 steam electric power generation plants using coal is 0–152,000 m^3/plant (mean 8440 m^3/plant), which is equal to 0–239 m^3/MW (mean 11.2 m^3/MW).
2. The cooling tower blowdown flow rate range of 120 steam electric power generation plants using gas is 0–10,900 m^3/plant (mean 1190 m^3/plant), which is equal to 0–99 m^3/MW (mean 11.6 m^3/MW).
3. The cooling tower blowdown flow rate range of 47 steam electric power generation plants using oil is 0–12,100 m^3/plant (mean 1040 m^3/plant), which is equal to 0–63.2 m^3/MW (mean 7.04 m^3/MW).

The data also indicate that there was a net increase from the influent concentration to the effluent concentration for the following compounds: trichlorofluoromethane, bromoform,

Table 1 Summary of Priority Pollutants in the Steam Electric Industry Once-Through Cooling Water

	Intake					Discharge				
Pollutant	Number of samples	Number of detections	Range of detections	Median of detections	Mean of detections	Number of samples	Number of detections	Range of detections	Median of detections	Mean of detections
Classical pollutants (mg/L)										
Total residual chlorine	11	6	<0.01–3	<0.01	<0.5	7	6	0.02–5.5	0.4	0.82
COD	11	1	35			7	0			
TDS	11	11	7–35,000	340	7,600	7	7	360–33,000	12,000	12,000
TSS	11	8	3–100	13	25	7	7	4–90	14	28
TOC	11	11	6–34	12	14	7	7	7.9–30	17	19
Free residual chlorine	10	1	500			7	1	190		
Phenolics	11	6	<0.005–0.015	0.010	0.009	7	5	0.007–0.26	0.02	0.06
Barium	11	5	0.01–0.06	0.024	0.029	6	0			
Calcium	11	5	0.084–51	45	37	6	0			
Manganese	11	5	0.053–0.2	0.066	0.096	6	0			
Magnesium	11	4	13–33	19	21	6	0			
Sodium	11	5	<15–49	21	24	6	0			
Iron	11	5	0.25–4	0.84	1.7	6	1	0.76		
Aluminum	11	4	0.28–2.4	1.2	1.3	6	0			
Boron	11	3	0.07–0.09	0.083	0.081	6	0			
Tin	11	2	0.03–0.036		0.033	6	0			
Titanium	11	3	0.018–0.051	0.040	0.036	6	0			
Molybdenum	11	1	0.009			6	0			
Cobalt	11	2	<0.005–0.01		0.0075	6	0			
Vanadium	11	0				6	0			
Toxic pollutants (μg/L)										
Toxic metals and inorganics										
Antimony	11	3	7–16	10	11	7	1	14		
Aresenic	11	3	3–5	4	4	7	0			
Chromium	11	6	7–24	13	14	6	1	14		
Copper	11	6	7–22	18	16	6	1	24		
Cyanide	11	1	10			7	0			
Lead	11	2	9–10		9.5	6	1	11		
Mercury	11	3	0.2–1.2	1.0	0.8	7	1	1		

Selenium	11	2	3.0–3.8		3.4	7	0	
Silver	11	1	30			6	1	36
Nickel	11	3	8–120	17	48	6	1	120
Zinc	11	4	32–340	79	130	5	1	24
Cadmium	11	2	13–40		26	6	1	16
Toxic organics								
Bromoform	9	0				11	1	31
Chlorodibromomethane	10	0				11	1	2.6
Bis(2-ethylhexyl) phthalate	11	4	<10–420	<10	110	10	0	
Gamma-BHC	10	0				11	1	<0.1
2,4-Dichlorophenol	11	0				11	1	6
1,2-Dichlorobenzene	11	0				11	1	30
1,4-Dichlorobenzene	11	1	18			10	0	
Benzene	10	3	<10–40	20	22	0		
2-Chloronapthalene	11	0				0		
Chloroform	10	1	<10			0		
1,1-Dichloroethylene	10	1	10			0		
Ethyl benzene	10	1	<10			0		
Methylene chloride	10	1	<10			0		
Phenol	11	3	9–26	<10	13	0		
Butyl benzyl phthalate	11	1	<10			0		
Di-*n*-butyl phthalate	11	4	<10– < 10	<10	<10	0		
Toluene	10	1	<10			0		
Trichloroethylene	10	1	<10			0		
1,1,1-Trichloroethane	11	0				0		
Pentachlorophenol	11	0				0		
Diethyl phthalate	11	1	50		50	0		
Tetrachloroethylene	11	1	<10			0		
Benzidene	11	0				0		
Methyl chloride	11	1	50			0		

Source: USEPA.

Table 2A Summary of Priority Pollutants in the Steam Electric Industry Recirculating Cooling Water

Pollutant	Intake				
	Number of samples	Number of detections	Range of detections	Median of detections	Mean of detections
Classical pollutant (mg/L)					
TDS	8	6	190–26,000	300	4,600
TSS	7	7	0.005–110	2	20
TOC	8	8	<1–34	15	16
Phenolics	7	5	0.002–0.02	0.007	0.008
Total residual chlorine	7	4	0.005–14	0.59	3.8
Sedium	8	5	17–6,000	95	1,300
Tin	8	2	0.06–0.3		0.18
Titanium	8	3	0.02–0.2	0.03	0.083
Iron	8	4	1.0–4.0	2	2.2
Vanadium	8	4	0.011–0.2	0.016	0.061
Barium	8	5	0.02–0.5	0.1	0.19
Boron	8	5	0.06–5.0	0.3	1.9
Calcium	8	6	6.9–340	56	120
Cobalt	8	3	0.008–0.01	0.01	0.009
Manganese	8	6	0.008–7.6	0.14	1.4
Magnesium	8	5	4.5–100	22	33
Molybdenum	8	4	0.02–0.08	0.02	0.035
Aluminum	8	2	0.7–2		1.4
Toxic pollutants (μg/L)					
Toxic metals and inorganics					
Antimony	8	2	4–7		5.5
Arsenic	8	0			
Cadmium	8	5	1.4–100	8	24
Chromium	8	6	7–440	71	140
Copper	8	8	10–700	30	110

Cyanide	8	2	4–15,000		7,500
Lead	8	5	6–500	11	120
Nickel	8	6	1.7–200	82	82
Silver	8	4	1.3–40	1.8	11
Selenium	8	2	2–2		2
Zinc	8	5	15–340	53	100
Thallium	8	1	20		
Mercury	8	1	0.5		
Toxic organics					
Benzene	8	2	1.2–2.4		1.8
Carbon tetrachloride	8	1	<1		
Chloroform	7	1	1.4		
1,2-Dichlorobenzene	8	2	5.3–18		12
Dichlorobromomethane	6	0			
Chlorodibromomethane	6	0			
Toluene	8	2	2–9.1		5.5
Trichloroethylene	8	1	4		
1,4-Dichlorobenzene	8	1	2.4		
2,4,6-Trichlorophenol	8	0			
2,4-Dichlorophenol	8	4	3–12	7	7.2
Pentachlorophenol	7	2	4–12		8
Bromoform	7	0			
			Discharge		
Classical pollutant (mg/L)					
TDS	8	6	430–34,000	1,600	11,000
TSS	8	8	2–460	30	84
TOC	8	8	8–76	10	28
Phenolics	7	4	2.5–20	8	9.6
Total residual chlorine	7	6	0.005–110	1.3	19
Sodium	8	5	33–7,000	210	1,500

Table 2B Summary of Priority Pollutants in the Steam Electric Industry Recirculating Cooling Water–Discharge Information

	Discharge				
Pollutant	Number of samples	Number of detections	Range of detections	Median of detections	Mean of detections
Tin	8	4	0.03–0.5	0.045	0.16
Titanium	8	2	0.02–0.2		0.11
Iron	8	4	0.3–4	1.5	1.8
Vanadium	8	5	0.01–0.2	0.02	0.054
Barium	7	3	0.02–0.2	0.1	0.11
Boron	8	5	0.06–4	0.07	1.2
Calcium	8	6	6.9–460	82	130
Cobalt	8	4	0.008–0.06	0.01	0.022
Manganese	8	5	0.05–0.1	0.1	0.076
Magnesium	8	5	4.9–57	20	28
Molybdenum	8	4	0.02–0.1	0.05	0.055
Aluminum	8	2	0.2–1		0.6
Toxic pollutant (μg/L)					
Toxic metals and inorganics					
Antimony	8	2	5–7		6
Arsenic	8	2	2–4		3
Cadmium	8	6	1–200	4.5	36
Chromium	8	8	2–550	42	150
Copper	8	8	42–3,800	55	560
Cyanide	8	2	3–5		4
Lead	8	4	3–800	85	240

Nickel	8	8	4–140	9	56
Silver	8	5	0.7–80	3	20
Selenium	8	0			
Zinc	8	5	38–780	200	270
Thallium	8	2	8–20		14
Mercury	8	2	0.2–1		0.6
Toxic organics					
Benzene	8	2	<1–1.5		1.0
Carbon tetrachloride	7	0			
Chloroform	8	2	<1–2.4		1.4
1,2-Dichlorobenzene	7	1	26		
Dichlorobromomethane	8	2	2.6–8.2		5.4
Chlorodibromomethane	8	2	<1–58		30
Toluene	7	1	24		
Trichloroethylene	8	1	4		
1,4-Dichlorobenzene	7	0			
2,4,6-Trichlorophenol	8	1	35		
2,4-Dichlorophenol	8	2	8–8		
Pentachlorophenol	8	2	4–4		8
Bromoform	8	1	150		4

Source: USEPA.

chlorodibromomethane, *bis*(2-ethylhexyl) phthalate, antimony, arsenic, cadmium, chromium, mercury, nickel, selenium, silver, thallium, benzene, tetrachloroethylene, toluene, copper, cyanide, lead, zinc, chloroform, phenol, asbestos, total dissolved solids, total suspended solids, total organic carbon, total residual chlorine, 1,2-dichlorobenzene, 2,4-dichlorophenol, boron, calcium, magnesium, molybdenum, total phenolics, sodium, tin, vanadium, cobalt, iron, chloride, 2,4,6-trichlorophenol, and pentachlorophenol. It must be recognized, however, that recirculating cooling systems tend to concentrate the dissolved solids present in the make-up water and, thus, a blowdown stream with many different compounds showing concentration increases is to be expected. Of the priority pollutants detected as net discharges, the concentration increase was greater than 10 ppb only for *bis*(2-ethylhexyl) phthalate, cadmium, chromium, nickel, selenium, silver, toluene, copper, cyanide, lead, zinc, phenol, 1,2-dichlorobenzene, total phenolics, and 2,4,6-trichlorophenol.

32.3.3 Ash Transport

The chemical compositions of both types of bottom ash, dry or slag, are quite similar. The major species present in bottom ash are silica (20–60 wt.% as SiO_2), alumina (10–35 wt.% as Al_2O_3), ferric oxides (5–35 wt.% as Fe_2O_3), calcium oxide (1–20 wt.% as CaO), magnesium oxide (0.3–0.4 wt.% as MgO), and minor amounts of sodium and potassium oxides (1–4 wt.%). In most instances, the combustion of coal produces more fly ash than bottom ash. Fly ash generally consists of very fine spherical particles, ranging in diameter from 0.5 to 500 microns. The major species present in fly ash are silica (30–50 wt.% as SiO_2), alumina (20–30 wt.% Al_2O_3), and titanium dioxide (0.4–1.3 wt.% as TiO_2). Other species that may be present include sulfur trioxide, carbon, boron, phosphorus, uranium, and thorium.

In addition to these major components, a number of trace elements are also found in bottom ash and fly ash. The trace elemental concentrations can vary considerably within a particular ash or between ashes. Generally, higher trace element concentrations are found in the fly ash than bottom ash; however, there are several cases where bottom ash exceeds fly ash concentration. Fly ash demonstrates an increased concentration trend with decreasing particle sizes.

During the verification sampling effort, the ash pond overflows of nine facilities were sampled to further quantify those effluent pollutants identified in the screening program. The data are presented in Table 3A and 3B for information of intake and discharge, respectively.

32.3.4 Low-Volume Wastes

Low-volume waste sources include water treatment processes that prevent scale formation such as clarification, filtration, lime/lime soda softening, ion exchange, reverse osmosis, and evaporation. Also included are drains and spills from floor and yard drains and laboratory streams.

Clarification Wastes

Clarification is the process of agglomerating the solids in a stream and separating them by settling. Chemicals that are commonly added to the clarification process do not contain any of the listed priority pollutants.

Table 3A Summary of Priority Pollutants in the Steam Electric Industry Ash Pond Overflow–Intake Information

	Intake				
Pollutant	Number of samples	Number of detections	Range of detections	Median of detections	Mean of detections
Classical pollutants (mg/L)					
Oil and grease	10	1	25		
TDS	11	8	130–530	260	300
TSS	9	8	0.005–170	13	41
TOC	10	7	5–21	10	9.4
Phenolics	10	4	0.006–0.04	0.01	0.02
Chloride	10	1	14		
Aluminium	11	5	0.2–2	0.5	0.78
Barium	11	7	0.017–0.06	0.03	0.032
Boron	11	6	0.06–0.1	0.075	0.077
Calcium	11	8	6.9–57	33	33
Cobalt	11	3	0.007–0.04	0.01	0.019
Manganese	11	8	0.04–0.8	0.082	0.19
Magnesium	11	8	4.5–23	6.9	13
Molybdenum	11	2	0.009–0.06		0.034
Sodium	11	8	15–57	26	28
Tin	11	3	0.01–0.036	0.03	0.025
Titanium	11	4	0.018–0.04	0.025	0.027
Iron	11	10	0.2–20	0.5	3
Vanadium	10	3	0.01–0.04	0.013	0.021
Yttrium	11	0			
Toxic pollutants (μg/L)					
Toxic metals and inorganics					
Antimony	11	3	3–7	4	4.7
Arsenic	11	2	2–7		2.5
Beryllium	11	0			
Cadmium	11	5	2.1–40	6.5	11
Chromium	11	9	3–4,000	10	460
Copper	11	11	8–700	20	84
Cyanide	11	2	4–15,000		7,500
Lead	11	9	1.7–20	10	11
Mercury	11	2	0.2–0.5		0.35
Nickel	11	10	1.7–2,000	18	210
Selenium	11	2	2–3		2.5
Silver	11	2	1.5–1.6		1.6
Thallium	11	1	1		
Zinc	11	7	15–88	32	4.3
Toxic organics					
Benzene	11	3	1.2–10	2.4	4.5
Carbon tetrachloride	11	1	1		
Chloroform	11	3	0.17–10	1.4	3.9
1,2-Dichlorobenzene	11	1	5.3		
Ethylbenzene	9	0			

(*continues*)

Table 3A *Continued*

Pollutant	Intake				
	Number of samples	Number of detections	Range of detections	Median of detections	Mean of detections
Toluene	11	2	2–9.1		5.5
Trichloroethylene	11	2	0.57–4		1.3
1,1-Dichloroethylene	11	0			
1,4-Dichlorobenzene	11	1	2.4		
Methylene chloride	10	1	10		
Phenol	11	1	9		
Bis(2-ethylhexyl) phthalate	11	1	10		
Butyl benzyl phthalate	11	1	10		
Di-*n*-butyl phthalate	11	1	10		
Diethyl phthalate	11	1	50		
Dimethyl phthalate	11	0	2.7		
Tetrachloroethylene	11	2	0.4–10		5.2
1,1,2,2-Tetrachloroethane	11	1	24		
1,1,1-Trichloroethane	11	1	0.68		
Pentachlorophenol	11	1	3.8		
Pesticides					
4,4′-DDD	11	1	1		

Table 3B Summary of Priority Pollutants in the Steam Electric Industry Ash Pond Overflow–Discharge Information

Pollutant	Discharge				
	Number of samples	Number of detections	Range of detections	Median of detections	Mean of detections
Classical pollutants (mg/L)					
Oil and grease	12	2	1–24		12
TDS	12	9	4–2,400	490	700
TSS	12	11	5–160	15	33
TOC	12	8	3–150	16	32
Phenolics	12	7	0.006–0.04	0.01	0.02
Chloride	12	2	37–37		37
Aluminium	12	6	0.06–5	0.55	1.4
Barium	12	8	0.04–0.2	0.06	0.083
Boron	12	8	0.08–2	0.6	0.95
Calcium	12	9	21–140	64	71
Cobalt	12	4	0.007–0.05	0.015	0.022
Manganese	12	8	0.01–1	0.1	0.24
Magnesium	12	9	5.6–20	9.5	11
Molybdenum	12	9	0.005–0.3	0.05	0.094
Sodium	12	9	15–70	32	34
Tin	12	6	0.007–0.036	0.025	0.022

(continues)

Table 3B *Continued*

	Discharge				
Pollutant	Number of samples	Number of detections	Range of detections	Median of detections	Mean of detections
Titanium	12	2	0.02–0.05		0.035
Iron	12	9	0.17–8	0.9	1.7
Vanadium	12	5	0.02–0.14	0.022	0.028
Yttrium	12	1	0.02		
Toxic pollutants (μg/L)					
Toxic metals and Inorganics					
Antimony	12	4	6–10	6.5	7.2
Arsenic	11	2	9–300		150
Beryllium	12	3	2–2.5	2.5	2.3
Cadmium	12	8	1–90	3.5	15
Chromium	12	12	4–1,000	12	120
Copper	12	11	8–80	25	43
Cyanide	12	2	22–22		22
Lead	12	8	1.2–120	8.5	25
Mercury	11	3	0.2–1.5	1.0	0.9
Nickel	12	12	5.2–470	21	65
Selenium	12	3	3–13	8	8
Silver	12	6	0.5–24	4.8	7.8
Thallium	12	0			
Zinc	12	8	15–1,200	100	260
Toxic organics					
Benzene	12	3	2–10	2	4.0
Carbon tetrachloride	11	0			
Chloroform	11	2	0.25–10		5.1
1,2-Dichlorobenzene	11	0			
Ethylbenzene	12	3	1–10		4.0
Toluene	12	2	3.5–3.5	1	3.5
Trichloroethylene	10	0			
1,1-Dichloroethylene	12	1	10		
1,4-Dichlorobenzene	12	1	2.4		
Methylene chloride	12	2	10–32		21
Phenol	12	1	4		
Bis(2-ethylhexyl)phthalate	12	1	10		
Butyl benzyl phthalate	12	0			
Di-*n*-butyl phthalate	12	1			
Diethyl phthalate	12	1	10		
Dimethyl phthalate	12	1	10		
Tetrachloroethylene	11	0	10		
1,1,2,2-Tetrachloroethane	11	0			
1,1,1-Trichloroethane	11	0			
Pentachlorophenol	12	1	6.5		
Pesticides					
4,4′-DDD					

Source: USEPA.

Ion Exchange Wastes

Ion exchange processes can be designed to remove all mineral salts in a one-unit operation and, as such, is the most common means of treating supply water. The process uses an organic resin that must be regenerated periodically by backwashing and releasing the solids. A regenerant solution is passed over the bed and it is subsequently washed.

The resulting exchange wastes are generally acidic or alkaline with the exception of sodium chloride solutions, which are neutral. While these wastes do not have significant amounts of suspended solids, certain chemicals such as calcium sulfate and calcium carbonate have extremely low solubilities and are often precipitated because of common ion effects.

Spent regenerant solutions, constituting a significant part of the total flow of wastewater from ion exchange regeneration, contains ions that are eluted from the ion exchange material plus the excess regenerant that is not consumed during regeneration. The eluted ions represent the chemical species that were removed from water during the service cycle of the process. Table 4A and 4B present a summary of ion exchange demineralizer regenerant wastes characterized in the surveillance and analysis study, for the information of intake and discharge, respectively.

Lime Softener Wastewater

Softening removes hardness using chemical precipitation. The two major chemicals used are calcium hydroxide and sodium carbonate, thus no priority pollutants will be introduced into the system.

Reverse Osmosis Wastewater

Reverse osmosis is a process used by some plants to remove dissolved salts. The waste stream from this process consists of reverse osmosis brine. In water treatment schemes reported by the industry, reverse osmosis was always used in conjunction with demineralizers, and sometimes with clarification, filtration, and ion exchange softening.

Floor and Yard Drain Wastewater

As a result of the numerous potential sources of wastewater from equipment drainage and leakage throughout a steam electric facility, the pollutants encountered in such wastewaters may be diverse. There have been little data reported for these waste streams; however, the pollutant parameters that may be of concern are oil and grease, pH, and suspended solids.

Laboratory Drain Wastewater

The wastes from the laboratories vary in quantity and constituents, depending on the use of the facilities and the type of power plant. The chemicals are usually present in extremely small quantities. It has been common practice to combine laboratory drains with other plant plumbing.

Boiler Blowdown

Boiler blowdown is generally of fairly high quality because the boiler feedwater must be maintained at high quality. Boiler blowdown having a high pH may contain a high dissolved solids concentration depending on boiler pressure. The sources of impurities in the blowdown are the intake water, internal corrosion of the boiler, and chemicals added to the boiler system. Impurities contributed by the intake water are usually soluble inorganic species (Na^+, K^+, Cl^-,

Table 4A Summary of Priority Pollutants in the Steam Electric Industry Demineralizer Regenerant–Intake Information

	Intake				
Pollutant	Number of samples	Number of detections	Range of detections	Median of detections	Mean of detections
Classical pollutants (mg/L)					
TDS	3	2	210–290		250
TSS	2	1	2.8		
TOC	3	2	2.3–9		5.6
Aluminium	3	1	0.50		
Barium	3	1	0.017		
Boron	3	0			
Calcium	3	1	49		
Manganese	3	1	0.065		
Magnesium	3	1	15		
Molybdenum	3	0			
Sodium	2	0			
Titanium	3	1	0.018		
Iron	3	2	0.001–0.84		0.42
Toxic pollutants (μg/L)					
Toxic metals and inorganics					
Antimony	3	0			
Arsenic	3	2	2–3		2.5
Cadmium	3	1	4		
Copper	3	3	9–22	22	18
Chromium	3	1	10		
Cyanide	3	0			
Lead	3	0			
Mercury	3	3	0.2–1.5	1	0.9
Nickel	3	1	8		
Selenium	3	1	1		
Silver	3	0			
Zinc	3	3	10–100	88	66
Thallium	3	0			
Toxic organics					
Benzene	3	1	<10		
Chloroform	3	2	4.4–68		36
1,1-Dichloroethylene	3	0			
Methylene chloride	3	1	<10		
Bromoform	3	1	23		
Dichlorobromomethane	3	1	0.87		
Chlorodibromomethane	3	1	0.17		
Phenol	3	2	4.2–9		6.6
Bis(2-ethylhexyl) Phthalate	3	1	<10		
Butyl benzyl phthalate	3	1	<10		
Di-*n*-butyl phthalate	3	1	<10		
Diethyl phthalate	3	1	50		
Tetrachloroethylene	3	1	<10		

(*continues*)

Table 4A *Continued*

Pollutant	Intake				
	Number of samples	Number of detections	Range of detections	Median of detections	Mean of detections
Trichloroethylene	3	2	0.13–<10		<5.1
Chlorobenzene	3	0			
1,1,2-Trichloroethane	3	1	0.23		
1,2-Dichlorobenzene	3	0			
1,3-Dichlorobenzene	3	0			
1,4-Dichlrorobenzene	3	0			
Nitrobenzene	3	0			

Table 4B Summary of Priority Pollutants in the Steam Electric Industry Demineralizer Regenerant–Discharge Information

Pollutant	Discharge				
	Number of samples	Number of detections	Range of detections	Median of detections	Mean of detections
Classical pollutant (mg/L)					
TDS	3	2	3,000–4,600		3,800
TSS	3	2	9.2–17		13
TOC	3	2	4.8–8		6.4
Aluminium	3	1	0.28		
Barium	3	0			
Boron	3	1	0.063		
Calcium	3	1	170		
Manganese	3	1	0.009		
Magnesium	3	1	17		
Molybdenum	3	1	0.015		
Sodium	3	1	160		
Iron	3	2	0.790–5		2.9
Toxic pollutant (μg/L)					
Toxic metals and inorganics					
Antimony	3	1	20		
Arsenic	1	0			
Cadmium	3	2	5–35		20
Copper	2	2	14–65		40
Chromium	3	2	14–26		20
Cyanide	3	2	0.04–47		24
Lead	3	1	24		
Mercury	2	2	1.6–6		3.8
Nickel	3	2	200–230		220
Selenium	2	1	4		
Silver	3	1	58		
Zinc	2	1	54		
Thallium	3	1	180		

(continues)

Table 4B *Continued*

Pollutant	Discharge				
	Number of samples	Number of detections	Range of detections	Median of detections	Mean of detections
Toxic organics					
Benzene	3	0			
Chloroform	3	3	1.8–140	38	60
1,1-Dichloroethylene	3	1	<10		
Methylene chloride	3	2	60–>220		>140
Bromoform	2	1	>10		
Dichlorobromomethane	3	1	70		
Chlorodibromomethane	3	1	30		
Phenol	3	2	3.8–4		3.9
Bis(2-ethylhexyl) Phthalate	3	1	<10		
Butyl benzyl phthalate	2	0			
Di-*n*-butyl phthalate	3	1	<10		
Diethyl phthalate	3	1	<10		
Tetrachloroethylene	3	1	<10		
Trichloroethylene	3	1	0.38		
Chlorobenzene	3	1	0.67		
1,1,2-Trichloroethane	3	1	0.68		
1,2-Dichlorobenzene	3	1	39		
1,3-Dichlorobenzene	3	1	0.3		
1,4-Dichlrorobenzene	3	1	5.2		
Nitrobenzene	3	1	81		

Source: USEPA.

SO_4^{2-}, etc.) and precipitates containing calcium/magnesium cations. Products of boiler corrosion are soluble and insoluble species of iron, copper, and other metals. A number of chemicals are added to the boiler feedwater to control scale formation, corrosion, pH, and solids deposition. Table 5 presents a summary of toxic and classical pollutants detected in verification analyses of boiler blowdown.

32.3.5 Metal Cleaning Wastes

Metal cleaning wastes include wastewater from chemical cleaning of boiler tubes, air preheater washwater, and boiler fireside washwater.

Chemical Cleaning of Boiler Tubes

The characteristics of waste streams emanating from the chemical cleaning of utility boilers are similar in many respects. The major constituents consists of boiler metals; that is, alloy metals used for boiler tubes, hot wells, pumps, and so on. Although waste streams from certain cleaning operations that are used to remove certain deposits, such as alkaline degreaser to remove oils and organics, do not contain heavy concentrations of metals, the primary purpose of the total boiler cleaning operation (all stages combined) is removal of heat

Table 5 Summary of Priority Pollutants in the Steam Electric Industry Boiler Blowdown

	Intake					Discharge				
Pollutant	Number of samples	Number of detections	Range of detections	Medium of detections	Mean of detections	Number of samples	Number of detections	Range of detections	Medium or detection	Mean or detections
Classical pollutants (mg/L)										
TDS	4	2	210–290		250	4	3	7–100	11	39
TSS	3	1	2.8			4	3	0.8–<5	<5	1.9
TOC	3	2	2.3–9		5.6	4	3	1.2–3	<3	1.9
Oil and grease	4	0				4	1	5		
Phenolics	4	2	4.2–<20		7.1	4	2	6.4–<20		8.2
Aluminum	4	0				4	1	0.21		
Calcium	4	1	49			4	2	<5–<5		<5
Manganese	4	1	0.065			4	0			
Magnesium	4	1	15			4	0			
Molybdenum	4	0				4	2	0.055–0.061		0.058
Sodium	3	0				4	1	<15		
Iron	4	2	0.01–0.84		0.43	3	1	0.06		
Titanium	4	1	0.018			4	0			
Toxic pollutants (μg/L)	4									
Toxic metals	4									
Antimony	4	0				4	3	6–20	10	12
Arsenic	4	2	2–3		2.5	4	2	2–2		2
Cadmium	4	1	4			4	1	5		
Chromium	4	1	10			4	1	6		
Copper	4	3	9–22	22	18	4	4	8–520	17	140
Lead	4	0				4	2	36–40		38
Mercury	4	3	0.2–1.5	1.0	0.9	3	1	1.7		
Nickel	4	1	8			4	1	1.3		
Selenium	4	0				4	1	5.7		
Zinc	4	3	10–100	88	66	4	3	10–72	68	50
Toxic organics	4									

Benzene	4	1	<10			4	2	30–290	160
1,1,1-Trichloroethane	4	0				4	1	<10	
1,1,2,2-Tetrachloroethane	4	0				4	1	<10	
Chloroform	4	3	4.4–6.8	<5	4.6	4	2	0.12–<10	2.6
1,1-Dichloroethylene	4	0				4	2	<10–<60	<35
Ethyl benzene	4	0				4	2	<10–<10	<10
Methylene chloride	4	1	<10			4	1	910	
Bis(2-thylhexyl)phthalate	4	1	<10			4	1	<10	
Butyl benzyl phthalate	4	1	<10			4	0		
Di-*n*-butyl phthalate	4	1	<10			4	2	<10–<10	<10
Diethyl phthalate	4	1	50			4	2	<10–<10	<10
Tetrachloroethylene	4	1	<10			4	2	<10–<10	<10
Toluene	4	0					2		
Trichloroethylene	4	2	0.13–<10			4	0		
Dichlorobromomethane	4	2	0.87–23		2.6	4	0		
Chlorodibromomethane	4	2	0.17–3.8		12	4	0		
1,1,2-Trichloroethane	4	1	0.23		2	4	0		
Bromoform		1	0.07			4	1	<10	
1,3-Dichloropropene	4	0				4	1	<10	
Phenol	4	0				4	1	10	

Source: USEPA.

transfer-retarding deposits, which consist mainly of iron oxides resulting from corrosion. This removal of iron is evident in all total boiler cleaning operations through its presence in boiler cleaning wastes.

Cleaning mixtures used include alkaline chelating rinses, proprietary chelating rinses, organic solvents, acid cleaning mixtures, and alkaline mixtures with oxidizing agents for copper removal. Wastes from these cleaning operations will contain iron, copper, zinc, nickel, chromium, hardness, and phosphates. In addition to these constituents, wastes from alkaline cleaning mixtures will contain ammonium ions, oxidizing agents, and high alkalinity; wastes from acid cleaning mixtures will contain fluorides, high acidity, and organic compounds; wastes from alkaline chelating rinses will contain high alkalinity and organic compounds; and wastes from most proprietary processes will be alkaline and will contain organic and ammonium compounds. Other waste constituents present in spent chemical cleaning solutions include wide ranges of pH, high dissolved solids concentrations, and significant oxygen demands (BOD and/or COD). The pH of spent solutions ranges from 2.5 to 11.0 depending on whether acidic or alkaline cleaning agents are employed.

Table 6 presents a summary of toxic and classical pollutants detected in three common cleansing solutions: ammoniacal sodium bromate, hydrochloric acid without copper complexer, and hydrochloric acid with copper complexer.

Boiler Fireside Wastewater

When boiler firesides are washed, the waste effluents produced contain an assortment of dissolved and suspended solids. Acid wastes are common for boilers fired with high-sulfur fuels. Sulfur oxides absorb onto fireside deposits, causing low pH and a high sulfate content in the waste effluent.

Air Preheater Wastewater

Fossil fuels with significant sulfur content will produce sulfur oxides that absorb on air preheater deposits. Water washing of these deposits produces an acidic effluent. Alkaline reagents are often added to washwater to neutralize acidity, prevent corrosion of metallic surfaces, and maintain an alkaline pH. Alkaline reagents might include soda ash (Na_2CO_3), caustic soda (NaOH), phosphates, and/or detergent. Preheater washwater contains suspended and dissolved solids, which include sulfates, hardness, and heavy metals including copper, iron, nickel, and chromium.

32.3.6 Ash Pile, Chemical Handling, and Construction Area Runoff

Runoff wastewater characteristics change all the time. No reliable data have been gathered. The readers are referred to the literature for similar technical information [4,5].

32.3.7 Coal Pile Runoff

No reliable data have been gathered for the wastewater characteristics of coal pile runoff. Example 3 (Section 32.6.3) presents the technical information on the characteristics of a combined wastewater (consisting of coal pile runoff, regeneration wastewater, and fly ash), and its treated effluent.

Table 6 Summary of Priority Pollutants in the Steam Electric Industry Metal Cleaning Wastes

	Ammoniated EDTA Solutions					Ammonia Sodium Bromide Solutions				
Pollutant	Number of samples	Number of detections	Range of detections	Median of detections	Mean of detections	Number of samples	Number of detections	Range of detections	Median of detections	Mean of detections
Classical pollutants (mg/L)										
TDS	2	2	60,000–74,000		67,000	3	3	340–1,400	1,000	920
TSS	1	1	24			3	3	8–77	71	52
COD						2	2	24–120		72
Oil and grease	1	1	41			2	2	<5–<5		<5
pH, pH Units	7	7	8.8–10	9.2	9.3	2	2	10–10		10
Phosphorous	1	1	260			2	2	10–30		20
Bromide						2	2	<5–52		28
Chloride						1	1	60		
Fluoride						2	2	1.5–6.1		3.0
Aluminum	1	1	31			2	2	<0.2–<0.2		<0.2
Calcium	2	2	21–45		33	3	2	0.4–3		1.7
Barium						2	2	<0.1–<0.1		<0.1
Sodium	1	1	370			3	3	3.7–59	15	26
Potassium						2	2	70–220		140
Tin						2	2	<1–<1		<1
Iron	7	7	2,200–8,300	6,900	6,300	5	4	0.15–4.9	1.8	2.2
Manganese	2	2	50–73		61	3	3	0.01–0.04	0.03	0.03
Magnesium	2	2	11–21		16	3	2	0.67–2.9		1.8

(*continues*)

Table 6 *Continued*

Pollutant	Ammoniated EDTA Solutions					Ammonia Sodium Bromide Solutions				
	Number of samples	Number of detections	Range of detections	Median of detections	Mean of detections	Number of samples	Number of detections	Range of detections	Median of detections	Mean of detections
Toxic pollutants (µg/L)										
Toxic metals										
Arsenic						3	3	2.5–310,000	40	100,000
Cadmium						2	2	<10–<10		<10
Chromium	3	3	10,000–26,000	12,000	16,000	2	2	<1–<20		<10
Copper	7	7	17–12,000,000	120,000	1,900,000	4	3	<5–<50	<5	<20
Lead						6	6	100,000–790,000	370,000	420,000
Mercury						3	3	<10–100	<10	37
Nickel	3	3	12,000–140,000	68,000	73,000	3	3	<0.2–15,000	<0.2	5,000
Selenium						5	4	80–260,000	1,500	57,000
Silver						3	3	<2–24,000	<2	8,000
Zinc	3	3	79,000–140,000	120,000	110,000	2	2	<10–<20		<15
						5	5	60–1,000	500	510

Source: USEPA

32.3.8 Wet Flue Gas Cleaning Blowdown

The readers are referred to another source for more detailed information regarding wet flue gas cleaning blowdown characteristics and treatment [2,3].

32.4 WASTE TREATMENT

32.4.1 End-of-Pipe Treatment Technologies

Wastewater effluents discharged to publicly owned treatment facilities are sometimes treated by physical or chemical systems to remove pollutants potentially hazardous to the POTW or which may be treated inadequately in the POTW. Such treatment methods are numerous, but they generally fall into one of three broad categories in accordance with their process objectives. These include pH control, removal of dissolved materials, and separation of phases.

The following is a summary of end-of-pipe treatment technologies commonly employed in the steam electric power generation industry, their objectives, equipment and processes required, and efficiency [12–22].

Neutralization

This is a process for pH adjustment, usually to within the range 6–9. Acid or base is used as required; this is usually in the form of sulfuric acid or lime.

Chemical Reduction

This is a process mainly used in power plants for reduction of hexavalent chromium to trivalent chromium. Sulfur dioxide, sodium bisulfite, sodium metabisulfite, and ferrous salts are common reducing agents to be used in the process. A pH range of 2–3 should be controlled. The process efficiency of removal is about 99.7%.

Precipitation

This is a process mainly used in power plants for removal of ions by forming insoluble salts. Common precipitating agents are lime, hydrogen sulfide, organic precipitants, and sods ash. Optimum pH depends on the ions to be removed. The removal efficiency for inorganic pollutants is as follows:

- Copper, 96.6%;
- Nickel, 91.7%;
- Chromium, 98.8 %;
- Zinc, 99.7%;
- Phosphate, 93.6%.

Ion Exchange

This is a process mainly used in power plants for removal of ions by sorption on the surface of a solid matrix. Synthetic cation and/or anion exchange resins are required depending on the pollutants to be removed. It may require pH adjustments. The removal efficiency for inorganic pollutants is as follows:

- Cyanide, 99%;
- Chromium, 98%;

- Copper, 95%;
- Iron, 100%;
- Cadmium, 92%;
- Nickel, 100%;
- Zinc, 75%;
- Phosphate, 90%;
- Sulfate, 97%;
- Aluminum, 98%.

Liquid/Liquid Extraction

This is a process mainly used in power plants for removal of soluble organics or chemically charged pollutants. The required chemicals are immiscible solvents that may contain chelating agents. It may require pH adjustments. The removal efficiency for inorganic pollutants is as follows:

- Phenol, 99%;
- Chromium, 99%;
- Nickel, 99%;
- Zinc, 99%;
- Fluoride, 68%;
- Iron, 99%;
- Molybdenum, 90%;

Disinfection

This is a process for destruction of microorganisms. Chlorine, hypochlorite salts, phenol, phenol derivatives, ozone, salts of heavy metals, chlorine dioxide, and so on are effective disinfectants. It may require pH adjustments.

Adsorption

This is a process mainly used in power plants for removal of sorbable contaminants. Activated carbon, synthetic sorbents are the common adsorbents to be used in the process. It may require pH adjustments. The process removal efficiency depends on the nature of the pollutants and the composition of the waste.

Chemical Oxidation

This is a process mainly used in power plants for destruction of cyanides using chlorine, hypochlorite salts, or ozone. The process removal efficiency is about 99.6% [12–19].

Distillation

This is a process mainly used in power plants for separation of dissolved matters by evaporation of the water. Multistage flash distillation, multiple-effect vertical long-tube vertical evaporation, submerged tube evaporation, and vapor compression are effective process equipment. It may require pH adjustment. The process removal efficiency is about 100%.

Reverse Osmosis (RO)

This is a process mainly used in power plants for separation of dissolved matter by filtration through a semipermeable membrane. Tubular membrane, hollow filter modules, or spiral-wound flat sheet membrane can be adopted for the RO process. Total dissolved solids (TDS) removal efficiency is about 93% [20].

Electrodialysis (ED)

This is a process mainly used in power plants for removal of dissolved polar compounds. Solute is exchanged between two liquids through a selective semipermeable membrane in response to differences in chemical potential between the two liquids. The process removal efficiency for TDS is about 62–96%.

Freezing

This is a process mainly used in power plants for separation of solute from liquid by crystallizing the solvent. Either direct refrigeration, or indirect refrigeration can be used. The process removal efficiency is over 99.5%.

32.4.2 Solid–Liquid Separation Technologies

The solid/liquid separation technologies commonly employed in the steam electric power generation industry include the following.

Skimming

This is a process for removal of floating solids from liquid wastes. It requires between about 1 and 15 minutes of retention time, and has a removal efficiency of 70–90%.

Clarification (Conventional)

This is a process for removal of suspended solids by settling. Typical examples are settling ponds and settling clarifiers. It requires 45 minutes to 2 hours retention time (RT), and can reduce TSS to 15 mg/L or below.

Flotation

This is an innovative separation process for removal of suspended solids and oil and grease by flotation followed by skimming. It requires very short RT (less than 30 minutes), and can achieve 90–99% removal efficiency [15,18].

Microstraining

This is mechanical separation process for removal of suspended solids by passing the wastewater through a microscreen. A removal efficiency for TSS is 50–80% depending on the pore size of the microscreen to be used.

Filtration

This is physical operation for removal of suspended solids by filtration through a bed of sand and gravel. TSS removal efficiency is 50–90% depending on the type of filter media used and the filtration rate.

Screening

This is a unit operation for removal of large solid matter by passing through screens. The efficiency for large solid removal is 50–99% depending on the type of coarse screen or bar screen to be used.

Thickening

This is a process for concentration of sludge by removing water. Either gravity thickening or dissolved air flotation thickening can be used. The thickening efficiency depends on the nature of sludge to be processed [15].

Pressure Filtration

This is a unit operation for separation of solid from liquid by passing through a semipermeable membrane or filter media under pressure. It requires 1 to 3 hours of RT, and reduces 50% of moisture content.

Heat Drying

This is a process for reducing the water content of sludge by heating. Flash drying, spray drying, rotary kiln drying, or multiple hearth drying can be used.

Ultrafiltration

This is a separation process for removal of macromolecules of suspended matter from the waste by filtration through a semipermeable membrane under pressure. Total solids removal of 95% and above can be achieved.

Sandbed Drying

This is a process for removal of moisture from sludge by evaporation and drainage through sands. The RT is as long as 1 to 2 days. It is practised extensively by industry due its low cost.

Vacuum Filtration

This is a process for solid–liquid separation by vacuum. It requires about 1 to 5 minutes RT, and can produce 30% solid in filter cake.

Centrifugation

This is a liquid/solid separation process using centrifugal force. The moisture of the sludge can be reduced to 65–70%.

Emulsion Breaking

This process is effective for separation of emulsified oil and water. It requires 2 to 8 hours of RT. Over 99% removal efficiency can be achieved if aluminum salts, iron salts, and other demulsifiers are used at optimum pH conditions. It is practised extensively by the industry [21,22].

22.4.3 Cleaner Production, Industrial Ecology, and Other Issues

Traditional industries operated in a one-way, linear fashion: natural resources from the environment are used for producing products for our society, and the generated wastes are dumped back into our environment. However, natural resources such as minerals and fossil fuels are present in finite amounts, and the environment has a limited capacity to absorb waste. The field of industrial ecology has emerged to address these issues in the power industry. The thermal energy generated from the power industry, for instance, may be reused for many domestic and industrial applications, for cost saving as well as thermal pollution control. The greenhouse gas, carbon dioxide, in flue gas is a pollutant, but can also be reused as a chemical agent in wastewater treatment [8]. An extension of the concept of sustainable manufacturing – industrial ecology – seeks to use resources efficiently and regards "wastes" as potential products [11,23].

All electric power plants should practise cleaner production and industrial ecology strategies.

Additional issues of air and thermal pollution for electric power generation are addressed in detail by Wisconsin Public Service Commission [9].

The World Nuclear Association [10] provides detailed technical information on nuclear electricity and related environmental, health and safety issues.

32.5 TREATMENT TECHNOLOGY COSTS

The investment cost, operating and maintenance costs, and energy costs for the application of control technologies to the wastewaters of the steam electric power generating industry have been analyzed. These costs were developed to reflect the conventional use of technologies in this industry. Several unit operation/unit process configurations have been analyzed for the cost of application of technologies and to select BPT and BAT levels of treatment. A detailed presentation of the applicable treatment technologies, cost methodology, and cost data are available in the literature [6,7].

32.6 PLANT-SPECIFIC EXAMPLES

32.6.1 Example 1

Plant 1226 is a bituminous coal-, oil-, and gas-fired electricity plant [1]. The recirculator cooling water system influent was sampled from a stream taken from the river and the effluent from the cooling tower blowdown stream. The effluent stream is used again in the ash sluice stream. Table 7 presents the data. The following additives are combined with the cooling tower influent:

- chlorine (biocide);
- calgon Cl-5 (corrosion inhibitor);
- sulfuric acid (scale prevention).

The addition is necessary for the control of pipe corrosion.

Table 7 Plant-Specific Treatment Data for Plant 1226 Recirculating Cooling Water

Pollutant	Influent	Effluent[a]
Classical pollutants (mg/L)		
TDS	190	1,000
TSS	14	8
TOC	10	11
Phenolics	0.01	0.008
TRC[b]	ND	<0.01
Aluminum	0.7	0.4
Barium	0.02	0.02
Boron	ND	0.06
Calcium	6.9	6.9
Cobalt	0.007	0.008
Manganese	0.2	0.1
Magnesium	4.5	4.9
Sodium	33	210
Titanium	0.02	0.02
Iron	2.0	3.0
Vanadium	ND	0.03
Flow (L/s)	745	630
Toxic pollutants (μg/L)		
Toxic metals		
Antimony	7	7
Arsenic	3	4
Cadmium	2.1	1.8
Chromium	7	20
Copper	10	48
Lead	11	3
Mercury	0.5	0.2
Nickel	14	6
Silver	1.3	0.7
Zinc	40	38
Toxic organics		
Chloroform	NA	<1
Bromoform	NA	150
Dichlorobromomethane	NA	8.2
Chlorodibromomethane	NA	58

[a]Percent removal not meaningful because water does not undergo any treatment.
[b]Total residual chorine.
Source: USEPA.

32.6.2 Example 2

Plant 1245 is an oil- and gas-fired electric generating facility. The samples chosen are the influent and effluent from a once-through cooling tower stream. The influent sample was taken from the makeup stream comprised of river water, with the effluent stream being a direct discharge from the condensers to the river. The cooling water does not undergo any treatment to remove pollutants. The data reflect the changes that may occur to such a stream due to evaporation and pipe corrosion. Table 8 presents plant-specific data for plant 1245.

Table 8 Plant-Specific Treatment Data for Plant 1245 Once-Through Cooling Water

Pollutant	Influent	Effluent[a]
Classical pollutants (mg/L)		
TDS	35,000	33,000
TSS	6	14
TOC	14	25
Phenolics	<5	<5
TRC[b]	<10	120
Flow (L/s)	4,380	4,380

[a]Percent removal is not meaningful due to the fact that the water does not undergo any treatment.
[b]Total residual chlorine.
Source: USEPA.

Table 9 Plant-Specific Treatment Data for Plant 3920 Fly Ash Pond Water

Pollutant	Influent	Effluent	Percent removal
Classical pollutant (mg/L)			
TDS	220	880	NM
TSS	12	73	NM
TOC	5	3	40
Phenolics	0.04	0.04	0
Barium	0.03	0.06	NM
Boron	0.08	1	NM
Calcium	28	120	NM
Cobalt	ND	0.007	NM
Manganese	0.05	0.3	NM
Magnesium	7.2	6.7	7
Molybdenum	ND	10	NM
Sodium	18	35	NM
Tin	ND	ND	NM
Aluminum	ND	5	NM
Iron	0.5	2	NM
Flow (L/s)	61.3	61.3	
Toxic pollutants (μg/L)			
Toxic metals			
Cadmium	ND	ND	NM
Chromium	11	30	NM
Copper	8	30	NM
Lead	20	8	60
Nickel	25	18	28
Zinc	ND	140	NM
Beryllium	ND	2	NM
Silver	ND	ND	NM

ND, not detected.
NM, not meaningful.
Source: USEPA.

Table 10 Plant-Specific Treatment Data for Plant 1742 Ash Pond Water and Once-Through Cooling Water

	Ash Pond			Once-Through Cooling Water		
Pollutant	Influent	Effluent	Percent removal	Influent	Effluent	Percent removal
Classical pollutant (mg/L)						
TDS	340	370	NM	340	1,200	NM
TSS	100	15	85	100	90	10
TOC	10	150	NM	10	9	10
Phenolics	0.006	0.01	NM	0.006	0.260	NM
TRC				NA	830	NM
Aluminum	2	ND	>99	2	NA	
Barium	0.06	0.05	17	0.06	NA	
Boron	0.09	0.2	NM	0.09	NA	
Calcium	51	51	NM	51	NA	
Cobalt	0.01	0.05	NM	0.01	NA	
Manganese	0.2	0.3	NM	0.2	NA	
Magnesium	23	20	13	23	NA	
Molybdenum	0.009	0.05	NM	0.009	NA	
Sodium	21	26	NM	21	NA	
Tin	0.03	0.03	0	0.03	NA	
Titanium	0.04	ND	>99	0.04	NA	
Iron	4	8	NM	4	NA	
Vanadium	ND	20	NM	ND	NA	
Flow (L/s)	4.38	4.38		1,440	1,440	
Toxic pollutants (μg/L)						
Toxic metals						
Cadmium	40	10	75	40	NA	
Chromium	22	1,000	NM	22	NA	
Copper	20	78	NM	20		
Lead	9	9	0	9	NA	
Nickel	17	470	NM	17	NA	
Silver	ND	ND	0	ND	NA	
Zinc	70	ND	>99	70	NA	
Mercury	ND	1.5	NM	NA	NA	

ND, not detected; NM, not meaningful; NA, not analyzed.
Source: USEPA.

Table 11 Plant-Specific Treatment Data for Plant 3001 Multiple Ash Ponds Water

Pollutant	Influent	Effluent	Percent removal
Classical pollutant (mg/L)			
TDS	530	490	8
TSS	170	30	82
Oil and grease	25	24	4
Phenolics	NA	0.01	NM
Aluminum	0.5	2	NM
Barium	0.04	0.2	NM
Boron	0.06	2	NM
Calcium	38	64	NM
Manganese	0.04	ND	>99
Cadmium	ND	0.008	NM
Magnesium	23	11	52
Molybdenum	ND	0.03	NM
Sodium	57	70	NM
Tin	ND	0.007	NM
Iron	0.2	ND	>99
Vanadium	ND	0.02	NM
Flow (L/s)	23.3	unknown	
Toxic pollutants (μg/L)			
Toxic metals			
Chromium	10	190	NM
Copper	10	ND	>99
Lead	ND	3	NM
Nickel	6	35	NM
Toxic organics			
1, 1, 2, 2-Tetrachloroethane	24	ND	>99

Analytic methods: V.7.3.31, Data set 2; ND, not detected; NM, not meaningful; NA, not analyzed.
Source: USEPA.

32.6.3 Example 3

Plant 3920 is a bituminous coal- and oil-fired plant with a generating capacity of 557 MW. This plant uses 1,220,000 Mg/year of coal. An ash settling pond was used to remove wastes from coal pile runoff, regeneration wastes, and fly ash. The influent data were obtained from the pond inlet whereas the effluent data were from the discharge stream to the river. The results of this treatment are shown in Table 9.

32.6.4 Example 4

Plant 1742 is a bituminous coal- and oil-fired plant producing 22 MW of electricity. Table 10 represents data that are from both the ash pond and the once-through cooling tower.

32.6.5 Example 5

Plant 3001 is a coal- and gas-fired facility with a generating capacity of 50 MW. The plant uses approximately 277,000 Mg/year of coal. The fly ash and bottom ash from the boiler are combined and put through a series of three settling ponds. The effluent from the ponds is discharged to the river. Table 11 shows the effectiveness of this treatment technology.

REFERENCES

1. USEPA. *Development Document for Effluent Limitations Guidelines and Standards for the Steam Electric Point Source Category*, EPA-440/1-80/029-b; US Environmental Protection Agency: Washington, DC, 1980; 597p.
2. Wang, L.K.; Taricska, J.; Hung, Y.T.; Eldridge, J.; Li, K. Wet and dry scrubbing. In *Air Pollution Control Engineering*; Wang, L.K., Pereira, N.C., Hung, Y.T., Eds.; Humana Press, Inc.: Totowa, NJ, 2004.
3. Wang, L.K.; Williford, C.; Chen, W.Y. Desulfurization and emission control. In *Advanced Air and Noise Pollution Control*; Wang, L.K.; Pereira, N.C., Hung, Y.T., Eds.; Humana Press, Inc.: Totowa, NJ, 2004.
4. Wang, L.K. Treatment of storm run-off by oil–water separation, flotation, filtration and adsorption, part A: wastewater treatment. In *Proceedings of the 44th Industrial Waste Conference*, Purdue University, Lafayette, IN, 1990; 655–666.
5. Wang, L.K. Treatment of storm run-off by oil–water separation, flotation, filtration and adsorption, part B: waste sludge management. In *Proceedings of the 44th Industrial Waste Conference*, Purdue University, Lafayette, IN, 1990; 667–673.
6. Wang, J.C.; Aulenbach, D.B.; Wang, L.K. Energy models and cost models for water pollution controls. In *Clean Production*; Misra, K.B., Ed.; Springer-Verlag: Berlin, Germany, 1996; 685–720.
7. Wang, L.K.; Chen, J.L.; Hung, Y.T. Performance and costs of air pollution control technologies. In *Advanced Air and Noise Pollution Control*; Wang, L.K., Pereira, N.C., Hung, Y.T., Eds.; Humana Press, Inc.: Totowa, NJ, 2004.
8. Wang, L.K.; Krouzek, J.V.; Kounitson, U. *Case Studies of Cleaner Production and Site Remediation*; United Nations Industrial Development Organization (UNIDO): Vienna, Austria, 1995; Training Manual No. DTT-5-4-95, 136.
9. Wisconsin Public Service Commission. *Air Quality Issues for Electric Power Generation*, Publication No. 6015B; Wisconsin Public Service Commission: State of Wisconsin, PO Box 7854, Madison, WI, 1998.
10. World Nuclear Association. Nuclear electricity. In *Nuclear Energy Made Simple*, Chapter 3; World Nuclear Association, 2003; www.world-nuclear.org/education/ne/ne3/htm.
11. Wang, L.K. Industrial ecology. In *Encyclopedia of Life Support Systems: Hazardous Waste Management*, Chapter 15; Grasso, D., Vogel, T., Smets, B., Eds.; Eolss Publishers Co., Ltd.: London, 2003; www.eolss.net/E-1-08-toc.aspx.
12. Wang, L.K. *Pretreatment and Ozonation of Cooling Tower Water, Part I*; U.S. Department of Commerce, National Technical Information Service: Springfield, VA, 1983; Technical Report No. PB84-192053, 34 p., April.
13. Wang, L.K. *Pretreatment and Ozonation of Cooling Tower Water, Part II*; U.S. Department of Commerce, National Technical Information Service: Springfield, VA, 1983; Technical Report No. PB84-192046, 29 p., Aug.
14. Wang, L.K. *Prevention of Airborne Legionairs' Disease by Formulation of A New Cooling Water For Use in Central Air Conditioning Systems*; U.S. Department of Commerce, National Technical Information Service, 1984; Technical Report No. PB85-215317/AS, 97 p., Aug.
15. Wang, L.K.; Krofta, M. Treatment of cooling tower water by dissolved air–ozone flotation. In *Proceedings of the Seventh Mid-Atlantic Industrial Waste Conference*, 1985; p. 207–216, June 1985.
16. Wang, L.K. *Recent Development in Cooling Water Treatment with Ozone*; Lenox Institute of Water Technology: Lenox, MA, 1988; Technical Report No. LIR/03-88/285, 237 p., March.
17. Wang, L.K. *Treatment of Cooling Tower Water with Ozone*; Lenox Institute of Water Technology: Lenox, MA, 1988; Technical Report No. LIR/05-88/303, 55 p., May.
18. Wang, L.K. *Analysis of Sludges Generated from Flotation Treatment of Storm Runoff Water*; U.S. Department of Commerce, National Technical Information Service: Springfield, VA, 1988; Technical Report No. PB88-20062I/AS, 20 p.

19. Wang, L.K.; and Krofta, M. *Treatment of Cooling Tower Water with Ozone*. Lenox Institute of Water Technology: Lenox, MA, 1988; Report No. LIR/05-88/303, 55 p., May.
20. Wang, L.K.; Kopko, S.P. *City of Cape Coral Reverse Osmosis Water Treatment Facility*; US Department of Commerce, National Technical Information Service: Springfield, VA, 1997; 15p., www.afssociety.org/publications, Association of Filtration Society.
21. Wang, L.K. *Evaluation and Development of Physical–Chemical Techniques for the Separation of Emulsified Oil from Water*, Project Report No. 189; Veridian Engineering (formerly Arvin Calspan Corp.): Buffalo, NY, 1973; 31 p., May.
22. Wang, L.K. Separation of emulsified oil from water. *Chem. Indust.* **1975**, 562–564.
23. Van Berkel, C.W.M. Cleaner production: a profitable road for sustainable development of Australian industry. *Clean Air* **1999**, *33* (*4*), 33–38.

Index